Teacher's Edition

Integrated Mathematics 3

Authors

Senior Authors

Rheta N. Rubenstein
Timothy V. Craine
Thomas R. Butts

Kerry Cantrell
Linda Dritsas
Valarie A. Elswick
Joseph Kavanaugh
Sara N. Munshin
Stuart J. Murphy
Anthony Piccolino
Salvador Quezada
Jocelyn Coleman Walton

McDougal Littell
A Houghton Mifflin Company
Evanston, Illinois   Boston Dallas Phoenix

Authors

Senior Authors

Rheta N. Rubenstein — Associate Professor of Education, University of Windsor, Windsor, Ontario

Timothy V. Craine — Assistant Professor of Mathematical Sciences, Central Connecticut State University, New Britain, Connecticut

Thomas R. Butts — Associate Professor of Mathematics Education, University of Texas at Dallas, Dallas, Texas

Kerry Cantrell — Mathematics Department Head, Marshfield High School, Marshfield, Missouri

Linda Dritsas — Mathematics Coordinator, Fresno Unified School District, Fresno, California

Valarie A. Elswick — Mathematics Teacher, Roy C. Ketcham Senior High School, Wappingers Falls, New York

Joseph Kavanaugh — Academic Head of Mathematics, Scotia-Glenville Central School District, Scotia, New York

Sara N. Munshin — Mathematics Teacher, Theodore Roosevelt High School, Los Angeles, California

Stuart J. Murphy — Visual Learning Specialist, Evanston, Illinois

Anthony Piccolino — Assistant Professor of Mathematics and Computer Science, Montclair State College, Upper Montclair, New Jersey

Salvador Quezada — Mathematics Teacher, Theodore Roosevelt High School, Los Angeles, California

Jocelyn Coleman Walton — Educational Consultant, Mathematics K-12, and former Mathematics Supervisor, Plainfield High School, Plainfield, New Jersey

The authors wish to thank **Jane Pflughaupt**, Mathematics Teacher, Pioneer High School, San Jose, California, and **Anita G. Morris**, Coordinator of Mathematics, Anne Arundel County Public Schools, Annapolis, Maryland, for their contributions to this Teacher's Edition.

ISBN: 0-395-64449-6

123456789 VH 99 98 97 96 95

Contents of the Teacher's Edition

PHILOSOPHY
of Integrated Mathematics

Goals of the Course

Integrated Mathematics has been written to prepare your students for success in college, and in their careers and daily lives in the 21st century, by helping them develop their abilities to:

- **Explore and solve mathematical problems**
- **Understand and apply mathematical concepts**
- **Think critically**
- **Work cooperatively with others**
- **Communicate ideas clearly**

Underlying Concept

This program is built on the idea that students develop better conceptual understanding of mathematics and stronger problem solving skills when they:

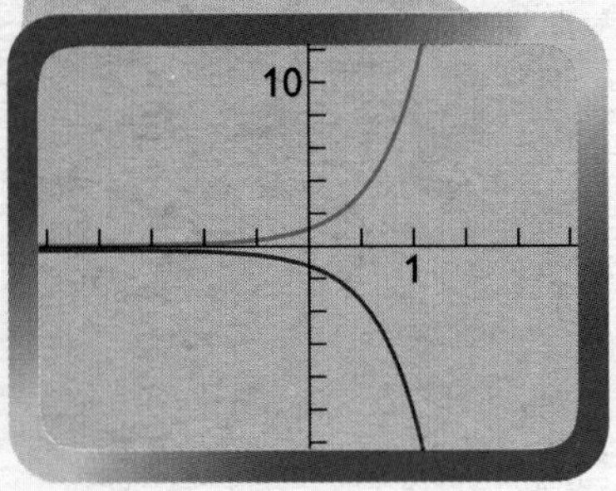

- **See the connections among different branches of mathematics**
- **Are actively involved in the learning process**
- **Study mathematics that is meaningful**
- **Continually build on prior learning because topics are spiraled**

Accessible and Inviting Mathematics

Integrated Mathematics was designed to make mathematics accessible and inviting. It opens the door to mathematics for more students by incorporating a variety of different teaching strategies, including:

- **Real-life applications**
- **Use of technology**
- **Visual and hands-on approaches**
- **Exploratory activities and projects**
- **Group work**
- **Open-ended problem solving**

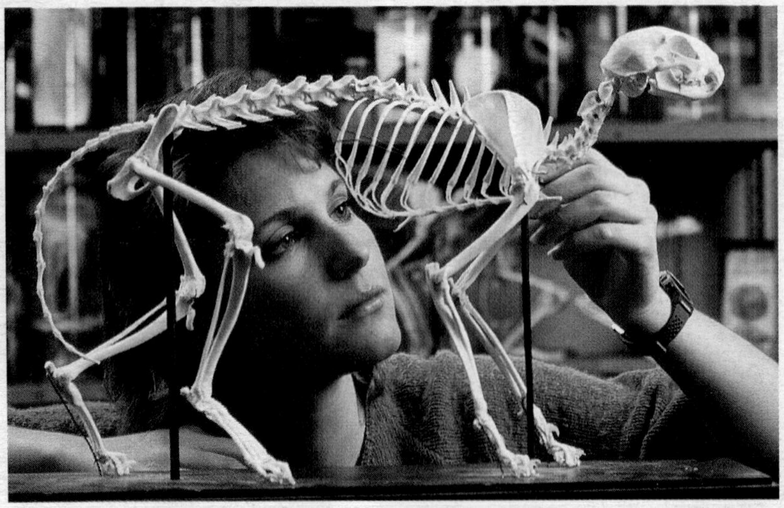

A Manageable Program

Integrated Mathematics makes it easy for you to manage these teaching strategies by incorporating them directly into the textbook. In addition, ongoing assessment that matches the instruction is included throughout the course.

Basis of the Curriculum

This program is based on the recommendations of the National Council of Teachers of Mathematics and other curriculum groups that emphasize problem solving, critical thinking, communication, and connections among mathematical topics and connections between mathematics and other subject areas.

Mathematical Content

Over a three-year period, *Integrated Mathematics* teaches the same mathematical topics as a contemporary Algebra 1/Geometry/Algebra 2 sequence. The difference is in the organization of the content. Instead of being divided into separate courses, algebra and geometry are taught in each of the three years. In addition, topics from logical reasoning, measurement, probability, statistics, discrete mathematics, and functions are interwoven throughout each year. The Topic Spiraling chart on page xiii of the textbook shows how mathematical concepts are spiraled over the three years of the program.

Integrated Mathematics provides a strong foundation for more advanced courses in mathematics and the many other fields that require quantitative reasoning and decision-making skills.

Advantages of an Integrated Approach

With an integrated approach, your students can:

- **Learn more mathematics**
- **Solve problems that are more realistic and more interesting**
- **Have better retention of what they have learned**

Field Testing

Preliminary versions of *Integrated Mathematics* were tried out by hundreds of teachers and thousands of students in many different types of classrooms nationwide. Their comments and suggestions have guided the development of this book. Here is what some teachers who piloted the book have said:

"With Integrated Mathematics, I can teach my students more mathematics and better mathematics in less time. The Precalculus teacher is looking forward to having my students next year."

"I am a person who does not change just for the sake of change. For me to switch from the traditional curriculum to the new integrated mathematics I needed to see that it really was a step forward. This approach is definitely enabling me to help produce more mathematically able students."

NAVIGATION

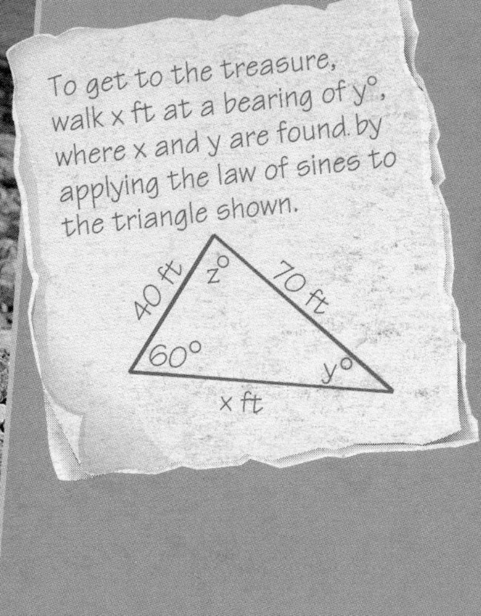
To get to the treasure, walk x ft at a bearing of y°, where x and y are found by applying the law of sines to the triangle shown.
40 ft
z°
70 ft
60°
y°
x ft

"I used to spend a lot of time scrambling around trying to find activities and applications to pull into my class. With this program, it has all been done for me."

Contents

Unit 1 In this unit, students investigate techniques for analyzing and solving real-world problems using algebra, discrete mathematics, statistics, and technology. Explorations, cooperative learning, and applications help students become actively involved.

Unit 2 Exploring and Applying Functions

Unit 2 The focus of this unit is modeling everyday situations using many kinds of functions. Connections to literature, transportation, medicine, physics, biology, and other topics underscore the relevance of mathematics.

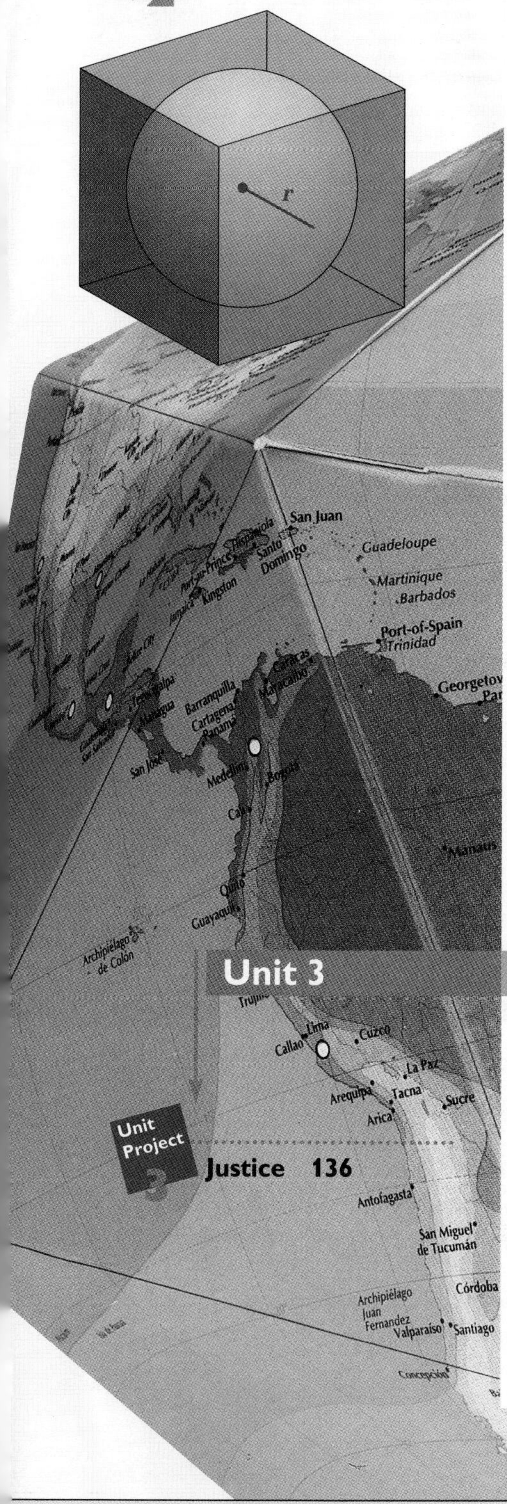

Unit 3 Logical Reasoning and Methods of Proof

Unit 3 In this unit, students extend their ability to apply logical reasoning by exploring important geometric concepts and methods of proof. Critical thinking skills are presented in geometric, algebraic, and everyday contexts to enable students to reach conclusions about the world around them.

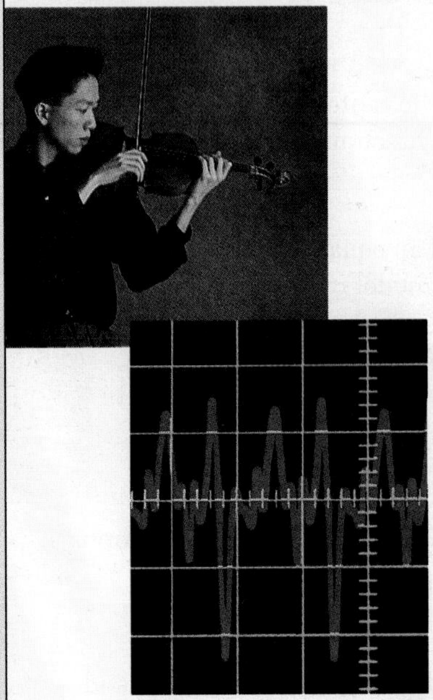

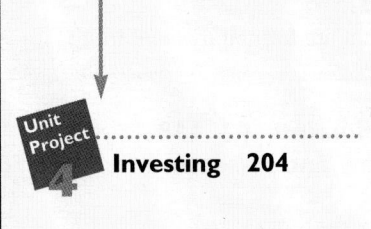

Unit 4 Sequences and Series

Unit 4 The focus of this unit is analyzing sequences and series using patterns, formulas, graphs, technology, and explorations. Throughout the unit, concepts are presented both algebraically and geometrically.

Unit 5 Exponential and Logarithmic Functions

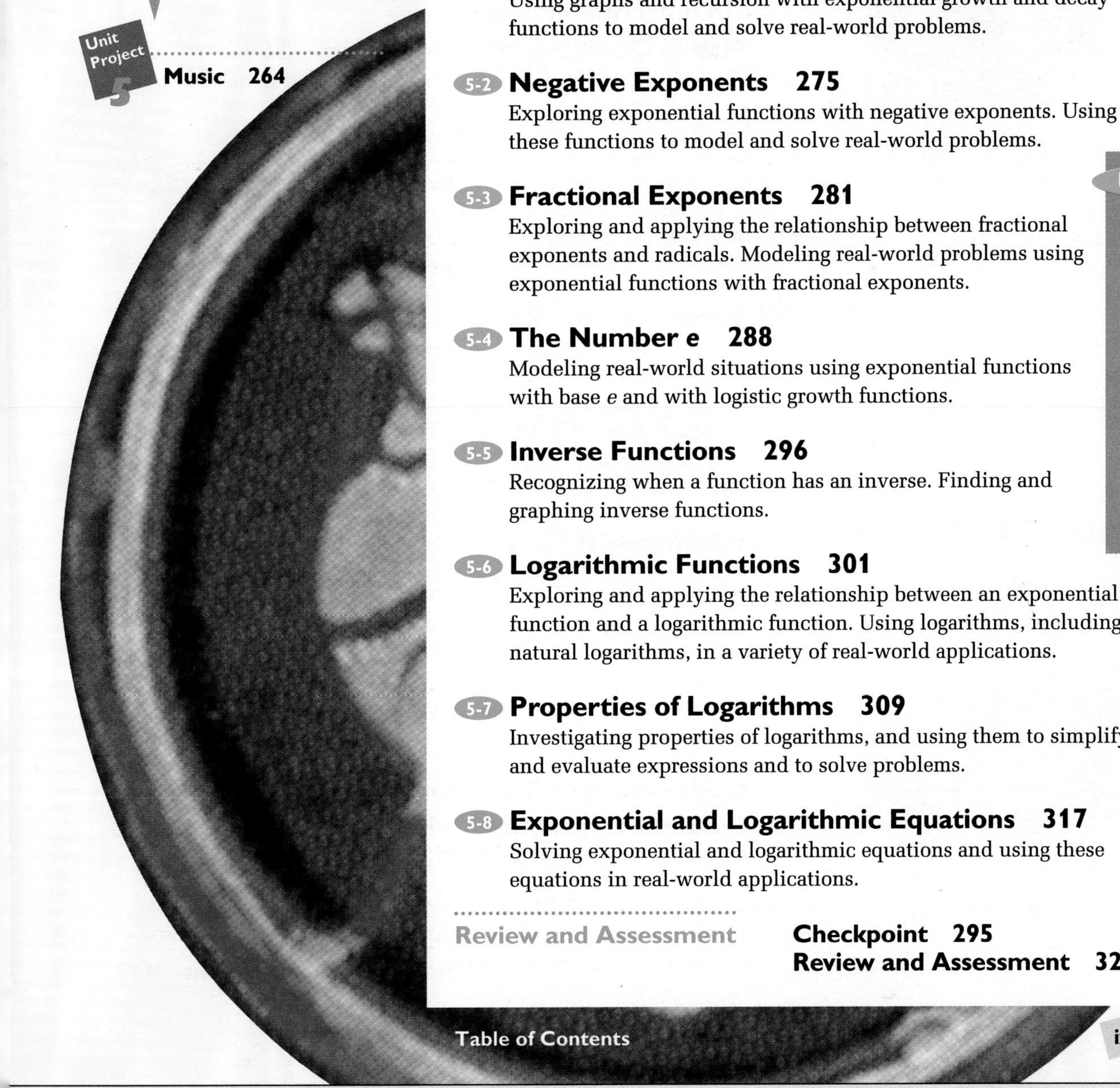

Unit 5 This unit extends the study of modeling real-world situations using functions begun in Unit 2. Exponential and logarithmic functions are the unifying theme linking algebra, geometry, discrete mathematics, and statistics with applications to business education, history, biology, music, art, and other areas.

Unit 6 Modeling and Analyzing Data

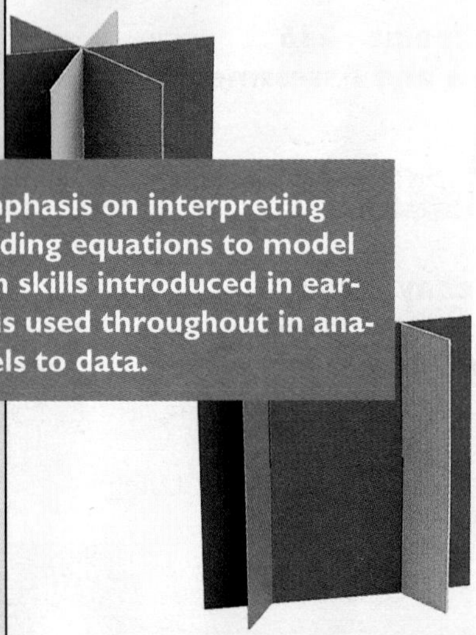

Unit 6 With its emphasis on interpreting real-world data and finding equations to model data, this unit builds on skills introduced in earlier units. Technology is used throughout in analyzing and fitting models to data.

Unit 7 Applying Probability Models

Unit 7 In this unit, students use simulation, formulas, diagrams, graphs, and area models to find probabilities. Frequent explorations, practical applications, and opportunities for cooperative learning motivate student interest and participation.

Unit 8 Angles, Trigonometry, and Vectors

Unit 8 This unit integrates real-world applications of trigonometry, geometry, and algebra. Connections to literature, geography, social studies, science, and other topics illustrate the importance of mathematics in everyday life.

Unit 9 Transformations of Graphs and Data

Unit 9 This unit extends the study of functions and data analysis introduced in previous units. Graphing technology is used to explore the relationship between changes in an equation and changes in its graph.

Unit 10 The study of periodic models in this unit links algebraic, geometric, and trigonometric concepts. Practical applications include finding equations to model real-world periodic data, periodicity in decorative patterns, and connections to science, literature, history, and health.

Unit 10 Periodic Models

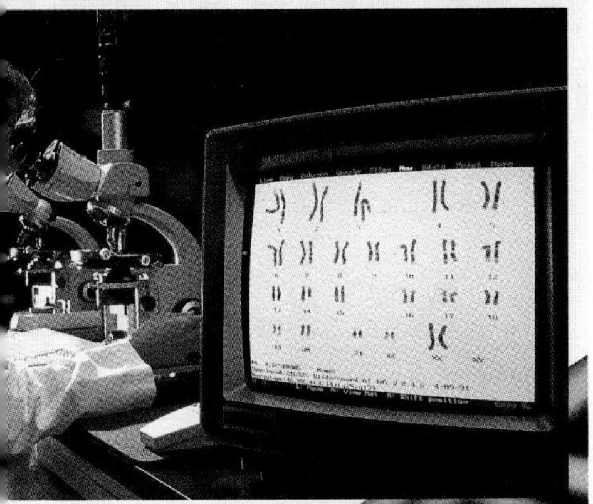

Student Resources

Integrated Mathematics Topic Spiraling

This chart shows how mathematical strands are spiraled over the three years of the *Integrated Mathematics* program.

	Course 1	Course 2	Course 3
Algebra	Linear equations Linear inequalities Multiplying binomials Factoring expressions	Quadratic equations Linear systems Rational equations Complex numbers	Polynomial functions Exponential and logarithmic functions Radical functions Parametric equations
Geometry	Angles, polygons, circles Perimeter, circumference Area, surface area Volume	Similar and congruent figures Geometric proofs Coordinate geometry Transformational geometry	Inscribed polygons Tangents to circles Transforming graphs Vectors, polar coordinates
Statistics, Probability	Analyzing data and displaying data Experimental and theoretical probability Geometric probability	Sampling methods Simulation Binomial distributions Compound events	Standard deviation Normal distribution Conditional probability z-scores
Logical Reasoning	Conjectures Counterexamples If-then statements	Inductive and deductive reasoning Valid and invalid reasoning Postulates and proof	Identities Contrapositive and inverse Comparing proof methods
Discrete Math	Discrete quantities Matrices to display data Lattices	Matrix operations Transformation matrices Counting techniques	Sequences and series Recursion Network diagrams
Trigonometry	Right-triangle ratios: sine, cosine, tangent Solving right triangles Tangent ratio as slope	Applying properties of special right triangles: 30° - 60° - 90° 45° - 45° - 90°	Circular trigonometric functions and graphs Modeling periodic situations Law of cosines, law of sines

Teachers today are being asked to teach more mathematics, and better mathematics, to more students.

How is this possible?

Integrated *Approach*

Integrated Mathematics interweaves mathematical topics and contemporary teaching strategies throughout the course. Key mathematical strands are spiraled through the units — and integrated within individual sections.

DATA ANALYSIS

Analyzing data to model and solve real-world problems is emphasized throughout this course.

ALGEBRA AND GEOMETRY

Both algebraic and geometric solutions are often presented to broaden students' problem solving skills and illustrate the relationship between these topics.

Section 1-7 **Maximizing and Minimizing**

Focus

Make quantities as large or as small as possible to solve real-life problems.

Peak Performance

Stickers on new cars give values for MPG (miles per gallon of gasoline) for city and highway driving.

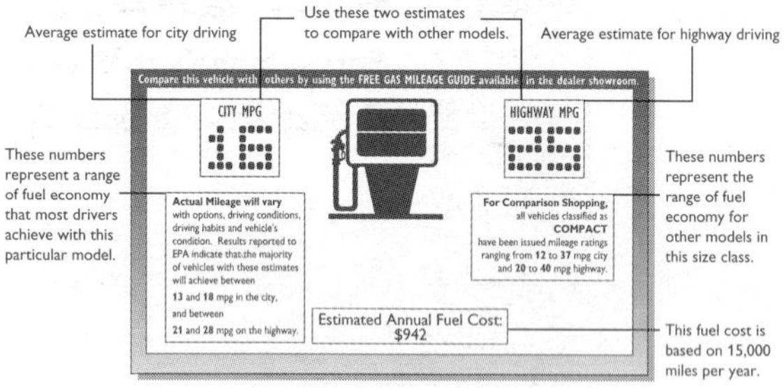

Average estimate for city driving
Use these two estimates to compare with other models.
Average estimate for highway driving
Compare this vehicle with others by using the FREE GAS MILEAGE GUIDE available in the dealer showroom.
CITY MPG
16
HIGHWAY MPG
25
These numbers represent a range of fuel economy that most drivers achieve with this particular model.
Actual Mileage will vary with options, driving conditions, driving habits and vehicle's condition. Results reported to EPA indicate that the majority of vehicles with these estimates will achieve between 13 and 18 mpg in the city, and between 21 and 28 mpg on the highway.
For Comparison Shopping, all vehicles classified as COMPACT have been issued mileage ratings ranging from 12 to 37 mpg city and 20 to 40 mpg highway.
These numbers represent the range of fuel economy for other models in this size class.
Estimated Annual Fuel Cost: $942
This fuel cost is based on 15,000 miles per year.

Talk it Over

The table gives the MPG for an average car at various speeds.

1. What happens to MPG as the speed of a car increases?
2. Do you think a car has a **maximum**, or greatest, MPG? Why or why not? If so, at what speed(s) do you think the maximum MPG occurs?
3. Do you think a car has a **minimum**, or least, MPG? Why or why not? If so, at what speed(s) do you think the minimum MPG occurs?

x = speed (mi/h)	y = MPG
15	21.0
25	30.0
35	33.6
45	33.5
55	30.3
65	24.9
75	20.0

In many situations it is important to make a quantity as large or as small as possible. For example, business managers want to make decisions that maximize income and minimize expenses. Drivers often want to maximize gas mileage or minimize the distance or time of a trip.

Sample Response

Student Resources Toolbox p. 651 *Graphing*

Method 1 Use a formula.

The function $y = -0.014x^2 + 1.2x + 6.9$ is a quadratic function, so its graph is a parabola. Since the coefficient of the x^2-term is negative, the parabola opens down. The vertex is on the line of symmetry of the parabola at the point where the value of the function is a maximum.

The speed that maximizes MPG is the x-coordinate of the vertex. To find this value, use the equation of the line of symmetry.

$$x = -\frac{b}{2a}$$ ← a and b are coefficients in the general quadratic function $y = ax^2 + bx + c$.

$$= -\frac{1.2}{(2)(-0.014)}$$ ← Substitute **1.2** for b and -0.014 for a.

$$\approx 42.9$$

A driving speed of about 43 mi/h maximizes MPG.

Method 2 Use a graph.

Graph $y = -0.014x^2 + 1.2x + 6.9$.

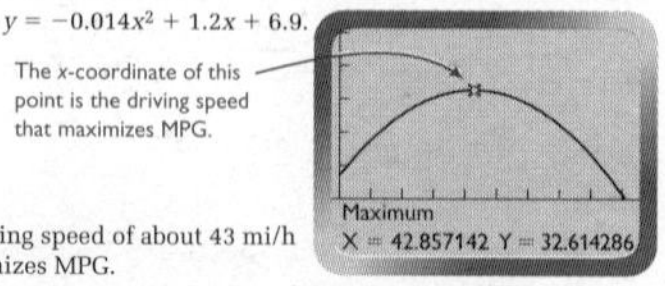

Now you are ready for: Exs. 1–11 on pp. 49–50

A driving speed of about 43 mi/h maximizes MPG.

Use Diagrams to Find Minimums

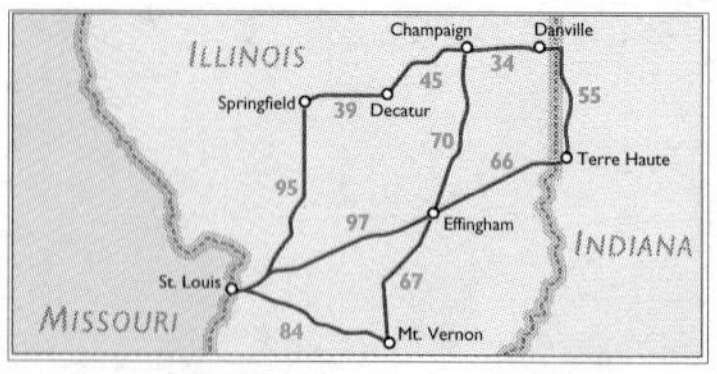

The map shows several cities in Illinois, Indiana, and Missouri, as well as the interstate highways connecting them. The driving distance between each pair of connected cities is labeled.

← All distances are given in miles.

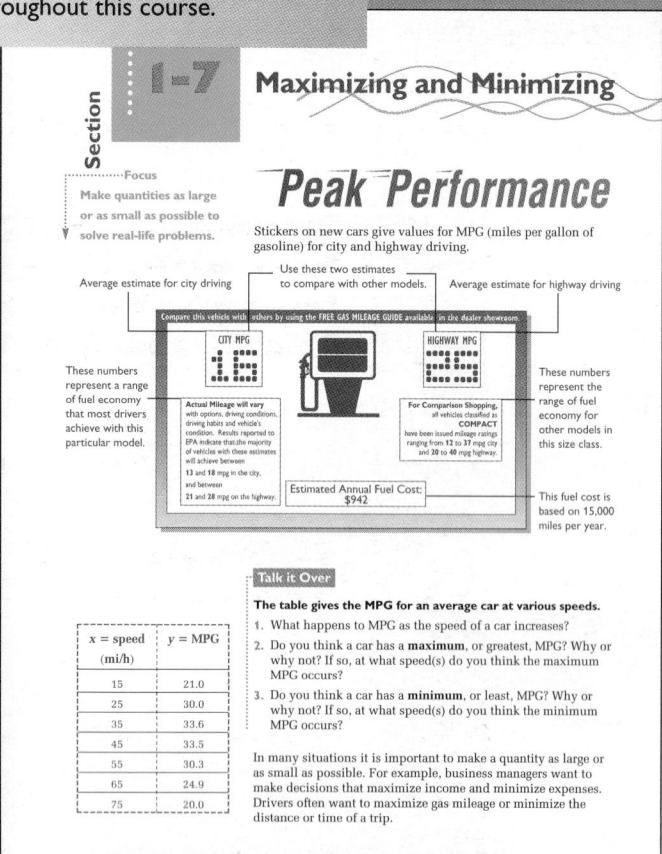

Topic Integration

TOPIC INTEGRATION

Topic spiraling enables students to continually build on prior learning. The list at the right illustrates how major topic strands are integrated into the sections of the book.

Talk it Over

4. Use the map on page 46. Suppose you want to travel from St. Louis to Danville using interstate highways.
   a. What do you think is the shortest route between the two cities? What is the length of this route?
   b. Describe the method you used in part (a) to find the route you think is shortest.

Sample 2 shows how to use an algorithm to find the minimum distance between two points.

Sample 2

Find the shortest route on interstate highways from St. Louis to Danville. Use the map on page 46.

Sample Response

Problem Solving Strategy: Use a diagram.

1. Draw a network diagram that models the map. Each vertex represents a city. Each edge represents an interstate highway. Distances do not need to be drawn to scale.

The red edges join labeled and unlabeled vertices.

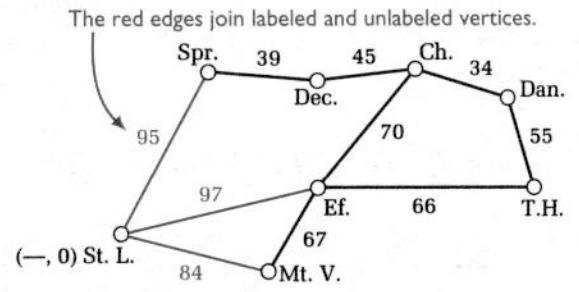

2. Label the starting point, St. Louis, with the ordered pair (—, 0).

3. For each edge that connects a labeled and an unlabeled vertex, find this sum:

$s =$ second entry of ordered pair for labeled vertex $+$ length of edge

Find *s* for each red edge.
St. L. to Spr.: 0 + 95 = 95
St. L. to Ef.: 0 + 97 = 97
St. L. to Mt. V.: 0 + 84 = 84 ⟵ This is the least sum.
Write the ordered pair (St. L., 84) next to "Mt. V."

4. Choose the edge from Step 3 that has the minimum sum *s*. Label the unlabeled vertex of that edge with this ordered pair:
(label of the other vertex of the edge, *s*)

5. Repeat Steps 3 and 4 until the vertex for Danville is labeled.

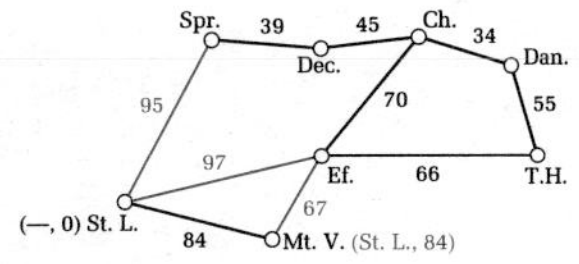

This is the least sum.
Find *s* for each red edge.
St. L. to Spr.: 0 + 95 = 95
St. L. to Ef.: 0 + 97 = 97

DISCRETE MATHEMATICS

Techniques of discrete mathematics, including network diagrams, matrices, and recursive formulas, receive special attention in this course.

6. When the vertex for Danville is labeled, use the ordered pairs to find the shortest route from St. Louis to Danville.

Starting from Danville, use the first coordinates of the ordered pairs to backtrack to St. Louis. This gives the shortest route.

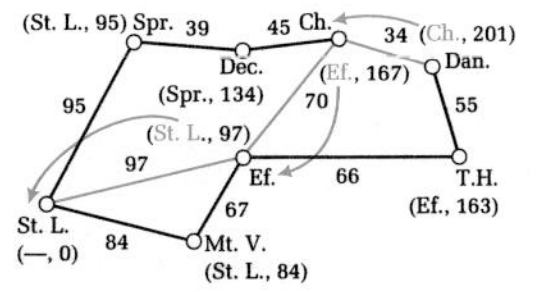

The shortest route from St. Louis to Danville on interstate highways is St. Louis to Effingham to Champaign to Danville.

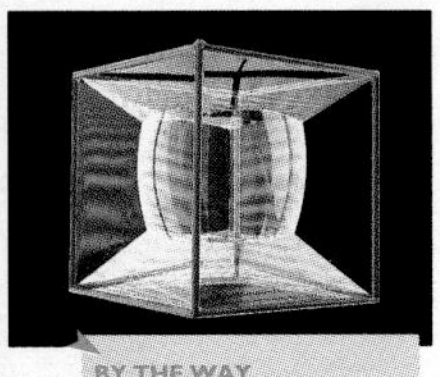

BY THE WAY...

Soap films continually form surfaces that minimize energy. Used with pegs and a map, soap films act as mechanical computers for constructing the shortest road network among a group of cities.

Talk it Over

Questions 5–8 are about Sample 2.

5. What is the length of the shortest route from St. Louis to Danville? What is the relationship between this length and the second entry of the ordered pair for Danville?
6. Why is the first entry of the ordered pair for St. Louis a dash instead of a vertex label?
7. How many times must you perform Steps 3 and 4 to find the shortest route from St. Louis to Danville?
8. Can you use the network diagram to find the shortest route from St. Louis to Terre Haute? from Springfield to Terre Haute? Why or why not?

Look Back

How can you use graphs of functions and network diagrams to maximize and minimize quantities?

Now you are ready for: Exs. 12–22 on pp. 50–52

ALGEBRA

Functions, equations, and graphs: linear, quadratic, piecewise, absolute value, composite, polynomial, rational, radical, exponential, logarithmic 1-7, Unit 2, Unit 5, 9-2, 9-4, 9-5

Systems of equations (inequalities): linear, matrix, parametric 1-5, 1-8, 6-5, 8-5

Modeling problem situations For example, 1-5, 2-6, 4-3, 5-1, 6-5, 8-5

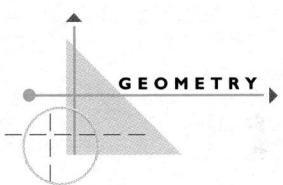

GEOMETRY

Logical reasoning and proof 3-1–3-7

Polygons, polyhedra, circles 3-4–3-9, 10-7

Transforming graphs and patterns 9-2, 9-4, 9-5, 10-4, 10-6

Vectors, polar coordinates 8-1–8-5

STATISTICS & PROBABILITY

Data analysis 1-3, 6-2, 9-1, 9-3, 9-5

Distribution of data 6-1–6-3, 9-6

Fitting models to data 6-4, 6-5

Probability of events 7-2–7-6

DISCRETE MATH

Network diagrams 1-6, 1-7

Sequences and series Unit 4

Recursion 4-3, 4-4

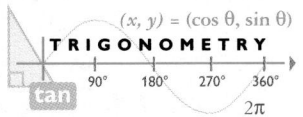

TRIGONOMETRY

Trigonometric ratios 8-2

Law of cosines, law of sines 8-6, 8-7

Circular functions and modeling 10-3–10-5

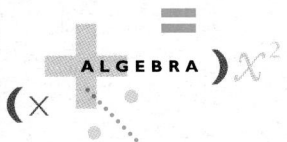

I want *all* my students to have the mathematical and problem solving skills they need. But some students just don't seem interested in learning math.

How can I get them involved?

Active Learning

Integrated Mathematics makes it easy for you to get your students actively involved because projects, explorations, activities, and discussion questions are built right into the book.

EXPLORATIONS

The Explorations get students involved in investigating math. They build strong conceptual understanding by helping students move from the concrete to the abstract.

Section 6-3 **Normal Distributions**

Focus
Recognize special properties of normal distributions.

DOES THIS RING A BELL?

EXPLORATION

How would you describe the distribution of heights of you and your classmates?

- **Materials: tape measures**
- **Work in a group of four students.**

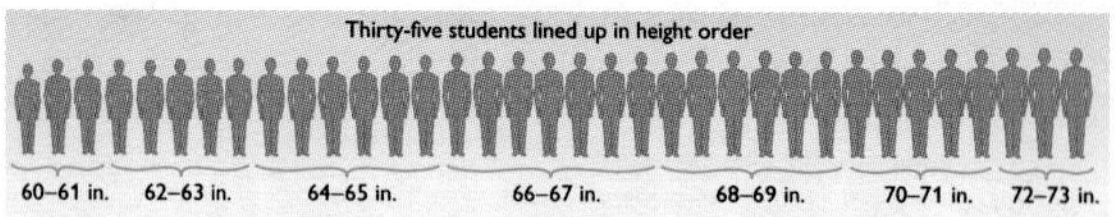

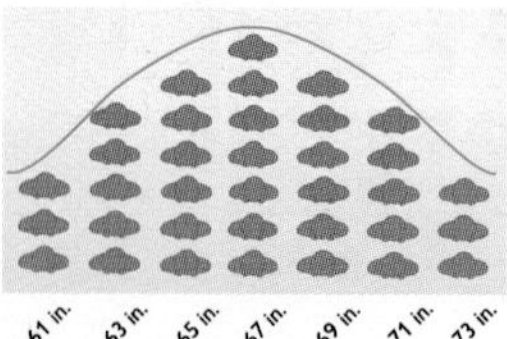

1. List the height of everyone in the class to the nearest inch. Measure students who do not know their heights. Make separate lists for males and females.
2. In your group, decide on a reasonable interval width and draw a relative frequency histogram for the heights of everyone in the class. What type of distribution have you drawn? Where does the mean fall? the median fall?
3. Draw two new relative frequency histograms, one for the male students only and one for the female students only. What type of distribution is shown in each histogram? How do the histograms compare with the histogram for the whole class? Where do the means and medians fall?
4. Suppose you drew a histogram for the heights of all the males or all the females in your school. What shape would you expect the new distribution to have?

Modeling and Analyzing Data

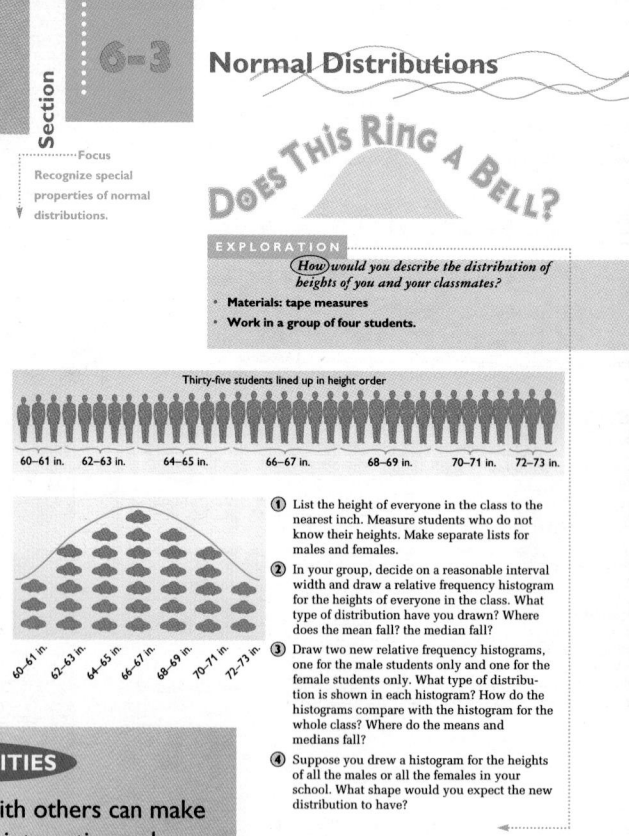

GROUP ACTIVITIES

Working with others can make math more interesting and more accessible for students.

UNIT PROJECTS

Each unit begins with a Unit Project. These projects give your students a chance to work on the types of open-ended, long-range problems that prepare them for future careers.

The Unit Projects

- put the mathematics in context
- unify related mathematical topics
- give all students an opportunity to participate and contribute

TALK IT OVER

The Talk it Over questions give students a chance to communicate their understanding.

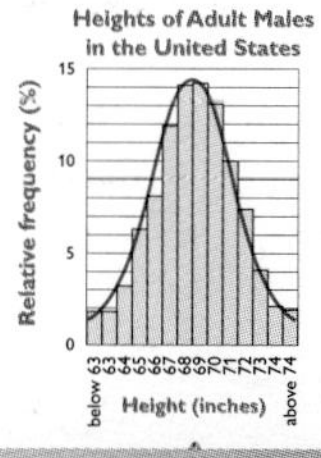

When you draw a smooth curve close to or through the tops of the rectangles of a histogram, you get a bell-shaped curve like the one shown. This bell curve is called a *normal curve*. A distribution with a histogram that follows a normal curve is called a **normal distribution.** In a normal distribution, the mean and median of the data are equal and fall at the line of symmetry for the curve.

Talk it Over

1. What do you think is the average height of men in the United States?
2. About what percent of men in the United States are between 66 in. and 71 in. tall?
3. What kind of distribution would you get if you made a histogram of the heights of all adults in the United States? Explain.
4. Normal distributions are always bell-shaped, but the shape of the bell can vary. Make a statement about the mean and standard deviation of these two normal distributions.

In a normal distribution, a known percent of the data falls within one, two, or three standard deviations of the mean.

68-95-99.7 RULE FOR NORMAL DISTRIBUTIONS

For a normal distribution:

- about 68% of the data are within one standard deviation of the mean.
- about 95% of the data are within two standard deviations of the mean.
- about 99.7% of the data are within three standard deviations of the mean.

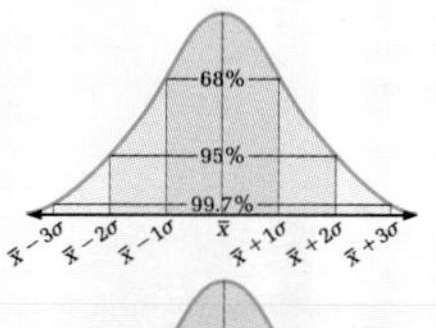

Since a normal distribution is symmetric, the 68-95-99.7 rule can be broken down in this way:

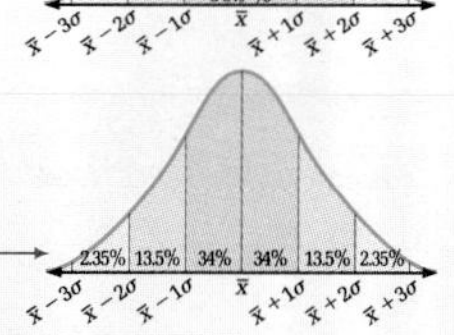

345

Talk it Over

5. Why must you know that data have a normal distribution before you can apply the 68-95-99.7 rule to the data?
6. If 95% of the data are within two standard deviations of the mean, what percent are outside two standard deviations?

BY THE WAY...

Seven states do not have lieutenant governors: Arizona, Maine, New Jersey, Oregon, Tennessee, West Virginia, and Wyoming.

Sample

The 1992–1993 annual salaries of the 43 state lieutenant governors in the United States have a normal distribution with $\bar{x} = \$57{,}000$ and $\sigma = \$22{,}000$.

a. Within what range do about 95% of the lieutenant governors' salaries fall?

b. About what percent of the lieutenant governors have an annual salary between \$13,000 and \$79,000?

Sample Response

a. The data have a normal distribution, so 95% of the salaries will fall within two standard deviations of the mean.

$$\bar{x} + 2\sigma = 57{,}000 + 2(22{,}000) = 101{,}000$$

$$\bar{x} - 2\sigma = 57{,}000 - 44{,}000 = 13{,}000$$

About 95% of the lieutenant governors have an annual salary between \$13,000 and \$101,000.

b. Draw a normal curve for the lieutenant governors' salaries and label what is known.

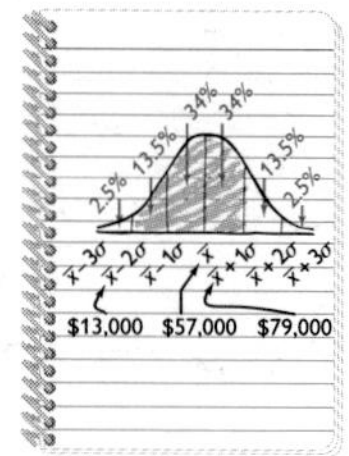

You know 13,000 is two standard deviations below the mean. Since 57,000 + 22,000 = 79,000, \$79,000 is one standard deviation above the mean.

So about 13.5% + 34% + 34% = 81.5% of the lieutenant governors have an annual salary between \$13,000 and \$79,000.

MODELING LEARNING

Modeling possible responses helps students learn, review, and apply concepts.

Now you are ready for: Exs. 6–16 on pp. 347–349

Look Back

What is a normal distribution?

346 Unit 6 Modeling and Analyzing Data

Unit Project 2

Make a Pedal Pitch

Your project is to give a presentation about bicycles to children in your community.

Each group member will choose one aspect of bicycling and plan to present information on it. For example, you may choose bicycle safety, bicycling for exercise, bike racing, or the physics of bicycles.

As you complete the "Working on the Unit Project" exercises in this unit, decide how you will use the information in your presentation. Save your work to support the data you present.

racing

Bike racing tracks are made of asphalt, clay, concrete, or wood. Most racers prefer the smooth finish of a wood track, like the one in the Major Taylor **Velodrome** in Indianapolis, named after the cyclist who was the first African-American world champion in any sport.

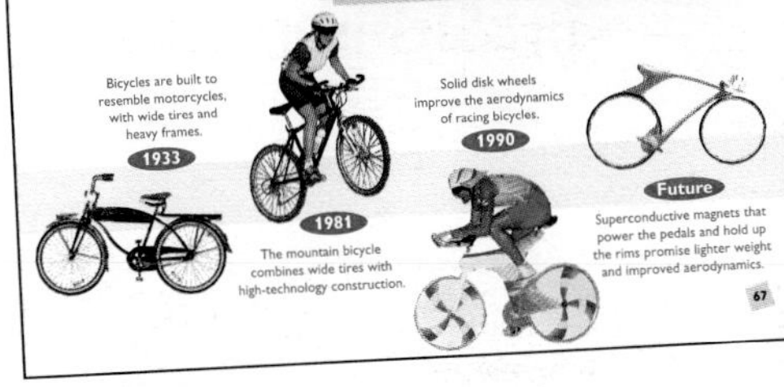

Bicycles are built to resemble motorcycles, with wide tires and heavy frames.
1933
1981
The mountain bicycle combines wide tires with high-technology construction.
Solid disk wheels improve the aerodynamics of racing bicycles.
1990
Future
Superconductive magnets that power the pedals and hold up the rims promise lighter weight and improved aerodynamics.

67

PROJECT EXERCISES

The Working on the Unit Project exercises in each section help students build the knowledge and the skills they need to complete the project.

Ongoing **ASSESSMENT**

22. **Open-ended** Alberta fills a glass with water, adds two ice cubes, and uses a straw to drink the water. Sketch a piecewise graph that you think could give the height of the water in the glass as a function of time. State the action associated with each piece of the graph.

Review **PREVIEW**

Write an equation for each function in the form $f(x) = mx + b$. *(Section 2-2)*

23. $f(-2) = -10$, $f(2) = 2$

24. $f(0) = 5$, $f(7) = \frac{3}{2}$

What name best describes each quadrilateral? *(Glossary)*

25.–27.

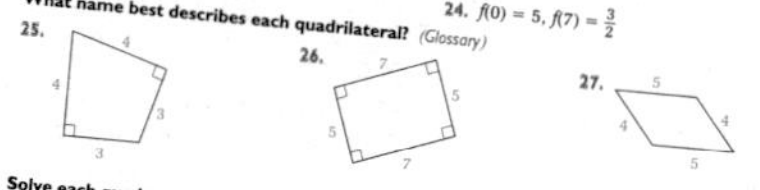

Solve each quadratic equation. *(Toolbox Skill 22)*

28. $x^2 + 3x + 1 = 0$

29. $4x^2 - 7 = 2x$

Working on the Unit Project

30. During the summer, Juanita likes to mountain bike. The graph shows the distance she travels on one trail as a function of time.

a. Write a piecewise function for the graph and give the function's domain and range.

b. How long does it take Juanita to complete the trail? How long does it take her to reach the halfway mark of the trail?

c. Is Juanita's speed constant? How can you tell by looking at the graph? What might be reasons for any change in speed?

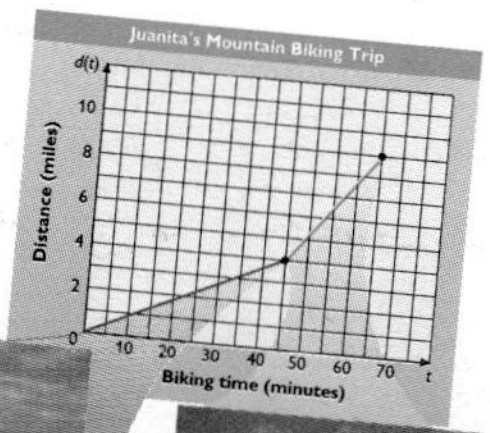

92

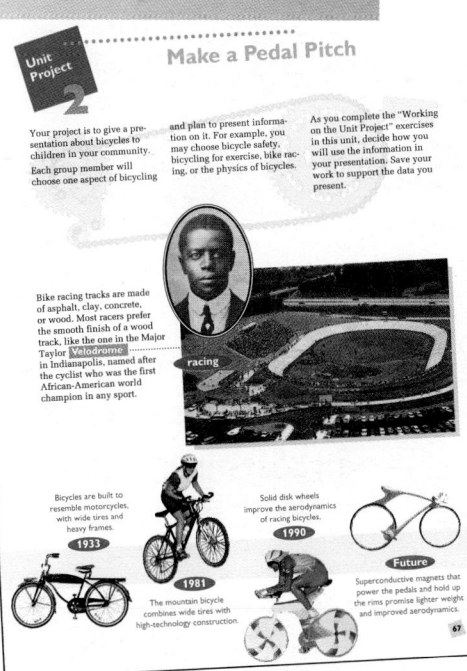

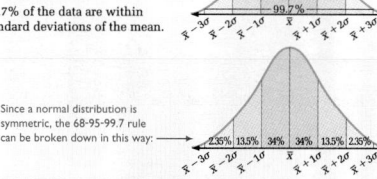

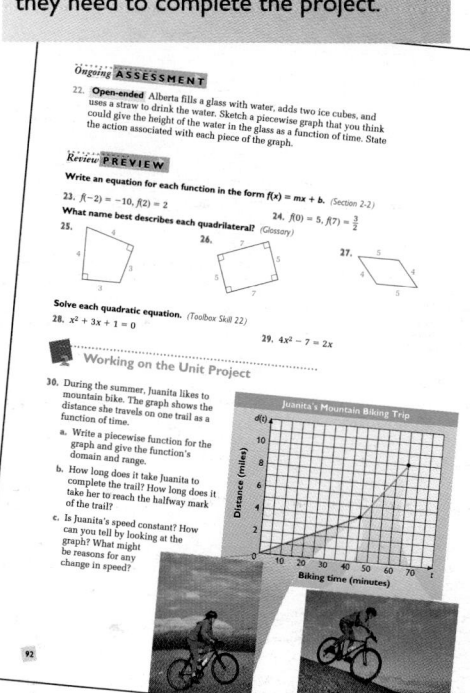

I'm tired of my students asking, "When am I ever going to use this?" I want them to realize that mathematics is useful and powerful.

How can I convince them?

Meaningful *Mathematics*

Integrated Mathematics focuses on important concepts in mathematics and shows how they can be applied to solve a wide variety of types of problems in daily life and in careers.

MATHEMATICAL MODELING

Modeling real-world situations is emphasized throughout the course. Using technology makes it easier to visualize problems and find solutions.

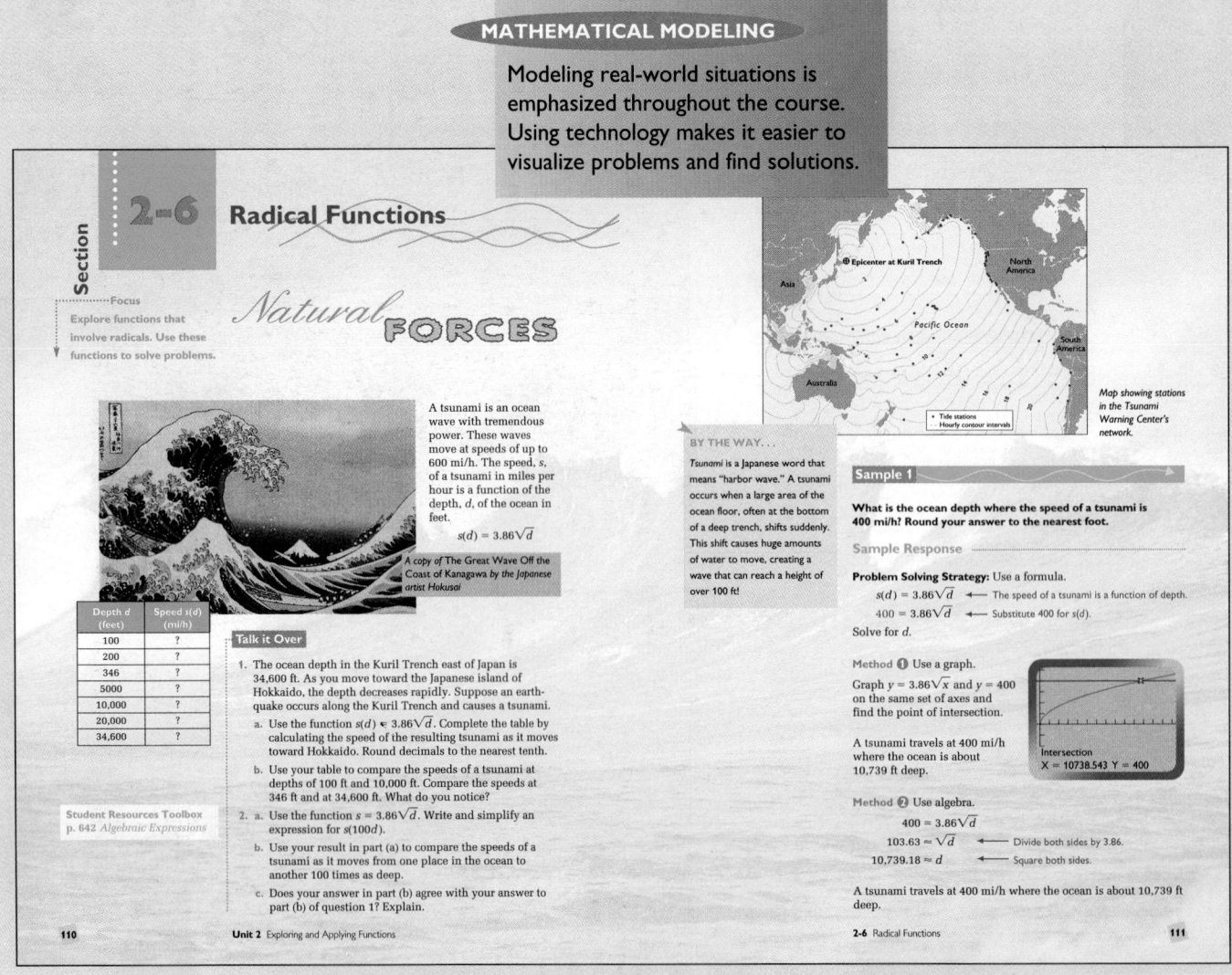

Section **2-6** **Radical Functions**

Focus
Explore functions that involve radicals. Use these functions to solve problems.

Natural FORCES

A copy of The Great Wave Off the Coast of Kanagawa by the Japanese artist Hokusai

A tsunami is an ocean wave with tremendous power. These waves move at speeds of up to 600 mi/h. The speed, s, of a tsunami in miles per hour is a function of the depth, d, of the ocean in feet.

$$s(d) = 3.86\sqrt{d}$$

Depth d (feet)	Speed $s(d)$ (mi/h)
100	?
200	?
346	?
5000	?
10,000	?
20,000	?
34,600	?

Talk it Over

1. The ocean depth in the Kuril Trench east of Japan is 34,600 ft. As you move toward the Japanese island of Hokkaido, the depth decreases rapidly. Suppose an earthquake occurs along the Kuril Trench and causes a tsunami.
   a. Use the function $s(d) = 3.86\sqrt{d}$. Complete the table by calculating the speed of the resulting tsunami as it moves toward Hokkaido. Round decimals to the nearest tenth.
   b. Use your table to compare the speeds of a tsunami at depths of 100 ft and 10,000 ft. Compare the speeds at 346 ft and at 34,600 ft. What do you notice?
2. a. Use the function $s = 3.86\sqrt{d}$. Write and simplify an expression for $s(100d)$.
   b. Use your result in part (a) to compare the speeds of a tsunami as it moves from one place in the ocean to another 100 times as deep.
   c. Does your answer in part (b) agree with your answer to part (b) of question 1? Explain.

Student Resources Toolbox
p. 642 *Algebraic Expressions*

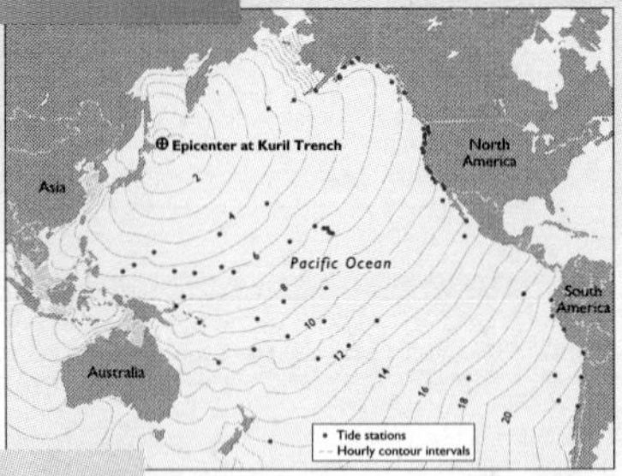

Map showing stations in the Tsunami Warning Center's network.

BY THE WAY...

Tsunami is a Japanese word that means "harbor wave." A tsunami occurs when a large area of the ocean floor, often at the bottom of a deep trench, shifts suddenly. This shift causes huge amounts of water to move, creating a wave that can reach a height of over 100 ft!

Sample 1

What is the ocean depth where the speed of a tsunami is 400 mi/h? Round your answer to the nearest foot.

Sample Response

Problem Solving Strategy: Use a formula.

$s(d) = 3.86\sqrt{d}$ ← The speed of a tsunami is a function of depth.

$400 = 3.86\sqrt{d}$ ← Substitute 400 for $s(d)$.

Solve for d.

Method 1 Use a graph.

Graph $y = 3.86\sqrt{x}$ and $y = 400$ on the same set of axes and find the point of intersection.

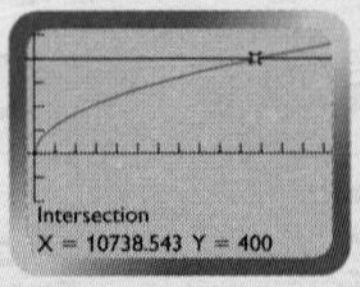

A tsunami travels at 400 mi/h where the ocean is about 10,739 ft deep.

Method 2 Use algebra.

$400 = 3.86\sqrt{d}$

$103.63 \approx \sqrt{d}$ ← Divide both sides by 3.86.

$10{,}739.18 \approx d$ ← Square both sides.

A tsunami travels at 400 mi/h where the ocean is about 10,739 ft deep.

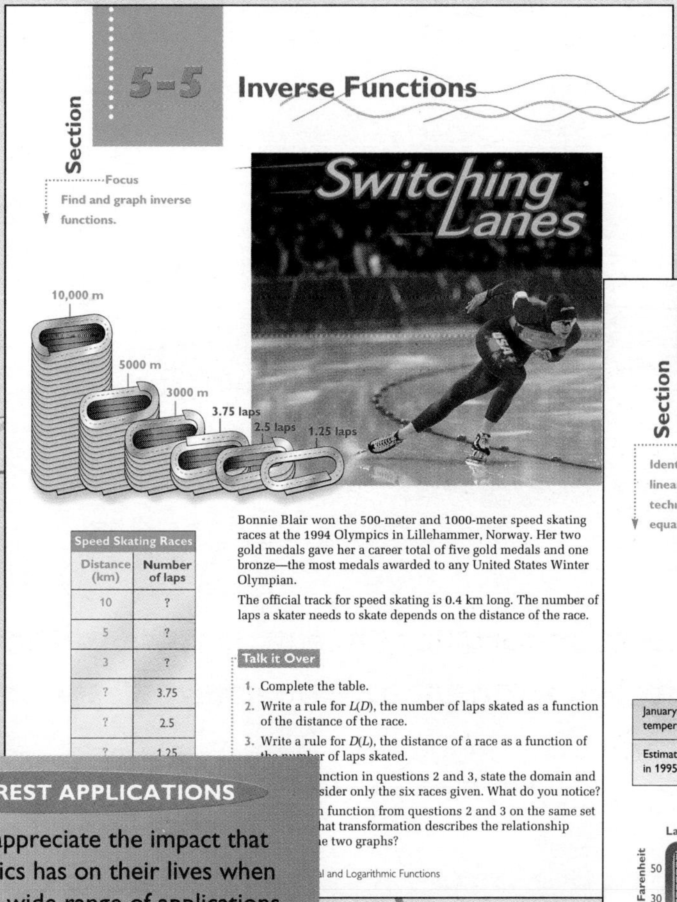

Section 5-5

Inverse Functions

Focus
Find and graph inverse functions.

Bonnie Blair won the 500-meter and 1000-meter speed skating races at the 1994 Olympics in Lillehammer, Norway. Her two gold medals gave her a career total of five gold medals and one bronze—the most medals awarded to any United States Winter Olympian.

The official track for speed skating is 0.4 km long. The number of laps a skater needs to skate depends on the distance of the race.

Speed Skating Races	
Distance (km)	Number of laps
10	?
5	?
3	?
?	3.75
?	2.5
?	1.25

Talk it Over

1. Complete the table.
2. Write a rule for $L(D)$, the number of laps skated as a function of the distance of the race.
3. Write a rule for $D(L)$, the distance of a race as a function of the number of laps skated.

...unction in questions 2 and 3, state the domain and ...sider only the six races given. What do you notice?

...n function from questions 2 and 3 on the same set ...hat transformation describes the relationship ...e two graphs?

...al and Logarithmic Functions

HIGH-INTEREST APPLICATIONS

Students appreciate the impact that mathematics has on their lives when they see a wide range of applications.

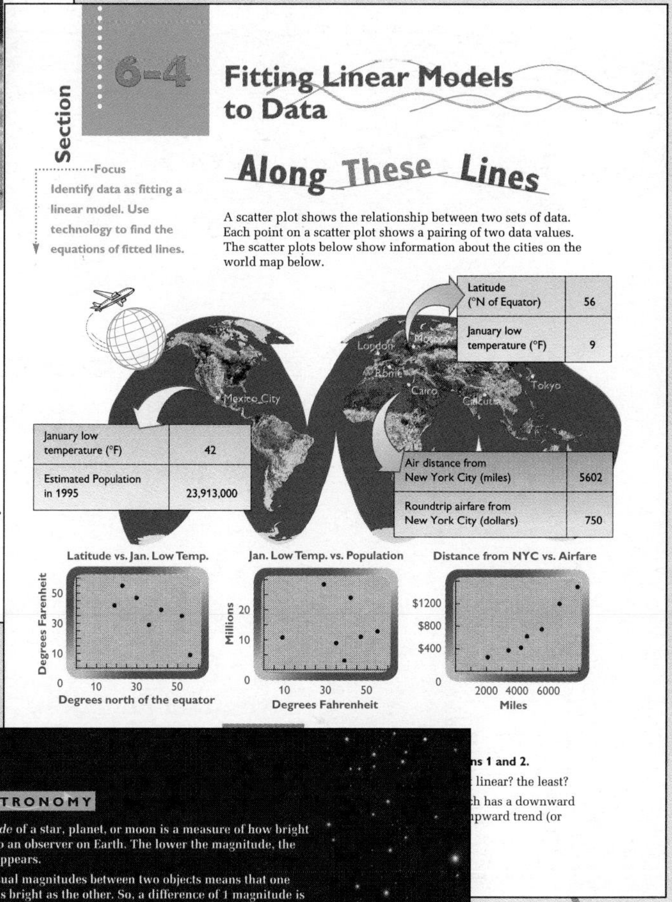

Section 6-4

Fitting Linear Models to Data

Focus
Identify data as fitting a linear model. Use technology to find the equations of fitted lines.

Along These Lines

A scatter plot shows the relationship between two sets of data. Each point on a scatter plot shows a pairing of two data values. The scatter plots below show information about the cities on the world map below.

...ns 1 and 2.
... linear? the least?
...h has a downward ...pward trend (or

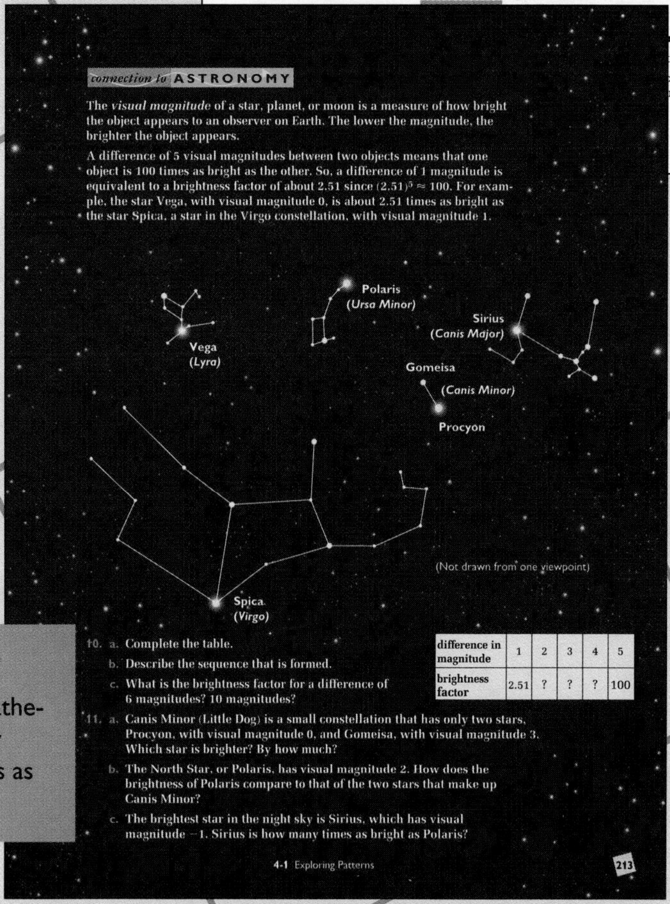

connection to ASTRONOMY

The *visual magnitude* of a star, planet, or moon is a measure of how bright the object appears to an observer on Earth. The lower the magnitude, the brighter the object appears.

A difference of 5 visual magnitudes between two objects means that one object is 100 times as bright as the other. So, a difference of 1 magnitude is equivalent to a brightness factor of about 2.51 since $(2.51)^5 \approx 100$. For example, the star Vega, with visual magnitude 0, is about 2.51 times as bright as the star Spica, a star in the Virgo constellation, with visual magnitude 1.

10. a. Complete the table.
    b. Describe the sequence that is formed.
    c. What is the brightness factor for a difference of 6 magnitudes? 10 magnitudes?

difference in magnitude	1	2	3	4	5
brightness factor	2.51	?	?	?	100

11. a. Canis Minor (Little Dog) is a small constellation that has only two stars, Procyon, with visual magnitude 0, and Gomeisa, with visual magnitude 3. Which star is brighter? By how much?
    b. The North Star, or Polaris, has visual magnitude 2. How does the brightness of Polaris compare to that of the two stars that make up Canis Minor?
    c. The brightest star in the night sky is Sirius, which has visual magnitude −1. Sirius is how many times as bright as Polaris?

4-1 Exploring Patterns 213

INTERDISCIPLINARY PROBLEMS

These theme exercises connect mathematics to other subject areas. They illustrate the power of mathematics as a problem solving tool.

I've heard that math educators need to make some changes. I'd like to try new teaching strategies, but I'm not sure how to get started.

How can I make change work for me?

Teaching Flexibility

The Integrated Mathematics program provides teaching flexibility that enables you to incorporate the changes that are right for you and your students — at a pace that is comfortable for you. This program makes it easy for you to try new approaches and accommodate different learning styles.

LEARNING STYLES

Integrated Mathematics addresses different learning styles by presenting concepts visually, verbally, and kinesthetically.

Section **2-5** Polynomial Functions

More POWER To You

Focus

Recognize, evaluate, and graph direct variation functions and polynomial functions.

EXPLORATION

How is the weight of a ball related to its diameter?

- **Materials: modeling clay, thin probe (for example, a straightened paper clip), ruler, spring or balance scale, graphics calculator**
- **Work in a group of four students.**

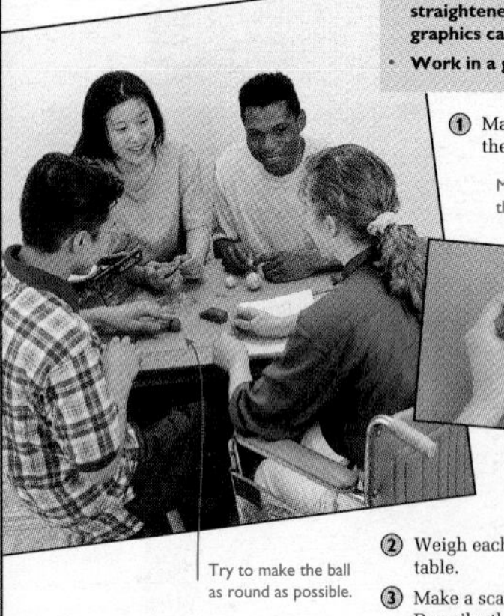

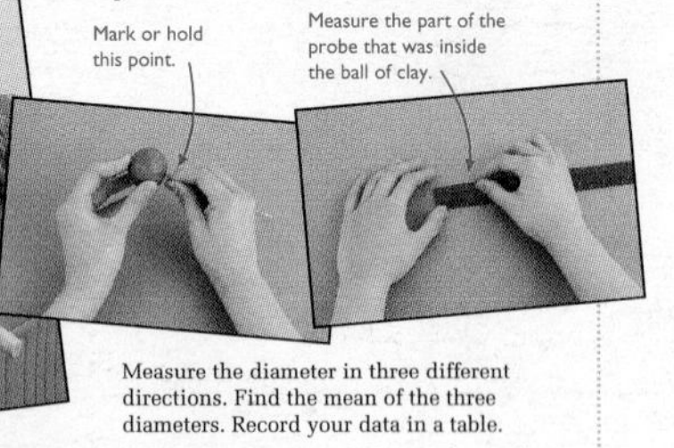
Mark or hold this point.
Measure the part of the probe that was inside the ball of clay.

Try to make the ball as round as possible.

1. Make eight balls of clay of different sizes. Use the probe to measure the diameter of each ball.

   Measure the diameter in three different directions. Find the mean of the three diameters. Record your data in a table.
2. Weigh each ball on the scale. Record the weights in the table.
3. Make a scatter plot of the ordered pairs (diameter, weight). Describe the shape of the graph.
4. Use the statistical functions of a graphics calculator to find an equation that models the data in the scatter plot. Use the calculator's "power regression" model. Write down the equation.
5. Suppose you could make perfect measurements in steps 1 and 2. The equation that you found in step 4 would have the form $f(x) = kx^3$. How close to 3 is the exponent in your equation? What is the value of k in your equation?

2-5 Polynomial Functions 101

Zeros of a Function

Sometimes polynomials are written in factored form. For example, you can write the volume of the prism below as the sum of the volumes of the four smaller prisms or as the product of the length, width, and height of the large prism.

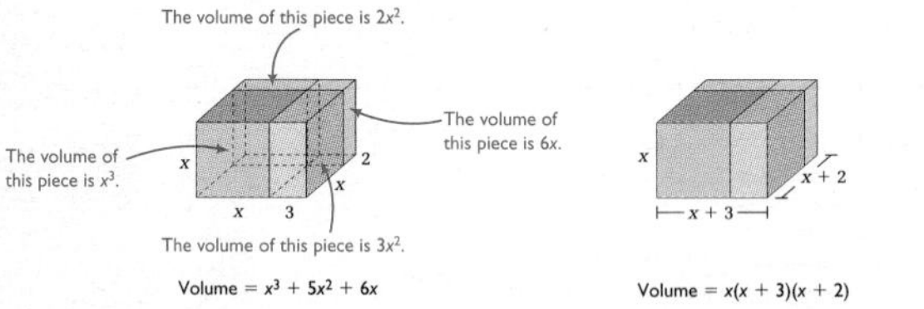
The volume of this piece is $2x^2$.
The volume of this piece is $6x$.
The volume of this piece is x^3.
The volume of this piece is $3x^2$.
x
2
x
x
3
x
x + 2
x + 3
Volume = $x^3 + 5x^2 + 6x$
Volume = $x(x + 3)(x + 2)$

The polynomial $V(x) = x(x + 3)(x + 2)$ is in factored form. The factored form of a polynomial function tells you the *zeros of the function*. A **zero of a function** is a value of x that makes $f(x)$ equal to zero. You can use the zeros to graph the function.

Sample 3

Graph the function $V(x) = x(x + 3)(x + 2)$.

Sample Response

Method 1 Graph the function by hand.

Step 1 Find the zeros of the function.

$V(x) = 0$ when $x = 0$, $x = -3$, or $x = -2$.
The graph crosses the x-axis at (0, 0), (−3, 0), and (−2, 0).

Step 2 Substitute values of x that are between and beyond the zeros to see the shape of the graph.

x	y
−4	−8
−2.5	0.625
−1	−2
1	12

Step 3 Graph all the points that you found and connect them with a smooth curve.

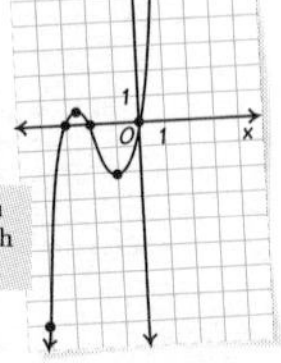
y
x
1
O
1

Unit 2 Exploring and Applying Functions

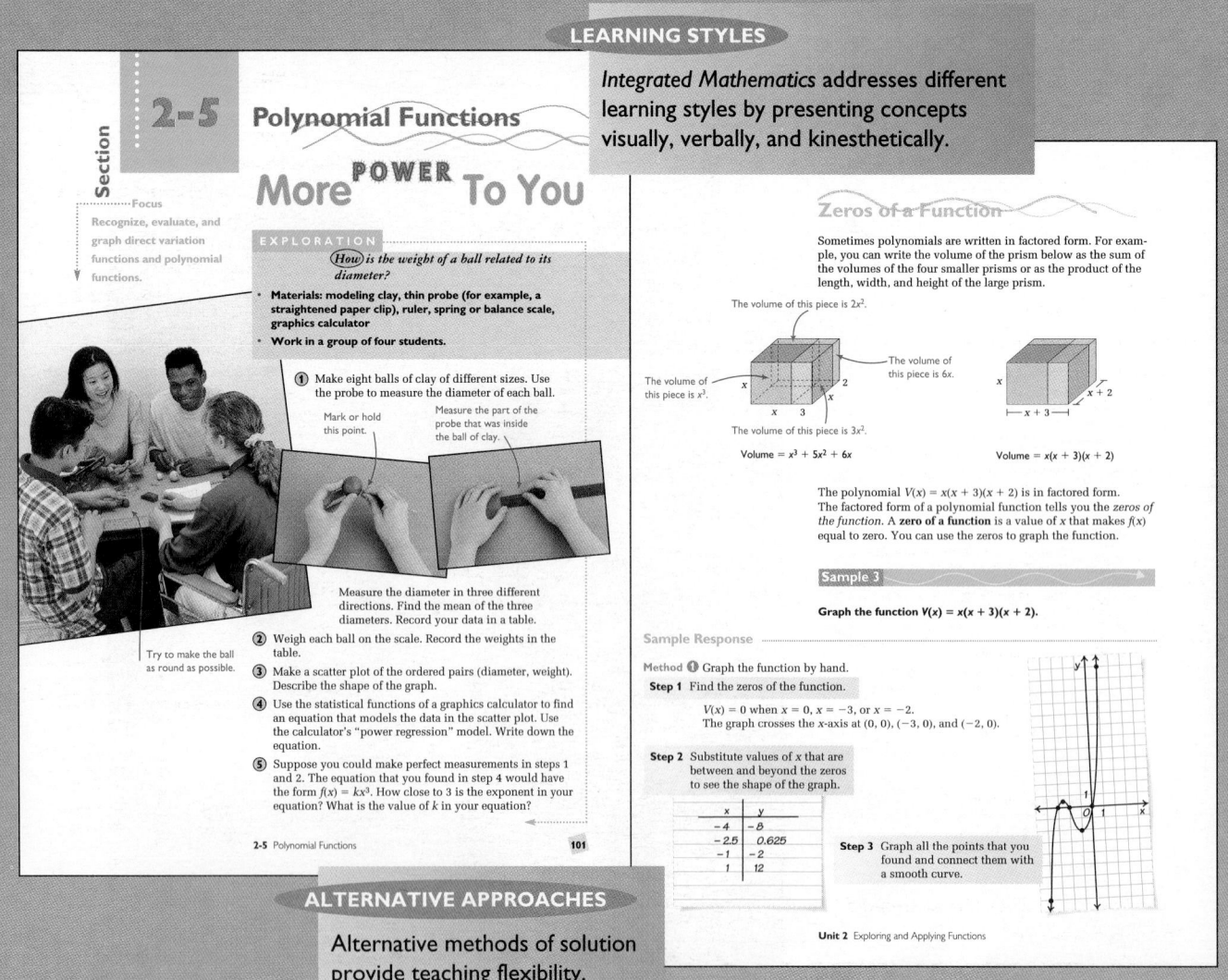

ALTERNATIVE APPROACHES

Alternative methods of solution provide teaching flexibility.

COMPLETE TEACHING SUPPORT

In the Teacher's Edition, each section has a planning list that helps you pace your lesson and use the support materials.

Method ❷ Use a graphics calculator or graphing software.
Enter the polynomial in factored form or as a sum:
$y = x(x + 3)(x + 2)$ or $y = x^3 + 5x^2 + 6x$.

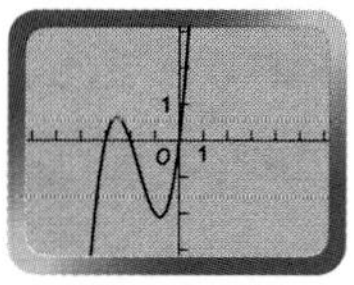

Talk it Over

9. The dimensions of a prism must be positive. Which points of the graph in Sample 3 represent this?
10. Describe how to use the graph in Sample 3 to find x when the volume of the prism is 40.

X Lab

POLYNOMIAL FUNCTIONS

A polynomial function has the form

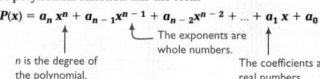

$P(x) = a_n x^n + a_{n-1}x^{n-1} + a_{n-2}x^{n-2} + \dots + a_1 x + a_0$

- n is the degree of the polynomial.
- The exponents are whole numbers.
- The coefficients are real numbers.

Example

$P(x) = 3x^5 + x^3 - 4x^2 + 1$

The factored form of a polynomial gives the zeros of the function. You can use the zeros to sketch a graph of the function.

Look Back

Use examples to explain why constant functions, linear functions, quadratic functions, and direct variation functions are types of polynomial functions.

Now you are ready for: Exs. 22–36 on pp. 107–109

2-5 Exercises and Problems

1. **Reading** What is the value of k for each of the direct variation functions in the graphs on page 102?

Tell whether each direct variation function is *one-to-one*.

2. $f(x) = 0.2x^3$
3. $g(x) = 1.2x^4$
4. $r(x) = 0.75x$

For Exercises 5–8, write an equation in the form $f(x) = kx^n$ for each direct variation function.

5. $k = 0.8$ and $n = 3$
6. $k = 1$ and $f(6) = 36$
7. $f(2) = 8$ and $n =$
8. $f(0.3) = 45$ and $f(x)$ varies directly with the square of x

9. Suppose a ball of clay with a diameter of 1 in. weighs 0.41 oz. Write a formula for the weight of the ball as a function of the diameter.

Student Resources T... p. 652 *Formulas*

10. **Driving** A driver's reaction distance, r, varies directly with the speed, s, of the car.
    a. Write an equation for r as a function of s. Use k as the constant.
    b. Suppose the reaction distance for a driver traveling 30 mi/h is 33 ft. Find the value of k. Rewrite your function using this value of k.
    c. For what speed is the reaction distance 50 ft?

11. **Trains** Taking curves at high speeds raises the average speed of a train, but passengers do not like how it feels. The force (F) you feel when a train goes around a curve depends on your mass (m), the radius of the curve (r), and the square of the train's velocity (v).

$$F = \frac{mv^2}{r}$$

(kg, m/s, newtons, meters)

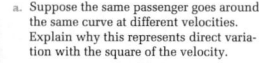

    a. Suppose the same passenger goes around the same curve at different velocities. Explain why this represents direct variation with the square of the velocity.
    b. Suppose the passenger has a mass of 60 kg and the radius of the turn is 9 m. Write a direct variation function that describes force in terms of velocity.
    c. Use your equation to find the force on the passenger when the velocity is 9 m/s.
    d. **Research** Find out what a *newton* is and after whom it is named.

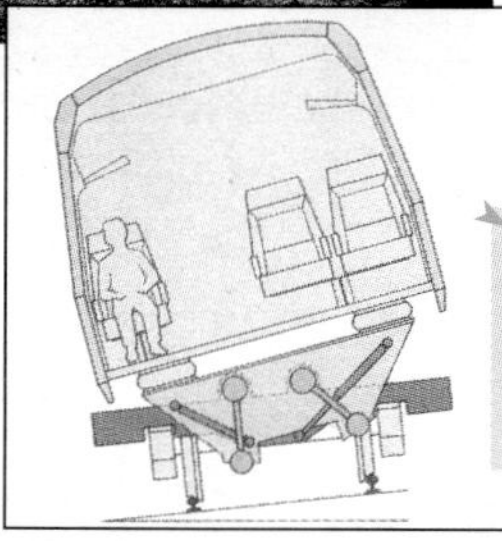

BY THE WAY...

New passenger cars for trains are designed to tilt as they round corners in order to cancel out 70% of the force experienced by riders.

Exercises continued

VARIED PRACTICE AND EXTENSIONS

The wide variety of exercises, problems, and activities in the textbook and in the support materials allows you to tailor the course to the needs of your students and your teaching preferences.

PLANNING

Objectives and Strands

See pages 66A and 66B.

Spiral Learning

See page 66B.

Materials List

- Modeling clay, thin probe
- Ruler, graph paper
- Spring or balance scale
- Graphing technology

Recommended Pacing

Section 2-5 is a two-day lesson.

Day 1

Pages 101–103: Exploration through Talk it Over 8, *Exercises 1–21*

Day 2

Pages 104–105: Zeros of a Function through Look Back, *Exercises 22–36*

Toolbox References

- **Toolbox Skill 29:** Formulas from Geometry

Extra Practice

See pages 610–612.

Warm-Up Exercises

- Warm-Up Transparency 2-5

Support Materials

- Practice Bank: Practice 14
- Activity Bank: Enrichment 13
- Study Guide: Section 2-5
- Problem Bank: Problem Set 4
- Explorations Lab Manual: Additional Exploration 3 Diagram Masters 2, 11
- Overhead Visuals: Folder 2
- Using TI-81 and TI-82 Calculators: Points of Symmetry for Cubic Polynomials
- Using Plotter Plus: Points of Symmetry for Cubic Polynomials
- Assessment Bank: Quiz 2-5,

Practice Bank

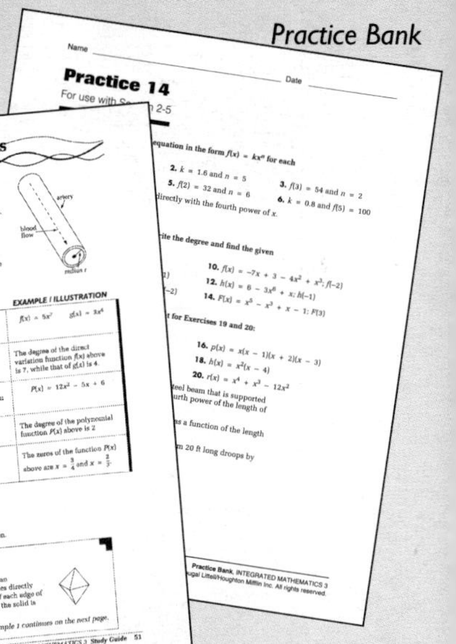

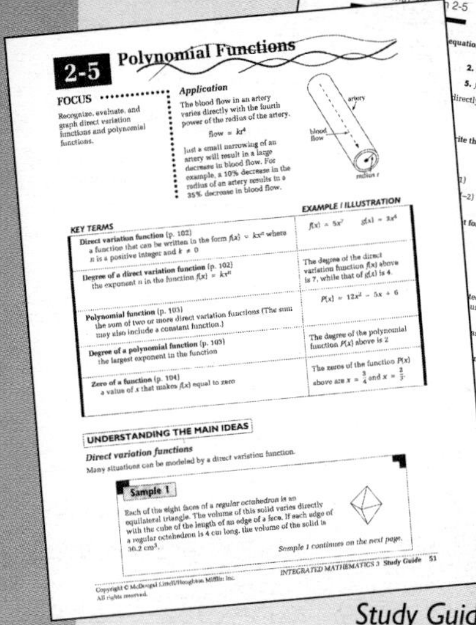

Study Guide

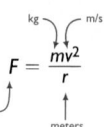

A Complete Program

Teaching Support

The *Integrated Mathematics* program supports the full range of teaching and learning needs.

TEACHER'S EDITION

- Student pages with answers and teaching commentary, planning information, assignment guide
- **Solution Key** with solutions to all problems also available

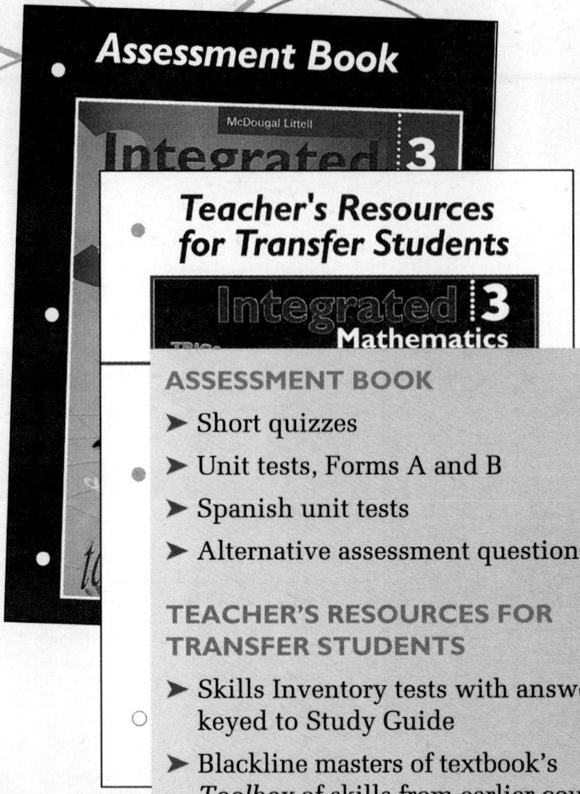

ASSESSMENT BOOK

- Short quizzes
- Unit tests, Forms A and B
- Spanish unit tests
- Alternative assessment questions

TEACHER'S RESOURCES FOR TRANSFER STUDENTS

- Skills Inventory tests with answers keyed to Study Guide
- Blackline masters of textbook's *Toolbox* of skills from earlier courses

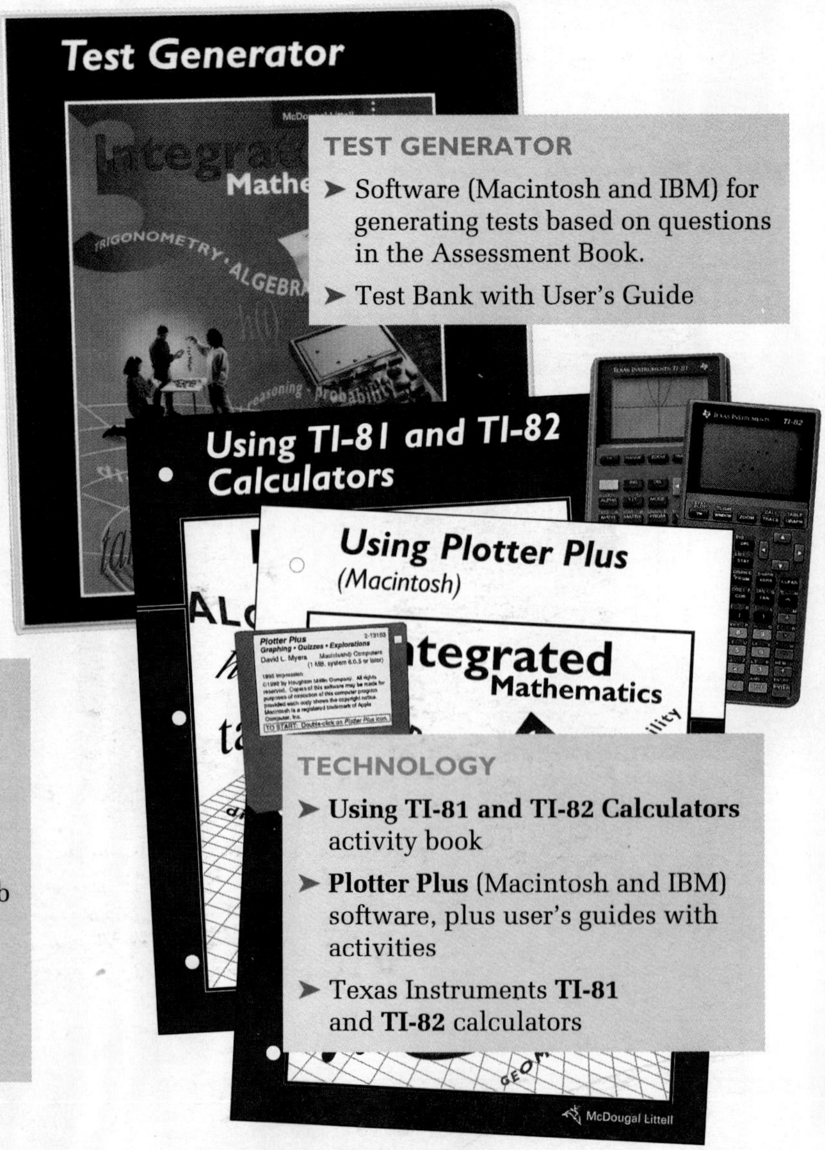

TEST GENERATOR

- Software (Macintosh and IBM) for generating tests based on questions in the Assessment Book.
- Test Bank with User's Guide

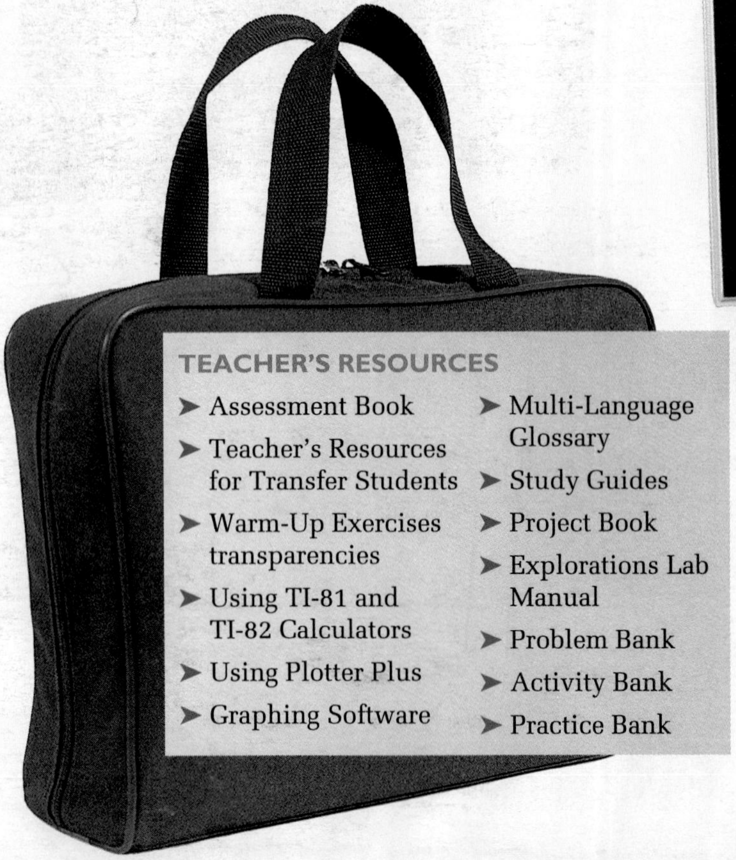

TEACHER'S RESOURCES

- Assessment Book
- Teacher's Resources for Transfer Students
- Warm-Up Exercises transparencies
- Using TI-81 and TI-82 Calculators
- Using Plotter Plus
- Graphing Software
- Multi-Language Glossary
- Study Guides
- Project Book
- Explorations Lab Manual
- Problem Bank
- Activity Bank
- Practice Bank

TECHNOLOGY

- **Using TI-81 and TI-82 Calculators** activity book
- **Plotter Plus** (Macintosh and IBM) software, plus user's guides with activities
- Texas Instruments **TI-81** and **TI-82** calculators

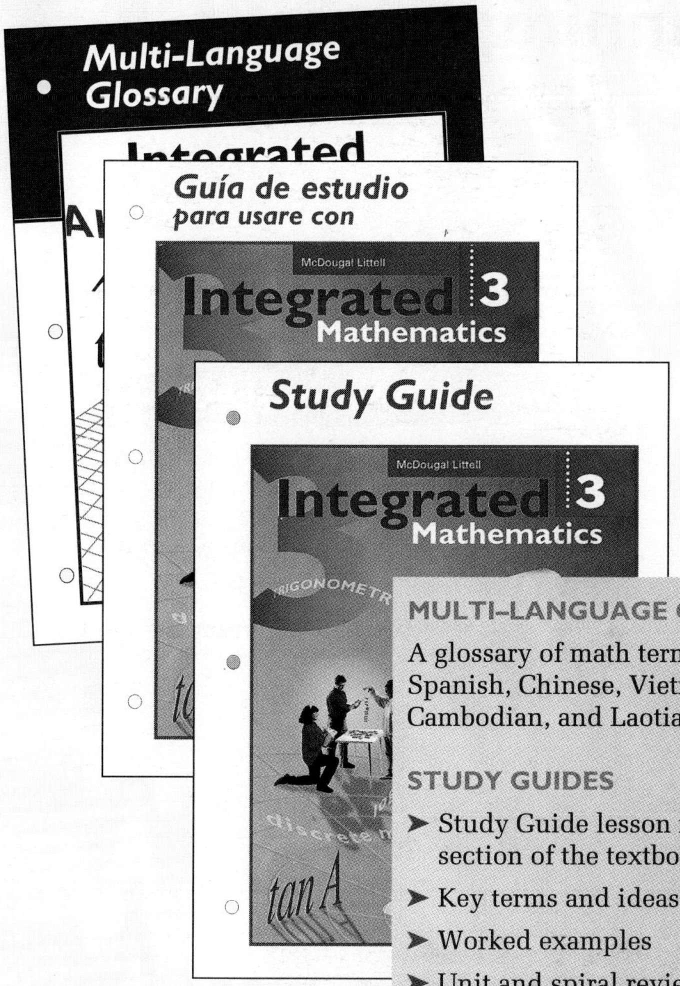

MULTI–LANGUAGE GLOSSARY

A glossary of math terms in English, Spanish, Chinese, Vietnamese, Cambodian, and Laotian

STUDY GUIDES

- Study Guide lesson for each section of the textbook
- Key terms and ideas
- Worked examples
- Unit and spiral reviews
- Complete Spanish edition

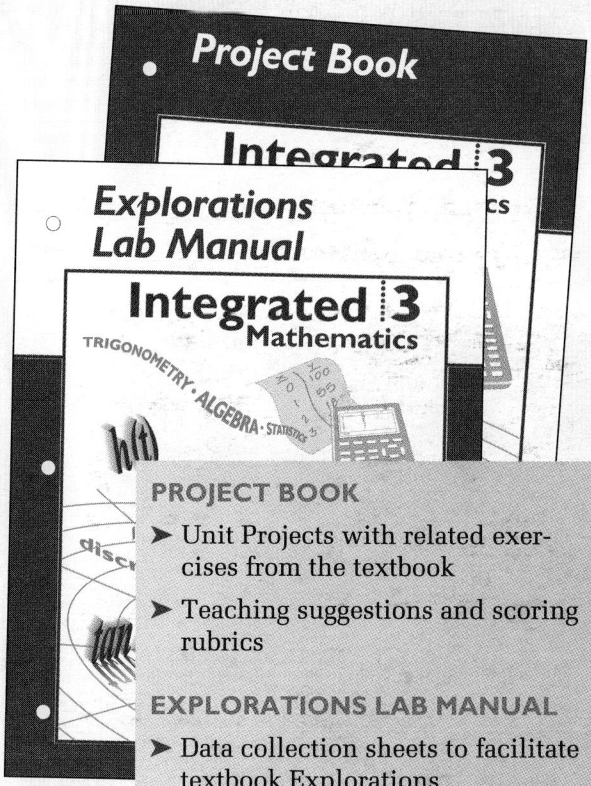

PROJECT BOOK

- Unit Projects with related exercises from the textbook
- Teaching suggestions and scoring rubrics

EXPLORATIONS LAB MANUAL

- Data collection sheets to facilitate textbook Explorations
- Additional Explorations to address different learning and teaching styles

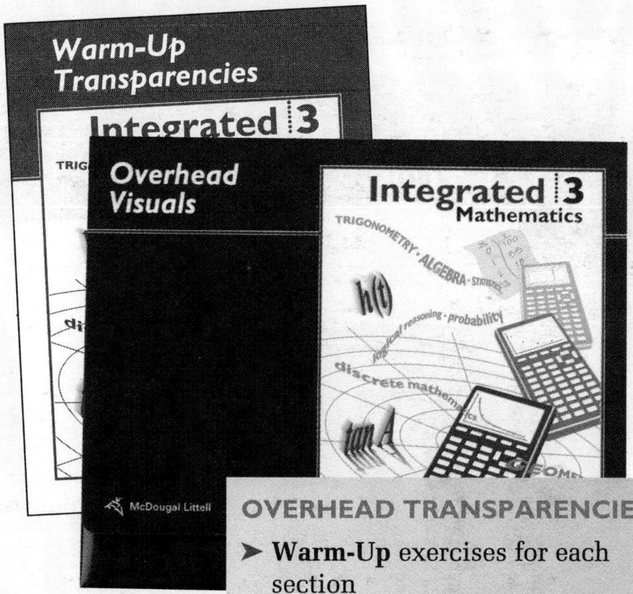

OVERHEAD TRANSPARENCIES

- **Warm-Up** exercises for each section
- Multi-color **Overhead Visuals** with teaching suggestions

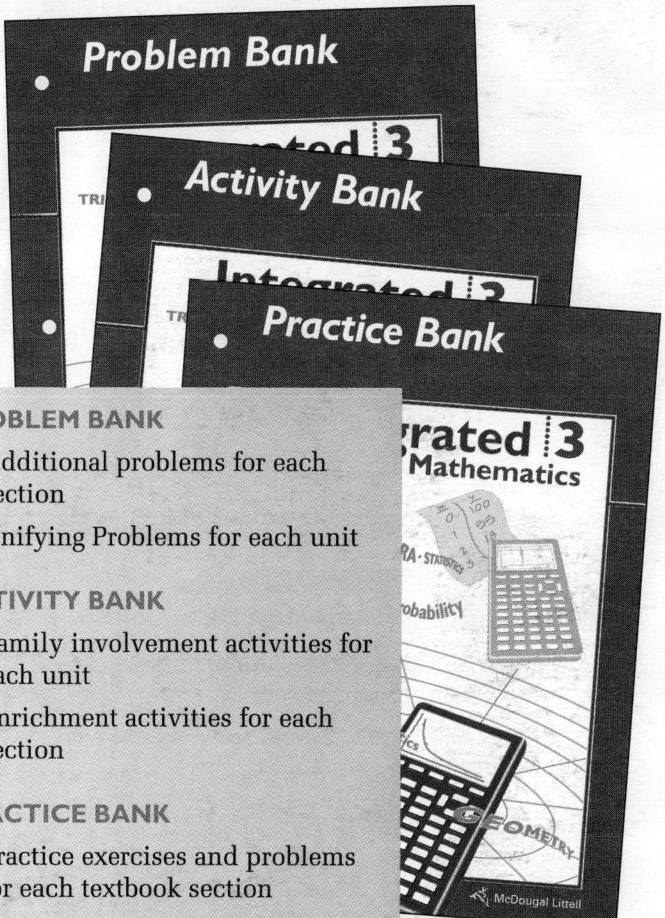

PROBLEM BANK

- Additional problems for each section
- Unifying Problems for each unit

ACTIVITY BANK

- Family involvement activities for each unit
- Enrichment activities for each section

PRACTICE BANK

- Practice exercises and problems for each textbook section
- Cumulative reviews for each unit

Using the TE

Special Planning Pages for Every Unit

Unit Overview

- **Overview** provides a summary of the mathematical topics, applications, use of technology, and problem solving strategies for each unit.
- **Unit Objectives** gives objectives and NCTM Standards for each section.

Topic Integration

- **Topic Spiraling** gives connections to past and future learning.
- **Integrating the Strands** shows the integration of mathematical strands throughout the unit.

Planning the Unit 4 — Sequences and Series

OVERVIEW

- Unit 4 covers the topics of sequences and series. Sequences are classified as arithmetic, geometric, or neither. Formulas are developed for sequences, both explicit and recursive. Subscript notation is used to represent the terms of a sequence.
- Series are introduced and students find sums for finite arithmetic and geometric series. Sums of infinite geometric series lead to the introduction of a limit of an infinite series. Sigma notation is introduced and used to write finite arithmetic series and finite geometric series.
- The theme of the **Unit Project** is to save for the future. Students learn how to make a small investment grow in order to pay for a large future expense. Students explore stocks, mutual funds, CDs, and other types of investments, using long-range planning for funding such needs as a college education.
- **Connections** to skydiving, investments, astronomy, fractals, genetics, knitting patterns, spider webs, and Greek myths are some of the topics integrated into the teaching materials and the exercises.
- **Graphics calculators** are used in Section 4-3 to find the limit of a sequence using recursion. **Computer software,** such as Plotter Plus, can be used in Sections 4-1 and 4-3 to work with spreadsheets.
- **Problem-solving strategies** used in Unit 4 include using patterns, graphs, and formulas.

Unit Objectives

Section	Objectives	NCTM Standards
4-1	• Describe and continue patterns.	
	• Graph sequences and find apparent limits.	1, 2, 3, 4, 12
4-2	• Use subscripts and formulas for sequences.	1, 2, 3, 4, 5, 12
4-3	• Write and use formulas for sequences in which each term is found by using the preceding term(s).	1, 2, 3, 4, 5, 12
4-4	• Identify sequences that have a common difference, a common ratio, or neither.	1, 2, 3, 4, 5, 12
	• Write explicit and recursive formulas.	
4-5	• Find the sum of a finite nongeometric series.	1, 2, 3, 4, 5, 12
4-6	• Use a formula to find the sum of a finite geometric series.	1, 2, 3, 4, 5, 12
4-7	• Find the sum of an infinite geometric series.	1, 2, 3, 4, 5, 12, 13

204A

Topic Spiraling

Section	Connections to Prior and Future Concepts
4-1	**Section 4-1** explores patterns that lead to a sequence of numbers. Students graph points on a coordinate plane and examine the graph for a limit of the sequence. Plotting points on a coordinate plane was first introduced in Section 4-1 of Book 1. The concept of limits is explored in Sections 4-7 and 5-4 of Book 3 and is an important concept in precalculus and calculus courses.
4-2	**Section 4-2** extends the concept of sequences begun in Section 4-1 of Book 3 to now include subscript notation and formulas to represent the sequences. Writing and understanding formulas for sequences is a skill used throughout the remainder of Unit 4 of Book 3.
4-3	**Section 4-3** explores sequences based on recursive formulas. Explicit and recursive formulas for the same sequence are compared. Recursive formulas are used again in Section 4-4 of Book 3.
4-4	**Section 4-4** explores sequences that are arithmetic, geometric, or neither. Explicit and recursive formulas are examined to find a common difference or ratio. Geometric and arithmetic means are introduced. Students use the explicit and recursive formulas first encountered in Sections 4-2 and 4-3 of Book 3.
4-5	**Section 4-5** extends the study of arithmetic sequences begun in Section 4-4 of Book 3. Students explore finding the sum of a finite arithmetic series. Sigma notation for summation is introduced, and is used in Section 4-6 of Book 3.
4-6	**Section 4-6** extends the study of geometric sequences begun in Section 4-4 of Book 3. Students explore finding the sum of a finite geometric series. The sigma notation introduced in Section 4-5 of Book 3 is used to represent the summation of a geometric series.
4-7	**Section 4-7** extends the work of Section 4-6 of Book 3. Students explore finding the sum of an infinite geometric series. The concept of a limit was informally introduced in Section 4-7 of Book 1, and the limit of a sequence was introduced in Section 4-1 of Book 3.

Integrating the Strands

Strands	Sections
Number	4-1, 4-2, 4-3, 4-4, 4-5, 4-6, 4-7
Algebra	4-1, 4-2, 4-3, 4-4, 4-5, 4-6, 4-7
Functions	4-1, 4-7
Geometry	4-1, 4-2, 4-4, 4-5, 4-6, 4-7
Trigonometry	4-2
Statistics and Probability	4-2
Discrete Mathematics	4-1, 4-2, 4-3, 4-4, 4-5, 4-6, 4-7
Logic and Language	4-1, 4-2, 4-3, 4-4, 4-5, 4-6, 4-7

204B

Planning the Unit 4

Sequences and Series

Section Planning Guide

- ➤ Essential exercises and problems are indicated in boldface.
- ➤ Ongoing work on the Unit Project is indicated in color.
- ➤ Exercises and problems that require student research, group work, manipulatives, or graphing technology are indicated in the column headed "Other."

Section	Materials	Pacing	Standard Assignment	Extended Assignment	Other
4-1	toothpicks, calculator, spreadsheet software	Day 1 Day 2	**2–8**, 12 **13–22**, 26–31, 32	**2–8**, 9–12 **13–22**, 24–31, 32	1 23, 32
4-2	spreadsheet software	Day 1	**1–9**, 10, 12, 15–21, 22	**1–9**, 10–21, 22	22
4-3	graphics calculator, spreadsheet software	Day 1	**1–13**, 19–22, 23, 24	**1–13**, 14–17, 19–22, 23, 24	18, 24
4-4	spreadsheet software	Day 1 Day 2	**1–22** **27–34**, 36–38, 39	**1–22**, 23, 24 25, 26, **27–34**, 36–38, 39	 35, 39
4-5	spreadsheet software	Day 1 Day 2	**1–6**, 7–11 **13–20**, 23–26, 27	**1–6**, 7–12 **13–20**, 21–26, 27	 27
4-6	spreadsheet software	Day 1	**1–15**, **20**, 21–26, 27	**1–15**, 16, 17, **20**, 21–26, 27	18, 19
4-7	large squares of paper, scissors, spreadsheet software	Day 1 Day 2	**1–8**, 9, 10 **11–23**, 25, 27–32, 33	**1–8**, 9, 10 **11–23**, 24–32, 33	 33
Review Test		**Day 1** **Day 2**	**Unit Review** **Unit Test**	**Unit Review** **Unit Test**	

Yearly Pacing	Unit 4 Total	Units 1–4 Total	Remaining	Total
	15 days (2 for Unit Project)	65 days	89 days	154 days

Support Materials

- ➤ See **Project Book** for notes on Unit 4 Project: Save for the Future.
- ➤ UPP and disk refer to **Using Plotter Plus** booklet and **Plotter Plus** disk.
- ➤ TI-81/82 refers to **Using TI-81 and TI-82 Calculators** booklet.
- ➤ Warm-up exercises for each section are available on **Warm-Up Transparencies.**

Section	Study Guide	Practice Bank	Problem Bank	Activity Bank	Explorations Lab Manual	Assessment Book	Visuals	Technology
4-1	4-1	Practice 29	Set 9	Enrich 26	Masters 1, 2, 12	Quiz 4-1		Statistics Spreadsheet (disk) Data Analyzer (disk)
4-2	4-2	Practice 30	Set 9	Enrich 27	Master 2	Quiz 4-2		
4-3	4-3	Practice 31	Set 9	Enrich 28		Quiz 4-3		Statistics Spreadsheet
4-4	4-4	Practice 32	Set 9	Enrich 29	Master 2	Quiz 4-4 Test 15		
4-5	4-5	Practice 33	Set 10	Enrich 30		Quiz 4-5		
4-6	4-6	Practice 34	Set 10	Enrich 31		Quiz 4-6		
4-7	4-7	Practice 35	Set 10	Enrich 32	Master 2	Quiz 4-7 Test 16		
Unit 4	Unit Review	Practice 36	Unifying Problem 4	Family Involve 4		Tests 17, 18		

204C

Teaching Information

- ➤ **Section Planning Guide** gives materials, pacing, and suggested assignments for each section.
- ➤ **Support Materials** lists all support materials for each section.

UNIT TESTS

Form A

Spanish versions of these tests are on pages 127–130 of the **Assessment Book.**

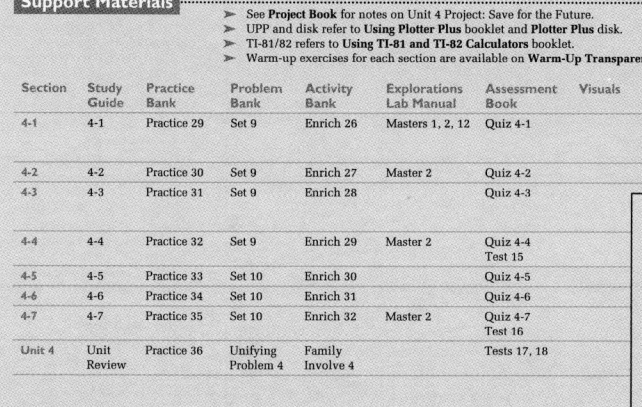

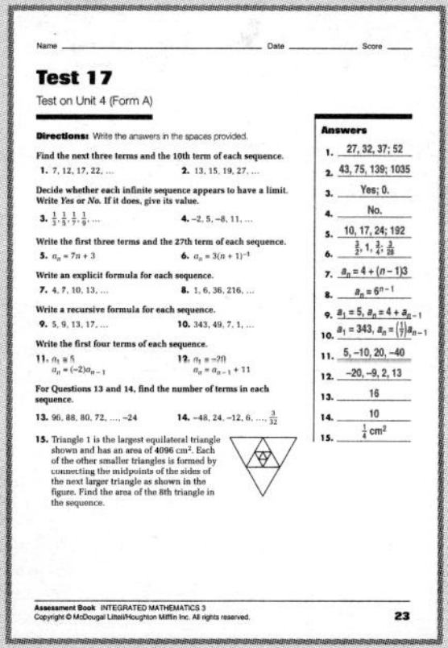

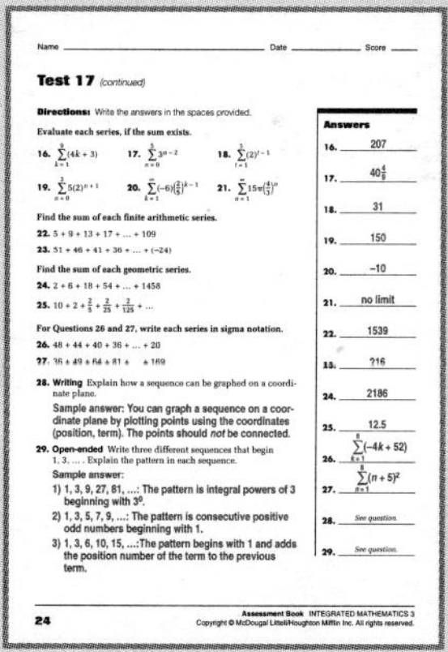

Form B

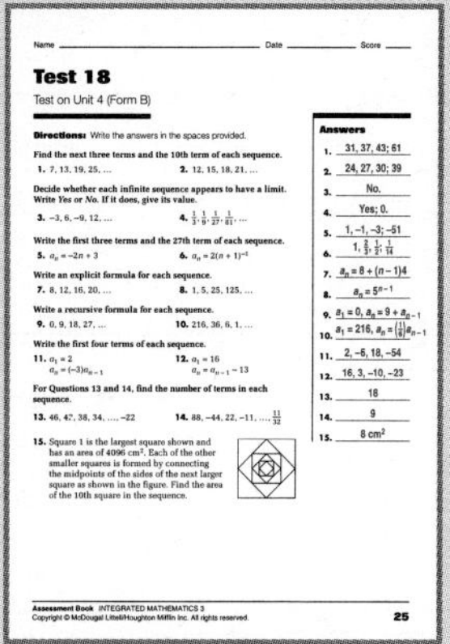

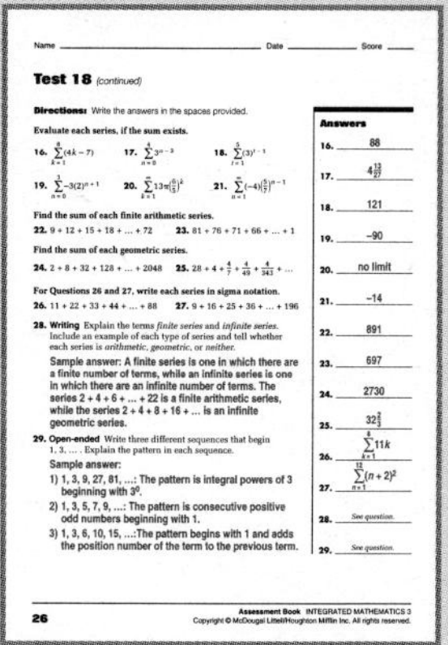

OUTSIDE RESOURCES

Books/Periodicals

Dence, Joseph B. and Thomas P. Dence. "A Rapidly Converging Recursive Approach to Pi." *Mathematics Teacher* (February 1993): pp. 121–124.

Schielack, Vincent P., Jr. "Tournaments and Geometric Sequences." *Mathematics Teacher* (February 1993): pp. 127–129.

HiMAP Module 2, *Recurrence Relations*, Section 3: "Counting Backwards: Carbon Dating": pp. 8–12.

Peitgen, Heinz-Otto, Hartmut Jurgens, and Dietmar Saupe. *Fractals for the Classroom* (Part one: Introduction to Fractals and Chaos; Part two: Complex Systems and Mandelbrot Set). NCTM, co-published with Springer-Verlag.

Activities/Manipulatives

Masalski, William J. "Topic: Compound Interest." *How to Use the Spreadsheet as a Tool in the Secondary Mathematics Classroom.* NCTM, 1990: pp. 16–19.

Jurgens, Hartmut, Evan Maletsky, Heinz-Otto Peitgen, Terry Perciante, Dietmar Saupe, and Lee Yunker. *Fractals for the Classroom: Strategic Activities, Vols. 1 & 2.* NCTM and Springer-Verlag.

Software

Exploring Chaos. Apple II series, Ver. 1.0. MECC, Minneapolis, MN.

Appleworks. Claris Corporation, 440 Clyde Avenue, Mountain View, CA.

Videos

Southern Illinois University at Carbondale. *World Population Review.* 1990.

204D

Teacher's Resources

- ➤ **Unit Tests** shows reduced facsimiles of Unit Tests, Forms A and B.
- ➤ **Outside Resources** lists books, periodicals, manipulatives and activities, software, and videos.

Facsimiles of the *Practice* masters appear in the side columns next to the section that they accompany.

A Teaching Plan for Every Section

Planning

A column referencing

- **Objectives and Strands**
- **Spiral Learning**
- **Materials List**
- **Recommended Pacing**
- **Toolbox References**
- **Extra Practice**
- **Warm-Up Exercises**
- **Support Materials**

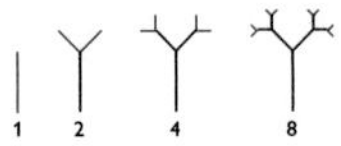

1 2 4 8

Look at the sequence of numbers of *new* branches that grow at each stage.

After the first term, each term is twice the term before it. You can use subscripts to express this relationship.

$a_n = 2a_{n-1}$ ← The term *before* the nth term is the $(n-1)$st term.
the nth term ↗

You can use the equation above to write a *recursive formula* for the sequence. A **recursive formula** for a sequence tells you how to find the nth term from the term(s) before it. A recursive formula has two parts:

$a_1 = 1$ ← The value(s) of the first term(s) are given.

$a_n = 2a_{n-1}$ ← A *recursion equation* shows how to find each term from the term(s) before it.

Sample 1

Write a recursive formula for the sequence 1, 2, 6, 24, ...

Sample Response

Write several terms of the sequence using subscripts. Then look for the relationship between each term and the term before it.

$a_1 = 1$
$a_2 = 2 = 2 \cdot 1$ —— Write a_2 in terms of a_1. → $a_2 = 2 \cdot a_1$
$a_3 = 6 = 3 \cdot 2$ —— Write a_3 in terms of a_2. → $a_3 = 3 \cdot a_2$
$a_4 = 24 = 4 \cdot 6$ —— Write a_4 in terms of a_3. → $a_4 = 4 \cdot a_3$
⋮

Write a recursion equation. → $a_n = na_{n-1}$ These values are always equal.

Use the value of the first term and the recursion equation to write a recursive formula for the sequence.

$a_1 = 1$
$a_n = na_{n-1}$

Talk it Over

4. a. Write an explicit formula for the sequence 1, 2, 6, 24, . . . in Sample 1.
   b. Which formula was easier to find? Do you think this is always the case? Why or why not?

4-3 Using Recursive Formulas 223

Answers to Talk it Over

4. a. $a_n = 1 \cdot 2 \cdot 3 \cdot \ldots \cdot n = n!$
   b. Answers may vary. An example is given. I think the recursive formula was easier to find in this case. I don't think this will always be the case. In general, I think no one method will always be easier.

TEACHING

Multicultural Note

Fractal images are very common in Islamic art. Islamic art seeks to represent the spirituality of nature and beings, rather than their physical characteristics. Traditional Islamic artists do not create realistic representations of people, animals, or objects from nature; instead, they represent these things in abstract fashion through complex patterns. Art in the Islamic world developed in this way in part because Islam does not allow images of people in any place of worship. This law, created many hundreds of years ago, is intended to allow people to worship without distractions.

Visual Thinking

Ask students to find examples of fractals in art, nature, computer graphics, and other areas. Ask them to highlight the fractals and to share examples of their findings with the class. This activity involves the visual skills of *recognition* and *communication.*

Additional Sample

S1 Write a recursive formula for the sequence 1, 4, 9, 16, 25, ...
Write several terms of the sequence using subscripts. Then look for the relationship between each term and the term before it.
$a_1 = 1$
$a_2 = 1 + 3 = a_1 + 3$
$a_3 = 4 + 5 = a_2 + 5$
$a_4 = 9 + 7 = a_3 + 7$
$a_5 = 16 + 9 = a_4 + 9$
Starting with the second term, the nth term, a_n, is a_{n-1} increased by the nth odd number, $2n - 1$. This information gives you a recursive formula.
$a_1 = 1$
$a_n = a_{n-1} + (2n - 1)$

223

PLANNING

Objectives and Strands
See pages 204A and 204B.

Spiral Learning
See page 204B.

Materials List
- Graphics calculator
- Spreadsheet software

Recommended Pacing
Section 4-3 is a one-day lesson.

Extra Practice
See pages 614–615.

Warm-Up Exercises
Warm-Up Transparency 4-3

Support Materials
- Practice Bank: Practice 31
- Activity Bank: Enrichment 28
- Study Guide: Section 4-3
- Problem Bank: Problem Set 9
- Using Mac Plotter Plus Disk: Data Analyzer
- Using IBM Plotter Plus Disk: Statistics Spreadsheet
- Assessment Book: Quiz 4-3, Alternative Assessment 3

Section 4-3 Using Recursiv

Focus
Write and use formulas for sequences in which each term is found by using the preceding term(s).

BRANCH

Many irregular shapes found in the
modeled by geometric shapes calle
show some of the early stages of th

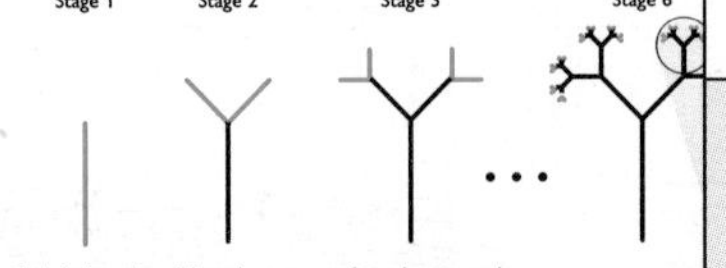

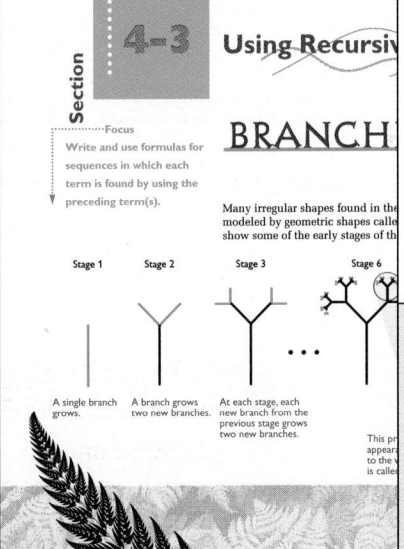

Stage 1 Stage 2 Stage 3 Stage 6

A single branch grows. A branch grows two new branches. At each stage, each new branch from the previous stage grows two new branches.

This pr appear to the v is calle

BY THE WAY...
Computer fractal programs can generate images from nature that are almost identical to the real thing. These images are called *fractal forgeries.*

Talk it Over

1. How are a fractal tree and a real tree alike? How are they different?
2. Write a sequence whose first four terms are the number of branches on a fractal tree at stages 1–4.
3. For the sequence in question 2, describe how each term after the first is related to the term before it.

222 Unit 4 Sequences and Series

Answers to Talk it Over

1. Summaries may vary. A fractal tree begins with a single branch; a tree has a trunk, which can be thought of as the first single branch. For both, new branches grow from existing branches. When you magnify any part of the fractal tree, what you see looks like the whole tree. This is not true of a real tree.
2. 1, 3, 7, ..., 63
3. Each term is a power of 2 added to the previous term. The power is the position number of the previous term.

222

Answers

Answers to Explorations, Talk it Over questions, Look Back questions, and Exercises and Problems are conveniently located at the bottom of each page.

In addition to the section side-column notes, a **Quick Quiz** is provided at each Checkpoint in the student book, as well as at the end of each unit.

Teaching

Notes on
- Explorations
- Talk it Over questions
- Additional Samples
- Teaching Tips
- Error Analysis
- Problem Solving
- Using Technology
- Using Manipulatives
- Communication
- Cooperative Learning
- Reasoning
- Mathematical Procedures
- Students Acquiring English
- Multicultural Information
- Visual Thinking
- Look Back questions

Method ❷ Use a spreadsheet.

Enter 650 in cell A1.

Key in =0.26*A1+650 in cell A2 and use the FILL DOWN command.

A12 = 0.26*A11 + 650 — This is the recursion equation for the 12th term.

Aspirin Levels

	A	B	C
1	650		
2	819		
3	862.94		
4	874.3644		
5	877.334744		
6	878.107033		
7	878.307829		
8	878.360035		
9	878.373609		
10	878.377138		
11	878.378056		
12	878.378295		

The sequence appears to approach a limit of about 878.

The amount of aspirin in his body will level off at about 878 mg.

Talk it Over

5. After which dose does the amount of aspirin first reach a level of 878 mg in Robert Nuttall's body? After how many hours does this happen?
6. In Sample 3, what will you key in differently if Robert Nuttall's first dose is 1000 mg and each dose after that is 650 mg?

Look Back

How are explicit and recursive formulas alike? How are they different?

4-3 Exercises and Problems

1. **Reading** What are the two parts of a recursive formula?

Write a recursive formula for each sequence.

2. 9, 14, 19, 24, . . .
3. −3, 9, −27, 81, . . .
4. $9, 3, 1, \frac{1}{3}, \ldots$
5. 8.32, 8.44, 8.56, 8.68, . . .
6. $2, \frac{1}{4}, 16, \frac{1}{256}, \ldots$
7. $\frac{1}{2}, \frac{2}{3}, \frac{4}{4}, \frac{8}{5}, \ldots$

Write the first six terms of each sequence.

8. $a_1 = 3$; $a_n = a_{n-1} - 5$
9. $a_1 = 0.5$; $a_n = (0.2)a_{n-1}$
10. $a_1 = 10$; $a_n = (a_{n-1})^2$
11. $a_1 = 5$; $a_n = \frac{a_{n-1}}{n}$
12. $a_1 = 5$; $a_n = (i)(a_{n-1})$
13. $a_1 = 1$; $a_2 = 2$; $a_n = (a_{n-1})(a_{n-2})$

Look Back

You may wish to have students write their responses to this question in their journals.

APPLYING

Suggested Assignment

Standard 1–13, 19–24

Extended 1–17, 19–24

Integrating the Strands

Number Exs. 1–19, 21–23

Algebra Exs. 1–23

Discrete Mathematics Exs. 1–19, 21–23

Logic and Language Exs. 1, 17, 24

Error Analysis

A higher than normal error rate in Exs. 8–13 may indicate that students are having difficulty with the notation. If this is the case, review the meaning of subscripts.

Using Technology

One way to enter sequences that are described by an explicit formula on the TI-82 is to use the List feature.

Consider the sequence defined by $b_n = n^2$. On the home screen, press [2nd][LIST]<seq(>[ALPHA][A][x²][,][A][,][1][,][10][,][1][)] to enter the following:

variable; position of 1st term; number of terms; increment

seq (A², A, 1, 10, 1)

When you press [ENTER], the calculator will display the sequence between braces (without commas). To store and view the terms in table form, press [STO▶] [2nd][L1] [ENTER]. To see L1 (list 1), press [STAT] <EDIT>. To display a particular term on the home screen, type L1 followed by the position number (in parentheses).

Answers to Exercises and Problems

1. the value of the first term and a recursion equation, a formula showing how to find each term from the term(s) before it
2. $a_1 = 9$; $a_n = a_{n-1} + 5$
3. $a_1 = -3$; $a_n = -3a_{n-1}$
4. $a_1 = 9$; $a_n = \frac{1}{3}(a_{n-1})$
5. $a_1 = 8.32$; $a_n = a_{n-1} + 0.12$
6. $a_1 = 2$; $a_n = (a_{n-1})^{-2}$
7. $a_1 = \frac{1}{2}$; $a_n = (2a_{n-1})\left(\frac{n}{n+1}\right)$
8. 3, −2, −7, −12, −17, −22
9. 0.5, 0.1, 0.02, 0.004, 0.0008, 0.00016
10. 10, 100, 10,000, 10^8, 10^{16}, 10^{32}
11. $5, \frac{5}{2}, \frac{5}{6}, \frac{5}{24}, \frac{1}{24}, \frac{1}{144}$
12. 5, 5*i*, −5, −5*i*, 5, 5*i*
13. 1, 2, 2, 4, 8, 32

Reasoning

You may wish to discuss how recursive formulas can be used to take a new look at some previous exercises. For example, in Section 4-1, Ex. 8 lends itself to using a recursive approach. When the *n*th cut is made, *n* new pieces of the circle are created. Suggest that students illustrate this with a diagram.

Teaching Tip

Students may find it especially interesting to discuss Ex. 14. The Fibonnaci sequence has a fascinating array of interesting properties. For example, write the first 10 terms of the Fibonnaci sequence on the board. Ask a student to come to the board and find the differences between successive terms.

1 1 2 3 5 8 ...

difference: 0 1 1 2 3 ...

The sequence of differences starts with 0 and then becomes the Fibonacci sequence itself!

There is an explicit formula for the Fibonacci sequence:

$$a_n = \frac{1}{\sqrt{5}}\left[\left(\frac{1+\sqrt{5}}{2}\right)^n - \left(\frac{1-\sqrt{5}}{2}\right)^n\right].$$

Have students discuss whether the explicit or recursive formula is easier to use.

Interdisciplinary Problems

The mathematics of sequences can be applied to analyze problem situations in many other disciplines. Exs. 14–16 illustrate this by examining situations in biology, medicine, and state lotteries.

connection to **BIOLOGY**

14. Male honeybees hatch from unfertilized eggs and therefore have a female parent, but no male parent. Look at the family tree of a male bee.
   a. Starting with the male honeybee, write the sequence of the number of bees in each preceding generation. This sequence is known as the *Fibonacci sequence* and occurs in many aspects of nature.
   b. Write a recursive formula for the Fibonacci sequence. (*Hint:* The recursion equation involves *two* previous terms.)

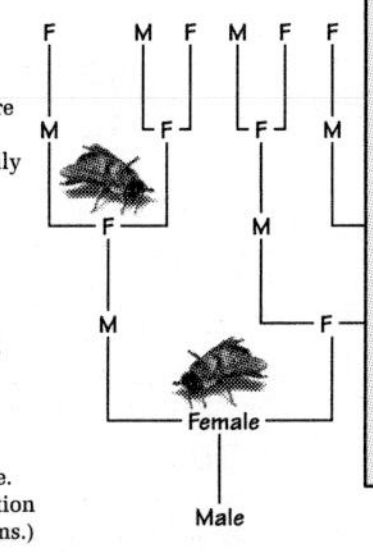

1 male honeybee

15. **Medicine** Naomi Pedraza has bronchitis. She takes 250 mg of an antibiotic every four hours. Only 33% of the antibiotic is left in her body by the time she takes a new dose. What will the amount of antibiotic in her body level off to?
16. **State Lotteries** In some states a $1,000,000 lottery winner receives $50,000 per year for 20 years. The winner receives $50,000 right away. The state then puts just enough money into an account to pay the prize winner $50,000 per year for the next 19 years, leaving a zero balance in the account after the 19th payment from the account.

   Suppose an account has a fixed annual interest rate of 6.33%. Let a_0 be the amount of money the state puts into the account and let a_n be the amount of money left in the account after the *n*th payment.
   a. Write a recursion equation for a_n.
   b. What is the value of a_{19}?
   c. Find a_{19} when a_0 = $950,000, when a_0 = $700,000, and when a_0 = $500,000. Do any of these values for a_0 come close to giving you the correct value for a_{19}?
   d. Solve your recursive formula in part (a) for a_{n-1}. Use this new equation as a recursive formula for the sequence that has a_{19} as the first term and a_0 as the last term. Will this method give you the correct value for a_0? Explain.
   e. What is the amount the state should put into the account?
   f. **Writing** Why do you think some states pay a lottery winner over many years rather than all at once?
17. **Writing** Explain how finding the 100th term of a sequence defined explicitly is different from finding the 100th term of the same sequence defined recursively.

Answers to Exercises and Problems

14. a. 1, 1, 2, 3, 5, 8, ...
   b. $a_1 = 1$; $a_2 = 1$; $a_n = a_{n-1} + a_{n-2}$
15. about 373 mg
16. a. $a_n = 1.0633a_{n-1} - 50{,}000$
   b. 0
   c. $1,303,791.65; $501,374.00; For $500,000, at a_{17} the amount becomes negative, so the formula no longer applies, since interest will no longer be paid; the value of a_0 that comes closest to giving the correct value for a_{19} is $500,000.
   d. $a_{n-1} = \frac{a_n + 50{,}000}{1.0633}$; Yes; for this sequence, the last term is the term ($543,792.69), which when used in the original sequence produces a last term of 0.
   e. $543,792.69
   f. Spreading the payments out over many years significantly decreases the cost to the state. In the case described in this exercise, the state pays out only about half as much by spreading out the payments rather than making a single payment of $1 million.
17. To find the 100th term of a sequence defined explicitly, you need only substitute the value 100 in the formula. To find the same term in a sequence defined recursively, you need to know the 99th term.

Applying

Notes on
- Suggested Assignments
- Integrating the Strands
- Problem Solving
- Using Technology
- Using Manipulatives
- Cooperative Learning
- Reasoning
- Multicultural Information
- Unit Projects
- Careers
- Applications
- Research
- Visual Thinking
- Interdisciplinary Problems
- Assessment

Pacing and Making Assignments

Pacing Chart

A yearly Pacing Chart and daily assignments are provided for two levels of courses—a standard course and an extended course. Both levels provide for 154 days, including days for using the Unit Openers, completing the Unit Project, and review and testing. The Pacing Chart below shows the number of days allotted for each unit of both courses. Semester and trimester divisions are indicated by a red rule and blue rules, respectively.

Unit	1	2	3	4	5	6	7	8	9	10
Standard Course	17	17	16	15	17	12	14	15	16	15
Extended Course	17	17	16	15	17	12	14	15	16	15

trimester semester trimester

Standard Course

The standard course is intended for students who enter with typical mathematical and problem solving skills. The course covers all ten units. The daily assignments include all the essential exercises and problems plus a number of other exercises that focus on higher-order thinking skills.

Extended Course

The extended course is intended for students who enter with strong mathematical and problem solving skills and who are able to understand new concepts quickly. The course covers all ten units. The daily assignments include all the essential exercises plus many other exercises that focus on higher-order thinking skills. It is recommended that these students be assigned some of the exercises that are listed in the Other column of the Section Planning Guide.

Helping Transfer Students

You may have students who enter your *Integrated Mathematics 3* class at the beginning of the year, or during the year, without having studied all of the mathematical topics covered in *Integrated Mathematics 1* and *Integrated Mathematics 2.* The supplementary publication *Teacher's Resources for Transfer Students* for *Integrated Mathematics 3* was developed to make it easier for you to help your transfer students catch up. This publication contains diagnostic Skills Inventory tests, blackline masters of the textbook's *Toolbox* of skills assumed from previous courses, and blackline masters of selected sections from the *Study Guide* for *Integrated Mathematics 2.* The Skills Inventories help you identify the concepts and skills that transfer students need to learn, and the *Toolbox* and *Study Guide* masters offer students a way to learn them.

Section Planning Guide

The Section Planning Guide for each unit is located on the interleaved pages preceding the unit. A part of the Section Planning Guide for Unit 4 is shown here. A key describing the exercises and problems for the assignments is given in each Section Planning Guide.

Section Planning Guide

- Essential exercises and problems are indicated in boldface.
- Ongoing work on the Unit Project is indicated in color.
- Exercises and problems that require student research, group work, manipulatives, or graphing technology are indicated in the column headed "Other."

Section	Materials	Pacing	Standard Assignment	Extended Assignment	Other
4-1	toothpicks, calculator, spreadsheet software	Day 1	**2–8**, 12	**2–8**, 9–12	1
		Day 2	**13–22**, 26–31, 32	**13–22**, 24–31, 32	23, 32
4-2	spreadsheet software	Day 1	**1–9**, 10, 12, 15–21, 22	**1–9**, 10–21, 22	22
4-3	graphics calculator, spreadsheet software	Day 1	**1–13**, 19–22, 23, 24	**1–13**, 14–17, 19–22, 23, 24	18, 24
4-4	spreadsheet software	Day 1	**1–22**	**1–22**, 23, 24	
		Day 2	**27–34**, 36–38, 39	25, 26, **27–34**, 36–38, 39	35, 39
4-5	spreadsheet software	Day 1	**1–6**, 7–11	**1–6**, 7–12	
		Day 2	**13–20**, 23–26, 27	**13–20**, 21–26, 27	27

Essential Exercises and Problems

These exercises and problems, indicated in boldface, are essential to understanding and applying the mathematical concepts presented in each section. They are listed in both the Standard and Extended Assignments and should be completed by all students.

Working on the Unit Project Exercises

These exercises provide students with ongoing work on the Unit Project. They are listed in color in both the Standard and Extended Assignments. Some of these exercises may involve group work, graphing technology, or manipulatives and may require time in class for completion. Others may involve student research and may require extended time outside of class for completion.

"Other" Exercises

These exercises require group work, use of technology, use of manipulatives, or student research. They have been placed in the "Other" category to alert you to the fact that those involving group work, technology, or manipulatives may require time in class for completion and those involving student research may require extended time outside of class for completion.

Support Materials

The extensive support materials available for *Integrated Mathematics 3* can provide an additional source for assignments that fit the needs of particular classes and teaching preferences. In particular, the *Practice Bank, Problem Bank,* and *Study Guide* can be used for this purpose. A complete list of ancillary materials available for each section can be found in the Support Materials chart that follows each Section Planning Guide on the interleaved pages.

Teaching Strategies

Effective Learning and Teaching

by **Gerlena R. Clark**
Mathematics Consultant
Jefferson City, Missouri

"All students can learn mathematics."

The Challenge

American businesses want their future employees to be able to work with others, to solve problems, to read and understand the principles of mathematics, and to communicate ideas. The primary question for teachers today is: How do I help students build a foundation of skills and information while simultaneously encouraging them to use their creative and intellectual abilities to solve real-world problems?

Recent research on how students learn suggests some strategies for accomplishing the challenging goal of preparing students for their lives as adults in the 21st century. These strategies are based on the assumption that:

> All students can learn mathematics if mathematics is taught in the way that students learn.

How Do Students Learn Mathematics?

Learning in the traditional manner is not sufficient in a world that demands attitudes that are conducive to creativity, as well as specific knowledge and skills. In the past students were expected to acquire facts through drill, practice, and memorization. The teacher was the giver of knowledge through lecturing and demonstrating.

Now we know that:

- Students must be actively involved in the learning process.
- Students learn best through dialogue, discussion, and interaction with others.
- Students benefit from reviewing, critiquing, and revising another's work as well as their own.
- What students learn is connected to how they learn.
- Students learn by experiencing tasks that are as closely aligned to real life as possible.
- Students learn by making connections to what they already know about the task and the real world.

We may summarize these statements by defining:

- **knowledge** as the result of individuals constructing meaning for themselves, by creating rules and hypotheses to explain what they've experienced.
- **intelligence** as a function of experiences. The brain learns best through first-hand experiences.

How Does This Learning Take Place in the Classroom?

If we put what we know about learning into action, what will be happening in the classroom?

In the classroom, students should:

- work with objects to represent mathematical models
- work in cooperative groups, or in pairs, as the task dictates
- write results, or outline strategies
- discuss mathematical ideas
- ask and answer each other's questions

In the classroom, the teacher should:

- allow time for students to think through problem formation and solution
- maintain an atmosphere of freedom for students' expressions
- encourage mathematical arguments with questions such as, Do you agree or disagree? Why?
- not focus on the "correct" answer, but allow discussion on alternative answers and solution procedures
- avoid paraphrasing what students say; ask students to clarify their own thinking
- model expected behavior for working in a group and for solving problems
- ask questions that will allow students to go beyond one-dimensional responses
- encourage students to go to each other for assistance
- encourage students to revise their written responses
- allow students to self-assess as well as assist in the assessment of others

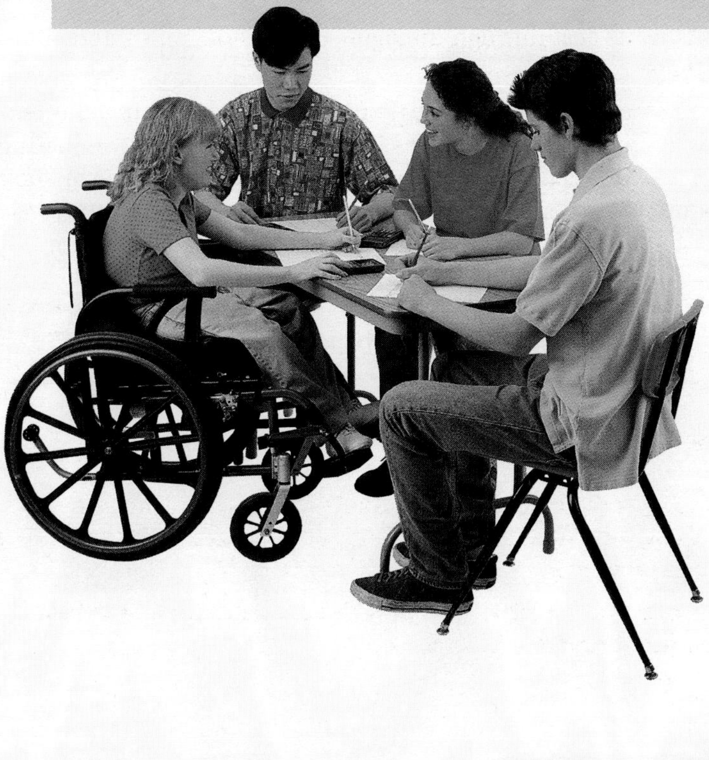

Meeting the Challenge

All of this implies a change in the way students learn and in the way teachers teach. Making this change may not be simple or easy at first, but it will become simpler and easier with time. And the result—students who know and use mathematics, who are mathematically empowered—will be of benefit to all.

In *Integrated Mathematics 3*
See student pages 32, 101, and 237 and side-column commentary on pages 17, 72, and 245.

Cooperative Learning

by ***Judith Collison***
Assistant Professor, Critical and Creative Thinking
University of Massachusetts/Boston

"Group work can decrease or eliminate math anxiety."

Benefits of Cooperative Learning

The NCTM *Curriculum and Evaluation Standards* stress the importance of developing skills of collaboration in mathematics teaching and learning. Developing skills of group participation is an important goal of all education, but it is especially useful in mathematics. Research has shown that group work in math classes has decreased or eliminated math anxiety; increased motivation, flexibility, confidence, self-esteem, curiosity, and perseverance; improved ability to solve problems and to communicate mathematically; and resulted in more positive attitudes towards mathematics.

When and how should cooperative strategies be used?

Teachers should use collaboration to create a sense of community and trust among students, as well as to create a deeper and more personal understanding of mathematics.

Cooperative forms are natural for problems that seem too big, too time-consuming, or too complex to be tackled by one person, and problems that require multiple perspectives, ability levels, or discussions.

Adequate preparations are key to the success of cooperative group work. The teacher needs to decide how to configure the groups, what type of group work will be used, how the problem will be divided among or within the groups, and how the work of the groups and of individual group members will be assessed.

Before students embark on their assignment, the teacher should verify that all groups and group members understand all instructions and expectations.

How should groups be formed?

The teacher selects the method by which group membership is determined. The composition of the group may be decided by the teacher or the students, according to some criteria, or randomly.

Types of group structure

Most typical cooperative structures include students working in *pairs,* in *small groups* made up of three or more members, or as a *whole class.* The teacher needs to choose the structure most useful for the activity or problem at hand. The following descriptions of the various types of cooperative structures are adapted from Neil Davidson's *Cooperative Learning in Mathematics: A Handbook for Teachers* (Addison-Wesley, 1990).

INTERVIEW (2-4 participants in each group) Most useful for getting students acquainted with each other in order to begin forming a sense of community. Members of the group ask each other questions dealing with either personal information or with applications of mathematics to their lives. They then share the information with the larger group or with the whole class.

Teacher's Choice
(by ability, social, psychological, or random grouping)

HOMOGENEOUS	HETEROGENEOUS	RANDOM
ability level talents/interests learning style social group psychological group	mixed levels complementary talents/interests combination of learning styles diverse ethnic representation	counting off according to height arbitrary numbering e.g., phone or social security numbers

Student's Choice

SELF-SELECTION
Students choose their own working partners.

MODIFIED SELF-SELECTION
Students list first, second, and third choices, and the teacher constructs groups based on these preferences.

Think-pair-share Useful for developing communication about concepts and procedures, and practice in problem solving. Students think about the problem alone, then discuss possible solutions with their partners and agree on the correct solution. They share their conclusions with the rest of the class. Having students work in pairs is probably the best cooperative format when using computers in problem solving.

Peer Practice and Drill There are several versions of this format. It provides opportunities to practice the mastery of skills or concepts.

a. **Partner drill** Students take turns asking their partners questions.

b. **Flashcards** Each student makes up flash cards with questions for a partner. The student presenting the cards should have the correct answer(s) for each question asked.

c. **Teams-Games-Tournaments** In every round, individual students from each team compete with each other. The team's score is determined by members of each team.

d. **Peer-pair-problem solving** (also called "pairs check") Each of the students in the pair has a unique role. One is the "solver" or "performer," the other is the "checker" or "coach." The "solver" works on solving the problem. The "checker" observes, gives hints, points out errors, and gives positive feedback and encouragement to the partner. The "checker" can only give suggestions, not part or all of the solution. Roles are reversed for the next problem to be solved.

Jigsaw Most useful for solving large or complex problems or doing an extended class project. The task is divided into several component parts, approaches, or topics.

a. Each group member is assigned to work on a different aspect of the problem. For example, if a problem requires students to make measurements, check and tabulate data, and report on results, a "measurer," "checker," "tabulator," and "reporter" may be designated.

b. Each group works on a different part of the problem or project. Each group's effort becomes part of the unified effort of the whole class, as pieces of a jigsaw puzzle fit together to form a larger, complete picture.

Roundtable This method is not highly interactive, but is excellent for brainstorming, for generating enthusiasm, and for generating a large number of answers to problems with multiple solutions. Each group has one piece of paper. After the teacher poses the question or problem, students write a solution or suggestion on the paper, then pass the paper to the next student. Within a given time frame, the paper continues to be circulated among the group (or the whole class). A student may choose to pass a round without penalty.

Whole class as a group The whole class as a group may decide on classroom procedures, divide work among the groups, decide on topics for class projects, brainstorm, present the results of their project to the school or the community, go over homework, review materials, or play mathematical games.

Is cooperative learning always an appropriate methodology?

Clearly, it is not always appropriate to use cooperative learning strategies in a mathematics class, nor should group work be the only strategy used. Students need to be engaged in a combination of individual and collaborative efforts. In addition to developing a common vocabulary and shared understanding of concepts, students need to develop a personal voice and a personal understanding of the ideas.

In ***Integrated Mathematics 3***
See student pages 51, 117, and 331 and side-column commentary on pages 35, 294, and 307.

Enhancing Mathematics Learning Using Graphing Technology

by ***Bill Leonard***
Mathematics Instructor
Shawnee Mission West High School; Shawnee Mission, Kansas

"Graphing technology helps students make connections between logic, symbols, and visualization."

Benefits of Graphing Technology

Developments in graphing software and graphics calculators have had a great impact on the teaching and learning of mathematics and have allowed teachers and students opportunities that were unavailable before. Teachers can present the same problem situation algebraically and geometrically within the same lesson. Students can visualize problems much more easily than in the past, and can attempt to solve problems that in the past might have been too complicated to deal with using only paper and pencil and algorithms.

Graphing technology can be used to facilitate students' many different individual learning styles, including verbal/linguistic, visual/spatial, logical, and kinesthetic. Graphing technology works as well with students whose learning style is interpersonal as with those who learn best working on their own. Perhaps the greatest strength of graphing technology is its facility in helping students to make connections between logic, symbols, and visualization.

Features that Enhance Different Learning Styles

Lines with the Same Slope

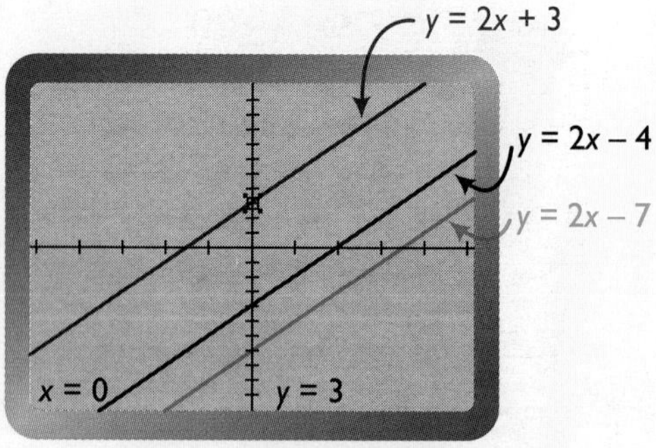

The relationship between two lines that have equal slope but different y-intercepts can quickly be seen by graphing the equations of two or more such lines using graphing software or a graphics calculator. The image on the screen helps all students, and especially visual learners, conclude that lines with equal slope but different y-intercepts are parallel. The effect of changes in the y-intercept for a line with given slope is also apparent, which helps make the slope-intercept form of an equation, $y = mx + b$, more meaningful. While an example such as this is particulary beneficial to the visual learner, the hands-on approach is also helpful to the kinesthetic learner.

Reflections

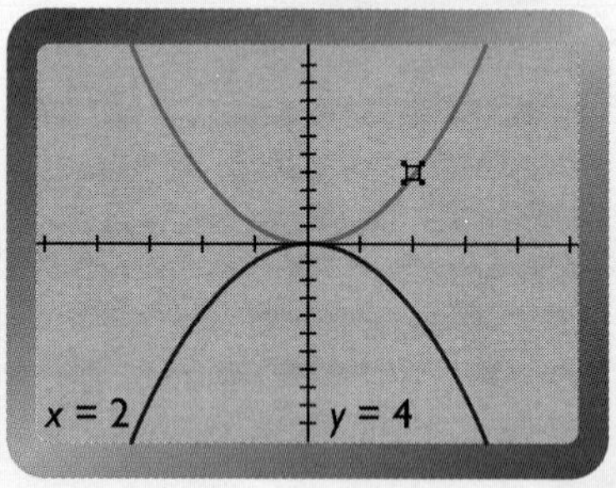

The relationship between changes in a quadratic equation and resulting translations of the graph can also be viewed quickly using graphing technology. The examples at the left and at the top of page T35 demonstrate how graphing technology can be used to show how changes in an equation affect its graph. For example, when the graphs of $y_1 = x^2$ and $y_2 = -x^2$ are displayed on the same screen, students can see that the effect of multiplying the coefficient of x^2 by -1 is to reflect the graph of the original equation in the x-axis. When 3 is added to the equation, the graph is translated up 3 units. When 3 is added to the x-term before squaring, the graph is shifted 3 units left.

Vertical Shift

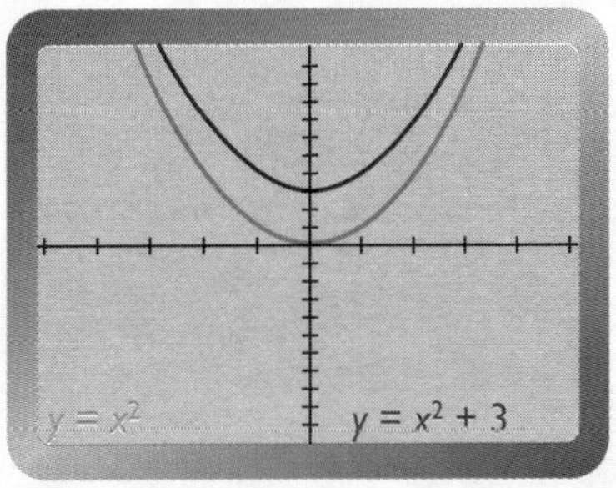

Horizontal Shift

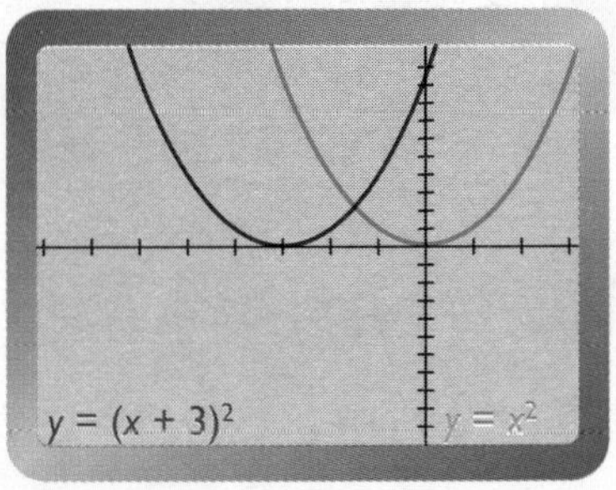

Solving a quadratic equation graphically by hand may be time-consuming, but, worse, may produce a graph from which it is extremely difficult to estimate solutions. However, using a graphics calculator or software produces the graph quickly. The student can use the TRACE and ZOOM features to estimate solutions quickly and accurately.

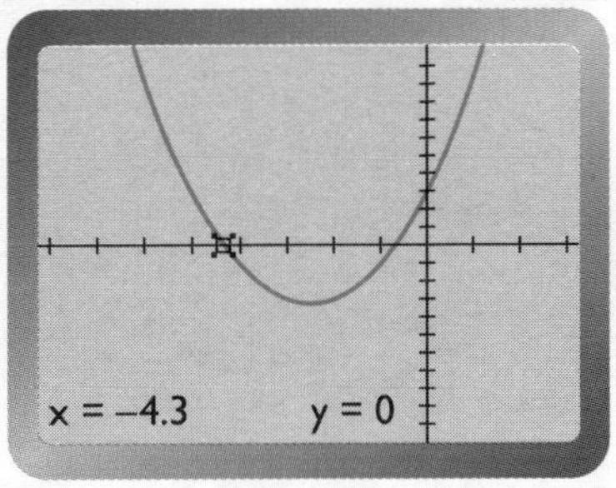

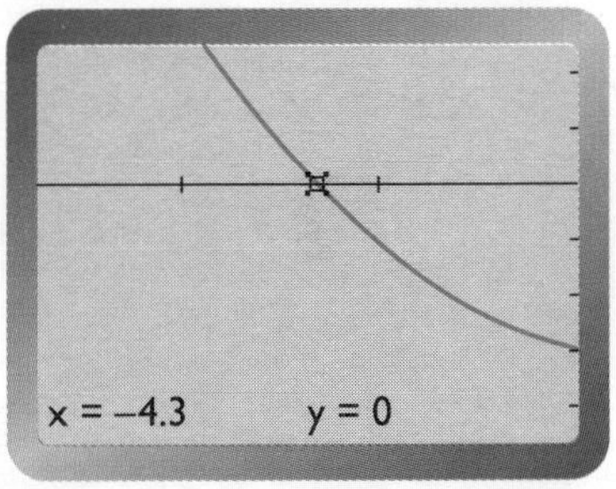

Graphing technology is also very useful for visualizing real-world data and fitting a line to the data points in order to predict unknown values.

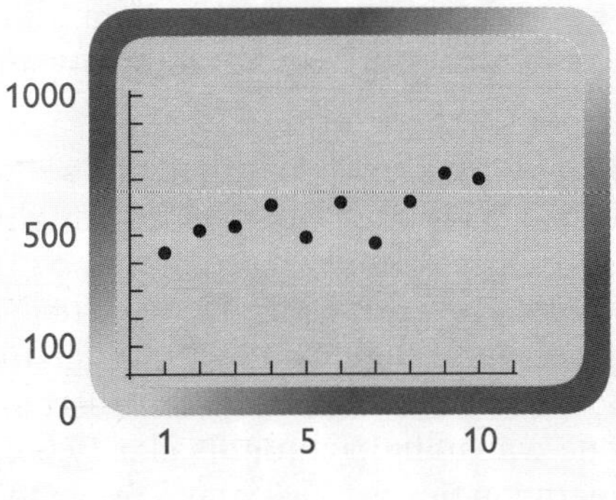

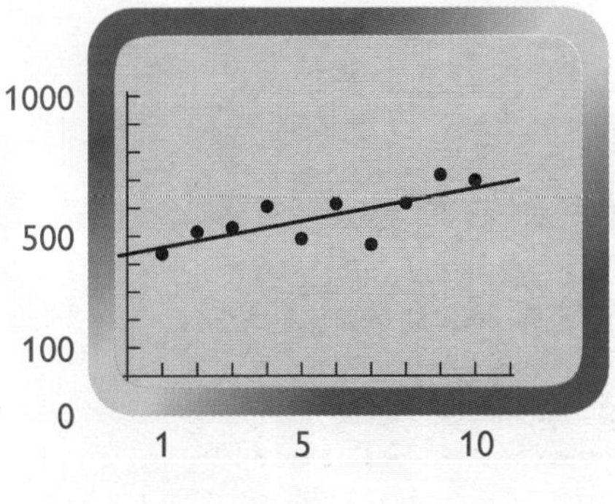

Meeting the Needs of Students for the Year 2000 and Beyond

As technology continues to become more and more a part of everyday life, students need to develop a familiarity and ease with using technology. They also need to recognize that math can be used to model and solve real-life problems. Graphics software and graphics calculators are successful tools for enabling students to achieve both of these goals. What may be just as important, they are fun to use!

In ***Integrated Mathematics 3***
See student pages 24, 268, and 355 and side-column commentary on pages 21, 129, and 220.

Writing in Mathematics

by ***Joan C. Countryman***
Head of School
Lincoln School; Providence, Rhode Island

"When students write they learn that mathematics is a human endeavor."

"Does your answer make sense?"

The student had come in for extra help and we were going over the homework problems. In the silence, I looked up, expecting some sort of defense for his solution to the exercise, but as our eyes met I realized that my question had startled him.

"Is it supposed to make sense?"

For too many students, studying mathematics has nothing to do with making sense. Practicing the steps, learning the rules, passing tests, adding, subtracting, multiplying, dividing—these are the activities of math class. If you get the right answer, the one in the back of the book, or the one your friend got, you can move on to the next task. If not, you must retrace your steps, find the mistake, do the calculations once more.

My student's quizzical look reminded me that, for some math students, none of it makes sense. I needed to find a way to help the young people enrolled in my classes build connections between what they already knew and what I wanted them to learn. I wanted to help them learn to stop and think about what we were doing in class. Why does the graph rise here and fall there? What does x represent in this example? Are squares of numbers always bigger than the numbers? Why not multiply and divide in this case?

I started asking my students to write about their work in mathematics because I thought that writing might help them move beyond a mechanical approach to learning, an approach that they found annoying but familiar. I wanted to help them discover the questions that are central to the discipline, the questions that mathematicians might pose. I also wanted them to think about themselves as learners. Over the years, I have found that one way to help students clarify, express, and reflect on their work in mathematics is to ask them to write to learn.

When I talk about writing in math I mean writing in its broadest sense. I ask students to take notes, make lists, record their observations and feelings, as well as to write essays, term papers, and stories. Having students write supports an active approach to teaching and learning. I expect my students to construct meaning. In order to make sense of the material, they must connect new information to what they already know. Writing helps them to learn to ask their own questions, and to explore some of the questions that I pose.

Many teachers use the writing process across the grades to help students construct mathematical knowledge. The examples at the left are from the math journals of tenth graders. The students recorded their observations about ellipses during a unit on analytic geometry. These brief comments, written at the end of class, provided insight for the teacher about how the students understood their work in conics.

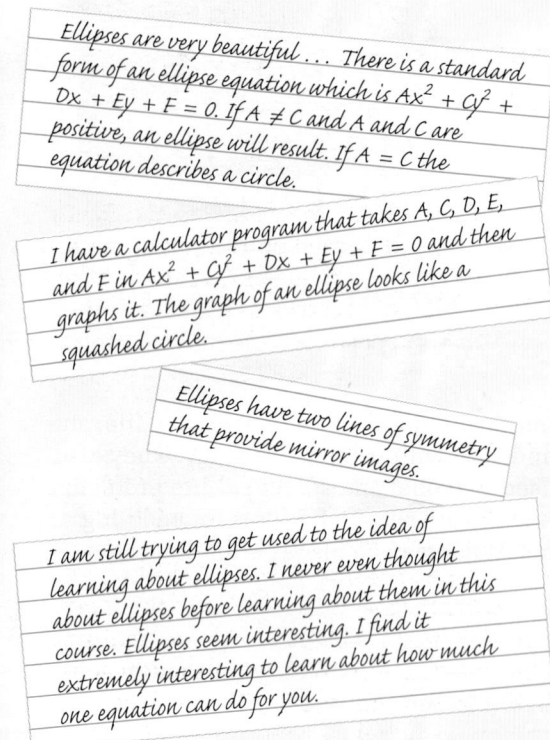

Ellipses are very beautiful . . . There is a standard form of an ellipse equation which is $Ax^2 + Cy^2 + Dx + Ey + F = 0$. If $A \neq C$ and A and C are positive, an ellipse will result. If $A = C$ the equation describes a circle.

I have a calculator program that takes A, C, D, E, and F in $Ax^2 + Cy^2 + Dx + Ey + F = 0$ and then graphs it. The graph of an ellipse looks like a squashed circle.

Ellipses have two lines of symmetry that provide mirror images.

I am still trying to get used to the idea of learning about ellipses. I never even thought about ellipses before learning about them in this course. Ellipses seem interesting. I find it extremely interesting to learn about how much one equation can do for you.

Keeping journals and writing word problems are just two of many activities that math teachers might require of their students. Autobiographies, the stories of their growth as math students, written in the first weeks of a math course, can serve to inform teachers about students' initial perceptions of mathematics and their own learning styles. Letters to parents or friends provide current accounts of coursework. Study guides, test questions, and lesson summaries can serve as excellent reviews. One way to get started is simply to ask students to write a comment, on the back of the homework, about the exercises they have completed. Which was the most difficult? Which provided the most insight on the material? Which ones were easy to complete?

Advocates of writing across the curriculum are not suggesting that all teachers assign essays and correct them as an English teacher might. Instead we imagine that teachers might ask students to think and write as essayists, scientists, historians, and mathematicians do, posing questions, and solving problems by writing and reflecting on the material of the discipline.

Most useful to us as teachers is the writing that provides insights on how our students think about their work. The following example is from a mathematics student who was also studying physics. For the teacher, the student's comments revealed the depth of his thinking about mathematical concepts of real phenomena.

> *Since I have been very interested and involved in physics, I wanted to find a topic for my final paper that would in some way investigate some of the principles we were studying in physics. I also wanted to study empirically one of these topics, that is, to take my own data, and develop equations based not on a textbook but on my own data-taking and analysis. For these reasons, I decided to examine the behavior of different balls as they bounce, specifically to establish a relationship between the height from which a ball is dropped and the height that it then bounces. While I worked on this project, I also became interested in the time over which a ball continues to bounce, and I added my investigations in this area to my paper.*

My hope is that one day students and teachers will write to learn freely. Pages of notes, stories, plays, lists, poems, sketches, and journal entries about math, language, literature, science, and history will help students make sense of the world in which they live. Teachers of all subjects will serve as coaches and experts about the learning process, knowledgeable about students—who they are, what they can do, what they know, and what they need to know.

What might students learn when they write about math? What can we as teachers learn from their writing? First, when students write they learn that mathematics is a human endeavor, one that comes not from the sky, but from the work of human minds. In fact, when students write some of the mathematics comes from their own minds. Second, when reading students' writing, teachers discover that learners, like mathematicians, must construct the mathematics for themselves. If we give them time to do this, they will succeed in constructing their understanding of the material. Finally, students and teachers will learn that meaning lies not in the words and symbols themselves, but in the ways that we use those words and symbols to make sense of information.

In ***Integrated Mathematics 3***
See student pages 5, 182, and 356 and side-column commentary on pages 124, 214, and 273.

Developing Good Problem Solvers

by ***Martha E. Wilson***
Preparatory Mathematics Specialist
Mathematical Sciences Teaching and Learning Center
University of Delaware; Newark, Delaware

"Teaching itself is a complex problem solving process."

The Goal of Mathematics Instruction

Mathematics educators have agreed for some time that problem solving is a very important, if not the most important, goal of mathematics instruction. Students who learn mathematics through drill in routine operations lose interest in the subject and miss the opportunity for intellectual development. Stimulating them to solve appropriately challenging problems, and helping them to solve those problems, provides students with interest in and tools for independent thinking.

Research in mathematics education has not been able to identify any one single way of teaching that is "best" for developing problem solving skills in mathematics for all students in all situations. There are, however, examples of good problem solving that point to actions that teachers might take to help students develop those skills.

Cooperative Learning

The National Council of Teachers of Mathematics *Curriculum and Evaluation Standards* and *Professional Standards for Teaching Mathematics* both call for a classroom where students have an opportunity to explore and investigate ideas, develop conjectures, and verify hypotheses. Cooperative learning in mathematics is often recommended as one way to help students because:

- they become active participants in the classroom
- they may be encouraged to discuss and communicate their ideas about mathematics in an environment that is less threatening than in whole-class discussions
- when groups of students struggle with an interesting problem, the outcome for many students is a new and refreshing view of doing mathematics

HIGHEST GEAR

Whole-Group Instruction

Research also suggests that an active, problem solving approach to learning can be accomplished in whole-group instruction, where the teacher directs the activities of the entire class. Students can become actively involved when the teacher selects and presents an interesting idea or problem and then leads students to a discovery of concepts and connections through a series of questions. Questions should develop a train of thought in logical sequence and should include these types:

- some moderately challenging, to stimulate thinking
- some factual, to bring out important facts or information
- some requiring considerable thought and formulation of a conclusion

Example

Not this: The formula $T(x) = 20 + 70e^{(-0.05x)}$ can be used to find the temperature (T) in degrees Celsius over the time (x) in minutes of soup that was heated to 90°C in a room with a temperature of 20°C. How long will it take for the soup to reach 50°C?

But this: Here is a plot that a class made from data from a cooling experiment. Soup was heated to 90°C and the temperature was recorded every second for 36 s. Discuss the model you would write for this data and how you might determine the temperature of the room.

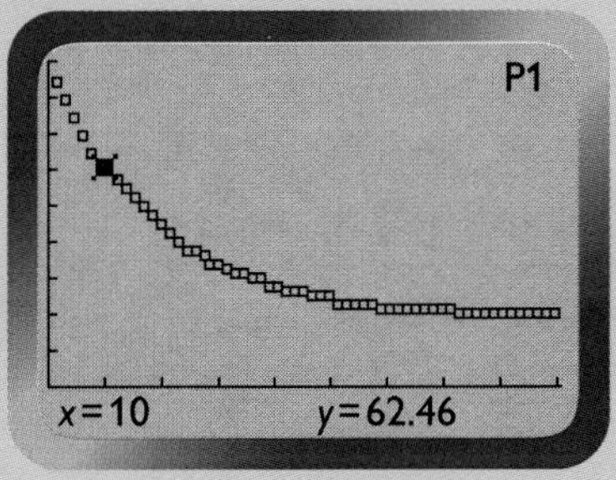

The Teacher's Role

The role of the teacher in preparing for either model of instruction becomes one of organizing for learning by selecting an engaging task or problem. An understanding of the background knowledge and interests of the students and thoughtful planning of future lessons is necessary for the design of this task. Good tasks prompt an interest in investigation whether they are presented for group work or for whole-class discussion.

Selecting Good Tasks

A task should be:

- set in a context that will engage the interest of students
- complex and difficult enough to challenge students' thinking, but not so difficult that they will give up quickly
- solvable by more than one method, so that a subsequent discussion can point out connections and the possibility of multiple approaches

In discussions, the teacher must prepare to ask the questions that lead to clear and concise mathematical conclusions and emphasize the connections that can be made.

Teachers as Problem Solvers

In ***Integrated Mathematics 3***
See student pages 11, 86, and 377 and side-column commentary on pages 123, 212, and 286.

Research on teaching and learning mathematics is still incomplete, and there is no indication that teaching mathematics must be done in a single prescribed way in order for students to become good mathematical problem solvers. Teachers may choose among a variety of styles, including teacher presentation, large-group activities, small-group cooperative learning, and combinations of these. As researchers have investigated several modes of teaching in search for the one that produces good problem solvers, they seem to have found that teaching itself is a complex problem solving process.

Sailing Yachts that race must meet design requirements. In 1989, the International America's Cup Class (IACC) rules included this formula:

$$\frac{L + 1.25\sqrt{S} - 9.8\sqrt[3]{D}}{0.388} \leq 42$$

41. Suppose you design a yacht with length 20.8 m and sail area 285 m^2. It displaces a volume of water equal to 21.4 m^3. Will your yacht meet the IACC requirements?
42. Suppose a designer maximizes the length of the yacht and the area of the sails. Estimate the minimum volume of water the yacht can displace.
43. Suppose a designer decides not to maximize the length of the yacht and the area of the sails. Does the minimum amount of water the yacht can displace go up or down? Explain.

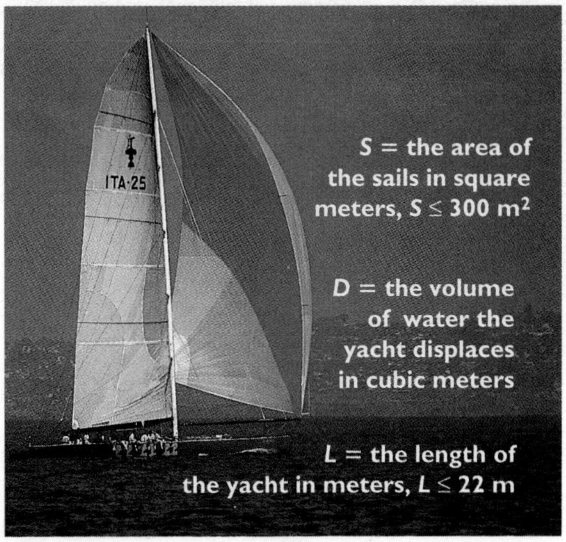

ITA-25
S = the area of the sails in square meters, S ≤ 300 m²
D = the volume of water the yacht displaces in cubic meters
L = the length of the yacht in meters, L ≤ 22 m

From ***Integrated Mathematics 3,*** page 117

Teaching Students Acquiring English

by ***Cesar Larriva***
Doctoral Student, Department of Education
Stanford University; Stanford, California

"The strategies for teaching students with limited English proficiency work well for all students."

Challenge

The educational backgrounds of LEP (limited English proficient) students in general are as varied as the nationalities they represent. Some students have received top-quality public or private education in their native countries while others have barely attended school. The students' first (native) language proficiency can also vary considerably and has a strong influence on student academic success in the second language setting. The LEP student generally comes from a culture where education is valued, and the school and the teacher are highly respected. These students are generally very motivated by their desire to succeed. They are also motivated by their desire to please the teacher and by peer pressure from other LEP students, which is largely achievement oriented.

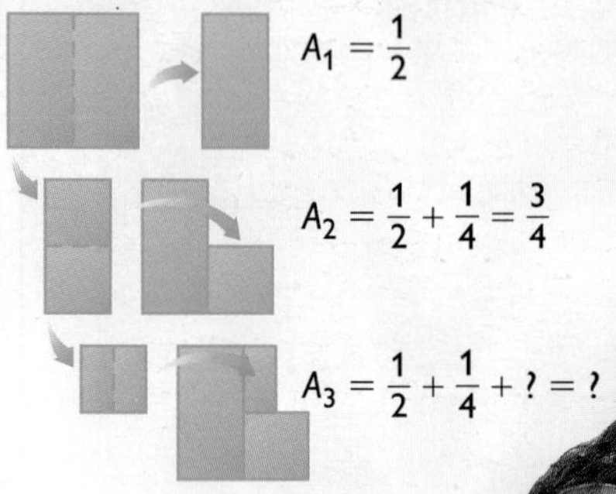

$A_1 = \frac{1}{2}$
$A_2 = \frac{1}{2} + \frac{1}{4} = \frac{3}{4}$
$A_3 = \frac{1}{2} + \frac{1}{4} + ? = ?$

The wide range of mathematics skills, English proficiency, and first language proficiency of LEP students within a classroom makes it challenging to meet the needs of all students. Meeting this challenge therefore requires special strategies.

Sheltering Techniques

Sheltering techniques are an important tool. Sheltered English instruction is a method of delivering subject matter (e.g., mathematics) instruction to LEP students using English as the medium for instruction. A language can be learned only if it is presented as comprehensible input. Information is comprehensible when the vocabulary and language used are familiar to the learner and the information is presented in a meaningful context.

In sheltered English instruction, the delivery of the message is simplified (sheltered), but subject matter remains challenging; material is not "watered down." Instruction in the sheltered English classroom is adjusted to ensure student comprehension. Input can be made more comprehensible by the following techniques.

Teacher's Speech

Language is simplified by avoiding compound sentences, by favoring simple grammatical structures such as the present tense, by limiting the vocabulary, and by avoiding use of idioms. A phrase such as "I want you to stay on top of things here" should be used only if its meaning is discussed first. Content is emphasized over grammatical accuracy. Student oral and written responses are evaluated based on content not grammatical accuracy. Important ideas are repeated several times for emphasis.

Providing Clues

Effort should constantly be made to provide contextual clues. These may be graphical representations such as photographs or graphs. The clues may be in written form, such as the posting of important vocabulary on a chalk or bulletin board; a written vocabulary word can be pointed to as the teacher uses it in the sentence during lecture. Clues may be physical, such as real objects and scaled models.

Acceptance

The classroom culture is supportive, motivational, and non-threatening so that the student's defense mechanism (which hinders participation) is low. LEP students experience a pre-speech stage or silent period during which active listening and learning occur without language production. Students will produce language when ready and should be encouraged but not pushed.

Manipulatives

Concepts are contextualized and communication is facilitated through the use of hands-on activities and manipulatives, such as algebra tiles. A great deal of emphasis is placed on making the abstract more concrete. For example, students can physically act out or model a problem. The diagram shows students physically plotting a linear equation on a tennis court or a school yard on which coordinate axes have been drawn using chalk.

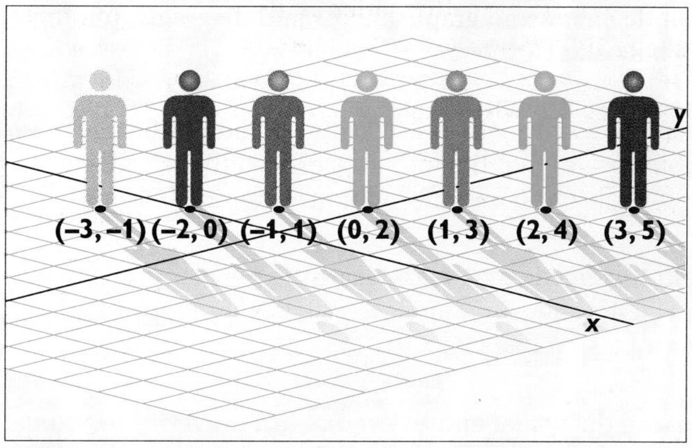

Students physically model the function $y = x + 2$.

Any technique like this that facilitates communication by decreasing the reliance on language is beneficial.

Prior Experience

Provide sufficient preparation and background when introducing a new topic. A lesson on probability using playing cards, for example, will definitely require an explanation of the playing cards themselves, since many LEP students have never seen these cards that are so familiar to us. New concepts should be presented in a context that is meaningful to the student.

It is important to recognize that LEP students have a wealth of prior skills and experience to draw upon. Many also have highly developed informal math skills. It is our job to help students utilize their prior knowledge. Since thinking skills transfer from one language to another, problems involving experiences such as travel, money, and school life can trigger students' interest and motivation. Classroom activities should encourage students to draw upon these.

Cooperative Learning

Related to the goal of encouraging LEP student participation is the goal of creating a positive feeling toward mathematics and the mathematics classroom. Maintaining a friendly non-threatening, supportive environment is an essential aspect of encouraging participation and thus learning. Additionally, fostering a feeling of community is essential in a classroom for students to acquire lasting knowledge. The community should generate and sustain a mathematics culture.

Cooperative learning groups play an important role in promoting the exchange of values and providing a forum for participation in the mathematics culture. The culture values and rewards inquiry, effort, and risk taking. It is the responsibility of the teacher to facilitate the development of such a culture within the classroom. Student conversational interaction is encouraged through cooperative learning activities. I am careful to seat LEP students next to others who can translate and/or provide support. Small groups provide a safer environment since many students are reluctant to ask questions or offer answers in a large class setting. Students pool their talents and strengths (e.g. language, computational, and problem solving abilities) to piece together solutions to problems.

The teacher should allow for a reasonable noise level since a class of 30 or 40 engaged in cooperative learning can generate considerable noise. Classroom management skills become important in maintaining the balance between the organized group debates which develop, and anarchy.

Language Development

Vocabulary (mathematical, technical, and general) is taught as part of subject matter instruction rather than in the traditional method that relied on vocabulary lists without connection to meaningful contexts. Therefore vocabulary is always presented as an incidental part of a lesson and is connected to real ideas and objects. In this way vocabulary learning takes on a new meaning for students; it has a purpose.

Students' language development is central to success for LEP students in the mathematics classroom and thus deserves added attention. Collaborative problem solving is an effective way to foster written and verbal communication between students. The student discussions create opportunities for students to hear themselves and other students using the language of mathematics. Students must be given opportunities to write and speak mathematically with each other as well as the teacher. Additional methods of promoting language development include the use of investigations and projects in collaborative group settings. Assigning portfolios and journal writing is also useful. Do not avoid assigning problems that require writing. Instead, use these problems as vehicles for students to develop oral and written language skills in cooperative group settings.

Effective Teaching

The instructional approaches recommended thus far for teaching mathematics to limited English proficient students are in fact nothing more than good teaching techniques that work well for all students. It is therefore possible to accommodate the needs of limited English proficient students and native English speakers concurrently without compromising the learning of either.

In *Integrated Mathematics 3*
See student pages 4, 155, and 251 and side-column commentary on pages 35, 99, and 113.

Teaching Strategies

Visual Learning Strategies

by ***Stuart J. Murphy***
Visual Learning Specialist
Evanston, Illinois

"Linking the visual and the verbal is a powerful teaching tool."

Our Visual Environment

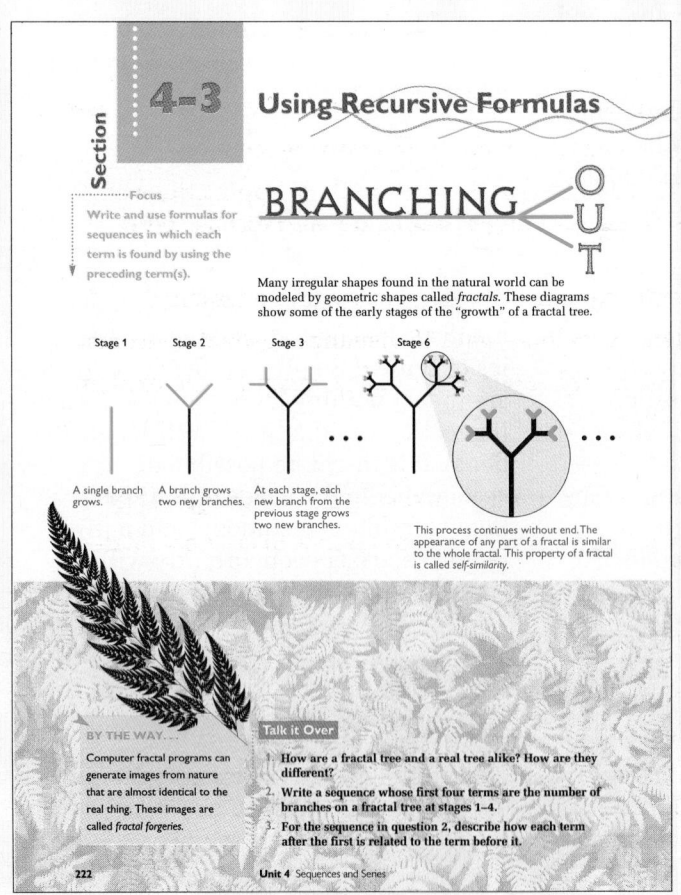

There is no question that we are living in an intensely visual environment. Whether from television and videos, or magazines and books, information regularly comes to us in a variety of formats. In addition to text, these formats include charts and graphs, maps and diagrams, photographs and illustrations, symbols and cartoons.

Even the way in which text is presented has become more varied to include a greater use of highlighted phrases, headlines, call-outs, and captions. The need to absorb more information—in more formats—has never been greater.

With this need come many learning opportunities. There is growing evidence that comprehension increases when verbal information is augmented by high-quality visual displays.

Linking the visual and the verbal is a powerful teaching tool. Such a link interests and motivates learners, provides more information, and reaches a broader audience than either method alone.

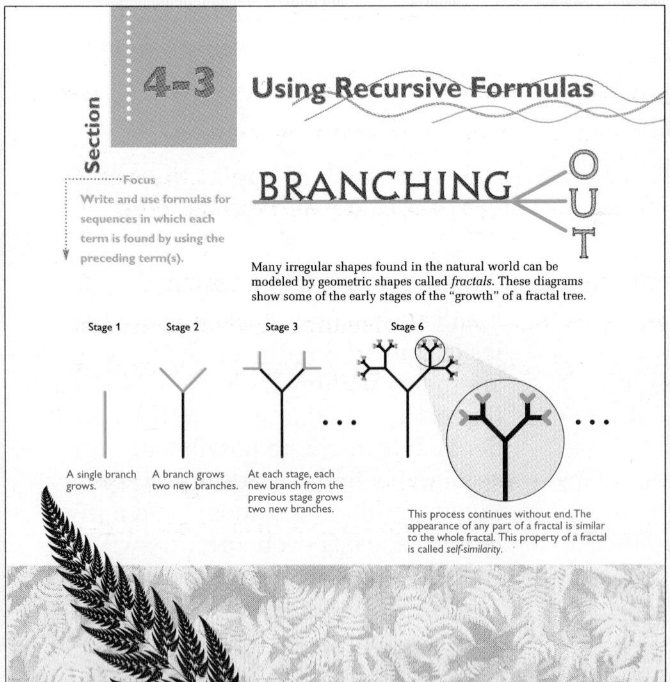
Section 4-3 Using Recursive Formulas

Focus: Write and use formulas for sequences in which each term is found by using the preceding term(s).

BRANCHING OUT

Many irregular shapes found in the natural world can be modeled by geometric shapes called *fractals.* These diagrams show some of the early stages of the "growth" of a fractal tree.

Stage 1 — A single branch grows.
Stage 2 — A branch grows two new branches.
Stage 3 — At each stage, each new branch from the previous stage grows two new branches.
Stage 6

This process continues without end. The appearance of any part of a fractal is similar to the whole fractal. This property of a fractal is called *self-similarity.*

BY THE WAY...
Computer fractal programs can generate images from nature that are almost identical to the real thing. These images are called *fractal forgeries.*

Talk it Over
1. How are a fractal tree and a real tree alike? How are they different?
2. Write a sequence whose first four terms are the number of branches on a fractal tree at stages 1–4.
3. For the sequence in question 2, describe how each term after the first is related to the term before it.

222 Unit 4 Sequences and Series

Visual Learning in Mathematics

In the study of mathematics, visual learning strategies play an especially important role. Understanding symbols—and the use of symbols within the concise language of mathematics—is critical to the ability to comprehend and express mathematical ideas.

Icons and symbols also play an important role in the use of technology in mathematical instruction. Calculators and computers use a carefully constructed symbolic language to provide direction and convey meaning to users.

Visual presentations help us to model mathematical ideas, to see patterns, and to understand relationships. Indeed, a basic understanding of many important mathematical concepts—concepts such as comparison, scale, dimension, translation, and perspective—depends upon the ability of the student to visualize.

A better understanding of how information is conveyed visually can also help us as we work with students who are visual/spatial learners, students who have limited English proficiency, and students who come from a variety of socio-economic and cultural backgrounds. In fact, using visual learning strategies can help us increase the learning potential of all students.

Integrated Mathematics includes a carefully planned visual learning strand to help students develop visual learning skills. Pages are designed to allow easy access to the material being presented and to provide multiple points of entry, including images, titles, diagrams, and call-outs.

Illustrations and applications help students connect mathematical ideas to real-world situations and to other lands and cultures.

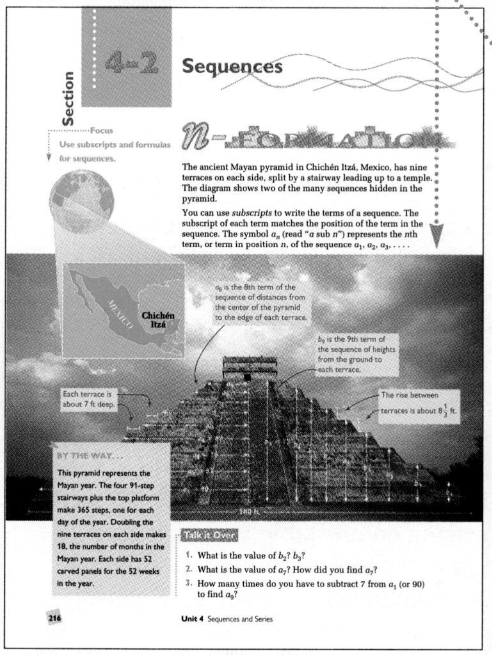

Section 4-2 Sequences

Focus: Use subscripts and formulas for sequences.

n-FORMATION

The ancient Mayan pyramid in Chichén Itzá, Mexico, has nine terraces on each side, split by a stairway leading up to a temple. The diagram shows two of the many sequences hidden in the pyramid.

You can use *subscripts* to write the terms of a sequence. The subscript of each term matches the position of the term in the sequence. The symbol a_n (read "a sub n") represents the nth term, or term in position n, of the sequence $a_1, a_2, a_3, \ldots$.

MEXICO
Chichén Itzá
a_8 is the 8th term of the sequence of distances from the center of the pyramid to the edge of each terrace.
b_9 is the 9th term of the sequence of heights from the ground to each terrace.
Each terrace is about 7 ft deep.
The rise between terraces is about $8\frac{1}{3}$ ft.
180 ft

BY THE WAY...

This pyramid represents the Mayan year. The four 91-step stairways plus the top platform make 365 steps, one for each day of the year. Doubling the nine terraces on each side makes 18, the number of months in the Mayan year. Each side has 52 carved panels for the 52 weeks in the year.

Talk it Over

1. What is the value of b_2? b_3?
2. What is the value of a_7? How did you find a_7?
3. How many times do you have to subtract 7 from a_1 (or 90) to find a_9?

216 Unit 4 Sequences and Series

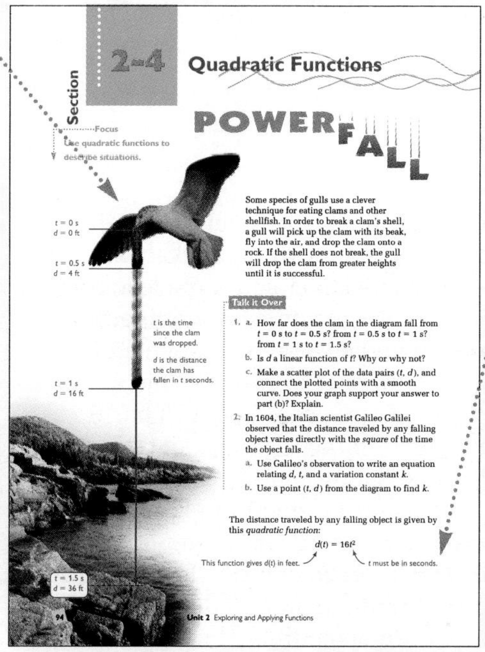

Section 2-4 Quadratic Functions

Focus: Use quadratic functions to describe situations.

POWER FALL

Some species of gulls use a clever technique for eating clams and other shellfish. In order to break a clam's shell, a gull will pick up the clam with its beak, fly into the air, and drop the clam onto a rock. If the shell does not break, the gull will drop the clam from greater heights until it is successful.

$t = 0$ s
$d = 0$ ft
$t = 0.5$ s
$d = 4$ ft
$t = 1$ s
$d = 16$ ft
$t = 1.5$ s
$d = 36$ ft
t is the time since the clam was dropped.
d is the distance the clam has fallen in t seconds.

Talk it Over

1. a. How far does the clam in the diagram fall from $t = 0$ s to $t = 0.5$ s? from $t = 0.5$ s to $t = 1$ s? from $t = 1$ s to $t = 1.5$ s?
   b. Is d a linear function of t? Why or why not?
   c. Make a scatter plot of the data pairs (t, d), and connect the plotted points with a smooth curve. Does your graph support your answer to part (b)? Explain.
2. In 1604, the Italian scientist Galileo Galilei observed that the distance traveled by any falling object varies directly with the *square* of the time the object falls.
   a. Use Galileo's observation to write an equation relating d, t, and a variation constant k.
   b. Use a point (t, d) from the diagram to find k.

The distance traveled by any falling object is given by this *quadratic function*:

$$d(t) = 16t^2$$

This function gives $d(t)$ in feet. t must be in seconds.

94 Unit 2 Exploring and Applying Functions

Special callouts explain mathematical concepts and thinking processes.

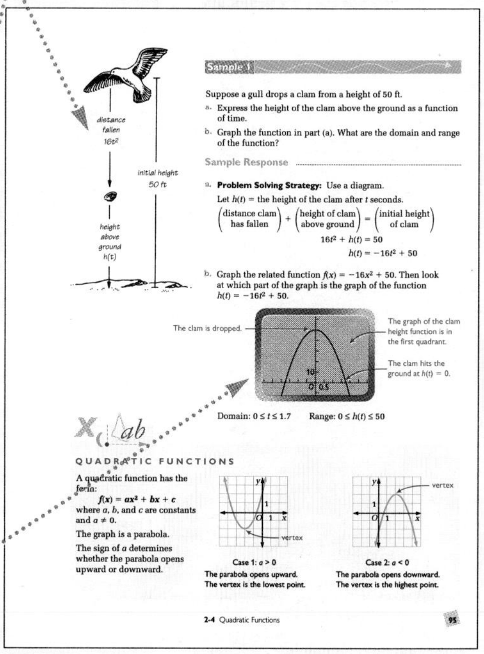

distance fallen $16t^2$
initial height 50 ft
height above ground $h(t)$

Sample 1

Suppose a gull drops a clam from a height of 50 ft.

a. Express the height of the clam above the ground as a function of time.
b. Graph the function in part (a). What are the domain and range of the function?

Sample Response

a. **Problem Solving Strategy:** Use a diagram.
Let $h(t)$ = the height of the clam after t seconds.

$$\begin{pmatrix}\text{distance clam}\\ \text{has fallen}\end{pmatrix} + \begin{pmatrix}\text{height of clam}\\ \text{above ground}\end{pmatrix} = \begin{pmatrix}\text{initial height}\\ \text{of clam}\end{pmatrix}$$

$$16t^2 + h(t) = 50$$

$$h(t) = -16t^2 + 50$$

b. Graph the related function $f(x) = -16x^2 + 50$. Then look at which part of the graph is the graph of the function $h(t) = -16t^2 + 50$.

The clam is dropped.
The graph of the clam height function is in the first quadrant.
The clam hits the ground at $h(t) = 0$.
10
O 0.5

Domain: $0 \le t \le 1.7$ Range: $0 \le h(t) \le 50$

QUADRATIC FUNCTIONS

A quadratic function has the form:

$$f(x) = ax^2 + bx + c$$

where a, b, and c are constants and $a \neq 0$.

The graph is a parabola.

The sign of a determines whether the parabola opens upward or downward.

vertex

Case 1: $a > 0$
The parabola opens upward.
The vertex is the lowest point.

Case 2: $a < 0$
The parabola opens downward.
The vertex is the highest point.

2-4 Quadratic Functions 95

Screen displays provide guidance for students with calculators and visual models for those without calculators.

Strategies for Developing Visual Skills

Here are some visual learning strategies that you can use on an ongoing basis.

- Display visual materials to interest, excite, and motivate students.
- Emphasize photos within the text that demonstrate real applications of mathematical concepts and ask students to consider and discuss other examples.
- Explain—or have students explain—the diagrams within a lesson.
- Develop—or have students develop—ways to visualize abstract concepts.
- When students are having trouble understanding a concept, try to explain the concept without using words.
- Encourage students to:
  - draw and sketch as part of their note-taking and journal practice
  - take photographs or clip photos from magazines to connect related ideas
  - demonstrate their thinking by mapping out the steps or acting out the process
  - show their understanding by drawing a concept map
  - construct charts, graphs, and diagrams to explain concepts

Using these strategies and the images that have been provided in *Integrated Mathematics* will help your students develop their visual learning skills—help them to link visual to verbal, process to concept, and learning to life.

In ***Integrated Mathematics 3***
See student pages 32, 159-160, and 207 and side-column commentary on pages 43, 95, and 230.

Assessment Methods

by **Karen S. Norwood**
Assistant Professor, Department of Mathematics and Science Education
North Carolina State University; Raleigh, North Carolina

"Assessment needs to be embedded in the instructional process."

Investigate the shape and the dimensions of the pen with the largest area that can be constructed with 36 feet of fencing.

Explain which is larger:

π^6 or 6^π

Do not use your calculator.

Write a paragraph explaining how the sine and the cosine ratios are alike and how they are different.

Demonstration
Tell the class everything you know about the graphs of

A $y = x^2 + 2x + 1$

and

B $y = -x^2 + 2$

Use a graphics calculator or computer software if you wish.

Assessment Goals

The purpose of assessment in mathematics is to improve and evaluate learning and teaching. In the teaching-learning process, it is imperative that assessment be used to broaden and inform, rather than restrict, the process. Assessment needs to be embedded in the instructional process, instead of being apart from it. This view was well stated in the NCTM's *Curriculum and Evaluation Standards for School Mathematics.* "In an instructional environment that demands a deeper understanding of mathematics, testing instruments that call for only the identification of single correct responses no longer suffice. Instead, our instruments must reflect the scope and intent of our instructional program to have students solve problems, reason, and communicate."

Traditional paper-and-pencil tests are incomplete measures of achievement. In fact, no single type of assessment can serve all the information needs of an educational institution. Using alternative assessment methods provides a more equitable measure of a student's mathematical progress, has less potential for bias, and encourages respect for diversity by modeling appreciation for varied approaches to a problem. The goals of alternative assessment are to:

- find out what the students already know
- evaluate the depth of the students' conceptual understanding and their ability to transfer this understanding to new and different situations
- evaluate the students' ability to communicate their understanding mathematically, make mathematical connections, and reason mathematically
- plan the mathematics instruction in order to achieve the objectives
- report individual student progress and show growth in mathematical maturity
- analyze the overall effectiveness of the mathematical instruction

When using alternative assessment, it is important to start slowly, so as not to become overwhelmed. Journal writing is a good place to start. Once you become comfortable with this technique, try to add another alternative assessment strategy to your repertoire. Don't try to use alternative assessment alone; involve colleagues, parents, and administrators.

Scoring

There are several ways to score alternative assessment assignments. One of the most simple methods is to divide papers into piles labeled "satisfactory" and "unsatisfactory." Then assign a grade from 0 to 3 based on the following criteria. Satisfactory papers are given a grade of 3 if the student gives a clear explanation with appropriate diagrams or graphs, and a score of 2 if the student's work is complete and shows understanding but contains computational errors or minor flaws in explanation. Unsatisfactory papers are given a 1 if the work is incomplete and contains serious conceptual errors along with flagrant computational errors. A score of 0 is given if little or no effort was made to complete the assignment.

Some people prefer to use a scale of 1 to 4, where 4 indicates excellent and 1 indicates unacceptable work. Five- and six-point scales are also used.

Alternative Assessment Formats

Several types of alternative assessment items are appropriate to the mathematics classroom.

JOURNALS Regular use of a journal encourages students to express complex mathematical concepts in words. Writing helps to make students aware of what they do and don't understand, what they can and cannot do. Reading a journal gives the teacher insight into the student's understanding.

RESEARCH PROJECTS Group or individual research projects allow students to investigate topics that encompass many mathematical concepts and their real-world applications. Examples of such projects are the Unit Projects in this book.

DEMONSTRATION/PERFORMANCE ASSESSMENT Teachers can assess their students' comprehension of a mathematical concept by asking them to explain the concept in their own words using such items as compasses, graph paper, calculators, and computers.

PROBLEM SOLVING Problem solving is considered to be the link between facts and algorithms and the real-life problem situations that we all face. Problem-solving activities include non-routine problems where the strategy necessary to solve the problem is not immediately apparent, and analysis and synthesis of previously learned knowledge are required.

PORTFOLIOS As artists and writers use portfolios to show off their best work, mathematics students can use portfolios to document their growth and the development of their mathematical power. Portfolios can be used to assess a student's mathematical reasoning, understanding, attitudes, and ability to communicate mathematically.

Both the teacher and the student should have input into selecting what will be included in a portfolio. For example, the teacher might determine how many pieces are to be included and the categories from which they will come. The student might be allowed to choose the pieces. The portfolio should include a table of contents and a cover letter. Each included work should be labeled with the date, a description of the task or problem, and the identity of the person who selected the work. A self-assessment should also be included.

The contents of a portfolio might include:

- open-ended questions, problems, and tasks, in which the student is asked to formulate hypotheses, explain a mathematical situation, make a generalization, and so on, either orally or in writing
- research projects
- presentations, discussions, and debates
- journal entries
- cooperative learning activities
- math logs: problems assigned by the teacher which require that the student not only show computations, but validate the solution
- problem solving
- investigations
- models and simulations
- interviews: students talk, individually or in groups, while the teacher listens and asks questions. The teacher may encourage students to further elaborate in an interview by using phrases such as, "I am interested in your thinking," or "I understand it better now, but..."
- photographs of items the student may have produced that are too bulky to fit in a portfolio
- work dealing with the same mathematical idea sampled at different times
- copies of awards or prizes

Ongoing **ASSESSMENT**

29\. a. **Open-ended** Write equations for at least ten different polynomial functions of varying degrees.

b. TECHNOLOGY Use a graphics calculator or graphing software to graph the functions. Record the degree of each polynomial function and how many *turning points* the function has.

c. Make a conjecture about the relationship between the number of turning points and the degree of the polynomial function.

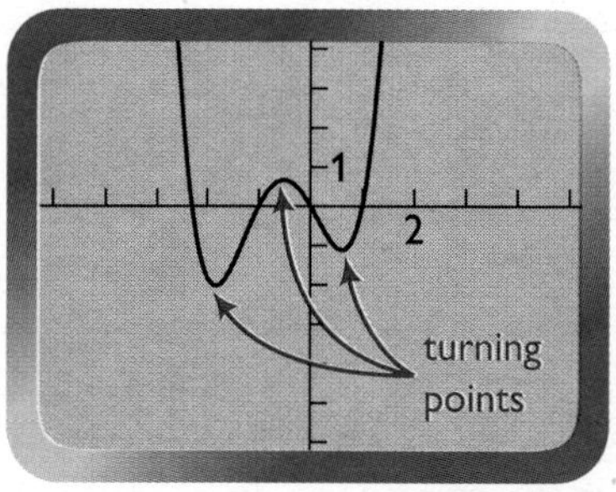

From ***Integrated Mathematics 3,*** page 108

In ***Integrated Mathematics 3***
See student pages 51, 227, and 337 and side-column commentary on pages 75, 117, and 271.

Teaching Discrete Mathematics and Data Analysis

by ***Loring Coes III***
Chair of the Mathematics Department
Rocky Hill School; East Greenwich, Rhode Island

"Discrete mathematics and data analysis are the basic skills for the future."

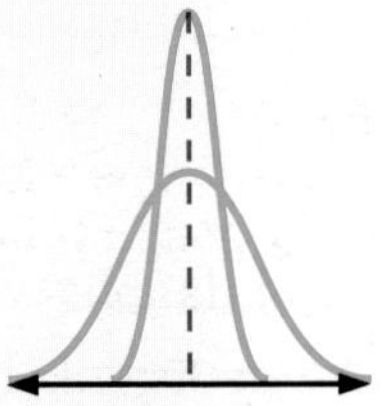

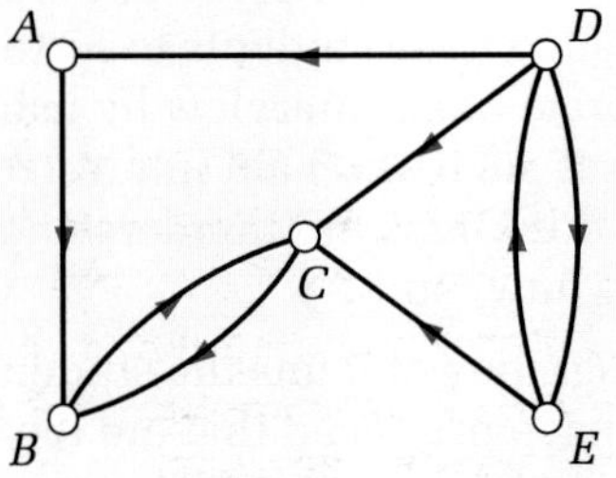

$$\begin{array}{c} \\ P \\ Q \\ R \\ S \end{array}\begin{array}{c} \begin{array}{cccc} P & Q & R & S \end{array} \\ \begin{bmatrix} 0 & 1 & 1 & 0 \\ 1 & 0 & 0 & 1 \\ 0 & 1 & 0 & 1 \\ 1 & 0 & 1 & 0 \end{bmatrix} \end{array}$$

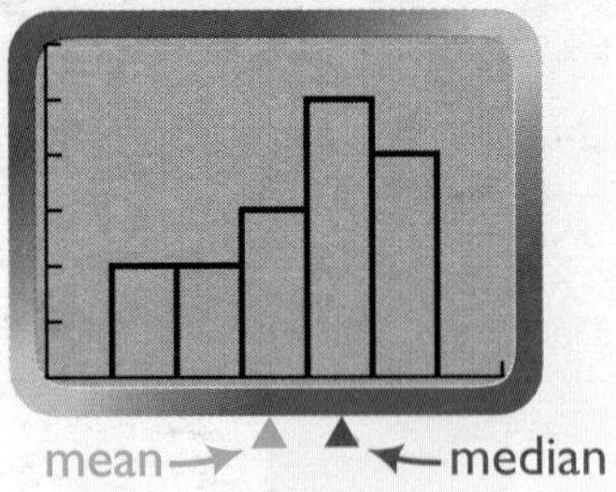

Why Teach Discrete Mathematics and Data Analysis?

Why do we need new topics like discrete mathematics and data analysis in the high school mathematics curriculum? Aren't the traditional topics good enough? What's wrong with the material we adults learned a generation ago? How can teachers justify these new ideas to parents and communities worried about basic skills? Are discrete mathematics and data analysis just fads, doomed to fade away in a few years? Are these topics important for future careers?

These are good questions, ones that swirl around in every conversation about changes in the mathematics curriculum, and these questions deserve good answers.

The Mathematics for Contemporary Life

Preparing for the Future

In the first place, there was nothing wrong with the mathematics of a generation ago—it reflected the needs of the time. The purpose of the current curriculum reform in mathematics education is not to find fault with the past but to help prepare students for their future as adults in the 21st century.

A Technological Environment

Today we live in an age of automated workplaces—and automated homes—where the mathematical skills of our parents and grandparents will leave our children wanting. More than ever before, we work, think, and play in a technological environment that becomes more sophisticated by the day. Rivers of information are at our fingertips, and worldwide communication is easy.

There are new challenges because of technology. How do we use computing power to our best advantage? How can we be sure that the machines are serving us, and not the other way around? How can we stand beside that flow of information and pick out what is most important to us? How can we be confident that we are reading the information well? Discrete mathematics and data analysis help us answer these questions.

Both discrete mathematics and data analysis have emerged hand in hand with technology. Rapid computation power has made problems in discrete mathematics and data analysis more accessible to us. On the other hand, understanding the mathematics helps us to use the technology more effectively.

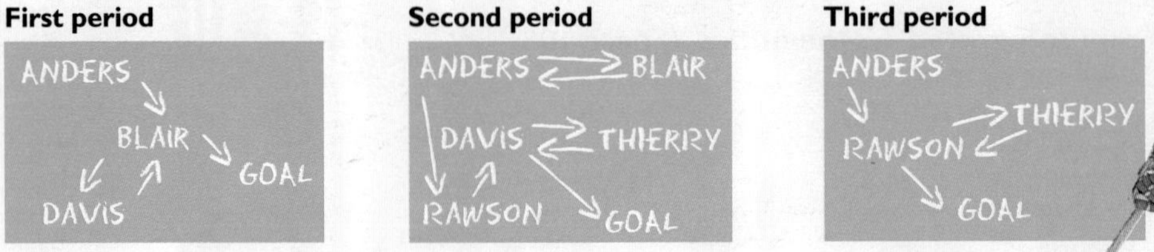

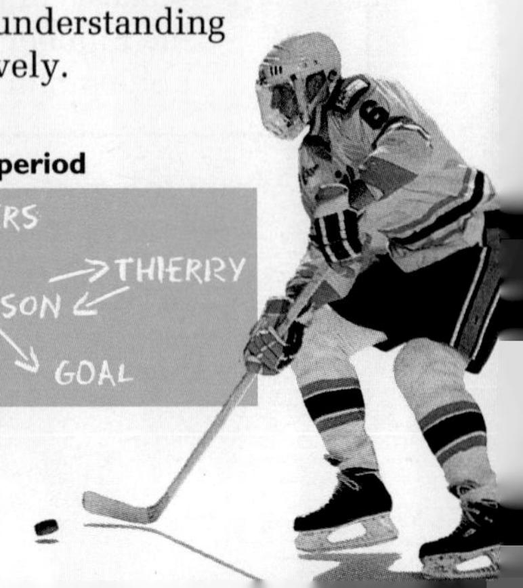

Data Analysis

In a sense, our parents never studied data analysis because they couldn't. The computations involved were just too long. In the world of paper-and-pencil mathematics, finding the mean height of 200 high school students would be an unreasonable task. Now, with a spreadsheet and a good plan, such a task would be an interesting one. On a larger scale, we expect people in business, science, and government to identify, collect, and analyze data relevant to their fields. Good decisions and good plans for all of us depend on accurate and competent understanding of data. The skills for this analysis must be taught in school.

Discrete Mathematics

Discrete mathematics has been called the mathematics of our time. It's a fitting title, because discrete structures are present at every level of our lives, from street maps to the electronic world of the Internet. The discrete steps in computer algorithms define the thinking and the power of these machines. Computer languages themselves use discrete mathematical design. Read any piece of computer code and you will see *recursion, counting,* and *decision nodes,* all important problem-solving tools in the world of discrete mathematics.

We use discrete mathematics every day whether we think about it or not. In choosing the roads we drive on the way to work, we're making discrete decisions at an informal, intuitive level. The phone call we make to a faraway city follows a set of discrete pathways and switching stations. Look at an airline map and you will see hubs and discrete pathways between cities. The structures of graph theory are a natural way to represent, to discuss, and to improve travel.

Mathematics for Informed Citizenship

In ***Integrated Mathematics 3***
See student pages 17, 40, and 351 and side-column commentary on pages 18, 238, and 379.

As citizens, we have a social responsibility to understand data discussions better than we have in the past, and our students will have this responsibility in the future. Most political arguments are laced with references to numbers, and it is important for all of us to be able to evaluate and interpret those arguments accurately. In the same way, voting methods and issues about decision-making—topics in discrete mathematics—need to be understood better by all citizens.

The Basic Skills for the Future

Discrete mathematics and data analysis will not fade away. They will grow in prominence as our world evolves. These topics, once the domain of a few specialists, will define some of the basic skills for the next century for all people.

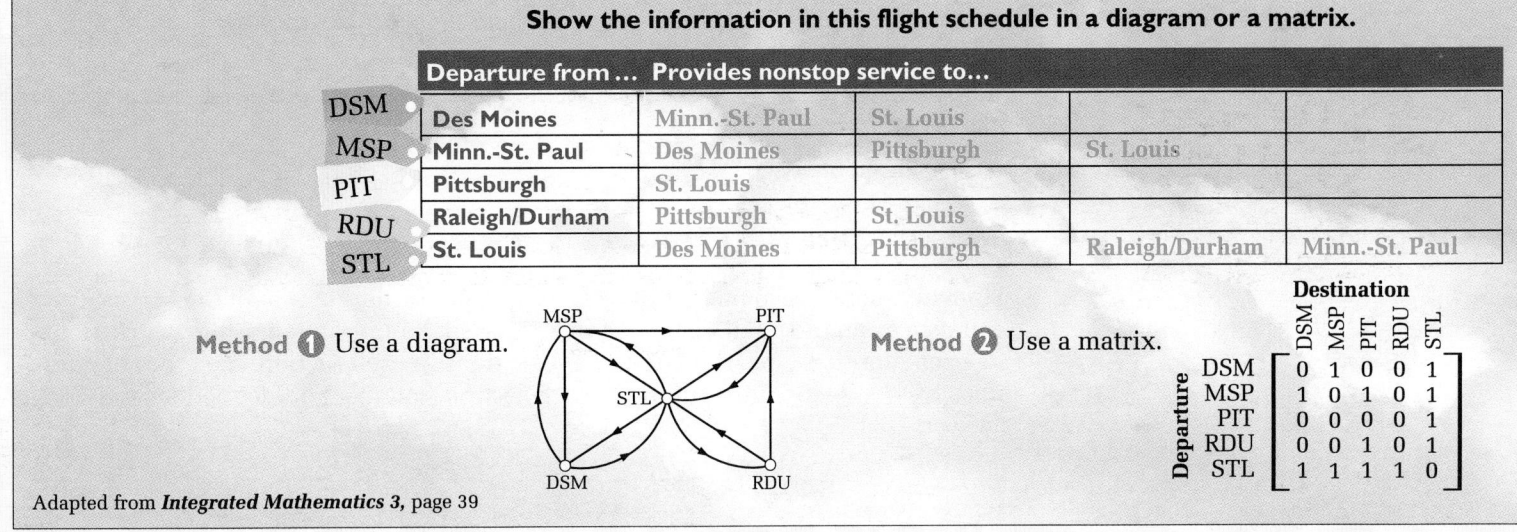

Show the information in this flight schedule in a diagram or a matrix.

	Departure from ...	Provides nonstop service to...			
DSM	**Des Moines**	Minn.-St. Paul	St. Louis		
MSP	**Minn.-St. Paul**	Des Moines	Pittsburgh	St. Louis	
PIT	**Pittsburgh**	St. Louis			
RDU	**Raleigh/Durham**	Pittsburgh	St. Louis		
STL	**St. Louis**	Des Moines	Pittsburgh	Raleigh/Durham	Minn.-St. Paul

Method 1 Use a diagram.

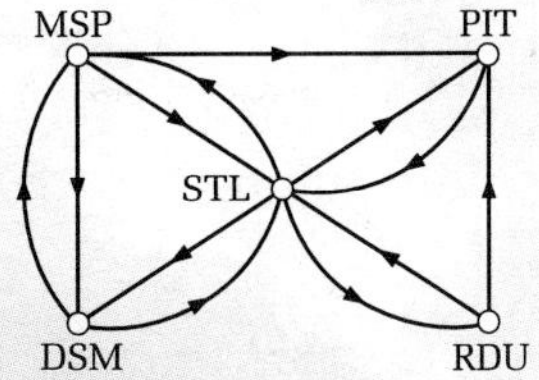

Method 2 Use a matrix.

Destination (columns) / **Departure** (rows)

	DSM	MSP	PIT	RDU	STL
DSM	0	1	0	0	1
MSP	1	0	1	0	1
PIT	0	0	0	0	1
RDU	0	0	1	0	1
STL	1	1	1	1	0

Adapted from ***Integrated Mathematics 3,*** page 39

Preparing for College Entrance Examinations

by ***Anita G. Morris***
Coordinator of Mathematics
Anne Arundel County Public Schools
Annapolis, Maryland

"Students need to acquire a solid base of knowledge in mathematics."

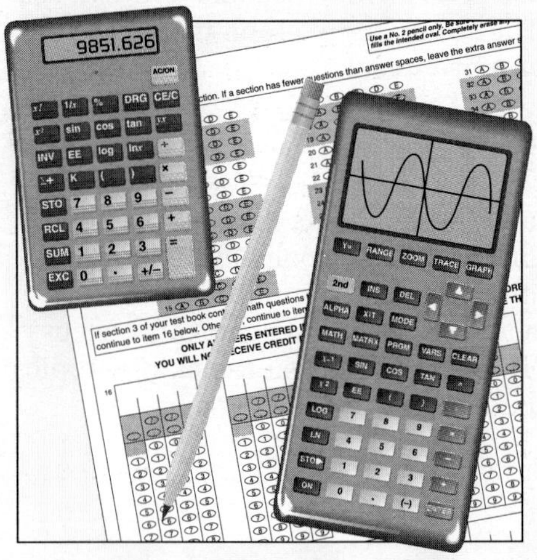

Purpose of College Entrance Examinations

College entrance examinations are intended to measure how well students will do in their first year in college. Test makers are very careful to state that the results of these tests are only predictors of success; they are not meant to measure a student's aptitude for academic work, knowledge, or intelligence. As education changes and the goals of colleges evolve to emphasize thinking rather than rote memorization, the format of college entrance exams will change to reflect a new emphasis on the ability to apply information and skills. The overall purpose of these tests, however, will remain the same.

Three-Tiered Approach to Test Preparation

The mathematics portion of college entrance exams, such as the SAT 1, generally tests three major areas in mathematics: arithmetic, algebra, and geometry. Student success on these tests rests on:

- acquiring a knowledge base in mathematics
- practicing general test-taking strategies
- practicing strategies specific to the mathematical items on the test

Acquiring a Knowledge Base in Mathematics

It is essential that students enroll in mathematics courses that will provide them with a substantive knowledge of arithmetic, algebra, and geometry. Advanced mathematics and formal geometry proofs are usually not included on college entrance exams; however, students should be encouraged to take as many mathematics courses in high school as possible. This not only prepares them for entrance exams but also better prepares them for success in college.

Practicing General Test-Taking Strategies

Test Structure and Format

Students should become familiar with the structure and format of the college entrance exam they will be taking, including such information as the instructions for each question type, the number of questions in each section, the types of questions in each section, the timing of each section, scoring, and deductions for wrong answers.

Key Concepts

Arithmetic

Simple computation

Operations with whole numbers, integers, and rational numbers

Number properties

Averages

Percents

Rates

Number lines

Ratio/proportion

Algebra

Verbal to algebraic translation

Substitution

Simplifying algebraic expressions

Factoring

Solving equations/inequalities

Exponents

Square roots

Quadratic equations

Algebraic symbol problems

Geometry

Parallel and perpendicular lines

Angles in geometric figures

Properties of triangles (right, isosceles, equilateral, 30-60-90)

Pythagorean theorem

Similarity

Properties of polygons, circles

Perimeter, circumference, area, and volume

Simple coordinate geometry

Attractive Distractors

Unlocking the test involves thinking like the test makers. Answer choices for multiple-choice sections will include incorrect answers that are based on common student errors. These are called "attractive distractors." Students should be aware of these distractors and not assume that because their answer appears among the answer choices that it is correct.

Deductions for Wrong Answers

Students may not have taken tests before on which credit is deducted for wrong answers. Test makers include these deductions to eliminate random guessing on multiple-choice items. A common procedure is the following: If there are five answer choices for a question, 1/4 of a point is deducted for a wrong answer; if there are four answer choices, 1/3 of a point is deducted for a wrong answer; 0 points are deducted for omitted answers. It has been found that if one can narrow the answer choices to two or three, taking an "educated guess" at an answer will increase one's scores substantially over the long run.

Difficulty of Questions and Timing

Some entrance exams are constructed so that the questions increase in difficulty over a section. It is extremely important for students to know whether the same number of points is awarded for answering easy questions as for answering hard questions. If so, on a timed test it is to students' advantage to concentrate on answering as many easy and medium questions as possible. They should avoid using a majority of testing time in answering difficult questions. Timing is essential in becoming a successful test taker.

Simulated Test Practice

Short-term test preparation should include simulated test practices, where students concentrate on general test-taking strategies, including:

- knowing the test structure
- knowing the format of each question type
- knowing the test directions
- knowing how to pace themselves
- learning when to guess
- concentrating on easy questions

Practicing Strategies Specific to Mathematical Items

Mathematical Topics

Within the three broad categories of arithmetic, algebra, and geometry, there are key concepts with which students should be familiar. The table at the left categorizes concepts that are often required. In addition, concepts of simple probability, logic, statistics, and other topics may be included on the test. Simply knowing these concepts will not assure success; students must be comfortable

Multiple Choice

If $3(x - 2) = 27$, then $x - 2$ is:

(A) 3
(B) 5
(C) 6
(D) 9
(E) 11

The correct answer is (D).

Quantitative Comparison

Column A	Column B
3^{12}	$3^{13} - 3^{12}$

Adding 3^{12} to both columns allows you to compare:

$2(3^{12})$	3^{13}
$2(3^{12})$	$3(3^{12})$

The answer is (B).

Student-Produced Response

If the lengths of two sides of a triangle are 4 cm and 7 cm, what is a possible value for the length of the third side?

Any value between 11 and 3 would be correct. If a student chose $9\frac{1}{2}$ as the solution, the correct gridded answer would be:

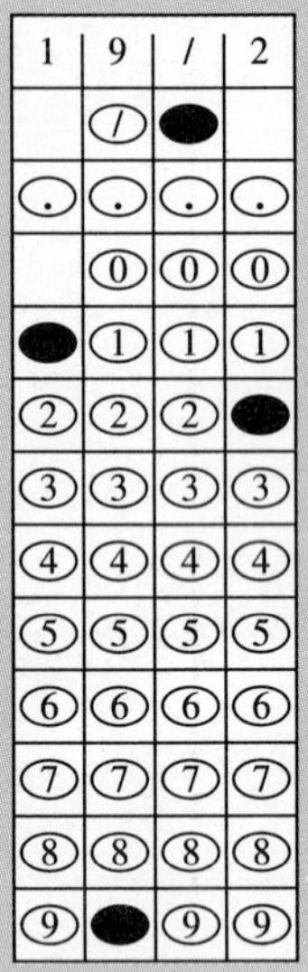

or

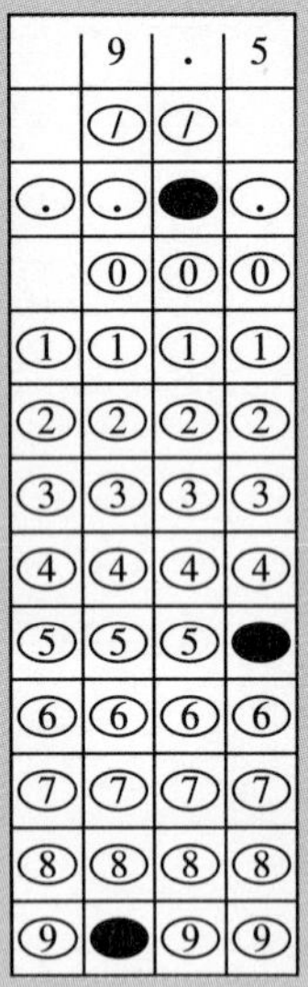

with applying them in a problem solving situation. Students should look over the list of required skills for the test they will be taking and categorize them as ones they know, ones they need to review, or ones that are difficult. Testing simulations, using actual items from previous tests, will help them decide where to concentrate their efforts. A long-term goal in test preparation should be increasing their ability to solve problems that apply these skills. This type of preparation will have the greatest effect on increasing their test scores and assuring their success in college.

Types of Questions

It is important that students be familiar with the question types typically found on college entrance exams. These include: five-choice Multiple-Choice questions, four-choice Quantitative Comparison questions, and Student-Produced Responses. An example of a Multiple-Choice question is shown at the left. In Quantitative Comparison questions, students are presented two quantities, one in Column A and one in Column B. They must determine whether the quantity in Column A is greater (choice A), the quantity in Column B is greater (choice B), the quantities are equal (choice C), or you cannot determine which is greater from the information given (choice D). An (E) response will not be scored. A typical Quantitative Comparison question appears at the left.

Student-Produced Responses require students to fill in the correct answer, rather than choose from a group of choices. Specific instructions for entering answers in the grid are provided by the test makers. A sample of this type of question, and two possible gridded answers for it, are shown at the left. With this type of question, only gridded answers are scored; handwritten answers at the top of the grid are not considered.

The Use of Calculators

Some college entrance exams now allow students to use calculators. However, calculators with a word processing unit, paper tape, or a typewriter keyboard are generally not allowed. Also forbidden are calculators that make noise or require a power outlet. Students are advised to bring a calculator with which they are familiar. It is particularly important that they know how their calculator handles order of operations. Students should not try to answer every question with the calculator. First, they should decide how to solve the problem; then if the calculator will help with the computation, they should use it. The test may be constructed so that every question can be answered without the use of a calculator, but appropriate calculator usage can give them an edge.

Integrated Mathematics 3

Authors

Senior Authors

Rheta N. Rubenstein
Timothy V. Craine
Thomas R. Butts

Kerry Cantrell
Linda Dritsas
Valarie A. Elswick
Joseph Kavanaugh
Sara N. Munshin
Stuart J. Murphy
Anthony Piccolino
Salvador Quezada
Jocelyn Coleman Walton

McDougal Littell
A Houghton Mifflin Company
Evanston, Illinois Boston Dallas Phoenix

Authors

Senior Authors

Rheta N. Rubenstein Associate Professor of Education, University of Windsor, Windsor, Ontario

Timothy V. Craine Assistant Professor of Mathematical Sciences, Central Connecticut State University, New Britain, Connecticut

Thomas R. Butts Associate Professor of Mathematics Education, University of Texas at Dallas, Dallas, Texas

Kerry Cantrell Mathematics Department Head, Marshfield High School, Marshfield, Missouri

Linda Dritsas Mathematics Coordinator, Fresno Unified School District, Fresno, California

Valarie A. Elswick Mathematics Teacher, Roy C. Ketcham Senior High School, Wappingers Falls, New York

Joseph Kavanaugh Academic Head of Mathematics, Scotia-Glenville Central School District, Scotia, New York

Sara N. Munshin Mathematics Teacher, Theodore Roosevelt High School, Los Angeles, California

Stuart J. Murphy Visual Learning Specialist, Evanston, Illinois

Anthony Piccolino Assistant Professor of Mathematics and Computer Science, Montclair State College, Upper Montclair, New Jersey

Salvador Quezada Mathematics Teacher, Theodore Roosevelt High School, Los Angeles, California

Jocelyn Coleman Walton Educational Consultant, Mathematics K-12, and former Mathematics Supervisor, Plainfield High School, Plainfield, New Jersey

All authors contributed to the planning and writing of the series. In addition to writing, the Senior Authors played a special role in establishing the philosophy of the program, planning the content and organization of topics, and guiding the work of the other authors.

Field Testing The authors give special thanks to the teachers and students in classrooms nationwide who used a preliminary version of this book. Their suggestions made an important contribution to its development.

ISBN: 0-395-64448-8 123456789 VH 99 98 97 96 95

Editorial Advisers and Reviewers

Carol A. Andrews Mathematics Teacher, Greece Arcadia High School, Rochester, New York

John G. Balzano Mathematics Teacher, Johanna Perrin Middle School, Fairport, New York

Janet M. Burt Mathematics and Science Supervisor, Clinton Central School, Clinton, New York

Barbara P. Chamberlain former Mathematics Teacher, Shorecrest High School, Seattle, Washington

Gerlena R. Clark Mathematics Consultant, Jefferson City, Missouri

John Fourcroy Mathematics Teacher, Lindhurst High School, Marysville, California

Charles V. Kruger Mathematics Teacher, Eden Valley-Watkins High School, Eden Valley, Minnesota

Terry Wozny Mathematics Educator, Port Washington School District, Port Washington, Wisconsin

Manuscript Reviewers

Cheryl Avalos Mathematics Consultant, Los Angeles County Office of Education, Los Angeles, California

Joan C. Countryman Head of School, Lincoln School, Providence, Rhode Island

Deborah K. Jenkins Mathematics Department Head, Pershing High School, Detroit, Michigan

Cesar Larriva Doctoral Student, Department of Education, Stanford University, Stanford, California

Les Leibovitch Mathematics Teacher, Van Nuys Middle School, Mathematics and Science Magnet, Sherman Oaks, California

Karen S. Norwood Assistant Professor, Department of Mathematics and Science Education, North Carolina State University, Raleigh, North Carolina

Jeri P. Philbrick Mathematics Department Chairperson, Oxnard High School, Oxnard, California

Matthias Vheru Secondary Mathematics Teacher, Locke High School/Los Angeles Unified School District, Los Angeles, California

Program Consultants

Joel L. Arougheti Mathematics Supervisor, High School of Art and Design, New York, New York

Sandrea Assam Assistant Principal for Math Instruction, Andrew Jackson High School, Queens, New York

William H. Harris Coordinator of Mathematics and Science, City School District of Troy, Troy, New York

Samuel Johnson, Jr. Director of Personnel, San Mateo Union High School District, San Mateo, California

Euclid Mejia Principal, George Washington High School, New York, New York

Richard P. Stucky Coordinator of Mathematics and Technology, Shoreline School District, Seattle, Washington

Barbara Griggs Wells UCLA Site Director, Mathematics Diagnostic Testing Project, UCLA, Los Angeles, California

Welcome to Integrated Mathematics 3!

A Strong Foundation

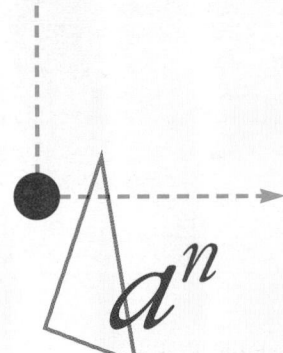

Integrated Mathematics 3 builds on the mathematical topics and problem solving techniques in *Integrated Mathematics 1* and *Integrated Mathematics 2*. This three-year series provides a strong foundation for more advanced courses in mathematics and the many other fields that require quantitative reasoning and decision-making skills.

Mathematical Strands

You can learn more with *Integrated Mathematics* because the mathematical topics are integrated. Over a three-year period, this program teaches all the essential topics in a contemporary Algebra 1/Geometry/Algebra 2 sequence, plus many other interesting topics.

- Algebra and Geometry are taught in each of the three years.
- Topics from Logical Reasoning, Measurement, Probability, Statistics, Discrete Mathematics, and Functions are interwoven throughout.

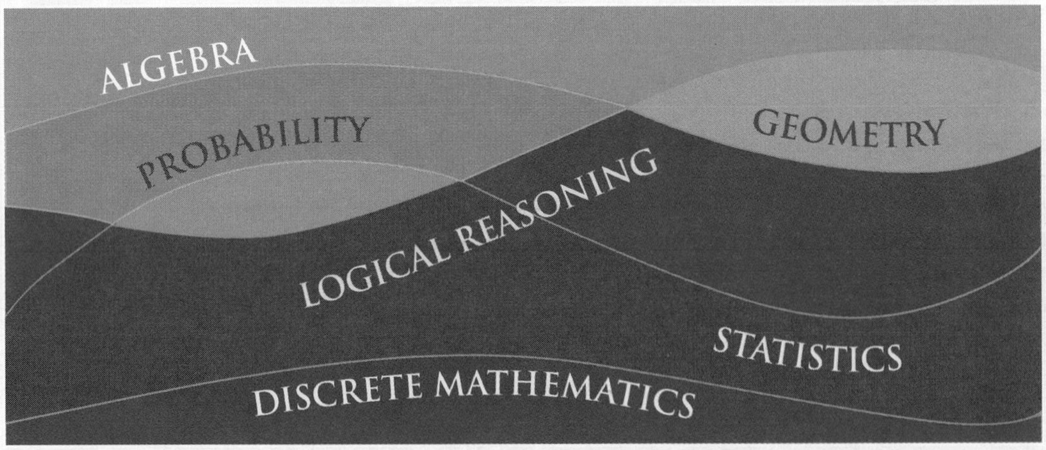

iv

$p \rightarrow q$

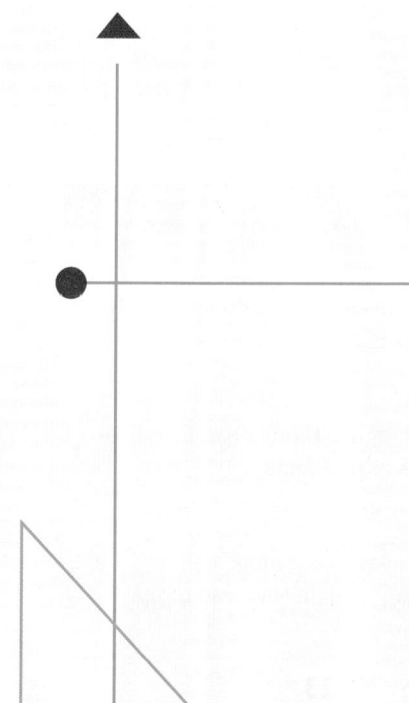

Course Goals

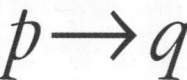

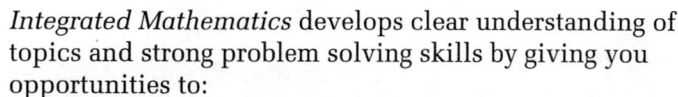

This new program has been written to prepare you for success in college, in careers, and in daily life in the 21st century.

It helps you develop the ability to:

- **Explore and solve mathematical problems**
- **Think critically**
- **Work cooperatively with others**
- **Communicate ideas clearly**

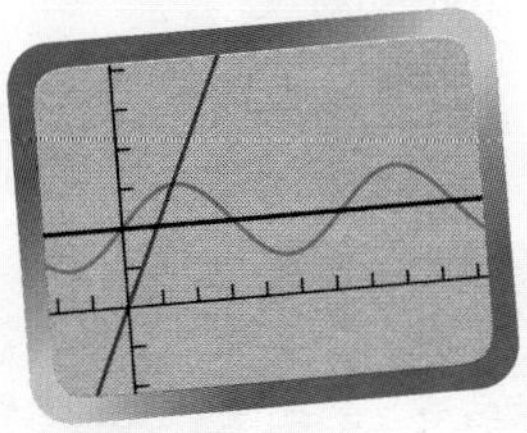

Advantages of this Program

Integrated Mathematics develops clear understanding of topics and strong problem solving skills by giving you opportunities to:

- **Get actively involved in learning**
- **Study meaningful mathematics**
- **See connections among different branches of mathematics**
- **Try a wide variety of types of problems, including real-world applications and long-term projects**
- **Use calculators and computers**

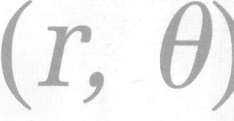

(r, θ)

v

Contents

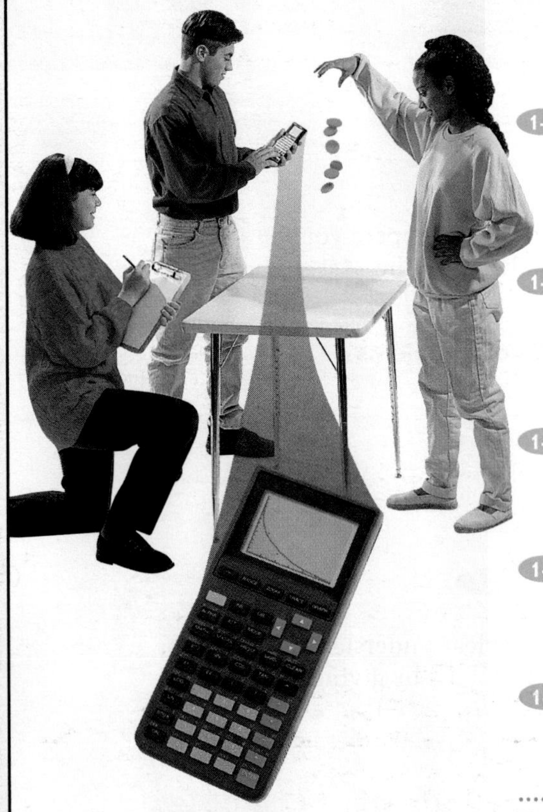

vii

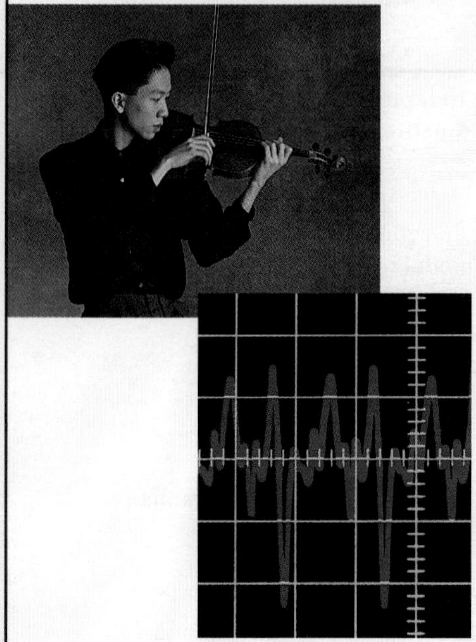

Unit 5 Exponential and Logarithmic Functions

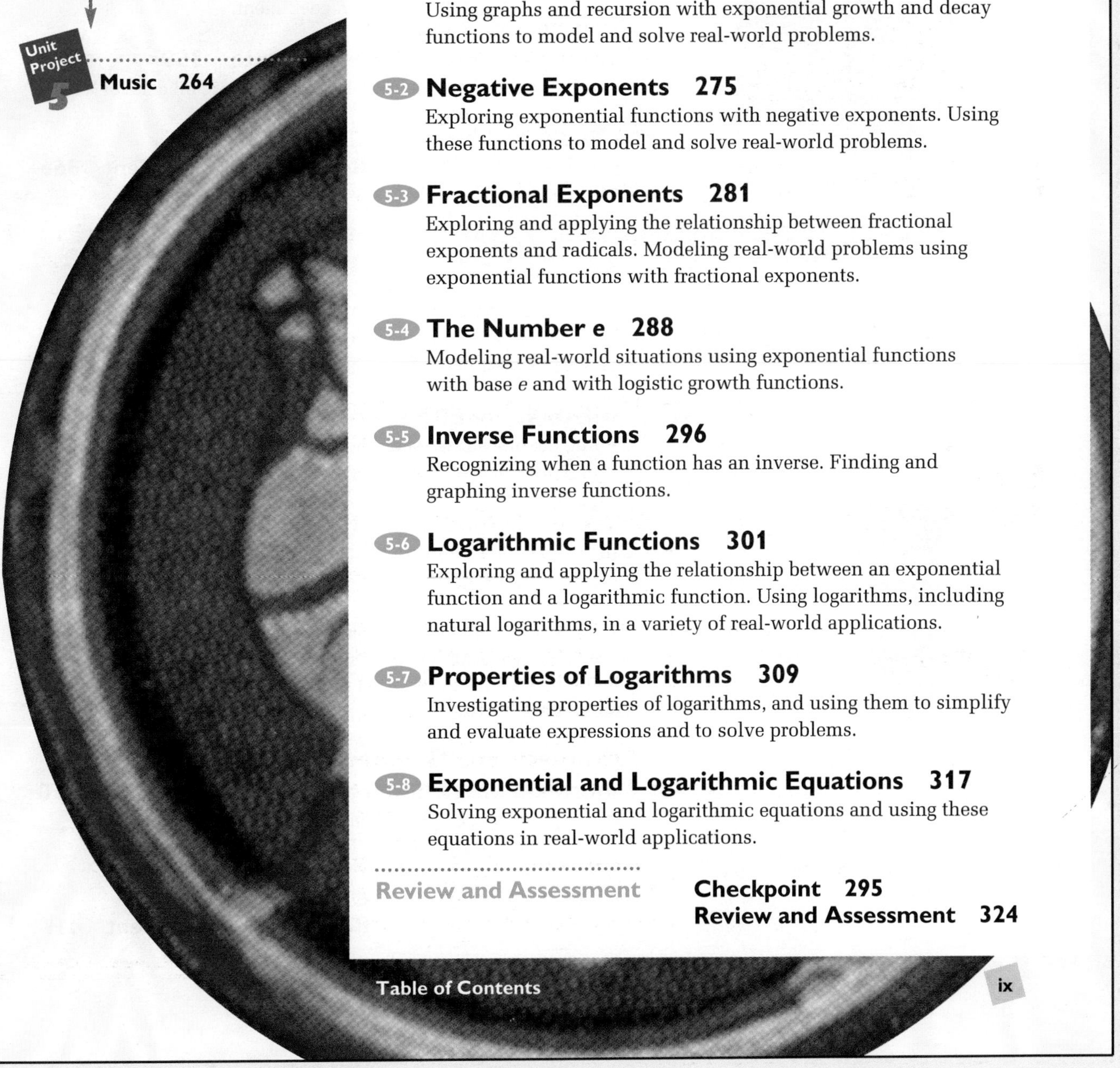

Unit 6 Modeling and Analyzing Data

Unit 7 Applying Probability Models

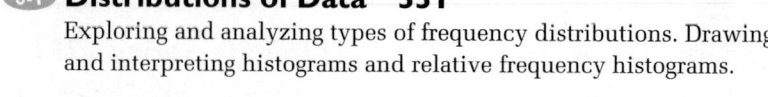

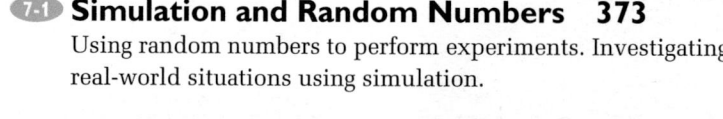

Unit 8 Angles, Trigonometry, and Vectors

Unit 9 Transformations of Graphs and Data

Unit 10 Periodic Models

xii

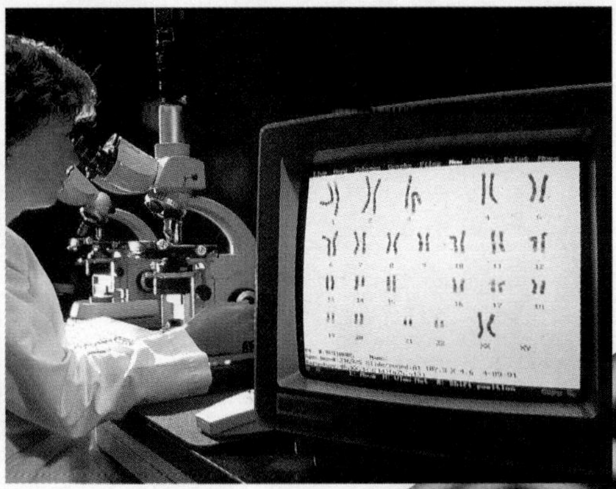

Student Resources

Integrated Mathematics Topic Spiraling

This chart shows how mathematical strands are spiraled over the three years of the *Integrated Mathematics* program.

	Course 1	Course 2	Course 3
Algebra	Linear equations Linear inequalities Multiplying binomials Factoring expressions	Quadratic equations Linear systems Rational equations Complex numbers	Polynomial functions Exponential and logarithmic functions Radical functions Parametric equations
Geometry	Angles, polygons, circles Perimeter, circumference Area, surface area Volume	Similar and congruent figures Geometric proofs Coordinate geometry Transformational geometry	Inscribed polygons Tangents to circles Transforming graphs Vectors, polar coordinates
Statistics, Probability	Analyzing data and displaying data Experimental and theoretical probability Geometric probability	Sampling methods Simulation Binomial distributions Compound events	Standard deviation Normal distribution Conditional probability z-scores
Logical Reasoning	Conjectures Counterexamples If-then statements	Inductive and deductive reasoning Valid and invalid reasoning Postulates and proof	Identities Contrapositive and inverse Comparing proof methods
Discrete Math	Discrete quantities Matrices to display data Lattices	Matrix operations Transformation matrices Counting techniques	Sequences and series Recursion Network diagrams
Trigonometry	Right-triangle ratios: sine, cosine, tangent Solving right triangles Tangent ratio as slope	Applying properties of special right triangles: 30° - 60° - 90° 45° - 45° - 90°	Circular trigonometric functions and graphs Modeling periodic situations Law of cosines, law of sines

What Students are Saying...

T<H T

Who has the **BEST IDEAS** about how mathematics should be taught? Students and teachers, of course! That is why preliminary versions of this book were tried out by thousands of students and their teachers in CLASSROOMS NATIONWIDE. The suggestions from these students and teachers have been incorporated into this book. Here is what some of the students who have already studied this course have said.

I enjoy doing the Explorations and solving application problems that relate to my life and the world around me.

I'm really involved in math class this year. I read and think about problems, and analyze and discuss solutions with other students.

cos C

xiv

Your Ideas are Important!

Get Involved

This course may be different from ones you have taken before. In this course you will be

- TALKING about mathematics
- working *together* to explore ideas
- gathering **DATA**
- looking for patterns
- making and testing **predictions**
- using ***calculators*** or **computers.**

Your ideas and viewpoints are important. Sharing them with others will help everyone learn more. So don't hold back. Jump right in and get involved.

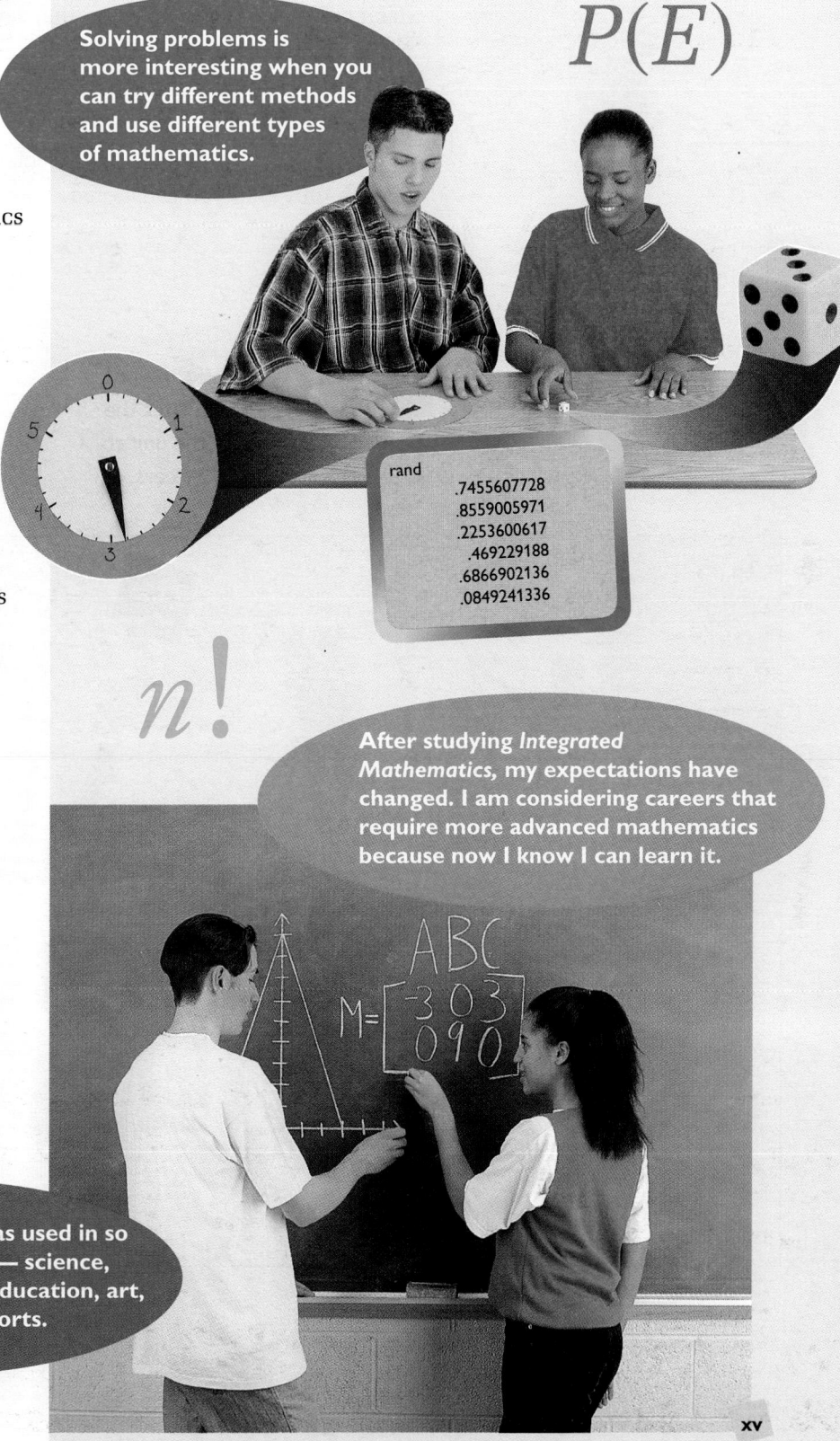

Solving problems is more interesting when you can try different methods and use different types of mathematics.

$P(E)$

$n!$

After studying *Integrated Mathematics,* my expectations have changed. I am considering careers that require more advanced mathematics because now I know I can learn it.

Guide to Your Course

The next ten pages will give you an overview of the organization of your book and a preview of what you will be learning in the course. They will help you get off to a good start.

I never realized math was used in so many different subjects — science, social studies, business education, art, music, literature, and sports.

xv

Unit Projects

Each unit begins with a project that sets the stage for the mathematics in the unit. The project gives you a chance to **apply** what you are learning right away. As you study the unit, you will gather the **INFORMATION** and develop the SKILLS to complete the project. The first three pages of each unit help you get started.

Project Theme

Each project has a theme, like music, that relates the mathematics of the unit to daily life and to careers.

unit 5

Exponential and Logarithmic Functions

If you wanted to write a song for the whole world to sing, what musical system would you use? In China most music is written in a five-note scale. Traditional European music is based on an eight-note octave. In India the raga is an important musical form. Ragas use scales with various sequences of five, six, or seven notes.

Different instruments sound different because of the blend of higher tones, or overtones, produced with each note. In 1965 the first electronic synthesizer to produce music with vibrating electric circuits was invented. Synthesizers can imitate all the instruments of the orchestra.

To uncover the secrets of the great violin makers of the past, scientist and violin maker Carleen M. Hutchins has handcrafted over 400 instruments in the last 40 years. When a violin is played, the vibrating strings cause both the wooden walls of the violin and the air inside to vibrate. By analyzing these vibrations, Carleen M. Hutchins has worked out basic rules for making violins.

264

Unit Project 5

Making Music with Mathematics

Your project is to give a lecture/performance that demonstrates how mathematics and music are related. Following the guidelines in the Working on the Unit Project exercises, your group should make two sets of panpipes and two simple guitars. You should describe how you built your instruments, use them to play music, and explain how they produce musical notes.

Describe at least two examples of the relationship between mathematics and music.

This wooden harp is from Zaire. African harps like this traditionally played a vital role in the lives of millions of people. Many harp songs tell of historical events or legends.

On a stringed instrument like a violin or a guitar, one way to raise the pitch, or produce a higher note, is to shorten or tighten a string. Another way is to use a thinner string.

Shortening the column of air that vibrates when a musician blows into a wind instrument like a flute or panpipe raises the pitch.

265

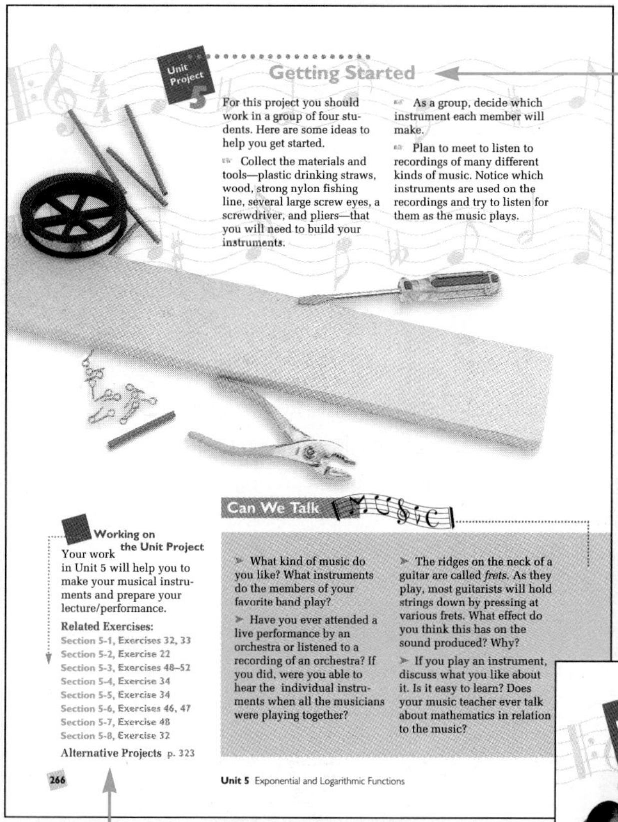

Unit Project 5

Getting Started

For this project you should work in a group of four students. Here are some ideas to help you get started.

- Collect the materials and tools—plastic drinking straws, wood, strong nylon fishing line, several large screw eyes, a screwdriver, and pliers—that you will need to build your instruments.
- As a group, decide which instrument each member will make.
- Plan to meet to listen to recordings of many different kinds of music. Notice which instruments are used on the recordings and try to listen for them as the music plays.

Working on the Unit Project

Your work in Unit 5 will help you to make your musical instruments and prepare your lecture/performance.

Related Exercises:

Section 5-1, Exercises 32, 33
Section 5-2, Exercise 22
Section 5-3, Exercises 48–52
Section 5-4, Exercise 34
Section 5-5, Exercise 34
Section 5-6, Exercises 46, 47
Section 5-7, Exercise 48
Section 5-8, Exercise 32

Alternative Projects p. 323

Can We Talk MUSIC

- What kind of music do you like? What instruments do the members of your favorite band play?
- Have you ever attended a live performance by an orchestra or listened to a recording of an orchestra? If you did, were you able to hear the individual instruments when all the musicians were playing together?
- The ridges on the neck of a guitar are called *frets*. As they play, most guitarists will hold strings down by pressing at various frets. What effect do you think this has on the sound produced? Why?
- If you play an instrument, discuss what you like about it. Is it easy to learn? Does your music teacher ever talk about mathematics in relation to the music?

266 Unit 5 Exponential and Logarithmic Functions

Starting the Project

At the beginning of each unit, there is a description of the project, hints for getting started, and questions to get you thinking and talking about the project.

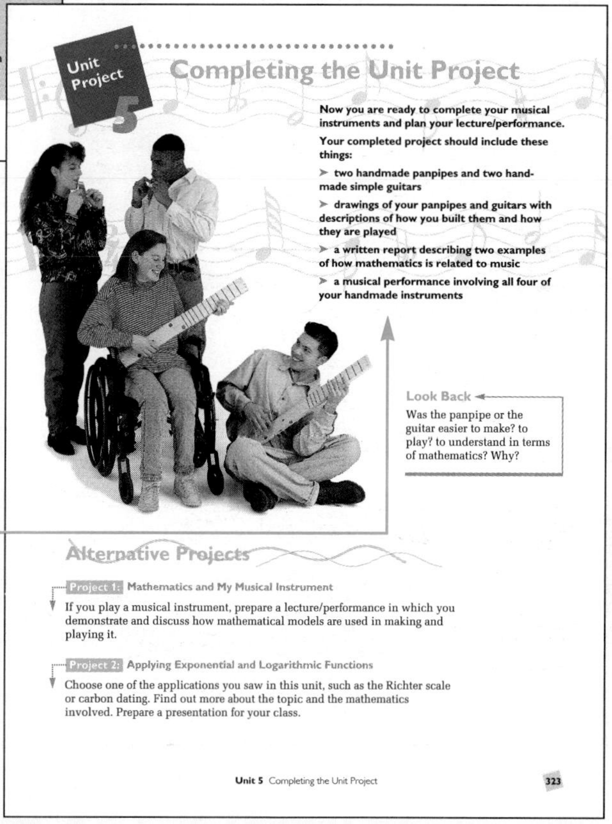

Unit Project 5

Completing the Unit Project

Now you are ready to complete your musical instruments and plan your lecture/performance.

Your completed project should include these things:

- **two handmade panpipes and two handmade simple guitars**
- **drawings of your panpipes and guitars with descriptions of how you built them and how they are played**
- **a written report describing two examples of how mathematics is related to music**
- **a musical performance involving all four of your handmade instruments**

Look Back

Was the panpipe or the guitar easier to make? to play? to understand in terms of mathematics? Why?

Alternative Projects

Project 1: Mathematics and My Musical Instrument

If you play a musical instrument, prepare a lecture/performance in which you demonstrate and discuss how mathematical models are used in making and playing it.

Project 2: Applying Exponential and Logarithmic Functions

Choose one of the applications you saw in this unit, such as the Richter scale or carbon dating. Find out more about the topic and the mathematics involved. Prepare a presentation for your class.

Unit 5 Completing the Unit Project 323

Working on the Unit Project

In each section of the unit, there are exercises to help you complete your work on the project.

Completing the Unit Project

After you complete the unit, you can finish the project and present your results. Then you are ready to look back over what you learned.

Unit Projects

Section Organization

The organization of material within sections is patterned after the way that you learn: Ideas are introduced. You EXPLORE them, **think** about them, and TALK about them with other students. You check that you UNDERSTAND them by working through some sample problems. Before going on, you pause and look back at what you have learned.

Application A problem situation connects the mathematics you will be studying to real-world applications.

Focus This is what you will be doing in the lesson.

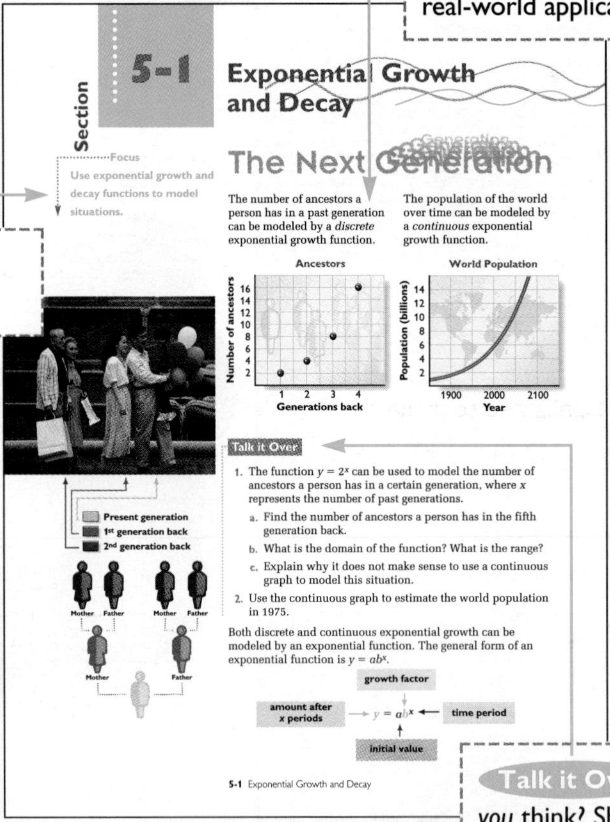

Section 5-1

Exponential Growth and Decay

Focus

Use exponential growth and decay functions to model situations.

The Next Generation

The number of ancestors a person has in a past generation can be modeled by a *discrete* exponential growth function.

The population of the world over time can be modeled by a *continuous* exponential growth function.

Talk it Over

1. The function $y = 2^x$ can be used to model the number of ancestors a person has in a certain generation, where x represents the number of past generations.
   a. Find the number of ancestors a person has in the fifth generation back.
   b. What is the domain of the function? What is the range?
   c. Explain why it does not make sense to use a continuous graph to model this situation.
2. Use the continuous graph to estimate the world population in 1975.

Both discrete and continuous exponential growth can be modeled by an exponential function. The general form of an exponential function is $y = ab^x$.

growth factor; amount after x periods → $y = ab^x$ ← time period; initial value

5-1 Exponential Growth and Decay

Talk it Over What do *you* think? Share your ideas. Talking them over can help you deepen your understanding.

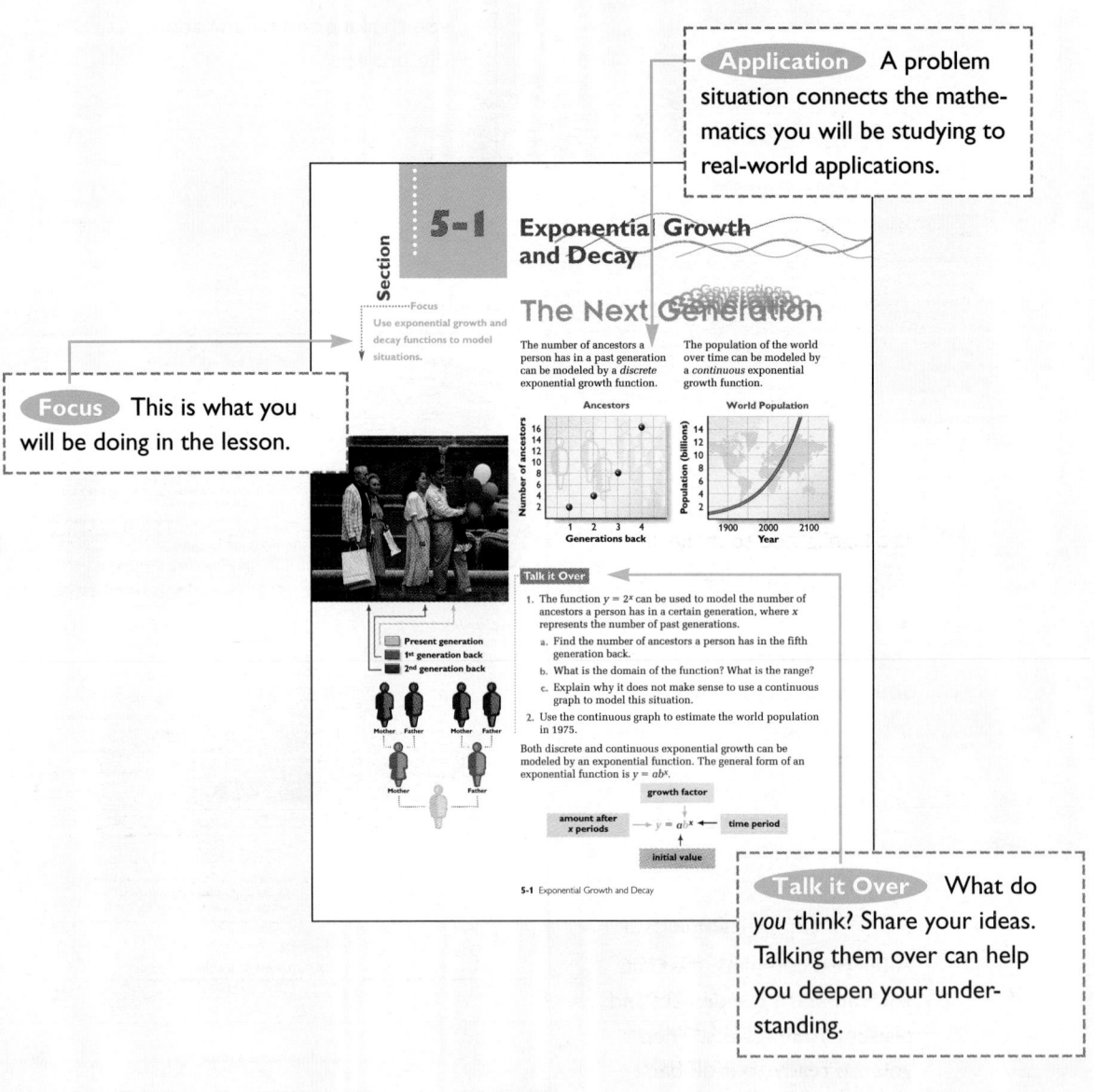

xviii

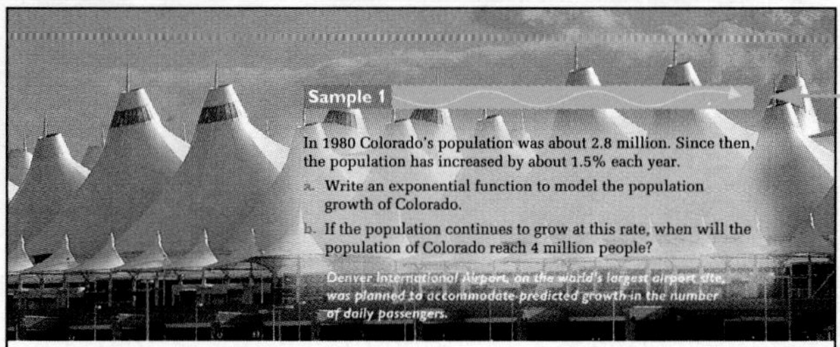

Sample 1

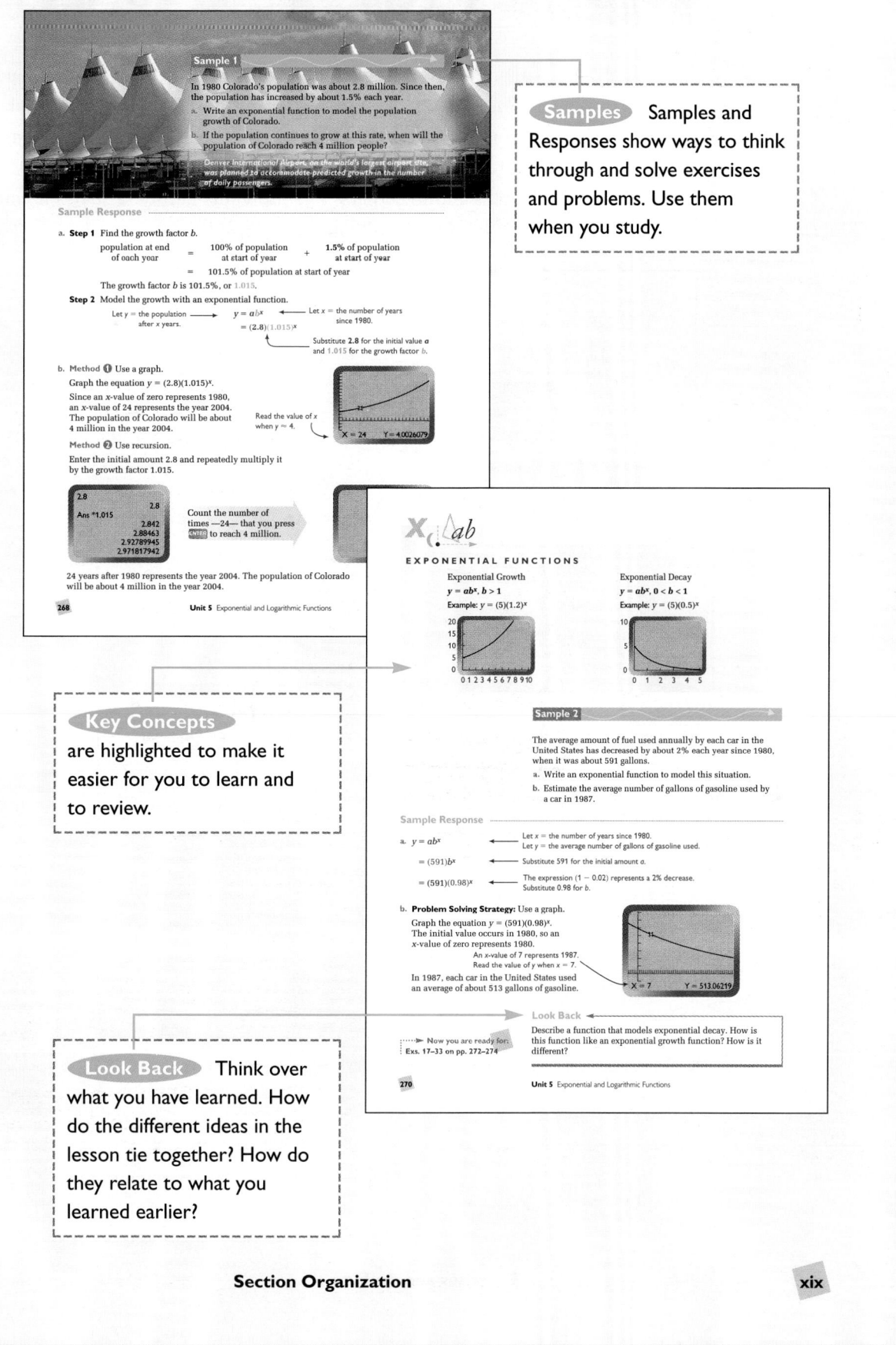

In 1980 Colorado's population was about 2.8 million. Since then, the population has increased by about 1.5% each year.

a. Write an exponential function to model the population growth of Colorado.

b. If the population continues to grow at this rate, when will the population of Colorado reach 4 million people?

Denver International Airport, on the world's largest airport site, was planned to accommodate predicted growth in the number of daily passengers.

Samples Samples and Responses show ways to think through and solve exercises and problems. Use them when you study.

Sample Response

a. **Step 1** Find the growth factor b.

population at end of each year = 100% of population at start of year + 1.5% of population at start of year

= 101.5% of population at start of year

The growth factor b is 101.5%, or 1.015.

Step 2 Model the growth with an exponential function.

Let y = the population after x years. → $y = ab^x$ ← Let x = the number of years since 1980.

$= (2.8)(1.015)^x$

Substitute **2.8** for the initial value a and 1.015 for the growth factor b.

b. **Method 1** Use a graph.

Graph the equation $y = (2.8)(1.015)^x$.

Since an x-value of zero represents 1980, an x-value of 24 represents the year 2004. The population of Colorado will be about 4 million in the year 2004.

Read the value of x when $y \approx 4$.

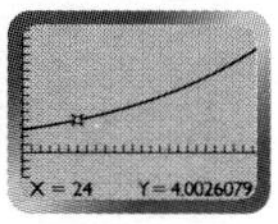

Method 2 Use recursion.

Enter the initial amount 2.8 and repeatedly multiply it by the growth factor 1.015.

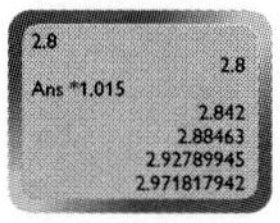

Count the number of times —24— that you press ENTER to reach 4 million.

24 years after 1980 represents the year 2004. The population of Colorado will be about 4 million in the year 2004.

EXPONENTIAL FUNCTIONS

Exponential Growth

$y = ab^x$, $b > 1$

Example: $y = (5)(1.2)^x$

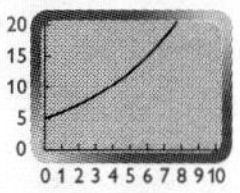

Exponential Decay

$y = ab^x$, $0 < b < 1$

Example: $y = (5)(0.5)^x$

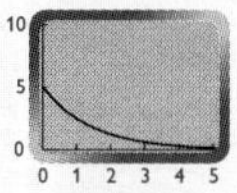

Key Concepts are highlighted to make it easier for you to learn and to review.

Sample 2

The average amount of fuel used annually by each car in the United States has decreased by about 2% each year since 1980, when it was about 591 gallons.

a. Write an exponential function to model this situation.

b. Estimate the average number of gallons of gasoline used by a car in 1987.

Sample Response

a. $y = ab^x$ ← Let x = the number of years since 1980. Let y = the average number of gallons of gasoline used.

$= (591)b^x$ ← Substitute 591 for the initial amount a.

$= (591)(0.98)^x$ ← The expression $(1 - 0.02)$ represents a 2% decrease. Substitute 0.98 for b.

b. **Problem Solving Strategy:** Use a graph.

Graph the equation $y = (591)(0.98)^x$.
The initial value occurs in 1980, so an x-value of zero represents 1980.

An x-value of 7 represents 1987. Read the value of y when $x = 7$.

In 1987, each car in the United States used an average of about 513 gallons of gasoline.

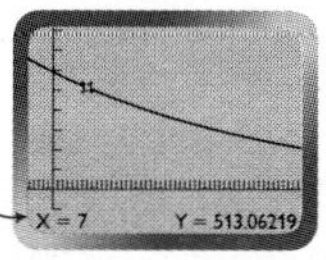

Look Back

Describe a function that models exponential decay. How is this function like an exponential growth function? How is it different?

Now you are ready for: Exs. 17–33 on pp. 272–274

Look Back Think over what you have learned. How do the different ideas in the lesson tie together? How do they relate to what you learned earlier?

Exercises and Problems

Each section has a wide variety of exercises and problems. Some **practice** and **EXTEND** the concepts and skills you have learned. Others apply the concepts to everyday situations and explore **connections** to other subject areas and to careers. The problems help you sharpen your **THINKING** and **problem solving** skills.

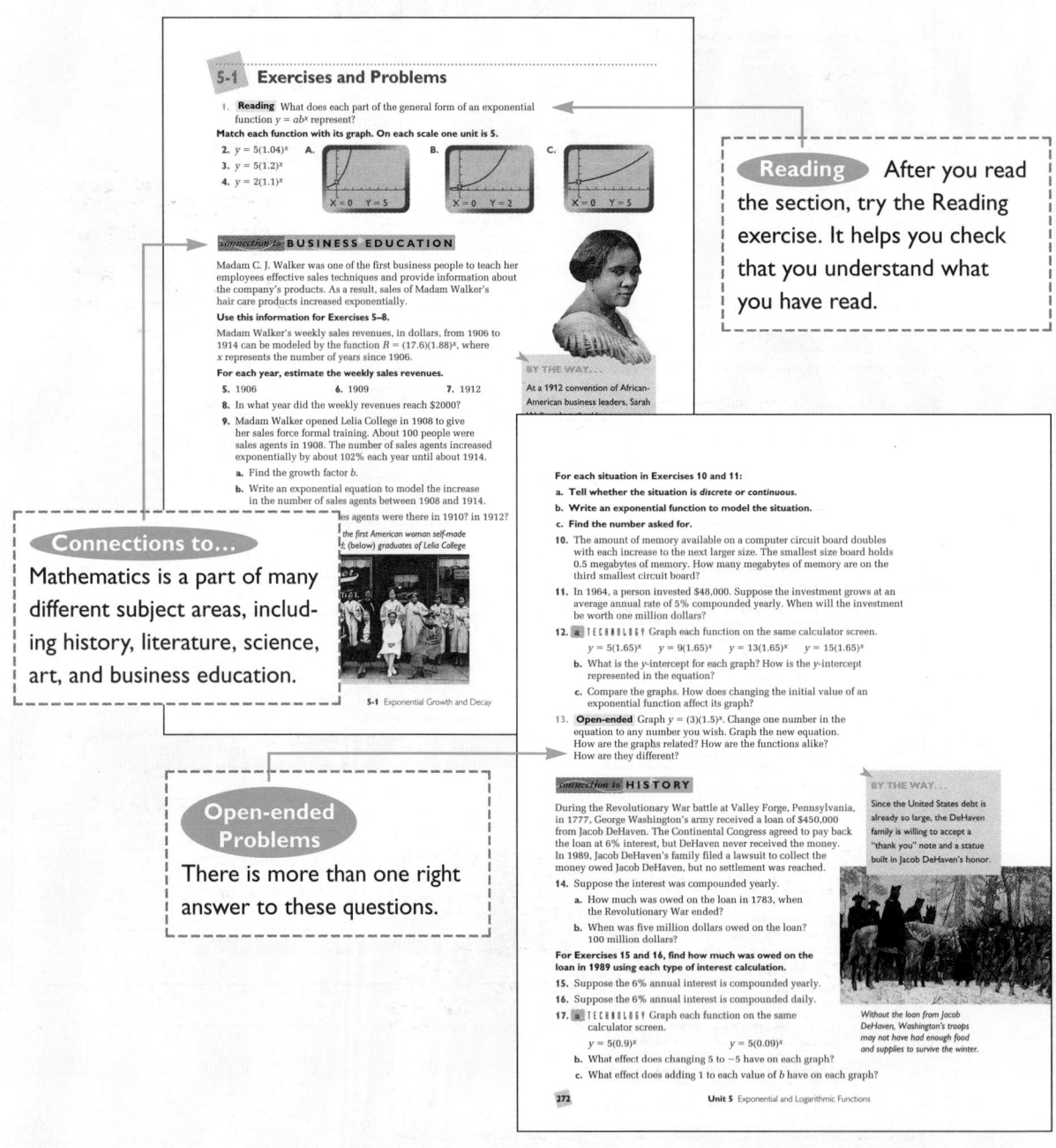

5-1 Exercises and Problems

1. **Reading** What does each part of the general form of an exponential function $y = ab^x$ represent?

Match each function with its graph. On each scale one unit is 5.

2. $y = 5(1.04)^x$
3. $y = 5(1.2)^x$
4. $y = 2(1.1)^x$

A.

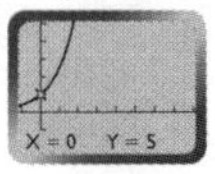

B.

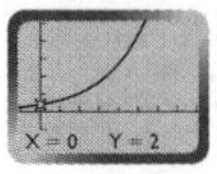

C.

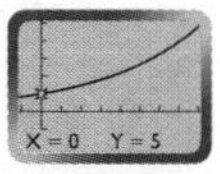

Reading After you read the section, try the Reading exercise. It helps you check that you understand what you have read.

connection to **BUSINESS EDUCATION**

Madam C. J. Walker was one of the first business people to teach her employees effective sales techniques and provide information about the company's products. As a result, sales of Madam Walker's hair care products increased exponentially.

Use this information for Exercises 5–8.

Madam Walker's weekly sales revenues, in dollars, from 1906 to 1914 can be modeled by the function $R = (17.6)(1.88)^x$, where x represents the number of years since 1906.

For each year, estimate the weekly sales revenues.

5. 1906
6. 1909
7. 1912
8. In what year did the weekly revenues reach $2000?
9. Madam Walker opened Lelia College in 1908 to give her sales force formal training. About 100 people were sales agents in 1908. The number of sales agents increased exponentially by about 102% each year until about 1914.
   a. Find the growth factor b.
   b. Write an exponential equation to model the increase in the number of sales agents between 1908 and 1914.
   ...es agents were there in 1910? in 1912?

BY THE WAY...

At a 1912 convention of African-American business leaders, Sarah

...the first American woman self-made ...d; (below) graduates of Lelia College

Connections to... Mathematics is a part of many different subject areas, including history, literature, science, art, and business education.

For each situation in Exercises 10 and 11:

a. Tell whether the situation is *discrete or continuous.*

b. Write an exponential function to model the situation.

c. Find the number asked for.

10. The amount of memory available on a computer circuit board doubles with each increase to the next larger size. The smallest size board holds 0.5 megabytes of memory. How many megabytes of memory are on the third smallest circuit board?
11. In 1964, a person invested $48,000. Suppose the investment grows at an average annual rate of 5% compounded yearly. When will the investment be worth one million dollars?
12. a. TECHNOLOGY Graph each function on the same calculator screen.
    $y = 5(1.65)^x$ $y = 9(1.65)^x$ $y = 13(1.65)^x$ $y = 15(1.65)^x$
    b. What is the y-intercept for each graph? How is the y-intercept represented in the equation?
    c. Compare the graphs. How does changing the initial value of an exponential function affect its graph?
13. **Open-ended** Graph $y = (3)(1.5)^x$. Change one number in the equation to any number you wish. Graph the new equation. How are the graphs related? How are the functions alike? How are they different?

Open-ended Problems There is more than one right answer to these questions.

connection to **HISTORY**

During the Revolutionary War battle at Valley Forge, Pennsylvania, in 1777, George Washington's army received a loan of $450,000 from Jacob DeHaven. The Continental Congress agreed to pay back the loan at 6% interest, but DeHaven never received the money. In 1989, Jacob DeHaven's family filed a lawsuit to collect the money owed Jacob DeHaven, but no settlement was reached.

BY THE WAY...

Since the United States debt is already so large, the DeHaven family is willing to accept a "thank you" note and a statue built in Jacob DeHaven's honor.

14. Suppose the interest was compounded yearly.
    a. How much was owed on the loan in 1783, when the Revolutionary War ended?
    b. When was five million dollars owed on the loan? 100 million dollars?

For Exercises 15 and 16, find how much was owed on the loan in 1989 using each type of interest calculation.

15. Suppose the 6% annual interest is compounded yearly.
16. Suppose the 6% annual interest is compounded daily.
17. a. TECHNOLOGY Graph each function on the same calculator screen.
    $y = 5(0.9)^x$ $y = 5(0.09)^x$
    b. What effect does changing 5 to −5 have on each graph?
    c. What effect does adding 1 to each value of b have on each graph?

Without the loan from Jacob DeHaven, Washington's troops may not have had enough food and supplies to survive the winter.

Ongoing Assessment

These problems pull together the different things you have studied in the section.

Ongoing **ASSESSMENT**

24. **Open-ended** Describe a situation that can be modeled by each of the following: *discrete exponential growth, continuous exponential growth, discrete exponential decay, continuous exponential decay.* Sketch a graph for each example.

Review/Preview

These exercises at the end of each section help you keep skills alive and get ready for the next section.

Review **PREVIEW**

Evaluate each series, if the sum exists. (Section 4-7)

25. $\sum_{n=0}^{\infty}(1.03)^n$

26. $\sum_{t=1}^{\infty}\left(\frac{1}{2}\right)^t$

27. $\sum_{n=1}^{\infty}-5(0.003)^{n-1}$

28. Can the sum of the interior angle measures of a regular polygon be 655°? Why or why not? (Section 3-4)

Simplify. Write answers with positive exponents. (Toolbox Skill 14)

29. $(3x^2)(5x^{-3})$

30. $5t(s^{-1}t^{-2})^{-2}$

31. $\frac{p^{-1}q^{-2}}{p^{-3}}$

Working on the Unit Project

The *pitch* of a note is how high or low the note sounds to a listener. Pitch depends on frequency. The frequencies of musical notes in a scale form a geometric sequence.

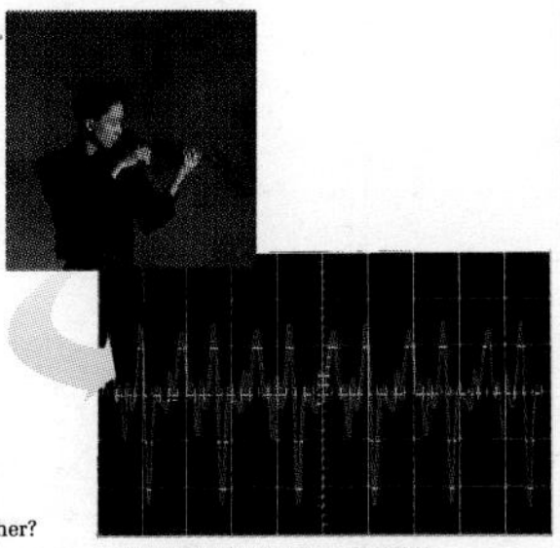

An electronic representation of a 440 Hz note produced by a vibrating violin string

32. Suppose there were a musical instrument whose notes are related to one another by the function $f(n) = 20(2)^n$, where n is the number of the note in the sequence, and the frequency of the note is measured in hertz (Hz).
    a. Find $f(1)$, the frequency of the note with the lowest pitch.
    b. Find the frequency for the next seven notes in the instrument's sequence.
    c. Graph the function. How does the frequency of a note change as the pitch of a note gets higher?
    d. The human ear can hear sounds in the range 20–20,000 Hz. What problem can you foresee in terms of the instrument described above?

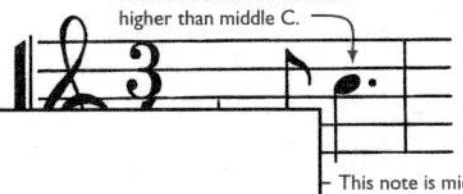

This C note is one octave higher than middle C.
This note is middle C.

33. In the music of Europe and the United States, the frequency of a note is double the frequency of the same note one *octave* lower. One of the C notes on a piano has a frequency of 130.8 Hz. Find the frequency of the C note that is one octave higher.

Writing

Problems that ask you to put your thoughts into words help you think and express your ideas clearly.

18. Suppose you repeat the Exploration using 30 dice and remove a die if a 6 lands face up.
    a. Write an equation to model the number of dice remaining after each toss.
    b. Write an equation to model the number of dice remaining after each toss if you remove a die if either a 1 or a 6 is face up.
    c. Graph each of the equations you wrote in parts (a) and (b) on the same axes. How are the graphs alike? How are they different?

19. a. **Writing** Compare the functions $y = (3)(1.25)^x$ and $y = (3)(0.75)^x$. How are the functions alike? How are they different?
    b. Graph each of the functions in part (a) on the same set of axes. Describe each graph. What point(s) do the graphs have in common?

Career After reaching a peak level, the concentration of a medication in a person's bloodstream decreases exponentially. Pharmacists can use an exponential model to determine the concentration level.

20. A medication for treating skin infections is available in three strengths. In each model, t represents the number of hours since the medication reached its peak concentration and y represents the concentration of the medication in the blood in micrograms per milliliter (μg/mL).

weakest strength	medium strength	greatest strength
$y = (0.8)(0.84)^t$	$y = (1.6)(0.84)^t$	$y = (3.2)(0.84)^t$

    a. Graph each equation on the same set of axes. What do the y-intercept values represent in this situation?
    b. How long does it take for each strength to reach a concentration of 0.1 μg/mL, the level at which the medication is considered cleared from a person's system?

21. The concentration of a stomach medication in a person's bloodstream in nanograms per milliliter (ng/mL) can be represented by the function $y = (492.5)(0.79)^t$, where t represents the number of hours since the medication reached its peak concentration.
    a. Graph the equation.
    b. Find the peak concentration.
    c. Find the time when half of the peak concentration remains. This is the medication's half-life.
    d. Use the information from the photo and your answer to part (c) to estimate when the stomach medication is cleared from a patient's blood stream.

Pharmacists multiply a medication's half-life by five to estimate when a medication is cleared from a patient's blood stream.

22. Repeat Exercise 21 for a hypertension medication whose concentration, in ng/mL, can be modeled by the equation $y = (115)(0.95)^t$.

23. Repeat Exercise 21 for an allergy medication whose concentration, in ng/mL, can be modeled by the equation $y = (263)(0.92)^t$.

Careers

Problems related to a variety of careers are included throughout the course to give you a preview of the job world.

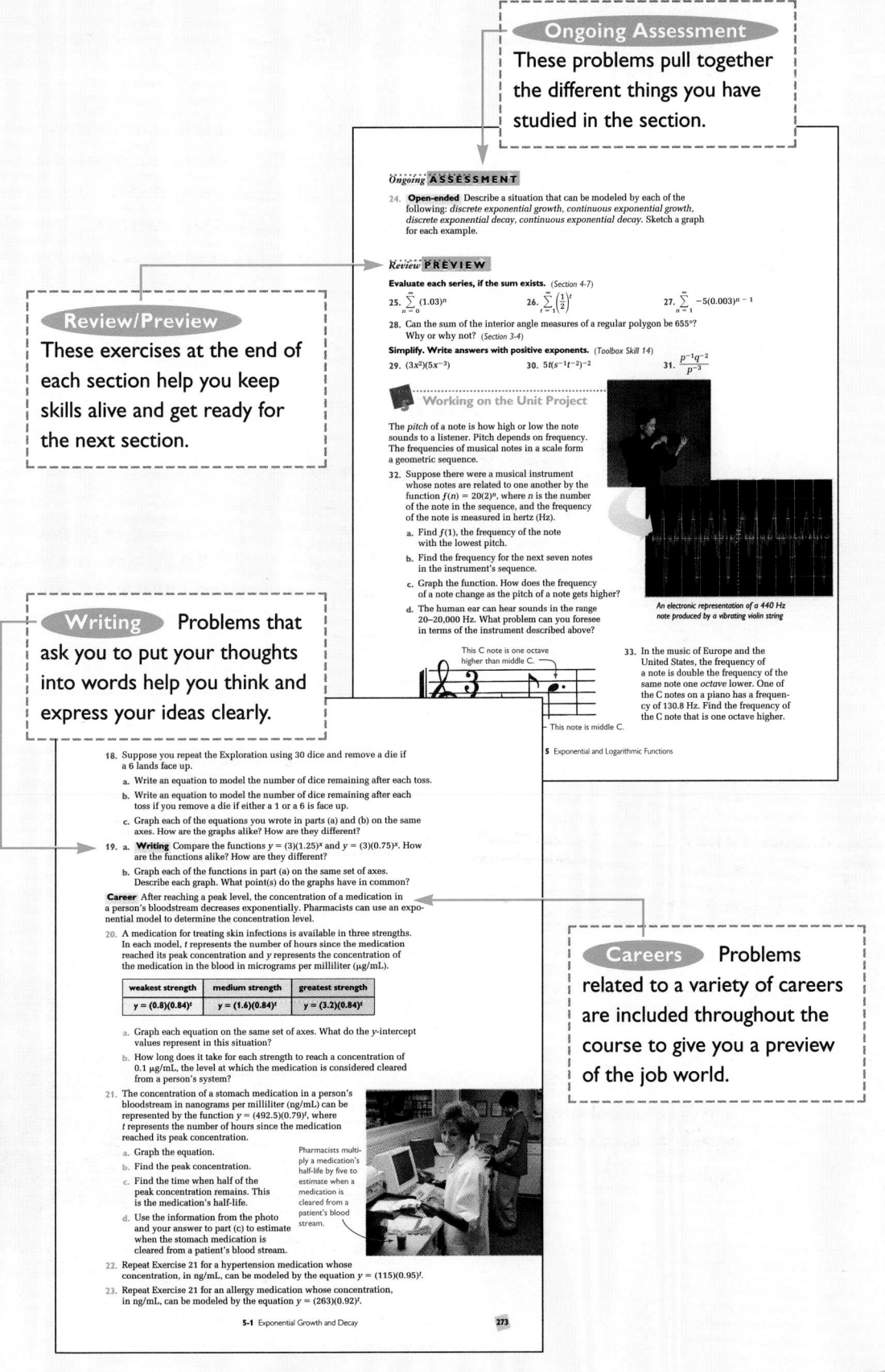

Exercises and Problems

Course Preview

Explorations

Explorations are an important part of this course. They will help you discover, understand, and connect mathematical ideas. In the Explorations, you will be gathering **data**, looking for **patterns**, and making **generalizations**. You will be working with others and sharing your ideas.

Section **5-4** The Number e

Focus
Model situations using exponential functions with base e.

In Your Be$t Intere$t

EXPLORATION

Does compounding interest more often give you unlimited returns?

- Materials: graphics calculators
- Work with a partner.

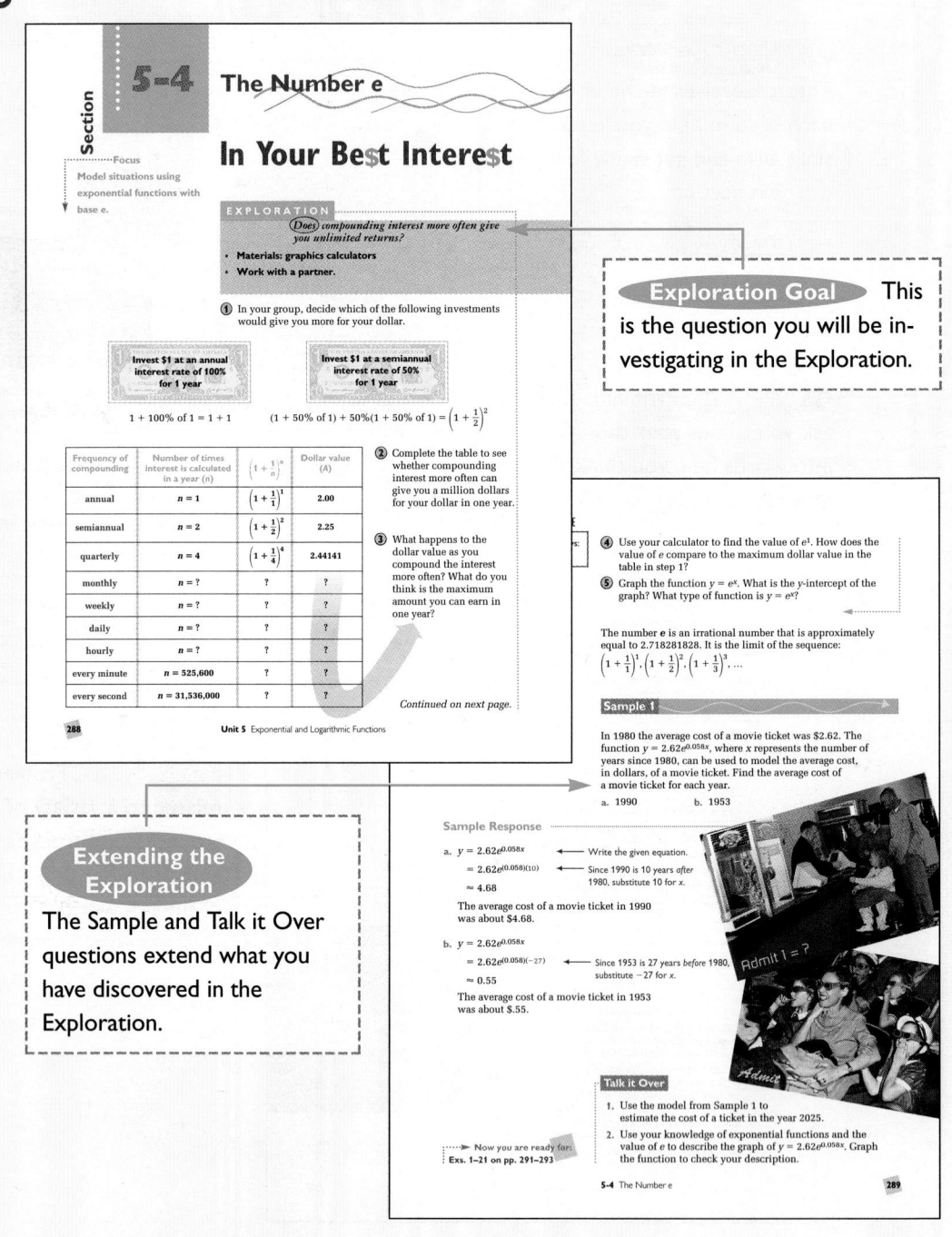

Exploration Goal This is the question you will be investigating in the Exploration.

① In your group, decide which of the following investments would give you more for your dollar.

Invest $1 at an annual interest rate of 100% for 1 year

$1 + 100\%$ of $1 = 1 + 1$

Invest $1 at a semiannual interest rate of 50% for 1 year

$(1 + 50\% \text{ of } 1) + 50\%(1 + 50\% \text{ of } 1) = \left(1 + \frac{1}{2}\right)^2$

Frequency of compounding	Number of times interest is calculated in a year (n)	$\left(1 + \frac{1}{n}\right)^n$	Dollar value (A)
annual	$n = 1$	$\left(1 + \frac{1}{1}\right)^1$	2.00
semiannual	$n = 2$	$\left(1 + \frac{1}{2}\right)^2$	2.25
quarterly	$n = 4$	$\left(1 + \frac{1}{4}\right)^4$	2.44141
monthly	$n = ?$	?	?
weekly	$n = ?$	?	?
daily	$n = ?$	?	?
hourly	$n = ?$	?	?
every minute	$n = 525,600$	?	?
every second	$n = 31,536,000$	?	?

② Complete the table to see whether compounding interest more often can give you a million dollars for your dollar in one year.

③ What happens to the dollar value as you compound the interest more often? What do you think is the maximum amount you can earn in one year?

Continued on next page.

④ Use your calculator to find the value of e^1. How does the value of e compare to the maximum dollar value in the table in step 1?

⑤ Graph the function $y = e^x$. What is the y-intercept of the graph? What type of function is $y = e^x$?

The number **e** is an irrational number that is approximately equal to 2.718281828. It is the limit of the sequence:
$\left(1 + \frac{1}{1}\right)^1, \left(1 + \frac{1}{2}\right)^2, \left(1 + \frac{1}{3}\right)^3, \ldots$

Sample 1

In 1980 the average cost of a movie ticket was $2.62. The function $y = 2.62e^{0.058x}$, where x represents the number of years since 1980, can be used to model the average cost, in dollars, of a movie ticket. Find the average cost of a movie ticket for each year.

a. 1990 b. 1953

Extending the Exploration
The Sample and Talk it Over questions extend what you have discovered in the Exploration.

Sample Response

a. $y = 2.62e^{0.058x}$ ← Write the given equation.
$= 2.62e^{(0.058)(10)}$ ← Since 1990 is 10 years *after* 1980, substitute 10 for x.
≈ 4.68

The average cost of a movie ticket in 1990 was about $4.68.

b. $y = 2.62e^{0.058x}$
$= 2.62e^{(0.058)(-27)}$ ← Since 1953 is 27 years *before* 1980, substitute -27 for x.
≈ 0.55

The average cost of a movie ticket in 1953 was about $.55.

Talk it Over

1. Use the model from Sample 1 to estimate the cost of a ticket in the year 2025.
2. Use your knowledge of exponential functions and the value of e to describe the graph of $y = 2.62e^{0.058x}$. Graph the function to check your description.

Now you are ready for: Exs. 1–21 on pp. 291–293

Review and Assessment

With this book you review and ASSESS your progress as you go along. In each unit there are one or two Checkpoints for self-assessment, plus a thorough Unit Review and Assessment at the end.

Assessment Matches Learning

The Checkpoint and Review and Assessment questions are like the ones in the unit.

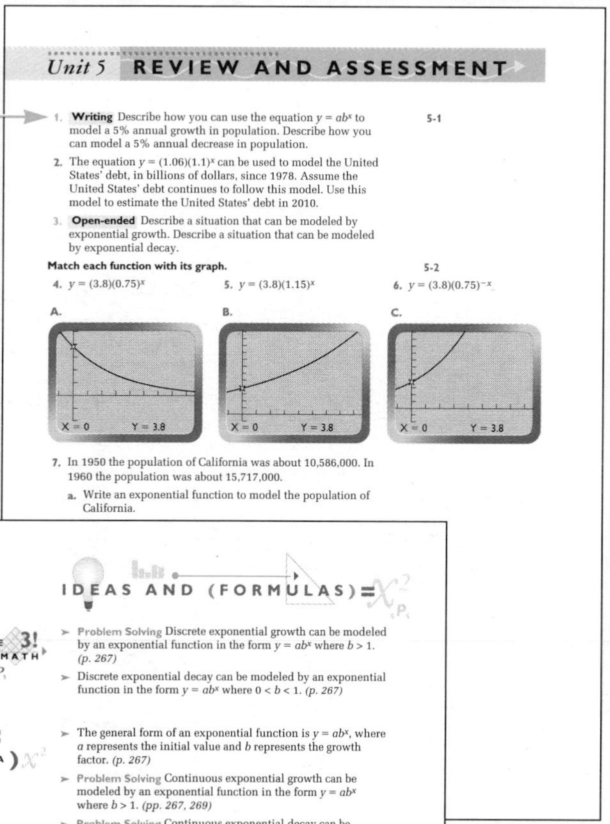

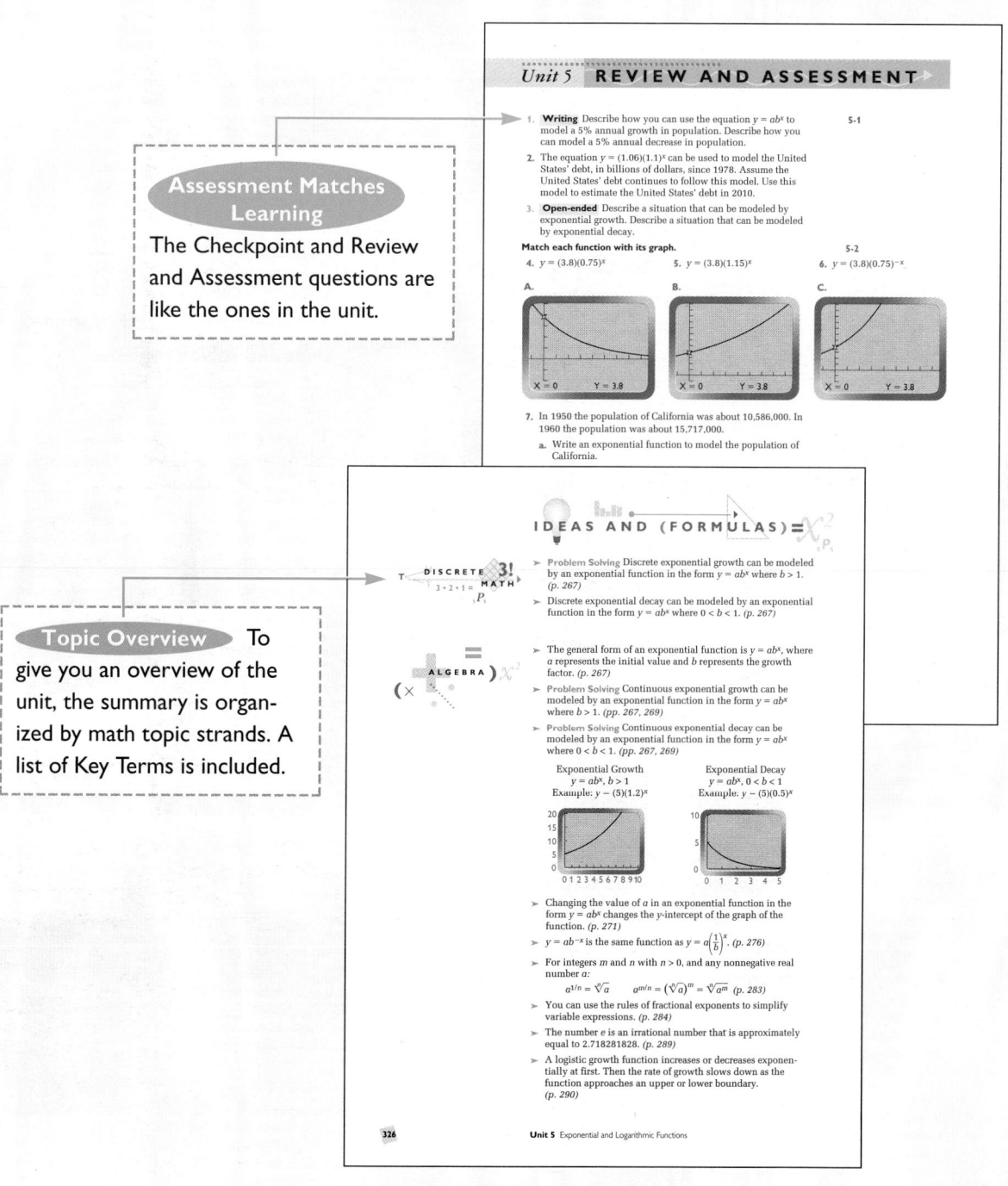

Unit 5 **REVIEW AND ASSESSMENT**

1. **Writing** Describe how you can use the equation $y = ab^x$ to model a 5% annual growth in population. Describe how you can model a 5% annual decrease in population. 5-1
2. The equation $y = (1.06)(1.1)^x$ can be used to model the United States' debt, in billions of dollars, since 1978. Assume the United States' debt continues to follow this model. Use this model to estimate the United States' debt in 2010.
3. **Open-ended** Describe a situation that can be modeled by exponential growth. Describe a situation that can be modeled by exponential decay.

Match each function with its graph. 5-2

4. $y = (3.8)(0.75)^x$
5. $y = (3.8)(1.15)^x$
6. $y = (3.8)(0.75)^{-x}$

A. B. C.

7. In 1950 the population of California was about 10,586,000. In 1960 the population was about 15,717,000.
   a. Write an exponential function to model the population of California.

Topic Overview To give you an overview of the unit, the summary is organized by math topic strands. A list of Key Terms is included.

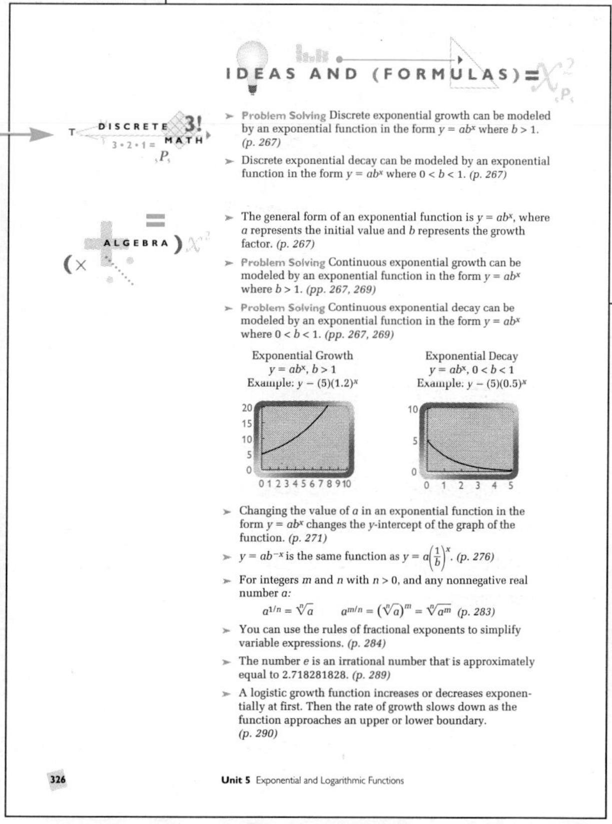

IDEAS AND (FORMULAS)=

DISCRETE MATH

- Problem Solving Discrete exponential growth can be modeled by an exponential function in the form $y = ab^x$ where $b > 1$. *(p. 267)*
- Discrete exponential decay can be modeled by an exponential function in the form $y = ab^x$ where $0 < b < 1$. *(p. 267)*

ALGEBRA

- The general form of an exponential function is $y = ab^x$, where a represents the initial value and b represents the growth factor. *(p. 267)*
- Problem Solving Continuous exponential growth can be modeled by an exponential function in the form $y = ab^x$ where $b > 1$. *(pp. 267, 269)*
- Problem Solving Continuous exponential decay can be modeled by an exponential function in the form $y = ab^x$ where $0 < b < 1$. *(pp. 267, 269)*

Exponential Growth	Exponential Decay
$y = ab^x$, $b > 1$	$y = ab^x$, $0 < b < 1$
Example: $y = (5)(1.2)^x$	Example: $y = (5)(0.5)^x$

- Changing the value of a in an exponential function in the form $y = ab^x$ changes the y-intercept of the graph of the function. *(p. 271)*
- $y = ab^{-x}$ is the same function as $y = a\left(\frac{1}{b}\right)^x$. *(p. 276)*
- For integers m and n with $n > 0$, and any nonnegative real number a:
  $a^{1/n} = \sqrt[n]{a}$ $a^{m/n} = (\sqrt[n]{a})^m = \sqrt[n]{a^m}$ *(p. 283)*
- You can use the rules of fractional exponents to simplify variable expressions. *(p. 284)*
- The number e is an irrational number that is approximately equal to 2.718281828. *(p. 289)*
- A logistic growth function increases or decreases exponentially at first. Then the rate of growth slows down as the function approaches an upper or lower boundary. *(p. 290)*

326 **Unit 5** Exponential and Logarithmic Functions

Course Preview

Technology

In this course you will see many different ways that CALCULATORS and **COMPUTERS** can make exploring ideas and solving problems easier.

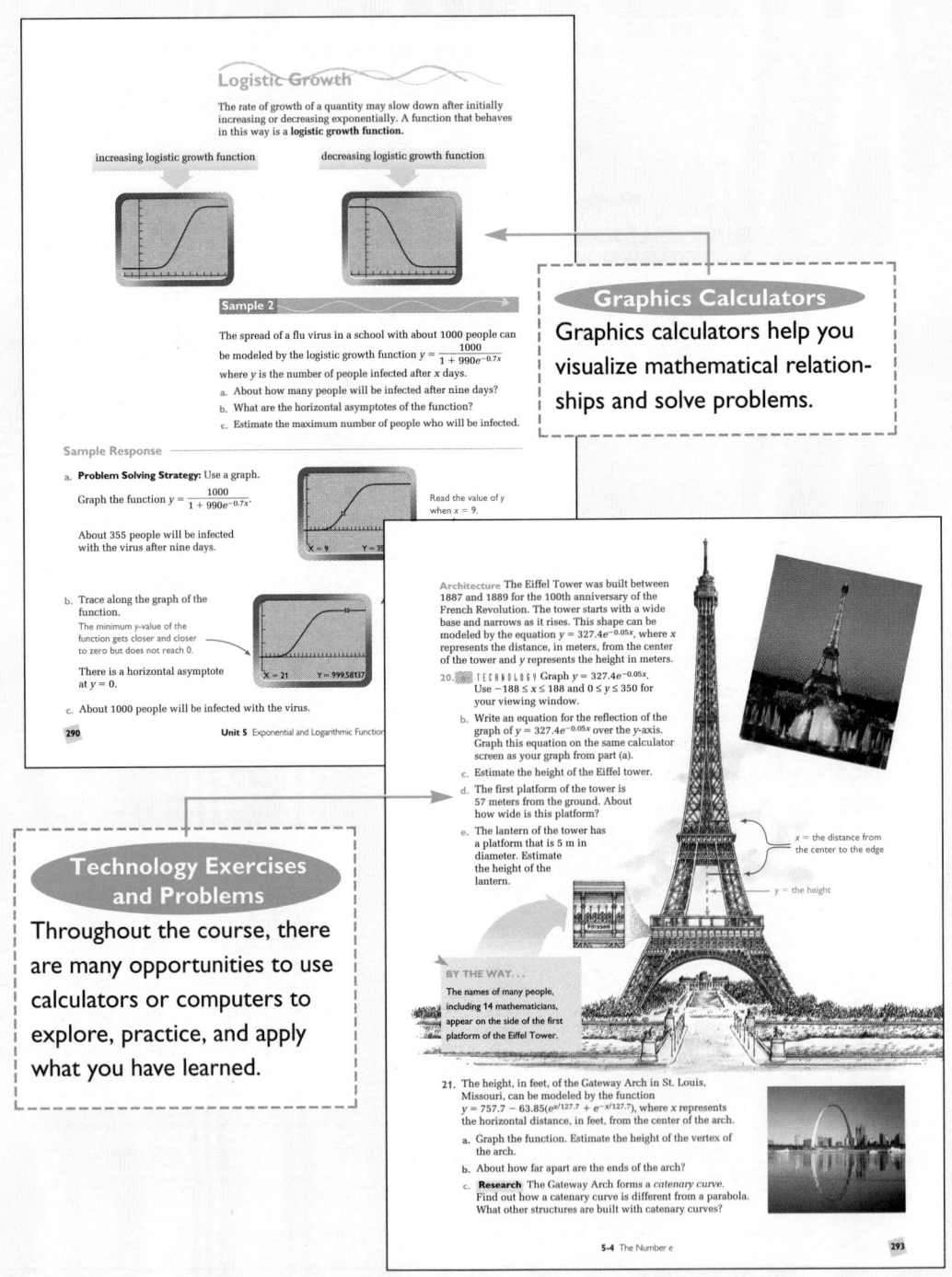

Logistic Growth

The rate of growth of a quantity may slow down after initially increasing or decreasing exponentially. A function that behaves in this way is a **logistic growth function.**

increasing logistic growth function

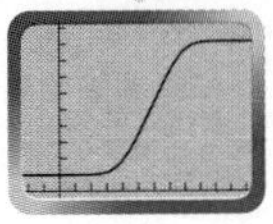

decreasing logistic growth function

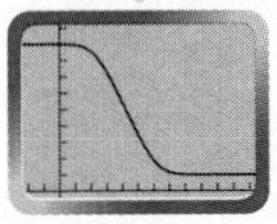

Graphics Calculators

Graphics calculators help you visualize mathematical relationships and solve problems.

Sample 2

The spread of a flu virus in a school with about 1000 people can be modeled by the logistic growth function $y = \frac{1000}{1 + 990e^{-0.7x}}$ where y is the number of people infected after x days.

a. About how many people will be infected after nine days?

b. What are the horizontal asymptotes of the function?

c. Estimate the maximum number of people who will be infected.

Sample Response

a. **Problem Solving Strategy:** Use a graph.

Graph the function $y = \frac{1000}{1 + 990e^{-0.7x}}$.

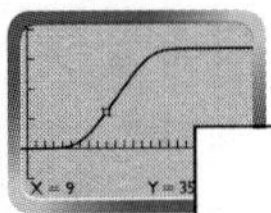

Read the value of y when $x = 9$.

About 355 people will be infected with the virus after nine days.

b. Trace along the graph of the function.

The minimum y-value of the function gets closer and closer to zero but does not reach 0.

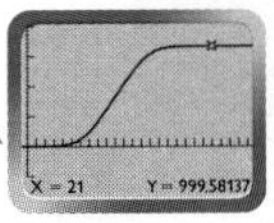

There is a horizontal asymptote at $y = 0$.

c. About 1000 people will be infected with the virus.

290 **Unit 5** Exponential and Logarithmic Function

Technology Exercises and Problems

Throughout the course, there are many opportunities to use calculators or computers to explore, practice, and apply what you have learned.

Architecture The Eiffel Tower was built between 1887 and 1889 for the 100th anniversary of the French Revolution. The tower starts with a wide base and narrows as it rises. This shape can be modeled by the equation $y = 327.4e^{-0.05x}$, where x represents the distance, in meters, from the center of the tower and y represents the height in meters.

20. a. TECHNOLOGY Graph $y = 327.4e^{-0.05x}$. Use $-188 \le x \le 188$ and $0 \le y \le 350$ for your viewing window.

b. Write an equation for the reflection of the graph of $y = 327.4e^{-0.05x}$ over the y-axis. Graph this equation on the same calculator screen as your graph from part (a).

c. Estimate the height of the Eiffel tower.

d. The first platform of the tower is 57 meters from the ground. About how wide is this platform?

e. The lantern of the tower has a platform that is 5 m in diameter. Estimate the height of the lantern.

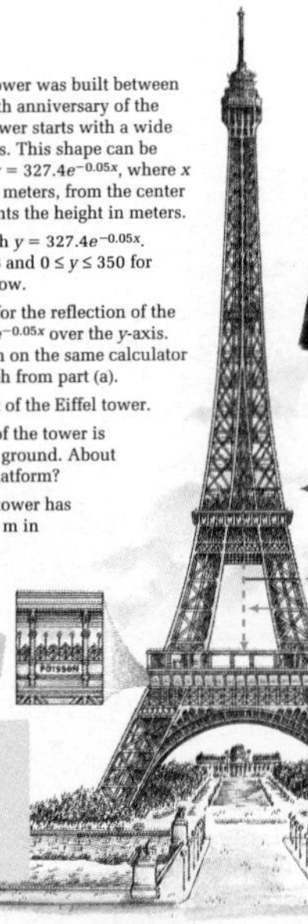

BY THE WAY...

The names of many people, including 14 mathematicians, appear on the side of the first platform of the Eiffel Tower.

21. The height, in feet, of the Gateway Arch in St. Louis, Missouri, can be modeled by the function $y = 757.7 - 63.85(e^{x/127.7} + e^{-x/127.7})$, where x represents the horizontal distance, in feet, from the center of the arch.

a. Graph the function. Estimate the height of the vertex of the arch.

b. About how far apart are the ends of the arch?

c. **Research** The Gateway Arch forms a *catenary curve.* Find out how a catenary curve is different from a parabola. What other structures are built with catenary curves?

5-4 The Number e 293

xxiv

Sample 1

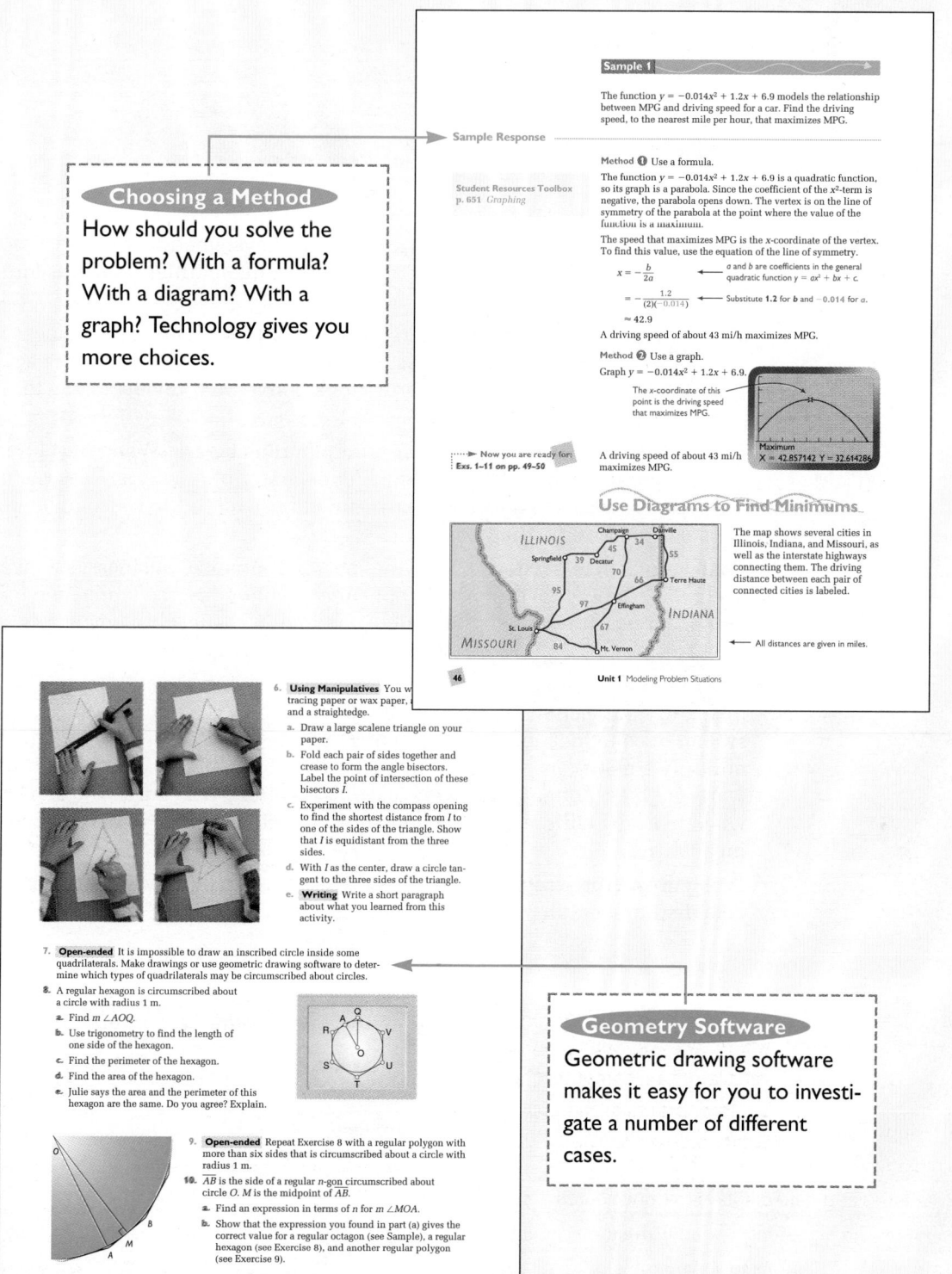

The function $y = -0.014x^2 + 1.2x + 6.9$ models the relationship between MPG and driving speed for a car. Find the driving speed, to the nearest mile per hour, that maximizes MPG.

Sample Response

Choosing a Method

How should you solve the problem? With a formula? With a diagram? With a graph? Technology gives you more choices.

Student Resources Toolbox p. 651 *Graphing*

Method ❶ Use a formula.

The function $y = -0.014x^2 + 1.2x + 6.9$ is a quadratic function, so its graph is a parabola. Since the coefficient of the x^2-term is negative, the parabola opens down. The vertex is on the line of symmetry of the parabola at the point where the value of the function is a maximum.

The speed that maximizes MPG is the x-coordinate of the vertex. To find this value, use the equation of the line of symmetry.

$x = -\frac{b}{2a}$ ← a and b are coefficients in the general quadratic function $y = ax^2 + bx + c$.

$= -\frac{1.2}{(2)(-0.014)}$ ← Substitute **1.2** for b and -0.014 for a.

≈ 42.9

A driving speed of about 43 mi/h maximizes MPG.

Method ❷ Use a graph.

Graph $y = -0.014x^2 + 1.2x + 6.9$.

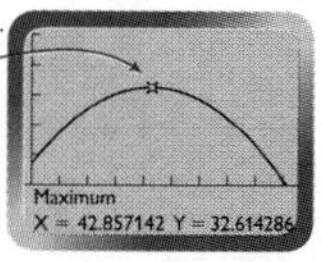

A driving speed of about 43 mi/h maximizes MPG.

Now you are ready for: Exs. 1–11 on pp. 49–50

Use Diagrams to Find Minimums

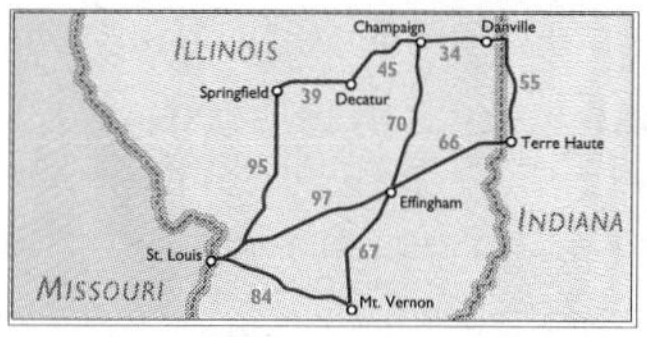

The map shows several cities in Illinois, Indiana, and Missouri, as well as the interstate highways connecting them. The driving distance between each pair of connected cities is labeled.

← All distances are given in miles.

6. **Using Manipulatives** You w... tracing paper or wax paper, ... and a straightedge.
   a. Draw a large scalene triangle on your paper.
   b. Fold each pair of sides together and crease to form the angle bisectors. Label the point of intersection of these bisectors *I*.
   c. Experiment with the compass opening to find the shortest distance from *I* to one of the sides of the triangle. Show that *I* is equidistant from the three sides.
   d. With *I* as the center, draw a circle tangent to the three sides of the triangle.
   e. **Writing** Write a short paragraph about what you learned from this activity.

7. **Open-ended** It is impossible to draw an inscribed circle inside some quadrilaterals. Make drawings or use geometric drawing software to determine which types of quadrilaterals may be circumscribed about circles.

Geometry Software

Geometric drawing software makes it easy for you to investigate a number of different cases.

8. A regular hexagon is circumscribed about a circle with radius 1 m.
   a. Find $m\angle AOQ$.
   b. Use trigonometry to find the length of one side of the hexagon.
   c. Find the perimeter of the hexagon.
   d. Find the area of the hexagon.
   e. Julie says the area and the perimeter of this hexagon are the same. Do you agree? Explain.

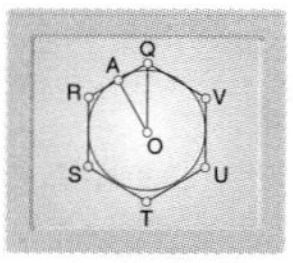

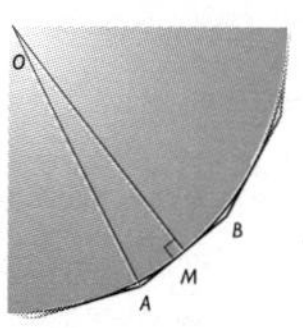

9. **Open-ended** Repeat Exercise 8 with a regular polygon with more than six sides that is circumscribed about a circle with radius 1 m.
10. $\overline{AB}$ is the side of a regular n-gon circumscribed about circle O. M is the midpoint of $\overline{AB}$.
    a. Find an expression in terms of n for $m\angle MOA$.
    b. Show that the expression you found in part (a) gives the correct value for a regular octagon (see Sample), a regular hexagon (see Exercise 8), and another regular polygon (see Exercise 9).

Planning the Unit

1 Modeling Problem Situations

OVERVIEW

- In **Unit 1,** methods of representing and analyzing real-world problems are presented. Students explore the use of algorithms, systematic lists, statistics, graphs and equations, diagrams, systems of equations, inequalities, matrices and networks.
- Box-and-whisker plots and scatter plots are used to examine trends and make predictions about trends concerning real-world data. A review of data measures is included in the **Student Resources Toolbox**, which can be found on pages 626–663. The Student Resources Toolbox provides a convenient review of key topics from previous courses.
- The **Unit Project** theme revolves around planning a park. When developing their park proposal, students need to consider a variety of issues, such as the needs of the community, where to locate the park, how to use math topics from the unit, and how to represent the final design.
- **Connections** to television scheduling, personal finance, business, science, interior decorating, newspaper reporting, consumer economics, recycling, music, and exercise machines are some of the topics included in the teaching materials and the exercises.
- **Graphics calculators** are used in Section 1-1 to graph the solution region of an inequality, in Section 1-3 to graph a scatter plot, in Sections 1-4 and 1-5 to solve a system of equations, in Section 1-5 to solve a matrix equation, in Section 1-6 to multiply a matrix by itself, and in Section 1-7 to graph a parabola. **Computer software**, such as Plotter Plus, can be used in Section 1-4 to graph lines, in Section 1-5 when working with matrices, in Section 1-7 with parabolas, and in Section 1-8 with inequalities.
- **Problem-solving strategies** used in Unit 1 include using algorithms, organized lists, data displays, graphs, equations, and diagrams.

Unit Objectives

Section	Objectives	NCTM Standards
1-1	• Write, apply, and compare step-by-step procedures for solving problems.	1, 2, 3, 4, 5, 12
1-2	• Use organized lists to solve problems.	1, 2, 3, 4, 5, 8, 12
1-3	• Use statistics and data displays to draw conclusions.	1, 2, 3, 4, 10, 12
1-4	• Use graphs and equations to model situations and make decisions.	1, 2, 3, 4, 5, 8
1-5	• Use systems of equations to solve problems.	1, 2, 3, 4, 5, 8, 12
1-6	• Use diagrams to solve problems.	1, 2, 3, 4, 12
1-7	• Make quantities as large or as small as possible to solve real-life problems.	1, 2, 3, 4, 5, 6, 8, 12
1-8	• Use systems of inequalities to model situations and find maximum and minimum values.	1, 2, 3, 4, 5, 8, 12

Topic Spiraling

Section	Connections to Prior and Future Concepts
1-1	**Section 1-1** formally introduces an algorithm as a means of solving a problem. Students analyze and write algorithms for both real-world and mathematical situations. Students have used algorithms throughout Books 1 and 2 to solve equations, inequalities, and so on, and will continue to use them in future mathematics courses.
1-2	**Section 1-2** continues the study of using systematic lists that was begun in Section 6-1 of Book 2. Students explore using systematic lists to make the best choices in solving real-world situations.
1-3	**Section 1-3** describes trends in data by examining box-and-whisker plots and scatter plots. Box-and-whisker plots were first introduced in Section 3-5 of Book 1, while scatter plots were first introduced in Section 4-5 of Book 1. The terms mean and median are reviewed. These terms were introduced in Section 3-2 of Book 1. Finding the mean will be an important skill in Section 6-2 of Book 3.
1-4	**Section 1-4** explores using a graph of two linear equations to model a situation and make a decision based on the graph. Graphing systems of linear equations was introduced in Section 8-5 of Book 1, reviewed in Section 3-1 of Book 2, and will be continued in Section 1-5 of Book 3. The concept is now extended to include break-even analysis.
1-5	**Section 1-5** continues the study of using systems of equations to solve problems. These topics are found in Section 8-5 of Book 1, Section 3-1 of Book 2, and Section 1-4 of Book 3. Matrix equations are reviewed, having been introduced in Section 3-8 of Book 2. Systems of three linear equations in three variables are introduced and solved by using matrices. Matrix methods to solve large systems of linear equations are studied in future courses in discrete mathematics.
1-6	**Section 1-6** introduces using networks and matrices to model problem situations. Students have used matrices to represent situations in Section 3-1 of Book 1. Network concepts are used in Section 1-7 of Book 3.
1-7	**Section 1-7** explores maximizing or minimizing real-life situations through the use of quadratic models and networks. Finding the vertex of a parabola was first introduced in Section 4-1 of Book 2. The concept of networks was first introduced in Section 1-6 of Book 3.
1-8	**Section 1-8** explores using linear programming to solve real-world problems. The graphing of linear inequalities used in linear programming was first introduced in Section 8-6 of Book 1 and extended to systems in Section 8-7 of Book 1.

Integrating the Strands

Strands	Sections
Number	1-1, 1-2
Algebra	1-1, 1-2, 1-3, 1-4, 1-5, 1-6, 1-7, 1-8
Functions	1-3, 1-7
Geometry	1-1, 1-2, 1-3, 1-4, 1-5, 1-6, 1-7, 1-8
Statistics and Probability	1-1, 1-3, 1-4, 1-8
Discrete Mathematics	1-1, 1-2, 1-3, 1-4, 1-6, 1-7, 1-8
Logic and Language	1-1, 1-2, 1-3, 1-4, 1-5, 1-6, 1-7, 1-8

Modeling Problem Situations

Section Planning Guide

- Essential exercises and problems are indicated in boldface.
- Ongoing work on the Unit Project is indicated in color.
- Exercises and problems that require student research, group work, manipulatives, or graphing technology are indicated in the column headed "Other."

Section	Materials	Pacing	Standard Assignment	Extended Assignment	Other
1-1	graphics calculator, scissors	Day 1	1, 2, **5–11**, 16–26, 27	1–3, **5–11**, 12–14, 16–26, 27	4, 15
1-2	toothpicks	Day 1 Day 2	**1–8, 12, 13** **15–19**, 20–29, 30	**1–8, 12, 13**, 14 **15–19**, 20–29, 30	9–11
1-3	graphing technology	Day 1	1, **2–9**, 11–13, 15–26, 27	1, **2–9**, 10–13, 15–26, 27	14
1-4	graphics calculator	Day 1	**1–16**, 18, 19, 23–30, 31	**1–16**, 17–30, 31	
1-5	graphics calculator	Day 1 Day 2	1, 2, **3–12** **13–15, 17–22**, 28–39, 40	1, 2, **3–12** **13–15, 17–22**, 23–39, 40	 16, 40
1-6	colored pencils, graphics calculator	Day 1 Day 2	**1–9**, 15, 16 **18–20**, 21–30, 31	**1–9**, 10–16 **18–20**, 21–30, 31	17
1-7	graphing technology, heavy-weight cardboard, nylon fishing line, metal washers, 3 identical mugs, protractor	Day 1 Day 2	1, **2–8**, 9 **12–14**, 16–21, 22	1, **2–8**, 9, 10 **12–14**, 16–21, 22	9c, 11 15
1-8		Day 1 Day 2	**1–6**, 7–9 **10–12**, 14–22, 23	**1–6**, 7–9 **10–12**, 14–22, 23	 13
Review Test		**Day 1** **Day 2**	**Unit Review** **Unit Test**	**Unit Review** **Unit Test**	

Yearly Pacing	Unit 1 Total	Remaining	Total
	17 days (2 for Unit Project)	137 days	154 days

Support Materials

- See **Project Book** for notes on Unit 1 Project: Plan a Park.
- UPP and disk refer to **Using Plotter Plus** booklet and **Plotter Plus** disk.
- TI-81/82 refers to **Using TI-81 and TI-82 Calculators** booklet.
- Warm-up exercises for each section are available on **Warm-Up Transparencies.**

Section	Study Guide	Practice Bank	Problem Bank	Activity Bank	Explorations Lab Manual	Assessment Book	Visuals	Technology
1-1	1-1	Practice 1	Set 1	Enrich 1		Quiz 1-1		TI-81/82, page 54
1-2	1-2	Practice 2	Set 1	Enrich 2	Master 10	Quiz 1-2		
1-3	1-3	Practice 3	Set 1	Enrich 3	Master 1	Quiz 1-3 Test 1		
1-4	1-4	Practice 4	Set 2	Enrich 4		Quiz 1-4		Line Plotter (disk)
1-5	1-5	Practice 5	Set 2	Enrich 5		Quiz 1-5	Folder 1	UPP, page 46 Matrix Reducer (disk) Matrix Calculator (disk)
1-6	1-6	Practice 6	Set 2	Enrich 6		Quiz 1-6		
1-7	1-7	Practice 7	Set 2	Enrich 7		Quiz 1-7		Parabola Plotter (disk)
1-8	1-8	Practice 8	Set 2	Enrich 8	Master 2	Quiz 1-8 Test 2		UPP, page 48 Inequality Plotter (disk)
Unit 1	Unit Review	Practice 9	Unifying Problem 1	Family Involve 1		Tests 3, 4		

Form A

Spanish versions of these tests are on pages 115–118 of the **Assessment Book.**

Name ____________ Date ________ Score ____

Test 3

Test on Unit 1 (Form A)

Directions: Write the answers in the spaces provided.

Graph each pair of equations and estimate the break-even point.

1. $E = 200 + 5x$
   $I = 20x$

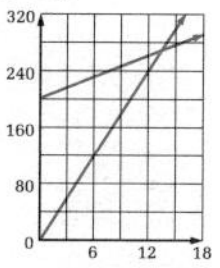

2. $E = 240 + 3.3x$
   $I = 15x$

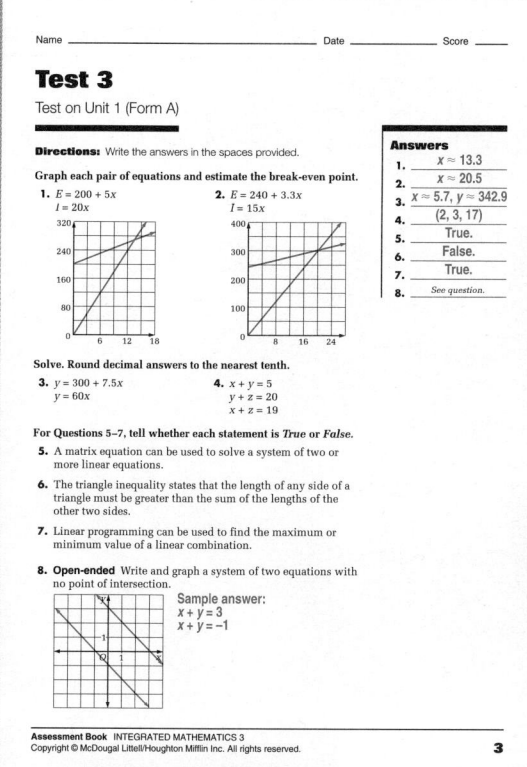

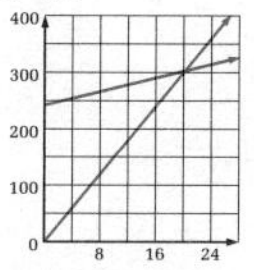

Solve. Round decimal answers to the nearest tenth.

3. $y = 300 + 7.5x$
   $y = 60x$

4. $x + y = 5$
   $y + z = 20$
   $x + z = 19$

For Questions 5–7, tell whether each statement is *True* or *False*.

5. A matrix equation can be used to solve a system of two or more linear equations.
6. The triangle inequality states that the length of any side of a triangle must be greater than the sum of the lengths of the other two sides.
7. Linear programming can be used to find the maximum or minimum value of a linear combination.
8. **Open-ended** Write and graph a system of two equations with no point of intersection.

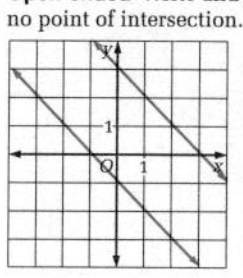

Sample answer:
$x + y = 3$
$x + y = -1$

Answers

1. $x \approx 13.3$
2. $x \approx 20.5$
3. $x \approx 5.7,\ y \approx 342.9$
4. (2, 3, 17)
5. True.
6. False.
7. True.
8. *See question.*

Name ____________ Date ________ Score ____

Test 3 *(continued)*

Directions: Write the answers in the spaces provided.

Use a matrix to represent the connections among the vertices of the diagram.

9. 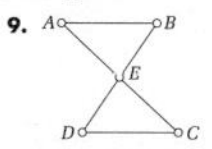

$$\begin{array}{c} \\ A \\ B \\ C \\ D \\ E \end{array}\begin{array}{c} \begin{array}{ccccc} A & B & C & D & E \end{array} \\ \begin{bmatrix} 0 & 1 & 0 & 0 & 1 \\ 1 & 0 & 0 & 0 & 1 \\ 0 & 0 & 0 & 1 & 1 \\ 0 & 0 & 1 & 0 & 1 \\ 1 & 1 & 1 & 1 & 0 \end{bmatrix} \end{array}$$

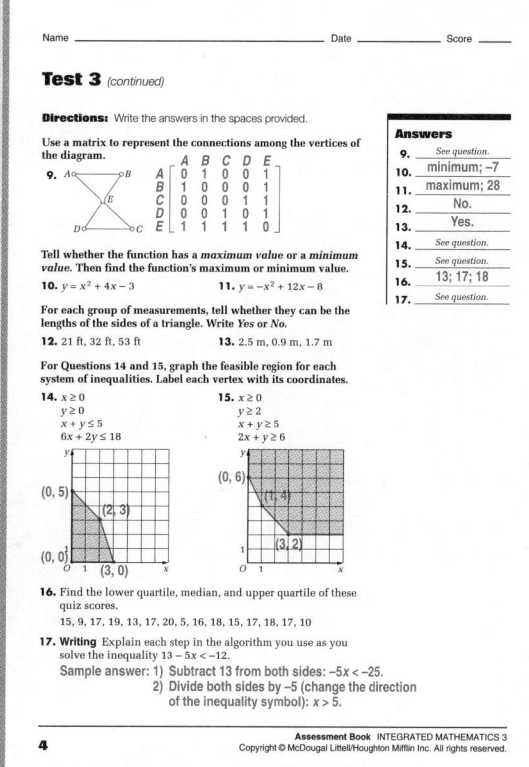

Tell whether the function has a *maximum value* or a *minimum value*. Then find the function's maximum or minimum value.

10. $y = x^2 + 4x - 3$
11. $y = -x^2 + 12x - 8$

For each group of measurements, tell whether they can be the lengths of the sides of a triangle. Write *Yes* or *No*.

12. 21 ft, 32 ft, 53 ft
13. 2.5 m, 0.9 m, 1.7 m

For Questions 14 and 15, graph the feasible region for each system of inequalities. Label each vertex with its coordinates.

14. $x \geq 0$
    $y \geq 0$
    $x + y \leq 5$
    $6x + 2y \leq 18$

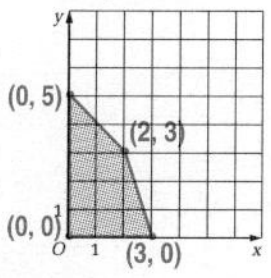

15. $x \geq 0$
    $y \geq 2$
    $x + y \geq 5$
    $2x + y \geq 6$

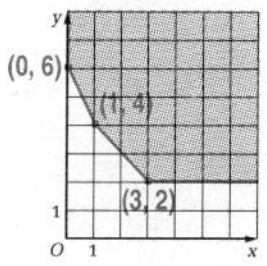

16. Find the lower quartile, median, and upper quartile of these quiz scores.
    15, 9, 17, 19, 13, 17, 20, 5, 16, 18, 15, 17, 18, 17, 10
17. **Writing** Explain each step in the algorithm you use as you solve the inequality $13 - 5x < -12$.
    Sample answer: 1) Subtract 13 from both sides: $-5x < -25$.
    2) Divide both sides by –5 (change the direction of the inequality symbol): $x > 5$.

Answers

9. *See question.*
10. minimum; –7
11. maximum; 28
12. No.
13. Yes.
14. *See question.*
15. *See question.*
16. 13; 17; 18
17. *See question.*

Form B

Name ____________ Date ________ Score ____

Test 4

Test on Unit 1 (Form B)

Directions: Write the answers in the spaces provided.

Graph each pair of equations and estimate the break-even point.

1. $E = 300 + 6x$
   $I = 32x$

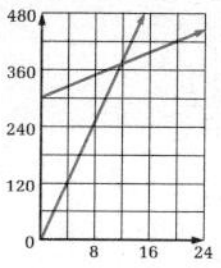

2. $E = 240 + 4.5x$
   $I = 15x$

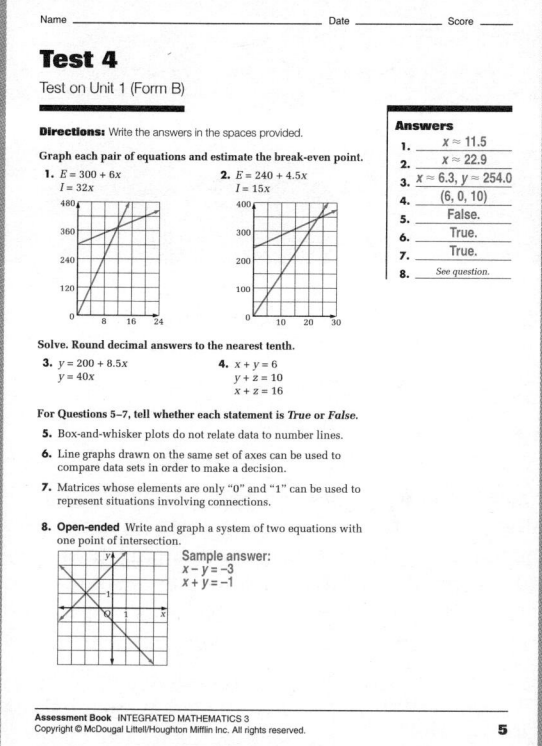

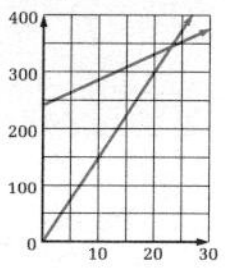

Solve. Round decimal answers to the nearest tenth.

3. $y = 200 + 8.5x$
   $y = 40x$

4. $x + y = 6$
   $y + z = 10$
   $x + z = 16$

For Questions 5–7, tell whether each statement is *True* or *False*.

5. Box-and-whisker plots do not relate data to number lines.
6. Line graphs drawn on the same set of axes can be used to compare data sets in order to make a decision.
7. Matrices whose elements are only "0" and "1" can be used to represent situations involving connections.
8. **Open-ended** Write and graph a system of two equations with one point of intersection.

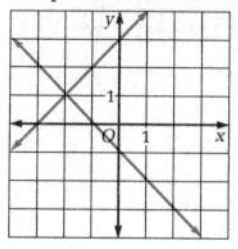

Sample answer:
$x - y = -3$
$x + y = -1$

Answers

1. $x \approx 11.5$
2. $x \approx 22.9$
3. $x \approx 6.3,\ y \approx 254.0$
4. (6, 0, 10)
5. False.
6. True.
7. True.
8. *See question.*

Name ____________ Date ________ Score ____

Test 4 *(continued)*

Directions: Write the answers in the spaces provided.

Use a matrix to represent the connections among the vertices of the diagram.

9. 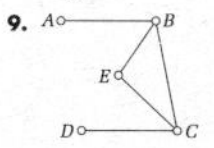

$$\begin{array}{c} \\ A \\ B \\ C \\ D \\ E \end{array}\begin{array}{c} \begin{array}{ccccc} A & B & C & D & E \end{array} \\ \begin{bmatrix} 0 & 1 & 0 & 0 & 0 \\ 1 & 0 & 1 & 0 & 1 \\ 0 & 1 & 0 & 1 & 1 \\ 0 & 0 & 1 & 0 & 0 \\ 0 & 1 & 1 & 0 & 0 \end{bmatrix} \end{array}$$

Tell whether the function has a *maximum value* or a *minimum value*. Then find the function's maximum or minimum value.

10. $y = -2x^2 + 8x - 3$
11. $y = x^2 - 2x - 8$

For each group of measurements, tell whether they can be the lengths of the sides of a triangle. Write *Yes* or *No*.

12. 13 cm, 31 cm, 22 cm
13. 3.6 in., 1.5 in., 2.1 in.

For Questions 14 and 15, graph the feasible region for each system of inequalities. Label each vertex with its coordinates.

14. $x \geq 0$
    $y \geq 0$
    $x + y \leq 4$
    $2x + y \leq 6$

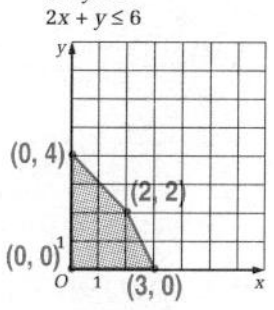

15. $x \geq 1$
    $y \geq 0$
    $x + y \geq 4$
    $5x + 3y \geq 18$

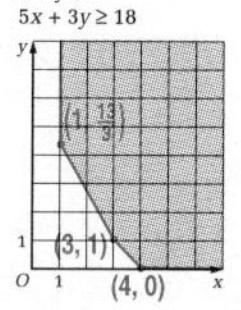

16. Find the lower quartile, median, and upper quartile of these quiz scores.
    12, 8, 18, 19, 14, 14, 18, 18, 20, 7, 11, 20, 18, 19, 15
17. **Writing** Explain each step in the algorithm you use as you solve the inequality $9 - 4x > -7$.
    Sample answer: 1) Subtract 9 from both sides: $-4x > -16$.
    2) Divide both sides by –4 (change the direction of the inequality symbol): $x < 4$.

Answers

9. *See question.*
10. maximum; 5
11. minimum; –9
12. Yes.
13. No.
14. *See question.*
15. *See question.*
16. 12; 18; 19
17. *See question.*

OUTSIDE RESOURCES

Books/Periodicals

Ballew, Hunter. "Sherlock Holmes, Master Problem Solver." *Mathematics Teacher* (November 1994): pp. 596–601.

Sandefur, James T. "Technology, Linear Equations, and Buying a Car." *Mathematics Teacher* (October 1992): pp. 562–567.

Copes, Wayne, William Sacco, Clifford Sloyer, and Robert Stark. "Reachability." *Contemporary Applied Mathematics, Graph Theory.* pp. 47–51. Janson Publications.

"Exploring Data" from the *Quantitative Literacy Series;* written by members of the Joint Committee on the Curriculum in Statistics and Probability of the American Statistical Association and the National Council of Teachers of Mathematics. Dale Seymour Publication.

Activities/Manipulatives

Wood, Eric. "Gas-Bill Mathematics." *Mathematics Teacher* (March 1995): pp. 214–218.

Software

Myers, David L. *Plotter Plus.* Boston, MA: Houghton Mifflin Company, 1995. Macintosh and MS-DOS (worksheets included).

Videos

Mathematical Eye. Program No. 5: Logic and Problem Solving. Journal Films, 1988.

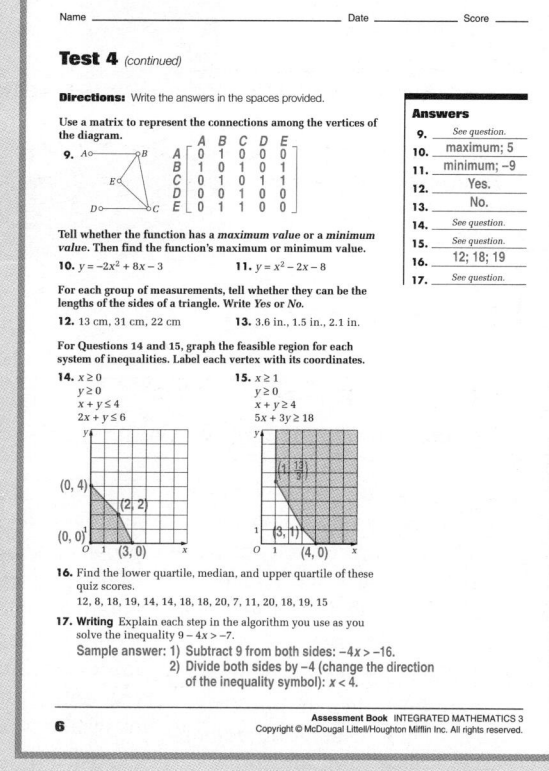

PROJECT GOALS

- Students plan a park and develop a proposal for funding and building it.
- Students survey the people in their community to find out what kind of park they want.
- Students use at least three math topics from this unit in their park design and proposal.

PROJECT PLANNING

Project Teams

Have students work on the project in groups of four. One way for the individuals in the group to distribute the work is as follows:

1. Surveyor: coordinates development and refinement of the survey questions, conducts the survey, and summarizes its results.
2. Designer: prepares a scale drawing, a CAD drawing, or a three-dimensional model of the park plan.
3. Mathematician: decides which math topics from the unit to include in the proposal and how they will be used.
4. Writer: writes the description of the park and how it meets the needs of the people in the community.

Support Materials

The ***Project Book*** contains information about the following topics for use with this Unit Project.

- Project Description
- Teaching Commentary
- Working on the Unit Project Exercises
- Completing the Unit Project
- Assessing the Unit Project
- Alternative Projects
- Outside Resources

unit 1

Modeling Problem Situations

COMMUNITY PLANNING

Picture some barren mud flats and marshes. To this land come some exiles chased out of their city by invaders from the north. Here they build a new life and a new city full of magnificent buildings, bridges, and squares. Business and culture thrive. This unlikely success story took place over 1400 years ago. The city is Venice.

Most cities grow gradually, without any plan. A few, like Washington, D.C., and Brasília are built to a **master plan.**

Washington, D.C.

This early plan of Washington features a triangle formed by the White House, the Capitol, and the Washington Monument.

Brasília

The Brazilian **architect** Oscar Niemeyer designed the main public buildings of Brasília, the capital of Brazil since 1960. The city was built in the shape of an airplane.

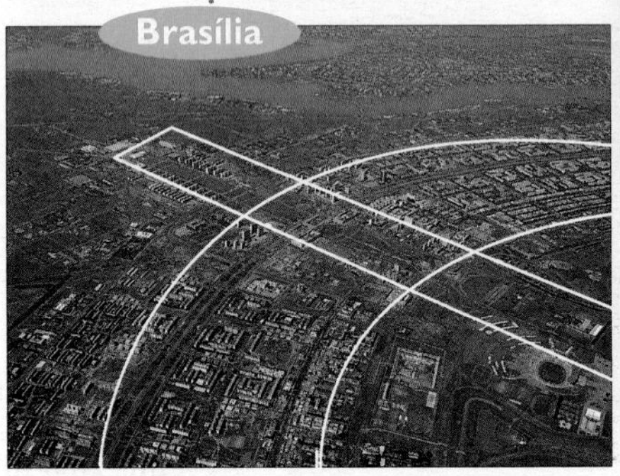

General Rubric for Unit Projects

Each unit project can be evaluated in many possible ways. The following rubric is just one way to evaluate these open-ended projects. It is based on a 4-point scale.

4 The student fully achieves all mathematical and project goals. The presentation demonstrates clear thinking and explanation.

3 The student substantially achieves the mathematical and project goals. The main thrust of the project and the mathematics behind it is understood, but there may be some minor misunderstanding of content, errors in computation, or weakness in presentation.

2 The student partially achieves the mathematical and project goals. A limited grasp of the main mathematical ideas or project requirements is demonstrated. Some of the work may be incomplete, misdirected, or unclear.

1 The student makes little progress toward accomplishing the goals of the project because of lack of understanding or lack of effort.

Unit Project

Plan a Park

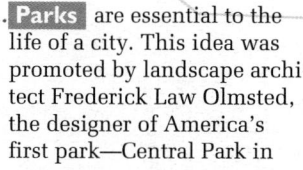

Your project is to plan a park and develop a proposal for funding and building it. First, you will need to survey the people in your community to find out what kind of park will meet the needs of young children through senior citizens, including people with disabilities.

Using the results of your survey, you should develop a park design and proposal. You should use at least three of these math topics from this unit:

- algorithms
- systematic lists
- data displays
- systems of equations
- network diagrams
- linear programming

You can use a scale drawing, CAD software, or a three-dimensional model to present your park design.

Parks are essential to the life of a city. This idea was promoted by landscape architect Frederick Law Olmsted, the designer of America's first park—Central Park in New York City. Olmsted's later work on the 1893 Chicago World's Fair helped set the direction for the emerging field of community planning.

Central Park

◄In 1791 Benjamin Banneker was a surveyor of the land selected for the new capital of the United States.

Mary Pat Hodge, seated, and Marsha Johnson served as contractors for a new kind of playground.

In Fort Worth, Texas, Debbie Bradshaw and other concerned citizens brought about the construction of a playground accessible to all children. The design was modeled on Janet Simpson's design for Fantasy Landing, in Dallas, the first public playground of this kind in the United States.

1

Students Acquiring English

The kinesthetic and visual nature of the park design project will enable students acquiring English to participate fully. You may want to suggest that they create questionnaires in their first language and survey people who speak their language to increase the diversity of people surveyed by their group.

ADDITIONAL BACKGROUND

Multicultural Note

Benjamin Banneker, Frederick Law Olmsted, and Oscar Niemeyer are three individuals whose contributions to the fields of design and architecture have had long-lasting effect. Banneker, an African-American, was an astronomer, mathematician, and surveyor who was well known for his skills. When the three-man surveying team for the design of Washington, D.C. was selected, Thomas Jefferson suggested Banneker as a member. After a year of planning, the chief surveyor suddenly resigned and took all plans with him; Banneker had to reproduce the plans from memory. The testament to his talents is the layout of Washington, D.C.'s governmental center. Frederick Law Olmsted, also an African-American, coined the term "landscape architect." He was the planner of New York City's Central Park, and was also the first person to improve a park through artistic designing. He went on to plan Prospect Park, Brooklyn; Fairmount Park, Philadelphia; South Park, Chicago; the grounds surrounding the nation's Capitol; and many others. Oscar Niemeyer, who helped plan and design Brasília, has said his designs were inspired by Brazilian social and climatic conditions and history. Brazil's capital city and its government buildings demonstrate both Niemeyer's architectural skill and dedication to his nation.

Suggested Rubric for Unit Project

4 The survey is successful in reaching people of different ages, including people with disabilities, in the community. The results of the survey are used to design the park. The math topics selected support the park design, as does the drawing or model used. The description of the park explains how the park meets the needs of the community.

3 Students' survey questions can be improved because they did not reach some members of the community. The written summary of the survey results has left out a few key points. The written description of the park does not explain in detail how the park meets the needs of the community.

2 Students' survey questions do not collect enough information to plan an adequate park for all people in the community. The drawings or models used are not clear and the written description shows a lack of effort.

1 Students have not developed an adequate survey and their implementation of the tasks needed to plan the park have many shortcomings. The final description is inadequate. Students should be encouraged to speak with the teacher as soon as possible to review their work and to make a new start on the project.

Modern Park Design

New parks are being designed today that meet the needs of all people, including people with disabilities. For example, resilient rubber surfacing on ramps into play areas allows for easier accessibility by people in wheelchairs. Special clusters of bricks along walkways serve as markers to other play areas for visually impaired children. Water fountains, restrooms, picnic areas, and parking lots are other facilities that need to be taken into consideration when designing a park that is fully accessible to all people.

ALTERNATIVE PROJECTS

Project 1, page 61

Management Science

Write a report on linear programming that describes several business applications, including airline scheduling and routing information transmitted over complex communications networks. Discuss the simplex and ellipsoid methods for finding the corner points representing maximum profit when all points cannot be tested.

Project 2, page 61

Secret Codes

Find out how to use Huffman codes to use network diagrams to send secret messages. Write an article explaining how to code and decode messages using Huffman codes.

Unit Project

Getting Started

For this project you should work in a group of four students. Here are some ideas to help you get started.

☞ Decide where to locate your park. You may make up a piece of land with any dimensions you choose or plan your park to fit a vacant lot in your community.

☞ Visit some parks in your area. Notice what kinds of equipment and facilities they contain and how they are arranged. Also notice who is in the park and what they are doing.

☞ Talk about what questions you should include on your survey. Brainstorm ways to reach people of different ages, cultures, and so on.

☞ Decide whether you will represent your park design with a paper-and-pencil drawing, a CAD drawing, or a three-dimensional model.

Working on the Unit Project

Your work in Unit 1 will help you plan a park.

Related Exercises:

Section 1-1, Exercise 27
Section 1-2, Exercise 30
Section 1-3, Exercise 27
Section 1-4, Exercise 31
Section 1-5, Exercise 40
Section 1-6, Exercise 31
Section 1-7, Exercise 22
Section 1-8, Exercise 23

Alternative Projects p. 61

Can We Talk COMMUNITY PLANNING

➤ How often do you use the parks in your area? What activities do you participate in when you go to a park?

➤ Some local parks and playgrounds are planned, funded, and built by volunteer community groups. Others are built by the parks department of the town or city, or the state. What are some advantages and disadvantages of both approaches?

➤ In the 17 years that it took to build New York City's Central Park, property values near the park went up by a factor of nine, but away from the park they only doubled. Studies show that property values increase when trees are planted on a city block. Why do you think this is?

➤ How are the streets in your community laid out? Do you think the area was built to a plan? Why or why not?

Answers to Can We Talk?

➤ Answers may vary. An example is given. I use the parks about every other day in the summer. I like to use the basketball and tennis courts.

➤ Answers may vary. An example is given. If volunteer groups did the work, the park may be designed to fit their specific needs and not the needs of all people in the community. Volunteer work is a good way to get people involved in their community. However, it is sometimes difficult to find enough volunteers and to decide who is in charge of different tasks. Fund raising can be difficult because revenues for upkeep of the park are not constant. If the city, town, or state builds the park, there are usually people paid to take care of it, and the funds for building the park and maintaining it are constant.

➤ Answers may vary. An example is given. People like trees. They can make a home or community look attractive and well established. They also provide other advantages, such as offering shade on a hot day.

➤ Answers may vary. An example is given. The area of my community was built to a plan because the main streets run north-south and east-west and are each one mile apart.

Section **1-1**

Algorithms

Step by Step

Focus

Write, apply, and compare step-by-step procedures for solving problems.

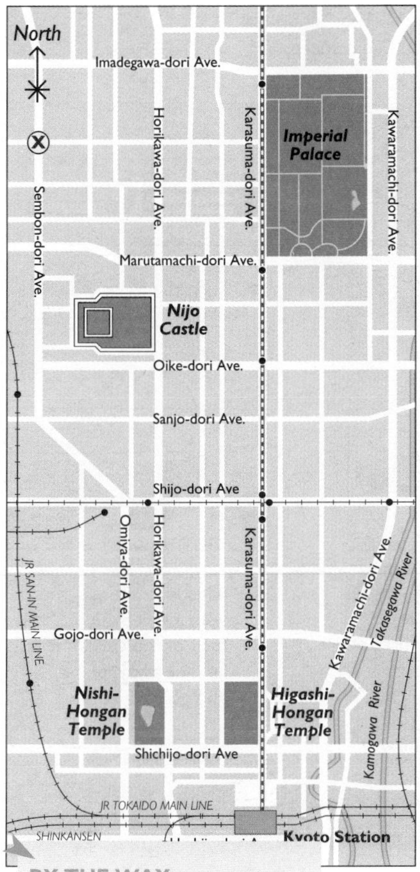

BY THE WAY...

Kyoto was the capital of Japan from 794 until 1868. It is still a cultural and artistic center.

Nijo Castle, built in 1603 by the Tokugawa shogun

Visitors to Kyoto, Japan, often use a map like this to find their way around the city.

Talk it Over

1. Give directions for walking from **Nijo Castle** to **Higashi-Hongan Temple.**
2. A tourist at point **X** walking north on Sembon-dori Avenue asked for directions to **Nishi-Hongan Temple.**

   Michiko gave these directions:
   - Turn right at Imadegawa-dori Avenue.
   - Turn right at Horikawa-dori Avenue.

   Jiro gave these directions:
   - Turn around and follow Sembon-dori Avenue past Nijo Castle to Omiya-dori Avenue.
   - Turn right at Omiya-dori Avenue.
   - Turn left at Gojo-dori Avenue.
   - Turn right at Horikawa-dori Avenue.

   a. Which directions are easier to state?
   b. Which are easier to follow?
   c. Which describe a shorter route?
   d. Which directions do you think are better? Why?

PLANNING

Objectives and Strands
See pages 1A and 1B.

Spiral Learning
See page 1B.

Materials List
- Graphics calculator
- Scissors
- Ruler
- Protractor

Recommended Pacing
Section 1-1 is a one-day lesson.

Extra Practice
See pages 608–610.

Warm-Up Exercises
Warm-Up Transparency 1-1

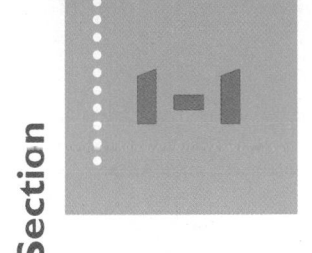

Support Materials
- Practice Bank: Practice 1
- Activity Bank: Enrichment 1
- Study Guide: Section 1-1
- Problem Bank: Problem Set 1
- Using TI-81 and TI-82 Calculators: The Euclidean Algorithm
- Assessment Book: Quiz 1-1, Alternative Assessment 1

Answers to Talk it Over

1. Answers may vary. An example is given. Walk south on Horikawa-dori Ave. to Shichijo-dori Ave. Turn left at Shichijo-dori Ave. and walk east.
2. Answers may vary. Examples are given.
   a. Michiko's
   b. Michiko's
   c. Jiro's
   d. (1) Jiro's; Michiko's directions give no idea of how long to walk along Horikawa-dori Ave. and I'd think I missed the temple. (2) Michiko's; I wouldn't have to make decisions at as many intersections as with Jiro's directions.

TEACHING

Talk it Over

Questions 1 and 2 introduce students to the idea of an algorithm by using a street map and directions.

Additional Sample

S1 Write an algorithm for placing a three-way telephone call to two friends who have agreed to be available between 9:00 P.M. and 9:30 P.M. Assume that your telephone is set up for three-way calling and that neither friend has call waiting and will be willing to take your call as soon as they get it.

Step 1. Phone friend 1.
Step 2. If no answer or busy, hang up and phone friend 2.
Step 3. If no answer or busy, hang up and repeat steps 1 and 2 until one person has answered.
Step 4. Tell the first person to hold on while you try the other person.
Step 5. Dial the other person's number.
Step 6. If the line is busy or there is no answer, return to the other person and tell him or her that you will call back as soon as you get the other person.
Step 7. Call the person who did not answer until you get through.
Step 8. Tell that person to hold while you get the person waiting for you to call back on the line.

Teaching Tip

After discussing Sample 1, ask students for other examples of algorithms, either from everyday life or from other disciplines. Ask which involve loops and which do not. Teaching strategies for the effective learning and teaching of mathematics are given on page T30 in the front of this Teacher's Edition.

A set of step-by-step directions for traveling from one place to another is an example of an *algorithm*. When you break down any process into steps, you are creating an **algorithm.**

Sample 1

Write an algorithm for parallel parking a car on the right side of the street.

Sample Response

1. Determine whether parking is permitted and whether your car will fit in the space between two cars. If the answer to either question is "No," continue looking for a parking space and repeat this step when you find a space.

2. Pull up next to the car in front of the space until the rear bumpers are side by side.

3. Shift into reverse.

4. Turn the front wheels all the way to the right.

5. Slowly back up into the parking space until your front wheels are opposite the rear bumper of the car in front.

6. Turn the steering wheel all the way to the left. Keep backing up until your car is parallel to the curb. If there is not enough room, pull out and go back to Step 2.

7. Check to see whether your car is within a foot of the curb. If it is farther away, pull out and go back to Step 2.

8. Straighten out the front wheels and center the car in the parking space.

Some algorithms, like travel directions, consist of a simple sequence of steps that you follow in order. Other algorithms, like the one for parallel parking, may ask you to make decisions or to complete a *loop*. A **loop** is a group of steps that you repeat either a certain number of times or until some condition is met.

BY THE WAY...

NASA engineer Christine Darden (center) wrote a new computer algorithm that researchers use to try to minimize sonic boom by altering the shape of the airplane.

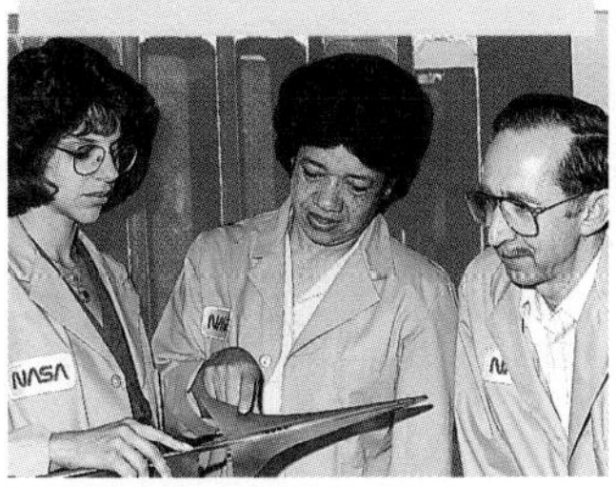

Talk it Over

3. In which steps of the algorithm in Sample 1 do you need to make a decision?
4. How do you know when to end each loop in Sample 1?
5. How would you change the algorithm in Sample 1 for parallel parking on the left side of a one-way street?

You can use algorithms to solve problems.

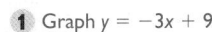

Sample 2

a. Use two different algorithms to solve the inequality $-3x + 9 < 4$.

b. **Writing** Compare the two algorithms you used in part (a).

Sample Response

a. **Method ①** Use algebra.

$-3x + 9 < 4$

$-3x < -5$ ⟵ Subtract 9 from both sides.

$x > \frac{5}{3}$ ⟵ Divide both sides by -3, and reverse the inequality symbol.

The solution is all real numbers greater than $\frac{5}{3}$.

Method ② Use a graph.

① Graph $y = -3x + 9$.

② Graph $y = 4$.

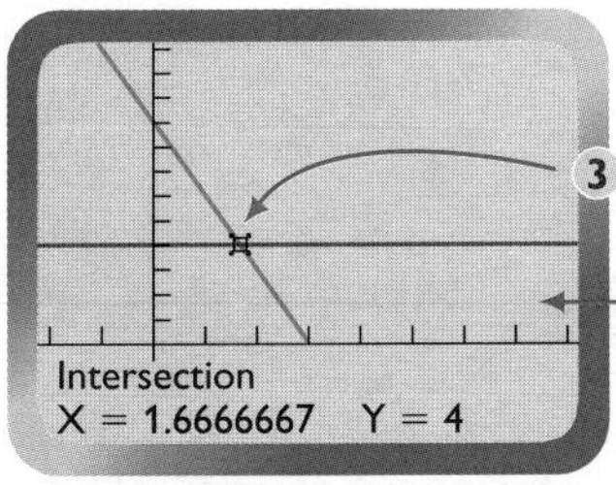

③ Estimate the x-coordinate of the intersection point.

The solutions of the inequality are the values of x for which the graph of $y = -3x + 9$ is *below* the graph of $y = 4$.

The solution is all real numbers greater than about 1.7.

Describe the advantages and disadvantages of each algorithm.

b. The algebraic method gives an exact solution, it is fast, and it is compact. However, I have to remember to reverse the inequality symbol when I divide both sides by a negative number.
The graphical method helps me see the solution region, but the boundary of the region can only be estimated.

Answers to Talk it Over

3. Steps 1, 6, and 7
4. End the first loop when a space is found that the car will fit in and in which it may be legally parked. End the second loop when there is enough space for the car to be backed in parallel to the curb. End the third loop when the car is parallel to the curb and less than 1 foot from it.
5. In Step 4, turn the front wheels all the way to the left. In Step 6, turn the steering wheel all the way to the right.

Talk it Over

Questions 3 and 4 ask students to analyze the algorithm in the Sample Response with regard to decision-making and loops. Question 5 has students alter the algorithm to perform a related procedure.

Additional Sample

S2 a. Use two different algorithms to find the measure of $\angle C$ of $\triangle ABC$, given that $m\angle A = 35°$ and $m\angle B = 98°$.

Method 1. Use the fact that the sum of the measures of the angles of a triangle is 180°. Write the equation $m\angle A + m\angle B + m\angle C = 180°$. Substitute 35° for $m\angle A$ and 98° for $m\angle B$, then solve for $m\angle C$.

$35° + 98° + m\angle C = 180°$

$133° + m\angle C = 180°$

$m\angle C = 47°$

The measure of $\angle C$ is 47°.

Method 2. Draw an angle of 35° by using a ruler and a protractor. Label it $\angle A$. Pick a point (other than A) on one side of $\angle A$ and label it B. Draw $\angle ABC$ so that $\overrightarrow{BC}$ intersects the other side of $\angle A$ at C and $m\angle ABC = 98°$.

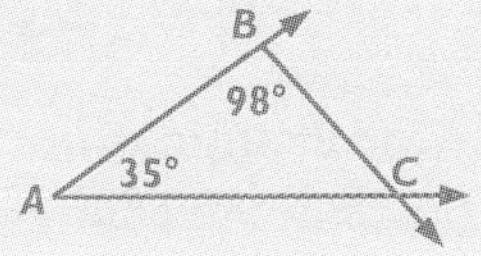

Use the protractor to measure $\angle ACB$. The measure is about 47° or 48°.

b. Compare the two algorithms in part (a).

The first method uses a known fact and an equation to obtain an exact answer. The second method, while visual, can only give an estimated answer.

Teaching Tip

After students have studied Sample 2, ask them to write an algorithm for solving any inequality of the form $ax \geq b$. Note that if $a = 0$ and $b \leq 0$, then all real numbers are solutions. If $a = 0$ and $b > 0$, then there are no solutions.

Talk it Over

Question 7 asks students to analyze whether the steps of an algorithm can be done in different orders. You may wish to have students discuss this question for Mathod 1 of Sample 2 as well.

Look Back

Students can respond to this question verbally. A volunteer can write the things suggested on the board.

APPLYING

Suggested Assignment

Standard 1, 2, 5–11, 16–27

Extended 1–3, 5–14, 16–27

Integrating the Strands

Number Exs. 11, 13–16

Algebra Exs. 6–10, 17–20

Geometry Ex. 4

Statistics and Probability Exs. 21–26

Discrete Mathematics Exs. 1–6, 11–16, 21–27

Logic and Language Exs. 1, 3, 12, 14, 16, 27

Communication: Writing

Ex. 3 involves students in writing directions as preparation for writing algorithms in Exs. 6 and 12. For Ex. 14, students compare two algorithms and identify some advantages and disadvantages of each. For Ex. 16, students again write and compare algorithms. Teaching strategies for writing in mathematics are given on page T36 in the front of this Teacher's Edition.

Talk it Over

6. What are the steps of an algorithm for using algebra to solve any linear inequality with one variable?
7. Can the steps of the algorithm in Method 2 of Sample 2 be done in a different order? If they can, what are the steps of the new algorithm?

Look Back

What are some things to keep in mind when you write a set of directions for solving a problem?

1-1 Exercises and Problems

1. **Reading** How can you tell whether an algorithm contains a loop?
2. **Clothing** A *sari* is a trditional form of dress in some parts of India and Sri Lanka. It is made with a piece of cloth 5 to 7 yd long and about 40 in. wide. Some of the steps in tying and draping a sari are shown out of order. Put these steps in the correct order.

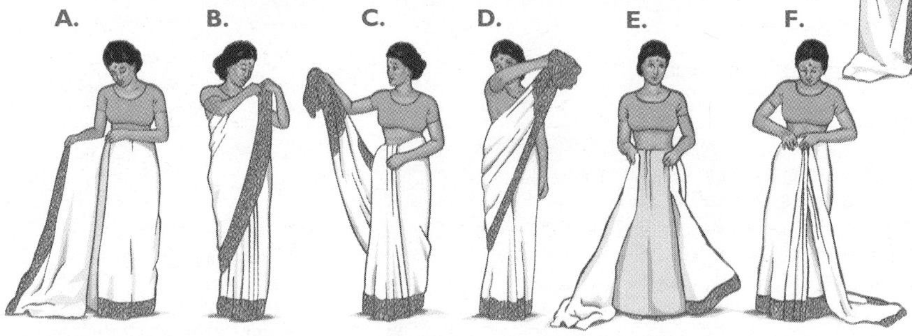

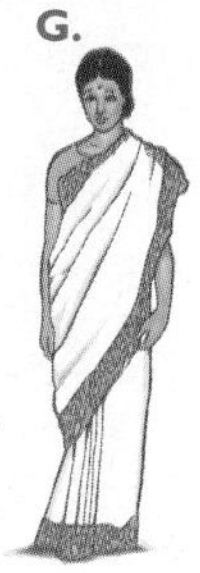

3. **Open-ended** Suppose it is the first day of school. A new student stops you outside your mathematics classroom and asks you for directions to the main office in your school. Write directions from the classroom to the main office.
4. **Using Manipulatives** Work with another student.
   a. Draw a large triangle on a sheet of paper and cut it out.
   b. Use paper folding to construct the three medians of the triangle.
   c. Write an algorithm for the procedure you used in part (b).
   d. Use your algorithm to construct the three medians of an obtuse triangle, an acute triangle, and a right triangle by paper folding.
   e. Use your results in part (d) to make a conjecture about the intersection of the three medians of any triangle.

A *median* of a triangle is a segment from a vertex to the midpoint of the opposite side.

Answers to Talk it Over

6. Answers may vary. An example is given for solving an inequality of the form $ax + b < c$. **Step 1:** If $b = 0$, go to Step 3. **Step 2:** Add the opposite of b to both sides of the inequality. **Step 3:** If $a = 1$, go to Step 6. **Step 4:** Multiply both sides of the inequality by the reciprocal of a. **Step 5:** If $a < 0$, reverse the inequality symbol. **Step 6:** Simplify if necessary.
7. Yes; **Step 1:** Graph $y = 4$. **Step 2:** Graph $y = -3x + 9$. **Step 3:** Estimate the x-coordinate of the intersection point.

Answers to Look Back

Answers may vary. Examples are given. Directions should be correct, in proper sequence, clear, concise, and easy to follow.

5. **Pottery** Santana Martinez makes coil pots in the traditional Pueblo style. The photographs show some of the key steps she follows. Put the steps in the correct order and write an algorithm for the process.

A.
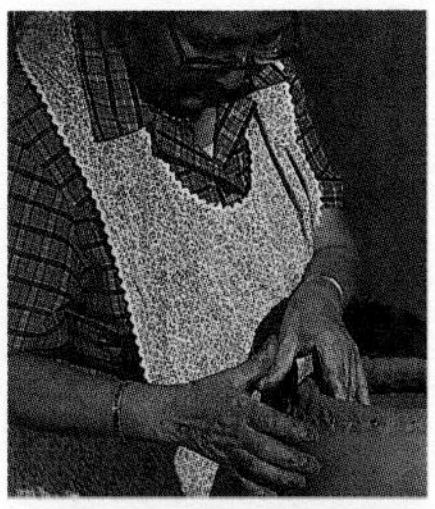
B.
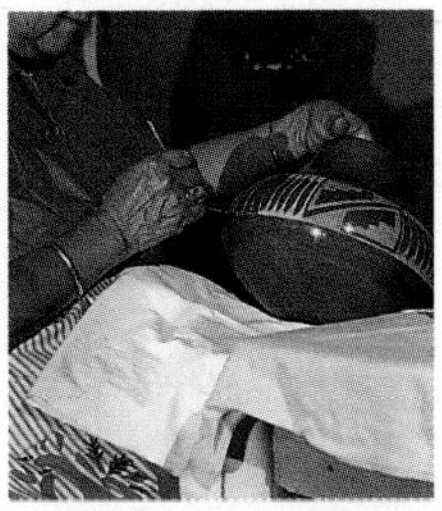
C.

D.

E.

F.
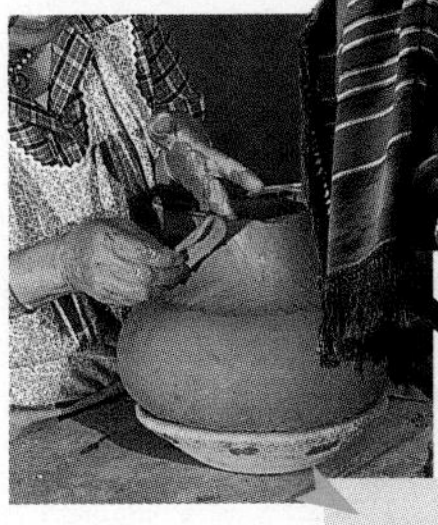

BY THE WAY...

Santana Martinez uses a technique developed by her mother-in-law, Maria Martinez, a Native American potter from New Mexico. In 1909 Maria Martinez re-created the technique that the ancient Pueblos used eight centuries earlier.

6. Write an algorithm to solve an equation of the form $ax + b = c$ for x using algebra.

Solve each inequality in two ways.

7. $\frac{x}{14} - 85 > 6$

8. $2x + 3 \leq 17$

9. $-5x - 18 \geq 4$

10. $23 > 12 - \frac{x}{9}$

11. Susan and Ryan use different algorithms for estimating a 15% tip for a server in a restaurant. In the state where they live, there is a 5% sales tax on restaurant meals.

Susan's Algorithm	Ryan's Algorithm
1. Move the decimal point one place to the left in the total cost of the meal. 2. Find half the amount in Step 1. 3. Add the amounts in Steps 1 and 2.	Multiply the sales tax by 3.

a. Compare Susan's and Ryan's algorithms. Which do you prefer? Why?

b. Use the fact that $\frac{1}{7} \approx 14\%$ or that $\frac{1}{6} \approx 17\%$ to write another algorithm for estimating a 15% tip.

c. Compare your algorithm from part (b) to Susan's and Ryan's algorithms. Which do you prefer? Why?

12. **Open-ended** Write an algorithm for an everyday activity. Include at least one decision step and one loop in your algorithm.

Using Manipulatives

Ex. 4 has students use manipultives and an algorithm, and their reasoning skills, to make a conjecture about the intersection of the three medians of any triangle. You may wish to extend this activity to angle bisectors.

Communication: Discussion

Students may wish to discuss the importance of algorithms for computers.

Integrating the Strands

The general concept of an algorithm permeates almost all strands of mathematics. Algorithms are used in calculations with numbers, solving algebraic equations, applying functions to solve problems, making measurements, and in solving certain problems in geometry, trigonometry, statistics, probability, discrete mathematics, and mathematical logic.

Answers to Exercises and Problems

1. The algorithm will tell you to repeat a step or steps either a certain number of times or until some condition is met.
2. E, F, A, C, D, B, G
3. Answers may vary.
4. a, b. Check students' work.
   c. Answers may vary. Example: **Step 1:** Fold the paper so that one sided of the triangle is folded in half. Pinch a fold at the midpoint. **Step 2:** Fold the opposite vertex to the midpoint of the side. Crease the fold. **Step 3:** Repeat Steps 1 and 2 until all medians have been made.
   d. Check students' work.
   e. Conjecture: the three medians of any triangle intersect in a point inside the triangle.
5. E, F, C, A, B, B
6. Answers may vary. An example is given. **Step 1:** If $b = 0$, go to Step 3. **Step 2:** Add the opposite of b to both sides of the equation. **Step 3:** If $a = 1$, go to Step 5. **Step 4:** Multiply both sides of the equation by the reciprocal of a. **Step 5:** Simplify if necessary.

7–10. Solution methods may vary. Solutions are given.

7. $x > 1274$
8. $x \leq 7$
9. $x \leq -4\frac{2}{5}$
10. $x > -99$
11. a. Answers may vary. An example is given. I prefer Ryan's algorithm because it requires fewer steps and calculations. If the server didn't add the sales tax to the check, then I'd use Susan's algorithm.
   b. Divide the bill, with tax, by 7. Divide the bill, without tax, by 6.
   c. Answers may vary. An example is given. I prefer Ryan's algorithm because multiplying is easier than dividing. My algorithm is less complicated than Susan's.
12. Answers may vary. An example is given. Baking a batch of cookies: **Step 1:** Find recipe. **Step 2:** Check to be sure I have ingredients. **Step 3:** Preheat oven. **Step 4:** Mix ingredients. **Step 5:** Fill cookie sheet with cookies. **Step 6:** Bake cookies. **Step 7:** If there is more cookie dough, repeat Steps 5 and 6.

Research

Exs. 13 and 14 present an interesting example of an ancient Egyptian algorithm for multiplication. Some students may wish to research other algorithms used by the Egyptians.

Students Acquiring English

For students acquiring English, being able to communicate that they understand the algorithms is more important than being able to demonstrate this understanding by writing a polished paragraph. For Ex. 14, consider offering these students the alternative of listing the advantages of each type of algorithm. Teaching strategies for satudents acquiring English are presented on page T40 of this Teacher's Edition.

Practice 1 For use with Section 1-1

Name ______________________ Date ____________

Practice 1

For use with Section 1-1

Solve each inequality in two ways.

1. $3x + 5 \leq -16$ $x \leq -7$
2. $-\frac{1}{2}x - 5 > 11$ $x < -32$
3. $-5x + 3 \geq -15$ $x \leq 3.6$
4. $\frac{4x}{3} + 7 < 3$ $x < -3$
5. $\frac{x}{6} - 5 > -10$ $x > -30$
6. $-8x + 9 \leq 7$ $x \geq 0.25$
7. $\frac{1}{3} - \frac{x}{6} < 4$ $x > -22$
8. $4 - \frac{3}{2}x \geq 17$ $x \leq -\frac{26}{3}$
9. The key steps of an algorithm for filling your car with gas at a self-service pump are out of order. Put these steps in the correct order. E, D, B, A, F, C
   - **A.** Place hose nozzle in gas pipe of car.
   - **B.** Turn on pump by lifting lever below hanger.
   - **C.** Pay cashier for the amount of gas you pumped.
   - **D.** Take nozzle off its hanger on pump.
   - **E.** If station requires a deposit before pumping, pay deposit.
   - **F.** Squeeze nozzle handle to pump; stops automatically.
10. Is there a decision in the algorithm in Exercise 9? If there is, which step involves the decision? Is there a loop in the algorithm in Exercise 9? If there is, describe the loop. yes; Step E; no
11. The following algorithm for finding the greatest common divisor (GCD) of two positive integers is called the Euclidean algorithm:

    Step 1: Divide the larger integer by the smaller integer.

    Step 2: Replace the larger of the two integers by the remainder from your division.

    Step 3: Repeat steps 1 and 2 with the integers you now have until the remainder is 0.

    Step 4: The last divisor is the GCD.

    a. Carry out the algorithm to find the GCD of 30 and 72. 6
    b. Carry out the algorithm to find the GCD of 28 and 72. 4
    c. Does the Euclidean algorithm contain a decision? If it does, describe the decision. no
    d. Does the Euclidean algorithm contain a loop? If it does, describe the loop. d. yes; Step 3 requires repetition of previous steps and contains a stopping condition.
12. *Open-ended* Write an algorithm to describe calling a friend on the phone. Include the possibilities that the line may be busy and that your friend may not be home. Answers may vary. Check students' work.

Practice Bank, INTEGRATED MATHEMATICS 3

1

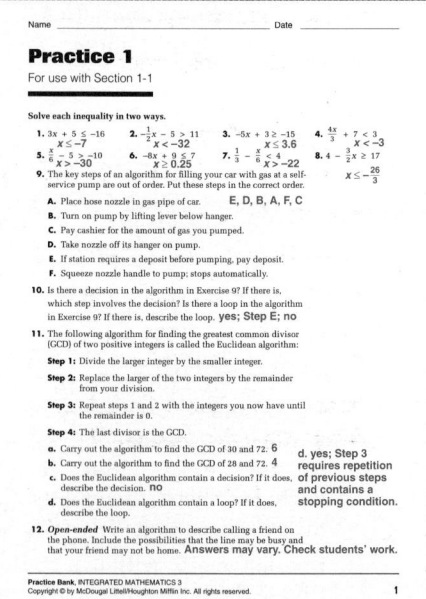

connection to **HISTORY**

Over 4000 years ago the Egyptians developed one of the earliest algorithms for multiplication. Today's computers use the idea on which the algorithm is based.

Examples

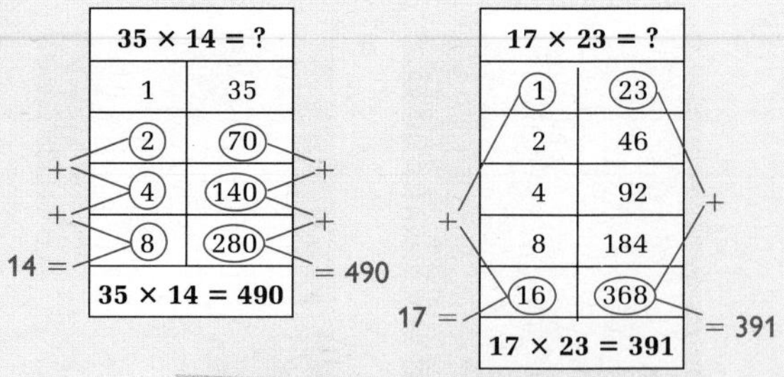

13. Explain how the Egyptian multiplication algorithm works.

14. **Writing** Compare the Egyptian multiplication algorithm to the multiplication algorithm you know. What are some advantages and disadvantages of each?

15 TECHNOLOGY Wenona used this algorithm with a graphics calculator.

a. What procedure does Wenona's algorithm perform?

b. Describe any loops in Wenona's algorithm.

Step 1 Enter a LIST.
Step 2 Set SUM and COUNT to 0.
Step 3 Read a number from LIST.
Step 4 Add the number in Step 3 to SUM.
Step 5 Add 1 to COUNT.
Step 6 Repeat Steps 3–5 until all the numbers in LIST have been read.
Step 7 Divide SUM by COUNT.

Ongoing ASSESSMENT

16. a. Write two different algorithms for making the change that a salesclerk gives to a customer who has made a purchase.

b. **Writing** Compare your algorithms.

Review PREVIEW

Factor. *(Toolbox Skill 17)*

17. $3x^2 + 4x - 7$ **18.** $3x^2 + 15x + 18$ **19.** $9x^2 - 64$ **20.** $16x^2 + 56x + 49$

Find each number of permutations. *(Toolbox Skill 3)*

21. 4 items arranged 3 at a time **22.** 6 items arranged 4 at a time **23.** 9 items arranged 5 at a time

Find each number of combinations. *(Toolbox Skill 4)*

24. 7 items chosen 2 at a time **25.** 12 items chosen 5 at a time **26.** 5 items chosen 4 at a time

Working on the Unit Project

27. The steps of an algorithm for planning a park are listed out of order. Put the steps in the correct order.

A. Build the park.
B. Choose the location of the park.
C. Operate the park.
D. Prepare a detailed design plan.
E. Identify the need for a park.
F. Identify sources of funds and make a budget.
G. Survey people in the area and use the results to develop a design concept.

Answers to Exercises and Problems

13. Summaries may vary. An example is given. A two-column table is set up. In one column, powers of 2 are listed. In the other, the multiples of the larger of the two factors and powers of 2 are listed. Powers of 2 are chosen so that their sum is the smaller factor. The related multiples of 2 are added to find the product. In the first example, $14 = 2^1 + 2^2 + 2^3 = 2 + 4 + 8$, so $14 \cdot 35 = (2 + 4 + 8)(35) = 2(35) + 4(35) + 8(35) = 70 + 140 + 280 = 490$.

14. Summaries may vary. An example is given. The algorithm I know also uses the distributive property, but is based on multiples of powers of 10. For example, $14 \cdot 35 = (4 + 10)(35) = 4(35) + 10(35)$. If the numbers are not too large, the Egyptian algorithm is fairly simple. However, if the numbers are very large, the Egyptian algorithm may be very time-consuming. It also may be confusing to write one of the factors as the sum of powers of 2.

15, 16. See answers in back of book.

17. $(3x + 7)(x - 1)$

18. $3(x + 3)(x + 2)$

19. $(3x - 8)(3x + 8)$

20. $(4x + 7)(4x + 7)$

21. 24 **22.** 360

23. 15,120 **24.** 21

25. 792 **26.** 5

27. E, F, B, G, D, A, C

Section **1-2**

Using Systematic Lists

Focus

Use organized lists to solve problems.

Get Organized

EXPLORATION

How can you find all the triangles that have a given perimeter?

- **Materials: 15 toothpicks**
- **Work with another student.**

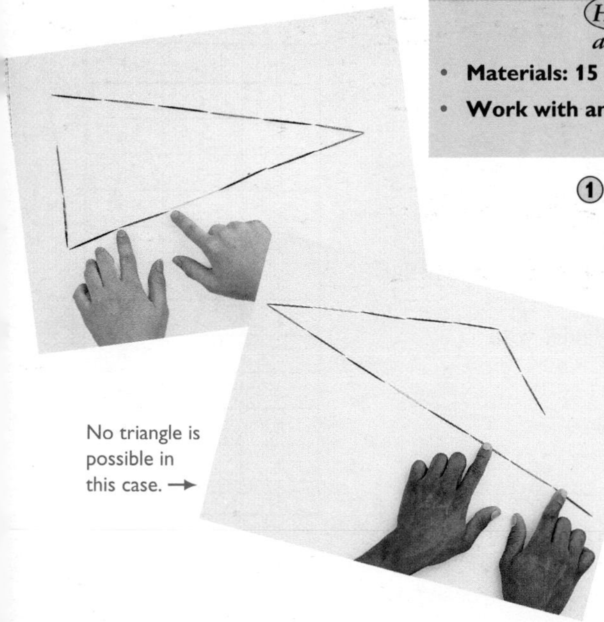

No triangle is possible in this case. →

① Try to build all the different triangles with a perimeter of 15 toothpicks. Record the number of triangles you built and the lengths of their sides. Describe any problem solving strategies and algorithms you used.

② Compare your results with those of another group. Did you leave out any triangles with side lengths that add up to 15? If you did, how can you improve your strategy?

In the Exploration, you probably discovered that you cannot build triangles with all combinations of side lengths. For example, you cannot build a triangle with sides that are 8 units, 5 units, and 2 units in length.

X(Lab

TRIANGLE INEQUALITY

The sum of the lengths of any two sides of a triangle is greater than the length of the third side.

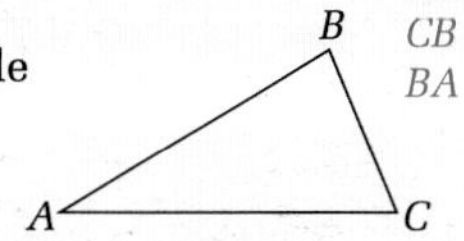

$AC + CB > AB$
$CB + BA > AC$
$BA + AC > CB$

Answers to Exploration

1. Answers may vary. There are 7 such triangles.

Side 1	Side 2	Side 3
7	1	7
7	2	6
7	3	5
7	4	4
6	3	6
6	4	5
5	5	5

2. Answers may vary.

PLANNING

Objectives and Strands

See pages 1A and 1B.

Spiral Learning

See page 1B.

Materials List

- Toothpicks

Recommended Pacing

Section 1-2 is a two-day lesson.

Day 1

Pages 9–10: Exploration through Talk it Over 4, *Exercises 1–14*

Day 2

Pages 11–12: Top of page 11 through Look Back, *Exercises 15–30*

Extra Practice

See pages 608–610.

Warm-Up Exercises

Warm-Up Transparency 1-2

Support Materials

- Practice Bank: Practice 2
- Activity Bank: Enrichment 2
- Study Guide: Section 1-2
- Problem Bank: Problem Set 1
- Explorations Lab Manual: Diagram Master 10
- Assessment Book: Quiz 1-2, Alternative Assessment 2

TEACHING

Exploration

This Exploration has two goals. First, using toothpicks, students discover that they cannot build triangles with all combinations of side lengths. This discovery provides a concrete and intuitive basis for understanding the meaning of the triangle inequality. The Exploration also motivates the need for using the problem-solving strategy of using a systematic list to be certain that all triangles have been found.

Additional Sample

S1 Find all the different scalene triangles with whole-number side lengths and a perimeter of 11.

Step 1. Make a systematic list of all possible combinations of side lengths that add up to 11.

	Side 1	Side 2	Side 3
X	1	1	9
X	1	2	8
X	1	3	7
X	1	4	6
X	1	5	5
X	2	2	7
X	2	3	6
	2	4	5
X	3	3	5
X	3	4	4

Step 2. Put an "X" next to each combination that does not satisfy the triangle inequality or is not scalene. The only triangle that is scalene, has whole-number side lengths, and has perimeter 11 is the 2-4-5 triangle.

Talk it Over

Students need to apply the triangle inequality to answer question 1. The other three questions direct students' attention to a deeper analysis of the algorithm used in Sample 1.

Systematic Lists

In the Exploration, you wanted to find all the possible combinations of side lengths that add up to 15. In situations like this, where there are many possibilities to explore, you can use a *systematic,* or organized, list to make sure that you do not leave out or repeat any combinations.

Sample 1

Find all the different triangles that have sides with whole-number lengths and a perimeter of 12.

Sample Response

Step 1 Make a systematic list of all possible combinations of side lengths that add up to 12.

Here is an algorithm that you can use to make a complete list in which no combinations repeat.

> Begin with 1 for the length of side 1. List in order all possible lengths for side 2. Subtract to find each length for side 3. Skip any duplicates.

	Side 1	Side 2	Side 3
X	1	1	10
X	1	2	9
X	1	3	8
X	1	4	7
X	1	5	6
X	2	2	8
X	2	3	7
X	2	4	6
	2	5	5
X	3	3	6
	3	4	5
	4	4	4

Step 2 Put an "X" next to all the combinations that do not satisfy the triangle inequality stated on page 9.

There are only three different triangles that have sides with whole-number lengths and a perimeter of 12:

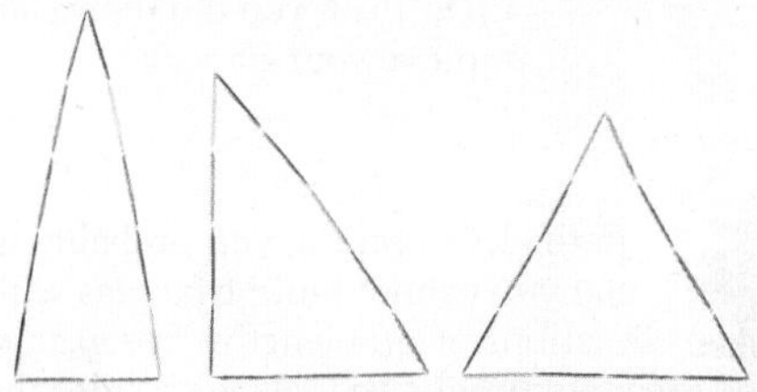

Talk it Over

1. Why are there no triangles that have a perimeter of 12 and at least one side of length 1?
2. Why does the algorithm used in Sample 1 produce all possible combinations of side lengths that add up to 12?
3. Can you use the same algorithm to list all the different orderings of 12 different letters? Why or why not?
4. Describe another algorithm that you can use to solve Sample 1.

Now you are ready for: Exs. 1–14 on p. 13

Answers to Talk it Over

1. If one side is 1 unit long, the other two sides have lengths x and $11 - x$ for some integer x. Because of the triangle inequality, all of the following must be true: $x + 1 > 11 - x$, $x + (11 - x) > 1$, and $1 + (11 - x) > x$. For all three to be true, x must be greater than 5 and less than 6. There is no such integer x.
2. The algorithm begins with the smallest possible length for one side and considers all possible lengths of each of the other two sides in order (from least to greatest for the second side and greatest to least for the third).
3. No; first, there is no relationship among letters similar to inequality with numbers. If you were to use "≤" to mean "is the same or comes before in alphabetical order" the algorithm would still not be successful. In the given algorithm, order does not matter. It would in producing different ordering of letters. Also, in Sample 1, side lengths can repeat in a single triangle; in deter-

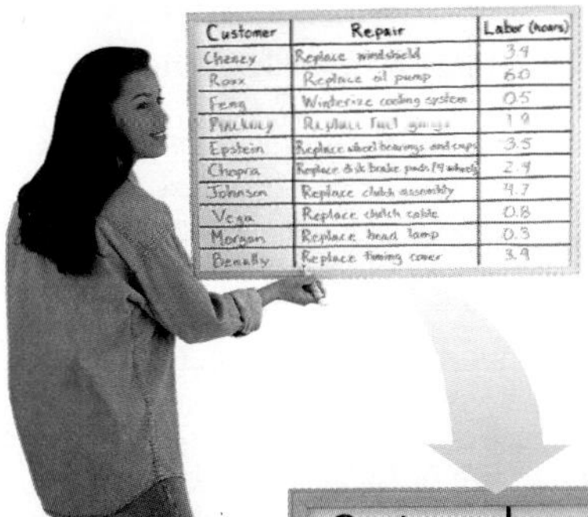

In many businesses, managers must make a schedule of tasks and must assign tasks to employees. Often the goal is to use the fewest number of people to complete the tasks.

Sample 2

Paula Hernandez is the manager of an automobile repair service. She lists the number of hours needed to complete the repair work on the cars waiting for service.

Customer	Repair	Labor (hours)
Cheney	Replace windshield	3.4
Roux	Replace oil pump	6.0
Feng	Winterize cooling system	0.5
Pinckney	Replace fuel gauge	1.9
Epstein	Replace wheel bearings and cups	3.5
Chopra	Replace disk brake pads (4 wheels)	2.4
Johnson	Replace clutch assembly	4.7
Vega	Replace clutch cable	0.8
Morgan	Replace head lamp	0.3
Benally	Replace timing cover	3.9

Each mechanic works 8 h per day and can make any repair. The repairs can be done in any order.

a. What is the least number of mechanics needed to complete all the repairs in one day?

b. How should Paula Hernandez assign the repairs so that she needs the least number of mechanics to complete all the repairs in one day?

Sample Response

a. Find the minimum number of mechanics.

sum of all the repair times ÷ **number of hours per day each mechanic works** = **number of mechanics needed**

(3.4 + 6.0 + 0.5 + 1.9 + 3.5 + 2.4 + 4.7 + 0.8 + 0.3 + 3.9) ÷ 8 ≈ 3.43

At least four mechanics are needed.

b. **Problem Solving Strategy:** Make a systematic list.

Try this algorithm for assigning the repair tasks:

Take each repair in the order listed and assign it to the first mechanic who has time to complete it.

1 Write the time for the first repair.

2 Start a second row, since 3.4 + 6.0 > 8.

3.4 + 0.5 + 1.9 + 0.8 + 0.3 = 6.9

6.0

3.5 + 2.4 = 5.9

4.7

3.9

3 Add the next two repairs to the first row, since 3.4 + 0.5 + 1.9 < 8.

4 Continue in the same way until all repairs are assigned.

The completed list contains five rows, so five mechanics are needed when the tasks are assigned this way.

Continued on next page.

Additional Sample

S2 In Sample 2, suppose the manager takes one last look at her estimates of how long it will take to complete the jobs. Suppose she decides that four of the estimates should change as follows:

Cheney	3.6 h
Pinckney	2.5 h
Epstein	4.0 h
Vega	1.0 h

a. Now what is the number of mechanics needed?
The total number of hours of work changes from 27.4 to 28.9. Since 28.9 ÷ 8 ≈ 3.61, it is still true that at least four mechanics are needed.

b. With the changes she has made in her estimates, how should she now assign the repair jobs?
Try both algorithms if necessary. The first algorithm yields the same assignments with the new estimates. It is desirable to use four mechanics instead of five, so try the second algorithm. This algorithm works just as before, but it gives new job assignments.
Mech. 1: Roux, Vega, Feng, Morgan
Mech. 2: Johnson, Pinckney
Mech. 3: Epstein, Benally
Mech. 4: Cheney, Chopra

Answers to Talk it Over

mining possible orders of letters, each letter would be used only once in each grouping.

4. Answers may vary. An example is given. Beginning with 10 for the length of side 1, list all possible lengths for sides 2 and 3 that add up to 12 and also satisfy this inequality: length of side 1 ≥ length of side 2 ≥ length of side 3. Cross out all the combinations that do not satisfy the triangle inequality.

Reasoning

Ask students whether they think the second algorithm in the Sample 2 response will always result in a job assignment that uses the minimum number of mechanics. Discuss why or why not.

Students may also wish to consider what might happen if the labor hours for the jobs changed. For example, if there were seven jobs of 4.1 h each and three jobs of 0.3 h each, what happens? (You would again conclude that at least four mechanics are needed. But when you begin planning job assignments, you quickly see that there is no way to make the assignments without using seven or more mechanics.)

Talk it Over

Questions 5–8 require that students think carefully about the two algorithms used in Sample 2. You may wish to have students work in cooperative groups to answer these questions.

Look Back

Students can begin by working independently on this activity. Then a brief class discussion of their answers would provide a valuable summary of the usefulness of the strategy of "Using a Systematic List."

APPLYING

Suggested Assignment

Day 1

Standard 1–8, 12, 13

Extended 1–8, 12–14

Day 2

Standard 15–30

Extended 15–30

The algorithm used on page 11 does not result in the least number of mechanics.

Try another algorithm:

1. Order the repair times from greatest to least.

   6.0, 4.7, 3.9, 3.5, 3.4, 2.4, 1.9, 0.8, 0.5, 0.3

2. Take each repair task in order and assign it to the first mechanic who has time to complete it.

When you use this algorithm to assign the repairs, only four mechanics are needed.

6.0 + 1.9 = **7.9** →

4.7 + 2.4 + 0.8 = **7.9** →

3.9 + 3.5 + 0.5 = **7.9** →

3.4 + 0.3 = **3.7** →

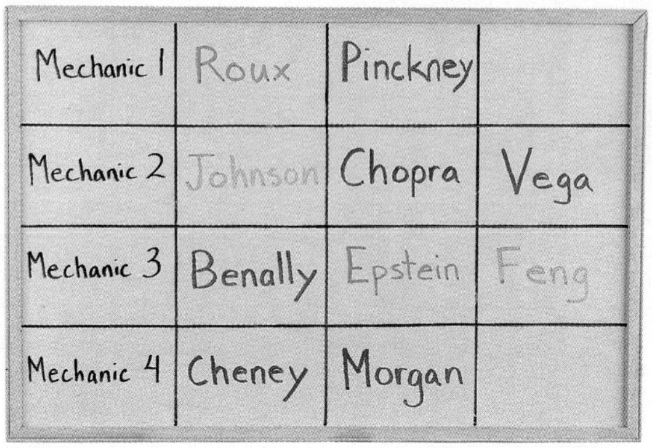
Mechanic 1
Roux
Pinckney
Mechanic 2
Johnson
Chopra
Vega
Mechanic 3
Benally
Epstein
Feng
Mechanic 4
Cheney
Morgan

Talk it Over

5. Describe how to make the list for the second algorithm in Sample 2.
6. Suppose in Sample 2 Paula Hernandez did *not* need to find the least number of mechanics. Compare the two algorithms used in part (b).
7. Think of another algorithm for assigning the repairs in Sample 2. Do you think this algorithm results in the least number of mechanics? Why or why not?
8. Why would Paula Hernandez *not* be able to use the methods shown in part (b) of Sample 2 if some of the customers have to pick up their cars by noon? if more than one mechanic is needed to do some of the repairs?

Now you are ready for: Exs. 15–30 on pp. 13–15

Look Back

Describe at least two kinds of situations in which "Using a Systematic List" is a useful problem solving strategy. Explain why this strategy is useful in each situation you describe.

Answers to Talk it Over

5. Assign the first task to a mechanic. The second task would take that mechanic past 8 h, so assign it to a new mechanic. Similarly, assign the third task to a third mechanic. For each succeeding task, determine if any of the first three mechanics (in order) can complete the task. If not, assign another mechanic.

6. Even if Paula Hernandez does not need to use the least number of mechanics, the second algorithm may be more reasonable, because each mechanic's time is used more efficiently. Three of the four will be working all day. With the first algorithm, four are working 6 h or less and two are working less than 5 h.

7, 8. See answers in back of book.

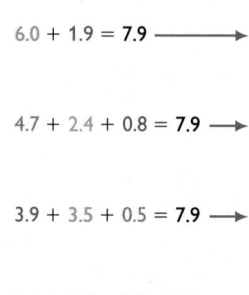

1-2 Exercises and Problems

1. **Reading** How are the lengths of the three sides of any triangle related?

For each group of measurements in Exercises 2–7, tell whether they can be the lengths of the sides of a triangle. Write *Yes* or *No*.

2. 2 ft, 7 ft, 9 ft
3. 4 in., 4 in., 6 in.
4. 12 cm, 7 cm, 15 cm
5. 7.4 m, 9 m, 14.2 m
6. 27 yd, 4 yd, 30 yd
7. $14\frac{1}{2}$ ft, 11 ft, $28\frac{1}{4}$ ft

8. Find all the sets of four whole numbers with a sum of 16.

Using Manipulatives For Exercises 9–11, use toothpicks to build all the triangles that have whole-number side lengths and each indicated perimeter.

9. 17
10. 24
11. 32

12. a. Find all the ways that you can make change for a $1 bill using quarters, dimes, and nickels.
    b. Which of the ways that you found in part (a) gives you 65¢ in exact change for a vending machine?
    c. Describe the algorithms you used in parts (a) and (b).

13. a. Find all isosceles triangles that have sides with whole-number lengths and a perimeter of 30.
    b. Which of the triangles you found in part (a) has the largest area?

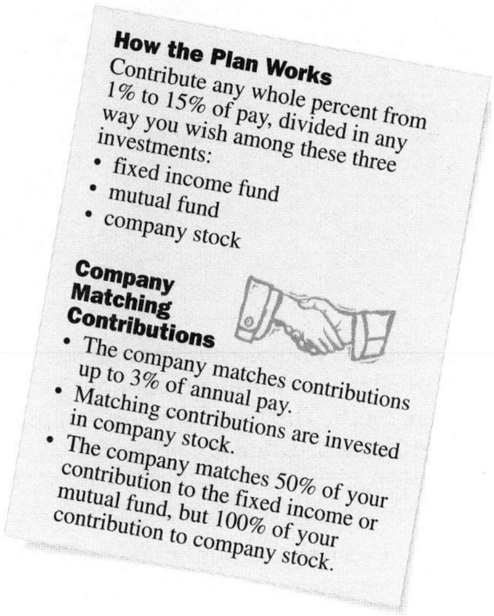
How the Plan Works
Contribute any whole percent from 1% to 15% of pay, divided in any way you wish among these three investments:
- fixed income fund
- mutual fund
- company stock

Company Matching Contributions
- The company matches contributions up to 3% of annual pay.
- Matching contributions are invested in company stock.
- The company matches 50% of your contribution to the fixed income or mutual fund, but 100% of your contribution to company stock.

14. Personal Finance Sri Iskandar is employed by a company that offers the employee savings plan described in this brochure. Find all the ways that she can contribute under 6% of her pay to the plan and still receive 3% in matching funds from the company.

15. Mel Walker collects objects that are made from cast iron. He wants to mail home the objects he bought for his collection while on vacation. The weights of the objects are:

34 lb 8 oz	26 lb 7 oz
25 lb 5 oz	43 lb 9 oz
38 lb 12 oz	30 lb 6 oz
57 lb	22 lb 8 oz
30 lb 7 oz	18 lb 14 oz

The maximum weight for a parcel-post package is 70 lb. The fewer packages he mails, the lower his mailing cost will be. How can he package the objects in the least number of parcel-post packages?

Integrating the Strands

Number Exs. 8, 12, 14, 25–28
Algebra Exs. 21–28
Geometry Exs. 1–7, 9–11, 13, 18, 25–28
Discrete Mathematics Exs. 12–20, 30
Logic and Language Exs. 1, 15–20, 29, 30

Mathematical Procedures

Students can solve Exs. 12 and 13 with greater confidence if they know in advance how many items they should have in their organized lists. Often this number can be determined by using prior knowledge or by beginning an organized list and looking for patterns that suggest how to predict the number of items in the final list. After they have completed the list, they can count to see if they do indeed have that many items.

Students Acquiring English

Students acquiring English may be unfamiliar with the banking terms in Ex. 14. You may want to explain the meanings of *mutual fund, fixed income,* and *company stock.*

Problem Solving

Ex. 15 may at first glance seem like a completely new problem. In fact, it is very much like Sample 2. In Sample 2, jobs of a certain size were assigned to mechanics until no further jobs could be assigned without exceeding 8 hours. Here, the weights are like the times needed to complete a job. The 70 lb weight limit is like the 8 h work limit. Students should learn to watch for such similarities so they can apply earlier methods to new situations. Teaching strategies for developing good problem solvers are given on page T38 in the front of this Teacher's Edition.

Answers to Look Back

Answers may vary. Examples: making a schedule, finding all possible combinations of sides of a triangle for a given perimeter. A systematic list is useful in finding all possible combinations when there are many possibilities to explore. Such a list helps ensure that no combinations are left out or repeated. A systematic list can also be used to assign tasks. Such a list helps determine the most efficient use of time or resources.

Answers to Exercises and Problems

1. The sum of the lengths of any two sides must be greater than the length of the third side. This relationship is called the triangle inequality.
2. No.
3. Yes.
4. Yes.
5. Yes.
6. Yes.
7. No.

8–14. See answers in back of book.

15. Answers may vary. An example is given. (1) 57 lb; (2) 43 lb 9 oz, 26 lb 7 oz; (3) 38 lb 12 oz, 30 lb 7 oz; (4) 34 lb 8 oz, 30 lb 6 oz; (5) 25 lb 5 oz, 22 lb 8 oz, 18 lb 14 oz

Visual Thinking

Ideas for helping students develop their visual learning skills are provided on page T42 in the front of this Teacher's Edition and at strategic points throughout the side-column notes. Visual learning skills include:

1. Observation
2. Identification
3. Recognition
4. Recall
5. Interpretation
6. Exploration
7. Correlation
8. Generalization
9. Inference
10. Perception
11. Communication
12. Self-Expression

Each Visual Thinking note indicates the visual learning skills that are involved in the suggested activity.

16. **Business** How many groomers does Ahmad Jackson need to schedule tomorrow? At this grooming business, the groomers are part-time workers who work 3 h 30 min per day. Treatment to remove fleas takes about 15 min.

1 h 25 min to groom a long-haired dog

Jackson's
Pet Grooming
Long-haired dogs to be groomed tomorrow:
Lhasa apso groom treat for fleas
Afghan hound groom
cocker spaniel groom treat for fleas
Old English sheep dog groom
Samoyed groom treat for fleas
Maltese groom
Tibetan terrier groom treat for fleas

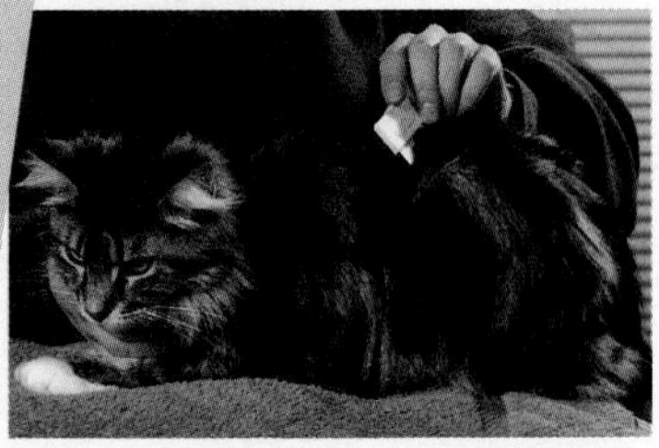

25 min to groom a cat

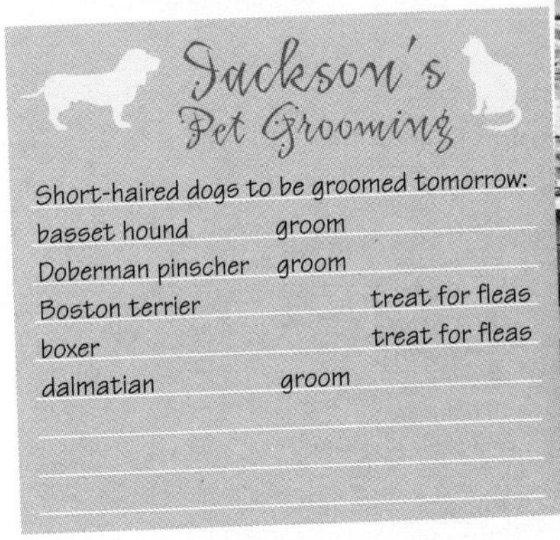

Jackson's
Pet Grooming
Short-haired dogs to be groomed tomorrow:
basset hound groom
Doberman pinscher groom
Boston terrier treat for fleas
boxer treat for fleas
dalmatian groom

20 min to groom a short-haired dog

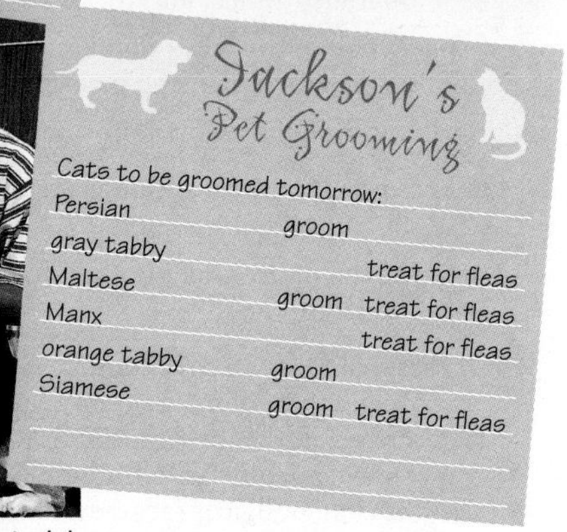

Jackson's
Pet Grooming
Cats to be groomed tomorrow:
Persian groom
gray tabby treat for fleas
Maltese groom treat for fleas
Manx treat for fleas
orange tabby groom
Siamese groom treat for fleas

17. **Television** Suppose a television station decides that the commercial breaks during a one-hour program cannot last longer than 2 min. The lengths of the commercials scheduled to run during the program are 60 s, 30 s, 30 s, 15 s, 30 s, 90 s, 60 s, 40 s, 30 s, 15 s, 20 s, 45 s, 60 s, 60 s, 30 s, and 90 s. The station can run the commercials at any time during the hour.
    a. How can the station schedule the commercials to minimize the number of commercial breaks during the program?
    b. **Writing** Describe the algorithm you used in part (a).

BY THE WAY. . .

In 1993 almost two thirds of TV network commercials and over four fifths of local TV station commercials were only 30 seconds long.

18. **Manufacturing** A sporting goods company packages baseballs in cubical boxes with edges that are 2.8 in. long. The company wants to ship the baseballs in cartons that hold four dozen baseballs.
    a. Find all possible dimensions of a shipping carton.
    b. What are the dimensions of the shipping carton that uses the least amount of material?

Answers to Exercises and Problems

16. 5 groomers

17. a. Six breaks are needed. Plans may vary. An example is given. Run two pairs of 60 s commercials, and two pairs of one 90 s and one 30 s commercials. Run the remaining three 30 s and the two 15 s commercials together. Run the 40 s, 20 s, and 45 s commercials together.
    b. Answers may vary. An example is given. I grouped numbers with a sum of 120. There were five such groups. I grouped the remaining commercials together, since their running time was less than 2 min.

18. a. 2.8 in. by 2.8 in. by 134.4 in.; 2.8 in. by 5.6 in. by 67.2 in.; 2.8 in. by 8.4 in. by 44.8 in.; 2.8 in. by 11.2 in. by 33.6 in.; 2.8 in. by 16.8 in. by 22.4 in.; 5.6 in. by 5.6 in. by 33.6 in.; 5.6 in. by 8.4 in. by 22.4 in.; 5.6 in. by 11.2 in. by 16.8 in.; 8.4 in. by 11.2 in. by 11.2 in.
    b. the 8.4 in. by 11.2 in. by 11.2 in. carton

19. She needs four boards. Plans may vary. An example is given. She could cut two 12 ft boards into three pieces each, one $6\frac{1}{2}$ ft long and the other two $2\frac{1}{2}$ ft long. The third board could be cut into four pieces, two 4 ft long and two 2 ft long. She would need a $6\frac{1}{2}$ ft piece from the fourth board.

19. Carpentry Alyssa wants to install shelves on one wall of her bedroom. She made this sketch. How can Alyssa cut the shelves from 12 ft boards in order to use the fewest boards?

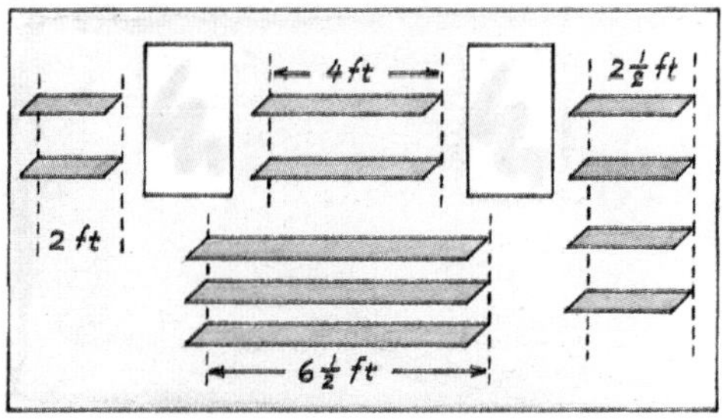

Ongoing ASSESSMENT

20. **Writing** The first algorithm used in part (b) of Sample 2 on page 11 is sometimes called the *first-fit* algorithm. The second algorithm used is sometimes called the *first-fit decreasing* algorithm. Do you think these are appropriate names for the algorithms? Why or why not?

Review PREVIEW

Solve each inequality in two ways. *(Section 1-1)*

21. $-6 + 4x < 11$
22. $\frac{x}{2} + 18 > -6$
23. $14 \le -3x + 8$
24. $-6x + 17 \le 2$

Find the value of *x* in each triangle. *(Toolbox Skill 30)*

25.
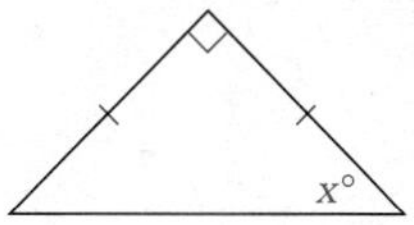

26.
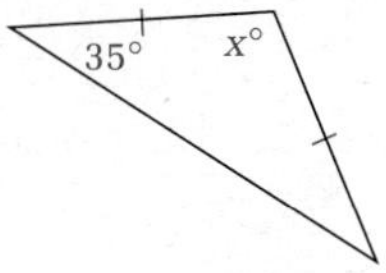

27.
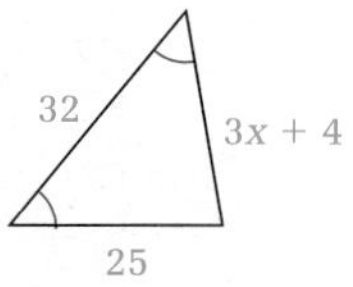

28.
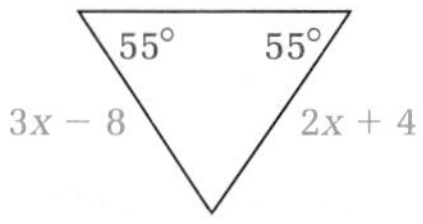

Use deductive reasoning to reach a conclusion. *(Toolbox Skill 35)*

29. If there is heavy traffic in the city, Corey will arrive at the theater after the curtain goes up. If Corey arrives at the theater after the curtain goes up, he will have to wait until the first act ends to be seated.

Working on the Unit Project

As you complete Exercise 30, think about how you will decide what kinds of equipment to install in the park you design.

30. A playground designer plans to include 15 pieces of equipment in a new playground. A survey of the children in the community and their parents shows that at least three swing sets, three slides, three whirligigs, and three sandboxes should be installed.
   a. Find all the possible combinations of swing sets, slides, whirligigs, and sandboxes that the designer can include.
   b. Suppose the survey shows that swings are twice as popular as the other types of equipment. How many of each type of equipment should the designer include?

Working on the Unit

Ex. 30 provides students with an opportunity to discuss in their project groups the kinds of equipment they may wish to install in the park they design. Each group should keep a written record of their answers to this exercise.

Practice 2 For use with Section 1-2

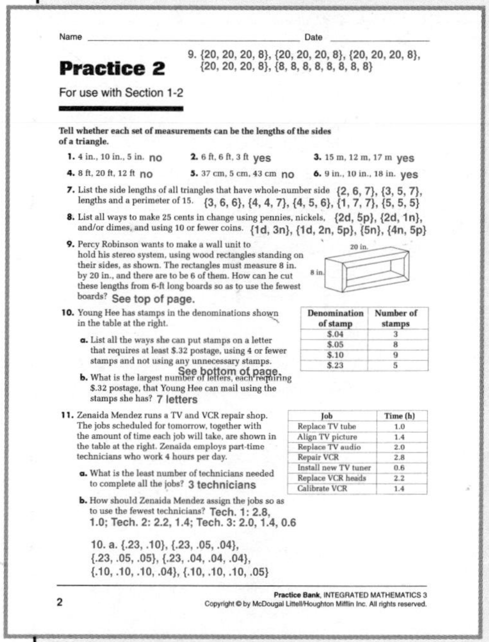

Name ______________ Date ______________

Practice 2

For use with Section 1-2

9. {20, 20, 20, 8}, {20, 20, 20, 8}, {20, 20, 20, 8}, {20, 20, 20, 8}, {8, 8, 8, 8, 8, 8, 8, 8}

Tell whether each set of measurements can be the lengths of the sides of a triangle.

1. 4 in., 10 in., 5 in. no
2. 6 ft, 6 ft, 3 ft yes
3. 15 m, 12 m, 17 m yes
4. 8 ft, 20 ft, 12 ft no
5. 37 cm, 5 cm, 43 cm no
6. 9 in., 10 in., 18 in. yes
7. List the side lengths of all triangles that have whole-number side lengths and a perimeter of 15. {2, 6, 7}, {3, 5, 7}, {3, 6, 6}, {4, 4, 7}, {4, 5, 6}, {1, 7, 7}, {5, 5, 5}
8. List all ways to make 25 cents in change using pennies, nickels, and/or dimes, and using 10 or fewer coins. {2d, 5p}, {2d, 1n}, {1d, 3n}, {1d, 2n, 5p}, {5n}, {4n, 5p}
9. Percy Robinson wants to make a wall unit to hold his stereo system, using wood rectangles standing on their sides, as shown. The rectangles must measure 8 in. by 20 in., and there are to be 6 of them. How can he cut these lengths from 6-ft long boards so as to use the fewest boards? See top of page.

20 in.
8 in.

10. Young Hee has stamps in the denominations shown in the table at the right.

Denomination of stamp	Number of stamps
\$.04	3
\$.05	8
\$.10	9
\$.23	5

   a. List all the ways she can put stamps on a letter that requires at least \$.32 postage, using 4 or fewer stamps and not using any unnecessary stamps. See bottom of page.
   b. What is the largest number of letters, each requiring \$.32 postage, that Young Hee can mail using the stamps she has? 7 letters

11. Zenaida Mendez runs a TV and VCR repair shop. The jobs scheduled for tomorrow, together with the amount of time each job will take, are shown in the table at the right. Zenaida employs part-time technicians who work 4 hours per day.

Job	Time (h)
Replace TV tube	1.0
Align TV picture	1.4
Replace TV audio	2.0
Repair VCR	2.8
Install new TV tuner	0.6
Replace VCR heads	2.2
Calibrate VCR	1.4

   a. What is the least number of technicians needed to complete all the jobs? 3 technicians
   b. How should Zenaida Mendez assign the jobs so as to use the fewest technicians? Tech. 1: 2.8, 1.0; Tech. 2: 2.2, 1.4; Tech. 3: 2.0, 1.4, 0.6

10. a. {.23, .10}, {.23, .05, .04}, {.23, .05, .05}, {.23, .04, .04, .04}, {.10, .10, .10, .04}, {.10, .10, .10, .05}

2 — Practice Bank, INTEGRATED MATHEMATICS 3

Answers to Exercises and Problems

20. Answers may vary. An example is given. Yes. The first-fit algorithm assigns each entry to the first category into which it will fit. The first-fit decreasing algorithm first orders the entries in decreasing order, then assigns each entry to the first category into which it will fit.

21. $x < 4\frac{1}{4}$

22. $x > -48$

23. $x \le -2$

24. $x \ge 2\frac{1}{2}$

25. 45

26. 110

27. 7

28. 12

29. If there is heavy traffic in the city, Corey will have to wait until the first act ends to be seated.

30. See answers in back of book.

PLANNING

Objectives and Strands

See pages 1A and 1B.

Spiral Learning

See page 1B.

Materials List

- Graph paper
- Graphics calculator or graphing software

Recommended Pacing

Section 1-3 is a one-day lesson.

Extra Practice

See pages 608–610.

Warm-Up Exercises

Warm-Up Transparency 1-3

Support Materials

- Practice Bank: Practice 3
- Activity Bank: Enrichment 3
- Study Guide: Section 1-3
- Problem Bank: Problem Set 1
- Explorations Lab Manual: Diagram Master 1
- Assessment Book: Quiz 1-3, Test 1, Alternative Assessment 3

Section 1-3

Using Statistics

TRENDS

Focus

Use statistics and data displays to draw conclusions.

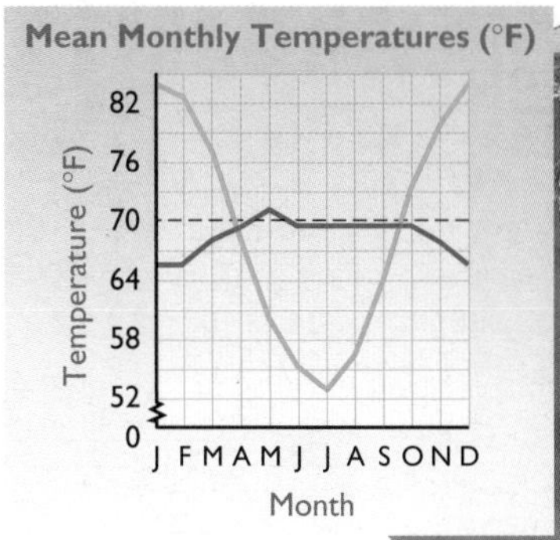

The graph for Caracas is red. The graph for Alice Springs is green.

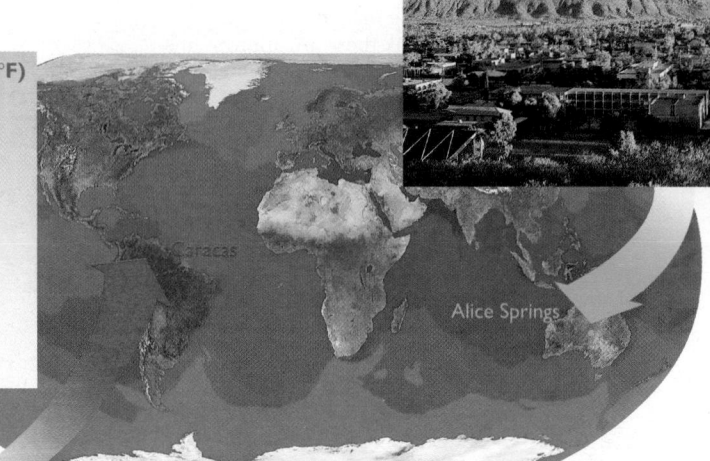

Both Alice Springs, Australia, and Caracas, Venezuela, have a **mean,** or average, annual temperature of about 70°F. The *line graph* compares their mean monthly temperatures.

Talk it Over

1. If you want temperatures near 70°F, any time of the year, which city should you visit?
2. If you want temperatures near 70°F, which city should you visit in January? in July? Why?
3. The mean annual rainfall is 63.6 in. for Alice Springs and 213.4 in. for Caracas.
   a. Do you think it rains more every month in Caracas than in Alice Springs?
   b. Does this statistic affect your decision to visit either city? Explain.

A statistic, such as the mean, summarizes a data set. A single statistic can be misleading. It is important to examine statistics and data displays carefully before you draw any conclusions.

Answers to Talk it Over

1. Caracas
2. Answers may vary. Examples are given. Caracas in both January and July; the temperature is at or near 70°F during both months in Caracas, but not in Alice Springs.
3. a. No; it means that, on average, Caracas is rainier than Alice Springs.
   b. Answers may vary. An example is given. Yes; if it rains a lot, then my activities could be limited.

You can use other data displays besides line graphs. A **box-and-whisker plot** is a data display that divides a data set into four parts. Each part represents about 25% of the data. The box contains the middle 50% of the data.

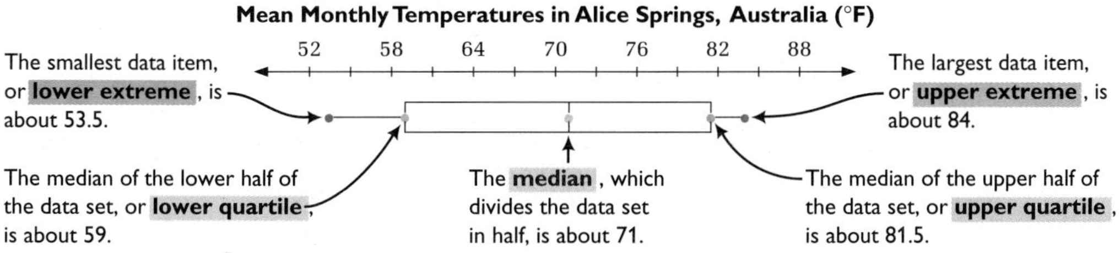

Mean Monthly Temperatures in Alice Springs, Australia (°F)
52 58 64 70 76 82 88
The smallest data item, or **lower extreme**, is about 53.5.
The largest data item, or **upper extreme**, is about 84.
The median of the lower half of the data set, or **lower quartile**, is about 59.
The **median**, which divides the data set in half, is about 71.
The median of the upper half of the data set, or **upper quartile**, is about 81.5.

Sample 1

The box-and-whisker plots show the gas mileage for some 1993 cars. *Gas mileage* is how far a car can travel on 1 gal of gasoline.

a. Which type(s) of cars have models with low gas mileage?

b. About what percent of compact cars have higher gas mileage than any large car? Explain.

c. If you want high gas mileage, which type of car should you buy based on the box-and-whisker plots? Why?

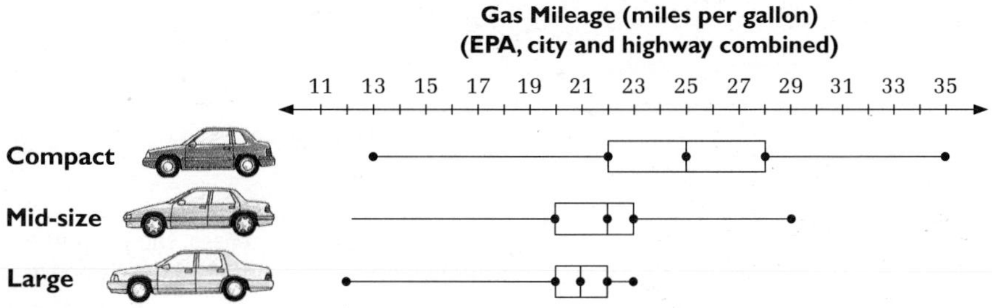

Gas Mileage (miles per gallon)
(EPA, city and highway combined)
11 13 15 17 19 21 23 25 27 29 31 33 35
Compact
Mid-size
Large

Sample Response

a. All three types of cars have models with low gas mileage. The lower extreme for all cars is about 12 or 13.

b. Between 50% and 75% of compact cars have higher gas mileage than any large car. The lower quartile for compact cars and the upper extreme for large cars are both about 23.

c. You should consider buying a compact car. Between 50% and 75% of compact cars have higher gas mileage than any large car. Since the upper quartile for compact cars and the upper extreme for mid-size cars are close together, about 25% of compact cars have higher gas mileage than any mid-size car.

TEACHING

Teaching Tip

You can use this diagram to help students remember what portions of the data items correspond to different parts of a box-and-whisker plot.

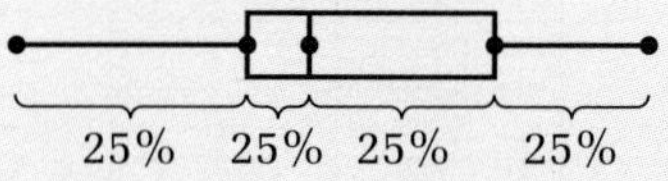

25% 25% 25% 25%

For additional information on discrete mathematics and data analysis, see page T46 in this Teacher's Edition.

Additional Sample

S1 The box-and-whisker plots show the time needed by three different groups of students to complete a particular math test. One group was allowed no calculators (NC), another allowed scientific calculators (SC), and the third allowed graphics calculators (GC).

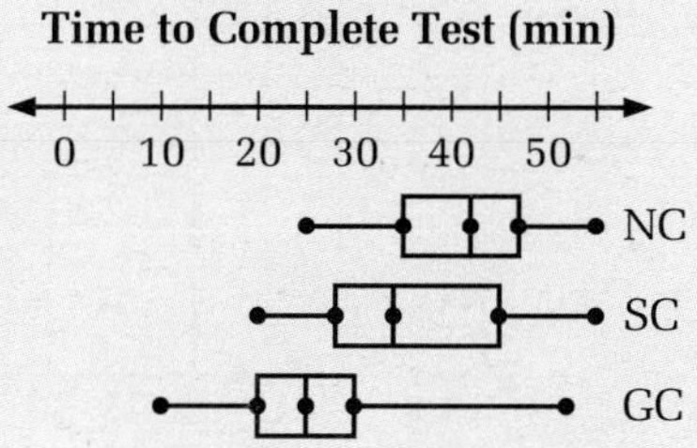

Time to Complete Test (min)
0 10 20 30 40 50
NC
SC
GC

a. In which group did about 75% of the students need 35 minutes or more to complete the test?
the no-calculator group

b. About what percent of students who used graphics calculators finished before any of the students who used scientific calculators? **25%**

c. Suppose the students scored about equally well in all three groups. What kind of calculators would you recommend the district purchase in the future? Why?
Answers may vary. An example is given. The district should try to purchase graphics calculators, since the graphics-calculator group seemed to work more efficiently.

Additional Sample

S2 A college compared scores on an algebra placement test and on a calculus final. The table shows the results for 10 students.

Algebra	Calculus
15	70
11	71
24	90
27	68
19	84
8	52
17	73
21	66
11	64
16	61

a. Make a scatter plot of the relationship between the algebra placement scores and the calculus exam scores.

Algebra Placement Scores and Calculus Exam Scores

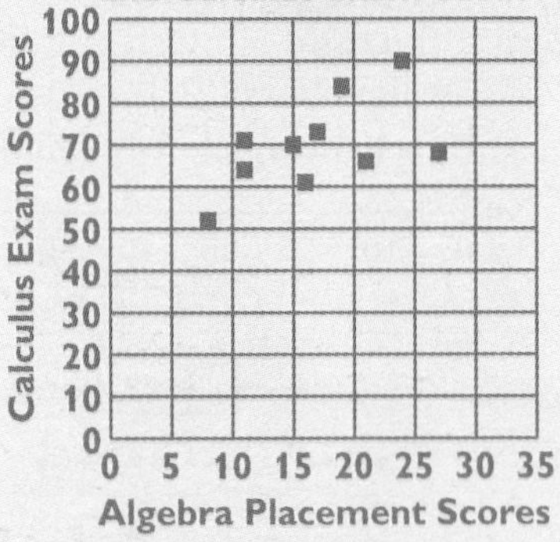

b. Describe any trends suggested by the scatter plot. The scatter plot rises to the right. It appears that as algebra placement scores increase, the calculus exam scores increase.

c. Explain how a student planning to take calculus can use this scatter plot. A person planning to take calculus might decide to master algebra.

Reasoning

Ask students if they could circle the dots in a part of the scatter plot for Sample 2 to show a *downward* trend. This activity will help students understand that showing only part of the data can lead to a different conclusion about trends than would be reached by examining the full data set.

Another type of data display is a *scatter plot.* A **scatter plot** is a graph that shows the relationship between two data sets.

Sample 2

The table shows data for 20 law schools in the United States.

a. Make a scatter plot of the relationship between out-of-state tuition and percent of graduates employed six months after graduation.

b. Describe any trends suggested by the scatter plot.

c. Explain how a person applying to law school can use this scatter plot.

1993 Law School Statistics

Law school	Out-of-state tuition (dollars)	% employed 6 months after graduation
A	18,548	100
B	17,750	97
C	20,186	98
D	19,095	98
E	19,740	97
F	19,400	97
G	12,111	84
H	18,170	94
I	18,995	90
J	15,678	96
K	12,202	87
L	12,544	96
M	19,156	88
N	10,190	93
O	11,529	89
P	7,280	85
Q	17,000	94
R	13,600	88
S	17,720	92
T	9,963	80

Sample Response

a. Plot the ordered pairs (18,548, 100), (17,750, 97), and so on.

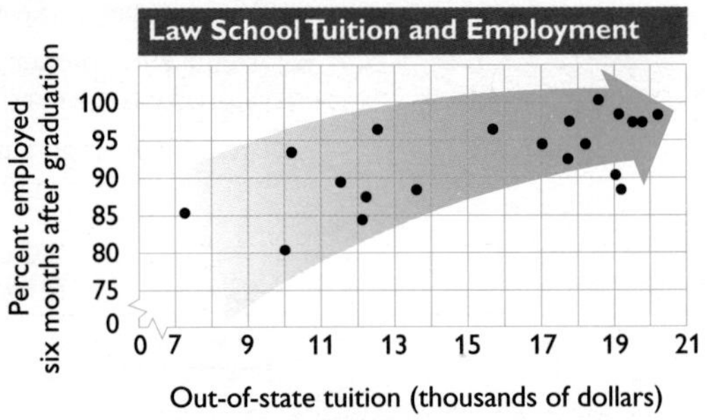

b. The scatter plot rises to the right. It appears that as out-of-state tuition increases, the percent of students employed six months after graduation also increases.

c. A person applying to law school may decide to attend a more expensive school if it means a better chance of getting a job after graduation.

BY THE WAY...

The first woman to graduate from a law school in the United States was Ada H. Kepley. She graduated from the Union College of Law in Chicago on June 30, 1870. Today, 30% to 40% of law students worldwide are women.

Talk it Over

4. Does it matter whether you graph the percent employed on the horizontal or vertical axis?
5. If the out-of-state tuition at law school Z is $15,000, about what percent of its graduates would you expect to be employed six months after graduation?

Answers to Talk it Over

4. No; the graph would still rise to the right, indicating the same relationship between increasing employment percentages and increasing tuition.
5. Estimates may vary. about 90%

Answers to Look Back

No; the data in a line graph may not be numerical. Yes; it is necessary to put the data in numerical order to make a box-and-whisker plot because you must determine quartiles. No; it is not necessary to put the data in numerical order in order to make a scatter plot, although it might help to avoid overlooking data points.

Look Back

Do you have to put the items of a data set in numerical order to make a line graph? a box-and-whisker plot? a scatter plot? Explain.

1-3 Exercises and Problems

Seasonal change in Shanghai, China

1. **Reading** How would a box-and-whisker plot for the mean monthly temperatures in Caracas differ from the one for Alice Springs? (See pages 16 and 17.)
2. Find the lower extreme, the lower quartile, the median, the upper quartile, and the upper extreme of these bowling scores.

   123, 254, 100, 121, 143, 93, 205,
   66, 111, 176, 89, 189, 205, 75, 100

Meteorology **For Exercises 3 and 4, use the table.**

Mean Monthly Temperature (°C)

	Jan.	Feb.	Mar.	Apr.	May	June	July	Aug.	Sept.	Oct.	Nov.	Dec.
Mexico City, Mexico	12	13	16	18	20	19	18	19	18	16	14	13
Shanghai, China	4	6	10	14	21	26	28	28	25	19	12	6

3. What is the mean annual temperature of each city?
4. a. Make a box-and-whisker plot for each city using the same number line.
   b. **Writing** Write three statements comparing the temperatures in the two cities. Use the box-and-whisker plots to justify your statements.
   c. **Open-ended** Do you think it is better to use a line graph to compare these data? Explain.

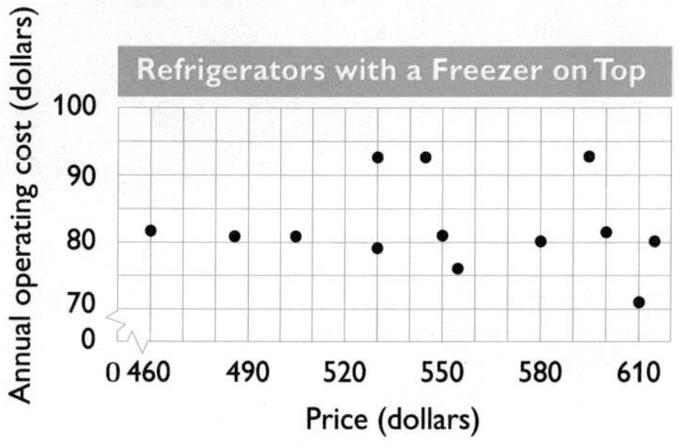

For Exercises 5 and 6, use the scatter plot.

5. a. Describe any trend(s) suggested by the scatter plot.
   b. Does every refrigerator represented on the scatter plot follow the trend(s) you found in part (a)? Explain.
6. About how much would you expect to pay each year to operate a refrigerator with a freezer on top that sells for $470?

Answers to Exercises and Problems

1. Answers may vary. An example is given. The box-and-whisker plot for Caracas would be much shorter than the plot for Alice Springs because the data for Caracas are much less spread out. The box for Caracas would be very short because the middle half of the data covers a spread of only about 1.5 degrees.

2–4. See answers in back of book.

5. a, b. Answers may vary. An example is given.
   a. Of the 13 points on the scatter plot, 9 appear to lie near the line $y = 80$, which indicates that the annual cost of operating a top-freezer refrigerator is fairly constant, about $80, no matter what the cost of the refrigerator.
   b. No; four of the data points lie fairly far from the line, indicating that four of the refrigerators cost somewhat more or less than $80 per year to operate.

Answers continued on next page.

Look Back

Have students explain their answers to these questions to the entire class. In so doing, any misunderstanding of how data are used to make graphs and plots may be determined and corrected.

APPLYING

Suggested Assignment

Standard 1–9, 11–13, 15–27

Extended 1–13, 15–27

Integrating the Strands

Algebra Exs. 20–26

Functions Exs. 20–22

Geometry Exs. 16–19

Statistics and Probability Exs. 1–15, 27

Discrete Mathematics Exs. 1–15

Logic and Language Exs. 1, 4, 10, 13, 15, 27

Communication: Reading

For Ex. 1, students first need to read the line graph on page 16 that shows mean monthly and annual temperatures for Caracas and Alice Springs. An analysis of the data for Caracas will help students answer this question.

Multicultural Note

Caracas, the capital of the Republic of Venezuela, is one of South America's most important cities. Located seven miles from the Caribbean coast, Caracas is Venezuela's cultural mecca, center of industry and commerce, and largest city. Founded in 1567 as Santiago de León de Caracas, the city is now home to over three million people, and boasts of a national library, an ultramodern theater, a number of major newspapers, beautiful stadiums, numerous historical sites, and skyscrapers that are among the tallest on the continent.

Mathematical Procedures

When data are not evenly distributed, students may find box-and-whisker plots difficult to understand. When students are constructing a box-and-whisker plot by hand, they should list all data items from least to greatest and use arrows to show where the data-list breaks into quartiles.

Career Note

Atmospheric scientists are specialists trained in a number of scientific disciplines. A background in chemistry is especially valuable because the atmosphere is composed of various gases such as oxygen, carbon dioxide, ozone, water vapor, nitrogen, and argon. Atmospheric scientists also understand the principles of physics and have excellent mathematical skills in order to make measurements and interpret data.

Television The A. C. Nielsen Company rated the prime-time television series of the four major commercial networks for the 1994 season. A higher rating indicates a larger percent of homes watching.

BY THE WAY...

"Nielsen families" agree to have all televisions metered for two years. Television viewers must sign on, respond to inquiries from the meter, and sign off *every* time they watch. Meters that reduce viewer responsibility are under development.

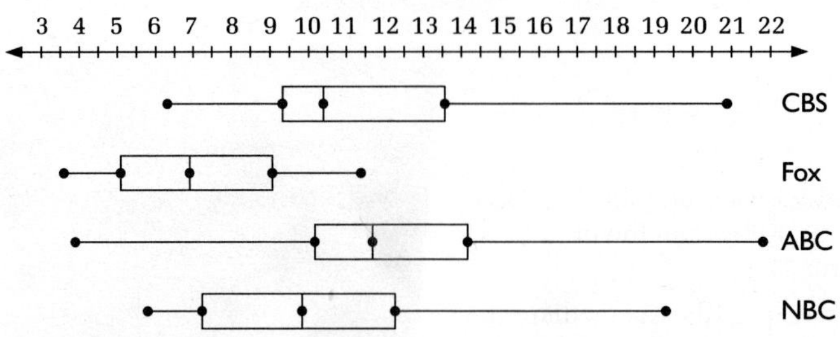

7. Which network had the lowest rated series in 1994?
8. About what percent of NBC series had higher ratings than all Fox series in 1994? Explain.
9. Which network had the best overall prime-time ratings in 1994? Explain.
10. **Open-ended** The Nielsen rating system includes both one-hour and half-hour prime-time series. Do you think one-hour shows should be rated separately from half-hour shows? Why or why not?

Career Atmospheric scientists in California measured the thickness of the ozone layer over a city with two instruments, one in a satellite in space and one at ground level. This scatter plot compares the two readings over 100 days.

11. What do you notice about the scales on the horizontal and vertical axes?
12. a. Since the instruments measure the same thing, what do you expect about the readings from each instrument?
    b. Does the scatter plot support your expectation?
13. **Open-ended** Why do you think scientists would want to compare the readings of two instruments measuring the same thing?

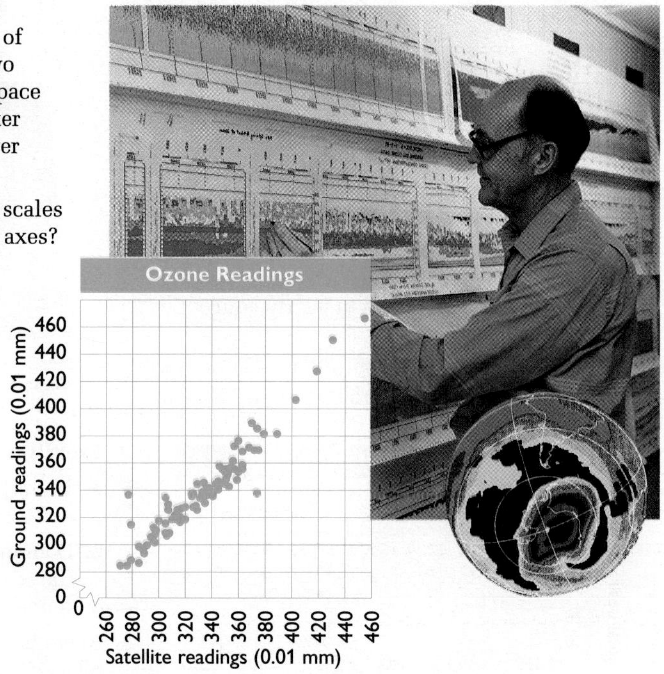

Answers to Exercises and Problems

6. about $80; Answers may vary. An example is given. Because the trend is fairly constant, I would expect to pay about the same as for most of the other refrigerators.
7. Fox
8. about 25%; the upper extreme for the Fox data is about equal to the upper quartile for the NBC data.
9. ABC; ABC has the highest upper extreme as well as the highest median.
10. Answers may vary. An example is given. I think it is not necessary, because even if one network has six half-hour shows in prime time while a second has only three one-hour shows, the networks are still sharing the same number of potential viewers and counting six separate shows does not give the first network any real advantage.
11. They are the same.
12. a. If both instruments are working properly and are not unduly influenced by other factors, I would expect the readings to be about the same.
    b. Yes.
13. Answers may vary. An example is given. Each instrument would work as a check on the other. If the readings were vastly different, the scientists would have to determine if one or both instruments were not working properly.
14. a, b. Check students' work.
    c. Answers may vary. An example is given. There is a slightly negative trend in the data. The greater the population density, the less willing people are to help strangers.
15. Answers may vary. An example is given. **Step 1:** Arrange the data in numerical order. **Step 2:** Determine the lower extreme, lower quartile, median, upper quartile, and upper extreme. **Step 3:** Draw a number line that includes all of the five numbers. Below the number line, locate points corresponding to each of the five numbers. **Step 4:** Draw segments between the dots for the lower extreme and lower quartile and between the upper quartile and upper extreme. **Step 5:** Draw short vertical segments through the quartiles and the median, then

14. **Group Activity** Work with another student.

Sociology In a recent study of 36 United States cities, researchers measured how willing people are to help people they do not know.

a. **Research** Find the population density of each city by dividing the population by the area of the city in square miles.

b. TECHNOLOGY Use a graphics calculator or graphing software.

Alternative Approach Work with another student.

Make a scatter plot of the relationship between ranking and population density for these cities.

c. Describe any trends suggested by the scatter plot.

Volunteers serving a free Thanksgiving dinner in San Antonio, Texas

Rank	City	Rank	City	Rank	City
1	Paterson, N.J.	13	San Francisco, Cal.	25	Columbus, Ohio
2	New York City, N.Y.	14	Buffalo, N.Y.	26	Indianapolis, Ind.
3	Los Angeles, Cal.	15	Bakersfield, Cal.	27	Chattanooga, Tenn.
4	Fresno, Cal.	16	Atlanta, Ga.	28	East Lansing, Mich.
5	Philadelphia, Pa.	17	Springfield, Mass.	29	Detroit, Mich.
6	Shreveport, La.	18	San Diego, Cal.	30	St. Louis, Mo.
7	Chicago, Ill.	19	San Jose, Cal.	31	Louisville, Ky
8	Providence, R.I.	20	Dallas, Texas	32	Knoxville, Tenn.
9	Boston, Mass.	21	Santa Barbara, Cal.	33	Memphis, Tenn
10	Salt Lake City, Utah	22	Worcester, Mass.	34	Nashville, Tenn.
11	Sacramento, Cal.	23	Kansas City, Mo.	35	Houston, Texas
12	Youngstown, Ohio	24	Canton, Ohio	36	Rochester, N.Y.

Least willing to help (rank 1)

Most willing to help (rank 36)

Ongoing ASSESSMENT

15. **Writing** Write an algorithm for drawing a box-and-whisker plot.

Review PREVIEW

Find the whole-number lengths of the sides of all triangles with each indicated perimeter. *(Section 1-2)*

16. 13 **17.** 14 **18.** 18 **19.** 26

For the graph of each function in Exercises 20–22: *(Toolbox Skill 28)*

a. Find an equation for the line of symmetry. **b. Find the y-intercept.**

20. $y = 4x^2 + 3x + 8$ **21.** $y = -3x^2 + 11x$ **22.** $y = 7x^2 - 5x + 1$

Solve each system of equations. *(Toolbox Skill 21)*

23. $3x - 2y = 12$, $y = 4x + 7$

24. $5x - 18 = 4y$, $y = 3x$

25. $x = 6y - 8$, $-5x + 7y = 11$

26. $x = 19 - 2y$, $x = 5y + 4$

Cooperative Learning

For Ex. 14, students work with a partner. Detailed suggestions on cooperative learning techniques are presented on page T32 in the front of this Teacher's Edition.

Using Technology

You can use the TI-82 to make box-and-whisker plots as well as scatter plots. Here is the basic procedure.

(1) Clear old data from list L_1 and enter the data you wish to plot in L_1.

(2) Sort the data in ascending order by using 2: SortA(on the EDIT menu, which you see when you press [STAT].

(3) Press [WINDOW] and set Xmin to the smallest number in the list L_1. Set Xmax to the largest number in L_1. If you wish no axes to show, go to the FORMAT menu for [WINDOW] and select Axes Off.

(4) Press [Y=] and clear the equations list. Press [2nd] [STAT PLOT]. Press 4 [ENTER] to clear any statistical graphs.

(5) Press [2nd] [STAT PLOT]1. On the screen where you see Plot 1 as the first line, select On, then |-[|]-|, then L_1 (on the line for Xlist), and 1 (on the line for Freq).

(6) Press [GRAPH] to see the plot. If you press [TRACE] and use [▶] and [◀], you will get the extremes and quartiles as readouts at the bottom of the screen.

Suggestions for enhancing mathematics learning by using graphing technology are given on page T34 in front of this Teacher's Edition.

Answers to Exercises and Problems

draw horizontal segments to create a box enclosing the region between the quartiles.

16.

Side 1	Side 2	Side 3
1	6	6
2	5	6
3	4	6
3	5	5
4	4	5

17.

Side 1	Side 2	Side 3
2	6	6
3	5	6
4	4	6
4	5	5

18, 19. See answers in back of book.

20. a. $x = -\frac{3}{8}$ b. 8

21. a. $x = 1\frac{5}{6}$ b. 0

22. a. $x = \frac{5}{14}$ b. 1

23. $\left(-5\frac{1}{5}, -13\frac{4}{5}\right)$

24. $\left(-2\frac{4}{7}, -7\frac{5}{7}\right)$

25. $\left(-\frac{10}{23}, 1\frac{6}{23}\right)$

26. $\left(14\frac{5}{7}, 2\frac{1}{7}\right)$

Working on the Unit Project

By completing Ex. 27, students are given experience in making data displays. They can apply this skill to analyzing the results of their surveys when completing the Unit Project.

Quick Quiz (1-1 through 1-3)

See page 65.

See also Test 1 in the *Assessment Book.*

Practice 3 For use with Section 1-3

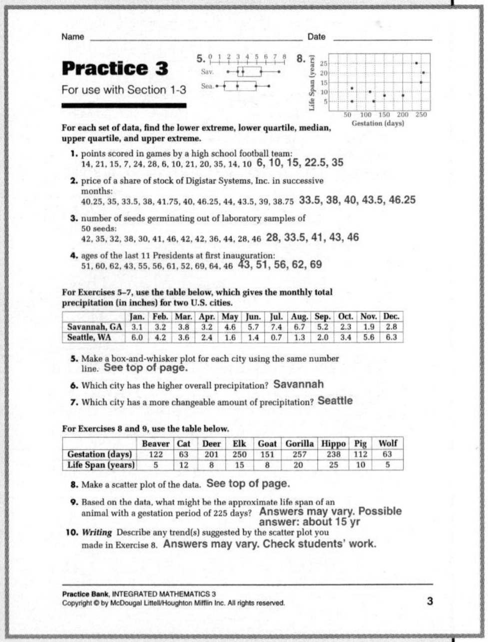

Name ______________________ Date ____________

Practice 3

For use with Section 1-3

For each set of data, find the lower extreme, lower quartile, median, upper quartile, and upper extreme.

1. points scored in games by a high school football team: 14, 21, 15, 7, 24, 28, 6, 10, 21, 20, 35, 14, 10 **6, 10, 15, 22.5, 35**
2. price of a share of stock of Digistar Systems, Inc. in successive months: 40.25, 35, 33.5, 38, 41.75, 40, 46.25, 44, 43.5, 39, 38.75 **33.5, 38, 40, 43.5, 46.25**
3. number of seeds germinating out of laboratory samples of 50 seeds: 42, 35, 32, 38, 30, 41, 46, 42, 42, 36, 44, 28, 46 **28, 33.5, 41, 43, 46**
4. ages of the last 11 Presidents at first inauguration: 51, 60, 62, 43, 55, 56, 61, 52, 69, 64, 46 **43, 51, 56, 62, 69**

For Exercises 5–7, use the table below, which gives the monthly total precipitation (in inches) for two U.S. cities.

	Jan.	Feb.	Mar.	Apr.	May	Jun.	Jul.	Aug.	Sep.	Oct.	Nov.	Dec.
Savannah, GA	3.1	3.2	3.8	3.2	4.6	5.7	7.4	6.7	5.2	2.3	1.9	2.8
Seattle, WA	6.0	4.2	3.6	2.4	1.6	1.4	0.7	1.3	2.0	3.4	5.6	6.3

5. Make a box-and-whisker plot for each city using the same number line. **See top of page.**
6. Which city has the higher overall precipitation? **Savannah**
7. Which city has a more changeable amount of precipitation? **Seattle**

For Exercises 8 and 9, use the table below.

	Beaver	Cat	Deer	Elk	Goat	Gorilla	Hippo	Pig	Wolf
Gestation (days)	122	63	201	250	151	257	238	112	63
Life Span (years)	5	12	8	15	8	20	25	10	5

8. Make a scatter plot of the data. **See top of page.**
9. Based on the data, what might be the approximate life span of an animal with a gestation period of 225 days? **Answers may vary. Possible answer: about 15 yr**
10. *Writing* Describe any trend(s) suggested by the scatter plot you made in Exercise 8. **Answers may vary. Check students' work.**

Practice Bank, INTEGRATED MATHEMATICS 3

3

Answers to Exercises and Problems

27. See answers in back of book.

Working on the Unit Project

As you complete Exercise 27, think about how you will use the results of your survey to make decisions about what facilities to include in your park plan.

27. **Writing** Suppose a park planner can include only three sports facilities in a new park. Which three sports should the park offer? Make at least two data displays that will help you answer. Explain your choices. The numbers in the table are in thousands.

People Who Play Selected Recreational Sports in the United States (1991)

Sport	Male	Female	Ages 7–17	Ages 18–34	Ages 35 and over
Baseball	13,640	2,908	10,040	4,312	2,197
Softball	11,878	7,768	6,644	8,997	4,006
Basketball	19,151	6,999	12,960	9,277	3,913
Soccer	6,657	3,935	7,322	2,010	660
Swimming	30,833	35,348	20,203	21,899	24,080
Tennis	9,207	7,495	4,053	7,645	5,005
Volleyball	11,495	11,091	6,863	11,146	4,579

Unit 1 CHECKPOINT

1. **Writing** Explain why box-and-whisker plots are useful for comparing two data sets.
2. Write an algorithm for choosing a movie to see with a group of friends. Include at least one loop. 1-1
3. Use this menu. Suppose a meal consists of a salad, a side order, and a beverage. Find all meals that cost less than $4. Assume that prices include tax. 1-2

Use these box-and-whisker plots.

4. About what percent of the countries represented used less than the equivalent of 400 million metric tons of coal each year? 1-3
5. Describe any trends shown by the graph.

Amount of Energy Used by the Top 20 Energy-Using Countries (equivalent of millions of metric tons of coal)

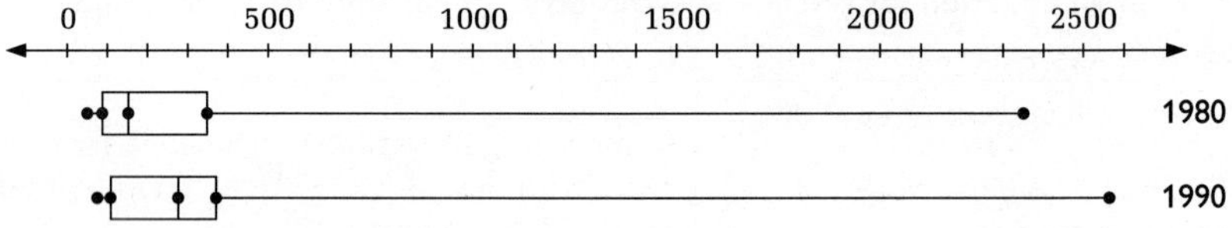

Answers to Checkpoint

1. Answers may vary. An example is given. Box-and-whisker plots let you compare the extremes and the quartiles. For example, you can determine which of the data sets contains the largest or smallest number. You can also compare which has half of its data grouped more closely about the median, or which is spread out more. You can also tell how the upper and lower quartiles of the data compare.
2. Answers may vary. An example is given. **Step 1:** Decide which movies are showing at the theater you plan to go to. **Step 2:** Decide which movie you would all like to see. If you do not agree on a movie, discuss your preferences, toss a coin , or compromise until a movie is selected. **Step 3:** Decide what times the movie is showing and whether any one time is suitable for all of you. If not, go back to the previous step to choose another movie.
3. garden salad, small cole slaw, small herbal tea
4. about 75% both years
5. Answers may vary. An example is given. Generally, usage is up slightly. Every specific point on the graph is higher in 1990 than it was in 1980.

Section

1-4 Using Graphs and Equations

Focus
Use graphs and equations to model situations and make decisions.

Making The Right Decision

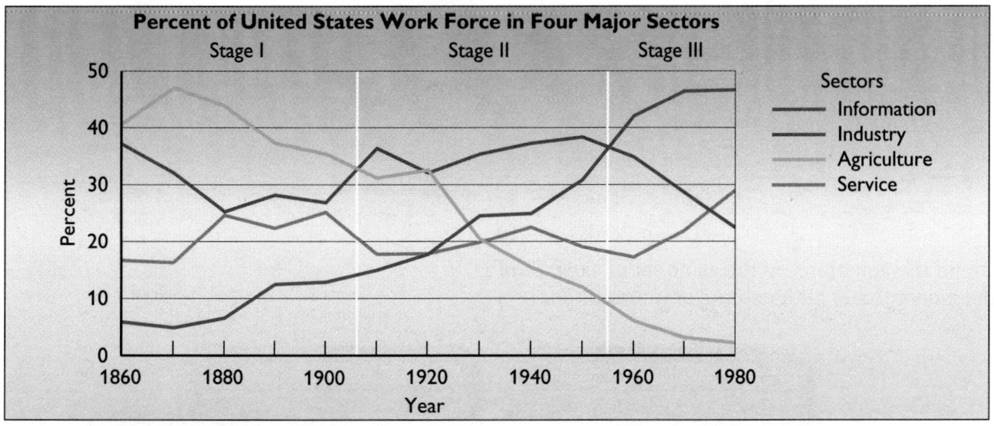

Agriculture in the United States in the last century and today

The graph shows how the percent of workers in four sectors of the United States economy changed from 1860 to 1980.

Talk it Over

1. Why do you think Stage I is called the *Agriculture Age*?
2. a. What happens to the Agriculture and Industry graphs at the end of Stage I?
   b. About what year did Stage I end?
3. a. What do you think Stage II is called?
   b. About what year did Stage II end?
4. What do you think Stage III is called? Why?

By comparing two graphs, you can determine when a quantity is greater than, less than, or equal to another quantity. This can help you make decisions.

PLANNING

Objectives and Strands
See pages 1A and 1B.

Spiral Learning
See page 1B.

Materials List
- Graphics calculator

Recommended Pacing
Section 1-4 is a one-day lesson.

Extra Practice
See pages 608–610.

Warm-Up Exercises
- Warm-Up Transparency 1-4

Support Materials
- Practice Bank: Practice 4
- Activity Bank: Enrichment 4
- Study Guide: Section 1-4
- Problem Bank: Problem Set 2
- Using IBM/Mac Plotter Plus Disk: Line Plotter
- Assessment Book: Quiz 1-4, Alternative Assessment 4

Answers to Talk it Over

1. There were more workers employed in agriculture than in any of the other three sectors.
2. a. They intersect.
   b. about 1905
3. a. the Industrial Age
   b. about 1954
4. the Information Age; More workers are employed in Information than in any of the other three sectors.

TEACHING

Talk it Over

Questions 1–4 help prepare students to analyze multiple line graphs to solve real-world problems.

Multicultural Note

Maggie Lena Walker (1867–1934) was born in Richmond, Virginia, the daughter of an African American who had been held in slavery. At fourteen, she joined the Independent Order of Saint Luke, a type of insurance company that helped African Americans living in the South pay for medical expenses and burial costs. In 1899, when she was made a top official of IOSL, its treasury had only $31.61 in it. By 1924, Walker opened a bank through IOSL (becoming in the process America's first female bank president) and increased that group's holdings to $3,480,540.19. Always arguing in favor of the capabilities and rights of both African Americans and women, Walker served on the boards of numerous activist organizations, including the NAACP.

Additional Samples

S1 There are just three pizza restaurants in Bensonville. All are within a block of each other. All will deliver to any location within the city limits. Delivery charges vary.
Louie's: $1.00 plus $.35 for each half mile;
Manny's: $2.00 plus $.25 for each half mile;
Leona's: $6.00
How should a person decide which place has the most reasonable delivery charge?

Model the information with equations.
Let x = number of half miles from restaurant.
Let y = total delivery charge.
Louie's: $y = 1 + 0.35x$
Manny's: $y = 2 + 0.25x$
Leona's: $y = 6$

BY THE WAY...

Maggie Lena Walker was the first woman to found a bank in the United States. In 1903, she opened the Saint Luke Penny Savings Bank in Richmond, Va.

Sample 1

Camille has saved $600 and wants to open her first checking account. Which bank offers the lowest monthly charge?

Monthly Charges

Central Bank:	$6.00 plus $.35 for each check you write
Main St. Bank:	$3.00 plus $.75 for each check you write
Co-op Bank:	$10.00, no matter how many checks you write

Sample Response

Problem Solving Strategy: Use a graph.

Model the information with equations.

Let x = the number of checks Camille writes each month.

Let y = the total monthly charge.

Central Bank: $y = 6 + 0.35x$
Main St. Bank: $y = 3 + 0.75x$
Co-op Bank: $y = 10$

Camille's monthly charge depends on the number of checks she expects to write.

Graph the equations on the same set of axes. Find the x-coordinate of the points of intersection.

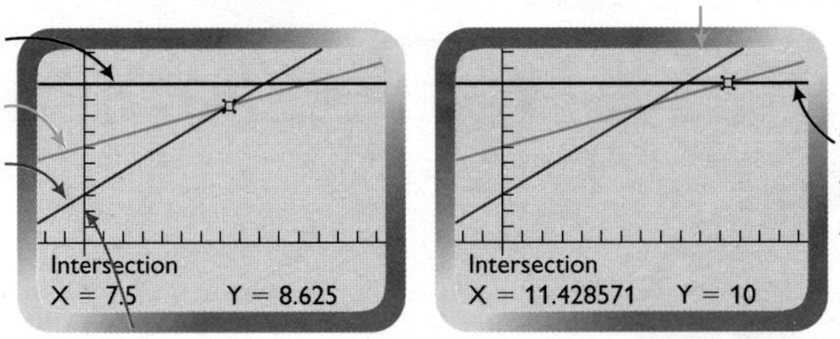

The Central Bank graph is below the others for 7.5 < x < 11.4.
y = 10
y = 6 + 0.35x
y = 3 + 0.75x
Intersection
X = 7.5
Y = 8.625
Intersection
X = 11.428571
Y = 10
The Co-op Bank graph is below the others for x > 11.4.

The Main St. Bank graph is below the others for $0 \le x < 7.5$. Since the number of checks must be a whole number, the monthly charges at Main St. Bank are the lowest for $x \le 7$.

Camille should choose:

Main St. Bank if she expects to write 7 or fewer checks each month.

Central Bank if she expects to write 8, 9, 10, or 11 checks each month.

Co-op Bank if she expects to write 12 or more checks each month.

Talk it Over

5. Why were only the x-coordinates of the points of intersection considered? What do the y-coordinates mean in this situation?
6. Why does Camille not have to consider the intersection of the graphs of $y = 3 + 0.75x$ and $y = 10$?

Answers to Talk it Over

5. The x-coordinates represent the number of checks Camille will write. The y-coordinates indicate the total monthly charge for the number of checks indicated by the x-coordinate.
6. At that point, both the graph of $y = 3 + 0.75x$ and the graph of $y = 10$ are above the graph of $y = 6 + 0.35x$. The monthly charges at Main St. Bank and Co-op Bank would be higher for that number of checks than for the same number of checks at Central Bank.

Break-Even Point

> **BY THE WAY...**
>
> A break-even chart or graph is one of the standard tools used by economists. It shows at a glance the point where loss ends and profit begins.

Until its income exceeds its expenses, a business loses money. When its income equals its expenses, a business has reached the *break-even point.*

Sample 2

Suppose the mortgage, taxes, and insurance for a 100-room motel in Joplin, Missouri, are $120,000 per year. Each occupied room costs $25 per day to maintain. Each unoccupied room costs $5 per day to maintain. The room rental charge is $35 per day.

The motel is open 365 days a year. What must the average number of occupied rooms be each day in order for this motel business to break even at the end of a year?

Sample Response

Step 1 Separate expenses from income.

Expenses: mortgage, taxes, insurance, maintenance
Income: room rental

Step 2 Write equations to model the situation.

Let r = the average number of occupied rooms each day.

Let E = the total annual expenses.

$$E = \text{mortgage, taxes, and insurance} + \text{annual maintenance, occupied rooms} + \text{annual maintenance, unoccupied rooms}$$

$$= 120{,}000 + 365 \cdot 25r + 365 \cdot 5(100 - r)$$

(number of unoccupied rooms: $100 - r$)

$$= 120{,}000 + 9125r + 182{,}500 - 1825r$$

$$= 302{,}500 + 7300r$$

Let I = annual income from room rental.

$$I = 365 \cdot 35r$$

$$= 12{,}775r$$

Step 3 Find the break-even point, when $I = E$.

Method ❶ Use a graph.

Graph the equations on the same set of axes.

The average number of occupied rooms each day should be about 55.3.

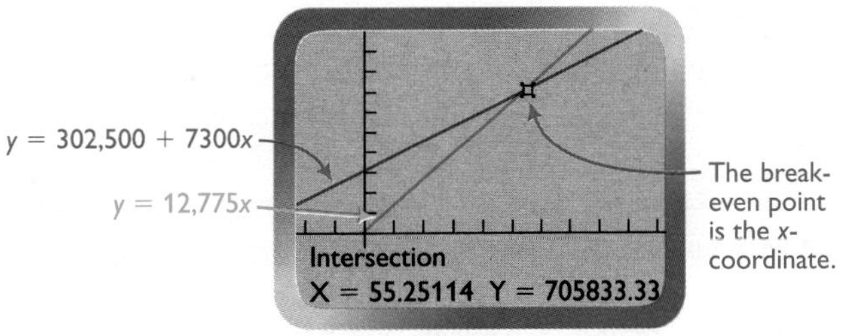

Method ❷ Use algebra.

$$302{,}500 + 7300r = 12{,}775r$$

$$55.3 \approx r$$

The average number of occupied rooms each day should be about 55.3.

Graph the equations on the same set of axes. Find the appropriate points of intersection.

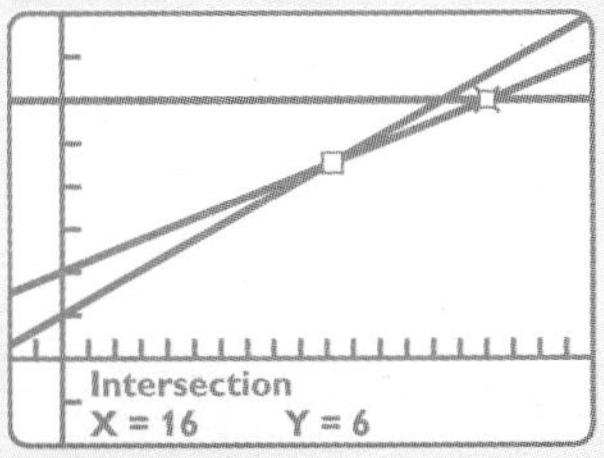

The intersection marked with a □ is (10, 4.5).

The graph for Louie's is below the graph for Manny's until you get to (10, 4.5). The graph for Manny's is below the graph for Leona's until you get to (16, 6). Recall that the first coordinate represents *half* miles.

A person should choose: Louie's if the distance is 5 or fewer miles; Manny's if the distance is between 5 and 8 miles; Leona's if the distance is more than 8 miles.

S2 Otis Turner has been thinking about creating menus for restaurants on a personal computer. He would like to buy a computer, color printer, and software and rent work space. He estimates software, equipment, and rent would cost $15,000. Materials for each job would cost about $20. He would charge restaurants $100 for a menu. How many jobs will he need before he can break even?

Step 1. Separate expenses from income. His expenses are the software, equipment, rent, and materials. His income is from the jobs he completes.

Step 2. Write equations to model the situation.

Let x = the number of jobs he completes.

Let E = expenses.

Let I = income from jobs.

$E = 15{,}000 + 20x$

$I = 100x$

Sample continued on next page.

Additional Sample (continued)

Step 3. Find the break-even point.

Method 1. Use a graph.

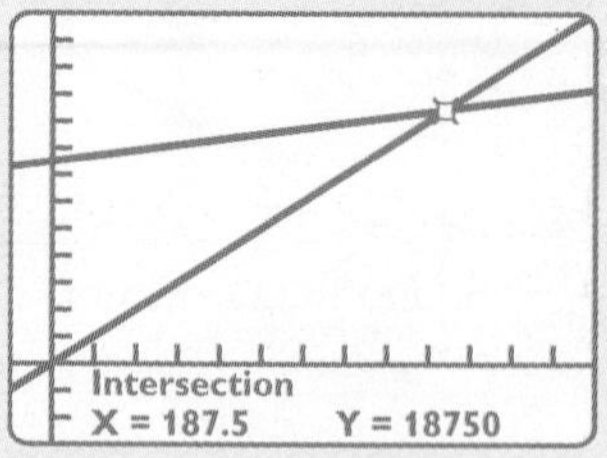

The graphs intersect at (187.5, 18,750). The number of jobs must be a whole number. He will need 188 jobs to break even.

Method 2. Use algebra.

$15{,}000 + 20x = 100x$

$15{,}000 = 80x$

$x = 187.5$

He will need 188 jobs to break even.

Talk it Over

Question 8 has students consider using algebra to solve Samples 1 and 2. You may wish to extend this question by asking students to suggest other problem-solving strategies. Students should be prepared to explain how their strategies would work.

Look Back

A thorough discussion of this question is necessary to ensure that students understand fully the significance of the intersection points.

Talk it Over

7. Why can you graph both the expenses and the income equations in Sample 2 on the same set of axes?
8. Explain how to solve Sample 1 with algebra. Which method, *algebra* or *a graph*, do you prefer for Sample 1? for Sample 2?

Look Back

Why are intersection points important when you make a decision based on a graph?

1-4 Exercises and Problems

1. **Reading** What can you find out by comparing graphs on the same set of axes?

Business The three graphs represent bids submitted to a large medical group practice by three firms that process patient bills.

- Physicians' Services (PS) bid a flat fee plus a charge per bill.
- Doctors' Service (DS) bid a flat fee less than that of PS, plus a charge per bill.
- Human Resources (HR) bid a charge per bill with no flat fee.

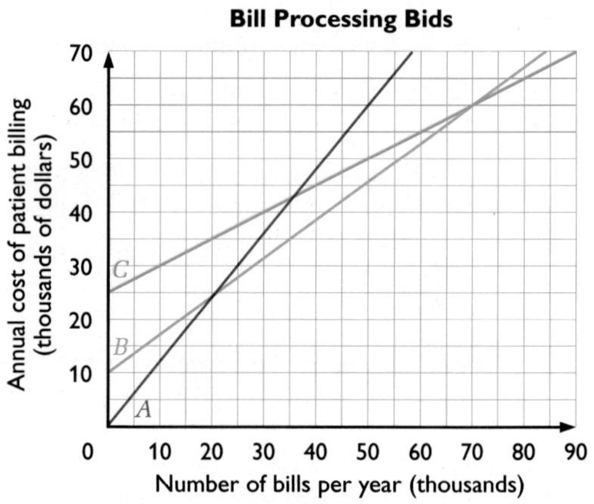

For Exercises 2–4, choose the letter of the graph that models each bid.

2. PS
3. DS
4. HR
5. Estimate the flat fee of the PS bid.
6. Which firm charges the most per bill? How do you know?
7. Which firm's bid is the lowest for processing 80,000 bills per year?
8. Let x = the number of bills that the medical group expects to process next year. Estimate the values of x for which the DS bid is the lowest.

Graph each pair of equations and estimate the break-even point.

9. $E = 500 + 3.5x$
   $I = 5x$
10. $E = 127 + 3x$
    $I = 25x$
11. $E = 0.52 + 0.04x$
    $I = 0.06x$
12. $E = 8200 + 73x$
    $I = 1436x$

Use algebra to find each break-even point. Round decimal answers to the nearest tenth.

13. $E = 220 + 4.3x$
    $I = 14x$
14. $E = 99 + 0.01x$
    $I = 0.28x$
15. $E = 725 + x$
    $I = 6.5x$
16. $E = 18 + 22x$
    $I = 24x$

Answers to Talk it Over

7. Both graphs involve the same control variable and independent variable.
8. Find the solutions of the three pairs of equations $y = 6 + 0.35x$ and $y = 3 + 0.75x$, $y = 6 + 0.35x$ and $y = 10$, and $y = 3 + 0.75x$ and $y = 10$. Check which bank is least expensive on each of the four intervals determined by the x-values of the three solutions. Answers may vary. Examples are given. The graphing solution to Sample 1 is easier to interpret. The algebra solution to Sample 2 is quicker.

Answers to Look Back

Intersection points represent the value at which two or more options provide the same results. Based on this, one can decide at which point one option is better than another. Intersection points can also tell where the break-even point for expenses and income occurs, so decisions can be made about a business opportunity.

Answers to Exercises and Problems

1. You can determine the intervals where the values of one function are less than, greater than, or equal to the values of another function.
2. *C*
3. *B*
4. *A*
5. $25,000 per year
6. Human Resources; Each graph is a linear equation with y-intercept the flat fee and slope the charge per bill. The graph for Human Resources has the steepest slope, therefore the highest charge per bill.
7. Physicians' Services
8. between 20,000 and 70,000 bills per year

9–12. Estimates may vary.

Gourmet

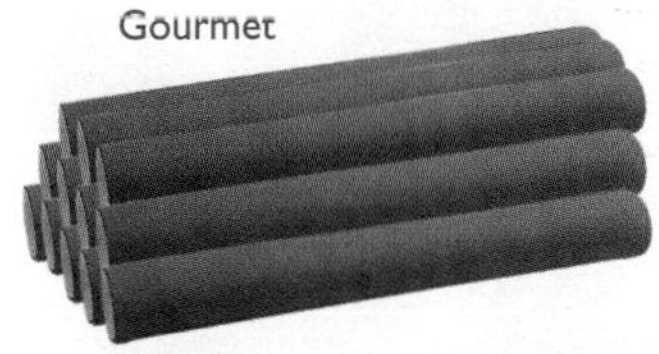

12 rolls at \$21 per roll

Autumn

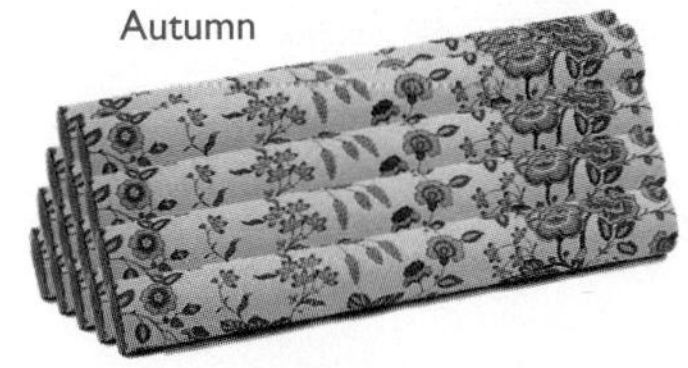

14 rolls at \$16 per roll

17. **Interior Decorating** Charlie Ruben wants to wallpaper his dining room. DJ Paperers charges \$18 per hour for labor to hang plain paper such as Gourmet. An extra \$5 per hour is added for patterned paper such as Autumn, since a helper is needed to finish the job in the same amount of time.

a. Write two equations to model the cost of wallpapering Charlie Ruben's dining room with each type of paper. Be sure to include the cost of materials.

b. Which paper is less expensive to hang if each job takes 4 h? 6 h?

18. **Consumer Economics** Anna Croft wants to rent a 15-foot truck for a day. She makes a list of rental charges at four companies.

U-Rent:	\$19.95 per day plus \$.59 per mile
MacRents:	\$44.95 per day plus \$.39 per mile
Wiley's:	\$79.95 per day plus \$.22 per mile
Zeke's:	\$150 per day with unlimited mileage

a. Write an equation that models each company's rental charges.

b. Which company should she choose? Why?

connection to **SCIENCE**

19. A rule of thumb for converting temperatures from Celsius to Fahrenheit is to double the Celsius reading and add 30. The formula is $F = 1.8C + 32$.

a. Write an equation that models the rule of thumb.

b. Graph both equations on the same set of axes.

c. Find the Celsius reading that gives the same Fahrenheit reading for both the rule of thumb and the formula.

d. Suppose that you consider the rule of thumb estimate good only if it is within 5°F of the actual reading. Use your graph from part (b) to find the values of C that meet this condition.

e. **Open-ended** What estimation error do you consider reasonable? What values of C meet your condition?

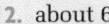

TECHNOLOGY NOTE

See the Technology Handbook, p. 600, for moving between graphs to see different y-values for the same x-value.

20. Alejandro earns extra money by typing papers and reports. He estimates that his per-hour cost for supplies is about \$1.25. He decides to invest in a new word processor that sells for \$559.

a. Write an equation that models Alejandro's expenses.

b. Alejandro charges \$12 per hour for typing. Write an equation that models Alejandro's income.

c. If Alejandro averages only five hours of typing per week, how many weeks will it take him to break even?

21. **Open-ended** Describe a situation that can be modeled by these expense and income equations:

$$E = 52.50 + 1.50x \qquad I = 3.25x$$

APPLYING

Suggested Assignment

Standard 1–16, 18, 19, 23–31

Extended 1–31

Integrating the Strands

Algebra Exs. 8–23

Geometry Exs. 26–29

Statistics and Probability Exs. 1–8, 24, 25

Discrete Mathematics Exs. 1–8, 17–22, 24, 25, 30

Logic and Language Exs. 1, 19, 21, 23, 31

Interdisciplinary Problems

Exs. 2–8, 17–19, and 22 apply the concepts and techniques of this section to real-world situations involving business, interior decorating, consumer economics, science, and recycling.

Reasoning

Students should be prepared to explain the reasoning behind their answers to Exs. 19(e) and 21. This can be done as a small group or full class discussion so that students may benefit from hearing different answers involving different reasoning.

Answers to Exercises and Problems

9. about 333.3

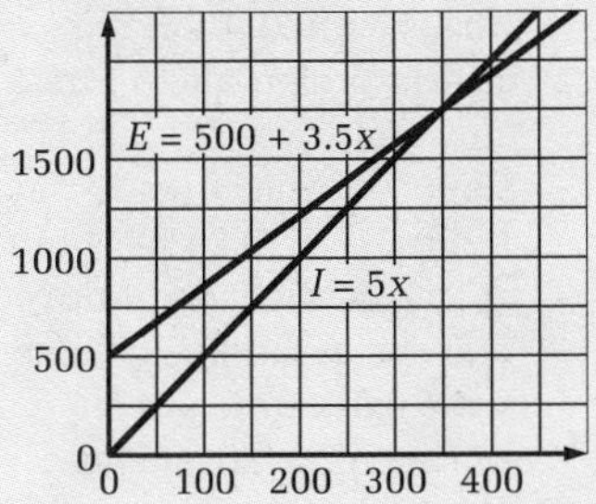

10. about 5.8

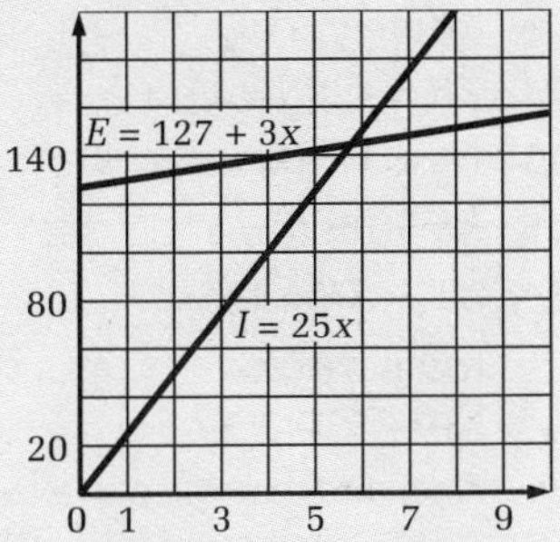

11. about 26

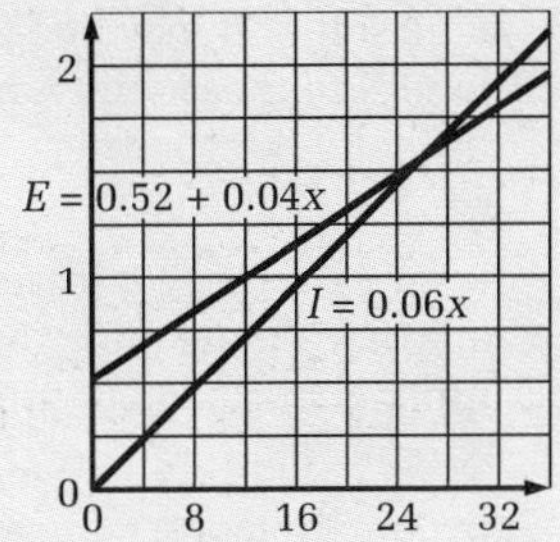

12. about 6

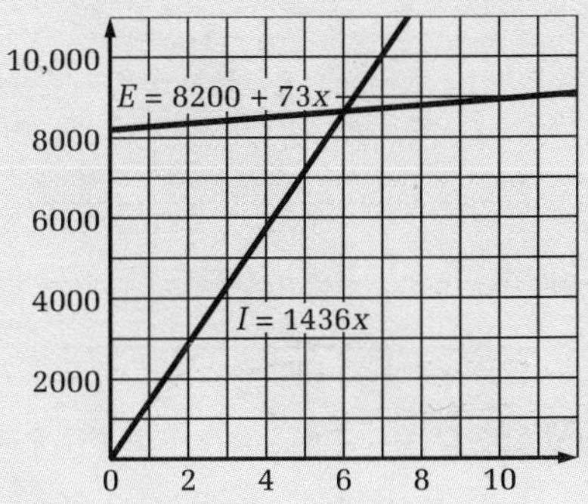

13. 22.7

14. 366.7

15. 131.8

16. 9

17. a. Let t be the total cost in dollars and h the number of hours needed. Autumn: $t = 252 + 18h$; Gourmet: $t = 224 + 23h$

b. Gourmet; Autumn

18–21. See answers in back of book.

Assessment: Open-ended

Students should answer the question: If a business has no break-even point, what can you deduce about the linear models for their income and expenses? Describe a situation in which the business is successful and one in which it is failing. For additional information on assessment methods, see page T44 in this Teacher's Edition.

Working on the Unit Project

Students can work in their project groups to discuss the situation presented in Ex. 31. Suggest that they keep notes about how to use graphs and equations to develop their own park proposal.

Practice 4 For use with Section 1-4

Name ______________________ Date ______________

Practice 4

For use with Section 1-4

Find the break-even point of each pair of equations. Round decimal answers to the nearest tenth.

1. $E = 26 + 3.2x$, $I = 4.5x$ **20**
2. $E = 225 + 1.8x$, $I = 6.3x$ **50**
3. $E = 90 + 0.6x$, $I = 4.2x$ **25**
4. $E = 7.5 + 1.9x$, $I = 8.0x$ **1.2**
5. $E = 145 + 22.5x$, $I = 44.5x$ **6.6**
6. $E = 16 + 3.4x$, $I = 7.2x$ **4.2**

Tomorrow, Inc. (TI) and Technologies Unlimited (TU) both charge a flat monthly fee to connect users to a global computer network, and both companies charge an hourly rate for a user's connect time, but TI's flat fee is higher. Info Services (IS) charges no flat fee, only an hourly rate. Use the graph at the right for Exercises 7–14.

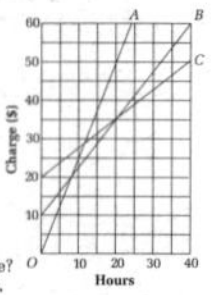

For Exercises 7–9, choose the letter of the graph that models each service.

7. TI **C**　8. TU **B**　9. IS **A**

10. Estimate the flat fee for TU. **about $10**

11. Which company charges the highest hourly rate for connect time? How do you know? **IS; Graph has the greatest slope.**

Exs. 12–14: Answers may vary.

For what range of hours of connect time is each company cheapest?

12. TI **$t > 20$**　13. TU **$8 < t < 20$**　14. IS **$0 < t < 8$**

15. Urmila bought a $350 lawn mower to use for her after-school lawn-mowing business. She estimates that fuel and maintenance on the mower will cost about $2.25 per hour of mowing.
    a. Write an expense equation that models this situation. Let t = number of hours mowing. **$E = 350 + 2.25t$**
    b. Urmila charges $12.50 per hour for mowing lawns. Write an income equation that models this situation. **$I = 12.5t$**
    c. Urmila averages 7 hours of mowing per week. How many weeks will it take her to break even? **5 weeks**

4　Practice Bank, INTEGRATED MATHEMATICS 3

22. **Recycling** Cardboard that is *baled* is easier to recycle and can be sold to earn money.
    a. Suppose it costs $290 a month to lease a baling machine. The cost of baling wire is $3.38 per ton of cardboard, and the cost of labor is $27.30 per ton. Write an equation that models monthly costs.
    b. Suppose a city receives $35 per baled ton of cardboard. Write an equation that models monthly income.
    c. How many tons of cardboard does the city in part (b) have to bale each month to break even?

Baled cardboard is compacted and bound with wire.

Ongoing **ASSESSMENT**

23. **Writing** How can you decide when to use a graph and when to use algebra to find a break-even point?

Review **PREVIEW**

For each data set, find the lower extreme, the lower quartile, the median, the upper quartile, and the upper extreme. *(Section 1-3)*

24. 25, 83, 44, 59, 26, 37, 49, 48, 68, 75, 39, 48

25. 80, 113, 140, 95, 87, 102, 117, 90, 94, 115

Find the exact measure of each unknown side in each triangle. *(Toolbox Skill 30)*

26.
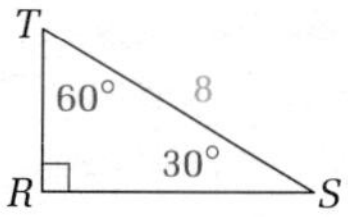

27.
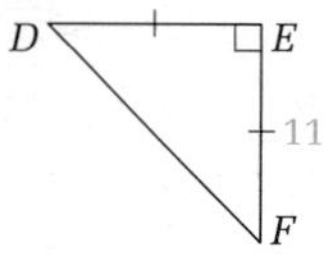

28.
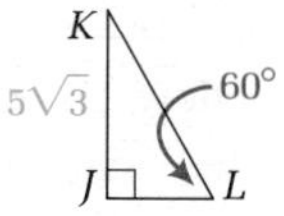

29.
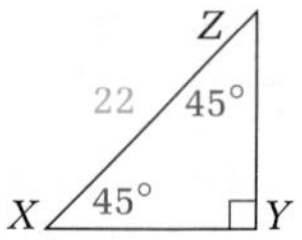

30. Add matrices B and C. *(Toolbox Skill 10)*　$B = \begin{bmatrix} 11 & 36 & 15 \\ 2 & 34 & 8 \end{bmatrix}$　$C = \begin{bmatrix} 1.6 & 42 & 13 \\ 8 & 36 & 1 \end{bmatrix}$

Working on the Unit Project

As you complete Exercise 31, think about how you can use graphs and equations to make the decisions needed to develop your park proposal.

31. A community group is writing a proposal for a new park. Since the park project began, the population in the area has grown. To handle the expected increase in park use, the group must add an additional employee to the proposed park staff or schedule 35 hours of overtime for the present staff. Should the group add another full-time employee or ask the present staff to work overtime? Explain your choice.

PARK WORKERS' WAGES AND HOURS

Hours: 35 hours per week

Regular hourly wages: Experienced—$8; New—$7

Overtime pay: Time-and-a-half (over 40 hours per week)

Benefits: 28% of earnings

Answers to Exercises and Problems

22. Let t be the number of tons of cardboard recycled each month.
    a. $E = 290 + (3.38 + 27.30)t$ or $E = 290 + 30.68t$
    b. $I = 35t$　c. about 67.1 tons

23. Answers may vary. An example is given. If an exact answer is needed, an algebraic solution may be preferable although sometimes a graphic solution may provide an exact answer. A graphic solution may help some people interpret the solution, while an algebraic solution may be more helpful to those who have difficulty interpreting graphs.

24. lower extreme: 25; lower quartile: 38; median: 48; upper quartile: 63.5; upper extreme: 83

25. lower extreme: 80; lower quartile: 90; median: 98.5; upper quartile: 115; upper extreme: 140

26. $TR = 4$; $RS = 4\sqrt{3}$

27. $DE = 11$; $DF = 11\sqrt{2}$

28. $KL = 10$; $JL = 5$

29. $XY = ZY = 11\sqrt{2}$

30. $\begin{bmatrix} 12.6 & 78 & 28 \\ 10 & 70 & 9 \end{bmatrix}$

31. Answers may vary. An example is given. If benefits paid out are 28% of all wages earned, then it is clearly preferable to hire a new employee at 1.28($7) = $8.96 per hour than to pay 1.28($8) = $10.24 per hour to an established employee. The cost would be greater if the extra time put a worker on overtime (at $15.36 per hour). I would hire a new employee.

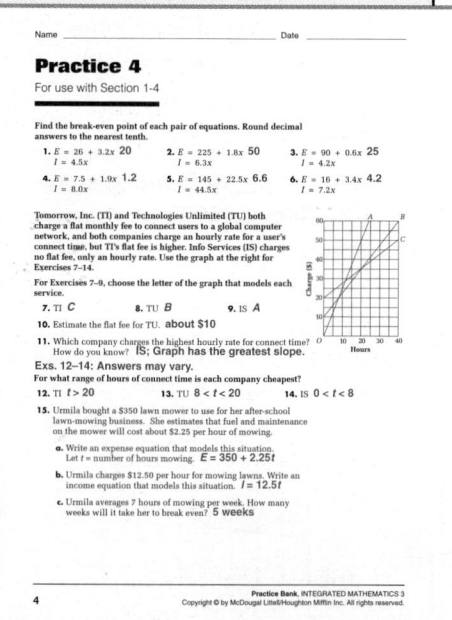

Section 1-5

Using Systems of Equations

Focus

Use systems of equations to solve problems.

Sound Systems

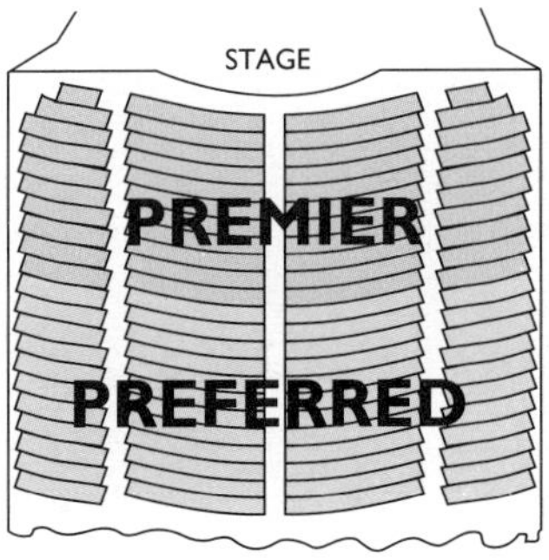

BY THE WAY...

In 1994, the record for the greatest number of tickets sold for a concert by a single band was about 180,000, for the concert given by Paul McCartney in 1990 in the Maracaña Stadium in Rio de Janeiro, Brazil.

Suppose a concert by a popular band involves these expenses:

- The band charges $20,000 plus 75% of the income from ticket sales.
- The costs for renting a 10,000-seat auditorium and paying for security, insurance, publicity, and so on are an additional $60,000.

The concert promoter wants to sell two types of tickets, preferred and premier.

Talk it Over

1. What limits the total number of tickets that can be sold?
2. Which type of ticket should be sold at a higher price? Why?
3. What other factors, besides the expenses given above, should the concert promoter consider when setting ticket prices?

PLANNING

Objectives and Strands

See pages 1A and 1B.

Spiral Learning

See page 1B.

Materials List

- Graphics calculator

Recommended Pacing

Section 1-5 is a two-day lesson.

Day 1

Pages 29–31: Opening paragraph through Sample 3, *Exercises 1–12*

Day 2

Pages 32–34: Systems of Linear Equations with Three Variables through Look Back, *Exercises 13–40*

Toolbox References

- **Toolbox Skill 8:** Representing a Data Set with a Matrix
- **Toolbox Skill 9:** Multiplying a Matrix by a Number
- **Toolbox Skill 10:** Adding and Subtracting Matrices
- **Toolbox Skill 11:** Multiplying Matrices

Extra Practice

See pages 608–610.

Warm-Up Exercises

Warm-Up Transparency 1-5

Support Materials

- Practice Bank: Practice 5
- Activity Bank: Enrichment 5
- Study Guide: Section 1-5
- Problem Bank: Problem Set 2
- Overhead Visuals: Folder 1
- Using Plotter Plus: Reducing Matrices to Solve Systems
- Using IBM Plotter Plus Disk: Matrix Reducer
- Using Mac Plotter Plus Disk: Matrix Calculator
- Assessment Book: Quiz 1-5

Answers to Talk it Over

1. the seating capacity of the auditorium
2. premier; These seats are closer to the stage.
3. Answers may vary. Examples: average ticket prices in the area; local sales of the band's music; how much of a profit the promoter can reasonably expect on each ticket

TEACHING

Additional Samples

S1 Refer to the information on page 29 and Sample 1. The promoter wonders whether it would be possible to increase profits to \$6000 by offering special star circle seats for \$60 per ticket. These would be in place of the \$36 premier tickets. Write two equations that model the new situation.

Let x = number preferred tickets. Let y = number of star circle tickets.

first equation: $x + y = 10,000$

second equation:

$28x + 60y - [20,000 + 0.75(28x + 60y) + 60,000] = 6000$

Simplify second equation.

$7x + 15y = 86,000$

These equations model the new situation.

$x + y = 10,000$

$7x + 15y = 86,000$

S2 Solve the system of equations:

$x + y = 10,000$

$7x + 15y = 86,000$

***Method 1.* Use algebra.**

Solve the first equation for y.

$y = 10,000 - x$

Substitute $10,000 - x$ for y in the second equation.

$7x + 15(10,000 - x) = 86,000$

$x = 8000$

To find y, substitute 8000 for x in the first equation.

$8000 + y = 10,000$

$y = 2000$

The promoter should sell 8000 preferred tickets and 2000 star circle tickets.

***Method 2.* Use a graph.**

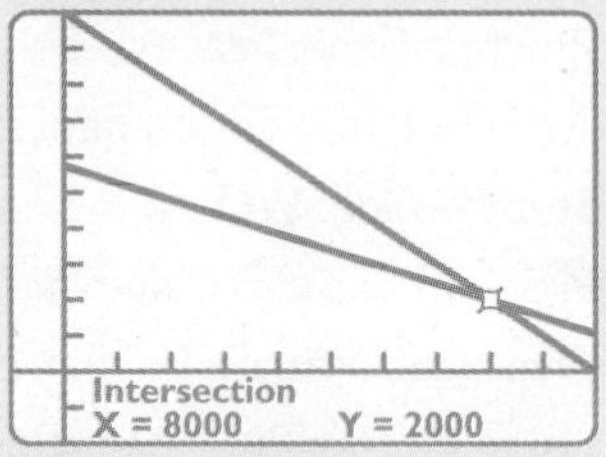

The promoter should sell 8000 preferred tickets and 2000 star circle tickets.

Sample 1

A concert promoter thinks a concert will be a sellout when tickets are priced as shown. The promoter needs to know how many of each type of ticket to sell to make a \$5000 profit. Write two equations that model this situation. Use the information on page 29 about the number of seats and the expenses.

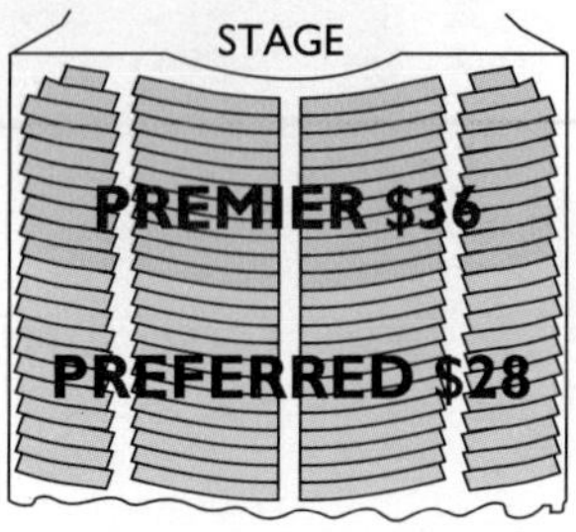

Sample Response

Let x = the number of preferred tickets. Let y = the number of premier tickets.

number of preferred tickets + number of premier tickets = number of tickets sold

$$x + y = 10,000$$ ← first equation

ticket income − expenses for band, auditorium, and so on = profit

$$28x + 36y - [20,000 + 0.75(28x + 36y) + 60,000] = 5000$$ ← second equation

(income from \$28 tickets: $28x$; income from \$36 tickets: $36y$)

These equations model the situation.

$x + y = 10,000$

$7x + 9y = 85,000$ ← Simplify the second equation.

A **system of equations** consists of two or more equations involving the same variables, as in Sample 1.

Sample 2

Solve the system of equations: $x + y = 10,000$
$7x + 9y = 85,000$

Sample Response

Method ❶ Use algebra.

$y = 10,000 - x$ ← Solve one equation for y.

$7x + 9(10,000 - x) = 85,000$ ← Substitute $10,000 - x$ for y in the other equation.

$x = 2500$

$x + y = 10,000$

$2500 + y = 10,000$ ← To find y, substitute 2500 for x in either of the original equations.

$y = 7500$

Method ❷ Use a graph.

Find the intersection point.

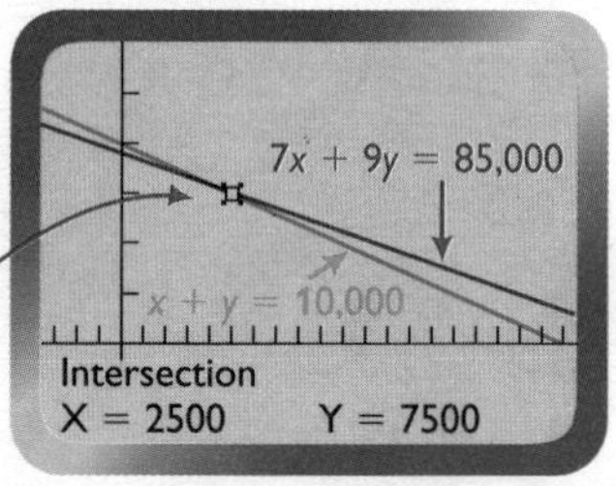

The promoter should sell 2500 preferred tickets and 7500 premier tickets.

Matrix Equations

Student Resources Toolbox
pp. 632–634 *Operations with Matrices*

You can write a system of linear equations as a *matrix equation*.

System of Equations	Matrix Equation
$ax + by = c$ $dx + ey = f$	coefficient matrix $\begin{bmatrix} a & b \\ d & e \end{bmatrix}\begin{bmatrix} x \\ y \end{bmatrix} = \begin{bmatrix} c \\ f \end{bmatrix}$ constant matrix

For a system with two variables, a matrix equation has the form: $A\begin{bmatrix} x \\ y \end{bmatrix} = B$

The solution of this matrix equation is $\begin{bmatrix} x \\ y \end{bmatrix} = A^{-1}B$.

inverse matrix

Sample 3

Write the system in Sample 1 as a matrix equation and use a graphics calculator to solve.

TECHNOLOGY NOTE

To use matrices, see Technology Handbook, p. 603.

Sample Response

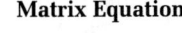

System of Equations	Matrix Equation	Solution of Matrix Equation
$x + y = 10,000$ $7x + 9y = 85,000$	$\begin{bmatrix} 1 & 1 \\ 7 & 9 \end{bmatrix}\begin{bmatrix} x \\ y \end{bmatrix} = \begin{bmatrix} 10,000 \\ 85,000 \end{bmatrix}$ Enter as matrix *A*. Enter as matrix *B*.	[A]⁻¹[B] [2500] [7500]

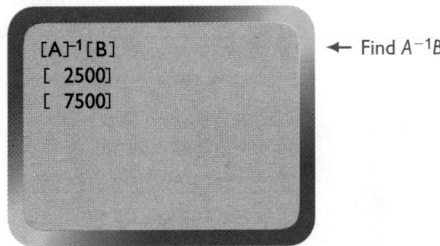

← Find $A^{-1}B$.

The resulting matrix tells you that the promoter should sell 2500 preferred tickets and 7500 premier tickets.

Now you are ready for:
Exs. 1–12 on p. 34

Additional Sample

S3 Write the system in Additional Sample S1 as a matrix equation and use a graphics calculator to solve.

System of Equations

$x + y = 10,000$
$7x + 15y = 86,000$

Matrix Equation

$\begin{bmatrix} 1 & 1 \\ 7 & 15 \end{bmatrix}\begin{bmatrix} x \\ y \end{bmatrix} = \begin{bmatrix} 10,000 \\ 86,000 \end{bmatrix}$

Enter as matrix A   Enter as matrix B

$[A]^{-1}[B]$
[8000]
[2000]

The resulting matrix tells you that the promoter should sell 8000 preferred tickets and 2000 star circle tickets.

Teaching Tip

Some students may feel uncomfortable solving the system in Sample 3 without seeing the inverse matrix $[A]^{-1}$ in standard form. You can always ask them to find $[A]^{-1}$ and then find $[A]^{-1}[B]$ by pressing 2nd [ANS] MATRX <[B]> ENTER (for the TI-82) or 2nd [ANS] 2nd [[B]] (for the TI-81).

Talk it Over

Questions 4 and 5 lead students to the conclusion that the solution to an equation or system of equations in three variables is an ordered triple.

Assessment: Open-ended

Ask students to consider these questions: Suppose a system of three equations does not have a solution. How would you describe geometrically the possible relationships between the planes? What happens when you try to solve this system using matrices?

BY THE WAY...

In 1683 the Japanese mathematician Seki Kowa developed a new algorithm for an ancient Chinese method of solving a system of linear equations. He used bamboo rods placed in squares on a table to represent the constants.

Systems of Linear Equations with Three Variables

A linear equation with two variables, x and y, can be written in the form

$$ax + by = c.$$

A **linear equation with three variables**, x, y, and z, can be written in the form

$$ax + by + cz = d.$$

Talk it Over

4. a. Is $(-1, 2)$ a solution of $3x + 4y = 5$? Explain.
   b. Is $(-1, 2, 3)$ a solution of $3x + 4y + 2z = 11$? Explain.
   c. Is $(-1, 2, -3)$ a solution of $3x + 4y + 2z = 11$? Explain.
5. a. Is $(-1, 2)$ a solution of this system? Explain.

$$3x + 4y = 5$$
$$2x + 5y = 8$$

   b. Is $(-1, 2, 3)$ a solution of this system? Why?

$$3x + 4y + 2z = 11$$
$$2x + 5y - z = 3$$
$$x - 2y + 4z = 6$$

This point is the graph of the *ordered triple* $(2, -3, 4)$.

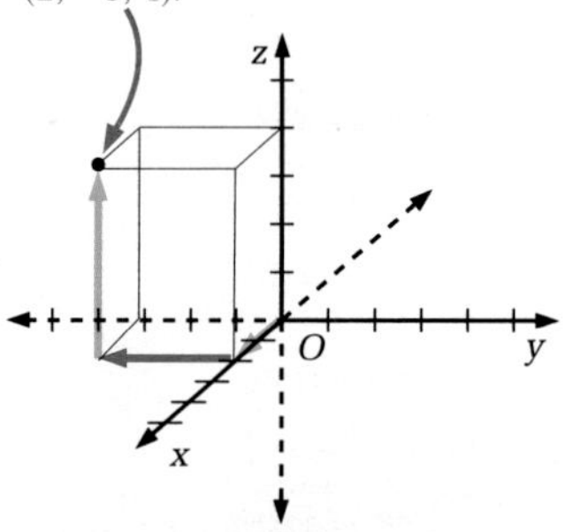

You can graph a solution of a linear equation with three variables in a three-dimensional coordinate system with an x-axis, a y-axis, and a z-axis that meet at right angles at the origin O.

The graph of *all* the solutions of a linear equation with three variables is a plane. The graph of a system of three linear equations with three variables is three planes. Only those points that are common to all three planes are solutions of the system.

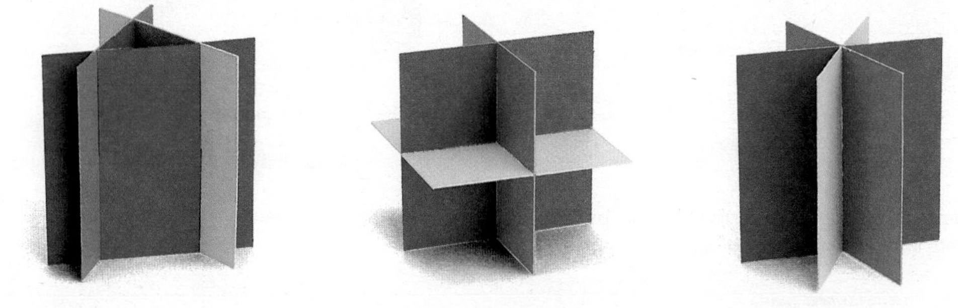

no common point of intersection — one common point of intersection — many common points of intersection

Answers to Talk it Over

4. a. Yes; when the given values of x and y are substituted in the equation, the result is a true statement. $3(-1) + 4(2) = -3 + 8 = 5$✓
   b. Yes; when the given values of x, y, and z are substituted in the equation, the result is a true statement. $3(-1) + 4(2) + 2(3) = -3 + 8 + 6 = 11$✓
   c. No; when the given values of x, y, and z are substituted in the equation, the result is not a true statement. $3(-1) + 4(2) + 2(-3) = -3 + 8 + (-6) = -1$; $-1 \neq 11$
5. a. Yes; $(-1, 2)$ is a solution of both equations.
   b. No; $(-1, 2, 3)$ is not a solution of the second equation or the third.

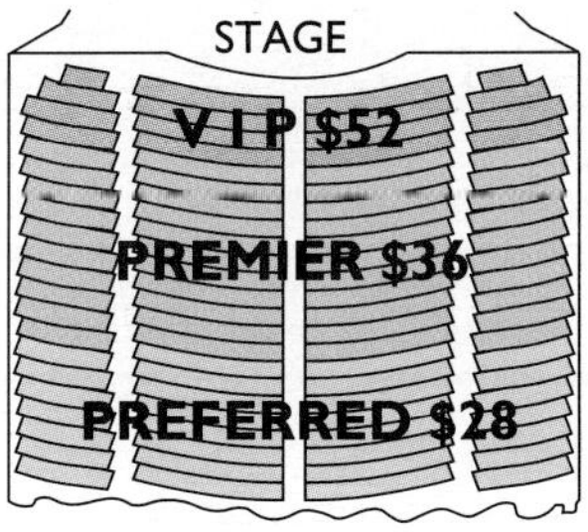

Sample 4

Suppose the promoter decides to sell three types of tickets for the concert described on page 29. The promoter wants the number of preferred tickets to equal the number of premier and VIP tickets combined. How many of each type of ticket must be sold in order to make a profit of $8000 when the concert is a sellout?

Sample Response

Let x = the number of preferred tickets, y = the number of premier tickets, and z = the number of VIP tickets.

1 Write three equations.

number of preferred tickets	+	number of premier tickets	+	number of VIP tickets	=	total number of tickets sold
x	+	y	+	z	=	10,000

ticket income − expenses for band, auditorium, and so on = profit

$$28x + 36y + 52z - [20{,}000 + 0.75(28x + 36y + 52z) + 60{,}000] = 8000$$

number of preferred tickets	=	number of premier tickets	+	number of VIP tickets
x	=	y	+	z

$$x + y + z = 10{,}000$$
$$7x + 9y + 13z = 88{,}000$$
$$x - y - z = 0$$

2 Rewrite each equation in the form $ax + by + cz = d$.

$$\begin{bmatrix} 1 & 1 & 1 \\ 7 & 9 & 13 \\ 1 & -1 & -1 \end{bmatrix}\begin{bmatrix} x \\ y \\ z \end{bmatrix} = \begin{bmatrix} 10{,}000 \\ 88{,}000 \\ 0 \end{bmatrix}$$

3 Write the matrix equation $A\begin{bmatrix} x \\ y \\ z \end{bmatrix} = B$.

$$\begin{bmatrix} x \\ y \\ z \end{bmatrix} = \begin{bmatrix} 5000 \\ 3000 \\ 2000 \end{bmatrix}$$

4 Use a graphics calculator to solve.

The promoter must sell 5000 preferred tickets, 3000 premier tickets, and 2000 VIP tickets to make a profit of $8000.

Talk it Over

6. a. The last equation in Sample 4 can be rewritten as $x = y + z$. Does substituting for x in the other two equations produce a system that you can solve with algebra? Explain.
   b. Is it possible to solve Sample 4 with a graphics calculator, as in Method 2 of Sample 2? Why or why not?

Answers to Talk it Over

6. a. Yes; methods may vary. An example is given. Substitute $y + z$ for x in the first equation to get $2y + 2z = 10{,}000$. Next, substitute $y + z$ for x in the second equation to get $16y + 20z = 88{,}000$. Then solve the system of equations in y and z to get $z = 2000$. Use this value to get $y = 3000$ in one of the two equations. Finally, use $z = 2000$ and $y = 3000$ in $x = y + z$ to get $x = 5000$.
   b. No; it is only possible to graph equations in two variables on a graphics calculator.

Additional Sample

S4 Refer to Sample 4. An advisor tells the promoter that there is a strong demand for concert tickets. She suggests making the total number of premier and VIP tickets four times the number of preferred tickets. How many tickets in each category must be sold to make a profit of $8000?

The advisor's suggestion would change the last equation of the system in Sample 4. If $y + z = 4x$, then the last equation will be $4x - y - z = 0$.

The matrix equation will then be

$$\begin{bmatrix} 1 & 1 & 1 \\ 7 & 9 & 13 \\ 4 & -1 & -1 \end{bmatrix}\begin{bmatrix} x \\ y \\ z \end{bmatrix} = \begin{bmatrix} 10{,}000 \\ 88{,}000 \\ 0 \end{bmatrix}.$$

Use a graphics calculator to solve the new matrix equation. The solution is

$$\begin{bmatrix} x \\ y \\ z \end{bmatrix} = \begin{bmatrix} 2000 \\ 7500 \\ 500 \end{bmatrix}.$$

The promoter must sell 2000 preferred tickets, 7500 premier tickets, and 500 VIP tickets.

Mathematical Procedures

For Sample 4, it is easy to solve the system by using the addition-or-subtraction method. You may wish to have students do this to review the method. Students should see that adding the first and third equations will quickly give a value for x. The rest of the work is straightforward.

Then have students consider a system for which the coefficients are more difficult to work with. For example:

$$17x - 23y + 9z = 103$$
$$-54x + 16y - 12z = 25$$
$$13x + 82y + 15z = 88$$

Students can discuss which method would be most efficient to solve this system.

Integrating the Strands

The mathematics of this section shows a beautiful integration of concepts from the strands of algebra, geometry, and discrete mathematics. Algebraic equations are displayed geometrically by their graphs and a system of equations is represented and solved by using a matrix equation.

Look Back

When discussing these questions, students should compare and contrast both the algebraic and geometric meanings involved with solving systems of equations.

APPLYING

Suggested Assignment

Day 1

Standard 1–12

Extended 1–12

Day 2

Standard 13–15, 17–22, 28–40

Extended 13–15, 17–40

Integrating the Strands

Algebra Exs. 1–39

Geometry Exs. 13–15

Logic and Language Exs. 1, 2, 16, 40

Error Analysis

Students may have misconceptions about solving systems of equations. Ex. 2 provides an excellent opportunity to check students' understanding of the methods for solving systems and to correct errors.

Application

Ex. 12 applies a system of two equations with two variables to a problem in aviation, while Exs. 23–25 apply a system of three equations to a problem in physics.

Now you are ready for: Exs. 13–40 on pp. 35–36

Look Back

How is solving a system of linear equations with three variables like solving a system of linear equations with two variables? How is it different?

1-5 Exercises and Problems

1. **Reading** Samples 2 and 3 show three methods for solving a system of equations. Which of these methods do you prefer? Why?
2. **Open-ended** Make a chart or a concept map about methods for solving systems of equations.

Solve. When necessary, round decimal answers to the nearest tenth.

3. $3x + 2y = 1$; $x + y = 1$
4. $x - y = 7$; $2x + y = 2$
5. $4x + 2y = 1$; $-6x + 4y = 9$
6. $3x - y = 12$; $2x + 3y = 74$
7. $x + y = 7000$; $5x + 7y = 39{,}000$
8. $2x - 3y = 5000$; $4x - 5y = 45{,}000$
9. $0.54x + 3.2y = 7.9$; $2.1x - 4.5y = 2.3$
10. $570x - 190y = 3400$; $320x + 780y = 4600$
11. $\frac{3}{4}x + \frac{1}{2}y = -1$; $\frac{5}{8}x + \frac{2}{5}y = 2$

12. **Aviation** In the atmosphere over the United States, winds generally blow from west to east. These winds affect the travel time of airplanes. A flight from San Francisco to New York may take 5.5 h, but a flight from New York to San Francisco in the same airplane may take an hour longer.
    a. The distance between San Francisco and New York is about 2800 mi. Use the formula

       distance = travel speed × time

       to write a system of two equations with two variables.
    b. Find the airplane's speed (without the wind). Round your answer to the nearest tenth.

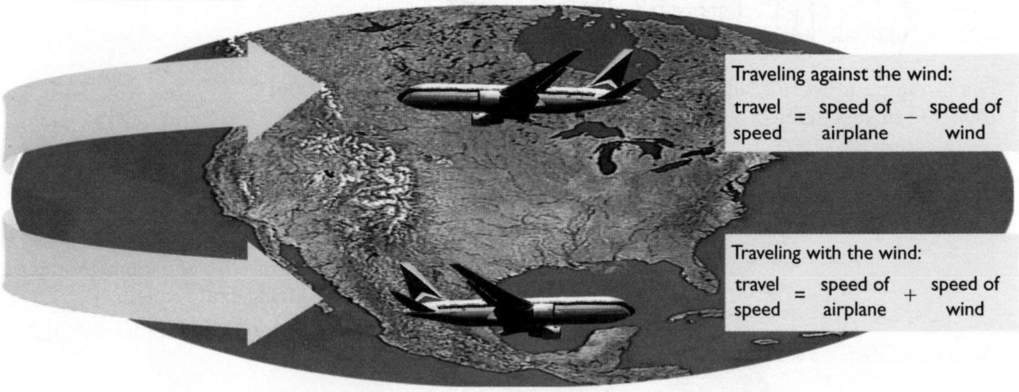

Traveling against the wind:
travel speed = speed of airplane − speed of wind
Traveling with the wind:
travel speed = speed of airplane + speed of wind

Answers to Look Back

In both cases, you need to find values of the variables that solve all the equations in the system. If there are two equations, the solution is an ordered pair. If there are three equations, the solution is an ordered triple.

Answers to Exercises and Problems

1. Answers may vary. An example is given. I prefer to use a graphics calculator to graph the equation if one is available because it is the easiest and fastest. If one is not available, then I like to use algebra. This method gives you an exact answer, although it may take some work to get to it.
2. Answers may vary. An example is given.

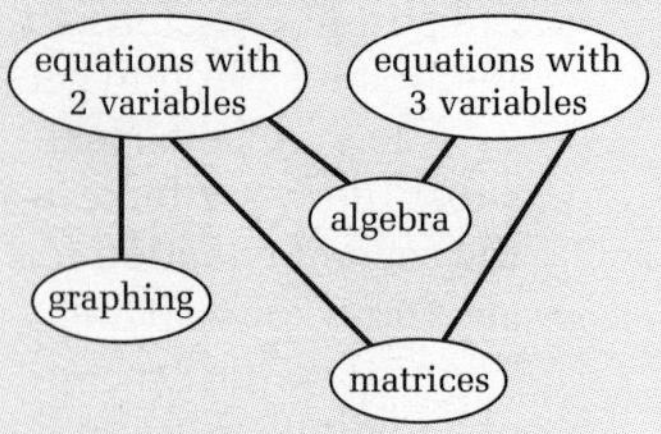

3. (–1, 2)
4. (3, –4)
5. (–0.5, 1.5)
6. (10, 18)

For Exercises 13–15, suppose planes *P*, *Q*, and *R* represent a system of equations. Tell whether each system has a solution. Explain.

13.

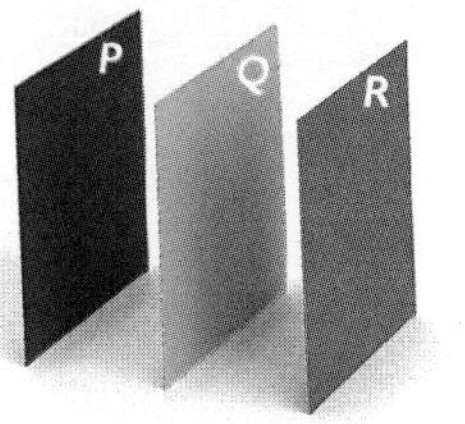

14.

15.

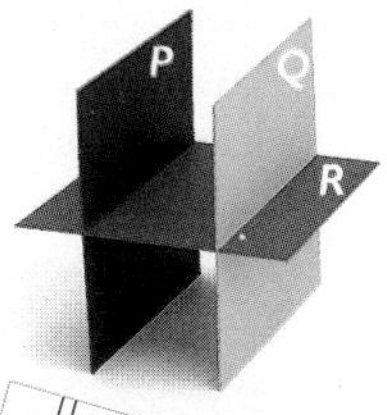

16. **Group Activity** Work with another student.

Writing Not having his graphics calculator handy, Dana decided to solve the system shown using algebra. He began by solving the first equation for z in terms of x and y, but he was not sure what to do next. Show Dana how to proceed, giving him explanations along the way.

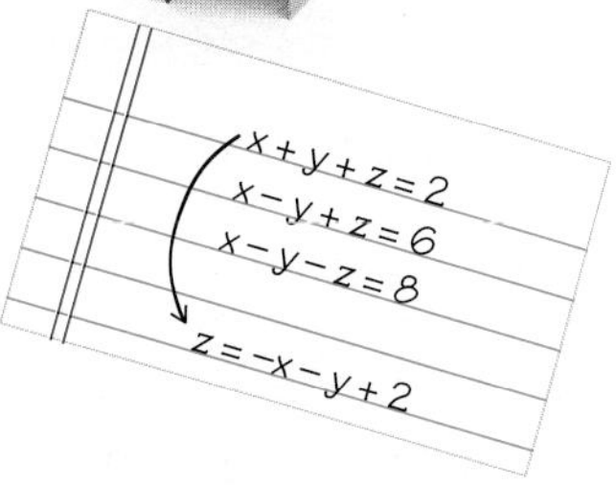

Solve. Round decimal answers to the nearest tenth.

17. $x + y = 7$
$y + z = 8$
$x + 2z = 2$

18. $2x - y = 7$
$6x + 3z = 9$
$x + 2y + 3z = -4$

19. $5x + y + 6z = 3$
$x - y + 10z = 9$
$5x + y - 2z = -9$

20. $7x + 2y + z = -10$
$5x + 5y + z = 13$
$2x - y + z = -5$

21. $3x + y - 4z = -3$
$5x + y - z = -6$
$10x + 4y + 5z = 15$

22. $-2x + 2y + 2z = 5$
$2x + 2y - z = 3$
$2x + 4y - z = 4$

connection to PHYSICS

Suppose an electric circuit has three resistors wired to an 18-volt battery. To find the currents I_A, I_B, and I_C (measured in amperes) across the resistors in this circuit, you can use these laws:

Current law:	The sum of the currents entering a node equals the sum of the currents leaving the node.
Voltage law:	Around any closed loop, the sum of the voltage drops across the resistors is equal to the voltage supplied by the battery.

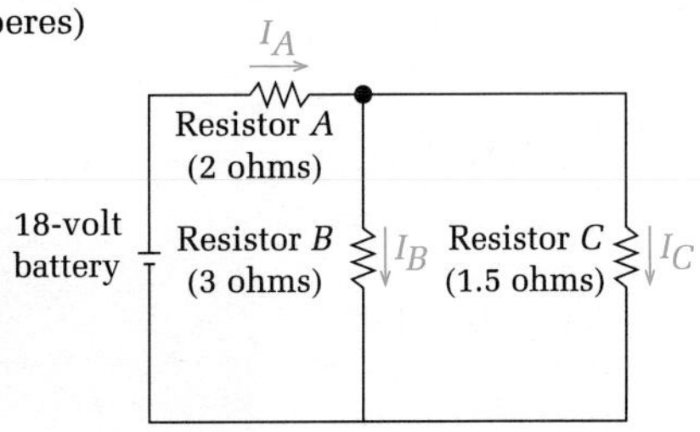

23. Use the current law to write an equation relating I_A, I_B, and I_C at the labeled node.

24. To calculate the voltage V (measured in volts) across a resistor, you use the formula $V = RI$, where the resistance R is measured in ohms and the current I is measured in amperes. Use this formula and the voltage law to write an equation for the closed loop containing resistors A and B.

25. Use the voltage law to write an equation for the closed loop running from the battery, through resistors A and C, and back to the battery.

26. Solve the system of three equations that you wrote in Exercises 23–25.

Answers to Exercises and Problems

7. (5000, 2000)
8. (55,000, 35,000)
9. (4.7, 1.7) 10. (7.0, 3.0)
11. (112, –170)
12. a. Let a = speed of the airplane and w = speed of the wind. $2800 = 5.5(a + w)$; $2800 = 6.5(a - w)$
b. about 470 mi/h
13. No; the planes have no common point of intersection.
14. Yes; the planes have many common points of intersection.
15. No; the planes have no common point of intersection.
16. See answers in back of book.
17. (0, 7, 1) 18. (–1, –9, 5)
19. (–2, 4, 1.5) 20. (–4, 5, 8)
21. (–3, 10, 1)
22. (4, 0.5, 6)
23. $I_A = I_B + I_C$
24. $2I_A + 3I_B = 18$
25. $2I_A + 1.5I_C = 18$
26. I_A = 6 amperes, I_B = 2 amperes, I_C = 4 amperes

Teaching Tip

If students make errors in Exs. 3–11 or 17–22, suggest that they solve each system by using two different methods and then check their work if the answers do not agree. Errors in paper-and-pencil solutions are likely to arise from computational mistakes, especially with the addition-or-subtraction method. Errors using the substitution method or with a calculator are often the result of incorrect solutions for y in terms of x. Matrix methods have the advantage that they require little or no advanced change in the coefficients of the system. However, students may apply the procedure incorrectly. They will get error messages if they enter $[A]^{-1}$ as [A]^–1 instead of $[A]^{-1}$ (with the x^{-1} key).

Students Acquiring English

For Ex. 16, consider offering students acquiring English the option of writing the equations and orally explaining the steps in the process of solving the equation.

Cooperative Learning

Ex. 16 provides an excellent opportunity for students to help one another to understand the procedure involved in solving systems algebraically. You may wish to have the groups share their explanations with the entire class.

Research

As background for Exs. 23–25, ask for volunteers to research the meanings of the following terms from physics: electric current, resistor, volt, ampere, and ohm. Ask the researchers to give a brief explanation of each term.

27. Business Koretta Jackson is the manager of *Koretta's Kitchen*, a new restaurant, which will be open from 6:00 A.M. to 11:00 P.M. daily. She must hire workers to cover three shifts. She needs 11 workers during the lunch rush from 11:00 A.M. to 2:00 P.M. and 16 workers during the dinner rush from 4:00 P.M. to 7:00 P.M. Her budget allows her to hire 20 workers. How many should she hire for each shift?

Shift I: 6:00 A.M.–2:00 P.M.
Shift II: 11:00 A.M.–7:00 P.M.
Shift III: 4:00 P.M.–12:00 midnight

Ongoing ASSESSMENT

28. **Open-ended** Write a system of three linear equations that has the solution (1, 5, 23). Check your answer by solving your system.

Review PREVIEW

Use algebra to find the break-even point when E = expenses and I = income. Round decimal answers to the nearest tenth. *(Section 1-4)*

29. $E = 187 + 5x$, $I = 5.6x$

30. $E = 260 + 3x$, $I = 7.34x$

31. $E = 112x - 3200$, $I = 55x$

32. $E = 1350 - 4x$, $I = 46x$

Solve each inequality. *(Section 1-1)*

33. $\frac{1}{2}x + 5 > 16$

34. $2(3 - x) \leq 9$

35. $2x + 2.5 > -3x$

Simplify. Write answers with positive exponents. *(Toolbox Skill 14)*

36. $2c^{-1/2}d^0$

37. $(-1 + x^0)y^{-1/3}$

38. $\frac{18n^{-5}}{3}$

39. $\frac{-3}{r^{-4}}$

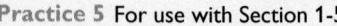

Working on the Unit Project

40. **Research** Write a report on the achievements of Frederick Law Olmsted. Include a discussion of some of his major park projects.

A section of Boston's "Emerald Necklace," a connected system of parks designed by Frederick Law Olmsted

Practice 5 For use with Section 1-5

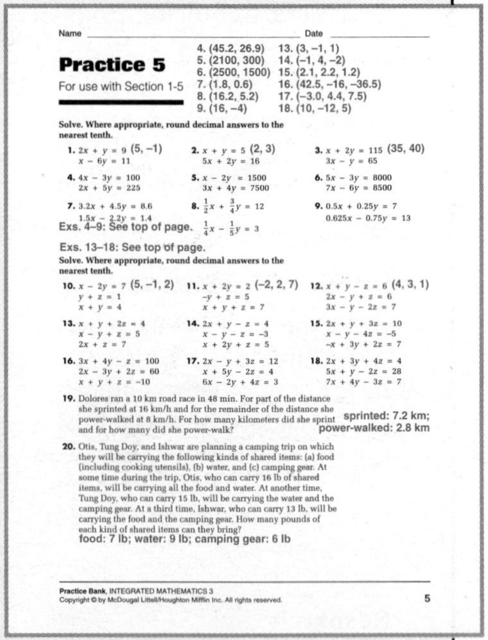

Name ______________________ Date ____________

Practice 5

For use with Section 1-5

4. (45.2, 26.9) 5. (2100, 300) 6. (2500, 1500) 7. (1.8, 0.6) 8. (16.2, 5.2) 9. (16, –4)
13. (3, –1, 1) 14. (–1, 4, –2) 15. (2.1, 2.2, 1.2) 16. (42.5, –16, –36.5) 17. (–3.0, 4.4, 7.5) 18. (10, –12, 5)

Solve. Where appropriate, round decimal answers to the nearest tenth.

1. $2x + y = 9$, $x - 6y = 11$ (5, –1)

2. $x + y = 5$, $5x + 2y = 16$ (2, 3)

3. $x + 2y = 115$, $3x - y = 65$ (35, 40)

4. $4x - 3y = 100$, $2x + 5y = 225$

5. $x - 2y = 1500$, $3x + 4y = 7500$

6. $5x - 3y = 8000$, $7x - 6y = 8500$

7. $3.2x + 4.5y = 8.6$, $1.5x - 2.2y = 1.4$

8. $\frac{1}{2}x + \frac{3}{4}y = 12$, $\frac{1}{4}x - \frac{1}{5}y = 3$

9. $0.5x + 0.25y = 7$, $0.625x - 0.75y = 13$

Exs. 4–9: See top of page.

Exs. 13–18: See top of page.

Solve. Where appropriate, round decimal answers to the nearest tenth.

10. $x - 2y = 7$, $y + z = 1$, $x + y = 4$ (5, –1, 2)

11. $x + 2y = 2$, $-y + z = 5$, $x + y + z = 7$ (–2, 2, 7)

12. $x + y - z = 6$, $2x - y + z = 6$, $3x - y - 2z = 7$ (4, 3, 1)

13. $x + y + 2z = 4$, $x - y + z = 5$, $2x + z = 7$

14. $2x + y - z = 4$, $x - y - z = -3$, $x + 2y + z = 5$

15. $2x + y + 3z = 10$, $x - y - 4z = -5$, $-x + 3y + 2z = 7$

16. $3x + 4y - z = 100$, $2x - 3y + 2z = 60$, $x + y + z = -10$

17. $2x - y + 3z = 12$, $x + 5y - 2z = 4$, $6x - 2y + 4z = 3$

18. $2x + 3y + 4z = 4$, $5x + y - 2z = 28$, $7x + 4y - 3z = 7$

19. Dolores ran a 10 km road race in 48 min. For part of the distance she sprinted at 16 km/h and for the remainder of the distance she power-walked at 8 km/h. For how many kilometers did she sprint and for how many did she power-walk? sprinted: 7.2 km; power-walked: 2.8 km

20. Otis, Tung Doy, and Ishwar are planning a camping trip on which they will be carrying the following kinds of shared items: (a) food (including cooking utensils), (b) water, and (c) camping gear. At some time during the trip, Otis, who can carry 16 lb of shared items, will be carrying all the food and water. At another time, Tung Doy, who can carry 15 lb, will be carrying the water and the camping gear. At a third time, Ishwar, who can carry 13 lb, will be carrying the food and the camping gear. How many pounds of each kind of shared items can they bring?
food: 7 lb; water: 9 lb; camping gear: 6 lb

Practice Bank, INTEGRATED MATHEMATICS 3

5

Answers to Exercises and Problems

27. She should hire 4 workers for Shift I, 7 workers for Shift II, and 9 workers for Shift III.

28. Answers may vary. An example is given. $x + y + z = 29$; $x + y - z = -17$; $x - y - z = -27$

29. 311.7

30. 59.9

31. 56.1

32. 27

33.

34.

35.

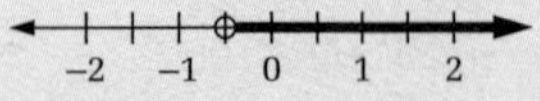

36. $\frac{2}{c^{1/2}}$

37. 0

38. $\frac{6}{n^5}$

39. $-3r^4$

40. Summaries may vary. Frederick Law Olmsted was an American landscape architect and author. His projects included Central Park in New York, Prospect Park in Brooklyn, New York, South Park in Chicago, Mt. Royal Park in Montreal, park systems in Buffalo and Boston, and the grounds of the World's Columbian Exposition in Chicago in 1893. These grounds later became Jackson Park.

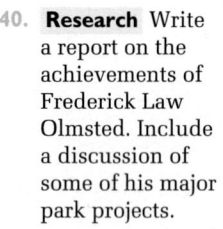

Section

1-6 Using Diagrams

MAKING CONNECTIONS

Focus
Use diagrams to solve problems.

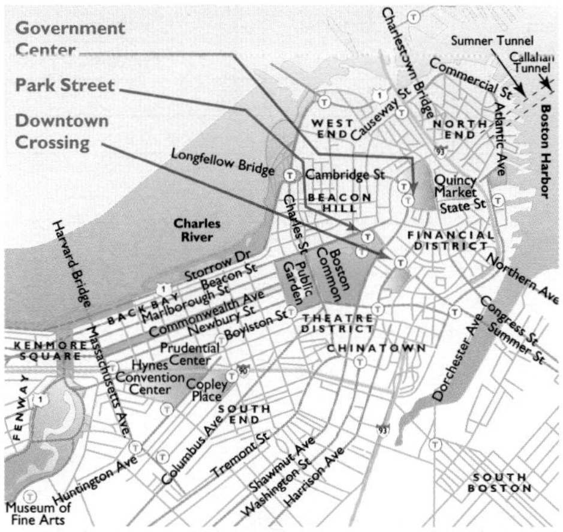

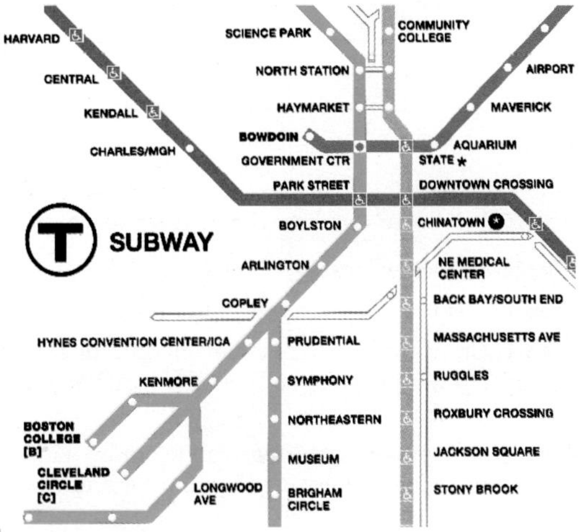

BY THE WAY...

The lines of the Boston public transit system are color-coded: Green Line for "Emerald Necklace" parks of Frederick Law Olmsted; Blue Line for Atlantic coast; Red Line for crimson of Harvard University; Orange Line for area once known as "Orange."

Talk it Over

1. Which map is easier to read? Why?
2. What do the dots represent on the transit map? What does it mean when a line is drawn to connect one dot to another? What does it mean when there is no line connecting two dots?
3. Describe a route from Museum station to Airport station on the public transit system. How many stations will you stop at before you reach Airport station?
4. The transit map is an example of a *network* diagram. A dot is a **vertex of a network**, and a line connecting two dots is an **edge of a network.** Name the vertices from Kenmore to Science Park along the Green Line. How many edges are there along this route?
5. On the transit map, the distance from Park Street to Government Center is equal to the distance from Park Street to Downtown Crossing. Are these distances the same on the street map?
6. How is the transit map like the street map? How is it different?

PLANNING

Objectives and Strands
See pages 1A and 1B.

Spiral Learning
See page 1B.

Materials List
- Colored pencils
- Graphics calculator

Recommended Pacing
Section 1-6 is a two-day lesson.

Day 1

Pages 37–39: Talk it Over 1 through Talk it Over 8, *Exercises 1–17*

Day 2

Pages 40–41: Coloring Network Diagrams through Look Back, *Exercises 18–31*

Extra Practice
See pages 608–610.

Warm-Up Exercises
- Warm-Up Transparency 1-6

Support Materials
- Practice Bank: Practice 6
- Activity Bank: Enrichment 6
- Study Guide: Section 1-6
- Problem Bank: Problem Set 2
- Assessment Book: Quiz 1-6, Alternative Assessments 5, 6

Answers to Talk it Over

1. The transportation map is easier to read. Summaries may vary. The labels are clearly written in horizontal lines. The labels on the street map are written within boundary lines in varying directions.
2. stations; There is a transportation line connecting the two stations. There is no transportation line directly connecting the two stations.
3. Answers may vary. An example is given. I would take the Green Line from the Museum station to Government Center, then take the Blue Line to the Airport. The Airport would be the twelfth stop.
4. Kenmore, Hynes, Copley, Arlington, Boylston, Park St., Government Center, Haymarket, North Station, Science Park; 9 edges
5. No.
6. Summaries may vary. Both maps indicate locations in the city of Boston. The street map shows the actual layout of the streets, with distances to scale. The transportation map shows transportation routes around the city. Distances are not to scale and streets are not shown. The street map would be helpful if you were walking or driving around the city. The transportation map would be of little help when walking and driving, other than indicating relative positions of some locations.

TEACHING

Talk it Over

Questions 1–6 introduce students to the concept of a network diagram and lead them to see how it is different from an actual map.

Additional Sample

S1 Saburo Sakato wants to assign the workers at his grocery store jobs they have done before. He has drawn up the following list.
Bagger: Linda, Miguel, Rama, Leon
Deli: Linda, Rama
Stockroom: Miguel, Leon
Checkout register: Miguel
How can people be assigned so that each person covers one job?

Use a diagram.
Step 1. List workers in one column and jobs in another.
Step 2. Draw a vertex for each worker and each job. Draw an edge to connect each worker to jobs he or she has done before.
Step 3. Assign workers to jobs by darkening the edge connecting the worker to the job. The vertex for "Checkout register" has only one edge, so it must be darkened. This leaves only Leon available for the stockroom job.

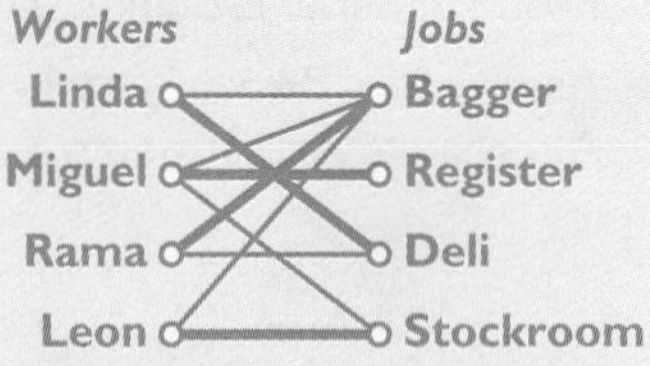

Rama and Linda can be assigned to the bagger and deli jobs in either of two ways. One way to assign workers to jobs is: Miguel to checkout register, Leon to stockroom, Linda to deli, and Rama to bagger.

Reasoning

After students have studied Sample 1, ask them why the first step is to look for vertices with only one edge. You may wish to ask what they would do in a situation where all vertices have more than one edge.

Sample 1

A high school newspaper staff is making assignments for the next edition of the paper. This list shows which reporters are available to cover each event. How can the stories be assigned so that each reporter covers one event?

Student council elections	Language club's international festival	Opening performance of the school play	Regional volleyball championship	Academic banquet
Kwan	Kwan	Epstein	Cárdenas	Cárdenas
Pinnoo	Cárdenas	Cárdenas	Kwan	Epstein
Cárdenas	Epstein		Johns	
			Pinnoo	

Sample Response

Problem Solving Strategy: Use a diagram.

Step 1 List the events in one column and the reporters in another.

Step 2 Draw a vertex for each event and for each reporter. Draw an edge to connect the vertices if a reporter can cover the event.

Step 3 Assign each reporter to an event by darkening the edge connecting the reporter and the event.

Watch Out! The point where two edges cross is not always a vertex.

1. Look for vertices that have only one edge. Johns has only one edge. Darken this edge.
2. Look at the vertices that have edges in common with "volleyball." Pinnoo has only two edges. Since Johns is covering volleyball, Pinnoo must cover the student council elections.

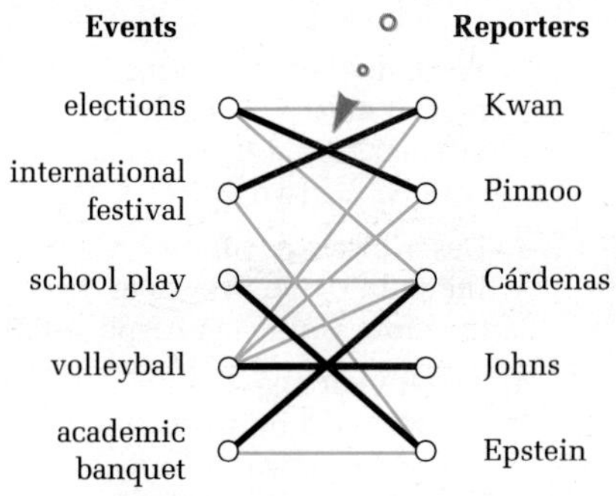

3. Look at the vertices that have edges in common with "elections." The only event that Kwan can cover is the international festival.
4. Cárdenas and Epstein can both cover the remaining stories.

This is one way to assign the reporters to cover the events:

Student council elections	International festival	School play	Volleyball tournament	Academic banquet
Pinnoo	Kwan	Epstein (or Cárdenas)	Johns	Cárdenas (or Epstein)

Sample 2

This table shows a nonstop flight schedule for a regional airline. Show this information in a diagram or a matrix.

	Departure from ...	Provides nonstop service to...			
DSM	**Des Moines**	Minn.-St. Paul	St. Louis		
MSP	**Minn.-St. Paul**	Des Moines	Pittsburgh	St. Louis	
PIT	**Pittsburgh**	St. Louis			
RDU	**Raleigh/Durham**	Pittsburgh	St. Louis		
STL	**St. Louis**	Des Moines	Pittsburgh	Raleigh/Durham	Minn.-St. Paul

Sample Response

Method 1 Use a diagram.

Draw a vertex for each city of departure. Connect the vertices with edges. Place arrowheads on the edges to show the direction of the flight.

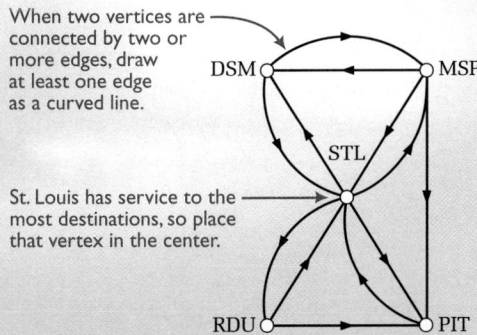

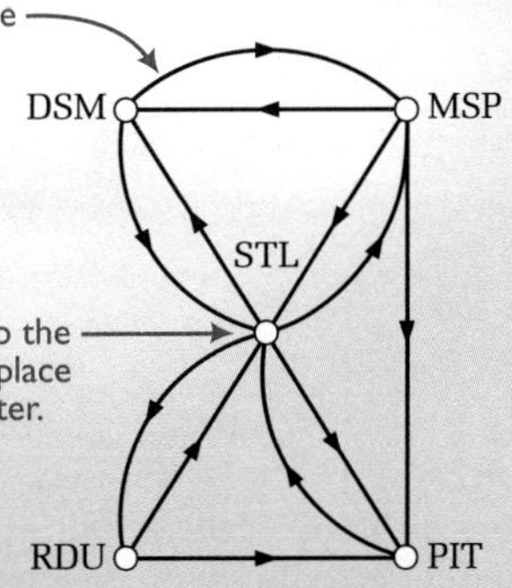

When two vertices are connected by two or more edges, draw at least one edge as a curved line.

Method 2 Use a matrix.

Let each row represent a city with nonstop service and each column represent a destination.

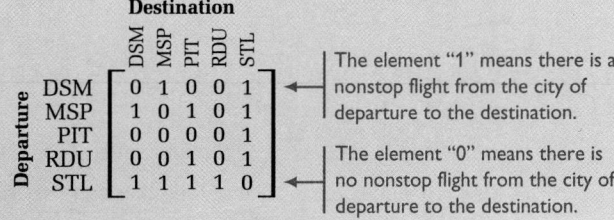

Destination (columns) / **Departure** (rows)

	DSM	MSP	PIT	RDU	STL
DSM	0	1	0	0	1
MSP	1	0	1	0	1
PIT	0	0	0	0	1
RDU	0	0	1	0	1
STL	1	1	1	1	0

The element "1" means there is a nonstop flight from the city of departure to the destination.

The element "0" means there is no nonstop flight from the city of departure to the destination.

Talk it Over

7. How many nonstop flights depart from Des Moines? from St. Louis? How do you get this information from the diagram? from the matrix?
8. Which representation is more helpful for finding the number of nonstop flights departing from a city? Which is more helpful for finding the number of roundtrip nonstop flights? Explain your choices.

Now you are ready for: Exs. 1–17 on pp. 41–43

Teaching Tip

You may find it helpful to draw the diagram for Sample 1 on an acetate sheet and use it with an overhead projector. Use a cloth to carefully *remove* edges as they are no longer needed.

Additional Sample

S2 A group of investors is exploring the possibility of starting an airline that serves only five major European capitals: Athens (A), Berlin (B), Madrid (M), Paris (P), and Rome (R). Their plan would call for daily nonstop flights as indicated by this flight list:

From	*To*
A	B, P
B	A, P
M	B, P
P	A, B, M, R
R	A, M, P

Show this information in a diagram or a matrix.

Method 1. **Use a diagram. Draw a vertex for each city, with Paris in the center. Connect vertices with edges that have arrowheads indicating the direction of the flight.**

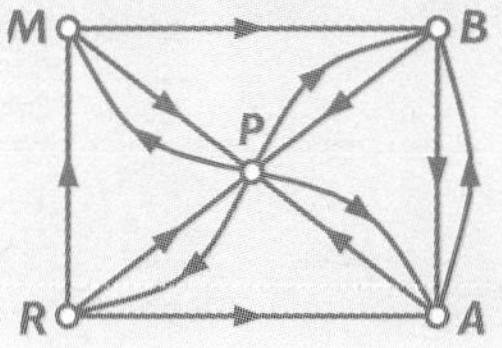

Method 2. **Use a matrix. Let each row represent a city of departure, and let each column represent a destination. The element "1" means that there is a nonstop flight from a certain city of departure to a certain destination. The element "0" means there is no such service.**

Destination (columns) / **Depart** (rows)

Depart	A	B	M	P	R
A	0	1	0	1	0
B	1	0	0	1	0
M	0	1	0	1	0
P	1	1	1	0	1
R	1	0	1	1	0

Talk it Over

Questions 7 and 8 have students compare the ease of finding information from the diagram and from the matrix.

Answers to Talk it Over

St. Louis has service to the most destinations, so place that vertex in the center.

7. 2 nonstop flights; 4 nonstop flights; Count the arrows coming out of the vertex for the indicated city; add the 1's in the row for the indicated city.
8. Answers may vary. An example is given. It is very easy to read the number of nonstop departing flights from either the diagram or the matrix. It is easier to read the number of roundtrip nonstop flights from the diagram because you only have to look for arrows going both ways between two vertices.

Additional Sample

S3 In Sample 3, suppose that Cárdenas decides in the second semester to join the science club as a third activity. Make a new activities schedule so that no activities with a member in common meet at the same time.

Step 1. **Draw a network diagram. Connect vertices that represent activities with members in common.**

Since Cárdenas has joined the science club, there are two new edges. They connect science club with school newspaper and honor society. See the diagram below.

Step 2. **Use the same procedure as that described in step 2 of the Sample 3 response. Instead of using color for the vertices, label the vertices with numbers. Labeling two vertices with the same number is like coloring them with the same color. If you begin labeling "school newspaper" with the number 1, then "drama club" can also be labeled 1. Label "science club" with 2. The only other vertex that can be labeled 2 is "student council." All other activities should be labeled 3.**

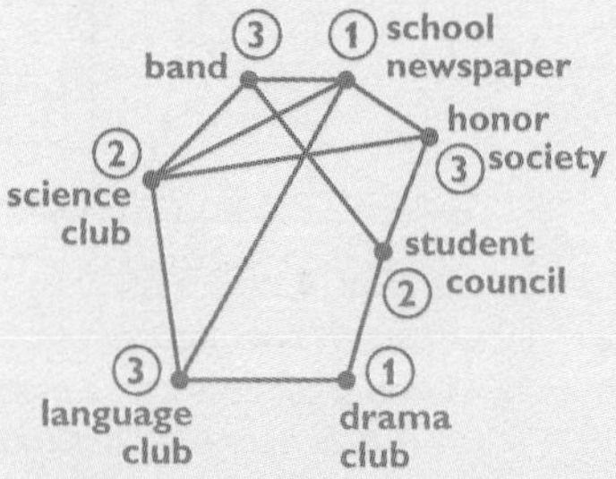

One possible schedule groups together activities represented by vertices with the same number.

Meeting time 1
school newspaper
drama club
Meeting time 2
science club
student council
Meeting time 3
band
honor society
language club

Coloring Network Diagrams

Sometimes it is helpful to color the vertices of a network diagram. Suppose you use different colors for vertices that are connected by edges. Then vertices that are the *same color* represent things, people, or ideas that can be grouped together.

Sample 3

Suppose an activities director at a high school is scheduling the meeting times for the school's activities. The director lists the students in each group who participate in more than one activity.

Make an activities schedule so that no activities with a member in common meet at the same time.

Band	Honor Society	Student Council	Drama Club	Language Club	Science Club	School Newspaper
Epstein	Volk	Medved	Medved	Mason	Hilbert	Kwan
Benally	Cárdenas	Volk	Mason	Hilbert	Benally	Johns
Kwan		Hayden		Johns		Cárdenas
Hayden						Epstein

Sample Response

Problem Solving Strategy: Use a diagram.

Step 1 Draw a network diagram for the activities. Connect vertices that represent activities with members in common.

Step 2 Color the diagram so that vertices that are connected by an edge have different colors. Use as few colors as you can.

1. To start, color any vertex (for example, language club ●).
2. Use the same color to color all vertices *not* connected to each other or the first vertex.
3. Use a second color for one of the other vertices and repeat Step 2.
4. Use a third color and repeat Steps 2 and 3 until all the vertices are colored.

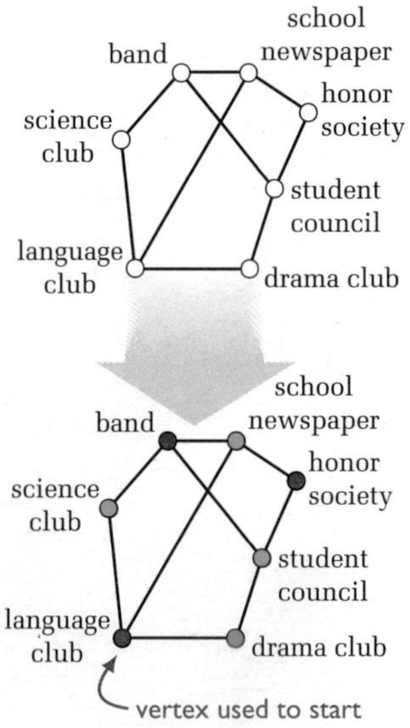

One possible schedule groups together activities represented by the vertices of the same color.

Meeting time A	Meeting time B	Meeting time C
band	student council	drama club
honor society	science club	
language club	school newspaper	

Answers to Talk it Over

9. Answers may vary. An example is given.

Meeting time A	Meeting time B	Meeting time C
band	student council	science club
honor society	language club	school newspaper
drama club		

10. Answers may vary. An example is given. Use each vertex as a starting point and find all possible colorings for the vertex. Record all resulting schedules and eliminate any duplicates.

Answers to Look Back

A diagram can show relationships and connections among information, for example, how cities are connected by airline flights, or how tasks can be assigned to individuals in a group. Coloring vertices in a diagram can help you to group the vertices, taking into account how they are connected.

Talk it Over

9. Find another activities schedule by coloring the diagram differently.
10. How do you think you can find all the possible schedules?

Look Back

How can a diagram help you organize information? How can coloring a diagram help you analyze information?

Now you are ready for: Exs. 18–31 on pp. 43–44

1-6 Exercises and Problems

1. **Reading** Which vertex in Sample 1 has the most edges connected to it? What does the number of edges connected to a vertex represent in this situation?

Use a matrix to represent the connections among the vertices of each diagram.

2. 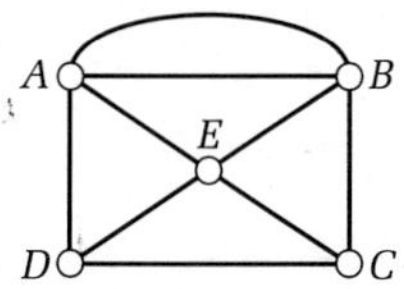

3. 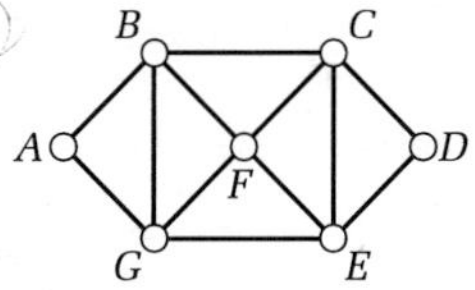

4. 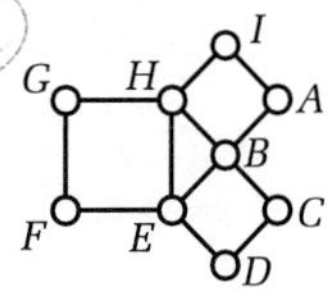

Draw a diagram to show how the points in each matrix are connected.

5. $$\begin{array}{c} \\ W \\ X \\ Y \\ Z \end{array}\begin{array}{c} \begin{matrix} W & X & Y & Z \end{matrix} \\ \begin{bmatrix} 0 & 1 & 1 & 1 \\ 1 & 0 & 0 & 1 \\ 1 & 0 & 0 & 0 \\ 1 & 1 & 0 & 0 \end{bmatrix} \end{array}$$

6. $$\begin{array}{c} \\ A \\ B \\ C \\ D \\ E \end{array}\begin{array}{c} \begin{matrix} A & B & C & D & E \end{matrix} \\ \begin{bmatrix} 0 & 1 & 1 & 0 & 1 \\ 1 & 0 & 0 & 1 & 0 \\ 1 & 0 & 0 & 1 & 0 \\ 0 & 1 & 1 & 0 & 0 \\ 1 & 0 & 0 & 0 & 0 \end{bmatrix} \end{array}$$

7. $$\begin{array}{c} \\ J \\ K \\ L \\ M \\ N \end{array}\begin{array}{c} \begin{matrix} J & K & L & M & N \end{matrix} \\ \begin{bmatrix} 0 & 0 & 0 & 1 & 1 \\ 0 & 0 & 1 & 1 & 1 \\ 0 & 1 & 0 & 1 & 0 \\ 1 & 1 & 1 & 0 & 1 \\ 1 & 1 & 0 & 1 & 0 \end{bmatrix} \end{array}$$

Each diagram matches a student with his or her preferences. Assign each task so that every student participates in one activity.

8. **Yearbook Staff Committees**

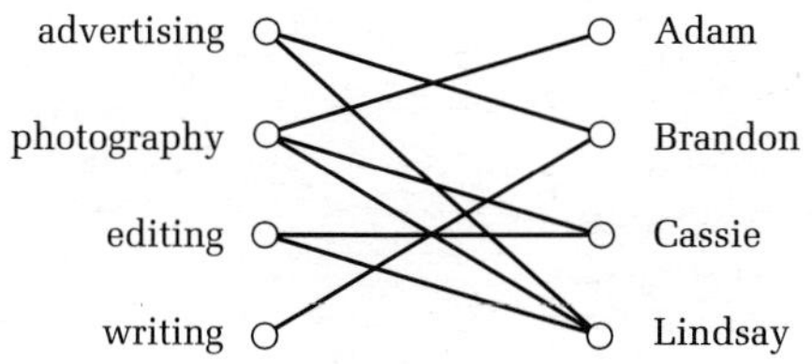

9. **Volunteer Club—Two Students per Location on Saturday**

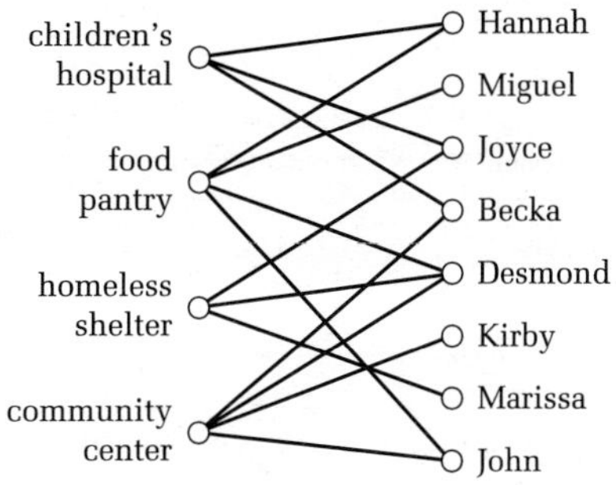

Look Back

A brief discussion of these questions will help students to summarize their understanding of how to use diagrams to organize and analyze information.

APPLYING

Suggested Assignment

Day 1

Standard 1–9, 15, 16

Extended 1–16

Day 2

Standard 18–31

Extended 18–31

Integrating the Strands

Algebra Exs. 2–7, 15, 17, 22–24, 31

Geometry Exs. 25–30

Discrete Mathematics Exs. 1–21, 31

Logic and Language Exs. 1, 15, 21, 31

Communication: Drawing

As students complete the exercises and problems for this section, they should begin to understand and appreciate the fact that network diagrams are powerful tools for communicating information.

Answers to Exercises and Problems

1. The vertices labeled "Cárdenas" and "volleyball" each have four edges connected to them. The number of edges connected to a story indicates the number of people who can cover it. The number of edges connected to a name indicate the number of stories the reporter can cover.

2–4. See answers in back of book.

5. 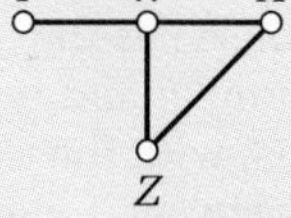

6. 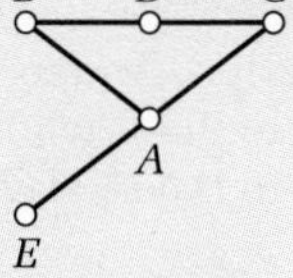

7. 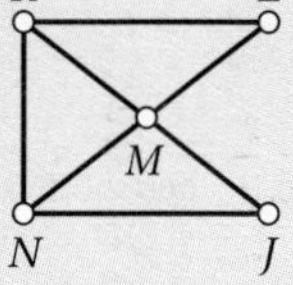

8. Adam: photography; Brandon: writing; Cassie: editing; Lindsay: advertising

9. Answers may vary. Examples are given. children's hospital: Joyce, Becka; food pantry: Miguel, Hannah; homeless shelter: Desmond, Marissa; community center: Kirby, John

Application

Network diagrams are a versatile problem-solving tool that can be applied to a variety of discrete mathematics situations. In Ex. 10, for example, architects may analyze a floor plan using a network diagram. Other examples of their use are in sports (Ex. 15) and in chemistry (Ex. 20).

Multicultural Note

In Africa, as in other large geographic regions, housing styles are as diverse as the peoples who live in them. African homes range in style from the rounded multi-unit grass thatch houses of the Zulu people to European-style homes in Nairobi, Kenya, to the multistory mud dwellings built by the Somolo people, to the Yoruba people's square-shaped homes centered on a courtyard. Experts have identified at least 20 distinct major groups of traditional architectural styles in Africa. Though there is no single type of "African" architecture, there are some characteristics common to many of the traditional dwellings. For example, homes in many African cultures are constructed to suit the needs of a family for a certain period of time, with the understanding that the housing needs of the next generation may be completely different. Also, traditional houses in many cultures are built as a collection of one-room units instead of as one building with many rooms.

Career Architects may draw a network diagram called an *access graph* to analyze the floor plan for a house or a building. In an access graph, each room is represented by a vertex and each edge connects two vertices when you can walk directly from one room to the other without passing through another room.

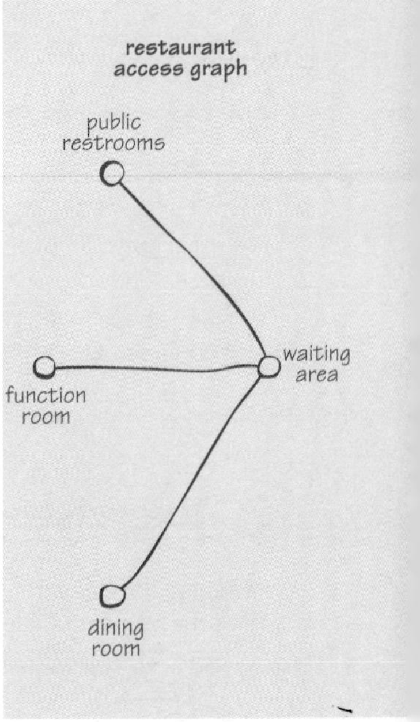

10. Suppose a client gives an architect the following preferences for the design of a restaurant.

Room	Has access to...
waiting area	dining room, function room, public restrooms
kitchen	dining room, employees' restrooms, function room
dining room	waiting area, kitchen, patio for outdoor dining
function room	waiting area, kitchen
patio for outdoor dining	dining room
employees' restrooms	kitchen
public restrooms	waiting area

a. Complete the access graph to show the connections between the rooms.

b. Do you think there is more than one possible graph? Explain.

For Exercises 11–13, draw an access graph for each plan of a traditional dwelling unit in Burkina Faso in west Africa.

11.

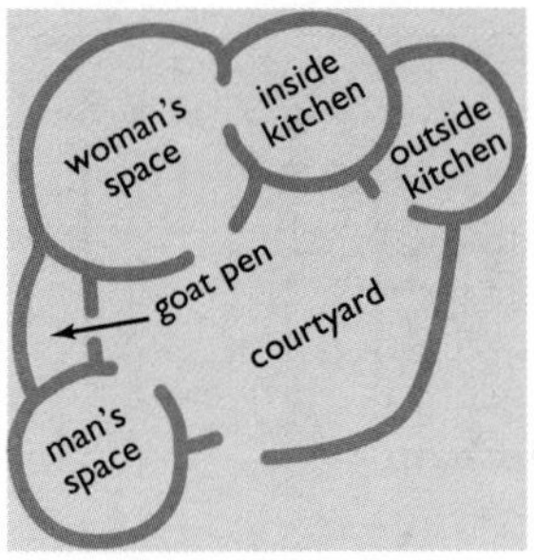

12.

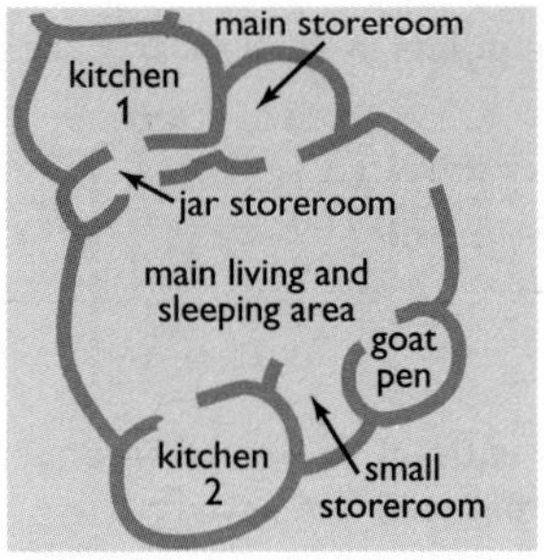

13.

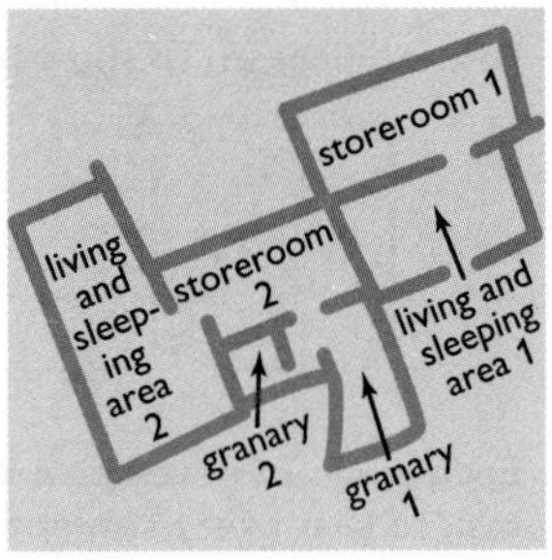

14. a. Draw an access graph for each house plan.

A.

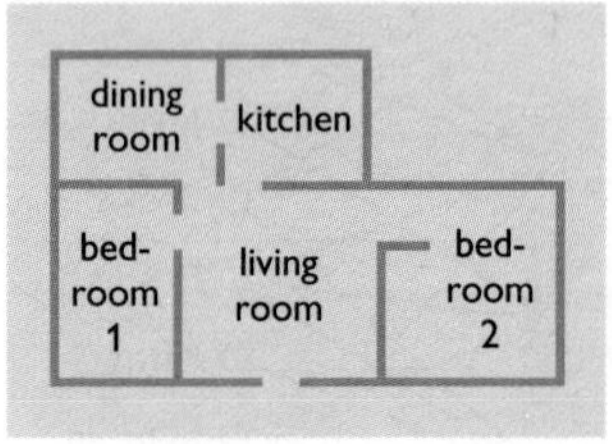

B.

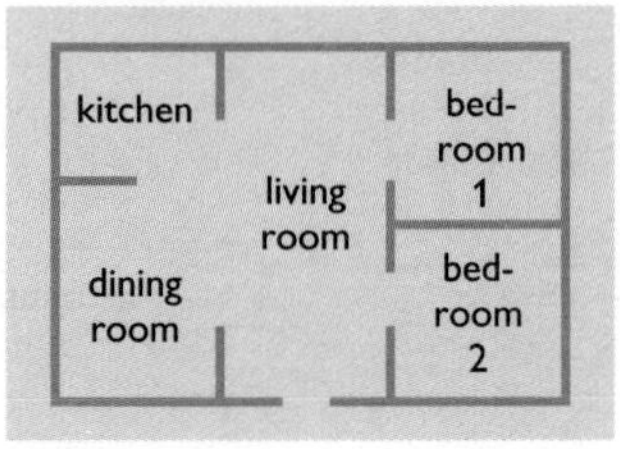

C.

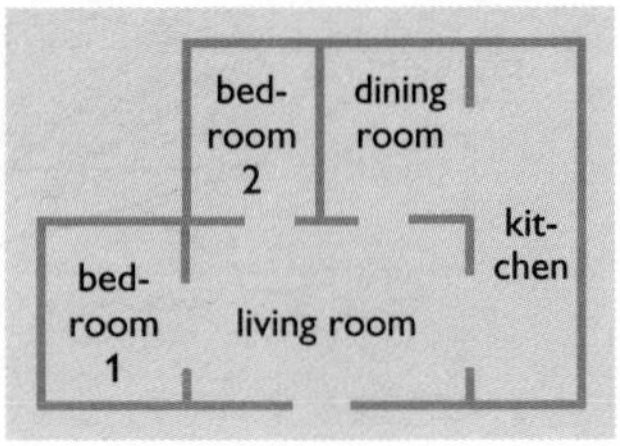

b. How are your access graphs alike? Do you see any patterns?

c. How are the house plans alike? How are they different?

Answers to Exercises and Problems

10. a.

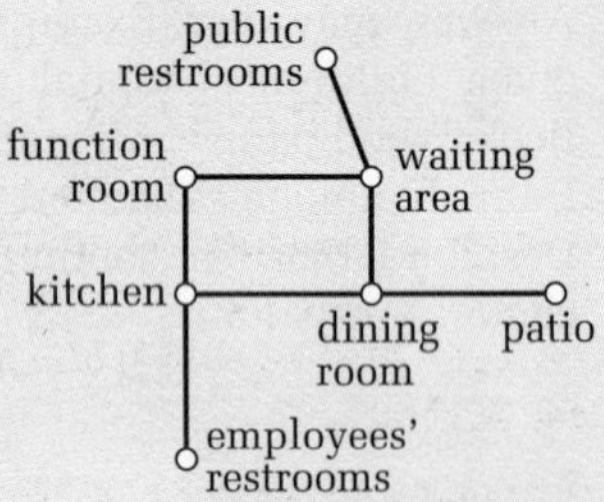

b. Yes; for example, there might be direct access from the function room to the public restrooms or to the dining room.

11–13. See answers in back of book.

14. a. A, B, C

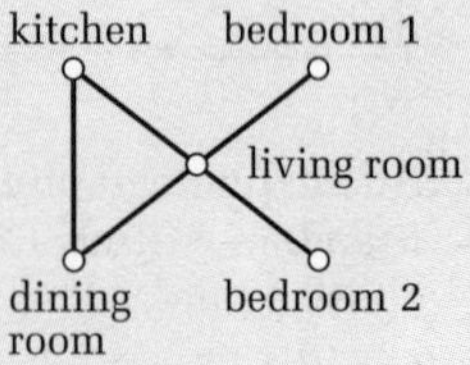

b. The access graphs are the same for each house. The living room connects to all other rooms in each house plan.

c. Each house has the same number and type of rooms, with the same access between rooms. The layouts differ.

15. a. First period: Anders passed to Blair, who passed to Davis. Davis passed the puck back to Blair, who scored. Second period: Anders passed to Blair, who passed back to Anders. Anders then passed to Rawson, Rawson passed to Davis, and Davis passed to Thierry. Thierry passed back to Davis, and Davis scored. Third period: Anders

connection to SPORTS

15. **Writing** Each diagram shows the passes that led to a goal during a hockey game.

First period

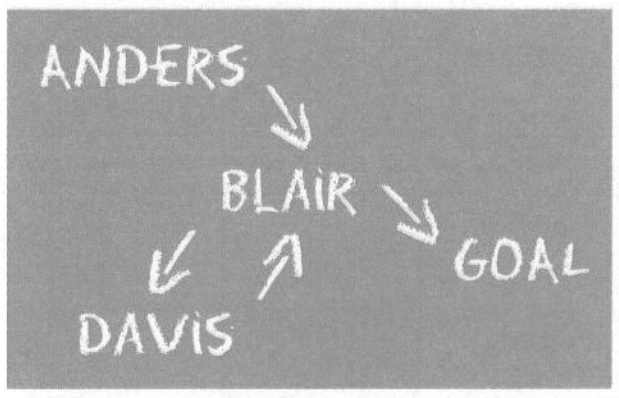

Second period

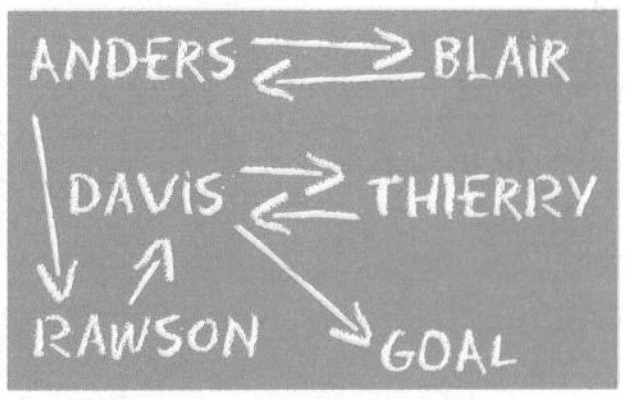

Third period

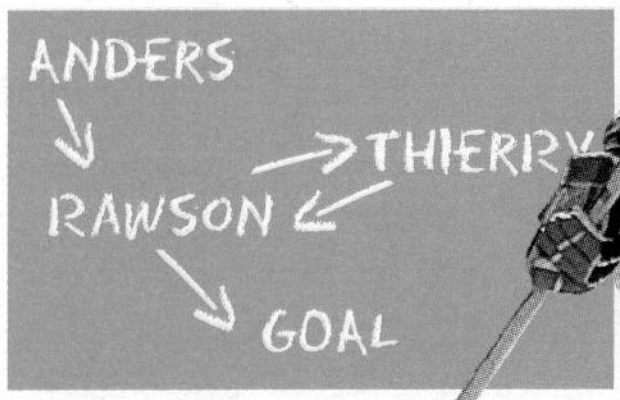

a. Describe how each goal was scored.

b. Show the information from each diagram in a matrix. For each matrix, explain what each row and each column represent.

c. Were any player(s) in all three plays? If so, which ones? Do you think it is easier to find out using the diagrams or the matrix? Why?

16. a. Draw a network diagram to represent the nonstop airline service shown.

b. How many nonstop flights are there?

c. Which airport is used the most by this airline?

Departure from ...	Provides nonstop service to ...		
Albuquerque	Salt Lake City	Denver	
Denver	Aspen	Cheyenne	Salt Lake City
Aspen	Denver		
Salt Lake City	Albuquerque	Denver	Cheyenne
Cheyenne	Denver	Aspen	

17. a. Use the flight schedule from Exercise 16. Make a matrix, M, to represent the nonstop service.

b TECHNOLOGY Use matrix multiplication to find M^2. This matrix will tell you the cities connected by a one-stop flight. A "1" means there is a one-stop, or 2-leg, flight. A "2" or "3" means there are two or three one-stop flights. A "0" means there are no one-stop flights. List all the one-stop flights. For each flight, tell where each stop occurs. (*Note*: Do not list flights that return to the city of departure.)

In Exercises 18 and 19, an edge is drawn between two vertices when two or more students are in the same class. How can exams be scheduled so that no student has a conflict?

18. **11th Grade Classes**

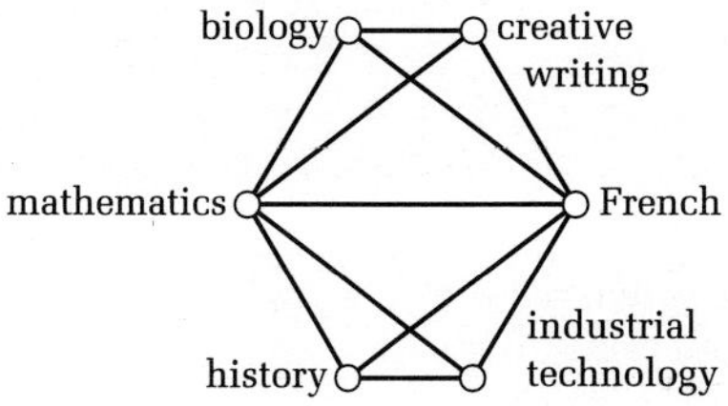

19. **12th Grade Classes**

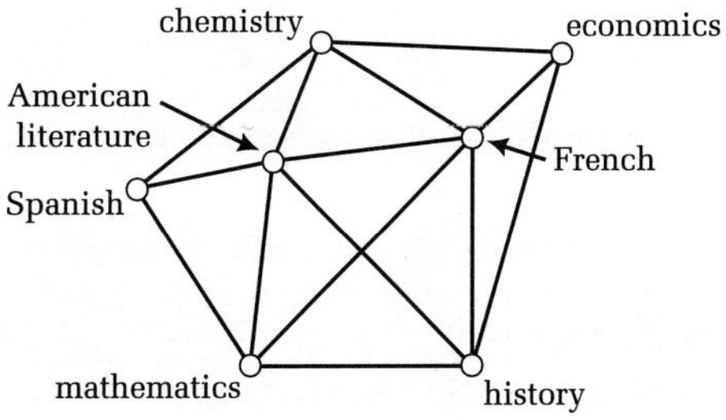

Visual Thinking

After doing Ex. 15, assign students to work in teams to create a diagram and matrix similar to those in the exercise and related to something in their own lives. Possibilities could include stops on different bus lines in their community, people in the group and their activities, and so on. Ask each group to explain its results to the class. This activity involves the visual skills of *exploration* and *communication.*

Answers to Exercises and Problems

passed to Rawson, who passed to Thierry. Thierry passed back to Rawson, who then scored.

b. The rows represent passes by the indicated player. The columns represent passes to the indicated player and the goal shot. The elements 1, 2, 3, and so on represent the order in which the passes were made; a 0 indicates no pass was made.

	Anders	Blair	Davis	Goal
Anders	0	1	0	0
Blair	0	0	2	4
Davis	0	3	0	0

	Anders	Blair	Davis	Rawson	Thierry	Goal
Anders	0	1	0	3	0	0
Blair	2	0	0	0	0	0
Davis	0	0	0	0	5	7
Rawson	0	0	4	0	0	0
Thierry	0	0	6	0	0	0

	Anders	Rawson	Thierry	Goal
Anders	0	1	0	0
Rawson	0	0	2	4
Thierry	0	3	0	0

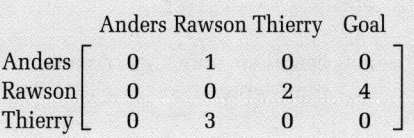

c. Anders; answers may vary.

16. a.

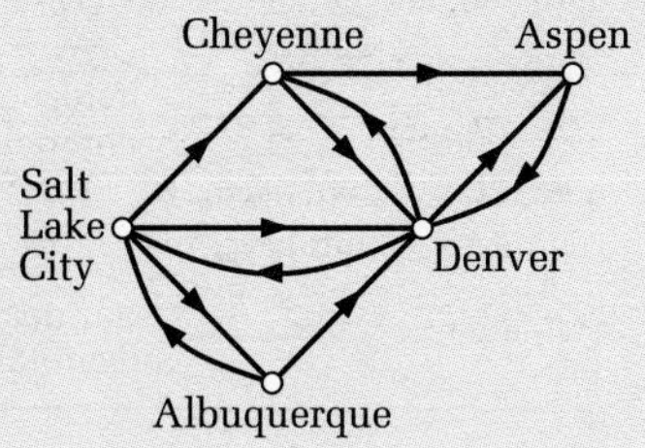

b. 11

c. Denver

17. See answers in back of book.

18, 19. Schedules may vary. Examples are given.

18. scheduling sequence: math; French; biology and history; creative writing and industrial technology

19. scheduling sequence: math; chemistry and history; economics and American literature; French and Spanish

Working on the Unit Project

When students develop their park design and proposal, they will need to use at least three math topics from this unit. As they work Ex. 31, they can record their thoughts about how to apply network diagrams to the design of a park.

Practice 6 For use with Section 1-6

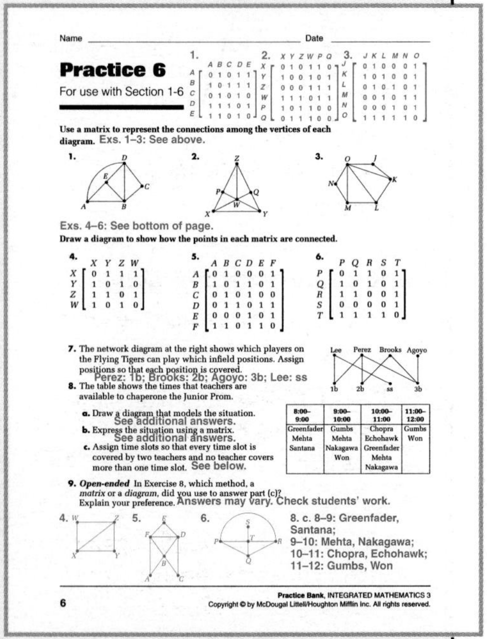

Name ______________________ Date ____________

Practice 6
For use with Section 1-6

1.

	A	B	C	D	E
A	0	1	0	1	1
B	1	0	1	1	1
C	0	1	0	1	0
D	1	1	1	0	1
E	1	1	0	1	0

2.

	X	Y	Z	W	P	Q
X	0	1	0	1	1	0
Y	1	0	0	1	0	1
Z	0	0	0	1	1	1
W	1	1	1	0	1	1
P	1	0	1	1	0	0
Q	0	1	1	1	0	0

3.

	J	K	L	M	N	O
J	0	1	0	0	0	1
K	1	0	1	0	0	1
L	0	1	0	1	0	1
M	0	0	1	0	1	1
N	0	0	0	1	0	1
O	1	1	1	1	1	0

Use a matrix to represent the connections among the vertices of each diagram. Exs. 1–3: See above.

1. 2. 3.

Exs. 4–6: See bottom of page.

Draw a diagram to show how the points in each matrix are connected.

4.

	X	Y	Z	W
X	0	1	1	1
Y	1	0	1	0
Z	1	1	0	1
W	1	0	1	0

5.

	A	B	C	D	E	F
A	0	1	0	0	0	1
B	1	0	1	1	0	1
C	0	1	0	1	0	0
D	0	1	1	0	1	1
E	0	0	0	1	0	1
F	1	1	0	1	1	0

6.

	P	Q	R	S	T
P	0	1	1	0	1
Q	1	0	1	0	1
R	1	1	0	0	1
S	0	0	0	0	1
T	1	1	1	1	0

7. The network diagram at the right shows which players on the Flying Tigers can play which infield positions. Assign positions so that each position is covered. Perez: 1b; Brooks: 2b; Agoyo: 3b; Lee: ss

8. The table shows the times that teachers are available to chaperone the Junior Prom.

   a. Draw a diagram that models the situation. See additional answers.
   b. Express the situation using a matrix. See additional answers.
   c. Assign time slots so that every time slot is covered by two teachers and no teacher covers more than one time slot. See below.

8:00–9:00	9:00–10:00	10:00–11:00	11:00–12:00
Greenfader Mehta Santana	Gumbs Mehta Nakagawa Won	Chopra Echohawk Greenfader Mehta Nakagawa	Gumbs Won

9. *Open-ended* In Exercise 8, which method, a *matrix* or a *diagram*, did you use to answer part (c)? Explain your preference. Answers may vary. Check students' work.

4. 5. 6.

8. c. 8–9: Greenfader, Santana; 9–10: Mehta, Nakagawa; 10–11: Chopra, Echohawk; 11–12: Gumbs, Won

6 Practice Bank, INTEGRATED MATHEMATICS 3

Chemistry In this table, a check (✔) means the chemicals in two groups can be stored together. A dash (—) means they cannot be stored together.

20. a. Represent the information in a network diagram.
    b. What does each vertex of your diagram represent? What does each edge represent?
    c. Find a way to group the chemicals for storage. How many storage areas are needed?

Can be stored with . . .

Chemical	Inorganic acids	Organic acids	Caustics	Amines and alkanolamines	Halogenated compounds	Alcohols and glycols	Phenols
Inorganic acids	✔	–	–	–	–	–	✔
Organic acids	–	✔	–	–	✔	✔	✔
Caustics	–	–	✔	✔	–	✔	–
Amines and alkanolamines	–	–	✔	✔	–	✔	–
Halogenated compounds	–	✔	–	–	✔	✔	✔
Alcohols and glycols	–	✔	✔	✔	✔	✔	✔
Phenols	✔	✔	–	–	✔	✔	✔

Ongoing ASSESSMENT

21. **Open-ended** Make an access graph of a building you are familiar with. Which rooms have access to the greatest number of other rooms? the least?

Review PREVIEW

Solve. *(Section 1-5)*

22. $-x - 3y - 2z = 1$
    $3x + 2y - z = -3$
    $-2x + y - 3z = 8$

23. $-2x - y - 3z = -10$
    $-x + 2y - z = -10$
    $4x - 3y - 2z = -5$

24. $-2x + y + z = -2$
    $-x + 5y - 3z = 13$
    $2x + 2y - z = 2$

Find each distance. *(Toolbox Skill 31)*

25. *DG*
26. *LP*
27. *BT*
28. *KM*
29. *TK*
30. *CP*

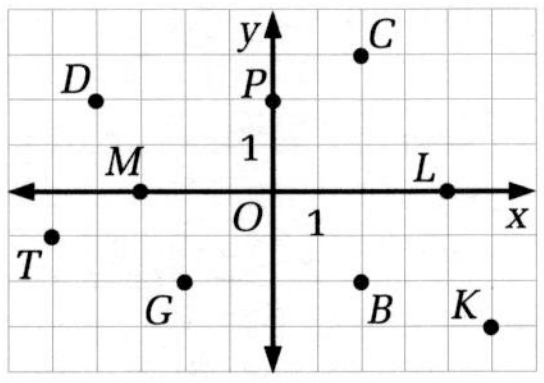

Working on the Unit Project

As you complete Exercise 31, think about how you can use network diagrams to design a park.

31. The table shows the amount of sunlight each type of flower needs.

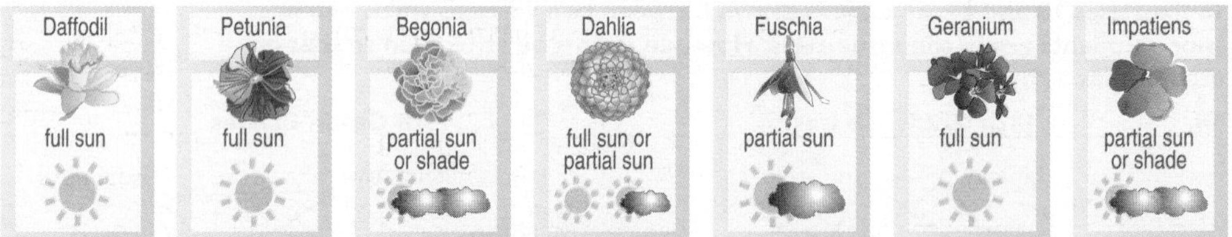

a. Show this information in a network diagram.
b. Show this information in a matrix.
c. Use your diagram or matrix to group together at least two types of flowers to plant in flower beds receiving each amount of sunlight.
d. **Writing** Do you think it is easier to see how to group the flowers in a diagram or in a matrix? Why?

Answers to Exercises and Problems

20. a. Diagrams may vary. An example is given.

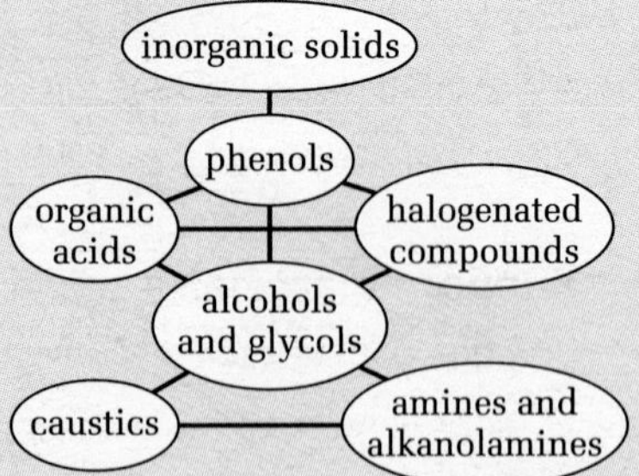

b. a chemical group; two chemical groups that can be stored together
c. Answers may vary. Check students' work. at least 3

21. Check students' work.
22. (–2, 1, –1)
23. (–1, –3, 5)
24. (–1, 0, –4)
25. $2\sqrt{5} \approx 4.5$
26. $2\sqrt{5} \approx 4.5$
27. $5\sqrt{2} \approx 7.1$
28. $\sqrt{73} \approx 8.5$
29. $2\sqrt{26} \approx 10.2$
30. $\sqrt{5} \approx 2.2$

31. a. Answers may vary. An example is given.

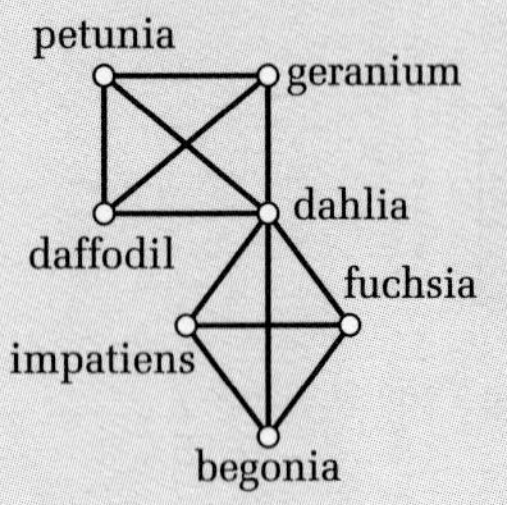

b.

	daffodil	petunia	begonia	dahlia	fuchsia	geranium	impatiens
daffodil	0	1	0	1	0	1	0
petunia	1	0	0	1	0	1	0
begonia	0	0	0	1	1	0	1
dahlia	1	1	1	0	1	1	1
fuchsia	0	0	1	1	0	0	1
geranium	1	1	0	1	0	0	0
impatiens	0	0	1	1	1	0	0

c. Answers may vary. An example is given. full sun: daffodils and geraniums; partial sun: fuchsias and dahlias; shade: begonias and impatiens
d. Answers may vary.

Section

1-7 Maximizing and Minimizing

Focus

Make quantities as large or as small as possible to solve real-life problems.

Peak Performance

Stickers on new cars give values for MPG (miles per gallon of gasoline) for city and highway driving.

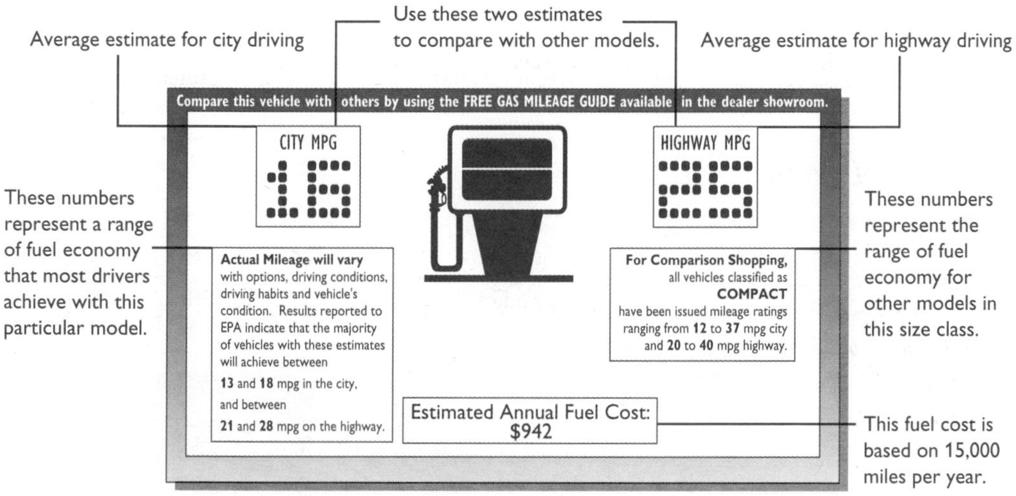

Average estimate for city driving
Use these two estimates to compare with other models.
Average estimate for highway driving
Compare this vehicle with others by using the FREE GAS MILEAGE GUIDE available in the dealer showroom.
CITY MPG
16
HIGHWAY MPG
25
These numbers represent a range of fuel economy that most drivers achieve with this particular model.
Actual Mileage will vary with options, driving conditions, driving habits and vehicle's condition. Results reported to EPA indicate that the majority of vehicles with these estimates will achieve between 13 and 18 mpg in the city, and between 21 and 28 mpg on the highway.
For Comparison Shopping, all vehicles classified as COMPACT have been issued mileage ratings ranging from 12 to 37 mpg city and 20 to 40 mpg highway.
These numbers represent the range of fuel economy for other models in this size class.
Estimated Annual Fuel Cost: $942
This fuel cost is based on 15,000 miles per year.

Talk it Over

The table gives the MPG for an average car at various speeds.

x = speed (mi/h)	y = MPG
15	21.0
25	30.0
35	33.6
45	33.5
55	30.3
65	24.9
75	20.0

1. What happens to MPG as the speed of a car increases?
2. Do you think a car has a **maximum**, or greatest, MPG? Why or why not? If so, at what speed(s) do you think the maximum MPG occurs?
3. Do you think a car has a **minimum**, or least, MPG? Why or why not? If so, at what speed(s) do you think the minimum MPG occurs?

In many situations it is important to make a quantity as large or as small as possible. For example, business managers want to make decisions that maximize income and minimize expenses. Drivers often want to maximize gas mileage or minimize the distance or time of a trip.

Answers to Talk it Over

1. It increases initially, reaches a maximum at about 35 mi/h, and then begins to decrease.
2. Yes; there must be a maximum. For example, there could be no speed at which a car achieves 200 MPG. The maximum appears to occur at about 35 mi/h.
3. Yes; clearly the MPG cannot be less than 0.

PLANNING

Objectives and Strands

See pages 1A and 1B.

Spiral Learning

See page 1B.

Materials List

- Graphics calculator or graphing software
- Smooth, heavy-weight cardboard
- Nylon fishing line
- Metal washer
- 3 identical plastic mugs
- Protractor

Recommended Pacing

Section 1-7 is a two-day lesson.

Day 1

Pages 45–46: Opening paragraph through Sample 1, *Exercises 1–11*

Day 2

Pages 46–48: Use Diagrams to Find Minimums through Look Back, *Exercises 12–22*

Toolbox References

- **Toolbox Skill 28:** Graphing Quadratic Functions

Extra Practice

See pages 608–610.

Warm-Up Exercises

Warm-Up Transparency 1-7

Support Materials

- Practice Bank: Practice 7
- Activity Bank: Enrichment 7
- Study Guide: Section 1-7
- Problem Bank: Problem Set 2
- Using IBM/Mac Plotter Plus Disk: Parabola Plotter
- Assessment Book: Quiz 1-7

Sample 1

The function $y = -0.014x^2 + 1.2x + 6.9$ models the relationship between MPG and driving speed for a car. Find the driving speed, to the nearest mile per hour, that maximizes MPG.

Sample Response

Student Resources Toolbox
p. 651 *Graphing*

Method ❶ Use a formula.

The function $y = -0.014x^2 + 1.2x + 6.9$ is a quadratic function, so its graph is a parabola. Since the coefficient of the x^2-term is negative, the parabola opens down. The vertex is on the line of symmetry of the parabola at the point where the value of the function is a maximum.

The speed that maximizes MPG is the x-coordinate of the vertex. To find this value, use the equation of the line of symmetry.

$$x = -\frac{b}{2a}$$ ← *a* and *b* are coefficients in the general quadratic function $y = ax^2 + bx + c$.

$$= -\frac{1.2}{(2)(-0.014)}$$ ← Substitute **1.2** for *b* and **−0.014** for *a*.

$$\approx 42.9$$

A driving speed of about 43 mi/h maximizes MPG.

Method ❷ Use a graph.

Graph $y = -0.014x^2 + 1.2x + 6.9$.

The x-coordinate of this point is the driving speed that maximizes MPG.

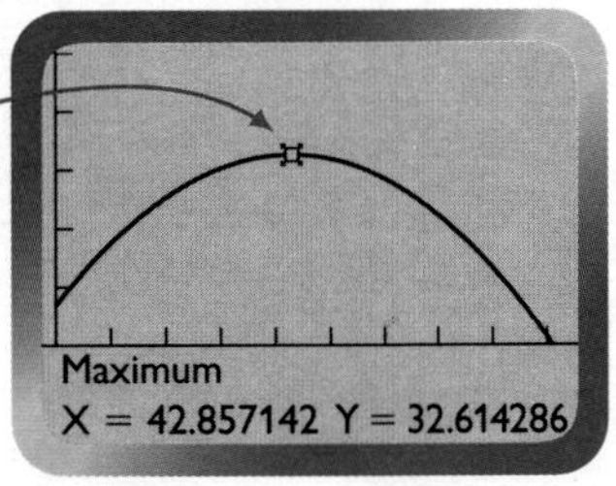

A driving speed of about 43 mi/h maximizes MPG.

Now you are ready for: Exs. 1–11 on pp. 49–50

Use Diagrams to Find Minimums

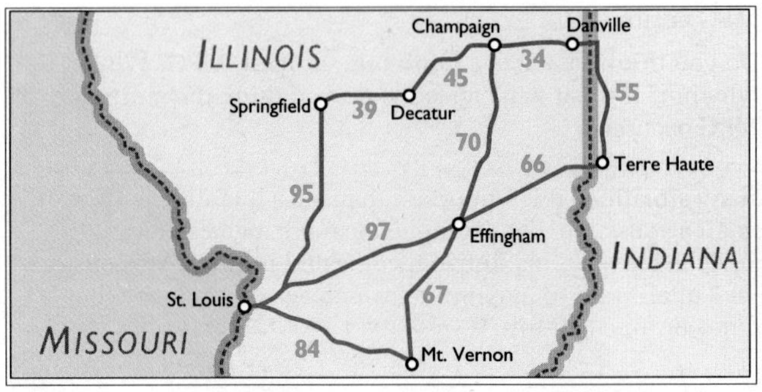

The map shows several cities in Illinois, Indiana, and Missouri, as well as the interstate highways connecting them. The driving distance between each pair of connected cities is labeled.

← All distances are given in miles.

TEACHING

Talk it Over

Questions 1–3 introduce students to the ideas of maximum and minimum values by using the gas mileage for a car.

Additional Sample

S1 If you ignore air resistance, the function $y = x - \frac{32}{3025}x^2$ models the relationship between the vertical height y (in feet) and the horizontal distance x (in feet) traveled by a projectile launched at a 45° angle with the ground and with an initial velocity of 55 ft/s. Find the horizontal distance of the projectile from the launch point when it reaches its maximum height.

Method 1. Use a formula. The graph is a parabola that opens down. The distance required is the x-coordinate of the vertex. $y = x - \frac{32}{3025}x^2$ is a quadratic function $y = ax^2 + bx + c$ for which $x = -\frac{b}{2a}$ is the x-coordinate of the vertex. In this particular case,

$$x = -\frac{1}{2\left(-\frac{32}{3025}\right)} = \frac{3025}{64} \approx 47.27.$$

The projectile reaches its maximum height when the horizontal distance from the launch point is about 47.27 ft.

Method 2. Use a graph. Graph $y = x - \frac{32}{3025}x^2$.

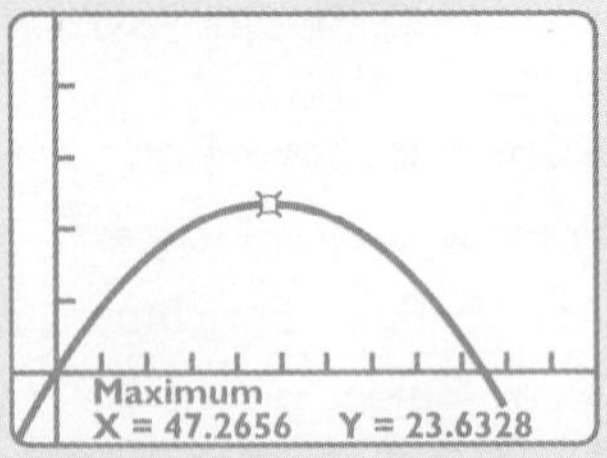

The projectile reaches its maximum height when the horizontal distance from the launch point is about 47.27 ft.

Talk it Over

4. Use the map on page 46. Suppose you want to travel from St. Louis to Danville using interstate highways.
   a. What do you think is the shortest route between the two cities? What is the length of this route?
   b. Describe the method you used in part (a) to find the route you think is shortest.

Sample 2 shows how to use an algorithm to find the minimum distance between two points.

Sample 2

Find the shortest route on interstate highways from St. Louis to Danville. Use the map on page 46.

Sample Response

Problem Solving Strategy: Use a diagram.

1 Draw a network diagram that models the map. Each vertex represents a city. Each edge represents an interstate highway. Distances do not need to be drawn to scale.

The red edges join labeled and unlabeled vertices.

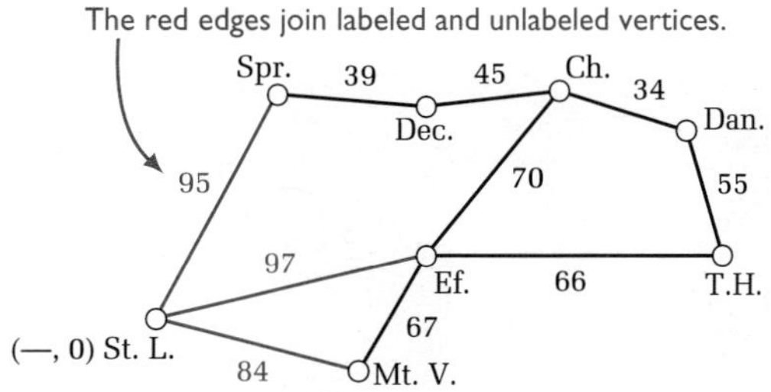

2 Label the starting point, St. Louis, with the ordered pair (—, 0).

3 For each edge that connects a labeled and an unlabeled vertex, find this sum:

$s =$ second entry of ordered pair for labeled vertex $+$ length of edge

Find s for each red edge.
St. L. to Spr.: 0 + 95 = 95
St. L. to Ef.: 0 + 97 = 97
St. L. to Mt. V.: 0 + 84 = 84 ⟵ This is the least sum.

Write the ordered pair (St. L., 84) next to "Mt. V."

4 Choose the edge from Step 3 that has the minimum sum s. Label the unlabeled vertex of that edge with this ordered pair:

(label of the other vertex of the edge, s)

5 Repeat Steps 3 and 4 until the vertex for Danville is labeled.

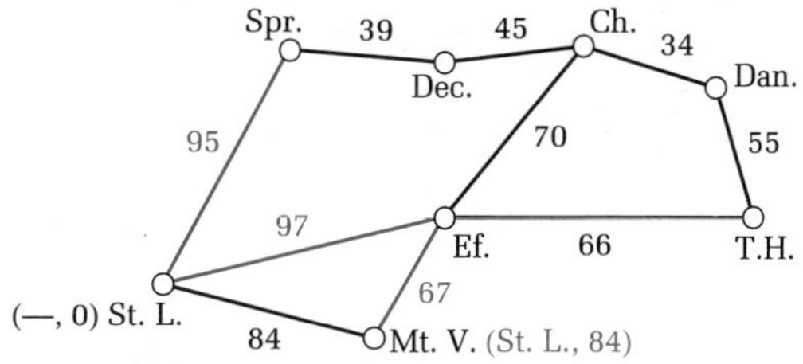

This is the least sum.
Find s for each red edge.
St. L. to Spr.: 0 + 95 = 95
St. L. to Ef.: 0 + 97 = 97
Mt. V. to Ef.: 84 + 67 = 151

Write the ordered pair (St. L., 95) next to "Spr."

Continued on next page.

Reasoning

In answering Talk it Over question 4, students will most likely come up with a number of different methods. Have students write their methods down on paper and then have them compare these methods to the algorithm in Sample 2.

Additional Sample

S2 This network diagram models roads connecting 10 cities. The cities are labeled with letters and the numbers indicate lengths of roads in miles. Find the shortest route from city A to city X.

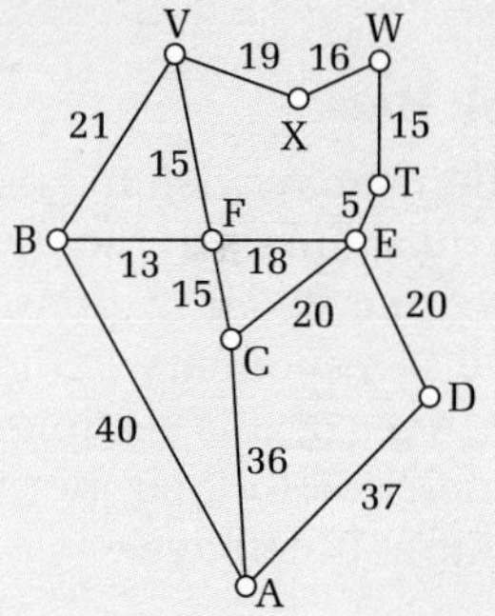

Label city A with the ordered pair (—, 0). Then follow steps 3 and 4 from Sample 2 until the vertex from city X is labeled.

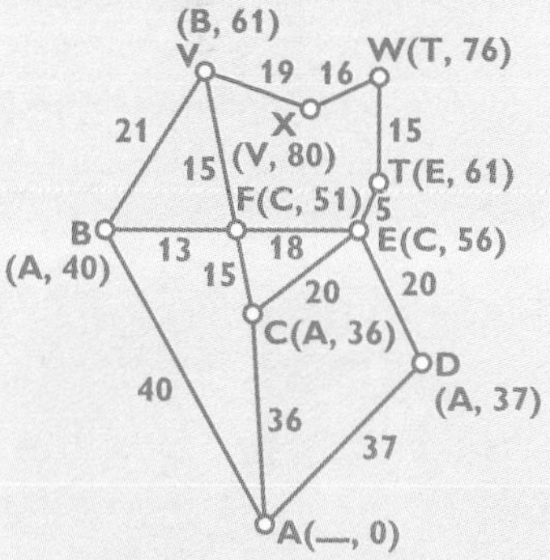

Starting from city X, use the first coordinates of the ordered pairs to backtrack to city A. The shortest route from city A to city X is A → B → V → X.

Answers to Talk it Over

4. a. St. Louis to Effingham to Champaign to Danville; 201 mi
   b. Answers may vary. An example is given. I used trial and error.

Talk it Over

Questions 5–7 have students analyze, and thus better understand, the algorithm presented in the Sample Response to Sample 2. Question 8 has students think about whether it is always possible to use a network diagram to find shortest routes.

Problem Solving

You may wish to have students discuss why making an organized list of possible routes from one vertex to another would or would not be a good approach for problems such as that of Sample 2.

Look Back

Students may wish to answer this question individually and in writing. Their responses can serve as a summary of the objectives of this section and be included in their journals for future reference.

6 When the vertex for Danville is labeled, use the ordered pairs to find the shortest route from St. Louis to Danville.

Starting from Danville, use the first coordinates of the ordered pairs to backtrack to St. Louis. This gives the shortest route.

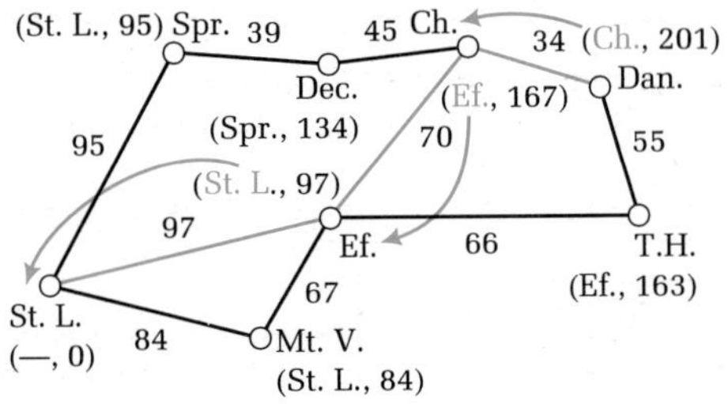

The shortest route from St. Louis to Danville on interstate highways is St. Louis to Effingham to Champaign to Danville.

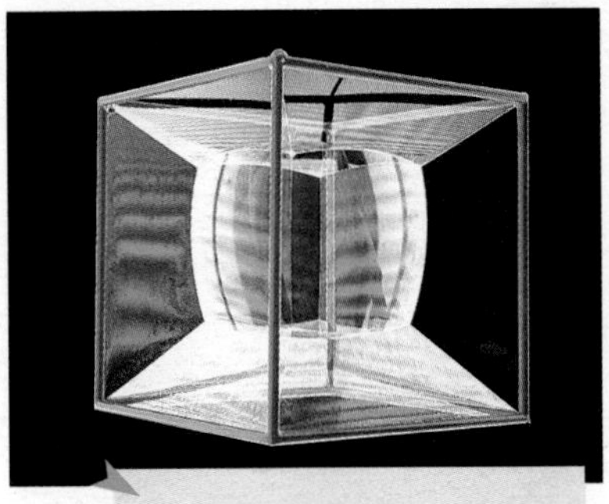

BY THE WAY...

Soap films continually form surfaces that minimize energy. Used with pegs and a map, soap films act as mechanical computers for constructing the shortest road network among a group of cities.

Talk it Over

Questions 5–8 are about Sample 2.

5. What is the length of the shortest route from St. Louis to Danville? What is the relationship between this length and the second entry of the ordered pair for Danville?
6. Why is the first entry of the ordered pair for St. Louis a dash instead of a vertex label?
7. How many times must you perform Steps 3 and 4 to find the shortest route from St. Louis to Danville?
8. Can you use the network diagram to find the shortest route from St. Louis to Terre Haute? from Springfield to Terre Haute? Why or why not?

Look Back

How can you use graphs of functions and network diagrams to maximize and minimize quantities?

Now you are ready for: Exs. 12–22 on pp. 50–52

Answers to Talk it Over

5. 201 mi; They are equal.
6. St. Louis is the starting point. There is no edge leading into this vertex.
7. 7 times
8. Yes. Yes. In both cases, you are able to do so, but it is unnecessarily complicated, since the shortest routes are fairly apparent.

Answers to Look Back

You can observe from the graph of a function where it achieves a maximum or minimum. You can use the algorithm described in Sample 2 to maximize or minimize quantities in network diagrams.

Answers to Exercises and Problems

1. Answers may vary. Examples are given. A business owner wants to maximize profit and minimize costs.
2. a. minimum
   b. –1
   c. –9
3. a. minimum
   b. 2
   c. –11
4. a. maximum
   b. 7
   c. 9
5. a. maximum
   b. –3
   c. 25.5
6. a. minimum
   b. –3.5
   c. 3.55

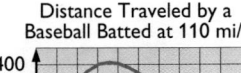

1-7 Exercises and Problems

1. **Reading** Describe one situation in which you want to maximize a quantity. Describe one situation in which you want to minimize a quantity.

For each quadratic function in Exercises 2–7:

a. Tell whether the function has a *maximum value* or a *minimum value*.

b. Find the value of x that maximizes or minimizes the function.

c. Find the function's maximum or minimum value.

2. $y = x^2 + 2x - 8$
3. $y = 3x^2 - 12x + 1$
4. $y = -x^2 + 14x - 40$
5. $y = -2.5x^2 - 15x + 3$
6. $y = 0.2x^2 + 1.4x + 6$
7. $y = -\frac{2}{3}x^2 + \frac{1}{6}x + \frac{1}{2}$

Distance Traveled by a Baseball Batted at 110 mi/h

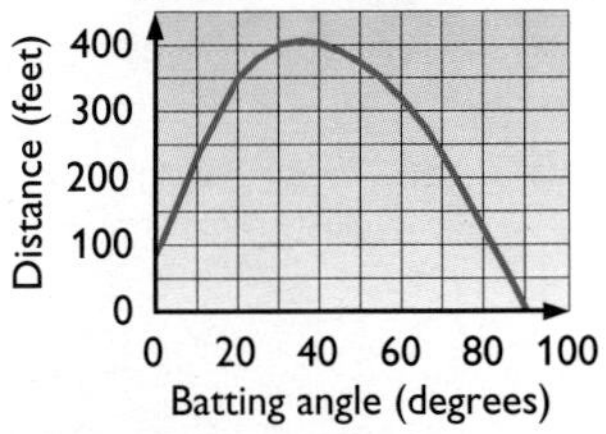

8. **Baseball** Use the graph.
   a. Describe what happens to the distance traveled by a baseball as the batting angle increases.
   b. What batting angle maximizes the distance a baseball travels? What is the maximum distance?
   c. **Writing** Why does the graph have a horizontal intercept at (90, 0)?

connection to **BIOLOGY**

Many birds migrate over long distances. Keeping energy use to a minimum during flights makes stored fat last longer, so the birds need to stop to feed less often. By controlling their flight speed, birds minimize the energy they use.

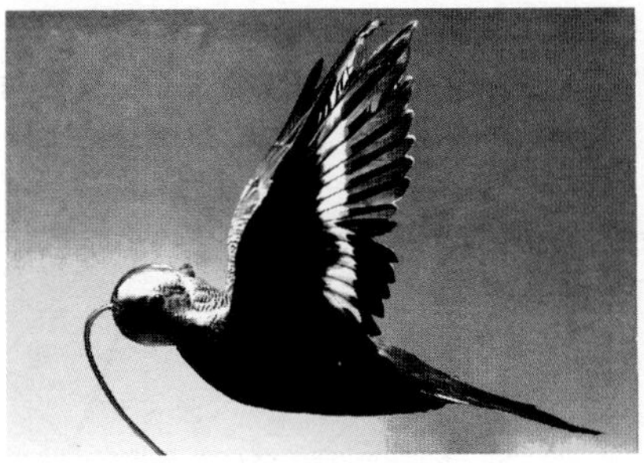

9. **Career** Zoologists gather data about the amount of oxygen birds use in flight as a measure of the amount of energy birds use. The amount of oxygen used by one type of parakeet from Australia each hour during flight is given by the function

$$f(s) = 1.96s^2 - 130s + 2920$$

where s is the flight speed in kilometers per hour (km/h) and $f(s)$ is measured in milliliters (mL).

   a. Find the flight speed that minimizes the amount of oxygen used each hour.
   b. **Writing** Explain why the average amount of oxygen this type of parakeet uses for every kilometer of distance it travels is given by the function

$$g(s) = \frac{f(s)}{s}$$

   where f is the function shown above.

   c. TECHNOLOGY Use a graphics calculator or graphing software. Graph $y = g(s)$. Is there a speed s that minimizes this function? Explain.

10. **Writing** Do you think a migrating bird should minimize the amount of oxygen it uses per unit of time or per unit of distance traveled? Explain.

APPLYING

Suggested Assignment

Day 1

Standard 1–9

Extended 1–10

Day 2

Standard 12–14, 16–22

Extended 12–14, 16–22

Integrating the Strands

Algebra Exs. 2–9, 11, 16–21

Functions Exs. 2–7, 9, 11

Geometry Exs. 11, 15

Discrete Mathematics Exs. 12, 13, 16, 22

Logic and Language Exs. 1, 8, 10, 14, 15

Career Note

Zoology is the science or branch of biology that deals with animals. A zoologist is a specialist in zoology. The animal kingdom is vast and ranges from single-celled animals called protozoa to fish, amphibians, reptiles, birds, and mammals. Students interested in animals can study zoology at a college or university. Employment opportunities are available as college teachers, in zoological gardens, where animals are kept for public exhibition, or in businesses and research laboratories, where animal life and behavior are studied.

Answers to Exercises and Problems

7. a. maximum
   b. $\frac{1}{8}$ c. $\frac{49}{96}$

8. a. The distance increases initially until it reaches a maximum, then decreases.
   b. about 35°; about 400 ft
   c. If the angle is 90°, the ball goes straight up and lands on the spot from which it was hit.

9. a. about 33.2 km/h
   b. $\frac{f(s)}{s} = \frac{\text{mL/h}}{\text{km/h}} = \frac{\text{mL}}{\text{km}}$
   c.

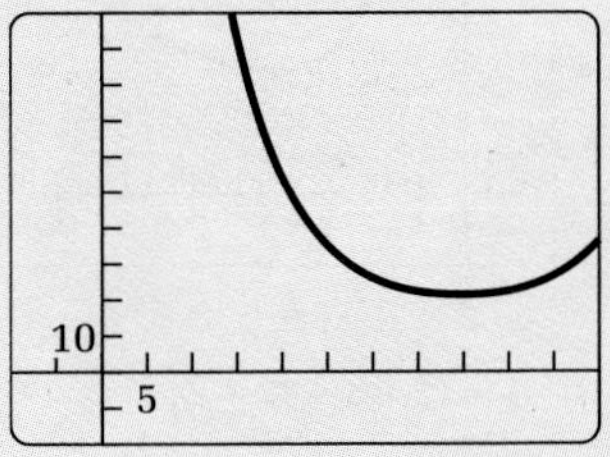

Yes; s and $g(s)$ must both be nonnegative and the first-quadrant portion of the graph has a lowest point at about (39.4, 21.3).

10. Answers may vary. An example is given. Since migrating birds stay in the air for very long periods of time, they should probably minimize oxygen used per unit of time.

Communication: Discussion

Ex. 14 will make for a good class discussion. All students who answer the question will think they have found the minimum number of trips, although actually they may not have. By discussing and comparing answers, students can see which algorithm or algorithms generate the minimum number of trips.

11. **Manufacturing** Suppose a food manufacturer is designing a can to hold 500 cm^3 of soup. To minimize tin and steel costs, the can's radius r and height h should be chosen so that the surface area of the can is as small as possible.

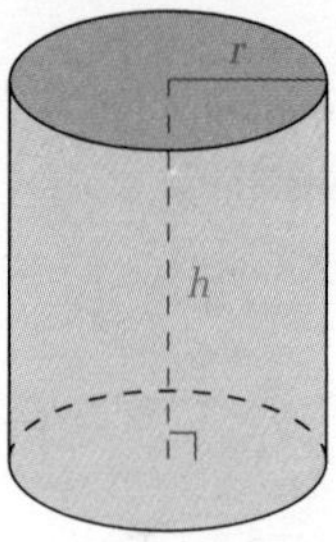

a. The volume V of a cylindrical can is given by the formula

$$V = \pi r^2 h.$$

Use the fact that $V = 500$ cm^3 to express h in terms of r.

b. The surface area S of a cylindrical can is given by the formula

$$S = 2\pi r^2 + 2\pi rh.$$

Use your expression for h from part (a) to write S as a function of r.

c. TECHNOLOGY Use a graphics calculator or graphing software. Graph the function you wrote in part (b). Find the radius that minimizes the surface area of the soup can. Then use your expression for h from part (a) to find the height that minimizes the surface area.

BY THE WAY...

Jean E. Taylor, a mathematics professor at Rutgers University, has solved several major problems involving the minimization of surface areas. She has written an algorithm for predicting the shapes that soap bubbles and crystals take in order to minimize energy.

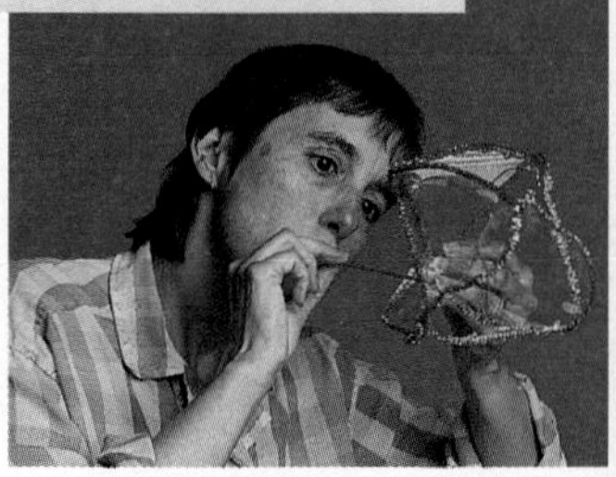

For each network diagram, find the shortest route from vertex *A* to vertex *Z*.

12.

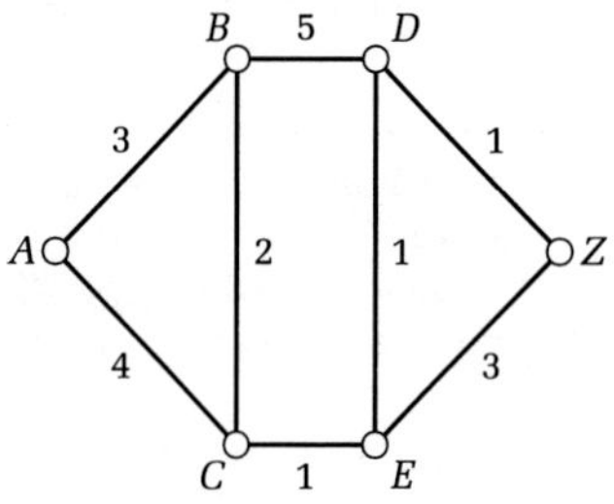

13.

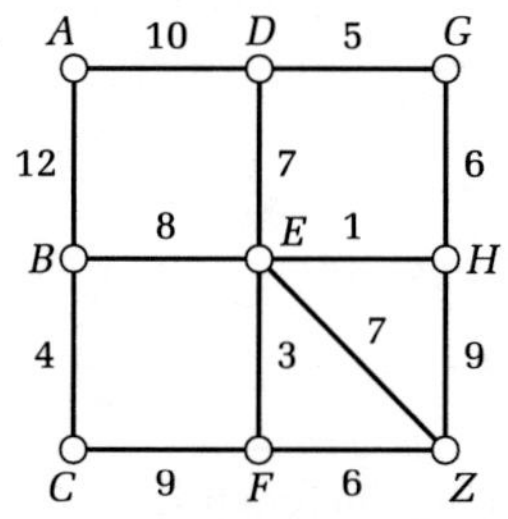

14. A classic puzzle problem asks you to suppose you need to transport a wolf, a goat, and a cabbage across a river by boat. You can take only one of them across at a time. The wolf and the goat cannot be left alone on the same side of the river. Also, the goat and the cabbage cannot be left alone on the same side. Write instructions for transporting the wolf, goat, and cabbage across the river so that the number of trips is minimized. How many trips are needed?

Answers to Exercises and Problems

11. a. $h = \frac{500}{\pi r^2}$

b. $S(r) = 2\pi r^2 + \frac{1000}{r}$

c.

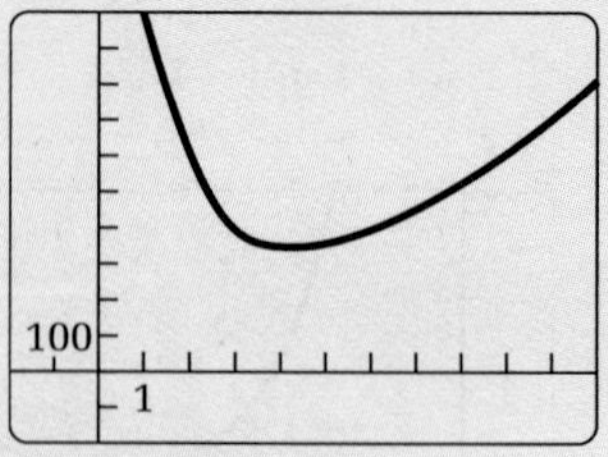

$r \approx 4.3$; $h \approx 8.6$

12. $A - C - E - D - Z$; 7

13. $A - D - E - Z$; 24

14. Take the goat. Go back and get the cabbage. Take it across the river, leave it there and bring the goat back. Leave the goat, and take the wolf across the river. Leave the wolf with the cabbage and go back to get the goat. Take the goat. This requires seven trips across the river.

Ongoing **ASSESSMENT**

15. **Group Activity** Work with another student. You will need a large piece of smooth, heavy-weight cardboard; nylon fishing line; a metal washer; and three identical plastic mugs.

   a. On the cardboard, mark three widely spaced points A, B, and C that are the vertices of an acute scalene triangle.

   b. Mark the point P that you think minimizes this sum:

$$d = d_A + d_B + d_C$$

   Find the value of d for the point P you chose.

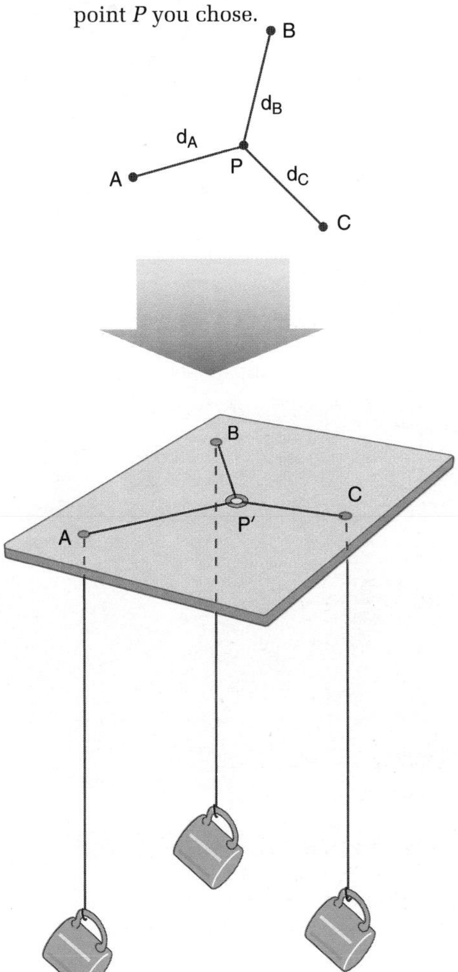

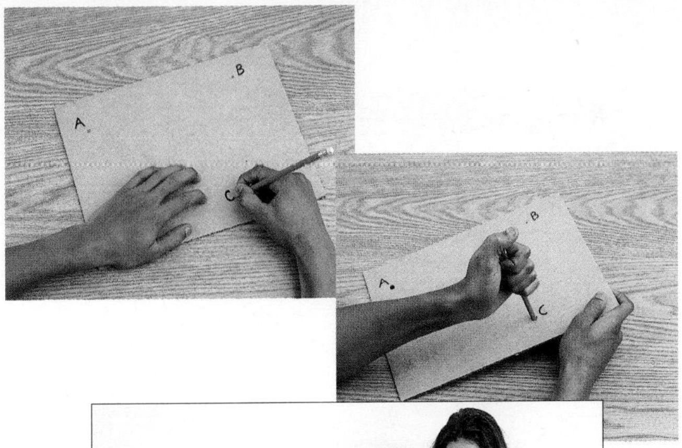

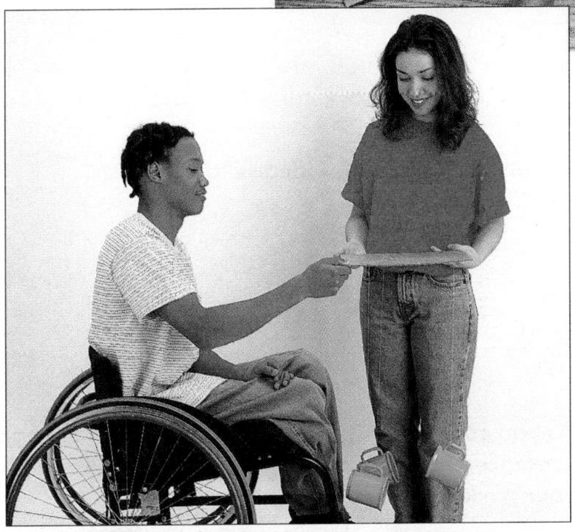

   c. Punch holes through the cardboard at points A, B, and C. Cut three equal lengths of nylon fishing line. Thread one piece through each hole and tie one end to a washer, as shown in the diagram. Tie the other end of each piece of string to one of the three mugs.

   d. Hold the piece of cardboard above the ground so that the mugs are hanging freely. Shake the cardboard gently until the washer settles down. Mark the point P' where the center of the washer ends up. This point minimizes the sum d in part (b).

   e. Find the value of d for your point P'. Compare this value of d with the value you found in part (b).

   f. Use a protractor to measure $\angle AP'B$, $\angle AP'C$, and $\angle BP'C$. What do you notice about the angle measures?

   g. **Open-ended** Describe a real-life problem that you can solve using this activity.

Using Manipulatives

Ex. 15 leads students to discover the fact that the point with the minimum total distance from the three vertices of a triangle is the one that forms three 120° angles with the vertices.

Answers to Exercises and Problems

15. a–f. Check students' work.

g. Answers may vary. An example is given. Where can you locate a warehouse that distributes merchandise to three outlet stores so that the sum of the distances between the warehouse and stores is minimized?

Review PREVIEW

16. Write a matrix that shows the connections among the vertices in this network diagram. *(Section 1-6)*

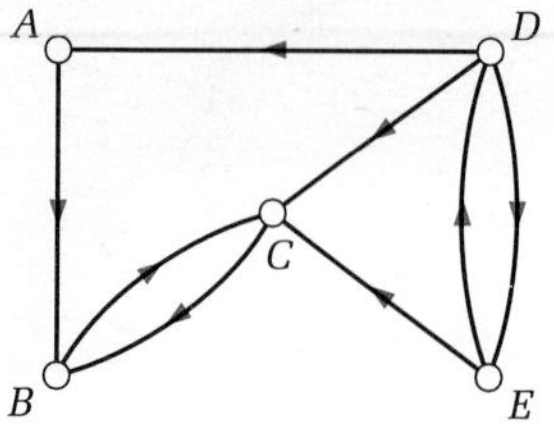

17. Use this matrix to draw a network diagram showing the connections among vertices P, Q, R, and S. *(Section 1-6)*

	P	Q	R	S
P	0	1	1	0
Q	1	0	0	1
R	0	1	0	1
S	1	0	1	0

Graph each system of inequalities. *(Toolbox Skill 27)*

18. $y \geq 2x$
$y \leq -3x - 1$

19. $x \geq 5$
$y < 2x - 4$

20. $y \leq 3x + 2$
$y \geq -2x + 7$

21. $y + \frac{1}{2}x < 1$
$x + 2y \leq 6$

Working on the Unit Project

As you complete Exercise 22, think about how you can use the strategies presented in this section to maximize and minimize quantities related to your park proposal.

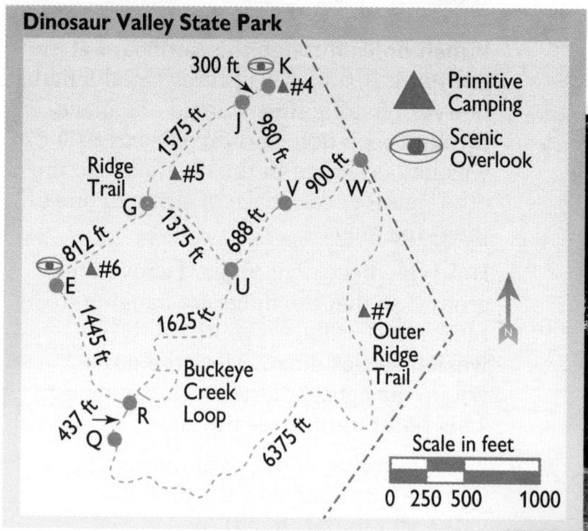

22. Use the map. All distances are given in feet. Find the shortest route from point Q to campsite #4.

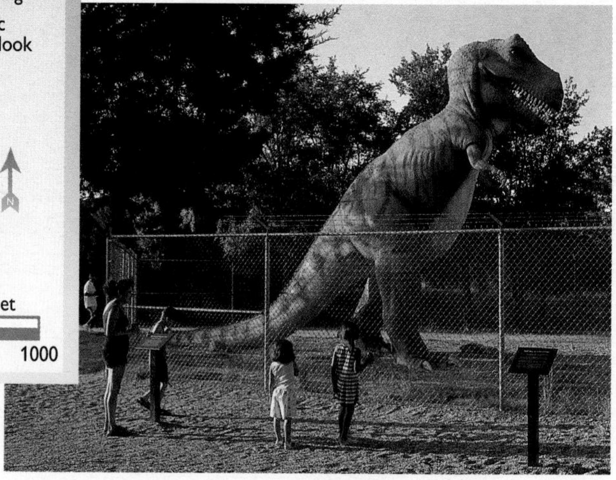

Practice 7 For use with Section 1-7

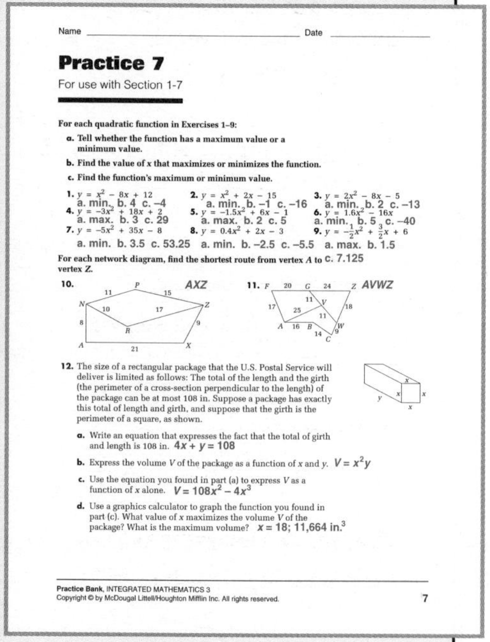

Name ________________ Date ________

Practice 7

For use with Section 1-7

For each quadratic function in Exercises 1–9:

a. Tell whether the function has a maximum value or a minimum value.

b. Find the value of x that maximizes or minimizes the function.

c. Find the function's maximum or minimum value.

1. $y = x^2 - 8x + 12$ a. min. b. 4 c. –4

2. $y = x^2 + 2x - 15$ a. min. b. –1 c. –16

3. $y = 2x^2 - 8x - 5$ a. min. b. 2 c. –13

4. $y = -3x^2 + 18x + 2$ a. max. b. 3 c. 29

5. $y = -1.5x^2 + 6x - 1$ a. max. b. 2 c. 5

6. $y = 1.6x^2 - 16x$ a. min. b. 5 c. –40

7. $y = -5x^2 + 35x - 8$ a. min. b. 3.5 c. 53.25

8. $y = 0.4x^2 + 2x - 3$ a. min. b. –2.5 c. –5.5

9. $y = -\frac{1}{2}x^2 + \frac{3}{2}x + 6$ a. max. b. 1.5 c. 7.125

For each network diagram, find the shortest route from vertex A to vertex Z.

10. AXZ

11. AVWZ

12. The size of a rectangular package that the U.S. Postal Service will deliver is limited as follows: The total of the length and the girth (the perimeter of a cross-section perpendicular to the length) of the package can be at most 108 in. Suppose a package has exactly this total of length and girth, and suppose that the girth is the perimeter of a square, as shown.

a. Write an equation that expresses the fact that the total of girth and length is 108 in. $4x + y = 108$

b. Express the volume V of the package as a function of x and y. $V = x^2y$

c. Use the equation you found in part (a) to express V as a function of x alone. $V = 108x^2 - 4x^3$

d. Use a graphics calculator to graph the function you found in part (c). What value of x maximizes the volume V of the package? What is the maximum volume? $x = 18$; 11,664 in.^{3}

Practice Bank, INTEGRATED MATHEMATICS 3

7

Answers to Exercises and Problems

16.

	A	B	C	D	E
A	0	1	0	0	0
B	0	0	1	0	0
C	0	1	0	0	0
D	1	0	1	0	1
E	0	0	1	1	0

17.

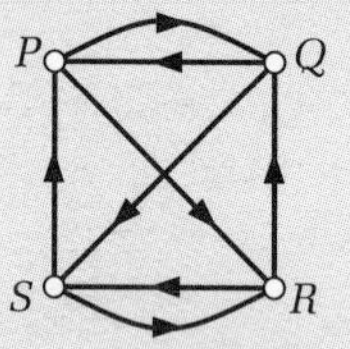

18.

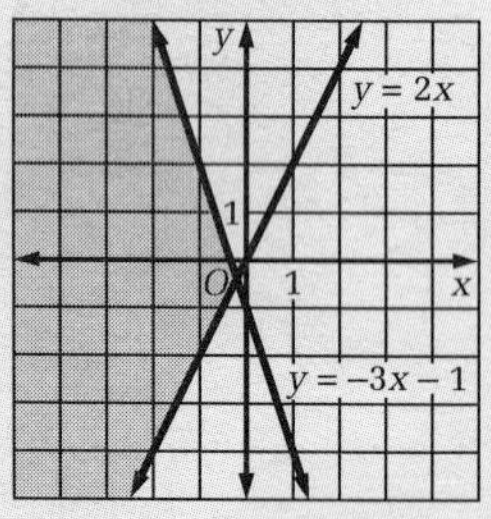

19.

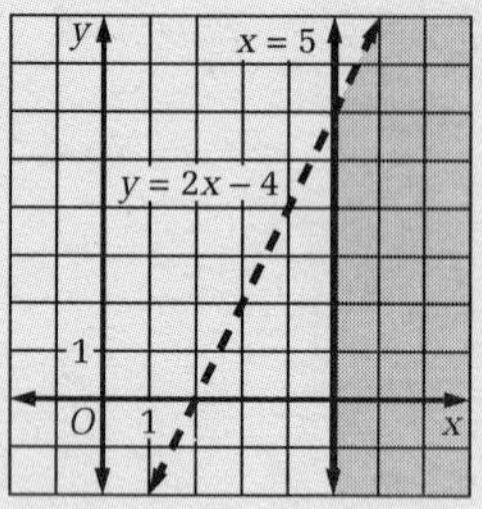

20.

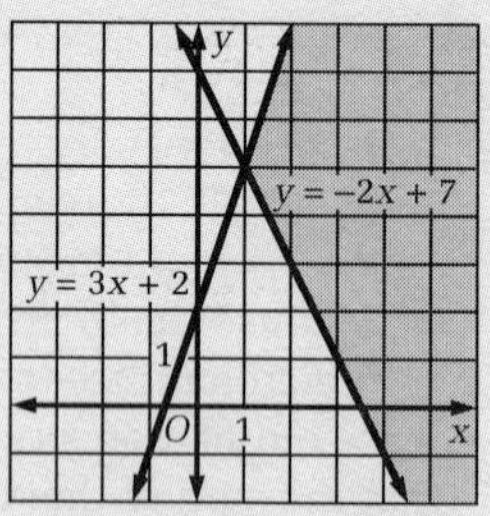

21.

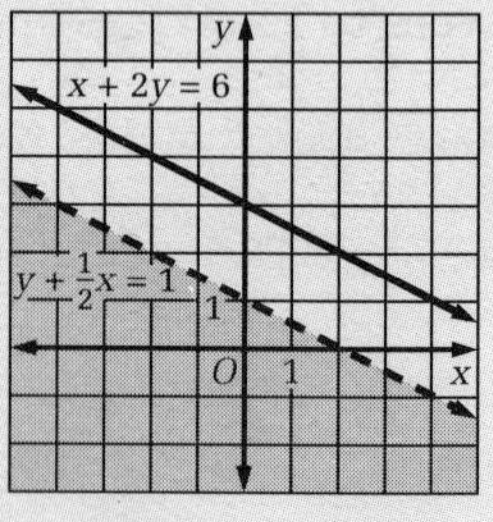

22. Q – R – U – V – J – K; 4030 ft

Section 1-8

Linear Programming

Focus
Use systems of inequalities to model situations and find maximum and minimum values.

A WINNING Combination

Advertise on WMMR	One 60-second ad costs ...
Morning (A.M.)	$200
Afternoon/evening (P.M.)	$50

Talk it Over

1. Look at WMMR's rate sheet. Why do you think A.M. ads are more expensive than P.M. ads?
2. Suppose you are the marketing manager for a store that sells casual clothing. Would you advertise a sale on WMMR in the morning? in the afternoon/evening? Why?
3. For each combination, find the total number of ads and the total cost of running the ads.
   a. ten A.M. ads and twelve P.M. ads
   b. six A.M. ads and ten P.M. ads
   c. ten A.M. ads and eight P.M. ads
   d. two A.M. ads and twenty P.M. ads
4. A **linear combination** is an expression in the form $ax + by$. Let x = the number of A.M. ads. Let y = the number of P.M. ads. Tell what each linear combination represents.
   a. $x + y$
   b. $200x + 50y$
5. You decide on two conditions. To which linear combination, $x + y$ or $200x + 50y$, does Condition 1 apply? Condition 2?
6. Which combination(s) of ads in question 3 meet(s) at least one of the conditions? Which meet(s) both conditions?
7. Can either x or y be negative in this situation? Why or why not?

Condition 1
You will spend at most $2200.
Condition 2
You will run at most 20 ads.

PLANNING

Objectives and Strands
See pages 1A and 1B.

Spiral Learning
See page 1B.

Materials List
- Graph paper

Recommended Pacing
Section 1-8 is a two-day lesson.

Day 1
Pages 53–55: Talk it Over 1 through Talk it Over 9, *Exercises 1–9*

Day 2
Pages 55–57: Middle of page 55 through Look Back, *Exercises 10–23*

Toolbox References
- **Toolbox Skill 21:** Solving a Linear System

Extra Practice
See pages 608–610.

Warm-Up Exercises
- Warm-Up Transparency 1-8

Support Materials
- Practice Bank: Practice 8
- Activity Bank: Enrichment 8
- Study Guide: Section 1-8
- Problem Bank: Problem Set 2
- Explorations Lab Manual: Additional Exploration 1, Diagram Master 2
- Using Plotter Plus: Linear Programming: the Corner-Point Principle
- Using IBM/Mac Plotter Plus Disk: Inequality Plotter
- Assessment Book: Quiz 1-8, Test 2

Answers to Talk it Over

1. Answers may vary. An example is given. I think more people listen to the radio in the morning than in the afternoon or evening, because people listen while getting ready for and going to work or school.
2. Answers may vary.
3. a. 22 ads; $2600
   b. 16 ads; $1700
   c. 18 ads; $2400
   d. 22 ads; $1400
4. a. the total number of ads
   b. the total cost of x A.M. ads and y P.M. ads
5. $200x + 50y$; $x + y$
6. b, c, and d; b
7. No; x and y represent numbers of ads, which are always nonnegative.

TEACHING

Talk it Over

Questions 1–7 introduce students to many of the concepts involved in linear programming, such as linear combinations and constraints, through a real-life situation. This situation is used throughout the section to explore linear programming.

Additional Sample

S1 A small company that makes animals for toy stores uses 1.5 m of yellow cloth to make a lion and 3 m to make a giraffe. The lion's legs require 1.5 m of a special plastic tubing, while the giraffe's legs require 6 m. The company has in stock 258 m of yellow cloth and 480 m of the plastic tubing. Assuming all other necessary materials are available, the number of the two animals that could be made without buying more cloth or tubing is the feasible region for the following system:

$x \geq 0$
$y \geq 0$
$1.5x + 3y \leq 258$
$1.5x + 6y \leq 480$

where x = number of lions and y = number of giraffes. Graph the feasible region for this system. Label each vertex with its coordinates.

Since $x \geq 0$ and $y \geq 0$, the feasible region lies in the first quadrant. Graph the equations $1.5x + 3y = 258$ and $1.5x + 6y = 480$. In the first quadrant, shade the region that consists of all points that lie below both. Also, mark the points on the line graphs that lie on the boundary of this region. Use algebra or a graphics calculator to find the coordinates of vertices of the feasible region.

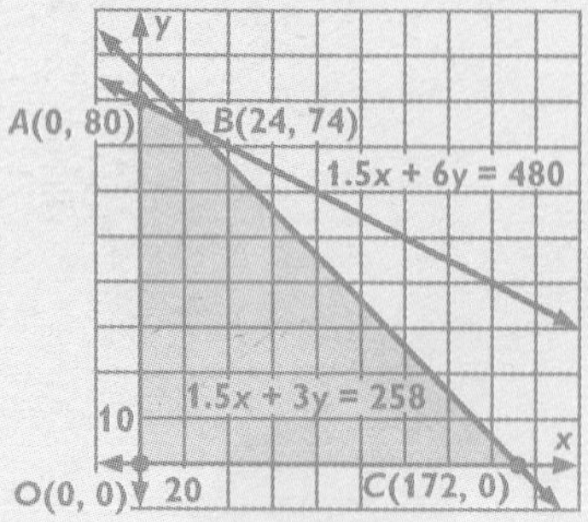

Any condition that must be met by a variable or by a linear combination of variables is a **constraint.**

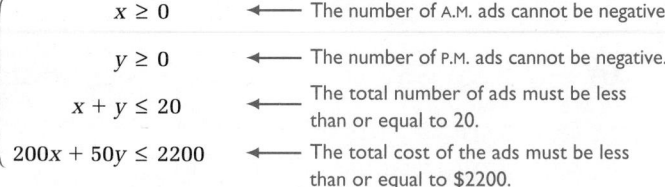

You can use a *system of linear inequalities* to express all the constraints on the numbers of ads.

$$\begin{cases} x \geq 0 \\ y \geq 0 \\ x + y \leq 20 \\ 200x + 50y \leq 2200 \end{cases}$$

- $x \geq 0$ ← The number of A.M. ads cannot be negative.
- $y \geq 0$ ← The number of P.M. ads cannot be negative.
- $x + y \leq 20$ ← The total number of ads must be less than or equal to 20.
- $200x + 50y \leq 2200$ ← The total cost of the ads must be less than or equal to $2200.

The graph of the solution of the system of inequalities includes all points that represent possible combinations of ads that meet all the constraints. The graph is called the **feasible region.**

Sample 1

Student Resources Toolbox
p. 644 *Solving Inequalities*

Graph the feasible region for the system of inequalities shown above. Label each vertex with its coordinates.

Sample Response

You must find the points that make all four inequalities true.

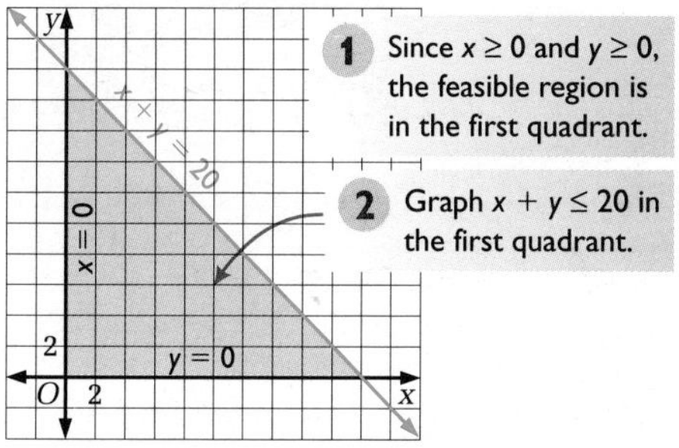

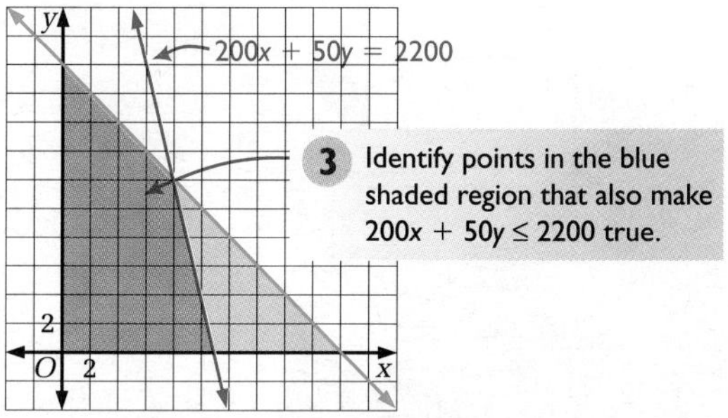

The feasible region consists of all points on or inside quadrilateral *ABCO*. You can find the coordinates of each vertex by solving a system of equations.

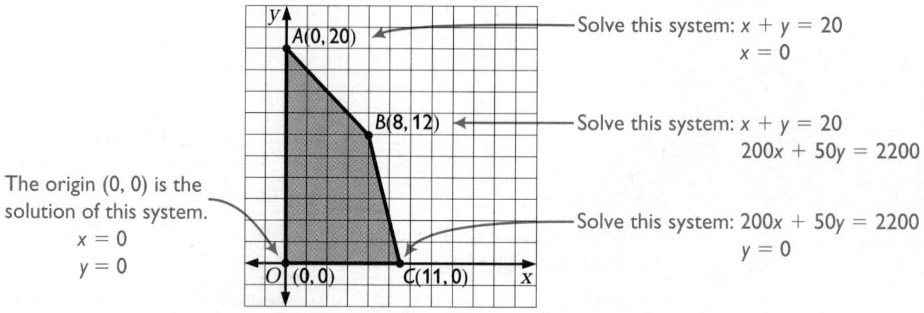

The vertex (0, 80) is the solution of the system $x = 0$, $1.5x + 6y = 480$. The vertex (24, 74) is the solution of the system $1.5x + 3y = 258$, $1.5x + 6y = 480$. The vertex (172, 0) is the solution of the system $y = 0$, $1.5x + 3y = 258$. The vertex (0, 0) is the solution of the system $x = 0$, $y = 0$.

Talk it Over

Questions 8 and 9 ask students to interpret the meaning of points on or inside the feasible region. Questions 10–13 lead students to discover the corner-point principle of linear programming.

Now you are ready for: Exs. 1–9 on pp. 57–58

Talk it Over

8. Find the total number of ads represented by each vertex of *ABCO* and the total cost of running the ads.
9. Does every point in the feasible region in Sample 1 represent a possible combination of ads? Explain.

The rate sheet on page 53 has another column of information that shows how many people, on average, are listening to WMMR at any one time.

Advertise on WMMR	One 60-second ad costs ...	and is heard by ...
Morning (A.M.)	\$200	90,000 people
Afternoon/evening (P.M.)	\$50	30,000 people

Given the original constraints, you want as many people as possible to hear the ads. You must find the *maximum-listeners point*. This is the point in the feasible region (see Sample 1) that represents the combination of ads that reaches the most listeners.

Talk it Over

10. Find the total number of listeners reached by the combination of ads represented by each point.
    a. (0, 10) b. (1, 7) c. (2, 4)

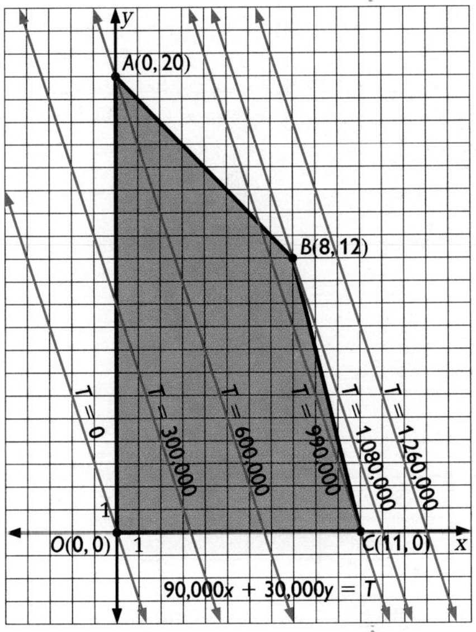

11. The graph shows the feasible region found in Sample 1 and the line $90{,}000x + 30{,}000y = T$ for several values of T.
    a. What does T represent?
    b. Where are all the points in question 10 located?
    c. Use the graph to find another combination of ads that reaches the same number of listeners as the combinations of ads in question 10.
12. Use the graph to find a combination of ads that reaches each number of listeners.
    a. 600,000 b. 990,000
    c. 1,080,000 d. 1,260,000
13. a. Describe what happens to the line $90{,}000x + 30{,}000y = T$ as the number of listeners increases.
    b. Explain why B is the maximum-listeners point.
    c. How many A.M. ads and how many P.M. ads should you run to reach the maximum number of listeners? What is the maximum number of listeners?

Answers to Talk it Over

8. O: 0 ads, \$0; A: 20 ads, \$1000; B: 20 ads, \$2200; C: 11 ads; \$2200
9. No; only points with both coordinates whole numbers.
10. a. 300,000 people
    b. 300,000 people
    c. 300,000 people
11. a. the total number of listeners
    b. on the line labeled $T = 300{,}000$ in the figure (the line with equation $90{,}000x + 30{,}000y = 300{,}000$)
    c. (3, 1)
12. a–c. Answers may vary. Examples are given.
    a. 4 A.M. ads, 8 P.M. ads
    b. 8 A.M. ads, 9 P.M. ads
    c. 8 A.M. ads, 12 P.M. ads
    d. The line $T = 1{,}260{,}000$ does not intersect the feasible region. None of the combinations will reach that many listeners.
13. a. It moves to the right and both intercepts increase.
    b. All other points in the feasible region fall below the graph of $90{,}000x + 30{,}000y = 1{,}080{,}000$. That is, for all other points in the feasible region, $T < 1{,}080{,}000$.
    c. 8 A.M. ads and 12 P.M. ads; 1,080,000 listeners

BY THE WAY. . .

In 1984 Narendra Karmarkar created a high-speed computer algorithm for linear programming that solves a problem with 800,000 variables in 10 hours.

The process used to find the combination of ads that maximizes the number of listeners is an example of **linear programming.**

In this situation, the maximum-listeners point is one of the vertices of the feasible region. This is an example of the *corner-point principle* of linear programming.

Sample 2

Use the WMMR rate sheet on page 55. Suppose that you choose these constraints for one day of radio advertisements.

Constraint 1	Constraint 2	Constraint 3
You will run at most twenty ads.	You will run at least as many A.M. ads as P.M. ads.	You want to reach at least 720,000 listeners.

How many A.M. ads and how many P.M. ads should you run in order to minimize the total cost? How much will the ads cost?

Sample Response

1 Represent the constraints with a system of linear inequalities.
Let x = the number of A.M. ads
Let y = the number of P.M. ads.

$x + y \leq 20$ ← Constraint 1

$x \geq y$ ← Constraint 2

$90{,}000x + 30{,}000y \geq 720{,}000$ ← Constraint 3

$x \geq 0$
$y \geq 0$ ← Include inequalities that show that x and y cannot be negative.

2 Graph the feasible region and find the coordinates of its vertices.

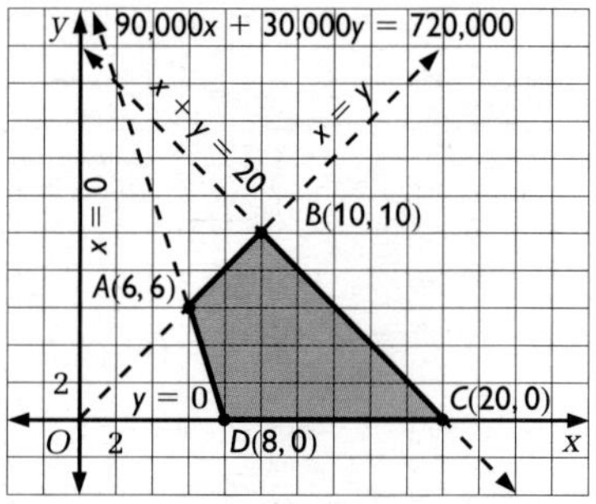

3 Write a linear combination that represents the total cost of the ads.

$200x + 50y$

4 Use the corner-point principle.

The minimum cost occurs at a vertex of the feasible region.

Find the total cost for the combination of ads represented by each vertex.

The minimum cost occurs at $A(6, 6)$. →

Vertex	Total cost (dollars)
(x, y)	$200x + 50y$
$A(6, 6)$	$200(6) + 50(6) = 1500$
$B(10, 10)$	$200(10) + 50(10) = 2500$
$C(20, 0)$	$200(20) + 50(0) = 4000$
$D(8, 0)$	$200(8) + 50(0) = 1600$

You should run six A.M. ads and six P.M. ads. The ads will cost $1500.

Additional Sample

S2 Refer to the situation in Additional Sample S1. The company is going to make more animals, but it decides to make no more than 75 animals in all and to keep the number of giraffes to no more than 5 more than the number of lions. The profit on each lion is $20 and the profit on each giraffe is $18. How many of each kind of animal should the company make to make the maximum profit?

Step 1. Represent the constraints with a system of linear inequalities.
Let x = number of lions.
Let y = number of giraffes.
The following system models the situation.
$x \geq 0$
$y \geq 0$
$x + y \leq 75$
$y \leq x + 5$

Step 2. Graph the feasible region and find the coordinates of its vertices.

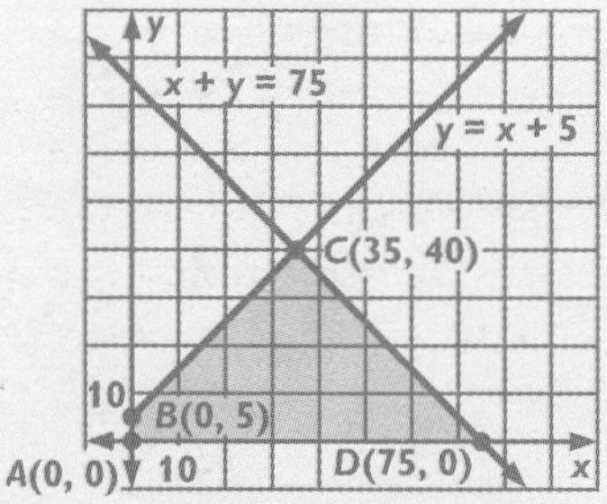

Step 3. Write a linear combination that represents the profit on x lions and y giraffes.
$20x + 18y$

Step 4. Use the corner-point principle. The maximum profit will occur at a vertex of the feasible region. Find the profit that corresponds to each vertex.

Vertex	Profit ($)
(x, y)	$20x + 18y$
$A(0, 0)$	0
$B(0, 5)$	90
$C(35, 40)$	1420
$D(75, 0)$	1500

The company will make the maximum profit, $1500, if it makes 75 lions and no giraffes.

LINEAR PROGRAMMING

You can use linear programming:

- when you can represent the constraints on the variables with a system of linear inequalities, and
- when the goal is to find the maximum or minimum value of a linear combination of the variables.

CORNER-POINT PRINCIPLE

Any maximum or minimum value of a linear combination of the variables will occur at one of the vertices of the feasible region.

Now you are ready for: Exs. 10–23 on pp. 59–60

Look Back

Explain the meaning of each of these terms in your own words: *linear combination, constraint,* and *feasible region.*

1-8 Exercises and Problems

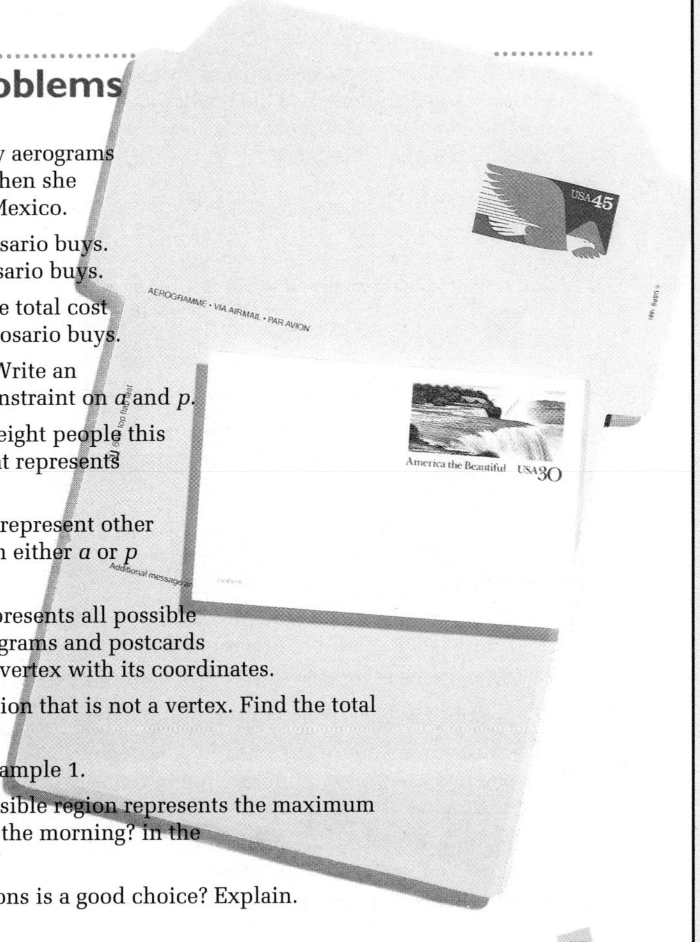

1. Rosario goes to the post office to buy aerograms and pre-stamped postcards to use when she writes to her friends and family in Mexico.

   Let a = the number of aerograms Rosario buys.
   Let p = the number of postcards Rosario buys.

   a. Write a linear combination for the total cost of the aerograms and postcards Rosario buys.

   b. Rosario has \$8.10 in her wallet. Write an inequality that represents this constraint on a and p.

   c. Rosario plans to write to at least eight people this weekend. Write an inequality that represents this constraint on a and p.

   d. Write two more inequalities that represent other constraints on a and p. (*Hint:* Can either a or p be negative in this situation?)

   e. Graph the feasible region that represents all possible combinations of numbers of aerograms and postcards that Rosario can buy. Label each vertex with its coordinates.

   f. Choose a point of the feasible region that is not a vertex. Find the total cost at that point.

2. Use the feasible region graphed in Sample 1.

   a. **Reading** Which vertex of the feasible region represents the maximum number of ads that can be run in the morning? in the afternoon/evening?

   b. Do you think either of these options is a good choice? Explain.

Look Back

Students can best complete this activity by writing down the meaning of each term.

APPLYING

Suggested Assignment

Day 1

Standard 1–9

Extended 1–9

Day 2

Standard 10–12, 14–23

Extended 10–12, 14–23

Integrating the Strands

Algebra Exs. 1–15

Geometry Exs. 19–22

Statistics and Probability Exs. 16–18

Discrete Mathematics Exs. 1–14, 16–18, 23

Logic and Language Exs. 2, 9, 10, 14, 23

Reasoning

Exs. 2 and 9 have students analyze feasible regions and then offer expanations as to some real-life choices that can be made using these regions.

Answers to Look Back

Summaries may vary. A linear combination is a sum of the form $ax + by$, where a and b are constants. A constraint is a condition that must be met by a variable or a linear combination of variables. The feasible region of a system of linear inequalities is the graph of the solution set of the system.

Answers to Exercises and Problems

1. a. $0.45a + 0.30p$

   b. $0.45a + 0.30p \leq 8.10$

   c. $a + p \geq 8$

   d. $a \geq 0$; $p \geq 0$

   e.

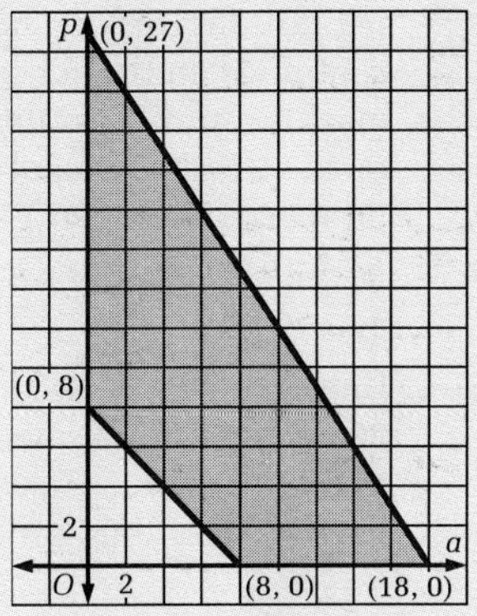

   f. Answers may vary. An example is given. For (10, 6), the total cost is \$6.30.

2. a. (11, 0); (0, 20)

   b. Answers may vary. An example is given. No; I think it is better to spread them out to reach a wider audience.

Interdisciplinary Problems

Many problems in the business world can be analyzed and solved by using linear programming techniques. This is illustrated by Exs. 7 and 10–12.

Graph the feasible region for each system of inequalities. Label each vertex with its coordinates.

3.	**4.**	**5.**	**6.**
$x \geq 0$	$1 \leq c \leq 4$	$a \geq 0$	$x \geq 0$
$y \geq 0$	$c + d \leq 14$	$2 \leq b \leq 8$	$y \geq 1$
$x + y \leq 6$	$d \geq 2c$	$10a + 5b \geq 30$	$x + y \geq 5$
$12x + 6y \leq 48$		$2a + b \leq 12$	$40x + 10y \geq 80$

7. **Manufacturing** A chemical company manufactures two silicon compounds using the raw materials shown in the table.

Raw material	Amount (kg) needed to make 1 kg of ... silicon carbide	silicon tetrachloride
silica	1.5	0.35
carbon	0.6	0.14
chlorine	0	0.84

a. For each raw material, write a linear combination that represents the total amount (in kilograms) of the raw material needed to manufacture c kg of silicon carbide and t kg of silicon tetrachloride.

b. The manufacturer has in stock 105 kg of silica, 84 kg of carbon, and 126 kg of chlorine. Write a system of inequalities that represents all the constraints in this situation.

BY THE WAY...

Some silicon compounds are used to make silicon chips for computers, calculators, radios, and televisions.

8. **Fitness** Emilio tries to meet several goals each time he works out on the ski machine and the treadmill. He drew this graph of the feasible region that represents all possible workouts that allow him to meet his goals.

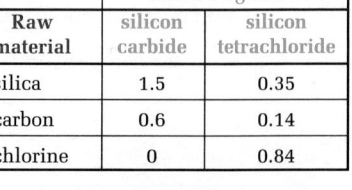

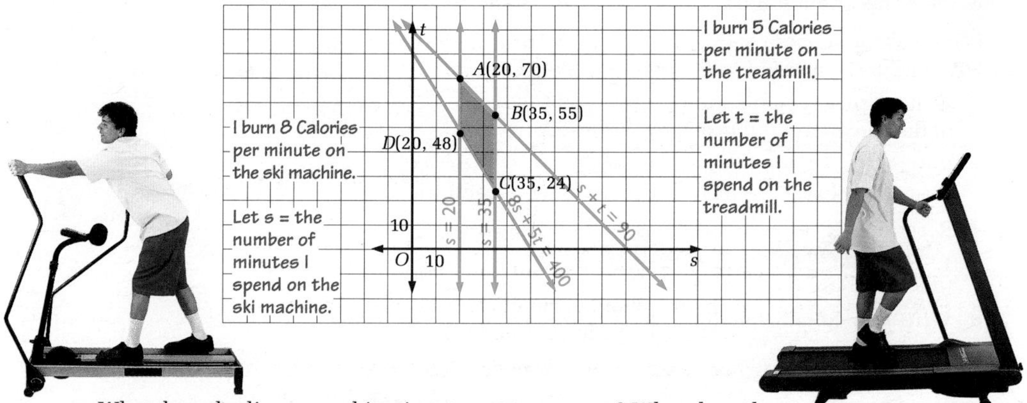

a. What does the linear combination $8s + 5t$ represent? What does the linear combination $s + t$ represent?

b. Write a system of inequalities that represents the feasible region.

c. Describe the goal (constraint) represented by each inequality in part (b).

9. a. For each workout represented by a vertex of $ABCD$, find the total length of the workout, the amount of time spent on each machine, and the number of Calories burned during the workout.

b. **Writing** Explain why Emilio might choose the workout represented by B over the workout represented by A. Explain why Emilio might choose the workout represented by C over the workout represented by D.

Answers to Exercises and Problems

3.
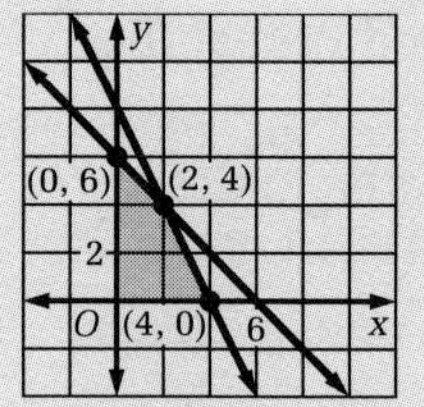

4.
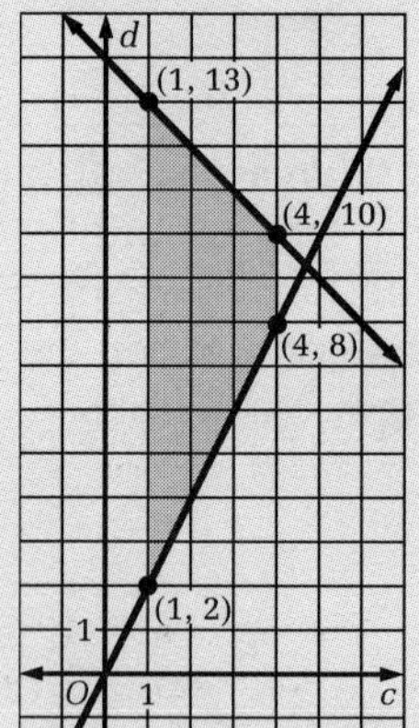

5.
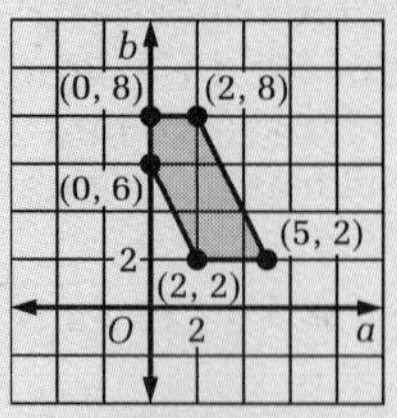

6.
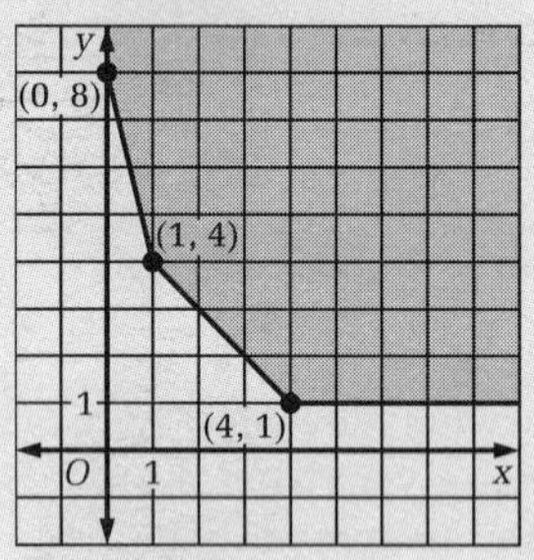

7. a. silica: $1.5c + 0.35t$; carbon: $0.6c + 0.14t$; chlorine: $0.84t$

b. $c \geq 0$; $t \geq 0$; $1.5c + 0.35t \leq 105$; $0.6c + 0.14t \leq 84$; $0.84t \leq 126$

8. a. the total number of Calories Emilio burns per minute working out s minutes on the ski machine and t minutes on the treadmill; total minutes of exercise

b. $20 \leq s \leq 35$; $8s + 5t \geq 400$; $s + t \leq 90$

c. He will spend between 20 and 35 minutes on the ski machine. He will burn at least 400 Calories. His total workout time on the ski machine and treadmill will be less than 90 minutes.

9. a. A: He works out for a total of 90 minutes, 20 minutes on the ski machine and 70 minutes on the treadmill, and burns 510 Calories. B: He works out for a total of 90 minutes, 35 minutes on the ski machine and 55 minutes on the treadmill, and burns 555 Calories. C: He works out for a total of 59 minutes, 35 minutes on the ski machine and 24 minutes on the treadmill, and burns 400 Calories. D: He works out for a total of 68 minutes, 20 minutes on the ski machine and 48 minutes on the treadmill, and burns 400 Calories.

10. Agriculture Rhonda Neilson will plant at most 100 acres of her farm with barley and corn. She can spend up to \$8400 to grow these two crops.

expenses: \$60 per acre
EXPECTED INCOME
\$125 per acre
Let b = the number of acres planted with barley.
expenses: \$140 per acre
EXPECTED INCOME
\$250 per acre
Let c = the number of acres planted with corn.

a. Write a system of inequalities that represents all the constraints on the number of acres that will be used for each crop on Rhonda Neilson's farm.

b. Graph the feasible region and find the coordinates of its vertices.

c. What does the linear combination $125b + 250c$ represent?

d. Graph the line $125b + 250c = 20{,}000$ on the same set of axes you used in part (b).

e. **Writing** What does your graph in part (d) tell you about the farm's total income from barley and corn? Explain how you know.

f. How many acres of each crop should Rhonda Neilson plant in order to maximize the income? What is the maximum income?

11. Use the situation described in Exercise 7. Suppose the company sells silicon carbide for \$320/kg and silicon tetrachloride for \$250/kg. How many kilograms of each product should the company manufacture in order to maximize income? What is the maximum income?

12. Business A small business produces oak chairs and walnut chairs.

- ➤ The business can make at most 20 chairs per week.
- ➤ Materials cost \$100 per oak chair and \$150 per walnut chair.
- ➤ The business has a budget of \$2400 per week for materials.
- ➤ Each oak chair sells for \$400 and each walnut chair for \$500.

How many of each type of chair should the business make each week to maximize income?

Application

Exs. 10–12 demonstrate the wide use of linear programming to maximize income and minimize cost. Businesses as diverse as farming, chemical manufacturing, and furniture making all employ this procedure. You may wish to have students suggest other businesses that use linear programming.

Answers to Exercises and Problems

b. Answers may vary. Examples are given. Emilio might choose B over A because he wants to burn more Calories in the same amount of time. He might choose C over D because he prefers to spend a shorter time to burn the same number of Calories.

10. a. $b \geq 0$; $c \geq 0$; $b + c \leq 100$; $60b + 140c \leq 8400$

b.

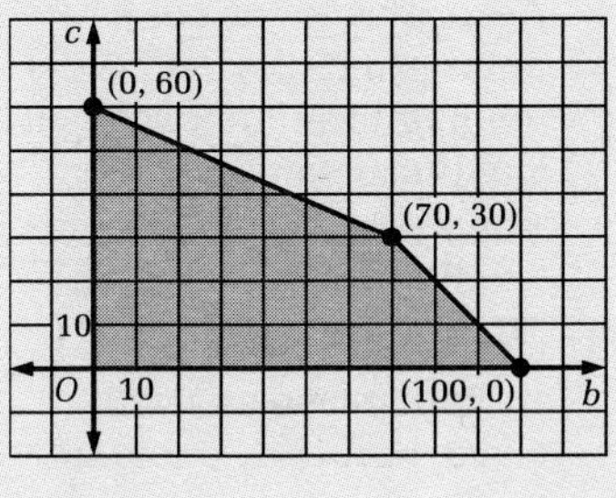

c. the total income from b acres of barley and c acres of corn

d.

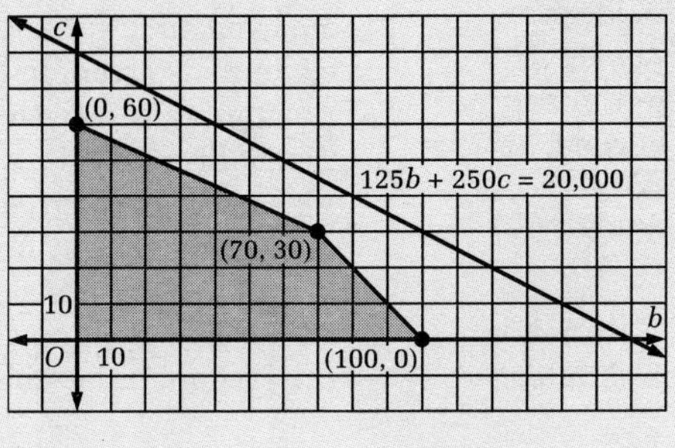

e. Rhonda Neilson's total income from barley and corn will be less than \$20,000. Every point in the feasible region is below the graph of $125b + 250c = 20{,}000$, so for every point (b, c) in the feasible region, $125b + 250c < 20{,}000$.

f. 70 acres of barley, 30 acres of corn; \$16,250

11. 35 kg of silicon carbide and 150 kg of silicon tetrachloride; \$48,700

12. 12 oak chairs, 8 walnut chairs (\$8800 income)

Working on the Unit Project

Students can work in their project groups to complete Ex. 23. Ideas about using linear programming to create a park design and proposal should be written down for possible use when completing the Unit Project.

Quick Quiz (1-4 through 1-8)

See page 62.

See also Test 2 in the *Assessment Book.*

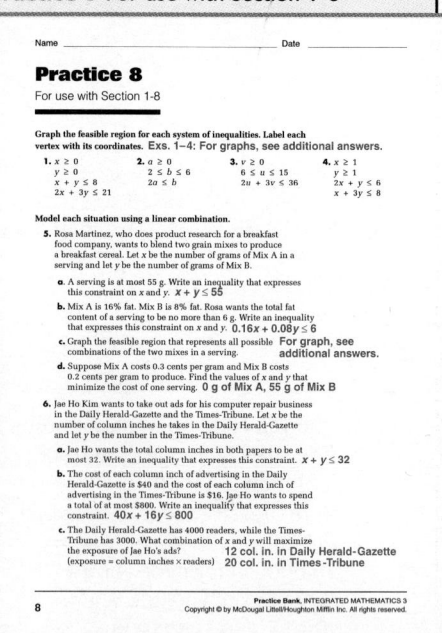

13. a. **Open-ended** Use the information in the tables to create and solve a linear programming problem about an exercise routine that consists of two types of exercise. Make up constraints on time and Calories burned.

Calories Burned During Exercise

Type of exercise	aerobics	swimming	squash	walking	jogging	bicycling
Cal/min	5	7	12	6	8	10

b. **Group Activity** Work with another student.

Trade the problems you wrote in part (a). Solve the problem you receive.

Ongoing ASSESSMENT

14. **Writing** Explain how systems of inequalities are used in linear programming.

Review PREVIEW

15. The function $y = -0.036x^2 + x + 6$ models the relationship between the height y of a soccer ball in feet and the horizontal distance x in feet that the ball travels after being struck at an angle of 45° and an initial speed of 30 ft/s. What is the maximum height of the ball and how far has it traveled horizontally when it reaches this height? Round answers to the nearest foot. *(Section 1-7)*

Suppose you roll one red die and one green die. Find each probability. *(Toolbox Skill 7)*

16. P(red 3 and green 3)
17. P(odd red and green 5)
18. P(odd red and odd green)

Find the surface area and volume of a sphere with each indicated radius or diameter. *(Toolbox Skill 29)*

19. radius: 4.5 cm
20. diameter: 72 in.
21. diameter: 18 m
22. radius: 13 ft

Working on the Unit Project

As you complete Exercise 23, think about how you can use linear programming in the creation of your park proposal.

23. **Writing** Suppose you plan to include in your park a large field that can be used for both touch football and field hockey games in the fall. Explain how you can use the information below to decide how many hours per week to schedule each sport in order to make the most profit from the use of the field.
    - hours per week that the field can be used
    - hourly maintenance costs for both games
    - weekly maintenance budget
    - hourly personnel costs for both games
    - weekly budget for personnel costs
    - admission charges for both games

Practice 8 For use with Section 1-8

Name ______________ Date ______________

Practice 8

For use with Section 1-8

Graph the feasible region for each system of inequalities. Label each vertex with its coordinates. Exs. 1–4: For graphs, see additional answers.

1. $x \geq 0$, $y \geq 0$, $x + y \leq 8$, $2x + 3y \leq 21$
2. $a \geq 0$, $2 \leq b \leq 6$, $2a \leq b$
3. $v \geq 0$, $6 \leq u \leq 15$, $2u + 3v \leq 36$
4. $x \geq 1$, $y \geq 1$, $2x + y \leq 6$, $x + 3y \leq 8$

Model each situation using a linear combination.

5. Rosa Martinez, who does product research for a breakfast food company, wants to blend two grain mixes to produce a breakfast cereal. Let x be the number of grams of Mix A in a serving and let y be the number of grams of Mix B.
   a. A serving is at most 55 g. Write an inequality that expresses this constraint on x and y. $x + y \leq 55$
   b. Mix A is 16% fat. Mix B is 8% fat. Rosa wants the total fat content of a serving to be no more than 6 g. Write an inequality that expresses this constraint on x and y. $0.16x + 0.08y \leq 6$
   c. Graph the feasible region that represents all possible combinations of the two mixes in a serving. For graph, see additional answers.
   d. Suppose Mix A costs 0.3 cents per gram and Mix B costs 0.2 cents per gram to produce. Find the values of x and y that minimize the cost of one serving. 0 g of Mix A, 55 g of Mix B
6. Jae Ho Kim wants to take out ads for his computer repair business in the Daily Herald-Gazette and the Times-Tribune. Let x be the number of column inches he takes in the Daily Herald-Gazette and let y be the number in the Times-Tribune.
   a. Jae Ho wants the total column inches in both papers to be at most 32. Write an inequality that expresses this constraint. $x + y \leq 32$
   b. The cost of each column inch of advertising in the Daily Herald-Gazette is \$40 and the cost of each column inch of advertising in the Times-Tribune is \$16. Jae Ho wants to spend a total of at most \$800. Write an inequality that expresses this constraint. $40x + 16y \leq 800$
   c. The Daily Herald-Gazette has 4000 readers, while the Times-Tribune has 3000. What combination of x and y will maximize the exposure of Jae Ho's ads? (exposure = column inches × readers) 12 col. in. in Daily Herald-Gazette 20 col. in. in Times-Tribune

8 Practice Bank, INTEGRATED MATHEMATICS 3

Answers to Exercises and Problems

13. a. Answers may vary. An example is given. Isaac has 50 minutes to exercise. He will bicycle for b minutes and jog for j minutes. He wants to jog at least 20 minutes. He wants to maximize Calories burned. Constraints: $b + j \leq 50$, $j \geq 20$, $b \geq 0$, $j \geq 0$, Maximum Calories burned: $10b + 8j = C$

    b. Answers may vary. Solution is given for part (a). He should jog for 20 minutes, bike for 30 minutes for a total of 460 Calories burned.

14. Answers may vary. An example is given. In linear programming, a system of inequalities (say in x and y) is used to describe constraints on x and y or on linear combinations of x and y. The solution set of the system, or feasible region, represents all possible combinations that satisfy all the constraints. According to the corner-point principle, the maximum or minimum value of any linear combination of x and y will occur at a vertex of the feasible region.

15. maximum height: about 13 ft; distance traveled: about 14 ft

16. $\frac{1}{36} \approx 0.03$

17. $\frac{1}{12} \approx 0.08$

18. $\frac{1}{4} \approx 0.25$

19. S.A. ≈ 254.5 cm^2; $V \approx$ 381.7 cm^3

20. S.A. ≈ 16,286 in.^{2}; $V \approx$ 195,432 in.^{3}

21. S.A. ≈ 1017.9 m^2; $V \approx$ 3053.6 m^3

Unit Project

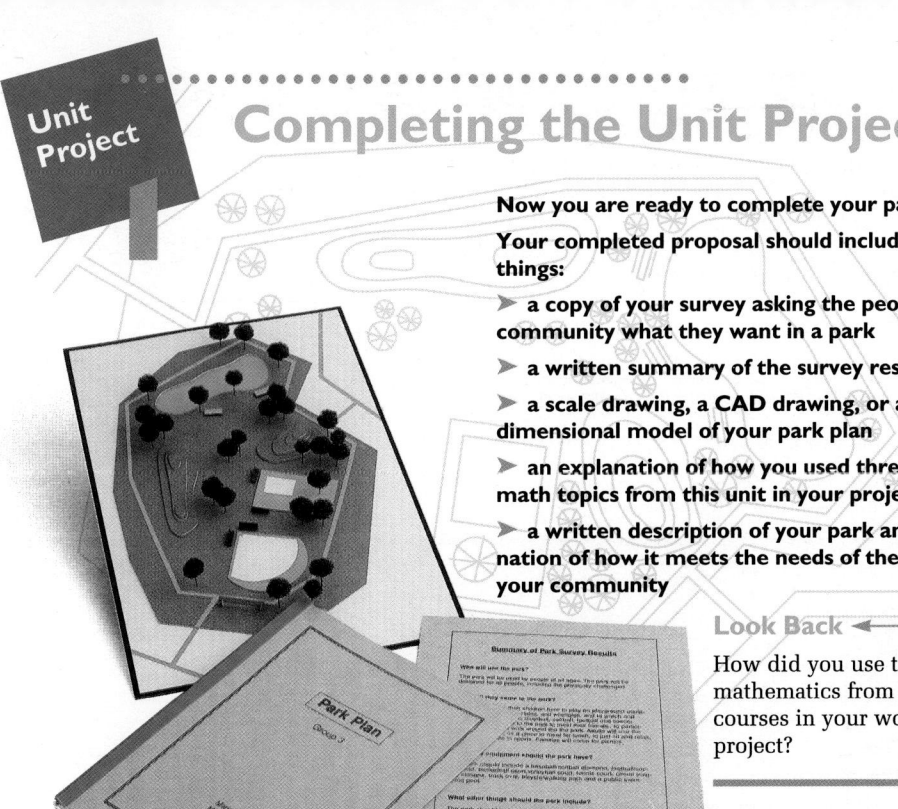

Completing the Unit Project

Now you are ready to complete your park proposal.

Your completed proposal should include these things:

- **a copy of your survey asking the people of your community what they want in a park**
- **a written summary of the survey results**
- **a scale drawing, a CAD drawing, or a three-dimensional model of your park plan**
- **an explanation of how you used three of the math topics from this unit in your project**
- **a written description of your park and an explanation of how it meets the needs of the people in your community**

Look Back

How did you use topics in mathematics from previous courses in your work on this project?

Alternative Projects

Project 1: Management Science

Write a report on linear programming. Include the contributions of George Dantzig, Narendra Karmarkar, and L. G. Khachian. Describe several business applications, including airline scheduling and the routing of information transmitted over complex communications networks. Also discuss the simplex and ellipsoid methods for finding the corner point that represents maximum profit in cases where there are so many corner points that you cannot test them all.

Project 2: Secret Codes

Huffman codes represent one way to use network diagrams to send messages in secret codes. Find out how to use Huffman codes. Write an article explaining how to code and decode messages with them.

Assessment

A scoring rubric for the Unit Project can be found on pages xxvi and 1 of this Teacher's Edition and also in the ***Project Book***.

Answers to Exercises and Problems

22. S.A. ≈ 2123.7 ft²; $V \approx 9202.8$ ft³

23. Answers may vary. An example is given. One constraint would require that the maintenance cost for x hours of football and y hours of field hockey be equal to or less than the proposed maintenance budget. Another would require that the personnel costs for x hours of football and y hours for field hockey be equal to or less than the proposed personnel budget. A third constraint is the total number of hours the field is available for football and field hockey, $x + y$. The objective would be to maximize the income from admission charges for the games played.

Unit Support Materials

- *Practice Bank:* Cumulative Practice 9
- *Study Guide:* Unit 1 Review
- *Problem Bank:* Unifying Problem 1
- *Assessment Book:* Unit Tests 3 and 4, Spanish Unit Tests, Alternative Assessment
- *Test Generator* software with Test Bank
- *Teacher's Resources for Transfer Students*

Quick Quiz (1-4 through 1-8)

1. Use algebra to find the break-even point when E = expenses and I = income. [1-4]
$E = 160 + 5x$
$I = 9x$
Break-even point is $x = 40$.

2. Is $(-1, 2, -3)$ a solution of $2x - 4y + 3z = -19$? [1-5]
Yes.

3. Write this system of equations as a matrix equation. [1-5]
$2x - 3y = 11$
$-5x + 8y = 13$

$$\begin{bmatrix} 2 & -3 \\ -5 & 8 \end{bmatrix}\begin{bmatrix} x \\ y \end{bmatrix} = \begin{bmatrix} 11 \\ 13 \end{bmatrix}$$

4. Complete this sentence. A dot is a __?__ of a network diagram, and a line connecting two dots is an __?__ of the network.
[1-6] vertex; edge

5. Tell whether the function $y = 0.4x^2 + 2.6x + 1$ has a maximum or minimum value and find the value of x that maximizes or minimizes the function. [1-7]
minimum value at $x = -3.25$

6. Graph the feasible region for this system of inequalities. [1-8]
$x \geq 0$
$y \leq 2$
$x + y \leq 6$
$50x + 20y \geq 80$

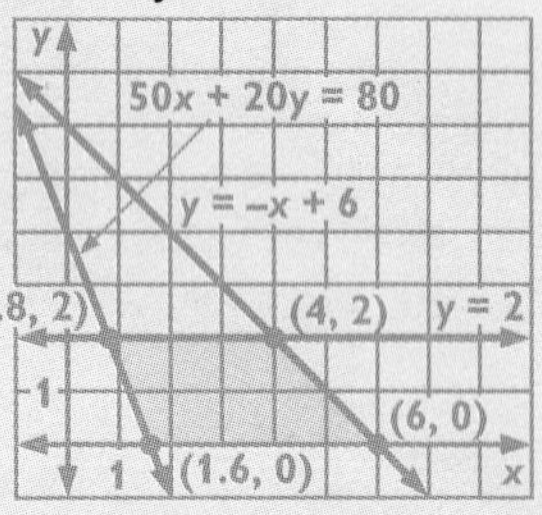

Unit 1 REVIEW AND ASSESSMENT

1. a. Write two algorithms for solving a system of two linear equations with two variables. 1-1

b. Writing Compare your algorithms.

2. Solve the inequality $3x > 7 + 10x$ in two ways.

3. What are the lengths of the sides of all the triangles that have sides whose lengths are whole numbers and a perimeter of 22? 1-2

4. a. What conclusion(s) can you draw from the scatter plot? 1-3

b. Does each conclusion you drew apply to every SLR zoom lens in the scatter plot? Why or why not?

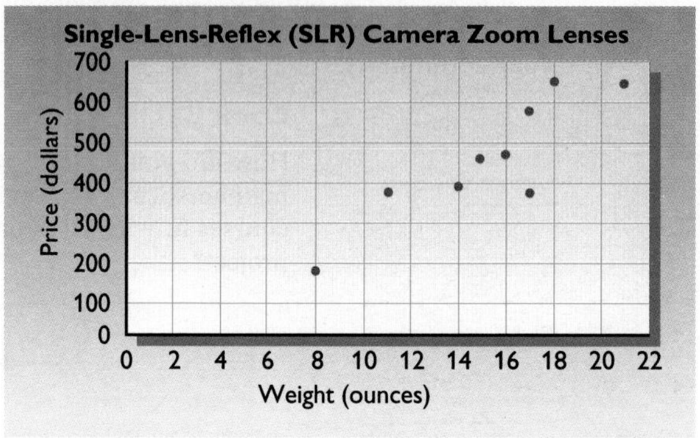

5. Robert DiNucci must select a monthly calling plan for local telephone service for his new apartment. He expects to limit local calls to Zone 1. 1-4

Measured Service	Unlimited Service	Measured Circle Service
\$6.39 per month plus 2.6¢ per min for local calls in Zone 1	\$12.59 per month for local calls in Zone 1	\$9.25 per month for up to two hours of local calls in Zone 1 plus 5.8¢ per min for local calls in Zone 1 after two hours

a. For how many minutes per month of local calls in Zone 1 is Measured Service the least expensive?

b. For how many minutes per month of local calls in Zone 1 is Unlimited Service the most expensive?

c. Which plan should Robert DiNucci choose in order to pay as little as possible for local phone service? Explain.

Solve. Round decimal answers to the nearest tenth. 1-5

6. $x - 3y - z = 1$
$2x + y + z = 3$
$x + 2y + 2z = 0$

7. $2x + z = 1$
$3y + 2z = -2$
$-3x + y - z = -3$

8. $x - y + 2z = 1$
$-x + 3y + z = 5$
$-2x + y - z = 4$

Answers to Unit 1 Review and Assessment

1. Answers may vary. Examples are given for two equations in x and y.

a. *Method 1:* **Step 1:** Solve for y in the first equation. **Step 2:** Substitute the resulting value for y in the second equation. **Step 3:** Solve the resulting equation for x. **Step 4:** Substitute the value of x in the first equation. **Step 5:** Solve the resulting equation for y. *Method 2:* **Step 1:** Graph both equations on a graphics calculator. **Step 2:** Use TRACE to determine the point of intersection and find x and y.

b. The results will be the same (although the first may result in a fractional answer, while the second gives a decimal approximation). Neither algorithm is significantly easier to use.

2. *Method 1:*
$3x > 7 + 10x$
$-7x > 7$ ← Subtract $10x$ from both sides.
$x < -1$ ← Divide both sides by -7, and reverse the inequality sign.

9. a. **Open-ended** Describe a situation that can be modeled by this network diagram. 1-6

b. Write a matrix that represents the connections in the diagram in part (a).

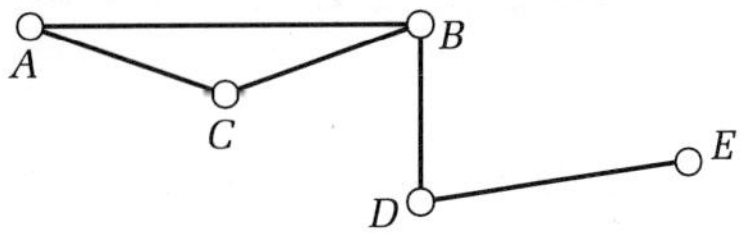

10. A landscape architect's plan for this flower bed states that the flowers in bordering sections are different colors. Show how to follow this plan using the least number of colors.

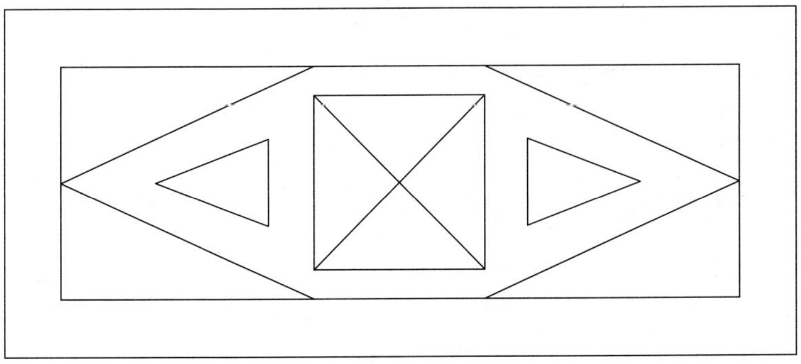

For each quadratic function in Questions 11–13: 1-7

a. Tell whether the function has a *maximum value* or a *minimum value*.

b. Find the value of *x* that maximizes or minimizes the function.

c. Find the function's maximum or minimum value.

11. $y = 2x^2 - 5x + 11$ 12. $y = -x^2 + 3x - 2$ 13. $y = -1.1x^2 - 4.5x + 2.3$

14. Katie Stein is a sales representative whose territory is Arizona and New Mexico. Her daily travel expenses average $120 in Arizona and $100 in New Mexico. She receives an annual travel allowance of $18,000. She must spend at least 50 days in Arizona and 60 days in New Mexico. If sales average $3000 per day in Arizona and $2800 per day in New Mexico, how many days should she spend in each state to maximize sales? 1-8

15. **Self-evaluation** Which of the problem-solving strategies presented in this unit have you used before? Which, if any, are new to you? Do you think your work in this unit has made you a better problem-solver? Explain.

16. **Group Activity** Work in a group of three students.

Use this system of equations.

$$16x - 2.4y = 147$$
$$5.8x + 9y = 258$$

a. One student should solve the system by graphing, another by algebra, and the third by matrices.

b. Compare your solutions. Explain any differences.

c. Compare the three algorithms the group used to solve the system.

Answers to Unit 1 Review and Assessment

Method 2: Graph $y = 3x$ and $y = 7 + 10x$ on the same set of axes.

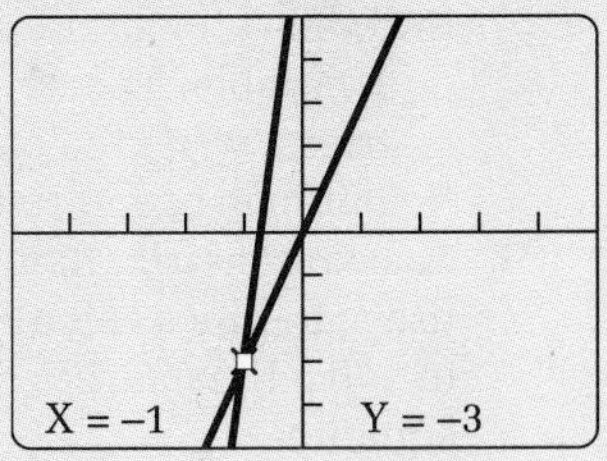

The solution of the inequality is all values of x for which the graph of $y = 7 + 10x$ is below the graph $y = 3x$, that is $x < -1$.

3.

Side 1	Side 2	Side 3
2	10	10
3	9	10
4	8	10
4	9	9
5	7	10
5	8	9
6	6	10
6	7	9
6	8	8
7	7	8

4. Answers may vary. Examples are given.

a. In general, as the weight of an SLR Camera zoom lens increases, the price increases.

b. No; although the points lie close to the line $y = 3x$, they do not all lie on the line. So, for example, there is a 17 oz lens that costs less than a 16 oz lens.

5. a. less than about 110 min or between about 128 min and about 238 min

b. less than about 178 min

c. If Robert expects to spend less than about 110 min or between about 128 min and about 238 min on local calls, he should get Measured Service. If he expects to spend between about 110 min and about 128 min on local calls, he should get Measured Circle Service. If he plans to spend more than 238 min on local calls, then he should get Unlimited Service.

6. (2, 1, −2)

7. (−2, −4, 5)

8. (−3, 0, 2)

9. a. Answers may vary. Example: The access graph might represent five towns and their connections by direct train service.

b.

	A	*B*	*C*	*D*	*E*
A	0	1	1	0	0
B	1	0	1	1	0
C	1	1	0	0	0
D	0	1	0	0	1
E	0	0	0	1	0

10. three colors

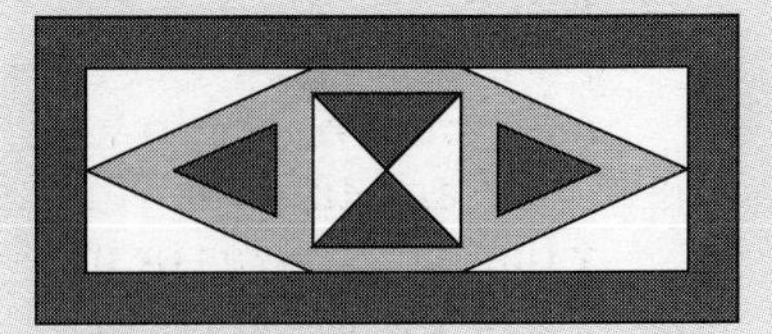

11. a. minimum

b. $1\frac{1}{4}$

c. $7\frac{7}{8}$

12. a. maximum

b. $1\frac{1}{2}$

c. $\frac{1}{4}$

13. a. maximum

b. about −2

c. about 6.9

Answers continued on next page.

IDEAS AND (FORMULAS) $= x^2$ ${}_5P_5$

ALGEBRA

- You can solve a linear inequality in one variable with algebra and with a graph. *(p. 5)*
- **Problem Solving** To solve certain problems modeled by two or more linear equations, you can find the intersection point of the graphs of the equations. *(p. 24)*
- To find a break-even point where income is equal to expenses, you can use algebra or a graph. *(p. 25)*
- You can use algebra or a graph to solve a system of two linear equations. *(p. 30)*

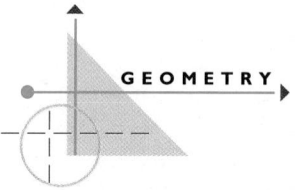

GEOMETRY

- The triangle inequality states that the sum of the lengths of any two sides of a triangle must be greater than the length of the third side. *(p. 9)*

$AC + CB > AB$
$CB + BA > AC$
$BA + AC > CB$

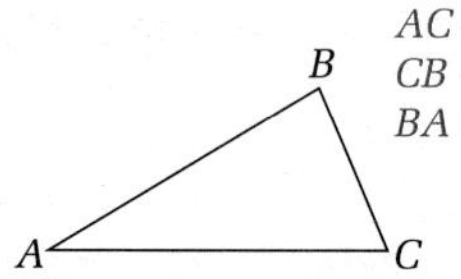

- Three planes can have no point of intersection, one point of intersection, or many points of intersection. *(p. 32)*

no common point of intersection | one common point of intersection | many common points of intersection

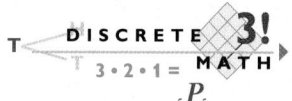

DISCRETE MATH

- Sometimes there is more than one possible algorithm for solving a problem. Algorithms may contain loops. *(p. 4)*
- **Problem Solving** You can use a systematic list to be sure that you do not leave out or repeat any possible solutions of a problem. *(p. 10)*
- A matrix equation can be used to solve a system of two or more linear equations. *(pp. 31, 33)*
- **Problem Solving** Network diagrams can be used to solve some problems that involve connections. The direction of the connections can also be represented in network diagrams. *(pp. 38, 39, 40)*

Answers to Unit 1 Review and Assessment

14. She should spend 50 days in Arizona and 120 days in New Mexico for a sales volume of \$486,000.

15. Answers may vary.

16. a. graphics calculator solution: (12.314789, 20.855263); hand graphing solution: about (12, 21); algebra solution: $\left(12\frac{393}{1316}, 20\frac{975}{1316}\right)$; matrix solution: (12.29863222, 20.74088146)

b. The algebraic solution is exact. The others are all approximations, the closest being the matrix solution.

c. To find a solution by graphing, find the coordinates of the points where the graphs of the equations intersect. To use algebra, find linear combinations that eliminate one variable. Solve for the other. Then use that value to find the value of the second variable. To use matrices, write the coefficient matrix and the constant matrix, then use a graphics calculator to solve the matrix equation.

- You can use matrices whose elements are only "0" or "1" to represent situations that involve connections. A "0" means there is no connection. A "1" means there is a connection. *(p. 39)*
- **Problem Solving** Sometimes it is useful to color the vertices of a network diagram in order to solve a problem that involves groupings. *(p. 40)*
- The region formed by the graphs of a system of linear inequalities that represent all the constraints on a situation is the feasible region that includes all possible solutions. *(p. 54)*
- You can find the maximum or minimum value of a linear combination by using linear programming. *(p. 56)*
- The corner-point principle states that for a linear programming problem, any maximum or minimum value of a linear combination $ax + by$ will occur at one of the vertices of the feasible region. *(p. 56)*

STATISTICS & PROBABILITY

- Line graphs drawn on the same set of axes can be used to compare data sets in order to make a decision. *(p. 16)*
- Box-and-whisker plots drawn under the same number line can be used to compare data sets in order to make a decision. *(p. 17)*
- A scatter plot that compares two data sets can be used to make a decision. *(p. 18)*

Key Terms

- **algorithm** (p. 4)
- **box-and-whisker plot** (p. 17)
- **lower extreme** (p. 17)
- **scatter plot** (p. 18)
- **inverse matrix, A^{-1}** (p. 31)
- **edge of a network** (p. 37)
- **linear combination** (p. 53)
- **linear programming** (p. 56)
- **loop** (p. 4)
- **median** (p. 17)
- **upper quartile** (p. 17)
- **system of equations** (p. 30)
- **linear equation with three variables** (p. 32)
- **maximum** (p. 45)
- **constraint** (p. 54)
- **mean** (p. 16)
- **upper extreme** (p. 17)
- **lower quartile** (p. 17)
- **matrix equation** (p. 31)
- **vertex of a network** (p. 37)
- **minimum** (p. 45)
- **feasible region** (p. 54)

Quick Quiz (1-1 through 1-3)

1. Explain the meaning of the term *algorithm.* [1-1]
An algorithm is a set of step-by-step directions that describes a particular process or accomplishes a specific task.

2. Write an algorithm for solving an inequality of the form $ax + b \geq c$, where $a \neq 0$. [1-1]
(a) Add $-b$ to each side of the inequality.
(b) Divide each side of the inequality in step (a) by a.
(c) If a is negative, reverse the direction of the inequality from step (b) from $\geq$ to $\leq$.
(d) If a is positive, do not reverse the direction of the inequality.

For each group of numbers, tell whether they can be the lengths of the sides of a triangle. [1-2]

3. 3, 6, 11 **No.**

4. 5, 9, 13 **Yes.**

5. Find the lower extreme, lower quartile, the median, the upper quartile, and the upper extreme of these temperatures. [1-3]
33, 47, 16, 22, 56, 30, 19, 44, 51, 27, 39, 46, 53
lower extreme: 16
lower quartile: 24.5
median: 39
upper quartile: 49
upper extreme: 56

Planning the Unit 2

Exploring and Applying Functions

OVERVIEW

- **Unit 2** explores a variety of functions and their graphs. Included in the study are linear, quadratic, piecewise, absolute value, polynomial, radical, and rational functions. The domain and range for the various functions are explored. Vertical and horizontal asymptotes are introduced. Students solve associated equations for some of these functions. Composite functions are introduced.
- The **Unit Project** is based on students making a presentation about bicycles or bicycling to children in their community. Students explore such topics as bicycle safety, bicycle design, the economics of bicycling, or the physics and mathematics of bicycling. They learn that the world of bicycling can be related to the mathematics of functions.
- **Connections** to surveying, heart rates, consumer economics, sports, road design, medicine, passenger trains, and space shuttles are some of the topics included in the teaching materials and the exercises.
- **Graphics calculators** are used in Section 2-2 to graph a fitted line, in Section 2-3 to graph piecewise and absolute functions, in Section 2-4 to graph quadratic functions, in Section 2-5 to graph polynomial functions, in Section 2-6 to graph radical functions, and in Section 2-7 to graph rational functions. **Computer software**, such as Plotter Plus, can be used in Section 2-1 when studying functions, in Section 2-2 to graph lines, and in Section 2-3 to graph parabolas.
- **Problem-solving strategies** used in Unit 2 include breaking the problem into parts, using a diagram, using a formula, using an equation, using a graph, and, in general, using various kinds of functions to solve problems.

Unit Objectives

Section	Objectives	NCTM Standards
2-1	• Recognize and describe functions.	1, 2, 3, 4, 5, 6
2-2	• Use linear functions to describe situations. • Find the domain and range of a linear function.	1, 2, 3, 4, 5, 6
2-3	• Use piecewise and absolute value functions to describe situations.	1, 2, 3, 4, 5, 6
2-4	• Use quadratic functions to describe situations.	1, 2, 3, 4, 5, 6
2-5	• Recognize, evaluate, and graph direct variation functions and polynomial functions.	1, 2, 3, 4, 5, 6
2-6	• Explore functions that involve radicals. • Use these functions to solve problems.	1, 2, 3, 4, 5, 6
2-7	• Understand how rational functions are different from other types of functions. • Use rational functions to solve problems.	1, 2, 3, 4, 5, 6
2-8	• Find a function of a function.	1, 2, 3, 4, 5, 6

Topic Spiraling

Section	Connections to Prior and Future Concepts
2-1	**Section 2-1** continues the study of functions and their graphs, including domain, range, and the vertical line test. The study of functions was introduced in Section 4-7 of Book 1, and developed further in Sections 2-1, 4-1, and 9-4 of Book 2. Additional types of functions will be studied throughout the remainder of this unit, in Units 5, 9, and 10, and in future mathematics courses.
2-2	**Section 2-2** concentrates on the study of linear functions. Students review slope and intercept, write equations, find values, and study the domain and range of these functions. Section 2-2 of Book 2 introduced linear functions. Linear functions are also used in Sections 2-3, 2-8, and 6-4 of Book 3.
2-3	**Section 2-3** introduces piecewise functions and absolute value functions. Both functions use linear graphing skills, studied in Section 2-2 of Book 2 and Section 2-2 of Book 3, and the concepts of domain and range, studied in Section 2-1 of Book 2 and Section 2-2 of Book 3.
2-4	**Section 2-4** explores using a quadratic function to model various situations. Students have studied quadratic functions in Unit 10 of Book 1 and in Unit 4 of Book 2. In Section 6-5 of Book 3, a quadratic model will be fit to data through regression.
2-5	**Section 2-5** introduces polynomial functions. Students recognize, evaluate, and graph direct variation functions and polynomial functions. Transformations of some of these graphs will be studied in Unit 9 of Book 3. Polynomial functions will play an important role in many future mathematics courses.
2-6	**Section 2-6** introduces radical functions. Solving radical equations is explored. Solving equations involving square roots was introduced in Section 2-9 and Section 9-2 of Book 1 and is used throughout Book 2.
2-7	**Section 2-7** introduces rational functions and discusses how they are different from other types of functions. Rational functions are graphed and vertical and horizontal asymptotes are explored. Concepts involving asymptotes will be used in Sections 4-1, 4-7, 5-4, and 9-2 of Book 3.
2-8	**Section 2-8** introduces composite functions. The concept was informally introduced in Section 10-2 of Book 1, where $y = x^2$ is replaced with $y = (x + 2)^2$, and in Section 4-2 of Book 2. Students use composite functions to solve problems.

Integrating the Strands

Strands	Sections
Algebra	2-1, 2-2, 2-3, 2-4, 2-5, 2-6, 2-7, 2-8
Functions	2-1, 2-2, 2-3, 2-4, 2-5, 2-6, 2-7, 2-8
Geometry	2-1, 2-3, 2-5, 2-6, 2-8
Trigonometry	2-3
Statistics and Probability	2-4, 2-6, 2-7, 2-8
Discrete Mathematics	2-1, 2-2, 2-8
Logic and Language	2-1, 2-2, 2-3, 2-4, 2-5, 2-6, 2-7, 2-8

Exploring and Applying Functions

Section Planning Guide

- Essential exercises and problems are indicated in boldface.
- Ongoing work on the Unit Project is indicated in color.
- Exercises and problems that require student research, group work, manipulatives, or graphing technology are indicated in the column headed "Other."

Section	Materials	Pacing	Standard Assignment	Extended Assignment	Other
2-1	graphics calculator	Day 1 Day 2	**1–3,** 6, 7, **8–13, 16–20** **21–32,** 33–39, 40	**1–3,** 4–7, **8–13,** 14, 15, **16–20** **21–32,** 33–39, 40	
2-2		Day 1 Day 2	**1–12,** 13–15, **16–19** **20–22, 27, 28,** 29–34, 35	**1–12,** 13–15, **16–19** **20–22,** 23–26, **27, 28,** 29–34, 35	
2-3		Day 1 Day 2	**1–8, 10** **12–17,** 19–29, 30	**1–8,** 9, **10,** 11 **12–17,** 18–29, 30	
2-4	graphics calculator	Day 1	**1–8,** 12, 13, **14,** 16–23, 24	**1–8,** 10–13, **14,** 16–23, 24	9, 15
2-5	modeling clay, thin probe, spring or balance scale, graphing technology	Day 1 Day 2	1, **2–10, 13–21** **22–25, 28,** 30–34, 35, 36	1, **2–10,** 12, **13–21** **22–25,** 26, 27, **28,** 30–34, 35, 36	11 29
2-6	graphics calculator	Day 1 Day 2	**1–7, 9–16** 18, **19–39,** 45–47, 48, 49	**1–7, 9–16** 18, **19–39,** 41–43, 45–47, 48, 49	8, 17 40, 44
2-7	graphics calculator	Day 1	**1–18,** 23–26, 27	**1–18,** 19–26, 27	27
2-8		Day 1	**1–8, 12, 15–18,** 20–22, 23–25	**1–8,** 9–11, **12,** 13, 14, **15–18,** 20–22, 23–25	19
Review Test		**Day 1** **Day 2**	**Unit Review** **Unit Test**	**Unit Review** **Unit Test**	

Yearly Pacing	Unit 2 Total	Units 1–2 Total	Remaining	Total
	17 days (2 for Unit Project)	34 days	120 days	154 days

Support Materials

- See **Project Book** for notes on Unit 2 Project: Make a Pedal Pitch.
- UPP and disk refer to **Using Plotter Plus** booklet and **Plotter Plus** disk.
- TI-81/82 refers to **Using TI-81 and TI-82 Calculators** booklet.
- Warm-up exercises for each section are available on **Warm-Up Transparencies.**

Section	Study Guide	Practice Bank	Problem Bank	Activity Bank	Explorations Lab Manual	Assessment Book	Visuals	Technology
2-1	2-1	Practice 10	Set 3	Enrich 9	Master 2	Quiz 2-1		Function Plotter (disk)
2-2	2-2	Practice 11	Set 3	Enrich 10	Masters 1, 2	Quiz 2-2		Line Plotter (disk)
2-3	2-3	Practice 12	Set 3	Enrich 11	Add. Expl. 2 Masters 1, 2	Quiz 2-3 Test 5		UPP, page 50 Parabola Plotter (disk)
2-4	2-4	Practice 13	Set 4	Enrich 12	Masters 1, 2	Quiz 2-4		
2-5	2-5	Practice 14	Set 4	Enrich 13	Add. Expl. 3 Masters 2, 11	Quiz 2-5	Folder 2	TI-81/82, page 56 UPP, page 51
2-6	2-6	Practice 15	Set 4	Enrich 14		Quiz 2-6 Test 6		
2-7	2-7	Practice 16	Set 5	Enrich 15	Master 2	Quiz 2-7		
2-8	2-8	Practice 17	Set 5	Enrich 16	Master 1	Quiz 2-8 Test 7		
Unit 2	Unit Review	Practice 18	Unifying Problem 2	Family Involve 2		Tests 8, 9		

UNIT TESTS

Form A

Spanish versions of these tests are on pages 119–122 of the **Assessment Book.**

Name ______________________ Date __________ Score ______

Test 8

Test on Unit 2 (Form A)

Directions: Write the answers in the spaces provided.

1. For $f(x) = \frac{-2}{5x-4}$, find $f(-1)$ and $f(5)$.

Tell whether each graph represents a function. Write *Yes* or *No*.

2.

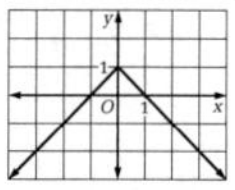

3.

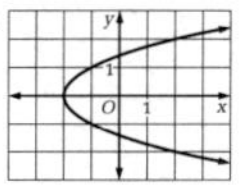

For Questions 4 and 5, write an equation for each function in the form $f(x) = mx + b$.

4. $f(-2) = 9, f(3) = -1$

5. $f(0) = -7, f(5) = 8$

6. Write a rule for a piecewise function of this graph.

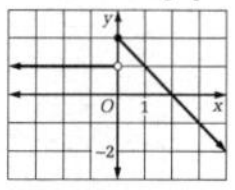

7. Graph this piecewise function.
$f(x) = \begin{cases} 2 & \text{if } x < 1 \\ x - 2 & \text{if } x \geq 1 \end{cases}$

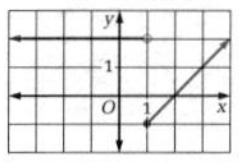

8. Graph the absolute value functions $y = |x - 2|$ and $y = |x| - 1$ on the same set of axes.

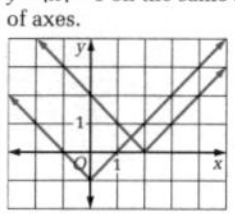

9. Graph the quadratic function $f(x) = (x + 1)^2$. Then find $f(-4)$ and $f(4)$.

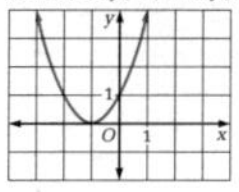

10. Which of the following are not quadratic functions?
A. $f(x) = -4x^2 + 5$ B. $f(x) = 0.3 + x$ C. $f(x) = \frac{-5}{x^2}$

Answers

1. $\frac{2}{9}, -\frac{2}{21}$
2. Yes.
3. No.
4. $f(x) = -2x + 5$
5. $f(x) = 3x - 7$
6. $y = \begin{cases} 1 & \text{if } x < 0 \\ 2 - x & \text{if } x \geq 0 \end{cases}$
7. *See question.*
8. *See question.*
9. 9; 25
10. B, C

10

Assessment Book INTEGRATED MATHEMATICS 3

Name ______________________ Date __________ Score ______

Test 8 *(continued)*

Directions: Write the answers in the spaces provided.

11. Which of the following is not a polynomial function?
A. $f(x) = \sqrt{3}x^4$ B. $f(x) = 0.5x^2 - 2x$ C. $f(x) = x^{-2} + 5x^{-1}$

12. Find the degree and the zeros of the polynomial function $f(x) = (x + 4)(x - 3)^2$.

13. Simplify $\sqrt{49x^2}$.

14. Solve $2x = \sqrt{8x + 21}$.

15. Graph the radical function $y = \sqrt{x + 1}$, and then find its domain and range.

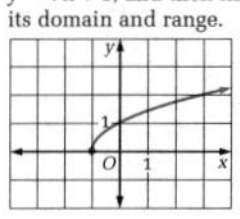

16. Graph the rational function $y = \frac{-2x}{x-1}$ and then find its domain and range.

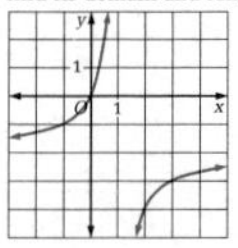

Let $f(x) = x^2 + 1$ and $g(x) = 3x^2$. Find each of the following.

17. $(f \circ g)(x)$ 18. $g(f(x))$ 19. $(g \circ g)(-2)$

For Questions 20 and 21, use the rational function $f(x) = \frac{x^2 + 9x + 8}{x + 2}$.

20. Find $f(-3)$.

21. Find x when $f(x) = 4$.

22. For the function $g(t) = \frac{3(t-3)}{t^2 - 5t + 6}$, find t when $g(t) = -6$.

23. **Open-ended** Give examples to show that the following statement is true. "All polynomial functions are rational functions, but not all rational functions are polynomial functions."
Sample answer: Any polynomial function, such as $f(x) = x^3 + 2x^2 + 1$, can be made into a rational function by writing the expression over a denominator of 1. A rational function, such as $f(x) = \frac{-2}{x-4}$, is not a polynomial function since there is a variable in the denominator.

24. **Writing** Define radical function. Give an example of a radical function.
Sample answer: A radical function is a function with a variable under a radical sign. The function $f(x) = \sqrt{x + 4}$ is a radical function.

Answers

11. C
12. 3; –4, 3
13. $7|x|$
14. $\frac{7}{2}$
15. D: $x \geq -1$; R: $y \geq 0$
16. D: $x \neq 1$; R: $y \neq -2$
17. $9x^4 + 1$
18. $3x^4 + 6x^2 + 3$
19. 432
20. 10
21. 0, –5
22. $\frac{3}{2}$
23. *See question.*
24. *See question.*

Assessment Book INTEGRATED MATHEMATICS 3

11

Form B

Name ______________________ Date __________ Score ______

Test 9

Test on Unit 2 (Form B)

Directions: Write the answers in the spaces provided.

1. For $f(x) = -2x^2 + 1$, find $f(-3)$ and $f(2)$.

Tell whether each graph represents a function. Write *Yes* or *No*.

2.

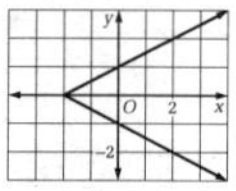

3.

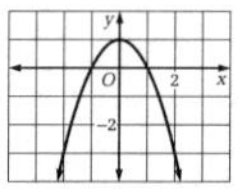

For Questions 4 and 5, write an equation for each function in the form $f(x) = mx + b$.

4. $f(5) = -1, f(-3) = 15$

5. $f(0) = -7, f(4) = 9$

6. Write a rule for a piecewise function of this graph.

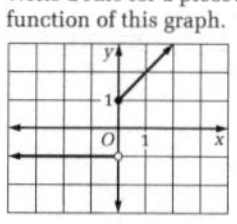

7. Graph this piecewise function.
$f(x) = \begin{cases} -1 & \text{if } x < -1 \\ 1 - x & \text{if } x \geq -1 \end{cases}$

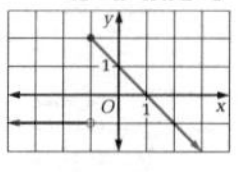

8. Graph the absolute value functions $y = |x + 2|$ and $y = |x| - 1$ on the same set of axes.

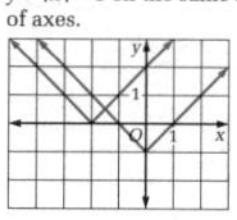

9. Graph the quadratic function $f(x) = (x - 1)^2$. Then find $f(-2)$ and $f(3)$.

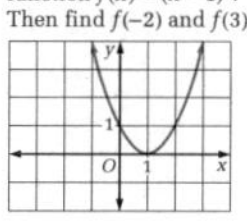

10. Which of the following are not quadratic functions?
A. $f(x) = 6 - 2x$ B. $f(x) = 4x^2 - 6$ C. $f(x) = \frac{x}{x+2}$

Answers

1. –17; –7
2. No.
3. Yes.
4. $f(x) = -2x + 9$
5. $f(x) = 4x - 7$
6. $y = \begin{cases} -1 & \text{if } x < 0 \\ x + 1 & \text{if } x \geq 0 \end{cases}$
7. *See question.*
8. *See question.*
9. 9; 4
10. A, C

12

Assessment Book INTEGRATED MATHEMATICS 3

Name ______________________ Date __________ Score ______

Test 9 *(continued)*

Directions: Write the answers in the spaces provided.

11. Which of the following is not a polynomial function?
A. $f(x) = \sqrt{7}x^3$ B. $f(x) = x^{-3} - 2x^{-2}$ C. $f(x) = 0.3x^2 - 6$

12. Find the degree and the zeros of the polynomial function $f(x) = x(x - 4)(x + 3)$.

13. Simplify $\sqrt{81x^2}$.

14. Solve $2x = \sqrt{5x + 9}$.

15. Graph the radical function $y = \sqrt{x - 1}$, and then find its domain and range.

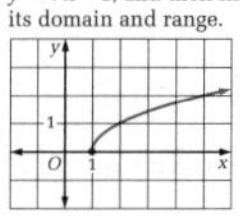

16. Graph the rational function $y = \frac{x}{x+2}$ and then find its domain and range.

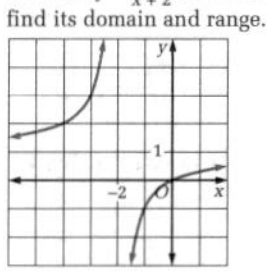

Let $f(x) = x^2 - 1$ and $g(x) = 2x^2$. Find each of the following.

17. $(g \circ f)(x)$ 18. $f(g(x))$ 19. $(g \circ g)(2)$

For Questions 20 and 21, use the rational function $f(x) = \frac{x^2 + 6x + 5}{x - 1}$.

20. Find $f(-4)$.

21. Find x when $f(x) = -5$.

22. For the function $h(t) = \frac{5(t-2)}{t^2 - 6t + 8}$, find t when $h(t) = -3$.

23. **Open-ended** Give examples to show that the following statement is true. "All quadratic functions are rational functions, but not all rational functions are quadratic functions."
Sample answer: Any quadratic function, such as $f(x) = x^2 + 2x + 1$, can be made into a rational function by writing the expression over a denominator of 1. A rational function, such as $f(x) = \frac{x-1}{x+2}$, is not a quadratic function because it cannot be rewritten in the form $f(x) = ax^2 + bx + c$.

24. **Writing** Define quadratic function. Give an example of a quadratic function.
Sample answer: A quadratic function is a function of the form $f(x) = ax^2 + bx + c$, where $a \neq 0$. The function $f(x) = 3x^2 - x + 5$ is a quadratic function.

Answers

11. B
12. 3; 0, 4, –3
13. $9|x|$
14. $\frac{9}{4}$
15. D: $x \geq 1$, R: $y \geq 0$
16. D: $x \neq -2$; R: $y \neq 1$
17. $2x^4 - 4x^2 + 2$
18. $4x^4 - 1$
19. 128
20. $\frac{3}{5}$
21. 0, –11
22. $\frac{7}{3}$
23. *See question.*
24. *See question.*

Assessment Book INTEGRATED MATHEMATICS 3

13

OUTSIDE RESOURCES

Books/Periodicals

"Are Graphs Just Pictures?" *The Language of Functions and Graphs,* Unit A2: pp. 74–81. The Shell Centre for Mathematical Education at the University of Nottingham, UK.

"Extending Some of the Sports Applications in This Volume." *Sourcebook of Applications of School Mathematics.* Prepared by Joint Committee of the MAA and NCTM: pp. 293–296. NCTM.

Activities/Manipulatives

Leiva, Miriam, Joan Ferrini-Mundy, and Loren P. Johnson. "Playing With Blocks: Visualizing Functions." *Mathematics Teacher* (November 1992): pp. 641–646.

Software

Myers, David L. *Plotter Plus.* Boston, MA: Houghton Mifflin Company, 1995. Macintosh and MS-DOS (worksheets included).

Videos

Videodisk: "Projectile Motion." *Science: Forces and Energy* (Side A). Macmillan/McGraw-Hill, 1993.

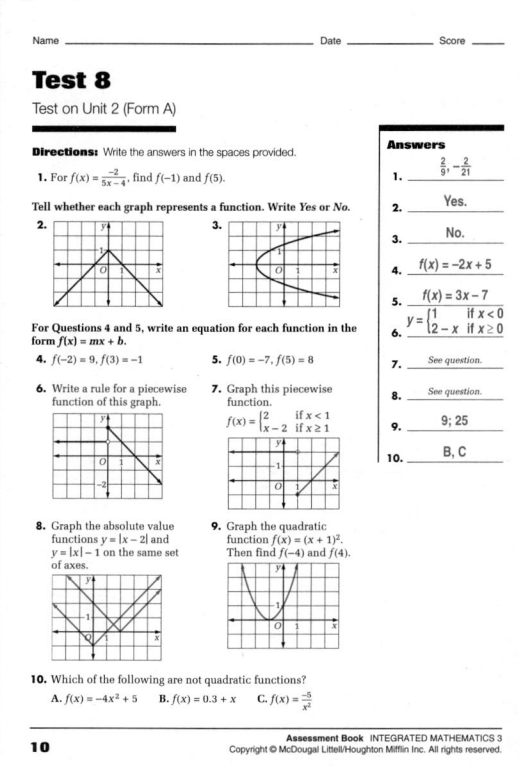

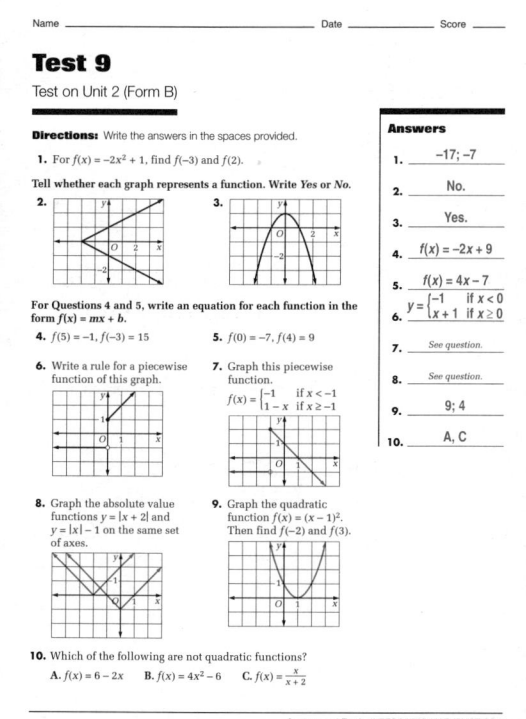

PROJECT GOALS

- Students give a presentation about bicycles to children in their community.
- Students work in a group of four to research their topics and plan a presentation.
- Each student in a group takes part in the presentation and uses posters, diagrams, or other props.

PROJECT PLANNING

Materials List

- Materials to make posters, charts, graphs, or other props

Project Teams

Have students work on the project in groups of four. Each group can look at the "Working on the Unit Project" exercises in the unit and each student in the group can choose one aspect of bicycling to research and report on.

Students should discuss how to coordinate their individual parts in the presentation and develop posters, diagrams, and other props to support their explanations. A preliminary outline should be developed at one of the first group meetings, which can then be revised as students begin to finalize their plans for the presentation.

unit 2 Functions

BICYCLING

What's a good way to avoid rush-hour traffic? Ride a bike! In China, Japan, Germany, and the Netherlands — countries with a high percent of bicycle owners — specially designed roads make it convenient for commuters to bike to work.

W&OD Railroad Park in Falls Church, Virginia

Cyclists are getting **on track**—railroad tracks! In 1963, naturalist May Theilgaard Watts had the idea of replacing abandoned railroad tracks with paths for walking and riding bicycles. Trails in almost every state now connect cities to towns and go through state and national forests.

1790 The first bicycles had no pedals. They were called "hobbyhorses".

1840 Pedals are added. Riders slide the pedals from front to back to move the rear wheel.

1870 The front wheel is made larger than the rear wheel so riders can go faster.

1885 A chain connects the pedals to the back wheel making a more efficient, safer bicycle.

66

Suggested Rubric for Unit Project

4 Each student in a group has chosen a specific aspect of bicycling and has researched it thoroughly. The outline of the presentation shows careful planning and each group member's role in it. Students' supporting props are neat and appropriately support their presentations. The presentations themselves are clear, complete, and interesting. All mathematical information is not only correct, but is also presented in a way that does not confuse the children in the audience.

3 Students have chosen different aspects of bicycling, but at least one group member has not done a very good job of researching it for background material. A preliminary outline has been written and revised, but each student's role in the presentation is not entirely clear. The props are good but could have been improved with more thought and effort. The presentation itself generally goes well, with some students doing a significantly better job than one or two others. Sufficient time is not allotted for a summary or for questions.

Unit Project 2

Make a Pedal Pitch

Your project is to give a presentation about bicycles to children in your community.

Each group member will choose one aspect of bicycling and plan to present information on it. For example, you may choose bicycle safety, bicycling for exercise, bike racing,or the physics of bicycles.

As you complete the "Working on the Unit Project" exercises in this unit, decide how you will use the information in your presentation. Save your work to support the data you present.

Bike racing tracks are made of asphalt, clay, concrete, or wood. Most racers prefer the smooth finish of a wood track, like the one in the Major Taylor **Velodrome** in Indianapolis, named after the cyclist who was the first African-American world champion in any sport.

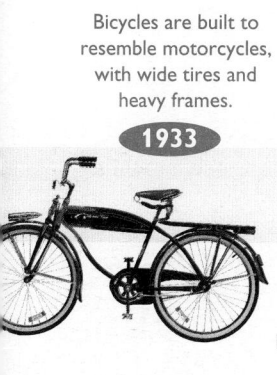

Bicycles are built to resemble motorcycles, with wide tires and heavy frames.

1933

1981

The mountain bicycle combines wide tires with high-technology construction.

Solid disk wheels improve the aerodynamics of racing bicycles.

1990

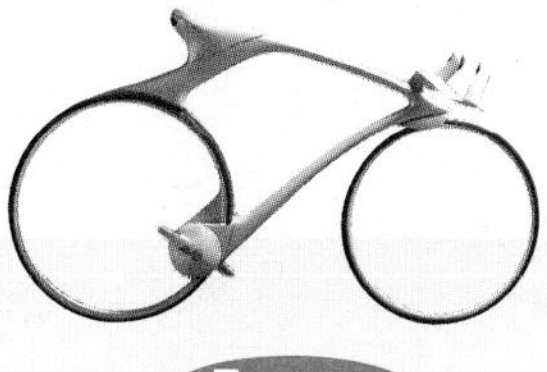

Future

Superconductive magnets that power the pedals and hold up the rims promise lighter weight and better aerodynamics.

67

Support Materials

The ***Project Book*** contains information about the following topics for use with this Unit Project.

- Project Description
- Teaching Commentary
- Working on the Unit Project Exercises
- Completing the Unit Project
- Assessing the Unit Project
- Alternative Projects
- Outside Resources

Students Acquiring English

You might want to encourage students acquiring English to create a miniglossary of cycling terms relevant to their particular cycling project.

ADDITIONAL BACKGROUND

Multicultural Note

Marshall W. "Major" Taylor was known as the "Fastest Bicycle Rider in the World" until his retirement in 1910. By the end of that year, he had won twenty-one races, placed second in thirteen, and came in third in eleven. Taylor is now recognized as the first African-American champion in any professional sport. Many millions of people throughout the world look to cycling not as a competitive sport but as a means of transportation and recreation today. In Shanghai, China, over four million people make the daily commute to work on bicycles.

Suggested Rubric for Unit Project

2 The students in the group do not fully carry out all aspects of the planning, research, prop-making, and outlining to deliver a well-coordinated presentation. Each phase of the project shows significant room for improvement. The presentation itself is often not clear and either too much or too little time is spent on either the overview, each member's part, the summary, or time for questions.

1 Group members do not work well together and fail to organize a coherent presentation. After students have chosen their topics, they do not follow up with appropriate research or outlines. Props are either not made or have been made so hastily that they are ineffective. The end result of the preliminary work by the group to develop a presentation is so poor that no presentation can actually be given. Students should be encouraged to speak with the teacher as soon as possible to review their work and to make a new start on the project.

The Zero Bike

A new bicycle design, based on emerging transportation technology being explored in Europe and Japan, has been created by Makota Makita and Hiroshi Tsuzaki, formally students at the Art Center College of Design in Los Angeles. This theoretical bicycle has wheel rims cradled by magnets and would be powered by magnetic pedals. The zero bike has a sculpted look and the wheels would turn without friction. The two designers think the technology they used has great potential for future bike design.

ALTERNATIVE PROJECTS

Project 1, page 132

Graphs

Look through published materials such as other books, magazines, and newspapers for graphs of polynomial and piecewise functions. Prepare a brief description of how each function is used in real-world situations and what it describes.

Project 2, page 132

Falling Objects

Find out how air resistance affects the velocity of a falling object by experimenting with different objects to find the constant of air resistance. Write a description of the results, using diagrams and function notation.

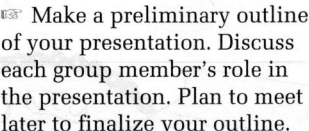

Unit Project 2 — Getting Started

For this project you should work in a group of four students. Here are some ideas to help you get started.

☞ Decide which topics you want to discuss. You may want to look at the "Working on the Unit Project" exercises at the end of each section for ideas. Decide which group member will present each topic.

☞ Make a preliminary outline of your presentation. Discuss each group member's role in the presentation. Plan to meet later to finalize your outline.

☞ In your group discuss how you can use posters, diagrams, and props to make your descriptions and explanations as clear and complete as possible.

☞ Plan to research your topics. You can use books about bicycles or talk to cyclists or bicycle experts in your community.

Working on the Unit Project

Your work in Unit 2 will help you prepare your presentation.

Related Exercises:

Section 2-1, Exercise 40
Section 2-2, Exercise 35
Section 2-3, Exercise 30
Section 2-4, Exercise 24
Section 2-5, Exercises 35, 36
Section 2-6, Exercises 48, 49
Section 2-7, Exercise 27
Section 2-8, Exercises 23–25

Alternative Projects p. 132

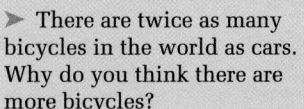

Can We Talk BICYCLING

➤ There are twice as many bicycles in the world as cars. Why do you think there are more bicycles?

➤ What are the advantages of commuting to school or work by bicycle? What are the disadvantages? What are some other uses of bicycles in your community?

➤ Why do you think there are special bicycles for racing and for traveling on dirt roads?

➤ What safety hazards do you need to watch for while riding a bicycle?

➤ What protective equipment do you think cyclists should wear? Why?

Answers to Can We Talk?

➤ Answers may vary. Examples are given. Bicycles are much less expensive to buy than cars and require much less maintenance. Bicycles take up less space and are easier to use in highly populated cities. Children and young teenagers can own a bicycle, whereas car owners must be at least 16 years of age in the United States and in many other countries.

➤ Answers may vary. Examples are given. Some advantages are that they provide good exercise, do not need an energy source to run, and do not require large parking areas. Some disadvantages are that they are slower, cannot be used in bad weather, and are difficult to ride wearing certain clothing. Bicycles can be used for delivering small objects and for recreation and exercise.

➤ Answers may vary. An example is given. Different uses require different bicycles. For example, a light-weight bicycle with a thin tire is used best for speed. On a dirt road, a sturdy bicycle with a thick tire is necessary.

➤ Answers may vary. Examples are given. vehicles, such as cars, trucks, or buses, construction areas, holes, or objects in the road

➤ Answers may vary. Examples are given. All cyclists should wear a safety helmet because head injuries are the most dangerous that can occur in bicycle accidents. Gloves for the hands and elbow and knee pads are useful because these are the parts of the body most likely to absorb the impact of a fall.

Section **2-1**

What Is a Function?

It All Depends

Focus

Recognize and describe functions.

Your fastest walking speed in miles per hour depends on your leg length in feet. Maximum speed = $\sqrt{32 \times \text{length}}$

inflcion
Fri·day [fráidi] *s.* viernes.
fried [fraid] *adj.* frito; freído; *p.p. de* **to fry.**
friend [frend] *s.* amigo, amiga.
friend·li·ness [fréndlinis] *s.* afabilidad; amistad.
friend·ly [fréndli] *adj.* amistoso, afable, amigable; …picio, favorable; *adv.* amistosamente.
…**ship** [fréndʃip] *s.* amistad.
fright·en [fráitn] *v.* espantar, asustar, atemorizar.

A dictionary pairs words with their meanings.

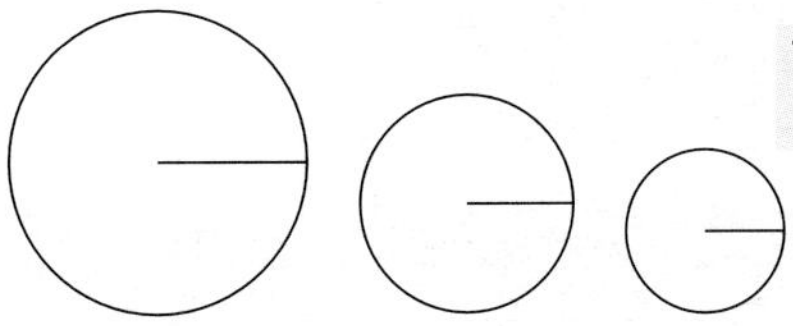

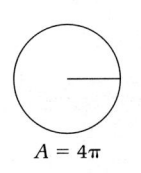

The area of a circle depends on its radius.

Everybody has a birthday.

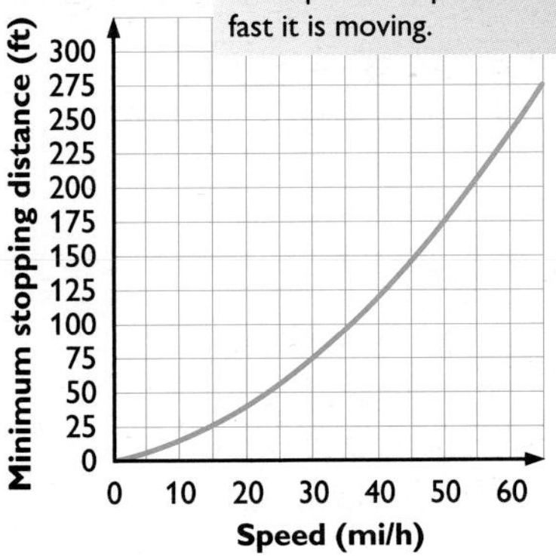

The minimum distance needed to stop a car depends on how fast it is moving.

Every real number except 0 has a reciprocal.

x	$\frac{1}{x}$
-2	-0.5
-1	-1
0	undefined
1	1
2	0.5

Each of these relationships has two variables. A variable that determines, or controls, another variable is called a *control variable.* A variable that is determined by, or depends on, another variable is called a *dependent variable.*

In the relationship between speed and stopping distance, the control variable is the speed of the car and the dependent variable is the distance needed to stop.

PLANNING

Objectives and Strands

See pages 66A and 66B.

Spiral Learning

See page 66B.

Materials List

- Graphics calculator
- Graph paper

Recommended Pacing

Section 2-1 is a two-day lesson.

Day 1

Pages 69–72: Opening paragraph through Talk it Over 6, *Exercises 1–20*

Day 2

Pages 72–73: Graphs and Functions through Look Back, *Exercises 21–40*

Extra Practice

See pages 610–612.

Warm-Up Exercises

- Warm-Up Transparency 2-1

Support Materials

- Practice Bank: Practice 10
- Activity Bank: Enrichment 9
- Study Guide: Section 2-1
- Problem Bank: Problem Set 3
- Explorations Lab Manual: Diagram Master 2
- Using IBM/Mac Plotter Plus Disk: Function Plotter
- Assessment Book: Quiz 2-1, Alternative Assessment 1

TEACHING

Integrating the Strands

The concept of a function is a powerful unifying idea that helps to integrate many of the diverse strands of mathematics. Notice that the definition of a function is given in very general terms as a *relationship* that pairs each value of the *control* variable with only one value of the *dependent* variable. Relationships between two variables can be found in number theory, algebra, geometry, trigonometry, statistics, probability, discrete mathematics, logic, and other branches of mathematics.

Students Acquiring English

Students acquiring English may benefit from developing a nonmathematical definition of the word *variable*. Discussing what things in nature and in everyday life can be characterized as *variable* will reinforce mathematical concepts and develop English learners' language skills.

Additional Sample

S1 Does this table represent a function? If so, find its domain and range.

Occupants	
1 person	\$85.00
2 people	\$105.00
3 people	\$120.00
4 people	\$130.00

This table represents a function. The domain is all values of the control variable: 1, 2, 3, 4. The range is all values of the dependent variable: 85, 105, 120, 130.

Talk it Over

Questions 1–3 check students' understanding of the concepts of a dependent variable and the domain and range of a function.

A *function* is a relationship for which each value of the control variable is paired with *only one* value of the dependent variable.

One word may have more than one meaning. This does not represent a function.

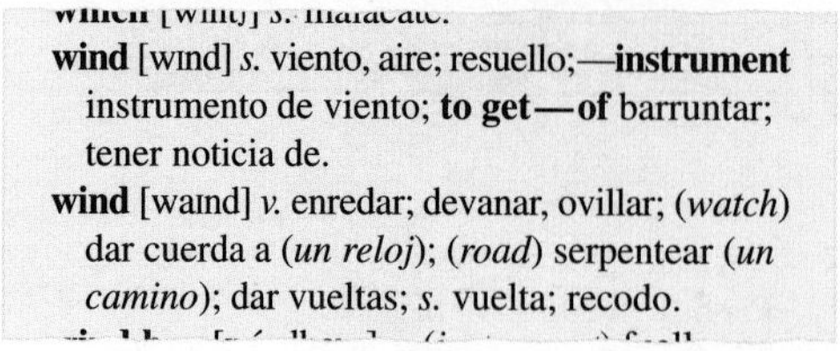
wind [wɪnd] *s.* viento, aire; resuello;—**instrument** instrumento de viento; **to get—of** barruntar; tener noticia de.
wind [waɪnd] *v.* enredar; devanar, ovillar; (*watch*) dar cuerda a (*un reloj*); (*road*) serpentear (*un camino*); dar vueltas; *s.* vuelta; recodo.

A person's birthday can be on only one date. This does represent a function.

Domain and Range

The **domain of a function** is all possible values of the control variable. The **range of a function** is all possible values of the dependent variable.

Sample 1

Does this table represent a function? If so, find its domain and range.

Tune-ups	
4 cylinders	\$49.95
6 cylinders	\$59.95
8 cylinders	\$69.95

Sample Response

Each value of the control variable (number of cylinders) is paired with only one value of the dependent variable (cost of tune-up). This table represents a function.

The domain is all values of the control variable: 4, 6, 8.

The range is all values of the dependent variable: 49.95, 59.95, 69.95.

Talk it Over

Questions 1–3 refer to the functions shown on page 69.

1. In the relationship between fastest walking speed and length of legs, which is the dependent variable?
2. What is the domain of the reciprocal function?
3. What is the range of the birthday function?

Answers to Talk it Over

1. walking speed
2. all real numbers except 0
3. all 366 days of the year

Notation for Functions

When a variable y is a function of a variable x, you can use **function notation:**

dependent variable (output) → $y = f(x)$ ← The notation $f(x)$ is read "f of x."

control variable (input) points to x

function name (Other letters can be used.) points to f

This means that y represents the value of the function f for a value of x.

When a function has a rule for pairing x-values with y-values, you can write:

$$y = f(x) = \text{rule}$$

For example, if the reciprocal function is named f,

$$y = f(x) = \frac{1}{x}.$$

Watch Out!
$f(x)$ does *not* mean f times x.

This statement is usually written in one of these ways:

$$y = f(x) \qquad y = \frac{1}{x} \qquad f(x) = \frac{1}{x}$$

The *values of a function* are the numbers in its range. One value of the reciprocal function is $f(2) = 0.5$.

Read "$f(2) = 0.5$" as → "f of 2 equals 0.5." → "the value of f at 2 is 0.5."

Sample 2

a. Use $f(x) = 3x^2 - 1$. Find $f(2)$ and $f(a)$.

b. Use $A(r) = \pi r^2$. Find $A(0.5)$ and $A(3 + x)$. Leave your answer in terms of π.

Sample Response

a. $f(x) = 3x^2 - 1$

$f(2) = 3(2)^2 - 1$ ← Substitute 2 for x.

$= 11$ ← Simplify.

$f(x) = 3x^2 - 1$

$f(a) = 3a^2 - 1$ ← Substitute a for x.

b. $A(r) = \pi r^2$

$A(0.5) = \pi(0.5)^2$ ← Substitute 0.5 for r.

$= \pi(0.25)$

$= 0.25\pi$

$A(r) = \pi r^2$

$A(3 + x) = \pi(3 + x)^2$ ← Substitute $3 + x$ for r.

$= \pi(9 + 6x + x^2)$

$= 9\pi + 6\pi x + \pi x^2$

Communication: Reading

You may wish to read the paragraph that introduces the notation for functions together with students in class. Take time to have students discuss the ideas presented because they provide the foundation for all subsequent work with functions.

Additional Sample

S2 a. Use $f(x) = x^2 + 2x$. Find $f(-2)$ and $f(a)$.

$f(x) = x^2 + 2x$

$f(-2) = (-2)^2 + 2(-2)$

$= 4 + (-4)$

$= 0;$

$f(x) = x^2 + 2x$

$f(a) = a^2 + 2a$

b. Use $V(r) = \frac{4}{3}\pi r^3$. Find $V(3)$ and $V(6k)$. Leave your answer in terms of π.

$V(r) = \frac{4}{3}\pi r^3$

$V(3) = \frac{4}{3}\pi(3)^3$

$= \frac{4}{3}\pi(27)$

$= 36\pi;$

$V(r) = \frac{4}{3}\pi r^3$

$V(6k) = \frac{4}{3}\pi(6k)^3$

$= \frac{4}{3}\pi(216k^3)$

$= 288\pi k^3$

Talk it Over

Questions 4–6 provide a check on students' understanding of function notation through the use of mathematical and real-world functions. Question 6(b) relates the concept of $f(0)$ to the graph of a function.

Teaching Tip

The word "many" in the term "many-to-one" simply means *two or more.* To ensure understanding, ask students to sketch graphs that would show various types of possibilities: one-to-one for the entire domain; one-to-one for negative values of x but many-to-one for positive values of x; one-to-one except for the values of x from 0 to 10, and so on. Students should understand that if the function is many-to-one for any part of its domain, then it is automatically considered to be a many-to-one function.

Additional Sample

S3 a. Is the function $f(x) = x^3$ *one-to-one* or *many-to-one*?

Write the function as $y = x^3$. Graph the function.

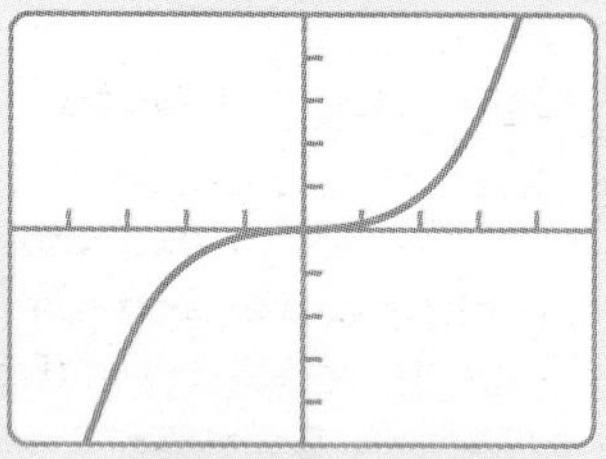

No horizontal line will intersect the graph in more than one point. The function is one-to-one.

b. Find the domain and range of f.

Any real number can be cubed. The domain is all real numbers. Any real number is the cube of another real number. The range is all real numbers.

Talk it Over

4. Write a rule for $C(r)$, the circumference of a circle as a function of the radius.
5. The birthday function is $B(\text{person}) = \text{date}$. What is $B(\text{you})$?
6. The minimum stopping distance function on page 69 is defined by the equation $y = x + \frac{x^2}{20}$.
   a. Choose a name for the function and write it in function notation.
   b. What is the value of the function at 0? How does the graph on page 69 show this?

Now you are ready for: Exs. 1–20 on pp. 73–75

Graphs and Functions

You can represent functions with graphs.

On a coordinate grid, the domain is on the x-axis and the range is on the y-axis.

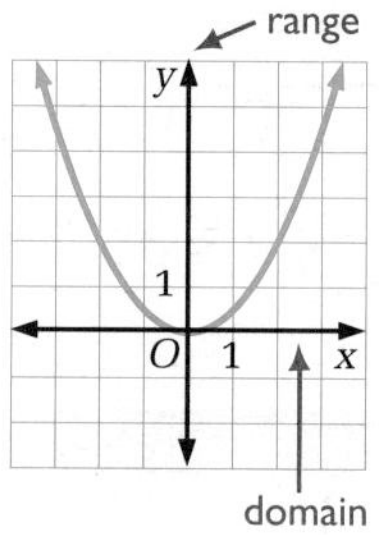

If no vertical line crosses a graph in two or more points, the graph represents a function.

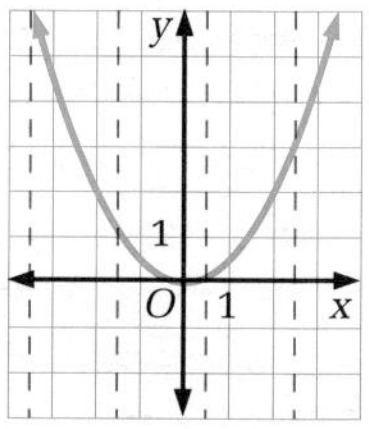

The minimum stopping distance function on page 69 is an example of a **one-to-one function.** Each member of the range is paired with exactly one member of the domain. If no horizontal line crosses the graph of a function in two or more points, the function is one-to-one.

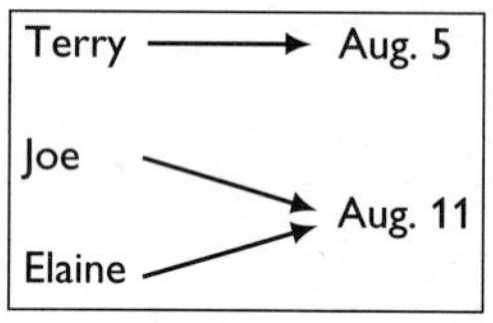

The birthday function is an example of a **many-to-one function.** A member of the range (a date) may be paired with more than one member of the domain (people).

Sample 3

a. Is the function $f(x) = x^2$ *one-to-one* or *many-to-one*?

b. Find the domain and range of f.

Answers to Talk it Over

4. $C(r) = 2\pi r$
5. Answers may vary. An example is given. $B(\text{George Washington}) = \text{February 22}$
6. a. Answers may vary. An example is given. Let m be the minimum stopping distance function; $m(x) = x + \frac{x^2}{20}$.
   b. 0; The point (0, 0) is on the graph. That is, for $x = 0$, $m(x) = 0$.

Answers to Look Back

Answers may vary. An example is given. You can tell if a graph represents a function by using the vertical-line test. If any two points of the graph lie on the same vertical line, the graph does not represent a function. In a table, determine whether any domain value (usually values in the first column) is paired with more than one range value (usually values in the second column). If so, the table does not represent a function.

Sample Response

a. Write the function as $y = x^2$. Graph the function.

The graph shows two different domain values, 2 and -2, paired with the same range value 4. The function is many-to-one.

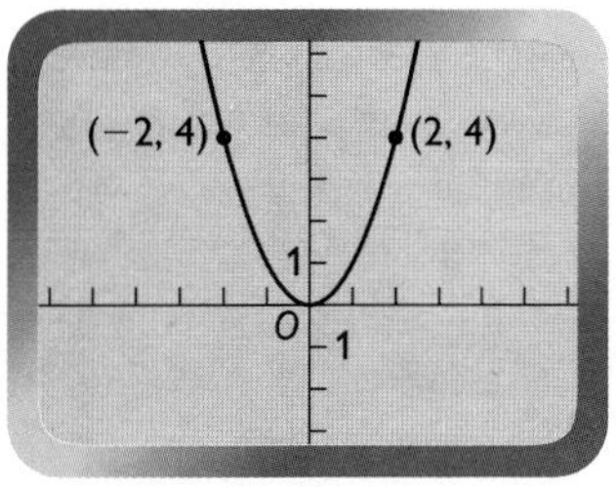

b. Any real number can be squared. The domain is all real numbers.

The square of any number is nonnegative. The range is all positive real numbers and zero. You can also describe the range by writing $y \geq 0$.

Look Back

How can you tell if a graph or a table of values represents a function?

Now you are ready for: Exs. 21–40 on pp. 75–76

2-1 Exercises and Problems

1. a. **Reading** Give an example of a relationship on pages 69–70 that is not a function.
   b. Explain how you can tell it is not a function.
2. a. Does this cast list represent a function? Explain why or why not.
   b. If so, what are the domain and range of the function?
3. Does this table give a person's phone number as a function of his or her name? Explain why or why not.

A Raisin in the Sun
a play by Lorraine Hansberry

The Cast, in order of appearance

Ruth Younger .... Libra Woods
Travis Younger .... Kyle Turner
Walter Lee Younger .... Warren Moore
Beneatha Younger .... Denise Loundes
Lena Younger .... Kenyatta Johnson
Joseph Asagai .... Steven Taylor
George Murchison .... Jerome Roberts
Karl Lindner .... Richard Wilson
Bobo .... Teritus Fortune

TECHNICAL CREW LIST

NAME	PHONE
Giles	555-5678 (workshop)
	555-7325 (house)
Isaac	555-2587
Kalisha	555-1234 (home)
	555-5432 (work)

Answers to Exercises and Problems

1. a. the relation that pairs words and meanings in the dictionary
   b. The word *friendly* on page 69, for example, is paired with several meanings.
2. a. Yes; no domain value (role) is paired with more than one range member (actor).
   b. The domain is the list of roles and the range is the list of actors.
3. No; the domain values Giles and Kalisha are both paired with two range values.

Look Back

This question can be used to check students' understanding of the meaning of a function. Students can respond verbally and ask questions among themselves until an understanding emerges from the discussion.

APPLYING

Suggested Assignment

Day 1

Standard 1–3, 6–13, 16–20

Extended 1–20

Day 2

Standard 21–40

Extended 21–40

Integrating the Strands

Algebra Exs. 8–15, 18, 19, 28, 29, 32, 34–36, 38, 39

Functions Exs. 1–33, 40

Geometry Exs. 5–7, 16, 17, 20, 40

Discrete Mathematics Exs. 16–20, 37

Logic and Language 1, 4, 25, 30, 33, 37, 40

Multicultural Note

Lorraine Hansberry (1930–1965) had a short but prolific career as a playwright, reporter, political activist, essayist, and screenwriter. In addition to writing *A Raisin in the Sun*, a play that has become an American classic, Hansberry wrote numerous articles on African American art, history, and women's rights. In 1959, Hansberry received the New York Drama Critics Circle award for Best Play of the Year. She was the first African American and one of the first women to gain this honor.

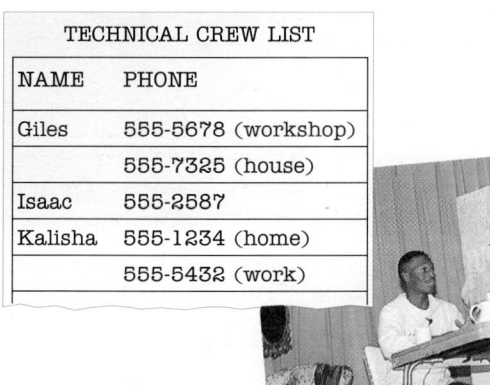

Error Analysis

Some students may interpret function notation as implied multiplication. If this is the case, have these students reread the discussion on page 71, paying particular attention to the Watch Out note.

Career Note

A surveyor is a person whose occupation is to make land surveys. A land survey shows the shape, boundaries, and position of a tract of land. Surveyors use the mathematics of geometry and trigonometry to calculate linear and angular measurements. Their main tool is a transit, which can be used to determine angles, bearings, and levels. It is mounted on a tripod. A weight called a plumb bob is suspended directly below the telescope of the transit to determine true vertical. Surveyors are generally employed by engineering firms that provide services for the construction of commercial buildings and roads, or with government agencies involved in surveying large tracts of land.

4. **Writing** The domain of a sales relationship is *merchandise items* and the range is their *prices*.
   a. Explain why, in any one store, customers expect the relationship to be a function.
   b. Explain why, in any one city, this relationship is probably not a function.
5. a. Use the diagram to explain why the area of an equilateral triangle is given by $A(s) = \frac{\sqrt{3}}{4}s^2$.
   b. Does this relationship represent a function?
   c. What is the control variable? the dependent variable?
   d. What are the domain and range of the function?

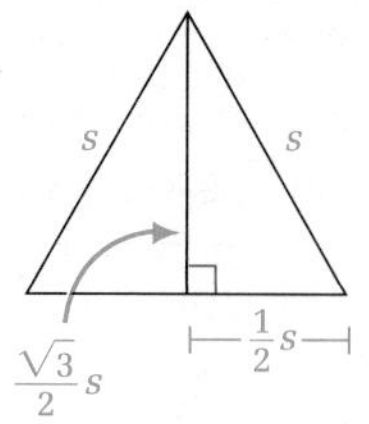

6. Write a rule for $S(e)$, the surface area of a cube as a function of the length of one edge.
7. Write a rule for $V(e)$, the volume of a cube as a function of the length of one edge.

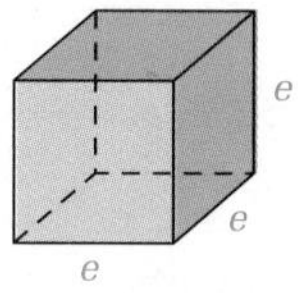

For each function:

a. Find $f(2)$. **b. Find $f(-4)$.** **c. Find $f(a + b)$.**

8. $f(x) = 6x^2$
9. $f(s) = \frac{1}{s+1}$
10. $f(x) = \frac{2x}{3+x}$
11. $f(t) = |t|$
12. $f(x) = 3x + 4$
13. $f(x) = 4x^2 - 3x + 1$

Career Very hot or very cold temperatures can affect the length of a steel tape measure. A surveyor can find out how the length of the tape measure changes by using a formula. For a tape calibrated at 68°F, the formula is:

$$C(T) = 0.00000645L(T - 68)$$

$C(T)$ is the change in the length of the tape, L is the original length of the tape, and T is the temperature at which the surveyor uses the tape.

14. Explain why the temperature is the control variable in this situation.
15. a. Write a formula for the change in the length of a tape originally 100 ft long.
    b. Suppose the tape is used at a temperature of 16°F. Use your formula to find the length of the tape.

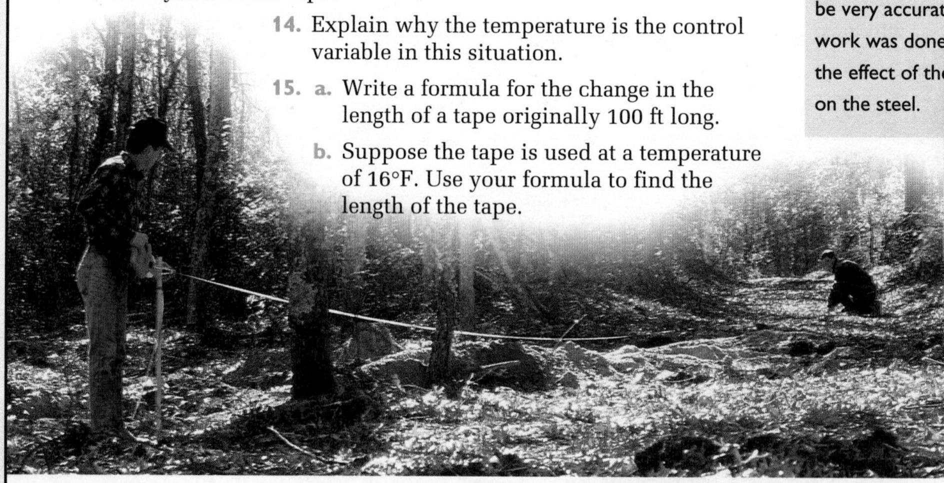

BY THE WAY...

The Gateway Arch in St. Louis is made of two triangular columns of steel. To make sure that the two columns would meet at the top, every measurement had to be very accurate. All survey work was done at night to avoid the effect of the heat of the sun on the steel.

Answers to Exercises and Problems

4. Answers may vary. An example is given.
   a. A customer would expect items that were exactly the same to have the same price all over the store; that is, no item for sale would have two different prices.
   b. The same item in two different stores might well have different prices.
5. a. Each altitude of an equilateral triangle divides the triangle into two 30°-60°-90° triangles. If the length of a side of the original triangle is s, the resulting right triangles have sides $\frac{1}{2}s$, s, and $\frac{\sqrt{3}}{2}s$. The area of the original triangle is $\frac{1}{2}$ × base × height = $\frac{1}{2}s \times \frac{\sqrt{3}}{2}s = \frac{\sqrt{3}}{4}s^2$.
   b. Yes; no two equilateral triangles with sides of the same length have different areas.
   c. length of a side, s; area $A(s)$
   d. all positive real numbers; all positive real numbers
6. $S(e) = 6e^2$
7. $V(e) = e^3$
8. a. 24
   b. 96
   c. $6a^2 + 12ab + 6b^2$
9. a. $\frac{1}{3}$ b. $-\frac{1}{3}$
   c. $\frac{1}{a+b+1}$
10. a. $\frac{4}{5}$ b. 8
    c. $\frac{2a+2b}{3+a+b}$
11. a. 2 b. 4
    c. $|a + b|$
12. a. 10 b. -8
    c. $3a + 3b + 4$
13. a. 11
    b. 77
    c. $4a^2 + 8ab + 4b^2 - 3a - 3b + 1$

16. The *truth function*, *T*, assigns the value 1 (true) or 0 (false) to statements. For example, *T*(All squares are rectangles.) = 1 and *T*(All circles are squares.) = 0. What are the domain and range of the truth function?

Find each value of the truth function, which is defined in Exercise 16.

17. *T*(All rectangles are parallelograms.)

18. *T*(If $-2x \geq -6$, then $x \geq 3$.)

19. *T*(If $x < 0$, then $\sqrt{x}$ is not a real number.)

20. *T*(All parallelograms are squares.)

Tell whether each graph represents a function.

21.

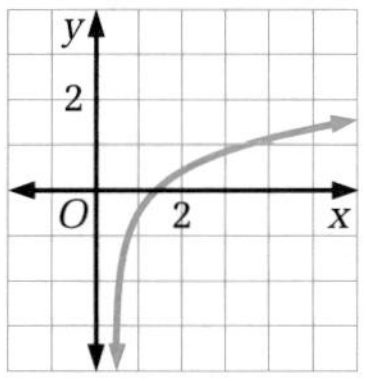

22.

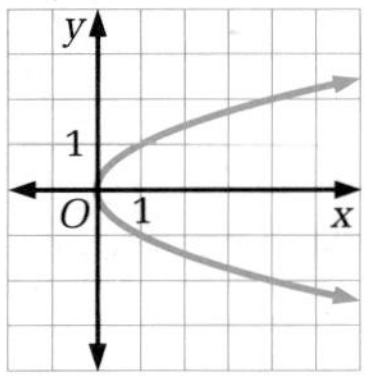

23.

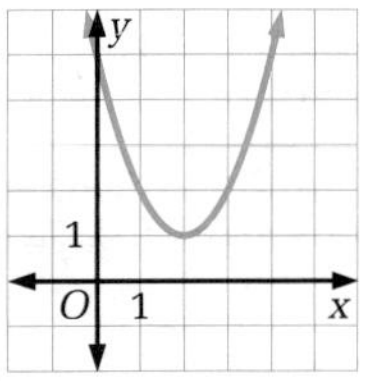

24. **Open-ended** Draw a graph that does not represent a function.

25. **Writing** A function can also be represented by a *mapping diagram,* in which two vertical number lines represent the domain and range. Akeem drew mapping diagrams for three relationships. Which is a *one-to-one function*, which is a *many-to-one function*, and which is *not a function*? Explain your answer.

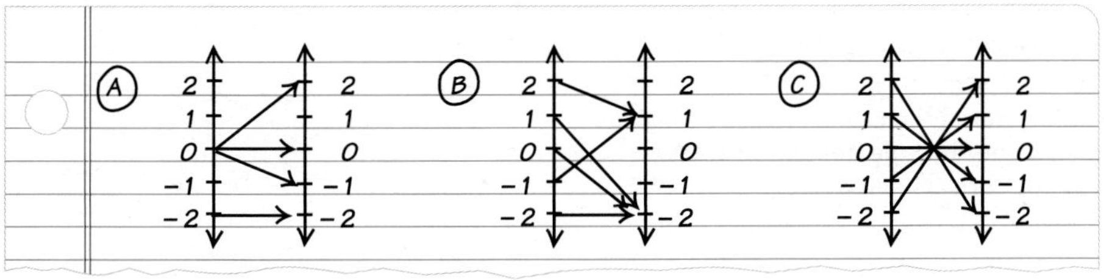

For Exercises 26–29, tell whether each function is *one-to-one* or *many-to-one*. Explain your answer.

26.

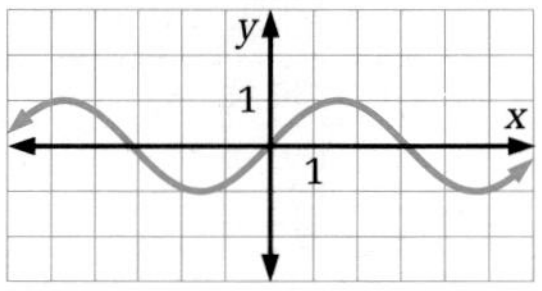

27.

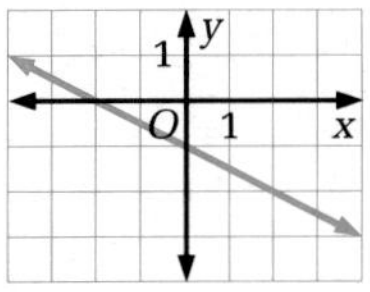

28. $g(x) = \frac{3}{5}x$

29. $c(x) = x^3$

30. **Reading** How does the graph in Sample 3 show that the range of *f* is the nonnegative real numbers?

*__31.__ Find the domain and range of the function shown in Exercise 23.

*__32.__ Find the domain and range of $f(x) = \sqrt{x}$.

Reasoning

Graphics calculators and computers operate in the same way as the truth function in Ex. 16. Students can experiment by working on the home screen, using inequalities that contain no variables. They can enter inequalities to see if they can deduce how the calculator evaluates them. Simple inequalities are easy. For example, since $3 < 2$ is false, it has the value 0. A calculator or computer will give the same result. When compound inequalities are considered, the situation changes. Ordinarily, we would consider $3 < 2 < 4$ to be false. A graphics calculator will deliver a value of 1 for $3 < 2 < 4$, meaning that it considers the expression to be true. Obviously, the calculator is "thinking" differently. Students who wish to understand more about this topic can consult the manuals for the calculators they are using.

Assessment: Open-ended

After completing Exs. 21–29, students should be able to draw a graph and a mapping diagram for a one-to-one function and a one-to-many function.

Answers to Exercises and Problems

14. The length of the tape depends on the temperature of its environment.

15. a. $C(T) = 0.000645(T - 68)$ or $C(T) = 0.000645T - 0.04386$

b. 99.96646 ft

16. The domain is all logical statements that can be determined to be true or false; the range is 0 and 1.

17. 1 **18.** 0 **19.** 1

20. 0 **21.** Yes. **22.** No.

23. Yes.

24. Answers may vary. An example is given.

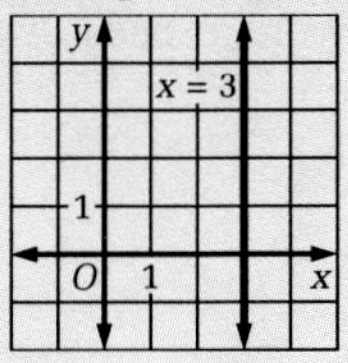

25. *B* and *C* are functions, since no member of either domain is paired with more than one member of the range. *B* is many-to-one, since two members of the domain are paired with the same member of the range. *C* is one-to-one, since every member of the range is paired with exactly one member of the domain. *A* is not a function, since 0 is paired with three different members of the range.

26. many-to-one; Any horizontal line $y = c$ for $-1 \leq c \leq 1$ intersects the graph in infinitely many points.

27. one-to-one; No horizontal line intersects the graph in more than one point.

28. one-to-one; Each value *y* of the range is paired with exactly one *x*-value.

29. one-to-one; Each value *y* of the range is paired with exactly one *x*-value.

30. No part of the graph lies in the third or fourth quadrants.

31. all real numbers; all real numbers greater than or equal to 1

32. all nonnegative real numbers; all nonnegative real numbers

Cooperative Learning

You may wish to have students work on Ex. 33 in small groups. This will provide them with an excellent opportunity to discuss key concepts associated with a function.

Working on the Unit Project

Students can work in their project groups to complete Ex. 40. They should save their answers to the questions for use in preparing the final presentation about bicycles to children in your community.

Practice 10 For use with Section 2-1

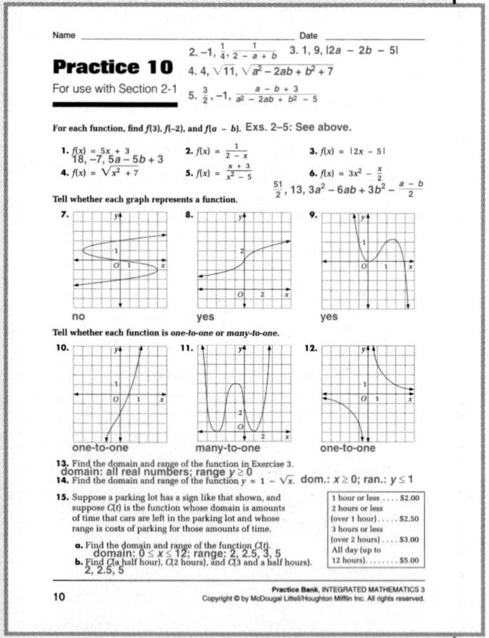
Name ______________________ Date ______________

Practice 10
For use with Section 2-1

2. $-1, \frac{1}{4}, \frac{1}{2-a+b}$ 3. 1, 9, $|2a - 2b - 5|$
4. 4, $\sqrt{11}$, $\sqrt{a^2 - 2ab + b^2 + 7}$
5. $\frac{3}{2}, -1, \frac{a-b+3}{a^2-2ab+b^2-5}$

For each function, find $f(3)$, $f(-2)$, and $f(a - b)$. Exs. 2–5: See above.

1. $f(x) = 5x + 3$ 18, −7, $5a - 5b + 3$
2. $f(x) = \frac{1}{2-x}$
3. $f(x) = |2x - 5|$
4. $f(x) = \sqrt{x^2 + 7}$
5. $f(x) = \frac{x+3}{x^2-5}$
6. $f(x) = 3x^2 - \frac{x}{2}$ $\frac{51}{2}$, 13, $3a^2 - 6ab + 3b^2 - \frac{a-b}{2}$

Tell whether each graph represents a function.

7. no
8. yes
9. yes

Tell whether each function is *one-to-one* or *many-to-one*.

10. one-to-one
11. many-to-one
12. one-to-one

13. Find the domain and range of the function in Exercise 3. domain: all real numbers; range $y \geq 0$
14. Find the domain and range of the function $y = 1 - \sqrt{x}$. dom.: $x \geq 0$; ran.: $y \leq 1$
15. Suppose a parking lot has a sign like that shown, and suppose $C(t)$ is the function whose domain is amounts of time that cars are left in the parking lot and whose range is costs of parking for those amounts of time.

1 hour or less	$2.00
2 hours or less (over 1 hour)	$2.50
3 hours or less (over 2 hours)	$3.00
All day (up to 12 hours)	$5.00

a. Find the domain and range of the function $C(t)$. domain: $0 \leq x \leq 12$; range: 2, 2.5, 3, 5
b. Find C(a half hour), C(2 hours), and C(3 and a half hours). 2, 2.5, 5

10 Practice Bank, INTEGRATED MATHEMATICS 3

Ongoing **ASSESSMENT**

33. **Writing** Describe a function that could represent the information in this mail-order catalog shipping chart. Include in your description a discussion of domain, range, and whether the function is one-to-one.

Shipping and Handling

The chart below shows how much your shipping and handling will be.

Order size	Shipping charges
$ 30.00 – 49.99	$ 7.35
$ 50.00 – 99.99	$ 8.60
$100.00 – 149.99	$10.30
$150.00 – 199.99	$13.10
$200.00 – 299.99	$16.45
$300.00 – 499.99	$22.80
$500.00 – 749.99	$30.65
$750.00 – 999.99	$48.45
$1,000 – 2,000.00	$58.65

Review **PREVIEW**

Graph the feasible region for each system of inequalities. Label each vertex with its coordinates. *(Section 1-8)*

34. $x \geq 0$
$y \geq 0$
$x - y \leq 4$

35. $y \leq 0$
$y \leq \frac{3}{2}x + 3$
$y \leq -x + 1$

36. $x \geq 2$
$2 \leq y \leq 5$
$y \leq -\frac{1}{2}x + 10$

37. Write an algorithm for preparing for your mathematics class. *(Section 1-1)*

38. Solve the inequality $8 < 3x - 10$. *(Section 1-1)*

39. Find the value of y in the equation $y = \frac{1}{4}x - 3$ when $x = -3$. *(Toolbox Skill 12)*

2 Working on the Unit Project

40. The diameter of the circle made by the front wheel of your bicycle depends on how much you turn the handlebars.
 a. Is the relationship between the angle of the turn and the diameter of the circle a function? Explain why or why not.
 b. **Open-ended** What do you think is a reasonable domain for the relationship? Explain your choice.

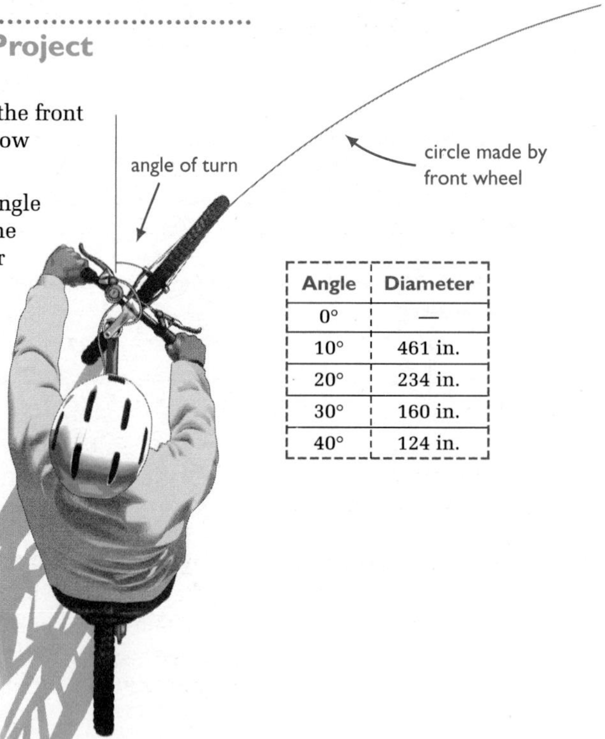

Angle	Diameter
0°	—
10°	461 in.
20°	234 in.
30°	160 in.
40°	124 in.

Answers to Exercises and Problems

33. Answers may vary. An example is given. The control variable is the size of the order and the dependent variable is the shipping charge. The domain is dollar amounts greater than or equal to $30 and less than or equal to $2000. The range is dollar amounts greater than or equal to $7.35 and less than or equal to $58.65. The function is many-to-one.

34.
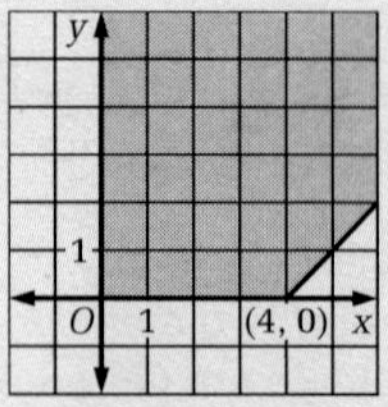

35.
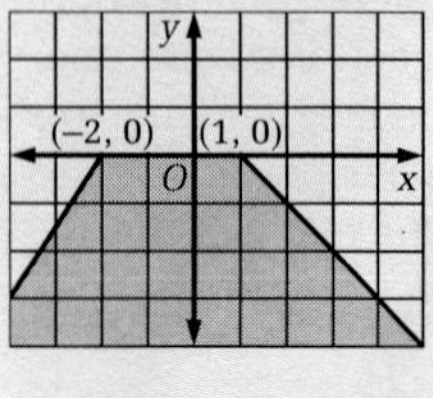

36.
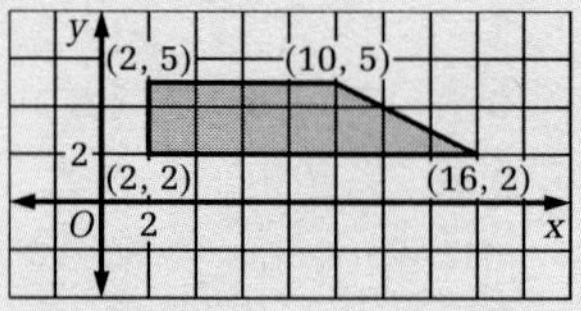

37. See answers in back of book.

38. $x > 6$

39. −3.75

40. a. Yes; for each angle, the diameter of the circle will be different.
 b. Answers may vary. An example is given. I think a reasonable domain is angle measures greater than or equal to 0° and less than about 75°. Explanations may vary.

Section 2-2

Linear Functions

Focus

Use linear functions to describe situations. Find the domain and range of a linear function.

STRAIGHT TALK

When you are camping high up in the Swiss Alps, water boils at a lower temperature than it does when you are camping in Death Valley. This happens because altitude affects the boiling temperature of water.

PLACE	Altitude (feet)	Boiling temperature of water (°F)
Cartagena, Colombia	Sea level (0 ft)	212
Fribourg, Switzerland	1929	209
Kimberley, South Africa	4013	205
Colorado Springs, United States	6012	201
Thimphu, Bhutan	7950	198

TECHNOLOGY NOTE

See the Technology Handbook, p. 605, for graphing a fitted line.

Talk it Over

1. Make a scatter plot of the data in the table. Use altitude as the control variable. What pattern do you see in the graph?
2. Does the graph represent a function? Explain why or why not.
3. a. Use your graph to estimate the boiling temperature of water when the altitude is 7000 ft.
   b. Draw a fitted line through the points on your graph. Use it to estimate the altitude when the boiling temperature is 207°F.
4. Where does the line cross the vertical axis? What does this point represent?

PLANNING

Objectives and Strands

See pages 66A and 66B.

Spiral Learning

See page 66B.

Materials List

- Graph paper

Recommended Pacing

Section 2-2 is a two-day lesson.

Day 1

Pages 77–79: Opening paragraph through Talk it Over 12, *Exercises 1–19*

Day 2

Pages 80–81: Domain and Range of a Linear Function through Look Back, *Exercises 20–35*

Extra Practice

See pages 610–612.

Warm-Up Exercises

Warm-Up Transparency 2-2

Support Materials

- Practice Bank: Practice 11
- Activity Bank: Enrichment 10
- Study Guide: Section 2-2
- Problem Bank: Problem Set 3
- Explorations Lab Manual: Diagram Masters 1, 2
- Using IBM/Mac Plotter Plus Disk: Line Plotter
- Assessment Book: Quiz 2-2

Answers to Talk it Over

1. See answers in back of book.
2. The graph appears to represent a function. It appears that no two points on the graph lie on the same vertical line.
3. a. Estimates may vary. Example: about 200°F
   b. Estimates may vary. Example: about 2800 ft

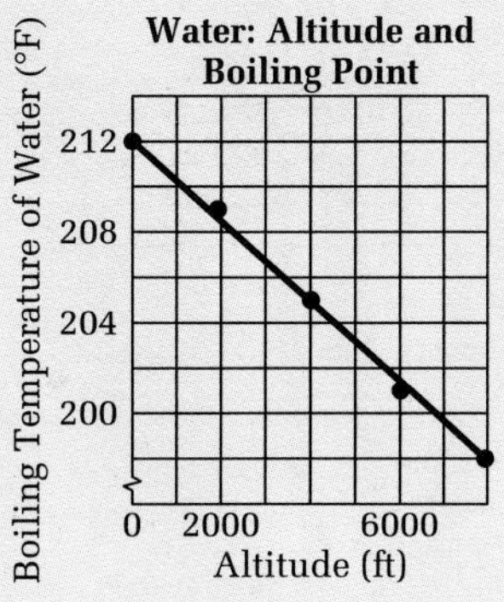

4. 212; the boiling temperature of water at an altitude of 0 ft, or sea level

TEACHING

Talk it Over

Questions 1–4 use real-world data, a scatter plot, and a fitted line to have students begin thinking about concepts associated with linear functions.

Teaching Tip

You may want to have students review what they learned about slope from *Integrated Mathematics 1* and *2*. Students should recall that slope is the ratio of rise over run and is the measure of the steepness of a line when it is drawn in the coordinate plane. Students should also recall that vertical lines have no slope and are not graphs of functions.

Additional Sample

S1 Yuen Shun rides his bicycle to school. The table gives data about how the time he has been riding relates to his distance from the school.

Time (min)	Distance from School (km)
5	5.25
8	4.8
10	4.5
20	3.0

a. Describe how the distance from school changes in relation to the time traveled.

Let t = time traveled (in min).
Let $d(t)$ = distance from school (in km) as a function of time.
Use any two pairs of data in the table. For example, use (5, 5.25) and (10, 4.5).
The slope describes how $d(t)$ changes as t changes.

$$m = \frac{\text{change in } d(t)}{\text{change in } t}$$

$$= \frac{4.5 - 5.25}{10 - 5}$$

$$= -0.15$$

If the time increases by 1 min, then the distance decreases by 0.15 km.

Some relationships, such as the one between altitude and boiling temperature, can be modeled by *linear functions*.

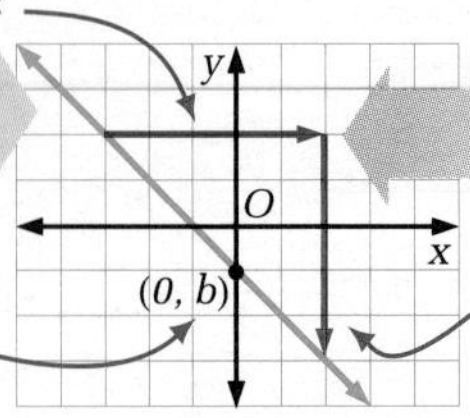
change in x
You can tell that a function is linear by looking at its graph. The graph is a line.
The value of b is the vertical intercept.
y
O
x
(0, b)
change in f(x)
You can also look at how f(x) changes when x changes by a constant amount: The change is constant.

$$m = \frac{\text{change in } f(x)}{\text{change in } x}$$

The ratio of the change in $f(x)$ to the change in x is the slope, m. Any **linear function** can be written in the form $f(x) = mx + b$.

Sample 1

a. Use the information in the table on the previous page. Describe how the boiling temperature of water changes in relation to altitude.

b. Write an equation for the boiling temperature of water as a function of altitude.

Sample Response

a. Let a = the altitude (in feet).

Let $T(a)$ = the temperature (in degrees Fahrenheit) as a function of altitude.

Use any two pairs of data in the table.
For example, use Fribourg (1929, 209) and Thimphu (7950, 198).

$$m = \frac{\text{change in } T(a)}{\text{change in } a}$$

The slope describes how $T(a)$ changes as a changes.

$$= \frac{198 - 209}{7950 - 1929}$$

$$= \frac{-11}{6021}$$

$$\approx -0.0018$$

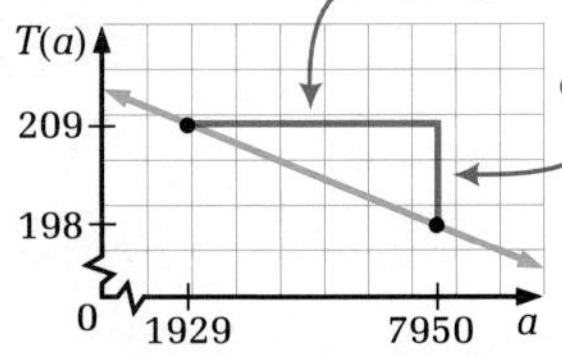
change in a
T(a)
change in T(a)
209
198
0
1929
7950
a

If the altitude *increases* by 1 ft, the boiling temperature of water *decreases* by about 0.0018°F.

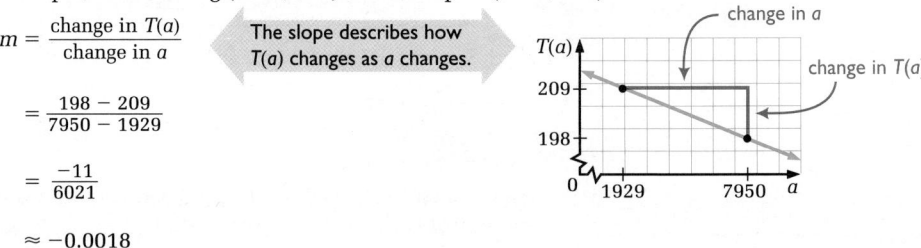

b. $\begin{pmatrix}\text{Boiling}\\ \text{temperature at}\\ \text{any altitude}\end{pmatrix} = \begin{pmatrix}\text{Boiling}\\ \text{temperature}\\ \text{at sea level}\end{pmatrix} + \begin{pmatrix}\text{Change in boiling}\\ \text{temperature per}\\ \text{foot of altitude}\end{pmatrix} \times \begin{pmatrix}\text{Altitude}\\ \text{(in ft)}\end{pmatrix}$

$$T(a) \quad = \quad 212 \quad + \quad (-0.0018)a$$

If the altitude is a, then the boiling point of water is $T(a) = -0.0018a + 212$.

Answers to Talk it Over

5. The coefficient of a is negative. The slope of the graph is negative, so as a increases, $T(a)$ decreases. As the values in the second column (altitudes) increase, the values in the third column (temperatures) decrease.
6. Cathie used the form of a linear function with the calculated value of m and

Talk it Over

5. How does the equation $T(a) = -0.0018a + 212$ in Sample 1 show that the boiling temperature of water *decreases* as the altitude *increases*? How does the graph show this? How does the table of data show it?
6. Cathie found an equation for the function in Sample 1 this way after finding the slope. Explain Cathie's method.
7. Choose two other data points from the table on page 77. Use them to find the equation for the boiling temperature of water as a function of altitude. Is your answer the same as the one in Sample 1?

$m = -0.0018$
$f(x) = mx + b$
$T(a) = -0.0018a + b$
$198 = -0.0018(7950) + b$
$198 = -14.31 + b$
$212 \approx b$
An equation of the function is $T(a) = -0.0018a + 212$.

Tell whether each function is linear. Explain how you know.

8. $f(x) = \frac{1}{2}x - 13$
9. $y = 4x$
10. $g(x) = x^2 + 0.6$

Sample 2

Use $f(x) = \frac{1}{3}x + 17$.

a. Find $f(9)$.

b. Find x when $f(x) = -1$.

Sample Response

a. Substitute 9 for x in the function.

$$f(x) = \frac{1}{3}x + 17$$
$$f(9) = \frac{1}{3}(9) + 17$$
$$= 20$$

b. Substitute -1 for $f(x)$ in the function and solve for x.

$$f(x) = \frac{1}{3}x + 17$$
$$-1 = \frac{1}{3}x + 17$$
$$-18 = \frac{1}{3}x \quad \leftarrow \text{Subtract 17 from both sides.}$$
$$-54 = x \quad \leftarrow \text{Multiply both sides by 3.}$$

Talk it Over

11. Describe how you can use the equation from Sample 1 to find the boiling temperature of water at an altitude of 10,000 ft. Find this temperature.
12. Describe how you can use the equation from Sample 1 to find the altitude of Shiraz, Iran, where the boiling temperature of water is 185°F. Find the altitude of Shiraz.

Now you are ready for: Exs. 1–19 on pp. 81–82

b. Write an equation for Yuen Shun's distance from school as a function of the time he has been riding.
Using the slope calculated, work back from the data point (5, 5.25) to see that he lives 6 km from school. Since (0, 6) is the vertical intercept of the graph, $d(t) = -0.15t + 6$.

Talk it Over

Question 5 asks students to analyze how the function, the graph, and the data all show that the boiling temperature decreases as the altitude increases. Question 6 leads students to examine another approach for finding the equation for the function $T(a) = -0.0018a + 212$. In Question 7, students go back to the table on page 77 to see if the equation that results from using different table values yields the same equation as Sample 1. Questions 8–10 check students' understanding of whether a particular function is linear or not.

Additional Sample

S2 Use $g(x) = \frac{4}{5}x - 8$.

a. Find $g(10)$.
Substitute 10 for x in the function.

$$g(x) = \frac{4}{5}x - 8$$
$$g(10) = \frac{4}{5}(10) - 8$$
$$= 8 - 8$$
$$= 0$$

b. Find x when $g(x) = 3$.
Substitute 3 for $g(x)$ in the function and solve for x.

$$g(x) = \frac{4}{5}x - 8$$
$$3 = \frac{4}{5}x - 8$$
$$11 = \frac{4}{5}x$$
$$13.75 = x$$

Answers to Talk it Over

substituted the a and $T(a)$ values from one of the pairs in the table. She solved the resulting equation for b.

7. Answers may vary. An example is given. For Kimberley and Colorado Springs, the value of m is $\frac{201 - 205}{6012 - 4013} = -\frac{4}{1999} \approx -0.002$ and the equation is $T(a) = -0.002a + 212$. The equation will vary only slightly, since for any two points in the table, the value of m rounded to three decimal places is -0.002.
8. Yes; the equation is written in the form $f(x) = mx + b$ with $m = \frac{1}{2}$ and $b = -13$.
9. Yes; the equation is written in the form $f(x) = mx + b$ with $m = 4$ and $b = 0$.
10. No; the equation cannot be written in the form $f(x) = mx + b$.
11. Substitute 10,000 for a; $T(10{,}000) = 212 + (-0.0018)(10{,}000) = 194$; 194°.
12. Substitute 185 for $T(a)$; $185 = 212 + (-0.0018)a$; $a = 15{,}000$; 15,000 ft.

Additional Sample

S3 Refer to Additional Sample S1. The library and a video store are along the route that Yuen Shun takes from home to school. He passes the library 12 minutes after he leaves home and the video store another 10 minutes after that. Use this information to find the domain and range of $d(t) = -0.15t + 6$ for the portion of his route from the library to the video store.

Yuen Shun passes the library when $t = 12$ and the video store when $t = 12 + 10 = 22$. So the domain for this portion of the route is all possible values of t for which $12 \le t \le 22$. The range is all values of $d(t)$ between $d(12)$ and $d(22)$.

When $t = 12$:

$$d(t) = (-0.15)(12) + 6 = 4.2$$

When $t = 22$:

$$d(t) = (-0.15)(22) + 6 = 2.7$$

The range is $2.7 \le d(t) \le 4.2$.

Mathematical Procedures

Students should be able to answer two basic questions about functions. First, how can you tell whether a particular number is in the domain of a function? Second, how can you tell whether a particular number is in the range of a function?

If the function is defined by an equation of the form $f(x)$ = <expression in x>, ask whether the number can be put in place of the variable to obtain a specific numerical result. If the answer is yes, then that number belongs to the domain. Otherwise, it does not.

To decide whether a certain number a is in the range of $f(x)$, replace $f(x)$ with a and examine the equation

a = <expression in x>.

Ask whether the equation can be solved for x. If there is at least one value of x that is a solution of the equation, then the number a is in the range of $f(x)$. Otherwise, it is not.

Domain and Range of a Linear Function

Any value of x can be substituted into $f(x) = mx + b$. This means that the domain of a linear function consists of all real numbers. Any value of $f(x)$ can be produced by the formula. This means that the range of a linear function also consists of all real numbers when $m \neq 0$.

Watch Out!
When a function represents an everyday situation, there may be limits on the domain and range.

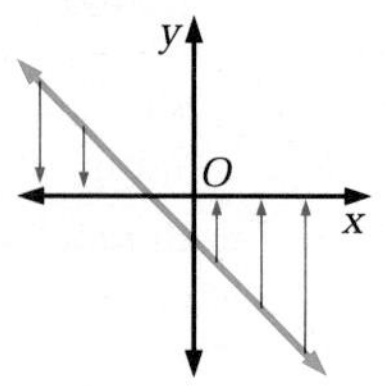

The graph of a linear function shows that the domain covers the entire x-axis.

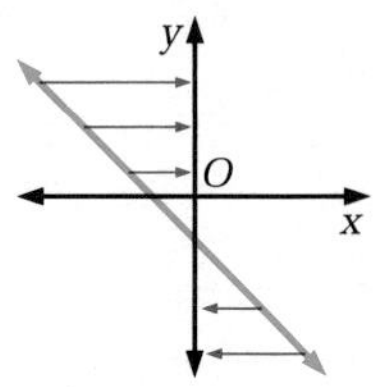

The graph also shows that the range covers the entire y-axis.

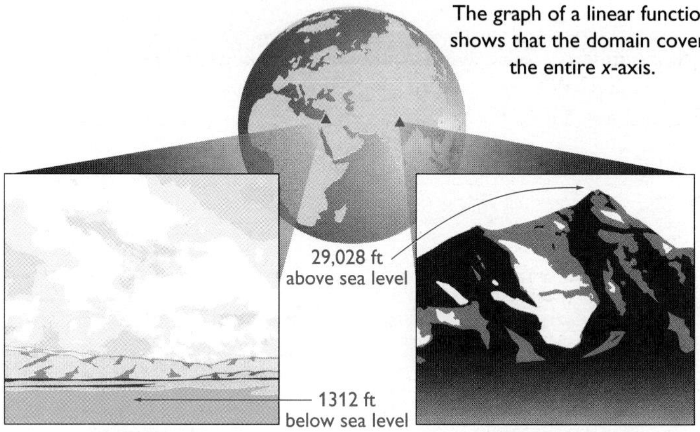

Dead Sea **Mount Everest**

Sample 3

The altitude of Mount Everest, the highest place on Earth, is 29,028 ft. The lowest place on the surface of Earth is the Dead Sea, 1312 ft below sea level.

Use this information to find the domain and range of $T(a) = -0.0018a + 212$ on the surface of Earth.

Sample Response

The domain is all possible values for a: $-1312 \le a \le 29{,}028$.

The range is all possible values for $T(a)$ over this domain.

When $a = -1312$:

$$T(a) = -0.0018(-1312) + 212 \approx 214.36$$

When $a = 29{,}028$:

$$T(a) = -0.0018(29{,}028) + 212 \approx 159.75$$

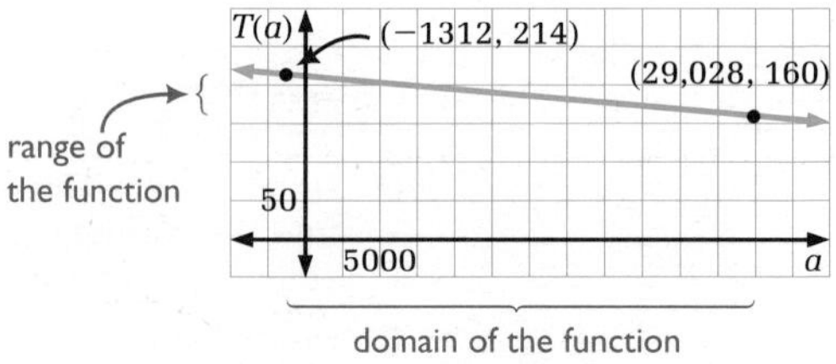

The range is about $160 \le T(a) \le 214$.

LINEAR FUNCTIONS

A linear function has the form $f(x) = mx + b$.

slope ↗ ↖ vertical intercept

$$m = \frac{\text{change in } f(x)}{\text{change in } x}$$

$$b = f(0)$$

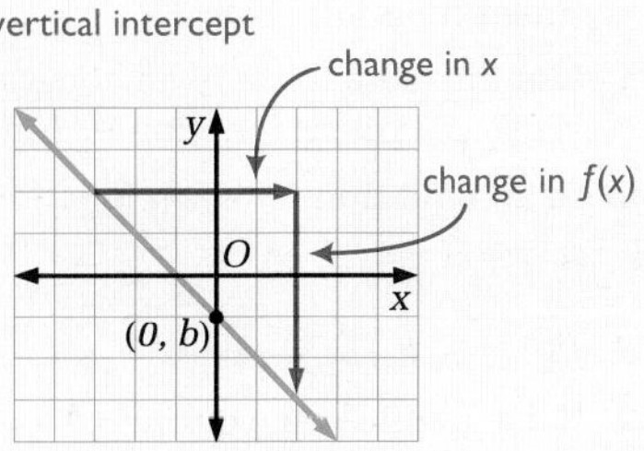

The graph of a linear function is a line.

Look Back

How many data points do you need to find the equation of a linear function? Why?

Now you are ready for: Exs. 20–35 on pp. 82–84

2-2 Exercises and Problems

1. **Reading** How can you tell that a function is linear?

Tell whether each graph represents a linear function.

2.
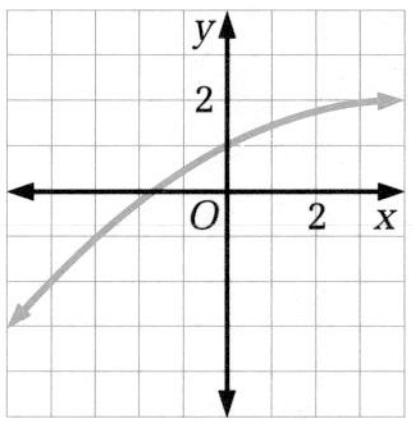

3.
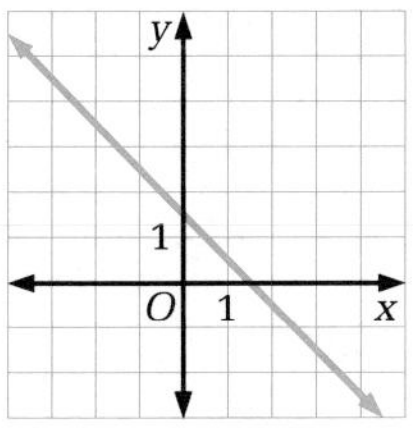

4.
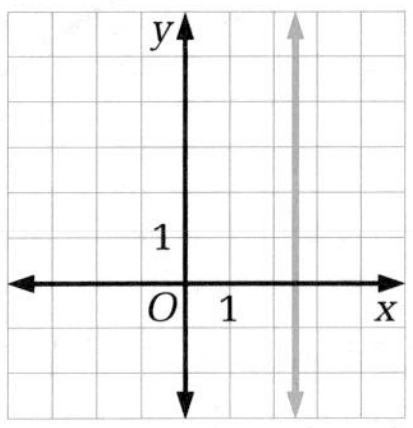

5.
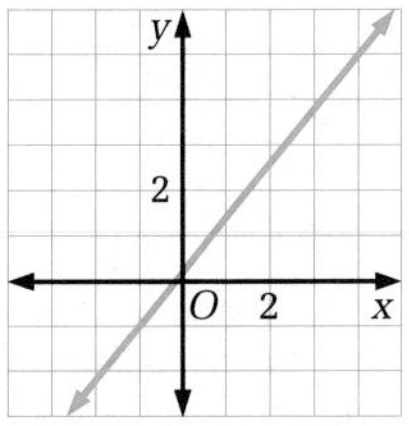

6. a. Which of the graphs in Exercises 2–5 represent a direct variation?
   b. Suppose the slope of a direct variation is m. Write an equation for this direct variation.

Student Resources
p. 672 *Visual Glossary of Functions*

Tell whether each function is linear. If it is, find $f(2)$, $f(-2)$, and $f(a + b)$.

7. $f(x) = 2 - x^2$
8. $f(x) = -3x - 6$
9. $f(a) = 4 + 10a$
10. $f(x) = x^3 - x$
11. $f(x) = \frac{3}{2}x + 4$
12. $f(d) = \frac{1}{d}$

Look Back

It is important that students be able to explain correctly why two data points are needed to find the equation of a linear function.

APPLYING

Suggested Assignment

Day 1

Standard 1–19

Extended 1–19

Day 2

Standard 20–22, 27–35

Extended 20–35

Integrating the Strands

Algebra Exs. 6–12, 15–22, 25–28, 30, 32–35

Functions Exs. 1–30, 35

Discrete Mathematics Ex. 31

Logic and Language Exs. 1, 20, 23, 24, 29, 31, 35

Answers to Look Back

You need two data points. Two data points allow you to determine the slope. You can then find the *y*-intercept using the slope, the coordinates of one point, and the slope-intercept form of an equation.

Answers to Exercises and Problems

1. Its graph is a line. Its equation can be written in the form $y = mx + b$.
2. No.
3. Yes.
4. No.
5. Yes.
6. a. the graph in Exercise 5
   b. $y = mx$ or $f(x) = mx$
7. No.
8. Yes; -12, 0, $-3a - 3b - 6$
9. Yes; 24, -16, $4 + 10a + 10b$
10. No.
11. Yes; 7, 1, $\frac{3}{2}a + \frac{3}{2}b + 4$
12. No.

Problem Solving

In modeling and solving real-world problems, it is important for students to remember that function equations must be used in a sensible manner. Almost always, the domain and range for a real-world situation is only a small portion of the domain and range for a function when it is viewed as a purely mathematical entity. Also, real-world data that display, say, a linear relationship over most of the domain, may suddenly become nonlinear at the extreme lower end or upper end of the domain.

Failure to keep these things in mind can lead to unrealistic conclusions. In Ex. 21, for example, $C(190) = 604$. But a cricket would not be alive for long at 190°F, let alone produce 604 chirps per minute.

Communication: Discussion

After students complete the exercises and problems in this section, discuss briefly what they should be able to do.

1. Recognize that functions of the form $y = mx + b$ are linear functions.
2. Evaluate a linear function $f(x)$ for given values of x.
3. Find a rule for a linear function $f(x)$ given any two ordered pairs $(x_1, f(x_1))$ and $(x_2, f(x_2))$.
4. Recognize that the domain and range of a linear function consist of real numbers.

Use the table of shoe sizes for Exercises 13–15.

Women's Shoe Sizes	
American	**European**
5	36
7	38
9	40
11	42

13. Suppose a woman's American shoe size is 8. What European size would you recommend in a walking shoe?
14. Do you think the relationship in the table is an example of a linear function? Why?
15. Use the table to write an equation for European women's shoe sizes as a linear function of American women's shoe sizes.

Write an equation for each function in the form $f(x) = mx + b$.

16. $f(3) = 9, f(5) = 15$
17. $f(-2) = 4, f(2) = -16$
18. $f(0) = 5, f(-5) = 10$
19. $f(10) = 4, m = \frac{1}{2}$
20. Each graph is an example of a *constant function*.

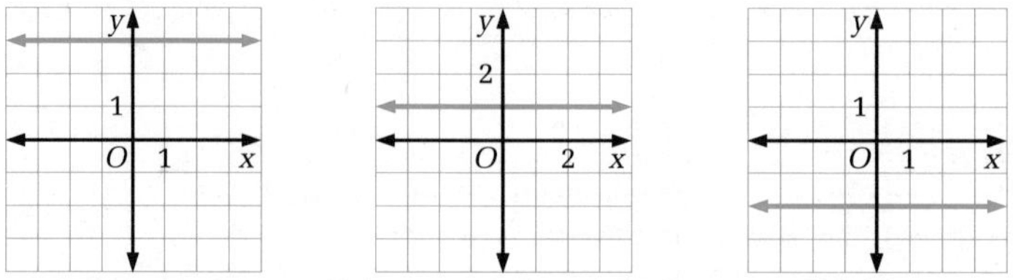

   a. Do the graphs represent linear functions?
   b. What is the slope of each graph?
   c. Suppose a constant function intersects the y-axis at $(0, c)$. Write an equation for this constant function.
   d. What is the domain of a constant function? What is the range of a constant function?
   e. **Open-ended** Give an everyday example of a constant function.
21. The number of chirps made by a snowy tree cricket can be modeled by a linear function of temperature. At 68°F, these crickets chirp 116 times per minute; at 80°F, they chirp 164 times per minute.
   a. Write an equation for $C(T)$, the number of chirps as a function of temperature.
   b. Find $C(90)$. What does it represent?
   c. Most crickets do not begin chirping until the temperature is about 39°F. Suppose that the highest temperature at which snowy tree crickets chirp is 105°F. Find the domain and range of $C(T)$.

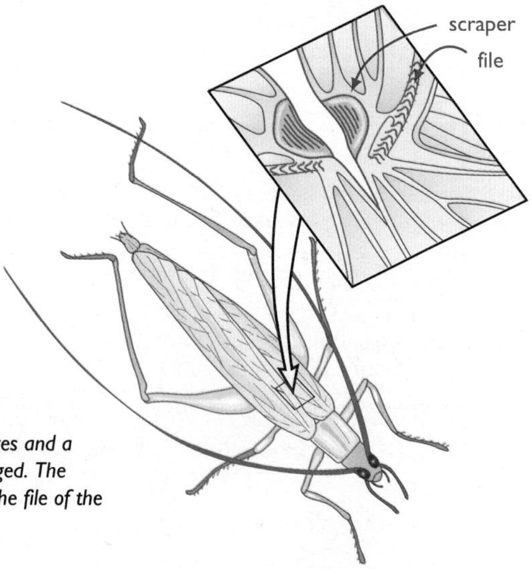

Each wing has a "file" with crossridges and a "scraper" that is hard and sharp-edged. The scraper of either wing rubs against the file of the other to produce a chirping sound.

Answers to Exercises and Problems

13. 39
14. Yes; for every increase of 2 in the first column, there is an increase of 2 in the second column, indicating the graph is a line with slope 1.
15. Let S be American shoe size and E European shoe size; $E(S) = S + 31$.
16. $f(x) = 3x$
17. $f(x) = -5x - 6$
18. $f(x) = -x + 5$
19. $f(x) = \frac{1}{2}x - 1$
20. a. Yes.
   b. 0
   c. $f(x) = c$
   d. all real numbers; the y-intercept
   e. Answers may vary. An example is given. The function that describes the income of a salaried employee as a function of the number of hours worked is a constant function. (A salaried employee is not paid an hourly wage.)

22. When you fill a car's gas tank, the amount of gasoline in the tank is a function of the time the pump has been on. Suppose a pump fills an empty tank at a rate of 3.9 gal/min and the tank holds up to 13.5 gal of gasoline.

a. Write an equation for the amount of gas in the tank as a function of time.

b. What are the domain and range in this situation?

***connection to* LITERATURE**

In 1861 Samuel White Baker and Florence von Sass Baker set out on a journey to search for the lake at high altitude that is the source of the Nile River. Their travels are recorded in Samuel Baker's journal, *The Albert N'Yanza, Great Basin of the Nile.*

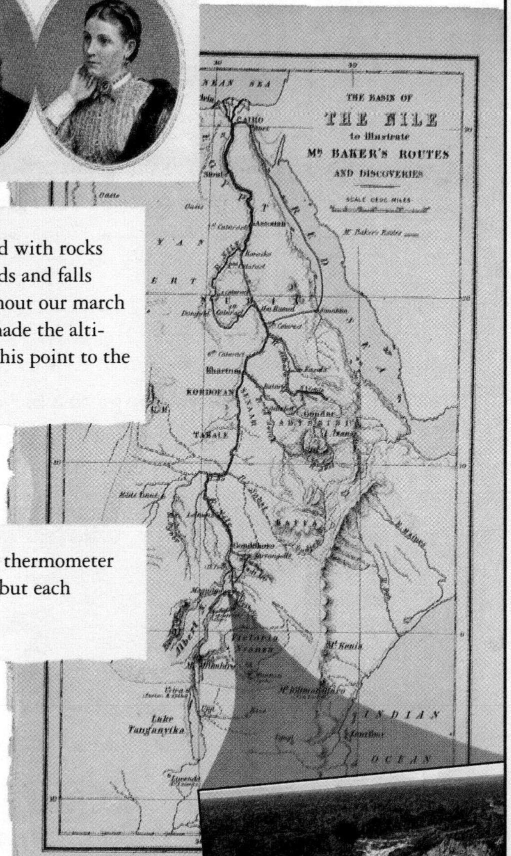

"The river was about 180 to 200 yards in width, but much obstructed with rocks and islands; the stream ran at about four miles per hour; and the rapids and falls were so numerous that the roar of water had been continuous throughout our march from Murchison's Falls. By observations of Casella's thermometer I made the altitude of the river level at the island of Patooan 3,195 feet; thus from this point to the level of the Albert Lake at Magungo there was a fall of 475 ft . . . ."

After recording the altitude at several points along the river, Samuel Baker comments:

"These observations were extremely satisfactory, and showed that the thermometer behaved well at every boiling, as there was no confusion of altitudes, but each corroborated the preceding."

For Exercises 23–25, use the formulas from Sample 1 on page 78.

23. Why do you think it was important for the Bakers to take measurements at different points along the river?

24. **Writing** Explain how the Bakers could use the boiling temperature of water to find the altitude.

25. For a decrease in altitude of 475 ft, what is the change in the boiling temperature of water?

26. In his journal, Samuel Baker states, "Thus Victoria is the first source; but from the Albert the river issues at once as the great White Nile." The altitude of Albert Lake is 2021 ft. Find the boiling temperature of water there.

Murchison's Falls

27. Use the function $g(x) = 3x + 2$.

a. Show that $g(y) + g(z) \neq g(y + z)$.

b. Show that $g(y) \cdot g(z) \neq g(y \cdot z)$.

Assessment: Standard

For Ex. 22, students should be able to decide that this is an example of direct variation. Ask how the problem could be reworded so that it is not an example of direct variation.

Interdisciplinary Problems

The excerpts from Samuel White Baker's journal and Exs. 23–26 demonstrate a real-world use of mathematics and science in helping to locate the source of the Nile River during the nineteenth century.

Error Analysis

Some students may assume that $f(a + b) = f(a) + f(b)$ and that $f(a \cdot b) = f(a) \cdot f(b)$. Exs. 27 and 28 will demonstrate that, in general, these relationships are not true.

Answers to Exercises and Problems

21. a. $C(T) = 4T - 156$

b. 204; the number of times per minute a snowy tree cricket chirps at 90°F

c. $39 \le T \le 105$; $0 \le C(T) \le 264$

22. a. Let t be the number of minutes the pump has been on and $g(t)$ be the amount of gas in the tank; $g(t) = 3.9t$.

b. $0 \le t \le 3.5$ (approximate); $0 \le g \le 13.5$

23. Answers may vary. An example is given. to confirm that the altitude increased as they went up river

24. They could use the function $T(a) = -0.0018a + 212$ to determine the altitude of the lake, given the temperature at which water boils there.

25. an increase of about 0.9°F

26. 208.4°F

27. a. $g(y) + g(z) = 3y + 2 + 3z + 2 = 3y + 3z + 4$; $g(y + z) = 3(y + z) + 2 = 3y + 3z + 2$

b. There are values of y and z for which $g(y) \cdot g(z) = g(y \cdot z)$. However, the statement is not true in general. Consider $y = 1$ and $z = 2$: $g(1) \cdot g(2) = (3(1) + 2)(3(2) + 2) = 40$; $g(1 \cdot 2) = g(2) = 3(2) + 2 = 8$

Working on the Unit Project

If students go to a bicycle shop to complete Ex. 35, suggest that they collect whatever information is available in the shop about bicycles. This research information may be very valuable in preparing the final presentation.

Practice 11 For use with Section 2-2

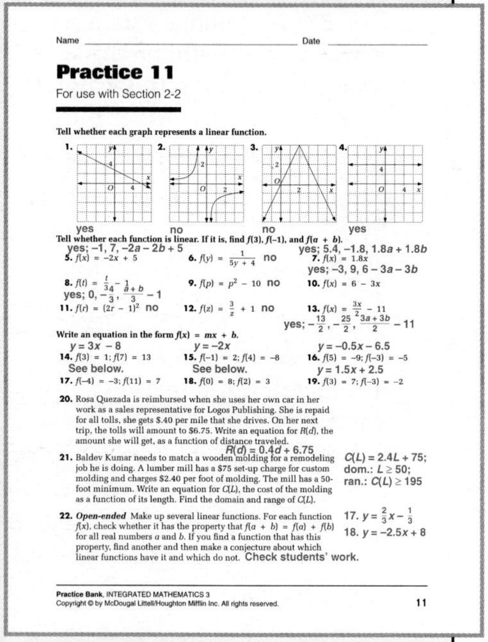

Name ______________ Date ______________

Practice 11

For use with Section 2-2

Tell whether each graph represents a linear function.

1. yes
2. no
3. no
4. yes

Tell whether each function is linear. If it is, find $f(3)$, $f(-1)$, and $f(a + b)$.

5. $f(x) = -2x + 5$ yes; $-1, 7, -2a - 2b + 5$
6. $f(y) = \frac{1}{5y + 4}$ no
7. $f(x) = 1.8x$ yes; $5.4, -1.8, 1.8a + 1.8b$
8. $f(t) = \frac{t}{3} - 1$ yes; $0, -\frac{4}{3}, \frac{a + b}{3} - 1$
9. $f(p) = p^2 - 10$ no
10. $f(x) = 6 - 3x$ yes; $-3, 9, 6 - 3a - 3b$
11. $f(r) = (2r - 1)^2$ no
12. $f(z) = \frac{3}{z} + 1$ no
13. $f(x) = \frac{3x}{2} - 11$ yes; $-\frac{13}{2}, -\frac{25}{2}, \frac{3a + 3b}{2} - 11$

Write an equation in the form $f(x) = mx + b$.

14. $f(3) = 1; f(7) = 13$ $y = 3x - 8$
15. $f(-1) = 2; f(4) = -8$ $y = -2x$
16. $f(5) = -9; f(-3) = -5$ $y = -0.5x - 6.5$
17. $f(-4) = -3; f(11) = 7$ See below.
18. $f(0) = 8; f(2) = 3$ See below.
19. $f(3) = 7; f(-3) = -2$ $y = 1.5x + 2.5$

20. Rosa Quezada is reimbursed when she uses her own car in her work as a sales representative for Logos Publishing. She is repaid for all tolls, she gets $.40 per mile that she drives. On her next trip, the tolls will amount to $6.75. Write an equation for $R(d)$, the amount she will get, as a function of distance traveled. $R(d) = 0.4d + 6.75$

21. Baldev Kumar needs to match a wooden molding for a remodeling job he is doing. A lumber mill has a $75 set-up charge for custom molding and charges $2.40 per foot of molding. The mill has a 50-foot minimum. Write an equation for $C(L)$, the cost of the molding as a function of its length. Find the domain and range of $C(L)$. $C(L) = 2.4L + 75$; dom.: $L \geq 50$; ran.: $C(L) \geq 195$

22. *Open-ended* Make up several linear functions. For each function $f(x)$, check whether it has the property that $f(a + b) = f(a) + f(b)$ for all real numbers a and b. If you find a function that has this property, find another and then make a conjecture about which linear functions have it and which do not. Check students' work.

17. $y = \frac{2}{3}x - \frac{1}{3}$
18. $y = -2.5x + 8$

Practice Bank, INTEGRATED MATHEMATICS 3

11

28. Use the function $f(x) = mx + b$.

  a. Show that $f(y) + f(z) \neq f(y + z)$. **b.** Show that $f(y) \cdot f(z) \neq f(y \cdot z)$.

Ongoing ASSESSMENT

29. Writing Does this table represent a linear function? Describe two ways you can tell.

x	y
−2	−8
0	−7
3	−5.5
7	−3.5
12	−1

Review PREVIEW

30. Use the function $f(x) = x^3 - x$. Find $f(5)$, $f(-1)$, and $f(2a)$. *(Section 2-1)*

31. Explain what the corner-point principle is. *(Section 1-8)*

Write an inequality to describe each graph. *(Toolbox Skill 27)*

32. 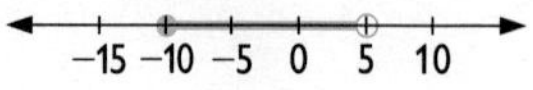

33. 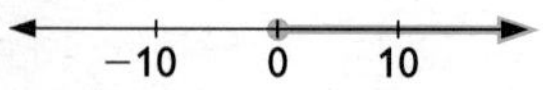

34. 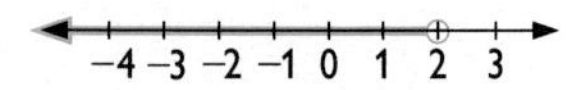

2 Working on the Unit Project

Use your group's bicycle or go to a bicycle shop.

35. The diagrams show the position of the chain in the highest and lowest gears of a multispeed bicycle.

IN HIGHEST GEAR

IN LOWEST GEAR

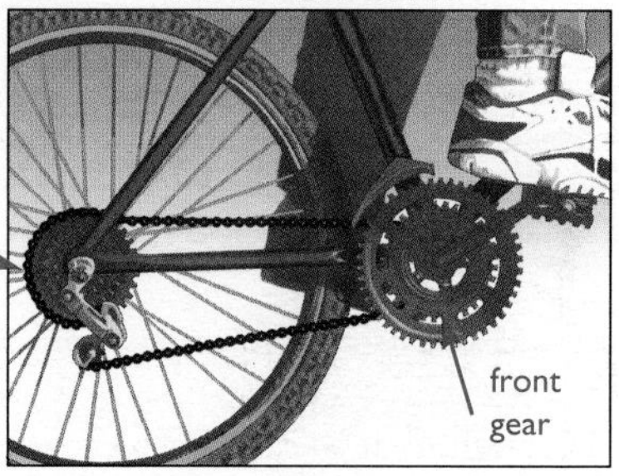

  a. Choose a gear on your bicycle. When your bicycle is in this gear, how many teeth are in the front gear? How many teeth are in the rear gear?

  b. The ratio $\frac{\text{teeth in front gear}}{\text{teeth in rear gear}}$ is the number of turns the rear wheel makes for one full turn of the pedals. Find this ratio for the gear that you chose in part (a).

  c. Write an equation for the number of turns of the back tire as a function of the number of turns of the pedals for the gear you chose.

  d. Describe how you could use the function you wrote for part (c) to find the speed of your bicycle as a function of the number of turns of the pedals. (*Hint:* The circumference of the rear tire is the distance your bicycle travels for one full turn of the rear wheel.)

Answers to Exercises and Problems

28. There are values of y and z for which $f(y) + f(z) = f(y) + f(z)$ and values for which $f(y) \cdot f(z) = f(y \cdot z)$. However neither statement is true in general. (The first statement is always true if $b = 0$. The second is always true if $m = 1$ and $b = 0$ or if $m = 0$ and $b = 1$.)
  a. For $b \neq 0$, $f(y) + f(z) = my + b + mz + b = my + mz + 2b$; $f(y + z) = m(y + z) + b = my + mz + b$.
  b. For $m \neq 0$ or 1 and $b \neq 0, 1$, or -1, consider $y = 0$ and $z = 0$: $f(0) \cdot f(0) = b \cdot b = b^2$; $f(0 \cdot 0) = f(0) = b$.

29. Yes. Answers may vary. Examples are given. Graph the points and use the vertical-line test. Notice that if there is such a function f, $f(0) = -7 = b$. Use two points to determine the slope, m, of the line between them. Then if the table represents a linear function, all the points in the table satisfy the equation $f(x) = 0.5x - 7$. Test each point.

30. 120; 0; $8a^3 - 2a$

31. When a linear programming problem is represented by a system of linear inequalities in x and y, any maximum or minimum value of a linear combination $ax + by$ will occur at one of the vertices of the feasible region.

32. $-10 \leq x < 5$

33. $x \geq 0$ 34. $x < 2$

35. a. Answers may vary. Example: front: 52, rear: 26
  b. 2
  c. Let p be the number of turns of the pedals and $n(p)$ the number of turns of the back tire; $n(p) = \frac{\text{teeth in front gear}}{\text{teeth in rear gear}} \cdot p$.
  d. To find the distance traveled per minute, determine the number of times the pedals are turned in one minute, then use the function in part (c) to find $n(p)$. If r is the radius of the rear tire in feet, the speed of the bicycle as a function of p is $2\pi r \cdot n(p)$ ft/min.

Section **2-3**

Piecewise and Absolute Value Functions

Focus

Use piecewise and absolute value functions to describe situations.

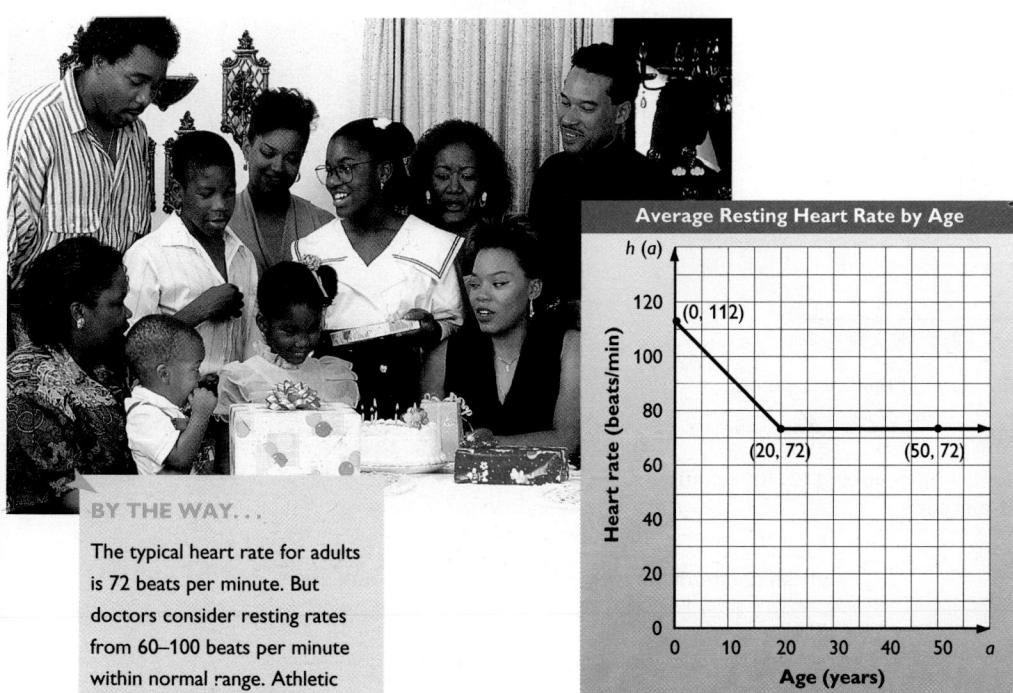

PIECE IT TOGETHER

BY THE WAY...

The typical heart rate for adults is 72 beats per minute. But doctors consider resting rates from 60–100 beats per minute within normal range. Athletic training slows the heart rate. Many athletes have resting rates from 40–60 beats per minute.

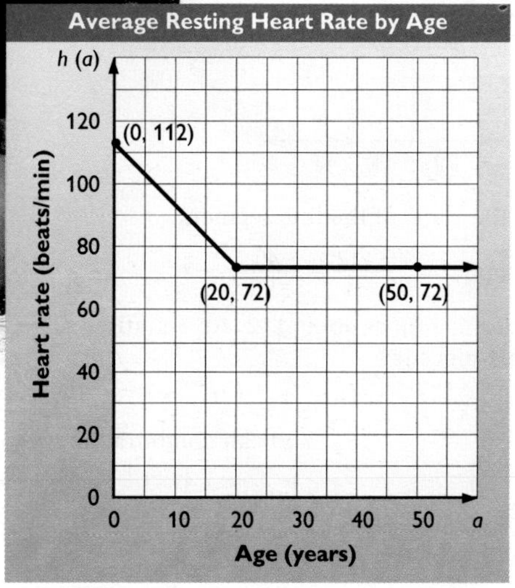

Talk it Over

1. Tell how a person's heart rate changes over the course of his or her lifetime. Refer to the graph.
2. Does the graph represent a function? Why or why not?
3. Do you think the graph can be described by a single equation? Explain.

Answers to Talk it Over

1. A person's heart rate decreases steadily over the first 20 years of life, then remains constant after the age of 20.
2. Yes; no two points have the same first coordinate and different second coordinates.
3. No; the first part of the graph clearly has an equation of the form $f(x) = mx + b$, where m is negative and b is positive. The second part of the graph clearly has an equation of the form $g(x) = c$, for a constant c.

PLANNING

Objectives and Strands

See pages 66A and 66B.

Spiral Learning

See page 66B.

Materials List

- Graph paper

Recommended Pacing

Section 2-3 is a two-day lesson.

Day 1

Pages 85–87: Talk it Over 1 through Sample 2, *Exercises 1–11*

Day 2

Pages 87–89: Absolute Value Functions through Look Back, *Exercises 12–30*

Toolbox References

- **Toolbox Skill 26:** Making a Table of Values to Graph

Extra Practice

See pages 610–612.

Warm-Up Exercises

- Warm-Up Transparency 2-3

Support Materials

- Practice Bank: Practice 12
- Activity Bank: Enrichment 11
- Study Guide: Section 2-3
- Problem Bank: Problem Set 3
- Explorations Lab Manual: Additional Exploration 2 Diagram Masters 1, 2
- Using Plotter Plus: Absolute Value Graphs
- Using IBM/Mac Plotter Plus Disk: Parabola Plotter
- Assessment Book: Quiz 2-3, Test 5, Alternative Assessment 2

TEACHING

Talk it Over

Questions 1–3 have students analyze a piecewise graph as to whether it represents a function and can be decribed by a single equation.

Additional Sample

S1 The graph shows a heating curve for an 18.0 g sample of water that is heated at a certain constant rate, starting with ice at –10°C and ending at the time when the last bit of ice has melted.

Heating Curve for Water

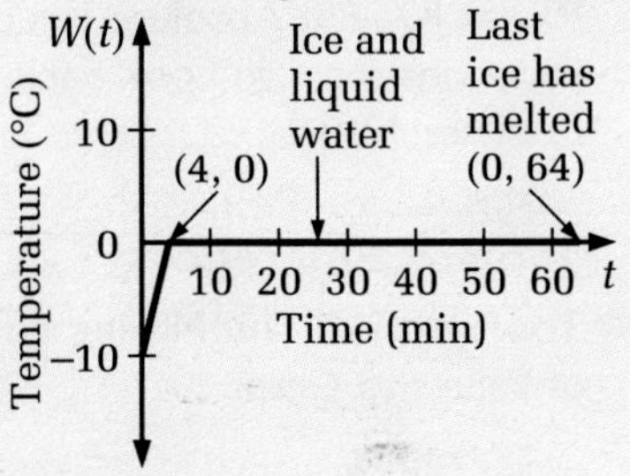

Express the temperature of the water sample as a function of time.

The slope of the line segment from the vertical intercept, (0, –10), to (4, 0) is $\frac{0-(-10)}{4-0} = \frac{10}{4} = 2.5$. An equation for this piece is $W(t) = 2.5t - 10$. The rest of the graph lies along the x-axis, which represents a constant function with equation $W(t) = 0$. The function can therefore be described as follows:

$$W(t) = \begin{cases} 2.5t - 10 & \text{if } 0 \le t \le 4 \\ 0 & \text{if } 4 < t \le 64 \end{cases}$$

Teaching Tip

Some students may feel uncomfortable with the fact that piecewise functions are not described by a single equation. Have students consider the graph of the following function.

$$y = \frac{(\sqrt{(x-3)^2} - x + 6)(\sqrt{x^2})}{(\sqrt{x})^2}$$

Students will agree that the equation is fairly complicated. But when you press GRAPH on a graphics calculator, you get a graph that is very similar to the graph in Sample 1. (Note

Sample 1

Express heart rate as a function of age.

Sample Response

Problem Solving Strategy: Break the problem into parts.

Separate the heart rate graph into two pieces. For each piece, write an equation and give the ages to which the equation applies.

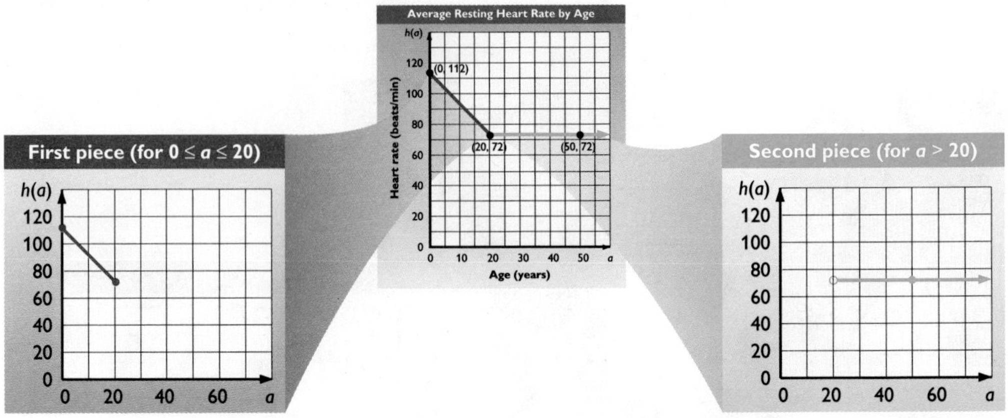

Find the slope of the line segment.

$$\text{slope} = \frac{72 - 112}{20 - 0} = \frac{-40}{20} = -2$$

The vertical intercept is 112. An equation for this piece is:

$h(a) = -2a + 112$

This piece represents a constant function. An equation for this piece is:

$h(a) = 72$

You can combine the two equations using this notation:

$$h(a) = \begin{cases} -2a + 112 & \text{if } 0 \le a \le 20 \\ 72 & \text{if } a > 20 \end{cases}$$

Talk it Over

4. Use the function from Sample 1 to predict the heart rate of a person 4 years old, 15 years old, and 50 years old.
5. a. Etenia's resting heart rate is 90 beats/min. What equation can you solve to estimate Etenia's age?
   b. Solve your equation from part (a).

The heart rate function in Sample 1 is an example of a *piecewise function.* A **piecewise function** is a function defined by two or more equations. Each equation applies to a different part of the function's domain.

Answers to Talk it Over

4. 104 beats/min; 82 beats/min; 72 beats/min
5. a. $90 = -2a + 112$
   b. 11; Etenia is 11 years old.

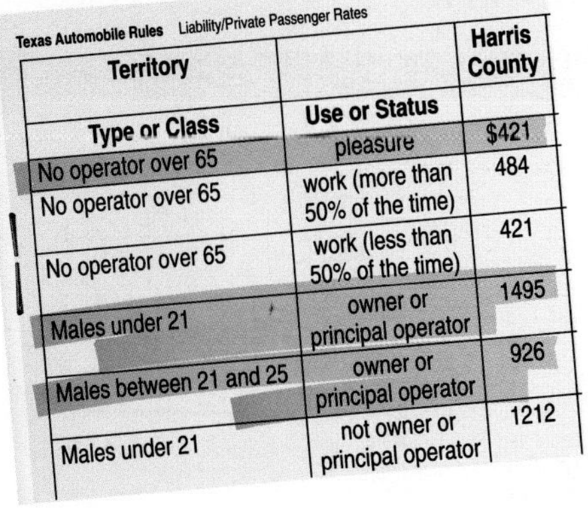

Texas Automobile Rules Liability/Private Passenger Rates

Territory		Harris County
Type or Class	**Use or Status**	$421
No operator over 65	pleasure	$421
No operator over 65	work (more than 50% of the time)	484
No operator over 65	work (less than 50% of the time)	421
Males under 21	owner or principal operator	1495
Males between 21 and 25	owner or principal operator	926
Males under 21	not owner or principal operator	1212

Sample 2

The table shows average annual car insurance premiums for male drivers living in Harris County, Texas. The minimum age for a regular driver's license is 16.

a. Express the premiums for the highlighted categories as a function of the age of a male driver.

b. Graph the function in part (a).

Sample Response

a. Let a be the age of the driver, and let $p(a)$ be the premium in dollars. The rule for $p(a)$ consists of one equation for each of the three categories.

$$p(a) = \begin{cases} 1495 & \text{if } 16 \le a < 21 \\ 926 & \text{if } 21 \le a \le 25 \\ 421 & \text{if } 25 < a \le 65 \end{cases}$$

b. Graph each equation in the rule for $p(a)$ over the appropriate part of the domain.

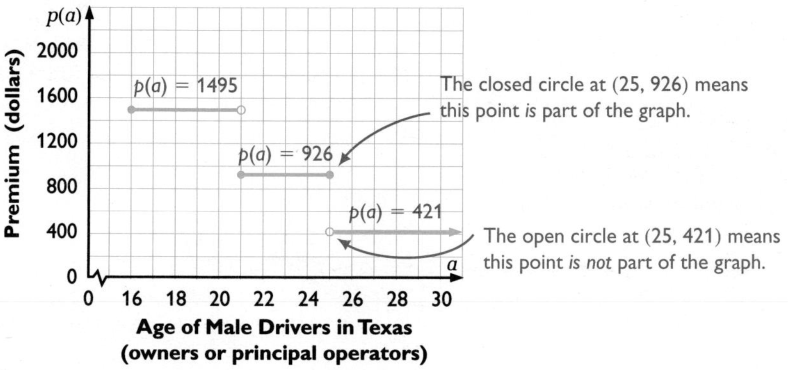

Now you are ready for: Exs. 1–11 on pp. 89–90

Absolute Value Functions

The function d in Exercise 10 on page 90 pairs each real number x with its *absolute value*, written $|x|$. The **absolute value** of a number can be defined by this piecewise rule:

$$|x| = \begin{cases} x & \text{if } x \ge 0 \\ -x & \text{if } x < 0 \end{cases}$$

← the original value if x is non-negative

← the opposite of the original value if x is negative

that 0 is not in the domain.) Ask students to describe this graph with a piecewise definition, as in Sample 1. Point out that piecewise functions often *can* be described by a single equation, but this may not be the most natural way to describe it.

Additional Sample

S2 The table shows what Jiffy Pak Messenger Services charges for Fast Track, a local delivery service that promises 2-hour delivery within the local area of any parcel of 2 pounds or less.

Weight	Charge
$\frac{1}{2}$ lb or under	$13
more than $\frac{1}{2}$ lb but not more than 1 lb	$20
more than 1 lb but not more than 2 lb	$25

a. Express the charge for a Fast Track parcel as a function of the weight of the parcel in pounds.

Let w be the weight of the parcel in pounds and let $C(w)$ be the charge in dollars for Fast Track delivery.

$$C(w) = \begin{cases} 13 \text{ if } 0 < w \le \frac{1}{2} \\ 20 \text{ if } \frac{1}{2} < w \le 1 \\ 25 \text{ if } 1 < w \le 2 \end{cases}$$

b. Graph the function.

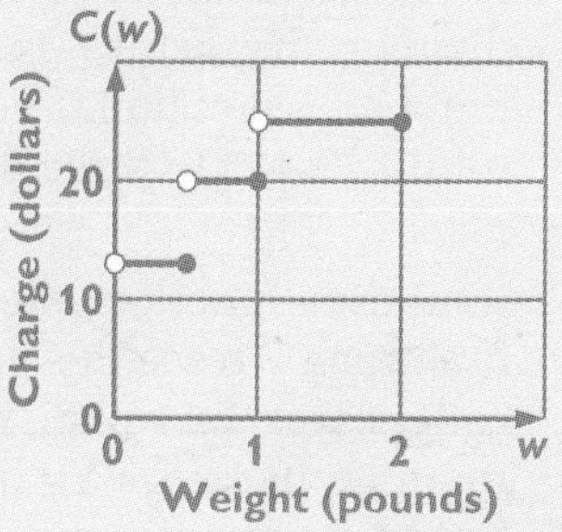

Additional Sample

S3 Graph $y = |x - 3|$.

Write $y = |x - 3|$ in piecewise form.

$$y = \begin{cases} x - 3 & \text{if } x - 3 \geq 0 \\ -(x - 3) & \text{if } x - 3 < 0 \end{cases} \text{ or}$$

$$y = \begin{cases} x - 3 & \text{if } x \geq 3 \\ -x + 3 & \text{if } x < 3 \end{cases}$$

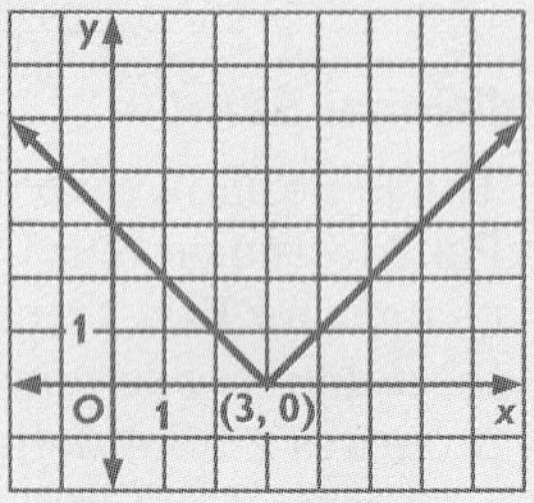

Using Technology

Most graphics calculators have certain piecewise functions as built-in functions. You may, for example, use the TI-81 or TI-82 to access the absolute value function. Press [2nd] [ABS]. To graph the function in Sample 3, press [Y=] and after Y1=, type the expression abs(X + 2). Press [GRAPH] to view the graph.

Other piecewise functions are available. One example is the rounding function. Press [MATH] [▶] to access it. The function round(lets you round any number to a specified number of decimal places. On the home screen, enter round(7.1415,3). When you press [ENTER], the calculator displays the result 7.142, the number you obtain if you round 7.1415 to three decimal places.

Rounding functions are step functions. To see this graphically, enter the equation $y =$ round(x, 0) on the Y= list. Use Dot Mode and the standard graphing window. Press [GRAPH] to view the graph.

Other piecewise functions on the MATH NUM menu are iPart (the integer part of any real number), fPart (the fractional part of any real number), and int (the greatest integer function).

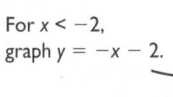

Sample 3

Graph $y = |x + 2|$.

Sample Response

Write $y = |x + 2|$ in piecewise form:

$$y = \begin{cases} x + 2 & \text{if } x + 2 \geq 0 \\ -(x + 2) & \text{if } x + 2 < 0 \end{cases} \quad \text{or} \quad y = \begin{cases} x + 2 & \text{if } x \geq -2 \\ -x - 2 & \text{if } x < -2 \end{cases}$$

Solve each inequality for x.

Graph $y = x + 2$ and $y = -x - 2$ over the appropriate parts of the x-axis:

For $x < -2$, graph $y = -x - 2$.

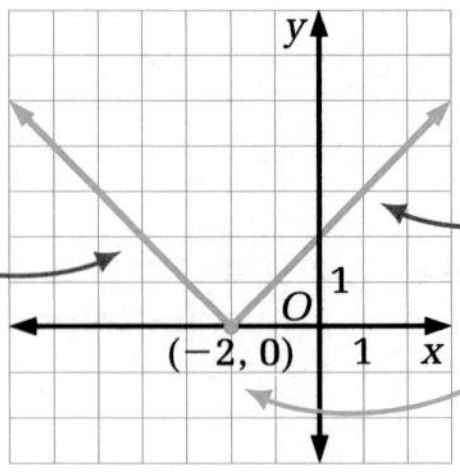

For $x \geq -2$, graph $y = x + 2$.

$y = |x + 2|$

The function in Sample 3 is called an *absolute value function.*

X(ab

ABSOLUTE VALUE FUNCTIONS

An **absolute value function** is a function that has a variable within the absolute value symbol.

Example

$y = 2|x + 1| - 4$

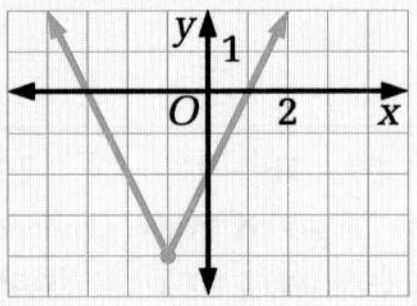

Talk it Over

Student Resources Toolbox p. 649 *Graphing*

6. a. Graph $y = |x| + 2$. Compare this graph with the graph of $y = |x + 2|$ in Sample 3.
   b. For which values of x is $|x + 2| = |x| + 2$?
7. Graph $y = |2x|$ and $y = 2|x|$ on the same set of axes. What do you notice about the two graphs? Compare the graphs with the graph of $y = |x|$.
8. Repeat question 7 for the graphs of $y = |-2x|$ and $y = -2|x|$.

Answers to Talk it Over

6. a.

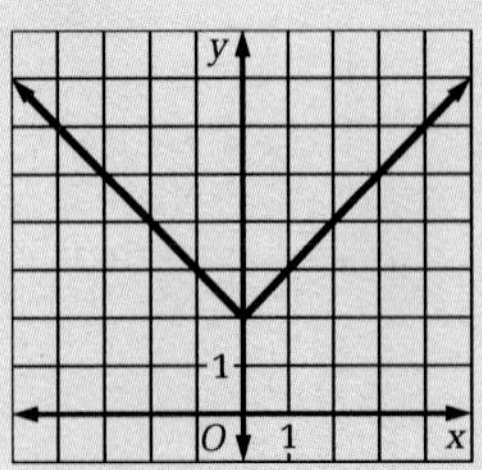

The graphs have the same shape; the graph of $y = |x| + 2$ is the graph of $y = |x + 2|$ shifted 2 units to the right and up 2 units.

b. all x greater than or equal to 0

7.

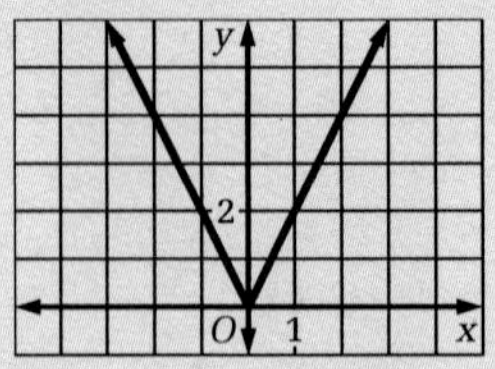

The graphs are identical. Each of the two sections of the graph of $y = |2x|$ and $y = 2|x|$ has slope twice that of the corresponding section of the graph of $y = |x|$.

8. See answers in back of book.

Now you are ready for:
Exs. 12–30 on pp. 90–92

Look Back

Explain why an absolute value function is a piecewise function.

2-3 Exercises and Problems

1. **Reading** Explain the meaning of this heart rate function from Sample 1:

$$h(a) = \begin{cases} -2a + 112 & \text{if } 0 \le a \le 20 \\ 72 & \text{if } a > 20 \end{cases}$$

Write a rule for a piecewise function for each graph.

2. 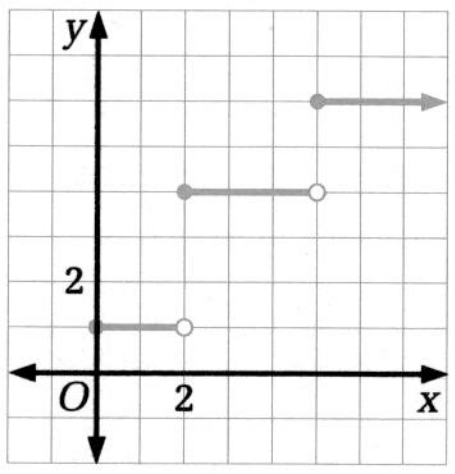

3. 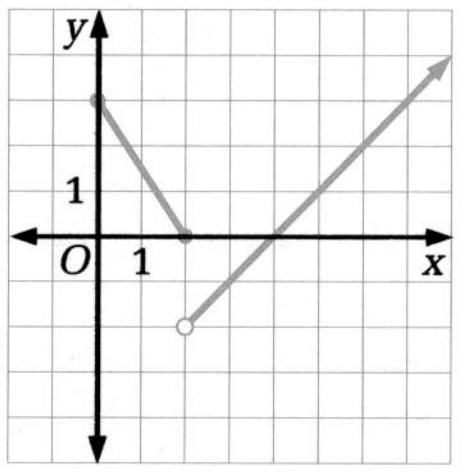

4. 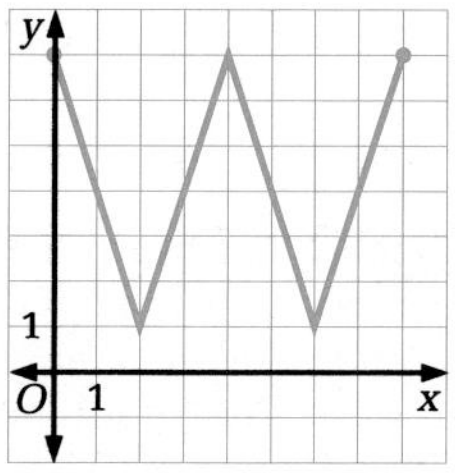

Graph each piecewise function.

5. $f(x) = \begin{cases} 4 & \text{if } 0 \le x < 3 \\ 8 & \text{if } x \ge 3 \end{cases}$

6. $g(t) = \begin{cases} 5 & \text{if } t \le -2 \\ -\frac{3}{2}t + 2 & \text{if } t > -2 \end{cases}$

7. $y = \begin{cases} x & \text{if } x < 0 \\ 2 & \text{if } x = 0 \\ x + 4 & \text{if } x > 0 \end{cases}$

8. $y = \begin{cases} 9 & \text{if } x < -3 \\ x^2 & \text{if } -3 \le x \le 3 \\ 9 & \text{if } x > 3 \end{cases}$

9. **Consumer Economics** The graph shows the monthly cost of Zachary's long-distance calling plan as a function of the length of time he talks on the telephone.
   a. Write a piecewise function for the graph.
   b. Find the cost of calling long-distance for 1.5 hours per month.
   c. Zachary wants his monthly long distance telephone bill to be no more than \$20. Find the maximum length of time he can call long-distance.

Monthly Cost of Long-Distance Calling

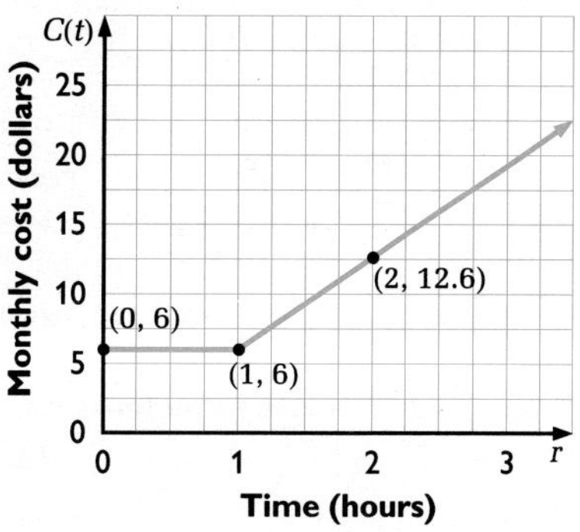

Talk it Over

Questions 6–8 lead students to discover some properties of the graphs of absolute value functions.

Look Back

After students have an opportunity to respond verbally with their explanations, suggest that they write a statement for inclusion in their journals.

APPLYING

Suggested Assignment

Day 1
Standard 1–8, 10
Extended 1–11

Day 2
Standard 12–17, 19–30
Extended 12–30

Integrating the Strands

Algebra Exs. 1–24, 28–30
Functions Exs. 1–24, 30
Geometry Exs. 25–27
Trigonometry Exs. 19–21
Logic and Language Exs. 1, 11, 30

Answers to Look Back

An absolute value function is a piecewise function because it has one rule for nonnegative values and another rule for negative values. An absolute value function is defined by two equations.

Answers to Exercises and Problems

1. Heart rate is a piecewise function of age. If a person's age is 20 or less, the heart rate is given by $h(a) = -2a + 112$. If his or her age is greater than 20, the heart rate is given by $h(a) = 72$.

2. $y = \begin{cases} 1 & \text{if } 0 \le x < 2 \\ 4 & \text{if } 2 \le x < 5 \\ 6 & \text{if } x \ge 5 \end{cases}$

3. $y = \begin{cases} -\frac{3}{2}x + 3 & \text{if } 0 \le x \le 2 \\ x - 4 & \text{if } x > 2 \end{cases}$

4. $y = \begin{cases} -3x + 7 & \text{if } 0 \le x < 2 \\ 3x - 5 & \text{if } 2 \le x < 4 \\ -3x + 19 & \text{if } 4 \le x < 6 \\ 3x - 17 & \text{if } 6 \le x \le 8 \end{cases}$

5. 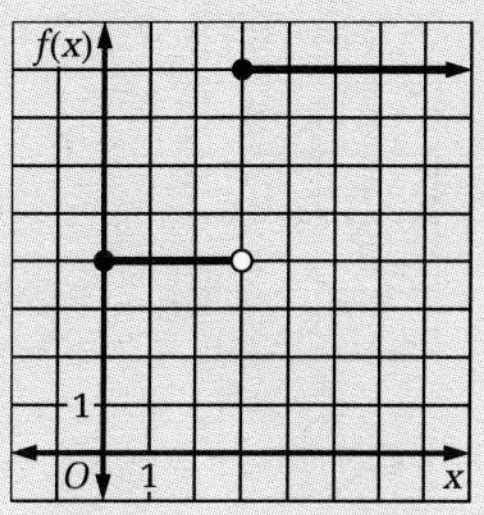

6. 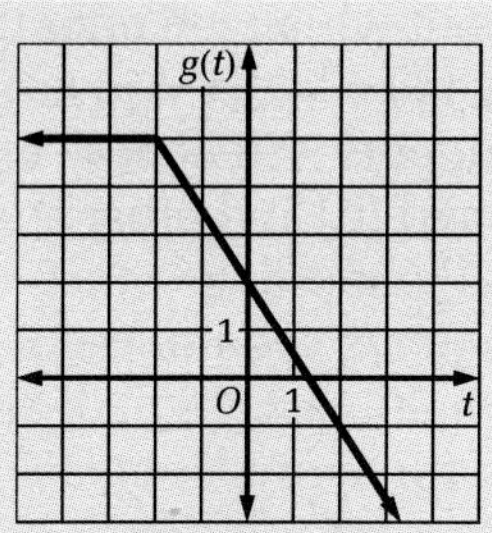

7. 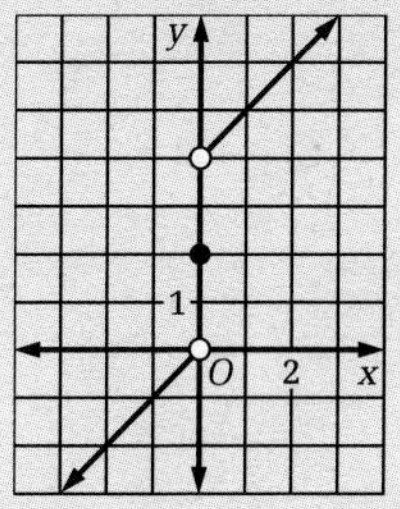

8, 9. See answers in back of book.

Integrating the Strands

Ex. 10 integrates the strands of algebra, geometry, and functions by having students look at the absolute value function in terms of distance from 0 on a number line. To be certain students make this connection, you may wish to ask them what function $d(x)$ really is.

Reasoning

If students enter the equation suggested in the Technology Note on a graphics calculator, they will indeed obtain the graph for the piecewise function described in the note. Be sure students understand *why* the equation
$y = 3x(x < 2) + (x - 1)(x \geq 2)$
gives the correct graph. If you replace x with a number a that is less than 2, the calculator will assign a value of 1 to $(a < 2)$ and a value of 0 to $(a \geq 2)$. The expression will be evaluated as follows:

$$\begin{aligned} &3a(a < 2) + (a - 1)(a \geq 2) \\ &= 3a(1) \quad + (a - 1)0 \\ &= 3a \quad + 0 \\ &= 3a \end{aligned}$$

Similarly, if you use a number b greater than or equal to 2, then $(b < 2)$ will be 0 and $(b \geq 2)$ will be 1. Thus, the result the calculator displays for $3b(b < 2) + (b - 1)(b \geq 2)$ will be equal in value to $b - 1$.

Communication: Discussion

Ex. 11 gives students an opportunity to explore, think, and write about which job offer they consider to be better. Since some students will choose one offer in part (d) and other students will choose the other offer, students should find it interesting to discuss their reasoning with their classmates.

10. a. Let d be a function that pairs a real number x with its distance from 0 on a number line. Copy and complete the table.

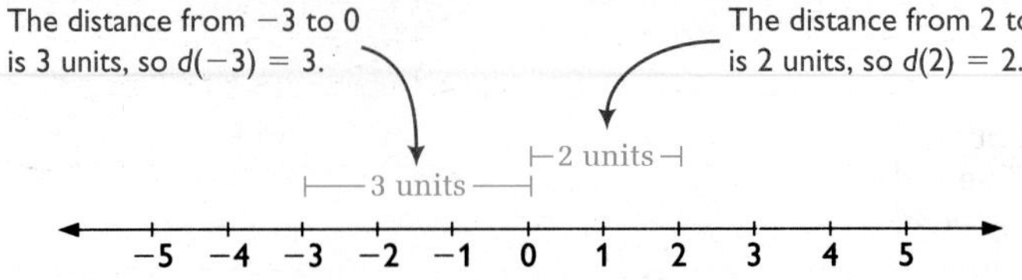

x	$d(x)$
−3	?
−2	?
−1	?
0	?
1	?
2	?
3	?

b. Use your completed table to graph $y = d(x)$. Describe the graph, and give the domain and range of d.

c. Complete each sentence with *equal* or *opposites.*

If x is a negative number, then x and $d(x)$ are __?__.

If x is a nonnegative number, then x and $d(x)$ are __?__.

d. Write a piecewise rule for d.

TECHNOLOGY NOTE

On many graphics calculators, you enter piecewise functions such as

$$y = \begin{cases} 3x & \text{if } x < 2 \\ x - 1 & \text{if } x \geq 2 \end{cases}$$

like this:

$y = 3x(x < 2) + (x - 1)(x \geq 2)$

11. Michelle Jacobson received job offers from two department stores, Clothing Galore and Fits You Right. Clothing Galore offered Michelle a 4% commission on all sales. Fits You Right offered her \$1000 per month plus a 2% commission on all sales over \$15,000.

a. Express Michelle's possible monthly earnings at Clothing Galore as a function of her sales.

b. Express Michelle's possible monthly earnings at Fits You Right as a function of her sales.

c. Graph the functions from parts (a) and (b) on the same set of axes.

d. **Writing** Based on your graphs, which job do you think Michelle should accept? Explain your reasoning.

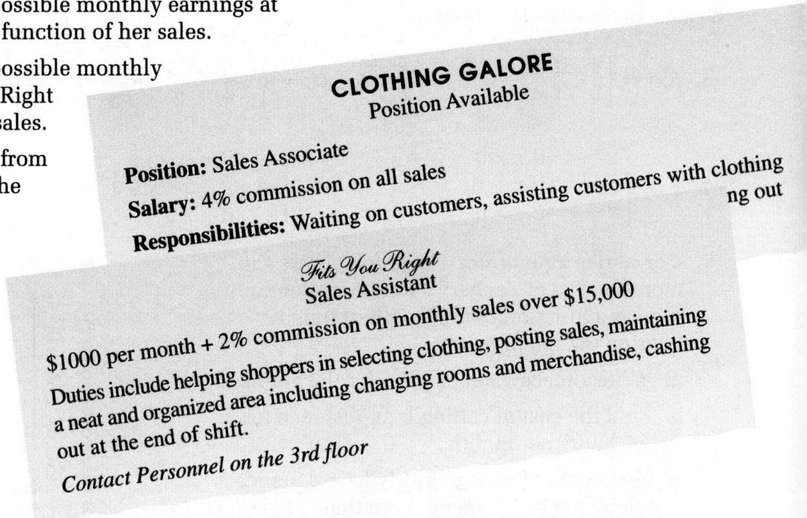

CLOTHING GALORE
Position Available
Position: Sales Associate
Salary: 4% commission on all sales
Responsibilities: Waiting on customers, assisting customers with clothing
ng out
Fits You Right
Sales Assistant
$1000 per month + 2% commission on monthly sales over $15,000
Duties include helping shoppers in selecting clothing, posting sales, maintaining a neat and organized area including changing rooms and merchandise, cashing out at the end of shift.
Contact Personnel on the 3rd floor

Graph each absolute value function.

12. $y = |x + 1|$

13. $y = |x - 1|$

14. $y = 2|x| + 3$

15. $y = 2|x + 3|$

16. $y = 3|x - 4| - 2$

17. $y = -4|x + 2| + 1$

Answers to Exercises and Problems

10. a.

x	$d(x)$
−3	3
−2	2
−1	1
0	0
1	1
2	2
3	3

b.

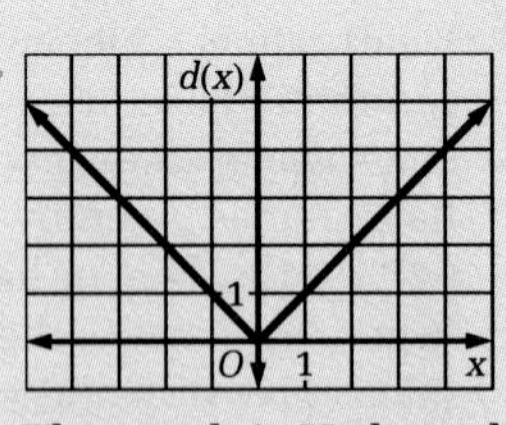

The graph is V-shaped, consisting of two rays, one with negative slope, one with positive slope. The domain is all real numbers, the range is all nonnegative real numbers.

c. opposites; equal

d. $d(x) = \begin{cases} x & \text{if } x \geq 0 \\ -x & \text{if } x < 0 \end{cases}$

11. Let x represent monthly sales in dollars and y represent monthly earnings.

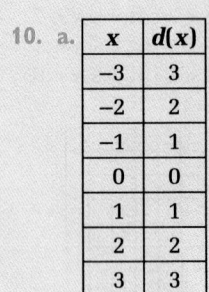

a. $y = 0.04x(x \geq 0)$

Depth finders Using an electronic device called a *depth finder*, the captain of a fishing boat measures the water depth to be 550 ft. Assume that the difference of the actual and measured depths is at most 5% of the actual depth.

18. a. The *absolute difference* between x and *any* fixed number a is given by the equation $y = |x - a|$. Write a function $f(x)$ that gives the absolute difference of the actual depth x and the measured depth.

b. You write "5% of the actual depth x" as $0.05x$. Graph the function from part (a) and $f(x) = 0.05x$ on the same axes.

c. What are the intersection points? What do these values mean?

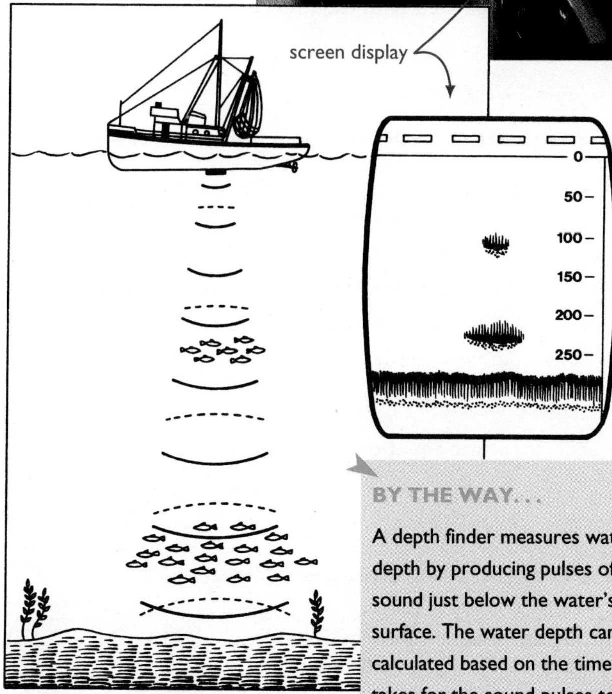

TECHNOLOGY NOTE

On many graphics calculators, you enter $y = |x - 550|$ as $y = \text{abs}(x - 550)$.

BY THE WAY...

A depth finder measures water depth by producing pulses of sound just below the water's surface. The water depth can be calculated based on the time it takes for the sound pulses to echo off the bottom of the body of water and return to the ship.

connection to **PHYSICS**

The path of the billiard ball in the diagram is described by this equation:

$$y = (\tan d°)\,|x - a|$$

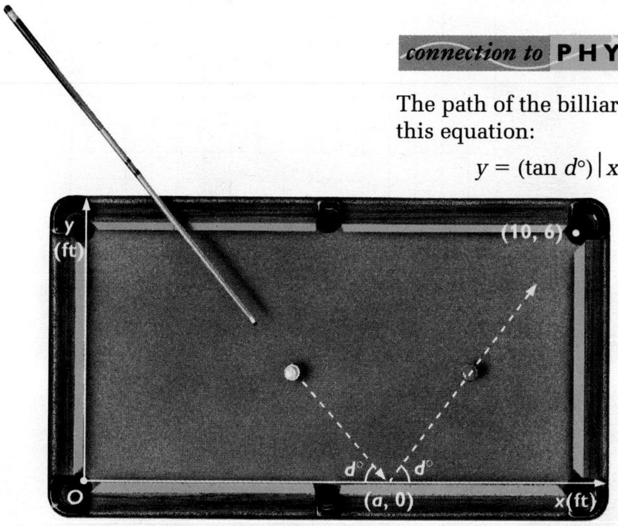

19. Suppose $a = 4.5$. For what value of d will the ball go into the corner pocket?

20. Suppose $d = 40$. For what value of a will the ball go into the corner pocket?

21. Suppose the ball is initially located at the point (2, 5). For what values of a and d will the ball go into the corner pocket?

Application

Exs. 19–21 provide a real-world application of functions and geometry to the game of billiards. You may wish to have any students who have played billiards describe how they attempt to make a shot like the one pictured in the diagram.

14.

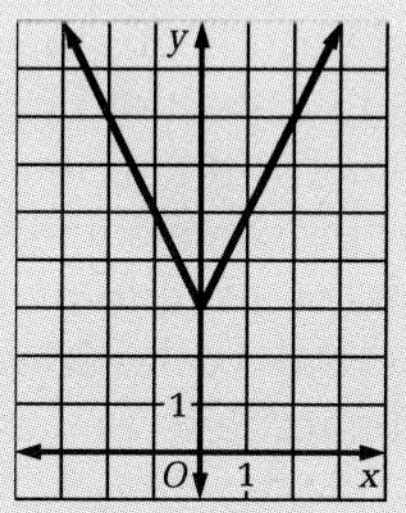

15.

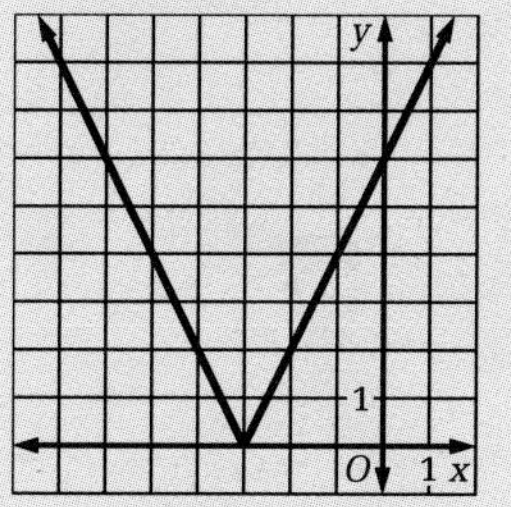

16.

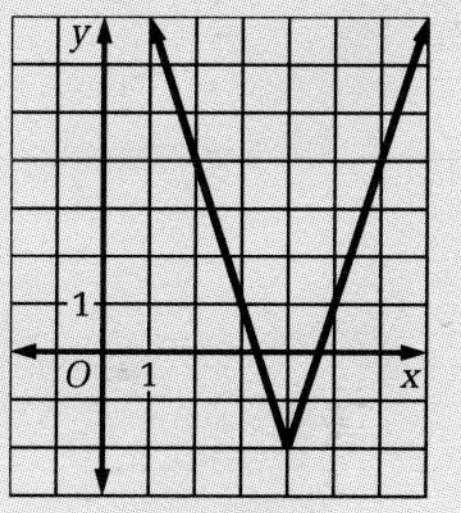

17.

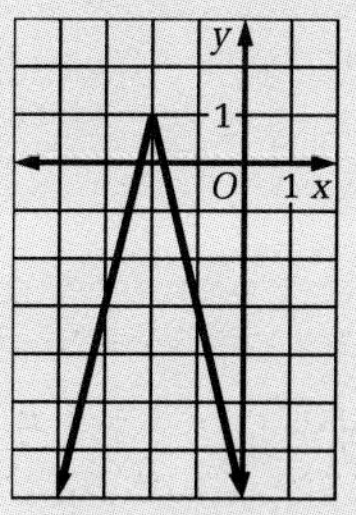

18. a. $f(x) = |x - 550|$

b.

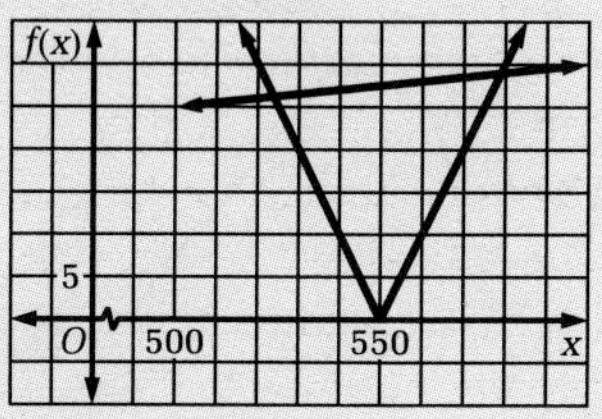

c. about (523.8, 26.2) and about (578.9, 28.9); These values mean that the actual depth is between about 524 ft and about 579 ft.

19. $d \approx 47°$

20. $a \approx 2.85$

21. $a \approx 5.6$; $d \approx 54°$

Answers to Exercises and Problems

b. $y = \begin{cases} 1000 & \text{if } 0 \le x < 15{,}000 \\ 1000 + 0.02(x - 15{,}000) & \text{if } x \ge 15{,}000 \end{cases}$

c.

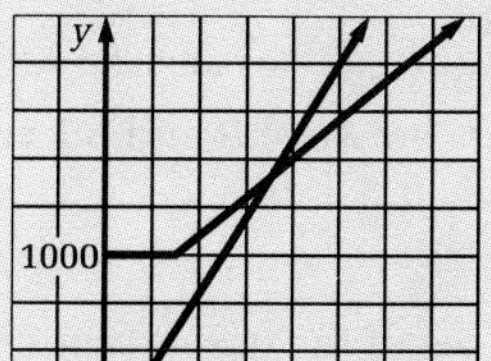

d. Answers may vary. An example is given. If Michelle is deciding only on the basis of income and if she is fairly sure her monthly sales will be less than $35,000, she should choose Fits You Right. If she is fairly sure she can maintain monthly sales of at least $35,000, she should choose Clothing Galore.

12. See answers in back of book.

13.

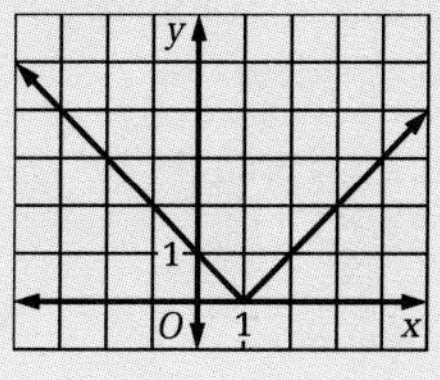

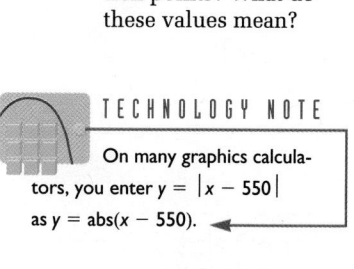

Cooperative Learning

Students can discuss Ex. 22 in small groups. They may even wish to perform the experiment described to see what happens to the height of the water as a function of time.

Working on the Unit Project

As students work Ex. 30, and subsequent Unit Project exercises, they should begin to think about an aspect of bicycling they would like to discuss in their presentations. Graphs could be used to make descriptions and explanations as clear and complete as possible.

Ongoing ASSESSMENT

22. **Open-ended** Alberta fills a glass with water, adds two ice cubes, and uses a straw to drink the water. Sketch a piecewise graph that you think could give the height of the water in the glass as a function of time. State the action associated with each piece of the graph.

Review PREVIEW

Write an equation for each function in the form $f(x) = mx + b$. *(Section 2-2)*

23. $f(-2) = -10$, $f(2) = 2$

24. $f(0) = 5$, $f(7) = \frac{3}{2}$

What name best describes each quadrilateral? *(Glossary)*

25. 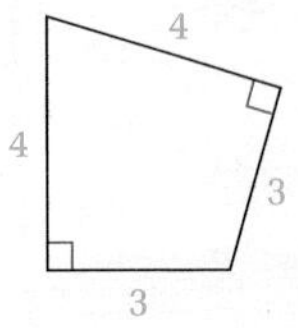

26. 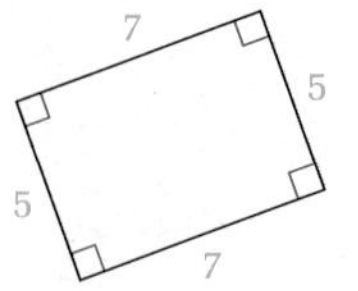

27. 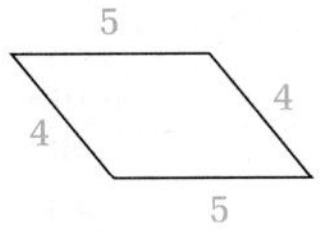

Solve each quadratic equation. *(Toolbox Skill 22)*

28. $x^2 + 3x + 1 = 0$

29. $4x^2 - 7 = 2x$

2 Working on the Unit Project

30. During the summer, Juanita likes to mountain bike. The graph shows the distance she travels on one trail as a function of time.
    a. Write a piecewise function for the graph and give the function's domain and range.
    b. How long does it take Juanita to complete the trail? How long does it take her to reach the halfway mark of the trail?
    c. Is Juanita's speed constant? How can you tell by looking at the graph? What might be reasons for any change in speed?

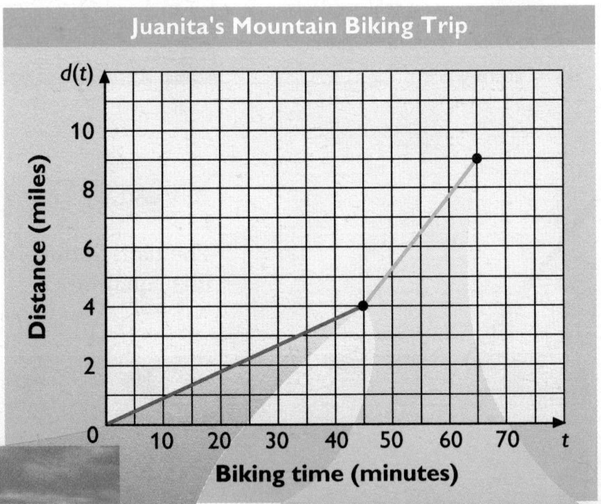

Answers to Exercises and Problems

22. 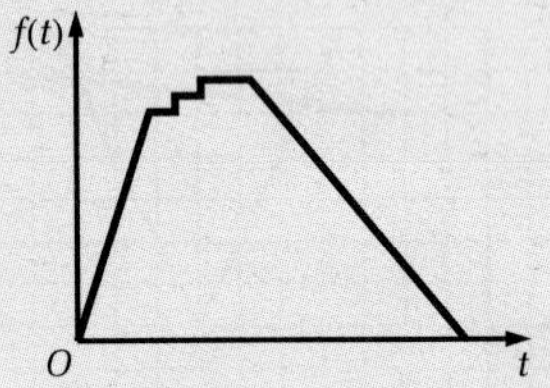

Sketches may vary. An example is given. At $t = 0$, there is no water in the glass. The first piece of the graph shows the water level increasing as Alberta fills the glass. The second shows the level for the time between when she fills the glass and when she starts drinking. The third piece shows the water level decreasing as Alberta drinks from the glass. (The shape of this piece of the graph may vary in the sketches, depending on whether Alberta draws steadily on the straw to drain the glass, or whether she sips sporadically.)

23. $f(x) = 3x - 4$

24. $f(x) = -\frac{1}{2}x + 5$

25. kite

26. rectangle

27. parallelogram

28. about –0.38; about –2.6

29. about 1.6; about –1.1

30. a. $d(t) = \begin{cases} 0.09t & \text{if } 0 \le t \le 45 \\ 0.25t - 7.25 & \text{if } 45 \le t \le 65 \end{cases}$
    The domain is all real numbers between 0 and 65; the range is all real numbers between 0 and 9.
    b. 65 min; about 47 min
    c. No. The slopes of the two pieces of the graph (which represent the rate at which Juanita travels) are not the same. Answers may vary. Example: The first 4 miles of the trip may be over more difficult terrain.

CHECKPOINT 1

1. **Writing** Describe how to use the vertical line test to determine whether a graph is a function. Why does this test work? 2-1

2. Does the table below represent a function? If so, what are its domain and range?

Soft Drinks

12 oz	\$.75
20 oz	\$1.00
32 oz	\$1.50

3. a. Write a rule for $V(r)$, the volume of a cone as a function of its radius. The height of the cone is 9 in.
   b. Find $V(2.5)$ and $V(7x)$.

4. Is the function $f(x) = x^2 + 5$ *one-to-one* or *many-to-one*? Explain your answer.

5. A cablevision company charges its subscribers a monthly fee of \$24.28 for its basic service. The company also offers an additional Pay-per-Watch movie service that costs \$4.99 per movie. 2-2
   a. Write a rule for the monthly cable fee as a function of the number of movies ordered.
   b. Suppose Julie orders 5 movies this month. How much will she pay this month for her cable service?
   c. Ferris budgets \$45 per month for his cable service. How many movies can he order each month?
   d. Suppose the cablevision company increases the monthly fee for basic service by \$3. How does that affect the number of movies Ferris can order?

6. $f(-6) = 7$ and $f(3) = 4$. Write an equation for $f(x)$ in the form $f(x) = mx + b$.

7. A supermarket has a discount on "family packs" of meat. Chicken costs \$2.00/lb for packages over 5 lb. Smaller packages are \$2.50/lb. Express the cost as a function of weight. 2-3

8. Graph the functions $y = |x - 3|$ and $y = |x| - 3$ on the same set of axes. How are the graphs related?

Quick Quiz (2-1 through 2-3)

See page 134.
See also Test 5 in the *Assessment Book.*

Practice 12 For use with Section 2-3

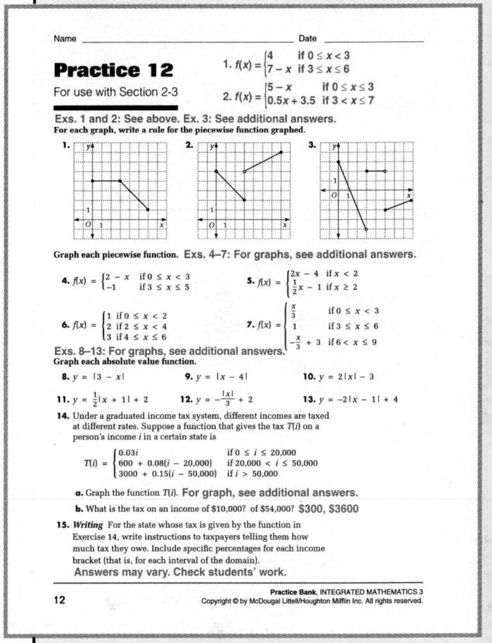

Name ______ Date ______

Practice 12

For use with Section 2-3

1. $f(x) = \begin{cases} 4 & \text{if } 0 \le x < 3 \\ 7 - x & \text{if } 3 \le x \le 6 \end{cases}$

2. $f(x) = \begin{cases} 5 - x & \text{if } 0 \le x \le 3 \\ 0.5x + 3.5 & \text{if } 3 < x \le 7 \end{cases}$

Exs. 1 and 2: See above. Ex. 3: See additional answers.

For each graph, write a rule for the piecewise function graphed.

1. 2. 3.

Graph each piecewise function. Exs. 4–7: For graphs, see additional answers.

4. $f(x) = \begin{cases} 2 - x & \text{if } 0 \le x < 3 \\ -1 & \text{if } 3 \le x \le 5 \end{cases}$

5. $f(x) = \begin{cases} 2x - 4 & \text{if } x < 2 \\ \frac{1}{2}x - 1 & \text{if } x \ge 2 \end{cases}$

6. $f(x) = \begin{cases} 1 & \text{if } 0 \le x < 2 \\ 2 & \text{if } 2 \le x < 4 \\ 3 & \text{if } 4 \le x \le 6 \end{cases}$

7. $f(x) = \begin{cases} \frac{x}{3} & \text{if } 0 \le x < 3 \\ 1 & \text{if } 3 \le x \le 6 \\ -\frac{x}{3} + 3 & \text{if } 6 < x \le 9 \end{cases}$

Exs. 8–13: For graphs, see additional answers.

Graph each absolute value function.

8. $y = |3 - x|$
9. $y = |x - 4|$
10. $y = 2|x| - 3$
11. $y = \frac{1}{2}|x + 1| + 2$
12. $y = -\frac{|x|}{3} + 2$
13. $y = -2|x - 1| + 4$

14. Under a graduated income tax system, different incomes are taxed at different rates. Suppose a function that gives the tax $T(i)$ on a person's income i in a certain state is

$$T(i) = \begin{cases} 0.03i & \text{if } 0 \le i \le 20{,}000 \\ 600 + 0.08(i - 20{,}000) & \text{if } 20{,}000 < i \le 50{,}000 \\ 3000 + 0.15(i - 50{,}000) & \text{if } i > 50{,}000 \end{cases}$$

a. Graph the function $T(i)$. For graph, see additional answers.
b. What is the tax on an income of \$10,000? of \$54,000? \$300, \$3600

15. ***Writing*** For the state whose tax is given by the function in Exercise 14, write instructions to taxpayers telling them how much tax they owe. Include specific percentages for each income bracket (that is, for each interval of the domain).
Answers may vary. Check students' work.

12

Practice Bank, INTEGRATED MATHEMATICS 3

Answers to Checkpoint

1. If any vertical line drawn through the graph intersects the graph in more than one point, the graph does not represent a function. If a vertical line intersects the graph in more than one point, there are two points on the graph with the same first coordinate and different second coordinates. Then the graph does not represent a function.

2. Yes. The domain consists of the numbers 12, 20, and 32. The range consists of the numbers 0.75, 1, and 1.5.

3. a. $V(r) = 3\pi r^2$
   b. about 59 units^3; about $462x^2$ units^3

4. many-to-one; For every real number x except 0, $x \neq -x$ but $f(x) = f(-x)$.

5. a. Let n be the number of movies ordered and $C(n)$ the cost in dollars; $C(n) = 4.99n + 24.28$.
   b. \$49.23
   c. 4 movies
   d. He will be able to order 3 movies per month. (If he carries over the \$2.75 he has left to the next month, he can order 4 movies every other month.)

6. $f(x) = -\frac{1}{3}x + 5$

7. Let w be the weight of the chicken in pounds and $c(w)$ be the cost in dollars;
$c(w) = \begin{cases} 2.5w & \text{if } 0 < w \le 5 \\ 2w & \text{if } w > 5 \end{cases}$

8. 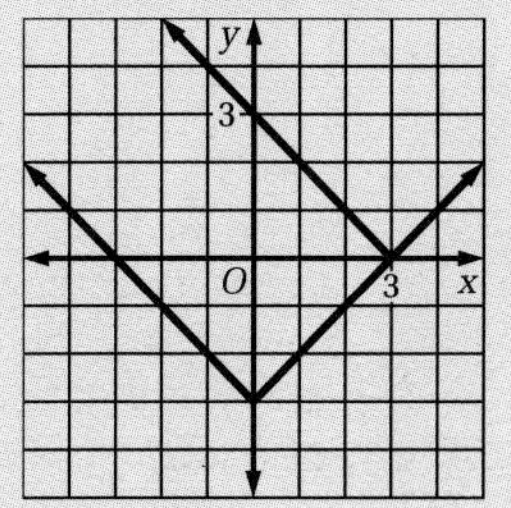

The graphs are translations of the graph of $y = |x|$. The graph of $y = |x - 3|$ is a translation 3 units to the right; the graph of $y = |x| - 3$ is a translation 3 units down.

PLANNING

Objectives and Strands
See pages 66A and 66B.

Spiral Learning
See page 66B.

Materials List
- Graph paper
- Graphics calculator

Recommended Pacing
Section 2-4 is a one-day lesson.

Extra Practice
See pages 610–612.

Warm-Up Exercises
- Warm-Up Transparency 2-4

Support Materials
- Practice Bank: Practice 13
- Activity Bank: Enrichment 12
- Study Guide: Section 2-4
- Problem Bank: Problem Set 4
- Explorations Lab Manual: Diagram Masters 1, 2
- Assessment Book: Quiz 2-4

Section 2-4 Quadratic Functions

POWER FALL

Focus
Use quadratic functions to describe situations.

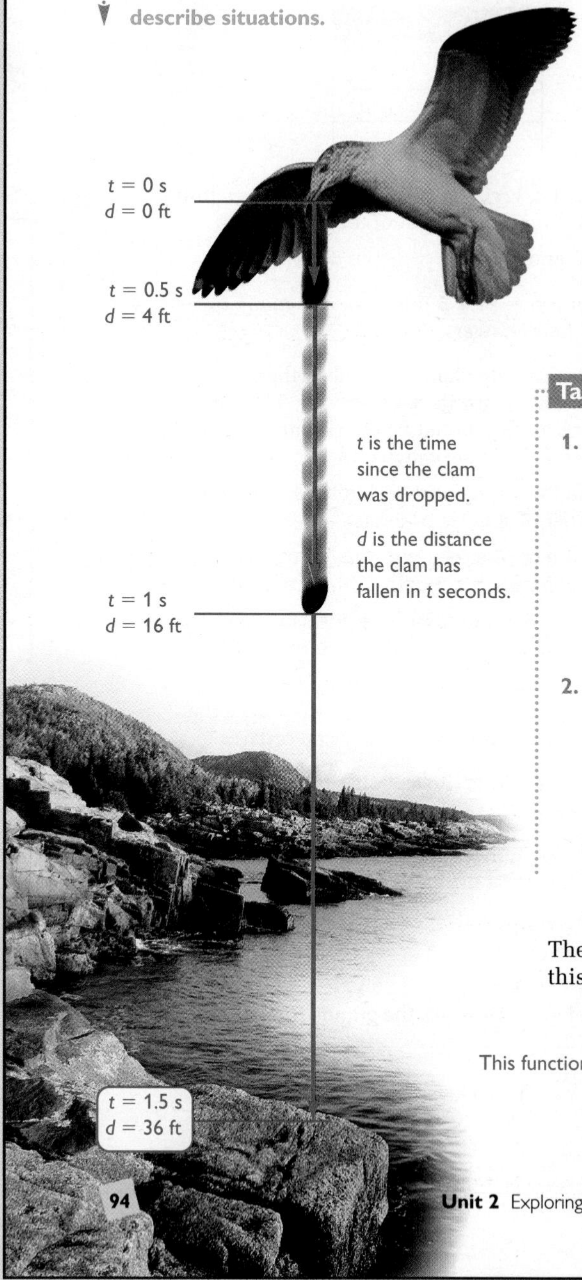

Some species of gulls use a clever technique for eating clams and other shellfish. In order to break a clam's shell, a gull will pick up the clam with its beak, fly into the air, and drop the clam onto a rock. If the shell does not break, the gull will drop the clam from greater heights until it is successful.

Talk it Over

1. a. How far does the clam in the diagram fall from $t = 0$ s to $t = 0.5$ s? from $t = 0.5$ s to $t = 1$ s? from $t = 1$ s to $t = 1.5$ s?
   b. Is d a linear function of t? Why or why not?
   c. Make a scatter plot of the data pairs (t, d), and connect the plotted points with a smooth curve. Does your graph support your answer to part (b)? Explain.
2. In 1604, the Italian scientist Galileo Galilei observed that the distance traveled by any falling object varies directly with the *square* of the time the object falls.
   a. Use Galileo's observation to write an equation relating d, t, and a variation constant k.
   b. Use a point (t, d) from the diagram to find k.

The distance traveled by any falling object is given by this *quadratic function*:

$$d(t) = 16t^2$$

This function gives $d(t)$ in feet. — t must be in seconds.

Answers to Talk it Over

1. a. 4 ft; 12 ft; 20 ft
   b. No. If d were a linear function of t, the clam would have fallen the same distance in every 0.5-second period.
   c. Yes; the graph is a curve, not a line.

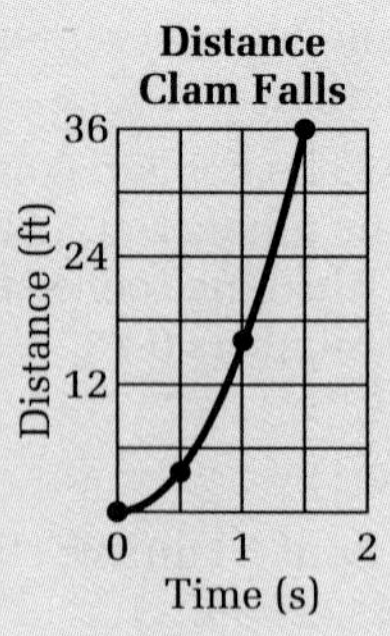

2. a. $d(t) = kt^2$
   b. Using the point (1, 16) gives $d(1) = 16$; $k \cdot 1^2 = 16$; $k = 16$.

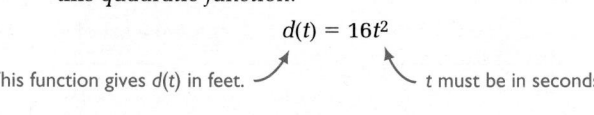

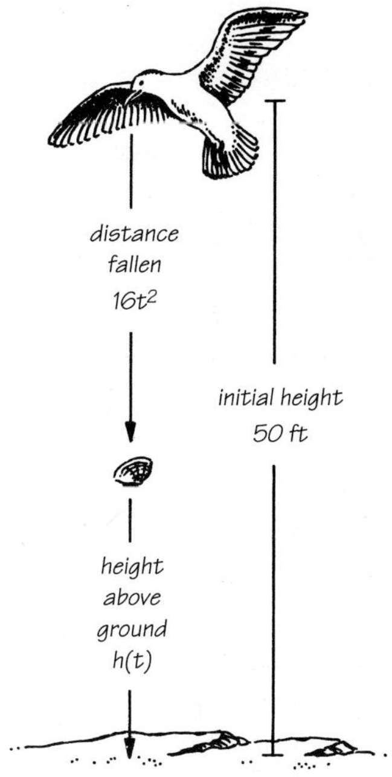

Sample 1

Suppose a gull drops a clam from a height of 50 ft.

a. Express the height of the clam above the ground as a function of time.

b. Graph the function in part (a). What are the domain and range of the function?

Sample Response

a. **Problem Solving Strategy:** Use a diagram.

Let $h(t)$ = the height of the clam after t seconds.

$$\begin{pmatrix}\text{distance clam}\\ \text{has fallen}\end{pmatrix} + \begin{pmatrix}\text{height of clam}\\ \text{above ground}\end{pmatrix} = \begin{pmatrix}\text{initial height}\\ \text{of clam}\end{pmatrix}$$

$$16t^2 + h(t) = 50$$

$$h(t) = -16t^2 + 50$$

b. Graph the related function $f(x) = -16x^2 + 50$. Then look at which part of the graph is the graph of the function $h(t) = -16t^2 + 50$.

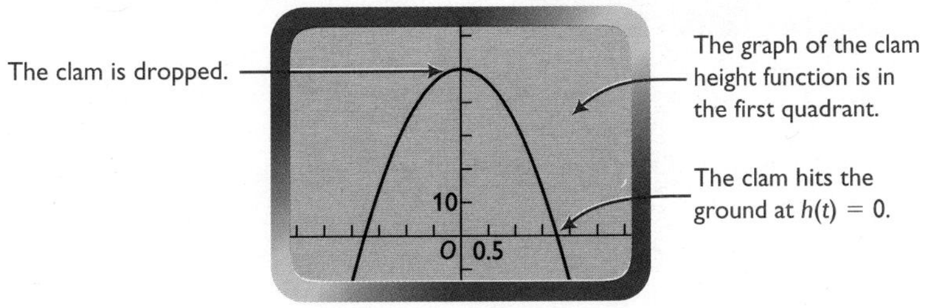

Domain: $0 \le t \le 1.7$ Range: $0 \le h(t) \le 50$

QUADRATIC FUNCTIONS

A quadratic function has the form:

$f(x) = ax^2 + bx + c$

where a, b, and c are constants and $a \neq 0$.

The graph is a parabola.

The sign of a determines whether the parabola opens upward or downward.

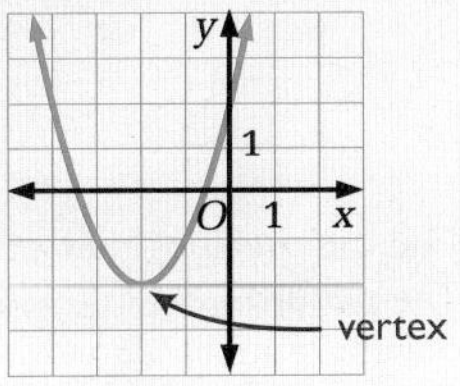

Case 1: $a > 0$

The parabola opens upward.
The vertex is the lowest point.

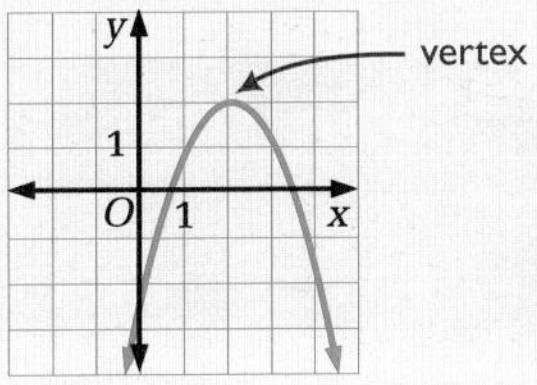

Case 2: $a < 0$

The parabola opens downward.
The vertex is the highest point.

TEACHING

Talk it Over

Questions 1 and 2 introduce students to a nonlinear function, that of an object in freefall.

Visual Thinking

When discussing Sample 1, ask students to select points on the diagram shown and indicate where those points occur on the graph. Ask them to select points on the graph and describe where they occur on the diagram. This activity involves the visual skills of *recognition* and *identification*.

Additional Sample

S1 On the moon, the distance traveled (in meters) by an object in freefall is given by $s(t) = 0.8t^2$, where t is the time (in seconds) that the object has been falling.

a. Suppose an astronaut tosses a small moon rock up and it reaches a height of 15 m before falling back to the surface of the moon. Express the height of the rock above the moon's surface as a function of t, the time it is in freefall from a height of 15 m.

Let $h(t)$ = height of the rock t seconds after it has started to fall back.
Then $h(t) = -0.8t^2 + 15$.

b. Graph the function in part (a). What are the domain and range of the function?

Graph the related function $f(x) = -0.8x^2 + 15$.

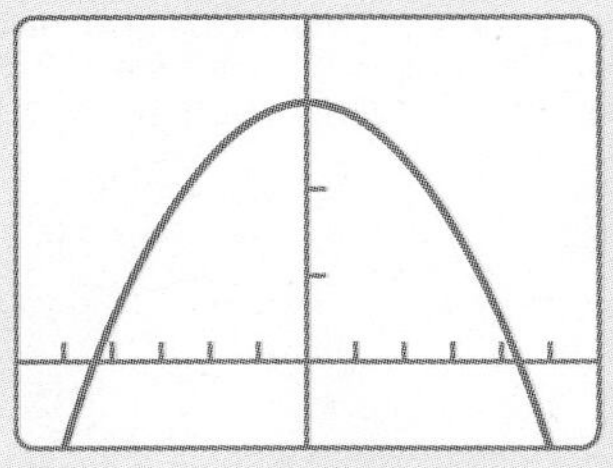

The graph for the height function for the rock is in the first quadrant. It reaches height $h(t) = 0$ (the surface) when $t \approx 4.33$.
Domain: $0 \le t \le 4.33$
Range: $0 \le h(t) \le 15$

Additional Sample

S2 When a baseball player hits a ball, the bat propels the ball from a height of 3 ft, at a speed of 150 ft/s, and at an angle of 53.1° with respect to the horizontal.

a. Express the height of the ball as a function of time.

Use the equation for the height of a thrown object.

$h(t) =$
$-16t^2 + (v_0 \sin A)t + h_0 =$
$-16t^2 + (150 \sin 53.1°)t + 3$
$\approx -16t^2 + 120t + 3$

b. When is the ball 59 ft above the ground?

Solve the equation when $h(t) = 59$.

Method 1. Use a graph. Graph $y = -16x^2 + 120x + 3$ and $y = 59$ on the same set of axes. Find the points of intersection.

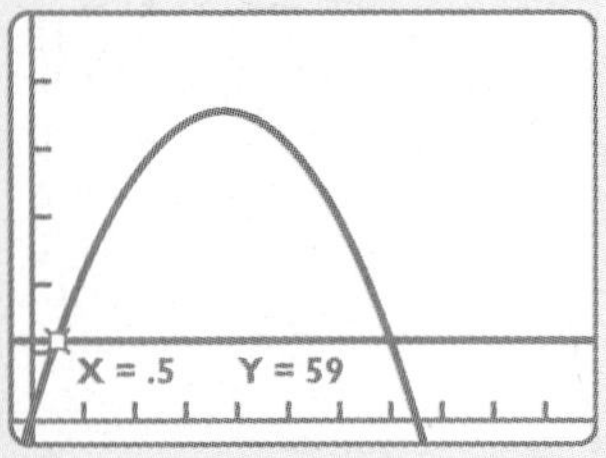

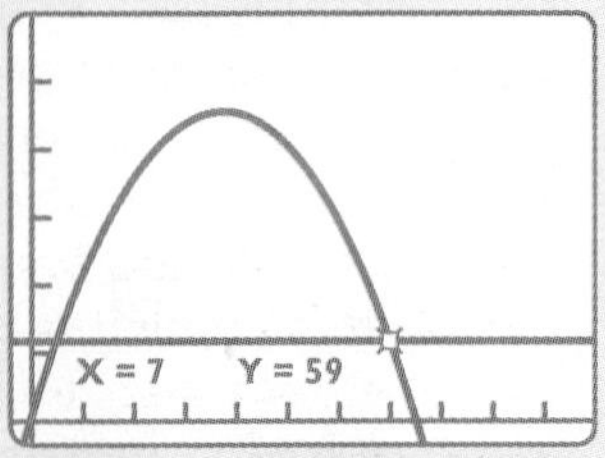

The x-coordinates of the points of intersection are about 0.5 and 7.0. The ball is 59 ft above the ground after about 0.5 s and 7 s.

Method 2. Use a formula. Write the equation in the form $at^2 + bt + c = 0$.

$-16t^2 + 120t + 3 = 59$
$-16t^2 + 120t - 56 = 0$

Solve the equation using the quadratic formula.

$$t = -\frac{b}{2a} \pm \frac{\sqrt{b^2 - 4ac}}{2a}$$
$$= -\frac{120}{2(-16)} \pm \frac{\sqrt{(120)^2 - 4(-16)(-56)}}{2(-16)}$$
$$= 3.75 \pm 3.25$$
$$= 0.5 \text{ or } 7$$

The ball is 59 ft above the ground after 0.5 s and 7 s.

The Height of a Thrown Object

In the function in Sample 1, $f(x) = -16x^2 + 50$, the initial height, h_0, is 50 feet.

If you throw an object instead of dropping it, the object's height in feet after t seconds is given by this equation:

$$h(t) = -16t^2 + (v_0 \sin A)t + h_0$$

h_0 is the object's initial height (in feet).

v_0 is the object's initial speed (in feet per second).

A is the angle at which the object is thrown with respect to the horizontal.

Sample 2

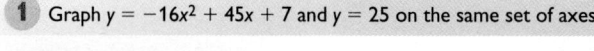

A lacrosse player uses a stick with a pocket on one end to throw a ball toward the opposing team's goal. Suppose the ball leaves a player's stick from an initial height of 7 ft, at a speed of 90 ft/s, and at an angle of 30° with respect to the horizontal.

a. Express the height of the ball as a function of time.

b. When is the ball 25 ft above the ground?

Sample Response

a. Use the equation for the height of a thrown object.

$$h(t) = -16t^2 + (v_0 \sin A)t + h_0$$
$$= -16t^2 + (90 \sin 30°)t + 7 \quad \leftarrow \text{Substitute } \mathbf{90} \text{ for } v_0,\ 30° \text{ for } A, \text{ and } 7 \text{ for } h_0.$$
$$= -16t^2 + 45t + 7$$

b. Solve the equation $h(t) = -16t^2 + 45t + 7$ when $h(t) = 25$.

Method 1

Problem Solving Strategy: Use a graph.

1 Graph $y = -16x^2 + 45x + 7$ and $y = 25$ on the same set of axes.

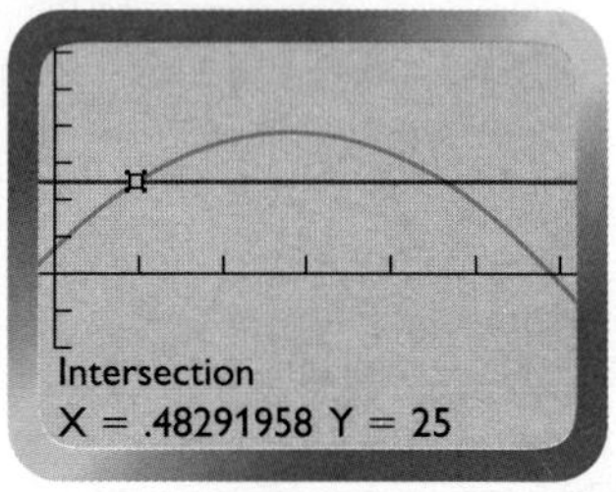

2 Find each point of intersection. The x-coordinates of these points are about 0.5 and 2.3.

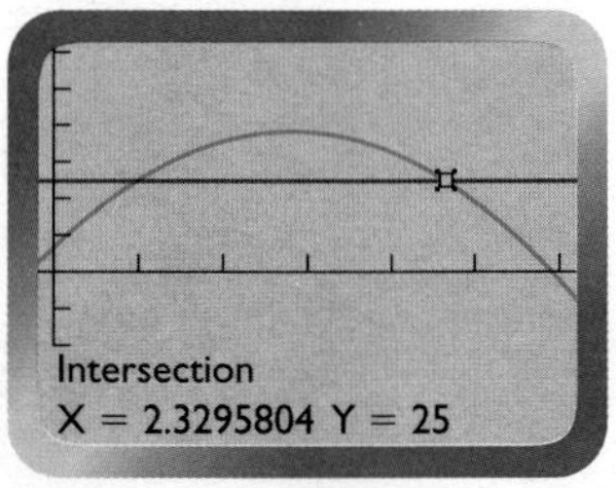

The ball is 25 ft above the ground after about 0.5 s and about 2.3 s.

Answers to Talk it Over

3. Answers may vary. An example is given. I like using a graphics calculator because it is simpler than using the quadratic formula.
4. after about 1.4 s; about 38.6 ft
5. No; it shows the height of the ball as a function of time and does not give any information about the angle of the ball with respect to the horizontal.
6. No; every height but the maximum is paired with two times.

Answers to Look Back

Answers may vary. Examples are given. The graph of a linear function is a line, while the graph of a quadratic function is a curve. The equation for a linear function contains only an x-term; the equation for a quadratic function contains an x^2-term. A linear function is one-to-one; a quadratic function is many-to-one.

Method ❷

Problem Solving Strategy: Use a formula.

$-16t^2 + 45t + 7 = 25$ ⟵ First write the equation in the form $at^2 + bt + c = 0$.

$-16t^2 + 45t - 18 = 0$ ⟵ Subtract 25 from both sides.

Then solve the equation using the quadratic formula.

$$t = -\frac{b}{2a} \pm \frac{\sqrt{b^2 - 4ac}}{2a}$$

$$= -\frac{45}{2(-16)} \pm \frac{\sqrt{(45)^2 - 4(-16)(-18)}}{2(-16)}$$ ⟵ Substitute −16 for *a*, 45 for *b*, and −18 for *c*.

$$= \frac{45}{32} \pm \frac{\sqrt{873}}{-32}$$

$$\approx 1.41 \pm (-0.92)$$

$$\approx 0.49 \text{ or } 2.33$$

The ball is 25 ft above the ground after about 0.5 s and about 2.3 s.

Talk it Over

3. In Sample 2, which method do you prefer? Why?
4. Use a graphics calculator to determine when the lacrosse ball reaches its maximum height. Find the maximum height.
5. Does the graph of $h(t) = -16t^2 + 45t + 7$ show the path of the lacrosse ball through the air? Explain.
6. Is the graph of the function one-to-one? Explain.

BY THE WAY...

Modern lacrosse is based on a game first played by the Iroquois people of New York and Canada. The Iroquois called their game *baggataway* or *tewaraathon*.

Look Back ⟵

List at least two ways in which quadratic functions differ from linear functions.

2-4 Exercises and Problems

1. **Reading** Look at the picture of the gull and clam on page 94. How far does the clam fall in 1.5 s?

Tell whether each function is a quadratic function. For each quadratic function:

a. Find $f(3)$, $f(-3)$, and $f(a + b)$.

b. Graph $f(x)$.

2. $f(x) = 1.5x^2$
3. $f(x) = 2x^2 + 3x - 2$
4. $f(x) = \frac{1}{2x^2 + 3x - 2}$
5. $f(x) = 5 - 7x^2$
6. $f(x) = (x - 4)^3$
7. $f(x) = (x - 4)^2$

Talk it Over

Questions 3–6 focus students' attention on Sample 2. For Question 5, students should understand *why* the graph of $h(t)$ does not show the path of the ball through the air.

Problem Solving

Talk it Over question 4 can be answered by using the ZOOM and TRACE features of the calculator. Another approach is possible for students who are using the TI-82. They can press [2nd] [CALC] and select maximum from the Calculate menu.

You may wish to ask if the Sample Response for Sample 2 provides information that could be used to check the answer for question 4. Students may recall that a parabola is symmetric with respect to the vertical line through its maximum or minimum point. Discuss how you can average the x-coordinates when the ball is 25 ft high to obtain the axis of symmetry, $x = 1.41$. Then evaluate $h(1.41)$ to find the maximum.

Look Back

You might ask students to respond to this question verbally and have one student make a list of the differences on the board.

APPLYING

Suggested Assignment

Standard 1–8, 12–14, 16–24

Extended 1–8, 10–14, 16–24

Integrating the Strands

Algebra Exs. 1–14, 16, 17, 19–24

Functions Exs. 1–11, 16, 17, 19, 20, 24

Statistics and Probability Ex. 18

Logic and Language Exs. 1, 10, 15, 17, 24

Answers to Exercises and Problems

1. 36 ft
2. a. quadratic; 13.5; 13.5; $1.5a^2 + 3ab + 1.5b^2$

   b. 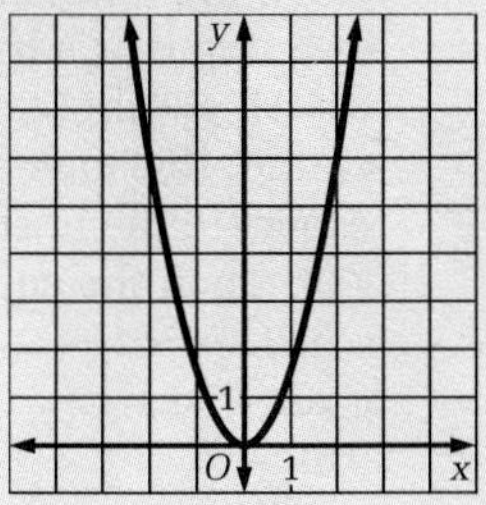

3. a. quadratic; 25; 7; $2a^2 + 4ab + 2b^2 + 3a + 3b - 2$

   b. 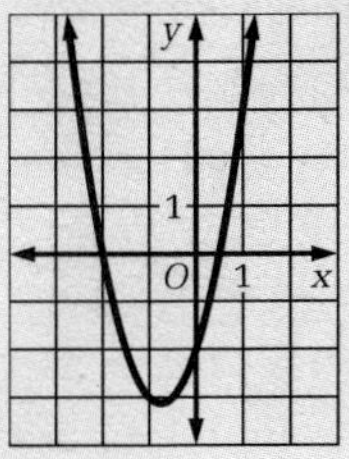

4. not quadratic
5. a. quadratic; −58; −58; $5 - 7a^2 - 14ab - 7b^2$

   b. 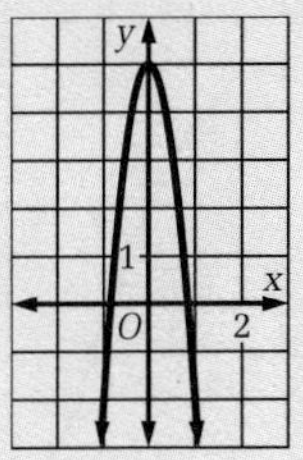

Answers continued on next page.

Career Note

Engineers who design roads usually hold a degree in civil engineering from a college or university. Civil engineers design public works such as roads, bridges, dams, canals, harbors, and so on, or they supervise the construction of such projects. When designing an actual road, the land for the road would first be surveyed to mark its position and surrounding boundaries.

Cooperative Learning

Working in small groups, students can discuss the derivation of the quadratic formula shown in Exs. 12–14. Various groups can then share their answers with the whole class.

8. Suppose a gull drops a clam from a height of 40 ft.

a. Express the height of the clam above the ground as a function of time.

b. Graph the function in part (a).

c. When does the clam hit the ground?

BY THE WAY...

Gulls shut down a driving range in Springfield, Massachusetts, when they mistook golf balls for clams. Players had to dodge balls dropped by gulls trying to break the golf balls open.

9. Suppose the lacrosse ball in Sample 2 is thrown from an initial height of 6 ft, at a speed of 120 ft/s, and at an angle of 10° with respect to the horizontal.

a. Express the height of the ball above the ground as a function of time.

b. When is the ball 5 ft above the ground?

c TECHNOLOGY Use a graphics calculator to determine when the ball reaches its maximum height. What is the maximum height?

10. Suppose an automatic pitching machine releases a baseball parallel to the ground at a height of 5 ft and at an initial speed of 70 mi/h. At the same instant, a second baseball is dropped from the same initial height.

a. Express the height of the pitching machine's baseball as a function of time. (Use the formula on page 96 and an angle of 0°.)

b. Express the height of the dropped baseball as a function of time. (Use the formula on page 94.)

c. **Writing** What do you notice about the functions from parts (a) and (b)? What does the relationship between the functions mean in this situation?

11. **Career** An engineer designing a curved road must make the curve's radius large enough so that car passengers are not pulled to one side as they go around the curve at the posted speed. The minimum radius for passenger comfort is given by this function:

$$r(s) = 0.334s^2$$

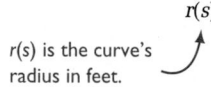

$r(s)$ is the curve's radius in feet.

s is the expected speed of traffic in miles per hour.

s (mi/h)	$r(s)$ (ft)
5	?
10	?
?	133.6
?	534.4

a. Copy and complete the table.

b. Calculate the ratios $\frac{r(10)}{r(5)}$, $\frac{r(20)}{r(10)}$, and $\frac{r(40)}{r(20)}$. Based on your calculations, by what factor does the minimum radius increase when the expected speed of traffic doubles?

c. Show that doubling the traffic speed *always* produces the result you observed in part (b). $\left(\textit{Hint: }\text{Use the ratio } \frac{r(2s)}{r(s)}.\right)$

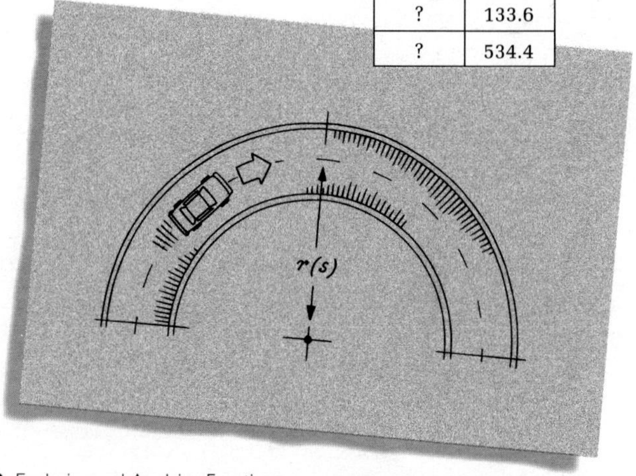

Answers to Exercises and Problems

6. not quadratic

7. a. quadratic; 1; 49; $a^2 + 2ab - 8a + b^2 - 8b + 16$

b. 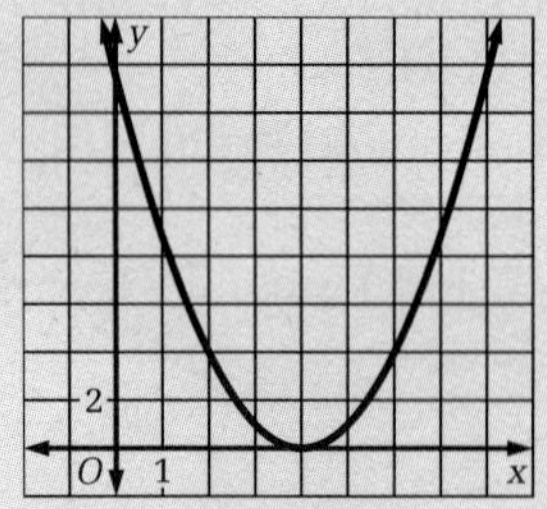

8. a. $h(t) = -16t^2 + 40$

b. 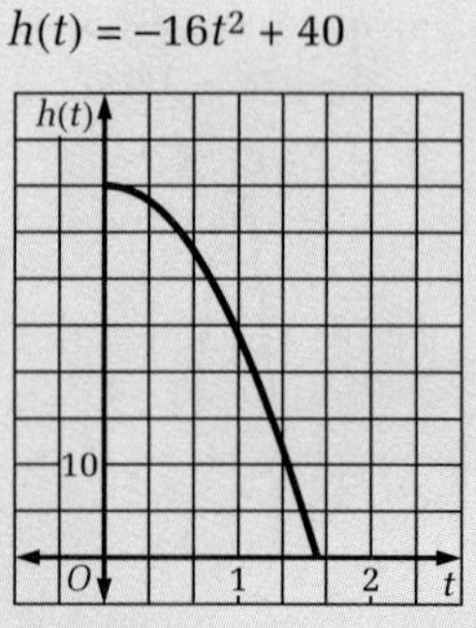

c. after about 1.6 s

9. a. $h(t) = -16t^2 + (120 \sin 10°)t + 6 = -16t^2 + 20.8t + 6$

b. after about 1.35 s

c. after about 0.65 s; about 12.8 ft

10. a. $h(t) = -16t^2 + 5$

b. $h(t) = -16t^2 + 5$

c. The functions are the same. This means that throwing a ball parallel to the ground has the same effect (on height as a function of time) as dropping it. If a ball is thrown parallel to the ground, the angle with respect to the horizontal is 0°. Then the equation is $h(t) = -16t^2 + h_0(\sin 0°)t + h_0 = -16t^2 + h_0$.

History of Mathematics The Indian mathematician Sridhara gave this derivation of the quadratic formula around A.D. 1025.

$$ax^2 + bx + c = 0 \quad \longleftarrow \text{Start with the general quadratic equation.}$$
$$ax^2 + bx = -c \quad \longleftarrow \text{Subtract } c \text{ from both sides.}$$
$$4a^2x^2 + 4abx = -4ac \quad \longleftarrow \text{Multiply both sides by } 4a.$$
$$4a^2x^2 + 4abx + b^2 = b^2 - 4ac \quad \longleftarrow \text{Add } b^2 \text{ to both sides.}$$
$$(2ax + b)^2 = b^2 - 4ac \quad \longleftarrow \text{Factor the left side of the equation.}$$
$$2ax + b = \pm\sqrt{b^2 - 4ac} \quad \longleftarrow \text{Take the square root of both sides.}$$
$$2ax = -b \pm \sqrt{b^2 - 4ac} \quad \longleftarrow \text{Subtract } b \text{ from both sides.}$$
$$x = \frac{-b \pm \sqrt{b^2 - 4ac}}{2a} \quad \longleftarrow \text{Divide both sides by } 2a.$$

12. Look at the fourth and fifth equations in Sridhara's derivation. Show that these equations are equivalent by expanding the product $(2ax + b)^2$.

13. Why is the condition $a \neq 0$ necessary in the quadratic formula?

14. Use the quadratic equation $2x^2 + 5x + 3 = 0$.
   a. What are the values of a, b, and c for this equation?
   b. Solve $2x^2 + 5x + 3 = 0$ by following the steps to the right of the equations in Sridhara's derivation.
   c. Solve $2x^2 + 5x + 3 = 0$ by substituting the values of a, b, and c directly into the quadratic formula. Are your solutions the same as those for part (b)?

15. **Research** Write a short report on Sridhara's life and mathematical work.

16. Cochita wants to build a rectangular corral for her horse. Three sides of the corral will be wooden fencing. The wall of the canyon will form the fourth side. Cochita has 200 ft of fencing and wants to build the largest corral possible.

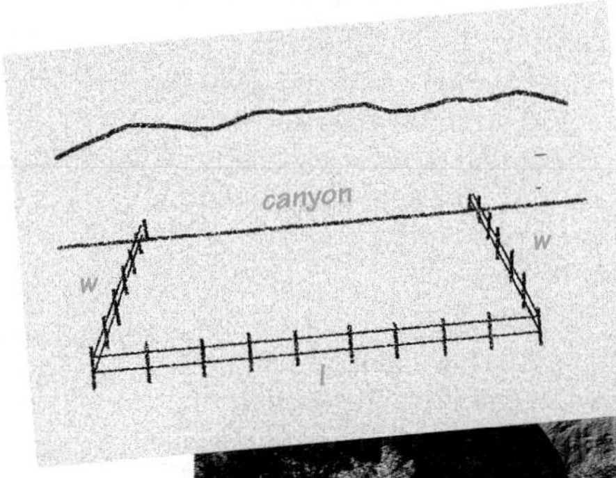

   a. Express the area A of Cochita's corral in terms of the length l and width w.
   b. Express l in terms of w.
   c. Express A as a function of w alone. Is the function $A = f(w)$ a quadratic function?
   d. Graph $A = f(w)$. What values of w and l maximize the corral's area? What is the maximum area?
   e. Suppose Cochita's corral does not have to be rectangular. Does the corral whose dimensions you found in part (d) still have the maximum possible area?

Students Acquiring English

For Ex. 13, you may want to allow students acquiring English to show why the condition $a \neq 0$ is necessary in the quadratic equation through examples rather than by writing an explanation in paragraph form.

Research

For Ex. 15, there are many excellent books that have been published on the history of mathematics. School and public libraries will most likely have some books that students can use for their research.

Error Analysis

In Ex. 16, some students may forget that Cochita's fencing is to be used on only three sides of the corral. Urge these students to reread the problem carefully and to refer to the diagram to point out the sides where fencing will be used.

Students can discuss the solution to this exercise in class. They can then look at how the function would change if all four sides required fencing. Students should find the solution for this situation and compare it to the solution for the original situation.

Answers to Exercises and Problems

11. a.

s (mi/h)	r(s) (ft)
5	8.35
10	33.4
20	133.6
40	534.4

b. 4; 4; 4; It increases by a factor of 4.

c. $\frac{r(2s)}{r(s)} = \frac{0.334(2s)^2}{0.334s^2} = \frac{4s^2}{s^2} = 4$

12. $(2ax + b)^2 = (2ax)^2 + 2(2ax)b + b^2 = 4a^2x^2 + 4axb + b^2$

13. to avoid division by 0; Also note that if $a = 0$, then the function is not a quadratic function. Part of the definition of a quadratic function is $a \neq 0$.

14. a. $a = 2$, $b = 5$, $c = 3$

b. $2x^2 + 5x = -3$; $16x^2 + 40x = -24$; $16x^2 + 40x + 25 = 1$; $(4x + 5)^2 = 1$; $4x + 5 = \pm 1$; $4x = -5 \pm 1$; $x = \frac{-5 \pm 1}{4} = -1$ or $-\frac{3}{2}$

c. $x = \frac{-5 \pm \sqrt{5^2 - 4 \cdot 2 \cdot 3}}{2 \cdot 2} = \frac{-5 \pm 1}{4} = -1$ or $-\frac{3}{2}$; Yes.

15. Answers may vary.

16. a. $A = lw$

b. $l = 200 - 2w$

c. $A(w) = w(200 - 2w)$ or $A(w) = 200w - 2w^2$; Yes.

d.

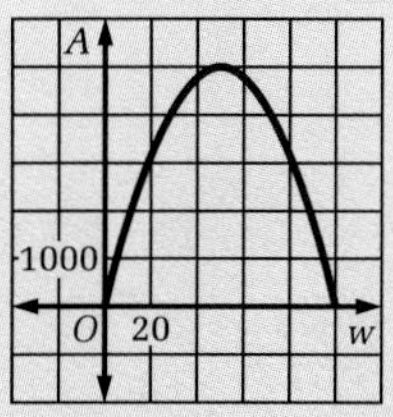

$w = 50$, $l = 100$; $A = 5000$ ft^2

e. No; the radius of a semicircular corral 200 ft around would be about 63.7 ft. The area of the corral would be about 6366 ft^2.

Working on the Unit Project

One aspect of bicycling that students may decide to give a presentation on is bicycle safety. The data in Ex. 24 can be supplemented with further research on safety issues associated with bicycling.

Practice 13 For use with Section 2-4

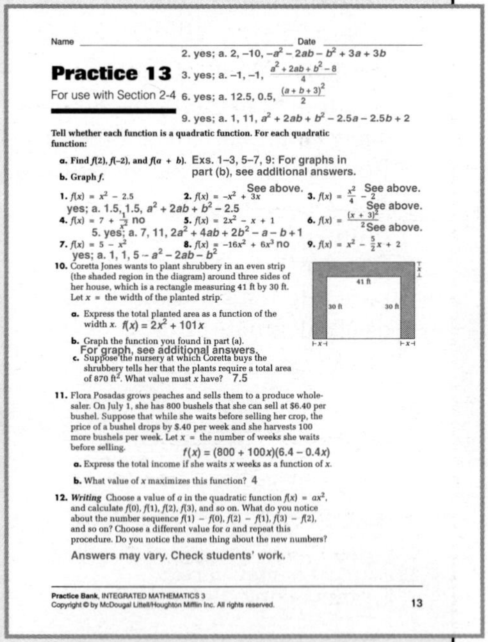

Name ______________________ Date ______________

Practice 13

For use with Section 2-4

2. yes; a. 2, −10, $-a^2 - 2ab - b^2 + 3a + 3b$
3. yes; a. −1, −1, $\frac{a^2 + 2ab + b^2 - 8}{4}$
6. yes; a. 12.5, 0.5, $\frac{(a+b+3)^2}{2}$
9. yes; a. 1, 11, $a^2 + 2ab + b^2 - 2.5a - 2.5b + 2$

Tell whether each function is a quadratic function. For each quadratic function:

a. Find $f(2)$, $f(-2)$, and $f(a + b)$. Exs. 1–3, 5–7, 9: For graphs in part (b), see additional answers.

b. Graph f.

1. $f(x) = x^2 - 2.5$ yes; a. 1.5, 1.5, $a^2 + 2ab + b^2 - 2.5$
2. $f(x) = -x^2 + 3x$ See above.
3. $f(x) = \frac{x^2}{4} - 2$ See above.
4. $f(x) = 7 + \frac{1}{x^2}$ no
5. $f(x) = 2x^2 - x + 1$ 5. yes; a. 7, 11, $2a^2 + 4ab + 2b^2 - a - b + 1$
6. $f(x) = \frac{(x+3)^2}{2}$ See above.
7. $f(x) = 5 - x^2$ yes; a. 1, 1, $5 - a^2 - 2ab - b^2$
8. $f(x) = -16x^2 + 6x^3$ no
9. $f(x) = x^2 - \frac{5}{2}x + 2$ See above.

10. Coretta Jones wants to plant shrubbery in an even strip (the shaded region in the diagram) around three sides of her house, which is a rectangle measuring 41 ft by 30 ft. Let x = the width of the planted strip.

   a. Express the total planted area as a function of the width x. $f(x) = 2x^2 + 101x$

   b. Graph the function you found in part (a). For graph, see additional answers.

   c. Suppose the nursery at which Coretta buys the shrubbery tells her that the plants require a total area of 870 ft². What value must x have? 7.5

41 ft
30 ft
30 ft
x

11. Flora Posadas grows peaches and sells them to a produce wholesaler. On July 1, she has 800 bushels that she can sell at $6.40 per bushel. Suppose that while she waits before selling her crop, the price of a bushel drops by $.40 per week and she harvests 100 more bushels per week. Let x = the number of weeks she waits before selling.

   a. Express the total income if she waits x weeks as a function of x. $f(x) = (800 + 100x)(6.4 - 0.4x)$

   b. What value of x maximizes this function? 4

12. *Writing* Choose a value of a in the quadratic function $f(x) = ax^2$, and calculate $f(0)$, $f(1)$, $f(2)$, $f(3)$, and so on. What do you notice about the number sequence $f(1) - f(0)$, $f(2) - f(1)$, $f(3) - f(2)$, and so on? Choose a different value for a and repeat this procedure. Do you notice the same thing about the new numbers?

Answers may vary. Check students' work.

13

Ongoing ASSESSMENT

17. **Writing** The function $h(t) = -16t^2 + (v_0 \sin A)t + h_0$ for the height of a thrown object is based on the assumption that there is no air resistance. Do you think the function will overestimate or underestimate an object's actual height at a given instant? Explain.

Review PREVIEW

18. Find the mean, median, and mode of these test scores: 85, 56, 89, 92, 84, 100, 75, 72, 81, 69. *(Toolbox Skill 1)*

Graph each function. *(Section 2-3)*

19. $y = \begin{cases} x + 3 & \text{if } x < 0 \\ 5 & \text{if } x = 0 \\ 2x - 1 & \text{if } x > 0 \end{cases}$

20. $f(x) = -3|x + 2| - 1$

Factor. *(Toolbox Skill 17)*

21. $x^2 + 2x - 24$

22. $6a^2 - a - 15$

23. $25x^3 - 100x$

2 Working on the Unit Project

24. In 1992, only about 20% of adults and 5% of children wore protective helmets while bicycling. In this exercise, you will investigate what happens if you fall without a helmet.

   a. When a person of average height sits on a bicycle, the person's head is about 5.3 ft above the ground. Suppose such a person falls from a bicycle at rest. Use the function $d(t) = 16t^2$ to estimate the time it takes for the person's head to hit the ground.

   b. The function $v(t) = 32t$ gives the speed $v(t)$ in feet per second of an object that falls for t seconds. Estimate the speed at which the head of the person in part (a) hits the ground.

   c. **Open-ended** A cyclist not wearing a helmet will generally have lasting brain damage if the cyclist's head hits the ground at more than 18 ft/s. Based on your answer to part (b), do you think wearing a helmet is a good idea? Explain.

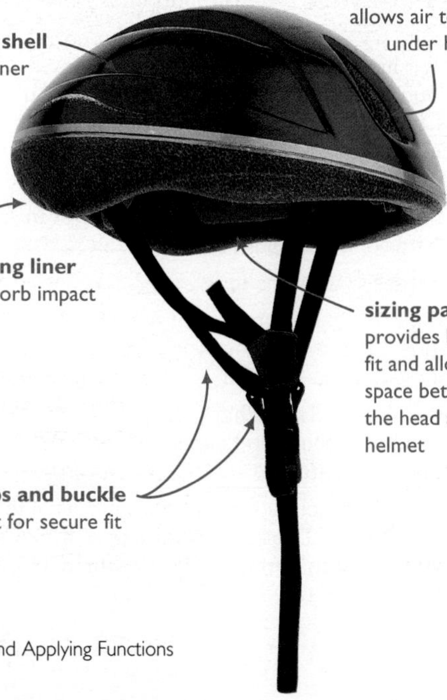

Answers to Exercises and Problems

17. The function will overestimate the height, since air resistance opposes the motion of the object.

18. 80.3; 82.5; none

19.

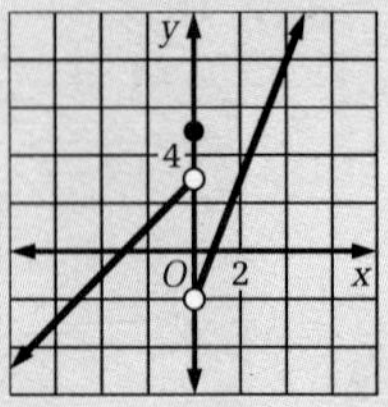

20.

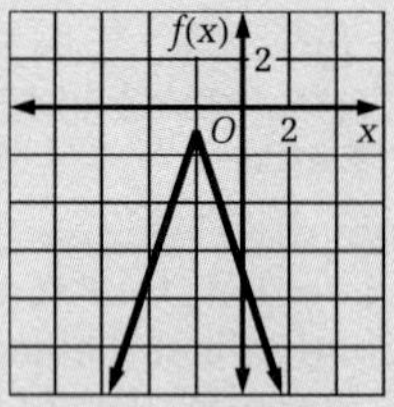

21. $(x + 6)(x - 4)$

22. $(3a - 5)(2a + 3)$

23. $25x(x + 2)(x - 2)$

24. a. about 0.58 s

   b. about 18.6 ft/s; The speed would increase if the bicycle were moving.

   c. Yes; even if the bicycle were not moving when the cyclist fell, he or she would probably incur lasting brain damage. The injuries would be worse if the bicycle were in motion when the cyclist fell.

Section

2-5 Polynomial Functions

More POWER To You

Focus

Recognize, evaluate, and graph direct variation functions and polynomial functions.

EXPLORATION

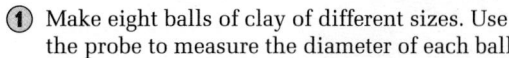

How is the weight of a ball related to its diameter?

- **Materials: modeling clay, thin probe (for example, a straightened paper clip), ruler, spring or balance scale, graphics calculator**
- **Work in a group of four students.**

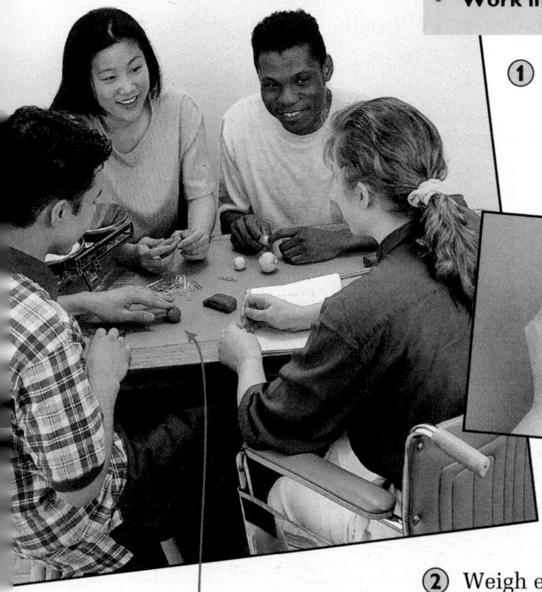

Try to make the ball as round as possible.

1. Make eight balls of clay of different sizes. Use the probe to measure the diameter of each ball.

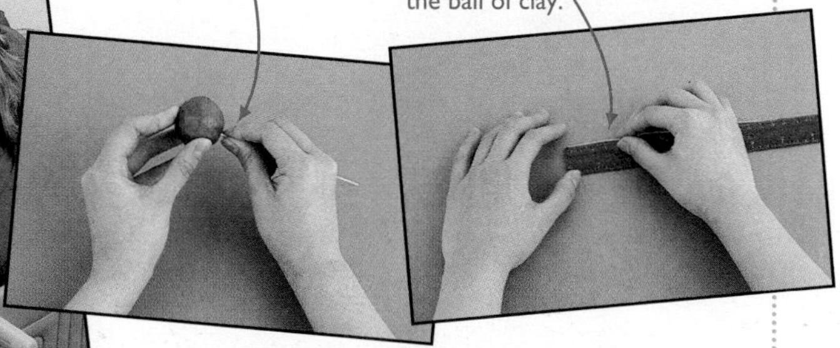

Mark or hold this point.
Measure the part of the probe that was inside the ball of clay.

Measure the diameter in three different directions. Find the mean of the three diameters. Record your data in a table.

2. Weigh each ball on the scale. Record the weights in the table.
3. Make a scatter plot of the ordered pairs (diameter, weight). Describe the shape of the graph.
4. Use the statistical functions of a graphics calculator to find an equation that models the data in the scatter plot. Use the calculator's "power regression" model. Write down the equation.
5. Suppose you could make perfect measurements in steps 1 and 2. The equation that you found in step 4 would have the form $f(x) = kx^3$. How close to 3 is the exponent in your equation? What is the value of k in your equation?

Answers to Exploration

1–5. Answers may vary. Check students' work.

PLANNING

Objectives and Strands

See pages 66A and 66B.

Spiral Learning

See page 66B.

Materials List

- Modeling clay, thin probe
- Ruler, graph paper
- Spring or balance scale
- Graphing technology

Recommended Pacing

Section 2-5 is a two-day lesson.

Day 1

Pages 101–103: Exploration through Talk it Over 8, *Exercises 1–21*

Day 2

Pages 104–105: Zeros of a Function through Look Back, *Exercises 22–36*

Toolbox References

- **Toolbox Skill 29:** Formulas from Geometry

Extra Practice

See pages 610–612.

Warm-Up Exercises

- Warm-Up Transparency 2-5

Support Materials

- Practice Bank: Practice 14
- Activity Bank: Enrichment 13
- Study Guide: Section 2-5
- Problem Bank: Problem Set 4
- Explorations Lab Manual: Additional Exploration 3 Diagram Masters 2, 11
- Overhead Visuals: Folder 2
- Using TI-81 and TI-82 Calculators: Point of Symmetry for Cubic Polynomials
- Using Plotter Plus: Point of Symmetry for Cubic Polynomials
- Assessment Bank: Quiz 2-5, Alternative Assessment 3

TEACHING

Exploration

The goal of the Exploration is to have students discover a specific direct variation function that describes the weight of a clay ball in terms of its diameter. In step 5, students learn that if they could make perfect measurements, then the equation they found would have the form $f(x) = kx^3$.

Additional Sample

S1 The weight, (w), of a ball bearing made from a particular kind of steel varies directly with the cube of its radius, (r). A ball bearing that has a radius of 0.4 cm has a weight of 2.1 g. Write a direct variation function that describes weight in terms of radius.

Use the general form for direct variation with the cube.
$w(r) = kr^3$
Substitute 0.4 for r and 2.1 for $w(r)$.

$$2.1 = k(0.4)^3$$
$$2.1 = k(0.064)$$
$$32.8125 = k$$

The function is $w(r) = 32.8125r^3$.

Talk it Over

Questions 1–3 have students investigate the graphs of the given functions concerning degree, symmetry, and whether they are one-to-one. Questions 4 and 5 introduce the concepts of odd and even functions.

X Lab

DIRECT VARIATION FUNCTIONS

A direct variation function has the form

$$f(x) = kx^n$$

where n is a positive integer and $k \neq 0$.

The exponent, n, is the degree of the function.

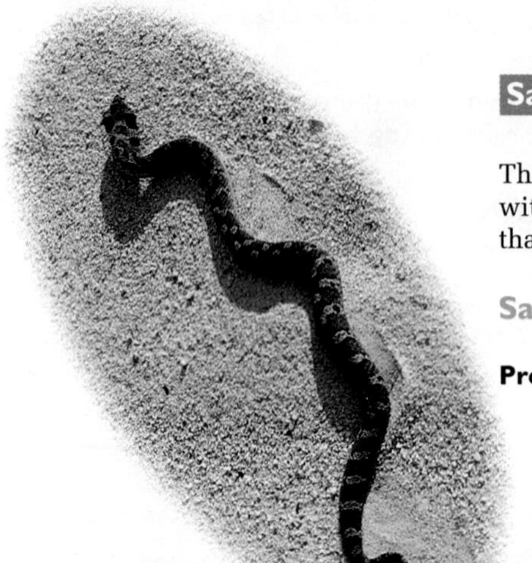

A snake that is 0.4 m long weighs 27.7 g.

Sample 1

The weight, w, of an adult male hognose snake varies directly with the cube of its length, l. Write a direct variation function that describes weight in terms of length.

Sample Response

Problem Solving Strategy: Use an equation.

$w(l) = kl^3$ ← Use the general form for direct variation with the cube.

$27.7 = k(0.4)^3$ ← Substitute the values you know.

$27.7 = k(0.064)$

$432.8 \approx k$

The function is $w(l) = 432.8l^3$.

Talk it Over

Use the graphs of $y = x$, $y = x^2$, $y = x^3$, $y = x^4$, and $y = x^5$.

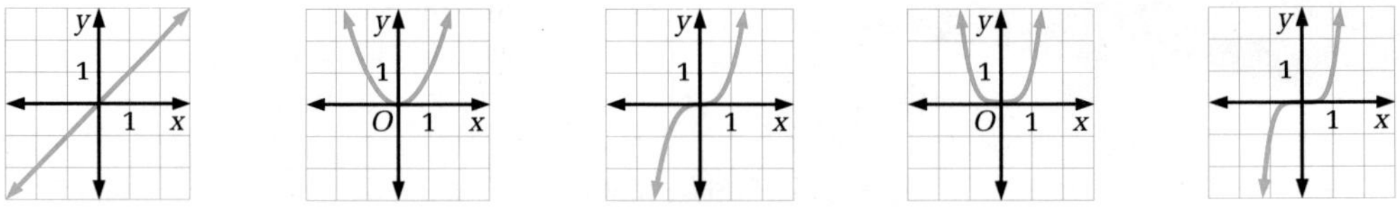

1. Explain why each function is a direct variation function. Give the degree of each function.
2. Which of the functions are one-to-one?
3. What kind of symmetry does each graph have?
4. The functions $f(x) = x$, $f(x) = x^3$, and $f(x) = x^5$ are *odd functions.* What do their graphs have in common?
5. The functions $f(x) = x^2$ and $f(x) = x^4$ are *even functions.* What do their graphs have in common?

Answers to Talk it Over

1. Each function has the form $f(x) = kx^n$; the degrees of the equations (from left to right) are 1, 2, 3, 4, and 5.
2. $y = x$; $y = x^3$; $y = x^5$
3. $y = x$, $y = x^3$, and $y = x^5$ have rotational symmetry of 180° about the origin; $y = x^2$ and $y = x^4$ have symmetry about the y-axis.
4. They are one-to-one and have 180° rotational symmetry about the origin. Their domain and range are all real numbers.
5. They are many-to-one and are symmetric about the y-axis. Their domain is all real numbers, and their range is all nonnegative real numbers.

Defining a Polynomial Function

The sum of one or more direct variation functions is a **polynomial function.** The sum may also include a constant function. An example of a polynomial function is

$$P(x) = x^4 + 16x^3 + 5x^2 - 13x + 6.$$

The coefficients of a polynomial function are real numbers, and the exponents are whole numbers. The **degree of a polynomial function** is the largest exponent.

Sample 2

Use the polynomial function $P(x) = 2x^4 + 0.2x^3 - x^2 + 1$.

a. Find $P(-2)$.

b. Find $P(3a)$.

Sample Response

a. $P(x) = 2x^4 + 0.2x^3 - x^2 + 1$

$P(-2) = 2(-2)^4 + 0.2(-2)^3 - (-2)^2 + 1$ ← Substitute −2 for *x*.

$= 2(16) + 0.2(-8) - 4 + 1$ ← Simplify.

$= 32 + (-1.6) - 4 + 1$

$= 27.4$

b. $P(x) = 2x^4 + 0.2x^3 - x^2 + 1$

$P(3a) = 2(3a)^4 + 0.2(3a)^3 - (3a)^2 + 1$ ← Substitute 3*a* for *x*.

$= 2(81a^4) + 0.2(27a^3) - 9a^2 + 1$ ← Simplify.

$= 162a^4 + 5.4a^3 - 9a^2 + 1$

Talk it Over

6. Use the function $f(x) = \frac{1}{2}x - 4$.
   a. Is *f* a polynomial function? What is its degree? What other type of function is this?
   b. Find *x* so that $f(x) = -1$.
7. Use the function $g(x) = -x^2 + 3x - 1$.
   a. Is *g* a polynomial function? What is its degree? What other type of function is this?
   b. Find *x* so that $g(x) = -1$.
8. Use the function in Sample 2. Suppose you wanted to find *x* so that $P(x) = -1$. What method could you use?

Now you are ready for: Exs. 1–21 on pp. 105–107

Additional Sample

S2 Use the polynomial function $P(x) = -3x^5 + 5x^2 - x + 4$.

a. Find $P(1)$.

$P(x) = -3x^5 + 5x^2 - x + 4$

$P(1) = -3(1)^5 + 5(1)^2 - 1 + 4$

$= -3 + 5 - 1 + 4$

$= 5$

b. Find $P(-2k)$.

$P(x) = -3x^5 + 5x^2 - x + 4$

$P(-2k) = -3(-2k)^5 + 5(-2k)^2 - (-2k) + 4$

$= -3(-32k^5) + 5(4k^2) + 2k + 4$

$= 96k^5 + 20k^2 + 2k + 4$

Error Analysis

Some students may think that linear functions and quadratic functions are not polynomial functions since they are already classified as linear or quadratic. Talk it Over questions 6 and 7 should help clear up this mistaken belief.

Answers to Talk it Over

6. a. Yes; 1; linear.
   b. 6
7. a. Yes; 2; quadratic.
   b. 0, 3
8. Answers may vary. An example is given. You could graph the function and see that there is no value for which $P(x) = -1$.

Additional Sample

S3 Graph the function $f(x) = \frac{1}{10}(x - 4)(x - 3)(x + 2)(x + 1)$.

Method 1. Graph the function by hand.

Step 1. Find the zeros of the function. $f(x) = 0$ when $x = 4$, $x = 3$, $x = -2$, or $x = -1$. The graph crosses the x-axis at (4, 0), (3, 0), (−2, 0), and (−1, 0).

Step 2. Substitute values of x that are between and beyond the zeros to see the shape of the graph.

x	y
−3	8.4
−1.5	−0.6
1	3.6
3.5	−0.6
5	8.4

Step 3. Graph all points and connect them with a smooth curve.

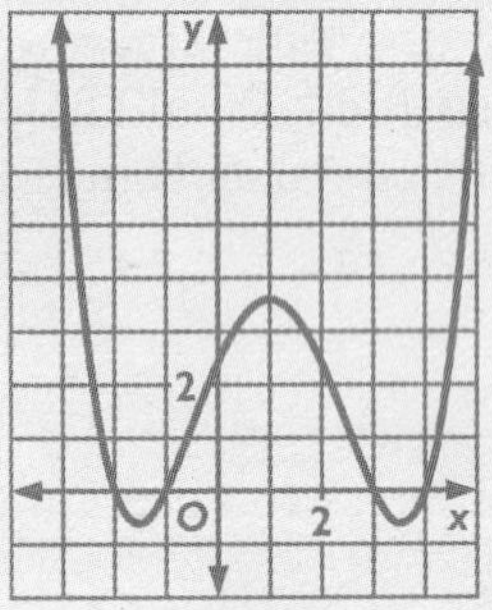

y
2
O
2
x

Method 2. Use a graphics calculator or graphing software. Enter the polynomial as $y = 0.1(x - 4)(x - 3)(x + 2)(x + 1)$ or $y = 0.1x^4 - 0.4x^3 - 0.7x^2 + 2.2x + 2.4$.

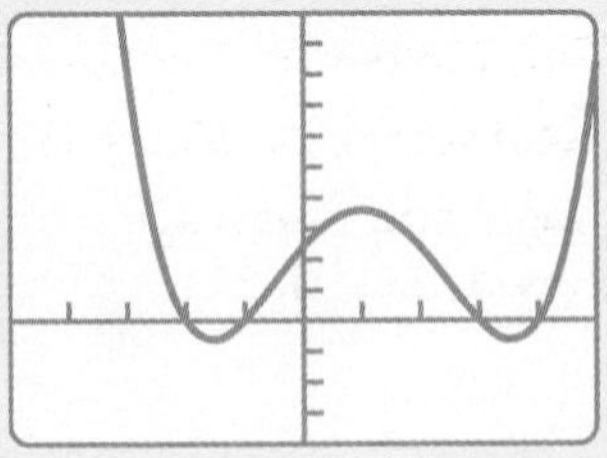

Zeros of a Function

Sometimes polynomials are written in factored form. For example, you can write the volume of the prism below as the sum of the volumes of the four smaller prisms or as the product of the length, width, and height of the large prism.

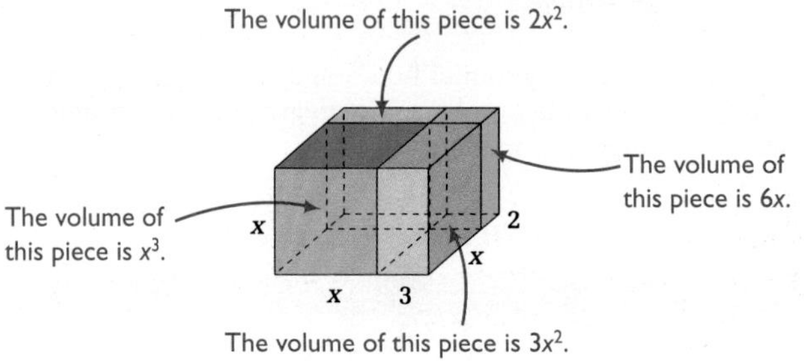

The volume of this piece is $2x^2$.
The volume of this piece is $6x$.
The volume of this piece is x^3.
x
2
x
x
3
The volume of this piece is $3x^2$.

Volume = $x^3 + 5x^2 + 6x$

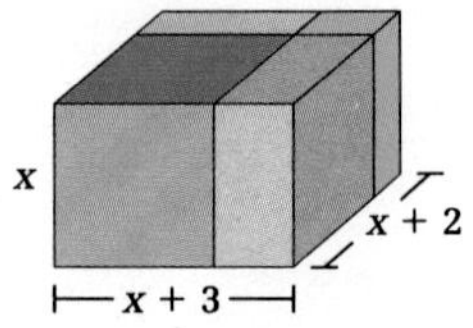

x
x + 2
x + 3

Volume = $x(x + 3)(x + 2)$

The polynomial $V(x) = x(x + 3)(x + 2)$ is in factored form. The factored form of a polynomial function tells you the *zeros of the function*. A **zero of a function** is a value of x that makes $f(x)$ equal to zero. You can use the zeros to graph the function.

Sample 3

Graph the function $V(x) = x(x + 3)(x + 2)$.

Sample Response

Method 1 Graph the function by hand.

Step 1 Find the zeros of the function.

$V(x) = 0$ when $x = 0$, $x = -3$, or $x = -2$.
The graph crosses the x-axis at (0, 0), (−3, 0), and (−2, 0).

Step 2 Substitute values of x that are between and beyond the zeros to see the shape of the graph.

x	y
−4	−8
−2.5	0.625
−1	−2
1	12

Step 3 Graph all the points that you found and connect them with a smooth curve.

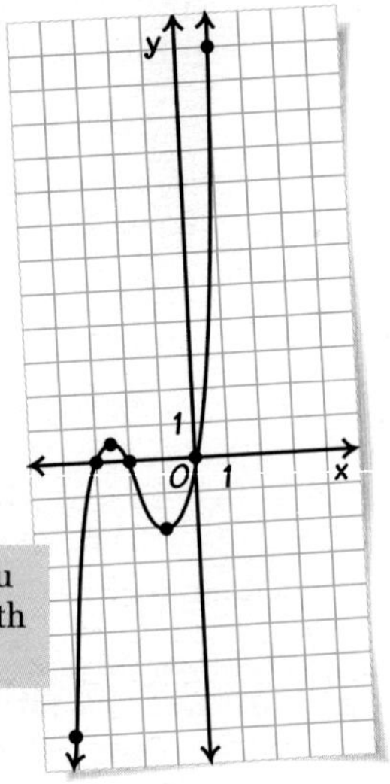

y
1
O
1
x

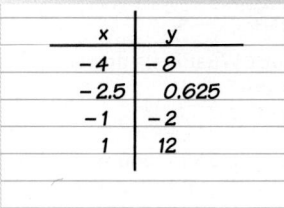

Method ❷ Use a graphics calculator or graphing software. Enter the polynomial in factored form or as a sum:

$y = x(x + 3)(x + 2)$ or $y = x^3 + 5x^2 + 6x$.

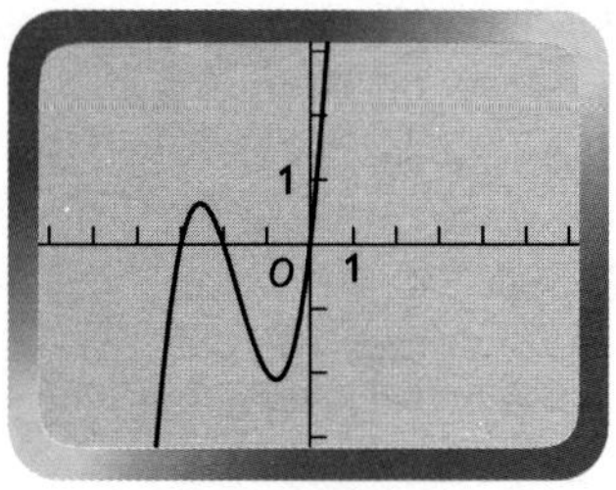

Talk it Over

9. The dimensions of a prism must be positive. Which points of the graph in Sample 3 represent this?
10. Describe how to use the graph in Sample 3 to find x when the volume of the prism is 40.

Xab

POLYNOMIAL FUNCTIONS

A polynomial function has the form

$P(x) = a_n x^n + a_{n-1}x^{n-1} + a_{n-2}x^{n-2} + \ldots + a_1 x + a_0$

- n is the degree of the polynomial.
- The exponents are whole numbers.
- The coefficients are real numbers.

Example

$P(x) = 3x^5 + x^3 - 4x^2 + 1$

The factored form of a polynomial gives the zeros of the function. You can use the zeros to sketch a graph of the function.

Look Back

Use examples to explain why constant functions, linear functions, quadratic functions, and direct variation functions are types of polynomial functions.

Now you are ready for: Exs. 22–36 on pp. 107–109

2-5 Exercises and Problems

1. **Reading** What is the value of k for each of the direct variation functions in the graphs on page 102?

Tell whether each direct variation function is *one-to-one*.

2. $f(x) = 0.2x^3$
3. $g(x) = 1.2x^4$
4. $r(x) = 0.75x$

Teaching Tip

In connection with Sample 3, you may wish to call attention to the relationship between the factored form of the polynomial and the zeros of the function. Challenge students to write a polynomial function in factored form that has zeros at three or four values of x that are supplied by members of the class. Students can check their results by graphing on a graphics calculator. Ask students to suggest ways to find an appropriate graphing window.

Look Back

Students can place some of their examples on the board in order to discuss the various types of functions listed.

APPLYING

Suggested Assignment

Day 1

Standard 1–10, 13–21

Extended 1–10, 12–21

Day 2

Standard 22–25, 28, 30–36

Extended 22–28, 30–36

Integrating the Strands

Algebra Exs. 1–30, 34–36

Functions Exs. 1–30, 35, 36

Geometry Exs. 28, 31–33

Logic and Language Exs. 1, 11, 29, 35

Answers to Talk it Over

9. the points in the first quadrant
10. Draw the graph of $y = 40$ on the same axes and determine the x-coordinate of the point where the graphs intersect. ($x = 2$)

Answers to Look Back

A polynomial function can be written in the form $P(x) = a_nx^n + a_{n-1}x^{n-1} + \ldots + a_1x + a_0$, where a_i is a real number and n is a whole number. A constant function such as $P(x) = 2$ has $n = 0$. A linear function such as $P(x) = 2x + 1$ has $n = 1$. A quadratic function such as $P(x) = 2x^2 + x + 1$ has $n = 2$. A direct variation function such as $f(x) = 2x^3$ has $n \geq 1$.

Answers to Exercises and Problems

1. 1
2. Yes.
3. No.
4. Yes.

Application

Polynomial functions often arise in solutions to real-world problems. Exs. 10–12 and 26–28 apply polynomial functions to a variety of practical problems.

Assessment: Standard

For Ex. 11, ask students these questions: Suppose you double the weight of the person on the train. How does this affect the force? Suppose you double the velocity. How does this affect the force?

For Exercises 5–8, write an equation in the form $f(x) = kx^n$ for each direct variation function.

5. $k = 0.8$ and $n = 3$
6. $k = 1$ and $f(6) = 36$
7. $f(2) = 8$ and $n = 4$
8. $f(0.3) = 45$ and $f(x)$ varies directly with the square of x

9. Suppose a ball of clay with a diameter of 1 in. weighs 0.41 oz. Write a formula for the weight of the ball as a function of the diameter.

Student Resources Toolbox
p. 652 *Formulas*

10. **Driving** A driver's reaction distance, r, varies directly with the speed, s, of the car.
    a. Write an equation for r as a function of s. Use k as the constant.
    b. Suppose the reaction distance for a driver traveling 30 mi/h is 33 ft. Find the value of k. Rewrite your function using this value of k.
    c. For what speed is the reaction distance 50 ft?

11. **Trains** Taking curves at high speeds raises the average speed of a train, but passengers do not like how it feels. The force (F) you feel when a train goes around a curve depends on your mass (m), the radius of the curve (r), and the square of the train's velocity (v).

$$F = \frac{mv^2}{r}$$

kg (m), m/s (v), newtons (F), meters (r)

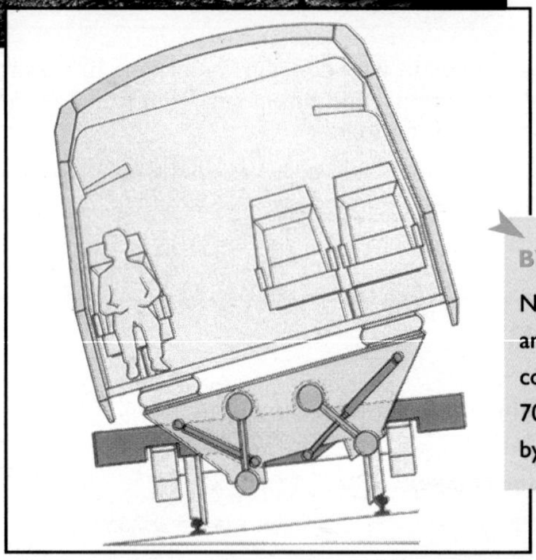

a. Suppose the same passenger goes around the same curve at different velocities. Explain why this represents direct variation with the square of the velocity.
b. Suppose the passenger has a mass of 60 kg and the radius of the turn is 9 m. Write a direct variation function that describes force in terms of velocity.
c. Use your equation to find the force on the passenger when the velocity is 9 m/s.
d. **Research** Find out what a *newton* is and after whom it is named.

BY THE WAY...

New passenger cars for trains are designed to tilt as they round corners in order to cancel out 70% of the force experienced by riders.

Answers to Exercises and Problems

5. $f(x) = 0.8x^3$
6. $f(x) = x^2$
7. $f(x) = \frac{1}{2}x^4$
8. $f(x) = 500x^2$
9. $w(d) = 0.41d^3$
10. a. $r = ks$
    b. 1.1; $r = 1.1s$
    c. about 45 mi/h
11. a. Since the same person and the same corner are involved, m and r are constants and the function rule is $F(v) = kv^2$ where $k = \frac{m}{r}$.
    b. $F(v) = \frac{60}{9}v^2 \approx 6.67v^2$
    c. 540 N
    d. a unit of force in the metric system (the force which, acting on a 1 kg mass produces an acceleration of 1 m/s²); Sir Isaac Newton, one of the greatest physicists of all time
12. a. $f(r) \approx 108.5r^4$
    b. about 59.35 mL/min
13. Yes.
14. No; $h(a) = 6^a$ is not a term of a polynomial because the exponent is a variable and not a constant.
15. No; in a polynomial function, the exponents are positive whole numbers.
16. a. 1; polynomial function of degree 1 and linear function
    b. 2; −7
    c. −10
17. a. 2; polynomial function of degree 2 and quadratic function
    b. 10; $1\frac{3}{4}$
    c. 0, −3
18. 3; −1
19. 4; 1
20. 5; 217
21. 4; 2
22. a. 0, 2, −3
    b.

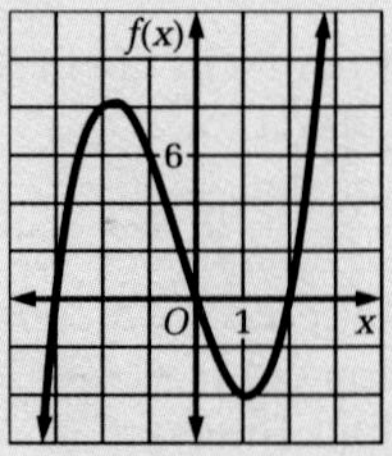

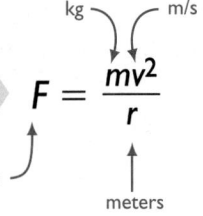

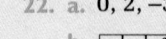

12. Medicine The blood flow in an artery varies directly with the fourth power of the radius of the artery.

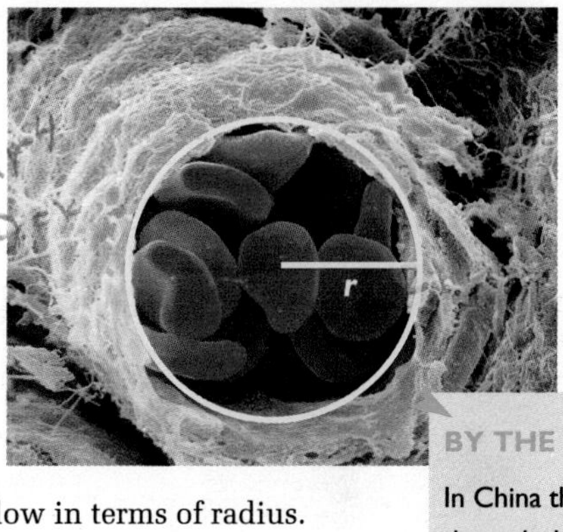

a. The radius of the aorta, a coronary artery, is about 1.2 cm at one point. The blood flow at that point is about 225 mL/min (milliliters per minute). Write a direct variation function that describes blood flow in terms of radius.

b. Farther away from the heart, the radius of the aorta is 0.86 cm. Find the blood flow at that point.

BY THE WAY...

In China the circulation of blood through the body was known and studied over 2000 years ago. Chinese physicians taught their students about blood flow and the heart by demonstrating with a system of bamboo tubes and bellows to pump liquid.

Tell whether each function is a polynomial function. If your answer is *No*, give a reason.

13. $g(x) = \frac{1}{2}x^3 - 5x$

14. $h(a) = 6^a - a^6$

15. $a(c) = c^{-2} + c^4 + 2c^5$

16. Use the function $f(x) = 1.5x - 7$.

a. What is the degree of f? What are two different names for this type of function?

b. Find $f(6)$ and $f(0)$.

c. Find x so that $f(x) = -22$.

17. Use the function $g(x) = x^2 + 3x$.

a. What is the degree of g? What are two different names for this type of function?

b. Find $g(-5)$ and $g\left(\frac{1}{2}\right)$.

c. Find x so that $g(x) = 0$.

For each polynomial function, write the degree and find the given value.

18. $f(x) = x^3 - 3x^2 + 2x - 1$; $f(2)$

19. $g(x) = 3x - x^2 + x^4 - 2x^3$; $g(1)$

20. $h(x) = x^5 - x^3 + 1$; $h(3)$

21. $f(x) = 2x^4 - x^2 + 1$; $f(-1)$

For Exercises 22–25:

a. Find the zeros of each function.

b. Graph each polynomial function.

22. $f(x) = x(x - 2)(x + 3)$

23. $g(x) = (x - 1)(x - 3)^2$

24. $h(x) = x(x^2 - x - 6)$

25. $r(x) = x^4 - 4x^3 + 4x^2$

Mathematical Procedures

For Exs. 22–25, it is not necessary to factor the polynomials when using a graphics calculator to produce the graph. If students are not using a calculator, then factoring can be useful, since it can give information about points where the graph may cross the x-axis.

Reasoning

Exs. 22–25 and Ex. 29 have to do with fundamental questions about the graphs of polynomial functions in general. How many zeros does a given polynomial function have? How many turning points does its graph have? While definitive answers are difficult to provide in this course, students can reason their way to plausible conjectures.

Here is one example of the kind of reasoning students can engage in. In Ex. 24, factor to write the function as $h(x) = x(x - 3)(x + 2)$. Since the solutions of $x(x - 3)(x + 2) = 0$ are -2, 0, and 3, the graph of $h(x)$ crosses the x-axis at $(-2, 0)$, $(0, 0)$, and $(3, 0)$. To find out what happens between these points, select a value of x between -2 and 0, say -1. Since $h(-1) = (-1)(-1 - 3)(-1 + 2) = 4$, the graph must have a turning point above the x-axis in the interval from $x = -2$ to $x = 0$. Consider $(0, 0)$ and $(3, 0)$. Since $h(1) = -6$, the graph must dip below the x-axis for the interval from $x = 0$ to $x = 3$. This is evidence of another turning point.

Answers to Exercises and Problems

23. a. 1, 3

b. 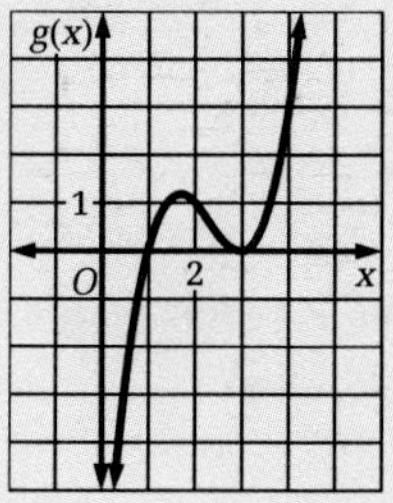

24. a. 0, −2, 3

b. 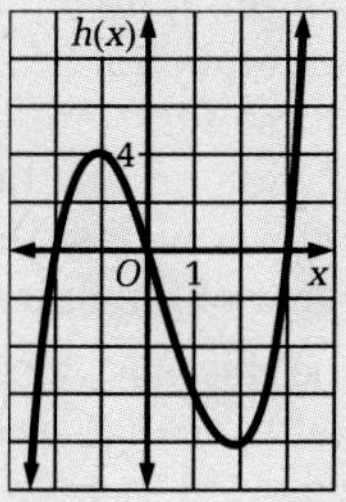

25. a. 0, 2

b. 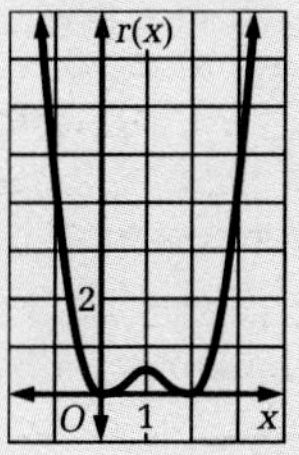

Integrating the Strands

Ex. 28 integrates the strands of algebra, geometry, and functions to provide a formula for finding the number of diagonals of a polygon.

Reasoning

Through the use of technology, Ex. 29 provides an opportunity for students to make a conjecture regarding the degree of a polynomial function and the number of turning points its graph has. You may wish to expand this reasoning process by having students use their graphs to make a conjecture about the degree of a polynomial function and its maximum number of real roots.

A vacuum cleaner makes a noise of approximately 80 decibels.

Noise Level A social scientist found that the percent (P) of people who are "highly annoyed" by noise from household appliances, traffic, airports, and so on depends on the decibel level (L) of the noise.

26. What percent of people are highly annoyed by a noise level of 80 decibels?

27. What percent of people are highly annoyed by a noise level of 0 decibels? What does this tell you about the graph of this polynomial?

28. Anne wrote a formula for the number of diagonals of a polygon.

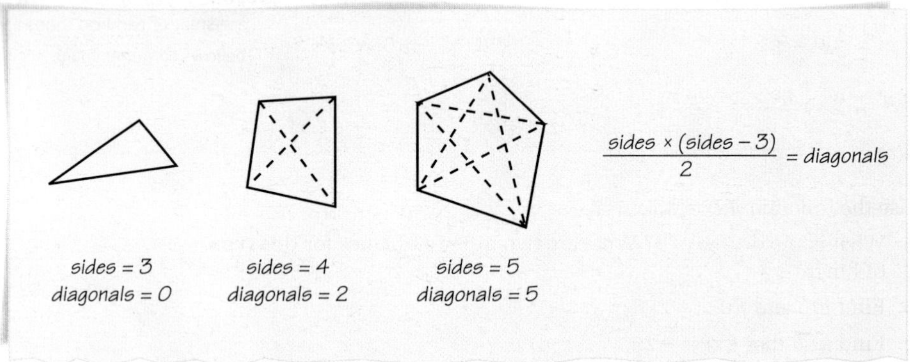

a. Write Anne's formula as a polynomial function for the number of diagonals in terms of the number of sides. What is its degree? What is its domain?

b. How many diagonals does an octagon have?

c. A polygon has 135 diagonals. How many sides does it have?

Ongoing **ASSESSMENT**

29. a. **Open-ended** Write equations for at least ten different polynomial functions of varying degrees.

b. TECHNOLOGY Use a graphics calculator or graphing software to graph the functions. Record the degree of each polynomial function and how many *turning points* the function has.

c. Make a conjecture about the relationship between the number of turning points and the degree of the polynomial function.

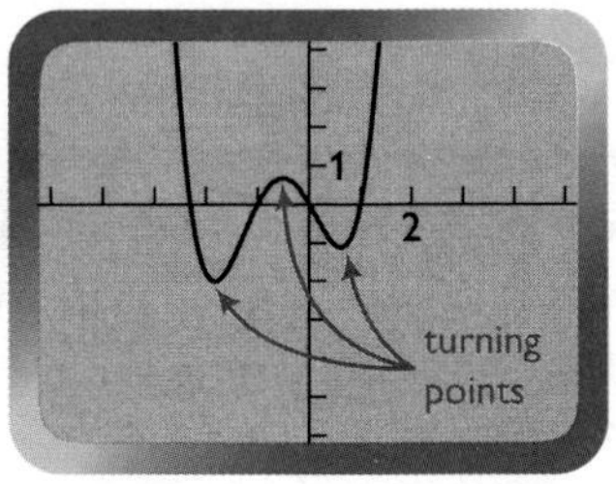

Answers to Exercises and Problems

26. about 52.4%

27. 0%; The graph intersects the origin.

28. a. $d(s) = \frac{1}{2}s^2 - \frac{3}{2}s$; 2; whole numbers greater than or equal to 3

b. 20 diagonals

c. 18 sides

29. a, b. Answers may vary.

c. Answers may vary. An example is given. I think the number of turning points of the graph of a polynomial function is at most one less than the degree of the function.

30. Let t be the time in seconds after the fall and $h(t)$ the height in feet of the pebble; $h(t) = -16t^2 + 75$.

31. No.

32. Yes.

33. Yes.

34. True.

35. a. about 382.1 W

b. about 38.6 W; about 343.5 W

Review PREVIEW

30. Suppose a pebble falls from a height of 75 ft. Express the height of the pebble above the ground as a function of time. *(Section 2-4)*

Tell whether each group of measurements can be the lengths of the sides of a triangle. Write *Yes* or *No*. *(Section 1-2)*

31. 7 mm, 8 mm, 19 mm **32.** 1 in., 0.5 in., 0.75 in. **33.** 10 yd, 10 yd, 10 yd

34. Is this statement *true* or *false*: $\sqrt{25 \cdot 16} = \sqrt{25} \cdot \sqrt{16}$? *(Toolbox Skill 19)*

2 Working on the Unit Project

If there were no friction from the air or the road, a cyclist would coast forever at a constant speed. A cyclist must supply energy to maintain a constant speed, however. The power (in watts) that a 75 kg rider on a 10 kg bicycle must supply to maintain a constant speed is $P = 0.2581v^3 + 3.509v$, where v is the speed in meters per second.

The first-degree term, 3.509v, is the power required to overcome rolling resistance.
The third-degree term, 0.2581v³, is the power required to overcome air resistance.

35. In a two-hour bicycle race, Bonnie rides at an average speed of 11 m/s (25 mi/h).

a. How much power must she supply to ride at this speed?

b. How much of the power is used to overcome rolling resistance? How much is used to overcome air resistance?

c. Which do you think is more important to a bicycle racer, rolling resistance or air resistance? Describe two things a racer could do in order to reduce the effect of the resistance.

36. When Arlean rides her bike to school, she supplies 0.1 horsepower (74.6 watts) to the bicycle in order to keep it moving at a constant speed. Use a graphics calculator to determine Arlean's speed when she rides to school.

Working on the Unit Project

The information presented in Exs. 35 and 36 can form the basis of a presentation in bike racing or the physics of bicycles. Including some mathematical information in the presentations would help to make them interesting and complete.

Practice 14 For use with Section 2-5

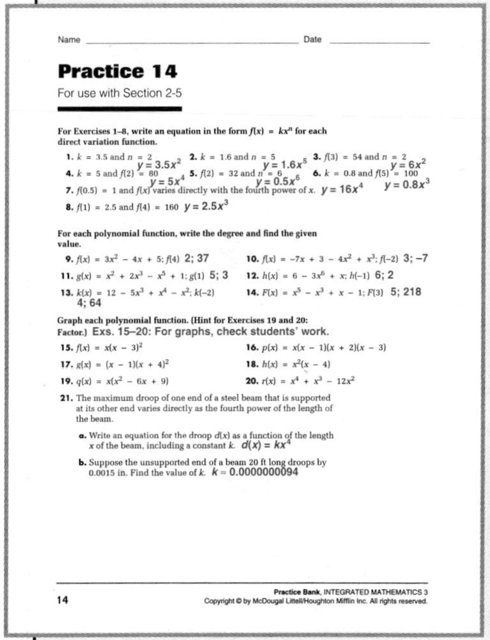

Name ______________ Date ______________

Practice 14

For use with Section 2-5

For Exercises 1–8, write an equation in the form $f(x) = kx^n$ for each direct variation function.

1. $k = 3.5$ and $n = 2$ $y = 3.5x^2$
2. $k = 1.6$ and $n = 5$ $y = 1.6x^5$
3. $f(3) = 54$ and $n = 2$ $y = 6x^2$
4. $k = 5$ and $f(2) = 80$ $y = 5x^4$
5. $f(2) = 32$ and $n = 6$ $y = 0.5x^6$
6. $k = 0.8$ and $f(5) = 100$ $y = 0.8x^3$
7. $f(0.5) = 1$ and $f(x)$ varies directly with the fourth power of x. $y = 16x^4$
8. $f(1) = 2.5$ and $f(4) = 160$ $y = 2.5x^3$

For each polynomial function, write the degree and find the given value.

9. $f(x) = 3x^2 - 4x + 5$; $f(4)$ 2; 37
10. $f(x) = -7x + 3 - 4x^2 + x^3$; $f(-2)$ 3; –7
11. $g(x) = x^2 + 2x^3 - x^5 + 1$; $g(1)$ 5; 3
12. $h(x) = 6 - 3x^6 + x$; $h(-1)$ 6; 2
13. $k(x) = 12 - 5x^3 + x^4 - x^2$; $k(-2)$ 4; 64
14. $F(x) = x^5 - x^3 + x - 1$; $F(3)$ 5; 218

Graph each polynomial function. (Hint for Exercises 19 and 20: Factor.) Exs. 15–20: For graphs, check students' work.

15. $f(x) = x(x - 3)^2$
16. $p(x) = x(x - 1)(x + 2)(x - 3)$
17. $g(x) = (x - 1)(x + 4)^2$
18. $h(x) = x^2(x - 4)$
19. $q(x) = x(x^2 - 6x + 9)$
20. $r(x) = x^4 + x^3 - 12x^2$
21. The maximum droop of one end of a steel beam that is supported at its other end varies directly as the fourth power of the length of the beam.
    a. Write an equation for the droop $d(x)$ as a function of the length x of the beam, including a constant k. $d(x) = kx^4$
    b. Suppose the unsupported end of a beam 20 ft long droops by 0.0015 in. Find the value of k. $k \approx 0.0000000094$

14 **Practice Bank**, INTEGRATED MATHEMATICS 3

Answers to Exercises and Problems

c. Answers may vary. An example is given. I think air resistance is more important because the power to overcome it is a function of the cube of the speed, while the power to overcome road resistance is a linear function of speed. I think it would help if the rider's clothing were tight-fitting but flexible and if the rider's position were low and close to the handlebars.

36. Estimates may vary. An example is given; about 5.93 m/s.

PLANNING

Objectives and Strands

See pages 66A and 66B.

Spiral Learning

See page 66B.

Materials List

- Graphics calculator
- Scissors
- Ruler

Recommended Pacing

Section 2-6 is a two-day lesson.

Day 1

Pages 110–113: Opening paragraph through Talk it Over 4, *Exercises 1–17*

Day 2

Pages 113–115: Radical Functions through Look Back, *Exercises 18–49*

Toolbox References

- **Toolbox Skill 19:** Simplifying Radical Expressions
- **Toolbox Skill 20:** Operating with Complex Numbers

Extra Practice

See pages 610–612.

Warm-Up Exercises

Warm-Up Transparency 2-6

Support Materials

- Practice Bank: Practice 15
- Activity Bank: Enrichment 14
- Study Guide: Section 2-6
- Problem Bank: Problem Set 4
- Assessment Book: Quiz 2-6, Test 6, Alternative Assessment 4

Section 2-6 Radical Functions

Focus

Explore functions that involve radicals. Use these functions to solve problems.

Natural FORCES

A copy of The Great Wave Off the Coast of Kanagawa *by the Japanese artist Hokusai*

A tsunami is an ocean wave with tremendous power. These waves move at speeds of up to 600 mi/h. The speed, s, of a tsunami in miles per hour is a function of the depth, d, of the ocean in feet.

$$s(d) = 3.86\sqrt{d}$$

Depth d (feet)	Speed $s(d)$ (mi/h)
100	?
200	?
346	?
5000	?
10,000	?
20,000	?
34,600	?

Talk it Over

1. The ocean depth in the Kuril Trench east of Japan is 34,600 ft. As you move toward the Japanese island of Hokkaido, the depth decreases rapidly. Suppose an earthquake occurs along the Kuril Trench and causes a tsunami.
   a. Use the function $s(d) = 3.86\sqrt{d}$. Complete the table by calculating the speed of the resulting tsunami as it moves toward Hokkaido. Round decimals to the nearest tenth.
   b. Use your table to compare the speeds of a tsunami at depths of 100 ft and 10,000 ft. Compare the speeds at 346 ft and at 34,600 ft. What do you notice?
2. a. Use the function $s = 3.86\sqrt{d}$. Write and simplify an expression for $s(100d)$.
   b. Use your result in part (a) to compare the speeds of a tsunami as it moves from one place in the ocean to another 100 times as deep.
   c. Does your answer in part (b) agree with your answer to part (b) of question 1? Explain.

Student Resources Toolbox
p. 642 *Algebraic Expressions*

Answers to Talk it Over

1. a.

Depth d (ft)	Speed $s(d)$ (mi/h)
100	38.6
200	54.6
346	71.8
5000	272.9
10,000	386
20,000	545.9
34,600	718

b. The speed at a depth of 10,000 ft is 10 times that at 100 ft. The speed at a depth of 34,600 ft is 10 times that at 346 ft.

2. a. $s(100d) = 3.86\sqrt{100d} = 38.6\sqrt{d}$

b. The speed is 10 times greater at a depth 100 times as deep.

c. Yes.

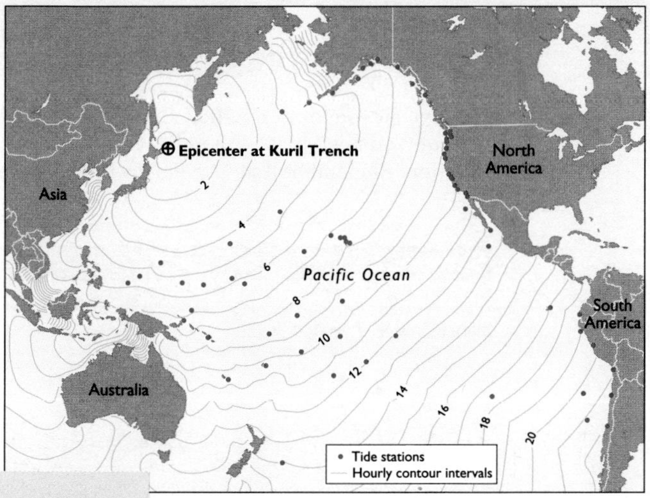

Map showing stations in the Tsunami Warning Center's network.

BY THE WAY...

Tsunami is a Japanese word that means "harbor wave." A tsunami occurs when a large area of the ocean floor, often at the bottom of a deep trench, shifts suddenly. This shift causes huge amounts of water to move, creating a wave that can reach a height of over 100 ft!

Sample 1

What is the ocean depth where the speed of a tsunami is 400 mi/h? Round your answer to the nearest foot.

Sample Response

Problem Solving Strategy: Use a formula.

$s(d) = 3.86\sqrt{d}$ ⟵ The speed of a tsunami is a function of depth.

$400 = 3.86\sqrt{d}$ ⟵ Substitute 400 for $s(d)$.

Solve for d.

Method ❶ Use a graph.

Graph $y = 3.86\sqrt{x}$ and $y = 400$ on the same set of axes and find the point of intersection.

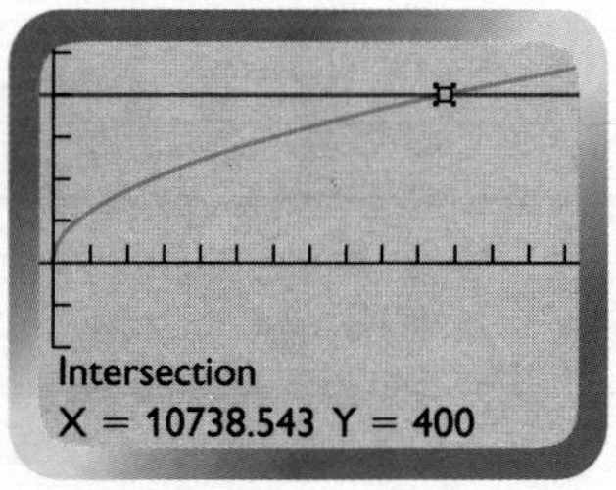

A tsunami travels at 400 mi/h where the ocean is about 10,739 ft deep.

Method ❷ Use algebra.

$400 = 3.86\sqrt{d}$

$103.63 \approx \sqrt{d}$ ⟵ Divide both sides by 3.86.

$10{,}739.18 \approx d$ ⟵ Square both sides.

A tsunami travels at 400 mi/h where the ocean is about 10,739 ft deep.

TEACHING

Talk it Over

Questions 1 and 2 lead students to understand how the speed of a tsunami wave is a function of the depth of the ocean. As the depth of the ocean decreases, the speed of the wave also decreases.

Additional Sample

S1 Suppose you make a small hole in the bottom of a cylindrical water tank that is open at the top. The velocity, v, of the water that squirts from the hole is a function of the depth, d, of the water in the tank: $v(d) = 8\sqrt{d}$.
Here, d is measured in feet and $v(d)$ in feet per second. What is the depth of the water in such a tank when water is squirting from the hole in the bottom at a velocity of 37 ft/s?

Use the formula $v(d) = 8\sqrt{d}$. Substitute 37 for $v(d)$.

$v(d) = 8\sqrt{d}$

$37 = 8\sqrt{d}$

Solve for d.

Method 1. Use a graph.

Graph $y = 8\sqrt{x}$ and $y = 37$ on the same set of axes and find the point of intersection.

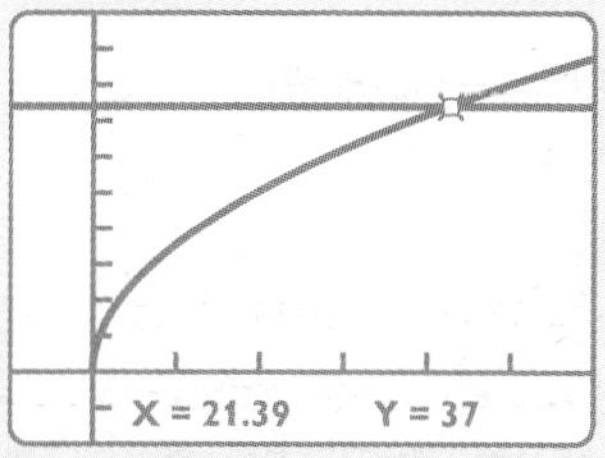

The depth of the water is about 21.4 ft.

Method 2. Use algebra.

$37 = 8\sqrt{d}$

$4.625 = \sqrt{d}$

$21.39 \approx d$

The depth of the water is about 21.4 ft.

Additional Sample

S2 Solve $\sqrt{2x+1} - x = -1$.

Method 1. Use algebra.

$$\sqrt{2x+1} - x = -1$$
$$\sqrt{2x+1} = x - 1$$
$$2x + 1 = x^2 - 2x + 1$$
$$0 = x^2 - 4x$$
$$0 = x(x - 4)$$

$x = 0$ *or* $x = 4$

Check.

$$\sqrt{2x+1} - x = -1$$
$$\sqrt{2(0)+1} - 0 \stackrel{?}{=} -1$$
$$\sqrt{1} - 0 \stackrel{?}{=} -1$$
$$1 - 0 \neq -1$$

0 is not a solution.

$$\sqrt{2x+1} - x = -1$$
$$\sqrt{2(4)+1} - 4 \stackrel{?}{=} -1$$
$$\sqrt{9} - 4 \stackrel{?}{=} -1$$
$$3 - 4 = -1 \checkmark$$

4 is a solution.

Method 2. Use a graph. Graph $y = \sqrt{2x+1} - x$ and $y = -1$ on the same set of axes and find all points of intersection.

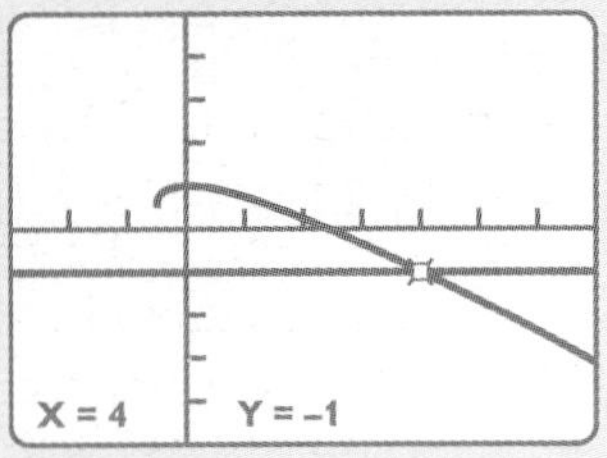

The solution is 4.

Mathematical Procedures

Graphical solutions are good checks on algebraic solutions of radical equations because the graphs do not give extraneous solutions. In Sample 2, the solution can be obtained graphically in either of two ways. The approach of Method 2 can be used, or you can graph $y = \sqrt{x+7} - 1 - x$ and find the x-intercept of the resulting graph.

The paragraph after the Sample Response observes that in the algebraic solution, the extraneous solution came about when the given equation was squared. This can be confirmed by graphing $y = x + 7$ and $y = x^2 + 2x + 1$, the left and right sides of the equation obtained after squaring. The graphs intersect at *two* points. It is important, therefore, not to change a radical equation algebraically before you solve it graphically.

Solving Radical Equations

To solve an equation in which a variable appears under a radical symbol, you need to get the radical alone on one side of the equation. Remember that $\sqrt{x}$ means the positive square root of x.

Sample 2

Solve $\sqrt{x+7} - 1 = x$.

Sample Response

Method ❶ Use algebra.

$\sqrt{x+7} - 1 = x$

$\sqrt{x+7} = x + 1$ ← Get the square root term alone on one side.

$x + 7 = x^2 + 2x + 1$ ← Square both sides.

$0 = x^2 + x - 6$ ← Subtract $(x + 7)$ from both sides.

$0 = (x + 3)(x - 2)$ ← Factor or use the quadratic formula.

$x + 3 = 0$ *or* $x - 2 = 0$ ← Set each factor equal to 0.

$x = -3$ *or* $x = 2$ ← Solve for x.

Check

$\sqrt{x+7} - 1 = x$ ← Use the *original* equation. → $\sqrt{x+7} - 1 = x$

$\sqrt{-3+7} - 1 \stackrel{?}{=} -3$ ← Substitute a solution for x. → $\sqrt{2+7} - 1 \stackrel{?}{=} 2$

$\sqrt{4} - 1 \stackrel{?}{=} -3$ | $\sqrt{9} - 1 \stackrel{?}{=} 2$

$2 - 1 \neq -3$ | $3 - 1 = 2 \checkmark$

−3 is not a solution. | 2 is a solution.

The solution is 2.

Method ❷ Use a graph.

Graph $y = \sqrt{x+7} - 1$ and $y = x$ on the same set of axes and find all points of intersection.

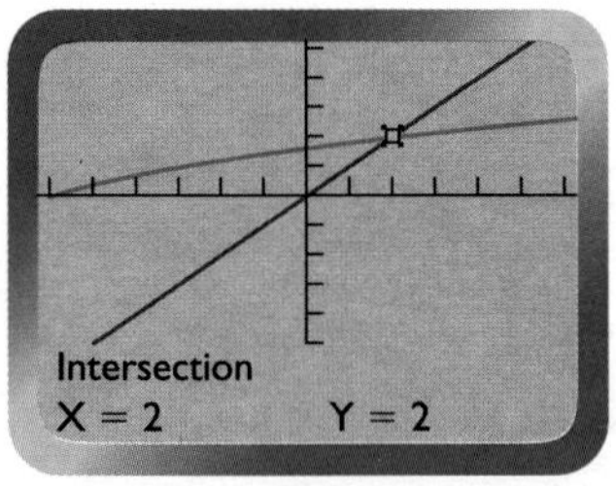

The solution is 2.

In Sample 2, squaring both sides of the equation produced a value for x that is not a solution of the original equation. Such an answer is called an **extraneous solution**.

Talk it Over

3. How does the graph in Method 2 of Sample 2 show that -3 is an extraneous solution?
4. Why does the graph of $y = \sqrt{x + 7} - 1$ stop at $(-7, -1)$?

Now you are ready for: Exs. 1–17 on pp. 115–116

Radical Functions

A function like $s(d) = 3.86\sqrt{d}$ is a *radical function.* A **radical function** is a function that has a variable under a radical symbol. In this book, assume that the domains and ranges of all functions are real numbers.

PROPERTIES OF RADICALS

		Examples
$\sqrt[n]{b} = x$ (You read this as "the *n*th root of *b*.")	when $x^n = b$	$\sqrt[4]{16} = 2$ because $2^4 = 16$
$\sqrt[n]{b^n} = b$	when *n* is odd	$\sqrt[5]{(-7)^5} = -7$
$\sqrt[n]{b^n} = \lvert b \rvert$	when *n* is even	$\sqrt[4]{(-7)^4} = \lvert -7 \rvert = 7$
$\sqrt[n]{ab} = \sqrt[n]{a} \cdot \sqrt[n]{b}$,	for all values of *a* and *b* when *n* is odd	$\sqrt[3]{8x^3} = \sqrt[3]{8} \cdot \sqrt[3]{x^3} = 2x$
$\sqrt[n]{ab} = \sqrt[n]{a} \cdot \sqrt[n]{b}$,	for positive values of *a* and *b* when *n* is even	$\sqrt[4]{256x^4} = \sqrt[4]{256} \cdot \sqrt[4]{x^4} = 4\lvert x \rvert$

Sample 3

Simplify.

a. $\sqrt[4]{(-3)^4}$ b. $\sqrt[3]{(t-5)^3}$ c. $\sqrt[6]{(2-x)^6}$

Sample Response

a. $\sqrt[4]{(-3)^4} = \lvert -3 \rvert = 3$ ← $\sqrt[n]{b^n} = \lvert b \rvert$ when *n* is even

b. $\sqrt[3]{(t-5)^3} = t - 5$ ← $\sqrt[n]{b^n} = b$ when *n* is odd

c. $\sqrt[6]{(2-x)^6} = \lvert 2 - x \rvert = \begin{cases} 2 - x & \text{when } x \le 2 \\ -(2-x) = x - 2 & \text{when } x > 2 \end{cases}$

Talk it Over

Questions 3 and 4 ask students to analyze the graph in Method 2 of Sample 2 with regard to the algebraic solution of Method 1 and to domain and range. Interpreting the information given by a graph is a valuable mathematical skill.

Students Acquiring English

Students acquiring English may need additional explanation of the phrase *nth root of x.* Explain that -th is a common ending for ordinal numbers, and that the letter *n* here has the same meaning it does in equations.

Error Analysis

A very common error is for students to think that $\sqrt[n]{b^n}$ is always equal to *b*. This error can be corrected by using instances in which the equality is true and instances in which it is not true. The examples in the Properties of Radicals box can help, as can Ex. 44 on page 117.

Additional Sample

S3 Simplify.

a. $\sqrt[5]{(-9)^5}$
As 5 is odd, $\sqrt[5]{(-9)^5} = -9$.

b. $\sqrt[4]{(x+2)^4}$
As 4 is even,
$\sqrt[4]{(x+2)^4} = \lvert x + 2 \rvert$
$= \begin{cases} x + 2 & \text{when } x \ge -2 \\ -(x+2) & \text{when } x < -2 \end{cases}$.

c. $\sqrt[7]{(3-x)^7}$
As 7 is odd,
$\sqrt[7]{(3-x)^7} = (3 - x) = -x + 3$.

Answers to Talk it Over

3. The graphs do not intersect the *x*-axis at $x = -3$.
4. $\sqrt{x + 7}$ is not a real number if $x < -7$.

Additional Sample

S4 Find the domain and range of each function.

a. $f(x) = -\sqrt{2x + 4}$

Graph the function.

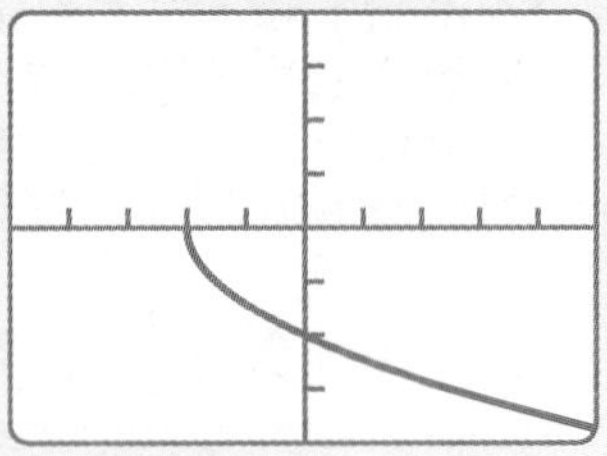

Domain: $x \geq -2$
Range: $y \leq 0$

b. $f(x) = \sqrt[3]{2 - x}$

Graph the function.

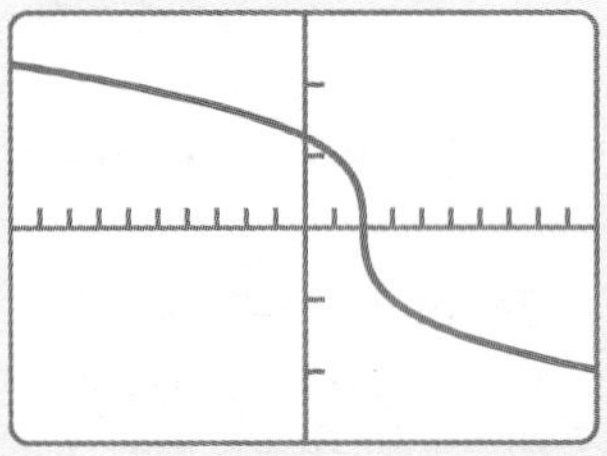

Domain: all real numbers
Range: all real numbers

Teaching Tip

Have students graph the functions from Sample 4 on their graphics calculators. Students should notice that when they trace along the graph for part (a), no y-values appear at the bottom of the screen for values of x less than –1. Have students discuss what this means in terms of the domain and range.

Suggest to students that they always use the TRACE feature when investigating domain and ranges. To see why, have students graph $y = \sqrt{x^2 - x}$. If they merely look at the graph, they might think that positive numbers are not in the domain of the function. When students use TRACE, they see that these numbers *are* in the domain. The mistaken impression arises because for $x \geq 0$, the graph coincides with the nonnegative portion of the x-axis.

Sample 4

Find the domain and range of each function.

a. $f(x) = \sqrt{x + 1}$ b. $f(x) = \sqrt[3]{x + 1}$

Sample Response

a. Graph $y = \sqrt{x + 1}$.

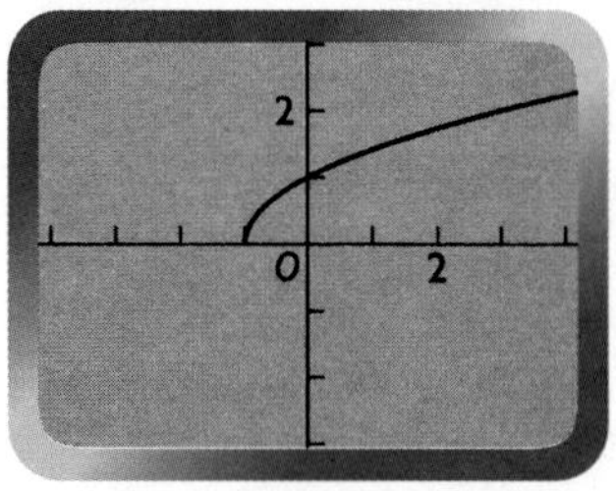

Range: $y \geq 0$

Domain: $x \geq -1$

b. Graph $y = \sqrt[3]{x + 1}$.

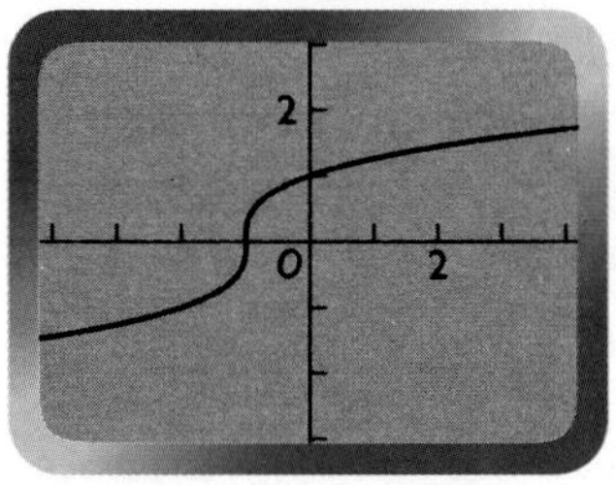

Range: all real numbers

Domain: all real numbers

TECHNOLOGY NOTE

To enter radicals, see p. 596 of the Technology Handbook.

Talk it Over

5. Jane found the domain of $f(x) = \sqrt{x + 1}$ in this way:

   $(x + 1)$ *cannot be negative.*

   $x + 1 \geq 0$

   $x \geq -1$

   Explain her reasoning.

6. Discuss the differences in the domains and ranges of the two functions of Sample 4.

RADICAL FUNCTIONS

A radical function is a function that has a variable under a radical symbol.

Examples

$f(x) = \sqrt{x}$

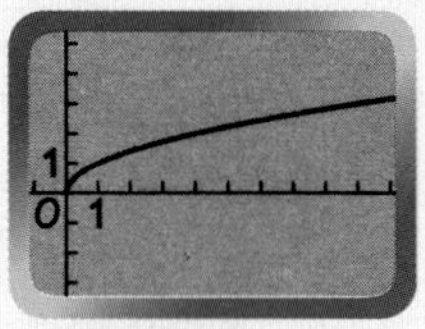

$f(x) = \sqrt[3]{x - 1}$

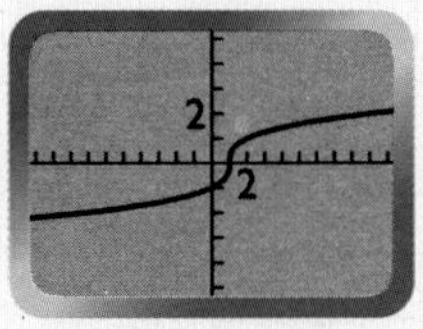

Answers to Talk it Over

5. The domain of the function is all possible values of the control variable. Jane knows that $x + 1$ must be nonnegative for $\sqrt{x + 1}$ to be a real number.

6. The first function's domain is limited by the fact that for $x < -1$, $\sqrt{x + 1}$ is not a real number. The range is limited by the fact that $\sqrt{x}$ is nonnegative for every real number x. For the second function, $\sqrt[3]{x + 1}$ is a real number for every real number x, and for every real number y, there is a real number z such that $y^3 = z$.

:····▶ Now you are ready for:
: Exs. 18–49 on pp. 116–118

Look Back ◀——

Describe how to use the expression under the radical symbol and the value of n to find the domain of a radical function.

2-6 Exercises and Problems

Find $f(x)$ when $x = 2$. Round decimals to the nearest tenth.

1. $f(x) = 2\sqrt{5x}$ **2.** $f(x) = -3\sqrt{x+4}$ **3.** $f(x) = -\sqrt{x-1} + 6$

Find s when $t = 5$.

4. $t + 1 = 4\sqrt{s}$ **5.** $t = \sqrt{-5s + 31}$ **6.** $-4t + 2 = -\sqrt{s+1}$

7. Make a table of values for $f(x) = \sqrt{25x}$ and $f(x) = 5\sqrt{x}$. What do you notice? Explain.

8. **Using Manipulatives** Work with another student.

a. Cut out a 6 in. by 2.5 in. piece of paper. Hold the piece of paper just above your partner's hand so that it will fall between your partner's thumb and forefinger. Drop the paper. Did your partner catch the paper?

b. Switch roles. Did you catch the paper?

c. Repeat parts (a) and (b) with a ruler. The zero mark on the ruler should be just between the catcher's fingers. Record the measurement at the point where the ruler is caught.

d. The time it takes to see the ruler falling and to catch it is called *reaction time*. It is modeled by the equation $t = \sqrt{\frac{d}{192}}$, where t = time in seconds and d = distance of fall in inches. Use your answers from part (c) to find the reaction times for you and your partner.

e. What is the maximum possible reaction time to still catch the paper in part (a)?

9. **Writing** Tell whether the statement $\sqrt{a+b} = \sqrt{a} + \sqrt{b}$ is *True* or *False*. Explain your answer.

10. **Reading** How can you tell if a solution of a radical equation is an extraneous solution?

Solve.

11. $\sqrt{x} = 2x$ **12.** $\sqrt{x} = x - 2$ **13.** $a = \sqrt{7a - 12}$

14. $\sqrt{5x - 6} = 2x$ **15.** $\sqrt{z+2} - 2 = z$ **16.** $x\sqrt{x + 13} - 1 = x$

Answers to Look Back

If n is odd, the domain of the function is all real numbers no matter what the expression under the radical symbol is. If n is even, the domain is all real numbers x for which the expression under the radical symbol is nonnegative.

Answers to Exercises and Problems

1. 6.3
2. −7.3
3. 5
4. 2.25
5. 1.2
6. 323
7. Table values may vary. The functions are the same. $\sqrt{25x} = \sqrt{25} \cdot \sqrt{x} = 5\sqrt{x}$
8. a–e. Answers may vary.
9. False. The statement is only true if $a = 0$ or $b = 0$.
10. Substitute the solution in the original equation.
11. 0, $\frac{1}{4}$
12. 4
13. 4, 3
14. no real solutions
15. −2, −1
16. about −12.15; about 0.37

Talk it Over

Question 5 leads students to think about how the domain of the function $f(x) = \sqrt{x+1}$ can be found by reasoning algebraically. Question 6 focuses students' attention on the differences between the domains and ranges of the two functions of Sample 4. If students generalize what they have discovered, they will be able to respond to the Look Back on this page.

APPLYING

Suggested Assignment

Day 1

Standard 1–7, 9–16

Extended 1–7, 9–16

Day 2

Standard 18–39, 45–49

Extended 18–39, 41–43, 45–49

Integrating the Strands

Algebra Exs. 1–45, 47–49

Functions Exs. 1–3, 7, 27–40, 44, 45, 47

Geometry Ex. 17

Statistics and Probability Ex. 46

Logic and Language Exs. 9, 10, 17

Using Manipulatives

Ex. 8 provides students with a first-hand look at a real-world activity that can be modeled with a radical function.

Error Analysis

Ex. 9 provides an excellent opportunity for students to self-correct a common error.

Mathematical Procedures

Ex. 18 can be used to demonstrate to students that some mathematical procedures that hold true for real numbers may not be true when imaginary numbers are involved.

Reasoning

Ex. 40 allows students to use technology to reach conclusions about the domain and range of radical functions based on whether the index of the radical is odd or even.

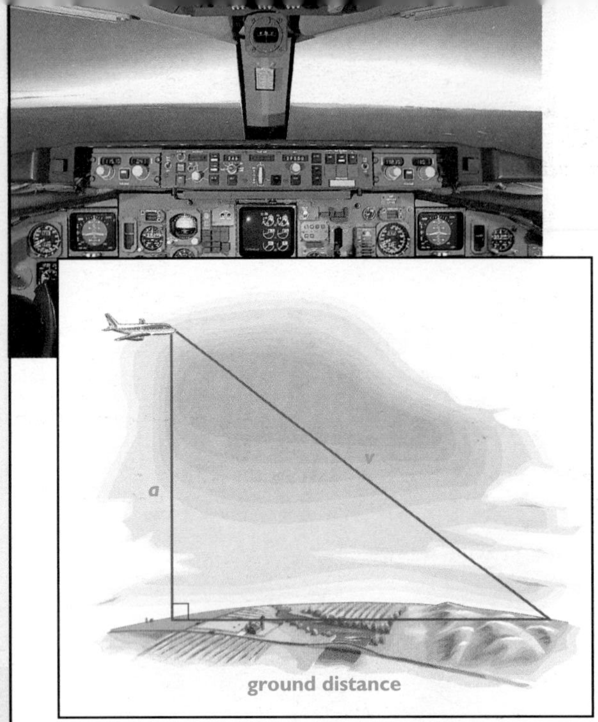

17. **Viewing Distance** When you are flying in an airplane, the distance you can see is modeled by the formula $v = \sqrt{1.5a}$, where v = the viewing distance to the horizon in miles and a = the altitude in feet.
    a. Suppose you are in an airplane and flying over your hometown at an altitude of 31,680 ft. What is the viewing distance to the horizon? Round your answer to the nearest mile.
    b. The ground distance along Earth's surface can be approximated by a straight line. Use the Pythagorean theorem to find the ground distance in miles from your hometown to the point you can see on the horizon. Round your answer to the nearest mile.
    c. **Research** Use an atlas or a map and the information from part (b) to find out what cities or towns are within your viewing distance at 31,680 ft over your hometown.

18. The statement $\sqrt{ab} = \sqrt{a} \cdot \sqrt{b}$ is true when a and b are positive.
    a. Show that $\sqrt{4 \cdot 16} = \sqrt{4} \cdot \sqrt{16}$.
    b. Show that $\sqrt{-4 \cdot 16} = \sqrt{-4} \cdot \sqrt{16}$.
    c. Compare the value of $\sqrt{-4 \cdot (-16)}$ with the value of $\sqrt{-4} \cdot \sqrt{-16}$.
    d. Tell whether the statement $\sqrt{ab} = \sqrt{a} \cdot \sqrt{b}$ is true when a and b are negative.

Student Resources Toolbox
p. 643 *Algebraic Expressions*

Simplify.

19. $\sqrt{64x^2}$
20. $\sqrt[4]{a^4}$
21. $\sqrt[3]{343p^3}$
22. $\sqrt[3]{-64a^3}$
23. $\sqrt[4]{(-2)^4}$
24. $\sqrt{(5-h)^2}$
25. $\sqrt{75x^2}$
26. $\sqrt[5]{-(x+1)^5}$

Graph each function, and then find the domain and range.

27. $f(a) = 3\sqrt{4a}$
28. $t = \sqrt{s} + 3.9$
29. $y = \sqrt{x-9}$
30. $y = \sqrt{2x}$
31. $y = \sqrt{2x+2}$
32. $y = \sqrt{2x} + 2$
33. Find the domain of $y = \sqrt{-5x}$ without graphing.

Find the domain and range of each function.

34. $f(x) = \sqrt[3]{4x}$
35. $y = 2\sqrt[3]{x-1.8}$
36. $y = \sqrt{-x}$
37. $s(t) = -3\sqrt{t} + 1$
38. $d = -\sqrt[3]{g+5}$
39. $f(x) = \sqrt[3]{x-7}$
40. TECHNOLOGY Graph $y = \sqrt[n]{x}$ for several values of n. What conclusion(s) can you make about the domain and range of radical functions when n is odd? when n is even?

Answers to Exercises and Problems

17. a. 218 mi
    b. 218 mi
    c. Answers may vary.
18. a. $\sqrt{4 \cdot 16} = \sqrt{64} = 8$;
    $\sqrt{4} \cdot \sqrt{16} = 2 \cdot 4 = 8$
    b. $\sqrt{-4 \cdot 16} = \sqrt{-64} = 8i$;
    $\sqrt{-4} \cdot \sqrt{16} = 2i \cdot 4 = 8i$
    c. $\sqrt{-4 \cdot (-16)} = \sqrt{64} = 8$;
    $\sqrt{-4} \cdot \sqrt{-16} = 2i \cdot 4i = 8i^2 = -8$
    d. No.
19. $8|x|$
20. $|a|$
21. $7p$
22. $-4a$
23. 2
24. $|5-h|$
25. $5\sqrt{3}|x|$
26. $-(x+1)$
27. 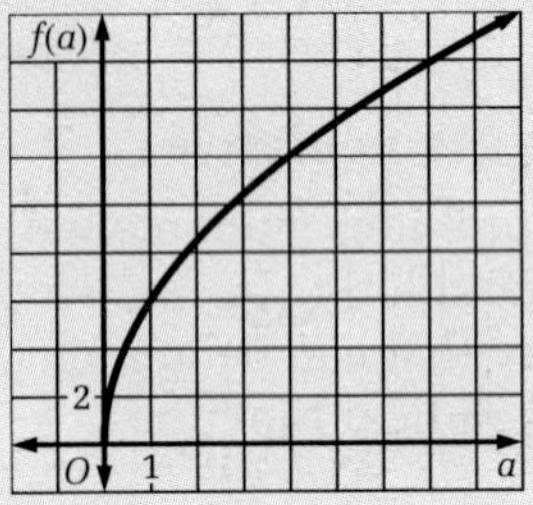

domain: all real numbers greater than or equal to 0; range: all real numbers greater than or equal to 0

28. 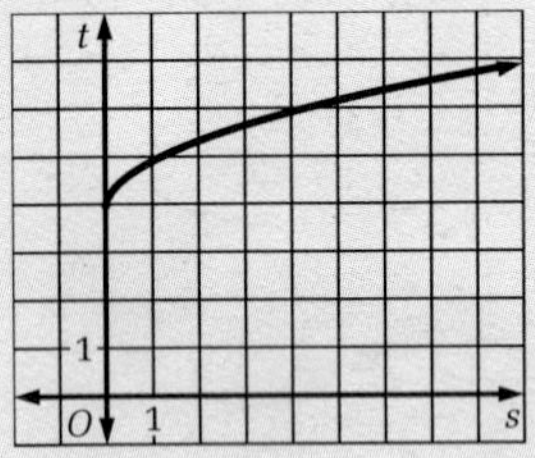

domain: all real numbers greater than or equal to 0; range: all real numbers greater than or equal to 3.9

29. 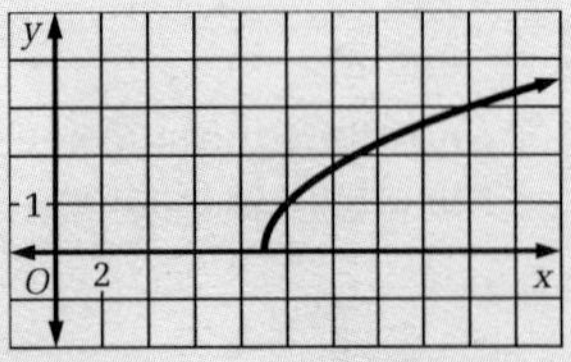

domain: all real numbers greater than or equal to 9; range: all real numbers greater than or equal to 0

30. 

domain: all real numbers greater than or equal to 0; range: all real numbers greater than or equal to 0

31. 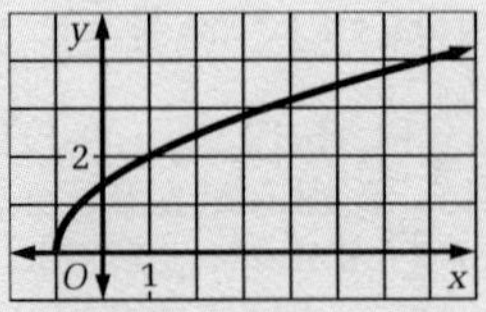

domain: all real numbers greater than or equal to –1; range: all real numbers greater than or equal to 0

Sailing Yachts that race must meet design requirements. In 1989, the International America's Cup Class (IACC) rules included this formula:

$$\frac{L + 1.25\sqrt{S} - 9.8\sqrt[3]{D}}{0.388} \leq 42$$

41. Suppose you design a yacht with length 20.8 m and sail area 285 m^2. It displaces a volume of water equal to 21.4 m^3. Will your yacht meet the IACC requirements?

42. Suppose a designer maximizes the length of the yacht and the area of the sails. Estimate the minimum volume of water the yacht can displace.

43. Suppose a designer decides not to maximize the length of the yacht and the area of the sails. Does the minimum amount of water the yacht can displace go up or down? Explain.

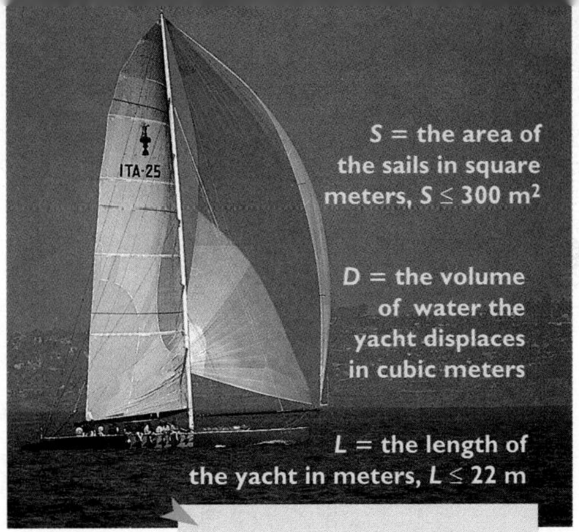

BY THE WAY...

America's Cup is an international sailing competition. The trophy, also called America's Cup, is named after the yacht, *America*, which won the first race held in 1851. It is the oldest trophy in international sports.

Ongoing ASSESSMENT

44. **Group Activity** Work in a group of three students.
    a. Without graphing, decide whether these functions are equivalent to $y = x$:
    $y = \sqrt{x^2}$ $\qquad$ $y = (\sqrt{x})^2$
    b. One of you should graph $y = x$. Another should graph $y = \sqrt{x^2}$. The third person should graph $y = (\sqrt{x})^2$.
    c. Compare your graphs. Was the decision you made in part (a) correct?
    d. Graph $y = \sqrt[3]{x^3}$ and $y = (\sqrt[3]{x})^3$. Compare these graphs to the graph of $y = x$.
    e. Make a conjecture about the functions $y = x$, $y = \sqrt[n]{x^n}$, and $y = (\sqrt[n]{x^n})^n$. Test your conjecture for different values of n.

Review PREVIEW

45. a. Graph the function $P(x) = x(x - 4)(x + 10)$. *(Section 2-5)*
    b. What are the zeros of the function in part (a)?

46. Find the lower extreme, the lower quartile, the median, the upper quartile, and the upper extreme of these batting averages. *(Section 1-3)*

    0.325, 0.249, 0.311, 0.188, 0.190, 0.234,
    0.214, 0.171, 0.202, 0.190, 0.298, 0.249

47. What is the only value of x not in the domain of $f(x) = \frac{1}{1 - x}$? Explain. *(Section 2-1)*

Assessment: Group Work

Have students compare the graphs of the following:
$f(x) = x^2$ and $g(x) = \sqrt{x}$
$f(x) = x^3$ and $g(x) = \sqrt[3]{x}$
$f(x) = x^4$ and $g(x) = \sqrt[4]{x}$
$f(x) = x^5$ and $g(x) = \sqrt[5]{x}$
Students should note the similarities and differences between even and odd powers of x and their radicals.

41. Yes.
42. Estimates may vary. An example is given. about 21.75 m^3
43. It goes down. Solving the given inequality for D yields $D \geq \left(\frac{0.388(42) - L - 125\sqrt{S}}{-9.8}\right)^3$. If L and S decrease from the maximums, the numerator decreases and the denominator stays the same, so D decreases.
44. a–e. Answers may vary. Examples are given.
    a. Neither is equivalent to $y = x$.
    b.

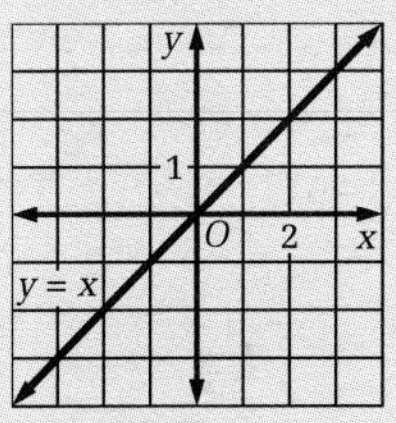

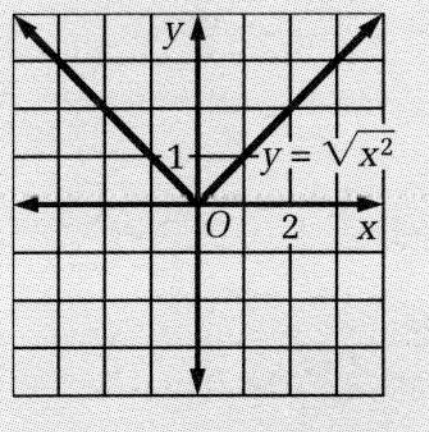

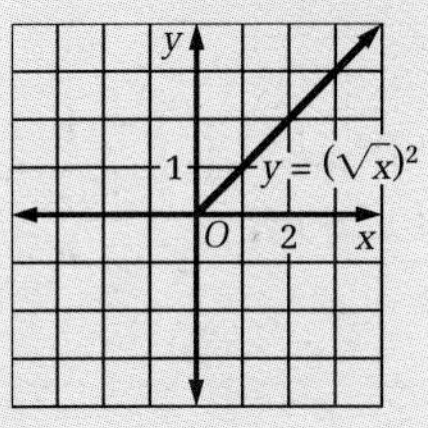

    c. Yes.
    d. The graphs are the same.

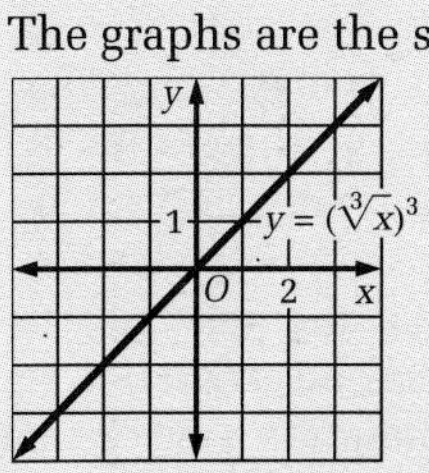

    e. If n is odd, $y = \sqrt[n]{x^n}$, $y = \left(\sqrt[n]{x}\right)^n$, and $y = x$ are the same function. If n is even, the three functions are different.

45–47. See answers in back of book.

Answers to Exercises and Problems

32.

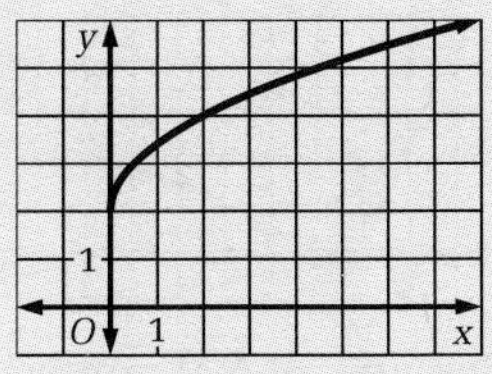

domain: all real numbers greater than or equal to 0; range: all real numbers greater than or equal to 2

33. all real numbers less than or equal to 0
34. all real numbers; all real numbers
35. all real numbers; all real numbers
36. all real numbers less than or equal to 0; all real numbers greater than or equal to 0
37. all real numbers greater than or equal to 0; all real numbers less than or equal to 1
38. all real numbers; all real numbers
39. all real numbers; all real numbers
40. Graphs will vary. If n is odd, the domain and range are all real numbers. If n is even, the domain is all real numbers that make the expression under the radical positive. The range is those numbers that satisfy the equation for the specified domain.

Working on the Unit Project

Students who plan to use data about bicycle racecourses with banked or sloped curves could illustrate their presentations using diagrams that show various curves. They can use Exs. 48 and 49 as a basis for designing their racetracks.

Quick Quiz (2-4 through 2-6)

See page 135.

See also Test 6 in the *Assessment Book*.

Practice 15 For use with Section 2-6

Name ______________________ Date ____________

Practice 15

For use with Section 2-6

Find y when $x = 5$. Round decimals to the nearest tenth.

1. $y = 6\sqrt{x+4}$ 18
2. $y = -3\sqrt{7x}$ −17.7
3. $y = 5 - \sqrt{x+3}$ 2.2
4. $y = 3\sqrt{x+8} + 1$ 11.8
5. $y - 2 = \sqrt{27 - 2x}$ 6.1
6. $y = 2\sqrt{4x-8} - 7$ −0.1

Find x when $y = 3$.

7. $y = 4\sqrt{x-1}$ $\frac{25}{16}$
8. $3y + 1 = \sqrt{x+5}$ 95
9. $7 - 2y = \sqrt{3x-1}$ $\frac{2}{3}$
10. $-y + 9 = 2\sqrt{x}$ 9
11. $-2y + 10 = \sqrt{8-x}$ −8
12. $4y - 7 = \sqrt{5x+3}$ $\frac{22}{5}$

Graph each function, and then find the domain and range.

Exs. 13–18: For graphs, see additional answers.

13. $f(x) = \sqrt{x+4}$ dom.: $x \ge -4$; ran.: $y \ge 0$
14. $y = 3\sqrt{x} - 5$ dom.: $x \ge 0$; ran.: $y \ge -5$
15. $f(x) = \frac{\sqrt{4-x}}{2}$ dom.: $x \le 4$; ran.: $y \ge 0$
16. $y = 6.5 - \sqrt{x}$ dom.: $x \ge 0$; ran.: $y \le 6.5$
17. $f(x) = \frac{1}{4}\sqrt{2x-1}$ dom.: $x \ge 0.5$; ran.: $y \ge 0$
18. $y = \sqrt{3x+6} - 7$ dom.: $x \ge -2$; ran.: $y \ge -7$

Simplify.

19. $\sqrt{81x}$ $9\sqrt{x}$
20. $\sqrt{48y}$ $4\sqrt{3y}$
21. $\sqrt{125w}$ $5\sqrt{5w}$
22. $\sqrt{98p}$ $7\sqrt{2p}$
23. $\sqrt{300r}$ $10\sqrt{3r}$
24. $\sqrt{180c}$ $6\sqrt{5c}$
25. $\sqrt[3]{1000t}$ $10\sqrt[3]{t}$
26. $\sqrt[3]{27u}$ $3\sqrt[3]{u}$
27. $\sqrt[3]{56n}$ $2\sqrt[3]{7n}$
28. $\sqrt[3]{250a}$ $5\sqrt[3]{2a}$
29. $\sqrt[3]{5000b}$ $10\sqrt[3]{5b}$
30. $\sqrt[3]{128k}$ $4\sqrt[3]{2k}$

Solve.

31. $3\sqrt{x} = x$ 0, 9
32. $\sqrt{x+2} = x$ 2
33. $\sqrt{2x-5} = x - 4$ 7
34. $\sqrt{3x+7} - 3 = x$ −2, −1
35. $\sqrt{6x+7} + 1 = 2x$ 3
36. $4\sqrt[3]{x} = x$ −8, 0, 8
37. The frequency f (in hertz) at which an electric guitar string vibrates is given by

$$f = \frac{1}{2L}\sqrt{\frac{T}{\mu}},$$

where L is the length of the string (in meters), T is the tension on the string (in newtons), and μ is the linear density of the string (in kilograms per meter). Solve the equation for T. $T = 4\mu f^2L^2$

Practice Bank, INTEGRATED MATHEMATICS 3

15

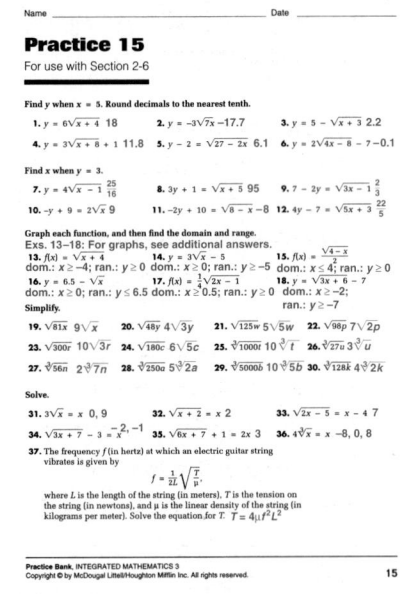

Answers to Exercises and Problems

48. about 9 m/s
49. a. about 9°
    b. about 120 m

2 Working on the Unit Project

Bicycle racecourses have *banked*, or sloped, curves. This helps keep the tires perpendicular to the road around a curve to prevent skidding. The approximate speed at which a bicyclist should enter a curve banked at $b°$ is given by the formula at the right.

$v = \sqrt{0.171rb}$

v = the speed in meters per second

r = the radius of the curve in meters

$b°$ = the measure of the angle at which the curve is banked

48. How fast should a bicyclist enter a 12° banked curve with a radius of 40 m?

49. a. Suppose you design a racetrack that has a curve with a radius of 80 m. You want the average speed on this part of the track to be 11 m/s. At what angle should you bank the curve?

b. You decide to design another curve and bank it at 7°. You want the average speed on this part of the track to be 12 m/s. With what radius should you design the curve?

Unit 2 CHECKPOINT 2

1. **Writing** Compare the graphs of quadratic functions with the graphs of square root functions. How are their shapes alike? How are they different?
2. Suppose a squirrel drops an acorn from a branch at a height of 60 ft. 2-4
   a. Express the height of the acorn above the ground as a function of time.
   b. Graph the function in part (a).
3. A baseball pitcher throws a ball from an initial height of 4 ft, at a speed of 120 ft/s, and at an angle of 1.5° with respect to horizontal.
   a. Use the equation $h(t) = -16t^2 + (v_0 \sin A)t + h_0$ to express the height of the ball as a function of time.
   b. When is the ball 1.5 ft above the ground?
4. Use the polynomial function $P(x) = x^3 - 9x^2 + 18x$. 2-5
   a. Find $P(-3)$ and $P(4b)$.
   b. Graph $P(x)$.
5. Use the function $f(x) = 4.2\sqrt{x}$. Solve for x when $f(x) = 200$.

Solve.

6. $\sqrt{2x-1} = x$
7. $\sqrt{11x+3} = 2x$

Graph each function, and then find the domain and range. 2-6

8. $f(x) = \sqrt{x-3}$
9. $f(b) = \sqrt[3]{b}$

Answers to Checkpoint

1. Summaries may vary. An example is given. The graphs of quadratic functions and square root functions are both curves. The graph of a quadratic function is a parabola. The graph of a square root function is a curve that resembles half of a parabola.
2. a. Let t be the time in seconds and $h(t)$ the height in feet. $h(t) = -16t^2 + 60$

   b.

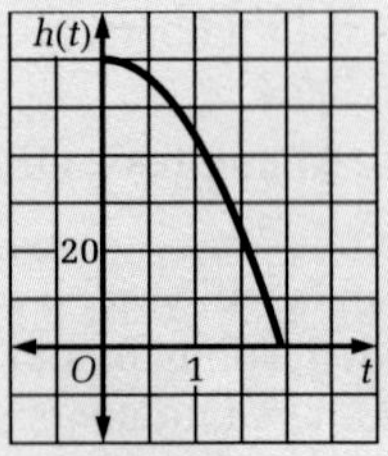

3. a. $h(t) \approx -16t^2 + 3.14t + 4$
   b. about 0.5 s
4. a. -162; $64b^3 - 144b^2 + 72b$

   b.

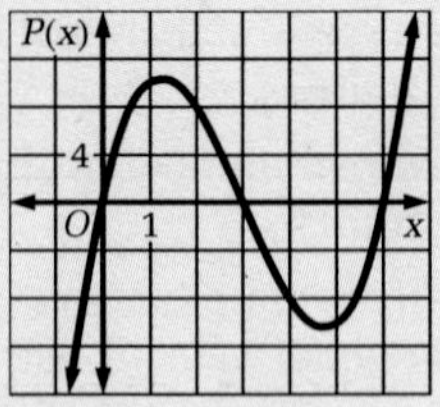

5. about 2267.57
6. 1
7. 3
8.

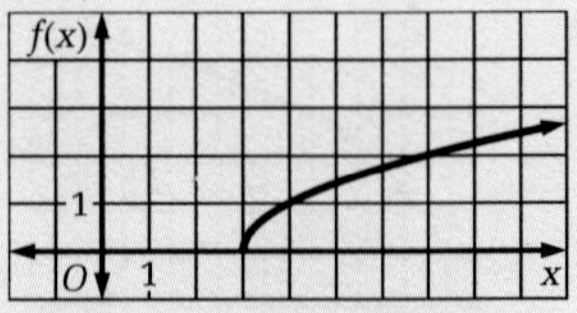

domain: all real numbers greater than or equal to 3; range: all real numbers greater than or equal to 0

Section **2-7**

Rational Functions

MAP IT OUT

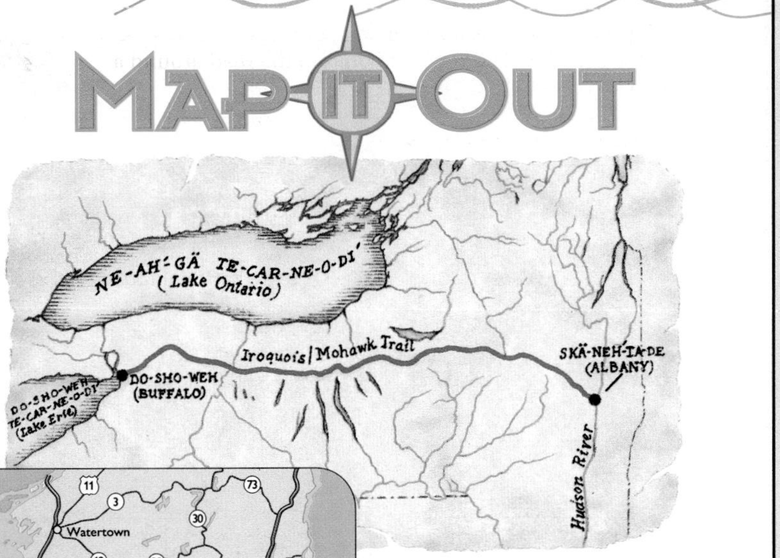

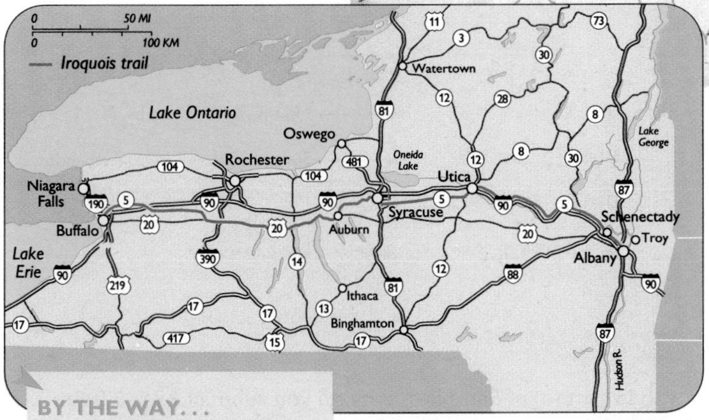

Focus

Understand how rational functions are different from other types of functions. Use rational functions to solve problems.

BY THE WAY...

The Iroquois/Mohawk trail was originally a footpath for the Iroquois Confederacy. Runners, traveling 100 mi a day, used the trail to carry messages between the Native American nations of the confederacy from Buffalo to Albany.

It is about a 300-mile trip across New York state from Buffalo to Albany along the Iroquois/Mohawk Trail.

Talk it Over

Speed s (miles per hour)	Time t (hours)
25	?
40	?
55	?
65	?

1. Complete the table for a 300-mile trip.
2. a. Write an equation that relates speed, time, and distance.
   b. Solve your equation for t. Since you know the distance for this trip, what does the time of the trip depend on?
   c. Why is the function in part (b) not a polynomial function?

PLANNING

Objectives and Strands

See pages 66A and 66B.

Spiral Learning

See page 66B.

Materials List

- Graph paper
- Graphics calculator

Recommended Pacing

Section 2-7 is a one-day lesson.

Toolbox References

Toolbox Skill 23: Solving Rational Equations

Extra Practice

See pages 610–612.

Warm-Up Exercises

Warm-Up Transparency 2-7

Support Materials

- Practice Bank: Practice 16
- Activity Bank: Enrichment 15
- Study Guide: Section 2-7
- Problem Bank: Problem Set 5
- Explorations Lab Manual: Diagram Master 2
- Assessment Book: Quiz 2-7

Answers to Checkpoint

9. all real numbers; all real numbers

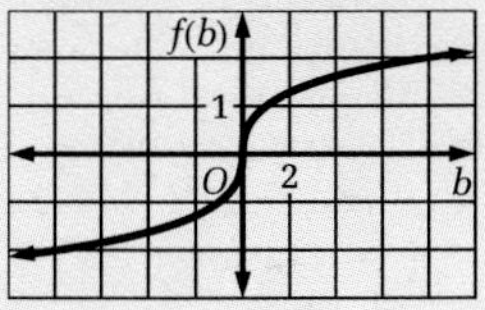

Answers to Talk it Over

1.

Speed s (miles per hour)	Time t (hours)
25	12
40	7.5
55	5.45
65	4.62

2. a. distance = speed × time or $d = st$
   b. $t = \frac{d}{s}$; the rate of speed
   c. In a polynomial function, exponents of the control variable are whole numbers in $t = ds^{-1}$, the exponent of s is a negative number.

TEACHING

Talk it Over

Questions 1 and 2 ask students to complete a table for a trip across New York state and to solve the relationship they discover between speed and time for the variable t. This function gives them an example of a rational function.

Additional Sample

S1 Use the rational function $f(x) = \frac{x^2 + x - 8}{x + 2}$.

a. Find $f(-4)$.

$$f(x) = \frac{x^2 + x - 8}{x + 2}$$
$$f(-4) = \frac{(-4)^2 + (-4) - 8}{(-4) + 2}$$
$$= \frac{16 - 4 - 8}{-4 + 2}$$
$$= -2$$

b. Find x when $f(x) = 2$.

$$f(x) = \frac{x^2 + x - 8}{x + 2}$$
$$2 = \frac{x^2 + x - 8}{x + 2}$$
$$2x + 4 = x^2 + x - 8$$
$$0 = x^2 - x - 12$$
$$0 = (x - 4)(x + 3)$$
$$x = 4 \text{ or } x = -3$$

Talk it Over

For Question 4, it is important for students to understand that the function in Sample 1 is not defined at $x = 1$.

Additional Sample

S2 Arlene Walters realized while she was making plans for her children's education that she was seven times as old as her youngest child, Mark, who is 5 years old. In how many years will Arlene be twice as old as Mark?

Let t = the number of years from now.

Let $r(t)$ = the ratio of Arlene's age to Mark's t years from now.

$r(t) = \frac{35 + t}{5 + t}$, so substitute 2 for $r(t)$.

$$2 = \frac{35 + t}{5 + t}$$
$$2(5 + t) = 35 + t$$
$$25 = t$$

Arlene will be twice as old as Mark in 25 years.

The function you wrote in part (b) of question 2 is an example of a *rational function*. A function that is defined by a *rational expression* is called a **rational function.** A **rational expression** is the quotient of two polynomials. Here are some other rational functions:

$$f(x) = \frac{1}{x} \qquad g(x) = \frac{x^2 + 4}{x + 2}$$

Sample 1

Use the rational function $f(x) = \frac{x^2 - 2x + 3}{x - 1}$.

a. Find $f(4)$.

b. Find x when $f(x) = 3$.

Sample Response

a.
$$f(x) = \frac{x^2 - 2x + 3}{x - 1}$$
$$f(4) = \frac{4^2 - 2(4) + 3}{4 - 1}$$
$$= \frac{11}{3}$$

b.
$$f(x) = \frac{x^2 - 2x + 3}{x - 1}$$
$3 = \frac{x^2 - 2x + 3}{x - 1}$ ← Substitute 3 for $f(x)$.

$3x - 3 = x^2 - 2x + 3$ ← Multiply both sides by $x - 1$.

$0 = x^2 - 5x + 6$ ← Subtract $3x - 3$ from both sides.

$0 = (x - 3)(x - 2)$ ← Factor or use the quadratic formula.

$x = 3$ or $x = 2$ ← Solve for x.

Student Resources Toolbox
p. 646 *Solving Equations and Inequalities*

Talk it Over

3. In part (b) of Sample 1, why do you subtract $3x - 3$ from both sides?
4. What happens when you try to find $f(1)$ for the function in Sample 1?

Sample 2

Roy took two tests and his average is 63. If he scores 100 on the rest of his tests, how many more tests does Roy need to take to raise his average to 85?

Answers to Talk it Over

3. to produce a quadratic equation in standard form
4. You get a fraction with denominator 0.

Sample Response

Roy's average is the sum of all his scores divided by the total number of tests.

Let t = the number of tests Roy needs.

Let $a(t)$ = the average of his test scores.

average of first two scores → $2 \cdot 63$; score on each of the remaining tests → $t \cdot 100$; total number of tests → $t + 2$

$$a(t) = \frac{2 \cdot 63 + t \cdot 100}{t + 2}$$

Substitute 85 for $a(t)$.

$$85 = \frac{2 \cdot 63 + t \cdot 100}{t + 2}$$

$$85(t + 2) = 126 + 100t \quad \leftarrow \text{Multiply both sides by } t + 2.$$

$$2.93 \approx t \quad \leftarrow \text{Solve for } t.$$

Roy must score 100 on three more tests to raise his average to 85.

Talk it Over

5. Why can you use the expression $2 \cdot 63$ for the sum of Roy's first two test scores even though you do not know his actual scores?
6. Suppose Roy has scores of 52 and 74 on his first two tests, but the teacher decides not to include the lowest score. How does the function in Sample 2 change?

Graphs of Rational Functions

There is not a single shape that describes the graphs of all rational functions. The shape of the graph of a rational function depends on the degrees of the polynomials in the numerator and in the denominator.

Sample 3

a. **Graph $f(x) = \frac{1}{x - 1}$.**

b. **Find the domain and range of $f(x)$.**

Sample Response

a. **Method 1** Sketch the graph by hand. Make a table of values.

Problem Solving Strategy: Break the problem into parts.

Step 1 When $x = 1$, $f(x)$ is undefined since the denominator is 0. No point on the graph of $f(x)$ lies on the line $x = 1$. Graph a dashed vertical line at $x = 1$ as a reminder.

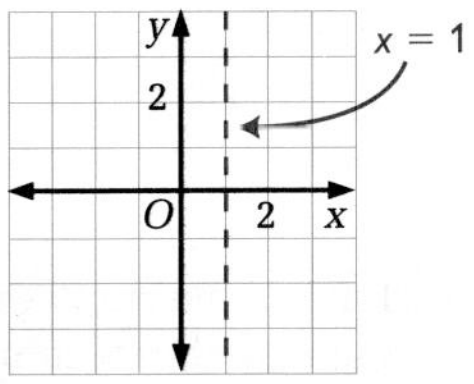

Continued on next page.

Answers to Talk it Over

5. The average is the sum divided by 2, so the sum is $2 \cdot 63$.
6. The function becomes $a(t) = \frac{74 + 100t}{t + 1}$.

Talk it Over

Questions 5 and 6 check students' understanding of the concept of an *average.*

Additional Sample

S3 a. Graph $f(x) = \frac{5}{x^2 + 1}$

Method 1. Sketch the graph by hand. Make a table of values.

Step 1. Since x^2 is greater than or equal to 0 for all values of x, the denominator, $x^2 + 1$, is always 1 or greater. So, the function is defined for all real numbers and is always positive.

Step 2. Look at $f(x)$ for positive values of x and for $x = 0$.

x	$f(x)$
0	5.0
1	2.5
2	1.0
5	0.1923
10	0.0495
50	0.0020

The farther x gets from 0 in the positive direction, the closer $f(x)$ gets to 0.

Step 3. Look at $f(x)$ for negative values of x.

x	$f(x)$
−1	2.5
−2	1.0
−5	0.1923
−10	0.0495
−50	0.0020

The farther x gets from 0 in the negative direction, the closer $f(x)$ gets to 0. Notice also that for any real number, x, and its opposite, $-x$, the values of $f(x)$ are the same. This means that the graph will be symmetric with respect to the y-axis.

Step 4. Graph the trend that steps 2 and 3 suggest.

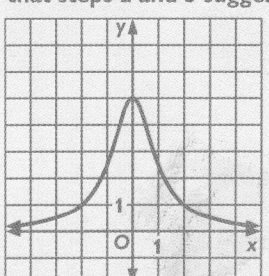

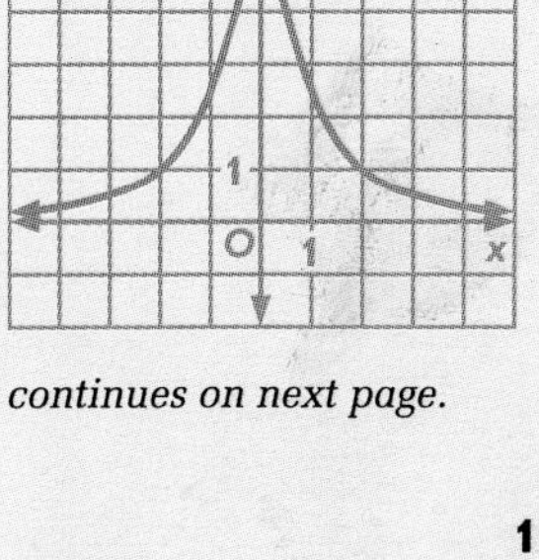

Sample continues on next page.

Additional Sample (continued)

Method 2. Use a graphics calculator. Enter the equation $y = \frac{5}{x^2 + 1}$. To see a smooth curve, use Connected mode.

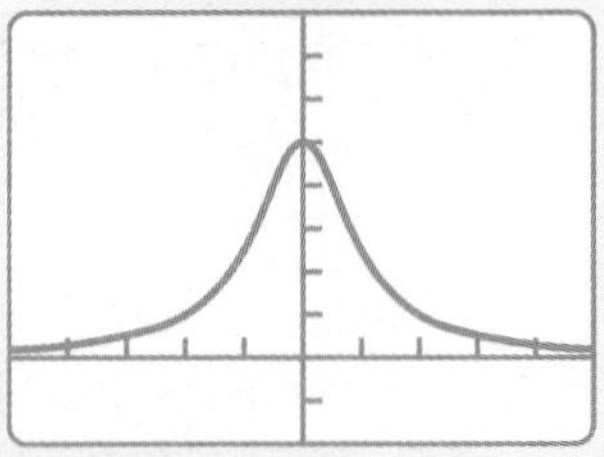

b. Find the domain and range of $f(x)$.

Look at the graph of $f(x)$. The domain is all real numbers. The range is all real numbers y such that $0 < y \leq 5$.

Step 2 Look at what happens to $f(x)$ as the value of x gets very close to 1.

	approach 1 from the left →			← approach 1 from the right		
x	0.99	0.999	0.9999	1.0001	1.001	1.01
$f(x)$	−100	−1000	−10,000	10,000	1000	100

As x approaches 1 from either direction, $|f(x)|$ becomes very large.

Step 3 Look at what happens to $f(x)$ as the value of x gets very far from 1.

	← get farther from 1 to the left			get farther from 1 to the right →		
x	−1,000,000	−100,000	−10,000	10,000	100,000	1,000,000
$f(x)$	−0.000001	−0.00001	−0.0001	0.0001	0.00001	0.000001

As x gets farther away from 1 in either direction, $|f(x)|$ approaches zero, but never equals zero.

Step 4 Graph the trend that the points in Steps 2 and 3 suggest.

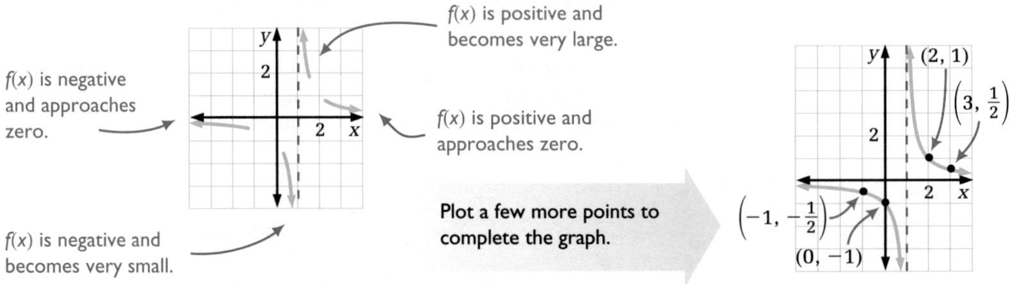
f(x) is positive and becomes very large.
f(x) is negative and approaches zero.
f(x) is positive and approaches zero.
f(x) is negative and becomes very small.
Plot a few more points to complete the graph.
(2, 1)
(3, 1/2)
(−1, −1/2)
(0, −1)

Method 2 Use a graphics calculator.

Enter the equation $y = \frac{1}{x - 1}$. (Some graphics calculators, when in Connected mode, will draw a line connecting the separate parts of a graph. To avoid this, use Dot mode.)

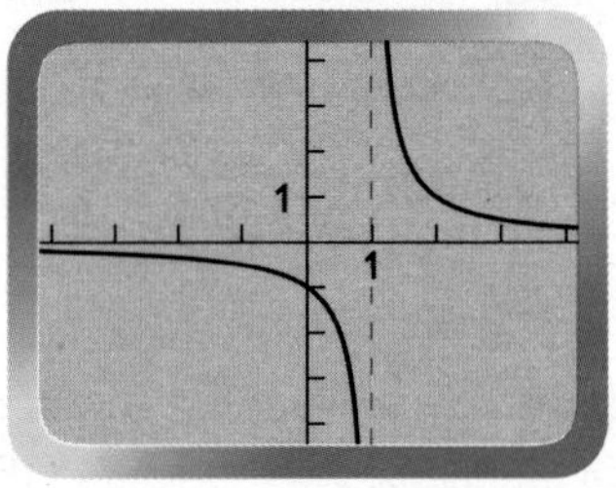

b. Look at the graph of $f(x)$. The domain is all real numbers except 1. The range is all real numbers except 0.

In Method 1 part (a) of Sample 3, the x-axis and the dashed line are called *asymptotes.* **Asymptotes** are lines that a graph approaches more and more closely. Asymptotes are helping lines used in graphing. They are usually graphed as dashed lines because they are not part of the graph.

Talk it Over

7. Explain why the value of the function in Sample 3 cannot be zero.
8. How can you find the domain of the function in Sample 3 without graphing?
9. The graph of the function in Sample 3 has both a vertical and a horizontal asymptote. What is an equation of the horizontal asymptote?
10. How are the asymptotes and the domain and range of a rational function related?

RATIONAL FUNCTIONS

A rational function is defined by the quotient of two polynomials.

Examples

$f(x) = \frac{x+6}{x+1}$ $\qquad$ $f(x) = \frac{5x^2}{x^2+1}$

Suppose the numerator and denominator have no common factor. Then the shape of the graph is a **hyperbola** whenever the numerator has degree 0 or 1 and the denominator has degree 1.

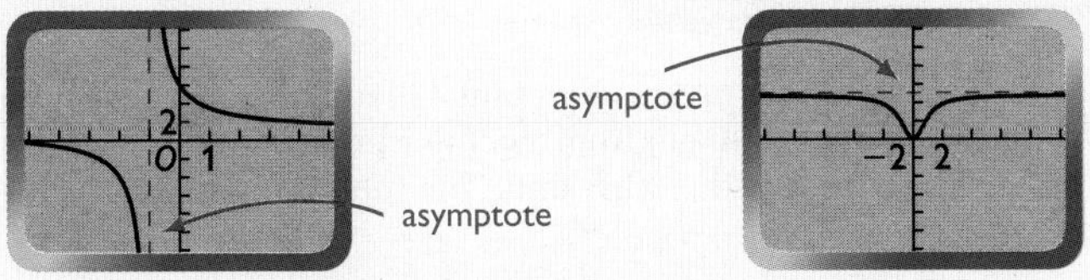

Look Back

Draw a concept map for these types of functions: *absolute value functions, rational functions, quadratic functions, radical functions,* and *polynomial functions.*

Talk it Over

Questions 7–10 have students analyze the function in Sample 3 in order to increase their understanding of the relationships between the domain, the range, and the horizontal and vertical asymptotes of the graph of a rational function.

Problem Solving

Students often think that any value of the independent variable that makes the denominator of a rational function 0 signals a vertical asymptote. Ask students to analyze if this is really the case.

A guess-and-check approach would probably work best. The guesses should start off with simple polynomials in the numerator and denominator. For $f(x) = \frac{x^2}{x}$ and $g(x) = \frac{x^3}{x^3 - x^2}$, the denominators are zero when $x = 0$. Hence, neither function is defined at $x = 0$. However, the y-axis is not a vertical asymptote.

Look Back

Students can share their concept maps in order to compare and contrast them, looking for meaningful differences. A fully developed and accurate map can be included in each student's journal.

Answers to Talk it Over

7. No fraction with numerator 1 is equal to 0.
8. Note that $\frac{1}{1-x}$ is not defined when $1 - x = 0$, that is, when $x = 1$.
9. $y = 0$
10. A vertical asymptote $x = k$ indicates that the domain of the function is all real numbers except k. A horizontal asymptote $y = n$ indicates that the range is all real numbers except n.

Answers to Look Back

Maps may vary. An example is given.

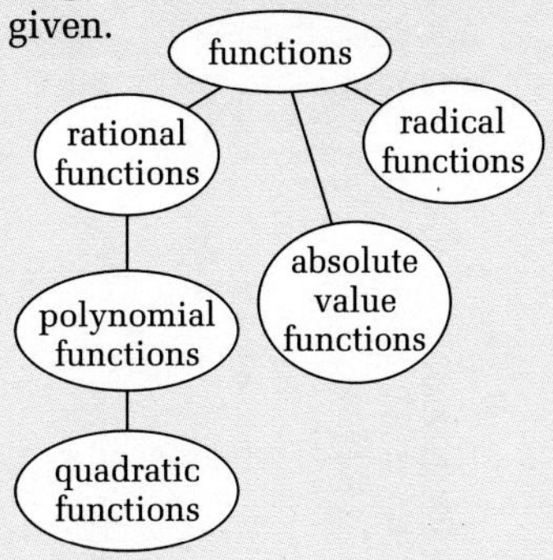

APPLYING

Suggested Assignment

Standard 1–18, 23–27

Extended 1–27

Integrating the Strands

Algebra Exs. 1–26

Functions Exs. 2–8, 10–22, 25, 26

Statistics and Probability Ex. 9

Logic and Language Exs. 1, 10, 22, 27

Reasoning

Ask students to analyze functions such as $f(x) = \frac{x(x-1)}{x}$ and $g(x) = x - 1$ as to whether or not they are exactly the same. Many students are under the impression that they are identical. They almost are but not quite. The function $f(x)$ does not have a value when $x = 0$, since division by 0 is undefined. Hence, 0 is not in the domain of $f(x)$. But all real numbers are in the domain of $g(x)$. For all values of x other than 0, $f(x)$ and $g(x)$ have the same value.

Interdisciplinary Problems

Ex. 21 presents an interesting connection to physics. Students may wish to explore further why the weight of an object depends on how far the object is from the center of Earth. (Weight is a function of the force of gravity. The farther away an object is from the center of gravity of Earth, the less the object weighs.) An object weighs less on the moon than on Earth because the moon is smaller than Earth and exerts less gravitational pull on the object. Students can also consider the distinction between the concepts of weight and mass.

2-7 Exercises and Problems

1. **Reading** Why are asymptotes usually graphed as dashed lines?

Find $f(x)$ when $x = -2$.

2. $f(x) = \frac{x-4}{2x}$
3. $f(x) = \frac{x^2+4x-7}{x^3+5}$
4. $f(x) = \frac{x^3-10}{x^2-x}$

Find t for the given value of $g(t)$.

5. $g(t) = \frac{t-1}{t-5}$; $g(t) = -3$
6. $g(t) = \frac{t^2-5}{t+3}$; $g(t) = -2$
7. $g(t) = \frac{t^2-2t+3}{t+2}$; $g(t) = 6$
8. $g(t) = \frac{2t^2+8t-40}{t-4}$; $g(t) = 5$
9. Anabel took two tests this semester and her average is 75. If she scores 100 on the rest of her tests, how many more tests does Anabel have to take to raise her average to 90?
10. **Writing** Explain why $f(x) = \frac{\sqrt{x-5}}{x}$ is not a rational function.

For Exercises 11–18:

a. Graph the function.

b. Write the equations of all asymptotes of the graph.

c. Find the domain and range of the function.

11. $f(x) = \frac{2}{x-1}$
12. $g(x) = \frac{x}{x-2}$
13. $y = \frac{2x-1}{x}$
14. $r(x) = \frac{6x^2}{x^2+1}$
15. $f(x) = \frac{1}{1-x^2}$
16. $f(x) = \frac{x^2}{3-x}$
17. $d(x) = \frac{-2x^2}{x^2+4}$
18. $f(x) = \frac{1}{x^2-x-6}$
19. **Open-ended** Write a rational function whose domain is all real numbers.

connection to **PHYSICS**

20. A *lever* is a bar that pivots at a point called the *fulcrum*. Suppose a mass of 20 grams is placed 30 cm from the fulcrum. A second mass, m, is placed d cm from the fulcrum at the other end. The function that models this situation is $d = \frac{600}{m}$.
    a. Suppose the second mass is also 20 g. How far from the fulcrum should the second mass be placed to balance the lever?
    b. Suppose the second mass is more than 20 g. Should this mass be placed more or less than 30 cm from the fulcrum to balance the lever?
    c. Suppose the second mass placed 20 cm from the fulcrum balances the lever. How many grams is this mass?

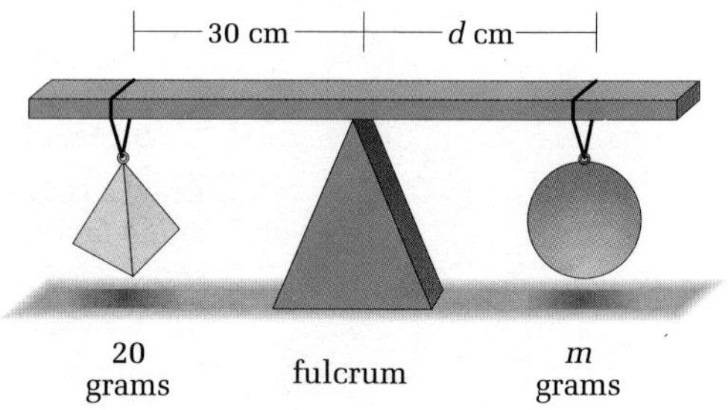

Answers to Exercises and Problems

1. to show that the graph may not cross those lines, they are not part of the graph
2. $\frac{3}{2}$
3. $\frac{11}{3}$
4. -3
5. 4
6. -1
7. 9, -1
8. $-4, \frac{5}{2}$
9. 3 tests
10. $f(x) = \frac{g(x)}{h(x)}$, where $g(x) = \sqrt{x-5}$ and $h(x) = x$. The function $g(x) = \sqrt{x-5}$ is not a polynomial function.

11–17. See answers in back of book.

18. a.

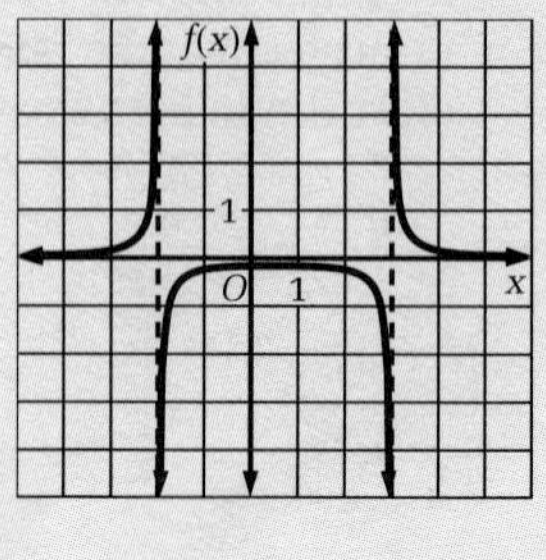

b. $x = 3$, $x = -2$, $y = 0$

c. domain: all real numbers except 3 and -2; range: all real numbers greater than 0 or less than or equal to -0.16

19. Answers may vary. An example is given. $f(x) = \frac{x}{x^2+4}$

connection to **SCIENCE**

21. The radius of Earth is about 4000 mi. On or near Earth, the weight, w, of an object depends on how far the object is from the center of Earth. The weight of an object is modeled by the function

$$w = \frac{(4000)^2 s}{(4000 + h)^2},$$

where h = the height above sea level in miles, and s = the weight on Earth at sea level in pounds.

   a. The Space Shuttle orbiter weighs about 190,000 lb at sea level. The orbiter is in orbit when it is about 300 mi above Earth. What does the orbiter weigh while in orbit?

   b. Suppose an astronaut inside the orbiter weighs 110 lb when the shuttle is in orbit. What is the astronaut's weight on Earth at sea level?

BY THE WAY...

Mae Jemison was the first African American woman to venture into space. As a Science Mission Specialist aboard the *Space Shuttle Endeavor*, she performed experiments on the effects of weightlessness and gravity.

Ongoing **ASSESSMENT**

22. **Writing** Describe how rational functions are different from linear functions. Include in your description a discussion of graphs and domain and range.

Review **PREVIEW**

Solve. *(Section 2-6)*

23. $\sqrt{9x - 20} = x$

24. $\sqrt{x + 3} = 2x$

25. Write $f(x) = |3x - 12|$ as a piecewise function. *(Section 2-3)*

26. Use $g(x) = 2x + 1$. Find $g\left(\frac{a}{2}\right)$. *(Section 2-2)*

2 Working on the Unit Project

27. **Research** Talk to an expert at a local bicycle store or read a book about bicycles to find out about shifting gears on a bicycle. What factors affect how often a rider should shift gears? What are the road or weather conditions that require high gears? low gears?

Communication: Writing

After students complete Ex. 22, ask a few volunteers to read their descriptions to the class. Students can then agree on a good description and enter it into their journals.

Working on the Unit Project

Students may wish to do this research independently. Then they can compile their data for use by the whole project group.

Practice 16 For use with Section 2-7

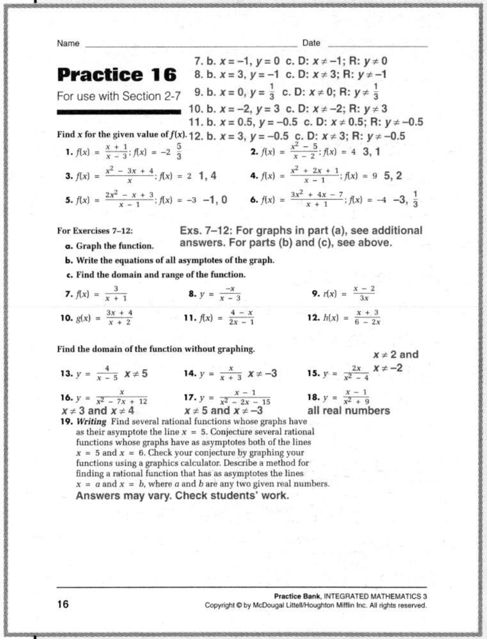

Name ______________ Date ______________

Practice 16
For use with Section 2-7

7. b. $x = -1, y = 0$ c. D: $x \neq -1$; R: $y \neq 0$
8. b. $x = 3, y = -1$ c. D: $x \neq 3$; R: $y \neq -1$
9. b. $x = 0, y = \frac{1}{3}$ c. D: $x \neq 0$; R: $y \neq \frac{1}{3}$
10. b. $x = -2, y = 3$ c. D: $x \neq -2$; R: $y \neq 3$
11. b. $x = 0.5, y = -0.5$ c. D: $x \neq 0.5$; R: $y \neq -0.5$
12. b. $x = 3, y = -0.5$ c. D: $x \neq 3$; R: $y \neq -0.5$

Find x for the given value of $f(x)$.

1. $f(x) = \frac{x+1}{x-3}$; $f(x) = -2$ $\frac{5}{3}$
2. $f(x) = \frac{x^2-5}{x-2}$; $f(x) = 4$ 3, 1
3. $f(x) = \frac{x^2-3x+4}{x}$; $f(x) = 2$ 1, 4
4. $f(x) = \frac{x^2+2x+1}{x-1}$; $f(x) = 9$ 5, 2
5. $f(x) = \frac{2x^2-x+3}{x-1}$; $f(x) = -3$ −1, 0
6. $f(x) = \frac{3x^2+4x-7}{x+1}$; $f(x) = -4$ −3, $\frac{1}{3}$

For Exercises 7–12:
a. Graph the function.
b. Write the equations of all asymptotes of the graph.
c. Find the domain and range of the function.

Exs. 7–12: For graphs in part (a), see additional answers. For parts (b) and (c), see above.

7. $f(x) = \frac{3}{x+1}$
8. $y = \frac{-x}{x-3}$
9. $r(x) = \frac{x-2}{3x}$
10. $g(x) = \frac{3x+4}{x+2}$
11. $f(x) = \frac{4-x}{2x-1}$
12. $h(x) = \frac{x+3}{6-2x}$

Find the domain of the function without graphing.

13. $y = \frac{4}{x-5}$ $x \neq 5$
14. $y = \frac{x}{x+3}$ $x \neq -3$
15. $y = \frac{2x}{x^2-4}$ $x \neq 2$ and $x \neq -2$
16. $y = \frac{x}{x^2-7x+12}$ $x \neq 3$ and $x \neq 4$
17. $y = \frac{x-1}{x^2-2x-15}$ $x \neq 5$ and $x \neq -3$
18. $y = \frac{x-1}{x^2+9}$ all real numbers
19. ***Writing*** Find several rational functions whose graphs have as their asymptote the line $x = 5$. Conjecture several rational functions whose graphs have as asymptotes both of the lines $x = 5$ and $x = 6$. Check your conjecture by graphing your functions using a graphics calculator. Describe a method for finding a rational function that has as asymptotes the lines $x = a$ and $x = b$, where a and b are any two given real numbers.
Answers may vary. Check students' work.

16 **Practice Bank**, INTEGRATED MATHEMATICS 3

23. 5, 4
24. 1
25. $f(x) = \begin{cases} 3x - 12 & \text{if } x \geq 4 \\ -3x + 12 & \text{if } x < 4 \end{cases}$
26. $a + 1$
27. Answers may vary. An example is given. When to shift gears depends on the terrain where you are biking. If the road is fairly flat, you might not need to change gears. But if the road is hilly, you would change gears for going up and down the hills and also for the flat road between hills. In rainy weather, you would probably want to ride in a high gear to get better traction on the wet road. A lower gear would not have as good traction because the wheel is rotated more.

Answers to Exercises and Problems

20. a. 30 cm
   b. less than 30 cm
   c. 30 g
21. a. about 164,413 lb
   b. about 127 lb
22. Summaries may vary. An example is given. Every linear function is a rational function, but not every rational function is a linear function. The graph of a linear function is a line. The domain and range of a linear function are all real numbers. The graph of a rational function depends on the degrees of the polynomials in the numerator and the denominator. The domain is all real numbers except those for which the polynomial in the denominator is 0. If the numerator of a rational function has degree 0 or 1 and the denominator has degree 1, the graph is a hyperbola and exclusions from the domain and range can be determined by noting the equations of the asymptotes.

PLANNING

Objectives and Strands

See pages 66A and 66B.

Spiral Learning

See page 66B.

Materials List

- Graph paper

Recommended Pacing

Section 2-8 is a one-day lesson.

Toolbox References

- **Toolbox Skill 13:** Adding and Subtracting Polynomials
- **Toolbox Skill 15:** Multiplying Polynomials
- **Toolbox Skill 17:** Factoring Polynomials
- **Toolbox Skill 18:** Simplifying Rational Expressions

Extra Practice

See pages 610–612.

Warm-Up Exercises

- Warm-Up Transparency 2-8

Support Materials

- Practice Bank: Practice 17
- Activity Bank: Enrichment 16
- Study Guide: Section 2-8
- Problem Bank: Problem Set 5
- Explorations Lab Manual: Diagram Master 1
- Assessment Book: Quiz 2-8, Test 7, Alternative Assessment 5

Section **2-8**

Combining Functions

Watt Comes Out?

Focus

Find a function of a function.

Sunshine on a rooftop in Santa Fe, New Mexico, may input 1000 watts of solar power to each square meter of solar panels. Suppose the solar panels can convert 12% of the solar power into electric power.

Diagram of a Solar Cell

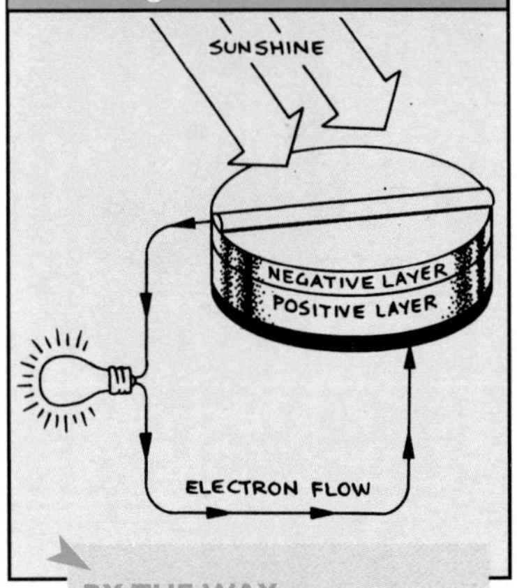

BY THE WAY...

In 1948, architect Eleanor Raymond and chemist/engineer Maria Telkes collaborated to build the first solar-heated house.

Talk it Over

1. Let A = the area of the solar panels in square meters. Let s = the total number of watts of solar power input to the panels. Write a formula for s as a function of A.
2. Let p = the total number of watts of electric power generated by solar panels on the rooftop. Write a formula for p as a function of s.
3. Using the results of questions 1 and 2, write a formula for p as a function of A.
4. How many watts of electric power can solar panels with an area of 2 m² generate on a sunny rooftop in Santa Fe?

You expressed p as a function of s, and s as a function of A. You combined the functions to find p as a function of A.

The new function $p(s(A))$ is called the **composite** of $p(s)$ and $s(A)$.

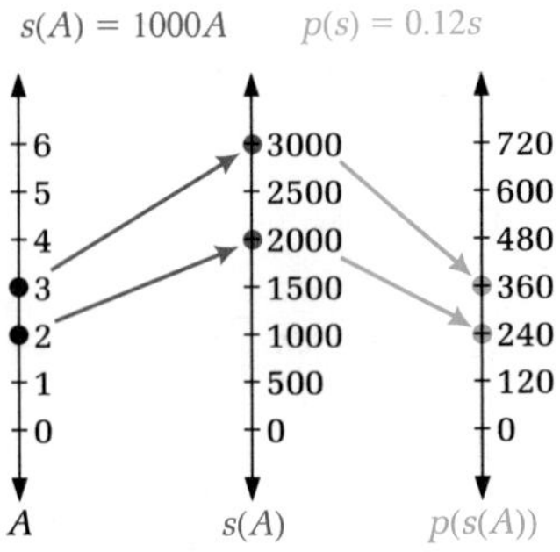

Answers to Talk it Over

1. $s(A) = 1000A$
2. $p(s) = 0.12s$
3. $p(A) = 120A$
4. 240 W

THE COMPOSITE OF TWO FUNCTIONS

The composite of two functions f and g is the function $f(g(x))$, which is read, "f of g of x."

You write $(f \circ g)(x)$, which is read "the composite of f and g."

To find $(f \circ g)(x)$, you find $g(x)$ first and then find $f(x)$ for that value.

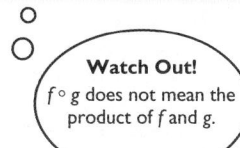

Watch Out!
$f \circ g$ does not mean the product of f and g.

Sample 1

Use the functions $h(x) = x^2 + x$ and $k(x) = 6x$.

a. **Find $(h \circ k)(x)$.**

b. **Find $(h \circ k)\left(\frac{1}{3}\right)$.**

Sample Response

a. $(h \circ k)(x) = h(k(x))$ ← Use the definition of the composite of two functions.

$= h(6x)$ ← Substitute $6x$ for $k(x)$.

$= (6x)^2 + 6x$ ← Substitute $6x$ for x in the rule for $h(x)$.

$= 36x^2 + 6x$ ← Simplify.

b. $(h \circ k)\left(\frac{1}{3}\right) = 36 \cdot \left(\frac{1}{3}\right)^2 + 6 \cdot \frac{1}{3}$ ← Substitute $\frac{1}{3}$ for x in the rule that you found in part (a).

$= 36 \cdot \frac{1}{9} + 2$

$= 6$

Talk it Over

5. In Sample 1, how are the range of $k(x)$ and the domain of $(h \circ k)(x)$ related?
6. Sandra's work is shown. Explain how she found $(h \circ k)\left(\frac{1}{3}\right)$.

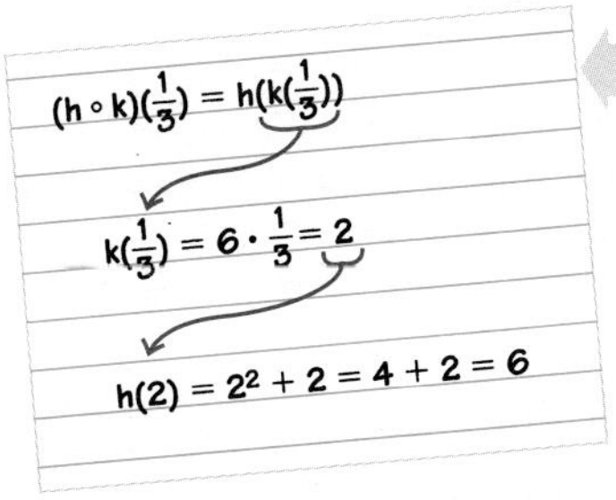

7. Use the functions $f(x) = x^2 - 1$ and $g(x) = 1 - 2x$.
   a. Find a function rule for $(f \circ g)(x)$.
   b. Find a function rule for $(g \circ f)(x)$.
   c. Do you think that $(f \circ g)(x)$ and $(g \circ f)(x)$ are always equal? Explain.

TEACHING

Talk it Over

Questions 1–3 lead students to combine two linear functions to arrive at a composite function which expresses electric power as a function of the area of solar panels.

Additional Sample

S1 Use the functions $f(x) = x^2 + 1$ and $g(x) = x^2 - 1$.

a. Find $(g \circ f)(x)$.

$(g \circ f)(x) = g(f(x))$
$= g(x^2 + 1)$
$= (x^2 + 1)^2 - 1$
$= x^4 + 2x^2 + 1 - 1$
$= x^4 + 2x^2$

b. Find $(g \circ f)(5)$.

$(g \circ f)(5) = 5^4 + 2(5)^2$
$= 625 + 50$
$= 675$

Talk it Over

Question 5 leads students to see the connection between the range of the function applied first, in this case $k(x)$, with the domain of the composite function. Question 6 provides an alternate procedure for finding values of composite functions. Question 7 points out the important idea that $f(g(x))$ does not necessarily equal $g(f(x))$.

Answers to Talk it Over

5. The domain of $(h \circ k)(x)$ is those numbers in the domain of $h(x)$ that are also in the range of $k(x)$.
6. Sandra found the value of $k\left(\frac{1}{3}\right)$ and substituted that value in the equation for h. She found $k\left(\frac{1}{3}\right) = 2$, then $h(2) = 6$.
7. a. $(f \circ g)(x) = 4x^2 - 4x$
   b. $(g \circ f)(x) = 3 - 2x^2$
   c. No; the given functions $f(x)$ and $g(x)$ provide a counterexample, since $4x^2 - 4x \neq 3 - 2x^2$ for all x.

Visual Thinking

Ask students to create a diagram like those shown in this section for the situation described in Sample 2. Encourage students to explain their diagrams and to consider other ways in which this information could be presented. This activity involves the visual skills of *interpretation* and *communication*.

Additional Sample

S2 Camila pays estimated income tax each quarter of the year. She pays 15% of her quarterly earnings and adds \$50 to be on the safe side. She earns \$90 a day. Write a function for her estimated tax payment for one quarter in terms of the number of days, d, that she worked during that quarter.

The amount she earns, $e(d)$, is a function of the number of days she worked: $e(d) = 90d$. The estimated tax payment she makes on earnings of x dollars will be a function, $E(x)$, of her earnings: $E(x) = 0.15x + 50$. For d days of work, her estimated tax payment will be

$$(E \circ e)(d) = E(e(d))$$
$$= E(90d)$$
$$= 0.15(90d) + 50$$
$$= 13.5d + 50$$

Talk it Over

Question 8 directs students' attention to identifying the control and dependent variable of each function. Question 9 requires an analysis of the domain of the cost function and its relationship to the range of the number of bottles function.

Look Back

Students should work through Samples 1 and 2 a second time if they cannot explain how to find the composite of two functions.

Sample 2

Jason is providing supplies for a morning canoe trip. He plans to have 2 bottles of flavored water for each person and 10 extra bottles. Each bottle costs \$.60. Write a function for the total cost of the water in terms of the number of people on the canoe trip.

Sample Response

1 Write the cost as a function of the number of bottles.

Let b = the number of bottles.
Let C = the total cost in dollars.
$C(b) = 0.6b$

2 Write the number of bottles as a function of the number of people on the trip.

Let p = the number of people on the trip.
$b(p) = 2p + 10$

3 Write the cost as a function of the number of people.

$$(C \circ b)(p) = C(b(p))$$
$$= C(2p + 10)$$
$$= 0.6(2p + 10)$$
$$= 1.20p + 6$$

Talk it Over

8. What are the control variable and the dependent variable of each function? Which functions have the same domain?
9. Leah says that the domain of the cost function is the range of the number-of-bottles function. Do you agree or disagree? Explain your answer.

Look Back

Explain how to find the composite of two functions.

2-8 Exercises and Problems

For Exercises 1–6, use the functions $f(x) = \sqrt{x}$, $g(x) = 2x$, and $h(x) = 3x^2 - 2$. Find each function rule or value.

1. $(f \circ g)(x)$
2. $(g \circ h)(x)$
3. $(h \circ g)(x)$
4. $(f \circ g)(8)$
5. $(g \circ h)(-2)$
6. $(h \circ g)(1)$

Answers to Talk it Over

8. the number of bottles and the cost; the number of people and the number of bottles; the number of people and the cost; $b(p)$ and $(C \circ b)(p)$
9. Answers may vary. An example is given. I disagree. The range of the number of bottles function is even numbers greater than 10. The domain of the cost function is positive whole numbers. However, every number in the range of the number of bottle functions is in the domain of the cost function.

Answers to Look Back

Answers may vary. An example is given. Finding the composite of two functions f and g, $f \circ g(x)$, involves finding the value of the function g at x and then the value of the function f at $g(x)$.

7. **Reading** How do you read $f(g(x))$ and $(f \circ g)(x)$?

8. a. Express the area A of a circle as a function of the radius r.
   b. Express the radius as a function of the circumference C.
   c. Write an expression for area as a function of the circumference of a circle.

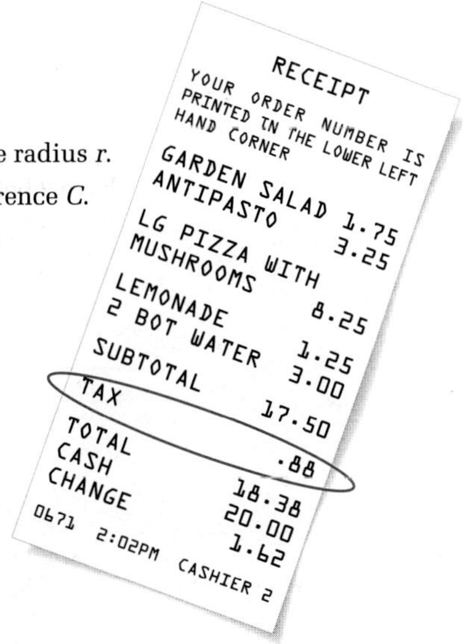

RECEIPT
YOUR ORDER NUMBER IS PRINTED IN THE LOWER LEFT HAND CORNER
GARDEN SALAD 1.75
ANTIPASTO 3.25
LG PIZZA WITH MUSHROOMS 8.25
LEMONADE 1.25
2 BOT WATER 3.00
SUBTOTAL 17.50
TAX .88
TOTAL 18.38
CASH 20.00
CHANGE 1.62
0671 2:02PM CASHIER 2

9. The tax on a restaurant bill is 5%. A customer leaves a tip that is three times the tax on the bill.
   a. Express the tax p as a function of the amount on the bill.
   b. Express the tip t as a function of the tax.
   c. Write a function for the tip in terms of the bill.

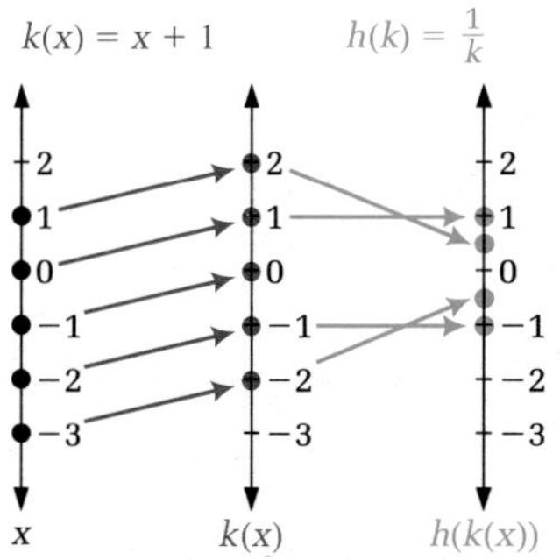

10. **Writing** This mapping diagram shows how to find the value of the composite of h and k. Choose one value of x from the first number line and explain how to find the corresponding value of $h(k(x))$.

11. Lily has a coupon for \$10 off at a stereo store. An ad for the store says that there is a 5% discount on any purchase.
    a. Let x be the price of an item. What does the function $c(x) = x - 10$ represent? What does the function $d(x) = 0.95x$ represent?
    b. Suppose Lily uses the 5% discount on an item, then uses the coupon on the same item. Write the price that she pays as a function of x.
    c. Suppose Lily uses the coupon first, then the discount. Write the price that she pays as a function of x.
    d. Find $(c \circ d)(x)$ and $(d \circ c)(x)$. Which of these is the function you wrote for part (b)? Which is the function you wrote for part (c)? Explain.

12. Use the functions $f(x) = \frac{1}{2}x + 4$ and $g(x) = 2x - 8$.
    a. Find $(f \circ g)(x)$ and $(g \circ f)(x)$. What do you notice?
    b. **Open-ended** The function $I(x) = x$ is called the *identity function.* Write rules for two other functions whose composite is the identity function.

APPLYING

Suggested Assignment

Standard 1–8, 15–18, 20–25

Extended 1–18, 20–25

Integrating the Strands

Algebra Exs. 1–21, 23–25

Functions Exs. 1–21, 23–25

Geometry Exs. 8, 22

Statistics and Probability Exs. 13, 14

Discrete Mathematics Exs. 13, 14

Logic and Language Exs. 7, 10, 12, 14, 25

Communication: Reading and Writing

Exs. 7–12 provide practice in reading and writing composite functions. Students should refer to the discussion of composite functions on page 127 if they need to review the notation used.

Using Technology

In Ex. 12, you can use the TI-81 or TI-82 to graph $(f \circ g)(x)$ and $(g \circ f)(x)$. Enter $\frac{1}{2}x + 4$ after Y_1= on the Y= list. Enter $2x - 8$ after Y_2=. After Y_3=, enter $Y_1(Y_2)$. To do this, press [2nd] [Y–VARS] <Function><Y1> [(] [2nd][Y–VARS] <Function> <Y2> [)]. Follow a similar procedure to enter $Y_2(Y_1)$ on the next line, after Y_4=.

To view all four graphs at the same time, press [GRAPH]. If viewing all graphs together is confusing, move the cursor to be on top of the equals sign for the equation whose graph you would like to suppress. Press [ENTER]. The equation is stored in the calculator, but its graph will not show on the screen.

If you view the graph of Y_3 while all other graphs are suppressed, you see that it is just like the graph of $y = x$. The same is true for Y_4. Since $Y_1(Y_2)$ and $Y_2(Y_1)$ correspond to $f \circ g$ and $g \circ f$, you can readily see that f and g are inverses of one another in this case.

Answers to Exercises and Problems

1. $\sqrt{2x}$
2. $6x^2 - 4$
3. $12x^2 - 2$
4. 4
5. 20
6. 10
7. f of g of x; the composite of f and g
8. a. $A(r) = \pi r^2$
   b. $r(C) = \frac{C}{2\pi}$
   c. $A(C) = \pi \cdot \left(\frac{C}{2\pi}\right)^2 = \frac{C^2}{4\pi}$
9. Let a be the amount on the bill.
   a. $p(a) = 0.05a$
   b. $t(p) = 3p$
   c. $t(a) = 0.15a$
10. Answers may vary. In each case, follow the arrow to find $k(x)$ on the second number line, then follow that arrow to find $h(k(x))$ on the third number line.
11. See answers in back of book.
12. a. $(f \circ g)(x) = x$; $(g \circ f)(x) = x$
    b. Answers may vary. An example is given. $f(x) = x + 1$ and $g(x) = x - 1$

Integrating the Strands

Exs. 13 and 14 apply the notion of combining functions to a biological situation involving the respiratory systems of a group of animals. Integrated into this situation are the strands of algebra, functions, statistics, discrete mathematics, and logic and language.

Multicultural Note

Some African animals that once abounded in the wild have recently been threatened with extinction. One example is the black rhinoceros, whose numbers are down from an estimated 65,000 in 1970 to fewer than 2500 today. Many rhinos are killed by poachers who are able to sell their horns for more than the price of gold. In fact, when smuggled into Asia, where crushed rhino horns are used by traditional healers in medication, the horns are worth about \$13,600 per pound (and rhino horns can weigh 20 pounds each). Many nations are making valiant efforts to preserve the rhino, sometimes taking drastic measures to ensure their safety. In Zimbabwe, park rangers patrol the game parks around the clock to capture poachers.

connection to **BIOLOGY**

The table contains data about the respiratory systems of eleven wild and domestic African animals.

Animal	Body mass (kg)	Lung volume (mL)	Maximum oxygen use $\left(\frac{\text{mL}}{\text{s}}\right)$
dwarf mongoose	0.418	27.4	0.89
banded mongoose	1.14	63.3	2.17
genet cat	1.415	86.8	2.41
suni	3	208.7	4.83
Grant's gazelle	10	562	9.03
wildebeest	102	7,678	75.7
waterbuck	126	9,383	102.6
eland	240	10,668	143.3
African goat	21.9	1,355	19.2
African sheep	22	1,662	15.3
Zebu cattle	151	8,036	68

genet cat

Grant's gazelle

wildebeest

waterbuck

13. a. Make a scatter plot of the data in the first two columns. Use body mass, m, as the control variable.

b. Draw a fitted line through the points on your graph.

c. Using your fitted line, write a rule for $v(m)$, lung volume as a function of body mass.

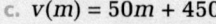

d. Repeat parts (a) and (b) for the last two columns of data. Use lung volume, v, as the control variable.

e. Using your fitted line from part (d), write a rule for $u(v)$, maximum oxygen use as a function of lung volume.

f. Write a rule for the composite function $u(v(m))$.

g. Use your rule from part (f) to estimate the maximum oxygen use of a camel whose body mass is 234.5 kg.

14. a. Ann Martin used the data in the table and data for eleven other African mammals. She found these relationships among body mass (m), lung volume (v), and maximum oxygen use (u):

$$u(v) = 0.009v + 3.36$$

$$v(m) = 58.19m + 106.66$$

Write a rule for the composite function $u(v(m))$.

b. **Writing** Use the rule you wrote in part (a) to estimate the maximum oxygen use for the eland and the dwarf mongoose in the table. Explain why Ann's equations may give better estimates of some data items than others. Compare the estimated values to the corresponding values in the table.

Answers to Exercises and Problems

13. Lines fitted to data may vary. Examples with related equations are given.

a, b.

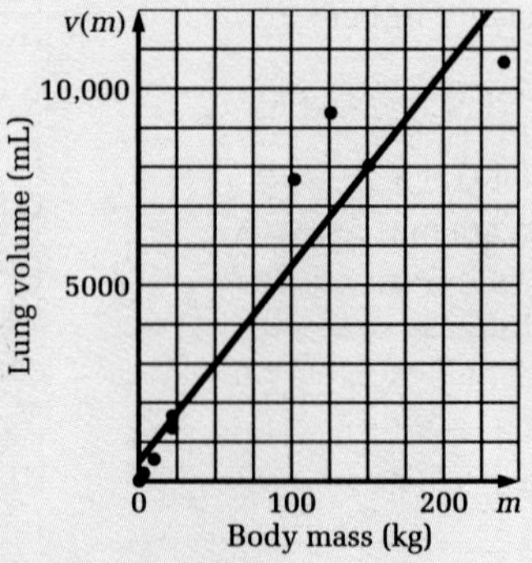

c. $v(m) = 50m + 450$

d.

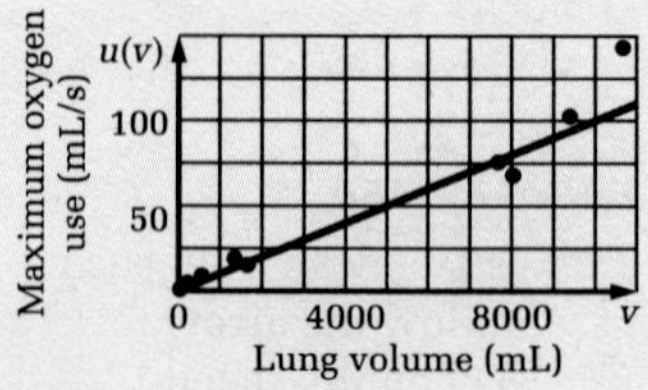

e. $u(v) = 0.01v$

f. $u(v(m)) = 0.5m + 4.5$

g. 121.75 mL/s

14. a. $u(v(m)) = 0.52m + 4.32$

b. 129.12 mL/s; 4.54 mL/s; The line Ann fitted to the data will be closer to some points that to others. The estimated value for the eland is closer to the table value than the estimated value for the dwarf mongoose is to its table value.

15. $x^2 - 3x + 2$

In Exercises 15–18, add, subtract, multiply, or divide the functions. Use $f(x) = x^2 - 5x + 4$, $g(x) = 2x - 2$, and $h(x) = x^2 - x$.

Student Resources Toolbox
p. 635 *Algebraic Expressions*

Example:

$$g(x) - f(x) = (2x - 2) - (x^2 - 5x + 4)$$
$$= 2x - 2 - x^2 + 5x - 4$$
$$= -x^2 + 7x - 6$$

Example:

$$\frac{f(x)}{h(x)} = \frac{x^2 - 5x + 4}{x^2 - x}$$
$$= \frac{(x-4)(x-1)}{x(x-1)} = \frac{x-4}{x};\ x \neq 1$$

15. $f(x) + g(x)$ **16.** $f(x) - h(x)$ **17.** $\frac{f(x)}{g(x)}$ **18.** $g(x) \cdot h(x)$

Ongoing ASSESSMENT

19. **Group Activity** Work with another student.
   a. Write rules for two different functions, $f(x)$ and $g(x)$.
   b. One student should find $f \circ g$ and the other should find $g \circ f$. Exchange papers and check each other's work.
   c. Which of the composite functions in part (b) was easier to find? Why?

Review PREVIEW

20. Find the domain and range of $k(x) = \frac{4}{2 - x}$. *(Section 2-7)*

21. Use the function $f(x) = x^2 - 3x + 2$. Find $f(3)$ and $f(2a)$. *(Section 2-1)*

22. Tell which of these sets of three measures provides enough information to prove that two triangles are congruent: SSS, SSA, ASA, AAS, AAA, SAS *(Postulates and Theorems, p. 670)*

2 Working on the Unit Project

The average speed v, in meters per second, of a rider in a 40 km bicycle race is given by the function $v(t) = \frac{40{,}000}{t}$, where t is the total time in seconds that it takes the rider to complete the course.

Racer	Time
Racer 1	1 h 12 min
Racer 2	1 h 14 min
Racer 3	1 h 17 min

Suppose the racecourse is flat and the cyclists ride at a constant speed. The power P in watts that a cyclist supplies to a bicycle is given by the function $P(v) = 0.2581v^3 + 3.509v$. Use the table.

23. What was Racer 1's average speed during the race?

24. How much power did Racer 1 supply to the bicycle?

25. a. Find $(P \circ v)(t)$. Explain what this function represents.
b. Use your formula for $P(t)$ to find the power that Racer 2 supplied to the bicycle.

Working on the Unit Project

Students can complete Exs. 23–25 in their project groups. The use of tables in presentations about bicycles to children could help to make the topics being discussed both interesting and clear.

Quick Quiz (2-7 through 2-8)

See page 133.

See also Test 7 in the *Assessment Book.*

Practice 17 For use with Section 2-8

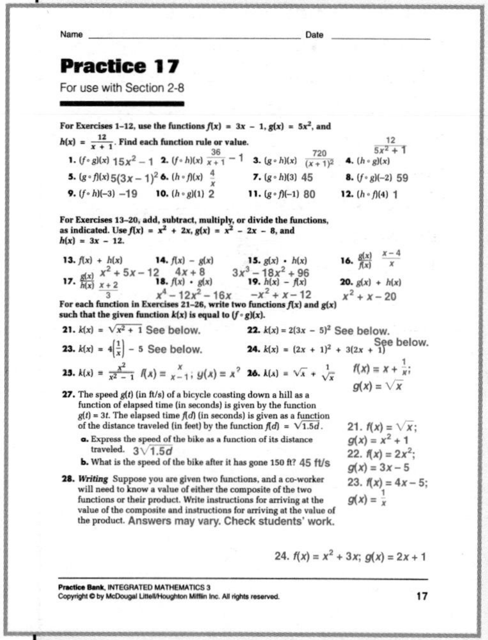

Name ______ Date ______

Practice 17

For use with Section 2-8

For Exercises 1–12, use the functions $f(x) = 3x - 1$, $g(x) = 5x^2$, and $h(x) = \frac{12}{x+1}$. Find each function rule or value.

1. $(f \circ g)(x)$ $15x^2 - 1$
2. $(f \circ h)(x)$ $\frac{36}{x+1} - 1$
3. $(g \circ h)(x)$ $\frac{720}{(x+1)^2}$
4. $(h \circ g)(x)$ $\frac{12}{5x^2+1}$
5. $(g \circ f)(x)$ $5(3x-1)^2$
6. $(h \circ f)(x)$ $\frac{4}{x}$
7. $(g \circ h)(3)$ 45
8. $(f \circ g)(-2)$ 59
9. $(f \circ h)(-3)$ -19
10. $(h \circ g)(1)$ 2
11. $(g \circ f)(-1)$ 80
12. $(h \circ f)(4)$ 1

For Exercises 13–20, add, subtract, multiply, or divide the functions, as indicated. Use $f(x) = x^2 + 2x$, $g(x) = x^2 - 2x - 8$, and $h(x) = 3x - 12$.

13. $f(x) + h(x)$ $x^2 + 5x - 12$
14. $f(x) - g(x)$ $4x + 8$
15. $g(x) \cdot h(x)$ $3x^3 - 18x^2 + 96$
16. $\frac{g(x)}{f(x)}$ $\frac{x-4}{x}$
17. $\frac{g(x)}{h(x)}$ $\frac{x+2}{3}$
18. $f(x) \cdot g(x)$ $x^4 - 12x^2 - 16x$
19. $h(x) - f(x)$ $-x^2 + x - 12$
20. $g(x) + h(x)$ $x^2 + x - 20$

For each function in Exercises 21–26, write two functions $f(x)$ and $g(x)$ such that the given function $k(x)$ is equal to $(f \circ g)(x)$.

21. $k(x) = \sqrt{x^2 + 1}$ See below.
22. $k(x) = 2(3x - 5)^2$ See below.
23. $k(x) = 4\left(\frac{1}{x}\right) - 5$ See below.
24. $k(x) = (2x + 1)^2 + 3(2x + 1)$ See below.
25. $k(x) = \frac{x^2}{x^2 - 1}$ $f(x) = \frac{x}{x-1}$; $g(x) = x^2$
26. $k(x) = \sqrt{x} + \frac{1}{\sqrt{x}}$ $f(x) = x + \frac{1}{x}$; $g(x) = \sqrt{x}$

27. The speed $g(t)$ (in ft/s) of a bicycle coasting down a hill as a function of elapsed time (in seconds) is given by the function $g(t) = 3t$. The elapsed time $f(d)$ (in seconds) is given as a function of the distance traveled (in feet) by the function $f(d) = \sqrt{1.5d}$.
   a. Express the speed of the bike as a function of its distance traveled. $3\sqrt{1.5d}$
   b. What is the speed of the bike after it has gone 150 ft? 45 ft/s

28. *Writing* Suppose you are given two functions, and a co-worker will need to know a value of either the composite of the two functions or their product. Write instructions for arriving at the value of the composite and instructions for arriving at the value of the product. Answers may vary. Check students' work.

21. $f(x) = \sqrt{x}$; $g(x) = x^2 + 1$
22. $f(x) = 2x^2$; $g(x) = 3x - 5$
23. $f(x) = 4x - 5$; $g(x) = \frac{1}{x}$
24. $f(x) = x^2 + 3x$; $g(x) = 2x + 1$

Practice Bank, INTEGRATED MATHEMATICS 3

17

Answers to Exercises and Problems

16. $-4x + 4$
17. $\frac{x-4}{2}$, $x \neq 1$
18. $2x^3 - 4x^2 + 2x$
19. a–c. Answers may vary. Check students' work.
20. domain: all real numbers except 2; range: all real numbers except 0
21. 2; $4a^2 - 6a + 2$
22. SSS, ASA, AAS, SAS
23. about 9.26 m/s
24. about 237.38 W
25. a. $P(v(t)) = \frac{1.65 \times 10^{13}}{t^3} + \frac{140{,}360}{t}$; the power in watts that a cyclist supplies to a bicycle as a function of the time t in seconds it takes the cyclist to finish a 40 km race
    b. about 220.12 W

Assessment

A scoring rubric for the Unit Project can be found on pages 66 and 67 of this Teacher's Edition and also in the ***Project Book***.

Unit Project 2

Completing the Unit Project

Now you are ready to organize your information and make your presentation.

If possible, schedule time at a local elementary school to make your presentation. Make sure that each group member has the information, props, diagrams, and so on needed to make the presentation. Decide on the order of the topics.

Your presentation should include these things:

- **an overview of the bicycle project**
- **each group member's presentation, including at least three posters, diagrams, or props**
- **a summary of what you have discussed**
- **time for questions from the audience**

Look Back

Write an evaluation of your presentation. What went well? What would you do differently? How would you change your presentation for a different age group?

Alternative Projects

Project 1: Graphs

Look through books, magazines, and newspapers for graphs of polynomial and piecewise functions. What situations do they describe? Collect several of them and prepare a brief description of how each function is used.

Project 2: Falling Objects

The functions in Section 2-4 that describe the height of an object do not include the force of air resistance. Find out how air resistance affects the velocity of a falling object. Experiment with different objects to find the constant of air resistance. Describe your results, using diagrams and function notation as needed.

Unit 2 REVIEW AND ASSESSMENT

Popcorn	
small	\$1.55
medium	\$1.95
large	\$2.45

2-1

1. Does this table represent a function? Explain why or why not.
2. Erin buys gasoline at a self-service station for \$1.17/gal.
   a. Write a rule for $C(g)$, the cost of buying g gallons of gasoline, as a function of price.
   b. Suppose the gasoline tank of Erin's car holds 12 gal. What are the domain and range of the function in part (a)?
3. Find $f(-2)$ and $f(a - b)$ for $f(z) = \frac{2}{z + 4}$.

Write an equation for each function in the form $f(x) = mx + b$. 2-2

4. $f(9) = 5$, $m = \frac{1}{9}$
5. $f(2) = 3$, $f(6) = -3$
6. Write a piecewise function for the graph. 2-3

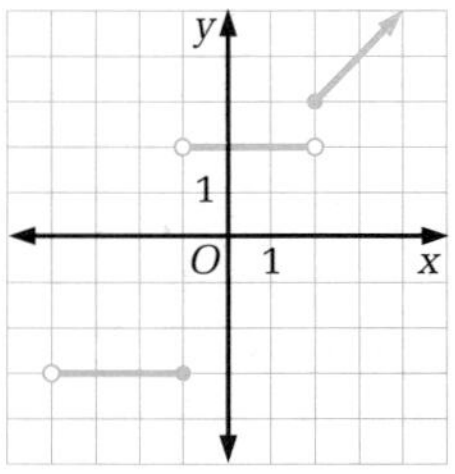

7. Jamelle drops a ball from an initial height of 4 ft. 2-4
   a. Use the model $h(t) = -16t^2 + h_0$ to express the height of the ball as a function of time.
   b. After 0.25 s, what is the height of the ball?
   c. When is Jamelle's ball 2 ft above the ground?
8. Find $f(2)$ and $f(-2)$ for $f(x) = x^2 + 5x - 14$.
9. The volume (V) of a sphere varies directly with the cube of its diameter (d^3). When $V = 288\pi$, $d = 12$. Write a direct variation function that expresses V as a function of d. 2-5
10. Use the function $f(x) = 7x^4 - 4x^2 - 12$.
    a. What is the degree of f? b. Find $f(1)$.
11. Simplify $\sqrt{12a^2}$. 2-6
12. Solve $x = \sqrt{2x - 1}$.
13. Use the function $f(x) = \frac{x^2 - x - 12}{5x + 9}$. 2-7
    a. Find $f(2)$. b. Find x when $f(x) = 10$.

Unit Support Materials

- *Practice Bank:* Cumulative Practice 18
- *Study Guide:* Unit 2 Review
- *Problem Bank:* Unifying Problem 2
- *Assessment Book:* Unit Tests 8 and 9, Spanish Unit Tests, Alternative Assessment
- *Test Generator* software with Test Bank
- *Teacher's Resources for Transfer Students*

Quick Quiz (2-7 through 2-8)

Use the rational function $f(x) = \frac{x^2 + 3x - 4}{x - 1}$.

1. Find $f(x)$ when $x = -4$. [2-7]
   $f(x) = 0$
2. Find x when $f(x) = 10$. [2-7]
   $x = 6$
3. Find the domain of the function $f(x) = \frac{-8x^2}{x^2 - 3x - 10}$ without graphing. [2-7]
   Domain is all real numbers except 5 and −2.
4. $f(x) = x^3 + x$ and $g(x) = 5x$. Find $(f \circ g)(x)$. [2-8]
   $(f \circ g)(x) = 125x^3 + 5x$

Use $f(x) = 3x^2 + 2x + 1$, $g(x) = 3x - 1$, and $h(x) = x^2 - 2$.

5. Find $f(x) + h(x)$. [2-8]
   $f(x) + h(x) = 4x^2 + 2x - 1$
6. Find $h(x) - g(x)$. [2-8]
   $h(x) - g(x) = x^2 - 3x - 1$

Answers to Unit 2 Review and Assessment

1. Yes; for every number in the domain, there is only one value in the range.
2. a. $C(g) = 1.17g$
   b. $0 \le g \le 12$; $0 \le C(g) \le 14.04$
3. 1; $\frac{2}{a - b + 4}$
4. $f(x) = \frac{1}{9}x + 4$
5. $f(x) = -\frac{3}{2}x + 6$
6. $f(x) = \begin{cases} -3 & \text{if } -4 < x \le -1 \\ 2 & \text{if } -1 < x < 2 \\ x + 1 & \text{if } x \ge 2 \end{cases}$
7. a. $h(t) = -16t^2 + 4$
   b. about 3 ft
   c. after about 0.35 s
8. 0; −20
9. $V = \frac{\pi}{6}d^3$
10. a. 4
    b. −9
11. $2\sqrt{3}|a|$
12. 1
13. a. $-\frac{10}{19}$
    b. about −1.93; about 52.93

Quick Quiz (2-1 through 2-3)

1. Use $f(x) = -4x^2 + 2$. Find $f(-2)$. [2-1] $f(-2) = -14$
2. Complete this sentence: The values of a function are the numbers in its __?__. [2-1] range
3. Use $f(x) = \frac{2}{3}x + 14$. Find x when $f(x) = -2$. [2-2] $x = -24$
4. Write an equation for this function in the form $y = mx + b$. $f(2) = 7, f(5) = 10$. [2-2] $y = x + 5$
5. How many equations are needed to define a piecewise function? [2-3] two or more
6. Is this statement *True* or *False*? Every absolute value function is a linear function. [2-3] False.

14. **Writing** Monica has played 19 basketball games and averaged 22 points per game. She has one game left and must average at least 23.2 points per game in order to win her league's scoring title. Do you think she can win the title? Explain.

15. $f(x) = 3x$ and $g(x) = \frac{1}{x}$. Find $(f \circ g)(x)$. **2-8**

16. a. Graph each function and find its domain and range.

$$f(x) = |x + 2| \qquad f(x) = 2\sqrt{x + 1} \qquad f(x) = \frac{10}{x - 5} \qquad f(x) = x(x + 4)(x + 6)$$

b. **Writing** Describe one way in which each graph is different from the others.

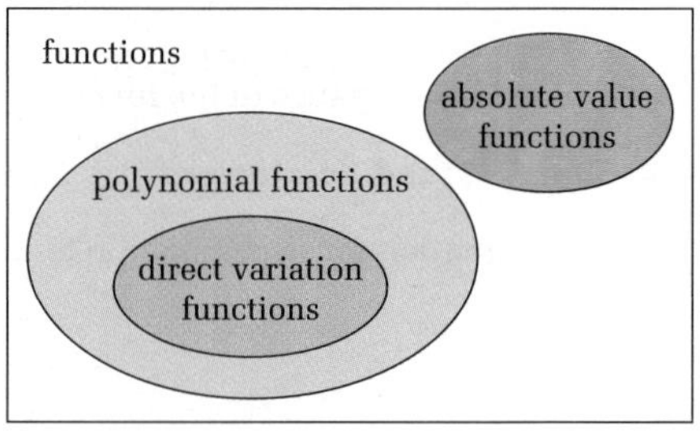

17. **Open-ended** Use this *Venn diagram* to make three statements about the relationships between some of the functions you have studied in this unit.

18. **Self-evaluation** Which topic in this unit would you spend some extra time reviewing? Why?

19. **Group Activity** Work in a group of four students.
Use the fact that the volume of a submerged object is equal to the volume of water that it displaces.
a. One pair of students should measure the volume of an object shaped like a sphere, such as a golf ball or softball, by submerging it in water in a straight-sided bucket or a plastic wastebasket. To estimate the volume of water displaced, you can use the formula $V = Bh$, where B is the area of the base of the bucket or wastebasket and h is the rise in the height of the water.
b. The other pair of students should measure the diameter of the spherical object by placing it between two books and measuring the distance between them. You can use the direct variation formula $V = \frac{4}{3}\pi r^3$, where V is the volume and r is the radius, to find the volume of the spherical object.
c. Repeat parts (a) and (b) for several spherical objects.
d. Compare results using the two methods. Which do you think is more accurate? Why?

IDEAS AND (FORMULAS) =

ALGEBRA

- The domain of a function is all possible values of the control variable. *(p. 70)*
- The range of a function is all possible values of the dependent variable. *(p. 70)*
- A function can be written in function notation, $y = f(x)$. *(p. 71)*
- A function may have the same output for many different inputs but may not have more than one output for one input. *(p. 72)*

Answers to Unit 2 Review and Assessment

14. No; she would need to score 46 points. It is unlikely she could do so, with an average of 22 points per game.

15. $(f \circ g)(x) = \frac{3}{x}$

16. a.

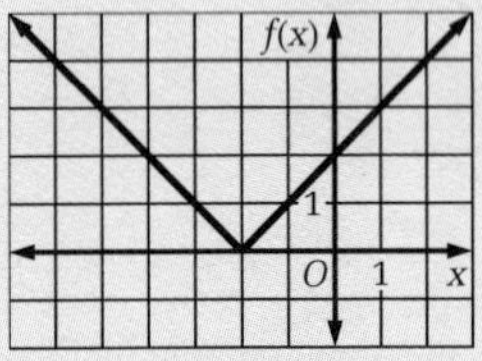

domain: all real numbers; range: all real numbers greater than or equal to 0

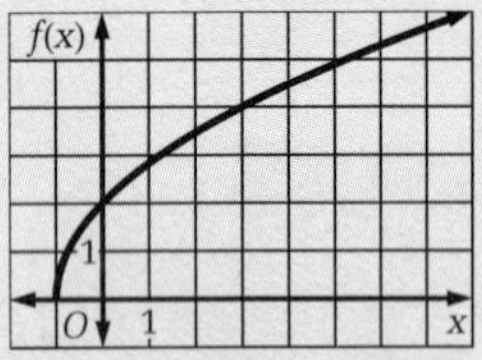

domain: all real numbers greater than or equal to –1; range: all real numbers greater than or equal to 0

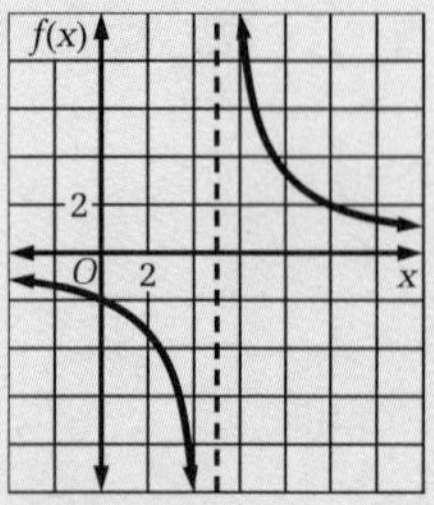

domain: all real numbers except 5; range: all real numbers except 0

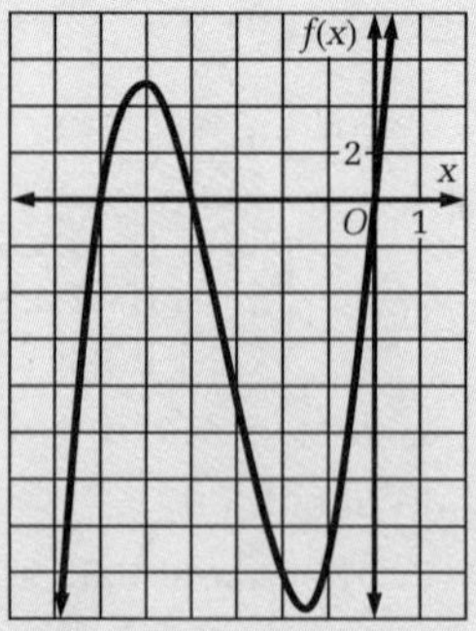

domain: all real numbers; range: all real numbers

b. Answers may vary. Examples are given. The first consists of two rays, the second is half of a parabola, the third is a hyperbola, and the fourth is an S-shaped curve.

- One-to-one functions have one input for each output. *(p. 72)*
- Linear functions have the form $f(x) = mx + b$. The slope is m and the vertical intercept is b. *(p. 78)*
- A piecewise function has different equations for different parts of its domain. *(p. 86)*
- An absolute value function is a special type of piecewise function. *(p. 87)*
- You can model the height of an object that is dropped or thrown with a quadratic function. *(pp. 94, 96)*
- Direct variation functions have the form $f(x) = kx^n$ where n is a positive integer and $k \neq 0$. *(p. 102)*
- Polynomial functions are the sum of one or more direct variation functions or a constant function. *(p. 103)*
- A zero of a function is a value of x that makes $f(x)$ equal to zero. *(p. 104)*
- Radical functions are functions with a variable under the radical symbol. *(p. 113)*
- Rational functions are defined by rational expressions. *(p. 120)*
- The composite of two functions is written $f \circ g$. You can find the composite by finding $f(g(x))$. *(p. 127)*

GEOMETRY

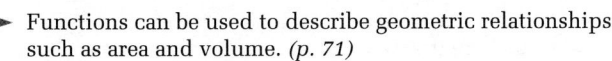

- Functions can be used to describe geometric relationships such as area and volume. *(p. 71)*
- The graph of a linear function is a line. *(p. 81)*
- The graph of a quadratic function is a parabola. *(p. 95)*
- Asymptotes are used as helping lines when graphing a rational function. *(p. 123)*
- The graphs of some rational functions are hyperbolas. *(p. 123)*

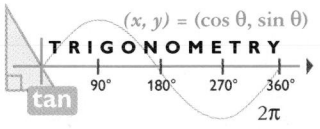

TRIGONOMETRY

- The angle at which an object is thrown affects the height of the object. *(p. 96)*

Key Terms

- **domain of a function** (p. 70)
- **one-to-one function** (p. 72)
- **piecewise function** (p. 86)
- **degree** (p. 103)
- **radical function** (p. 113)
- **asymptote** (p. 123)
- **range of a function** (p. 70)
- **many-to-one function** (p. 72)
- **absolute value function** (p. 88)
- **zero of a function** (p. 104)
- **rational function** (p. 120)
- **hyperbola** (p. 123)
- **function notation ($f(x)$)** (p. 71)
- **linear function** (p. 78)
- **polynomial function** (p. 103)
- **extraneous solution** (p. 112)
- **rational expression** (p. 120)
- **composite** (p. 126)

Quick Quiz (2-4 through 2-6)

1. Write the general form of a quadratic function. [2-4]
   $f(x) = ax^2 + bx + c$
2. Use $f(x) = 5 - 4x + 2x^2$. Find $f(-2)$. [2-4] **$f(-2) = 21$**
3. Write the general form of a direct variation function. [2-5] **$f(x) = kx^n$**
4. Define a polynomial function. [2-5]
   A polynomial function is the sum of one or more direct variation functions or a constant function.
5. Solve $\sqrt{x+9} - 3 = x$. [2-6]
   $x = 0$
6. Simplify $\sqrt[3]{(x-4)^3}$. [2-6]
   $x - 4$

Answers to Unit 2 Review and Assessment

17. Answers may vary. Examples are given. Every direct variation function is a polynomial function. No absolute value function is a direct variation function. Some rational functions are polynomial functions.
18. Answers may vary.
19. Answers may vary.

Logical Reasoning and Methods of Proof

OVERVIEW

- **Unit 3** covers writing proofs, both synthetic and coordinate, and introduces the technique of indirect proof. Students compare and contrast coordinate and synthetic methods of proof and learn to understand the need for clear definitions. Coordinate geometry is reviewed in the **Student Resources Toolbox**.
- Theorems are introduced that relate to interior and exterior angle measures in polygons. Inscribed polygons and circumscribed polygons are studied. Angle and arc relationships are explored for central angles and inscribed angles. The five regular polyhedra are introduced and their properties examined.
- The **Unit Project** is based on using logical reasoning to prepare a legal brief. Students plan and write logical arguments for situations, choose questions and decide in which order to ask them, and explore alibis as a means of indirect proof in everyday contexts.
- **Connections** to geography, social studies, architecture, coin-collecting, navigation, puzzles, communications, obtaining a restaurant license, and choosing theater seats are some of the topics included in the teaching materials and the exercises.
- **Graphics calculators** or **computer spreadsheet software** are used in Section 3-8 to find the semiperimeters and areas of regular n-gons circumscribed about a circle.
- **Problem-solving strategies** used in Unit 3 are based upon writing synthetic proofs, coordinate proofs, and indirect proofs.

Unit Objectives

Section	Objectives	NCTM Standards
3-1	• Review proofs and explore the relationship between implications and their converses, inverses, and contrapositives.	1, 2, 3, 4, 7, 14
3-2	• Compare and contrast coordinate and synthetic methods of proof.	1, 2, 3, 4, 5, 7, 8, 14
3-3	• Understand the need for clear definitions. • Reconsider the definition of trapezoid.	1, 2, 3, 4, 5, 7, 8, 14
3-4	• Find formulas for the sums of interior and exterior angle measures in polygons.	1, 2, 3, 4, 5, 7, 8, 14
3-5	• Examine relationships between inscribed polygons and circles.	1, 2, 3, 4, 5, 7, 8, 14
3-6	• Examine relationships between inscribed angles and arcs of circles.	1, 2, 3, 4, 5, 7, 8, 14
3-7	• Use indirect proof in geometric, algebraic, and everyday contexts. • Examine the properties of tangents to circles.	1, 2, 3, 4, 5, 7, 8, 14
3-8	• Examine relationships between circumscribed polygons and circles.	1, 2, 3, 4, 5, 7, 8, 14
3-9	• Recognize the five regular polyhedra and examine properties of regular and semiregular polyhedra.	1, 2, 3, 4, 5, 7, 8, 14

Topic Spiraling

Section	Connections to Prior and Future Concepts
3-1	**Section 3-1** reviews writing proofs, first introduced in Units 7 and 8 of Book 2. Implications, converses, inverses, and contrapositives are explored. Logical reasoning and proof are important concepts that will be used in many future mathematics courses.
3-2	**Section 3-2** explores the differences and similarities between coordinate and synthetic proofs. Formulas for midpoint and distance, first introduced in Sections 5-2 and 5-3 of Book 2, are reviewed.
3-3	**Section 3-3** explores the need for clear definitions. Inclusive and exclusive definitions are studied and compared as trapezoids are linked to parallelograms. The importance of using clear definitions was first introduced in Section 7-4 of Book 2. The use of clear and concise definitions is an important part of all mathematics courses.
3-4	**Section 3-4** covers the sum of the measures of the interior and of the exterior angles of a polygon. Angle relationships in a triangle were first explored in Section 2-5 of Book 1 and reviewed in Section 8-2 of Book 2.
3-5	**Section 3-5** explores relationships between inscribed polygons and circles. The relationship between central angles and their arcs is extended to inscribed angles in the next section.
3-6	**Section 3-6** examines relationships between inscribed angles and arcs of a circle. Skills relating to inscribed triangles learned in Section 3-5 of Book 3 are used here.
3-7	**Section 3-7** introduces indirect proof. Students write indirect proofs of geometry theorems and number theory facts. Theorems about tangents to circles are written. Indirect proof is a powerful tool that will be used in many future mathematics courses.
3-8	**Section 3-8** explores relationships between circumscribed polygons and circles, thus reversing the positions of the polygons and circles studied in Section 3-5. Two methods for finding the area of a circumscribed polygon are explored.
3-9	**Section 3-9** extends the study of three-dimensional figures to include the five regular polyhedra. Properties of regular and semiregular polyhedra are explored. The study of three-dimensional figures was first encountered in Sections 9-5 through 9-8 of Book 1 and continued in Unit 10 of Book 2.

Integrating the Strands

Strands	Sections
Number	3-3, 3-4, 3-6, 3-7
Algebra	3-1, 3-2, 3-3, 3-4, 3-5, 3-6, 3-7, 3-8, 3-9
Functions	3-1, 3-3, 3-4, 3-7, 3-8, 3-9
Measurement	3-2, 3-9
Geometry	3-1, 3-2, 3-3, 3-4, 3-5, 3-6, 3-7, 3-8, 3-9
Trigonometry	3-5, 3-7, 3-8
Discrete Mathematics	3-3
Logic and Language	3-1, 3-2, 3-3, 3-4, 3-5, 3-6, 3-7, 3-8, 3-9

Planning the Unit 3

Logical Reasoning and Methods of Proof

Section Planning Guide

- Essential exercises and problems are indicated in boldface.
- Ongoing work on the Unit Project is indicated in color.
- Exercises and problems that require student research, group work, manipulatives, or graphing technology are indicated in the column headed "Other."

Section	Materials	Pacing	Standard Assignment	Extended Assignment	Other
3-1	straws or stirrers, scissors	Day 1 Day 2	1, **4–15** **16–20,** 23–26, **27–29,** 31–35, 36	1–3, **4–15** **16–20,** 23–26, **27–29,** 31–35, 36	 21, 22, 30
3-2		Day 1	**2–11, 13,** 14, 16–19, 20	1, **2–11,** 12, **13,** 14–19, 20	
3-3		Day 1	1, **2–7, 11–18,** 19–23, 24	1, **2–7,** 8–10, **11–18,** 19–23, 24	
3-4		Day 1	**1–13,** 15, 17–21, 22	**1–13,** 14, 15, 17–21, 22	16
3-5	compass	Day 1 Day 2	**1–5** **6–10,** 11, **12–14,** 18–21, 22	**1–5** **6–10,** 11, **12–14,** 15, 16, 18–21, 22	 17
3-6	protractor, scissors, compass, straightedge	Day 1	**1–4, 6–8, 11,** 16–20, 21	**1–4, 6–8,** 10, **11,** 13–20, 21	5, 9, 12
3-7		Day 1 Day 2	**1–3,** 4, **6, 7** **11, 12, 14,** 15–21, 22	**1–3,** 4, 5, **6, 7,** 8–10 **11, 12, 14,** 15–21, 22	 13
3-8	tracing paper, compass, straightedge, spreadsheet or graphics calculator	Day 1	**1–5, 8, 10,** 12–15, 16	**1–5,** 7, **8,** 9, **10,** 12–15, 16	6, 11
3-9	polyhedral nets, toothpicks, gum drops or marshmallows	Day 1	**1, 5, 8–13, 15–19,** 23–27, 28	**1,** 4, **5,** 6, 7, **8–13,** 14, **15–19,** 20, 23–27, 28	2, 3, 21, 22
Review Test		**Day 1** **Day 2**	**Unit Review** **Unit Test**	**Unit Review** **Unit Test**	

Yearly Pacing	Unit 3 Total	Units 1–3 Total	Remaining	Total
	16 days (2 for Unit Project)	50 days	104 days	154 days

Support Materials

- See **Project Book** for notes on Unit 3 Project: State Your Case.
- UPP and disk refer to **Using Plotter Plus** booklet and **Plotter Plus** disk.
- TI-81/82 refers to **Using TI-81 and TI-82 Calculators** booklet.
- Warm-up exercises for each section are available on **Warm-Up Transparencies.**

Section	Study Guide	Practice Bank	Problem Bank	Activity Bank	Explorations Lab Manual	Assessment Book	Visuals	Technology
3-1	3-1	Practice 19	Set 6	Enrich 17		Quiz 3-1		
3-2	3-2	Practice 20	Set 6	Enrich 18		Quiz 3-2		
3-3	3-3	Practice 21	Set 6	Enrich 19		Quiz 3-3 Test 10		
3-4	3-4	Practice 22	Set 7	Enrich 20	Add. Expl. 4	Quiz 3-4		
3-5	3-5	Practice 23	Set 7	Enrich 21		Quiz 3-5		
3-6	3-6	Practice 24	Set 7	Enrich 22		Quiz 3-6 Test 11		
3-7	3-7	Practice 25	Set 8	Enrich 23		Quiz 3-7		TI-81/82, page 58 UPP, page 53
3-8	3-8	Practice 26	Set 8	Enrich 24		Quiz 3-8		
3-9	3-9	Practice 27	Set 8	Enrich 25		Quiz 3-9 Test 12	Folder 3	
Unit 3	Unit Review	Practice 28	Unifying Problem 3	Family Involve 3		Tests 13, 14		

Form A

Spanish versions of these tests are on pages 123–126 of the **Assessment Book.**

Name ______ Date ______ Score ______

Test 13

Test on Unit 3 (Form A)

Directions: Write the answers in the spaces provided.

1. Write the inverse of the following statement. Then determine whether the inverse is *True* or *False*.
"If a quadrilateral is a rectangle, then it is a parallelogram."
If a quadrilateral is not a rectangle, then it is not a parallelogram; false.
2. Is the implication given in Question 1 *True* or *False*?
3. Show that quadrilateral *ABCD* with vertices *A*(0, 0), *B*(6, 0), *C*(5, 2), and *D*(1, 2) is an isosceles trapezoid.
$AB = \sqrt{(6-0)^2 + (0-0)^2} = 6$; $BC = \sqrt{(5-6)^2 + (2-0)^2} = \sqrt{5}$; $CD = \sqrt{(1-5)^2 + (2-0)^2} = 4$; $AD = \sqrt{(1-0)^2 + (2-0)^2} = \sqrt{5}$; slope of $\overline{AB}$: $\frac{0-0}{6-0} = 0$; slope of $\overline{CD}$: $\frac{2-2}{1-5} = 0$; Since the slopes of $\overline{AB}$ and $\overline{CD}$ are equal, *ABCD* is a trapezoid. Since *BC* = *AD*, *ABCD* is an isosceles trapezoid.

For Questions 4 and 5, tell whether the word one means "exactly one" or "at least one."

4. You may enter our contest if you purchase one of our products.
5. You are still eligible for co-curricular activities if you have one D.
6. Find the measure of each interior angle of a regular octagon.
7. Find the measure of each exterior angle of a regular pentagon.
8. **Writing** Explain how the formula for finding the sum of the interior angles of a polygon can be derived. Include a drawing to illustrate your explanation.
Sample answer: By drawing all the diagonals from one vertex of a polygon, the polygon is divided into two fewer triangles than there are sides of the polygon (that is, (*n* – 2) triangles if *n* = the number of sides). Since the sum of the measures of the angles of a triangle is 180°, and since the angles of the polygon are formed by combinations of the angles of these triangles, the formula is *S* = (*n* – 2)180°.

6 sides
6 – 2 = 4 triangles

Answers	
1.	See question.
2.	True.
3.	See question.
4.	at least one
5.	exactly one
6.	135°
7.	72°
8.	See question.

Assessment Book INTEGRATED MATHEMATICS 3

17

Name ______ Date ______ Score ______

Test 13 *(continued)*

Directions: Write the answers in the spaces provided.

Find the measure of each arc or angle.

9. ∠*NQA* 10. ∠*M*
11. ∠*MNQ* 12. ∠*NRQ*
13. $\overset{\frown}{NQ}$ 14. $\overset{\frown}{MRN}$

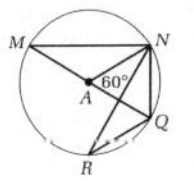

For Questions 15–20, find the measure of each angle. In circle *O*, $\overrightarrow{PA}$ and $\overrightarrow{PB}$ are tangents, $\overrightarrow{PD}$ bisects ∠*BPA*, and *m* ∠*AOP* = 40°.

15. ∠*BOC* 16. ∠*BOD* 17. ∠*PBO*
18. ∠*APO* 19. ∠*OAD* 20. ∠*DAP*

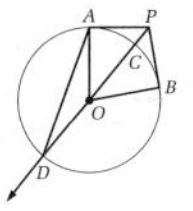

21. Find the area of circumscribed polygon *RSTU* if the radius of circle *O* is 5.25 in. and the polygon is a kite with *ST* = 9 in. and *RU* = 12 in.

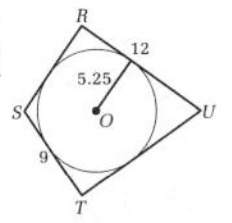

The base of a pyramid is a regular pentagon and each of its other faces is an equilateral triangle. Use this information for Questions 22–25.

22. How many faces does the pyramid have?
23. How many edges does the pyramid have?
24. How many vertices does the pyramid have?
25. Find the defect at the vertex where the triangular faces meet.
26. **Open-ended** Write an indirect proof to show that the measure of each interior angle of a regular hexagon is not 150°.
Sample answer: Assume that the measure of each interior angle of a regular hexagon is 150°. Then the sum of the measures is 6(150°) = 900°. So, (*n* – 2)180° = 900° and thus *n* – 2 = 5, or *n* = 7. But this contradicts the fact that a hexagon has 6 sides. Therefore, the assumption is false and it must be true that the measure of each interior angle of a regular hexagon is not 150°.

Answers	
9.	60°
10.	30°
11.	90°
12.	30°
13.	60°
14.	240°
15.	40°
16.	140°
17.	90°
18.	50°
19.	20°
20.	110°
21.	110.25 in.^{2}
22.	6
23.	10
24.	6
25.	60°
26.	See question.

18
Assessment Book INTEGRATED MATHEMATICS 3

Form B

Name ______ Date ______ Score ______

Test 14

Test on Unit 3 (Form B)

Directions: Write the answers in the spaces provided.

1. Write the contrapositive of the following statement. Then determine whether the inverse is *True* or *False*.
"If a quadrilateral is a square, then it is a rhombus."
If a quadrilateral is not a rhombus, then it is not a square; true.
2. Is the implication given in Question 1 *True* or *False*?
3. Show that quadrilateral *JKLM* with vertices *J*(0, 0), *K*(3, 0), *L*(6, 2), and *M*(–3, 2) is an isosceles trapezoid.
$JK = \sqrt{(3-0)^2 + (0-0)^2} = 3$; $KL = \sqrt{(6-3)^2 + (2-0)^2} = \sqrt{13}$; $LM = \sqrt{(-3-6)^2 + (2-2)^2} = 9$; $JM = \sqrt{(-3-0)^2 + (2-0)^2} = \sqrt{13}$; slope of $\overline{JK}$: $\frac{0-0}{3-0} = 0$; slope of $\overline{LM}$: $\frac{2-2}{-3-6} = 0$; Since the slopes of $\overline{JK}$ and $\overline{LM}$ are equal, *JKLM* is a trapezoid. Since *KL* = *JM*, *JKLM* is an isosceles trapezoid.

For Questions 4 and 5, tell whether the word one means "exactly one" or "at least one."

4. One hour of tutoring will help you improve your grade.
5. You may enter the competition with one partner.
6. Find the measure of each interior angle of a regular pentagon.
7. Find the measure of each exterior angle of a regular octagon.
8. **Writing** In circle *O* at the right, draw and label a central angle and an inscribed angle that intercept the same arc. Explain the relationship between the arc measure and the angle measures.
Sample answer: As shown in the figure, when a central angle and an inscribed angle intercept the same arc, the measure of the central angle is the same as the arc measure and the measure of the inscribed angle is half that of the arc (and also of the central angle).

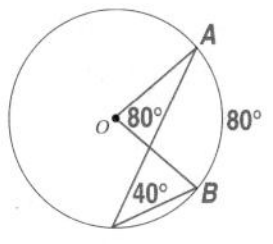

Answers	
1.	See question.
2.	True.
3.	See question.
4.	at least one
5.	exactly one
6.	108°
7.	45°
8.	See question.

Assessment Book INTEGRATED MATHEMATICS 3

19

Name ______ Date ______ Score ______

Test 14 *(continued)*

Directions: Write the answers in the spaces provided.

Find the measure of each arc or angle.

9. ∠*MQP* 10. ∠*QMN*
11. ∠*QAM* 12. ∠*MNQ*
13. $\overset{\frown}{QM}$ 14. $\overset{\frown}{QPN}$

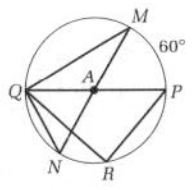

For Questions 15–20, find the measure of each angle. In circle *O*, $\overrightarrow{PA}$ and $\overrightarrow{PB}$ are tangents, $\overrightarrow{PD}$ bisects ∠*BPA*, and *m* ∠*AOP* = 50°.

15. ∠*OAD* 16. ∠*BOC* 17. ∠*BOD*
18. ∠*PBO* 19. ∠*APO* 20. ∠*DAP*

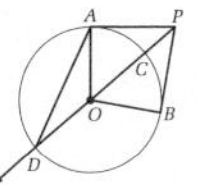

21. Find the area of circumscribed polygon *RSTU* if the radius of circle *O* is 4.25 in. and the polygon is a kite with *ST* = 7 in. and *RU* = 10 in.

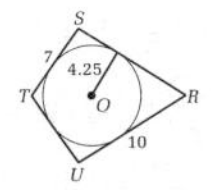

The base of a pyramid is a regular hexagon and each of its other faces is an isosceles triangle with base angles measuring 70°. Use this information for Questions 22–25.

22. How many faces does the pyramid have?
23. How many edges does the pyramid have?
24. How many vertices does the pyramid have?
25. Find the defect at the vertex where the triangular faces meet.
26. **Open-ended** Write an indirect proof to show that the measure of each interior angle of a regular decagon is 144°.
Sample answer: Assume that the measure of each interior angle of a regular decagon is not 144°. Then, since a decagon has 10 sides, the sum of the measures of the interior angles is not 10(144°), or 1440°. So, (*n* – 2)180° ≠ 1440° and thus *n* – 2 ≠ 8, and *n* ≠ 10. But this contradicts the fact that a decagon does have 10 sides. Therefore, the assumption is false and it is true that the measure of each interior angle of a regular decagon is 144°.

Answers	
9.	30°
10.	30°
11.	120°
12.	60°
13.	120°
14.	300°
15.	25°
16.	50°
17.	130°
18.	90°
19.	40°
20.	115°
21.	72.25 in.^{2}
22.	7
23.	12
24.	7
25.	120°
26.	See question.

20
Assessment Book INTEGRATED MATHEMATICS 3

OUTSIDE RESOURCES

Books/Periodicals

Serra, Michael. Chapter 1: "Inductive Reasoning": pp. 39–69. Chapter 6: "Arc Length": pp. 282–287. *Discovering Geometry: An Inductive Approach.* Key Curriculum Press.

Brandell, Joseph L. "Helping Students Write Paragraph Proofs in Geometry." *Mathematics Teacher* (October 1994): pp. 498–502.

Activities/Manipulatives

Van Dyke, Frances. "A Concrete Approach to Mathematical Induction." *Mathematics Teacher* (April 1995): pp. 302–307.

Miller, William A. and Robert G. Clason. "Golden Triangles, Pentagons, and Pentagrams." *Mathematics Teacher* (May 1994): pp. 338–341.

Software

Baulac, I., F. Bellemain, and J. M. Laborde. Cabri: The Interactive Geometry Notebook. Pacific Grove, CA: Brooks/Cole Publishing Co., 1992.

Videos

The Platonic Solids. Animated video illustrating properties of regular polyhedra, 17 minutes. Key Curriculum Press.

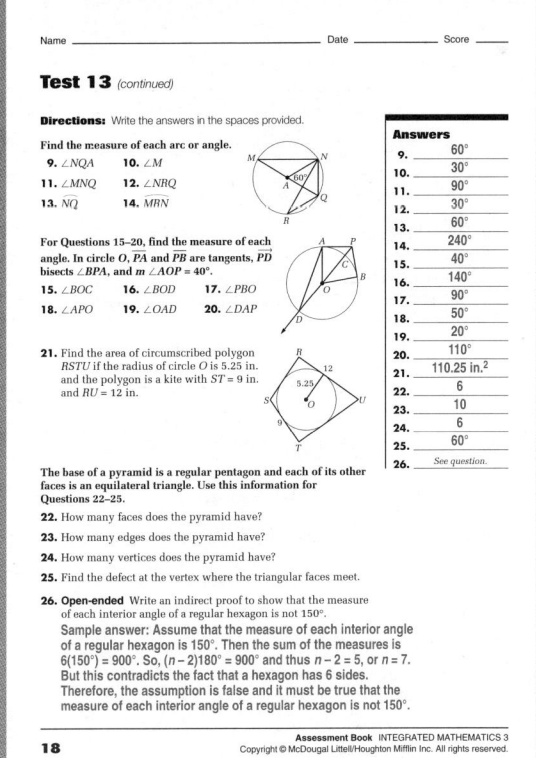

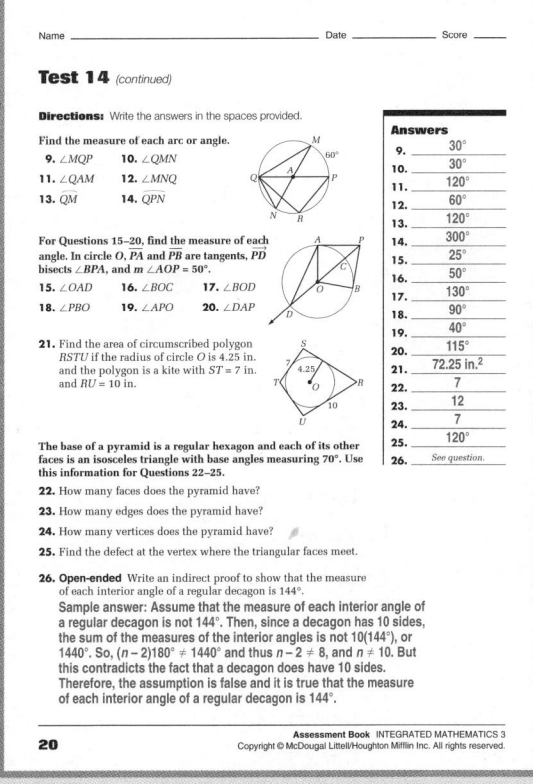

PROJECT GOALS

- Students prepare a legal brief for a case involving a girl who is not allowed to play on a boys' high school soccer team.
- Students prepare a legal brief that supports either the girl's position or the school's position.

PROJECT PLANNING

Project Teams

Have students work on the project in groups of four. Students should read and discuss the law that applies to the case. They should write a "witness statement" in which they look at the evidence the witness might offer to support the plaintiff or the defense. Students should examine the strengths and weaknesses of their witnesses and consider if any other witnesses could be used to support their case. After visiting a real courtroom or watching a TV trial, students should discuss the kinds of reasoning that lawyers use to convince people of their cases.

Support Materials

The ***Project Book*** contains information about the following topics for use with this Unit Project.

- Project Description
- Teaching Commentary
- Working on the Unit Project Exercises
- Completing the Unit Project
- Assessing the Unit Project
- Alternative Projects
- Outside Resources

unit 3

Logical Reasoning and Methods of Proof

Should a girl be allowed to play on a boys' team in a sport where there is no girls' team, or vice versa? Would you go to court to sue for the right to try out for a school team?

You will use the witness statements on pages 136–138 to help you develop your case for or against the Springfield school system.

Justice

Sarah Wilkins, a junior, played soccer for many years before moving to Springfield last year. When she learned there was no girls' team at Springfield High School, she went to the tryouts for the boys' team, but the soccer coach would not let her try out.

Sarah decided to file a suit against the Springfield school system for sex discrimination. She cited the law shown below.

Title IX of the 1972 Educational Amendments states, in part:

No person in the United States shall, on the basis of sex, be excluded from participation in, be denied the benefits of, or be subjected to discrimination under any education program or activity receiving Federal financial assistance . . .

"I have coached boys' soccer for eight years. In previous years, a sign-up sheet for girls interested in playing soccer did not get enough signatures to support developing a girls' team. I know that boys play rough and think they may assume a girl is weaker than they are. I think some boys would be embarrassed by being on a team with a girl, or ashamed of losing to a girl."

Springfield High School Boys' Soccer Coach

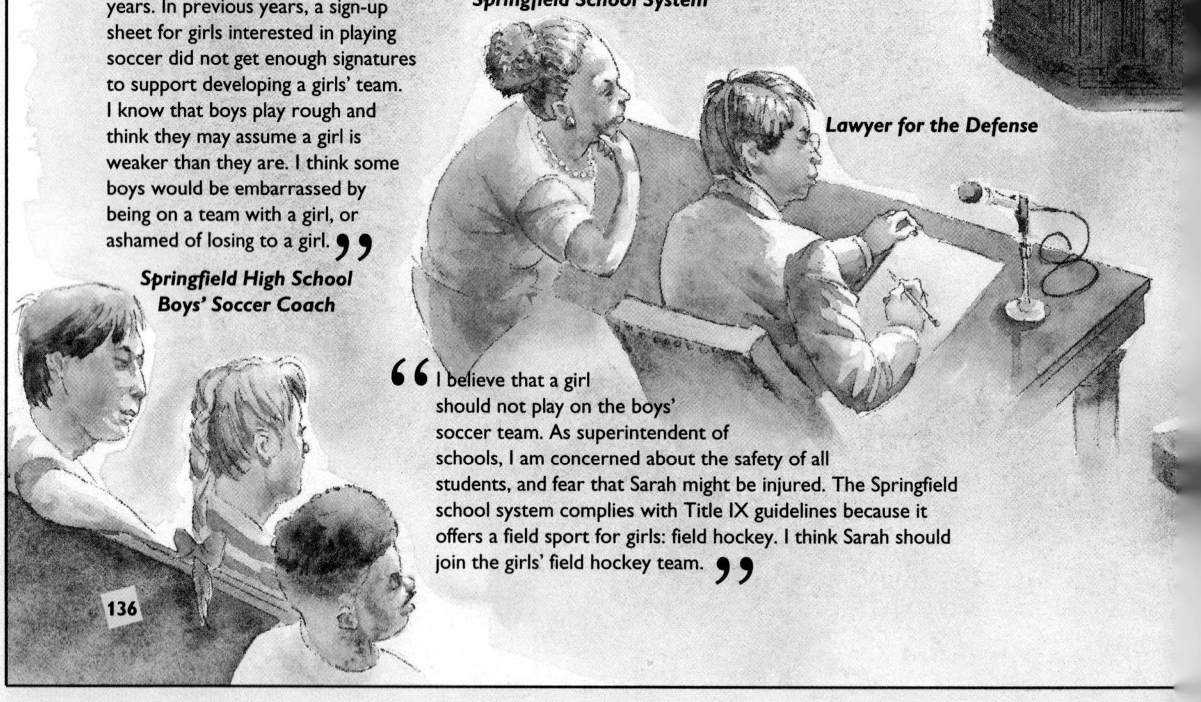

"I believe that a girl should not play on the boys' soccer team. As superintendent of schools, I am concerned about the safety of all students, and fear that Sarah might be injured. The Springfield school system complies with Title IX guidelines because it offers a field sport for girls: field hockey. I think Sarah should join the girls' field hockey team."

Suggested Rubric for Unit Project

4 Students' briefs include all of the things listed in the textbook on page 137. The brief is well written, clearly organized, and complete. The questions for witnesses are asked in an order that makes good sense and is effective. Definitions of key terms in the law are precise. The reasoning used to justify the group's position demonstrates a correct understanding of the mathematical reasoning processes presented in the unit. The closing arguments use the evidence presented and are convincing.

3 Students' briefs include all of the things listed in the textbook, but some of them are not as complete as they could be. The order of questions for witnesses can be improved and some of the questions do not do an effective job of cross-examination. The reasoning used is satisfactory but is not entirely convincing. Counterarguments do not cover all the points made by the opponent's case. Closing arguments have some shortcomings.

Unit Project 3

State Your Case

Your project is to prepare a legal brief for the case of Wilkins vs. Springfield School System. A **brief** is an outline that lists your main lines of reasoning in a case, along with supporting evidence. It is sometimes filed by an attorney before arguing a case in court.

Your brief should include these things:

BRIEF

- an opening statement
- questions you plan to ask witnesses
- questions you plan to ask witnesses for the opposing side during cross-examination
- the law as it applies to your case
- the reasoning you will use to convince others that the evidence presented justifies your position according to the law
- counterarguments you will make to your opponent's case
- closing arguments

Sarah Wilkins
Junior, Springfield High School

"I began playing soccer at age 10 at an all-girls school. I hoped to continue playing in college, and feel that I must play in high school to get an athletic scholarship. I am a good player and am not afraid of playing with and against boys. I think I should be able to try out for the boys' soccer team. I do not think it is fair to ask me to play a different sport."

Sarah Wilkins's father

"Soccer has been a way for my daughter to excel. Her attitude toward studying and doing homework improved when she began playing soccer. I know that playing soccer is important to her. I am concerned that boys might tease her if she joined the boys' team, but I am more concerned about the bad effects that not playing might have on her."

137

Students Acquiring English

This project has several writing exercises that require polished language skills. You may want to make sure that groups have both students acquiring English and students fluent in English so that group members can help each other with those exercises.

ADDITIONAL BACKGROUND

Multicultural Note

In recent years, a number of professional women athletes have become highly successful and internationally famous as a result of their accomplishments. Billie Jean King, Chris Evert, Martina Navratilova, Aranxta Sanchez Vicario, and Monica Seles are among those who have made names for themselves in professional tennis. Althea Gibson, a tennis star in the 1950s, later became a professional golfer; she was the first African-American woman to win both tennis and golf championships. Nancy Lopez and Ayako Okamoto are two other highly successful women golfers. In general, professional women athletes have been restricted to solo sports since team sports, including baseball, football, and soccer, have very few professional women's teams. Manon Rheaume, a minor league ice hockey goalie, recently became one of the first women ever to play professionally in a men's league.

Suggested Rubric for Unit Project

2 Students' briefs include all of the things a brief should, but the content of many of the items is superficial. There are some errors in the use of logical arguments in both drawing conclusions from the evidence presented and in making counterarguments to the opponent's case. Closing arguments are weak and not convincing.

1 Students' briefs are incomplete and poorly organized. Important information has been omitted and frequent errors in reasoning render the brief useless. Counter-arguments and closing arguments are ineffective. Students should be encouraged to speak with the teacher as soon as possible to review their work and to make a new start on the project.

Rules of Evidence

In a real courtroom, elaborate rules of evidence have to be followed by the attorneys to ensure that both parties receive a fair trial. If a question is raised about a rule being violated, the judge decides whether to allow the evidence or to exclude it from the record. Some rules of evidence apply to the form of questions that can be asked of witnesses. For example, an attorney cannot ask a witness a leading question. A leading question suggests to the witness the desired answer. If a leading question is asked, the burden is on the opposing counsel to object to the judge. In the absence of an objection, the evidence will most likely be allowed by the judge. Thus, the attorneys in a trial need to know the rules of evidence in order to properly represent their clients.

ALTERNATIVE PROJECTS

Project 1, page 199

Research a Famous Case

Research a famous court case and prepare a written or oral report about it. Describe the reasoning used in the case and tell the outcome.

Project 2, page 199

Put on a Mock Trial

Simulate a trial in the classroom. Students play the roles of judge, jurors, attorneys, and so forth.

Unit Project 3

Getting Started

Sarah's Former Soccer Coach

"I was Sarah's soccer coach at another school for the past seven years. I have watched Sarah improve her soccer playing and was happy to present her with a Most Valuable Player award. I think Sarah is as good a player as most boys her age, and as good as many college-aged women players. These next two years are important ones for Sarah to continue developing her skills."

Senior and Captain of the Boys' Soccer Team

"I believe that adding a girl to my soccer team will ruin team spirit. Teamwork requires cooperation, and not all boys will support a girl team member. Part of the experience of being on a team—analyzing our mistakes together—will be ruined by the presence of a girl. I also expect other teams in the league may make fun of my team."

For this project you should work in a group of four students. Half the groups will be assigned to work for the plaintiff, and half for the defense. Here are some ideas to help you get started.

☞ Read the law that applies to the case. Discuss how you might convince a jury or judge that your case is supported by the law.

☞ Look at the witness statements on this page and pages 136 and 137. Discuss the evidence you think the witnesses might offer to support the plaintiff or the defendant.

☞ Discuss the case with your team. What are the strengths of the witnesses you have? What are their weaknesses? What other witnesses, if any, do you think you can use to support your case?

☞ Visit a courtroom during a trial that is open to the public, or watch a trial on TV. Talk about the kinds of reasoning that lawyers use to convince a judge or a jury.

Working on the Unit Project

Your work in Unit 3 will help you see how reasoning is used in arguing a court case.

Related Exercises:

Section 3-1, Exercise 36
Section 3-2, Exercise 20
Section 3-3, Exercise 24
Section 3-4, Exercise 22
Section 3-5, Exercise 22
Section 3-6, Exercise 21
Section 3-7, Exercise 22
Section 3-8, Exercise 16
Section 3-9, Exercise 28

Alternative Projects p. 199

Can We Talk Justice

➤ Why do you think a person might not be selected as a juror for this case by the lawyers for the defense? by the lawyers for the plaintiff?

➤ Attorneys use reasoning to try to convince a jury. What role, if any, do you think emotions play in a courtroom?

➤ What do you think the outcome of this case should be? What do you think would happen if this case occurred at your school?

➤ Discuss the impact of pretrial publicity and/or cameras in the courtroom on the outcome of a trial that has been in the news recently.

Answers to Can We Talk?

➤ Answers may vary. Examples are given. A person should not be selected as a juror if he or she was involved in a sport at Springfield High School, if there was any relationship with the defense or the plaintiff, or if the person is not a resident of Springfield.

➤ Answers may vary. An example is given. Emotions can be used to help a jury understand a person's reason for doing something or a person's state of mind. They can also be used by attorneys to try to sway a jury to their point of view or to emphasize a particular line of reasoning.

➤ Answers may vary. An example is given. I think Sarah should be allowed to try out for the team. If she is a good soccer player, then she should be allowed to try out for the boys' team. If this occurred at my school, I think Sarah would have been allowed to try out for the team.

➤ Answers may vary depending upon cases in the news recently.

Section

3-1 Implications and Proof

Focus

Review proofs and explore the relationship between implications and their converses, inverses, and contrapositives.

ON THE DIAGONAL

EXPLORATION

How do properties of diagonals determine quadrilaterals?

- **Materials: straws or stirrers cut to unequal lengths**
- **Work in a group of four students.**

1. Take two straws of unequal length. Find their midpoints. Make an x so the straws intersect at their midpoints. Draw the figure you get by joining the endpoints. What kind of quadrilateral is it?
2. Repeat step 1 using a different angle at the point of intersection.
3. Compare your results with other members of your group. Make a conjecture about what you observe.
4. Repeat steps 1 and 2 using two straws of equal length. Make a conjecture about the kind of quadrilaterals you can draw by joining the endpoints.
5. Using the two straws of equal length from step 4, make an x so the straws do not intersect at their midpoints. Draw the figure you get by joining the endpoints. Do you have a special quadrilateral?

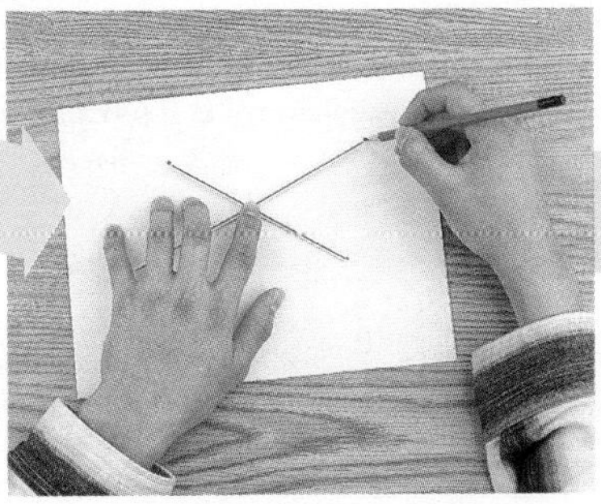

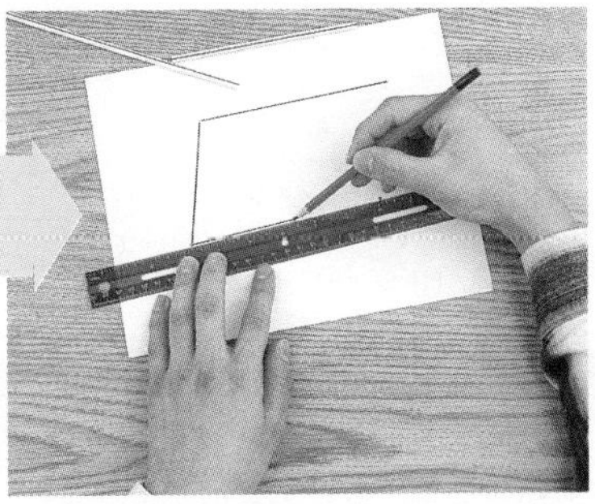

PLANNING

Objectives and Strands

See pages 136A and 136B.

Spiral Learning

See page 136B.

Materials List

- Straws or stirrers
- Scissors

Recommended Pacing

Section 3-1 is a two-day lesson.

Day 1

Pages 139–141: Exploration through Bisecting Diagonals Theorem, *Exercises 1–15*

Day 2

Pages 142–143: Implications through Look Back, *Exercises 16–36*

Toolbox References

- **Toolbox Skill 33:** Recognizing Congruent Figures
- **Toolbox Skill 34:** Understanding Implications

Extra Practice

See pages 612–613.

Warm-Up Exercises

- Warm-Up Transparency 3-1

Support Materials

- Practice Bank: Practice 19
- Activity Bank: Enrichment 17
- Study Guide: Section 3-1
- Problem Bank: Problem Set 6
- Assessment Book: Quiz 3-1, Alternative Assessment 1

Answers to Exploration

1. parallelogram
2. parallelogram
3. If the diagonals of a quadrilateral intersect at their midpoints (that is, bisect each other), the quadrilateral is a parallelogram.
4. The resulting figure is a rectangle. If the diagonals of a quadrilateral are equal in measure and bisect each other, then the quadrilateral is a rectangle.
5. In general, no. (The resulting figure *may* be a kite.)

TEACHING

Exploration

The goal of the Exploration is to have students discover two properties of quadrilaterals that are determined by their diagonals: (1) If the two diagonals of a quadrilateral bisect each other, then the quadrilateral is a parallelogram; and (2) if the two diagonals are equal in length and bisect each other, then the quadrilateral is a rectangle.

Talk it Over

Questions 1 and 2 review the concept of the converse of an if-then statement, and question 2 reminds students that the converse of a true statement is not necessarily true.

Student Resources Toolbox
p. 657 *Logical Reasoning*

Talk it Over

Remember that the *converse* of an if-then statement is formed by interchanging the hypothesis (the "if" part) and conclusion (the "then" part). State the converse of each statement. Do you think each converse is true? Why or why not?

1. If a quadrilateral is a parallelogram, then the diagonals bisect each other.
2. If a quadrilateral is a rectangle, then the diagonals are equal in length.

Writing Proofs

The converse of a true statement is not necessarily true. To be sure that a converse is true, you must prove it.

Synthetic proof is a proof built using a system of postulates and theorems in which the properties of figures, but not their actual measurements, are studied.

JUSTIFICATIONS FOR SYNTHETIC PROOF

To prove a theorem, you may use the following as justifications:

- given statements
- definitions
- postulates
- previously proved theorems

A list of postulates and previously proved theorems is on page 668–671.

Student Resources Toolbox
p. 656 *Relationships*

Sample 1

Prove that if the two diagonals of a quadrilateral bisect each other, then the quadrilateral is a parallelogram.

Sample Response

Given $AE = CE$, $BE = DE$

Prove $ABCD$ is a parallelogram.

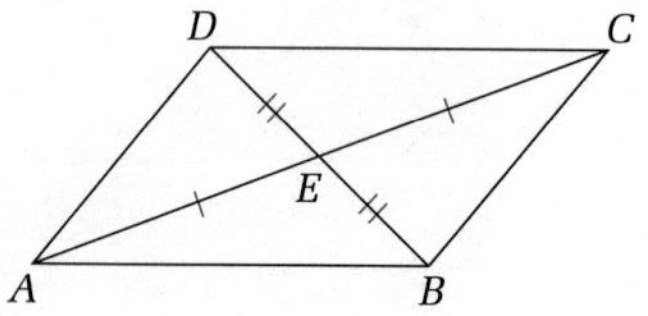

Answers to Talk it Over

1. If the diagonals of a quadrilateral bisect each other, then the quadrilateral is a parallelogram; True. Since vertical angles are equal in measure, two pairs of congruent triangles are formed (SAS). If two lines are intersected by a transversal and alternate interior angles are equal in measure, the lines are parallel. This fact can be used with the pairs of congruent triangles to show that both pairs of opposite sides of the quadrilateral are parallel.
2. If the diagonals of a quadrilateral are equal in length, then the quadrilateral is a rectangle; False. The diagonals must also bisect each other.

Plan Ahead

Show that $\triangle ABE$ and $\triangle CDE$ are congruent, so $m\angle 3 = m\angle 4$, which means that $\overline{AB} \parallel \overline{CD}$.

Repeat with $\triangle ADE$ and $\triangle CBE$, to show that $m\angle 7 = m\angle 8$, which means that $\overline{AD} \parallel \overline{BC}$.

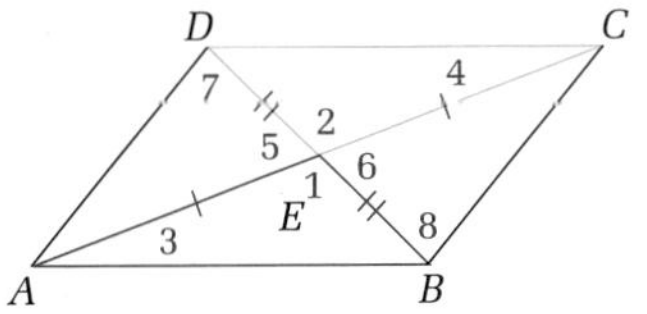

Show Your Reasoning

Statements	Justifications
1. $AE = CE$ and $BE = DE$	1. Given
2. $m\angle 1 = m\angle 2$	2. Vertical ∠s are = in measure.
3. $\triangle ABE \cong \triangle CDE$	3. SAS (Steps 1 and 2)
4. $m\angle 3 = m\angle 4$	4. Corresponding parts of ≅ △s are = in measure.
5. $\overline{AB} \parallel \overline{CD}$	5. If two lines are intersected by a transversal and alternate interior ∠s are = in measure, then the lines are ∥.
6. $m\angle 5 = m\angle 6$	6. Vertical ∠s are = in measure.
7. $\triangle ADE \cong \triangle CBE$	7. SAS (Steps 1 and 6)
8. $m\angle 7 = m\angle 8$	8. Corresponding parts of ≅ △s are = in measure.
9. $\overline{AD} \parallel \overline{BC}$	9. If two lines are intersected by a transversal and alternate interior ∠s are = in measure, then the lines are ∥.
10. $ABCD$ is a parallelogram.	10. Definition of parallelogram (Steps 5 and 9)

Talk it Over

3. In the plan for the proof, one goal is to show that $\overline{AB} \parallel \overline{CD}$. Which steps in the proof show this?
4. Another goal in the plan for the proof is to show that $\overline{AD} \parallel \overline{BC}$. Which steps in the proof show this?

BISECTING DIAGONALS THEOREM

Theorem 3.1 If the two diagonals of a quadrilateral bisect each other, then the quadrilateral is a parallelogram.

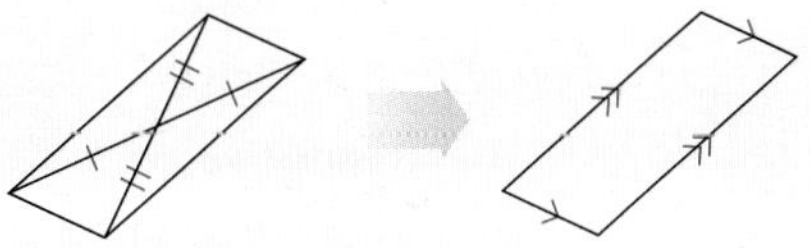

Now you are ready for: Exs. 1–15 on pp. 143–144

Additional Sample

S1 *Given:* For quadrilateral $ABCD$, M is the midpoint of diagonals $\overline{AC}$ and $\overline{BD}$. $AC = BD$

Prove: $ABCD$ is a rectangle.

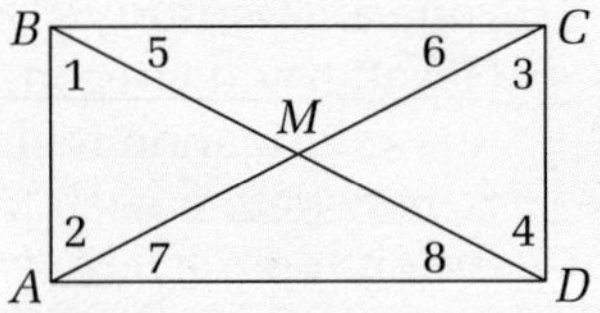

Plan: Show that because M is the midpoint of $\overline{AC}$ and $\overline{BD}$ and $AC = BD$, $\triangle BMA$ and $\triangle CMD$ are congruent and isosceles. Likewise, prove that $\triangle BMC$ and $\triangle AMD$ are congruent and isosceles. Since the base angles of isosceles triangles are equal in measure, $m\angle 1 = m\angle 2 = m\angle 3 = m\angle 4$ and $m\angle 5 = m\angle 6 = m\angle 7 = m\angle 8$. It follows that $m\angle ABC = m\angle BAD$ and $m\angle BCD = m\angle CDA$. By the result proved in Sample 1, $ABCD$ is a parallelogram. Thus, the co-interior angles ABC and BAD are supplementary. Since they are congruent, each has a measure of 90° and is a right angle. Likewise, angles BCD and CDA are right angles. Therefore, $ABCD$ is a rectangle by the definition of *rectangle*.

Talk it Over

Questions 3 and 4 help students to analyze and understand the structure of the proof given in Sample 1.

Answers to Talk it Over

3. Steps 1–5
4. Steps 1 and 6–9

Additional Sample

S2 The lower 48 states of the United States are divided into four time zones: Eastern, Central, Mountain, and Pacific. Some of the states in the Pacific time zone are California, Nevada, and Oregon. Use this information to determine whether each implication is *True* or *False*.

a. If you live in the Pacific time zone, then you live in Portland, Oregon. ($p \to q$)
False. A resident of Nevada lives in the Pacific time zone but not in Portland, Oregon.

b. If you live in Portland, Oregon, then you live in the Pacific time zone. ($q \to p$)
True. The state of Oregon and the city of Portland are in the Pacific time zone.

c. If you do not live in Portland, Oregon, then you do not live in the Pacific time zone. (not $q \to$ not p)
False. If you live in Reno, Nevada, you do not live in Portland, Oregon. Nevertheless, you live in the Pacific time zone.

d. If you do not live in the Pacific time zone, then you do not live in Portland, Oregon. (not $p \to$ not q)
True. You cannot live in Portland, Oregon without living in the Pacific time zone.

Teaching Tip

Students may need to review the conditions under which an implication, $p \to q$, is true or false. They can use T for *true*, F for *false*, and construct a truth table.

p	q	$p \to q$
T	T	T
T	F	F
F	T	T
F	F	T

BY THE WAY...

Over 66 million people live in Madhya Pradesh, a large state in India. The majority of them speak Hindi, but some also speak Urdu and other languages.

Implications

If-then statements are called **implications** and can be represented in symbols. $p \to q$ is read "p implies q" or "if p, then q."

Sample 2

Use the map to determine whether each implication is *True* or *False*.

a. If you live in Madhya Pradesh, then you live in India. ($p \to q$)

b. If you live in India, then you live in Madhya Pradesh. ($q \to p$)

c. If you do not live in Madhya Pradesh, then you do not live in India. (not $p \to$ not q)

d. If you do not live in India, then you do not live in Madhya Pradesh. (not $q \to$ not p)

Sample Response

a. True. The map shows that any person, X, who lives in Madhya Pradesh must also live in India, because Madhya Pradesh is a state in India.

b. False. Many residents of India do not live in Madhya Pradesh. This is the *converse* of the first implication.

c. False. Many people do not live in Madhya Pradesh but still live in India. This is the *inverse* of the original implication.

d. True. Any person, Y, who lives outside India cannot possibly live inside Madhya Pradesh. This is the *contrapositive* of the original implication.

IMPLICATIONS

Original Implication
$p \to q$ If p, then q.

Contrapositive
not $q \to$ not p If not q, then not p.

An implication and its contrapositive are either *both true* or *both false*. They are logically equivalent.

Converse
$q \to p$ If q, then p.

Inverse
not $p \to$ not q If not p, then not q.

The converse and inverse of an implication are not necessarily true just because the implication is true.

Answers to Look Back

Converse: If a quadrilateral is a parallelogram, then the diagonals of the quadrilateral bisect each other; True.

Inverse: If the diagonals of a quadrilateral do not bisect each other, then the quadrilateral is not a parallelogram; True.

Contrapositive: If a quadrilateral is not a parallelogram, then the diagonals of the quadrilateral do not bisect each other; True.

Look Back

State the converse, inverse, and contrapositive of the theorem in this section. Determine which of these are true.

Now you are ready for: Exs. 16–36 on pp. 144–146

3-1 Exercises and Problems

1. In the Exploration and in Sample 1, you used inductive and deductive reasoning. Give an example of each.
2. **Reading** The word *synthetic* has a special meaning in geometry. What is it? Give another meaning for the word *synthetic* in a different context.
3. **Writing** Describe how this object demonstrates something you discovered in the Exploration.

For each pair of triangles in Exercises 4–7:

a. Can you use the given information to prove that two triangles are congruent? Write *Yes* or *No*.

b. If your answer to part (a) is *Yes*, what postulate or theorem would you use to prove the triangles are congruent?

4.
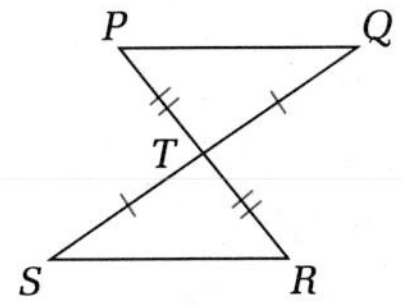

5.
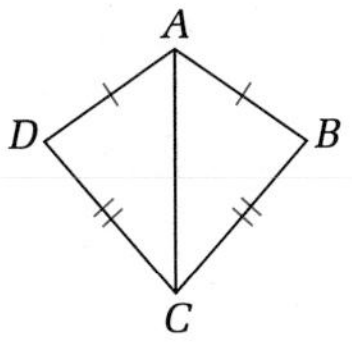

6.
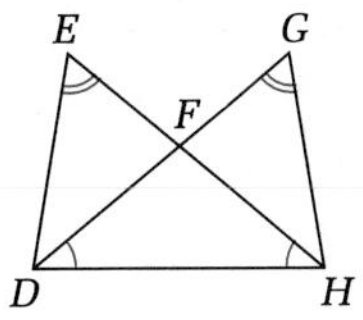

7.
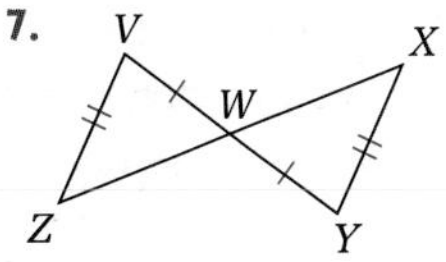

For each postulate or theorem in Exercises 8–10:

a. State the converse.

b. Tell whether the converse is *True* or *False*.

c. If your answer to part (b) is *False*, give a counterexample.

8. If two parallel lines are intersected by a transversal, then corresponding angles are equal in measure.
9. If two lines are intersected by a transversal and alternate interior angles are equal in measure, then the lines are parallel.
10. If a quadrilateral is a rhombus, then its diagonals are perpendicular to each other.

APPLYING

Suggested Assignment

Day 1

Standard 1, 4–15

Extended 1–15

Day 2

Standard 16–20, 23–29, 31–36

Extended 16–20, 23–29, 31–36

Integrating the Strands

Algebra Exs. 31, 35

Functions Ex. 31

Geometry Exs. 1–15, 19, 20, 23, 32–35

Logic and Language Exs. 1–3, 8–30, 36

Answers to Exercises and Problems

1. Answers may vary. An example is given. In the Exploration, conjectures were based on several observations. That is an example of inductive reasoning. The proof in Sample 1 used facts, definitions, and accepted principles and properties to prove a general statement. That is an example of deductive reasoning.
2. In geometry, "synthetic" refers to proofs based on a system of postulates and theorems in which the properties of figures, but not their actual measurements, are used. "Synthetic" may also mean artificial, as in synthetic fur or leather.
3. The seat of the chair is supported by two sets of diagonals of equal length intersecting at their midpoints. These diagonals indicate that corresponding vertices form a rectangle. So, the seat of the chair is parallel to the floor.
4. a. Yes. b. SAS
5. a. Yes. b. SSS
6. a. Yes. b. AAS
7. a. No.
8. a. If two lines are intersected by a transversal and corresponding angles are equal in measure, the lines are parallel.
   b. True.
9. a. If two parallel lines are intersected by a transversal, then alternate interior angles are equal in measure.
   b. True.
10. a. If the diagonals of a quadrilateral are perpendicular to each other, then the quadrilateral is a rhombus.
    b. False.
    c. kite

Communication: Discussion

Students can best review proofs by trying to write them independently and then by discussing them with their classmates. The use of small groups gives students an opportunity to analyze other's proofs and exchange ideas. A whole-class discussion can be used to identify and correct any major problems and to bring closure to the proofs being discussed.

Interdisciplinary Problems

The relationship between implications and their converses, inverses, and contrapositives apply not only to mathematics but also to other disciplines. Exs. 16–18 involve maps as used in geography, while Exs. 25 and 26 use a connection to social studies.

Answers to
Exercises and Problems

11. ❶ definition of parallelogram; ❷ If two ∥ lines are intersected by a transversal, then alternate interior ∠s are equal in measure; ❸ Reflexive property of equality; ❹ ASA; ❺ Corres. parts of ≅ △s are = in measure.

12–15. See answers in back of book.

16–18. Answers may vary. Examples are given.

16. a. If you are at the South Pole, then you are in Antarctica.
b. If you are not in Antarctica, then you are not at the South Pole.

17. a. If you are in Queensland, then you are in Australia.
b. If you are not in Australia, then you are not in Queensland.

18. a. If you are in Bolivia, then you are in South America.
b. If you are not in South America, then you are not in Bolivia.

19. Converse: If two lines form right angles, then the lines are perpendicular; True. Inverse: If two lines are not perpendicular, then the lines do not form right angles; True. Contrapositive: If two lines do not form right angles, then the lines are not perpendicular; True.

11. Give a justification for each step in this flow proof of the statement: If a quadrilateral is a parallelogram, then opposite sides are equal in measure.

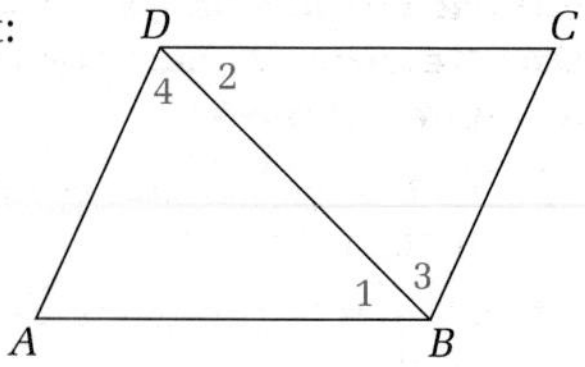

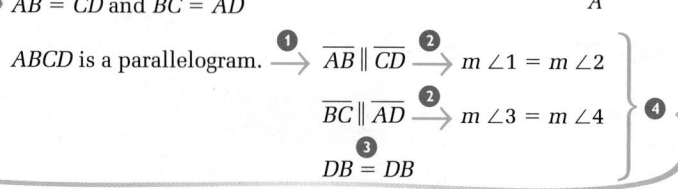

Given $ABCD$ is a parallelogram.

Prove $AB = CD$ and $BC = AD$

$ABCD$ is a parallelogram. $\xrightarrow{❶}$ $\overline{AB} \parallel \overline{CD}$ $\xrightarrow{❷}$ $m\angle 1 = m\angle 2$

$\overline{BC} \parallel \overline{AD}$ $\xrightarrow{❷}$ $m\angle 3 = m\angle 4$

❸ $DB = DB$

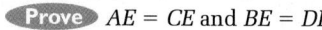

$\xrightarrow{❹}$ $\triangle ABD \cong \triangle CDB$ $\xrightarrow{❺}$ $AB = CD$ and $BC = AD$

12. Rewrite the flow proof in Exercise 11 in two-column form.

13. a. State the converse of the theorem proved in Exercise 11.
b. Prove the converse in a two-column proof. (*Hint:* Use the same pair of congruent triangles and SSS.)

14. Prove that the diagonals of a parallelogram bisect each other.

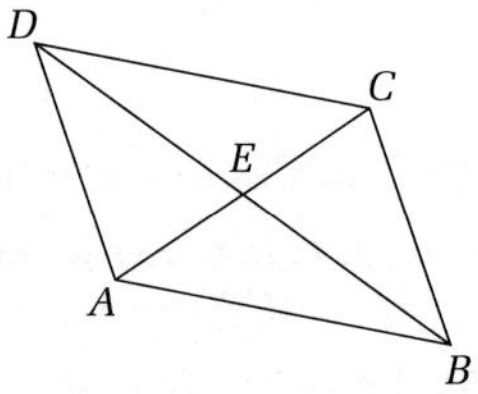

Given Parallelogram $ABCD$ with diagonals intersecting at E

Prove $AE = CE$ and $BE = DE$

Plan Ahead

Prove $\triangle AEB \cong \triangle CED$ by using ASA and the theorem proved in Exercise 11.

15. Prove that if a parallelogram is a rectangle, then its diagonals are equal in measure. Use the theorem proved in Exercise 11.

Geography **For each map in Exercises 16–18:**

a. Write an implication suggested by the map.

b. Write the contrapositive of your implication.

16.

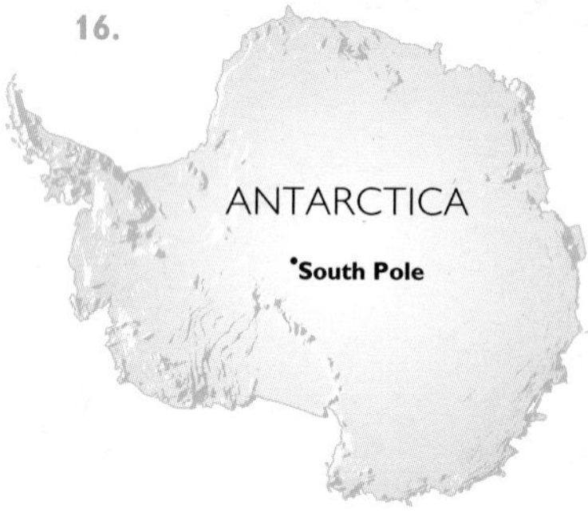

17.

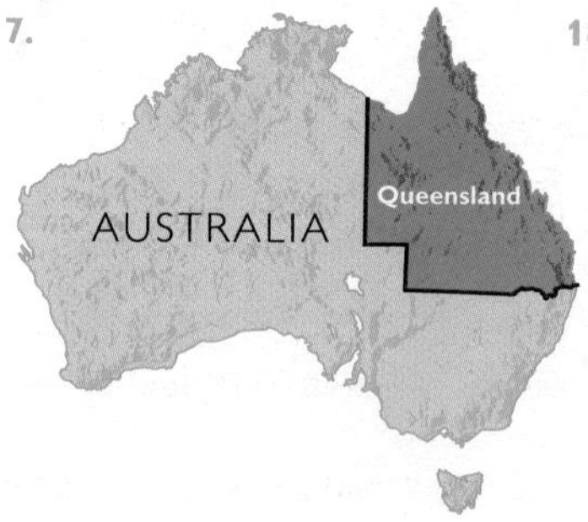

18.

20. Converse: If two triangles are similar, then they are congruent; False; any two similar triangles for which the ratio of the lengths of two sides is not 1:1 are not congruent. Inverse: If two triangles are not congruent, then they are not similar; False; a pair of triangles such as those described in the preceding counterexample are not congruent, but are similar. Contrapositive: If two triangles are not similar, then they are not congruent; True.

21. Converse: If a compound contains carbon, then it is a sugar; False; carbon monoxide contains carbon, but is not a sugar. Inverse: If a compound is not a sugar, then it does not contain carbon; False; carbon dioxide is not a sugar, but contains carbon. Contrapositive: If a compound does not contain carbon, then it is not a sugar; True.

22. Converse: If a football team adds six points to its score, then the team scores a touchdown; False; for example, the team might score two field goals. Inverse: If a football team does not score a touchdown, then the team does not add six points to its score; False; for example, the team might score two

For each implication, state its converse, inverse, and contrapositive. Determine which are *True* and which are *False*. If *False*, give a counterexample.

19. If two lines are perpendicular, then they form right angles.

20. If two triangles are congruent, then they are similar.

21. **Research** If a compound is a sugar, then it contains carbon.

22. **Research** If a football team scores a touchdown, then the team adds six points to its score.

23. **Open-ended** State three theorems that have been previously proved. Make three new statements by forming the contrapositives of the proved theorems. How do you know that the three new statements are also true?

24. **Writing** A friend insists that the converse and inverse of any implication are either both true or both false. Explain why you agree or disagree.

Margaret Chase Smith

BY THE WAY...

More than 150 women have served in Congress. Four women have served in *both* the Senate and House of Representatives:
Margaret Chase Smith of Maine,
Barbara A. Mikulski of Maryland,
Olympia Snowe of Maine, and
Barbara Boxer of California.

connection to **SOCIAL STUDIES**

Article I Section 3 of the United States Constitution states:

No Person shall be a Senator who shall not have attained to the Age of Thirty years.

25. a. Complete this implication: "If a person has not attained to the age of thirty years, then __?__."

b. State the contrapositive of the implication.

26. If a person is a Representative in Congress, then he or she must be at least twenty-five years old. Rewrite this statement in the language used in the United States Constitution.

For Exercises 27–29, let $p \rightarrow q$ *be* the implication "If today is Monday, then tomorrow is Tuesday."

27. Show that the inverse of the converse of the implication $p \rightarrow q$ is the same as the contrapositive of $p \rightarrow q$.

28. Find the converse of the contrapositive of the implication $p \rightarrow q$. What is the relationship of your answer to the original implication $p \rightarrow q$?

29. Find the inverse of the inverse of the implication $p \rightarrow q$. What do you notice?

Error Analysis

If there are an unusual number of errors in Exs. 27–29, check to see whether students are confused by the terminology. Students can be led through one exercise step by step. For example, for Ex. 27:

(1) inverse of (converse of ($p \rightarrow q$))

$q \rightarrow p$

not $q \rightarrow$ not p

(2) contrapositive of ($p \rightarrow q$)

not $q \rightarrow$ not p

For both (1) and (2), the resulting implication is not $q \rightarrow$ not p.

Answers to Exercises and Problems

field goals. Contrapositive: If a football team does not add six points to its score, then the team does not score a touchdown; True.

23. Answers may vary. Examples are given. If two sides of a triangle are equal in measure, then the angles opposite those sides are equal in measure. If the angles opposite two sides of a triangle are not equal in measure, then the two sides are not equal in measure. If two lines are intersected by a transversal and alternate interior angles are equal in measure, then the lines are parallel. If two lines are not parallel, then alternate interior angles formed when the lines are intersected by a transversal are not equal in measure. If two angles form a linear pair, then they are supplementary. If two angles are not supplementary, then they do not form a linear pair. The new statements are all true because an implication and its contrapositive are logically equivalent; that is, either both are true or both are false.

24. The friend is right. The converse and inverse of an implication are logically equivalent. The inverse is the contrapositive of the converse. Implication: $p \rightarrow q$; Converse: $q \rightarrow p$; Inverse: not $p \rightarrow$ not q; Contrapositive of the converse: not $p \rightarrow$ not q

25. a. the person shall not be a Senator

b. If a person is a Senator, then that person is at least thirty years old.

26. No Person shall be a Representative in Congress who shall not have attained to the Age of Twenty-five years.

27. Implication: $p \rightarrow q$ (If today is Monday, then tomorrow is Tuesday.) Converse: $q \rightarrow p$ (If tomorrow is Tuesday, then today is Monday.) Inverse of the converse: not $q \rightarrow$ not p (If tomorrow is not Tuesday, then today is not Monday.) Contrapositive: not $q \rightarrow$ not p (If tomorrow is not Tuesday, then today is not Monday.)

28. Contrapositive: not $q \rightarrow$ not p (If tomorrow is not Tuesday, then today is not Monday.) Converse of the contrapositive: not $p \rightarrow$ not q (If today is not Monday, then tomorrow is not Tuesday.) The converse of the contrapositive is the inverse.

29. Inverse: not $p \rightarrow$ not q (If today is not Monday, then tomorrow is not Tuesday.) Inverse of the inverse: not (not p) $\rightarrow$ not (not q) or $p \rightarrow q$ (If today is Monday, then tomorrow is Tuesday.) The inverse of the inverse is the original implication.

Working on the Unit Project

Students can work independently on Ex. 36. Then, in their project groups of four, they can compare their answers and decide in part (b) which side could use the statement.

Assessment: Portfolio

As students complete the first part of their unit projects, they should note that many times a converse of a statement is used incorrectly as a fact and that this leads to illogical thinking. Students should enter their findings on Ex. 36 into their working portfolios for use at the end of the unit.

Practice 19 For use with Section 3-1

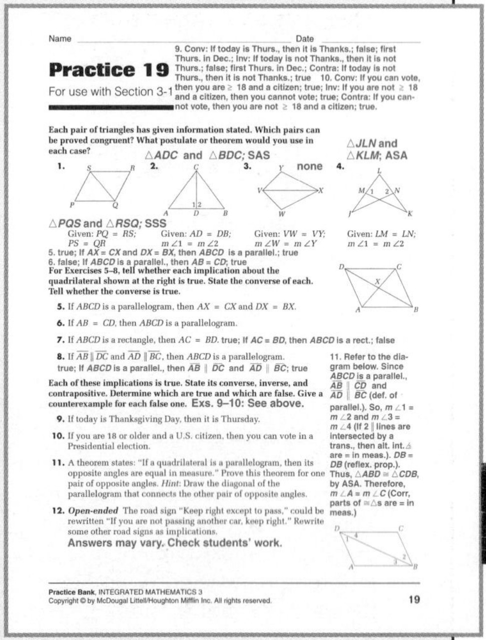

Name ______________________ Date ____________

Practice 19
For use with Section 3-1

9. Conv: If today is Thurs., then it is Thanks.; false; first Thurs. in Dec.; Inv: If today is not Thanks., then it is not Thurs.; false; first Thurs. in Dec.; Contra: If today is not Thurs., then it is not Thanks.; true 10. Conv: If you can vote, then you are ≥ 18 and a citizen; true; Inv: If you are not ≥ 18 and a citizen, then you cannot vote; true; Contra: If you cannot vote, then you are not ≥ 18 and a citizen; true.

Each pair of triangles has given information stated. Which pairs can be proved congruent? What postulate or theorem would you use in each case?

1. $\triangle PQS$ and $\triangle RSQ$; SSS
Given: $PQ = RS$; $PS = QR$

2. $\triangle ADC$ and $\triangle BDC$; SAS
Given: $AD = DB$; $m\angle 1 = m\angle 2$

3. none
Given: $VW = VY$; $m\angle W = m\angle Y$

4. $\triangle JLN$ and $\triangle KLM$; ASA
Given: $LM = LN$; $m\angle 1 = m\angle 2$

5. true; If $AX = CX$ and $DX = BX$, then $ABCD$ is a parallel.; true
6. false; If $ABCD$ is a parallel., then $AB = CD$; true

For Exercises 5–8, tell whether each implication about the quadrilateral shown at the right is true. State the converse of each. Tell whether the converse is true.

5. If $ABCD$ is a parallelogram, then $AX = CX$ and $DX = BX$.
6. If $AB = CD$, then $ABCD$ is a parallelogram.
7. If $ABCD$ is a rectangle, then $AC = BD$. true; If $AC = BD$, then $ABCD$ is a rect.; false
8. If $\overline{AB} \parallel \overline{DC}$ and $\overline{AD} \parallel \overline{BC}$, then $ABCD$ is a parallelogram. true; If $ABCD$ is a parallel., then $\overline{AB} \parallel \overline{DC}$ and $\overline{AD} \parallel \overline{BC}$; true

Each of these implications is true. State its converse, inverse, and contrapositive. Determine which are true and which are false. Give a counterexample for each false one. Exs. 9–10: See above.

9. If today is Thanksgiving Day, then it is Thursday.
10. If you are 18 or older and a U.S. citizen, then you can vote in a Presidential election.
11. A theorem states: "If a quadrilateral is a parallelogram, then its opposite angles are equal in measure." Prove this theorem for one pair of opposite angles. *Hint*: Draw the diagonal of the parallelogram that connects the other pair of opposite angles.
12. *Open-ended* The road sign "Keep right except to pass," could be rewritten "If you are not passing another car, keep right." Rewrite some other road signs as implications.
Answers may vary. Check students' work.

11. Refer to the diagram below. Since $ABCD$ is a parallel., $\overline{AB} \parallel \overline{CD}$ and $\overline{AD} \parallel \overline{BC}$ (def. of parallel.). So, $m\angle 1 = m\angle 2$ and $m\angle 3 = m\angle 4$ (If 2 ∥ lines are intersected by a trans., then alt. int. ∠s are = in meas.). $DB = DB$ (reflex. prop.). Thus, $\triangle ABD \cong \triangle CDB$, by ASA. Therefore, $m\angle A = m\angle C$ (Corr. parts of ≅△s are = in meas.)

Practice Bank, INTEGRATED MATHEMATICS 3

19

Ongoing **ASSESSMENT**

30. **Group Activity** Work in a group of three students. Rotate roles so that you get to play each role.

Student 1: Write an implication but do not show it to the others.

Student 2: Pick one of the three words *converse*, *inverse*, or *contrapositive* and announce it to everyone.

Student 1: Write the converse, inverse, or contrapositive as chosen by Student 2 and give it to Student 3.

Student 3: State the original implication.

Review **PREVIEW**

31. Use the functions $f(x) = x + 1$ and $g(x) = x^2$. Find $(f \circ g)(x)$. *(Section 2-8)*

Find the reflection of each point about the y-axis. *(Toolbox Skill 38)*

32. (4, 6)
33. (−3, 0)
34. (a, b)

35. Find the distance between points $P(6, 0)$ and $Q(0, 3)$. *(Toolbox Skill 31)*

3 Working on the Unit Project

36. A blood sample is often a critical piece of evidence in a trial. Suppose blood found at the scene of a burglary presumably belongs to the burglar. The police announce to the press: "If this blood came from the suspect, then the suspect's blood group matches the blood group of the sample."

a. Write the converse and the contrapositive of the above implication.

b. Which statement, the converse or the contrapositive, could be used most convincingly in a court of law? Which side, the prosecution or the defense, could use that statement?

DNA fingerprinting laboratory

Answers to Exercises and Problems

30. Check students' work.

31. $(f \circ g)(x) = x^2 + 1$

32. (−4, 6) 33. (3, 0)

34. $(-a, b)$ 35. $3\sqrt{5} \approx 6.7$

36. a. Converse: If the suspect's blood group matches the blood group of the sample, then this blood came from the suspect. Contrapositive: If the suspect's blood group does not match the blood group of this sample, then this blood did not come from the suspect.

b. Answers may vary. An example is given. The contrapositive could be used more effectively because it is true, while the converse is false. (Millions of people have the same blood group.) The defense would use the contrapositive (assuming they know that the defendant's blood group does not match the blood group of the sample).

Section **3-2**

Coordinate and Synthetic Proof

Focus

Compare and contrast coordinate and synthetic methods of proof.

WILL YOUR PROOF HOLD UP?

If a roof is large, it may begin to sag under its own weight. There are many ways to support a roof. One is called a *scissors truss.*

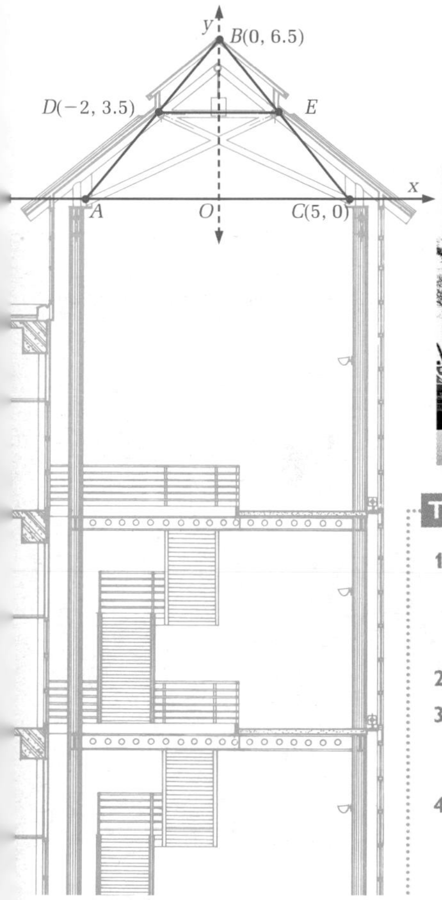

Talk it Over

1. Figure *ADEC* is a trapezoid where $\overline{DE} \parallel \overline{AC}$. Both $\triangle ABC$ and trapezoid *ADEC* have a line of symmetry. $\triangle ABC$ is an isosceles triangle. Do you think trapezoid *ADEC* should be called an *isosceles trapezoid?* Why or why not?
2. What are the coordinates of points *E* and *A*?
3. Support beams $\overline{DC}$ and $\overline{AE}$ are the diagonals of trapezoid *ADEC*. How could you use coordinates to find out whether support beams $\overline{DC}$ and $\overline{AE}$ are the same length?
4. A **median of a triangle** is a segment joining a vertex to the midpoint of the opposite side. How could you use coordinates to determine whether $\overline{DC}$ and $\overline{AE}$ are medians of $\triangle ABC$?

Student Resources Toolbox
p. 654 *Formulas and Relationships*

You can use *coordinate proof* or synthetic proof to show that the medians to the legs of an isosceles triangle are equal in measure. **Coordinate proof** is a proof based on a coordinate system in which all points are represented by ordered pairs of numbers.

PLANNING

Objectives and Strands

See pages 136A and 136B.

Spiral Learning

See page 136B.

Recommended Pacing

Section 3-2 is a one-day lesson.

Toolbox References

- **Toolbox Skill 31:** Using the Distance and Midpoint Formula

Extra Practice

See pages 612–613.

Warm-Up Exercises

Warm-Up Transparency 3-2

Support Materials

- Practice Bank: Practice 20
- Activity Bank: Enrichment 18
- Study Guide: Section 3-2
- Problem Bank: Problem Set 6
- Assessment Book: Quiz 3-2, Alternative Assessment 2

Answers to Talk it Over

1. Answers may vary. An example is given. Yes. I think *ADEC* should be called an isosceles trapezoid because it appears to have two sides that are equal in measure, just like an isosceles triangle.
2. *E*(2, 3.5); *A*(−5, 0)
3. Use the distance formula and the coordinates of *A*, *C*, *D*, and *E*.
4. Use the midpoint formula and the coordinates of *A* and *B* to find the coordinates of the midpoint of $\overline{AB}$ to determine if they are the same as the coordinates of *D* (they are not). Repeat for $\overline{CB}$ and *E*. $\overline{DC}$ and $\overline{AE}$ are not medians.

TEACHING

Talk it Over

Questions 1–4 review the concept of symmetry, introduce isosceles trapezoids and the median of a triangle, and lead students to think about how to use coordinates to establish certain facts.

Additional Sample

Show that if the midpoints of the sides of a square are joined in order, then the resulting quadrilateral is a square.

Method 1. Write a coordinate proof.

Step 1. Every square has four right angles. Place the square in the first quadrant so that two sides intersect at the origin. Choose coordinates for the vertices so as to simplify calculations.

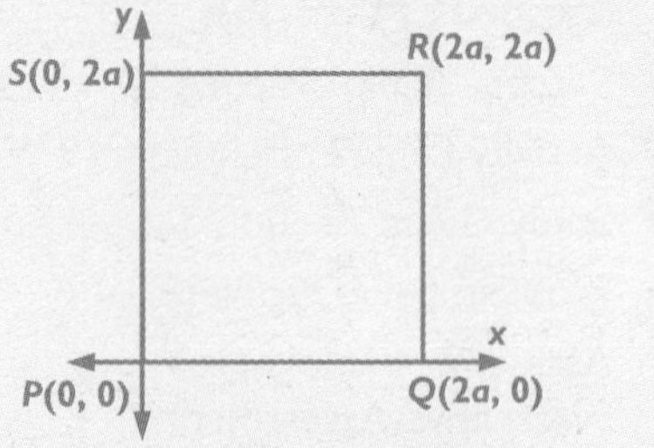

Step 2. Use the formula midpoint $= \left(\frac{x_1 + x_2}{2}, \frac{y_1 + y_2}{2}\right)$ to find the midpoints of the sides of square *PQRS*. The results are shown in the figure below. Join the midpoints in order.

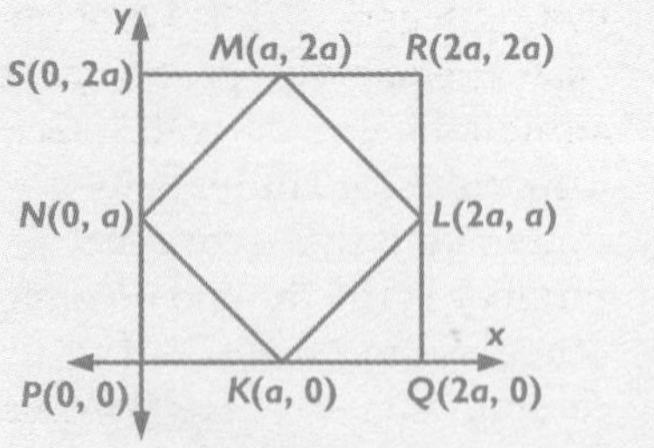

Step 3. Calculate the slope and length of each side of *KLMN*.

slope $\overline{KL} = \frac{a - 0}{2a - a} = 1$

slope $\overline{LM} = \frac{2a - a}{a - 2a} = -1$

slope $\overline{MN} = \frac{a - 2a}{0 - a} = 1$

slope $\overline{NK} = \frac{0 - a}{a - 0} = -1$

Since the product of the slopes of each pair of consecutive sides is −1, these four pairs of sides are perpendicular. Thus, all angles of *KLMN* are right angles.

$KL = \sqrt{(2a - a)^2 + (a - 0)^2} = a\sqrt{2}$

$LM = \sqrt{(a - 2a)^2 + (2a - a)^2} = a\sqrt{2}$

X(ab

JUSTIFICATIONS FOR COORDINATE PROOF

In addition to the justifications allowed for synthetic proofs, you may use the following as justifications in coordinate proofs:

- the distance and midpoint formulas
- Parallel lines have the same slope.
- Perpendicular lines have slopes that are negative reciprocals of each other.
- A geometric figure may be placed anywhere in the coordinate plane.

Sample

Show that the medians drawn to the legs of an isosceles triangle are equal in measure.

Sample Response

Method ❶ Write a coordinate proof.

Step 1 Since every isosceles triangle has a line of symmetry, you can place the triangle on coordinate axes so that the *y*-axis is the line of symmetry. Use variables to describe the coordinates of each vertex.

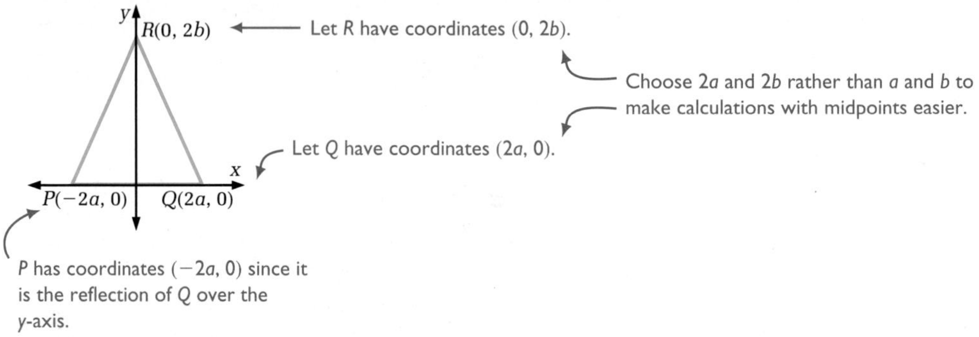

Step 2 Use the midpoint formula to find the midpoints of the legs.

$$\text{Midpoint} = \left(\frac{x_1 + x_2}{2}, \frac{y_1 + y_2}{2}\right)$$

$$\text{Midpoint of } \overline{PR} = T\left(\frac{-2a + 0}{2}, \frac{0 + 2b}{2}\right) = T(-a, b)$$

$$\text{Midpoint of } \overline{QR} = S\left(\frac{2a + 0}{2}, \frac{0 + 2b}{2}\right) = S(a, b)$$

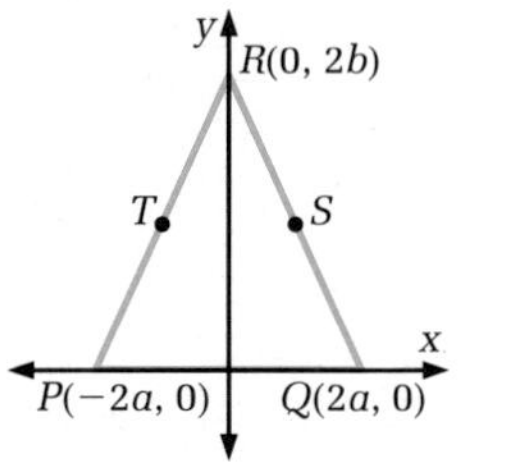

Step 3 The medians are $\overline{QT}$ and $\overline{PS}$. Use the distance formula to show that $QT = PS$.

Distance $= \sqrt{(x_2 - x_1)^2 + (y_2 - y_1)^2}$

$QT = \sqrt{(-a - 2a)^2 + (b - 0)^2} = \sqrt{9a^2 + b^2}$

$PS = \sqrt{[a - (-2a)]^2 + (b - 0)^2} = \sqrt{9a^2 + b^2}$

Therefore $QT = PS$ and the two medians are equal in measure.

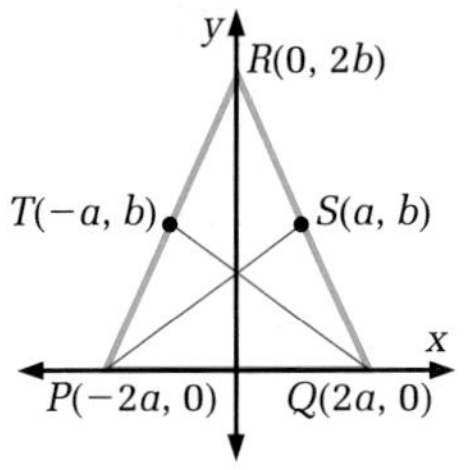

Method ❷ Write a synthetic proof.

Given Isosceles $\triangle PQR$ with $PR = QR$ and medians $\overline{PS}$ and $\overline{QT}$

Prove $PS = QT$

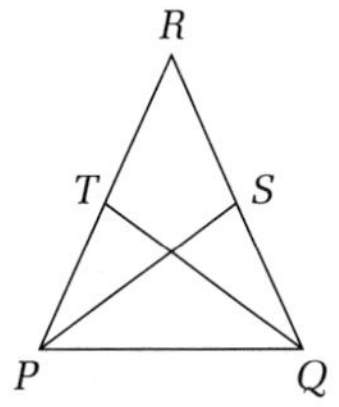

Plan Ahead

Show in a flow proof that $\triangle PQS \cong \triangle QPT$, and then use the fact that corresponding parts of congruent triangles are equal in measure.

Show Your Reasoning

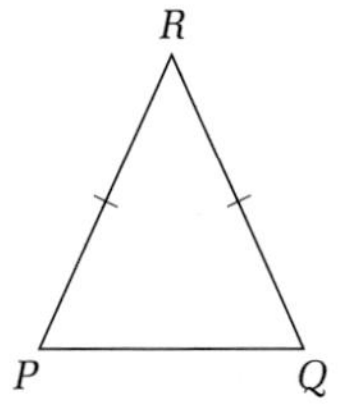

$PR = QR$

→ ❶

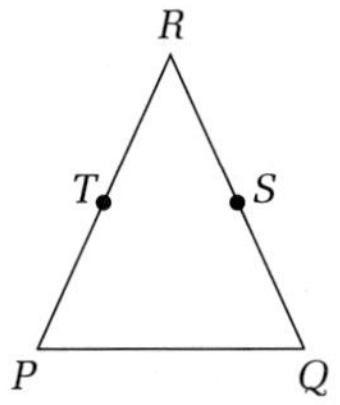

$\frac{PR}{2} = \frac{QR}{2}$

❷ $PT = \frac{PR}{2}$

❷ $SQ = \frac{QR}{2}$

} ❸ →

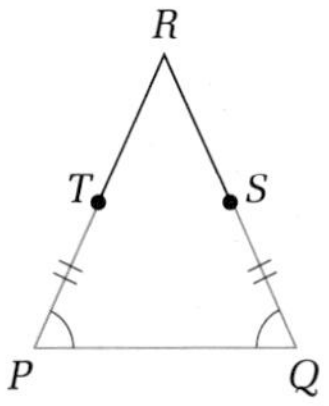

$PT = SQ$

❹ $m\angle P = m\angle Q$

❺ $PQ = PQ$

} ❻ →

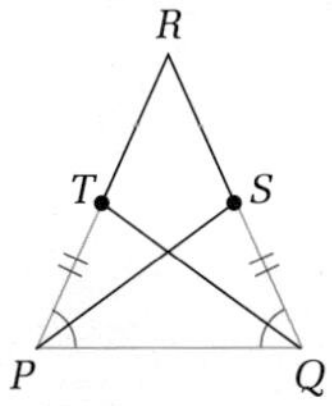

$\triangle PQS \cong \triangle QPT$

→ ❼

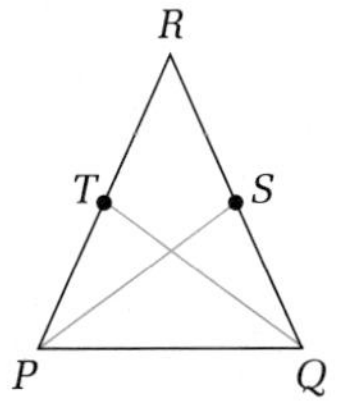

$PS = QT$

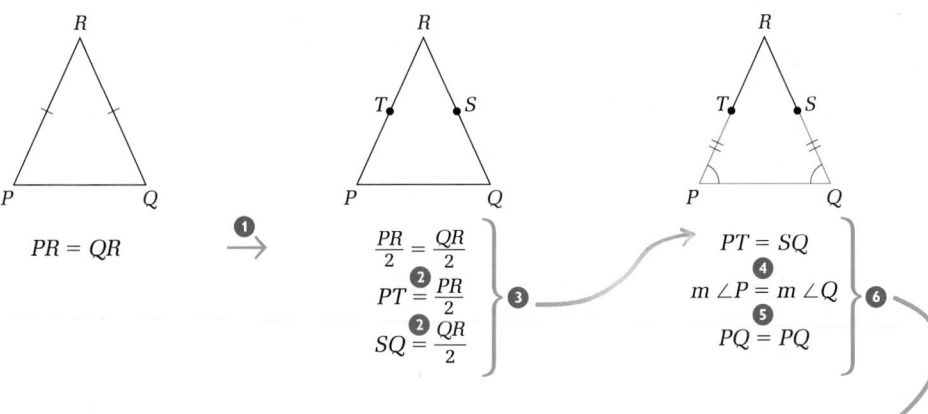

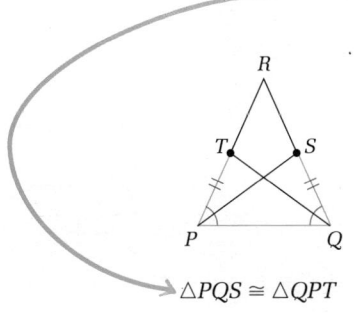

You will provide justifications in Exercise 5.

$MN = \sqrt{(0 - a)^2 + (a - 2a)^2} = a\sqrt{2}$

$NK = \sqrt{(0 - a)^2 + (a - 0)^2} = a\sqrt{2}$

Since all angles of *KLMN* are right angles and all sides have the same length, *KLMN* is a square.

Method 2. Write a synthetic proof.

Given: *PQRS* is a square. *K, L, M,* and *N* are midpoints.

Prove: *KLMN* is a square.

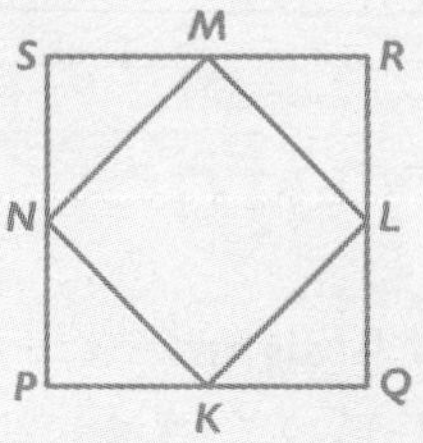

Plan: Prove that $\triangle PKN$, $\triangle QKL$, $\triangle RML$, and $\triangle SNM$ are congruent isosceles right triangles. Conclude that the acute angles of these triangles are 45° angles. Hence, $\angle NKL$, $\angle KLM$, $\angle LMN$, and $\angle MNK$ are right angles. Since $\overline{KL}$, $\overline{LM}$, $\overline{MN}$, and $\overline{NK}$ are corresponding sides of congruent triangles, *KLMN* has four sides of equal measure and four right angles. Hence, *KLMN* is a square.

Teaching Tip

With regard to the note at the right of the diagram in Method I, Step 1 of the Sample, ask students where the use of $2a$ and $2b$ leads to the greatest simplification in calculations. (step 3) There may be a few students who think that $2a$, $-2a$, and $2b$ signify coordinates that are even integers. If that were the case, the analytic proof would suffer a loss of generality. It is important for students to understand that every real number is the double of some other real number. For this reason, the use of these forms for the coordinates is purely a matter of convenience and involves no loss of generality whatsoever.

Reasoning

In the synthetic proof for the Sample, $\triangle PQS$ and $\triangle QPT$ were used to show that $PS = QT$. Ask students if it is possible to use other triangles. (Yes, $\triangle PRS$ and $\triangle QRT$.) What postulate or theorem guarantees that these triangles are congruent? (SAS)

Look Back

Students can respond to these questions verbally as a volunteer lists the responses at the board. When the questions have been answered completely, students should copy the lists from the board into their journals.

APPLYING

Suggested Assignment

Standard 2–11, 13, 14, 16–20

Extended 1–20

Integrating the Strands

Algebra Exs. 1–4, 6–16

Measurement Exs. 1, 3, 7, 8, 10, 11, 15, 16

Geometry Exs. 1–19

Logic and Language Exs. 1, 12, 14, 17–20

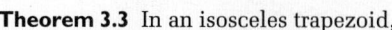

XLab

ISOSCELES TRIANGLES AND ISOSCELES TRAPEZOIDS

Theorem 3.2 In an isosceles triangle, the medians drawn to the legs are equal in measure.

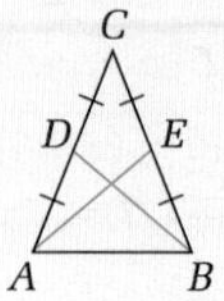

Definition An isosceles trapezoid is a trapezoid with a line of symmetry that passes through the midpoints of the bases.

Theorem 3.3 In an isosceles trapezoid,

(1) the legs are equal in measure,

(2) the diagonals are equal in measure, and

(3) the two angles at each base are equal in measure.

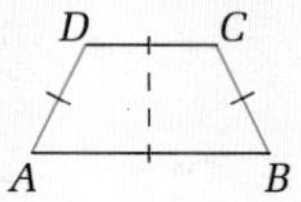

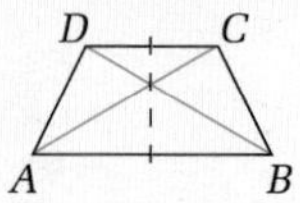

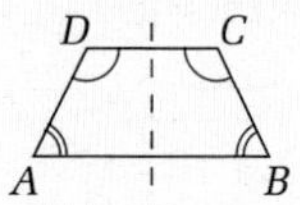

You will use both coordinate and synthetic methods in Exercise 11 to prove Theorem 3.3.

Look Back

In what ways are isosceles triangles and isosceles trapezoids alike? In what ways are they different?

3-2 Exercises and Problems

1. **Reading** How can you use coordinate geometry to show that two segments are equal in measure?
2. In pentagon *GHJKL*, the *y*-axis is a line of symmetry. Find the coordinates of *H* and *L*.

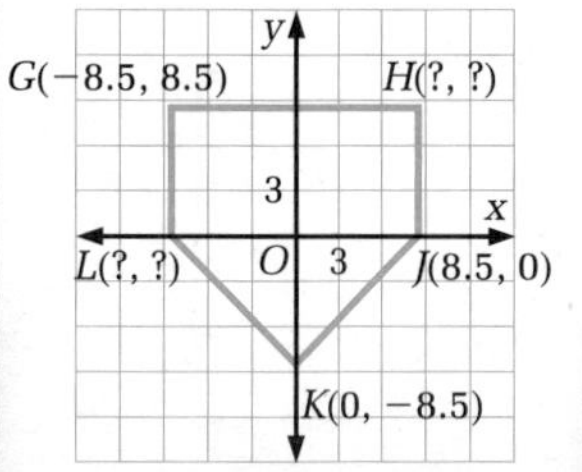

(Measurements are in inches.)

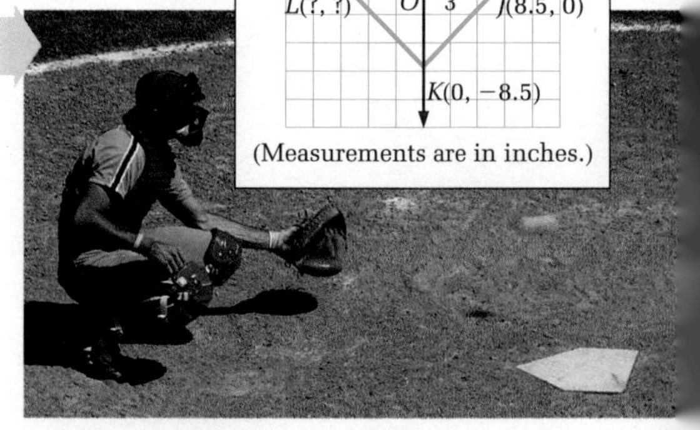

Architecture Use the illustration of the scissors truss on page 147.

3. Use coordinates to determine whether support beams $\overline{DC}$ and $\overline{AE}$ are the same length.
4. Use coordinates to determine whether $\overline{DC}$ and $\overline{AE}$ are medians of $\triangle ABC$.
5. Provide justifications for the statements of the synthetic proof in Method 2 of the Sample on page 149.

Answers to Look Back

Answers may vary. Examples are given. Isosceles triangles and isosceles trapezoids both have a line of symmetry and at least two sides equal in measure. An isosceles triangle has one pair of base angles that are equal in measure. An isosceles trapezoid has two pairs of base angles that are equal in measure. An isosceles trapezoid has two diagonals that are equal in measure. An isosceles triangle has no diagonals. An isosceles triangle has three sides, while an isosceles trapezoid has four.

Answers to Exercises and Problems

1. Use the coordinates of the given points and the distance formula.
2. *H*(8.5, 8.5); *L*(−8.5, 0)
3. $\overline{DC}$ has endpoints (−2, 3.5) and (5, 0) and length $\sqrt{(5-(2))^2 + (0-3.5)^2} = \sqrt{61.25}$. *AE* has endpoints (2, 3.5) and (−5, 0) and length $\sqrt{(2-(-5))^2 + (3.5-0)^2} = \sqrt{61.25}$. The supports are the same length.
4. By the distance formula, $\overline{BD} = \sqrt{13}$ and $\overline{AD} = \sqrt{21.25}$, so *D* is not the midpoint of $\overline{BA}$ and $\overline{DC}$ is not a median. $\overline{BE} = \sqrt{13}$ and $\overline{EC} = \sqrt{21.25}$, so *E* is not the midpoint of $\overline{BC}$ and $\overline{AE}$ is not a median.
5. ❶ Division property of equality; ❷ Definition of median; ❸ Substitution property; ❹ Base ∠s of an isosceles triangle are = in measure; ❺ Reflexive property of equality; ❻ SAS; ❼ Corres. parts of ≅ △s are = in measure.
6. Yes; if *PQRS* were folded over the *y*-axis, one half would fit exactly over the other.

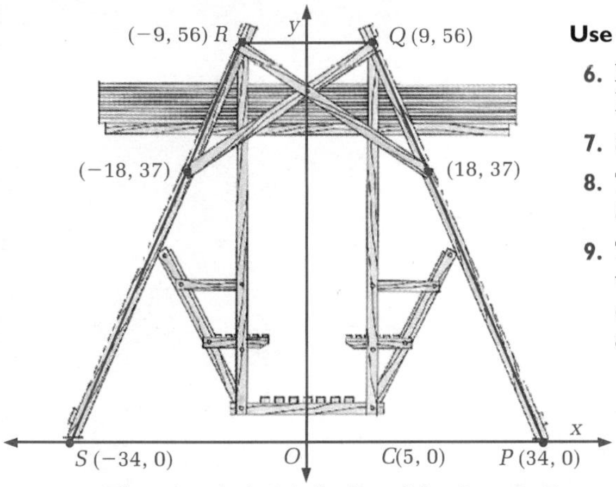

Use this picture of a swing.

6. Does the frame of the swing at the left have a line of symmetry? Justify your answer.
7. Show that the sides are the same length.
8. The diagonal supports keep the frame steady. Show that the supports are the same length.
9. The frame would be sturdier if the supports were at the diagonals of trapezoid $PQRS$. Why do you think that the supports were not placed at the diagonals?

10. The triangle in Method 1 of the Sample Response on page 148 has coordinates $P(-2a, 0)$, $Q(2a, 0)$, and $R(0, 2b)$. Show that this triangle is isosceles by showing that $PR = QR$.

11. In isosceles trapezoid $PQRS$ the y-axis is a line of symmetry that passes through the midpoints of the bases.
 a. Use the coordinates of isosceles trapezoid $PQRS$ to show that $\overline{PS}$ and $\overline{QR}$ are equal in measure. (Part (1) of Theorem 3.3)
 b. Show that the diagonals, $\overline{PR}$ and $\overline{QS}$, are equal in measure. (Part (2) of Theorem 3.3)
 c. Use pairs of overlapping congruent triangles in a synthetic proof to show that $m\angle SPQ = m\angle RQP$ and $m\angle QRS = m\angle PSR$. (Part (3) of Theorem 3.3.)

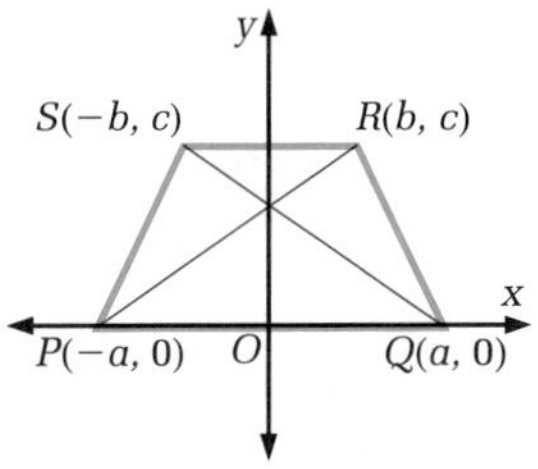

12. **Writing** Explain why it would be difficult to write a coordinate proof of part (3) of Theorem 3.3.

Start with general coordinates for a quadrilateral.

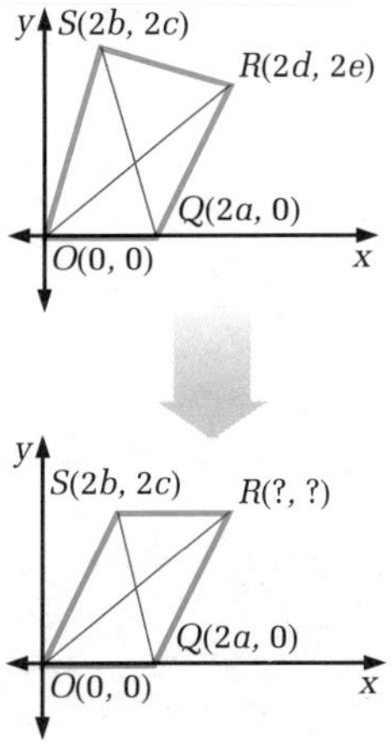

13. In this exercise you will write a coordinate proof of Theorem 3.1: If the two diagonals of a quadrilateral bisect each other, then the quadrilateral is a parallelogram.
 a. Find the coordinates of the midpoint of $\overline{OR}$.
 b. Find the coordinates of the midpoint of $\overline{QS}$.
 c. $\overline{OR}$ bisects $\overline{QS}$, so their midpoints have the same coordinates. Complete:

 $d = \underline{\ ?\ }$ and $e = \underline{\ ?\ }$.

 d. Rewrite the coordinates of R in terms of a, b, and c. Then show that quadrilateral $OQRS$ is a parallelogram. (*Hint:* Show that opposite sides are parallel because they have the same slope.)

Mathematical Procedures

For Ex. 11, it may help students to reflect on how analytic proofs and synthetic proofs compare with one another. In analytic proofs, the "given" is embodied in the way the figure is placed in the coordinate plane and coordinates are chosen. Once these decisions are made, the course of the proof is usually predictable. You perform the calculations to establish the desired relationships. For synthetic proofs, there are often alternate approaches. This is one reason why a brief statement of a plan for a proof is helpful. The plan focuses attention on the particular approach to be used and helps the person writing the proof to work through all the details.

Integrating the Strands

Coordinate proofs integrate the strands of algebra, geometry, measurement, and logic and language.

Answers to Exercises and Problems

7. One side has endpoints (−9, 56) and (−34, 0) and length $\sqrt{(-9-(-34))^2 + (56-0)^2} = \sqrt{3761}$. The other side has endpoints (9, 56) and (34, 0) and length $\sqrt{(9-34)^2 + (56-0)^2} = \sqrt{3761}$. The sides are the same length.

8. One support has endpoints (−9, 56) and (18, 37) and length $\sqrt{(-9-18)^2 + (56-37)^2} = \sqrt{1090}$. The other support has endpoints (−18, 37) and (9, 56) and length $\sqrt{(-18-9)^2 + (37-56)^2} = \sqrt{1090}$. The supports are the same length.

9. It would make it difficult, if not impossible, to get into the swing.

10. $PR = \sqrt{(0-(-2a))^2 + (2b-0)^2} = \sqrt{4a^2 + 4b^2} = 2\sqrt{a^2 + b^2}$; $QR = \sqrt{(0-2a)^2 + (2b-0)^2} = \sqrt{4a^2 + 4b^2} = 2\sqrt{a^2 + b^2}$

11, 12. See answers in back of book.

13.

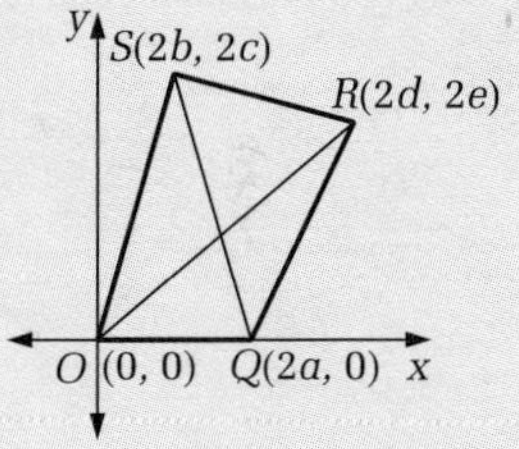

a. $\left(\frac{2d+0}{2}, \frac{2e+0}{2}\right) = (d, e)$

b. $\left(\frac{2a+2b}{2}, \frac{2c}{2}\right) = (a+b, c)$

c. $a + b$; c

d.

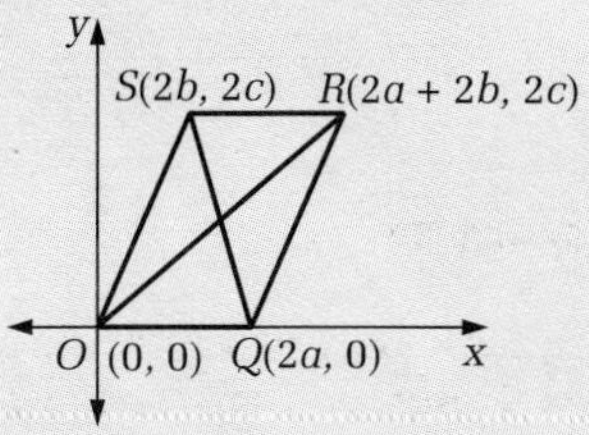

The coordinates of R are $(2a + 2b, 2c)$. $\overline{OQ}$ is on the x-axis, so the slope of $\overline{OQ}$ is 0. The slope of $\overline{RS}$ is $\frac{2c-2c}{2a+2b-2b} = \frac{0}{2a} = 0$, so $\overline{OQ} \parallel \overline{RS}$. The slope of $\overline{SO}$ is $\frac{2c-0}{2b-0} = \frac{c}{b}$. The slope of $\overline{RQ}$ is $\frac{2c-0}{2a+2b-2b} = \frac{c}{b}$. Then $\overline{SO} \parallel \overline{RQ}$ and $OQRS$ is a parallelogram.

Communication: Writing

When comparing and contrasting coordinate and synthetic proofs, as in Ex. 14, students will begin to appreciate the fact that sometimes one method of proof may be much easier to use than another.

Multicultural Note

The Kpelle people form the single largest ethnic group in contemporary Liberia. When Liberia was founded in 1847 by freed African Americans from the United States, the Kpelle people already resided in a large portion of the land. They are thought to have moved into the region from the north in the sixteenth century. Of the approximate 2,580,000 people living in Liberia today, more than 10% are Kpelle.

Practice 20 For use with Section 3-2

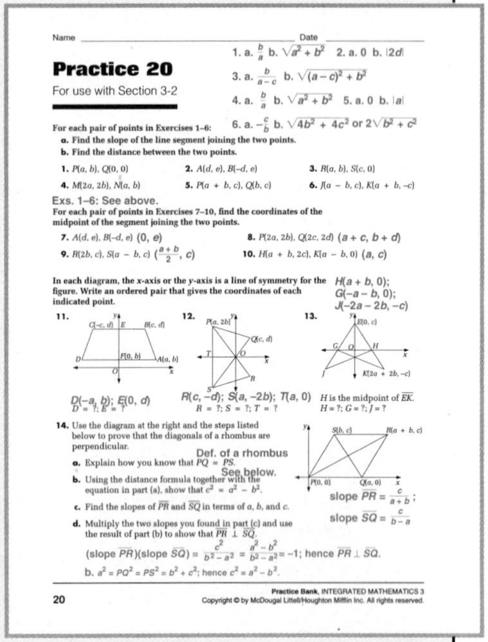

Name ______________ Date ______________

Practice 20

For use with Section 3-2

1. a. $\frac{b}{a}$ b. $\sqrt{a^2 + b^2}$ 2. a. 0 b. $|2d|$
3. a. $\frac{b}{a-c}$ b. $\sqrt{(a-c)^2 + b^2}$
4. a. $\frac{b}{a}$ b. $\sqrt{a^2 + b^2}$ 5. a. 0 b. $|a|$
6. a. $-\frac{c}{b}$ b. $\sqrt{4b^2 + 4c^2}$ or $2\sqrt{b^2 + c^2}$

For each pair of points in Exercises 1–6:
a. Find the slope of the line segment joining the two points.
b. Find the distance between the two points.

1. $P(a, b)$, $Q(0, 0)$ 2. $A(d, e)$, $B(-d, e)$ 3. $R(a, b)$, $S(c, 0)$
4. $M(2a, 2b)$, $N(a, b)$ 5. $P(a + b, c)$, $Q(b, c)$ 6. $J(a - b, c)$, $K(a + b, -c)$

Exs. 1–6: See above.

For each pair of points in Exercises 7–10, find the coordinates of the midpoint of the segment joining the two points.

7. $A(d, e)$, $B(-d, e)$ $(0, e)$ 8. $P(2a, 2b)$, $Q(2c, 2d)$ $(a + c, b + d)$
9. $R(2b, c)$, $S(a - b, c)$ $(\frac{a+b}{2}, c)$ 10. $H(a + b, 2c)$, $K(a - b, 0)$ (a, c)

In each diagram, the x-axis or the y-axis is a line of symmetry for the figure. Write an ordered pair that gives the coordinates of each indicated point.

11. $D = ?$; $E = ?$ — $D(-a, b)$; $E(0, d)$
12. $R = ?$; $S = ?$; $T = ?$ — $R(c, -d)$; $S(a, -2b)$; $T(a, 0)$
13. H is the midpoint of $\overline{EK}$. $H = ?$; $G = ?$; $J = ?$ — $H(a + b, 0)$; $G(-a - b, 0)$; $J(-2a - 2b, -c)$

14. Use the diagram at the right and the steps listed below to prove that the diagonals of a rhombus are perpendicular.
   a. Explain how you know that $PQ = PS$. Def. of a rhombus
   b. Using the distance formula together with the equation in part (a), show that $c^2 = a^2 - b^2$. See below.
   c. Find the slopes of $\overline{PR}$ and $\overline{SQ}$ in terms of a, b, and c. slope $\overline{PR} = \frac{c}{a+b}$; slope $\overline{SQ} = \frac{c}{b-a}$
   d. Multiply the two slopes you found in part (c) and use the result of part (b) to show that $\overline{PR} \perp \overline{SQ}$.

(slope $\overline{PR}$)(slope $\overline{SQ}$) $= \frac{c^2}{b^2 - a^2} = \frac{a^2 - b^2}{b^2 - a^2} = -1$; hence $\overline{PR} \perp \overline{SQ}$.
b. $a^2 = PQ^2 = PS^2 = b^2 + c^2$; hence $c^2 = a^2 - b^2$.

20 Practice Bank, INTEGRATED MATHEMATICS 3

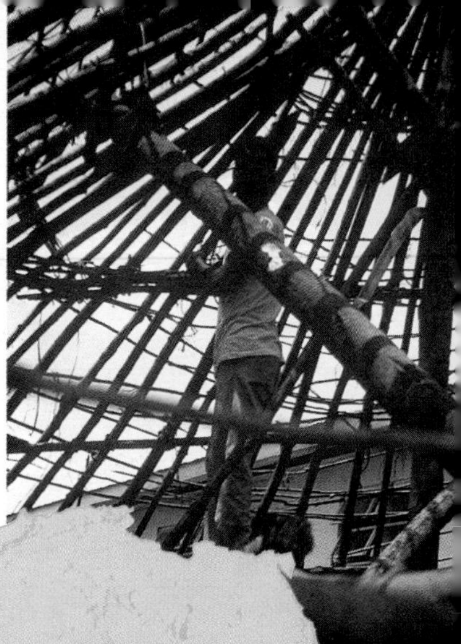

14. **Writing** Which proof of Theorem 3.1 do you prefer, the synthetic proof on pages 140 and 141 of Section 3-1, or the coordinate proof in Exercise 13? Why?

15. Architecture The Kpelle people of Liberia build round houses and rectangular houses of wood. To make a rectangle, they lay two poles of equal length on the ground, one across the other intersecting at their midpoints. The ends of the poles are the corners of a rectangle.

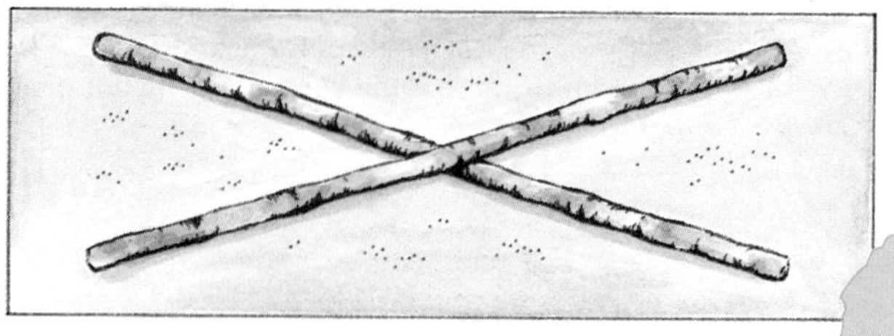

   a. Write a conjecture that justifies the Kpelle construction method.
   b. Prove your conjecture from part (a).

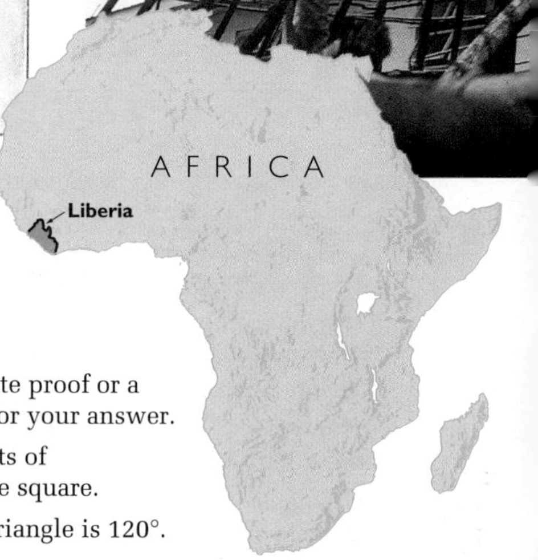

Ongoing ASSESSMENT

16. **Open-ended** Decide whether you would use a coordinate proof or a synthetic proof to prove each conjecture. Give a reason for your answer.
   a. In every square, a segment that connects the midpoints of two opposite sides has the same length as a side of the square.
   b. The measure of each exterior angle of an equilateral triangle is 120°.

Review PREVIEW

17. Write the contrapositive of the following statement: "If a quadrilateral is a square, then it is a rectangle." *(Section 3-1)*

18. Write true sentences of the form "All _?_ are _?_" using the following pairs of words: *(Toolbox Skill 34, Glossary)*
   a. rhombuses, squares
   b. parallelograms, quadrilaterals
   c. parallelograms, rectangles

19. Write a definition for the word *parallelogram* using the phrase "if and only if." *(Toolbox Skill 34, Glossary)*

3 Working on the Unit Project

20. In a courtroom, as in mathematics, there is often more than one way to argue a case. Outline two or more plans for how you can argue your case.

Answers to Exercises and Problems

14. Answers may vary. An example is given. I prefer the synthetic proof because I don't like working with coordinates.
15. See answers in back of book.
16. Answers may vary. Examples are given.
   a. I think I could use coordinates to prove this because I could use the standard coordinates for a square, then use the midpoint formula to express the coordinates of the endpoints of the segment in terms of the coordinates of the vertices of the square. Then I could use the distance formula to show that the lengths of the sides and the segment are equal.
   b. I would use a synthetic proof. I don't think I could use coordinates to prove this, because I wouldn't know how to use coordinates to express the measure of either the interior or exterior angles of the triangle.
17. If a quadrilateral is not a rectangle, then it is not a square.
18. a. All squares are rhombuses.
   b. All parallelograms are quadrilaterals.
   c. All rectangles are parallelograms.
19. Answers may vary. An example is given. A quadrilateral is a parallelogram if and only if both pairs of opposite sides are parallel.
20. Answers may vary.

Section

3-3 Another Look at Definitions

Focus
Understand the need for clear definitions. Reconsider the definition of trapezoid.

On Your Own Terms

To attract customers, the manager of a bookstore sets up a small refreshment counter and sells hot drinks, lemonade, and pastry. The manager has provided tables and chairs in a corner of the store where customers can eat a snack and enjoy a book.

> No commercial establishment may sell meals to be consumed on its premises without having obtained a restaurant license from the Department of Public Health.

The City Health Inspector visits the bookstore and tells the manager to apply for a restaurant license. The inspector cites a city law.

The manager tells the inspector that a bookstore operator does not need a restaurant license. They take the matter to court and argue over how the city law should be interpreted.

Talk it Over

1. Read the section of the city law cited above. List the terms whose definition you must know to decide whether the bookstore manager must apply for a restaurant license.
2. Write a definition for each of the terms listed in question 1.
3. On the basis of your definitions, decide whether the bookstore manager must apply for a restaurant license.
4. Compare your decision with the decisions of your classmates.

PLANNING

Objectives and Strands
See pages 136A and 136B.

Spiral Learning
See page 136B.

Recommended Pacing
Section 3-3 is a one-day lesson.

Extra Practice
See pages 612–613.

Warm-Up Exercises
Warm-Up Transparency 3-3

Support Materials
- Practice Bank: Practice 21
- Activity Bank: Enrichment 19
- Study Guide: Section 3-3
- Problem Bank: Problem Set 6
- Assessment Book: Quiz 3-3, Test 10

Answers to Talk it Over

1–3. Answers may vary. Examples are given.

1. Among the terms that may need to be defined in order to interpret the city ordinance are: *commercial establishment, meals,* and *premises.*
2. A commercial establishment could be defined as any place of business that offers goods or services for sale. The bookstore is a commercial establishment. A meal could be defined as any food consumed in one sitting or as a substantial serving of food as opposed to a snack or light refreshment. Premises could be defined as land owned by a business and the buildings on the land, or as the building occupied by a business.
3. If a meal is defined as any food consumed in one sitting and premises are defined as land owned by a business and the buildings on the land, then the bookstore manager must apply for a license. If a meal is defined as a substantial serving of food as opposed to a snack or light refreshment or if premises are defined as only the building occupied by a business, then the owner does not have to apply for a license.
4. Answers may vary.

TEACHING

Students Acquiring English

You may want to invite students acquiring English to define the verbs *include* and *exclude* in order to highlight the definitions of the mathematical concepts *inclusive* and *exclusive.*

Talk it Over

Questions 1–4 lead students to appreciate the fact that it is necessary to agree upon the definitions of key terms.

Questions 5–7 introduce the inherent ambiguity in the original definition of trapezoid. Students will see that two interpretations are possible. This leads to an inclusive definition (at least one pair of parallel sides) and an exclusive definition (exactly one pair of parallel sides).

For questions 8–11, students apply the inclusive definition of trapezoid.

Mathematical Procedures

The use of inclusive definitions has practical and theoretical value in mathematics. It helps prevent proliferation of special theorems and stresses connections between categories of mathematical objects.

Teaching Tip

Ask students for other examples of inclusive definitions they have encountered in their study of mathematics. Possible examples are isosceles triangle (to include equilateral triangles) and rectangle (to include squares). Ask if you get an inclusive or exclusive definition if you define real numbers to be numbers that can be represented by infinite decimals. Does the definition include rational numbers that can be written as terminating decimals? (Yes, since terminating decimals such as 1.37 cannot only be represented as 1.370000... but also as 1.369999..., with a nonzero repeating digit.)

Inclusive and Exclusive Definitions

Definitions are important in interpreting matters of law. In mathematics, definitions are equally important. A good mathematical definition should be clear and easy to interpret.

A trapezoid is a quadrilateral with one pair of parallel sides.

In quadrilateral *MNRS*, $\overline{MN} \parallel \overline{SR}$. Is *MNRS* a trapezoid? The discussion of the students below shows that there may not be a simple answer.

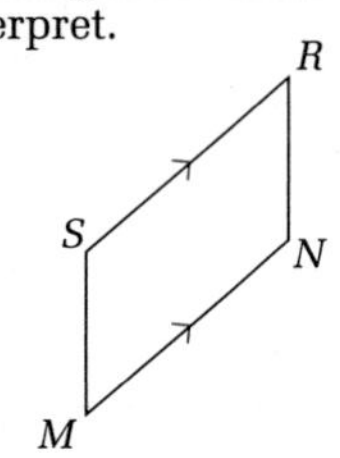

MNRS must be a trapezoid, because it does have one pair of parallel sides, $\overline{MN}$ and $\overline{SR}$.
But MNRS could have another pair of parallel sides, $\overline{MS}$ and $\overline{NR}$. I'd call *MNRS* a parallelogram.

Talk it Over

5. Do you agree with what Silas says about *MNRS*?
6. Do you agree with Carly's response?
7. Silas tells Carly, "A parallelogram has *at least* one pair of parallel sides, so it is a trapezoid." Do you agree or disagree? Explain.

Using Silas's definition of *trapezoid*, which of these statements are true?

8. All parallelograms are trapezoids.
9. All rectangles are trapezoids.
10. All squares are trapezoids.
11. All kites are trapezoids.

Carly interprets "one" to mean "exactly one," but Silas says it means "at least one." To settle the dispute you must agree on which interpretation to use. This book will use this definition:

> A **trapezoid** is a quadrilateral with *at least one* pair of parallel sides.

With this definition, you can conclude that *MNRS* must be a trapezoid, regardless of whether $\overline{MS}$ and $\overline{NR}$ are also parallel.

Because the new definition of trapezoid *includes* parallelograms as trapezoids, it is called an **inclusive definition.** Carly's definition *excludes* parallelograms and is called an **exclusive definition.**

Answers to Talk it Over

5–7. Answers may vary. Examples are given.

5. Yes; *MNRS* fits the given definition of trapezoid.
6. Yes; the definition does not exclude the possibility of two pairs of parallel sides.
7. disagree; I don't think a quadrilateral can be both a parallelogram and a trapezoid.
8. True.
9. True.
10. True.
11. False.

A Quadrilateral Chart

With the inclusive definition of trapezoid, a parallelogram may be considered as one kind of trapezoid. In the previous section you saw examples of another special trapezoid, the isosceles trapezoid.

Sample

Explain why this statement is true: If a quadrilateral is a rectangle, then it is an isosceles trapezoid.

Sample Response

Suppose the rectangle is named *PQRS*. All rectangles are parallelograms, and all parallelograms are trapezoids. So *PQRS* is a trapezoid.

You can place *PQRS* so that the *y*-axis is a line of symmetry through the midpoints of $\overline{PQ}$ and $\overline{SR}$. So *PQRS* is an isosceles trapezoid.

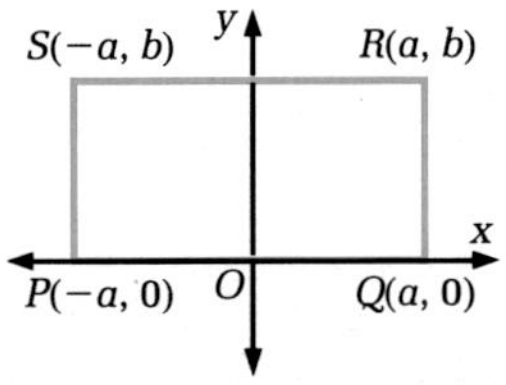

This chart shows the relationships among the quadrilaterals you have now studied. Figures in the quadrilateral chart inherit all the properties of figures above them to which they are linked.

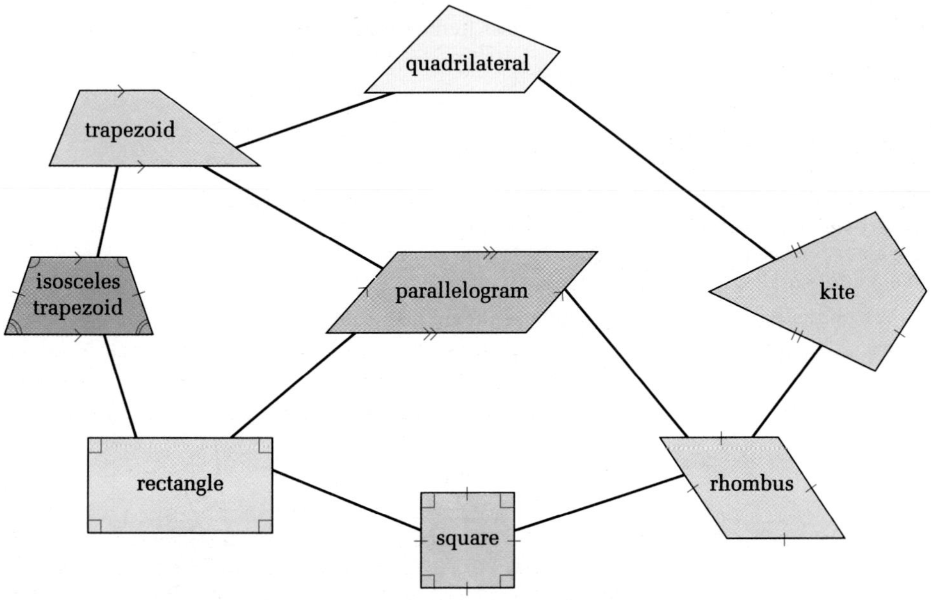

Additional Sample

Explain why this statement is true: If a quadrilateral is a square, then it is an isosceles trapezoid.

Every square is a rectangle. The Sample Response shows that every rectangle is an isosceles trapezoid. Therefore, every square is an isosceles trapezoid.

Reasoning

In connection with the Sample, students can consider what is meant by the bases of a trapezoid. Ask them whether it is possible for a trapezoid to have more than one pair of bases. (Yes.) If so, what kind of trapezoid would this be? (parallelogram)

Visual Thinking

Check students' understanding by asking them to verbally explain the diagram on this page. Ask them to identify which of the figures contain right, acute, and obtuse angles. This activity involves the visual skills of *recognition* and *interpretation*.

Look Back

A class discussion of this question with examples will help all students gain a deeper understanding of the fundamental importance of having clear definitions.

APPLYING

Suggested Assignment

Standard 1–7, 11–24

Extended 1–24

Integrating the Strands

Number Ex. 5

Algebra Exs. 4, 21

Functions Ex. 10

Geometry Exs. 6, 7, 11–21, 23

Discrete Mathematics Ex. 22

Logic and Language Exs. 1–20, 22, 24

Interdisciplinary Problems

The need for clear definitions exists in all subject matter disciplines as well as in other areas of life where verbal and written communications are essential. Exs. 1–3, 8, 9, and 24 provide nonmathematical situations for students to consider.

Look Back

Why is it important to have clear definitions? Give mathematical and nonmathematical examples to explain your answer.

3-3 Exercises and Problems

1. **Driver's Licenses** Suppose the law where you live reads as follows:

   - No person shall be eligible for a driver's license if he or she is less than 16 years old.
   - No person under the age of 18 shall be eligible for a driver's license without having successfully completed a state-approved course in driver education, consisting of at least 30 hours of instruction, under the supervision of a qualified instructor.

   a. Identify the words or phrases in this law that you feel need to be well defined in order for this law to be interpreted.

   b. Write a good definition for one of the words or phrases you listed in part (a).

For each sentence in Exercises 2–6, tell whether the word *one* means "exactly one" or "at least one."

2. To graduate from high school you must pass one art course.
3. Each senior is entitled to one free guest pass to the senior luncheon.
4. The equation $2x + 5 = 9$ has one solution.
5. An even number has one factor that is a power of two.
6. Through a point not on a line, one line may be drawn perpendicular to the given line.
7. This book uses an *inclusive* definition of trapezoid: A trapezoid is a quadrilateral with *at least one* pair of parallel sides. Some other books use an *exclusive* definition: A trapezoid is a quadrilateral with *exactly one* pair of parallel sides.

   a. Which figures are trapezoids if you use the inclusive definition?

   b. Which figures are trapezoids if you use the exclusive definition?

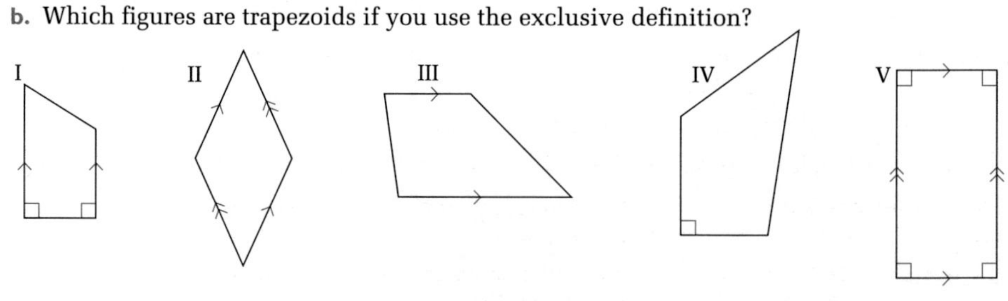

Answers to Look Back

Explanations may vary. Examples are given. It is important that words be clearly defined because it is difficult to exchange ideas and information or to follow rules if the words we are using do not have clear and generally accepted meanings. For example, suppose a sign at a swimming pool reads "Caps must be worn with long hair." How is "long hair" defined? Is it below the ears or is it shoulder length or longer? In mathematics, if you want to prove a property of all trapezoids, it is important that you know which of the definitions given on page 154 is appropriate.

Answers to Exercises and Problems

1. Answers may vary. Examples are given.

   a. Among the terms or phrases which may need to be defined in order to interpret the law are: *successfully completed, state-approved course in driver's instruction,* and *qualified instructor.*

   b. A qualified instructor is a person who satisfies the requirements established by the state department of motor vehicles for a driver-education instructor and is licensed by the state.

2. at least one
3. exactly one
4. exactly one

Exercises 8 and 9 contain applications of inclusive and exclusive definitions in two situations.

8. **Movie Theaters** A movie theater charges one rate for children and another rate for teenagers. Children are defined as people who are at least 6 years old but under 13. Teenagers are from age 13 through age 19. Does the definition of "child" include or exclude "teenager"?

9. **Open-ended** Write an inclusive definition for "packaging that uses recycled material." Then write an exclusive definition.

10. **Reading** Review the definitions of *polynomial function* and *rational function* from Unit 2. Does the set of rational functions include or exclude the set of polynomial functions?

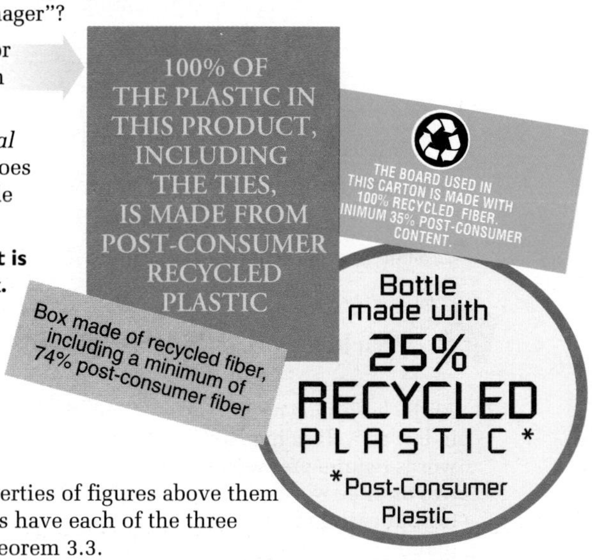

100% OF THE PLASTIC IN THIS PRODUCT, INCLUDING THE TIES, IS MADE FROM POST-CONSUMER RECYCLED PLASTIC
THE BOARD USED IN THIS CARTON IS MADE WITH 100% RECYCLED FIBER. MINIMUM 35% POST-CONSUMER CONTENT.
Box made of recycled fiber, including a minimum of 74% post-consumer fiber
Bottle made with 25% RECYCLED PLASTIC*
*Post-Consumer Plastic

For Exercises 11–14, tell whether each statement is *True* or *False*, using the definitions from this book.

11. All squares are isosceles trapezoids.

12. Every trapezoid is a parallelogram.

13. If a kite is a rhombus, then it is a square.

14. If a quadrilateral is a rectangle, then it is a trapezoid.

15. Figures in the quadrilateral chart inherit properties of figures above them to which they are linked. Show that rectangles have each of the three properties of isosceles trapezoids listed in Theorem 3.3.

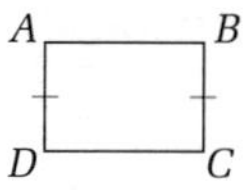

A
B
D
C

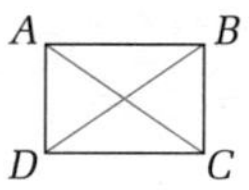

A
B
D
C

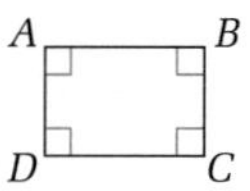

A
B
D
C

For Exercises 16–18, write a definition for each type of polygon using a biconditional statement. For example, a quadrilateral is a parallelogram *if and only if* it has two pairs of parallel sides.

16. trapezoid

17. isosceles trapezoid

18. kite

Ongoing **ASSESSMENT**

For Exercises 19 and 20, use this information.

Sometimes two different definitions can describe exactly the same figure. Here are two definitions of *rectangle*.

Definition 1: A rectangle is a quadrilateral with four right angles.

Definition 2: A rectangle is a parallelogram with at least one right angle.

19. Using Definition 1, prove this statement in a paragraph proof: If a quadrilateral is a rectangle, then it is a parallelogram.

20. Using Definition 2, prove this statement in a paragraph proof: If a quadrilateral is a rectangle, then it has four right angles.

Answers to Exercises and Problems

5. at least one
6. exactly one
7. a. I, II, III, V
   b. I, III
8. exclude
9. Answers may vary. Examples are given. Inclusive: Packaging that uses recycled material must contain some recycled components. Exclusive: Packaging that uses recycled material must consist of only recycled components.
10. A polynomial function is the sum of one or more direct variation functions. (The sum may include a constant function.) A rational function is a function with the form $f(x) = \frac{p(x)}{q(x)}$, where $p(x)$ and $q(x)$ are both polynomials and $q(x)$ cannot be equal to zero; includes.
11. True.
12. False.
13. False.
14. True.

15–20. See answers in back of book.

Error Analysis

For Exs. 16–18, some students may write erroneous definitions as well as definitions that are "workable" but not official. These errors may be corrected by having students refer to the Glossary or Index to locate the definition in question.

Communication: Discussion

Discuss the difference between an erroneous definition and a workable-but-not-official definition. An erroneous definition frequently includes a figure that should not be included or excludes a figure that should not be excluded. For example, it is erroneous to define a kite as a figure with two pairs of consecutive congruent sides. This would include figures such as the one shown below.

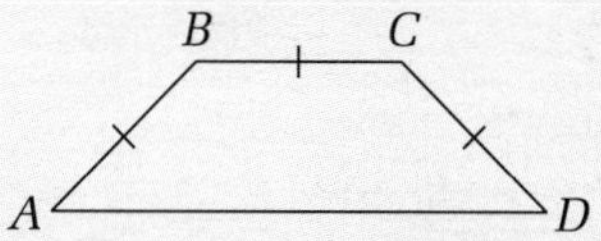

B
C
A
D

An example of a workable-but-not-official definition of a parallelogram might be that a quadrilateral is a parallelogram if and only if it has a pair of parallel, congruent sides. This is a workable definition because every figure that is a parallelogram by this definition is also a parallelogram in accordance with the "official" definition (a quadrilateral with two pairs of parallel sides). The reverse is also true. In this sense, these definitions of *parallelogram* are equivalent and interchangeable. In a deductive system, however, it is important to agree on the official definition of the term, since that determines what must be proved when trying to show that a figure is a parallelogram.

Assessment: Standard

Students should be able to write two definitions for a square, one using the fact that all squares are rhombuses and one using the fact that all squares are rectangles. They should also be able to write a biconditional statement for each definition.

Working on the Unit Project

Students can work in their project groups to discuss Ex. 24. They may wish to record their explanations in their journals for future reference.

Quick Quiz (3-1 through 3-3)

See page 202.

See also Test 10 in the *Assessment Book*.

Practice 21 For use with Section 3-3

Name ______________ Date ______________

Practice 21
For use with Section 3-3

10. A quad. is a rhom. if and only if all sides are = in meas. 11. A quad. is a kite if and only if at least one diag. is a line of symmetry. 12. A quad. is a parallel. if and only if both pairs of opp. sides are ∥. 13. A quad. is a sq. if and only if it is a rect. and a rhom.

Using the definitions stated in the text, determine whether each statement is *True* or *False*.

1. All rectangles are right trapezoids. true
2. If a quadrilateral is a parallelogram, then it is a trapezoid. true
3. All isosceles trapezoids are kites. false
4. If a quadrilateral is a rhombus, then it is an isosceles trapezoid. false

For Exercises 5–8, determine whether the word "one" means "exactly one" or "at least one."

5. The senior class will choose one person as president, one person as secretary, and one person as treasurer. exactly one
6. In order to use the car pool lane, a car must contain one person besides the driver. at least one
7. For good health, you should eat one serving of fruit each day. at least one
8. Each citizen of voting age should cast one vote on election day. exactly one
9. Suppose you define a rectangle as follows: "A rectangle is a quadrilateral whose angles are all right angles and whose four sides are not all equal in measure."
   a. Is this definition *inclusive* or *exclusive*? exclusive
   b. According to this definition, is "All squares are rectangles" a true statement? no
   c. According to this definition, is "Some rectangles are squares" a true statement? no
   d. According to this definition, is "No rectangles are squares" a true statement? yes

Exs. 10–13: See top of page. Definitions may vary. Samples are given.

Write a definition for each polygon using a biconditional statement.

10. rhombus 11. kite 12. parallelogram 13. square

14. ***Open-ended*** Write a definition of some familiar term, such as *window*. Give your definition to a friend, and ask your friend to identify what you are defining. If he or she cannot do so, change your definition. If he or she can identify it, ask whether he or she can think of any other things that fit the definition. If so, change your definition until it actually defines what you have in mind.
Answers may vary. Check students' work.

Practice Bank, INTEGRATED MATHEMATICS 3

21

Answers to Exercises and Problems

21. $PQ = \sqrt{(-1-0)^2 + (-2-0)^2} = \sqrt{5}$; $QR = \sqrt{(2-0)^2 + (1-0)^2} = \sqrt{5}$; $RS = \sqrt{(1-2)^2 + (-1-1)^2} = \sqrt{5}$; $PS = \sqrt{(1-(-1))^2 + (-1-(-2))^2} = \sqrt{5}$; Since $PQ = QR = RS = PS$, $PQRS$ is a rhombus.

22. Answers may vary. An example is given. **Step 1:** Determine the author's last name. **Step 2:** Look in the card catalog under the author's last name and find the reference number. **Step 3:** If you know the general location of the book, go to Step 5. **Step 4:** Find the general location of the book. If necessary, ask the librarian. **Step 5:** Use the reference number to locate the book.

23, 24. See answers in back of book.

Review PREVIEW

21. Show that the quadrilateral $PQRS$ with vertices $P(-1, -2)$, $Q(0, 0)$, $R(2, 1)$, and $S(1, -1)$ is a rhombus. *(Section 3-2)*

22. Write an algorithm for finding a specific book at a library. *(Section 1-1)*

23. a. Use quadrilateral $ABCD$ to complete each statement.

    $m\angle A + m\angle 1 + m\angle 3 = \underline{\ ?\ }$

    $m\angle C + m\angle 2 + m\angle 4 = \underline{\ ?\ }$

    $m\angle 1 + \underline{\ ?\ } = m\angle CDA$

    $m\angle 3 + m\angle 4 = \underline{\ ?\ }$

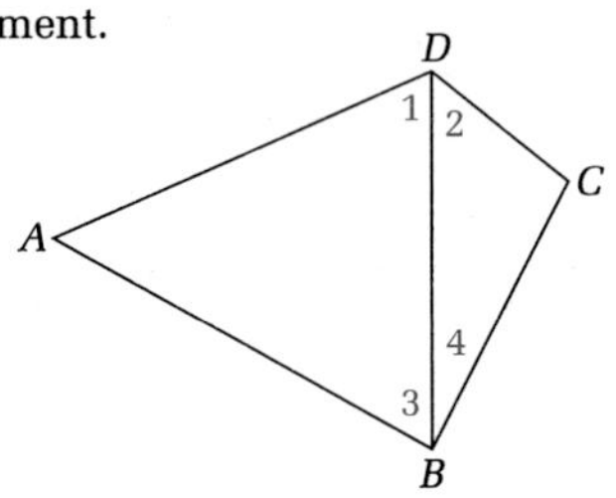

    b. Use the results in part (a) to show that $m\angle A + m\angle ABC + m\angle C + m\angle CDA = 360°$.

    c. State the theorem that part (b) proves. *(Theorems, pp. 668-671)*

Working on the Unit Project

24. At the beginning of this section, you saw that in law as well as in mathematics, certain words require precise definitions. Choose two or three words that are relevant to your case and explain why their definitions may help determine the outcome.

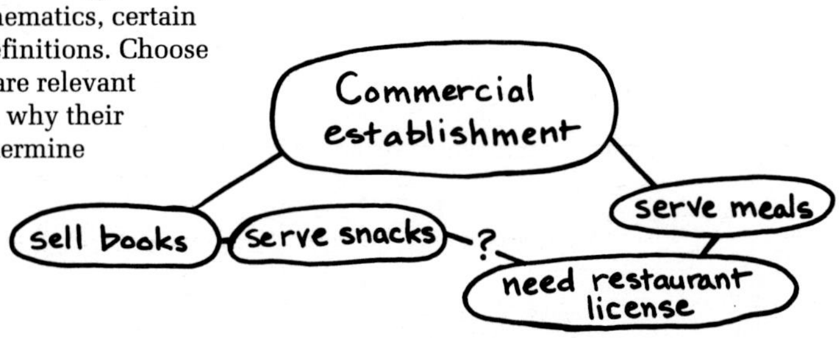

Unit 3 CHECKPOINT 1

1. **Writing** Explain how to tell the difference between the inverse and the contrapositive of an implication. 3-1

2. State the inverse of this implication: If a quadrilateral is a rectangle, then its diagonals are equal in measure. Is the inverse true?

Tell whether each triangle with the given vertices is isosceles. Use both the distance formula and symmetry to justify your answer. 3-2

3. $D(-6, 0)$, $E(5, 0)$, $F(0, 3)$
4. $J(-4, 5)$, $K(0, -2)$, $L(4, 5)$

Tell whether each statement is *True* or *False*. 3-3

5. All rectangles are isosceles trapezoids.
6. All isosceles trapezoids have two lines of symmetry.

Answers to Checkpoint

1. In the inverse of an implication, the "if" and the "then" are in the same order as in the original implication. In the contrapositive, they are in reverse order. In both cases, the "if" and the "then" are negated.

2. If a quadrilateral is not a rectangle, then its diagonals are not equal in measure; False.

3. $FD = \sqrt{(0-(-6))^2 + (3-0)^2} = 3\sqrt{5}$; $FE = \sqrt{(0-5)^2 + (3-0)^2} = \sqrt{34}$; $ED = \sqrt{(5-(-6))^2 + (0-0)^2} = 11$. $\triangle DEF$ is not isosceles, since no two sides are equal in measure. $\triangle DEF$ has no line of symmetry, so $\triangle DEF$ is not isosceles.

4. $JK = \sqrt{(-4-0)^2 + (5-(-2))^2} = \sqrt{65}$; $LK = \sqrt{(4-0)^2 + (5-(-2))^2} = \sqrt{65}$; Since two sides of $\triangle JKL$ are equal in measure, $\triangle JKL$ is isosceles. $\triangle JKL$ is an isosceles triangle because it has a line of symmetry, the y-axis.

5. True.
6. False.

Section

3-4 Angles in Polygons

Focus

Find formulas for the sums of interior and exterior angle measures in polygons.

One Thing Leads To Another

You know a theorem of geometry that states:

The sum of the measures of the angles of a triangle is 180°.

You can use this fact to find the sum of the angle measures of any polygon.

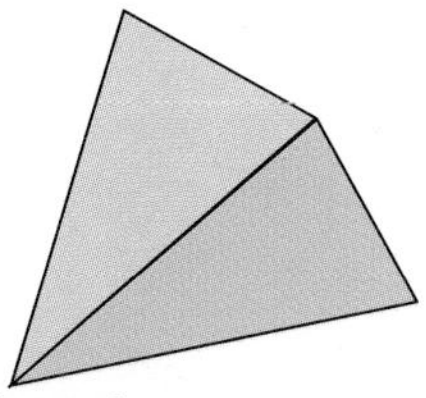

quadrilateral

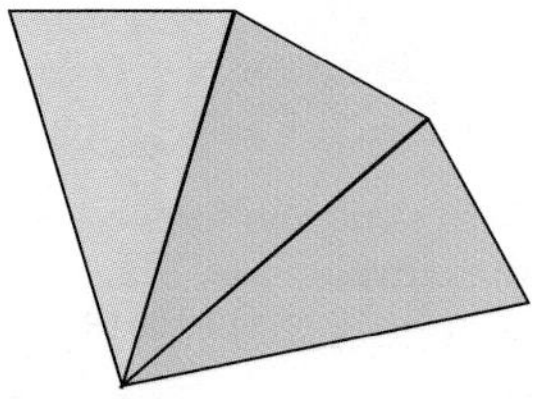

pentagon

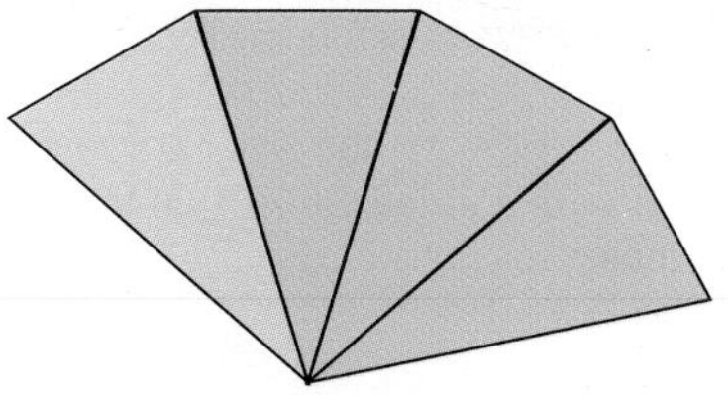

hexagon

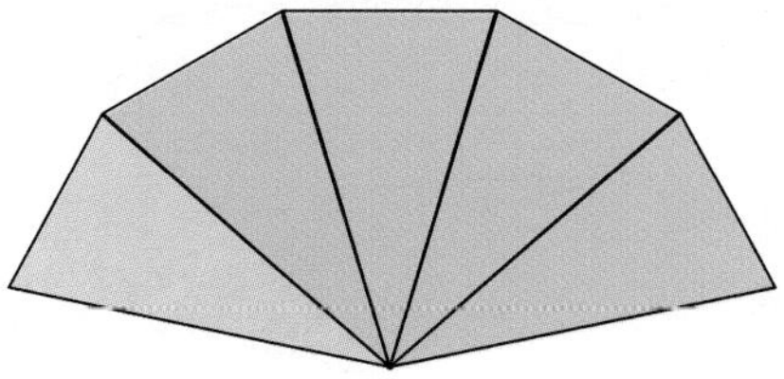

heptagon

Student Resources Toolbox
p. 668 *Postulates and Theorems*

Talk it Over

Use the drawings to answer question 1.

1. What is the sum of the angle measures of any quadrilateral? any pentagon? any hexagon? any heptagon?
2. Complete the chart. Look for patterns.

Name of polygon	Number of sides, n	Sum of angle measures $S(n)$
Triangle	3	180° = (180°)
Quadrilateral	4	_?_ = _?_ (180°)
Pentagon	5	_?_ = _?_ (180°)
Hexagon	6	_?_ = _?_ (180°)
Heptagon	7	_?_ = _?_ (180°)
Octagon	8	_?_ = _?_ (180°)

3. Find a general formula for the function $S(n)$, the sum of the angle measures of a polygon with n sides. Explain why you think this formula will work for *all* polygons.

PLANNING

Objectives and Strands

See pages 136A and 136B.

Spiral Learning

See page 136B.

Recommended Pacing

Section 3-4 is a one-day lesson.

Extra Practice

See pages 612–613.

Warm-Up Exercises

Warm-Up Transparency 3-4

Support Materials

- Practice Bank: Practice 22
- Activity Bank: Enrichment 20
- Study Guide: Section 3-4
- Problem Bank: Problem Set 7
- Explorations Lab Manual: Additional Exploration 4
- Assessment Book: Quiz 3-4

Answers to Talk it Over

1. 360°; 540°; 720°; 900°

2.

Name of polygon	Number of sides, n	Sum of angle measures, $S(n)$
Triangle	3	180° = (180°)
Quadrilateral	4	360° = 2(180°)
Pentagon	5	540° = 3(180°)
Hexagon	6	720° = 4(180°)
Heptagon	7	900° = 5(180°)
Octagon	8	1080° = 6(180°)

3. $S(n) = (n - 2)(180°)$ Answers may vary. An example is given. Every time another side is added, another triangle can be formed, so 180° is added to the sum of the angle measures. For example, the formula is true for octagons, $S(8) = (8 - 2)(180°) = 6(180°) = 1080°$. When a ninth side is added, 180° is added and the sum of the measures of the angles is $1260° = 7(180°) = (9 - 2)(180°)$.

TEACHING

Talk it Over

Questions 1–3 lead students to discover the general formula $S(n) = (n - 2)180°$ for the sum of the measures of the angles of a polygon with n sides.

Additional Sample

S1 Suppose that the formula $D(n) = \frac{n(n-3)}{2}$ gives the number of diagonals for a polygon with k sides. Show that the formula gives the number of diagonals for a polygon with $k + 1$ sides.

Let $A_1, A_2, \ldots, A_k, A_{k+1}$ be the vertices of the polygon with $k + 1$ sides. All diagonals of the k-gon are also diagonals of the $(k + 1)$-gon. The segments obtained by connecting A_{k+1} to each of the other vertices (k vertices in all) are diagonals, except for $A_{k+1}A_1$ and $A_{k+1}A_k$. This gives an additional $k - 2$ diagonals, and the segment A_1A_k, which is a side of the k-gon $A_1A_2 \ldots A_k$, is a diagonal of the $(k + 1)$-gon. The total number of diagonals for the $(k + 1)$-gon is therefore $\frac{k(k-3)}{2} + (k - 1)$.

$$\frac{k(k-3)}{2} + (k-1)$$
$$= \frac{k^2 - 3k + 2(k-1)}{2}$$
$$= \frac{k^2 - k - 2}{2}$$
$$= \frac{(k+1)(k-2)}{2}$$
$$= \frac{(k+1)[(k+1) - 3]}{2}$$

This last expression is equal to $\frac{n(n-3)}{2}$ when $n = k + 1$. Thus, if the formula works for the k-gon, it also works for the $(k + 1)$-gon.

The Sum of the Measures of the Angles of a Polygon

You may have conjectured that in a polygon with n sides, the sum of the measures of the angles is given by the formula

$$S(n) = (n - 2)180°.$$

This formula works for polygons with 3, 4, 5, 6, and 7 sides, but how can you be sure that it will work for *all* values of n? A proof of the conjecture uses a technique called *mathematical induction.*

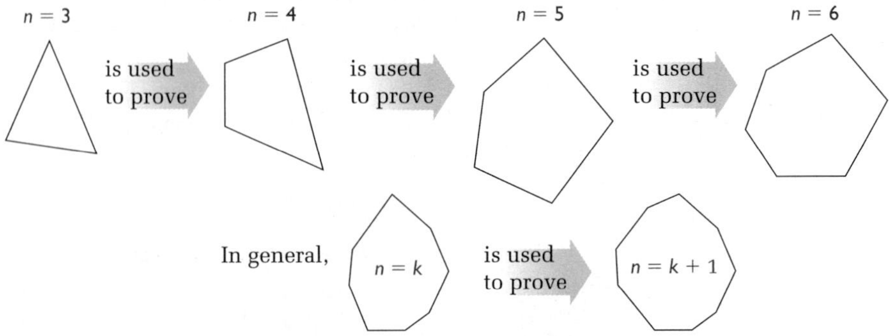

To show that the pattern will continue forever, you must show that you can always use the current case, $S(k) = (k - 2)180°$, to prove the next case, $S(k + 1) = ([k + 1] - 2)180°$.

Sample 1

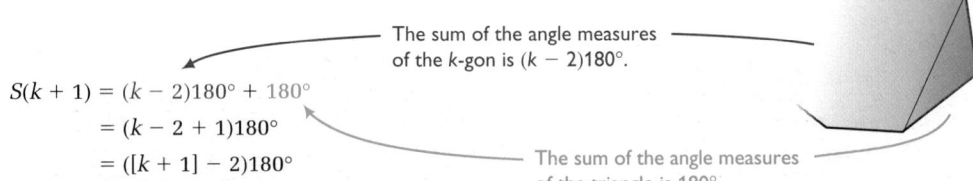

Suppose that the formula $S(n) = (n - 2)180°$ gives the sum of the angle measures for a polygon with k sides. Show that the formula gives the sum of the angle measures for a polygon with $k + 1$ sides.

Sample Response

Draw a diagonal of the $(k + 1)$-gon to form a triangle and a k-gon.

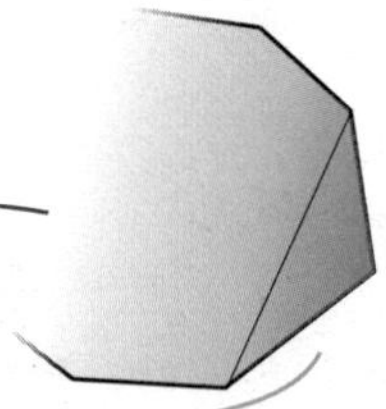

$$S(k+1) = (k-2)180° + 180°$$
$$= (k - 2 + 1)180°$$
$$= ([k+1] - 2)180°$$

The last expression is equal to $(n - 2)180°$ when $n = k + 1$, so the formula $S(n) = (n - 2)180°$ works for the $(k + 1)$-gon if it works for the k-gon.

Talk it Over

4. Explain how $(k - 2)180° + 180°$ can be rewritten as $(k - 2 + 1)180°$ in Sample 1 using the distributive property.
5. In Sample 1, the case of $n = k$ is used to prove the case of $n = k + 1$. How does Sample 1 convince you that $S(n) = (n - 2)180°$ for all integral values of n greater than 3?

XLab

INTERIOR AND EXTERIOR ANGLE MEASURES IN POLYGONS

Theorem 3.4 The sum of the angle measures of an n-gon is given by the formula $S(n) = (n - 2)180°$.

Theorem 3.5 The sum of the exterior angle measures of an n-gon, one angle at each vertex, is 360°.

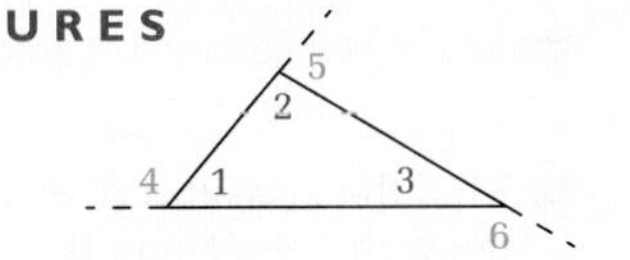

$m\angle 1 + m\angle 2 + m\angle 3 = 180°$
$m\angle 4 + m\angle 5 + m\angle 6 = 360°$

You will prove Theorem 3.5 in Exercise 11.

Regular Polygons

A polygon is a **regular polygon** if and only if all its sides are equal in measure and all its angles are equal in measure.

Sample 2

Find the measure of each interior and each exterior angle of the regular octagon.

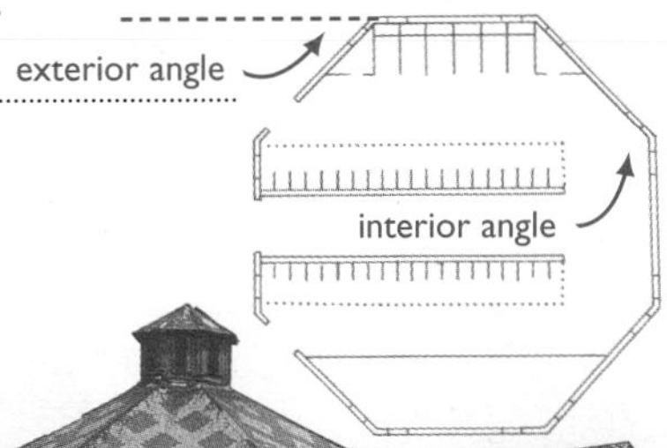

Sample Response

First find the sum of the interior angle measures.

$S(n) = (n - 2)180° = (8 - 2)180° = 6(180°) = 1080°$

Since the eight interior angles are equal in measure, the measure of each angle is $\frac{1080°}{8} = 135°$.

The sum of the exterior angles is 360°. Since each exterior angle is a supplement of one of the interior angles, the eight exterior angles are equal in measure. The measure of each exterior angle is $\frac{360°}{8} = 45°$. (Also, $180° - 135° = 45°$.)

Octagonal barn dating from 1873

Talk it Over

Question 5 has students analyze the basic idea behind mathematical induction and has them offer their own explanation about why it is a convincing argument. You may wish to have students discuss their answers to this question either in groups or as a whole class discussion.

Error Analysis

Some students may miss the concept "one angle at each vertex" in Theorem 3.5. This can lead to errors in finding the number and measures of exterior angles. Stress to students that there is only one exterior angle at each vertex just as their is only one interior angle.

Additional Sample

S2 Find the measure of each interior and each exterior angle of a regular 16-gon.

The sum of the interior angle measures is $S(n) = (n - 2)180° = (16 - 2)180° = 14(180°) = 2520°$. Therefore, each interior angle has a measure of $\frac{2520°}{16} = 157.5$ °. The sum of the measures of the exterior angles is 360°. So, each exterior angle has a measure of $\frac{360°}{16} = 22.5°$.

Answers to Talk it Over

4. $(k - 2)180° + 180° =$
$(k - 2)180° + (1)180° =$
$(k - 2 + 1)180°$
5. Answers may vary. An example is given. I know the formula is true for $n = 3$. Since I can show that whenever it is true for n, it is true for $n + 1$, I am convinced that it is true for all values of n greater than 3.

Look Back

Students may give slightly different answers to how they would find the measure of an exterior angle. A brief discussion should convince them that two methods are possible.

APPLYING

Suggested Assignment

Standard 1–13, 15, 17–22

Extended 1–15, 17–22

Integrating the Strands

Number Ex. 15

Algebra Exs. 4, 11, 13, 15, 16, 20

Functions Exs. 4, 12, 13, 15, 16

Geometry Exs. 1–13, 17–19, 21

Logic and Language Exs. 1, 14, 16, 17, 22

Teaching Tip

Some students may be coin collectors. Ask these students if they have any unusually shaped coins they would want to bring to class. If any of these coins are regular polygons, they can be used to extend Exs. 5–10.

Look Back

Explain how to find the measure of an interior angle of a regular polygon. How do you find the measure of an exterior angle?

3-4 Exercises and Problems

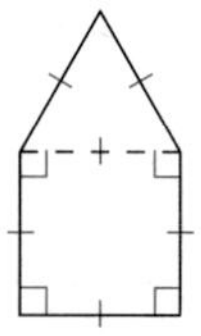

1. **Reading** Is a square a regular polygon? Explain.
2. a. Explain why this pentagon is not a regular polygon.
   b. Find the measures of the angles of the pentagon and show that their sum is 540°.
3. Can each figure be a regular polygon? Explain.
   a. rectangle b. rhombus c. trapezoid
4. Use the formula $S(n) = (n - 2)180°$ for the sum of the angle measures of a polygon.
   a. Explain why $n = 2$ and $n = 6.5$ are not in the domain of $S(n)$.
   b. Find the domain and range of $S(n)$.
   c. What kind of function is $S(n)$?

Coins Find the measures of each interior and each exterior angle of the regular polygon on each coin.

5.

7-gon from Sierra Leone

6.

8-gon from Netherlands Antilles

7.

9-gon from Portugal

8.

10-gon from Tanzania

9.

11-gon from Canada

10.

12-gon from South Vietnam

Answers to Look Back

First, use the number of sides, n, to determine the sum of the measures of the interior angles, $(n - 2)180°$. Next, use the fact that all of the angles are equal in measure. Divide the sum of the measures of the angles by the number of sides to find the measure of an interior angle: $\frac{(n-2)180°}{n}$. To find the measure of an exterior angle, divide 360° by n or subtract the measure of an interior angle from 180°.

Answers to Exercises and Problems

1. Yes. All the sides of a square are equal in measure, and all four angles are right angles and thus equal in measure. So a square is a regular polygon.
2. a. The angles are not all equal in measure.
   b. There are two 90° angles, two 150° angles, and one 60° angle. 90° + 90° + 150° + 150° + 60° = 540°
3. a. Yes; a square is a rectangle that is a regular polygon.
   b. Yes; a square is a rhombus that is a regular polygon.
   c. Yes; a square is a trapezoid that is a regular polygon.
4. a. There are no polygons with 2 sides and no polygons with 6.5 sides.
   b. domain: all integers greater than or equal to 3; range: $180n$ for all integers $n \geq 1$
   c. one-to-one, linear
5. about 128.6°; about 51.4°
6. 135°; 45°
7. 140°; 40°
8. 144°; 36°
9. about 147.3°; about 32.7°
10. 150°; 30°

11. Complete this proof of Theorem 3.5: The sum of the exterior angle measures of an n-gon, one angle at each vertex, is 360°.

Given a polygon with n sides

$S(n)$ = the sum of the interior angle measures.

$E(n)$ = the sum of the exterior angle measures.

Prove $E(n) = 360°$

a. At each vertex, the sum of the measures of the interior angle and one exterior angle is 180°. Justification: _?_

b. Based on part (a), what is the combined sum of interior and exterior angle measures in a polygon with n vertices? ($S(n) + E(n) =$ _?_)

c. Substitute $(n - 2)180°$ for $S(n)$ in the equation you wrote in part (b) and solve for $E(n)$. $E(n) =$ _?_

12. **a.** Find the domain and range of $E(n)$ in Exercise 11.

b. What kind of function is $E(n)$?

13. Let $X(n)$ represent the measure of one exterior angle of a regular polygon of n sides. Let $I(n)$ represent the measure of one interior angle of a regular polygon of n sides.

a. Find algebraic expressions for $X(n)$ and $I(n)$.

b. What kind of functions are $X(n)$ and $I(n)$?

c. Sketch graphs of $X(n)$ and $I(n)$.

14. **Writing** Explain how these dominos are like a proof that uses mathematical induction.

15. The diagrams suggest that the sum of the first n odd integers is n^2.

a. Draw the next two pictures in the sequence.

b. The sum of the first n odd integers is given by the function $F(n) = 1 + 3 + 5 + \ldots + (2n - 1)$. What does $2n - 1$ represent?

c. Suppose $F(k) = k^2$ is a formula for the sum of the first k odd integers. Show that $F(k + 1) = (k + 1)^2$.

d. Do you think the statement $F(n) = n^2$ is true for all positive integers n? Explain.

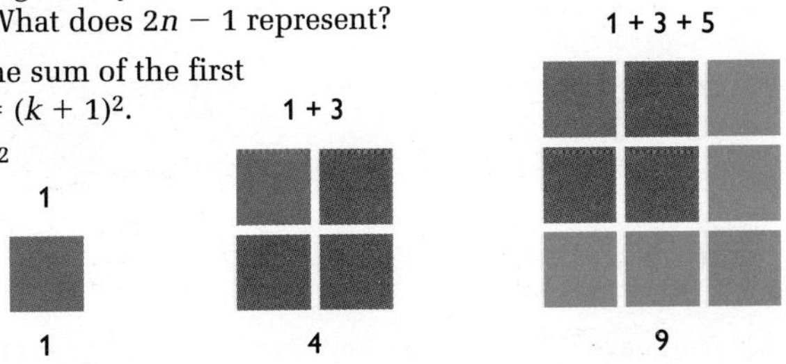

Mathematical Procedures

In connection with Ex. 14, students should understand that any result to be proved by mathematical induction must be known to be true in at least one initial case. It is true that if one domino falls, then the next one must also fall. Still, it takes a fall by an initial domino for the rest of the action to occur.

Students Acquiring English

For Ex. 14, you may want to pair students acquiring English with students fluent in English for the writing of the explanation. This is a good exercise to illustrate visually the logical progression of proofs.

Answers to Exercises and Problems

11. a. At each vertex, the interior angle and one exterior angle form a linear pair. Angles that form a linear pair are supplementary.

b. $n(180°)$

c. $(n - 2)180° + E(n) = n(180°)$; $E(n) = n(180°) - (n - 2)180° = 2(180°) = 360°$

12. a. all integers greater than or equal to 3; 360

b. constant

13. a. $X(n) = \dfrac{360°}{n}$; $I(n) = \dfrac{(n-2)180°}{n}$

b. rational

c.

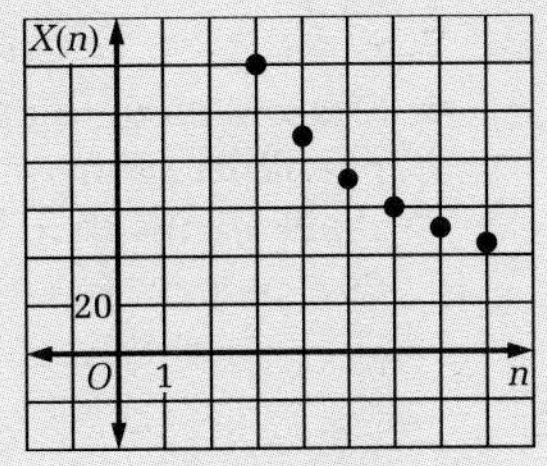

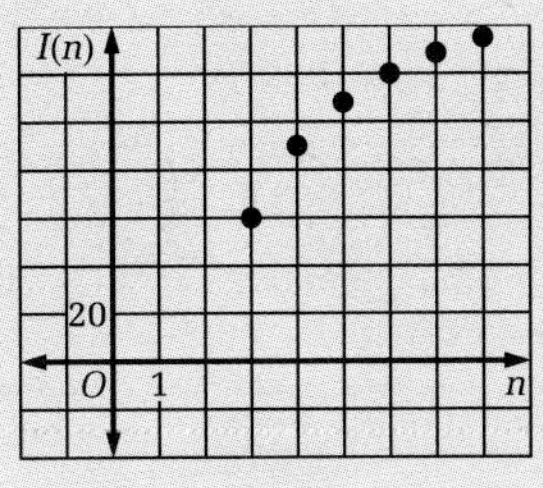

14. Answers may vary. An example is given. As each domino falls, it strikes the next one, allowing it to fall and strike the next one. The process continues as long as there are dominoes to be struck. In a mathematical induction proof, each step that is proved allows the next step to be proved.

15. a.

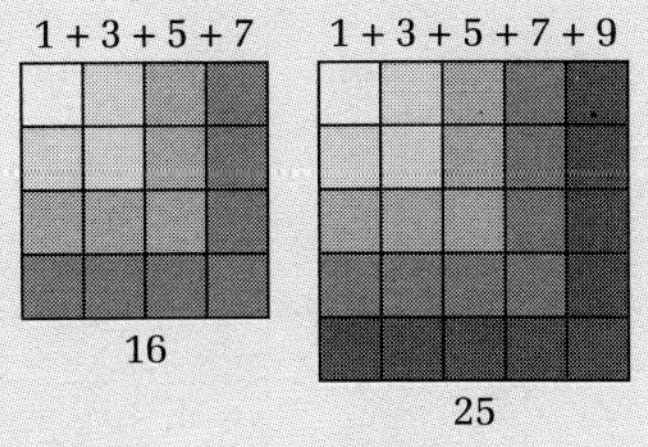

b. the nth odd number

c. $F(k + 1) = 1 + 3 + 5 + \ldots + (2k - 1) + 2k + 1 = F(k) + 2k + 1 = k^2 + 2k + 1 = (k + 1)^2$

d. Yes; by mathematical induction, whenever $F(k)$ is true, $F(k + 1)$ is true.

Communication: Drawing

Diagrams can help students establish a connection between the situations in Additional Sample S1 and Ex. 16. Suggest that students draw polygons and their diagonals. Ask what happens if they think of each segment as corresponding to a handshake. What do the vertices of the polygon represent in terms of the handshake problem? (the people who are shaking hands)

Working on the Unit Project

Students should keep a written record of their examples to use when completing the Unit Project.

Practice 22 For use with Section 3-4

Name ______________________ Date ____________

Practice 22
For use with Section 3-4

13. d. $\frac{k^2-3k}{2} + k - 1 = \frac{k^2-3k+2k-2}{2}$
$= \frac{(k+1)^2 - 3(k+1)}{2}$

Find the sum of the measures of the angles and the sum of the measures of the exterior angles, one angle at each vertex, of each polygon.

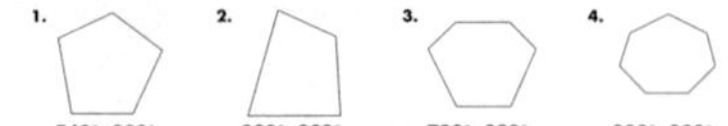

1. 540°, 360° 2. 360°, 360° 3. 720°, 360° 4. 900°, 360°

Find the measure of each angle and each exterior angle of a regular polygon with the given number of sides.

5. 3 60°, 120° 6. 8 135°, 45° 7. 10 144°, 36°
8. 9 140°, 40° 9. 12 150°, 30° 10. 15 156°, 24°

11. Kin Yui measures four of the angles of a pentagon and finds that their measures are 85°, 110°, 125°, and 105°. What is the measure of the fifth angle of Kin Yui's pentagon? 115°

12. A hexagon has a line of symmetry that passes through the midpoints of two sides. Two of the three angles on one side of the line of symmetry have measures 102° and 128°, respectively. What is the measure of the third angle on the same side of the line of symmetry? 130°

13. In this exercise, you will use mathematical induction to prove that a polygon with n sides has $\frac{n^2-3n}{2}$ diagonals.

Number of sides n	Number of diagonals	$\frac{n^2-3n}{2}$
4	2	2
5	5	5
6	9	9

a. Complete the table at the right by sketching each polygon and counting its diagonals.

b. Is the statement true for $n = 4$, 5, and 6? yes

c. Suppose $F(k) = \frac{k^2-3k}{2}$ is a formula for the number of diagonals of a polygon with k sides. Suppose one side is replaced by two (adjacent) sides, so that the polygon now has $k+1$ sides. How many new diagonals will be created? (Don't forget the one that used to be a side!) $k-1$

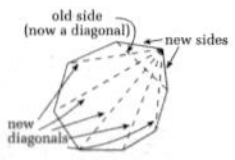

d. Use the result of part (c) to show that $F(k+1) = \frac{(k+1)^2 - 3(k+1)}{2}$. See top of page.

22 Practice Bank, INTEGRATED MATHEMATICS 3

16. **Group Activity** Work in a group of five students.

a. Complete the table showing the number of handshakes if everyone shakes hands once with every person.

Number of people	Number of handshakes
p	$H(p)$
2	?
3	?
4	?
5	?

b. Show that $H(p) = \frac{p(p-1)}{2}$ for $p = 2, 3, 4,$ and 5.

c. Suppose a sixth person joins your group. How many more handshakes will there be as you welcome the sixth person?

d. Suppose k people all shake hands with each other, for a total of $H(k) = \frac{k(k-1)}{2}$ handshakes. One more person walks into the room and shakes hands with everyone. Show that $H(p) = \frac{p(p-1)}{2}$ when $p = k + 1$.

e. **Writing** How many handshakes will there be if everyone in your class shakes hands once with everyone else? Explain.

Ongoing ASSESSMENT

17. **Writing** Explain why it is possible to create a regular polygon with exterior angles that measure 15°, but it is not possible to create a regular polygon with exterior angles that measure 25° or 35°.

Review PREVIEW

Tell whether each statement is *True* or *False*. *(Section 3-3)*

18. Every square is a trapezoid.

19. If a quadrilateral is a rhombus, then it is a trapezoid.

20. Solve $\sqrt{-x + 13} = x - 1$. *(Section 2-6)*

21. Which of the following show(s) a perpendicular bisector of $\overline{PQ}$? Explain. *(Glossary, Toolbox Skill 33)*

a.
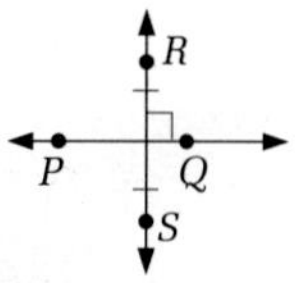

b.
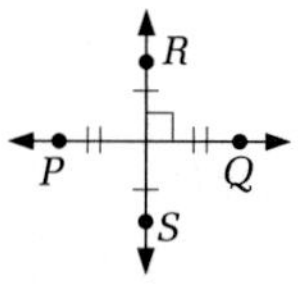

c.
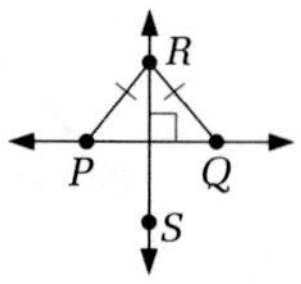

d.
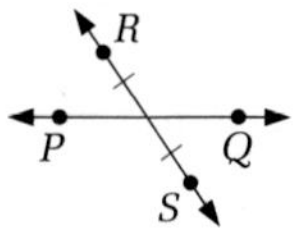

Working on the Unit Project

22. You have learned to use both inductive and deductive reasoning. Give examples of how both types of reasoning can be used in your brief.

Answers to Exercises and Problems

16. a.

Number of people	Number of handshakes
p	$H(p)$
2	1
3	3
4	6
5	10

b. For $p = 2$, $\frac{2(2-1)}{2} = 1 = H(2)$.
For $p = 3$, $\frac{3(3-1)}{2} = 3 = H(3)$.
For $p = 4$, $\frac{4(4-1)}{2} = 6 = H(4)$.
For $p = 5$, $\frac{5(5-1)}{2} = 10 = H(5)$.

c. 5

d. Since each of the k people in the room will shake hands with the new person, $H(k+1) = H(k) + k = \frac{k(k-1)}{2} + k = \frac{k(k-1)+2k}{2} = \frac{k^2-k+2k}{2} = \frac{k^2+k}{2} = \frac{(k+1)k}{2}$ or $\frac{(k+1)[(k+1)-1]}{2}$.

e. Using mathematical induction, you know that if there are x people in your class, there will be $\frac{x(x-1)}{2}$ handshakes.

17. Since 15 is a factor of 360, it is possible to create a regular polygon with exterior angles that measure 15°. The polygon will have $\frac{360}{15} = 24$ sides. 25 and 35 are not factors of 360, so it is not possible to create regular polygons with exterior angles that measure 25° or 35°.

18. True.

19. True.

20. 4

21. See answers in back of book.

22. Answers may vary.

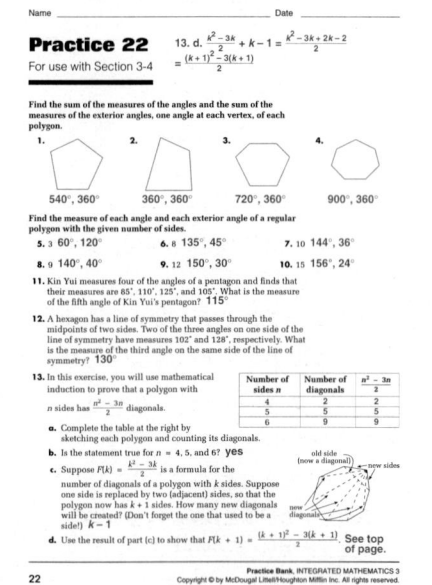

Section **3-5**

Inscribed Polygons

STRIKE A CHORD

Focus
Examine relationships between inscribed polygons and circles.

EXPLORATION

Can a circle be drawn through the vertices of any triangle?

- **Materials: rulers, compasses**
- **Work in a group of four students.**

(1) On a sheet of paper, draw an acute scalene triangle and label it $\triangle ABC$.

(2) Fold the paper so that A lies directly on top of C. The line formed by the crease of the paper is the perpendicular bisector of $\overline{AC}$.

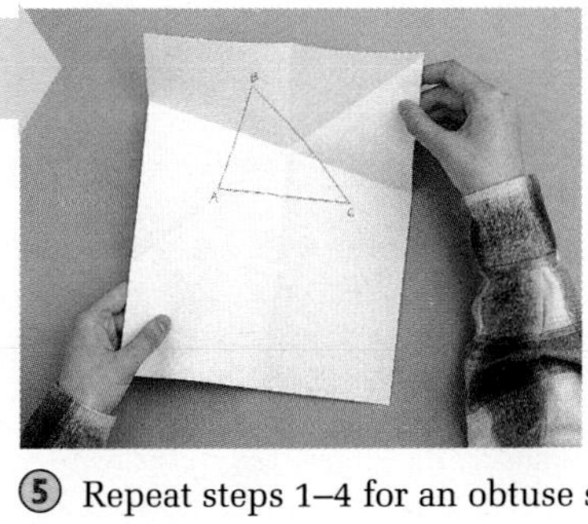

(3) Fold the paper again to find the perpendicular bisectors of $\overline{BC}$ and $\overline{AB}$.

(4) Label the point O where the three perpendicular bisectors intersect. With a compass, see if you can draw a circle with center O that passes through A, B, and C.

(5) Repeat steps 1–4 for an obtuse scalene triangle.

(6) Discuss these questions in your group.

a. Why does the crease made in step 2 produce the perpendicular bisector of $\overline{AC}$?

b. Make a conjecture about where the center of the circle would lie if $\triangle ABC$ were a right triangle.

c. If $\triangle ABC$ is scalene, none of the perpendicular bisectors passes through a vertex. Would this be so with an isosceles triangle? with an equilateral triangle?

PLANNING

Objectives and Strands
See pages 136A and 136B.

Spiral Learning
See page 136B.

Materials List
- Ruler
- Compass

Recommended Pacing

Section 3-5 is a two-day lesson.

Day 1

Pages 165–166: Exploration through Perpendicular Bisector of a Chord, *Exercises 1–5*

Day 2

Pages 167–168: Top of page 167 through Look Back, *Exercises 6–22*

Extra Practice
See pages 612–613.

Warm-Up Exercises
Warm-Up Transparency 3-5

Support Materials
- Practice Bank: Practice 23
- Activity Bank: Enrichment 21
- Study Guide: Section 3-5
- Problem Bank: Problem Set 7
- Assessment Book: Quiz 3-5, Alternative Assessment 3

Answers to Exploration

1–5. Check students' work.

6. a. Answers may vary. An example is given. The perpendicular bisector of a segment is a line perpendicular to the segment at its midpoint. Since A lies directly on top of C, the point where the fold intersects $\overline{AC}$ is the midpoint of $\overline{AC}$. The two angles that the fold makes with $\overline{AC}$ are equal in measure and are a linear pair, so both are right angles. The crease then forms a line that is perpendicular to $\overline{AC}$ and intersects it at its midpoint.

b. at the midpoint of the hypotenuse

c. No. If the isosceles triangle is not equilateral, exactly one perpendicular bisector (that of the base) passes through a vertex. If the triangle is equilateral, all three perpendicular bisectors pass through vertices.

TEACHING

Exploration

The goal of the Exploration is to lead students to discover that the perpendicular bisectors of the chords of a circle all pass through the center of the circle. Students also explore where the three perpendicular bisectors of various types of triangles meet and relate this to circumscribed circles.

Visual Thinking

Ask students to sketch a triangle with both an inscribed and a circumscribed circle, and a circle with both an inscribed and a circumscribed triangle. This activity involves the visual skills of *interpretation* and *correlation.*

Additional Sample

S1 Prove that the bisector of the angle opposite the base of an isosceles triangle inscribed in a circle passes through the center of the circle.

Given: **$\triangle ABC$ with $AB = BC$ inscribed in circle O. $\overrightarrow{BP}$ is the bisector of $\angle ABC$.**

Prove: **$\overrightarrow{BP}$ passes through O.**

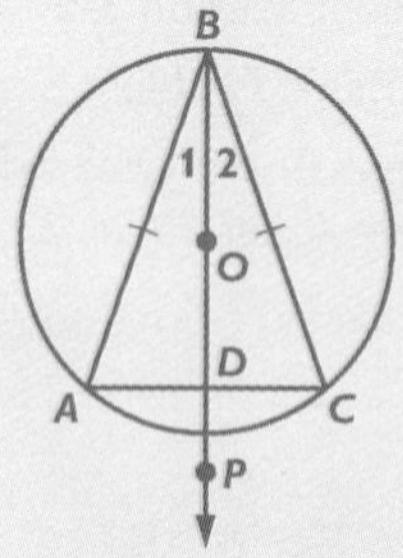

Plan: Let D be the point where $\overrightarrow{BP}$ intersects $\overline{AC}$. Since $\overrightarrow{BP}$ bisects $\angle ABC$, then $m\angle 1 = m\angle 2$. $\overline{BD}$ is a side common to $\triangle ABD$ and $\triangle CBD$ and $AB = BC$. Thus, $\triangle ABD \cong \triangle CBD$ (SAS) and corresponding parts $\overline{AD}$ and $\overline{CD}$ are equal in measure. This implies D is the midpoint of chord $\overline{AC}$. Since corresponding angles $\angle ADB$ and $\angle CDB$ are equal in measure, then they are right angles and $\overrightarrow{BP} \perp \overline{AC}$. From the result in Sample 1, $\overrightarrow{BP}$ passes through O.

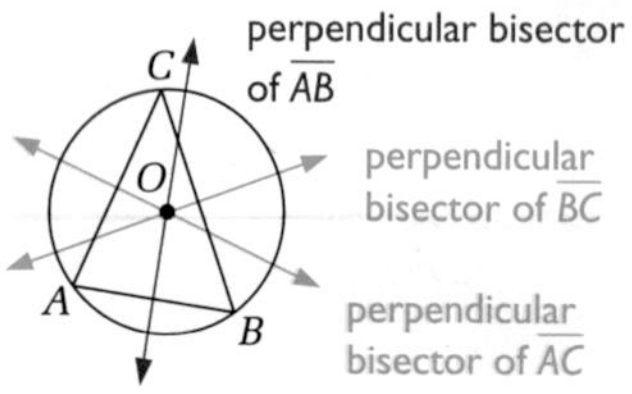

$\overline{AB}$ is a chord of circle O.

Inscribed Triangles

The triangles in the Exploration are **inscribed** in the circles you drew. The circles are **circumscribed** about the triangles. The sides of the triangles are segments joining two points on the circle. Such segments are called **chords** of the circle.

The Exploration suggests it is always possible to circumscribe a circle about a triangle. The point where the perpendicular bisectors of the sides meet is the center of the circumscribed circle.

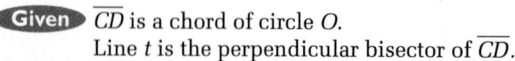

Sample 1

Prove that the perpendicular bisector of a chord of a circle passes through the center of the circle.

Sample Response

Given $\overline{CD}$ is a chord of circle O.
Line t is the perpendicular bisector of $\overline{CD}$.

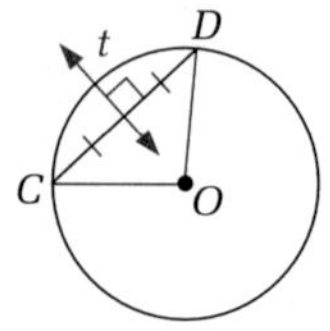

Prove O lies on the perpendicular bisector of $\overline{CD}$.

Plan Ahead

Show that $OC = OD$, and therefore O lies on the perpendicular bisector of $\overline{CD}$.

C and D both lie on circle O, since $\overline{CD}$ is a chord of circle O.

Since O is the center of the circle, $\overline{OC}$ and $\overline{OD}$ are radii of circle O.

By definition, all radii of a circle are equal in measure, so O is the same distance from points C and D.

The perpendicular bisector theorem (p. 669) states that if a point is the same distance from both endpoints of a segment, then it lies on the perpendicular bisector of the segment. Therefore O lies on the perpendicular bisector of $\overline{CD}$.

PERPENDICULAR BISECTOR OF A CHORD

Theorem 3.6 The perpendicular bisector of a chord of a circle passes through the center of the circle.

$\overline{CD}$ is a perpendicular bisector of $\overline{AB}$.

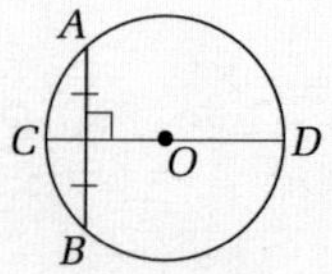

Now you are ready for:
Exs. 1–5 on pp. 168–169

Although every triangle may be inscribed in a circle, not all polygons have this property.

Any regular polygon, however, can be inscribed in a circle. An example is regular pentagon *DEFGH* inscribed in circle *O*.

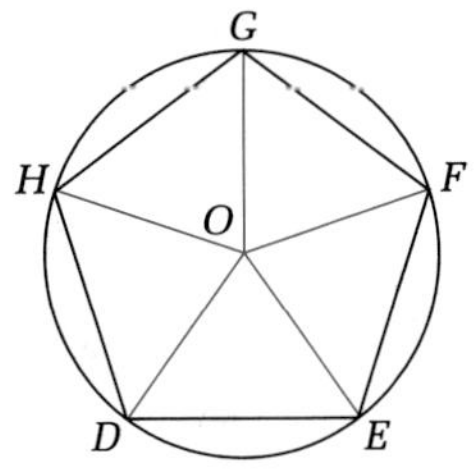

Talk it Over

1. In the regular pentagon shown, which triangles are congruent? How do you know that they are congruent?
2. Find the measures of the angles at center *O*.

Central Angles and Arcs

A **central angle** of a circle is an angle with its vertex at the center of the circle. The measure of an arc *intercepted* (cut off) by a central angle is equal to the measure of that central angle.

Arcs may be classified according to their measures in degrees. A major arc is named with three letters to distinguish it from the minor arc with the same endpoints.

CLASSIFICATIONS OF ARCS

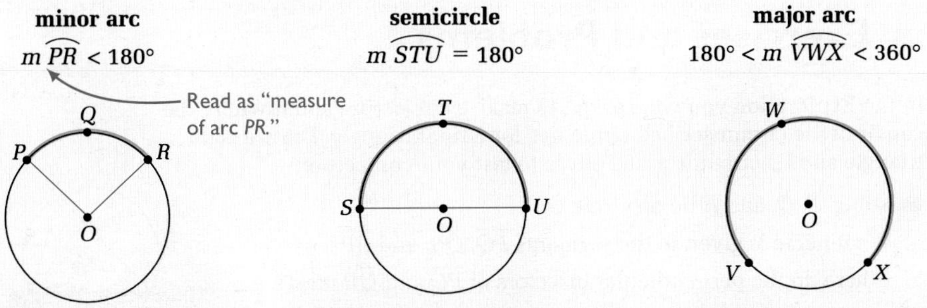

minor arc: $m\overset{\frown}{PR} < 180°$. Name with two or three letters: $\overset{\frown}{PR}$ or $\overset{\frown}{PQR}$. Equal in measure to central angle *POR*.

semicircle: $m\overset{\frown}{STU} = 180°$. Name with three letters: $\overset{\frown}{STU}$. $\overline{SU}$ is a diameter.

major arc: $180° < m\overset{\frown}{VWX} < 360°$. Name with three letters: $\overset{\frown}{VWX}$.

Talk it Over

Questions 1 and 2 demonstrate that when a regular polygon is inscribed in a circle, the central angles formed are equal in measure. This leads to a discussion of how the degree measure of an arc is related to the degree measure of its central angles.

Answers to Talk it Over

1. $\triangle ODE$, $\triangle OEF$, $\triangle OFG$, and $\triangle OGH$, and $\triangle OHD$ ($\triangle OED$, $\triangle OFE$, $\triangle OGF$, $\triangle OHG$, and $\triangle ODH$ are all congruent to the five given triangles as well.) All ten triangles are congruent by SSS. The sides that are radii of the circle are equal in measure. The bases that are sides of the regular polygon are equal in measure as well.
2. 72°

Additional Sample

S2 Find the measure of each arc or angle. ($\overline{AB}$ and $\overline{CD}$ are diameters.)

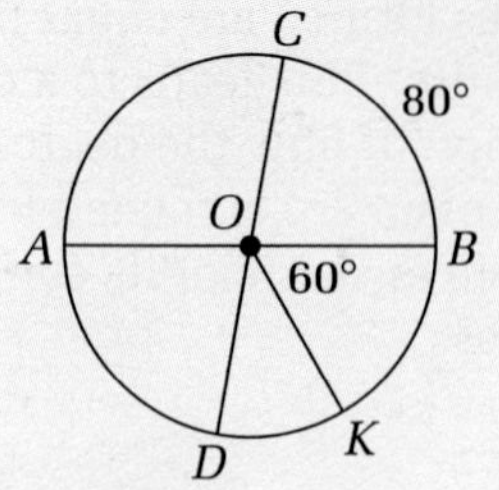

a. $\angle DOK$

Since $m\overset{\frown}{BC} = 80°$, $m\angle BOC = 80°$. Therefore, $m\angle KOC = 60° + 80° = 140°$. Since $\angle DOK$ and $\angle KOC$ are supplementary, $m\angle DOK = 180° - 140° = 40°$.

b. $\overset{\frown}{KAC}$

Since $m\overset{\frown}{KAC} + m\overset{\frown}{CBK}$ is 360° and $m\overset{\frown}{BK} = 60°$, the measure of $\overset{\frown}{KAC}$ is $360° - (80° + 60°) = 360° - 140° = 220°$.

c. $\overset{\frown}{AC}$

The angles AOC and COB are supplementary. It follows that $m\angle AOC = 180° - 80° = 100°$. The measure of $\overset{\frown}{AC}$ is equal to the measure of its central angle, 100°.

d. $\overset{\frown}{KCD}$

$m\angle DOK = 40°$ (from part (a)). Therefore, $m\overset{\frown}{KCD} = 360° - 40° = 320°$.

APPLYING

Suggested Assignment

Day 1

Standard 1–5

Extended 1–5

Day 2

Standard 6–14, 18–22

Extended 6–16, 18–22

Integrating the Strands

Algebra Exs. 4, 5, 21

Geometry Exs. 1–22

Trigonometry Ex. 16

Logic and Language Exs. 2, 3, 11, 18, 22

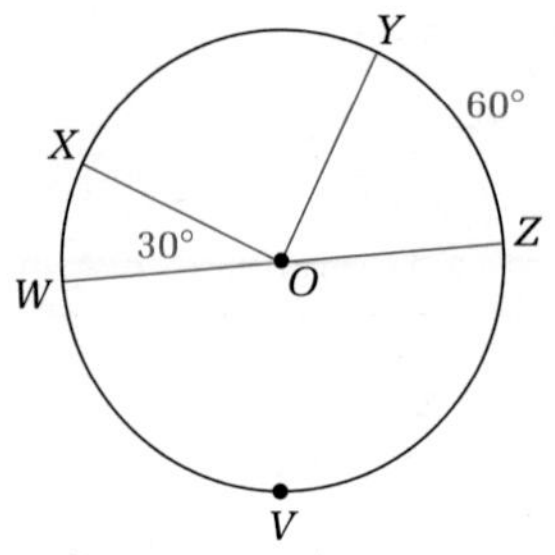

Sample 2

Find the measure of each arc or angle.

a. $\overset{\frown}{WX}$

b. $\angle YOZ$

c. $\overset{\frown}{WVZ}$

d. $\overset{\frown}{XY}$

Sample Response

a. The measure of a minor arc is equal to the measure of its central angle, so $m\overset{\frown}{WX} = 30°$.

b. The measure of a central angle is equal to the measure of the arc it intercepts, so $m\angle YOZ = 60°$.

c. The measure of a semicircle is 180°, so $m\overset{\frown}{WVZ} = 180°$.

d. The sum of the arcs of a circle is 360°.
$m\overset{\frown}{XY} = 360° - (30° + 180° + 60°) = 90°$

Now you are ready for: Exs. 6–22 on pp. 169–171

Look Back

Based on what you have learned in this section, what types of polygons may be inscribed in a circle?

3-5 Exercises and Problems

1. In the Exploration you were asked to make a conjecture about where the center of the circumscribed circle lies for a right triangle. Draw a right triangle and its circumscribed circle to test your conjecture.

2. **Reading** P, Q, and R lie on circle O.

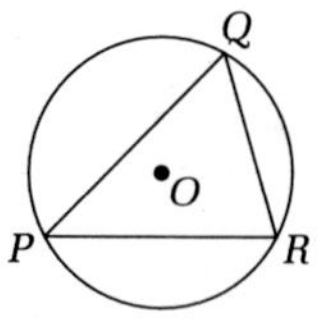

   a. What name is given to the segments $\overline{PQ}$, $\overline{QR}$, and $\overline{RP}$?

   b. Where do the perpendicular bisectors of $\overline{PQ}$ and $\overline{QR}$ meet?

3. **Writing** Explain why you agree or disagree with this statement: "For every triangle there is a point that is equidistant from all three vertices."

Answers to Look Back

all triangles and all regular polygons

Answers to Exercises and Problems

1.

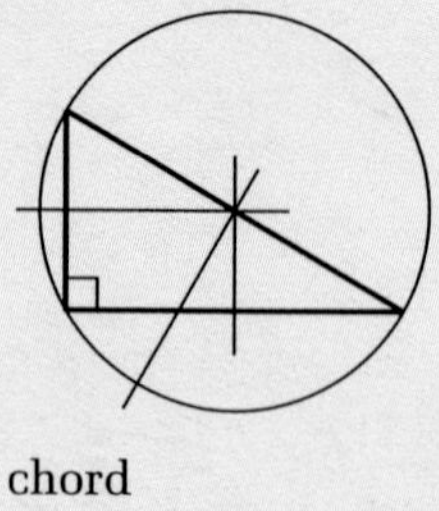

2. a. chord

   b. at O

3. agree; The statement is true because every triangle has a circumscribed circle, the center of which is equidistant from all three vertices.

4. a. $10^2 + 0^2 = 100$, so R lies on the circle. $6^2 + 8^2 = 100$, so S lies on the circle.

   b. (8, 4)

4. A circle with equation $x^2 + y^2 = 100$ has its center at the origin O and radius 10.

   a. Show that the points $R(10, 0)$ and $S(6, 8)$ lie on the circle.

   b. Find the midpoint of $\overline{RS}$.

   c. Find the slope of $\overline{RS}$ and the slope of a line perpendicular to $\overline{RS}$.

   d. Find an equation of the perpendicular bisector of $\overline{RS}$.

   e. Show that the perpendicular bisector of $\overline{RS}$ passes through the center of the circle (Theorem 3.6).

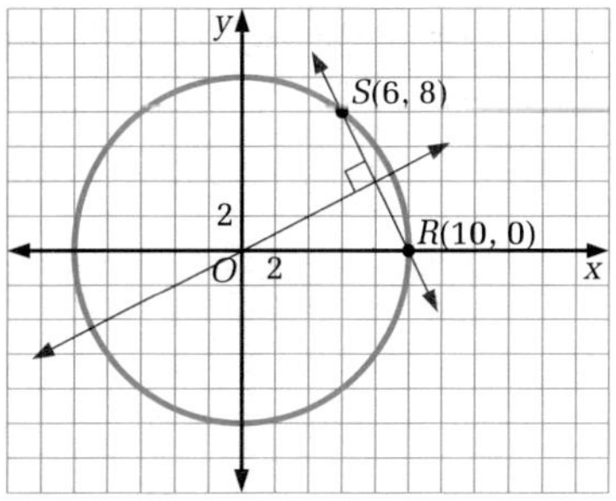

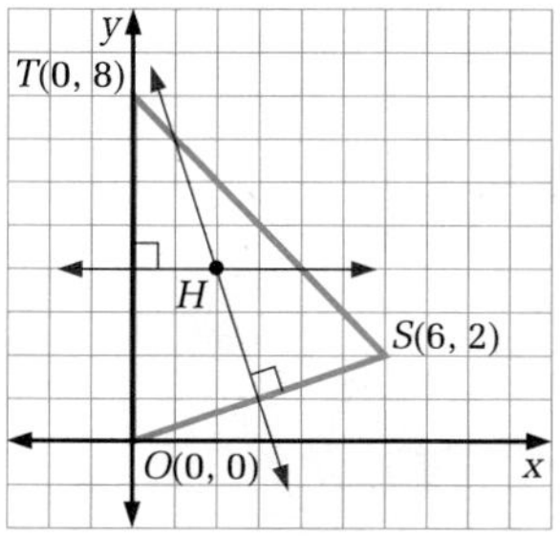

5. In coordinate geometry it is possible to inscribe a triangle in a circle by finding the equation of the circle. Start with $\triangle OST$ with vertices $O(0, 0)$, $S(6, 2)$, and $T(0, 8)$.

   a. Find the midpoints of $\overline{OT}$ and $\overline{OS}$.

   b. The perpendicular bisector of $\overline{OT}$ is a horizontal line. Find its equation.

   c. Find an equation of the perpendicular bisector of $\overline{OS}$.

   d. Find the coordinates of H, the point where these two perpendicular bisectors intersect.

   e. Find the distance from H to each vertex and the equation of the circle with center H passing through O, S, and T.

   f. Find an equation of the perpendicular bisector of $\overline{ST}$ and show that it also passes through H.

Find the measure of each arc or angle.

6. $\overset{\frown}{CD}$
7. $\angle AOB$
8. $\overset{\frown}{AED}$

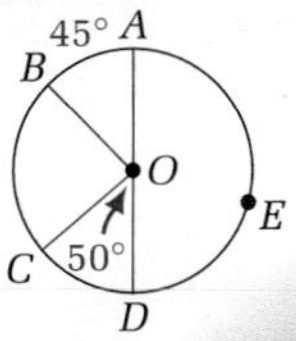

connection to GEOGRAPHY

9. Distances traveled by ships are usually measured in *nautical miles* rather than the more familar *statute* (land) *miles*.

   One nautical mile is defined as the length of $\frac{1}{60}°$ of arc of Earth's circumference.

   a. Find Earth's circumference in nautical miles.

   b. Use the fact that one nautical mile is about 1.15 statute miles to find Earth's circumference and diameter in statute miles.

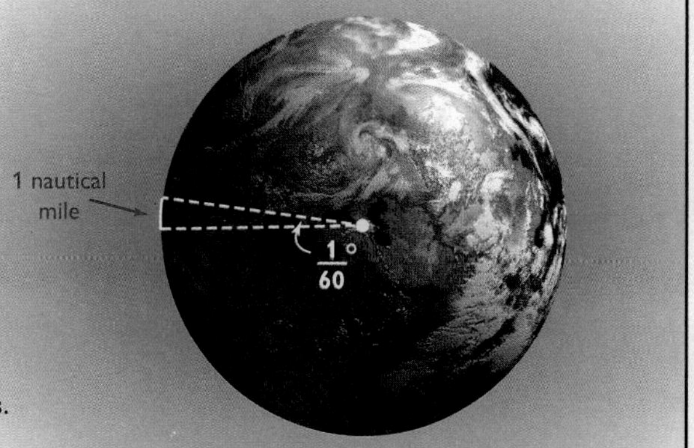

(Not drawn to scale)

Answers to Exercises and Problems

c. $-2; \frac{1}{2}$

d. $y = \frac{1}{2}x$

e. The center of the circle is (0, 0). $0 = \frac{1}{2}(0)$, so the perpendicular bisector of $\overline{SR}$ passes through the center of the circle.

5. a. (0, 4); (3, 1)

   b. $y = 4$

   c. $y = -3x + 10$

   d. (2, 4)

   e. $2\sqrt{5} \approx 4.5$; $(x - 2)^2 + (y - 4)^2 = 20$

   f. $y = x + 2$; $4 = 2 + 2$, so the perpendicular bisector of $\overline{ST}$ passes through H.

6. 50°

7. 45°

8. 180°

9. a. 21,600 nautical miles

   b. about 24,840 statute miles; about 7907 statute miles

Using Technology

The instance of Theorem 3.6 that students verify in Ex. 4 uses simple coordinates for all points. For more complicated coordinates, you can use a graphics calculator such as the TI-81 or TI-82.

First, graph Y1 = √(100–X²) and Y2 = –√(100–X²). Press [ZOOM] and select menu item 5 to "square-up" the graph.

You need to select endpoints (*A*, *B*) and (*C*, *D*) for a chord. Press [TRACE] and move the cursor to the point you would like to use for (*A*, *B*). Then press [2nd] [QUIT] to go to the home screen. Press [ALPHA] [X] [STO▶] [ALPHA] [A] [ENTER]. (For the TI-81, there is no need to press [ALPHA] after you press [STO▶].) This stores the *x*-coordinate of the point as the value of *A*. Press [ALPHA] [Y] [STO▶] [ALPHA] [B] [ENTER] to store the *y*-coordinate as the value of *B*. Follow a similar procedure to select and store the coordinates of a second point (*C*, *D*) (not on the same vertical line as (*A*, *B*)).

Now, calculate and store the coordinates (*E*, *F*) of the midpoint of the segment from (*A*, *B*) to (*C*, *D*). Work on the home screen. Type (A+C)/2→E and press [ENTER]. Next, type (B+D)/2→F and press [ENTER]. Finally, find the product of the slope of the chord through (*A*, *B*) and (*C*, *D*), and the slope of the line through (*E*, *F*) and (0, 0). To do this, type ((D–B)/(C–A))*F/E. When you press [ENTER], you get –1.

Variations on this procedure can be used to verify instances of a wide variety of geometric theorems. The variables for coordinates can be used with items from the DRAW menu to show any segments you want to display with the graph.

Interdisciplinary Problems

Exs. 9 and 15 connect the mathematics of this section to problems involving geography and social studies.

Multicultural Note

The Cherokee numbered about 29,000 and lived in the southern Appalachian mountains when Europeans first arrived in the Americas. The 17th and 18th centuries saw many Cherokee skirmishes with colonists and with other Native American peoples, but the early 19th century was a time of comparative peace and prosperity. In 1821, the Cherokee Sequoyah developed an alphabet of 85 characters to represent the Cherokee language in order to preserve his people's history and culture in writing in their native language. Eight years later, gold was discovered on Cherokee land and the peace was broken: the Cherokee were forcibly removed from their traditional land and made to march 800 miles to the west to settle in what is now Oklahoma. Today, there are an estimated 135,000 Cherokee descendants, most of whom live in Oklahoma or North Carolina.

10. Using number lines like the one shown, graph the sets of measures (in degrees) for:

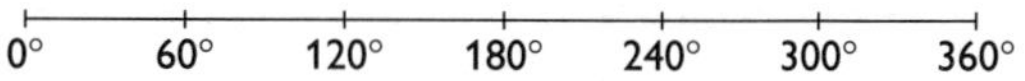

a. all possible major arcs

b. all possible minor arcs

11. Make a concept map for what you have learned about central angles and arcs.

12. A regular polygon with 15 sides is inscribed in a circle. Find the degree measure of each intercepted arc.

13. Equilateral $\triangle ABC$ is inscribed in circle O. Find each measure.

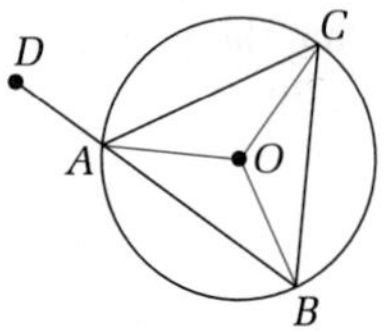

a. $m\angle AOB$ **b.** $m\widehat{AB}$

c. $m\angle OAB$ **d.** $m\angle DAC$

14. A regular polygon with n sides is inscribed in a circle. Prove that the measure of a central angle is equal to the measure of one of the exterior angles of the polygon.

connection to SOCIAL STUDIES

15. The Cherokee people had a system of government based on seven clans. Their council house consisted of seven sections, one for each clan, and its shape was a regular *heptagon*. Find the measure of each central angle of this figure.

BY THE WAY...

This design symbolizes the seven clans of the Cherokee people. The number 7 is sacred to the Cherokee. Besides having 7 clans, they use 7 kinds of wood to kindle the fire in their 7-sided council house, and they have 7 yearly festivals.

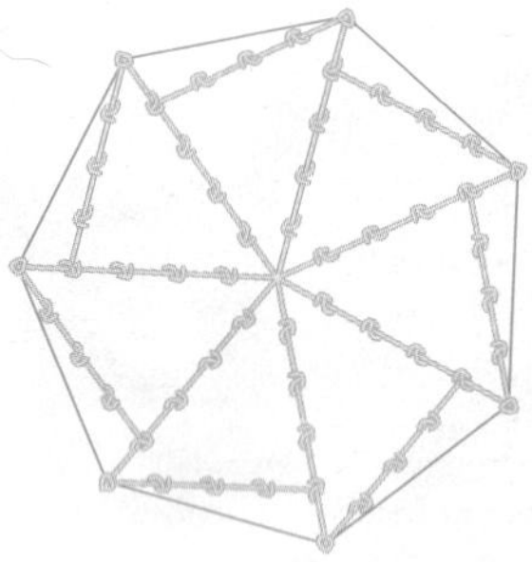

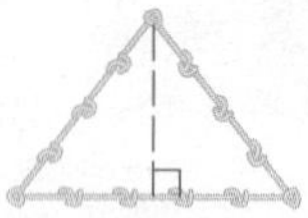

16. An almost-regular heptagon can be constructed using a rope with 13 equally-spaced knots. Form isosceles triangles with sides measuring 4 units, 4 units, and 5 units and a common central vertex. The outer vertices of the triangles mark the vertices of a heptagon.

Show that this method produces a central angle which is within one degree of the angle needed for a true regular heptagon. (*Hint:* Draw a right triangle and use one of the trigonometric ratios.)

Answers to Exercises and Problems

10. a.

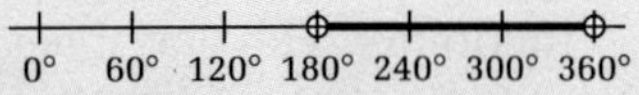

b.

11. Diagrams may vary. An example is given.

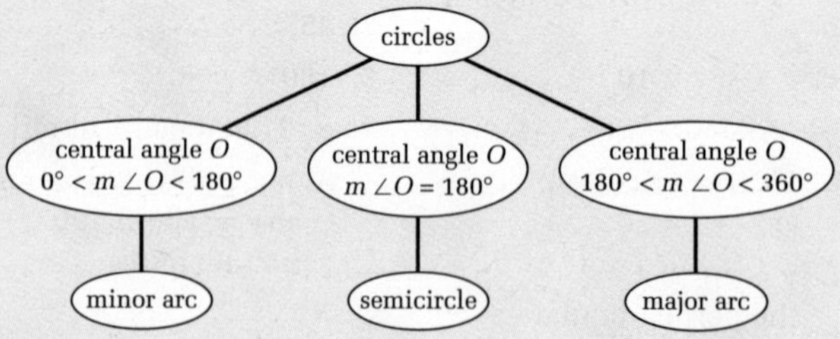

12. 24°

13. a. 120° **b.** 120°

c. 30° **d.** 120°

14. Since there are n central angles, all equal in measure, the measure of each is $\frac{360°}{n}$. The measure of each exterior angle of a regular polygon with n sides is also $\frac{360°}{n}$.

17. **Using Manipulatives** A regular hexagon can be constructed by compass and straightedge alone. Draw a circle and choose a point A on the circle. With the same radius as the circle, draw arcs at points B, C, D, E, and F on the circle. Join these points to form hexagon $ABCDEF$. Connect all points to the center O. Show that each central angle formed is 60° and that $ABCDEF$ is a regular hexagon.

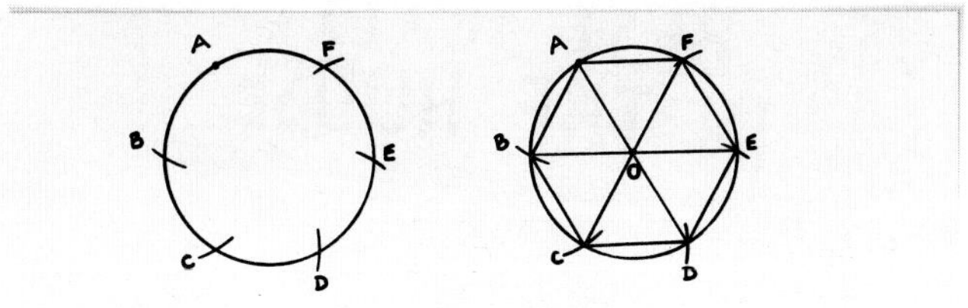

Ongoing ASSESSMENT

18. **Open-ended** Some quadrilaterals can be inscribed in a circle, but others cannot, as this parallelogram demonstrates. Make drawings or use geometric drawing software to determine which types of quadrilaterals can be inscribed in circles.

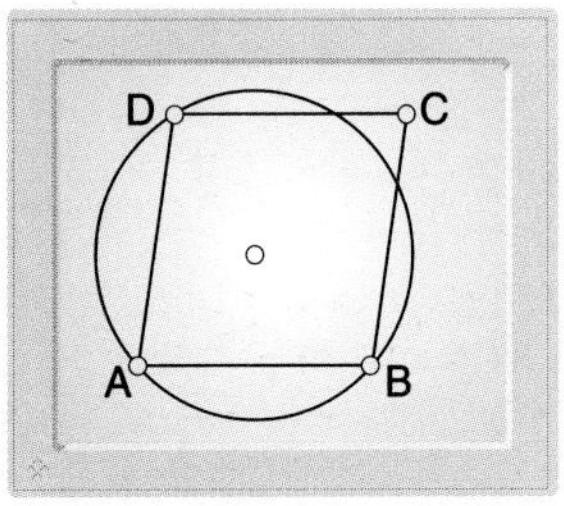

Review PREVIEW

19. What is the measure of an interior angle in a regular 13-gon? *(Section 3-4)*

20. Triangles ABC and DEF are similar triangles. What must be true about their corresponding angles? about their corresponding sides? *(Toolbox Skill 32)*

21. Find the value of each unknown angle measure. *(Toolbox Skill 30)*

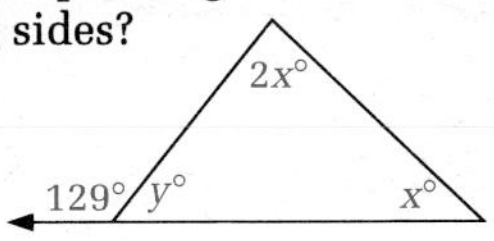

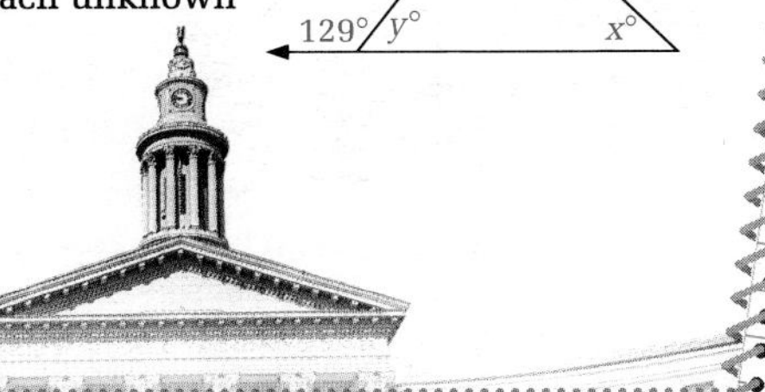

3 Working on the Unit Project

22. Read these simplified rules of evidence that will be used in deciding your case. Are these rules more like postulates or theorems in mathematics? Explain.

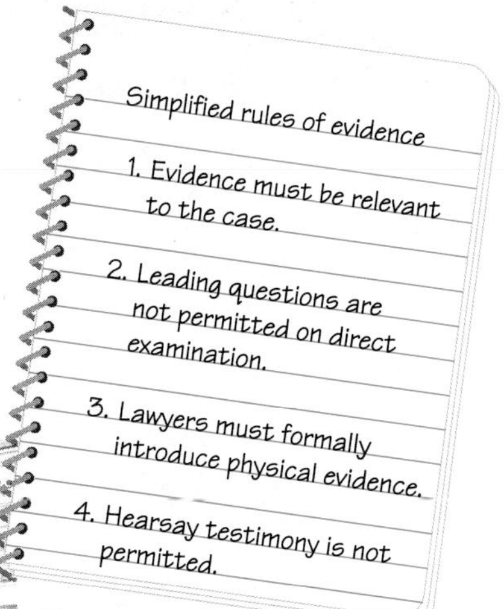

Simplified rules of evidence
1. Evidence must be relevant to the case.
2. Leading questions are not permitted on direct examination.
3. Lawyers must formally introduce physical evidence.
4. Hearsay testimony is not permitted.

The complete rules of evidence can be found in legal periodicals, historical documents, statutes, and trial records.

Assessment: Open-ended

For Ex. 18, students have found that a nonrectangular parallelogram cannot be inscribed in a circle. Have them draw parallelogram $ABCD$ with $A(0, 0)$, $B(2, 4)$, $C(6, 4)$, and $D(4, 0)$. They should find the equation of the circle which passes through A, B, and C and relocate D so that it lies on the circle.

Working on the Unit Project

Students should meet in their project groups to discuss Ex. 22. In order to prepare a sound legal brief, they will need to understand the law as it applies to their case.

Practice 23 For use with Section 3-5

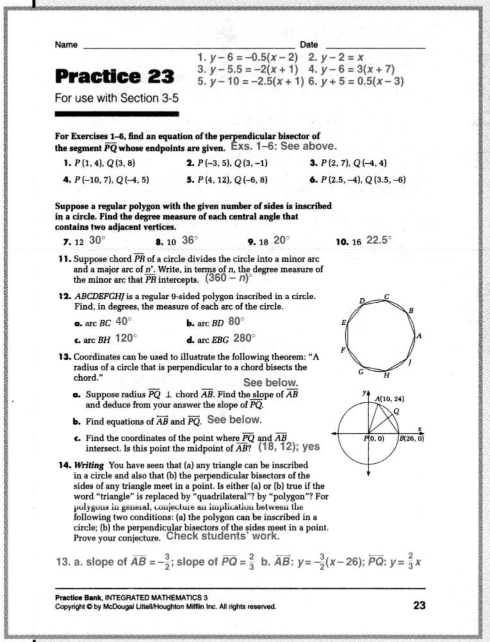

Name ____________ Date ________

Practice 23

For use with Section 3-5

1. $y - 6 = -0.5(x - 2)$ 2. $y - 2 = x$
3. $y - 5.5 = -2(x + 1)$ 4. $y - 6 = 3(x + 7)$
5. $y - 10 = -2.5(x + 1)$ 6. $y + 5 = 0.5(x - 3)$

For Exercises 1–6, find an equation of the perpendicular bisector of the segment $\overline{PQ}$ whose endpoints are given. Exs. 1–6: See above.

1. $P(1, 4)$, $Q(3, 8)$
2. $P(-3, 5)$, $Q(3, -1)$
3. $P(2, 7)$, $Q(-4, 4)$
4. $P(-10, 7)$, $Q(-4, 5)$
5. $P(4, 12)$, $Q(-6, 8)$
6. $P(2.5, -4)$, $Q(3.5, -6)$

Suppose a regular polygon with the given number of sides is inscribed in a circle. Find the degree measure of each central angle that contains two adjacent vertices.

7. 12 30°
8. 10 36°
9. 18 20°
10. 16 22.5°

11. Suppose chord $\overline{PR}$ of a circle divides the circle into a minor arc and a major arc of $n°$. Write, in terms of n, the degree measure of the minor arc that $\overline{PR}$ intercepts. $(360 - n)°$

12. $ABCDEFGHJ$ is a regular 9-sided polygon inscribed in a circle. Find, in degrees, the measure of each arc of the circle.
   a. arc BC 40°
   b. arc BD 80°
   c. arc BH 120°
   d. arc EBG 280°

D C B A J H G F E

13. Coordinates can be used to illustrate the following theorem: "A radius of a circle that is perpendicular to a chord bisects the chord." See below.
   a. Suppose radius $\overline{PQ} \perp$ chord $\overline{AB}$. Find the slope of $\overline{AB}$ and deduce from your answer the slope of $\overline{PQ}$.
   b. Find equations of $\overleftrightarrow{AB}$ and $\overleftrightarrow{PQ}$. See below.
   c. Find the coordinates of the point where $\overline{PQ}$ and $\overline{AB}$ intersect. Is this point the midpoint of $\overline{AB}$? (18, 12); yes

y A(10, 24) Q x P(0, 0) B(26, 0)

14. ***Writing*** You have seen that (a) any triangle can be inscribed in a circle and also that (b) the perpendicular bisectors of the sides of any triangle meet in a point. Is either (a) or (b) true if the word "triangle" is replaced by "quadrilateral"? by "polygon"? For polygons in general, conjecture an implication between the following two conditions: (a) the polygon can be inscribed in a circle; (b) the perpendicular bisectors of the sides meet in a point. Prove your conjecture. Check students' work.

13. a. slope of $\overline{AB} = -\frac{3}{2}$; slope of $\overline{PQ} = \frac{2}{3}$ b. $\overleftrightarrow{AB}$: $y = -\frac{3}{2}(x - 26)$; $\overleftrightarrow{PQ}$: $y = \frac{2}{3}x$

Practice Bank, INTEGRATED MATHEMATICS 3

23

Answers to Exercises and Problems

15. about 51.4°

16. When the altitude is drawn to the base of one of the isosceles triangles, two right triangles are formed. The cosine ratio can be used to find the measure of one of the acute angles, which is also a central angle of the circle in which the heptagon is inscribed. The cosine of the angle is $\frac{2.5}{4} = 0.625$; the measure of the angle is about 51.3°. The central angle formed by a true heptagon would be $\frac{360°}{7} \approx 51.4°$.

17. The chords formed by any two consecutive arcs are equal in measure to the radius of the circle, so the process produces six equilateral triangles. Each central angle is an angle of one of those triangles, and so has measure 60°. Each angle of the hexagon produced is the sum of two angles of equilateral triangles, and so has measure 120°. Since all the sides and all the angles of $ABCDEF$ are equal in measure, $ABCDEF$ is a regular hexagon by definition.

18. A square, a rectangle, a non-rectangular isosceles trapezoid, or a kite with two right angles can be inscribed in a circle. Various irregular quadrilaterals can also be inscribed in a circle.

19. about 152.3°

20. Corresponding angles are equal in measure and corresponding sides are in proportion.

21. 43°; 86°; 51°

22. Answers may vary. An example is given. The rules of evidence are more like postulates than theorems because the rules of evidence are basic assumptions accepted without proof.

PLANNING

Objectives and Strands

See pages 136A and 136B.

Spiral Learning

See page 136B.

Materials List

- Protractor
- Scissors
- Compass
- Straightedge

Recommended Pacing

Section 3-6 is a one-day lesson.

Extra Practice

See pages 612–613.

Warm-Up Exercises

Warm-Up Transparency 3-6

Support Materials

- Practice Bank: Practice 24
- Activity Bank: Enrichment 22
- Study Guide: Section 3-6
- Problem Bank: Problem Set 7
- Assessment Book: Quiz 3-6, Test 11, Alternative Assessment 4

Section 3-6 Arcs and Angles

Focus

Examine relationships between inscribed angles and arcs of circles.

BEST SEATS IN THE HOUSE

Helen and Tim are trying to decide where to sit in a theater.

"I want to sit close enough to get a good view of the stage," says Helen.

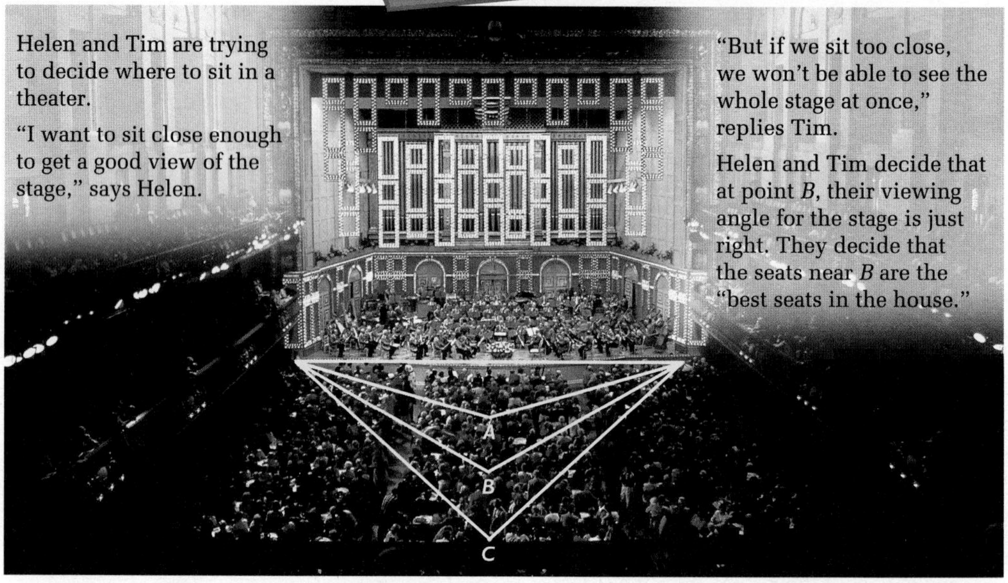

"But if we sit too close, we won't be able to see the whole stage at once," replies Tim.

Helen and Tim decide that at point B, their viewing angle for the stage is just right. They decide that the seats near B are the "best seats in the house."

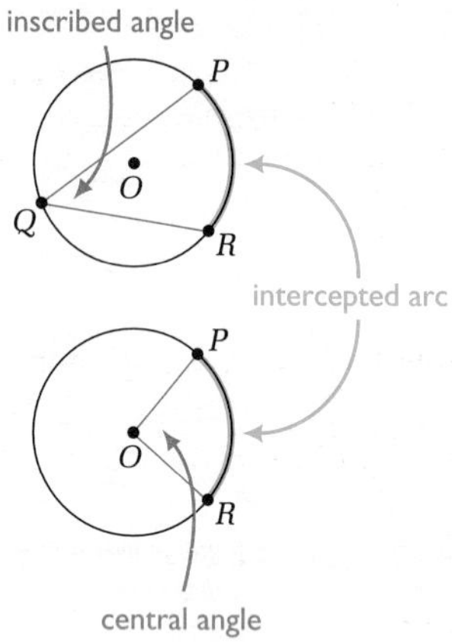

Talk it Over

1. Which point, A, B, or C, has the largest viewing angle for the stage? Which point has the smallest?
2. Are there any other places Helen and Tim can sit in the theater with the same viewing angle for the stage as point B? If so, where do you think these points are located?

The answer to question 2 is related to a theorem about inscribed angles in circles. An **inscribed angle** is an angle formed by two chords that intersect at a point on a circle. The arc that lies within an inscribed angle is called its **intercepted arc.**

You have learned that a minor arc is measured by finding its central angle, the angle formed by two radii from the center to the endpoints of the arc. The degree measure of an arc may be written near the center of the circle or along the arc itself.

Answers to Talk it Over

1. A; C
2. Yes; let X and Y be the endpoints of the stage. Consider a circle of which $\overline{XY}$ is a chord and B a point on the circle. The points with the same viewing angle are on $\overparen{XBY}$.

Sample 1

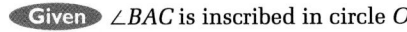

Prove that the measure of an inscribed angle of a circle is equal to half the measure of its intercepted arc.

Sample Response

Given $\angle BAC$ is inscribed in circle O.

Prove $m\angle BAC = \frac{1}{2}m\,\widehat{BC}$

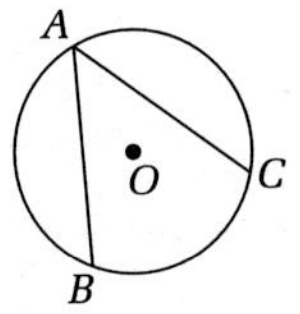

Plan Ahead

Show that $m\angle BAC = \frac{1}{2}m\angle BOC$. Then use the fact that $m\angle BOC = m\,\widehat{BC}$ and use the substitution property.

Show Your Reasoning

Draw $\overline{BC}$ and radii $\overline{OA}$, $\overline{OB}$, and $\overline{OC}$.
$\triangle AOB$, $\triangle BOC$, and $\triangle COA$ are isosceles.
Mark equal angles using x°, y°, and z°.

In $\triangle ABC$, $2x + 2y + 2z = 180$.
In $\triangle BOC$, $2x + m\angle BOC = 180$.
So, $2x + m\angle BOC = 2x + 2y + 2z$.
Using algebra and substitution,
$m\angle BOC = 2y + 2z$.

Since $m\angle BAC = y + z$ and $m\angle BOC = 2y + 2z$,
$m\angle BOC = 2\,m\angle BAC$.

Since the measure of an arc is equal to the measure of its central angle,
$m\,\widehat{BC} = m\angle BOC = 2\,m\angle BAC$.
Therefore, $\frac{1}{2}m\,\widehat{BC} = m\angle BAC$.

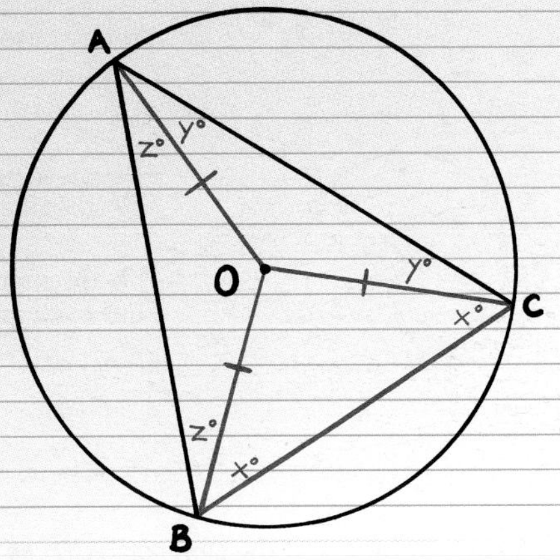

This proof assumes that point O lies inside $\angle BAC$. The theorem is also true when point O lies on a side of the angle or outside the angle. You will prove this in Exercises 7 and 8.

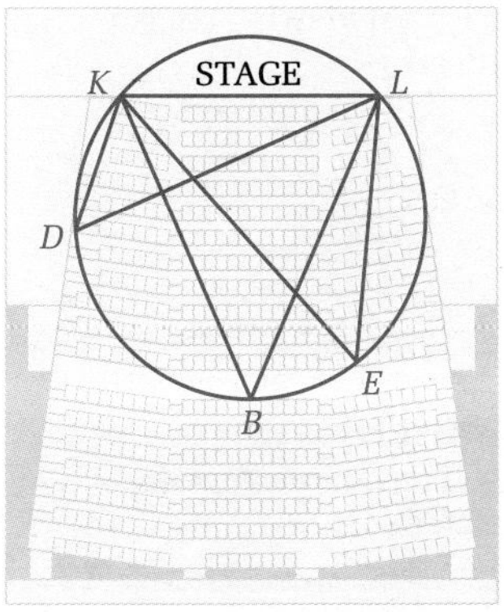

Talk it Over

3. In the diagram in Sample 1, what is the intercepted arc for $\angle BCA$?
4. What kind of arc is intercepted by an inscribed angle of 90°?
5. In the figure, why do $\angle D$, $\angle B$, and $\angle E$ all have the same measure?

Answers to Talk it Over

3. $\widehat{AB}$
4. a semicircle
5. They are inscribed angles that intercept the same arc, so each has measure $\frac{1}{2}\widehat{KL}$.

TEACHING

Talk it Over

Questions 1 and 2 prepare students for understanding the concepts of an inscribed angle and its intercepted arc and the theorems on page 174.

Additional Sample

S1 Use the result from Sample 1 to prove that the opposite angles of a quadrilateral inscribed in a circle are supplementary.

Given: Inscribed quadrilateral $ABCD$.

Prove: $\angle A$ and $\angle C$ are supplementary. $\angle B$ and $\angle D$ are supplementary.

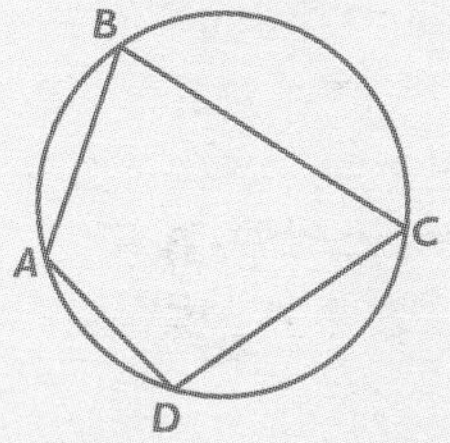

Proof: By the result from Sample 1, $m\angle A = \frac{1}{2}m\,\widehat{BCD}$ and $m\angle C = \frac{1}{2}m\,\widehat{BAD}$.
Thus, $m\angle A + m\angle C$
$= \frac{1}{2}m\,\widehat{BCD} + \frac{1}{2}m\,\widehat{BAD}$
$= \frac{1}{2}(m\,\widehat{BCD} + m\,\widehat{BAD})$.
Since $\widehat{BCD}$ and $\widehat{BAD}$ make up the whole circle, the sum of their measures is 360°. Therefore, $m\angle A + m\angle C = \frac{1}{2}(360°) = 180°$. Thus, $\angle A$ and $\angle C$ are supplementary.
A similar argument shows that $\angle B$ and $\angle D$ are supplementary.

Reasoning

It is important for students to understand that proofs must cover all possibilities in order to be valid. In proving the result in Sample 1, it is easy to overlook the role of the position of points A, B, and C in relation to the center O. Sample 1 can be used to illustrate the importance of examining the reasoning in a proof to be sure that it works for all cases. If it does not, then either the reasoning must be changed, or the problem must be analyzed in terms of special cases that cover all possibilities.

Additional Sample

S2 O is the center of the circle below. $\overline{BE}$ is a diameter. Find each angle or arc measure.

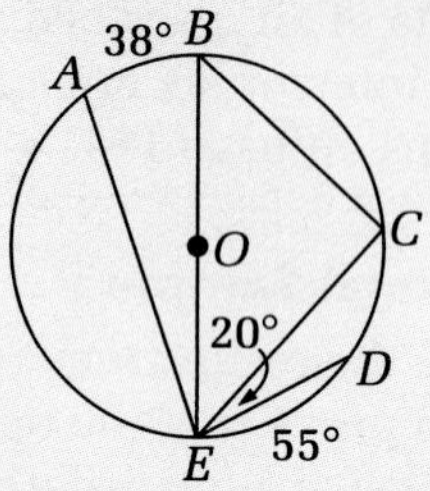

a. $m\angle BEC$

Since $\overline{BE}$ is a diameter, $\widehat{BCE}$ is a semicircle and, hence, has measure 180°.

By Theorem 3.7, $m\widehat{CD}$ must be 40°. Therefore, $m\widehat{BC} = 180° - (55° + 40°) =$ 85°, and $m\angle BEC = \frac{1}{2}m\widehat{BC} = 42.5°$.

b. $m\angle AED$

By Theorem 3.7, $m\angle AEB = 19°$. From part (a), $m\angle BEC = 42.5°$. Therefore, $m\angle AED = 19° + 42.5° + 20° = 81.5°$.

c. $m\widehat{AE}$

$m\widehat{AE} = m\widehat{EAB} - m\widehat{AB} = 180° - 38° = 142°$.

d. $m\angle EBC$

Since $\angle C$ intercepts a semicircle, $m\angle C = 90°$. From part (a), $m\angle BEC = 42.5°$. Since the sum of the measures of the angles of $\triangle EBC$ is 180°, $m\angle EBC = 180° - 90° - 42.5° = 47.5°$.

Look Back

These questions offer a good check on students' understanding of some of the concepts of this section. You may wish to have students discuss their answers with each other.

INSCRIBED ANGLE THEOREMS

Theorem 3.7 The measure of an inscribed angle of a circle is equal to half the measure of its intercepted arc.

Theorem 3.8 An inscribed angle whose intercepted arc is a semicircle is a right angle.

Theorem 3.9 If two inscribed angles in the same circle intercept the same arc, then they are equal in measure.

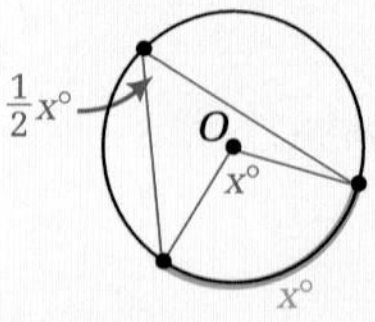

Theorem 3.7

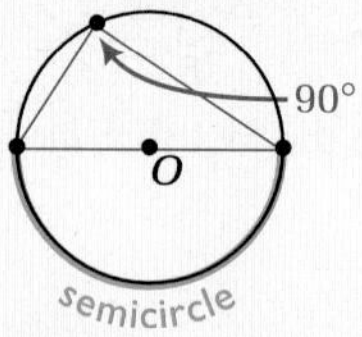

Theorem 3.8

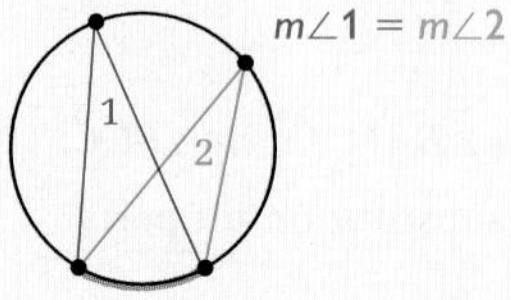

Theorem 3.9

You can use the theorems above to find angle measures and arc measures.

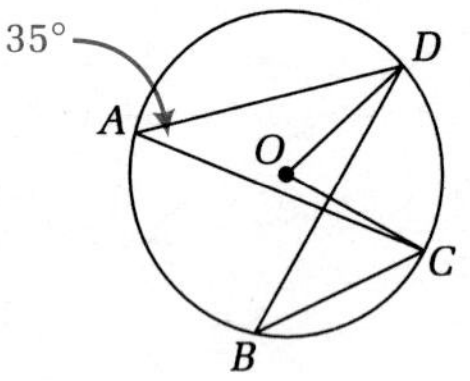

Sample 2

O is the center of the circle at the left. In circle O, $m\angle A = 35°$. Find each angle or arc measure.

a. $m\angle B$

b. $m\widehat{CD}$

c. $m\widehat{DAC}$

d. $m\angle COD$

Sample Response

a. $\angle A$ and $\angle B$ intercept the same arc, $\widehat{CD}$, so $m\angle A = m\angle B = 35°$.

b. The measure of an inscribed angle is half the measure of its intercepted arc, so $m\widehat{CD} = 2\,m\angle A = 70°$.

c. $\widehat{DAC}$ is a major arc. Since the measure of the entire circle is 360°, $m\widehat{DAC} = 360° - m\widehat{CD} = 360° - 70° = 290°$.

d. $\angle COD$ is a central angle and equal in measure to its intercepted arc, so $m\angle COD = m\widehat{CD} = 70°$.

Look Back

When are two inscribed angles in a circle equal in measure? When a central angle and an inscribed angle intercept the same arc, how are their measures related?

Answers to Look Back

Two inscribed angles in a circle are equal in measure when they intercept arcs that are equal in measure. The measure of the central angle is twice that of the inscribed angle.

3-6 Exercises and Problems

1. **Reading** What is the difference between an inscribed angle and a central angle?

Find all the missing measures of angles and arcs.

2. 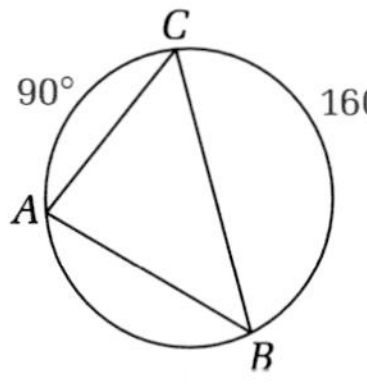

3. 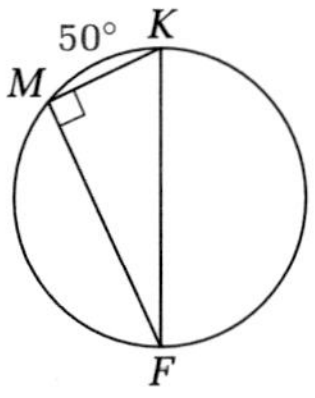

4. 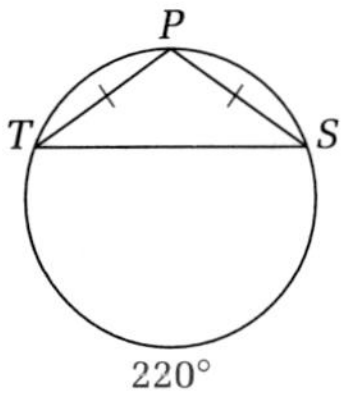

5. **Using Manipulatives** You can simulate the "best seat in the house" problem with a paper model of the theater and viewing angle.
   a. On one sheet of paper draw a segment to represent the stage.
   b. On another sheet of paper draw a viewing angle of 30° and cut it out.
   c. Place the viewing angle at many different positions, each time with the endpoints of the stage lying on the sides of the angle. Draw a dot next to the vertex of the viewing angle each time. What do you notice?

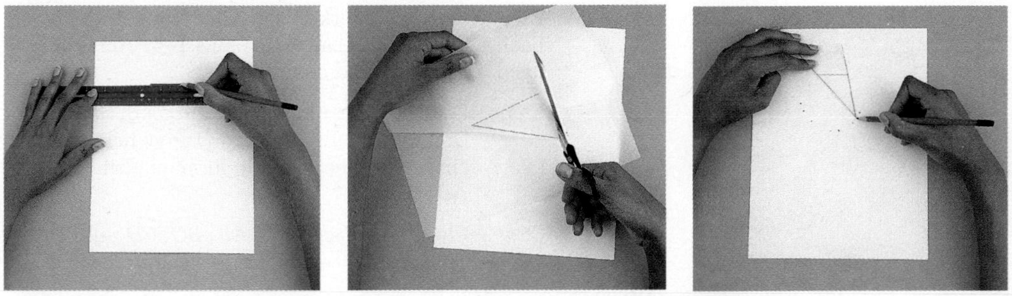

   d. Predict the result if you use a viewing angle of 45° instead of 30°. Cut out a viewing angle of 45° and check your prediction.

6. *DEFGH* is a regular pentagon inscribed in a circle.
   a. Find the measures of $\overset{\frown}{DE}$, $\overset{\frown}{EF}$, and $\overset{\frown}{FG}$.
   b. Use the results of part (a) to find $m\angle DHG$.
   c. Find $m\angle DHG$ by another method.

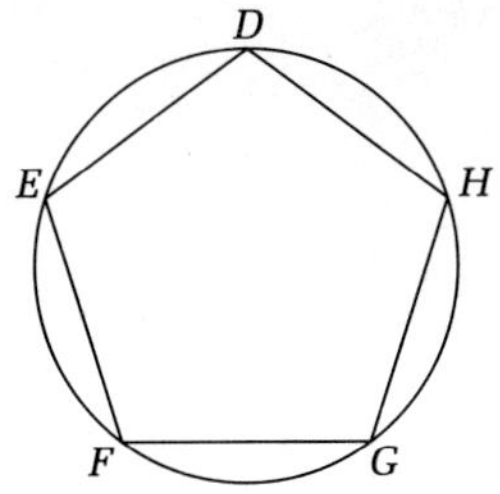

APPLYING

Suggested Assignment
Standard 1–4, 6–8, 11, 16–21
Extended 1–4, 6–8, 10, 11, 13–21

Integrating the Strands
Number Ex. 20
Algebra Exs. 7, 8, 10, 11
Geometry Exs. 1–17
Logic and Language Exs. 1, 16, 18, 19, 21

Error Analysis
A common error that some students make is to equate the measure of an inscribed angle with the measure of its intercepted arc. This error can be corrected by reminding students that in order for an angle and its intercepted arc to have the same measure, the angle must be a *central* angle; that is, one with its vertex at the center of the circle.

Answers to Exercises and Problems

1. The vertex of an inscribed angle of a circle is on the circle, so the sides of an inscribed angle are chords of the circle. The vertex of a central angle is at the center of the circle, so the sides of a central angle are radii of the circle.
2. $m\overset{\frown}{AB} = 110°$; $m\angle B = 45°$; $m\angle C = 55°$; $m\angle A = 80°$
3. $m\overset{\frown}{KF} = 180°$; $m\overset{\frown}{MF} = 130°$; $m\angle K = 65°$; $m\angle F = 25°$
4. $m\overset{\frown}{TP} = m\overset{\frown}{PS} = 70°$; $m\angle T = m\angle S = 35°$; $m\angle P = 110°$
5. a, b. Check students' work.
   c. Check students' work. The dots lie on a circle with radius equal to the measure of the segment representing the stage.
   d. Predictions may vary. The dots lie on a circle whose radius is the measure of the segment representing the stage divided by $\sqrt{2}$.
6. a. $m\overset{\frown}{DE} = m\overset{\frown}{EF} = m\overset{\frown}{FG} = 72°$
   b. 108°
   c. Methods may vary. An example is given. $\angle DHG$ is an angle of a regular pentagon, so $m\angle DHG = \frac{(5-2)(180°)}{5} = 108°$.

Cooperative Learning

After students complete Ex. 9, you may wish to have the groups state their conjectures to the class and present their proofs at the board.

7. Here is a diagram for a proof of Theorem 3.7 for the case where the center of the circle lies *on one of the sides* of the angle. Write a proof.

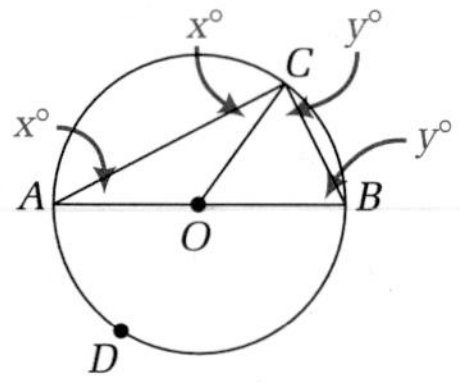

Given $\angle BAC$ is inscribed in circle O.

Prove $m\angle BAC = \frac{1}{2} m\,\widehat{BC}$

Plan Ahead

Show that $m\angle AOC = 2y$, $m\angle BOC = 2x$, and $x + y = 90°$.

8. Here is a diagram for a proof of Theorem 3.7 for the case where the center of the circle lies *outside* the angle.

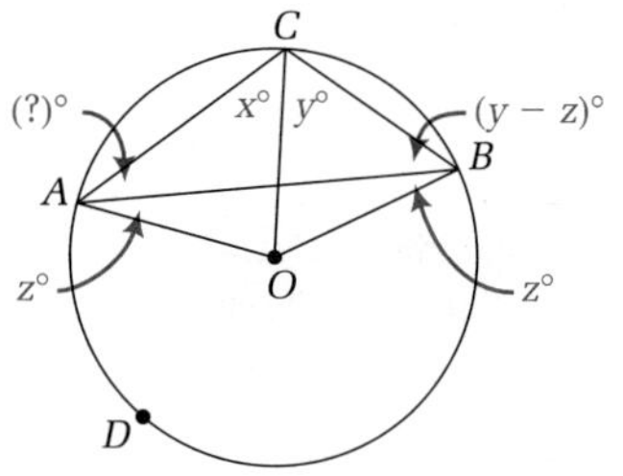

a. Write $m\angle BAC$ in terms of x and z.

b. Prove that $m\angle BAC = \frac{1}{2} m\widehat{BC}$.

Plan Ahead

Use $\triangle BOC$ to show that $2y + m\angle BOC = 180°$. Then write an equation based on the sum of the angle measures in $\triangle ABC$. Show that $m\angle BOC = 2x - 2z = 2m\angle BAC$.

9. Group Activity Work in a group of four students. Each of you should carry out steps 1–4. You will need protractors and compasses.

Step 1: Draw a circle of any size and mark its center. Draw points A, B, C, and D anywhere on the circle. Connect them to form quadrilateral $ABCD$.

Step 2: Find the measures of $\widehat{AB}$, $\widehat{BC}$, $\widehat{CD}$, and $\widehat{DA}$. The sum of these measures should be 360°.

Step 3: Find the measures of $\angle DAB$ and $\angle DCB$.

Step 4: Find the measures of $\angle ABC$ and $\angle ADC$.

What relationship do you find between opposite angles of your quadrilaterals? State your conjecture as a theorem and prove it.

10. Theorem 3.8 can be illustrated using coordinates of points on the circle with equation $x^2 + y^2 = 25$.

a. Show that each of these points lies on the circle: $P(-5, 0)$, $Q(5, 0)$, $A(-4, 3)$, $B(-3, 4)$, $C(0, 5)$, $D(3, 4)$, $E(4, 3)$.

b. Find the slopes of $\overline{PA}$, $\overline{PB}$, $\overline{PC}$, $\overline{PD}$, and $\overline{PE}$.

c. Find the slopes of $\overline{QA}$, $\overline{QB}$, $\overline{QC}$, $\overline{QD}$, and $\overline{QE}$.

d. Use slopes to explain why $\angle PAQ$, $\angle PBQ$, $\angle PCQ$, $\angle PDQ$, and $\angle PEQ$ are all right angles.

Answers to Exercises and Problems

7.

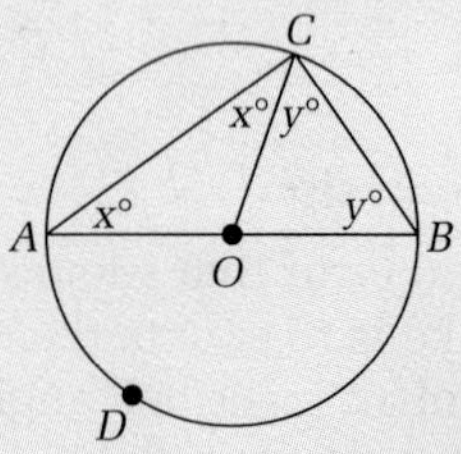

In $\triangle ABC$, $2x + 2y = 180°$. In $\triangle BOC$, $2y + m\angle BOC = 180°$. So $m\angle BOC = 2x$. Consequently, $m\angle BOC = 2m\angle BAC$. Since the measure of an arc is equal to the measure of its central angle, $m\widehat{BC} = m\angle BOC = 2m\angle BAC$. Therefore, $m\angle BAC = \frac{1}{2}m\widehat{BC}$.

8.

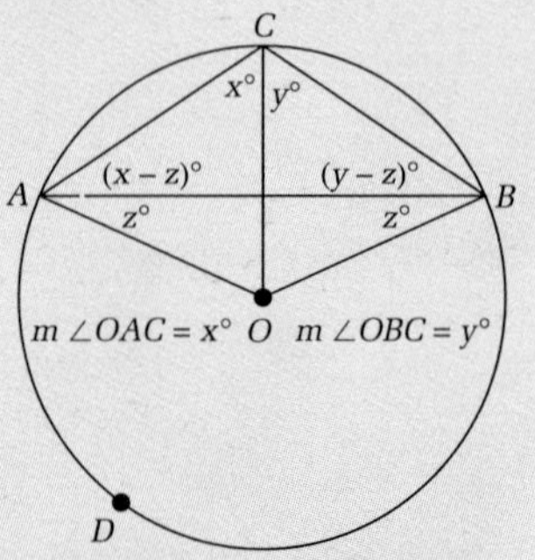

a. $(x - z)°$

b. $\overline{OA}$, $\overline{OC}$, and $\overline{OB}$ are all radii of circle O, so all three are equal in measure. Then $\triangle OAC$, $\triangle OCB$, and $\triangle OAB$ are isosceles triangles and x, y, and z are correctly marked on the diagram. In $\triangle ABC$, $2x + 2y - 2z = 180°$. In $\triangle BOC$, $2y + m\angle BOC = 180°$. So $m\angle BOC = 2x - 2z$. Consequently, $m\angle BOC = 2m\angle BAC$. Since the measure of an arc is equal to the measure of its central angle, $m\widehat{BC} = m\angle BOC = 2m\angle BAC$. Therefore, $m\angle BAC = \frac{1}{2}m\widehat{BC}$.

9. $\angle DAB$ and $\angle DCB$ are supplementary. $\angle ABC$ and $\angle ADC$ are also supplementary. Theorem: If quadrilateral $ABCD$ is inscribed in a circle, then opposite angles of $ABCD$ are supplementary. Proof: $m\angle DAB = \frac{1}{2}m\widehat{BCD}$ and $m\angle DCB = \frac{1}{2}m\widehat{DAB}$, so $m\angle DAB + m\angle DCB = \frac{1}{2}(m\widehat{BCD} + m\widehat{DAB}) = \frac{1}{2}(360°) = 180°$. Then $\angle DAB$ and $\angle DCB$ are supplementary. $m\angle ABC = \frac{1}{2}m\widehat{ADC}$ and $m\angle ADC = \frac{1}{2}m\widehat{ABC}$, so $m\angle ABC + m\angle ADC = \frac{1}{2}(m\widehat{ADC} + m\widehat{ABC}) = \frac{1}{2}(360°) = 180°$. Then $\angle ABC$ and $\angle ADC$ are supplementary.

10. a. P: $(-5)^2 + 0^2 = 25$; Q: $5^2 + 0^2 = 25$; A: $(-4)^2 + 3^2 = 16 + 9 = 25$; B: $(-3)^2 + 4^2 = 9 + 16 = 25$; C: $0^2 + 5^2 = 25$; D: $3^2 + 4^2 = 25$; E: $4^2 + 3^2 = 25$

b. slope of $\overline{PA}$: $\frac{3-0}{-4-(-5)} = 3$; slope of $\overline{PB}$: $\frac{4-0}{-3-(-5)} = 2$; slope of $\overline{PC}$: $\frac{5-0}{0-(-5)} = 1$; slope of $\overline{PD}$: $\frac{4-0}{3-(-5)} = \frac{1}{2}$; slope of $\overline{PE}$: $\frac{3-0}{4-(-5)} = \frac{1}{3}$

11. Write a coordinate proof of Theorem 3.8: An inscribed angle whose intercepted arc is a semicircle is a right angle.

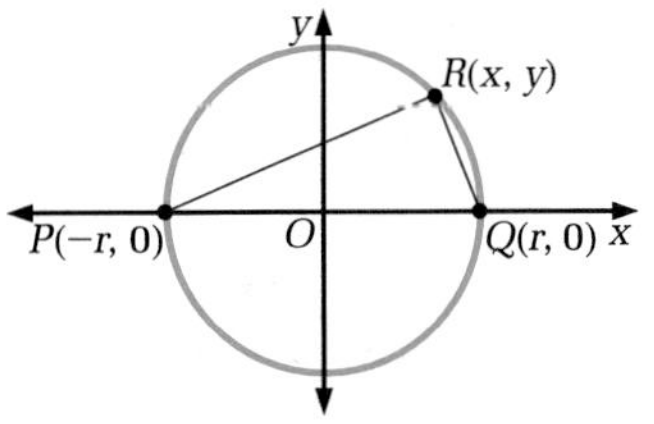

Given $P(-r, 0)$, $Q(r, 0)$, and $R(x, y)$; $x^2 + y^2 = r^2$

Prove $\angle PRQ$ is a right angle.

Plan Ahead

Show that the slope of $\overline{PR}$ is $\frac{y}{x + r}$ and the slope of $\overline{QR}$ is $\frac{y}{x - r}$. Then show that the product of these slopes is -1, using the fact that $x^2 - r^2 = -y^2$.

12. Two chords, $\overline{AB}$ and $\overline{CD}$, intersect inside a circle at point E.

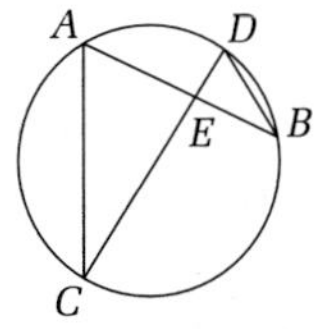

a. **Using Manipulatives** Draw a large circle and any two chords that intersect inside it. Label the triangles formed as in the diagram. Fold your drawing once through point E so that the similarity of the triangles is clear.

b. Prove that $\triangle ACE \sim \triangle DBE$.

Navigation Boat pilots have to be careful to avoid dangerous areas. Navigation charts often provide information about the "danger angle." Two landmarks on the shore at L and M are identified. A circle is drawn through L and M that encloses the danger area. $\angle LCM$ is called the danger angle.

13. Explain why any point C on major arc $\overset{\frown}{LNM}$ will give the same danger angle.

14. The pilot of the boat at B wants to stay away from the danger area. Should the pilot try to keep $m\angle LBM$ greater than or less than $m\angle LCM$? Explain.

15. **Open-ended** Find or draw a lake or ocean map with a danger area near the shore. Draw a circle that passes through two landmarks on the shore and that completely encloses the danger area. Measure the danger angle a boat pilot should use to avoid the danger area.

Assessment: Standard

For Ex. 12, students should be able to write proportions using the lengths of the sides of the similar triangles: $\frac{AE}{DE} = \frac{CE}{EB}$. They can then show that $AE \cdot EB = CE \cdot DE$ and state this as a theorem.

Application

Exs. 13–15 apply the mathematics of inscribed angles and arcs of circles to a practical problem that involves the navigation of a boat.

Answers to Exercises and Problems

c. slope of $\overline{QA}$: $\frac{3-0}{-4-5} = -\frac{1}{3}$; slope of $\overline{QB}$: $\frac{4-0}{-3-5} = \frac{1}{2}$; slope of $\overline{QC}$: $\frac{5-0}{0-5} = -1$; slope of $\overline{QD}$: $\frac{4-0}{3-5} = -2$; slope of $\overline{QE}$: $\frac{3-0}{4-5} = -3$

d. $\overline{PA} \perp \overline{QA}$, $\overline{PB} \perp \overline{QB}$, $\overline{PC} \perp \overline{QC}$, $\overline{PD} \perp \overline{QD}$, and $\overline{PE} \perp \overline{QE}$, since the slopes of each pair of lines are negative reciprocals. So the given angles are all right angles.

11. The slope of $\overline{PR}$ is $\frac{y-0}{x-(-r)} = \frac{y}{x+r}$. The slope of $\overline{QR}$ is $\frac{y-0}{x-r} = \frac{y}{x-r}$. Since $x^2 - r^2 = -y^2$, the product of the slopes is $\left(\frac{y}{x+r}\right)\left(\frac{y}{x-r}\right) = \frac{y^2}{x^2-r^2} = \frac{y^2}{-y^2} = -1$. Then $\overline{PR} \perp \overline{QR}$ and $\angle PRQ$ is a right angle.

12. a. Fold the drawing along the bisector of $\angle AED$.

b. $m\angle A = m\angle D$, since both inscribed angles intercept the same arc. $m\angle C = m\angle B$, since both inscribed angles intercept the same arc. Then $\triangle ACE \sim \triangle DBE$ by AA.

13. The danger angle for any point C on $\overset{\frown}{LNM}$ intercepts $\overset{\frown}{LM}$ and has measure $\frac{1}{2} m\overset{\frown}{LM}$.

14. less than; The farther from the danger area, the less the measure of $\angle LBM$ will be.

15. Answers may vary. An example is shown.

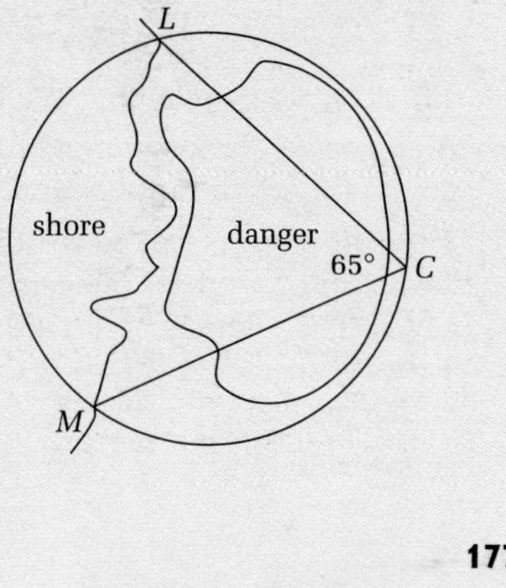

Working on the Unit Project

Each student should work Ex. 21 independently. Then they can compare their questions and explanations as a group activity.

Quick Quiz (3-4 through 3-6)

See page 203.

See also Test 11 in the *Assessment Book.*

Practice 24 For use with Section 3-6

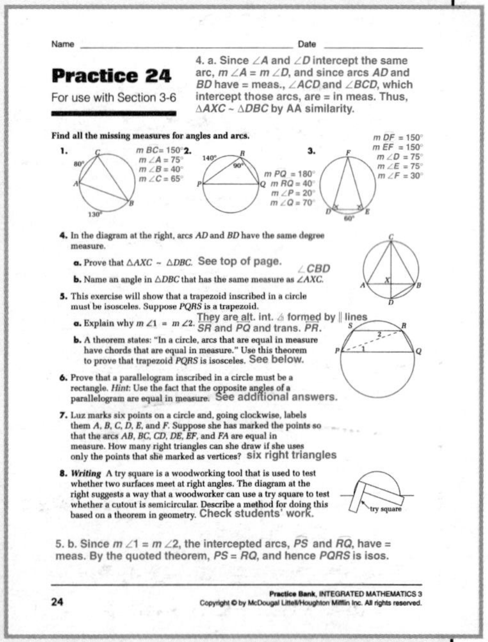

Name ______________________ Date ____________

Practice 24

For use with Section 3-6

4. a. Since $\angle A$ and $\angle D$ intercept the same arc, $m\angle A = m\angle D$, and since arcs AD and BD have = meas., $\angle ACD$ and $\angle BCD$, which intercept those arcs, are = in meas. Thus, $\triangle AXC \sim \triangle DBC$ by AA similarity.

Find all the missing measures for angles and arcs.

1. $m\,BC = 150°$; $m\angle A = 75°$; $m\angle B = 40°$; $m\angle C = 65°$
2. $m\,PQ = 180°$; $m\,RQ = 40°$; $m\angle P = 20°$; $m\angle Q = 70°$
3. $m\,DF = 150°$; $m\,EF = 150°$; $m\angle D = 75°$; $m\angle E = 75°$; $m\angle F = 30°$

4. In the diagram at the right, arcs AD and BD have the same degree measure.
   a. Prove that $\triangle AXC \sim \triangle DBC$. See top of page.
   b. Name an angle in $\triangle DBC$ that has the same measure as $\angle AXC$. $\angle CBD$
5. This exercise will show that a trapezoid inscribed in a circle must be isosceles. Suppose $PQRS$ is a trapezoid.
   a. Explain why $m\angle 1 = m\angle 2$. They are alt. int. ∠s formed by ∥ lines $\overline{SR}$ and $\overline{PQ}$ and trans. $\overline{PR}$.
   b. A theorem states: "In a circle, arcs that are equal in measure have chords that are equal in measure." Use this theorem to prove that trapezoid $PQRS$ is isosceles. See below.
6. Prove that a parallelogram inscribed in a circle must be a rectangle. *Hint*: Use the fact that the opposite angles of a parallelogram are equal in measure. See additional answers.
7. Luz marks six points on a circle and, going clockwise, labels them A, B, C, D, E, and F. Suppose she has marked the points so that the arcs AB, BC, CD, DE, EF, and FA are equal in measure. How many right triangles can she draw if she uses only the points that she marked as vertices? six right triangles
8. ***Writing*** A try square is a woodworking tool that is used to test whether two surfaces meet at right angles. The diagram at the right suggests a way that a woodworker can use a try square to test whether a cutout is semicircular. Describe a method for doing this based on a theorem in geometry. Check students' work.

5. b. Since $m\angle 1 = m\angle 2$, the intercepted arcs, $\widehat{PS}$ and $\widehat{RQ}$, have = meas. By the quoted theorem, $PS = RQ$, and hence $PQRS$ is isos.

24 Practice Bank, INTEGRATED MATHEMATICS 3

Ongoing ASSESSMENT

16. **Writing** Explain how the types of angles in a triangle are related to the types of arcs they intercept when the triangle is inscribed in a circle.

Review PREVIEW

17. Draw a regular octagon inscribed in a circle. What is the measure of the arc intercepted by one side of the polygon? *(Section 3-5)*

For Exercises 18 and 19, what conclusion can you reach from each set of conditions? *(Toolbox Skill 35)*

18. Tanya eats either taco salad or pizza for lunch. She does not eat pizza.
19. Akio filled out an application with either a black pen, a blue pen, or a pencil. He did not use a pencil.
20. Show that if two integers are odd, then their sum is even. *(Toolbox Skills 13 and 17)*

3 Working on the Unit Project

21. In writing a proof, the order of the steps is important. In questioning a witness, the order in which questions are asked is important.

    Pick one of your witnesses and write four questions you plan to ask. Put them in order. Explain why you chose that order.

Unit 3 CHECKPOINT 2

1. **Writing** Which has larger interior angles, a regular 18-gon or a regular 19-gon? Which has larger exterior angles? Explain. 3-4
2. Describe how you can find the center of a circle that circumscribes a given triangle. 3-5

Find all the missing measures of angles and arcs. 3-6

3.

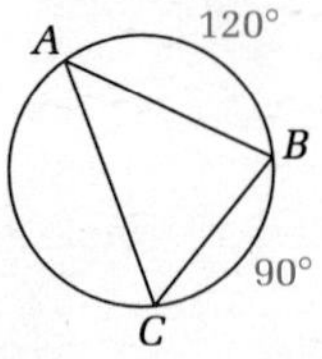

4.

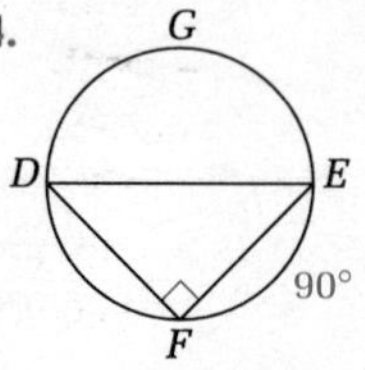

5.

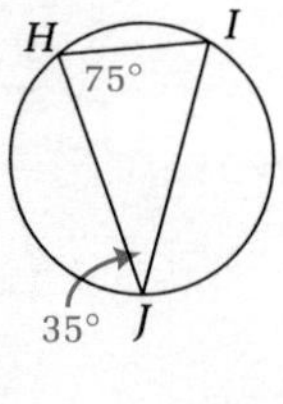

Answers to Exercises and Problems

16. An acute angle intercepts a minor arc. A right angle intercepts a semicircle. An obtuse angle intercepts a major arc.
17. 45°
18. Tanya eats taco salad for lunch.
19. Akio used a black pen or a blue pen to fill out the application.
20. The odd integers can be written $2k + 1$ and $2n + 1$ for integers k and n. Their sum is $2k + 1 + 2n + 1 = 2k + 2n + 2 = 2(k + n + 1)$, which is even.
21. Answers may vary.

Answers to Checkpoint

1. a regular 19-gon; a regular 18-gon; The measure of an interior angle of a regular n-gon is $\frac{(n-2)180°}{n} = 180° - \frac{360°}{n}$, which increases as n increases. The measure of an exterior angle of a regular n-gon is $\frac{360°}{n}$, which decreases as n increases.
2. Construct the perpendicular bisectors of two of the sides. The point where the bisectors intersect is the center of the circumscribed circle. The radius is the distance from the center to any vertex of the triangle.
3. $m\,\widehat{AC} = 150°$; $m\angle A = 45°$; $m\angle B = 75°$; $m\angle C = 60°$
4. $m\,\widehat{DGE} = 180°$; $m\,\widehat{DF} = 90°$; $m\angle D = 45°$; $m\angle E = 45°$
5. $m\angle I = 70°$; $m\,\widehat{JI} = 150°$; $m\,\widehat{HJ} = 140°$; $m\,\widehat{HI} = 70°$

Section 3-7

Indirect Proof

Focus

Use indirect proof in geometric, algebraic, and everyday contexts. Examine the properties of tangents to circles.

Say It Isn't So

Here is a puzzle. Two opposite corners of a checkerboard have been removed. You have some dominoes, each of which covers two squares of the board. Can you cover the board with dominoes? No dominoes may overlap or spill over the edge of the checkerboard.

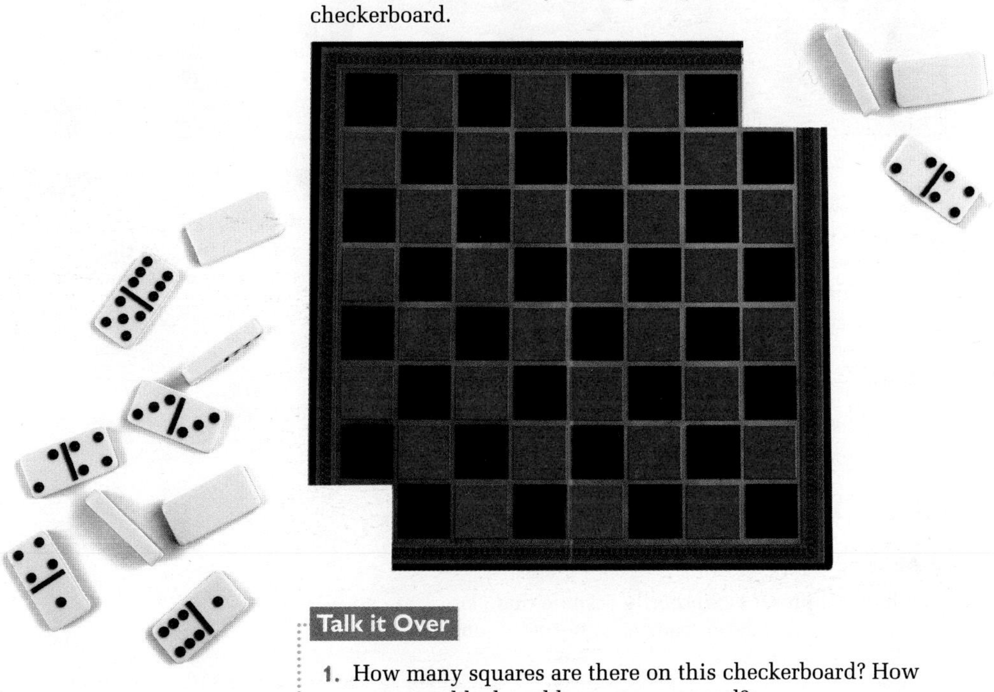

Talk it Over

1. How many squares are there on this checkerboard? How many are black and how many are red?
2. How many dominoes would it take to cover the checkerboard?
3. If one half of a domino covers a black square, what color must the other half cover?
4. If the correct number of dominoes covers the checkerboard, how many black squares will be covered? How many red squares?
5. Is it possible to solve the domino-checkerboard puzzle? Explain.

PLANNING

Objectives and Strands

See pages 136A and 136B.

Spiral Learning

See page 136B.

Recommended Pacing

Section 3-7 is a two-day lesson.

Day 1

Pages 179–181: Opening paragraph through Sample 2, *Exercises 1–10*

Day 2

Pages 181–183: Tangents and Secants through Look Back, *Exercises 11–22*

Extra Practice

See pages 612–613.

Warm-Up Exercises

Warm-Up Transparency 3-7

Support Materials

- Practice Bank: Practice 25
- Activity Bank: Enrichment 23
- Study Guide: Section 3-7
- Problem Bank: Problem Set 8
- Using TI-81 and TI-82 Calculators: Tangent Lines to a Circle
- Using Plotter Plus: Tangent Lines to a Circle
- Assessment Book: Quiz 3-7, Alternative Assessment 5

Answers to Talk it Over

1. 62 squares; 32 black squares, 30 red squares
2. 31 dominoes
3. red
4. 31 black; 31 red
5. No; there are only 30 red squares to be covered. There are 32 black squares.

TEACHING

Talk it Over

Questions 1–5 use a puzzle involving a checkerboard to introduce students to the concept of indirect proof.

Teaching Tip

Ask students for examples of indirect proof from everyday situations. You may find it especially helpful to consider arguments like those that lawyers use in courtroom situations.

Additional Sample

S1 Prove that if l, m, and n are three different lines in the same plane such that $l \parallel m$ and $l \parallel n$, then $m \parallel n$.

Given: Lines l, m, and n are three different lines. Lines l, m, and n lie in the same plane. $l \parallel m$, $l \parallel n$

Prove: $m \parallel n$

Proof: Assume that $m \nparallel n$. Then m and n intersect in one point P. There are then two lines that pass through P and are parallel to line l. This contradicts Postulate 12, which states that through a point not on a given line, there is one and only one line parallel to the given line. Thus, the assumption must be false, which means that $m \parallel n$.

Assuming the Opposite

The domino-checkerboard puzzle illustrates *indirect proof.* In an **indirect proof,** you show that if you temporarily assume the opposite of what you want to prove, you are led to an impossible situation or a contradiction of a known fact.

You don't believe me? Suppose I could manage to cover all the squares. Each domino would cover one black square and one red square.
Sure, that's obvious.
Altogether I would have covered 31 red squares and 31 black squares. But I cut out two red squares so I have only 30 red squares left.
So what we just assumed could be done is impossible!
Right. And that's why we can be sure that we'll never be able to cover the board with 31 dominoes.

Sample 1

Prove that if two lines lie in the same plane and they are perpendicular to the same line, then they are parallel.

Sample Response

Given Line j is perpendicular to line m at A.
Line k is perpendicular to line m at B.
Lines j, k, and m lie in the same plane.

Prove $j \parallel k$

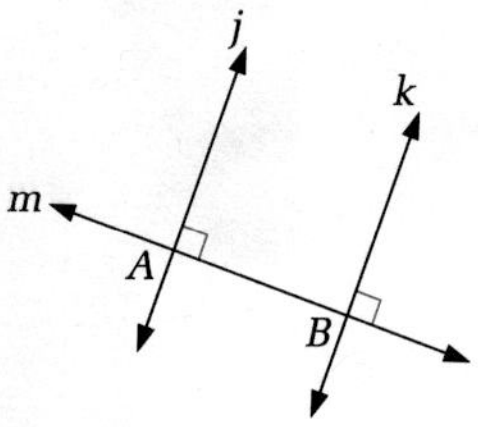

Plan Ahead

Use an indirect proof. Temporarily assume that j and k are not parallel. Show that this assumption leads to a contradiction.

Show Your Reasoning

Suppose that j and k are not parallel. Then since they lie in the same plane, they must intersect. Label as C the point at which they intersect.

Then in $\triangle ABC$, $\angle CAB$ is a right angle and $\angle CBA$ is a right angle, since perpendicular lines form right angles. The sum of the measures of the three angles in the triangle is $90° + 90° + m\angle ACB$. This sum must be greater than $180°$.

This contradicts the Triangle Sum Theorem, so the assumption that j and k are not parallel must be false. Therefore $j \parallel k$.

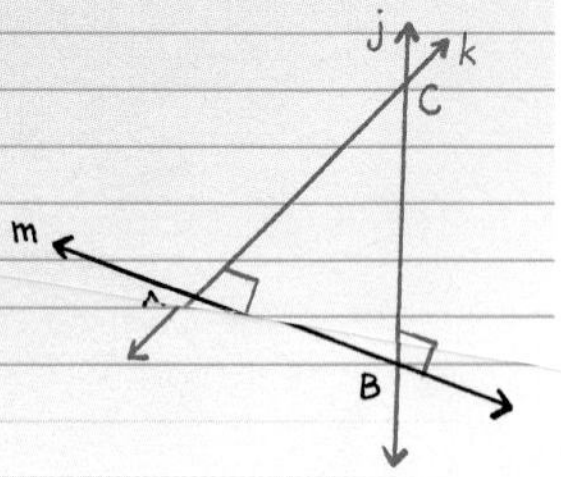

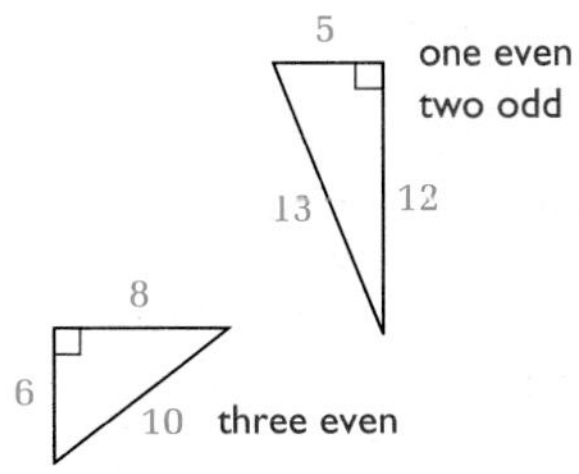

Sample 2

Three positive integers, a, b, and c, that satisfy the equation $a^2 + b^2 = c^2$ are called a *Pythagorean triple.* Prove that every Pythagorean triple contains at least one even number. Use an indirect proof.

Sample Response

Suppose there is a Pythagorean triple in which none of the numbers is even. That means a, b, and c are all odd numbers.

Then a^2 is an odd number, because $a^2 = a \cdot a$, and the product of two odd numbers is odd. For the same reason, b^2 and c^2 must be odd.

Then $a^2 + b^2$ is the sum of two odd numbers, which is even. But $a^2 + b^2 = c^2$, and c^2 is odd (Exercise 20, page 178).

This is a contradiction, because $a^2 + b^2$ cannot be even and odd at the same time, so the original assumption that there is a Pythagorean triple with three odd numbers must be false.

Therefore, every Pythagorean triple contains at least one even number.

Now you are ready for:
Exs. 1–10 on pp. 183–185

Tangents and Secants

If a line intersects a circle, it does so in either one point or two points.

TANGENTS AND SECANTS

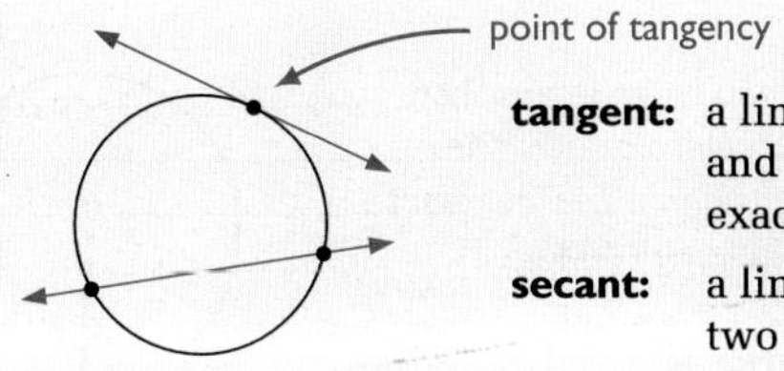

tangent: a line in the plane of a circle and intersecting the circle in exactly one point

secant: a line intersecting the circle in two points

Additional Sample

S2 Prove that if a, b, and c are a Pythagorean triple, then no number of the triple can be twice one of the other numbers.

Proof: Suppose that one of the numbers, say b, is twice one of the other numbers, say, a. Then the numbers of the triple are a, $2a$, and c. One of the numbers must have a square equal to the sum of the squares of the other two. It cannot be a, since the square of $2a$ will be greater than the square of a. Therefore, $(2a)^2 = a^2 + c^2$ or $c^2 = a^2 + (2a)^2$. If the first equation is solved for c, then $c = a\sqrt{3}$. If the second equation is solved for c, then $c = a\sqrt{5}$. But $\sqrt{3}$ and $\sqrt{5}$ are not rational numbers, so neither $a\sqrt{3}$ nor $a\sqrt{5}$ can be an integer. On the other hand, c is an integer, since Pythagorean triples must be integers. This is a contradiction. So the initial assumption is false. Therefore, no number of the triple can be twice one of the others.

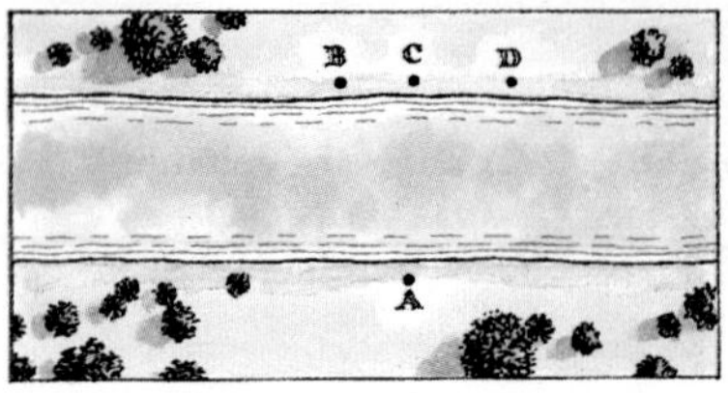

Talk it Over

6. Which point on a tangent line is closest to the center of the circle?
7. A city planner wants to build a bridge to cross the river from point A. To which point on the opposite shore should the planner build to make the shortest bridge?

SHORTEST PATH POSTULATE

A segment can be drawn perpendicular to a given line from a point not on the line. The length of this segment is the shortest distance from the point to the line.

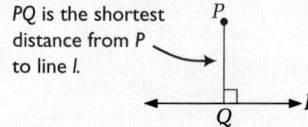

PQ is the shortest distance from P to line l.

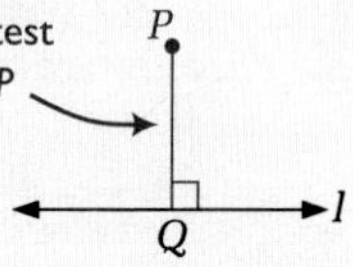

Sample 3

Prove that if a line is tangent to a circle, then the line is perpendicular to the radius drawn from the center to the point of tangency.

Sample Response

Given Line t is tangent to circle O at P.

Prove $\overline{OP}$ is perpendicular to t.

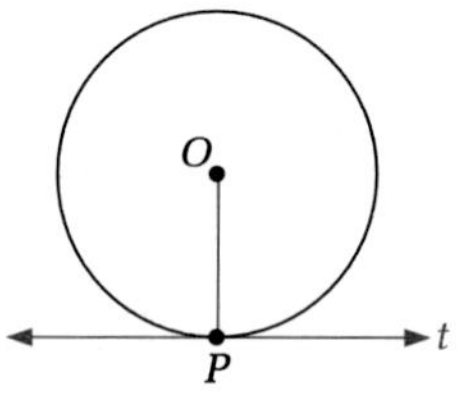

Plan Ahead

Use the Shortest Path Postulate in an indirect proof. Temporarily assume that $\overline{OP}$ is not perpendicular to t.

Show Your Reasoning

Suppose that $\overline{OP}$ is not perpendicular to t. Then there is another segment $\overline{OM}$ from O to line t which is perpendicular to t. By the Shortest Path Postulate, $OM < OP$. Because a tangent line intersects a circle in only one point, M lies outside the circle and $OM > OP$.

It is impossible for $\overline{OM}$ to be both shorter and longer than the radius $\overline{OP}$. So the assumption that $\overline{OP}$ is not perpendicular to t is false. Therefore, $\overline{OP}$ is perpendicular to t.

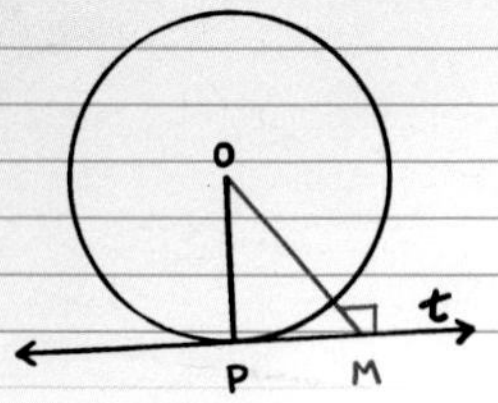

Answers to Talk it Over

6. the point of tangency
7. C

Talk it Over

Questions 6 and 7 lead students to think about two situations that illustrate the Shortest Path Postulate.

Additional Sample

S3 Prove that if circle O and circle Q are tangent to line t at the same point P, and points O and Q are on opposite sides of line t, then P is the only point where circles O and Q intersect.

Given: Circle O is tangent to line t at P. Circle Q is tangent to line t at P. Points O and Q are on opposite sides of line t.

Prove: P is the only point where circles O and Q intersect.

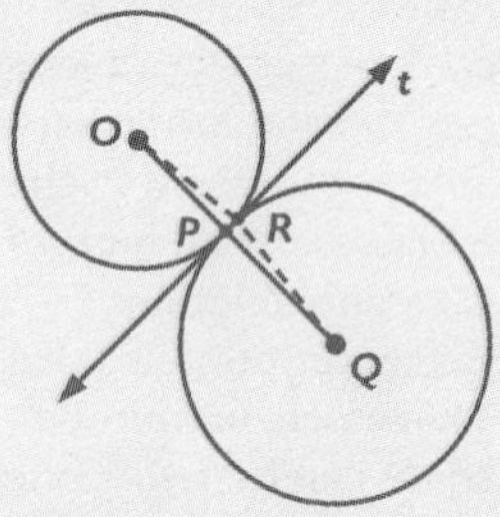

Proof: Circles O and Q intersect at P, since it is given that each circle is tangent to t at P. Suppose that there is a second point, R, at which the circles intersect. From the result in Sample 3, $\overline{OP} \perp t$ and $\overline{QP} \perp t$. Therefore, $\angle OPQ$ is a straight angle. Thus, OPQ is the shortest path from O to Q. But the length of this path is OP + PQ, and the length of ORQ is OR + RQ = OP + PQ, since OP and OR are radii of circle O and PQ and RQ are radii of circle Q. Thus, ORQ has the same length as the shortest path from O to Q. Since there is only one shortest path from O to Q, R must lie on $\overline{OQ}$ and be the same point as P. This contradicts the assumption that R is a *second* point of intersection of the two circles. Therefore, circle O and circle Q intersect only at P.

THEOREMS ABOUT TANGENTS

Theorem 3.10 If a line is tangent to a circle, then the line is perpendicular to the radius drawn from the center to the point of tangency.

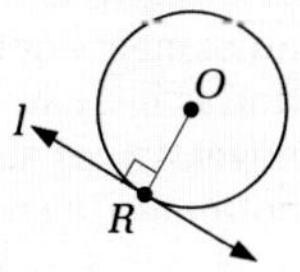

Theorem 3.11 If a line in the plane of a circle is perpendicular to a radius at its outer endpoint, then the line is tangent to the circle.

Theorem 3.12 If two tangent segments are drawn from the same point to the same circle, then they are equal in measure.

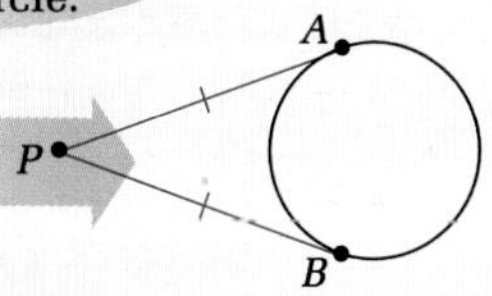

Now you are ready for: Exs. 11–22 on pp. 185–186

Look Back

Describe the technique of indirect proof.

3-7 Exercises and Problems

1. **Reading** There are four indirect proofs in this section. Complete this table to summarize them.

	To prove	Temporarily assume the opposite
Domino-Checkerboard Puzzle	The board cannot be covered.	It can be covered.
Sample 1	line $j \parallel$ line k	?
Sample 2	All three numbers in a Pythagorean triple cannot be odd.	?
Sample 3	?	$\overline{OP}$ is not perpendicular to line t.

2. Fill in the blanks to complete this proof of the fact that no triangle has two obtuse angles.

   Suppose ? . Let the two obtuse angles be $\angle A$ and $\angle B$. Then $m\angle A > 90°$ and $m\angle B$? .

   The sum, $m\angle A + m\angle B + m\angle C >$? .

   This contradicts the ? theorem. Therefore ? .

3. Use an indirect proof to show that no quadrilateral has four acute angles.

Look Back

A discussion of the technique of indirect proof would help those students who may not yet fully understand this procedure.

APPLYING

Suggested Assignment

Day 1

Standard 1–4, 6, 7

Extended 1–10

Day 2

Standard 11, 12, 14–22

Extended 11, 12, 14–22

Integrating the Strands

Number Exs. 7, 10, 13

Algebra Exs. 10, 13, 14, 20

Functions Ex. 20

Geometry Exs. 1–3, 6, 7, 11, 12, 14–17, 19, 21

Trigonometry Ex. 15

Logic and Language Exs. 1, 4, 5, 8, 9, 18, 22

Integrating the Strands

The exercises in this section apply the technique of indirect proof in geometric, algebraic, and real-world situations. As students write indirect proofs, they begin to develop an understanding of indirect reasoning. Number theory facts, such as Exs. 10 and 13 on page 185, often involve the use of algebra in their proofs.

Answers to Look Back

Indirect proof is a method of proof in which you show that if you temporarily assume the opposite of what you want to prove, you are led to an impossible situation or a contradiction. It follows that what you temporarily assumed to be true could not be true, and the opposite is true.

Answers to Exercises and Problems

1. Sample 1: Line j is not parallel to line k; Sample 2: All three numbers in a Pythagorean triple are odd; Sample 3: $\overline{OP}$ is perpendicular to line t.

2. a triangle has two obtuse angles; > 90°; 180°; triangle sum; the assumption that a triangle has two obtuse angles must be false. No triangle has two obtuse angles.

3. Suppose quadrilateral $ABCD$ has four acute angles. Then $m\angle A < 90°$, $m\angle B < 90°$, $m\angle C < 90°$, and $m\angle D < 90°$. Then $m\angle A + m\angle B + m\angle C + m\angle D < 360°$. This contradicts the fact that the sum of the measures of the interior angles of a quadrilateral is 360°. So the assumption that a quadrilateral can have four acute angles must be false. Therefore, no quadrilateral can have four acute angles.

Application

Indirect proof is a general way of reasoning that can be applied to both mathematical and nonmathematical situations. Exs. 4, 5, 8, 9, and 22 provide students with examples of the use of indirect proof in everyday contexts.

Mathematical Procedures

Indirect proofs are frequently used to prove the uniqueness, the existence, or the nonexistence of a mathematical "object." Mathematicians prefer direct proofs to indirect proofs and use indirect proofs only when a direct proof cannot be found.

4. Elisa Rodriguez is a sales representative. She has scheduled a one-hour meeting with a customer in Indianapolis at 9:00 A.M. On the same day she must meet another customer in Evansville at noon. Her only transportation is her car. The distance between the cities is 168 mi. Elisa Rodriguez never drives above the 65 mi/h speed limit. Prove that she cannot attend both meetings.

5. At Sunnydale High School the following schedule of classes is available for seniors.

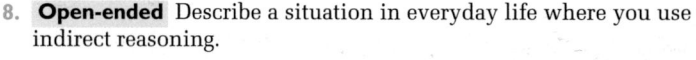

PERIOD	COURSES OFFERED
1	PHYSICS, ACCOUNTING, ADVANCED FRENCH
2	DISCRETE MATHEMATICS, ENGLISH, CALCULUS
3	AMERICAN GOVERNMENT, ENGLISH, PHYSICS
4	AMERICAN GOVERNMENT, ADVANCED SPANISH, BIOLOGY
5	INTERNATIONAL RELATIONS, MUSIC THEORY, COMPUTER SCIENCE
6	ACCOUNTING, STUDIO ART, DISCRETE MATHEMATICS

All seniors are required to take English and American Government. Prove that no senior can take both Calculus and Biology.

6. Prove that if two lines in the same plane are parallel to the same line, then they are parallel to each other. Use the postulate: Through a point not on a line there is exactly one line parallel to a given line.

7. Prove that there is no Pythagorean triple with two even numbers and one odd number.

8. **Open-ended** Describe a situation in everyday life where you use indirect reasoning.

9. **Map making** On a map, regions that share a border are not the same color.

Here is a section of a map, showing six European countries. Using this section of the map, write an indirect proof of this statement:

Some maps require at least four colors.

(*Hint:* Start by assuming this map can be colored with only three colors.)

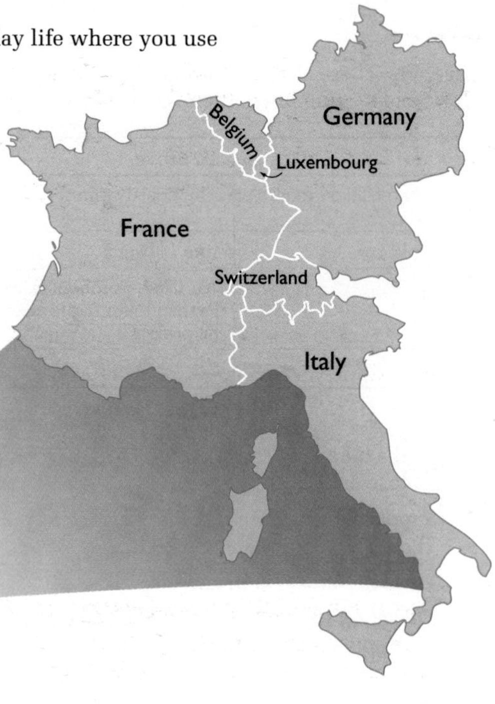

Answers to Exercises and Problems

4. Suppose Elisa can attend both meetings. She must leave Indianapolis at 10:00 A.M. and arrive in Evansville, 168 miles away, two hours later. She will have to travel at 84 mi/h. This contradicts the given information that Elisa never drives more than 65 mi/h. The assumption that Elisa can attend both meetings must be false. Therefore, Elisa cannot attend both meetings.

5. Suppose that a senior is taking both Calculus and Biology. The student must take Calculus second period and Biology fourth period. Then the senior must be taking both English and American Government third period since all seniors are required to take both those courses and that is the only time either does not conflict with Calculus or Biology. A student cannot take two courses in one period. The assumption that a senior is taking both Calculus and Biology must be false. Therefore, no senior can take both Calculus and Biology.

6. *Given:* lines j, k, and l in a plane; $j \parallel l$; $k \parallel l$
Prove: $j \parallel k$

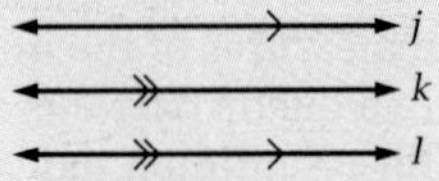

Proof: Suppose that j is not parallel to k. Let P be the point where j and k intersect. Since $j \parallel l$ and $k \parallel l$, P is not on l. Then j and k are both lines through P that are parallel to l. This contradicts the postulate that states that through a point not on a line, there is exactly one line parallel to the given line. The assumption that j is not parallel to k must be false. Therefore, $j \parallel k$.

7. Suppose there is a Pythagorean triple with two even numbers and one odd number. There are two possibilities. One, either a or b is odd, and two, c is odd. Suppose a is odd and b and c are even. Then a^2 is odd and b^2 is even, so $a^2 + b^2 = c^2$ is odd, since it is the sum of an even number and an odd number. This contradicts the assumption that c is even. Suppose a and b are even and c is odd. Then a^2 and b^2 are even, so $a^2 + b^2 = c^2$ is even, since it is the sum of two even numbers. This contradicts the assumption that c is odd. The assumption that there is a Pythagorean triple with two even numbers and one odd number must be false. Therefore, there is no Pythagorean triple with two even numbers and one odd number.

10. Puzzles In a *magic square* the sum of the numbers in each row, column, and diagonal is the same. The classical 3-by-3 magic square uses the first nine positive integers.

Use an indirect proof to explain why there is no magic square using only the first nine prime numbers: 2, 3, 5, 7, 11, 13, 17, 19, and 23.

Detail from Albrecht Dürer's Melancholia

11. Copy and complete this indirect proof of Theorem 3.11: If a line in the plane of a circle is perpendicular to a radius at its outer endpoint, then the line is tangent to the circle.

Given $\overline{OA}$ is a radius of circle O. Line m is in the plane of circle O and is perpendicular to $\overline{OA}$.

Prove Line m is a tangent to circle O.

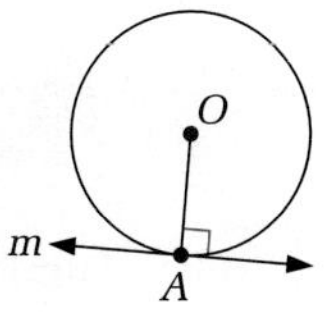

Since line m intersects the circle at A, it is either a tangent or a secant. Suppose m is not a tangent to the circle. Then it is a secant and it intersects the circle at another point B.

Draw $\overline{OB}$. Then, since $\overline{OA}$ and $\overline{OB}$ are both radii, $OA = OB$ and $\triangle OAB$ is isosceles. (Continue the indirect proof from here.)

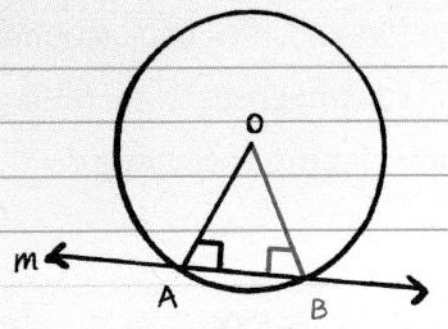

12. Copy and complete this proof of Theorem 3.12: If two tangent segments are drawn from the same point to the same circle, then they are equal in measure.

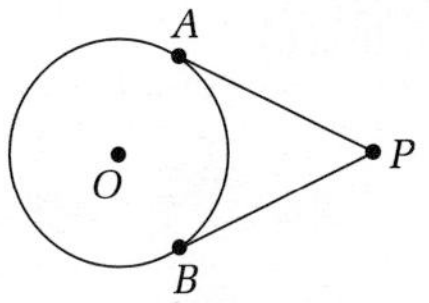

Given $\overline{PA}$ and $\overline{PB}$ are both tangents to circle O at A and B.

Prove $PA = PB$

Plan Ahead

Draw $\overline{OA}$, $\overline{OB}$, and $\overline{OP}$. Use theorem 3.10 to show that $\triangle OAP$ and $\triangle OBP$ are both right triangles. Use the Pythagorean theorem to find PA and PB in terms of h and r.

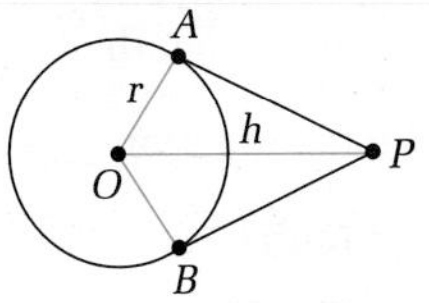

13. Research Find an indirect proof that $\sqrt{2}$ is an irrational number and present it to the class.

14. a. Solve the system of equations: $x + y = 4$, $x^2 + y^2 = 8$

b. Show that this system represents a circle and a line tangent to the circle. Find the coordinates of the point of tangency.

c. Use coordinate geometry to show that the radius to the point of tangency is perpendicular to the tangent line.

Answers to Exercises and Problems

8. Answers may vary. An example is given. A secretary returns from lunch and notices the message light on the telephone answering machine is not lit. The secretary concludes there are no messages, because if there were, the light would be lit.

9. Suppose that the map shown can be colored with only three colors, say, red, blue, and yellow. Belgium, Luxembourg, and Germany must be three different colors, since any two of the countries share a border. Color Belgium red, Luxembourg blue, and Germany yellow. There is no color left to color France, which borders on these three differently-colored countries. The original assumption that the map can be colored with only three colors must be false. Therefore, there are some maps that require at least four colors.

10. Suppose there is a magic square using the first nine prime numbers. The sum of each row, column, and diagonal must be the same. The sum of any three rows or three columns must be the sum of the nine numbers, 100. Then the sum of any one row or column must be $\frac{100}{3} = 33\frac{1}{3}$. This contradicts the fact that the sum of three integers is an integer. The assumption that there is a magic square using the first nine prime numbers must be false. Therefore, there is no magic square using the first nine prime numbers.

11. Since $\overline{OA} \perp m$, $\angle A$ is a right angle. Since $\triangle OAB$ is isosceles, its base angles are equal in measure and $m\angle B = 90°$. Then since $m\angle AOB > 0$, $m\angle AOB + m\angle A + m\angle B > 180°$. This contradicts the triangle sum theorem. The assumption that m is not tangent to the circle must be false. Therefore, m is tangent to circle O.

12. Draw $\overline{OA}$, $\overline{OB}$, and $\overline{OP}$. $\overline{PA} \perp \overline{OA}$ and $\overline{PB} \perp \overline{OB}$. (If a line is tangent to a circle, then it is perpendicular to the radius from the point of tangency.) Then $\triangle OAP$ and $\triangle OBP$ are right triangles. so by the Pythagorean theorem, $PA = \sqrt{h^2 - r^2}$ and $PB = \sqrt{h^2 - r^2}$. Then $PA = PB$.

13. Suppose that $\sqrt{2}$ is a rational number. Let a and b be the integers such that $\sqrt{2} = \frac{a}{b}$ and $\frac{a}{b}$ is in lowest terms; that is, a and b have no common factors other than 1. a^2 and b^2 are integers and $\frac{a^2}{b^2} = 2$, so $a^2 = 2b^2$ and a^2 is even. Then a is even, so $a = 2n$ for some integer n. But $2b^2 = a^2 = (2n)^2 = 4n^2$ and $b^2 = 2n^2$, so b^2 and b are also both even. This contradicts the given information that a and b have no common factors other than 1. The assumption that $\sqrt{2}$ is rational must be false. Therefore, $\sqrt{2}$ is irrational.

14. a. (2, 2)

b. $x^2 + y^2 = 8$ is the graph of a circle with center (0, 0), and radius $\sqrt{8} = 2\sqrt{2}$. Since the line $x + y = 4$ intersects the circle at only one point (the system has only one solution), the line is tangent to the circle. The point of tangency is (2, 2).

c. The point of tangency is (2, 2). The slope of the radius to the point of tangency is $\frac{2-0}{2-0} = 1$. The slope of the tangent line is −1. Since the slopes of the two lines are negative reciprocals, the radius to the point of tangency is perpendicular to the tangent line.

Communication: Writing

Ex. 18 provides an excellent opportunity to check students' understanding of how to write an indirect proof. Students who have trouble answering this exercise should review this section.

Working on the Unit Project

Ex. 22 allows students to make a connection between the mathematical concept of indirect proof and the legal concept of an alibi defense.

Practice 25 For use with Section 3-7

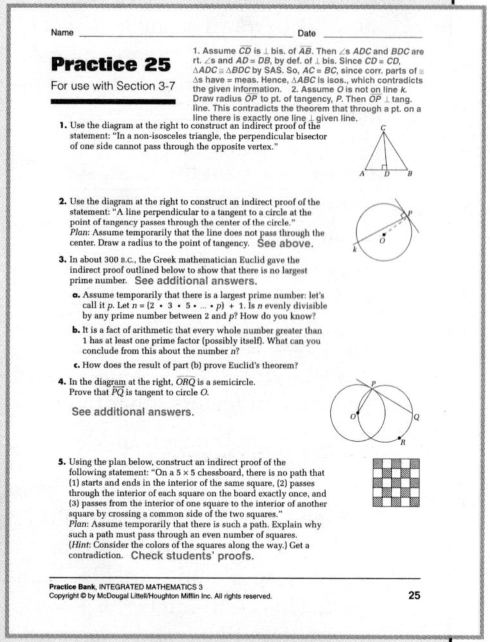
Name ____________ Date ____________

Practice 25
For use with Section 3-7

1. Assume $\overline{CD}$ is $\perp$ bis. of $\overline{AB}$. Then $\angle$s ADC and BDC are rt. $\angle$s and $AD = DB$, by def. of $\perp$ bis. Since $CD = CD$, $\triangle ADC \cong \triangle BDC$ by SAS. So, $AC = BC$, since corr. parts of $\cong$ $\triangle$s have = meas. Hence, $\triangle ABC$ is isos., which contradicts the given information. 2. Assume O is not on line k. Draw radius $\overline{OP}$ to pt. of tangency, P. Then $\overline{OP} \perp$ tang. line. This contradicts the theorem that through a pt. on a line there is exactly one line $\perp$ given line.

1. Use the diagram at the right to construct an indirect proof of the statement: "In a non-isosceles triangle, the perpendicular bisector of one side cannot pass through the opposite vertex."
2. Use the diagram at the right to construct an indirect proof of the statement: "A line perpendicular to a tangent to a circle at the point of tangency passes through the center of the circle." *Plan*: Assume temporarily that the line does not pass through the center. Draw a radius to the point of tangency. See above.
3. In about 300 B.C., the Greek mathematician Euclid gave the indirect proof outlined below to show that there is no largest prime number. See additional answers.
   a. Assume temporarily that there is a largest prime number: let's call it p. Let $n = (2 \cdot 3 \cdot 5 \cdot \ldots \cdot p) + 1$. Is n evenly divisible by any prime number between 2 and p? How do you know?
   b. It is a fact of arithmetic that every whole number greater than 1 has at least one prime factor (possibly itself). What can you conclude from this about the number n?
   c. How does the result of part (b) prove Euclid's theorem?
4. In the diagram at the right, $\overparen{ORQ}$ is a semicircle. Prove that $\overrightarrow{PQ}$ is tangent to circle O.

   See additional answers.
5. Using the plan below, construct an indirect proof of the following statement: "On a 5 × 5 chessboard, there is no path that (1) starts and ends in the interior of the same square, (2) passes through the interior of each square on the board exactly once, and (3) passes from the interior of one square to the interior of another square by crossing a common side of the two squares." *Plan*: Assume temporarily that there is such a path. Explain why such a path must pass through an even number of squares. (*Hint*: Consider the colors of the squares along the way.) Get a contradiction. Check students' proofs.

Practice Bank, INTEGRATED MATHEMATICS 3

25

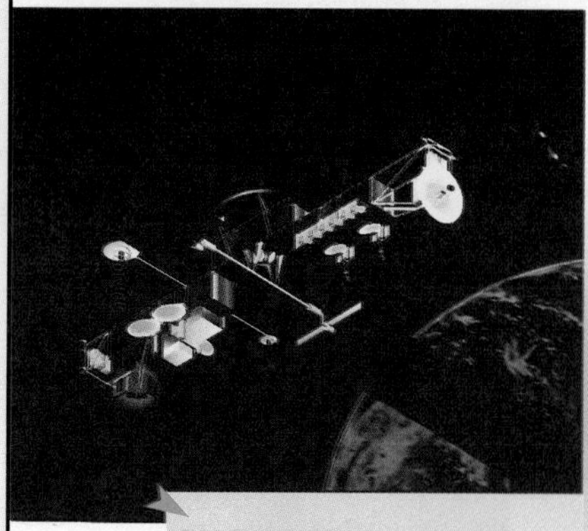

BY THE WAY...

The Milstar satellite provides global communications. Systems director Dr. Wanda Austin provided engineering support to the Milstar. Her background includes degrees in mathematics and systems engineering.

Global Communications Communications satellites are used to transmit radio and television signals around the world. They circle Earth in a stationary orbit at a height of 22,300 mi above the surface. At this height they travel just fast enough to keep up with Earth's rotation. The radius of Earth is about 4000 mi.

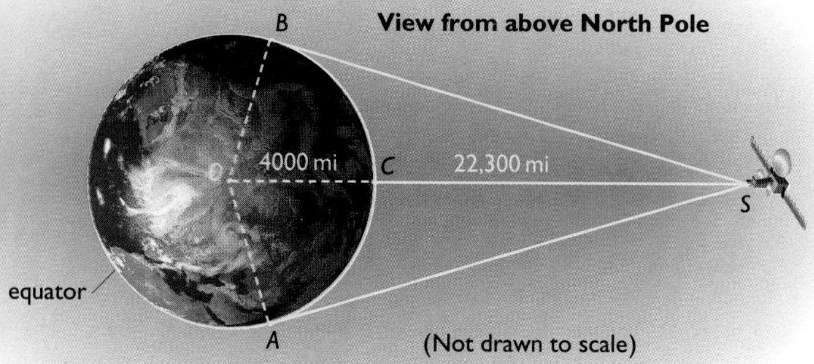

15. A communications satellite is placed in orbit at point S. Use trigonometry to find the degree measure of $\overparen{ACB}$.
16. Use proportions to find the length of $\overparen{ACB}$, the number of miles along the equator that the satellite covers.
17. Suppose three satellites are placed in orbit around the equator, 120° apart. Would the range of these satellites include the entire equator? the entire surface of Earth? Explain.

Ongoing **ASSESSMENT**

18. **Writing** Explain to a friend who missed class how to write an indirect proof.

Review **PREVIEW**

19. Draw any quadrilateral with vertices on a circle. Draw its diagonals. Label your diagram and name *two* pairs of similar triangles. Explain how you can prove that the triangles are similar. *(Section 3-6)*
20. Use the function $f(x) = \dfrac{x^2 + 4x + 1}{x + 1}$.
    a. Find $f(-4)$.
    b. Find x when $f(x) = 3$.
21. Find the circumference and the area of a circle with radius 4.5 m. *(Toolbox Skill 29)*

3 Working on the Unit Project

22. What is an alibi defense? How can an alibi be used as an indirect proof? Give an example of a case you have read about in which an alibi played a significant role.

Answers to Exercises and Problems

15. about 162.5°
16. about 11,345 mi
17. Yes; each satellite would cover an arc of about 162.5° at the equator, and together they would cover the entire equator. They would not cover the entire surface of Earth. Each satellite covers a region of the globe consisting of a circle and the region above it. (Picture a circular shadow cast on Earth.) The three regions determined by the satellites would not completely cover Earth. For example, the poles would not be covered.
18. Answers may vary. An example is given. First assume the opposite of what you want to prove. Show that this assumption leads to a contradiction, so that what you assumed to be true could not be, and the opposite (what you hoped to prove) is true.
19. See answers in back of book.
20. a. $-\frac{1}{3}$ b. 1, −2
21. circumference: 9π m, or about 28.3 m; area: 20.25π m^2, or about 63.6 m^2
22. An alibi defense is an argument that the defendant was somewhere else when the crime was committed. It is like an indirect proof in that, if we suppose the defendant is guilty, it follows that the defendant was at the scene when the crime was committed. This contradicts the testimony that he or she was somewhere else at the time. Then the assumption that the defendant is guilty must be false. Therefore, the defendant is innocent. Examples may vary.

Section 3-8

Circumscribed Polygons

Focus
Examine relationships between circumscribed polygons and circles.

GETTING AROUND

At the right, a circle is inscribed in a square. The square is circumscribed about the circle. The sides of a circumscribed polygon are tangents to the inscribed circle.

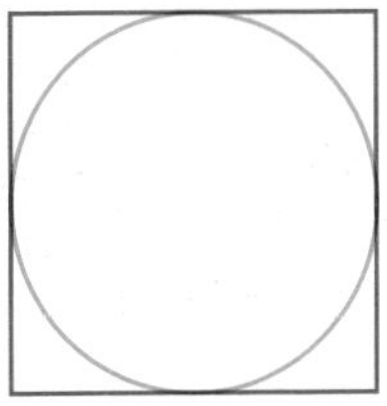

A regular polygon with n sides may be circumscribed about a circle by drawing tangents at points on the circle that form n arcs of equal measure.

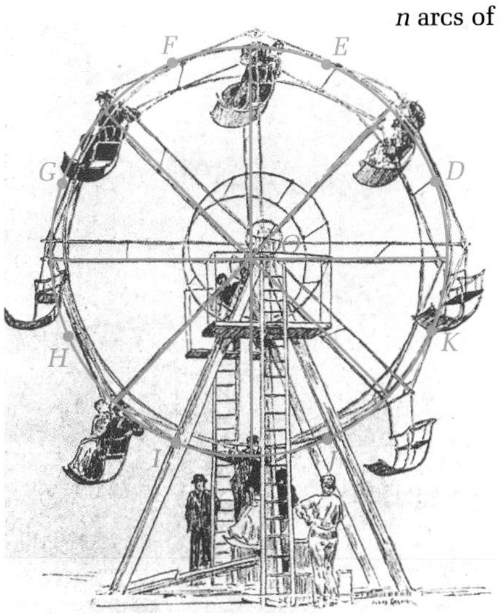

Pleasure wheel from about 1860

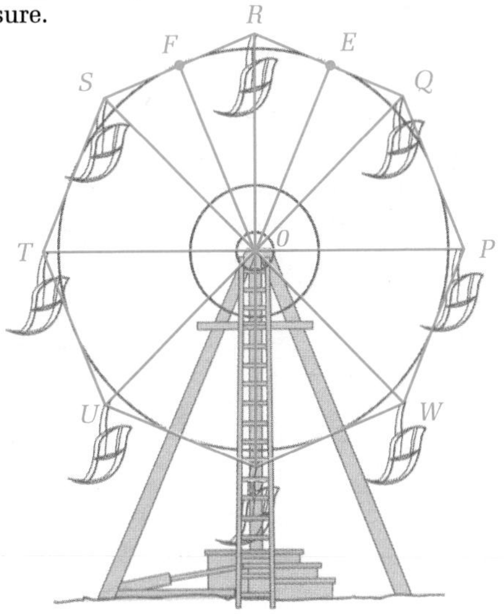

Talk it Over

1. In the diagrams above, $D, E, F, G, H, I, J,$ and K are spaced equally around circle O. Find $m\,\widehat{EF}$. What other arcs have the same measure?
2. Tangents are drawn at the eight points on the circle to form octagon $PQRSTUVW$. Explain why $\triangle OER$ and $\triangle OFR$ are congruent.
3. Find $m\angle EOR$ and $m\angle FOR$. What other angles must have the same measures?
4. Explain why $\triangle SOR$ and $\triangle ROQ$ are congruent.
5. Explain why $PQRSTUVW$ is a regular octagon.

PLANNING

Objectives and Strands
See pages 136A and 136B.

Spiral Learning
See page 136B.

Materials List
- Tracing paper or wax paper
- Compass
- Straightedge
- Spreadsheet software or graphics calculator

Recommended Pacing
Section 3-8 is a one-day lesson.

Extra Practice
See pages 612–613.

Warm-Up Exercises
- Warm-Up Transparency 3-8

Support Materials
- Practice Bank: Practice 26
- Activity Bank: Enrichment 24
- Study Guide: Section 3-8
- Problem Bank: Problem Set 8
- Assessment Book: Quiz 3-8, Alternative Assessment 6

Answers to Talk it Over

1. 45°; all arcs whose endpoints are consecutive points on the circle O
2. Answers may vary. An example is given. $OR = OR$ (The measure of a segment is equal to itself.) $OF = OE$ (All radii of a circle are equal in measure.) $FR = FE$ (If two tangent segments are drawn from the same point to the same circle, then they are equal in measure.) Then $\triangle OER \cong \triangle OFR$ by SSS.
3. $22\frac{1}{2}°$; If $\overline{OG}, \overline{OH}, \overline{OI}, \overline{OJ}, \overline{OK}$, and $\overline{OD}$ were drawn, the sixteen central angles produced would all have measure $22\frac{1}{2}°$.
4. $\triangle EOR \cong \triangle FOR$, so $FR = ER$ and $m\angle ORF = m\angle ORE$. (Corresponding parts of congruent triangles are equal in measure.) $OR = OR$ (Reflexive property of equality) Then $\triangle SOR \cong \triangle ROQ$ by ASA.
5. Using the same methods as in questions 1–4, you can prove that $\triangle SOR \cong \triangle ROQ \cong \triangle QOP \cong \triangle POW \cong \triangle WOV \cong \triangle VOU \cong \triangle UOT \cong \triangle TOS$. Then the sides of $PQRSTUVW$ are corresponding parts of congruent triangles and are equal in measure. Using the same property along with addition of angle measures, you can show that all the angles of $PQRSTUVW$ are equal in measure.

TEACHING

Talk it Over

Questions 1–5 give students the opportunity to explore a number of concepts and properties of inscribed regular polygons.

Additional Sample

Suppose the radius of circle O is 30 cm. Find the area of the regular circumscribed pentagon $ABCDE$.

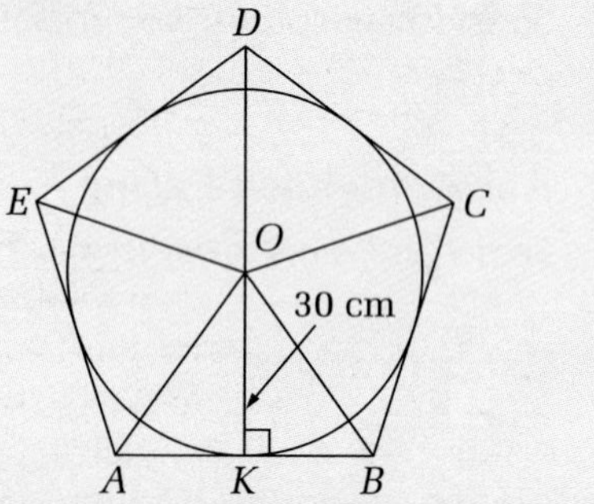

Step 1. Use the tangent ratio to find KB in $\triangle KOB$.

$m\angle AOB = \frac{360°}{5} = 72°$, so

$m\angle KOB = \frac{1}{2} m\angle AOB = 36°$.

In right $\triangle KOB$, $\tan 36° = \frac{KB}{KO} = \frac{KB}{30}$.

$KB \approx 21.7963$.

Hence, $AB = 2(KB) \approx 43.5926$ cm.

Step 2. Use the formula $A = \frac{1}{2}bh$.

Area of $\triangle AOB = \frac{1}{2}(AB)(KO)$

$\approx \frac{1}{2}(43.5926)(30)$

$= 653.889 \text{ cm}^2$

Step 3. Multiply the area of $\triangle AOB$ by 5:

$5(653.889) \approx 3269.45$

The area of $ABCDE$ is about 3269.45 cm^2.

Talk it Over

Question 6 introduces the concept of semiperimeter. Question 7 leads students to the area formula given at the bottom of the page.

Sample

Suppose the radius of circle O is 1 m. Find the area of regular octagon *PQRSTUVW*.

Sample Response

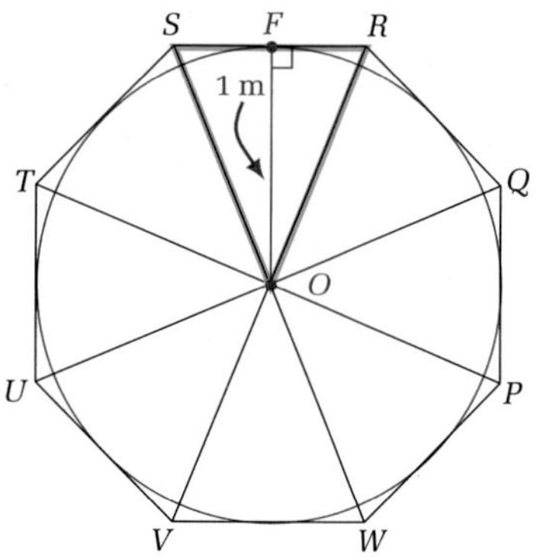

Since the octagon is made of 8 congruent triangles, you can find the area of one triangle and multiply it by 8.

To find the area of $\triangle SOR$, you need to find the length RS.

Step 1 Use the tangent ratio to find RF in right $\triangle FOR$.

$m\angle SOR = \frac{360°}{8} = 45°$; $m\angle FOR = \frac{1}{2} \cdot m\angle SOR = 22.5°$

In right $\triangle FOR$, $\frac{RF}{OF} = \frac{RF}{1} = \tan 22.5°$

$RF \approx 0.414$

$RS = 2(RF) \approx 0.828$ (m)

Step 2 Use the formula $A = \frac{1}{2}bh$.

Area of $\triangle SOR = \frac{1}{2}(RS)(OF)$

$\approx \frac{1}{2}(0.828)(1) = 0.414 \text{ (m}^2)$

Step 3 Multiply the area of $\triangle SOR$ by 8: $8(0.414) \approx 3.31$

The area of octagon $PQRSTUVW$ is about 3.31 m^2.

Talk it Over

6. The **semiperimeter** of a polygon is half the perimeter. Find the semiperimeter of octagon *PQRSTUVW*.
7. Show that the area of an octagon may be found by multiplying the radius of the inscribed circle and the semiperimeter.
8. Find the area of the circle. How does it compare to the area of the octagon?

The area of any circumscribed polygon may be found when the radius of the inscribed circle and the sides are known. This is true for nonregular as well as regular polygons.

AREA OF A CIRCUMSCRIBED POLYGON

The area of any circumscribed polygon is the product of the radius (r) of the inscribed circle and the semiperimeter (s).

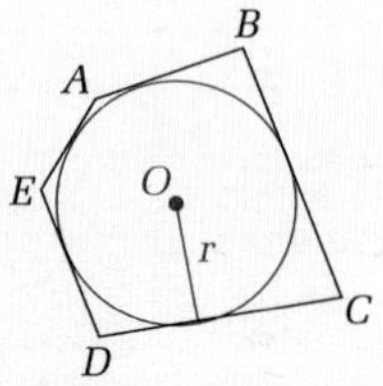

$A = rs$

$\frac{AB + BC + CD + DE + AE}{2}$ = semiperimeter

Answers to Talk it Over

6. about 3.312 m
7. The octagon consists of eight congruent triangles. Let p be the perimeter of the octagon. Each triangle has height h (which is also the radius of the circle) and base $\frac{p}{8}$. The area of one triangle is $\frac{1}{2}h\left(\frac{p}{8}\right) = h \cdot \frac{p}{16}$. The area of the octagon is $8 \cdot h\left(\frac{p}{16}\right) = h \cdot \frac{p}{2}$. So the area of the octagon is equal to the radius times one half of the perimeter, which is the semiperimeter.
8. The area of the circle is about 3.14 m^2; the area of the circle is slightly less than that of the octagon.
9. a. $s = \pi r$

   b. $A = sr$ or $\frac{s^2}{\pi}$

As the number of sides of a regular polygon increases, its area approaches that of the inscribed circle.

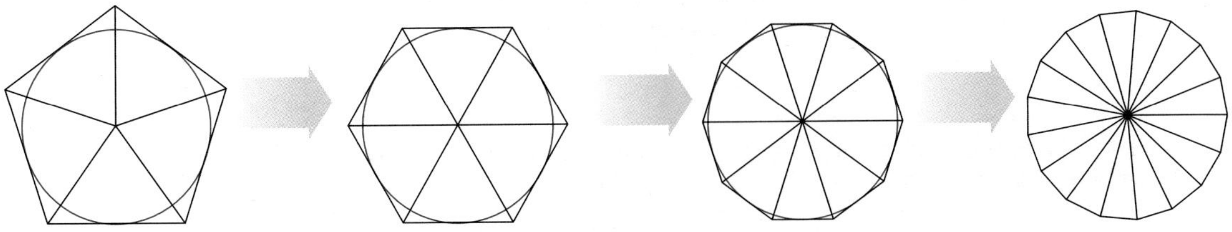

Talk it Over

9. a. The semiperimeter of a circle is half of its circumference. Write a formula for the semiperimeter s of a circle in terms of r.
   b. Use your answer to part (a) to write a formula for the area of a circle in terms of s.

Look Back

What is the relationship between an inscribed circle and its circumscribed polygon?

3-8 Exercises and Problems

1. **Reading** What is a semiperimeter?
2. Circle O is inscribed in $\triangle ABC$. Find the perimeter of the triangle.

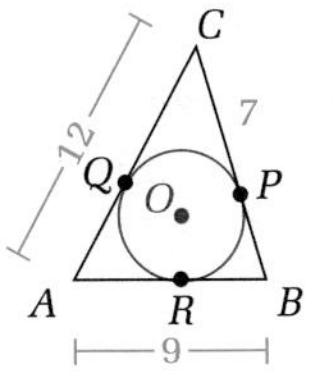

3. Circle O is inscribed in a regular pentagon. Find $m\angle AOP$ and $m\angle APO$.

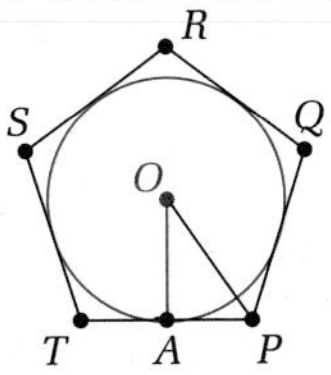

4. $\triangle BCD$ has an inscribed circle whose radius is $4\frac{1}{3}$ cm. Find the area of the triangle.

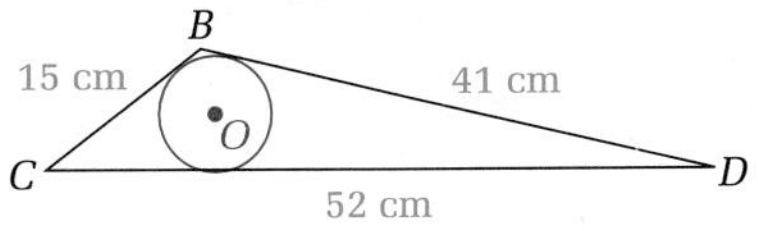

5. Circle O is inscribed in a triangle. $AB = 25$ ft, $AC = 63$ ft, and $BC = 52$ ft. The radius of the circle is 9 ft. Use two methods to find the area of the triangle.

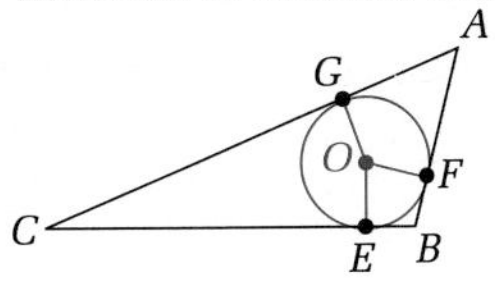

Using Technology

Students can use a graphics calculator to see that as the number of sides of a regular polygon increases, its area approaches that of the inscribed circle. Suppose the radius of the inscribed circle is 1. The area of this circle is π. Using the method of the Sample, it can be shown that the area of the circumscribed n-gon is given by $A(n) = n \tan \frac{180°}{n}$.

Set the calculator to degree mode. Enter the equation $y = x \tan\left(\frac{180}{x}\right)$ on the Y= list. Use Xmin = 0, Xmax = 100, Ymin = –1, Ymax = 5. Press [GRAPH] and trace to the right. Notice that as the x-coordinate increases, the y-coordinate is fairly close to π. To see what happens when the circumscribed polygon has a very large number of sides, multiply the Xmax setting by 10, 100, or 1000.

The Table feature of the TI-82 can be used to answer Ex. 11 on page 191. Press [2nd] [TblSet] to set TblMin = 3 and ΔTbl = 1. Students can then enter the equations for columns 2 through 5 of the spreadsheet on the Y= list. View the table by pressing [2nd] [TABLE]. Change ΔTbl to 2 for part (b). Remind students that the *smallest* x-value that is meaningful in this situation is $x = 3$.

APPLYING

Suggested Assignment

Standard 1–5, 8, 10, 12–16

Extended 1–5, 7–10, 12–16

Integrating the Strands

Algebra Ex. 15

Functions Ex. 15

Geometry Exs. 1–14

Trigonometry Exs. 8, 9

Logic and Language Exs. 1, 6, 11, 12, 16

Answers to Look Back

Answers may vary. An example is given. The sides of the circumscribed polygon are tangents to the inscribed circle. The area of the circumscribed polygon is equal to the product of its semiperimeter and the circle's radius.

Answers to Exercises and Problems

1. A semiperimeter of a figure is half the perimeter.
2. 32 units
3. 36°; 54°
4. 234 cm²
5. Method 1: $A = rs = 9\left(\frac{25 + 63 + 52}{2}\right) = 630$ ft²; Method 2: Draw $\overline{OA}$, $\overline{OB}$, and $\overline{OC}$; A = area of $\triangle AOB$ + area of $\triangle AOC$ + area of $\triangle BOC = \frac{1}{2} \cdot 9 \cdot 25 + \frac{1}{2} \cdot 9 \cdot 63 + \frac{1}{2} \cdot 9 \cdot 52 = 630$ ft²

Using Manipulatives

The activity of Ex. 6 allows students to explore some properties of circumscribed triangles. As an extension of this exercise, you may wish to have students compare and contrast their results with those of the Exploration on page 165 that involved inscribed triangles.

Communication: Discussion

A discussion of Exs. 8–10 will provide students with an opportunity to share their results and bring closure to these exercises.

6. **Using Manipulatives** You will need tracing paper or wax paper, a compass, and a straightedge.
   a. Draw a large scalene triangle on your paper.
   b. Fold each pair of sides together and crease to form the angle bisectors. Label the point of intersection of these bisectors *I*.
   c. Experiment with the compass opening to find the shortest distance from *I* to one of the sides of the triangle. Show that *I* is equidistant from the three sides.
   d. With *I* as the center, draw a circle tangent to the three sides of the triangle.
   e. **Writing** Write a short paragraph about what you learned from this activity.

7. **Open-ended** It is impossible to draw an inscribed circle inside some quadrilaterals. Make drawings or use geometric drawing software to determine which types of quadrilaterals may be circumscribed about circles.

8. A regular hexagon is circumscribed about a circle with radius 1 m.
   a. Find $m\angle AOQ$.
   b. Use trigonometry to find the length of one side of the hexagon.
   c. Find the perimeter of the hexagon.
   d. Find the area of the hexagon.
   e. Julie says the area and the perimeter of this hexagon are the same. Do you agree? Explain.

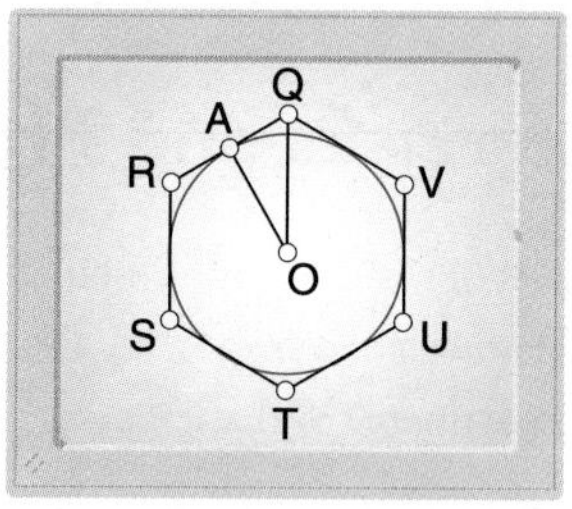

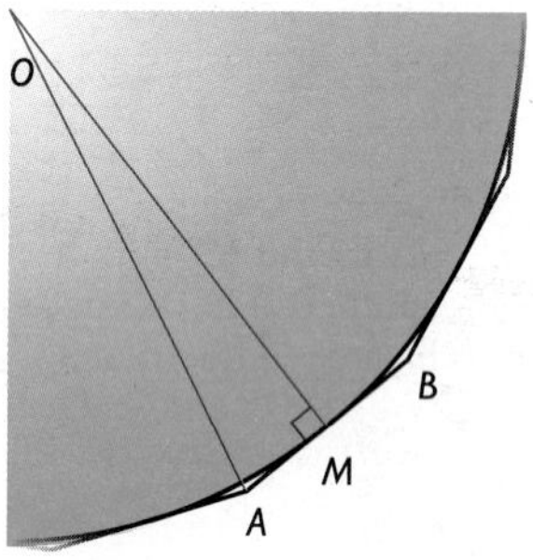

9. **Open-ended** Repeat Exercise 8 with a regular polygon with more than six sides that is circumscribed about a circle with radius 1 m.

10. $\overline{AB}$ is the side of a regular *n*-gon circumscribed about circle *O*. *M* is the midpoint of $\overline{AB}$.
    a. Find an expression in terms of *n* for $m\angle MOA$.
    b. Show that the expression you found in part (a) gives the correct value for a regular octagon (see Sample), a regular hexagon (see Exercise 8), and another regular polygon (see Exercise 9).

Answers to Exercises and Problems

6. a–d. Check students' work.
   e. The angle bisectors of a triangle intersect in a point that is equidistant from the sides of the triangle. A circle can be inscribed in any triangle. The center is the intersection of the angle bisectors and the radius is the distance to a side.

7. Check students' work. Squares, trapezoids, rhombuses, kites, and various irregular quadrilaterals can be circumscribed about circles.

8. a. 30°
   b. about 1.15 m
   c. about 6.92 m
   d. about 3.46 m^2
   e. No. The numerical values of the *semiperimeter* and the area are the same, but the units are different.

9. Answers may vary. Results for a decagon are given.
   a. 18°
   b. about 0.65 m
   c. about 6.5 m
   d. about 3.25 m^2

11 TECHNOLOGY Use spreadsheet software or a graphics calculator to find the semiperimeters and areas of regular n-gons circumscribed about a circle with radius 1 unit. Use the degree mode, or convert from degrees to radians, if necessary.

a. **Writing** Start with $n = 3$ and increase n by 1 each time. Extend the pattern to $n = 25$. Describe any patterns you see in your chart.

Semiperimeters, areas

n	$m\angle MOA$	tan MOA	Length of side	Semiperimeter	Area
3	60°	1.7320499361	3.4640998722	5.1961498083	5.1961498083
4	45°	1	2	4	4
5	36°	0.7265423282	1.4530846566	3.6327116414	3.6327116414

b. Change the spreadsheet so that n doubles each time. Extend this pattern until you get an area that is about 3.14159. What is n?

c. What happens to the ratio of the area of the circumscribed polygon to the area of the circle as the number of sides increases? What happens to the ratio of the perimeter to the circumference?

Ongoing ASSESSMENT

12. **Writing** Explain why the formula $A = rs$ can be used to find the area of any nonregular polygon that is circumscribed about a circle.

Review PREVIEW

13. Write an indirect proof to show that $m\angle MOA$ in the figure for Exercise 10 can never be 25° if the polygon is regular. *(Section 3-7)*

14. Identify these space figures by name. *(Toolbox Skill 29)*

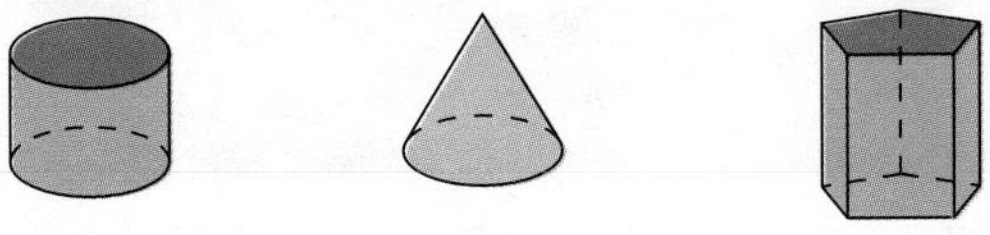

15. Tell whether the function $f(x) = \frac{4}{5}x - 2$ is linear. Find $f(-5)$, $f(0)$, and $f(20)$. *(Section 2-2)*

3 Working on the Unit Project

16. As an attorney you want to discredit your opponent's arguments.
 a. List the arguments you think the opposing attorneys will make. List your counterarguments to their arguments.
 b. Make a list of questions you intend to ask in cross-examining witnesses for the other side.

"A unique and stirring plea, counsellor."

Problem Solving

For Ex. 11, students can discuss the approach they used. What planning had to be done? What formulas had to be developed? Are there any results that help provide a check of the formulas used? Students should understand that as n increases, the numbers in the area column should be getting closer and closer to π.

Working on the Unit Project

As students work on Ex. 16, they should save the list of questions for use in writing their legal briefs.

Practice 26 For use with Section 3-8

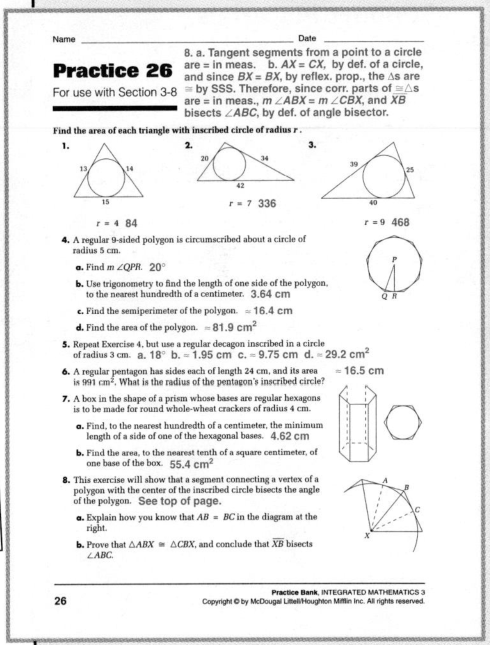

Name ______ Date ______

Practice 26
For use with Section 3-8

8. a. Tangent segments from a point to a circle are = in meas. b. $AX = CX$, by def. of a circle, and since $BX = BX$, by reflex. prop., the △s are ≅ by SSS. Therefore, since corr. parts of ≅△s are = in meas., $m\angle ABX = m\angle CBX$, and $\overline{XB}$ bisects $\angle ABC$, by def. of angle bisector.

Find the area of each triangle with inscribed circle of radius r.

1. 13, 14, 15; $r = 4$ 84
2. 20, 34, 42; $r = 7$ 336
3. 39, 25, 40; $r = 9$ 468

4. A regular 9-sided polygon is circumscribed about a circle of radius 5 cm.
 a. Find $m\angle QPR$. 20°
 b. Use trigonometry to find the length of one side of the polygon, to the nearest hundredth of a centimeter. 3.64 cm
 c. Find the semiperimeter of the polygon. ≈ 16.4 cm
 d. Find the area of the polygon. ≈ 81.9 cm²
5. Repeat Exercise 4, but use a regular decagon inscribed in a circle of radius 3 cm. a. 18° b. ≈ 1.95 cm c. ≈ 9.75 cm d. ≈ 29.2 cm²
6. A regular pentagon has sides each of length 24 cm, and its area is 991 cm². What is the radius of the pentagon's inscribed circle? ≈ 16.5 cm
7. A box in the shape of a prism whose bases are regular hexagons is to be made for round whole-wheat crackers of radius 4 cm.
 a. Find, to the nearest hundredth of a centimeter, the minimum length of a side of one of the hexagonal bases. 4.62 cm
 b. Find the area, to the nearest tenth of a square centimeter, of one base of the box. 55.4 cm²
8. This exercise will show that a segment connecting a vertex of a polygon with the center of the inscribed circle bisects the angle of the polygon. See top of page.
 a. Explain how you know that $AB = BC$ in the diagram at the right.
 b. Prove that $\triangle ABX \cong \triangle CBX$, and conclude that $\overline{XB}$ bisects $\angle ABC$.

26 **Practice Bank**, INTEGRATED MATHEMATICS 3

Answers to Exercises and Problems

10. a. $\frac{180°}{n}$
 b. octagon: $\frac{180°}{8} = 22\frac{1}{2}°$; hexagon: $\frac{180°}{6} = 30°$; Answers may vary. For decagon: $\frac{180°}{10} = 18°$

11. See answers in back of book.

12. By drawing segments from the center of the circle to the vertices of the n-gon, you divide the n-gon into n triangles each with height r, the radius of the circle. Let s_k be the length of one of the sides of the n-gon. The area of the n-gon, then, is $\frac{1}{2}rs_1 + \frac{1}{2}rs_2 + \ldots + \frac{1}{2}rs_n = \frac{1}{2}r(s_1 + s_2 + \ldots + s_n) = \frac{1}{2}r$ (perimeter) $= rs$.

13. Suppose that $m\angle MOA = 25°$. Then the figure is an n-gon where $\frac{180}{n} = 25$ and $n = 7.2$. This contradicts the fact that n must be a whole number. The assumption that $m\angle MOA = 25°$ must be false. Therefore, $m\angle MOA$ can never be 25°.

14. cylinder; cone; pentagonal prism

15. Yes; –6; –2; 14

16. Check students' work.

PLANNING

Objectives and Strands
See pages 136A and 136B.

Spiral Learning
See page 136B.

Materials List
- Polyhedral nets
- Toothpicks
- Gum drops, raisins, or miniature marshmallows

Recommended Pacing
Section 3-9 is a one-day lesson.

Extra Practice
See pages 612–613.

Warm-Up Exercises
- Warm-Up Transparency 3-9

Support Materials
- Practice Bank: Practice 27
- Activity Bank: Enrichment 25
- Study Guide: Section 3-9
- Problem Bank: Problem Set 8
- Overhead Visuals: Folder 3
- Assessment Book: Quiz 3-9, Test 12, Alternative Assessment 7

Section 3-9 Polyhedra

A GAP IN THE NET

Focus

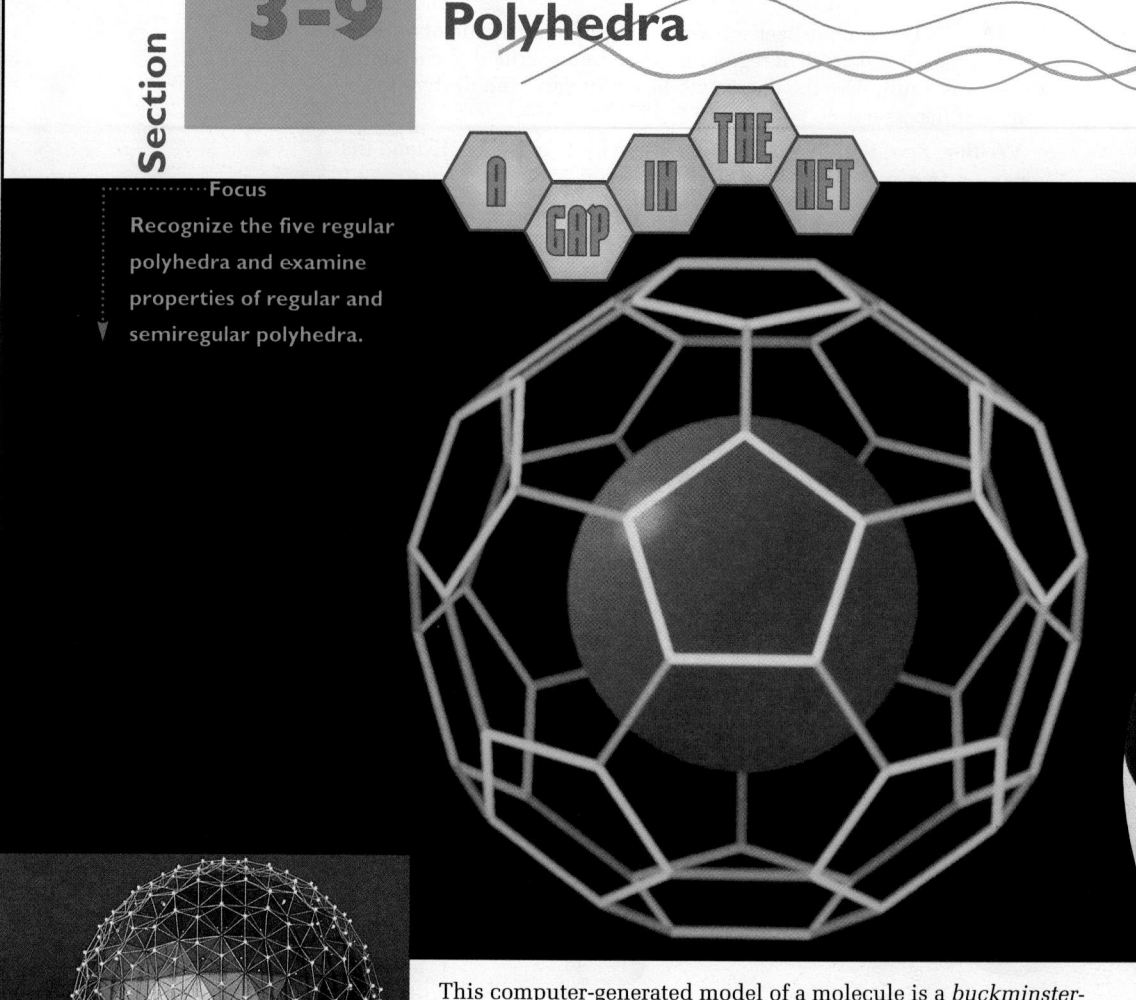

Recognize the five regular polyhedra and examine properties of regular and semiregular polyhedra.

BY THE WAY...

The buckminsterfullerene molecule was discovered in 1985. It is named after R. Buckminster Fuller, the architect who invented geodesic domes.

This computer-generated model of a molecule is a *buckminsterfullerene,* or *buckyball* for short. It has 60 atoms of carbon joined as 12 regular pentagons and 20 regular hexagons. Crystallized buckyballs may form a substance harder than diamonds.

Buckyballs have the same structure as the pattern on soccer balls. You can visualize the structure of a buckyball by imagining an atom of carbon at each of the vertices on a soccer ball.

Talk it Over

1. How many hexagons and how many pentagons meet at each vertex of a buckyball?
2. How many degrees are there in each angle at each vertex?
3. Find the sum of the measures of the angles at each vertex.

Answers to Talk it Over

1. two hexagons, one pentagon
2. two 120° angles, one 108° angle
3. 348°

Regular and Semiregular Polyhedra

A space figure whose faces are all polygons is a **polyhedron** (plural: polyhedra). A buckyball is an example of a *semiregular polyhedron*. The faces of a **semiregular polyhedron** are all regular polygons, and at each vertex there is the same number of faces of each type.

Since the faces include both regular pentagons and regular hexagons, the buckyball is *not* a regular polyhedron. In a **regular polyhedron** only one type of regular polygon is used for the faces, and there is the same number of faces at each vertex.

Semiregular Polyhedra

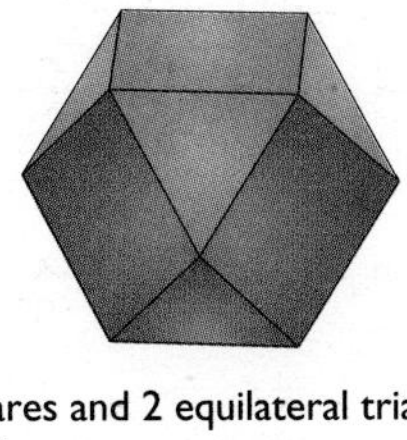

2 squares and 2 equilateral triangles at each vertex

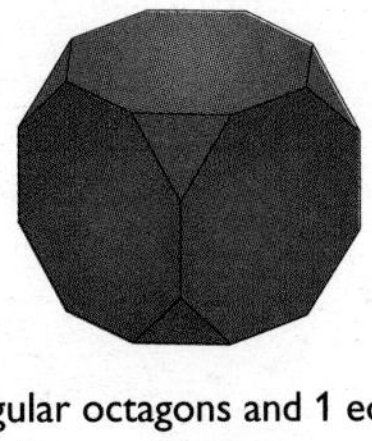

2 regular octagons and 1 equilateral triangle at each vertex

Regular Polyhedra

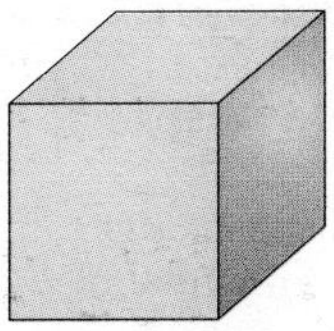

3 squares at each vertex

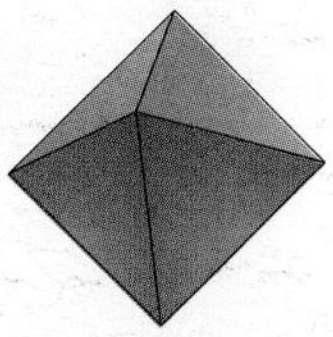

4 equilateral triangles at each vertex

The first model of the buckyball was made by cutting regular pentagons and regular hexagons out of paper and pasting them together. On the left is part of a **net** of polygons that can be used to make this model. The existence of a gap allows a net to "fold up" to make a polyhedron.

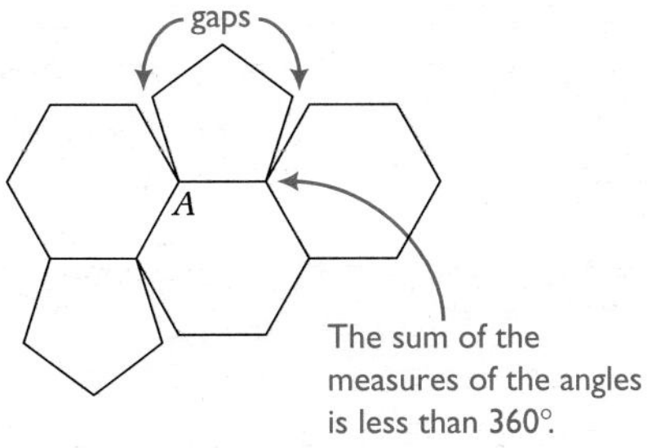
gaps
A
The sum of the measures of the angles is less than 360°.

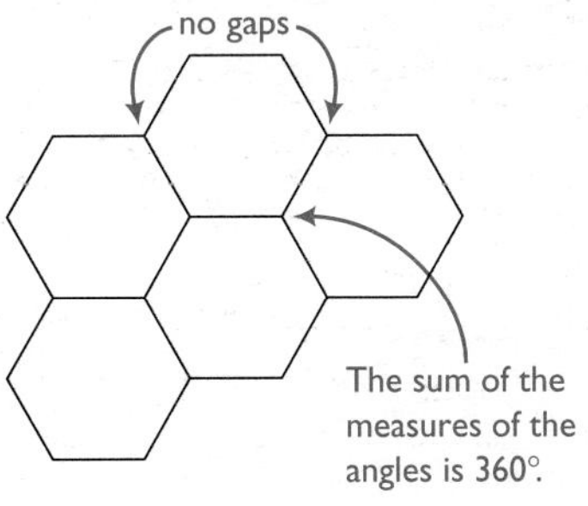
no gaps
The sum of the measures of the angles is 360°.

TEACHING

Talk it Over

Questions 1–3 use the buckyball to introduce students to the fact that in a convex polyhedron, the sum of the angles at each vertex is less than 360°.

Teaching Tip

Most, if not all, of the terms in this section will be new to students. You may wish to have students make a list of these newly defined terms with their definitions, and perhaps with an illustration, as a personal reference list.

Using Manipulatives

The Convex Polyhedron Postulate will be easy for students to see intuitively if they make a few simple models from construction paper. Suggest the following procedure. Cut an angular wedge from one side of a rectangular piece of paper. From the vertex, draw 5 or 6 arbitrarily selected dashed lines that form acute angles.

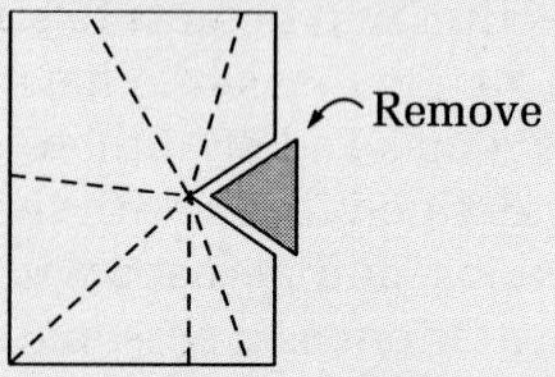

Fold along the dotted lines and join the edges where the wedge was removed. The sum of the acute angles at the vertex of the resulting three-dimensional model is obviously less than 360° (due to the removal of the wedge). This model does not give a complete polyhedron, but it illustrates clearly what happens at one vertex.

Talk it Over

Questions 4 and 5 lead students to understand the difference between a tessellation and a polyhedral net.

Additional Sample

Find the sum of the defects for all the vertices of a regular octahedron.

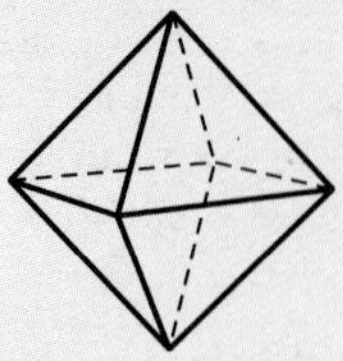

At each vertex there are four 60° angles, for a total of 240°.

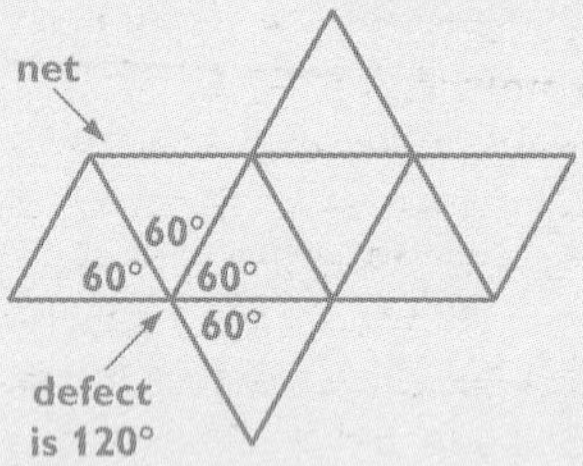

The defect is 360° – 240° = 120°. Since a regular octahedron has six vertices, the sum of the defects is 6(120°) = 720°.

Talk it Over

4. The angle measure of the gap in a net for a polyhedron is called the **defect** for the vertex. Find the defect at vertex A in the net for the buckyball on page 193.
5. How is the tessellation of hexagons different from the net for the buckyball?

The polyhedra you study in this course are convex. Prisms and pyramids are convex polyhedra. All regular and semiregular polyhedra are convex.

CONVEX POLYHEDRON POSTULATE

In any convex polyhedron, the sum of the measures of the angles at each vertex is less than 360°.

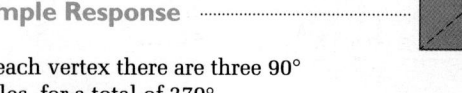

BY THE WAY...

Grace Chisholm Young (1868–1944) was the first woman to receive an official doctorate in mathematics in Germany. In her *First Book of Geometry* she included nets, believing they would help students become familiar with three-dimensional figures.

Sample

Find the sum of the defects for all the vertices of a cube.

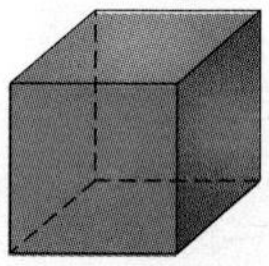

Sample Response

At each vertex there are three 90° angles, for a total of 270°.

The defect at each vertex is 360° − 270° = 90°.

Since a cube has eight vertices, the sum of the defects is 8(90°) = 720°.

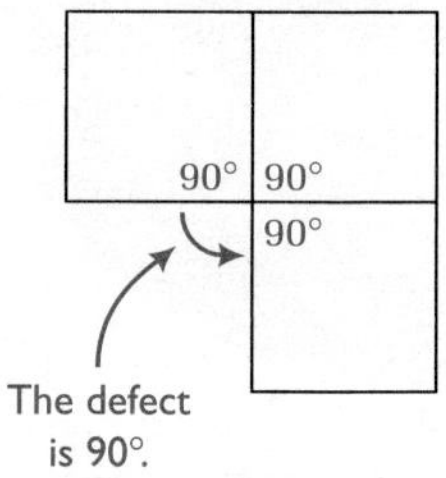

The Five Regular Polyhedra

Starting with triangles, there are regular polygons with any number of sides. There are regular quadrilaterals, pentagons, hexagons, heptagons, and so on. You might conjecture that there are infinitely many types of regular polyhedra also.

Answers to Talk it Over

4. 12°
5. The tessellation of hexagons has no defects; the sum of the measures of the angles at each vertex is 360°.

BY THE WAY...

The five regular polyhedra are also called the "Platonic Solids" after the Greek philosopher Plato.

Talk it Over

6. A regular tetrahedron has 3 equilateral triangles at each vertex. Is it possible to have a regular polyhedron with 4 equilateral triangles at each vertex? with 5? with 6? with more than 6?
7. Is it possible to have a regular polyhedron with more than 3 squares at each vertex? with more than 3 pentagons at each vertex?
8. Is it possible to make a regular polyhedron with hexagons? Why or why not?

The *Talk it Over* questions demonstrate that it is impossible to have more than five regular polyhedra.

REGULAR POLYHEDRA

Theorem 3.13 There are exactly five regular polyhedra.

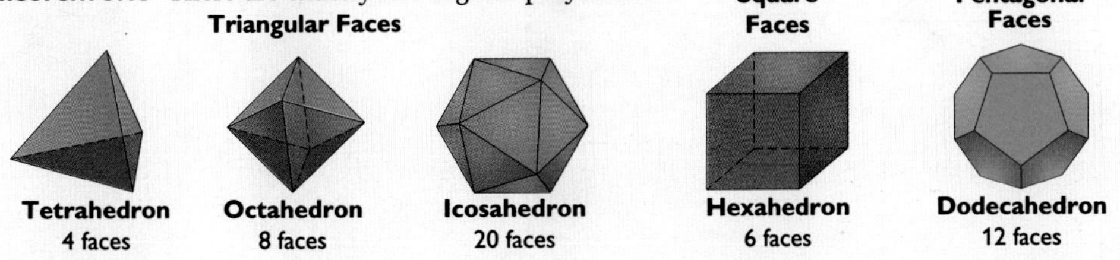

Look Back

Explain why a regular polyhedron cannot be made using polygons with seven or more sides.

3-9 Exercises and Problems

1. **Reading** What is a semiregular polyhedron?
2. **Using Manipulatives** Construct the five regular polyhedra. Use polyhedral nets provided by your teacher or use toothpicks and gum drops.
3. **Using Manipulatives** Construct three congruent regular hexagons and show that they will not "fold up" to make the corner of a polyhedron.
4. A *regular tessellation* is a tiling of the plane made up of only one type of regular polygon. Show that only three regular polygons can form regular tessellations.

Talk it Over

Questions 6–8 lead students through an informal proof of the fact that there are exactly five regular polyhedra.

Look Back

You may wish to call upon some students to give a verbal explanation to this question. Students can benefit from hearing how their classmates think, and making verbal statements in class gives students an opportunity to use and improve upon their communication skills.

APPLYING

Suggested Assignment

Standard 1, 5, 8–13, 15–19

Extended 1, 4–20, 23–28

Integrating the Strands

Algebra Exs. 24–27

Functions Ex. 27

Measurement Exs. 19, 23

Geometry Exs. 1–23

Logic and Language Exs. 1, 15–18, 21, 28

Answers to Talk it Over

6. Yes; Yes; No; No. The sum of the measures of the angles at each vertex must be less than 360°, so there can be no more than five equilateral triangles at each vertex.
7. No; No. The sum of the measures of the angles at each vertex must be less than 360°, so there can be no more than three squares or three pentagons at each vertex.
8. No. The measure of each angle of a regular hexagon is 120°. The sum of the measures of three such angles is 360°.

Answers to Look Back

Let n be the number of sides of the polygons forming a regular polyhedron. There must be at least three polygons at each vertex. The sum of the measures of the angles at each vertex must be less than 360°. Then $3\frac{(n-2)180}{n} < 360$, or $n < 6$.

Answers to Exercises and Problems

1. a space figure with faces that are all regular polygons of at least two types, with the same number of faces of each type at each vertex

2, 3. Check students' work.

4. The sum of the measures of the angles at each vertex must be 360°, and there must be at least three polygons at each vertex. Therefore, the greatest possible angle measure is 120°, which is the measure of each angle of a hexagon. Since the polygons are regular, the measure of each angle must be a factor of 360°. Since the measure of each angle of a regular pentagon is 108°, a tessellation cannot consist of regular pentagons. The only possibilities are equilateral triangles, squares, and regular hexagons.

Using Manipulatives

Physical models of the five regular polyhedra are available. If possible, have a number of these models in the classroom for students to use as a guide in doing Ex. 2.

Problem Solving

In connection with Ex. 6, ask students if they can prove that Euler's formula holds for all prisms. Students should be able to draw diagrams that are helpful in answering questions that point the way to a proof: How is the number of lateral faces related to the number of sides of a polygonal base of the prism? How is the number of vertices of the prism related to the number of sides of a polygonal base?

5. For each of the regular polyhedra, find the number of faces, vertices, and edges. Complete the table to show that Euler's formula, $F + V = E + 2$, holds.

	Number of Faces (F)	Number of Vertices (V)	Number of Edges (E)
tetrahedron	?	?	?
hexahedron	?	?	?
octahedron	?	?	?
dodecahedron	?	?	?
icosahedron	?	?	?

6. A *prism* is a polyhedron formed when two congruent polygons in parallel planes, with corresponding sides parallel, are joined at corresponding vertices. Prisms are classified according to the shape of the base. Show that Euler's formula, $F + V = E + 2$, holds for each of the prisms below.

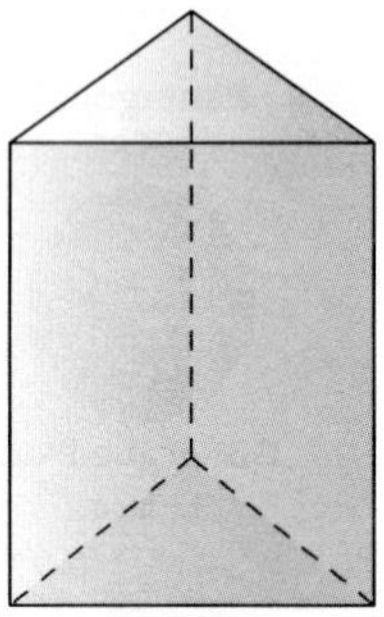
triangular prism

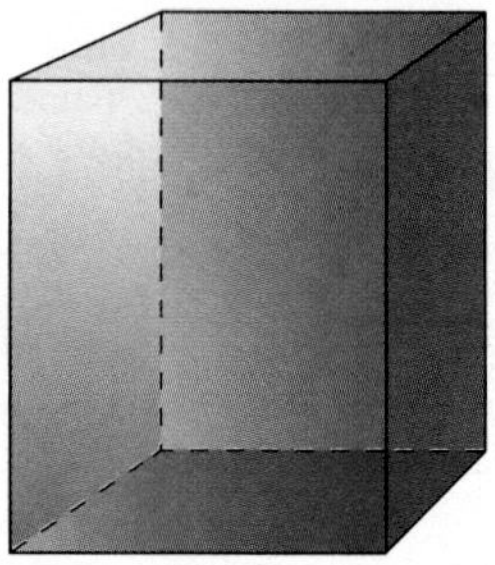
rectangular prism

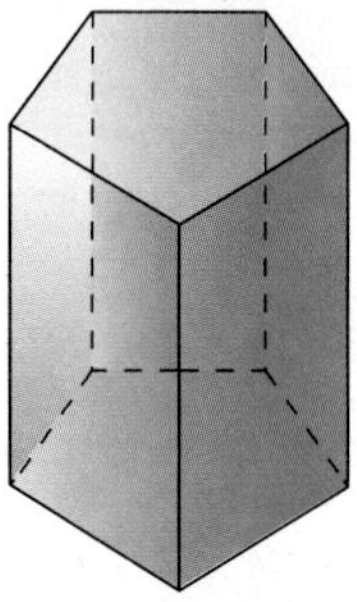
pentagonal prism

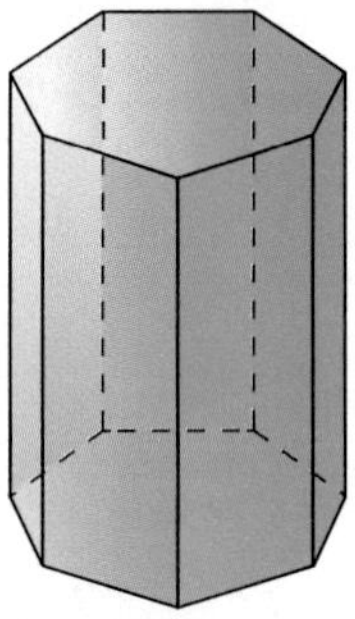
heptagonal prism

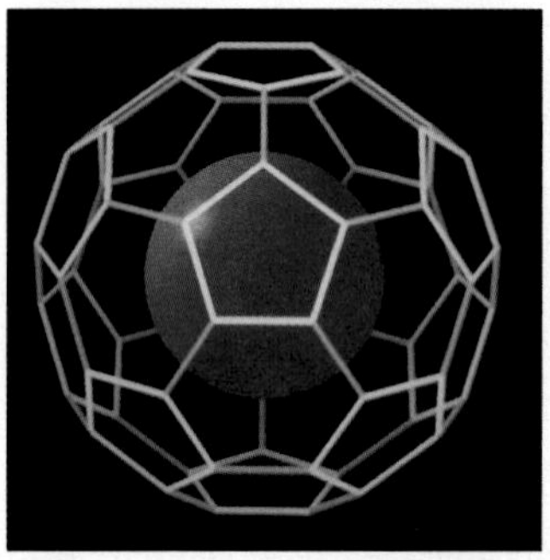

7. Find the number of edges in a buckyball using three methods.

a. There are 60 vertices. Each vertex connects three edges. That gives a total of _?_ edges. But you have counted each edge how many times? Divide to find the answer.

b. Each edge is the side of a face. There are 12 pentagonal faces and 20 hexagonal faces. How many sides is that altogether? You have now counted each edge how many times? Divide to find the answer.

c. You know the number of faces (F) and the number of vertices (V). Use Euler's formula, $F + V = E + 2$, to find the number of edges.

Find the sum of the defects for all the vertices of each polyhedron.

8. regular tetrahedron **9.** regular octahedron **10.** buckyball

11. prism with regular pentagonal base

Answers to Exercises and Problems

5. See answers in back of book.
6. triangular prism: 5 faces, 6 vertices, 9 edges; $5 + 6 = 9 + 2$; rectangular prism: 6 faces, 8 vertices, 12 edges; $6 + 8 = 12 + 2$; pentagonal prism: 7 faces, 10 vertices, 15 edges; $7 + 10 = 15 + 2$; heptagonal prism: 9 faces, 14 vertices, 21 edges; $9 + 14 = 21 + 2$
7. a. 180; twice; 90 edges
   b. 180 sides; twice; 90 edges
   c. $32 + 60 = E + 2$; $E = 90$
8. 720° 9. 720°
10. 720° 11. 720°
12. The sum of the defects at each vertex of any convex polyhedron is 720°. Examples will vary.
13. Suppose that there is a regular polyhedron with faces that have more than five sides. That is, the faces are regular n-gons with $n \geq 6$. The measure of each angle of the n-gon is $\frac{(n-2)180°}{n} = 180° - \frac{360°}{n}$. Since $n \geq 6$, the measure of each angle is greater than or equal to 120°. There are at least three n-gons at each vertex of the polyhedron, so the sum of the measures of the angles at each vertex is greater than or equal to 360°. This contradicts the convex polyhedron postulate. The assumption that there is a regular polyhedron with faces that have more than five sides must be false. Therefore, there is no regular polyhedron with faces that have more than five sides.

12. Make a conjecture based on your answers to Exercises 8–11 and the Sample. Test your conjecture for several other polyhedra.

13. Write an indirect proof to show that no regular polyhedron has faces with more than five sides.

14. **Open-ended** Make a chart similar to the quadrilateral chart to show relationships among polyhedra. Include these categories in your chart: regular polyhedron, semiregular polyhedron, prism, cube, buckyball, and others that you choose.

Tell whether *exactly* or *at least* fits best in each sentence.

15. Every polyhedron has (exactly/at least) three faces at each vertex.

16. Each edge of a polyhedron is shared by (exactly/at least) two faces.

17. A prism has (exactly/at least) one pair of parallel faces.

18. There are (exactly/at least) five regular polyhedra.

19. Each edge of the cube measures 2 m. One sphere is circumscribed about the cube. The other sphere is inscribed in the cube.

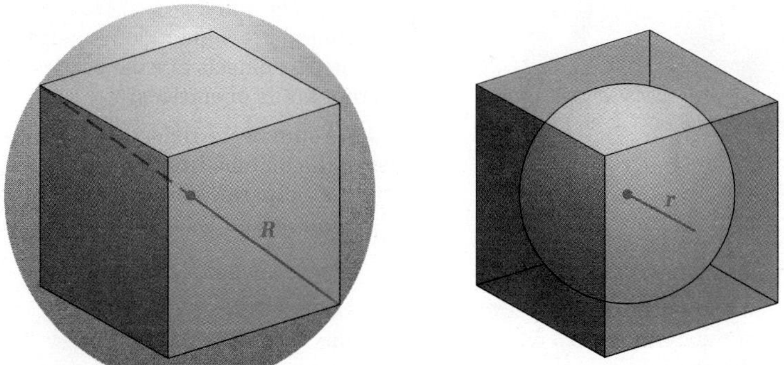

Remember:
Volume of sphere $= \frac{4}{3}\pi r^3$
Surface area $= 4\pi r^2$

a. Find the radii r and R of the two spheres. (*Hint:* A diameter of the circumscribed sphere is equal in length to a diagonal of the inscribed cube.)

b. Find the surface area of each sphere and compare it with the surface area of the cube.

c. Find the volume of each sphere and compare it with the volume of the cube.

20. This "globe" by R. Buckminster Fuller is made by circumscribing a regular icosahedron about Earth. Which is greater, the surface area of Earth or the surface area of the icosahedron?

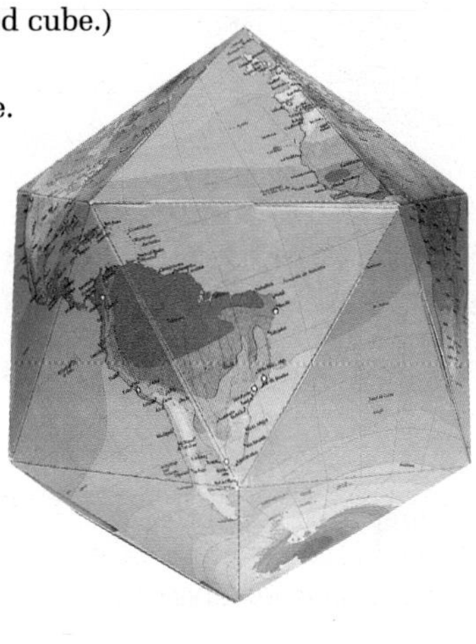

21. **Research** Learn how the structures of different crystals are related to polyhedra.

Reasoning

Exs. 12 and 13 allow students to use their thinking skills to make and test a conjecture about the sum of the defects for all the vertices of various polyhedra and to write an indirect proof showing that no regular polyhedron has faces with more than five sides.

Research

Some students may wish to research the career of R. Buckminster Fuller and report on his life, work, and achievements.

Communication: Writing

Students can write a report based on their research notes for Ex. 21. Some students may wish to include the reports in their portfolios.

Answers to Exercises and Problems

14.

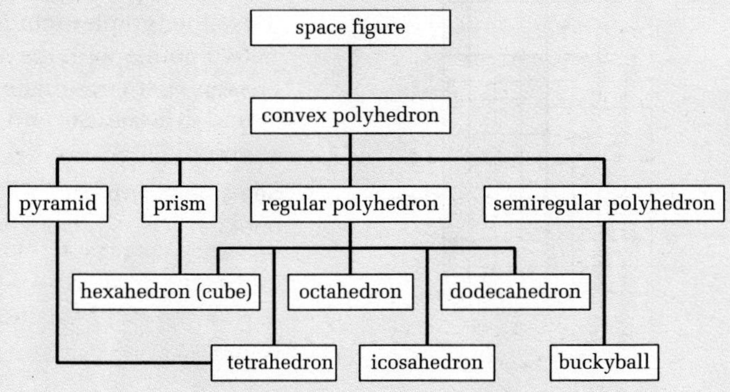

15. at least
16. exactly
17. at least
18. exactly
19. a. $r = 1$ m; $R = \sqrt{3}$ m $\approx$ 1.7 m
b. The inner sphere has surface area 4π m^2 or about 12.6 m^2. The outer sphere has surface area 12π m^2 or about 37.7 m^2. The cube has surface area 24 m^2.
c. The inner sphere has volume $\frac{4}{3}\pi$ m^3 or about 4.2 m^3. The outer sphere has volume $4\sqrt{3}\pi$ m^3 or about 21.8 m^3. The cube has volume 8 m^3.

20. the surface area of the icosahedron

21. Answers may vary. An example is given. Crystals have regular shapes due to the arrangements of their atoms. Crystals have flat surfaces that intersect at edges and are classified according to their axes of symmetry. Cubes and octahedra are said to have isometric systems of symmetry, for example, having three axes of equal length that are perpendicular to each other.

Quick Quiz (3-7 through 3-9)

See page 200.

See also Test 12 in the *Assessment Book.*

Practice 27 For use with Section 3-9

Name ______________________ Date ____________

Practice 27

For use with Section 3-9

Each regular polyhedron has the given number of vertices. Find the number of edges of each polyhedron.

1. octahedron; 6 vertices *Hint*: Each vertex is the intersection of 4 edges. 12
2. icosahedron; 12 vertices *Hint*: Each vertex is the intersection of 5 edges. 30
3. dodecahedron; 20 vertices *Hint*: Each vertex is the intersection of 3 edges. 30
4. Write an indirect proof that shows there is no regular polyhedron with faces that are regular hexagons.

4. Assume there is a regular polyhedron whose faces are regular hexagons. Since at least 3 faces must share each vertex, the sum of the angle measures at each vertex is at least $3 \cdot 120° = 360°$. This contradicts the Convex Polyhedron Postulate.

Find the sum of the defects for all the vertices of each polyhedron.

5. regular icosahedron; 12 vertices 720°
6. regular dodecahedron; 20 vertices 720°

Three regular polyhedra are shown below. List *all* of the polyhedra shown to which each statement applies.

dodecahedron icosahedron octahedron

7. Each vertex is part of five different faces. icosahedron
8. The defect at each vertex is 36°. dodecahedron
9. Each edge is shared by exactly 2 faces. all
10. At least one cross-section of the polyhedron is a square. octahedron
11. ***Open-ended*** Design a net for a prism and cut your net out of paper or cardboard. Find the sum of the defects for all the vertices of your prism. Does your net actually fold up into the prism, as you thought it would? If not, what went wrong? Describe a way of constructing other nets for the same prism.
Answers may vary. Check students' work.

Practice Bank, INTEGRATED MATHEMATICS 3 27

Ongoing ASSESSMENT

22. **Group Activity** Work in a group of four students.

Using toothpicks as edges, and gumdrops, raisins, or miniature marshmallows as vertices, build as many different convex polyhedra as you can that use *only equilateral triangles* as faces. They do not have to be regular polyhedra.

a. These polyhedra are called *deltahedra.* Make a systematic list of the deltahedra you build, including the number of vertices, edges, and faces for each.

b. Check to see if Euler's formula, $F + V = E + 2$, holds for your deltahedra.

c. Choose one of your deltahedra. How many vertices join 3 edges? How many join 4 edges? How many join 5 edges? What are the defects at each of these different kinds of vertices?

d. Find the sum of the defects at the vertices of the deltahedron you chose in part (c). Compare your answer with others in your group. What do you notice?

Review PREVIEW

23. A prism whose base is a regular hexagon is circumscribed about a cylinder. The radius of the cylinder is 5 cm, the height of the cylinder is 10 cm, and the semiperimeter of the hexagon is $10\sqrt{3}$ cm.

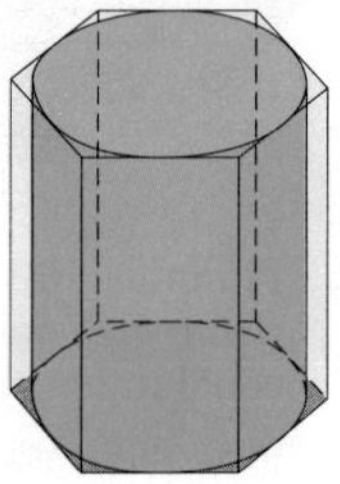

a. Find the volume of both the prism and the cylinder.

b. Suppose a prism whose base is a regular 12-gon is circumscribed about the same cylinder. How does its volume compare with that of the cylinder? with that of the hexagonal prism? *(Section 3-8)*

Solve. *(Toolbox Skills 22 and 23)*

24. $\frac{x-5}{2} + 3 = 7$

25. $10 < -5(x + 2)$

26. $2x^2 + 6 = 7x$

27. a. Graph $y = 3x + 4$. *(Section 2-1)*

b. On a separate set of axes, plot the points (−1, 1), (0, 4), (1, 7), (2, 10).

c. Are both graphs examples of functions? Explain.

3 Working on the Unit Project

28. As you finish preparing your brief, describe in what ways a legal brief is like a mathematical proof and in what ways it is different. Describe the most interesting part of preparing your brief.

Answers to Exercises and Problems

22. Answers may vary. Examples are given.

a. 6 faces, 5 vertices, and 9 edges; 10 faces, 7 vertices, and 15 edges;
12 faces, 8 vertices, and 18 edges

b. Euler's formula holds;
$6 + 5 = 9 + 2$; $10 + 7 = 15 + 2$; $12 + 8 = 18 + 2$

c. For the third deltahedron listed in part (a), no vertices join three edges, four vertices join four edges, and four vertices join five edges. At a vertex that joins four edges, the defect is 120°. At a vertex that joins five edges, the defect is 60°.

d. 720°; The sum of the defects is always 720°.

23. a. prism: $500\sqrt{3}$ cm³ ≈ 866 cm³; cylinder: 250π cm³ ≈ 785.4 cm³

b. Its volume is greater than that of the cylinder but less than that of the hexagonal prism.

24. 13

25. $x < -4$

26. 2; 1.5

27. a.

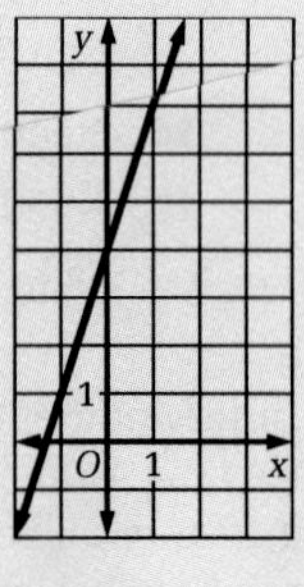

b.

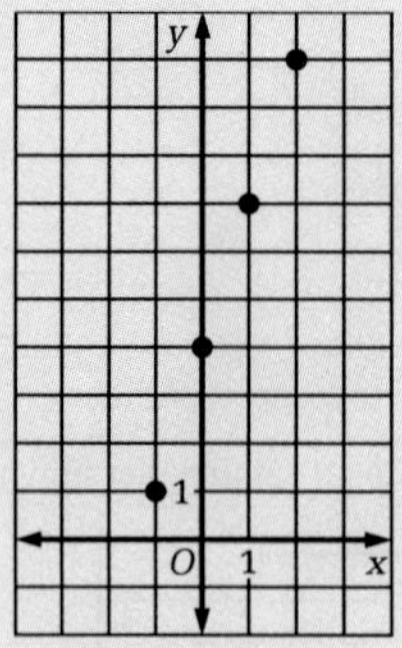

c. Yes; answers may vary. An example is given. Neither graph includes two points with the same first coordinate and different second coordinates.

28. See answers in back of book.

Unit Project 3

Completing the Unit Project

Now you are ready to finish your brief. Your finished project should include:

- **an opening statement**
- **questions you plan to ask witnesses**
- **questions you plan to ask witnesses for the opposing side during cross-examination**
- **the law as it applies to your case**
- **the reasoning you will use to convince others that the evidence presented justifies your position according to the law**
- **counterarguments you will make to your opponent's case**
- **closing arguments**

Look Back

How did you use reasoning in constructing your brief? Do you think your arguments are convincing?

Alternative Projects

Project 1: Research a Famous Case

Research a famous court case. Prepare a written or oral report about the case. Describe the reasoning used by the attorneys during the trial, and tell the outcome.

Project 2: Put on a Mock Trial

Simulate a trial in your classroom. Have students act as judge, jurors, bailiff, witnesses, defense, plaintiff, attorneys, and so forth. You may want to act out the case of Wilkins vs. Springfield School System.

Assessment

A scoring rubric for the Unit Project can be found on pages 136 and 137 of this Teacher's Edition and also in the ***Project Book.***

Unit Support Materials

- *Practice Bank:* Cumulative Practice 28
- *Study Guide:* Unit 3 Review
- *Problem Bank:* Unifying Problem 3
- *Assessment Book:* Unit Tests 13 and 14; Spanish Unit Tests; Alternative Assessment
- *Test Generator* software with Test Bank
- *Teacher's Resources for Transfer Students*

Quick Quiz (3-7 through 3-9)

Complete.

1. In writing an indirect proof, you begin by temporarily ___?___. Then you write a deductive proof that arrives at ___?___. Therefore, you conclude that ___?___. [3-7]
 assuming the opposite of what you want to prove; a contradiction of a known fact; the temporary assumption is false
2. Describe how to find the area of a polygon circumscribed about a circle. [3-8]
 Multiply the radius of the inscribed circle and the semiperimeter of the polygon.
3. What are the names of the five regular polyhedra? [3-9]
 tetrahedron, hexahedron, octahedron, dodecahedron, icosahedron

Unit 3 REVIEW AND ASSESSMENT

1. $ABCD$ is a parallelogram; $\overline{AC} \perp \overline{BD}$. 3-1
 a. Prove that $\triangle DEC \cong \triangle BEC$.
 b. Explain in a short paragraph how you can use the proof from part (a) to complete the proof of the following statement:
 If the diagonals of a parallelogram are perpendicular, then the parallelogram is a rhombus.

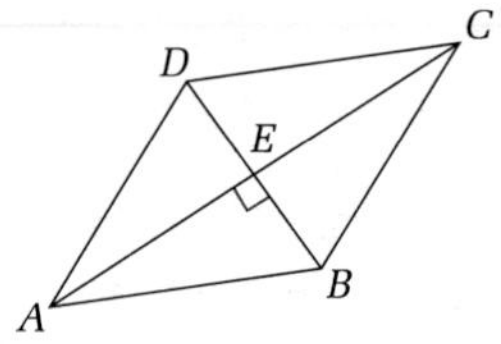

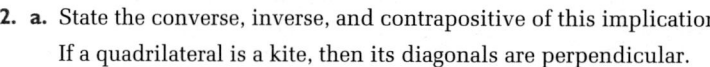

2. a. State the converse, inverse, and contrapositive of this implication:
 If a quadrilateral is a kite, then its diagonals are perpendicular.
 b. Tell whether each statement in part (a) is *True* or *False*. Give a counterexample for each false statement.
3. The y-axis is a line of symmetry for hexagon $QRSTUV$. 3-2
 a. Find the coordinates of points Q, S, and U.
 b. Show that this hexagon can be formed by joining two isosceles trapezoids.

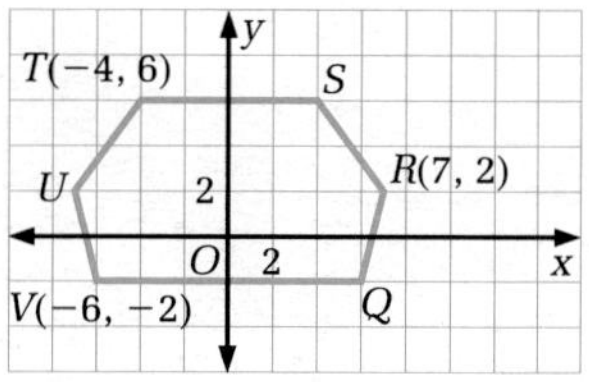

4. **Writing** Suppose you use an exclusive rather than an inclusive definition of trapezoid. How does the quadrilateral chart shown on page 155 change? 3-3
5. **Open-ended** Identify at least four words or phrases that must be clearly defined in order to properly interpret these bylaws:

> The treasurer shall be elected by majority vote of the executive board. To serve as treasurer, a person must be a member in good standing for three years. The treasurer is responsible for ensuring that all club funds are deposited in an appropriate account at a reliable financial institution.

For each polygon described in Questions 6 and 7, determine the number of sides. 3-4

6. Each interior angle of a regular polygon measures 135°.
7. Each exterior angle of a regular polygon measures 30°.
8. Q is the midpoint of $\overline{TS}$, and R lies on the perpendicular bisector of $\overline{TS}$. $RS = 17$ cm and $RQ = 15$ cm. Find QS, TQ, and TR. 3-5

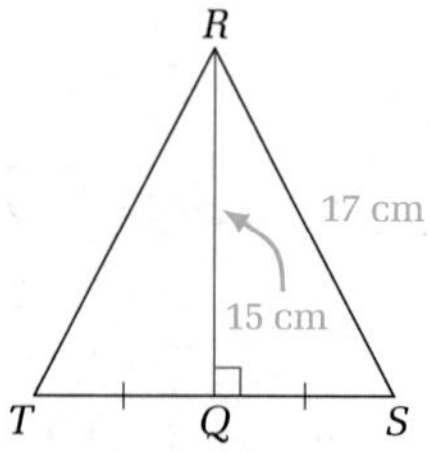

Answers to Unit 3 Review and Assessment

1. a. $ABCD$ is a parallelogram and $\overline{AC} \perp \overline{BD}$ (Given); $\angle DEC$ and $\angle BEC$ are right angles (Def. of $\perp$ lines); $\angle DEC$ and $\angle BEC$ are equal in measure (Def. of right angles); $\overline{AC}$ and $\overline{BD}$ bisect each other (The diagonals of a parallelogram bisect each other.); $DE = BE$ (Def. of segment bisector); $CE = CE$ (Reflexive property of equality); $\triangle DEC \cong \triangle BEC$ (SAS)
 b. Answers may vary. An example is given. I would use the proof to show that $\overline{DC}$ and $\overline{BC}$ are corresponding parts of congruent triangles and are equal in measure. Since $ABCD$ is a parallelogram, we already know that $\overline{DA}$ and $\overline{CB}$ are equal in measure and $\overline{AB}$ and $\overline{DC}$ are equal in measure. It would then follow that all four sides of $ABCD$ are equal in measure and $ABCD$ is a rhombus.
2. a. Converse: If the diagonals of a quadrilateral are perpendicular, then the quadrilateral is a kite. Inverse: If a quadrilateral is not a kite, then its diagonals are not perpendicular. Contrapositive: If the diagonals of a quadrilateral are not perpendicular, then the quadrilateral is not a kite.
 b. The converse and the inverse are false. Counterexample: Consider a quadrilateral whose diagonals are perpendicular, but whose intersec-

9. Find all the missing measures for angles and arcs. 3-6

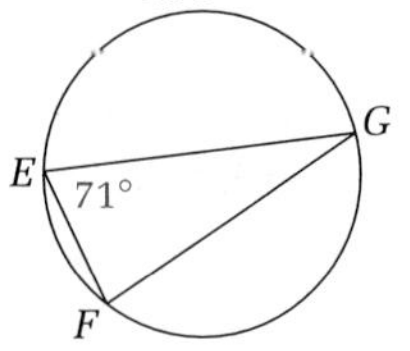

10. **Open-ended** Eight points, labeled A through H, are equally spaced around a circle. List all the measures of the inscribed angles you can make using just these points. Give an example of an inscribed angle for each angle measure.

11. Write an indirect proof to show that no regular polygon has an exterior angle measuring 150°. 3-7

12. An astronaut in a spacecraft at point R observes the horizon of Earth at point H.
   a. If the spacecraft is 100 mi above the surface of Earth, find the distance from R to H. Earth's radius is about 4000 mi.
   b. State any theorems you used to answer part (a).
   c. Find a formula for the distance from R to H when the spacecraft is d mi above the surface of Earth.

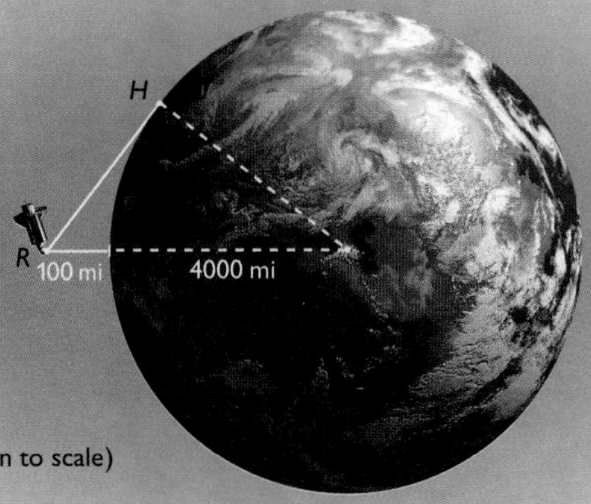

(Not drawn to scale)

13. A circle has a radius of 5 in. Find the area of a square inscribed in the circle and the area of a square circumscribed about the circle. Is the area of the circle closer to the area of the inscribed square or the circumscribed square? 3-8

14. A pyramid with a square base has equilateral triangles as its four other faces. 3-9
   a. Explain why the pyramid is not a regular polyhedron.
   b. Find the sum of the defects at each vertex.

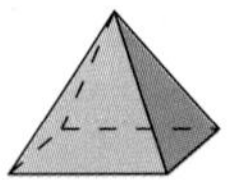

15. **Self-evaluation** Which kind of proof do you prefer to write, a synthetic proof or a coordinate proof? Why?

Answers to Unit 3 Review and Assessment

tion is not the midpoint of either diagonal. The contrapositive is true.

3. a. $Q(6, -2)$; $S(4, 6)$; $U(-7, 2)$
   b. $QRSTUV$ can be formed by joining $QVUR$ and $UTSR$, which are isosceles trapezoids, since both have one pair of parallel sides ($\overline{TS}$, $\overline{UR}$, and $\overline{VQ}$ are all parallel to the x-axis) and both have a line of symmetry (the y-axis).

4. There are no longer segments joining parallelograms to trapezoids or rectangles to trapezoids and isosceles trapezoids.

5. Answers may vary. Among the terms or phrases which may need to be defined in order to interpret the by-laws are *member, in good standing, appropriate account,* and *reliable financial institution.* Sample definitions are given. A member in good standing is one who has paid dues for the last membership year and has been absent from fewer than half of all meetings (unless those absences were due to illness or for other causes found reasonable by the executive board). A reliable financial institution is one which is licensed by the state to do business and is insured by the Federal Deposit Insurance Corporation.

6. 8 sides

7. 12 sides

8. 8 cm; 8 cm; 17 cm

9. $m\,\overset{\frown}{GF} = 142°$; $m\,\overset{\frown}{EF} = 54°$; $m\angle G = 27°$; $m\angle F = 82°$

10. 22.5° (for example, $\angle BAC$); 45° ($\angle BAD$); 67.5° ($\angle BAE$); 90° ($\angle BAF$); 112.5° ($\angle BAG$); 135° ($\angle BAH$)

11. Suppose that P is a regular n-gon with an exterior angle measuring 150°. P has n such exterior angles and their sum is 360°. (The sum of the exterior angles of an n-gon is 360°.) Then $150n = 360$ or $n = \frac{360}{150} = 2.4$. This contradicts the fact that n must be a whole number. So the assumption that a regular polygon can have an exterior angle measuring 150° must be false. Therefore, no regular polygon has an exterior angle measuring 150°.

12. a. about 900 mi
   b. If a line is tangent to a circle, then it is perpendicular to the radius from the center to the point of tangency. In any right triangle, the square of the length of the hypotenuse is equal to the sum of the squares of the lengths of the legs.
   c. Let h be the distance from R to H; $h = \sqrt{(4000 + d)^2 - (4000)^2} = \sqrt{8000d + d^2}$.

13. 50 in.^{2}; 100 in.^{2}; the circumscribed square

14. a. In a regular polyhedron, only one type of regular polygon is used for the faces.
   b. 720° (There are 4 vertices with defects of 150° and one vertex with a defect of 120°.

15. Answers may vary. An example is given. I think it is easier to write a coordinate proof, because I know I usually have to begin with the distance formula, the midpoint formula, or slope relationships.

Quick Quiz (3-1 through 3-3)

1. For the implication *If p, then q,* write each of the following. [3-1]
   a. the contrapositive
      If not q, then not p.
   b. the inverse
      If not p, then not q.
   c. the converse
      If q, then p.
2. Write a statement that is logically equivalent to this statement. [3-1]
   If a quadrilateral is a parallelogram, then its two diagonals bisect each other.
   If the two diagonals of a quadrilateral do not bisect each other, then the quadrilateral is not a parallelogram.
3. What four justifications may be used in coordinate proofs? [3-2]
   (1) the distance and midpoint formula
   (2) Parallel lines have the same slope.
   (3) Perpendicular lines have slopes that are negative reciprocals of each other.
   (4) A geometric figure may be placed anywhere in the coordinate plane.
4. Tell whether this statement is *True* or *False.* [3-3]
   All parallelograms are trapezoids. **True.**

16. **Group Activity** Work in a group of three students.
    a. Draw a circle and a circumscribed triangle.
    b. Pass your paper to the student on your right. Measure the radius of the circle and the sides of the triangle on your new paper. Find the area of the triangle using the formula $A = rs$.
    c. Again pass your paper to the student on your right. Use a different method to find the area of the triangle on your new paper.
    d. Compare your result from part (c) with the result found in part (b) by the student on your left.

IDEAS AND (FORMULAS)

GEOMETRY

- If the two diagonals of a quadrilateral bisect each other, then the quadrilateral is a parallelogram. *(p. 141)*
- In an isosceles triangle, the medians drawn to the legs are equal in measure. *(p. 150)*
- In an isosceles trapezoid, (1) the legs are equal in measure, (2) the diagonals are equal in measure, and (3) the two angles at each base are equal in measure. *(p. 150)*
- The sum of the interior angle measures of an n-gon is given by the formula $S(n) = (n - 2)180°$. *(p. 161)*
- The sum of the exterior angle measures of an n-gon, one at each vertex, is 360°. *(p. 161)*
- The perpendicular bisector of a chord of a circle passes through the center of the circle. *(p. 166)*
- The measure of an inscribed angle of a circle is equal to half the measure of its intercepted arc. *(p. 174)*
- An inscribed angle whose intercepted arc is a semicircle is a right angle. *(p. 174)*
- If two inscribed angles in the same circle intercept the same arc, then they are equal in measure. *(p. 174)*
- A segment can be drawn perpendicular to a given line from a point not on the line. The length of this segment is the shortest distance from the point to the line. *(p. 182)*
- If a line is tangent to a circle, then the line is perpendicular to the radius from the center to the point of tangency. *(p. 183)*

Answers to Unit 3 Review and Assessment

16. Check students' work.

- If a line in the plane of a circle is perpendicular to a radius at its outer endpoint, then the line is tangent to the circle. *(p. 183)*
- If two tangent segments are drawn from the same point to the same circle, then they are equal in measure. *(p. 183)*
- The area of any circumscribed polygon may be found by multiplying the radius (r) of the inscribed circle and the semiperimeter (s). *(p. 188)*
- In any convex polyhedron, the sum of the measures of the angles at each vertex is less than 360°. *(p. 194)*
- There are exactly five regular polyhedra. *(p. 195)*

LOGICAL REASONING ($p \leftrightarrow q$, *if - then*)

- The contrapositive of a true implication is always true. The converse and inverse of a true implication are not necessarily true. *(p. 142)*
- In mathematics, as well as in other aspects of life, precise definitions are often essential. *(p. 154)*
- The formula for the sum of the measures of the interior angles of an n-gon can be proved using mathematical induction. *(p. 160)*
- An indirect proof begins by temporarily assuming the opposite of what is to be proved and showing that this leads to an impossible situation or to a contradiction. *(p. 180)*

ALGEBRA

- Coordinate proofs can be used as well as synthetic proofs to demonstrate geometric theorems. *(p. 147)*

Key Terms

synthetic proof (p. 140)	**implication** (p. 142)	**converse** (p. 142)
inverse (p. 142)	**contrapositive** (p. 142)	**median of a triangle** (p. 147)
coordinate proof (p. 147)	**isosceles trapezoid** (p. 150)	**trapezoid** (p. 154)
inclusive definition (p. 154)	**exclusive definition** (p. 154)	**regular polygon** (p. 161)
inscribed (pp. 166, 187)	**circumscribed** (pp. 166, 187)	**chord** (p. 166)
central angle (p. 167)	**minor arc** (p. 167)	**semicircle** (p. 167)
major arc (p. 167)	**inscribed angle** (p. 172)	**intercepted arc** (p. 172)
indirect proof (p. 180)	**tangent** (p. 181)	**secant** (p. 181)
semiperimeter (p. 188)	**polyhedron** (p. 193)	**semiregular polyhedron** (p. 193)
regular polyhedron (p. 193)	**net** (p. 193)	**defect** (p. 194)

Quick Quiz (3-4 through 3-6)

1. Find the sum of the measures of the interior angles of a regular polygon with 20 sides. [3-4] **3240°**
2. Find the measure of each interior and each exterior angle of a regular polygon with 30 sides. [3-4] **168°; 12°**
3. What fact is true about the perpendicular bisector of a chord of a circle? [3-5] **It passes through the center of the circle.**
4. Complete each of the following. [3-5]
   For any minor arc AB, $m\overset{\frown}{AB}$ __?__ 180°. **<**
   For any semicircle XYZ, $m\overset{\frown}{XYZ}$ __?__. **= 180°**
   For any major arc RST, 180° __?__ $m\overset{\frown}{RST}$ __?__ 360° **<; <**
5. A central angle of a circle intercepts an arc of 120°. What is the measure of an inscribed angle that intercepts the same arc? [3-6] **60°**

Planning the Unit 4

Sequences and Series

OVERVIEW

- **Unit 4** covers the topics of sequences and series. Sequences are classified as arithmetic, geometric, or neither. Formulas are developed for sequences, both explicit and recursive. Subscript notation is used to represent the terms of a sequence.
- Series are introduced and students find sums for finite arithmetic and geometric series. Sums of infinite geometric series lead to the introduction of a limit of an infinite series. Sigma notation is introduced and used to write finite arithmetic series and finite geometric series.
- The theme of the **Unit Project** is to save for the future. Students learn how to make a small investment grow in order to pay for a large future expense. Students explore stocks, mutual funds, CDs, and other types of investments, using long-range planning for funding such needs as a college education.
- **Connections** to skydiving, investments, astronomy, fractals, genetics, knitting patterns, spider webs, and Greek myths are some of the topics integrated into the teaching materials and the exercises.
- **Graphics calculators** are used in Section 4-3 to find the limit of a sequence using recursion. **Computer software,** such as Plotter Plus, can be used in Sections 4-1 and 4-3 to work with spreadsheets.
- **Problem-solving strategies** used in Unit 4 include using patterns, graphs, and formulas.

Unit Objectives

Section	Objectives	NCTM Standards
4-1	• Describe and continue patterns. • Graph sequences and find apparent limits.	1, 2, 3, 4, 12
4-2	• Use subscripts and formulas for sequences.	1, 2, 3, 4, 5, 12
4-3	• Write and use formulas for sequences in which each term is found by using the preceding term(s).	1, 2, 3, 4, 5, 12
4-4	• Identify sequences that have a common difference, a common ratio, or neither. • Write explicit and recursive formulas.	1, 2, 3, 4, 5, 12
4-5	• Find the sum of a finite nongeometric series.	1, 2, 3, 4, 5, 12
4-6	• Use a formula to find the sum of a finite geometric series.	1, 2, 3, 4, 5, 12
4-7	• Find the sum of an infinite geometric series.	1, 2, 3, 4, 5, 12, 13

Topic Spiraling

Section	Connections to Prior and Future Concepts
4-1	**Section 4-1** explores patterns that lead to a sequence of numbers. Students graph points on a coordinate plane and examine the graph for a limit of the sequence. Plotting points on a coordinate plane was first introduced in Section 4-1 of Book 1. The concept of limits is explored in Sections 4-7 and 5-4 of Book 3 and is an important concept in precalculus and calculus courses.
4-2	**Section 4-2** extends the concept of sequences begun in Section 4-1 of Book 3 to now include subscript notation and formulas to represent the sequences. Writing and understanding formulas for sequences is a skill used throughout the remainder of Unit 4 of Book 3.
4-3	**Section 4-3** explores sequences based on recursive formulas. Explicit and recursive formulas for the same sequence are compared. Recursive formulas are used again in Section 4-4 of Book 3.
4-4	**Section 4-4** explores sequences that are arithmetic, geometric, or neither. Explicit and recursive formulas are examined to find a common difference or ratio. Geometric and arithmetic means are introduced. Students use the explicit and recursive formulas first encountered in Sections 4-2 and 4-3 of Book 3.
4-5	**Section 4-5** extends the study of arithmetic sequences begun in Section 4-4 of Book 3. Students explore finding the sum of a finite arithmetic series. Sigma notation for summation is introduced, and is used in Section 4-6 of Book 3.
4-6	**Section 4-6** extends the study of geometric sequences begun in Section 4-4 of Book 3. Students explore finding the sum of a finite geometric series. The sigma notation introduced in Section 4-5 of Book 3 is used to represent the summation of a geometric series.
4-7	**Section 4-7** extends the work of Section 4-6 of Book 3. Students explore finding the sum of an infinite geometric series. The concept of a limit was informally introduced in Section 4-7 of Book 1, and the limit of a sequence was introduced in Section 4-1 of Book 3.

Integrating the Strands

Strands	Sections
Number	4-1, 4-2, 4-3, 4-4, 4-5, 4-6, 4-7
Algebra	4-1, 4-2, 4-3, 4-4, 4-5, 4-6, 4-7
Functions	4-1, 4-7
Geometry	4-1, 4-2, 4-4, 4-5, 4-6, 4-7
Trigonometry	4-2
Statistics and Probability	4-2
Discrete Mathematics	4-1, 4-2, 4-3, 4-4, 4-5, 4-6, 4-7
Logic and Language	4-1, 4-2, 4-3, 4-4, 4-5, 4-6, 4-7

Planning the Unit 4

Sequences and Series

Section Planning Guide

➢ Essential exercises and problems are indicated in boldface.
➢ Ongoing work on the Unit Project is indicated in color.
➢ Exercises and problems that require student research, group work, manipulatives, or graphing technology are indicated in the column headed "Other."

Section	Materials	Pacing	Standard Assignment	Extended Assignment	Other
4-1	toothpicks, calculator, spreadsheet software	Day 1 Day 2	**2–8**, 12 **13–22**, 26–31, 32	**2–8**, 9–12 **13–22**, 24–31, 32	1 23, 32
4-2	spreadsheet software	Day 1	**1–9**, 10, 12, 15–21, 22	**1–9**, 10–21, 22	22
4-3	graphics calculator, spreadsheet software	Day 1	**1–13**, 19–22, 23, 24	**1–13**, 14–17, 19–22, 23, 24	18, 24
4-4	spreadsheet software	Day 1 Day 2	**1–22** **27–34**, 36–38, 39	**1–22**, 23, 24 25, 26, **27–34**, 36–38, 39	 35, 39
4-5	spreadsheet software	Day 1 Day 2	**1–6**, 7–11 **13–20**, 23–26, 27	**1–6**, 7–12 **13–20**, 21–26, 27	 27
4-6	spreadsheet software	Day 1	**1–15**, **20**, 21–26, 27	**1–15**, 16, 17, **20**, 21–26, 27	18, 19
4-7	large squares of paper, scissors, spreadsheet software	Day 1 Day 2	**1–8**, 9, 10 **11–23**, 25, 27–32, 33	**1–8**, 9, 10 **11–23**, 24–32, 33	 33
Review Test		**Day 1** **Day 2**	**Unit Review** **Unit Test**	**Unit Review** **Unit Test**	

Yearly Pacing	Unit 4 Total	Units 1–4 Total	Remaining	Total
	15 days (2 for Unit Project)	65 days	89 days	154 days

Support Materials

➢ See **Project Book** for notes on Unit 4 Project: Save for the Future.
➢ UPP and disk refer to **Using Plotter Plus** booklet and **Plotter Plus** disk.
➢ TI-81/82 refers to **Using TI-81 and TI-82 Calculators** booklet.
➢ Warm-up exercises for each section are available on **Warm-Up Transparencies.**

Section	Study Guide	Practice Bank	Problem Bank	Activity Bank	Explorations Lab Manual	Assessment Book	Visuals	Technology
4-1	4-1	Practice 29	Set 9	Enrich 26	Masters 1, 2, 12	Quiz 4-1		Statistics Spreadsheet (disk) Data Analyzer (disk)
4-2	4-2	Practice 30	Set 9	Enrich 27	Master 2	Quiz 4-2		
4-3	4-3	Practice 31	Set 9	Enrich 28		Quiz 4-3		Statistics Spreadsheet (disk) Data Analyzer (disk)
4-4	4-4	Practice 32	Set 9	Enrich 29	Master 2	Quiz 4-4 Test 15		
4-5	4-5	Practice 33	Set 10	Enrich 30		Quiz 4-5		
4-6	4-6	Practice 34	Set 10	Enrich 31		Quiz 4-6		
4-7	4-7	Practice 35	Set 10	Enrich 32	Master 2	Quiz 4-7 Test 16		TI-81/82, page 60
Unit 4	Unit Review	Practice 36	Unifying Problem 4	Family Involve 4		Tests 17, 18		

UNIT TESTS

Form A

Spanish versions of these tests are on pages 127–130 of the **Assessment Book.**

Name ______ Date ______ Score ______

Test 17

Test on Unit 4 (Form A)

Directions: Write the answers in the spaces provided.

Find the next three terms and the 10th term of each sequence.

1. 7, 12, 17, 22, ...
2. 13, 15, 19, 27, ...

Decide whether each infinite sequence appears to have a limit. Write *Yes* or *No*. If it does, give its value.

3. $\frac{1}{3}, \frac{1}{5}, \frac{1}{7}, \frac{1}{9}, \ldots$
4. −2, 5, −8, 11, ...

Write the first three terms and the 27th term of each sequence.

5. $a_n = 7n + 3$
6. $a_n = 3(n+1)^{-1}$

Write an explicit formula for each sequence.

7. 4, 7, 10, 13, ...
8. 1, 6, 36, 216, ...

Write a recursive formula for each sequence.

9. 5, 9, 13, 17, ...
10. 343, 49, 7, 1, ...

Write the first four terms of each sequence.

11. $a_1 = 5$, $a_n = (-2)a_{n-1}$
12. $a_1 = -20$, $a_n = a_{n-1} + 11$

For Questions 13 and 14, find the number of terms in each sequence.

13. 96, 88, 80, 72, ..., −24
14. $-48, 24, -12, 6, \ldots, \frac{3}{32}$

15. Triangle 1 is the largest equilateral triangle shown and has an area of 4096 cm^2. Each of the other smaller triangles is formed by connecting the midpoints of the sides of the next larger triangle as shown in the figure. Find the area of the 8th triangle in the sequence.

Answers

1. 27, 32, 37; 52
2. 43, 75, 139; 1035
3. Yes; 0.
4. No.
5. 10, 17, 24; 192
6. $\frac{3}{2}, 1, \frac{3}{4}; \frac{3}{28}$
7. $a_n = 4 + (n-1)3$
8. $a_n = 6^{n-1}$
9. $a_1 = 5, a_n = 4 + a_{n-1}$
10. $a_1 = 343, a_n = \left(\frac{1}{7}\right)a_{n-1}$
11. 5, −10, 20, −40
12. −20, −9, 2, 13
13. 16
14. 10
15. $\frac{1}{4}$ cm^2

Assessment Book INTEGRATED MATHEMATICS 3

23

Name ______ Date ______ Score ______

Test 17 *(continued)*

Directions: Write the answers in the spaces provided.

Evaluate each series, if the sum exists.

16. $\sum_{k=1}^{9}(4k+3)$
17. $\sum_{n=0}^{5} 3^{n-2}$
18. $\sum_{t=1}^{5}(2)^{t-1}$
19. $\sum_{n=0}^{3} 5(2)^{n+1}$
20. $\sum_{k=1}^{\infty}(-6)\left(\frac{2}{5}\right)^{k-1}$
21. $\sum_{n=1}^{\infty} 15\pi\left(\frac{4}{3}\right)^{n}$

Find the sum of each finite arithmetic series.

22. 5 + 9 + 13 + 17 + ... + 109
23. 51 + 46 + 41 + 36 + ... + (−24)

Find the sum of each geometric series.

24. 2 + 6 + 18 + 54 + ... + 1458
25. $10 + 2 + \frac{2}{5} + \frac{2}{25} + \frac{2}{125} + \ldots$

For Questions 26 and 27, write each series in sigma notation.

26. 48 + 44 + 40 + 36 + ... + 20
27. 36 + 49 + 64 + 81 + ... + 169

28. **Writing** Explain how a sequence can be graphed on a coordinate plane.
    Sample answer: You can graph a sequence on a coordinate plane by plotting points using the coordinates (position, term). The points should *not* be connected.

29. **Open-ended** Write three different sequences that begin 1, 3, ... Explain the pattern in each sequence.
    Sample answer:
    1) 1, 3, 9, 27, 81, ...: The pattern is integral powers of 3 beginning with 3^0.
    2) 1, 3, 5, 7, 9, ...: The pattern is consecutive positive odd numbers beginning with 1.
    3) 1, 3, 6, 10, 15, ...:The pattern begins with 1 and adds the position number of the term to the previous term.

Answers

16. 207
17. $40\frac{4}{9}$
18. 31
19. 150
20. −10
21. no limit
22. 1539
23. 216
24. 2186
25. 12.5
26. $\sum_{k=1}^{8}(-4k+52)$
27. $\sum_{n=1}^{8}(n+5)^2$
28. *See question.*
29. *See question.*

24

Assessment Book INTEGRATED MATHEMATICS 3

Form B

Name ______ Date ______ Score ______

Test 18

Test on Unit 4 (Form B)

Directions: Write the answers in the spaces provided.

Find the next three terms and the 10th term of each sequence.

1. 7, 13, 19, 25, ...
2. 12, 15, 18, 21, ...

Decide whether each infinite sequence appears to have a limit. Write *Yes* or *No*. If it does, give its value.

3. −3, 6, −9, 12, ...
4. $\frac{1}{3}, \frac{1}{9}, \frac{1}{27}, \frac{1}{81}, \ldots$

Write the first three terms and the 27th term of each sequence.

5. $a_n = -2n + 3$
6. $a_n = 2(n+1)^{-1}$

Write an explicit formula for each sequence.

7. 8, 12, 16, 20, ...
8. 1, 5, 25, 125, ...

Write a recursive formula for each sequence.

9. 0, 9, 18, 27, ...
10. 216, 36, 6, 1, ...

Write the first four terms of each sequence.

11. $a_1 = 2$, $a_n = (-3)a_{n-1}$
12. $a_1 = 16$, $a_n = a_{n-1} - 13$

For Questions 13 and 14, find the number of terms in each sequence.

13. 46, 42, 38, 34, ..., −22
14. $88, -44, 22, -11, \ldots, \frac{11}{32}$

15. Square 1 is the largest square shown and has an area of 4096 cm^2. Each of the other smaller squares is formed by connecting the midpoints of the sides of the next larger square as shown in the figure. Find the area of the 10th square in the sequence.

Answers

1. 31, 37, 43; 61
2. 24, 27, 30; 39
3. No.
4. Yes; 0.
5. 1, −1, −3; −51
6. $1, \frac{2}{3}, \frac{1}{2}; \frac{1}{14}$
7. $a_n = 8 + (n-1)4$
8. $a_n = 5^{n-1}$
9. $a_1 = 0, a_n = 9 + a_{n-1}$
10. $a_1 = 216, a_n = \left(\frac{1}{6}\right)a_{n-1}$
11. 2, −6, 18, −54
12. 16, 3, −10, −23
13. 18
14. 9
15. 8 cm^2

Assessment Book INTEGRATED MATHEMATICS 3

25

Name ______ Date ______ Score ______

Test 18 *(continued)*

Directions: Write the answers in the spaces provided.

Evaluate each series, if the sum exists.

16. $\sum_{k=1}^{8}(4k-7)$
17. $\sum_{n=0}^{4} 3^{n-3}$
18. $\sum_{t=1}^{5}(3)^{t-1}$
19. $\sum_{n=0}^{3} -3(2)^{n+1}$
20. $\sum_{k=1}^{\infty} 13\pi\left(\frac{6}{5}\right)^{k}$
21. $\sum_{n=1}^{\infty}(-4)\left(\frac{5}{7}\right)^{n-1}$

Find the sum of each finite arithmetic series.

22. 9 + 12 + 15 + 18 + ... + 72
23. 81 + 76 + 71 + 66 + ... + 1

Find the sum of each geometric series.

24. 2 + 8 + 32 + 128 + ... + 2048
25. $28 + 4 + \frac{4}{7} + \frac{4}{49} + \frac{4}{343} + \ldots$

For Questions 26 and 27, write each series in sigma notation.

26. 11 + 22 + 33 + 44 + ... + 88
27. 9 + 16 + 25 + 36 + ... + 196

28. **Writing** Explain the terms *finite series* and *infinite series*. Include an example of each type of series and tell whether each series is *arithmetic*, *geometric*, or *neither*.
    Sample answer: A finite series is one in which there are a finite number of terms, while an infinite series is one in which there are an infinite number of terms. The series 2 + 4 + 6 + ... + 22 is a finite arithmetic series, while the series 2 + 4 + 8 + 16 + ... is an infinite geometric series.

29. **Open-ended** Write three different sequences that begin 1, 3, ... Explain the pattern in each sequence.
    Sample answer:
    1) 1, 3, 9, 27, 81, ...: The pattern is integral powers of 3 beginning with 3^0.
    2) 1, 3, 5, 7, 9, ...: The pattern is consecutive positive odd numbers beginning with 1.
    3) 1, 3, 6, 10, 15, ...:The pattern begins with 1 and adds the position number of the term to the previous term.

Answers

16. 88
17. $4\frac{13}{27}$
18. 121
19. −90
20. no limit
21. −14
22. 891
23. 697
24. 2730
25. $32\frac{2}{3}$
26. $\sum_{k=1}^{8} 11k$
27. $\sum_{n=1}^{12}(n+2)^2$
28. *See question.*
29. *See question.*

26

Assessment Book INTEGRATED MATHEMATICS 3

OUTSIDE RESOURCES

Books/Periodicals

Dence, Joseph B. and Thomas P. Dence. "A Rapidly Converging Recursive Approach to Pi." *Mathematics Teacher* (February 1993): pp. 121–124.

Schielack, Vincent P., Jr. "Tournaments and Geometric Sequences." *Mathematics Teacher* (February 1993): pp. 127–129.

HiMAP Module 2, *Recurrence Relations*, Section 3: "Counting Backwards: Carbon Dating": pp. 8–12.

Peitgen, Heinz-Otto, Hartmut Jurgens, and Dietmar Saupe. *Fractals for the Classroom* (Part one: Introduction to Fractals and Chaos; Part two: Complex Systems and Mandelbrot Set). NCTM, co-published with Springer-Verlag.

Activities/Manipulatives

Masalski, William J. "Topic: Compound Interest." *How to Use the Spreadsheet as a Tool in the Secondary Mathematics Classroom.* NCTM, 1990: pp. 16–19.

Jurgens, Hartmut, Evan Maletsky, Heinz-Otto Peitgen, Terry Perciante, Dietmar Saupe, and Lee Yunker. *Fractals for the Classroom: Strategic Activities, Vols. 1 & 2.* NCTM and Springer-Verlag.

Software

Exploring Chaos. Apple II series, Ver. 1.0. MECC, Minneapolis, MN.

Appleworks. Claris Corporation, 440 Clyde Avenue, Mountain View, CA.

Videos

Southern Illinois University at Carbondale. *World Population Review.* 1990.

PROJECT GOALS

- Students learn to make a small investment grow over a period of 20 to 25 years in order to pay for a large future expense.
- Students learn to trace the effect of inflation by predicting the increase in the purchase price of an item.
- Students use computer spreadsheet software to explore changes in the value of their investments and in the price of items purchased over many years.

PROJECT PLANNING

Materials List

- Computer spreadsheet software
- Business sections of newspapers

Project Teams

Have students work on the project in groups of four. Students can meet and decide what the group will plan to save for. They should then research the present and past costs of their savings choice. Students with spreadsheet software experience should demonstrate its use to the group. The group can then do library research together, looking through business sections of newspapers for investment opportunities in stock tables, mutual funds, or other investment vehicles.

unit 4

Sequences and Series

INVE$TING

The incredible shrinking dollar . . . **Sixty years ago, you could treat yourself and three friends to a movie for a dollar! Today, your own ticket may cost $7 or more.**

. . . And how to stretch it. **Now imagine that sixty years ago you had invested a dollar at 5% annual interest compounded quarterly. By now your dollar would have grown to $20 — enough to buy tickets for you and three friends. (But you may have to go to the bargain matinee!)**

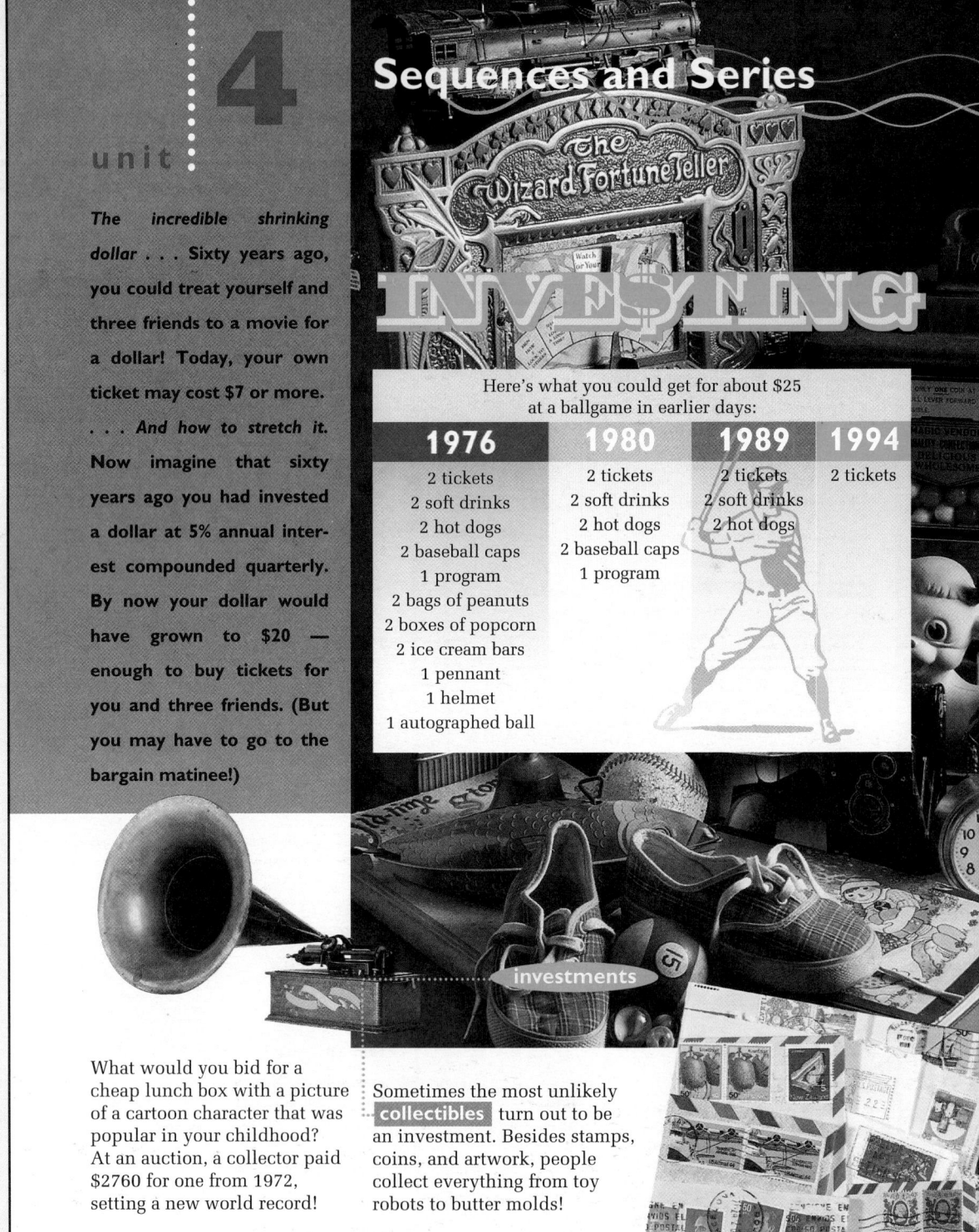

Here's what you could get for about $25 at a ballgame in earlier days:

1976	1980	1989	1994
2 tickets 2 soft drinks 2 hot dogs 2 baseball caps 1 program 2 bags of peanuts 2 boxes of popcorn 2 ice cream bars 1 pennant 1 helmet 1 autographed ball	2 tickets 2 soft drinks 2 hot dogs 2 baseball caps 1 program	2 tickets 2 soft drinks 2 hot dogs	2 tickets

What would you bid for a cheap lunch box with a picture of a cartoon character that was popular in your childhood? At an auction, a collector paid $2760 for one from 1972, setting a new world record!

Sometimes the most unlikely **collectibles** turn out to be an investment. Besides stamps, coins, and artwork, people collect everything from toy robots to butter molds!

204

Suggested Rubric for Unit Project

4 Students decide what their group will save for and collect accurate information regarding current costs and how its price has changed over the past 20 years. They use spreadsheet software to trace the effect of inflation on the purchase price of the item they chose to save for. Students select an investment and learn to make it grow in order to pay for their future purchase. All work is done accurately and the printouts of the spreadsheet calculations and the data displays based on the spreadsheets demonstrate that students have learned to make their investment grow at a rate that will pay for their future purchase. The written report provides a clear and accurate summary of the results of the project and describes how sequences and series were used to achieve the results obtained.

3 Students choose a future item to purchase and an investment vehicle to pay for it. They have some difficulties using the spreadsheet software, and their calculations regarding the effects of inflation on the future purchase price contain some errors. The investment chosen does not grow at a rate that is sufficient to pay for the item in 20–25 years, but it is close. The written report does not contain all of the spreadsheet calculations and the use of sequences and series in developing the results is somewhat sketchy.

2 Students select both a future item to purchase and an investment to pay for it. They do not, however, fully understand the goals of the project,

Unit Project 4

Save for the Future

Your project is to learn how to make a small investment grow in order to pay for a college education, a down payment on a home, a trip around the world, or any other high-cost item 20–25 years in the future.

You will trace the effect of inflation by predicting the increase in the purchase price of the item you choose.

You will use computer spreadsheet software to explore the changes in both the value of your investment and the price of your purchase over many years. Spreadsheet software stores information in matrices, allowing you to quickly and easily see the effect of a change in the interest or inflation rate.

People around the world invest in stocks to try to offset inflation.

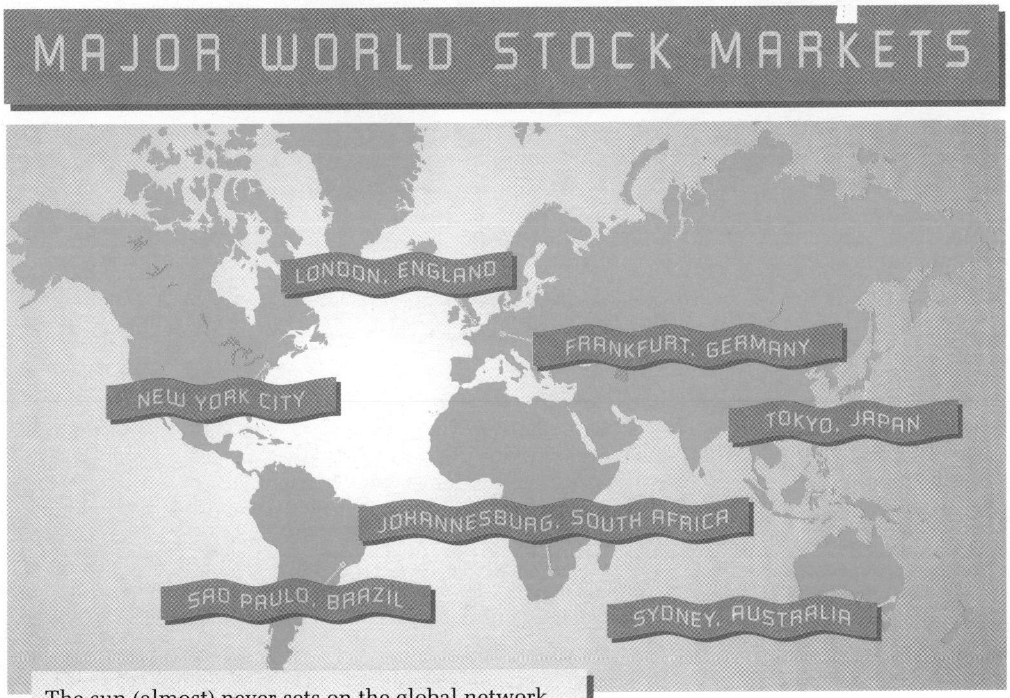

The sun (almost) never sets on the global network of stock exchanges. Three hours after the New York Stock Exchange closes for the day, the one in Tokyo is just opening. Three hours after the Tokyo exchange closes, the one in London opens.

205

Support Materials

The *Project Book* contains information about the following topics for use with this Unit Project.

- Project Description
- Teaching Commentary
- Working on the Unit Project Exercises
- Completing the Unit Project
- Assessing the Unit Project
- Alternative Projects
- Outside Resources

Students Acquiring English

The unit opener has several terms that may be difficult for some students acquiring English. You may want to make sure students are familiar with terms related to investing.

ADDITIONAL BACKGROUND

Multicultural Note

The stock exchanges in Buenos Aires, Mexico City, Tokyo, Hong Kong, and London are among those that are of global importance today. The Buenos Aires Stock Exchange, founded in 1854, is one of the most important South American exchanges. Since 1989, its dollar volume trade has jumped from about one million dollars a day to fifty or sixty million. The Buenos Aires Stock Exchange is only one of several exchanges in Argentina: minor exchanges include the Rosario Stock Exchange, the Córdoba Stock Exchange, and the Mercado Abierto Electrónico.

Suggested Rubric for Unit Project

and their use of the spreadsheet software does not help them to achieve the desired results. The investment does not grow at a rate that can pay for the projected future price. The effects of inflation on the price of the item may have been calculated incorrectly. The written report attempts to explain what has happened. Sequences and series are either not used, are used incorrectly, or their use contains errors that invalidate the results.

1 Students attempt to understand and trace the effects of inflation on the price of an item, but their work is incorrect. There are problems trying to use spreadsheet software. An investment is chosen, but attempts to make it grow over a period of 20–25 years contain many errors and are unsuccessful. Students should be encouraged to speak with the teacher as soon as possible to review their work and to make a new start on the project.

Investment Clubs

Financial surveys have consistently shown that most people are scared of the volatility of the stock market. Consequently, they tend to place their money in bank savings accounts or certificates of deposit, which protect their principal from declining and pay a given rate of interest. Over time, however, inflation will erode most of the gain on these types of savings. To make money grow over time, only the stock market has consistently yielded rates of return that beat the inflation rate. Recognizing this fact, many small investors have joined or started investment clubs that consist of a dozen or so members. The members pool moderate amounts of money to buy individual stocks that they hope will double in value within five years. Clubs generally meet once a month to discuss potential new stock purchases. In the early 1990s, the typical established club was about eight years old and had an annual rate of return of over 13%.

ALTERNATIVE PROJECTS

Project 1, page 259

Sequences and Series in Everyday Things

Look for sequences and series in the real world. Include examples that are visual and some that can only be heard. Tell whether the progression is a sequence or series, arithmetic or geometric, finite or infinite, and if infinite, whether limits or sums exist.

Project 2, page 259

Fractal Patterns

Research fractal shapes discussed in the unit. Describe how each shape is generated and find related sequences and series. Analyze the sequences and series discovered.

Unit Project 4

Getting Started

For this project you should work in a group of four students. Here are some ideas to help you get started.

☞ Decide what your group will plan to save for. Discuss how you can find out how much it costs today and how its price has changed over the past 20 years.

☞ Look through the business sections of newspapers. Examine the stock tables, mutual fund listings, and information about other investment opportunities.

☞ Have members of the group with experience using spreadsheets demonstrate their use.

☞ Plan to meet again later to schedule library research.

Working on the Unit Project

Your work in Unit 4 will help you save for the future.

Related Exercises:

Section 4-1, Exercise 32
Section 4-2, Exercise 22
Section 4-3, Exercises 23, 24
Section 4-4, Exercise 39
Section 4-5, Exercise 27
Section 4-6, Exercise 27
Section 4-7, Exercise 33

Alternative Projects p. 259

Can We Talk INVE$TING

➤ In the last few years, have you noticed any changes in the prices of the things you buy? Have prices increased or decreased? Why do you think this has happened? Have price changes affected how you spend your money?

➤ Do you collect stamps, baseball cards, or something else, or do you know someone who does? Do you know the value of the collection?

➤ Can you think of anything that was once a good investment but is a poor one now? Can you think of something that was once a poor investment but is a good one now? What happened to bring about any changes?

➤ Economists sometimes use the term "purchasing power" when they talk about the value of a dollar. What do you think this term means and why is it useful?

Answers to Can We Talk?

➤ Answers may vary. Examples are given. Overall, prices have increased in the last few years. This has happened due to the increased cost of materials, transportation, and wages for employees. Yes; I am more cautious about how and when I spend my money.

➤ Answers may vary, depending upon the various collections.

➤ Answers may vary. Check students' work.

➤ Answers may vary. An example is given. "Purchasing power" means how much of something you can buy for a dollar or for a specified amount of money. It is useful because it can be used to compare the value of money today with its value in the past or its projected value in the future.

Section **4-1**

Exploring Patterns

What's next?

Focus
Describe and continue patterns. Graph sequences and find apparent limits.

Stage 1

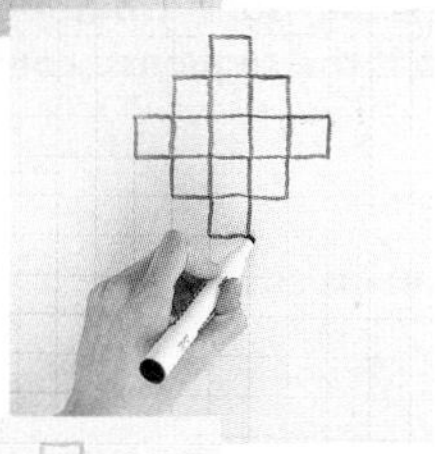

Stage 2

Stage 3

EXPLORATION

***How** can you describe and extend patterns?*

- **Materials: graph paper**
- **Work with another student.**

(1) **a.** Use graph paper and draw the shapes shown in Stages 1–4. Draw Stage 2 from Stage 1, Stage 3 from Stage 2, and Stage 4 from Stage 3.

b. Describe any patterns you used to draw each stage from the stage before it.

c. Continue the squares pattern by drawing a Stage 5 from Stage 4.

d. Copy and complete this table.

Stage number	1	2	3	4	5
Number of squares	5	?	?	?	?

e. Suppose you continue drawing more stages of the squares pattern. How many squares will be in Stage 10? Explain how you found your answer.

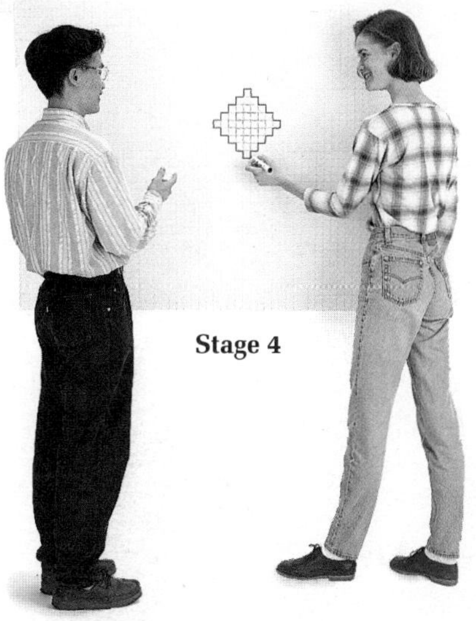

Stage 4

(2) Use the list of numbers $\frac{1}{2}, \frac{2}{3}, \frac{3}{4}, \frac{4}{5}, \ldots$.

a. Describe any patterns you see in the list.

b. Use the patterns you see to find the next three numbers in the list.

c. Find the tenth number in the list.

(3) Repeat Step 2 for each list of numbers.

List A 96, 48, 24, 12, . . .

List B 2, 5, 8, 11, 14, . . .

List C 1, −1, 1, −1, . . .

PLANNING

Objectives and Strands

See pages 204A and 204B.

Spiral Learning

See page 204B.

Materials List

- Graph paper
- Toothpicks
- Calculator
- Spreadsheet software

Recommended Pacing

Section 4-1 is a two-day lesson.

Day 1

Pages 207–209: Exploration through Talk it Over 1, *Exercises 1–12*

Day 2

Pages 209–211: Graphs and Limits of Sequences through Look Back, *Exercises 13–32*

Extra Practice

See pages 614–615.

Warm-Up Exercises

Warm-Up Transparency 4-1

Support Materials

- Practice Bank: Practice 29
- Activity Bank: Enrichment 26
- Study Guide: Section 4-1
- Problem Bank: Problem Set 9
- Explorations Lab Manual: Diagram Masters 1, 2, 12
- Using Mac Plotter Plus Disk: Data Analyzer
- Using IBM Plotter Plus Disk: Statistics Spreadsheet
- Assessment Book: Quiz 4-1, Alternative Assessment 1

Answers to Exploration

1. See answers in back of book.

2. a. Answers may vary. Examples are given. The numerators are the consecutive positive integers beginning with 1. The denominators are the consecutive positive integers beginning with 2. Also, the nth fraction is $\frac{n}{n+1}$.

b. $\frac{5}{6}, \frac{6}{7}, \frac{7}{8}$ c. $\frac{10}{11}$

3. Answers may vary. Examples are given.

A. a. Each number is half that of the number before it.

b. 6, 3, $\frac{3}{2}$ c. $\frac{3}{16}$

B. a. Each number is 3 more than the number before it.

b. 17, 20, 23

c. 29

C. a. All the entries in odd-numbered positions in the list are 1. All the entries in even-numbered positions in the list are −1.

b. 1, −1, 1

c. −1

TEACHING

Exploration

The goal of the Exploration is to introduce students to the concept of a sequence by having them examine patterns occurring in sequences of geometric shapes and in sequences of numbers. In each case, students use the patterns they observe to extend the sequences.

Additional Sample

S1 A large, square plaza will be covered with the same size white and gray stone squares to make a pattern, as shown in the diagram.

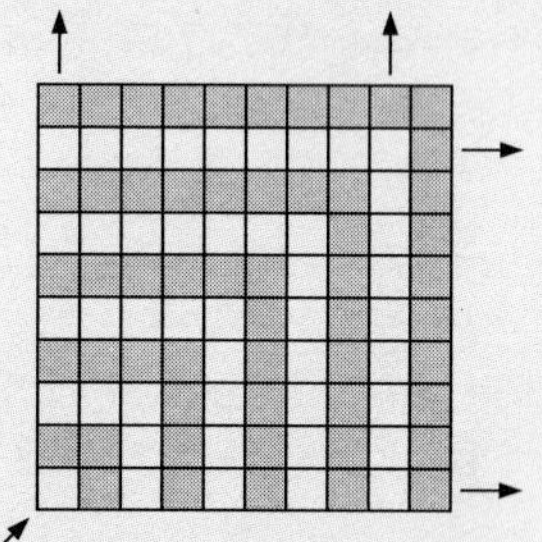

Start

The pattern starts at the lower left corner and expands as shown. Find the number of white stone squares in the eighth band of white squares.

The sequence is 1, 5, 9, ...

***Method 1.* Look for a relationship between the band number (terms) and the number of squares (position).**

position:	1	2	3 ...	8
	1+(0·4)	1+(1·4)	1+(2·4)	1+(7·4)
	↓	↓	↓	↓
term:	1	5	9 ...	29

The number of white stone squares in the eighth band is 29.

***Method 2.* Look for a relationship between each term and the term before it.**

position:	1	2	3	4	5	6	7	8
term and increase:	1	5	9	13	17	21	25	29
	+4	+4	+4					

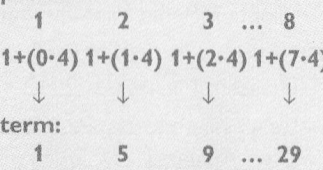

The number of white stone squares in the eighth band is 29.

The lists of numbers in the Exploration are called *sequences*. A **sequence** is an ordered list of numbers, called the **terms** of the sequence. A sequence with a last term is a **finite sequence.** A sequence with no last term is an **infinite sequence.** A counting number identifies each term's position in the sequence.

positions of terms →	1	2	3	4	
terms of a sequence →	**2,**	**5,**	**8,**	**11,**	...

11 is the *fourth term* of the sequence.

These three dots show that the sequence continues in the same pattern. This sequence is infinite.

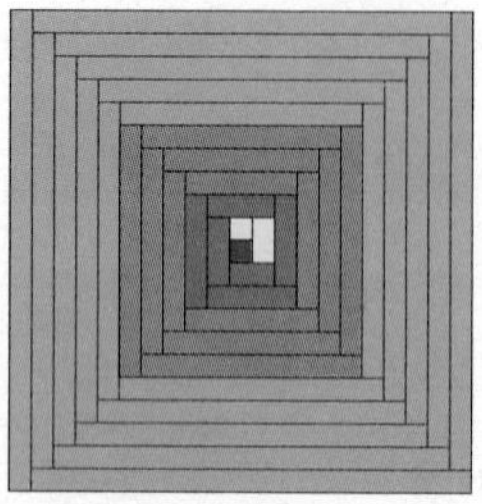

Sample 1

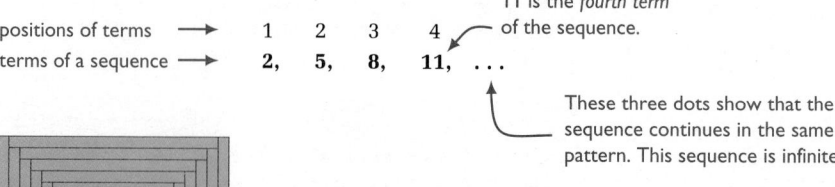

Write the sequence of the number of cloth strips in each color used to make this square for a log cabin quilt. Do not count the middle square. Find the 10th term if the sequence continued.

Sample Response

There are 2 yellow strips, 6 blue strips, 12 pink strips, and 20 green strips shown. The sequence is 2, 6, 12, 20, ...

Method 1

Look for a relationship between each term and its position number.

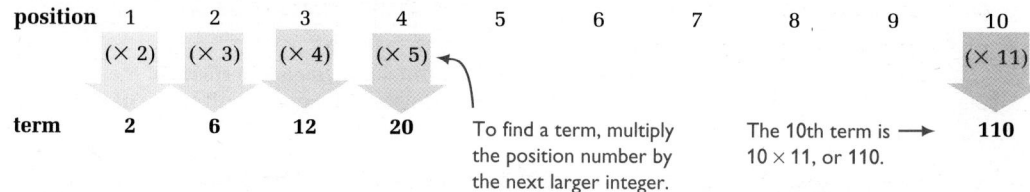

position	1	2	3	4	5	6	7	8	9	10
	(× 2)	(× 3)	(× 4)	(× 5)						(× 11)
term	**2**	**6**	**12**	**20**						**110**

To find a term, multiply the position number by the next larger integer.

The 10th term is → 10 × 11, or 110.

Method 2

Look for a relationship between each term and the term before it.

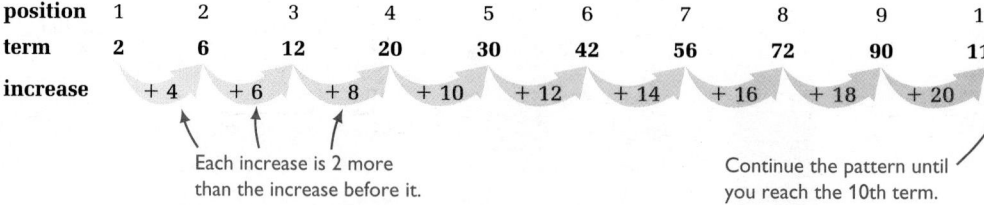

position	1	2	3	4	5	6	7	8	9	10
term	**2**	**6**	**12**	**20**	**30**	**42**	**56**	**72**	**90**	**110**
increase	+ 4	+ 6	+ 8	+ 10	+ 12	+ 14	+ 16	+ 18	+ 20	

Each increase is 2 more than the increase before it.

Continue the pattern until you reach the 10th term.

Talk it Over

1. a. Find a pattern in the sequence 1, 8, 27, 64, ...
   b. What is the 10th term of the sequence in part (a)?
   c. Which method shown in Sample 1 did you use to find the 10th term?

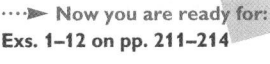

Now you are ready for:
Exs. 1–12 on pp. 211–214

Graphs and Limits of Sequences

Each term of a sequence is paired with the number that gives its position in the sequence. By plotting points with coordinates (*position*, *term*), you can graph a sequence on a coordinate plane.

Sample 2

Graph each sequence on a coordinate plane. Plot at least six points.

a. 2, 5, 8, 11, . . .

b. $1, \frac{1}{2}, \frac{1}{3}, \frac{1}{4}, \ldots$

Sample Response

a. Find the pattern.

Starting with the second term, each term is 3 more than the term before it. Continue the pattern for the 5th and 6th terms.

position	1	2	3	4	5	6
term	2	5	8	11	14	17

Plot the points (*position*, *term*).

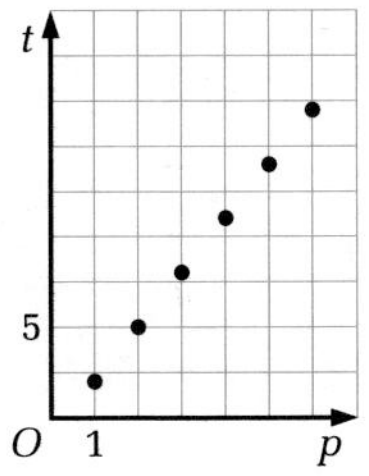

b. Find the pattern.

Each term is the reciprocal of its position number. Continue the pattern for the 5th and 6th terms.

position	1	2	3	4	5	6
term	1	$\frac{1}{2}$	$\frac{1}{3}$	$\frac{1}{4}$	$\frac{1}{5}$	$\frac{1}{6}$

Plot the points (*position*, *term*).

Watch Out!
Do not connect the points on the graph of a sequence.

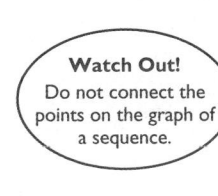

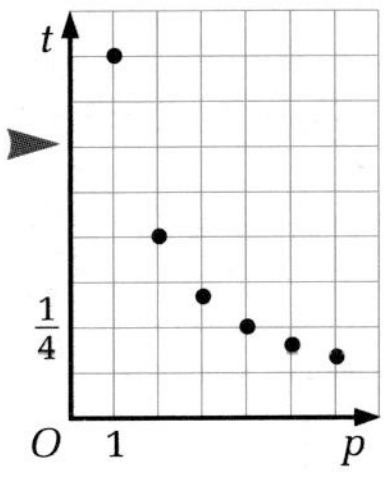

Talk it Over

Question 1 provides students with an opportunity to look for a pattern in a sequence of numbers and to apply what they have learned from Sample 1.

Additional Sample

S2 Graph each sequence on a coordinate plane. Plot at least six points.

a. 1, 3, 5, 7, ...

Starting with the second term, each term is 2 more than the term before it.

position	1	2	3	4	5	6
term	1	3	5	7	9	11

Plot the points (*position, term*).

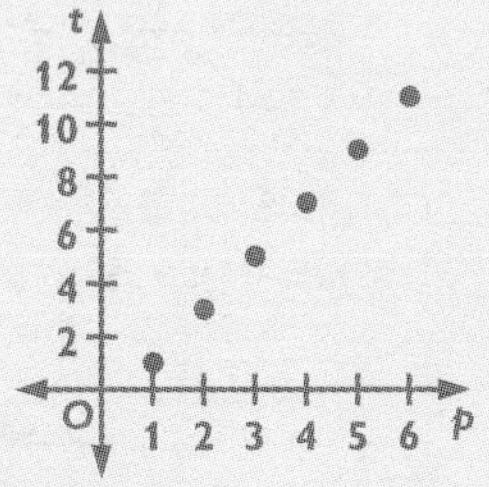

b. $\frac{1}{2}, \frac{2}{3}, \frac{3}{4}, \frac{4}{5}, \ldots$

The position number tells the numerator of the fraction for that position, and the denominator is one greater than the numerator.

position	1	2	3	4	5	6
term	$\frac{1}{2}$	$\frac{2}{3}$	$\frac{3}{4}$	$\frac{4}{5}$	$\frac{5}{6}$	$\frac{6}{7}$

Plot the points (*position, term*).

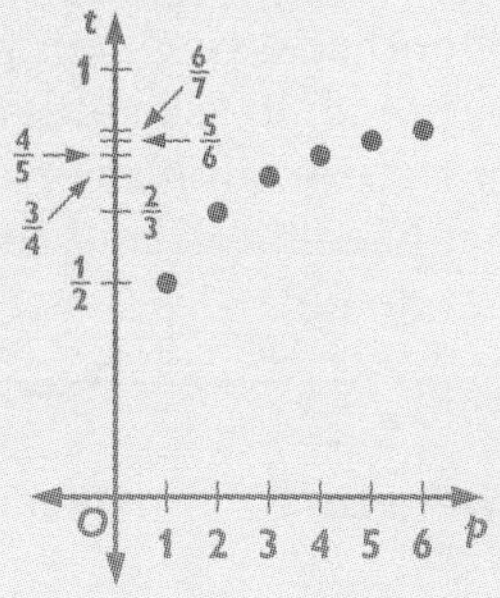

Answers to Talk it Over

1. a. Each term is the cube of the position number.
   b. 1000
   c. Method 1

Talk it Over

Questions 2–4 connect the idea of a sequence to that of a function. Students consider why the graphs in Sample 2 are functions, whether the graph of any sequence represents a function, and why the graph of a sequence is a discrete function. Question 5 leads students to the concept of a limit of a sequence.

Additional Sample

S3 Decide whether each infinite sequence appears to have a limit. If it does, give its value.

a. 1, 0, 1, 0, 0, 1, 0, 0, 0, ...
The sequence does not have a limit. Each group of 0's has one more 0 than the preceding group, but it is followed by a 1. As a result, there is no single fixed number that the terms of the sequence get close to.

b. $\frac{1}{1}, \frac{1}{4}, \frac{1}{9}, \frac{1}{16}, \ldots$
The denominator of each fraction can be found by squaring the position number for the fraction. All the numerators are 1. The 100th term will be $\frac{1}{100^2} = \frac{1}{10,000}$. The terms get very small very rapidly. The sequence appears to have a limit of 0.

Teaching Tip

After discussing Sample 3, you may wish to ask students to make up examples of sequences that have limits and sequences that do not.

Talk it Over

2. Do the graphs in Sample 2 represent functions? Why or why not?
3. Will the graph of *every* sequence represent a function? If so, what is the domain of the function? What is the range?
4. Are the graphs of sequences *continuous* or *discrete*? Explain.
5. Use the sequences in Sample 2.
   a. Suppose you continue listing the terms of each sequence. What number, if any, will the terms get closer to the farther out you go in each sequence?
   b. Suppose you continue plotting more points of each sequence. Describe what the graph will look like the farther out you go in each sequence.

The terms of the sequence in part (b) of Sample 2 appear to approach 0. When the terms of an infinite sequence get closer and closer to a single fixed number L, then L is called the **limit of the sequence.**

Not all infinite sequences have limits. The terms of the sequence in part (a) of Sample 2 do not get close to a single fixed number. This sequence does not have a limit.

Sample 3

Decide whether each infinite sequence appears to have a limit. If it does, give its value.

a. $\frac{1}{2}, \frac{5}{4}, \frac{7}{8}, \frac{17}{16}, \frac{31}{32}, \frac{65}{64}, \ldots$

b. 1, 2, 3, 1, 2, 3, 1, 2, 3, . . .

Sample Response

a. **Step 1** Find a trend.

Look at the decimal value of each term or look at the graph of the sequence.

Even-numbered terms decrease toward 1.

position	1	2	3	4	5	6
term	0.5	1.25	0.875	1.0625	0.96875	1.015625

Odd-numbered terms increase toward 1.

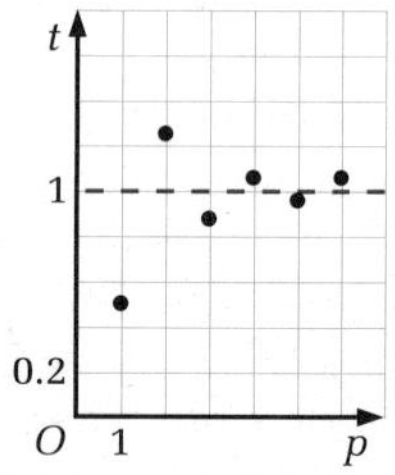

The points get closer and closer to the line $t = 1$.

Answers to Talk it Over

2. Yes. Answers may vary. An example is given. The graphs pass the vertical-line test.
3. Yes; the positive integers; ranges may vary.
4. The graph of a sequence is discrete because the domain is a set of distinct points.
5. a. (a) The sequence does not get close to any number.
   (b) 0
   b. (a) The terms will increase steadily as you go farther out in the sequence. The points lie on the line with equation $y = 3x - 1$.
   (b) The graph will get closer and closer to, but never intersect, the x-axis as you go farther out in the sequence. The points lie on the curve with equation $y = \frac{1}{x}$.
6. 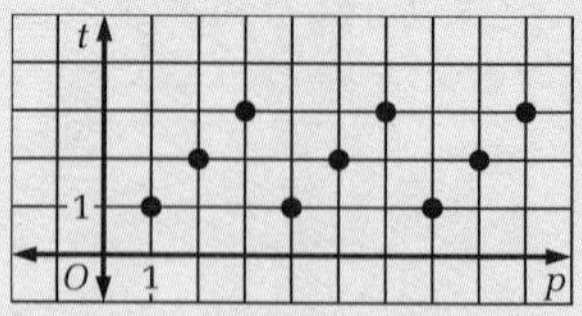

There is no single number to which the terms get close.
7. Answers may vary. An example is given. The limit of a sequence is a number that the sequence approaches but never reaches. The speed limit is a speed that the car approaches, reaches, and exceeds.

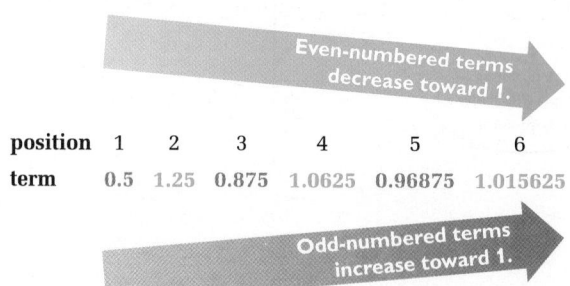

Step 2 When possible, check a term farther out in the sequence to see if it follows the same trend.

Find the pattern of the sequence. Look at the 20th term.

The numerators in the even positions are one *more* than the denominators.

$\frac{1}{2}, \frac{5}{4}, \frac{7}{8}, \frac{17}{16}, \frac{31}{32}, \frac{65}{64}, \ldots$

The denominators are powers of two. The power matches the position number.

20th term $\rightarrow \frac{2^{20}+1}{2^{20}} \approx 1.000000954$

The 20th term follows the trend from Step 1.

The sequence $\frac{1}{2}, \frac{5}{4}, \frac{7}{8}, \frac{17}{16}, \frac{31}{32}, \frac{65}{64}, \ldots$ appears to have a limit of 1.

b. Since the numbers 1, 2, and 3 will continue to repeat, there is no *single* fixed number that the terms of the sequence get close to. The sequence 1, 2, 3, 1, 2, 3, 1, 2, 3, . . . has no limit.

Talk it Over

6. Graph the sequence in part (b) of Sample 3. Explain how the graph shows that the sequence has no limit.
7. Explain how the limit of a sequence is different from a speed limit.

Look Back

Describe at least two methods you can use to look for a pattern in the terms of a sequence. Describe at least two methods you can use to decide whether an infinite sequence appears to have a limit.

Now you are ready for: Exs. 13–32 on pp. 214–215

4-1 Exercises and Problems

1. **Using Manipulatives** Use toothpicks.
   a. Build each of these shapes. Use one toothpick for each side of the smallest square.
   b. Describe any patterns you used to build each stage from the one before it.
   c. How many toothpicks will be in Stage 10?

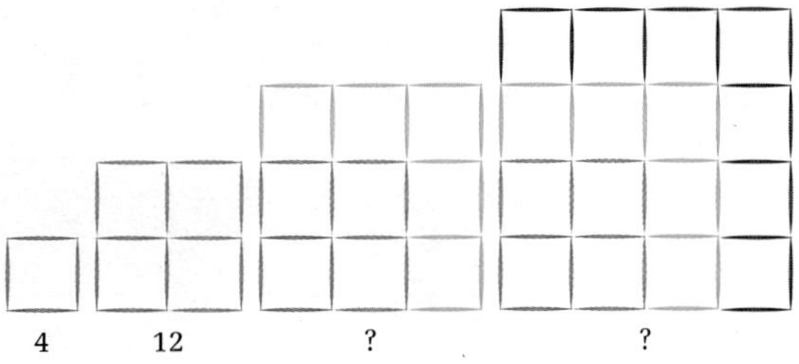

Using Technology

Parametric equations were first introduced in Unit 9 of *Integrated Mathematics 2.* You can graph a finite sequence of up to six terms on the TI-82 by changing the mode settings to Dot and Par (for parametric). For the graphing window, use Tmin = 1, Tmax = 6, Tstep = 1, and whatever axis settings are appropriate for the finite sequence you are graphing.

Press [Y=] and enter the position number and term for each point on the graph. If the first term is 1 and the second term is $\frac{1}{2}$, the equations list should show

$X_{1T} = 1$ $Y_{1T} = 1$
$X_{2T} = 2$ $Y_{2T} = 1/2$

and so on. When you have entered all position numbers and terms, press [GRAPH].

Look Back

You may wish to involve students in a class discussion of this Look Back. In so doing, students will learn how their classmates look for patterns and limits of sequences, which may be quite different from their own approaches.

APPLYING

Suggested Assignment

Day 1
Standard 2–8, 12
Extended 2–12

Day 2
Standard 13–22, 26–32
Extended 13–22, 24–32

Integrating the Strands

Number Exs. 1–5, 8–11, 13–26
Algebra Exs. 28–30
Functions Ex. 31
Geometry Ex. 27
Discrete Mathematics Exs. 1–26, 32
Logic and Language Exs. 9, 12, 25, 26, 32

Answers to Look Back

Summaries may vary. To find patterns, look for a relationship between each term and its position number or look for a relationship between each term and the term before it. To determine whether an infinite sequence has a limit, find a trend or look at the graph of the sequence. Then check a term farther out in the sequence to see if the trend holds or if the graph continues to approach an asymptote.

Answers to Exercises and Problems

1. a. Check students' work. 24, 40
   b. Answers may vary. An example is given. At each stage, I let the existing square be the lower left corner of the next square, which I built by adding squares along the top and the right side.
   c. 220 toothpicks

Problem Solving

The pattern for the sequence in Ex. 8 may be difficult for some students to discover. One reason is that the parts of the circle diagrams become very small very rapidly as more lines are added. You can suggest that students use a full sheet of paper and imagine that it is part of the inside of a large circle. It is important that all points of intersection of the segments be visible to make the parts countable.

Discovering the pattern may take several diagrams. These questions may help some students.

(1) Does the diagram you have at each stage give the *maximum* number of parts? How can you be sure?

(2) How can you go from one stage to the next? When you add one more segment, how can you be sure that you have the *maximum* number of parts for the next stage?

(3) When you add a segment to go from one stage to the next, by how many parts do you increase the total?

For each sequence in Exercises 2–8:

a. Find the next three terms.

b. Find the 12th term.

c. Describe the method you used to find the 12th term.

2. 3, 6, 9, 12, . . . **3.** 3, 6, 12, 24, . . . **4.** 3, 6, 10, 15, . . . **5.** 3, 6, 18, 72, . . .

6. A, AB, ABC, ABCD, . . . **7.** triangle, square, pentagon, hexagon, . . .

8. The sequence of the maximum number of pieces, not necessarily equal, a circle can be cut into with one cut, two cuts, three cuts, four cuts, and so on.

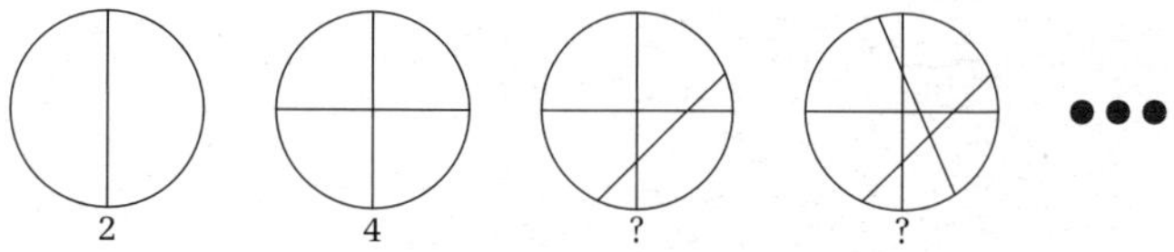

9. Skydiving A group of free-falling skydivers may join together to form a geometric-shaped formation before opening their parachutes. This is called *relative work skydiving.*

a. **Open-ended** Write the first three terms of a sequence that you see in the rings of skydivers. Explain the pattern in your sequence.

b. Find the 4th, 5th, and 10th terms of your sequence.

BY THE WAY. . .

The parachute that a skydiver wears today is made of nylon. In the 12th century, the Chinese experimented with rigid, umbrella-like devices.

Answers to Exercises and Problems

2. a. 15, 18, 21 b. 36
 c. Answers may vary. An example is given. I noticed that the term is three times the position number.

3. a. 48, 96, 192 b. 6144
 c. Answers may vary. An example is given. I noticed that each term is twice the term before it. I continued the pattern to find the 12th term.

4. a. 21, 28, 36
 b. 91
 c. Answers may vary. An example is given. I noticed that each increase is 1 more than the one before it. I continued the pattern to find the 12th term.

5. a. 360, 2160, 15,120
 b. 1,437,004,800
 c. Answers may vary. An example is given. I noticed that each term is the product of the term before it and the position number. I continued the pattern to find the 12th term.

6. a. ABCDE, ABCDEF, ABCDEFG
 b. ABCDEFGHIJKL
 c. I noticed that the terms are the first n letters of the alphabet in alphabetical order, where n is the position number. The twelfth letter of the alphabet is L.

7. a. heptagon, octagon, nonagon
 b. 14-gon
 c. I noticed each term is the name of a figure with $n + 2$ sides, where n is the position number.

8. a. 7, 11, 16 b. 79
 c. I noticed that each increase is n, where n is the position number. I continued the pattern to find the 12th term.

connection to **ASTRONOMY**

The *visual magnitude* of a star, planet, or moon is a measure of how bright the object appears to an observer on Earth. The lower the magnitude, the brighter the object appears.

A difference of 5 visual magnitudes between two objects means that one object is 100 times as bright as the other. So, a difference of 1 magnitude is equivalent to a brightness factor of about 2.51 since $(2.51)^5 \approx 100$. For example, the star Vega, with visual magnitude 0, is about 2.51 times as bright as the star Spica, a star in the Virgo constellation, with visual magnitude 1.

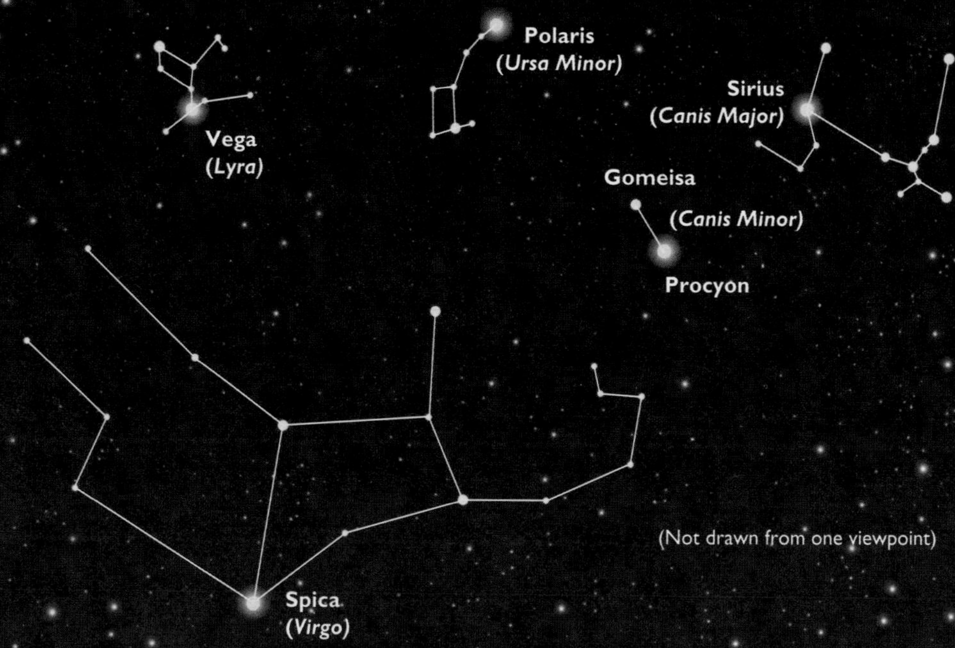

10. a. Complete the table.
   b. Describe the sequence that is formed.
   c. What is the brightness factor for a difference of 6 magnitudes? 10 magnitudes?

difference in magnitude	1	2	3	4	5
brightness factor	2.51	?	?	?	100

11. a. Canis Minor (Little Dog) is a small constellation that has only two stars, Procyon, with visual magnitude 0, and Gomeisa, with visual magnitude 3. Which star is brighter? By how much?
   b. The North Star, or Polaris, has visual magnitude 2. How does the brightness of Polaris compare to that of the two stars that make up Canis Minor?
   c. The brightest star in the night sky is Sirius, which has visual magnitude −1. Sirius is how many times as bright as Polaris?

Interdisciplinary Problems

Exs. 10 and 11 apply the concept of a sequence to the visual magnitude of a star, planet, or moon.

Research

Some students may find astronomy a fascinating subject, and you may wish to ask a small group of these students to research the stars and constellations referred to in Exs. 10 and 11. They can then provide some background information to the class when these exercises are discussed.

Answers to Exercises and Problems

9. a. Answers may vary. An example is given. From the innermost ring outward, the number of skydivers form the sequence 5, 10, 15, …. For each ring, the number of skydivers increases by 5.
   b. 4th term: $15 + 5 = 20$; 5th term: $20 + 5 = 25$; 10th term: $5(10) = 50$

10. a. Answers may vary due to rounding. These answers are based on 2.51.

difference in magnitude	1	2	3	4	5
brightness factor	2.51	6.30	15.8	39.7	100

   b. Each term is about 2.51 times the term before it.
   c. about 250; about 9925

11. a. Procyon; about 15.8 times brighter
   b. Procyon is about 6.30 times brighter than Polaris. Polaris is about 2.51 times brighter than Gomeisa.
   c. about 15.8 times as bright

Communication: Writing

For Ex. 25, part (c), students should try to write a clear and concise explanation of why they prefer one method over the other. You may wish to collect the written explanations and read a few of them to the class for further discussion.

Answers to Exercises and Problems

12. The domain of a sequence is the positive integers.

13. a. 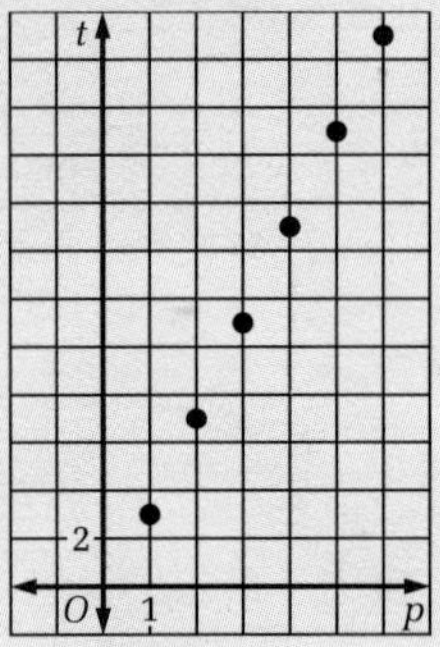

b. No; each term is 4 more than the term before it. The terms increase without limit.

14. a. 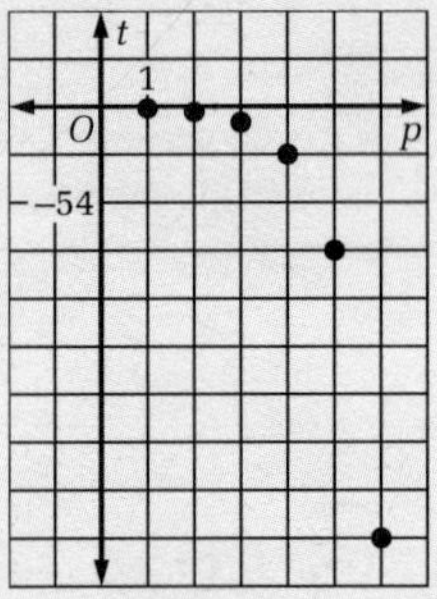

b. No; each term is 3 times the term before it. The terms decrease without limit.

15. a. 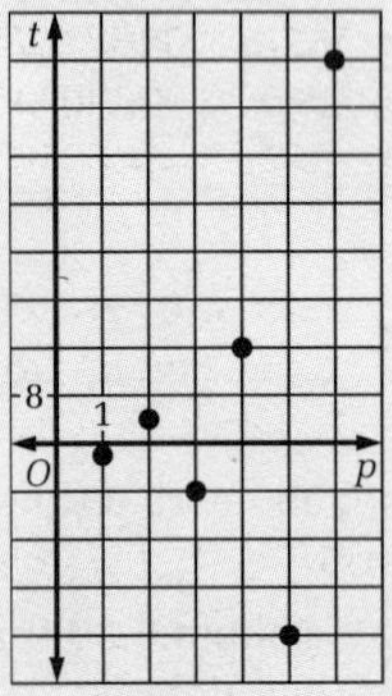

b. No; each term is –2 times the term before it. Odd-numbered terms decrease without limit, while even-numbered terms increase without limit.

12. **Reading** Why are the points of a graph of a sequence not connected?

For Exercises 13–18:

a. Graph each sequence. Plot at least six points.

b. Decide whether each sequence appears to have a limit.

13. $3, 7, 11, 15, \ldots$

14. $-1, -3, -9, -27, \ldots$

15. $-2, 4, -8, 16, \ldots$

16. $-1, \frac{1}{2}, -\frac{1}{3}, \frac{1}{4}, \ldots$

17. $10, 8, 6, 4, \ldots$

18. $1, 8, 27, 64, \ldots$

For Exercises 19–22, tell whether each sequence appears to have a limit. Justify your answer.

19. $\frac{1}{3}, \frac{1}{9}, \frac{1}{27}, \frac{1}{81}, \ldots$

20. $3, 6, 9, 12, \ldots$

21. $-0.5, 0.25, -0.125, 0.0625, \ldots$

22. $-5, 25, -125, 625, \ldots$

23. a. TECHNOLOGY Use a calculator to evaluate the first 20 terms of the sequence

$$\sqrt{3}, \sqrt{\sqrt{3}}, \sqrt{\sqrt{\sqrt{3}}}, \sqrt{\sqrt{\sqrt{\sqrt{3}}}}, \ldots.$$

b. What appears to be the limit of this sequence?

For Exercises 24 and 25, use the sequence $\frac{3}{4}, \frac{8}{9}, \frac{15}{16}, \frac{24}{25}, \frac{35}{36}, \ldots$.

24. Kimiko says that each of these numbers is a term of the sequence.

$\frac{99}{100}$ $\frac{399}{400}$ $\frac{899}{900}$

a. Describe a pattern Kimiko may have used to find the three terms.

b. Find the decimal value of each of Kimiko's terms.

c. What number appears to be the limit of this sequence?

25. a. The 1st term is $\frac{1}{4}$ less than 1. How much less than 1 is the 2nd term? the 3rd term? the 4th term? the 5th term?

b. What number appears to be the limit of the sequence of differences in part (a)?

c. **Writing** Compare the method used in part (a) to find the terms of the sequence with the method described in Exercise 24. Do both methods give you the same answer? Which method do you prefer? Why?

Ongoing ASSESSMENT

26. **Open-ended** Write at least four sequences that start with 2, 4, ... Describe a pattern in each sequence.

Review PREVIEW

27. Show that the sum of the defects at the vertices of a regular octahedron is 720°. *(Section 3-9)*

16. a. 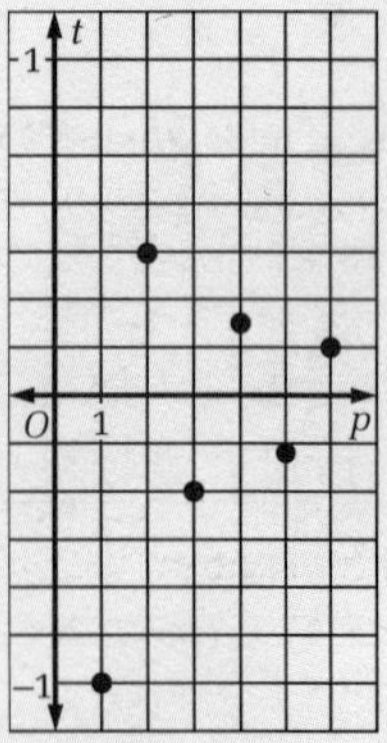

b. Yes; odd-numbered terms increase toward 0; even-numbered terms decrease toward 0.

17. a. 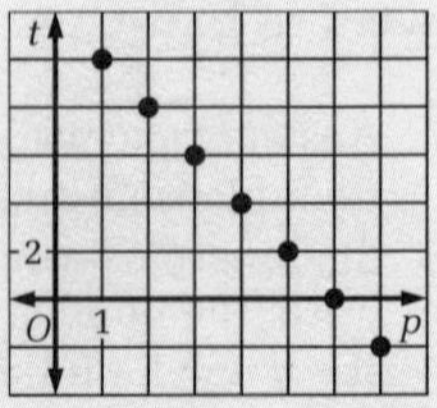

b. No; each term is 2 less than the term before it. The terms decrease without limit.

Simplify. *(Toolbox Skill 14)*

28. 2^{-5} **29.** $(-4)^0$ **30.** 7^{-3}

31. Find $f(0)$, $f(-3)$, and $f(12)$ when $f(x) = 4x - 5$. *(Section 2-1)*

4 Working on the Unit Project

TECHNOLOGY NOTE

Technology Handbook

See p. 595.

32. a. Research Find the most recent edition of the *Digest of Educational Statistics.* Find and record the data on the total annual tuition, room, and board costs at four-year universities, both public and private, for the last 29 years.

b. Make a five-column spreadsheet as shown.

Digest for Educational Statistics

	A	B	C	D	E
1	Year	Public U. Cost	% increase	Private U. Cost	% increase
2	1964	1051		2202	
3	1965	1105	= (B3 − B2)/B2	2316	
4	1966	1171		2456	

Columns C and E show the percent of increase in cost from the previous year. Format them to display percents with two decimal places.

c. Below the last row of the percent of increase columns (C and E), find the average percent of increase for each type of university.

Digest for Educational Statistics

	A	B	C	D	E
29	1991	6051	8.34%	17779	7.73%
30	1992	6449	6.58%	18892	6.26%
31					
32		Average % inc.	= AVERAGE (C3:C30)	Average % inc.	
33					

= AVERAGE (E3:E30)

d. Use the rates from part (b) to continue your spreadsheet by predicting the total annual cost of each type of university for the next 25 years.

Digest for Educational Statistics

	A	B	C	D	E
30	1992	6449	6.58%	18892	6.26%
31					
32		Average % inc.	6.73%	Average % inc.	8.01%
33					
34	1993	= B30*1.0673			
35	1994				

Project into the future by multiplying the last year for which you have data by 1 + the average percent of increase (expressed as a decimal).

Format columns B and D to display whole numbers.

Digest for Educational Statistics

	A	B	C	D	E
30	1992	6449	6.58%	18892	6.26%
31					
32		Average % inc.	6.73%	Average % inc.	8.01%
33					
34	1993	6883			
35	1994	= B34*1.0673			
36	1995				

Since the data in column B skips a few lines, enter the formula again before you FILL DOWN.

e. How much do you predict it will cost to attend each type of university 25 years from now?

Working on the Unit Project

Students should work in their project groups of four to complete Ex. 32. Students can share the tasks of recording the data, using the spreadsheet software to complete the spreadsheet, and making predictions for the future.

Practice 29 For use with Section 4-1

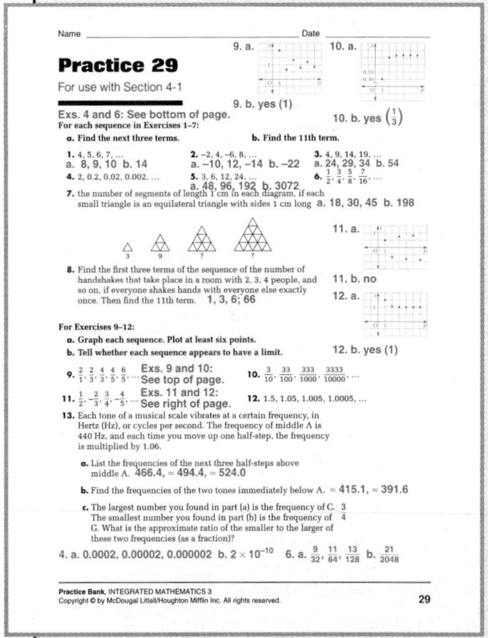

Name ______ Date ______

Practice 29

For use with Section 4-1

9. a. 10. a.

9. b. yes (1)

Exs. 4 and 6: See bottom of page.

10. b. yes $\left(\frac{1}{3}\right)$

For each sequence in Exercises 1–7:

a. Find the next three terms. b. Find the 11th term.

1. 4, 5, 6, 7, … a. 8, 9, 10 b. 14

2. −2, 4, −6, 8, … a. −10, 12, −14 b. −22

3. 4, 9, 14, 19, … a. 24, 29, 34 b. 54

4. 2, 0.2, 0.02, 0.002, …

5. 3, 6, 12, 24, … a. 48, 96, 192 b. 3072

6. $\frac{1}{2}, \frac{3}{4}, \frac{5}{8}, \frac{7}{16}, \ldots$

7. the number of segments of length 1 cm in each diagram, if each small triangle is an equilateral triangle with sides 1 cm long a. 18, 30, 45 b. 198

3 9 ? ?

11. a.

8. Find the first three terms of the sequence of the number of handshakes that take place in a room with 2, 3, 4 people, and so on, if everyone shakes hands with everyone else exactly once. Then find the 11th term. 1, 3, 6; 66

11. b. no

12. a.

For Exercises 9–12:

a. Graph each sequence. Plot at least six points.

b. Tell whether each sequence appears to have a limit.

12. b. yes (1)

9. $\frac{2}{1}, \frac{2}{3}, \frac{4}{3}, \frac{4}{5}, \frac{6}{5}, \ldots$ Exs. 9 and 10: See top of page.

10. $\frac{3}{10}, \frac{33}{100}, \frac{333}{1000}, \frac{3333}{10000}, \ldots$

11. $\frac{1}{2}, -\frac{2}{3}, \frac{3}{4}, -\frac{4}{5}, \ldots$ Exs. 11 and 12: See right of page.

12. 1.5, 1.05, 1.005, 1.0005, …

13. Each tone of a musical scale vibrates at a certain frequency, in Hertz (Hz), or cycles per second. The frequency of middle A is 440 Hz, and each time you move up one half-step, the frequency is multiplied by 1.06.

a. List the frequencies of the next three half-steps above middle A. 466.4, ≈ 494.4, ≈ 524.0

b. Find the frequencies of the two tones immediately below A. ≈ 415.1, ≈ 391.6

c. The largest number you found in part (a) is the frequency of C. The smallest number you found in part (b) is the frequency of G. What is the approximate ratio of the smaller to the larger of these two frequencies (as a fraction)? $\frac{3}{4}$

4. a. 0.0002, 0.00002, 0.000002 b. 2×10^{-10} 6. a. $\frac{9}{32}, \frac{11}{64}, \frac{13}{128}$ b. $\frac{21}{2048}$

Practice Bank, INTEGRATED MATHEMATICS 3

29

23. a. 1.732050808; 1.316074013; 1.14720269; 1.071075483; 1.034927767; 1.017313996; 1.008619847; 1.004300676; 1.002148031; 1.001073439; 1.000536576; 1.000268252; 1.000134117; 1.000067056; 1.000033528; 1.000016764; 1.000008382; 1.000004191; 1.000002095; 1.000001048

b. 1

24. a. The denominators are 2^2, 3^2, 4^2, and so on. The numerators are $2^2 - 1$, $3^2 - 1$, $4^2 - 1$, and so on.

b. 0.99, 0.9975, 0.9988888889

c. 1

25–27. See answers in back of book.

28. $\frac{1}{32}$

29. 1

30. $\frac{1}{343}$

31. −5; −17; 43

32. a–e. Check students' work.

Answers to Exercises and Problems

18. a.

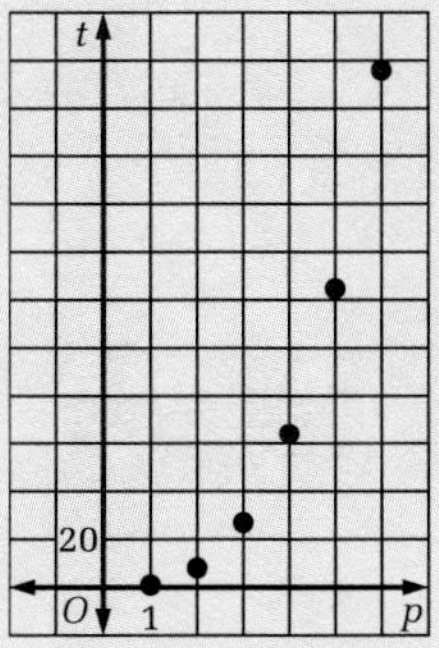

b. No; each term is the cube of the position number. The terms increase without limit.

19. Yes; the terms decrease toward 0.

20. No; each term is 3 times the term before it. The terms increase without limit.

21. Yes; odd-numbered terms increase toward 0, while even-numbered terms decrease toward 0.

22. No; each term is −5 times the term before it. Odd-numbered terms decrease without limit, while even-numbered terms increase without limit.

PLANNING

Objectives and Strands
See pages 204A and 204B.

Spiral Learning
See page 204B.

Materials List
- Graph paper
- Spreadsheet software

Recommended Pacing
Section 4-2 is a one-day lesson.

Toolbox References
- **Toolbox Skill 14:** Using Rules of Exponents

Extra Practice
See pages 614–615.

Warm-Up Exercises
- Warm-Up Transparency 4-2

Support Materials
- Practice Bank: Practice 30
- Activity Bank: Enrichment 27
- Study Guide: Section 4-2
- Problem Bank: Problem Set 9
- Explorations Lab Manual: Diagram Master 2
- Assessment Book: Quiz 4-2, Alternative Assessment 2

Section 4-2

Sequences

Focus
Use subscripts and formulas for sequences.

n-FORMATION

The ancient Mayan pyramid in Chichén Itzá, Mexico, has nine terraces on each side, split by a stairway leading up to a temple. The diagram shows two of the many sequences hidden in the pyramid.

You can use *subscripts* to write the terms of a sequence. The subscript of each term matches the position of the term in the sequence. The symbol a_n (read "a sub n") represents the nth term, or term in position n, of the sequence $a_1, a_2, a_3, \ldots$.

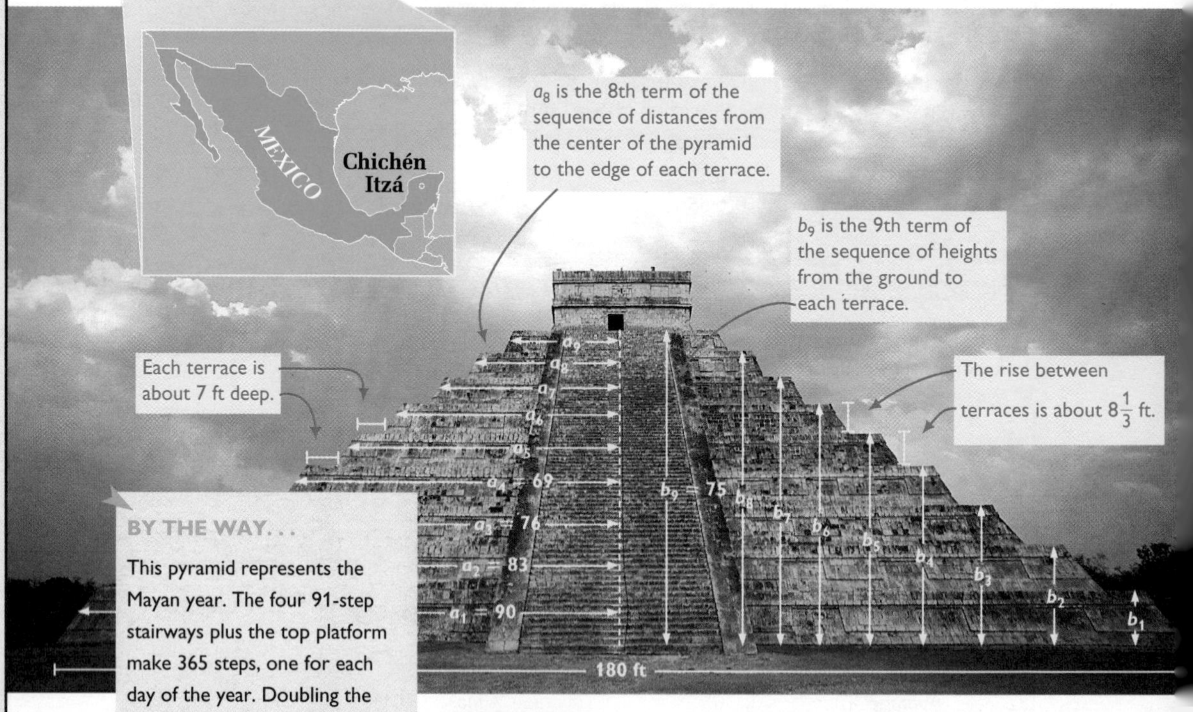

BY THE WAY...
This pyramid represents the Mayan year. The four 91-step stairways plus the top platform make 365 steps, one for each day of the year. Doubling the nine terraces on each side makes 18, the number of months in the Mayan year. Each side has 52 carved panels for the 52 weeks in the year.

Talk it Over

1. What is the value of b_2? b_3?
2. What is the value of a_7? How did you find a_7?
3. How many times do you have to subtract 7 from a_1 (or 90) to find a_9?

Answers to Talk it Over

1. $16\frac{2}{3}$ ft; 25 ft
2. 48 ft; the width of the first terrace is 90 ft; that is, $a_1 = 90$. Each succeeding terrace is 7 ft narrower, so the width of the seventh terrace is 42 ft less, that is, $a_7 = 90 - 42 = 48$.
3. eight times

Sample 1

Find a formula for the sequence of distances from the center of the pyramid to the edge of each terrace.

Sample Response

Step 1 Find a pattern. The pattern is a repeated subtraction of 7.

Step 2 Write the first few terms. Show how you found each term.

$a_1 = 90$

$a_2 = 83 = 90 - 1(7)$ ← Subtract 7 from 90 once.

$a_3 = 76 = 90 - 2(7)$ ← Subtract 7 from 90 twice.

$a_4 = 69 = 90 - 3(7)$ ← Subtract 7 from 90 three times.

Step 3 Express the pattern in terms of n.

For each term, a_n, the number of times you subtract 7 from 90 is one less than the subscript, n. The formula for the sequence is:

$a_n = 90 - (n - 1)(7)$ ← Subtract 7 from 90 a total of $(n - 1)$ times.

The formula found in Sample 1 is an example of an **explicit formula** for a sequence because it gives the value of any term a_n in terms of n.

Talk it Over

4. Explain why the explicit formula $a_n = 90 - (n - 1)(7)$ is a rule for a function. What is the control variable? What is the dependent variable? What is the domain? What is the range?
5. Find an explicit formula for the sequence of heights from the ground to each terrace of the pyramid.

Sample 2

Write the first three terms and the 12th term of each sequence.

a. $a_n = 3n - 5$

b. $b_n = 2^{1-n}$

Sample Response

Substitute 1, 2, 3, and 12 for n in each explicit formula.

a. $a_1 = 3(1) - 5 = -2$
$a_2 = 3(2) - 5 = 1$
$a_3 = 3(3) - 5 = 4$
$a_{12} = 3(12) - 5 = 31$

b. $b_1 = 2^{1-1} = 2^0 - 1$
$b_2 = 2^{1-2} = 2^{-1} = \frac{1}{2}$
$b_3 = 2^{1-3} = 2^{-2} = \frac{1}{4}$
$b_{12} = 2^{1-12} = 2^{-11} = \frac{1}{2048}$

Student Resources Toolbox
p. 637 *Algebraic Expressions*

Answers to Talk it Over

4. For each value of n, there is a single value of $90 - (n - 1)(7)$. The control variable is the position number and the dependent variable is the term. The domain is the positive integers. The range is the set of numbers given by the sequence 90, 83, 76, 69, ...
5. $b_n = \frac{25n}{3}$

TEACHING

Talk it Over

Questions 1–3 provide students with an opportunity to find the value of specific terms of the two sequences associated with a Mayan pyramid. An understanding of subscript notation is necessary to answer these questions correctly.

Additional Sample

S1 The year 2000 will be a leap year. The next leap year will be 2004. Find the formula that gives the sequence of leap years, starting with 2000.

Leap years occur every fourth year. The first four leap years can be found by using this information.

$L_1 = 2000$
$L_2 = 2004 = 2000 + 1(4)$
$L_3 = 2008 = 2000 + 2(4)$
$L_4 = 2012 = 2000 + 3(4)$

To find the nth term of the sequence, find the product of $n - 1$ and 4 and add it to 2000. The formula is $L_n = 2000 + (n - 1)4$.

Talk it Over

Question 4 relates the explicit formula from Sample 1 to the concept of function and to ideas related to a function. Question 5 requires that students find an explicit formula using the data for the pyramid.

Additional Sample

S2 Write the first three terms and the 10th term of each sequence.

a. $a_n = 2n + 1$

Substitute 1, 2, 3, and 10 for n in the formula.

$a_1 = 2(1) + 1 = 3$
$a_2 = 2(2) + 1 = 5$
$a_3 = 2(3) + 1 = 7$
$a_{10} = 2(10) + 1 = 21$

b. $b_n = 1 - \frac{1}{n}$

Substitute 1, 2, 3, and 10 for n in the formula.

$b_1 = 1 - \frac{1}{1} = 0$
$b_2 = 1 - \frac{1}{2} = \frac{1}{2}$
$b_3 = 1 - \frac{1}{3} = \frac{2}{3}$
$b_{10} = 1 - \frac{1}{10} = \frac{9}{10}$

Additional Sample

S3 Alejandro Reyna has a bacteria culture that increases by 10% each day. He estimates the current bacteria count to be 10,000.

a. Find a formula for the bacteria count after n days.

With a starting count of P, the number of bacteria after 1 day can be found as follows:

$$\text{new count} = P + 0.1P = P(1 + 0.1) = P(1.1)$$

After 1 day, any bacteria count P is multiplied by 1.1. Thus, the counts on successive days are multiplied by successive powers of 1.1.

$c_0 = 10{,}000$ (initial count)

$c_1 = 10{,}000(1.1)$ (count after 1 day)

$c_2 = 10{,}000(1.1)(1.1) = 10{,}000(1.1)^2$ (count after 2 days)

$c_3 = 10{,}000(1.1)^2(1.1) = 10{,}000(1.1)^3$ (count after 3 days)

$\vdots$

$c_n = 10{,}000(1.1)^n$ (count after n days)

b. Find the approximate bacteria count after 20 days.

Use the formula from part (a).

$c_{20} = 10{,}000(1.1)^{20} \approx 67{,}275$

After 20 days, there are approximately 67,275 bacteria.

Mathematical Procedures

The Sample Response for Sample 3 illustrates a procedure that can be important in finding solutions to mathematical problems: using a pattern of specific cases to generate a general formula.

When the first term of a sequence represents a starting value before any change occurs, the subscript 0 is often used. In a sequence of monthly bank account balances, the first term v_0 would be the initial deposit, the next term v_1 would be the balance after the *first* month's interest, and so on.

Sample 3

Samara Reese invests $1000 to open a savings account that earns 3% annual interest, compounded monthly.

a. Find a formula for the value of her account after n months.

b. Find the value of her account after one year.

Sample Response

a. **1** Find the monthly interest rate.

$$\frac{1}{12} \text{ of } 3\% = \frac{0.03}{12} = 0.0025$$

2 Look at what happens after one month to an account with starting value v.

$\text{new value} = v + 0.0025v$	← The value of the account increases by $0.0025v$.
$= v(1 + 0.0025)$	← Factor out v.
$= v(1.0025)$	← The value of the account is multiplied by 1.0025.

3 Write several terms of the sequence of values of the account. Look for a pattern. To see the pattern, leave each value unsimplified.

$v_0 = 1000$	← starting value
$v_1 = 1000(1.0025)$	← value at the end of one month
$v_2 = v_1(1.0025)$	← Multiply the value at the end of one month by 1.0025.
$= \mathbf{1000(1.0025)}(1.0025)$	← Substitute $1000(1.0025)$ for v_1.
$= 1000(1.0025)^2$	← value at the end of two months
$v_3 = v_2(1.0025)$	← Multiply the value at the end of two months by 1.0025.
$= \mathbf{1000(1.0025)^2}(1.0025)$	← Substitute $1000(1.0025)^2$ for v_2.
$= 1000(1.0025)^3$	← value at the end of three months
$\vdots$	
$v_n = 1000(1.0025)^n$	← value at the end of n months

b. Use the formula from part (a). One year is 12 months, so find v_{12}.

$v_{12} = 1000(1.0025)^{12}$ ← Substitute 12 for n.

≈ 1030.42

After one year, the value of Samara's account will be $1030.42.

Answers to Talk it Over

6. The domain of the explicit formula is the nonnegative integers rather than the positive integers.
7. about $1161.62

Answers to Look Back

In an explicit formula, the value of each term is given in terms of the subscripted position number, that is, a_n is given in terms of n.

Talk it Over

6. How does using 0 instead of 1 as the subscript for the first term of a sequence affect the domain of its explicit formula?
7. In Sample 3, find the value of Samara's investment after five years.

Look Back

Explain the role of subscripts when finding an explicit formula of a sequence.

4-2 Exercises and Problems

1. **Reading** When would you use the subscript 0 for the first term of a sequence?

Write the first four terms and the 24th term of each sequence.

2. $b_n = 5n + 12$
3. $a_n = \frac{n-1}{2}$
4. $t_n = \frac{n}{2n+3}$
5. $a_n = \frac{3^{2-n}}{n}$

For Exercises 6–9, write an explicit formula for each sequence.

6. 11, 17, 23, 29, . . .
7. 3, 9, 27, 81, . . .
8. $\frac{1}{2}, \frac{2}{3}, \frac{3}{4}, \frac{4}{5}, \ldots$
9. The number of diagonals from *one* vertex of a regular polygon.

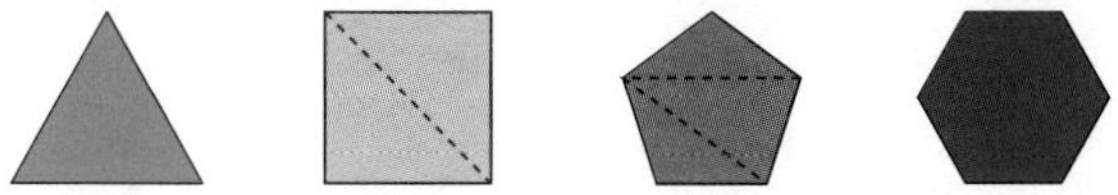

$d_1 = 0$ 0 diagonals | $d_2 = 1$ 1 diagonal | $d_3 = 2$ 2 diagonals | $d_4 = ?$

10. Use the formula $b_n = 2^{1-n}$ from part (b) of Sample 2.
    a. Find b_{100}, b_{250}, and b_{500}.
    b. Do you think this sequence has a limit? Explain.
    c. **Writing** Explain how you can use an explicit formula for an infinite sequence to decide if the sequence has a limit.
11. In Japan, a child is considered to be one year old at birth. Let y_n be the sequence of ages for a Japanese person, where y_0 = age at birth.
    a. Explain why the subscript 0 is used for the first term.
    b. The subscript does not match the position in this sequence. What does the subscript match?
    c. Find an explicit formula for this sequence.

Talk it Over

Question 6 directs students' attention to the subscript of the first term of a sequence (0 or 1) so they can see the effect on the domain of its explicit formula. The domain will be counting numbers if the sequence starts with zero and whole numbers if it starts with a one.

Look Back

You may wish to ask a few students to respond to this Look Back so that the class can hear different explanations. This approach would benefit any students having difficulty understanding the role of subscripts when finding an explicit formula of a sequence.

APPLYING

Suggested Assignment

Standard 1–10, 12, 15–22

Extended 1–22

Integrating the Strands

Number Exs. 1–10, 13, 15, 16, 22

Algebra Exs. 1–12, 14, 15, 18–21

Geometry Exs. 9, 14

Trigonometry Ex. 14

Statistics and Probability Ex. 17

Discrete Mathematics Exs. 1–17, 22

Logic and Language Exs. 1, 10, 13, 15, 22

Students Acquiring English

You may wish to pair students acquiring English with English-proficient partners for the two written responses in Ex. 10.

Answers to Exercises and Problems

1. when the initial value of a sequence is the value before any changes take place
2. 17, 22, 27, 32; 132
3. $0, \frac{1}{2}, 1, \frac{3}{2}; \frac{23}{2}$
4. $\frac{1}{5}, \frac{2}{7}, \frac{1}{3}, \frac{4}{11}; \frac{24}{51} = \frac{8}{17}$
5. $3, \frac{1}{2}, \frac{1}{9}, \frac{1}{36}; \frac{1}{3^{22} \cdot 24} \approx$ 0.000000000001328
6. $a_n = 11 + (n-1)6 = 6n + 5$
7. $a_n = 3^n$
8. $a_n = \frac{n}{n+1}$
9. $d_n = n - 3$, where n is the number of vertices of the polygon
10. a. about $1.578 \cdot 10^{-30}$; about $1.105 \cdot 10^{-75}$; about 0
    b. Answers may vary. An example is given. I think the limit of this sequence is 0. The value of b_{500} is so small the calculator gives the value 0 or an overflow error. Also, no matter how large n gets, 2^{1-n} will never be negative.
    c. Answers may vary. An example is given. You can look at what will happen when n is very large.
11. a. so the sequence of ages will begin with 1 rather than 2
    b. the age after n years of life
    c. $y_n = n + 1$

Using Technology

There are various ways to use the TI-82 to generate the terms of a sequence for which you have an explicit formula. The approach described here will work on the TI-81 and TI-82.

Use the Dot and Parametric mode settings. On the TI-81, parametric mode appears on the mode menu as Param. On the TI-82, it is Par. After you have fixed the mode settings, press RANGE (TI-81) or WINDOW (TI-82). Set Tmin to 1 and Tmax to the greatest position number you think you will need. (For example, if you think you will never want to go beyond the 100th term of the sequence, use Tmax = 100). Set Tstep to 1. If you want to view the graph, choose appropriate values for Xmin, Xmax, Ymin, and Ymax. Next, press Y=. On the first line, enter XIT=T. (To enter the variable T, which represents the position number, press X|T on the TI-81 or X T θ on the TI-82.) On the second line, enter the expression from the explicit formula, using T for the position variable. For example, for Ex. 4, the line that you enter should read YIT=T/(2T+3).

On the TI-81 and TI-82, you can view terms of the sequence by pressing GRAPH. Once graphing is completed, press TRACE. The readouts at the bottom of the screen show the position number as the value of T and XIT. The value of the term is displayed as the value of YIT.

Application

Exs. 12–14 provide students with three applications of sequences to situations that involve investing money, personal finance, and geometry.

12. Maxim Gutte invests \$500 to open a money market account that earns 4.2% annual interest, compounded monthly.

a. Find an explicit formula for the value of his account after n months.

b. Find the value of his account after two years.

13. a. Personal Finance In Sample 3, Samara Reese's account increased from \$1000 to \$1030.42 after one year. By what percent did the account increase? This percent is called the *effective annual yield.*

b. How does the effective annual yield compare to the annual interest rate compounded monthly for the problem in Sample 3?

c. What is the value of Samara's savings account after one year when the 3% annual interest is compounded yearly? when it is compounded daily?

d. **Writing** Find the effective annual yield of Samara's account when the interest is compounded yearly and daily. Discuss the effect different compounding periods have on the effective annual yield.

14. In the last unit, you used the perimeter of a regular polygon circumscribed about a circle to estimate the circumference of the circle. You can also use a regular polygon inscribed in a circle to estimate the circumference. The more sides the polygon has, the better the estimate will be.

BY THE WAY...

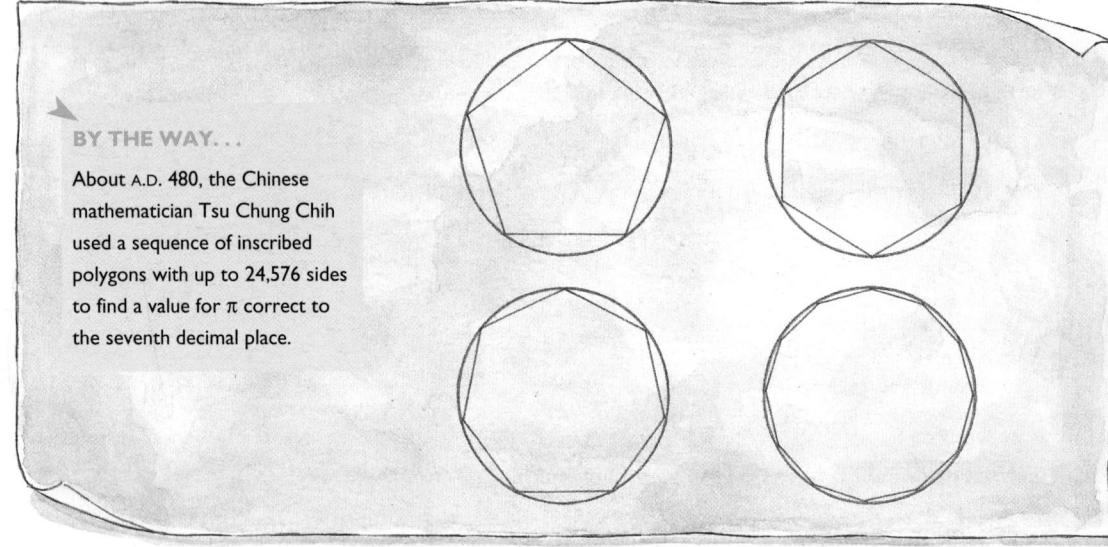

About A.D. 480, the Chinese mathematician Tsu Chung Chih used a sequence of inscribed polygons with up to 24,576 sides to find a value for π correct to the seventh decimal place.

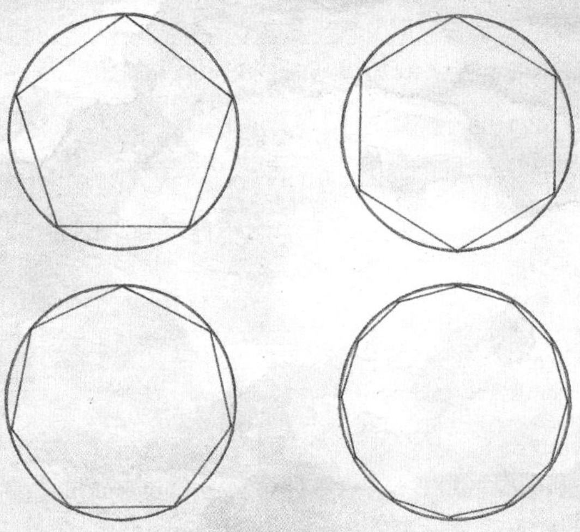

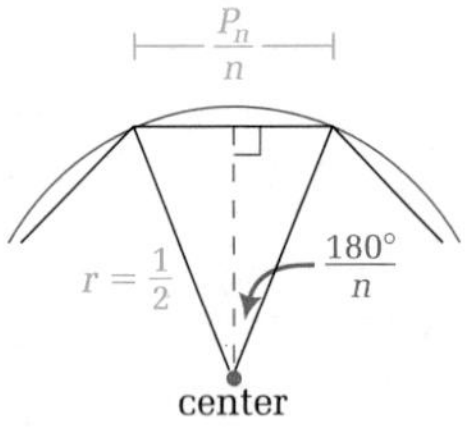

Let P_n = the perimeter of a regular n-gon inscribed in a circle of diameter 1. Use the diagram at the left.

a. Explain why the measure of the labeled angle is $\frac{180°}{n}$.

b. Complete this statement to find an explicit formula for the sequence of perimeters P_n:

$$\sin\left(\frac{180°}{n}\right) = \underline{\ ?\ }$$

c. What is the apparent limit of the sequence of perimeters? Why?

Answers to Exercises and Problems

12. a. $v_n = 500(1.0035)^n$
b. \$543.73

13. a. 3.042%
b. The effective annual yield is very slightly higher than the annual interest rate. (about 3.04%, compared to 3%)
c. \$1030; \$1030.45
d. yearly: 3%; daily: 3.045%; The more times per year the interest is compounded, the higher the effective annual yield. However, the differences are very small.

14. a. The triangle is isosceles, so the perpendicular to the base through the vertex angle divides the triangle into two congruent triangles (AAS). Since corresponding parts of congruent triangles are equal in measure, the measure of the labeled angle is $\frac{360°}{n} \div 2 = \frac{180°}{n}$.
b. $\frac{P_n}{2n} \div \frac{1}{2} = \frac{P_n}{n}$; $P_n = n\sin\left(\frac{180°}{n}\right)$

Ongoing ASSESSMENT

15. a. **Open-ended** Write the first five terms of a sequence with a pattern that involves repeated division by the same value.
    b. Write an explicit formula for your sequence in part (a).

Review PREVIEW

16. a. Graph this sequence. *(Section 4-1)*

$$-1, 1, -\frac{1}{2}, \frac{1}{2}, -\frac{1}{3}, \frac{1}{3}, \ldots$$

    b. Does the sequence in part (a) appear to have a limit?

17. Use Pascal's triangle to find the number of possible ways to choose two representatives from six students. *(Toolbox Skill 5)*

Simplify. *(Toolbox Skill 20)*

18. $(7 + 2i) + (4 + i)$ 19. $(2i)(3i)$ 20. $(5 + i)^2$ 21. $4i(1 - 2i)$

4 Working on the Unit Project

22. One way of investing money is with long-term time deposit accounts for 2.5 years or more. The interest rates for these accounts tend to be higher because the bank has more time to put the money to work.
    a. **Research** Find out the current rate for a long-term deposit account at a local bank.
    b. Use your answers to part (d) of Exercise 32 on page 215. Choose a lump sum amount of money that, if you invest it for 25 years at the rate in part (a), it will be enough to pay the cost of a public university education in 25 years. Choose another lump sum for a private university. Set up a spreadsheet to test your choice.
    c. Continue to choose lump sums and repeat part (b) until the return on your investment is within a dollar of the university cost needed.
    d. Choose the highest rate from the table and repeat part (c). How does your new lump sum compare with the lump sum from part (c)?
    e. Choose the lowest rate from the table and repeat part (c). How does your new lump sum compare with the lump sum from part (c)?

This table shows average interest rates for long-term accounts in the period 1983–1993.

Year	Rate
1983	10.48
1984	10.80
1985	8.73
1986	6.61
1987	7.86
1988	8.39
1989	7.86
1990	7.52
1991	5.56
1992	4.77
1993	4.29

Multicultural Note

Tsu Chung Chih inscribed his sequence of polygons inside a circle ten feet across to determine the value of pi as 3.1415929203. The Greek Ptolemy had already computed the value of pi to four decimal places, but Tsu Chung Chih and his son refined that value to seven places more than eleven hundred years before any European would. The written records of Tsu Chung Chih and his son Tsu Keng Chih have disappeared, but their findings were included in historical records from the period.

Practice 30 For use with Section 4-2

Name ______________ Date ______________

Practice 30
For use with Section 4-2

15. a. $1500(1.005)^n$ b. \$1690.74
16. a. $1200(1.0025)^n$ b. \$1274.11
17. a. $2400(1.00625)^n$ b. \$2787.10

For Exercises 1–8, write an explicit formula for each sequence.

1. 2, 7, 12, 17, … $a_n = -3 + 5n$
2. 5, 25, 125, 625, … $a_n = 5^n$
3. 0.2, 0.04, 0.008, 0.0016, … $a_n = (0.2)^n$
4. 8, 3, −2, −7, … $a_n = 13 - 5n$
5. $\frac{3}{1}, \frac{4}{2}, \frac{5}{3}, \frac{6}{4}, \ldots$ $a_n = \frac{n+2}{n}$
6. 2, 5, 10, 17, … $a_n = n^2 + 1$
7. the number of squares in each configuration below, and so on $a_n = (n + 1)^2 - 1$

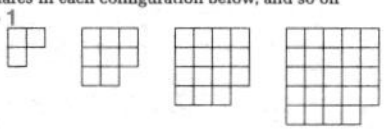

8. the number of handshakes that occur in a room containing 1 person, 2 people, 3 people, and so on, if each person shakes hands with every other person exactly once $a_n = \frac{n(n-1)}{2}$

Find the first four terms and the 12th term of each sequence.

9. $a_n = 3n - 1$ 2, 5, 8, 11; 35
10. $b_n = 2n^2 - n$ 1, 6, 15, 28; 276
11. $t_n = \frac{n+1}{n}$ $\frac{2}{1}, \frac{3}{2}, \frac{4}{3}, \frac{5}{4}; \frac{13}{12}$
12. $b_n = 2^n - 1$ 1, 3, 7, 15; 4095
13. $a_n = 3 - 5n$ −2, −7, −12, −17; −57
14. $t_n = \frac{1}{(n+1)(n+2)}$ $\frac{1}{6}, \frac{1}{12}, \frac{1}{20}, \frac{1}{30}; \frac{1}{182}$

For Exercises 15–17, assume the given amount of money *P* is invested at the given annual rate *r*, compounded monthly.

a. Find a formula for the value of the investment after *n* months.
b. Find the value of the investment after 2 years. Exs. 15–17: See top of page.

15. $P = \$1500$; $r = 6\%$ 16. $P = \$1200$; $r = 3\%$ 17. $P = \$2400$; $r = 7.5\%$

18. a. Let $y_3, y_4, y_5, \ldots$ be the sequence of measures of one angle of a regular polygon with 3, 4, 5, … sides. Why does it make sense to start this sequence with y_3? A polygon cannot have fewer than 3 sides.
    b. Give an explicit formula for y_n. (*Hint:* Recall Theorem 3.4.) $y_n = \frac{180(n-2)}{n}$
    c. Find the measures of one angle of the regular polygons with 30, 40, and 100 sides. Do you think the sequence has a limit? If so, what is it? 168, 171, 176.4; yes; 180

19. ***Open-ended*** Make up several explicit formulas for sequences. Without calculating, predict the 20th term of each sequence. Then calculate the actual value of the 20th term, using your formula. Was your prediction accurate? Answers may vary. Check students' work.

30 **Practice Bank**, INTEGRATED MATHEMATICS 3

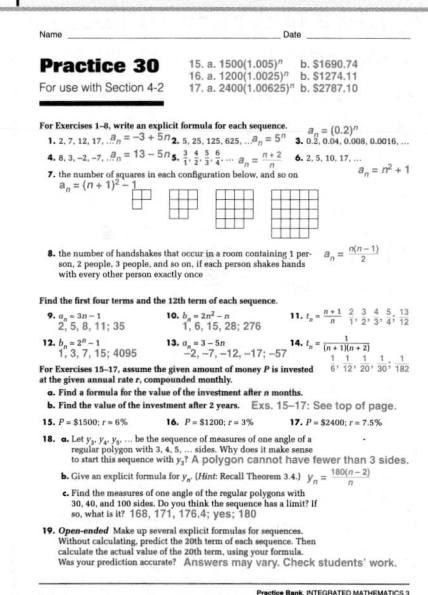

Answers to Exercises and Problems

c. π; As n increases, the product $n \sin\left(\frac{180°}{n}\right)$ gets closer and closer to π, which is the circumference of the circle.

15. Answers may vary. An example is given.
    a. 8, 4, 2, 1, $\frac{1}{2}$
    b. $a_n = \frac{16}{2^n}$

16. a.

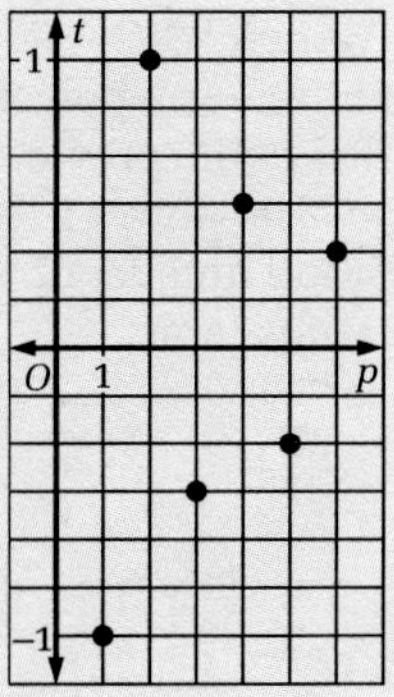

    b. Yes; odd-numbered terms increase toward 0, while even-numbered terms decrease toward 0.

17. 15 ways
18. $11 + 3i$
19. −6
20. $24 + 10i$
21. $4i + 8$
22. a–e. Answers may vary. Check students' work.

PLANNING

Objectives and Strands
See pages 204A and 204B.

Spiral Learning
See page 204B.

Materials List
- Graphics calculator
- Spreadsheet software

Recommended Pacing
Section 4-3 is a one-day lesson.

Extra Practice
See pages 614–615.

Warm-Up Exercises
- Warm-Up Transparency 4-3

Support Materials
- Practice Bank: Practice 31
- Activity Bank: Enrichment 28
- Study Guide: Section 4-3
- Problem Bank: Problem Set 9
- Using Mac Plotter Plus Disk: Data Analyzer
- Using IBM Plotter Plus Disk: Statistics Spreadsheet
- Assessment Book: Quiz 4-3, Alternative Assessment 3

Section 4-3 Using Recursive Formulas

Focus
Write and use formulas for sequences in which each term is found by using the preceding term(s).

BRANCHING OUT

Many irregular shapes found in the natural world can be modeled by geometric shapes called *fractals*. These diagrams show some of the early stages of the "growth" of a fractal tree.

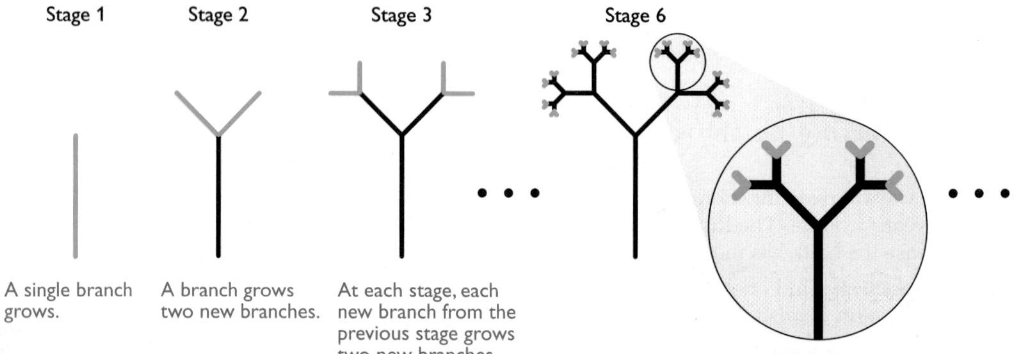

This process continues without end. The appearance of any part of a fractal is similar to the whole fractal. This property of a fractal is called *self-similarity*.

BY THE WAY...

Computer fractal programs can generate images from nature that are almost identical to the real thing. These images are called *fractal forgeries*.

Talk it Over

1. **How are a fractal tree and a real tree alike? How are they different?**
2. **Write a sequence whose first four terms are the number of branches on a fractal tree at stages 1–4.**
3. **For the sequence in question 2, describe how each term after the first is related to the term before it.**

Answers to Talk it Over

1. Summaries may vary. A fractal tree begins with a single branch; a tree has a trunk, which can be thought of as the first single branch. For both, new branches grow from existing branches. When you magnify any part of the fractal tree, what you see looks like the whole tree. This is not true of a real tree.
2. 1, 3, 7, ..., 63
3. Each term is a power of 2 added to the previous term. The power is the position number of the previous term.

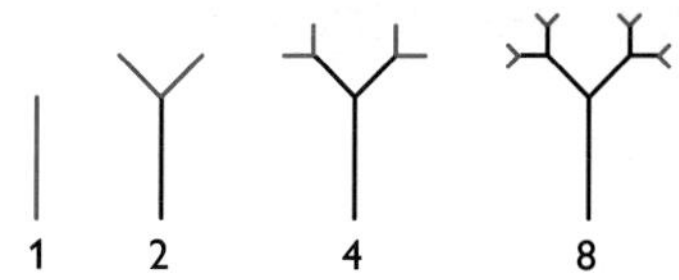

Look at the sequence of numbers of *new* branches that grow at each stage.

After the first term, each term is twice the term before it. You can use subscripts to express this relationship.

$$a_n = 2a_{n-1}$$

the nth term → a_n; a_{n-1} ← The term *before* the nth term is the $(n-1)$st term.

You can use the equation above to write a *recursive formula* for the sequence. A **recursive formula** for a sequence tells you how to find the nth term from the term(s) before it. A recursive formula has two parts:

$a_1 = 1$ ← The value(s) of the first term(s) are given.

$a_n = 2a_{n-1}$ ← A *recursion equation* shows how to find each term from the term(s) before it.

Sample 1

Write a recursive formula for the sequence 1, 2, 6, 24, ...

Sample Response

Write several terms of the sequence using subscripts. Then look for the relationship between each term and the term before it.

$a_1 = 1$

$a_2 = 2 = 2 \cdot 1$ — Write a_2 in terms of a_1. → $a_2 = 2 \cdot a_1$

$a_3 = 6 = 3 \cdot 2$ — Write a_3 in terms of a_2. → $a_3 = 3 \cdot a_2$

$a_4 = 24 = 4 \cdot 6$ — Write a_4 in terms of a_3. → $a_4 = 4 \cdot a_3$

⋮ ⋮

Write a recursion equation. → $a_n = na_{n-1}$ (These values are always equal.)

Use the value of the first term and the recursion equation to write a recursive formula for the sequence.

$$a_1 = 1$$
$$a_n = na_{n-1}$$

Talk it Over

4. a. Write an explicit formula for the sequence 1, 2, 6, 24, . . . in Sample 1.
   b. Which formula was easier to find? Do you think this is always the case? Why or why not?

Answers to Talk it Over

4. a. $a_n = 1 \cdot 2 \cdot 3 \cdot \ldots \cdot n = n!$
   b. Answers may vary. An example is given. I think the recursive formula was easier to find in this case. I don't think this will always be the case. In general, I think no one method will always be easier.

TEACHING

Multicultural Note

Fractal images are very common in Islamic art. Islamic art seeks to represent the spirituality of nature and beings, rather than their physical characteristics. Traditional Islamic artists do not create realistic representations of people, animals, or objects from nature; instead, they represent these things in abstract fashion through complex patterns. Art in the Islamic world developed in this way in part because Islam does not allow images of people in any place of worship. This law, created many hundreds of years ago, is intended to allow people to worship without distractions.

Visual Thinking

Ask students to find examples of fractals in art, nature, computer graphics, and other areas. Ask them to highlight the fractals and to share examples of their findings with the class. This activity involves the visual skills of *recognition* and *communication.*

Additional Sample

S1 Write a recursive formula for the sequence 1, 4, 9, 16, 25, ...

Write several terms of the sequence using subscripts. Then look for the relationship between each term and the term before it.

$a_1 = 1$
$a_2 = 1 + 3 = a_1 + 3$
$a_3 = 4 + 5 = a_2 + 5$
$a_4 = 9 + 7 = a_3 + 7$
$a_5 = 16 + 9 = a_4 + 9$

Starting with the second term, the nth term, a_n, is a_{n-1} increased by the nth odd number, $2n - 1$. This information gives you a recursive formula.

$a_1 = 1$
$a_n = a_{n-1} + (2n - 1)$

Additional Samples

S2 Write the first four terms of each sequence.

a. $a_1 = 1$

$a_n = n \cdot a_{n-1}$

$a_1 = 1$

Substitute 2, then 3, then 4 in the recursive formula.

$a_2 = 2 \cdot a_1 = 2 \cdot 1 = 2$

$a_3 = 3 \cdot a_2 = 3 \cdot 2 = 6$

$a_4 = 4 \cdot a_3 = 4 \cdot 6 = 24$

b. $b_1 = 0$

$b_2 = 2$

$b_n = b_{n-1} - 2b_{n-2}$

$b_1 = 0$

$b_2 = 2$

Substitute 3, then 4 in the recursive formula.

$b_3 = b_2 - 2b_1 = 2 - 2(0) = 2$

$b_4 = b_3 - 2b_2 = 2 - 2(2) = -2$

S3 Jun Chen gets a weekly paycheck that comes to exactly \$525 after all taxes and payroll deductions. She has opened a no-charge checking account and plans to deposit her check in the account each week. If she has 30% of the previous week's amount in her account each time she deposits her new check, what will her balance be in several weeks?

Write a recursive formula to model what she has in her account after each new deposit.

$a_1 = 525$

$a_n = 0.3a_{n-1} + 525$

Method 1. Use a graphics calculator.

Enter a_1. Next, type .3Ans+525 and press ENTER repeatedly. The calculator will display the numbers from a_1 through a_{12} (See Method 2). The amount in the account will level off to about \$750.

Method 2. Use a spreadsheet. Enter 525 in cell A1. Key in =0.3*A1+525 in cell A2 and use the FILL DOWN command.

	A	B
1	525	
2	682.5	
3	729.75	
4	743.925	
5	748.1775	
6	749.45325	
7	749.835975	
8	749.9507925	
9	749.9852378	
10	749.9955713	
11	749.9986714	
12	749.9996014	

The amount in the account will level off to about \$750.

Sample 2

Write the first four terms of each sequence.

a. $a_1 = -1$

$a_n = 3a_{n-1} + 4$

b. $b_1 = 5$

$b_2 = 1$

$b_n = b_{n-1} - b_{n-2}$

Sample Response

a. $a_1 = -1$

Substitute 2 first, then 3, and then 4 for n in the recursion equation.

$a_2 = 3a_1 + 4 = 3(-1) + 4 = 1$

$a_3 = 3a_2 + 4 = 3(1) + 4 = 7$

$a_4 = 3a_3 + 4 = 3(7) + 4 = 25$

b. $b_1 = 5$

$b_2 = 1$

Substitute 3 first and then 4 for n in the recursion equation.

$b_3 = b_2 - b_1 = 1 - 5 = -4$

$b_4 = b_3 - b_2 = -4 - 1 = -5$

Sample 3

Robert Nuttall injured his shoulder. For the pain, his doctor prescribed 650 mg of aspirin to be taken every 6 h. Only 26% of the aspirin remains in his body by the time he takes a new dose. What will happen to the amount of aspirin in his body if he takes aspirin for several days?

Sample Response

You can write a recursion equation to model the amount of aspirin in Robert Nuttall's body after every dose.

amount of aspirin after nth dose	=	26% of the amount of aspirin after previous dose	+	new dose of 650 mg
a_n	=	$(0.26)(a_{n-1})$	+	650

Look at what happens to a_n the farther out you go in the sequence.

Method ❶ Use a graphics calculator.

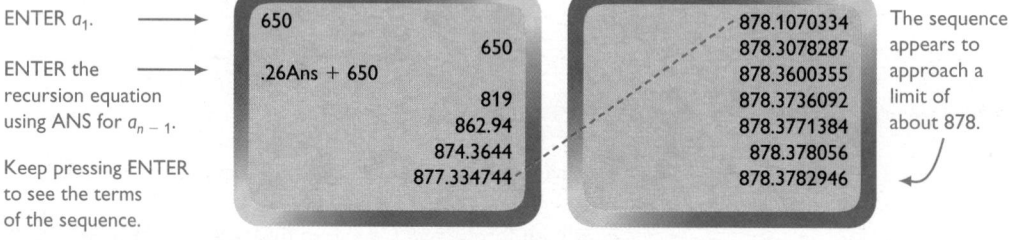

ENTER a_1. →

ENTER the recursion equation using ANS for a_{n-1}. →

Keep pressing ENTER to see the terms of the sequence.

```
650
                650
.26Ans + 650
                819
             862.94
           874.3644
         877.334744
```

```
878.1070334
878.3078287
878.3600355
878.3736092
878.3771384
 878.378056
878.3782946
```

The sequence appears to approach a limit of about 878.

The amount of aspirin in his body will level off at about 878 mg.

Answers to Talk it Over

5. the sixth dose; 30 h
6. The first term would be 1000 rather than 650. The recursion equation would be the same.

Answers to Look Back

Both give a formula for each term in a sequence. A recursive formula tells you how to find the nth term of the sequence from the terms before it. An explicit formula tells you how to find the nth term by using the position number n.

Method ❷ Use a spreadsheet.

Enter 650 in cell A1.

Key in =0.26*A1+650 in cell A2 and use the FILL DOWN command.

A12 = 0.26*A11 + 650 (This is the recursion equation for the 12th term.)

Aspirin Levels

	A	B	C
1	650		
2	819		
3	862.94		
4	874.3644		
5	877.334744		
6	878.107033		
7	878.307829		
8	878.360035		
9	878.373609		
10	878.377138		
11	878.378056		
12	878.378295		

The sequence appears to approach a limit of about 878.

The amount of aspirin in his body will level off at about 878 mg.

Talk it Over

5. After which dose does the amount of aspirin first reach a level of 878 mg in Robert Nuttall's body? After how many hours does this happen?
6. In Sample 3, what will you key in differently if Robert Nuttall's first dose is 1000 mg and each dose after that is 650 mg?

Look Back

How are explicit and recursive formulas alike? How are they different?

4-3 Exercises and Problems

1. **Reading** What are the two parts of a recursive formula?

Write a recursive formula for each sequence.

2. 9, 14, 19, 24, . . .
3. −3, 9, −27, 81, . . .
4. $9, 3, 1, \frac{1}{3}, \ldots$
5. 8.32, 8.44, 8.56, 8.68, . . .
6. $2, \frac{1}{4}, 16, \frac{1}{256}, \ldots$
7. $\frac{1}{2}, \frac{2}{3}, \frac{4}{4}, \frac{8}{5}, \ldots$

Write the first six terms of each sequence.

8. $a_1 = 3$; $a_n = a_{n-1} - 5$
9. $a_1 = 0.5$; $a_n = (0.2)a_{n-1}$
10. $a_1 = 10$; $a_n = (a_{n-1})^2$
11. $a_1 = 5$; $a_n = \frac{a_{n-1}}{n}$
12. $a_1 = 5$; $a_n = (i)(a_{n-1})$
13. $a_1 = 1$; $a_2 = 2$; $a_n = (a_{n-1})(a_{n-2})$

Answers to Exercises and Problems

1. the value of the first term and a recursion equation, a formula showing how to find each term from the term(s) before it
2. $a_1 = 9$; $a_n = a_{n-1} + 5$
3. $a_1 = -3$; $a_n = -3a_{n-1}$
4. $a_1 = 9$; $a_n = \frac{1}{3}(a_{n-1})$
5. $a_1 = 8.32$; $a_n = a_{n-1} + 0.12$
6. $a_1 = 2$; $a_n = (a_{n-1})^{-2}$
7. $a_1 = \frac{1}{2}$; $a_n = (2a_{n-1})\left(\frac{n}{n+1}\right)$
8. 3, −2, −7, −12, −17, −22
9. 0.5, 0.1, 0.02, 0.004, 0.0008, 0.00016
10. 10, 100, 10,000, 10^8, 10^{16}, 10^{32}
11. $5, \frac{5}{2}, \frac{5}{6}, \frac{5}{24}, \frac{1}{24}, \frac{1}{144}$
12. 5, $5i$, −5, $-5i$, 5, $5i$
13. 1, 2, 2, 4, 8, 32

Look Back

You may wish to have students write their responses to this question in their journals.

APPLYING

Suggested Assignment

Standard 1–13, 19–24

Extended 1–17, 19–24

Integrating the Strands

Number Exs. 1–19, 21–23

Algebra Exs. 1–23

Discrete Mathematics Exs. 1–19, 21–23

Logic and Language Exs. 1, 17, 24

Error Analysis

A higher than normal error rate in Exs. 8–13 may indicate that students are having difficulty with the notation. If this is the case, review the meaning of subscripts.

Using Technology

One way to enter sequences that are described by an explicit formula on the TI-82 is to use the List feature.

Consider the sequence defined by $b_n = n^2$. On the home screen, press [2nd] [LIST] <seq(> [ALPHA] [A] [x²] [,] [A] [,] 1 [,] 10 [,] 1 [)] to enter the following:

variable; position of 1st term; number of terms; increment

seq (A², A, 1, 10, 1)

When you press [ENTER], the calculator will display the sequence between braces (without commas). To store and view the terms in table form, press [STO▶] [2nd] [L1] [ENTER]. To see L1 (list 1), press [STAT] <EDIT>. To display a particular term on the home screen, type L1 followed by the position number (in parentheses).

Reasoning

You may wish to discuss how recursive formulas can be used to take a new look at some previous exercises. For example, in Section 4-1, Ex. 8 lends itself to using a recursive approach. When the nth cut is made, n new pieces of the circle are created. Suggest that students illustrate this with a diagram.

Teaching Tip

Students may find it especially interesting to discuss Ex. 14. The Fibonnaci sequence has a fascinating array of interesting properties. For example, write the first 10 terms of the Fibonnaci sequence on the board. Ask a student to come to the board and find the differences between successive terms.

$$\begin{array}{lcccccc} & 1 & 1 & 2 & 3 & 5 & 8 \ldots \\ \text{difference:} & 0 & 1 & 1 & 2 & 3 & \ldots \end{array}$$

The sequence of differences starts with 0 and then becomes the Fibonacci sequence itself!

There is an explicit formula for the Fibonnaci sequence:

$$a_n = \frac{1}{\sqrt{5}}\left[\left(\frac{1+\sqrt{5}}{2}\right)^n - \left(\frac{1-\sqrt{5}}{2}\right)^n\right].$$

Have students discuss whether the explicit or recursive formula is easier to use.

Interdisciplinary Problems

The mathematics of sequences can be applied to analyze problem situations in many other disciplines. Exs. 14–16 illustrate this by examining situations in biology, medicine, and state lotteries.

connection to **BIOLOGY**

14. Male honeybees hatch from unfertilized eggs and therefore have a female parent, but no male parent. Look at the family tree of a male bee.

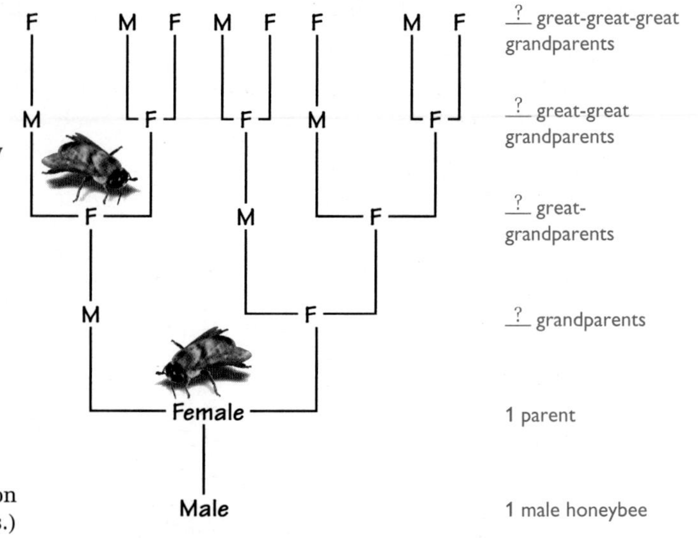

a. Starting with the male honeybee, write the sequence of the number of bees in each preceding generation. This sequence is known as the *Fibonacci sequence* and occurs in many aspects of nature.

b. Write a recursive formula for the Fibonacci sequence. (*Hint:* The recursion equation involves *two* previous terms.)

15. **Medicine** Naomi Pedraza has bronchitis. She takes 250 mg of an antibiotic every four hours. Only 33% of the antibiotic is left in her body by the time she takes a new dose. What will the amount of antibiotic in her body level off to?

16. **State Lotteries** In some states a \$1,000,000 lottery winner receives \$50,000 per year for 20 years. The winner receives \$50,000 right away. The state then puts just enough money into an account to pay the prize winner \$50,000 per year for the next 19 years, leaving a zero balance in the account after the 19th payment from the account.

Suppose an account has a fixed annual interest rate of 6.33%. Let a_0 be the amount of money the state puts into the account and let a_n be the amount of money left in the account after the nth payment.

a. Write a recursion equation for a_n.

b. What is the value of a_{19}?

c. Find a_{19} when a_0 = \$950,000, when a_0 = \$700,000, and when a_0 = \$500,000. Do any of these values for a_0 come close to giving you the correct value for a_{19}?

d. Solve your recursive formula in part (a) for a_{n-1}. Use this new equation as a recursive formula for the sequence that has a_{19} as the first term and a_0 as the last term. Will this method give you the correct value for a_0? Explain.

e. What is the amount the state should put into the account?

f. **Writing** Why do you think some states pay a lottery winner over many years rather than all at once?

17. **Writing** Explain how finding the 100th term of a sequence defined explicitly is different from finding the 100th term of the same sequence defined recursively.

Answers to Exercises and Problems

14. a. 1, 1, 2, 3, 5, 8, ...
b. $a_1 = 1$; $a_2 = 1$; $a_n = a_{n-1} + a_{n-2}$

15. about 373 mg

16. a. $a_n = 1.0633a_{n-1} - 50{,}000$
b. 0
c. \$1,303,791.65; \$501,374.00; For \$500,000, at a_{17} the amount becomes negative, so the formula no longer applies, since interest will no longer be paid; the value of a_0 that comes closest to giving the correct value for a_{19} is \$500,000.
d. $a_{n-1} = \dfrac{a_n + 50{,}000}{1.0633}$; Yes; for this sequence, the last term is the term (\$543,792.69), which when used in the original sequence produces a last term of 0.
e. \$543,792.69
f. Spreading the payments out over many years significantly decreases the cost to the state. In the case described in this exercise, the state pays out only about half as much by spreading out the payments rather than making a single payment of \$1 million.

17. To find the 100th term of a sequence defined explicitly, you need only substitute the value 100 in the formula. To find the same term in a sequence defined recursively, you need to know the 99th term.

Ongoing ASSESSMENT

18. Group Activity Work with another student.

This fractal is called a *Koch snowflake.* The first four stages in the formation of a Koch snowflake are shown. At each stage, the middle third of each side is replaced with an equilateral triangle.

Stage 1 Stage 2 Stage 3 Stage 4

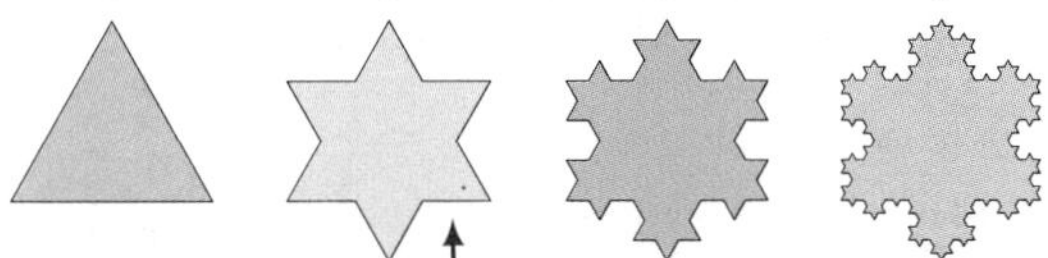

The stage 2 figure has 12 sides.

Each side measures 3 cm.

a. Suppose each side of the original equilateral triangle is 9 cm. Complete the table.

b. The table in part (a) shows the first three terms of three sequences. Write a recursive formula for each of the sequences.

c. Use your formulas from part (b) to find the 12th term of each sequence.

d. What appears to be the limit, if there is one, of each sequence?

	Number of segments	Length of each segment	Perimeter
Stage 1	3	9	27
Stage 2	?	?	?
Stage 3	?	?	?

Review PREVIEW

19. Find the first four terms and the 9th term of the sequence $b_n = \frac{3^n}{n+5}$. *(Section 4-2)*

20. Solve $x = \sqrt{1 - \frac{3}{2}x}$. *(Section 2-6)*

For each sequence: *(Section 4-1)*

a. Write the next three terms. b. Write the 10th term.

c. Describe the method you used to write the 10th term.

21. 5, 10, 20, 40, . . .

22. 5, 10, 15, 20, . . .

4 Working on the Unit Project

23. Repeat parts (a)–(e) of Exercise 22 on page 221, except, instead of choosing a lump sum, this time choose an amount of money to make equal annual deposits into a long-term deposit account for 25 years.

24. Research Read about the many different ways you can invest money. Here are some books that advise people how and where to invest their money.

Sylvia Porter's Money Book: How To Earn It, Spend It, Save It, Invest It, Borrow It, And — Use It To Better Your Life by Sylvia Field Porter

Susan Lee's ABZs of Money & Finance by Susan Lee

Everyone's Money Book by Jane Bryant Quinn

Assessment: Open-ended

For Ex. 14, ask students if they can write an explicit formula for the Fibonacci sequence.

Working on the Unit Project

For Ex. 23, students may want to choose one of the lump sums used in Ex. 22 on page 221, but now divide it into equal parts and invest it over time in equal annual deposits. They can then compare the results at the end of 25 years.

Practice 31 For use with Section 4-3

Name ____________ Date ____________

Practice 31

For use with Section 4-3

14. 1.2, 1.02, 1.002, 1.0002, 1.00002, 1.000002

For Exercises 1–9, write a recursive formula for each sequence.

1. −1, 3, 7, 11, ... **2.** 48, 24, 12, 6, ... **3.** 3, −3, 3, −3, ...

4. 4, 5, 7, 10, ... **5.** 1, 5, 14, 30, ... **6.** 2, 5, 11, 23, ...

7. $1, \frac{1}{2}, \frac{1}{6}, \frac{1}{24}, \ldots$ **8.** $\frac{1}{1}, \frac{2}{10}, \frac{3}{100}, \frac{4}{1000}, \ldots$ **9.** $\frac{1}{5}, -1, 5, -25, \ldots$

Exs. 1–9: See additional answers.

For Exercises 10–18, list the first six terms of each sequence. Ex. 14: See top of page.

10. $a_1 = -5$, $a_n = a_{n-1} + 4$ −5, −1, 3, 7, 11, 15

11. $a_1 = 3$, $a_n = -2a_{n-1}$ 3, −6, 12, −24, 48, −96

12. $a_1 = 2$, $a_n = na_{n-1}$ 2, 4, 12, 48, 240, 1440

13. $a_1 = 1$, $a_n = 3a_{n-1} - 1$ 1, 2, 5, 14, 41, 122

14. $a_1 = 1.2$, $a_n = \frac{1}{10}(a_{n-1} - 1) + 1$

15. $a_1 = 3$, $a_n = \frac{na_{n-1}}{n-1}$ 3, 6, 9, 12, 15, 18

16. $a_1 = 1$, $a_2 = 3$, $a_n = a_{n-1} + a_{n-2}$ 1, 3, 4, 7, 11, 18

17. $a_1 = 2$, $a_2 = 1$, $a_n = a_{n-1} - 2a_{n-2}$ 2, 1, −3, −5, 1, 11

18. $a_1 = 1$, $a_2 = 2$, $a_n = \frac{a_{n-1}}{a_{n-2}}$ 1, 2, 2, 1, $\frac{1}{2}$, $\frac{1}{2}$

19. The first three stages in the formation of a space-filling curve are shown. At stage n, the original square is divided into 4^n smaller squares, and a path is drawn through their centers.

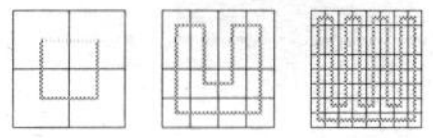

a. If each side of the original square has length 16 cm, complete the table for the first three stages of the space-filling curve. ("Segment" means a segment joining the centers of adjacent squares.) See table.

	Number of segments	Length of each seg.	Total len. of path
Stage 1	4	8 cm	32 cm
Stage 2	16	4 cm	64 cm
Stage 3	64	2 cm	128 cm

b. Write recursive formulas for the sequences in the columns of the table in part (a). Then use your formulas to find the 10th term of each sequence.

$a_1 = 4$, $a_n = 4a_{n-1}$; $b_1 = 8$, $b_n = \frac{1}{2}b_{n-1}$; $c_1 = 32$, $c_n = 2c_{n-1}$; 1,048,576; 0.015625; 16,384

Practice Bank, INTEGRATED MATHEMATICS 3

Answers to Exercises and Problems

18. a.

	Number of segments	Length of each segment	Perimeter
Stage 1	3	9	27
Stage 2	12	3	36
Stage 3	48	1	48

b. number of segments: $a_1 = 3$; $a_n = 4a_{n-1}$; length of each segment: $a_1 = 9$; $a_n = \frac{a_{n-1}}{3}$; perimeter: $a_1 = 27$; $a_n = \frac{4a_{n-1}}{3}$

c. 12,582,912; $\approx 5.08 \times 10^{-5}$; ≈ 639.28

d. none; 0; none

19. $\frac{3}{6}, \frac{9}{7}, \frac{27}{8}, \frac{81}{9}; \frac{19{,}683}{14}$

20. $\frac{1}{2}$

21. a. 80, 160, 320

b. 2560

c. Answers may vary. An example is given. I used an explicit formula, $a_n = 5 \cdot 2^{n-1}$; $a_{10} = 5 \cdot 2^9 = 2560$.

22. a. 25, 30, 35

b. 50

c. Answers may vary. An example is given. I used an explicit formula, $a_n = 5 \cdot n$; $a_{10} = 5 \cdot 10 = 50$.

23. Check students' work.

24. Answers may vary.

PLANNING

Objectives and Strands

See pages 204A and 204B.

Spiral Learning

See page 204B.

Materials List

- Graph paper
- Spreadsheet software

Recommended Pacing

Section 4-4 is a two-day lesson.

Day 1

Pages 228–231: Opening paragraph through Sample 2, *Exercises 1–24*

Day 2

Pages 231–233: General Formulas for Geometric Sequences through Look Back, *Exercises 25–39*

Extra Practice

See pages 614–615.

Warm-Up Exercises

Warm-Up Transparency 4-4

Support Materials

- Practice Bank: Practice 32
- Activity Bank: Enrichment 29
- Study Guide: Section 4-4
- Problem Bank: Problem Set 9
- Explorations Lab Manual: Diagram Master 2
- Assessment Book: Quiz 4-4, Test 15, Alternative Assessment 4

Section 4-4

Arithmetic and Geometric Sequences

Focus

Identify sequences that have a common difference, a common ratio, or neither. Write explicit and recursive formulas.

What's the Difference

Sequences are found in many everyday situations—in fuel tank readings, in calendar dates, and in the heights of a bouncing ball. Although these situations seem to have nothing in common, they are sometimes described mathematically using the same kind of sequences.

Talk it Over

Use these two groups of sequences.

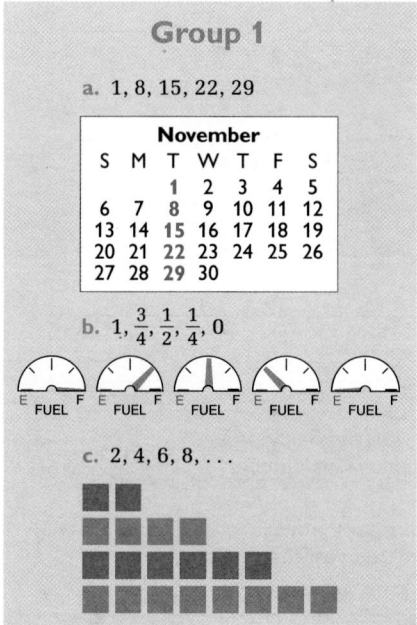

Group 1

a. 1, 8, 15, 22, 29

November

S	M	T	W	T	F	S
		1	2	3	4	5
6	7	8	9	10	11	12
13	14	15	16	17	18	19
20	21	22	23	24	25	26
27	28	29	30			

b. $1, \frac{3}{4}, \frac{1}{2}, \frac{1}{4}, 0$

c. 2, 4, 6, 8, . . .

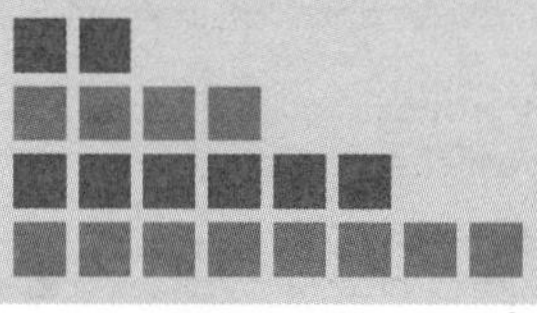

Group 2

a. 32, 24, 18, 13.5, . . .

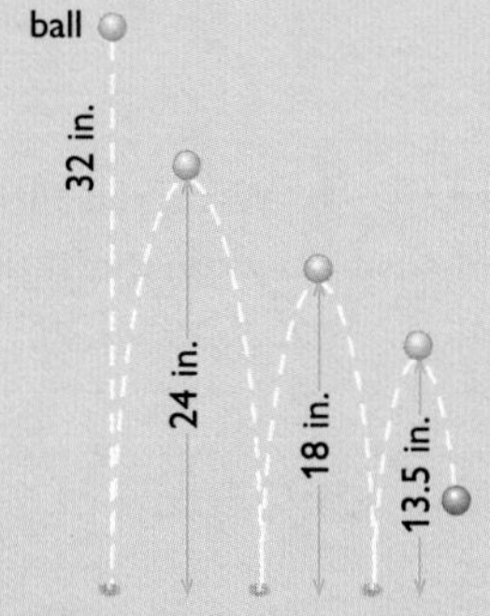

b. $1, \frac{1}{2}, \frac{1}{4}, \frac{1}{8}, \ldots$

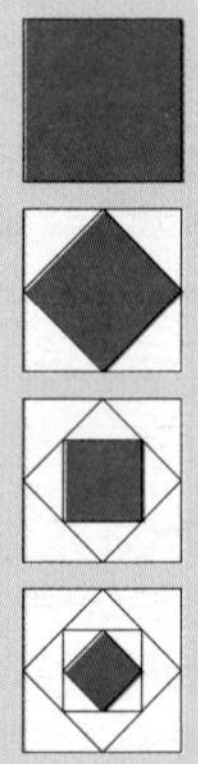

c. 2, 4, 8, 16, . . .

1. Write a recursive formula for each sequence in Group 1.
2. Write a recursive formula for each sequence in Group 2.
3. Describe any similarities within each group. Describe any differences between the two groups.

Answers to Talk it Over

1. a. $a_1 = 1$; $a_n = a_{n-1} + 7$
   b. $a_1 = 1$; $a_n = a_{n-1} - \frac{1}{4}$
   c. $a_1 = 2$; $a_n = a_{n-1} + 2$
2. a. $a_1 = 32$; $a_n = \frac{3}{4}a_{n-1}$
   b. $a_1 = 1$; $a_n = \frac{1}{2}a_{n-1}$
   c. $a_1 = 2$; $a_n = 2a_{n-1}$
3. Answers may vary. An example is given. The recursion equations for all the sequences in Group 1 involve adding the same number to the previous term. The recursion equations for all the sequences in Group 2 involve multiplying the previous term by the same number. The graphs of Group 1 functions are linear; the graphs of Group 2 functions are curves.

A sequence in which the difference between any term and the term before is a constant is an **arithmetic sequence.**

2, 4, 6, 8, . . .

+ 2 | + 2 | + 2

The value $a_n - a_{n-1}$ is a constant. This value is called the **common difference.**

A sequence in which the ratio of any term to the term before it is a constant is a **geometric sequence.**

2, 4, 8, 16, . . .

× 2 | × 2 | × 2

The value $\frac{a_n}{a_{n-1}}$ is a constant. This value is called the **common ratio.**

Sample 1

Tell whether each sequence is *arithmetic, geometric,* or *neither.*

a. 27, 9, 3, 1, . . . b. 1, 3, 7, 15, . . . c. 12, 9.5, 7, 4.5, . . .

Sample Response

Look for a common difference or a common ratio.

a. sequence 27, 9, 3, 1, . . .

difference → −18, −6, −2 — There is no common difference.

ratio → $\frac{1}{3}$, $\frac{1}{3}$, $\frac{1}{3}$ — The common ratio is $\frac{1}{3}$.

The sequence is geometric.

b. sequence 1, 3, 7, 15, . . .

difference → 2, 4, 8 — There is no common difference.

ratio → 3, $\frac{7}{3}$, $\frac{15}{7}$ — There is no common ratio.

The sequence is neither arithmetic nor geometric.

c. sequence 12, 9.5, 7, 4.5, . . .

difference → −2.5, −2.5, −2.5 — The common difference is −2.5.

The sequence is arithmetic.

Talk it Over

4. a. Write an explicit formula for the arithmetic sequence in Sample 1.
   b. To find the *n*th term of an arithmetic sequence, how many times do you add the common difference to the first term?
5. a. Write an explicit formula for the geometric sequence in Sample 1.
   b. To find the *n*th term of a geometric sequence, how many times do you multiply the first term by the common ratio?

Answers to Talk it Over

4. a. $a_n = 12 - 2.5(n - 1)$ or $a_n = 14.5 - 2.5n$
   b. $(n - 1)$ times
5. a. $a_n = 27\left(\frac{1}{3}\right)^{n-1}$ or $a_n = \frac{81}{3^n}$
   b. $(n - 1)$ times

TEACHING

Talk it Over

Questions 1–3 ask students to compare and contrast sequences with a common difference and sequences with a common ratio, thus helping students to see the difference between arithmetic and geometric sequences.

Additional Sample

S1 Tell whether each sequence is *arithmetic, geometric,* or *neither.*

a. 2, 5, 9, 17, ...

seq. 2, 5, 9, 17, ...
diff. → 3 4 8
ratio → $\frac{5}{2}$ $\frac{9}{5}$ $\frac{17}{9}$

The sequence is neither. There is neither a common difference nor a common ratio.

b. 6, 1, −4, −9, ...

seq. 6, 1, −4, −9, ...
diff. → −5 −5 −5
ratio → $\frac{1}{6}$ −4 $\frac{9}{4}$

The sequence is arithmetic, with a common difference of −5.

c. 6, 4, $\frac{8}{3}$, $\frac{16}{9}$, ...

seq. 6, 4, $\frac{8}{3}$, $\frac{16}{9}$, ...
diff. → −2 $-\frac{4}{3}$ $-\frac{8}{9}$
ratio → $\frac{2}{3}$ $\frac{2}{3}$ $\frac{2}{3}$

The sequence is geometric, with a common ratio of $\frac{2}{3}$.

Talk it Over

Questions 4 and 5 lead students from the formulas for the specific cases in Sample 1 to the general formulas presented on pages 230 and 231.

Assessment: Open-ended

On page 230, the explicit formula for an arithmetic sequence, $a_n = a_1 + (n - 1)d$, can be written as $a_n = nd + (a_1 - d)$. Ask students to explain what each term in the formula means. Then ask if the sequence 3, 7, 11, 15, ... is arithmetic. Write the explicit formula for this sequence in the form $a_n = nd + (a_1 - d)$.

Visual Thinking

Encourage students to discuss sequences in the environment, such as pillars, windows, and other architectural features; fabric and graphic designs; and events such as their class schedules and athletic activities. Ask them to create diagrams of these sequences and to demonstrate them to the class. This activity involves the visual skills of *generalization* and *communication*.

Additional Sample

S2 A gardener is going to plant daffodils in rows of various lengths in one section of a large park. He plans to use 150 bulbs in the first row, 142 in the second row, 126 in the fourth row, and so on. He will stop when he gets to the row with 30 bulbs. How many rows will he plant?

Step 1. Decide whether the sequence is arithmetic, geometric, or neither.

The sequence is arithmetic, and $d = -8$.

Step 2. Use a formula. Write the general explicit formula for an arithmetic sequence.

$a_n = a_1 + (n - 1)d$

Substitute 30 for a_n, 150 for a_1, and -8 for d.

$$30 = 150 + (n - 1)(-8)$$
$$-120 = (n - 1)(-8)$$
$$15 = n - 1$$
$$16 = n$$

The gardener will plant 16 rows of bulbs.

Talk it Over

6. a. Graph five terms of each sequence and compare.
   $2, 4, 6, 8, \ldots$ $\qquad$ $2, 4, 8, 16, \ldots$
   b. Graph five terms of each sequence and compare.
   $1, \frac{3}{4}, \frac{1}{2}, \frac{1}{4}, 0$ $\qquad$ $1, \frac{1}{2}, \frac{1}{4}, \frac{1}{8}, \ldots$
   c. Use your answers to parts (a) and (b) to make a conjecture about the graphs of arithmetic and geometric sequences.

GENERAL FORMULAS FOR ARITHMETIC SEQUENCES

For the sequence $a_1, a_2, a_3, a_4, \ldots$ with $d = a_n - a_{n-1}$

Example: $3, 5, 7, 9, \ldots$

Explicit formula

$\mathbf{a_n = a_1 + (n - 1)d}$ (nth term; 1st term; common difference)

Example: $a_n = 3 + (n - 1)2$

Recursive formula

$\mathbf{a_1}$ **= value of first term**

$\mathbf{a_n = a_{n-1} + d}$ (nth term; previous term; common difference)

Example: $a_1 = 3$, $a_n = a_{n-1} + 2$

Graph:

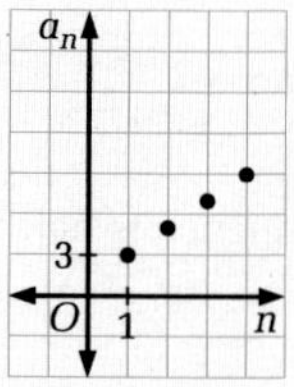

You can use an explicit formula to find out how many terms are in a finite sequence that is arithmetic or geometric.

Sample 2

Etenia is knitting a sweater with a repeating triangle pattern. The pattern repeat for each triangle is to knit 33 stitches, purl 29 stitches, knit 25 stitches, purl 21 stitches, and so on, ending with one knit stitch. How many rows are there in each triangle?

Answers to Talk it Over

6. a.

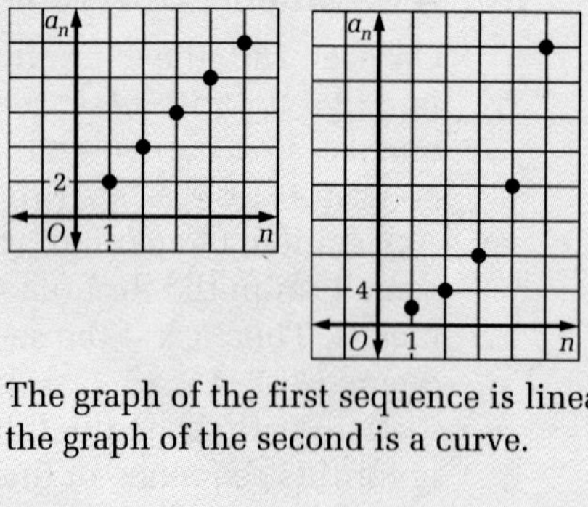

The graph of the first sequence is linear; the graph of the second is a curve.

b.

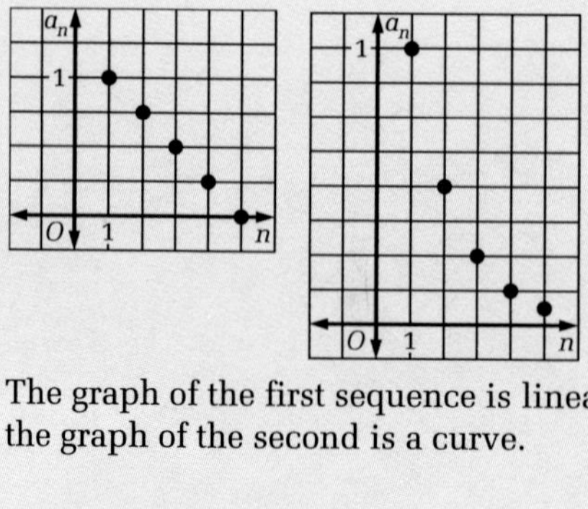

The graph of the first sequence is linear; the graph of the second is a curve.

c. The graph of an arithmetic sequence is part of a line. The graph of a geometric sequence is part of a curve.

Sample Response

Step 1 Decide whether the sequence is arithmetic, geometric, or neither.

$a_1, \quad a_2, \quad a_3, \quad a_4, \ldots, a_n$

$\downarrow \quad \downarrow \quad \downarrow \quad \downarrow \quad\quad \downarrow$

33, 29, 25, 21, . . . , 1

−4 −4 −4 ⟵ The sequence is arithmetic and $d = -4$.

Step 2 Problem Solving Strategy: Use a formula.

Write the general explicit formula for an arithmetic sequence.

$a_n = a_1 + (n - 1)d$

Substitute **1** for a_n, 33 for a_1, and −4 for d.

$1 = 33 + (n - 1)(-4)$

$1 = 33 - 4n + 4$

$-36 = -4n$

$9 = n$

There are 9 rows in each triangle.

Now you are ready for:
Exs. 1–24 on pp. 233–234

GENERAL FORMULAS FOR GEOMETRIC SEQUENCES

For the sequence $a_1, a_2, a_3, a_4, \ldots$ with $r = \dfrac{a_n}{a_{n-1}}$:

Example

3, 6, 12, 24, . . .

Explicit formula

$\mathbf{a_n = a_1 r^{n-1}}$

nth term, 1st term, common ratio

$a_n = 3 \cdot 2^{n-1}$

Recursive formula

$\mathbf{a_1}$ **= value of first term**

$\mathbf{a_n = (a_{n-1})r}$

nth term, previous term, common ratio

$a_1 = 3$

$a_n = (a_{n-1}) \cdot (2)$

Graph:

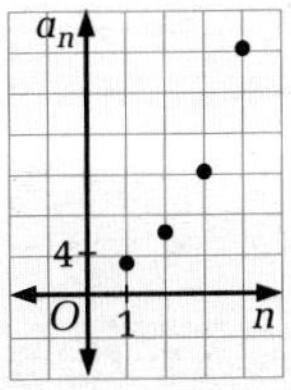

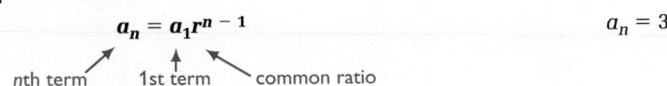

Error Analysis

Students have just learned four now formulas, two explicit and two recursive, for arithmetic and geometric sequences. Some students may confuse parts of one formula with parts of another or be confused as to when to use each formula. To help students learn the formulas and when to use them, they should work examples involving the formulas. This can be done in small groups, thus allowing students to help one another.

Additional Samples

S3 Find the 12th term of the geometric sequence 3, –15, 75, –375, ...

The common ratio is –5.

$a_n = a_1 r^{n-1}$

$a_{12} = 3(-5)^{12-1}$

$= 3(-48{,}828{,}125)$

$= -146{,}484{,}375$

S4 Find the value of x in the geometric sequence –2, x, –8, ...

The ratios $\frac{x}{-2}$ and $\frac{-8}{x}$ are both equal to the common ratio and hence are equal to each other.

$\frac{x}{-2} = \frac{-8}{x}$

$x^2 = 16$

$x = \pm\sqrt{16}$

$x = \pm 4$

x is 4 or –4.

Reasoning

The following questions require that students think about the definitions of arithmetic and geometric sequences.
(1) Is it possible for a sequence to be both arithmetic *and* geometric? (Yes; the terms are all the same nonzero number.)
(2) Can two *nonconsecutive* terms of a geometric sequence be equal? (Yes; consider 4, –4, 4, –4, ..., where the common ratio is –1.)
(3) Can two *nonconsecutive* terms of an arithmetic sequence be equal? (Yes; any such sequence has a common difference of 0, which means that all of its terms are equal.)

Talk it Over

Questions 7 and 8 point out that sequences with unknown terms, such as the one in Sample 4, can be either geometric or arithmetic. Question 8 introduces the concept of arithmetic mean.

Sample 3

Find the ninth term of the geometric sequence −2, 6, −18, 54, ...

Sample Response

The common ratio is −3.

$a_n = a_1 r^{n-1}$ ← Use the general explicit formula for a geometric sequence.

$a_9 = -2(-3)^{9-1}$ ← Substitute −2 for a_1, −3 for r, and 9 for n.

$= -2(-3)^8$

$= -2(6561)$

$= -13{,}122$

You can find an unknown term between any two known terms of an arithmetic or geometric sequence.

Sample 4

Find the value of x in the geometric sequence 3, x, 18, ...

Sample Response

In a geometric sequence, $\frac{a_n}{a_{n-1}}$ is a constant.

$\frac{x}{3} = \frac{18}{x}$ ← $\frac{a_2}{a_1} = \frac{a_3}{a_2}$

$x^2 = 54$ ← Use cross products.

$x = \pm\sqrt{54}$ ← Take the square root of both sides.

$= \pm 3\sqrt{6}$ ← Simplify.

x is $3\sqrt{6}$ or $-3\sqrt{6}$.

Watch Out!
A positive number has both a positive and a negative square root.

In Sample 4, the *positive* value of x is known as the *geometric mean* of 3 and 18. The **geometric mean** of any two positive numbers a and b is $\sqrt{ab}$ since the sequence a, $\sqrt{ab}$, b is geometric.

Talk it Over

7. a. How many different geometric sequences are possible from the sequence in Sample 4? List the first three terms of each possible sequence.
   b. Find the common ratio of each sequence in part (a).
   c. What is the next term of each sequence in part (a)?

Answers to Talk it Over

7. a. 2 sequences; 3, $3\sqrt{6}$, 18; 3, $-3\sqrt{6}$, 18
   b. $\sqrt{6}$; $-\sqrt{6}$
   c. $18\sqrt{6}$; $-18\sqrt{6}$
8. a. 10.5
   b. 1 sequence; 7.5
   c. 10.5
   d. The arithmetic mean and the mean are the same.

Answers to Look Back

Answers may vary. An example is given. In either an arithmetic or geometric sequence, any two consecutive terms are related in the same way, either by a common difference or a common ratio. That is, for all values of k and n, a_{n-1} and a_n are related in the same way that a_{k-1} and a_k are related. In an arithmetic sequence, any two consecutive terms have the same difference. In a geometric sequence, any two consecutive terms have the same ratio.

8. a. Suppose 3, x, 18, ... is an arithmetic sequence. Find x.
   b. How many different sequences are possible in part (a)? What is the common difference for each sequence?
   c. Find the mean of 3 and 18.
   d. The value of x in part (a) is the *arithmetic mean* of 3 and 18. How is the arithmetic mean of two numbers related to the mean of the numbers?

Look Back

How are arithmetic and geometric sequences alike? How are they different?

Now you are ready for: Exs. 25–39 on pp. 234–236

4-4 Exercises and Problems

Tell whether each sequence is *arithmetic, geometric,* or *neither.*

1. 5, 8, 11, 14, ...
2. 24, 11, −2, −15, ...
3. 1, 11, 111, 1111, ...
4. 1, 10, 100, 1000, ...
5. 7, 8, 7, 8, ...
6. $1, \frac{1}{5}, \frac{1}{25}, \frac{1}{125}, \ldots$

Tell whether the sequence defined by each explicit formula is *arithmetic, geometric,* or *neither.*

7. $a_n = 2n + 1$
8. $d_n = 3(2)^n$
9. $a_n = \frac{n+1}{n+2}$
10. $t_n = \frac{(-2)^n}{8}$
11. $a_n = 5 - 3n$
12. $d_n = 4n - \frac{1}{2}$

For each sequence in Exercises 13–18:

a. Decide whether the sequence is *arithmetic, geometric,* or *neither.*

b. Write a recursive formula.

13. 64, 16, 4, 1, ...
14. 3, 5, 9, 17, ...
15. 5, 3, 1, −1, ...
16. −3, −12, −21, −30, ...
17. the distance traveled after n hours by a jet flying at 550 mi/h, assuming there are no head or tail winds
18. the amount of money in a savings account after n months, if the initial amount is \$500 and the annual interest rate is 3.6% compounded monthly

Find the number of terms in each sequence.

19. 15, 12, 9, 6, ..., −36
20. −21, −9, 3, 15, ..., 219
21. 72, 56, 40, 24, ..., −440
22. 8, 15, 22, 29, ..., 99

Teaching Tip

After Talk it Over question 8, you may wish to ask whether knowing two terms (not necessarily consecutive) and their positions in an arithmetic sequence will allow you to find the first term and the common difference. Let students investigate this question by picking the values of the terms and their positions. You can then consider the analogous question for geometric sequences. In this connection, see the Reasoning note on page 232.

Look Back

A brief discussion of these questions should help students understand clearly the distinguishing characteristics of arithmetic and geometric sequences.

APPLYING

Suggested Assignment

Day 1

Standard 1–22

Extended 1–24

Day 2

Standard 27–34, 36–39

Extended 25–34, 36–39

Integrating the Strands

Number Exs. 1–22, 26–31, 34–36

Algebra Exs. 1–38

Geometry Exs. 24, 25, 32, 33, 37

Discrete Mathematics Exs. 1–31, 34, 35, 39

Logic and Language Exs. 23, 26, 34, 39

Answers to Exercises and Problems

1. arithmetic
2. arithmetic
3. neither
4. geometric
5. neither
6. geometric
7. arithmetic
8. geometric
9. neither
10. geometric
11. arithmetic
12. arithmetic
13. a. geometric
    b. $a_1 = 64$; $a_n = \frac{1}{4}a_{n-1}$
14. a. neither
    b. $a_1 = 3$; $a_n = a_{n-1} + 2^{n-1}$
15. a. arithmetic
    b. $a_1 = 5$; $a_n = a_{n-1} - 2$
16. a. arithmetic
    b. $a_1 = -3$; $a_n = a_{n-1} - 9$
17. a. arithmetic
    b. $a_1 = 550$; $a_n = a_{n-1} + 550$
18. a. geometric
    b. $a_1 = 500$; $a_n = a_{n-1}(1.003)$
19. 18 terms
20. 21 terms
21. 33 terms
22. 14 terms

Research

Exs. 24 and 25 demonstrate that arithmetic sequences can be found in nature. You may wish to have students research other natural phenomena that involve arithmetic or geometric sequences, or the Fibonacci sequence.

Communication: Reading

Ex. 26 presents an excellent check on whether students understand that a geometric mean must be positive.

23. The *ywegale* is the smallest measure of weight in Myanmar and weighs about 0.026 g. Two ywegale equal one *ywegi,* two ywegi equal one *pai,* two pai equal one *moo,* two moo equal one *mat,* two mat equal one *ngamu,* and two ngamu equal one *tical.*
    a. Express every measure of weight in terms of ywegale.
    b. What type of sequence does the weight system in Myanmar form? What is the common difference or ratio?
    c. **Open-ended** Name another weight system that forms the same type of sequence as the one in Myanmar. What is the common difference or ratio?

connection to **BIOLOGY**

Use this information for Exercises 24 and 25.

Scientists measure every part of a spider's web to understand the web's construction. When spinning a web, a spider makes several radii and then connects them with a spiral. The lengths of the cross threads between any two adjacent radii form an arithmetic sequence.

24. The shortest cross thread between two adjacent radii in a web measures 3.3 mm. The longest cross thread measures 46.5 mm. The difference in length between adjacent cross threads is 1.2 mm. How many cross threads are between the two radii?
25. The first cross thread between two radii in another web measures 3.0 mm. The last cross thread measures 42.6 mm. There are 13 cross threads in this section. What is the common difference in this sequence of cross threads?
26. **Reading** Why is $-3\sqrt{6}$ *not* a geometric mean of 3 and 18?
27. Use the geometric sequence 5, x, 7, ...
    a. Find the value of x. b. Find the common ratio. c. Find the next three terms.
28. Suppose $a_3 = 15$ and $a_5 = 2$ in a geometric sequence.
    a. Find the value of a_4. b. Find the common ratio. c. Find the value of a_1.

For each sequence in Exercises 29–31:

a. Find the value of x when the sequence is arithmetic.

b. Find the value of x when the sequence is geometric.

29. $-20, x, -4, \ldots$
30. $21, x, 10, \ldots$
31. $4, x, 16, \ldots$

Answers to Exercises and Problems

23. a. One tical equals sixty-four ywegale, one ngamu equals thirty-two ywegale, one mat equals sixteen ywegale, one moo equals eight ywegale, one pai equals four ywegale, and one ywegi equals two ywegale.
    b. geometric; The ratio is 2 if the first term is considered to be the ywegale; the ratio is $\frac{1}{2}$ if the first term is considered to be the tical.
    c. Answers may vary. An example is given. the metric system of mass; The common ratio is 10. (one centigram = ten milligrams; one decigram = ten centigrams, and so on)
24. 36 cross threads
25. 3.3 mm
26. The geometric mean of two numbers is positive.
27. a. $\sqrt{35}$ or $-\sqrt{35}$
    b. $\frac{\sqrt{35}}{5}$ or $-\frac{\sqrt{35}}{5}$
    c. $\frac{7\sqrt{35}}{5}, \frac{49}{5}, \frac{49\sqrt{35}}{25}$ or $-\frac{7\sqrt{35}}{5}, \frac{49}{5}, -\frac{49\sqrt{35}}{25}$
28. a. $\sqrt{30}$ or $-\sqrt{30}$
    b. $\frac{\sqrt{30}}{15}$ or $-\frac{\sqrt{30}}{15}$
    c. 112.5
29. a. -12
    b. $4\sqrt{5}$ or $-4\sqrt{5}$
30. a. 15.5
    b. $\sqrt{210}$ or $-\sqrt{210}$
31. a. 10
    b. 8 or -8
32. $2\sqrt{14} \approx 7.5$
33. $8\sqrt{3} \approx 13.9$
34. The numbers 10 and -20 do not have a geometric mean. That is, there is no real number x for which $x = \sqrt{10(-20)} = \sqrt{-200}$.
35. a. A: $-\frac{1}{2}$; B: -4; C: $-\frac{1}{4}$; D: -3

Use this theorem for Exercises 32 and 33.

If the altitude is drawn to the hypotenuse of a right triangle, then the measure of the altitude is the geometric mean between the measures of the parts of the hypotenuse.

Find the length of the altitude of each right triangle.

32.

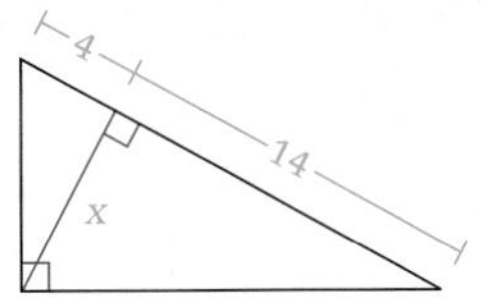

33.

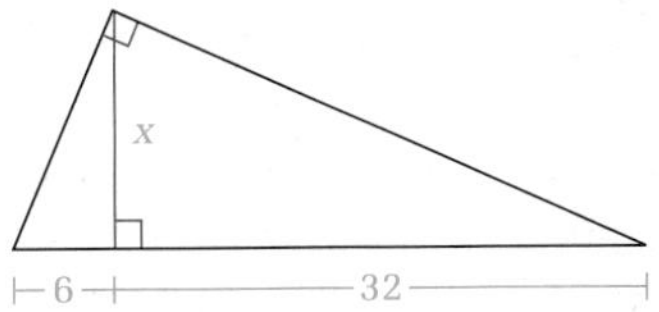

34. **Writing** Explain why there is no real number x for which $10, x, -20, \ldots$ is a geometric sequence.

Ongoing ASSESSMENT

35. **Group Activity** Work with another student.

Use these geometric sequences:

A. $6, -3, \frac{3}{2}, -\frac{3}{4}, \ldots$

B. $-5, 20, -80, 320, \ldots$

C. $\frac{1}{2}, -\frac{1}{8}, \frac{1}{32}, -\frac{1}{128}, \ldots$

D. $1, -3, 9, -27, \ldots$

a. What is the common ratio for each sequence?

b. Graph each sequence on a different set of axes.

c. What conjecture can you make about the graphs of geometric sequences with a negative common ratio?

d. Tell whether each sequence appears to have a limit.

e. What conjecture can you make about the limits of geometric sequences with a negative common ratio?

Review PREVIEW

36. Write the first six terms of this sequence. *(Section 4-3)*

$b_1 = 15$

$b_n = b_{n-1} + (2n - 1)$

37. Use the triangle. Find the value of x. *(Toolbox Skill 30)*

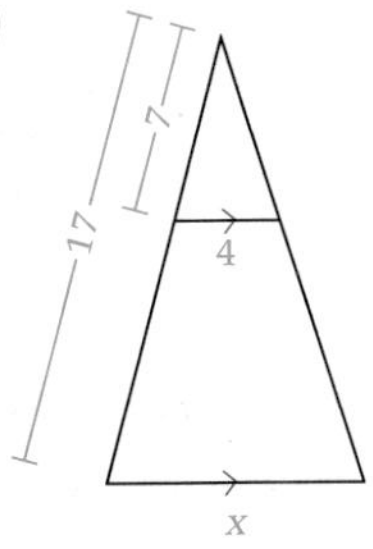

38. Find the value of A in the formula

$A = \frac{1}{2}(b_1 + b_2)h$, when $b_1 = 5$, $b_2 = 9$, and $h = 6$. *(Toolbox Skill 29)*

Cooperative Learning

Ex. 35 has students work with a partner to explore geometric sequences with a negative common ratio, make conjectures about their graphs, and decide whether or not they have limits.

Answers to Exercises and Problems

b. A:

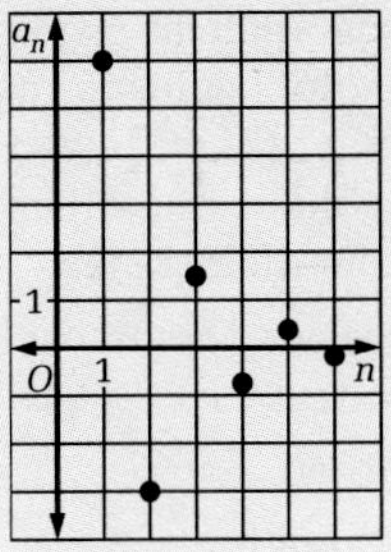

B:

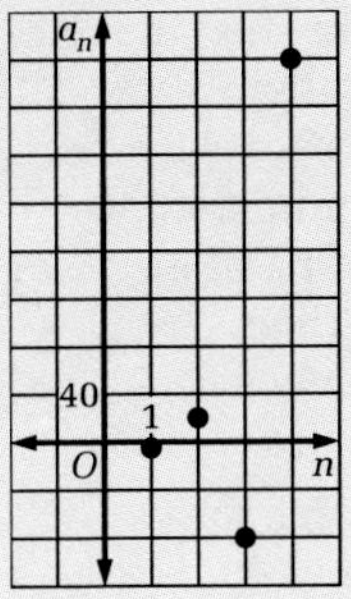

C:

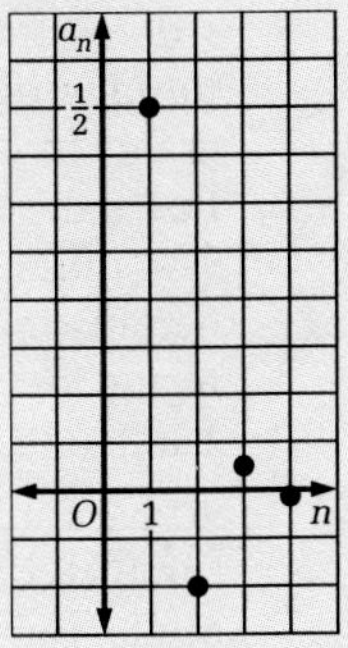

D:

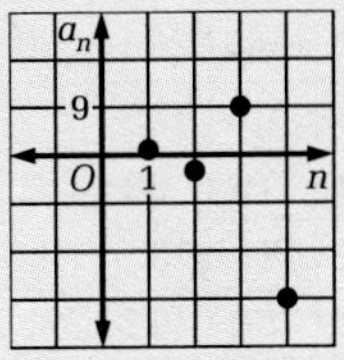

c. Answers may vary. An example is given. The graphs of such sequences lie on two curves. If the absolute value of the common ratio is less than 1, the curves approach the x-axis. If the absolute value of the common ratio is greater than 1, the curves increase without limit.

d. A: Yes, 0; B: No; C: Yes, 0; D: No.

e. If the absolute value of the common ratio is less than 1, the limit of the sequence is 0. If the absolute value of the common ratio is greater than 1, the sequence has no limit.

36. 15, 18, 23, 30, 39, 50

37. $\frac{68}{7} \approx 9.7$

38. 42

Quick Quiz (4-1 through 4-4)

See page 263.

See also Test 15 in the *Assessment Book*.

4 Working on the Unit Project

The Consumer Price Index (CPI) is a measure of the inflation rate in the United States or the change in the cost of goods and services from the previous year. "Cost of living" salary increases are based on the CPI.

39\. a. **Research** Find the most recent edition of the *Statistical Abstract of the United States.* Find and record the data in the Annual Percent Changes in Consumer Prices table for the last 30 years or so. Find the mean and median of the data. Which do you think better represents the data? Why?

b. **Research** Find the average annual entry-level salary for a career that interests you. Assume annual salary increases that match the more representative statistic in part (a).

c. Choose a percent of salary from part (b) to save for a public university education. Choose another percent to save for a private university. Use the same interest rate that you found in part (a) of Exercise 22 on page 221. Use a spreadsheet to see how much is saved over 25 years.

d. Repeat part (c), choosing different percents of salary to go to savings to find what percent of salary should be saved for a public education and for a private education.

Unit 4 CHECKPOINT

1\. Kiyoshi is sewing a sunburst quilt made up of parallelograms. 4-1

a. **Writing** Describe any pattern you see in the quilt.

b. Use the pattern to find the number and color of parallelograms she will need for a fifth ring and a sixth ring.

For each sequence in Exercises 2 and 3:

a. Plot at least six points of the sequence.

b. Decide whether the sequence appears to have a limit.

2\. $3, -4, 5, -6, \ldots$

3\. $\frac{1}{2}, \frac{1}{4}, \frac{1}{8}, \frac{1}{16}, \ldots$

Write an explicit formula for each sequence. 4-2

4\. $1, \frac{4}{5}, \frac{7}{9}, \frac{10}{13}, \ldots$

5\. $-2, \frac{1}{2}, 1\frac{1}{3}, 1\frac{3}{4}, \ldots$

Write a recursive formula for each sequence. 4-3

6\. $6, 13, 20, 27, \ldots$

7\. $1, 4, 9, 16, \ldots$

Tell whether each sequence is *arithmetic, geometric,* or *neither.* 4-4

8\. $3, -6, 12, -24, \ldots$

9\. $1, 3, 7, 15, \ldots$

10\. $\frac{7}{8}, \frac{3}{4}, \frac{5}{8}, \frac{1}{2}, \ldots$

11\. $2, 3, 5, 8, \ldots$

Practice 32 For use with Section 4-4

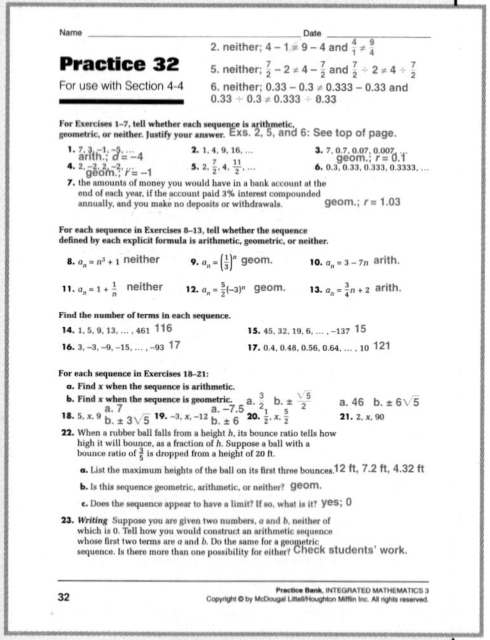

Name ______________ Date ______________

Practice 32

For use with Section 4-4

2\. neither; $4 - 1 \neq 9 - 4$ and $\frac{4}{1} \neq \frac{9}{4}$

5\. neither; $\frac{7}{2} - 2 \neq 4 - \frac{7}{2}$ and $\frac{7}{2} \div 2 \neq 4 \div \frac{7}{2}$

6\. neither; $0.33 - 0.3 \neq 0.333 - 0.33$ and $0.33 \div 0.3 \neq 0.333 \div 0.33$

For Exercises 1–7, tell whether each sequence is arithmetic, geometric, or neither. Justify your answer. Exs. 2, 5, and 6: See top of page.

1\. $7, 3, -1, -5, \ldots$ arith.; $d = -4$

2\. $1, 4, 9, 16, \ldots$

3\. $7, 0.7, 0.07, 0.007, \ldots$ geom.; $r = 0.1$

4\. $2, -2, 2, -2, \ldots$ geom.; $r = -1$

5\. $2, \frac{7}{2}, 4, \frac{11}{2}, \ldots$

6\. $0.3, 0.33, 0.333, 0.3333, \ldots$

7\. the amounts of money you would have in a bank account at the end of each year, if the account paid 3% interest compounded annually, and you make no deposits or withdrawals. geom.; $r = 1.03$

For each sequence in Exercises 8–13, tell whether the sequence defined by each explicit formula is arithmetic, geometric, or neither.

8\. $a_n = n^3 + 1$ neither

9\. $a_n = \left(\frac{1}{3}\right)^n$ geom.

10\. $a_n = 3 - 7n$ arith.

11\. $a_n = 1 + \frac{1}{n}$ neither

12\. $a_n = \frac{5}{2}(-3)^n$ geom.

13\. $a_n = \frac{3}{4}n + 2$ arith.

Find the number of terms in each sequence.

14\. $1, 5, 9, 13, \ldots, 461$ 116

15\. $45, 32, 19, 6, \ldots, -137$ 15

16\. $3, -3, -9, -15, \ldots, -93$ 17

17\. $0.4, 0.48, 0.56, 0.64, \ldots, 10$ 121

For each sequence in Exercises 18–21:

a. Find x when the sequence is arithmetic.

b. Find x when the sequence is geometric.

18\. $5, x, 9$ a. 7 b. $\pm 3\sqrt{5}$

19\. $-3, x, -12$ a. -7.5 b. ± 6

20\. $\frac{1}{2}, x, \frac{5}{2}$ a. $\frac{3}{2}$ b. $\pm \frac{\sqrt{5}}{2}$

21\. $2, x, 90$ a. 46 b. $\pm 6\sqrt{5}$

22\. When a rubber ball falls from a height h, its bounce ratio tells how high it will bounce, as a fraction of h. Suppose a ball with a bounce ratio of $\frac{3}{5}$ is dropped from a height of 20 ft.

a. List the maximum heights of the ball on its first three bounces. 12 ft, 7.2 ft, 4.32 ft

b. Is this sequence geometric, arithmetic, or neither? geom.

c. Does the sequence appear to have a limit? If so, what is it? yes; 0

23\. *Writing* Suppose you are given two numbers, a and b, neither of which is 0. Tell how you would construct an arithmetic sequence whose first two terms are a and b. Do the same for a geometric sequence. Is there more than one possibility for either? Check students' work.

32 Practice Bank, INTEGRATED MATHEMATICS 3

Answers to Exercises and Problems

39\. a. Data are given for the years 1960 through 1992.

Year	'60	'61	'62	'63	'64	'65	'66	'67	'68	'69	'70
% change	1.7	1.0	1.0	1.3	1.3	1.6	2.9	3.1	4.2	5.5	5.7
Year	'71	'72	'73	'74	'75	'76	'77	'78	'79	'80	'81
% change	4.4	3.2	6.2	11.0	9.1	5.8	6.5	7.6	11.3	13.5	10.3
Year	'82	'83	'84	'85	'86	'87	'88	'89	'90	'91	'92
% change	6.2	3.2	4.3	3.6	1.9	3.6	4.1	4.8	5.4	4.2	3.0

The mean is 4.9 and the median is 4.2. Answers may vary. An example is given. I think the mean and median are so close that neither is more typical of the data.

b–d. Check students' work.

Answers to Checkpoint

1\. a. If the inside ring is considered the first ring, the number of parallelograms in each ring is 8 times the ring number. The number of parallelograms in the rings forms the sequence 8, 16, 24, 32, ...

b. 40 parallelograms for the fifth ring and 48 parallelograms for the sixth ring, for a total of 88 parallelograms; If the color begins to repeat, Kiyoski will need 40 golden and 48 violet parallelograms.

2–8. See answers in back of book.

9\. neither

10\. arithmetic

11\. neither

Section

4-5 Arithmetic Series and Sigma Notation

Focus

Find the sum of a finite nongeometric series.

THAT SPECIAL SUM-THING

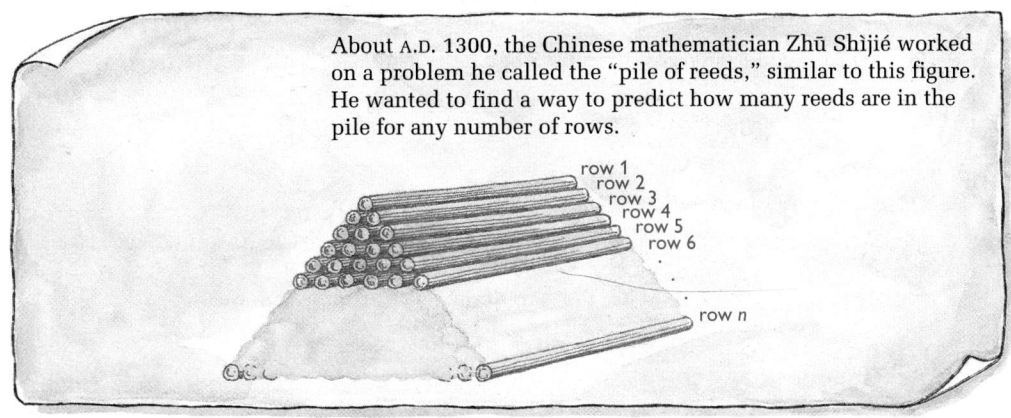

About A.D. 1300, the Chinese mathematician Zhū Shìjié worked on a problem he called the "pile of reeds," similar to this figure. He wanted to find a way to predict how many reeds are in the pile for any number of rows.

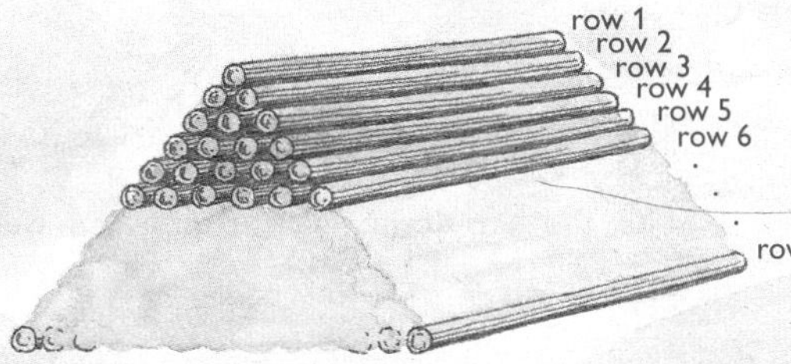

BY THE WAY...

Zhū Shìjié was a wandering teacher for more than 20 years. He also wrote two works. His second work, *Precious Mirror of the Four Elements*, began with what is now called Pascal's triangle. This scheme of numbers was considered an old one even then.

Talk it Over

1. Write a sequence for the number of reeds in each row.
2. How many reeds are in the whole pile when there are six rows?
3. Explain how your answer to question 2 is related to the first six terms of your sequence in question 1.
4. Use your sequence in question 1 to write an expression for the number of reeds in the pile when there are n rows.
5. To find out how many reeds are in Zhū Shìjié's pile when there are 100 rows, you have to find this sum.

$$1 + 2 + 3 + 4 + \cdots + 97 + 98 + 99 + 100$$

   a. What is the sum of 1 + 100? 2 + 99? 3 + 98? 4 + 97?
   b. How many sums like the ones you found in part (a) are in $1 + 2 + 3 + 4 + \cdots + 97 + 98 + 99 + 100$? Explain.
   c. Use the information from parts (a) and (b) to find the number of reeds in 100 rows of Zhū Shìjié's pile.
6. Do you think the method described in question 5 will work if the difference between terms is not 1? if the first term is not 1? if the last term is not 100? if there is an odd number of terms being added?

PLANNING

Objectives and Strands

See pages 204A and 204B.

Spiral Learning

See page 204B.

Materials List

- Spreadsheet software

Recommended Pacing

Section 4-5 is a two-day lesson.

Day 1

Pages 237–238: Opening paragraph through Sample 1, *Exercises 1–12*

Day 2

Pages 238–241: Sigma Notation through Look Back, *Exercises 13–27*

Extra Practice

See pages 614–615.

Warm-Up Exercises

- Warm-Up Transparency 4-5

Support Materials

- Practice Bank: Practice 33
- Activity Bank: Enrichment 30
- Study Guide: Section 4-5
- Problem Bank: Problem Set 10
- Assessment Book: Quiz 4-5, Alternative Assessments 5, 6

Answers to Talk it Over

1. 1, 2, 3, 4, ..., n
2. 21 reeds
3. It is the sum of the first six terms of the sequence.
4. $1 + 2 + 3 + 4 + \ldots + n$
5. a. 101; 101; 101; 101
   b. 50 sums; There are 100 numbers altogether and they are grouped in pairs.
   c. $\frac{100}{2} \cdot 101 = 5050$
6. a. Yes; Yes; Yes; Yes, except that the middle term will not be paired with another term.

TEACHING

Talk it Over

Questions 1–4 lead students intuitively to the concept of an arithmetic series. Questions 5 and 6 lead students to discover a method for finding the sum of the first n terms of an arithmetic series.

Additional Sample

S1 Find the sum of the finite arithmetic series 3 + 13 + 23 + 33 + ... + 93.

Step 1. Find the number of terms in the series.

$a_n = a_1 + (n - 1)d$
$93 = 3 + (n - 1)10$
$93 = 3 + 10n - 10$
$100 = 10n$
$10 = n$

Step 2. Use the formula for the sum of a finite arithmetic series.

$S = \frac{n(a_1 + a_n)}{2}$
$= \frac{10(3 + 93)}{2}$
$= 480$

The sum of the series is 480.

Reasoning

When a finite sequence is given by an explicit formula, you can often decide whether the sequence and related series are arithmetic, geometric, or neither without computing actual terms. For example, in Sample 2, part (a), the nth term is $5 - 2n$, the $(n + 1)$st term is $5 - 2(n + 1)$. Simplify the difference $[5 - 2(n + 1)] - [5 - 2n]$ and the result is -2. Since this result is independent of the value of n, all the common differences are -2, which means that the series is arithmetic.

When the terms of a sequence are added, the indicated sum of the terms is called a **series.** For example:

sequence → 1, 2, 3, 4, 5, 6

series → 1 + 2 + 3 + 4 + 5 + 6

A *finite series* has a last term, but an *infinite series* does not. The finite series above is an **arithmetic series** since the terms form an arithmetic sequence.

SUM OF A FINITE ARITHMETIC SERIES

The general formula for the sum S of a finite arithmetic series with n terms is

$$S = \frac{n(a_1 + a_n)}{2}.$$

Sample 1

Find the sum of the finite arithmetic series 7 + 12 + 17 + 22 + · · · + 52.

Sample Response

Step 1 Find the number of terms in the series.

$a_n = a_1 + (n - 1)d$ ← Use the general explicit formula for an arithmetic sequence.

$52 = 7 + (n - 1)5$ ← Substitute **52** for a_n, **7** for a_1, and **5** for d.

$52 = 7 + 5n - 5$ ← Solve for n.

$50 = 5n$

$10 = n$

Step 2 Use the formula for the sum of a finite arithmetic series.

$S = \frac{n(a_1 + a_n)}{2}$

$= \frac{10(7 + 52)}{2}$ ← Substitute **10** for n, **7** for a_1, and **52** for a_n.

$= 295$

Now you are ready for: Exs. 1–12 on pp. 241–242

Sigma Notation

A series can be written in a compact form using *sigma notation.* **Sigma notation** uses the *summation symbol* Σ, which is the Greek letter *sigma.*

In sigma notation, the series from Sample 1, $7 + 12 + 17 + 22 + \cdots + 52$, looks like this.

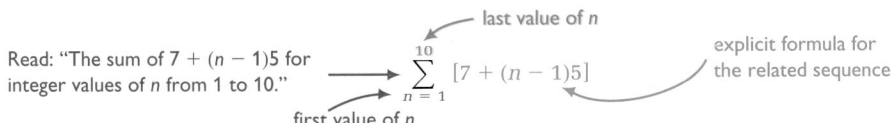

Read: "The sum of $7 + (n - 1)5$ for integer values of n from 1 to 10."

$$\sum_{n=1}^{10} [7 + (n - 1)5]$$

last value of n (10); first value of n ($n = 1$); explicit formula for the related sequence ($[7 + (n - 1)5]$)

When you substitute the values of n into the formula, you write the series in *expanded form.*

$$\sum_{n=1}^{10} [7 + (n - 1)5] = [7 + (\mathbf{1} - 1)5] + [7 + (\mathbf{2} - 1)5] + \cdots + [7 + (\mathbf{10} - 1)5]$$
$$= 7 + 12 + \cdots + 52$$

Sample 2

Evaluate.

a. $\displaystyle\sum_{n=1}^{8} (5 - 2n)$ b. $\displaystyle\sum_{n=2}^{7} |4 - n|$

Sample Response

a. **Step 1** Write the series in expanded form.

$$\sum_{n=1}^{8} (5 - 2n) = [5 - 2 \cdot 1] + [5 - 2 \cdot 2] + [5 - 2 \cdot 3] + [5 - 2 \cdot 4] + [5 - 2 \cdot 5] + [5 - 2 \cdot 6] + [5 - 2 \cdot 7] + [5 - 2 \cdot 8]$$
$$= 3 + 1 + (-1) + (-3) + (-5) + (-7) + (-9) + (-11)$$

Step 2 Find the sum of the series.

The series is arithmetic since there is a common difference of -2.

$$S = \frac{n(a_1 + a_n)}{2}$$ ← Use the formula for the sum of a finite arithmetic series.

$$= \frac{8(3 + (-11))}{2}$$ ← Substitute **8** for n, **3** for a_1, and **−11** for a_n.

$$= -32$$

b. **Step 1** Write the series in expanded form.

$$\sum_{n=2}^{7} |4 - n| = |4 - 2| + |4 - 3| + |4 - 4| + |4 - 5| + |4 - 6| + |4 - 7|$$

Step 2 Find the sum of the series.

$$S = 2 + 1 + 0 + 1 + 2 + 3$$
$$= 9$$

Additional Sample

S2 Evaluate.

a. $\displaystyle\sum_{n=1}^{6} (3n + 10)$

Step 1. Write the series in expanded form.

$$\sum_{n=1}^{6} (3n + 10)$$
$$= [3 \cdot 1 + 10] + [3 \cdot 2 + 10] + [3 \cdot 3 + 10] + [3 \cdot 4 + 10] + [3 \cdot 5 + 10] + [3 \cdot 6 + 10]$$
$$= 13 + 16 + 19 + 22 + 25 + 28$$

Step 2. Find the sum of the series. Use the formula for the sum of the finite arithmetic series.

$$S = \frac{n(a_1 + a_n)}{2}$$
$$= \frac{6(13 + 28)}{2}$$
$$= 123$$

b. $\displaystyle\sum_{n=3}^{8} (-1)^n(2n)$

Step 1. Write the series in expanded form.

$$\sum_{n=3}^{8} (-1)^n(2n)$$
$$= (-1)^3(6) + (-1)^4(8) + (-1)^5(10) + (-1)^6(12) + (-1)^7(14) + (-1)^8(16)$$

Step 2. Find the sum of the series.

$$S = (-6) + 8 + (-10) + 12 + (-14) + 16$$
$$= 6$$

Reasoning

Ask students under what conditions can they add two expressions involving sigma notation. Use examples such as $\displaystyle\sum_{n=1}^{8} (5 - 2n) + \sum_{n=1}^{8} (3n + 9)$ and $\displaystyle\sum_{n=1}^{10} (5 - 2n) + \sum_{n=2}^{10} (3n + 9)$ to generate discussion.

Talk it Over

Questions 7 and 8 ask students to consider when the use of the formula for the sum of a finite arithmetic series is appropriate.

Additional Sample

S3 Write each series in sigma notation.

a. the total number of square tiles needed to build these six models

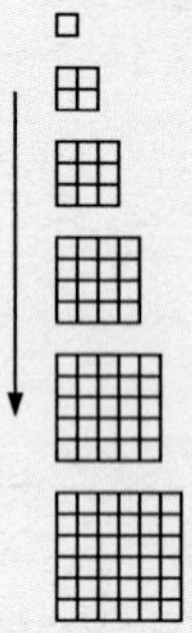

Step 1. Find an explicit formula for the sequence of areas.
$A_1 = 1 = 1^2$
$A_2 = 4 = 2^2$
$A_3 = 9 = 3^2$
An explicit formula is $A_n = n^2$.
Step 2. Find the first and last values of n. There are six terms in the series. The first value of n is 1, and the last value of n is 6.
Step 3. Write the series in sigma notation.
$\sum_{n=1}^{6} n^2$

b. $15 + 8 + 1 + (-6) + \ldots + (-90)$

Step 1. Find an explicit formula for the related sequence 15, 8, 1, –6, ..., –90. The sequence is arithmetic since there is a common difference of –7.
$a_n = a_1 + (n - 1)d$
$= 15 + (n - 1)(-7)$
Step 2. Find the first and last values of n.
The first value of n is 1. Substitute –90 for a_n to find the last value of n.
$-90 = 15 + (n - 1)(-7)$
$-90 = 15 - 7n + 7$
$16 = n$
Step 3. Write the series in sigma notation.
$\sum_{n=1}^{16} (-7n + 22)$

Talk it Over

7. In part (a) of Sample 2, is using the sum formula easier than adding up the terms? Do you think the sum formula is always easier to use? Why or why not?
8. In part (b) of Sample 2, why can you not use the formula for the sum of a finite arithmetic series?

Sample 3

Write each series in sigma notation.

a. the total number of unit cubes needed to build these five models

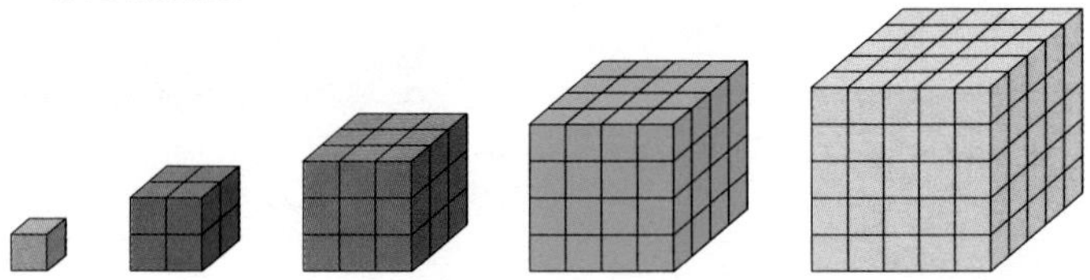

b. $7 + 10 + 13 + 16 + \cdots + 64$

Sample Response

a. **Step 1** Find an explicit formula for the related sequence.

The total number of unit cubes in each model is equal to its volume. Write several terms of the sequence of the volumes of the models.

$V_1 = 1^3$
$V_2 = 2^3$
$V_3 = 3^3$ ← Each term is the third power of its position number.

An explicit formula for the sequence is $V_n = n^3$.

Step 2 Find the first and last values of n.

There are five terms in the series. The first value of n is 1, and the last value is 5.

Step 3 Write the series in sigma notation.

$$\sum_{n=1}^{5} n^3$$

b. **Step 1** Find an explicit formula for the related sequence 7, 10, 13, 16, . . . , 64.

The series is arithmetic since there is a common difference of 3.

$a_n = a_1 + (n - 1)d$ ← Use the explicit formula for an arithmetic sequence.
$= 7 + (n - 1)3$ ← Substitute **7** for a_1 and 3 for d.

Answers to Talk it Over

7. Answers may vary. An example is given. I think it was easier to use the formula. I think deciding which method is easier depends on several things, including the number of terms in the series and the types of numbers. For example, if the terms involved fractions or both positive and negative numbers, I think it would be easier to use the formula.
8. The series is not arithmetic.

Step 2 Find the first and last values of n.

To find the last value of n, find out how many terms the series has.

$a_n = 7 + (n - 1)3$ ← Use the explicit formula for the related sequence.

$64 = 7 + (n - 1)3$ ← Substitute 64 for a_n.

$64 = 7 + 3n - 3$

$60 = 3n$

$20 = n$

The first value of n is 1, and the last value is 20.

Step 3 Write the series in sigma notation.

$$\sum_{n=1}^{20} [7 + (n - 1)3]$$

Now you are ready for: Exs. 13–27 on pp. 242–243

Look Back

Describe the relationship between a series and its related sequence.

4-5 Exercises and Problems

1. **Reading** What three things must you know about a finite arithmetic series in order to use the sum formula?

For Exercises 2–6, find the sum of each finite arithmetic series.

2. $6 + 11 + 16 + 21 + \cdots + 116$
3. $98 + 94 + 90 + 86 + \cdots + 6$
4. $1 + 1.2 + 1.4 + 1.6 + \cdots + 5.0$
5. $13 + 12.25 + 11.5 + 10.75 + \cdots + 1$
6. the sum of all positive three-digit multiples of 5
7. **Festivals** *Kwanzaa* is an African-American festival. Each candle represents one of the seven principles associated with Kwanzaa. On the first night, one candle is lit and then blown out. Each night after that, one new candle is lit, and all candles from the previous nights are lit again and then blown out.
   a. Write the series for the sum of the candles that are lit *again* during the festival and find the sum.
   b. What is the total number of lightings and re-lightings of candles for the festival?

Using Technology

The TI-82 can be used to find the sum of a finite series provided you have an explicit formula for the terms of the series and the series has 99 or fewer terms. Press [2nd] [LIST]MATH <sum>. The word *sum* will appear on the home screen. Then enter the related sequence as described in the Using Technology note in Section 4-3, page 225. For example, to find the value of $\sum_{n=1}^{8}(5 - 2n)$, enter the expression sum seq (5–2A,A,1,8,1) on the home screen. When [ENTER] is pressed, the calculator will display the sum.

If you would like to find the sum of more terms, edit the preceding expression. For example, if you want the sum of the first 20 terms of the preceding sequence, press [2nd] [ENTRY]. Then move the cursor to 8 and change 8 to 20 to obtain sum seq (5–2A,A,1,20,1). When you press [ENTER], the new sum will be displayed.

APPLYING

Suggested Assignment

Day 1

Standard 1–11

Extended 1–12

Day 2

Standard 13–20, 23–27

Extended 13–27

Integrating the Strands

Number Exs. 1–27

Algebra Exs. 1–22, 24

Geometry Exs. 12, 21, 22

Discrete Mathematics Exs. 1–27

Logic and Language Exs. 1, 21, 23, 27

Answers to Talk it Over

A series with n terms is the indicated sum of the first n terms of its related sequence.

Answers to Exercises and Problems

1. the first term, the last term, and the number of terms
2. 1403
3. 1248
4. 63
5. 119
6. 98,550
7. a. $0 + 1 + 2 + 3 + 4 + 5 + 6$; 21
   b. 28

Integrating the Strands

Ex. 12 integrates ideas from the strands of number, algebra, geometry, and discrete mathematics.

Career Note

The Unit 4 Project activities help students learn how to save for the future to pay for major expenditures such as a college education or a down payment on a home. A major concern of many working adults today is how to save enough money to finance their retirement years, which may last for as long as 30 years after they stop working. To avoid making what could be costly mistakes, some people employ the service of a *financial planner* or *consultant* to analyze their financial situation and to chart an investment program that will build their net worth for retirement. Financial planners are trained to invest money for long-term growth and hold a Certified Financial Planner (CFP) certificate. They work for a variety of financial businesses, such as brokerage firms, banks, financial-planning firms, or are self-employed. Students can look in the yellow pages of their local telephone books to identify the financial planners living in their area.

Open-ended **For each number, write an arithmetic series that has the number as its sum. Each series should have at least four terms.**

8. 25
9. 49
10. 15
11. 36

connection to **BIOLOGY**

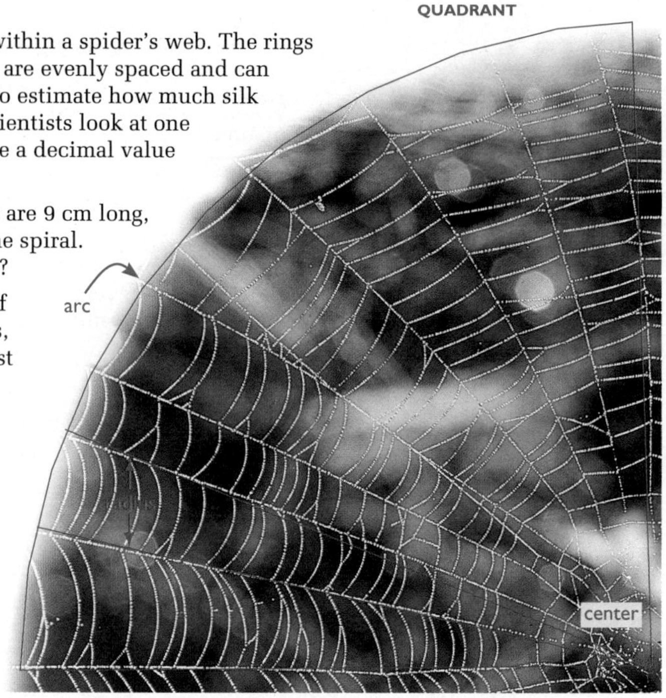

12. There are sequences hidden within a spider's web. The rings in the spiral of a spider's web are evenly spaced and can be approximated by circles. To estimate how much silk a spider uses for the spiral, scientists look at one section of the web. (Do not use a decimal value for π until part (d).)
    a. Suppose the radii in a web are 9 cm long, and there are 36 rings in the spiral. How far apart are the rings?
    b. Write the first four terms of the sequence of arc lengths, starting with the one closest to the center of the web. (*Hint:* What is the formula for circumference?)
    c. What is the length of the last arc?
    d. How much silk is used to make the spiral in one section of the web?
    e. How much silk is used to make the spiral for the whole web?

Evaluate.

13. $\sum_{k=1}^{8} 4k$
14. $\sum_{n=5}^{10} (2n-5)$
15. $\sum_{n=0}^{5} (-2)^n$
16. $\sum_{t=-2}^{4} (3+t)^2$

Write each series in sigma notation.

17. $28 + 23 + 18 + 13 + \cdots + (-7)$
18. $(-13) + (-2) + 9 + 20 + \cdots + 108$
19. $\frac{1}{2^3} + \frac{1}{2^4} + \frac{1}{2^5} + \cdots + \frac{1}{2^{10}}$
20. $\frac{1}{3} + \frac{2}{3} + 1 + \frac{4}{3} + \frac{5}{3} + \cdots + 33$

21. Shelley Tres wants to paint the side of a 16-step stairway. She needs to find the area in order to know how much paint to buy.

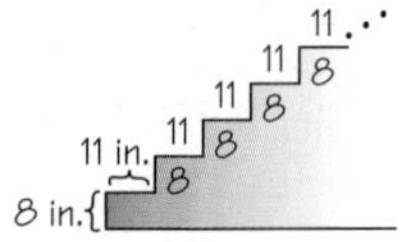

    a. **Writing** Explain how the side of the stairway can be divided into a series of rectangles.
    b. Write a series in sigma notation for the areas of the rectangles.
    c. A gallon of paint will cover about 300 square feet. Shelley Tres has about a half gallon of paint and wants to apply two coats. Should she buy more paint? Explain.

Answers to Exercises and Problems

8–11. Answers may vary. Examples are given.
8. $1 + 3 + 5 + 7 + 9$
9. $-2 + 1 + 4 + 7 + 10 + 13 + 16$
10. $1 + 2 + 3 + 4 + 5$
11. $1 + 3 + 5 + 7 + 9 + 11$
12. a. 0.25 cm
    b. $\frac{\pi}{8}, \frac{\pi}{4}, \frac{3\pi}{8}, \frac{\pi}{2}$
    c. $\frac{9\pi}{2}$
    d. $\frac{333\pi}{4}$ cm $\approx$ 261.54 cm
    e. 333π cm $\approx$ 1046.15 cm $\approx$ 10.5 m
13. 144
14. 60
15. −21
16. 140
17. $\sum_{n=1}^{8}[28 + (n-1)(-5)] =$ $\sum_{n=1}^{8}(33 - 5n)$
18. $\sum_{n=1}^{12}[-13 + (n-1)(11)] =$ $\sum_{n=1}^{12}(11n - 24)$
19. $\sum_{n=3}^{10} \frac{1}{2^n}$
20. $\sum_{n=1}^{99} \frac{n}{3}$
21. a. Answers may vary. An example is given. The side can be divided into a series of rectangles with width 11 and lengths consecutive multiples of 8.

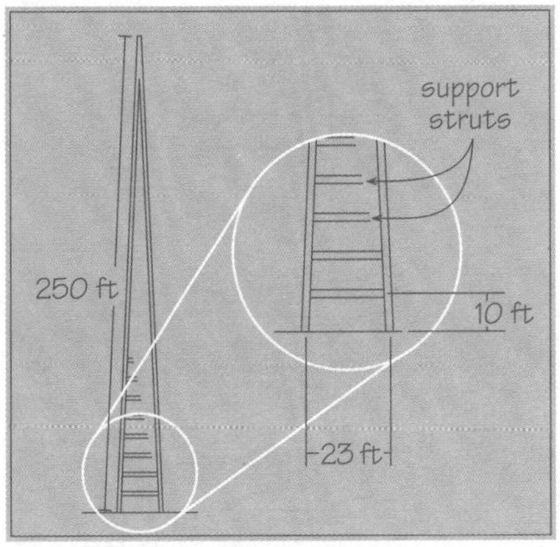

22. Rosa Bonillas is designing a radio tower with four identical sides. One side is shown. The length of each leg is 250 ft. Each pair of adjacent legs is 23 ft apart on the ground. The tower needs a horizontal supporting *strut*, or steel beam, every 10 ft along the leg.
   a. Use similar triangles to find the lengths of the bottom four struts on one side.
   b. Write a sequence for the lengths of the struts. Is the sequence *arithmetic, geometric,* or *neither*?
   c. Write a series in sigma notation for the minimum length of steel beam Rosa will need for the struts on one side of the tower.
   d. Find the minimum length of steel beam Rosa needs for the struts for the whole tower.

Ongoing ASSESSMENT

23. **Open-ended** Without using a formula, write to a seventh-grader explaining how to use a shortcut to find the sum of the whole numbers from 1 to 50.

Review PREVIEW

24. Find the value of x in the geometric sequence 12, x, 1, ... *(Section 4-4)*
25. Find AB and BA. *(Toolbox Skill 11)*

$$A = \begin{bmatrix} 0 & 3 & -2 \\ 1 & 2 & 1 \end{bmatrix} \qquad B = \begin{bmatrix} \frac{1}{2} & -5 \\ 4 & 2 \\ -1 & 2 \end{bmatrix}$$

26. a. What type of sequence is 3, 6, 12, 24, 48, 96? *(Section 4-4)*
   b. Find the sum of the terms of the sequence in part (a).

4 Working on the Unit Project

27. **Group Activity** Work in a group of four students.
   a. **Research** Each student should research one of these types of investments: Certificates of Deposit (CDs), United States Savings Bonds, mutual funds, and zero-coupon bonds. Find out a typical rate of return, whether there are any tax advantages, and what kind of risk is involved with each investment type. Summarize your results with a chart and report.
   b. Set up a spreadsheet to calculate the results when your annual investment remains the same but is split between two different investment types.

Students Acquiring English

Ex. 23 is an excellent project for students acquiring English. You may wish to pair students acquiring English with English-proficient partners to facilitate the project.

Working on the Unit Project

Students researching mutual funds should begin by choosing one of the major categories of funds, such as aggressive growth funds, growth funds, balanced funds, income funds, or bond funds. Typical rates of returns for various categories of funds can vary greatly.

Practice 33 For use with Section 4-5

Name ______________________ Date ______________

Practice 33

For use with Section 4-5

11. $\sum_{k=1}^{35}(3k+7)$ 12. $\sum_{k=1}^{12}(22-9k)$ 13. $\sum_{k=1}^{9}(3+k)^3$
14. $\sum_{k=1}^{11}\frac{2^k}{3}$ 15. $\sum_{k=1}^{11}\frac{7}{4+k}$ 16. $\sum_{k=1}^{10}\frac{1}{k(k+1)}$

For Exercises 1–6, find the sum of each finite arithmetic series.

1. $8 + 11 + 14 + 17 + \ldots + 95$ 1545
2. $48 + 43 + 38 + 33 + \ldots + (-12)$ 234
3. $2 + 13 + 24 + 35 + \ldots + 123$ 750
4. $\frac{1}{2} + \frac{5}{2} + \frac{9}{2} + \frac{13}{2} + \ldots + \frac{53}{2}$ 189
5. $0.25 + 2 + 3.75 + 5.5 + \ldots + 30$ 272.25
6. $32.5 + 31.9 + 31.3 + 30.7 + \ldots + 19.9$ 576.4

Evaluate.

7. $\sum_{k=1}^{6}(3k+1)$ 69
8. $\sum_{k=5}^{12}(1-2k)$ −128
9. $\sum_{k=2}^{7}\left(\frac{1}{2}\right)^k$ $\frac{63}{128}$
10. $\sum_{k=-3}^{5}(k^2+1)$ 78

Write each series in sigma notation. Exs. 11–16: See top of page.

11. $10 + 13 + 16 + 19 + \ldots + 112$
12. $13 + 4 + (-5) + (-14) + \ldots + (-86)$
13. $4^3 + 5^3 + 6^3 + 7^3 + \ldots + 12^3$
14. $\frac{2}{3} + \frac{2^2}{3} + \frac{2^3}{3} + \frac{2^4}{3} + \ldots + \frac{2^{11}}{3}$
15. $\frac{7}{5} + \frac{7}{6} + 1 + \frac{7}{8} + \ldots + \frac{7}{15}$
16. $\frac{1}{1 \cdot 2} + \frac{1}{2 \cdot 3} + \frac{1}{3 \cdot 4} + \ldots + \frac{1}{10 \cdot 11}$
17. Melor Ramón estimates that her peach orchard will yield 35 bushels of peaches during the first week of the harvest, and that her yield will increase by 5 bushels per week for the next 6 weeks. How many bushels can she expect to harvest during that time? 350 bushels
18. Lisle Washington has two job offers. The starting salary for both jobs is \$10,000 for 6 months' work, or \$20,000 per year. In the first job, Lisle would get a raise of \$500 (in his salary for 6 months) at the end of every 6 months. In the second, he would get a raise of \$1500 (in his yearly salary) at the end of every year. Calculate his total earnings after 5 years on each job. Which job pays more for those 5 years? \$122,500; \$115,000; the first job
19. A roll of paper used in the production of books contains the equivalent of 3750 concentric cylinders of paper. The smallest cylinder (wound on a core) has a diameter of 3 in. and each successive cylinder is 0.004 in. larger in diameter than the preceding one. What is the total length of paper on such a roll? about 124,000 in.

Practice Bank, INTEGRATED MATHEMATICS 3

33

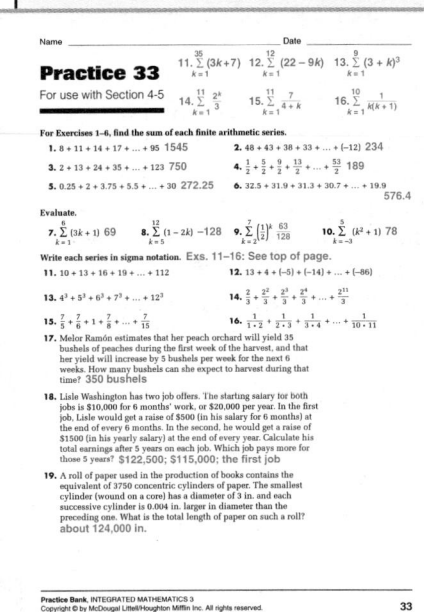

Answers to Exercises and Problems

b. $\sum_{n=1}^{16} 11(8n)$

c. 11,968; 11,968 in.$^2 \approx$ 83 ft^2; Yes; she needs enough paint to cover about 166 ft^2. The half gallon she has will cover only about 150 ft^2.

22. a. 22.08 ft; 21.16 ft; 20.24 ft; 19.32 ft

b. $a_n = 22.08 - 0.92(n - 1)$ or $a_n = 23 - 0.92n$; arithmetic

c. $\sum_{n=1}^{24}(23 - 0.92n)$

d. 1104 ft

23. Summaries may vary. Look at the numbers in the list: 1 + 2 + 3 + ... + 48 + 49 + 50. Pair the numbers, beginning with the first and last: 1 + 50; 2 + 49, 3 + 48, and so on. Notice that all the sums equal 51. Since there are 50 numbers, there are 25 such pairs. Then the sum of the 50 numbers is 25(51) = 1275.

24. $2\sqrt{3}$ or $-2\sqrt{3}$

25. $AB = \begin{bmatrix} 14 & 2 \\ 7.5 & 1 \end{bmatrix}$; $BA = \begin{bmatrix} -5 & -8.5 & -6 \\ 2 & 16 & -6 \\ 2 & 1 & 4 \end{bmatrix}$

26. a. geometric
   b. 189

27. a, b. Check students' work.

PLANNING

Objectives and Strands
See pages 204A and 204B.

Spiral Learning
See page 204B.

Materials List
- Spreadsheet software

Recommended Pacing
Section 4-6 is a one-day lesson.

Extra Practice
See pages 614–615.

Warm-Up Exercises
- Warm-Up Transparency 4-6

Support Materials
- Practice Bank: Practice 34
- Activity Bank: Enrichment 31
- Study Guide: Section 4-6
- Problem Bank: Problem Set 10
- Assessment Book: Quiz 4-6

Section 4-6

Geometric Series

Use a formula to find the sum of a finite geometric series.

"SUM" MORE

The NCAA Women's Basketball Tournament involves 64 teams. In each round, each team plays one other team and only the winners go on to the next round. The chart shows some results for the 1994 East-West regional games.

BY THE WAY...

Lynette Woodard, a University of Kansas All American, became the leading career scorer in the history of women's college basketball in 1981. After graduating, she captained the United States Olympic Basketball team that won a gold medal in 1984. Lynette was the first woman to play with the Harlem Globetrotters.

Talk it Over

1. a. Write the sequence for the number of games played in each round of a tournament involving 64 teams. What kind of sequence have you formed?
   b. Write the series related to your sequence in part (a).
   c. How many games are played in the tournament?

Answers to Talk it Over

1. a. 32, 16, 8, 4, 2, 1; geometric
   b. 32 + 16 + 8 + 4 + 2 + 1
   c. 63 games

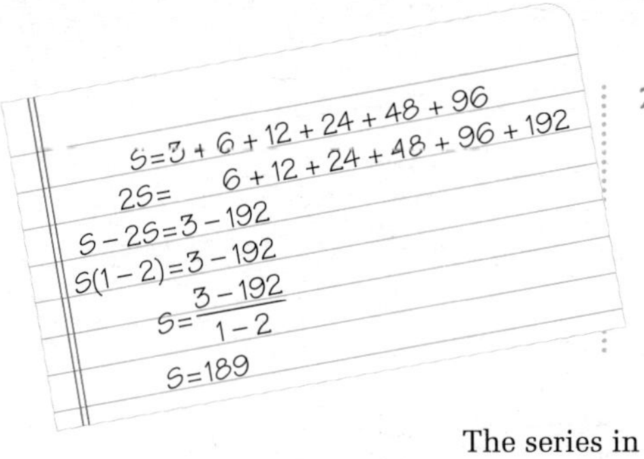

2. Elena used this method to find the sum of the series 3 + 6 + 12 + 24 + 48 + 96.

   a. Explain her method. Is Elena's answer correct?

   b. Use Elena's method to find the sum of your series in part (b) of question 1. Did you get the same answer as before?

The series in question 2 and in part (b) of question 1 are finite *geometric series*. The terms of a **geometric series** form a geometric sequence. There is a formula for the sum of a finite geometric series.

Look closely at Elena's next-to-last step in question 2.

first term → last term × common ratio

$$S = \frac{3 - 192}{1 - 2}$$

common ratio

SUM OF A FINITE GEOMETRIC SERIES

The general formula for the sum S of a finite geometric series with n terms and common ratio r is

$$S = \frac{a_1 - a_n r}{1 - r}.$$

Sample 1

Find the sum of the series (−1) + 3 + (−9) + 27 + · · · + 2187.

Sample Response

Step 1 Decide whether the series is arithmetic, geometric, or neither.

There is a common ratio of −3, so the series is geometric.

Step 2 Use the formula for the sum of a finite geometric series.

$$S = \frac{a_1 - a_n r}{1 - r}$$

$$S = \frac{-1 - 2187(-3)}{1 - (-3)}$$ ← Substitute −1 for a_1, 2187 for a_n, and −3 for r.

$$= \frac{6560}{4}$$

$$= 1640$$

The sum of the series is 1640.

TEACHING

Talk it Over

Questions 1 and 2 use specific examples to lead students to the general formula for the sum of a finite geometric series.

Teaching Tip

Elena's method in Talk it Over question 2 is important and you may wish to have students solve two or three similar problems by the same method. The method involves multiplying both sides of the first equation by the common ratio and then subtracting to get a simpler equation that can be solved for S. This method is the basis from which the general formula is derived. If students understand the method, they can use it if they have trouble remembering the formula. The same is true for the method used to find the sum of an arithmetic series. (See Section 4-5, page 237, Talk it Over questions 1–6).

Additional Sample

S1 Find the sum of the series 7 + 14 + 28 + 56 + ... + 7168.

Step 1. **Decide whether the series is arithmetic, geometric, or neither. There is a common ratio of 2, so the series is geometric.**

Step 2. **Use the formula for the sum of a finite geometric series.**

$$S = \frac{a_1 + a_n r}{1 - r}$$

$$= \frac{7 - 7168(2)}{1 - 2}$$

$$= 14{,}329$$

The sum of the series is 14,329.

Answers to Talk it Over

2. a. She multiplied each term of the series by the common ratio, 2. The new series has the same terms as the original series except that it does not include the first term of the original and has a new last term, 192. When she subtracts, all but the first term of the original series and the last term of the new series cancel out, leaving $S - rS = a_1 - ra_n$. She then divides both sides by $1 - r$. Her answer is correct.

   b. $S = 32 + 16 + 8 + 4 + 2 + 1$; $\frac{1}{2}S = 16 + 8 + 4 + 2 + 1 + \frac{1}{2}$; $S - \frac{1}{2}S = 32 - \frac{1}{2}$; $S = 64 - 1 = 63$; Yes.

Talk it Over

Questions 3–5 ask students to consider features and limitations of the formula for the sum of a finite geometric series.

Additional Sample

S2 Evaluate $\sum_{n=1}^{6}(-4)3^n$.

Step 1. Decide whether the series is arithmetic, geometric, or neither.

$\sum_{n=1}^{6}(-4)3^n$
$= (-4)3^1 + (-4)3^2 + (-4)3^3 + (-4)3^4 + (-4)3^5 + (-4)3^6$

The series is geometric with a common ratio of 3.

Step 2. Find the values of the first and last terms.

$a_1 = (-4)3^1 = -12$ $\quad a_6 = (-4)3^6 = -2916$

Step 3. Use the formula for the sum of a finite geometric series.

$S = \frac{a_1 + a_n r}{1 - r}$
$= \frac{-12 - (-2916)3}{1 - 3}$
$= -4368$

The sum of the series is −4368.

Talk it Over

3. In Sample 1, why is the sum of the series less than the last term?
4. What three things do you need to know about a geometric series in order to find its sum using the formula?
5. a. Write a geometric series having six terms and a common ratio of 1.
   b. Explain why you cannot use the formula for the sum of a finite geometric series to find the sum of your series.
   c. Explain how to find the sum of your series without adding up the terms.

Using Σ

Geometric series can also be written in sigma notation. To write the geometric series for the number of games played in the NCAA Tournament in sigma notation, you must find an explicit formula for the related geometric sequence.

$a_n = a_1 r^{n-1}$ ← Use the general explicit formula for a geometric sequence.

$a_n = 32\left(\frac{1}{2}\right)^{n-1}$ ← Substitute **32** for a_1 and $\frac{1}{2}$ for r.

There are six terms in the series. In sigma notation, the series looks like this.

$$\sum_{n=1}^{6} 32\left(\frac{1}{2}\right)^{n-1}$$

Sample 2

Evaluate $\sum_{n=1}^{8} 13(2)^n$.

Sample Response

Step 1 Decide whether the series is arithmetic, geometric, or neither.

Write the series in expanded form.

$$\sum_{n=1}^{8} 13(2)^n = 13(2)^1 + 13(2)^2 + 13(2)^3 + 13(2)^4 + 13(2)^5 + 13(2)^6 + 13(2)^7 + 13(2)^8$$

The series is geometric with a common ratio of 2.

Answers to Talk it Over

3. Alternating terms of the sequence are negative.
4. the first and last terms and the common ratio
5. a. Answers may vary. An example is given. $3 + 3 + 3 + 3 + 3 + 3$
   b. Since $r = 1$, $1 - r = 0$ and $\frac{a_1 - a_n r}{1 - r}$ is not defined.
   c. Since all the terms are the same, multiply the first term by the number of terms. $3 \cdot 6 = 18$

Answers to Look Back

Summaries may vary. The sum of a finite series is the quotient that results when the common ratio multiplied by the last term is subtracted from the first term and the result is divided by the difference between 1 and the common ratio.

Step 2 Find the values of the first and last terms.

$a_1 = 13(2)^1 = 26$ $\qquad a_8 = 13(2)^8 = 3328$

Step 3 Use the formula for the sum of a finite geometric series.

$$S = \frac{a_1 - a_n r}{1 - r}$$

$$S = \frac{26 - 3328(2)}{1 - 2}$$ ← Substitute **26** for a_1, 3328 for a_n, and **2** for r.

$$= 6630$$

The sum of the series is 6630.

Look Back

Write the formula for the sum of a finite geometric series in words without using any symbols.

4-6 Exercises and Problems

1. **Reading** What values do you need to know to write a geometric series in sigma notation?

Find the sum of each series.

2. $3 + 15 + 75 + 375 + \cdots + 9375$
3. $4 + (-8) + 16 + (-32) + \cdots + 1024$
4. $5 + 1 + 0.2 + 0.04 + \cdots + 0.000064$
5. $10 + \frac{20}{3} + \frac{40}{9} + \frac{80}{27} + \cdots + \frac{640}{729}$

Evaluate.

6. $\sum_{n=1}^{17} 5(-1)^n$
7. $\sum_{t=0}^{7} \frac{2}{5}(2)^t$
8. $\sum_{n=1}^{8} (-1)\left(-\frac{2}{3}\right)^{n-1}$
9. $\sum_{s=1}^{10} 3(0.5)^{s-1}$

For each series in Exercises 10–15:

a. Decide whether the series is *arithmetic, geometric,* or *neither*.

b. Find the sum.

10. $0 + 2 + 6 + 12 + \cdots + 56$
11. $(-3) + 3 + 9 + 15 + \cdots + 33$
12. $\sum_{t=0}^{4} 7(0.1)^t$
13. $\frac{1}{5} + \left(-\frac{1}{10}\right) + \frac{1}{20} + \left(-\frac{1}{40}\right) + \cdots + \frac{1}{320}$
14. $\sum_{n=0}^{10} -\frac{1}{4}(n + 1)$
15. $\sum_{n=1}^{6} \frac{1}{2}(3)^{n-1}$

Look Back

Students can work independently on this writing activity. Then you may wish to have volunteers read aloud what they have written. Some students may need to revise their original statements before entering them into their journals.

APPLYING

Suggested Assignment

Standard 1–15, 20–27

Extended 1–17, 20–27

Integrating the Strands

Number Exs. 2–17, 19–22, 25–27

Algebra Exs. 2–17, 19–24, 27

Geometry Exs. 23, 24

Discrete Mathematics Exs. 1–17, 19–22, 25–27

Logic and Language Exs. 1, 17–21

Answers to Exercises and Problems

1. the number of terms and an explicit formula for the related geometric sequence, for which you need to know the first term and the common ratio
2. 11,718
3. 684
4. 6.249984
5. about 28.24
6. –5
7. 102
8. about –0.58
9. about 5.99
10. a. neither
    b. 168
11. a. arithmetic
    b. 105
12. a. geometric
    b. 7.7777
13. a. geometric
    b. 0.134375
14. a. arithmetic
    b. –16.5
15. a. geometric
    b. 182

Cooperative Learning

Exs. 16–20 provide interesting situations that students can explore together working in small groups. Specific suggestions pertaining to these exercises are contained in the teaching notes on this page and page 249.

Interdisciplinary Problems

Exs. 16–18 relate the use of sequences to a story from Greek mythology.

Communication: Writing

As they write their responses to Ex. 17, part (c), students should come to the realization that this series is infinite and continuously increasing, and, therefore, has no sum. The concept of infinite series will be discussed in Section 4-7.

connection to **LITERATURE**

In a Greek myth, Herakles performed twelve dangerous tasks to become immortal, or be allowed to live forever. The second task was to defeat the Hydra, a monster with nine heads.

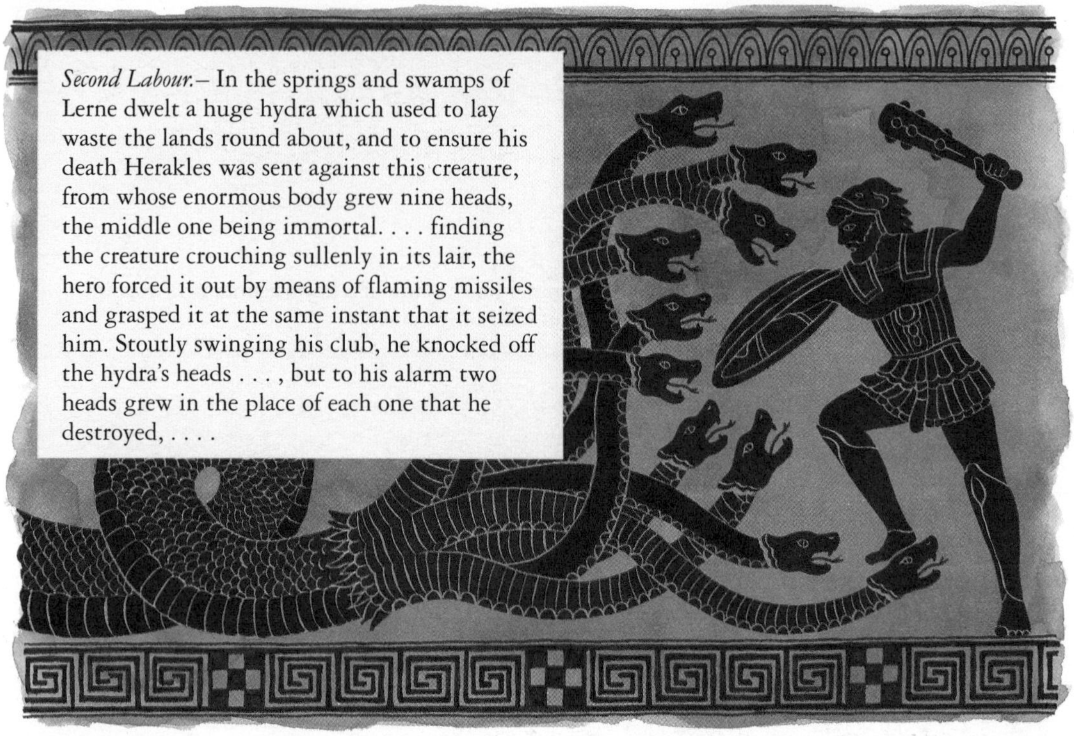
Second Labour.– In the springs and swamps of Lerne dwelt a huge hydra which used to lay waste the lands round about, and to ensure his death Herakles was sent against this creature, from whose enormous body grew nine heads, the middle one being immortal. . . . finding the creature crouching sullenly in its lair, the hero forced it out by means of flaming missiles and grasped it at the same instant that it seized him. Stoutly swinging his club, he knocked off the hydra's heads . . . , but to his alarm two heads grew in the place of each one that he destroyed, ...

16. Suppose with one swing of his club, Herakles can knock off all the Hydra's heads, except the immortal one, before any new heads replace them.

a. Write a series in sigma notation for the total number of heads that are knocked off after five swings.

b. Is the series in part (a) *arithmetic, geometric,* or *neither*?

c. Find the sum of your series.

17. Suppose Herakles can knock off only one head with each swing, and the head is replaced by two new heads before he can swing his club again.

a. Write the first five terms of the sequence for the number of heads that the Hydra has after each swing.

b. Is the sequence in part (a) *arithmetic, geometric,* or *neither*?

c. **Writing** Explain why the sum of the series related to the sequence in part (a) has no meaning.

18. **Research** Find a copy of this myth. How did Herakles finally defeat the Hydra?

Answers to Exercises and Problems

16. a. $\sum_{n=1}^{5} 8 \cdot 2^{n-1}$

b. geometric

c. 248

17. a. $9 + 10 + 11 + 12 + 13$

b. arithmetic

c. The terms of the sequence indicate the number of heads the Hydra has after the indicated swing. Adding the number of heads the Hydra has at two different times makes no sense. For example, after the second swing, the Hydra does not have $9 + 10 = 19$ heads.

18. Herakles burned the neck after cutting off each head.

19. a. 1, 5, 25, 125, ...

b. \$1562.50

c. 13 levels

d. Answers may vary. An example is given. Such letters are inherently fraudulent. A person would get no money until six stages after he or she had mailed out letters (until his or her name moved up to the top of the list). That means only those in the first seven levels (a total

connection to HISTORY

BY THE WAY...

The "Prosperity Club" chain letter is believed to have begun in Denver in April, 1935. As a result of its popularity, the volume of first class mail rose so high and so suddenly in Denver, there was talk of a breakdown in the postal system.

19. In 1935 the famous "Prosperity Club" chain letter caused quite a stir throughout the United States. The letter contained a list of six names and addresses and asked you to send a dime to the person at the top of the list and remove his or her name. You were then to add your name to the bottom of the list and pass the letter on to five friends. These five friends made up the second level in the chain.

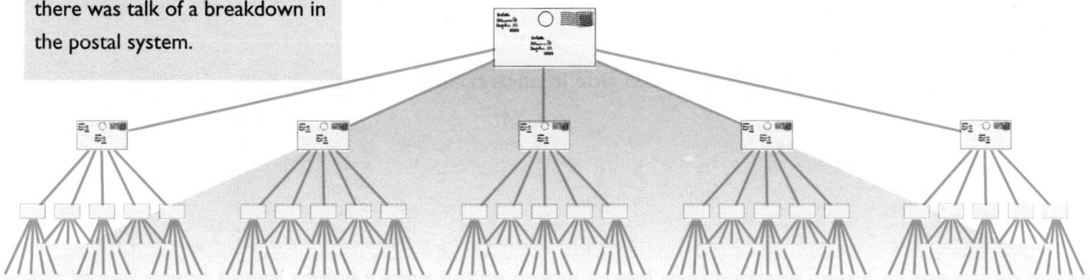

a. Write several terms of the sequence of the number of people at each level of the chain.

b. The letter claimed that if you did not break the chain, you would receive a certain amount of money. How much was this?

c. The population of the United States in 1935 was about 127,250,000. Assuming no duplications, how many levels would it take for the entire population of the United States to receive the chain letter?

d. **Writing** Use your answer from part (c) to give a reason why it is illegal to send chain letters that request money.

e. **Research** Explain how events in the United States during the 1930s might account for the popularity of this chain letter.

20. A poem that is over 200 years old reads:

As I was going to St. Ives,
I met a man with seven wives,
Each wife had seven sacks,
Each sack had seven cats,
Each cat had seven kits:
Kits, cats, sacks, and wives,
How many were there going to St. Ives?

a. Write a series for the number of people and objects the author met.

b. Find the sum of your series.

c. **Open-ended** There are several possible answers to the question in the last line. Find an answer and justify it.

Problem Solving

When students solve problems, they should be aware of all the important conditions and assumptions and be able to explain why each condition or assumption is necessary. For example, in Ex. 19, part (c), the assumption is that no duplications occur. You may wish to discuss why this assumption is important. How would the problem and its solution change if this condition were removed?

Communication: Discussion

You may wish to have students discuss their answers to Ex. 20, part (c). The discussion should help students realize that the answer to a question is related to how one interprets the question.

Answers to Exercises and Problems

of 19,531 people) would get any money. The people who added their names from level 8 on would never recieve any money. Only a fraction of the population could benefit, while the majority would lose not only their 10¢, but also the postage for the letters they mailed.

e. Answers may vary. An example is given. In the mid-1930s, the United States was in the middle of the Great Depression, a time of serious economic hardship. The hope of getting a large return on a small investment would probably have been very tempting to those who did not understand the problems with such a scheme.

20. a. $1 + 7 + 49 + 343 + 2401$

b. $\sum_{n=1}^{5} 7^{n-1} = 2801$

c. Answers may vary. Examples are given. The most frequently given answer is 0, justified by the fact that the speaker is going to St. Ives. Therefore, those he or she meets are not going to, but coming from St. Ives. The number could be 1 (the speaker). If the speaker meets no others on their way to St. Ives, the answer might be 2800 (if kits, cats, sacks, and wives are counted), 2801 if the man who is met is counted as well, or 2802 if the speaker is also counted.

Working on the Unit Project

Students can work independently on Ex. 27 within their project groups. Each student in a group may wish to select a different type of investment. Then they can compare the results obtained.

Practice 34 For use with Section 4-6

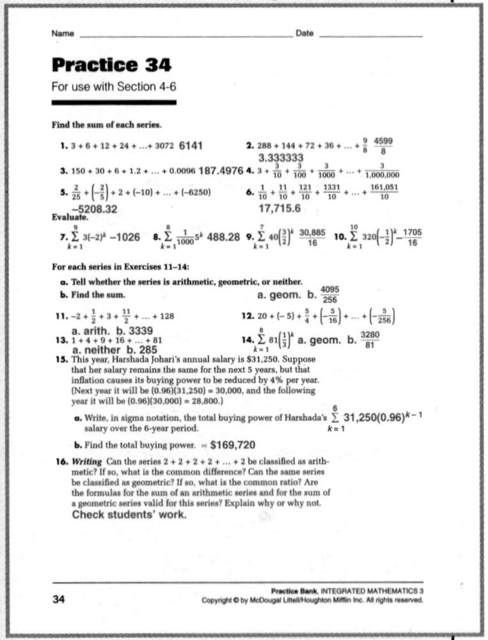

Name ______________ Date ______________

Practice 34

For use with Section 4-6

Find the sum of each series.

1. $3 + 6 + 12 + 24 + \dots + 3072$ 6141
2. $288 + 144 + 72 + 36 + \dots + \frac{9}{8}$ $\frac{4599}{8}$
3. $150 + 30 + 6 + 1.2 + \dots + 0.0096$ 187.4976
4. $3 + \frac{3}{10} + \frac{3}{100} + \frac{3}{1000} + \dots + \frac{3}{1,000,000}$ 3.333333
5. $\frac{2}{25} + \left(-\frac{2}{5}\right) + 2 + (-10) + \dots + (-6250)$ −5208.32
6. $\frac{1}{10} + \frac{11}{10} + \frac{121}{10} + \frac{1331}{10} + \dots + \frac{161,051}{10}$ 17,715.6

Evaluate.

7. $\sum_{k=1}^{9} 3(-2)^k$ −1026
8. $\sum_{k=1}^{8} \frac{1}{1000}5^k$ 488.28
9. $\sum_{k=1}^{7} 40\left(\frac{3}{2}\right)^k$ $\frac{30,885}{16}$
10. $\sum_{k=1}^{10} 320\left(-\frac{1}{2}\right)^k$ $-\frac{1705}{16}$

For each series in Exercises 11–14:

a. Tell whether the series is arithmetic, geometric, or neither.

b. Find the sum.

11. $-2 + \frac{1}{2} + 3 + \frac{11}{2} + \dots + 128$ a. arith. b. 3339
12. $20 + (-5) + \frac{5}{4} + \left(-\frac{5}{16}\right) + \dots + \left(-\frac{5}{256}\right)$ a. geom. b. $\frac{4095}{256}$
13. $1 + 4 + 9 + 16 + \dots + 81$ a. neither b. 285
14. $\sum_{k=1}^{8} 81\left(\frac{1}{3}\right)^k$ a. geom. b. $\frac{3280}{81}$
15. This year, Harshada Johari's annual salary is $31,250. Suppose that her salary remains the same for the next 5 years, but that inflation causes its buying power to be reduced by 4% per year. (Next year it will be (0.96)(31,250) = 30,000, and the following year it will be (0.96)(30,000) = 28,800.)
    a. Write, in sigma notation, the total buying power of Harshada's salary over the 6-year period. $\sum_{k=1}^{6} 31,250(0.96)^{k-1}$
    b. Find the total buying power. = $169,720
16. ***Writing*** Can the series 2 + 2 + 2 + 2 + ... + 2 be classified as arithmetic? If so, what is the common difference? Can the same series be classified as geometric? If so, what is the common ratio? Are the formulas for the sum of an arithmetic series and for the sum of a geometric series valid for this series? Explain why or why not. Check students' work.

34 Practice Bank, INTEGRATED MATHEMATICS 3

Copyright © by McDougal Littell/Houghton Mifflin Inc. All rights reserved.

Ongoing ASSESSMENT

21. **Open-ended** Suppose that your birthday is April 30, and that a wealthy relative gives you this choice of gifts.
    A. $1,000,000 on your birthday
    B. $.01 on April 1, $.02 on April 2, $.04 on April 3, $.08 on April 4, and so on with the last payment on your birthday

    Which gift would you choose? Justify your answer.

Review PREVIEW

22. Find the sum of the series $(-1) + 6 + 13 + 20 + \cdots + 111$. *(Section 4-5)*

Find the exact measure of each unknown side in each triangle. *(Toolbox Skill 30)*

23.

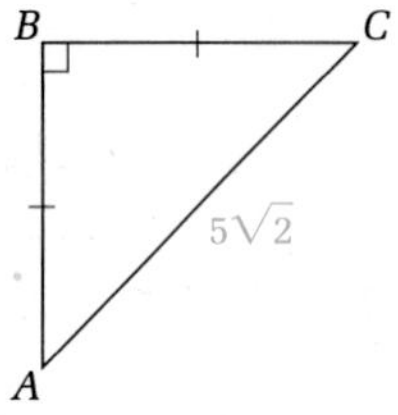

24.

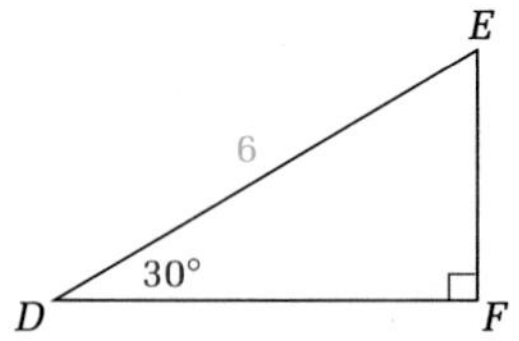

Decide whether each sequence appears to have a limit. Justify your answer. *(Section 4-1)*

25. 5, 11, 17, 23, . . .

26. 7, −3.5, 1.75, −0.875, . . .

4 Working on the Unit Project

27. Suppose you have only six years to save for a university education. Select one or more investment types. Use a spreadsheet to find the annual amount, within a dollar, that you will need to invest.

    Assume you continue to invest during the four years it takes to earn a degree. Be sure to show the layout of your spreadsheet and the formulas used to make the calculations.

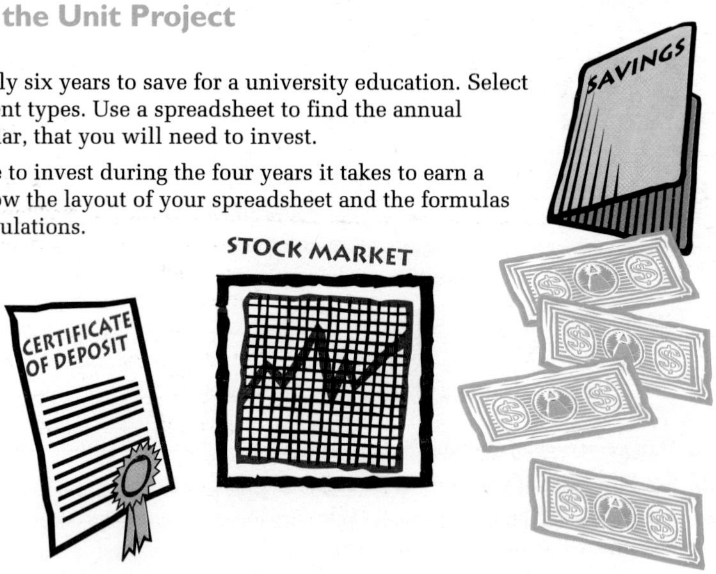

Answers to Exercises and Problems

21. Answers may vary. An example is given. If I were guaranteed that my relative would not change his or her mind or run out of money before April 30, I would choose B. The amount received over the thirty days would be $\sum_{n=1}^{5} 0.01(2^{n-1}) =$ $10,737,418.24.
22. 935
23. $AB = BC = 5$
24. $EF = 3$; $DF = 3\sqrt{3}$
25. No; there is no single number to which the terms of the sequence get close.
26. Yes; the odd-numbered terms decrease toward 0 and the even-numbered terms increase toward 0.
27. Check students' work.

Section 4-7

Infinite Series

TAKE IT TO THE LIMIT

Focus

Find the sum of an infinite geometric series.

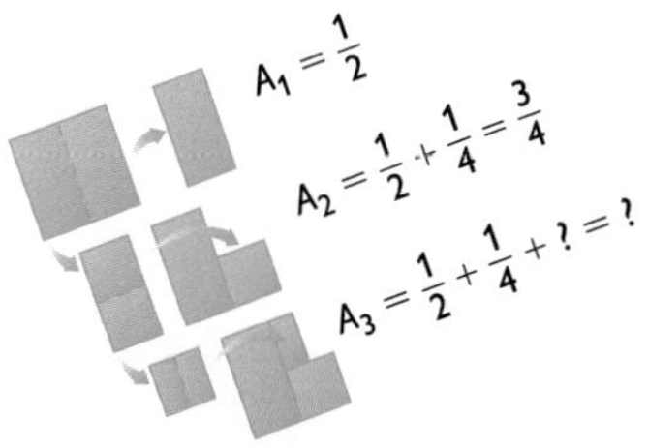

$A_1 = \frac{1}{2}$
$A_2 = \frac{1}{2} + \frac{1}{4} = \frac{3}{4}$
$A_3 = \frac{1}{2} + \frac{1}{4} + ? = ?$

EXPLORATION

Can you find the sum of an infinite series?

- **Materials: large squares of paper, scissors**
- **Work with another student.**

Let the area of a square be one square unit.

1. Fold a square in half and tear or cut along the fold line. Set aside one half. What is the area A_1 of this piece?
2. Fold the leftover piece in half and tear or cut it along the fold line. Place one half next to the piece from step 1. What is the combined area A_2 of the pieces?
3. Keep repeating step 2 until the leftover piece is too small to tear or cut. Be sure to find the combined area of the pieces you set aside at each step.
4. Write several terms of the sequence $A_1, A_2, A_3, A_4, \ldots$. What limit does this sequence appear to have?
5. **a.** Suppose you can continue halving the paper forever. Write a series that represents the combined area of the pieces that are being put aside.
   b. Is your series *finite* or *infinite*? Is it *arithmetic* or *geometric*? What is the common difference or ratio?
   c. Look at the area you are forming with the pieces that are being put aside. What must the sum of your series be? Why?
6. Compare your answers to steps 4 and 5(c). What do you notice?

BY THE WAY...

Sonya Kovalevsky (1850–1891) was a Russian mathematician who worked with infinite series. In 1874 she became the first woman since the Renaissance to receive a doctorate in mathematics.

PLANNING

Objectives and Strands

See pages 204A and 204B.

Spiral Learning

See page 204B.

Materials List

- Large squares of paper
- Scissors
- Graph paper
- Spreadsheet software

Recommended Pacing

Section 4-7 is a two-day lesson.

Day 1

Pages 251–253: Exploration through Talk it Over 2, *Exercises 1–10*

Day 2

Pages 253–255: Infinite Geometric Series through Look Back, *Exercises 11–33*

Extra Practice

See pages 614–615.

Warm-Up Exercises

Warm-Up Transparency 4-7

Support Materials

- Practice Bank: Practice 35
- Activity Bank: Enrichment 32
- Study Guide: Section 4-7
- Problem Bank: Problem Set 10
- Explorations Lab Manual: Diagram Master 2
- Using TI-81 and TI-82 Calculators: Contrasting Two Infinite Series
- Assessment Book: Quiz 4-7, Test 16, Alternative Assessment 7

Answers to Exploration

1. $\frac{1}{2}$
2. $\frac{3}{4}$
3. Check students' work.
4. $A_3 = \frac{7}{8}, A_4 = \frac{15}{16}, A_5 = \frac{31}{32}, A_6 = \frac{63}{64}, A_7 = \frac{127}{128}, A_8 = \frac{255}{256}, A_9 = \frac{511}{512}$; the limit is 1.
5. a. $\frac{1}{2} + \frac{1}{4} + \frac{1}{8} + \frac{1}{16} + \frac{1}{32} + \ldots + \frac{1}{2^n}$
   b. infinite; geometric; $\frac{1}{2}$
   c. 1; The area of the original sheet of paper is 1.
6. The limit of the sequence $A_1, A_2, A_3, \ldots$ and the sum of the series are the same.

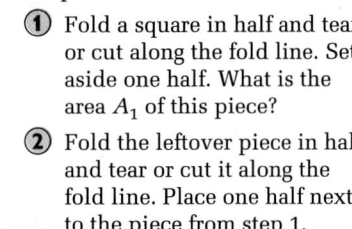

TEACHING

Exploration

The goal of the Exploration is to use a manipulative activity to lead students to discover the rule for finding the sum of an infinite series.

Additional Sample

S1 Tell whether each series has a sum. If it does, give its value.

a. $1 + (-1) + 2 + (-2) + \ldots$

This series is neither geometric nor arithmetic.

$S_1 = 1$
$S_2 = 0$
$S_3 = 2$
$S_4 = 0$
$S_5 = 3$
$S_6 = 0$

Graph the partial sums.

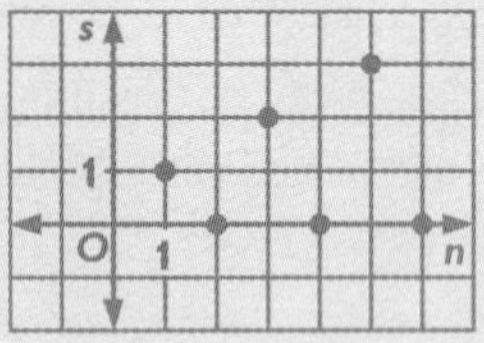

The series does not have a sum.

b. $1 + 1.1 + 1.21 + 1.331 + \ldots$

This is a geometric series with common ratio 1.1.

$S_1 = 1$
$S_2 = 2.1$
$S_3 = 3.31$
$S_4 = 4.641$
$S_5 = 6.1051$
$S_6 = 7.71561$

Graph the partial sums.

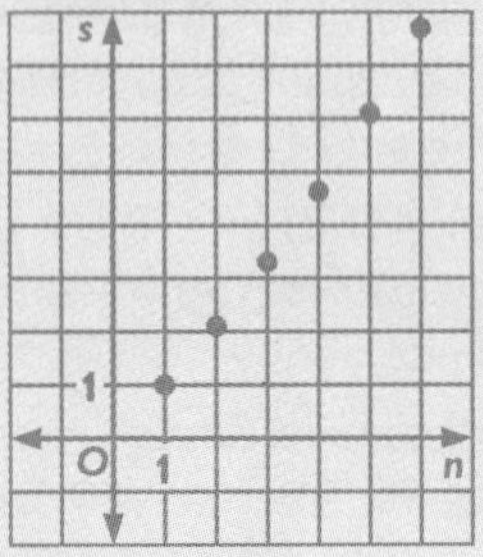

The series does not have a sum.

The sum of the first n terms of an infinite series is called a **partial sum** of the series. In the Exploration, the combined areas $\frac{1}{2}, \left(\frac{1}{2} + \frac{1}{4}\right), \left(\frac{1}{2} + \frac{1}{4} + \frac{1}{8}\right), \cdots$ form a sequence of partial sums for the infinite series $\frac{1}{2} + \frac{1}{4} + \frac{1}{8} + \cdots$ that represents the sum of the areas of the pieces put aside.

SUM OF AN INFINITE SERIES

If the sequence of partial sums of an infinite series has a limit, then that limit is the sum of the series.

Sample 1

Tell whether each series appears to have a sum. If it does, give its value.

a. $3 + (-1.5) + 0.75 + (-0.375) + \cdots$

b. $0.5 + (-1) + 2 + (-4) + \cdots$

c. $2 + 5 + 8 + 11 + \cdots$

Sample Response

Write several terms. Find the first few partial sums to see whether there is a trend.

a. This is a geometric series. The common ratio is -0.5.

$3 + (-1.5) + 0.75 + (-0.375) + 0.1875 + (-0.09375) + \cdots$

$S_1 = 3$

$S_2 = 3 + (-1.5) = 1.5$

$S_3 = 1.5 + 0.75 = 2.25$

$S_4 = 2.25 + (-0.375) = 1.875$

$S_5 = 1.875 + 0.1875 = 2.0625$

$S_6 = 2.0625 + (-0.09375) = 1.96875$

Each partial sum after the first is the sum of the previous partial sum and the next term in the series.

Graph the partial sums.

The limit of the sequence of partial sums appears to be 2.

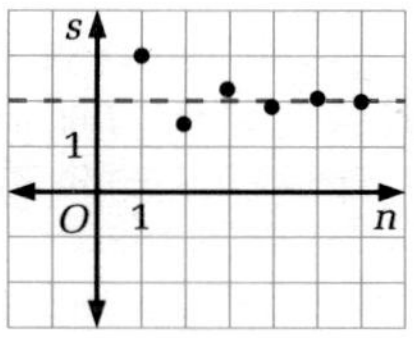

The sum of the series appears to be 2.

b. This is a geometric series. The common ratio is -2.

$0.5 + (-1) + 2 + (-4) + 8 + (-16) + \cdots$

$S_1 = 0.5$

$S_2 = 0.5 + (-1) = -0.5$

$S_3 = -0.5 + 2 = 1.5$

$S_4 = 1.5 + (-4) = -2.5$

$S_5 = -2.5 + 8 = 5.5$

$S_6 = 5.5 + (-16) = -10.5$

The series does not have a sum.

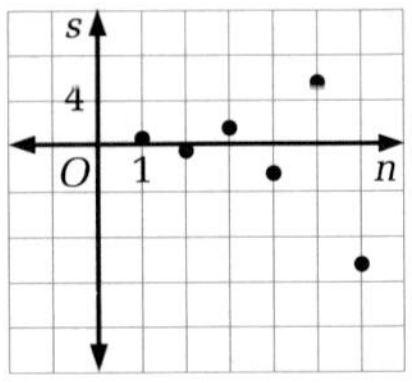

The partial sums do not get close to any number.

c. This is an arithmetic series. The common difference is 3.

$2 + 5 + 8 + 11 + 14 + 17 + \cdots$

$S_1 = 2$

$S_2 = 2 + 5 = 7$

$S_3 = 7 + 8 = 15$

$S_4 = 15 + 11 = 26$

$S_5 = 26 + 14 = 40$

$S_6 = 40 + 17 = 57$

The series does not have a sum.

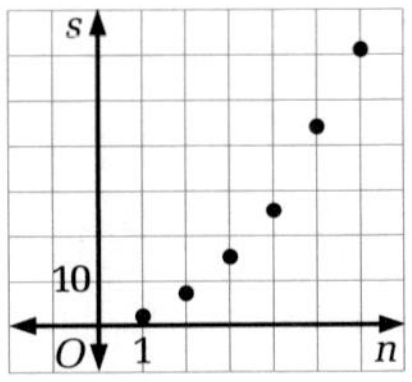

The partial sums increase and do not get close to any number.

Talk it Over

1. Does the method for finding the partial sums of the three series in Sample 1 involve recursion? Explain.
2. a. In part (c) of Sample 1, the arithmetic series with a positive common difference did not have a sum. Write any infinite arithmetic series with a negative common difference. Does it appear to have a sum?

   b. Do you think every infinite arithmetic series has a sum? Why or why not?

Now you are ready for: **Exs. 1–10 on pp. 255–256**

Infinite Geometric Series

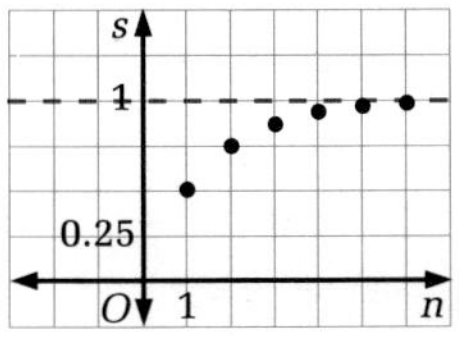

This graph shows that the series in the Exploration with $r = 0.5$ appears to have a sum.

The graph in part (a) of Sample 1 shows that a geometric series with $r = -0.5$ appears to have a sum.

In both of these cases, $|r| < 1$.

c. $4 + 0.8 + 0.16 + 0.032 + 0.0064 + \ldots$

This is a geometric series with common ratio 0.2.

$S_1 = 4$

$S_2 = 4.8$

$S_3 = 4.96$

$S_4 = 4.992$

$S_5 = 4.9984$

$S_6 = 4.99968$

Graph the partial sums.

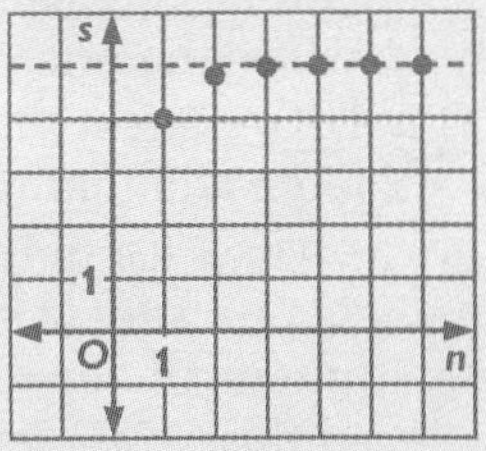

The sum of the series appears to be 5.

Talk it Over

Question 2 asks students to expand upon part (c) of Sample 1 and then to make a conjecture about whether an infinite arithmetic series will have a sum.

Answers to Talk it Over

1. Yes; each partial sum S_n is defined in terms of S_{n-1}.
2. a. Answers may vary. An example is given. For the sequence 1, 0, –1, –2, –3, ..., the common difference is –1 and the sequence of partial sums is 1, 1, 0, –2, –5, ... The series has no sum. (This will be true for any such example.)

   b. No. No arithmetic sequence with $d \neq 0$ will have a sum. The sequence of partial sums will increase (if $d > 0$) or decrease (if $d < 0$) without limit.

Error Analysis

Students may use the word *sequence* when they should use the word *series.* This is a comparatively harmless error when the sequence or series is finite. It can be a serious error when the sequence and related series are infinite. It is entirely possible for a sequence to have a limit, while the related series does not. To correct the error, ask students whether the situation involves a *list* of terms (a sequence) or a *sum* (a series). After students understand why they used the wrong term, ask them to restate what they said earlier using the correct term.

Talk it Over

Question 3 asks students to apply the formula for the sum of an infinite geometric series. Questions 4 leads students to think about whether or not any infinite geometric series with $|r| > 1$ has a sum.

Additional Sample

S2 Decide whether the series $3 + (-2) + \frac{4}{3} + \left(-\frac{8}{9}\right) + \ldots$ has a sum. If it does, give its value.

Step 1. The series is geometric with $r = -\frac{2}{3}$. Since $\left|-\frac{2}{3}\right| < 1$, the series has a sum.

Step 2. Use the formula for the sum of an infinite geometric series.

$$S = \frac{a_1}{1-r} = \frac{3}{1-\left(-\frac{2}{3}\right)} = \frac{3}{\frac{5}{3}} = \frac{9}{5}$$

The sum of the series is $\frac{9}{5}$.

For any *infinite* geometric series with $|r| < 1$, the terms get closer and closer to zero. This suggests substituting zero for a_n in the formula for the sum of a finite geometric series.

$$S = \frac{a_1 - a_n r}{1 - r}$$

$$= \frac{a_1 - 0 \cdot r}{1 - r} \quad \longleftarrow \text{Substitute 0 for } a_n.$$

$$= \frac{a_1}{1 - r}$$

SUM OF AN INFINITE GEOMETRIC SERIES

The general formula for the sum S of an infinite geometric series $a_1 + a_2 + \cdots$ with common ratio r where $|r| < 1$ is

$$\mathbf{S = \frac{a_1}{1 - r}}.$$

Talk it Over

3. Find the sum of the series in part (a) of Sample 1 by using the formula. Did you get the same answer as before?
4. a. In part (b) of Sample 1, the infinite geometric series with $r < -1$ did not have a sum. Write any infinite geometric series with $r > 1$. Does your series appear to have a sum? Explain.
   b. Do you think every infinite geometric series with $|r| > 1$ has a sum?

Sample 2

Decide whether the series $4 + \frac{4}{5} + \frac{4}{25} + \frac{4}{125} + \cdots$ has a sum. If it does, give its value.

Sample Response

Step 1 The series is geometric with $r = \frac{1}{5}$. Since $\left|\frac{1}{5}\right| < 1$, the series has a sum.

Step 2 $S = \frac{a_1}{1 - r}$ ← Use the formula for the sum of an infinite geometric series.

$= \frac{4}{1 - \frac{1}{5}}$ ← Substitute 4 for a_1 and $\frac{1}{5}$ for r.

$= 5$ ← Simplify.

The sum of the series is 5.

Answers to Talk it Over

3. $\frac{3}{1-\left(-\frac{1}{2}\right)} = 2$; Yes.
4. a. Answers may vary. An example is given. For the sequence 2, 4, 8, 16, …, the common ratio is 2 and the sequence of partial sums is 2, 6, 14, 30, … . The series has no sum. (This will be true for any such example.)
   b. No. No geometric series with $|r| > 1$ will have a sum. The sequence of partial sums behaves as did the one in Sample 1(b) if $r < -1$ and increases without limit if $r > 1$.

Answers to Look Back

Determine whether the sequence of partial sums appears to have a limit. If it does, that limit is the sum of the series.

To write an infinite series in sigma notation, use the symbol ∞ for "infinity." In sigma notation, the infinite series

$$\left(\frac{1}{2}\right)^1 + \left(\frac{1}{2}\right)^2 + \left(\frac{1}{2}\right)^3 + \left(\frac{1}{2}\right)^4 + \cdots$$

looks like this:

$$\sum_{n=1}^{\infty} \left(\frac{1}{2}\right)^n$$ ← Read: "The sum of $\left(\frac{1}{2}\right)^n$ for integer values of n from one to infinity."

Sample 3

Evaluate $\sum_{n=1}^{\infty} (-0.25)^{n-1}$ if the sum exists.

Sample Response

Step 1 To see whether the sum exists, write the series in expanded form.

$$\sum_{n=1}^{\infty} (-0.25)^{n-1} = (-0.25)^0 + (-0.25)^1 + (-0.25)^2 + (-0.25)^3 + \cdots$$

The series is geometric with $r = -0.25$. Since $|-0.25| < 1$, the series has a sum.

Step 2 Find the sum.

$$S = \frac{a_1}{1-r}$$ ← Use the formula for the sum of an infinite geometric series.

$$= \frac{1}{1-(-0.25)}$$ ← Substitute **1** for a_1 and **−0.25** for r.

$$= \frac{1}{1.25} = 0.8$$

$$\sum_{n=1}^{\infty} (-0.25)^{n-1} = 0.8.$$

Look Back

How can you use partial sums to find the sum of an infinite series?

Now you are ready for: Exs. 11–33 on pp. 257–258

4-7 Exercises and Problems

For each series in Exercises 1–4:

a. Graph the first six partial sums.

b. Decide whether the series appears to have a sum. If it does, give its value.

1. $8 + 1 + (-6) + (-13) + \cdots$
2. $\frac{8}{5} + \frac{8}{25} + \frac{8}{125} + \frac{8}{625} + \cdots$
3. $5 + (-7.5) + 11.25 + (-16.875) + \cdots$
4. $(-6) + 3 + \left(-\frac{3}{2}\right) + \frac{3}{4} + \cdots$

Additional Sample

S3 Evaluate $\sum_{n=1}^{\infty} (0.4)^{n-1}$ if the sum exists.

Step 1. **Write the series in expanded form.**

$\sum_{n=1}^{\infty} (0.4)^{n-1}$
$= (0.4)^0 + (0.4)^1 + (0.4)^2 + (0.4)^3 + \ldots$

The series is geometric with $r = 0.4$. Since $|0.4| < 1$, the series has a sum.

Step 2. **Find the sum. Use the formula for the sum of an infinite geometric series.**

$S = \frac{a_1}{1-r}$
$= \frac{1}{1-0.4}$
$= \frac{5}{3}$

$\sum_{n=1}^{\infty} (0.4)^{n-1} = \frac{5}{3}.$

APPLYING

Suggested Assignment

Day 1
Standard 1–10
Extended 1–10

Day 2
Standard 11–23, 25, 27–33
Extended 11–33

Integrating the Strands

Number Exs. 1–30
Algebra Exs. 1–30
Functions Ex. 32
Geometry Exs. 10, 26, 31
Discrete Mathematics Exs. 1–30, 33
Logic and Language Exs. 9, 28, 32, 33

Answers to Exercises and Problems

1. a.

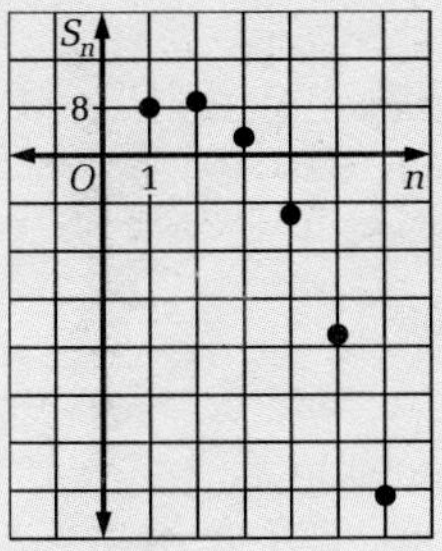

b. No.

2. See answers in back of book.

3. a.

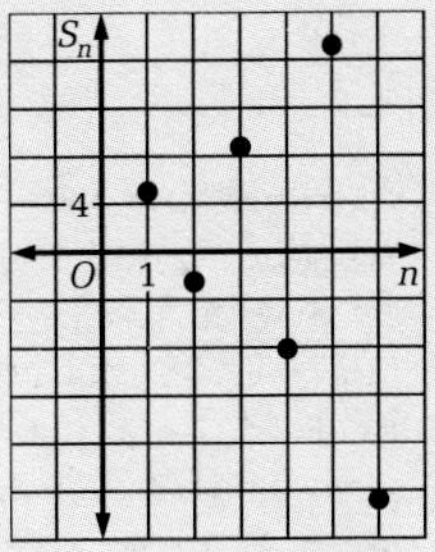

b. No.

4. a.

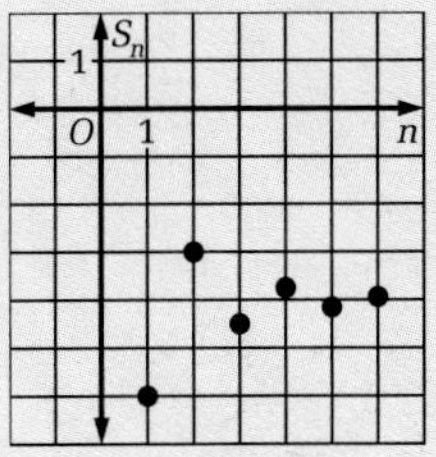

b. Yes; –4.

Interdisciplinary Problems

Sequences and series, both finite and infinite, can be found in a variety of mathematical and real-world situations. Exs. 9 and 10 relate sequences and fractals. Ex. 24 relates infinite geometric series to repeating decimals. Ex. 25 relates an infinite series to a physics problem. Ex. 26 makes a connection to geometry.

Decide whether each series appears to have a sum. If it does, give its value.

5. $12 + (-6) + 3 + (-1.5) + \cdots$

6. $(-4) + 0 + 4 + 8 + \cdots$

7. $\frac{2}{3} + (-2) + 6 + (-18) + \cdots$

8. $9 + \frac{9}{4} + \frac{9}{16} + \frac{9}{64} + \cdots$

9. Suppose the branches in each stage of a fractal tree are half the length of the branches in the previous stage.

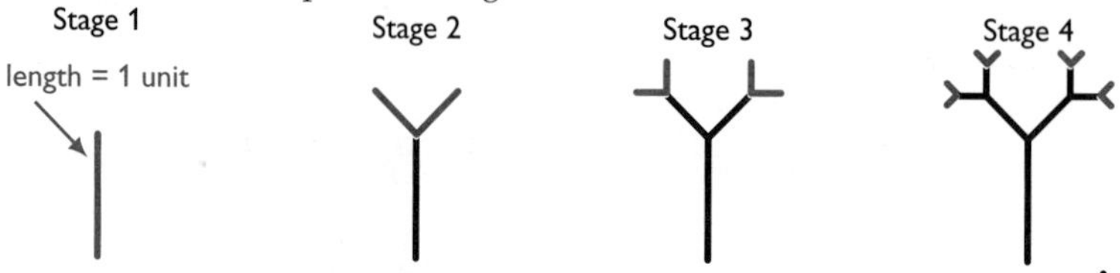

a. Write a sequence of partial sums for the total length of one path of the fractal tree.

b. Decide whether the sequence in part (a) appears to have a limit. If it does, give its value.

c. Write a sequence of partial sums for the total length of all the branches of a fractal tree, if the length of the first branch is 1.

d. Decide whether the sequence in part (c) appears to have a limit. If it does, give its value.

e. **Writing** Explain how the length of a path of the tree can be finite, when the length of all branches of the tree is infinite.

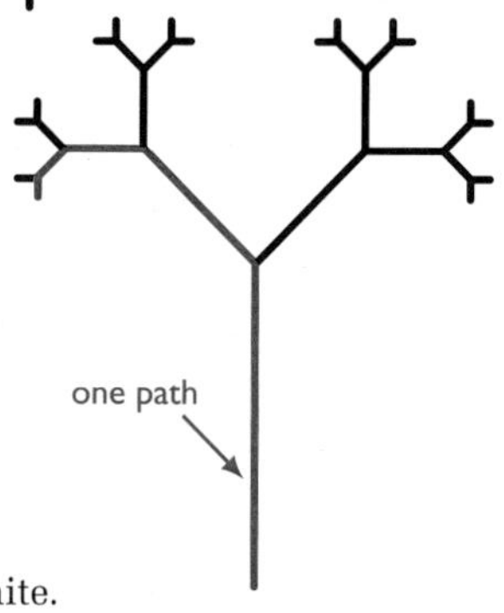

10. Another example of a fractal is Sierpinski's triangle. Start with a triangle with area 1. In Stage 1, the triangle is divided into four congruent triangles by connecting the midpoints of the sides, and the center triangle is removed. In Stage 2, the three remaining triangles are divided as before and each center triangle is removed. At each successive stage, the remaining triangles are divided as before and each center triangle is removed.

a. Write a sequence of partial sums for the area of the removed triangles.

b. What limit does the sequence appear to have, if any?

Answers to Exercises and Problems

5. Yes; 8.
6. No.
7. No.
8. Yes; 12.
9. a. $1, \frac{3}{2}, \frac{7}{4}, \frac{15}{8}, \ldots$
   b. Yes; 2.
   c. 1, 2, 3, 4, ...
   d. No.
   e. There are an infinite number of branches.
10. a. $\frac{1}{4}, \frac{7}{16}, \frac{37}{64}, \frac{175}{256}, \ldots$
   b. 1
11. those for which $|r| < 1$
12. Yes; $-\frac{64}{7} \approx -9.14$.
13. Yes; 40.5.
14. No.
15. Yes; –2.
16. Yes; 100.
17. No.
18. 1.5
19. 66
20. no sum
21. no sum
22. about -1×10^{-12}
23. 12.5

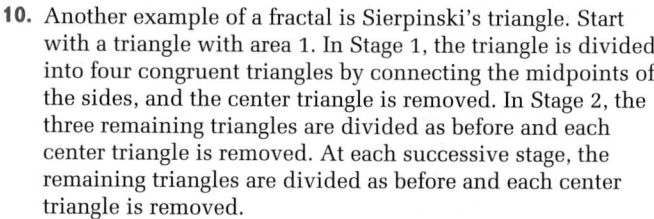

11. **Reading** Which geometric series have sums?

Tell whether each series has a sum. If it does, give its value.

12. $(-16) + 12 + (-9) + 6\frac{3}{4} + \cdots$

13. $13\frac{1}{2} + 9 + 6 + 4 + \cdots$

14. $0.5 + 3.5 + 6.5 + 9.5 + \cdots$

15. $(-1) + (-0.5) + (-0.25) + (-0.125) + \cdots$

16. $90 + 9 + 0.9 + 0.09 + \cdots$

17. $\sqrt{2} + 2 + 2\sqrt{2} + 4 + \cdots$

Evaluate each series, if the sum exists.

18. $\sum_{n=1}^{\infty} \left(\frac{1}{3}\right)^{n-1}$

19. $\sum_{t=0}^{\infty} 12\left(\frac{9}{11}\right)^{t}$

20. $\sum_{n=1}^{\infty} \left(\frac{1}{2} + 2n\right)$

21. $\sum_{n=1}^{\infty} 17\left(\frac{5}{4}\right)^{n-1}$

22. $\sum_{k=4}^{\infty} -(0.001)^{k}$

23. $\sum_{t=0}^{\infty} 10(0.2)^{t}$

24. The repeating decimal 0.3333 . . . can be expressed as the infinite geometric series: $0.333333\ldots = 0.3 + 0.03 + 0.003 + 0.0003 + \cdots$.
   a. Find the sum of the series. Give your answer as a fraction.
   b. What fraction is equal to 0.454545 . . . ?
   c. Show that $0.999999\ldots = 1$.

connection to **PHYSICS**

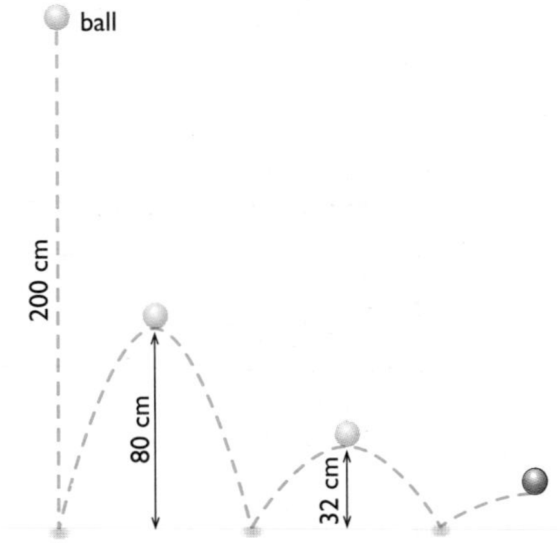

25. Use the diagram. Suppose a ball dropped from a height of 200 cm will bounce forever. Each bounce height is 0.4 of the height of the bounce before it.
   a. Write a series for the vertical distance the ball travels only when it is *falling*.
   b. Write a series for the vertical distance the ball travels only when it is *bouncing upward*.
   c. Find the total vertical distance the ball travels.

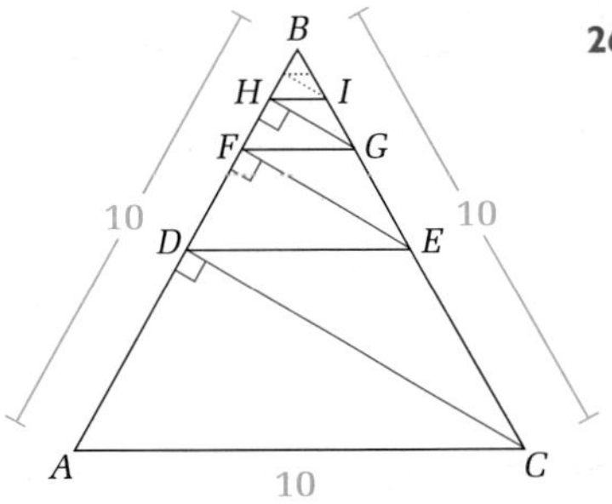

For Exercise 26, use these theorems and the diagram.

If a segment joins the midpoints of two sides of a triangle, then the segment is parallel to the third side and half as long. A median of an equilateral triangle is also an altitude.

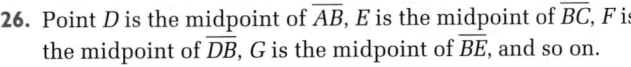

26. Point D is the midpoint of $\overline{AB}$, E is the midpoint of $\overline{BC}$, F is the midpoint of $\overline{DB}$, G is the midpoint of $\overline{BE}$, and so on.
   a. Find the sum of the lengths of the horizontal segments of the zigzag line from C to D to E to F to G to H, and so on.
   b. Find the sum of the lengths of the diagonal segments of the zigzag line.
   c. What is the length of the zigzag line?
   d. What is the length of the zigzag line for an equilateral triangle with sides of length 20 units? 30 units? n units?

Using Technology

The Using Technology note in Section 4-5 on page 241 explains how to use the TI-82 to find the sum for a series of 99 terms or less. If you wish to find a sum for more than 99 terms, the calculation can be separated into parts. For example, to find the partial sum for the first 150 terms of $\sum_{n=1}^{\infty}(0.5)^n$, enter the expression sum seq (.5^A,A,1,99,1) on the home screen. When you press [ENTER], the calculator displays the sum of the first 99 terms. Write down the answer. Then press [2nd] [ENTRY] and edit the previous expression to read sum seq (.5^A,A,100,150,1). When you press [ENTER], the calculator will display the sum of terms 100 through 150. Add the previous sum to this result to get the sum of the first 150 terms.

Assessment: Standard

For Ex. 26, suppose the sides of the equilateral triangle measure 1 unit. Ask students to find the sum of the lengths of the segments joining the midpoints of each successive triangle.

Answers to Exercises and Problems

24. a. $\frac{1}{3}$
   b. $\frac{45}{99} = \frac{5}{11}$
   c. $0.999999\ldots = \sum_{n=1}^{\infty} 9(0.1)^n = \frac{0.9}{1-0.1} = 1$

25. a. $\sum_{n=0}^{\infty} 200(0.4)^n$ or $\sum_{n=1}^{\infty} 200(0.4)^{n-1}$
   b. $\sum_{n=0}^{\infty} 80(0.4)^n$ or $\sum_{n=1}^{\infty} 80(0.4)^{n-1}$
   c. $466.\overline{6}$ cm

26. a. 10 units
   b. $10\sqrt{3}$ units
   c. $10 + 10\sqrt{3}$ units
   d. $20 + 20\sqrt{3}$ units; $30 + 30\sqrt{3}$ units; $n + n\sqrt{3}$ units

Quick Quiz (4-5 through 4-7)

See page 260.

See also Test 16 in the *Assessment Book.*

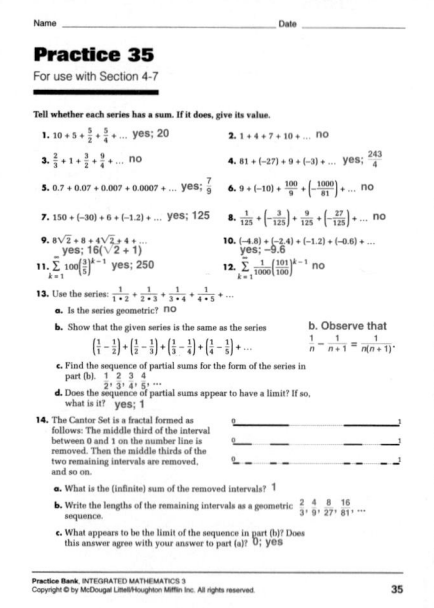

Practice 35 For use with Section 4-7

Name ______________________ Date ____________

Practice 35

For use with Section 4-7

Tell whether each series has a sum. If it does, give its value.

1. $10 + 5 + \frac{5}{2} + \frac{5}{4} + \ldots$ yes; 20
2. $1 + 4 + 7 + 10 + \ldots$ no
3. $\frac{2}{3} + 1 + \frac{3}{2} + \frac{9}{4} + \ldots$ no
4. $81 + (-27) + 9 + (-3) + \ldots$ yes; $\frac{243}{4}$
5. $0.7 + 0.07 + 0.007 + 0.0007 + \ldots$ yes; $\frac{7}{9}$
6. $9 + (-10) + \frac{100}{9} + \left(-\frac{1000}{81}\right) + \ldots$ no
7. $150 + (-30) + 6 + (-1.2) + \ldots$ yes; 125
8. $\frac{1}{125} + \left(-\frac{3}{125}\right) + \frac{9}{125} + \left(-\frac{27}{125}\right) + \ldots$ no
9. $8\sqrt{2} + 8 + 4\sqrt{2} + 4 + \ldots$ yes; $16(\sqrt{2} + 1)$
10. $(-4.8) + (-2.4) + (-1.2) + (-0.6) + \ldots$ yes; -9.6
11. $\sum_{k=1}^{\infty} 100\left(\frac{3}{5}\right)^{k-1}$ yes; 250
12. $\sum_{k=1}^{\infty} \frac{1}{1000}\left(\frac{101}{100}\right)^{k-1}$ no
13. Use the series: $\frac{1}{1 \cdot 2} + \frac{1}{2 \cdot 3} + \frac{1}{3 \cdot 4} + \frac{1}{4 \cdot 5} + \ldots$
    a. Is the series geometric? no
    b. Show that the given series is the same as the series $\left(\frac{1}{1} - \frac{1}{2}\right) + \left(\frac{1}{2} - \frac{1}{3}\right) + \left(\frac{1}{3} - \frac{1}{4}\right) + \left(\frac{1}{4} - \frac{1}{5}\right) + \ldots$ b. Observe that $\frac{1}{n} - \frac{1}{n+1} = \frac{1}{n(n+1)}$.
    c. Find the sequence of partial sums for the form of the series in part (b). $\frac{1}{2}, \frac{2}{3}, \frac{3}{4}, \frac{4}{5}, \ldots$
    d. Does the sequence of partial sums appear to have a limit? If so, what is it? yes; 1
14. The Cantor Set is a fractal formed as follows: The middle third of the interval between 0 and 1 on the number line is removed. Then the middle thirds of the two remaining intervals are removed, and so on.

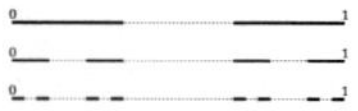

    a. What is the (infinite) sum of the removed intervals? 1
    b. Write the lengths of the remaining intervals as a geometric sequence. $\frac{2}{3}, \frac{4}{9}, \frac{8}{27}, \frac{16}{81}, \ldots$
    c. What appears to be the limit of the sequence in part (b)? Does this answer agree with your answer to part (a)? 0; yes

Practice Bank, INTEGRATED MATHEMATICS 3

35

27. Suppose the bob of a pendulum travels 90% as far on each swing as on the previous swing and will swing forever. When the bob travels 50 cm on the first swing, what is the total distance the bob will swing?

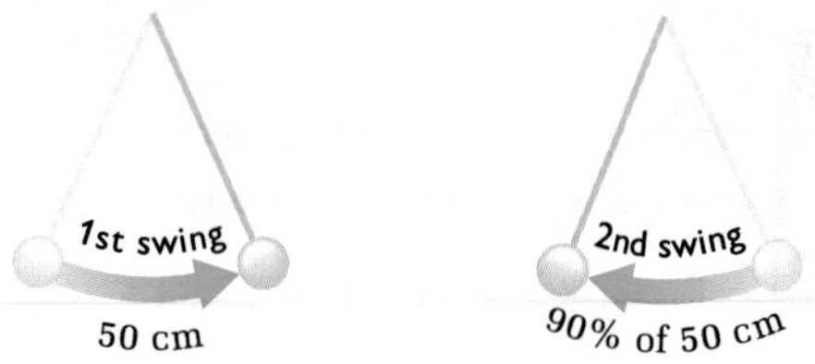

Ongoing **ASSESSMENT**

28. Use this true statement: If an infinite sequence does not have a limit, then the related series does not have a sum.

a. Does the infinite sequence $\frac{1}{2}, \frac{2}{3}, \frac{3}{4}, \frac{4}{5}, \cdots$ appear to have a limit? If so, what is the value?

b. Does the series $\frac{1}{2} + \frac{2}{3} + \frac{3}{4} + \frac{4}{5} + \cdots$ appear to have a sum? Explain.

c. **Writing** What is the inverse of the original statement? Is it also true? Explain.

Review **PREVIEW**

Find the sum of each series. *(Section 4-6)*

29. $500 + 50 + 5 + 0.5 + \cdots + 0.00005$

30. $0.2 + 0.4 + 0.8 + 1.6 + \cdots + 25.6$

31. A circle has radius 4 cm. Find the length of the longest chord of the circle. Explain your reasoning. *(Section 3-5)*

32. a. **Writing** Explain why this graph represents a function. *(Section 2-1)*

b. Find the domain and range of the function.

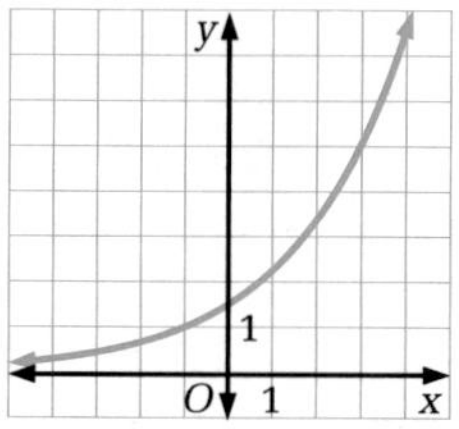

Working on the Unit Project

33. a. **Research** Find these things: the current rate for student and personal loans, limitations on amounts borrowed, and the terms of the repayment agreements. Summarize your findings.

b. Using the information from part (a), set up a spreadsheet showing how to finance the cost of a university education 25 years from now through both student and personal loans without any savings. Show a repayment plan spread out over ten years.

c. **Writing** Discuss other options for making a four-year university education affordable in 25 years. Be as specific as you can about the amount of savings you are suggesting.

Answers to Exercises and Problems

27. 500 cm

28. a. Yes; 1.

b. No. As n increases, S_n increases without limit.

c. If an infinite sequence appears to have a limit, then the related series has a sum. No; the series in this exercise provides a counterexample.

29. 555.55555

30. 51

31. 8 cm; The longest chord of a circle is a diameter. The length of a diameter is twice the length of a radius.

32. a. No vertical line intersects the graph in more than one point.

b. all real numbers; all positive real numbers

33. a–c. Answers may vary. Check students' work.

Unit Project 4

Completing the Unit Project

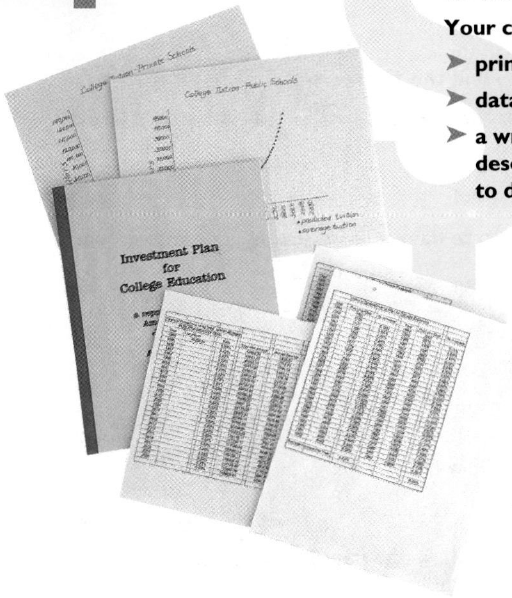

Now you are ready to complete your savings plan for the future.

Your completed plan should include these things:

- **printouts of your spreadsheet calculations**
- **data displays based your spreadsheets**
- **a written report summarizing your results and describing how you used sequences and series to develop them**

Look Back

Did any of your results surprise you? Why or why not?

Alternative Projects

Project 1: **Sequences and Series in Everyday Things**

Look for sequences and series in the world around you. Include at least two examples each from architecture, science, music, art, and business. Include some progressions in time and some in space. Find some sequences and series that are visual and some that can only be heard. In each case, tell whether the progression is a *sequence* or a *series,* whether it is *arithmetic, geometric,* or *neither,* and whether it is *finite* or *infinite.* For infinite progressions, tell whether sequences appear to have limits and whether series have sums.

Project 2: **Fractal Patterns**

Research fractal shapes, such the simple fractal tree in Section 4-3, the Koch snowflake (see Exercise 18 of Section 4-3) and other Koch curves, the Sierpinski triangle (see Exercise 10 in Section 4-7), the Sierpinski carpet, and Peano curves. Describe how each fractal shape is generated and find as many related sequences and series as you can. Analyze the sequences and series you discover.

Assessment

A scoring rubric for the Unit Project can be found on pages 204 and 205 of this Teacher's Edition and also in the ***Project Book.***

Unit Support Materials

- *Practice Bank:* Cumulative Practice 36
- *Study Guide:* Unit 4 Review
- *Problem Bank:* Unifying Problem 4
- *Assessment Book:* Unit Tests 17 and 18, Spanish Unit Tests, Alternative Assessment
- *Test Generator* software with Test Bank
- *Teacher's Resources for Transfer Students*

Quick Quiz (4-5 through 4-7)

1. Find the sum of the series $5 + 10 + 15 + 20 + \ldots + 100$. [4-5] **1050**
2. Evaluate. [4-5] $\sum_{n=1}^{5}(5n-2)$ **65**
3. Find the sum of the series $10 + 20 + 40 + \ldots + 2560$. [4-6] **5110**
4. Evaluate. [4-6] $\sum_{n=1}^{5}10(3)^n$ **3630**

Find the sum of the infinite geometric series, if it exists. [4-7]

5. $0.1 + 0.02 + 0.004 + \ldots$ **0.125**
6. $8 - 20 + 50 - \ldots$ **This series does not have a sum.**

For each sequence in Questions 1–3: 4-1

a. Find the next three terms.

b. Find the 10th term.

c. Writing Describe the method you used to find the 10th term.

1. $1, 4, 13, 40, \ldots$
2. $3, 8, 18, 38, \ldots$
3. AA, AB, AC, AD, . . .

For each sequence in Questions 4–6:

a. Graph the sequence. Plot at least six points.

b. Writing Tell whether the sequence appears to have a limit. Justify your answer.

4. $\frac{1}{3}, \frac{3}{5}, \frac{5}{7}, \frac{7}{9}, \ldots$
5. $7, 14, 21, 28, \ldots$
6. $7, \left(-\frac{7}{2}\right), \frac{7}{4}, \left(-\frac{7}{8}\right), \ldots$

Write an explicit formula for each sequence. 4-2

7. $9, 3, 1, \frac{1}{3}, \ldots$
8. $\frac{1}{3}, \frac{1}{4}, \frac{1}{5}, \frac{1}{6}, \ldots$

Find the first four terms and the 17th term of each sequence.

9. $b_n = |n - 3|$
10. $d_n = \frac{18 - n}{n}$
11. Moira paid for a car stereo on an installment plan. She paid a \$200 down payment, and then paid \$30 per month for 12 months.
    a. Write an explicit formula for the sequence of the total amount she has paid at the end of n months.
    b. The stereo cost \$500. How much extra did Moira pay in interest?

Write the first five terms. 4-3

12. $a_1 = 3$, $a_n = -(a_{n-1} - 7)$
13. $a_1 = 6$, $a_n = (-0.6)a_{n-1}$
14. Write a recursive formula for the sequence 11, 24, 37, 50, ...
15. **Archery** An archery target has a center, or bull's-eye, with a radius of 4 cm, and nine rings that are each 4 cm wide.
    a. Find the sequence of the areas of the rings.
    b. Write a recursive formula and an explicit formula for this sequence.
    c. Find the area of the ninth ring.

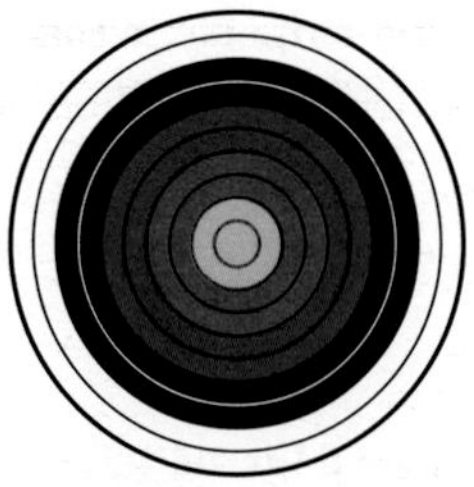

Answers to Unit 4 Review and Assessment

1. a. 121, 364, 1093
   b. 29,524
   c. Answers may vary. An example is given. I noticed that the terms increase by powers of 3. I continued the pattern to find the 10th term.
2. a. 78, 158, 318
   b. 2558
   c. Answers may vary. An example is given. I noticed that the second term is 5 more than the first, and each increase after that is twice the previous increase. I continued the pattern to find the 10th term.
3. a. AE, AF, AG
   b. AJ
   c. Answers may vary. An example is given. I noticed that the terms follow a pattern. The first letter is always A. The second letters are in alphabetical order. The tenth letter of the alphabet is J.
4. a.

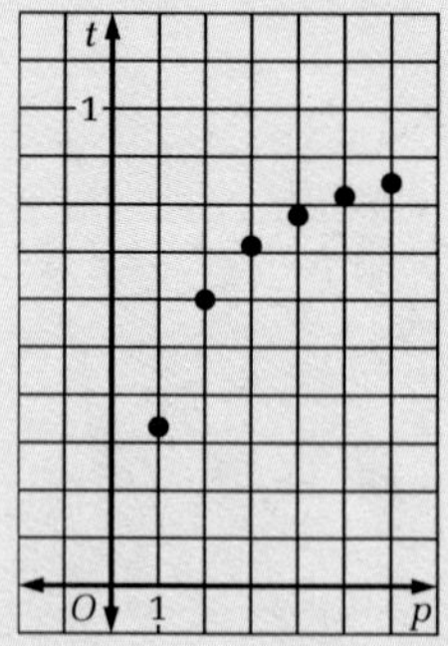

   b. Yes; the terms appear to approach 1.
5. a.

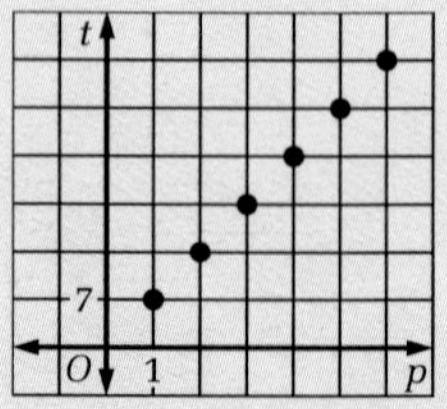

   b. No. The terms appear to increase without limit.

For each sequence in Questions 16–18: 4-4

a. Tell whether the sequence is *arithmetic*, *geometric*, or *neither*.

b. Write a recursive formula.

16. 10, 21, 43, 87, . . .

17. $\frac{2}{3}, \frac{4}{9}, \frac{16}{81}, \frac{256}{6561}, \ldots$

18. $\frac{1}{8}, \frac{1}{2}, \frac{7}{8}, 1\frac{1}{4}, \ldots$

19. Find the sum of the series 8 + 11 + 14 + 17 + · · · + 98. 4-5

20. Evaluate $\sum_{n=1}^{7} 2n + 4$.

21. Write the series 95 + 90 + 85 + 80 + · · · + 0 in sigma notation.

22. **Writing** Draw a concept map that summarizes what you have learned in this unit. Your map should include the relationships among a series, its related sequence, the sequence of partial sums of the series, and the sum of the series.

23. Find the sum of the series 300 + 30 + 3 + 0.3 + · · · + 0.00003. 4-6

24. Evaluate $\sum_{k=0}^{8} 4\left(\frac{1}{4}\right)^k$.

For each series in Questions 25 and 26: 4-7

a. Graph the first six partial sums.

b. Tell whether each series appears to have a sum. If it does, give its value.

25. 500 + 100 + 20 + 4 + · · ·

26. 0.003 + 0.03 + 0.3 + 3 + · · ·

Evaluate each series, if the sum exists.

27. $\sum_{n=1}^{\infty} 11\left(\frac{3}{2}\right)^{n-1}$

28. $\sum_{t=0}^{\infty} -3\left(\frac{5}{8}\right)^t$

29. **Open-ended** Write two infinite geometric series, only one of which has a sum.

30. **Self-evaluation** Do you think it is easier to find an explicit formula or a recursive formula for a sequence? Do you find one type of formula more useful than the other? Explain.

31. **Group Activity** Work with another student.

a. One of you should write four terms of an arithmetic sequence.

b. The other should write an explicit formula for an arithmetic sequence.

c. Trade papers. The student who wrote the terms should now write the first four terms of the sequence defined by the other student's formula. The student who wrote the formula should write a formula for the other student's sequence.

d. Repeat parts (a)–(c) with a geometric sequence.

Answers to Unit 4 Review and Assessment

6. a.

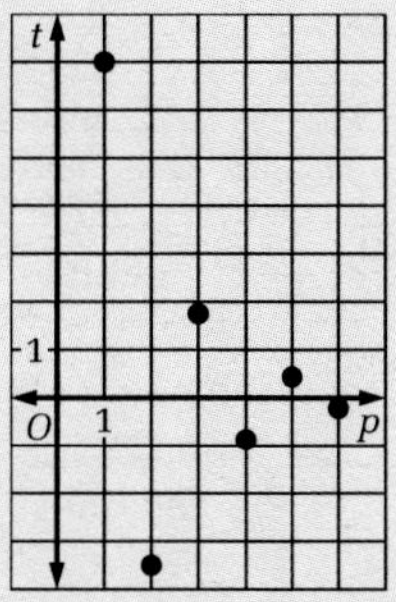

b. Yes. the terms appear to approach 0.

7. $a_n = 3^{3-n}$

8. $a_n = \frac{1}{n+2}$

9. 2, 1, 0, 1; 14

10. 17, 8, 5, $\frac{7}{2}$; $\frac{1}{17}$

11. a. $a_n = 200 + 30n$

b. $60

12. 3, 4, 3, 4, 3

13. 6, −3.6, 2.16, −1.296, 0.7776

14. $a_1 = 11$; $a_n = a_{n-1} + 13$

15. a. $48\pi, 80\pi, 112\pi, 144\pi, 176\pi, 208\pi, 240\pi, 272\pi, 304\pi$

b. recursive: $a_1 = 48\pi$, $a_n = a_{n-1} + 32\pi$; explicit: $a_n = 48\pi + 32\pi(n-1)$

c. 304π cm^2 $\approx$ 955 cm^2

16. a. neither

b. $a_1 = 10$; $a_n = 2a_{n-1} + 1$

17. a. neither

b. $a_1 = \frac{2}{3}$; $a_n = (a_{n-1})^2$

18. a. arithmetic

b. $a_1 = \frac{1}{8}$; $a_n = a_{n-1} + \frac{3}{8}$

19. 1643

20. 84

21. $\sum_{n=1}^{20}(100 - 5n)$

22. See answer on next page.

23. 333.33333

24. about 5.33

25. a.

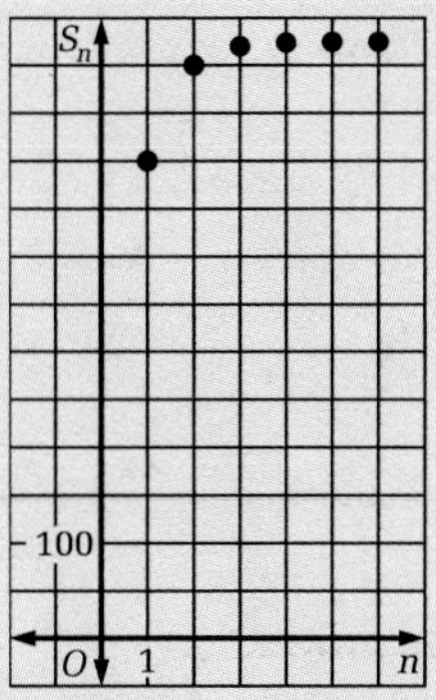

b. Yes; 625.

26. a.

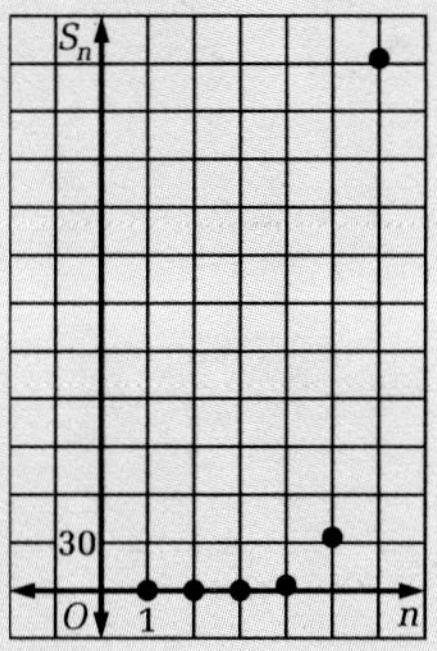

b. No.

27. no sum

28. −8

29. Answers may vary. Examples are given.

$\sum_{n=1}^{\infty}\left(\frac{1}{2}\right)^n$ has a sum. (The sum is 1.)

$\sum_{n=1}^{\infty} 2^n$ has no sum.

30. Answers may vary. Examples are given. I usually find it easier to find a recursive formula. Both types are useful. In finding sums, the explicit formula is more useful.

31. a–d. Check students' work.

IDEAS AND (FORMULAS)

- Three dots (· · ·) show that some terms are not shown but the sequence or series continues in the same pattern. *(p. 208)*
- The general explicit formula for an arithmetic sequence where d is the common difference is

$$a_n = a_1 + (n - 1)d. \text{ (p. 230)}$$

- The general explicit formula for a geometric sequence with common ratio r is

$$a_n = a_1 r^{n-1}. \text{ (p. 230)}$$

- The formula for the sum S of a finite arithmetic series with n terms is

$$S = \frac{n(a_1 + a_n)}{2}. \text{ (p. 238)}$$

- The symbol Σ (sigma) can be used to express both finite and infinite series. *(pp. 238, 246, 255)*
- When you substitute the values of n into a series in sigma notation, you write the series in expanded form. *(p. 239)*
- The formula for the sum S of a finite geometric series with n terms and common ratio r is

$$S = \frac{a_1 - a_n r}{1 - r}. \text{ (p. 245)}$$

- When the sequence of partial sums of an infinite series has a limit, then that limit is the sum of the series. *(p. 252)*
- When $|r| < 1$, an infinite geometric series has a sum. *(p. 254)*
- The formula for the sum of an infinite geometric series with common ratio r where $|r| < 1$ is

$$S = \frac{a_1}{1 - r}. \text{ (p. 254)}$$

- The number identifying each term's position in a sequence is usually a counting number. *(p. 208)*
- You can graph a sequence on a coordinate plane by plotting points with the coordinates (*position, term*). The points on a graph of a sequence should not be connected. *(p. 209)*
- The graph of any sequence represents a function whose domain is the counting numbers only. *(p. 209)*
- The symbol a_n is used to represent the nth term of the sequence $a_1, a_2, a_3, \ldots$. *(p. 216)*
- Sometimes the first term of a sequence is called a_0, rather than a_1, to show the starting value before any change occurs. *(p. 218)*

Answers to Unit 4 Review and Assessment

22. Maps will vary. An example is given.

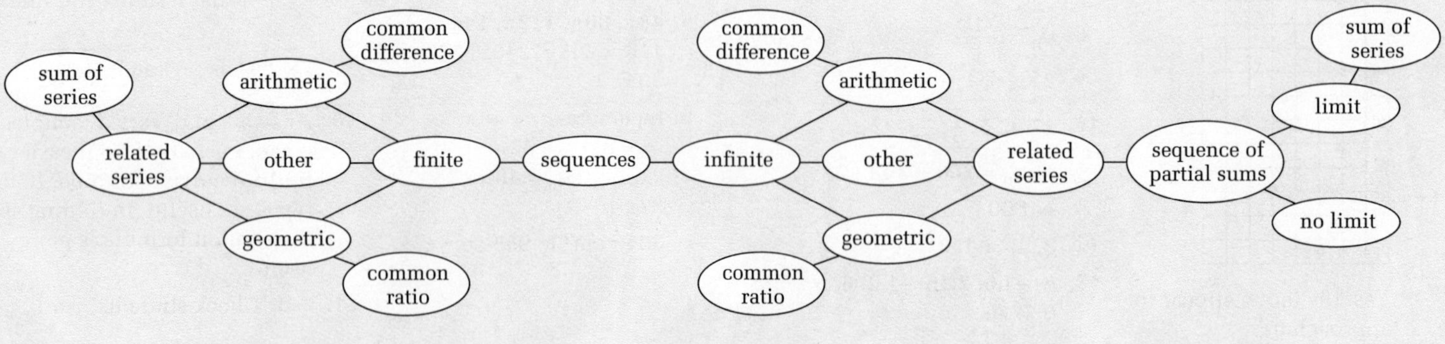

- A graphics calculator or a spreadsheet can help you find the nth term of a sequence described by a recursive formula. *(pp. 224–225)*
- The recursive formula for an arithmetic sequence where d is the common difference is:

  a_1 = value of the first term
  $a_n = a_{n-1} + d$ *(p. 230)*

- The graph of an arithmetic sequence is a discrete linear function. *(p. 230)*
- The recursive formula for a geometric sequence where r is the common ratio is:

  a_1 = value of the first term
  $a_n = (a_{n-1})r$ *(p. 231)*

- The graph of a geometric sequence is a discrete curve. *(p. 231)*

$p \leftrightarrow q$ LOGICAL *if - then* REASONING

- There are two ways to describe the pattern of a sequence. You can look for a relationship between each term and its position number (that is, an explicit formula), or you can look for a relationship between each term and the term(s) before it (that is, a recursive formula). *(p. 208)*
- To determine whether a sequence appears to have a limit, look at the graph of the sequence or the decimal value of each term to see a trend. Verify your observation with a term farther out in the sequence. *(p. 210)*

Key Terms

- **sequence** (p. 208)
- **infinite sequence** (p. 208)
- **recursive formula** (p. 223)
- **geometric sequence** (p. 229)
- **series** (p. 238)
- **Σ** (p. 238)
- **∞** (p. 255)
- **term of sequence** (p. 208)
- **limit of a sequence** (p. 210)
- **arithmetic sequence** (p. 229)
- **common ratio** (p. 229)
- **arithmetic series** (p. 238)
- **geometric series** (p. 245)
- **finite sequence** (p. 208)
- **explicit formula** (p. 217)
- **common difference** (p. 229)
- **geometric mean** (p. 232)
- **sigma notation** (p. 238)
- **partial sum** (p. 252)

Quick Quiz (4-1 through 4-4)

1. What is the eighth term of this sequence? [4-1]
   $\frac{5}{2}, 1, -\frac{1}{2}, -2, \ldots$ **−8**
2. Does this sequence appear to have a limit? If it does, give its value. [4-1]
   $\frac{1}{2}, \frac{2}{3}, \frac{3}{4}, \frac{4}{5}, \frac{5}{6}, \ldots$ **Yes; 1.**
3. Write the first three terms and the tenth term of this sequence. [4-2]
   $a_n = 2n - 1$ **1, 3, 5; 19**
4. Write a recursive formula for the sequence 3, 11, 27, 59, ... [4-3]
   $a_1 = 3$
   $a_n = 2a_{n-1} + 5$

Tell whether each sequence is *arithmetic, geometric,* or *neither.* [4-4]

5. $4, \frac{1}{4}, 4, \frac{1}{4}, 4, \frac{1}{4}, \ldots$ **neither**
6. $-18, -11, -4, 3, \ldots$
   arithmetic
7. $\frac{1}{2}, \frac{3}{2}, \frac{9}{2}, \frac{27}{2}, \frac{81}{2}, \ldots$ **geometric**

Planning the Unit

5 Exponential and Logarithmic Functions

OVERVIEW

- **Unit 5** explores exponential and logarithmic functions. Both types of functions are graphed, and rules relating to simplifying expressions involving each are presented. Expressions with exponents are extended from whole number exponents to negative exponents to rational exponents.
- The irrational number *e* is introduced, as is the natural logarithm. Students convert exponential expressions to logarithmic expressions, and vice versa. Exponential and logarithmic equations are solved. Reflecting graphs can be reviewed in the **Student Resources Toolbox**.
- The theme of the **Unit Project** involves music and mathematics. Students make four musical instruments and give a lecture/performance using their instruments. They consider the connections between mathematics and music and give two examples.
- **Connections** to history, medicine, music, cellular phones, biology, populations, banking, architecture, and prices of movie tickets are integrated into the teaching materials and exercises.
- **Graphics calculators** are used in Sections 5-1 and 5-2 to solve problems, in Section 5-3 to find fractional exponents, in Section 5-4 to approximate *e* and to graph functions, in Section 5-5 to graph an inverse function, and in Section 5-6 to evaluate a logarithm. **Computer software**, such as Plotter Plus, can be used in Sections 5-1, 5-2, 5-4, and 5-5 to graph functions.
- **Problem-solving strategies** used in Unit 5 include using a graph and using a formula.

Unit Objectives

Section	Objectives	NCTM Standards
5-1	• Use exponential growth and decay functions to model situations.	1, 2, 3, 4, 5, 6
5-2	• Use exponential functions with negative *x*-values to model situations.	1, 2, 3, 4, 5, 6
5-3	• Use expressions involving fractional exponents or radicals. • Model situations using exponential functions with fractional exponents.	1, 2, 3, 4, 5, 6
5-4	• Model situations using exponential functions with base *e*.	1, 2, 3, 4, 5, 6
5-5	• Find and graph inverse functions.	1, 2, 3, 4, 5, 6
5-6	• Recognize and evaluate logarithmic functions. • Use logarithmic functions to solve problems.	1, 2, 3, 4, 5, 6
5-7	• Learn about the properties of logarithms. • Use these properties to solve problems.	1, 2, 3, 4, 5, 6
5-8	• Use exponential and logarithmic equations to solve real-life problems.	1, 2, 3, 4, 5, 6

Topic Spiraling

Section	Connections to Prior and Future Concepts
5-1	**Section 5-1** uses exponential growth and decay functions to model real-world situations. These exponential functions were introduced in Section 2-7 of Book 2 and are explored further in Sections 5-2, 5-3, and 5-4 of Book 3.
5-2	**Section 5-2** extends using exponential functions as models to exponential functions having negative exponents. Functions whose base is greater than 1 and functions whose base is between 0 and 1 are both examined.
5-3	**Section 5-3** introduces fractional exponents for exponential functions. This extends work in exponential functions developed in the previous two sections. A number with a fractional exponent is related to a radical. Radicals were covered in Sections 2-9 and 9-2 of Book 1 and were used throughout Book 2.
5-4	**Section 5-4** introduces the number e as a base of an exponential function. A limit of a sequence is used to define e. Limits were introduced in Section 4-1 of Book 3. The number e is used in connection with the natural logarithm in Section 5-6 of Book 3. Both the functions e^x and $\ln x$ will be used in future mathematics courses.
5-5	**Section 5-5** explores graphs of functions and their inverses. Graphs have been used extensively throughout Books 1 and 2, and the graphs of functions were covered in detail in Unit 2 of Book 3. Reflections over $y = x$ are used to graph inverses. Reflections over lines were first encountered in Section 10-1 of Book 1.
5-6	**Section 5-6** introduces logarithmic functions as the inverse of exponential functions, building on the skills learned in the last section. Logarithmic forms and exponential forms of expressions are compared. Exponential and logarithmic equations are used in Section 5-8 of Book 3.
5-7	**Section 5-7** introduces the properties of logarithms that can be used to simplify expressions. These properties are used to solve equations in Section 5-8 of Book 3.
5-8	**Section 5-8** explores solving exponential and logarithmic equations that model real-life situations. The skills learned in the prior sections of Unit 5 are brought together in this section.

Integrating the Strands

Strands	Sections
Number	5-1, 5-2, 5-3, 5-4, 5-5, 5-6, 5-7,
Algebra	5-1, 5-2, 5-3, 5-4, 5-5, 5-6, 5-7, 5-8
Functions	5-1, 5-2, 5-3, 5-4, 5-5, 5-6, 5-7, 5-8
Measurement	5-4
Geometry	5-1, 5-5
Statistics and Probability	5-7, 5-8
Discrete Mathematics	5-1, 5-6, 5-7, 5-8
Logic and Language	5-1, 5-2, 5-3, 5-4, 5-5, 5-6, 5-7, 5-8

Planning the Unit 5
Exponential and Logarithmic Functions

Section Planning Guide

- Essential exercises and problems are indicated in boldface.
- Ongoing work on the Unit Project is indicated in color.
- Exercises and problems that require student research, group work, manipulatives, or graphing technology are indicated in the column headed "Other."

Section	Materials	Pacing	Standard Assignment	Extended Assignment	Other
5-1	graphics calculator, objects such as pennies, container	Day 1	**1–4,** 5–9, **10, 11, 13–16**	**1–4,** 5–9, **10, 11, 13–16**	12
		Day 2	**18–21,** 24–31, 32, 33	**18–21,** 22–31, 32, 33	17
5-2	graphics calculator	Day 1	1, **2–7,** 11, 13, 15–21, 22	1, **2–7,** 8, 9, 11–21, 22	10, 22
5-3	graphics calculator	Day 1	**1–18,** 19, 20, 22	**1–18,** 19, 20, 22	21
		Day 2	**23–30,** 31–35, **37–39,** 41–47, 48–52	**23–30,** 31–35, **37–39,** 40–47, 48–52	36
5-4	graphics calculator, 8 plastic straws, scissors, masking tape, centimeter ruler	Day 1	**1–8, 12–15,** 19	**1–8,** 10, 11, **12–15,** 19	9, 16–18, 20, 21
		Day 2	26–33, 34	25–33, 34	22–24
5-5	graphing technology	Day 1	**1–8, 10–21, 23–25,** 29–33, 34	**1–8,** 9, **10–21,** 22, **23–25,** 29–33, 34	26–28, 34
5-6	graphics calculator	Day 1	**1,** 2, **3–25**	**1,** 2, **3–25**	
		Day 2	**26–32,** 35, 40–45, 46, 47	**26–32,** 33–38, 40–45, 46, 47	39
5-7	graphics calculator	Day 1	**2–20**	1, **2–20**	
		Day 2	**23–28, 32–35,** 38–47, 48	21, **23–28, 32–35,** 36–47, 48	22, 29–31
5-8	graphing technology	Day 1	**1–7,** 8–10, **11–16,** 24–31, 32	**1–7,** 8–10, **11–16,** 21, 22, 24–31, 32	17–20, 23
Review		**Day 1**	**Unit Review**	**Unit Review**	
Test		**Day 2**	**Unit Test**	**Unit Test**	

Yearly Pacing	Unit 5 Total	Units 1–5 Total	Remaining	Total
	17 days (2 for Unit Project)	82 days	72 days	154 days

Support Materials

- See **Project Book** for notes on Unit 5 Project: Making Music with Mathematics.
- UPP and disk refer to **Using Plotter Plus** booklet and **Plotter Plus** disk.
- TI-81/82 refers to **Using TI-81 and TI-82 Calculators** booklet.
- Warm-up exercises for each section are available on **Warm-Up Transparencies.**

Section	Study Guide	Practice Bank	Problem Bank	Activity Bank	Explorations Lab Manual	Assessment Book	Visuals	Technology
5-1	5-1	Practice 37	Set 11	Enrich 33	Master 13	Quiz 5-1		Function Plotter (disk)
5-2	5-2	Practice 38	Set 11	Enrich 34		Quiz 5-2		Function Plotter (disk)
5-3	5-3	Practice 39	Set 11	Enrich 35		Quiz 5-3		TI-81/82, page 62 UPP, page 55
5-4	5-4	Practice 40	Set 11	Enrich 36	Add. Expl. 5 Master 14	Quiz 5-4 Test 19		Function Plotter (disk)
5-5	5-5	Practice 41	Set 12	Enrich 37	Add. Expl. 6 Master 2	Quiz 5-5	Folder 4	Function Plotter (disk)
5-6	5-6	Practice 42	Set 12	Enrich 38		Quiz 5-6		
5-7	5-7	Practice 43	Set 12	Enrich 39	Master 2	Quiz 5-7		
5-8	5-8	Practice 44	Set 12	Enrich 40		Quiz 5-8 Test 20		TI-81/82, page 63 UPP, page 56
Unit 5	Unit Review	Practice 45	Unifying Problem 5	Family Involve 5		Tests 21–23		

Form A

Spanish versions of these tests are on pages 131–134 of the **Assessment Book.**

Name ______________________ Date __________ Score _____

Test 21

Test on Unit 5 (Form A)

Directions: Write the answers in the spaces provided.

1. Rewrite the function $y = (3.1)\left(\frac{2}{3}\right)^{-x}$ in the form $y = ab^x$.

For Questions 2–5, simplify.

2. $36^{5/2}$ **3.** $300^{1/2}$

4. $\sqrt[3]{81y^{12}}$ **5.** $\sqrt[4]{x^{-1}} \cdot \sqrt[4]{16x^{-3}}$

6. Find the value of b when $f(x) = 5b^x$ and $f\left(\frac{2}{3}\right) = 80$.

The formula $A = 5000e^{rt}$ can be used to find the dollar value of an investment of $5000 after t years when the interest is compounded continuously at a rate of r percent.

7. Find the value of the investment after 10 years if the interest rate is 6%.

8. Find the value of the investment after 10 years if the interest rate is 9%.

Graph each function and its reflection over the line $y = x$. Is the reflection the graph of a function? Write *Yes* or *No*.

9. $y = 1.5^x$

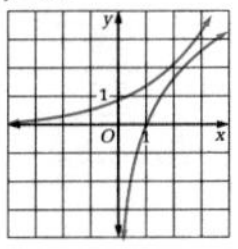

10. $y = 2x^2 - 3$

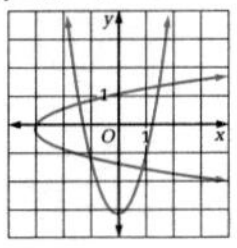

Tell whether each statement is *True* or *False*.

11. The graph of $y = ab^x$ is a reflection about the y-axis of the graph of $y = ab^{-x}$.

12. For any integer $a \neq 0$, $a^{1/n} = \frac{1}{a^n}$.

13. The number e is a rational number equal to 2.71.

14. $\log \frac{1}{5} = \log 1 \div \log 5$

Answers

1.	$y = (3.1)\left(\frac{3}{2}\right)^x$
2.	7776
3.	$10\sqrt{3}$
4.	$3y^4\sqrt[3]{3}$
5.	$\frac{2}{\lvert x \rvert}$
6.	64
7.	$9,110.59
8.	$12,298.02
9.	Yes.
10.	No.
11.	True.
12.	False.
13.	False.
14.	False.

Name ______________________ Date __________ Score _____

Test 21 *(continued)*

Directions: Write the answers in the spaces provided.

Write an equation of the inverse for each function.

15. $y = 2.3x - 4$ **16.** $y = x^{3/2}$ **17.** $y = 3x^3 + 1$

Evaluate each logarithm. Round decimal answers to the nearest hundredth.

18. $\log \frac{1}{1000}$ **19.** $\ln 12$ **20.** $\log_8 512$

Solve. Round decimal answers to the nearest hundredth.

21. $\log_x 343 = 3$ **22.** $e^{2x} = \frac{1}{e^{-6}}$ **23.** $10^x = 28$

Write each expression in terms of $\log_5 M$ and $\log_5 N$.

24. $\log_5 M^3N$ **25.** $\log_5 \frac{M}{\sqrt[4]{N}}$

Solve.

26. $\log_2 x + \log_2 (x + 2) = 3$

27. $\log_3 (7m) = 1 + \log_3 (m + 2)$

For Questions 28–30, evaluate to three decimal places.

28. $\log_2 42$ **29.** $\log_7 24$ **30.** $\log_5 \frac{3}{4}$

31. Simplify $\log_4 4 + \log_4 32 - \log_4 8$.

32. Open-ended Explain the connection between an exponential equation and a logarithmic equation. Support your explanation with an example that demonstrates the connection.

Sample answer: Exponential and logarithmic equations are inverses of each other. For example, to find the inverse of the exponential function $y = 3^x$, we interchange x and y to obtain $x = 3^y$. Taking the logarithm base-3 of both sides to solve for y yields $\log_3 x = y$, which is a logarithmic equation.

33. Writing Describe a logistic growth function.

Sample answer: A logistic growth function is one that increases or decreases exponentially at first, but then the rate of growth slows as the value of the function approaches its upper or lower boundary.

Answers

15.	$y = \frac{x+4}{2.3}$
16.	$y = x^{2/3}$
17.	$y = \sqrt[3]{\frac{x-1}{3}}$
18.	−3
19.	2.48
20.	3
21.	7
22.	3
23.	1.45
24.	$3 \log_5 M + \log_5 N$
25.	$\log_5 M - \frac{1}{4} \log_5 N$
26.	2
27.	$\frac{3}{2}$
28.	5.392
29.	1.633
30.	−0.179
31.	2
32.	*See question.*
33.	*See question.*

Form B

Name ______________________ Date __________ Score _____

Test 22

Test on Unit 5 (Form B)

Directions: Write the answers in the spaces provided.

1. Rewrite the function $y = (4.7)\left(\frac{5}{2}\right)^{-x}$ in the form $y = ab^x$.

For Questions 2–5, simplify.

2. $64^{3/2}$ **3.** $162^{1/2}$

4. $\sqrt[3]{16y^{12}}$ **5.** $\sqrt[4]{x^{-3}} \cdot \sqrt[4]{256x^{-1}}$

6. Find the value of b when $f(x) = 3b^x$ and $f\left(\frac{4}{3}\right) = 243$.

The formula $A = 3000e^{rt}$ can be used to find the dollar value of an investment of $3000 after t years when the interest is compounded continuously at a rate of r percent.

7. Find the value of the investment after 10 years if the interest rate is 6%.

8. Find the value of the investment after 10 years if the interest rate is 12%.

Graph each function and its reflection over the line $y = x$. Is the reflection the graph of a function? Write *Yes* or *No*.

9. $y = 3x^2 + 1$

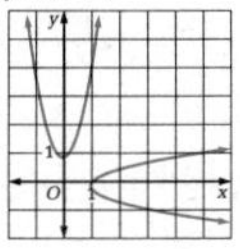

10. $y = 1.8^x$

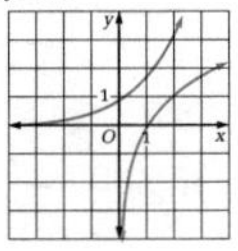

Tell whether each statement is *True* or *False*.

11. The graph of $y = ab^x$ is a reflection about the line $y = x$ of the graph of $y = ab^{-x}$.

12. For any integer $n > 0$ and any positive real number a, $a^{1/n} = \sqrt[n]{a}$.

13. The number e is an irrational number; $e \approx 2.718281828$.

14. $\log \frac{3}{4} = \log 3 - \log 4$

Answers

1.	$y = (4.7)\left(\frac{2}{5}\right)^x$
2.	512
3.	$9\sqrt{2}$
4.	$2y^4\sqrt[3]{2}$
5.	$\frac{4}{\lvert x \rvert}$
6.	27
7.	$5466.36
8.	$9960.35
9.	No.
10.	Yes.
11.	False.
12.	True.
13.	True.
14.	True.

Name ______________________ Date __________ Score _____

Test 22 *(continued)*

Directions: Write the answers in the spaces provided.

Write an equation of the inverse for each function.

15. $y = 4.9x + 2$ **16.** $y = x^{5/3}$ **17.** $y = 3x^3 - 2$

Evaluate each logarithm. Round decimal answers to the nearest hundredth.

18. $\log \frac{1}{100{,}000}$ **19.** $\ln 32$ **20.** $\log_5 3125$

Solve. Round decimal answers to the nearest hundredth.

21. $\log_x 216 = 3$ **22.** $e^{x+1} = \frac{1}{e^{-4}}$ **23.** $10^x = 32$

Write each expression in terms of $\log_5 M$ and $\log_5 N$.

24. $\log_5 \frac{M^3}{N}$ **25.** $\log_5 M^4\sqrt{N}$

Solve.

26. $\log_3 x + \log_3 (x - 6) = 3$

27. $\log_4 (5m) = 1 + \log_4 (m + 3)$

For Questions 28–30, evaluate to three decimal places.

28. $\log_2 124$ **29.** $\log_5 22$ **30.** $\log_4 \frac{3}{5}$

31. Simplify $\log_5 5 + \log_5 25 - \log_5 125$.

32. Open-ended Explain how the inverse of a function is determined algebraically. Give an example to explain the steps you follow.

Sample answer: The inverse of a function is found by interchanging the control and dependent variables and then solving the resulting equation for the dependent variable. For example, to find the inverse of the function $y = 2x + 1$, first switch the variables: $x = 2y + 1$. Then solve for y: $x - 1 = 2y$ and $y = \frac{x-1}{2}$.

33. Writing Explain the difference between a common logarithm and a natural logarithm.

Sample answer: A common logarithm is base-10, and the notation "log" is used to indicate a common logarithm. A natural logarithm is base-e, and the notation "ln" is used.

Answers

15.	$y = \frac{x-2}{4.9}$
16.	$y = x^{3/5}$
17.	$y = \sqrt[3]{\frac{x+2}{3}}$
18.	−5
19.	3.47
20.	5
21.	6
22.	3
23.	1.51
24.	$3 \log_5 M - \log_5 N$
25.	$4 \log_5 M + \frac{1}{2} \log_5 N$
26.	9
27.	12
28.	6.954
29.	1.921
30.	−0.368
31.	0
32.	*See question.*
33.	*See question.*

OUTSIDE RESOURCES

Books/Periodicals

"Looking at Exponential Functions." *The Language of Functions and Graphs,* Unit B3: pp. 120–125. The Shell Centre for Mathematical Education at the University of Nottingham, UK.

Jacobs, Harold R. *Mathematics: A Human Endeavor.* (Chapter 4: Large Numbers and Logarithms.) San Francisco, CA: W.H. Freeman and Company, 1994.

Jones, Graham A. "Mathematical Modeling in a Feast of Rabbits." *Mathematics Teacher* (December 1993): pp. 770–773.

Activities/Manipulatives

Kincaid, Charlene, Guy Mauldin, and Deanna Mauldin. "The Marble Sifter: A Half-Life Simulation." *Mathematics Teacher* (December 1993): pp. 748–759.

Software

Myers, David L. *Plotter Plus.* Boston, MA: Houghton Mifflin Company, 1995. Macintosh and MS-DOS (worksheets included).

Videos

Southern Illinois University at Carbondale. *World Population Review,* 1990.

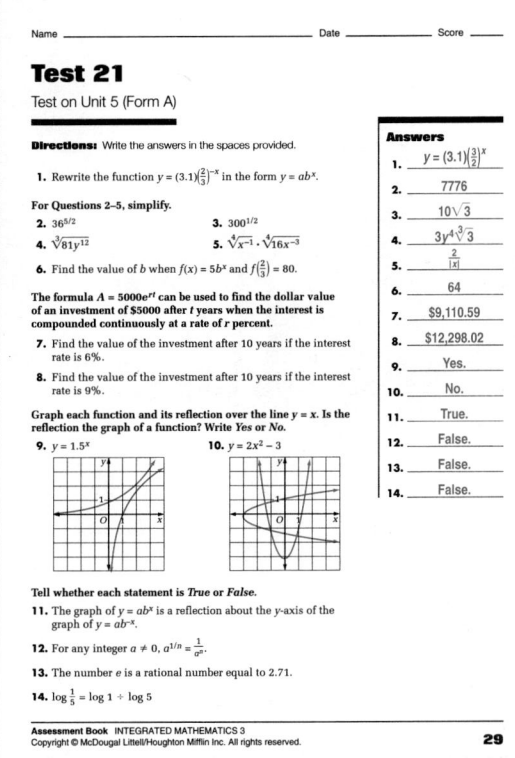

PROJECT GOALS

- Students give a lecture/performance that demonstrates how mathematics and music are related.
- Students make four musical instruments, two sets of panpipes and two simple guitars.
- Students describe at least two examples of the relationship between mathematics and music.

PROJECT PLANNING

Materials List

- Plastic drinking straws
- Wood
- Nylon fishing line
- Large screw eyes
- Screwdriver, Pliers
- Centimeter rule
- Calculator, Scissors
- Masking tape
- Marking pen
- **Note:** Students should use care in working with these materials and should do so under the supervision of a teacher.

Project Teams

Have students work on the project in groups of four. Students can begin by collecting the materials and tools they need to build their instruments. Each student should decide which instrument to make. They should then plan to listen to recordings of different kinds of music together, notice which instruments are being played, and then try to listen for them as the music is being played. Students should then proceed to make their instruments and write a report which contains drawings of the instruments and descriptions of how they were built. Two examples of how mathematics is related to music should also be included in the report.

unit 5

Exponential and Logarithmic Functions

If you wanted to write a song for the whole world to sing, what musical system would you use? In China most music is written in a five-note scale. Traditional European music is based on an eight-note octave. In India the raga is an important musical form. Ragas use scales with various sequences of five, six, or seven notes.

handcrafted

To uncover the secrets of the great **violin makers** of the past, scientist and violin maker Carleen M. Hutchins has handcrafted over 400 instruments in the last 40 years. When a violin is played, the vibrating strings cause both the wooden walls of the violin and the air inside to vibrate. By analyzing these vibrations, Carleen M. Hutchins has worked out basic rules for making violins.

Different instruments sound different because of the blend of higher tones, or overtones, produced with each note. In 1965 the first **electronic** synthesizer to produce music with vibrating electric circuits was invented. Synthesizers can imitate all the instruments of the orchestra.

digital

Suggested Rubric for Unit Project

4 Students successfully make their instruments following the guidelines in the "Working on the Unit Project Exercises." They write a report that contains clear and accurate drawings of their instruments with detailed descriptions of how they were built and how they are played. The report contains two examples of how mathematics is related to music. Students give an interesting lecture and entertaining musical performance that involves all four instruments.

3 Students make their instruments following the textbook's guidelines, but one or two of the instruments are not very well made. A report is written, but the descriptions of how the instruments were made and how they are played is sketchy in parts. One of the examples of the relationship between mathematics and music is weak. The lecture/performance is good, but more practice would have made it better.

2 Some students have difficulty making their instruments and the finished product does not work well. The written

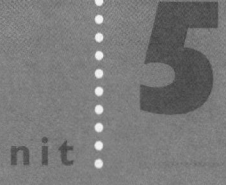

Unit Project 5

Making Music with Mathematics

Your project is to give a lecture/performance that demonstrates how mathematics and music are related. Following the guidelines in the Working on the Unit Project exercises, your group should make two sets of panpipes and two simple guitars. You should describe how you built your instruments, use them to play music, and explain how they produce musical notes.

Describe at least two examples of the relationship between mathematics and music.

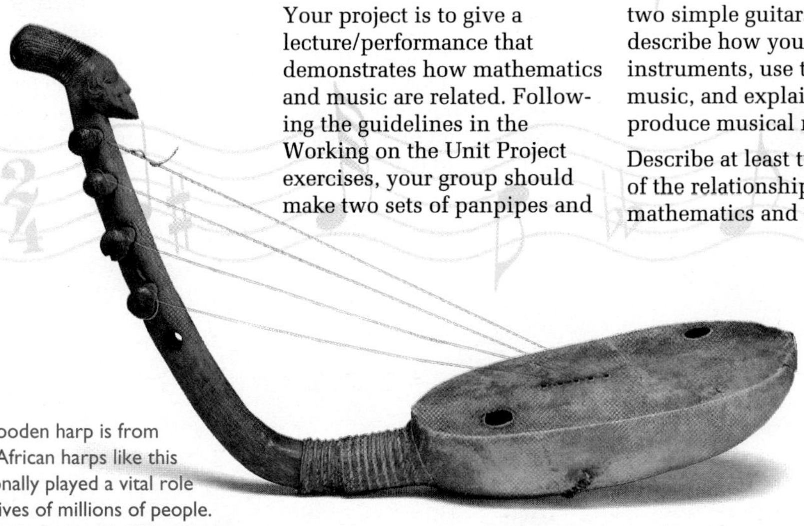

This wooden harp is from Zaire. African harps like this traditionally played a vital role in the lives of millions of people. Many harp songs tell of historical events or legends.

On a **stringed instrument** like a violin or a guitar, one way to raise the pitch, or produce a higher note, is to shorten or tighten a string. Another way is to use a thinner string.

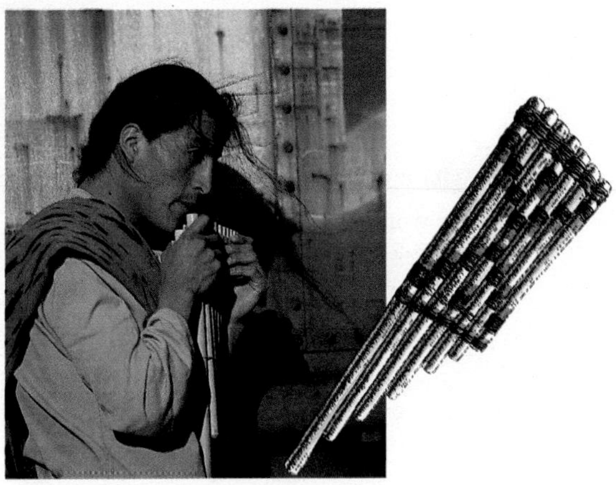

Shortening the column of air that vibrates when a musician blows into a **wind instrument** like a flute or panpipe raises the pitch.

Support Materials

The ***Project Book*** contains information about the following topics for use with this Unit Project.

- Project Description
- Teaching Commentary
- Working on the Unit Project Exercises
- Completing the Unit Project
- Assessing the Unit Project
- Alternative Projects
- Outside Resources

Students Acquiring English

Constructing musical instruments is an excellent activity for students acquiring English in that it draws on kinesthetic, auditory, and mathematical skills. You may want to ask students to describe musical instruments popular in their countries of origin and to speculate on the connection between those instruments and mathematics.

ADDITIONAL BACKGROUND

Multicultural Note

Most traditional cultures construct musical instruments of various sorts. In Thailand, one common instrument is the *rang*, a percussion instrument with twenty-one bamboo "keys" and a stand made in the form of a Thai river boat. The Nigerian *agbe* is a rattle made from the gourd of a calabash tree. The gourd is cleaned and covered with a string net woven with beads, bamboo strips, or cowrie shells. In the highlands of Bolivia and Peru, panpipes made of fired clay and wood have been part of the musical tradition for centuries. The *ombis* or *n'goms* of Uganda, Kenya, and the Congo are arched harps. These have resonator boxes covered with animal hides, and strings to pluck.

Suggested Rubric for Unit Project

report has some serious omissions involving the drawings of the instruments or the required descriptions. The examples of how mathematics is related to music contain some errors and misunderstandings on a conceptual level. The lecture/performance is somewhat disorganized and not pleasing to the ear.

1 Students are unable to complete the project because they have not been able to make all of the instruments. A report is attempted, but it is incomplete. Drawings, descriptions, and examples are either inaccurate or missing. No lecture/performance is given. Students should be encouraged to speak with the teacher as soon as possible to review their work and to make a new start on the project.

Physics, Math, and Music

The physics of music involves an understanding of physical processes. Musical instruments produce sound waves, which are described by physicists as the propagation of a disturbance of air particles, creating a wave moving with a definite speed. These waves, when striking the ear within a frequency range of 20 to 20,000 vibrations per second, are called sound waves. Using mathematics as a tool to express their ideas, physicists write formulas to understand the properties of sound waves. Some properties of interest are the amplitude, velocity, and intensity of sound waves. One thing that scientists do not study, however, is the beauty of particular sounds made by musicians. Musicians are artists and beauty is a subjective concept that is outside the realm of scientific inquiry. All people, however, can enjoy the beauty of the sound that is called music.

ALTERNATIVE PROJECTS

Project 1, page 323

Mathematics and My Musical Instrument

Students who play a musical instrument prepare a lecture/performance in which they demonstrate and discuss how mathematical models are used in making and playing it.

Project 2, page 323

Applying Exponential and Logarithmic Functions

Choose one application from this unit, such as the Richter scale or carbon dating. Find out more about the topic, and the mathematics involved. Prepare a presentation for the class.

Unit Project 5

Getting Started

For this project you should work in a group of four students. Here are some ideas to help you get started.

☞ Collect the materials and tools—plastic drinking straws, wood, strong nylon fishing line, several large screw eyes, a screwdriver, and pliers—that you will need to build your instruments.

☞ As a group, decide which instrument each member will make.

☞ Plan to meet to listen to recordings of many different kinds of music. Notice which instruments are used on the recordings and try to listen for them as the music plays.

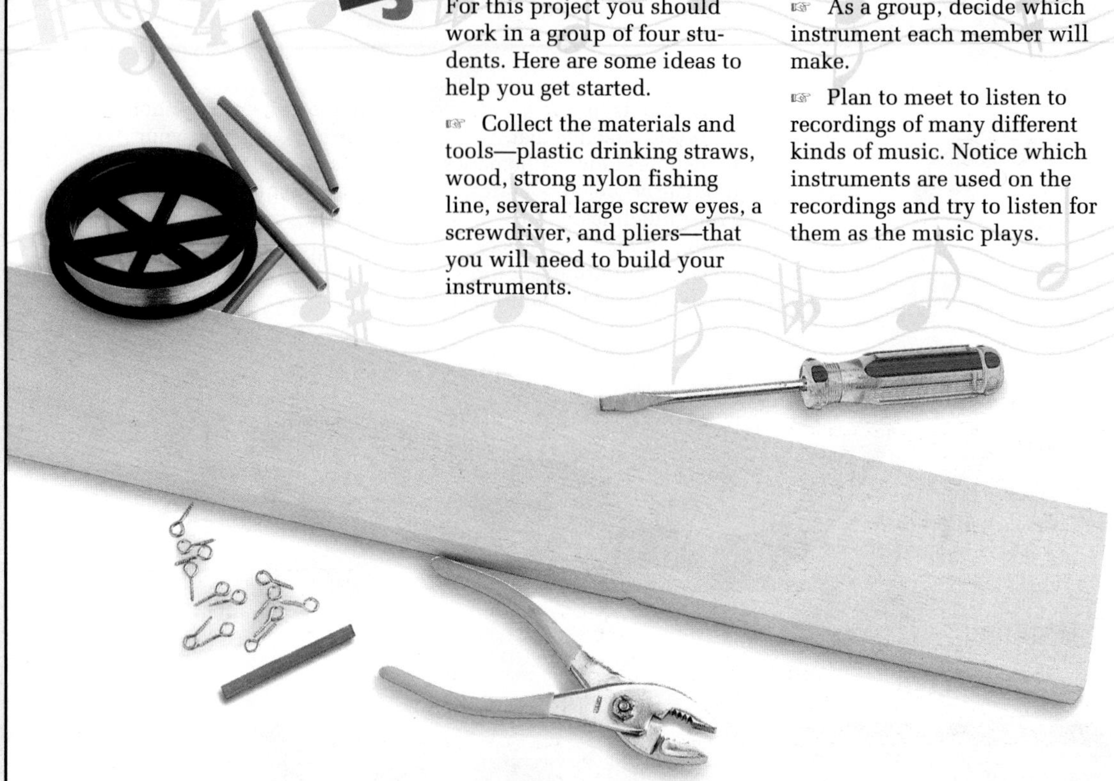

Working on the Unit Project

Your work in Unit 5 will help you to make your musical instruments and prepare your lecture/performance.

Related Exercises:

Section 5-1, Exercises 32, 33
Section 5-2, Exercise 22
Section 5-3, Exercises 48–52
Section 5-4, Exercise 34
Section 5-5, Exercise 34
Section 5-6, Exercises 46, 47
Section 5-7, Exercise 48
Section 5-8, Exercise 32

Alternative Projects p. 323

Can We Talk MUSIC

➤ What kind of music do you like? What instruments do the members of your favorite band play?

➤ Have you ever attended a live performance by an orchestra or listened to a recording of an orchestra? If you did, were you able to hear the individual instruments when all the musicians were playing together?

➤ The ridges on the neck of a guitar are called *frets*. As they play, most guitarists will hold strings down by pressing at various frets. What effect do you think this has on the sound produced? Why?

➤ If you play an instrument, discuss what you like about it. Is it easy to learn? Does your music teacher ever talk about mathematics in relation to the music?

Answers to Can We Talk?

➤ Answers may vary. An example is given. I like jazz music. My favorite band has a piano, electric guitar, drums, trumpet, and saxophone.

➤ Answers may vary. An example is given. Yes. Yes, at certain times you can hear the individual instruments. If you listen carefully and look for a particular instrument, you can sometimes hear it separately from the other instruments.

➤ Answers may vary. An example is given. I think pressing on various frets changes the note or pitch of the guitar because the length of the string that vibrates to make a sound is changed.

➤ Answers may vary. Check students' work.

Section **5-1**

Exponential Growth and Decay

Focus

Use exponential growth and decay functions to model situations.

The Next Generation

The number of ancestors a person has in a past generation can be modeled by a *discrete* exponential growth function.

The population of the world over time can be modeled by a *continuous* exponential growth function.

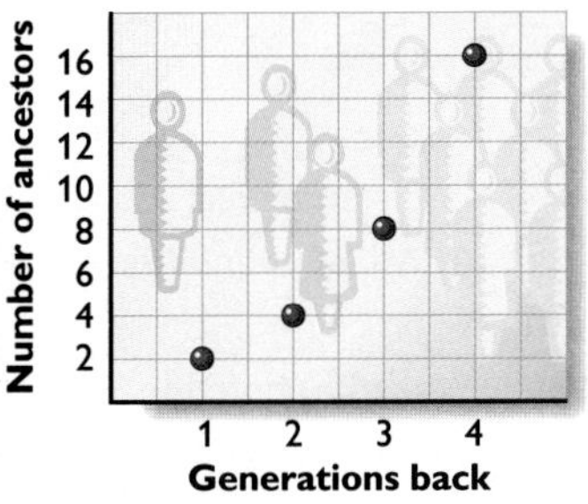

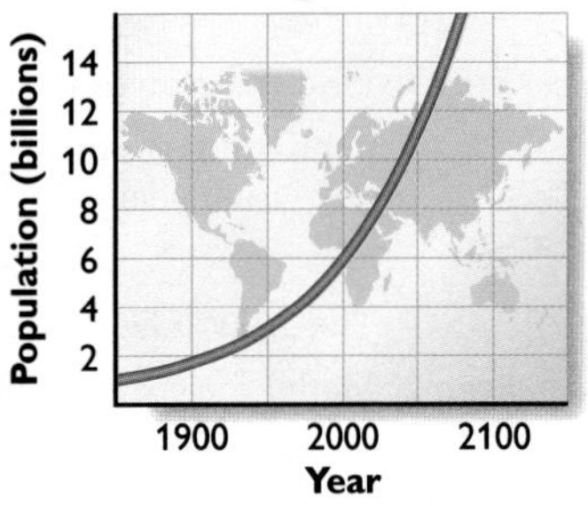

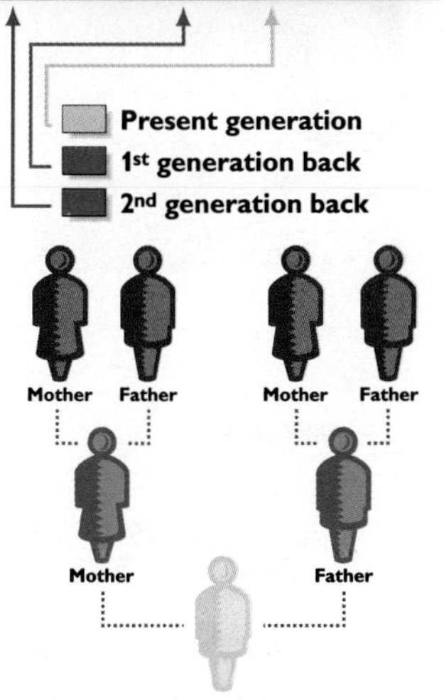

Talk it Over

1. The function $y = 2^x$ can be used to model the number of ancestors a person has in a certain generation, where x represents the number of past generations.
   a. Find the number of ancestors a person has in the fifth generation back.
   b. What is the domain of the function? What is the range?
   c. Explain why it does not make sense to use a continuous graph to model this situation.
2. Use the continuous graph to estimate the world population in 1975.

Both discrete and continuous exponential growth can be modeled by an exponential function. The general form of an exponential function is $y = ab^x$.

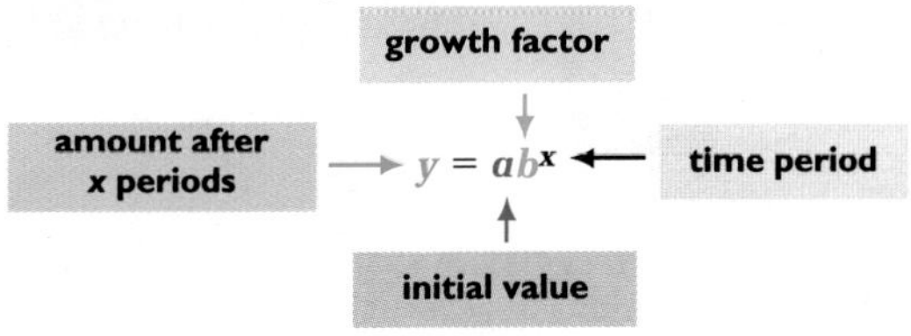

PLANNING

Objectives and Strands

See pages 264A and 264B.

Spiral Learning

See page 264B.

Materials List

- Graphics calculator
- 50 objects such as pennies
- Container

Recommended Pacing

Section 5-1 is a two-day lesson.

Day 1

Pages 267–269: Opening paragraph through Talk it Over 4, *Exercises 1–16*

Day 2

Pages 269–270: Exploration through Look Back, *Exercises 17–33*

Extra Practice

See pages 615–617.

Warm-Up Exercises

Warm-Up Transparency 5-1

Support Materials

- Practice Bank: Practice 37
- Activity Bank: Enrichment 33
- Study Guide: Section 5-1
- Problem Bank: Problem Set 11
- Explorations Lab Manual: Diagram Master 13
- Using IBM/Mac Plotter Plus Disk: Function Plotter
- Assessment Book: Quiz 5-1, Alternative Assessments 1, 2

Answers to Talk it Over

1. a. 32 ancestors
   b. the positive integers; all numbers 2^n for n an integer and $n \geq 1$
   c. The number of generations and the number of ancestors must both be positive whole numbers.
2. Estimates may vary. An example is given; about 4 billion people.

TEACHING

Talk it Over

Questions 1 and 2 have students analyze and use discrete and continuous models of exponential growth. Though the population of the world is discrete, a continuous model can be used to model population growth over time.

Additional Sample

S1 Monthly benefits for Social Security in May 1992 were $23,307 million. Since then, benefits have increased about 5.4% per year.

a. Write an exponential function to model the growth of monthly Social Security benefits paid each year.
Step 1. **Find the growth factor *b*.**
(benefits at end of each year) = (100% of benefits at start of year) + (5.4% of benefits at start of year)
= 105.4% of benefits at start of year
b = 105.4% or 1.054
Step 2. **Model the growth with an exponential function. Let y = the benefits (in millions) after x years. Let x = the number of years since 1992. Substitute 23,307 for the initial value a and 1.054 for the growth factor b into $y = ab^x$.**
$y = (23,307)(1.054)^x$

b. If benefits continue to grow at this rate, when will the monthly Social Security benefits reach $50,000 million?
Method 1. **Use a graph. Graph $y = (23,307)(1.054)^x$**

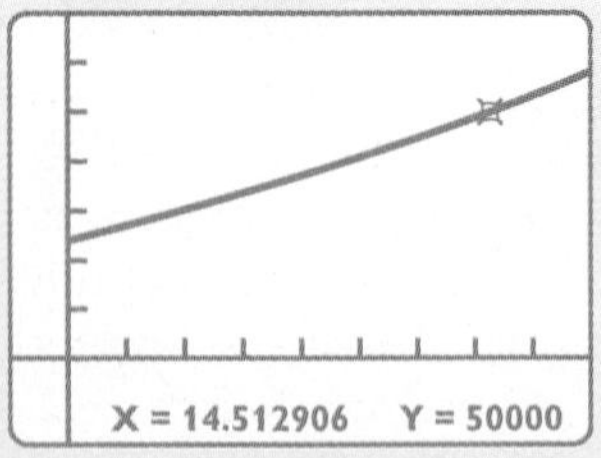

Since an x-value of zero represents 1992, an x-value of 15 represents the year 2007. The benefits will reach $50,000 million in the year 2007.

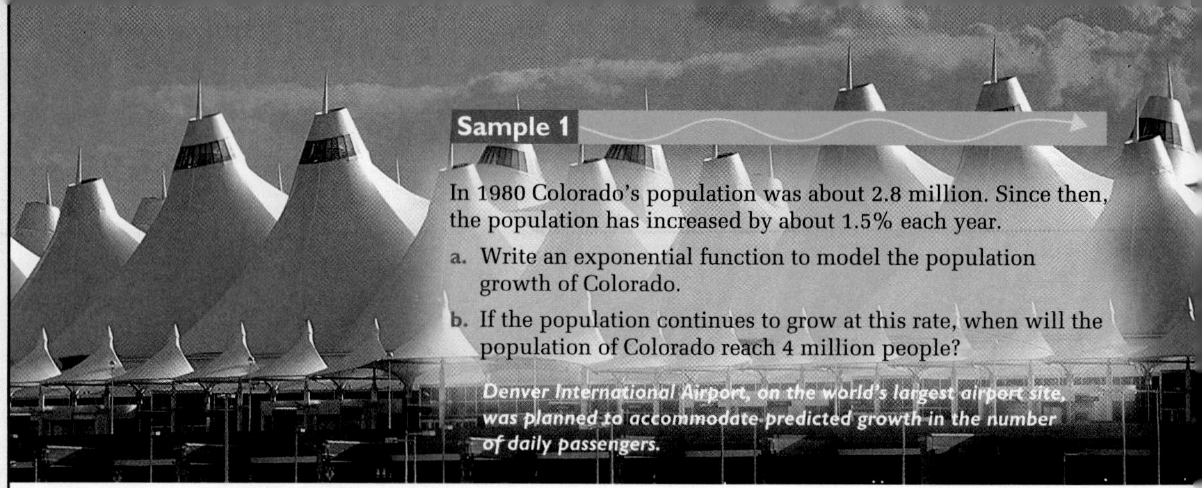

Sample 1

In 1980 Colorado's population was about 2.8 million. Since then, the population has increased by about 1.5% each year.

a. Write an exponential function to model the population growth of Colorado.

b. If the population continues to grow at this rate, when will the population of Colorado reach 4 million people?

Denver International Airport, on the world's largest airport site, was planned to accommodate predicted growth in the number of daily passengers.

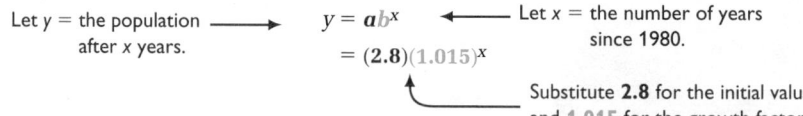

Sample Response

a. **Step 1** Find the growth factor b.

population at end of each year = 100% of population at start of year + **1.5%** of population at start of year
= 101.5% of population at start of year

The growth factor b is 101.5%, or 1.015.

Step 2 Model the growth with an exponential function.

Let y = the population after x years. → $y = ab^x$ ← Let x = the number of years since 1980.

$= (2.8)(1.015)^x$

Substitute **2.8** for the initial value a and **1.015** for the growth factor b.

b. **Method 1** Use a graph.

Graph the equation $y = (2.8)(1.015)^x$.

Since an x-value of zero represents 1980, an x-value of 24 represents the year 2004. The population of Colorado will be about 4 million in the year 2004.

Read the value of x when $y \approx 4$.

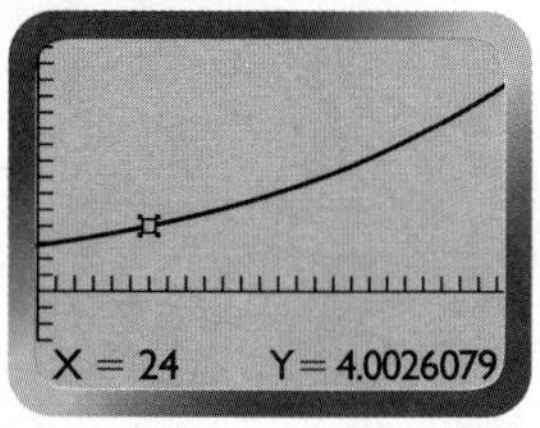

Method 2 Use recursion.

Enter the initial amount 2.8 and repeatedly multiply it by the growth factor 1.015.

```
2.8
            2.8
Ans *1.015
          2.842
        2.88463
     2.92789945
     2.971817942
```

Count the number of times —24— that you press ENTER to reach 4 million.

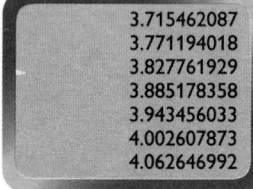

```
3.715462087
3.771194018
3.827761929
3.885178358
3.943456033
4.002607873
4.062646992
```

24 years after 1980 represents the year 2004. The population of Colorado will be about 4 million in the year 2004.

Answers to Talk it Over

3. The model assumes that the growth rate will remain unchanged during the time period in question. The growth rate may not stay the same. For example, it might increase (for example, in response to an economic boom in the state) or decrease (for example, in response to a decline in available natural resources).

4. For the same value of a, the greater the value of b, the steeper is the graph of $y = ab^x$.

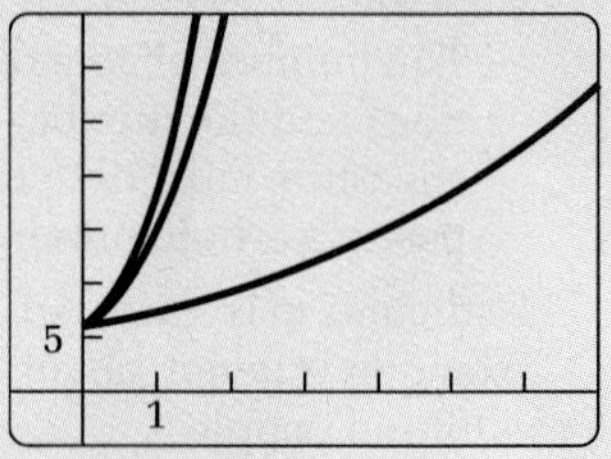

Now you are ready for: Exs. 1–16 on pp. 271–272

Talk it Over

3. What information does the model in Sample 1 assume? What may affect the accuracy of the model's estimate for Colorado's population?
4. Graph each function on the same calculator screen. How does changing the value of b in $y = ab^x$ affect the graph of an exponential function?

   $y = (6)(3.1)^x$ $\quad y = (6)(2.5)^x$ $\quad y = (6)(1.25)^x$

EXPLORATION

Can an exponential model show a decreasing amount?

- **Materials: 50 objects with 2 differently marked sides (for example, pennies), a container large enough to hold your 50 objects, graphics calculators**
- **Work in a group of four students.**

Toss	Number of objects remaining
0	50
1	?
2	?
3	?
⋮	?

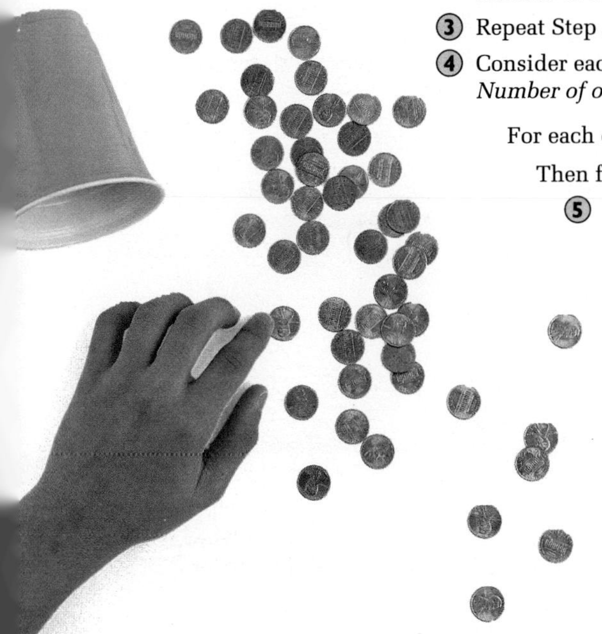

1. In this Exploration, you will remove objects that land "face up." Decide which side of your object will be the face.
2. Shake the objects in a container and spill them onto a flat surface. Remove the ones that land face up. Record the number of objects that remain.
3. Repeat Step 2 until there are no objects left.
4. Consider each data value listed in the row of data labeled *Number of objects remaining* as a term a_n in a sequence.

   For each data value less than 50, find the ratio $\frac{a_n}{a_{n-1}}$.

   Then find the average of these ratios.
5. What number represents the *initial value* in this situation? Use the initial value and the average ratio you found in Step 4 to write an exponential equation to model this situation. Graph your equation.
6. Compare your equations and graphs with those of other groups. Describe any patterns you see.
7. Besides representing the decay factor in the number of objects remaining after each toss, what do you think the value of b represents in this situation? Explain your reasoning.

Answers to Exploration

1. Answers may vary.

2, 3. Answers may vary. On average, the number of objects remaining after each toss should be about half the number before the given toss.

4. Answers may vary. The average of the ratios should be about $\frac{1}{2}$.

5. 50; $y = 50(0.5)^x$

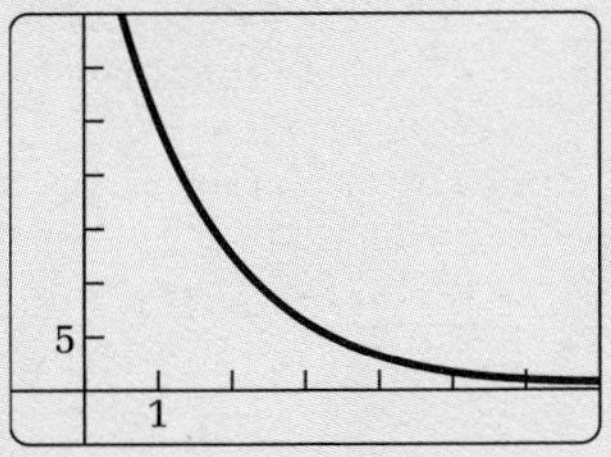

6. a should be 50 for all groups; b should be about 0.5.

7. b represents the probability that an object will land face up (or will not land face up). On the first toss, the probability that a coin will not land face up (and therefore, remain for the second toss) is $\frac{1}{2}$, so the number left after the first toss should be about 50(0.5). Similarly, the number left after the second toss should be about 50(0.5)(0.5) or $50(0.5)^2$.

Method 2. Use recursion. Enter the initial amount 23,307 and repeatedly multiply it by the growth factor of 1.054. Count the number of times–15–you press ENTER to reach 50,000. 15 years after 1992 represents the year 2007. The benefits will reach $50,000 million in the year 2007.

Mathematical Procedures

Many models can be represented fairly accurately by an equation, in this case, by an exponential equation. A regression coefficient can be computed that indicates how accurately an equation fits actual data. Students should realize that, as a model, any results are only estimates of a real situation.

Using Technology

In Method 2 of Sample 1, students can also use a spreadsheet. Have students enter the value 2.8 in cell B2 and B2*1.015 in B3. Have students copy the formula in several more cells of the column. Students can count the number of times it takes until a population of 4,000,000 is reached.

Talk it Over

Question 3 has students think about the real-world factors that may affect the accuracy of a mathematical model. Question 4 has students explore the effect of different values of b on the graph of the function with form ab^x.

Exploration

The goal of the Exploration is to have students discover, through the use of manipulatives, an exponential decay equation. By graphing their equations and comparing the equations and graphs with those of other groups, students are led to the exponential decay function presented at the top of page 270.

EXPONENTIAL FUNCTIONS

Exponential Growth

$y = ab^x$, $b > 1$

Example: $y = (5)(1.2)^x$

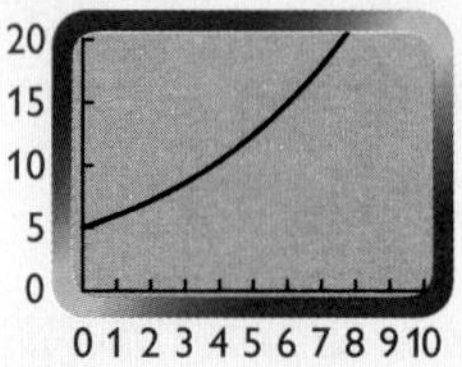

Exponential Decay

$y = ab^x$, $0 < b < 1$

Example: $y = (5)(0.5)^x$

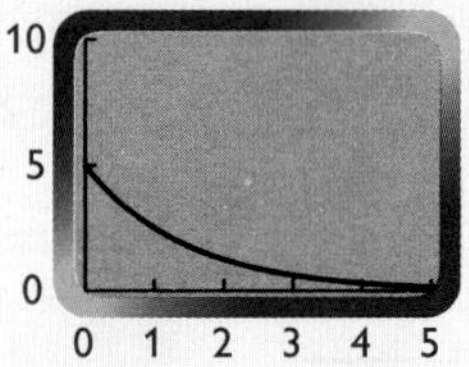

Sample 2

The average amount of fuel used annually by each car in the United States has decreased by about 2% each year since 1980, when it was about 591 gallons.

a. Write an exponential function to model this situation.

b. Estimate the average number of gallons of gasoline used by a car in 1987.

Sample Response

a. $y = ab^x$ ← Let x = the number of years since 1980. Let y = the average number of gallons of gasoline used.

$= (591)b^x$ ← Substitute 591 for the initial amount a.

$= (591)(0.98)^x$ ← The expression $(1 - 0.02)$ represents a 2% decrease. Substitute 0.98 for b.

b. **Problem Solving Strategy:** Use a graph.

Graph the equation $y = (591)(0.98)^x$. The initial value occurs in 1980, so an x-value of zero represents 1980.

An x-value of 7 represents 1987. Read the value of y when $x = 7$.

In 1987, each car in the United States used an average of about 513 gallons of gasoline.

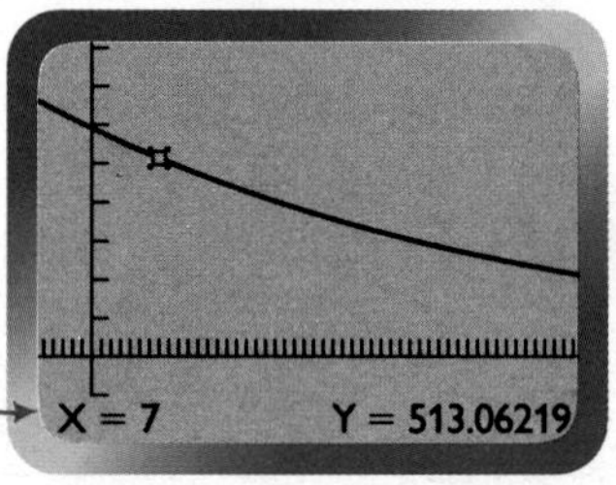

Look Back

Describe a function that models exponential decay. How is this function like an exponential growth function? How is it different?

Now you are ready for: Exs. 17–33 on pp. 272–274

Additional Sample

S2 In 1984, funds for the Emergency Food Assistance program were about $1,075 million. Since 1984, this fund has decreased about 19% per year.

a. Write an exponential function to model this situation.

$y = ab^x$

Let x = the number of years since 1984.

Let y = the funds available (in millions) for the Emergency Food Assistance program.

Substitute 1075 for the initial amount.

$y = (1075)b^x$

The expression $(1 - 0.19)$ represents a 19% decrease. Substitute 0.81 for b.

$y = (1075)(0.81)^x$

b. Estimate the funds available for the Emergency Food Assistance program in 1990.

Graph the equation $y = (1075)(0.81^x)$.

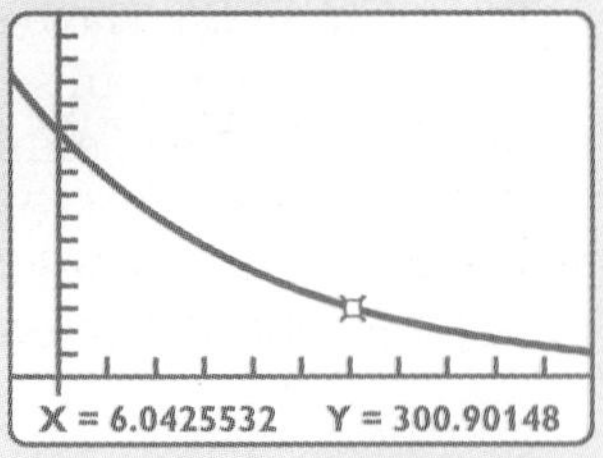

The initial value occurs in 1984, so an x-value of 6 represents 1990. Read the value of y when $x = 6$. In 1990, the funds available for the Emergency Food Assistance program were about $301 million.

Look Back

This Look Back would make a good journal entry. Students could then discuss their answers with a classmate.

Answers to Look Back

A function that models exponential decay has the form $y = ab^x$, where $0 < b < 1$. A function that models exponential growth has the same form, except that $b > 1$.

5-1 Exercises and Problems

1. **Reading** What does each part of the general form of an exponential function $y = ab^x$ represent?

Match each function with its graph. On each scale one unit is 5.

2. $y = 5(1.04)^x$
3. $y = 5(1.2)^x$
4. $y = 2(1.1)^x$

A.
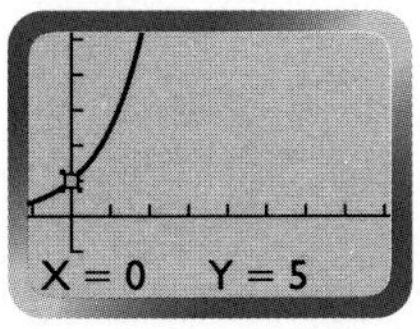

B.
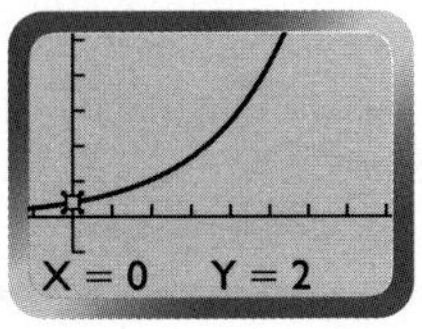

C.
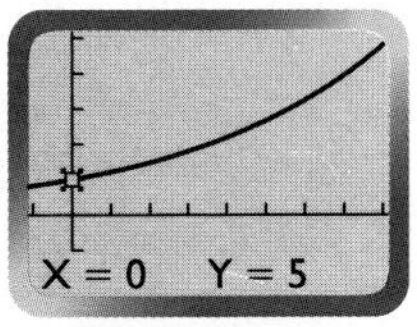

connection to **BUSINESS EDUCATION**

Madam C. J. Walker was one of the first business people to teach her employees effective sales techniques and provide information about the company's products. As a result, sales of Madam Walker's hair care products increased exponentially.

Use this information for Exercises 5–8.

Madam Walker's weekly sales revenues, in dollars, from 1906 to 1914 can be modeled by the function $R = (17.6)(1.88)^x$, where x represents the number of years since 1906.

For each year, estimate the weekly sales revenues.

5. 1906
6. 1909
7. 1912
8. In what year did the weekly revenues reach $2000?
9. Madam Walker opened Lelia College in 1908 to give her sales force formal training. About 100 people were sales agents in 1908. The number of sales agents increased exponentially by about 102% each year until about 1914.
   a. Find the growth factor b.
   b. Write an exponential equation to model the increase in the number of sales agents between 1908 and 1914.
   c. About how many sales agents were there in 1910? in 1912?

Madam C. J. Walker (right), the first American woman self-made millionaire, in the car she loved; (below) graduates of Lelia College

BY THE WAY. . .

At a 1912 convention of African-American business leaders, Sarah Walker described her success:

*I am a woman who came from the cotton fields of the South. I was promoted from there to the washtub. Then I was promoted to the cook kitchen, and from there **I promoted myself** into the business of manufacturing hair goods and preparations. I have built my own factory on my own ground. My object in life is not simply to make money for myself or to spend it on myself. I love to use a part of what I make in trying to help others.*

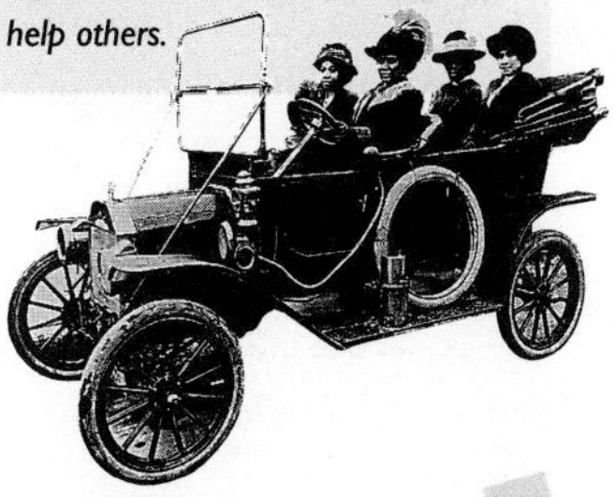

APPLYING

Suggested Assignment

Day 1

Standard 1–11, 13–16

Extended 1–11, 13–16

Day 2

Standard 18–21, 24–33

Extended 18–33

Integrating the Strands

Number Exs. 25–27

Algebra Exs. 1–27, 29–33

Functions Exs. 1–24, 32, 33

Geometry Ex. 28

Discrete Mathematics Exs. 10, 11, 25–27

Logic and Language Exs. 1, 13, 19, 24

Assessment: Portfolio

Students should collect the various types of functions encountered throughout this unit, including their graphs and applications.

Communication: Reading

Ex. 1 provides an excellent check of students' understanding of the parts of $y = ab^x$.

Answers to Exercises and Problems

1. a represents the initial amount, b represents the rate of change, x represents the time interval, and y represents the amount after x time intervals.
2. C
3. A
4. B
5. $17.60
6. $116.95
7. $777.07
8. 1914
9. a. 2.02
   b. Let n be the number of sales agents; $n = 100(2.02)^x$, where x is the number of years since 1908.
   c. 408 sales agents; 1665 sales agents

Reasoning

Exs. 12 and 17 have students use graphics calculators to explore how changing various values in the general exponential function $y = ab^x$ affects the graphs of the functions.

Interdisciplinary Problems

Ex. 14 presents an interesting connnection to history and economics, and also allows students to see how quickly an exponential function increases.

For each situation in Exercises 10 and 11:

a. Tell whether the situation is *discrete* or *continuous*.

b. Write an exponential function to model the situation.

c. Find the number asked for.

10. The amount of memory available on a computer circuit board doubles with each increase to the next larger size. The smallest size board holds 0.5 megabytes of memory. How many megabytes of memory are on the third smallest circuit board?

11. In 1964, a person invested \$48,000. Suppose the investment grows at an average annual rate of 5% compounded yearly. When will the investment be worth one million dollars?

12. a. TECHNOLOGY Graph each function on the same calculator screen.

    $y = 5(1.65)^x$ $\quad y = 9(1.65)^x$ $\quad y = 13(1.65)^x$ $\quad y = 15(1.65)^x$

    b. What is the y-intercept for each graph? How is the y-intercept represented in the equation?

    c. Compare the graphs. How does changing the initial value of an exponential function affect its graph?

13. **Open-ended** Graph $y = (3)(1.5)^x$. Change one number in the equation to any number you wish. Graph the new equation. How are the graphs related? How are the functions alike? How are they different?

connection to **HISTORY**

During the Revolutionary War battle at Valley Forge, Pennsylvania, in 1777, George Washington's army received a loan of \$450,000 from Jacob DeHaven. The Continental Congress agreed to pay back the loan at 6% interest, but DeHaven never received the money. In 1989, Jacob DeHaven's family filed a lawsuit to collect the money owed Jacob DeHaven, but no settlement was reached.

BY THE WAY...

Since the United States debt is already so large, the DeHaven family is willing to accept a "thank you" note and a statue built in Jacob DeHaven's honor.

Without the loan from Jacob DeHaven, Washington's troops may not have had enough food and supplies to survive the winter.

14. Suppose the interest was compounded yearly.

    a. How much was owed on the loan in 1783, when the Revolutionary War ended?

    b. When was five million dollars owed on the loan? 100 million dollars?

For Exercises 15 and 16, find how much was owed on the loan in 1989 using each type of interest calculation.

15. Suppose the 6% annual interest is compounded yearly.

16. Suppose the 6% annual interest is compounded daily.

17. a. TECHNOLOGY Graph each function on the same calculator screen.

    $y = 5(0.9)^x$ $\qquad y = 5(0.09)^x$

    b. What effect does changing 5 to -5 have on each graph?

    c. What effect does adding 1 to each value of b have on each graph?

Answers to Exercises and Problems

10. a. discrete
    b. $y = 0.5(2)^x$
    c. 2 megabytes

11. a. discrete
    b. $y = 48{,}000(1.05)^x$
    c. in 2027

12. a.

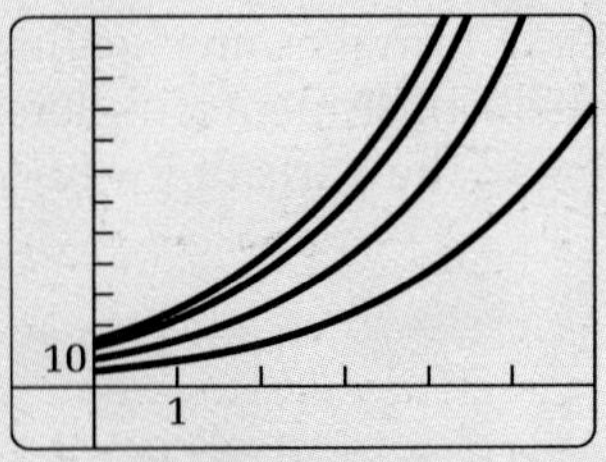

    b. 5, 9, 13, 15; The y-intercept is the initial value a in the expression ab^x.

    c. For a fixed value of b, increasing the value of a changes the y-intercept of the graph and increases the steepness of the graph.

13. Answers will vary.

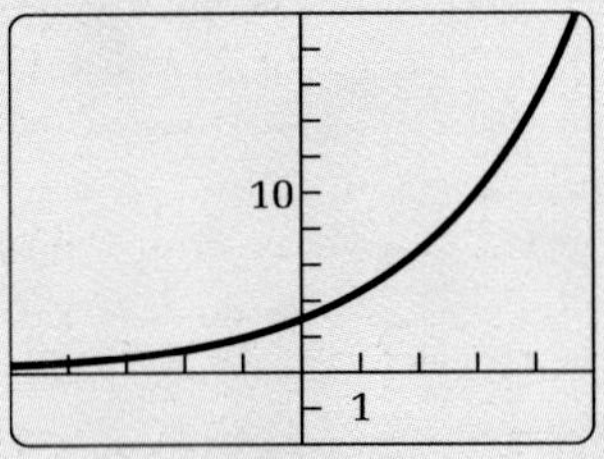

14. a. about \$638,334
    b. 1819; 1870

15. about \$104,245,000,000

16. based on 365 days/year, about \$150,500,000,000

17. a.

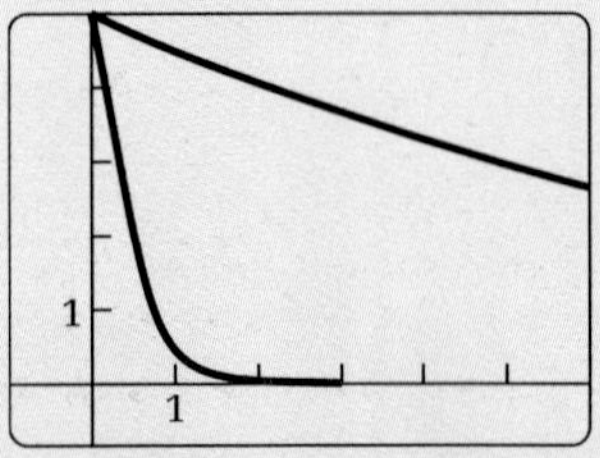

    b. Both contain (0, 5). The graph of $y = 5(0.09)^x$ declines sharply as x increases, while the graph of $y = 5(0.9)^x$ declines slowly.

    c. The graphs are reflected in the x-axis.

    d. The graphs rise as x increases instead of declining as x increases.

18. a. $y = 30\left(\frac{5}{6}\right)^x$

    b. $y = 30\left(\frac{2}{3}\right)^x$

    c.

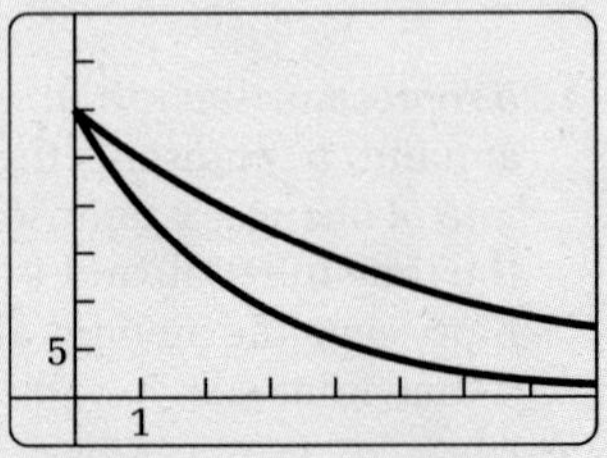

Both curves are exponential and model exponential decay. The rate of decay is greater for the first function; the

18. Suppose you repeat the Exploration using 30 dice and remove a die if a 6 lands face up.
   a. Write an equation to model the number of dice remaining after each toss.
   b. Write an equation to model the number of dice remaining after each toss if you remove a die if either a 1 or a 6 is face up.
   c. Graph each of the equations you wrote in parts (a) and (b) on the same axes. How are the graphs alike? How are they different?

19. a. **Writing** Compare the functions $y = (3)(1.25)^x$ and $y = (3)(0.75)^x$. How are the functions alike? How are they different?
   b. Graph each of the functions in part (a) on the same set of axes. Describe each graph. What point(s) do the graphs have in common?

Career After reaching a peak level, the concentration of a medication in a person's bloodstream decreases exponentially. Pharmacists can use an exponential model to determine the concentration level.

20. A medication for treating skin infections is available in three strengths. In each model, t represents the number of hours since the medication reached its peak concentration and y represents the concentration of the medication in the blood in micrograms per milliliter (μg/mL).

weakest strength	medium strength	greatest strength
$y = (0.8)(0.84)^t$	$y = (1.6)(0.84)^t$	$y = (3.2)(0.84)^t$

   a. Graph each equation on the same set of axes. What do the y-intercept values represent in this situation?
   b. How long does it take for each strength to reach a concentration of 0.1 μg/mL, the level at which the medication is considered cleared from a person's system?

21. The concentration of a stomach medication in a person's bloodstream in nanograms per milliliter (ng/mL) can be represented by the function $y = (492.5)(0.79)^t$, where t represents the number of hours since the medication reached its peak concentration.
   a. Graph the equation.
   b. Find the peak concentration.
   c. Find the time when half of the peak concentration remains. This is the medication's half-life.
   d. Use the information from the photo and your answer to part (c) to estimate when the stomach medication is cleared from a patient's blood stream.

Pharmacists multiply a medication's half-life by five to estimate when a medication is cleared from a patient's blood stream.

22. Repeat Exercise 21 for a hypertension medication whose concentration, in ng/mL, can be modeled by the equation $y = (115)(0.95)^t$.

23. Repeat Exercise 21 for an allergy medication whose concentration, in ng/mL, can be modeled by the equation $y = (263)(0.92)^t$.

Communication: Writing

Ex. 19 has students compare an exponential growth function with an exponential decay function. This will help students solidify their understanding of the differences and similarties of the two functions.

Career Note

Exs. 20–23 introduce students to the way pharmacists use exponential decay models to determine the concentration of medication in a person's bloodstream. Students wishing to become pharmacists must pursue a five-year course of study at a school of pharmacy within a university. The science of dealing with the preparation, use, and effects of drugs is called pharmacology.

Communication: Drawing

As an extension of Ex. 24 on page 274, students could choose one of the types of exponential models and present the model and its graph in poster form to the class.

Multicultural Note

Ex. 32 on page 274 discusses the pitch of a note. The jazz *blue note* is a pitch that has no written representation in Western music. The pitch of a blue note is generally the third or seventh scale tone flattened approximately one quarter step. This pitch change helps give jazz a soulful sound different from many other types of music. The blue note was not a part of American mainstream music until the early twentieth century, when jazz became popular for a wide variety of audiences. What became known as the blue note, however, was part of African American music long before jazz became popular. Bessie Smith, Blind Lemon Jefferson, and Louis Armstrong are some of the early jazz greats who helped to popularize the blue note.

Answers to Exercises and Problems

graph of the first function declines more steeply than that of the second.

19. a. Both functions are exponential functions, that is, both have the form $y = ab^x$. For the first equation, $b > 1$, so the first function models exponential growth. For the second equation, $0 < b < 1$, so the second function models exponential decay.

b.

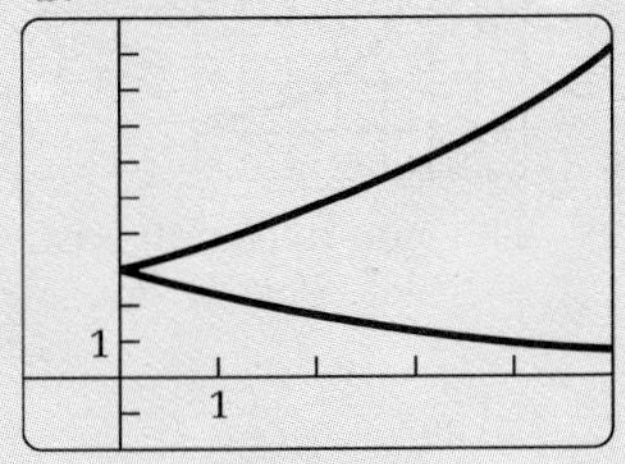

Both curves are exponential. The first models exponential growth; the second models exponential decay. The graphs have the point (0, 3) in common.

20–23. See answers in back of book.

Ongoing ASSESSMENT

24. **Open-ended** Describe a situation that can be modeled by each of the following: *discrete exponential growth, continuous exponential growth, discrete exponential decay, continuous exponential decay.* Sketch a graph for each example.

Review PREVIEW

Evaluate each series, if the sum exists. (*Section 4-7*)

25. $\sum_{n=0}^{\infty} (1.03)^n$

26. $\sum_{t=1}^{\infty} \left(\frac{1}{2}\right)^t$

27. $\sum_{n=1}^{\infty} -5(0.003)^{n-1}$

28. Can the sum of the interior angle measures of a regular polygon be 655°? Why or why not? (*Section 3-4*)

Simplify. Write answers with positive exponents. (*Toolbox Skill 14*)

29. $(3x^2)(5x^{-3})$

30. $5t(s^{-1}t^{-2})^{-2}$

31. $\frac{p^{-1}q^{-2}}{p^{-3}}$

5 Working on the Unit Project

The *pitch* of a note is how high or low the note sounds to a listener. Pitch depends on frequency. The frequencies of musical notes in a scale form a geometric sequence.

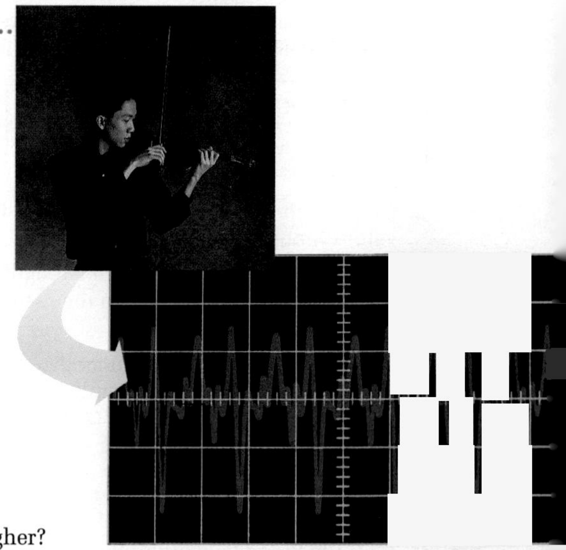

An electronic representation of a 440 Hz note produced by a vibrating violin string

32. Suppose there were a musical instrument whose notes are related to one another by the function $f(n) = 20(2)^n$, where n is the number of the note in the sequence, and the frequency of the note is measured in hertz (Hz).
    a. Find $f(1)$, the frequency of the note with the lowest pitch.
    b. Find the frequency for the next seven notes in the instrument's sequence.
    c. Graph the function. How does the frequency of a note change as the pitch of a note gets higher?
    d. The human ear can hear sounds in the range 20–20,000 Hz. What problem can you foresee in terms of the instrument described above?

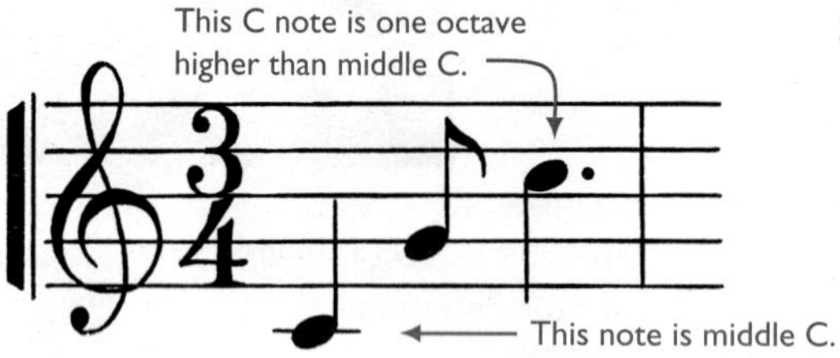
This C note is one octave higher than middle C.
This note is middle C.

33. In the music of Europe and the United States, the frequency of a note is double the frequency of the same note one *octave* lower. One of the C notes on a piano has a frequency of 130.8 Hz. Find the frequency of the C note that is one octave higher.

Working on the Unit Project

Exs. 32 and 33 introduce students to the connection between the pitch and frequency of a musical note. Students also learn that the frequencies of musical notes in a scale form a geometric sequence.

Practice 37 For use with Section 5-1

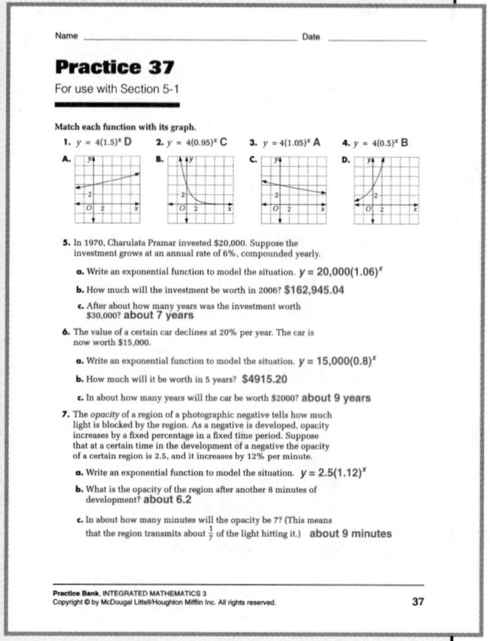

Name ______________ Date ______________

Practice 37

For use with Section 5-1

Match each function with its graph.

1. $y = 4(1.5)^x$ D
2. $y = 4(0.95)^x$ C
3. $y = 4(1.05)^x$ A
4. $y = 4(0.5)^x$ B

5. In 1970, Charulata Pramar invested $20,000. Suppose the investment grows at an annual rate of 6%, compounded yearly.
    a. Write an exponential function to model the situation. $y = 20{,}000(1.06)^x$
    b. How much will the investment be worth in 2006? $162,945.04
    c. After about how many years was the investment worth $30,000? about 7 years
6. The value of a certain car declines at 20% per year. The car is now worth $15,000.
    a. Write an exponential function to model the situation. $y = 15{,}000(0.8)^x$
    b. How much will it be worth in 5 years? $4915.20
    c. In about how many years will the car be worth $2000? about 9 years
7. The *opacity* of a region of a photographic negative tells how much light is blocked by the region. As a negative is developed, opacity increases by a fixed percentage in a fixed time period. Suppose that at a certain time in the development of a negative the opacity of a certain region is 2.5, and it increases by 12% per minute.
    a. Write an exponential function to model the situation. $y = 2.5(1.12)^x$
    b. What is the opacity of the region after another 8 minutes of development? about 6.2
    c. In about how many minutes will the opacity be 7? (This means that the region transmits about $\frac{1}{7}$ of the light hitting it.) about 9 minutes

Practice Bank, INTEGRATED MATHEMATICS 3

37

Answers to Exercises and Problems

24. Answers may vary. Examples are given. An amount is deposited in a savings account which has interest compounded monthly. The value of the account represents discrete exponential growth.

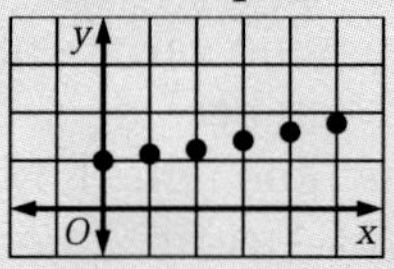

A bacteria is cultured in a growing medium. The growth of the bacteria represents continuous exponential growth.

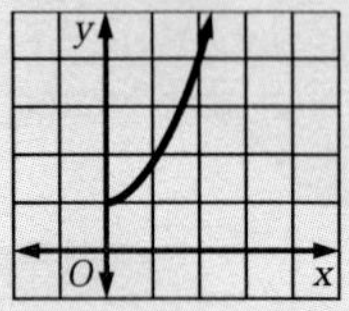

The number of teams remaining in a single-elimination tournament (such as the NCAA basketball championship tournament) represents discrete exponential decay.

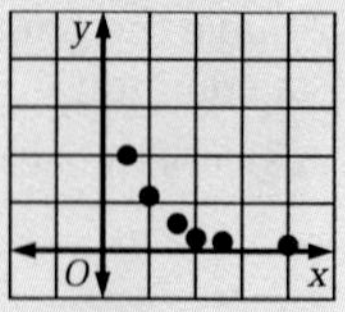

The amount of radioactive material left in a radioactive substance over a period of time represents continuous exponential decay.

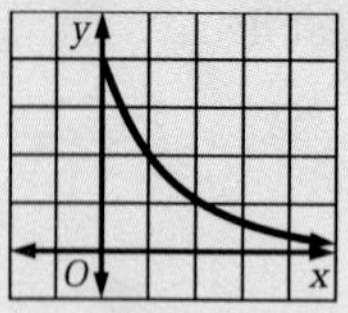

25. The sum does not exist.

26. 1

27. $-\frac{5}{0.997} \approx -5.015$

28. No; if such a polygon existed, the number, n, of sides of the polygon would satisfy the equation $(n - 2)180° = 655$. The solution to this equation is not a whole number.

29. $\frac{15}{x}$

30. $5s^2t^2$

31. $\frac{p^2}{q^2}$

32, 33. See answers in back of book.

Section

5-2 Negative Exponents

Focus

Use exponential functions with negative x-values to model situations.

NATIONAL PASTIME

Talk it Over

The function $y = (283)(1.2)^x$ can be used to model the average salary for a professional baseball player in the United States, in thousands of dollars, since 1984.

1. What year is represented by an x-value of 0? of 3? of 8?
2. What x-value should you use to find the average salary in 1991?
3. Use the x-value you found in question 2 to estimate the average salary in 1991.
4. What x-value should you use to find the average salary in 1983?

Use the function $y = (283)(1.2)^x$ to estimate the average salary in each year.

5. 1979
6. 1973
7. 1967

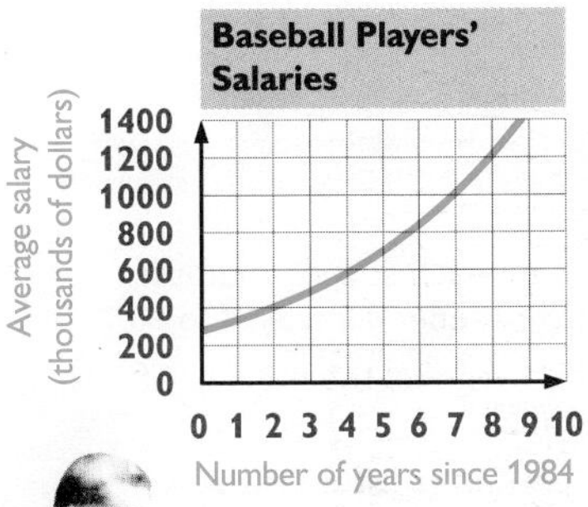

Mickey Cochrane, the best hitting, fastest running catcher up to that time, in action with the 1930 Philadelphia Athletics

Sometimes an exponential function models a change in a quantity over a certain period of time. To evaluate the function at an *earlier* time, you use negative x-values. In the model of baseball players' average salary, you use negative x-values to find the average salary in a year before 1984.

PLANNING

Objectives and Strands

See pages 264A and 264B.

Spiral Learning

See page 264B.

Materials List

- Graphics calculator

Recommended Pacing

Section 5-2 is a one-day lesson.

Toolbox References

- **Toolbox Skill 14:** Using Rules of Exponents

Extra Practice

See pages 615–617.

Warm-Up Exercises

- Warm-Up Transparency 5-2

Support Materials

- Practice Bank: Practice 38
- Activity Bank: Enrichment 34
- Study Guide: Section 5-2
- Problem Bank: Problem Set 11
- Using IBM/Mac Plotter Plus Disk: Function Plotter
- Assessment Book: Quiz 5-2

Answers to Talk it Over

1. 1984; 1987; 1992
2. 7
3. about $1,014,000
4. –1
5. about $114,000
6. about $38,000
7. about $12,750

TEACHING

Talk it Over

Questions 1–3 ask students to think about various positive x-values for the given function in relation to their real-world meaning. Questions 4–7 extend this reasoning to have students consider negative x-values and their relationship to the real-world data.

Exploration

The goal of the Exploration is for students to investigate the effect negative exponents have on an exponential function and to discover the relationship between $y = ab^{-x}$ and $y = a\left(\frac{1}{b}\right)^x$.

Teaching Tip

You may wish to point out that the Sample problem is asking students to *extrapolate* information, since the year 1986 is *outside* of the range of data from 1989 to 1990. If students are asked to find a value *within* the given range of data, then they interpolate to find the approximate answer. Extrapolation is often a less accurate estimate than interpolation.

Additional Sample

The population of Wisconsin was about 4,700,000 in 1980, and about 4,900,000 in 1990.

a. Write an exponential function to model the population of Wisconsin over time.

Step 1. Find the growth factor b.

$\frac{\text{population in 1990}}{\text{population in 1980}} = \frac{4{,}900{,}000}{4{,}700{,}000} \approx 1.043$

Step 2. Model the situation with an exponential function. Let x = the number of 10-year periods since 1980 and let y = the population of Wisconsin in millions.

Using $y = ab^x$, substitute 4.7 for the initial value a and 1.043 for the growth factor b.

$y = (4.7)(1.043)^x$

EXPLORATION

How does a negative exponent affect an exponential function?

- **Materials: graphics calculators**
- **Work with another student.**

① Graph $f(x) = 3 \cdot 2^x$. Include both positive and negative values for x. What are the domain and the range of $f(x)$?

② Write an equation for the image when $f(x) = 3 \cdot 2^x$ is reflected over the y-axis.

③ Graph $g(x) = 3 \cdot 2^{-x}$ and $h(x) = (3)\left(\frac{1}{2}\right)^x$ on the same set of axes. What do the graphs tell you about the relationship between $g(x)$ and $h(x)$?

④ Show that $(2)\left(\frac{1}{4}\right)^x = (2)(4)^{-x}$.

Student Resources Toolbox p. 637 *Algebraic Expressions*

Sample

In 1989 there were 24.1 students for each school computer in the United States. By 1990 this number decreased to 20.7.

a. Write an exponential function to model the number of students per computer over time.

b. Use the model you wrote in part (a) to estimate the number of students for each computer in 1986.

Sample Response

a. **Step 1** Find the decay factor b.

$\frac{\text{no. of students per computer in 1990}}{\text{no. of students per computer in 1989}} = \frac{20.7}{24.1} \approx 0.86$ ← The value of b must be less than 1 since the function is decreasing.

Step 2 Model the situation with an exponential function.

$y = ab^x$ ← Let x = the number of years since 1989.

$= (24.1)(0.86)^x$ ← Substitute **24.1** for the initial value a and **0.86** for the decay factor b.

b. Since the value $x = 0$ represents 1989, the value $x = -3$ represents 1986.

Method ❶ Use a graph.

Graph $y = (24.1)(0.86)^x$.

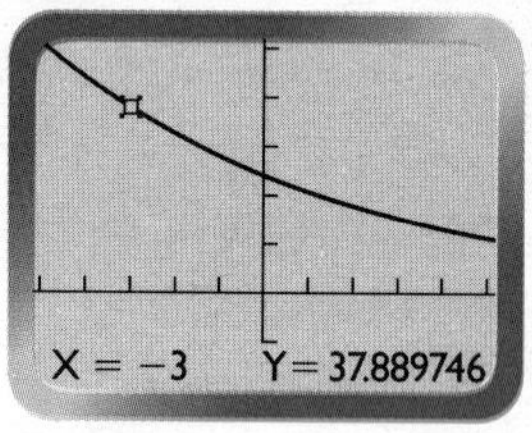

Read the value of y when $x = -3$.

Method ❷ Use an equation.

$y = (24.1)(0.86)^x$

$= (24.1)(0.86)^{-3}$ ← Substitute −3 for x.

≈ 37.9

There were about 38 students for each computer.

Answers to Exploration

1.

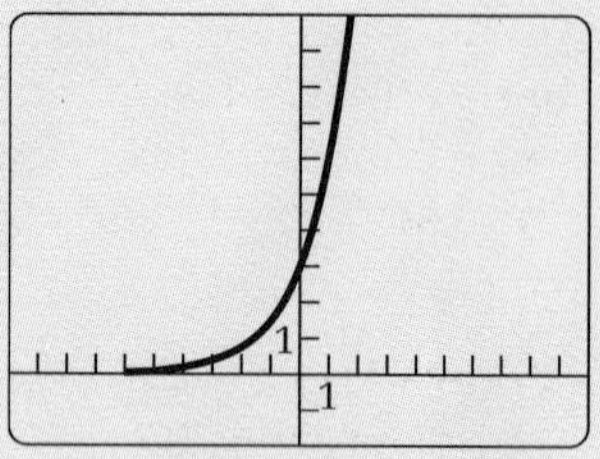

the real numbers; the positive real numbers

2–4. See answers in back of book.

Answers to Look Back

Answers may vary. Examples are given. A quantity, such as salaries, that models exponential growth relative to a base year can be modeled by an equation of the form $y = ab^x$. Negative values of x have meaning; they indicate years before the base year, while positive values of x indicate years after the base year. If the exponential growth is relative to a base year before which the quantity did not exist, negative values of x have no meaning. One example would be the growth in sales of a newly invented product.

Look Back

Give an example of a situation that can be modeled by an equation of the form $y = ab^x$ where negative values of x have meaning. Describe a situation in which negative values of x would not have meaning.

5-2 Exercises and Problems

1. **Reading** What other value can you use for a in the equation $y = ab^x$ to model the situation in the Sample? What x-value represents 1986 in this model?

Rewrite each function in the form $y = ab^x$.

2. $y = (25)\left(\frac{5}{6}\right)^{-x}$
3. $y = (0.5)(8)^{-x}$
4. $y = (8)(0.125)^{-x}$

For each graph, answer parts (a)–(c).

5.

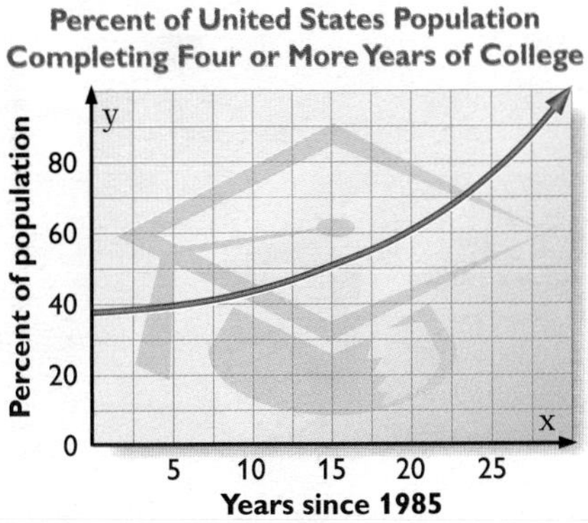

6.

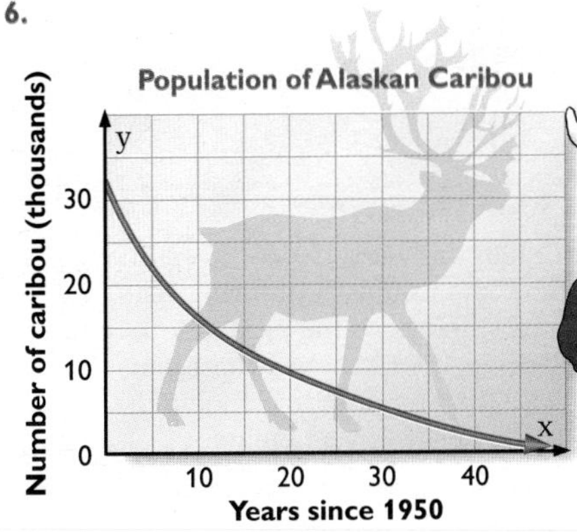

a. What year is represented by an x-value of 0? of -5? of 15?

b. What happens to the graph as the x-values become negative? positive?

c. **Writing** What happens to the y-values as the x-values increase? Do all y-values make sense in this situation? Why or why not?

7. **Cellular Phones** The first commercial cellular phone system in the United States began service in 1983. The function $y = (91.6)(1.77)^x$, where x represents the number of years since 1983, shows how many thousands of cellular phone subscribers there are in the United States.

a. Estimate the number of subscribers in 1984.

b. Without calculating, tell how many subscribers there were in 1982.

c. **Writing** The domain of the function is $x \geq 0$. Explain why it does not make sense to assign a negative value to x in this situation.

Answers to Exercises and Problems

1. 20.7; −4
2. $y = (25)(1.2)^x$
3. $y = (0.5)(0.125)^x$
4. $y = (8)(8)^x = 8^{x+1}$
5. a. 1985; 1980; 2000
   b. The graph declines and moves toward the x-axis; the graph rises and moves away from the x-axis.
   c. y-values increase; Yes; a percent can be any real number.
6. a. 1950; 1945; 1965
   b. The graph rises and moves away from the x-axis; the graph declines and moves toward the x-axis
   c. y-values decrease; No; the number of caribou must be a positive integer.
7. a. about 162,132
   b. 0
   c. For years represented by negative values of x, there was no cellular phone system, and so no subscribers.

b. Use the model you wrote in part (a) to estimate the population of Wisconsin in 1970.

Since $x = 0$ represents 1980, the value −1 represents one 10-year period before 1980, or the year 1970.

Method 1. Use a graph.

Graph $y = (4.7)(1.043)^x$

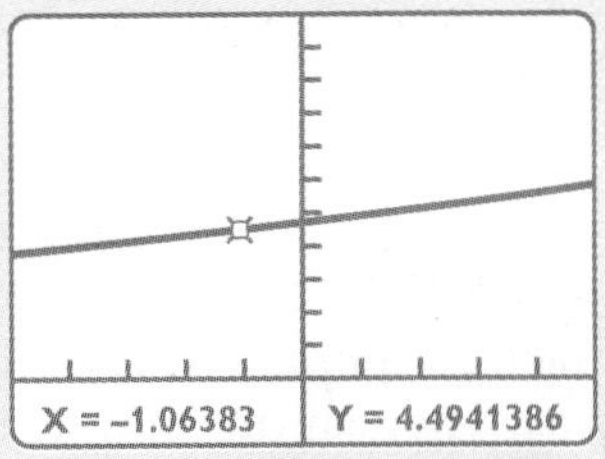

Read the value of y when $x = -1$. There were about 4,500,000 people in Wisconsin in 1970.

Method 2. Use an equation.

$y = (4.7)(1.043)^x$

$= (4.7)(1.043)^{-1}$

≈ 4.5

There were about 4,500,000 people in Wisconsin in 1970.

Look Back

You may wish to have every student write an answer to this Look Back and then share them in small groups.

APPLYING

Suggested Assignment

Standard 1–7, 11, 13, 15–22

Extended 1–9, 11–22

Integrating the Strands

Number Exs. 2–4, 17, 18, 21

Algebra Exs. 1–21

Functions Exs. 1–16

Logic and Language Exs. 1, 5–7, 10, 12, 15, 22

Students Acquiring English

In Exs. 1–7 there are three items that require a significant amount of writing. Consider pairing students acquiring English with English proficient partners for these exercises in order to make the writing assignments more accessible.

Reasoning

Ex. 7 on page 277 points out that not every situation makes sense for negative values of x. The feasible range for a variable must be considered before evaluating. Just because a numerical value can be found does not mean that it is a valid answer for a problem.

Application

As Exs. 8–13 demonstrate, exponential models are often used for population studies. A year is given the x-value of zero, then positive x-values are used to estimate populations after that date, and negative x-values are used to estimate populations before that date.

Multicultural Note

The Gold Rush of 1848–1850 attracted thousands of people to San Francisco and changed the site from a town of 800 inhabitants in 1848 to a city of many thousands in 1849. Significant numbers of gold seekers came from China, Mexico, Chile, Australia, Italy, and France, as well as from the eastern states of the United States. Even today, many of San Francisco's neighborhoods show ethnic origins established in that period. The city's famed Chinatown was first formed during the Gold Rush, as was the traditional Italian neighborhood known as North Beach.

connection to **HISTORY**

Although the population of the United States has increased, different areas have experienced various rates of growth at different times. These rates are often determined by economic events.

8. Use the equation $y = (46.6)(1.23)^x$ to model the population of the United States in millions, where x is the number of decades after 1890.
   a. Which year is represented by an x-value of 10? of 0? of -2?
   b. What was the approximate population of the United States in 1790? in 1840? in 1940?

9. The function $y = (2988)(1.132)^x$, where x represents the number of months since March of 1849, models the population of San Francisco between 1847 and 1849.
   a. Find the approximate population of San Francisco in January of 1848, when gold was discovered in the Sierra Nevada.
   b. Find the approximate population of San Francisco in January of 1849, after President James K. Polk addressed Congress and announced the discovery of gold.

10. In the 1960s and early 1970s, the silicon chip was developed in Santa Clara County, California. The function $y = (1.5)(1.026)^x$ can be used to model the population, in millions, of this region from 1960 to 1992, where x represents the number of years since 1990. Use this function to estimate the population of Santa Clara County during each year below.
    a. In 1969 a chip becomes the main component of a computer.
    b. In 1971 the world's first "computer on a chip" is developed.
    c. In 1974 the phrase "Silicon Valley" is used to describe this area where many electronics and computer corporations are located.
    d. **Research** Use the function to estimate the present population of Santa Clara County. Then look up the present population. Give a possible reason for any differences in the two numbers.

Answers to Exercises and Problems

8. a. 1990; 1890; 1870
   b. about 5,880,000; about 16,550,000; about 131,190,000
9. a. about 530
   b. about 2300
10. a. about 875,000
    b. about 921,000
    c. about 995,000
    d. Answers may vary. An example is given. about 1,700,000; about 1,500,000; Differences might be due to changes in the computer industry and the economy in general.
11. a. $y = 993{,}678(1.5786)^x$, where x is the number of decades since 1920
    b. Estimates may vary; about 457,280.
    c. Estimates may vary; about 574,590.
12. a. a graph showing exponential decay

11. During the first half of the twentieth century, many people moved to Detroit to work on the assembly line at the Ford Motor Company. The company offered wages of $25 per week, about $12 more than the average salary of that time.

   a. Use the populations of Detroit during 1920 and 1930 to write an exponential function that models the population of Detroit.

   b. Estimate the population of Detroit in 1903, when the Ford Motor Company was founded. (*Hint:* Since 10 years is represented by 1 unit, 1 year is represented by 0.1 unit.)

   c. Repeat part (b) for 1908, the year the Model T was invented.

Year	Population
1920	993,678
1930	1,568,662

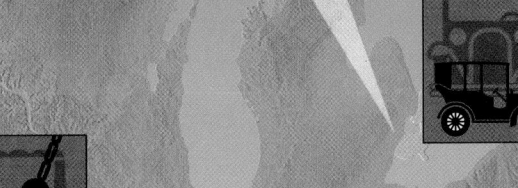

12. Since 1970, manufacturing companies in Gary, Indiana, have become more modernized. Fewer workers are needed at these companies and people have left to take jobs in other cities.

   a. **Writing** Describe a graph that could represent the change in Gary's population.

   b. The population of Gary, Indiana, can be modeled by the function $y = (218{,}378)(0.82)^x$. Graph this function. How does the model compare with your graph from part (a)?

13. After the Industrial Revolution, people began moving out of rural areas and into cities. In 1890 only 64.9% of people in the United States lived in a rural area as opposed to 71.8% in 1880.

   a. Write an exponential function to model the rural population as a percent of the total United States population.

   b. Estimate the percent of people living in rural areas in 1870.

   c. Estimate the percent of people living in rural areas in 1840. Tell whether you think this is a reasonable estimate and why.

Answers to Exercises and Problems

b. 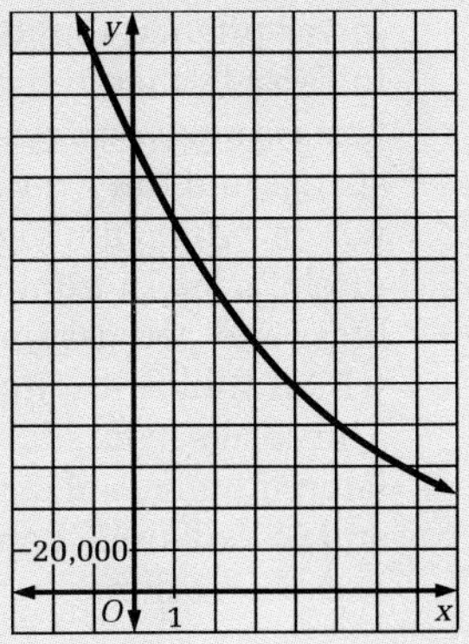

Answers may vary. The graph should be as described in part (a).

13. a. $y = 0.718(0.904)^x$, where x is the number of decades since 1880

b. Estimates may vary; about 79.4%.

c. Estimates may vary; about 107.5%. This is not a reasonable estimate. Even 100% would not be a reasonable estimate, since there were cities in the United States in 1840.

Communication: Writing

Ex. 15 presents an excellent opportunity to check students' understanding of the relationship between exponential functions with positive and negative exponents.

Practice 38 For use with Section 5-2

Name ______________ Date ______________

Practice 38

For use with Section 5-2

1. $y = 15\left(\frac{3}{2}\right)^x$ 2. $y = 4\left(\frac{4}{3}\right)^x$
3. $y = 48\left(\frac{2}{5}\right)^x$ 4. $y = 100\left(\frac{20}{21}\right)^x$
5. $y = (-14)(4)^x$ 6. $y = 22(5)^x$

Rewrite each function in the form $y = ab^x$.

1. $y = (15)\left(\frac{2}{3}\right)^{-x}$ 2. $y = (4)(0.75)^{-x}$ 3. $y = (48)\left(\frac{5}{2}\right)^{-x}$
4. $y = (100)(1.05)^{-x}$ 5. $y = (-14)(0.25)^{-x}$ 6. $y = (22)\left(\frac{1}{5}\right)^{-x}$
7. The approximate population of San Diego, California between 1960 and 1990 can be modeled by the function $y = (875{,}538)(1.024)^x$, where x represents the number of years since 1980. Use this function to estimate the population of San Diego in each year.
   a. 1990 1,109,876 b. 1970 690,678 c. 1965 613,445 d. 1960 544,849
8. Magnesium-27, an isotope of the element magnesium, decays radioactively in such a way that at the end of every minute 7% less magnesium-27 is present than was present at the start of that minute. Suppose a sample contains 30.0 mg of magnesium-27 at 1:00 P.M.
   a. Write an equation that models the amount y of magnesium-27 present in the sample at x minutes after 1:00 P.M. $y = 30(0.93)^x$
   b. Find the amount of magnesium-27 present at 1:15 P.M. about 10.1 mg
   c. Find the amount of magnesium-27 present at 12:52 P.M. about 53.6 mg
   d. When was the amount of magnesium-27 in the sample about 40 mg? at about 12:56 P.M.
9. A certain bacteria culture increases at the rate of 4% per minute. At 6:00 A.M., there were 150,000 bacteria in the culture.
   a. Write an equation that models the number y of bacteria in the culture x minutes after 6:00 A.M. $y = 150{,}000(1.04)^x$
   b. How many bacteria were present in the culture at 5:45 A.M.? about 83,290 bacteria
   c. At what time were there about 90,000 bacteria present? at about 5:47 A.M
10. ***Writing*** Suppose Quantity A is increasing by 10 units per minute and Quantity B is increasing by 5% per minute, and suppose both A and B stand at 100 units at 12:00 noon. Describe the relative sizes of the two quantities before and after 12:00 noon. In particular, were they ever equal before 12:00 noon, and will they ever be equal again after 12:00 noon? Before 12:00 noon, A < B. For about 26 min 35 s after noon, A > B. After that, B remains greater than A. The quantities are equal at 12:00 noon and 26 min 35 s after noon.

38 **Practice Bank**, INTEGRATED MATHEMATICS 3

14. Mark and Chandra's data from a chemistry lab shows that the amount of a material left after each experiment depends on the length of the experiment.

Length of experiment (min)	7	8	9	10	11
Amount of material left (mL)	187.5	75	30	12	4.8

Mark Owens

$\frac{75}{187.5} = 0.4$

$f(x) = (187.5)(0.4)^x$

Chandra Ray

$\frac{4.8}{12} = 0.4$

$g(x) = (12)(0.4)^x$

a. Explain the method each person used to find an equation to model the data.

b. How long an experiment does an x-value of zero represent in each of the functions?

c. Use Chandra's function to estimate the amount of material that is left after each length of time.

3 min 5 min
12 min 15 min

d. Repeat part (c) using Mark's function. How do your answers compare with your answers from part (c)?

e. Use the rules of exponents to show that $g(x) = f(x + 3)$. What does this tell you about the relationship between the graphs of $g(x)$ and $f(x)$? Graph the functions to check your answer.

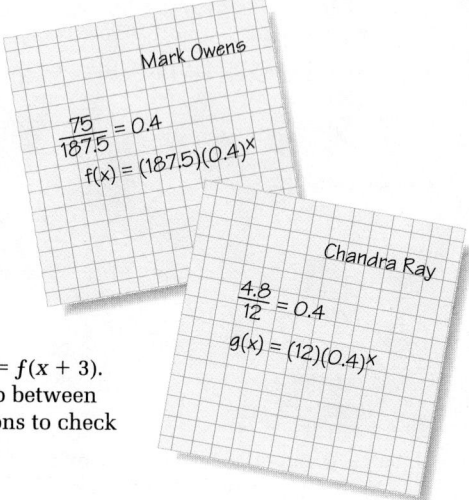

Ongoing **ASSESSMENT**

15. **Writing** Explain the relationship between the graphs of the functions $y = ab^x$ and $y = ab^{-x}$.

Student Resources Toolbox
p. 637 *Algebraic Expressions*

Review **PREVIEW**

16. In 1940 the population of the United States was about 132 million. Since then, it has increased at an average rate of 1.3% each year. Write an exponential function to model the population of the United States. (Section 5-1)

Simplify. (Section 2-6)

17. $\sqrt[3]{(-2)^3}$ 18. $\sqrt[4]{(-1)^4}$ 19. $\sqrt[3]{27d^9}$ 20. $\sqrt[6]{64x^6}$

21. Use the geometric sequence 41, x, 369, ... (Section 4-4)
    a. Find the value of x.
    b. Find the common ratio.
    c. Find the next three terms.

5 Working on the Unit Project

22. **Research** Find out how many of each instrument are included in a typical orchestra and where each is located in the orchestra. How do you think an instrument's position affects the performance?

Answers to Exercises and Problems

14. a. Mark used the values 187.5 and 75 from the table to find the decay factor and 187.5 as the initial value. Chandra used the values 12 and 4.8 from the table to find the decay factor and 12 as the initial value.
    b. For $f(x) = 187.5(0.4)^x$, an x-value of 0 represents an experiment of 7 min; for $g(x) = 12(0.4)^x$, an x-value of 0 represents an experiment of 10 min.
    c. 3 min: about 7324.2 mL; 5 min: about 1171.9 mL; 12 min: about 1.92 mL; 15 min: about 0.1 mL
    d. 3 min: about 7324.2 mL; 5 min: about 1171.9 mL; 12 min: about 1.92 mL; 15 min: about 0.1 mL; They are the same.
    e. $f(x + 3) = 187.5(0.4)^{x+3} = 187.5(0.4)^x(0.4)^3 = 187.5(0.064)(0.4)^x = 12(0.4)^x = g(x)$; The graph of $g(x)$ is the graph of $f(x)$ shifted left 3 units. Check students' graphs.

15. See answers in back of book.
16. $y = (132{,}000{,}000)(1.013)^x$, where x is the number of years since 1940
17. -2 18. 1
19. $3d^3$ 20. $2|x|$
21. a. ± 123
    b. 3
    c. ± 1107, 3321, ± 9963
22. Answers may vary. A typical modern orchestra usually has 90 to 110 members divided into four sections: the strings (violins, violas, cellos, double basses), with 40 to 60 members; the woodwinds (flutes, piccolos, clarinets, oboes, English horn, bassoons), with 8 to 16 members; the brass (French horns, trumpets, trombones, tubas), with 7 to 16 members; and percussion (timpani, cymbals, triangle, drums, gongs, xylophone, and so on), with 2 to 6 members. Since lower-pitched sounds tend to carry farther, the orchestra is arranged so that the strings and the woodwinds, which are higher-pitched, are in front of the brass and percussion, which are lower-pitched. This also accounts for the larger numbers of strings and woodwinds than brass and percussion.

Section

5-3 Fractional Exponents

Focus

Use expressions involving fractional exponents or radicals. Model situations using exponential functions with fractional exponents.

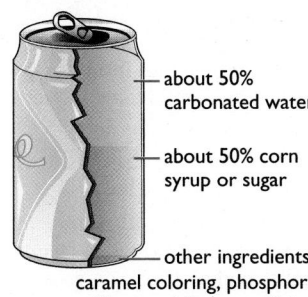

Getting To The Root Of It

An average can of a soft drink may contain about 40 mg of caffeine. In an adult, the half-life of caffeine in the bloodstream is about 5 hours. This is how long it takes for half the amount of caffeine to be eliminated from the bloodstream.

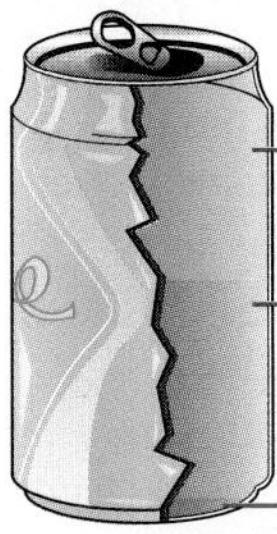

Sample 1

a. Suppose an adult drinks a soft drink containing 40 mg of caffeine. The amount of caffeine in the bloodstream of an adult is a function of the number of 5-hour periods after the beverage is consumed. Write an equation to model this function.

b. How much caffeine is left in the bloodstream 1 hour after an adult drinks a soft drink with 40 mg of caffeine?

Sample Response

a. Find an equation to model the amount of caffeine in the bloodstream after x periods of 5 hours.

After one half-life (5 hours) $(40)\left(\frac{1}{2}\right) = 20$ mg of caffeine remaining

After two half-lives (2 · 5 hours) $(40)\left(\frac{1}{2}\right)\left(\frac{1}{2}\right) = 10$ mg of caffeine remaining

This pattern can be modeled by the function $f(x) = 40\left(\frac{1}{2}\right)^x$, where x represents the number of 5-hour periods after an adult consumes a soft drink.

b. One hour is $\frac{1}{5}$ of a 5-hour period.

$$f(x) = 40\left(\frac{1}{2}\right)^x$$

$$f(1) = (40)\left(\frac{1}{2}\right)^{1/5} \quad \leftarrow \text{Substitute } \tfrac{1}{5} \text{ for } x.$$

$$\approx 34.82$$

There are about 35 mg of caffeine left in the bloodstream.

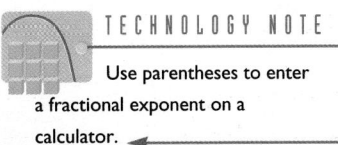

TECHNOLOGY NOTE

Use parentheses to enter a fractional exponent on a calculator.

PLANNING

Objectives and Strands

See pages 264A and 264B.

Spiral Learning

See page 264B.

Materials List

- Graphics calculator

Recommended Pacing

Section 5-3 is a two-day lesson.

Day 1

Pages 281–283: Opening paragraph through Talk it Over 4, *Exercises 1–22*

Day 2

Page 284: Top of page 284 through Look Back, *Exercises 23–52*

Toolbox References

- **Toolbox Skill 14:** Using Rules of Exponents

Extra Practice

See pages 615–617.

Warm-Up Exercises

- Warm-Up Transparency 5-3

Support Materials

- Practice Bank: Practice 39
- Activity Bank: Enrichment 35
- Study Guide: Section 5-3
- Problem Bank: Problem Set 11
- Using TI-81 and TI-82 Calculators: The Shape of $y = x^k$
- Using Plotter Plus: The Shape of $y = x^k$
- Assessment Book: Quiz 5-3

TEACHING

Additional Sample

S1 a. Suppose your school holds an outdoor all-day crafts festival on Saturday. Due to weather during the afternoon, half of the people at the festival leave every half hour. If 1200 people first attend, write an equation to model the number of people remaining.

After one half-life $\left(\frac{1}{2}\text{ hour}\right)$, $1200\left(\frac{1}{2}\right) = 600$ people remain. After two half-lives, $1200\left(\frac{1}{2}\right)\left(\frac{1}{2}\right) =$ 300 people remain. This pattern can be modeled by the function $f(x) = 1200\left(\frac{1}{2}\right)^x$, where x represents the number of half-hour periods.

b. How many people remain after 3 hours?

Three hours is 6 half-hour periods.

$f(x) = 1200\left(\frac{1}{2}\right)^x$

$f(6) = (1200)\left(\frac{1}{2}\right)^6$

≈ 18.75

About 19 people remain after 3 hours.

Talk it Over

Question 1 provides students with an opportunity to apply what they learned in Sample 1. Question 2 checks students' understanding of the concept of half-life.

Talk it Over

1. The half-life of caffeine in the bloodstream of a child is about 2 hours. Rewrite the function in Sample 1 to represent the amount of caffeine in the bloodstream of a child after t hours.
2. What does it mean to say that carbon 14 has a half-life of 5780 years?

EXPLORATION

***What** does a fractional exponent mean?*

- **Work with another student.**

$y = 2^x$	
x	y
0	1
1	?
2	?
3	?

(1) Complete the first table. Notice that the x-values form an arithmetic sequence. What type of sequence do the y-values form?

(2) Choose a different arithmetic sequence of integers for the x-values. What type of sequence do the y-values form?

(3) The x-values in the second table also form an arithmetic sequence. Assume that you can expect the y-values to form a geometric sequence. What number makes 1, __?__, 2 a geometric sequence? Copy and complete the table with exact y-values.

(4) Use your results from step 3 to complete each equation.

$2^{1/2} =$ __?__ $\quad 2^{3/2} =$ __?__

Confirm these results by using a calculator to find decimal values for both sides of each equation.

(5) What do you think is the value of $2^{1/3}$? of $2^{2/3}$? (*Hint:* Complete the third table.)

Answers to Talk it Over

1. $f(x) = (40)\left(\frac{1}{2}\right)^x$, where x is the number of 2-hour periods since the beverage was consumed
2. It takes 5780 years for half of a given amount of carbon 14 to decay.

Answers to Exploration

1.

x	y
0	1
1	2
2	4
3	8

geometric

2. Integer sequences may vary; geometric

3. $\sqrt{2}$

x	y
0	1
$\frac{1}{2}$	$\sqrt{2}$
1	2
$\frac{3}{2}$	$2\sqrt{2}$
2	4

4. $\sqrt{2}$; $(\sqrt{2})^3 = 2\sqrt{2}$;
$2^{1/2} = \sqrt{2} \approx 1.414213562$;
$2^{3/2} = (\sqrt{2})^3 \approx 2.828427125$

5. $\sqrt[3]{2}$; $(\sqrt[3]{2})^2$

x	0	$\frac{1}{3}$	$\frac{2}{3}$	1
y	1	$\sqrt[3]{2}$	$(\sqrt[3]{2})^2$	2

Raising any nonnegative number a to the $\frac{1}{2}$ power is the same as finding its square root.

$(a^{1/2})^2 = a^{(1/2)(2)} = a^1 = a$ $\qquad$ $(\sqrt{a})^2 = a$

These are equal.

RULES FOR FRACTIONAL EXPONENTS

For integers m and n with $n > 0$, and any nonnegative real number a:

		Examples
$a^{1/n} = \sqrt[n]{a}$	← Both expressions mean the nth root of a.	$4^{1/3} = \sqrt[3]{4}$
$a^{m/n} = \left(\sqrt[n]{a}\right)^m = \sqrt[n]{a^m}$		$5^{3/4} = \left(\sqrt[4]{5}\right)^3 = \sqrt[4]{5^3}$

Fractional exponents follow the same rules as integral exponents.

Student Resources Toolbox
p. 637 *Algebraic Expressions*

Sample 2

Simplify.

a. $9^{5/2}$ $\qquad$ b. $4^{3/5}$

Sample Response

a. $9^{5/2} = \left(\sqrt{9}\right)^5$ or $\sqrt{9^5}$ ← Use the rule $a^{m/n} = \left(\sqrt[n]{a}\right)^m = \sqrt[n]{a^m}$.

$= 3^5$ ← Use $\left(\sqrt{9}\right)^5$, since $\sqrt{9} = 3$.

$= 243$

b. $4^{3/5} = \left(\sqrt[5]{4}\right)^3$ or $\sqrt[5]{4^3}$ ← Use the rule $a^{m/n} = \left(\sqrt[n]{a}\right)^m = \sqrt[n]{a^m}$.

$= \sqrt[5]{64}$ ← Use $\sqrt[5]{4^3}$, since $4^3 = 64$.

$= \sqrt[5]{32 \cdot 2}$ ← Factor 64 as $32 \cdot 2$, since $32 = 2^5$.

$= \sqrt[5]{2^5 \cdot 2}$ ← Write 32 as 2^5.

$= \sqrt[5]{2^5} \cdot \sqrt[5]{2}$ ← Use the rule $\sqrt[n]{a} \cdot \sqrt[n]{b} = \sqrt[n]{ab}$.

$= 2\sqrt[5]{2}$ ← $\sqrt[n]{b^n} = b$ when n is odd.

Now you are ready for:
Exs. 1–22 on pp. 285–286

Talk it Over

3. Write $\sqrt[5]{6^2}$ as a power with a fractional exponent.
4. Use the fact that $18 = 2 \cdot 3^2$ to simplify $18^{3/2}$.

Exploration

The goal of the Exploration on page 282 is to have students discover the meaning of fractional exponents. By applying their knowledge of arithmetic and geometric sequences, students are led to the conclusion that fractional exponents represent roots.

Additional Sample

S2 Simplify.

a. $16^{3/2}$

$16^{3/2} = \left(\sqrt{16}\right)^3$ or $\sqrt{16^3}$

Use $\left(\sqrt{16}\right)^3$.

$\left(\sqrt{16}\right)^3 = 4^3$

$= 64$

b. $9^{2/3}$

$9^{2/3} = \left(\sqrt[3]{9}\right)^2$ or $\sqrt[3]{9^2}$

Use $\sqrt[3]{9^2}$.

$\sqrt[3]{9^2} = \sqrt[3]{81}$

$= \sqrt[3]{27 \cdot 3}$

$= 3\sqrt[3]{3}$

Talk it Over

Questions 3 and 4 give students practice with the rules for fractional exponents.

Answers to Talk it Over

3. $6^{2/5}$
4. $18^{3/2} = (2 \cdot 3^2)^{3/2} = 2^{3/2} \cdot 3^{2 \cdot 3/2} = 2^{1 + 1/2} \cdot 3^3 = 2 \cdot 2^{1/2} \cdot 3^3 = 54\sqrt{2}$

Additional Samples

S3 Simplify.

a. $\sqrt[4]{m^9}$

$\sqrt[4]{m^9} = \sqrt[4]{m^8 \cdot m}$
$= \sqrt[4]{(m^2)^4} \cdot \sqrt[4]{m}$
$= m^2\sqrt[4]{m}$

b. $\sqrt[3]{27t^6s^5}$

$\sqrt[3]{27t^6s^5} = \sqrt[3]{27} \cdot \sqrt[3]{t^6} \cdot \sqrt[3]{s^5}$
$= 3t^2 \cdot \sqrt[3]{s^3 \cdot s^2}$
$= 3t^2s\sqrt[3]{s^2}$

S4 Find the value of b when $f(x) = 4b^x$ and $f\left(\frac{3}{4}\right) = 32$.

$f(x) = 4b^x$
$32 = 4b^{3/4}$
$8^{4/3} = (b^{3/4})^{4/3}$
$2^4 = b^1$
$16 = b$

Teaching Tip

Another possible solution for Sample 3, part (b) exists. $\sqrt[4]{d^2} = d^{2/4}$, and $d^{2/4}$ can be written as $d^{1/2}$. Caution students that d must be nonnegative in order for $\sqrt[4]{d^2}$ to be written as $\sqrt{d}$. This can be accomplished by writing $3c^2|d|\sqrt[4]{d^2}$ as $3c^2|d|\sqrt{|d|}$.

Using Technology

The TI-82 has a menu item for finding nth roots. It is item 5 of the MATH menu. To find the 7th root of 128, go to the home screen and type 7 [MATH] [$\sqrt[x]{\ }$] 128. The expression will be displayed as $7\sqrt[x]{128}$. Press [ENTER] to evaluate the expression. The result is 2.

Look Back

Ask a few students to respond verbally. Their explanations will provide a review for the entire class.

You can use the rules of fractional exponents to simplify variable expressions.

Sample 3

Simplify.

a. $\sqrt[3]{x^7}$

b. $\sqrt[4]{81c^8d^6}$

Sample Response

a. $\sqrt[3]{x^7} = \sqrt[3]{x^6 \cdot x}$
$= \sqrt[3]{(x^2)^3} \cdot \sqrt[3]{x}$ ← $x^6 = (x^2)^3$ and $\sqrt[n]{a} \cdot \sqrt[n]{b} = \sqrt[n]{ab}$
$= x^2\sqrt[3]{x}$

b. $\sqrt[4]{81c^8d^6} = \sqrt[4]{81} \cdot \sqrt[4]{c^8} \cdot \sqrt[4]{d^6}$ ← $\sqrt[n]{a} \cdot \sqrt[n]{b} = \sqrt[n]{ab}$
$= 3 \cdot c^2 \cdot \sqrt[4]{d^4 \cdot d^2}$ ← $c^8 = (c^2)^4$
$= 3c^2 \cdot \sqrt[4]{d^4} \cdot \sqrt[4]{d^2}$ ← $\sqrt[n]{a} \cdot \sqrt[n]{b} = \sqrt[n]{ab}$
$= 3c^2\,|d|\,\sqrt[4]{d^2}$ ← $\sqrt[n]{a^n} = |a|$ when n is even.

TECHNOLOGY NOTE

You can use fractional exponents to find nth roots on scientific calculators. However, you might get an error message if you use this method to find the nth root of a negative number.

You can solve equations like $x^3 = 1000$ by finding the cube root of both sides, which is the same as raising both sides to the reciprocal power $\frac{1}{3}$. This technique can be used with fractional exponents as well.

Sample 4

Find the value of b when $f(x) = 3b^x$ and $f\left(\frac{3}{2}\right) = 24$.

Sample Response

$f(x) = 3b^x$
$24 = 3b^{3/2}$ ← Substitute **24** for $f(x)$ and $\frac{3}{2}$ for x.
$8 = b^{3/2}$ ← Divide both sides by 3.
$8^{2/3} = (b^{3/2})^{2/3}$ ← Raise each side to the reciprocal of the exponent 3/2.
$(\sqrt[3]{8})^2 = b^1$
$4 = b$

Now you are ready for: Exs. 23–52 on pp. 286–287

Look Back

Explain how fractional exponents and radicals are related.

Answers to Look Back

Raising a positive real number to the $\frac{1}{n}$ power is the same as taking its nth root.

5-3 Exercises and Problems

1. **Reading** State the rules for fractional exponents. Give your own examples.

Write each expression in radical form.

2. $33^{1/2}$
3. $13^{1/5}$
4. $5^{2/3}$
5. $15^{5/3}$

Write each expression as a power with a fractional exponent.

6. $\sqrt[3]{10}$
7. $\sqrt[4]{12}$
8. $\sqrt{7^3}$
9. $\sqrt[3]{21^4}$
10. In Sample 2, one method is shown in part (a) and another in part (b). Which method would you use to simplify $16^{4/3}$? Explain.

Simplify.

11. $8^{3/4}$
12. $81^{3/4}$
13. $3^{1/2} \cdot 3^{1/3}$
14. $100^{5/2} \cdot 10^{1/4}$
15. $\left(\frac{64}{25}\right)^{3/2}$
16. $\frac{7^{4/5}}{7^{2/5}}$
17. $20^{5/2}$
18. $250^{2/3}$

connection to BIOLOGY

Biologists sometimes grow bacteria in a liquid medium. A reasonable range for the mass of a bacteria culture in 1 L of a liquid is 0.1 g to 10 g. For Exercises 19 and 20, suppose a biologist plans to start with 0.1 g of a bacteria in 1 L of a liquid.

19. a. Write an exponential function to represent the mass of the bacteria after *t* hours when the growth rate of the bacteria is 35% per hour.
    b. What will be the mass of the bacteria after 20 min? after 75 min?
    c. What is the domain of the function you wrote in part (a)? Are the function values in part (b) reasonable?
    d. When will the mass of the population have doubled?
20. Suppose the doubling period of another type of bacteria is 2 h. Use an exponential function to find the mass of the bacteria after 5 h and after 10 h. Are these values reasonable? Why or why not?

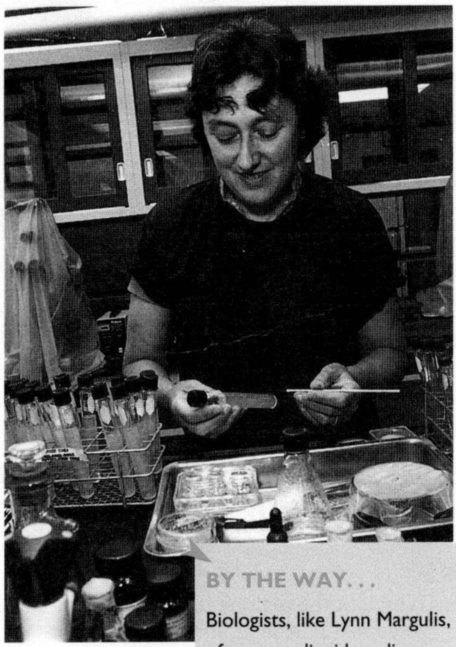

BY THE WAY...

Biologists, like Lynn Margulis, often use a liquid medium instead of a solid medium to grow bacteria because the bacteria grow more quickly. Dr. Margulis's research has led her to the theory that parts of animal and plant cells originated as free-living bacteria.

APPLYING

Suggested Assignment

Day 1

Standard 1–20, 22

Extended 1–20, 22

Day 2

Standard 23–35, 37–39, 41–52

Extended 23–35, 37–52

Integrating the Strands

Number Exs. 2–18, 21

Algebra Exs. 1–20, 22–52

Functions Exs. 19, 20, 31–39, 44–52

Logic and Language Exs. 1, 10, 22, 36, 40, 41

Communication: Discussion

Ex. 10 asks students to analyze the best way to simplify $16^{4/3}$. Since some students will choose the method of part (a) and others the method of part (b), a class discussion of the reasons involved in making the choice should prove enlightening to all students.

Interdisciplinary Problems

Many students think of mathematics as being connected only to physics and chemistry. Exs. 19 and 20 demonstrate how a biologist might use an exponential model to study and predict the growth of bacteria.

Answers to Exercises and Problems

1. For integers m and n with $n > 0$ and any positive real number a, $a^{1/n} = \sqrt[n]{a}$, and $a^{m/n} = (\sqrt[n]{a})^m = \sqrt[n]{a^m}$. Examples may vary.
2. $\sqrt{33}$
3. $\sqrt[5]{13}$
4. $\sqrt[3]{5^2}$ or $\sqrt[3]{25}$
5. $\sqrt[3]{15^5}$ or $\sqrt[3]{759{,}375}$ or $15\sqrt[3]{225}$
6. $10^{1/3}$
7. $12^{1/4}$
8. $7^{3/2}$
9. $21^{4/3}$
10. Choices may vary. The first method is inappropriate because the cube root of 16 is not rational. The second method can be used. $16^{4/3} = \sqrt[3]{16^4} = \sqrt[3]{16^3 \cdot 16} = 16\sqrt[3]{16} = 16\sqrt[3]{2^3 \cdot 2} = 32\sqrt[3]{2}$
11. $4\sqrt[4]{2}$
12. 27
13. $\sqrt[6]{243}$
14. $100{,}000\sqrt[4]{10}$
15. $\frac{512}{125} = 4\frac{12}{125}$
16. $\sqrt[5]{49}$
17. $800\sqrt{5}$
18. $25\sqrt[3]{4}$
19. a. $f(t) = 0.1(1.35)^t$
    b. about 0.11 g; about 0.15 g
    c. nonnegative real numbers; Yes.
    d. after about 2.3 h
20. $f(t) = 0.1(2)^{t/2}$; about 0.57 g; about 3.2 g; Yes. Both values are reasonable, since they both fall within the range given for a reasonably sized culture.

Using Technology

If students are using a TI-82 graphics calculator, the expressions in Ex. 21 can be evaluated by using fractional exponents or root functions from the MATH menu. Students should not overlook the warning in the Technology Note that accompanies Sample 3 on page 284. They may want to see how their calculators handle powers of negative numbers. If they get unexpected or strange results, explain that most calculators are not designed to deal with non-integer powers of negative numbers. For instance, (–1)^(1/3) will be evaluated to give –1. On the other hand, (–1)^(2/3) will result in an error message.

Problem Solving

Ex. 36 emphasizes the fact that the results of a problem are only as good as the model used to represent the set of data. Students should realize that models do not give exact results but should give as close an approximation as possible.

Assessment: Investigation

For Ex. 36, students could also research the population growth in their region since 1970 and graph these results. A scatter plot of growth in house prices versus population growth could determine whether there is a positive correlation between the two.

21. TECHNOLOGY Use a calculator to find $\sqrt[3]{64}$ and $\sqrt[3]{-64}$. Tell which keys or menu items you used. If you get an error message, does it mean that there is no answer or that your calculator will not perform that operation?

22. **Writing** Explain in words or in symbols why $(\sqrt[n]{a})^m$ is the same as $\sqrt[n]{a^m}$. Give specific examples.

Simplify.

23. $\sqrt{x^3}$	24. $\sqrt{x^9y^8}$	25. $\sqrt{49v^5}$	26. $\sqrt[3]{24c^3}$
27. $\sqrt{36z^{-4}}$	28. $\sqrt[4]{16p^6q^2}$	29. $\sqrt[3]{d^6}\cdot\sqrt[3]{d^{-6}}$	30. $\sqrt{x^3}\cdot\sqrt[3]{x}$

Real Estate The exponential curves in the diagram below approximate the median sales prices of new one-family houses from 1970 to 1992.

Median Sales Prices of New One-Family Houses, 1970 to 1992

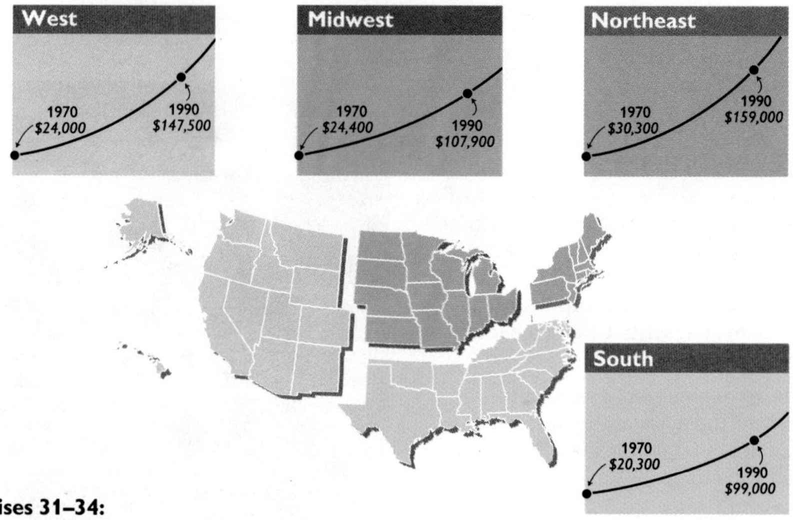

For Exercises 31–34:

a. Find the growth factor for the prices for each region.

b. Write the function represented by each exponential curve. (*Hint:* Use number of years since 1970 as the control variable.)

31. West 32. Midwest 33. Northeast 34. South

35. a. Which of the regions in Exercises 31–34 had the greatest median sales price in 1970? in 1990?
    b. Which had the greatest growth factor from 1970 to 1992?

36. a. **Research** Find the most recent figure available for the median sales price of new one-family houses in the West.
    b. Suppose housing prices continued to grow at the same rate after 1992. Use the function you wrote in Exercise 31 to estimate the median sales price in the year from part (a).
    c. **Open-ended** Compare the results of parts (a) and (b). Is the actual number close to your estimate? If not, suggest some ways that you could improve the model.

Answers to Exercises and Problems

21. Results may vary. On the TI-82, 64 [^] [(] 1 [÷] 3 [)] = 4; [(–)] 64 [^] [(] 1 [÷] 3 [)] = –4. Some calculators may not perform the operation for the root of a negative number, but will instead give an error message. This does not always indicate that there is no answer. In this case, it is known that $(-4)^3 = -64$.

22. $(\sqrt[n]{a})^m = (a^{1/n})^m = a^{(1/n)m} = a^{m(1/n)} = \sqrt[n]{a^m}$; Examples may vary.

23. $|x|\sqrt{x}$ 24. $x^2y^2\sqrt[4]{x}$

25. $7v^2\sqrt{v}$ 26. $2c\sqrt[3]{3}$

27. $6z^{-2}$

28. $2|p|\sqrt[4]{p^2q^2} = 2|p|\sqrt{|pq|}$

29. 1 30. $x\sqrt[6]{x^5}$

31. a. $\left(\frac{147.5}{24}\right)^{1/20} \approx 1.095$
    b. $f(x) = 24(1.095)^x$

32. a. 1.077
    b. $f(x) = (24.4)(1.077)^x$

33. a. 1.086
    b. $f(x) = (30.3)(1.086)^x$

34. a. 1.082
    b. $f(x) = (20.3)(1.082)^x$

35. a. the Northeast; the Northeast
    b. the West

36. a–c. Answers may vary.

37. 1,000,000,000

38. $3\sqrt{3}$ 39. $\frac{16}{25} = 0.64$

40. a. Kerri used a rule for fractional exponents to write $\sqrt[3]{x^7}$ as $x^{7/3}$. He then rewrote the exponent: $\frac{7}{3} = \frac{6}{3} + \frac{1}{3} = 2 + \frac{1}{3}$, and he used the product of powers rule to write

Solve each function for *b*.

37. $f(x) = \frac{1}{5}b^x, f\left(\frac{1}{3}\right) = 200$ **38.** $g(x) = 9b^x, g\left(\frac{2}{3}\right) = 27$ **39.** $h(x) = 64b^x, h\left(-\frac{3}{2}\right) = 125$

40. a. Writing Describe Kerri's method for simplifying $\sqrt[3]{x^7}$.

b. Use Kerri's method to simplify $\sqrt{x^5}$.

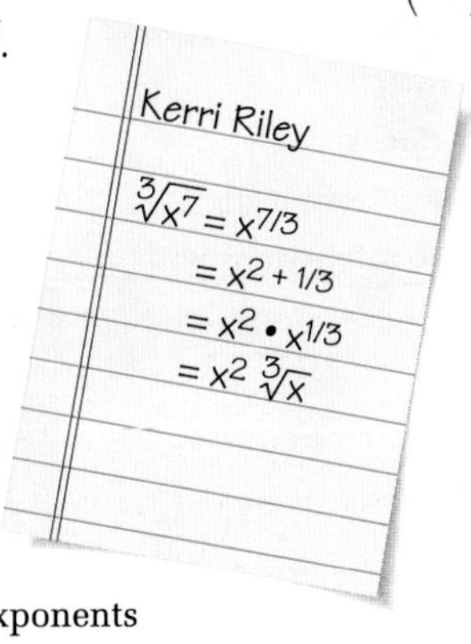

Ongoing ASSESSMENT

41. Writing Write an algorithm to find the growth factor of an exponential function given the initial value and one other value of the function.

Review PREVIEW

42. Show two different ways to write $\left(\frac{1}{b}\right)^x$. Use the rules of exponents to show that the expressions you wrote are equal. *(Section 5-2)*

43. Write an algorithm for multiplying two binomials. *(Section 1-1, Toolbox Skill 15)*

Tell whether each function is an example of *exponential growth* or *exponential decay*. *(Section 5-1)*

44. $y = (3.2)(3.5)^{-2x}$ **45.** $y = (0.5)(1.8)^x$ **46.** $y = (1.8)(0.5)^x$ **47.** $y = (1.8)(0.5)^{-x}$

5 Working on the Unit Project

An octave spans the notes from any note to a note with the same name that is eight full tones above or below the original note. The frequency of the last note in an octave is double that of the first note. The frequencies of the notes of the white and black keys of a piano form a geometric sequence.

If you number the 88 keys of a piano from left to right beginning with 1, then the function

$$f(n) = (27.5)(2)^{(n-1)/12}$$

represents the frequency of the *n*th key.

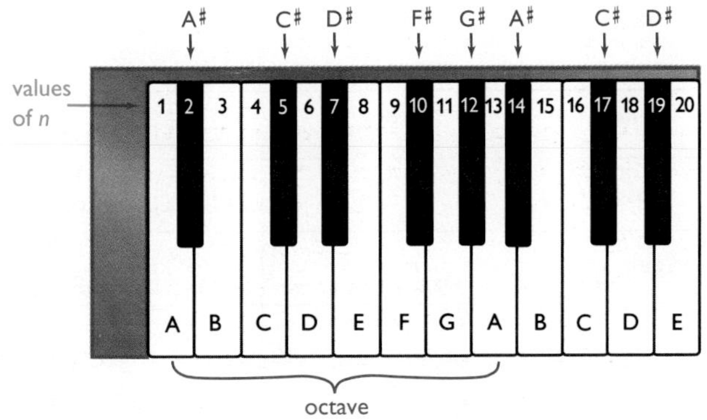

For Exercises 48–50, find the frequency of each given note.

48. the 5th lowest note **49.** the 20th lowest note **50.** the 28th highest note

51. Graph the function for the frequency of the notes on a piano. What is the domain and range of this function? Explain your reasoning.

52. Use the rules for exponents to rewrite $f(n)$ in two different ways.

Mathematical Procedures

The method of simplifying a radical expression in Ex. 40 is an important one as it emphasizes the definition of a rational exponent. The use of absolute value symbols is still important if students prefer this method of simplifying even roots.

Working on the Unit Project

Exs. 48–52 familiarize students with the exponential function used to calculate the frequencies of the notes on a piano.

Practice 39 For use with Section 5-3

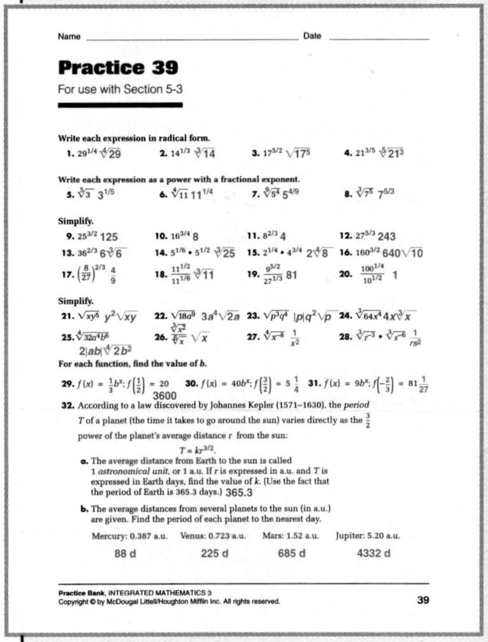
Name ______________ Date ______________

Practice 39

For use with Section 5-3

Write each expression in radical form.

1. $29^{1/4}$ $\sqrt[4]{29}$ 2. $14^{1/3}$ $\sqrt[3]{14}$ 3. $17^{5/2}$ $\sqrt{17^5}$ 4. $21^{3/5}$ $\sqrt[5]{21^3}$

Write each expression as a power with a fractional exponent.

5. $\sqrt[5]{3}$ $3^{1/5}$ 6. $\sqrt[4]{11}$ $11^{1/4}$ 7. $\sqrt[9]{5^4}$ $5^{4/9}$ 8. $\sqrt[3]{7^5}$ $7^{5/3}$

Simplify.

9. $25^{3/2}$ 125 10. $16^{3/4}$ 8 11. $8^{2/3}$ 4 12. $27^{5/3}$ 243

13. $36^{2/3}$ $6\sqrt[3]{6}$ 14. $5^{1/6} \cdot 5^{1/2}$ $\sqrt[3]{25}$ 15. $2^{1/4} \cdot 4^{3/4}$ $2\sqrt[4]{8}$ 16. $160^{3/2}$ $640\sqrt{10}$

17. $\left(\frac{8}{27}\right)^{2/3}$ $\frac{4}{9}$ 18. $\frac{11^{1/2}}{11^{1/6}}$ $\sqrt[3]{11}$ 19. $\frac{9^{5/2}}{27^{1/3}}$ 81 20. $\frac{100^{1/4}}{10^{1/2}}$ 1

Simplify.

21. $\sqrt{xy^5}$ $y^2\sqrt{xy}$ 22. $\sqrt{18a^9}$ $3a^4\sqrt{2a}$ 23. $\sqrt{p^3q^4}$ $|p|q^2\sqrt{p}$ 24. $\sqrt[3]{64x^4}$ $4x\sqrt[3]{x}$

25. $\sqrt[4]{32a^4b^6}$ $2|ab|\sqrt[4]{2b^2}$ 26. $\frac{\sqrt[3]{x^2}}{\sqrt[6]{x}}$ $\sqrt{x}$ 27. $\sqrt[4]{x^{-8}}$ $\frac{1}{x^2}$ 28. $\sqrt[3]{r^{-3}} \cdot \sqrt[3]{s^{-6}}$ $\frac{1}{rs^2}$

For each function, find the value of *b*.

29. $f(x) = \frac{1}{3}b^x; f\left(\frac{1}{2}\right) = 20$ 3600 30. $f(x) = 40b^x; f\left(\frac{3}{2}\right) = 5$ $\frac{1}{4}$ 31. $f(x) = 9b^x; f\left(-\frac{2}{3}\right) = 81$ $\frac{1}{27}$

32. According to a law discovered by Johannes Kepler (1571–1630), the *period* T of a planet (the time it takes to go around the sun) varies directly as the $\frac{3}{2}$ power of the planet's average distance r from the sun:

$$T = kr^{3/2}.$$

a. The average distance from Earth to the sun is called 1 *astronomical unit*, or 1 a.u. If r is expressed in a.u. and T is expressed in Earth days, find the value of k. (Use the fact that the period of Earth is 365.3 days.) 365.3

b. The average distances from several planets to the sun (in a.u.) are given. Find the period of each planet to the nearest day.

Mercury: 0.387 a.u.	Venus: 0.723 a.u.	Mars: 1.52 a.u.	Jupiter: 5.20 a.u.
88 d	225 d	685 d	4332 d

Practice Bank, INTEGRATED MATHEMATICS 3

39

Answers to Exercises and Problems

$x^{2 + 1/3}$ as $x^2 \cdot x^{1/3}$. Finally, he used a rule for fractional exponents to write $x^{1/3}$ as $\sqrt[3]{x}$, so that $x^2 \cdot x^{1/3} = x^2\sqrt[3]{x}$.

b. $\sqrt{x^5} = x^{5/2} = x^{2 + 1/2} = x^2 \cdot x^{1/2} = x^2\sqrt{x}$

41. Summaries may vary. Let $(0, a)$ be the initial value and (t, v) the other value. Divide v by a and find the tth root of the quotient. The rate of change b is $\left(\frac{v}{a}\right)^{1/t}$.

42. $\left(\frac{1}{b}\right)^x = b^{-x}; \left(\frac{1}{b}\right)^x = \frac{1^x}{b^x} = \frac{1}{b^x}$

43. Methods may vary. An example is given.

Step 1: Find the product of the two first terms.

Step 2: Find the product of the two outside terms.

Step 3: Find the product of the two inside terms.

Step 4: Find the product of the two last terms.

Step 5: Add your results from Steps 1 through 4.

Step 6: Simplify if necessary.

44. exponential decay

45. exponential growth

46. exponential decay

47. exponential growth

48. about 34.6 Hz

49. about 82.4 Hz

50. about 830 Hz

51.

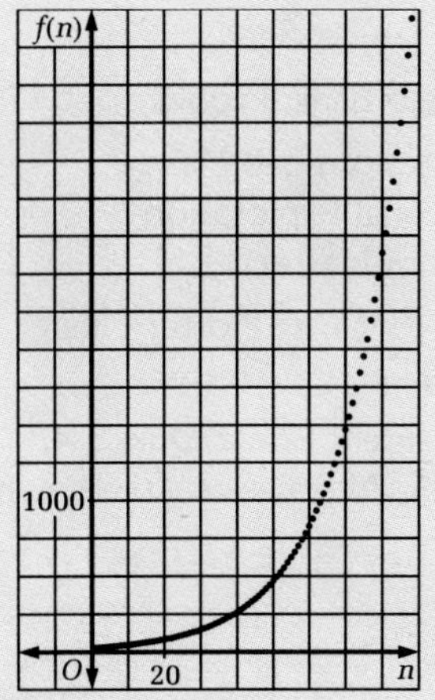

domain: integers greater than or equal to 1 and less than or equal to 88 (there are 88 keys on a piano); range: the 88 different frequencies, which are between 27.5 and 4186

52. $f(n) = 27.5(\sqrt[12]{2^{n-1}})$; $f(n) = 27.5(\sqrt[12]{2})^{n-1} \approx (26.0)(1.0595)^n$

Objectives and Strands
See pages 264A and 264B.

Spiral Learning
See page 264B.

Materials List
- Graphics calculator
- 8 plastic straws
- Scissors
- Masking tape
- Centimeter ruler

Recommended Pacing
Section 5-4 is a two-day lesson.

Day 1

Pages 288–289: Exploration through Talk it Over 2, *Exercises 1–21*

Day 2

Pages 290–291: Logistic Growth through Look Back, *Exercises 22–34*

Extra Practice
See pages 615–617.

Warm-Up Exercises
Warm-Up Transparency 5-4

Support Materials
- Practice Bank: Practice 40
- Activity Bank: Enrichment 36
- Study Guide: Section 5-4
- Problem Bank: Problem Set 11
- Explorations Lab Manual: Additional Exploration 5 Diagram Master 14
- Using IBM/Mac Plotter Plus Disk: Function Plotter
- Assessment Book: Quiz 5-4, Test 19, Alternative Assessment 3

Section **5-4**

The Number e

In Your Be$t Intere$t

Focus

Model situations using exponential functions with base e.

EXPLORATION

Does compounding interest more often give you unlimited returns?

- **Materials: graphics calculators**
- **Work with a partner.**

① In your group, decide which of the following investments would give you more for your dollar.

Invest $1 at an annual interest rate of 100% for 1 year

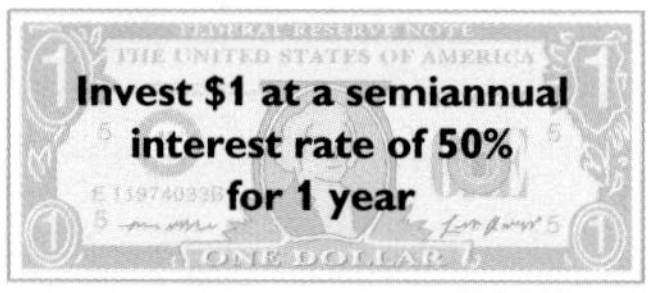

Invest $1 at a semiannual interest rate of 50% for 1 year

$1 + 100\%\text{ of }1 = 1 + 1$

$(1 + 50\%\text{ of }1) + 50\%(1 + 50\%\text{ of }1) = \left(1 + \frac{1}{2}\right)^2$

Frequency of compounding	Number of times interest is calculated in a year (n)	$\left(1 + \frac{1}{n}\right)^n$	Dollar value (A)
annual	$n = 1$	$\left(1 + \frac{1}{1}\right)^1$	2.00
semiannual	$n = 2$	$\left(1 + \frac{1}{2}\right)^2$	2.25
quarterly	$n = 4$	$\left(1 + \frac{1}{4}\right)^4$	2.44141
monthly	$n = ?$	?	?
weekly	$n = ?$	?	?
daily	$n = ?$	?	?
hourly	$n = ?$	?	?
every minute	$n = 525,600$	?	?
every second	$n = 31,536,000$	?	?

② Complete the table to see whether compounding interest more often can give you a million dollars for your dollar in one year.

③ What happens to the dollar value as you compound the interest more often? What do you think is the maximum amount you can earn in one year?

Continued on next page.

Answers to Exploration

1. $1 + 1 = 2$; $\left(1 + \frac{1}{2}\right)^2 = 2.25$; the second investment
2. See answers in back of book.
3. The dollar value increases, but by smaller and smaller amounts. It appears that the maximum amount is about $2.72.
4. 2.718281828; Answers may vary due to the number of decimal places in the table value.
5. 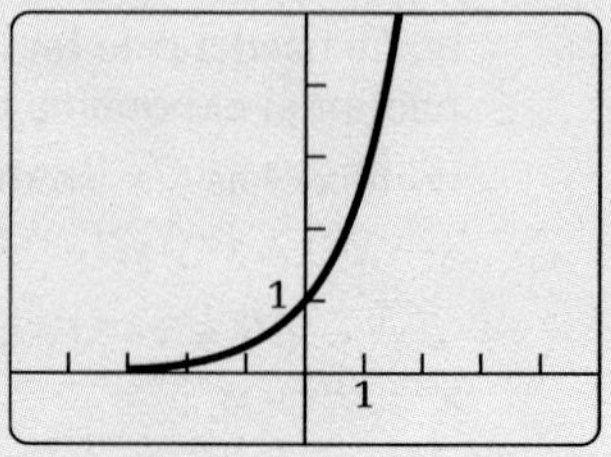

1; exponential

TECHNOLOGY NOTE

To find e^1, press these keys:

2nd e^x 1 ENTER

(4) Use your calculator to find the value of e^1. How does the value of e compare to the maximum dollar value in the table in step 1?

(5) Graph the function $y = e^x$. What is the y-intercept of the graph? What type of function is $y = e^x$?

The number **e** is an irrational number that is approximately equal to 2.718281828. It is the limit of the sequence:

$$\left(1 + \frac{1}{1}\right)^1, \left(1 + \frac{1}{2}\right)^2, \left(1 + \frac{1}{3}\right)^3, \ldots$$

Sample 1

In 1980 the average cost of a movie ticket was \$2.62. The function $y = 2.62e^{0.058x}$, where x represents the number of years since 1980, can be used to model the average cost, in dollars, of a movie ticket. Find the average cost of a movie ticket for each year.

a. 1990 b. 1953

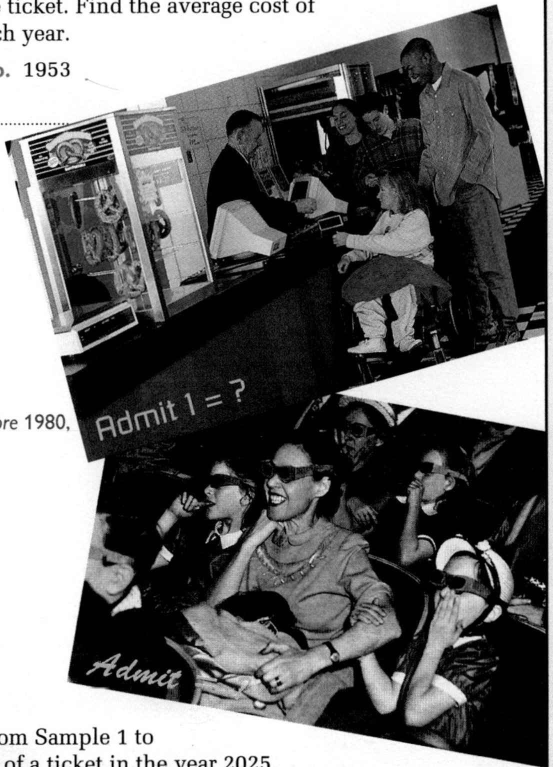

Sample Response

a. $y = 2.62e^{0.058x}$ ← Write the given equation.

$= 2.62e^{(0.058)(10)}$ ← Since 1990 is 10 years *after* 1980, substitute 10 for x.

≈ 4.68

The average cost of a movie ticket in 1990 was about \$4.68.

b. $y = 2.62e^{0.058x}$

$= 2.62e^{(0.058)(-27)}$ ← Since 1953 is 27 years *before* 1980, substitute -27 for x.

≈ 0.55

The average cost of a movie ticket in 1953 was about \$.55.

Talk it Over

1. Use the model from Sample 1 to estimate the cost of a ticket in the year 2025.
2. Use your knowledge of exponential functions and the value of e to describe the graph of $y = 2.62e^{0.058x}$. Graph the function to check your description.

Now you are ready for: Exs. 1–21 on pp. 291–293

TEACHING

Exploration

The goal of the Exploration is to have students discover the number e by using a graphics calculator to explore the effects of compounding 100% interest more and more frequently.

Using Technology

The terms of the sequence $a_n = \left(1 + \frac{1}{n}\right)^n$ approach e rather slowly. You also can obtain e as the sum of the infinite series $1 + \frac{1}{1!} + \frac{1}{2!} + \frac{1}{3!} + \ldots$. Since $n!$ grows rapidly, you would expect the first 15 terms of this series to give a good approximation to e. Students can test this by applying ideas from the Using Technology notes in Unit 4. Since 0! is defined as 1, the series can be written as $\sum_{A=0}^{\infty} \frac{1}{A!}$. To find the sum of the first 15 terms, use the TI-82 to evaluate sum seq(1/(A!),A,0,14,1). The calculator gives an approximate value of 2.718281828. (On the TI-81, enter the expression 1/0!+1/1!+1/2!+...+1/14! and press ENTER.)

Additional Sample

S1 In 1970, one high school sold student lunches for \$.55. The function $y = 0.55e^{0.05x}$, where x represents the number of years since 1970, can be used to model the average cost, in dollars, of a lunch. Find the average cost of a lunch for each year.

a. 1995

$y = 0.55e^{0.05x}$

Since 1995 is 25 years after 1970, substitute 25 for x.

$y = 0.55e^{0.05(25)} \approx 1.92$

The average cost in 1995 is about \$1.92.

b. 1960

$y = 0.55e^{0.05x}$

Since 1960 is 10 years before 1970, substitute -10 for x.

$y = 0.55e^{0.05(-10)} \approx 0.33$

The average cost in 1960 was about \$.33.

Answers to Talk it Over

1. about \$35.63
2. Answers may vary. An example is given. Since the rate of change is greater than 1, the function models exponential growth. The graph lies completely above the x-axis, approaching the x-axis as x decreases and increasing exponentially as x increases. It has a y-intercept of 2.62.

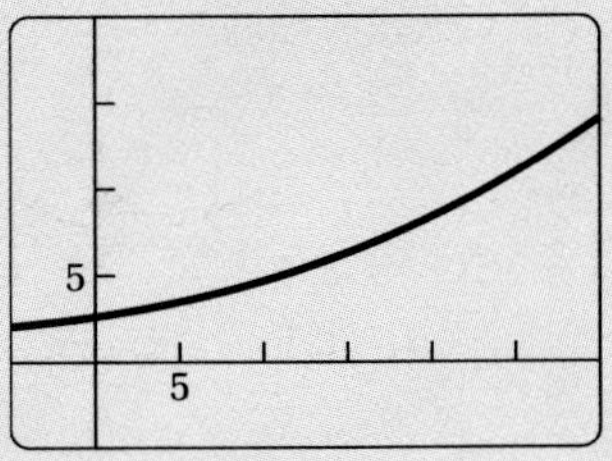

Logistic Growth

The rate of growth of a quantity may slow down after initially increasing or decreasing exponentially. A function that behaves in this way is a **logistic growth function.**

increasing logistic growth function

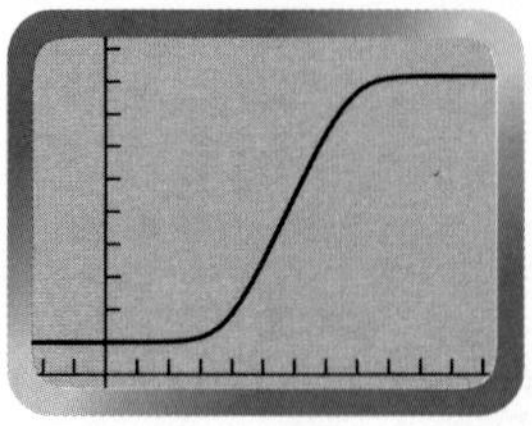

decreasing logistic growth function

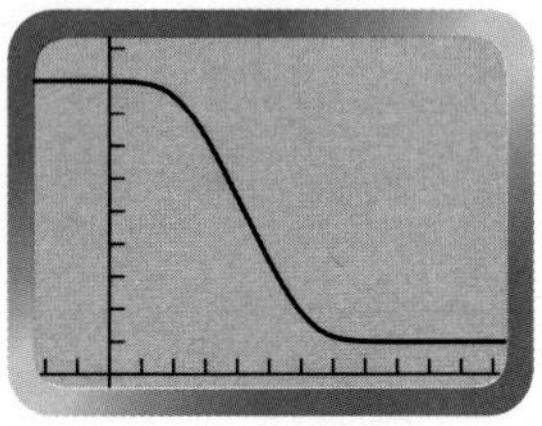

Sample 2

The spread of a flu virus in a school with about 1000 people can be modeled by the logistic growth function $y = \frac{1000}{1 + 990e^{-0.7x}}$, where y is the number of people infected after x days.

a. About how many people will be infected after nine days?

b. What are the horizontal asymptotes of the function?

c. Estimate the maximum number of people who will be infected.

Sample Response

a. **Problem Solving Strategy:** Use a graph.

Graph the function $y = \frac{1000}{1 + 990e^{-0.7x}}$.

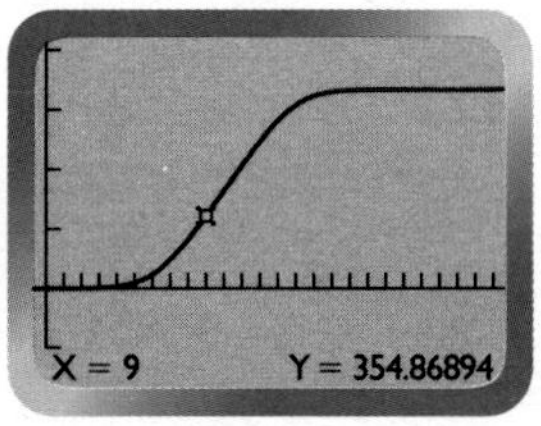

Read the value of y when $x = 9$.

About 355 people will be infected with the virus after nine days.

b. Trace along the graph of the function.

The minimum y-value of the function gets closer and closer to zero but does not reach 0.

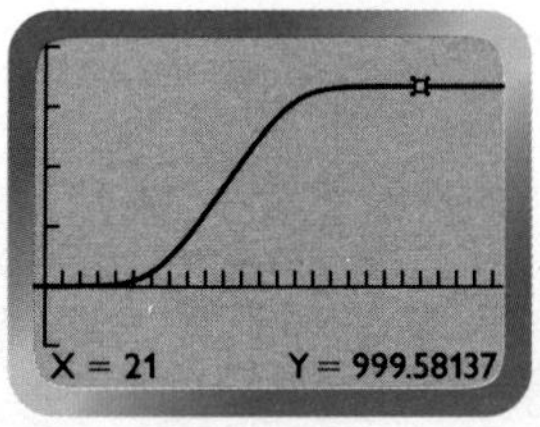

The maximum y-value of the function gets closer and closer to 1000 but does not reach 1000.

There is a horizontal asymptote at $y = 0$.

There is a horizontal asymptote at $y = 1000$.

c. About 1000 people will be infected with the virus.

Visual Thinking

Check students' understanding of a logistic growth function by asking them to explain the two graphs shown. Encourage them to describe what is happening at the beginning and at the end of each curve. Ask them what would have to change in order for the curve to change. This activity involves the visual skills of *correlation* and *interpretation*.

Additional Sample

S2 The spread of a rumor in a town of 5000 people can be modeled by the logistic growth function $y = \frac{5000}{1 + 800e^{-0.8x}}$, where y is the number of people who have heard the rumor after x days.

a. About how many people will have heard the rumor after eight days?
Use a graph. Graph the function $y = \frac{5000}{1 + 800e^{-0.8x}}$.

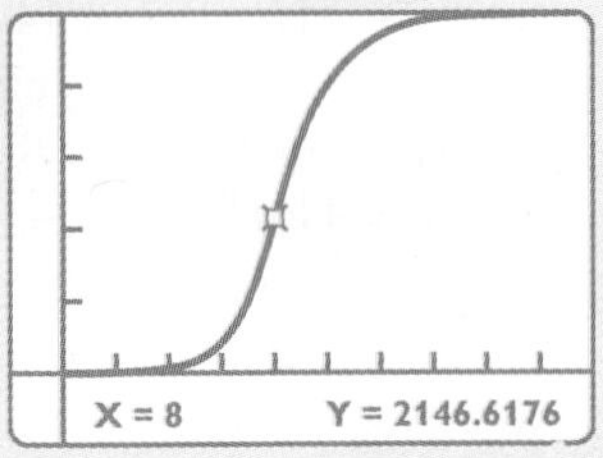

Read the value of y when $x = 8$.
About 2147 people will have heard the rumor in eight days.

b. What are the horizontal asymptotes of the function?
Trace along the graph of the function. The minimum y-value gets closer and closer to 0. The maximum y-value gets closer and closer to 5000. The horizontal asymptotes are $y = 0$ and $y = 5000$.

c. Estimate the maximum number of people who will hear the rumor.
About 5000 people will hear the rumor.

Answers to Look Back

Over part of its domain, a logistic growth function behaves like an exponential function. However, the growth rate eventually slows down and as domain values increase, the graph approaches an asymptote, rather than increasing or decreasing exponentially.

Look Back

How is a logistic growth function like an exponential function? How is it different?

Now you are ready for: Exs. 22–34 on pp. 294–295

5-4 Exercises and Problems

1. **Reading** What kind of number is e? What is the approximate value of e?

Without using a calculator, choose the letter that represents the best estimate for each expression.

2. e^{-1}
3. $2e$
4. $e - 1$
5. e^2

A. 7
B. 1.7
C. 5.5
D. $\frac{1}{3}$

6. How is the number e like π? Which number is larger?
7. Find $f(3.6)$ when $f(x) = 5.3e^{-4x}$.
8. Find $f(14.5)$ when $f(x) = \dfrac{125{,}000}{(1 + 124e^{-1.5x})}$.

connection to SCIENCE

The percent of available surface light, L, that filters down through bodies of water can be modeled by the exponential function $L(x) = 100e^{kx}$, where k is a measure of the murkiness of the water, and x is the depth below the surface in meters.

9. In clear water, the value of k in $L(x) = 100e^{kx}$ is about -0.2.
   a. TECHNOLOGY Use a graphics calculator or graphing software. Graph an equation that models the percent of available light in clear water. What percent of available surface light does a diver have 4 m below the surface?
   b. How far below the surface can a diver swim in clear water before only 50% of the surface light is available?
10. Repeat Exercise 9 for murky water. Use the function $L(x) = 100e^{-0.3x}$.
11. Repeat Exercise 9 for muddy water. Use the function $L(x) = 100e^{-0.5x}$.

APPLYING

Suggested Assignment

Day 1

Standard 1–8, 12–15, 19

Extended 1–8, 10–15, 19

Day 2

Standard 26–34

Extended 25–34

Integrating the Strands

Number Exs. 1–6, 26, 27

Algebra Exs. 7–24, 28–34

Functions Exs. 7–11, 16–24, 31–33

Measurement Ex. 34

Logic and Language Exs. 1, 18, 21, 25

Communication: Discussion

Exs. 1–6 can be used to discuss the number e. If students' estimates differ for Exs. 2–5, they should be able to explain their reasoning for choosing a particular estimate.

Answers to Exercises and Problems

1. irrational; about 2.718281828
2. D
3. C
4. B
5. A
6. e and π are both irrational numbers; π is larger.
7. about $2.95 \cdot 10^{-6}$
8. about 125,000

9. a.

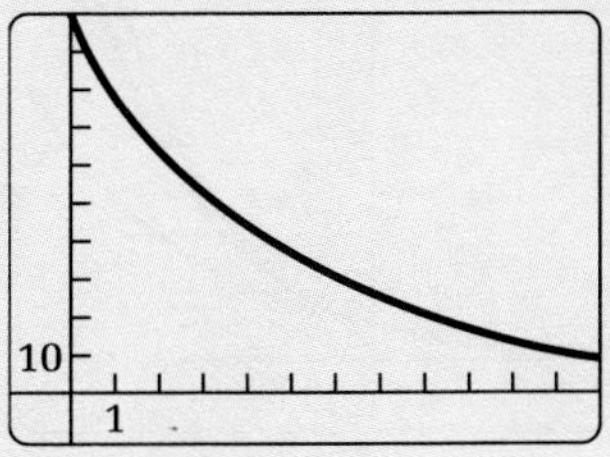

about 45%

b. about 3.5 m

10. a.

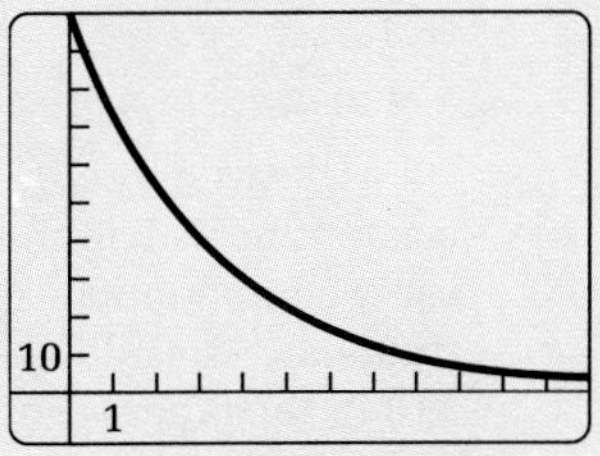

about 30%

b. about 2.3 m

11. a.

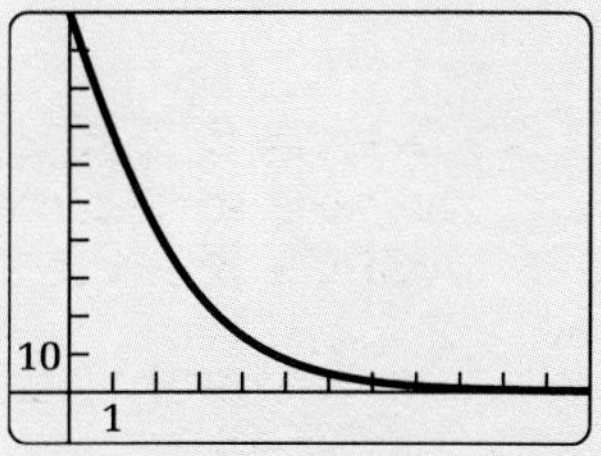

about 14%

b. about 1.4 m

Application

Exs. 12–15 involve the important concept of compound interest. Students should understand that the power of compound interest can make a small investment grow into a large amount of money over a long period of time. For example, an investment of \$100 a month earning 10% interest a year will grow to one million dollars in 46 years (assuming tax-free compounding). Thus, a 20-year-old person who follows this investment strategy would be a millionaire at a retirement age of 66. The total amount invested would have been \$100 × 12 × 46, or only \$55,200!

Students Acquiring English

Some students acquiring English may be unfamiliar with the banking terms found in Exs. 12–15. Consider having volunteers explain each of the following terms: *investment, interest, compounded continuously, compounded yearly, compounded monthly, compounded quarterly, bonus, accounts,* and *initial deposits.*

Reasoning

Exs. 16 and 17 have students explore the graph of these functions to see how they are alike and how they are different. Students also see how changing the value of the exponent affects each graph. For Ex. 18, students see how the graph of $y = e^x$ can be translated in the plane by adding or subtracting a constant to the function $y = e^x$.

Banking The formula $A = P\left(1 + \frac{r}{n}\right)^{nt}$ can be used to find the dollar value of an investment when interest is compounded n times a year for t years at a rate of r percent on a beginning balance of P dollars.

Becky Quezada invests \$100 at 6% interest. Find the value of her investment after one year with each type of compound interest.

12. yearly
13. monthly
14. daily

15. Bob Drake sold an antique stuffed bear for \$2000. He decides to invest the money in a bank account until he retires in 25 years.
    a. The formula $A = Pe^{rt}$ can be used to find the dollar value of an investment of P dollars when interest is compounded continuously for t years at a rate of r percent. Find the amount that he will earn in a bank that pays 5.25% interest compounded continuously.
    b. Find the amount that he will earn in a bank that adds \$10 to the initial deposit as a bonus for new accounts, and pays 5.25% interest compounded quarterly.
    c. Should he invest in the account with *continuous compounding* or the one with *quarterly compounding*? Why?

BY THE WAY...

When tulips were first introduced in Holland during the 1590s, some bulbs cost only a few florins. In the 1630s, a farmer paid 2500 florins for one Viceroy tulip bulb (shown here by Dutch painter Judith Leyster). Payment took the form of 160 bushels of wheat, 320 bushels of rye, four fat oxen, eight pigs, a dozen sheep, two oxheads of beverage, four tons of butter, a thousand pounds of cheese, a bed, some clothing, and a silver beaker. All that for one tulip bulb!

BY THE WAY...

A 16th century Leonardo da Vinci notebook containing sketches and theories about why the moon shines and why the sky is blue sold for \$30.8 million in 1994. The notebook last sold for \$5.6 million in 1980.

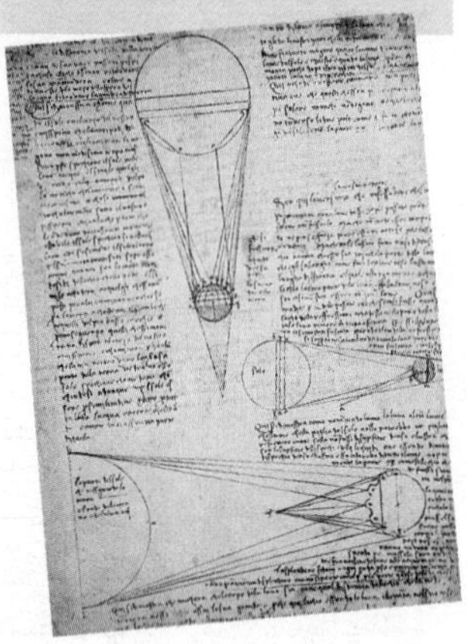

16. a TECHNOLOGY Graph the functions $y = 2^x$, $y = e^x$, and $y = 3^x$ on the same screen. How are the graphs alike? How are they different?
    b. For which values of x is e^x greater than 2^x? For which values of x is e^x greater than 3^x?

17. TECHNOLOGY Repeat Exercise 16 using the functions $y = 2^{-x}$, $y = e^{-x}$, and $y = 3^{-x}$.

18. a TECHNOLOGY Graph the functions $y = e^x$ and $y = e^x + 4$ on the same screen. Describe how the functions are alike and how they are different.
    b. **Open-ended** Think of a situation that can be modeled by each function in part (a). Explain how the situations are different.
    c. Use your answer from part (a) to sketch the graph $y = e^x - 2$.

19. Between 1888 and 1988, the United States Post Office issued new stamps at an exponential rate. The number of new stamps issued, S, can be modeled by the function $S = 83e^{0.024t}$, where t is the number of years since 1888.
    a. About how many new stamps were issued in 1900?
    b. When will the number of new United States stamps issued reach 2000?
    c. According to the model, what is the growth rate?

Answers to Exercises and Problems

12. \$106 13. \$106.17
14. \$106.18
15. a. \$7430.90 b. \$7404.56
    c. He will earn more in the account in which interest is compounded continuously; the added \$10 in the other account does not make up for the interest lost by compounding quarterly rather than continuously.

16. a.

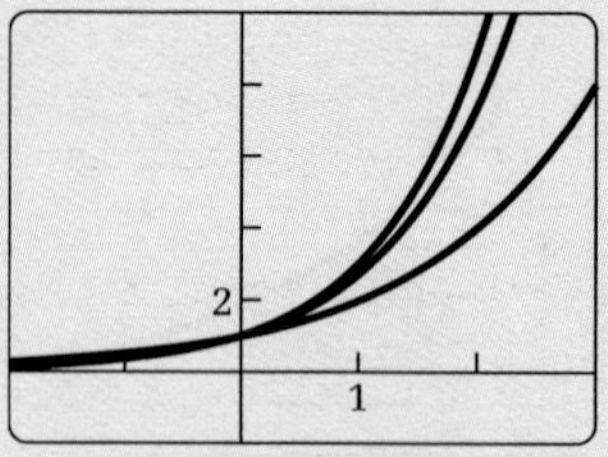

All three graphs are exponential and have the same y-intercept. The steepnesses of the graphs differ.

b. $x > 0$; $x < 0$

17. a.

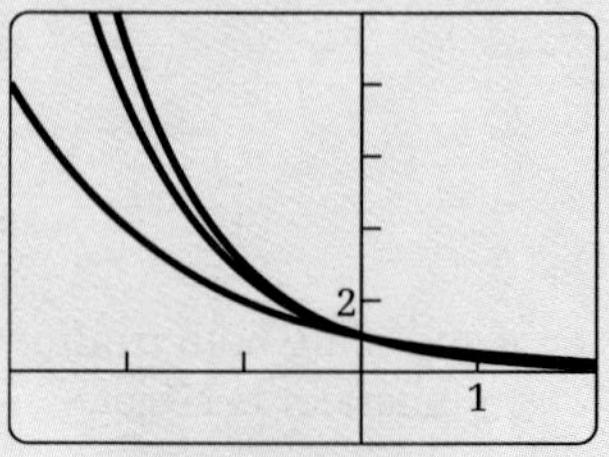

All three graphs are exponential and have the same y-intercept.

Architecture The Eiffel Tower was built between 1887 and 1889 for the 100th anniversary of the French Revolution. The tower starts with a wide base and narrows as it rises. This shape can be modeled by the equation $y = 327.4e^{-0.05x}$, where x represents the distance, in meters, from the center of the tower and y represents the height in meters.

20. a. TECHNOLOGY Graph $y = 327.4e^{-0.05x}$. Use $-188 \le x \le 188$ and $0 \le y \le 350$ for your viewing window.
    b. Write an equation for the reflection of the graph of $y = 327.4e^{-0.05x}$ over the y-axis. Graph this equation on the same calculator screen as your graph from part (a).
    c. Estimate the height of the Eiffel tower.
    d. The first platform of the tower is 57 meters from the ground. About how wide is this platform?
    e. The lantern of the tower has a platform that is 5 m in diameter. Estimate the height of the lantern.

BY THE WAY...

The names of many people, including 14 mathematicians, appear on the side of the first platform of the Eiffel Tower.

21. The height, in feet, of the Gateway Arch in St. Louis, Missouri, can be modeled by the function $y = 757.7 - 63.85(e^{x/127.7} + e^{-x/127.7})$, where x represents the horizontal distance, in feet, from the center of the arch.
    a. Graph the function. Estimate the height of the vertex of the arch.
    b. About how far apart are the ends of the arch?
    c. **Research** The Gateway Arch forms a *catenary curve.* Find out how a catenary curve is different from a parabola. What other structures are built with catenary curves?

Answers to Exercises and Problems

The steepnesses of the graphs differ.

b. $x < 0$; $x > 0$

18. a.

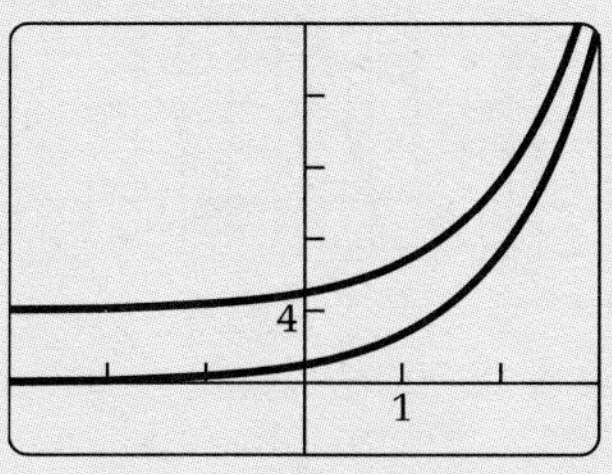

Both graphs are exponential. The graph of $y = e^x + 4$ is the graph of $y = e^x$ translated up 4 units.

b. Examples may vary.

c.

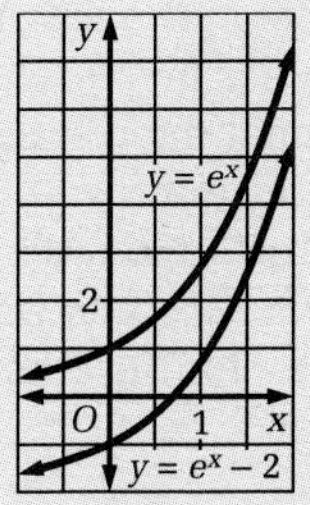

19. a. about 111 stamps
    b. 2021
    c. about 2.4% per year

20. a, b.

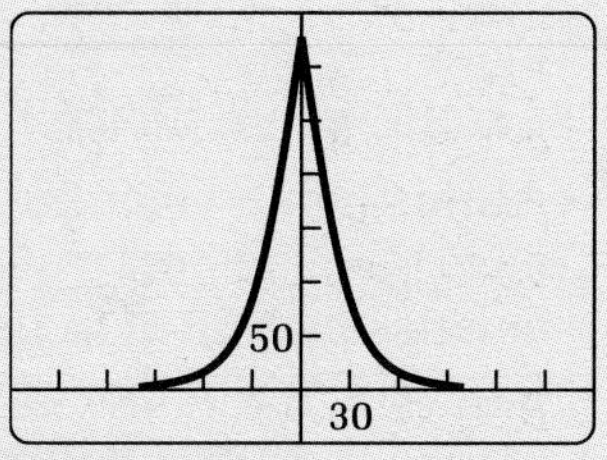

b. $y = 327.4e^{0.05x}$
c. about 327 m
d. about 70 m
e. about 289 m

21. a.

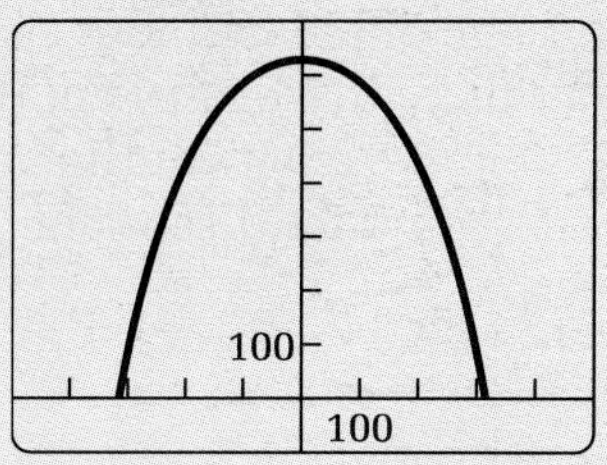

about 630 ft

b. about 630 ft

c. Answers may vary. An example is given. A catenary is the curve theoretically formed by a perfectly flexible, uniformly dense and inextensible cable suspended from its endpoints. The equation of a catenary is $y = \frac{a}{2}(e^{x/a} + e^{-x/a})$, while the equation of a parabola is of the form $y = ax^2$. Some bridges are built with catenary curves.

Career Note

To earn a medical degree, a student must pursue eight years of higher education. During the first four years, students typically earn a bachelor's degree in a course of study called premedical, which involves a heavy concentration in science courses such as biology and organic chemistry. Admission to medical school is very competitive, and only students with an excellent academic record are accepted. After four years of medical school, all new medical doctors must then serve an internship at a hospital for one or more years. During this time, they learn to treat patients under the supervision of experienced doctors. Many doctors then still continue their medical education by specializing in a particular branch of medicine such as surgery, internal medicine, or radiology. You may wish to ask students to identify the special types of doctors they know.

Cooperative Learning

At this point, you may wish to have students review the mathematics presented in the first four sections of this unit. Cooperative learning groups of four provide a good method for doing this. Within a group, each student is responsible for making a one-page review outline for one of the first four sections of this unit. Members of the group then share outlines with each other for discussion.

TECHNOLOGY For Exercises 22 and 23, graph each function on a graphics calculator. Sketch the screen. Identify the horizontal asymptotes.

22. $f(x) = \dfrac{43}{(1 + 2e^{-x})}$ 23. $f(x) = \dfrac{1}{(1 + 0.7e^{-2x})}$

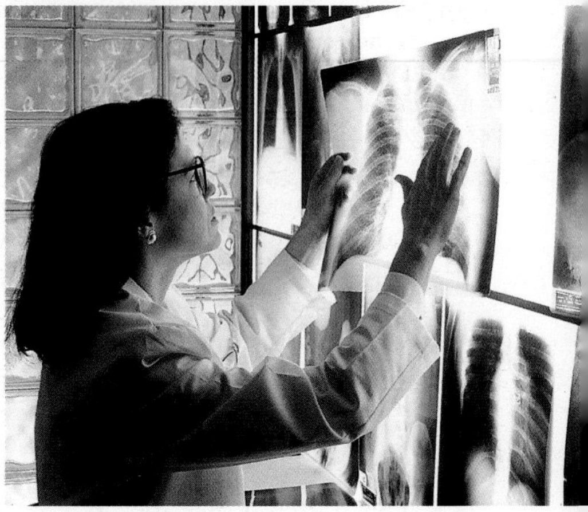

24. **Higher Education** The number of medical degrees awarded to women each year between 1960 and 1990 can be modeled by the logistic function $f(x) = \dfrac{5300}{(1 + 100e^{-0.27x})}$, where x is the number of years since 1960.

a. Graph the function. About how many women earned a medical degree in 1960? in 1976?

b. Between what years did the number of women earning medical degrees appear to increase exponentially? What may have been a cause?

c. In what year did the number of women earning medical degrees begin to level off?

d. The number of law degrees awarded to women each year between 1960 and 1990 can be modeled by the logistic function $f(x) = \dfrac{15{,}000}{(1 + 403e^{-0.33x})}$. Repeat parts (a)–(c) for the number of women obtaining law degrees.

e. Graph each function on the same calculator screen. When was the number of women awarded medical degrees greater than the number awarded law degrees? When were these numbers equal?

f. By 1990, about how many more law degrees than medical degrees were awarded to women?

Ongoing ASSESSMENT

25. **Writing** Explain why it does not make sense for banks to offer an account in which interest is compounded hourly.

Review PREVIEW

Simplify. *(Section 5-3)*

26. $125^{1/3}$ 27. $1000^{4/3}$ 28. $\sqrt{98x^3}$ 29. $\sqrt[4]{256b^{12}c^{10}}$

30. Graph the feasible region for this system of inequalities. Label each vertex with its coordinates. *(Section 1-8)* $x \geq 0$, $y \geq x$, $y \leq 6$

For each function, graph the function and its reflection over the line y = x. *(Toolbox Skill 38)*

31. $y = 3x + 2$ 32. $y = x^2 + 4x - 5$ 33. $y = 2^x$

Answers to Exercises and Problems

22.

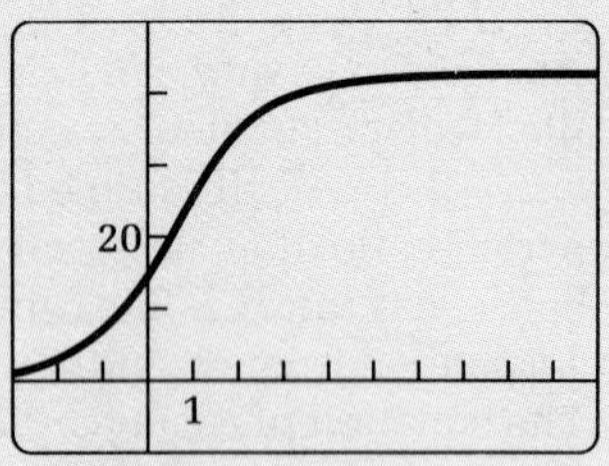

$y = 0$; $y = 43$

23, 24. See answers in back of book.

25. Answers may vary. An example is given. The additional interest earned by the account would be insignificant, but maintaining such accounts would be costly for the bank.

26. 5

27. 10,000

28. $7|x|\sqrt{2x}$

29. $4|b^3|c^2\sqrt[4]{c^2}$ or $4|b^3|c^2\sqrt{|c|}$

30.

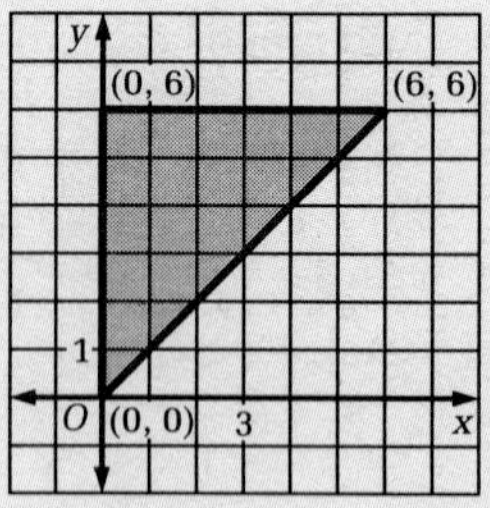

31.

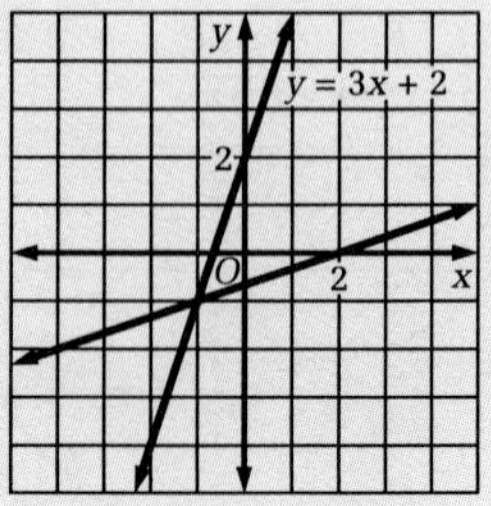

Working on the Unit Project

34. To make a panpipe, you will need 8 plastic straws, scissors, masking tape, a centimeter ruler, a calculator, and an instrument to produce a middle C note, such as a piano.

   a. Blow across the opening of one of the straws and compare the note with middle C. Cut off short segments of the straw to raise the pitch until the note of the straw sounds like middle C. Measure the length of the straw to the nearest 0.1 cm. Record this length as l_1.

   b. Use your value for l_1 from part (a) and the equation $l_n = l_1 e^{-0.057(n-1)}$, where n is the number of the note, to find the lengths of the straws, in centimeters, needed to produce the remaining notes for your panpipe.

   c. Use your measurements from part (b) to cut the straws for the remaining notes. Compare each note to the note produced by the musical instrument. Label each straw with the name of the note it plays.

   d. Place the straws on a flat surface in order from largest to smallest. Align the top edges. Put masking tape across the set. Pick up the set and wrap the rest of the tape around it.

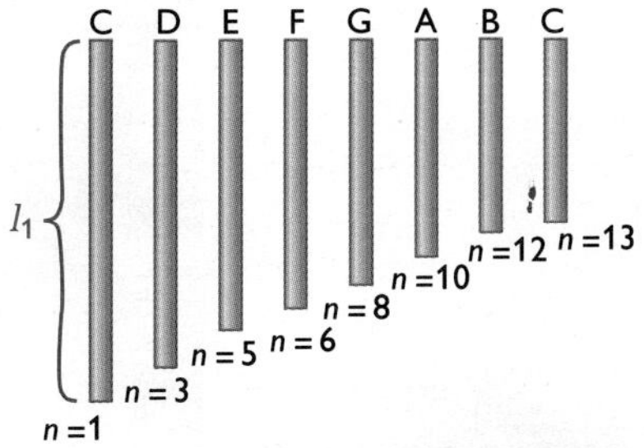

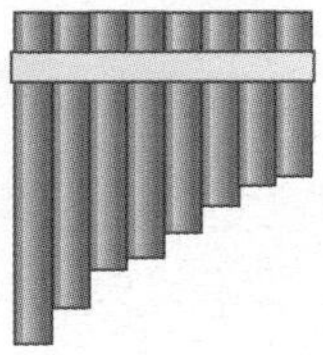

Working on the Unit Project

In Ex. 34, students work in groups of four to make an actual instrument using plastic straws. Once the first straw is constructed and measured, group members can divide up the construction of the remaining straws.

Quick Quiz (5-1 through 5-4)

See page 327.

See also Test 19 in the *Assessment Book.*

Unit 5 CHECKPOINT

1. **Writing** How is an exponential growth function like an exponential decay function? How are they different? 5-1

2. The number of golfers in the United States has grown exponentially since 1970. 5-2

Year	Number of golfers
1980	15,112,000
1985	17,520,000

   a. Write a function to model the number of golfers in the United States.

   b. Estimate the number of golfers in the year 2025.

   c. Estimate the number of golfers in the year 1970.

Simplify. 5-3

3. $27^{2/3}$ 4. $81^{5/4}$ 5. $32^{3/5}$

Sona Doshi invested $250 at 7% interest. Find the value of her investment after two years with each type of compounding. 5-4

6. yearly 7. monthly 8. daily

Practice 40 For use with Section 5-4

Name ______________ Date ______________

Practice 40
For use with Section 5-4

1. 57.2577; 4.6069 2. 0.0282; 89.5518
3. 0.0042; 0.0015 4. 0.0356; 0.0049
5. 8334.4189; 15.2238 6. 0.0183; 0.4997

For each function, find $f(2.5)$ and $f(-0.02)$. Round answers to the nearest ten-thousandth.

1. $f(x) = 4.7e^x$ 2. $f(x) = 84e^{-3.2x}$ 3. $f(x) = \frac{1}{1 + 640e^{-0.4x}}$

4. $f(x) = \frac{1}{1 + 200e^{-0.8x}}$ 5. $f(x) = 7.6(e^{2.8x} + e^{-2.8x})$ 6. $f(x) = \frac{1}{e^{1.6x} + e^{-1.6x}}$

7. Linh Luong invested $4000 for 3 years in a bank that pays 5% interest compounded monthly.

   a. What was the value of her investment at the end of that period? $4645.89

   b. What would have been the value of her investment if she had put the money in another bank paying the same interest, compounded daily? $4647.29

8. The median age of first marriage for men in the U.S. underwent logistic growth between 1955 and 1990, with growth function

$$f(x) = 22.6 + \frac{4}{1 + 130e^{-0.2x}}$$

where x is the number of years after 1955.

   a. Graph the function. What was the median age of first marriage for men in 1970? in 1980? See below.

   b. Between what years did the median age of first marriage for men rise exponentially? See below.

   c. In what year did the median age of first marriage for men start its rise? about 1949

9. The height of a rope bridge above the bottom of a gorge is given by the function

$$f(x) = 60 + 10(e^{0.01x} + e^{-0.01x}),$$

where x is the horizontal distance from the center of the bridge.

   a. Graph the function. How high is the center of the bridge above the bottom of the gorge? For graph, see additional answers.; 80 ft

   b. Each end of the bridge is 90 ft above the bottom of the gorge. How wide is the gorge where the bridge crosses it? about 192 ft

8. a. For graph, see additional answers.; about 23.1 yr; about 24.7 yr
b. Answers may vary. A sample answer is given: between about 1949 and 1983.

40 Practice Bank, INTEGRATED MATHEMATICS 3

Answers to Exercises and Problems

32. 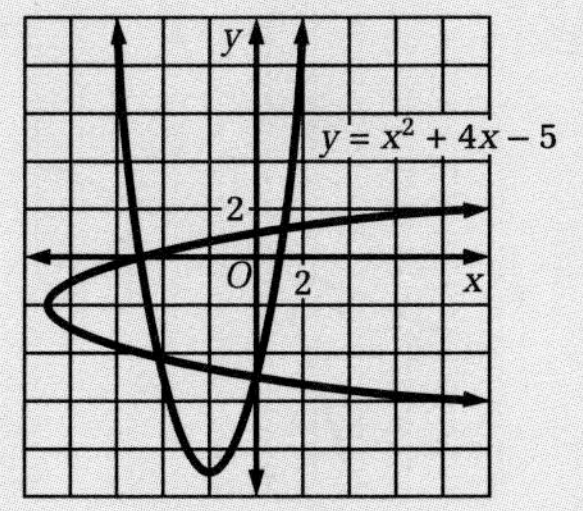

33. 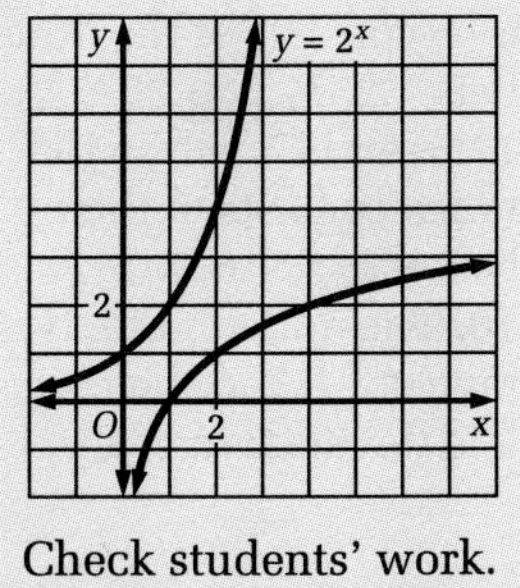

34. Check students' work. The straw length for a middle C note should be about 16 cm.

Answers to Checkpoint

1. Exponential growth functions and exponential decay functions both have the form $y = ab^x$. For an exponential growth function, $b > 1$. For an exponential decay function, $0 < b < 1$.

2. a. $y = 15{,}112{,}000(1.16)^x$, where x is the number of 5-year periods after 1980

   b. about 27,360,000

   c. about 11,231,000

3. 9
4. 243
5. 8
6. $286.23
7. $287.45
8. $287.57

PLANNING

Objectives and Strands

See pages 264A and 264B.

Spiral Learning

See page 264B.

Materials List

- Graph paper
- Graphics calculator or graphing software

Recommended Pacing

Section 5-5 is a one-day lesson.

Extra Practice

See pages 615–617.

Warm-Up Exercises

- Warm-Up Transparency 5-5

Support Materials

- Practice Bank: Practice 41
- Activity Bank: Enrichment 37
- Study Guide: Section 5-5
- Problem Bank: Problem Set 12
- Explorations Lab Manual: Additional Exploration 6 Diagram Master 2
- Overhead Visuals: Folder 4
- Using IBM/Mac Plotter Plus Disk: Function Plotter
- Assessment Book: Quiz 5-5, Alternative Assessments 4, 5

Section 5-5 Inverse Functions

Focus

Find and graph inverse functions.

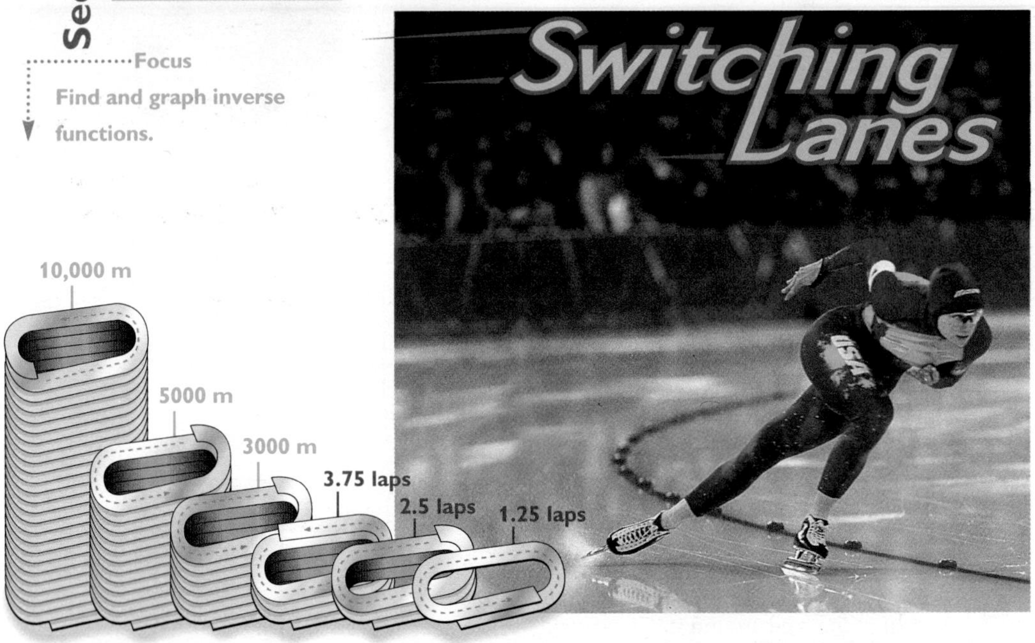

Speed Skating Races	
Distance (km)	Number of laps
10	?
5	?
3	?
?	3.75
?	2.5
?	1.25

Bonnie Blair won the 500-meter and 1000-meter speed skating races at the 1994 Olympics in Lillehammer, Norway. Her two gold medals gave her a career total of five gold medals and one bronze—the most medals awarded to any United States Winter Olympian.

The official track for speed skating is 0.4 km long. The number of laps a skater needs to skate depends on the distance of the race.

Talk it Over

1. Complete the table.
2. Write a rule for $L(D)$, the number of laps skated as a function of the distance of the race.
3. Write a rule for $D(L)$, the distance of a race as a function of the number of laps skated.
4. For each function in questions 2 and 3, state the domain and range. Consider only the six races given. What do you notice?
5. Graph each function from questions 2 and 3 on the same set of axes. What transformation describes the relationship between the two graphs?

Answers to Talk it Over

1.

Distance (km)	Number of laps
10	25
5	12.5
3	7.5
1.5	3.75
1	2.5
0.5	1.25

2. $L(D) = \frac{D}{0.4} = 2.5D$
3. $D(L) = 0.4L$
4. The domain of L consists of the numbers 0.5, 1, 1.5, 3, 5, and 10, while the range consists of the numbers 1.25, 2.5, 3.75, 7.5, 12.5, and 25. The domain of D consists of the numbers 1.25, 2.5, 3.75, 7.5, 12.5, and 25, while the range consists of the numbers 0.5, 1, 1.5, 3, 5, and 10. The domain of L is the range of D and the range of L is the domain of D.
5. 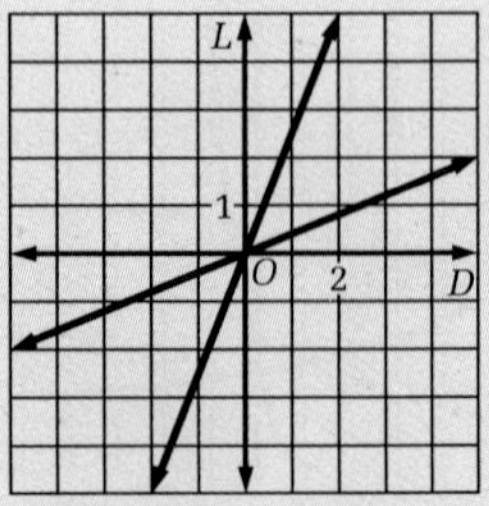

Each graph is the image of the other after reflection over the line $L = D$.

Two functions f and g are **inverse functions** if $g(b) = a$ whenever $f(a) = b$.

The graph of the inverse is the reflection of the graph of the function over the line $y = x$.

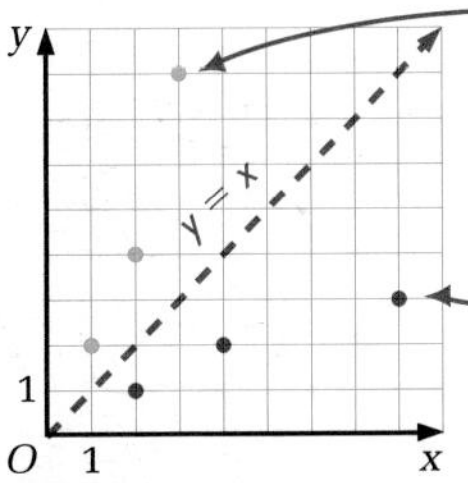

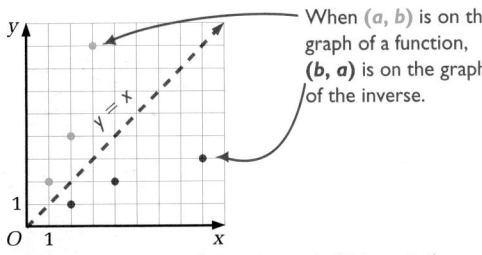

When (a, b) is on the graph of a function, (b, a) is on the graph of the inverse.

The inverse of a function $f(x)$ can be written $f^{-1}(x)$ or f^{-1}.

An inverse is a function. If the reflection of a function over the line $y = x$ is not a function, then the inverse does not exist.

Watch Out!
Although the exponent -1 usually means the reciprocal, $f^{-1}(x)$ is *not* equal to $\frac{1}{f(x)}$.

Sample 1

Graph each function and its reflection over the line y = x. Is the reflection the graph of a function?

a. $y = 2^x$

b. $y = x^2$

Sample Response

Step 1 Plot points to graph each function. Then interchange the coordinates of the ordered pairs and plot these points. Draw smooth curves through each set of points.

a. $(1, 2) \rightarrow (2, 1)$

$\left(-1, \frac{1}{2}\right) \rightarrow \left(\frac{1}{2}, -1\right)$

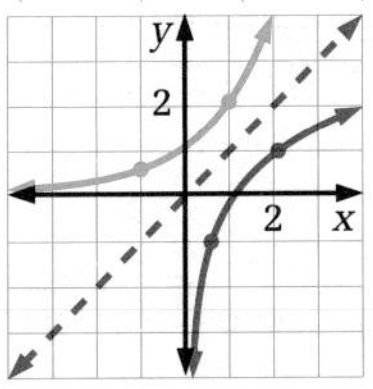

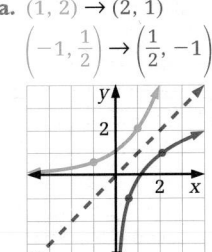

b. $(2, 4) \rightarrow (4, 2)$

$(-2, 4) \rightarrow (4, -2)$

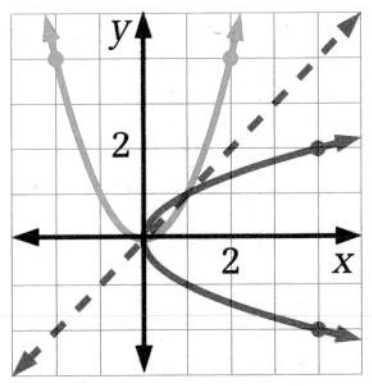

Step 2 Use the vertical line test to decide whether the reflection is the graph of a function.

The reflection of the graph of $y = 2^x$ represents a function.

The reflection of the graph of $y = x^2$ does *not* represent a function. The function $y = x^2$ does *not* have an inverse.

Just as you use the vertical line test to tell whether a graph is a function, you can use the horizontal line test to tell whether a function has an inverse.

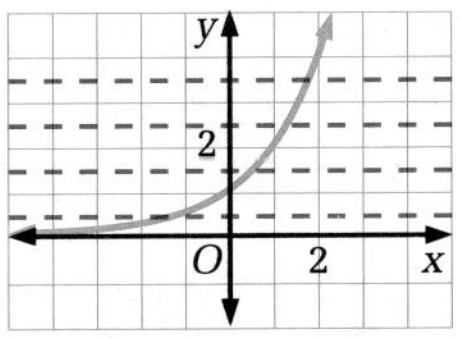

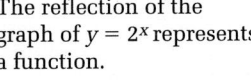

No two points lie on the same horizontal line. The function has an inverse.

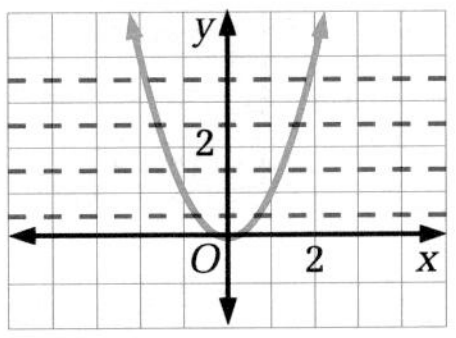

Two points lie on the same horizontal line. The function does not have an inverse.

TEACHING

Talk it Over

In questions 1–3, students construct a function and its inverse. Question 4 helps students discover the relationship of domains and ranges for inverse functions, and question 5 helps students discover that a graph and its inverse are reflections of each other over the line $y = x$.

Integrating the Strands

Graphs of inverse functions can be found by using geometry. Given a point P on the function, find its corresponding point P' on the inverse by drawing a line m through P perpendicular to the line $y = x$. Let Q be the intersection of line m with $y = x$. P' will be the point on m across the line $y = x$ from P such that $PQ = QP'$. Note that the line $y = x$ is the perpendicular bisector of $\overline{PP'}$.

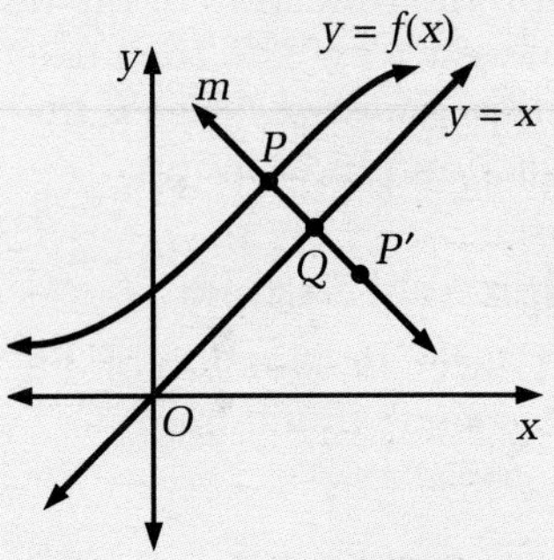

Additional Sample

S1 Graph each function and its reflection over the line $y = x$. Is the reflection the graph of a function?

a. $y = 3^x$

Step 1. Plot points to graph $y = 3^x$. Then interchange the ordered pairs and plot these points. Draw smooth curves through each set of points.

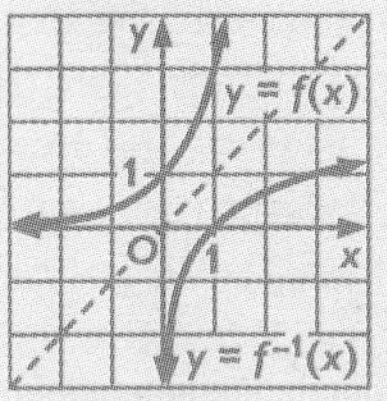

Step 2. Use the vertical line test. Yes, the reflection is a function.

Sample continued on next page.

Additional Sample (continued)

b. $y = |2x|$

Step 1. Plot points to graph $y = |2x|$. Then interchange the ordered pairs and plot these points.

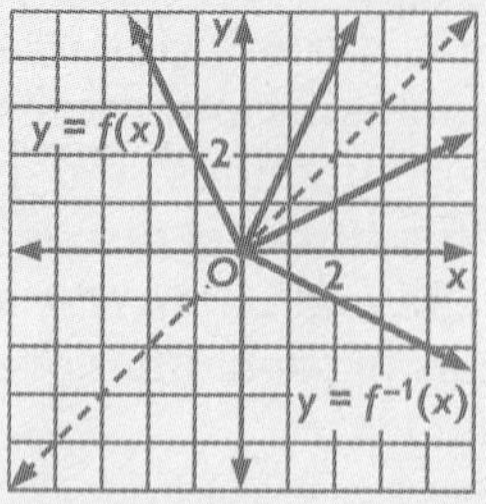

Step 2. Use the vertical line test. No, the reflection is not a function.

Talk it Over

When doing questions 6–8, students may become confused as to when to use the vertical line test and when to use the horizontal line test. Students should realize that they can use the horizontal line test on the original function to see if it has an inverse. Or, they can graph the reflection of the function over $y = x$ and then use the vertical line test to see if the reflection is a function.

Additional Sample

S2 Write an equation of the inverse function of $f(x) = \frac{3}{4}x + 5$.

$$f(x) = \frac{3}{4}x + 5$$
$$y = \frac{3}{4}x + 5$$
$$x = \frac{3}{4}y + 5$$
$$x - 5 = \frac{3}{4}y$$
$$\frac{4}{3}x - \frac{20}{3} = y$$
$$f^{-1}(x) = \frac{4}{3}x - \frac{20}{3}$$

Error Analysis

Finding the equation of an inverse function is a simple procedure, yet many students have problems with it. One problem area is switching x and y. Students may try to solve for x. Emphasize that the variables are merely changing positions; that no solving is being done.

Talk it Over

Tell whether each graph represents a function. If so, does it have an inverse? Explain.

6.

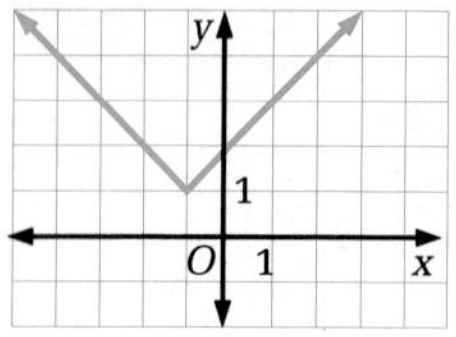

7.

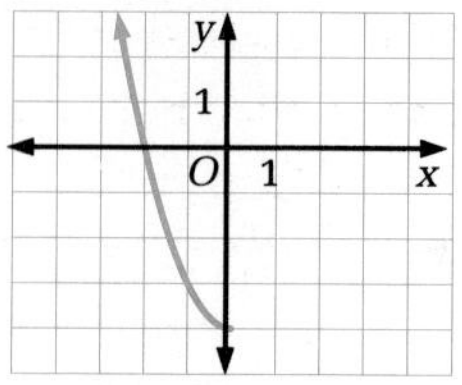

8.

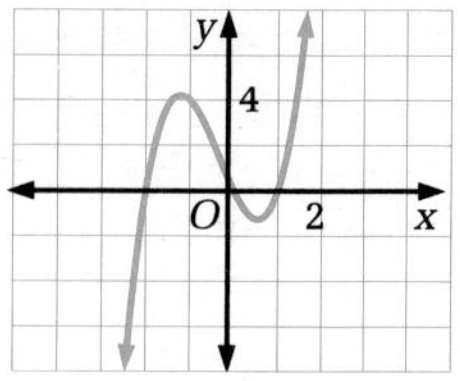

9. Suppose the function $y = x^2$ represents the area (y) of a square as a function of the length (x) of one side.
   a. What are the domain and the range of this area function?
   b. What part of the graph in Sample 1(b) is the graph of this area function?
   c. Does the function have an inverse? Explain.

Sample 2

Write an equation of the inverse function of $f(x) = \frac{2}{3}x - 4$.

Sample Response

$f(x) = \frac{2}{3}x - 4$	← Write the original equation.
$y = \frac{2}{3}x - 4$	← Write y for $f(x)$.
$x = \frac{2}{3}y - 4$	← Interchange the positions of x and y.
$x + 4 = \frac{2}{3}y$	← Solve for y.
$\frac{3}{2}x + 6 = y$	← y represents $f^{-1}(x)$.
$f^{-1}(x) = \frac{3}{2}x + 6$	

Look Back

In Sample 2 the third step was to interchange the variables x and y. Explain why this method works.

Answers to Talk it Over

6. Yes; No; the inverse does not pass the horizontal line test.
7. Yes; Yes; the inverse passes the horizontal line test.
8. Yes; No; the inverse does not pass the horizontal line test.
9. a. nonnegative real numbers; nonnegative real numbers
   b. the portion of the graph in the first quadrant
   c. Yes; the function $y = x^2$ with the domain restricted to $x \geq 0$ has an inverse. Its graph passes the horizontal line test. The graph of the inverse is the first-quadrant portion of the inverse function shown in Sample 1(b).

Answers to Look Back

For every point (x, y) on the graph of $f(x)$, (y, x) is on the graph of $f^{-1}(x)$.

5-5 Exercises and Problems

1. **Reading** How are the domain and the range of a function related to the domain and the range of its inverse?
2. Suppose $(-1, 3)$ is a solution of $f(x)$. Find a solution of $f^{-1}(x)$.

For each function, tell whether the inverse exists. If it does, graph the inverse. If it does not exist, give a domain so that the function has an inverse.

3. 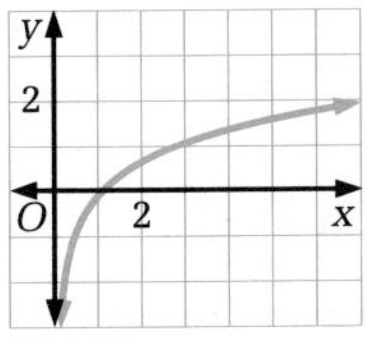

4. 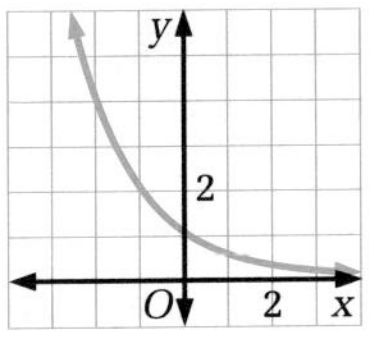

5. 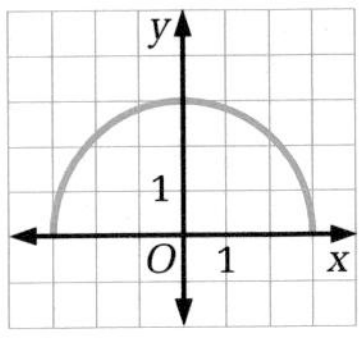

Graph each function and its reflection over the line y = x. Is the reflection the graph of a function?

6. $y = x^{1/3}$
7. $y = e^x$
8. $y = x^4 + \frac{1}{2}$

9. The traditional Japanese calendar is 660 years ahead of the calendar developed in Europe and used today in the United States. For example, when the European year was 1000, the Japanese year was 1660.
   a. Write an equation for the function that converts the European year to the Japanese year.
   b. Write an equation for the inverse function that converts the Japanese year to the European year.
   c. What is the slope of each equation?
   d. **Writing** Will the graphs of the equations in parts (a) and (b) intersect? Give both a mathematical reason and a practical reason for your answer.

BY THE WAY...

According to legend, Jimmu Tenno became the first emperor of Japan in 660 B.C. The Japanese calendar is based on the start of Jimmu's reign—placing the Japanese year 660 years ahead of the European calendar.

Write an equation of the inverse of each function.

10. $y = \frac{1}{3}x - 5$
11. $y = -x + 6$
12. $y = x - 3\frac{1}{2}$
13. $y = x$
14. $y = -2x^2, x \geq 0$
15. $y = 2x^5 - 3$
16. $y = x^{2/5}$
17. $y = 2|x|, x \leq -1$
18. $y = mx + b$
19. $y = x + b$
20. $y = ax^b$
21. $y = x^{a/b}$

22. **Travel** Leanne used this equation to exchange her United States dollars for Indian rupees: $r = 31.15d$, where r is the number of rupees with the same value as d dollars.
   a. Write an equation for the inverse of the function. Can you use the inverse function to exchange rupees for dollars? Explain.
   b. What are the domain and the range of the function? of the inverse function?

Answers to Exercises and Problems

1. The domain of $f(x)$ is the range of $f^{-1}(x)$ and the range of $f(x)$ is the domain of $f^{-1}(x)$.
2. (3, –1)
3. Yes. 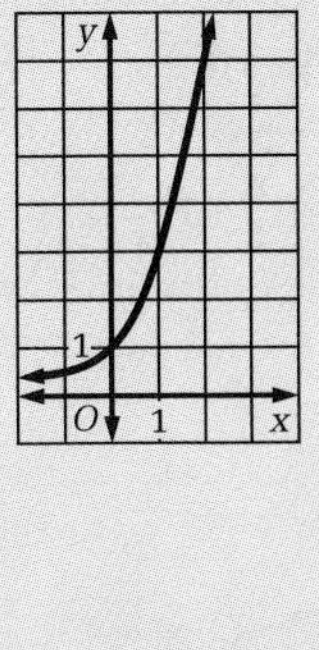

4. Yes. 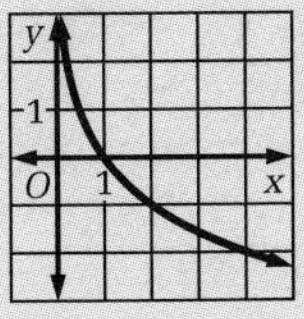

5. No. Answers may vary. An example is given. 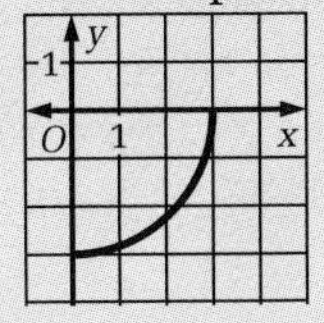

6. Yes. 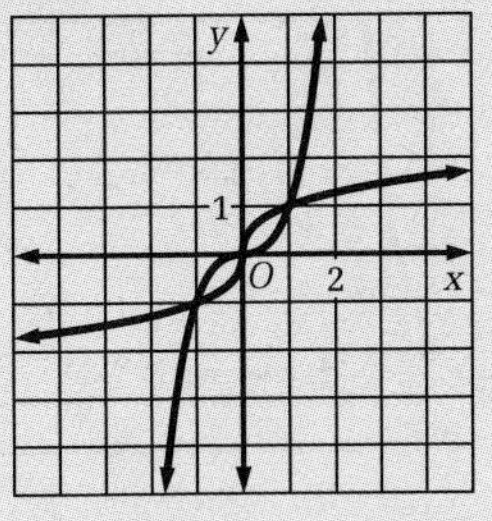

7–22. See answers in back of book.

APPLYING

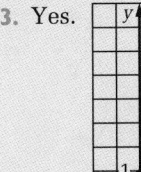

Suggested Assignment

Standard 1–8, 10–21, 23–25, 29–34

Extended 1–25, 29–34

Integrating the Strands

Number Exs. 31–33

Algebra Exs. 1–29

Functions Exs. 1–29

Geometry Ex. 30

Logic and Language Exs. 1, 9, 26, 29, 34

Using Technology

As noted in Ex. 27 on page 300, parametric equations can be used to examine functions and the relations that are their inverses. (For a review of parametric equations, students can refer to the Toolbox or Using TI-81 and TI-82 Calculators booklet.)

Consider the function $y = 2^x$ in Sample 1, part (a). Set the calculator to parametric mode. On the Y= list, enter the following equations.

for $y = 2^x$:
X1T=T
Y1T=2^T
for the inverse relation:
X2T=2^T
Y2T=T
for the line $y = x$:
X3T=T
X3T=T

Use the following initial window settings: Tmin = –5, Tmax = 5, Tstep = 0.1, Xmin = –5, Xmax = 5, Ymin = –5, Ymax = 5. Press ZOOM and select menu item 5 to "square up" the graph. The original function and the inverse relation both appear to pass the vertical line test. Hence, both are functions and each function is the inverse of the other.

Assessment: Investigation

Students should be given a list of various functions and an accompanying set of their graphs, for example, $f(x) = x^2$, $f(x) = \sqrt{x}$, $x > 0$, $f(x) = |x + 1|$, $f(x) = 3x + 2$, $f(x) = 3x^3 - 2x^2 + 3x - 1$. Before determining whether each function has an inverse, students should determine whether it is one-to-one or many-to-one. They can reflect the graphs about the line $y = x$ to determine whether an inverse exists. Students should write a conclusion about one-to one functions and inverses. They should then find the inverses of the functions, if they exist.

Practice 41 For use with Section 5-5

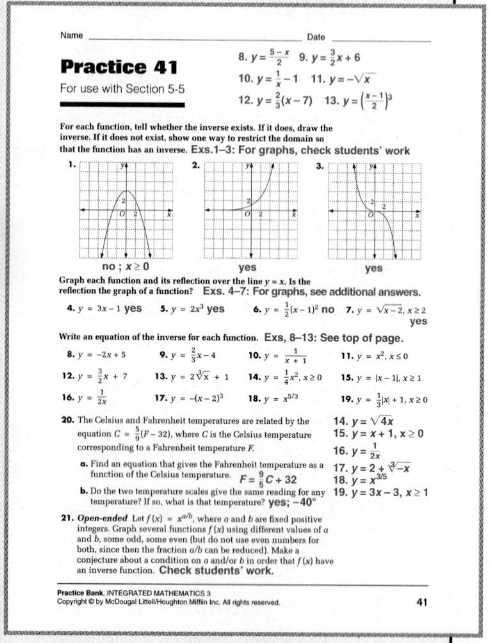

Name ______________________ Date ______________

Practice 41

For use with Section 5-5

8. $y = \frac{5-x}{2}$ 9. $y = \frac{3}{2}x + 6$
10. $y = \frac{1}{x} - 1$ 11. $y = -\sqrt{x}$
12. $y = \frac{2}{3}(x - 7)$ 13. $y = \left(\frac{x-1}{2}\right)^3$

For each function, tell whether the inverse exists. If it does, draw the inverse. If it does not exist, show one way to restrict the domain so that the function has an inverse. Exs.1–3: For graphs, check students' work

1. no ; $x \geq 0$
2. yes
3. yes

Graph each function and its reflection over the line $y = x$. Is the reflection the graph of a function? Exs. 4–7: For graphs, see additional answers.

4. $y = 3x - 1$ yes
5. $y = 2x^3$ yes
6. $y = \frac{1}{2}(x-1)^2$ no
7. $y = \sqrt{x-2}$, $x \geq 2$ yes

Write an equation of the inverse for each function. Exs, 8–13: See top of page.

8. $y = -2x + 5$
9. $y = \frac{2}{3}x - 4$
10. $y = \frac{1}{x+1}$
11. $y = x^2$, $x \leq 0$
12. $y = \frac{3}{2}x + 7$
13. $y = 2\sqrt[3]{x} + 1$
14. $y = \frac{1}{4}x^2$, $x \geq 0$
15. $y = |x - 1|$, $x \geq 1$
16. $y = \frac{1}{2x}$
17. $y = -(x-2)^3$
18. $y = x^{5/3}$
19. $y = \frac{1}{3}|x| + 1$, $x \geq 0$

20. The Celsius and Fahrenheit temperatures are related by the equation $C = \frac{5}{9}(F - 32)$, where C is the Celsius temperature corresponding to a Fahrenheit temperature F.
 a. Find an equation that gives the Fahrenheit temperature as a function of the Celsius temperature. $F = \frac{9}{5}C + 32$
 b. Do the two temperature scales give the same reading for any temperature? If so, what is that temperature? yes; −40°

14. $y = \sqrt{4x}$
15. $y = x + 1$, $x \geq 0$
16. $y = \frac{1}{2x}$
17. $y = 2 + \sqrt[3]{-x}$
18. $y = x^{3/5}$
19. $y = 3x - 3$, $x \geq 1$

21. ***Open-ended*** Let $f(x) = x^{a/b}$, where a and b are fixed positive integers. Graph several functions $f(x)$ using different values of a and b, some odd, some even (but do not use even numbers for both, since then the fraction a/b can be reduced). Make a conjecture about a condition on a and/or b in order that $f(x)$ have an inverse function. Check students' work.

Practice Bank, INTEGRATED MATHEMATICS 3

41

In Exercises 23–25, tell whether each function has an inverse.

23. $f(x)$ is not one-to-one.
24. $f(x)$ is linear with slope -1.
25. $f(x) = k$

26. a TECHNOLOGY Use a graphics calculator or graphing software. Find where the graphs of $f(x)$ and $f^{-1}(x)$ intersect.

➤ $f(x) = 4x - 12$, $f^{-1}(x) = 0.25x + 3$
➤ $f(x) = \frac{1}{x}$, $f^{-1}(x) = \frac{1}{x}$
➤ $f(x) = x^3$, $f^{-1}(x) = \sqrt[3]{x}$

b. **Writing** Make a conjecture about where the graph of a function intersects the graph of its inverse. Why do you think this is true?

27 TECHNOLOGY Use a graphics calculator or graphing software.

a. Set $x_1 = t$ and $y_1 = 2t - 7$. How should you set x_2 and y_2 in order to graph the reflection of the graph over the line $y = x$?

b. Set $x_1 = t$ and $y_1 = 2e^t$ to graph $y = 2e^x$. Then use the method from part (a) to graph its reflection over the line $y = x$.

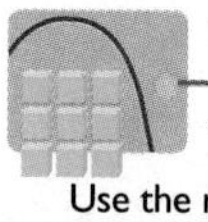

TECHNOLOGY NOTE

Work in parametric mode. Use the range key to set the t, x, and y values between -10 and 10.

Ongoing ASSESSMENT

28. **Group Activity** Work in a group of four students. After completing each part, give your paper to the person on your right.

a. Each student should write an equation for a function that has an inverse. Do not write an exponential function.

b. Write the equation for the inverse of the function you receive.

c. On the same set of axes, graph the original function and the inverse function you receive.

d. Check the equation and the graphs you receive. Discuss any errors.

Review PREVIEW

29. The equation $y = 3.4e^{0.12x}$ can be used to model the number of ATM transactions, in billions, where x represents the number of years since 1985. Find the number of transactions in 1990. *(Section 5-4)*

30. Write a coordinate proof to show that $OPQR$ is a kite. *(Section 3-2 and Toolbox Skill 31)*

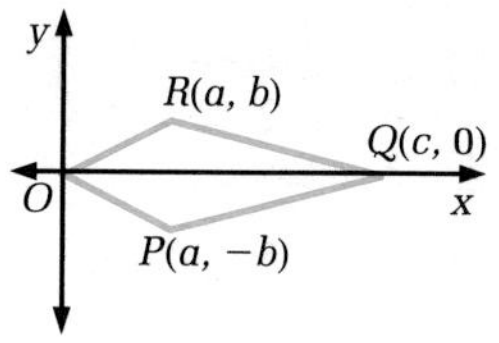

Write each number in scientific notation. *(Toolbox Skill 14)*

31. 0.00000291
32. 47,000,000
33. 482,900

Working on the Unit Project

34. **Research** Find a book of music for beginners that you can use to play songs on your instruments.

Answers to Exercises and Problems

23. No.
24. Yes.
25. No.
26. a. (4, 4); The graphs coincide; (0, 0), (1, 1), and (–1, –1).
 b. at a point on the line $y = x$; For points on this line, interchanging the coordinates results in the same point.
27. a. $x_2 = 2t - 7$, $y_2 = t$
 b.

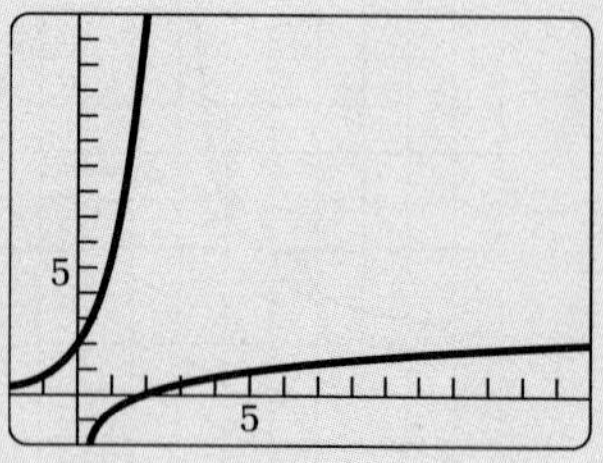

28. Check students' work.
29. about 6.2 billion transactions
30. Answers may vary. An example is given. Use the distance formula to find OR, OP, RQ, and QP. $OR = \sqrt{(a-0)^2 + (b-0)^2} = \sqrt{a^2 + b^2}$, $OP = \sqrt{(a-0)^2 + (b-0)^2} = \sqrt{a^2 + b^2}$, $RQ = \sqrt{(c-a)^2 + (0-b)^2} = \sqrt{(c-a)^2 + b^2}$, and $QP = \sqrt{(c-a)^2 + (0+b)^2} = \sqrt{(c-a)^2 + b^2}$. Therefore, $OR = OP$ and $RQ = QP$. By the definition of a kite, $OPQR$ is a kite.
31. $2.91 \cdot 10^{-6}$
32. $4.7 \cdot 10^7$
33. $4.829 \cdot 10^5$
34. Check students' work.

Section

5-6 Logarithmic Functions

Recording of a 5.1 quake

Focus

Recognize and evaluate logarithmic functions. Use logarithmic functions to solve problems.

Richter scale number	Relative size	
6	1,000,000	10^6
5	100,000	10^5
2	100	10^2
1	10	10^1

Epicenter of 1989 earthquake at Los Gatos, California

On January 19, 1994, an earthquake measuring 5.1 on the Richter scale struck California. A one-point increase in the Richter scale number indicates a ten-fold increase in the relative size of the earthquake.

Talk it Over

1. What relationship do you notice between a Richter number and the relative size of the earthquake?
2. What is the relative size of the California earthquake that measured 5.1 on the Richter scale?
3. Is an earthquake that measures 8 on the Richter scale four times the size of an earthquake that measures 2 on the scale? Explain.
4. Does the sequence of Richter scale numbers represent an *arithmetic sequence* or a *geometric sequence*? Does the sequence of relative size numbers represent an *arithmetic sequence* or a *geometric sequence*?
5. The relative size of an earthquake is a function of its Richter scale number x: $f(x) = 10^x$. Look at the graphs below. Does $f(x)$ have an inverse function? Why or why not?

$y = 10^x$ — Interchange x and y. — $x = 10^y$

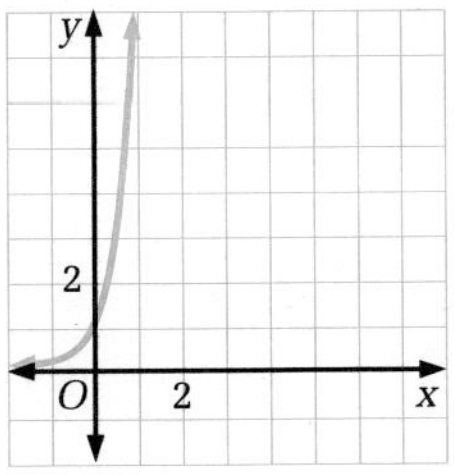

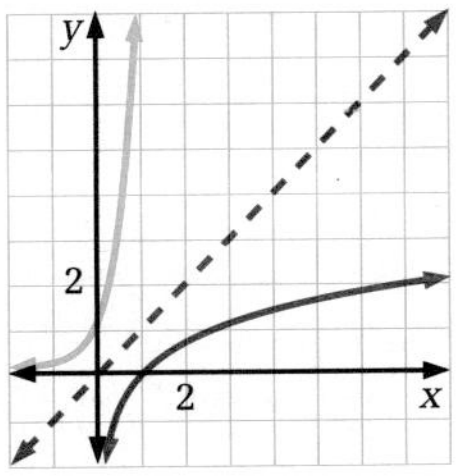

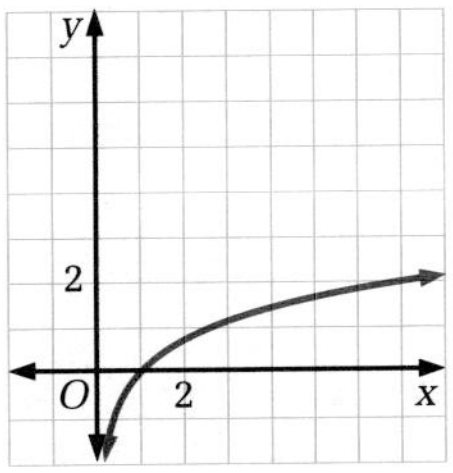

PLANNING

Objectives and Strands

See pages 264A and 264B.

Spiral Learning

See page 264B.

Materials List

- Graphics calculator

Recommended Pacing

Section 5-6 is a two-day lesson.

Day 1

Pages 301–303: Opening paragraph through Sample 1, *Exercises 1–25*

Day 2

Pages 303–305: Sample 2 through Look Back, *Exercises 26–47*

Extra Practice

See pages 615–617.

Warm-Up Exercises

Warm-Up Transparency 5-6

Support Materials

- Practice Bank: Practice 42
- Activity Bank: Enrichment 38
- Study Guide: Section 5-6
- Problem Bank: Problem Set 12
- Assessment Book: Quiz 5-6, Alternative Assessment 6

Answers to Talk it Over

1. The scale number is the same as the number of zeroes in the relative size.
2. about 125,893
3. No; the relative size of an earthquake that measures 8 on the Richter scale is 100,000,000. The relative size of an earthquake that measures 2 on the Richter scale is 100. The size of the larger earthquake is a million times the size of the smaller quake.
4. arithmetic; geometric
5. Yes; the graph passes the horizontal line test.

TEACHING

Talk it Over

Questions 1–5 use the Richter scale to have students analyze a real-life exponential function and to reach the conclusion that exponential functions have inverses, thus setting the stage for the introduction of the logarithmic function.

Visual Thinking

Have students work in groups to research the relative size of recent earthquakes. Encourage each group to select the locations of the quakes that it will study. Have them create tables and graphs of their quakes like those shown on page 301. Ask each group to use these displays to explain the Richter scale numbers and relative size data to the class. This activity involves the visual skills of *exploration* and *communication.*

Communication: Reading

Have a student read aloud the line, "A logarithm of any positive real number x is the exponent a when you write x as a power of a base b." It is helpful for students to think of logarithms as exponents. For example, when students are finding log 10,000, they are looking for the exponent n in the equation $10^n = 10{,}000$.

Teaching Tip

Discuss changing exponential form to logarithmic form, and vice versa. Many problems students encounter will begin in one form but are solved by converting to the other form.

Talk it Over

In question 6, students should observe that a logarithm with base 1 will not give a definitive answer by considering the corresponding inverse function $y = 1^x$.

The exponential function $f(x) = 10^x$ has an inverse function called a **logarithmic function** that is written as $f^{-1}(x) = \log_{10} x$. A **logarithm** of any positive real number x is the exponent a when you write x as a power of a base b.

$$\log_b x = a \text{ when } x = b^a$$

Read as "the base-*b* logarithm of *x*" or "log base *b* of *x*."

The base of a logarithm can be any positive number except 1. A logarithm with base 10 is called a **common logarithm.** You write the common logarithm of x as $\log x$. A logarithm with base e is called a **natural logarithm.** It is usually written as $\ln x$.

Logarithmic and Exponential Form

An exponential function and a logarithmic function are related in the following way:

$$x = b^y \qquad y = \log_b x$$

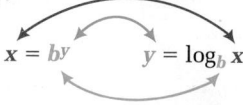

TECHNOLOGY NOTE

You can use LOG or LN to find a common logarithm or a natural logarithm on your calculator.

Exponential Form: $x = b^y$	**Logarithmic Form: $y = \log_b x$**
$1000 = 10^3$	$3 = \log 1000$
$16 = 4^2$	$2 = \log_4 16$
$\frac{1}{2} = 2^{-1}$	$-1 = \log_2 \frac{1}{2}$
$5 \approx e^{1.6}$	$1.6 \approx \ln 5$

Talk it Over

6. The base of a logarithm can be any positive number except 1. Why do you think 1 cannot be a base of a logarithm?
7. Write each equation in exponential form.
   a. $\log_5 25 = 2$ b. $\log_4 4 = 1$ c. $\log (0.01) = -2$
8. Write each equation in logarithmic form.
   a. $10^5 = 100{,}000$ b. $4^{3/2} = 8$ c. $7^0 = 1$

LOGARITHMS

When x and b are positive numbers ($b \neq 1$):

$$\log_b x = a \text{ if and only if } x = b^a$$

common logarithm: $\log x = a$ when $10^a = x$

natural logarithm: $\ln x = a$ when $e^a = x$

Examples

$\log_2 16 = \log_2 2^4 = 4$

$\log 100{,}000 = \log_{10} 10^5 = 5$

$\ln 8 \approx \log_e e^{2.08} \approx 2.08$

Answers to Talk it Over

6. because 1 raised to any power is 1
7. a. $5^2 = 25$
   b. $4^1 = 4$
   c. $10^{-2} = 0.01$
8. a. $\log (100{,}000) = 5$
   b. $\log_4 8 = \frac{3}{2}$
   c. $\log_7 1 = 0$

Sample 1

Evaluate each logarithm. Round decimal answers to the nearest hundredth.

a. $\log_4 64$ b. $\log \frac{1}{10{,}000}$ c. $\ln 5.3$ d. $\log 145$

Sample Response

a. $\log_4 64 = \log_4 4^3$ ← Write 64 as a power of the base 4.

$= 3$

b. $\log \frac{1}{10{,}000} = \log 10^{-4}$ ← Rewrite $\frac{1}{10{,}000}$ as a power of the base 10.

$= -4$

c. $\ln 5.3 \approx 1.67$ ← Use a calculator to find the natural logarithm.

d. $\log 145 \approx 2.16$ ← Use a calculator to find the common logarithm.

Now you are ready for: Exs. 1–25 on p. 305

Sample 2

Solve $\log_x 81 = 2$.

Sample Response

$\log_x 81 = 2$

$x^2 = 81$ ← Rewrite as an exponential equation.

$x = 9$ ← Find the positive square root of both sides, since the base of a logarithm is positive.

Sample 3

Solve. Round decimal answers to the nearest hundredth.

a. $10^x = 15$ b. $e^x = 29$

Sample Response

a. $10^x = 15$

$x = \log 15$ ← Rewrite as a logarithmic equation.

≈ 1.18 ← Use a calculator to find the common logarithm.

b. $e^x = 29$

$x = \ln 29$ ← Rewrite as a logarithmic equation.

≈ 3.37 ← Use a calculator to find the natural logarithm.

Additional Samples

S1 Evaluate each logarithm. Round decimal answers to the nearest hundredth.

a. $\log_5 125$

Write 125 as a power of the base 5.

$\log_5 125 = \log_5 5^3 = 3$

b. $\log \frac{1}{100}$

Rewrite $\frac{1}{100}$ as a power of the base 10.

$\log \frac{1}{100} = \log 10^{-2} = -2$

c. $\ln 18.9$

Use a calculator to find the natural logarithm.

$\ln 18.9 \approx 2.94$

d. $\log 87$

Use a calculator to find the common logarithm.

$\log 87 \approx 1.94$

S2 Solve $\log_x 64 = 3$.

Rewrite as an exponential equation.

$\log_x 64 = 3$

$x^3 = 64$

$x = 4$

S3 Solve. Round decimal answers to the nearest hundredth.

a. $10^x = 34$

Rewrite as a logarithmic equation. Then use a calculator to find the common logarithm.

$x = \log 34$

≈ 1.53

b. $e^x = 47$

Rewrite as a logarithmic equation. Then use a calculator to find the natural logarithm.

$x = \ln 47$

≈ 3.85

Additional Sample

S4 a. If $[H+] = 10^{-10}$ for an antacid tablet, what is the pH level for the antacid?

$pH = -\log [H+]$
$pH = -\log 10^{-10}$
$= -(-10)$
$= 10$

b. The concentration of hydrogen ions in a solution is 3.7×10^{-5}. What is the pH level of this particular solution?

$pH = -\log [H+]$
$pH = -\log (3.7 \times 10^{-5})$
$\approx -(-4.4)$
≈ 4.4

pH Values of Some Common Substances

Substance	pH
lemon juice	2.1–2.3
tomato juice	4.1
milk	6.6
pure water	7.0
eggs	7.6–8.0
baking soda	8.5
ammonia	11.9

Applying Logarithms

Acidity is a measure of the concentration of hydrogen ions, $[H^+]$, measured in moles/L. Since $[H^+]$ is usually a very large or very small number, it is converted to a pH number between 0 and 14 that shows relative acidity. A solution that has a pH less than 7 is acidic and a solution that has a pH greater than 7 is basic. A solution with a pH of 7 is neutral. The conversion formula is $pH = -\log [H^+]$.

Talk it Over

9. Normal drinking water has a pH between 6.3 and 6.6. Is this water more acidic or less acidic than pure water?
10. Coffee has a pH of 5.0. When milk is added, does the pH level of the mixture become more acidic or less acidic than the coffee alone?

Sample 4

Piki bread is a Native American food that is made with blue corn flour and water. To release the B-vitamin niacin from the flour mixture, limewater or ash water needs to be added to the batter.

The batter changes color from a deep blue to a grayish blue when limewater or ash water is added. The Hopi and Zuni people use the color of the batter to determine when the bread will taste the best and be the most nutritious.

a. Suppose the color of the batter changes to a grayish blue when $[H^+] = 10^{-8}$. Find the pH level of the grayish blue batter.

b. Suppose the concentration of hydrogen ions in blue corn flour is about 1.6×10^{-7}. What is the pH level of blue corn flour? Is the flour *acidic* or *basic*?

Sample Response

a. **Problem Solving Strategy:** Use a formula.

$pH = -\log [H^+]$ ← Write the conversion formula.
$pH = -\log 10^{-8}$ ← Substitute for $[H^+]$.
$= -(-8)$
$= 8$

The grayish blue batter has a pH of 8.

b. **Problem Solving Strategy:** Use a formula.

$pH = -\log [H^+]$
$pH = -\log (1.6 \times 10^{-7})$
$\approx -(-6.8)$ ← Use a calculator to find the common logarithm.
≈ 6.8

The blue corn flour has a pH of about 6.8. The flour is acidic.

Answers to Talk it Over

9. more acidic
10. less acidic

Answers to Look Back

The common and natural logarithms of a number x are both exponents when x is written as a power of a base. The common logarithm of x is the power of 10 that is equal to x. The natural logarithm of x is the power of e that is equal to x.

Look Back

How are common logarithms and natural logarithms alike? How are they different?

Now you are ready for: Exs. 26–47 on pp. 305–308

5-6 Exercises and Problems

1. **Reading** How is the inverse of an exponential function related to a logarithmic function?

2. a. **Geology** On July 11, 1993, an earthquake that registered 6.1 on the Richter scale struck the coast of northern Chile. This earthquake is how many times greater in size than an earthquake that registers 5.1 on the Richter scale?

   b. On January 15, 1993, an earthquake that registered 7.1 on the Richter scale struck the Hokkaido region in Japan. This earthquake is how many times greater in size than an earthquake that registers 5.1 on the Richter scale?

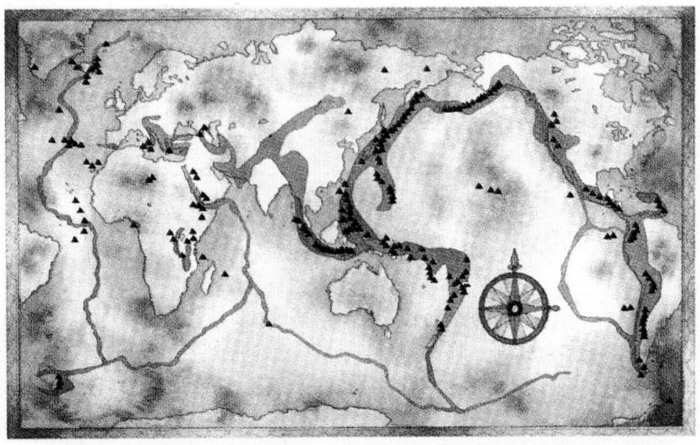

About 80% of all earthquakes occur along the borders of the Pacific Ocean. This zone is called the Ring of Fire because of its earthquakes (red areas) and volcanoes (black triangles).

Write each equation in logarithmic form.

3. $y = 7^x$
4. $a = \left(\frac{1}{2}\right)^b$
5. $e^1 = e$
6. $3^3 = 27$
7. $2^{-3} = \frac{1}{8}$
8. $10^2 = 100$
9. $8^0 = 1$
10. $\sqrt{36} = 6$

Write each equation in exponential form.

11. $\log_6 1 = 0$
12. $\log_4 64 = 3$
13. $\log 10{,}000 = 4$
14. $\log_3 \sqrt{81} = 2$
15. $\ln 2.718 \approx 1$
16. $\ln 8 \approx 2.08$

Evaluate each logarithm. Round decimal answers to the nearest hundredth.

17. $\log 100$
18. $\log_5 625$
19. $\log 10$
20. $\log \frac{1}{1000}$
21. $\ln 10$
22. $\ln 100$
23. $\ln 1$
24. $\log 625$
25. a. $\log_2 8 = 3$ and $\log_2 16 = 4$. Estimate the value of $\log_2 10$.

    b. **Writing** Explain your reasoning in part (a).

Solve. Round decimal answers to the nearest hundredth.

26. $\log_x 9 = 2$
27. $\log_x 8 = -3$
28. $e^x = \frac{1}{e^2}$
29. $10^x = 21$
30. $\log 1 = x$
31. $e^x = 10$
32. Acid rain is rain that has a pH below 5.6. Suppose the rain's concentration of hydrogen ions, $[H^+]$, is 10^{-4}. Is this acid rain? (Use the formula on page 304.)

Look Back

You may wish to extend this question by having students graph $y = \log x$ and $y = \ln x$ and compare the graphs. Students may also be interested in researching why these two logarithmic functions have the names that they do.

APPLYING

Suggested Assignment

Day 1

Standard 1–25

Extended 1–25

Day 2

Standard 26–32, 35, 40–47

Extended 26–38, 40–47

Integrating the Strands

Number Exs. 2, 5–25, 32, 33

Algebra Exs. 1–45, 47

Functions Exs. 1, 39, 40, 45

Discrete Mathematics Exs. 33, 34

Logic and Language Exs. 1, 25, 38

Reasoning

Most students will mentally perform a linear interpolation to answer Ex. 25 and give an answer of 3.25. Have students use the x^y key on a scientific calculator to improve their estimates. (A very good approximation is 3.322.) Ask students to explain why a linear process does not produce as good an approximation as one would think.

Answers to Exercises and Problems

1. For any positive number b, $f(x) = \log_b x$ and $g(x) = b^x$ are inverse functions.
2. a. 10 times

   b. 100 times
3. $\log_7 y = x$
4. $\log_{1/2} a = b$
5. $\ln e = 1$
6. $\log_3 27 = 3$
7. $\log_2 \left(\frac{1}{8}\right) = -3$
8. $\log 100 = 2$
9. $\log_8 1 = 0$
10. $\log_{36} 6 = \frac{1}{2}$
11. $6^0 = 1$
12. $4^3 = 64$
13. $10^4 = 10{,}000$
14. $3^2 = \sqrt{81}$
15. $e^1 \approx 2.718$
16. $e^{2.08} \approx 8$
17. 2
18. 4
19. 1
20. −3
21. 2.30
22. 4.61
23. 0
24. 2.80
25. a. Estimates may vary; about 3.25.

    b. Answers may vary. An example is given. 10 is about one-quarter of the way between 8 and 16.
26. 3
27. $\frac{1}{2}$
28. −2
29. 1.32
30. 0
31. 2.30
32. Yes; the pH is 4.

Career Note

Many professional photographers developed an interest in photography in their teenage years, when taking pictures of family or friends. They later developed their skills into a career in order to earn a living. Photographers perform a useful service at significant events in peoples' lives, as they take pictures to record the events. Some photographers, however, have raised their skills to an art and their work is exhibited in galleries throughout the world. Students may not realize that the first photographic lens was made in 1812 and that the first photographic image on paper was made in 1838. Photographic film was developed in 1889 by George Eastman.

connection to **ART**

Photographers Ansel Adams and Fred Archer developed the *Zone System*, a system photographers can use to adjust the exposure to lighten or darken selected portions of the images in a black-and-white photograph.

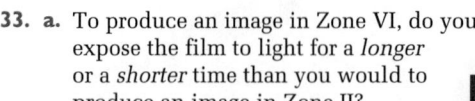

To make an image whose color matches Zone III look like it is in Zone V, the photographer adjusts the exposure time by two settings.

Zone	O	I	II	III	IV	V	VI	VII	VIII	IX
Gray scale										
Exposure time (seconds)	2	1	$\frac{1}{2}$	$\frac{1}{4}$	$\frac{1}{8}$	$\frac{1}{16}$	$\frac{1}{32}$	$\frac{1}{64}$	$\frac{1}{128}$	$\frac{1}{256}$

33. **a.** To produce an image in Zone VI, do you expose the film to light for a *longer* or a *shorter* time than you would to produce an image in Zone II?

b. What type of sequence is formed by the exposure times listed in the table? Write an equation to find the exposure time given the Zone number.

34. Film is made of tiny crystals of light-sensitive silver. When light reaches the film in a camera and the film is developed, the density of silver on the film changes—more silver is left where the scene was bright and less silver is left where it was dark.

a. Would an image exposed in Zone II leave *more* or *less* silver on the film than an image exposed in Zone VII?

b. The formula $d = \log \frac{1}{t}$ can be used to find the density of silver, d, on the film after an exposure time, t. Find the density of silver remaining on the film after each exposure time shown in the Zone System table. Round your answers to the nearest tenth. What type of sequence do these values form?

Moon and Half Dome, Yosemite National Park, California, 1960, *by Ansel Adams. How many zones can you see in this photograph?*

Answers to Exercises and Problems

33. a. shorter time

b. geometric; $f(z) = 2\left(\frac{1}{2}\right)^z$

34. a. less silver

b. −0.3, 0, 0.3, 0.6, 0.9, 1.2, 1.8, 2.1, 2.4; arithmetic

35. **Earthquakes** Many scientists prefer to use a scale that measures the total amount of energy released by an earthquake. A formula that relates the number on the Richter scale to the energy of an earthquake is

$$r = 0.67 \log E - 7.6,$$

where r is the number on the Richter scale and E is the energy in ergs.

a. What is the Richter number of an earthquake that releases 3.9×10^{15} ergs of energy?

b. What is the Richter number of an earthquake that releases 2.5×10^{20} ergs of energy?

Marine Biology Rain can deposit acidic pollutants in lakes, rivers, and streams. This causes the water chemistry to change.

36. Fresh-water clams require a pH level that is above 6.0. Can a lake where the concentration of hydrogen ions is 10^{-7} support clams? Why or why not?

37. When the pH level in a lake drops below 4.5, fish cannot survive. Can a lake where the concentration of hydrogen ions is 10^{-5} support fish? Why or why not?

38. a. **Writing** The pH value for a fresh-water aquarium with tropical fish should be between 6.0 and 6.8, and should never drop below 5.5. Would it be safe for the fish to be put in water where the concentration of hydrogen ions is 1.2×10^{-7}? Explain.

b. What is the greatest concentration of hydrogen ions that is safe for fish? Is this *acidic* or *basic*?

BY THE WAY...

Some fish are very sensitive to the pH of the water they live in. For instance, a Scleropages requires water with a pH between 5.5 and 6. A bumblebee goby can survive in water with a more varied pH, between 7 and 9.

Ongoing **ASSESSMENT**

39. **Group Activity** Work with another student.

a. Each student should write two equations in exponential form on a sheet of paper. On the back of your paper, write each equation in logarithmic form.

b. Trade papers with your partner and write each equation in the alternate form.

c. Return each paper to its owner. Check the equations and discuss any differences between the equations on the reverse side of the paper and the equations written by your partner.

Review **PREVIEW**

40. Find the equation of the inverse function of $y = 4x + 1$. *(Section 5-5)*

Simplify. Write each answer without negative exponents. *(Toolbox Skill 14)*

41. $(8x^3)(2xy)(y^{-2})$ 42. $(15t^3)(3t^2)^{-1}$ 43. $\left(\frac{6a^2}{4a^5b^6}\right)^2$ 44. $\left(\frac{mn^{-5}}{n^{-7}}\right)^{-3}$

45. Describe the graph of $f(x) = -16x^2 + 20$. *(Section 2-4)*

Application

When using the formula in Ex. 35 that relates Richter scale values to the energy of an earthquake, the energy values are given in ergs. An erg is the amount of work done by a force of one dyne over a distance of one centimeter. A dyne is a unit of force that would give one gram of mass an acceleration of one centimeter per second per second. One erg is a fairly small amount of energy, which is why the energy from an earthquake is measured in ergs.

Cooperative Learning

Ex. 39 provides an opportunity for students to help one another strengthen their understanding of the relationship between logarithms and exponents.

Answers to Exercises and Problems

35. a. about 2.8
b. about 6.1

36. Yes; the pH is 7.

37. Yes; the pH is 5.

38. a. No; the pH is about 6.92, which is above the recommended level.
b. 3.16×10^{-6}; acidic

39. Check students' work.

40. $y = \frac{1}{4}(x - 1)$

41. $\frac{16x^4}{y}$

42. $5t$

43. $\frac{9}{4a^6b^{12}}$

44. $\frac{1}{m^3n^6}$

45. Answers may vary. An example is given. The graph of $y = -16x^2 + 20$ is a parabola that opens down. Its vertex is at (0, 20).

Working on the Unit Project

To make a stringed instrument, you will need a piece of wood about 30 in. × 3 in. × $\frac{3}{4}$ in., nylon fishing line cut into four equal lengths (about 85 cm each), 8 large screw eyes, a screwdriver, pliers, and a marking pen.

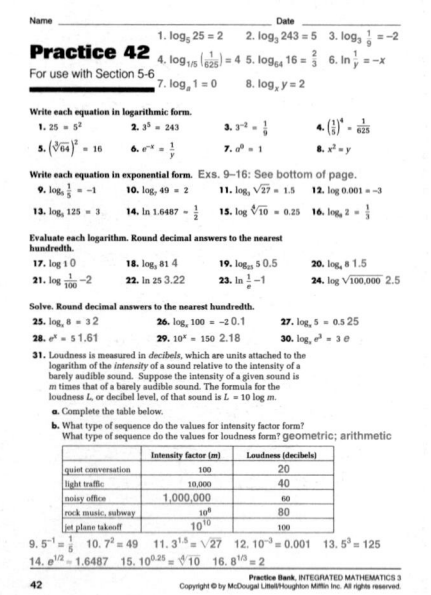

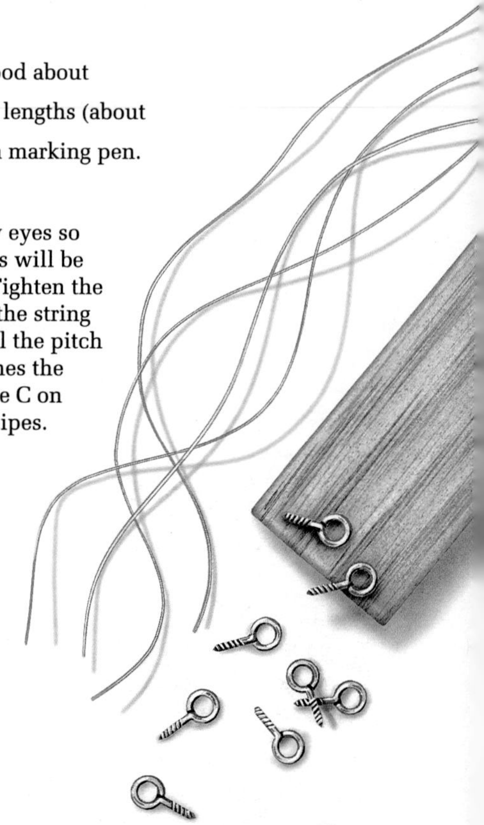

46. **a.** Position the screw eyes so that all four strings will be the same length. Tighten the first string. Pluck the string and tighten it until the pitch of the string matches the pitch of the middle C on your group's panpipes.

b. Repeat the tightening and tuning procedure for the remaining three strings for the notes E, G, and upper C.

47. You can use the equation $l_n = l_1(0.9439)^{n-1}$ to find the distance, in centimeters, from one end of the strings to the position of the *frets* that will allow you to play other notes.

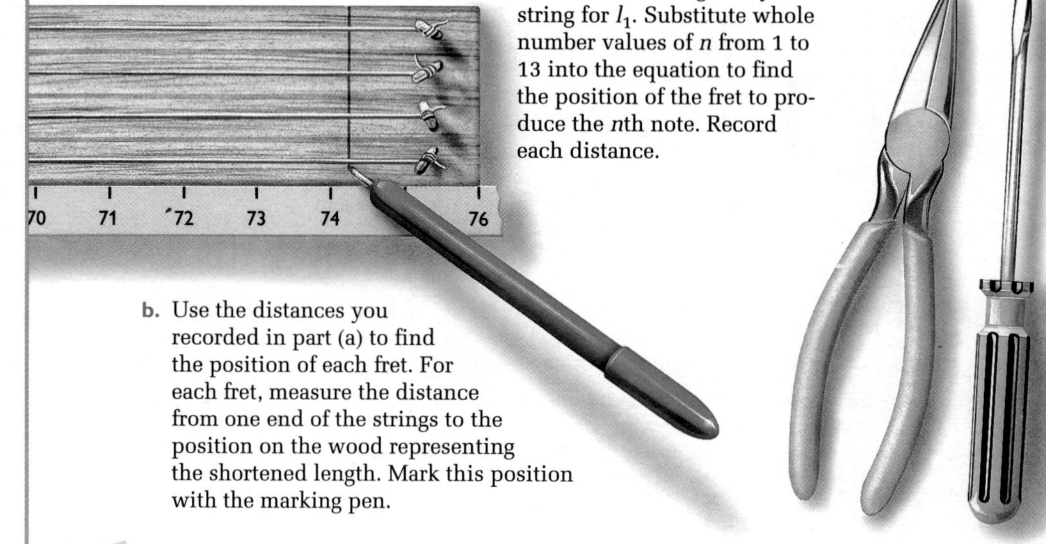

a. Substitute the length of your string for l_1. Substitute whole number values of n from 1 to 13 into the equation to find the position of the fret to produce the nth note. Record each distance.

b. Use the distances you recorded in part (a) to find the position of each fret. For each fret, measure the distance from one end of the strings to the position on the wood representing the shortened length. Mark this position with the marking pen.

Practice 42 For use with Section 5-6

Name ______________________ Date ____________

Practice 42
For use with Section 5-6

1. $\log_5 25 = 2$ 2. $\log_3 243 = 5$ 3. $\log_3 \frac{1}{9} = -2$
4. $\log_{1/5}\left(\frac{1}{625}\right) = 4$ 5. $\log_{64} 16 = \frac{2}{3}$ 6. $\ln \frac{1}{y} = -x$
7. $\log_a 1 = 0$ 8. $\log_x y = 2$

Write each equation in logarithmic form.

1. $25 = 5^2$ **2.** $3^5 = 243$ **3.** $3^{-2} = \frac{1}{9}$ **4.** $\left(\frac{1}{5}\right)^4 = \frac{1}{625}$

5. $\left(\sqrt[3]{64}\right)^2 = 16$ **6.** $e^{-x} = \frac{1}{y}$ **7.** $a^0 = 1$ **8.** $x^2 = y$

Write each equation in exponential form. Exs. 9–16: See bottom of page.

9. $\log_5 \frac{1}{5} = -1$ **10.** $\log_7 49 = 2$ **11.** $\log_3 \sqrt{27} = 1.5$ **12.** $\log 0.001 = -3$

13. $\log_5 125 = 3$ **14.** $\ln 1.6487 = \frac{1}{2}$ **15.** $\log \sqrt[4]{10} = 0.25$ **16.** $\log_8 2 = \frac{1}{3}$

Evaluate each logarithm. Round decimal answers to the nearest hundredth.

17. $\log 1$ 0 **18.** $\log_3 81$ 4 **19.** $\log_{25} 5$ 0.5 **20.** $\log_4 8$ 1.5

21. $\log \frac{1}{100}$ −2 **22.** $\ln 25$ 3.22 **23.** $\ln \frac{1}{e}$ −1 **24.** $\log \sqrt{100{,}000}$ 2.5

Solve. Round decimal answers to the nearest hundredth.

25. $\log_x 8 = 3$ 2 **26.** $\log_x 100 = -2$ 0.1 **27.** $\log_x 5 = 0.5$ 25

28. $e^x = 5$ 1.61 **29.** $10^x = 150$ 2.18 **30.** $\log_x e^3 = 3$ e

31. Loudness is measured in *decibels*, which are units attached to the logarithm of the *intensity* of a sound relative to the intensity of a barely audible sound. Suppose the intensity of a given sound is m times that of a barely audible sound. The formula for the loudness L, or decibel level, of that sound is $L = 10 \log m$.

a. Complete the table below.

b. What type of sequence do the values for intensity factor form? What type of sequence do the values for loudness form? geometric; arithmetic

	Intensity factor (m)	Loudness (decibels)
quiet conversation	100	20
light traffic	10,000	40
noisy office	1,000,000	60
rock music, subway	10^8	80
jet plane takeoff	10^{10}	100

9. $5^{-1} = \frac{1}{5}$ 10. $7^2 = 49$ 11. $3^{1.5} = \sqrt{27}$ 12. $10^{-3} = 0.001$ 13. $5^3 = 125$
14. $e^{1/2} = 1.6487$ 15. $10^{0.25} = \sqrt[4]{10}$ 16. $8^{1/3} = 2$

42 Practice Bank, INTEGRATED MATHEMATICS 3

Answers to Exercises and Problems

46, 47. Check students' work.

Section **5-7**

Properties of Logarithms

Focus

Learn about the properties of logarithms. Use these properties to solve problems.

SPLITTING LOGS

EXPLORATION

How can you rewrite the logarithm of a product, a quotient, or a power?

- **Materials: graphics calculators**
- **Work with another student.**

A. $y = 4 \log x$

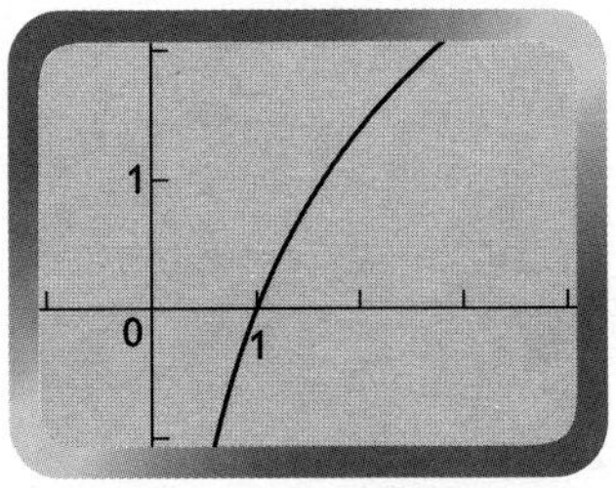

B. $y = (\log 4)(\log x)$

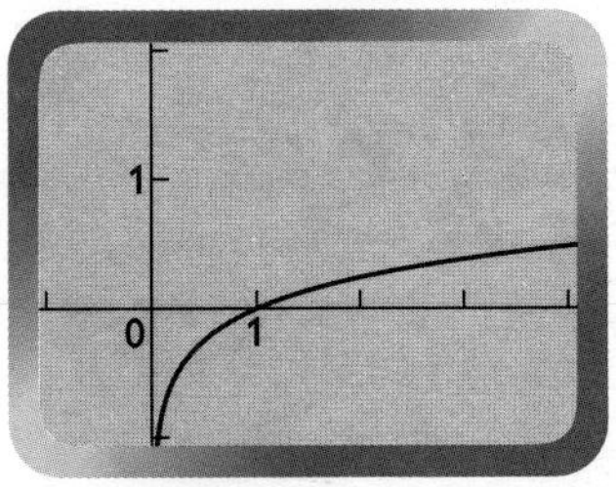

C. $y = \log 4 + \log x$

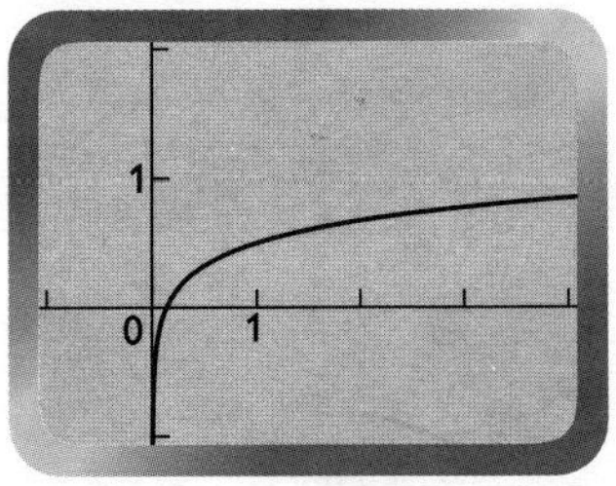

1 **a.** Graph $y = \log 4x$. Choose the function at the left that appears to have the same graph.

b. Check your answer to part (a) by graphing the function you chose and $y = \log 4x$ on the same screen. Are the two graphs the same?

c. Complete this statement: When $x > 0$, $\log 4x = \underline{?}$.

2 For parts (a)–(c), graph the two functions on the same screen, and tell whether the graphs are the same. Be sure to enter $\log \frac{x}{2}$ as $\log \left(\frac{x}{2}\right)$.

a. $y = \log \frac{x}{2}$, $y = \frac{\log x}{\log 2}$

b. $y = \log \frac{x}{2}$, $y = \log x - \log 2$

c. $y = \log \frac{x}{2}$, $y = \frac{1}{2} \log x$

3 Use your results from step 2 to complete this statement: When $x > 0$, $\log \frac{x}{2} = \underline{?}$.

4 For parts (a)–(c), graph the two functions on the same screen, and tell whether the graphs are the same.

a. $y = \log x^3$, $y = \log 3x$

b. $y = \log x^3$, $y = (\log x)^3$

c. $y = \log x^3$, $y = 3 \log x$

5 Use your results from step 4 to complete this statement: When $x > 0$, $\log x^3 = \underline{?}$.

PLANNING

Objectives and Strands

See pages 264A and 264B.

Spiral Learning

See page 264B.

Materials List

- Graphics calculator
- Graph paper

Recommended Pacing

Section 5-7 is a two-day lesson.

Day 1

Pages 309–310: Exploration through Sample 2, *Exercises 1–20*

Day 2

Pages 311–313: Sample 3 through Look Back, *Exercises 21–48*

Extra Practice

See pages 615–617.

Warm-Up Exercises

Warm-Up Transparency 5-7

Support Materials

- Practice Bank: Practice 43
- Activity Bank: Enrichment 39
- Study Guide: Section 5-7
- Problem Bank: Problem Set 12
- Explorations Lab Manual: Diagram Master 2
- Assessment Book: Quiz 5-7, Alternative Assessment 7

Answers to Exploration

1. a. C
   b. Yes.
   c. $\log 4 + \log x$
2. a. No.
   b. Yes.
   c. No.
3. $\log x - \log 2$
4. a. No.
   b. No.
   c. Yes.
5. $3 \log x$

TEACHING

Exploration

The goal of the Exploration is to have students discover various properties of exponents by comparing graphs of logarithmic functions.

Teaching Tip

This is a good time to remind students about the difference between inductive and deductive reasoning. By seeing an example in which these properties of logarithms are true, students can use inductive reasoning to conjecture that the properties are always true. However, the properties are proved to be true using deductive reasoning. These properties are fairly simple to prove. They are given as Exs. 32–34 on page 314.

Additional Samples

S1 Write each expression in terms of log M and log N.

a. $\log M^4N^2$

$\log M^4N^2$
$= \log M^4 + \log N^2$
$= 4 \log M + 2 \log N$

b. $\log \frac{\sqrt[4]{N}}{M^2}$

$\log \frac{\sqrt[4]{N}}{M^2}$
$= \log \sqrt[4]{N} - \log M^2$
$= \log N^{1/4} - \log M^2$
$= \frac{1}{4} \log N - 2 \log M$

S2 Write as a single logarithm: $\ln 25 + 4 \ln 5 - \ln 125$

$\ln 25 + 4 \ln 5 - \ln 125$
$= \ln 25 + \ln 5^4 - \ln 125$
$= \ln 25 + \ln 625 - \ln 125$
$= \ln 25 + \ln \frac{625}{125}$
$= \ln 25 + \ln 5$
$= \ln (25 \cdot 5)$
$= \ln 125$

Error Analysis

When simplifying a logarithmic expression, students may forget to change powers and roots to coefficients and vice versa. This mistake is especially common when roots are involved. The Samples and Additional Samples on this page should be helpful in eliminating this error.

Xab

PROPERTIES OF LOGARITHMS

M, N, and b are positive numbers with $b \neq 1$, and k is any real number.

Product of Logarithms Property	$\log_b MN = \log_b M + \log_b N$
Quotient of Logarithms Property	$\log_b \frac{M}{N} = \log_b M - \log_b N$
Power of Logarithms Property	$\log_b M^k = k \log_b M$

Watch Out!
There are no properties that let you simplify $\log_b (M + N)$ or $\log_b (M - N)$.

Sample 1

Write each expression in terms of log *M* and log *N*.

a. $\log M^2N^3$ b. $\log \frac{\sqrt[3]{M}}{N^4}$

Sample Response

a. $\log M^2N^3 = \log M^2 + \log N^3$ ← Use the product property.
$= 2 \log M + 3 \log N$ ← Use the power property.

b. $\log \frac{\sqrt[3]{M}}{N^4} = \log \sqrt[3]{M} - \log N^4$ ← Use the quotient property.
$= \log M^{1/3} - \log N^4$ ← Write the radical in exponential form.
$= \frac{1}{3} \log M - 4 \log N$ ← Use the power property.

Sample 2

Write as a single logarithm:
ln 18 − 2 ln 3 + ln 4

Sample Response

$\ln 18 - 2 \ln 3 + \ln 4 = \ln 18 - \ln 3^2 + \ln 4$ ← Use the power property.
$= \ln 18 - \ln 9 + \ln 4$
$= \ln \frac{18}{9} + \ln 4$ ← Use the quotient property.
$= \ln 2 + \ln 4$
$= \ln (2 \cdot 4)$ ← Use the product property.
$= \ln 8$

Now you are ready for:
Exs. 1–20 on p. 313

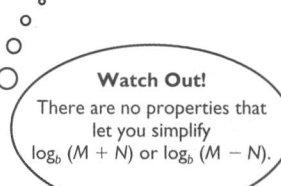

Sample 3

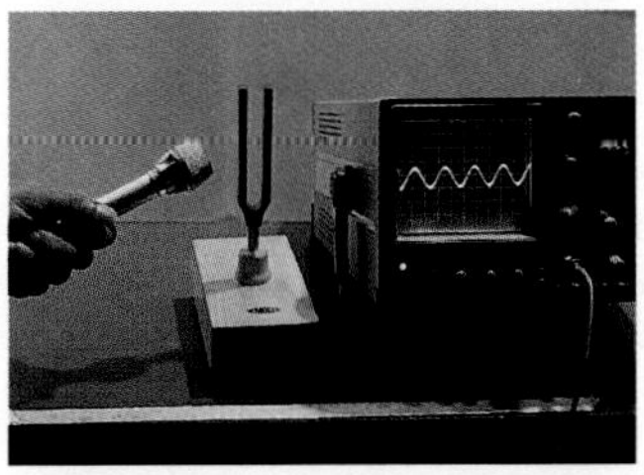

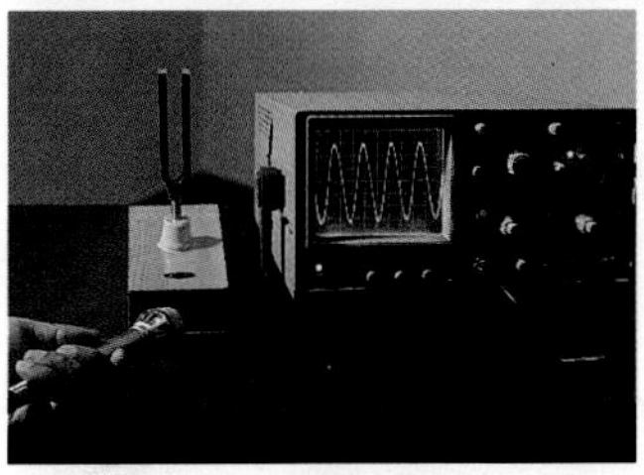

The *loudness* L of a sound is related to the sound's *intensity* I by the equation

$$L = 10 \log \frac{I}{I_0}$$

where L is measured in decibels, I is measured in watts per square meter, and I_0 is the intensity of a barely audible sound. Suppose the intensity of the sound from a tuning fork is doubled. By how many decibels does the loudness increase?

An oscilloscope screen shows an electronic representation of the sound waves produced when the prongs of a tuning fork vibrate. The harder a tuning fork is struck, the more intense the sound will be. This is shown on an oscilloscope as an increase in the height of the graph.

Sample Response

Let L_1 = the original loudness, and let L_2 = the loudness after the intensity is doubled.

increase in loudness $= L_2 - L_1$

$= 10 \log \frac{2I}{I_0} - 10 \log \frac{I}{I_0}$

$= 10 \log \left(2 \cdot \frac{I}{I_0}\right) - 10 \log \frac{I}{I_0}$ ← Rewrite $\frac{2I}{I_0}$.

$= 10\left(\log 2 + \log \frac{I}{I_0}\right) - 10 \log \frac{I}{I_0}$ ← Use the product property.

$= 10 \log 2 + 10 \log \frac{I}{I_0} - 10 \log \frac{I}{I_0}$ ← Use the distributive property.

$= 10 \log 2$ ← Simplify.

≈ 3

The loudness increases by about 3 decibels.

Talk it Over

1. In general, does doubling the intensity of a sound cause the loudness to double? Explain.
2. Explain how to use the quotient property to solve the problem in Sample 3 by using this as the third step:

$$\text{increase in loudness} = 10\left(\log \frac{2I}{I_0} - \log \frac{I}{I_0}\right)$$

BY THE WAY...

The decibel is named after Alexander Graham Bell, the inventor of the telephone.

Additional Sample

S3 Suppose being in your classroom is about 5 times as loud as being in your living room with the TV and stereo off. By how many decibels does the loudness increase?

Let L_1 = the loudness in your living room.
Let L_2 = the loudness in your classroom.

$L_2 - L_1$

$= 10 \log \frac{5I}{I_0} - 10 \log \frac{I}{I_0}$

$= 10 \log \left(5 \cdot \frac{I}{I_0}\right) - 10 \log \frac{I}{I_0}$

$= 10\left(\log 5 + \log \frac{I}{I_0}\right) - 10 \log \frac{I}{I_0}$

$= 10 \log 5$

≈ 7

The loudness increases by about 7 decibels.

Talk it Over

Question 2 asks students to do the problem in Sample 3 using the quotient property of logarithms rather than the product property. This helps students to realize that sometimes problems can be done in different ways.

Answers to Talk it Over

1. No; as the sample shows, doubling the intensity of a sound increases the loudness by about 3 decibels. The loudness would double only if the original loudness were 3 decibels.
2. increase in loudness =
$L_2 - L_1 =$
$10 \log \frac{2I}{I_0} - 10 \log \left(\frac{I}{I_0}\right) =$
$10\left(\log \frac{2I}{I_0} - \log \frac{I}{I_0}\right) =$
$10\left(\log\left(\frac{2I}{I_0} \div \frac{I}{I_0}\right)\right) =$
$10 \log \left(\frac{2I}{I_0} \cdot \frac{I_0}{I}\right) =$
$10 \log 2 \approx 3$

Communication: Reading

The change-of-base property is extremely valuable for the reasons given in the first paragraph on this page. Be certain students read this paragraph and understand the importance of the property in terms of using technology with logarithms whose bases are not 10 or *e*.

Additional Samples

S4 Evaluate $\log_4 15$ to three decimal places.

Use the change-of-base property to write $\log_4 15$ in terms of common logarithms.

$\log_4 15 = \frac{\log 15}{\log 4} \approx 1.953$

S5 Use a graphics calculator to graph $y = \log_5 x$.

Use the change-of-base property.

$y = \frac{\log x}{\log 5}$

Graph $y = \frac{\log x}{\log 5}$.

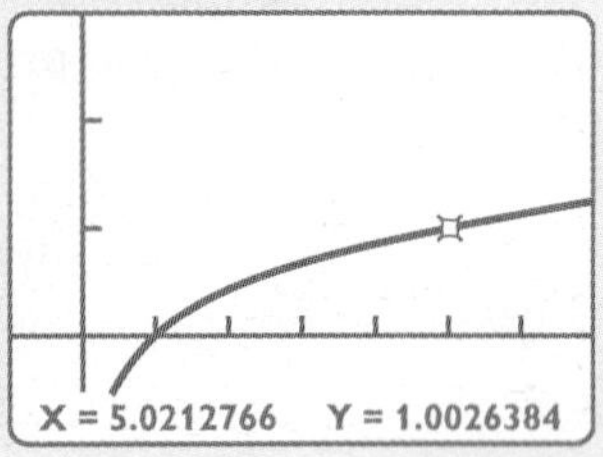

Check.

You know that $\log_5 5 = 1$. Using TRACE confirms this.

Changing the Base of a Logarithm

Most calculators have keys for evaluating common and natural logarithms like log 7 and ln 7. You can use the following property to evaluate a logarithm whose base is not 10 or *e*. You will prove this property in Exercise 35.

CHANGE-OF-BASE PROPERTY OF LOGARITHMS

For all positive numbers with $b \neq 1$ and $c \neq 1$,

$$\log_b M = \frac{\log_c M}{\log_c b}.$$

Sample 4

Evaluate $\log_3 8$ to three decimal places.

Sample Response

$\log_3 8 = \frac{\log 8}{\log 3}$ ← Use the change-of-base property to write $\log_3 8$ in terms of common logarithms.

≈ 1.893 ← Simplify.

Sample 5

Use a graphics calculator to graph $y = \log_6 x$.

Sample Response

1 Use the change-of-base property to write $y = \log_6 x$ in terms of common logarithms:

$y = \frac{\log x}{\log 6}$

2 Graph $y = \frac{\log x}{\log 6}$.

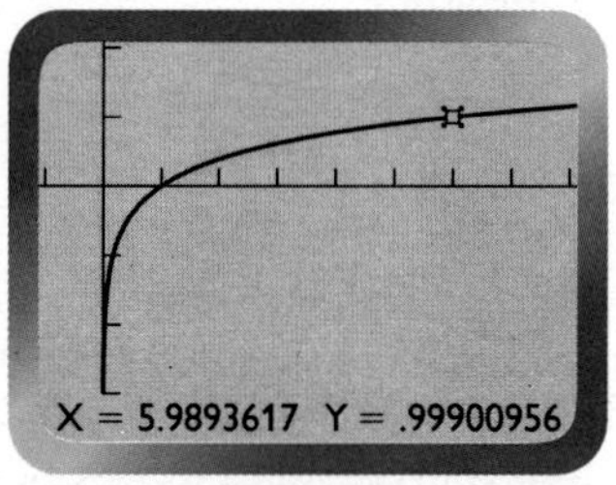

3 Check

You know that $\log_6 6 = 1$. Using TRACE confirms this.

Answers to Talk it Over

3. Find $3^{1.893}$.
4. $\log_3 8 = \frac{\ln 8}{\ln 3} \approx \frac{2.0794}{1.0986} \approx 1.893$
5. Graph $y = \frac{\ln x}{\ln 6}$. This graph is identical to the graph of $\frac{\log x}{\log 6}$ from Sample 5.

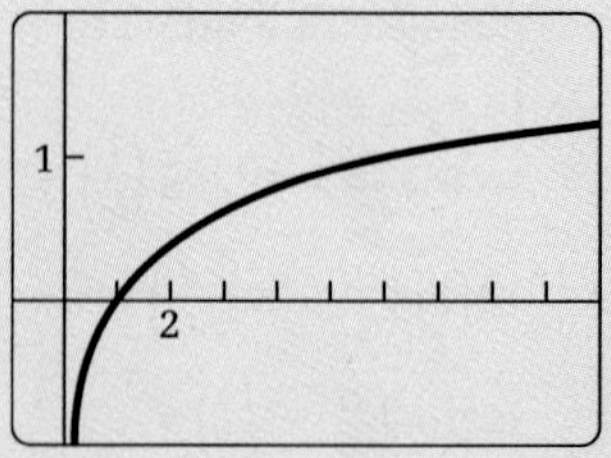

Answers to Look Back

$\log_b x$ is defined only for positive real numbers *x*.

Talk it Over

3. How can you use the power key on a calculator to check the answer to Sample 4?
4. Use natural logarithms to evaluate the logarithm in Sample 4.
5. Use natural logarithms to graph the function in Sample 5.

Look Back

Look back at the properties of logarithms on pages 310 and 312. Explain why M and N must be positive numbers.

Now you are ready for: Exs. 21–48 on pp. 313–316

5-7 Exercises and Problems

1. **Reading** In part (b) of Sample 1, why was it necessary to write the radical in exponential form?
2. Complete each statement.
   a. The logarithm of the product of two numbers equals the _?_ of the logarithms of the numbers.
   b. The logarithm of the _?_ of two numbers equals the difference of the logarithms of the numbers.

Write each expression in terms of log *M* and log *N*.

3. $\log M^3$
4. $\log MN^2$
5. $\log \frac{M^7}{N^5}$
6. $\log \frac{1}{N^2}$
7. $\log \sqrt[5]{M^4}$
8. $\log M\sqrt{N}$

Write each expression as a single logarithm.

9. $\ln 2 - 3 \ln 7$
10. $\ln 16 + 2 \ln 5 - \ln 4$
11. $\log_4 7 + \log_4 5$
12. $7 \log x + \frac{1}{4} \log x^8$
13. $3 \ln 2 + \frac{1}{2} \ln 49 - \ln 14$
14. $4 \log_6 t - 8 \log_6 u + 5 \log_6 v$

Evaluate.

15. $\log 5 + \log 2$
16. $2 \log_6 12 - \log_6 4$
17. $\log_5 6 - \log_5 150$

Let $x = \log_3 2$ and $y = \log_3 10$. Write each expression in terms of *x* and *y*.

18. $\log_3 20$
19. $\log_3 \frac{1}{8}$
20. $\log_3 250$
21. **Acoustics** Use the information in Sample 3 on page 311.
   a. Suppose the intensity of the sound from the tuning fork is tripled. By how many decibels does the loudness increase?
   b. Suppose the intensity of the sound from the tuning fork is increased by a factor of n. By how many decibels does the loudness increase? (Your answer will involve n.)

Answers to Exercises and Problems

1. In order to use the power property, $\sqrt[3]{M}$ had to be written as $M^{1/3}$.
2. a. sum
   b. quotient
3. $3 \log M$
4. $\log M + 2 \log N$
5. $7 \log M - 5 \log N$
6. $-2 \log N$
7. $\frac{4}{5} \log M$
8. $\log M + \frac{1}{2} \log N$
9. $\ln \frac{2}{343}$
10. $\ln 100$
11. $\log_4 35$
12. $\log x^9$
13. $\ln 4$
14. $\log_6 \frac{t^4 v^5}{u^8}$
15. 1
16. 2
17. −2
18. $x + y$
19. $-3x$
20. $3y - 2x$
21. a. by about 4.77 decibels
    b. $10 \log n$ decibels

Talk it Over

In questions 4 and 5, students verify that the change of base property works with natural logarithms.

APPLYING

Suggested Assignment

Day 1

Standard 2–20

Extended 1–20

Day 2

Standard 23–28, 32–35, 38–48

Extended 21, 23–28, 32–48

Integrating the Strands

Number Exs. 15–17, 23–28, 42–44

Algebra Exs. 1–48

Functions Exs. 1, 29–38

Statistics and Probability Exs. 36, 37

Discrete Mathematics Exs. 36, 37, 42–44

Logic and Language Exs. 1, 2, 22, 32–35, 38

Mathematical Procedures

Ex. 2 provides a quick and accurate check on whether students understand the procedures involved in the product and quotient properties for logarithms.

Problem Solving

In part (b) of Ex. 21, students write an expression for the increase in decibels in terms of n. Have students compute how many times louder an urban street (80 decibels) would have to be to reach the threshold of pain (120 decibels). (10,000 times as loud)

Research

Exs. 21 and 22 provide two real-world applications of logarithmic functions. You may wish to have students research other applications of natural and common logarithmic functions.

Reasoning

Exs. 32–35 have students use deductive reasoning to prove the properties of logarithms presented in this section. Since these demonstrations use properties of exponents, students are once again reminded of the relationship between logarithms and exponents.

22. **Aviation** A *pressure altimeter* is an instrument that finds the height of an airplane above sea level based on the pressure of the surrounding air. The height h and pressure p are related by the equation

$$h = -26{,}400 \ln\left(\frac{p}{2120}\right)$$

where h is measured in feet and p is measured in pounds per square foot.

a. TECHNOLOGY Graph the equation above on a graphics calculator. Does an increase in air pressure indicate an increase or decrease in altitude?

b. Suppose that an airplane's altimeter measures a 20% drop in air pressure. How has the altitude of the airplane changed?

c. **Writing** Look at the photograph of the airplane. Why would the reading on a pressure altimeter *not* equal the airplane's height above the ground?

Evaluate to three decimal places.

23. $\log_2 11$
24. $\log_5 8$
25. $\log_7 6$
26. $\log_6 7$
27. $\log_9 \frac{4}{5}$
28. $\log_{1/4} 0.92$

TECHNOLOGY **Graph each function on a graphics calculator.**

29. $y = \log_2 x$
30. $y = \log_5 x$
31. $y = \log_{1/3} x$
32. Show that $\log_b MN = \log_b M + \log_b N$. (*Hint*: Let $\log_b M = x$ and $\log_b N = y$. Then $M = b^x$ and $N = b^y$.)
33. Show that $\log_b M^k = k \log_b M$. (*Hint*: Let $\log_b M = x$. Then $M = b^x$.)
34. Use the product and power properties to show that $\log_b \frac{M}{N} = \log_b M - \log_b N$. (*Hint*: Write $\frac{M}{N}$ as MN^{-1}.)
35. Show that $\log_b M = \frac{\log_c M}{\log_c b}$. (*Hint*: Let $\log_b M = x$. Then $M = b^x$.)

Answers to Exercises and Problems

22. a.

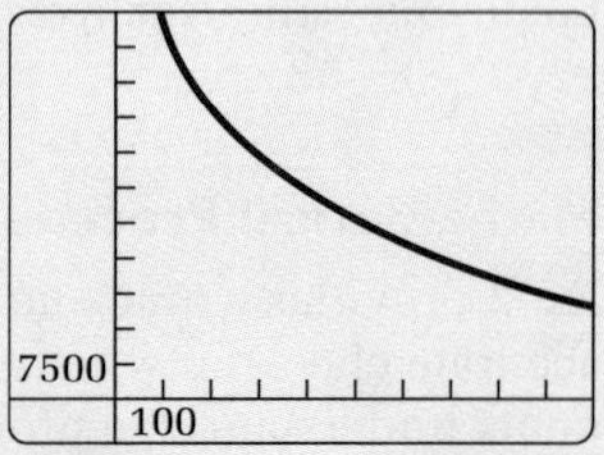

decrease

b. increased by about 5891 ft

c. The reading on the pressure altimeter measures the height above sea level, not above the ground.

23. 3.459
24. 1.292
25. 0.921
26. 1.086
27. −0.102
28. 0.060

29.

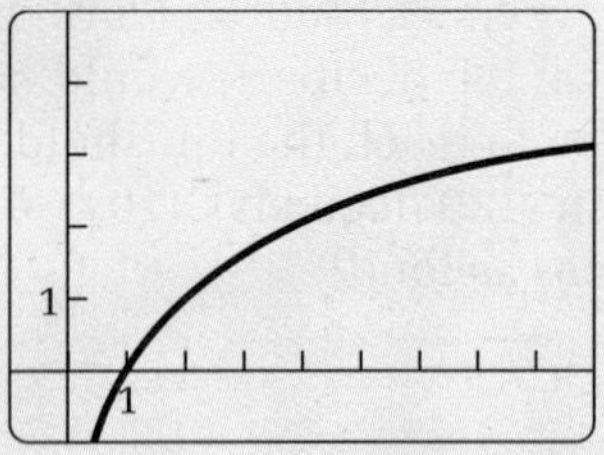

30.

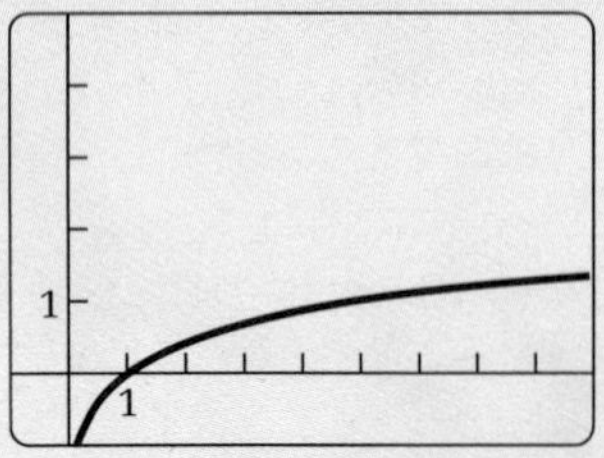

31.

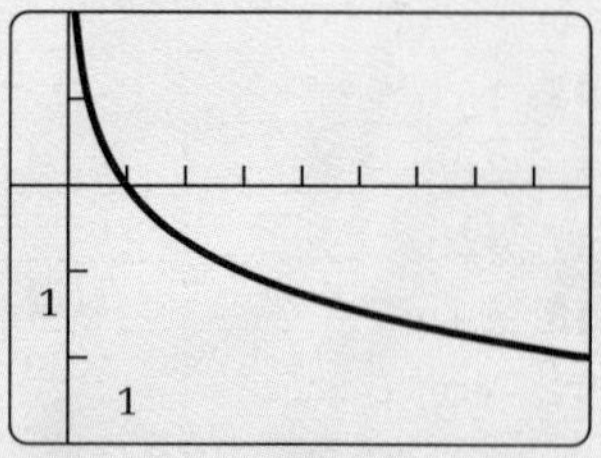

32. Let $\log_b M = x$ and $\log_b N = y$. Then $M = b^x$ and $N = b^y$, so that $\log_b MN = \log_b (b^x \cdot b^y) = \log_b b^{x+y} = x + y = \log_b M + \log_b N$

33. Let $\log_b M = x$. Then $M = b^x$, so that $\log_b M^k = \log_b (b^x)^k = \log_b b^{kx} = kx = k \log_b M$.

34. $\log_b \frac{M}{N} = \log_b MN^{-1} = \log_b M + \log_b N^{-1} = \log_b M + (-1)(\log_b N) = \log_b M - \log_b N$

35. Let $\log_b M = x$. Then $M = b^x$, so that $\log_c M = \log_c b^x = x \log_c b = \log_b M \cdot \log_c b$; so $\log_b M = \frac{\log_c M}{\log_c b}$.

connection to BIOLOGY

The pictures show some typical weights and wingspans for different types of birds. Let x = the weight and y = the wingspan.

36. a. Make a scatter plot of the data pairs (x, y). Draw a smooth curve that passes as close as possible to each plotted point.

b. Writing Use the data or the graph to answer these questions: Does wingspan appear to be a linear function of weight? an exponential function of weight? Give a reason for your answer.

37. a. Copy and complete the table below. Give each logarithm to two decimal places.

b. Make a scatter plot of the data pairs (u, v). What do you notice about the plotted points?

c. Draw a fitted line for your scatter plot. Write an equation of your fitted line in terms of u and v.

d. Write the equation you wrote in part (c) in terms of x and y. (*Hint*: Replace u with $\ln x$ and v with $\ln y$.)

e. Show that your equation from part (d) can be written in the form $y = ax^b$, where a and b are constants. A function of this form is called a *power function*.

f. Use the power function you wrote in part (e) to estimate the wingspan of a golden eagle weighing 8.2 lb.

Bird Weights and Wingspans

Type of bird	$u = \ln x$	$v = \ln y$
cuckoo	?	?
sparrow hawk	?	?
curlew	?	?
graylag goose	?	?
griffon vulture	?	?

Cuckoo
0.23 lb
1.90 ft

Sparrow hawk
0.49 lb
2.46 ft

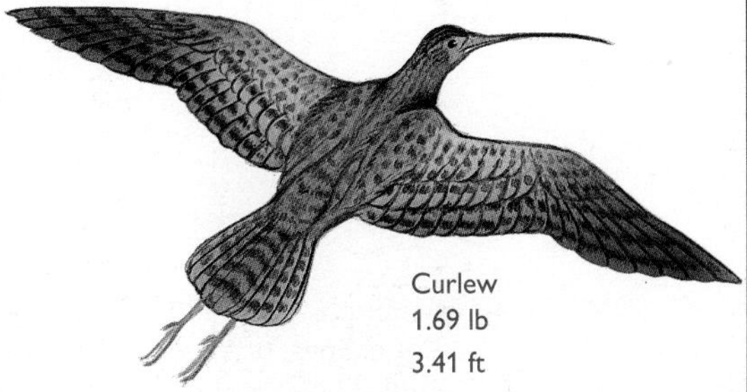
Curlew
1.69 lb
3.41 ft

Graylag goose
6.76 lb
5.35 ft

Griffon vulture
16.03 lb
8.40 ft

Integrating the Strands

The mathematics involved in doing Exs. 36 and 37 integrates the strands of algebra, functions, statistics, and discrete mathematics.

Answers to Exercises and Problems

36. a.

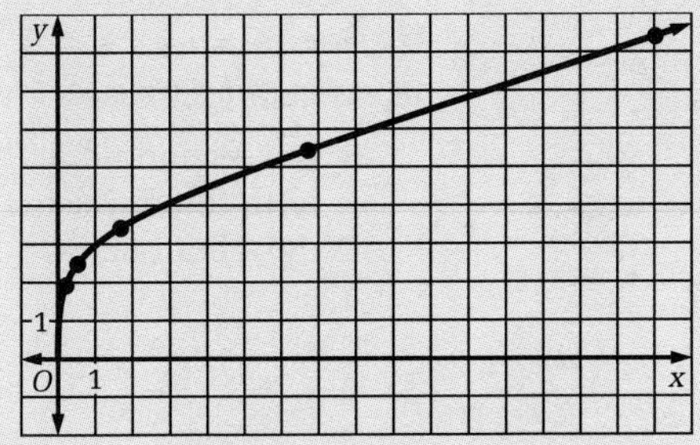

b. No; No; the points do not lie on a line or on an exponential curve.

37. a.

Type of bird	$u = \ln x$	$v = \ln y$
cuckoo	−1.47	0.64
sparrow hawk	−0.71	0.90
curlew	0.52	1.23
graylag goose	1.91	1.68
griffon vulture	2.77	2.13

b. The points appear to represent a linear relationship.

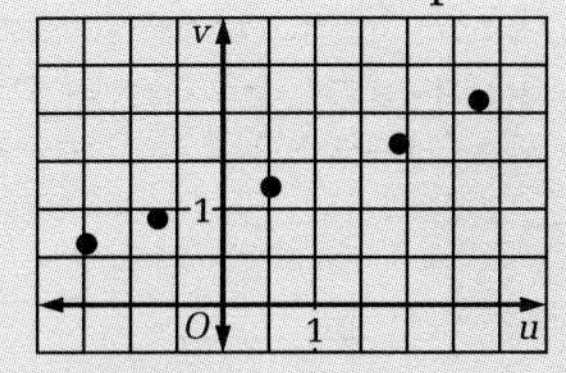

c. Answers may vary. An example is given. $v = 0.38u + 1.11$

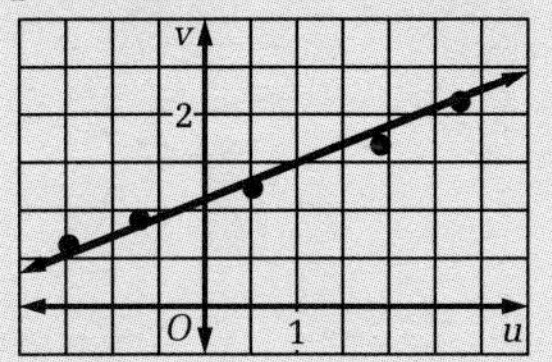

d–f. Answers may vary. Examples based on the equation given in part (c) above are given.

d. $\ln y = 0.38 \ln x + 1.11$

e. $\ln y = \ln x^{0.38} + \ln 3.03 = \ln (3.03x^{0.38})$; $y = 3.03x^{0.38}$

f. about 6.7 ft

Students Acquiring English

For Ex. 38, you may wish to pair students acquiring English with English-proficient partners to help students form written answers.

Working on the Unit Project

Ex. 48 introduces students to the formula for reverberation time at the Royal Albert Hall in London. This formula contains a natural logarithm. Students can do this exercise individually at home and compare their answers the next day.

Ongoing ASSESSMENT

38. **Writing** Suppose $f(x) = \log \sqrt{x}$ and $g(x) = \frac{1}{2} \log x$. Explain how to convince someone that $f(x) = g(x)$ for $x > 0$ using the properties of logarithms and using graphs.

Review PREVIEW

Write each equation in exponential form. *(Section 5-6)*

39. $\log_3 81 = 4$ **40.** $\ln 5 \approx 1.61$ **41.** $\log_2 64 = 6$

Evaluate each series. *(Sections 4-5, 4-7)*

42. $\sum_{k=1}^{10} (3k - 2)$ **43.** $\sum_{n=0}^{\infty} 4\left(\frac{5}{7}\right)^n$ **44.** $\sum_{m=1}^{5} \frac{1}{4}(2^{m-1})$

Solve. *(Section 2-6)*

45. $\sqrt{x} = 3x$ **46.** $z = \sqrt{z + 8} - 2$ **47.** $x = x\sqrt{x - 2}$

Working on the Unit Project

When a musician plays an instrument in a concert hall, the sound *reverberates*, or echoes, off the walls of the hall even after the musician has stopped playing. The time it takes for the sound to fade away is called the *reverberation time.*

48. At the Royal Albert Hall in London, England, the approximate reverberation time T (in seconds) is given by the function

$$T = 5.24 - 0.408 \ln f$$

where f is the frequency of the sound in hertz (Hz).

a. Suppose a violinist plays a middle C in the Royal Albert Hall. Estimate the reverberation time. Use 261.6 Hz as the frequency of middle C.

b. Suppose the violinist in part (a) plays two notes. The frequency of the second note is half the frequency of the first note. Which note has the longer reverberation time? By how many seconds do the reverberation times differ?

Practice 43 For use with Section 5-7

Name ______________________ Date __________

Practice 43

For use with Section 5-7

23.b.

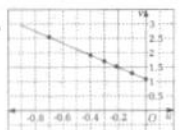

Let $x = \log_5 3$ and $y = \log_5 12$. Write each expression in terms of x and y.

1. $\log_5 9$ $2x$ **2.** $\log_5 36$ $x + y$ **3.** $\log_5 4$ $y - x$

4. $\log_5 \frac{1}{144}$ $-2y$ **5.** $\log_5 108$ $2x + y$ **6.** $\log_5 15$ (*Hint:* $15 = 3 \cdot 5$) $x + 1$

Write each expression as a single logarithm.

7. $\log 6 + \log 11$ $\log 66$ **8.** $\log_3 250 - 2\log_3 10$ $\log_3 2.5$ **9.** $\frac{1}{3}\log 8 + \log 15$ $\log 30$

10. $\frac{1}{2}(\log_5 9 + \log_5 4)$ $\log_5 6$ **11.** $5\log x - \log x^3$ $\log x^2$ **12.** $2\log 10 - \frac{1}{2}\log 25 + \log 9$ $\log 180$

Evaluate.

13. $\log_3 36 - \log_3 4$ 2 **14.** $\log 125 + \log 8$ 3 **15.** $\log_5 12 - \log_5 60$ -1

Write each expression in terms of log M and log N.

16. $\log M^4 N$ $4\log M + \log N$ **17.** $\log \frac{\sqrt{M}}{N}$ $\frac{1}{2}\log M - \log N$ **18.** $\log \sqrt[3]{MN^2}$ $\frac{1}{3}(\log M + 2\log N)$

Evaluate to three decimal places.

19. $\log_4 7$ 1.404 **20.** $\log_8 \frac{3}{2}$ 0.195 **21.** $\log_7 \sqrt{2}$ 0.178 **22.** $\log_5 \frac{1}{3}$ -0.683

23. The intensities of illumination of a surface when the surface is placed at various distances from the same light source are given in the top table shown here.

a. Copy and complete the second table.

b. Make a scatter plot of the data pairs (u, v). See top of page.

c. Draw a fitted line for your scatter plot. Write the equation of the fitted line in terms of u and v. Sample: $v = -1.91u + 1.18$

d. Use the equation of the fitted line to get a relationship between x and y of the form $y = ax^b$. Sample: $y = 15x^{-2}$

e. Use the power function you wrote in part (d) to estimate the intensity with the surface 2 m from the source. about 3.75 candelas

x = Distance (in meters)	y = Intensity (in candelas)
1.00	15
0.80	23
0.60	43
0.50	54
0.40	85
0.20	333

$u = \log x$	$v = \log y$
0	1.1761
−0.0969	1.3617
−0.2218	1.6335
−0.3010	1.7324
−0.3979	1.9294
−0.6990	2.5224

24. ***Writing*** Explain how tables giving the values of logarithms could be used to carry out computations such as multiplication and finding roots. Check students' work.

Practice Bank, INTEGRATED MATHEMATICS 3

43

Answers to Exercises and Problems

38. For $x > 0$, $\sqrt{x} > 0$ and $\log \sqrt{x} = \log x^{1/2} = \frac{1}{2} \cdot \log x$ by the power property of logarithms. That is, for every $x > 0$, $f(x) = g(x)$. Also, for $x > 0$, the graphs of $f(x)$ and $g(x)$ are identical, which can be seen by graphing both by hand, or with a graphics calculator or graphing software.

39. $3^4 = 81$

40. $e^{1.61} \approx 5$

41. $2^6 = 64$

42. 145

43. 14

44. $7\frac{3}{4}$

45. $0; \frac{1}{9}$

46. 1

47. 3

48. a. about 2.97 s

b. the second note (the one with the lower frequency); about 0.28 s

Section

5-8 Solving Exponential and Logarithmic Equations

Focus

Use exponential and logarithmic equations to solve real-life problems.

Some of the carbon in the air we breathe and in the food we eat is a harmless radioactive form called *carbon 14.* Over time, carbon 14 breaks down due to *radioactive decay.*

When a plant or animal dies, the amount of carbon 14 in the cells begins to decrease exponentially.

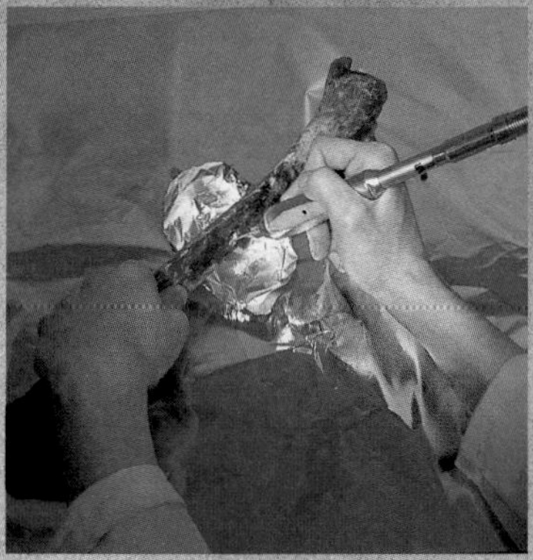

$$A(t) = A_0(0.883)^t$$

- amount of carbon 14 present t thousand years after death → $A(t)$
- amount of time since death (in thousands of years) → t
- amount of carbon 14 present at time of death → A_0

In 1968 a bison bone was found in Calgary, Canada. Scientists measured its carbon 14 content using the equation above. They estimated that the bison died 8000 years ago. This process is called *carbon-14 dating.*

PLANNING

Objectives and Strands

See pages 264A and 264B.

Spiral Learning

See page 264B.

Materials List

- Graphics calculator or graphing software

Recommended Pacing

Section 5-8 is a one-day lesson.

Extra Practice

See pages 615–617.

Warm-Up Exercises

- Warm-Up Transparency 5-8

Support Materials

- Practice Bank: Practice 44
- Activity Bank: Enrichment 40
- Study Guide: Section 5-8
- Problem Bank: Problem Set 12
- Using TI-81 and TI-82 Calculators: Comparing x^k and 2^x
- Using Plotter Plus: Comparing x^k and 2^x
- Assessment Book: Quiz 5-8, Test 20

TEACHING

Additional Sample

S1 An archeologist found a bone that contained 48% of the carbon 14 present when the animal died. Estimate the age of the bone.

$A_0(0.883)^t = A(t)$

Substitute $0.48A_0$ for $A(t)$.

$A_0(0.883)^t \approx 0.48A_0$

$(0.883)^t \approx 0.48$

$\log (0.883)^t \approx \log 0.48$

$t \log 0.883 \approx \log 0.48$

$t \approx \frac{\log 0.48}{\log 0.883}$

≈ 6

The bone is about 6000 years old.

Talk it Over

Questions 1–3 have students analyze how to solve an exponential equation based on the procedure presented in Sample 1. If students can provide a clear explanation to question 3, they should have little difficulty in solving exponential equations. For question 4, students may wish to use their method to solve Sample 1 graphcially.

Additional Sample

S2 Solve $\log_4 x + \log_4 (x - 6) = 2$.

$\log_4 x + \log_4 (x - 6) = 2$

$\log_4 x(x - 6) = 2$

Write the equation in exponential form.

$x(x - 6) = 4^2$

$x^2 - 6x = 16$

$x^2 - 6x - 16 = 0$

$(x - 8)(x + 2) = 0$

$x - 8 = 0$ or $x + 2 = 0$

$x = 8$ $\quad x = -2$

$\log_4 (-2)$ is not defined.

The solution is 8.

Talk it Over

Questions 5 and 6 give students experience in solving exponential or logarithmic problems using a graphics calculator.

Sample 1

The bison bone found in 1968 contained about 37% of the carbon 14 present when the bison died. Estimate the age of the bone.

Sample Response

Problem Solving Strategy: Use a formula.

$A_0(0.883)^t = A(t)$

$A_0(0.883)^t \approx 0.37A_0$ ← Substitute $0.37A_0$ for $A(t)$.

$(0.883)^t \approx 0.37$ ← Divide both sides by A_0.

$\log (0.883)^t \approx \log 0.37$ ← Take the common logarithm of both sides.

$t \log 0.883 \approx \log 0.37$ ← Use the power property of logarithms.

$t \approx \frac{\log 0.37}{\log 0.883}$ ← Divide both sides by log 0.883.

$t \approx 8$

The bone is about 8000 years old.

Chemical analysis proved this ax to be copper, making the Iceman (see below) the only mummy found from the Copper Age.

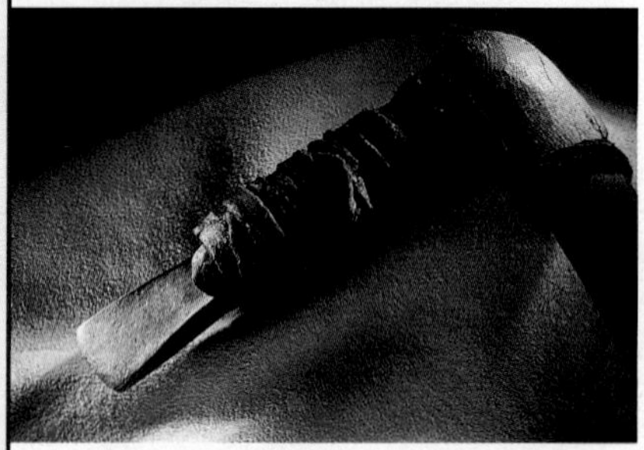

Talk it Over

1. Describe the method used in Sample 1 to solve the exponential equation $(0.883)^t = 0.37$.
2. To solve the equation $(0.883)^t = 0.37$, can you take the natural logarithm of both sides instead of the common logarithm? Explain.
3. Explain how to solve $5^x = 27$ using the method shown in Sample 1.
4. How can you solve the problem in Sample 1 with a graph?

BY THE WAY...

On September 19, 1991, the mummified body of a man buried in a glacier was discovered in the Tyrolean Alps. Scientists used carbon-14 dating to determine that he lived around 3000 B.C. Known as the Iceman, he was five feet two inches tall, had wavy, medium length hair and a beard, and was between 25 and 40 years old.

Answers to Talk it Over

1. Take the common logarithms of both sides of the equation to express the equation in logarithmic form. Use the power property of logarithms to express $\log (0.883)^t$ as $t \log 0.883$. Finally, divide both sides of the equation by log 0.883.
2. Yes. Using logarithms with any base will result in the same answer.
3. Take the common logarithms of both sides of the equation. Use the power property of logarithms to express $\log 5^x$ as $x \log 5$. Divide both sides of the equation by log 5; $x \approx 2.05$.
4. Graph the equation $y = 0.883^t$ and use TRACE to find the t-value when $y = 0.37$.
5. Graph $y = \frac{\log x + \log (x - 2)}{\log 2}$ and $y = 3$ on the same screen and use TRACE to find the x-coordinate of the intersection point.
6. Graph both equations on the same screen and use TRACE to find the x-coordinate of the intersection point; $x \approx 1.9$, $y \approx 5.2$.

In Sample 1, the power property of logarithms was used to solve an exponential equation. You can also use properties of logarithms to solve logarithmic equations.

Sample 2

Solve $\log_2 x + \log_2 (x - 2) = 3$.

Sample Response

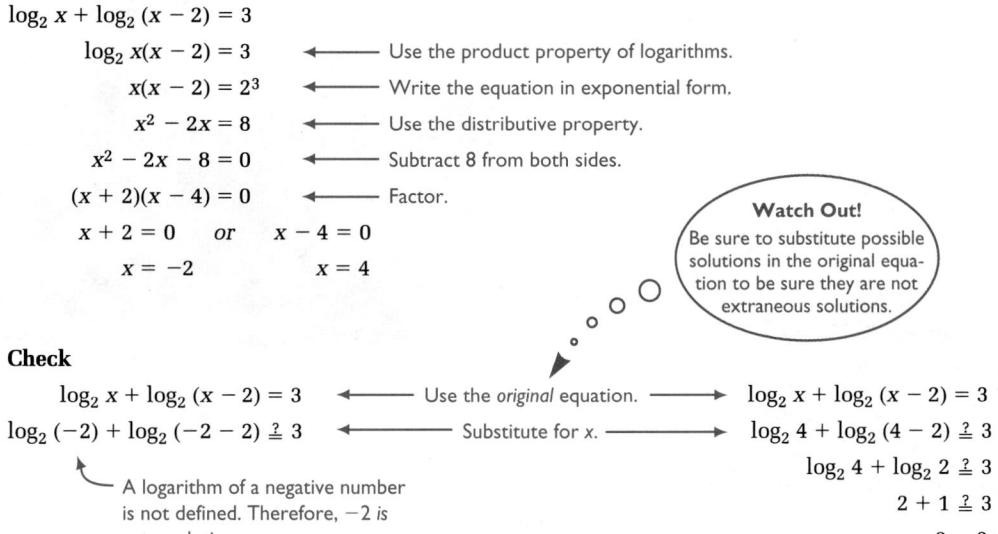

$\log_2 x + \log_2 (x - 2) = 3$

$\log_2 x(x - 2) = 3$ ← Use the product property of logarithms.

$x(x - 2) = 2^3$ ← Write the equation in exponential form.

$x^2 - 2x = 8$ ← Use the distributive property.

$x^2 - 2x - 8 = 0$ ← Subtract 8 from both sides.

$(x + 2)(x - 4) = 0$ ← Factor.

$x + 2 = 0$ *or* $x - 4 = 0$

$x = -2$ $\quad$ $x = 4$

Watch Out!
Be sure to substitute possible solutions in the original equation to be sure they are not extraneous solutions.

Check

$\log_2 x + \log_2 (x - 2) = 3$ ← Use the *original* equation. → $\log_2 x + \log_2 (x - 2) = 3$

$\log_2 (-2) + \log_2 (-2 - 2) \stackrel{?}{=} 3$ ← Substitute for x. → $\log_2 4 + \log_2 (4 - 2) \stackrel{?}{=} 3$

A logarithm of a negative number is not defined. Therefore, -2 *is not* a solution.

$\log_2 4 + \log_2 2 \stackrel{?}{=} 3$

$2 + 1 \stackrel{?}{=} 3$

$3 = 3$ ✓

The solution is 4.

Talk it Over

5. Solve the equation in Sample 2 by using a graphics calculator to graph $y = \log_2 x + \log_2 (x - 2)$ and $y = 3$ on the same screen. (*Hint:* Use the change-of-base property to write the first function in terms of common logarithms.)
6. Explain how to solve this system of equations by using a graphics calculator. $\quad y = 4(1.14)^x$, $\quad y = 2(1.65)^x$

Look Back

Explain how you can use the properties of logarithms to solve exponential and logarithmic equations.

Answers to Look Back

Summaries may vary. The properties of logarithms can be used to rewrite exponential equations so the variable is no longer an exponent and to rewrite logarithmic equations as exponential equations. For example, an exponential equation such as $5^x = 7$ can be rewritten as $\log 5^x = \log 7$. Then using the power property of logarithms, the new equation can be written $x \log 5 = \log 7$. Then $x = \frac{\log 7}{\log 5} = 1.209$. The logarithmic equation $\log_2 x - \log_2 (x - 3) = 2$ can be rewritten using the quotient property of logarithms: $\log_2 \left(\frac{x}{x-3}\right) = 2$. The exponential form of the equation is $\frac{x}{x-3} = 4$; $x = 4x - 12$; $3x = 12$; $x = 4$.

Using Technology

Students should be careful when they solve logarithmic equations graphically. In Talk it Over question 5, students can easily use the standard graphing window to solve the system of equations.

After discussing question 5, ask students to change the second equation from $y = 3$ to $y = -6$. Keep the first equation the same ($y = \frac{\log x}{\log 2} + \frac{\log (x - 2)}{\log 2}$, after using the change-of-base property). When the graphs are viewed on the standard window, it appears that they do not intersect. This is deceptive and occurs because the x-values for all pixels to the left of the bottommost point displayed for the logarithmic function are outside the domain of the function.

To study the situation more closely, have students use the following procedure. Trace to the bottommost point for the logarithmic function. Once there, press [ZOOM] and select menu item 1 to draw a box. Press [◄] twice, press [▲] twice, then press [ENTER].

Next, press [►] four times.

Finally, hold down [▼] to draw a box that goes all the way to the bottom of the window; once there, press [ENTER]. The portion of the graph now displayed shows points that were not visible on the original graph. The new graph shows that the graphs of the logarithmic function and $y = -6$ do indeed intersect.

Look Back

This question will help students think about how and when to use the methods presented in this section. A class discussion of this Look Back would give students the opportunity to vocalize and review these methods.

APPLYING

Suggested Assignment

Standard 1–16, 24–32

Extended 1–16, 21, 22, 24–32

Integrating the Strands

Algebra Exs. 2–28, 32

Functions Exs. 8–10, 17–23, 32

Discrete Mathematics Exs. 30, 31

Statistics and Probability Exs. 30, 31

Logic and Language Exs. 1, 22, 24

Communication: Reading

Ex. 1 provides an opportunity for students to think about extraneous solutions. As with radical equations, extraneous solutions can, and often do, occur when solving logarithmic equations. Students should pay particular attention to the Watch Out note on page 319.

Research

The famous prehistoric cave paintings in Lascaux, France were also dated using carbon-14 dating. Some students may wish to find out more about these paintings and how scientists used carbon-14 dating to discover how old they were.

Assessment: Open-ended

The half-life of carbon 14 is approximately 5570 years. In other words, in any material containing carbon 14, there will be only half as many of these atoms present 5570 years from now. Ask students how this verifies the model for carbon-14 dating used on page 317, $A(t) = A_0(0.883)^t$.

5-8 Exercises and Problems

1. **Reading** Why is it not possible for -2 to be a solution of the logarithmic equation in Sample 2?

Solve.

2. $2^x = 3$
3. $12^y = 4$
4. $100(1.05)^t = 350$
5. $8(0.32)^t = 1$
6. $5^{3n} = 9$
7. $e^{-2x} = 6$

Archaeology **For Exercises 8–10, use the function $A(t) = A_0(0.883)^t$ described on page 317.**

8. In 1950 some wood in Zoser's Step Pyramid in Saqqara, Egypt, contained about 59% of its original carbon 14. Estimate the age of the wood in 1950.
9. In the summer of 1993, scientists were exploring a burial mound of the ancient Pazyryk culture in southern Siberia. In one of the tombs they found a mummified woman. Research showed that the mummy contained 75% of the body's original carbon. Estimate the age of the mummy.
10. Find the half-life of carbon 14. (*Hint:* You want the value of t for which $A(t) = \frac{A_0}{2}$.)

Solve.

11. $\log_2 x + \log_2 (x - 4) = 5$
12. $\log_7 (x - 2) + \log_7 (x - 8) = 1$
13. $\log_3 (2y - 1) + \log_3 (y + 7) = 3$
14. $\log_5 m = 1 - \log_5 (m + 2)$
15. $\log_2 5x - \log_2 (x + 6) = 3$
16. $2 \log_3 z - \log_3 (z - 2) = 2$

TECHNOLOGY Use a graphics calculator or graphing software to solve each system of equations. Write the coordinates of each solution (x, y) to the nearest hundredth.

17. $y = 4(1.15)^x$; $y = 2(1.65)^x$
18. $y = 3 \ln x$; $y = \ln 3x$
19. $y = 10 \cdot 2^x$; $4x + 2y = 160$

Use the diagram at the right. Assume that the rate of population growth remains constant in both countries.

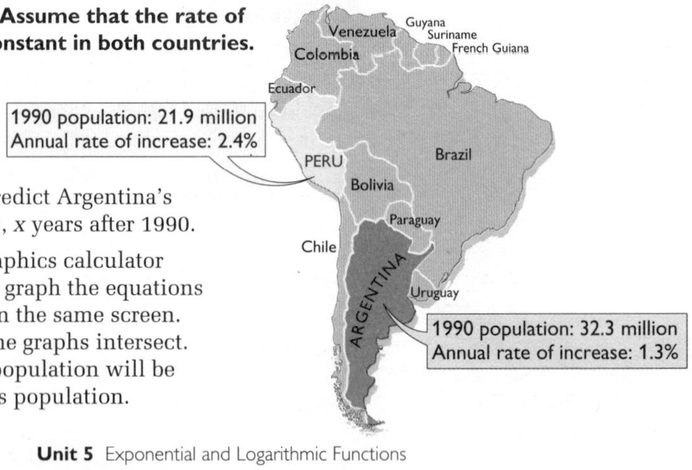

20. a. Write an equation to predict Peru's population, in millions, x years after 1990.
    b. Write an equation to predict Argentina's population, in millions, x years after 1990.
    c. TECHNOLOGY Use a graphics calculator or graphing software to graph the equations from parts (a) and (b) on the same screen. Find the point where the graphs intersect. Estimate when Peru's population will be greater than Argentina's population.

Answers to Exercises and Problems

1. $\log_2 x + \log_2 (x - 2)$ is not defined for $x = -2$.

2–7. Answers are given to the nearest hundredth.

2. 1.58
3. 0.56
4. 25.68
5. 1.82
6. 0.46
7. −0.90
8. about 4200 years
9. about 2300 years
10. about 5600 years
11. 8
12. 9
13. 2
14. about 1.45
15. no solution
16. 6, 3
17. about (1.92, 5.23)
18. about (1.73, 1.65)
19. about (2.89, 74.21)
20. a. $y = 21.9(1.024)^x$
    b. $y = 32.3(1.013)^x$

connection to LITERATURE

In Michael Crichton's *The Andromeda Strain*, a group of scientists tries to stop the spread of deadly bacteria from outer space. At one point, the scientists discuss how quickly bacteria can multiply.

> The mathematics of uncontrolled growth are frightening. A single cell of the bacterium *E. coli* would, under ideal circumstances, divide every twenty minutes. That is not particularly disturbing until you think about it, but the fact is that bacteria multiply geometrically: one becomes two, two become four, four become eight, and so on. In this way, it can be shown that in a single day, one cell of *E. coli* could produce a supercolony equal in size and weight to the entire planet earth.

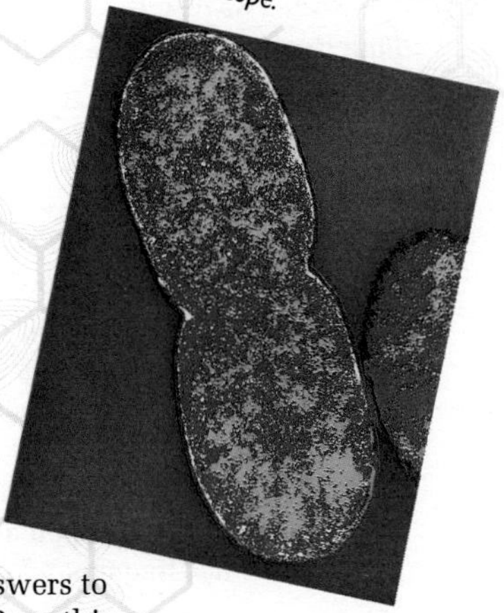

E. coli dividing, viewed through an electron microscope.

21. Suppose a single cell of *E. coli* exists under ideal circumstances.

a. Write an equation that gives the number N of cells after t twenty-minute periods.

b. How long does it take to produce one million cells of *E. coli*?

22. a. How many cells can a single cell of *E. coli* produce in one day?

b. The radius of Earth is about 2×10^7 ft. Find Earth's volume. (*Hint:* Use the formula $V = \frac{4}{3}\pi r^3$ for the volume of a sphere.)

c. Based on the excerpt from *The Andromeda Strain* and your answers to parts (a) and (b), what is the volume of a single cell of *E. coli*? Does this volume seem reasonable? Explain.

23. Personal Finance When you take out a loan to buy a car, you must repay a portion of the loan with interest each month. The amount $A(x)$ you still owe after making x monthly payments is given by the formula

$$A(x) = \left(A_0 - \frac{P}{r}\right)(1 + r)^x + \frac{P}{r}$$

where A_0 is the original amount of the loan, P is the monthly payment, and r is the monthly interest rate expressed as a decimal. (*Hint*: The monthly rate is the annual rate divided by 12.)

a. Alvin Thompson borrows \$10,800 at 12% annual interest to buy a new car. His monthly payment is \$240.24. Write a function that gives the amount Alvin still owes after making x monthly payments.

b. As soon as Alvin drives his car off the dealership lot, the car's resale value drops to about \$8600. The resale value continues to decrease by about 1.1% each month. Write a function that gives the resale value $R(x)$ after x months.

c. TECHNOLOGY Use a graphics calculator or graphing software to graph the functions from parts (a) and (b) on the same screen. When should Alvin sell his car?

Reasoning

Exs. 21 and 22 pose an interesting situation. Students first find the terms of a geometric progression, then find the sum of these terms. The answers to these two exercises may convince some students that one of the scientists overstated the deadliness of the bacteria.

Problem Solving

Refer to Ex. 23. People often make additional payments each month that can be applied directly to the principal owed. Suppose Alvin pays an additional \$20 per month. Have students use a spreadsheet to see how this affects the amount owed after x months. Ask students to write an equation that represents this situation.
$\left[A(x) = \left(A_0 - \frac{P}{r}\right)(1 + r)^x + \frac{P}{r} - 20x\right]$

Answers to Exercises and Problems

c. in about 36 yr, or 2026

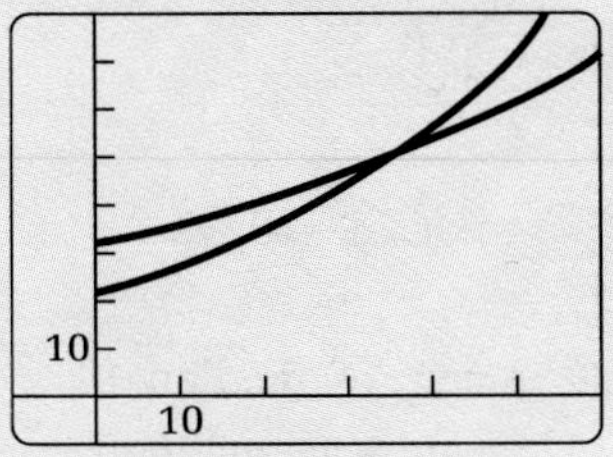

21. a. $N(t) = 2^t$

b. about 20 twenty-minute periods, or about $6\frac{2}{3}$ h

22. a. about $(4.72 \cdot 10^{21})$ Z cells

b. about $(3.35 \cdot 10^{22})$ ft^3

c. about 0.14 ft^3; No. Explanations will vary.

23. a. $A(x) = -13{,}224(1.01)^x + 24{,}024$

b. $R(x) = 8600(0.989)^x$

c. in about 2.5 yr

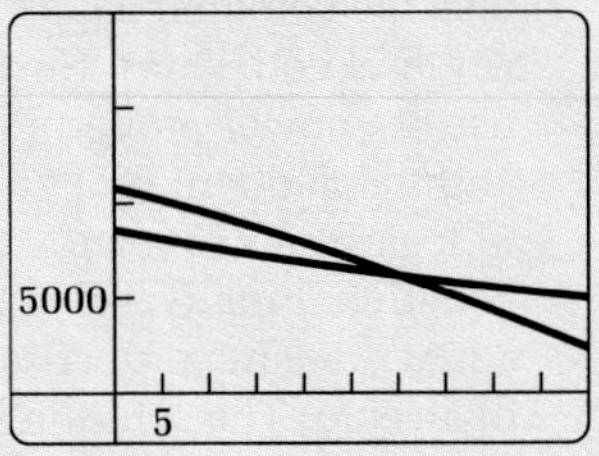

Working on the Unit Project

Ex. 32 shows students that functions can be used in the most unusual places. These questions can be answered as a group or each student can work on them individually in class or at home.

Quick Quiz (5-5 through 5-8)

See page 324.

See also Test 20 in the *Assessment Book.*

Practice 44 For use with Section 5-8

Name ______________________ Date ____________

Practice 44

For use with Section 5-8

Solve. Give answers to three decimal places.

1. $5^x = 12$ 1.544
2. $7^y = 3$ 0.565
3. $1000(1.08)^t = 2000$ 9.006
4. $14(0.88)^t = 3$ 12.050
5. $3^{2x+1} = 5$ 0.232
6. $15e^{x-1} + 2 = 7$ −0.099

Solve.

7. $\log_4 x + \log_4 (x + 6) = 2$ 2
8. $\log_3 x + \log_3 (x - 2) = 1$ 3
9. $\log_2 (x - 9) + \log_2 (x - 2) = 3$ 10
10. $2 \log_7 y - \log_7 (3y + 10) = 0$ 5
11. $\log_5 (6z - 1) - 2 \log_5 z = 1$ 1, 0.2
12. $\log_2 (x - 3) + \log_2 (x + 3) = 4$ 5
13. $2 \log_6 v - \log_6 (v + 12) = 1$ 12
14. $2 \log_3 x = \log_3 (2x - 5) + 2$ 3, 15
15. The population of a certain town was 2500 in 1993 and has been increasing at the rate of 5% per year since then.
    a. Write an equation that gives the population of the town as a function of time t in years after 1993. $p = 2500(1.05)^t$
    b. During what year will the population of the town reach 3600? 2000
16. In 1990, Lien Hua Kim invested $10,000 at 8% annual interest. At the same time, Kenesha Jenkins invested $11,000 at an annual rate of 6%.
    a. Express the values of the two investments as functions of time t in years after 1990. $v = 10{,}000(1.08)^t$; $v = 11{,}000(1.06)^t$
    b. In how many years will the two investments be worth the same amount? in about 5.1 yr
17. ***Writing*** A *mortgage* is a loan that is paid off in equal monthly installments M, given by the formula
$$M = \frac{rP}{1 - (1 + r)^{-t}},$$
where r is the monthly rate of interest, P is the amount borrowed, and t is the number of monthly payments. Suppose you know r, P, and M. Describe a method for finding t.
Check students' work.

44 **Practice Bank,** INTEGRATED MATHEMATICS 3

Ongoing ASSESSMENT

24. **Writing** List the steps needed to solve an exponential equation of the form $ab^x = c$ where a, b, and c are positive numbers. What is the solution in terms of a, b, and c?

Review PREVIEW

Write each expression as a single logarithm. *(Section 5-7)*

25. $2 \log_5 6 - \log_5 4$
26. $3 \log_b p + 7 \log_b p^2 - \log_b q$

Simplify. *(Toolbox Skill 15)*

27. $w(w + 1)(w - 1)$
28. $7(2a^2 - 3) - 2(5a^2 - a + 1)$
29. $\frac{x^3 - 2x^2 + x}{x - 1}$

Find the mean and the median of each set of numbers. *(Toolbox Skill 1)*

30. 85, 100, 58, 77, 92, 81, 39
31. 2.22, 0.98, 5.74, 3.48, 1.53, 4.35

5 Working on the Unit Project

32. Notice the long crest on top of the dinosaur's head. This *Parasaurolophus* made hooting sounds by forcing air through nasal passages inside the crest.
    a. The sound made by air blowing through a tube open at both ends has frequency f given by
    $$f = \frac{170}{l}$$
    where f is measured in hertz and l is the length of the tube in meters. Use this formula to find the frequency of the sound made by *Parasaurolophus.* The total length of its nasal passages was about 2 m.
    b. You can estimate the pitch of a sound made by *Parasaurolophus* by playing the note on a piano having about the same frequency. The frequency of the note made by the nth piano key is given by
    $$f = (27.5)(2)^{(n - 1)/12}$$
    where $n = 1$ corresponds to the leftmost key. Find the number of the piano key whose sound is closest in pitch to a sound made by *Parasaurolophus.*

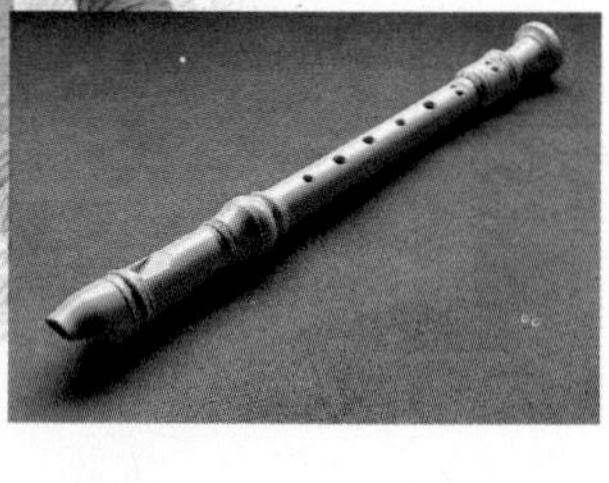

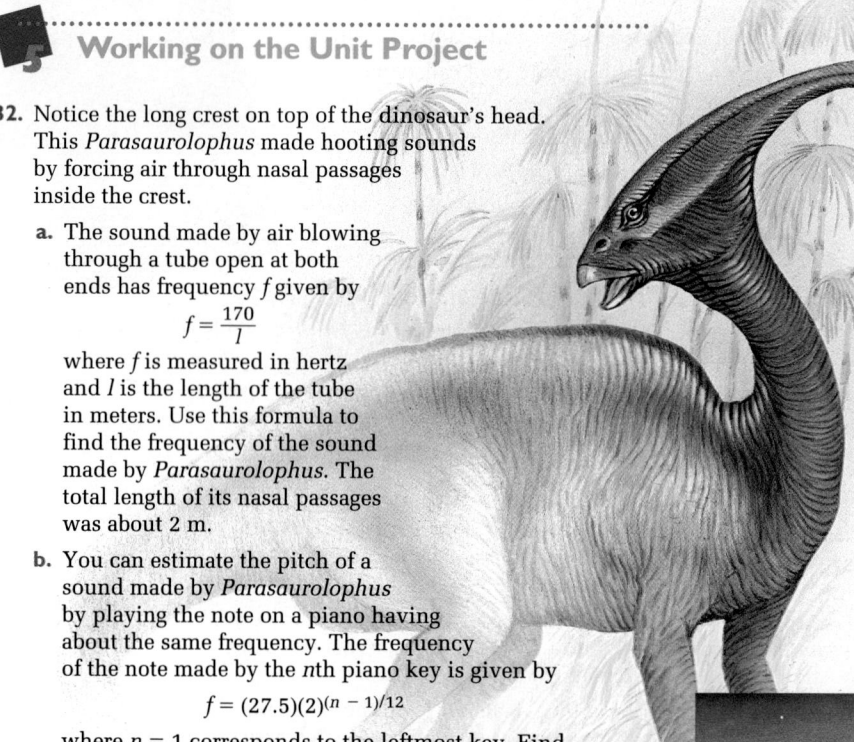

A recorder is also a tube open at both ends. Sound is made by blowing air through the tube.

Answers to Exercises and Problems

24. First, divide both sides of the equation by a. Then take the common logarithm of both sides. Next, use the power property and the quotient property of logarithms to rewrite the equation. Finally, solve the equation using properties of algebra. The solution is $x = \frac{\log c - \log a}{\log b}$; $ab^x = c$; $b^x = \frac{c}{a}$; $x \log b = \log \frac{c}{a}$; $x \log b = \log c - \log a$; $x = \frac{\log c - \log a}{\log b}$.
25. $\log_5 9$
26. $\log_b \frac{p^{17}}{q}$
27. $w^3 - w$
28. $4a^2 + 2a - 23$
29. $x^2 - x$
30. mean: 76; median: 81
31. mean: 3.05; median: 2.85
32. a. 85 Hz
    b. 21

Unit Project 5

Completing the Unit Project

Now you are ready to complete your musical instruments and plan your lecture/performance.

Your completed project should include these things:

- **two handmade panpipes and two handmade simple guitars**
- **drawings of your panpipes and guitars with descriptions of how you built them and how they are played**
- **a written report describing two examples of how mathematics is related to music**
- **a musical performance involving all four of your handmade instruments**

Look Back

Was the panpipe or the guitar easier to make? to play? to understand in terms of mathematics? Why?

Alternative Projects

Project 1: Mathematics and My Musical Instrument

If you play a musical instrument, prepare a lecture/performance in which you demonstrate and discuss how mathematical models are used in making and playing it.

Project 2: Applying Exponential and Logarithmic Functions

Choose one of the applications you saw in this unit, such as the Richter scale or carbon dating. Find out more about the topic and the mathematics involved. Prepare a presentation for your class.

Assessment

A scoring rubric for the Unit Project can be found on pages 264 and 265 of this Teacher's Edition and also in the ***Project Book***.

Unit Support Materials

- *Practice Bank:* Cumulative Practice 45
- *Study Guide:* Unit 5 Review
- *Problem Bank:* Unifying Problem 5
- *Assessment Book:* Unit Tests 21 and 22 Cumulative Test 23 Spanish Unit Tests Alternative Assessment
- *Test Generator* software with Test Bank
- *Teacher's Resources for Transfer Students*

Quick Quiz (5-5 through 5-8)

1. Write an equation for the inverse of $y = -9(x + 5)$. [5-5] $y = -\frac{1}{9}x - 5$
2. Solve for x. $\log_x 9 = -2$ [5-6] $\frac{1}{3}$
3. Write the expression in terms of $\log x$ and $\log y$. $\log \frac{\sqrt{x}}{y^2}$ [5-7] $\frac{1}{2}\log x - 2\log y$
4. Evaluate $\log_3 12$ to the nearest hundredth. [5-7] 2.26
5. A biologist found that the number of bacteria in his sample increased exponentially according to the formula $A(t) = A_0(1.021)^t$, where t is the time in minutes. How long will it take the sample to double? [5-8] about 33 min

Unit 5 REVIEW AND ASSESSMENT

1. **Writing** Describe how you can use the equation $y = ab^x$ to model a 5% annual growth in population. Describe how you can model a 5% annual decrease in population. 5-1
2. The equation $y = (1.06)(1.1)^x$ can be used to model the United States' debt, in billions of dollars, since 1978. Assume the United States' debt continues to follow this model. Use this model to estimate the United States' debt in 2010.
3. **Open-ended** Describe a situation that can be modeled by exponential growth. Describe a situation that can be modeled by exponential decay.

Match each function with its graph. 5-2

4. $y = (3.8)(0.75)^x$
5. $y = (3.8)(1.15)^x$
6. $y = (3.8)(0.75)^{-x}$

A.

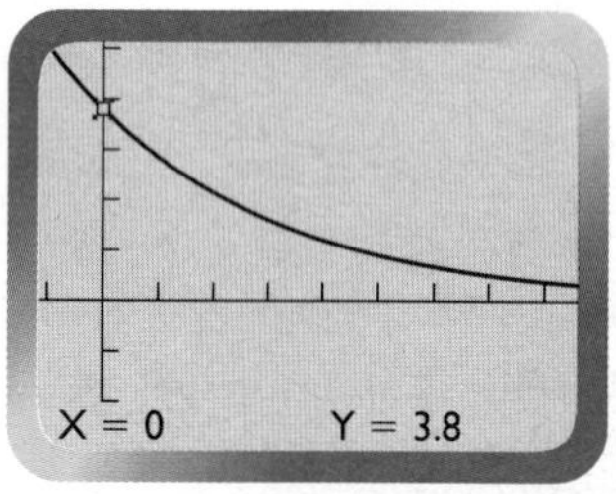

B.

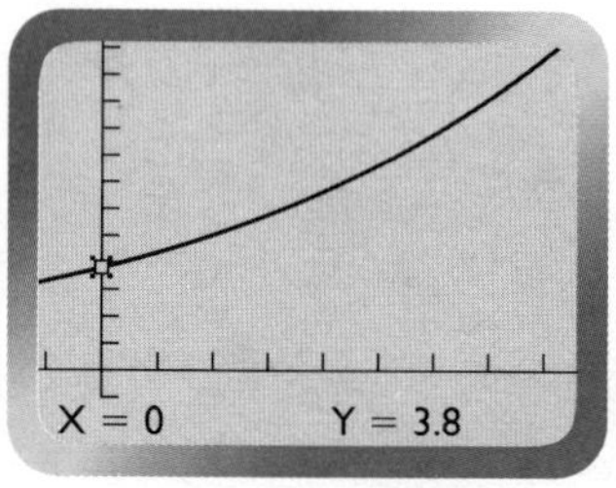

C.

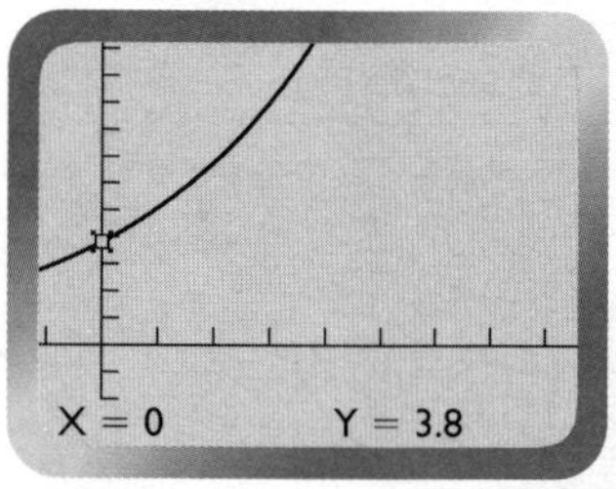

7. In 1950 the population of California was about 10,586,000. In 1960 the population was about 15,717,000.
   a. Write an exponential function to model the population of California.
   b. Use your function to estimate the population in 1900.
8. Rewrite $y = (20)\left(\frac{1}{5}\right)^{-x}$ in the form $y = ab^x$.

Simplify. 5-3

9. $25^{3/2}$
10. $2000^{1/3}$
11. $\sqrt[4]{64x^{16}}$
12. Find b when $f(x) = 2b^x$ and $f\left(\frac{4}{3}\right) = 162$.
13. The breathing rate of a cyclist in liters of oxygen per minute can be modeled by the equation $y = 0.25e^{0.11x}$, where x represents the speed of the bicycle in miles per hour. 5-4
    a. Find the breathing rate for a cyclist riding at 12 mi/h.
    b. Suppose the cyclist in part (a) wants to double the breathing rate. How fast does the cyclist have to ride?
14. TECHNOLOGY Graph the function $f(x) = \frac{5}{1 + 0.5e^{-2x}}$. What are the horizontal asymptotes of the graph?

Answers to Unit 5 Review and Assessment

1. The growth factor would be 1.05, so the equation would be $y = a(1.05)^x$. The decay factor would be 0.95, so the equation would be $y = a(0.95)^x$.
2. about 22 billion dollars
3. Answers may vary. Examples are given. value of an investment at a fixed rate of interest over a period of time; amount of carbon 14 left in organic material over time
4. A
5. B
6. C
7. a. $y = 10586000(1.485)^x$, where x is the number of decades since 1950
   b. about 1,466,000
8. $y = (20)(5^x)$
9. 125
10. $10\sqrt[3]{2}$
11. $2x^4\sqrt[4]{4}$
12. 27
13. a. 0.94 L/min
    b. about 18 mi/h
14. $y = 0$; $y = 5$

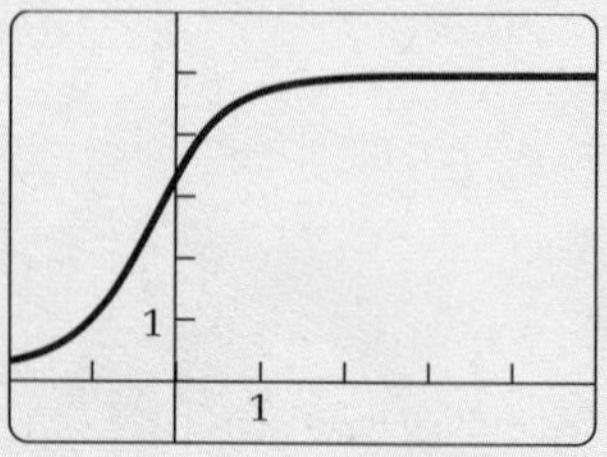

15. Yes.

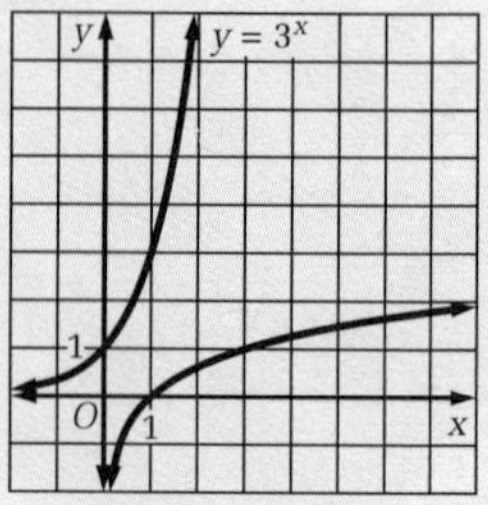

Graph each function and its reflection over the line y = x. Determine whether each function has an inverse. 5-5

15. $y = 3^x$ **16.** $y = 4x^2 + 1$

17. Tell whether the inverse of $y = \frac{1}{4}x + 2$ exists. If it exists, write its equation.

Evaluate each logarithm. Round decimal answers to the nearest hundredth. 5-6

18. $\log \frac{1}{100{,}000}$ **19.** $\ln 6$ **20.** $\log_6 216$

Solve. Round decimal answers to the nearest hundredth.

21. $\log_x 121 = 2$ **22.** $e^x = \frac{1}{e^4}$ **23.** $10^x = 35$

For Questions 24–26, write each expression in terms of log *M* and log *N*. 5-7

24. $\log N^7$ **25.** $\log MN^5$ **26.** $\log \frac{\sqrt{M}}{N^6}$

27. Simplify $\log_3 4 + \log_3 20 - \log_3 6$.

28. Find the value of $\log_3 17$ to three decimal places.

29. Solve $\log_2 x + \log_2 (x - 4) = 5$. 5-8

For Questions 30 and 31, use the formula $A = Pe^{rt}$ and the information below. 5-4, 5-8

In 1983 the United States government agreed to pay the Native American Pequot people for land they owned in 1856. The Pequots asked for \$900,000. A government official suggested that the Pequots be paid the value of the land in 1856 (\$8091.17) plus interest.

30. How much did the government official offer the Pequots in 1983, considering the average 1983 interest rate of 10% compounded continuously?

31. Find the interest rate, compounded continuously, requested by the Pequots.

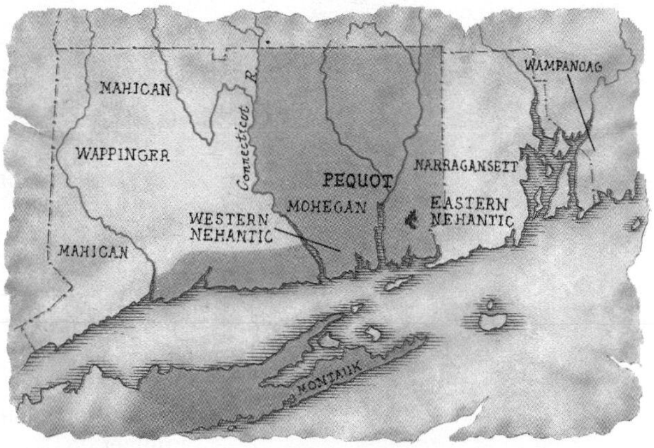

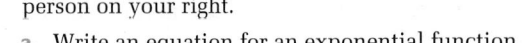

The region of greatest expansion of the Pequot people in the 1600s is shown in orange. The small purple region is all the land they owned in 1856.

32. **Self-evaluation** Do you prefer to solve problems like *Question 7* or like *Question 26*? Why?

33. **Group Activity** Work in a group of four students. After completing each part, pass your paper to the person on your right.

a. Write an equation for an exponential function.

b. Write the equation for the inverse of the function you receive.

c. Graph the original function and the inverse function you receive on the same set of axes.

d. Check the equation and graphs you receive. Discuss any errors you find.

Answers to Unit 5 Review and Assessment

16. No.

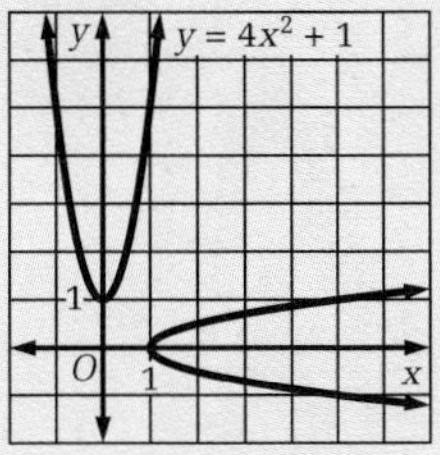

17. Yes; $y = 4x - 8$.

18. −5

19. 1.79

20. 3

21. 11

22. −4

23. 1.54

24. $7 \log N$

25. $\log M + 5 \log N$

26. $\frac{1}{2} \log M - 6 \log N$

27. $\log_3 \left(13\frac{1}{3}\right)$

28. 2.579

29. 8

30. about 2.65 billion dollars

31. about 3.7%

32. Answers may vary.

33. Check students' work.

IDEAS AND (FORMULAS)

DISCRETE MATH

- **Problem Solving** Discrete exponential growth can be modeled by an exponential function in the form $y = ab^x$ where $b > 1$. *(p. 267)*
- Discrete exponential decay can be modeled by an exponential function in the form $y = ab^x$ where $0 < b < 1$. *(p. 267)*

ALGEBRA

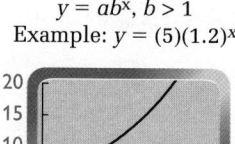

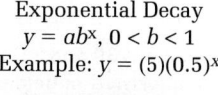

- The general form of an exponential function is $y = ab^x$, where a represents the initial value and b represents the growth factor. *(p. 267)*
- **Problem Solving** Continuous exponential growth can be modeled by an exponential function in the form $y = ab^x$ where $b > 1$. *(pp. 267, 269)*
- **Problem Solving** Continuous exponential decay can be modeled by an exponential function in the form $y = ab^x$ where $0 < b < 1$. *(pp. 267, 269)*

Exponential Growth
$y = ab^x$, $b > 1$
Example: $y = (5)(1.2)^x$

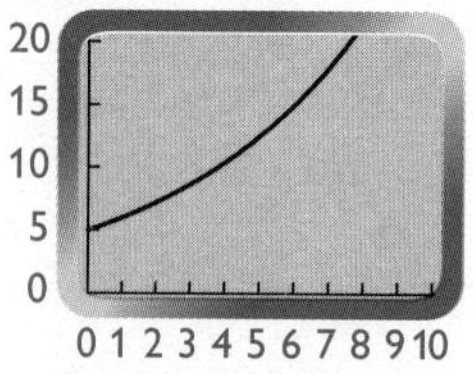

Exponential Decay
$y = ab^x$, $0 < b < 1$
Example: $y = (5)(0.5)^x$

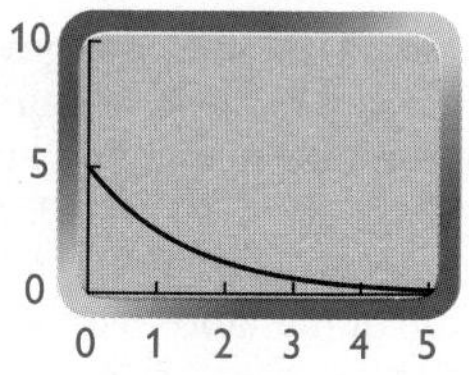

- Changing the value of a in an exponential function in the form $y = ab^x$ changes the y-intercept of the graph of the function. *(p. 271)*
- $y = ab^{-x}$ is the same function as $y = a\left(\frac{1}{b}\right)^x$. *(p. 276)*
- For integers m and n with $n > 0$, and any nonnegative real number a:

  $a^{1/n} = \sqrt[n]{a}$ $\qquad$ $a^{m/n} = \left(\sqrt[n]{a}\right)^m = \sqrt[n]{a^m}$ *(p. 283)*
- You can use the rules of fractional exponents to simplify variable expressions. *(p. 284)*
- The number e is an irrational number that is approximately equal to 2.718281828. *(p. 289)*
- A logistic growth function increases or decreases exponentially at first. Then the rate of growth slows down as the function approaches an upper or lower boundary. *(p. 290)*

- Two functions f and g are inverse functions if $g(b) = a$ whenever $f(a) = b$. *(p. 297)*
- The inverse of a function can be written $f^{-1}(x)$ or f^{-1}. *(p. 297)*
- If the reflection of the graph of a function over the line $y = x$ is not a function, then the function does not have an inverse. *(p. 297)*
- You can use the horizontal line test to tell whether a function has an inverse. *(p. 297)*
- The base b logarithm of any positive real number x is the exponent a when you write x as a power of a base b. *(p. 302)*

  $\log_b x = a$ when $x = b^a$
- A natural logarithm of any positive real number x is the exponent a when you write x as a power of e. *(p. 302)*

  $\ln x = a$ when $e^a = x$
- Let M, N, b, and c be positive numbers with $b \neq 1$ and $c \neq 1$. *(pp. 310, 312)*

  $\log_b MN = \log_b M + \log_b N$ (product of logarithms property)

  $\log_b \frac{M}{N} = \log_b M - \log_b N$ (quotient of logarithms property)

  $\log_b M^k = k \log_b M$ (power of logarithms property)

  $\log_b M = \frac{\log_c M}{\log_c b}$ (change-of-base property of logarithms)
- You can use properties of logarithms to solve exponential and logarithmic equations. Some solutions may be extraneous. *(p. 319)*

GEOMETRY

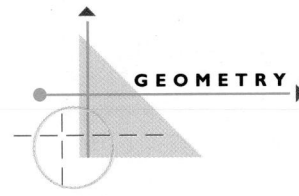

- The graph of $y = ab^x$ is a reflection over the y-axis of the graph of $y = ab^{-x}$. *(p. 276)*
- The graph of the inverse of a function is the reflection of the graph of the function over the line $y = x$. *(p. 297)*

Key Terms

- **logistic growth function** (p. 290)
- **common logarithm** (p. 302)
- **inverse functions** (p. 297)
- **natural logarithm** (p. 302)
- **logarithm** (p. 302)
- **logarithmic function** (p. 302)

Quick Quiz (5-1 through 5-4)

1. John Harth plans to stay at his new job until he retires in 10 years. He is offered a choice of salary: either \$20,000 with a 5% annual raise or \$20,000 with a \$1000 annual raise. Which salary should he choose and why? [5-1]
   the first choice; After the second year, both his salary and his raises are greater.
2. Give an example of a situation that can be modeled by a function of the form $y = ab^x$, where negative values of x make sense. [5-2]
   Answers will vary. An example is given. Let y be the number of tourists to visit an amusement park and x the years after 1995. A negative value of x would represent any year before 1995.
3. Explain why the function $y = 2(3)^{-x}$ is an exponential decay function. [5-2]
   It is equivalent to $y = 2\left(\frac{1}{3}\right)^x$.
4. Simplify. $\sqrt[4]{32a^8b^{13}}$ [5-3]
   $2a^2b^3\sqrt[4]{2b}$
5. Give the horizontal asymptotes of the graph of $f(x) = \frac{3.7}{1 + 2.1e^{-4x}}$. [5-4]
   $y = 0$ and $y = 3.7$

Planning the Unit

6 Modeling and Analyzing Data

OVERVIEW

- **Unit 6** introduces frequency distributions and the different shapes that are common to a distribution. The standard deviation is introduced as a measure of variability in statistics, then related to the shape of a normal distribution. Students explore various ranges of data as they relate to the normal distribution.
- Statistical modeling is explored. Linear models are derived from linear regression and by finding the least-squares line and the median-median line. Data are fitted through quadratic regression and by solving a system of linear equations. A comparison is made between linear regression and quadratic regression to see which best fits a set of data. Measures of central tendency are reviewed in the **Student Resources Toolbox**.
- The **Unit Project** revolves around the idea of creating a profile of a typical car. Students conduct a survey to determine all the features that people typically choose in a car and use statistics to analyze the results of the survey. They complete the project by creating a visual display showing the features of their car, along with a written analysis of the survey data.
- **Connections** to medicine, quality control of manufactured products, demographics, warranties, pumpkin sizes, and world travel are some of the topics included in the teaching materials and the exercises.
- **Graphics calculators** are used in Section 6-2 to find means and standard deviations, in Section 6-4 to find the equation of a median-median line and a least-squares line, and in Section 6-5 to graph a least-squares line and to find the least-squares parabola. **Computer software**, such as Plotter Plus, can be used in Sections 6-1 and 6-3 to draw histograms.
- **Problem-solving strategies** used in Unit 6 include using frequency distributions, using formulas, using scatter plots, and using technology.

Unit Objectives

Section	Objectives	NCTM Standards
6-1	• Identify different types of frequency distribution.	1, 2, 3, 4, 10, 12
6-2	• Use mean, range, and standard deviation to compare data sets with technology.	1, 2, 3, 4, 5, 10, 12
6-3	• Recognize special properties of normal distributions.	1, 2, 3, 4, 5, 10, 12
6-4	• Identify data as fitting a linear model. • Use technology to find the equations of fitted lines.	1, 2, 3, 4, 5, 10, 12
6-5	• Decide whether a quadratic model fits data. • Use technology and matrices to find quadratic models.	1, 2, 3, 4, 5, 10, 12

Topic Spiraling

Section	Connections to Prior and Future Concepts
6-1	**Section 6-1** explores different types of frequency distributions—uniform, mound-shaped, skewed, and bimodal. The histograms used in these distributions were first introduced in Section 3-4 of Book 1. Various distributions of both discrete and continuous data are studied in future courses in statistics.
6-2	**Section 6-2** introduces standard deviation as a measure of variability in statistics. The mean and range are also used to compare sets of data. The mean and range were first covered in Section 3-2 of Book 1. Standard deviation is used again in Sections 6-3 and 9-6 of Book 3.
6-3	**Section 6-3** explores the normal distribution and the bell-shaped curve. Familiarity with the mean is necessary, having been covered in Section 3-2 of Book 1 and Section 6-2 of Book 3. Familiarity with the standard deviation is also necessary, having been covered in Section 6-2 of Book 3. The normal distribution is used in Section 9-6 of Book 3, where it is transformed into the standard normal distribution.
6-4	**Section 6-4** covers fitting raw data to a linear model by means of linear (least-squares line and median-median line) regression. The concept of regression for modeling is encountered again in Section 6-5 of Book 3.
6-5	**Section 6-5** introduces fitting raw data to a quadratic model. Students use a calculator to determine quadratic regression. The concept of regression to determine equations was first introduced in Section 6-4 of Book 3. From only three points, students write and solve systems of equations by using inverse matrices. This process was first introduced in Section 3-8 of Book 2.

Integrating the Strands

Strands	Sections
Number	6-1
Algebra	6-1, 6-2, 6-3, 6-4, 6-5
Functions	6-1, 6-3, 6-5
Geometry	6-4, 6-5
Statistics and Probability	6-1, 6-2, 6-3, 6-4, 6-5
Discrete Mathematics	6-1, 6-2, 6-3, 6-4, 6-5
Logic and Language	6-1, 6-2, 6-3, 6-4, 6-5

Planning the Unit 6

Modeling and Analyzing Data

Section Planning Guide

- Essential exercises and problems are indicated in boldface.
- Ongoing work on the Unit Project is indicated in color.
- Exercises and problems that require student research, group work, manipulatives, or graphing technology are indicated in the column headed "Other."

Section	Materials	Pacing	Standard Assignment	Extended Assignment	Other
6-1	clock with a second hand	Day 1 Day 2	**1–6,** 9 **10–13,** 14–17, 18	**1–6,** 7–9 **10–13,** 14–17, 18	
6-2	calculator	Day 1	**1–6,** 10–13, 14	**1–6,** 7–13, 14	
6-3	tape measure	Day 1 Day 2	**1–3,** 4 **6–9,** 13–16, 17	**1–3,** 4, 5 **6–9,** 10–16, 17	
6-4	graphics calculator or statistical software	Day 1 Day 2	**1–4** **11–13,** 14–17, 18	**1–4** **11–13,** 14–17, 18	5–10 12d, 18
6-5	graphics calculator or statistical software, small beans or other counters, 3 jar lids of different sizes	Day 1	**1–3, 5–8, 10,** 15, 16, 17	**1–3, 5–8, 10,** 15, 16, 17	4, 9, 11–14
Review Test		**Day 1** **Day 2**	**Unit Review** **Unit Test**	**Unit Review** **Unit Test**	

Yearly Pacing	Unit 6 Total	Units 1–6 Total	Remaining	Total
	12 days (2 for Unit Project)	94 days	60 days	154 days

Support Materials

- See **Project Book** for notes on Unit 6 Project: The Mean Machine.
- UPP and disk refer to **Using Plotter Plus** booklet and **Plotter Plus** disk.
- TI-81/82 refers to **Using TI-81 and TI-82 Calculators** booklet.
- Warm-up exercises for each section are available on **Warm-Up Transparencies.**

Section	Study Guide	Practice Bank	Problem Bank	Activity Bank	Explorations Lab Manual	Assessment Book	Visuals	Technology
6-1	6-1	Practice 46	Set 13	Enrich 41	Master 1	Quiz 6-1		Statistics Spreadsheet (disk) Data Analyzer (disk)
6-2	6-2	Practice 47	Set 13	Enrich 42	Master 1	Quiz 6-2		UPP, page 58
6-3	6-3	Practice 48	Set 13	Enrich 43	Master 1	Quiz 6-3 Test 24	Folder 5	TI-81/82, page 65
6-4	6-4	Practice 49	Set 14	Enrich 44	Add. Expl. 7 Master 2	Quiz 6-4		Statistics Spreadsheet (disk) Data Analyzer (disk)
6-5	6-5	Practice 50	Set 14	Enrich 45	Masters 1, 2	Quiz 6-5 Test 25		
Unit 6	Unit Review	Practice 51	Unifying Problem 6	Family Involve 6		Tests 26, 27		

Form A

Spanish versions of these tests are on pages 135–138 of the **Assessment Book.**

Name ______________ Date ________ Score ____

Test 26

Test on Unit 6 (Form A)

Directions: Write the answers in the spaces provided.

Answers
1. about 248
2. about 72
3. *See question.*
4. *See question.*
5. *See question.*
6. $y = 2x^2 + 5$

A group of high school students were asked how many hours they spent on homework each week. The survey results are shown in the histogram below. Use this information for Questions 1 and 2.

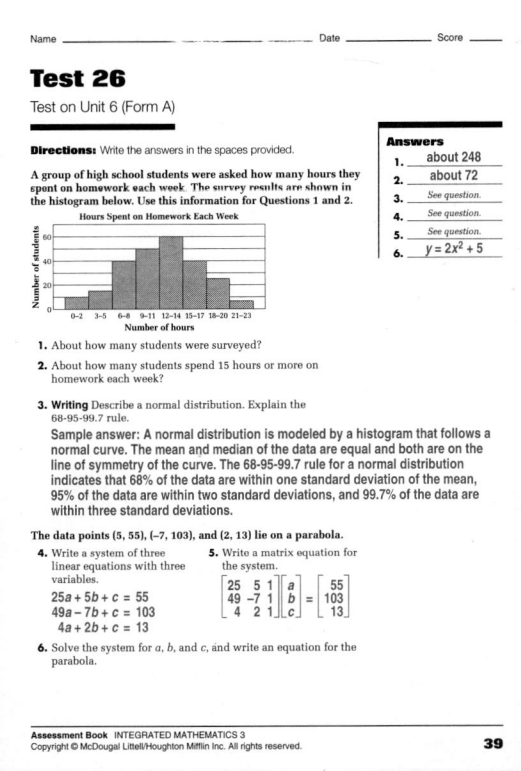

Hours Spent on Homework Each Week

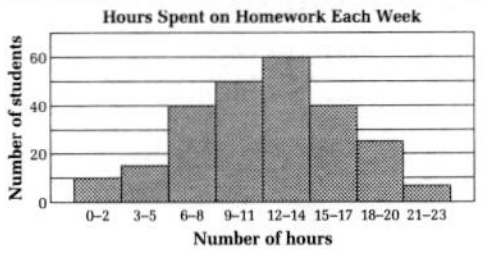

1. About how many students were surveyed?
2. About how many students spend 15 hours or more on homework each week?
3. **Writing** Describe a normal distribution. Explain the 68-95-99.7 rule.
   Sample answer: A normal distribution is modeled by a histogram that follows a normal curve. The mean and median of the data are equal and both are on the line of symmetry of the curve. The 68-95-99.7 rule for a normal distribution indicates that 68% of the data are within one standard deviation of the mean, 95% of the data are within two standard deviations, and 99.7% of the data are within three standard deviations.

The data points (5, 55), (–7, 103), and (2, 13) lie on a parabola.

4. Write a system of three linear equations with three variables.
   $25a + 5b + c = 55$
   $49a - 7b + c = 103$
   $4a + 2b + c = 13$
5. Write a matrix equation for the system.
   $$\begin{bmatrix} 25 & 5 & 1 \\ 49 & -7 & 1 \\ 4 & 2 & 1 \end{bmatrix}\begin{bmatrix} a \\ b \\ c \end{bmatrix} = \begin{bmatrix} 55 \\ 103 \\ 13 \end{bmatrix}$$
6. Solve the system for a, b, and c, and write an equation for the parabola.

Assessment Book INTEGRATED MATHEMATICS 3

39

Name ______________ Date ________ Score ____

Test 26 *(continued)*

Directions: Write the answers in the spaces provided.

Answers
7. 20.7; 28; ≈8.58
8. 60%
9. *See question.*
10. $y = 2.04x + 18.7$
11. $y = x + 24.8$
12. *See question.*

Last year the Wolverine football team scored the following number of points in its 10 games.

Number of points: 17, 7, 28, 21, 24, 35, 14, 10, 31, 20

7. Find the mean, the range, and the standard deviation of the data.
8. Find the percent of the data within one standard deviation of the mean.

This year the Wolverine football team scored the following number of points in its 10 games. Use this information for Questions 9–11.

Game	1	2	3	4	5	6	7	8	9	10
Points	27	17	24	14	34	38	41	28	45	31

9. Make a scatter plot of the 10 data points. Let x = the game number and y = the number of points scored during that game.

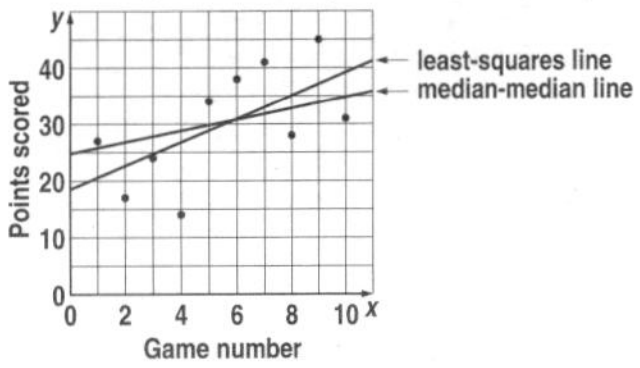

10. Find an equation of the least-squares line. Graph the line on the scatter plot above. Label the line.
11. Find an equation of the median-median line. Graph the line on the scatter plot above. Label the line.
12. **Open-ended** Create a situation, including data, that demonstrates a bimodal distribution.
    Sample answer: A survey of high school students showed that 55% of the freshmen, 15% of the sophomores, 20% of the juniors, and 65% of the seniors wanted to have a holiday dance.

40
Assessment Book INTEGRATED MATHEMATICS 3

Form B

Name ______________ Date ________ Score ____

Test 27

Test on Unit 6 (Form B)

Directions: Write the answers in the spaces provided.

Answers
1. about 337
2. about 62
3. *See question.*
4. *See question.*
5. *See question.*
6. $y = 4.5x^2 + 7$

A group of high school students were asked how many hours they worked at a job each week. The survey results are shown in the histogram below. Use this information for Questions 1 and 2.

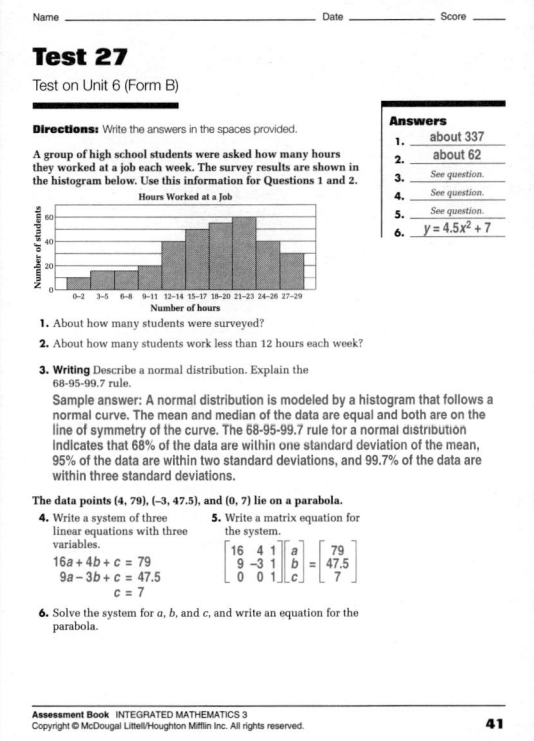

Hours Worked at a Job

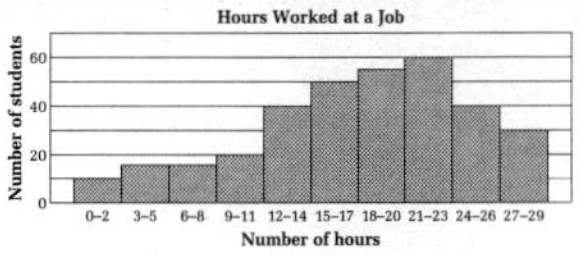

1. About how many students were surveyed?
2. About how many students work less than 12 hours each week?
3. **Writing** Describe a normal distribution. Explain the 68-95-99.7 rule.
   Sample answer: A normal distribution is modeled by a histogram that follows a normal curve. The mean and median of the data are equal and both are on the line of symmetry of the curve. The 68-95-99.7 rule for a normal distribution indicates that 68% of the data are within one standard deviation of the mean, 95% of the data are within two standard deviations, and 99.7% of the data are within three standard deviations.

The data points (4, 79), (–3, 47.5), and (0, 7) lie on a parabola.

4. Write a system of three linear equations with three variables.
   $16a + 4b + c = 79$
   $9a - 3b + c = 47.5$
   $c = 7$
5. Write a matrix equation for the system.
   $$\begin{bmatrix} 16 & 4 & 1 \\ 9 & -3 & 1 \\ 0 & 0 & 1 \end{bmatrix}\begin{bmatrix} a \\ b \\ c \end{bmatrix} = \begin{bmatrix} 79 \\ 47.5 \\ 7 \end{bmatrix}$$
6. Solve the system for a, b, and c, and write an equation for the parabola.

Assessment Book INTEGRATED MATHEMATICS 3

41

Name ______________ Date ________ Score ____

Test 27 *(continued)*

Directions: Write the answers in the spaces provided.

Answers
7. 68.2; 23; ≈6.87
8. 70%
9. *See question.*
10. $y = 2.07x + 71.5$
11. $y = 2.57x + 68.2$
12. *See question.*

Last year the Wolverine basketball team scored the following number of points in its first 10 games.

Number of points: 72, 61, 64, 58, 63, 75, 72, 81, 64, 72

7. Find the mean, the range, and the standard deviation of the data.
8. Find the percent of the data within one standard deviation of the mean.

This year the Wolverine basketball team scored the following number of points in its first 10 games. Use this information for Questions 9–11.

Game	1	2	3	4	5	6	7	8	9	10
Points	73	72	83	82	79	84	85	91	86	94

9. Make a scatter plot of the 10 data points. Let x = the game number and y = the number of points scored during that game.

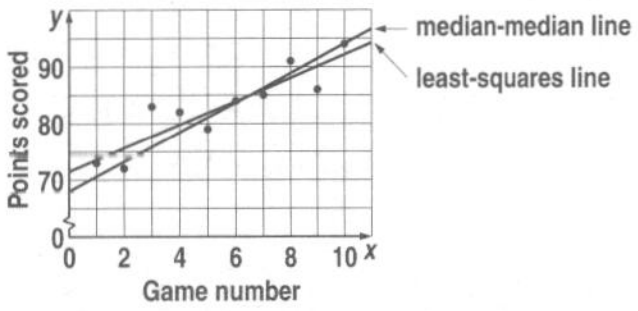

10. Find an equation of the least-squares line. Graph the line on the scatter plot above. Label the line.
11. Find an equation of the median-median line. Graph the line on the scatter plot above. Label the line.
12. **Open-ended** Create a situation, including data, that demonstrates a skewed distribution.
    Sample answer: A survey of high school students showed that 15% of the freshmen, 55% of the sophomores, 20% of the juniors, and 65% of the seniors wanted to have a holiday dance.

42
Assessment Book INTEGRATED MATHEMATICS 3

OUTSIDE RESOURCES

Books/Periodicals

Kullman, David E. "Patterns of Postage-Stamp Production." *Mathematics Teacher* (March 1992): pp. 188–189.

"Exploring Data" from the *Quantitative Literacy Series*; written by members of the Joint Committee on the Curriculum in Statistics and Probability of the American Statistical Association and the National Council of Teachers of Mathematics. Dale Seymour publication.

Sandefur, James T. "Technology, Linear Equations, and Buying a Car." *Mathematics Teacher* (October 1992): pp. 562–567.

Activities/Manipulatives

Baker, Patricia Cooper. "Supply and Demand—An Application of Linear Equations." *Mathematics Teacher* (October 1991): pp. 554–556.

Brunner, Regina Baron and Carl E. Brunner. "How Much Does Camouflage Help?" *Mathematics Teacher* (December 1994): pp. 676–681.

Wallace, Edward C. "Exploring Regression with a Graphing Calculator." *Mathematics Teacher* (December 1993): pp. 741–743.

Software

f(g)Scholar. IBM. Southhampton, PA: Future Graph, 1992.

Statistics Workshop. Macintosh family. Scotts Valley, CA: Wings for Learning, 1991.

Videos

Futures with Jaime Escalante. Program No. 1: Statistics. PBS, 1990.

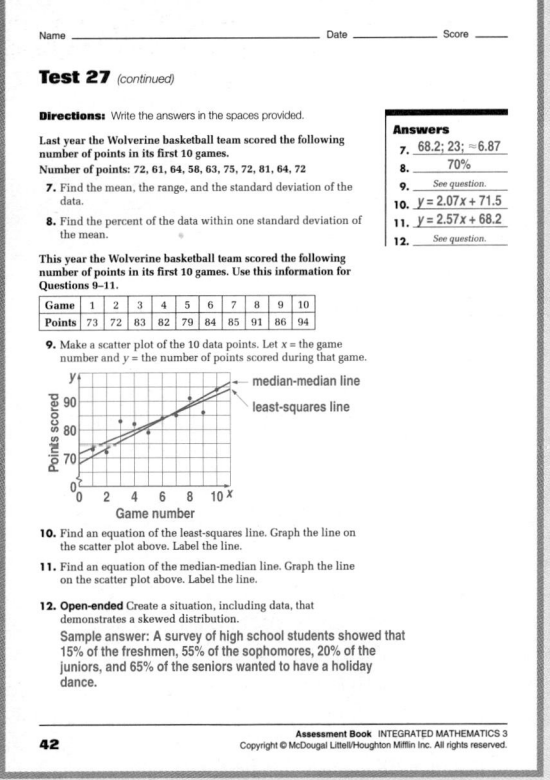

PROJECT GOALS

- Students create a profile of a typical car, one that has all the features that people typically choose.
- Students conduct a survey of at least 30 people to collect information about their preferences in a new car.
- Students analyze the data collected using concepts and skills learned in this unit.

PROJECT PLANNING

Materials List

- Magazines and newspapers that contain advertisements for cars

Project Teams

Have students work on the project in groups of four. One way for the individuals in the group to distribute the work is as follows:

1. Researcher: finds advertisements for cars in magazines and newspapers and makes a list of the major features and options of the cars advertised.
2. Interviewer: visits a car dealership to interview salespeople and customers regarding features.
3. Surveyor: coordinates development and refinement of the survey questions, conducts the survey, and summarizes its results.
4. Artist: makes a visual display showing the features of the car.

unit 6

Modeling and Analyzing Data

CARS

September brings an end to summer, the beginning of a new school year — and the unveiling of next year's new car models. Every year for the past decade, there have been over 500 models to choose from. For a couple of years in the late 1980s, there were more than 600!

Color is so important to new-car buyers that some car makers use computer-aided design (CAD) systems for designing color as well as shape and construction. Color designers use the systems to create colors that people associate with "feeling" words like elegant and dynamic. The software even adjusts the color display to show how a color looks under different weather conditions.

There are two types of car **safety features**: those that help prevent accidents and those that protect passengers in case an accident happens.

preventive Antilock braking systems (ABS) and all-wheel drive (AWD) help stop skidding and increase traction.

protective Airbags, seat belts, and reinforced frames help prevent injury.

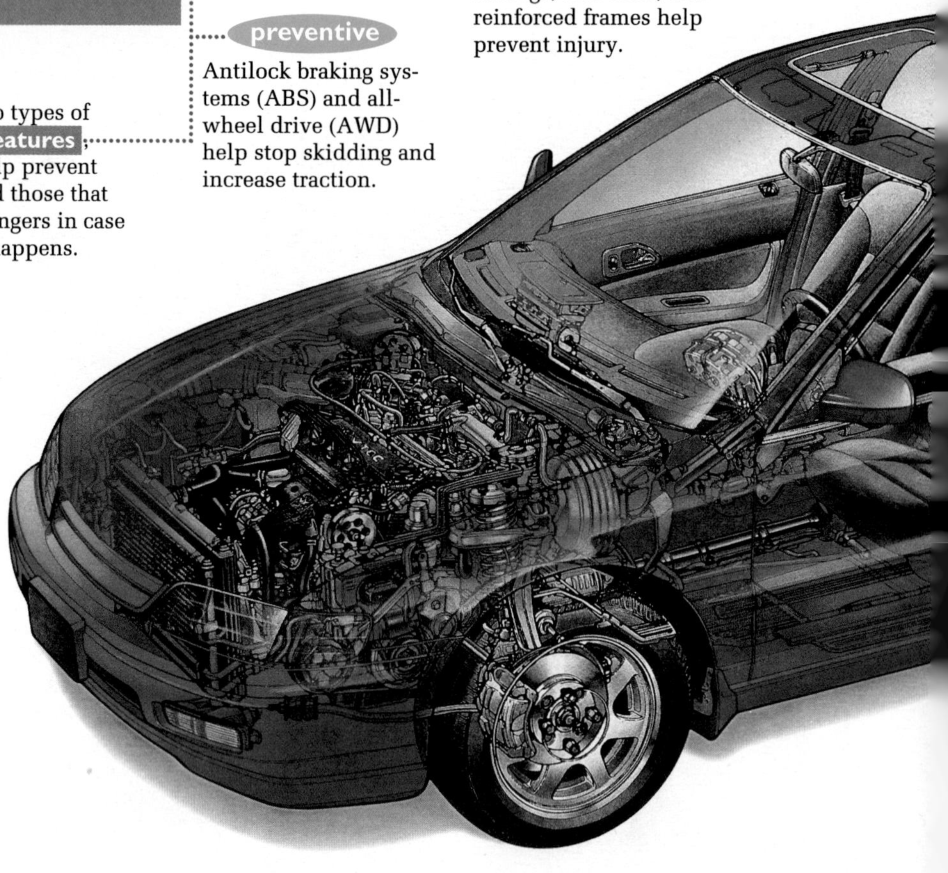

328

Suggested Rubric for Unit Project

4 Students select a sample of people to survey and collect data to create a profile of a typical car. They write a report that presents a complete data table with a histogram of the survey results. The work is accurate and neatly presented. An interesting visual display effectively shows the features of their typical car. The written analysis of the survey data accurately uses the mathematics of the unit. Means and standard deviations are correctly calculated. The conclusions arrived at are supported by the data.

3 Students collect data using a small sample of at least 30 people. The written report contains a data table, histogram, and visual display of the car's features, but the work could be organized in a clearer fashion. The written analysis of the survey data contains some errors and all of the features assembled are not supported by the results of the survey.

2 Students' surveys are incomplete and significant information about the features people want in a new car is missing. Students' written reports are carelessly prepared

Unit Project 6

The Mean Machine

Your project is to create a profile of a typical car, one that has all the features that people typically choose. To collect information about preferences, you should survey at least 30 people. Ask them to suppose they are planning to buy a new car that suits their present lifestyle and is priced within their present budget. Using the concepts and skills you learn in this unit, you should analyze the data collected and assemble the features that make up the typical car. Present your conclusions in a visual display combined with a written report.

environment

To reduce air pollution, engines will burn **cleaner** and cars made of lighter materials will burn less fuel.

In Brazil some cars use ethanol made from sugar cane as an alternative to gasoline.

new technology

Intelligent Vehicle Highway Systems (IVHS) will make the car of the future safer and cleaner, automatically:

- electronic linkage of steering and braking
- radar for avoiding collisions and changing lanes
- infrared imaging for night driving
- sensors that monitor weather conditions

329

Support Materials

The ***Project Book*** contains information about the following topics for use with this Unit Project.

- Project Description
- Teaching Commentary
- Working on the Unit Project Exercises
- Completing the Unit Project
- Assessing the Unit Project
- Alternative Projects
- Outside Resources

ADDITIONAL BACKGROUND

Multicultural Note

During World War II, significant numbers of women came to automobile factories and other heavy machinery operations to work. By September of 1943, ten million American men had gone to war, and women were called upon to replace them as factory workers, producing war supplies such as automobiles, planes, and tanks. In Detroit, where many automobile factories were converted to war production, over 80% of the new hirees in September and October of 1942 were women. Many women were employed as factory line workers; others enlisted in machine shop courses to become skilled operators. In November of 1942, *Newsweek* reported that "depending on the industry, women today make up from 10 to 88 percent of total personnel in most war plants."

Suggested Rubric for Unit Project

and omit some of the things the completed project should contain. The written analysis of the survey data contains some mathematical errors. The profile developed does not accurately portray people's preferences for features in a new car.

1 Students do an inadequate job in surveying people about the features they want in a new car. The information collected is not organized into a meaningful report. Omissions and mathematical errors make the conclusions invalid. Students should be encouraged to speak with the teacher as soon as possible to review their work and to make a new start on the project.

New Technology

Electronic-based safety systems have the potential to vastly reduce auto-related fatalities. Such systems take time to develop and test, however, and can take as long as 10 to 15 years before they fully penetrate the marketplace. New technology will include keyless-entry systems with remote-control starts and programmable position controls, integrated traction control and antilock brake systems, intelligent vehicle highway systems, and collision-avoidance systems, using radar and automatic pilot systems. On a global scale, these types of safety systems will take time to implement, and some of them may not be mandatory until the late 21st century.

ALTERNATIVE PROJECTS

Project 1, page 365

TV Ratings

Find out how television rating services collect data and how television networks analyze the data to make decisions about programming. Write a report describing what has been learned. The report should discuss the reliability of both the data and the decisions based on it.

Project 2, page 365

Is it Normal?

List at least three kinds of data sets that are normally distributed and three that are not normally distributed. Obtain samples of all six types of data sets. For each set, find the mean(s) and, if possible, the standard deviation. Describe each distribution and determine which, if any, are normally distributed.

Unit Project 6

Getting Started

For this project you should work in a group of four students. Here are some ideas to help you get started.

☞ Discuss what kind of sampling method you should use for selecting a sample for your survey.

☞ Look at advertisements for cars in magazines and newspapers. Make a list of some of the major built-in features and options of the cars advertised.

☞ Plan to visit a car dealership. Try to interview some of the salespeople and some of the customers. Find out what features new-car buyers want.

Can We Talk CARS

Working on the Unit Project

Your work in Unit 6 will help you create a profile of a typical car.

Related Exercises:

Section 6-1, **Exercise 18**
Section 6-2, **Exercise 14**
Section 6-3, **Exercise 17**
Section 6-4, **Exercise 18**
Section 6-5, **Exercise 17**

Alternative Projects p. 365

➤ Suppose you are planning to buy a new car tomorrow. What color will you choose? Why? What colors appear to be popular this year? What colors were popular last year? the year before?

➤ How do you think factors besides personal taste affect color choice in a new car? Consider security, visibility, weather (temperature, rain, snow, amount of sun), frequency of cleaning, and so on.

➤ Which of the safety features named on page 328 do you think are the most effective? Why?

➤ Suppose you ask a random sample of 10,000 people how much money they are willing to spend on a new car. Describe what you think the shape of a histogram that displays the data will be. What prices do you think are typical for the sample—very high, very low, or in-between? How did you decide?

Answers to Can We Talk?

➤ Answers may vary, depending upon personal preference and year.

➤ Answers may vary. Examples are given. Red cars are the most common color of stolen cars. White cars and black cars can show dirt and scratches more than other colors. Black cars absorb the light (making them hotter), while white cars reflect the light (making them cooler).

➤ Answers may vary. An example is given. I think seat belts are the most effective because they work no matter where a car is hit or what type of accident a car is involved in.

➤ Answers may vary. An example is given. I think the histogram will be bell-shaped. I think the typical prices will be the average prices of all the cars made, not very high or very low but in-between because the sample is random and is large enough to get a good distribution.

Section **6-1**

Distributions of Data

Focus
Identify different types of frequency distributions.

How Does It SHAPE Up?

EXPLORATION

What do survey results "look" like?

- **Materials: clock with a second hand**
- **Work in a group of four students.**

Student Resources Toolbox
p. 627 *Making a Histogram*

You take a pulse by placing your fingers on the inside of the wrist above the thumb. Count the number of beats in 15 seconds, then multiply by 4.

1. The whole class should answer these questions with the results shown on the board.
   - In what month were you born?
   - What is the last digit of your telephone number?
   - How many letters are in your full first name?
   - How many times does your heart beat per minute while you are sitting at your desk?

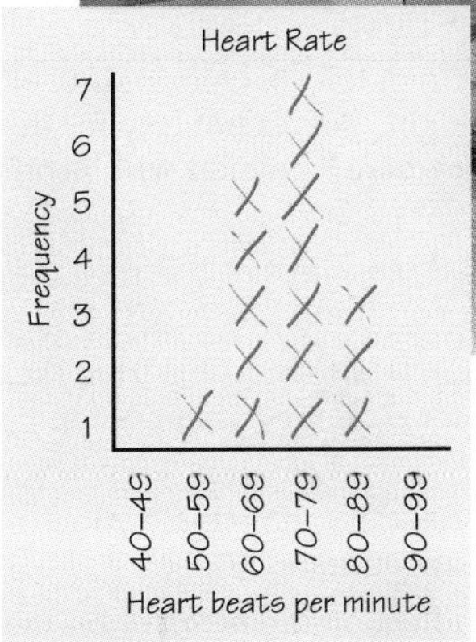

The horizontal axis shows the *intervals*.

2. A **histogram** shows how many data values are in each interval. Your group should make histograms showing the results of each survey question.
3. Compare the shapes of your histograms. Does the shape of each histogram look the way you expected?
4. How many students were born during the summer? How many have a telephone number with a last digit less than 4? How many have first names with three to five letters? How many have heart rates of over 70 beats per minute?
5. Discuss what you think is the "average" answer to each survey question. Write down your group's reasons.

Answers to Exploration

1–5. Answers may vary.

PLANNING

Objectives and Strands
See pages 328A and 328B.

Spiral Learning
See page 328B.

Materials List
- Clock with a second hand
- Graph paper

Recommended Pacing
Section 6-1 is a two-day lesson.

Day 1
Pages 331–332: Exploration through Talk it Over 4, *Exercises 1–9*

Day 2
Pages 333–334: Sample 1 through Look Back, *Exercises 10–18*

Toolbox References
- **Toolbox Skill 1:** Finding Mean, Median, Mode, and Range
- **Toolbox Skill 2:** Making a Histogram

Extra Practice
See pages 617–619.

Warm-Up Exercises
Warm-Up Transparency 6-1

Support Materials
- Practice Bank: Practice 46
- Activity Bank: Enrichment 41
- Study Guide: Section 6-1
- Problem Bank: Problem Set 13
- Explorations Lab Manual: Diagram Master 1
- Using Mac Plotter Plus Disk: Data Analyzer
- Using IBM Plotter Plus Disk: Statistics Spreadsheet
- Assessment Book: Quiz 6-1, Alternative Assessment 1

TEACHING

Exploration

The goal of the Exploration is to have students learn how to represent data values in a histogram and to gain some experience in reading and interpreting histograms.

Talk it Over

Question 2 leads students to discover a method for estimating the median on a histogram. In question 4, students learn what an outlier is and that the mean is more affected by outliers than the median.

Students Acquiring English

You may wish to pair students acquiring English with English-proficient partners to discuss the Talk it Over questions because of the many technical terms involved.

Multicultural Note

Dorothy Crowfoot Hodgkin was born in Cairo, Egypt in 1910 and educated in England. Her deciphering of the structure of vitamin B_{12}, for which she won the 1964 Nobel Prize for chemistry, was done with a method called X-ray crystallographic analysis. In this type of analysis, the planes of an atom diffract X-ray beams in different patterns as the beams pass through. Knowing the structure of vitamin B_{12} has given scientist a greater understanding of the body's production of red blood cells, which in turn has helped them prevent pernicious anemia. In addition to determining the structures of vitamin B_{12} and insulin, Crowfoot Hodgkin determined the molecular structures of cholesterol iodide and penicillin.

A histogram is a graph of a **frequency distribution.** A frequency distribution can be categorized by the general shape of its histogram. Notice where the mean and median fall on each histogram.

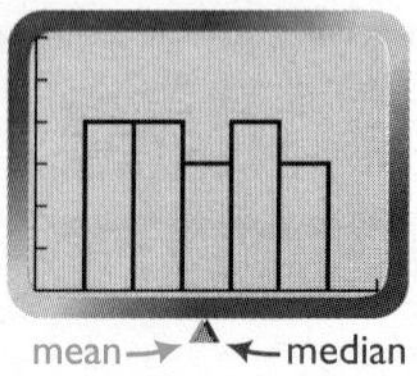

In a **uniform distribution,** all the intervals have approximately equal frequencies.

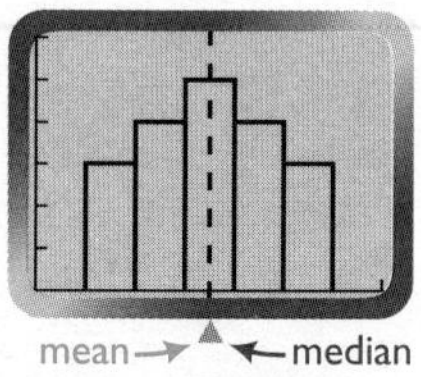

A **mound-shaped distribution** is approximately symmetrical about a line passing through the interval or between the intervals with the greatest frequency.

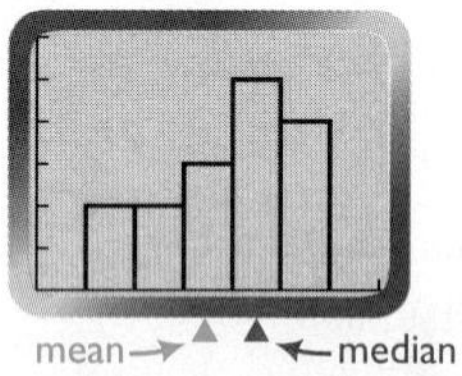

In a **skewed distribution,** the interval or group of intervals that contains the greatest frequencies is near one end.

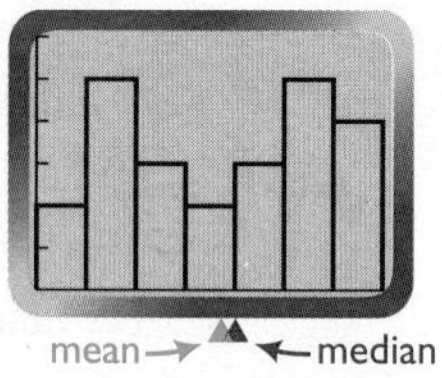

A **bimodal distribution** has two distinct intervals or groups of intervals that contain the greatest frequencies.

Student Resources Toolbox
p. 626 *Mean, Median, and Mode*

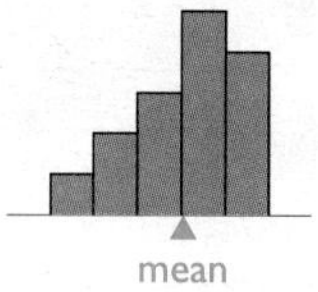

Now you are ready for:
Exs. 1–9 on pp. 335–336

Talk it Over

1. Tell whether the shape of each of your survey histograms is most like a *uniform, mound-shaped, skewed,* or *bimodal* distribution or a combination of these.
2. The mean of a data set would fall at the "balance point" of its histogram if the bars had weight. This is not true for the median. How you can estimate where a median will be on a histogram?
3. The mean and median are called *measures of central tendency* of a data set. Why do you think this is so?
4. In a data set, an **outlier** is a value whose distance from the center of the data is much greater than the distances for other data values.
   a. How do you think an outlier appears on a histogram? Does a histogram always show outliers?
   b. Will outliers have a greater effect on the *mean* or on the *median* of a data set? Why?

Answers to Talk it Over

1. Answers may vary. An example is given. The telephone-digit data set shows a uniform distribution.
2. Estimate the total heights of all the bars and divide by 2. The median will be in the bar that contains the item of data that corresponds to this quotient.
3. Answers may vary. An example is given. I think measures of central tendency describe how numbers relate to a central, or average, value.
4. a. An outlier might show as a bar separated from the rest of the bars. If the intervals are large enough, the histogram might not show the outliers at all.
   b. The median is determined by the numerical order of the numbers in a data list, not by how large or small they are, so an outlier has no effect on the median. For example, for the data 3, 5, 6, 7, 7, 7, 8, 12, if the last data point were 1750 instead of 12, the median would still be 7. Since the outlier may significantly change the sum of the data, the mean would be affected.

Sample 1

Blood sugar readings: July

before breakfast						
129	78	161	150	102	167	132
90	113	218	63	170	124	193
216	68	93	134	163	240	209
101	190	196	78	104	122	224
140	106	120				
before lunch						
56	190	191	213	41	57	49
119	71	96	41	51	40	40
57	44	114	47	44	83	157
64	100	78	51	54	138	61
66	39	133				

Tricia Barry has diabetes. She tries to keep her blood glucose (sugar) level at 80 to 120 mg of glucose per deciliter of blood, which is the normal level. Her diabetes is considered in poor control if her glucose level consistently goes below 40 or above 240.

a. Draw a histogram for each data set and identify the type of distribution.

b. Compare where the mean and median of the data fall on each histogram.

c. **Writing** Make a conclusion based on the histograms.

Sample Response

a. Tricia needs to know how often her readings are below 40 or above 240. The lowest possible reading is zero, so let the intervals be 0–39, 40–79, 80–119, and so on until the last interval includes the highest reading. Count the frequencies of the readings for each interval.

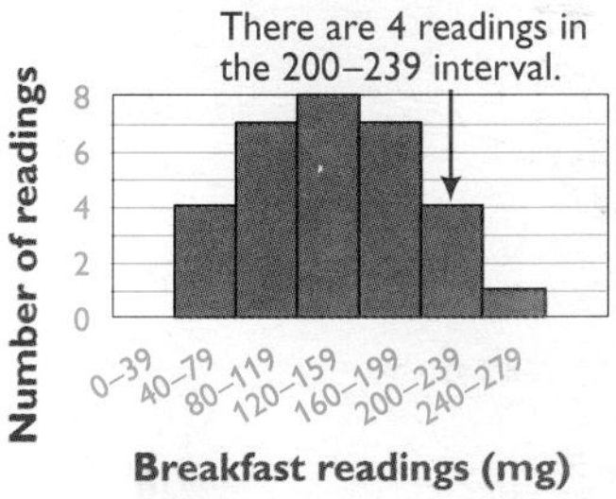

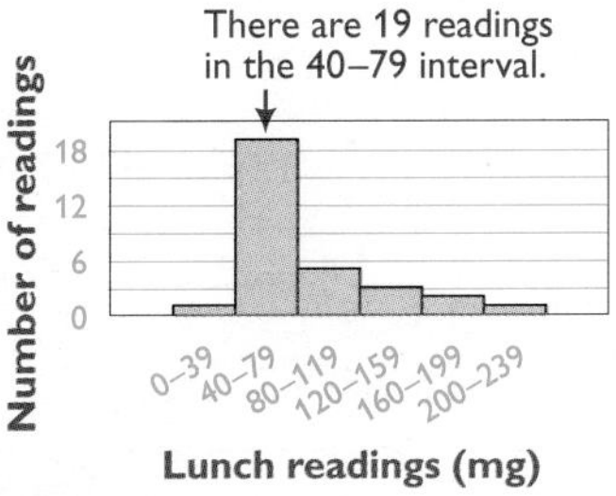

The breakfast distribution is basically mound-shaped.
The lunch distribution is skewed.

b.

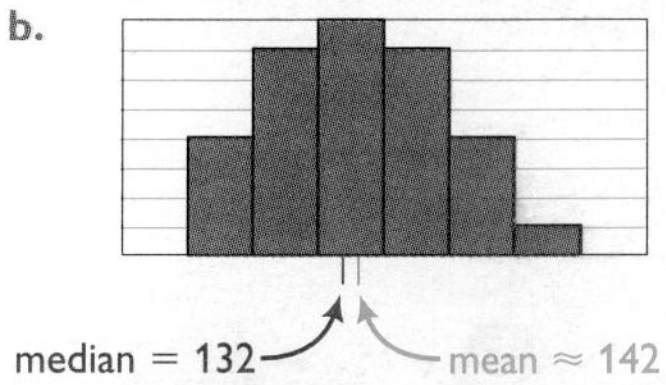

The mean and median of the breakfast data are close together and fall within the same interval.

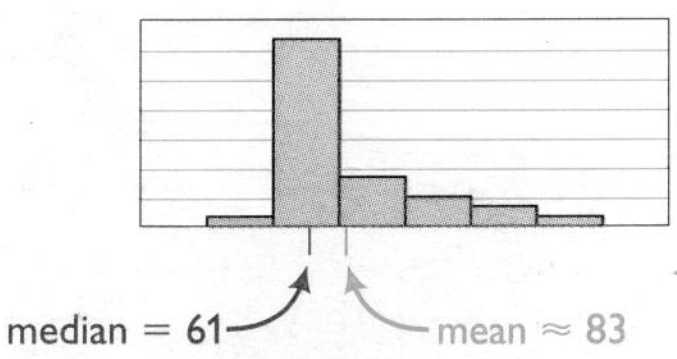

The mean and median of the lunch data are farther apart and fall in separate intervals.

c. Tricia's blood glucose readings tend to be higher than normal at breakfast and lower than normal at lunch. However, since she has gone below 40 only once and above 240 only once at either meal time, her diabetes seems in good control.

BY THE WAY. . .

The structure of insulin, the drug used to control diabetes, was discovered in 1969 by Dorothy Crowfoot Hodgkin. She won the Nobel Prize in 1964 for discovering the structure of vitamin B_{12}.

Additional Sample

S1 George Maynard has high blood pressure. He must try to keep his systolic pressure between 110 and 140 mmHg and his diastolic pressure between 75 and 90 mmHg. The doctor does not want his blood pressure to go over 150/95.

Blood Pressure Readings, March

Systolic:

118 146 132 128 121 125
134 117 121 138 142 96
128 113 127 122 134 140
167 158 139 128 118 121
132 114 126 118 137 132

Diastolic:

83 85 89 87 82 84 85 90 78
77 82 84 85 80 78 85 86 90
98 96 95 89 88 87 82 78 80
81 78 80

a. Draw a histogram for each data set and identify the type of distributions. **George needs to know how often his readings are above 150 and above 95. Count the frequencies of the reading for each interval. (See graphs below.) The systolic distribution is basically mound-shaped. The diastolic distribution is slightly skewed.**

b. Compare where the mean and median of the data fall on each histogram. **Systolic: mean = 129, median = 128; The mean and median are close together and fall within the same interval. Diastolic: mean = 84.7, median = 84.5; The mean and median are close together and fall within the same interval.**

c. Make a conclusion based on the histograms. **There were only 2 systolic readings over 150 and 2 diastolic readings over 95. George's blood pressure seems to be under control. The only high readings for both occurred at the same time, which may have been due to an extenuating circumstance.**

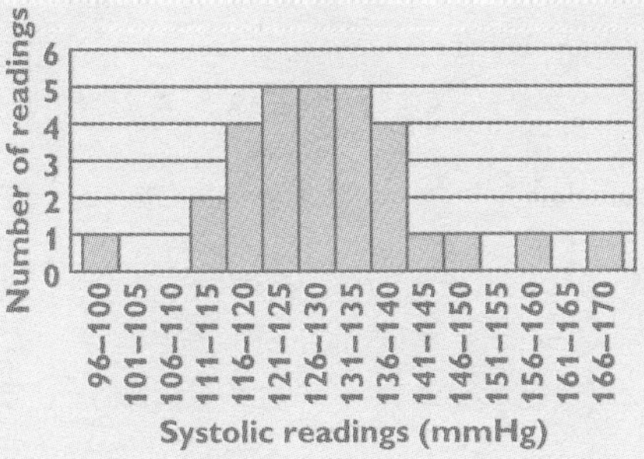

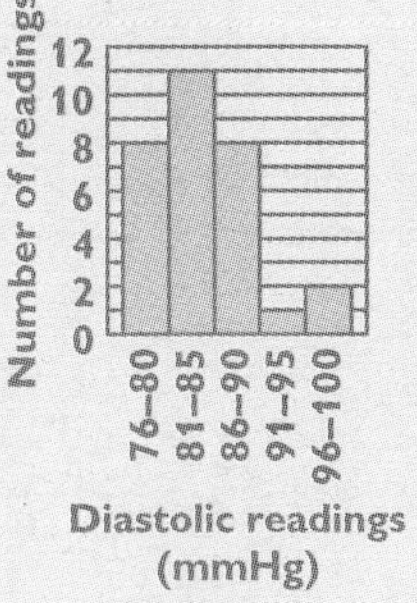

Additional Sample

S2 Draw a relative frequency histogram for each data set in Additional Sample S1.

There are 30 readings for each blood pressure. Divide the number of readings in the interval by the total number of readings, then multiply by 100 to find the relative frequency as a percent.

Systolic:

Interval	No. of readings	Percent of total readings
96–100	1	3.3
101–105	0	0
106–110	0	0
111–115	2	6.7
116–120	4	13.3
121–125	5	16.7
126–130	5	16.7
131–135	5	16.7
136–140	4	13.3
141–145	1	3.3
146–150	1	3.3
151–155	0	0
156–160	1	3.3
161–165	0	0
166–170	1	3.3
Total	30	99.9

Diastolic:

Interval	No. of readings	Percent of total readings
76–80	8	26.7
81–85	11	36.7
86–90	8	26.7
91–95	1	3.3
96–100	2	6.7
Total	30	100.1

(See graphs below.)

Look Back

You may wish to have a student volunteer answer this Look Back in class. Students might also discuss how a relative frequency histogram can be made from a frequency histogram.

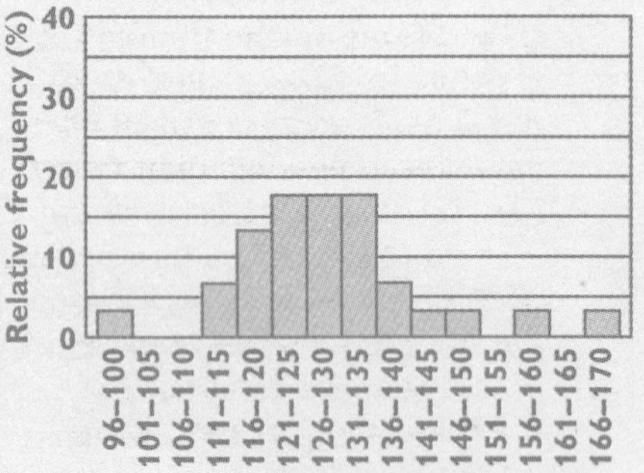

Systolic readings (mmHg)

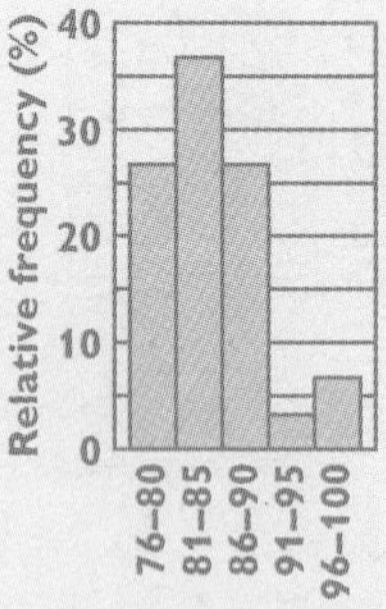

Diastolic readings (mmHg)

Relative Frequency Histograms

The scale on the vertical axis of each histogram in Sample 1 tells you how many data values are in each interval. In a **relative frequency histogram,** the scale on the vertical axis tells you what fraction or what percent of the data values are in each interval.

Sample 2

Draw a relative frequency histogram for each data set in Sample 1.

Sample Response

There are 31 readings for each meal time. Divide the number of readings in the interval by the total number of readings, then multiply by 100 to find the relative frequency as a percent.

INTERVAL	BREAKFAST Number of readings	BREAKFAST Percent of total readings	LUNCH Number of readings	LUNCH Percent of total readings
0–39	0	0	1	3.2
40–79	4	12.9	19	61.3
80–119	7	22.6	5	16.1
120–159	8	25.8	3	9.7
160–199	7	22.6	2	6.5
200–239	4	12.9	1	3.2
240–279	1	3.2	0	0
Total	31	100	31	100

$4 \div 31 \times 100 \approx 12.9$

12.9% of the readings are in the 40–79 interval.

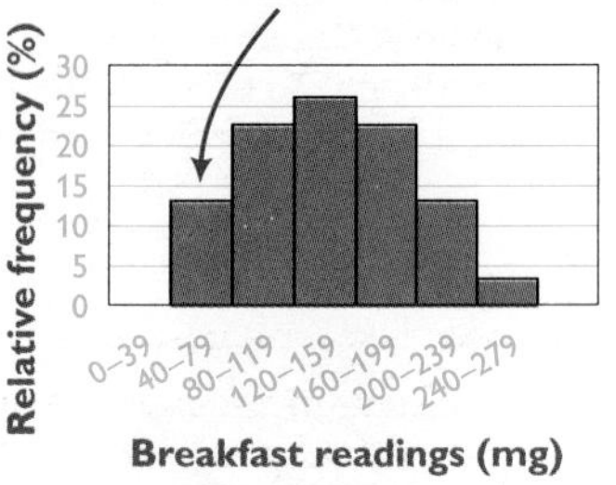

Breakfast readings (mg)

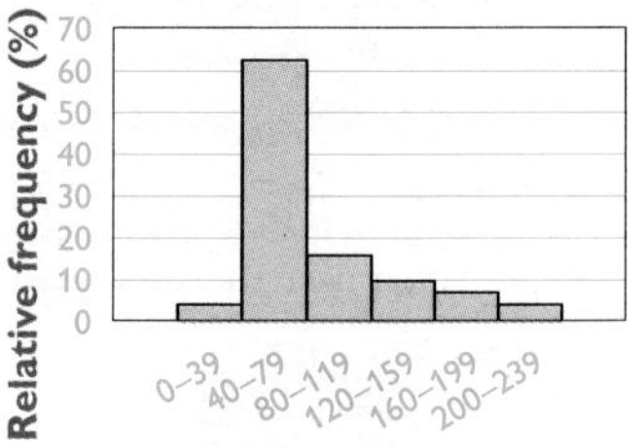

Lunch readings (mg)

Talk it Over

5. Compare the histograms in Sample 2 with those in Sample 1. How are they alike? How are they different?
6. What percent of breakfast readings were 160 or higher? What percent of lunch readings were 159 or lower?

Now you are ready for:
Exs. 10–18 on pp. 336–337

Look Back

Explain how to make a relative frequency histogram.

6-1 Exercises and Problems

1. **Reading** How is a skewed distribution different from a mound-shaped distribution?
2. **Writing** Explain how you chose the intervals for the histograms in the Exploration on page 331.
3. Use the data in Sample 1. Change the size of the intervals and draw a new histogram for the breakfast or the lunch data. What do you notice about the shape of the new histogram?

For Exercises 4–7, use these histograms based on a survey given to a high school junior class.

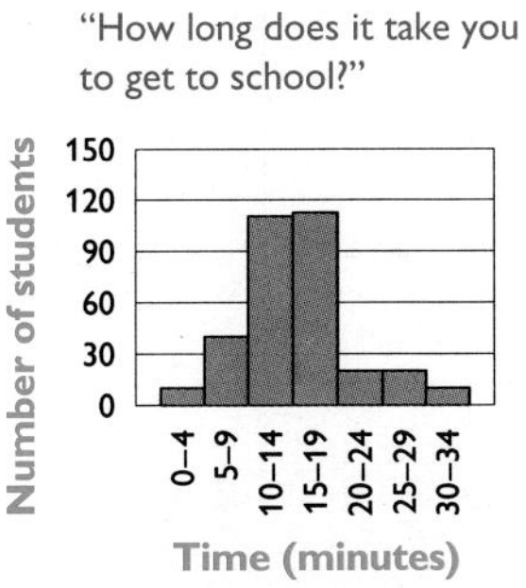

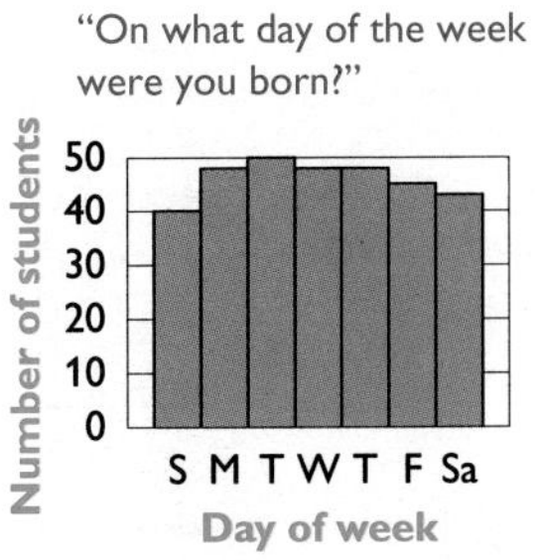

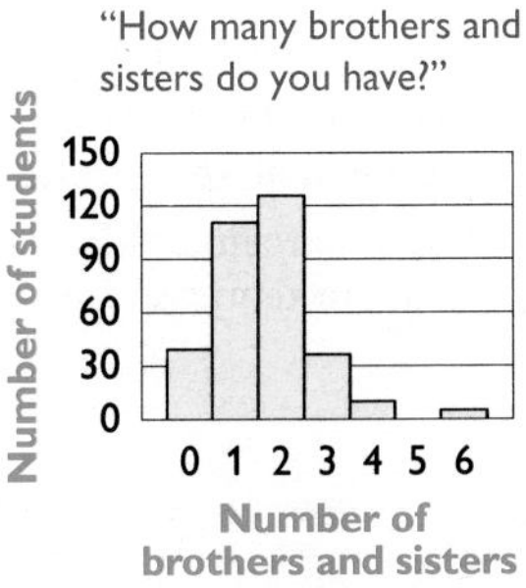

4. About how many students were surveyed?
5. a. About how many students were born on each weekday? on each weekend day?
   b. Does the histogram for birthdays look the way you expected? Why or why not?
6. Do most of these students have many brothers and sisters? Explain.
7. **Writing** What do you think is the average number of minutes it takes these students to get to school? What do you think is the average number of brothers and sisters? Explain your reasoning.
8. a. **Ornithology** Draw a histogram of the bird nest temperature data.
   b. Identify the type of distribution. Justify your answer.

The air temperature in the nest of a common tern is 36.0°C.

Air Temperature in Birds' Nests

Temperature (°C)	Number of species
30.0–30.9	1
31.0–31.9	2
32.0–32.9	3
33.0–33.9	7
34.0–34.9	7
35.0–35.9	9
36.0–36.9	3
37.0–37.9	1

APPLYING

Suggested Assignment

Day 1
Standard 1–6, 9
Extended 1–9

Day 2
Standard 10–18
Extended 10–18

Integrating the Strands

Number Ex. 17
Algebra Exs. 15, 16
Functions Ex. 16
Statistics and Probability Exs. 1–14, 18
Discrete Mathematics Exs. 1–14, 17, 18
Logic and Language Exs. 1, 2–6, 9, 10, 12–14, 18

Communication: Writing

Exs. 2 and 7 provide students with an opportunity to express their ideas in writing. For Ex. 7, after students have written their reasoning, you may wish to call upon two or three students to read their explanations to the class.

Application

Histograms can be applied to many different kinds of real-world situations as is illustrated by Exs. 4–14. These exercises use histograms to analyze data involving junior high school students, ornithology, language, baseball salaries, doctors' salaries, and voting and registration patterns of different groups of people.

Answers to Talk it Over

5. Each histogram in Sample 2 is identical to the corresponding one in Sample 1 except for the scale on the vertical axis.
6. 38.7% (about 40%); 90.3% (about 90%)

Answers to Look Back

Summaries may vary. Choose reasonable intervals for the data and count the frequencies for each interval. Determine how many numbers are in the data set, divide each frequency by the total, and multiply by 100 to determine the relative frequency. Use the relative frequencies to choose an appropriate scale for the vertical axis.

Answers to Exercises and Problems

1. In a skewed distribution, the interval or intervals that contain the greatest frequencies are at one end. In a mound-shaped distribution, the interval with the greatest frequency is in the middle of the histogram.
2. Answers may vary. An example is given. For the question involving month of birth, I used a separate interval for each month. For the question involving telephone numbers, I used a separate interval for each possible digit from 0 to 9. For the question involving number of letters in a first name, I used a separate interval for each number of letters, ranging from 2 up to the highest response. For the heart rate question, I used intervals of 10 heart beats.

3–8. See answers in back of book.

Research

In Ex. 9, histograms of word lengths are used to study language. Histograms involving words can also be used to study the authorship of certain written documents. Some students may wish to read the article "Deciding Authorship" by Frederick Mosteller and David Wallace in the book *Statistics: A Guide to the Unknown*, published by Holden-Day Co.

Using Technology

For Ex. 12, students can use the TI-82 to help sort the data. The process is simple. Press STAT and select item 4 (ClrList) on the EDIT menu. On the home screen, after ClrList, type L_1,L_2. Press ENTER. This ensures that you have cleared space for entering data. Press STAT and select item 1 (Edit) on the EDIT menu. Enter the data for the National League as list L_1. Enter the data for the American League as list L_2. Press STAT and select item 2 (Sort A( ) on the EDIT menu. (Note: On this menu, A stands for *ascending* and D stands for *descending.*) On the home screen, after Sort A(, type L_1,L_2). Press ENTER. The calculator sorts each list in ascending order. If you return to examine the data, they are now easy to group in whatever way you wish in order to make a frequency table and to decide on intervals.

Assessment: Open-ended

For Ex. 13, students should be able to find a way to compare the two sets of data on one graph. One possible way would be to construct a frequency polygon for each set of data by connecting the midpoints of the bars of the histograms. They could then show each frequency polygon on one graph. Students should be able to write at least three conclusions based on the salary groupings about the comparative salaries of female and male doctors.

9. a. Draw a histogram for the lengths of words in English. Draw a histogram for the lengths of words in Italian.
   b. Identify the type of distribution for each histogram. Justify your answers.
   c. Find the mean and median for each data set and show them on the histograms.
   d. **Open-ended** What can you say about the lengths of words used in English and Italian?

10. **Reading** How is a histogram different from a relative frequency histogram?

11. a. Make a relative frequency histogram for the data in Exercise 8.
    b. What percent of bird species have nests with air temperatures of 35.0°C or greater?

Lengths of words in the first paragraph of *At Bertram's Hotel* by Agatha Christie

Number of letters	Number of words English	Italian
1	1	3
2	5	4
3	6	8
4	7	4
5	4	6
6	6	4
7	4	4
8	2	3
9	1	0
10	0	2
11	0	0
12	1	1
13	0	1

1993 Average Annual Baseball Players' Salaries by Team

National League	1,453,538	1,371,692	1,584,302
	327,926	653,484	1,072,981
	1,448,200	543,401	1,431,099
	1,039,897	850,580	877,048
	1,264,315	935,501	
American League	1,078,359	1,329,414	923,511
	1,304,747	548,156	1,331,414
	1,503,247	811,852	978,462
	1,578,192	1,430,493	1,052,694
	1,046,661	1,708,000	

12. a. **Salaries** Make a histogram for the data for each league.
    b. Identify each type of distribution.
    c. Estimate the mean and median of each data set and compare where they fall on the histograms.
    d. **Writing** Use the histograms to compare the salaries in the two leagues. What conclusion(s) can you make? Explain your reasoning.

13. a. **Salaries** Make a relative frequency histogram for each data set.
    b. Identify the type of distribution for each histogram.
    c. What percent of doctors in each group had an annual salary between $100,000 and $299,999 in 1992?
    d. **Writing** What conclusion(s) can you make based on the histograms? Explain your reasoning.

1992 United States Annual Doctors' Salaries

Salary ($)	Female (%)	Male (%)
0–99,999	52	25
100,000–199,999	36	43
200,000–299,999	10	21
300,000–399,999	1	6
400,000 +	1	5

Answers to Exercises and Problems

9. Answers may vary depending on intervals chosen. Examples are given.

a, c.

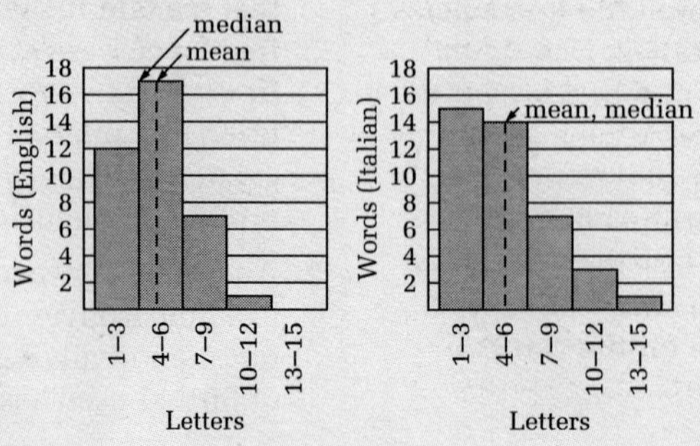

b. Both distributions are basically skewed; each has the interval with the greatest frequency at or close to one end.

c. 4.8, 4; 5.05, 5

d. Answers may vary. An example is given. I think Italian words tend to be slightly longer than English words.

10–13. See answers in back of book.

Ongoing **ASSESSMENT**

14. **Open-ended** Make at least five conclusions based on these voting and registration histograms. Justify your conclusions.

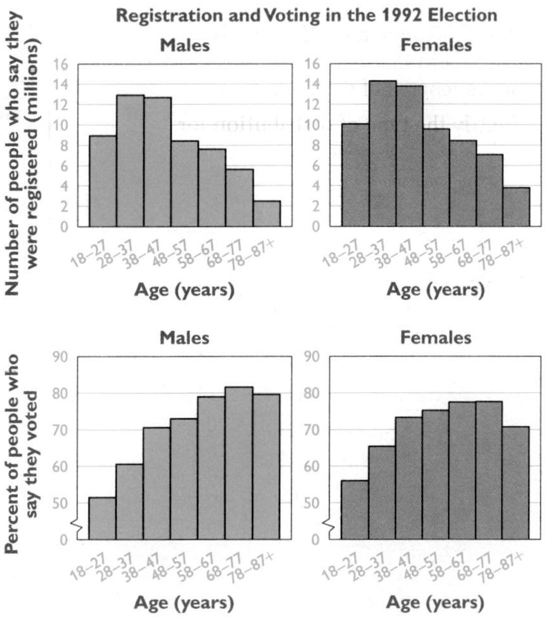

Review **PREVIEW**

15. Solve $4^x = 20$. *(Section 5-8)*

16. Tell whether the inverse of $y = -\frac{1}{2}x^2$ exists when $x \geq 0$. If it exists, write an equation of the inverse function. *(Section 5-5)*

17. Evaluate $\sum_{n=0}^{6} (3 - 4n)$. *(Section 4-5)*

Working on the Unit Project

18. a. Create a survey to give to 30 people. Ask them to suppose they plan to buy a new car that suits their present lifestyle and is priced within their present budget. Here are some questions you can ask.
   - ➤ How much money will you pay for the car?
   - ➤ How many miles per gallon of gasoline will it average?
   - ➤ About how many miles per year will you drive it?
   - ➤ What color will it be?
   - ➤ How many doors will it have?
   - ➤ How many options and safety features will it have? What are they?
   - ➤ How long do you plan to keep the car?

   b. Make a histogram for the responses to as many survey questions as you can. Can you make a histogram for every set of responses? Explain.

   c. Identify the type of distribution for each histogram in part (b). Show where the mean and median fall on your histograms.

Working on the Unit Project

Students can work in their project groups to create the survey and discuss how to sample. After the surveys are completed, the group can compile the results and each student can do one or two of the histograms. The group should reconvene to discuss Ex. 18(b) and (c). This information should be saved by each student for use in completing the Unit Project.

Practice 46 For use with Section 6-1

Name ______________________ Date ____________

Practice 46

For use with Section 6-1

4. a.

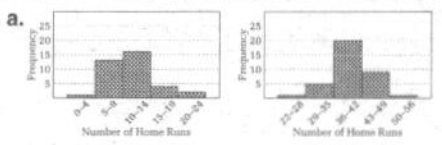

For Exercises 1–3: Exs. 1–3: For histograms in part (a), see additional answers.

a. Draw a histogram of the data.

b. Identify the type of distribution. 1. b. mound-shaped 2. b. skewed

1.

Heights of Active Volcanoes in Asia	
Height (1000 ft)	Number of Volcanoes
over 13	1
12-13	2
11-12	1
10-11	2
9-10	4
8-9	3
7-8	5
6-7	5
5-6	5
4-5	6
3-4	2
2-3	3

2.

Congressional Representation (1980)	
Number of Reps	Number of States
1-3	17
4-6	11
7-9	8
10-12	5
13-15	1
16-18	1
19-21	3
22-24	1
25-27	0
28-30	1
31-33	1
over 33	1

3.

Length of Last Names of U.S. Presidents	
Number of Letters in Name	Number of Presidents
4	4
5	6
6	9
7	8
8	8
9 or more	6

b. mound-shaped or uniform

4. a. Make a histogram for each data set of home run leaders' totals. See top of page.

b. Identify each type of distribution. both mound-shaped

c. Compare where the mean and median of the data fall on each histogram. See below.

d. What conclusion(s) can you make based on the histogram? Distributions are similar.

5. ***Open-ended*** Give some possible reasons for the difference in the distribution of home run totals for the two periods of time shown in your histograms for Exercise 4. Check students' work.

Home Run Leader's Totals (National and American Leagues)				
1901-1918	16	6	9	9
	9	12	10	12
	7	10	21	14
	19	19	24	12
	12	8	4	16
	13	10	8	12
	8	7	9	10
	11	10	12	9
	7	12	9	11
1971-1988	48	40	44	36
	38	38	52	40
	48	48	31	37
	40	36	37	37
	49	39	33	37
	32	32	36	32
	39	46	45	41
	22	39	39	43
	40	40	49	42

4. c. first: med. = 10, mean = 11.31; second: med. = 39, mean = 39.58

46 **Practice Bank**, INTEGRATED MATHEMATICS 3

Answers to Exercises and Problems

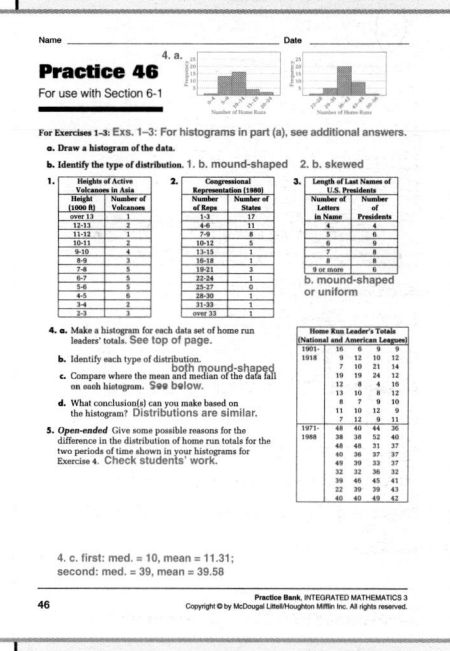

14. Answers may vary. Examples are given. The number of people, both male and female, who say they are registered decreased with age, since both histograms are skewed to the left. While the numbers are similar, slightly more women in each age category say they are registered to vote. The number of people, both male and female, who say they voted increased with age, since both of those histograms are skewed to the right. Both histograms in each category have the same shape for males and females, so the age distributions are similar for both sexes. Potential voters between 18 and 27 years of age do not take their right to vote very seriously, since less than 10% are registered to vote, and only about 50% of those registered do actually vote.

15. about 2.16

16. No.

17. –63

18. a. Answers may vary.

   b. Check students' work. Some responses are not numerical and, thus, cannot be displayed in a histogram.

   c. Check students' work.

PLANNING

Objectives and Strands
See pages 328A and 328B.

Spiral Learning
See page 328B.

Materials List
- Calculator
- Graph paper

Recommended Pacing
Section 6-2 is a one-day lesson.

Extra Practice
See pages 617–619.

Warm-Up Exercises
Warm-Up Transparency 6-2

Support Materials
- Practice Bank: Practice 47
- Activity Bank: Enrichment 42
- Study Guide: Section 6-2
- Problem Bank: Problem Set 13
- Explorations Lab Manual: Diagram Master 1
- Using Plotter Plus: Exploring the Standard Deviation
- Assessment Book: Quiz 6-2, Alternative Assessment 2

Section 6-2 Standard Deviation

Focus
Use mean, range, and standard deviation to compare data sets with technology.

How Far Off?

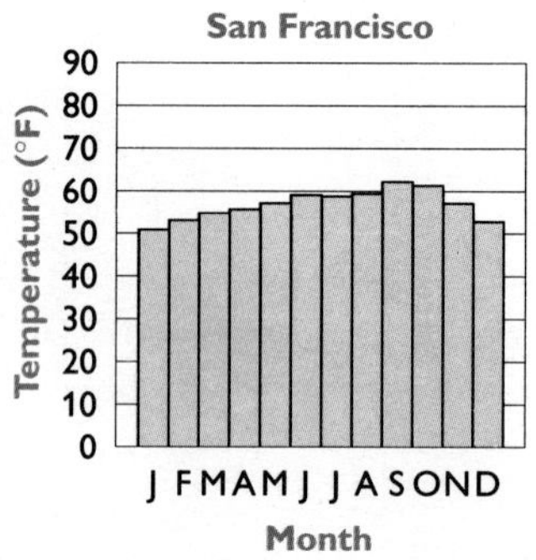

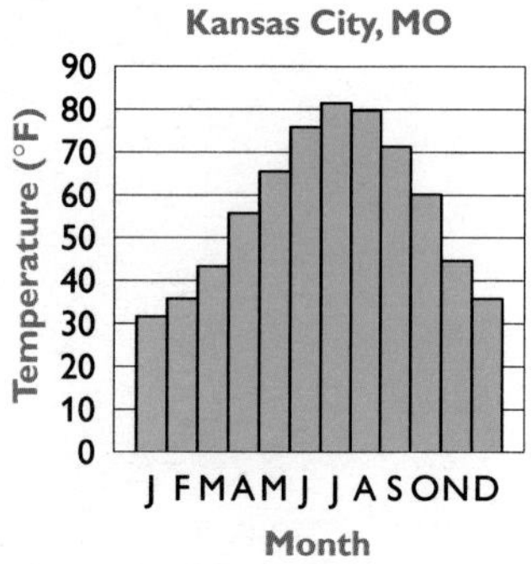

Monthly Mean Temperatures (°F)												
	J	F	M	A	M	J	J	A	S	O	N	D
San Francisco	50.7	53.0	54.7	55.7	57.4	59.1	58.8	59.4	62.0	61.4	57.4	52.5
Kansas City, MO	31.7	35.8	43.3	55.7	65.6	75.9	81.5	79.8	71.3	60.2	44.6	35.8

Talk it Over

1. What do the graphs show about the temperature pattern in each city?
2. Find the mean annual temperature for each city. Is the mean alone a good way to compare the temperatures in these cities? Explain.
3. The **range** of a data set is the difference between the greatest and least data values. What is the range of temperatures for each city?
4. Are the mean and the range a good way to compare the temperatures in these cities? Explain.
5. When you subtract the mean from a data value, you find the value's *deviation from the mean.* For both cities, find the deviation from the mean for each temperature. What is the average of these deviations?

For example, in the San Francisco data, the deviation from the mean for February is 53.0 − 56.8 = −3.8.

Answers to Talk it Over

1. The San Francisco distribution is fairly uniform, while the Kansas City distribution is basically mound-shaped.
2. about 56.8° for both cities; No. Although the histograms show that the monthly mean temperatures for the two cities are very different, the mean annual temperature is the same for both cities.
3. San Francisco: 11.3°; Kansas City: 49.8°
4. Answers may vary. An example is given. I think the mean and range together are better than the mean alone. Given both, I can tell that there is a greater variation in the temperature in Kansas City.
5. San Francisco: −6.1, −3.8, −2.1, −1.1, 0.6, 2.3, 2, 2.6, 5.2, 4.6, 0.6, −4.3; Kansas City: −25.1, −21, −13.5, −1.1, 8.8, 19.1, 24.7, 23, 14.5, 3.4, −12.2, −21; San Francisco average: 0; Kansas City average: 0

The mean is one statistic you can use to describe a data set, but it describes only the center. The range is an example of a *measure of variability* because it measures how variable or spread out the data are. Another measure of variability uses the square of the deviation of each data value from the mean.

STANDARD DEVIATION

Read as "x bar." ⟶ For a data set $x_1, x_2, x_3, \ldots, x_n$, where $\bar{x}$ = the mean of the data values and n = the number of data values, the **standard deviation** σ is

Read as "sigma." ⟶

$$\sigma = \sqrt{\frac{(x_1 - \bar{x})^2 + (x_2 - \bar{x})^2 + \cdots + (x_n - \bar{x})^2}{n}}$$

$$= \sqrt{\frac{\sum_{i=1}^{n} (x_i - \bar{x})^2}{n}}$$

To find the standard deviation:

1. Find the mean, $\bar{x}$.
2. Find the deviations from the mean, $x_i - \bar{x}$.
3. Find the squares of the deviations, $(x_i - \bar{x})^2$.
4. Find the sum of the squares, then divide by the number of data values, n.
5. Take the positive square root.

Sample 1

Find the standard deviation of the monthly mean temperatures in San Francisco.

Sample Response

Method ❶ Use a formula.

standard deviation ⟶ $\sigma = \sqrt{\frac{\sum_{i=1}^{12} (x_i - 56.8)^2}{12}}$ mean: $\bar{x} = \frac{50.7 + 53.0 + \cdots + 52.5}{12} \approx 56.8$

$$= \sqrt{\frac{(50.7 - 56.8)^2 + (53.0 - 56.8)^2 + \cdots + (52.5 - 56.8)^2}{12}}$$

≈ 3.4

Method ❷ Use a calculator.

Put your calculator in statistics mode or select the statistics menu. Enter the data. Have the calculator compute and display the mean, usually shown as $\bar{x}$, and the standard deviation, usually σ_x or σ_n.

1–Var Stats
$\bar{x}$ = 56.84166667
Σx = 682.1
Σx^2 = 38912.41
Sx = 3.576554547
σx = 3.424290275
↓n = 12

TEACHING

Talk it Over

Questions 1–5 lead students to discover the limitations of the mean alone in describing a set of data. Students learn that the range and the deviation from the mean can be used with the mean to get a better understanding of the data.

Communication: Reading

It is important for students to be able to read complex mathematical expressions such as the one for standard deviation. You may want to pair students and have them read the expression aloud to each other, making sure they use the proper names for the various symbols.

Additional Sample

S1 Find the standard deviation of the price in dollars of the same bag of groceries being sold at ten different supermarkets.

19.10	17.60	18.30	17.50
18.10	20.30	21.00	19.80
20.20	20.60		

Method 1. See below.

Method 2. Use a calculator. Put your calculator in statistics mode or select the statistics menu. Enter the data. Have the calculator compute and display the mean, usually shown as $\bar{x}$, and the standard deviation, usually σ_x or σ_n.

1-Var Stats
$\bar{x}$ = 19.25
Σx = 192.5
Σx^2 = 3720.85
Sx = 1.300640868
σx = 1.233896268
↓n = 10

Method 1: Use a formula.

mean:

$$\bar{x} = \frac{19.10 + 17.60 + \ldots + 20.60}{10} = 19.25$$

$$\sigma = \sqrt{\frac{\sum_{i=1}^{10}(x_i - 19.25)^2}{10}} =$$

$$\sqrt{\frac{(19.10 - 19.25)^2 + (17.60 - 19.25)^2 + \ldots + (20.60 - 19.25)^2}{10}} \approx 1.2$$

Talk it Over

In question 7, students will see that although the standard deviations for the two data sets are quite different, the percentage of data that lies within one standard deviation from the mean in each case is fairly similar.

Multicultural Note

The English name *popcorn* derives from the Middle English word *poppe*, meaning "explosive sound," but popcorn was unknown by Europeans until they arrived in the Americas. Popcorn had been a part of life in the Americas for at least five thousand years. Native Americans developed a variety of ways to pop corn, and popcorn was used as food, as a material for jewelry, and even in amulets for Aztec religious ceremonies. Hernando Cortés was given popcorn by the Aztecs in 1510, and at their first Thanksgiving dinner in 1621, the pilgrims ate popcorn given to them by their Wampanoag neighbors.

Additional Sample

S2 A company that makes raisin bread advertises that in each slice of bread there are at least 10 raisins. Sample loaves are tested 20 times per day, and a slice of bread from each loaf is examined. If there are less than 8 raisins in the slice, the loaf is considered not salable. How is each baker performing this day?

Number of raisins per slice

Baker 1
8 12 10 13 14 7 11 8 15 10 14 6 8 11 13 14 10 12 4 8

Baker 2
10 12 13 8 13 14 5 12 10 9 14 11 10 9 11 10 12 8 9 14

Find the mean and standard deviation for the number of raisins per slice of bread from each baker.

Calculations for Kansas City Temperatures

1–Var Stats
$\bar{x} = 56.76666667$
$\Sigma x = 681.2$
$\Sigma x^2 = 42316.9$
$Sx = 18.20950468$
$\sigma x = 17.4342734$
$\downarrow n = 12$

Talk it Over

6. Which city has more variability in temperature? Explain.
7. For both cities, find $\bar{x} + \sigma$ and $\bar{x} - \sigma$. What *percent* of temperatures in each city were between these values? These values are within one standard deviation of the mean.

Standard deviation is used in business for quality control. Machines that manufacture products are designed to minimize error, but their performance must be tested often.

Sample 2

Writing Puff Pop Company produces microwave popcorn and has two machines that package the kernels. Sample packages are collected from each machine 20 times per day. The popcorn from each package is popped. If there are more than 15 unpopped kernels, the package is considered not salable. How is each machine performing this day?

Number of unpopped kernels

Machine 1	9, 8, 10, 12, 12, 10, 11, 16, 11, 13, 11, 9, 10, 14, 11, 14, 12, 15, 11, 15
Machine 2	8, 9, 9, 14, 15, 9, 8, 10, 18, 9, 10, 16, 18, 10, 8, 17, 20, 10, 8, 12

Sample Response

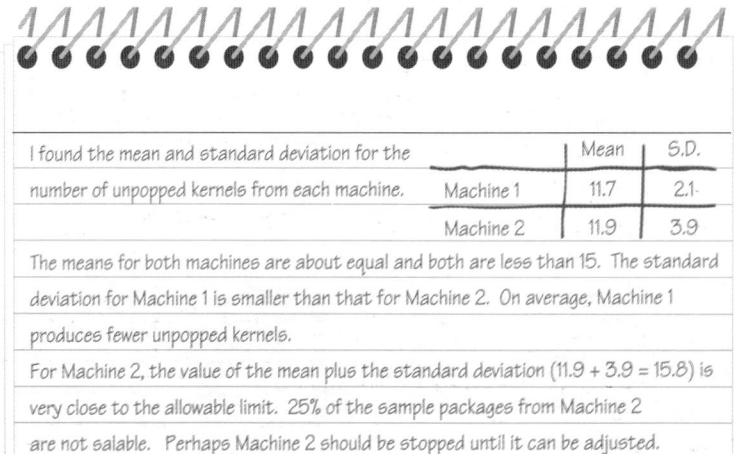

I found the mean and standard deviation for the number of unpopped kernels from each machine.

	Mean	S.D.
Machine 1	11.7	2.1
Machine 2	11.9	3.9

The means for both machines are about equal and both are less than 15. The standard deviation for Machine 1 is smaller than that for Machine 2. On average, Machine 1 produces fewer unpopped kernels.

For Machine 2, the value of the mean plus the standard deviation (11.9 + 3.9 = 15.8) is very close to the allowable limit. 25% of the sample packages from Machine 2 are not salable. Perhaps Machine 2 should be stopped until it can be adjusted.

Machine 1

1–Var Stats
$\bar{x} = 11.7$
$\Sigma x = 234$
$\Sigma x^2 = 2830$
$Sx = 2.202868943$
$\sigma x = 2.147091055$
$\downarrow n = 20$

Machine 2

1–Var Stats
$\bar{x} = 11.9$
$\Sigma x = 238$
$\Sigma x^2 = 3138$
$Sx = 4.011824628$
$\sigma x = 3.910242959$
$\downarrow n = 20$

Answers to Talk it Over

6. Kansas City; The standard deviation for Kansas City, about 17.4, is greater than that for San Francisco, about 3.4. Also, the range for Kansas City, 49.8, is greater than the range for San Francisco, 11.3.

7. San Francisco: 60.2, 53.4; 58.3% of the temperatures are within one standard deviation of the mean; Kansas City: 74.2, 39.4; 50% of the temperatures are within one standard deviation of the mean.

Answers to Look Back

The mean describes only the center of a data set. The range and standard deviation describe how spread out the data are. You might want to use measures of variability to standardize products and for quality control.

	Mean	SD
Baker 1	10.4	2.96
Baker 2	10.7	2.30

The means are about equal and both are greater than 10. The standard deviation for Baker 1 is greater than that for Baker 2. On average, Baker 2 produces fewer slices of bread with less than 8 raisins. For Baker 1, just one standard deviation below the mean is under the eight raisin limit. Baker 1 might stir the batter a little more in order to more evenly distribute the raisins throughout the loaf.

Look Back

Why might you want to use other statistical measures besides the mean to compare data sets?

6-2 Exercises and Problems

1. **Reading** How many data items do you use to find the range? How many data items do you use to find the standard deviation?
2. a. Find the mean of the credit card interest rates in each state.

Interest Rates of Credit Cards Issued by New Hampshire and Iowa Banks

NH	10.9,	13.9,	14.4,	14.4,	14.4,	14.5,	14.9,	14.9,	14.9,	14.9,	17.9,	17.9		
IA	13.4,	13.4,	13.4,	13.4,	13.4,	13.4,	13.4,	15.0,	15.0,	15.0,	15.0,	15.0,	15.0,	15.0

b. Find the range of the credit card interest rates in each state.

c. **Writing** Suppose you know only the mean and the range of the data. In which state would you expect to find the bank offering the lowest interest rate? Explain.

For Exercises 3–5:

a. Find the mean. **b. Find the range.** **c. Find the standard deviation.**

d. Find the percent of the data within one standard deviation of the mean.

3. 1984–1993 U.S. Women's Open winning golf scores 290, 280, 287, 285, 277, 278, 284, 283, 280, 280
4. 1984–1993 U.S. Men's Open winning golf scores 276, 279, 279, 277, 278, 278, 280, 282, 285, 272
5. Length of 1991 and 1992 Space Shuttle flights (days) 8, 6, 9, 9, 7, 8, 8, 8, 13, 7, 7, 9, 7
6. **Writing** Compare the means and ranges of the data sets in Exercises 3 and 4. What conclusion(s) can you make?
7. a. **Demographics** Find the mean and the standard deviation of these two data sets. What conclusion(s) can you make?

b. Find the mean and the standard deviation of the South Atlantic data set *without* the outlier. Compare these new values to the mean and the standard deviation of the New England data set. What conclusion(s) can you make? Are they the same conclusion(s) you made in part (a)?

c. **Writing** Discuss the effects of an outlier on the mean and the standard deviation of a data set.

Number of Physicians per 10,000 People in 1992

South Atlantic Region		New England Region	
Delaware	22.3	Maine	20.9
Maryland	34.0	New Hampshire	20.8
Washington, D.C.	64.5	Vermont	26.4
Virginia	21.7	Massachusetts	34.1
West Virginia	19.1	Rhode Island	27.5
N. Carolina	19.8	Connecticut	31.2
S. Carolina	17.3		
Georgia	18.4		
Florida	22.2		

Look Back

A response to this Look Back can be placed in students' journals. A few students may wish to share their responses with the class.

APPLYING

Suggested Assignment

Standard 1–6, 10–14

Extended 1–14

Integrating the Strands

Algebra Exs. 9, 12

Statistics and Probability Exs. 1–11, 13, 14

Discrete Mathematics Exs. 1–11, 13, 14

Logic and Language Exs. 1, 2, 6–8, 10, 14

Assessment: Investigation

For the data in Ex. 7, students should investigate the median as a measure of central tendency and discuss how outliers affect this measure.

Answers to Exercises and Problems

1. 2, the least and the greatest; all the data items
2. Answers are given to the nearest tenth.
   a. New Hampshire: 14.8; Iowa: 14.2
   b. New Hampshire: 7; Iowa: 1.6
   c. Answers may vary. An example is given. I think the bank offering the lowest rate is in New Hampshire. The mean is lower for Iowa, but the range is greater for New Hampshire, so I think the lowest rate for New Hampshire is probably lower than the lowest rate for Iowa.
3. a. 282.4 b. 13 c. 3.9 d. 60%
4. a. 278.6 b. 13 c. 3.3 d. 80%
5. a. 8.2 b. 7 c. 1.7 d. 84.6%
6. Conclusions may vary. Men's and women's scores have the same range, but, on average, the men's scores are lower.
7. See answers in back of book.

Application

Ex. 8 illustrates how a manufacturing business might use mean and standard deviation in quality control.

Visual Thinking

Have students work in groups to consider a real business application of standard deviation. The applications might involve quality control or cost deviations. They might be business applications within the school, such as sales of items at sporting events, or from the outside world. Ask students to collect enough data to create a histogram. Encourage them to discuss their histograms with the class, describing the mean, the range, the standard deviation, and the real-world meaning of the data. This activity involves the visual skills of *interpretation* and *communication*.

Career Note

Quality control of new products is an extremely important issue in today's international world of business. Companies that manufacture products that do not meet the standards of consumers quickly find out their competitors are able to take business away from them. Successful products meet the needs of consumers by being trouble free. Thus, within every business, there are individuals who are responsible for establishing quality control procedures and seeing to it that the procedures are followed. Very often these people have backgrounds in engineering or in industrial management. They understand the business, its customers, and how to design and build quality products.

Communication: Discussion

Students should discuss the results of Ex. 10 in order to make clear and reinforce the effects of transforming data.

8. **Quality Control** Many companies test a sample of their product several times a day in order to detect a problem as soon as possible. The companies may use a tool called a *quality control chart.* The diameter of a penny must be between 18.95 and 19.15 mm. The inspector at a United States Mint measures the diameters of four pennies to the nearest 0.0001 mm about three times a day to see on average how much the diameters vary.

Newly minted pennies being checked for size at the United States Mint

a. Complete the table by finding the mean diameter of these samples to the nearest 0.0001 mm.

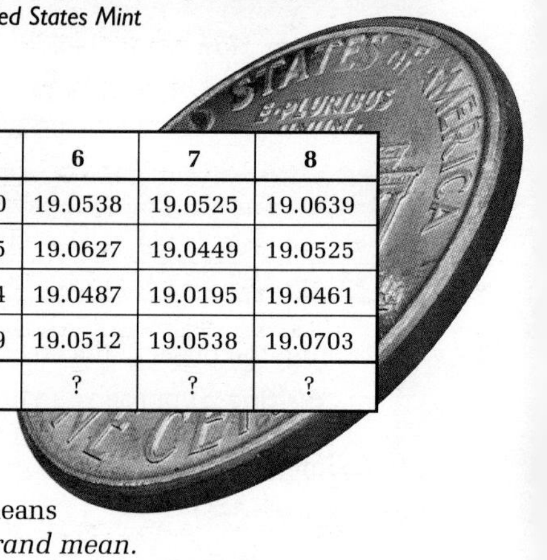

Batch	1	2	3	4	5	6	7	8
Penny 1	19.0627	19.0398	19.0639	19.0703	19.0500	19.0538	19.0525	19.0639
Penny 2	19.0639	19.0296	19.0512	19.0512	19.0525	19.0627	19.0449	19.0525
Penny 3	19.0449	19.0284	19.0639	19.0601	19.0474	19.0487	19.0195	19.0461
Penny 4	19.0703	19.0220	19.0474	19.0703	19.0639	19.0512	19.0538	19.0703
Sample mean	19.0605	?	?	?	?	?	?	?

(19.0627 + 19.0639 + 19.0449 + 19.0703) ÷ 4 = 19.0605

b. Find the mean and the standard deviation of the sample means in the table. The mean of the sample means is called the *grand mean.*

c. The machine's performance is allowed to vary as much as three standard deviations from the grand mean. The *upper control limit* is the measurement that is three standard deviations above the grand mean. The *lower control limit* is the measurement that is three standard deviations below the grand mean. Find the upper and lower control limits.

d. Copy the Penny Control Chart. Find the value of the grand mean in the left column. Draw a heavy horizontal line across the chart at this value. Draw heavy horizontal lines across the chart at the upper control limit and the lower control limit.

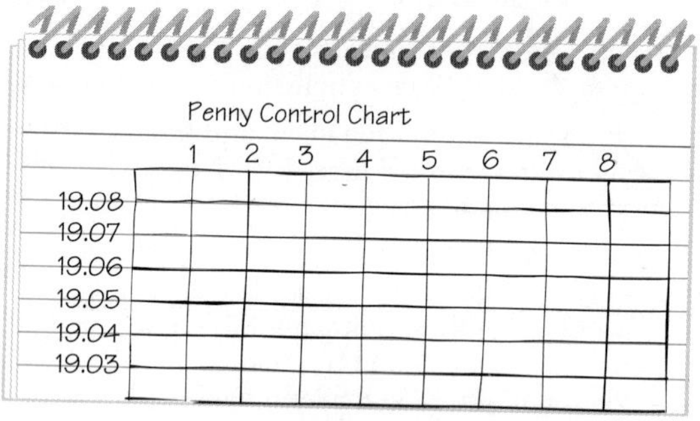

e. **Writing** Below are some sample means of penny diameters measured a week later. Plot each sample mean as a point on your control chart and connect the points. What conclusion(s) can you make? Explain your reasoning.

Batch	1	2	3	4	5	6	7	8
Sample mean	19.0521	19.0491	19.0544	19.0532	19.0500	19.0528	19.0535	19.0497

Answers to Exercises and Problems

8. a. 2: 19.0300; 3: 19.0566; 4: 19.0630; 5: 19.0535; 6: 19.0541; 7: 19.0427; 8: 19.0582

b. 19.0523; 0.0102

c. 19.0217; 19.0829

d, e.

Penny Control Chart

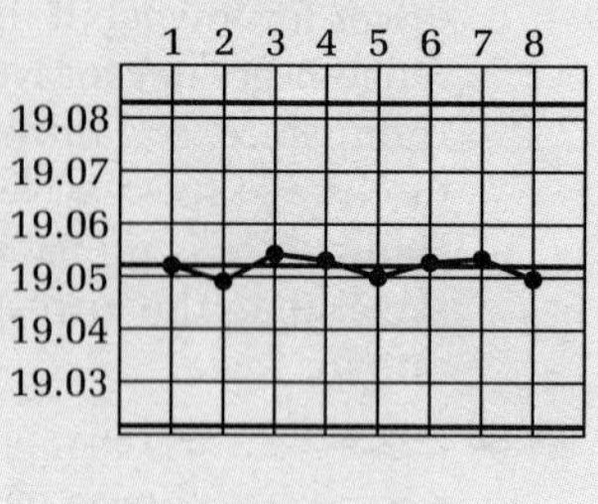

e. The machine's performance is well within operating limits. The 8 sample coins are all within one standard deviation of the grand mean.

9. a. The proportion of data that lies within 2 standard deviations of the means is at least $1 - \frac{1}{2^2} = 1 - \frac{1}{4} = \frac{3}{4} = 75\%$.

9. For any distribution, the fraction of the data that lies within k standard deviations of the mean is always at least $1 - \frac{1}{k^2}$, where $k > 1$.

a. Show that for any distribution at least 75% of the data lie within two standard deviations of the mean.

b. For a distribution with $\overline{x} = 80$ and $\sigma = 7$, at least what percent of the data lies between 59 and 101?

Ongoing **ASSESSMENT**

10. a. **Open-ended** Write a ten-number data set without any outliers. Find the mean, range, and standard deviation of your data set.

b. Add 4 to each data value in your data set. Find the mean, range, and standard deviation of your new data set.

c. Multiply each data value in your original data set by 4. Find the mean, range, and standard deviation of this new data set.

d. **Writing** What conclusion(s) can you make based on the results of parts (a)–(c)? Explain.

Review **PREVIEW**

11. Make a relative frequency histogram for the data set given for machine 1 in Sample 2 on page 340. *(Section 6-1)*

12. Solve $\sqrt{2x + 10} - 5 = x$. *(Section 2-6)*

13. a. What type of distribution is shown? *(Section 6-1)*

b. About what percent of students received a score between 70 and 79?

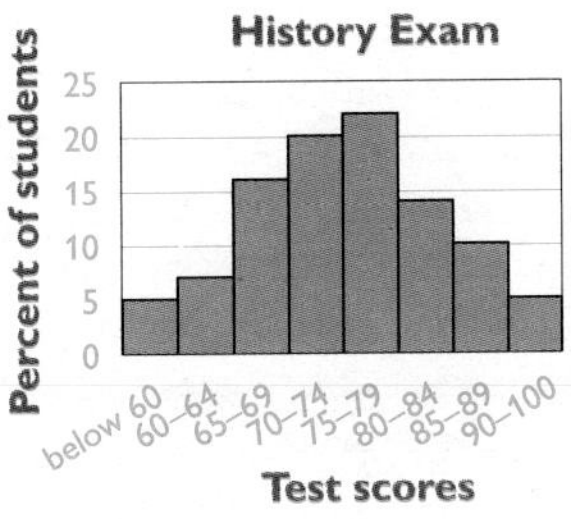

Working on the Unit Project

Use the results of your survey from Exercise 18 on page 337.

14. a. Find the range and the standard deviation of the responses to as many survey questions as you can. How variable are the responses?

b. Is there greater or less variety in the answers than you expected?

c. **Writing** Would the mean of a set of responses with greater variability or less variability tend to be on the list of features for the typical car? Explain.

Students Acquiring English

Sections 6-1 and 6-2 introduce many terms vital for understanding problems in this unit. Have students create a list of important terms and their definitions in these two sections, and add to the list as the unit continues. Some of the terms for the first two sections to include are *frequency distribution, uniform distribution, mound-shaped distribution, skewed distribution, bimodal distribution, relative frequency histogram, measure of variability,* and *standard deviation.*

Working on the Unit Project

Working individually, students can answer Ex. 14 about one or two of the survey questions. The group can then meet to discuss parts (b) and (c).

Practice 47 For use with Section 6-2

Name ______________ Date ______________

Practice 47
For use with Section 6-2

1. a. 263.66 b. 50.7 c. ≈ 18.7 d. 60%
2. a. 241.35 b. 42.3 c. ≈ 13.8 d. 60%
3. a. 1366 b. 1260 c. ≈ 375.6 d. ≈ 76.9%
4. a. 38.875 b. 56.1 c. ≈ 17.7 d. ≈ 58.3%
5. a. ≈ 49.24 b. 70.1 c. ≈ 24.1 d. 55.6%
6. a. ≈ 5329.33 b. 3213 c. ≈ 966.8 d. ≈ 66.7%

For Exercises 1–6: See above.

a. Find the mean. **b.** Find the range. **c.** Find the standard deviation.

d. Find the percent of the data within one standard deviation of the mean.

1. Women's Olympic 400-Meter Freestyle Times 1956-1992 (seconds)
294.6, 290.6, 283.3, 271.8, 259.4, 249.9, 248.8, 247.1, 243.9, 247.2

2. Men's Olympic 400-Meter Freestyle Times 1956-1992 (seconds)
267.3, 258.3, 252.2, 249.0, 240.3, 231.9, 231.3, 231.2, 227.0, 225.0

3. U.S. Silver Production in Every Fifth Year 1930-1990 (tons)
1578, 1428, 2164, 904, 1347, 1134, 1120, 1238, 1400, 1087, 1006, 1227 2125

4. U.S. Population per Square Mile 1880-1990 (every 10 years)
14.2, 17.8, 21.5, 26.0, 29.9, 34.7, 37.2, 42.6, 50.6, 57.7, 64.0, 70.3

5. Airplane Passengers by Airline in 1993 for Top Nine Carriers (millions)
84.8, 82.5, 69.7, 53.7, 44.1, 37.5, 37.3, 18.9, 14.7

6. Per Pupil Expenditure on Public Education by State 1993 (S. Atlantic Region)
7387, 6028, 6040, 5198, 5352, 4653, 4242, 4174, 4890

7. Audio Concepts, Inc. produces a headphone set that can be adjusted to accommodate different-sized heads. The company's survey of vertical distances between the top of the head and the ears produced the following data (in inches):

4.5, 4.8, 6.2, 5.5, 5.6, 5.4, 5.8, 6.0, 5.8, 6.2, 4.6, 5.0, 5.4, 6.2, 5.8

a. Find the range and the mean of the data. 1.7, 5.52

b. Find the standard deviation of the data. ≈ 0.55

c. Suppose the company decides to accommodate three standard deviations from the mean in designing their headphones. Find the range of distances that must be accommodated. 3.87 in. to 7.17 in.

8. The following figures represent the governors' salaries, in thousands of dollars to the nearest thousand dollars, in the ten most populous states of the U.S.:

114, 130, 99, 105, 103, 110, 98, 112, 85, 94

a. Find the mean and the standard deviation of the data. 105; 11.8

b. Find the mean and the standard deviation without the 130 figure. Would you say that this figure is an outlier? ≈ 102.2; ≈ 8.8; yes

Practice Bank, INTEGRATED MATHEMATICS 3

47

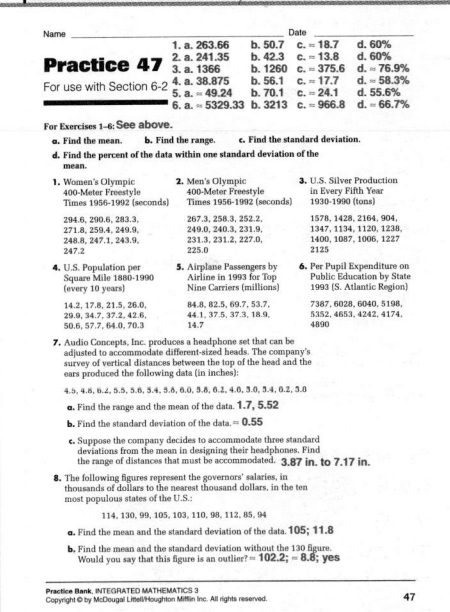

Answers to Exercises and Problems

b. The data between 59 and 101 are within 3 standard deviations of the mean; $1 - \frac{1}{3^2} = 1 - \frac{1}{9} = \frac{8}{9} \approx 89\%$.

10. a–c. Answers may vary. An example is given.

a. data: 15, 15, 17, 18, 20, 21, 21, 21, 23, 25; mean: 19.6; range: 10; standard deviation: 3.14

b. data: 19, 19, 21, 22, 24, 25, 25, 25, 27, 29; mean: 23.6; range: 10; standard deviation: 3.14

c. data: 60, 60, 68, 72, 80, 84, 84, 84, 92, 100; mean: 78.4; range: 40; standard deviation: 12.55

d. If the same number is added to each number in a data set, the mean is increased by that number, while the range and standard deviation are unchanged. If each number in a data set is multiplied by the same number, the mean, range, and standard deviation are all multiplied by that number.

11. Answers may vary depending on intervals chosen. In the example shown, the data are not grouped into intervals.

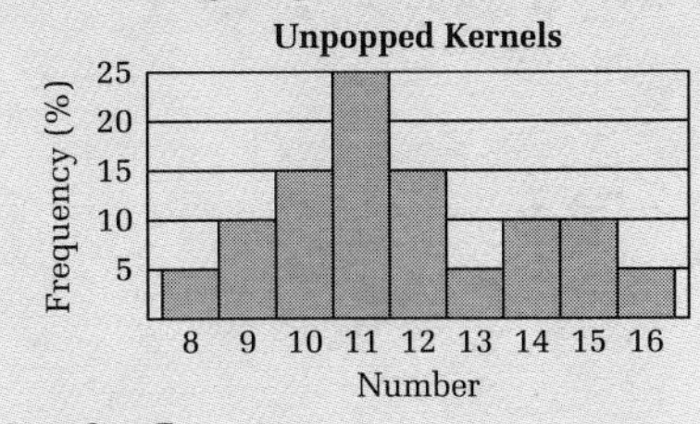

12. −3, −5

13, 14. See answers in back of book.

PLANNING

Objectives and Strands

See pages 328A and 328B.

Spiral Learning

See page 328B.

Materials List

- Tape measure
- Graph paper

Recommended Pacing

Section 6-3 is a two-day lesson.

Day 1

Pages 344–345: Exploration through Talk it Over 4, *Exercises 1–5*

Day 2

Pages 345–346: Rule for Normal Distribution through Look Back, *Exercises 6–17*

Extra Practice

See pages 617–619.

Warm-Up Exercises

- Warm-Up Transparency 6-3

Support Materials

- Practice Bank: Practice 48
- Activity Bank: Enrichment 43
- Study Guide: Section 6-1
- Problem Bank: Problem Set 13
- Explorations Lab Manual: Diagram Master 1
- Overhead Visuals: Folder 5
- Using TI-81 and TI-82 Calculators: The Distribution of Sample Means
- Assessment Book: Quiz 6-3, Test 24, Alternative Assessment 3

Section 6-3 Normal Distributions

Focus

Recognize special properties of normal distributions.

EXPLORATION

How would you describe the distribution of heights of you and your classmates?

- **Materials: tape measures**
- **Work in a group of four students.**

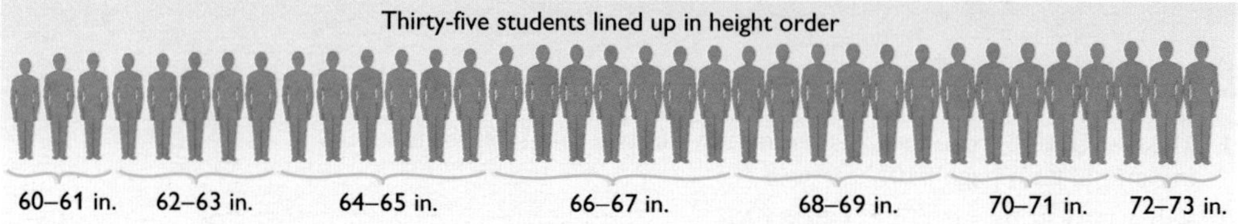

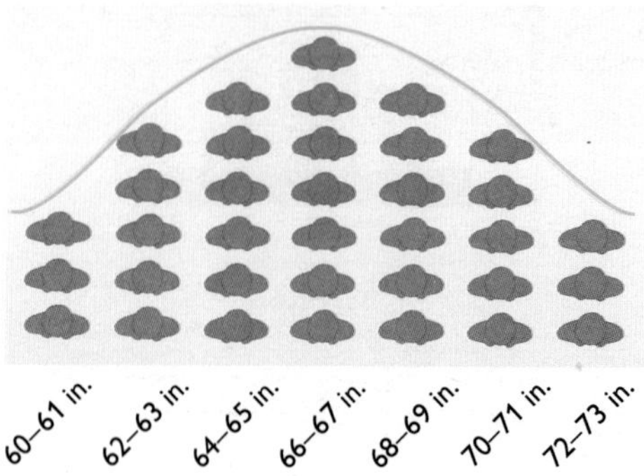

1. List the height of everyone in the class to the nearest inch. Measure students who do not know their heights. Make separate lists for males and females.
2. In your group, decide on a reasonable interval width and draw a relative frequency histogram for the heights of everyone in the class. What type of distribution have you drawn? Where does the mean fall? the median fall?
3. Draw two new relative frequency histograms, one for the male students only and one for the female students only. What type of distribution is shown in each histogram? How do the histograms compare with the histogram for the whole class? Where do the means and medians fall?
4. Suppose you drew a histogram for the heights of all the males or all the females in your school. What shape would you expect the new distribution to have?

Answers to Exploration

1. Check students' work.
2. Answers may vary. The distribution should be mound-shaped, with the mean and median approximately at the line of symmetry.
3. The distributions should be similar to the original distribution, mound-shaped with the mean and median approximately at the line of symmetry. The mean and median for the boys will probably be higher than those for the class in general, while those for the girls will probably be lower than those for the class in general.
4. The shape of the distribution should be similar to the original.

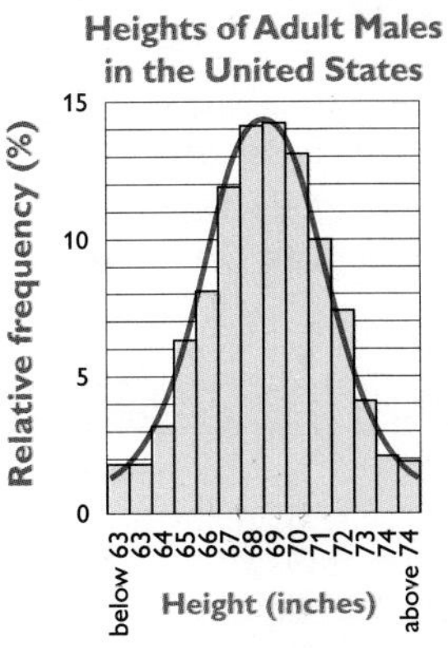

When you draw a smooth curve close to or through the tops of the rectangles of a histogram, you get a bell-shaped curve like the one shown. This bell curve is called a *normal curve.* A distribution with a histogram that follows a normal curve is called a **normal distribution.** In a normal distribution, the mean and median of the data are equal and fall at the line of symmetry for the curve.

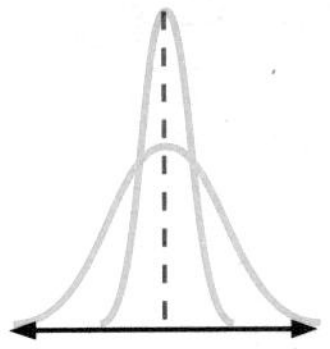

Talk it Over

1. What do you think is the average height of men in the United States?
2. About what percent of men in the United States are between 66 in. and 71 in. tall?
3. What kind of distribution would you get if you made a histogram of the heights of all adults in the United States? Explain.
4. Normal distributions are always bell-shaped, but the shape of the bell can vary. Make a statement about the mean and standard deviation of these two normal distributions.

Now you are ready for: Exs. 1–5 on p. 347

In a normal distribution, a known percent of the data falls within one, two, or three standard deviations of the mean.

68-95-99.7 RULE FOR NORMAL DISTRIBUTIONS

For a normal distribution:

- about 68% of the data are within one standard deviation of the mean.
- about 95% of the data are within two standard deviations of the mean.
- about 99.7% of the data are within three standard deviations of the mean.

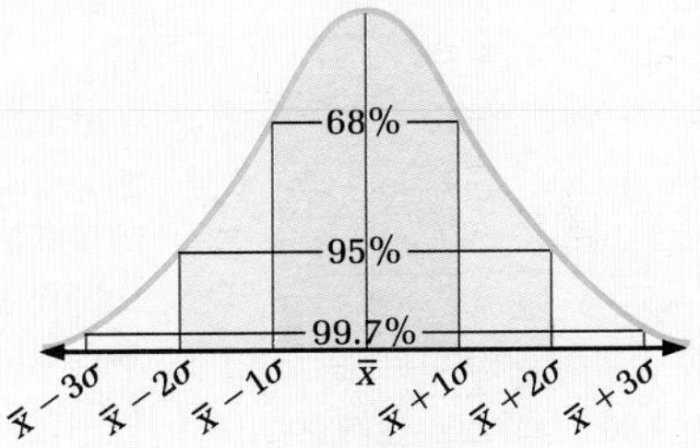

Since a normal distribution is symmetric, the 68-95-99.7 rule can be broken down in this way:

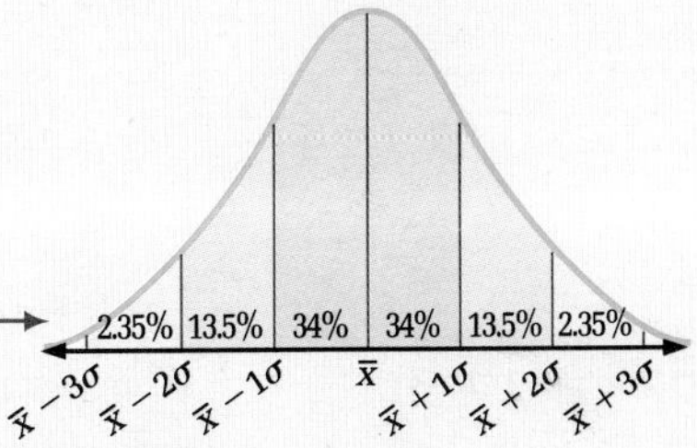

TEACHING

Exploration

The goal of the Exploration is to lead students to the concept of a normal distribution by having them draw relative frequency histograms using data that are normally distributed.

Talk it Over

Question 4 encourages students to think about how the shape of a normal curve is affected by the standard deviation.

Visual Thinking

Check students' understanding of a normal distribution by having them discuss the meaning of the *converse* of each of the statements in the Rule for Normal Distributions. If 68% of the data are within one standard deviation of the mean, what about the other 32%? What does that suggest in the graph at the top of the page? This activity involves the visual skills of *recognition* and *perception.*

Error Analysis

Some students may misinterpret the 68-95-99.7 rule by interpreting it to mean "within two standard deviations total" rather than "within two standard deviations in each direction." Suggest that these students add the percents in the graph shown at the bottom of this page.

Answers to Talk it Over

1. about 68.5 in.
2. about 71%
3. normal; Most heights and weights approximate normal distributions.
4. Both distributions have the same mean, but the distribution described by the shorter, wider bell-shaped curve has a greater standard deviation than the distribution described by the narrower curve.

Talk it Over

In question 6, students see the different ways the 68-95-99.7 rule can be applied.

Additional Sample

In a large school, the heights of female students are approximately normally distributed with $\bar{x} = 64.5$ inches and $\sigma = 2.5$ inches.

a. Within what range do about 95% of the female students' heights fall?

The data have a normal distribution, so 95% of the heights will fall within two standard deviations of the mean.

$\bar{x} + 2\sigma = 64.5 + 2(2.5)$
$= 69.5$
$\bar{x} - 2\sigma = 64.5 - 5.0$
$= 59.5$

About 95% of the female students have heights between 59.5 and 69.5 inches.

b. About what percent of the female students have heights between 62 and 67 inches?

Draw a normal curve for the female students' heights and label what is known. You know that 62 and 67 are both one standard deviation from the mean. So about 34% + 34% = 68% of the female students have heights between 62 and 67 inches.

Teaching Tip

For the Sample Response, part (b), refer students to the normal distribution on page 345. Students should become familiar with the 2.35%-13.5%-34%-34%-13.5%-2.35% spread of data.

Look Back

Students can answer this question by listing the properties of the normal distribution. The class could work on this together while one student writes the properties on the board.

Talk it Over

5. Why must you know that data have a normal distribution before you can apply the 68-95-99.7 rule to the data?
6. If 95% of the data are within two standard deviations of the mean, what percent are outside two standard deviations?

BY THE WAY...

Seven states do not have lieutenant governors: Arizona, Maine, New Jersey, Oregon, Tennessee, West Virginia, and Wyoming.

Sample

The 1992–1993 annual salaries of the 43 state lieutenant governors in the United States have a normal distribution with $\bar{x} = \$57,000$ and $\sigma = \$22,000$.

a. Within what range do about 95% of the lieutenant governors' salaries fall?

b. About what percent of the lieutenant governors have an annual salary between $13,000 and $79,000?

Sample Response

a. The data have a normal distribution, so 95% of the salaries will fall within two standard deviations of the mean.

$$\bar{x} + 2\sigma = 57,000 + 2(22,000)$$
$$= 101,000$$
$$\bar{x} - 2\sigma = 57,000 - 44,000$$
$$= 13,000$$

About 95% of the lieutenant governors have an annual salary between $13,000 and $101,000.

b. Draw a normal curve for the lieutenant governors' salaries and label what is known.

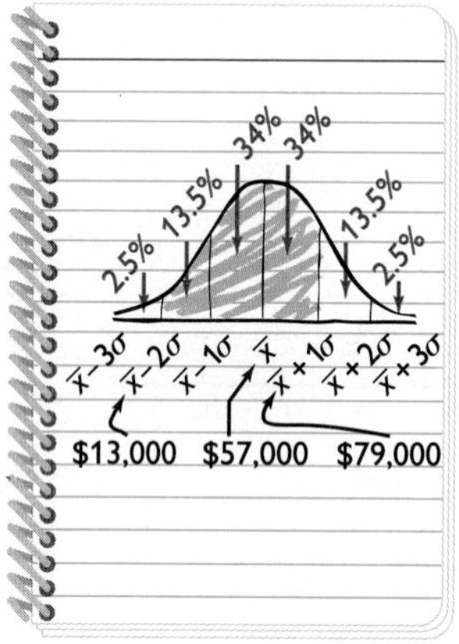

You know 13,000 is two standard deviations below the mean. Since 57,000 + 22,000 = 79,000, $79,000 is one standard deviation above the mean.

So about 13.5% + 34% + 34% = 81.5% of the lieutenant governors have an annual salary between $13,000 and $79,000.

Now you are ready for: Exs. 6–16 on pp. 347–349

Look Back

What is a normal distribution?

Answers to Talk it Over

5. The 68-95-99.7 rule applies only to normal distributions.
6. 5%

Answers to Look Back

A normal distribution is a symmetrical mound-shaped distribution that keeps its shape the larger the data set gets no matter how small the interval size is. In a true normal distribution, the mean and median are equal and fall at the line of symmetry.

6-3 Exercises and Problems

For each data set in Exercises 1–3:

a. Make a histogram.

b. Show the mean and median on the histogram.

c. Tell whether the data appear to have a normal distribution.

1. Time 50 people took to complete the same test (minutes)

   30, 29, 24, 28, 27, 40, 38, 23, 39, 33, 27, 32, 19, 40, 35, 25, 28, 30, 28, 27, 36, 25, 26, 30, 34, 34, 37, 37, 24, 24, 25, 34, 34, 26, 27, 33, 30, 26, 32, 38, 24, 29, 29, 31, 33, 38, 33, 33, 24, 34

2. Weights of 25 men given pre-employment physical exams (pounds)

   188, 188, 168, 200, 185, 185, 175, 175, 185, 190, 192, 183, 180, 205, 195, 195, 170, 170, 190, 180, 185, 180, 190, 182, 190

3. Lifetime of 30 100-watt light bulbs (hours)

   1013, 1000, 914, 910, 947, 1075, 1044, 1000, 1122, 1029, 982, 940, 994, 1038, 1036, 950, 950, 1044, 1081, 872, 900, 973, 1097, 1025, 970, 1022, 979, 1050, 984, 1000

4. a. Make a new data set of the last digit of each weight in Exercise 2. Draw a histogram for this data set. Give each digit its own interval.

   b. **Writing** Explain why the histogram in part (a) might convince someone that the men were not actually weighed.

5. a. **Open-ended** Sketch two normal distributions where the standard deviations are equal but the means are not equal.

   b. **Open-ended** Sketch two normal distributions where the standard deviations are not equal and the means are not equal.

6. **Reading** What is the 68-95-99.7 rule?

7. **Health** The blood cholesterol readings of a group of women have a normal distribution with $\overline{x} = 172$ and $\sigma = 14$.

   a. Within what range do about 95% of the readings fall?

   b. About what percent of the women have readings between 158 and 172?

   c. Readings higher than 200 are considered undesirable. About what percent of the readings are undesirable?

8. The amount of juice dispensed from a machine is normally distributed with a mean of 10.50 oz and a standard deviation of 0.75 oz.

   a. About 68% of the amounts dispensed fall within what range?

   b. About what percent of the time will the machine overfill a 12 oz cup?

9. **Manufacturing** The ICY Company offers a ten-year warranty for free repair of the major parts of a refrigerator. The life span of an ICY refrigerator shows a normal distribution with a mean of 17 years and a standard deviation of 3.3 years. About what percent of ICY refrigerators should the company expect to have to repair free of charge?

APPLYING

Suggested Assignment

Day 1

Standard 1–4

Extended 1–5

Day 2

Standard 6–9, 13–17

Extended 6–17

Integrating the Strands

Algebra Exs. 14–16

Functions Exs. 14, 15

Statistics and Probability Exs. 1–13, 17

Discrete Mathematics Exs. 1–13, 16, 17

Logic and Language Exs. 4, 6, 10, 11, 12, 17

Mathematical Procedures

Mathematicians analyze data by studying how they are distributed over intervals. The most important data distribution is the normal distribution. This is so because measurements of many things in the real world give rise to distributions that are approximately normal.

Reasoning

In Ex. 8(b), students should realize that two standard deviations above the mean implies that the cup has been filled to the brim, but has not yet overflowed.

Answers to Exercises and Problems

1–4. Answers may vary depending on interval size chosen. Examples are given.

1. a, b.

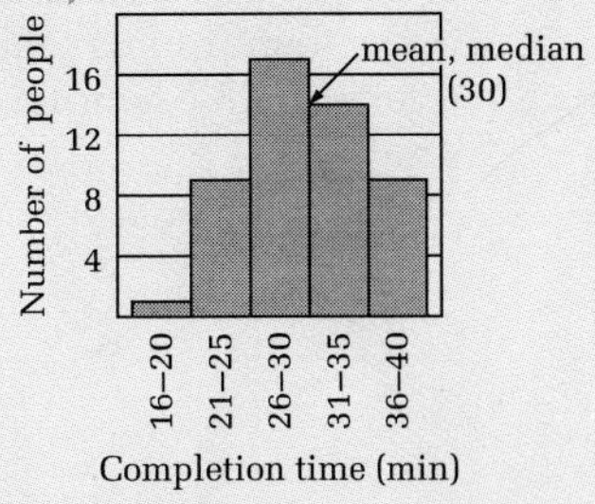

c. No.

2. a, b.

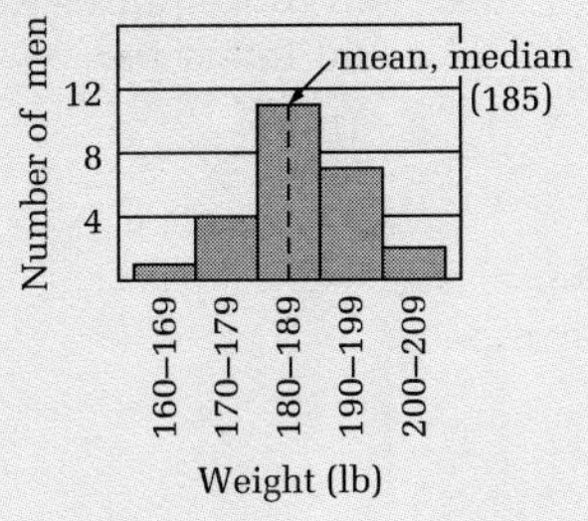

c. Yes.

3. a, b.

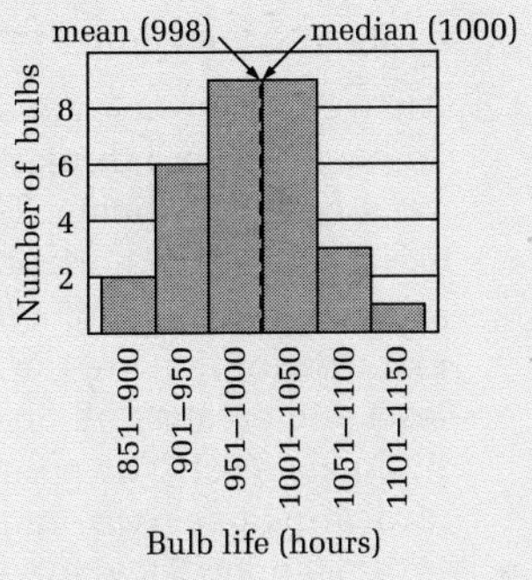

c. Yes.

4–6. See answers in back of book.

7. a. 144 and 200

   b. about 34%

   c. about 2.5%

8. a. 9.75 oz and 11.25 oz

   b. about 2.5%

9. about 2.5%

Interdisciplinary Problems

Exs. 7–11 provide some interesting data that are normally distributed. Many characteristics of animal populations, such as heights, weights, and foot sizes are normally distributed. Biologists can then use the properties of normal curves to make predictions about those characteristics in a population.

Communication: Discussion

You may wish to ask those students who have read the book *Jurassic Park* or have seen the movie to tell the class something about it.

Communication: Writing

In responding to Ex. 12 in writing, students should give a mathematical explanation of their answer to the question.

connection to LITERATURE

In *Jurassic Park* by Michael Crichton, Dr. Henry Wu uses dinosaur DNA to create dinosaurs for a zoo-like park. A mathematician, Dr. Ian Malcolm, claims that there are more dinosaurs in the park than there should be.

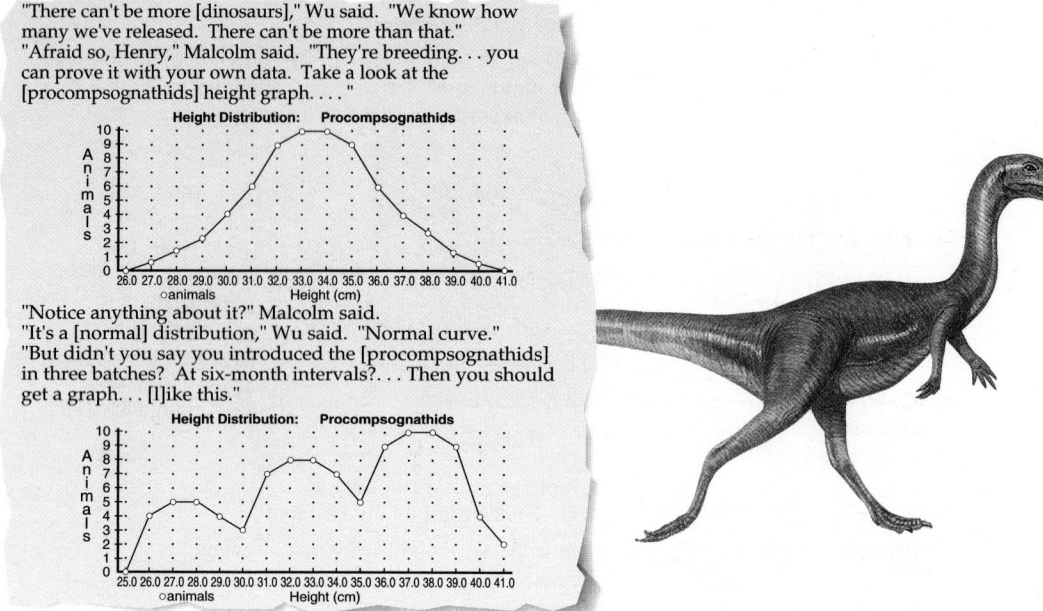

"There can't be more [dinosaurs]," Wu said. "We know how many we've released. There can't be more than that."
"Afraid so, Henry," Malcolm said. "They're breeding. . . you can prove it with your own data. Take a look at the [procompsognathids] height graph. . . . "

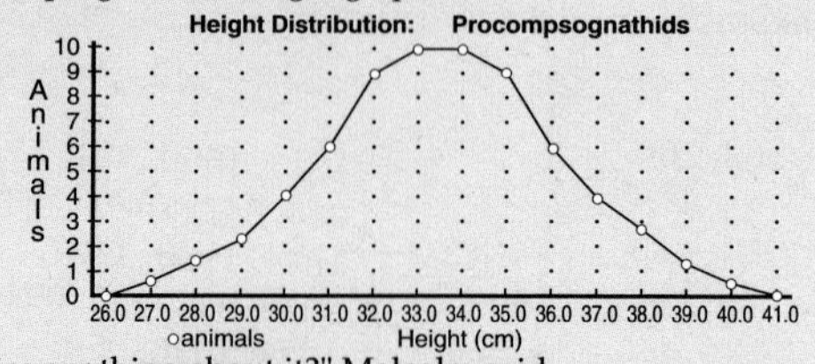

"Notice anything about it?" Malcolm said.
"It's a [normal] distribution," Wu said. "Normal curve."
"But didn't you say you introduced the [procompsognathids] in three batches? At six-month intervals?. . . Then you should get a graph. . . [l]ike this."

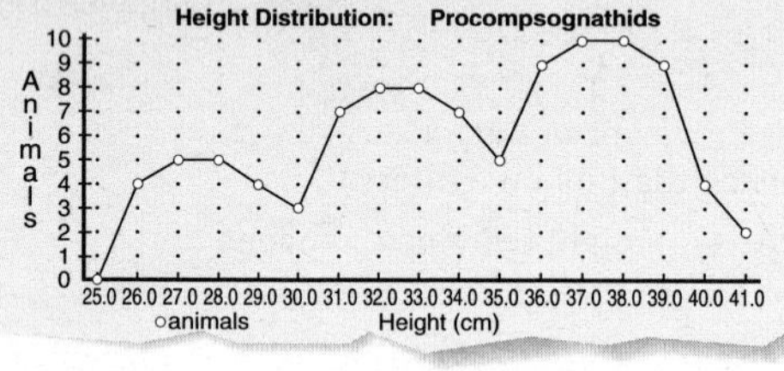

10. a. Suppose the only procompsognathids in the park are the ones that Dr. Wu introduced. For the second graph to be accurate, what must be true about the procompsognathids in each batch that was introduced?

b. Suppose the procompsognathids were all introduced as adults. What would be the expected height distribution?

11. a. Why does a normal height distribution convince Dr. Malcolm that the dinosaurs are breeding?

b. What is the approximate mean of the procompsognathids' heights?

c. The normal distribution graph above shows the heights of 65 procompsognathids. About how many have a height within one standard deviation of the mean? within two standard deviations of the mean?

Ongoing ASSESSMENT

12. Writing There is a rule of thumb that says that in a normal distribution $\sigma \approx \frac{\text{range}}{6}$. Do you think this formula gives a good estimate of σ? Explain.

Answers to Exercises and Problems

10. a. The procompsognathids were all young or, at least, still growing when they were introduced.

b. normal distribution

11. a. The heights are supposedly for three distinct age groups, so you would not expect a normal distribution. If the distribution is normal, then there must be procompsognathids other than the ones introduced.

b. about 33.5 cm

c. about 44; about 62

12. Answers may vary. An example is given. I think it is a reasonable estimate. The endpoints of the horizontal scale for the graph of a normal distribution are the least and greatest numbers in the data set. The points indicating one, two, and three standard deviations of the mean roughly divide the horizontal axis between those points into six equal intervals.

13. 13.4°; 10.5°

14. $y = 0.5x - 1$

15. $y = 4x + 3$

16. $\begin{bmatrix} 2\frac{2}{3} \\ -5\frac{2}{3} \end{bmatrix}$

17. a. Answers may vary.

b. Answers may vary. Sometimes a larger sample size will show a normal distribution that is not evident from a relatively smaller sample size.

c. Answers may vary.

Review **PREVIEW**

13. Find the mean and the standard deviation of these temperatures (°F) taken at 6 A.M. on ten consecutive days in January. *(Section 6-2)*

$-2, -1, 10, 15, 12, 16, 10, 12, 32, 30$

Write an equation for each function in the form $y = mx + b$. *(Section 2-2)*

14. $f(4) = 1$, $m = 0.5$

15. $f(0) = 3$, $f(2) = 11$

16. Solve $\begin{bmatrix} 3 & 0 \\ 1 & -2 \end{bmatrix}\begin{bmatrix} x \\ y \end{bmatrix} = \begin{bmatrix} 8 \\ 14 \end{bmatrix}$. *(Section 1-5)*

Working on the Unit Project

Use the histograms from Exercise 18 on page 337.

17. a. Do any of your survey histograms appear to show a normal distribution? How can you decide?

b. Open-ended Do you think more of the data would show a normal distribution if your survey involved more than 30 people? Explain.

c. For all normally distributed survey responses, within what range do 68% of the responses fall? Within what range do 95% fall?

Unit 6 CHECKPOINT

CEO Salaries at Major U.S. Corporations in 1992 (thousands of dollars)

Financial institutions	Communications companies
2010	1434
1381	2258
2250	908
2185	1573
1164	1851
2433	1349
776	1547
1782	1086
1227	1705
1039	
1318	
2752	

1. Writing How is a normal distribution different from the other types of distributions presented in this unit?

2. a. Salaries Make a relative frequency histogram for each data set in the table of salaries of chief executive officers (CEOs). **6-1**

b. Identify each type of distribution.

c. What percent of executives in each group made between \$750,000 and \$1,500,000 in 1992? between \$1,500,000 and \$2,500,000?

d. Make a comparison based on the histograms.

3. The main span length of ten notable suspension bridges is given in feet. Find the mean, the range, and the standard deviation of the lengths. **6-2**

4626; 4260; 4200; 3800; 3609;
3576; 3524; 3500; 3323; 1595

4. Suppose the test scores on a Spanish exam show a normal distribution with $\overline{x} = 75$ and $\sigma = 8$. Within what range do about 68% of the scores fall? **6-3**

Working on the Unit Project

Ex. 17 helps students to see how the characteristics of a normal distribution apply to their survey histograms. Students may be able to use the results of this exercise when they write their final report for the Unit Project.

Quick Quiz (6-1 through 6-3)

See page 369.

See also Test 24 in the *Assessment Book.*

Practice 48 For use with Section 6-3

Name ______ Date ______

Practice 48

For use with Section 6-3

1. a.

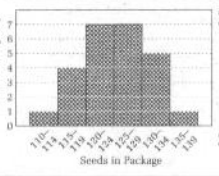

2. a.

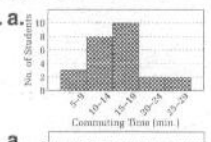

3. a.

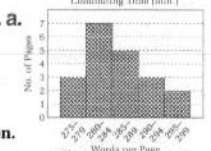

For Exercises 1–3:

a. Make a histogram for each data set.

b. Plot the mean and median on the histogram.

c. Tell whether the data appear to show a normal distribution.

1. Number of Sunflower Seeds in a Small Package
120, 130, 118, 125, 128, 130, 122, 110, 127, 135, 115, 118, 124, 124, 130, 132, 126, 128, 123, 122, 126, 125, 120, 116, 131
b. mean = 124.2; med. = 125 c. yes

2. Commuting time of Students in the Same Math Class (minutes)
15, 27, 18, 20, 15, 12, 16, 18, 11, 8, 17, 9, 9, 10, 14, 22, 25, 16, 15, 10, 13, 14, 16, 17, 14
b. mean = 15.24; med. = 15 c. no

3. Number of Words on Each of the First 20 Pages of a Novel
285, 280, 299, 278, 290, 283, 288, 292, 280, 280, 284, 295, 286, 275, 291, 282, 286, 289, 280, 277
b. mean = 284.9; med. = 284.5 c. no

4. The widths of maple boards from a sawmill have a normal distribution with a mean width of 7.2 in. and a standard deviation of 0.9 in.

a. What percent of the boards have widths between 6.3 in. and 8.1 in.? 68%

b. Within what range of widths do 95% of the boards fall? between 5.4 in. and 9.0 in.

c. A certain lumber yard will not accept boards of width less than 5.4 in. What percent of the output of the sawmill will not be accepted by this lumber yard? 2.5%

5. Toshio Toyama is training for a marathon race. The times of his practice runs form a normal distribution with $\overline{x} = 2$ h 33 min and $\sigma = 3.5$ min.

a. About 68% of his times fall within what range? between 2 h 29.5 min and 2 h 36.5 min

b. What percent of his times are between 2 h 26 min and 2 h 33 min? 47.5%

c. In order to qualify for a certain marathon, a runner must have an official time of 2:40 or less in a previous race. Once Toshio chooses a qualifying race, what is the probability that he will qualify for the marathon? 97.5%

6. *Writing* Choose some measurement that you can make on many different objects of the same kind, or a measurement of some variable quantity that you can make at many different times. Make a histogram of your data. Tell whether you think the data form a normal distribution and why or why not. Answers may vary. Check students' work.

48 **Practice Bank**, INTEGRATED MATHEMATICS 3

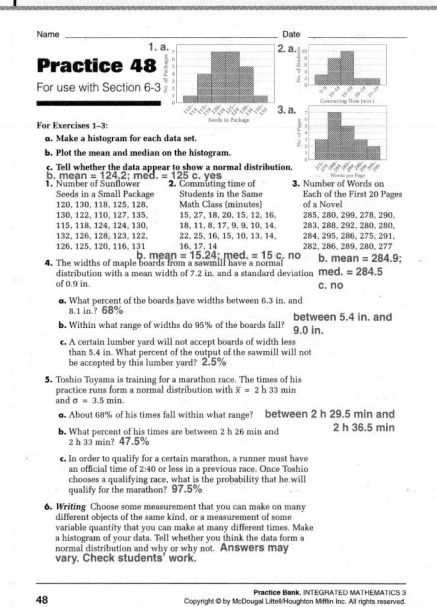

Answers to Checkpoint

1. See answers in back of book.

2. a. Answers may vary depending on the intervals chosen. Examples are given. See histograms at the right.

Financial Institutions CEO Salaries (1992)

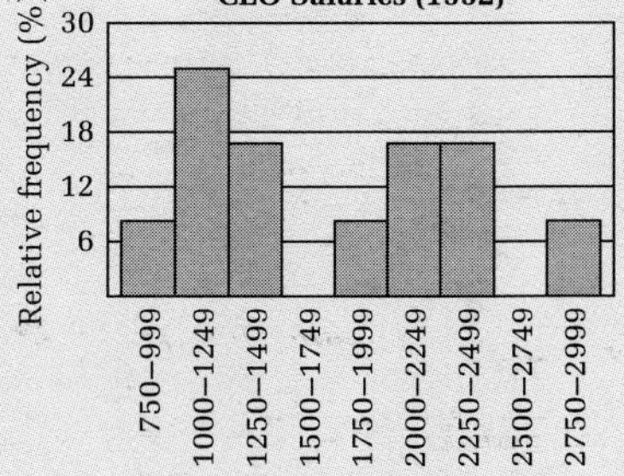

Communications Companies CEO Salaries (1992)

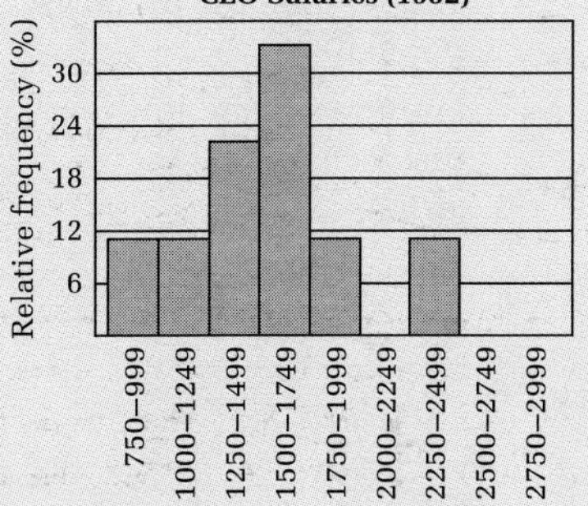

b. financial: basically bimodal; communications: basically mound-shaped

c. financial: 50%; about 42%; communications: about 44.5%; about 55.5%

d. The range of salaries for financial institutions is greater than that for communications companies, so there is more variation for financial institutions. The two groups appear to have similar median salaries, but it appears that, overall, salaries are higher in financial institutions.

3. about 3601; 3031; about 775

4. between 67 and 83

PLANNING

Objectives and Strands
See pages 328A and 328B.

Spiral Learning
See page 328B.

Materials List
- Graph paper
- Graphics calculator or statistical software

Recommended Pacing

Section 6-4 is a two-day lesson.

Day 1

Pages 350–352: Opening paragraph through top of page 352, *Exercises 1–10*

Day 2

Pages 352–353: Median-median Line through Look Back, *Exercises 11–18*

Toolbox References
- **Toolbox Skill 1:** Finding Mean, Median, Mode, and Range

Extra Practice
See pages 617–619.

Warm-Up Exercises
- Warm-Up Transparency 6-4

Support Materials
- Practice Bank: Practice 49
- Activity Bank: Enrichment 44
- Study Guide: Section 6-4
- Problem Bank: Problem Set 14
- Explorations Lab Manual: Additional Exploration 7, Diagram Master 2
- Using Mac Plotter Plus Disk: Data Analyzer
- Using IBM Plotter Plus Disk: Statistics Spreadsheet
- Assessment Book: Quiz 6-4, Alternative Assessment 4

Section

6-4 Fitting Linear Models to Data

Focus

Identify data as fitting a linear model. Use technology to find the equations of fitted lines.

Along These Lines

A scatter plot shows the relationship between two sets of data. Each point on a scatter plot shows a pairing of two data values. The scatter plots below show information about the cities on the world map below.

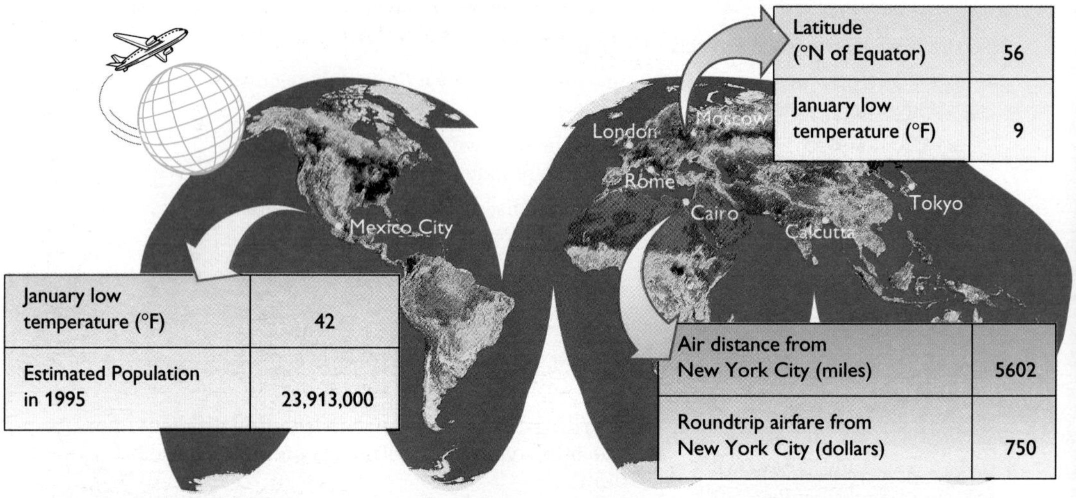

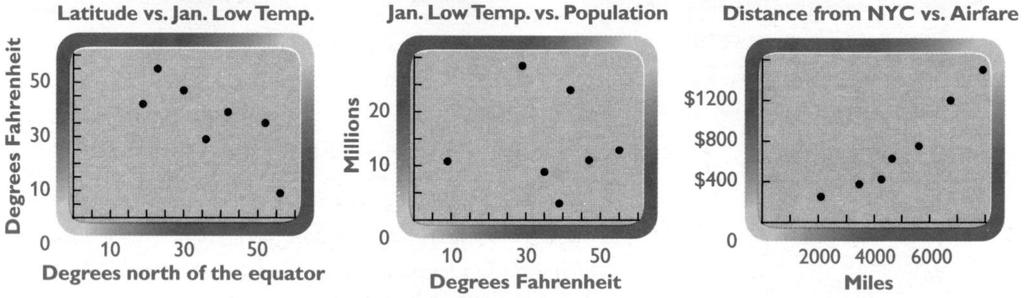

Talk it Over

Use the three scatter plots above in questions 1 and 2.

1. Which of the scatter plots looks the most linear? the least?
2. Of the scatter plots that look linear, which has a downward trend (or negative slope)? which has an upward trend (or positive slope)?

Answers to Talk it Over

1. distance from NYC vs. airfare; Jan. low temp. vs. population
2. latitude vs. Jan. low temp.; distance from NYC vs. airfare

The least-squares line minimizes the sum of the squares of the vertical distances shown in red.

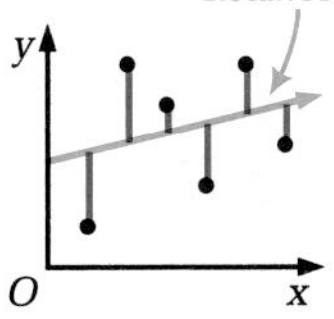

Least-Squares Line

The process of fitting a curve to a set of data is called *regression.* Although several linear regression models may be used, the *least-squares line* is most common. You can use technology to find an equation of the least-squares line for a set of data and use the equation to predict unknown values.

Sample 1

Over a number of years the weights of prize-winning pumpkins at the annual Topsfield Fair in Topsfield, Massachusetts, have tended to increase. Predict the winning pumpkin weight for the year 2005.

BY THE WAY...

The Pilgrims found pumpkins easy to grow, as is shown by this verse from a song.

For pottage, and puddings,
and custards, and pies,
Our pumpkins, and parsnips,
are common supplies;
We have pumpkin at morning,
and pumpkin at noon,
If it was not for pumpkin
we should soon be undoon.

Winning Pumpkin Weights — Topsfield Fair

Year	1984	1985	1986	1987	1988	1989	1990	1991	1992	1993
Weight (lb)	433	515.5	530.5	604.5	490	614	470	617.5	718	698

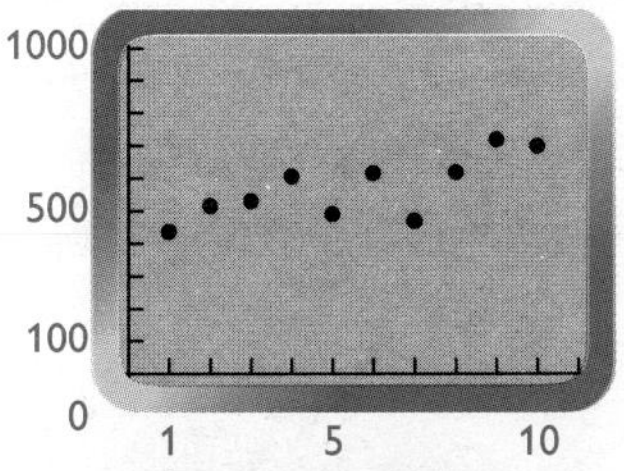

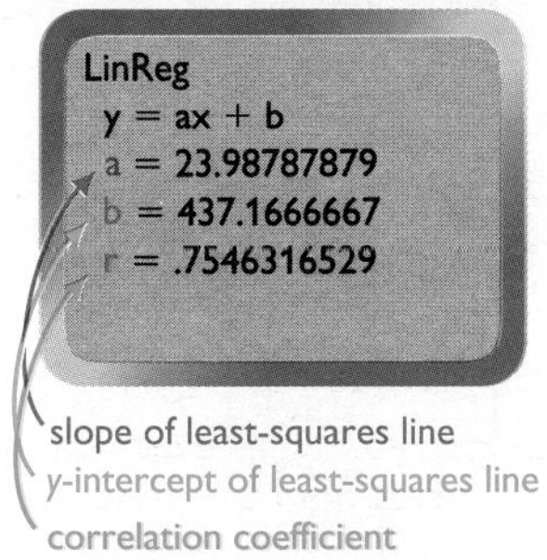

Sample Response

Step 1 Make a scatter plot of the ten data points. Let x represent the number of years after 1983 and y represent the winning pumpkin weight that year. Use technology to perform a linear regression.

Record the regression model for the data. It should be about the same as this equation: $y = 23.99x + 437.2$.

Step 2 Use the equation of the least-squares line to predict the unknown value.

$y = 23.99x + 437.2$

$= 23.99(22) + 437.2$ ← 2005 is 22 years after 1983. Substitute 22 for x.

$= 964.98$

According to this model, the winning pumpkin weight for the year 2005 will be about 965 lb.

TEACHING

Talk it Over

In questions 1 and 2, students analyze scatter plots to determine if a linear relationship exists and if so, to identify whether it has a downward or upward trend.

Communication: Discussion

Discuss the illustration at the top left of this page with students. They should realize that the distances represent the difference between the actual y-value and the y-value predicted by the model for a given x-value.

Integrating the Strands

The least-squares line integrates the strands of algebra, geometry, and statistics. The statistical data that are collected are modeled using an algebraic linear relationship whose graph is a straight line. The strength of the linear model is then measured using a statistic called the correlation coefficient, which is defined on page 352.

Additional Samples

S1 The Kumar family planted a tree when they moved into their new home in 1985. Predict the height of the tree in 2008.

Height of Tree (inches)

Year	Height
1985	47
1986	56
1987	72
1988	85
1989	93
1990	110
1991	118
1992	136
1993	164
1994	185

Step 1. Make a scatter plot of the ten data points. Let x represent the number of years after 1984 and y represent the height of the tree that year. Use technology to perform a linear regression.

Sample continued on next page.

Additional Samples (continued)

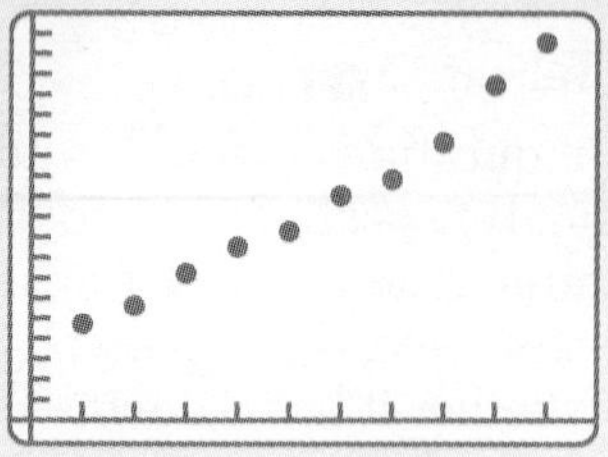

Record the regression model for the data. It should be about the same as this equation: y = 14.75x + 25.5.

```
LinReg
 y = ax + b
 a = 14.75151515
 b = 25.46666667
 r = .987542017
```

Step 2. Use the equation of the least-squares line to predict the unknown value.
2008 is 24 years after 1984.
$y = 14.75x + 25.5$
$= 14.75(24) + 25.5$
$= 379.5$
According to this model, the height of the tree in the year 2008 will be about 380 inches.

S2 Find an equation of the median-median line for the height of the tree in Additional Sample 1.
Step 1. Make a scatter plot of the data.

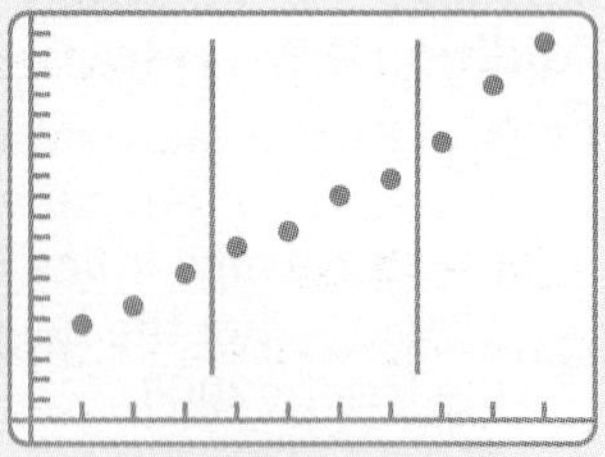

Step 2. On the scatter plot, draw two vertical lines that divide the points into three equal or nearly equal groups. In this case, the ten data points are divided into groups of three, four, and three.
Step 3. Find a summary point for each of the three groups. The summary point for each of the three groups has these coordinates: (median x-value, median y-value).
For the three groups, the summary points are (2, 56), (5.5, 101.5), and (9, 164).
Step 4. Calculate the slope of the line through the outermost summary points.
$m = \frac{164 - 56}{9 - 2} \approx 15.43$

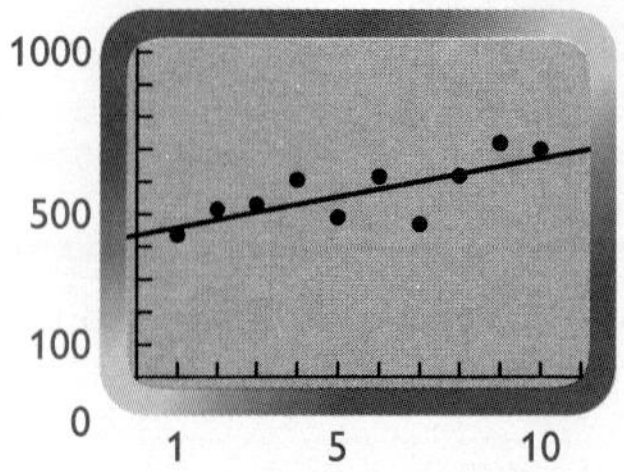

Now you are ready for:
Exs. 1–10 on pp. 354–356

The **correlation coefficient** for a data set describes how well the data points can be modeled by a line. The correlation is a number r between -1 and 1, inclusive. The closer $|r|$ is to 1, the more linear a graph of the data looks. A random distribution of data has a correlation close to 0.

For the data in Sample 1, $r \approx 0.75$. This positive correlation is due to the fact that the weights have tended to increase over time. The scattering of points around the least-squares line is why the correlation is not closer to 1.

Median-median Line

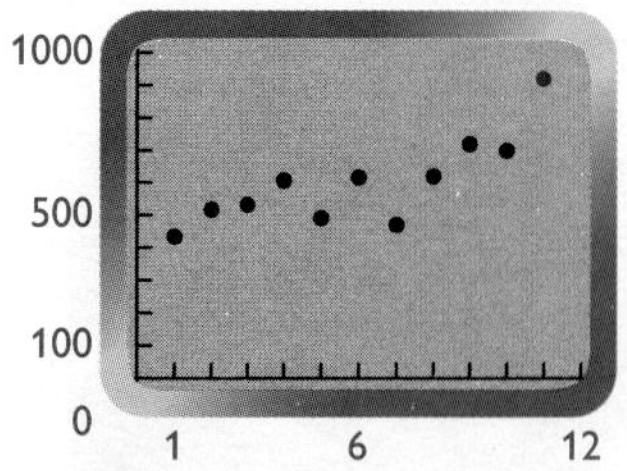

A graphics calculator uses every data point, even outliers, in performing a least-squares linear regression. The red point in the scatter plot represents the winning pumpkin weight of 914 lb at the 1994 Topsfield Fair. Another linear model, called the **median-median line,** is less influenced by outliers such as (11, 914). This line uses *summary points* calculated using medians.

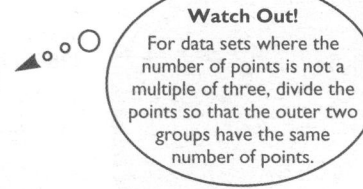

Student Resources Toolbox
p. 626 *Discrete Mathematics*

Sample 2

Find an equation of the median-median line for the winning pumpkin weights at the Topsfield Fair for 1984–1994.

Sample Response

Step 1 Make a scatter plot of the data.

Step 2 On the scatter plot, draw two vertical lines that divide the points into three equal or nearly equal groups. In this case, the eleven data points are divided into groups of four, three, and four.

Watch Out!
For data sets where the number of points is not a multiple of three, divide the points so that the outer two groups have the same number of points.

Step 3 Find a summary point for each of the three groups. The summary point for each of the three groups has these coordinates:

(median x-value, median y-value)

For the three groups, the summary points are (2.5, 523), (6, 490), and (9.5, 708).

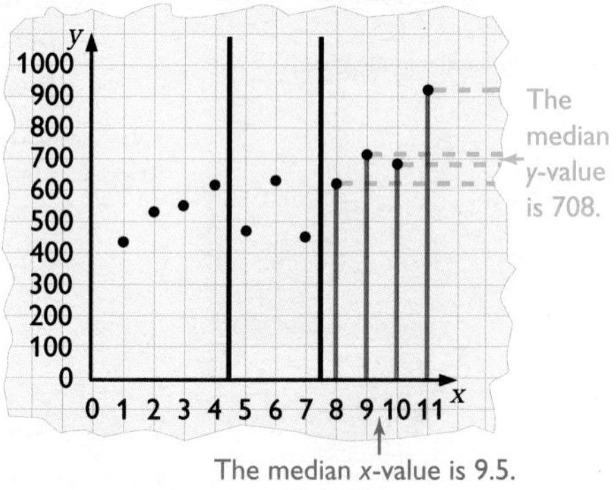

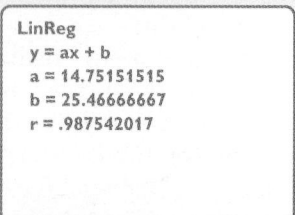

Step 4 Calculate the slope of the line through the outermost summary points.

$$m = \frac{708 - 523}{9.5 - 2.5} \approx 26.43$$

Step 5 Find the point whose coordinates are the means of the x- and y-coordinates of all three summary points.

$$\left(\frac{2.5 + 6 + 9.5}{3}, \frac{523 + 490 + 708}{3}\right), \text{ or about } (6, 573.7)$$

Step 6 The median-median line has the slope found in step 4 and passes through the point found in step 5.

$$y = mx + b$$

$$y = 26.43x + b$$ ← Substitute 26.43 for m from step 4.

$$573.7 = (26.43)(6) + b$$ ← Substitute 6 for x and 573.7 for y from step 5.

$$573.7 = 158.58 + b$$

$$415.1 \approx b$$

An equation of the median-median line is $y = 26.43x + 415.1$.

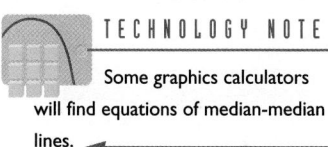

TECHNOLOGY NOTE

Some graphics calculators will find equations of median-median lines.

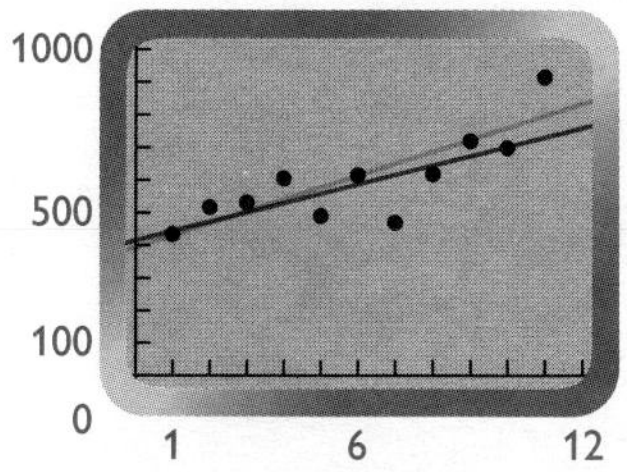

least-squares line:
$y = 33.67x + 398.4$

median-median line:
$y = 26.43x + 415.1$

Talk it Over

The graph shows both the least-squares line and the median-median line for the Topsfield Fair pumpkin data for 1984–1994 (from Exercise 6 and Sample 2).

3. Explain why the point (11, 914) may be an outlier.
4. Which of the lines has a greater slope? Give a reason why the lines have different slopes.
5. Would you expect the data point for the year 1995, (12, ?), to be closer to the median-median line or to the least-squares line? Why?
6. For a data set that includes outliers, which line is better for making predictions, the *least-squares line* or the *median-median line*?

Now you are ready for: Exs. 11–18 on pp. 357–358

Look Back

How do you determine whether to use the least-squares line or the median-median line for a given data set?

Step 5. **Find the point whose coordinates are the means of the x- and y-coordinates of all three summary points.**

$$\left(\frac{2 + 5.5 + 9}{3}, \frac{56 + 101.5 + 164}{3}\right)$$

or (5.5, 107.2)

Step 6. **The median-median line has the slope found in step 4 and passes through the point found in step 5.**

$$y = mx + b$$
$$y = 15.43x + b$$
$$107.2 = 15.43(5.5) + b$$
$$22.3 \approx b$$

An equation of the median-median line is $y = 15.43x + 22.3$.

Reasoning

Students should understand that a correlation close to negative one would represent a relationship where the dependent variable decreases as the independent variable increases.

Teaching Tip

Students should understand that $|r|$ close to 1 means r is close to 1 or close to –1. The negative indicates negative slope only. A correletion close to –1 indicates a strong linear relationship. You might have students graph six points on the line $y = -3x + 2$, and find the correlation coefficient of these points.

Talk it Over

Questions 4–6 lead students to compare how a least-squares line and a median-median line are affected by outliers.

Look Back

Students should write their responses to this Look Back in their journals.

Answers to Talk it Over

3. The point is not near either the least-squares line or the median-median line.
4. the least-squares line; The slope of the least-squares line is determined using actual data points, while the slope of the median-median line is determined using points whose coordinates are means of the coordinates of summary points.
5. Answers may vary. An example is given. I think the point will be closer to the median-median line. That line is less affected by the point (11, 914) than is the least-squares line.
6. the median-median line

Answers to Look Back

The median-median line should be used when a data set includes apparent outliers, which would influence the least-squares line.

APPLYING

Suggested Assignment

Day 1
Standard 1–4
Extended 1–4

Day 2
Standard 11–18
Extended 11–18

Integrating the Strands

Algebra Exs. 4–13, 16, 17
Geometry Ex. 13
Statistics and Probability Exs. 1–15, 18
Discrete Mathematics Exs. 15–17
Logic and Language Exs. 1, 5, 6, 8–12, 14, 18

6-4 Exercises and Problems

TECHNOLOGY **For the technology exercises in this section, you can use a graphics calculator or statistical software.**

1. **Reading** Would you fit a linear model to data with a correlation coefficient that is close to zero? Why or why not?

2. Which of the scatter plots on page 350 do you think has each correlation coefficient? Why?
   a. $r = 0.89$ b. $r = -0.04$ c. $r = -0.77$

3. **Demographics** Suppose you have a scatter plot with the population of some cities along the horizontal axis and the area of the same cities along the vertical axis.
   a. In which range do you expect the correlation coefficient r for the data set to be: $-1 \le r < -\frac{1}{2}$, $-\frac{1}{2} \le r \le \frac{1}{2}$, or $\frac{1}{2} < r \le 1$?
   b. Make a scatter plot of the data below. Look at the distribution of the data points. Does your scatter plot support your prediction in part (a)? Explain.

Cities Around the World

	Mexico City	Cairo	London	Tokyo	Calcutta	Moscow	Rome
Estimated population in 1995 (thousands)	23,913	11,155	8,897	28,447	12,885	10,769	3,079
Area (square miles)	522	104	874	1089	209	379	69

4. You can calculate the correlation coefficient r for a set of data points $(x_1, y_1), (x_2, y_2), \ldots$ by using the following formula:

mean of the products of the x- and y-values

product of the mean x-value and the mean y-value

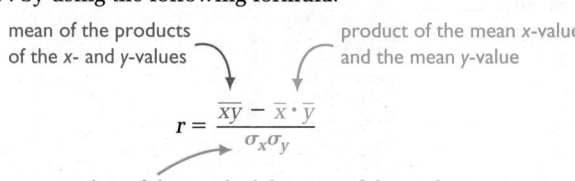

$$r = \frac{\overline{xy} - \overline{x} \cdot \overline{y}}{\sigma_x \sigma_y}$$

product of the standard deviation of the x-values and the standard deviation of the y-values

   a. Make a scatter plot of these points:
   $(-2, -3), (0, 1.5), (1, 2), (3, 3)$
   b. Do the data points appear close to some line? Estimate the correlation coefficient.
   c. Find $\overline{xy}$, $\overline{x}$, $\overline{y}$, σ_x, and σ_y.
   d. Use the formula above to calculate the correlation coefficient.
   e. How close was your estimate in part (b)?

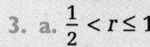

Answers to Exercises and Problems

1. No; if the correlation coefficient of a data set is close to 0, the distribution of the data set is nearly random. A line would not model the data well.
2. a. distance from NYC vs. airfare; It is the data set with a positive correlation.
   b. Jan. low temp. vs. population; Correlation is negative and the graph of the data does not appear linear, so the absolute value of the correlation should be close to 0.
   c. latitude vs. Jan. low temp.; The correlation is negative and the graph of the data appears more linear than that for the other negative correlation.
3. a. $\frac{1}{2} < r \le 1$
   b.

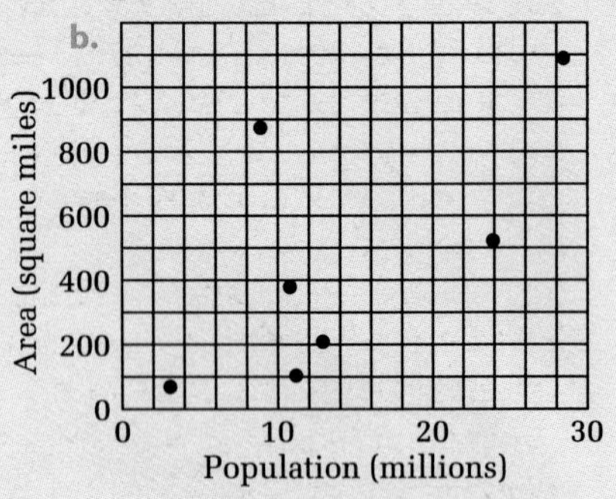

Yes; the data show a general upward trend, but the data points are somewhat scattered.

4. a.

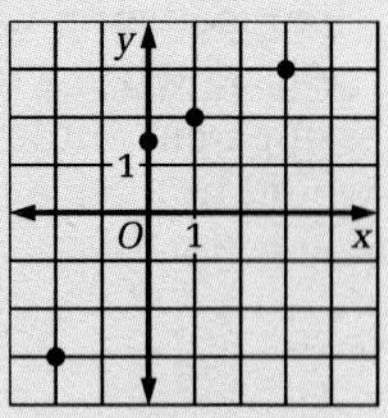

   b. Yes. Estimates will vary. An example is given. between $\frac{1}{2}$ and 1
   c. 4.25; 0.5; 0.875; 1.8; 2.3
   d. about 0.92
   e. Answers may vary.
5. See answers in back of book.
6. a. $y = 33.67x + 398.4$, where x is the number of years since 1983
   b.

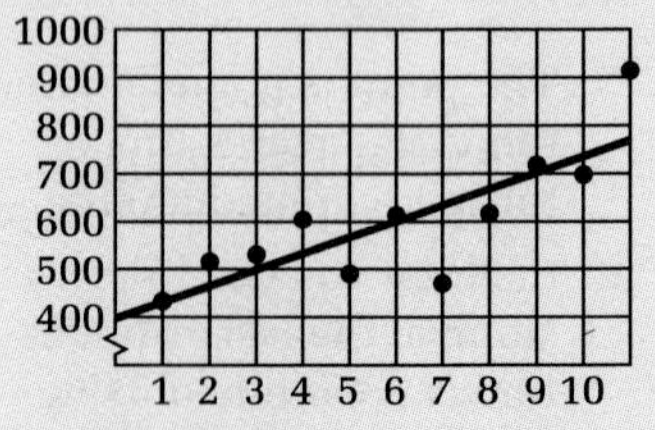

5 TECHNOLOGY The table at the right shows somo information about canned soups prepared with an equal volume of water. The portions are each 100 g.

TYPE OF SOUP	Water (percent)	Food energy (Calories)
Beef noodle	93.2	28
Celery, cream of	92.3	26
Chicken, cream of	91.9	39
Minestrone	89.5	43
Mushroom, cream of	89.6	56
Tomato	90.5	36
Vegetable beef	91.9	32
Vegetable with beef broth	91.7	32

a. Make a scatter plot of the data. Let x = the percent of water (including the water content of the ingredients) and y = the number of Calories.

b. Find an equation of the least-squares line. Graph the line.

c. **Writing** Explain what a negative correlation coefficient means in this case.

d. Split pea soup is 85.4% water. What would you expect to be the number of Calories in 100 g of split pea soup?

6 TECHNOLOGY The record pumpkin weight in 1994 at the Topsfield Fair was 914 lb. Use this fact and the data from Sample 1 on page 351.

a. Find an equation of the least-squares line for the pumpkin weight data for 1984–1994.

b. Make a scatter plot and graph the least-squares line.

c. **Writing** Compare the line in parts (a) and (b) with the line in Sample 1. Explain why the slope of the line for 1984–1994 is greater than the slope of the line for 1984–1993.

BY THE WAY...

Pumpkins are eaten for good fortune in Japan for *Tōji* (the Winter Solstice), especially in rural areas.

7 TECHNOLOGY Japan joined the International Pumpkin Association's World Championship in 1987, competing against the United States, Great Britain, and Canada. The top weights in Japan for the years 1987–1994 are shown below.

Top Pumpkin Weights — Japan

Year	1987	1988	1989	1990	1991	1992	1993	1994
Weight (lb)	326.5	313	473	439	606.5	587.3	556.7	690

a. Make a scatter plot of the data.

b. Find an equation of the least-squares line. (*Hint:* You may want to use "Years since 1986" for the x-values.)

c. Use the equation of the least-squares line to predict the top pumpkin weight in Japan for the year 2000. How does this compare to the prediction for Topsfield in Sample 1?

Answers to Exercises and Problems

c. The y-intercept of the line for 1984–1994 is less than the y-intercept of the line for 1984–1993. The slope of the line for 1984–1994 is greater than the slope of the line for 1984–1993 because of the point (11, 914).

7. a.

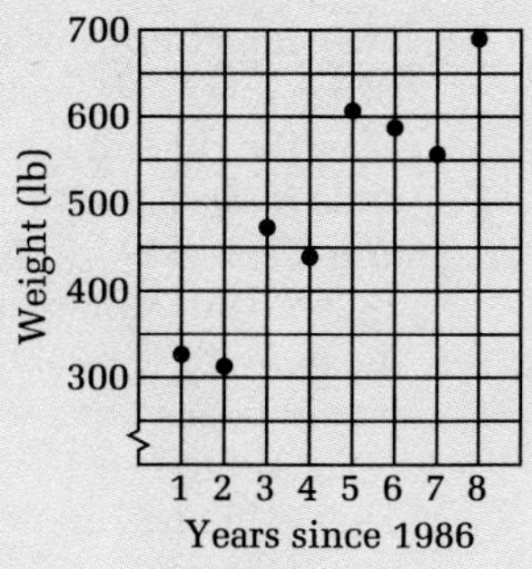

b. $y = 50.9x + 270$, where x is the number of years since 1986

c. 982.6 lb; This prediction is much greater than the prediction for Topsfield.

Multicultural Note

Sample 1 on page 351, Ex. 6, and Ex. 7 discuss pumpkins. Pumpkins were probably first cultivated between 5000 B.C. and 2500 B.C. in southeastern Mexico. Over the centuries, they came to be a staple in the diet of many North American peoples. In the southwest, the Cochiti pueblo emphasized the importance of pumpkins by naming one of their two major subdivisions Pumpkin (the other was Turquoise). The northeastern Native Americans generally cultivated pumpkins along with corn and kidney beans in small mounds place at three foot intervals. When the Pilgrims arrived in North America, the cultivation of pumpkins was one of the many valuable skills they learned from their Native American neighbors.

Using Technology

In Ex. 7, students are asked to use the least-squares line to make a prediction. The safest way to do this is to enter the equation with full calculator accuracy for the coefficients. On the TI-82, first find the coefficients. Next, press [Y=]. Enter the equation on the Y1= line as follows. Press [VARS], select item 5 (Statistics), press [▶] to get the EQ menu. Once this menu is displayed, select item 7 (RegEQ). The equation is transferred to the Y= list. Press [GRAPH] to view the graph. (For the TI-81, consult the manual.)

Students often need to round coefficients in equations for least-squares lines. To see how rounding affects the graph (and, hence, predictions made from the graph), it is best to graph the equation that uses nonrounded coefficients together with the equation that has rounded coefficients. By using the TRACE feature and the [▼] and [▲] keys, you can gauge how much y-values differ from one equation to the other.

Mathematical Procedures

Ex. 9 introduces the term *extrapolate*. Values found through extrapolation are often less accurate than values found through interpolation, or estimating values *within* the given information.

Cooperative Learning

Ex. 10 can be extended into a cooperative project. Partners can first make a poster showing their data and graphs. The least-squares line can then be drawn on the poster and students can present their results to the class.

8 TECHNOLOGY The table shows the number of Calories you might use while in-line skating at a comfortable pace for different periods of time.

In-Line Skating

Time (minutes)	3	7	12	18	25	33
Energy used (Calories)	29	67	115	172	238	314

a. Make a scatter plot of the data points. Find an equation of the least-squares line and find the correlation coefficient.

b. **Writing** State a conjecture suggested by the graph.

c. Find the mean for the skating time, $\overline{x}$, and the mean for the energy used, $\overline{y}$.

d. Show that $(\overline{x}, \overline{y})$ is a point on the least-squares line.

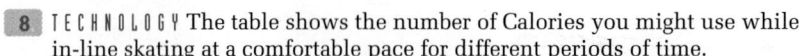

9. The National Youth Physical Fitness Program establishes qualifying standards for the Presidential Physical Fitness Awards. The table below shows the standards for boys ages 9–17 on the one-mile run test.

One-mile Run Test

Age (years)	9	10	11	12	13	14	15	16	17
Time (seconds)	511	477	452	431	410	386	380	368	366

a TECHNOLOGY Find an equation of the least-squares line.

b. What is the meaning of the slope and the y-intercept of the equation of the least-squares line in this situation?

c. **Writing** When a model is used to predict results beyond the given data, you *extrapolate* from the model. Tell whether you think this model would be useful for extrapolating the one-mile run time for a 25-year-old man.

10. **Group Activity** Work with another student.

a. **Open-ended** Collect data about your classmates or friends. Some suggested topics are given below, or you may decide to gather information on some other topic.

- heights of students and heights of same-sex adults in the family
- the number of hours spent on homework each week and the number of hours spent watching TV each week

b TECHNOLOGY Make a scatter plot of student height versus same-sex adult height, or of hours of homework versus hours of TV, or of the other information you gathered.

c. **Writing** Make a conjecture about the relationship between the two sets of data in your scatter plot.

d TECHNOLOGY Does your scatter plot look linear? If so, find an equation of the least-squares line.

Answers to Exercises and Problems

8. a.

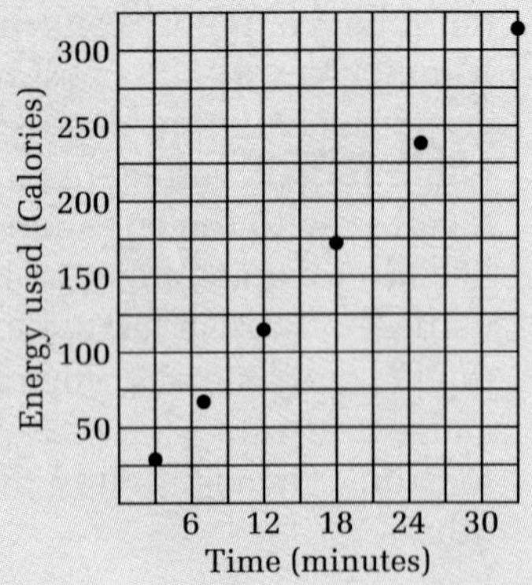

$y = 9.5x + 0.7$; $r = 0.999997$

b. The number of Calories used while skating is a linear function of the time in minutes spent skating.

c. $\frac{49}{3}$; $\frac{935}{6}$

d. $9.5\left(\frac{49}{3}\right) + 0.7 = \frac{4676}{30} \approx \frac{4675}{30} = \frac{935}{6}$

9. a. $y = -18.3x + 657.6$

b. The slope indicates that as age increases by 1 year, the time in seconds it takes to run 1 mile decreases by about 18.3 s. The y-intercept is the number of seconds it would take a person 0 years old to run 1 mile. This value has no meaning in the real world.

c. With this model, the time for a 25-year-old man is 3 min 20 s. This would establish a new world record, so the model is not useful for extrapolating for the average 25-year-old man.

10. Check students' work.

11. **Health Care** The table shows that health care spending on a per person basis is higher in the United States than in many other countries.

Health Care Spending in 1991		
COUNTRY	GDP (dollars per person)	Spending on Health Care (dollars per person)
Greece	7,775	450
Portugal	9,191	624
Spain	12,719	848
New Zealand	13,883	1047
Netherlands	16,530	1360
France	18,227	1650
Canada	19,178	1915
United States	22,204	2867

The gross domestic product (GDP) is the total value of goods and services produced entirely within a country in one year.

a. Make a scatter plot of the data with GDP on the horizontal axis.

b. Draw vertical lines to divide the points into three equal or nearly equal groups.

c. Plot or circle the summary points.

d. Use the method from Sample 2 to find an equation of the median-median line. Graph the line.

e. **Writing** Tell whether or not you think the median-median line is a good fit for the health care spending data. Would the least-squares line be a better fit? Give a reason for your answer.

12. **Demographics** Use the information about population and area for the metropolitan regions in the table.

Cities Around the World	Mexico City	Cairo	London	Tokyo	Calcutta	Moscow	Rome
Estimated population in 1995 (thousands)	23,913	11,155	8,897	28,447	12,885	10,769	3,079
Area (square miles)	522	104	874	1089	209	379	69

a. **Writing** Which linear model would you expect to be a better fit for the data, the *least-squares line* or the *median-median line*? Explain your choice.

b. Find an equation of the model you chose and draw the line.

c. The estimated mid-year population in 1995 for the city of Chicago is 6,541,000. Use your model from part (b) to predict the area of Chicago.

d. **Research** Find the actual area of the Chicago metropolitan region from an almanac or encyclopedia. How do the actual value and your prediction compare?

Career Note

Exs. 3 and 12 discuss demographics. A demographer is a person who uses statistics to study human populations. Demographers are particularly interested in the size, density, distribution, and other characteristics of a population, such as births, deaths, marriages, diseases, and so on. Every 10 years, the U.S. Bureau of the Census counts the number of people in the United States and also collects other information such as ages and income. Statistical methods are then used to organize and analyze the data collected so it can be interpreted and understood . The science of vital and social statistics is called demography.

Answers to Exercises and Problems

11. a–c.

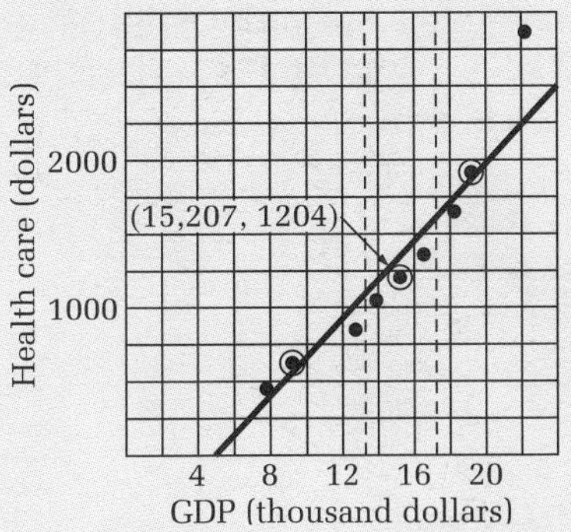

d. Answers may vary due to rounding; $y = 0.13x - 639.54$.

e. Answers may vary. An example is given. I think the median-median line is a better fit than the least-squares line. The data point for the United States appears to be an outlier.

12. a. Answers may vary. An example is given. I think the median-median line would be a better fit because the data point for London appears to be an outlier.

b. Equations may vary due to rounding. least-squares line: $y = 0.028x + 61.23$; median-median line: $y = 0.017x + 257.13$

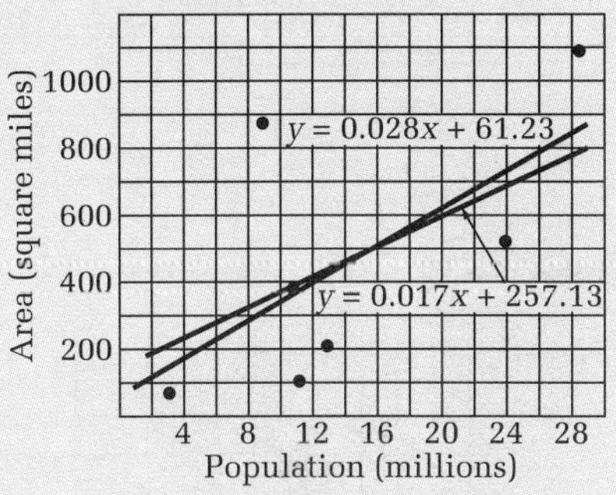

c. Predictions may vary due to rounding. least-squares line: about 244 mi^2; median-median line: about 368 mi^2

d. Comparisons may vary based on whether population and area are given for the city proper or for the metropolitan area.

Integrating the Strands

Ex. 13 shows how geometry and statistics can be integrated when finding the median-median line. The point found by averaging the x- and y-coordinates of the summary points is also the centroid of the triangle formed by connecting these summary points.

Working on the Unit Project

For Ex. 18, students examine their survey data to see if linear relationships exist between any two sets of responses. Students can divide up the work for part (a) among group members and then discuss part (b) together.

Practice 49 For use with Section 6-4

Name ______________________ Date ______________

Practice 49

For use with Section 6-4

1. The table shows the population and the cost of living in 10 selected American cities.

City	1990 Population (100,000's)	1993 Cost of Living Index (Percent)
Boston, MA	5.74	137.1
San Diego, CA	11.10	128.5
Minneapolis, MN	3.68	102.4
Houston, TX	16.30	99.4
Omaha, NE	3.32	92.9
Dallas, TX	10.06	105.2
Seattle, WA	5.16	119.7
Toledo, OH	3.33	101.6
Denver, CO	4.68	105.1
Atlanta, GA	3.94	101.5

a. Suppose you had a scatter plot of the data in the table, showing the population along the horizontal axis and the cost of living index along the vertical axis. In which range would you expect the correlation coefficient r for the data set to be:

$-1 \le r \le -\frac{1}{2}$, $-\frac{1}{2} \le r \le \frac{1}{2}$, or $\frac{1}{2} \le r \le 1$? $-\frac{1}{2} \le r \le \frac{1}{2}$

b. Make a scatter plot of the data. Look at the distribution of the data points. Does your scatter plot support your answer for part (a)? See additional answers.

2. The table below shows the times spent by nine participants in a Community Action Day, and the amounts of money they raised.

Time Spent (h)	1.5	2.3	2.4	2.8	1.8	2.0	3.2	4.0	4.5
Money Raised ($)	50	75	70	80	65	75	90	95	100

a. Find an equation of the regression line for the data. $y = 14.55x + 38.17$

b. Make a scatter plot of the data and graph the regression line. See additional answers.

c. Use the regression line to predict how much a student who worked 5 h might raise. about $111

3. The table below shows the levels of blood cholesterol of a group of women and the average number of hours per week they spend exercising.

Exercise Time (h)	1.5	2.0	1.0	3.0	2.5	3.5	5.0	4.0	4.5
Cholesterol Level	200	175	190	160	165	180	155	170	175

a. Make a scatter plot of the data. See additional answers.

b. Draw vertical lines to divide the data points into three equal groups. See additional answers.

c. Find an equation of the median-median line. $y = -6.67x + 195$

d. Use the equation of the median-median line to estimate the blood cholesterol level of a woman who exercises 5.0 h per week. about 162

4. ***Writing*** Do you think there is a cause-and-effect relationship between the two quantities graphed in Exercise 3? Why or why not? What other factors could account for the high correlation between the two quantities?
Answers may vary. Check students' work.

Practice Bank, INTEGRATED MATHEMATICS 3

49

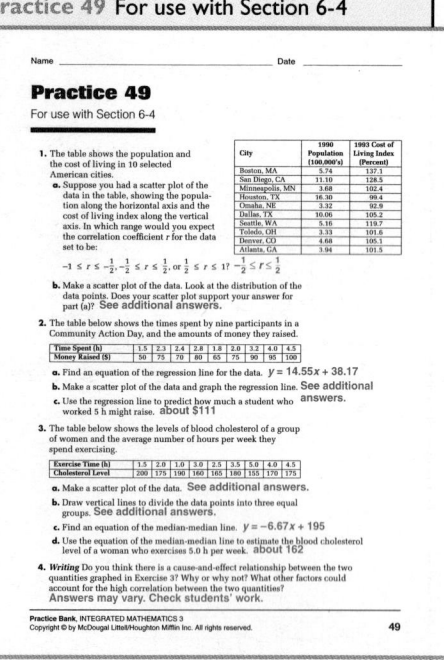

13. In finding the median-median line, the first step is to find the three summary points. If these three points do not lie on the same line, they are the vertices of a triangle. The median-median line passes through a special point called the *centroid* of this triangle.

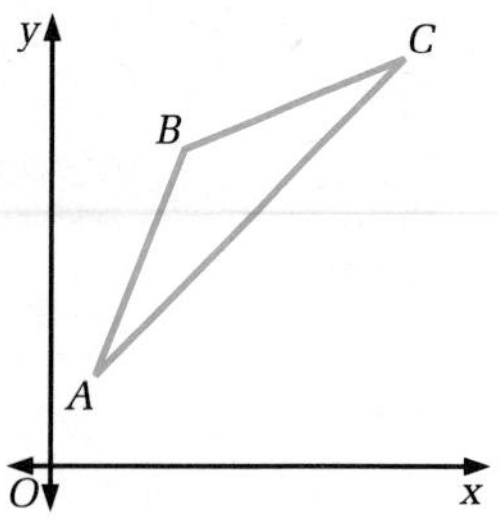

a. On graph paper, locate these three summary points: $A(2, 4)$, $B(6, 14)$, and $C(16, 18)$. Draw $\triangle ABC$.

b. Find the midpoints of each side of $\triangle ABC$.

c. Draw the medians of $\triangle ABC$ by joining each vertex to the midpoint of the opposite side. Find the coordinates of the point where the three medians meet. This is the centroid of the triangle.

d. Show that the coordinates of the centroid can be found by averaging the x-coordinates and averaging the y-coordinates of the summary points.

Ongoing ASSESSMENT

14. **Writing** Compare the median-median line and the least-squares line. How are they alike? How are they different?

Review PREVIEW

15. The weight of fish in Central Pond have a normal distribution with $\bar{x} = 28$ lb and $\sigma = 9$ lb. About what percent of the fish in the pond weigh between 10 lb and 37 lb? *(Section 6-3)*

16. Write the first five terms of the sequence: $a_1 = -3$ *(Section 4-3)*
$a_n = 2a_{n-1} + 5$

17. Write a matrix equation for the system: *(Section 1-5)*

$$2x + 3y - z = 16$$
$$-x + 5y + 8z = 16$$
$$5x - y = -7$$

6 Working on the Unit Project

Use the results of your survey from Exercise 18 on page 337.

18 TECHNOLOGY

a. Make several scatter plots comparing two sets of responses from your survey. Do any of the scatter plots suggest a linear model?

b. Are you surprised at the number of scatter plots that are linear? Explain.

c. Choose the scatter plot that looks most linear and find an equation of the least-squares line.

Answers to Exercises and Problems

13. a–c.

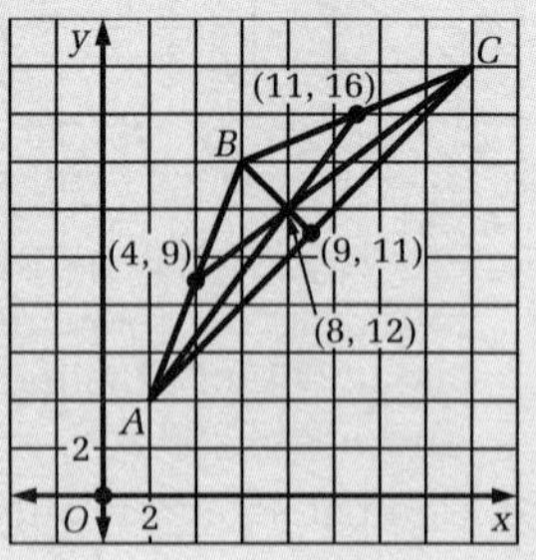

d. $\left(\frac{4 + 11 + 9}{3}, \frac{9 + 16 + 11}{3}\right) =$ (8, 12)

14. Answers may vary. An example is given. Both are lines whose equations provide linear models for a set of data and can be used to predict future behavior. The least-squares equation is based on all the data points, while the equation of the median-median line is based on points that summarize the data. If the data set includes outliers at its extremes, the median-median line may model the data better than the least-squares line.

15. about 81.5%

16. −3, −1, 3, 11, 27

17. $\begin{bmatrix} 2 & 3 & -1 \\ -1 & 5 & 8 \\ 5 & -1 & 0 \end{bmatrix} \begin{bmatrix} x \\ y \\ z \end{bmatrix} = \begin{bmatrix} 16 \\ 16 \\ -7 \end{bmatrix}$

18. Answers may vary. Check students' work.

Section

6-5 Fitting Quadratic Models to Data

Focus

Decide whether a quadratic model fits data. Use technology and matrices to find quadratic models.

STRAIGHT OR CURVED

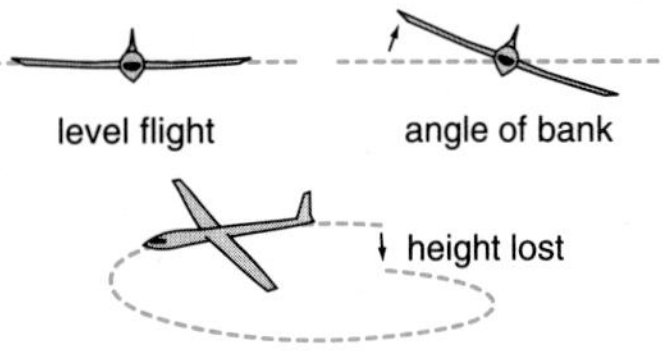

Talk it Over

The table below shows the height loss in feet for a glider completing a circle at the given angle of bank.

Angle of bank (degrees)	20	25	30	35	40	45	50
Height loss in 360° turn (feet)	108	91	81	74	71	70	71

x	y-values Actual	y-values Model	Deviation d	d^2
20	108	?	?	?
25	91	?	?	?
30	81	?	?	?
35	74	?	?	?
40	71	?	?	?
45	70	?	?	?
50	71	?	?	?
				$\Sigma d^2 = ?$

1. Copy and complete the table at the left, using parts (a)–(e).
   a. Use a graphics calculator to find an equation of the least-squares line for the data. Let x represent the angle of bank and y represent the height loss in feet.
   b. Substitute the x-values from the data points to find the y-values of corresponding points on the least-squares line.
   c. Find the vertical deviation d between each data point and the least-squares line. Use the formula
      $$d = \text{actual } y\text{-value} - \text{model's } y\text{-value}.$$
      The deviation may be positive, negative, or zero.
   d. Find the square of each of the deviations, d^2.
   e. Find the sum of the squares of the deviations, Σd^2. Statisticians use this value to measure how well a model fits a data set.
2. Look at the scatter plot shown. The data points at each end are above the least-squares line, and those in the middle are below the line. Do the points seem to lie on a curve rather than on a line? If so, what type of curve? Explain.

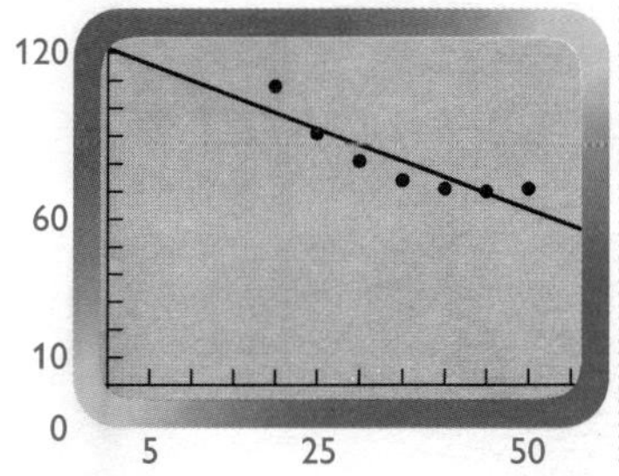

PLANNING

Objectives and Strands

See pages 328A and 328B.

Spiral Learning

See page 328B.

Materials List

- Graph paper
- Graphics calculator or statistical software
- Small beans or other counters
- 3 jar lids of different sizes

Recommended Pacing

Section 6-5 is a one-day lesson.

Extra Practice

See pages 617–619.

Warm-Up Exercises

- Warm-Up Transparency 6-5

Support Materials

- Practice Bank: Practice 50
- Activity Bank: Enrichment 45
- Study Guide: Section 6-5
- Problem Bank: Problem Set 14
- Explorations Lab Manual: Diagram Masters 1, 2
- Assessment Book: Quiz 6-5, Test 25, Alternative Assessments 5, 6

Answers to Talk it Over

1. a. Answers may vary due to rounding; $y = -1.16x + 121.6$.

b–d.

x	y-values Actual	y-values Model	Deviation d	d^2
20	108	98.4	9.6	92.16
25	91	92.6	–1.6	2.56
30	81	86.8	–5.8	33.64
35	74	81	–7	49
40	71	75.2	–4.2	17.64
45	70	69.4	0.6	0.36
50	71	63.6	7.4	54.76

e. 250.12

2. Yes; a parabola.

TEACHING

Additional Sample

S1 The table below shows the median prices of houses for sale in a certain town.

Year	Price
1984	$42,500
1985	45,000
1986	48,000
1987	62,000
1988	65,000
1989	68,000
1990	67,500
1991	80,000
1992	81,000
1993	79,000
1994	82,500

a. Use a graphics calculator to find an equation of the least-squares line for the data.

Let x = the number of years since 1984 and y = the median price of a house. The equation of the least-squares line is about $y = 4304.55x + 43{,}977$.

b. Find the least-squares parabola for the house price data and compare it with the least-squares line.

Step 1. Use technology to perform a quadratic regression. A parabola that fits the data is $y = -273.31x^2 + 7037.6x + 39{,}878$.

Step 2. Decide whether a linear model or a quadratic model fits the data better. Find which model has the smaller *sum of squares*. (See below).

The quadratic model has a smaller sum of squares, so the parabola is a better fit for the data than the line.

Linear Model

x	Actual y-value	$y = 4304.55x + 43{,}977$	Deviation d	d^2
0	42,500	43,977	–1477	2,181,529
1	45,000	48,282	–3282	10,771,524
2	48,000	52,586	–4586	21,031,396
3	62,000	56,891	5109	26,101,881
4	65,000	61,195	3805	14,478,025
5	68,000	65,500	2500	6,250,000
6	67,500	69,804	–2304	5,308,416
7	80,000	74,109	5891	34,703,881
8	81,000	78,413	2587	6,692,569
9	79,000	82,718	–3718	13,823,524
10	82,500	87,023	–4523	20,457,529
				$\sum d^2 = 161{,}800{,}274$

Quadratic Model

x	Actual y-value	$y = -273.31x^2 + 7037.6x + 39{,}878$	Deviation d	d^2
0	42,500	39,878	2622	6,874,884
1	45,000	46,642	–1642	2,696,164
2	48,000	52,860	–4860	23,619,600
3	62,000	58,531	3469	12,033,961
4	65,000	63,655	1345	1,809,025
5	68,000	68,233	–233	54,289
6	67,500	72,264	–4764	22,695,696
7	80,000	75,749	4251	18,071,001
8	81,000	78,687	2313	5,349,969
9	79,000	81,078	–2078	4,318,084
10	82,500	82,923	–423	178,929
				$\sum d^2 = 97{,}701{,}602$

The points in the scatter plot on page 359 seem to lie closer to a parabola than to a line. In this case it is better to model the glider data by a **least-squares parabola,** which you can obtain by performing a quadratic regression using technology.

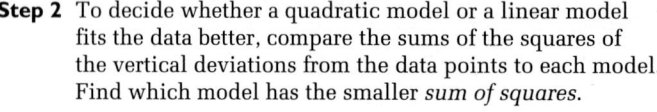

Sample 1

Find the least-squares parabola for the glider data and compare it to the least-squares line.

Sample Response

Step 1 Recall that the general equation for a parabola is $y = ax^2 + bx + c$.

Use technology to perform a quadratic regression and find a, b, and c. A parabola that fits the data is $y = 0.068x^2 - 5.9x + 198$.

With technology, graph this quadratic model along with the scatter plot and the linear model.

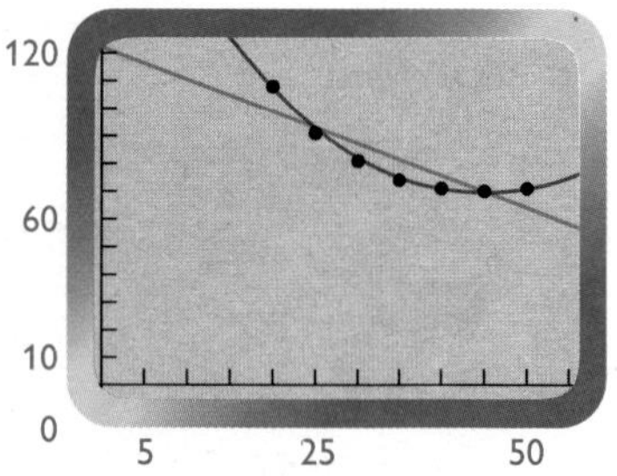

Step 2 To decide whether a quadratic model or a linear model fits the data better, compare the sums of the squares of the vertical deviations from the data points to each model. Find which model has the smaller *sum of squares.*

x	Actual y-value	$y = 0.068x^2 - 5.9x + 198$	Deviation d	d^2
20	108	107.2	0.8	0.64
25	91	93	−2	4
30	81	82.2	−1.2	1.44
35	74	74.8	−0.8	0.64
40	71	70.8	0.2	0.04
45	70	70.2	−0.2	0.04
50	71	73	−2	4
				$\sum d^2 = 10.8$

The sum of the squares of the deviations from the data points to the quadratic model is 10.8. From *Talk it Over* question 1, you should have found that the sum of the squares of the deviations for the linear model is about 300.

The quadratic model has a smaller sum of squares, so the parabola is a better fit for the data set than the line.

The glider data in Sample 1 was modeled using quadratic regression. You can also use a system of equations to fit a quadratic model to data. For this method, you need only three data points.

Sample 2

Use a system of equations to find a quadratic model to fit the data points (20, 108), (35, 74), and (50, 71).

Sample Response

1 To fit a quadratic model to the data, you need to find a, b, and c in the equation $y = ax^2 + bx + c$. Substitute the three data points.

$400a + 20b + c = 108$ ← Substitute 20 for x and 108 for y.

$1225a + 35b + c = 74$ ← Substitute 35 for x and 74 for y.

$2500a + 50b + c = 71$ ← Substitute 50 for x and 71 for y.

2 This is a system of three equations with three variables. The matrix equation representing this system is:

$$\begin{bmatrix} 400 & 20 & 1 \\ 1225 & 35 & 1 \\ 2500 & 50 & 1 \end{bmatrix} \begin{bmatrix} a \\ b \\ c \end{bmatrix} = \begin{bmatrix} 108 \\ 74 \\ 71 \end{bmatrix}$$

← coefficient matrix · variable matrix = constant matrix

3 Use a graphics calculator to multiply both sides by the inverse of the coefficient matrix to find the solution matrix.

$$\begin{bmatrix} a \\ b \\ c \end{bmatrix} = \begin{bmatrix} 0.06889 \\ -6.0556 \\ 201.556 \end{bmatrix}$$

← solution matrix

A quadratic model for the given data is $y = 0.069x^2 - 6.1x + 200$.

Talk it Over

3. Why are three linear equations needed to fit a quadratic model to data?
4. Why is the quadratic model found in Sample 1 slightly different from the model found in Sample 2?
5. Would choosing a different group of three data points from the data set result in the same quadratic model as in Sample 2? Explain.
6. How could you decide which quadratic model fits a data set better?

Look Back

Describe which types of data sets may be best modeled by a linear least-squares equation, by a median-median equation, or by a quadratic least-squares equation.

Additional Sample

S2 Use a system of equations to find a quadratic model to fit the data points (8, 110), (13, 305), and (4, 26).

Find a, b, and c in the equation $y = ax^2 + bx + c$. Substitute the three data points.

$64a + 8b + c = 110$
$169a + 13b + c = 305$
$16a + 4b + c = 26$

This is a system of three equations with three variables. The matrix equation representing this system is:

$$\begin{bmatrix} 64 & 8 & 1 \\ 169 & 13 & 1 \\ 16 & 4 & 1 \end{bmatrix} \begin{bmatrix} a \\ b \\ c \end{bmatrix} = \begin{bmatrix} 110 \\ 305 \\ 26 \end{bmatrix}$$

Use a graphics calculator to multiply both sides by the inverse of the coefficient matrix to find the solution matrix.

$$\begin{bmatrix} a \\ b \\ c \end{bmatrix} = \begin{bmatrix} 2 \\ -3 \\ 6 \end{bmatrix}$$

A quadratic model for the given data is $y = 2x^2 - 3x + 6$.

Integrating the Strands

Sample 2 integrates the strands of algebra, functions, statistics, and discrete mathematics. Matrices are used to solve the algebraic system of equations. The solution to the system is the quadratic function, which models the given data.

Answers to Talk it Over

3. There are three unknowns: a, b, and c.
4. In Sample 1, all the points were used to find the least-squares parabola. In Sample 2, a curve that passes through 3 of the points was found.
5. No; only if all the points actually are on the same curve as the points in Sample 2.
6. Choose the quadratic equation for which the sum of the squares of the distances from the curve is the least.

Answers to Look Back

A data set in which the points appear to lie on or near a straight line may best be modeled by a linear least-squares equation. If the data set includes outliers, the data may best be modeled by the median-median equation. Data points whose scatter plot resembles a parabolic curve may best be modeled by a quadratic least-squares equation.

APPLYING

Suggested Assignment

Standard 1–3, 5–8, 10, 15–17

Extended 1–3, 5–8, 10, 15–17

Integrating the Strands

Algebra Exs. 2, 5–9

Functions Exs. 3, 11, 12, 14

Geometry Ex. 15

Statistics and Probability Exs. 1–14, 16, 17

Discrete Mathematics Exs. 5–9

Logic and Language Exs. 1, 2, 4, 9, 10

6-5 Exercises and Problems

TECHNOLOGY **For the technology exercises in this section, you can use a graphics calculator or statistical software.**

1. **Reading** How would you determine whether a linear model or a quadratic model fits a set of data better?

2. a. Use the least-squares quadratic equation from Sample 1 to predict the height loss for an angle of bank of 60°.

   b. The actual height loss is 80 feet. Explain why this value and the value in part (a) might be different.

   c. At about what angle of bank is there least height loss?

3. **Automobiles** An advertisement for a sports car states that the car reaches its maximum power, measured in horsepower (hp), at a lower engine speed, measured in revolutions per minute (rpm), than other cars. Some estimated values on which this claim was made are shown here.

Engine speed (rpm)	1000	2000	3000	4000	5000	6000
Engine power (hp)	16	35	55	72	77	68

   a. Make a scatter plot of the data.

   b. Find a quadratic model for the data.

   c. At about what engine speed does the maximum engine power occur?

4. TECHNOLOGY Look at the scatter plot on page 350 of Section 6-4 that shows distance and roundtrip airfare. The data are shown below.

WORLD TRAVEL	Mexico City	Cairo	London	Tokyo	Calcutta	Moscow	Rome
Air distance from New York City (miles)	2094	5602	3458	6740	7918	4665	4281
Roundtrip airfare from New York City (dollars)	250	750	375	1200	1500	624	520

   a. Perform a linear regression and graph the least-squares line for the data.

   b. Find a quadratic model for the data points.

   c. Graph the parabola on the same scatter plot as the least-squares line.

   d. **Writing** Which model seems to fit the data better, the *linear* model or the *quadratic* model? Explain.

Answers to Exercises and Problems

1. Graph points to see if they appear to lie on a line or on a curve.

2. a. about 88.8 ft

   b. The equation is a model and may not fit the data exactly. Also, all the values are rounded.

   c. about 43.4°

3. a.

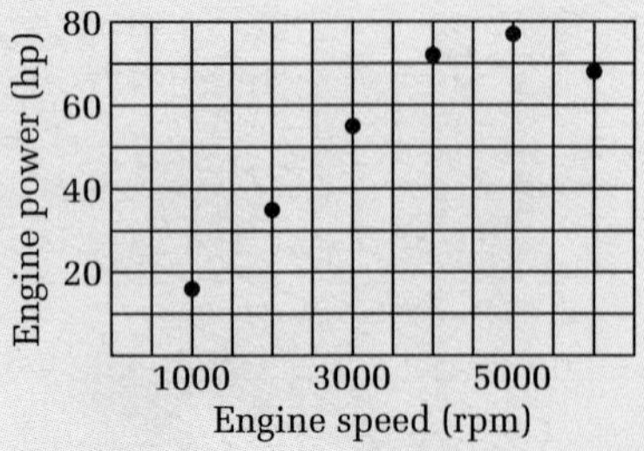

   b. Answers may vary depending on whether the matrix method or technology is used. An example is given; $y = (-4.13 \times 10^{-6})x^2 + 0.039x - 19.2$.

   c. about 4718 rpm

4. See answers in back of book.

5. a. $4a + 2b + c = 20$; $25a + 5b + c = 56$; $9a - 3b + c = 0$

   b. $\begin{bmatrix} 4 & 2 & 1 \\ 25 & 5 & 1 \\ 9 & -3 & 1 \end{bmatrix} \cdot \begin{bmatrix} a \\ b \\ c \end{bmatrix} = \begin{bmatrix} 20 \\ 56 \\ 0 \end{bmatrix}$

   c. $y = x^2 + 5x + 6$

6. a. $c = -4$; $0.36a + 0.6b + c = -2.2$; $0.64a - 0.8b + c = -0.8$

   b. $\begin{bmatrix} 0 & 0 & 1 \\ 0.36 & 0.6 & 1 \\ 0.64 & -0.8 & 1 \end{bmatrix} \cdot \begin{bmatrix} a \\ b \\ c \end{bmatrix} = \begin{bmatrix} -4 \\ -2.2 \\ -0.8 \end{bmatrix}$

   c. $y = 5x^2 - 4$

7. a. $a + b + c = 16$; $a - b + c = 16$; $4a - 2b + c = -32$

   b. $\begin{bmatrix} 1 & 1 & 1 \\ 1 & -1 & 1 \\ 4 & -2 & 1 \end{bmatrix} \cdot \begin{bmatrix} a \\ b \\ c \end{bmatrix} = \begin{bmatrix} 16 \\ 16 \\ -32 \end{bmatrix}$

   c. $y = -16x^2 + 32$

8. a. $c = 9$; $a + b + c = 4$; $9a + 3b + c = 0$

   b. $\begin{bmatrix} 0 & 0 & 1 \\ 1 & 1 & 1 \\ 9 & 3 & 1 \end{bmatrix} \cdot \begin{bmatrix} a \\ b \\ c \end{bmatrix} = \begin{bmatrix} 9 \\ 4 \\ 0 \end{bmatrix}$

In Exercises 5–8, the data points lie on a parabola.

a. Write a system of three linear equations with three variables.

b. Write a matrix equation for the system.

c. Solve the system for *a*, *b*, and *c*, and write an equation for the parabola.

5. (2, 20), (5, 56), (−3, 0)
6. (0, −4), (0.6, −2.2), (−0.8, −0.8)
7. (1, 16), (−1, 16), (−2, −32)
8. (0, 9), (1, 4), (3, 0)

9. a. Trace this scatter plot.

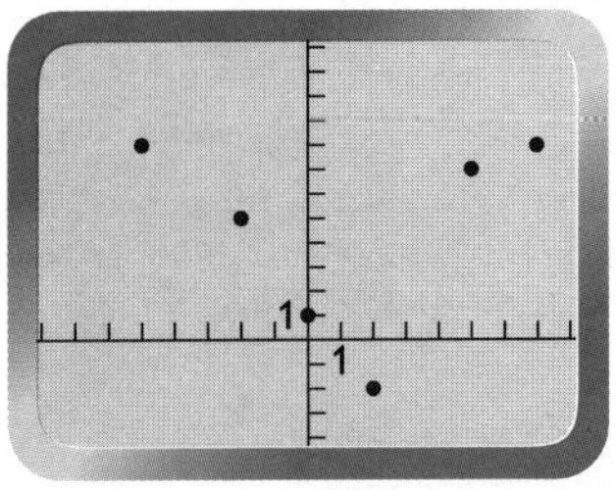

b. Sketch a parabola that you think fits the data points well.

c. TECHNOLOGY Find a quadratic regression equation using technology or choose three points and use a matrix equation to find a quadratic model.

d. Graph your equation from part (c) on the same set of axes as your sketch in part (b).

e. **Writing** How do the graphs in part (d) compare?

Experience (years)	Salary (dollars)
1	35,000
3	40,000
5	44,000
7	48,000
9	52,000
11	55,000
13	58,000
15	60,000
17	63,000
19	65,000
21	67,000
23	68,500
25	70,000
27	71,000
29	72,000

Average Annual Salaries

Salaries **The table gives the average annual salaries of engineers with master's degrees in the United States. Use the table for Exercises 10–12.**

10. a. Make a scatter plot of the data.

b. **Writing** How does the appearance of the scatter plot suggest a quadratic model?

11. TECHNOLOGY

a. Find a quadratic model for the data.

b. Graph the quadratic model on the same set of axes as the scatter plot.

c. What is the meaning of the *y*-intercept in this situation?

12. TECHNOLOGY

a. What is the approximate maximum salary for an engineer with a master's degree according to the model from Exercise 11?

b. After how many years of employment does an engineer with a master's degree reach the maximum salary?

c. According to the model, what happens after the maximum salary is reached?

BY THE WAY...

Engineering is a profession that puts scientific knowledge to practical use. Some of the branches of engineering are aerospace, biomedical, chemical, civil, electrical, environmental, industrial, mechanical, and nuclear.

Communication: Drawing

For Exs. 5–8, students can sketch what they think the parabola should look like based on the given points and discuss how they determined whether their parabolas open upward or downward. Do their sketches match the equation for each exercise?

Using Technology

Graphics calculators often permit students to find models other than linear and quadratic models. For exercises such as Ex. 10, you may want to encourage students to explore and compare other options that the calculator offers for regression equations.

Problem Solving

In Ex. 12, the graph shows a maximum salary while the table does not. Students can discuss why maximum salaries are necessary in any field. Ask students to examine the equation to determine why the graph shows a minimum value.

Answers to Exercises and Problems

c. $y = x^2 - 6x + 9$

9. a, b. Sketches may vary. An example is given.

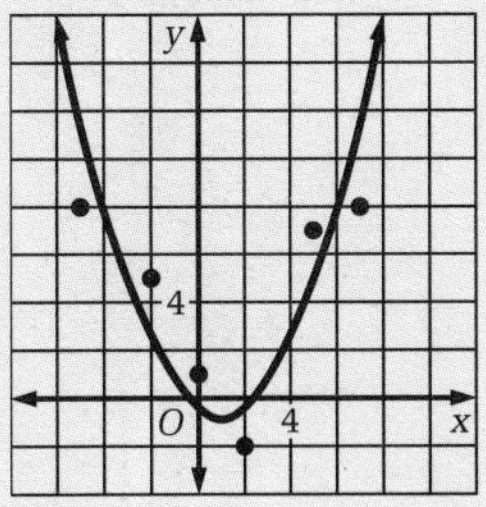

c. Answers may vary depending on whether the matrix method or technology is used. Answers derived from the matrix method will vary slightly depending on the points chosen. An example is given based on using the matrix method; $y = 0.32x^2 - 1.27x + 1$.

d. Answers may vary. An example is given.

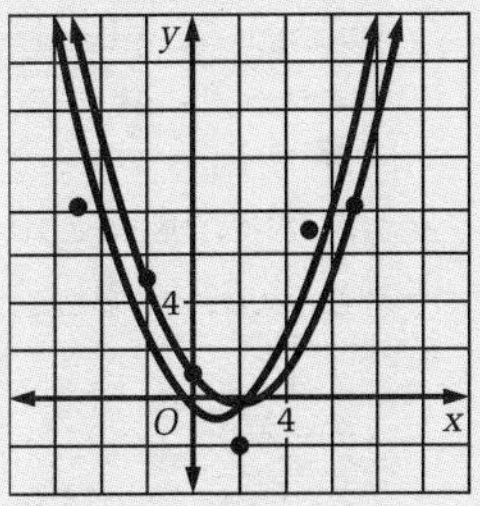

e. Answers may vary.

10. a.

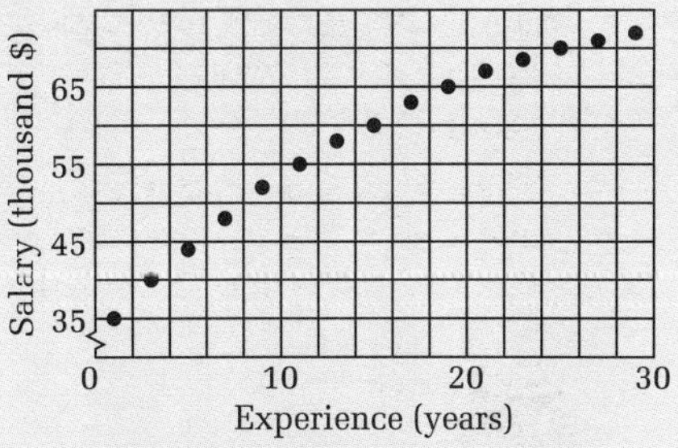

b. The points of the scatter plot appear to lie on a parabolic curve.

11. See answers in the back of book.

12. a. Estimates may vary; about \$72,500.

b. about 33 years

c. Salaries begin to fall.

Quick Quiz (6-4 through 6-5)

See page 366.

See also Test 25 in the *Assessment Book.*

Ongoing ASSESSMENT

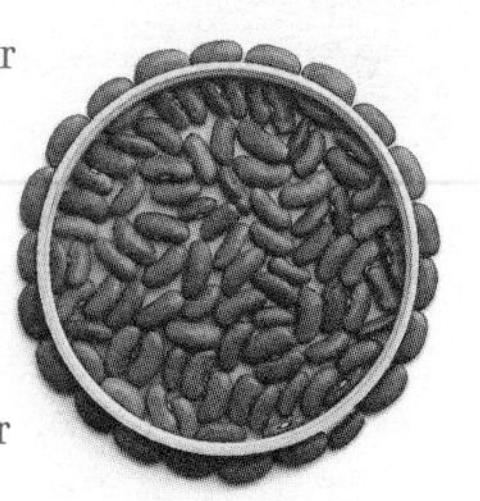

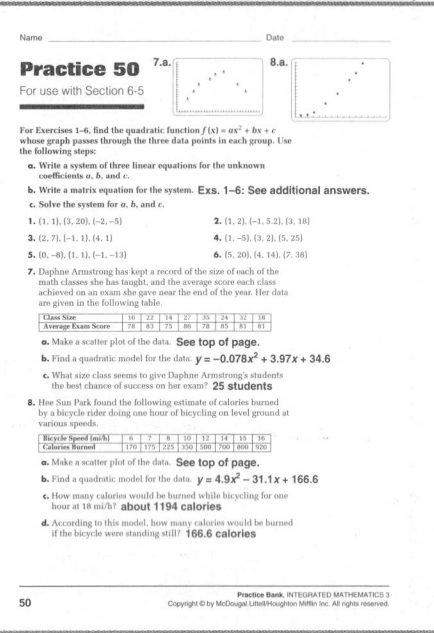

13. Group Activity Work with another student. You will need small beans or other counters, and three jar lids of different sizes.

a. **Using Manipulatives** Use three different-sized jar lids. Using small beans, collect two types of data from each lid—the number of beans needed to fill the lid one layer deep and the number of beans, laid end-to-end, needed to surround the lid.

b. Combine your data with the data of one other group and make a scatter plot of the six points.

c. TECHNOLOGY Find a model that you can use to predict the number of beans needed to fill a lid if you know the number it takes to surround the lid.

d. Compare your model with those of other groups. Discuss similarities and differences in the models.

Review PREVIEW

14. Use the table. *(Section 6-4)*

Tuition at Selected Universities in Illinois

Resident (dollars)	3025	2533	2455	3006	2590	2937
Nonresident (dollars)	7381	6228	6135	7506	6042	7425

a. Draw a scatter plot of the data.

b. Which kind of function appears to model the data? Why?

c TECHNOLOGY Find an equation to model this situation.

15. In $\triangle ABC$, $m\angle A = 100°$. Use an indirect proof to show that $\angle B$ is not a right angle. *(Section 3-7)*

16. Suppose you roll a die. What is the probability that you get an odd number? *(Toolbox Skill 6)*

6 Working on the Unit Project

Use your scatter plots from Exercise 18 on page 358.

17. a. Do any of your scatter plots suggest a quadratic model? Are you surprised at the number of scatter plots that are quadratic? Explain.

b. Choose the scatter plot that looks most quadratic and find a quadratic model for it.

Practice 50 For use with Section 6-5

Name ______________________ Date ____________

Practice 50

For use with Section 6-5

7.a.

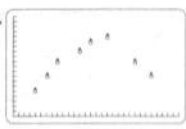

8.a.

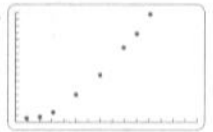

For Exercises 1–6, find the quadratic function $f(x) = ax^2 + bx + c$ whose graph passes through the three data points in each group. Use the following steps:

a. Write a system of three linear equations for the unknown coefficients *a*, *b*, and *c*.

b. Write a matrix equation for the system. **Exs. 1–6: See additional answers.**

c. Solve the system for *a*, *b*, and *c*.

1. (1, 1), (3, 20), (–2, –5)

2. (1, 2), (–1, 5.2), (3, 18)

3. (2, 7), (–1, 1), (4, 1)

4. (1, –5), (3, 2), (5, 25)

5. (0, –8), (1, 1), (–1, –13)

6. (5, 20), (4, 14), (7, 38)

7. Daphne Armstrong has kept a record of the size of each of the math classes she has taught, and the average score each class achieved on an exam she gave near the end of the year. Her data are given in the following table.

Class Size	16	22	14	27	35	24	32	18
Average Exam Score	78	83	75	86	78	85	81	81

a. Make a scatter plot of the data. **See top of page.**

b. Find a quadratic model for the data. $y = -0.078x^2 + 3.97x + 34.6$

c. What size class seems to give Daphne Armstrong's students the best chance of success on her exam? **25 students**

8. Hee Sun Park found the following estimate of calories burned by a bicycle rider doing one hour of bicycling on level ground at various speeds.

Bicycle Speed (mi/h)	6	7	8	10	12	14	15	16
Calories Burned	170	175	225	350	500	700	800	920

a. Make a scatter plot of the data. **See top of page.**

b. Find a quadratic model for the data. $y = 4.9x^2 - 31.1x + 166.6$

c. How many calories would be burned while bicycling for one hour at 18 mi/h? **about 1194 calories**

d. According to this model, how many calories would be burned if the bicycle were standing still? **166.6 calories**

50 **Practice Bank**, INTEGRATED MATHEMATICS 3

Answers to Exercises and Problems

13. Answers will vary. Check students' work.

14. See answers in back of book.

15. Suppose that $\angle B$ is a right angle. Then, since $m\angle C > 0$, $m\angle A + m\angle B + m\angle C = 100° + 90° + m\angle C > 190°$. This is not possible, since the sum of the measures of the angles of a triangle is 180°. So the assumption that $\angle B$ is a right angle is false. Therefore, $\angle B$ is not a right angle.

16. 0.5

17. Answers may vary.

Unit Project 6

Completing the Unit Project

Now you are ready to complete your profile of a typical car.

Your completed project should include these things:

- **a data table containing your survey results**
- **histograms of your survey results**
- **a visual display showing the features of your typical car**
- **a written analysis of your survey data, including a description of each distribution and its mean and standard deviation, as well as an explanation of how you selected the features of your typical car**

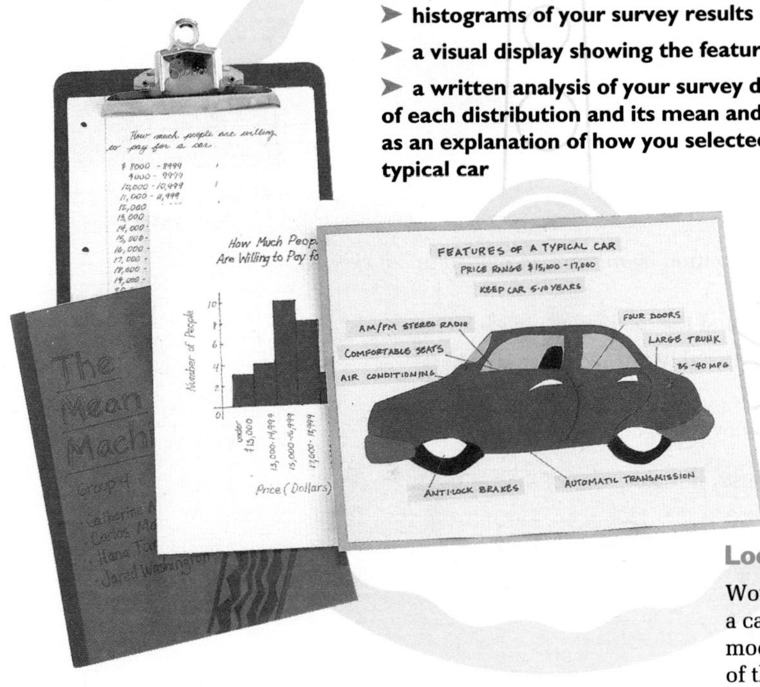

Look Back

Would you recommend that a car maker design a car model based on your profile of the typical car? Why or why not?

Alternative Projects

Project 1: TV Ratings

Find out how television rating services such as the A.C. Nielsen Company collect data. Also find out how television networks analyze these data in order to make decisions about programming. Write a report describing what you have learned. Include a discussion of the reliability of the data and of the decisions based on it.

Project 2: Is It Normal?

List at least three kinds of data sets that you think are normally distributed and three kinds that you think are not normally distributed. Obtain samples of all six types of data sets. For each data set, find the mean(s) and, if possible, the standard deviation. Describe each distribution and determine which, if any, of the data sets are normally distributed.

Assessment

A scoring rubric for the Unit Project can be found on pages 328 and 329 of this Teacher's Edition and also in the ***Project Book***.

Unit Support Materials

- *Practice Bank:* Cumulative Practice 51
- *Study Guide:* Unit 6 Review
- *Problem Bank:* Unifying Problem 6
- *Assessment Book:* Unit Tests 26 and 27, Spanish Unit Tests, Alternative Assessment
- *Test Generator* software with Test Bank
- *Teacher's Resources for Transfer Students*

Quick Quiz (6-4 through 6-5)

1. A data set has a correlation of –0.9. What conclusions can you make about the data? [6-4] **The two variables have a strong linear relationship with a negative slope.**
2. The three summary points for a median-median line are (3.1, 21), (5.9, 37.8), and (9.2, 51.6). Find the equation of the median-median line. [6-4] $y = 5.02x + 6.35$
3. What does the sum of the squares of the deviations tell you about a data set? [6-5] **how closely a particular model fits the data; It is used to compare one model to another.**
4. What system of equations would you use to find a quadratic model that goes through the three points (1, 7), (7, –23), (15, –175)? [6-5]
   $7 = a + b + c$,
   $-23 = 49a + 7b + c$,
   $-175 = 225a + 15b + c$

Unit 6 REVIEW AND ASSESSMENT

1. These histograms show the distribution of the ages of students in three swimming classes. **6-1**

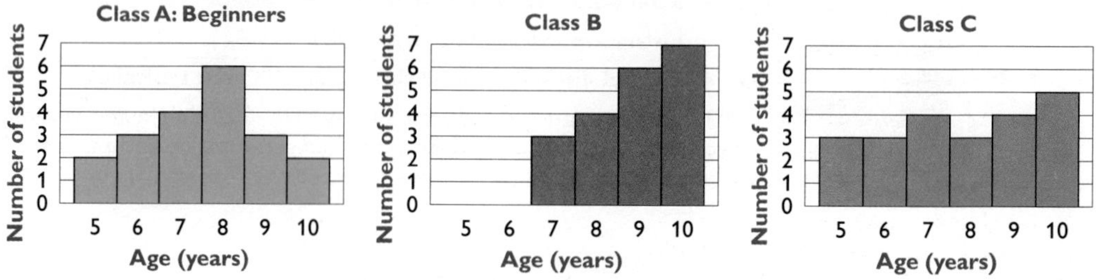

a. Tell whether each distribution is *uniform, mound-shaped, skewed,* or *bimodal.*

b. What are the mean age and the median age in each class?

c. **Writing** Which class probably has the best swimmers? Justify your answer.

2. a. Make a relative frequency histogram of this data.

Number of Counties in Each State of the United States
67, 12, 15, 75, 58, 63, 8, 3, 67, 159, 4, 44, 102, 92, 99, 105, 120, 64, 16, 23, 14, 83, 87, 82, 114, 56, 93, 16, 10, 21, 33, 62, 100, 53, 88, 77, 36, 67, 5, 46, 67, 95, 254, 29, 14, 95, 39, 55, 72, 23

b. Identify any outliers in the data set.

c. Find the mean and the median of the data and show them on your histogram.

d. What percent of states have between 20 and 49 counties?

3. a. Find the mean and the range of the number of sons and of the number of daughters of Presidents of the United States. **6-2**

Children of Presidents of the United States

n	0	1	2	3	4	5	6	7	8	9	10	11	12	13	14
No. of Presidents with *n* sons	13	4	8	7	4	2	1	1	1	0	0	0	0	0	0
No. of Presidents with *n* daughters	12	12	11	2	1	2	1	0	0	0	0	0	0	0	0
No. of Presidents with *n* children	6	2	8	6	7	3	5	1	1	0	1	0	0	0	1

Thirteen Presidents had no sons.
Eleven Presidents had two daughters.
President John Tyler had 14 children: 8 sons and 6 daughters.

Answers to Unit 6 Review and Assessment

1. a. A: mound-shaped; B: skewed; C: uniform
   b. A: 7.55, 8; B: 8.85, 9; C: 7.77, 8
2. a. Intervals may vary; bimodal.

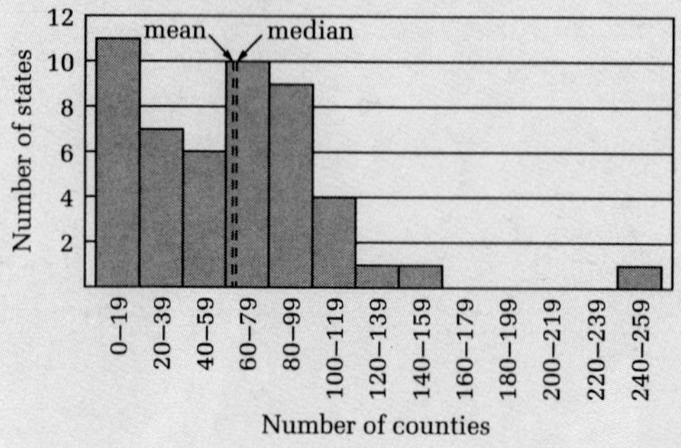

   b. 254
   c. mean: 61.64; median: 62.5; See graph.
   d. 18%
3. a. sons: 2.15, 8; daughters: 1.46, 6
   b. 2.05; 1.47

b. Find the standard deviation of the number of sons and of the number of daughters.

c. **Writing** The mean of the number of total children is about 3.6 with a standard deviation of about 2.8. Compare this to the values you found for sons and for daughters. What do you notice? Explain why you think the differences exist.

4. The age at inauguration for Presidents of the United States shows a normal distribution with a mean of 55 years and a standard deviation of 6 years. 6-3

a. About 95% of the Presidents were within what age range?

b. What percent of the Presidents were between the ages of 55 and 67 at inauguration? between the ages of 49 and 55?

5. a. Copy the sketch of a normal curve. Show the mean of the distribution.

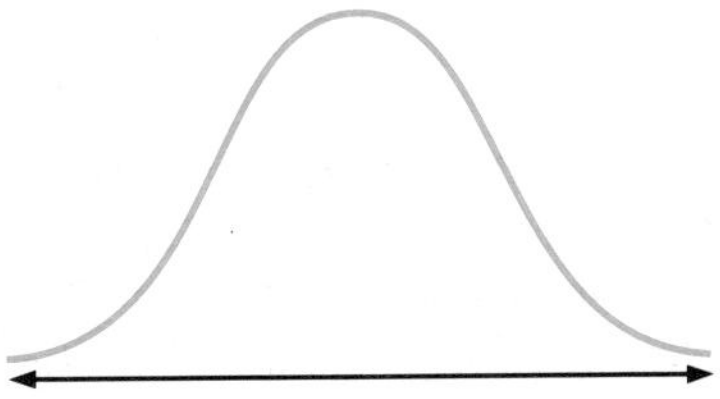

b. **Open-ended** On the copy, sketch another normal curve with the same mean but a different standard deviation.

c. **Open-ended** On the same drawing, sketch another normal curve with a different mean but the same standard deviation as the given curve.

d. **Writing** Explain how a normal curve changes when the mean and the standard deviation change.

6. The price of a pizza depends upon both the size (area) of the pizza and the fixed cost of making the pizza. 6-4

The area of a circle is
$A = \pi \cdot (\text{radius})^2$.
Use $\pi \approx 3.14$.

Plain Cheese Pizzas at Central Pizza

Size (diameter)	Personal (7 in.)	Small (10 in.)	Medium (12 in.)	Large (14 in.)	X-Large (16 in.)
Area (in.^{2})	38.48	78.54	113.10	153.94	201.06
Price ($)	2.49	6.49	8.49	11.49	14.49

a. Find a linear equation that models the price of a pizza as a function of its area.

b. Find the area of a pizza that is 20 in. in diameter and predict its price.

Answers to Unit 6 Review and Assessment

c. The mean is the sum of the means; $\frac{\text{number of children}}{\text{number of presidents}} = \frac{\text{number of daughters}}{\text{number of presidents}} + \frac{\text{number of sons}}{\text{number of presidents}}$. The standard deviation is greater when the numbers are combined because the presidents had so many more sons than daughters.

4. a. 43 yr to 67 yr

b. 47.5%; 34%

5. a–c. Sketches may vary.

a, b.

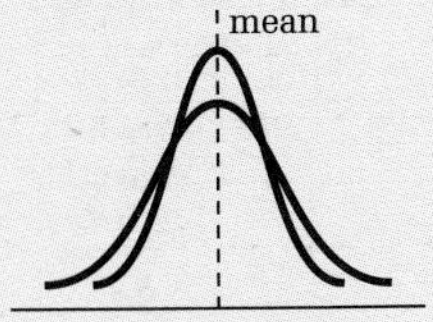

c.

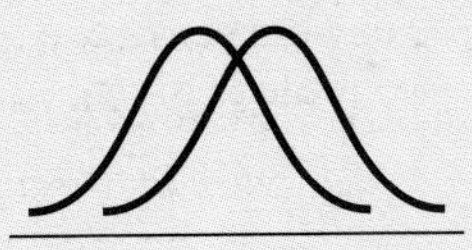

d. If the standard deviation is unchanged and the mean is increased (or decreased), the curve retains the same shape but shifts to the right (or left). If the mean is unchanged and the standard deviation increases (or decreases), the line of symmetry of the curve is unchanged, but the curve becomes flatter and wider (or steeper and narrower).

6. a. $P = 0.0723A + 0.23$

b. 314 in.^{2}; $22.93

Median Household Income

Years since 1986	Income (dollars)
1	26,061
2	27,225
3	28,906
4	29,943
5	30,126

7. **Household Income** The median household income for all households in the United States for the years 1987–1991 is shown in the table.
   a. Use quadratic least-squares regression or a system of equations to find a quadratic model for the situation.
   b. According to the model, what was the median household income in 1986?

8. **Self-evaluation** You see statistics every day in newspapers and magazines. Do you feel you understand the statistics quoted in the media better now that you have completed this unit? Explain.

9. **Group Activity** Work with another student.
   a. Make a scatter plot of the data.
   b. One student should find an equation of the least-squares line for the data while the other student finds a quadratic model to fit the data.
   c. Discuss and decide which model seems to fit the data better. Justify your answer.

Average Years of Life Left (1990)

Age (years)	Life left (years)	Age (years)	Life Left (years)
0	75.4	45	33.4
5	71.2	50	29.0
10	66.3	55	24.8
15	61.3	60	20.8
20	56.6	65	17.2
25	51.9	70	13.9
30	47.2	75	10.9
35	42.6	80	8.3
40	38.0	85	6.1

IDEAS AND (FORMULAS)

- Making a scatter plot of paired data helps you to see if the data have a linear relationship, a quadratic relationship, or neither. *(pp. 350, 360)*
- Technology allows you to find a linear or quadratic model of a scatter plot easily. *(pp. 351, 360)*
- You can use linear or quadratic models to make predictions about a data set. *(pp. 351, 360)*
- You can use matrices to solve a system of three linear equations with three variables to find a quadratic model for data. *(p. 361)*

Answers to Unit 6 Review and Assessment

7. Answers may vary. An example is given.
   a. $y = -186.14x^2 + 2201.7x + 23{,}895$, where x is the number of years since 1986
   b. $23,895
8. Answers may vary.
9. a.

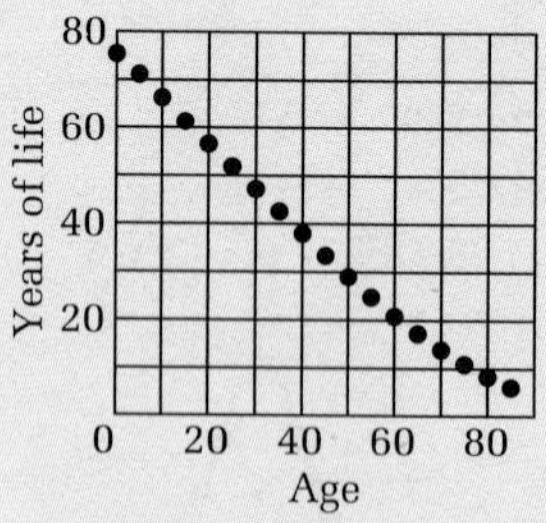

   b. Answers may vary. Examples are given. $y = -0.84x + 73.4$; $y = 0.0029x^2 - 1.09x + 76.67$
   c. Both models fit the data very well, but the quadratic model given is a better fit than the linear model because $\Sigma\, d^2$ is less for the quadratic model.

LOGICAL REASONING *if - then* $p \leftrightarrow q$

- Comparing the shapes of histograms allows you to make decisions regarding data sets. *(p. 333)*
- Range and standard deviation can help you make decisions when comparing data sets. *(pp. 338, 339)*

STATISTICS & PROBABILITY

- A frequency distribution can be categorized by the shape of its histogram. *(p. 332)*
- You can estimate where the mean and median of a data set will fall on a histogram. *(p. 332)*
- Histograms can show you how many data values are in each interval or what percent of data values are in each interval. *(pp. 333, 334)*
- Measures of variability tell you how spread out the data are. *(p. 339)*
- Technology allows you to find the mean and standard deviation of a data set easily. *(p. 339)*
- In a normal distribution, about 68% of the data are within one standard deviation of the mean, about 95% of the data are within two standard deviations of the mean, and about 99.7% of the data are within three standard deviations of the mean. *(p. 345)*
- You can find linear or quadratic models to fit data using technology and other methods. *(pp. 351, 360)*
- A correlation coefficient that is close to 1 or −1 shows that a set of data points lies close to the least-squares line. A coefficient that is close to 0 shows that a line is not a good model for the data set. *(p. 352)*
- The median-median line may be a better fit than the least-squares line when the data set has outliers. *(p. 353)*

Key Terms

- **histogram** (p. 331)
- **mound-shaped distribution** (p. 332)
- **outlier** (p. 332)
- **frequency distribution** (p. 332)
- **skewed distribution** (p. 332)
- **uniform distribution** (p. 332)
- **bimodal distribution** (p. 332)
- **relative frequency histogram** (p. 334)
- **range** (p. 338)
- **$\bar{x}$** (p. 339)
- **normal distribution** (p. 345)
- **median-median line** (p. 352)
- **σ** (p. 339)
- **least-squares line** (p. 351)
- **least-squares parabola** (p. 360)
- **standard deviation** (p. 339)
- **correlation coefficient** (p. 352)

Quick Quiz (6-1 through 6-3)

Jana asked the other students in her class how many cousins they had. She made a histogram of the results.

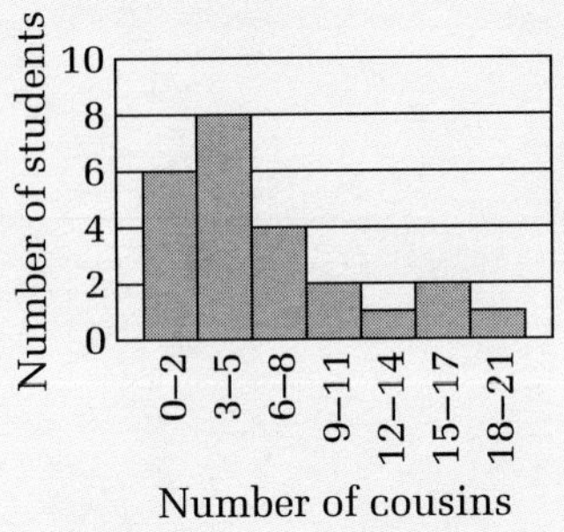

1. What type of distribution is this? [6-1] **skewed**
2. Estimate the mean and the median number of cousins. [6-1] **Answers will vary. The mean should be higher than the median; mean about 6, median about 5.**

Below are the heights in inches of the 12 players on two different basketball teams.

Team A: 73, 69, 77, 78, 71, 77, 75, 82, 79, 77, 84, 78

Team B: 77, 67, 81, 83, 76, 74, 69, 70, 88, 84, 73, 74

3. Find the mean and standard deviation for each team. [6-2] **Team A: about 76.7, about 4.1; Team B: about 76.3, about 6.2**
4. Compare the means and standard deviations of the two teams. What conclusion(s) can you make? [6-2] **Answers will vary. An example is given. The means of the two teams were about the same, but the standard deviation for team B was higher than for team A. This suggests that the players on team A had heights close to the mean, while team B had some shorter and taller players.**
5. Assume that each team's heights follow a normal distribution. Use the mean and standard deviation to determine about what percentage of the players on team A should have heights above 80.8 inches. [6-3] **16%**

Applying Probability Models

OVERVIEW

- **Unit 7** covers several aspects of probability. Students explore generating random numbers and using them to simulate an experiment. Finding an expected value is applied to deciding whether a game is fair or not.
- Probability problems include finding the probability that either of two events will occur. Mutually exclusive and complementary events are reviewed. Area models from geometry are used to solve probability problems. Independent and dependent events are explored, as is conditional probability. Binomial experiments are extended to multiple events and significance levels are used to decide if an outcome is due to chance. Counting techniques used in the unit are reviewed in the **Student Resources Toolbox**.
- The **Unit Project** revolves around studying the Native American game of da-un-dah-qua and changing a rule to create a new game. Students apply the mathematics of the unit to analyze both the older game and the new game. They also demonstrate how the new game is played.
- **Connections** to airline ticket sales, bus routes, gravity, cash register sales, allergies, market research, game shows, and genetics are some of the topics included in the teaching materials and the exercises.
- **Graphics calculators** are used to generate random numbers in Sections 7-1, 7-3, 7-4, and 7-6.
- **Problem-solving strategies** used in Unit 7 include using simulations, diagrams, area models, and formulas.

Unit Objectives

Section	Objectives	NCTM Standards
7-1	• Use simulations with random numbers to solve problems.	1, 2, 3, 4, 10, 12
7-2	• Find probabilities in situations with continuous outcomes and with discrete outcomes. • Find the probability that either of two events occurs.	1, 2, 3, 4, 5, 11, 12
7-3	• Use area models with coordinates to solve probability problems.	1, 2, 3, 4, 5, 11, 12
7-4	• Find conditional probabilities. • Recognize independent and dependent events.	1, 2, 3, 4, 5, 11, 12
7-5	• Find the average gain or loss in a situation.	1, 2, 3, 4, 5, 11, 12
7-6	• Find probabilities in binomial experiments. • Use significance levels in statistical decision-making.	1, 2, 3, 4, 5, 10, 11, 12

Topic Spiraling

Section	Connections to Prior and Future Concepts
7-1	**Section 7-1** explores using simulation to solve problems. Random numbers are introduced and are used to set up simulation experiments. Simulation was first introduced in Section 1-2 of Book 2.
7-2	**Section 7-2** introduces the probability of either of two events occurring. The difference between continuous and discrete outcomes is explored. Probability was covered in Sections 6-2 and 9-4 of Book 1 and in Unit 6 of Book 2, and will be developed further in Unit 7 of Book 3.
7-3	**Section 7-3** extends the concept of using geometric models for finding probability. Students use area models and coordinates to solve probability problems. Geometric probability was introduced in Section 9-4 of Book 1.
7-4	**Section 7-4** covers conditional probability and independent and dependent events. Independent and dependent events were first explored in Section 6-4 of Book 2. Students explore conditional probability using both tree diagrams and the formula.
7-5	**Section 7-5** explores the expected value of a situation. Students use expected value to decide whether or not a game is fair. Skills in probability from Unit 6 of Book 2 and Unit 7 of Book 3 are used.
7-6	**Section 7-6** extends the study of the probability of binomial experiments that was begun in Sections 6-7 and 6-8 of Book 2. The topic of decision-making based on probability is introduced. Binomial experiments and properties of the binomial distribution are studied in great detail in future courses in probability and statistics.

Integrating the Strands

Strands	Sections
Number	7-1, 7-2, 7-4
Algebra	7-1, 7-2, 7-3, 7-4, 7-5, 7-6
Functions	7-2, 7-3
Geometry	7-1, 7-2, 7-3, 7-4, 7-5
Statistics and Probability	7-1, 7-2, 7-3, 7-4, 7-5, 7-6
Discrete Mathematics	7-4, 7-6
Logic and Language	7-1, 7-2, 7-3, 7-4, 7-5, 7-6

Planning the Unit 7

Applying Probability Models

Section Planning Guide

- Essential exercises and problems are indicated in boldface.
- Ongoing work on the Unit Project is indicated in color.
- Exercises and problems that require student research, group work, manipulatives, or graphing technology are indicated in the column headed "Other."

Section	Materials	Pacing	Standard Assignment	Extended Assignment	Other
7-1	calculator with random number function, 6 two-sided disks, dish	Day 1 Day 2	1, **2–9,** 10 **11–16,** 20, 22–24, 25	1, **2–9,** 10 **11–16,** 17–20, 22–24, 25	 21
7-2	calculator	Day 1 Day 2	**1–8** **9–12, 14, 16,** 20–26, 27, 28	**1–8** **9–12,** 13, **14,** 15, **16,** 17, 18, 20–26, 27, 28	 19
7-3	calculator with random number function, tracing paper	Day 1	2, **3–7,** 9–16, 17	1, 2, **3–7,** 9–16, 17	8, 17
7-4	spinner, calculator with random number function	Day 1 Day 2	**2–9, 11,** 13 **16–22,** 26, 29–32, 33–36	1, **2–9,** 10, **11,** 13, 14 **16–22,** 23–32, 33–36	12, 15
7-5		Day 1	**4–8,** 10, **14–17,** 19–21, 22	1–3, **4–8,** 9–13, **14–17,** 19–21, 22	18
7-6	calculator with random number function, 12 playing cards	Day 1 Day 2	**2–11,** 12, 13 **14–20,** 21, 25–29, 30	1, **2–11,** 12, 13 **14–20,** 21–23, 25–29, 30	 24
Review Test		**Day 1** **Day 2**	**Unit Review** **Unit Test**	**Unit Review** **Unit Test**	

Yearly Pacing	Unit 7 Total	Units 1–7 Total	Remaining	Total
	14 days (2 for Unit Project)	108 days	46 days	154 days

Support Materials

- See **Project Book** for notes on Unit 7 Project: Create a New Game from an Older Game.
- UPP and disk refer to **Using Plotter Plus** booklet and **Plotter Plus** disk.
- TI-81/82 refers to **Using TI-81 and TI-82 Calculators** booklet.
- Warm-up exercises for each section are available on **Warm-Up Transparencies.**

Section	Study Guide	Practice Bank	Problem Bank	Activity Bank	Explorations Lab Manual	Assessment Book	Visuals	Technology
7-1	7-1	Practice 52	Set 15	Enrich 46	Add. Expl. 8 Masters 5, 15	Quiz 7-1	Folder 6	
7-2	7-2	Practice 53	Set 15	Enrich 47	Master 2	Quiz 7-2		
7-3	7-3	Practice 54	Set 15	Enrich 48	Masters 1, 2, 16	Quiz 7-3 Test 28		
7-4	7-4	Practice 55	Set 16	Enrich 49		Quiz 7-4	Folder 6	
7-5	7-5	Practice 56	Set 16	Enrich 50		Quiz 7-5		
7-6	7-6	Practice 57	Set 16	Enrich 51		Quiz 7-6 Test 29		
Unit 7	Unit Review	Practice 58	Unifying Problem 7	Family Involve 7		Tests 30, 31		

UNIT TESTS

Form A

Spanish versions of these tests are on pages 139–142 of the **Assessment Book.**

Name ______ Date ______ Score ______

Test 30

Test on Unit 7 (Form A)

Directions: Write the answers in the spaces provided.

1. Design and test a formula for a calculator to produce a set of random integers from 1 to 50.

Tell whether each situation appears to produce a sequence of random numbers. Write *Yes* or *No*.

2. You ask 50 people to tell you the amount of change they have in their pockets.
3. You ask 75 students to choose a number between 1 and 100.

For each event, tell whether the outcomes are *discrete* or *continuous*, and then find the probability of the event.

4. When you look at a clock, the hour hand is between 8 and 9.
5. You draw a heart at random from a well-shuffled standard deck of playing cards.

For Questions 6–8, tell whether each statement is *True* or *False*. Carmella, Kim, Jared, and Alejandro are the members of their school's debate team.

6. Selecting Carmella and selecting one of the other three persons are mutually exclusive events.
7. Selecting Carmella and selecting Jared or Alejandro are complementary events.
8. Selecting Kim and not selecting Kim are complementary events.
9. **Open-ended** Generate your own sequence of 10 random numbers. Explain why they are random.
   Answers will vary.
10. Suppose two friends both say they will call you tonight between 7:30 and 8:30. You plan to talk to each friend for 20 min. Use coordinates to find the probability that one friend will try to call while you are talking to the other.
   probability: $\frac{5}{9}$

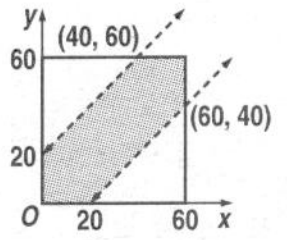

Answers

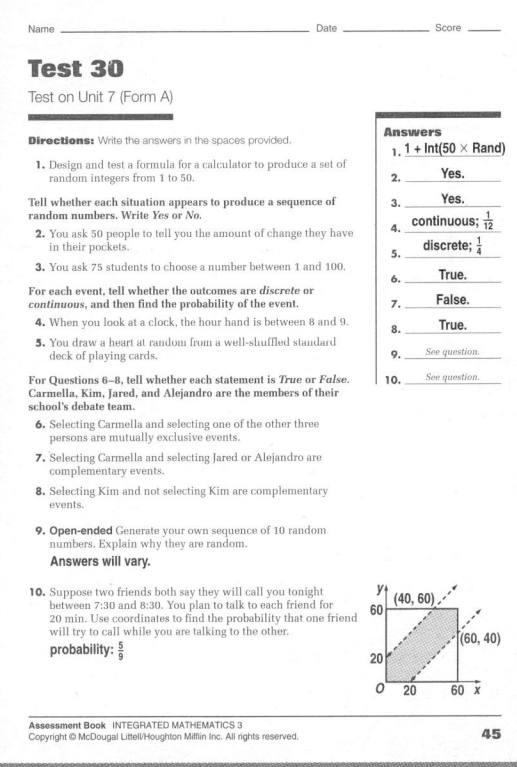

1. 1 + Int(50 × Rand)
2. Yes.
3. Yes.
4. continuous; $\frac{1}{12}$
5. discrete; $\frac{1}{4}$
6. True.
7. False.
8. True.
9. *See question.*
10. *See question.*

Name ______ Date ______ Score ______

Test 30 *(continued)*

Directions: Write the answers in the spaces provided.

The table below lists the science courses taken by the students at a community college. Suppose a student is selected at random. Find each probability.

	Females	Males	Total
Biology	205	185	390
Chemistry	240	250	490
Physiology	160	140	300
Total	605	575	1180

11. *P*(female | student takes Physiology)
12. *P*(male | student takes Biology)
13. *P*(student takes Chemistry | male)

A card is drawn at random from a standard deck of playing cards. You win $1.50 if you draw a face card and $20 if you draw an ace. Otherwise, you lose $2.

14. What is the expected value of the game?
15. Suppose you are the operator of the game. What should you charge in order to make the game fair?

A survey of the households in a city showed that 24% have two or more VCRs and that 57% of those with two or more VCRs reported watching at least 7 movies a week. However, only 28% of those with fewer than two VCRs reported watching at least 7 movies a week. Use this information for Questions 16 and 17.

16. Is the number of VCRs owned independent of the number of movies watched each week?
17. What is the probability that a household with two or more VCRs will watch fewer than 7 movies each week?
18. Eight cars are in a parking lot on a very cold day. Suppose the probability of any one of them not starting is 0.12. What is the probability that exactly three of the cars do not start?
19. **Writing** Explain how to find the expected value of an event.
   Multiply the value of each outcome by its probability and then add all the products.

Answers

11. $\frac{8}{15}$
12. $\frac{37}{78}$
13. $\frac{10}{23}$
14. $0.50 gain
15. $0.50
16. No.
17. 0.1032
18. ≈0.051
19. *See question.*

Form B

Name ______ Date ______ Score ______

Test 31

Test on Unit 7 (Form B)

Directions: Write the answers in the spaces provided.

1. Design and test a formula for a calculator to produce a set of random numbers from 1 to 40.

Tell whether each situation appears to produce a sequence of random numbers. Write *Yes* or *No*.

2. You ask 100 people to tell you the last digit of their telephone number.
3. You ask 50 students to choose a number between 1 and 200.

For each event, tell whether the outcomes are *discrete* or *continuous*, and then find the probability of the event.

4. You draw a king from a well-shuffled standard deck of playing cards.
5. When you look at a clock, the hour hand is between 5 and 7.

For Questions 6–8, tell whether each statement is *True* or *False*. Pindhi, Guadalupe, Amel, Lakim, and Brian are the members of their school's debate team.

6. Selecting Guadalupe and selecting Brian are mutually exclusive events.
7. Selecting Amel and not selecting Amel are complementary events.
8. Selecting Pindhi and selecting Lakim or Guadalupe are complementary events.
9. **Open-ended** Generate your own sequence of 10 random numbers. Explain why they are random.
   Answers will vary.
10. Suppose two friends both say they will call you tonight between 7:30 and 9:00. You plan to talk to each friend for 30 min. Use coordinates to find the probability that one friend will try to call while you are talking to the other.
   probability: $\frac{5}{9}$

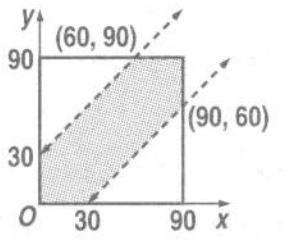

Answers

1. 1 + Int(40 × Rand)
2. Yes.
3. Yes.
4. discrete; $\frac{1}{13}$
5. continuous; $\frac{1}{6}$
6. True.
7. True.
8. False.
9. *See question.*
10. *See question.*

Name ______ Date ______ Score ______

Test 31 *(continued)*

Directions: Write the answers in the spaces provided.

The table below lists the results of a survey of the students at a large high school. Suppose a student is selected at random. Find each probability.

Favorite Music	Females	Males	Total
Country	125	100	225
Rock	200	375	575
New Wave	250	75	325
Total	575	550	1125

11. *P*(female | student likes New Wave)
12. *P*(male | student likes Rock)
13. *P*(student likes Country | female)

A card is drawn at random from a standard deck of playing cards. You win $5 if you draw a face card and $25 if you draw an ace. Otherwise, you lose $3.

14. What is the expected value of the game?
15. Suppose you are the operator of the game. What should you charge in order to make the game fair?

A survey of the households in a city showed that 37% have two or more cars and that 47% of those with two or more cars reported driving more than 500 miles per week. Only 18% of those with fewer than two cars reported driving more than 500 miles per week. Use this information for Questions 16 and 17.

16. Is the number of cars owned independent of the number of miles driven each week?
17. What is the probability that the persons in a household with two or more cars drive less than 500 miles per week?
18. Six cars are in a parking lot on a very cold day. Suppose the probability of any one of them not starting is 0.15. What is the probability that exactly two of the cars do not start?
19. **Writing** List the four characteristics of a binomial experiment.
   1) There are exactly two outcomes for each trial. 2) The trials are independent. 3) Each trial has the same probability of success. 4) The experiment has a fixed number of trials.

Answers

11. $\frac{10}{13}$
12. $\frac{15}{23}$
13. $\frac{5}{23}$
14. $1.00 gain
15. $1.00
16. No.
17. 0.1961
18. ≈0.176
19. *See question.*

OUTSIDE RESOURCES

Books/Periodicals

Ekeland, Ivar. *The Broken Dice and Other Mathematical Tales of Chance.* University of Chicago Press, 1993.

Floyd, Jeffrey K. "A Discrete Analysis of Final Jeopardy." *Mathematics Teacher* (May 1994): pp. 328–331.

Watson, James M. "Conditional Probability: Its Place in the Mathematics Curriculum." *Mathematics Teacher* (January 1995): pp. 12–17.

Copes, Wayne, William Sacco, Clifford Sloyer, and Robert Stark. "Queues." *Contemporary Applied Mathematics, Graph Theory,* Part IV, "Variable Arrivals and Service Times—Simulation": pp. 16–34. Janson Publications.

Quantitative Literacy Series. The Art and Techniques of Simulation. "Applications for Estimating Probability through Simulations." Dale Seymour publication.

Software

Geometric Probability. IBM PC or compatible with 256K memory. Accompanied by materials including 19 problems and activities. NCTM.

PROJECT GOALS

- Students create a new game by changing at least one of the rules of the Native American game da-un-dah-qua.
- Students use the ideas of probability presented in the unit to study how their rule change affects the new game.
- Students demonstrate how their new game is played.

PROJECT PLANNING

Materials List

- Six two-sided disks
- Dish or flat bowl

Project Teams

Have students work on the project in groups of four. Da-un-dah-qua is a game for two players, so the members of the group can take turns playing the game. As the groups solve the "Working on the Unit Project" exercises throughout the unit, they should keep a written record of their answers for use in completing the project. Group members can share responsibility for writing the answers to these exercises. They should work together, however, to discuss all of the exercises and arrive at group decisions regarding how to change at least one of the rules of da-un-dah-qua to create a new game. They should also work together to develop the final written set of rules for their new game.

unit 7

Applying Probability Models

Changing the Rules

About 100 years ago, football emerged as a distinct game in the United States and Canada. It was the result of decades of rules changes in the English game of rugby.

Ongoing rules changes in each country have turned American and Canadian football into quite different games. With a larger field and fewer chances to gain 10 yards before giving up the ball, Canadian football features more passing, more touchdowns, and a faster pace.

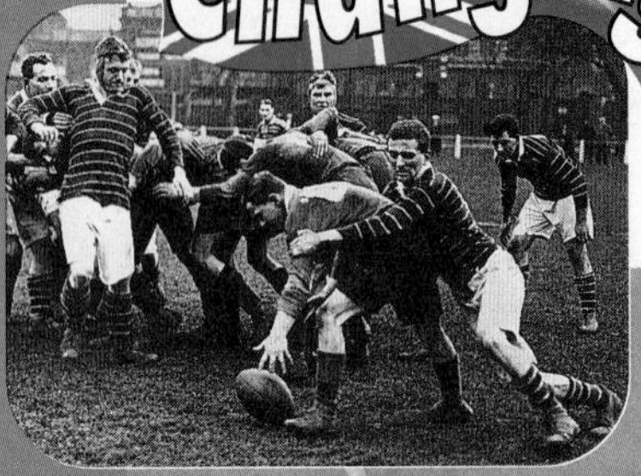

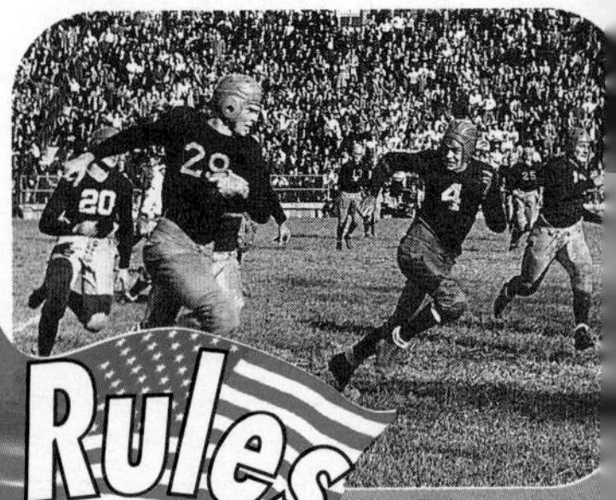

Sports Rules

Game Theory

Competition among businesses is a "game" or rivalry in which each company tries to develop a winning strategy. Some economists use a strategy called *game theory* to help businesses predict how their competitors will react to a decision like an increase in prices. The theory assigns probabilities to the other companies' responses and counter-responses.

370

Suggested Rubric for Unit Project

4 Students learn and play the game da-un-dah-qua. They apply the mathematics of probability correctly to analyze the game. They successfully create a new game, develop a written set of rules for it, and are able to analyze the game using probability concepts. They also discuss key questions about the new game, such as is it fair. All mathematical work is free of errors. Students demonstrate how to play their new game.

3 Students learn and play the game of da-un-dah-qua. They are able to change one of its rules and develop a new game. Their probability analysis of the new game and the original game, however, has some inaccuracies but is substantially correct. The rules for the new game are clearly written and a demonstration of how to play the game is given.

2 Students learn and play the game of da-un-dah-qua but have difficulty creating a new game. The rules for the new game are somewhat confusing and it appears the game is not fair. The probability analysis of

Unit Project 7

Rules of the Game

Your project is to create a new game by changing at least one of the rules of the Native American game *da-un-dah-qua*. You will then use the ideas of probability presented in this unit to study your new game.

Da-un-dah-qua is a traditional game played at midwinter and fall festivals by the Cayuga, a Native American people from western New York State and the Lake Ontario region of Canada. The rules of da-un-dah-qua appear with the "Working on the Unit Project" exercises in Section 7-1.

As you complete the "Working on the Unit Project" exercises, you will apply the ideas of probability to analyze da-un-dah-qua. A written set of rules for your new game, a probability analysis of both games, and a demonstration of your new game will complete your project.

The unwritten rules that govern social customs change as you move across national borders. Cultural tips from travel books and **cultural consultants** guide business people in their dealings around the globe.

The traditional greeting in Japan is a bow.

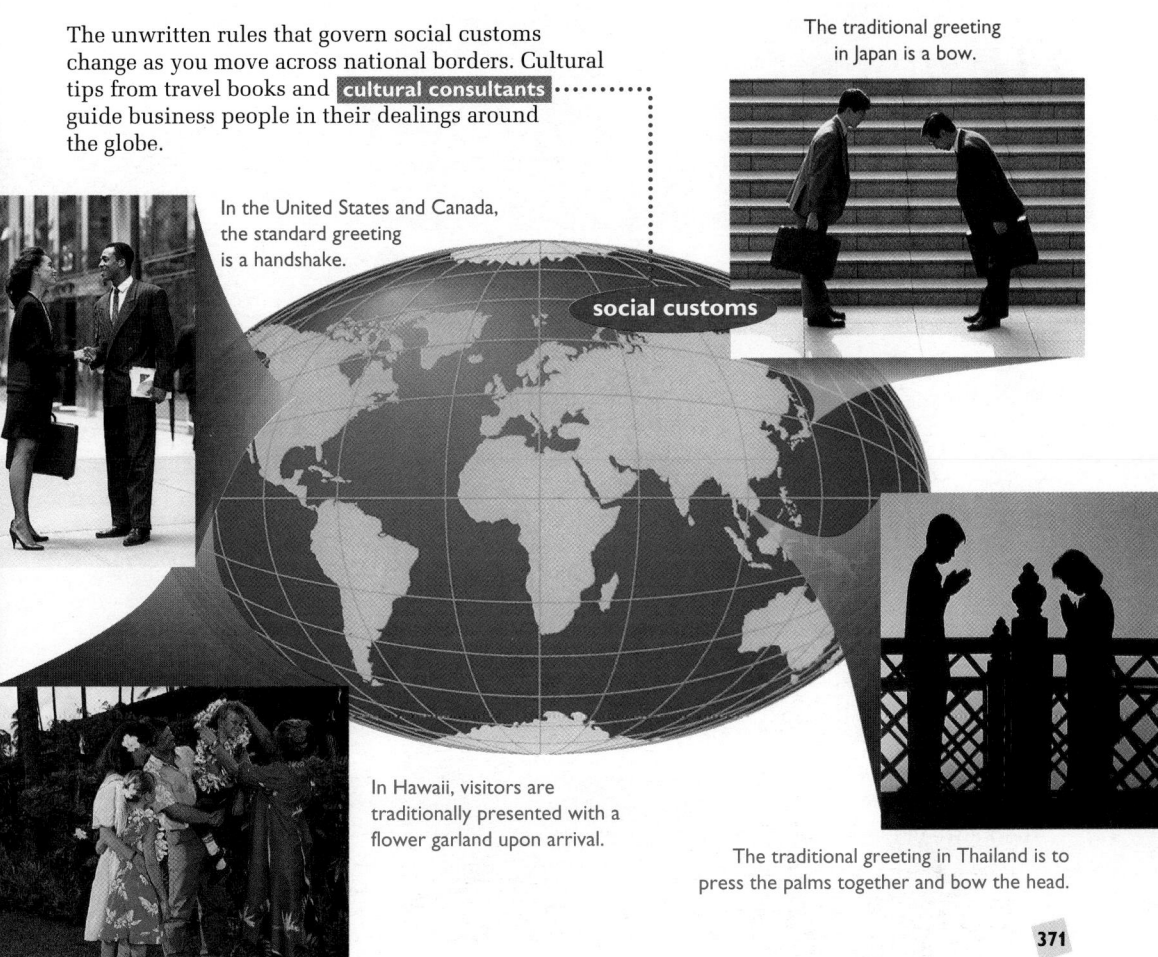

In the United States and Canada, the standard greeting is a handshake.

In Hawaii, visitors are traditionally presented with a flower garland upon arrival.

The traditional greeting in Thailand is to press the palms together and bow the head.

Support Materials

The ***Project Book*** contains information about the following topics for use with this Unit Project.

- Project Description
- Teaching Commentary
- Working on the Unit Project Exercises
- Completing the Unit Project
- Assessing the Unit Project
- Alternative Projects
- Outside Resources

Students Acquiring English

Creating a new game from da-un-dah-qua is an excellent activity for students acquiring English since it requires both creativity and an understanding of mathematical probabilities. You may want to ask students to talk about games of chance that are popular in their countries of origin.

ADDITIONAL BACKGROUND

Multicultural Note

The Cayugas lived along the Cayuga River (named for them) in what is now New York State. One of the original tribes in the Iroquois League, a confederation of related tribes, the Cayugas share much of their culture and history with other Iroquois. The Cayugas had at least thirteen major villages, many of which were located near wetlands; accordingly, the Cayugas were known as the People of the Marsh. Like all Iroquois, the Cayugas were master hunters, skilled at trapping deer (for meat and clothes), beavers (for robes and mittens), bears (for robes), and porcupines (for wampum belts to record treaties). They were also capable farmers, and relied heavily on corn, beans, and squash for sustenance. The Cayugas and other Iroquois were matrilineal, and women chose the tribal chiefs and league representatives.

Suggested Rubric for Unit Project

both the original game and the new game contains many errors. A demonstration of how the new game is played raises many questions by other students who think it is not a very good game.

1 Students learn and play the original game but are not successful in creating a new game. The new rules are confusing, and the probability analyses are incorrect. Students cannot demonstrate how the new game is played. Students should be encouraged to speak with the teacher as soon as possible to review their work and to make a new start on the project.

Game Theory

Game theory is a mathematical subject that studies situations in which two or more players are able to act freely to select various options. First developed in 1944, the theory was applied to economic situations in which perfect competition could not be assumed to be the guiding principal. Since that time, game theory has grown to be a powerful tool in analyzing and understanding not only economic situations, but also interactions among businesses, nations, and even biological species. In general, concepts of game theory can be applied in any situation where competition and/or cooperation arise.

ALTERNATIVE PROJECTS

Project 1, page 418

Consumer Economics and Maintenance Agreements

Use the ideas of probability to determine whether maintenance agreements that consumers buy for electronic items and home appliances are a good buy. Collect data on manufacturers' warranties as well as probabilities on failures of items. Estimate unavailable probabilities and describe assumptions made.

Project 2, page 418

Analyzing Advertising Claims

Advertisements are often based on claims that use statistics. Select at least three such claims, obtain the statistical data for each claim, and analyze the claim using the ideas of probability to see if it is justified by the data.

Unit Project 7

Getting Started

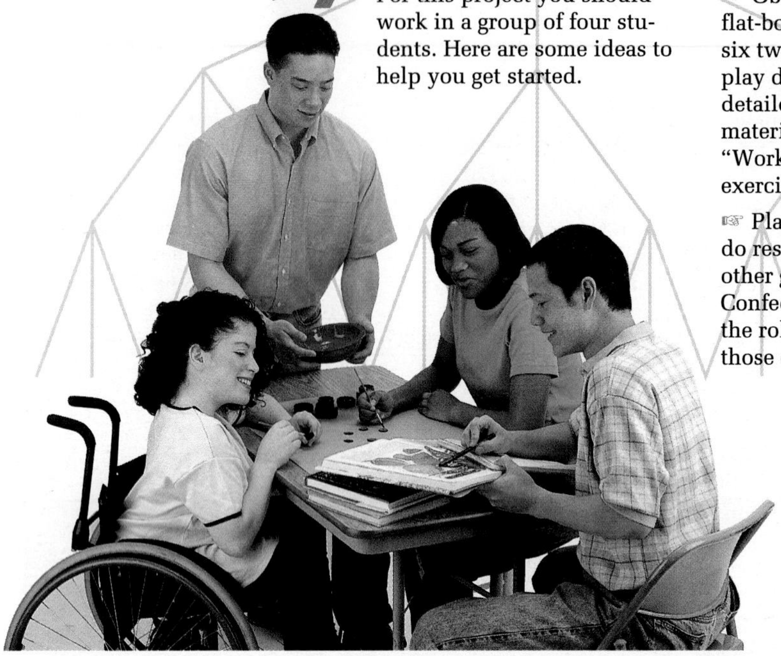

For this project you should work in a group of four students. Here are some ideas to help you get started.

☞ Obtain or make the flat-bottomed bowl and the six two-sided pieces used to play da-un-dah-qua. For a detailed description of the materials needed, see the "Working on the Unit Project" exercises on page 381.

☞ Plan to go to the library to do research on the Cayuga and other groups of the Iroquois Confederacy. Learn more about the role of da-un-dah-qua in those cultures.

Working on the Unit Project

Your work in Unit 7 will help you to create a new game based on da-un-dah-qua and to analyze your new game.

Related Exercises:

Section 7-1, Exercises 25–27
Section 7-2, Exercises 27, 28
Section 7-3, Exercise 17
Section 7-4, Exercises 33–36
Section 7-5, Exercise 22
Section 7-6, Exercise 30

Alternative Projects p. 418

Can We Talk Changing the Rules

When rules are changed, the results of those rules also change.

➤ Imagine changing one of the postulates listed in Unit 3. Predict the effect this has on the theorems in that unit.

➤ Your school has rules. Choose one and discuss the possible effects of a change in that rule.

➤ What changes in rules have occurred in a sport that you follow? How have the changes affected the sport?

➤ Do you think it is good that the United States constitution can be amended? Why or why not?

➤ Scientists develop theories and laws that model how the world works. Think of a law from a science course. Discuss what would happen if you made a change in the law.

➤ Some years ago a state passed a law declaring that the value of pi is exactly 3. Do you think this kind of rule change makes sense? Why or why not?

Answers to Can We Talk?

➤ Answers may vary. An example is given. If one of the postulates were changed, then the theorems that used the postulate would also change.

➤ Answers may vary. An example is given. School begins at 8:15 A.M. If this time were changed to, say 7:30 A.M., the bus schedule would have to change, as would the times for lunches and classes.

➤ Answers may vary. An example is given. The three-point shot has been added to basketball. This makes the game more exciting and allows teams that are far behind to catch up quickly.

➤ Answers may vary. An example is given. Yes, because as our society changes, its laws can be amended to help people live better lives.

➤ Answers may vary. An example is given. The Law of Conservation of Energy: Energy can never be created or destroyed. If this law were not true, then it would be possible to create new energy and there would never be an energy shortage in the future.

➤ Answers may vary. An example is given. No, because pi (π) is defined as the ratio of the circumference of a circle to its diameter, which is a constant value for all circles.

Section **7-1**

Simulation and Random Numbers

Focus
Use simulations to solve problems.

PICK A LETTER

There is a scratch-off circle on each ticket for an event in a series of concerts by international artists. One of the letters M, U, S, I, and C is hidden behind the circle. Anyone who holds tickets that spell "MUSIC" wins a free T-shirt. The letters are all equally likely to appear and their order is random.

Talk it Over

1. What is the probability of getting an "I" on the first ticket you buy?
2. What is the minimum number of tickets a person needs in order to win a T-shirt? Do you think there is a maximum number of tickets a person needs in order to win a T-shirt?
3. What do you think is the average number of tickets a person needs in order to win a T-shirt?

PLANNING

Objectives and Strands
See pages 370A and 370B.

Spiral Learning
See page 370B.

Materials List
- Calculator with random number function
- 6 disks
- Dish

Recommended Pacing

Section 7-1 is a two-day lesson.

Day 1

Pages 373–375: Opening paragraph through Talk it Over 7, *Exercises 1–10*

Day 2

Pages 376–378: Exploration through Look Back, *Exercises 11–27*

Extra Practice
See pages 619–620.

Warm-Up Exercises
- Warm-Up Transparency 7-1

Support Materials
- Practice Bank: Practice 52
- Activity Bank: Enrichment 46
- Study Guide: Section 7-1
- Problem Bank: Problem Set 15
- Explorations Lab Manual: Additional Exploration 8, Diagram Masters 5, 15
- Overhead Visuals: Folder 6
- Assessment Book: Quiz 7-1, Alternative Assessment 1

Answers to Talk it Over

1. 20%, or $\frac{1}{5}$
2. 5; If an infinite number of tickets were printed, then no number of tickets would guarantee winning a T-shirt. But in this example, only a finite number of tickets will be printed. Since it is equally likely that each letter will appear, there must be an equal number of tickets, n, with each letter. Since there are 5 different letters, buying $4n + 1$ tickets guarantees getting one of each letter, thus winning a T-shirt.
3. Answers may vary.

TEACHING

Talk it Over

Questions 1–3 start students thinking about probability and the limits of theoretical probability. Students will investigate the minimum, maximum, and average number of tickets needed using simulation in the Exploration on page 376.

Reasoning

When discussing random numbers, students need to understand that each event should have an equally likely chance to occur, as is the case when using the spinner shown in the text. However, the events are not equally likely if the spinner looks like this one.

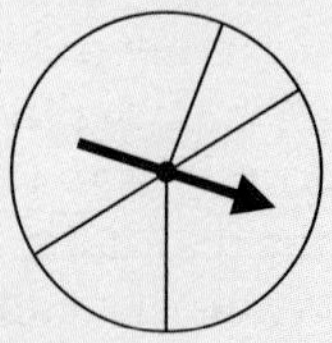

Research

An interesting research project for some students would be to find out why calculators may not give completely random numbers as is mentioned in the Watch Out! A written report for this project would make an excellent contribution to a student's portfolio.

You can investigate the T-shirt problem from page 373 with a *simulation*. A **simulation** is an experiment that is enough like a real event that the event can be studied without the cost, time, materials, or risk of the real event. Scientists, business people, engineers, and others use simulation to test new materials and procedures.

Producing Random Numbers

Simulations sometimes involve *randomness*. Just as raindrops fall on a sidewalk in no apparent pattern and each point on the sidewalk has an equally likely chance of being hit, a sequence of **random numbers** has no pattern and each number has an equally likely chance of appearing in the sequence.

You can generate random numbers with a spinner. If conditions are perfect, you cannot predict the results from one spin to the next.

Watch Out!
Physical conditions are never perfect (pointers bend and surfaces may not be perfectly flat). Even calculators may not give completely random numbers.

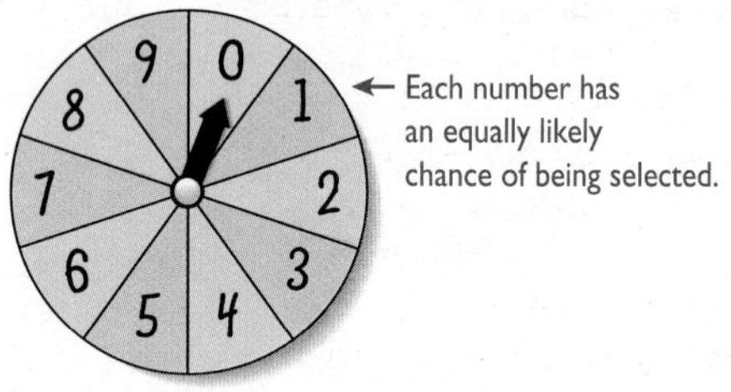

Calculators can also be used to generate random numbers. Many calculators have a random number function that gives a number between 0 and 1.

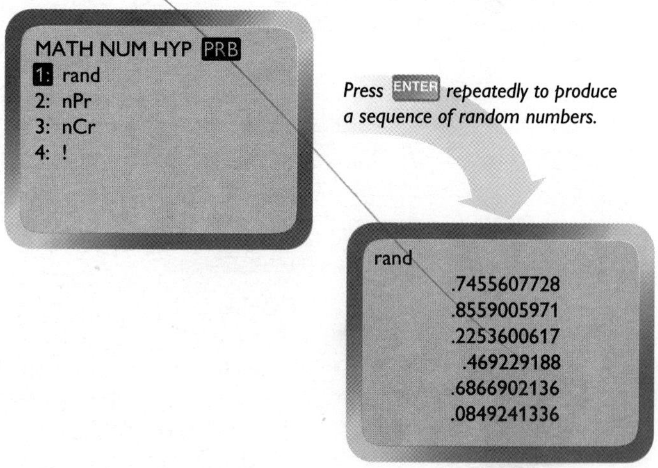

TECHNOLOGY NOTE
The random number function on your calculator may be a key or a menu option. On some calculators it is called "RND."

To produce a random number in any interval, you can use a formula to change the calculator's output.

Sample 1

Use a random number function on a calculator to simulate the outcomes of a roll of one die.

Sample Response

There are six equally likely outcomes. The digits 1, 2, 3, 4, 5, 6 are the six possible numbers.

	Formula		Example
Step 1 Use the "rand" function to get a random number between 0 and 1.	rand	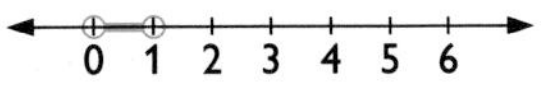	0.3714226
Step 2 Multiply by 6 to get a random number between 0 and 6.	6 • rand	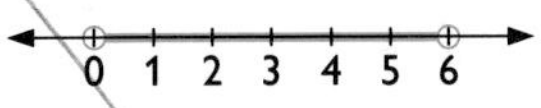	2.2285356
Step 3 Use the greatest integer function to get a random integer between 0 and 5.	int(6 • rand)		2
Step 4 Add 1 to get a random integer between 1 and 6.	int(6 • rand) + 1	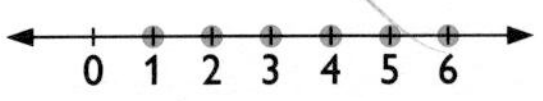	3

The calculator screen shows a random sequence produced by the formula in Step 4.

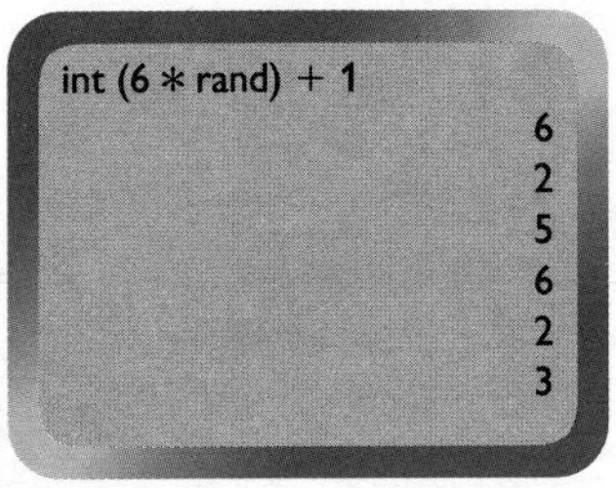

randint

Talk it Over

4. Suppose you change the coefficient in the formula in Step 2 of Sample 1 from 6 to 9. How does this change the results?
5. Suppose you change the coefficient to 9 and the number added in Step 4 from 1 to 3. How does this change the results?
6. Design and test a formula for a calculator to produce random integers from 10 to 29.
7. Design and test a formula to simulate tossing a coin.

Now you are ready for: Exs. 1–10 on p. 378

Answers to Talk it Over

4. The random sequence produced will consist of integers between 0 and 9.
5. The random sequence produced will consist of integers between 3 and 11.
6. int (20*rand) + 10
7. int (2*rand) + 1

Additional Sample

S1 Use a random number function on a calculator to simulate the outcomes of a roll of one 12-sided die.

There are 12 equally likely outcomes: 1, 2, 3, 4, 5, 6, 7, 8, 9, 10, 11, and 12.
Step 1. **rand yields a random decimal value between 0 and 1.**
Step 2. **12 • rand yields a random decimal value between 0 and 12.**
Step 3. **Using the greatest integer function, int(12 • rand) yields a random integer between 0 and 11.**
Step 4. **Adding 1 results in a random integer between 1 and 12, inclusive. The function needed is int(12 • rand) + 1.**

Talk it Over

Questions 4 and 5 involve students in Sample 1 in order to help them better understand the process of creating a formula that can produce random integers. Questions 6 and 7 give students an opportunity to apply what they have just learned.

Multicultural Note

Dice predate the earliest written records. In ancient times, dice (or objects similar to dice) were used as divining tools in religious ceremonies in many parts of the world. These early dice were made of many different materials, including plum stones, seeds, buffalo bone, deer horn, pebbles, pottery, beaver teeth, seashells, and walnut shells. Some time prior to 3000 B.C., women and children in Asia began to play a game of skill using rectangular dice made from the leg bones of sheep or goats, the first known use of dice for non-religious purposes. Archaeological and written evidence shows that after 3000 B.C. the Egyptians, Thebans, Greeks, Indians, Romans, and many other peoples also began to use dice for playing games. Dice remained popular throughout the Middle Ages, and continue to be widely used today in board games and other games of chance.

Teaching Tip

As students do the Exploration, they may need to review the meaning of the terms *range, mode, median, mean, histogram, relative frequency,* and *experimental probability.*

Exploration

The goal of the Exploration is to lead students to an understanding of the concept of a probability distribution. This is done by having each group of students conduct 10 trials of a simulation experiment using a calculator with a random number function. Data collected from all groups in the class can then be compared, analyzed, and represented graphically in a histogram, which represents a probability distribution.

Integrating the Strands

Algorithms integrate logic and language with other areas of mathematics. They are logical step-by-step instructions that can be used to solve complex multistep problems. An algorithm is used in Sample 2 to design a simulation. On page 413, an algorithm is used in statistical decision-making. Algorithms are also commonly used in algebra and geometry.

EXPLORATION

How many tickets do you need in order to win a T-shirt?

- **Materials: calculators with a random number function**
- **Work in a group of five students.**

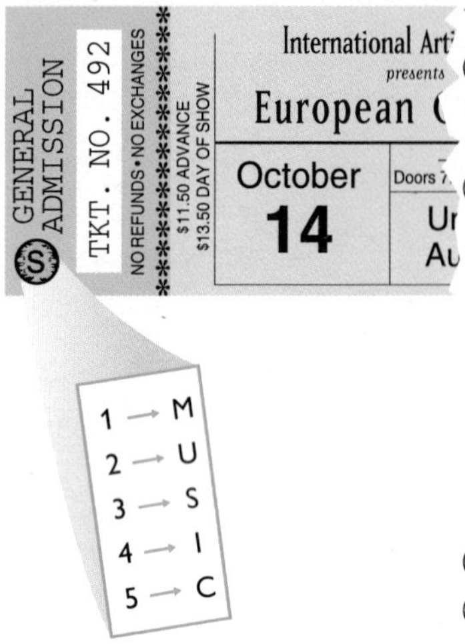

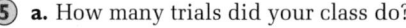

You will investigate the T-shirt problem on page 373.

1. Use the numbers 1 through 5 to represent the letters on the tickets. Design a calculator formula to generate random numbers from 1 to 5.
2. For each random number, record the letter. Continue until you have all five letters. This is one trial of the experiment. How many tickets (numbers) did you need to get all five letters? Record this value.

Example

Random numbers: 5 3 2 3 5 4 3 1

Letters represented: C S U S C I S M

Number of tickets needed: 8

3. Repeat step 2 nine times.
4. Compare your results with the results of other groups. Make a line plot of all the results.

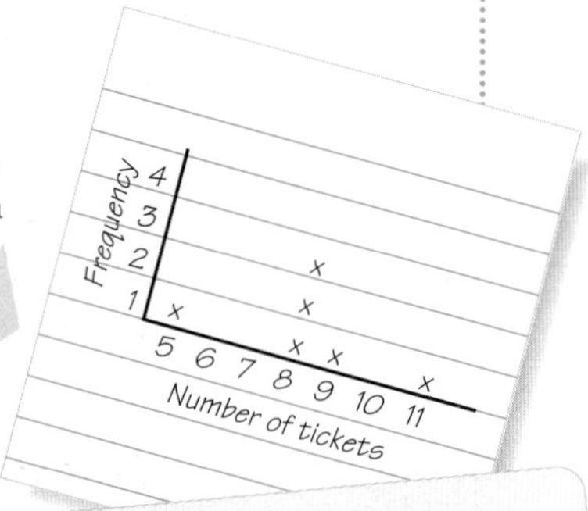

5. a. How many trials did your class do?
   b. What is the range of the number of tickets needed?
   c. What is the mode of the number of tickets needed? What is the median? What is the mean?
   d. Suppose the class repeats this experiment. How do you think the results will be different? How will they be the same?
6. Make a relative frequency histogram to show the distribution of results from the ticket simulation for the class. You can think of the relative frequencies on the vertical axis as experimental probabilities, so the histogram represents a **probability distribution.**
7. Is the distribution for your class *skewed, uniform, mound-shaped,* or *bimodal?*

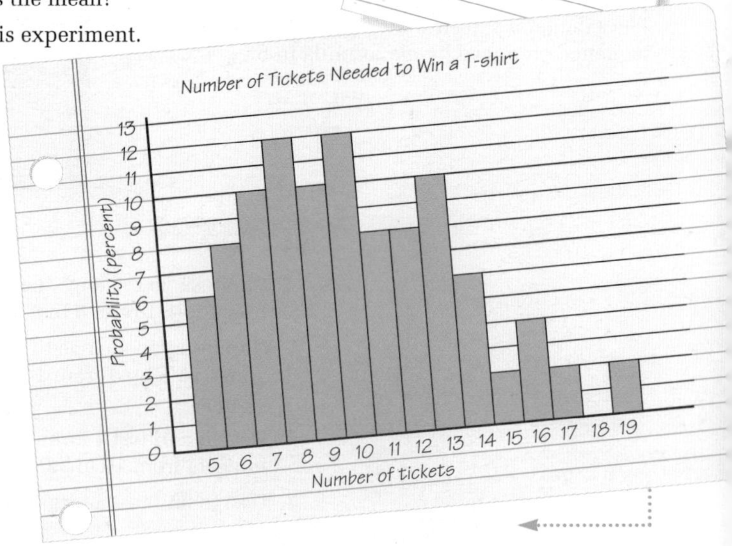

Answers to Exploration

1. Answers may vary. An example is given. int (5*rand) + 1

2–5(c). Answers may vary. Examples are given for a series of 10 trials, which resulted in the following numbers of tickets needed to get all 5 letters: 8, 8, 9, 12, 10, 12, 7, 6, 8, 10

2. 8

3. 8, 8, 9, 12, 10, 12, 7, 6, 8, 10

4. Graph shows the trials for one group.

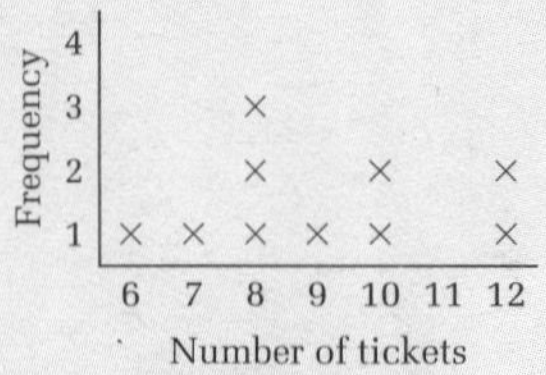

5. a. 10 trials per group
   b. 6 c. 8; 8.5; 9
   d. Answers may vary. An example is given. I think that with a greater number of trials there will be more outliers, but I think the measures of central tendency will be similar to those found with fewer trials.

6, 7. Answers may vary. Check students' work.

A Simulation Algorithm

In experiments involving chance, the results of specific trials are not predictable, but patterns may appear after many repeated trials. A simulation model can help you find these patterns.

Sample 2

Design a simulation for this situation.

Suppose a commuter plane holds 12 people. People buy tickets in advance. On average, 15% of the people who buy tickets do not take the flight. The airline sells 14 tickets for each flight. How often are flights overbooked because more than 12 people with tickets show up for a flight?

Sample Response

Use this algorithm.

1 State the problem clearly.	Find out how often 13 or 14 people show up for a flight.
2 Define the key question.	Select a passenger. Does the passenger show up for the flight?
3 State the underlying assumptions.	The probability that a person does not show up is 0.15. Each person's arrival has no effect on any other person's arrival.
4 Choose a model to answer the key question.	Generate random numbers from 00 to 99. A number from 01 to 15 inclusive means that the person does not show up. In all other cases, the person shows up.
5 Define and conduct a trial.	Pick 14 two-digit random numbers.
6 Record the observation of interest.	How many people showed up for the flight? Is it overbooked?
7 Repeat the last two steps several times.	Pick 14 two-digit random numbers repeatedly.
8 Summarize the information and draw conclusions.	Make a table or graph. Analyze and interpret the results.

Additional Sample

S2 Design a simulation for this situation.
A small hotel has 35 rooms available for overnight use. People make reservations in advance, but on the average, 9% of them do not show up. The hotel takes reservations for 40 rooms every night. How often would a customer be turned away due to overbooking?

Use this algorithm.
(1) State the problem clearly. Find out how often 36, 37, 38, 39, or 40 people show up with a reservation.
(2) Define the key question. Select a customer. Does that customer show up for the reserved room?
(3) State the underlying assumptions. The probability that a person does not show up is 0.09. Each person's arrival for the reservation has no effect on any other person's arrival.
(4) Choose a model to answer the key question. Generate random integers from 00 to 99. A number from 00 to 08 inclusive means that the person does not show up. In all other cases, the person shows up.
(5) Define and conduct a trial. Pick 40 two-digit random numbers.
(6) Record the observation of interest. How many people showed up with reservations? Is the hotel overbooked?
(7) Repeat the last two steps several times. Pick 40 random two-digit numbers repeatedly.
(8) Summarize the information and draw conclusions. Make a table or graph. Analyze and interpret the results.

Talk it Over

Question 9 uses the T-shirt problem to give students practice in using the simulation algorithm presented in Sample 2.

Students Acquiring English

For the Look Back, students acquiring English might be given the option of using random numbers to simulate an event by means of an example rather than by describing the simulation in a paragraph.

Look Back

After deciding how they would simulate their events, some students may wish to present their experiments to the class.

APPLYING

Suggested Assignment

Day 1

Standard 1–10

Extended 1–10

Day 2

Standard 11–16, 20, 22–25

Extended 11–20, 22–25

Integrating the Strands

Number Exs. 2–9

Algebra Ex. 22

Geometry Ex. 23

Statistics and Probability Exs. 1–21, 24–27

Logic and Language Exs. 1, 10, 11

Assessment: Open-ended

For Ex. 6, students are asked to design a calculator formula to produce a set of random integers from 0 to 9. Ask students how they can use a telephone book to produce the same type of random numbers. If only the integers from 3 through 8, inclusive, are wanted, can a telephone book still be used? How?

Talk it Over

8. For Sample 2, Sally used the numbers from 85 to 99 to represent passengers who did not show up for a flight, and numbers from 00 to 84 to represent passengers who did. Is this model the same as the model used in the Sample Response? Why or why not?
9. What are the eight steps of the simulation algorithm for the T-shirt experiment on page 376?

Now you are ready for: Exs. 11–27 on pp. 378–381

Look Back

Think of a real event that you can simulate by doing an experiment with random numbers. Describe how to use random numbers to simulate the event.

7-1 Exercises and Problems

1. **Reading** What are some benefits of simulating experiments?
2. Which of these numbers has a greater chance of appearing in a set of three-digit random numbers: *888*, *567*, or *351*? Why?

Tell whether each situation appears to produce a sequence of random numbers. Give a reason for your answer.

3. You ask each person in your class to think of a number and then to say it out loud to you.
4. You ask ten people for the last digit of their phone number.
5. You remove the face cards from a deck of cards, shuffle well, and use the number on the top card.

For Exercises 6–9, design and test a formula for a calculator to produce each set of random numbers.

6. integers from 0 to 9
7. integers from 1 to 100
8. integers from 5 to 8
9. integers from 11 to 29

10. **Writing** Give three advantages of using simulation instead of actually buying tickets to investigate the T-shirt problem.
11. In the Exploration, suppose that the probability that a ticket contains an "M" is $\frac{1}{25}$, and the probability for each of the other letters is $\frac{6}{25}$.
    a. How do you think this will affect the number of tickets needed to spell "MUSIC"?
    b. **Writing** Describe how to simulate the experiment with the new probabilities.

BY THE WAY...

To protect the privacy of people who use automatic teller machines, banks use random numbers to code transactions. A computer decodes the transaction using a mathematical procedure called the Data Encryption Standard.

Answers to Talk it Over

8. Yes; 15 of the random numbers represent passengers who do not take the flight, and 85 represent passengers who do.
9. See answers in back of book.

Answers to Look Back

Answers may vary. An example is given. The probability that a particular genetic trait will be passed on to any given offspring is 25%. If a couple gives birth to four children, what is the probability that none of the children will inherit the trait? To simulate the experiment, use the function int(4*rand) + 1 to generate random numbers between 1 and 4 inclusive. Let the number 1 represent a child that inherits the trait and the numbers 2, 3, and 4 represent children that do not. For each trial, generate a sequence of four numbers.

connection to SOCIAL STUDIES

Silvana's government project is about the population of the United States. She made this graph to show the distribution of ages.

Ages of United States Population (1992)

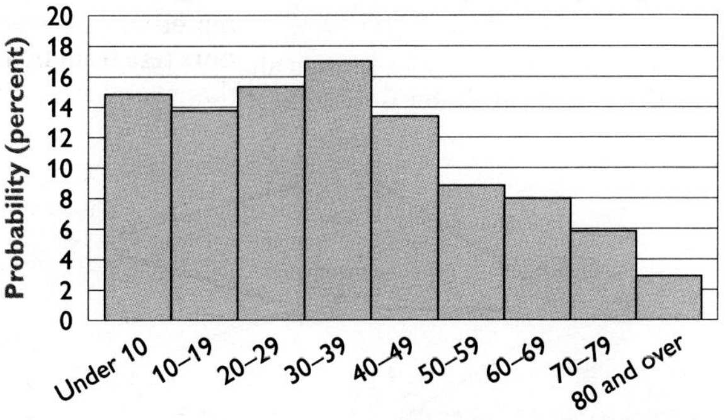

12. If you select a person at random, what age group is he or she most likely to be in?

13. Suppose you select a person at random. Estimate the probability that he or she is younger than 30.

14. Suppose the distribution of ages were uniform. What would the graph look like?

BY THE WAY...

The United States is expected to experience *population aging* over the next 50 years. The Census Bureau predicts that in 2050, almost 30% of the United States population will be 60 or older.

Design and carry out a simulation to investigate each situation.

15. Suppose you randomly stop students in the hall and ask them what month they were born in. On average, how many people do you have to ask before you find someone born in the same month as you? (Assume that birth months are equally likely.)

16. Mina Parekh has a stack of résumés for a position she wants to fill. All the résumés are from qualified candidates. Suppose the probability that an applicant is an excellent candidate is $\frac{1}{5}$ and the probability that an applicant is an outstanding candidate is $\frac{1}{10}$. On average, how many applicants must Mina Parekh interview before she finds an excellent or outstanding candidate?

Design a simulation to investigate each situation.

17. A group of equally qualified candidates for a job contains 30 men and 20 women. Five people are hired. What is the probability that four or five of the people hired are women?

18. Change the airplane simulation in Sample 2 so that the plane holds the number of people shown in the diagram, the airline sells 14 tickets, and there is a 25% chance that a person does not take the flight. Then how often is the plane overbooked?

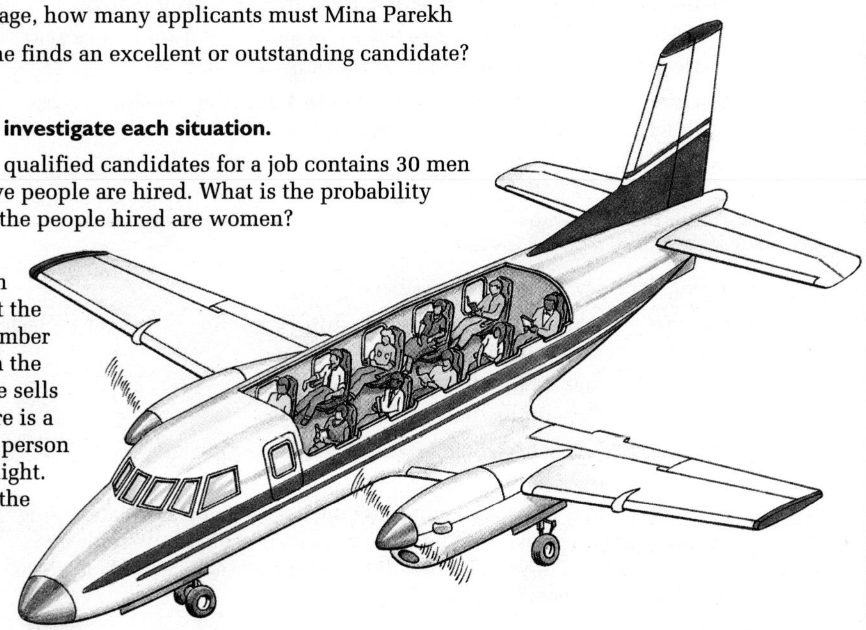

Using Technology

In connection with Exs. 6–9, have students explore the graph of y = rand. Suggest a graphing window with Xmin = 0, Xmax = 94, Ymin = –1, and Ymax = 2. View the graph with the calculator set to dot mode. Students will notice some rather unexpected behavior if they trace back and forth along the graph.

Students can use parametric and dot mode to explore the situations in Exs. 6–9. Get their thinking started by examining the situation in Sample 1. Enter X_{1T}=T and Y_{1T}=int(6*rand)+1. For the graphing window, use Tmin = 1, Tmax = 94, Tstep = 1, Xmin = 0, Xmax = 94, Ymin = –3, and Ymax = 10. After students have studied the graph, ask how they might use it to find an experimental probability for rolling the same number three times in a row with a single die. You can get a new graph to simulate more rolls by simply reentering X_{1T} = T.

Communication: Writing

Exs. 10 and 11 provide an opportunity for students to express their ideas about simulations in writing. You may wish to select a small random sample of students' responses for evaluation in order to provide constructive feedback as to the accuracy of the work and the quality of the writing.

Cooperative Learning

Students may enjoy working on Exs. 15–20 in small groups of three or four. In so doing, they have an opportunity to use their skills by discussing each situation and sharing ideas.

Application

Ask students if they think the method of Ex. 17 can be used to determine if job discrimination is being practiced by a particular employer. Students should support their answers with mathematical reasons.

Answers to Exercises and Problems

1. Answers may vary. An example is given. Simulating an experiment can help to save on costs, time, and materials, and avoid risks that might be involved in a real event.
2. All three numbers have an equal chance of appearing in a set of three-digit random numbers. None of the digits has any effect on the others, so that for example, the fact that the first digit of the first number is 8 does not make it any more or less likely that the second and third digits will be 8 as well.
3. No. Since there is no restriction on the range of numbers, a random distribution cannot be achieved. Responses would tend to be positive one- or two-digit numbers. Also, by saying the numbers out loud, some students' responses might be affected by others' responses.
4. Yes (no two people in the group have the same phone number); each of the digits 0–9 is equally likely to occur.

5–18. See answers in back of book.

Communication: Discussion

After the groups complete Ex. 21, you may wish to have some of them present their results to the class. All students can benefit from seeing how the different groups designed and carried out their simulations.

19. Design a simulation to investigate this situation: Suppose a water system has five pumps that work independently of each other.

At any time, the probability that a pump does not work is $\frac{1}{4}$. For water to flow from A to B, both pumps along at least one path must be working. How often can water flow from A to B?

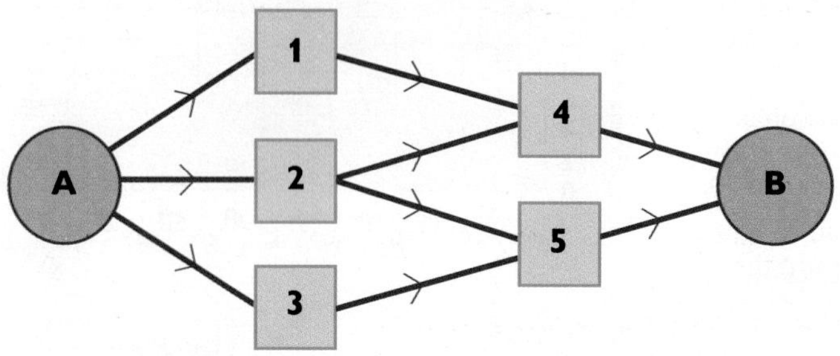

20. The tickets for a school fundraiser are numbered from 1 to n, the number of tickets sold. They will be used in a drawing for a door prize. Three friends arrive at the fundraiser separately (and randomly). Their ticket numbers are 43, 67, and 105. The friends estimate the number of tickets sold by using this algorithm:

- Find the mean of the differences between any two of their numbers.
- Add the mean to the greatest of their numbers.

a. What is their estimate of the total number of people at the fundraiser?

b. Estimate the probability that one of the friends wins the prize.

c. Suppose another friend joins the group and has ticket number 148. How does this change their estimate of n?

d. Choose three random numbers between 1 and 100. Use the process described above to estimate n (which in this case is 100). Do this several times. Is the mean of the estimates close to 100?

Ongoing **ASSESSMENT**

21. Group Activity Work with another student.

Suppose you can get to school by two different routes. You have collected this data over a period of several weeks:

Route A: 70% of the time it takes 5 minutes.
30% of the time it takes 8 minutes.

Route B: 20% of the time it takes 3 minutes.
80% of the time it takes 6 minutes.

Design and carry out a simulation in which one of you simulates using Route A for five weeks and the other simulates using Route B for five weeks (25 school days). Find the average time in each case. On average, which route is faster?

Answers to Exercises and Problems

19. Methods of simulation and results may vary. Use the function int(4*rand) + 1. Let the number 1 indicate a non-working pump and the numbers 2, 3, and 4 indicate a working pump. For each trial, generate a sequence of five random numbers. A sequence of 1's appearing as in the following patterns represents a situation in which the water will not flow from A to B:
1, 1, 1, –, –;
–, –, –, 1, 1;
1, 1, –, –, 1;
1, 1, 1, 1, 1;
–, 1, 1, 1, 1;
1, –, 1, 1, 1;
1, 1, –, 1, 1.
Record the number of sequences and the number that have any of the above patterns. Analyze the data.

20. a. about 146
b. $\frac{3}{146}$, or about 2%
c. about 236
d. Answers will vary.

21. Methods of simulation and results may vary. An example is given. For each route, use the function int(10*rand) + 1. For each trial for Route A, generate a sequence of 25 random numbers and let the numbers 1–7 represent a time of 5 minutes and 8–10 represent a time of 8 minutes. For Route B, generate a sequence of 25 random numbers and let the numbers 1 and 2 represent a time of 3 minutes and the numbers 3–10 represent a time of 6 minutes. On average, Route B is faster.

22. $y = x^2 + 2x - 8$

23. 160°; 20°

24. $\frac{1}{8}$, or 12.5%

25. a. No; the event "6 disks land same color up" has only two outcomes. The event "5 disks land same color up" has twelve.

Review PREVIEW

22. A parabola passes through the points $(-4, 0)$, $(-2, -8)$, and $(1, -5)$. Find an equation for the parabola. *(Section 6-5)*

23. Find the measures of each interior and each exterior angle of a regular 18-gon. *(Section 3-4)*

24. Suppose you toss three coins. What is the probability of getting all heads? *(Toolbox Skill 6)*

Working on the Unit Project

Da-un-dah-qua is a game for two players.

- Place six disks in a dish.
- Mark the disks so that the two faces are different from each other. For example, one side can be black and the other side green.
- One player shakes or rattles the dish so that the disks jump and resettle.

Rules of DA-UN-DAH-QUA

- If the disks land with all the same color showing, the player scores 5 points and gets another turn.
- If five of the disks land with the same color showing, the player scores 1 point and gets another turn.
- For all other results, the player scores no points and the other player gets a turn.
- The first player who scores 40 points wins the game.

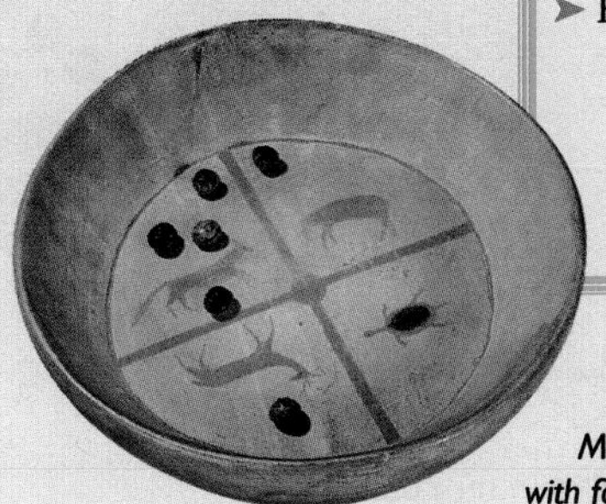

The Iroquois people used this decorated wooden bowl to play da-un-dah-qua at the Midwinter Ceremonial. The bowl is decorated with four clan symbols. The disks are peach pits burned black on one side.

25. a. Do you think that the events "6 disks land same color side up" and "5 disks land same color side up" are equally likely? Why or why not?
    b. Make a list of all the possible outcomes. Which event is most likely?

26. How can you simulate the toss of one disk? of six disks?

27. **Group Activity** Work with another student.
    Play two or three rounds of da-un-dah-qua. Record the outcome of each turn.

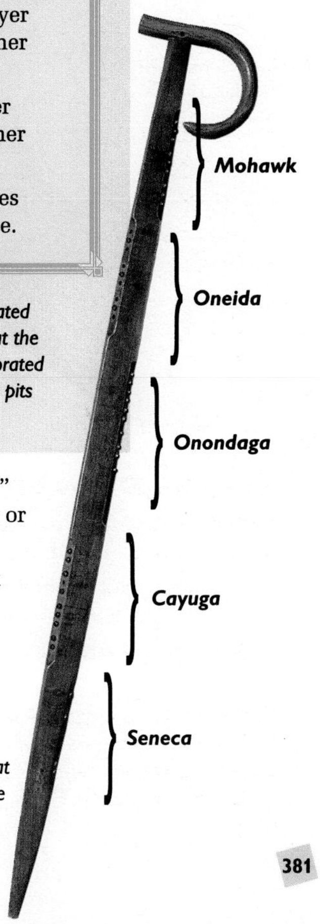

This maple cane was used to call the roll of the chiefs at meetings of the Iroquois League of Peace. The cane has five sections, one for each of the five nations that made up the league.

Working on the Unit Project

Since students will need to do a probability analysis of the game of da-un-dah-qua when completing the Unit Project, they should keep a written record of their answers to all exercises from this section throughout the unit.

Practice 52 For use with Section 7-1

Name ______________________ Date __________

Practice 52
For use with Section 7-1

11. Use 2 spinners like those shown below. Spin both for each trial. Tally trials and tally successes (paper arrives in time). Conduct at least 50 trials. Divide no. of successes by total no. of trials.

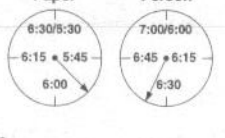

Tell whether each situation appears to produce a sequence of random numbers. Give a reason for your answer.

1. You ask each of 100 people the first letter of their family name. no; Letters such as "Q" and "X" are less likely.
2. In a parking lot, you record the last three digits of those license numbers that contain at least three digits. yes
3. You ask members of your math class to tell their house numbers. yes
4. You throw two dice and record the sum of the numbers showing. no; A sum of 3 is less likely than a sum of 7.
5. You look at a clock 20 times during a day and record the number of seconds after the minute shown by the second hand. yes

Design and test a formula for a calculator to produce each set of random numbers.

6. integers from 0 to 8 int(9 * rand)
7. integers from 1 to 12 int(12 * rand) + 1
8. integers from 9 to 15 int(7 * rand) + 9
9. even numbers from 2 to 16 2 * int(8 * rand) + 2

Exs. 10–12: Simulations may vary. Sample answers are given.

For Exercises 10–12, design a simulation to investigate the situation.

10. The probability that an egg will have cracked by the time it reaches the supermarket shelf is $\frac{1}{8}$. On average, how many cartons of 12 eggs must you open to find a carton with 12 unbroken eggs? Generate rand. nos. bet. 1 and 8. Let 1 stand for a broken egg. Count sets of 12 nos. until no 1 occurs.
11. Eduardo Sánchez leaves home for the office at a random time between 6:00 A.M. and 7:00 A.M. His newspaper is delivered at a random time between 5:30 A.M. and 6:30 A.M. What is the probability that the paper arrives before he leaves? See top of page.
12. A group of 5 students is to be chosen from the dance committee to make arrangements for the winter prom. The committee contains 6 boys and 8 girls. How many groups must be chosen at random before a group is chosen that contains at least 4 girls? 12. Flip a coin in sets of 5 flips until one such set contains at least 4 heads.
13. *Open-ended* Describe a real-life situation that could be modeled by the following experiment: Tags numbered from 1 to 20 are placed in a bag. A group of four tags is drawn at random. The tags are replaced in the bag, and more groups of four tags are drawn and replaced. Answers may vary. Check students' work.

52 Practice Bank, INTEGRATED MATHEMATICS 3

Answers to Exercises and Problems

b. In the list, A represents one color and B represents the other color.

AAAAAA, AAAAAB, AAAABA, AAAABB, AAABAA, AAABAB, AAABBA, AAABBB, AABAAA, AABAAB, AABABA, AABABB, AABBAA, AABBAB, AABBBA, AABBBB, ABAAAA, ABAAAB, ABAABA, ABAABB, ABABAA, ABABAB, ABABBA, ABABBB, ABBAAA, ABBAAB, ABBABA, ABBABB, ABBBAA, ABBBAB, ABBBBA, ABBBBB, BAAAAA, BAAAAB, BAAABA, BAAABB, BAABAA, BAABAB, BAABBA, BAABBB, BABAAA, BABAAB, BABABA, BABABB, BABBAA, BABBAB, BABBBA, BABBBB, BBAAAA, BBAAAB, BBAABA, BBAABB, BBABAA, BBABAB, BBABBA, BBABBB, BBBAAA, BBBAAB, BBBABA, BBBABB, BBBBAA, BBBBAB, BBBBBA, BBBBBB

Answers may vary. An example is given. The most likely outcomes are those with two of one color and four of the other color showing.

26. Use the function int(2*rand) to generate a random number, with 0 representing "color A up" and 1 representing "color B up." Use the same function to generate a sequence of 6 random numbers. Again, let 0 represent "color A up" and 1 represent "color B up."

27. Check students' results.

PLANNING

Objectives and Strands

See pages 370A and 370B.

Spiral Learning

See page 370B.

Materials List

- Graph paper
- Calculator

Recommended Pacing

Section 7-2 is a two-day lesson.

Day 1

Pages 382–383: Talk it Over 1 through Sample 1, *Exercises 1–8*

Day 2

Pages 384–386: Probability with *or* through Look Back, *Exercises 9–28*

Toolbox References

- **Toolbox Skill 3:** Counting Permutations
- **Toolbox Skill 6:** Finding the Probability of an Event

Extra Practice

See pages 619–620.

Warm-Up Exercises

- Warm-Up Transparency 7-2

Support Materials

- Practice Bank: Practice 53
- Activity Bank: Enrichment 47
- Study Guide: Section 7-2
- Problem Bank: Problem Set 15
- Explorations Lab Manual: Diagram Master 2
- Assessment Book: Quiz 7-2, Alternative Assessment 2

Section 7-2 Probability of Either of Two Events

Focus

Find probabilities in situations with continuous outcomes and with discrete outcomes. Find the probability that either of two events occurs.

TO BE OR NOT TO BE

Talk it Over

1. Answer each question for rolling the die and for spinning the pointer on the dial shown.

All the positive *real numbers* less than 6 are around the dial.
Each face of a die contains one of the *integers* 1–6.
0
1
2
3
4
5
The pointer spins freely.

   a. What are all the possible outcomes?
   b. Are the outcomes *discrete* or *continuous?* Explain your choice.

2. Suppose you roll the die. What is the probability of the event "a number less than or equal to 5"? Explain how you found the answer.

3. Charlene found the probability that the pointer on the dial will stop at a number less than or equal to 5 this way:

   The length of the arc from 0 to 5 is $\frac{5}{6}$ of the circumference of the entire circle, so $P(\text{number} \le 5) = \frac{5}{6}$.

   Explain why her method works.

Student Resources Toolbox p. 631 *Probability*

Answers to Talk it Over

1. a. die: 1, 2, 3, 4, 5, 6; dial: all real numbers greater than 0 and less than 6
   b. die: discrete (exactly six possible outcomes); dial: continuous (the number of outcomes cannot be counted)
2. $\frac{5}{6} \approx 0.83$; $\frac{\text{number of favorable outcomes}}{\text{number of possible outcomes}} = \frac{5}{6}$
3. Since the outcomes are continuous, measurements can be used to compare the ratio of favorable outcomes to possible outcomes.

Student Resources Toolbox
p. 628 *Using Counting Techniques*

To find the probability of an event with discrete outcomes, you can find the number of outcomes by direct counting or by using the multiplication counting principle or permutations. To find the probability of an event with continuous outcomes, you can measure lengths, areas, and so on.

$$P(\text{event}) = \frac{\text{size of favorable outcomes}}{\text{size of possible outcomes}}$$

"Size" can be a count or a measurement.

Sample 1

For each situation, find the probability of the event "the sum of both outcomes is less than or equal to 5."

a. Suppose you roll two dice.

b. Suppose you spin the pointers on two dials like the one on page 382.

Sample Response

a. When you roll two dice, the outcomes are discrete. There are 6 • 6 = 36 possible outcomes.

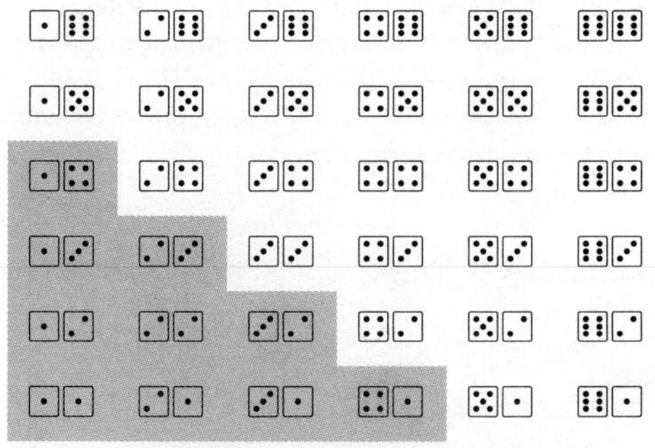

There are 10 outcomes for which sum ≤ 5.

$$P(\text{sum} \le 5) = \frac{\text{number of favorable outcomes}}{\text{number of possible outcomes}}$$

$$= \frac{10}{36} \approx 0.28$$

b. When you spin the pointers on two dials, the outcomes are continuous.
Let x = all possible outcomes on one dial.
Let y = all possible outcomes on the other dial.

Points (x, y) in the square represent all the possible outcomes for two dials.

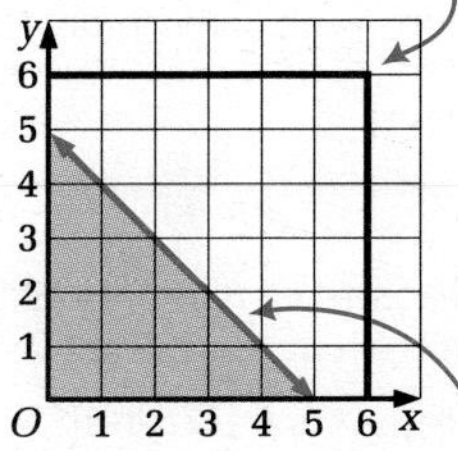

Points in the shaded region represent outcomes for which $x + y \le 5$.

$$P(\text{sum} \le 5) = \frac{\text{area of shaded triangle}}{\text{area of 6-by-6 square}}$$

$$= \frac{12.5}{36} \approx 0.35$$

Now you are ready for:
Exs. 1–8 on p. 387

TEACHING

Talk it Over

Question 1 leads students to consider the difference between discrete and continuous outcomes. Question 3 introduces students to a way of computing probabilities geometrically for outcomes that are continuous.

Additional Sample

S1 For each situation, find the probability of the event "the sum of both outcomes is between 3 and 6 inclusive."

a. Suppose you roll two dice.

The 36 possible outcomes of rolling two dice are discrete. There are 14 outcomes for which the sum is between 3 and 6 inclusive.

$P(3 \le \text{sum} \le 6) = \dfrac{\text{number of favorable outcomes}}{\text{number of possible outcomes}}$

$= \dfrac{14}{36} \approx 0.39$

b. Suppose you spin the pointers on two dials like the one on page 382.

The outcomes of spinning the two dials are continuous. Let x = all possible outcomes on one dial. Let y = all possible outcomes on the other dial.

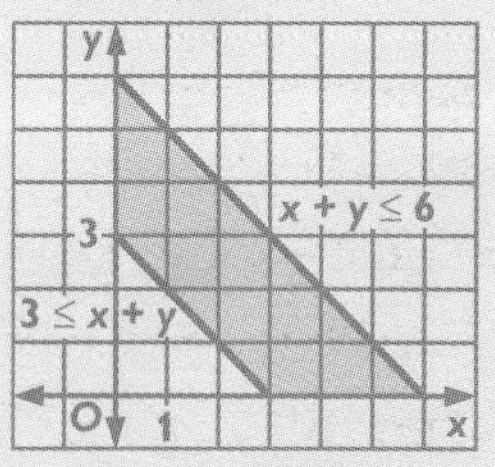

Points in the shaded region represent outcomes for which $3 \le x + y \le 6$.

$P(3 \le x + y \le 6) = \dfrac{\text{area of the shaded region}}{\text{area of 6-by-6 square}} = \dfrac{13.5}{36} \approx 0.38$

Talk it Over

Students use Venn diagrams in questions 4 and 5 to discover a way to find the probability of either of two events, the rule for which is given on page 385.

Problem Solving

Venn diagrams can be used to analyze probability problems. Ask students to recall how they have used Venn diagrams in the past. A few samples could be drawn on the board to review the basic ideas.

Additional Sample

S2 Refer to the schedule in Sample 2. Find the probability that Renata Baines waits no more than 8 min for the T-184 bus or no more than 8 min for the T-172 bus.

Step 1. Find all possible waiting times for both buses. Let x = the waiting time for the T-184 bus and let y = the waiting time for the T-172 bus. The points inside the rectangle formed by $0 \le x \le 20$ and $0 \le y \le 15$ represent all possible waiting times for the buses. The area of this rectangle is $15 \cdot 20 = 300$.

Step 2. Find all waiting times in which a bus arrives within 8 min. Graph the inequalities $0 \le x \le 8$ and $0 \le y \le 8$.

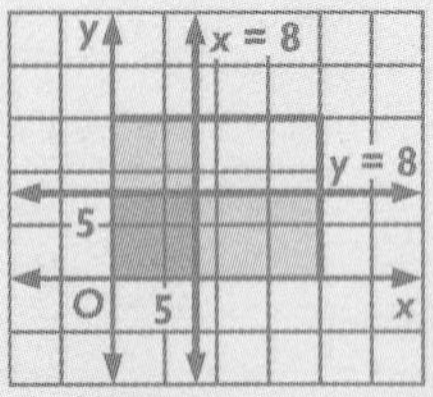

area of the shaded region = sum of the areas of the two shaded rectangles − area of the doubly shaded region = $(8 \cdot 15 + 8 \cdot 20) - 8 \cdot 8 = 216$

Step 3. Find the ratio of the areas you found in steps 1 and 2.

$P(0 \le x \le 8 \text{ or } 0 \le y \le 8) = \frac{\text{area of the shaded region}}{\text{area of 15-by-20 rectangle}} = \frac{216}{300} \approx 0.72$

Probability with *or*

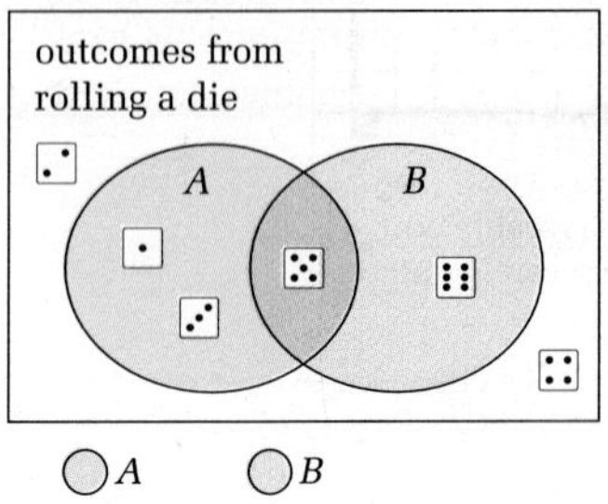

Recall that logical statements in mathematics use the inclusive *or*. The statement "either event A occurs or event B occurs" is true when *at least one* of the events occurs. It is true:

- when *only* event A occurs,
- when *only* event B occurs,
- when *both* events A and B occur.

Talk it Over

4. Use the Venn diagram. Event A is "odd number." Event B is "number greater than 4." Find each probability.
   a. $P(A)$
   b. $P(B)$
   c. $P(A \text{ and } B)$
5. How can you use your answers to question 4 to find $P(A \text{ or } B)$?

Sample 2

Renata Baines rides the bus to work. She can take either the T-184 bus or the T-172 bus. During the morning and afternoon rush hours, the T-184 bus runs every 20 min, while the T-172 bus runs every 15 min. What is the probability that she will wait no more than 5 min for a bus during rush hour?

Sample Response

Find the probability that she waits no more than 5 min for the T-184 bus or no more than 5 min for the T-172 bus.

Problem Solving Strategy: Use a diagram.

Step 1 Find all the possible waiting times for both buses.

- Let x = the waiting time for the T-184 bus.
- Let y = the waiting time for the T-172 bus.

The points inside the rectangle formed by $0 \le x \le 20$ and $0 \le y \le 15$ represent all possible waiting times for the buses.

The area of this rectangle is $15 \cdot 20 = 300$.

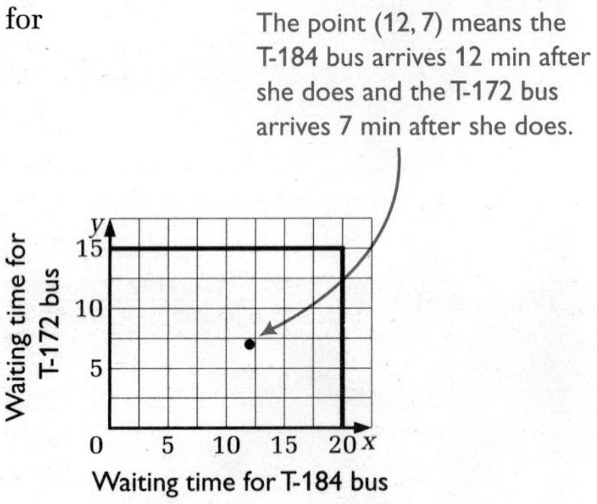

Answers to Talk it Over

4. a. $\frac{1}{2} = 0.5$
   b. $\frac{1}{3} \approx 0.33$
   c. $\frac{1}{6} \approx 0.17$
5. To find $P(A \text{ or } B)$ for the given situation, you can add $P(A)$ and $P(B)$ and subtract $P(A \text{ and } B)$ so no outcomes are counted twice. $P(A \text{ or } B) = \frac{1}{2} + \frac{1}{3} - \frac{1}{6} = \frac{2}{3} \approx 0.67$

Step 2 Find all waiting times in which a bus arrives within 5 min.

Graph the inequalities $0 \le x \le 5$ and $0 \le y \le 5$.

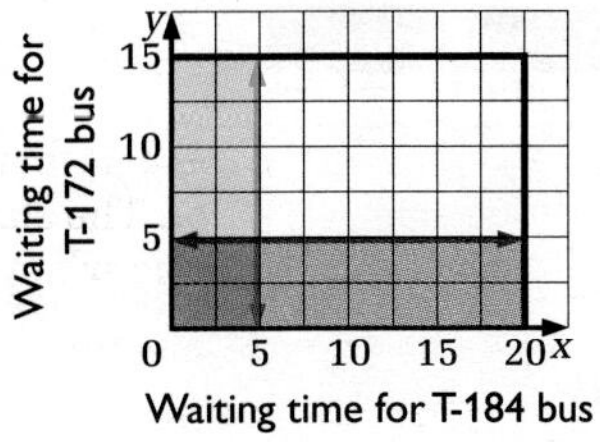

area of the shaded region	=	sum of the areas of the two shaded rectangles	−	area of the doubly shaded region
	=	$(5 \cdot 15 + 5 \cdot 20)$	−	$5 \cdot 5$
	= 150			

Step 3 Find the ratio of the areas you found in steps 1 and 2.

$$P(0 \le x \le 5 \text{ or } 0 \le y \le 5) = \frac{\text{area of the shaded region}}{\text{area of 15-by-20 rectangle}}$$

$$= \frac{150}{300} = 0.5$$

The probability that Renata Baines will wait no more than 5 min is 0.5.

PROBABILITY OF EITHER OF TWO EVENTS

To find the probability that either of two events occurs, you can add the probabilities of the events and subtract the probability that both events occur.

For any two events A and B:

$$\mathbf{P(A \text{ or } B) = P(A) + P(B) - P(A \text{ and } B)}$$

Talk it Over

6. Describe the region in Sample 2 that contains the points that represent this event:

   The waiting time for the T-184 bus is no more than 5 min *and* the waiting time for the T-172 bus is no more than 5 min.

7. Suppose you roll a die. Find each probability.

   a. *P*(4)
   b. *P*(odd number)
   c. *P*(4 or odd number)
   d. *P*(4 and odd number)

8. Suppose you replace "odd number" with "not 4" in question 7. How do your answers to parts (b)–(d) change?

Answers to Talk it Over

6. the square region bounded by the x- and y-axes and the lines $x = 5$ and $y = 5$

7. a. $\frac{1}{6} \approx 0.17$

   b. $\frac{3}{6} = \frac{1}{2} = 0.5$

   c. $\frac{4}{6} = \frac{2}{3} \approx 0.67$

   d. 0

8. Part (b) becomes $\frac{5}{6}$, or about 0.83; part (c) becomes 1; part (d) remains 0.

Visual Thinking

Ask students to explain in their own words what is taking place in relation to the buses in each of the shaded areas of the diagram at the top of this page. This activity involves the visual skills of *interpretation* and *communication.*

Students Acquiring English

Some students acquiring English may have difficulty gaining an understanding of the meaning of *Probability of Either of Two Events* from the text alone. You may want to clarify the situation to which the theorem applies as follows: There are two events that could happen. You want to find the probability that one or the other (but not both) will happen.

Talk it Over

Question 8 prepares students to understand how the rule for the probability of either of two events is affected in the case of mutually exclusive events.

Communication: Drawing

Students should be able to draw a Venn diagram to illustrate the formula for finding the probability of either of two events.

Error Analysis

Some students seem to have a difficult time distinguishing the differences in the situations and formulas used for determining probabilities of mutually exclusive events and probabilities of either of two events. Students should discuss the two types of situations and ask themselves which formula to apply.

Teaching Tip

Students should understand that the terms *mutually exclusive* and *complementary* do not have the same meaning. When rolling a die, $P(3)$ and $P(4)$ are mutually exclusive events, but they are not complementary.

Look Back

In answering this Look Back, students tie together the two parts of this section. You may wish to have students respond in writing. A brief class discussion would ensure that all students understand the main ideas of the section.

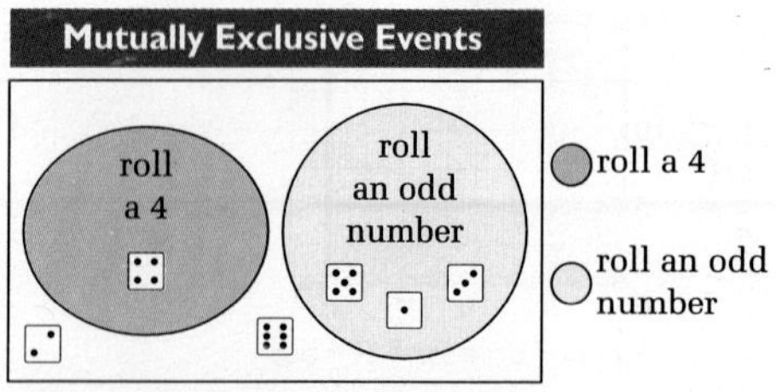

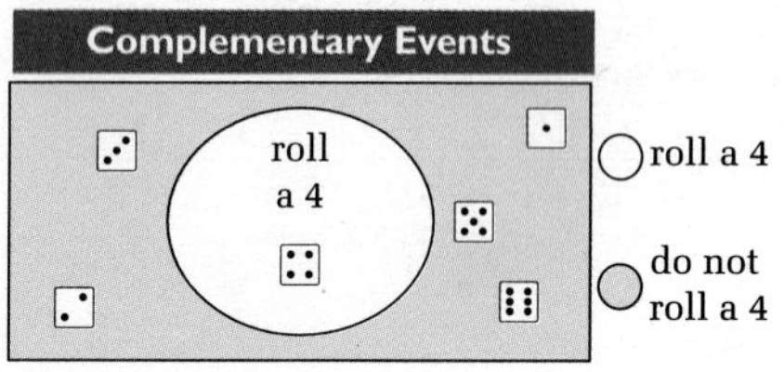

All the outcomes not in "roll a 4" are in its complement.

In Sample 2, both buses can arrive at the bus stop in 5 min or less from the time Renata Baines begins to wait. However, some events *cannot* occur at the same time. Events like these are called **mutually exclusive events.** When you roll a die, the events "roll a 4" and "roll an odd number" are mutually exclusive.

Two mutually exclusive events that account for all possibilities are called **complementary events.** When you roll a die, the events "roll a 4" and "do not roll a 4" are complementary events.

PROBABILITY OF MUTUALLY EXCLUSIVE EVENTS

When two events are mutually exclusive, you can add to find the probability that either one occurs. For any two mutually exclusive events A and B:

$$\mathbf{P(A \text{ or } B) = P(A) + P(B)}$$

Complementary Events

The sum of the probabilities of complementary events is 1. For any event A:

$$\mathbf{P(\text{not } A) = 1 - P(A)}$$

Talk it Over

9. Explain how to use the general formula for $P(A \text{ or } B)$ to derive the formula for the probability of mutually exclusive events.
10. Explain how to use the general formula for $P(A \text{ or } B)$ to derive the formula for the probability of complementary events.

Look Back

Compare finding the probability of either of two events when the outcomes of both events are continuous and when the outcomes of both events are discrete.

Now you are ready for: Exs. 9–28 on pp. 387–389

Answers to Talk it Over

9. If A and B are mutually exclusive events, then $P(A \text{ and } B) = 0$, so $P(A \text{ or } B) = P(A) + P(B) - P(A \text{ and } B) = P(A) + P(B) - 0 = P(A) + P(B)$.

10. A and not A are complementary events, so A and not A are mutually exclusive, which means that $P(A \text{ and not } A) = 0$. Also, A and not A account for all possibilities, so $P(A \text{ or not } A) = 1$. Then $P(A \text{ or not } A) = P(A) + P(\text{not } A) - P(A \text{ and not } A)$. Then $1 = P(A) + P(\text{not } A) - 0$ and $P(\text{not } A) = 1 - P(A)$.

Answers to Look Back

Probability is found by using the ratio $\frac{\text{size of favorable outcomes}}{\text{size of possible outcomes}}$, in both cases. If the outcomes are continuous, the sizes are measurements. If the outcomes are discrete, the sizes are counts.

7-2 Exercises and Problems

1. **Reading** How can you find the probability of an event with continuous outcomes?

For each event in Exercises 2–8:

a. Tell whether the outcomes are *discrete* or *continuous*.

b. Find the probability of the event.

2. The freely spinning pointer on a dial numbered 0–8 stops at a value between 1 and 4.
3. You roll a 9 on a 12-sided die that has the numbers 1 through 12 on its faces.
4. You choose an ace at random from a well-shuffled standard deck of cards.

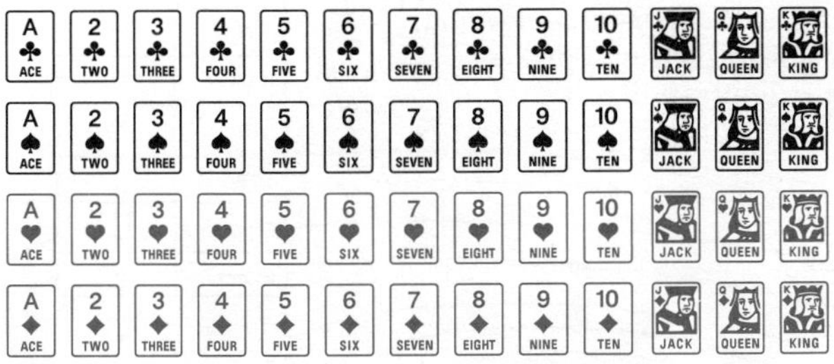

Each suit has 3 face cards — jack, queen, and king.

5. When you look at a clock, the second hand is between 6 and 7.
6. Suppose every board in a pile of 8-foot pine boards has a knot in it somewhere. A board is chosen at random and the knot is within 1 ft of one end of the board.
7. When you spin the pointers on the dial for Exercise 2 and the dial on page 382, the sum of the numbers where the pointers stop is greater than 10.
8. When you spin the pointers on the dial for Exercise 2 and the dial on page 382, the sum of the numbers where the pointers stop is between 2 and 4.

Find the probability of each event.

9. When you spin the pointers on two dials like the one on page 382, the sum of the numbers is less than 2 or greater than 3.
10. You choose an ace or a heart at random from a well-shuffled standard deck of playing cards.
11. You choose a face card or a red card at random from a well-shuffled standard deck of playing cards.
12. When you roll two dice, the sum of the outcomes is greater than 6 or you get a 4 on one of the dice.
13. **Writing** You use probability to solve the problem in Sample 2 because the outcomes in this situation are random. What is the source of randomness in this situation? Explain.

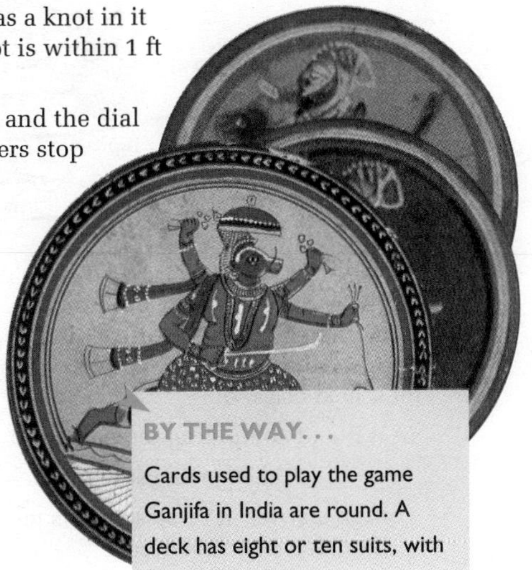

BY THE WAY...

Cards used to play the game Ganjifa in India are round. A deck has eight or ten suits, with 12 cards in each suit.

APPLYING

Suggested Assignment

Day 1

Standard 1–8

Extended 1–8

Day 2

Standard 9–12, 14, 16, 20–28

Extended 9–18, 20–28

Integrating the Strands

Number Exs. 20–22

Algebra Exs. 7–9, 18, 23–25

Functions Exs. 23–25

Geometry Exs. 7–9, 18

Statistics and Probability Exs. 1–22, 26–28

Logic and Language Exs. 1–8, 13, 16, 17, 19, 27

Using Technology

The exercises on this page deal with theoretical probabilities. Students can use the random number capabilities of the TI-81 and TI-82 to perform simulations. For example, in Ex. 12, students can enter the expression (int(6*rand)+1) + (int(6*rand)+1) on the home screen. If they press ENTER repeatedly, they will generate sums for successive rolls of two dice. They can keep a tally of the sums and use it when they are ready to calculate a probability for the simulation. Students should discuss why the expression used for the simulation cannot be simplified by using 2(int(6*rand)+1).

Answers to Exercises and Problems

When necessary, probabilities are rounded to the nearest hundredth.

1. Use measurements such as length or area to determine the sizes of favorable outcomes and possible outcomes.
2. a. continuous
   b. $\frac{3}{8} = 0.375$
3. a. discrete
   b. $\frac{1}{12} \approx 0.083$
4. a. discrete
   b. $\frac{4}{52} = \frac{1}{13} \approx 0.077$
5. a. continuous
   b. $\frac{1}{12} \approx 0.083$
6. a. continuous
   b. $\frac{2}{8} = \frac{1}{4} = 0.25$
7. a. continuous
   b. $\frac{8}{48} = \frac{1}{6} \approx 0.17$
8. a. continuous
   b. $\frac{6}{48} = \frac{1}{8} \approx 0.13$
9. $\frac{67}{72} \approx 0.93$
10. $\frac{16}{52} = \frac{4}{13} \approx 0.31$
11. $\frac{32}{52} = \frac{8}{13} \approx 0.62$
12. $\frac{25}{36} \approx 0.69$
13. Renata's arrival time at the bus stop is random. If she arrived at the same time every day, she would take the same bus and have exactly the same waiting time. It would not be a matter of probability.

Career Note

Astronomy is the study of the objects in the universe. Astronomers sometimes work with probes that travel into space to collect data. The data from these probes are often used to understand how the stars, planets, and other objects in the universe were formed, and also how the universe itself was formed. All astronomers need a strong background in mathematics and science, especially physics and chemistry. Those astronomers and other scientists who are interested in learning how the universe was formed, and what its ultimate fate may be, are also called cosmologists. Many cosmologists are essentially mathematical physicists. Their theories about the universe are expressed in highly abstract form using mathematical models.

Communication: Discussion

Ex. 16 can be used as a class discussion activity. Students whose answers to the various questions are different can explain their reasoning to support their responses.

Visual Thinking

Ask students to create diagrams of the real-life situations that they have described in response to Ex. 17. Their responses may take the form of Venn diagrams or other types of diagrams that are pictured in this section. Ask them to explain their diagrams to the class. This activity involves the visual skills of *correlation* and *communication.*

connection to **SCIENCE**

Use the Venn diagram for Exercise 14.

Suppose you select a planet at random. Find each probability.

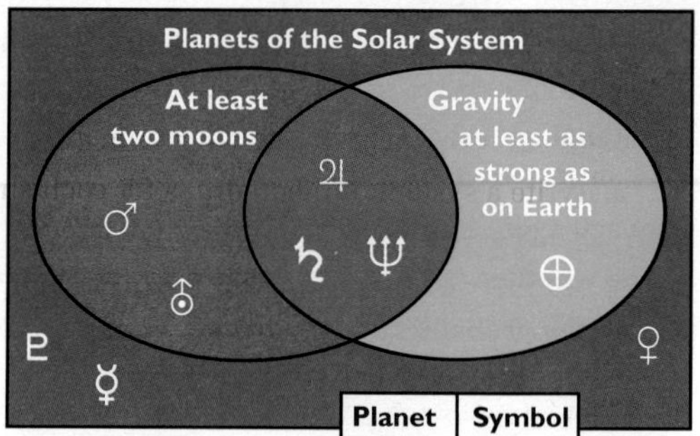

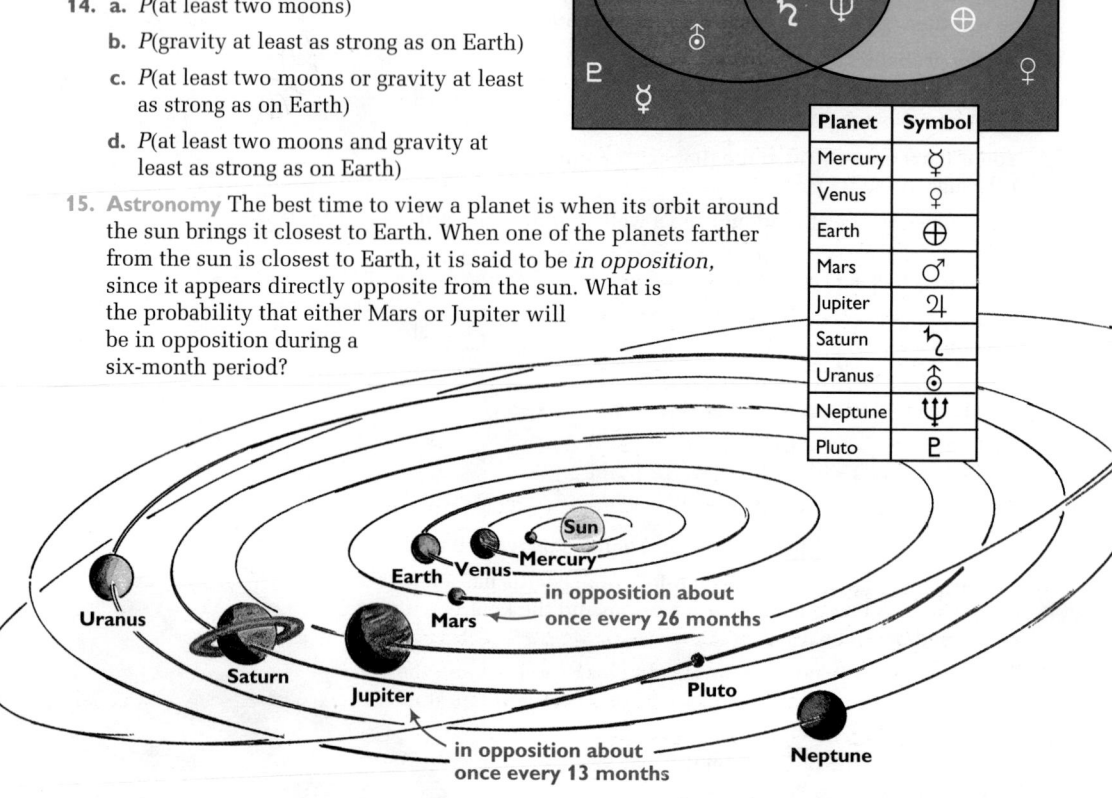

Planet	Symbol
Mercury	☿
Venus	♀
Earth	⊕
Mars	♂
Jupiter	♃
Saturn	♄
Uranus	⛢
Neptune	♆
Pluto	♇

14. **a.** *P*(at least two moons)

b. *P*(gravity at least as strong as on Earth)

c. *P*(at least two moons or gravity at least as strong as on Earth)

d. *P*(at least two moons and gravity at least as strong as on Earth)

15. **Astronomy** The best time to view a planet is when its orbit around the sun brings it closest to Earth. When one of the planets farther from the sun is closest to Earth, it is said to be *in opposition,* since it appears directly opposite from the sun. What is the probability that either Mars or Jupiter will be in opposition during a six-month period?

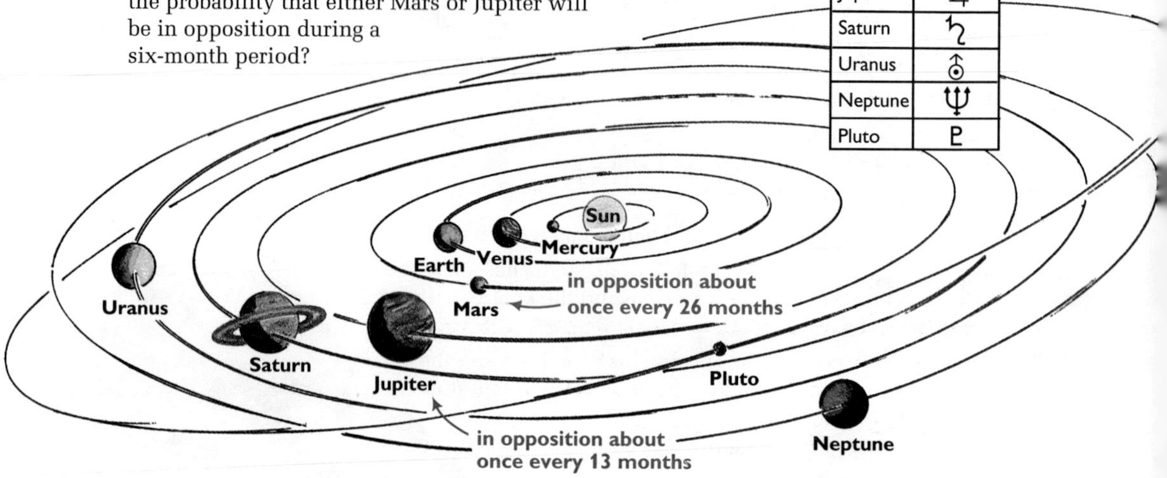

16. Suppose some students meet in the student center between classes. They take off their almost-identical denim jackets and pile them on a chair. Suppose they grab their jackets without looking when the bell rings for their next class.

a. Without computing the probability, tell whether you think

P(at least one student picks up the correct jacket)

is greater when *three* students meet or when *two* students meet. Explain your reasoning.

b. Find *P*(at least one student picks up the right jacket) in each situation described in part (a).

c. Does your answer to part (b) surprise you? Why or why not?

d. What do you think the probability is in this situation when four students meet? Check your answer by computing the probability.

Answers to Exercises and Problems

14. **a.** $\frac{5}{9} \approx 0.56$

b. $\frac{4}{9} \approx 0.44$

c. $\frac{2}{3} \approx 0.67$

d. $\frac{1}{3} \approx 0.33$

15. $\frac{99}{169} \approx 0.59$

16. See answers in back of book.

17. Answers may vary. An example is given. When you are in school, the events "you are in math class" and "you are in English class" are mutually exclusive but not complementary.

18. **a.** $0 \le x \le 5$, $0 \le y \le 15$, $y \ge x$; $0 \le x \le 20$, $0 \le y \le 5$, $y \le x$

b. The trapezoid with vertical bases represents the event that she waits longer for the T-172 bus than the T-184 bus, but no more than 5 minutes. The trapezoid with horizontal bases represents the event that she waits longer for the T-184 bus than the T-172 bus, but no more than 5 minutes.

c. Yes; No. **d.** $\frac{1}{2} = 0.5$

17. **Open-ended** Describe a real-life situation in which two events are mutually exclusive but are not complementary.

18. Amir drew this diagram to model the situation in Sample 2.
    a. Write a system of inequalities for each shaded trapezoid.
    b. What event does each trapezoid represent?
    c. Are the events represented by the two shaded trapezoids mutually exclusive? Are the events represented by the two shaded rectangles in Step 2 of Sample 2 mutually exclusive?
    d. Use Amir's diagram to solve Sample 2.

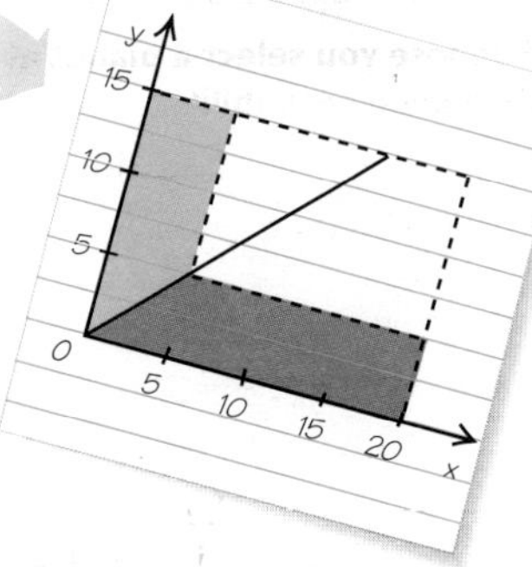

Ongoing ASSESSMENT

19. **Group Activity** Work with another student.
    a. Each of you should describe a possible event when three dice are rolled.
    b. Together, you should perform a simulation to find the experimental probability of either of the two events.
    c. Separately, you should calculate the theoretical probability of either of the two events.
    d. Compare your answers in part (c) to your results in part (b).

Review PREVIEW

Design and test a formula for a calculator to produce each set of random numbers. *(Section 7-1)*

20. integers from 0 to 12
21. integers from 1 to 35
22. integers from 47 to 62

For Exercises 23–25, graph each absolute value function. *(Section 2-3)*

23. $y = |x|$
24. $y = |x + 2|$
25. $y = 3|x| - 1$

26. Suppose a data set has a normal distribution with $\overline{x} = 384$ and $\sigma = 29$. *(Section 6-3)*
    a. Within what range do about 95% of the data items fall?
    b. About what percent of the data values are between 384 and 413?

7 Working on the Unit Project

Use the rules for the Cayuga game of da-un-dah-qua, which are given in the "Working on the Unit Project" exercises in Section 7-1.

27. Are the outcomes of the game *discrete* or *continuous*? Explain your choice.
28. Find the probability of each possible combination of two different colors.

Working on the Unit Project

In Ex. 28, students find the probabilities for each possible outcome on one turn of da-un-dah-qua. These probabilities should be written in students' journals for later reference in completing the Unit Project.

Practice 53 For use with Section 7-2

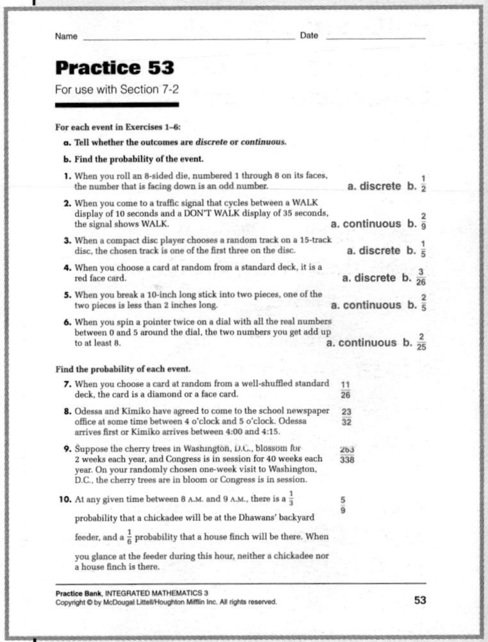

Name ______________ Date ______________

Practice 53

For use with Section 7-2

For each event in Exercises 1–6:

a. Tell whether the outcomes are *discrete* or *continuous*.

b. Find the probability of the event.

1. When you roll an 8-sided die, numbered 1 through 8 on its faces, the number that is facing down is an odd number. a. discrete b. $\frac{1}{2}$
2. When you come to a traffic signal that cycles between a WALK display of 10 seconds and a DON'T WALK display of 35 seconds, the signal shows WALK. a. continuous b. $\frac{2}{9}$
3. When a compact disc player chooses a random track on a 15-track disc, the chosen track is one of the first three on the disc. a. discrete b. $\frac{1}{5}$
4. When you choose a card at random from a standard deck, it is a red face card. a. discrete b. $\frac{3}{26}$
5. When you break a 10-inch long stick into two pieces, one of the two pieces is less than 2 inches long. a. continuous b. $\frac{2}{5}$
6. When you spin a pointer twice on a dial with all the real numbers between 0 and 5 around the dial, the two numbers you get add up to at least 8. a. continuous b. $\frac{2}{25}$

Find the probability of each event.

7. When you choose a card at random from a well-shuffled standard deck, the card is a diamond or a face card. $\frac{11}{26}$
8. Odessa and Kimiko have agreed to come to the school newspaper office at some time between 4 o'clock and 5 o'clock. Odessa arrives first or Kimiko arrives between 4:00 and 4:15. $\frac{23}{32}$
9. Suppose the cherry trees in Washington, D.C., blossom for 2 weeks each year, and Congress is in session for 40 weeks each year. On your randomly chosen one-week visit to Washington, D.C., the cherry trees are in bloom or Congress is in session. $\frac{263}{338}$
10. At any given time between 8 A.M. and 9 A.M., there is a $\frac{1}{3}$ probability that a chickadee will be at the Dhawans' backyard feeder, and a $\frac{1}{6}$ probability that a house finch will be there. When you glance at the feeder during this hour, neither a chickadee nor a house finch is there. $\frac{5}{9}$

Practice Bank, INTEGRATED MATHEMATICS 3

53

Answers to Exercises and Problems

19. Check students' work.
20. int(13*rand)
21. int(35*rand) + 1
22. int(16*rand) + 47
23.

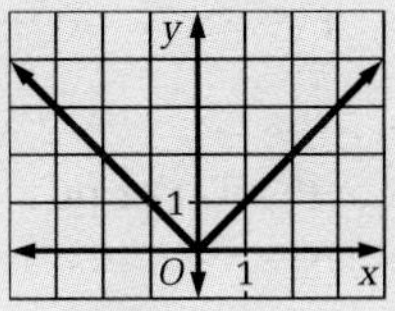

24.

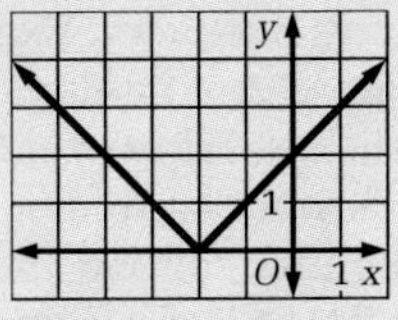

25.

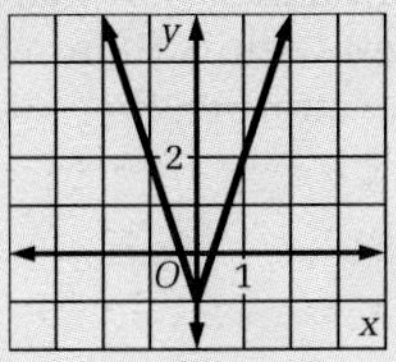

26. a. between 326 and 442
    b. about 34%
27. discrete; Each outcome can be identified by the number of disks that land color A up or the number of disks that land color B up, in each case, a number between 0 and 6.
28. $P(0 \text{ color A, } 6 \text{ color B}) = \frac{1}{64} \approx 0.02$; $P(1 \text{ color A, } 5 \text{ color B}) = \frac{6}{64} = \frac{3}{32} \approx 0.09$; $P(2 \text{ color A, } 4 \text{ color B}) \approx \frac{15}{64} \approx 0.23$; $P(3 \text{ color A, } 3 \text{ color B}) = \frac{20}{64} = \frac{5}{16} \approx 0.31$; $P(4 \text{ color A, } 2 \text{ color B}) = \frac{15}{64} \approx 0.23$; $P(5 \text{ color A, } 1 \text{ color B}) = \frac{6}{64} = \frac{3}{32} \approx 0.09$; $P(6 \text{ color A, } 0 \text{ color B}) = \frac{1}{64} \approx 0.02$

PLANNING

Objectives and Strands
See pages 370A and 370B.

Spiral Learning
See page 370B.

Materials List
- Calculator with random number function
- Graph paper
- Tracing paper

Recommended Pacing
Section 7-3 is a one-day lesson.

Extra Practice
See pages 619–620.

Warm-Up Exercises
- Warm-Up Transparency 7-3

Support Materials
- Practice Bank: Practice 54
- Activity Bank: Enrichment 48
- Study Guide: Section 7-3
- Problem Bank: Problem Set 15
- Explorations Lab Manual: Diagram Masters 1, 2, 16
- Assessment Book: Quiz 7-3, Test 28

Section **7-3**

Geometric Models for Probability

Focus
Use area models with coordinates to solve probability problems.

WAITING AREA

EXPLORATION

What is the probability that two friends will meet?

- **Materials: calculators with a random number function**
- **Work with another student.**

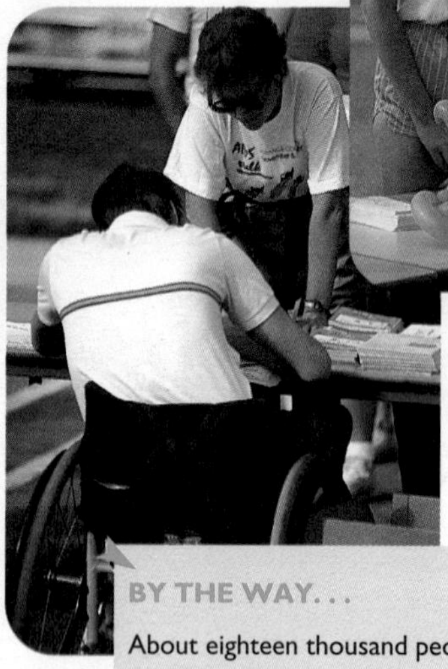

BY THE WAY...
About eighteen thousand people participated in an AIDS Walk through Golden Gate Park in San Francisco on July 17, 1994. The walk raised $2.8 million, a record for a single fundraising event in the Bay Area.

Suppose two friends take part in a fundraiser walk. They agree to arrive for registration between 8:00 A.M. and 8:30 A.M. They decide that they will wait for each other for 10 minutes, but no longer. What is the probability that they will meet?

Use simulation to find out.

1. Use random numbers to represent the friends' times in minutes after 8:00. Generate numbers from 0 to 29. Each trial is represented by two random numbers (one for each person).
2. Decide whether the times are within 10 minutes of each other. If so, the friends will meet. Record whether or not they will meet.

Example:

TRIAL	One friend's arrival time	The other friend's arrival time	Did they meet?
1	23	5	no (18 min apart)
2	6	13	yes (7 min apart)

3. Repeat the trial several times and record your results.
4. Combine your data with the data collected by the rest of the class. Using all the trials from the entire class, what is the experimental probability that the friends will meet?
5. What assumptions did you make in this simulation?

Answers to Exploration

1. Use the function int(30*rand).

2, 3. See answers in back of book.

4. Answers may vary. Based on the sample results, the probability of meeting is $\frac{9}{20}$ or 45%.

5. Answers may vary. Examples are given. I assumed that the arrival times of the friends were random and independent of one another, and that if the friends arrived within 10 minutes of one another, they would meet.

In the Exploration, you used simulation to find out whether the friends will meet. Another way to find out is to use coordinates. Then you can compare the areas of two regions, as you did in Section 7-2.

Sample

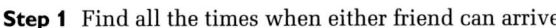

Use coordinates to find the theoretical probability of the event described in the Exploration.

Sample Response

Step 1 Find all the times when either friend can arrive.

Let x and y be the two arrival times in minutes after 8:00.

Either friend can arrive any time during a 30-minute period, so $0 \le x \le 30$ and $0 \le y \le 30$.

The area of this square is $30 \cdot 30 = 900$.

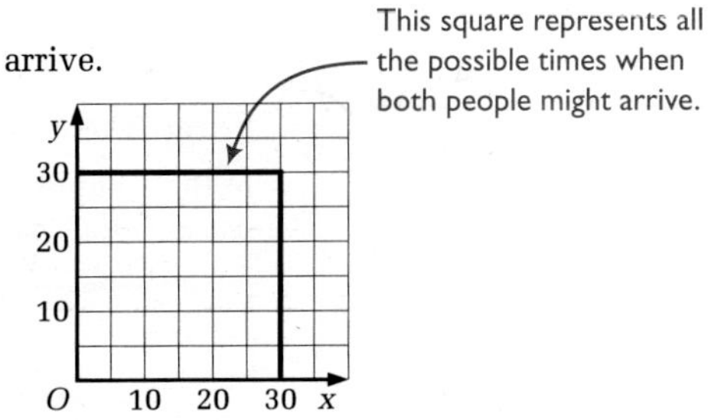

Step 2 Find out when both friends will be there.

One friend arrives x minutes after 8:00.

The number of minutes after 8:00 that the other friend arrives, y, must be less than $x + 10$ and greater than $x - 10$.

Graph the region that the inequalities $y < x + 10$ and $y > x - 10$ represent.

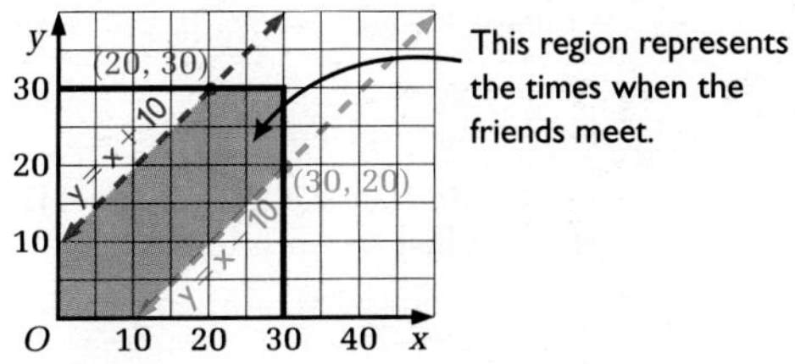

The area of the shaded region is

(the area of the square) − (the area of the two triangles in the corners).

Area of each triangle $= \frac{1}{2}(20)(20) = 200$ ⟵ $A = \frac{1}{2}bh$

Area of shaded region $= 900 - 2(200) = 500$

Step 3 Find the ratio of the two areas.

$$P(\text{friends meet}) = \frac{\text{area of shaded region}}{\text{area of square}}$$

$$= \frac{500}{900} \approx 0.56$$

The theoretical probability is about 0.56.

TEACHING

Exploration

The goal of this Exploration is to have students work with a partner to solve a probability problem using simulation. This same problem will be solved using a geometric model in the Sample on page 391, allowing students an opportunity to compare the two methods.

Additional Sample

Given the information from the Exploration, use coordinates to determine the theoretical probability that the two friends will meet if they agree to wait for each other for 15 minutes.

Step 1. **Find the times when either friend can arrive. Let x and y be the two arrival times in minutes after 8:00. Either friend can arrive any time during a 30-minute period, so $0 \le x \le 30$ and $0 \le y \le 30$. The area of the square is $30 \cdot 30 = 900$.**

Step 2. **Find out when both friends will be there. If one friend arrives x minutes after 8:00, then the number of minutes after 8:00 that the other friend arrives, y, must be less than $x + 15$ minutes and greater than $x - 15$ minutes. Graph the region that the inequalities $y < x + 15$ and $y > x - 15$ represent.**

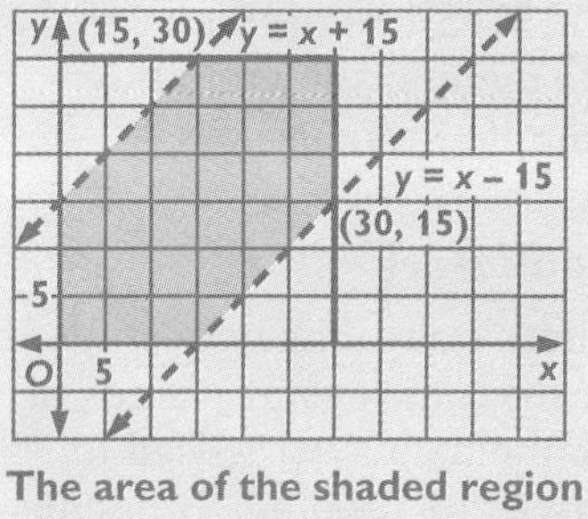

The area of the shaded region is (area of the square) − (area of the two triangles in the corners).

Area of each triangle = $\left(\frac{1}{2}\right)(15)(15) = 112.5$

Area of shaded region = $900 - 2 \cdot 112.5 = 675$

Step 3. **Find the ratio of the two areas.**

$P(\text{friends meet}) = \frac{\text{area of shaded region}}{\text{area of square}} = \frac{675}{900} = 0.75$

The theoretical probability is 0.75.

Integrating the Strands

The use of a geometric model to solve a probability problem, as was done in the Sample, is often called geometric probability. Geometric probability is a wonderful example of how two seemingly unrelated branches of mathematics can be used together to solve complex problems. Algebra was also used in the Sample to find and graph the equations that form the boundaries of the shaded region.

Talk it Over

Question 1 leads students to explore the differences between an experimental and a theoretical solution to the problem introduced in the Exploration.

Teaching Tip

Students may suggest a "short-cut" of just counting the squares of a doubly shaded region, either when examining the Sample Response or when answering Talk it Over question 3. This method works well when relatively small scales are used for the coordinate system. Students should be aware, however, that there will be times when counting is inappropriate, and it will be necessary to use the method shown in the Sample Response.

Look Back

These questions help students to see that several strands of mathematics were used to solve one problem. Students could discuss this Look Back with a partner.

Talk it Over

1. How close is the probability you found in step 4 of the Exploration to the result in the Sample?
2. May says that the shaded region in the Sample represents the friends' arriving at the same time ($y = x$) and all times within ten minutes before or after that. Do you agree with this statement? Why or why not?
3. Describe another way to find the area of the shaded region in the Sample.
4. Suppose the friends agree to wait 20 min for each other. How does this change the graph in the Sample? How does it change the probability?

Look Back

In the Sample, where did you use probability? Where did you use algebra? Where did you use geometry?

7-3 Exercises and Problems

1. **Reading** Suppose the friends decide to meet between 8 A.M. and 9 A.M. How would you change the simulation in the Exploration?
2. **Writing** What does the expression $x - 10 < y < x + 10$ mean? Why does it model the situation in the Exploration?
3. Find the probability that the friends meet if they agree to wait just five minutes.
4. Suppose you and your mother open a joint checking account. You each deposit \$50 and agree that no one will write a check over \$60. You each write one check.
   a. What does the square region represent?
   b. What does the line $x + y = 100$ represent?
   c. What are the coordinates of the vertices of the shaded region? What is the area of the shaded region?
   d. What is the probability that the account is overdrawn?

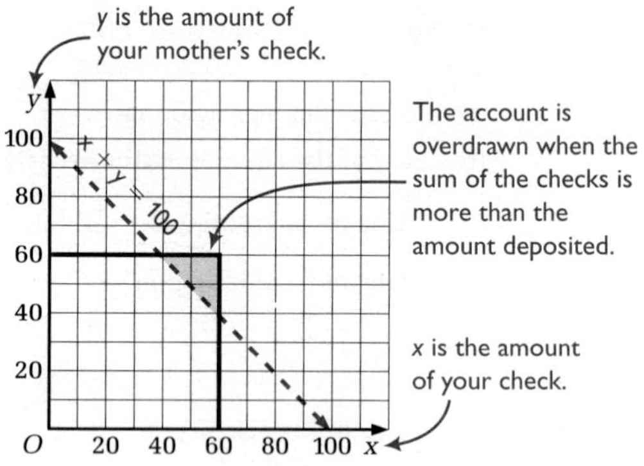

y is the amount of your mother's check.
The account is overdrawn when the sum of the checks is more than the amount deposited.
x is the amount of your check.
x + y = 100
y
100
80
60
40
20
O
20 40 60 80 100 x

5. Suppose two sisters open a joint checking account. One sister deposits \$30 and the other deposits \$20. They agree that neither of them will write a check over \$40. They each write one check. Use coordinates to find the probability that they overdraw the account.

Answers to Talk it Over

1. Answers may vary.
2. May is right. She describes $y = x$ and the two boundary lines. The region *does* represent the friends arriving at the same time ($y = x$) and all times *within* 10 minutes of that event, for the region inside the box.
3. Answers may vary. An example is given. Divide the shaded region into a rectangle and two congruent triangles. Then the area of the shaded region is $10\sqrt{2}(20\sqrt{2}) + 2\left(\frac{1}{2}\right)(10)(10)$ or 500 square units.
4. The region of the graph would be bounded by the inequalities $y < x + 20$ and $y > x - 20$. The probability would be about 0.89.

Answers to Look Back

Probability is used when areas are used to compare the sizes of favorable outcomes and possible outcomes. Algebra is used when arrival times and their relationships are described by inequalities. Geometry is used when the problem is described in terms of coordinates and when areas are calculated.

6. Suppose two friends both say that they will call you tonight between 7:00 and 7:30. You plan to talk to each friend for 15 minutes. Use coordinates to find the probability that their calls will overlap.

7. Janelle and Abby work in a clothing store that has only one cash register. It takes two minutes to ring up a sale. Suppose they each ring up a sale at some time in a ten-minute period. Use coordinates to find the probability that one of them will have to wait to use the register.

8. You can estimate the area of an irregularly shaped region with a geometric probability model. Use this algorithm with tracing paper and graph paper:

1 Trace the outline of the irregularly shaped region on tracing paper.

2 Place the tracing over graph paper on which you have drawn x- and y-axes. Draw a rectangle that encloses the shape. Find the dimensions of the rectangle in terms of the coordinate system.

Use any conveniently sized grid.

Put the origin at a corner.

3 Produce random ordered pairs with coordinates (x, y), where x is a positive number less than or equal to the length of the rectangle and y is a positive number less than or equal to the width of the rectangle.

4 For each ordered pair, score 1 point if it is in the region and 0 points if it is not.

5 After at least 100 trials, find the ratio of the sum of the points in the region to the total number of trials.

6 Use the map scale to find the area that the rectangle represents. Multiply the ratio by the area of the rectangle to estimate the area.

a. Use this method to estimate the area of Mauritania.

b. Explain why this algorithm works.

c. **Research** Find the area of Mauritania in an atlas or an almanac. How close is the estimate you made in part (a) to the actual area? Describe two ways to make a better estimate.

APPLYING

Suggested Assignment

Standard 2–7, 9–17

Extended 1–7, 9–17

Integrating the Strands

Algebra Exs. 2–8, 14, 15

Functions Exs. 14, 15

Geometry Exs. 3–9

Statistics and Probability Exs. 1, 3–13, 16

Logic and Language Exs. 1, 2, 8, 9, 17

Communication: Reading

Exs. 1 and 2 require students to read and rethink the problem in the Exploration using both the simulation and geometric methods.

Research

Students may extend their work in Ex. 8 to other geographic areas. Consider allowing students to choose a different country, state, or county. They could then follow the format of Ex. 8 to estimate the area of their region.

Answers to Exercises and Problems

1. You would generate random numbers between 0 and 59.

2. Answers may vary. An example is given. The inequality says that the absolute value of the difference between x and y is less than 10. It represents the condition that the difference between the two arrival times is less than 10 minutes if the two friends are to meet.

3. $\frac{275}{900} \approx 0.31$

4. a. all the possible sums of the amounts of the two checks

b. the amounts for which the sum of the two checks is the entire balance of the account

c. (40, 60), (60, 40), and (60, 60); 200

d. $\frac{200}{3600} = \frac{1}{18} \approx 0.056$

5. 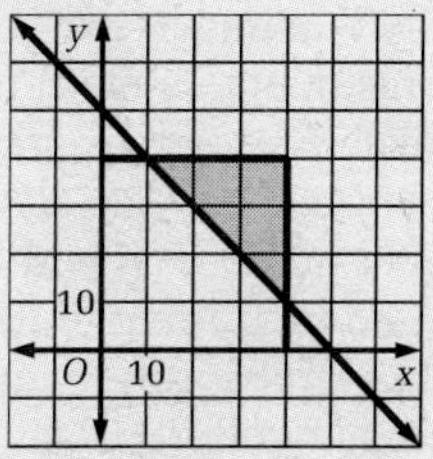

x is the amount of one sister's check; y is the amount of the other sister's check; the square represents all possible sums of the two checks written; the account is overdrawn if $x + y > 50$, that is, if the sum of the checks is more than was deposited.
$P(\text{overdrawn check}) = \frac{450}{1600} = \frac{9}{32} \approx 0.28$

6, 7. See answers in back of book.

8. a. Answers may vary.

b. The probability that a randomly chosen ordered pair inside the rectangle is inside the indicated region is equal to the area of the indicated region divided by the area of the rectangle.

c. about 400,000 mi²; Answers may vary. One way to get a better estimate is to increase the number of randomly chosen ordered pairs. Another is to divide the figure on the coordinate system into regions whose areas you can calculate and add to find the area of the region.

Quick Quiz (7-1 through 7-3)

See page 421.

See also Test 28 in the *Assessment Book*.

Practice 54 For use with Section 7-3

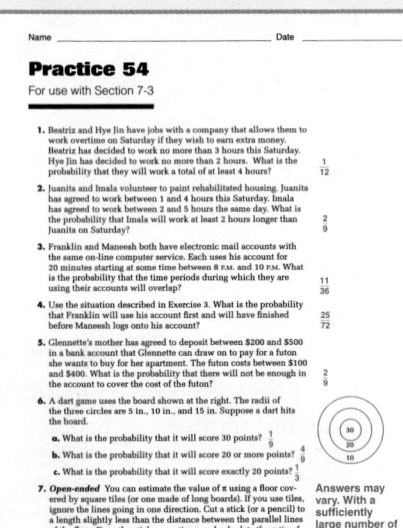

Name ________________ Date ________

Practice 54

For use with Section 7-3

1. Beatriz and Hye Jin have jobs with a company that allows them to work overtime on Saturday if they wish to earn extra money. Beatriz has decided to work no more than 3 hours this Saturday. Hye Jin has decided to work no more than 2 hours. What is the probability that they will work a total of at least 4 hours? $\frac{1}{12}$
2. Juanita and Imala volunteer to paint rehabilitated housing. Juanita has agreed to work between 1 and 4 hours this Saturday. Imala has agreed to work between 2 and 5 hours the same day. What is the probability that Imala will work at least 2 hours longer than Juanita on Saturday? $\frac{2}{9}$
3. Franklin and Maneesh both have electronic mail accounts with the same on-line computer service. Each uses his account for 20 minutes starting at some time between 8 P.M. and 10 P.M. What is the probability that the time periods during which they are using their accounts will overlap? $\frac{11}{36}$
4. Use the situation described in Exercise 3. What is the probability that Franklin will use his account first and will have finished before Maneesh logs onto his account? $\frac{25}{72}$
5. Glennette's mother has agreed to deposit between \$200 and \$500 in a bank account that Glennette can draw on to pay for a futon she wants to buy for her apartment. The futon costs between \$100 and \$400. What is the probability that there will not be enough in the account to cover the cost of the futon? $\frac{2}{9}$
6. A dart game uses the board shown at the right. The radii of the three circles are 5 in., 10 in., and 15 in. Suppose a dart hits the board.

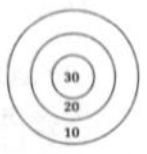

   a. What is the probability that it will score 30 points? $\frac{1}{9}$
   b. What is the probability that it will score 20 or more points? $\frac{4}{9}$
   c. What is the probability that it will score exactly 20 points? $\frac{1}{3}$
7. ***Open-ended*** You can estimate the value of π using a floor covered by square tiles (or one made of long boards). If you use tiles, ignore the lines going in one direction. Cut a stick (or a pencil) to a length slightly less than the distance between the parallel lines of the floor. Drop the stick many times and calculate the ratio of the number of drops for which the stick lands touching a line to the total number of drops. Compare this ratio to the value $\frac{2}{\pi}$ obtained from a calculator. Answers may vary. With a sufficiently large number of trials, the ratio should be approximately $\frac{2}{\pi}$.

54 **Practice Bank**, INTEGRATED MATHEMATICS 3

Ongoing **ASSESSMENT**

9. **Open-ended** Think of a problem that you can solve with a geometric probability model. Sketch a model that represents the problem and solve it.

Review **PREVIEW**

Two dice are thrown. Find the probability of each event. *(Section 7-2)*

10. The sum is 7 or 5.
11. The sum is less than 4 or greater than 8.
12. The sum is greater than 9 or equal to 3.
13. The sum is less than 6 or equal to 2.

Write an equation for each function in the form $f(x) = mx + b$. *(Section 2-2)*

14. $f(4) = 3, f(-1) = 13$
15. $f(7) = 0, f(11) = 0$
16. A person chooses a bill at random from a wallet that contains four \$1 bills, five \$5 bills, two \$10 bills, and one \$20 bill. *(Toolbox Skill 7)*
    a. Find P(\$1 bill).
    b. Find P(bill worth less than \$10).

Working on the Unit Project

17. **Research** Other Native American groups have different names and rules for the game of da-un-dah-qua. The first Europeans to see the game called it "Dish." Find out about the different ways people play this game and what its role is in Native American society.

Unit 7 CHECKPOINT

1. **Writing** Name two places in this unit where you used a formula. How are the formulas used? How are they related to probability?

For Exercises 2 and 3, design and test a formula for a calculator to produce each set of random numbers. 7-1

2. integers from 0 to 19
3. integers from 6 to 12
4. Design a simulation for this situation: Suppose a florist receives 60 orders for bouquets each weekend. Since, on average, 5% of the orders are cancelled later, the florist prepares 58 bouquets. How often does the florist have at least one extra bouquet?

For Exercises 5 and 6, suppose you draw a card from a well-shuffled standard deck. Find each probability. 7-2

5. P(diamond or club)
6. P(diamond or jack)
7. It takes 30 seconds to be served at a pretzel stand. Two people arrive randomly within a two-minute period. Use coordinates to find the probability that one of them will have to wait to be served. 7-3

Answers to Exercises and Problems

9. Answers may vary. An example is given.

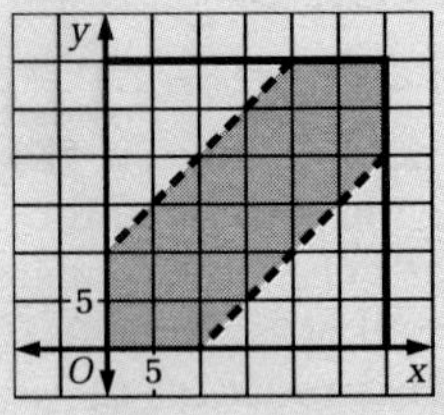

area of square = 900
area of shaded region = 500

The Lams' daughters both shower in the morning between 6:30 and 7:00. Each showers for 10 minutes. What is the probability one will have to wait? Answer: $\frac{500}{900} \approx 0.56$

10. $\frac{10}{36} = \frac{5}{18} \approx 0.28$
11. $\frac{13}{36} \approx 0.36$
12. $\frac{8}{36} = \frac{2}{9} \approx 0.22$
13. $\frac{10}{36} = \frac{5}{18} \approx 0.28$
14. $f(x) = -2x + 11$
15. $f(x) = 0$
16. a. $\frac{4}{12} = \frac{1}{3} \approx 0.33$
    b. $\frac{9}{12} = \frac{3}{4} = 0.75$
17. See answers in back of book.

Answers to Checkpoint

1. Answers may vary. Examples are given. I used a formula in Exercise 10 on page 387 to find the probability that one of two events that are not mutually exclusive will occur. The formula $P(A \text{ or } B) = P(A) + P(B) - P(A \text{ and } B)$ is used by finding the individual probabilities and subtracting the probability that both events occur. I also used the formula for the area of a triangle in Exercise 6 on page 393. Areas are used when finding probability of an event with continuous outcomes.
2. int(20*rand)
3. int(7*rand) + 6
4. For each trial, use the function int(20*rand) + 1 to generate a sequence of 60 random numbers between 1 and 20. Let the number 1 represent a cancelled order and the numbers 2–20 represent orders that are not cancelled. Analyze the number of sequences for which the number 1 appears three or more times.

5–7. See answers in back of book.

Section **7-4**

Conditional Probability and Independent Events

Focus

Find conditional probabilities. Recognize independent and dependent events.

Healthy Conditions?

Allergic
Not allergic
Total
Male students
12
7
19
Female students
9
8
17
Total
21
15
36

Simone noticed that in the fall and spring, many of her classmates frequently sneezed or coughed. She wondered whether their symptoms were due to allergies. The school nurse surveyed Simone's class for allergies to airborne substances. The results are shown in the table.

Sample 1

Find the probability that a student selected at random in Simone's class is not allergic, given that the student is a male.

Sample Response

Problem Solving Strategy: Use a diagram.

Make a tree diagram to represent the allergy data for Simone's class.

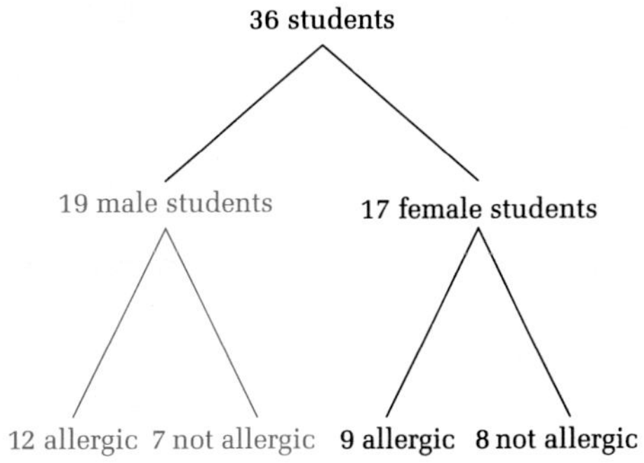

Look at the part of the diagram that is shown in color. There are 19 male students. Of the 19 males, 7 are not allergic, so the probability that a male selected at random from the class is not allergic is $\frac{7}{19}$, or about 0.37.

PLANNING

Objectives and Strands

See pages 370A and 370B.

Spiral Learning

See page 370B.

Materials List

- Spinner with 10 equal sectors
- Calculator with random number function

Recommended Pacing

Section 7-4 is a two-day lesson.

Day 1

Pages 395–397: Opening paragraph through Conditional Probability, *Exercises 1–15*

Day 2

Pages 397–398: Dependent and Independent Events through Look Back, *Exercises 16–36*

Extra Practice

See pages 619–620.

Warm-Up Exercises

- Warm-Up Transparency 7-4

Support Materials

- Practice Bank: Practice 55
- Activity Bank: Enrichment 49
- Study Guide: Section 7-4
- Problem Bank: Problem Set 16
- Overhead Visuals: Folder 6
- Assessment Book: Quiz 7-4, Alternative Assessment 3

TEACHING

Teaching Tip

In Sample 1, students examine a real-life situation that can be analyzed using probabilities and a tree diagram. The content of this Sample leads into the definition of conditional probability.

Additional Sample

S1 Find the probability that a student selected at random from Gregory's class owns her own CD player, given that the student is a female.

Students	Own	Do not own	Total
Male	8	14	22
Female	5	11	16
Total	13	25	38

Make a tree diagram to represent the data for Gregory's class.

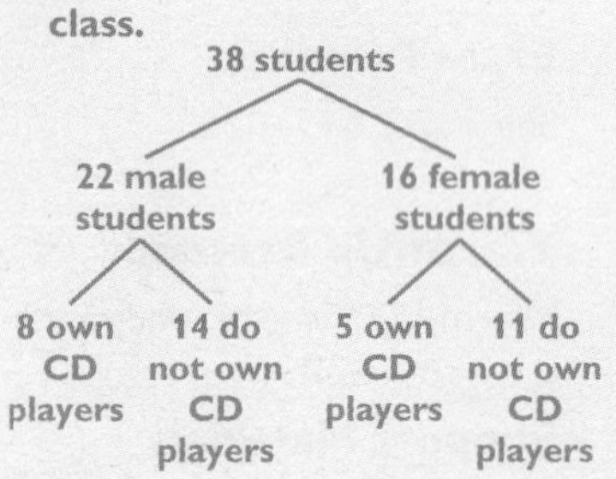

There are 16 female students. Of the 16 female students, 5 own a CD player, so the probability that a female student selected at random owns a CD player is $\frac{5}{16}$, or about 0.31.

Error Analysis

In Sample 1, some students may have difficulty understanding why the probability of not allergic, given a male student, is not $\frac{7}{36}$. Have these students list the original sample space using two-letter combinations to represent each student. Use F = female, M = male, A = allergic, N = not allergic. Have students cross off any outcome that does not meet the given. This will eliminate all outcomes of "F." Students can then clearly see that the sample space they have to work with has changed, so the denominator of the probability ratio can no longer be 36.

Talk it Over

1. In Sample 1, why is the probability $\frac{7}{19}$, not $\frac{7}{36}$?
2. What is the probability that a student is allergic, given that the student is a female?

In situations like that in Sample 1, you need to find a probability under restricted conditions. In Sample 1, the condition is that the student is a male. This is called a **conditional probability.** In these cases the number of possible outcomes is reduced to reflect the restrictions.

You write the conditional probability in Sample 1 as

P(not allergic | male),

which is read, "the probability that a student is not allergic given that the student is a male."

You can use a *probability tree diagram* to show the probabilities that events will occur. These two partially completed diagrams are based on the allergy data from Simone's class.

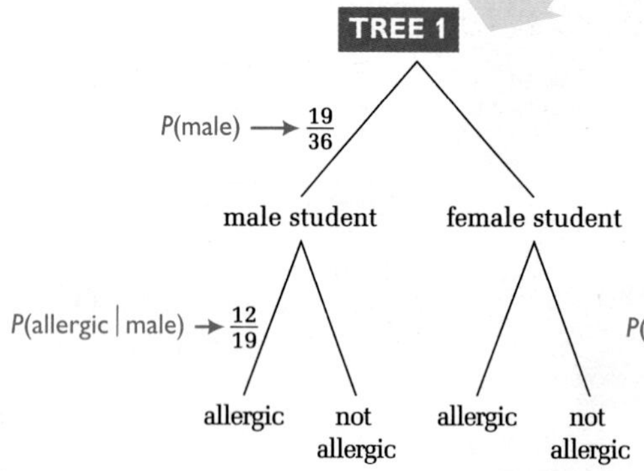

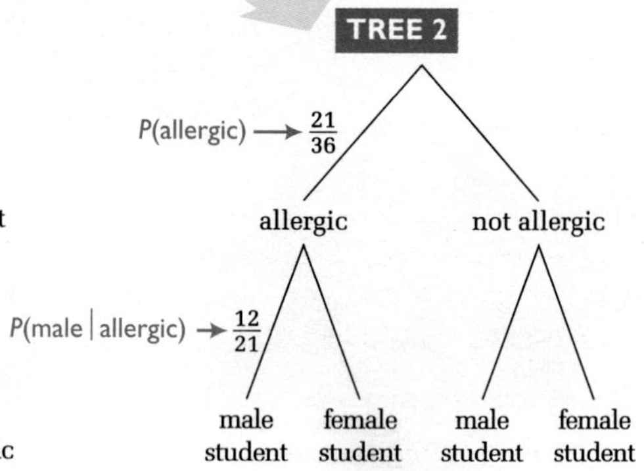

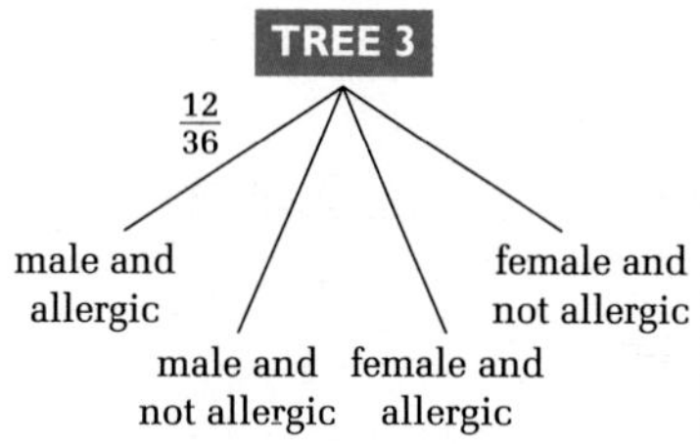

Talk it Over

3. Compare Tree 1 and Tree 2. Describe any differences.
4. Tree 3 shows the probabilities that a randomly chosen student is a male or a female *and* is or is not allergic.
   a. Look at the branches of Trees 1, 2, and 3 that have been completed. Do you see any relationship among the probabilities in Tree 1 and Tree 3? among the probabilities in Tree 2 and Tree 3?
   b. Choose corresponding branches from Trees 1–3 and find the probabilities. Does the relationship from part (a) still hold?

Answers to Talk it Over

1. It was given that the student selected is male. There are 19 male students.
2. $\frac{9}{17} \approx 0.53$
3. Tree 1 determines probabilities such as *P*(allergic | male), while Tree 2 determines probabilities such as *P*(male | allergic).
4. a. *P*(male and allergic) = *P*(male) · *P*(allergic | male); *P*(male and allergic) = *P*(allergic) · *P*(male | allergic)
   b. Answers may vary. An example is given. *P*(female) = $\frac{17}{36}$; *P*(not allergic | female) = $\frac{8}{17}$; *P*(not allergic) = $\frac{15}{36}$; *P*(female | not allergic) = $\frac{8}{15}$; *P*(female and not allergic) = $\frac{8}{36}$. The relationship holds.

CONDITIONAL PROBABILITY

For any events A and B:

$$P(A \text{ and } B) = P(A) \cdot P(B \mid A) \quad \text{and} \quad P(B \mid A) = \frac{P(A \text{ and } B)}{P(A)}$$

You read this as "the probability of B given A."

Now you are ready for: Exs. 1–15 on pp. 399–400

Dependent and Independent Events

In some situations, the outcome of one event changes the probability of the outcome of another event.

Sample 2

Two green balls and three orange balls are placed in a jar. One ball is randomly chosen and its color is recorded. Then a second ball is chosen and its color is recorded. What is the probability of getting a green ball on the first draw and an orange ball on the second draw in each situation?

a. The first ball is replaced before the second is drawn.

b. The first ball is not replaced before the second is drawn.

Sample Response

Problem Solving Strategy: Use a diagram.

a. The first ball is replaced, so the probabilities are the same for each draw.

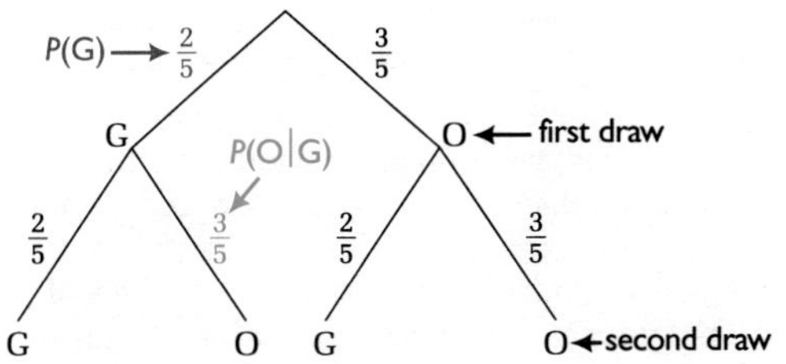

$P(G \text{ and } O) = P(G) \cdot P(O \mid G)$

$= \frac{2}{5} \cdot \frac{3}{5}$

$= \frac{6}{25} = 0.24$

b. The first ball is not replaced, so the probabilities change for the second draw.

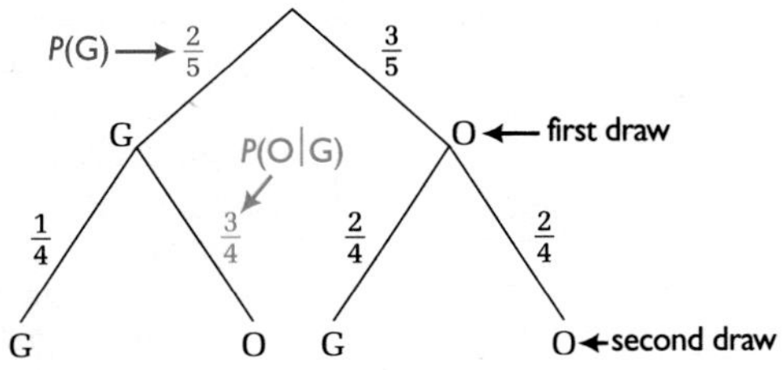

$P(G \text{ and } O) = P(G) \cdot P(O \mid G)$

$= \frac{2}{5} \cdot \frac{3}{4}$

$= \frac{6}{20} = 0.3$

Additional Sample

S2 Shelley places 8 plain beads and 5 flowered beads in a box. One bead is randomly chosen and it is noted whether it is plain or flowered. Then a second bead is drawn and its design is noted. What is the probability of getting a plain bead on the first draw and a flowered bead on the second draw in each situation?

a. The first bead is replaced before the second is drawn.

Use a diagram. The first bead is replaced, so the probabilities are the same for each draw.

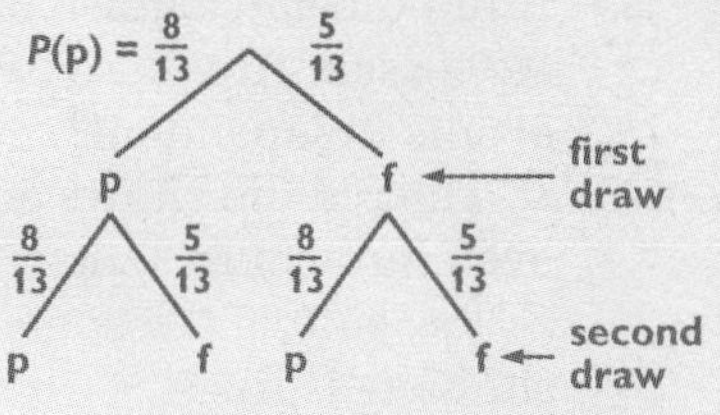

$P(p \text{ and } f) = P(p) \cdot P(f \mid p) = \frac{8}{13} \cdot \frac{5}{13} = \frac{40}{169} \approx 0.24$

b. The first bead is not replaced before the second is drawn.

Use a diagram. The first bead is not replaced, so the probabilities change for the second draw.

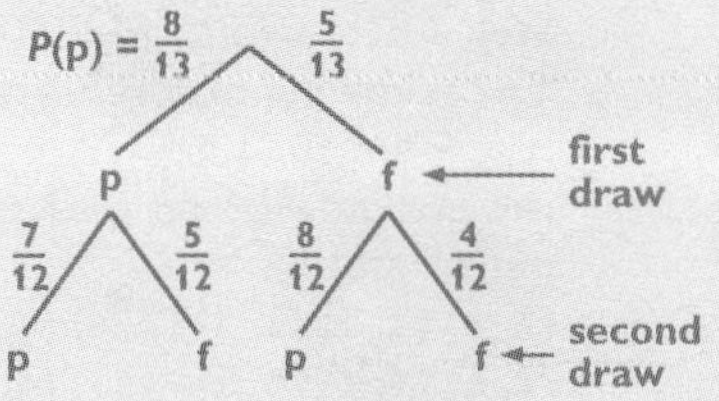

$P(p \text{ and } f) = P(p) \cdot P(f \mid p) = \frac{8}{13} \cdot \frac{5}{12} = \frac{10}{39} \approx 0.26$

Using Manipulatives

If some students are having difficulty understanding the concepts of dependent and independent events, have them consider this situation. What if two cards are drawn from a deck by first drawing one card, recording it, putting it back into the deck and then drawing a second card? Will the probability be the same as drawing a first card, keeping it, and then drawing a second card? Students can use actual cards to compute the probabilities of drawing various cards, with and without replacement.

Additional Sample

S3 A red die and a green die are tossed. Are these two events independent?
R: The red die shows an even number of spots.
S: The sum of the spots showing is five.

Use the definitions of independent events: If $P(S \mid R) = P(S)$, then the events are independent.

Step 1. Determine $P(S)$. The following ordered pairs represent a sum of five when the dice are tossed (red die, green die): (1, 4), (2, 3), (3, 2), (4, 1).

$P(S) = \frac{4}{36} = \frac{1}{9}$

Step 2. Determine $P(S \mid R)$. The following ordered pairs represent the possibilities of the red die with an even number of spots.

(2, 1) (2, 2) (2, 3) (2, 4) (2, 5) (2, 6) (4, 1) (4, 2) (4, 3) (4, 4) (4, 5) (4, 6) (6, 1) (6, 2) (6, 3) (6, 4) (6, 5) (6, 6)

R has 18 equally likely outcomes. Of these, 2 have a sum of five.

$P(S) = \frac{2}{18} = \frac{1}{9}$

The values are equal. The two events are independent.

Talk it Over

5. For each of the two cases in Sample 2, find the probability that the two balls drawn are *both* green.
6. In which situation in Sample 2 can you say that the probability of the event "the second ball is orange" depends on the probability of the event "the first ball is green"?

Two events are **independent** if and only if the occurrence of one does not affect the probability of the occurrence of the other.

INDEPENDENT EVENTS

Events *A* and *B* are independent if and only if $P(B \mid A) = P(B)$.

Sample 3

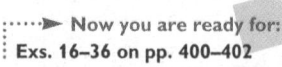

Use the allergy data on page 395. Are the events "female student" and "allergic" independent?

Sample Response

Use the definition of independent events.

If $P(\text{allergic} \mid \text{female}) = P(\text{allergic})$, then "female" and "allergic" are independent events.

$$P(\text{allergic}) = \frac{21}{36} \approx 0.58$$

$$P(\text{allergic} \mid \text{female}) = \frac{9}{17} \approx 0.53$$

The values are not equal. The events are not independent.

Slightly more than half of the female students (about 0.53) in Simone's class have allergies, yet females make up slightly less than half of the class (about 0.47). If "allergic" were independent of "female," then you could expect allergies to be distributed among male and female students in proportion to the sizes of these two groups. That does not happen here, so the events are *dependent*.

Now you are ready for: Exs. 16–36 on pp. 400–402

Look Back

Give examples of events you think are independent. How can you test mathematically whether they are independent?

Answers to Talk it Over

5. with replacement: $\frac{4}{25}$ = 0.16; without replacement: 0.1
6. situation (b), in which the first ball was not replaced

Answers to Look Back

Answers may vary. An example is given. I think that eye color and gender are independent events. To test this, I could record the gender and eye color of a large, randomly-chosen group. Using these results, I could calculate the probability of each eye color for each gender. If these characteristics are independent, then the probability of a given eye color should be the same for both males and females.

7-4 Exercises and Problems

1. **Reading** Suppose you are tutoring another student. Explain the meaning of *conditional probability.*

This table gives majors of students at a small technical college. Use the data for Exercises 2–9.

	Freshmen	Sophomores	Juniors	Seniors	Total
Architecture	50	40	45	25	160
Engineering	30	25	35	40	130
Business	20	25	20	45	110
Total	100	90	100	110	400

Suppose a student is selected at random. Find each probability.

2. *P*(freshman)
3. *P*(architecture major)
4. *P*(freshman and architecture major)
5. *P*(sophomore | business major)
6. *P*(business major | sophomore)
7. *P*(sophomore or business major)
8. Show that these expressions are equal:

   $P(\text{senior} \mid \text{engineering major})$ and $\dfrac{P(\text{senior and engineering major})}{P(\text{engineering major})}$

9. **Open-ended** Create a conditional probability question based on the table above. Answer the question.
10. **Reading** How is the second formula in the box at the top of page 397 related to the first formula?
11. A person is chosen at random from shoppers at a mall. Suppose the probability that the person has red hair and freckles, *P*(red hair and freckles), is $\frac{3}{25}$, and $P(\text{freckles}) = \frac{9}{25}$. What is *P*(red hair | freckles)?
12. **Research** Speak to your principal or to a person in the school administration office.
    a. Find out how many students are in your school. Find out how many students are in one grade.
    b. Find out how many students in that grade have last names that begin with a certain letter.
    c. What is the probability that a student chosen at random from that grade at your school has a last name beginning with that letter?
13. A road has stop lights at two consecutive intersections.
    ➤ The probability of passing through the first intersection without stopping is 0.5.
    ➤ The probability of passing through the second intersection without stopping, given that you made it through the first without stopping, is 0.8.

    What is the probability of passing through both intersections without stopping?

Mathematical Procedures

Students can use the definition of conditional probability to verify the value for *P*(allergy | female) in the Sample 3 Response.

$P(\text{allergic} \mid \text{female}) = \dfrac{P(\text{allergic and female})}{P(\text{female})} = \dfrac{\frac{9}{36}}{\frac{17}{36}} = \dfrac{9}{36} \bullet \dfrac{36}{17} = \dfrac{9}{17} \approx 0.53$

Look Back

This Look Back may be used for a class discussion. One student can write the class' suggestions for independent events on the board. Then the whole class can discuss how to test them mathematically.

APPLYING

Suggested Assignment

Day 1

Standard 2–9, 11, 13

Extended 1–11, 13, 14

Day 2

Standard 16–22, 26, 29–36

Extended 16–36

Integrating the Strands

Number Ex. 15

Algebra Exs. 30, 31

Statistics and Probability Exs. 1–30, 32–36

Discrete Mathematics Ex. 32

Logic and Language Exs. 1, 10, 14, 15, 23, 24, 26, 29

Visual Thinking

Assign students to work in groups to create probability tree diagrams of the information presented in the table for Exs. 2–9. For each of the events described in Exs. 2–7, encourage the groups to explain their diagrams to the class and to discuss the probability of each event. This activity involves the visual skills of *exploration* and *communication.*

Answers to Exercises and Problems

1. Answers may vary. An example is given. Conditional probability is a probability of related events. The conditional probability of event *A* given event *B* is the probability that *A* occurs given that *B* does occur.
2. $\frac{100}{400} = \frac{1}{4} = 0.25$
3. $\frac{160}{400} = \frac{2}{5} = 0.4$
4. $\frac{50}{400} = \frac{1}{8} = 0.125$
5. $\frac{25}{110} = \frac{5}{22} \approx 0.23$
6. $\frac{25}{90} = \frac{5}{18} \approx 0.28$
7. $\frac{175}{400} = \frac{7}{16} = 0.4375$
8. $P(\text{senior} \mid \text{engineering major}) = \frac{40}{130} \approx 0.31$; $\dfrac{P(\text{senior and engineering major})}{P(\text{engineering major})} = \dfrac{40/400}{130/400} = \dfrac{40}{130} \approx 0.31$
9. Answers may vary. An example is given. A student is chosen. What is the probability that the student is an architecture major given that he or she is a freshman? *P*(architecture | freshman) = 0.5

10–13. See answers in back of book.

Cooperative Learning

Allow time for students to work in pairs on Ex. 15. Students should think about the strategy for the hints they give. Each group should write a summary of their ideas for parts (b) and (c). Groups could then present and share their strategies.

Career Note

Market researchers test new products and other consumer goods, such as movies, by first introducing them to small groups of people. Their surveys or other tests are designed to see if people will buy the product and if it appeals to any particular group, such as teenagers, for example. The data collected are analyzed using probability and statistics.

Market researchers help businesses decide whether to produce new or revised products. They also help businesses decide how and when to advertise a product. A common form of market research today is the use of focus groups. A market researcher leads a small group of potential customers through a series of questions to determine their opinions and reactions to a new product. The discussion is generally observed behind one-way mirrors by the product designers, who can see and hear directly how people in the group respond.

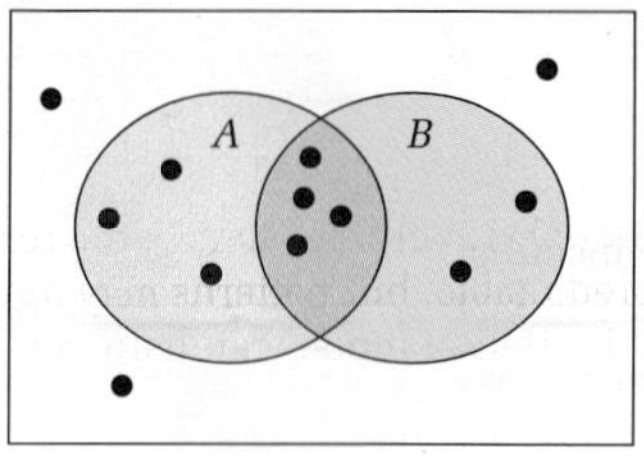

14. **Writing** Samuel drew this diagram to represent $P(B \mid A)$. He said, "We know that A has occurred, so look inside A. The probability of B given A is the ratio of the overlap of A and B to all of A. So we have $P(B \mid A) = \frac{\text{number in both } A \text{ and } B}{\text{number in } A}$."

    Do you agree with Samuel's argument? Why or why not?

15. **Group Activity** Work with another student. You will need a spinner with ten equal sectors or a calculator with a random number function.

    Have one student try to guess which random digit from 0 to 9 has been produced by the other student with a spinner or a calculator. The student producing the random number should give hints like, "It is odd," or "It is prime," or "It is less than 5."

    a. Play the game 30 times with a partner. Take turns generating the random digit. The player with the most correct guesses wins.

    b. **Writing** Which hints give more information than others? Give examples.

    c. **Writing** How does conditional probability play a role in this game?

Market Research A movie studio conducted a test screening to get audience reactions to a new film. Some results are shown.

The researchers used a random sample of 50 people so that predictions can be made from the data.

40% of the people surveyed were males.
60% of the males liked the film.
70% of the females liked the film.

Use the test screening results for Exercises 16–25.

16. a. Make a table to summarize the results.

    b. Use the table to complete each tree diagram.

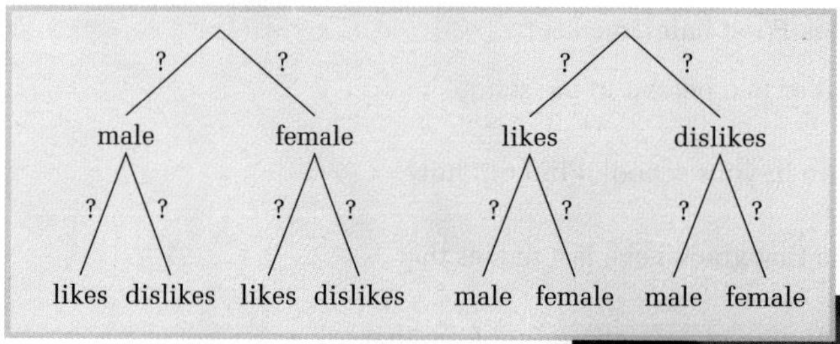

BY THE WAY...

Movie studios usually begin screening a movie three to four months before it is released. The test screenings, along with other market research, are used to determine whether there is enough awareness of the movie and how strongly the audience wants to see it.

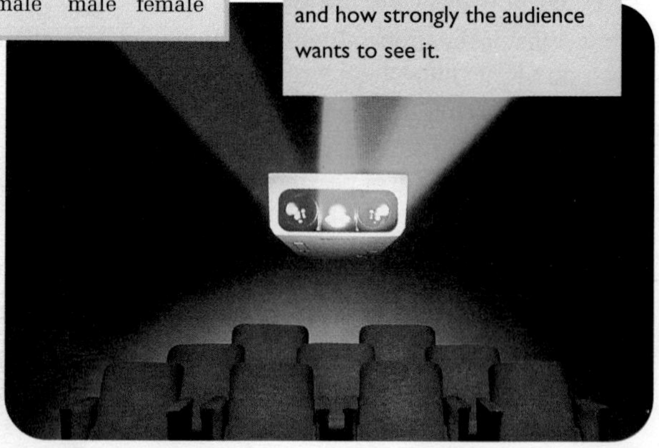

Suppose a person is chosen at random from the market research study. Find each probability.

17. P(female)
18. P(female | person likes the film)
19. P(person likes the film | male)
20. P(male | person who likes the film)
21. P(female likes the film)
22. P(male likes the film)

Answers to Exercises and Problems

14. Yes; by the formula for conditional probability: $P(B \mid A) = \frac{P(A \text{ and } B)}{P(A)} = \frac{\text{number in both } A \text{ and } B}{\text{total number}} \div \frac{\text{number in } A}{\text{total number}} = \frac{\text{number in both } A \text{ and } B}{\text{number in } A}$.

15. a. Answers may vary.

    b. Answers may vary. Examples are given. The hint "It is odd" eliminates five possibilities, while the hint "It is prime" eliminates six. A hint such as "It is a multiple of 7" eliminates eight possibilities.

    c. Initially, without a hint, we know only that the number is a digit from 0 to 9, and each digit is equally likely. With a hint, we have a conditional probability, P(digit is d | information given by hint). The probability of guessing correctly will be higher when hints are given.

16. See answers in back of book.

17. $\frac{30}{50} = \frac{3}{5} = 0.6$

18. $\frac{21}{33} = \frac{7}{11} \approx 0.64$

19. $\frac{12}{20} = \frac{3}{5} = 0.6$

20. $\frac{12}{33} = \frac{4}{11} \approx 0.36$

21. $\frac{21}{50} = 0.42$

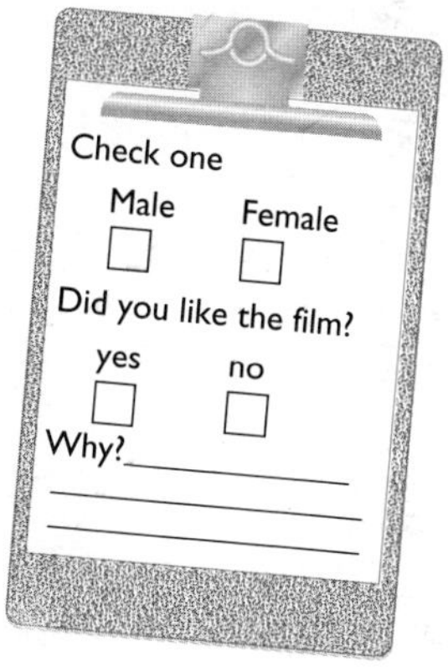

23. **Writing** To answer Exercises 17–22, did you use a formula or a tree diagram from Exercise 16, part (b)? Explain why you prefer one method more than the other.

24. **Writing** Are the events "likes the film" and "male" independent? Explain.

25. **Open-ended** Make a table of possible outcomes of a survey where "likes the film" and "female" are independent.

26. Four boys and five girls are on a committee to organize a school dance. They are having trouble scheduling a time to meet. At each meeting, a random member of the committee is absent.

  a. Find the probability that a boy is absent at one meeting and then a girl is absent at the next meeting.

  b. Suppose students are removed from the committee if they miss a meeting. Find the probability that a boy is absent at one meeting and then a girl is absent at the next meeting.

  c. **Writing** In which case, part (a) or part (b), are the events "boy is absent" and "girl is absent" independent? Explain.

27. Use the formulas for conditional probability and independent events to show that if events A and B are independent, then $P(A \text{ and } B) = P(A) \cdot P(B)$.

28. **Weather** Before the development of radar, weather satellites, computers, and all the other technology used by today's forecasters, people used the behavior of animals and plants to predict the weather.

The reliability of these natural weather predictors varies greatly. In England, people used a plant called the scarlet pimpernel to predict whether it would rain.

  a. When the pimpernel's flowers open, the chance that it will not rain that day is 92%. Express this probability as a conditional probability.

  b. When the flowers do not open, the chance of rain is 17%. Express this probability as a conditional probability.

Communication: Writing

Exs. 24 and 26(c) give students an opportunity to analyze an event to determine whether it is independent. This should help students to develop their ability to tell whether events are independent or not without using the mathematical definition. Students can try to determine from the language if the events are independent and then check their answers using the mathematical definition.

Answers to Exercises and Problems

22. $\frac{12}{50} = \frac{6}{25} = 0.24$

23. Answers may vary. An example is given. I prefer to use a tree diagram, because it helps me see the relationships among the events.

24. No; $P(\text{male} \mid \text{likes film}) = 0.36$; $P(\text{male}) = 0.4$.

25. Answers may vary. An example is given.

	Likes Film	Dislikes Film	Total
Females	21	9	30
Males	14	6	20
Total	35	15	50

26. a. $\frac{20}{81} \approx 0.25$

  b. $\frac{20}{72} = \frac{5}{18} \approx 0.28$

  c. in part (a); In part (b), $P(\text{boy is absent} \mid \text{girl is absent}) = \frac{1}{2}$, while $P(\text{boy is absent}) = \frac{4}{9}$.

27. If A and B are independent events, then $P(B \mid A) = P(B)$. Then $P(A \text{ and } B) = P(A) \cdot P(B \mid A) = P(A) \cdot P(B)$.

28. a. $P(\text{no rain} \mid \text{pimpernel's flowers open}) = 0.92$

  b. $P(\text{rain} \mid \text{pimpernel's flowers do not open}) = 0.17$

Working on the Unit Project

Exs. 33–36 give students an opportunity to apply conditional probabilities, tree diagrams, and dependent and independent events to the game of da-un–dah-qua. All of the concepts in the section must be used to answer this group of questions. Students should keep a record of this work in their journals for later reference.

Practice 55 For use with Section 7-4

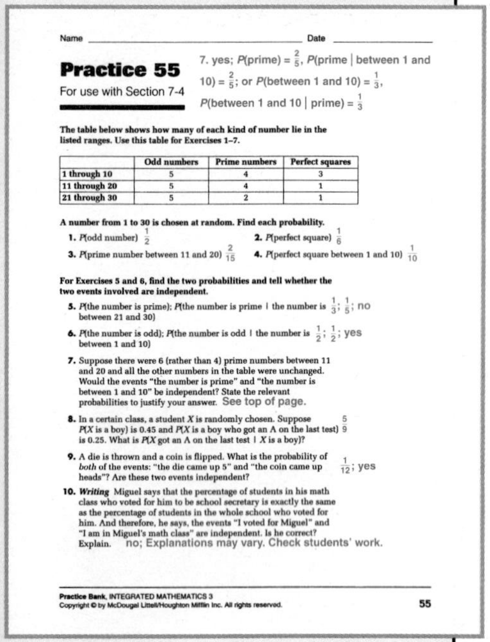

Name ____________________ Date __________

Practice 55

For use with Section 7-4

7. yes; $P(\text{prime}) = \frac{2}{5}$, $P(\text{prime} \mid \text{between 1 and 10}) = \frac{2}{5}$; or $P(\text{between 1 and 10}) = \frac{1}{3}$, $P(\text{between 1 and 10} \mid \text{prime}) = \frac{1}{3}$

The table below shows how many of each kind of number lie in the listed ranges. Use this table for Exercises 1–7.

	Odd numbers	Prime numbers	Perfect squares
1 through 10	5	4	3
11 through 20	5	4	1
21 through 30	5	2	1

A number from 1 to 30 is chosen at random. Find each probability.

1. P(odd number) $\frac{1}{2}$
2. P(perfect square) $\frac{1}{6}$
3. P(prime number between 11 and 20) $\frac{2}{15}$
4. P(perfect square between 1 and 10) $\frac{1}{10}$

For Exercises 5 and 6, find the two probabilities and tell whether the two events involved are independent.

5. P(the number is prime); P(the number is prime | the number is between 21 and 30) $\frac{1}{3}$; $\frac{1}{5}$; no
6. P(the number is odd); P(the number is odd | the number is between 1 and 10) $\frac{1}{2}$; $\frac{1}{2}$; yes
7. Suppose there were 6 (rather than 4) prime numbers between 11 and 20 and all the other numbers in the table were unchanged. Would the events "the number is prime" and "the number is between 1 and 10" be independent? State the relevant probabilities to justify your answer. See top of page.
8. In a certain class, a student *X* is randomly chosen. Suppose P(*X* is a boy) is 0.45 and P(*X* is a boy who got an A on the last test) is 0.25. What is P(*X* got an A on the last test | *X* is a boy)? $\frac{5}{9}$
9. A die is thrown and a coin is flipped. What is the probability of *both* of the events: "the die came up 5" and "the coin came up heads"? Are these two events independent? $\frac{1}{12}$; yes
10. ***Writing*** Miguel says that the percentage of students in his math class who voted for him to be school secretary is exactly the same as the percentage of students in the whole school who voted for him. And therefore, he says, the events "I voted for Miguel" and "I am in Miguel's math class" are independent. Is he correct? Explain. no; Explanations may vary. Check students' work.

Practice Bank, INTEGRATED MATHEMATICS 3

55

Ongoing ASSESSMENT

29. **Writing** Mutually exclusive events are always dependent. Explain why or give an example.

Review PREVIEW

30. Suppose you go to a mall with a friend. You agree to shop on your own and then meet at the fountain between 7:00 P.M. and 8:30 P.M. You agree to wait for each other for 15 min before going home. What is the probability that you and your friend meet? *(Section 7-3)*
31. Write the expression log M^2N^3 in terms of log M and log N. *(Section 5-7)*
32. How many different groups of three class representatives can be chosen from a class of 22 students? *(Toolbox Skill 4)*

7 Working on the Unit Project

Use the description of da-un-dah-qua on page 381.

33. a. Complete the tree diagram for what can happen when a player has two consecutive tosses in the game of da-un-dah-qua.
    b. What property did you use to complete the diagram?

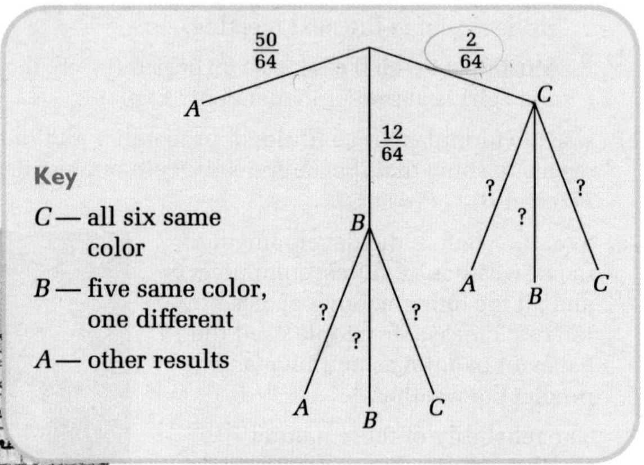

The Cayuga lived in longhouses made of bark over a frame of bent trees. They grew corn and other vegetables. To celebrate the first corn of the season, the Cayuga held a festival at which they played da-un-dah-qua.

34. Find each probability for two tosses.
    a. P(all six same color | five same color, one different)
    b. P(all green | all green)
35. Find the probability of getting all six of the same color on each of two consecutive tosses.
36. Suppose you have two consecutive tosses. What is the probability of getting all six the same color on one toss and getting exactly five the same color on the other toss?

Answers to Exercises and Problems

29. Answers may vary. An example is given. Two events are independent if and only if the occurrence of one does not affect the probability of the occurrence of the other. If two events are mutually exclusive, the occurrence of either makes it impossible for the other to occur.

30. $\frac{2475}{8100} \approx 0.31$

31. $2 \log M + 3 \log N$

32. 1540 groups

33. a.

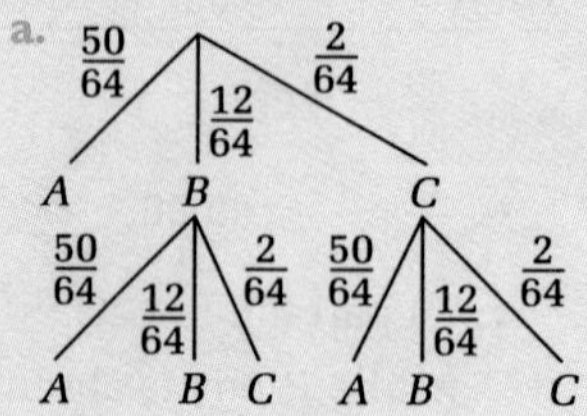

b. If events *A* and *B* are independent, then $P(B \mid A) = P(B)$.

34. a. $\frac{2}{64} = \frac{1}{32} = 0.03125$

b. $\frac{1}{64} = 0.015625$

35. $\frac{2}{64} \cdot \frac{2}{64} = \frac{4}{4096} \approx 0.00098$

36. $\frac{2}{64} \cdot \frac{12}{64} = \frac{24}{4096} \approx 0.0059$

Expected Value

Focus

Find the average gain or loss in a situation.

What's a Fair Price?

How do insurance companies decide what is a fair rate to charge? They use data from past experience. For example, suppose insurance companies have found that, on average, in 10,000 homeowners' policies the following claims have occurred in the past.

Average value of claim	Number of claims per 10,000 policies
\$200,000	1
\$20,000	10
\$2,000	200
\$0	9789

BY THE WAY...

According to legend, merchants along the Yangtze River in ancient China "insured" their cargo by dividing it equally among all the boats. If a boat overturned in the rapids, each merchant shared equally in the loss.

How much does an insurance company expect to pay in one year on 10,000 policies in this situation?

$$\$200{,}000 \cdot 1 + \$20{,}000 \cdot 10 + \$2000 \cdot 200 + \$0 \cdot 9789 = \$800{,}000$$

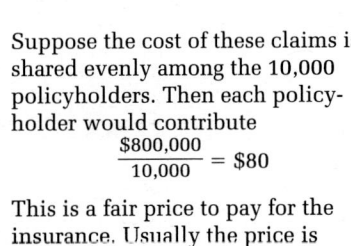

Suppose the cost of these claims is shared evenly among the 10,000 policyholders. Then each policyholder would contribute

$$\frac{\$800{,}000}{10{,}000} = \$80$$

This is a fair price to pay for the insurance. Usually the price is greater because the insurance company has additional costs and wants to make a profit.

PLANNING

Objectives and Strands

See pages 370A and 370B.

Spiral Learning

See page 370B.

Recommended Pacing

Section 7-5 is a one-day lesson.

Extra Practice

See pages 619–620.

Warm-Up Exercises

Warm-Up Transparency 7-5

Support Materials

- Practice Bank: Practice 56
- Activity Bank: Enrichment 50
- Study Guide: Section 7-5
- Problem Bank: Problem Set 16
- Assessment Book: Quiz 7-5, Alternative Assessment 4

TEACHING

Talk it Over

Question 1 provides an opportunity for students to analyze the procedures presented for finding a fair price for insurance. This will help students to understand the expected value formula.

Additional Sample

S1 A community organization is selling raffle tickets to benefit a local scholarship fund. Ten thousand tickets are sold at $1.00 each. There are two $500 prizes, four $100 prizes, and six $20 prizes. On the average, how much can a person who buys one ticket expect to win or lose?

Use a formula.

***Step 1.* Find the probability of winning each prize.**

$P(\$500) = \frac{2}{10{,}000} = 0.0002$

$P(\$100) = \frac{4}{10{,}000} = 0.0004$

$P(\$20) = \frac{6}{10{,}000} = 0.0006$

$P(\$0) = \frac{9{,}988}{10{,}000} = 0.9988$

***Step 2.* Find the expected value if the raffle were free.**

($500)(0.0002) + ($100)(0.0004) + ($20)(0.0006) + ($0)(0.9988) = $0.152

***Step 3.* Find the expected value for the raffle, including the cost of the ticket.**

$.152 – $1 = –$.848

A person who buys one raffle ticket should expect to lose $.848, or about 85 cents.

You can calculate a fair price to pay for insurance in the situation on page 403 another way. You can find the probability of each claim by dividing each number of claims by 10,000.

Average value of claim	Probability
$200,000	0.0001
$20,000	0.001
$2,000	0.02
$0	0.9789

$\frac{1}{10{,}000} = 0.0001$

$\frac{10}{10{,}000} = 0.001$

$\frac{200}{10{,}000} = 0.02$

$\frac{9789}{10{,}000} = 0.9789$

Then you can find a fair price to pay for a policy this way:

$$(\$200{,}000)(0.0001) + (\$20{,}000)(0.001) + (\$2000)(0.02) + (\$0)(0.9789) = \$80$$

Talk it Over

1. Compare the two methods for finding a fair price.
   a. How are they alike? How are they different?
   b. Which do you prefer? Why?
   c. Use the distributive property to explain why the two methods are equivalent.
2. Does it make sense for an insurance company's policyholders to share the cost of everyone's insurance evenly? Why or why not?

The second method for finding a fair price can be used in many situations involving probabilities and gains or losses. The result is called **expected value,** which is the average gain or loss in repeated occurrences of a situation.

EXPECTED VALUE

To find an expected value, multiply the value of each outcome, x_i, by its probability, $P(x_i)$, and then add all the products.

Expected value $= x_1 \cdot P(x_1) + x_2 \cdot P(x_2) + x_3 \cdot P(x_3) + \cdots + x_n \cdot P(x_n)$

Sample 1

A nonprofit organization is selling lottery tickets for $1 each. There is one $500 prize, two $100 prizes, and ten $5 prizes. Two thousand tickets are sold. On average, how much can a person who buys a ticket expect to win or lose?

Answers to Talk it Over

1. a. Answers may vary. An example is given. In both cases, each average claim is multiplied by the number of such claims and divided by 10,000 and the results are added. The order of the operations is different for the two methods.
   b. Answers may vary.
   c. $\left(\frac{1}{10{,}000}\right)[\$200{,}000(1) + (\$20{,}000)(10) + (\$2000)(200) + (\$0)(9789)] = \frac{\$200{,}000(1)}{10{,}000} + \frac{(\$20{,}000)(10)}{10{,}000} + \frac{(\$2000)(200)}{10{,}000} + \frac{(\$0)(9789)}{10{,}000}$
2. Answers may vary. An example is given. I think it makes sense to share the cost evenly unless a policyholder has repeated claims. Then I think that policyholder should pay more, being more of a risk.

Sample Response

Problem Solving Strategy: Use a formula.

Step 1 Find the probability of winning each prize.

$P(\$500) = \frac{1}{2000} = 0.0005 \qquad P(\$100) = \frac{2}{2000} = 0.001$

$P(\$5) = \frac{10}{2000} = 0.005 \qquad P(\$0) = \frac{1987}{2000} = 0.9935$

Step 2 Find the expected value if the game were free.

$(\$500)(0.0005) + (\$100)(0.001) + (\$5)(0.005) + (\$0)(0.9935) = \$.375$

Step 3 Find the expected value including the cost of the ticket.

$\$.375 - \$1 = -\$.625$ ← Subtract the cost of a ticket.

A person who buys this lottery ticket for $1 can expect to lose $.625. When a person plays the lottery repeatedly, or when a number of people each play the lottery once, the *average* of all the gains and losses is a loss of 62.5 cents for every ticket bought.

In Sample 1, if each person paid $.375 instead of $1 to play the lottery, the expected value would be $0. In that case the lottery would be considered a *fair game.* Any game or situation with an expected value of 0 is considered **fair**. Lotteries are fundraisers, so they are designed to have a negative expected value.

Sample 2

What price makes the following a fair game?

Roll three dice. You win as many dollars as there are "aces" (single dots) showing.

Sample Response

Problem Solving Strategy: Use a diagram.

Step 1 Make a probability tree diagram to show all the possible outcomes.

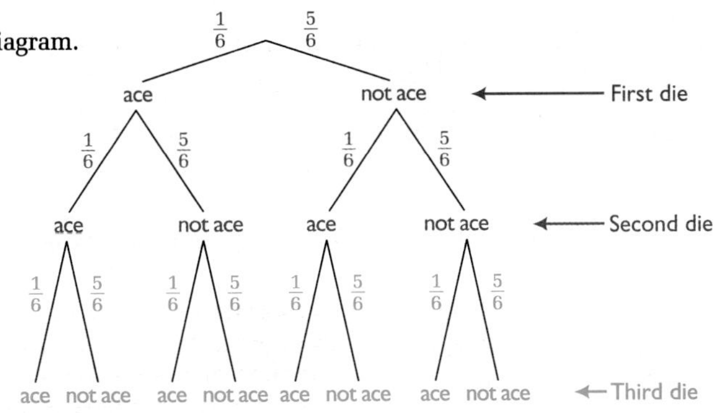

Continued on next page.

Additional Sample

S2 What price makes the following a fair game?
Draw three cards from a standard deck of cards. Each draw is done without replacement. Getting exactly one heart is worth $1, exactly two hearts is worth $5, and exactly three hearts is worth $10.

Use a diagram.

***Step 1.* Make a probability tree diagram of the outcomes.**

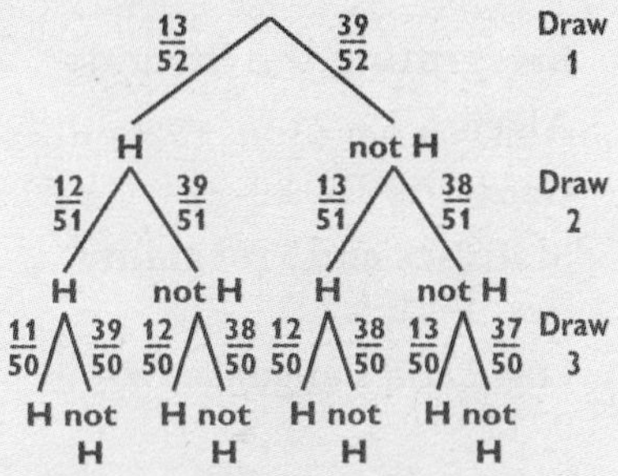

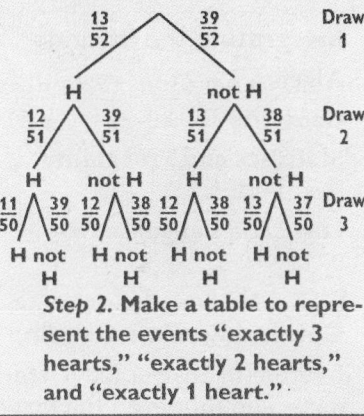

***Step 2.* Make a table to represent the events "exactly 3 hearts," "exactly 2 hearts," and "exactly 1 heart."**

Outcome	You win	Probability
exactly 3 hearts	$10	$\frac{13}{52} \cdot \frac{12}{51} \cdot \frac{11}{50} = \frac{11}{850}$
exactly 2 hearts	$5	$\frac{13}{52} \cdot \frac{12}{51} \cdot \frac{39}{50} + \frac{13}{52} \cdot \frac{39}{51} \cdot \frac{12}{50} + \frac{39}{52} \cdot \frac{13}{51} \cdot \frac{12}{50} = \frac{117}{850}$
exactly 1 heart	$1	$\frac{13}{52} \cdot \frac{39}{51} \cdot \frac{38}{50} + \frac{39}{52} \cdot \frac{13}{51} \cdot \frac{38}{50} + \frac{39}{52} \cdot \frac{38}{51} \cdot \frac{13}{50} = \frac{741}{1700}$

***Step 3.* Find the expected value if the game were free.**

$\frac{11}{850}(\$10) + \frac{117}{850}(\$5) + \frac{741}{1700}(\$1) = \$1.25$

A price of $1.25 makes the game fair. Then the expected value is zero.

Look Back

This question can lead to a lively discussion of what students had previously considered a fair game.

APPLYING

Suggested Assignment

Standard 4–8, 10, 14–17, 19–22

Extended 1–17, 19–22

Integrating the Strands

Algebra Ex. 21

Geometry Ex. 20

Statistics and Probability Exs. 1–19, 22

Logic and Language Exs. 3, 9, 11

Communication: Reading

In order to answer Ex. 3, students should reread the text on page 403. This can lead to a brief discussion of the need for a business to be able to pay its bills (costs) and to make a profit.

Step 2 Make a table to represent the events "exactly 3 aces," "exactly 2 aces," "exactly 1 ace," and "no aces."

Outcome	You win	Probability
exactly 3 aces	\$3	$\frac{1}{6} \cdot \frac{1}{6} \cdot \frac{1}{6} = \frac{1}{216}$
exactly 2 aces	\$2	$\frac{1}{6} \cdot \frac{1}{6} \cdot \frac{5}{6} \cdot 3 = \frac{15}{216}$
exactly 1 ace	\$1	$\frac{1}{6} \cdot \frac{5}{6} \cdot \frac{5}{6} \cdot 3 = \frac{75}{216}$
no aces	\$0	$\frac{5}{6} \cdot \frac{5}{6} \cdot \frac{5}{6} = \frac{125}{216}$

In these two cases, there are three ways that the outcome can occur.

Step 3 Find the expected value if the game were free.

$$\frac{1}{216} \cdot \$3 + \frac{15}{216} \cdot \$2 + \frac{75}{216} \cdot \$1 + \frac{125}{216} \cdot \$0 = \frac{\$3 + \$30 + \$75 + \$0}{216} = \$.50$$

A price of \$.50 makes the game fair. Then the expected value is zero.

Look Back

What makes a game fair?

7-5 Exercises and Problems

Suppose the insurance company described on page 403 finds that the numbers of claims change to the ones in this table.

Average value of claim	Number of claims per 10,000 policies
\$200,000	1
\$20,000	20
\$2,000	300
\$0	9679

1. What does the company expect to pay in this situation?
2. What is a fair price when each of the 10,000 customers shares the cost evenly?
3. **Reading** Is the answer to Exercise 2 the amount the insurance company charges its customers? Why or why not?

A fair die is rolled. You win a dollar for every dot showing on the top when the die lands.

4. What is the expected value for the game?
5. Suppose you are the operator of the game. What should you charge to make the game fair?

Answers to Look Back

A game is fair if its expected value is 0.

Answers to Exercises and Problems

1. \$1,200,000
2. \$120
3. No; the company has costs other than claims and needs to make a profit as well.
4. \$3.50
5. \$3.50
6. a. 2, 3, 5; 1, 4, 6

   b. $\frac{-\$1}{6}$, or about –17 cents
7. No; the person running the game.
8. The person can expect to win 26 cents, which does not cover the \$1 cost of a ticket. Overall, the person *loses* 74 cents.
9. The expected value for any situation is not necessarily one of the outcomes. The expected value is found by

For Exercises 6 and 7, suppose you roll a fair die. You win $1 for every dot showing on the top if the result is a prime number. You pay $1 per dot for any other result.

6. a. For which number(s) from 1 to 6 do you add to your winnings? For which number(s) do you subtract from your winnings?

   b. What is the expected value for this game?

7. Is this a fair game? If not, who is favored, the player or the person running the game?

8. A club is selling lottery tickets for $1 each. There is one $400 prize, two $50 prizes, and ten $2 prizes. Two thousand tickets are sold. On average, what can a person who buys a ticket expect to win or lose?

9. **Writing** In Sample 1, the expected value (–$.625) is not one of the possible outcomes ($500, $100, $5, or $0). Explain.

10. A food service company wants to make a bid for all the cafeterias in a large school district. The research and planning needed to make the bid costs $3000. If the bid is accepted, the company makes $23,000 over the life of the contract. Suppose the bid has a 25% chance of being accepted. Should the company make the bid? Why or why not?

Shenitra Richards

"Prize"	Net award	Number awarded	Probability	Product*
$500	$499	1	$\frac{1}{2000}$	$\frac{499}{2000}$
$100	$99	2	$\frac{2}{2000}$	$\frac{198}{2000}$
$5	$4	10	$\frac{10}{2000}$	$\frac{40}{2000}$
$0	– $1	1987	$\frac{1987}{2000}$	$-\frac{1987}{2000}$

Expected value $= -\frac{1250}{2000}$

$= -0.625$

* Product of Net award and Probability

11. **Reading** Shenitra has another way to do Sample 1. Here is her work.

    a. Explain Shenitra's method.

    b. Compare it with the one in Sample 1.

    c. Why do both methods give the same answer?

    d. Which do you prefer? Why?

12. Discount Shopping As customers enter a store one day, they receive discount coupons for their purchases that day. The amount of the discount on the coupon is not known until the cashier scratches the coupon. The store claims that for every 100 coupons, one reads "50% off," ten read "30% off," twenty read "20% off," and the rest read "10% off." On average, what discount can a customer expect to receive?

Communication: Writing

Ex. 9 provides an excellent check on whether students understand the concept of expected value.

Interdisciplinary Problems

Many businesses, including insurance companies, use probability and statistics to help them make decisions. In the opening example on page 403, an insurance company is using probability and expected value to determine how much to charge for a homeowner's policy. In Ex. 10, students see how a food service company uses expected value to decide if they will bid for a particular contract.

Mathematical Procedures

Ex. 11 gives students an opportunity to analyze another method for finding expected value. Some students may prefer to use this method in their work rather than the one presented in Sample 1.

Answers to Exercises and Problems

subtracting the expected payoff from the cost of playing.

10. Answers may vary. In the long run, a company would be likely to make money if it made many bids for contracts in many situations similar to this. However, there is a 75% chance that the company will lose $3000. If this were an uncommon situation, or if the company would place itself in financial jeopardy by losing $3000, it might not want to take the risk of placing the bid.

11. a. Shenitra subtracted the cost of the ticket before calculating the expected payoff.

    b. The methods are similar, except that Shenitra figured the cost of the ticket into the expected payoff, rather than computing the expected payoff, then subtracting the ticket cost to find the expected value.

    c. the distributive property

    d. Answers may vary. An example is given. I prefer the sample method because it gives the expected payoff and the expected value, and the arithmetic for Shenitra's method is a little more complicated.

12. 14.4%

Cooperative Learning

Students can work on Ex. 13 in class with a partner. Two or three groups could present a review of their results to the entire class.

Communication: Listening

You might want to have a student read the selection from "The Promise" aloud to the class either before students begin to work on Exs. 14–17 or prior to the discussion of the answers to these questions.

Students Acquiring English

Some students acquiring English may find Chaim Potok's description of the carnival game difficult to follow. These students may benefit from being paired with students fluent in English for the exercises.

Multicultural Note

Chaim Potok, author of *The Promise,* was born in New York City in 1929. An ordained rabbi, Potok worked as a university instructor, an editor for various Jewish publications, a U.S. Army lieutenant in Korea, and a novelist. Although he decided to become a novelist after reading Evelyn Waugh's *Brideshead Revisited* when he was sixteen, Potok did not publish any of his major written works until 1967 when *The Chosen,* a novel about the lives of two generations of Hasidic Jews in Brooklyn, was published. *The Chosen* was followed by *The Promise* in 1969. Some of Potok's other works include *My Name is Asher Lev* and *In the Beginning.*

13. Game Shows A contestant on a game show has won \$1000 and a chance to spin the wheel. The wheel has the values shown. (Hitting the "bankrupt" sector means that the contestant loses the \$1000 winnings.)

a. Design a simulation in which 20 people are each in the same situation as this contestant. Keep track of the results.

b. On average, what is a contestant's gain (or loss) in part (a)?

c. Calculate the expected value from spinning the wheel. Compare this result to your answer from part (b).

connection to LITERATURE

In *The Promise*, by Chaim Potok, a pitchman at a fair interests three friends in a carnival game in the early 1950s.

> The counter that fronted the booth was painted gold. Upon it lay a white wooden board and a brown-leather dice cup. The board contained ten rows of holes, ten holes in each row, one hundred holes in all. A number ranging from one to ten had been painted in black beneath each hole. The numbers appeared to be in random sequence and stood out sharp and black beneath the holes, the insides of which were bright red. . . .
>
> . . .The dice cup contained eight silver balls. The balls were tossed onto the board. The numbers below the holes into which they fell were added together. A card, which (the pitchman) now brought out and placed on the counter next to the board, was then consulted to determine the value of the total. As soon as the player's separate throws added up to ten, he was given a (prize).
>
> "Can't be any simpler than that," the pitchman said. . . .
>
> "How much is it to play?" I heard Michael ask.
>
> "Twenty-five cents a toss."

The pitchman in the story keeps changing the rules of the game as people play. Suppose, however, that a number of people play the game as described. The histogram shows the distribution of the number of tosses each player needs in order to score 10 points and win the prize.

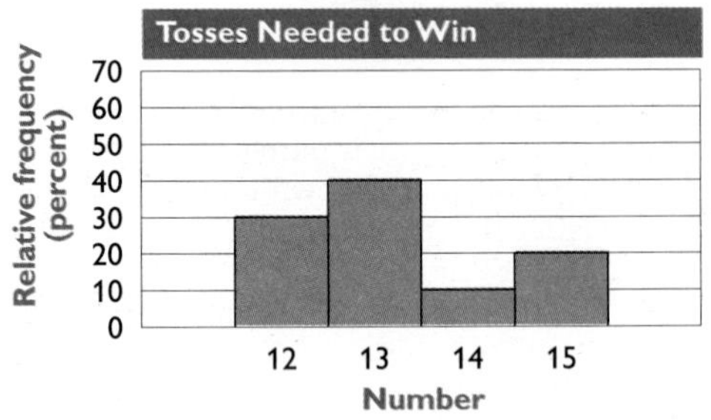

14. What is the expected number of tosses needed?

15. What is the expected cost of the game?

16. What is the expected value for the game when the prize is worth \$2?

17. Is the game fair? Why or why not?

Answers to Exercises and Problems

13. a. Simulation method may vary. For each trial, use the function int(6*rand) to generate a sequence of 20 random numbers between 0 and 5. Let 0 represent bankrupt, 1 represent a \$100 win, 2 and 3 represent a \$200 win, 4 represent a \$300 win, and 5 represent a \$500 win.

b. Answers may vary.

c. \$50; Answers may vary.

14. 13.2 tosses

15. \$3.30 16. –\$1.30

17. No; if the game were fair, the expected value would be 0.

18. a, b.

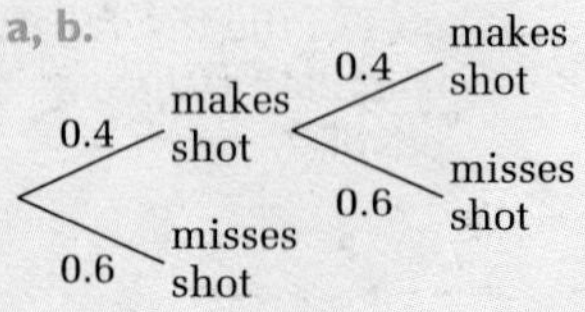

c. 2 points; 1 point; 0 points

d. 0.76

19. $\frac{2}{3}$ 20. 90°

21. a. $(a + b)^3 = 1a^3 + 3a^2b + 3ab^2 + 1b^3$; $(a + b)^4 = 1a^4 + 4a^3b + 6a^2b^2 + 4ab^3 + 1b^4$

b. Answers may vary. Examples are given. The

Ongoing ASSESSMENT

18. **Group Activity** Work with another student.

A basketball player is in a *one-and-one situation.* This means the player has a chance to shoot a free throw for one point. If she makes a basket, she has a chance to shoot a second free throw. The player's free-throw record is 40%.

a. Make a tree diagram with labels to represent the possible outcomes of the free throw(s).

b. What is the probability that the player makes a basket on a free throw? What is the probability that she does not make a basket? Fill in the probabilities along the branches of your tree diagram.

c. How many points will the player score if she makes a basket on both free throws? if she makes a basket on the first free throw only? if she does not make a basket on the first free throw?

d. What is the expected value for this situation?

Review PREVIEW

19. Find $P(B \mid A)$ when $P(A \text{ and } B) = \frac{2}{5}$ and $P(A) = \frac{3}{5}$. *(Section 7-4)*

20. What is the measure of an inscribed angle of a circle whose intercepted arc is a semicircle? *(Section 3-6)*

21. a. Complete this chart of the first five powers of $(a + b)$. *(Toolbox Skill 16)*

b. Describe any patterns you see.

$(a + b)^1 = 1a + 1b$
$(a + b)^2 = 1a^2 + 2ab + b^2$
$(a + b)^3 = ?$
$(a + b)^4 = ?$
$(a + b)^5 = 1a^5 + 5a^4b + 10a^3b^2 + 10a^2b^3 + 5ab^4 + 1b^5$

Working on the Unit Project

The table shows the point values that the Cayuga use in the game of da-un-dah-qua.

22. a. Complete the table.

b. Find the expected value (points) for one toss.

c. How are the points the Cayuga use related to the probabilities of the various outcomes?

d. **Open-ended** Design a scoring system that is more closely related to the probabilities of different outcomes.

Outcomes	Probability	Points
6 Blue	?	5
5B and 1G	?	1
4B and 2G	?	0
3B and 3G	?	0
2B and 4G	?	0
1B and 5G	?	1
6 Green	?	5

Working on the Unit Project

In Ex. 22, students apply their knowledge of expected value and the related probabilities to the game of da-un-dah-qua. Ask a few students to share their design of a scoring system for part (d) with the class.

Practice 56 For use with Section 7-5

Name ____________ Date ____________

Practice 56

For use with Section 7-5

1. The table at the right shows the prizes to be awarded in a contest.

Value of prize	Number of prizes
$10,000	1
$1,000	5
$500	10
$200	50

a. Suppose 50,000 people enter the contest. What is a fair price to enter the contest? $.60

b. Suppose 40,000 people enter. Do you think the fair price is higher or lower than in part (a)? Calculate the fair price in this situation. higher; $.75

2. A fair die is tossed. Suppose you win $1 for every spot showing when an odd number comes up and you win $0 when an even number comes up.

a. What is your expected value for the game? $1.50

b. Suppose you are using this game to make money for charity and you want to be able to contribute to the charity $.50 of what each player pays to play the game. How much should you charge each player? $2.00

3. Two marbles are drawn in succession from a bag containing 3 red marbles and 5 blue marbles. The first marble is returned to the bag before the second is drawn. You win $5 if both of the marbles drawn are red, $3 if both are blue. You lose $1.50 if the two marbles drawn are of different colors. What price makes this a fair game? $1.17

4. Use a tree diagram to answer the question in Exercise 3, if the first marble is *not* returned to the bag after being drawn. $.80

5. At a fundraising fair, a ticket for 3 rubber balls to throw at wooden figures costs $5. The probabilities of hitting the three kinds of figures with one ball are given in the table. Suppose 240 such tickets are sold. How much money can the fair expect to raise? $516

	Probability of hitting figure	Payoff for hitting figure
Small figure	$\frac{1}{10}$	$3
Medium-sized figure	$\frac{1}{5}$	$2
Large figure	$\frac{1}{4}$	$1

6. ***Open-ended*** Suppose you were asked to help organize a fund-raising fair. Describe an event you would propose, along with probabilities of success for participants and the price of a ticket. Calculate the funds raised from each ticket sold. Answers may vary. Check students' work.

56 **Practice Bank**, INTEGRATED MATHEMATICS 3

Answers to Exercises and Problems

coefficients of the expanded form of $(a + b)^n$ are the numbers in row n of Pascal's triangle. The powers of a in the expanded form of $(a + b)^n$ begin at n and decrease to 0, while the powers of b begin at 0 and increase to n. The sum of the powers of a and b in each term equals n.

22. a.

Outcomes	Probability	Points
6B	0.015625	5
5B and 1G	0.09375	1
4B and 2G	0.234375	0
3B and 3G	0.3125	0
2B and 4G	0.234375	0
1B and 5G	0.09375	1
6G	0.015625	5

b. 0.34375

c. The two outcomes with the lowest probability earn the highest number of points. The two outcomes with the next higher probability earn a lesser number of points. The other 50 possible outcomes earn no points.

d. Answers may vary. An example is given. 6B or 6G: 60 points; 5B and 1G or 5G and 1B: 10 points; 4B and 2G or 4G and 2B: 4 points; 3B and 3G: 3 points

PLANNING

Objectives and Strands
See pages 370A and 370B.

Spiral Learning
See page 370B.

Materials List
- Calculator with random number function
- 12 playing cards

Recommended Pacing
Section 7-6 is a two-day lesson.

Day 1

Pages 410–412: Opening paragraph through Probability in a Binomial Experiment, *Exercises 1–13*

Day 2

Pages 412–414: Paragraph in middle of page through Look Back, *Exercises 14–30*

Extra Practice
See pages 619–620.

Warm-Up Exercises
Warm-Up Transparency 7-6

Support Materials
- Practice Bank: Practice 57
- Activity Bank: Enrichment 51
- Study Guide: Section 7-6
- Problem Bank: Problem Set 16
- Assessment Book: Quiz 7-6, Test 29

Section 7-6 Binomial Experiments and Decision-Making

Focus
Find probabilities in binomial experiments.
Use significance levels in statistical decision-making.

Most *Likely* To SUCCEED

To determine whether the outcome in a real-life situation can be due to chance alone, you can use a simulation.

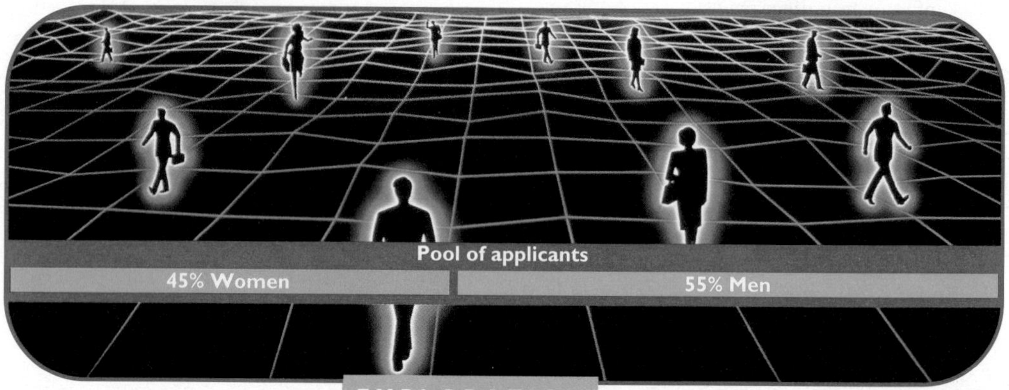

EXPLORATION

Is the outcome due to chance alone?

- **Materials: calculators with a random number function**
- **Work in a group of four students.**

In a very large pool of equally qualified candidates for a job, 55% are men and 45% are women. Ten people are hired. Eight of them are men. Assume that people are hired at random and that the selection of one person does not affect the ratio of men and women in the pool.

1. Design a simulation to investigate this problem. For each trial, record the number of men hired.
2. Combine your results with those of the other groups. Make a probability distribution of all the results.
3. Use the results from your class to estimate the probability of each event.
   a. Eight of the people hired are men.
   b. Nine of the people hired are men.
   c. Ten of the people hired are men.
   d. Eight or more of the people hired are men.

Answers to Exploration

1. For each trial, use the function int(100*rand) + 1 to generate a sequence of ten numbers between 1 and 100. Let the numbers 1–55 represent men and the numbers 56–100 represent women. Count the number of sequences generated and the number in which eight of the numbers are less than 56.

2. Results may vary. Sample results for 50 trials are given. Results of trials (number of men hired): 6, 7, 9, 6, 3, 4, 6, 6, 7, 6, 5, 7, 6, 3, 4, 2, 5, 6, 3, 1, 4, 2, 4, 5, 4, 4, 4, 6, 6, 5, 7, 8, 6, 6, 5, 4, 7, 7, 5, 7, 5, 8, 5, 6, 6, 6, 5, 6, 5, 6

n	0	1	2	3	4	5	6	7	8	9	10
$P(n)$	0	0.02	0.04	0.06	0.16	0.20	0.32	0.14	0.04	0.02	0

3. Answers may vary. Experimental probabilities based on the sample results from step 2 are given.
   a. 0.04
   b. 0.02
   c. 0
   d. 0.06

The experiment in the Exploration has several important characteristics. These are the characteristics of a **binomial experiment.** You can analyze a binomial experiment with a simulation, as you did in the Exploration, or with probability theory.

CHARACTERISTICS OF BINOMIAL EXPERIMENTS

	Examples
➤ There are exactly two outcomes for each trial.	Either a man or a woman is hired.
➤ The trials are independent.	The ratio of men and women in a very large pool of candidates does not change.
➤ Each trial has the same probability of success.	P(man is hired) = 0.55 for each trial.
➤ The experiment has a fixed number of trials.	Ten people are hired.

The "success" of a trial is defined arbitrarily. Here, "man is hired" is considered a "success."

Sample 1

In the hiring situation in the Exploration, what is the theoretical probability that exactly 8 of the 10 people hired are men?

Sample Response

Step 1 Find the number of favorable outcomes.

When a person is hired, there are two possible outcomes: a man (M) is hired or a woman (W) is hired. When 10 people are hired, outcomes such as MMWWMMMMMM and MWMMMMMMWM are considered favorable.

The number of favorable outcomes is the number of combinations of 10 items taken 8 at a time.

$$_{10}C_8 = \frac{10!}{8!2!} = \frac{10 \cdot 9}{2 \cdot 1} = 45$$

Step 2 Find the probability of each outcome.

The trials are independent, so you can multiply to find the probability of each outcome.

P(man is hired) P(woman is hired)

$$P(\text{one combination of } 8 \text{ men and } 2 \text{ women}) = (0.55)^8(0.45)^2$$

Step 3 Find the sum of the probabilities of the 45 mutually exclusive outcomes.

$$P(\text{exactly 8 men are hired}) = 45(0.55)^8(0.45)^2 \approx 0.076$$

Since each outcome has the same probability, you can multiply.

The probability that exactly 8 men are hired is about 7.6%.

TEACHING

Exploration

The goal of this Exploration is to give students experience with a binomial situation. This problem is solved using simulation; in Sample 1, it will be solved using probability theory.

Mathematical Procedures

Students should recall that a selection from a group of items when order is not important is a combination.

$$_nC_r = \frac{_nP_r}{r!} = \frac{n!}{(n-r)!r!}$$

Additional Sample

S1 Determine the theoretical probability that exactly 8 of the 10 people hired in the Exploration are women.

Step 1. Find the number of favorable outcomes. This is the number of combinations of 10 items taken 8 at a time.

$$_{10}C_8 = \frac{10!}{8!2!} = \frac{10 \cdot 9}{2 \cdot 1} = 45$$

Step 2. Find the probability of each outcome. The trials are independent. P(one combination of 8 women and 2 men) = $(0.45)^8(0.55)^2$

Step 3. Find the sum of the probabilities of the 45 mutually exclusive outcomes.

P(exactly 8 women hired) = $45(0.45)^8(0.55)^2 \approx 0.023$

The probability that exactly 8 women are hired is about 2.3%.

Using Technology

Students can use a graphics calculator to compute permutations and combination. To compute $_5P_2$, enter 5. Then press MATH. Select $_nP_r$ to compute permutations. Press ENTER. Then press 2 ENTER. The display reads 20. Select $_nC_r$ from the MATH menu to compute combinations.

Talk it Over

In order to answer Question 2, students need to recall that they can use Pascal's triangle to find combinations. The symbol ${}_{10}C_8$ represents the item in the 10th row on the 8th diagonal. Question 4 leads students through another calculation of a binomial probability. This problem is a good introduction to the formula that follows.

Teaching Tip

Ask students to make some general observations about the histogram of probability distributions before they consider Talk it Over questions 5–7 on page 413. For example, what do they know, and what can they assume by looking at the histogram?

Talk it Over

1. Compare the experimental probability that you found in the Exploration with the result in Sample 1. How close was your estimate to the theoretical probability?
2. Describe how to find ${}_{10}C_8$ without using a formula.
3. Suppose all 10 of the people hired are men. What is the probability of that happening by chance?
4. Suppose that 30% of the candidates are men and 70% are women.
   a. Which step(s) of Sample 1 change? How?
   b. Find the probability that exactly 8 men are hired in this situation. How does this probability compare with the one in Sample 1?

PROBABILITY IN A BINOMIAL EXPERIMENT

In a binomial experiment with n trials, the probability of exactly r successes is:

$$P(r) = {}_nC_r \cdot p^r q^{n-r}$$

number of possible outcomes ($_nC_r$); probability of success in each trial (p); $1 - p$, or probability of failure (q)

Now you are ready for: Exs. 1–13 on p. 415

This probability distribution shows the probabilities of all the possible outcomes in the hiring problem in the Exploration, using the formula above.

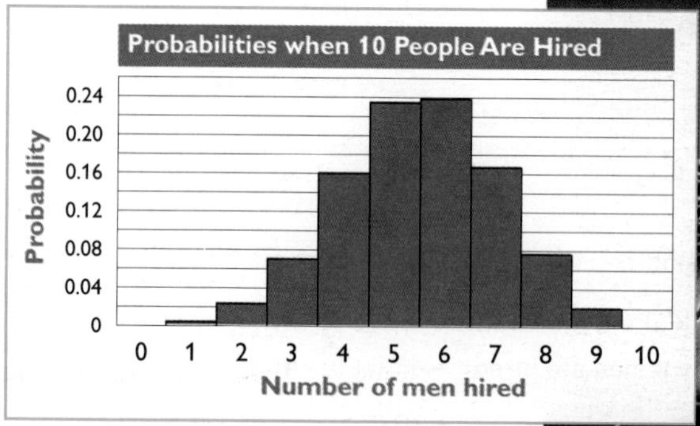

Answers to Talk it Over

1. Answers may vary. Responses are given based on the sample results from step 2 of the Exploration. The results predict a 4% probability, while theoretically the probability is 7.6%.
2. Answers may vary. Examples are given. Use a systematic list, or use Pascal's triangle. The number in row 10 and diagonal 8 of Pascal's triangle is ${}_{10}C_8$.
3. 0.55^{10}, or about 0.25%
4. a. step 2 and step 3 since $P(\text{man}) = 0.3$ and $P(\text{woman}) = 0.7$
   b. about 0.14%; This probability is about $\frac{1}{50}$ of the probability found in Sample 1.

Talk it Over

Use the histogram on page 412.

5. What is the most likely outcome of the experiment in the Exploration?
6. Describe the shape of the probability distribution.
7. Use the histogram to estimate the probability of each event.
   a. Exactly seven men are hired.
   b. Seven or more men are hired.
   c. Explain why you can add probabilities to answer part (b).

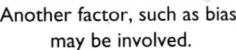

Decision-Making with Probability

One of the important uses of probability is to make decisions like, "Can an outcome be due to chance alone?" To make a decision like this, researchers complete these steps.

Create a probability model based on chance alone.

↓

Decide which probabilities to consider unusual if the results were due to chance.

Probabilities often used are 0.10, 0.05, and 0.01, each of which is called a **significance level.**

↓

Compare the model with the actual results.

↓

Is the probability that a result is due to chance less than 0.05?

In situations like the hiring problem, researchers often use a 0.05 or 5% significance level.

YES → Another factor, such as bias, may be involved.

NO → The outcome can be due to chance alone.

Talk it Over

8. Use the results of the Exploration. Is there reason to suspect bias if eight or more men are hired?
9. Use the probability distribution on page 412. For which outcome(s) is it reasonable to suspect bias?
   a. No men are hired.
   b. One man is hired or no men are hired.
   c. Two or fewer men are hired.
   d. Three or fewer men are hired.

Reasoning

Sometimes outcomes are based on chance and sometimes bias is involved. Students need to examine probability distributions and recognize when bias is a factor. If the probability is greater than the significance level, the outcome may be due to chance, and the bias is considered to be minimal. If the probability is less than the significance level, there may have been bias.

Application

Environmentalists and medical researchers often use statistical decision-making to show that the concentration of a certain disease in a particular area may not be due to chance alone. If this is the case, they may work together to see if there is an environmental cause for the disease.

Talk it Over

Questions 8 and 9 give students an opportunity to try the decision-making algorithm on both their simulated and theoretical probabilities. It also leads students to realize that having the data in a histogram can help them make quick decisions about the likelihood of a particular outcome.

Answers to Talk it Over

5. Six men are hired.
6. mound-shaped
7. a. about 0.17
   b. about 0.265
   c. The events are mutually exclusive.
8. No.
9. a. Yes.
   b. Yes.
   c. Yes.
   d. No.

Additional Sample

S2 Use the hiring situation of the Exploration and a significance level of 0.05. Suppose that 8 or more women are hired. Can the outcome be due to chance alone?

P(8 or more women are hired) = P(exactly 8 hired) + P(exactly 9 hired) + P(exactly 10 hired) = $45(0.45)^8(0.55)^2 + 10(0.45)^9(0.55)^1 + (0.45)^{10} \approx 0.023 + 0.004 + 0.0003 \approx 0.0273$

This probability is less than 0.05, so there may have been bias.

Teaching Tip

Some problems may be easier to solve if complements are used. For example, if 8 people are to be chosen from 10 available, and students are asked to find the probability of *at least* 3 women being chosen, they may find it easier to compute $1 - P(\text{no women}) - P(1\text{ woman}) - P(2\text{ women})$ than it would be to compute the probabilities of exactly 3 through exactly 8, and adding these.

Reasoning

The paragraph about statistical decision-making (immediately before the Look Back) should be discussed in class to ensure that all students understand its meaning.

Look Back

Students should write their response to this Look Back in their journals and then discuss it with a partner.

Sample 2

Use the hiring situation on page 410.

a. Suppose that 8 or more men are hired. At the 0.05 significance level, can the outcome be due to chance alone?

b. Suppose that 9 or more men are hired. At the 0.05 significance level, can the outcome be due to chance alone?

Sample Response

a. The outcome "8 or more men are hired" means that exactly 8 men are hired, exactly 9 men are hired, or exactly 10 men are hired.

$$\begin{aligned} P(\text{8 or more men are hired}) &= P(\text{exactly 8 men are hired}) + P(\text{exactly 9 men are hired}) \\ &\quad + P(\text{exactly 10 men are hired}) \\ &= 45(0.55)^8(0.45)^2 + 10(0.55)^9(0.45)^1 + (0.55)^{10} \\ &\approx 0.076 + 0.021 + 0.003 \\ &\approx 0.1 \end{aligned}$$

This probability is greater than 0.05, so the outcome can be due to chance alone.

b.
$$\begin{aligned} P(\text{9 or more men are hired}) &= P(\text{exactly 9 men are hired}) + P(\text{exactly 10 men are hired}) \\ &= 10(0.55)^9(0.45)^1 + (0.55)^{10} \\ &\approx 0.021 + 0.003 \\ &\approx 0.023 \end{aligned}$$

This probability is less than 0.05, so there may have been bias.

Statistical decision-making *never proves or disproves* a claim. It does not have the certainty of deductive reasoning. Statistical analysis uses inferences that give reason to *support or deny* a claim.

Now you are ready for: Exs. 14–30 on pp. 416–417

Look Back

Explain how to use significance levels in statistical decision-making.

Answers to Look Back

Create a probability model based on chance alone and compare the model with the actual results. Choose an appropriate significance level that will be used to determine whether it is likely that bias is involved in the actual situation or if the results can be due to chance alone.

Answers to Exercises and Problems

1. Let the numbers 1–25 represent men and the numbers 26–100 represent women. Count the number of sequences generated in which eight of the numbers are less than 26.
2. about 0.00039
3. about 0.00000525
4. about 0.00000095
5. about 0.00042
6. about 0.056
7. There are exactly two outcomes for each trial. The trials are independent. Each trial has the same probability of success. The experiment has a fixed number of trials.
8. n is the number of trials, r is the number of successes, p is the probability of

7-6 Exercises and Problems

For Exercises 1–6, suppose 25% of the people are male and 75% are female in the hiring situation on page 410.

1. What changes would you make to the simulation in the Exploration?

Find the probability of each event.

2. Exactly 8 men are hired.
3. Exactly 9 men are hired.
4. Exactly 10 men are hired.
5. At least 8 men are hired.
6. What is the probability that all ten of the people hired are women?
7. **Reading** What are the characteristics of a binomial experiment?
8. **Reading** Tell what each variable represents in this formula for an exact number of successes, r, in a binomial experiment:

$$P(r) = {}_nC_r \cdot p^r q^{n-r}$$

9. Suppose a group of job candidates contains exactly 11 men and 9 women.
   a. What is the probability that the first person hired is a man?
   b. What is the probability that the second person hired is a man?
   c. Why is the situation *not* a binomial experiment?
10. Perform an experiment by tossing a coin until it lands "heads." Explain why the binomial experiment formula does *not* apply in this case.
11. There are five true-false questions on a quiz. A student guesses randomly. Find the probability of each event.
    a. Exactly three answers are correct.
    b. Exactly four answers are correct.
    c. Exactly five answers are correct.
    d. At least three answers are correct.
12. In a school community-service club, 58% of the students are female and 42% are male. Twenty students are selected to attend a conference. What is the probability that exactly 12 of them are female?
13. **Gardening** Suppose that in a package of pepper seeds, the probability that a given seed will sprout is 0.9. Find the probability of each event when five pepper seeds are planted.
    a. Exactly four seeds sprout.
    b. Five seeds sprout.

APPLYING

Suggested Assignment

Day 1

Standard 2–13

Extended 1–13

Day 2

Standard 14–21, 25–30

Extended 14–23, 25–30

Integrating the Strands

Algebra Exs. 27–29

Statistics and Probability Exs. 1–26, 30

Discrete Mathematics Ex. 26

Logic and Language Exs. 7–10, 15, 21

Assessment: Standard

For Ex. 6, students should determine the level at which they would suspect bias: Would it be if 6 or more women were hired? 7 or more? 8 or more? 9 or more? 10 or more? Ask students how they would know, without computing, that if 9 or more women were hired there may be bias. (Since 55% were men and 45% were women and there was bias for hiring 9 or more men, with fewer women in the pool there would be bias for hiring 9 or more women.)

Answers to Exercises and Problems

success for each trial, and q is the probability of failure for each trial.

9. a. $\frac{11}{20} = 0.55$
   b. $\frac{10}{19} \approx 0.53$ if the first person hired is male, and $\frac{11}{19} \approx 0.58$ if the first person hired is female
   c. The trials are not independent, so the trials do not all have the same probability of success.
10. The experiment does not have a fixed number of trials.
11. a. 0.3125
    b. 0.15625
    c. 0.03125
    d. 0.50
12. about 0.18
13. a. 0.32805
    b. 0.59049

Communication: Writing

Ex. 21 provides an opportunity for students to clarify their understanding of significance levels.

Interdisciplinary Problems

Probability theory can be used to analyze genetic properties, as is shown in Ex. 22. Remind students that genetics is a subspecialty of biology that deals with the science of heredity.

The principal of a school randomly selects six juniors and six seniors to answer a survey about after-school jobs. In their school, P(junior has a job) = 0.5 and P(senior has a job) = 0.75. These graphs show the probabilities of the possible outcomes.

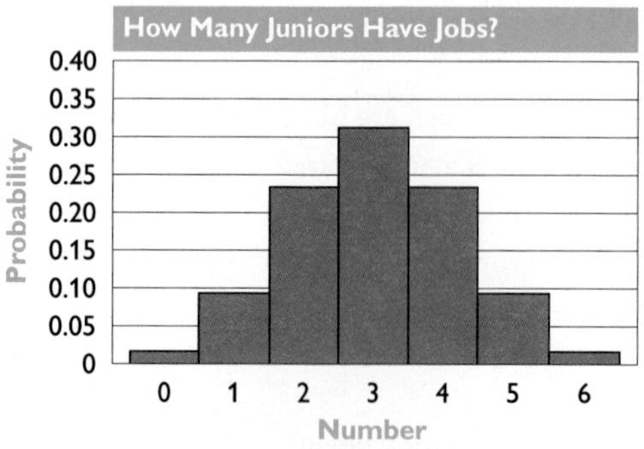

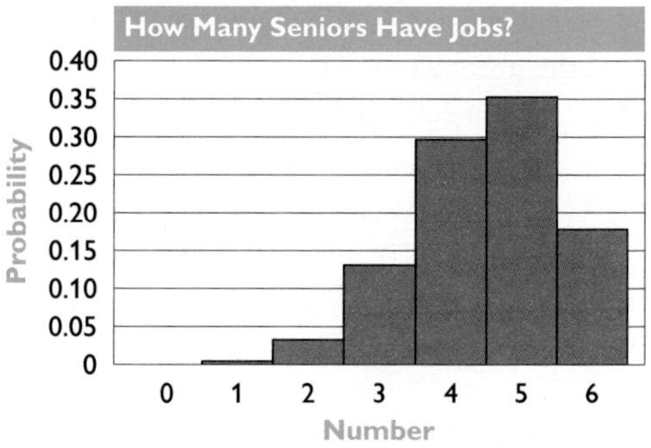

Use the graphs for Exercises 14–20.

14. Compare the shapes of the graphs. In each case, what is the most likely outcome?

15. Estimate P(exactly four seniors selected have jobs) and P(exactly four juniors selected have jobs). Explain why the probabilities are different.

Estimate the probability of each event.

16. Exactly two juniors have jobs.

17. Two or fewer juniors have jobs.

18. Exactly three seniors have jobs.

19. Three or more seniors have jobs.

20. Explain how to use the graph to find the expected value for the number of seniors selected who have jobs.

21. **Writing** What does a 0.01 significance level mean?

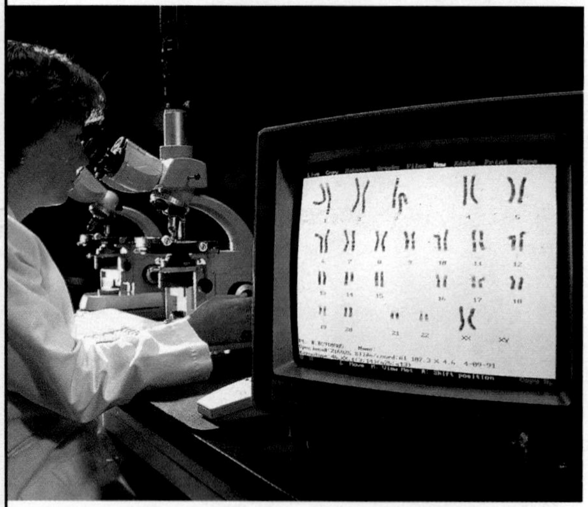

22. **Genetics** Children inherit genes independently from each parent. In order to have type O blood, a child must inherit two O genes, one from each parent. When both parents carry one O gene and one A gene, each child has a 0.25 probability of having blood type O.

Find the probability that in this situation, three or more children have blood type O when there are five children in the family.

Genetics researcher analyzing human chromosomes, the carriers of the genes

Answers to Exercises and Problems

14. The graph for juniors is mound-shaped, with 3 the most likely outcome. The graph for seniors is more skewed, with 5 the most likely outcome.

15–19. Estimates may vary.

15. about 0.29; about 0.23; The values are different because P(senior has a job) > P(junior has a job).

16. about 0.23

17. about 0.33

18. about 0.13

19. about 0.96

20. Multiply each outcome 0–6 by the probability of the outcome, and add these results to get the expected number of students with jobs.

21. If an experiment yields a result whose probability due to chance is less than 1%, the experimenters will assume that another factor, such as bias, is involved.

22. about 0.104

23. about 0.023

24. **a, b.** Answers may vary.

c. $n = 12$; $p = 0.25$; $q = 0.75$; r is any number from 0 to 12.

23. Five cars are in a parking lot on a very cold day. Suppose the probability of any one of them not starting is 0.05. What is the probability that at least two of the cars do not start?

Ongoing ASSESSMENT

24. **Group Activity** Work with another student. You will test your mind-reading skills. Use twelve playing cards, three of each of the four suits, or make four sets of three cards each using four different symbols (for example, circle, star, and so on).

   a. One of you should shuffle and hold the cards. He or she should look at the top card. The other student should guess what suit (or symbol) it is. The first student should record whether the guess is correct but keep the answer secret until after the trial is completed. Continue until all the cards are used. Let r = the number of cards correctly identified.

   b. Trade roles and repeat the experiment.

   c. For this experiment, give the value of each variable in the formula $P(r) = {}_nC_r \cdot p^r q^{n-r}$.

   d. For each value of r, find the probability that r cards were correctly identified by chance alone. At the 0.05 significance level, do your answers support the idea that either of you can read minds?

Cards like these were used in early experiments to test for the existence of extrasensory perception (ESP).

Review PREVIEW

25. Two hundred raffle tickets are sold to raise money for the library. The grand prize is a quilt worth \$500, and the second prize is a \$50 gift certificate from a book store. Each ticket costs \$3. Find the expected value for the raffle. *(Section 7-5)*

26. Find the mean and the standard deviation of the number of Calories in six different flavors of instant soup: 150, 210, 180, 120, 140, 225. *(Section 6-2)*

Find the distance between each pair of points. *(Toolbox Skill 31)*

27. (0, 0) and (8, −6)

28. (0, 0) and (−5, −12)

29. (0, 0) and (−5, 5)

Working on the Unit Project

30. Suppose two people are playing da-un-dah-qua. Find the probability that one person scored exactly 6 points on each of 3 of her first 8 plays.

Working on the Unit Project

In Ex. 30, students are given the opportunity to look at the game of da-un-dah-qua as a binomial experiment. They have now used the ideas of probability presented in the unit to analyze the game.

Quick Quiz (7-4 through 7-6)

See page 419.

See also Test 29 in the *Assessment Book.*

Practice 57 For use with Section 7-6

Name ______________________ Date ____________

Practice 57

For use with Section 7-6

1. A fair coin is tossed 6 times. Find the probability of each event.

   a. Exactly 4 tosses are heads. $\frac{15}{64}$

   b. Exactly 5 tosses are heads. $\frac{3}{32}$

   c. Exactly 6 tosses are heads. $\frac{1}{64}$

   d. At least 4 tosses are heads. $\frac{11}{32}$

2. A die tossed 10 times shows "1" on five of the tosses and other numbers on the other five tosses. Using a 0.05 significance level, would you conclude that the die is not fair? **yes**

3. A multiple-choice test has 4 choices for each question.

   a. What is the probability of correctly answering a question by randomly guessing? What is the probability of incorrectly answering a question? $\frac{1}{4}$; $\frac{3}{4}$

   b. On a 5-question test, what is the probability of getting exactly 3 correct answers? exactly 4 correct answers? **about 0.0879; about 0.0146**

4. Leon Brooks has a free-throw percentage of 70%. This means that the probability of his making a given free throw is 0.7. What is the probability that Leon will make exactly 8 of his next 10 free throws? (Assume that Leon has taken so many free throws that his free-throw percentage does not change significantly as he shoots these 10.) **about 0.233**

5. A compact disc player can be programmed to play random tracks on a disk. Suppose there are 8 tracks on a disk and the player chooses track five 3 times during a run of 7 plays and other tracks for the rest of the plays. Using a 0.02 significance level, would you say that the compact disc player is working correctly? **yes**

6. The U.S. Weather Service has predicted a 60% chance of rain for a particular region during the next 8 days.

   a. What is the probability that exactly 3 of the 8 days will be rainy in the region? **about 0.124**

   b. What is the probability that there will be 3 rainy days in a row? (*Hint*: The events "days 1, 2, and 3 are rainy," "days 2, 3, and 4 are rainy," and so on, are mutually exclusive.) **about 0.0133**

Practice Bank, INTEGRATED MATHEMATICS 3

57

Answers to Exercises and Problems

d. $P(0 \text{ successes}) \approx 0.0317$; $P(1 \text{ success}) \approx 0.127$; $P(2 \text{ successes}) \approx 0.232$; $P(3 \text{ successes}) \approx 0.258$; $P(4 \text{ successes}) \approx 0.194$; $P(5 \text{ successes}) \approx 0.103$; $P(6 \text{ successes}) \approx 0.040$; $P(7 \text{ successes}) \approx 0.0115$; $P(8 \text{ successes}) \approx 0.00239$; $P(9 \text{ successes}) \approx 0.00035$; $P(10 \text{ successes}) \approx 0.000035$; $P(11 \text{ successes}) \approx 0.0000022$; $P(12 \text{ successes}) \approx 0.00000006$; Answers may vary. An example is given. 6 or more correct guesses suggest that some factor other than chance may be involved.

25. −25¢

26. 170.8; 37.7

27. 10

28. 13

29. $5\sqrt{2} \approx 7.07$

30. about 0.000083

Assessment

A scoring rubric for the Unit Project can be found on pages 370 and 371 of this Teacher's Edition and also in the ***Project Book.***

Unit Project 7

Completing the Unit Project

Now you are ready to demonstrate your new game based on da-un-dah-qua.

Your completed project should include these things:

- **a probability analysis of the game of da-un-dah-qua based on your answers to the "Working on the Unit Project" exercises throughout the unit**
- **a written set of rules for your new game**
- **a probability analysis of your new game, including a discussion of these questions — Is the game fair? Will playing the game take more or less time than playing da-un-dah-qua? Is the game easier than da-un-dah-qua? more difficult?**
- **a demonstration of how your new game is played**

Look Back

How has your study of the rules of a game changed the way you think about games?

Alternative Projects

Project 1: **Consumer Economics and Maintenance Agreements**

Consumers can buy maintenance agreements for many electronics items and home appliances. Use the ideas of probability to determine whether maintenance agreements are a good buy. You will need to find out the terms of manufacturers' warranties, as well as the probabilities that the items fail after the warranty expires or that the failures are not covered by the warranty. Estimate probabilities that are unavailable and describe any assumptions you made.

Project 2: **Analyzing Advertising Claims**

Advertisements on television, radio, and in newspapers and magazines are full of claims based on statistics — for example, taste tests. Select at least three claims of this type. Obtain the statistical data that are the basis of each claim. Using the ideas of probability, analyze each claim to determine whether it is justified by the data.

Answers to Unit 7 Review and Assessment

1. If the numbers given to the students were originally assigned in a random manner (for example, by putting all the names in a hat and assigning 1 to the first name, and so on), then the procedure described would give a random sampling. If the numbers were not originally assigned at random (for example, if the numbers were given out in sequence by someone who simply visited all the areas of the school), then the sequence produced would not be random, since it is likely that friends would have numbers assigned at the same time and would arrive together, as would groups of athletes, band members, and so on.
2. int(100*rand)
3. int(10*rand) + 1
4. a. Add the frequencies; divide each frequency by the sum.
   b. Answers may vary. Examples are given. If a call is selected at random, the probability that it was received between 4 P.M. and midnight is about 55%. If a call is selected at random, it is most likely to be in the 8 P.M.–midnight category.
5. For each trial, use the function int(100*rand) + 1 to generate a sequence of 10 random numbers. Let the numbers 1–60 represent people with previous experience and the numbers 61–100 represent people with no previous experience. Count the number of sequences generated and

1. Suppose you give each student at your school a number from 1 to 786. You write down the numbers of the first 25 people who enter the auditorium for a pep rally. Will this produce a random sequence? Explain. 7-1

For Questions 2 and 3, design and test a formula for a calculator to produce each set of random numbers.

2. integers from 0 to 99
3. integers from 1 to 10

4. a. **Writing** Describe how to change this frequency distribution to a probability distribution.

   b. **Open-ended** Make two conclusions about the probability distribution.

Incoming Calls at a Taxi Company in One Day

Time period	Number of calls
midnight–4 A.M.	0
4 A.M.–8 A.M.	28
8 A.M.–noon	126
noon–4 P.M.	97
4 P.M.–8 P.M.	148
8 P.M.–midnight	153

5. In a very large pool of job applicants, 60% have previous experience for the job and 40% do not. Of the 10 people hired, 7 have no experience. Design a simulation to investigate the situation.

For each event in Questions 6–8, find the probability. 7-2

6. Two dice are tossed. The sum of the numbers is 9 or greater.
7. You look at a clock at a random moment. The minute hand is between 1 and 3.
8. You look at a clock at a random moment. The minute hand is between 12 and 4 or it is between 2 and 5.
9. **Writing** Give an example of events that are mutually exclusive but not complementary. Justify your response.
10. Arnie, Bettina, Cal, Dave, and Ellen are on a work crew. One crew member is selected for a special job. Tell whether each statement is *True* or *False*.

    a. Selecting Arnie and selecting Bettina are mutually exclusive events.

    b. Selecting Arnie and selecting Bettina are complementary events.

11. You agree with a friend that you will begin an on-line conversation by electronic mail some time between 7 P.M. and 9 P.M. You each agree to be on the computer for half an hour some time in that interval. What is the probability that you connect? 7-3

12. a. **Writing** What does it mean in everyday words for events to be independent? 7-4

    b. What does it mean mathematically for events to be independent?

Answers to Unit 7 Review and Assessment

the number in which 7 numbers greater than 60 appear. Analyze the data.

6. $\frac{5}{18} \approx 0.28$
7. $\frac{1}{6} \approx 0.17$
8. $\frac{5}{12} \approx 0.42$
9. Answers may vary. Examples are given. Two dice are rolled. Their sum is less than 5 or greater than 7. Both events cannot occur at the same time, so they are mutually exclusive. However, they account for only 21 of the 36 possible outcomes, so the events are not complementary.
10. a. True.

    b. False.
11. $\frac{7}{16} = 0.4375$
12. a. Two events are independent if the occurrence of one of the events does not affect the probability of the occurrence of the other.

    b. Events A and B are independent if the probability that B occurs given that A has occurred is exactly the same as the probability of B alone.

Unit Support Materials

- *Practice Bank:* Cumulative Practice 58
- *Study Guide:* Unit 7 Review
- *Problem Bank:* Unifying Problem 7
- *Assessment Book:* Unit Tests 30 and 31, Spanish Unit Tests, Alternative Assessment
- *Test Generator* software with Test Bank
- *Teacher's Resources for Transfer Students*

Quick Quiz (7-4 through 7-6)

A drawer contains 6 blue socks and 8 white socks. Carlotta will reach in and randomly draw one sock and then another without replacing the first sock.

1. Make a tree diagram to represent this situation. Include the probability for each branch of the diagram. [7-4]

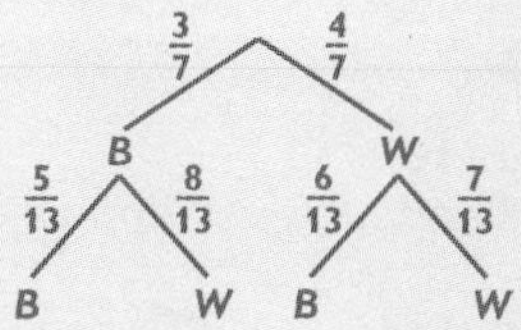

2. Find the probability of getting a blue sock and then another blue sock. [7-4] $\frac{15}{91}$
3. Are drawing the two blue socks independent or dependent events? Explain. [7-4] dependent events; $P(\text{blue}) = \frac{3}{7}$ and $P(\text{blue} \mid \text{blue}) = \frac{5}{13}$. These probabilities would have to be equal for the events to be independent.
4. Lottery tickets for a special fund-raiser are selling for fifty cents each. Suppose 12,000 tickets are sold. There are 10 prizes of $100 and 20 prizes of $50. What is the expected value of this lottery? Would you buy a ticket? Explain. [7-5] about 17 cents; Answers will vary. A sample answer is given. No; you pay fifty cents and expect to win only about 17 cents.

Quick Quiz continued on next page.

Quick Quiz (continued)

A 10-question, multiple-choice quiz is given. Each question has 4 choices for answers. A student claims that she knew nothing on the test and only choose answers at random.

5. Find the probability that she got exactly 8 answers correct. [7-6] **about 0.0004**

6. Using a significance level of 0.05, do you believe that the student knew nothing about the test if she had 8 correct answers? Explain. [7-6] **No; the probability of getting 8 correct answers with random guessing is less than 0.05 so that, most likely, the student's quiz results were not due to chance alone. She probably knew some of the material .**

For Questions 13–15, use this table. A student is selected at random. Find each probability.

	Male	Female	Total
In the band	8	7	15
Not in the band	12	5	17
Total	20	12	32

13. P(in the band | female) **14.** P(female | in the band)

15. Are the events "in the band" and "female" independent? Explain.

16. A survey of households in a small community showed that 26% have three or more television sets, 42% of those with three or more television sets reported watching over 100 hours per week, and 35% of those with fewer than three television sets reported watching over 100 hours per week.

a. Make a tree diagram or a table for this data.

b. Is owning three or more television sets independent of watching over 100 hours per week? Explain.

17. In a state lottery, players buy cards for $1 each. Players mark any three digits (for example, 004 or 029) on a card. Players who select the winning number win $600. Find the expected value for this lottery. **7-5**

18. The table shows the distribution of the Calories in 100 low-Calorie frozen dinners.

a. What is the mean of this probability distribution?

b. Suppose you choose one of the 100 dinners at random. What is the expected value of the number of Calories?

Number of Calories	Percent of dinners
250	10
260	20
270	20
280	30
300	0
310	10
320	10

19. Market Research An ad for cat food claims that "4 out of 5 cats prefer Kitty Nibbles." Suppose that each cat was offered two bowls of food, Kitty Nibbles and Brand X, and chose randomly between them. Find the probability that 4 out of 5 cats chose Kitty Nibbles. **7-6**

20. Twenty students are chosen at random for a crowd scene in a school play. Twenty percent of the students in the school are seniors. What is the probability that the crowd has 75% or more seniors?

21. **Self-evaluation** After completing this unit, do you think you will be better able to understand and think critically about claims involving probability and chance? Why or why not?

22. **Group Activity** Work with another student.
A pool of citizens for a jury has 25 men and 15 women. The pool was chosen at random.

a. Is selecting 12 people at random from this pool an example of a binomial experiment? Why or why not?

b. Design and carry out a simulation to estimate the probability that of the 12 people selected, 8 are women.

Answers to Unit 7 Review and Assessment

13. $\frac{7}{12} \approx 0.58$

14. $\frac{7}{15} \approx 0.47$

15. No. $P(\text{female}) = \frac{20}{32} = \frac{5}{8}$; $P(\text{female} \mid \text{in the band}) = \frac{7}{15}$; $\frac{5}{8} \neq \frac{7}{15}$

16. a.

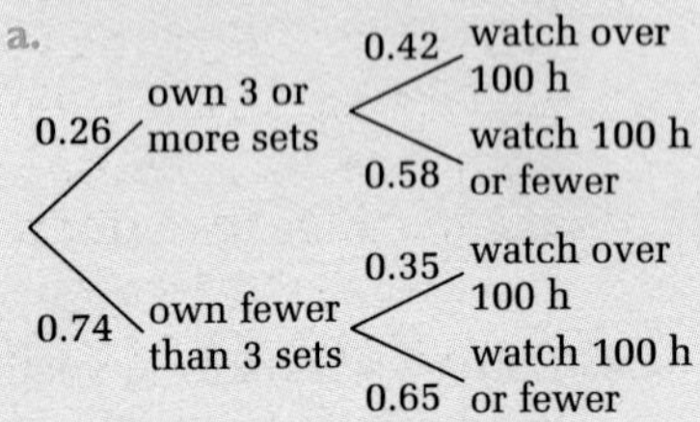

	Watch over 100 h	Watch 100 h or less	Total
Own 3 or more sets	0.109	0.151	0.26
Own fewer than 3 sets	0.259	0.481	0.74
Total	0.368	0.632	1

b. No; P(over 100 hours per week | three or more televisions) ≈ 0.11; P(over 100 hours per week) ≈ 0.37

17. −40¢

18. a. 278 b. 278 Calories

19. ≈ 0.16

20. ≈ 0.00000018

21. Answers may vary.

22. a. No; The trials are not independent and each trial does not have the same probability of success. (The pool is too small. Suppose the first two jurors chosen are women. For the first trial, P(woman) = 0.375. For the second trial, P(woman) ≈ 0.359.)

b. Answers may vary. An example is given. Using a deck of cards, let black cards represent men, and remove one black card from the deck so that 25 remain. Let red cards represent women and remove 11 red cards from the deck so 15 remain. Shuffle the remaining cards, and select 12. Perform this trial 100 times and analyze the data.

IDEAS AND (FORMULAS) $= \chi^2$ ${}_5P_5$

STATISTICS & PROBABILITY

- You can use simulation to study a real event without the time, cost, risk, or materials needed for the real event. (*p. 374*)
- A sequence of random numbers has no pattern, and each number has an equal chance of appearing in the sequence. (*p. 374*)
- The probability of any event is $\frac{\text{size of favorable outcomes}}{\text{size of possible outcomes}}$.

  "Size" can be a count or a measurement. (*p. 383*)
- For any two events A and B: $P(A \text{ or } B) = P(A) + P(B) - P(A \text{ and } B)$ (*p. 385*)
- You can use tree diagrams to investigate conditional probabilities. (*p. 396*)
- The probability that event B occurs given that event A occurred is written $P(B \mid A)$. For any two events A and B,

  $P(A \text{ and } B) = P(A) \cdot P(B \mid A)$ and $P(B \mid A) = \frac{P(A \text{ and } B)}{P(A)}$.

  (*p. 397*)
- Two events A and B are independent if and only if $P(B \mid A) = P(B)$. (*p. 398*)
- To find the expected value of an event, multiply the value of each outcome, x_i, by its probability, $P(x_i)$, and then add all the products. (*p. 404*)

  Expected value $= x_1 \cdot P(x_1) + x_2 \cdot P(x_2) + \cdots + x_n \cdot P(x_n)$
- In a binomial experiment with n trials, where the probability of success in each trial is p, and the probability of failure is $q = 1 - p$, the probability of exactly r successes is: (*p. 412*)

  $$P(r) = {}_nC_r \cdot p^r q^{n-r}$$

GEOMETRY

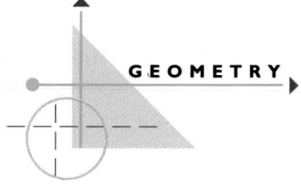

- You can use area models with coordinates to find probabilities. (*p. 391*)

Key Terms

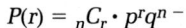

- **simulation** (p. 374)
- **random numbers** (p. 374)
- **probability distribution** (p. 376)
- **mutually exclusive events** (p. 386)
- **complementary events** (p. 386)
- **conditional probability** (p. 396)
- **independent events** (p. 398)
- **expected value** (p. 404)
- **fair (game)** (p. 405)
- **binomial experiment** (p. 411)
- **significance level** (p. 413)

Quick Quiz (7-1 through 7-3)

1. Design a random number simulation to investigate the number of rainy days possible in a 10-day stretch if the probability of rain on any one day is 15%. [7-1] **Answers will vary. A sample answer is given. Generate random numbers from 00 to 99. Let numbers from 01 to 15 inclusive mean that it rained on that day. All other numbers mean that it did not rain. Pick 10 two-digit random numbers repeatedly and determine how many times it rained in each trial.**
2. On a highway, there are 10 miles between Exits 1 and 2, 12 miles between Exits 2 and 3, and 17 miles between Exits 3 and 4. If accidents are equally likely to happen on any part of the highway, what is the probability of an accident occurring between Exits 2 and 3? [7-2] $\frac{4}{13}$
3. Jordon has 5 pennies, 1 nickel, 2 dimes, and 3 quarters in his pocket. If he draws one coin out at random, what is the probability that he has at least ten cents or a penny? [7-2] $\frac{10}{11}$
4. It takes 8 minutes for an emergency room receptionist to process a patient. Suppose two people arrive randomly within a 30-minute period. If you were to use coordinates to find the probability that one of them will have to wait to be helped, what system of inequalities would form the boundaries of the shaded region? [7-3] $x > 0$, $y > 0$, $x < 30$, $y < 30$, $y < x + 8$, $y > x - 8$

Planning the Unit

8 Angles, Trigonometry, and Vectors

OVERVIEW

- **Unit 8** introduces polar coordinates for locating points in the plane. The relationship between polar coordinates and rectangular coordinates is explored. Vectors are introduced and represented both geometrically and algebraically. Parametric equations are used to represent both linear and quadratic models.
- The study of trigonometry is extended to angles greater than 90°. Trigonometric functions are used to relate polar and rectangular coordinates. The law of cosines and the law of sines is used to find missing measures in triangles. The law of sines is also used to determine the ambiguous case (SSA) when solving a triangle.
- The **Unit Project** theme is to plan a treasure hunt. Students create a treasure map, a set of numbered clues, and a solution map. The clues can involve vectors, parametric equations, the law of sines, the law of cosines, and polar coordinates. Students trade their maps to hunt for a treasure.
- **Connections** to the game of chess, a marching band, an airplane's flight, tobogganing, physics, skateboard ramps, baseball, gardening, surveying, and astronomy are some of the topics included in the teaching materials and the exercises.
- **Graphics calculators** are used in Section 8-2 to convert polar coordinates to rectangular coordinates and in Section 8-5 to graph parametric equations and parametric equations involving trigonometric functions. **Computer software**, such as Plotter Plus, can be used in Section 8-1 to plot polar graphs.
- **Problem-solving strategies** include using a formula, using algebra, using vector diagrams, using parametric equations, and using trigonometry.

Unit Objectives

Section	Objectives	NCTM Standards
8-1	• Use polar coordinates to locate points.	1, 2, 3, 4, 5
8-2	• Convert from polar to rectangular coordinates and vice versa. • Extend the definitions of cosine, sine, and tangent. • Use a Pythagorean identity.	1, 2, 3, 4, 5, 9
8-3	• Use drawings of vectors to solve problems.	1, 2, 3, 4, 5
8-4	• Use algebra to solve problems involving vectors.	1, 2, 3, 4, 5, 9
8-5	• Use parametric equations to solve problems.	1, 2, 3, 4, 5, 9
8-6	• Derive the law of cosines and use it to find measures of sides and angles in triangles.	1, 2, 3, 4, 5, 8, 9
8-7	• Use the law of sines to find measures of sides and angles in triangles.	1, 2, 3, 4, 5, 8, 9

Topic Spiraling

Section	Connections to Prior and Future Concepts
8-1	**Section 8-1** introduces polar coordinates to locate points. Polar graph paper was first introduced in Section 4-4 of Book 1. Polar coordinates are used in Sections 8-2 and 8-3 of Book 3 and in a number of future mathematics courses.
8-2	**Section 8-2** introduces converting polar coordinates to rectangular coordinates, and vice versa. Skills with sine, cosine, and tangent functions are required. These functions were first introduced in Sections 6-7 and 7-1 of Book 1. The Pythagorean theorem was first encountered in Section 9-1 of Book 1. Polar coordinates and trigonometric functions are combined into a Pythagorean identity.
8-3	**Section 8-3** introduces the graphical representation of a vector, and uses these representations to solve problems. Polar and rectangular coordinates, from Section 8-2 of Book 3, are used to represent vectors. The graphical method of adding vectors is explored. Vectors are explored further in Section 8-4 of Book 3.
8-4	**Section 8-4** introduces algebraic techniques for solving problems involving vectors. Vectors written in rectangular form are added.
8-5	**Section 8-5** uses the concept of parametric equations, first introduced for lines in Section 9-6 of Book 2, and extends it to parabolas. The sine and cosine functions, introduced in Section 6-7 of Book 1 and reviewed in Section 8-2 of Book 3, are used as one model of a parametric equation. Parametric equations resulting in both lines and parabolas are graphed. Linear graphs have been studied in Unit 8 of Book 1, and reviewed in Section 2-2 of Book 2 and Sections 1-4 and 2-2 of Book 3. Graphs of parabolas were introduced in Section 10-7 of Book 1, and reviewed in Section 4-1 of Book 2 and Section 2-4 of Book 3.
8-6	**Section 8-6** develops the law of cosines and applies it to solving triangles other than right triangles. Using trigonometry to solve a right triangle was first explored in Section 6-7 of Book 1 and extended in Section 8-8 of Book 2.
8-7	**Section 8-7** develops the law of sines and applies it to solving triangles other than right triangles. This section covers triangles other than ones that could be solved in Section 8-6 of Book 3.

Integrating the Strands

Strands	Sections
Algebra	8-2, 8-3, 8-4, 8-5, 8-6, 8-7
Functions	8-2
Measurement	8-1, 8-2, 8-3, 8-4, 8-6
Geometry	8-1, 8-2, 8-3, 8-4, 8-5, 8-6, 8-7
Trigonometry	8-1, 8-2, 8-3, 8-4, 8-5, 8-6, 8-7
Statistics and Probability	8-1, 8-3, 8-7
Discrete Mathematics	8-2, 8-3, 8-4, 8-5, 8-7
Logic and Language	8-1, 8-2, 8-3, 8-4, 8-5, 8-6, 8-7

Planning the Unit 8 Angles, Trigonometry, and Vectors

Section Planning Guide

- Essential exercises and problems are indicated in boldface.
- Ongoing work on the Unit Project is indicated in color.
- Exercises and problems that require student research, group work, manipulatives, or graphing technology are indicated in the column headed "Other."

Section	Materials	Pacing	Standard Assignment	Extended Assignment	Other
8-1	polar graph paper	Day 1	**2–7, 16–30,** 35–38, 39	1, **2–7,** 8–15, **16–30,** 31–33, 35–38, 39	34
8-2	calculator, polar graph paper, protractor	Day 1 Day 2 Day 3	**1–3, 5–10** **12–16, 22–41** **42–55,** 57–59, 60	**1–3,** 4, **5–10,** 11 **12–16,** 17–21, **22–41** **42–55,** 57–59, 60	 56
8-3	protractor, calculator	Day 1	**5–16,** 20–23, 24	2–4, **5–16,** 17–23, 24	1
8-4	graphics calculator, protractor	Day 1	**2–7, 9–16,** 22–25, 26	1, **2–7,** 8, **9–16,** 17, 19, 20, 22–25, 26	18, 21
8-5	graphics calculator	Day 1	**5–11, 13–32,** 34–38, 39	1–4, **5–11,** 12, **13–32,** 34–38, 39	**13–28,** 33
8-6	centimeter ruler, protractor, compass	Day 1 Day 2	**1–4, 10–12** **17, 18, 22–24,** 29–31, 32	**1–4,** 5–9, **10–12,** 13–16 **17, 18,** 19–21, **22–24,** 25–27, 29–31, 32	 28
8-7	protractor, compass, calculator	Day 1 Day 2	**3–11** **16–33,** 35–37, 38	1, 2, **3–11,** 12–14 15, **16–33,** 35–37, 38	 34
Review Test		**Day 1** **Day 2**	**Unit Review** **Unit Test**	**Unit Review** **Unit Test**	

Yearly Pacing	Unit 8 Total	Units 1–8 Total	Remaining	Total
	15 days (2 for Unit Project)	123 days	31 days	154 days

Support Materials

- See **Project Book** for notes on Unit 8 Project: Plan a Treasure Hunt.
- UPP and disk refer to **Using Plotter Plus** booklet and **Plotter Plus** disk.
- TI-81/82 refers to **Using TI-81 and TI-82 Calculators** booklet.
- Warm-up exercises for each section are available on **Warm-Up Transparencies.**

Section	Study Guide	Practice Bank	Problem Bank	Activity Bank	Explorations Lab Manual	Assessment Book	Visuals	Technology
8-1	8-1	Practice 59	Set 17	Enrich 52	Master 8	Quiz 8-1	Folder 7	Polar Graph Plotter (disk)
8-2	8-2	Practice 60	Set 17	Enrich 53	Master 8	Quiz 8-2		
8-3	8-3	Practice 61	Set 17	Enrich 54	Masters 2, 17	Quiz 8-3	Folder 8	
8-4	8-4	Practice 62	Set 17	Enrich 55	Add. Expl. 9 Master 2	Quiz 8-4 Test 32		
8-5	8-5	Practice 63	Set 18	Enrich 56	Master 2	Quiz 8-5		TI-81/82, page 67
8-6	8-6	Practice 64	Set 18	Enrich 57	Masters 18, 19	Quiz 8-6		
8-7	8-7	Practice 65	Set 18	Enrich 58	Master 20	Quiz 8-7 Test 33		
Unit 8	Unit Review	Practice 66	Unifying Problem 8	Family Involve 8		Tests 34, 35		

Form A

Spanish versions of these tests are on pages 143–146 of the **Assessment Book.**

Name ______ Date ______ Score ______

Test 34

Test on Unit 8 (Form A)

Directions: Write the answers in the spaces provided.

For Questions 1–4, name the point having each pair of polar coordinates.

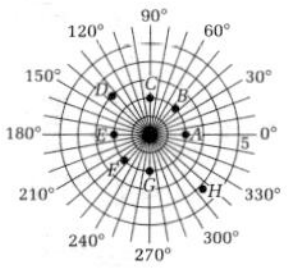

1. (4.2, 315°) **2.** (–2, 90°)

3. (3, 135°) **4.** (–2, 45°)

5. Convert (4, –90°) to rectangular coordinates.

6. Convert (3, 110°) to rectangular coordinates.

7. Convert (–2, 5) to polar coordinates.

8. Find all angles θ between 0° and 360° for which $\sin \theta = -0.707$.

9. Find $\cos \theta$ when $\sin \theta = -0.3951$ and θ is in Quadrant III.

10. Find the rectangular form of the vector sum (2, –5) + (–3, 1).

11. Find the polar form of the vector sum (2, 75°) + (4, –20°).

For Questions 12 and 13, use the law of cosines to calculate the length of the missing side in $\triangle ABC$.

12. $a = 25.3$, $b = 16.8$, $m\angle C = 52°$

13. $m\angle A = 103°$, $b = 10.7$, $c = 8.5$

14. Find the length of each side of a regular pentagon inscribed in a circle of radius 8 in.

15. Open-ended Write two vectors in rectangular form and show how to find their sum algebraically. Then show the sum graphically and verify that the resultant has the coordinates found algebraically.

Sample answer: Let v = (1, 2) and u = (5, 3). Then w = v + u = (1, 2) + (5, 3) = (1 + 5, 2 + 3) = (6, 5).

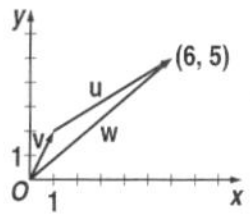

Answers

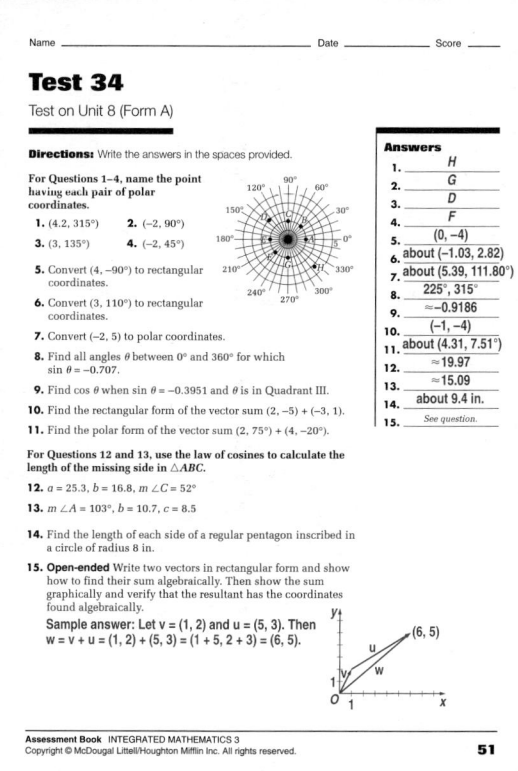

1. H
2. G
3. D
4. F
5. (0, –4)
6. about (–1.03, 2.82)
7. about (5.39, 111.80°)
8. 225°, 315°
9. ≈–0.9186
10. (–1, –4)
11. about (4.31, 7.51°)
12. ≈19.97
13. ≈15.09
14. about 9.4 in.
15. *See question.*

Assessment Book INTEGRATED MATHEMATICS 3

51

Name ______ Date ______ Score ______

Test 34 *(continued)*

Directions: Write the answers in the spaces provided.

For Questions 16 and 17, combine the two parametric equations into a single equation relating y to x.

16. $x = 4.5t$, $y = 2.25t$

17. $x = 5 \cos t$, $y = 5 \sin t$

18. Describe the graph of the parametric equations given in Question 17.

a circle of radius 5 with center at (0, 0)

For each angle measure and pair of side lengths, tell how many triangles can be formed.

19. $m\angle A = 48°$, $a = 5$, $b = 6$

20. $m\angle A = 105°$, $a = 22$, $b = 15$

Find the unknown measures in each triangle.

21.

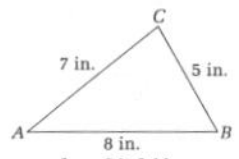

$m\angle A \approx 38.21°$, $m\angle B = 60°$, $m\angle C \approx 81.79°$

22.

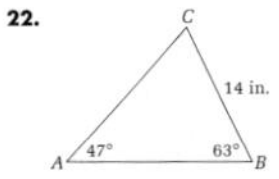

$b \approx 17.06$ in., $c \approx 17.99$ in., $m\angle C = 70°$

A 50-ft ramp makes an angle of 4.9° with the horizontal. To meet new accessibility guidelines, the ramp must be lowered so it makes an angle of 2.7° with the horizontal. Use this information for Questions 23 and 24.

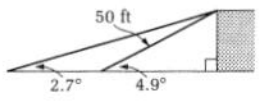

23. What will be the length of the new ramp?

24. How much farther along the ground will the new ramp extend?

25. Writing State the law of sines and explain when you can use it to solve a triangle.

$\frac{a}{\sin A} = \frac{b}{\sin B} = \frac{c}{\sin C}$; Sample explanation: you can use the law of sines whenever you are given ASA or AAS information for a triangle; you can use the law when given SSA information, but there may not be a unique solution.

Answers

16. $y = \frac{1}{2}x$
17. $\frac{x^2}{25} + \frac{y^2}{25} = 1$
18. *See question.*
19. 2 triangles
20. 1 triangle
21. *See question.*
22. *See question.*
23. ≈90.66 ft
24. ≈40.75 ft
25. *See question.*

52

Assessment Book INTEGRATED MATHEMATICS 3

Form B

Name ______ Date ______ Score ______

Test 35

Test on Unit 8 (Form B)

Directions: Write the answers in the spaces provided.

For Questions 1–4, name the point having each pair of polar coordinates.

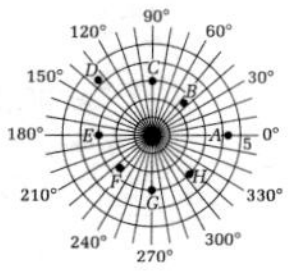

1. (4.3, 135°) **2.** (–3, 270°)

3. (3, 315°) **4.** (–2.5, 225°)

5. Convert (3, 180°) to rectangular coordinates.

6. Convert (5, 70°) to rectangular coordinates.

7. Convert (5, –3) to polar coordinates.

8. Find all angles θ between 0° and 360° for which $\sin \theta = -0.866$.

9. Find $\cos \theta$ when $\sin \theta = 0.5782$ and θ is in Quadrant II.

10. Find the rectangular form of the vector sum (1, –3) + (–3, 1).

11. Find the polar form of the vector sum (4, 55°) + (3, –80°).

For Questions 12 and 13, use the law of cosines to calculate the length of the missing side in $\triangle ABC$.

12. $a = 35.4$, $b = 46.7$, $m\angle C = 42°$

13. $m\angle A = 123°$, $b = 20.3$, $c = 16.5$

14. Find the length of each side of a regular octagon inscribed in a circle of radius 10 in.

15. Open-ended Make up an example for which the law of sines will not work and explain why it will not work. Include a drawing to illustrate your example.

Sample answer: Let $m\angle A = 40°$, $b = 20$ in., and $a = 5$ in. Using the law of sines, $\frac{5}{\sin 40°} = \frac{20}{\sin B}$ and thus $\sin B = \frac{20 \sin 40°}{5}$, or $\sin B = 4 \sin 40°$. But $\sin B \neq 4 \sin 40°$ since $4 \sin 40° > 1$. (When trying to construct the triangle, side BC is too short to form a triangle.)

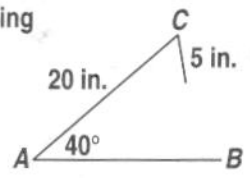

Answers

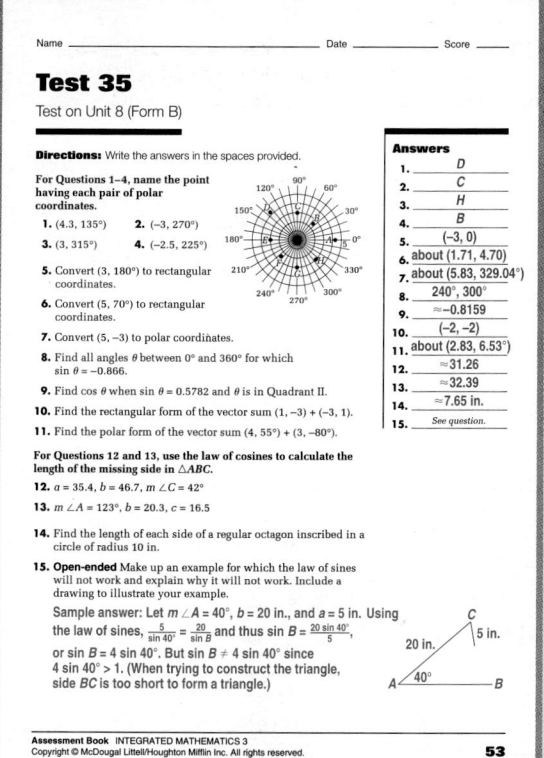

1. D
2. C
3. H
4. B
5. (–3, 0)
6. about (1.71, 4.70)
7. about (5.83, 329.04°)
8. 240°, 300°
9. ≈–0.8159
10. (–2, –2)
11. about (2.83, 6.53°)
12. ≈31.26
13. ≈32.39
14. ≈7.65 in.
15. *See question.*

Assessment Book INTEGRATED MATHEMATICS 3

53

Name ______ Date ______ Score ______

Test 35 *(continued)*

Directions: Write the answers in the spaces provided.

For Questions 16 and 17, combine the two parametric equations into a single equation relating y to x.

16. $x = 12t$, $y = 4t$

17. $x = 4 \cos t$, $y = 4 \sin t$

18. Describe the graph of the parametric equations given in Question 17.

a circle of radius 4 with center at (0, 0)

For each angle measure and pair of side lengths, tell how many triangles can be formed.

19. $m\angle A = 112°$, $a = 18$, $b = 14$

20. $m\angle A = 34°$, $a = 8$, $b = 10$

Find the unknown measures in each triangle.

21.

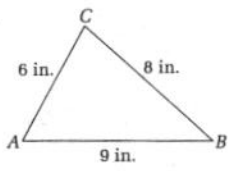

$m\angle A \approx 60.61°$, $m\angle B \approx 40.80°$, $m\angle C \approx 78.59°$

22.

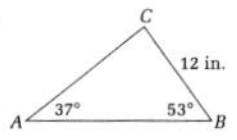

$b \approx 15.92$ in., $c \approx 19.94$ in., $m\angle C = 90°$

The roof of a house is being reconstructed toaccomodate heavy snows. The current 32-ft roofline makes an 18.2° angle with the horizontal. The owner has decided to construct the new roof so that it makes a 50° angle with the horizontal. Use this information for Questions 23 and 24.

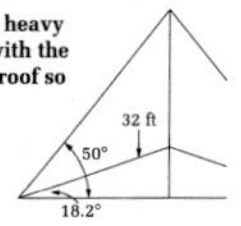

23. What will be the length of the new roofline?

24. How much higher will the new roof be?

25. Writing State one case of the law of cosines and explain when you can use it to solve a triangle.

Sample answer: $c^2 = a^2 + b^2 - 2ab \cos C$; this case of the law of cosines is most useful when a, b, and $m\angle C$ are known, or when a, b, and c are known.

Answers

16. $y = \frac{1}{3}x$
17. $\frac{x^2}{16} + \frac{y^2}{16} = 1$
18. *See question.*
19. 1 triangle
20. 2 triangles
21. *See question.*
22. *See question.*
23. ≈47.29 ft
24. ≈26.23 ft
25. *See question.*

54

Assessment Book INTEGRATED MATHEMATICS 3

OUTSIDE RESOURCES

Books/Periodicals

Germain-McCarthy, Yvelyne. "Circular Graphs: Vehicles for Conic and Polar Connections." *Mathematics Teacher* (January 1995): pp. 26–28.

Esty, Warren W. "Finding Points of Intersection of Polar-Coordinate Graphs." *Mathematics Teacher* (September 1991): pp. 472–477.

The Madison Project. "Matrices and Space Capsules": pp. 372–380. *Explorations in Math.*

Activities/Manipulatives

Hurwitz, Marsha. "Discovering the Law of Sines." *Mathematics Teacher* (November 1991): pp. 634–635.

Software

deLange, Jan. *Gliding.* Wings for Learning/Sunburst, 1992.

Johnson, Nancy Anne. *GyroGraphics.* IBM compatible, includes 15-page booklet. Cipher Systems, Stillwater, OK, 1990.

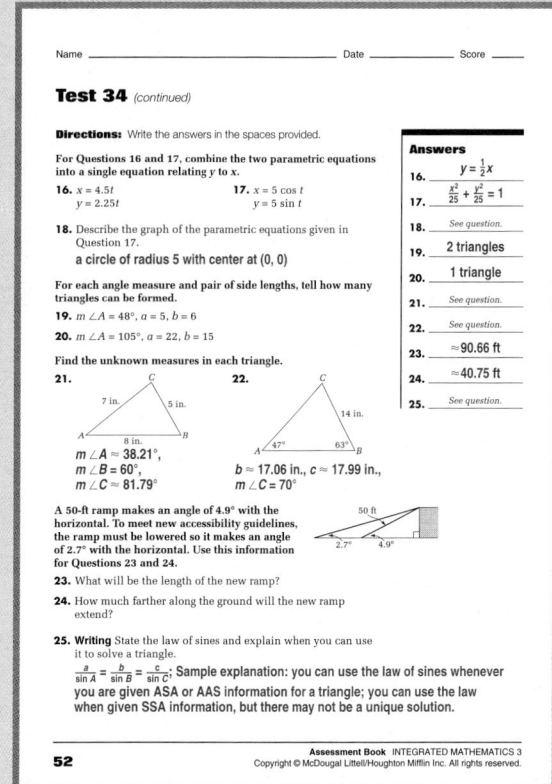

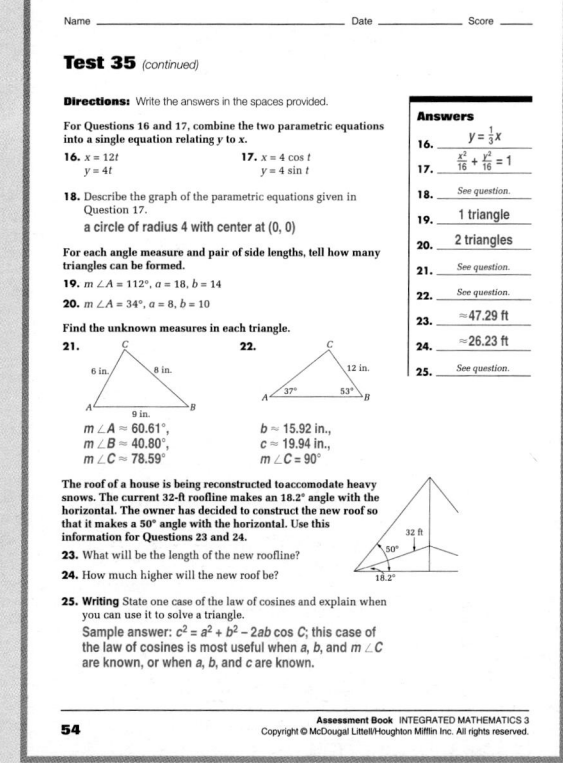

PROJECT GOALS

- Students plan a treasure hunt that can be solved with pencil and paper using a protractor and a ruler.
- Students create a treasure map, a set of numbered clues, and a solution map.
- Students' clues apply at least three mathematical topics from the unit.

PROJECT PLANNING

Materials List

- Protractor
- Ruler

Project Teams

Have students work on the project in groups of four. Groups can meet and begin by selecting a location for their treasure hunt. They should try to obtain a map of the location or make one if none is available, and then decide where to locate the buried treasure. Creating the treasure map is closely tied to the mathematics of this unit. Students should work together as a group to complete all of the "Working on the Unit Project" exercises. These exercises will guide students in using the mathematics of the unit to create their maps.

Support Materials

The ***Project Book*** contains information about the following topics for use with this Unit Project.

- Project Description
- Teaching Commentary
- Working on the Unit Project Exercises
- Completing the Unit Project
- Assessing the Unit Project
- Alternative Projects
- Outside Resources

unit **8**

Angles, Trigonometry, and Vectors

NAVIGATION

Over four thousand years ago, experienced navigators from Southeast Asia set off across the Pacific in wooden canoes. They had no charts or instruments. Navigating by the stars, the wind, the waves, the birds, and the clouds, they settled the tiny islands scattered over a triangular area covering millions of square miles of open ocean — now Polynesia.

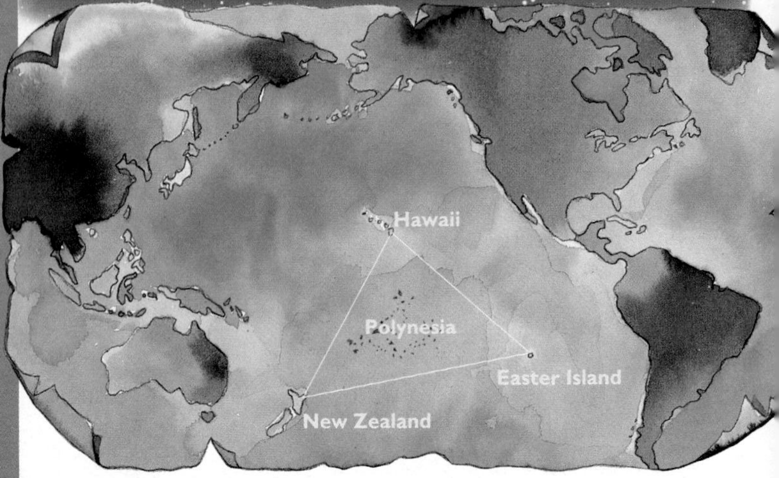

All modern navigators use maps. But before maps were drawn, they were carried in memory. The mental maps of the ancient Polynesians were based on time rather than distance and included up to 80 islands spread out over open ocean.

The people of the Marshall Islands in the Pacific made **stick-charts** to show directions of currents. The shells mark islands.

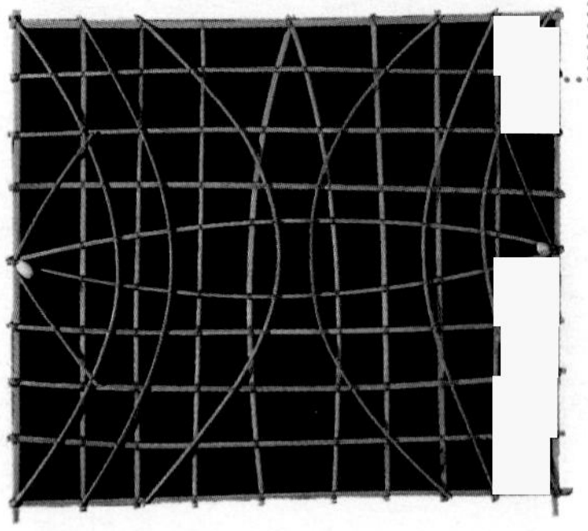

migration

The migratory flight of the **arctic tern** may be the greatest navigational feat among animals. Each year this bird travels 22,000 miles — from the North Pole to the South Pole and back again. Migrating birds, insects, bats, fishes, whales, and land animals use many of the navigational clues used by the ancient navigators.

422

Suggested Rubric for Unit Project

4 Students create a treasure map, a set of numbered clues, and a solutions map showing how the clues led to the treasure. The maps are clearly drawn and the clues apply three or more topics from the unit. The mathematics is used correctly and the clues are not too difficult to follow. The written critique of the treasure hunt by the group that tries it is very favorable.

3 Students complete all aspects of the project. There are, however, some minor mathematical problems with a few of the clues. The mathematics used in the clues contains at least one error. The group using the solutions map is able to locate the treasure and their written critique points out how improvements can be made in the solutions map.

2 Students complete their treasure map, clues, and solutions map, but the group that uses them has difficulty locating the treasure. The clues are open to more than one interpretation and at least one clue

Unit Project 8

Plan a Treasure Hunt

Your project is to plan a treasure hunt that can be solved with pencil and paper using a protractor and a ruler.

Your group should create a treasure map, a set of numbered clues, and a solution map showing how the clues lead to the treasure.

Your clues should apply at least three of the topics from the unit:

- polar coordinates
- vectors
- law of cosines and law of sines
- parametric equations

Some animals have a sixth sense referred to as *magnetic direction finding*. Bees, robins, and some fish can feel the directed lines of force of Earth's magnetic field. Lacking this sixth sense, people use a **magnetic compass** which was developed separately in ancient China and in Europe.

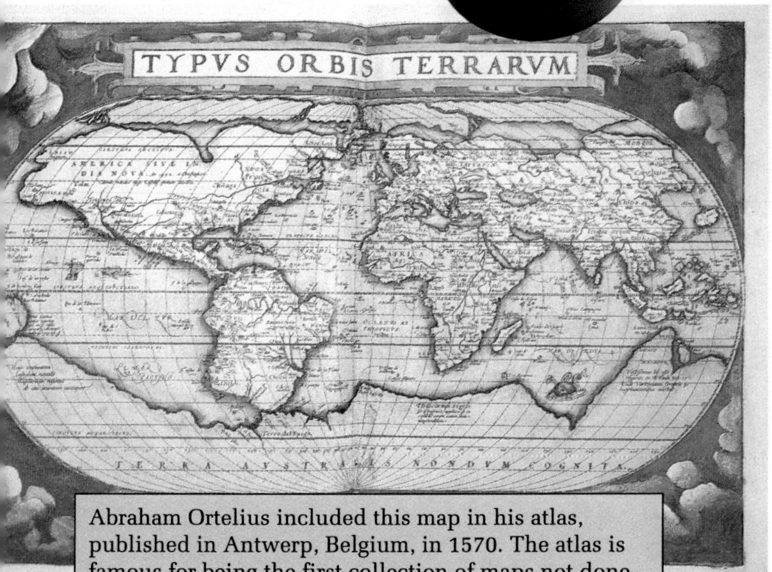

Abraham Ortelius included this map in his atlas, published in Antwerp, Belgium, in 1570. The atlas is famous for being the first collection of maps not done by the original mapmaker. He credited 87 others.

In Greenland, the native people carved driftwood to make three-dimensional maps. The earliest printed maps date back over 2000 years in China.

ADDITIONAL BACKGROUND

Multicultural Note

Not only the Chinese, but also the ancient Egyptians, Babylonians, and Hellenistic Greeks used maps to set down in usable form the geographic knowledge they acquired. Rough maps drawn by Inuit (Eskimos) helped Western explorers through the Arctic. Montagnais and Naskapi Native Americans drew maps of the Labrador Peninsula for explorer Henry Youle Hind. Montezuma, king of the Aztecs, provided Spanish conquistador Hernando Cortés with a painted map of the Gold Coast. Australian Aborigines, Tahitians, and Marshall Islanders all aided voyagers by drawing maps for them. The Incas, too, were accomplished mapmakers. In Europe, cartography was almost nonexistent during the Middle Ages but resurfaced in the 1300s. By the 1400s, maps were in wide use and were becoming more accurate, though without any indication of the Americas. After the sixteenth century, Gerhard Kremer, called Mercator, developed the first modern map projections.

Suggested Rubric for Unit Project

does not apply the mathematics correctly. The written critique is not favorable because the treasure could not be found without difficulty.

1 Students create a treasure map but do not understand how to create mathematical clues. The clues they write do not lead to the treasure because of errors and omissions. The group searching for the treasure is confused and frustrated. Students should be encouraged to speak with the teacher as soon as possible to review their work and to make a new start on the project.

Magnetic Direction Finding

The magnetic field generated by Earth, which acts as an enormous bar magnet, seems to be used by some animals for navigational purposes. Experiments have been done with honey bees that show the bees have a remarkable sensitivity to magnetic fields that they use in their navigation. A number of bird species also seem to use Earth's magnetic field to navigate. The European robin, the indigo bunting, the bobolink, pigeons, and ring-billed gulls all seem to use a magnetic compass during flight. Magnetic effects have also been observed in some fish. Biologists suspect that long-range oceanic migrators, such as salmon, tuna, and eels find their way by perceiving Earth's magnetic field directly.

ALTERNATIVE PROJECTS

Project 1, page 475

Ocean Navigation

Write a research report on ocean navigation. Find out how boat pilots use bearings and maps to stay on course by using information about currents, tides, and so on.

Project 2, page 475

Orienteering

Find out about the sport of orienteering. Create a classroom display that shows how participants use maps and magnetic compasses to plot courses between control points.

Unit Project 8

Getting Started

For this project you should work in a group of four students. Here are some ideas to help you get started.

- In your group select a location for your treasure hunt. You may want to consider using the school yard, an open field, a park, or a beach.
- Obtain a map of the location you selected, or plan to make one if none is available.
- Decide where to locate the buried treasure.

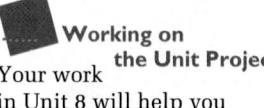

Working on the Unit Project

Your work in Unit 8 will help you plan your treasure hunt.

Related Exercises:

Section 8-1, **Exercise 39**
Section 8-2, **Exercise 60**
Section 8-3, **Exercise 24**
Section 8-4, **Exercise 26**
Section 8-5, **Exercise 39**
Section 8-6, **Exercise 32**
Section 8-7, **Exercise 38**

Alternative Projects p. 475

Can We Talk NAVIGATION

➤ There are four major ways to navigate:

(1) celestial navigation, or using the sun, the stars, and the planets

(2) dead reckoning, or keeping track of how long and how far you have traveled

(3) piloting, or using landmarks

(4) with instruments, such as a magnetic compass

Do you ever use any of these techniques? If you do, tell how you apply them in various situations.

➤ In many cultures, people learn special "tricks" for determining direction. In North America, people used the fact that tree rings are thicker on the north side. In the Tropics, people know that coconut palms lean into the prevailing wind. In northern Australia, the aboriginal people use the fact that termite mounds are oriented north-south.

Have you ever used any special "tricks" to find your way around or avoid getting lost? How do you think you can improve your sense of direction?

Answers to Can We Talk?

➤ Answers may vary. An example is given. I use piloting and instruments most often. Some landmarks I use in piloting are road signs and buildings. Some instruments I use are speedometers and odometers. I apply these techniques when traveling from one place to another in my car.

➤ Answers may vary. An example is given. I have used the position of the sun to help me understand in what direction I am traveling.

Section **8-1**

Polar Coordinates

Get Your Bearings

Focus
Use polar coordinates to locate points.

Turquoise Lake is an alpine glacial lake just outside Leadville, Colorado. To hike from the picnic area to the entrance of Baby Doe campground, you walk about 1.7 km at a *bearing* of 29°.

Leadville, Colorado, was the site of a gold and silver rush in the 1870s and 1880s.

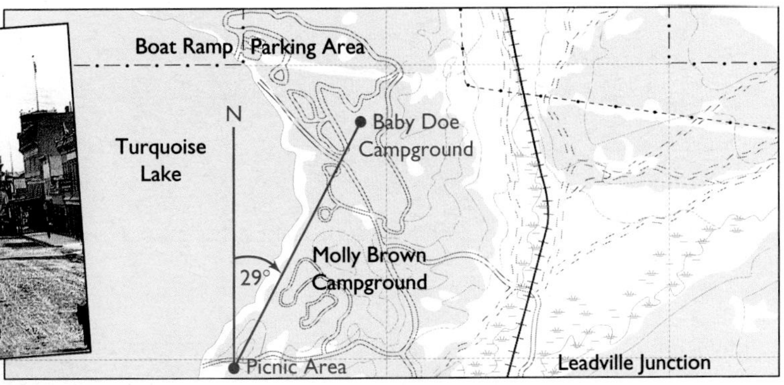

A **bearing** is an angular direction given by a magnetic compass. Bearings are used in orienteering and navigation.

In bearings, 0° is at the north and angles are measured clockwise.

From the picnic area, the bearing of Baby Doe campground is 29°.

Polar coordinates use both a distance from a reference point and an angle to describe locations. They are used in mathematics, science, and technology.

In polar coordinates, 0° is along a horizontal ray to the right of the reference point. Angles are measured counterclockwise. Polar coordinates are written as an ordered pair.

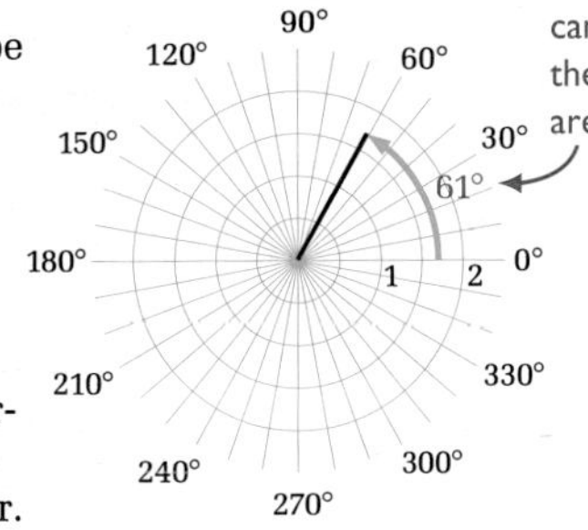

Polar coordinates of Baby Doe campground from the picnic area are (1.7, 61°).

(r, θ)

r (for radius) is the polar distance.

θ (*theta*) is the measure of the polar angle.

PLANNING

Objectives and Strands
See pages 422A and 422B.

Spiral Learning
See page 422B.

Materials List
- Polar graph paper

Recommended Pacing
Section 8-1 is a one-day lesson.

Extra Practice
See pages 621–622.

Warm-Up Exercises
- Warm-Up Transparency 8-1

Support Materials
- Practice Bank: Practice 59
- Activity Bank: Enrichment 52
- Study Guide: Section 8-1
- Problem Bank: Problem Set 17
- Explorations Lab Manual: Diagram Master 8
- Overhead Visuals: Folder 7
- Using Mac Plotter Plus Disk: Polar Graph Plotter
- Assessment Book: Quiz 8-1, Alternative Assessment 1

TEACHING

Students Acquiring English

Make sure students acquiring English are clear on the difference between clockwise and counterclockwise, since those terms are used repeatedly in the section. Consider using a clock to clarify the difference between the two.

Additional Sample

S1 For the new design of Compass Center, a city planner proposes a system of circular streets, closed to traffic, spaced one block apart but having a common center. Major public buildings would be located at locations marked with dots and letters. Straight streets which intersect the circular streets would be open to traffic.

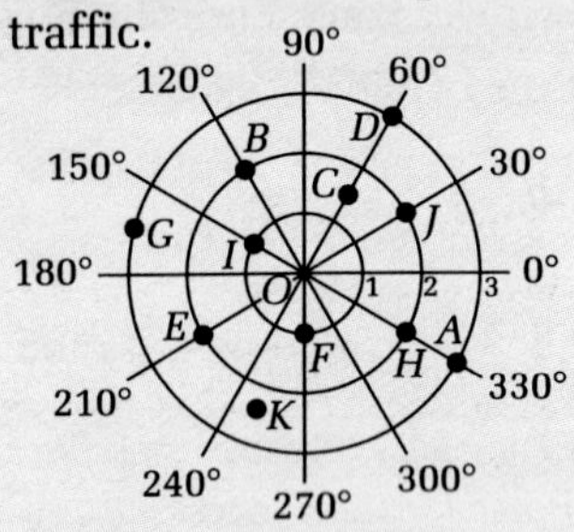

Use the map to name the site with each pair of polar coordinates.

a. (2, 210°)
From the center O, move right 2 blocks. Then move counterclockwise from 0° to 210°. The site at this location is building *E*.

b. (3, 60°)
Move right from center O a distance of 3 blocks. Then move counterclockwise from 0° to 60°. The site at this location is building *D*.

Talk it Over

Questions 1–3 lead students to find other pairs of polar coordinates for the same point by changing the angle. Students also investigate how these pairs are related to each other.

Sample 1

Use the map to name the site with each pair of polar coordinates.

a. (0.8, 175°) **b.** (0.4, 330°)

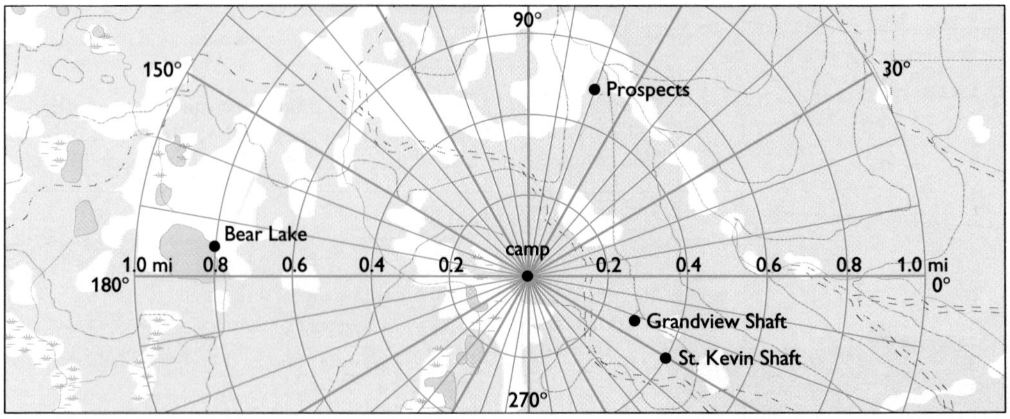

Sample Response

a. Move out 0.8 mi to the right from the campsite along the horizontal ray, and then travel counterclockwise around the circle from 0° to 175°.

The site at this location is the shore of Bear Lake.

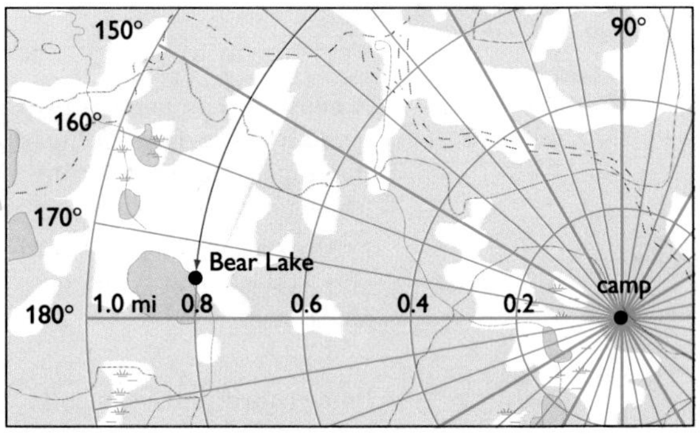

b. Move out 0.4 mi to the right from the campsite, and then travel counterclockwise around the circle to 330°.

The site at this location is St. Kevin Shaft.

Negative angles are measured clockwise.

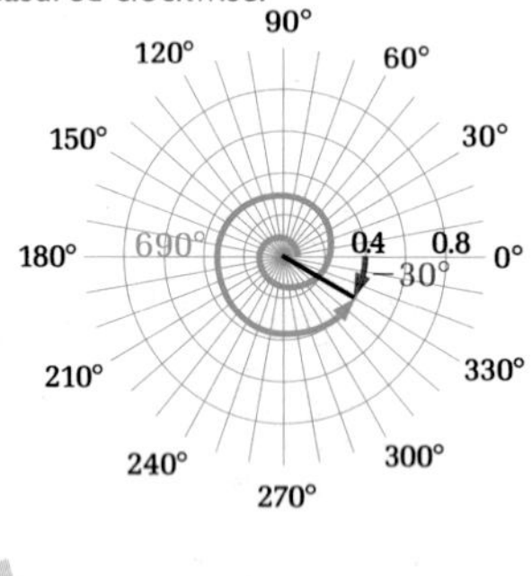

Talk it Over

When working with polar coordinates, angles do not need to have values between 0° and 360°.

1. Two other ways of giving coordinates for St. Kevin Shaft are (0.4, −30°) and (0.4, 690°). How are these coordinate pairs related to the coordinates given in Sample 1, part (b)?
2. Name two other pairs of polar coordinates that give the location of Bear Lake.
3. Describe how to find other pairs of polar coordinates for any point (r, θ).

Answers to Talk it Over

1. For each pair, $r = 0.4$ and $\theta = 330° + n(360°)$ for some integer n. For −30°, $n = -1$; for 690°, $n = 1$.
2. Answers may vary. Examples are given. (0.8, −185°); (0.8, 535°)
3. Answers may vary. An example is given. Use the same r-coordinate and find $\theta + n(360°)$ for integer values of n.
4. Answers may vary. An example is given. I think it is easier to use negative angles when the negative angles are small.
5. You can measure distances using a ruler or compass and measure angles using a protractor. You must first designate a reference point.

Sample 2

Use the map in Sample 1 to write polar coordinates for each location.

a. Prospects   b. Grandview Shaft

Sample Response

Draw a circle through each site, centered at the camp. Then use the markings of the polar grid to estimate r and θ.

a.

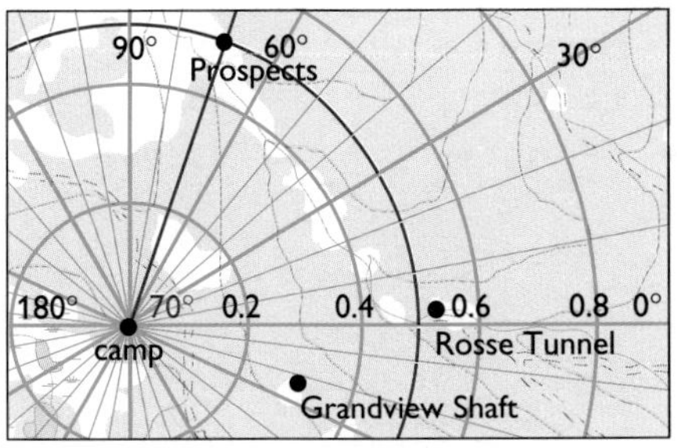

Polar coordinates for Prospects are about (0.5, 70°).

b.

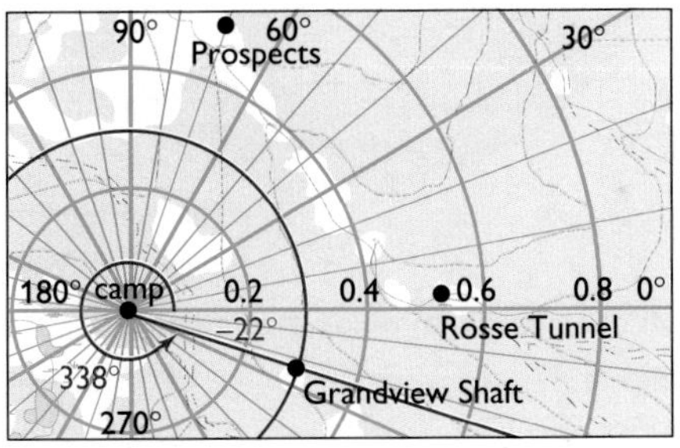

Polar coordinates for Grandview Shaft are about (0.3, 338°) or (0.3, −22°).

Talk it Over

4. Two pairs of polar coordinates are given in Sample 2, part (b). Which one do you prefer? Why?
5. How can you find polar coordinates on a map that does not contain a polar grid? What tools do you need?

Look Back

Describe some differences between a rectangular coordinate ordered pair (x, y) and a polar coordinate ordered pair (r, θ).

8-1 Exercises and Problems

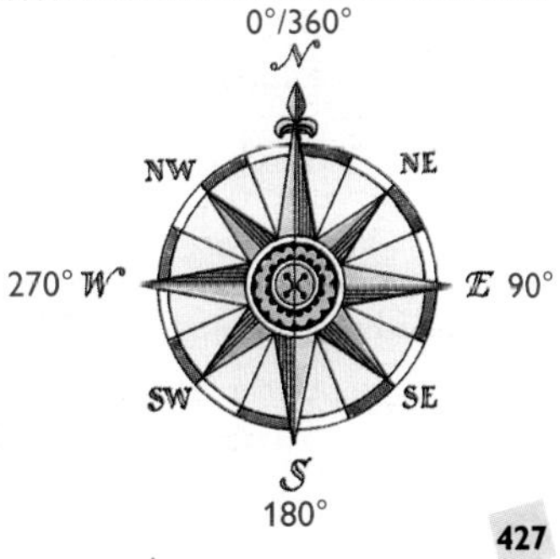

1. **Reading** Describe the similarities and differences between bearings and the angles used in polar coordinates.

For each compass direction, find the bearing and an angle used in polar coordinates.

2. west
3. south
4. northwest
5. northeast
6. southeast
7. southwest

Answers to Look Back

A rectangular coordinate ordered pair uses distances along the coordinate axes to describe locations. A polar coordinate ordered pair uses a distance from a reference point and an angle to describe locations.

Answers to Exercises and Problems

1. Both are angles measured from a fixed direction. In bearings, 0° is at the north and angles are measured clockwise. In polar coordinates, 0° is along a horizontal ray to the right of the reference point and angles are measured counterclockwise.
2. 270°; 180°
3. 180°; 270°
4. 315°; 135°
5. 45°; 45°
6. 135°; 315°
7. 225°; 225°

Additional Sample

S2 Use the diagram in Additional Sample S1 to find polar coordinates for each building.

a. building C

Draw a circle with center O and passing through C. Then estimate r and θ for the polar coordinates. Polar coordinates for C are about (1.5, 60°).

b. building K

Follow the procedure for part (a). Polar coordinates for K are about (2.4, 250°).

Reasoning

Sample 2 shows that the polar angle for a point may be positive or negative. Ask students if they think the polar distance in a pair of polar coordinates can be negative. (Yes.) Students should then consider how they might locate a point whose polar coordinates are (–5, 60°).

Visual Thinking

Ask students to create a rough sketch that shows the location of an ordered pair on a rectangular coordinate plane. Encourage them to sketch a polar coordinate grid over their rectangular grid. Then ask them to describe the same location in polar terms. This activity involves the visual skills of *correlation* and *interpretation*.

APPLYING

Suggested Assignment

Standard 2–7, 16–30, 35–39

Extended 1–33, 35–39

Integrating the Strands

Measurement Exs. 1–35, 38, 39

Geometry Exs. 1–35, 38, 39

Trigonometry Ex. 38

Statistics and Probability Exs. 36, 37

Logic and Language Exs. 1, 34, 39

Career Note

Throughout this unit, many types of maps are used. The science of making maps is known as cartography. A person who makes maps is called a cartographer. Cartographers use information from land surveyors, aerial photographs and other sources to make their maps. Increasingly, cartographers are using computers.

Interdisciplinary Problems

Many of the exercises in this section make connections to other disciplines. Exs. 8–15 enable students to see how angles can be used by astronomers to locate stars. In a connection to literature, polar coordinates are used in Exs. 22–27 to locate places on a map. In Exs. 31–34, students play a modified version of the game of chess by using a circular chessboard and polar coordinates.

Using Technology

In connection with Exs. 16–21, students can use the TI-82 to explore polar coordinates. Set the calculator to degree mode and polar mode. Press WINDOW and use these settings: θmin = 0, θmax = 360, θstep = 5. For x and y, use $-9.4 \le x \le 9.4$ and $-6.2 \le y \le 6.2$, with a scale of 1. After making these settings, move from the WINDOW menu to the FORMAT menu. Select PolarGC.

Next, press Y= and enter the equations $r_1 = 1$, $r_2 = 2$, $r_3 = 3$, …, $r_6 = 6$. Press GRAPH. The calculator will draw six concentric circles centered at the origin. Have students press TRACE and use the arrow keys to move the cursor to various locations. They should observe the polar coordinates at the bottom of the screen.

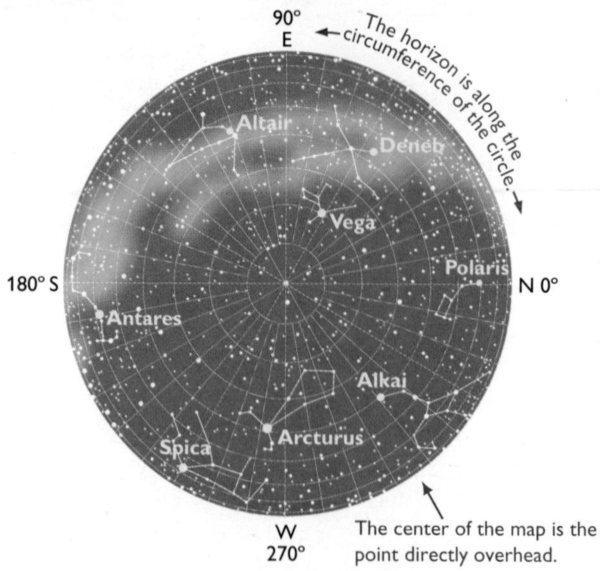

connection to SCIENCE

Astronomers sometimes locate a star by measuring its *azimuth*, the angle along the horizon from North. The star map shows the night sky in Shanghai, China, on June 5, 1996.

Estimate the azimuth of each star to the nearest five degrees.

8. Vega **9.** Polaris

10. Deneb **11.** Altair

Name a star with each azimuth.

12. 240° **13.** 190°

14. 310° **15.** 263°

Graph each point on polar graph paper.

16. $A(3, 45°)$ **17.** $B(2, -42°)$ **18.** $C(2.5, 790°)$

19. $D(2, -1000°)$ **20.** $E(3.2, 250°)$ **21.** $F(2.4, 640°)$

connection to LITERATURE

J. R. R. Tolkien created a world of fantasy in his trilogy *The Lord of the Rings.* The story takes place in an invented land called "Middle Earth."

Name the place in Middle Earth with each pair of polar coordinates.

22. (280, 138°)

23. (640, 29°)

24. (740, −32°)

Estimate polar coordinates for each place in Middle Earth.

25. Weather Hills

26. Gladden Fields

27. Dead Marshes

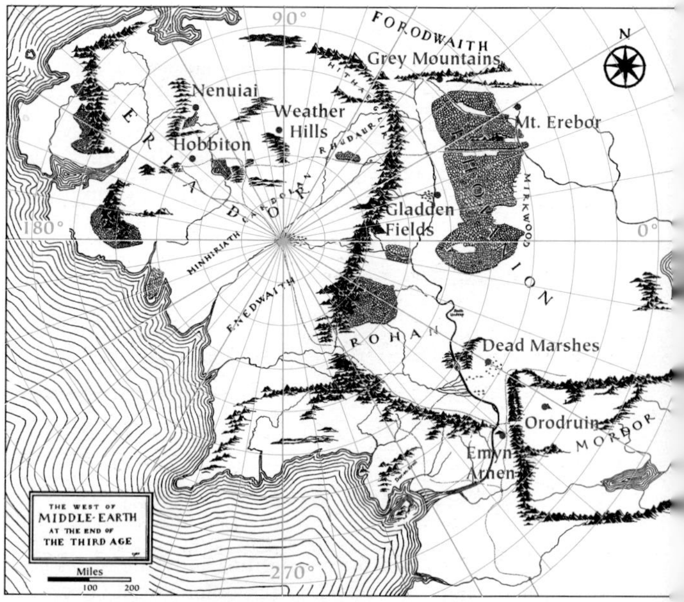

Open-ended **Draw an example of each regular polygon centered at the origin and write polar coordinates for its vertices.**

28. equilateral triangle **29.** regular pentagon **30.** regular octagon

Answers to Exercises and Problems

8. 60°
9. 0°
10. 55°
11. 110°
12. Spica
13. Antares
14. Alkai
15. Arcturus

16–21.

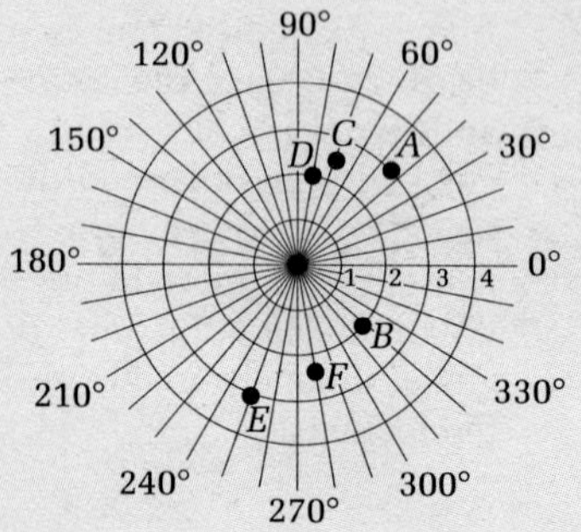

22. Hobbiton **23.** Mt. Erebor
24. Orodruin
25–27. Estimates may vary.
25. (about 260, about 92°)
26. (about 390, about 18°)
27. (about 560, about −30°)
28–30. Answers may vary. Examples are given.
28. $(k, 90°)$, $(k, 210°)$, $(k, 330°)$ for any positive k

Chess Each square on a standard chessboard is labeled by its *file* (a–h) and *rank* (1–8). The player with the white pieces is called "White," and the player with the black pieces is called "Black." White's queen is at d1.

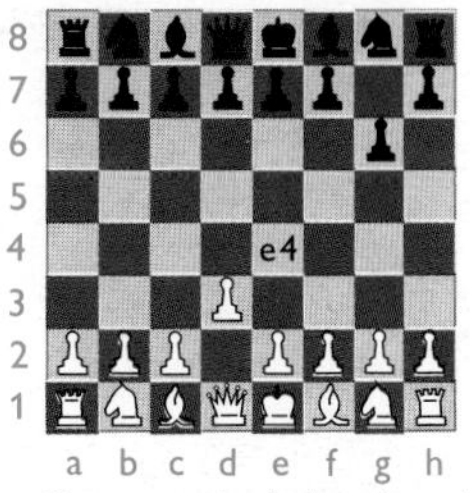

On a standard chessboard, a bishop can move along a diagonal to an edge.

Use the circular chessboard shown. The ranks are wrapped into circles. White's queen is at (1, 112.5°).

31. White starts a game by moving a pawn from d2 to d3. Write polar coordinates of the piece before and after the move.

32. Black responds by moving a pawn from g7 to g6. Write polar coordinates of the piece before and after the move.

33. Black announces "Check" because the bishop at f8 threatens White's king at e1. Write polar coordinates of each position along the indicated path from the bishop to the king.

34. a. **Research** Find out how pieces move in a standard chess game. Describe how they should move on the circular chessboard.

   b. Find at least three ways for White to block Black's bishop by moving a piece into its path. Write polar coordinates of the pieces before and after the moves. Which move do you think is best?

   c. **Group Activity** Work with another student. Play a game of chess on a circular chessboard. At the end of the game, write the polar coordinates of the remaining pieces.

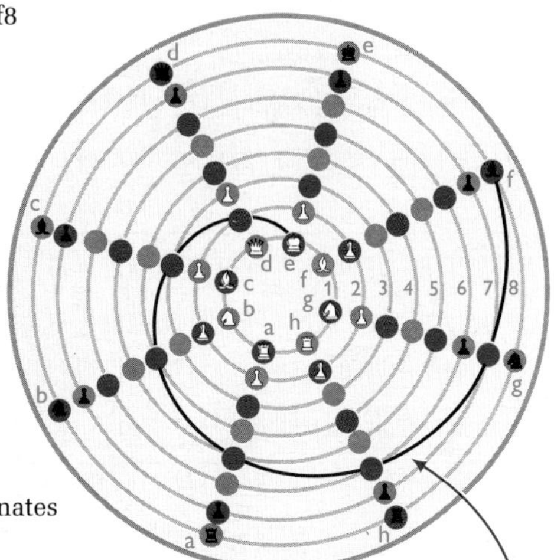

A bishop can move along a spiral until it reaches the inner or outer circle.

Ongoing ASSESSMENT

35. **Open-ended** Create a map of a real or an imaginary place. Use polar coordinates to give references to at least four locations.

Review PREVIEW

36. What is the probability of getting 6 or more correct answers when you guess randomly on a quiz of 10 true-or-false questions? *(Section 7-6)*

37. A die is rolled. Find the probability of the event "less than 2 or greater than 4." *(Section 7-2)*

38. Explain how to find the cosine and sine of an acute angle in a right triangle by measuring the sides of the triangle. *(Toolbox Skill 36)*

8 Working on the Unit Project

39. Choose a starting point on your treasure-hunt map and write a clue that uses polar coordinates to describe the location of the next clue.

The θ-coordinate that students see at the bottom of the screen will be positive because the window setting was for $0 \le \theta \le 360$. This is easy to change. Have students try $-180 \le \theta \le 180$. Next, have them try $-360 \le \theta \le 720$.

Cooperative Learning

If there are any chess players in the class, you may wish to organize small groups around each player so he or she can demonstrate how the game of chess is played on a standard chessboard. Then students can pair up to play a game on a circular chessboard, as is suggested in Ex. 34(c).

Practice 59 For use with Section 8-1

Name ______________ Date ______________

Practice 59

For use with Section 8-1

Exs. 1–12: Sample angle measures are given.
Give a pair of polar coordinates for each labeled point.

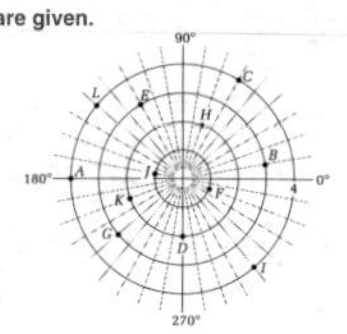

1. A (4, 180°) 2. B (3, 10°)
3. C (4, 60°) 4. D (2, 270°)
5. E (3, 120°) 6. F (1, –20°)
7. G (3, –140°) 8. H (2, 70°)
9. I (4, –50°) 10. J (1, 170°)
11. K (2, –160°) 12. L (4, 140°)

Graph each point on polar graph paper.

13. (3, 90°) 14. (2, –10°) 15. (4, 140°)
16. (1.5, 210°) 17. (2.5, –45°) 18. (3.2, –60°)
19. (2, 280°) 20. (3, 225°) 21. (4, 165°)

Exs. 13–21: For graphs, see additional answers.

Give the bearing of each direction. Note that "north-northwest" means a direction halfway between north and northwest.

22. north-northeast 22.5° 23. west-northwest 292.5° 24. east-northeast 67.5°
25. south-southeast 157.5° 26. north-northwest 337.5° 27. west-southwest 247.5°

Imagine a polar coordinate system drawn on the face of a clock, with the origin at the pivot point of the hands. Suppose the hour hand is 3 in. long and the minute hand is 5 in. long. Find polar coordinates for the outer tips of each of the two hands at each time. (Use inches as units.)

28. three o'clock (3, 0°); (5, 90°) 29. five o'clock (3, –60°); (5, 90°) 30. one-thirty (3, 45°); (5, –90°)
31. seven-thirty (3, –135°); (5, –90°) 32. nine-twenty (3, 170°); (5, –30°) 33. ten-forty (3, 130°); (5, –150°)

Practice Bank, INTEGRATED MATHEMATICS 3

59

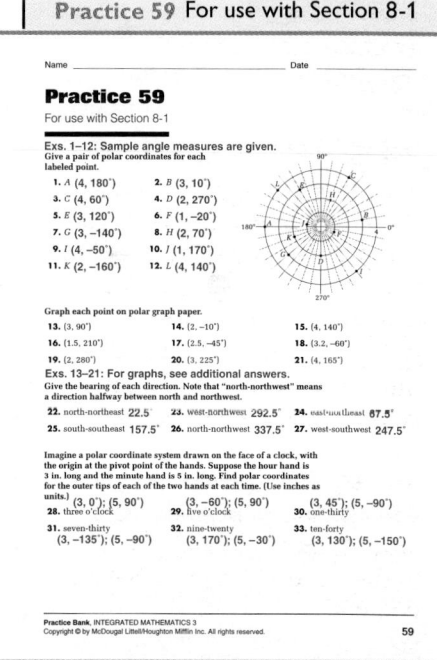

Answers to Exercises and Problems

29. (*k*, 90°); (*k*, 162°); (*k*, 234°); (*k*, 306°); (*k*, 378°) for any positive *k*

30. (*k*, 0°); (*k*, 45°); (*k*, 90°); (*k*, 135°); (*k*, 180°); (*k*, 225°); (*k*, 270°); (*k*, 315°) for any positive *k*

31. (2, 112.5°); (3, 112.5°)

32. (7, 337.5°); (6, 337.5°)

33. (8, 22.5°); (7, 337.5°); (6, 292.5°); (5, 247.5°); (4, 202.5°); (3, 157.5°); (2, 112.5°); (1, 67.5°)

34. See answers in back of book.

35. Answers may vary. Check students' work.

36. about 0.38

37. $\frac{1}{2} = 0.5$

38. Measure the length of the hypotenuse. To find the cosine of an acute angle, measure the length of the leg adjacent to the angle and divide by the length of the hypotenuse. To find the sine of the acute angle, measure the length of the leg opposite the angle and divide by the length of the hypotenuse.

39. Answers may vary. Check students' work.

PLANNING

Objectives and Strands
See pages 422A and 422B.

Spiral Learning
See page 422B.

Materials List
- Calculator
- Polar graph paper
- Protractor

Recommended Pacing

Section 8-2 is a three-day lesson.

Day 1

Pages 430–431: Opening paragraph through Talk it Over 2, *Exercises 1–11*

Day 2

Pages 432–433: Exploration through Talk it Over 5, *Exercises 12–41*

Day 3

Pages 434–435: Converting Rectangular Coordinates through Look Back, *Exercises 42–60*

Extra Practice
See pages 621–622.

Warm-Up Exercises
- Warm-Up Transparency 8-2

Support Materials
- Practice Bank: Practice 60
- Activity Bank: Enrichment 53
- Study Guide: Section 8-2
- Problem Bank: Problem Set 17
- Explorations Lab Manual: Diagram Master 8
- Assessment Book: Quiz 8-2, Alternative Assessment 2

Section **8-2**

Converting Coordinates

One Way Or Another

Focus

Convert from polar to rectangular coordinates, and vice versa; extend the definitions of cosine, sine, and tangent; and use a Pythagorean identity.

A call to a hospital in Michigan says that help is needed at an accident at the intersection of Seven Mile Road and Farmington. The hospital sends two response teams, one by helicopter and one by ambulance.

The polar coordinates and rectangular coordinates locate the same point.

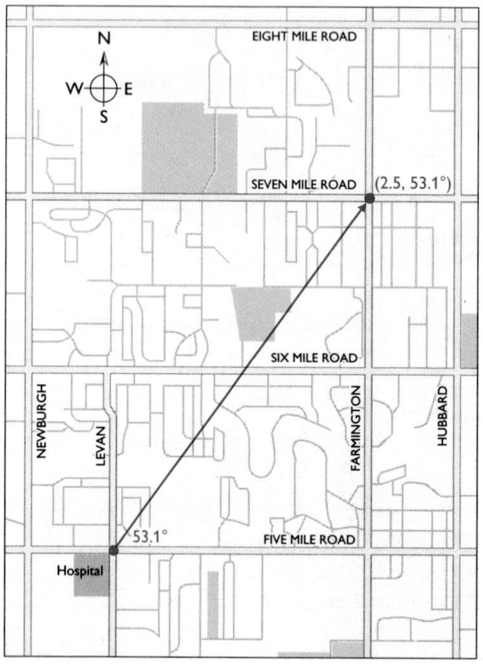

The helicopter flies 2.5 mi at an angle of about 53.1° north of east.

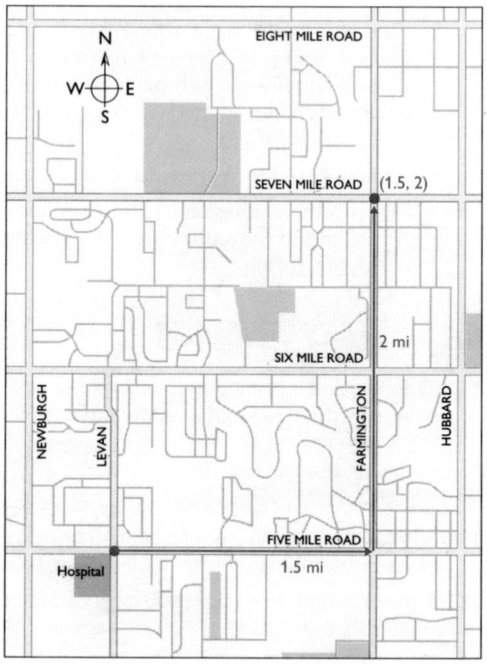

The ambulance travels 1.5 mi east on Five Mile Road, then 2 mi north on Farmington.

Both response teams arrive at the accident site, but they use different coordinate systems to get there.

Converting Polar Coordinates to Rectangular Coordinates

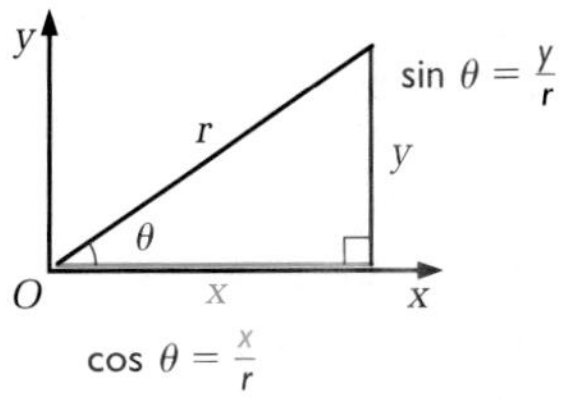

The coordinate system you have used most often is called a *rectangular coordinate system*. You can use what you know about trigonometry to convert polar coordinates in the first quadrant to rectangular coordinates.

$$x = r\cos\theta \qquad y = r\sin\theta$$

$$(x, y) = (r\cos\theta, r\sin\theta)$$

Sample 1

Convert the polar coordinates (38, 82°) to rectangular coordinates. Round decimal answers to the nearest hundredth.

Sample Response

Method 1 Use a formula.

$x = r\cos\theta$

$= 38\cos 82°$ ← Substitute **38** for *r* and **82°** for *θ*.

≈ 5.289

$y = r\sin\theta$

$= 38\sin 82°$ ← Substitute **38** for *r* and **82°** for *θ*.

≈ 37.630

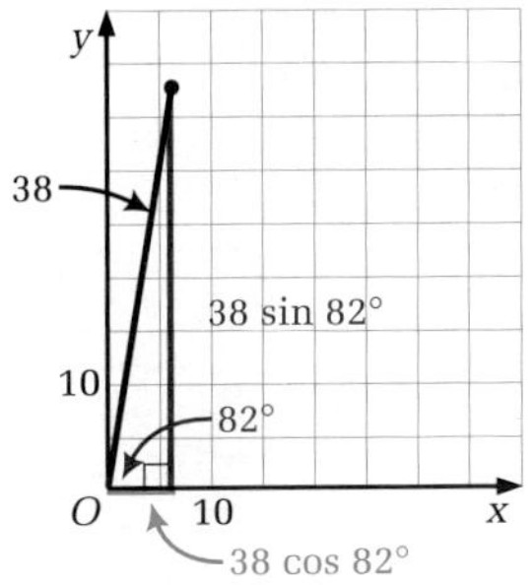

The rectangular coordinates for the point are about (5.29, 37.63).

Method 2 Use a calculator that converts polar coordinates to rectangular coordinates.

Be sure the calculator is in degree mode.

The rectangular coordinates for the point are about (5.29, 37.63).

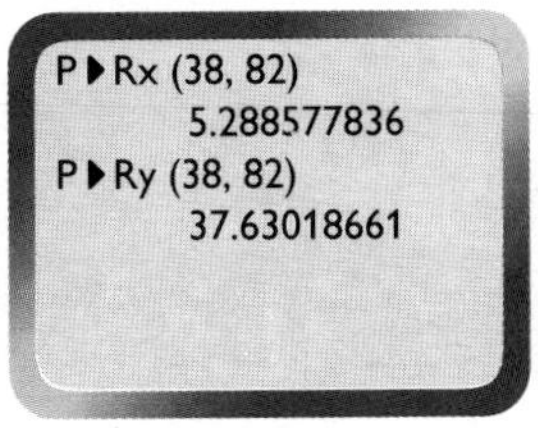

Talk it Over

1. The accident location described on page 430 has polar coordinates (2.5, 53.1°). Show that its rectangular coordinates are about (1.5, 2).
2. a. How are the points with polar coordinates (38, 82°), (38, 442°), and (38, −278°) related?
   b. Convert (38, 442°) and (38, −278°) to rectangular coordinates.

Now you are ready for: Exs. 1–11 on p. 436

TEACHING

Using Technology

The TI-81 and TI-82 use different procedures for converting between rectangular and polar coordinates. On the TI-82, press [2nd] [ANGLE]. Items 5 and 6 of the ANGLE menu are used to convert from rectangular to polar coordinates. Items 7 and 8 are for converting from polar to rectangular. On the TI-81, press [MATH]. Item 1 on the MATH menu is used to convert from rectangular to polar. Select item 1 and type the rest of the coordinate pair on the home screen, then press [ENTER] to get the *r*-coordinate. Once this coordinate is displayed, press [ALPHA] [θ] [ENTER] to get the *θ*-coordinate. Use an analogous procedure for converting from polar to rectangular. Use menu item 2 to get the *x*-coordinate. Once that is displayed, press [ALPHA] [Y] [ENTER] to get the *y*-coordinate.

Additional Sample

S1 Convert the polar coordinates (47, 55°) to rectangular coordinates. Round decimal answers to the nearest hundredth.

Method 1. **Use a formula.**

$x = r\cos\theta$

$= 47\cos 55$

≈ 26.96

$y = r\cos\theta$

$= 47\sin 55$

≈ 38.50

The rectangular coordinates for the point are about (26.96, 38.50).

Method 2. **Use a calculator. The rectangular coordinates for the point are about (26.96, 38.50).**

Talk it Over

Question 2 will help students to realize that there is not a unique set of polar coordinates for a given pair of rectangular coordinates.

Answers to Talk it Over

1. $r = 2.5$ and $\theta = 53.1°$; $x = r\cos\theta \approx 2.5(0.6) = 1.5$ and $y = r\sin\theta \approx 2.5(0.8) = 2$
2. a. The three pairs of polar coordinates describe the same point.
   b. (5.3, 37.6); (5.3, 37.6)

Exploration

The goal of the Exploration is to have students understand that the trigonometric ratios can be defined for angles of any measure. Also, students will discover some properties of these ratios that will help them to evaluate sines, cosines, and tangents of nonacute angles.

Reasoning

Some students may notice that the definitions of cos θ, sin θ, and tan θ all involve x and y. Challenge students to write a relationship involving the three trigonometric ratios based on this observation.

EXPLORATION

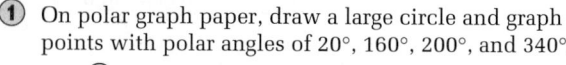

What are the cosine, sine, and tangent of angles less than 0° or greater than 90°?

- **Materials: polar graph paper, scientific calculators**
- **Work with another student.**

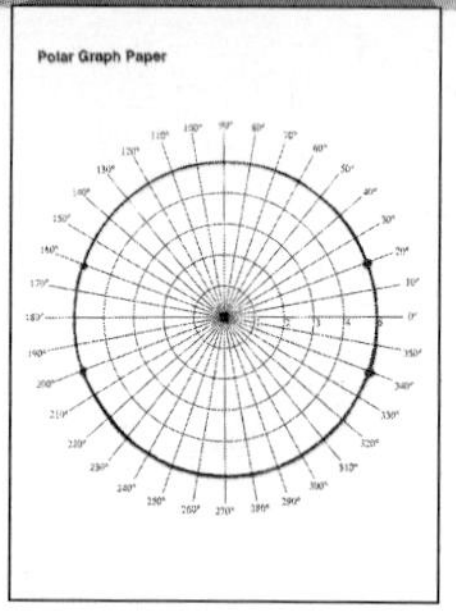

1. On polar graph paper, draw a large circle and graph points with polar angles of 20°, 160°, 200°, and 340°.
2. Describe the symmetry in your diagram.
3. Use a calculator to find the cosine of each angle. What do you notice?
4. Use a calculator to find the sine of each angle. What do you notice?
5. Repeat steps 1–4 for angles of 42°, 138°, −42°, and −138°.
6. Make and test a conjecture about the cosines of angles like 42° and −42° whose measures are opposites.
7. Test a conjecture about the sines of supplementary angles.
8. In which quadrants is the cosine of an angle positive? negative? In which quadrants is the sine of an angle positive? negative?
9. In which quadrants is the tangent of an angle positive? negative?

Expanding the Definitions of Cosine, Sine, and Tangent

The Exploration suggests that you can define the trigonometric ratios for angles of any measure.

COSINE, SINE, AND TANGENT OF ANY ANGLE

The trigonometric ratios are defined as

$$\cos\theta = \frac{x}{r} \qquad \sin\theta = \frac{y}{r} \qquad \tan\theta = \frac{y}{x} \quad (\text{when } x \neq 0)$$

where

- x and y represent the x- and y-coordinates of a point with polar coordinates (r, θ).
- x and y can be positive, negative, or zero (except as indicated).
- r is positive.

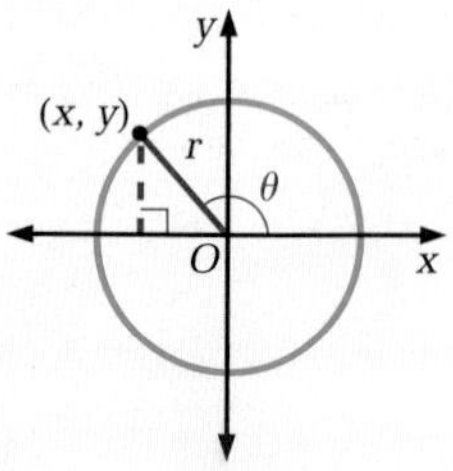

Answers to Exploration

1.

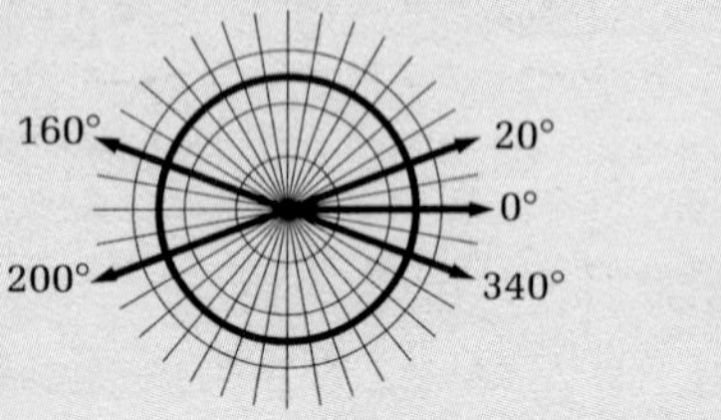

2. All four points are 20° away from the horizontal line through the reference point.
3. cos 20° = 0.9397; cos 160° = −0.9397; cos 200° = −0.9397; cos 340° = 0.9397; cos 340° = cos 20° = −cos 200° = −cos 160°
4. sin 20° = 0.3420; sin 160° = 0.3420; sin 200° = −0.3420; sin 340° = −0.3420; sin 160° = sin 20° = −sin 200° = −sin 340°
5. step 1:

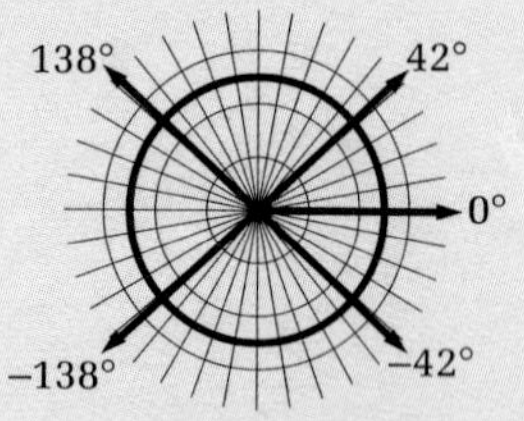

step 2: All four points are 42° away from the horizontal line through the reference point.

step 3: cos 42° = 0.7431; cos 138° = −0.7431; cos (−42°) = 0.7431; cos (−138°) = −0.7431; cos 42° = cos (−138°) = −cos (−42°) = −cos 138°

step 4: sin 42° = 0.6691; sin 138° = 0.6691; sin (−42°) = −0.6691; sin (−138°) = −0.6691; sin 42° = sin 138° = −sin (−42°) = −sin (−138°)

6. cos θ = cos $(-\theta)$
7. sin θ = sin $(180 - \theta)$
8. I and IV; II and III; I and II; III and IV
9. I and III; II and IV

To find the measure of an angle when you know its sine, cosine, or tangent, you can use the *inverse sine* ($\sin^{-1}$), *inverse cosine* ($\cos^{-1}$), or *inverse tangent* feature of a graphics calculator. Entering $\sin^{-1} 0.3420$ into a calculator is like asking, "What is an angle whose sine is 0.3420?"

You saw in the Exploration that there are *two* angles between 0° and 360° whose sine is 0.3420. When you use the *inverse sine* feature on a calculator, the calculator gives only 20° as the answer. You need to use what you know about symmetry to find the other angle, 160°.

Sample 2

a. Find all angles θ between 0° and 360° for which $\sin \theta = 0.4000$.

b. Find all angles θ between 0° and 360° for which $\cos \theta = -0.2567$.

Sample Response

a. $\sin^{-1} 0.4000 \approx 23.58°$ ← Use SIN⁻¹.

This angle is in Quadrant I.

There is also an angle in Quadrant II whose sine is 0.4000.

By symmetry, the other angle is its supplement, about 180° − 23.58°, or about 156.42°.

Two angles with a sine of 0.4000 are about 23.58° and about 156.42°.

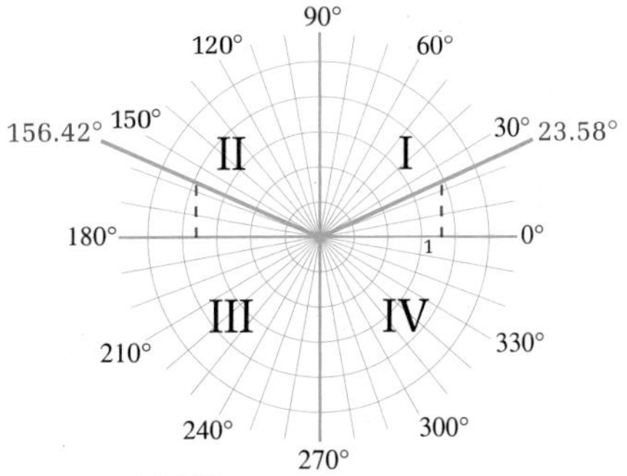

b. $\cos^{-1}(-0.2567) \approx 104.87°$ ← Use COS⁻¹.

This angle is in Quadrant II.

There is also an angle in Quadrant III whose cosine is −0.2567.

By symmetry, this angle has the opposite measure, about −104.87°, or about 255.13°.

Two angles with a cosine of −0.2567 are about 104.87° and about 255.13°.

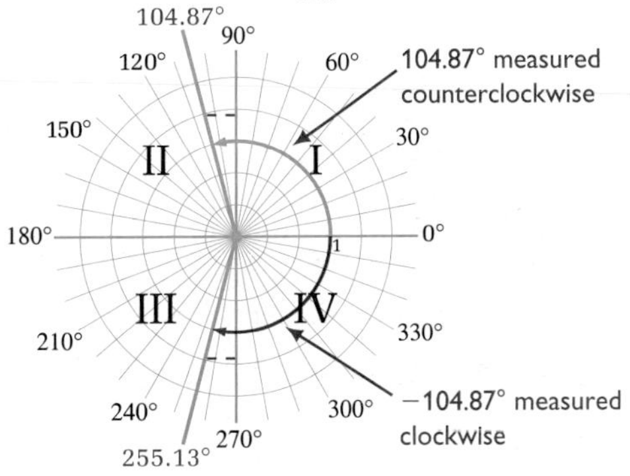

Talk it Over

3. Find all angles θ between 0° and 360° for which $\cos \theta = 0$.
4. Find all angles θ between 0° and 360° for which $\sin \theta = 0$.
5. Find all angles θ between 0° and 360° for which $\tan \theta = -2$.

Now you are ready for: Exs. 12–41 on pp. 436–438

Communication: Discussion

A discussion of how to use the inverse trigonometric feature of a graphics calculator may be worthwhile for some students. They will most likely have questions about the notation and why the term *inverse* is used.

Additional Sample

S2 a. Find all angles θ between 0° and 360° for which $\sin \theta = -0.5000$.

$\sin^{-1}(-0.5000) = -30°$ This angle is in Quadrant IV and corresponds to a positive angle of 330°. By symmetry, there is a positive angle in Quadrant III for which $\sin \theta = -0.5000$ and it is 180° + 30°, or 210°. Two angles with a sine of −0.5000 are 210° and 330°.

b. Find all angles θ between 0° and 360° for which $\cos \theta = 0.5222$.

$\cos^{-1} 0.5222 \approx 58.52°$ This angle is in Quadrant I. By symmetry, there is an angle in Quadrant IV whose cosine is 0.5222. This angle is 360° − 58.52°, or 301.48°. Two angles with a cosine of 0.5222 are 58.52° and 301.48°.

Problem Solving

Observing symmetry in mathematical situations may be the simplest and most direct way to answer certain questions. Sample 2 is a good illustration of this fact.

Answers to Talk it Over

3. 90°; 270°
4. 0°; 180°; 360°
5. 116.6°; 296.6°

Additional Sample

S3 Convert the rectangular coordinates (2, –3) to polar coordinates, where θ is an angle between 0° and 360°.

Method 1. **Use a formula.**

Step 1. **Find *r*. Use the Pythagorean theorem.**

$r^2 = x^2 + y^2$

$r = \sqrt{x^2 + y^2}$

$= \sqrt{(2)^2 + (-3)^2}$

$= \sqrt{13}$

≈ 3.606

Step 2. **Find the polar angle θ. Use a calculator to find $\cos^{-1}\left(\frac{x}{r}\right)$.**

$\cos^{-1}\left(\frac{x}{r}\right) = \cos^{-1}\left(\frac{2}{\sqrt{31}}\right)$

$\approx 56.310°$

The polar angle 56.310° has the correct cosine value, but it lies in Quadrant I. The point *A*(2, –3) is in Quadrant IV. By symmetry, the polar angle for *A*(2, –3) is about –56.310°. The positive angle with the same cosine value is 360° – 56.310°, or about 303.69°. Polar coordinates for (2, –3) are about (3.606, 303.69°).

Method 2. **Use a calculator that converts rectangular coordinates to polar coordinates.**

R ▶ Pr (2, –3)
3.605551275
R ▶ Pθ (2, –3)
–56.30993247

The positive angle with the same cosine value is 360° – 56.310°, or about 303.69°. Polar coordinates for (2, –3) are about (3.606, 303.69°).

Integrating the Strands

Converting rectangular coordinates to polar coordinates using Method 1 of the Sample 3 Response integrates the strands of algebra, functions, geometry, and trigonometry. Students use algebra to find *r*. The angle is found using an inverse trigonometric function together with coordinate geometry.

Converting Rectangular Coordinates to Polar Coordinates

To convert from rectangular coordinates to polar coordinates, you can use the Pythagorean theorem and the inverse trigonometric features of a calculator.

Sample 3

Convert the rectangular coordinates (−3, −7) to polar coordinates, where θ is an angle between 0° and 360°.

Sample Response

Method 1 Use a formula.

1 Find *r*.

$r^2 = x^2 + y^2$ ← Use the Pythagorean theorem.

$r = \sqrt{x^2 + y^2}$

$= \sqrt{(-3)^2 + (-7)^2}$

$= \sqrt{58}$

≈ 7.616

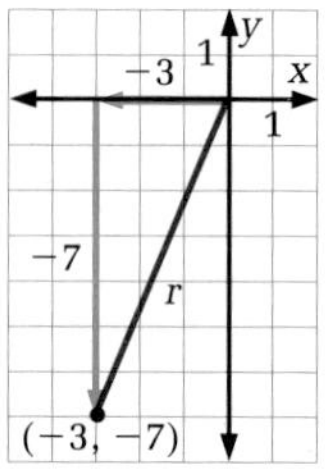

2 Find the polar angle θ.

You can use the inverse trigonometric feature of a calculator.

$\cos^{-1}\frac{x}{r} = \cos^{-1}\frac{-3}{7.616}$ ← Substitute −3 for *x* and **7.616** for **r**.

$= \cos^{-1}(-0.3939)$

$\approx 113.20°$

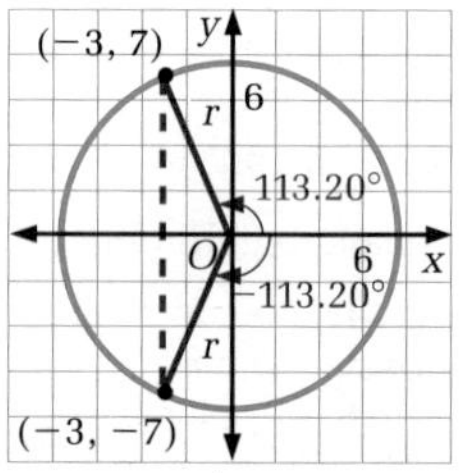

The polar angle 113.20° has the correct cosine value, but it lies in Quadrant II. The angle with the opposite measure, −113.20°, lies in Quadrant III and has the same cosine value.

The positive angle with the same cosine value is 360° − 113.20°, or 246.80°.

Polar coordinates for (−3, −7) are about (7.62, 246.80°).

Method 2 Use a calculator that converts rectangular coordinates to polar coordinates.

The positive angle with the same cosine value is 360° − 113.20°, or 246.80°.

R▶Pr (−3, −7)
7.615773106
R▶Pθ (−3, −7)
−113.1985905

Polar coordinates for (−3, −7) are about (7.62, 246.80°).

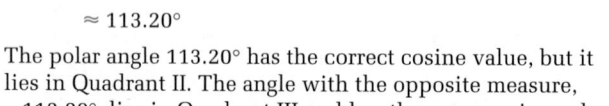

Answers to Talk it Over

6. $\sin^{-1}\frac{y}{r} = \sin^{-1}\frac{-7}{7.616} =$ $\sin^{-1}(-0.9191) = -66.8°$; The polar angle in Quadrant III with this sine value is 246.8°.

7. $\tan^{-1}\frac{y}{x} = \tan^{-1}\frac{-7}{-3} =$ $\tan^{-1}(2.333) = 66.8°$; The polar angle in Quadrant III with this tangent value is 246.8°.

8. Method 1: Use the Pythagorean theorem to find *r* and use the inverse trigonometric feature of a calculator to find θ. $r^2 = \sqrt{5^2 + (-8)^2}$; $r \approx 9.434$; $\cos^{-1}\frac{x}{r} = \cos^{-1}\frac{5}{9.434} \approx 58°$. The polar angle 57.99° has the correct cosine value but is in the first quadrant. The polar angle in Quadrant IV with this cosine is 302.01°. The polar coordinates of (5, –8) are about (9.43, 302.01). Method 2: Use a graphics calculator.

Watch Out! When you use the inverse trigonometric functions, the calculator may not give the polar angle that is in the same quadrant as the given point.

Talk it Over

6. How can you find the polar angle of $(-3, -7)$ using $\sin^{-1}\frac{y}{r}$?
7. How can you find the polar angle of $(-3, -7)$ using $\tan^{-1}\frac{y}{x}$?
8. Explain how to convert $(5, -8)$ to polar coordinates.

A Pythagorean Identity

Applying the Pythagorean theorem to the right triangle shown leads to an important relationship between $\cos\theta$ and $\sin\theta$ for any angle θ:

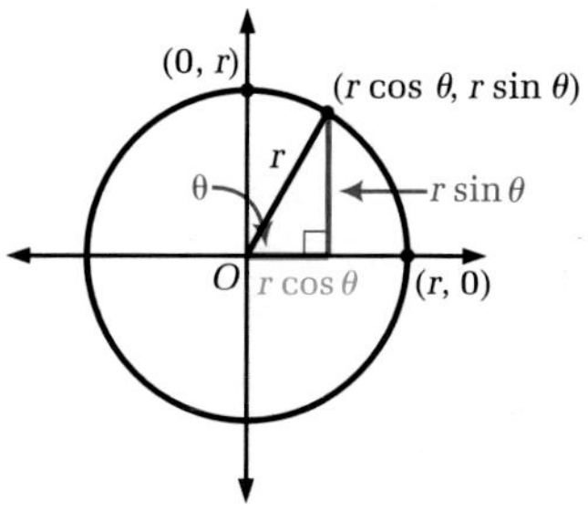

$(r\cos\theta)^2 + (r\sin\theta)^2 = r^2$ ← Use the Pythagorean theorem.

$r^2(\cos\theta)^2 + r^2(\sin\theta)^2 = r^2$ ← Use the fact that $(ab)^2 = a^2b^2$.

$\cos^2\theta + \sin^2\theta = 1$ ← Divide both sides by r^2.

"$(\cos\theta)^2$" is written "$\cos^2\theta$."

This result is sometimes called a **Pythagorean identity.** It shows how $\sin\theta$ and $\cos\theta$ are related and allows you to find one when the other is known.

Sample 4

Find $\cos\theta$ when $\sin\theta = -0.2955$ and θ is in Quadrant III.

Sample Response

$\cos^2\theta + \sin^2\theta = 1$ ← Use this Pythagorean identity.

$\cos^2\theta + (-0.2955)^2 = 1$ ← Substitute -0.2955 for $\sin\theta$.

$\cos\theta = \pm\sqrt{1-(-0.2955)^2}$ ← Solve for $\cos\theta$.

$\cos\theta \approx \pm 0.9553$ ← Simplify.

In Quadrant III, $\cos\theta$ is negative, so $\cos\theta \approx -0.9553$.

Look Back

Make a graph or chart showing where $\sin\theta$, $\cos\theta$, and $\tan\theta$ are positive, negative, zero, or undefined.

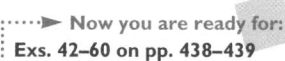

Now you are ready for: Exs. 42–60 on pp. 438–439

Talk it Over

Questions 6 and 7 lead students to understand that polar angles can be found by using any of the three trigonometric ratios.

Teaching Tip

If the radius of the circle shown on this page is 1, then the x- and y-coordinates of the point on the circle are $\cos\theta$ and $\sin\theta$. The equation of a circle of radius 1 with center at the origin is $x^2 + y^2 = 1$. Hence, $\cos^2\theta + \sin^2\theta = 1$.

Additional Sample

S4 Find $\sin\theta$ when $\cos\theta = 0.2308$ and θ is in Quadrant IV.

Use the Pythagorean identity $(\cos\theta)^2 + (\sin\theta)^2 = 1$.

$\cos^2\theta + \sin^2\theta = 1$

$(0.2308)^2 + \sin^2\theta = 1$

$\sin\theta = \pm\sqrt{1-(0.2308)^2}$

$\sin\theta \approx \pm 0.9730$

In Quadrant IV, $\sin\theta$ is negative, so $\sin\theta \approx -0.9730$.

Cooperative Learning

You may wish to have students work in groups of four to answer this Look Back. Each student could answer the question for one quadrant of the coordinate plane and then explain his or her graph or chart to the group.

Answers to Look Back

	Quadrant				Angle measure			
	I	II	III	IV	0°, 360°	90°	180°	270°
cosine	+	−	−	+	+	0	−	0
sine	+	+	−	−	0	+	0	−
tangent	+	−	+	−	0	not defined	0	not defined

APPLYING

Suggested Assignment

Day 1

Standard 1–3, 5–10

Extended 1–11

Day 2

Standard 12–16, 22–41

Extended 12–41

Day 3

Standard 42–55, 57–60

Extended 42–55, 57–60

Integrating the Strands

Algebra Exs. 11, 13, 42–55, 58–60

Functions Ex. 56

Measurement Exs. 2, 4, 9, 11

Geometry Exs. 1–12, 14–51, 57, 60

Trigonometry Exs. 1, 3–8, 10–12, 14–56, 60

Discrete Mathematics Ex. 58

Logic and Language Exs. 1, 13–25, 56, 60

Communication: Discussion

The reading, writing, and open-ended exercises on this page provide excellent opportunities for students to discuss their results verbally. You may wish to have some students explain their answers at the board.

8-2 Exercises and Problems

1. **Reading** Explain why the equations $x = r\cos\theta$ and $y = r\sin\theta$ can be used to find rectangular coordinates when you know polar coordinates of a point.

For Exercises 2–4, use the map on page 430.

2. Suppose a helicopter from the hospital flies 2.24 mi at an angle of 63.4° north of east to reach the site of another emergency. What is the nearest intersection?
3. Convert (2.24, 63.4°) to rectangular coordinates. Describe how an ambulance can drive to the site of the emergency.
4. a. **Open-ended** Choose another location somewhere northeast of the hospital. Use a protractor and a ruler to estimate polar coordinates of the location using the hospital as the origin. (Assume that the distance between Five Mile Road and Six Mile Road is one mile.)
   b. Convert the polar coordinates from part (a) to rectangular coordinates.

Convert each pair of polar coordinates to rectangular coordinates. Round decimal answers to the nearest hundredth.

5. (1, 0°) 6. (1, 30°) 7. (1, 45°) 8. (1, 60°)

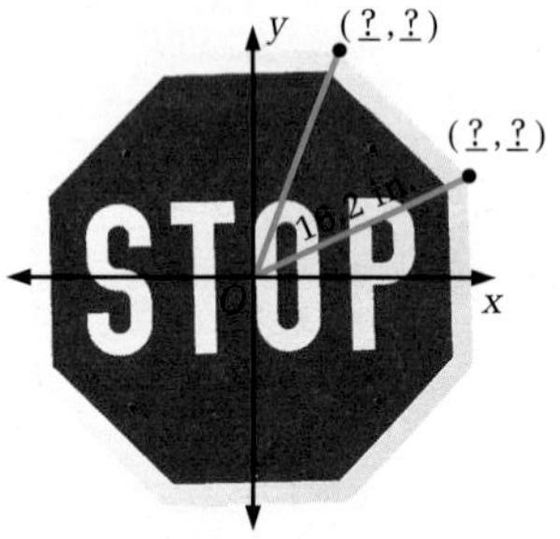

For Exercises 9 and 10, use this stop sign.

9. Two vertices of the stop sign lie in Quadrant I. What are their polar coordinates?
10. Convert the polar coordinates of the vertices in Quadrant I to rectangular coordinates.
11. **Open-ended** On polar graph paper, draw a polygon in Quadrant I with at least four vertices.
    a. Write polar coordinates for each vertex.
    b. Convert the polar coordinates you wrote in part (a) to rectangular coordinates.
    c. Find the lengths of all sides of the polygon.
12. a. Polar coordinates for the vertices of a rectangle are (6, 30°), (6, 150°), (6, −150°), (6, −30°). What are the rectangular coordinates of the vertices?
    b. **Open-ended** Use polar coordinates to describe the vertices of a rectangle. What are the rectangular coordinates of the vertices?
13. **Writing** Explain why $\tan\theta = \dfrac{\sin\theta}{\cos\theta}$.

Answers to Exercises and Problems

1. If a perpendicular is constructed from the point with polar coordinates (r, θ) to the x-axis and a line is drawn from the origin to the point, a right triangle is formed with acute angle θ and hypotenuse r. The horizontal leg has length x and the vertical leg has length y, so $\cos\theta = \frac{x}{r}$ and $\sin\theta = \frac{y}{r}$. Then $x = r\cos\theta$ and $y = r\sin\theta$.
2. Answers may vary. An example is given. The nearest street intersection is 7 blocks west of the site of the first emergency.
3. (1, 2); Drive 1.5 mi east on Five Mile Rd., 2 mi north on Farmington, and then 0.5 mi west on Seven Mile Rd.
4. Answers may vary. Check students' work.
5. (1, 0) 6. (0.87, 0.5)
7. (0.71, 0.71) 8. (0.5, 0.87)
9. (16.2, 22.5°) and (16.2, 67.5°)
10. (14.97, 6.20) and (6.20, 14.97)
11. Answers may vary. An example is given, with answers rounded to the nearest tenth.

    a.

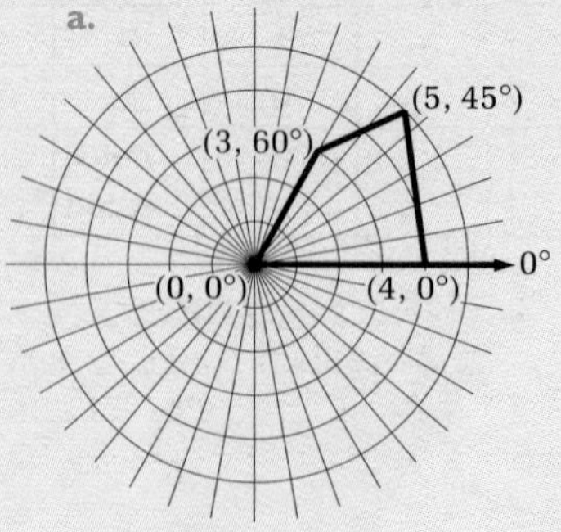

    b. (0, 0), (4, 0), (3.5, 3.5), (1.5, 2.6)
    c. 4, 3.5, 2.2, 3
12. a. (5.20, 3), (−5.20, 3), (−5.20, −3), (5.20, −3)
    b. Answers may vary. An example is given for a rectangle in the first quadrant with one vertex at the origin and sides of lengths a and b.

connection to GEOGRAPHY

At the South Pole, every direction is north. To talk about locations, scientists at the United States' Amundsen-Scott station near the South Pole sometimes use a coordinate system in which the Greenwich Meridian is called "north."

A polar grid with 0° at "east" is shown on the map of Antarctica.

Convert the polar coordinates of each scientific research station to rectangular coordinates. Distances are in miles. Describe the location of each station in words.

14. Syowa (Japan) (1440, 50°)

15. Belgrano II (Argentina) (840, 125°)

16. Scott Base (New Zealand) (840, 283°)

17. Vostok (Russia) (790, −17°)

18. Mawson (Australia) (1540, 27°)

19. Troll (Norway) (1240, 88°)

20. Byrd Surface Camp (U.S.) (690, 209°)

21. Dumont D'Urville (France) (1600, −50°)

Example

The Bay of Whales is at about (820, 254°). Its rectangular coordinates are about (820 cos 254°, 820 sin 254°), or about (−230, −790).

The Bay of Whales is about 230 mi "west" and about 790 mi "south" of the South Pole.

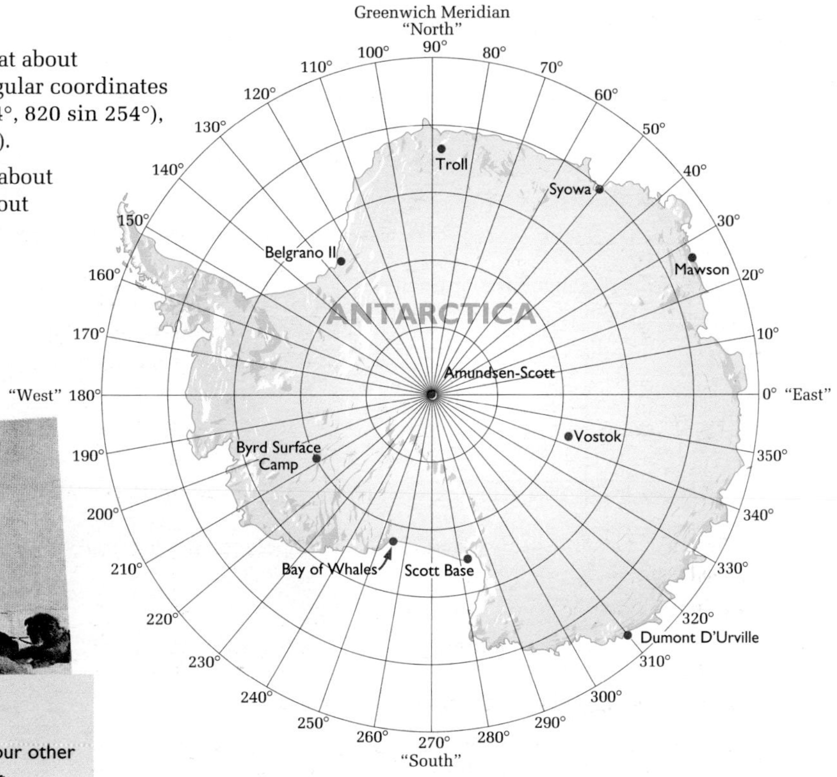

BY THE WAY...

Roald Amundsen and four other Norwegians began the first successful trip to the South Pole from the Bay of Whales in 1911. Their round-trip journey took 99 days.

Application

Positions or locations on Earth can be determied using *latitude,* north and south of the equator, and *longitude,* east and west of the prime (first) meridian. As Earth rotates, the sun crosses each meridian once each day. When this happens, all locations on the meridian have the same time. Thus, when it is noon at one meridian, it is midnight on the opposite side of Earth and a new day is beginning. The Greenwich meridian passes through Greenwich, England.

Answers to Exercises and Problems

The polar coordinates are (0, 0°), (*a*, 0°), $\left(\sqrt{a^2 + b^2}, \tan^{-1}\frac{b}{a}\right)$, and (*b*, 90°). The rectangular coordinates are (0, 0), (*a*, 0), (*a*, *b*), and (0, *b*).

13. $\tan\theta = \frac{y}{x} = \frac{y/r}{x/r} = \frac{\sin\theta}{\cos\theta}$

14. about (926, 1103); Syowa is about 926 mi "east" and 1103 mi "north" of the South Pole.

15. about (−482, 688); Belgrano II is about 482 mi "west" and 688 mi "north" of the South Pole.

16. about (189, −818); Scott Base is about 189 mi "east" and 818 mi "south" of the South Pole.

17. about (755, −231); Vostok is about 755 mi "east" and 231 mi "south" of the South Pole.

18. about (1372, 699); Mawson is about 1372 mi "east" and 699 mi "north" of the South Pole.

19. about (43, 1239); Troll is about 43 mi "east" and 1239 mi "north" of the South Pole.

20. about (−603, −335); Byrd Surface Camp is about 603 mi "west" and 335 mi "south" of the South Pole.

21. about (1028, −1226); Dumont D'Urville is about 1028 mi "east" and 1226 mi "south" of the South Pole.

Error Analysis

If students' work for Exs. 26–41 contains errors, it may be because they are trying to supply answers without thinking carefully about symmetry. Drawing a sketch will help students in their thinking.

Complete each statement with *opposite* or *the same*.

22. Supplementary angles have _?_ sines.

23. Supplementary angles have _?_ cosines.

24. Angles with opposite measures have _?_ sines.

25. Angles with opposite measures have _?_ cosines.

For each angle, find another angle between 0° and 360° with the same cosine.

26. 60° **27.** 145° **28.** 200° **29.** −10°

For each angle, find another angle between 0° and 360° with the same sine.

30. 100° **31.** −15° **32.** 260° **33.** 25°

For each angle, find another angle between 0° and 360° with the same tangent.

34. 45° **35.** 170° **36.** −150° **37.** −80°

Find all angles θ between 0° and 360° that fit each condition.

38. $\sin\theta = 0.5$ **39.** $\cos\theta = -0.89$ **40.** $\cos\theta = 0.25$ **41.** $\sin\theta = -0.75$

Convert each pair of rectangular coordinates to polar coordinates, where θ is an angle between 0° and 360°.

42. (12, −5) **43.** (4, 4) **44.** (−7.5, −7.5)

45. (−27.8, 3) **46.** (0, 7) **47.** (−4.3, 0)

48. (4, −3) **49.** $(-1, \sqrt{3})$ **50.** $(4\sqrt{3}, -4)$

51. Convert the rectangular coordinates of the pentagon to polar coordinates.

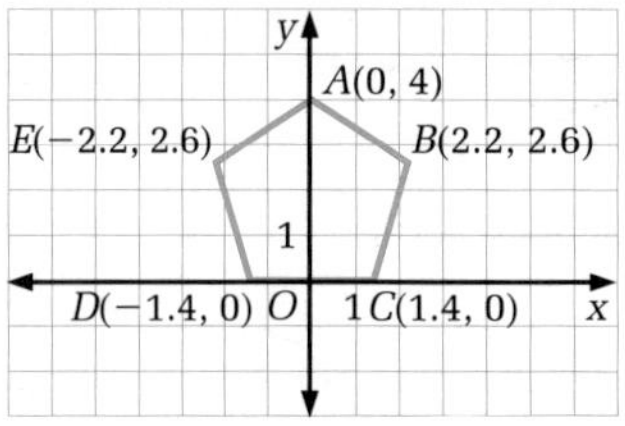

Use the Pythagorean identity $\cos^2\theta + \sin^2\theta = 1$.

52. Find all possible values of $\sin\theta$ when $\cos\theta = \frac{2}{3}$.

53. Find $\cos\theta$ when $\sin\theta = \frac{3}{5}$ and θ is in Quadrant II.

54. Find $\sin\theta$ when $\sin\theta < 0$ and $\cos\theta = -\frac{5}{13}$.

55. Find $\cos\theta$ when $\cos\theta < 0$ and $\sin\theta = 0.2121$.

Answers to Exercises and Problems

22. the same
23. opposite
24. opposite
25. the same
26. 300°
27. 215°
28. 160°
29. 10°, 350°
30. 80°
31. 195°, 345
32. 280°
33. 155°
34. 225°
35. 350°
36. 30°, 210°
37. 100°, 280°
38. 30°, 150°
39. 152.87°, 207.13°
40. 75.52°, 284.48°
41. 311.41°, 228.59°
42. (13, −22.62°)
43. (5.7, 45°)
44. (10.6, −135°)
45. (28.0, 173.84°)
46. (7, 90°)
47. (4.3, 180°)
48. (5, −36.87°)

Ongoing ASSESSMENT

Group Activity Work in a group of four students.

56. a. Copy and complete the table for all the angles θ between 0° and 360° whose measures are multiples of 10.
    b. Graph the ordered pairs.
    c. Connect the points. Describe the shape.
    d. Does the graph represent a function?
    e. Describe any connections between this activity and the Exploration on page 432.

$m\angle\theta$	$\cos\theta = x$	$\sin\theta = y$	(x, y)
0°	1	0	(1, 0)
10°	?	?	?
20°	?	?	?
⋮	⋮	⋮	⋮
350°	?	?	?
360°	?	?	?

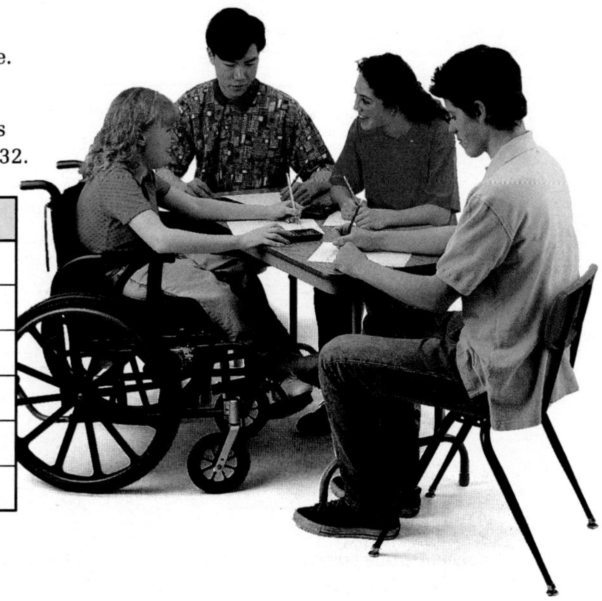

Review PREVIEW

57. Polar coordinates of two vertices of an equilateral triangle are (0, 0°) and (2, 60°). Find polar coordinates of all the points that can be the third vertex. *(Section 8-1)*

58. Write the first six terms of the sequence defined by the recursive formula $a_1 = 2000$, $a_n = 1.07a_{n-1} + 2000$. *(Section 4-3)*

59. Find the distance from (−1, 1) to (−4, 5). *(Toolbox Skill 31)*

8 Working on the Unit Project

60. Use polar coordinates to locate the next spot on your map. Write a clue that uses mathematics from this section.

To find the next clue, travel the same distance as you would if you went 40 m north and 22 m east. The polar angle you should use has a sine of −0.2 and lies in Quadrant III.

Cooperative Learning

Part (e) of Ex. 56 asks students to make connections between this activity and the Exploration on page 432. You may wish to form the groups of four here by pairing two of the pairs of students who worked together on the Exploration.

Working on the Unit Project

You may wish to suggest that each member of the project group write one clue using as many concepts from this section as possible. Then the group could meet and choose the best clue to use for the project.

Practice 60 For use with Section 8-2

Name ______________________ Date ______________

Practice 60

For use with Section 8-2

9. a. 280° b. 100° c. 260°
10. a. 235° b. 55° c. 305°
11. a. 14° b. 194° c. 166°
12. a. 140° b. 320° c. 40°
13. a. 165° b. 345° c. 15°
14. a. 110° b. 290° c. 70°
15. a. 338° b. 158° c. 202°
16. a. 60° b. 240° c. 120°

Convert the polar coordinates to rectangular coordinates. Give answers to the nearest hundredth.

1. (2, 20°) (1.88, 0.68)
2. (3, 65°) (1.27, 2.72)
3. (1, 135°) (−0.71, 0.71)
4. (2, 210°) (−1.73, −1)
5. (5.2, −40°) (3.98, −3.34)
6. (4.5, 160°) (−4.23, 1.54)
7. (2.4, 78°) (0.50, 2.35)
8. (7.6, 285°) (1.97, −7.34)

For each given angle, find (a) another angle between 0° and 360° with the same cosine, (b) another angle between 0° and 360° with the same sine, and (c) another angle between 0° and 360° with the same tangent.

9. 80°
10. 125°
11. −14°
12. 220°
13. 195°
14. −110°
15. 22°
16. 300°

Exs. 9–16: See top of page.

Find all angles θ between 0° and 360° that fit each condition. Round answers to the nearest tenth of degree.

17. $\cos\theta = 0.22$ 77.3°, 282.7°
18. $\sin\theta = 0.58$ 35.5°, 144.5°
19. $\sin\theta = -0.62$ 321.7°, 218.3°
20. $\cos\theta = -0.87$ 150.5°, 209.5°
21. $\sin\theta = 0.45$ 26.7°, 153.3°
22. $\cos\theta = -0.18$ 100.4°, 259.6°
23. $\cos\theta = 0.73$ 43.1°, 316.9°
24. $\sin\theta = -0.55$ 326.6°, 213.4°

Convert each set of rectangular coordinates to polar coordinates. Round coordinates to the nearest tenth. Sample angle measures are given.

25. (−6, 8) (10, 126.9°)
26. (0, −4) (4, −90°)
27. (−5, −5) (7.1, −135°)
28. (3, −3) (4.2, −45°)
29. $(-\sqrt{3}, 1)$ (2, 150°)
30. (8, −15) (17, −61.9°)
31. (2.4, 1.8) (3, 36.9°)
32. $(-7\sqrt{2}, 7\sqrt{2})$ (14, 135°)
33. (−4.2, −3.6) (5.5, −139.4°)
34. (5.7, −0.6) (5.7, −6.0°)
35. $(-9, 9\sqrt{3})$ (18, 120°)
36. (−3.5, −10.5) (11.1, −108.4°)

Use the Pythagorean identity $\cos^2\theta + \sin^2\theta = 1$. Express answers to the nearest hundredth.

37. Find the possible values of $\cos\theta$ if $\sin\theta = \frac{1}{3}$. ±0.94
38. Find $\sin\theta$ if θ is in Quadrant IV and $\cos\theta = \frac{7}{25}$. −0.96
39. Find $\sin\theta$ if $\sin\theta > 0$ and $\cos\theta = -0.26$. 0.97
40. ***Open-ended*** Suppose you were designing a theater. Describe a design for which it would be best to use polar coordinates to designate the seat locations and a design for which rectangular coordinates would be best. Give reasons for your choices. Answers may vary. Check students' work.

60 **Practice Bank,** INTEGRATED MATHEMATICS 3

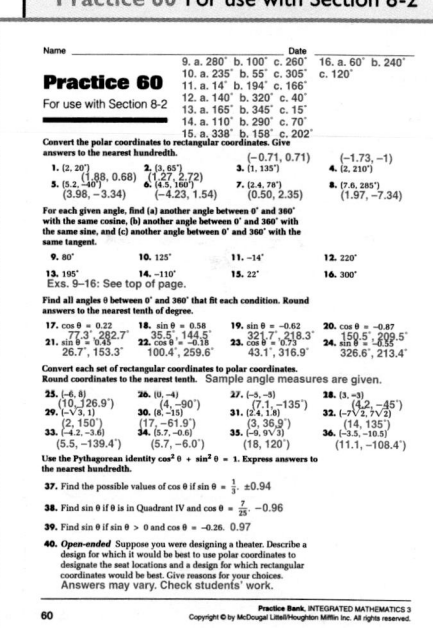

Answers to Exercises and Problems

49. (2, 120°)
50. (8, −30°)
51. $C(1.4, 0°)$, $B(3.4, 49.8°)$, $A(0, 90°)$, $E(3.4, 130.2°)$, $D(1.4, 180°)$
52. $\frac{\sqrt{5}}{3}, -\frac{\sqrt{5}}{3}$
53. $-\frac{4}{5}$
54. $-\frac{12}{13}$
55. −0.9772
56. See answers in back of book.
57. (2, 0°) or (2, 120°)
58. 2000, 4140, 6429.8, 8879.886, 11,501.47802, 14,306.58148
59. 5
60. Answers may vary.

PLANNING

Objectives and Strands
See pages 422A and 422B.

Spiral Learning
See page 422B.

Materials List
- Graph paper
- Protractor
- Ruler
- Calculator

Recommended Pacing
Section 8-3 is a one-day lesson.

Extra Practice
See pages 621–622.

Warm-Up Exercises
- Warm-Up Transparency 8-1

Support Materials
- Practice Bank: Practice 61
- Activity Bank: Enrichment 54
- Study Guide: Section 8-3
- Problem Bank: Problem Set 17
- Explorations Lab Manual: Diagram Master 2, 17
- Overhead Visuals: Folder 8
- Assessment Book: Quiz 8-3, Alternative Assessment 3

Section **8-3**

The Geometry of Vectors

Focus
Use drawings of vectors to solve problems.

Straight and *Arrow*

EXPLORATION

How can you use distances and angles to travel from place to place?

- **Materials: rulers, protractors, graph paper**
- **Work with another student.**

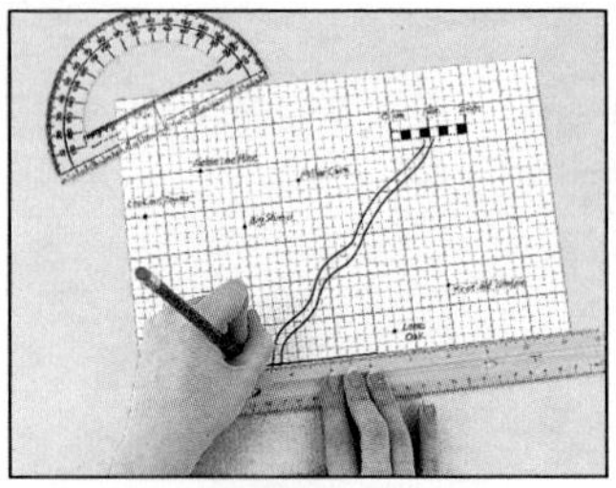

Draw a horizontal line through the origin, or starting point.

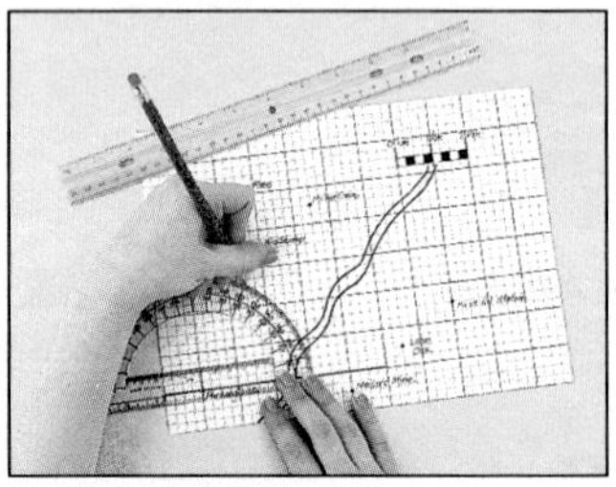

Measure a 60° angle as on a polar coordinate system.

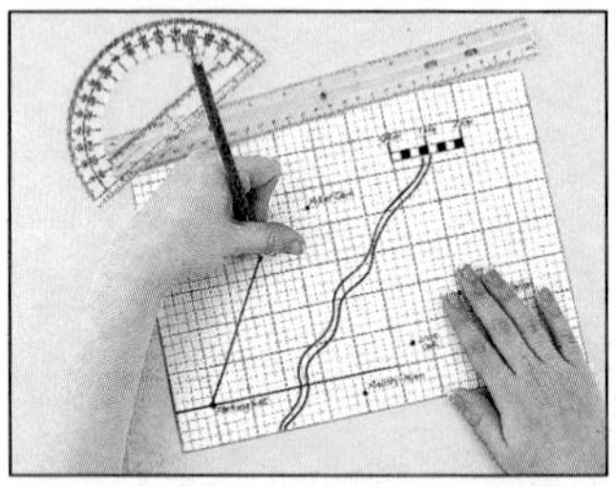

Draw a point 4.0 km from the starting point.

① Make one copy of the map on graph paper.

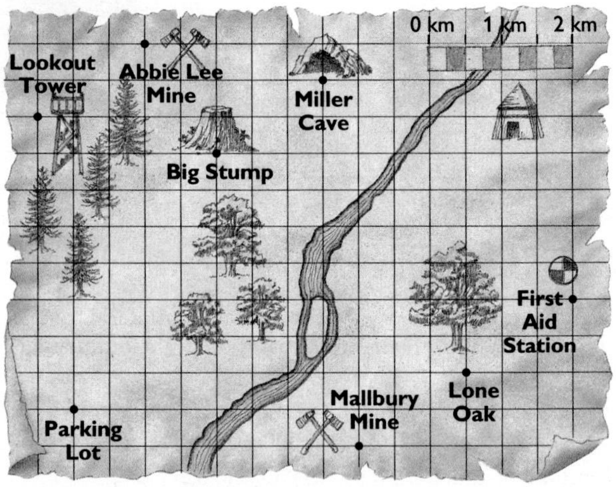

② Take turns following the directions below. Use each point along the way as a new origin.
- Begin at the parking lot and travel 4.0 km at an angle of 60°. (The diagrams at the left show how to complete this instruction.)
- From this point travel 4.6 km at −41°.
- Now travel 1.8 km at 34°.
- Finally, travel 4.6 km at 139°.

③ You should now be at Miller Cave. If you are not, go back and check each other's work.

④ Suppose you walk in a straight line from the parking lot to Miller Cave. About how far will you walk and at what angle?

Answers to Exploration

1–3. Check students' work.
4. about 5.8 km at 52°

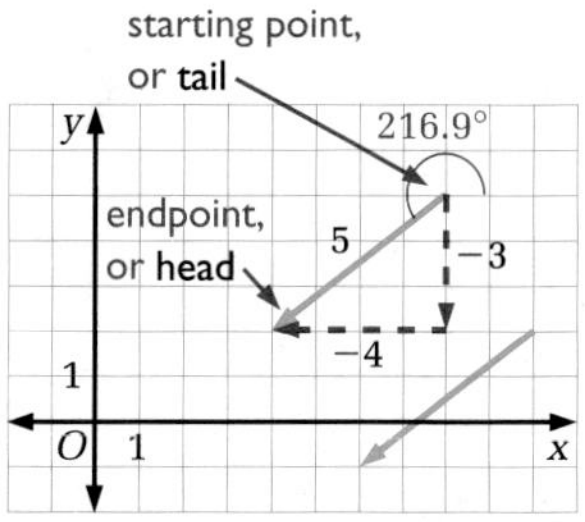

The instructions that guided your moves in the Exploration, such as "4.0 km at 60°," are examples of vectors. A **vector** is a quantity that has both a **magnitude,** or size, and a direction.

Vectors are commonly named with a bold letter, such as **v.** Here are some ways to represent a vector **v**:

- Draw an arrow, as shown in the graph. Each blue arrow represents the same vector, since each describes the same magnitude and direction. You can draw a vector anywhere in a plane.
- Use polar form: $\mathbf{v} = (5, 216.9°)$
- Use rectangular form: $\mathbf{v} = (-4, -3)$

Talk it Over

1. Discuss the difference between the two uses of $(-4, -3)$ in this diagram.

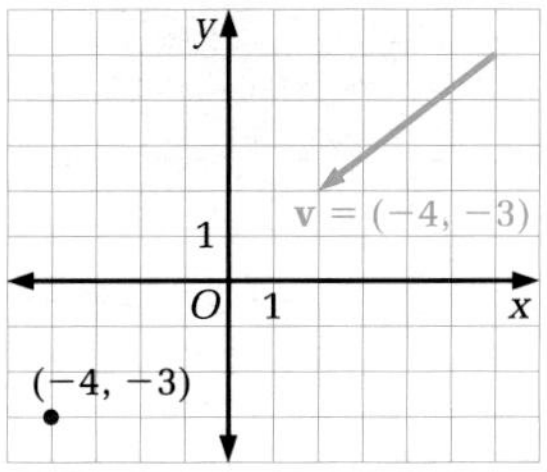

For questions 2 and 3, use this diagram.

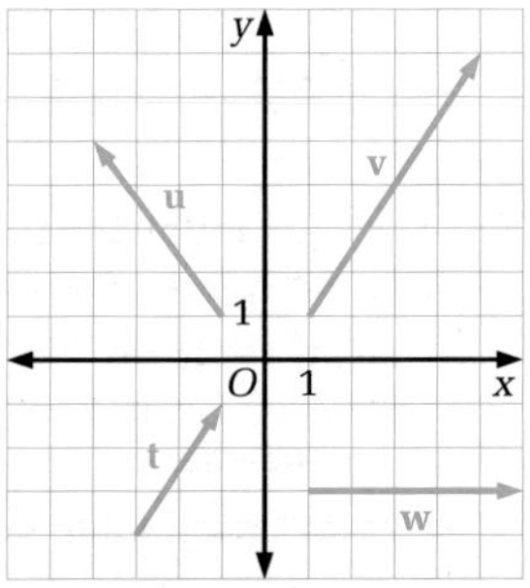

2. a. Write **t** in rectangular form.
   b. Write **u** in polar form.
3. a. Which two vectors have the same magnitude?
   b. Which two vectors have the same direction?

TEACHING

Exploration

The goal of the Exploration is to introduce students to the concept of a vector by having them use distances and angles to travel from place to place on a map.

Mathematical Procedures

Students may wonder why vectors are represented in both polar form and rectangular form. The reason has to do in part with the applications of vector concepts in physics and geometry. The polar form of a vector shows both the magnitude of the vector and its direction, and this itself can be useful. The algebra of vectors, however, is somewhat complicated when the coordinates are in polar form. Vectors in polar form cannot be added or subtracted by adding or subtracting their corresponding components. The rectangular form makes vector addition and subtraction easy, but does not show the magnitude of the vector. The magnitude of a vector (a, b) is easy to calculate since it is simply $\sqrt{a^2 + b^2}$.

Talk it Over

Question 1 directs students' attention to the difference between a vector and a point. Question 3 provides students with an opportunity to differentiate between the concepts of magnitude and direction.

Answers to Talk it Over

1. The (–4, –3) on the left of the diagram indicates the rectangular coordinates of the point in the plane that is 4 units to the left of the *y*-axis and 3 units below the *x*-axis. The (–4, –3) on the right indicates the vector with horizontal shift –4 and vertical shift –3.
2. a. (2, 3)
   b. (5, 126.9°)
3. a. **u** and **w**
   b. **t** and **v**

Additional Sample

S1 Refer to the diagram for Sample 1. Suppose a third student, Leona, also starts from the same corner as Jamal and Ja-Wen. Leona's path can be represented by the two vectors in polar form: (36 paces, 30°) then (20 paces, 60°). Draw a vector that represents the path that Leona follows.

Draw the first vector for Leona's path, with its tail at the origin. From the head of that vector, draw the vector (20 paces, 60°). The head of the second vector is at the same point where Jamal's and Ja-Wen's paths end.

Visual Thinking

Ask students to create their own rough sketches of the diagram shown in Sample 1. Encourage them to draw the vectors described on their sketches, to label the parts, and to describe the shifts using both rectangular and polar form. This activity involves the visual skills of *exploration* and *interpretation*.

Talk it Over

In question 4, students see that vector addition is commutative. In question 5, students are led to understand the concept of scalar multiplication.

Sample 1

Jamal and Ja-Wen are in the marching band's half-time show at a football game. They start from the same corner of the field and march to the same point. Jamal's path can be represented by two vectors in polar form: (20 paces, 60°) then (36 paces, 30°). Ja-Wen marches in a straight line. Draw a vector that represents the path Ja-Wen follows.

Sample Response

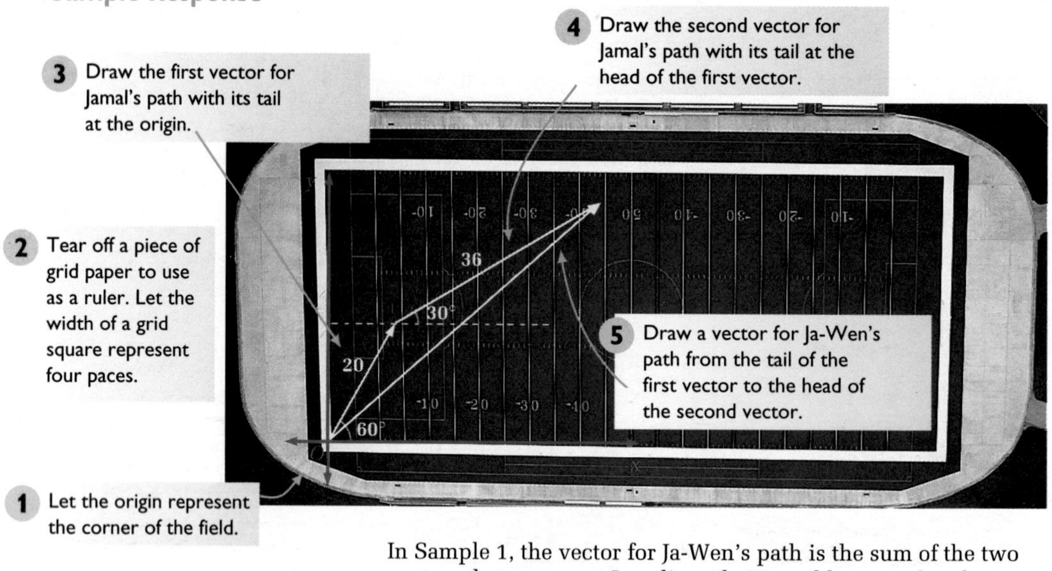

4 Draw the second vector for Jamal's path with its tail at the head of the first vector.
3 Draw the first vector for Jamal's path with its tail at the origin.
2 Tear off a piece of grid paper to use as a ruler. Let the width of a grid square represent four paces.
5 Draw a vector for Ja-Wen's path from the tail of the first vector to the head of the second vector.
1 Let the origin represent the corner of the field.
36
30°
20
60°

In Sample 1, the vector for Ja-Wen's path is the sum of the two vectors that represent Jamal's path. You add vectors by placing them head to tail. The sum is a vector from the tail of the first vector to the head of the second. It is called a **resultant vector.**

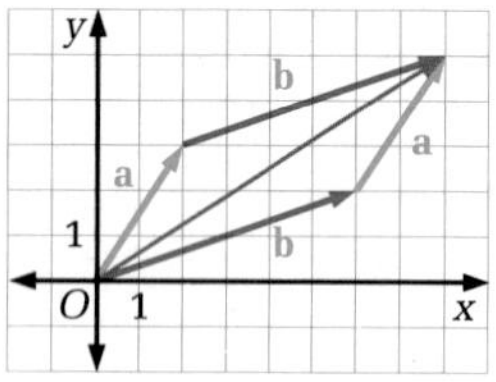

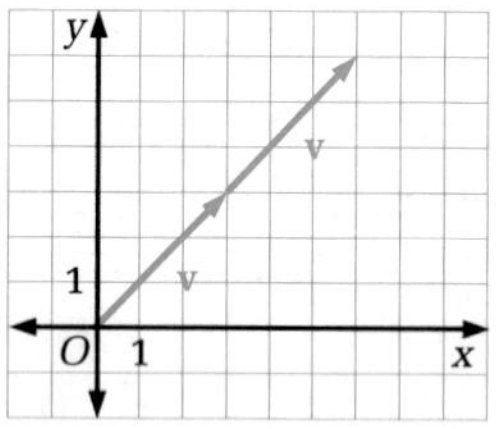

Talk it Over

4. a. In Sample 1, suppose the order of Jamal's vectors is reversed. Does this affect Ja-Wen's vector?
   b. Use the diagram to explain why the order in which you add two vectors does not affect the sum.
5. a. Let $\mathbf{u} = \mathbf{v} + \mathbf{v}$. How does $\mathbf{u}$ compare with $\mathbf{v}$ in magnitude and direction?
   b. Another way to write $\mathbf{u}$ is $2\mathbf{v}$. Describe the vector $\mathbf{w} = 3\mathbf{v}$. Describe $\mathbf{t} = \frac{1}{2}\mathbf{v}$.
   c. Let $(-1)\mathbf{v}$ be the vector $\mathbf{v}$ with its head and tail interchanged. How does $(-1)\mathbf{v}$ compare with $\mathbf{v}$ in magnitude and direction?

Answers to Talk it Over

4. a. No.
   b. Answers may vary. An example is given. The figure shows the vectors added in both orders, beginning at the same point. The resulting figure is a parallelogram whose diagonal is the resultant vector from both additions, $\mathbf{a} + \mathbf{b}$ and $\mathbf{b} + \mathbf{a}$.
5. a. The magnitude of $\mathbf{u}$ is twice that of $\mathbf{v}$; the direction is the same.
   b. $\mathbf{w}$ has magnitude three times that of $\mathbf{v}$ and the same direction as $\mathbf{v}$; $\mathbf{t}$ has magnitude half that of $\mathbf{v}$ and the same direction as $\mathbf{v}$.
   c. $(-1)\mathbf{v}$ has the same magnitude as $\mathbf{v}$. Its direction is opposite that of $\mathbf{v}$.

A vector $\mathbf{v} = k\mathbf{u}$ is called a **scalar multiple** of **u**. For $k > 0$, **v** has the same direction as **u** and a magnitude k times as great. For $k < 0$, **v** has the opposite direction as **u** and a magnitude $|k|$ times as great.

A special case is $\mathbf{v} = (-1)\mathbf{u}$, where **v** has the same magnitude as **u** but the opposite direction. In this case, **v** can be written as $-\mathbf{u}$, and **v** and **u** are called **opposite vectors.** The sum of two opposite vectors is the zero vector.

Sample 2

Draw the vector w = u − 2v, with u = (8, 83°) and v = (3.5, 144°). Use a protractor and a ruler to draw each sum. Write the resultant vector in polar form and in rectangular form.

Sample Response

Think of **w** as the sum of the vectors **u** and -2**v**.

1. Start at the origin. Draw **u**.
2. Draw **v** from the head of **u**.
3. Draw -2**v** from the head of **u**. It is a vector with twice the magnitude of **v** and the opposite direction.

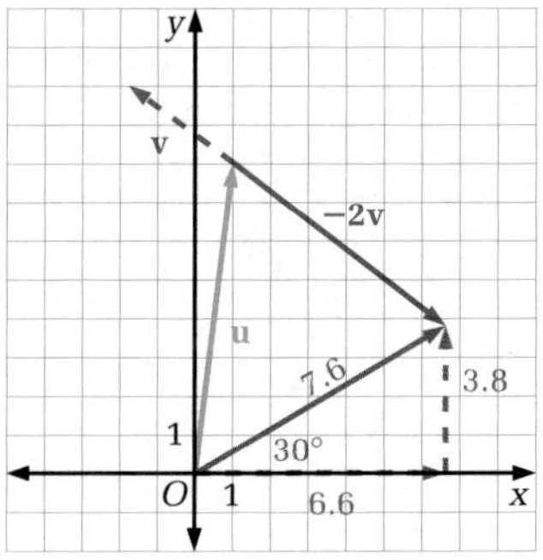

4. Draw the resultant vector **u** + (-2**v**) from the tail of **u** to the head of -2**v**.
5. To write the vector in polar form, measure to find the magnitude and direction of the resultant vector.
6. To write the vector in rectangular form, estimate the horizontal shift and the vertical shift from the tail to the head of the resultant vector.

The polar form of the resultant vector is about (7.6, 30°).
The rectangular form is about (6.6, 3.8).

Look Back

Describe a geometric method for adding two vectors.

Answers to Look Back

Draw the first vector with its tail at the origin. Draw the second vector with its tail at the head of the first vector. Draw a vector from the tail of the first vector to the head of the second. This vector is the resultant or sum of the vectors.

Additional Sample

S2 Draw the vector $\mathbf{t} = \mathbf{r} - \frac{1}{2}\mathbf{s}$, with $\mathbf{r} = (3.2, 37°)$ and $\mathbf{s} = (4, 0°)$. Use a protractor and a ruler to draw each sum. Write the resultant vector in polar form and in rectangular form.

Start at the origin. Draw r. From the head of r, draw s. From the head of r, draw $-\frac{1}{2}$s, the vector in the opposite direction from s and with half the magnitude. Draw the resultant $\mathbf{r} - \frac{1}{2}\mathbf{s}$ from the tail of r to the head of $-\frac{1}{2}$s.

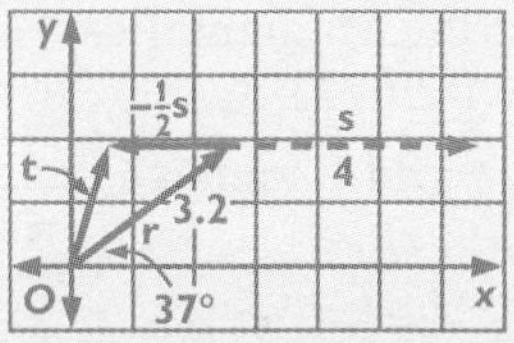

Measure to find the approximate magnitude and direction of the resultant vector. The polar form is about (2.0, 74°). Estimate the horizontal shift and vertical shift from the tail to the head of the resultant vector. The rectangular form is about (0.6, 1.9).

Communication: Drawing

Students need to understand the importance of making accurate drawings for vector problems, such as in Sample 2. Students should also understand that their answers will be approximations.

Students Acquiring English

For the Look Back, you might consider allowing students acquiring English to use diagrams to show a geometric method for adding two vectors instead of describing a method in a paragraph.

Look Back

You may wish to have one student answer this Look Back, while another writes the response on the board. The class could then consider a method for subtracting two vectors.

APPLYING

Suggested Assignment

Standard 5–16, 20–24

Extended 2–24

Integrating the Strands

Algebra Exs. 11–17, 20, 21, 24

Measurement Exs. 1, 3–20, 24

Geometry Exs. 1–21, 23

Trigonometry Exs. 21, 23

Statistics and Probability Ex. 22

Discrete Mathematics Exs. 1–20, 24

Logic and Language Exs. 1, 2, 17, 19, 20, 24

Teaching Tip

For Exs. 5–10, you may wish to save students some time by providing them with copies of the map of landmarks on an island.

Communication: Writing

For Ex. 17, ask some students to demonstrate their results at the board. Looking at many examples will convince students that the relationship is always true, although they should understand that inductive reasoning of this kind is certainly not a proof.

Application

Exs. 18 and 19 show students how vectors can be used to navigate a canoe from one island to another.

8-3 Exercises and Problems

1. **Group Activity** Work with another student. Use the map on page 440.
   a. Create your own set of instructions for traveling from one point on the map to another. Use at least four steps in your instructions.
   b. Follow each other's instructions. Did you each arrive at the correct destination? If not, go back and check each other's work.
2. **Reading** Which instructions in the Exploration on page 440 can be represented by opposite vectors?
3. Janice and Malcolm are dance partners. During a dance, Janice walks from Malcolm at a 45° angle for eight steps and then continues at a −25° angle for another eight steps. When she stops, Malcolm walks in a straight line to join her. Draw a vector that represents the path Malcolm follows.
4. Suppose you are given these instructions in a treasure hunt: Walk five paces from the doorstep at a 30° angle and then 10 paces at a 120° angle. Draw a vector that represents your new position relative to the doorstep.

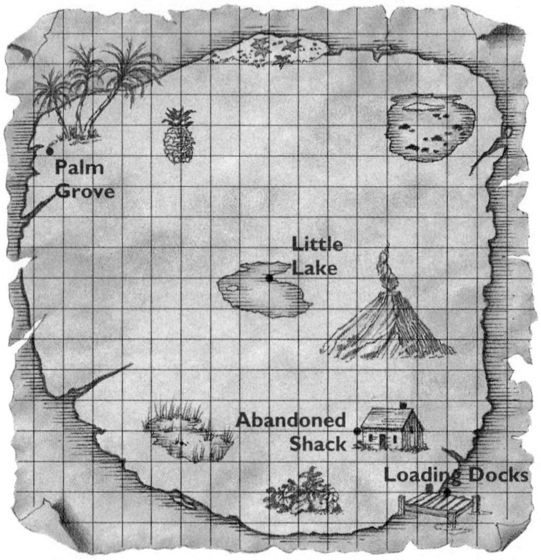

For Exercises 5–10, copy the map of landmarks on an island. Use Little Lake as the origin. Use the given coordinates to locate each landmark.

5. Slate Lagoon is at (3.3, 5.3).
6. Hermit's Cave is at (4.7, −40°).
7. Cool Spring: (12, −36°) from Palm Grove
8. Black Cliffs: (−7, 4.7) from the Abandoned Shack
9. Thinking Spot: (−9, −2) from Slate Lagoon
10. Feather Falls: (10, 100°) from the Loading Docks

Let t = (1, 3), u = (−7, 3), v = (3, 60°), and w = (4, 90°). Use a protractor and a ruler to draw each sum. Write the resultant vector in polar form and in rectangular form. Estimate answers to the nearest tenth.

11. **t** + **u**
12. **t** − **u**
13. −**t** + 2**u**
14. **v** + 3**w**
15. −**v** + (−**w**)
16. −(**v** + **w**)

17. **Writing** Compare the results of Exercises 15 and 16. Make a conjecture about the relationship between −**a** + (−**b**) and −(**a** + **b**). Test your conjecture with another pair of vectors.

Answers to Exercises and Problems

1. Answers may vary. Check students' work.
2. 4.6 km at −41° and 4.6 km at 139°
3.

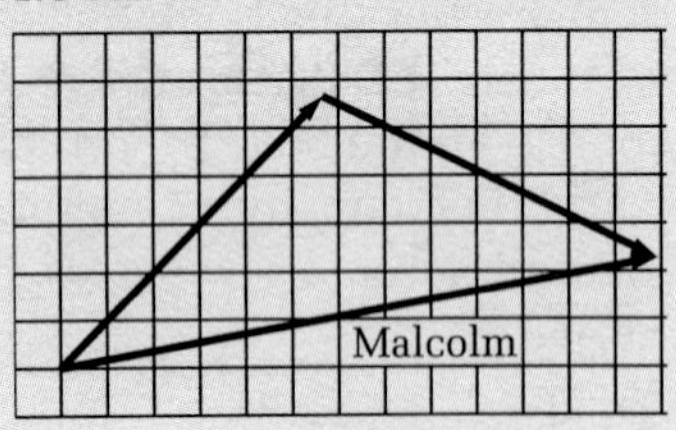

4.

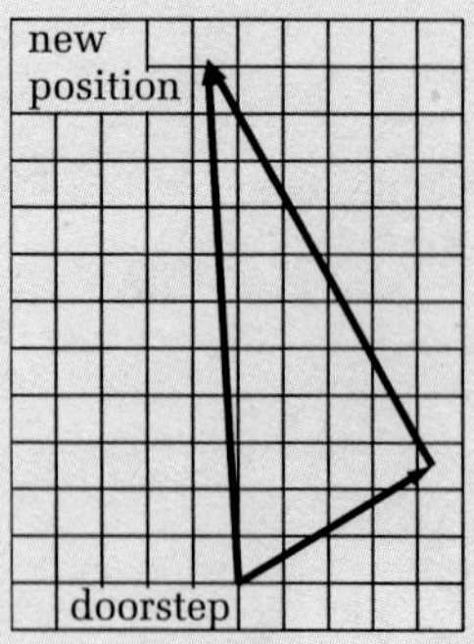

5–14. See answers in back of book.

15.

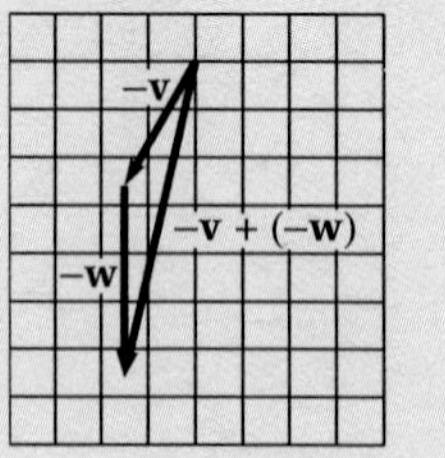

(6.8, −102.8°); (−1.5, −6.6)

16.

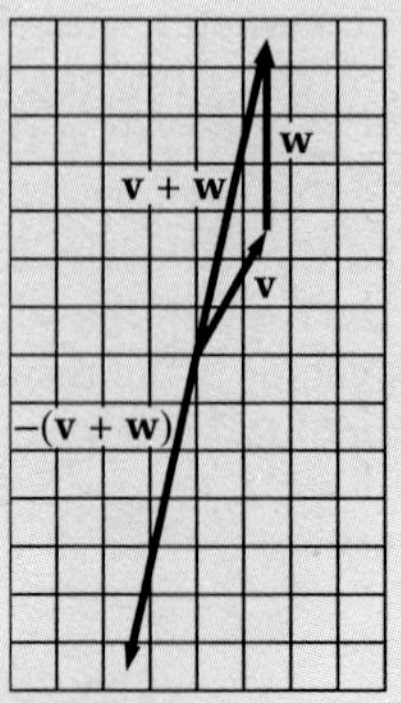

(6.8, −102.8°); (−1.5, −6.6)

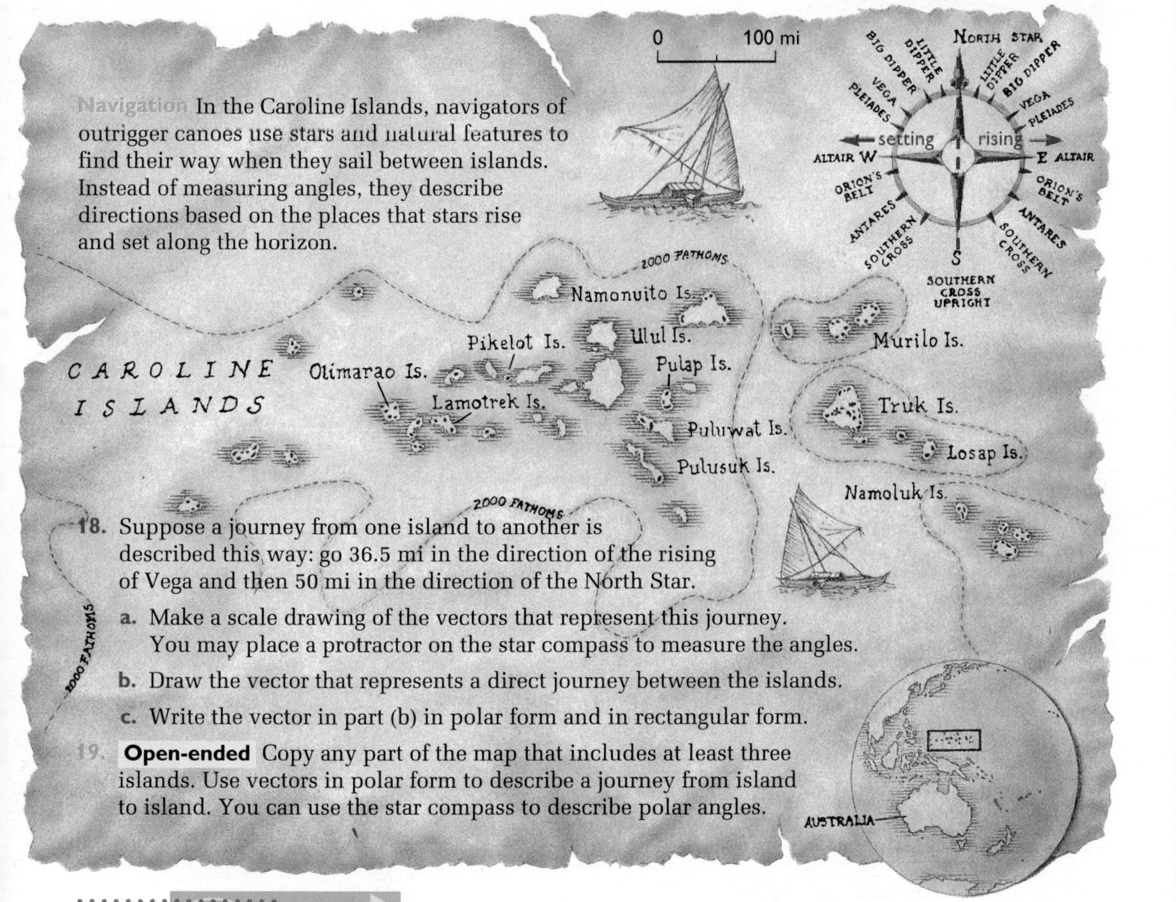

Navigation In the Caroline Islands, navigators of outrigger canoes use stars and natural features to find their way when they sail between islands. Instead of measuring angles, they describe directions based on the places that stars rise and set along the horizon.

18. Suppose a journey from one island to another is described this way: go 36.5 mi in the direction of the rising of Vega and then 50 mi in the direction of the North Star.
 a. Make a scale drawing of the vectors that represent this journey. You may place a protractor on the star compass to measure the angles.
 b. Draw the vector that represents a direct journey between the islands.
 c. Write the vector in part (b) in polar form and in rectangular form.

19. **Open-ended** Copy any part of the map that includes at least three islands. Use vectors in polar form to describe a journey from island to island. You can use the star compass to describe polar angles.

Ongoing ASSESSMENT

20. a. **Open-ended** Draw any three vectors **a**, **b**, and **c**.
 b. Draw (**a** + **b**) and (**b** + **c**).
 c. Draw (**a** + **b**) + **c** and **a** + (**b** + **c**).
 d. What is the relationship between the two vectors in part (c)?

Review PREVIEW

21. Convert the rectangular coordinates (7.2, −4.8) to polar coordinates. *(Section 8-2)*
22. Three students are chosen at random from 8 juniors and 10 seniors. What is the probability that at least 2 seniors are chosen? *(Section 7-4)*
23. Convert the polar coordinates (12.8, −83°) to rectangular coordinates. *(Section 8-2)*

8 Working on the Unit Project

24. Write a clue for your treasure hunt that requires someone to draw the sum of two vectors to get to the place where the next clue is.

Multicultural Note

The Caroline Islands are a group of islands located in the South Pacific. Home to many different peoples, the islands were probably settled by Asian voyagers in the first century A.D. Today, the islands are divided into four groups that together make up the Federated States of Micronesia. The capital group is Pohnpei, whose largest urban center, Kolonia Town, has about 6000 inhabitants.

Working on the Unit Project

Students should work in their project group to do Ex. 24. After the clue is written, at least one student in the group should check its mathematical accuracy.

Practice 61 For use with Section 8-3

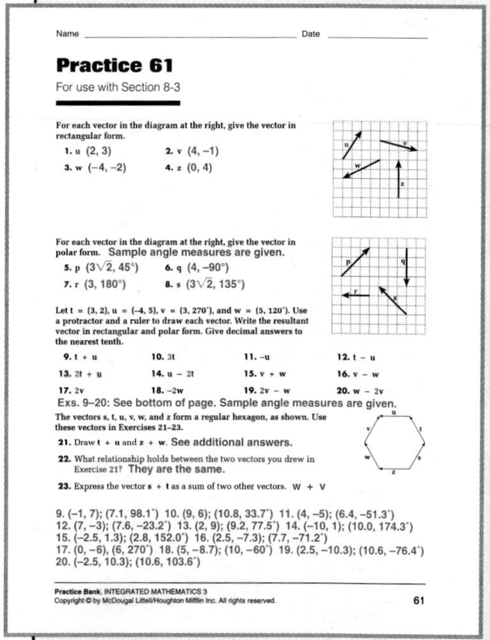
Name ______ Date ______

Practice 61
For use with Section 8-3

For each vector in the diagram at the right, give the vector in rectangular form.
1. u (2, 3) 2. v (4, −1)
3. w (−4, −2) 4. z (0, 4)

For each vector in the diagram at the right, give the vector in polar form. Sample angle measures are given.
5. p $(3\sqrt{2}, 45°)$ 6. q (4, −90°)
7. r (3, 180°) 8. s $(3\sqrt{2}, 135°)$

Let t = (3, 2), u = (−4, 5), v = (3, 270°), and w = (5, 120°). Use a protractor and a ruler to draw each vector. Write the resultant vector in rectangular and polar form. Give decimal answers to the nearest tenth.

9. t + u 10. 3t 11. −u 12. t − u
13. 2t + u 14. u − 2t 15. v + w 16. v − w
17. 2v 18. −2w 19. 2v − w 20. w − 2v

Exs. 9–20: See bottom of page. Sample angle measures are given.

The vectors s, t, u, v, w, and z form a regular hexagon, as shown. Use these vectors in Exercises 21–23.

21. Draw t + u and z + w. See additional answers.
22. What relationship holds between the two vectors you drew in Exercise 21? They are the same.
23. Express the vector s + t as a sum of two other vectors. w + v

9. (−1, 7); (7.1, 98.1°) 10. (9, 6); (10.8, 33.7°) 11. (4, −5); (6.4, −51.3°) 12. (7, −3); (7.6, −23.2°) 13. (2, 9); (9.2, 77.5°) 14. (−10, 1); (10.0, 174.3°) 15. (−2.5, 1.3); (2.8, 152.0°) 16. (2.5, −7.3); (7.7, −71.2°) 17. (0, −6), (6, 270°) 18. (5, −8.7); (10, −60°) 19. (2.5, −10.3); (10.6, −76.4°) 20. (−2.5, 10.3); (10.6, 103.6°)

Practice Bank, INTEGRATED MATHEMATICS 3

61

Answers to Exercises and Problems

17. The resultant vectors have the same magnitude and direction; −**a** + (−**b**) = −(**a** + **b**). Examples may vary.

18. a.
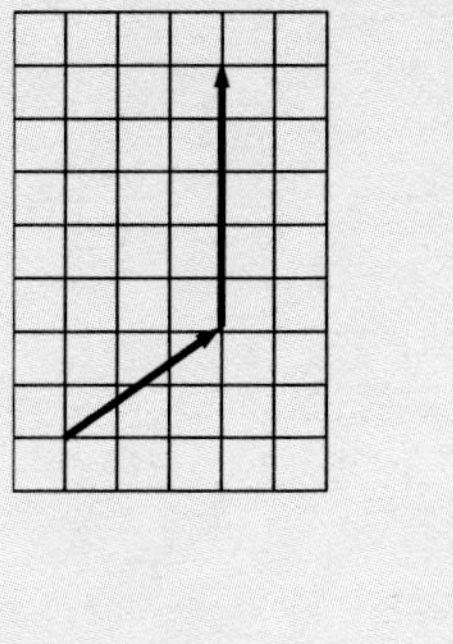

b.
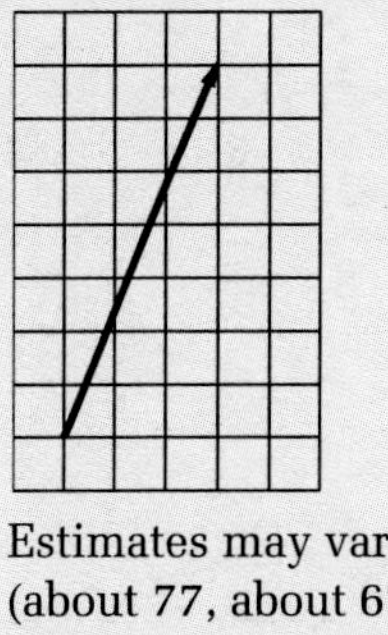

c. Estimates may vary; (about 77, about 67°); (about 30, about 71).

19. Answers may vary. Check students' work.
20. a–c. Drawings may vary. Check students' work.
 d. **a** + (**b** + **c**) and (**a** + **b**) + **c** are the same vector.
21. (8.7, −33.69°)
22. 0.588
23. (1.56, −12.70)
24. Answers may vary. Check students' work.

PLANNING

Objectives and Strands
See pages 422A and 422B.

Spiral Learning
See page 422B.

Materials List
- Graphics calculator
- Graph paper
- Ruler
- Protractor

Recommended Pacing
Section 8-4 is a one-day lesson.

Extra Practice
See pages 621–622.

Warm-Up Exercises
- Warm-Up Transparency 8-4

Support Materials
- Practice Bank: Practice 62
- Activity Bank: Enrichment 55
- Study Guide: Section 8-4
- Problem Bank: Problem Set 17
- Explorations Lab Manual: Additional Exploration 9 Diagram Master 2
- Assessment Book: Quiz 8-4, Test 32, Alternative Assessment 4

Section 8-4 The Algebra of Vectors

Focus
Use algebra to solve problems involving vectors.

Up Up And Away

When an airplane takes off, it moves horizontally and vertically at the same time.
The speed of the airplane in the vertical direction is called the *rate of climb.*
200 mi/h
15°
The speed of the airplane in the horizontal direction is called the *ground speed.*

Sample 1

A passenger jet takes off at an angle of 15°. Suppose the speed of the jet is 200 mi/h after the jet takes off.

a. Find the jet's ground speed to the nearest mile per hour.

b. Find the jet's rate of climb to the nearest mile per hour.

Sample Response

Use a vector **v** to represent the speed and direction of the jet. **v** is the sum of a horizontal vector $\mathbf{v_x}$ and a vertical vector $\mathbf{v_y}$.

a. The ground speed is the magnitude of $\mathbf{v_x}$. Use trigonometry to find this magnitude.

$$\cos 15° = \frac{\text{magnitude of } \mathbf{v_x}}{200}$$

$$200 \cos 15° = \text{magnitude of } \mathbf{v_x}$$

$$193.2 \approx \text{magnitude of } \mathbf{v_x}$$

The ground speed is about 193 mi/h.

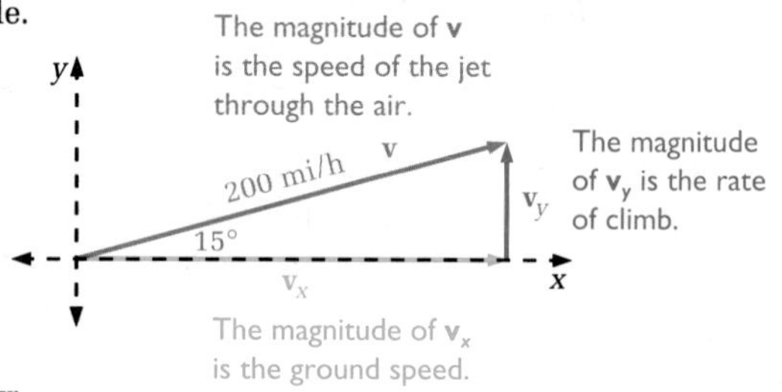

b. The rate of climb is the magnitude of $\mathbf{v_y}$. You can use trigonometry to show that the magnitude of $\mathbf{v_y} = 200 \sin 15° \approx 51.8$.

The rate of the climb is about 52 mi/h.

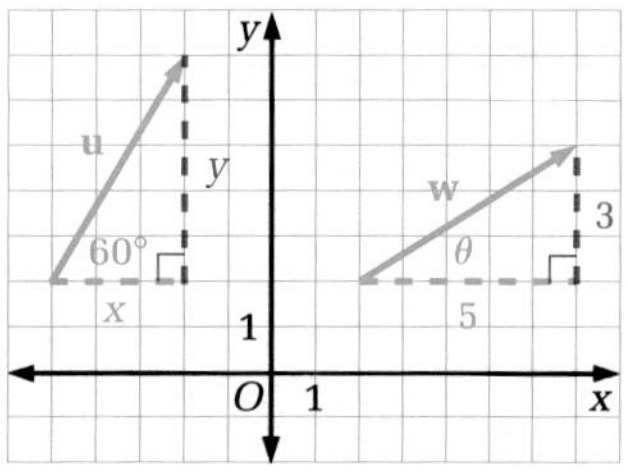

Talk it Over

1. Explain why it makes sense to use *ground speed* and *rate of climb* for an airplane's horizontal and vertical speeds.
2. a. What is a polar form of the vector **v** in Sample 1? What information about the jet does the polar form give?
   b. What is the rectangular form of **v**? What information does the rectangular form give?
3. a. Complete this statement: The rectangular form (x, y) of **u** in the diagram is **u** = ($\underline{?}$ cos $\underline{?}$, $\underline{?}$ sin $\underline{?}$).
   b. Complete this statement: You can find a polar form (r, θ) of **w** in the diagram using the equations $r^2 = (\underline{?})^2 + (\underline{?})^2$ and $\tan \theta = \underline{?}$.
4. Compare converting between polar and rectangular forms of vectors with converting between polar and rectangular coordinates.

Adding Vectors in Rectangular Form

The vectors **u** = (5, 2) and **v** = (3, 4) are shown, along with their sum **u** + **v**. Notice that:

$$\mathbf{u} + \mathbf{v} = (5, 2) + (3, 4) = (5 + 3, 2 + 4) = (8, 6)$$

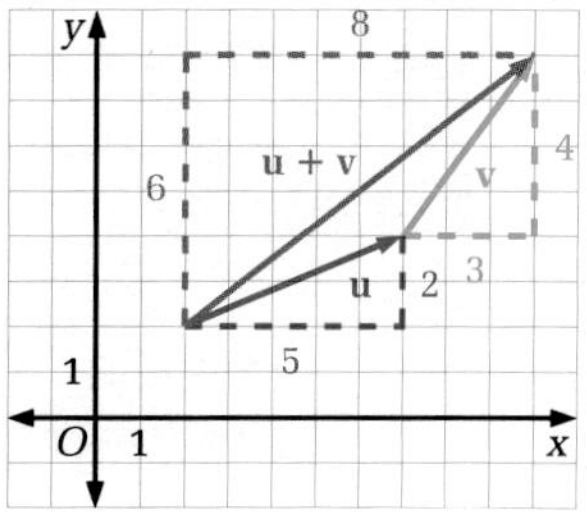

This result illustrates the following rule.

ADDING TWO VECTORS IN RECTANGULAR FORM

For any two vectors $\mathbf{u} = (x_1, y_1)$ and $\mathbf{v} = (x_2, y_2)$:

$$\mathbf{u} + \mathbf{v} = (x_1, y_1) + (x_2, y_2) = (x_1 + x_2, y_1 + y_2)$$

Example

$$(3, -2) + (4, 6) = (3 + 4, -2 + 6) = (7, 4)$$

Watch Out!
When two vectors are in polar form, you *cannot* use the above rule. You must convert the vectors to rectangular form before adding.

Answers to Talk it Over

1. The horizontal component of the airplane's speed is its speed relative to the ground, or *ground speed.* The vertical component of the airplane's speed is the rate at which it is climbing, or *rate of climb.*
2. a. (200, 15°); its actual speed through the air and its angle of ascent
   b. (193.19, 51.8); the horizontal and vertical distances of the jet from its takeoff point after 1 h
3. a. $\sqrt{34}$; 60; $\sqrt{34}$; 60
   b. 5; 3; $\frac{3}{5}$
4. The methods are the same.

TEACHING

Additional Sample

S1 A cyclist is going up a straight section of road that is inclined at a 5° angle with the horizontal. The cyclist is maintaining a steady speed of 12 mi/h.

a. Find the cyclist's horizontal ground speed.

Use trigonometry.

$$\cos 5° = \frac{\text{horizontal speed}}{\text{speed on road}}$$

$$\cos 5° = \frac{\text{horizontal speed}}{12}$$

12 cos 5° = horizontal speed

Since 12 cos 5° ≈ 11.95, the cyclist's horizontal ground speed is about 11.95 mi/h, almost the same as his speed going along the road.

b. Find the cyclist's rate of climb.

Use trigonometry.

$$\sin 5° = \frac{\text{rate of climb}}{\text{speed on road}}$$

$$\sin 5° = \frac{\text{rate of climb}}{12}$$

12 sin 5° = rate of climb

Since 12 sin 5° ≈ 1.05, the cyclist's rate of climb is about 1.05 mi/h.

Talk it Over

Question 2 provides students with an opportunity to contrast the polar and rectangular forms of vectors. Question 3 and 4 lead students to develop a method for converting between the polar and rectangular forms of vectors.

Teaching Tip

A class discussion about the double role played by ordered pairs such as (3, –5) and (7, 30°) would be worthwhile. On the one hand, they are used to specify locations of points in the plane. On the other hand, they name vectors, which can be thought of as directed segments in the plane.

Mathematical Procedures

In print, it is common practice to boldface a single letter used to name a vector. In handwritten material, it is not easy to use boldface, so many mathematicians draw an arrow over a letter that represents a vector (for example, $\vec{v}$). You may wish to recommend this practice to your students.

Additional Sample

S2 An airplane is flying at a speed of 325 mi/h and at a bearing of 50°. The airplane meets a wind that has a speed of 48 mi/h and a bearing of 330°. Find the resultant speed and bearing of the airplane.

Let a = speed and direction of the airplane with no wind. Let w = speed and direction of the wind. The polar angles that correspond to bearings of 50° and 330° are 40° and 120°, respectively.

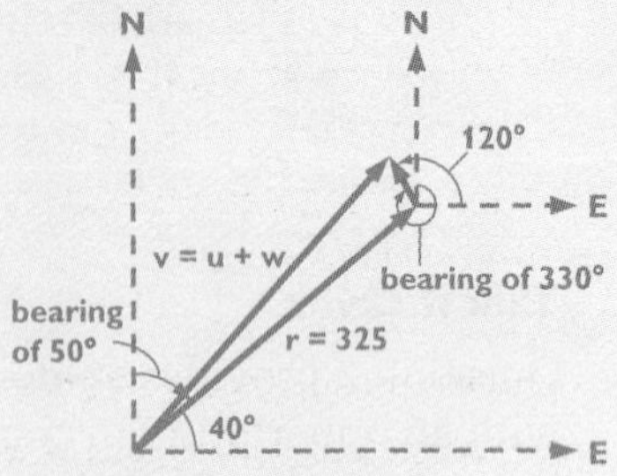

Step 1. Write a and w in polar form.
a = (325, 40°)
w = (48, 120°)

Step 2. Convert a and w to rectangular form.
Use a graphics calculator and round to the nearest hundredth.
a ≈ (248.96, 208.91)
w ≈ (–24, 41.57)

Step 3. Let v be the vector for the resulting speed of the airplane.
v = a + w ≈ (248.96, 208.91) + (–24, 41.57)
≈ (224.96, 250.48)

Step 4. Convert v to polar form. Use a graphics calculator.
v ≈ (336.67, 48.07°)

The polar angle 48.07° corresponds to a bearing of 41.93°. So, the speed of the airplane as it flies through the wind is about 337 mi/h and its bearing is about 41.9°.

Sample 2

An airplane is flying at a speed of 300 mi/h and at a bearing of 40° at a constant altitude. The airplane meets a wind whose speed is 60 mi/h and whose bearing is 155°. Find the resulting speed and bearing of the airplane as it flies through the wind.

Sample Response

Let **a** = the speed and direction of the airplane with no wind.
Let **w** = the speed and direction of the wind.
Then **v** = **a** + **w** represents the resulting speed and direction of the airplane as it flies through the wind.
Find **v** in polar form.

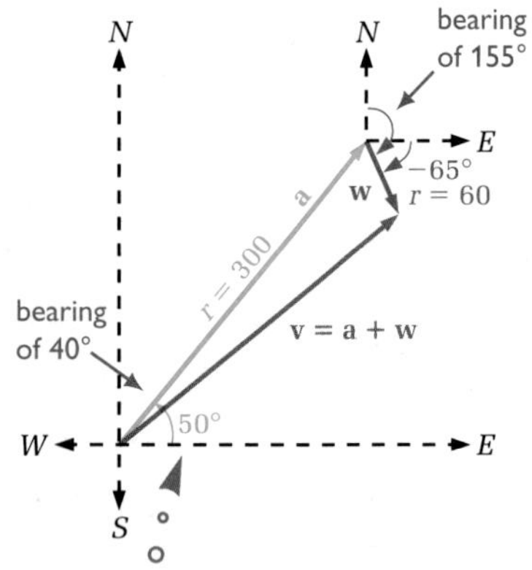

Step 1 Write **a** and **w** in polar form.

a = (300, 50°) ← The measure of the polar angle is 90° − 40° = 50°.

w = (60, −65°) ← The measure of the polar angle is 90° − 155° = −65°.

Step 2 Convert **a** and **w** to rectangular form.

```
P▶Rx (300, 50)
      192.8362829
P▶Ry (300, 50)
      229.8133329
```

```
P▶Rx (60, −65)
      25.3570957
P▶Ry (60, −65)
      −54.37846722
```

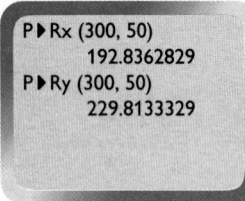

a ≈ (192.84, 229.81) **w** ≈ (25.36, −54.38)

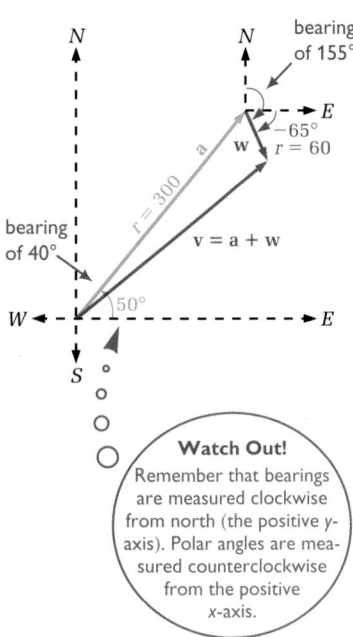

Watch Out! Remember that bearings are measured clockwise from north (the positive *y*-axis). Polar angles are measured counterclockwise from the positive *x*-axis.

Step 3 Find **v** in rectangular form.

v = **a** + **w**
≈ (192.84, 229.81) + (25.36, −54.38)
≈ (192.84 + 25.36, 229.81 + (−54.38))
≈ (218.20, 175.43)

Step 4 Convert **v** to polar form.

v ≈ (279.98, 38.80°)

```
R▶Pr (218.20, 175.43)
      279.9766506
R▶Pθ (218.20, 175.43)
      38.79883507
```

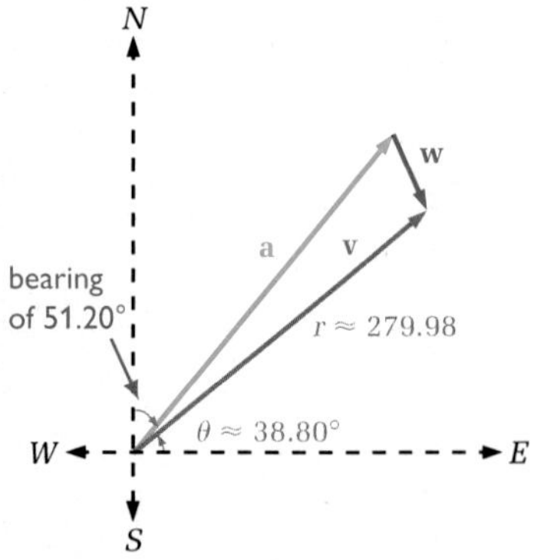

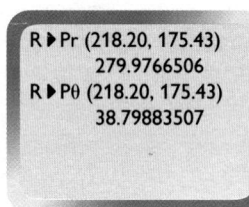

The resulting speed of the airplane as it flies through the wind is the magnitude of **v**, about 279.98 mi/h.

The airplane's bearing is about 90° − 38.80°, or about 51.20°.

Talk it Over

Questions 5–7 refer to Sample 2 on page 448.

5. Why was it necessary to convert **a** and **w** to rectangular form in Step 2?
6. Why was it necessary to convert **v** to polar form in Step 4?
7. a. Does the wind increase or decrease the airplane's speed?
   b. Suppose the wind's bearing is 30°. Does this change your answer to part (a)? Explain.
   c. What wind bearing maximizes the airplane's speed? What wind bearing minimizes the airplane's speed?

Look Back

How do you add two vectors given in rectangular form? How do you add two vectors given in polar form?

8-4 Exercises and Problems

1. **Reading** What are the missing steps in the solution in part (b) of Sample 1?

Convert each vector to polar form.

2. (3, 5)
3. (−6, 2)
4. (−4, −9)

Convert each vector to rectangular form.

5. (7, 60°)
6. (1, 143°)
7. (2.5, −35°)
8. **Writing** Janet and Alma were asked to express the sum of **u** = (9, 10°) and **v** = (6, 82°) in rectangular form. Write a paragraph comparing their solutions. Be sure to identify any mistakes.

Janet Kazdan

$u + v = (9, 10°) + (6, 82°)$

$= (9 + 6, 10° + 82°)$

$= (15, 92°)$

$= (15 \cos 92°, 15 \sin 92°)$

$\approx (-0.52, 14.99)$

Alma Van Dorn

$u + v = (9, 10°) + (6, 82°)$

$= (9 \cos 10°, 9 \sin 10°) + (6 \cos 82°, 6 \sin 82°)$

$\approx (8.86, 1.56) + (0.84, 5.94)$

$\approx (8.86 + 0.84, 1.56 + 5.94)$

$\approx (9.70, 7.50)$

Talk it Over

Question 5 directs students' attention to the fact that the formula for adding two vectors given on page 447 is only for vectors in rectangular form.

Look Back

Students should write their response to this Look Back in their journals.

APPLYING

Suggested Assignment

Standard 2–7, 9–16, 22–26

Extended 1–17, 19, 20, 22–26

Integrating the Strands

Algebra Exs. 1–4, 8–26

Measurement Exs. 18, 21

Geometry Exs. 1–8, 12–21

Trigonometry Exs. 1–8, 12–21, 25

Discrete Mathematics Exs. 2–22, 26

Logic and Language Exs. 1, 8, 19, 21, 26

Communication: Writing

Ex. 8 provides an excellent opportunity to check students' understanding of vector addition.

Answers to Talk it Over

5. so the vectors could be added by adding coordinates
6. to find the bearing of the airplane
7. a. decreases
   b. Yes; with a bearing of 30°, the wind would increase the airplane's speed.
   c. 40°; 220°

Answers to Look Back

Add the *x*-coordinates and the *y*-coordinates. Convert the vectors to rectangular form and add the coordinates.

Answers to Exercises and Problems

1. $\sin 15° = \frac{\text{magnitude of } \mathbf{v}_y}{\text{magnitude } \mathbf{v}} = \frac{\text{magnitude of } \mathbf{v}_y}{200}$; magnitude of $\mathbf{v}_y = 200 \sin 15° \approx 51.76$
2. (5.83, 59.04°)
3. (6.32, 161.57°)
4. (9.85, −113.96°)
5. (3.5, 6.06)
6. (−0.80, 0.60)
7. (2.05, −1.43)
8. Janet added the vectors by adding the polar coordinates, and then converted to rectangular coordinates. Her method is incorrect. Alma's method is correct. She converted the polar coordinates to rectangular coordinates, and then added the coordinates.

Interdisciplinary Problems

Vectors are used extensively in physics. In Exs. 15–17, vectors are used to determine velocities. In Exs. 19 and 20, students work with vectors to represent forces.

Teaching Tip

Students may wish to discuss examples of vector quantities from physics. They can also discuss some quantities that are not vectors. Force, velocity, and momentum are important examples of vector quantities because they involve both magnitude and direction. Volume, temperature, mass, and area are not vector quantities, because they involve magnitude but *not* direction.

Find each vector sum in rectangular form.

9. (4, 0) + (0, 3) **10.** (9, 2) + (5, −11) **11.** $(3\sqrt{2}, -1) + (-\sqrt{2}, 2)$

Find each vector sum in polar form.

12. (1, 70°) + (2, 10°) **13.** (85, −98°) + (6, −7°) **14.** (5, 50°) + (10, −50°)

15. An airplane takes off at an angle of 10° and at a speed of 250 mi/h. Find the airplane's ground speed and rate of climb.

16. An airplane is flying at a speed of 180 mi/h and at a bearing of 350°. The airplane meets a wind whose speed is 45 mi/h and whose bearing is 15°. Find the speed and bearing of the airplane as it flies through the wind.

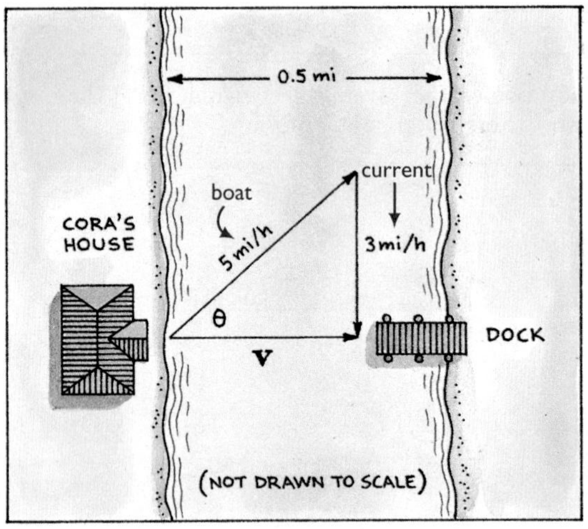

17. Cora Turner wants to row her boat to a dock directly across a river from her house. Cora can row at 5 mi/h. The speed of the current is 3 mi/h.

a. At what angle θ should she point the boat in order to reach the dock?

b. How long will it take her to reach the dock? (*Hint:* Find the magnitude of **v**, the vector representing the combined speed and direction of the boat and current.)

18. **Group Activity** Work with another student. You will need a ruler and graph paper.

a. Draw the vector 3**u** where **u** = (1, 2). Use your drawing to find the rectangular form of 3**u**. Complete:

$$3\mathbf{u} = (\underline{?} \cdot 1, \underline{?} \cdot 2)$$

b. What do you think is the rectangular form of k**u** where **u** = (x, y) and k is any real number? Justify your answer.

c. Draw the vector **u** − **v** where **u** = (4, 5) and **v** = (1, 3). Use your drawing to find the rectangular form of **u** − **v**. Complete:

$$\mathbf{u} - \mathbf{v} = (\underline{?} - \underline{?}, \underline{?} - \underline{?})$$

d. What do you think is the rectangular form of **u** − **v** where **u** = (x_1, y_1) and **v** = (x_2, y_2)? Justify your answer.

e. Find the rectangular form of 2**u** − 3**v** for the vectors in part (c).

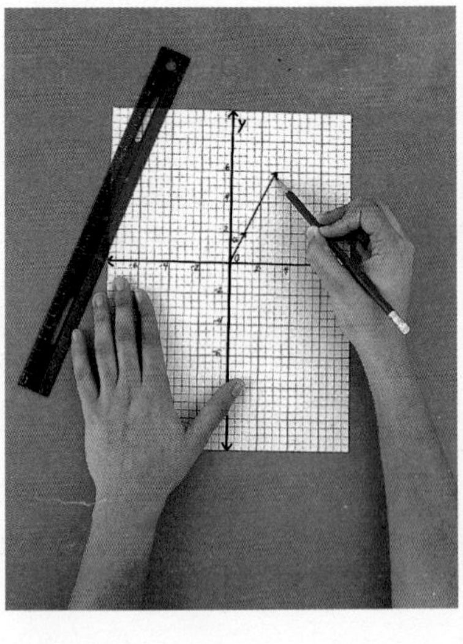

Answers to Exercises and Problems

9. (4, 3) 10. (14, −9)

11. $(2\sqrt{2}, 1)$

12. (2.65, 29.11°)

13. (85.10, −93.96°)

14. (10.37, −21.66°)

15. about 246.2 mi/h; about 43.41 mi/h

16. 221.6 mi/h; 354.92°

17. a. about 36.87° or 37° N of E, or a bearing of 53°

b. 7.5 min

18. a. (3, 6); 3; 3

b. (kx, ky); The magnitude of k**u** is $|k|$ times the magnitude of **u** or $|k|(x^2 + y^2) = |k|x^2 + |k|y^2$, which is the magnitude of the vector (kx, ky). The direction of k**u** is the same as **u** if k is positive and opposite to **u** if k is negative. The direction of (kx, ky) is the same as **u** if k is positive and opposite to **u** if k is negative.

c. (3, 2); 4; 1; 5; 3

d. $(x_1 - x_2, y_1 - y_2)$; $\mathbf{u} - \mathbf{v} = \mathbf{u} + (-\mathbf{v}) = \mathbf{u} + (-1)\mathbf{v} = (x_1, y_1) + (-1)(x_2, y_2) = (x_1, y_1) + (-x_2, -y_2) = (x_1 - x_2, y_1 - y_2)$

e. (5, 1)

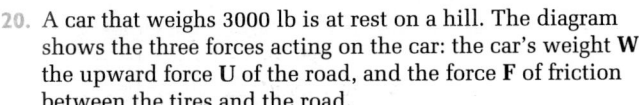

connection to PHYSICS

A *force* is a push or a pull. Physicists use vectors to represent forces.

19. Etenia is pulling her brother Ben in a sled. Etenia pulls with a force of 7.5 lb at an angle of 39° with respect to the horizontal.

a. What amount of force pulls the sled forward? (*Hint:* Find the magnitude of $\mathbf{F_x}$.)

b. **Writing** How can Etenia increase the amount of force that pulls the sled forward *without* increasing the total amount of force she exerts?

BY THE WAY...

The Ojibwa people of North America and Canada made flat wooden sleds called *nobugidabans* to carry people and supplies over the snow. British traders mispronounced this word as *toboggans.*

20. A car that weighs 3000 lb is at rest on a hill. The diagram shows the three forces acting on the car: the car's weight **W**, the upward force **U** of the road, and the force **F** of friction between the tires and the road.

a. Write **W** in rectangular form. (*Hint:* First write **W** in polar form.)

b. Let f be the magnitude of **F**. Write **F** in rectangular form. (Your answer will involve f.)

c. Let u be the magnitude of **U**. Write **U** in rectangular form. (Your answer will involve u.)

d. Since the car is at rest, the sum of the forces acting on the car is zero:

$$\mathbf{W} + \mathbf{U} + \mathbf{F} = (0, 0)$$

Use this fact to find the values of u and f.

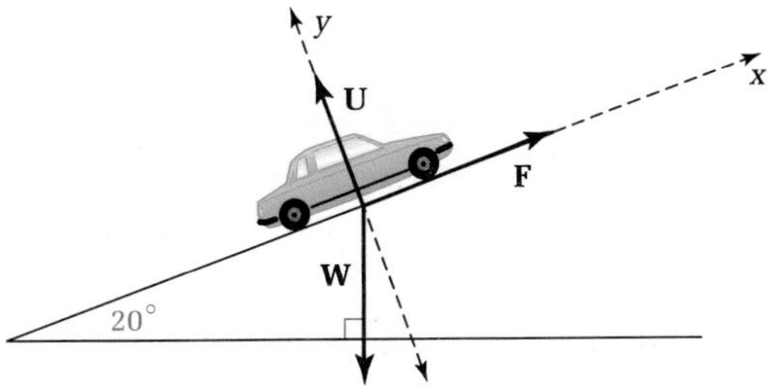

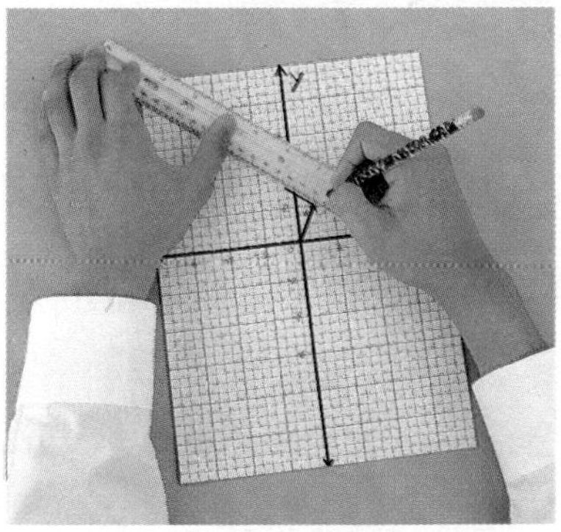

Ongoing **ASSESSMENT**

21. **Group Activity** Work with another student. You will need a ruler, a protractor, and graph paper.

a. Draw a vector **u** on a coordinate grid. Draw a second vector **v** whose tail is at the head of **u**.

b. Use the ruler and protractor to find a polar form (r_1, θ_1) of **u**. Then find a polar form (r_2, θ_2) of **v**.

c. Draw the vector $\mathbf{w} = (r_1 + r_2, \theta_1 + \theta_2)$ on the same coordinate grid you used in part (a). Based on your drawing, does $\mathbf{w} = \mathbf{u} + \mathbf{v}$? What does this tell you about adding vectors in polar form?

Application

To understand the problems in Exs. 19 and 20, students should understand the meaning of the term *force* as used in physics. When a force is applied to an object, the effect is determined by two things: the *magnitude* of the push or pull and its *direction.* Therefore, forces are represented by vectors rather than scalars.

Assessment: Open-ended

For Ex. 19, ask students to answer these questions. If Etenia doubles the amount of force with which she pulls the sled, will the forward force be doubled? If she pulls the sled at half the angle to the ground, will the forward force be doubled? What do you think $\mathbf{F_y}$ is in real-life terms? Calculate $\mathbf{F_y}$.

Answers to Exercises and Problems

19. a. about 58.3 lb

b. Decrease the angle of the force. For example, if she pulled with a force of 75 lb at an angle of 30°, there would be a 70 lb force pulling the toboggan forward. Etenia could do this by using a longer rope.

20. a. (–1026.06, –2819.08)

b. $(f, 0)$

c. $(0, u)$

d. $u = 1026.06$; $f = 2819.08$

21. a, b. Answers may vary.

c. No; vectors in polar form cannot be added by adding coordinates.

Working on the Unit Project

After completing Ex. 26, students should test the clues they have written so far. Each member of a group can follow all four clues to check that they all arrive at the same place.

Quick Quiz (8-1 through 8-4)

See page 479.

See also Test 32 in the *Assessment Book.*

Practice 62 For use with Section 8-4

Name ______________________ Date ____________

Practice 62

For use with Section 8-4

19. about (5.9, 53.4°)
20. about (11.0, 106.3°)
21. about (19.1, −133.0°)
22. about (12.6, 15.5°)
23. about (18.5, −5.1°)
24. about (5.8, 169.0°)

Convert each vector to polar form. Round decimal answers to the nearest tenth. Sample angle measures are given.

1. (2, 7) (7.3, 74.1°) 2. (3, −5) (5.8, −59.0°) 3. $(-4, 4\sqrt{3})$ (8, 120°)
4. (−6, −8) (10, −126.9°) 5. (4, 9) (9.8, 66.0°) 6. (−3, 7) (7.6, 113.2°)

Convert each vector to rectangular form. Round decimal answers to the nearest tenth.

7. (5, 150°) (−4.3, 2.5) 8. (3, 225°) (−2.1, −2.1) 9. (8, 38°) (6.3, 4.9)
10. $(\sqrt{2}, -45°)$ (1, −1) 11. (9, 295°) (3.8, −8.2) 12. (10, 165°) (−9.7, 2.6)

Find each vector sum in rectangular form.

13. (−7, 2) + (0, 5) (−7, 7) 14. (4, 5) + (6, −12) (10, −7) 15. (−1, −3) + (−7, 9) (−8, 6)
16. (3, 11) + (−10, 6) (−7, 17) 17. (2, −9) + (−5, 4) (−3, −5) 18. (2.5, −1.4) + (6, 8) (8.5, 6.6)

Find each vector sum in polar form. Exs. 19–24: See top of page.

19. (2, 40°) + (4, 60°) 20. (14, 125°) + (5, −10°) 21. (12, 200°) + (10, −100°)
22. (3, 155°) + (15, 8°) 23. (24, −50°) + (17, 80°) 24. (3, 110°) + (5, −160°)

25. A traffic light is suspended from two poles by wires that meet at point *X* in the diagram, $m\angle\theta = 38°$ and $m\angle\sigma = 52°$. Wire *XA* pulls at point *X* with a force of magnitude 25 lb. Let *f* be the magnitude of the pull at *X* by wire *XB*.

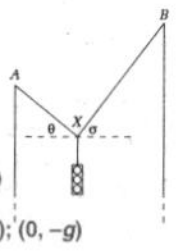

a. Write the force vectors **u** and **v** for the two wires in polar form (using *f* to express **v**). u = (25, 142°); v = (*f*, 52°)

b. Write **u** and **v** in rectangular form. u = (−19.7, 15.4); v = (*f* cos 52°, *f* sin 52°)

c. Let **w** be the weight of the traffic light. If the magnitude of **w** is *g*, write **w** in polar and rectangular form, using *g*. (*g*, 270°); (0, −*g*)

d. Since the light is stationary, **u** + **v** + **w** = (0, 0). Use this equation to find *f* and *g*. *f* = 32; *g* = 40.6

26. ***Open-ended*** Suspend an object from a string, like the traffic light in Exercise 25, but allow the object to slide along the string. Measure the weight of the object and the angles θ and σ. Use these measurements to calculate the magnitudes of the forces exerted by the two ends of the string. What do you notice about these angles and these forces? Answers may vary. Check students' work.

62 Practice Bank, INTEGRATED MATHEMATICS 3

Review PREVIEW

22. Let **u** = (2, −5). Write −3**u** in rectangular form. *(Section 8-3)*

Solve. *(Section 5-8)*

23. $200(1.08)^t = 500$

24. $\log_2 (x - 3) + \log_2 (x - 7) = 5$

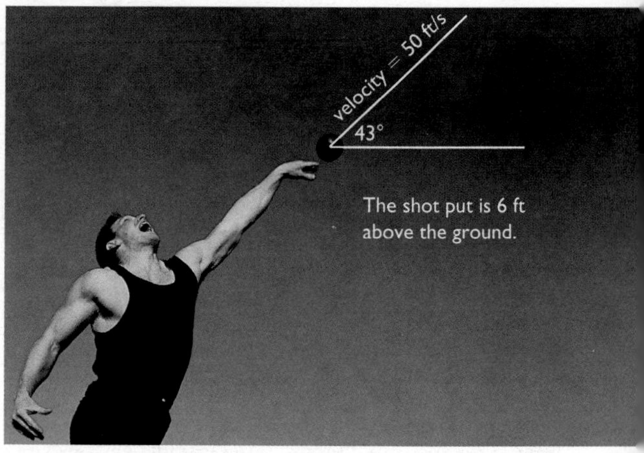

25. **Sports** Mark puts a shot as shown. *(Section 2-4)*

a. Use the equation

$$h(t) = -16t^2 + (v_0 \sin \theta)t + h_0$$

to express the height of the shot put above the ground as a function of time.

b. When does the shot put hit the ground?

8 Working on the Unit Project

26. Write a clue for your treasure hunt that requires someone to add two or more vectors in rectangular form in order to locate the next clue.

Unit 8 CHECKPOINT

1. **Writing** Suppose the magnitude of **u** is r_1 and the magnitude of **v** is r_2. Is it always true that the magnitude of **u** + **v** is $r_1 + r_2$? Explain your answer.

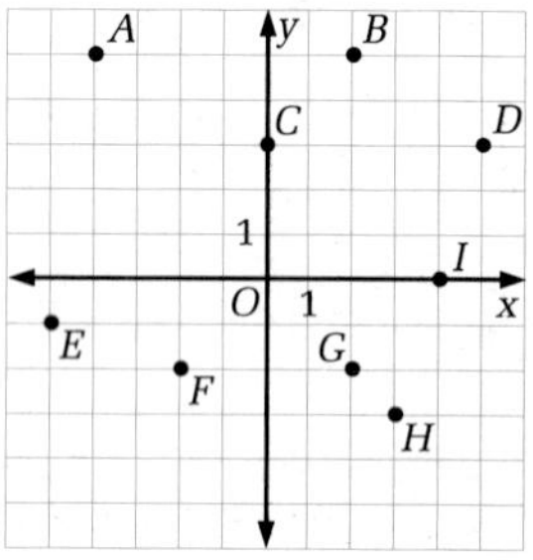

Name the point with each given pair of polar coordinates. 8-1

2. (3, 90°) 3. (2.8, −135°) 4. (2.8, −45°)

Find polar coordinates for each point.

5. *A* 6. *H* 7. *I*

8. Convert (8, 110°) to rectangular coordinates. 8-2

9. Convert (3, −5) to polar coordinates.

10. Find all angles θ between 0° and 360° for which cos θ = 0.6243.

11. Find sin θ when cos θ = −0.4892 and θ is in Quadrant II.

12. Draw the vector **w** = 3**u** − **v** where **u** = (2, 50°) and **v** = (3, −70°). 8-3

13. Find **u** = (7, 12) + (−4, 2) in rectangular form. 8-4

14. Find **v** = (5, 41°) + (9, −100°) in polar form.

Answers to Exercises and Problems

22. (−6, 15)
23. about 11.9
24. 11
25. a. $h(t) = -16t^2 + 34t + 6$
b. after about 2.3 s
26. Answers may vary.

Answers to Checkpoint

1. No; the magnitude of **u** + **v** will be $r_1 + r_2$ only if **u** and **v** have the same direction.
2. *C* 3. *F* 4. *G*
5. (6.4, 128.66°)
6. (4.24, −45°)
7. (4, 0°)
8. (−2.74, 7.52)
9. (5.83, −59.04°)
10. 51.37°; 308.63°
11. 0.8722
12.

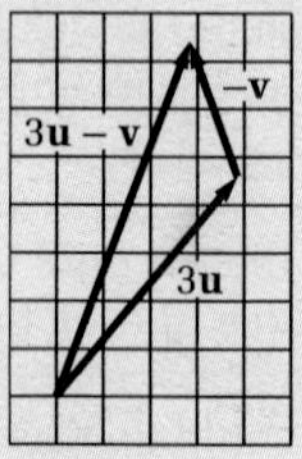

13. (3, 14)
14. (6.0, −68.4°)

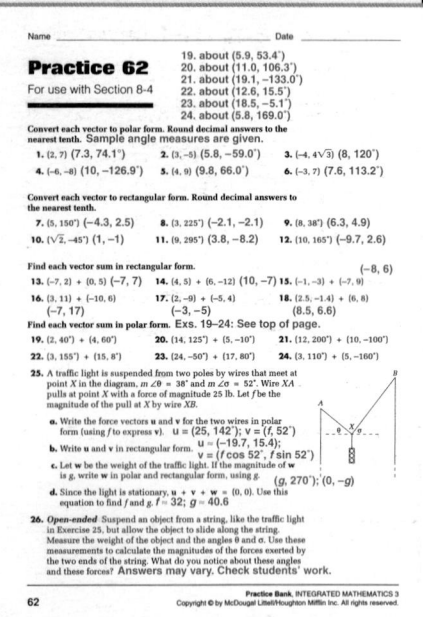

Section **8-5**

Parametric Equations

TIMED TRAVEL

Focus

Use parametric equations to solve problems.

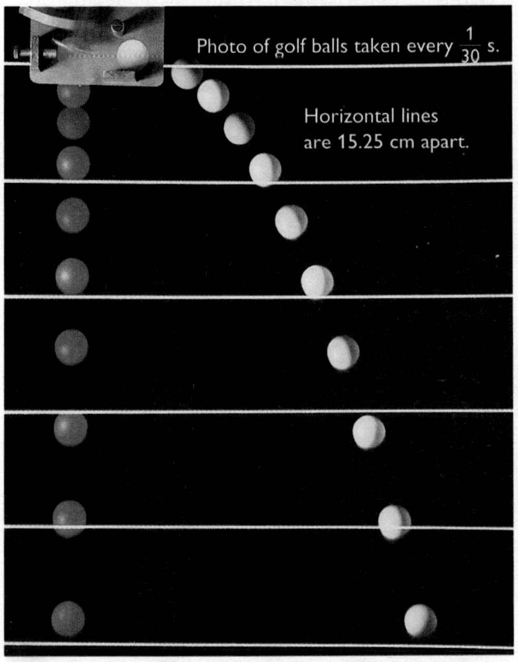

Photo of golf balls taken every $\frac{1}{30}$ s.
Horizontal lines are 15.25 cm apart.

Talk it Over

The photo shows two golf balls falling from the same starting point. The ball on the right was given an initial horizontal velocity of 2 m/s.

1. a. Compare the horizontal motion of the two golf balls. What do you notice?
   b. Compare the vertical motion of the two golf balls. What do you notice?
   c. Would you say that the second golf ball's horizontal motion has an effect on its vertical motion? Explain.
2. a. When an object falls, its vertical position varies as the square of the time it falls:
   $$y = at^2$$
   The vertical position of each golf ball at the bottom of the photograph is about −0.784 m after 0.4 s. Use this to determine the constant of variation a.
   b. The second ball traveled the horizontal distances given in this table. Find an equation giving x as a function of t.

Time t (seconds)	Horizontal position x (meters)
0.1	0.2
0.2	0.4
0.3	0.6

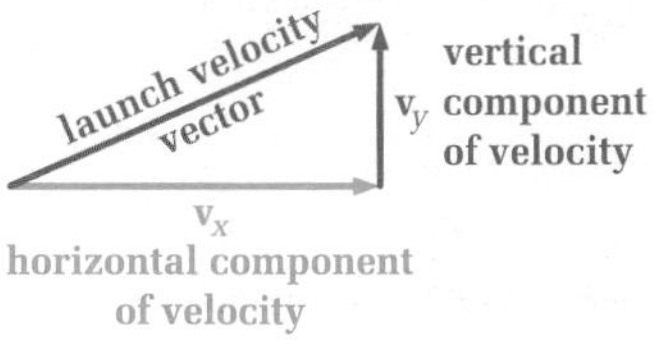

When an object falls, whether or not it is given an initial push, its horizontal and vertical movements are independent of each other.

Suppose an object is given a launch velocity **v** with horizontal and vertical components as shown. Then at time t the object's horizontal position x and vertical position y are given by these equations, where g, the acceleration due to gravity, is about 9.8 m/s² or $g \approx 32$ ft/s². The variables x and y each depend on a third variable, t. The equations relating x and y to t are called **parametric equations** and t is called a **parameter**.

$$x = v_x t$$
(v_x: magnitude of vector $\mathbf{v}_x$)

$$y = v_y t - \frac{1}{2}gt^2$$
(v_y: magnitude of vector $\mathbf{v}_y$)

PLANNING

Objectives and Strands

See pages 422A and 422B.

Spiral Learning

See page 422B.

Materials List

- Graphics calculator
- Graph paper

Recommended Pacing

Section 8-5 is a one-day lesson.

Extra Practice

See pages 621–622.

Warm-Up Exercises

- Warm-Up Transparency 8-5

Support Materials

- Practice Bank: Practice 63
- Activity Bank: Enrichment 56
- Study Guide: Section 8-5
- Problem Bank: Problem Set 18
- Explorations Lab Manual: Diagram Master 2
- Using TI-81 and TI-82 Calculators: Lissajous Figures
- Assessment Book: Quiz 8-5, Alternative Assessment 5

Answers to Talk it Over

1. a. The golf ball on the left has no horizontal motion. The golf ball on the right moves horizontally at a steady rate as it falls.
   b. The golf balls move vertically at the same rate.
   c. No; the vertical motion of the two golf balls is the same.
2. a. $a \approx -4.9$
   b. $x = 2t$

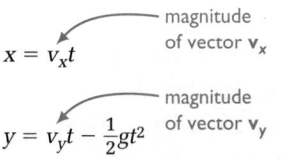

TEACHING

Additional Sample

S1 A stunt driver goes off the top of an inclined ramp at a speed of 120 mi/h and at an angle of 25°.

a. Express the horizontal and vertical positions, relative to the launch point, using parametric equations. Express positions in feet.

120 mi/h is equivalent to 176 ft/s, so use (176, 25°) for the velocity vector. Convert this vector to rectangular form. (176, 25°) = (159.51, 74.38) = (v_x, v_y) Substitute 159.51 for v_x and 74.38 for v_y in the parametric equations from page 453. The parametric equations for the horizontal and vertical positions are $x = 159.51t$ and $y = 74.38t - 16t^2$.

b. Use a graphics calculator to graph the path of the car. Use this graph to find the position of the car 1.5 s after it leaves the top of the ramp.

Use a viewing window with $0 \le x \le 1000$ and $-50 \le y \le 100$. Set the parameter t to go from 0 to 6 with a t-step of 0.1.

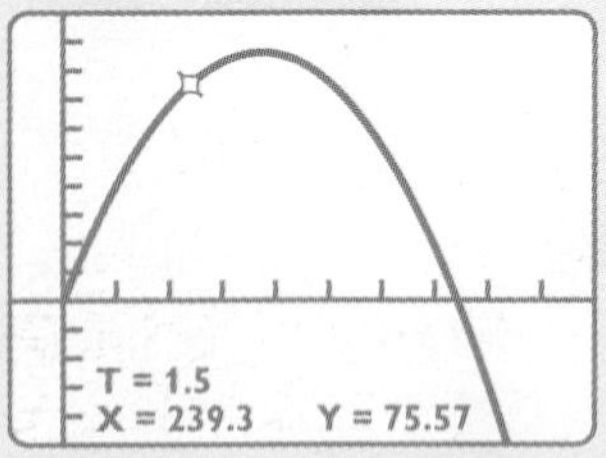

After 1.5 s, the car is about 239.3 ft to the right of the ramp and about 75.57 ft above the end of the ramp.

c. Combine the parametric equations into one equation relating y to x.

Solve $x = 159.51t$ for t and substitute in the equation for y.

$$t = \frac{x}{159.51}$$

$$y = 74.38\left(\frac{x}{159.51}\right) - 16\left(\frac{x}{159.51}\right)^2$$

$$\approx 0.467x - 0.00063x^2$$

An equation of the path is $y = 0.467x - 0.00063x^2$.

Sample 1

Suppose that when a skateboarder flies off a ramp, the skateboarder is moving at an angle of 41° with a velocity of 11 ft/s.

a. Express the horizontal and vertical positions, relative to the launch point of the skateboarder, using parametric equations.

b. Use a graphics calculator to graph the path of the moving skateboarder. Use this graph to find the skateboarder's position 0.2 s after the jump.

c. Combine the two parametric equations into a single equation relating y to x.

Sample Response

a. Find v_x and v_y.

(11, 41°) ≈ (8.3, 7.2) ← Convert the velocity vector from polar to rectangular form.

(11, 41°) is the velocity vector; 8.3 is v_x; 7.2 is v_y.

$$x = v_x t \qquad y = v_y t - \frac{1}{2}(32)t^2$$

$$= 8.3t \qquad = 7.2t - 16t^2$$

← Substitute **8.3** for v_x and **7.2** for v_y.

The parametric equations for the horizontal and vertical positions are $x = 8.3t$, and $y = 7.2t - 16t^2$.

b. An appropriate viewing window is $0 \le x \le 7$ and $-3 \le y \le 3$. Set the parameter t to go from 0 to 1 with a t-step of 0.01.

After 0.2 s, the skateboarder is 1.66 ft to the right of and 0.8 ft above the end of the ramp.

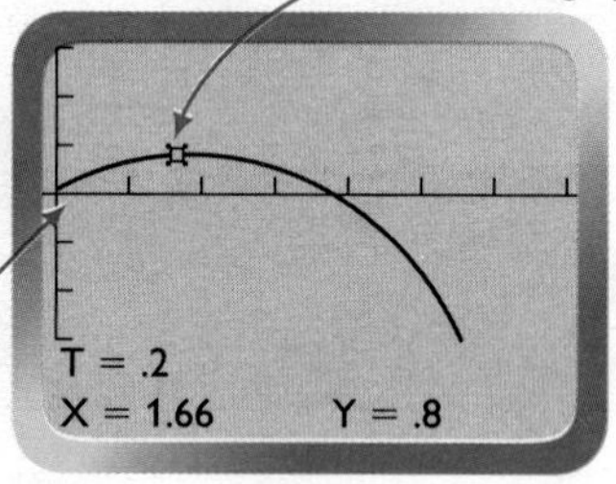

c. **1** Solve one of the equations for t.

$$x = 8.3t$$

$$\frac{x}{8.3} = t$$

2 Then substitute for t in the other equation.

$$y = 7.2t - 16t^2$$

$$= 7.2\left(\frac{x}{8.3}\right) - 16\left(\frac{x}{8.3}\right)^2$$

← Substitute $\frac{x}{8.3}$ for t.

$$= 0.87x - 0.24x^2$$

An equation of the path is $y = 0.87x - 0.24x^2$.

TECHNOLOGY NOTE

When graphing parametric equations, make these settings:

t-min ← the minimum value of t

t-max ← the maximum value of t

t-step ← the increment of time in each step from t-min to t-max

You also need to set the calculator mode to Parametric.

Talk it Over

3. a. What is the *y*-value when the skateboard hits the ground?
   b. When does the skateboard hit the ground?
4. Graph $y = 0.87x - 0.24x^2$. Do you think it is the same as the graph of the parametric equations?
5. How does changing the launch vector to (11, 20°) affect the path of the skateboard?

Sample 2

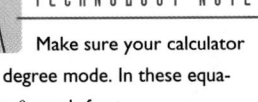

TECHNOLOGY NOTE

Make sure your calculator is in degree mode. In these equations, θ stands for *t*.

Suppose an object's motion is described by these parametric equations.

$$x = 2\cos\theta \qquad y = 2\sin\theta \qquad \text{where } 0° \le \theta \le 360°$$

a. Plot the path of this object.

b. Combine the two parametric equations into a single equation in terms of *x* and *y*.

Sample Response

a. An appropriate viewing window is $-6 \le x \le 6$, and $-4 \le y \le 4$. Set the parameter to go from 0 to 360 with a step of 1.

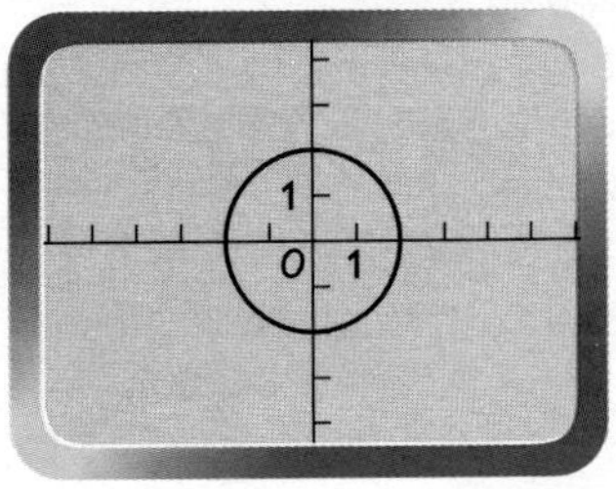

The path appears to be a circle of radius 2 centered at the origin.

b. Solve $x = 2\cos\theta$ and $y = 2\sin\theta$ for $\cos\theta$ and $\sin\theta$.

$$\cos\theta = \frac{x}{2}$$

$$\sin\theta = \frac{y}{2}$$

$$\cos^2\theta + \sin^2\theta = 1 \quad \longleftarrow \text{Use this Pythagorean identity.}$$

$$\left(\frac{x}{2}\right)^2 + \left(\frac{y}{2}\right)^2 = 1 \quad \longleftarrow \text{Substitute } \frac{x}{2} \text{ for } \cos\theta \text{ and } \frac{y}{2} \text{ for } \sin\theta.$$

$$\frac{x^2}{4} + \frac{y^2}{4} = 1 \quad \longleftarrow \text{Simplify.}$$

$$x^2 + y^2 = 4$$

An equation of the path is $x^2 + y^2 = 4$.

Look Back

Describe the effect of launch velocity on a falling object.

Additional Sample

S2 Suppose an object's motion is described by these parametric equations.
$x = 2\cos\theta$, $y = 3\sin\theta$, where $0° \le \theta \le 360°$

a. Plot the path of this object.
Use a viewing window with $-4.7 \le x \le 4.7$ and $-3.1 \le y \le 3.1$. Set the parameter to go from 0 to 360 with a step of 1.

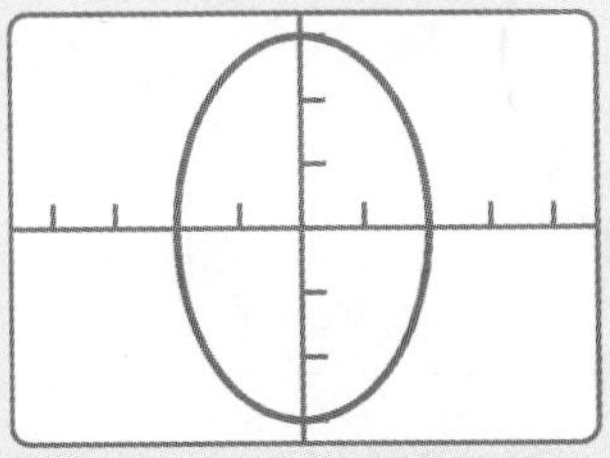

The path appears to be an ellipse centered at the origin, with a horizontal axis of 4 and a vertical axis of 6.

b. Combine the two parametric equations into a single equation.
Solve each of the parametric equations for the term involving θ.

$$\frac{x}{2} = \cos\theta,\ \frac{y}{3} = \sin\theta$$

Use the Pythagorean identity, substituting $\frac{x}{2}$ for $\cos\theta$ and $\frac{y}{3}$ for $\sin\theta$.

$$\cos^2\theta + \sin^2\theta = 1$$

$$\left(\frac{x}{2}\right)^2 + \left(\frac{y}{3}\right)^2 = 1$$

$$\frac{x^2}{4} + \frac{y^2}{9} = 1$$

The equation of the path is $9x^2 + 4y^2 = 36$.

Communication: Drawing

When sketching the graph of parametric equations such as those in Sample 2, you may wish to label some points with *t*-values to demonstrate the path of an object following the equations.

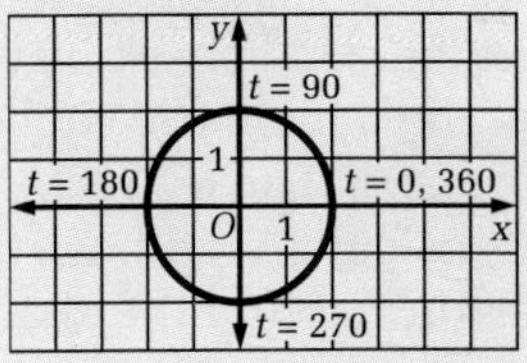

Answers to Talk it Over

3. a. −3
   b. after about 0.71 s
4. 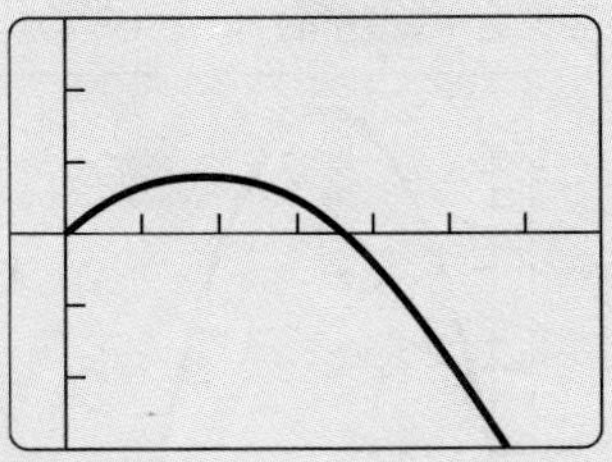

   Yes.
5. Answers may vary. An example is given. The skateboard will not achieve the same height and will hit the ground slightly sooner.

Answers to Look Back

If an object is given an initial velocity **v** with horizontal component v_x and vertical component v_y, its horizontal position at time *t* is $x = v_x t$ and its vertical position at time *t* is $y = v_y t - \frac{1}{2}gt^2$, where *g* is the acceleration due to gravity.

APPLYING

Suggested Assignment

Standard 5–11, 13–32, 34–39

Extended 1–32, 34–39

Integrating the Strands

Algebra Exs. 1–24, 26–33, 35–39

Geometry Exs. 4, 11–24, 26–33, 37–39

Trigonometry Exs. 4, 9–24, 26–33

Discrete Mathematics Exs. 35, 36

Logic and Language Exs. 1, 11, 12, 25, 34, 39

Reasoning

A discussion of Ex. 1 can be used to inquire why the equation for the horizontal component is linear and not quadratic. The reasoning is clear for students who are familiar with Newton's first law of motion. This law states that every object remains at rest or in motion in a straight line at constant speed unless acted upon by an unbalanced force. If air resistance is ignored, this law describes the horizontal component for the motion considered in Sample 1 on page 454.

Teaching Tip

Exs. 1–3 can be used to ascertain whether students understand the concept of velocity.

Problem Solving

In Ex. 11(b), students can obtain a single equation in x and y by using the algebraic procedure of Sample 1, part (c) on page 454. Ask if they can think of another way to find the equation. One possibility is to graph the parametric equations for part (a) on a TI-82 graphics calculator and use the Trace feature to find the x- and y-coordinates of three or more points of the graph. Then enter these coordinates as statistical data and perform a quadratic regression.

8-5 Exercises and Problems

1. **Reading** What does g represent in the equation for the vertical position of a falling object as a function of time? What is the value of g?
2. What is the relationship between these two different equations for y?

$$y = at^2 \quad \text{and} \quad y = v_y t - \frac{1}{2}gt^2$$

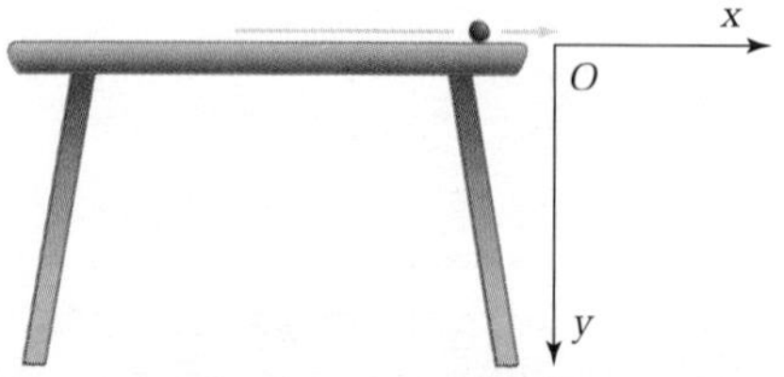

3. A billiard ball with a horizontal velocity of 3 m/s and a vertical velocity of 0 m/s rolls off of a table that is 1 m high.
   a. Write parametric equations to describe the position of the billiard ball at any time t.
   b. Complete the following table using the equations.

TIME (seconds)	Horizontal position x	Vertical position y	Ordered pair
0.0	?	?	?
0.1	?	?	?
0.2	?	?	?
0.3	?	?	?
0.4	?	?	?
0.5	?	?	?

   c. Estimate the time it takes for the billiard ball to hit the floor. Explain your answer.

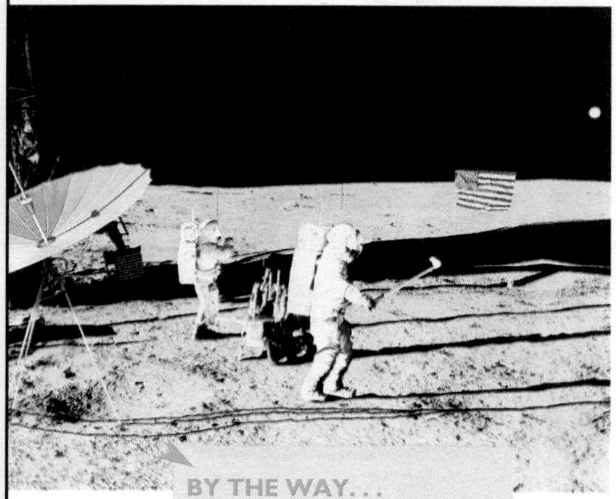

BY THE WAY...

During the 1971 Apollo 14 mission, astronaut Alan B. Shepard surprised Mission Control when he played golf on one of his moon walks. He estimated that his best shot traveled 400 yd.

4. **Golf** Suppose you hit a golf ball on Earth at an angle of 10° with a velocity of 235 ft/s.
   a. Write parametric equations that describe the path of the ball.
   b. Suppose you play golf on the moon, where $g = 2.7$ ft/s². How far will the ball travel?

Combine each pair of parametric equations into a single equation relating y to x.

5. $x = 2.5t,\ y = 3.4t$
6. $x = 53.9t,\ y = -4.9t^2$
7. $x = 9.1t,\ y = 3.23t - 4.9t^2$
8. $x = 4.03t,\ y = 4.03t - 16.1t^2$
9. $x = 3.6 \cos t,\ y = 3.6 \sin t$
10. $x = (6 \cos t) + 3,\ y = (6 \sin t) - 2$

Answers to Exercises and Problems

1. acceleration due to gravity; $g \approx -9.8$ m/s² in the metric system and $g \approx -32$ ft/s² in the customary system.
2. The first equation, $y = at^2$, is for an object with a launch velocity of 0. The second is for an object with a nonzero launch velocity.
3. See answers in back of book.
4. a. $x = 231.4t$; $y = 40.8t - 16t^2$
   b. about 6993 ft (or about 1.3 mi)
5. $y = 1.36x$
6. $y = -0.002x^2$
7. $y = 0.35x - 0.06x^2$
8. $y = x - 0.99x^2$
9. $x^2 + y^2 = 12.96$ or $y = \pm\sqrt{12.96 - x^2}$
10. $(x - 3)^2 + (y + 2)^2 = 36$ or $y = -2 \pm \sqrt{36 - (x - 3)^2}$
11. a. $x = 33.88t$; $y = 26.47t - 4.9t^2 + 1$

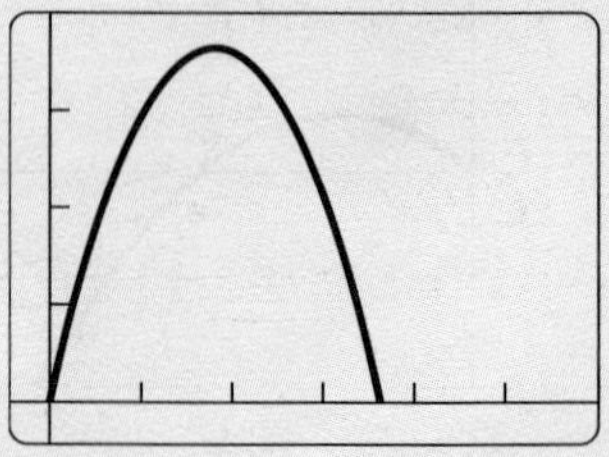

11. Baseball A batter hits a baseball with a velocity of 43 m/s and at an angle of 38°. The ball is 1 m above the ground when it is hit.

a. Write and graph parametric equations to describe the path of the ball.

b. Convert the parametric equations into a single equation.

c. Graph the single equation.

d. Compare the graphs.

e. What are the advantages and disadvantages of each type of equation?

BY THE WAY...

The Silver Bullets professional women's baseball team was founded in 1994. It is the first all-female team to be recognized by the National Association of Professional Baseball Leagues and compete against all-male minor-league teams.

12. Gardening To be sure her whole garden gets water, Wynona changes the angle of the hose so the water sprays farther. The water flows out of her garden hose at a velocity 7.8 m/s, and the end of the hose is about 1.5 m above the ground.

8 m

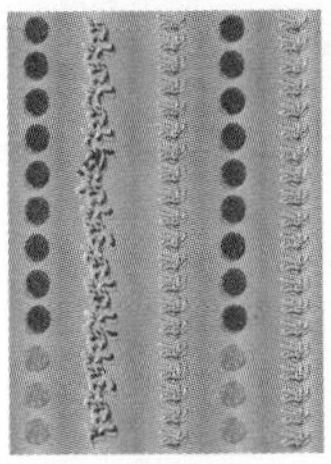

5 m

a. Write parametric equations that describe the path of the water when the hose is held at an angle of 30°, 40°, 50°, and 60°.

b. Graph the equations.

c. **Writing** Can Wynona water the whole garden while standing in one place? Does it matter where she stands? Explain your reasoning.

TECHNOLOGY **In the parametric equations for Exercises 13–32, the variable t stands for θ. Use the following settings on a graphics calculator:**

$0 \le t \le 360$ t-step = 1 $-12 \le x \le 12$ $-8 \le y \le 8$

For Exercises 13–15, describe the graph of each pair of parametric equations.

13. $x = 8 \cos t$, $y = 8 \sin t$

14. $x = 5 \cos t$, $y = 5 \sin t$

15. $x = 7.5 \cos t$, $y = 7.5 \sin t$

16. Write parametric equations for a circle with radius 3 centered at the origin.

17. Graph the equations $x = (4 \cos t) + 4$ and $y = (4 \sin t) - 3$.

18. Write parametric equations for a circle with radius 7 centered at the point $(-3, -2)$.

19. What can you conclude about the graph of the parametric equations $x = (a \cos t) + b$ and $y = (a \sin t) + c$ where $a > 0$?

For Exercises 20–22, graph Exercise 13. Then change each setting on the calculator as indicated. Describe how the graph changes.

20. Change t-step to 10.

21. Change t-min to 90.

22. Change t-max to 285.

23. Write parametric equations for a semicircle with radius 6 units in quadrants III and IV.

24. Write parametric equations for a circular arc with measure 47° and radius 7 units starting at 115°.

25. **Open-ended** Graph a pair of parametric equations. Change the t-step setting on the calculator. Predict the change in the graph. Check it.

Multicultural Note

During World War II, many male baseball players went overseas to fight. Philip Wrigley, then the owner of the Chicago Cubs, decided to help maintain the national interest in baseball by creating a women's league, which he named the All-American Girls Professional Baseball League. In May 1943, the first tryouts were held, and four women's teams, the Rockford Peaches, Racine Belles, Kenosha Comets, and South Bend Blue Sox, were quickly formed. The league later expanded to include six other teams. Many of the players were excellent athletes: Joanne Weaver of the Fort Wayne Daisies hit .429 one season, Sophie Kurys of the Racine Belles stole 201 bases in 203 attempts one year, and Jean Faut of the South Bend Blue Sox pitched two perfect games and won three pitching championships. Soon after World War II ended, the league was disbanded.

Integrating the Strands

Exs. 13–33 integrate the strands of algebra, trigonometry, and geometry. These exercises lead students to discover how circles and ellipses can be represented using parametric equations that involve trigonometric functions.

Using Technology

Some students may be interested in attempting to solve the graphing problem in Ex. 17 by using a single *polar* equation rather than a pair of parametric equations. They will need to be familiar with the ideas in the Using Technology note on page 428 in order to do this.

Answers to Exercises and Problems

b. $y = 0.78x - 0.0043x^2 + 1$

c.

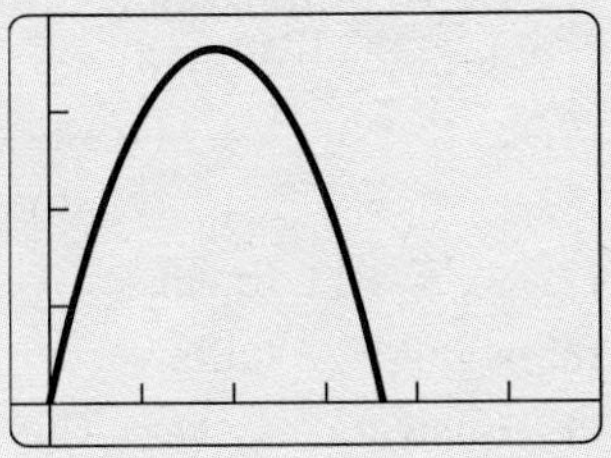

d. The graphs are essentially the same. They may not be identical due to rounding.

e. Answers may vary. An example is given. The parametric equations allow you to determine the horizontal and vertical positions of the ball at any given time t. The single equation allows you to determine the height of the ball at any given horizontal distance x from the point where the ball was hit.

12–25. See answers in back of book.

Assessment: Standard

For Ex. 29, students should be able to state that in $x = a \cos t + b$ and $y = c \sin t + d$ if $a = c$, the graph of the equations is a circle with center at (b, d) and radius a. If $a \neq c$, the graph is an ellipse with center (b, d) minor (or major) axis of c and major (or minor) axis of a.

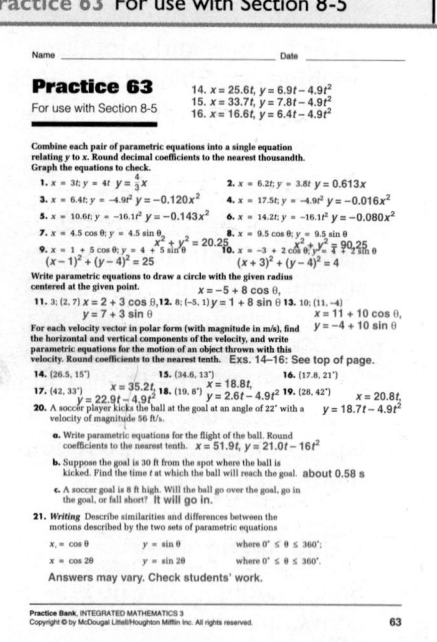

TECHNOLOGY A geometric shape called an *ellipse* can be graphed using parametric equations. Use the following settings on a graphics calculator:

$0 \le t \le 360$ t-step = 1 $-12 \le x \le 12$ $-8 \le y \le 8$

For Exercises 26–28, describe the graph of each pair of parametric equations.

26. $x = 8 \cos t$
$y = 6 \sin t$

27. $x = (5.5 \cos t) - 1$
$y = 8 \sin t$

28. $x = (4 \cos t) + 3$
$y = (7 \sin t) - 1$

29. What can you conclude about the graph of the parametric equations $x = (a \cos t) + b$ and $y = (c \sin t) + d$ where $a > 0$ and $c > 0$?

Write parametric equations for each graph.

30.

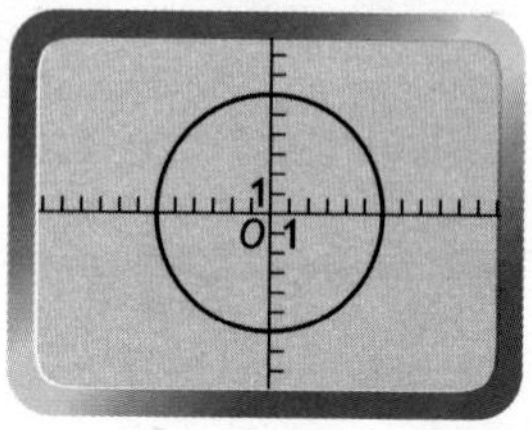

31.

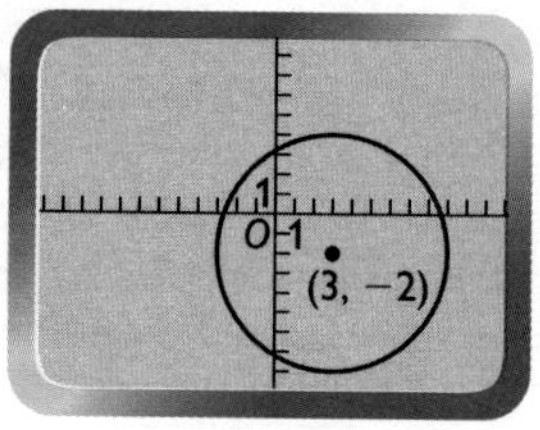

32.

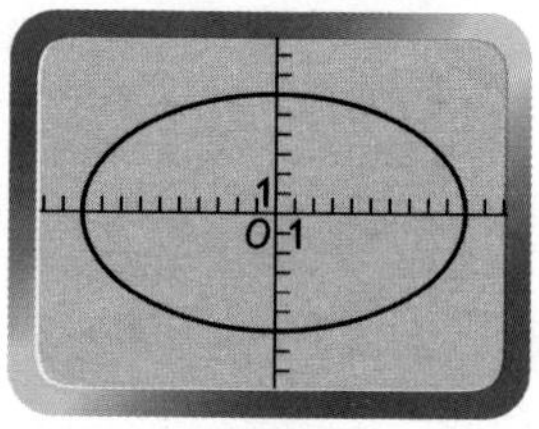

Ongoing ASSESSMENT

33. **Group Activity** Work with another student.
   a. Write parametric equations for a circle or an arc.
   b. Trade equations with your partner. Sketch the graph of the equations without using a graphics calculator.

34. **Writing** Describe a situation that can be modeled using parametric equations. Write three questions that can be answered using the parametric equations.

Review PREVIEW

For Exercises 35 and 36: *(Section 8-4)*

a. Find v + u. **b. Find u − v.**

35. $\mathbf{v} = (1, 5)$, $\mathbf{u} = (2, -7)$

36. $\mathbf{v} = (-2, 3)$, $\mathbf{u} = (5, -6)$

37. Find the length of the hypotenuse of a right triangle whose legs are 10 cm and 5 cm long. *(Toolbox Skill 30)*

38. Find the distance between (3, 7) and (5, 0). *(Toolbox Skill 31)*

8 Working on the Unit Project

39. Plot two more locations on your treasure hunt map: the location of the last clue and the location of the treasure. You will use these in Sections 8-6 and 8-7.

Practice 63 For use with Section 8-5

Name ______________ Date ______________

Practice 63
For use with Section 8-5

14. $x = 25.6t$, $y = 6.9t - 4.9t^2$
15. $x = 33.7t$, $y = 7.8t - 4.9t^2$
16. $x = 16.6t$, $y = 6.4t - 4.9t^2$

Combine each pair of parametric equations into a single equation relating y to x. Round decimal coefficients to the nearest thousandth. Graph the equations to check.

1. $x = 3t$; $y = 4t$ $y = \frac{4}{3}x$
2. $x = 6.2t$; $y = 3.8t$ $y = 0.613x$
3. $x = 6.4t$; $y = -4.9t^2$ $y = -0.120x^2$
4. $x = 17.5t$; $y = -4.9t^2$ $y = -0.016x^2$
5. $x = 10.6t$; $y = -16.1t^2$ $y = -0.143x^2$
6. $x = 14.2t$; $y = -16.1t^2$ $y = -0.080x^2$
7. $x = 4.5 \cos \theta$; $y = 4.5 \sin \theta$ $x^2 + y^2 = 20.25$
8. $x = 9.5 \cos \theta$; $y = 9.5 \sin \theta$ $x^2 + y^2 = 90.25$
9. $x = 1 + 5 \cos \theta$; $y = 4 + 5 \sin \theta$ $(x - 1)^2 + (y - 4)^2 = 25$
10. $x = -3 + 2 \cos \theta$; $y = 4 + 2 \sin \theta$ $(x + 3)^2 + (y - 4)^2 = 4$

Write parametric equations to draw a circle with the given radius centered at the given point.

11. 3; (2, 7) $x = 2 + 3 \cos \theta$, $y = 7 + 3 \sin \theta$
12. 8; (−5, 1) $x = -5 + 8 \cos \theta$, $y = 1 + 8 \sin \theta$
13. 10; (11, −4) $x = 11 + 10 \cos \theta$, $y = -4 + 10 \sin \theta$

For each velocity vector in polar form (with magnitude in m/s), find the horizontal and vertical components of the velocity, and write parametric equations for the motion of an object thrown with this velocity. Round coefficients to the nearest tenth. Exs. 14–16: See top of page.

14. (26.5, 15°)
15. (34.6, 13°)
16. (17.8, 21°)
17. (42, 33°) $x = 35.2t$, $y = 22.9t - 4.9t^2$
18. (19, 8°) $x = 18.8t$, $y = 2.6t - 4.9t^2$
19. (28, 42°) $x = 20.8t$, $y = 18.7t - 4.9t^2$

20. A soccer player kicks the ball at the goal at an angle of 22° with a velocity of magnitude 56 ft/s.
   a. Write parametric equations for the flight of the ball. Round coefficients to the nearest tenth. $x = 51.9t$, $y = 21.0t - 16t^2$
   b. Suppose the goal is 30 ft from the spot where the ball is kicked. Find the time t at which the ball will reach the goal. about 0.58 s
   c. A soccer goal is 8 ft high. Will the ball go over the goal, go in the goal, or fall short? It will go in.

21. ***Writing*** Describe similarities and differences between the motions described by the two sets of parametric equations

$x = \cos \theta$ $y = \sin \theta$ where $0° \le \theta \le 360°$;
$x = \cos 2\theta$ $y = \sin 2\theta$ where $0° \le \theta \le 360°$.

Answers may vary. Check students' work.

Practice Bank, INTEGRATED MATHEMATICS 3
 63

Answers to Exercises and Problems

26. an ellipse with its center at the origin, 12 units high, and 16 units wide

27. an ellipse with its center at (−1, 0), 11 units wide, and 16 units high on the line $x = -1$

28. an ellipse with its center at (3, −1), 8 units wide on the line $y = -1$, and 14 units high on the line $x = 3$

29. If $a \neq c$, the graph is an ellipse centered at (b, d) with one axis $2a$ units high on the line $x = b$, and one axis $2c$ units wide on the line $y = d$. If $a = c$, the graph is a circle with radius a and center (b, d).

30. $x = 6 \cos t$; $y = 6 \sin t$

31. $x = 6 \cos t + 3$;
$y = 6 \sin t - 2$

32. $x = 10 \cos t$; $y = 6 \sin t$

33. Answers may vary. Check students' work.

34. Answers may vary. An example is given. A ball is tossed out a window 5 ft above ground at an angle of 20° with the horizontal and an initial speed of 3 ft/s. Equations: $x = 2.82t$, $y = 1.03t - 16t^2 + 5$. Questions: What is the ball's maximum height? How long does it take for the ball to hit the ground? How far from the window does the ball land?

35. a. (3, −2) b. (1, −12)

36. a. (3, −3) b. (7, −9)

37. $5\sqrt{5} \approx 11.18$ cm

38. about 7.28 units

39. Answers may vary. Check students' work.

Section **8-6**

The Law of Cosines

Focus
Derive the law of cosines and use it to find measures of sides and angles in triangles.

Devils Tower National Monument looms above the plains in the northeastern corner of Wyoming.

The map shows distance and angle measurements to the base of the tower from three viewpoints along the hiking trail that circles the tower.

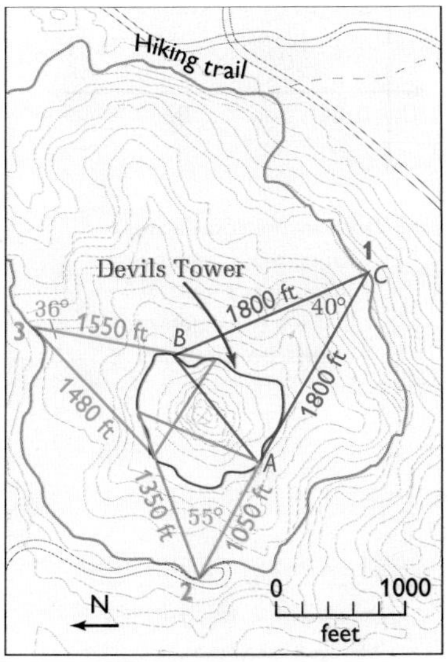

EXPLORATION

How can you find the length of the third side of a triangle from the lengths of the other two sides and the measure of their included angle?

- **Materials: centimeter rulers, protractors**
- **Work in a group of three students.**

① Choose one of the three viewpoints along the trail.

② Use the side-length and angle measures in the table below to make a scale drawing of $\triangle ABC$, where C is your viewpoint, and A and B are points on Devils Tower. Use the scale 1 cm = 100 ft.

③ Measure AB to the nearest millimeter on your scale drawing and record it in a table like the one below. How many feet does this represent through Devils Tower?

④ Draw an x-axis through points C and B, and a y-axis through point C. Find rectangular coordinates for point A.

⑤ Find AB using the distance formula. Record your results in the table. Compare your answers with the measurements on your scale drawing.

	Find *AB* using a scale drawing.				Find *AB* using the distance formula.			
	CB	$m\angle C$	*CA*	*AB*	*C*	*B*	*A*	*AB*
1	18 cm	40°	18 cm	?	(0, 0)	(18, 0)	(?, ?)	?
2	10.5 cm	55°	13.5 cm	?	(0, 0)	(10.5, 0)	(?, ?)	?
3	14.8 cm	36°	15.5 cm	?	(0, 0)	(14.8, 0)	(?, ?)	?

PLANNING

Objectives and Strands
See pages 422A and 422B.

Spiral Learning
See page 422B.

Materials List
- Centimeter ruler
- Protractor
- Compass

Recommended Pacing
Section 8-6 is a two-day lesson.

Day 1
Pages 459–461: Exploration through Talk it Over 3, *Exercises 1–16*

Day 2
Pages 462–463: Exploration through Look Back, *Exercises 17–32*

Extra Practice
See pages 621–622.

Warm-Up Exercises
Warm-Up Transparency 8-6

Support Materials
- Practice Bank: Practice 64
- Activity Bank: Enrichment 57
- Study Guide: Section 8-6
- Problem Bank: Problem Set 18
- Explorations Lab Manual: Diagram Masters 18, 19
- Assessment Book: Quiz 8-6, Alternative Assessment 6

Answers to Exploration

1, 2. Answers may vary. Check students' work.

3–5. Answers may vary due to measurement and rounding. Examples are given.

3. 1: 12.3 cm (1230 ft);
2: 11.4 cm (1140 ft);
3: 9.4 cm (940 ft)

4. 1: (13.8, 11.6);
2: (7.7, 11.1);
3: (12.5, 9.1)

5.

	Find *AB* using a scale drawing.			
	CB	$m\angle C$	*CA*	*AB*
1	18 cm	40°	18 cm	12.3 cm
2	10.5 cm	55°	13.5 cm	11.4 cm
3	14.8 cm	36°	15.5 cm	9.4 cm

	Find *AB* using the distance formula.			
	C	*B*	*A*	*AB*
1	(0, 0)	(18, 0)	(13.8, 11.6)	12.3 cm
2	(0, 0)	(10.5, 0)	(7.7, 11.1)	11.4 cm
3	(0, 0)	(14.8, 0)	(12.5, 9.1)	9.4 cm

The answers are the same.

TEACHING

Exploration

The goal of the Exploration is to motivate the derivation of the law of cosines, which is given on this page.

Reasoning

Ask students whether the derivation of the law of cosines on this page assumes anything about the size of $\angle C$. Does it work if $\angle C$ is a right angle? an obtuse angle? Students should be able to explain their thinking.

Assessment: Standard

Have students verify that the law of cosines is true for equilateral triangles and right isosceles triangles.

Error Analysis

As with the Pythagorean theorem, students may forget to take the square root of c^2 (or a^2 or b^2) at the conclusion of using the law of cosines. Pointing out the large square root symbol in the last line of step 4 on this page may help to lessen this error.

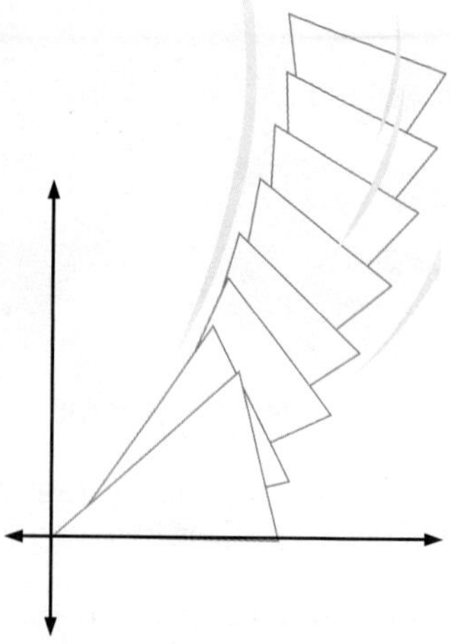

The triangles in the Exploration on page 459 are SAS (side-angle-side) constructions. These constructions yield unique triangles.

It is useful to be able to find the length of the third side without doing the construction and measuring. If the triangle is a right triangle, you can use the Pythagorean theorem. Otherwise, you can use the *law of cosines,* which is derived from the distance formula like this:

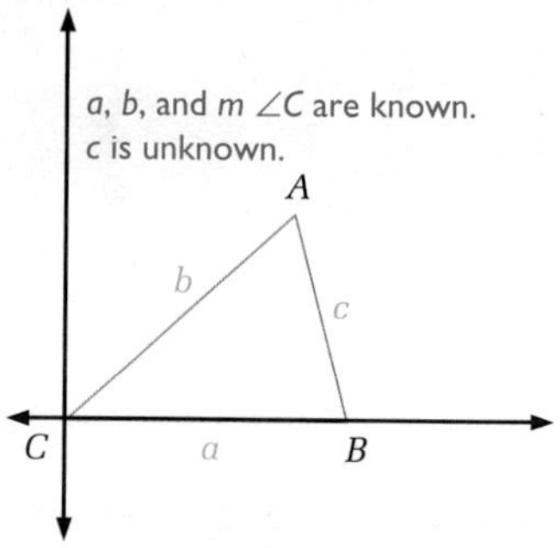

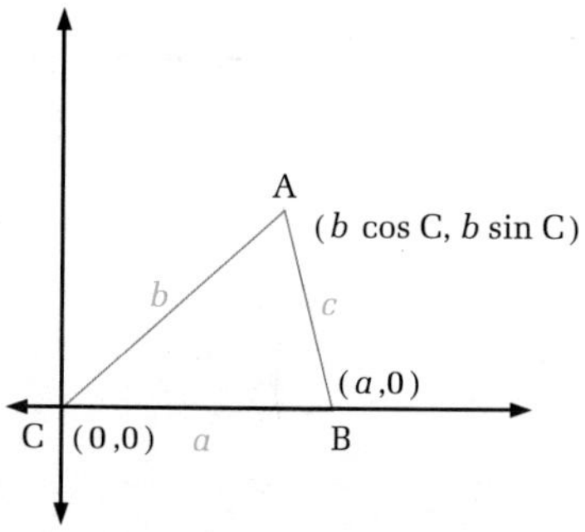

1 Given a triangle, place it so that one vertex is at the origin and one side lies on the *x*-axis.

2 Label the triangle. It is customary to label length *a* opposite $\angle A$, length *b* opposite $\angle B$, and length *c* opposite $\angle C$.

3 Identify the coordinates of each vertex in terms of the known measurements. You can use trigonometry to find the coordinates of A.

4 Use the distance formula to find *c*.

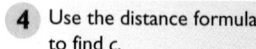

$d = \sqrt{(x_2 - x_1)^2 + (y_2 - y_1)^2}$	← Write the distance formula.
$c = \sqrt{(a - b\cos C)^2 + (0 - b\sin C)^2}$	← Substitute **c** for **d**, **(a, 0)** for (x_2, y_2), and **(b cos C, b sin C)** for (x_1, y_1).
$c^2 = (a - b\cos C)^2 + (0 - b\sin C)^2$	← Square both sides.
$= a^2 - 2ab\cos C + b^2\cos^2 C + b^2\sin^2 C$	← Simplify.
$= a^2 - 2ab\cos C + b^2(\cos^2 C + \sin^2 C)$	← Factor b^2 from the last two terms.
$= a^2 - 2ab\cos C + b^2(1)$	← Substitute 1 for $\cos^2 C + \sin^2 C$.
$= a^2 + b^2 - 2ab\cos C$	← Simplify and reorder terms.
$c = \sqrt{a^2 + b^2 - 2ab\cos C}$	← Take the positive square root of both sides.

The result resembles the Pythagorean theorem for right triangles, but it works for all triangles. You can use it to find the length of the third side of a triangle when you know the lengths of two sides and the measure of the included angle. If you put vertex *A* or *B* at the origin instead of vertex *C*, you can derive the related expressions for a^2 or b^2.

LAW OF COSINES

In any $\triangle ABC$:

$$a^2 = b^2 + c^2 - 2bc \cos A$$

$$b^2 = a^2 + c^2 - 2ac \cos B$$

unknown side length → $c^2 = a^2 + b^2 - 2ab \cos C$

(known side length: a; known side length: b; measure of the included angle: C)

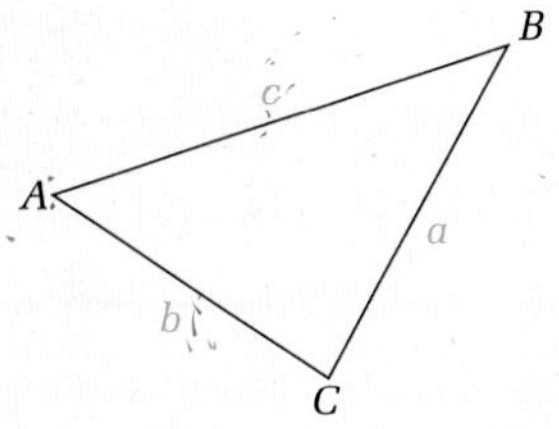

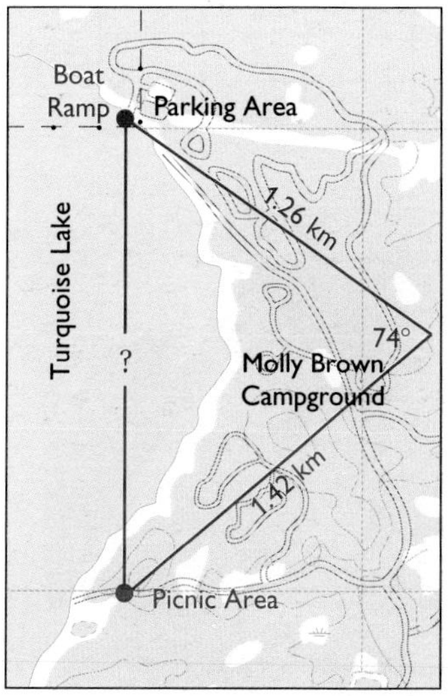

Sample 1

Find the distance from the picnic area to the boat ramp across Turquoise Lake.

Sample Response

The map shows two sides of a triangle that have lengths 1.26 km and 1.42 km, and an included angle that measures 74°.

$c^2 = a^2 + b^2 - 2ab \cos C$ ← Use the law of cosines.

$= (1.26)^2 + (1.42)^2 - 2(1.26)(1.42) \cos 74°$ ← Substitute **1.26** for a, 1.42 for b, and 74° for C

≈ 2.62

$c \approx 1.62$ ← Take the positive square root of both sides.

The boat ramp is about 1.6 km from the picnic area.

Talk it Over

1. In the derivation of the law of cosines, why can you substitute 1 for the expression $\cos^2 C + \sin^2 C$?
2. Use the law of cosines to calculate AB in the triangle you drew in the Exploration on page 459. Compare your answer to the results you found in the Exploration.
3. Can you use the law of cosines to find the hypotenuse of a right triangle? How is the Pythagorean theorem related to the law of cosines?

→ Now you are ready for: Exs. 1–16 on pp. 463–465

Additional Sample

S1 In an older section of downtown Bensonville, three streets intersect to form a triangular region ABC.

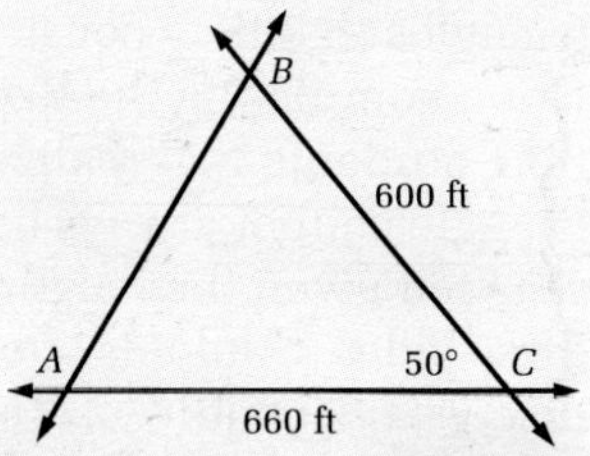

Streets $\overline{AC}$ and $\overline{CB}$ make an angle of 50° with each other. If $\overline{AC}$ and $\overline{CB}$ have the lengths shown in the diagram, what is the length of the third side of the triangular region?

Use the law of cosines.

$c^2 = a^2 + b^2 - 2ab \cos C$

$c^2 = (600)^2 + (660)^2 - 2(600)(660) \cos 50°$

$c^2 \approx 286{,}512.21$

$c \approx 535.3$

The length of the third side is about 535.3 ft.

Problem Solving

Ask students how they could use the law of cosines to decide, without using a calculator, whether a triangle whose side lengths are given is acute, right, or obtuse.

Talk it Over

In question 3, students discover that when trying to find the hypotenuse of a right triangle, the law of cosines reduces to the Pythagorean theorem.

Answers to Talk it Over

1. because of the Pythagorean identity, which states that for every angle θ, $\cos^2 \theta + \sin^2 \theta = 1$
2. Answers may vary due to measurement or rounding; 1: 1230 ft; 2: 1140 ft; 3: 940 ft; They are the same.
3. Yes; since the term $2ab \cos C = 0$, the formula simplifies to $a^2 + b^2 = c^2$. The Pythagorean theorem is a special case of the law of cosines for $\theta = 90°$.

Exploration

The goal of the Exploration is to have students construct triangles using side-side-side constructions and to find the measures of the angles of the triangles that they construct. Upon completing the Exploration, students will realize that a triangle cannot always be constructed given the lengths of three sides. Students should discuss the conditions that result in no triangle being constructed.

Additional Sample

S2 A triangle has sides of lengths 3.7 m, 5.4 m, and 5.9 m. Calculate the measure of the angle opposite the longest side of the triangle.

Let $\angle A$ be the angle opposite the longest side. Use the law of cosines to find cos *A*.

$$\cos A = \frac{b^2 + c^2 - a^2}{2bc}$$

$$\cos A = \frac{(3.7)^2 + (5.4)^2 - (5.9)^2}{2(3.7)(5.4)}$$

$\cos A \approx 0.2012$

Use the inverse cosine feature of a calculator.

$\cos^{-1} 0.2012 \approx 78.4°$

The measure of the angle opposite the longest side is about 78°.

EXPLORATION

How can you find the measures of the angles of a triangle when you know the lengths of the three sides?

- **Materials: rulers, compasses, protractors**
- **Work in a group of three students.**

① Construct the triangles whose side lengths are given in rows 2, 3, and 4 of the table below.

	AB	*AC*	*BC*	$m\angle A$	$m\angle B$	$m\angle C$
1	8.6 cm	6.4 cm	4.1 cm	27°	45°	108°
2	6.7 cm	7.2 cm	4.4 cm	?	?	?
3	8.2 cm	3.5 cm	6.5 cm	?	?	?
4	4.2 cm	6.9 cm	3.6 cm	?	?	?
5	8.7 cm	3.7 cm	4.0 cm	?	?	?

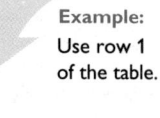

Example:
Use row 1 of the table.

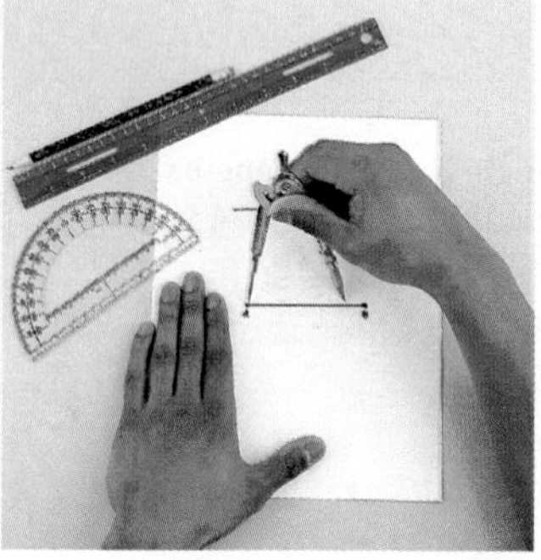

Draw $\overline{AB}$. From point *A*, draw an arc with radius *AC*.

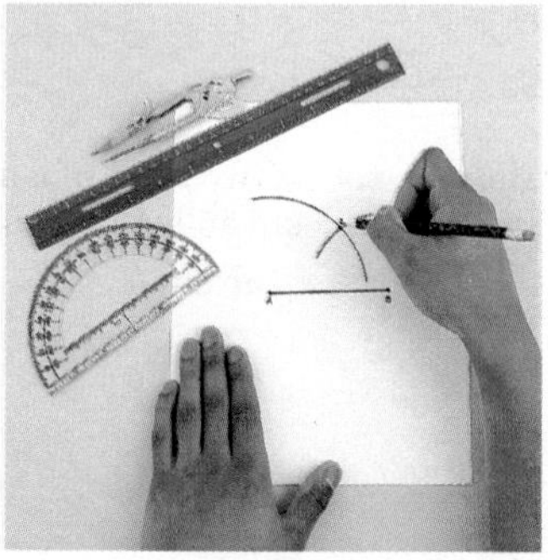

From point *B*, draw an arc with radius *BC*. Label the intersection of the arcs as point *C*.

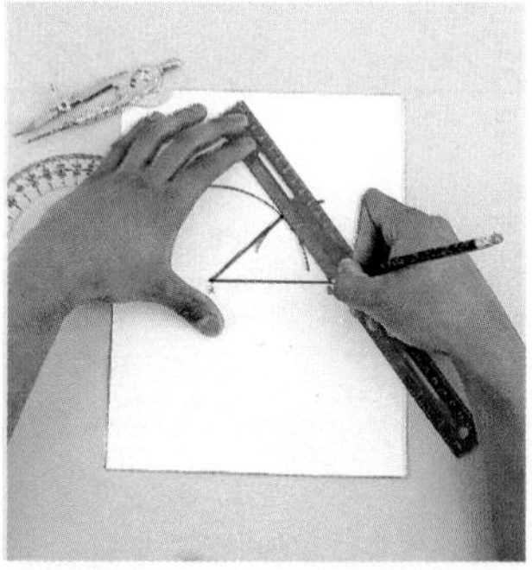

Draw $\overline{AC}$ and $\overline{BC}$ to complete the triangle.

② Measure the angles of your triangles with a protractor. Record them in a table like the one above.

③ Is the sum of the angle measures in each triangle 180°? Why or why not?

④ What happens when you try to construct a triangle using the lengths given in row 5 of the table above?

The triangles in the Exploration are SSS (side-side-side) constructions. If the sides can form a triangle, it is a unique triangle.

Answers to Exploration

1. Check students' work.
2.

	AB	*AC*	*BC*	$m\angle A$	$m\angle B$	$m\angle C$
1	8.6 cm	6.4 cm	4.1 cm	27°	45°	108°
2	6.7 cm	7.2 cm	4.4 cm	37°	78°	65°
3	8.2 cm	3.5 cm	6.5 cm	50°	24°	106°
4	4.2 cm	6.9 cm	3.6 cm	26°	124°	30°
5	8.7 cm	3.7 cm	4.0 cm	—	—	—

3. The sum of the measues of the angles should be 180°, but may not be exactly 180° due to rounding of measurements.
4. It cannot be done. The given side lengths do not form a triangle, since $8.7 > 3.7 + 4.0$.

Sample 2

Verify the measure of $\angle A$ for the triangle in row 1 of the table in the Exploration on page 462.

Sample Response

$a^2 = b^2 + c^2 - 2bc \cos A$ ← Write the law of cosines that involves $\angle A$.

$\cos A = \dfrac{b^2 + c^2 - a^2}{2bc}$ ← Solve for cos *A*.

$= \dfrac{(6.4)^2 + (8.6)^2 - (4.1)^2}{2(6.4)(8.6)}$ ← Substitute **6.4** for ***b***, **8.6** for ***c***, and **4.1** for ***a***.

≈ 0.8913 ← Simplify.

$m\angle A \approx 26.96°$ ← Use the inverse cosine feature of a calculator.

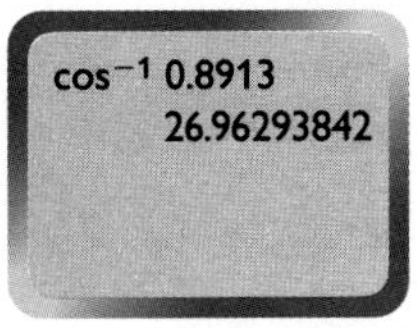

The measure of $\angle A$ is about 27°.

Look Back

Under what conditions can you use the law of cosines to find an unknown side length or an unknown angle measure in a triangle?

Now you are ready for: Exs. 17–32 on pp. 465–466

8-6 Exercises and Problems

1. **Reading** What measurements of a triangle do you need to know to use the law of cosines to find an unknown side length?

For Exercises 2–4, use the given information to draw $\triangle ABC$. Measure the length of the third side. Then use the law of cosines to calculate the length of the missing side. Compare your results.

2. $a = 8.3$ cm, $b = 11.3$ cm, $m\angle C = 37°$
3. $b = 4.7$ cm, $c = 7.1$ cm, $m\angle A = 112°$
4. $c = 9.2$ cm, $a = 3.9$ cm, $m\angle B = 67°$

5. **Writing** Place any $\triangle ABC$ on a coordinate grid so that $\angle B$ is at the origin. Using the method on page 460, derive this form of the law of cosines:

$$b^2 = a^2 + c^2 - 2ac \cos B$$

Look Back

The Look Back focuses students' attention on the conditions when it is appropriate to use the law of cosines.

APPLYING

Suggested Assignment

Day 1

Standard 1–4, 10–12

Extended 1–16

Day 2

Standard 17, 18, 22–24, 29–32

Extended 17–27, 29–32

Integrating the Strands

Algebra Exs. 1–32

Measurement Exs. 2–4, 7

Geometry Exs. 1–17, 19–30, 32

Trigonometry Exs. 1–17, 19–26, 28–32

Logic and Language Exs. 1, 5, 21, 32

Error Analysis

Some students may have errors in their work for Exs. 2–4 or Ex. 28 because they did not apply the law of cosines correctly. Students can correct their own work by making sketches to ensure they are using the appropriate equation from the law of cosines and are substituting correctly. Discuss ideas students may have for checking results. For example, do the angle measures add up to 180°? Does a scale drawing agree with the algebraic results?

Answers to Look Back

You can use the law of cosines if you know the lengths of two sides of a triangle and the measure of the included angle. When you know the lengths of all three sides but you do not know the measures of any of the angles, you can use an alternate form which has been solved for cos *A*, cos *B*, or cos *C*.

Answers to Exercises and Problems

1. the lengths of two sides and the measure of the included angle

2–4. Check students' drawings. Measurements may vary slightly. Calculated values are given to the nearest tenth.

2. 6.8 cm 3. 9.9 cm

4. 8.5 cm

5. Let *ABC* be a triangle on the coordinate plane with *B* at the origin and *C* on the *x*-axis; that is, $B = (0, 0)$ and $C = (a, 0)$. The coordinates of *A* are $(c \cos B, c \sin B)$. Then, by the distance formula, $b = \sqrt{(c \cos B - a)^2 + (c \sin B - 0)^2}$; $b^2 = c^2 (\cos^2 B) - 2ac \cos B + a^2 + c^2 (\sin^2 B) = c^2 ((\cos^2 B) + (\sin^2 B)) + a^2 - 2ac \cos B = a^2 + c^2 - 2ac \cos B$.

Mathematical Procedures

Many students associate the law of cosines with triangles only. Exs. 7 and 10–12 show how the law of cosines can be used with other polygons.

Application

Exs. 8 and 9 illustrate how the law of cosines can be used by chemists to model molecules.

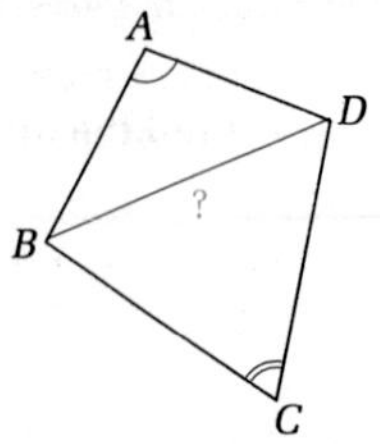

6. Suppose two sides of a triangle are 3 in. long and the measure of their included angle is 60°. What is the length of the third side?

7. **Open-ended** Draw any large quadrilateral $ABCD$. Measure the sides and $\angle A$ and $\angle C$. Find BD using the law of cosines with two different triangles. Check your answers by measuring.

connection to **CHEMISTRY**

Chemists model molecules by measuring bond angles and the distances between atoms. Molecules are in constant motion, so the measurements are constantly changing. The distances and bond angles given are average values. (Note: 1 picometer (pm) = 10^{-12} m.)

Find the distance between two hydrogen atoms in each molecule.

8. water (H_2O)

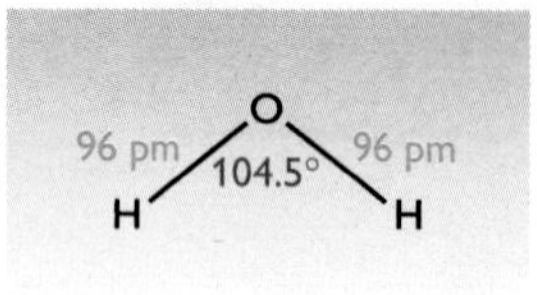

9. methane (CH_4)

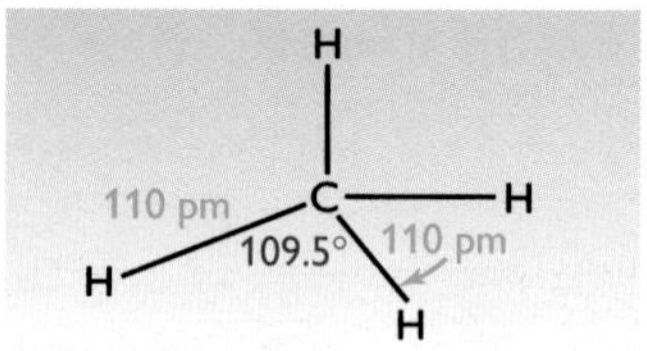

Find the length of a side of each inscribed regular polygon.

10.

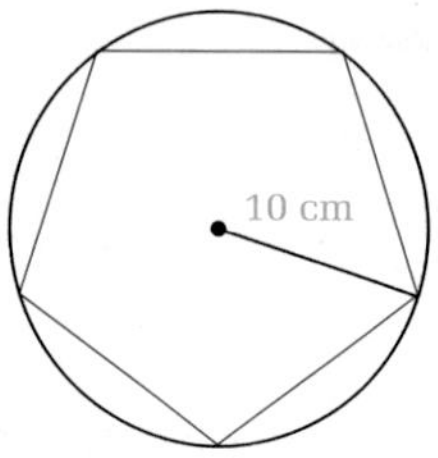

11.

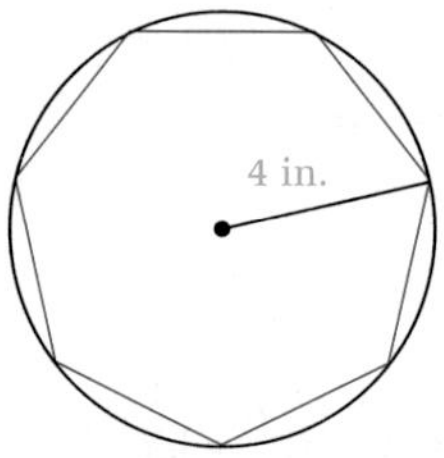

12.

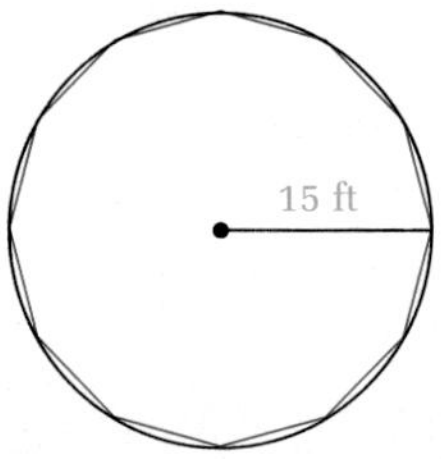

13. Dome-shaped tents often have regular hexagons as a base.

a. One popular dome-shaped tent has a base shaped like a regular hexagon with sides of length 48 in. How tall is the tallest person who can sleep stretched out full length in this tent along the line shown?

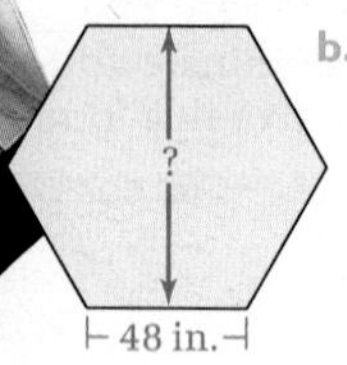

b. **Open-ended** How long must each side of the hexagonal base of a dome-shaped tent be for you to sleep stretched out full length?

Answers to Exercises and Problems

6. 3 in.

7. Answers may vary. Check students' work.

8, 9. Answers are given to the nearest whole number.

8. about 152 pm

9. about 180 pm

10–12. Answers are given to the nearest tenth.

10. 11.8 cm

11. 3.5 in.

12. 7.8 ft

13. a. about 83 in. or 6 ft 11 in.

b. Answers may vary. For example, a person 5 ft 6 in. tall would require a tent with sides at least 38 in. long.

14. 438.14 ft

15. about 2.3°

16. 4085.5 ft, or a little more than three-quarters of a mile

17. The angle at the vertex marked by the Lincoln Memorial is 44.4°; the third angle is 10.4°.

18. a. $c^2 - a^2 - b^2 = -2ab \cos C$; $\frac{c^2 - a^2 - b^2}{-2ab} = \frac{-2ab \cos C}{-2ab}$; $\cos C = \frac{c^2 - a^2 - b^2}{-2ab} = \frac{a^2 + b^2 - c^2}{2ab}$

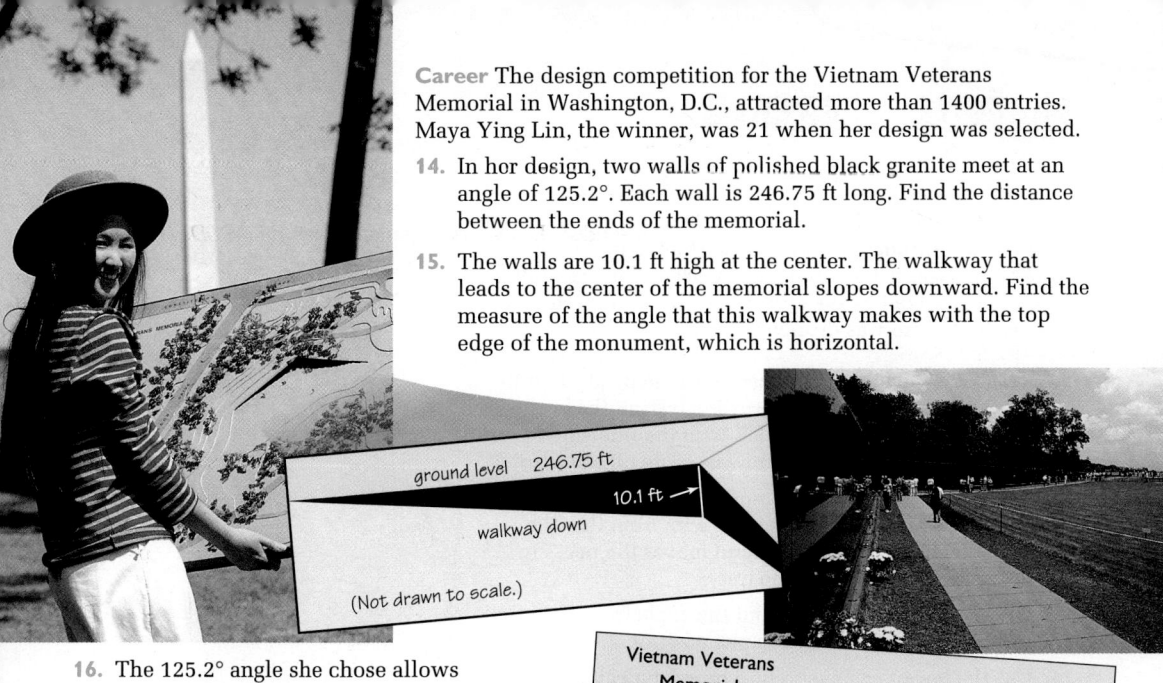

Career The design competition for the Vietnam Veterans Memorial in Washington, D.C., attracted more than 1400 entries. Maya Ying Lin, the winner, was 21 when her design was selected.

14. In her design, two walls of polished black granite meet at an angle of 125.2°. Each wall is 246.75 ft long. Find the distance between the ends of the memorial.

15. The walls are 10.1 ft high at the center. The walkway that leads to the center of the memorial slopes downward. Find the measure of the angle that this walkway makes with the top edge of the monument, which is horizontal.

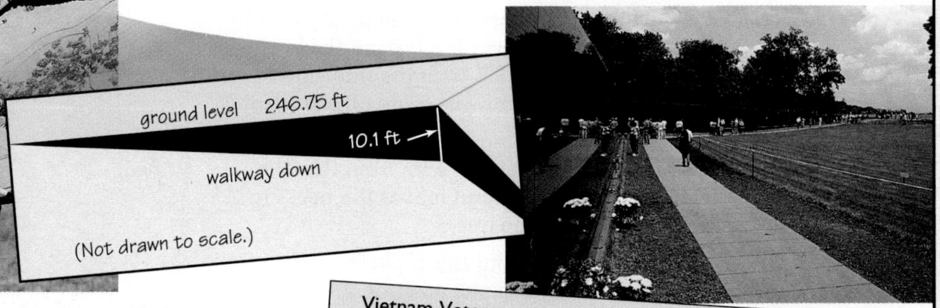

16. The 125.2° angle she chose allows the Vietnam Veterans Memorial to point to two other nearby landmarks.

Find the distance between the Lincoln Memorial and the Washington Monument.

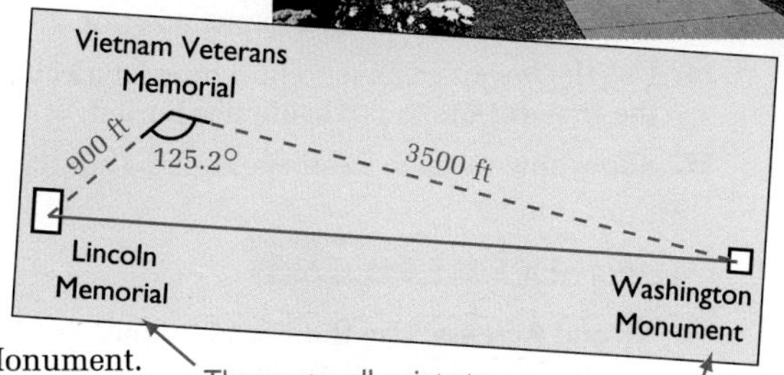

17. Find the other two angles in the triangle formed by the Vietnam Veterans Memorial, the Lincoln Memorial, and the Washington Monument.

18. a. Solve the equation $c^2 = a^2 + b^2 - 2ab \cos C$ for $\cos C$.
 b. Solve the equation $b^2 = a^2 + c^2 - 2ac \cos B$ for $\cos B$.

19. Use the law of cosines to verify the measures of $\angle B$ and $\angle C$ in row 1 of the table on page 462.

20. The side lengths in row 5 of the table on page 462 do not form a triangle. What happens if you try to use these lengths and the law of cosines to find the angles of a triangle?

21. **Writing** Explain how you can use the law of cosines several times to find the length of the fourth side of a quadrilateral when you know the lengths of three sides and the measures of all four angles.

A *dihedral angle* of a polyhedron is the angle between adjacent faces. Use the labeled triangles to find the measure of a dihedral angle in each regular polyhedron.

22. tetrahedron

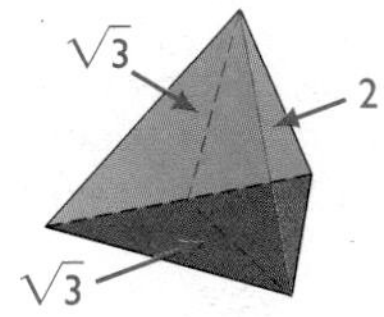

23. octahedron

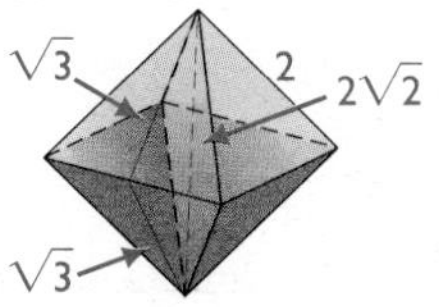

24. icosahedron

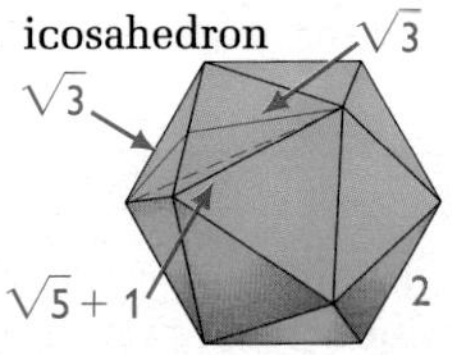

Research

Some students may want to learn more about Maya Ying Lin. What was her training and what has she done since designing the Vietnam Veterans Memorial?

Teaching Tip

After completing Ex. 18, students may wish to record the formulas for cos A (from page 463), cos B, and cos C in their journals for future use.

Communication: Writing

Ex. 21 offers an excellent opportunity to check students' understanding of how to use the law of cosines.

Answers to Exercises and Problems

b. $b^2 - a^2 - c^2 = -2ac \cos B$; $\dfrac{b^2 - a^2 - c^2}{-2ac} = \dfrac{-2ac \cos B}{-2ac}$; $\cos B = \dfrac{b^2 - a^2 - c^2}{-2ac} = \dfrac{a^2 + c^2 - b^2}{2ac}$

19. $\cos B = \dfrac{a^2 + c^2 - b^2}{2ac} = \dfrac{4.1^2 + 8.6^2 - 6.4^2}{2(4.1)(8.6)} = 0.7063$; $m\,\angle B \approx 45°$; $\cos C = \dfrac{a^2 + b^2 - c^2}{2ab} = \dfrac{4.1^2 + 6.4^2 - 8.6^2}{2(4.1)(6.4)} = -0.3085$; $m\,\angle C \approx 108°$

20. The cosines calculated for $\angle A$ and $\angle B$ are greater than 1. The cosine calculated for $\angle B$ is less than -1. No such angles are possible. For every angle θ, $-1 \le \cos\theta \le 1$.

21. Suppose that $ABCD$ is a quadrilateral for which you know the measures of all four angles and the lengths AB, BC, and CD. Draw diagonal $\overline{AC}$ and use the law of cosines to find AC. Use the alternate form of the law of cosines with $\triangle ABC$ to find $m\,\angle ACB$. Subtract to find $m\,\angle ACD$ and use the law of cosines again to find AD.

22–26. Answers are given to the nearest tenth.

22. 70.5°

23. 109.5°

24. 138.2°

Students Acquiring English

You may wish to pair students for the "Connection to Literature" exercises since some of the vocabulary, such as pirate, peg, and charm, may not be familiar.

Working on the Unit Project

When doing Ex. 32, students will need to remember the conditions under which the law of cosines can be applied.

Practice 64 For use with Section 8-6

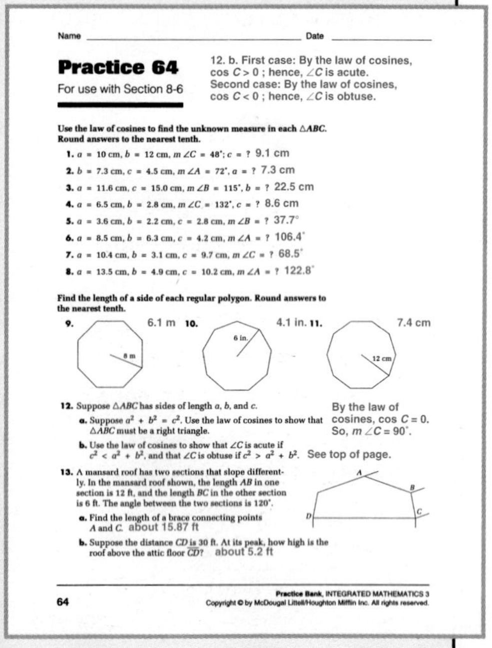

Name ____________ Date ____________

Practice 64

For use with Section 8-6

12. b. First case: By the law of cosines, $\cos C > 0$; hence, $\angle C$ is acute. Second case: By the law of cosines, $\cos C < 0$; hence, $\angle C$ is obtuse.

Use the law of cosines to find the unknown measure in each $\triangle ABC$. Round answers to the nearest tenth.

1. $a = 10$ cm, $b = 12$ cm, $m\angle C = 48°$; $c = ?$ 9.1 cm
2. $b = 7.3$ cm, $c = 4.5$ cm, $m\angle A = 72°$, $a = ?$ 7.3 cm
3. $a = 11.6$ cm, $c = 15.0$ cm, $m\angle B = 115°$, $b = ?$ 22.5 cm
4. $a = 6.5$ cm, $b = 2.8$ cm, $m\angle C = 132°$, $c = ?$ 8.6 cm
5. $a = 3.6$ cm, $b = 2.2$ cm, $c = 2.8$ cm, $m\angle B = ?$ 37.7°
6. $a = 8.5$ cm, $b = 6.3$ cm, $c = 4.2$ cm, $m\angle A = ?$ 106.4°
7. $a = 10.4$ cm, $b = 3.1$ cm, $c = 9.7$ cm, $m\angle C = ?$ 68.5°
8. $a = 13.5$ cm, $b = 4.9$ cm, $c = 10.2$ cm, $m\angle A = ?$ 122.8°

Find the length of a side of each regular polygon. Round answers to the nearest tenth.

9. (8 m) 6.1 m
10. (6 in.) 4.1 in.
11. (12 cm) 7.4 cm

12. Suppose $\triangle ABC$ has sides of length a, b, and c.
   a. Suppose $a^2 + b^2 = c^2$. Use the law of cosines to show that $\triangle ABC$ must be a right triangle. By the law of cosines, $\cos C = 0$. So, $m\angle C = 90°$.
   b. Use the law of cosines to show that $\angle C$ is acute if $c^2 < a^2 + b^2$, and that $\angle C$ is obtuse if $c^2 > a^2 + b^2$. See top of page.
13. A mansard roof has two sections that slope differently. In the mansard roof shown, the length AB in one section is 12 ft, and the length BC in the other section is 6 ft. The angle between the two sections is 120°.
   a. Find the length of a brace connecting points A and C. about 15.87 ft
   b. Suppose the distance CD is 30 ft. At its peak, how high is the roof above the attic floor $\overline{CD}$? about 5.2 ft

64 **Practice Bank**, INTEGRATED MATHEMATICS 3

connection to **LITERATURE**

In Edgar Allan Poe's story "The Gold-Bug," William Legrand decodes a pirate's clue to buried treasure. The clue leads him to a skull hidden in a tree. An assistant drops a gold charm through one eye of the skull, and Legrand places a peg where the charm lands. When he digs 50 ft away, as instructed by the clue, he finds nothing. He realizes the charm was dropped through the wrong eye.

> "This mistake made a difference ... in the position of the peg...; of course the error, however trivial in the beginning, increased as we proceeded with the line, and by the time we had gone fifty feet threw us quite off the scent."

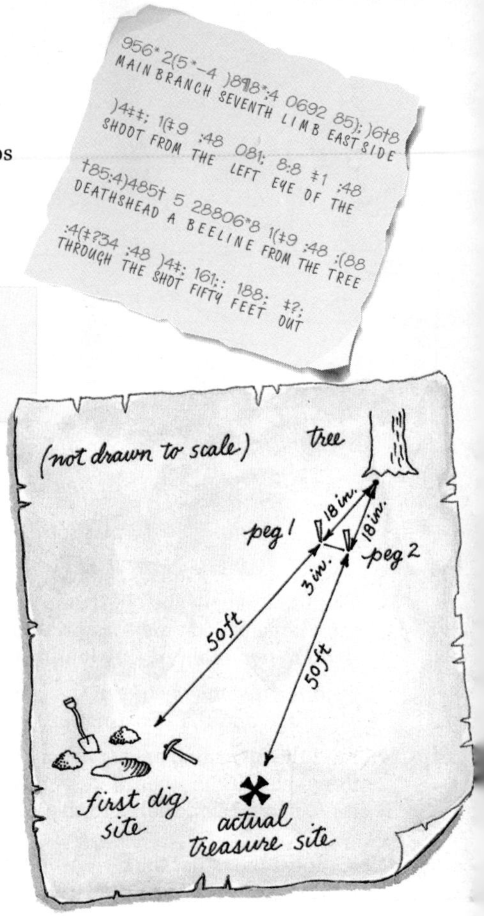

25. Suppose the gold charm lands 18 in. from the tree. To correct the mistake, Legrand moves the peg 3 in. but keeps it 18 in. from the tree.

    Use the law of cosines to find the angle from peg position 1 to the tree to peg position 2.
26. Use the law of cosines to find how far away from the first dig site the treasure lies buried.
27. Show how to solve Exercise 26 using similar triangles.

Ongoing **ASSESSMENT**

28. **Group Activity** Use the law of cosines to fill in the unknown lengths and angle measures for the triangles in the table.

a	b	c	$m\angle A$	$m\angle B$	$m\angle C$
129	43	?	?	?	76°
78	?	53	?	157°	?
52	17	60	?	?	?

Review **PREVIEW**

29. Write parametric equations for a circle that has a radius of 5 and its center at the origin. *(Section 8-5)*
30. Tell whether each statement is *True* or *False*? *(Toolbox Skill 36)*

    a. $h = a \sin C$ b. $h = c \sin A$

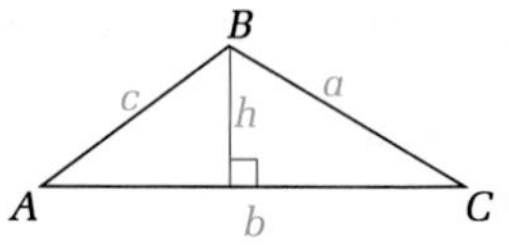

31. Solve the proportion $\frac{7}{\sin 30°} = \frac{196}{x}$. *(Toolbox Skill 37)*

8 Working on the Unit Project

32. Write a clue for your treasure hunt that requires the use of the law of cosines.

Answers to Exercises and Problems

25. 9.6°
26. 8.33 ft
27. The triangle defined by the tree and the two peg positions is similar to the triangle defined by the tree and the two dig sites. The ratio of the lengths of the corresponding sides is $\frac{18 \text{ in.}}{50 \text{ ft}} = \frac{18}{600} = \frac{3}{100}$. Then the distance between the two sites is 100 in. or about 8.3 ft.
28. Answers are given to the nearest whole number. Answers may vary slightly and the sum of the angle measures may not equal 180° due to rounding.

a	b	c	$m\angle A$	$m\angle B$	$m\angle C$
129	43	126	84°	19°	76°
78	128	53	14°	157°	9°
52	17	60	54°	15°	110°

29. $x = 5 \cos \theta$; $y = 5 \sin \theta$
30. a. True.
    b. True.
31. 14
32. Answers may vary.

Section 8-7

The Law of Sines

Going to GREAT Lengths

Focus

Use the law of sines to find measures of sides and angles in triangles.

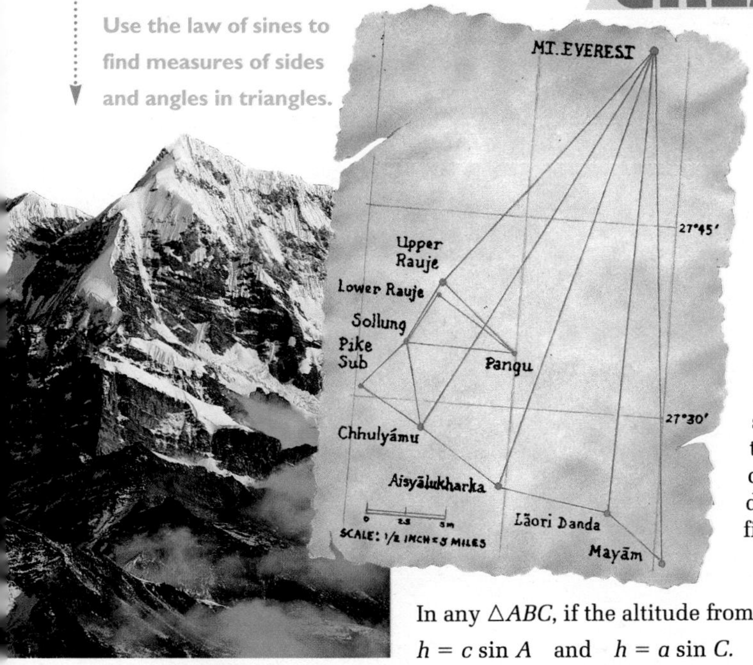

The Great Trigonometric Survey of India in 1852 declared Mt. Everest to be the highest mountain in the world.

In the 1950s, surveyors measured angles of elevation from eight different sites. These angles are useful for finding the height only if you know the distance of each observation site from Mt. Everest. These distances cannot be measured directly, but the *law of sines*, developed below, can be used to find them.

In any $\triangle ABC$, if the altitude from B has length h, then $h = c \sin A$ and $h = a \sin C$.

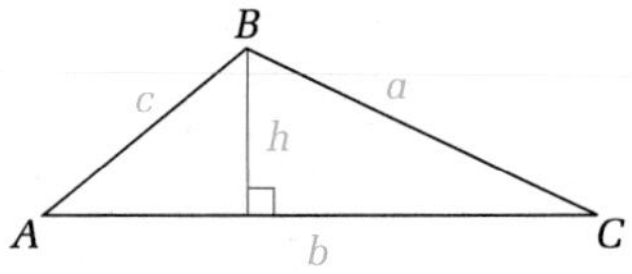

$$c \sin A = a \sin C$$ ← The two expressions for h are equal.

$$\frac{c \sin A}{\sin A \sin C} = \frac{a \sin C}{\sin A \sin C}$$ ← Divide both sides by $\sin A \sin C$.

$$\frac{c}{\sin C} = \frac{a}{\sin A}$$ ← Simplify.

By using other altitudes, you can show that in any $\triangle ABC$, the ratio $\frac{b}{\sin B}$ has the same value as $\frac{c}{\sin C}$ and as $\frac{a}{\sin A}$.

LAW OF SINES

In any $\triangle ABC$, the lengths of the sides are in proportion with the sines of the opposite angles.

$$\frac{a}{\sin A} = \frac{b}{\sin B} = \frac{c}{\sin C}$$

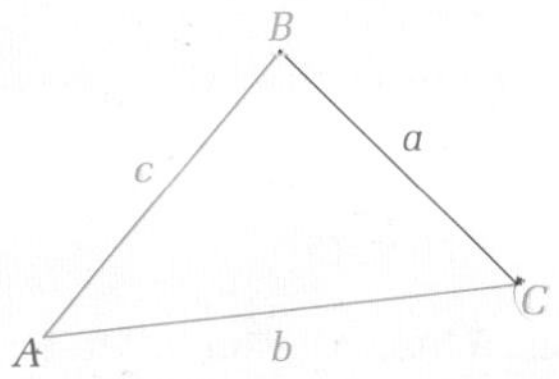

PLANNING

Objectives and Strands

See pages 422A and 422B.

Spiral Learning

See page 422B.

Materials List

- Protractor
- Ruler
- Compass
- Calculator

Recommended Pacing

Section 8-7 is a two-day lesson.

Day 1

Pages 467–469: Opening paragraph through Talk it Over 2, *Exercises 1–14*

Day 2

Pages 469–471: Exploration through Look Back, *Exercises 15–38*

Extra Practice

See pages 621–622.

Warm-Up Exercises

Warm-Up Transparency 8-7

Support Materials

- Practice Bank: Practice 65
- Activity Bank: Enrichment 58
- Study Guide: Section 8-7
- Problem Bank: Problem Set 18
- Explorations Lab Manual: Diagram Master 20
- Assessment Book: Quiz 8-7, Test 33, Alternative Assessment 7

TEACHING

Multicultural Note

Located in south-central Asia, Tibet is often called the "roof of the world" since its mountain ranges and plateaus are the highest in the world. Many of Tibet's people are nomadic shepherds and herders who travel through the northern uplands of the country. Most non-nomadic residents of Tibet live in the south of the country, where Lhasa, the capital, is located. Barley flour and yak products are the staple foods in Tibet, the yak providing milk, cheese, butter, and meat. Yaks are also used as transportation, as beasts of burden, and as a source of fiber with which to make cloth. Tibet was in former years an independent theocracy governed by Buddhist lamas; in recent years, it has been under the control of China.

Additional Sample

S1 Part of a sunken vessel is spotted from two points A and B that are 73 meters apart on a straight shoreline.

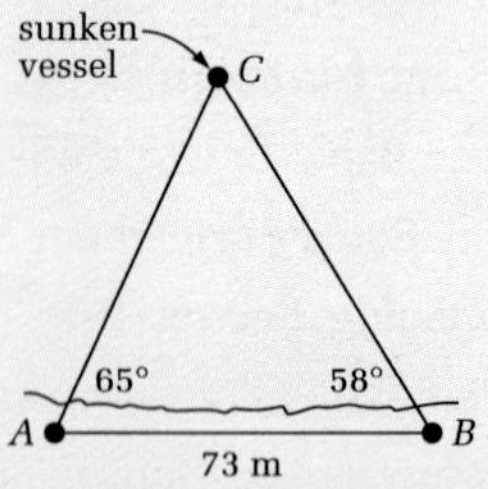

A civil engineer measures the angles formed by $\overline{AB}$ and the two sight lines shown in the diagram. Find the distance from point B to the sunken vessel.

Use the law of sines.

$$\frac{BC}{\sin A} = \frac{AB}{\sin C}.$$

$$m\angle C = 180° - (65° + 58°) = 57°$$

Solve for BC.

$$\frac{BC}{\sin 65°} = \frac{73}{\sin 57°}$$

$$BC = \frac{73 \sin 65°}{\sin 57°} \approx 78.9$$

The distance from point B to the sunken vessel is about 78.9 m.

BY THE WAY...

The Tibetan name for Mt. Everest is Chomolangma, the name of a bird sanctuary that includes the peak.

Sample 1 shows that you can use the law of sines to find unknown measures in a triangle given the measures of two angles and the length of an included side (ASA).

Sample 2 shows that you can also use this law to find unknown measures in a triangle given the measures of two angles and the length of a non-included side (AAS).

Sample 1

Find the distance from Aisyālukharka to Mt. Everest.

Sample Response

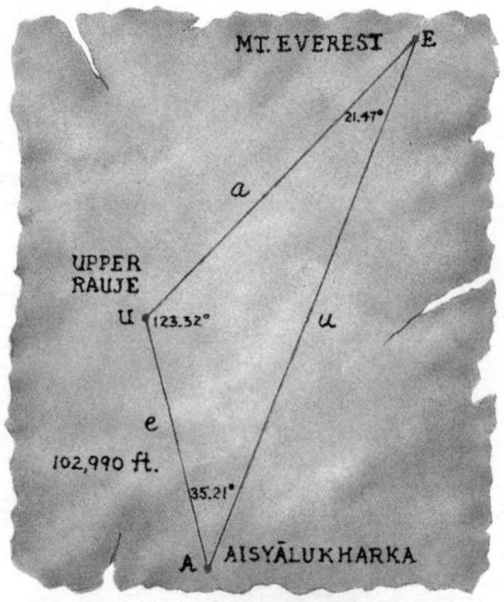

On the map, $m\angle U = 123.32°$, $m\angle E = 21.47°$, and $e = 102{,}990$.

$$\frac{u}{\sin U} = \frac{e}{\sin E}$$ ← Use the law of sines for △UEA.

$$\frac{u}{\sin 123.32°} = \frac{102{,}990}{\sin 21.47°}$$ ← Substitute **123.32°** for **U**, **102,990** for *e*, and **21.47°** for **E**.

$$u = \frac{102{,}990 \sin 123.32°}{\sin 21.47°}$$ ← Solve for *u*.

$$\approx 235{,}128$$ ← Simplify.

The distance from Aisyālukharka to Mt. Everest is about 235,130 ft.

Sample 2

Find the distance from Mayām to Mt. Everest.

Sample Response

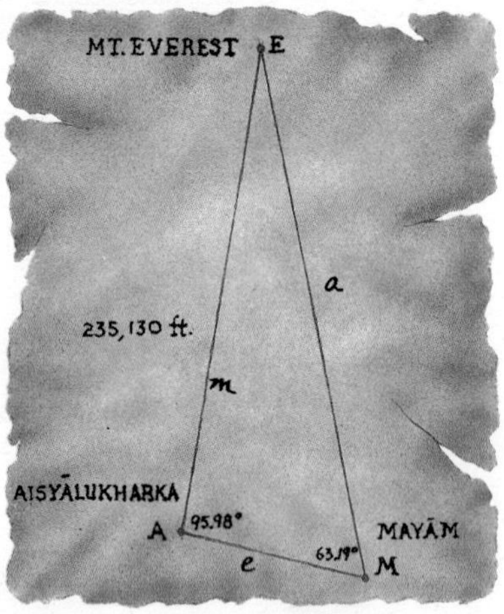

On the map, $m\angle A = 95.98°$, $m\angle M = 63.19°$, and $m = 235{,}130$.

$$\frac{a}{\sin A} = \frac{m}{\sin M}$$ ← Use the law of sines for △AEM.

$$\frac{a}{\sin 95.98°} = \frac{235{,}130}{\sin 63.19°}$$ ← Substitute **95.98°** for **A**, **235,130** for *m*, and **63.19°** for **M**.

$$a = \frac{235{,}130 \sin 95.98°}{\sin 63.19°}$$ ← Solve for *a*.

$$\approx 262{,}015$$ ← Simplify.

The distance from Mayām to Mt. Everest is about 262,020 ft.

Answers to Talk it Over

1. from Aisyālukharka: about 29,030 ft; from Mayām: about 28,720 ft
2. SAS

BY THE WAY...

Mt. Everest is so massive that its gravitational field makes liquid surfaces tilt away from the horizontal. Surveyors must adjust their angle measurements to take this into account.

Now you are ready for: Exs. 1–14 on pp. 471–472

Talk it Over

1. Use the tangent ratio and the information in the table to estimate the height of Mt. Everest measured from either Aisyālukharka or Mayām.

Observation site	Station altitude	Distance to Mt. Everest	Angle of elevation
Aisyālukharka	8670 ft	235,130 ft	4.95°
Mayām	10,948 ft	262,020 ft	3.88°

2. Sample 1 uses the law of sines with ASA information. Sample 2 uses the law of sines with AAS information. What other triangle information can you use with the law of sines?

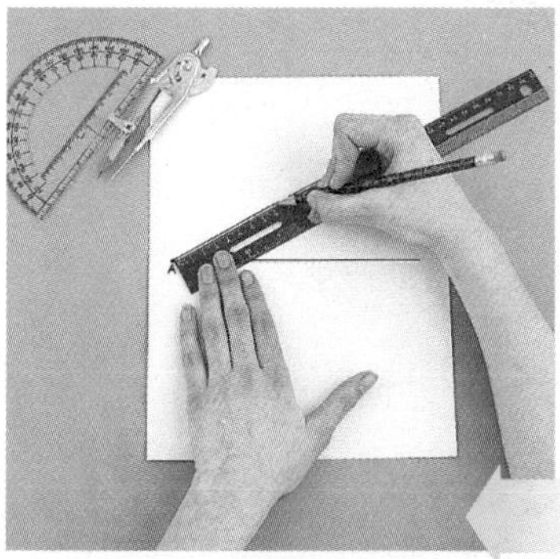

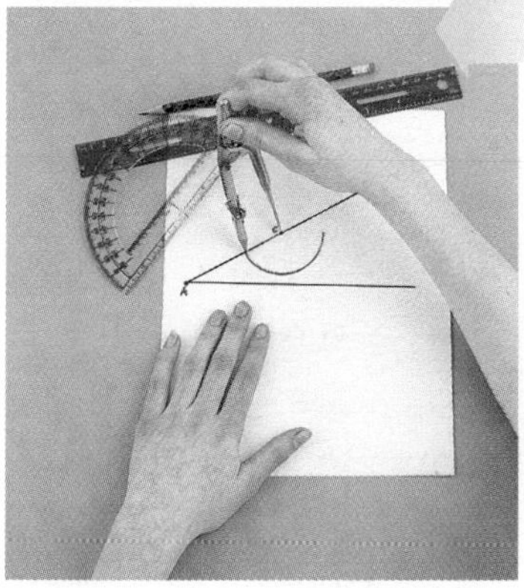

EXPLORATION

How many triangles can you construct if you are given SSA information?

- **Materials: protractors, rulers, compasses**
- **Work in a group of four students.**

Each student should follow steps 1–6 to complete one row of the table below.

1. Draw a long baseline. Label the left endpoint A.
2. Use a protractor to draw a 30° angle at point A.
3. Measure 8 cm from point A and mark point C as shown.
4. Put one end of the compass at C and draw an arc with radius a.
5. If your arc intersects the baseline, label the point(s) of intersection B.
6. List the number of triangles you can construct using the SSA information in each row. Then tell whether the triangles formed are *acute, right,* or *obtuse.*

a	b	$m\angle A$	Number of triangles	Type(s) of triangles
3 cm	8 cm	30°	?	?
4 cm	8 cm	30°	?	?
6 cm	8 cm	30°	?	?
10 cm	8 cm	30°	?	?

7. Write a conjecture about how the relationship between a, b, and $b \sin A$ influences the number of triangles formed.

Additional Sample

S2 A TV station has its broadcast tower in a wooded area close to the community of Pine Level. A service road 2.7 km long goes from Pine Level to the tower. The station management is considering a second service road from Tompkins, a community close to Pine Level.

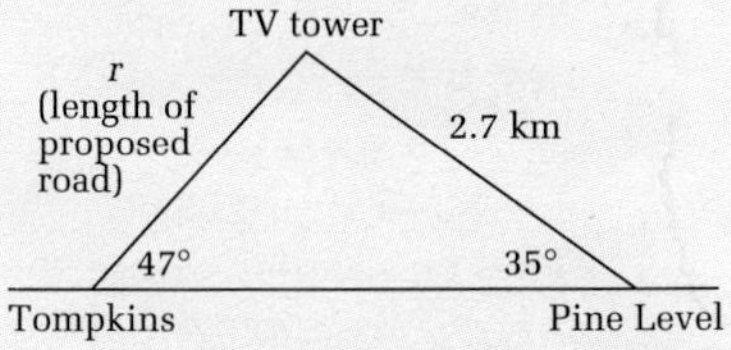

Find the length of the proposed service road. Assume all roads are straight.

Use the law of sines.

$$\frac{r}{\sin 35°} = \frac{2.7}{\sin 47°}$$

$$r = \frac{2.7 \sin 35°}{\sin 47°}$$

$$\approx 2.12$$

The length of the proposed road is about 2.12 km.

Reasoning

Ask students whether they can solve a triangle, given all the side lengths, by using *only* the law of cosines. This question can be discussed in class or as an open-ended portfolio activity.

Talk it Over

Question 2 directs students' attention to the conditions under which the law of sines can be applied.

Exploration

The goal of the Exploration is for students to gain experience constructing triangles when given two sides and the nonincluded angle. Students should realize that two triangles can sometimes be made under these conditions. After completing this Exploration, students may wish to compare their results with those obtained from the Exploration on page 462.

Answers to Exploration

1–5. See table results below.

6.

a	b	$m\angle A$	Number of triangles	Types of triangles
3 cm	8 cm	30°	0	—
4 cm	8 cm	30°	1	right
6 cm	8 cm	30°	2	obtuse
10 cm	8 cm	30°	1	acute

7. Based on the Exploration, it would be appropriate to conjecture that given an acute angle A, if $a < b \sin A$, then no triangle can be formed. If $a = b \sin A$ or if $a \geq b$, one triangle can be formed. If $a > b \sin A$ and $a < b$, two triangles can be formed.

Additional Sample

S3 Find the unknown measures for $\triangle ABC$, with $a = 9$ m, $b = 13$ m, and $m\angle A = 42°$.

Step 1. **Find the measure of one of the unknown angles.**

Since $9 > 13 \sin 42°$ and $9 < 13$, two triangles are formed.

$$\frac{a}{\sin A} = \frac{b}{\sin B}$$

$$\frac{9}{\sin 42°} = \frac{13}{\sin B}$$

$$\sin B = \frac{13 \sin 42°}{9}$$

$$\approx 0.9665$$

$m\angle B \approx 75°$ or $105°$

The two angles whose sine is 0.9665 are supplementary.

Step 2. **Find $m\angle C$ and c for each of two cases.**

Case 1: $m\angle B \approx 75°$

In this case, $m\angle C \approx 63°$.

$$\frac{c}{\sin C} = \frac{a}{\sin A}$$

$$\frac{c}{\sin 63°} = \frac{9}{\sin 42°}$$

$$c = \frac{9 \sin 63°}{\sin 42°}$$

$$\approx 11.98 \text{ m}$$

Case 2: $m\angle B \approx 105°$

In this case, $m\angle C \approx 33°$.

$$\frac{c}{\sin C} = \frac{a}{\sin A}$$

$$\frac{c}{\sin 33°} = \frac{9}{\sin 42°}$$

$$c = \frac{9 \sin 33°}{\sin 42°}$$

$$\approx 7.33 \text{ m}$$

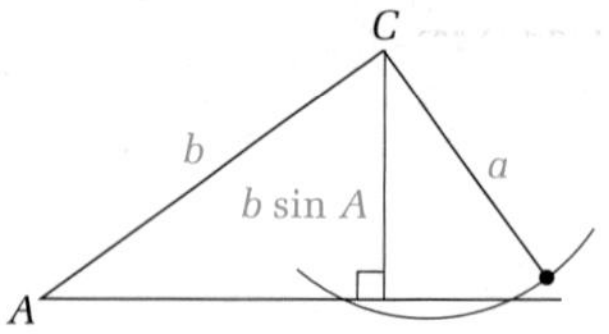

Two triangles can be formed from the same SSA information when two conditions are met:

$a > b \sin A$	$a < b$
The arc from C is long enough to intersect the baseline twice.	The arc is not long enough to intersect the baseline to the left of A outside the triangle.

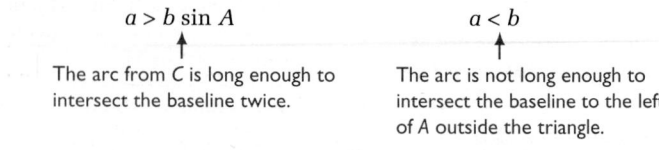

This situation is called the **ambiguous case.** This means you cannot tell whether one, two, or no triangles are formed.

Sample 3

Find the unknown measures for $\triangle ABC$ with $a = 4$ cm, $b = 7.2$ cm, and $m\angle A = 30°$.

Sample Response

Step 1 Find the measure of one of the unknown angles.

Since $4 > 7.2 \sin 30°$ and $4 < 7.2$, two triangles can be formed.

$\frac{a}{\sin A} = \frac{b}{\sin B}$ ← Use the law of sines to find $m\angle B$.

$\frac{4}{\sin 30°} = \frac{7.2}{\sin B}$ ← Substitute **4** for ***a***, **30°** for ***A***, and **7.2** for ***b***.

$\sin B = \frac{7.2 \sin 30°}{4}$ ← Solve for sin *B*.

$= 0.9$ ← Simplify.

$m\angle B \approx 64°$ or $116°$ ← Use the inverse sine feature of a calculator. The two angles whose sine is 0.9 are supplementary.

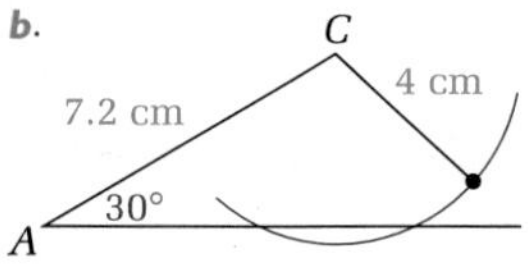

Step 2 Find $m\angle C$ and c for each of the two cases.

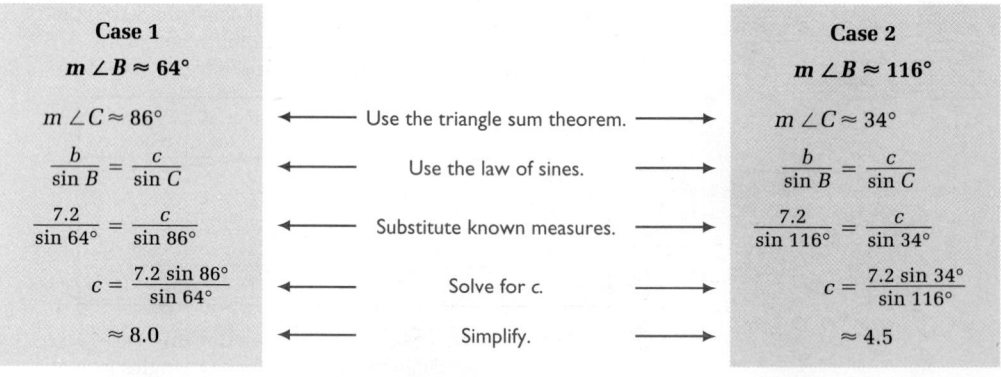

Case 1: $m\angle B \approx 64°$		Case 2: $m\angle B \approx 116°$
$m\angle C \approx 86°$	← Use the triangle sum theorem. →	$m\angle C \approx 34°$
$\frac{b}{\sin B} = \frac{c}{\sin C}$	← Use the law of sines. →	$\frac{b}{\sin B} = \frac{c}{\sin C}$
$\frac{7.2}{\sin 64°} = \frac{c}{\sin 86°}$	← Substitute known measures. →	$\frac{7.2}{\sin 116°} = \frac{c}{\sin 34°}$
$c = \frac{7.2 \sin 86°}{\sin 64°}$	← Solve for *c*. →	$c = \frac{7.2 \sin 34°}{\sin 116°}$
≈ 8.0	← Simplify. →	≈ 4.5

Answers to Look Back

SSA; As demonstrated in the Exploration, given two sides a and b and a nonincluded angle A, no triangle, one triangle, or two triangles may be formed depending on the relationships among a, b, and $b \sin A$.

Answers to Exercises and Problems

1. The distances to the mountain can be found using the law of sines.
2. from Aisyālukharka: about 29,030 ft; from Mayām: about 28,720 ft; Answers may vary. Differences may be due to inexact measurements and rounding. The estimates could be improved by more exact measurements and by rounding only at the final stage.

3–9. Measurements are given to the nearest tenth.

3. $m\angle Z = 57°$; $x = 16.7$ m; $y = 21.3$ m
4. $m\angle T = 33°$; $r = 162.4$ ft; $s = 110.8$ ft
5. $m\angle A = 104°$; $a = 40.3$ m; $b = 21.4$ m

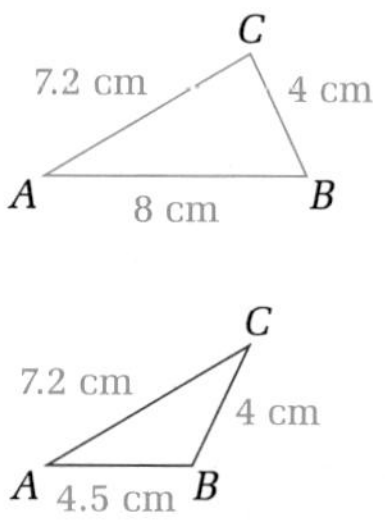

Measurements for the two triangles that meet the SSA conditions in Sample 3 are summarized in this table.

	$m\angle A$	$m\angle B$	$m\angle C$	a	b	c
Case 1	30°	64°	86°	4 cm	7.2 cm	8.0 cm
Case 2	30°	116°	34°	4 cm	7.2 cm	4.5 cm

Look Back

Of all the combinations of side lengths and angle measures used to construct triangles, which do not always produce a unique triangle? Why?

Now you are ready for:
Exs. 15–38 on pp. 473–474

8-7 Exercises and Problems

1. **Reading** How can surveyors use trigonometry to find the height of Mt. Everest without measuring distances to the mountain?
2. Find the height of Mt. Everest using the observation site you did not choose in *Talk it Over* question 1. Compare your answer with the other height you found. Explain any difference in the answers. How can you get a better estimate of the height of Mt. Everest?

Find the unknown measures in each triangle.

3. 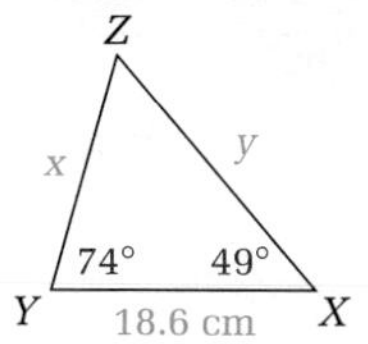

4. 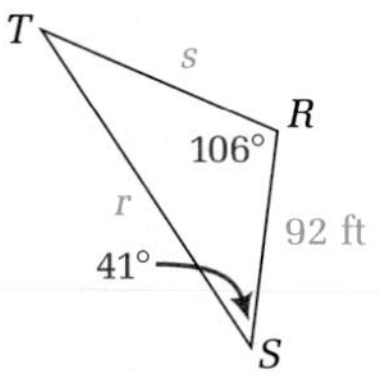

5. 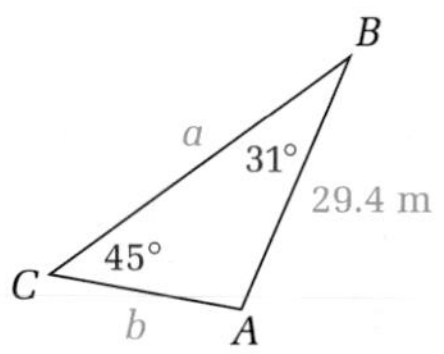

Copy and complete the table. Sketch each triangle.

	$m\angle A$	$m\angle B$	$m\angle C$	a	b	c
6.	?	28°	13°	?	6.2	?
7.	109°	?	25°	58	?	?
8.	52°	?	74°	?	?	62
9.	90°	38°	?	?	19	?

10. The diagram shows measurements a surveyor has made. Find x and y, the two distances to the flag.

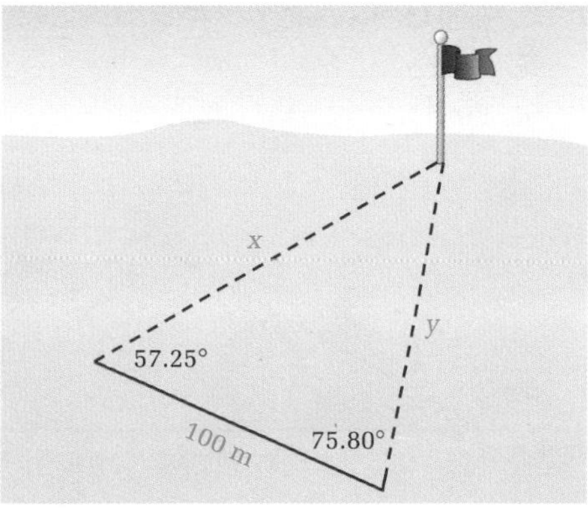

Answers to Exercises and Problems

	$m\angle A$	$m\angle B$	$m\angle C$	a	b	c
6.	139°	28°	13°	8.7	6.2	3.0
7.	109°	46°	25°	58	44.1	25.9
8.	52°	54°	74°	50.8	52.2	62
9.	90°	38°	52°	30.9	19	24.3

10. 132.7 m; 115.1 m

APPLYING

Suggested Assignment

Day 1

Standard 3–11

Extended 1–14

Day 2

Standard 16–33, 35–38

Extended 15–33, 35–38

Integrating the Strands

Algebra Exs. 3–15, 18–33, 38

Geometry Exs. 3–36, 38

Trigonometry Exs. 1–15, 18–33, 36, 38

Statistics and Probability Ex. 37

Discrete Mathematics Ex. 37

Logic and Language Exs. 1, 16, 38

Application

The subject of trigonometry has many applications to the real world. Exs. 1, 2, and 10–15 illustrate some practical applications involving indirect measurement, surveying, carpentry, access standards, angular separation of the sun and a planet, and astronomical units.

Error Analysis

Errors in solving problems such as those in Exs. 3–10 may be due to careless substitution. Urge students to double check the way in which they substituted. It is easy to make errors in paper-and-pencil calculations or in keying in numbers on calculators.

In exercises that involve SSA information, students may omit one possible solution. This can be corrected by urging students to be alert to cases where the given information is SSA. They should then check to see whether there will be no solution, one solution, or two solutions.

Research

Trigonometry is one of the oldest branches of mathematics and its origins are obscure. Land measurements and astronomy have most likely played a role in the early development of trigonometry. Some students may be interested in researching the history of trigonometry and writing a report for inclusion in their portfolios.

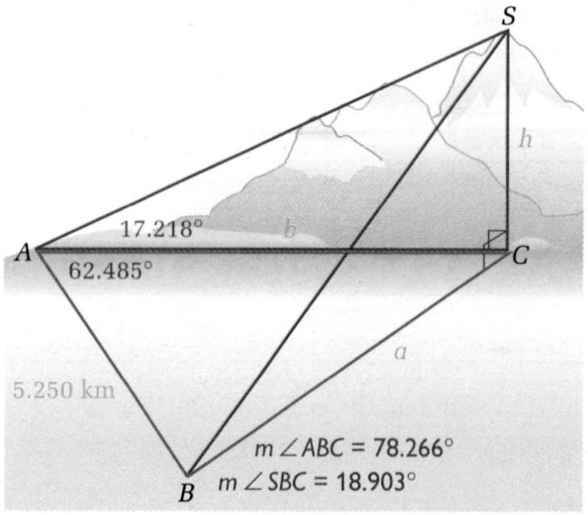

11. **Surveying** The diagram shows measurements taken by a surveyor at sea level to determine the height of a mountain.
    a. Find $m\angle ACB$.
    b. Use the law of sines to find a and b.
    c. Find h using right $\triangle ACS$.
    d. Find h using right $\triangle BCS$.
    e. What is the average of the two heights in parts (c) and (d)?
    f. Repeat parts (a) through (e), rounding all angle measurements to whole numbers. How does this reduction in accuracy affect the estimate of the mountain's height?

12. **Astronomy** When you hold a finger in front of your face and look at your finger with one eye at a time, your finger appears to shift against the background. This principle, known as *parallax*, has been used to estimate the distance to nearby stars, using the diameter of Earth's orbit (about 3.0×10^8 km) as the base of a triangle.
    a. Suppose an isosceles triangle models the position of Alpha Centauri, a bright star near Earth. The angle at Alpha Centauri is 0.000417°. Find the distance from Alpha Centauri to Earth.
    b. Light travels 9.46×10^{12} km in one year. How many years does it take light to travel from Alpha Centauri to Earth?

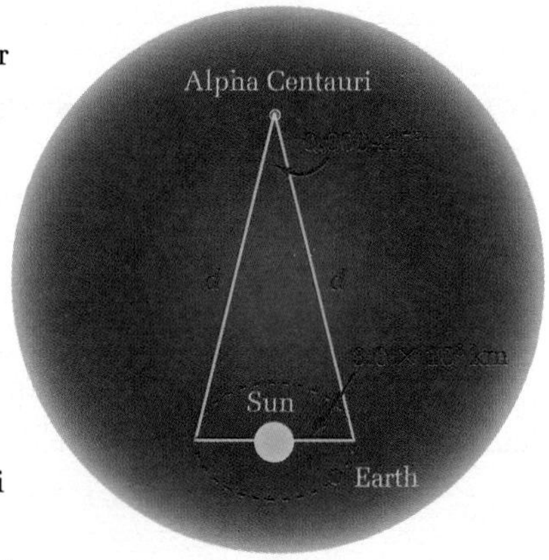

13. **Carpentry** Part of a roof has a slope of $\frac{4}{12}$, and the other part of the roof has a slope of $\frac{6}{12}$.
    a. Find the measures of $\angle A$ and $\angle B$.
    b. Find the measure of the angle at the peak of the roof.
    c. Find a and b.
    d. Find h.

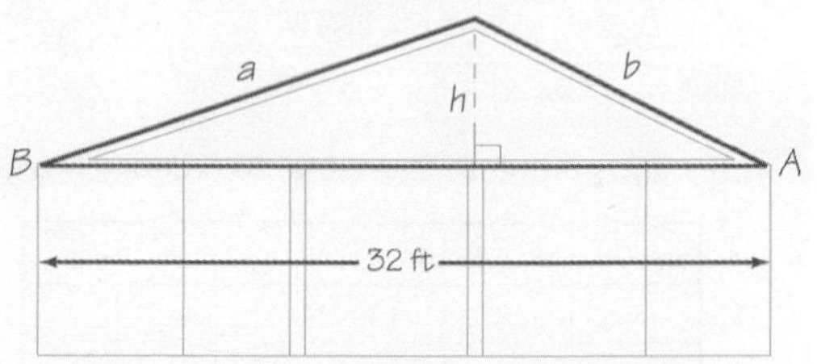

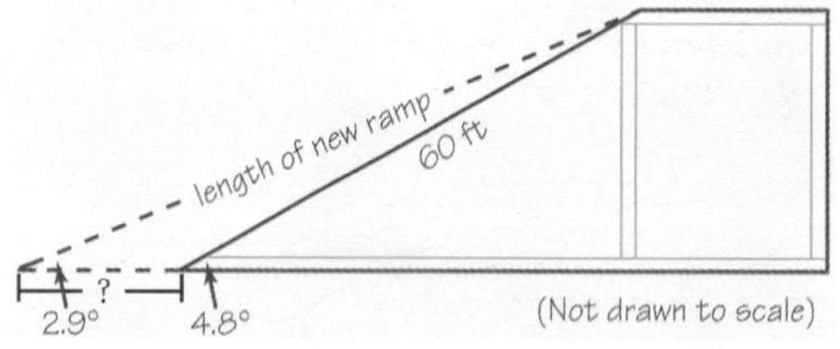

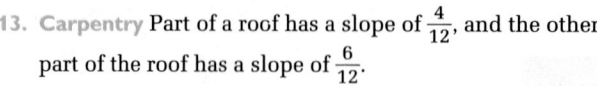

14. **Access Standards** A 60 ft ramp makes an angle of 4.8° with the ground. To meet new accessibility guidelines, the ramp must make an angle of 2.9° with the ground.
    a. How long will the new ramp be?
    b. How much farther along the ground will the new ramp extend?

Answers to Exercises and Problems

11. a. 39.249°
    b. 7.359 km; 8.124 km
    c. 2.518 km
    d. 2.520 km
    e. 2.519 km
    f. (a) 40° (b) 7.21 km; 8.000 km (c) 2.443 km (d) 2.483 km (e) 2.463 km; The estimate is decreased about 2.2%.

12. a. 4.12×10^{13} km
    b. 4.36 yr

13. a. 26.6°; 18.4°
    b. 135°
    c. 20.3 ft; 14.3 ft
    d. 6.4 ft

14. a. about 99.2 ft
    b. about 39.3 ft

15. See answers in back of book.

16. a. If a perpendicular were drawn from the given point to the other side of $\angle A$, its length would be $b \sin A$. If $a < b \sin A$, the segment is too short to reach the other side of A. If $a = b \sin A$, the segment just reaches the other side. If $a > b \sin A$, the segment will intersect the line containing the other side of $\angle A$

connection to SCIENCE

15. The *angular separation* of the sun and a planet is the angle formed by the sun, Earth, and the planet. The average distance from the sun to Earth is called 1 *astronomical unit* (a.u.).

On January 5, 1997, the angular separation of the sun and Venus is 21.147°. The distance from the sun to Earth on that day is 0.983 a.u. and from the sun to Venus is 0.724 a.u. On the same day, the angular separation of the sun and Mercury is 8.206°. The distance from the sun to Mercury is 0.323 a.u.

a. Find the unknown angle measures and side length in $\triangle V_1SE$ and $\triangle V_2SE$, which represent the possible positions of Venus, the sun, and Earth.

b. Find the unknown angle measures and side length in $\triangle M_1SE$ and $\triangle M_2SE$, which represent the possible positions of Mercury, the sun, and Earth.

c. Suppose a high-powered telescope reveals the sun shining on the planets as shown.

Is Venus at V_1 or V_2? 

Is Mercury at M_1 or M_2? 

d. Using your answers from parts (a)–(c), make a scale drawing showing the positions of the sun, Earth, Venus, and Mercury on January 5, 1997. Label all the distances among them. (*Hint:* Use the law of cosines to find the distance from Mercury to Venus.)

BY THE WAY...

Because the orbit of Venus is inside Earth's orbit, Venus appears near the Sun at dawn or dusk. That is why it is called the "morning star" or "evening star."

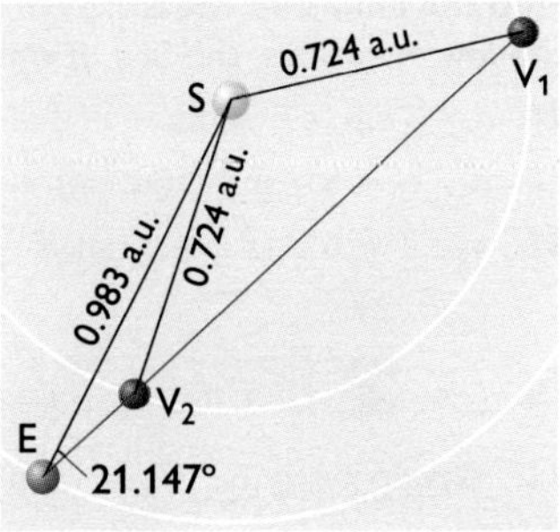

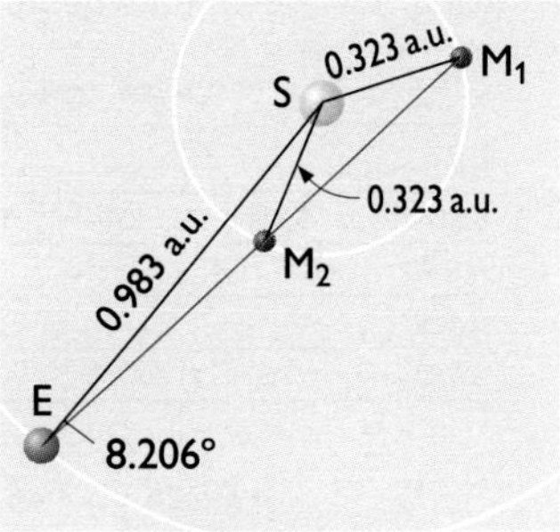

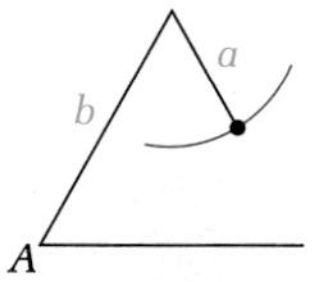

$a < b \sin A$

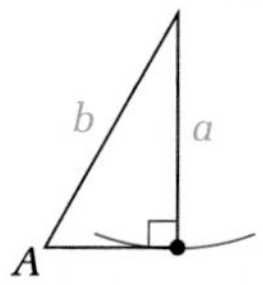

$a = b \sin A$

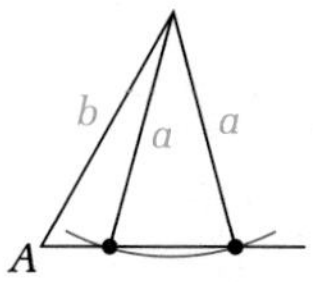

$a > b \sin A$
$a < b$

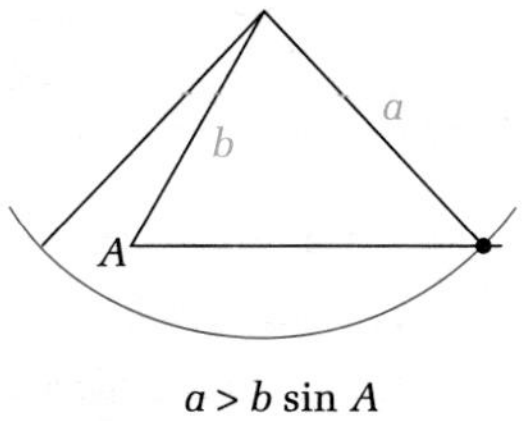

$a > b \sin A$
$a > b$

16. Four constructions that contain SSA information are shown.

a. For each construction, find the number of triangles formed and explain how you found the answer.

b. How many triangles will be formed when $a > b \sin A$ and $a = b$? Explain.

17. Draw diagrams, similar to those in Exercise 16, showing all the possibilities when A is an obtuse angle.

Mathematical Procedures

After completing Exs. 28–33, students should realize that in many cases involving the law of cosines, it is easier to use the law of cosines once and then switch to the law of sines to complete the problems.

Cooperative Learning

Ex. 34 on page 474 provides a useful summary of ways to solve a given triangle. Students can expand on this table to show the most efficient approach.

Answers to Exercises and Problems

twice, both times on the opposite side of $\angle A$ if $a < b$ (producing two triangles) and once on the opposite side of $\angle A$ if $a > b$ (producing one triangle).

b. If $a > b \sin A$ and $a = b$, one isosceles triangle would be formed. Since $a > b \sin A$, the segment is long enough to reach the opposite side of $\angle A$.

17.

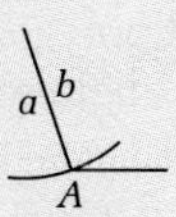

$a = b$

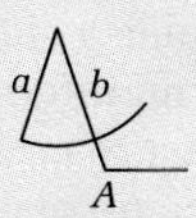

$a < b$

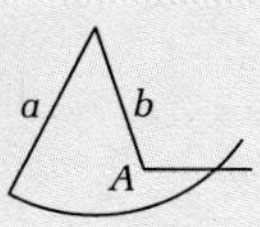

$a > b$

If $a \le b$, no triangle is formed.
If $a > b$, one triangle is formed.

Working on the Unit Project

After completing Ex. 38, each student in the group should follow all the clues they have written for their treasure hunt. Then they can compare their work to make sure that all members arrive at the same place.

Quick Quiz (8-5 through 8-7)

See page 476.

See also Test 33 in the *Assessment Book.*

Practice 65 For use with Section 8-7

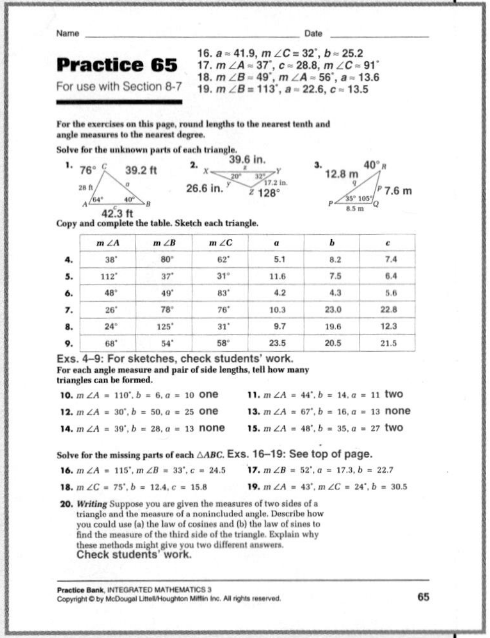
Name ______________ Date ______________

Practice 65
For use with Section 8-7

16. $a \approx 41.9$, $m\angle C = 32°$, $b \approx 25.2$
17. $m\angle A \approx 37°$, $c \approx 28.8$, $m\angle C \approx 91°$
18. $m\angle B \approx 49°$, $m\angle A \approx 56°$, $a \approx 13.6$
19. $m\angle B = 113°$, $a \approx 22.6$, $c \approx 13.5$

For the exercises on this page, round lengths to the nearest tenth and angle measures to the nearest degree.

Solve for the unknown parts of each triangle.

1. 76° C 39.2 ft; 28 ft; a; A 64°; 40° B; c 42.3 ft
2. 39.6 in.; X z Y; 20° 32°; 26.6 in. y; 17.2 in.; Z 128°
3. 12.8 m; 40° R; q; p 7.6 m; P 35° 105° Q; 8.5 m

Copy and complete the table. Sketch each triangle.

	$m\angle A$	$m\angle B$	$m\angle C$	a	b	c
4.	38°	80°	62°	5.1	8.2	7.4
5.	112°	37°	31°	11.6	7.5	6.4
6.	48°	49°	83°	4.2	4.3	5.6
7.	26°	78°	76°	10.3	23.0	22.8
8.	24°	125°	31°	9.7	19.6	12.3
9.	68°	54°	58°	23.5	20.5	21.5

Exs. 4–9: For sketches, check students' work.

For each angle measure and pair of side lengths, tell how many triangles can be formed.

10. $m\angle A = 110°$, $b = 6$, $a = 10$ one
11. $m\angle A = 44°$, $b = 14$, $a = 11$ two
12. $m\angle A = 30°$, $b = 50$, $a = 25$ one
13. $m\angle A = 67°$, $b = 16$, $a = 13$ none
14. $m\angle A = 39°$, $b = 28$, $a = 13$ none
15. $m\angle A = 48°$, $b = 35$, $a = 27$ two

Solve for the missing parts of each $\triangle ABC$. Exs. 16–19: See top of page.

16. $m\angle A = 115°$, $m\angle B = 33°$, $c = 24.5$
17. $m\angle B = 52°$, $a = 17.3$, $b = 22.7$
18. $m\angle C = 75°$, $b = 12.4$, $c = 15.8$
19. $m\angle A = 43°$, $m\angle C = 24°$, $b = 30.5$

20. ***Writing*** Suppose you are given the measures of two sides of a triangle and the measure of a nonincluded angle. Describe how you could use (a) the law of cosines and (b) the law of sines to find the measure of the third side of the triangle. Explain why these methods might give you two different answers.
Check students' work.

Practice Bank, INTEGRATED MATHEMATICS 3

65

For each angle measure and pair of side lengths, tell how many triangles can be formed.

18. $m\angle A = 47°$, $b = 6$, $a = 7$
19. $m\angle A = 112°$, $b = 11$, $a = 20$
20. $m\angle A = 32°$, $b = 9$, $a = 6$
21. $m\angle A = 57°$, $b = 11$, $a = 4$
22. $m\angle A = 12°$, $b = 192$, $a = 40$
23. $m\angle A = 149°$, $b = 27$, $a = 24$

Find the unknown measures in each $\triangle ABC$.

24. $m\angle A = 72°$, $m\angle C = 19°$, $b = 15.1$
25. $m\angle A = 111°$, $a = 12$, $c = 9$
26. $m\angle C = 76°$, $a = 18$, $c = 12.4$
27. $m\angle B = 108°$, $b = 10.9$, $c = 11.6$

Find the unknown measures in each $\triangle RST$. Tell whether you used the *law of cosines*, the *law of sines*, or *both*.

28. $r = 8$ cm, $s = 4$ cm, $t = 9$ cm
29. $m\angle R = 62°$, $m\angle T = 49°$, $s = 13$ km
30. $m\angle S = 81°$, $r = 16.2$ cm, $t = 9.1$ cm
31. $m\angle S = 119°$, $m\angle T = 15°$, $t = 4$ in.
32. $m\angle T = 57°$, $r = 6.3$ cm, $s = 17$ cm
33. $m\angle R = 17°$, $m\angle T = 91°$, $r = 3.2$ km

Ongoing ASSESSMENT

34. **Group Activity** Work with another student.

Write *triangle sum theorem, law of sines,* or *law of cosines* to complete each row. Then tell whether it is easier to find *angles* first or *sides* first.

GIVEN	To find sides, use...	To find angles, use...	Example
SAS	?	?	8–6 Sample 1, p. 461
SSS	all sides known	?	8–6 Sample 2, p. 463
ASA	?	?	8–7 Sample 1, p. 468
AAS	?	?	8–7 Sample 2, p. 468
SSA	?	?	8–7 Sample 3, p. 470

Review PREVIEW

35. Write a form of the law of cosines for a triangle with sides of length x, y, and z and angles X, Y, and Z. *(Section 8-6)*
36. Write an indirect proof to show that no regular polygon has an interior angle whose measure is 80°. *(Section 3-7)*
37. **Open-ended** Write an eight-number data set. Find the standard deviation of the data set. *(Section 6-2)*

8 Working on the Unit Project

38. Write a last clue for your treasure hunt that leads to the treasure and that can be answered only by applying the law of sines. An example is shown.

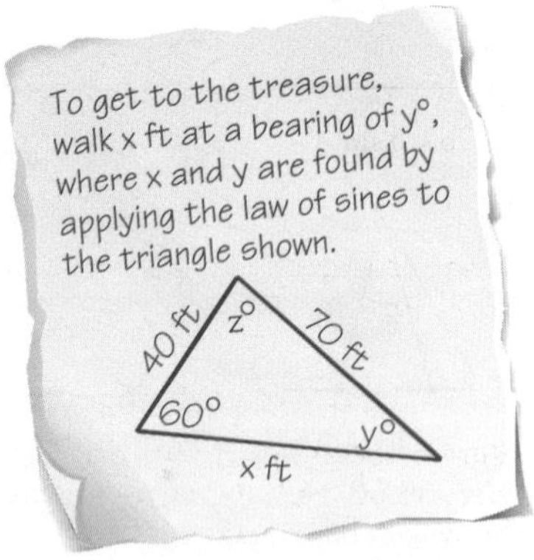

To get to the treasure, walk x ft at a bearing of y°, where x and y are found by applying the law of sines to the triangle shown.
40 ft
z°
70 ft
60°
y°
x ft

Answers to Exercises and Problems

18. 1 triangle
19. 1 triangle
20. 2 triangles
21. no triangle
22. 2 triangles
23. no triangle

24–27. Answers are given to the nearest tenth. Answers may vary slightly due to rounding or to solution method.

24. $m\angle B = 89°$; $b = 14.4$; $c = 4.9$
25. $m\angle B = 24.6°$; $m\angle C = 44.4°$; $b = 5.4$
26. not possible
27. not possible

28–33. Solution methods may vary; examples are given. Answers are given to the nearest tenth. Answers may vary slightly due to rounding or to solution method.

28. $m\angle R = 62.7°$; $m\angle S = 26.4°$; $m\angle T = 90.9°$; law of cosines
29. $m\angle S = 69°$; $r = 12.3$ km; $s = 10.5$ km; law of sines
30. $s = 17.3$ cm; $m\angle T = 31.3°$; $m\angle R = 67.7°$; both
31. $m\angle R = 46°$; $r = 11.1$ in.; $s = 13.5$ in.; law of sines
32. $t = 14.6$ cm; $m\angle R = 21.2°$; $m\angle S = 101.8°$; both
33. $m\angle S = 72°$; $s = 10.4$ km; $t = 10.9$ km; law of sines
34. See answers in back of book.
35. $x^2 = y^2 + z^2 - 2yz \cos X$; $y^2 = x^2 + z^2 - 2xz \cos Y$; $z^2 = x^2 + y^2 - 2xy \cos Z$
36. Methods may vary. An example is given. Suppose a regular n-sided polygon has an interior angle whose measure is 80°. Then the measure of an exterior angle is 180° – 80° = 100°. The measure of each exterior angle of a regular n-sided polygon is $\frac{360°}{n}$, so $100° \cdot n = 360°$ and $n = 3.6$. This contradicts the fact that n must be an integer. The assumption that there is a regular polygon with an interior angle whose measure is 80° is false. Therefore, no regular polygon has an interior angle whose measure is 80°.
37. Answers may vary. An example is given. data: 75, 66, 84, 82, 88, 71, 72, 87; standard deviation: 7.67
38. Answers may vary.

Unit Project 8

Completing the Unit Project

Now you are ready to share your map and clues with another group so that they can search for the buried treasure.

Your completed project should include these things:

- **a treasure map**
- **numbered clues written on separate pieces of paper**
- **a solutions map on which you write the clues at the points where searchers stop to pick up new clues**
- **a written critique of your treasure hunt by the group that tries it**

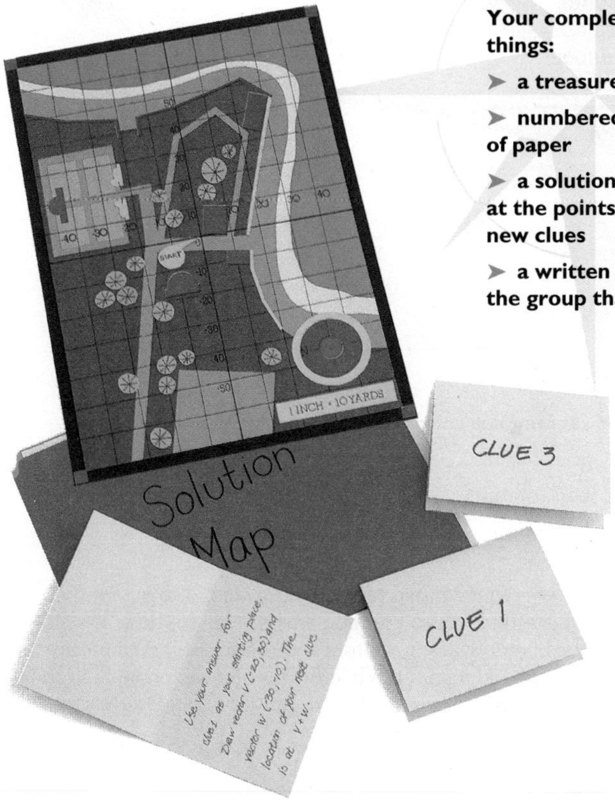

Look Back

Were the members of another group able to locate the buried treasure from your group's clues? Were any of the clues too imprecise? too easy or too difficult to follow?

Alternative Projects

Project 1: Ocean Navigation

Write a research report on ocean navigation. Find out how boat pilots use bearings and maps to stay on course by using information about currents, tides, and so on.

Project 2: Orienteering

Find out about the sport of orienteering. Create a classroom display that shows how participants use maps and magnetic compasses to plot courses between *control points*.

Assessment

A scoring rubric for the Unit Project can be found on pages 422 and 423 of this Teacher's Edition and also in the ***Project Book.***

Unit Support Materials

- *Practice Bank:* Cumulative Practice 66
- *Study Guide:* Unit 8 Review
- *Problem Bank:* Unifying Problem 8
- *Assessment Book:* Unit Tests 34 and 35, Spanish Unit Tests, Alternative Assessment
- *Test Generator* software with Test Bank
- *Teacher's Resources for Transfer Students*

Quick Quiz (8-5 through 8-7)

A toy rocket is launched at an angle of 64° with an initial velocity of 17 ft/s.

1. Express the horizontal and vertical positions, relative to the launch point of the rocket, using parametric equations. [8-5] $x = 7.5t$, $y = 15.3t - 16t^2$
2. Combine the two parametric equations of the rocket's position into a single equation. [8-5] $y = 2.04x - 0.28x^2$
3. Write parametric equations for a circle with radius 7 centered at (1, –2). [8-5] $x = 7\cos t + 1$, $y = 7\sin t - 2$
4. Find $m\angle A$ in $\triangle ABC$ if $a = 2.1$, $b = 3.7$, and $c = 3.2$. [8-6] about 34.5°
5. Find the unknown measures for $\triangle ABC$ with $a = 4.7$, $b = 6.5$, and $m\angle A = 43°$. [8-7] $m\angle B \approx 71°$, $m\angle C \approx 66$, $c \approx 6.3$ or $m\angle B \approx 109°$, $m\angle C \approx 28°$, and $c \approx 3.2$

Unit 8 REVIEW AND ASSESSMENT

For Questions 1–13, use this diagram.

Carol Grant Gould and James L. Gould studied the behavior of honey bees to find out how they navigate and communicate. In one study, a bee was trained to find food at three stations about 180 m from the hive.

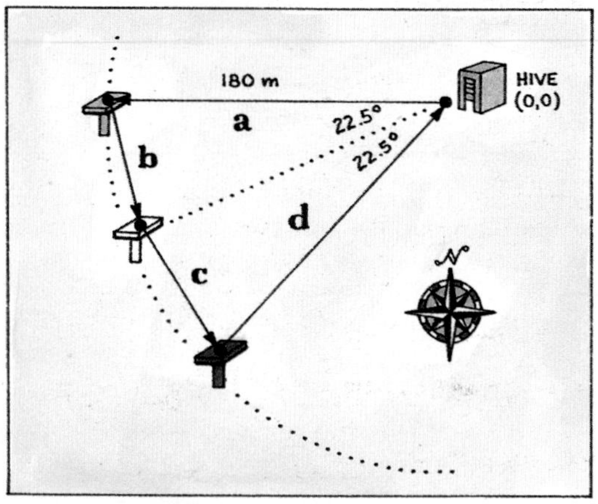

Find polar coordinates for each station. Use the hive as the origin. 8-1

1. blue station
2. white station
3. red station

Bee landing at hive entrance

BY THE WAY...

This study was done in 1976. The bee was marked with a dab of paint matching the color of each station it reached. It was known as the "bicentennial bee."

Find rectangular coordinates for each station. 8-2

4. blue station
5. white station
6. red station

The bee was first trained to fly along vector **a** to get food at the blue station. When the food was moved to the white station, the bee flew along vectors **a** and **b** to get it. When the food was moved again, the bee flew along vectors **a**, **b**, and **c** to get it. The bee returned directly to the hive along vector **d**.

Draw and label each vector. 8-3

7. **a** + **b**
8. **b** + **c**
9. **a** + **b** + **c**
10. the vector that the bee traveled to return to the hive from the red station

Find each vector sum in rectangular form. Use the rectangular coordinates you found in Questions 4–6. 8-4

11. **a** + **b**
12. **a** + **b** + **c**
13. **a** + **b** + **c** + **d**

Describe the graph of each pair of parametric equations. Use these settings: $0 \le t \le 360$, t-step = 1, $-12 \le x \le 12$, and $-8 \le y \le 8$. 8-5

14. $x = \cos t$, $y = \sin t$
15. $x = \cos t + 4$, $y = \sin t - 5$
16. $x = 4\cos t$, $y = 5\sin t$

Answers to Unit 8 Review and Assessment

1. (180, 180°)
2. (180, 202.5°)
3. (180, 225°)
4. (–180, 0)
5. about (–166.3, –68.9)
6. about (–127.3, –127.3)
7. 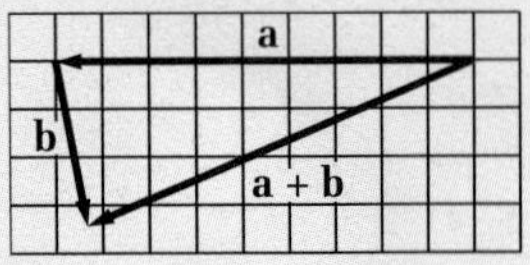

8. 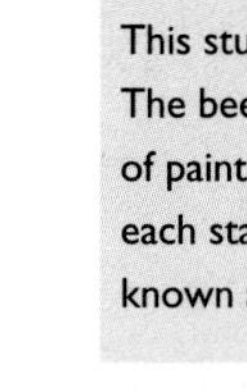

9. 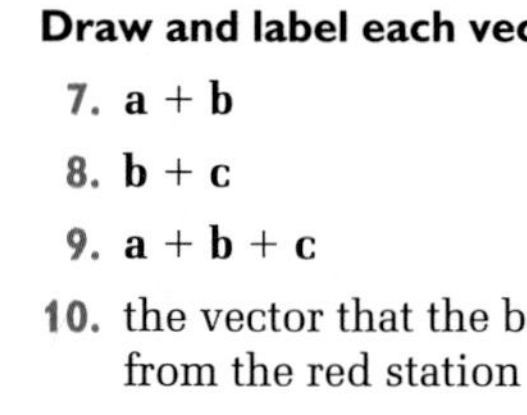

10. 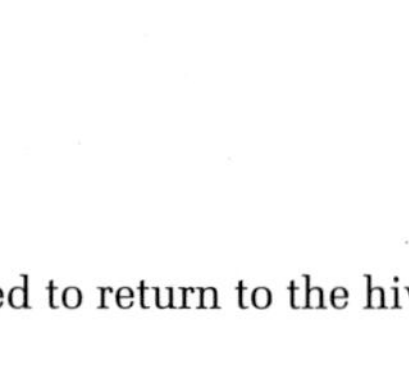

11. (–166.3, –68.9)
12. (–127.3, –127.3)
13. (0, 0)
14. a circle with radius 1 centered at the origin

17. Find the distance between the tips of the stunt kite's wings.

18. Sketch a regular pentagon inscribed in a circle of radius 2 in. What is the length of each side?

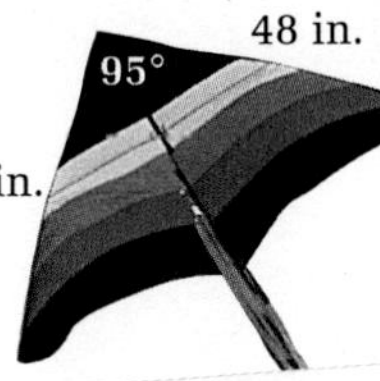

48 in.
95°
48 in.

8-6

Use the information from Sarah's journal.

8-7

19. Draw a sketch of the area and Sarah's hike back to camp.

20. How many paces is it from the rocks to camp?

21. What is the distance in paces across the lake from the cottonwood tree to camp?

22. Suppose Sarah walked at a rate of 100 paces/min and rested at the rocks for 17 min. At what time did she return to camp?

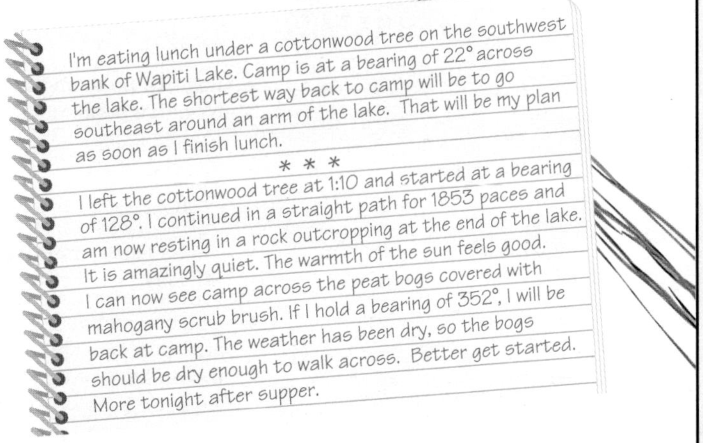

I'm eating lunch under a cottonwood tree on the southwest bank of Wapiti Lake. Camp is at a bearing of 22° across the lake. The shortest way back to camp will be to go southeast around an arm of the lake. That will be my plan as soon as I finish lunch.
* * *
I left the cottonwood tree at 1:10 and started at a bearing of 128°. I continued in a straight path for 1853 paces and am now resting in a rock outcropping at the end of the lake. It is amazingly quiet. The warmth of the sun feels good. I can now see camp across the peat bogs covered with mahogany scrub brush. If I hold a bearing of 352°, I will be back at camp. The weather has been dry, so the bogs should be dry enough to walk across. Better get started. More tonight after supper.

23. **Self-evaluation** What dimensions of a triangle do you need to know in order to use the law of cosines? the law of sines? How can you be sure you remember the laws and know when to use them?

24. **Group Activity** Work in a group of four students.

a. Copy and complete the table of angle measures for the five types of triangles in this geodesic dome. Round answers to the nearest hundredth of a degree.

b. **Open-ended** Choose a vertex of the dome and show that the sum of the measures of the angles at the vertex is less than 360°.

	Side lengths			Angle measures		
I	*a*	*a*	*b*	?	?	?
II	*b*	*c*	*c*	?	?	?
III	*c*	*d*	*f*	?	?	?
IV	*d*	*d*	*e*	?	?	?
V	*e*	*e*	*e*	60°	60°	60°

This geodesic dome is based on an icosahedron whose faces are each subdivided into 16 triangles.

Each vertex is 15 ft from the center of the dome.

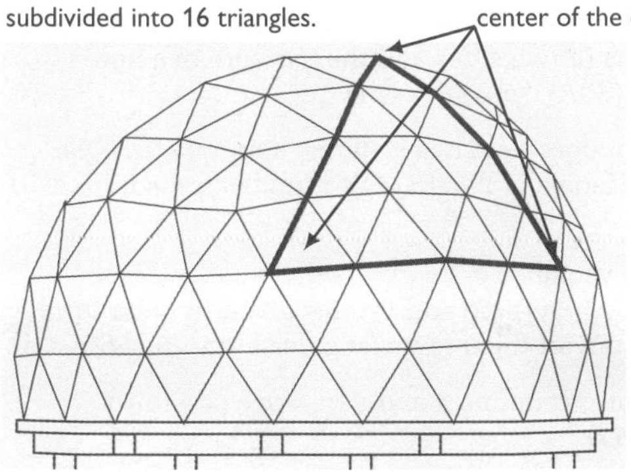

The triangles in the geodesic dome appear to be equilateral, but most of them are not.

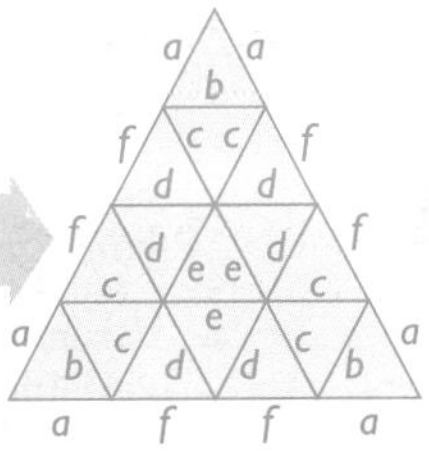

Side	Length (in.)
a	45.57
b	53.14
c	53.02
d	56.32
e	58.49
f	53.75

Answers to Unit 8 Review and Assessment

15. a circle with radius 1 centered at (4, –5)

16. an ellipse with its center at the origin, 8 units wide, 10 units high

17. about 78.6 in.

18. 2.35 in.

19.

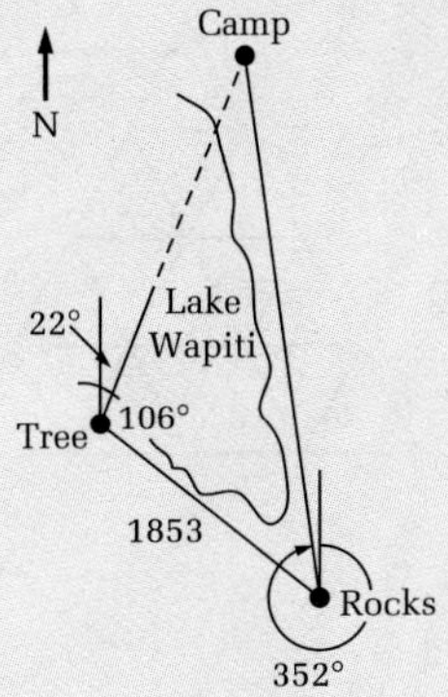

N
Camp
22°
Lake Wapiti
Tree
106°
1853
Rocks
352°

20. about 3562 paces

21. about 2574 paces

22. about 2:28

23. the lengths of two sides and the measure of the included angle or the lengths of all three sides; the measures of two angles and the length of the included side or the lengths of two sides and the measure of a non-included angle; Answers may vary.

24. Angle measures may vary slightly due to solution method or rounding. The sum of the measures of the angles may not be equal to 180° due to rounding.

Answers continued on next page.

IDEAS AND (FORMULAS) $= x^2$ ${}_5P_5$

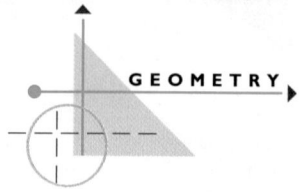

GEOMETRY

- Polar coordinates (r, θ) describe the location of a point using a distance r from a reference point and an angle θ measured from 0° along a horizontal to the right of the reference point. *(p. 425)*
- Positive polar angles are measured counterclockwise from 0°. Negative polar angles are measured clockwise from 0°. *(pp. 425, 426)*
- The rectangular coordinates of a point with polar coordinates (r, θ) are $(r \cos \theta, r \sin \theta)$. *(p. 431)*
- Vectors are quantities with both magnitude and direction. *(p. 441)*
- You can add vectors by placing them in a head-to-tail arrangement. The sum is the vector with its tail at the tail of the first vector and its head at the head of the last vector. *(p. 442)*
- The opposite of a vector **v** is a vector with the same magnitude as **v** but the opposite direction. *(p. 443)*
- The vector $k\mathbf{v}$ is a scalar multiple of vector **v**. It has the same direction as **v** and a magnitude that is k times the magnitude of **v**. *(p. 443)*
- You can use the law of cosines to find the length of the third side of a triangle when you know the lengths of two sides of a triangle and the measure of their included angle (SAS). *(p. 461)*
- You can use the law of cosines to find any angle in a triangle when you know the lengths of all three sides (SSS). *(p. 463)*
- You can use the law of sines to find unknown dimensions in a triangle when you know the measures of two angles in the triangle and the length of one side (ASA or AAS) or when you know the lengths of two sides and the measure of a non-included angle (SSA). *(pp. 468, 470)*
- SSA information does not always define a unique triangle. Sometimes no triangle or two triangles can be formed. *(p. 470)*

ALGEBRA

- The sum of two vectors $\mathbf{u} = (x_1, y_1)$ and $\mathbf{v} = (x_2, y_2)$ is the vector $(x_1 + x_2, y_1 + y_2)$. To add vectors given in polar form, you must first convert them to rectangular form. *(p. 447)*
- Parametric equations define x and y in terms of a third variable t. *(p. 453)*

Answers to Unit 8 Review and Assessment

a.

	Side lengths			Angle measures		
I	a	a	b	54.33°	54.33°	71.33°
II	b	c	c	60.15°	59.93°	59.93°
III	c	d	f	57.54°	63.67°	58.80°
IV	d	d	e	58.72°	58.72°	62.57°
V	e	e	e	60°	60°	60°

b.

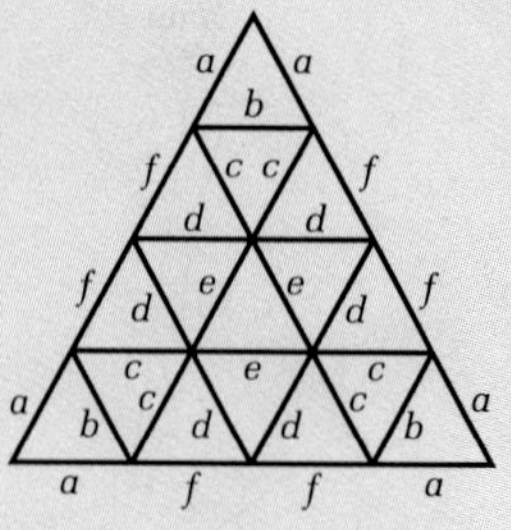

Answers may vary. An example is given. For the vertex indicated in the figure, the sum of the measures of the angles is 60.15° + 58.80° + 58.80° + 58.72° + 58.72° + 60° = 355.19°, which is less than 360°.

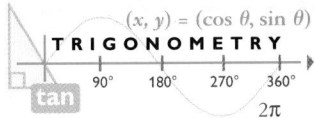

$(x, y) = (\cos\theta, \sin\theta)$

TRIGONOMETRY

90° 180° 270° 360°

tan 2π

- The cosine, sine, and tangent ratios can be defined for any angle. The values of *x* and *y* used are the coordinates of a point with polar coordinates (r, θ). *(p. 432)*

$$\cos\theta = \frac{x}{r} \quad \sin\theta = \frac{y}{r} \quad \tan\theta = \frac{y}{x} \quad (\text{when } x \neq 0)$$

- Supplementary angles have the same sine. *(p. 433)*
- Angles with opposite measures have the same cosine. *(p. 433)*
- To convert rectangular coordinates to polar coordinates, use the Pythagorean theorem to find *r*. Then use the inverse trigonometric feature of a calculator to find the measure of θ. The calculator may not display the correct polar angle. Use what you know about the quadrant the point lies in to find the correct polar angle. *(p. 434)*
- A Pythagorean identity relates the sine and cosine of any angle. *(p. 435)*

$$\cos^2\theta + \sin^2\theta = 1$$

- The law of cosines is related to the Pythagorean theorem but it applies to all triangles. *(p. 461)*

$$a^2 = b^2 + c^2 - 2bc\cos A$$

- The law of sines states that lengths of the sides of any triangle are in proportion with the sines of the opposite angles. *(p. 467)*

$$\frac{a}{\sin A} = \frac{b}{\sin B} = \frac{c}{\sin C}$$

Key Terms

- **bearing** (p. 425)
- **vector** (p. 441)
- **tail of a vector** (p. 441)
- **polar coordinates** (p. 425)
- **magnitude** (p. 441)
- **resultant vector** (p. 442)
- **Pythagorean identity** (p. 435)
- **head of a vector** (p. 441)
- **scalar multiple of a vector** (p. 443)
- **opposite vectors** (p. 443)
- **law of cosines** (p. 461)
- **parametric equations** (p. 453)
- **law of sines** (p. 467)
- **parameter** (p. 453)
- **ambiguous case** (p. 470)

Quick Quiz (8-1 through 8-4)

1. Jacob walks 2.5 miles southwest. Write two different pairs of polar coordinates to represent Jacob's location after his walk. [8-1]
Answers may vary. A sample answer is given. (2.5, –135°), (2.5, 225°)

2. Convert (21, 162°) to rectangular coordinates. [8-2]
about (–20.0, 6.5)

3. Convert (3, –4) to polar coordinates. [8-2]
Answers may vary. A sample answer is given. about (5, –53°)

4. Find all angles θ between 0° and 360° for which $\tan\theta = 1.2349$. [8-2]
51° and 231°

5. Describe how to add two vectors geometrically. [8-3]
Place them head to tail. The sum is a vector from the tail of the first vector to the head of the second vector.

6. Find **w** = (3, 15°) + (6, 210°) in polar form. [8-4]
about (3.2, –136.3°)

Planning the Unit 9

Transformations of Graphs and Data

OVERVIEW

- **Unit 9** covers transformations of graphs and data. Graphs are translated, stretched, and reflected. Combinations of transformations are performed.
- Transformations on data include adding a constant to data, and multiplying data by a constant. The standard normal distribution and the standard *z*-score are obtained and used to compare sets of data.
- The **Unit Project** is about a job interview. Students prepare a list of questions to use in interviewing a person who represents one career field. Students also explore qualification requirements for a job, working conditions, and salaries for different careers. They conduct an actual interview and take notes to use in a subsequent discussion.
- **Connections** to musical composition, football, toll roads, income tax, biology, parking lot prices, telephone rates, literature, and swimming are some of the topics included in the teaching materials and the exercises.
- **Graphics calculators** are used in Sections 9-1 and 9-3 to compute a standard deviation for a data set, in Section 9-2 to graph equations, in Section 9-3 to find an equation of the least-squares line for a data set, and in Section 9-7 to recognize the effect of a reflection on a graph and on the graph's equation. **Computer software**, such as Plotter Plus, can be used in Sections 9-1 and 9-3 to draw histograms.
- **Problem-solving strategies** used in Unit 9 include using technology, using a formula, using a graph, and using a diagram.

Unit Objectives

Section	Objectives	NCTM Standards
9-1	• Recognize the effect that adding a constant to data values has on the mean, median, range, and standard deviation.	1, 2, 3, 4, 5, 10
9-2	• Recognize the relationship between the graph of $y - k = f(x - h)$ and the parent graph of $y = f(x)$.	1, 2, 3, 4, 5, 6
9-3	• Recognize the effect that multiplying data values by a constant has on the mean, median, and standard deviation.	1, 2, 3, 4, 5, 10
9-4	• Recognize the effects of horizontal and vertical stretches on sets of points and graphs of functions.	1, 2, 3, 4, 5, 6
9-5	• Recognize the effects of multiple transformations on sets of data and graphs of functions.	1, 2, 3, 4, 5, 6, 10
9-6	• Use *z*-scores and the standard normal distribution to compare data.	1, 2, 3, 4, 5, 10
9-7	• Recognize the effect of a reflection on a graph and on the graph's equation.	1, 2, 3, 4, 5, 6, 10

Topic Spiraling

Section	Connections to Prior and Future Concepts
9-1	**Section 9-1** explores the effect that adding a constant has on the measures of central tendency and the standard deviation of a set of data. Measures of central tendency were first covered in Section 3-2 of Book 1, and reviewed in Section 1-3 of Book 3. Measures of variability were covered in Section 6-2 of Book 3.
9-2	**Section 9-2** explores translations of a parent graph. The parent graphs were covered originally in Unit 2 and in Sections 5-1, 5-2, and 5-6 of Book 3.
9-3	**Section 9-3** explores the effects on measures of central tendency and on the standard deviation when data are multiplied by a constant. Measures of central tendency were first covered in Section 3-2 of Book 1, and reviewed in Section 1-3 of Book 3. Measures of variability were covered in Section 6-2 of Book 3.
9-4	**Section 9-4** explores the effects of horizontal and vertical stretches on graphs. Matrix multiplication is used to represent some stretches. Multiplying matrices was first covered in Section 3-7 of Book 2, and is reviewed in the **Student Resources Toolbox**.
9-5	**Section 9-5** explores multiple transformations to a set of data and to a function. These transformations incorporate skills from the previous sections of Unit 9 of Book 3.
9-6	**Section 9-6** introduces standardizing the normal distribution, first introduced in Section 6-3 of Book 3. Students calculate *z*-scores for a set of data, and use the *z*-scores to compare sets of data. A *z*-score requires the standard deviation, first covered in Section 6-2 of Book 3. Percentiles are introduced.
9-7	**Section 9-7** explores reflecting a graph over the *x*-axis or the *y*-axis. Equations of graphs are obtained from a graph that has been reflected, translated, and/or stretched. Transformations were first covered in Sections 4-3, 6-6, 10-1, and 10-2 of Book 1, and were reviewed and expanded upon in Sections 3-6 and 4-2 of Book 2.

Integrating the Strands

Strands	Sections
Algebra	9-1, 9-2, 9-3, 9-4, 9-5, 9-6
Functions	9-1, 9-2, 9-3, 9-4, 9-5, 9-6, 9-7
Geometry	9-1, 9-2, 9-4, 9-5, 9-7
Trigonometry	9-1, 9-4
Statistics and Probability	9-1, 9-2, 9-3, 9-4, 9-5, 9-6, 9-7
Discrete Mathematics	9-1, 9-2, 9-3, 9-4, 9-5, 9-6, 9-7
Logic and Language	9-1, 9-2, 9-3, 9-4, 9-5, 9-6, 9-7

Planning the Unit 9

Transformations of Graphs and Data

Section Planning Guide

- ➤ Essential exercises and problems are indicated in boldface.
- ➤ Ongoing work on the Unit Project is indicated in color.
- ➤ Exercises and problems that require student research, group work, manipulatives, or graphing technology are indicated in the column headed "Other."

Section	Materials	Pacing	Standard Assignment	Extended Assignment	Other
9-1	calculator	Day 1	**2–8,** 9–17, 20–27, 28	1, **2–8,** 9–27, 28	
9-2	graphing technology	Day 1 Day 2	**1–9,** 10–13 **15–21,** 22, 23, 25–30, 31	**1–9,** 10–14 **15–21,** 22–30, 31	 31
9-3	graphics calculator	Day 1	**2, 3, 5–7,** 9–18, 19	1, **2, 3,** 4, **5–7,** 9–18, 19	8
9-4	graphing technology	Day 1 Day 2	**1–6, 9–15** 16, **17, 23–30, 34–37,** 39–44, 45, 46	**1–6,** 7, 8, **9–15** 16, **17,** 22, **23–30,** 31–33, **34–37,** 39–44, 45, 46	 18–21, 38
9-5		Day 1 Day 2	**1–4, 7–9** **10–16, 18–23,** 25–27, 28	**1–4,** 5, 6, **7–9** **10–16,** 17, **18–23,** 25–27, 28	 24
9-6		Day 1 Day 2	**1–4** **5–13,** 17–20, 21	**1–4** **5–13,** 14–20, 21	
9-7	graphing technology, MIRA® transparent mirror	Day 1 Day 2	**1–7** **12–17,** 19–21, 22	**1–7** 8–11, **12–17,** 19–21, 22	 8c, 18, 22c
Review Test		**Day 1** **Day 2**	**Unit Review** **Unit Test**	**Unit Review** **Unit Test**	

Yearly Pacing	Unit 9 Total	Units 1–9 Total	Remaining	Total
	16 days (2 for Unit Project)	139 days	15 days	154 days

Support Materials

- ➤ See **Project Book** for notes on Unit 9 Project: Job Interview.
- ➤ UPP and disk refer to **Using Plotter Plus** booklet and **Plotter Plus** disk.
- ➤ TI-81/82 refers to **Using TI-81 and TI-82 Calculators** booklet.
- ➤ Warm-up exercises for each section are available on **Warm-Up Transparencies.**

Section	Study Guide	Practice Bank	Problem Bank	Activity Bank	Explorations Lab Manual	Assessment Book	Visuals	Technology
9-1	9-1	Practice 67	Set 19	Enrich 59	Masters 1, 2	Quiz 9-1		Statistics Spreadsheet (disk) Data Analyzer (disk)
9-2	9-2	Practice 68	Set 19	Enrich 60	Add. Expl. 10 Master 2	Quiz 9-2	Folder 9	
9-3	9-3	Practice 69	Set 19	Enrich 61	Masters 1, 2	Quiz 9-3		Statistics Spreadsheet (disk) Data Analyzer (disk)
9-4	9-4	Practice 70	Set 19	Enrich 62	Master 2	Quiz 9-4 Test 36		
9-5	9-5	Practice 71	Set 20	Enrich 63	Master 2	Quiz 9-5		
9-6	9-6	Practice 72	Set 20	Enrich 64	Master 2	Quiz 9-6		
9-7	9-7	Practice 73	Set 20	Enrich 65	Add. Expl. 11 Master 2	Quiz 9-7 Test 37		TI-81/82, page 69 UPP, page 59
Unit 9	Unit Review	Practice 74	Unifying Problem 9	Family Involve 9		Tests 38, 39		

Form A

Spanish versions of these tests are on pages 147–150 of the **Assessment Book.**

Name ________________ Date ________ Score ____

Test 38

Test on Unit 9 (Form A)

Directions: Write the answers in the spaces provided.

The table at the right shows the price of an individual pizza at five different airports.

Chicago (O'Hare)	$4.75
San Francisco	$4.25
New York (JFK)	$5.25
Los Angeles	$4.50
Denver	$5.00

1. Find the mean and median of the pizza prices.

2. Find the range and standard deviation of the pizza prices.

3. Find the total cost of each pizza if 6% sales tax is added to the price of each pizza.

Chi: \$5.04; SF: \$4.51; NY: \$5.57; LA: \$4.77; D: \$5.30

4. Use what you know about multiplying each value in a data set by a constant to find the mean, median, range, and standard deviation for the total cost of the pizzas.

≈\$5.04; ≈\$5.04; \$1.06; ≈\$0.37

For each equation, find the parent function and the translation vector.

5. $y = \sqrt{x + 2}$ **6.** $y + 5 = \log(x - 6)$

Without using graphing technology, sketch each parent graph and translate it to obtain a graph of each equation.

7. $y - 4 = |x + 2|$ **8.** $y + 2 = (x - 3)^2$

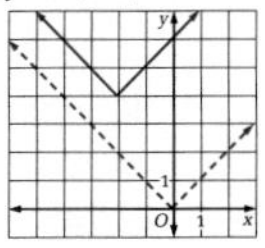

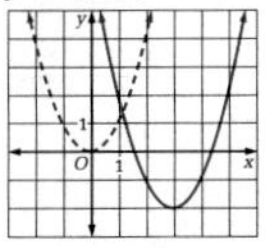

Pong's score on a test was 92. Deidre's score on the same test was 85. The class average was 87, and the standard deviation was 8.2. Use this information for Questions 9 and 10.

9. Find Pong's z-score. **10.** Find Deidre's z-score.

Answers

1. \$4.75; \$4.75
2. \$1.00; ≈\$0.35
3. *See question.*
4. *See question.*
5. $y = \sqrt{x}$; (−2, 0)
6. $y = \log x$; (6, −5)
7. *See question.*
8. *See question.*
9. ≈0.61
10. ≈−0.24

Assessment Book INTEGRATED MATHEMATICS 3

57

Name ________________ Date ________ Score ____

Test 38 *(continued)*

Directions: Write the answers in the spaces provided.

11. Writing Explain what happens when each value in a data set is multiplied by the same positive constant.

The mean, median, range, and standard deviation are multiplied by the constant.

Write an equation for each graph. Assume that each graph is a transformation of the graph of the given parent function.

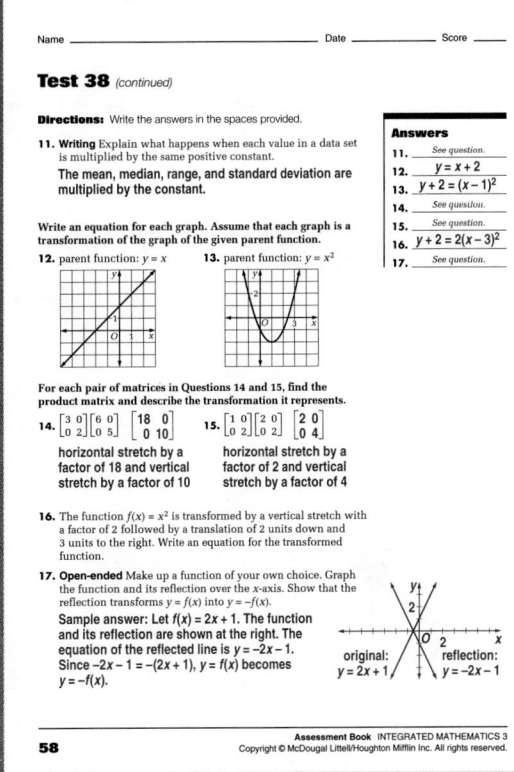

12. parent function: $y = x$ **13.** parent function: $y = x^2$

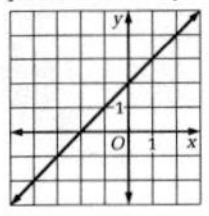

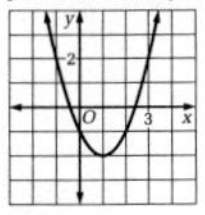

For each pair of matrices in Questions 14 and 15, find the product matrix and describe the transformation it represents.

14. $\begin{bmatrix} 3 & 0 \\ 0 & 2 \end{bmatrix}\begin{bmatrix} 6 & 0 \\ 0 & 5 \end{bmatrix}$ $\begin{bmatrix} 18 & 0 \\ 0 & 10 \end{bmatrix}$

horizontal stretch by a factor of 18 and vertical stretch by a factor of 10

15. $\begin{bmatrix} 1 & 0 \\ 0 & 2 \end{bmatrix}\begin{bmatrix} 2 & 0 \\ 0 & 2 \end{bmatrix}$ $\begin{bmatrix} 2 & 0 \\ 0 & 4 \end{bmatrix}$

horizontal stretch by a factor of 2 and vertical stretch by a factor of 4

16. The function $f(x) = x^2$ is transformed by a vertical stretch with a factor of 2 followed by a translation of 2 units down and 3 units to the right. Write an equation for the transformed function.

17. Open-ended Make up a function of your own choice. Graph the function and its reflection over the x-axis. Show that the reflection transforms $y = f(x)$ into $y = -f(x)$.

Sample answer: Let $f(x) = 2x + 1$. The function and its reflection are shown at the right. The equation of the reflected line is $y = -2x - 1$. Since $-2x - 1 = -(2x + 1)$, $y = f(x)$ becomes $y = -f(x)$.

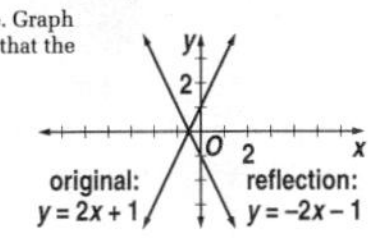

Answers

11. *See question.*
12. $y = x + 2$
13. $y + 2 = (x - 1)^2$
14. *See question.*
15. *See question.*
16. $y + 2 = 2(x - 3)^2$
17. *See question.*

58
Assessment Book INTEGRATED MATHEMATICS 3

Form B

Name ________________ Date ________ Score ____

Test 39

Test on Unit 9 (Form B)

Directions: Write the answers in the spaces provided.

The table at the right shows the price of a round trip ticket from Seattle to Omaha on five different airlines.

United	$175
American	$199
TWA	$187
Alaska Air	$167
Northwest	$191

1. Find the mean and median of the ticket prices.

2. Find the range and standard deviation of the ticket prices.

3. Find the total cost of each ticket if 7% sales tax is added to the price of each ticket.

U: \$187.25; Am: \$212.93; T: \$200.09; Al: \$178.69; N: \$204.37

4. Use what you know about multiplying each value in a data set by a constant to find the mean, median, range, and standard deviation for the total cost of each ticket.

≈\$196.67; \$200.09; \$34.24; ≈\$12.23

For each equation, find the parent function and the translation vector.

5. $y = \sqrt{x - 3}$ **6.** $y - 3 = \ln(x + 4)$

Without using graphing technology, sketch each parent graph and translate it to obtain a graph of each equation.

7. $y + 3 = |x + 5|$ **8.** $y - 4 = (x + 1)^2$

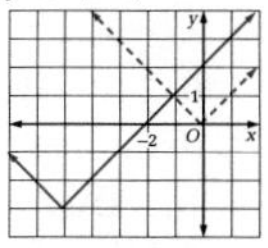

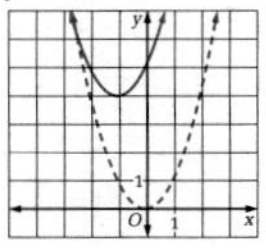

Akira's score on a test was 95. Lee's score on the same test was 84. The class average was 86, and the standard deviation was 5.6. Use this information for Questions 9 and 10.

9. Find Akira's z-score. **10.** Find Lee's z-score.

Answers

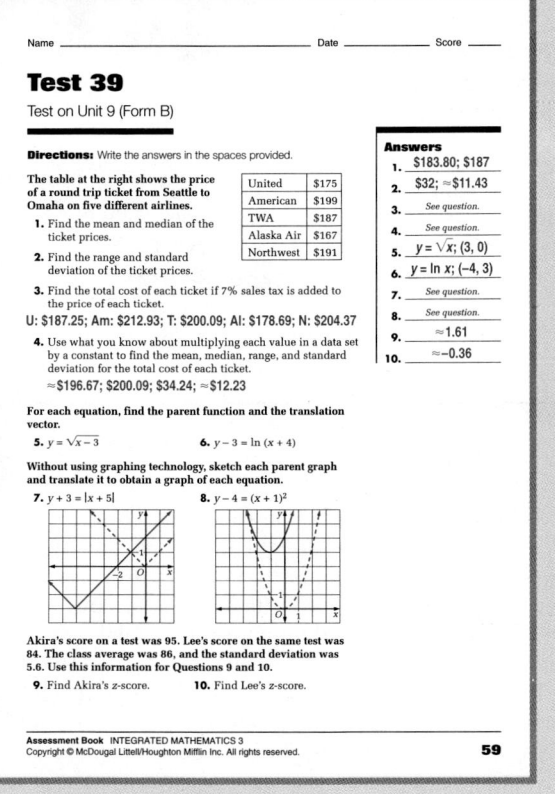

1. \$183.80; \$187
2. \$32; ≈\$11.43
3. *See question.*
4. *See question.*
5. $y = \sqrt{x}$; (3, 0)
6. $y = \ln x$; (−4, 3)
7. *See question.*
8. *See question.*
9. ≈1.61
10. ≈−0.36

Assessment Book INTEGRATED MATHEMATICS 3

59

Name ________________ Date ________ Score ____

Test 39 *(continued)*

Directions: Write the answers in the spaces provided.

11. Writing Explain what happens when the graph of $y = f(x)$ is stretched horizontally by a positive factor a and vertically by a positive factor b. Sample answer: The x-coordinate of each point on the graph is multiplied by a while the y-coordinate of each point is multiplied by b.

Write an equation for each graph. Assume that each graph is a transformation of the graph of the given parent.

12. parent function: $y = x$ **13.** parent function: $y = x^2$

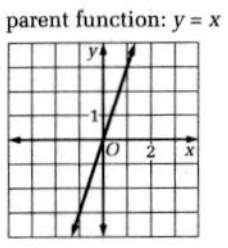

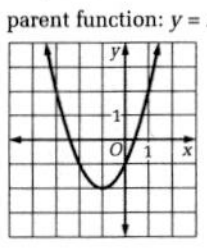

For each pair of matrices in Questions 14 and 15, find the product matrix and describe the transformation it represents.

14. $\begin{bmatrix} 2 & 0 \\ 0 & 3 \end{bmatrix}\begin{bmatrix} 5 & 0 \\ 0 & 6 \end{bmatrix}$ $\begin{bmatrix} 10 & 0 \\ 0 & 18 \end{bmatrix}$

horizontal stretch by a factor of 10 and vertical stretch by a factor of 18

15. $\begin{bmatrix} 2 & 0 \\ 0 & 1 \end{bmatrix}\begin{bmatrix} 3 & 0 \\ 0 & 3 \end{bmatrix}$ $\begin{bmatrix} 6 & 0 \\ 0 & 3 \end{bmatrix}$

horizontal stretch by a factor of 6 and vertical stretch by a factor of 3

16. The function $f(x) = x^2$ is transformed by a vertical stretch with a factor of 3 followed by a translation of 2 units up and 3 units to the left. Write an equation for the transformed function.

17. Open-ended Make up a function of your own choice. Graph the function and its reflection over the y-axis. Show that the reflection transforms $y = f(x)$ into $y = f(-x)$.

Sample answer: Let $f(x) = 2x + 2$. The function and its reflection are shown at the right. The equation of the reflected line is $y = -2x + 2$. Since $-2x + 2 = 2(-x) + 2$, $y = f(x)$ becomes $y = f(-x)$.

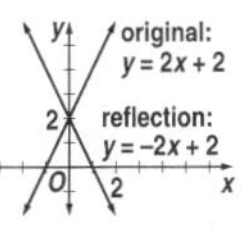

Answers

11. *See question.*
12. $y = 3x$
13. $y + 2 = (x + 1)^2$
14. *See question.*
15. *See question.*
16. $y - 2 = 3(x + 3)^2$
17. *See question.*

60
Assessment Book INTEGRATED MATHEMATICS 3

OUTSIDE RESOURCES

Books/Periodicals

Jacobs, Harold R. *Mathematics: A Human Endeavor.* Chapter 9: Measures of Variability: pp. 554–563. San Francisco, CA: W.H. Freeman and Company, 1994.

Ohio State University Calculator and Computer Precalculus Project. *Precalculus Mathematics: A Graphing Approach,* Section 3.3: "Quadratic Functions and Geometric Transformation": pp. 177–189. Addison-Wesley.

Software

Myers, David L. *Plotter Plus.* Boston, MA: Houghton Mifflin Company, 1995. Macintosh and MS-DOS (worksheets included).

Parker, David B. *AcroSpin.* Salt Lake City, UT, 1989. IBM PC compatible, 2 disks plus manual. Allows the user to rotate, translate, enlarge, or shrink an image.

Keir, Marilyn L. and Roy A. Keir. *AlgeBrush.* Salt Lake City, UT: MareWare, 1993.

f(g)Scholar. IBM. Southhampton, PA: Future Graph, 1992.

PROJECT GOALS

- Students develop a list of questions to use in interviewing a person representing a career field.
- Students discuss career fields they might be interested in exploring and choose one field for their group to investigate.
- Students interview a career representative in class and develop a set of notes to review and analyze what they have learned in the interview.

PROJECT PLANNING

Project Teams

Have students work on the project in groups of four. They can begin by discussing career fields they might find interesting, how to find someone to interview, and by asking the guidance office for help in finding information on careers. The class as a whole may then wish to discuss the career choices of each group. In order to provide as much career information as possible to students, a decision may be made to have each group investigate a different career. All group members should contribute questions for the interview and then edit the list to obtain a final set of questions.

Support Materials

The ***Project Book*** contains information about the following topics for use with this Unit Project.

- Project Description
- Teaching Commentary
- Working on the Unit Project Exercises
- Completing the Unit Project
- Assessing the Unit Project
- Alternative Projects
- Outside Resources

unit 9

Transformations of Graphs and Data

Suppose you could travel back through time. In the 4th millennium B.C. in Mesopotamia, you meet a wheelwright making wheels out of three planks clamped together. In the 1600s you meet a village blacksmith forging wagon parts in colonial America. You tell the wheelwright and the blacksmith about the role of computer graphics in the career of a 20th century automobile designer. They never heard of the job.

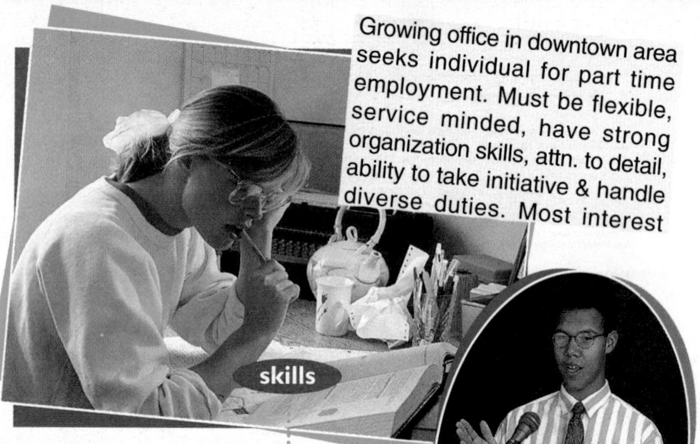

Growing office in downtown area seeks individual for part time employment. Must be flexible, service minded, have strong organization skills, attn. to detail, ability to take initiative & handle diverse duties. Most interest
skills

interests
The sucessful candidate must be self-motivated and possess strong communication skills and enjoy customer contact. We offer

When you go job hunting, you bring a **résumé** of your schooling, job experiences, honors, and special skills. Employers may also be looking for traits you take for granted, traits that followed you from childhood into adulthood.

480

Suggested Rubric for Unit Project

4 Students select a career to investigate, develop a list of questions for an interview, find someone to interview, and actually conduct the interview. Their questions cover all of the topics listed on page 533 and incorporated mathematical ideas from the Working on the Unit Project exercises. The interview with a career representative in class is informative and interesting. Students take notes during the interview and organize them in a way that can be used to review and analyze what was learned in the interview. The whole job interview process from beginning to end is well organized and effectively implemented.

3 Students complete the project successfully. There are, however, some areas that could be improved with more thought and work. The list of questions leaves out a few important topics or covers them only superficially. The role of mathematics on the job is not pursued in detail. The interview itself goes well but could have been longer or shorter. Only some of the students take notes during the interview, so the subsequent

Unit Project 9

Job Interview

A job interview is an opportunity for you to ask questions about a company and the job it is offering, as well as to answer questions about yourself. For your project, your group will plan to invite someone in one career field to visit your class and answer questions.

As you work through the Working on the Unit Project exercises, you will understand some of the topics to bring up in a job interview, such as the role and scoring of any special tests needed for entry and the expected availability of job openings in the future.

Your completed project will include a list of questions to use in interviewing a career representative, a completed interview, and an analysis of the information gained in the interview.

Statistics show that an adult working in the United States will probably change jobs six or seven times in a lifetime. In today's **global** economy, your special interests and skills may take you far from home to a part of the world that has the right job for you.

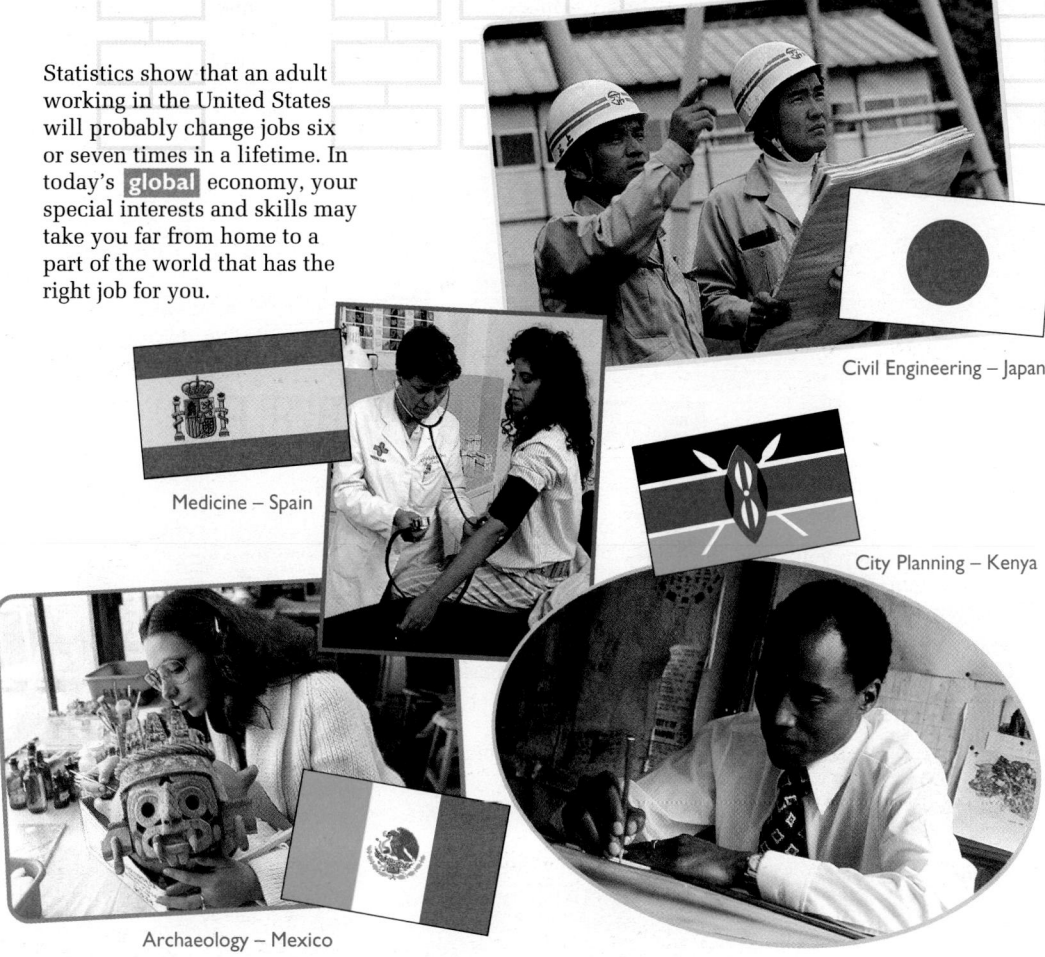

Civil Engineering – Japan

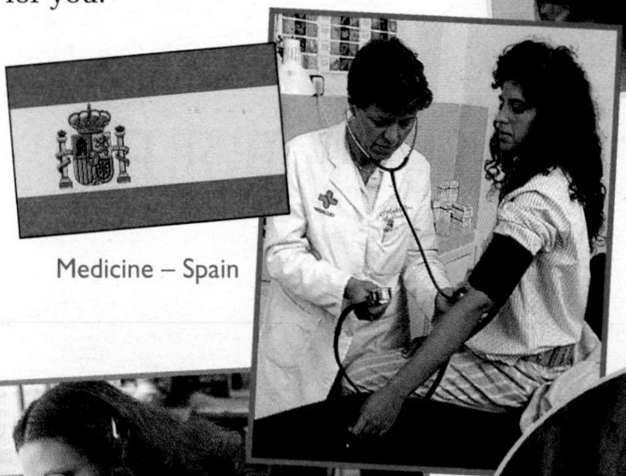

Medicine – Spain

City Planning – Kenya

Archaeology – Mexico

Suggested Rubric for Unit Project

review and analysis of what has been learned is incomplete.

2 Students complete the project but do not follow all of the guidelines stated in the textbook. The list of interview questions is incomplete. Important issues are not covered. There is little or no understanding of how mathematics plays a role in the career selected. The interview is not interesting, and the career representative is, at times, confused by students' questions because they do not make much sense. The follow-up notes for the interview are not well organized and students are not sure what they have learned.

1 Students are unable to complete the project. They select a career, but their list of questions is disorganized and shows a lack of effort in collecting the data required to conduct a successful interview. No career representative is invited to class. Students should be encouraged to speak with the teacher as soon as possible to review their work and to make a new start on the project.

Students Acquiring English

This unit project may be difficult for some students acquiring English since most of the project requires writing in English. You may want to consider allowing students to write the job interview questions and to conduct the interview in their native language. As an alternative, you may want to make sure that groups that have students acquiring English also have students fluent in English. Students can then ask their peers for assistance, if needed.

ADDITIONAL BACKGROUND

Multicultural Note

Kenya is a nation with a diverse population and a breathtaking natural landscape. About forty different ethnic groups are represented in the population of thirty million. The Kikuyu, numbering about two million, is the largest group; the Kalenjin, Kamba, Luhya, and Luo each have more than one million members. The Kenyan way of life is also varied: some Kenyans are nomads, some are urban homeowners, and some are rural farmers. Kenya's landscape is as different as its people, divided into three major regions: a tropical coastal strip, a dry plains area, and a fertile highland. Both the highlands and the coast are densely populated. The plains cover seventy-five percent of the country and are sparsely populated. The wildlife habitat in Kenya's plains supports hundreds of species of animals, including giraffes, elephants, lions, cheetahs, ostriches, rhinoceroses, zebras, antelope, buffalo, leopards, and hippopotamuses.

A Changing World

The world of jobs and careers is constantly changing as new technologies emerge that affect how things are done, how products are made, or even what products are made. Change is occurring on an international level as the world becomes more interrelated economically and culturally. Young people entering the job market today will need to be able to adapt to change and cope with it over a period of decades until retirement. Specific knowledge skills are essential for performing any job successfully. But dealing with change and building a lifelong career requires other skills and attitudes such as being able to learn new things, being able to solve problems, being able to communicate and work with others, and being flexible.

ALTERNATIVE PROJECTS

Project 1, page 533

Re-scaling the SAT Test

Use articles from newspapers, magazines, and literature from the College Board to research why the mean score of the verbal and math sections for the Scholastic Aptitude Test was re-scaled to 500. Write a paper explaining the College Board's decision and how students and parents should interpret the new scores.

Project 2, page 533

Transformations of Functions

Find at least six examples of transformed functions that model real-world applications. Use the parent functions studied in the text. Make a poster of the applications along with graphs of the transformed functions.

Unit Project 9

Getting Started

For this project you should work in a group of four students. Here are some ideas to help you get started.

☞ Ask your guidance counselor for the *Kuder General Interest Survey.* This survey will give you a profile of your interests and a rating in nine career fields.

☞ In your group, discuss career fields you might be interested in exploring.

☞ Ask parents, neighbors, teachers, and friends for the names of individuals working in the career you have chosen to investigate. You may wish to ask some of them about working conditions and job satisfaction.

☞ The class as a whole may wish to discuss career choices so that each group investigates a different career.

☞ Plan to meet with your group later to contact a career resource person and prepare questions for the interview.

Working on the Unit Project

Your work in Unit 9 will help you plan questions for the interview.

Related Exercises:

Section 9-1, Exercise 28
Section 9-2, Exercise 31
Section 9-3, Exercise 19
Section 9-4, Exercises 45, 46
Section 9-5, Exercise 28
Section 9-6, Exercise 21
Section 9-7, Exercise 22

Alternative Projects p. 533

Can We Talk CAREERS

➤ What jobs exist today that were unheard of when your great grandparents were looking for their first jobs?

➤ Do you see any shifts in the nature of the jobs available from the 1800s to the 1900s? Can you give some reasons for these changes?

➤ Do you prefer a 9-to-5 job, or one with flexible hours? working alone or with a team? working outdoors or indoors? Do you want a job that requires extensive travel?

➤ Have you ever been interviewed for a job? What kinds of questions were you asked? Did any surprise you?

➤ What are some jobs that require mathematical skills? What are those skills?

➤ Many employers expect to train new employees in the details of a job, but they want people to have problem solving skills. List characteristics that you think make a person a good problem solver.

Answers to Can We Talk?

➤ Answers may vary. Examples are given. jobs in the telecommunications, computer, packaging, and public relations fields

➤ Answers may vary. An example is given. The shift is from hands-on work and manufacturing to paper-and-pencil or computer work. The development of technology is one reason for the change.

➤ Answers may vary. An example is given. Yes. Questions related to my school work and attitudes toward work in general were asked. The question about how many days of school I missed was surprising.

➤ Answers may vary, depending upon students' preferences.

➤ Answers may vary. Examples are given. Almost all jobs require mathematical skills. Some jobs that require a great deal of mathematics are engineering, architecture, surveying, and carpentry.

➤ Answers may vary. An example is given. Some characteristics are being able to plan a good problem-solving strategy, being able to define a problem, being able to use graphs and diagrams, and being able to persist in working on a problem until it is solved.

Section **9-1**

Adding a Constant to Data

Focus

Recognize the effect that adding a constant to data values has on the mean, median, range, and standard deviation.

NEED A LIFT?

People who plaster ceilings sometimes walk on stilts. This allows the plasterers to reach high ceilings without having to move a ladder. The table shows the heights of a group of plasterers without stilts and with stilts.

Heights of Plasterers

Name	Without stilts (inches)	With stilts (inches)
Bob	72	86
Karl	65	79
Portia	63	77
Kendra	66	80
Ralph	74	88
Marcus	68	82
Jenny	64	78
Jordan	66	80
Sandra	62	76

Talk it Over

1. How are the numbers in the second and third columns of the table related?
2. Find the mean of the heights without stilts. Find the mean of the heights with stilts. How do the means compare?
3. Repeat question 2 using the median.
4. Repeat question 2 using the range.
5. The histograms show the distribution of heights for the plasterers without stilts and with stilts. Describe the relationship between the two histograms.

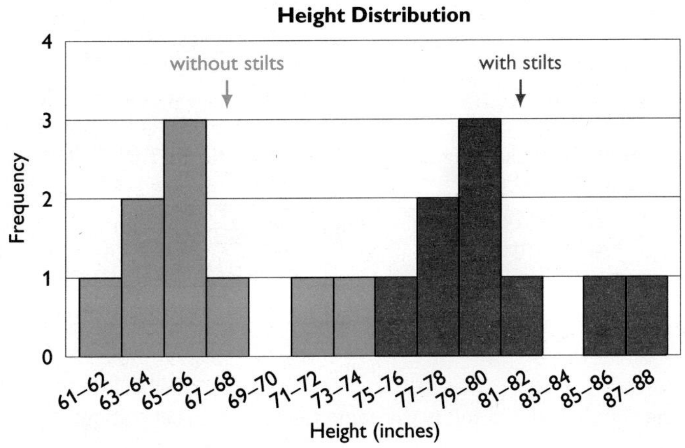

PLANNING

Objectives and Strands

See pages 480A and 480B.

Spiral Learning

See page 480B.

Materials List

- Calculator
- Graph paper

Recommended Pacing

Section 9-1 is a one-day lesson.

Extra Practice

See pages 622–624.

Warm-Up Exercises

Warm-Up Transparency 9-1

Support Materials

- Practice Bank: Practice 67
- Activity Bank: Enrichment 59
- Study Guide: Section 9-1
- Problem Bank: Problem Set 19
- Explorations Lab Manual: Diagram Masters 1, 2
- Using Mac Plotter Plus Disk: Data Analyzer
- Using IBM Plotter Plus Disk: Statistics Spreadsheet
- Assessment Book: Quiz 9-1

Answers to Talk it Over

1. The number in each row of the third column is 14 more than the number in the corresponding row of the second column.
2. about 66.7; about 80.7; The mean of the heights with stilts is 14 more than the mean of the heights without stilts.
3. 66; 80; The median of the heights with stilts is 14 more than the median of the heights without stilts.
4. 12; 12; The ranges are the same.
5. The histograms have exactly the same shape. The second is the first shifted to the right.

TEACHING

Talk it Over

Questions 1–5 lead students to recognize how the mean, median, range and distribution of a data set are affected when a constant is added to the data values.

Additional Sample

S1 The table shows the weights of the people going on a camping trip without backpacks and with them. All the backpacks weigh 12 lb.

Weights of Campers		
Name	Without backpack (pounds)	With backpack (pounds)
Ben	132	144
Diedra	135	147
Carla	128	140
Yori	136	148
Hana	127	139
Juan	131	143
Hoen	133	145
Otis	129	141
Dora	130	142
Mavis	145	157

Find the standard deviation of the weights without backpacks and with them. Compare the values.

Method 1: Use a calculator.
(without backpack)

1 – Var Stats
$\overline{x} = 132.6$
$\Sigma x = 1326$
$\Sigma x^2 = 176074$
$Sx = 5.232377832$
$\sigma x = 4.963869458$
$\downarrow n = 10$

The standard deviation without backpacks is about 4.96 lb.
(with backpack)

1 – Var Stats
$\overline{x} = 144.6$
$\Sigma x = 1446$
$\Sigma x^2 = 209338$
$Sx = 5.232377832$
$\sigma x = 4.963869458$
$\downarrow n = 10$

The standard deviation with backpacks is about 4.96 lb. The standard deviation is the same for both data sets, about 4.96 lb.

Sample 1

Use the table on page 483. Find the standard deviation of the heights without stilts and of the heights with stilts. Compare the values.

Sample Response

Method ❶ Use a calculator.

Put your calculator in statistics mode or select the statistics menu. Enter the data and have the calculator compute the standard deviation for each data set.

1–Var Stats
$\overline{x} = 66.66666667$
$\Sigma x = 600$
$\Sigma x^2 = 40130$
$Sx = 4.031128874$
$\sigma x = 3.80058475$ ← standard deviation without stilts
$\downarrow n = 9$

1–Var Stats
$\overline{x} = 80.66666667$
$\Sigma x = 726$
$\Sigma x^2 = 58694$
$Sx = 4.031128874$
$\sigma x = 3.80058475$ ← standard deviation with stilts
$\downarrow n = 9$

The standard deviation is the same for both data sets, about 3.8 in.

Method ❷ Use a formula.

$$\text{standard deviation} = \sqrt{\frac{\sum_{i=1}^{n}(x_i - \overline{x})^2}{n}}$$

$\overline{x}$: This represents the mean height.
n: This represents the number of heights.

Compute the mean height without stilts as you did in *Talk it Over* question 2 on page 483: $\overline{x} \approx 66.7$.

$$\text{standard deviation without stilts} \approx \sqrt{\frac{(72 - 66.7)^2 + (65 - 66.7)^2 + \cdots + (62 - 66.7)^2}{9}}$$

$$\approx \sqrt{\frac{5.3^2 + (-1.7)^2 + \cdots + (-4.7)^2}{9}}$$

$$\approx 3.80$$

Compute the mean height with stilts as you did in *Talk it Over* question 2 on page 483: $\overline{x} \approx 80.7$.

$$\text{standard deviation with stilts} \approx \sqrt{\frac{(86 - 80.7)^2 + (79 - 80.7)^2 + \cdots + (76 - 80.7)^2}{9}}$$

$$\approx \sqrt{\frac{5.3^2 + (-1.7)^2 + \cdots + (-4.7)^2}{9}}$$

$$\approx 3.80$$

The standard deviation is the same for both data sets, about 3.8 in.

Answers to Talk it Over

6. Answers may vary. An example is given. The standard deviation is a measure of spread. The data in both cases are spread out in exactly the same manner. Let x_i represent the first data points and X_i represent the second. For every i, $X_i = x_i + 14$, so $X_i - \overline{X} = (x_i + 14) - (\overline{x} + 14) = x_i - \overline{x}$). Since

ADDING A CONSTANT TO A DATA SET

When a constant value is added to every value in a data set,

(1) the graph of the distribution, the mean, and the median are translated by the constant value, and

(2) the range and the standard deviation (σ) remain the same.

Example

Add 8 to each data value.

$\bar{x} = 3$
median = 3.5
range = 3
$\sigma \approx 1.1$

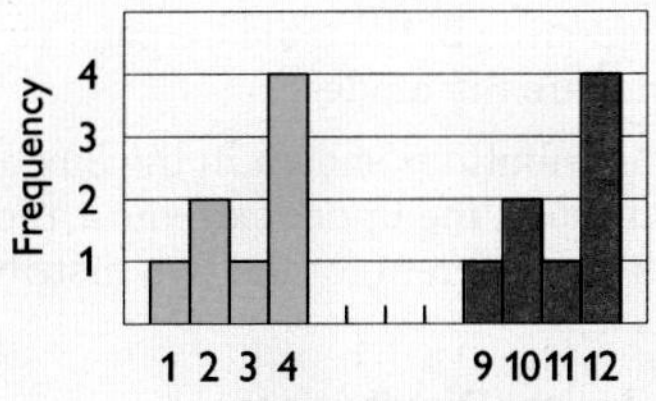

$\bar{x} = 3 + 8 = 11$
median = 3.5 + 8 = 11.5
range = 3
$\sigma \approx 1.1$

Talk it Over

6. Explain why the standard deviations in Sample 1 are the same with or without stilts.
7. Suppose the same constant negative value is added to every value in a data set. Tell whether each statistical measure will *increase*, *decrease*, or *stay the same*. Explain your choice.
   a. mean
   b. median
   c. range
   d. standard deviation
8. For the situation in question 7, will a histogram of the data be translated to the left or to the right? Explain your choice.

Sample 2

Suppose the stilts that the plasterers wear in the situation on pages 483 and 484 are 18 in. stilts. Find the mean and the standard deviation of the plasterers' heights with 18 in. stilts.

Sample Response

Use the mean of the heights without stilts from Sample 1: about 66.7. Add 18 to find the mean with stilts: 66.7 + 18 = 84.7. The mean of the plasterers' heights with 18 in. stilts is about 84.7 in.

The standard deviation of the heights with 18 in. stilts is the same as it is without stilts in Sample 1: about 3.8 in.

Method 2: **Use a formula. Compute the mean weight, $\bar{x}$, without backpack: $\bar{x} = 132.6$. Then use the formula for standard deviation.**

std. dev. (without backpack)

$$= \sqrt{\frac{(132 - 132.6)^2 + \ldots + (145 - 132.6)^2}{10}}$$

$$= \sqrt{\frac{(-0.6)^2 + \ldots + 12.4^2}{10}} \approx 4.96$$

Compute the mean weight, $\bar{x}$, with backpack: $\bar{x} = 144.6$.

std. dev. (with backpack)

$$= \sqrt{\frac{(144 - 144.6)^2 + \ldots + (157 - 144.6)^2}{10}}$$

$$= \sqrt{\frac{(-0.6)^2 + \ldots + 12.4^2}{10}} \approx 4.96$$

The standard deviation is the same for both data sets, about 4.96 lb.

Talk it Over

Question 6 provides an opportunity for students to think about the meaning of a standard deviation in order to explain why it is unaffected when adding a constant to every value in a data set.

Additional Sample

S2 Suppose that the backpacks for the situation in Additional Sample S1 weigh 13.5 lb. Find the mean and the standard deviation of the campers' weights with 13.5 lb. backpacks.

Use the mean without backpacks from Additional Sample 1: 132.6. Add 13.5 to find the mean with backpacks. The mean of the campers' weights with 13.5 lb backpacks is 132.6 + 13.5, or 146.1 lb. The standard deviation with 13.5 lb backpacks is the same as it is without backpacks: about 4.96 lb.

Answers to Talk it Over

the sum of the deviations is the same, the variance and standard deviation are the same.

7. a. decrease; When the same constant value is added to each value in a data set, the mean is translated by the constant value, in this case, a negative number.

   b. decrease; When the same constant value is added to each value in a data set, the median is translated by the constant value, in this case, a negative number.

   c. stay the same; When the same constant value is added to each value in a data set, the range remains the same.

   d. stay the same; When the same constant value is added to each value in a data set, the standard deviation remains the same.

8. left; When the same constant value is added to each value in a data set, the graph of the distribution is translated by the constant value, in this case, a negative number.

Look Back

A class response to this Look Back can serve as a check on students' understanding of the focus of this section.

APPLYING

Suggested Assignment

Standard 2–17, 20–28

Extended 1–28

Integrating the Strands

Algebra Exs. 20–23, 27

Functions Exs. 24–26

Geometry Ex. 20

Trigonometry Ex. 20

Statistics and Probability Exs. 1–8, 16–19, 27, 28

Discrete Mathematics Exs. 1–8, 16–19, 21–23, 27, 28

Logic and Language Exs. 1, 9–15, 27, 28

Multicultural Note

Stilts have been a part of work, play, and ceremony throughout the world for thousands of years. Archaeological evidence of stiltwalkers has been found in ancient civilizations in Nigeria, China, Italy, Tanzania, India, Tahiti, New Zealand, and Mexico. Today stilts are used in many ways by many people. In Newchwang, China, each New Year's parade features a dragon and other performers, all on stilts. New Zealand Maoris tell a traditional fable about supernatural thieves who walk on stilts so as to never leave a footprint. In the fruit orchards of California, stilts are sometimes used by people who pick peaches, plums, and apricots. A popular Tahitian game requires players to stand on stilts and try to knock each other down. Boys often play on stilts in Nigeria, trying to outdo each other with stunts. And in Trinidad, stilt dancers called *Moco Jumbie* participate in Carnival festivals.

Look Back

Identify three effects of adding a constant value to every value in a data set.

9-1 Exercises and Problems

1. **Reading** Which statistical measures are not affected by adding a constant value to every value in a data set?
2. Find the mode of the heights without stilts in the table on page 483 and the mode of the heights with stilts. Compare the modes.
3. Compare the medians for the two distributions shown in the box-and-whisker plots. Do the same for the lower extremes, the upper extremes, and the ranges. Which of these statistical measures are different for the two distributions? Which are the same?

Height Distributions

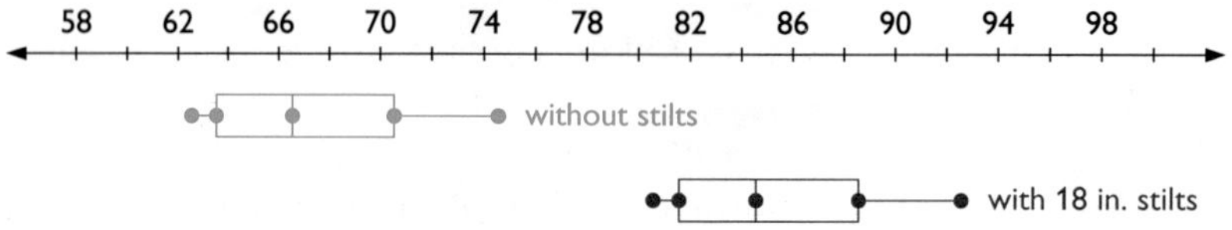

For Exercises 4–8, use the table.

4. Find the mean and the median of the Calories per serving for these cereals without milk.
5. Find the range and the standard deviation of the Calories per serving for these cereals without milk.
6. If you use a $\frac{1}{2}$ cup of skim milk, this adds 44 Cal. Predict the mean and the standard deviation of the Calories per serving for these cereals served with $\frac{1}{2}$ cup of skim milk.
7. Test your prediction from Exercise 6. Add 44 Cal to each entry in the Calorie column of the table. Then compute the mean and the standard deviation. Do you get the results you predicted? If not, why not?
8. **a.** Draw a box-and-whisker plot of cereal prices.
   b. Suppose there is a $.25 coupon for each box of cereal. Draw a box-and-whisker plot that shows how this will affect the amount customers will pay for the cereal.

BRAND	Calories per serving without milk	Price of a 15-oz box (dollars)
A	160	2.19
B	80	1.89
C	70	3.20
D	90	2.99
E	110	1.65
F	110	1.99
G	130	3.10
H	210	2.89
I	110	2.19
J	120	1.89
K	110	2.79
L	50	1.80

Answers to Look Back

When the same constant value is added to each value in a data set, the graph of the distribution, the mean, and the median are translated by the common value.

Answers to Exercises and Problems

1. the range and the standard deviation
2. 66 and 80; The mode of the heights with stilts is 14 more than the mode of the heights without stilts.
3. The median, lower extremes, and upper extremes for the heights with 18-in. stilts are each 18 more than the corresponding values for the heights without stilts. The ranges are the same.
4. 112.5; 110
5. 160; about 40.2
6. 156.5; about 40.2
7. 156.5; 40.2; The results should be the same.

connection to MUSIC

A musical composition can be transposed by raising or lowering the pitch of each note by the same amount. On a piano, intervals between notes are measured in half steps. The symbol ♯ indicates a half step up and the symbol ♭ represents a half step down.

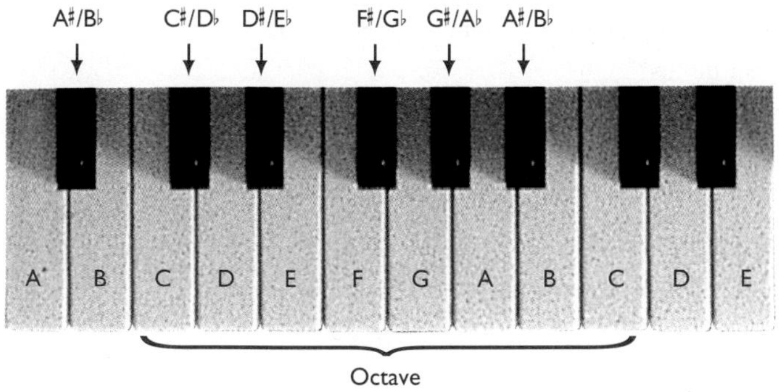

Suppose a composition is transposed up two half steps (one whole step). What does each note become?

9. D **10.** F♯ **11.** A♭ **12.** B

13. Suppose you want a composition that starts on a G to start lower, on a C. What must you do to each note in the composition?

14. What would you have to do to a composition in order to transpose it up an octave?

15. **Writing** When a piece of music is transposed, you can still recognize the melody. Why is that? How does this relate to what you have learned in this section?

Football The 1993 Dallas Cowboys football team had 58 players. The table shows the distribution of weights. All weights are given to the nearest pound. The mean weight for the team was 238.3 lb.

16. Draw a histogram for the data. Locate the mean on your histogram.

17. The clothes and equipment that football players wear for a game add about 10 lb to each player's weight.

a. How does adding 10 lb to each player's weight affect the mean weight of the team?

b. How does adding 10 lb to each player's weight affect the position, size, and shape of the histogram you drew in Exercise 16?

Distribution of Weights of 1993 Dallas Cowboys Players

Weight range in pounds	Frequency	Weight range in pounds	Frequency
180–189	8	260–269	2
190–199	5	270–279	4
200–209	7	280–289	6
210–219	3	290–299	0
220–229	5	300–309	2
230–239	4	310–319	1
240–249	5	320–329	3
250–259	3		

Communication: Reading

Exs. 1–3 provide students with an opportunity to read, interpret, compare, and analyze statistical data.

Reasoning

Ask students to make a conjecture about how the box-and-whisker plots for two sets of data compare if one set of data differs from the other by a constant. Students should be able to explain their thinking.

Error Analysis

Students who use calculators may have errors when they use 1-VarStats from the STAT CALC menu to compute means, standard deviations, and so on. They should check that they have entered all data correctly. If there are data in more than one of the STAT lists, they should be careful, after selecting 1-VarStats, to specify the list for which the statistical calculations are to be performed.

Interdisciplinary Problems

Exs. 9–15 relate the mathematical concepts of this section to the transposition of a musical composition. You may wish to ask those students who can read music to explain or elaborate upon any musical ideas in these exercises that other students may be curious about or may not understand.

Answers to Exercises and Problems

8. a. Cereal Prices (dollars)

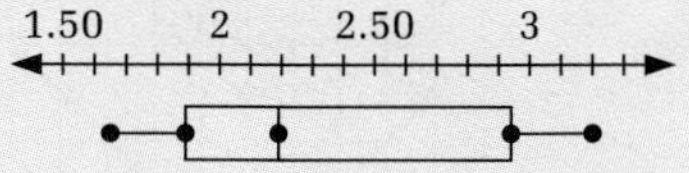

b. Cereal Prices (dollars) with coupon

9. E **10.** G♯ **11.** B♭ **12.** C

13. Transpose it $3\frac{1}{2}$ steps down.

14. Transpose each note up 6 full steps.

15. Answers will vary. An example is given. The notes still have the same relationship to each other. This compares to the fact that the standard deviation of a set of data does not change when the same constant value is added to each value in a data set.

16. Distribution of Weights of 1993 Dallas Cowboy Players

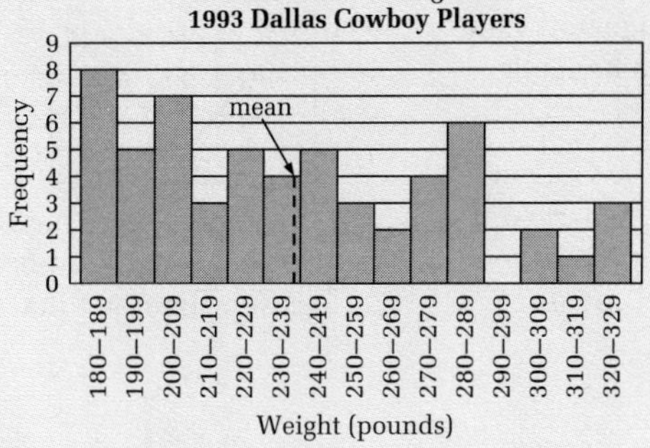

17. a. The mean is increased by 10.

b. The size and shape do not change; the position shifts to the right.

Communication: Writing

Ex. 27 gives students experience writing a nongeometric proof. Encourage students to write a plan for their proof before attempting to write the proof itself. Challenge students to use sigma notation in their proofs.

Mathematical Procedures

In connection with Ex. 27, you may wish to ask students to justify algebraically the fact that data sets whose data items differ by the same constant will have standard deviations that differ by that same constant.

Practice 67 For use with Section 9-1

Name ______________________ Date ____________

Practice 67

For use with Section 9-1

Liang is designing the upper deck of a sports stadium. In order to plan the heights of rows, he has taken measurements of the eye-level heights (in inches) of a sample of seated people. His data are displayed below. Use these data for Exercises 1–5.

35, 36, 38, 38, 39, 40, 41, 42, 42, 42, 43, 43, 44, 44, 44, 45, 45, 46, 46, 46, 47, 48, 48, 48, 49, 50, 52, 52, 54, 54

1. Find the mean and the standard deviation of the data. 44.7; about 4.92
2. The railing in front of the first row of the upper deck is 3 ft high. How much higher should the first row be if there is to be a mean of 1 ft between the top of the railing and the viewers' eye levels? 3.3 in.
3. Liang wants the mean eye-level height of the second row of the upper deck to be 10 in. above the mean eye-level height of the first row. Rewrite the data with the eye-level heights of the second row as measured from the floor of the first row. Then find the mean and the standard deviation of the data. 45, 46, 48, 48, 49, 50, 51, 52, 52, 52, 53, 53, 54, 54, 54, 55, 55, 56, 56, 56, 57, 58, 58, 58, 59, 60, 62, 62, 64, 64; $\bar{x} = 54.7$; $\sigma \approx 4.92$
4. Make up frequency tables for the original data and for the data from Exercise 3, using "35–37," "38–40," and so on as intervals. For frequency tables, see additional answers.
5. Draw histograms based on the tables you made in Exercise 4. For histograms, see additional answers.

The scores of a group of students on a math test are shown below. Use these scores in Exercises 6–8.

55, 58, 62, 64, 65, 68, 72, 74, 74, 76, 78, 80, 82, 84, 85, 85, 87, 91, 93, 94

6. Find the mean, the median, the range, and the standard deviation of the data. 76.35; 77; 39; about 11.30
7. Suppose the teacher who gave the test decides to raise everyone's grade by 5 points. Rewrite the data using the revised scores. 60, 63, 67, 69, 70, 73, 77, 79, 79, 81, 83, 85, 87, 89, 90, 90, 92, 96, 98, 99
8. Find the mean, median, range, and standard deviation of the revised scores. Then draw a box-and-whisker plot comparing the original test scores with the revised scores. See bottom of page.
9. ***Open-ended*** Make up some data about sales at a supermarket, a clothing store, or a record store. Find the mean, the median, and the standard deviation of the data. Revise the sales data by adding a constant. Find the mean, median, and standard deviation of the revised data. Answers may vary. Check students' work.

8. 81.35; 82; 39; about 11.30; For plot, see additional answers.

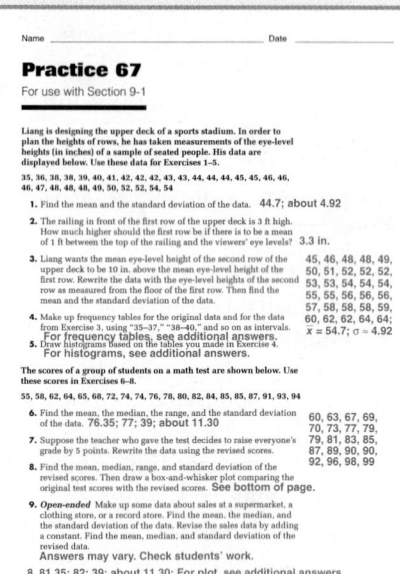

18. Suppose the mean annual contribution to an environmental organization is \$50. The organization spends \$3 per year for a newsletter to each contributor. What effect does it have on the mean, the median, the range, and the standard deviation of the annual contributions when you subtract the cost of the newsletters from the contributions?

19. Annual wages for hourly workers at Fit-Rite Corporation have a mean of \$21,300 and a standard deviation of \$3500. All hourly workers receive the same end-of-year bonus that increases the mean to \$22,100.

a. How much is the end-of-year bonus?

b. What is the standard deviation of the hourly workers' annual wages at Fit-Rite when the bonus is included?

Review PREVIEW

20. The measures of two angles of a triangle are 30° and 120°. The length of the side opposite the 30° angle is 12 cm. Find the length of the side opposite the 120° angle. *(Section 8-7)*

Suppose u = (8, −3) and v = (−1, 2). Find each sum. *(Section 8-4)*

21. **u** + **v**   **22.** 2**u** + **v**   **23.** 2**u** − **v**

For each function, draw a graph and find the domain and the range. *(Sections 2-1, 2-3)*

24. $f(x) = |x|$ **25.** $f(x) = |x| + 3$ **26.** $f(x) = |x + 3|$

Ongoing ASSESSMENT

27. **Writing** Write a paragraph proof of this statement:

If a constant number is added to each item in a data set, then the mean of the data set is increased by that number.

Given n data values $x_1, x_2, x_3, \ldots, x_n$ with mean $\bar{x}$

Prove The mean of the n data values $x_1 + k, x_2 + k, x_3 + k, \ldots, x_n + k$ is $\bar{x} + k$.

9 Working on the Unit Project

28. a. Draw a histogram for the data.

b. Draw a histogram for the same group 10 years later.

c. Describe the relationship between the graphs in parts (a) and (b).

d. Suppose that 65 is retirement age. Predict the number of faculty who will retire within the next 10 years.

e. **Open-ended** What factors besides retirement might affect the number of job openings for new college faculty over the next 10 years?

College Faculty in 1990

Ages in years	Number of faculty
20–24	5,000
25–29	45,000
30–34	55,000
35–39	93,000
40–44	85,000
45–49	80,000
50–54	99,000
55–60	150,000
60–64	100,000

Answers to Exercises and Problems

18. The mean and the median decrease by 3; the range and standard deviation are unchanged.

19. a. \$800

b. \$3500

20. about 20.78 cm

21. (7, −1)

22. (15, −4)

23. (17, −8)

24.

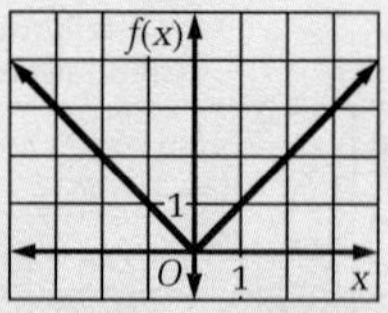

domain: all real numbers; range: all real numbers greater than or equal to 0

25.

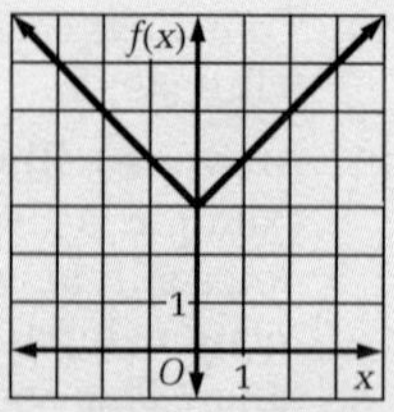

domain: all real numbers; range: all real numbers greater than or equal to 3

26.

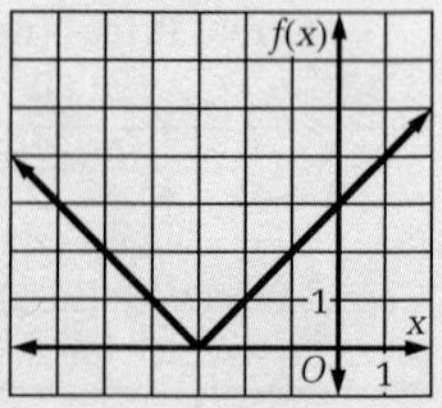

domain: all real numbers; range: all real numbers greater than or equal to 0

27, 28. See answers in back of book.

Section

9-2 Translating Graphs

ON THE MOVE

Focus

Recognize the relationship between the graph of $y - k = f(x - h)$ and the parent graph of $y = f(x)$.

EXPLORATION

How do h *and* k *affect the graph of* $y - k = f(x - h)$?

- **Materials: graphics calculators or graphing software**
- **Work in a group of four students.**

parent function: $y = x^2$

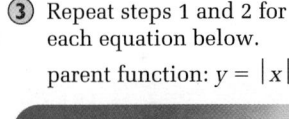

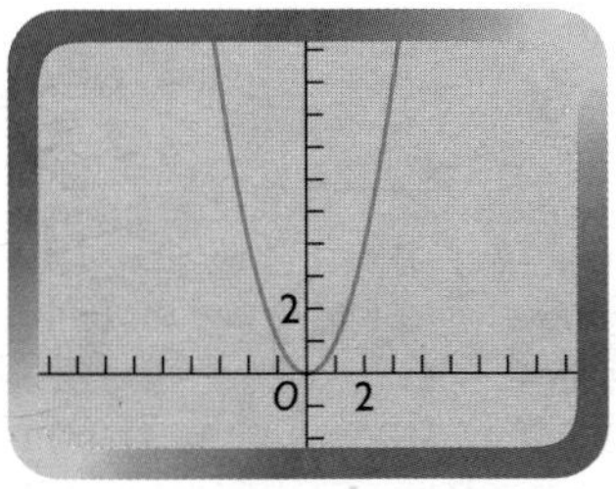

(1) Graph these equations and the *parent* function on the same set of axes. You may have to solve for y first. Use a window that will give you equal unit spacing on the two axes.

a. $y - 1 = x^2$

b. $y = (x - 2)^2$

c. $y - 1 = (x - 2)^2$

(2) Describe how the graph of each equation in step 1 is related to the graph of the parent function.

(3) Repeat steps 1 and 2 for each equation below.

parent function: $y = |x|$

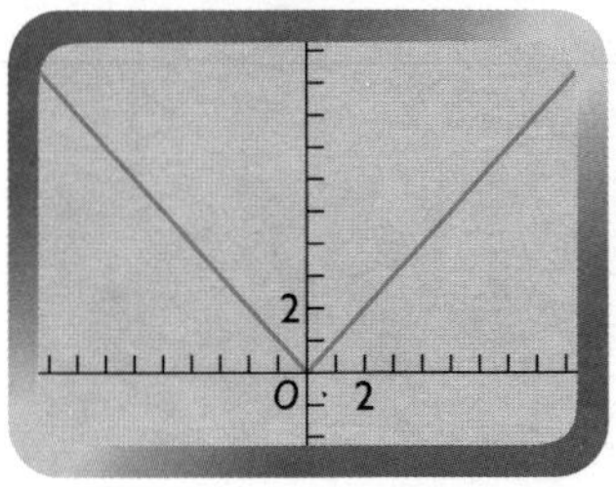

a. $y - 1 = |x|$

b. $y = |x - 2|$

c. $y - 1 = |x - 2|$

(4) Repeat steps 1 and 2 for each equation below.

parent function: $y = \sqrt{x}$

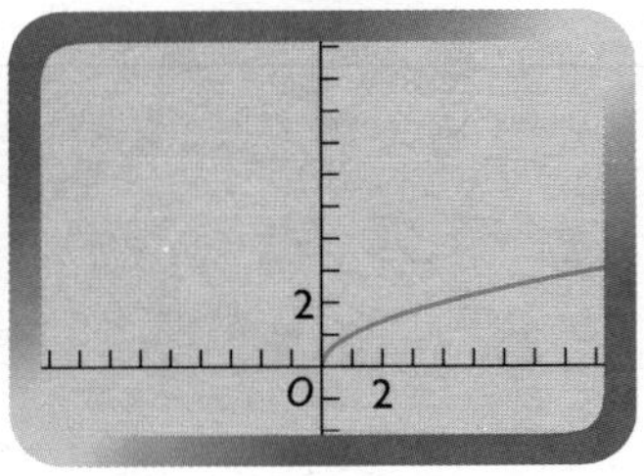

a. $y - 1 = \sqrt{x}$

b. $y = \sqrt{x - 2}$

c. $y - 1 = \sqrt{x - 2}$

(5) Repeat steps 1 and 2 for each equation below.

parent function: $y = 10^x$

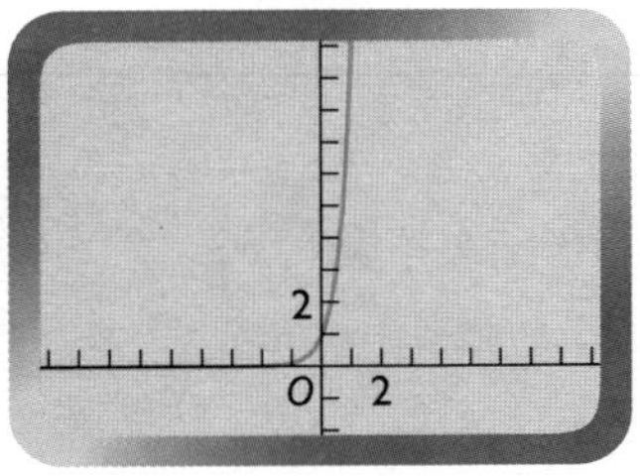

a. $y - 1 = 10^x$

b. $y = 10^{x - 2}$

c. $y - 1 = 10^{x - 2}$

(6) Make a conjecture about how the graph of $y - k = f(x - h)$ is related to the graph of the parent function $y = f(x)$.

PLANNING

Objectives and Strands

See pages 480A and 480B.

Spiral Learning

See page 480B.

Materials List

- Graphics calculator or graphing software
- Graph paper

Recommended Pacing

Section 9-2 is a two-day lesson.

Day 1

Pages 489–490: Exploration through Sample 1, *Exercises 1–14*

Day 2

Pages 491–492: Finding Equations for Graphs through Look Back, *Exercises 15–31*

Extra Practice

See pages 622–624.

Warm-Up Exercises

- Warm-Up Transparency 9-2

Support Materials

- Practice Bank: Practice 68
- Activity Bank: Enrichment 60
- Study Guide: Section 9-2
- Problem Bank: Problem Set 19
- Explorations Lab Manual: Additional Exploration 10 Diagram Master 2
- Overhead Visuals: Folder 9
- Assessment Book: Quiz 9-2, Alternative Assessment 1

Answers to Exploration

1. a.

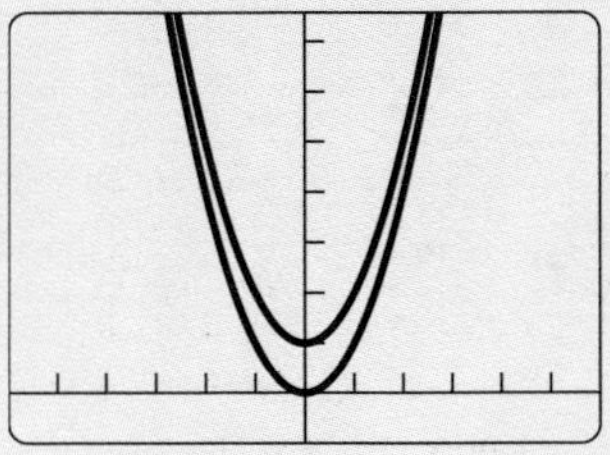

b.

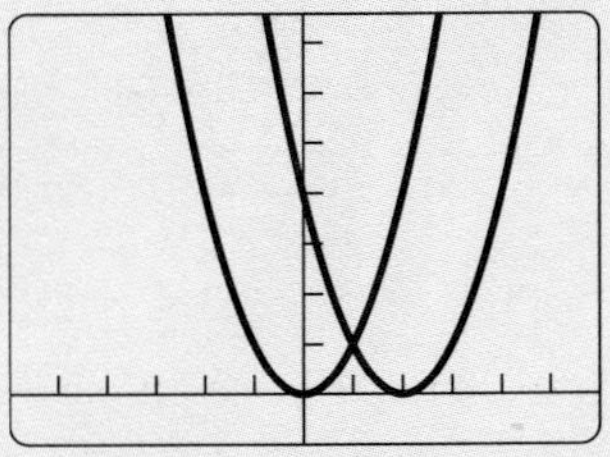

c.

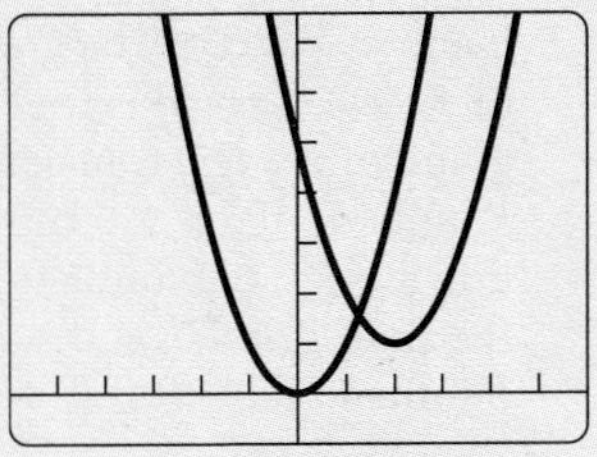

2. The graph of $y - 1 = x^2$ is the graph of $y = x^2$ translated 1 unit up. The graph of $y = (x - 2)^2$ is the graph of $y = x^2$ translated 2 units to the right. The graph of $y - 1 = (x - 2)^2$ is the graph of $y = x^2$ translated 1 unit up and 2 units to the right.

3–6. See answers in back of book.

TEACHING

Exploration

The goal of the Exploration is to have students discover how changes in the equation of a parent function translate the graph of the parent function in the coordinate plane.

Talk it Over

Questions 1 and 2 help students to solidify their understanding of how the graph of a parent function can be translated in the coordinate plane.

Integrating the Strands

The concept of a translation is used throughout the strands of algebra, geometry, functions, discrete mathematics, statistics and probability. Translations can also be represented in different ways by using notations from different strands, as Talk it Over question 3 demonstrates.

Additional Sample

S1 Sketch the graph of $y + 5 = (x - 7)^4$.

Find the parent function, h, and k. The parent function is a power function with exponent 4: $f(x) = x^4$.
$y - k = (x - h)^4$ is
$y - (-5) = (x - 7)^4$, so
$h = -5$ and $k = 7$.
Sketch the graph of the parent function $y = x^4$. Then translate the graph by the vector $(7, -5)$.

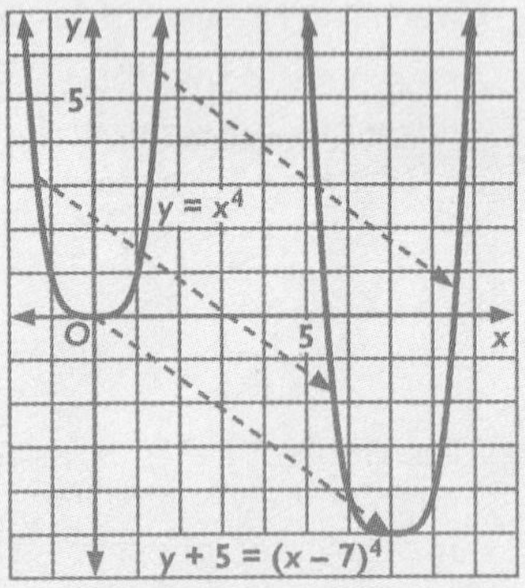

Xlab

TRANSLATING GRAPHS

The graph of $y - k = f(x - h)$ can be found by translating the graph of the parent function $y = f(x)$ by the vector (h, k).

Example

$f(x) = \log x$; $(h, k) = (-5, 2)$

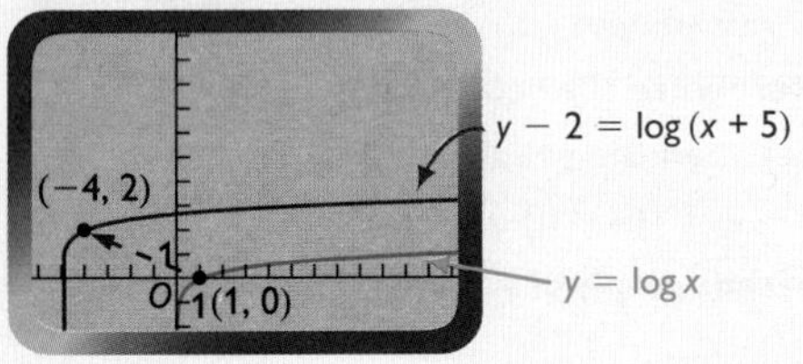

Talk it Over

1. In step 1 of the Exploration, what is the vertex of the parent graph? What is the vertex of the graph of $y - 1 = (x - 2)^2$? What vector transforms the first vertex into the second?
2. Which graphs in the Exploration were horizontal translations of the parent graph? Which were vertical translations? How can you identify horizontal and vertical translations if you are given values for h and k?
3. Complete each rule for the translation $P \to P'$ that transforms the graph of the parent function $y = f(x)$ into the graph of $y - k = f(x - h)$.

   a. $P(x, y) \to P'(\underline{?}, \underline{?})$

   b. $\overset{P}{\begin{bmatrix} x \\ y \end{bmatrix}} + \begin{bmatrix} ? \\ ? \end{bmatrix} = \overset{P'}{\begin{bmatrix} x' \\ y' \end{bmatrix}}$

Sample 1

Sketch the graph of $y - 3 = (x + 2)^3$.

Sample Response

1 Find the parent function, h, and k.

The parent function is a cubic: $f(x) = x^3$
$y - k = (x - h)^3$ is
$y - 3 = (x + 2)^3$, so
$k = 3$ and $h = -2$.

2 Sketch the graph of the parent function $y = x^3$.

3 Translate the graph by the vector $(-2, 3)$.

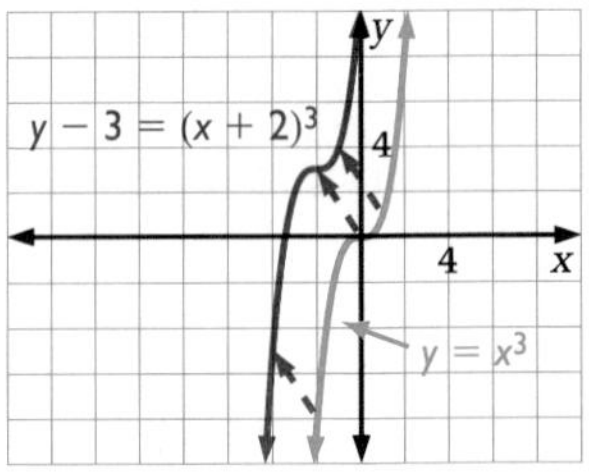

Now you are ready for: Exs. 1–14 on pp. 492–493

Answers to Talk it Over

1. (0, 0); (2, 1); (2, 1)
2. b and c; a and c; The graph of $y - k = f(x - h)$ is the graph of $y = f(x)$ translated k units up if $k > 0$, k units down if $k < 0$, h units to the right if $h > 0$, and h units to the left if $h < 0$.
3. a. $x + h$; $y + k$

   b. $\begin{bmatrix} h \\ k \end{bmatrix}$

Finding Equations for Graphs

You can use what you have learned about translating graphs to find an equation of a graph. To do this, you first have to decide which parent function has a graph with the same shape. There is a Visual Glossary of Functions on page 672.

Sample 2

Write an equation for each graph.

a. 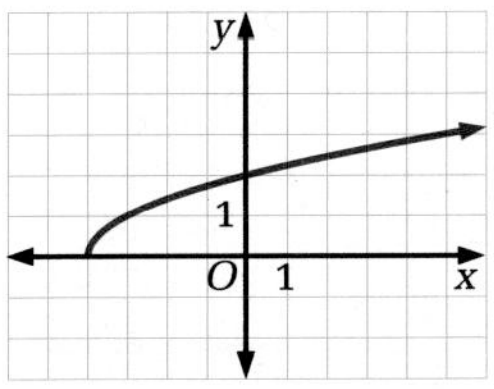

b. 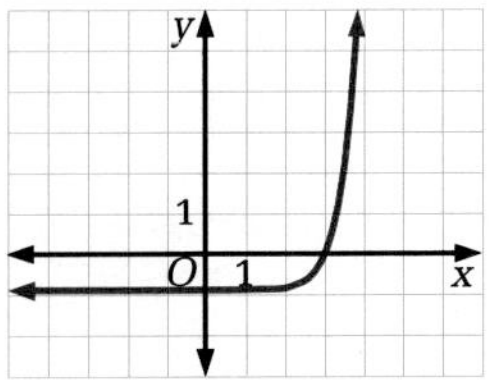

Sample Response

Step 1 Look at the shape of each graph to identify its parent function.

a. The parent function is the square root function.

b. The parent function is an exponential function.

Step 2 Identify the translation vector by finding a key point on the parent graph and its image on the translated graph.

a. 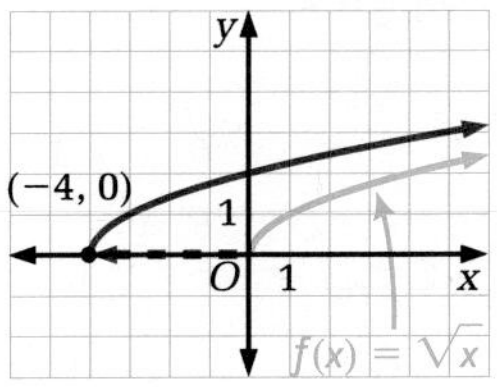

$O(0, 0) \rightarrow O'(-4, 0)$

The translation vector (h, k) is $(-4, 0)$.

b. 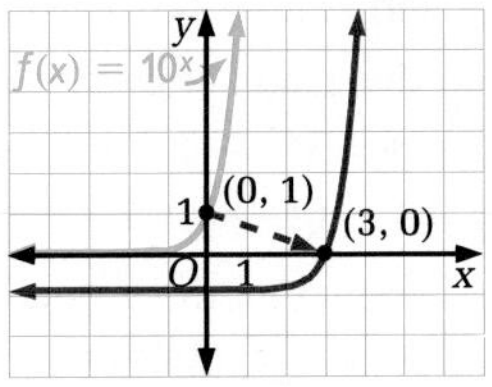

$P(0, 1) \rightarrow P'(3, 0)$

The translation vector (h, k) is $(3, -1)$.

Step 3 Write an equation of the form $y - k = f(x - h)$ for each graph.

a. $f(x) = \sqrt{x}$, $h = -4$, $k = 0$

$$y - k = f(x - h)$$
$$y - 0 = \sqrt{x - (-4)}$$
$$y = \sqrt{x + 4}$$

b. $f(x) = 10^x$, $h = 3$, $k = -1$

$$y - k = f(x - h)$$
$$y - (-1) = 10^{x - 3}$$
$$y + 1 = 10^{x - 3}$$

Additional Sample

S2 Write an equation for each graph.

a.

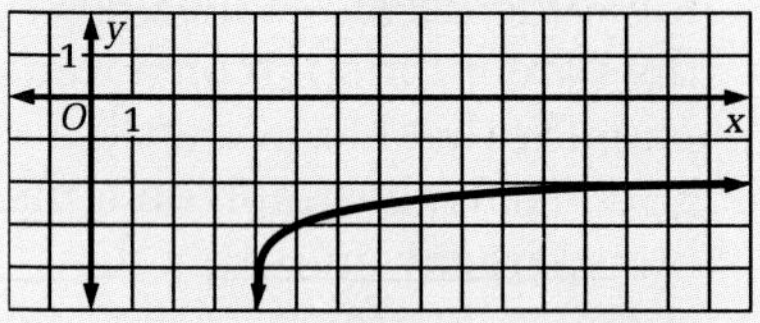

Step 1. **Look at the shape to identify the parent function. The parent function is the logarithmic function.**

Step 2. **Identify the translation vector by finding a key point on the parent graph and its image on the translated graph.**

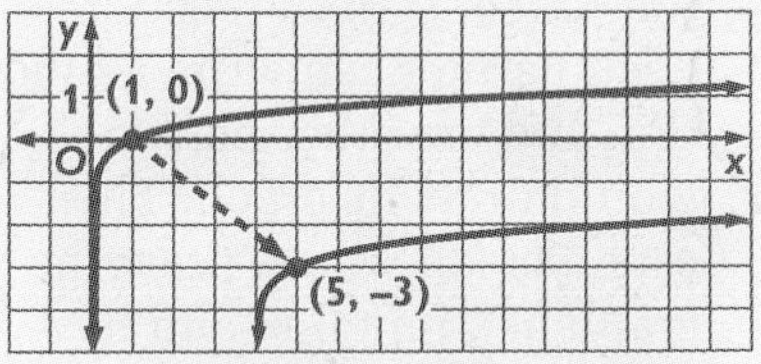

$P(1, 0) \rightarrow P'(5, -3)$, so the translation vector (h, k) is $(4, -3)$.

Step 3. **Write an equation of the form $y - k = f(x - h)$ for the graph.**

$$y - k = f(x - h)$$
$$y - (-3) = \log (x - 4)$$
$$y + 3 = \log (x - 4)$$

b.

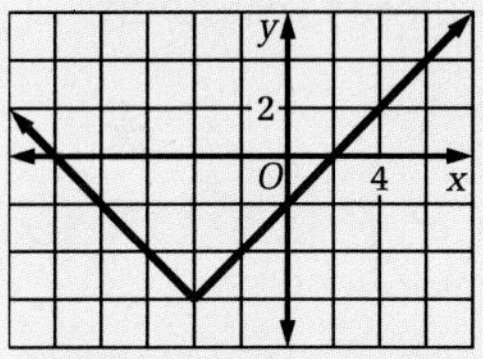

Step 1. **The parent function is the absolute value function.**

Step 2. **Identify the translation vector.**

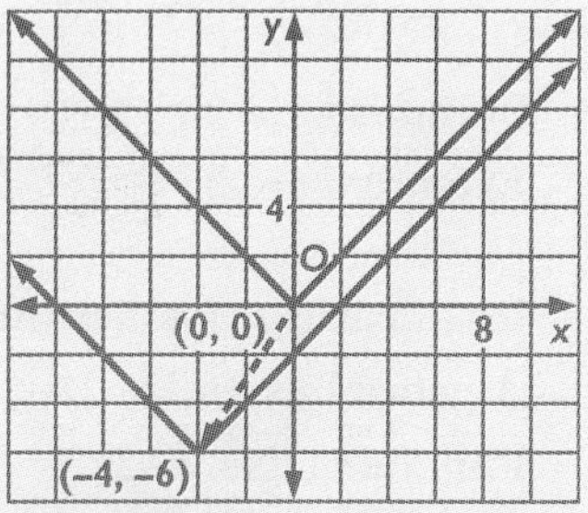

$O(0, 0) \rightarrow O'(-4, -6)$, so the translation vector (h, k) is $(-4, -6)$.

Step 3. **Write an equation.**

$$y - k = f(x - h)$$
$$y - (-6) = |x - (-4)|$$
$$y + 6 = |x + 4|$$

Teaching Tip

You may wish to use transparencies to make ideas about translating graphs more vivid for students. Draw the graph of the parent function on a transparency that has a coordinate grid. Trace the graph on a plain transparency and slide the copy to illustrate translations by a number of different vectors.

Cooperative Learning

After answering the Look Back, you may wish to have students work with a partner to review the concepts of the section. Each partner can write a function and the other person would draw its graph. Then each person would draw the graph of a function and have their partner write its equation.

Look Back

Students can respond to these questions verbally or in writing. Written responses can be read aloud to check for accuracy and understanding of the concepts involved.

APPLYING

Suggested Assignment

Day 1

Standard 1–13

Extended 1–14

Day 2

Standard 15–23, 25–31

Extended 15–31

Integrating the Strands

Algebra Exs. 9–14, 22–25

Functions Exs. 1–24

Geometry Exs. 9, 25

Statistics and Probability Ex. 26

Discrete Mathematics Exs. 1–3, 9, 26–30

Logic and Language Exs. 4, 12–14, 24, 25, 31

Talk it Over

4. The point (4, 9) is on the graph given in part (b) of Sample 2. Show that this ordered pair is a solution of the equation found in step 3 of part (b).
5. In part (a) of Sample 2, the origin was the key point used to find the translation. Why was the origin not used in part (b)?
6. How is the domain of the function in part (a) of Sample 2 related to the domain of its parent function?
7. How is the asymptote for the graph in part (b) of Sample 2 related to the asymptote of its parent graph?

Look Back

How can you use translations and parent functions to sketch the graph of a function? How can you use translations and parent graphs to write an equation of a graph?

Now you are ready for: Exs. 15–31 on pp. 494–495

9-2 Exercises and Problems

For Exercises 1–3, the equations are in the form $y - k = f(x - h)$. For each equation, find the parent function and the translation vector.

1. $y - 7 = (x + 2)^2$
2. $y - 2 = \log (x - 1)$
3. $y = 10^{x + 9}$
4. **Reading** Choose the letter of the translation that transforms the graph of $y = f(x)$ into the graph of $y - 2 = f(x + 5)$.

   A. left 2, up 5　B. right 5, down 2　C. left 5, down 2　D. left 5, up 2

Sketch the graph of each function.

5. $y + 9 = \sqrt{x - 4}$
6. $y + 1 = |x - 3|$
7. $y - 4 = \log x$
8. $y = (x + 6)^3$
9. a. Show that the vertices of trapezoid *ABCD* are on the graph of $y = x^2$.

   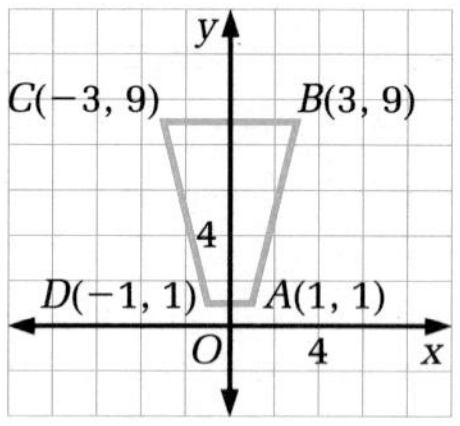

   b. Find the coordinates of the vertices of trapezoid *ABCD* after the translation $P \to P'$.

   $$\overset{P}{\begin{bmatrix} x \\ y \end{bmatrix}} + \begin{bmatrix} -1 \\ -8 \end{bmatrix} = \overset{P'}{\begin{bmatrix} x' \\ y' \end{bmatrix}}$$

   c. Suppose you transform the graph of $y = x^2$ using the translation in part (b). Write an equation of the new graph.

   d. Show that each vertex you found in part (b) is on the parabola whose equation you found in part (c).

Answers to Talk it Over

4. $9 + 1 = 10^{4-3}$, $10 = 10$✓
5. The origin is not part of the graph of $y = 10^x$.
6. The domain of the parent function is all real numbers greater than or equal to 0. The domain of the function in part (a) is all real numbers greater than or equal to −4. If you think of the domain of the parent function as an interval of the *x*-axis, the domain of the second function is the interval with the endpoint shifted 4 units to the left.
7. The asymptote of the graph in part (b) is the asymptote of the parent graph translated 1 unit down.

Answers to Look Back

To graph the function $y + k = f(x + h)$, graph the parent function $y = f(x)$ and then use the translation vector (h, k) to translate the graph. To use translations and parent graphs to write an equation of a graph, first look at the shape of the graph to identify its parent function. Then find a key point on the parent graph and find its image on the translated graph to determine the translation vector (h, k). Use the values of *h* and *k* to write the equation.

Toll Roads The Kansas Turnpike is a toll road. It begins at the border of Oklahoma and Kansas and continues for 236 mi to the border of Kansas and Missouri, near Kansas City.

Suppose a car enters the turnpike at South Topeka.

Let x = the number on the mile marker where the car exits the turnpike.

Let y = the toll in dollars.

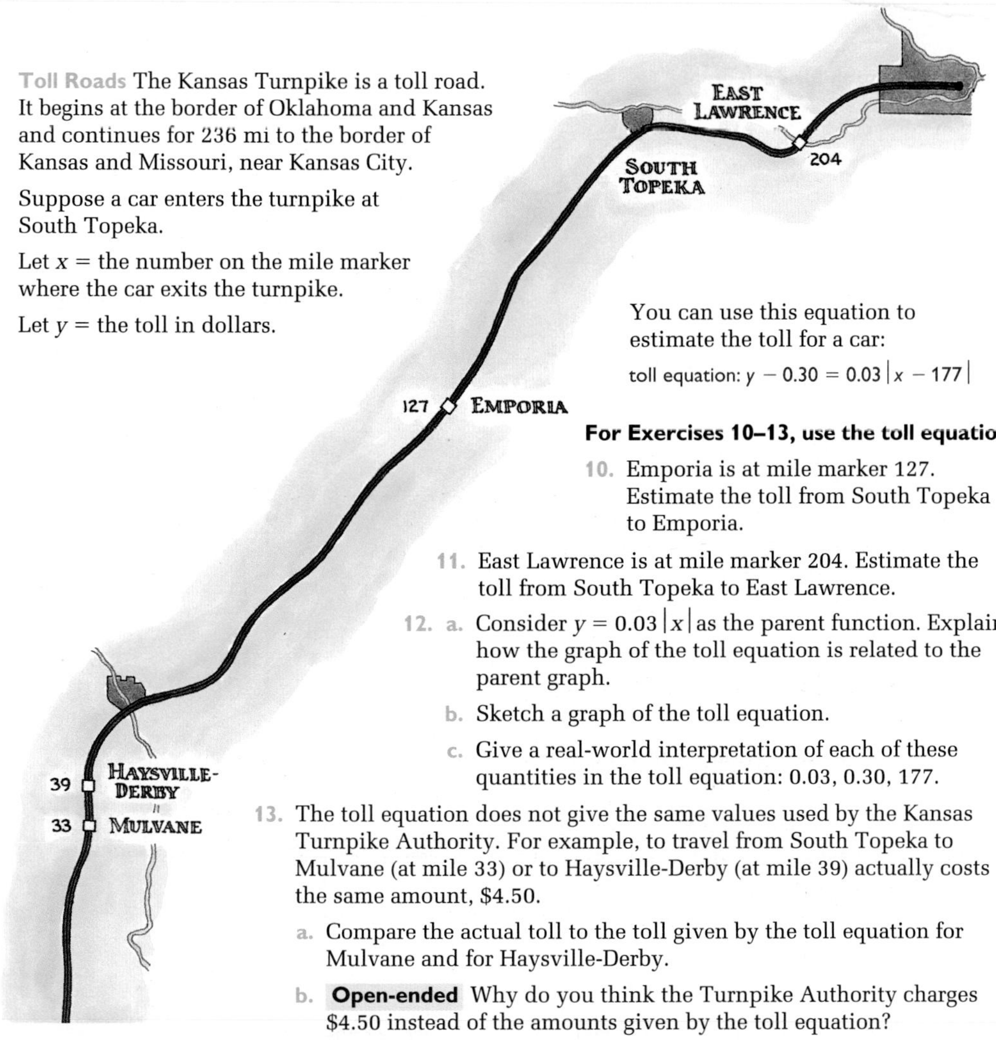

You can use this equation to estimate the toll for a car:

toll equation: $y - 0.30 = 0.03|x - 177|$

For Exercises 10–13, use the toll equation.

10. Emporia is at mile marker 127. Estimate the toll from South Topeka to Emporia.

11. East Lawrence is at mile marker 204. Estimate the toll from South Topeka to East Lawrence.

12. a. Consider $y = 0.03|x|$ as the parent function. Explain how the graph of the toll equation is related to the parent graph.
    b. Sketch a graph of the toll equation.
    c. Give a real-world interpretation of each of these quantities in the toll equation: 0.03, 0.30, 177.

13. The toll equation does not give the same values used by the Kansas Turnpike Authority. For example, to travel from South Topeka to Mulvane (at mile 33) or to Haysville-Derby (at mile 39) actually costs the same amount, \$4.50.
    a. Compare the actual toll to the toll given by the toll equation for Mulvane and for Haysville-Derby.
    b. **Open-ended** Why do you think the Turnpike Authority charges \$4.50 instead of the amounts given by the toll equation?

14. Income Tax In 1993 a single person who was self-supporting and had no other dependents paid no federal income tax on the first \$6050 of *earned* income. The tax rate for the first \$22,100 of *taxable* income was 15%.
    a. Let x = taxable income. Let y = income tax. Sketch a graph of the function $y = 0.15x$.
    b. On the same set of axes, let x = earned income. Let y = income tax. Sketch the graph of the function $y = 0.15(x - 6050)$.
    c. Consider the graph in part (a) as a parent graph. Describe the translation that transforms it into the graph in part (b).
    d. **Writing** Explain why both graphs, if properly interpreted, give correct values for the income tax of a single person with no other dependents.
    e. Do either of the equations in parts (a) and (b) apply to a single person earning \$3500? Explain.

Application

Exs. 10–14 apply the concept of a parent function and the mathematics of translating graphs to practical problems involving the use of toll roads and calculating income tax. On pages 494 and 495, Exs. 22–24 use parent functions and their translations to examine a situation involving a spacelab module above the surface of Earth.

Answers to Exercises and Problems

1. $y = x^2$; $(-2, 7)$
2. $y = \log x$; $(1, 2)$
3. $y = 10^x$; $(-9, 0)$
4. D
5. 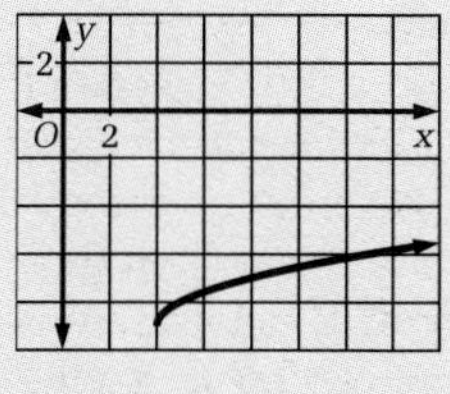
6. 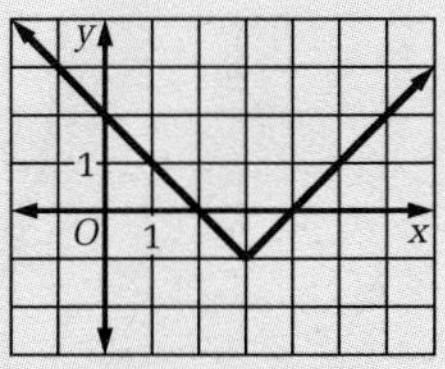
7. 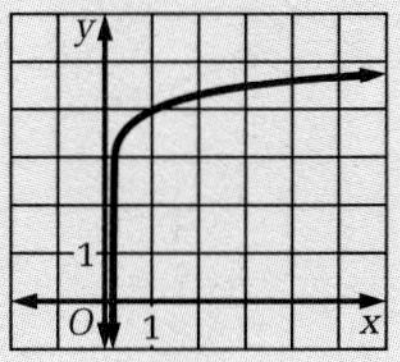
8. 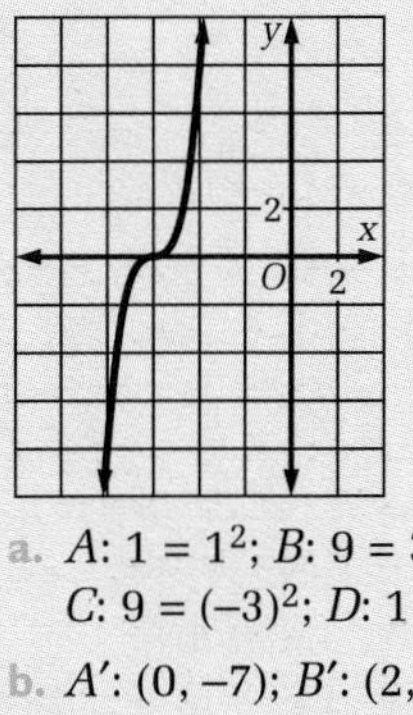
9. a. A: $1 = 1^2$; B: $9 = 3^2$; C: $9 = (-3)^2$; D: $1 = (-1)^2$
   b. A': $(0, -7)$; B': $(2, 1)$; C': $(-4, 1)$; D': $(-2, -7)$
   c. $y + 8 = (x + 1)^2$
   d. A': $-7 + 8 = (0 + 1)^2$; B': $1 + 8 = (2 + 1)^2$; C': $1 + 8 = (-4 + 1)^2$; D': $-7 + 8 = (-2 + 1)^2$
10. \$1.80
11. \$1.11
12. a. The graph is translated by the vector (177, 0.30); that is, the graph is translated 177 units right and 0.30 units up.
    b. 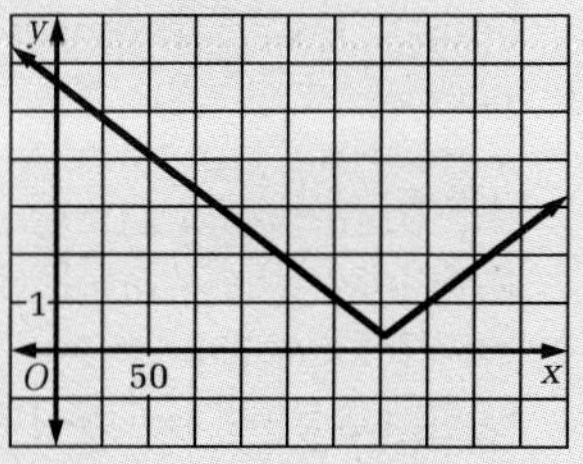
    c. 0.03 represents a charge of 3¢ per mile; 0.30 represents a base charge of 30¢, and 177 represents the mile-marker reading at South Topeka.
13. a. For Mulvane, the toll given by the toll equation is \$4.62, 12¢ more than the actual toll. For Haysville-Derby, the toll given by the toll equation is \$4.44, 6¢ less than the actual toll.
    b. Answers may vary. An example is given. I think tolls are rounded to make it easier to make change and to save time at the toll booths.
14. See answers in back of book.

Mathematical Procedures

The equations for Exs. 19 and 20 are not too difficult to obtain because the translation vector is vertical in one case and horizontal in the other. Exponential and logarithmic functions whose graphs result from "diagonal" translations are more difficult to work with. In those cases, it may help students to make a copy of the parent graph on tracing paper. They can mark some key points and label them with their coordinates. Students can then translate the parent graph by sliding the tracing paper so that its graph fits on top of the other graph.

Communication: Discussion

You may wish to discuss how graphs such as those in Exs. 19 and 20 can be distinguished from a single branch of a hyperbola (Ex. 21). Students should consider how much information they have when they are shown only part of a graph.

Multicultural Note

Sally Ride, who in 1983 went into orbit aboard the space shuttle *Challenger,* was the first American woman in space. Two Russian women preceded her: in 1963, Valentina Tereshkova became the first woman in space, and nineteen years later, Svetlana Savitskaya journeyed into space. Ride's participation in the *Challenger*'s ninth trip was a historic event for American women; the shuttle's take-off on June 18, 1983 drew a crowd of 500,000. In 1992, Dr. Mae Jemison became the first African-American woman astronaut when she traveled into space aboard the shuttle *Endeavor.*

Write an equation for each graph.

15.

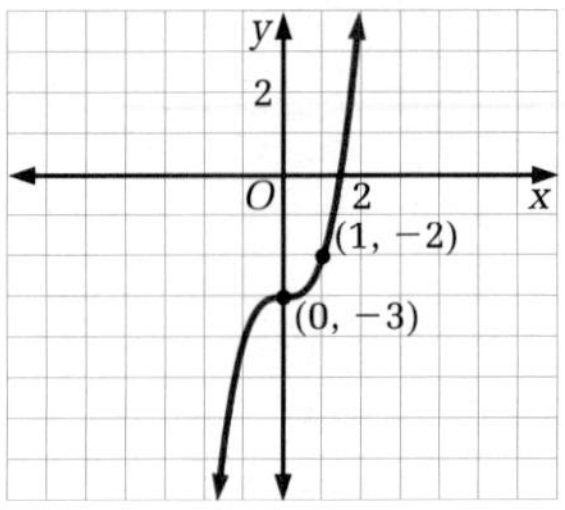

16.

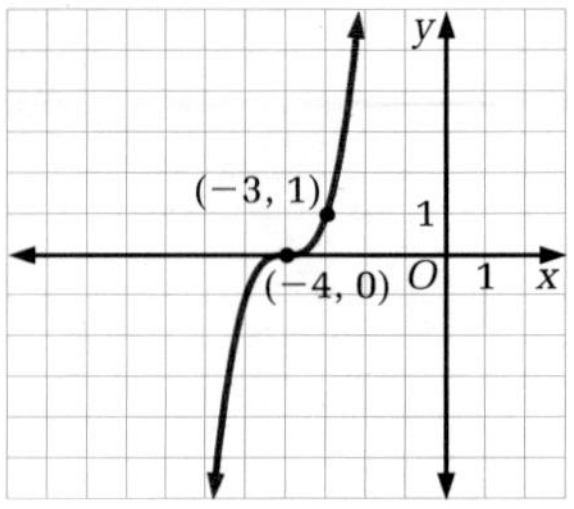

17.

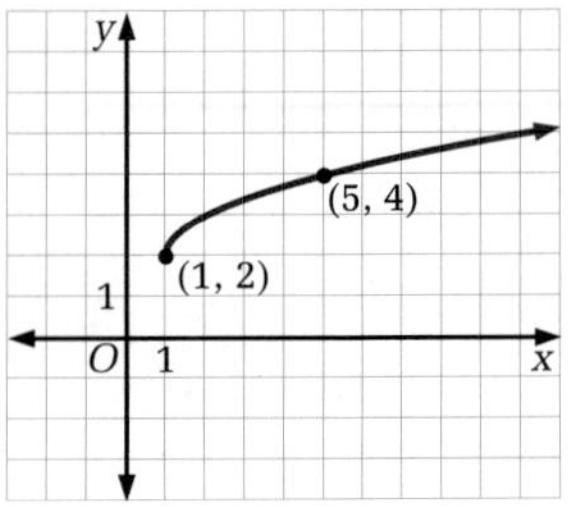

18.

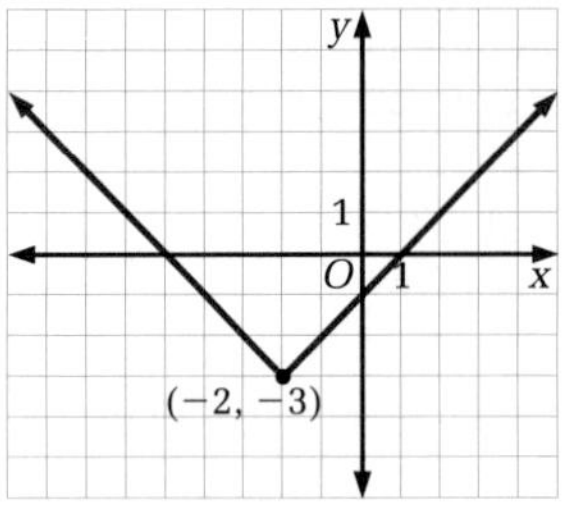

19.

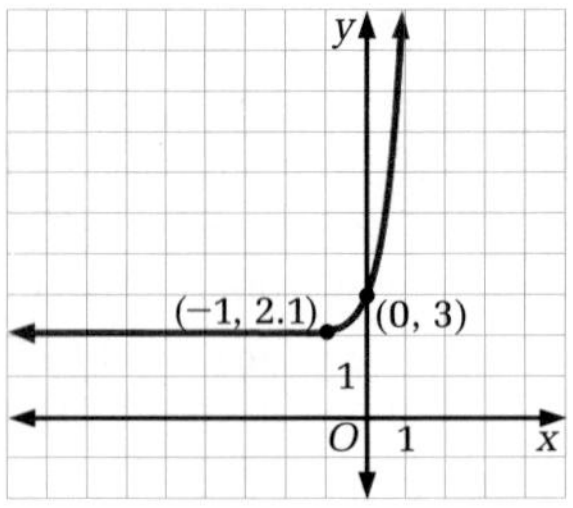

20.

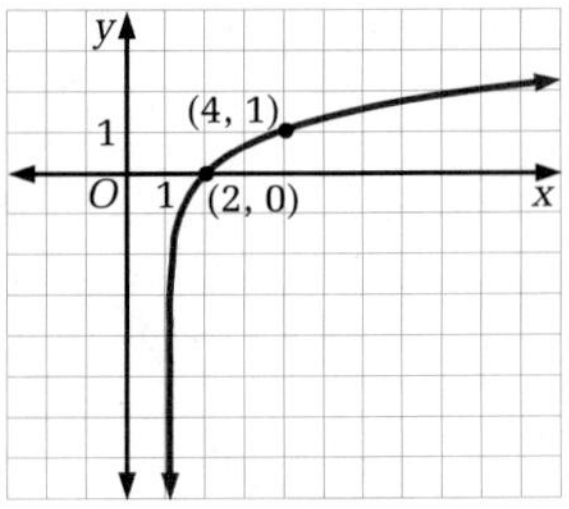

21. For the graph shown at the right, $y = \frac{1}{x}$ is the parent function.

a. Find the values of h and k.

b. Find the equations of the asymptotes for the translated graph.

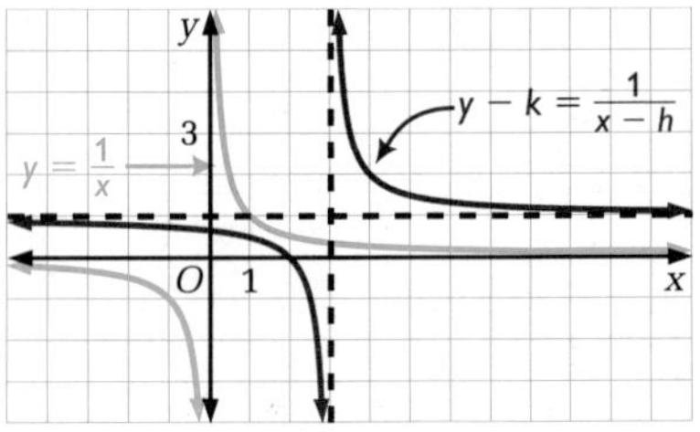

Space Exploration In July of 1994, surgeon Chiaki Mukai became the first Japanese woman in space. She was the payload specialist on an international mission aboard the space shuttle *Columbia*.

The Spacelab module in the payload bay held most of the experiments performed during the mission. The following equation approximates the module's weight w (in pounds) at a distance d (in miles) above the surface of Earth.

$$w = \frac{3.68 \times 10^{11}}{(4000 + d)^2}$$

For Exercises 22–24, use the equation above.

22. About how much does the Spacelab module weigh on the surface of Earth?

23. *Columbia*'s altitude while in orbit was 184 mi. About how much did the Spacelab module weigh while in orbit?

Answers to Exercises and Problems

15. $y + 3 = x^3$
16. $y = (x + 4)^3$
17. $y - 2 = \sqrt{x - 1}$
18. $y + 3 = |x + 2|$
19. $y - 2 = 10^x$
20. $y = \log (x - 1)$
21. a. $h = 3$; $k = 1$
    b. $x = 3$; $y = 1$
22. about 23,000 lb
23. about 21,022 lb

24. a, b.

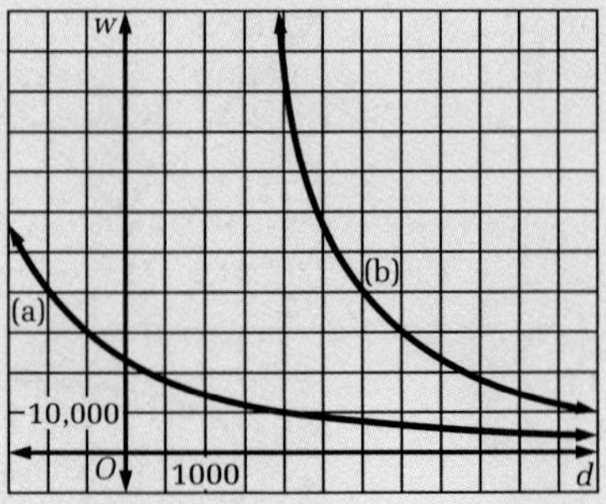

24. a. Graph the function at the bottom of page 494.

b. Graph the parent function $w = \frac{3.68 \times 10^{11}}{d^2}$ on the same set of axes.

c. Compare the graphs in parts (a) and (b).

d. Give a real-world interpretation of the function in part (b). (*Hint:* The radius of Earth is about 4000 mi.)

e. What values of d would you substitute into the function in part (b) to answer the questions asked in Exercises 22 and 23?

BY THE WAY...

On July 17, 1975, astronauts from the United States shook hands with Russian cosmonauts in outer space as part of the Apollo-Soyuz mission. This was the first time that spacecraft from two different countries met in orbit.

Ongoing ASSESSMENT

25. **Open-ended** Choose a radius. Write an equation for a circle that has this radius and is centered at the origin. Write an equation for a congruent circle with its center in the fourth quadrant. What translation transforms the original circle into the congruent circle?

Review PREVIEW

26. The mean of a set of test scores is 72 and the standard deviation is 9. The highest score is 90. Suppose 10 points are added to each score. What will be the mean and the standard deviation of the new scores? *(Section 9-1)*

27. Evaluate $\sum_{n=1}^{4} \left(\frac{2}{3}\right)^n$. *(Section 4-6)*

Multiply each pair of matrices. *(Toolbox Skill 11)*

28. $\begin{bmatrix} 5 & 0 \\ 0 & 2 \end{bmatrix} \begin{bmatrix} 1 & 0 \\ 0 & 4 \end{bmatrix}$

29. $\begin{bmatrix} 2 & 0 \\ 0 & \frac{1}{4} \end{bmatrix} \begin{bmatrix} \frac{4}{5} & 0 \\ 0 & 2 \end{bmatrix}$

30. $\begin{bmatrix} 3 & 0 \\ 0 & 2 \end{bmatrix} \begin{bmatrix} 2 & 3 & 2 & 4 \\ 1 & 0 & 2 & 1 \end{bmatrix}$

Working on the Unit Project

31. **Research** Read about the qualifications, salaries, working conditions, and job opportunities for the careers you are investigating. Here are some books that contain information:

- *Occupational Outlook Handbook* by the U.S. Bureau of Labor Statistics
- *Statistical Abstract of the United States* by the U.S. Bureau of the Census
- *The American Almanac of Jobs and Salaries* by John W. Wright
- *What Color Is Your Parachute?* by Richard N. Bolles

Reasoning

Ex. 25 leads students to the general equation for a circle, $(x - h)^2 + (y - k)^2 = r^2$. You may wish to have students write equations for other congruent circles with centers in the first, second, and third quadrants as well.

Assessment: Portfolio

Students should keep a list in their portfolios of the parent functions (and relations) discussed so far. These may include $y = x^2$, $y = x^3$, $y = \sqrt{x}$, $y = 10^x$, $y = |x|$, $y = \log x$, $y = \frac{1}{x}$, and $x^2 + y^2 = r^2$. They should show graphical examples and how a translation by the vector (h, k) affects the graph.

Practice 68 For use with Section 9-2

Name ____________ Date ____________

Practice 68

For use with Section 9-2

1. $y = x^3$; (3, −1) 2. $y = \sqrt{x}$; (2, 5)
3. $y = 2^x$; (0, 4) 4. $y = \frac{1}{x}$; (−2, 3)
13. $y + 9 = (x - 5)^4$ 14. $y - 8 = \log(x - 3)$
15. $y + 6 = 10^{x-1}$ 16. $y + 1 = \sqrt[3]{x + 2}$
17. $y - 10 = |x + 7|$ 18. $y - 3 = \frac{1}{x - 2}$

For Exercises 1–6, the equations are in the form $y - k = f(x - h)$. For each equation, find the parent function and the translation vector. Exs. 1–4: See above.

1. $y + 1 = (x - 3)^3$
2. $y - 5 = \sqrt{x - 2}$
3. $y - 4 = 2^x$
4. $y - 3 = \frac{1}{x + 2}$
5. $y = \log(x - 7)$ $y = \log x$; (7, 0)
6. $y + 6 = |x + 8|$ $y = |x|$; (−8, −6)

Sketch the graph of each function. Exs. 7–12: See additional answers.

7. $y - 2 = (x + 1)^2$
8. $y + 3 = \log(x - 2)$
9. $y + 5 = 10^{x-2}$
10. $y + 3 = |x - 1|$
11. $y - 2 = (x + 1)^3$
12. $y - 4 = \sqrt{x - 3}$

For each parent function, find the equation of the graph after translation by the given vector v. Exs. 13–18: See top of page.

13. $y = x^4$; $\mathbf{v} = (5, -9)$
14. $y = \log x$; $\mathbf{v} = (3, 8)$
15. $y = 10^x$; $\mathbf{v} = (1, -6)$
16. $y = \sqrt[3]{x}$; $\mathbf{v} = (-2, -1)$
17. $y = |x|$; $\mathbf{v} = (-7, 10)$
18. $y = \frac{1}{x}$; $\mathbf{v} = (2, 3)$

For each graph, the equation of the parent function is given. Find the translation vector and the equation of the graph shown.

19. $y = x^3$ See below.

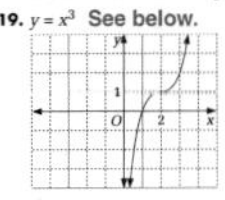

20. $y = \log x$ See below.

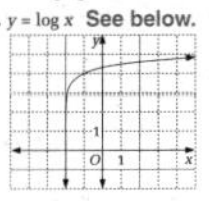

21. $y = \sqrt{x}$ (−3, −1); $y + 1 = \sqrt{x + 3}$

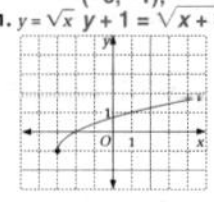

22. The height y (in feet) of a ball thrown straight upward with a velocity of 40 ft/s is given, as a function of the time t (in seconds) after the ball is thrown, by the equation $y = 40t - 16t^2$.

a. Suppose the ball is thrown from the top of a building 50 ft high. Rewrite the equation so that it gives the ball's height above the ground in this new situation. $y = 50 + 40t - 16t^2$

b. Suppose the ball is thrown at 12:00 noon and t represents the time after 11:59:58 (2 seconds before noon). Rewrite the given equation so that it gives the ball's height as a function of the new values of t. $y = 40(t - 2) - 16(t - 2)^2$

19. (2, 1); $y - 1 = (x - 2)^3$
20. (−2, 4); $y - 4 = \log(x + 2)$

68 Practice Bank, INTEGRATED MATHEMATICS 3

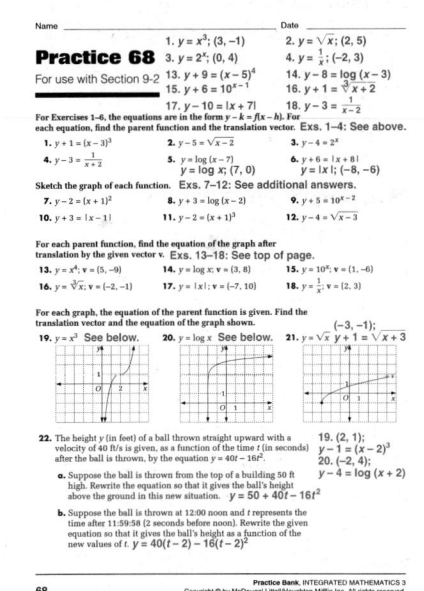

Answers to Exercises and Problems

c. The graph in part (a) is the parent graph (b) translated 4000 units left.

d. The function approximates the weight in pounds of the Spacelab module at a distance d miles from the center of Earth.

e. 4000, 4184

25. Answers may vary. An example is given. Let the radius be 5 and the original equation be $x^2 + y^2 = 25$. The circle with equation $(x - 3)^2 + (y + 3)^2 = 25$ is congruent and has its center in the fourth quadrant. The vector (3, −3) transforms the original circle into the congruent circle.

26. 82; 9

27. about 1.6

28. $\begin{bmatrix} 5 & 0 \\ 0 & 8 \end{bmatrix}$

29. $\begin{bmatrix} 1\frac{3}{5} & 0 \\ 0 & \frac{1}{2} \end{bmatrix}$

30. $\begin{bmatrix} 6 & 9 & 6 & 12 \\ 2 & 0 & 4 & 2 \end{bmatrix}$

31. Answers may vary.

PLANNING

Objectives and Strands
See pages 480A and 480B.

Spiral Learning
See page 480B.

Materials List
- Graphics calculator
- Graph paper

Recommended Pacing
Section 9-3 is a one-day lesson.

Extra Practice
See pages 622–624.

Warm-Up Exercises
Warm-Up Transparency 9-3

Support Materials
- Practice Bank: Practice 69
- Activity Bank: Enrichment 61
- Study Guide: Section 9-3
- Problem Bank: Problem Set 19
- Explorations Lab Manual: Diagram Masters 1, 2
- Using Mac Plotter Plus Disk: Data Analyzer
- Using IBM Plotter Plus Disk: Statistics Spreadsheet
- Assessment Book: Quiz 9-3

Section 9-3

Multiplying Data by a Constant

CONSTANT CHANGES

Focus

Recognize the effect that multiplying data values by a constant has on the mean, median, range, and standard deviation.

The French unit of money is the *franc*. The number of dollars equivalent to one franc is called the *exchange rate*. The table shows the numbers of francs spent by students in a high school French club during a trip to France. It also shows the equivalent numbers of dollars spent.

STUDENT	Francs spent	Equivalent dollars
Abdul	1630	326
Adriana	1285	257
Carrie	1480	296
Hamid	1870	374
Hannah	1190	238
Jared	1705	341
Kimiko	1570	314
Matthew	1120	224
Rosa	1860	372
Tanya	1550	310
Teodoro	1240	248

Talk it Over

1. Find the number that makes this equation true:
$$\text{number of francs spent} \times \underline{\ ?\ } = \text{number of dollars spent}$$
What was the exchange rate at the time of the trip?
2. a. Find the mean of the francs data.
   b. Find the mean of the dollars data.
   c. Find the number that makes this equation true:
$$\text{mean of francs data} \times \underline{\ ?\ } = \text{mean of dollars data}$$
3. Repeat question 2 using the median.
4. Repeat question 2 using the range.
5. The histograms show the distributions of the francs and dollars data. How are the histograms alike? different?

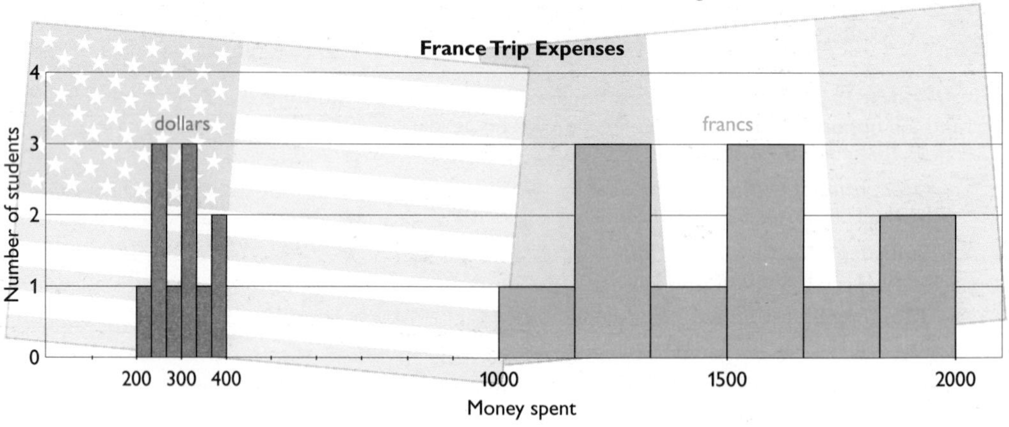

Answers to Talk it Over

1. 0.2; 0.2 dollars to 1 franc
2. a. 1500
   b. 300
   c. 0.2
3. a. 1550
   b. 310
   c. 0.2
4. a. 750
   b. 150
   c. 0.2
5. The columns are the same height. The columns in the second histogram are five times as wide as the columns in the first. The endpoints of each column in the second histogram are 5 times the endpoints of the corresponding column in the first.

Sample 1

Use the table on page 496. Find the standard deviation of the francs data and of the dollars data.

Sample Response

Enter the francs and dollars data on a calculator. Have the calculator compute the standard deviation for each data set.

standard deviation of francs data →

1–Var Stats
$\bar{x} = 1500$
$\Sigma x = 16500$
$\Sigma x^2 = 25437550$
$Sx = 262.2117465$
$\sigma x = 250.0090907$
$\downarrow n = 11$

standard deviation of dollars data →

1–Var Stats
$\bar{x} = 300$
$\Sigma x = 3300$
$\Sigma x^2 = 1017502$
$Sx = 52.4423493$
$\sigma x = 50.00181815$
$\downarrow n = 11$

The standard deviation of the francs data is about 250 francs. The standard deviation of the dollars data is about \$50.

Talk it Over

6. Use the results of Sample 1. Find the number that makes this equation true:

$$\text{standard deviation of francs data} \times \underline{\ ?\ } = \text{standard deviation of dollars data}$$

7. Suppose each of these data values is multiplied by 4:

6, 10, 7, 1, 16

Predict how the mean, the median, the range, and the standard deviation change. Check your prediction by computing these statistics for the original and transformed data sets.

Answers to Talk it Over

6. 0.2
7. Predictions will vary. Each will be multiplied by 4. For the original data, the mean is 8, the median is 7, the range is 15, and the standard deviation is about 4.94. For the transformed data, the mean is 32, the median is 28, the range is 60, and the standard deviation is about 19.76.

TEACHING

Additional Sample

S1 A new branch of a city public library requested a printout of the number of books one week overdue by cardholders and the amount of money due from each cardholder. The printout gave the following number of books (names of cardholders omitted): 4, 7, 2, 3, 4, 1, 5, 2, 1, 3, 3, 5, 4, 3, 2, 4, 2, 3, 4, 4.
The printout of amounts (in dollars) due from the cardholders with these books was as follows: 1.20, 2.10, .60, .90, 1.20, .30, 1.50, .60, .30, .90, .90, 1.50, 1.20, .90, .60, 1.20, .60, .90, 1.20, 1.20.
Find the mean and the standard deviation of the data on the number of books overdue by one week and on the amounts due on these books.

Enter the books and amounts due on a calculator. Have the calculator compute the mean and the standard deviation for each data set.

Books one week overdue:

1 – Var Stats
$\bar{x} = 3.3$
$\Sigma x = 66$
$\Sigma x^2 = 258$
$Sx = 1.454575359$
$\sigma x = 1.417744688$
$\downarrow n = 20$

The mean number of books one week overdue is 3.3, and the standard deviation is about 1.42 books.

Amount due:

1 – Var Stats
$\bar{x} = .99$
$\Sigma x = 19.8$
$\Sigma x^2 = 23.22$
$Sx = .4363726076$
$\sigma x = .4253234064$
$\downarrow n = 20$

The mean amount of money due from cardholders with books one week overdue is \$.99, and the standard deviation is about \$.43.

Visual Thinking

Have students create histograms of the situations described in Sample 2. Ask them to present their histograms to the class and describe the similarities and differences. This activity involves the visual skills of *identification* and *communication.*

Additional Sample

S2 The first day that a new credit union opened for business, seventy-five employees started savings accounts. The median deposit was $465. The range of the deposits was $620. Interest is 5.9% compounded quarterly. If these employees made no withdrawals or additional deposits for one year, what would be the median amount in the accounts and the range of amounts in the accounts after one year?

The amount in each account would be multiplied by a factor of $\left(1 + \frac{0.059}{4}\right)^4$, or about 1.06. The new median amount would be $465 × 1.06, or $492.90. The range of amounts in the accounts after one year would be $620 × 1.06, or $657.20.

Look Back

The answer to the Look Back provides a summary of the focus of this section.

APPLYING

Suggested Assignment

Standard 2, 3, 5–7, 9–19

Extended 1–7, 9–19

Integrating the Strands

Algebra Exs. 8, 9, 13–18

Functions Exs. 10–12

Statistics and Probability Exs. 1–9, 19

Discrete Mathematics Exs. 1–9, 19

Logic and Language Exs. 1, 7, 8, 9, 19

X Lab

MULTIPLYING DATA VALUES BY A CONSTANT

When each value in a data set is multiplied by the same positive constant a, then the mean, the median, the range, and the standard deviation are also multiplied by a.

Sample 2

The mean annual salary for employees at a tree nursery is $18,000. The standard deviation of the salaries is $7000. Suppose the nursery owner gives all employees a 4% raise. Find the mean and the standard deviation of the new salaries.

Sample Response

The effect of a 4% raise is that each salary is multiplied by 1.04. Therefore, the mean and the standard deviation are also multiplied by 1.04.

$$\text{new mean} = 1.04 \times \text{old mean} = (1.04)(18{,}000) = 18{,}720$$

$$\text{new standard deviation} = 1.04 \times \text{old standard deviation} = (1.04)(7000) = 7280$$

The mean of the new salaries is $18,720. The standard deviation is $7280.

Look Back

Identify at least three effects of multiplying each value in a data set by a positive constant.

9-3 Exercises and Problems

1. **Reading** Explain how you can use the histograms on page 496 to compare the ranges of the dollars and francs data.

For Exercises 2 and 3:

a. Find the mean, the median, the range, and the standard deviation of the given data.

b. Repeat part (a) for the data obtained by multiplying each value by *a*.

c. Make histograms of the original and transformed data. Draw your histograms on the same set of axes.

2. data: 93, 71, 96, 62, 85, 99, 75
   $a = 0.5$

3. data: 7.3, 9.2, 9.7, 9.6, 5.8, 5.5, 6.1, 5.2
   $a = 3$

Answers to Look Back

When each value in a data set is multiplied by the same positive number, the mean, the median, the range, and the standard deviation are multiplied by the same number.

Answers to Exercises and Problems

1. Compare the difference between the last endpoint and the first for the two histograms.
2. a. 83; 85; 37; about 13
   b. 41.5; 42.5; 18.5; about 6.5
   c. Data intervals may vary.

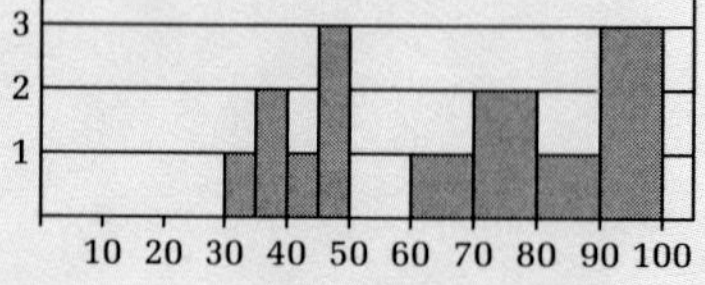

3. a. 7.3; 6.7; 4.5; about 1.8
   b. 21.9; 20.1; 13.5; about 5.4
   c. Data intervals may vary.

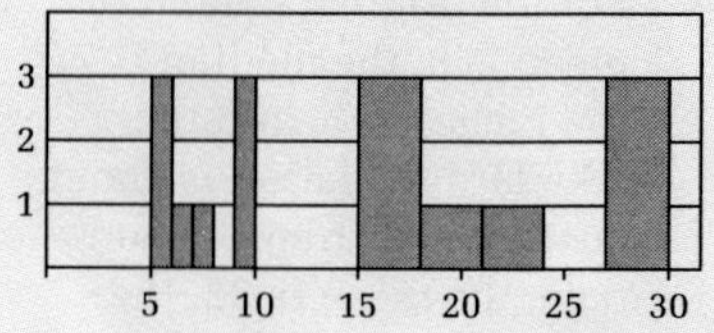

4. **Consumer Economics** At an electronics store, the mean price of a VCR is \$330. The standard deviation of the prices is \$45. During a clearance sale, all merchandise in the store is discounted by 15%. Find the mean and standard deviation of the new VCR prices.

connection to SCIENCE

The table shows the diameters of the nine planets in Earth's solar system.

PLANET	Diameter in miles	Diameter as a multiple of Earth's diameter
Mercury	3,030	0.38
Venus	7,520	0.95
Earth	7,930	1.00
Mars	4,200	0.53
Jupiter	88,700	11.19
Saturn	74,600	9.41
Uranus	31,600	3.98
Neptune	30,800	3.88
Pluto	1,420	0.18

5. Let d be the diameter of a planet expressed as a multiple of Earth's diameter. Complete each sentence with an inequality involving d.
   a. If a planet is larger than Earth, then __?__.
   b. If a planet is smaller than Earth, then __?__.
6. Find the mean, the median, the range, and the standard deviation of the data in the second column.
7. a. Find the number that makes this equation true:

$$\text{diameter in miles} \times \underline{?} = \text{diameter as a multiple of Earth's diameter}$$

   b. Find the mean, the median, the range, and the standard deviation of the data in the third column.
   c. **Writing** Aaron looked at the median from part (b) and concluded that Earth is an average-sized planet. Yolanda looked at the mean and concluded that Earth is much smaller than average. Explain why both conclusions are reasonable.

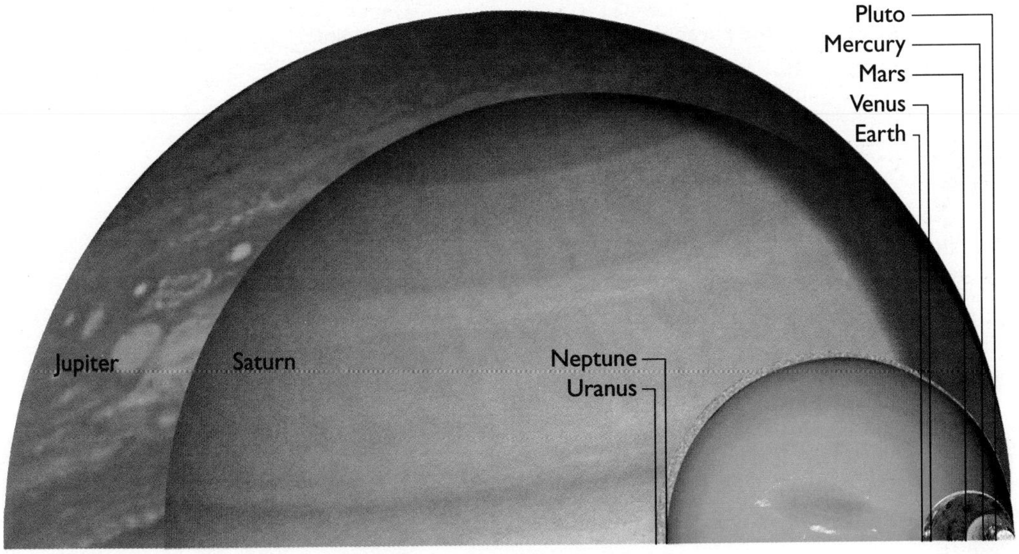

Using Technology

The quickest way to perform the calculations for Exs. 2 and 3 is to do the work on a TI-82 graphics calculator. Once students have entered the given data as statistical list L_1, they can multiply the data by the given constant by using the command $aL_1 \rightarrow L_2$, where a is the constant. When this command is put on the home screen and ENTER is pressed, the calculator multiplies all the items in L_1 by a and stores the resulting products as statistical list L_2.

Application

In Ex. 4, students should understand that a discounted price results when the amount of discount is subtracted from the original price. With a 15% discount, 1 – 0.15 or 0.85 of the original mean of \$330 is the mean of the new prices. The same constant should be used to find the standard deviation.

Communication: Writing

After students write their explanations for Ex. 7 (c), they can choose a partner and read their answers to one another. Modifications and revisions in answers should be made if necessary.

Answers to Exercises and Problems

4. \$280.50; \$38.25
5. a. $d > 1$
   b. $d < 1$
6. about 27,756; 7930; 87,280; about 30,903
7. a. about 0.000126
   b. Answers are rounded to the nearest tenth. 3.5; 1.0; 11.0; 3.9
   c. Aaron used the fact that the value of d for Earth is the median; Yolanda used the fact that the value of d for Earth is significantly less than the mean.

Reasoning

You may wish to discuss Ex. 9 in class and then invite students to explain the algebraic reasoning that shows what happens to the standard deviation. The reasoning for the standard deviation makes use of the result proved in this exercise.

Practice 69 For use with Section 9-3

Name ______________________ Date ____________

Practice 69
For use with Section 9-3

1. a. 30.4, 30, 18, 5.4 b. 60.8, 60, 36, 10.9
2. a. 4.8, 4.7, 3.8, 1.1, b. 14.5, 14.1, 11.4, 3.3
3. a. 8.6, 8.6, 3.8, 1.0 b. 43, 43, 19, 5.0
4. a. 159.3, 156, 46, 14.1 b. 39.8, 39, 11.5, 3.5

Exs. 1–4: For parts (a) and (b), see above.
For Exercises 1–4: For histograms for part (c), see additional answers.

a. Find the mean, median, range, and standard deviation of the given data. Round decimal answers to the nearest tenth.

b. Repeat part (a) for the data obtained by multiplying each value by *a*.

c. Make histograms of the original and transformed data. Draw your histograms on the same axis.

1. data: 25, 27, 32, 24, 35, 38, 26, 22, 28, 30, 32, 40, 36; $a = 2$

2. data: 4.2, 3.6, 5.8, 5.2, 4.8, 7.0, 6.4, 4.0, 4.6, 5.4, 3.2, 3.8; $a = 3$

3. data: 8.6, 7.8, 6.4, 9.4, 10.2, 9.8, 8.6, 8.4, 7.6, 9.0, 9.6, 8.2; $a = 5$

4. data: 144, 160, 152, 148, 140, 156, 168, 172, 182, 186, 148, 156; $a = 0.25$

5. The table at the right gives the numbers by which you would have to multiply your weight on Earth in order to find your weight on other planets. Suppose the weights of a group of people have a mean of 147 lb and a standard deviation of 35 lb. Find the mean and the standard deviation of the weights of the group on each planet.

Planet	Weight factor
Mercury	0.37
Venus	0.88
Earth	1.00
Jupiter	2.64
Saturn	1.15

a. Mercury 54.39; 12.95 **b.** Venus 129.36; 30.8

c. Jupiter 388.08; 92.4 **d.** Saturn 169.05; 40.25

6. On a standard clarinet or trumpet, the note that is called C has a frequency that is 0.89 times the frequency of the note called C on the piano. The same is also true of all the other notes that these two instruments can play.

a. Find the frequency of each note on a clarinet for the given frequency of the corresponding note on the piano:
E: 329.6 Hz G#: 415.3 Hz B-flat: 466.2 Hz ≈293.3 Hz ≈369.6 Hz ≈414.9 Hz

b. Find the frequency of each note on the piano for the given frequency of the corresponding note on the clarinet:
F#: 329.6 Hz D: 261.6 Hz B: 440 Hz ≈370.3 Hz ≈293.9 Hz ≈494.4 Hz

7. ***Writing*** Suppose that the company you work for is considering two kinds of raises: (1) raising each employee's salary by a fixed amount each year, or (2) raising each employee's salary by a fixed percent each year. Discuss the advantages and disadvantages of each system, taking into consideration starting salaries, amounts of fixed raises, and percents for percentage raises.
Answers may vary. Check students' work.

Practice Bank, INTEGRATED MATHEMATICS 3

69

8. **Biology** The table shows the recommended average weights for men of different heights.

   a. TECHNOLOGY Use a graphics calculator to find an equation of the least-squares line for the data.
   b. Make a new table giving the heights in centimeters and the weights in kilograms. (*Hint:* 1 in. = 2.54 cm and 1 lb ≈ 0.45 kg)
   c. TECHNOLOGY Use a graphics calculator to find an equation of the least-squares line for the centimeter/kilogram data in part (b).
   d. **Writing** What is the relationship between the vertical intercepts of the least-squares lines in parts (a) and (c)? What is the relationship between the slopes of the lines?

Height in inches	Weight in pounds
64	145
66	153
68	161
70	170
72	179
74	188
76	199

Ongoing **ASSESSMENT**

9. **Writing** Write a paragraph proof of this statement: If each value in a data set is multiplied by a, then the mean is also multiplied by a. (*Hint:* Let $x_1, x_2, \ldots, x_n$ be the original data values, and let $\bar{x}$ be their mean. You need to show that the mean of $ax_1, ax_2, \ldots, ax_n$ is $a\bar{x}$.)

Review **PREVIEW**

Without using graphing technology, sketch the graph of each function. *(Section 9-2)*

10. $y = |x - 3|$ **11.** $y - 2 = x^2$ **12.** $y + 4 = \sqrt{x - 5}$

Evaluate to three decimal places. *(Section 5-7)*

13. $\log_2 5$ **14.** $\log_7 3$ **15.** $\log_{1/3} 9.68$ **16.** $\log_{11} \frac{3}{4}$

Write an equation for the parabola passing through the given points. *(Section 6-5)*

17. (1, 8), (2, 3), (4, 10) **18.** (−5, −2), (3, 1), (6, −0.5)

Working on the Unit Project

19. The table shows average starting salaries for college graduates entering certain occupations in 1993.
    a. Display these data on a bar graph.
    b. Suppose that starting salaries will be 34% greater in 2003. What will the average starting salaries be then? Display these data on a bar graph.
    c. Describe the transformation that relates the graphs you drew in parts (a) and (b).

OCCUPATION	Average starting salary
teaching	\$22,505
sales/marketing	\$28,536
business administration	\$27,564
engineering	\$35,004
computer science	\$31,164

Answers to Exercises and Problems

8. a. a and b are rounded to the nearest tenth; $w = 4.5h - 141.8$.

   b.

Height in centimeters	Weight in kilograms
162.56	65.25
167.64	68.85
172.72	72.45
177.8	76.5
182.88	80.55
187.96	84.6
193.04	89.55

   c. a and b are rounded to the nearest tenth; $w = 0.8h - 63.8$.

   d. The vertical intercept of the second least-squares line is about 0.45 times that of the first. The slope of the second least-squares line is the slope of the first multiplied by the ratio of 0.45 to 2.54 (≈ 0.18).

9. Let $x_1, x_2, x_3, \ldots, x_n$ be the original data values, and let $\bar{x}$ be their mean. The transformed data values are $ax_1, ax_2, ax_3, \ldots, ax_n$. Their mean is $\frac{ax_1 + ax_2 + ax_3 + \ldots + ax_n}{n} = \frac{a(x_1 + x_2 + x_3 + \ldots + x_n)}{n} = a\bar{x}$.

10.

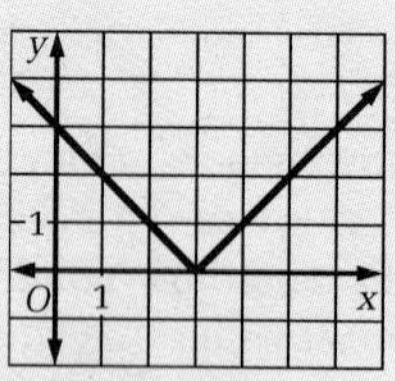

11.

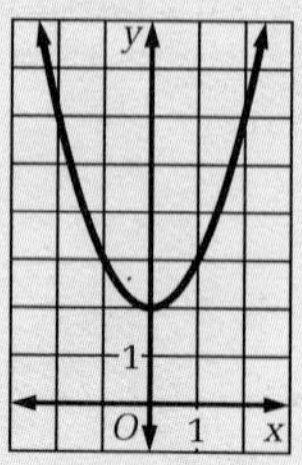

12.

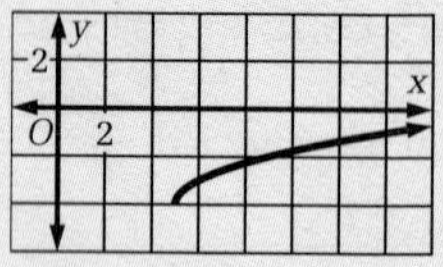

13. 2.322 14. 0.565

15. −2.066 16. −0.120

17. $y \approx 2.8x^2 - 13.5x + 18.7$ or $y = \frac{17}{6}x^2 - \frac{27}{2}x + \frac{56}{3}$

18. $y \approx -0.08x^2 + 0.2x + 1.1$ or $y = \frac{-7}{88}x^2 + \frac{19}{88}x + \frac{47}{44}$

19. See answers in back of book.

Section

9-4 Stretching Graphs

Focus

Recognize the effects of horizontal and vertical stretches on sets of points and graphs of functions.

Pulled In Many Directions

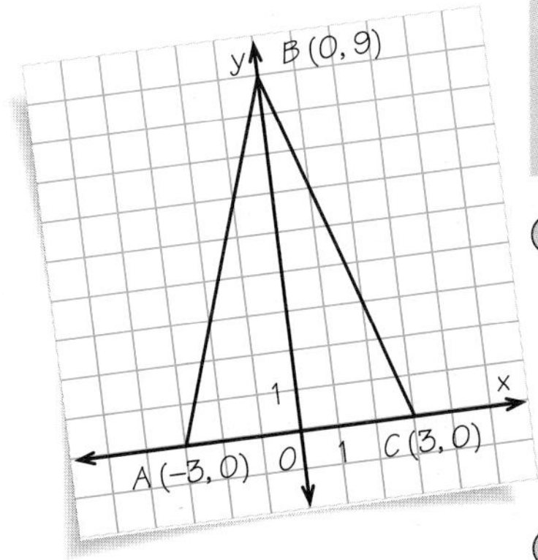

EXPLORATION

What happens to a figure when you multiply its coordinates by positive constants a and b?

- **Materials: graph paper, graphics calculators (optional)**
- **Work in a group of three students.**

1. You can use a **transformation matrix** to represent a transformation. Use the transformation matrix $T = \begin{bmatrix} 3 & 0 \\ 0 & 1 \end{bmatrix}$ and the matrix M shown. Find the product matrix $M' = TM$.

$$M = \begin{array}{c} \begin{matrix} A & B & C \end{matrix} \\ \begin{bmatrix} -3 & 0 & 3 \\ 0 & 9 & 0 \end{bmatrix} \end{array}$$ ← Matrix M represents the vertices of the triangle.

2. Draw triangle $A'B'C'$ represented by the matrix M'. Describe the effect of transformation T on the original triangle.

3. Repeat steps 1 and 2 for each transformation matrix.

a. $T = \begin{bmatrix} 1 & 0 \\ 0 & 2 \end{bmatrix}$ **b.** $T = \begin{bmatrix} 3 & 0 \\ 0 & 1 \end{bmatrix}\begin{bmatrix} 1 & 0 \\ 0 & 2 \end{bmatrix} = \begin{bmatrix} 3 & 0 \\ 0 & 2 \end{bmatrix}$ **c.** $T = \begin{bmatrix} \frac{1}{3} & 0 \\ 0 & \frac{1}{2} \end{bmatrix}$

4. Suppose a and b can be any positive numbers. Make a conjecture about the effects of the transformation matrix $T = \begin{bmatrix} a & 0 \\ 0 & b \end{bmatrix}$.

5. Write an equation of the parabola that passes through the vertices of the original triangle. Graph the equation.

6. Use the form $\frac{y}{b} = -\left(\frac{x}{a}\right)^2 + 9$.

a. Write an equation of the parabola that passes through the vertices of the triangle you drew in step 2. What is the value of a? of b?

b. Repeat part (a) for the triangle you drew in part (a) of step 3.

c. Graph your equations from parts (a) and (b). Compare the graphs with your graph from step 5.

Student Resources Toolbox
p. 634 *Matrices*

Now you are ready for:
Exs. 1–15 on pp. 504–505

PLANNING

Objectives and Strands

See pages 480A and 480B.

Spiral Learning

See page 480B.

Materials List

- Graph paper
- Graphics calculator or graphing software

Recommended Pacing

Section 9-4 is a two-day lesson.

Day 1

Page 501: Exploration, *Exercises 1–15*

Day 2

Pages 502–503: Top of page 502 through Look Back, *Exercises 16–46*

Extra Practice

See pages 622–624.

Warm-Up Exercises

Warm-Up Transparency 9-4

Support Materials

- Practice Bank: Practice 70
- Activity Bank: Enrichment 62
- Study Guide: Section 9-4
- Problem Bank: Problem Set 19
- Explorations Lab Manual: Diagram Master 2
- Assessment Book: Quiz 9-4, Test 36

Answers to Exploration

1. $\begin{bmatrix} -9 & 0 & 9 \\ 0 & 9 & 0 \end{bmatrix}$

2. 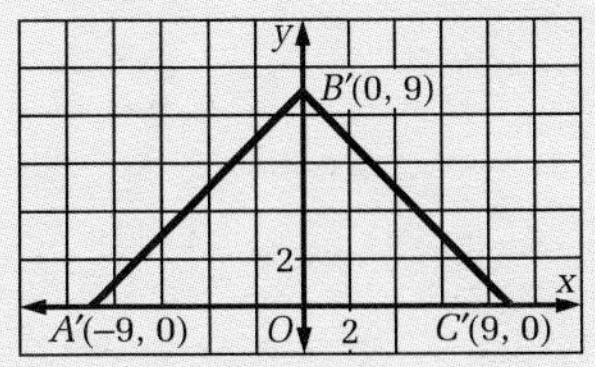

The graph is stretched horizontally by a factor of 3.

3. See answers in back of book.

4. The transformation stretches the graph horizontally by a factor of a and vertically by a factor of b.

5. $y = -x^2 + 9$

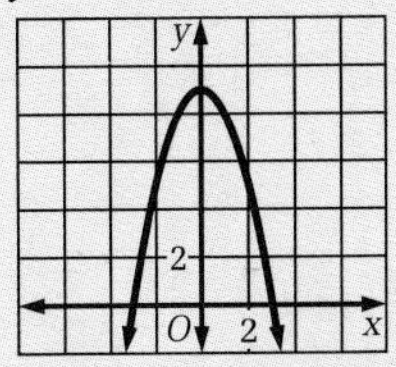

6. See answers in back of book.

TEACHING

Exploration

This Exploration has two goals. The first is to introduce students to the transformation matrix and have them discover the effects of this matrix. The second goal is for students to discover how the equation and graph of a transformed parabola are related to the original parabola.

Communication: Drawing

When sketching the graph of a vertically stretched function, students should draw the parent function first. Then the points on the x-axis will be the same for the transformed graph. Next, they can apply the stretch to the maximum and minimum values of the parent function to produce the maximum and minimum values of the transformed function. The rest of the graph can then be sketched easily.

Additional Sample

S1 The parent function $y = -2|x + 3|$ and the function $y = -2\left|\frac{x}{2} + 3\right|$ are graphed on the same set of axes.

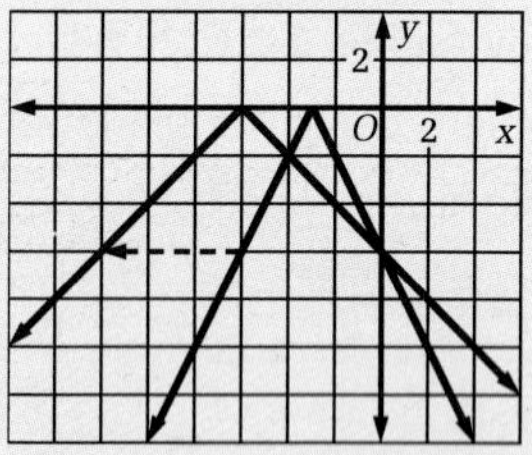

Describe the relationship between the two graphs.

The parent graph can be stretched horizontally by a factor of 2 to transform it into the graph of $y = -2\left|\frac{x}{2} + 3\right|$. The point (0, −6), where the parent graph crosses the y-axis, remains fixed in the new graph.

When you transformed the coordinates of the triangles in the Exploration, the triangles stretched.

$$\begin{bmatrix} a & 0 \\ 0 & b \end{bmatrix}\begin{bmatrix} x_1 & x_2 \\ y_1 & y_2 \end{bmatrix} = \begin{bmatrix} ax_1 + 0y_1 & ax_2 + 0y_2 \\ 0x_1 + by_1 & 0x_2 + by_2 \end{bmatrix} = \begin{bmatrix} ax_1 & ax_2 \\ by_1 & by_2 \end{bmatrix}$$

The segment with endpoints (x_1, y_1) and (x_2, y_2) is stretched by a factor of a in the horizontal direction.

The segment is stretched by a factor of b in the vertical direction.

You can stretch the graphs of functions as well.

STRETCHING GRAPHS

When the graph of $y = f(x)$ is stretched horizontally by the positive factor a,

(1) all points where the original graph crosses the y-axis remain fixed, and

(2) an equation of the new graph is $y = f\left(\frac{x}{a}\right)$.

When the graph of $y = f(x)$ is stretched vertically by the positive factor b,

(1) all points where the original graph crosses the x-axis remain fixed, and

(2) an equation of the new graph is $\frac{y}{b} = f(x)$.

Sample 1

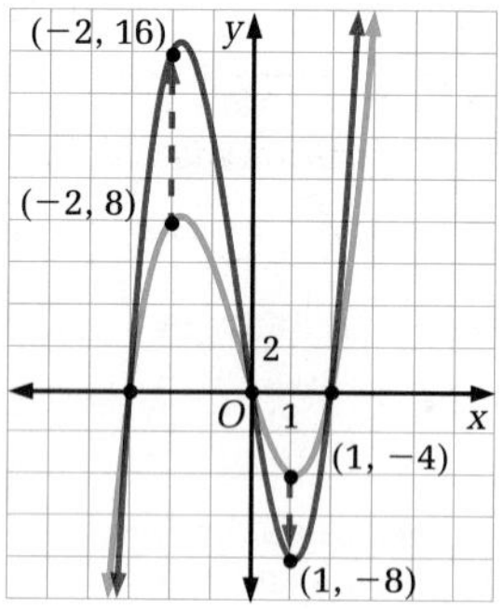

Writing The parent function $y = x(x - 2)(x + 3)$ and the function $y = 2x(x - 2)(x + 3)$ are graphed on the same set of axes. Describe the relationship between the two graphs using the language of transformations.

Sample Response

The function $y = 2x(x - 2)(x + 3)$ is related to the parent function in this way:

$$\frac{y}{2} = x(x - 2)(x + 3) = f(x)$$

The parent graph can be stretched vertically by the factor 2 to transform it into the graph of $y = 2x(x - 2)(x + 3)$.

The points where the parent graph crosses the x-axis remain fixed in the new graph: (0, −3), (0, 0), and (0, 2).

Talk it Over

1. Use the parent function $y = x(x-2)(x+3)$ from Sample 1.
   a. Find the stretch factor and the direction of stretch for the graph of the function $y = \frac{x}{2}\left(\frac{x}{2} - 2\right)\left(\frac{x}{2} + 3\right)$.
   b. Sketch the graph of the function in part (a).
2. Suppose you stretch the graph of the parent function horizontally by the factor $\frac{1}{2}$ and vertically by the factor 3.
   a. Find an equation of the new graph.
   b. Which points on the original graph will remain fixed?

Watch Out!
When the stretch factor in a given direction is between 0 and 1, the graph *shrinks* in that direction.

Sample 2

Andy Martinez manages a parking lot. He originally charged \$3 for the first hour or fraction of an hour plus \$2 for each additional hour or fraction of an hour. He is increasing his rates.

a. Use the graphs. Describe how the rates are changing.

b. Suppose the function that represents the original rates is $y = f(x)$. Write a function that represents the new rates.

Original Rates

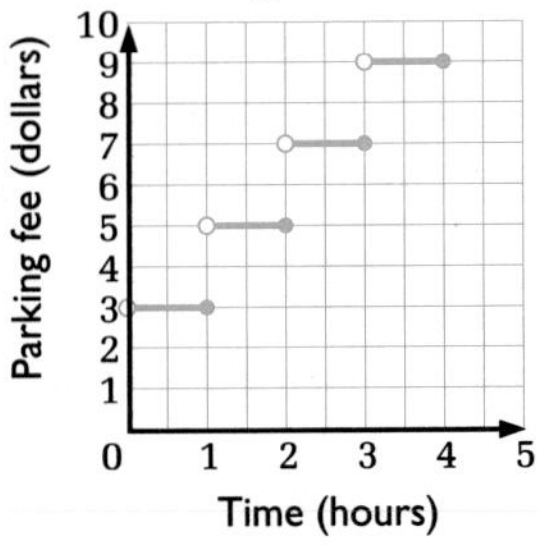

New Rates

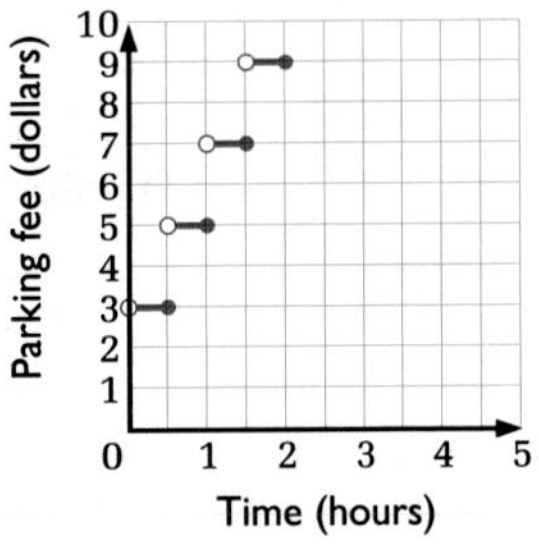

Sample Response

a. The time intervals are shrinking by 50%, to half hours instead of hours, but the fees per interval stay the same.

b. The graph shrinks only in the horizontal direction, so $y = f\left(\frac{x}{a}\right)$. Since $a = \frac{1}{2}$, $y = f\left(\frac{x}{\frac{1}{2}}\right)$. This can also be written $y = f(2x)$.

Look Back

For what values of a is the graph of $\frac{y}{b} = f\left(\frac{x}{a}\right)$ produced by stretching the graph of $y = f(x)$? In what direction?

Now you are ready for: Exs. 16–46 on pp. 505–509

Students Acquiring English

Some students acquiring English may have difficulty understanding what *stretch* means in this context, especially when "stretching" a graph by any number between 0 and 1 shrinks the graph. You may want to make sure that all students understand this use of *stretch*, since stretching is an important part of several Unit 9 sections.

Additional Sample

S2 A bicycle rental shop charges a basic fee and adds an additional charge based on the time a customer has the bicycle. During the warmer months, the rates change to take into account increased demand. The graph shows the rates for October through April and for May through September.

Bicycle Rental Rates

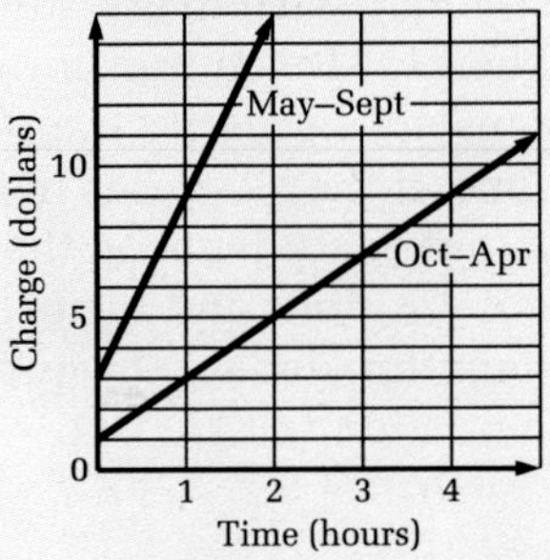

a. Use the graphs. Describe how the rates change for the warmer months.
The change in y-intercept indicates that the basic fee has increased by a factor of 3. The change in slope indicates the same change in the hourly rate.

b. Suppose the function that represents the October through April charges is $y = f(x)$. Write a function that represents the May through September charges.
The graph is stretched only in the vertical direction, so the function can be represented by $y = 3f(x)$.

Answers to Talk it Over

1. a. The parent graph is stretched horizontally by a factor of 2.
   b. See graph at right
2. a. $\frac{y}{3} = 2x(2x-2)(2x+3)$
   b. the origin

1. b.

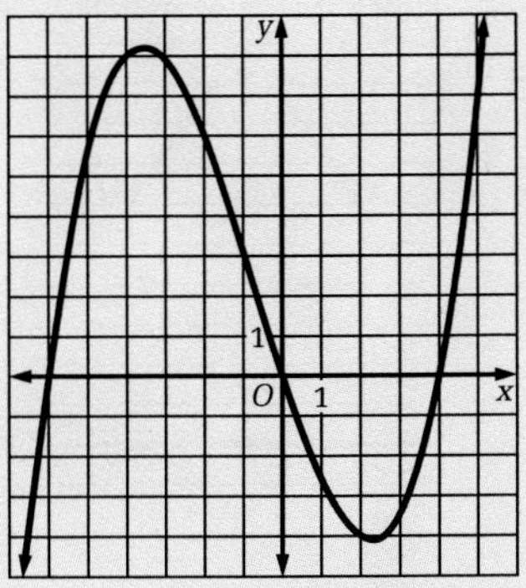

Answers to Look Back

for positive values of a; horizontally

APPLYING

Suggested Assignment

Day 1

Standard 1–6, 9–15

Extended 1–15

Day 2

Standard 16, 17, 23–30, 34–37, 39–46

Extended 16, 17, 22–37, 39–46

Integrating the Strands

Algebra Exs. 13–15, 17, 22, 30–33, 38

Functions Exs. 17–38, 44–46

Geometry Exs. 1–4, 7, 9–15, 30

Trigonometry Exs. 40–43

Statistics and Probability Ex. 39

Discrete Mathematics Exs. 1–14, 39

Logic and Language Exs. 7, 16

Reasoning

You may wish to ask students whether stretch transformations are similarity transformations. If the answer is not apparent to students on the basis of definitions and logic, have them think about the question in relation to Exs. 1–4.

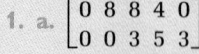

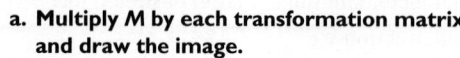

9-4 Exercises and Problems

Matrix M represents the polygon $OPQRS$.

$$M = \begin{array}{c} \begin{matrix} O & P & Q & R & S \end{matrix} \\ \begin{bmatrix} 0 & 8 & 8 & 4 & 0 \\ 0 & 0 & 6 & 10 & 6 \end{bmatrix} \end{array}$$

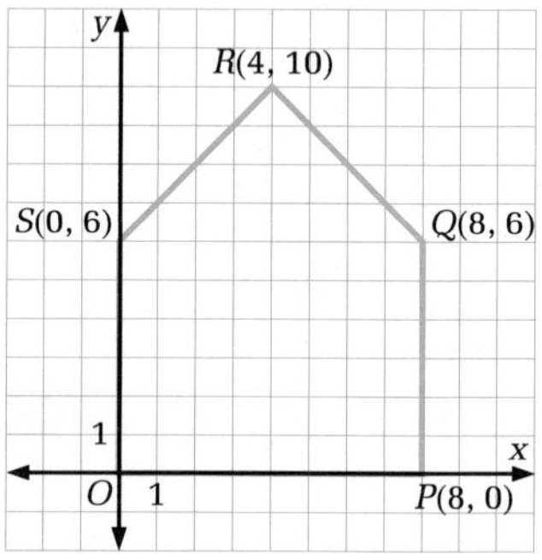

a. Multiply M by each transformation matrix and draw the image.

b. Describe the effect of each transformation on the polygon.

1. $\begin{bmatrix} 1 & 0 \\ 0 & \frac{1}{2} \end{bmatrix}$

2. $\begin{bmatrix} \frac{3}{2} & 0 \\ 0 & 1 \end{bmatrix}$

3. $\begin{bmatrix} 1 & 0 \\ 0 & 2 \end{bmatrix}$

4. $\begin{bmatrix} \frac{3}{2} & 0 \\ 0 & 2 \end{bmatrix}$

For each pair of matrices in Exercises 5 and 6:

a. Describe the transformation represented by each matrix.

b. Find the product matrix and describe the transformation it represents.

5. $\begin{bmatrix} 4 & 0 \\ 0 & 1 \end{bmatrix}\begin{bmatrix} 1 & 0 \\ 0 & 3 \end{bmatrix}$

6. $\begin{bmatrix} 1 & 0 \\ 0 & \frac{1}{3} \end{bmatrix}\begin{bmatrix} \frac{3}{2} & 0 \\ 0 & 1 \end{bmatrix}$

7. Open-ended Draw a polygon on a coordinate plane and use a matrix to represent its coordinates. Then choose two 2×2 matrices, one representing a horizontal stretch and one a vertical stretch. Find the image of your polygon after each of the transformations represented by your matrices.

8. a. Identify the set of points that remain unchanged after the transformation represented by $T = \begin{bmatrix} a & 0 \\ 0 & 1 \end{bmatrix}$, $a > 0$.

b. Identify the set of points that remain unchanged after the transformation represented by $T = \begin{bmatrix} 1 & 0 \\ 0 & b \end{bmatrix}$, $b > 0$.

Matrix M represents $\triangle ABC$ shown on page 501. Apply each transformation to matrix M. Then describe the effect.

$$M = \begin{array}{c} \begin{matrix} A & B & C \end{matrix} \\ \begin{bmatrix} -3 & 0 & 3 \\ 0 & 9 & 0 \end{bmatrix} \end{array}$$

9. $\begin{bmatrix} -3 & 0 \\ 0 & 1 \end{bmatrix}$

10. $\begin{bmatrix} 1 & 0 \\ 0 & -\frac{1}{2} \end{bmatrix}$

11. $\begin{bmatrix} -3 & 0 \\ 0 & -\frac{1}{2} \end{bmatrix}$

12. $\begin{bmatrix} 1 & 0 \\ 0 & -3 \end{bmatrix}$

Answers to Exercises and Problems

1. a. $\begin{bmatrix} 0 & 8 & 8 & 4 & 0 \\ 0 & 0 & 3 & 5 & 3 \end{bmatrix}$

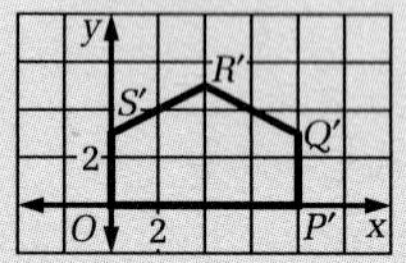

b. The polygon is stretched vertically by a factor of $\frac{1}{2}$.

2. a. $\begin{bmatrix} 0 & 12 & 12 & 6 & 0 \\ 0 & 0 & 6 & 10 & 6 \end{bmatrix}$

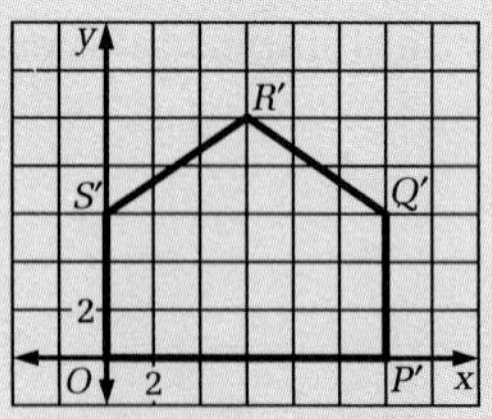

b. The polygon is stretched horizontally by a factor of $\frac{3}{2}$.

3. a. $\begin{bmatrix} 0 & 8 & 8 & 4 & 0 \\ 0 & 0 & 12 & 20 & 12 \end{bmatrix}$

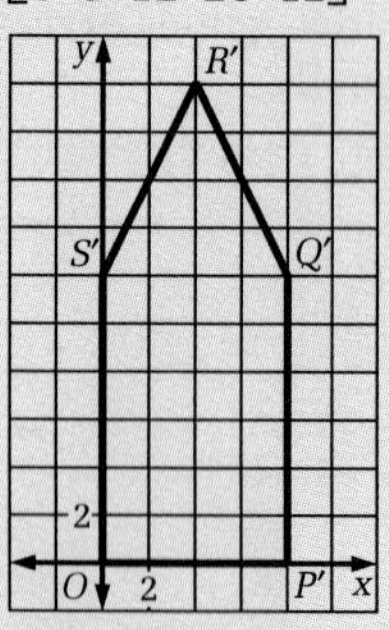

b. The polygon is stretched vertically by a factor of 2.

4. a. $\begin{bmatrix} 0 & 12 & 12 & 6 & 0 \\ 0 & 0 & 12 & 20 & 12 \end{bmatrix}$

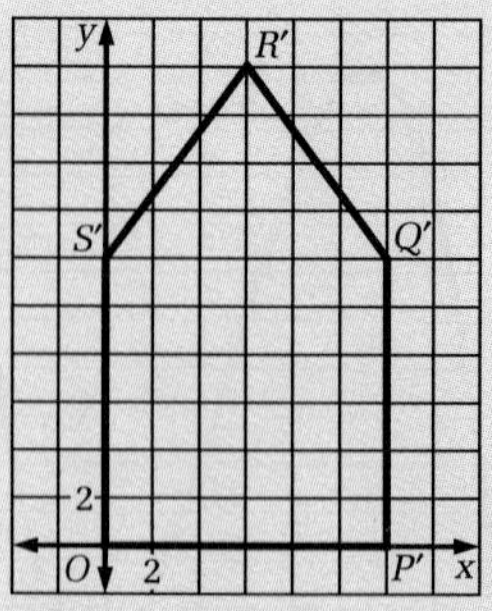

b. The polygon is stretched horizontally by a factor of $\frac{3}{2}$ and vertically by a factor of 2.

5–12. See answers in back of book.

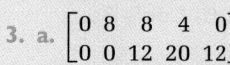

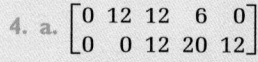

For Exercises 13–15, use matrix *M* and the circle.

A stretched circle is called an *ellipse.*

$$M = \begin{bmatrix} & A & B & C & D & E & F \\ & 5 & 4 & 0 & -4 & -5 & 0 \\ & 0 & 3 & 5 & 3 & 0 & -5 \end{bmatrix}$$

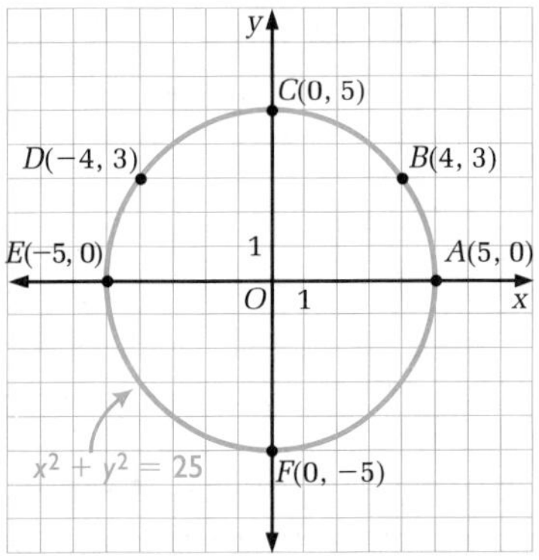

13. a. Find the image of matrix *M* after the transformation represented by the matrix $\begin{bmatrix} 1 & 0 \\ 0 & 2 \end{bmatrix}$.

b. Plot the image points and draw the ellipse formed by this transformation.

c. Show that the image points are solutions of this equation:

$$x^2 + \left(\frac{y}{2}\right)^2 = 25$$

14. a. Find the image of matrix *M* after the transformation represented by the matrix $\begin{bmatrix} 3 & 0 \\ 0 & 1 \end{bmatrix}$.

b. Plot the image points and draw the ellipse formed by this transformation.

c. Show that the image points are solutions of this equation:

$$\left(\frac{x}{3}\right)^2 + y^2 = 25$$

15. Sketch the graph of the ellipse whose equation is $\left(\frac{x}{3}\right)^2 + \left(\frac{y}{2}\right)^2 = 25$.

16. Reading What set of points remains fixed when you stretch a graph horizontally by the factor *a*? vertically by the factor *b*?

17. For the situation in Sample 2 on page 503, suppose Andy Martinez had decided to double his fees for each hour instead of shrinking the time intervals by 50%.

a. Sketch a graph to represent the new rates.

b. Suppose the function that represents the original rates is $y = f(x)$. Write a function that represents the new rates.

For Exercises 18–21, use the parent function shown.

a TECHNOLOGY Use a graphics calculator or graphing software to graph each function on the same set of axes as the parent function.

b. Describe the transformation that transforms the parent graph into the graph of each function.

18. $y = 3x(x - 2)(x + 2)$

19. $y = \frac{1}{3}x(x - 2)(x + 2)$

20. $y = 3x(3x - 2)(3x + 2)$

21. $y = \frac{x}{3}\left(\frac{x}{3} - 2\right)\left(\frac{x}{3} + 2\right)$

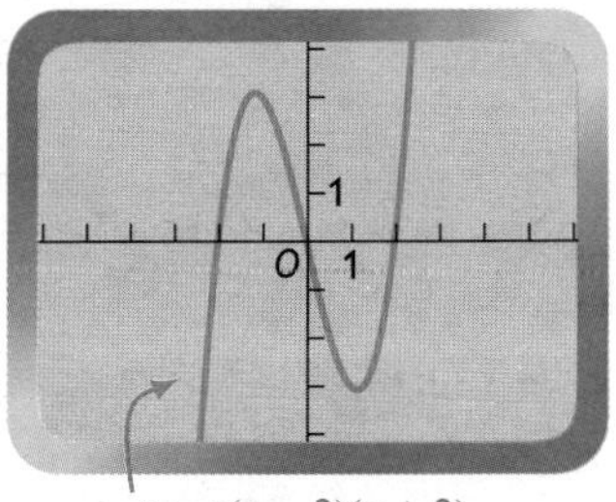

$y = x(x - 2)(x + 2)$

Communication: Writing

Exs. 1–6 and 9–12 require students to write a description of the effects of various transformations. You may wish to have some students read their descriptions to the class. This will provide an excellent review of the basic ideas of this section and an opportunity to discuss them further.

Answers to Exercises and Problems

13. a. $\begin{bmatrix} 5 & 4 & 0 & -4 & -5 & 0 \\ 0 & 6 & 10 & 6 & 0 & -10 \end{bmatrix}$

b.

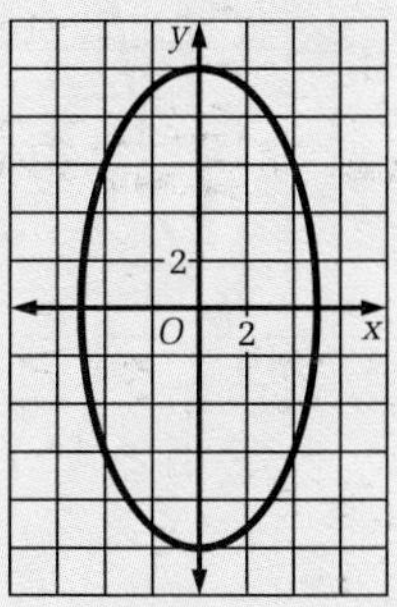

c. $(5, 0)$: $5^2 + \left(\frac{0}{2}\right)^2 = 25$;

$(4, 6)$: $4^2 + \left(\frac{6}{2}\right)^2 =$

$16 + 9 = 25$;

$(0, 10)$: $0^2 + \left(\frac{10}{2}\right)^2 = 25$;

$(-4, 6)$: $(-4)^2 + \left(\frac{6}{2}\right)^2 =$

$16 + 9 = 25$;

$(-5, 0)$: $(-5)^2 + \left(\frac{0}{2}\right)^2 = 25$;

$(0, -10)$: $0^2 + \left(\frac{-10}{2}\right)^2 = 25$

14, 15. See answers in back of book.

16. all points where the original graph intersects the *y*-axis; all points where the original graph intersects the *x*-axis

17. a.

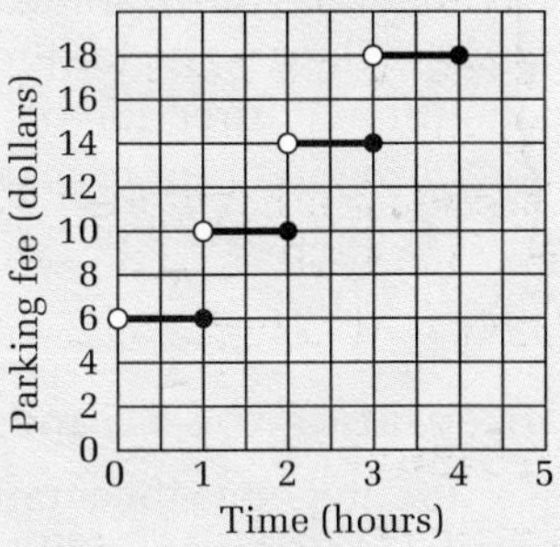

b. $\frac{y}{2} = f(x)$ or $y = 2(f(x))$

18. a.

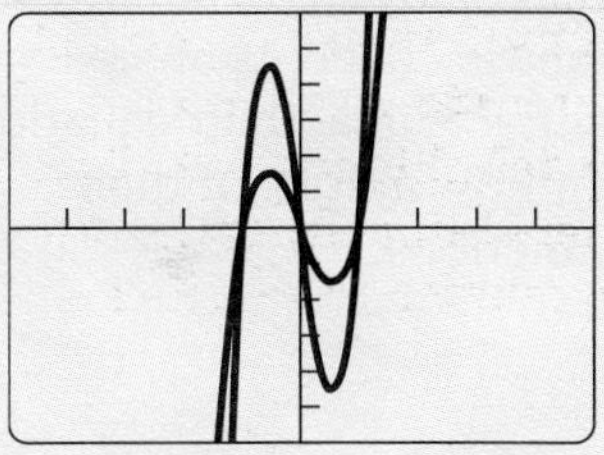

b. The graph is stretched vertically by a factor of 3.

19. a.

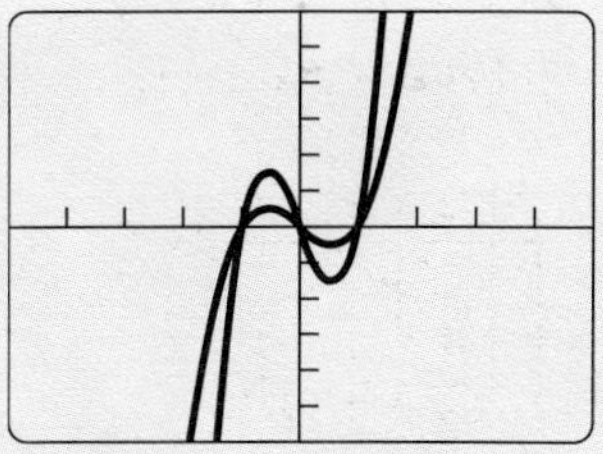

b. The graph is stretched vertically by a factor of $\frac{1}{3}$.

20. a.

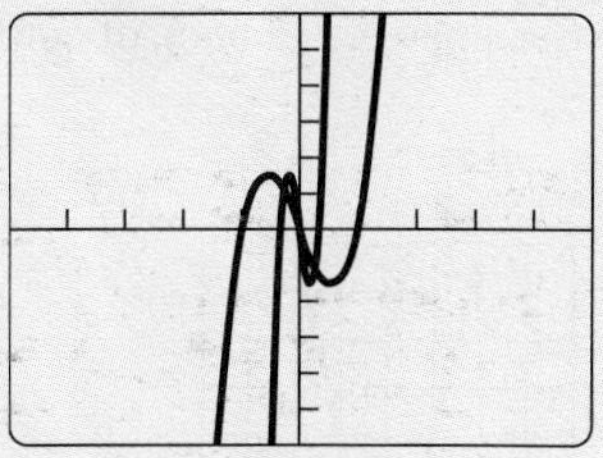

b. The graph is stretched horizontally by a factor of $\frac{1}{3}$.

21. See answers in back of book.

Application

The rates imposed by businesses who charge customers a per unit fee, such as the telephone company in Ex. 22 and the parking lot in Sample 2, can often be modeled by a step function. Have students compare and contrast these two problems. Why does Sample 2 lead to a horizontal stretch and Ex. 22 lead to a vertical stretch? Ask students to give examples of other real-world situations that can be modeled by step functions.

Reasoning

Exs. 23–26 require students to analyze the graphs for the amount and direction of stretching in order to make a correct match with an equation.

Integrating the Strands

Ex. 30 integrates the strands of algebra, geometry, and functions. Students can easily see the relationship among the geometric concepts of area and surface area, their equations, and the graphs of their functions.

22. **Telephone Rates** Between 8:00 A.M. and 5:00 P.M. a telephone company charges these weekday rates for a long distance call made between Boston and Worcester:

    $.29 for the first minute (or fraction of a minute)

    $.12 for each additional minute (or fraction of a minute)

    a. Let t = the length of a call in minutes. Let c = the cost of a call in dollars. Sketch a graph of the relationship between c and t for values of t in the domain $0 \le t \le 10$.

    b. Evening rates apply between 5:00 P.M. and 11:00 P.M. They are 60% of the weekday rates. Sketch a graph of the evening rates between the two cities on the same set of axes as in part (a).

    c. Describe the relationship between the graphs you drew in parts (a) and (b) using the language of transformations.

    d. Suppose the function that represents the weekday rates is $y = f(x)$. Write a function that represents the evening rates.

For Exercises 23–26, the parent function is $y = -x^2 + 4$. Match each transformed graph with its equation.

23.
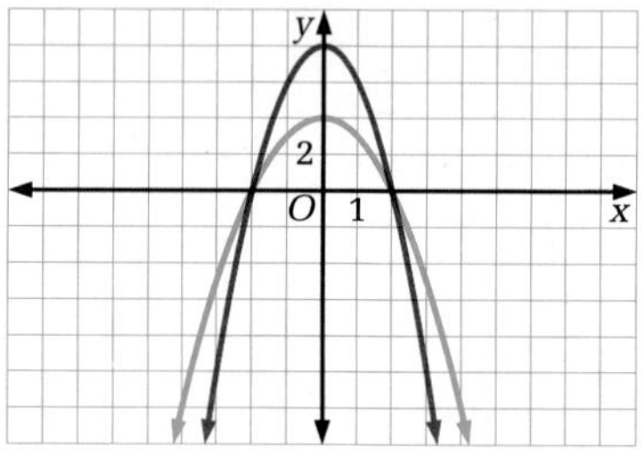

24.
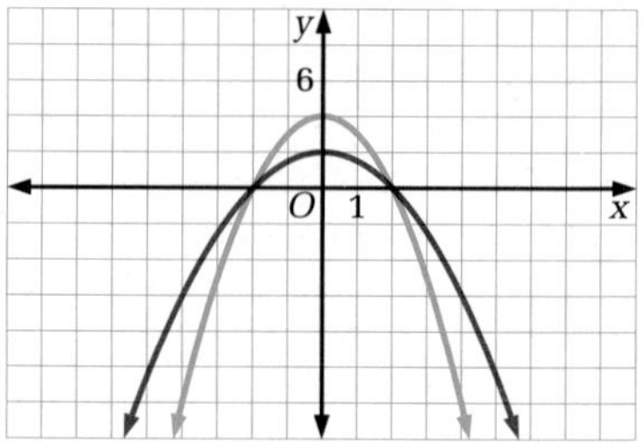

25.
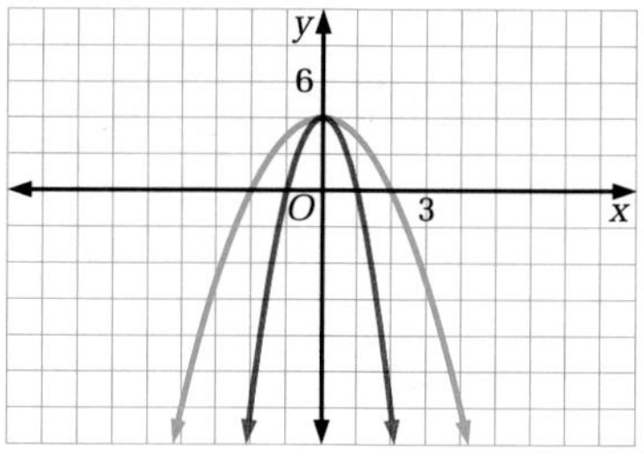

26.
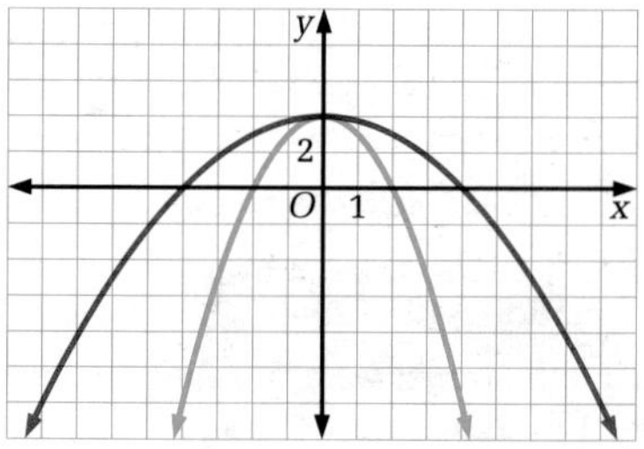

A. $\frac{y}{2} = -x^2 + 4$ **B.** $y = -\left(\frac{x}{2}\right)^2 + 4$ **C.** $\frac{y}{\frac{1}{2}} = -x^2 + 4$ **D.** $y = -\left(\frac{x}{\frac{1}{2}}\right)^2 + 4$

Without using graphing technology, sketch the graph of each function.

27. $y = 4|x|$ 28. $y = 10^{x/3}$ 29. $y = 5 \log x$

30. Let x represent the length of a side of a square and an edge of a cube.

    a. Graph the area of the square as a function of x.

    b. On the same set of axes, graph the surface area of the cube as a function of x.

    c. **Writing** Describe the relationship between the two graphs using the language of transformations.

Answers to Exercises and Problems

22. See answers in back of book.

23. A 24. C

25. D 26. B

27.
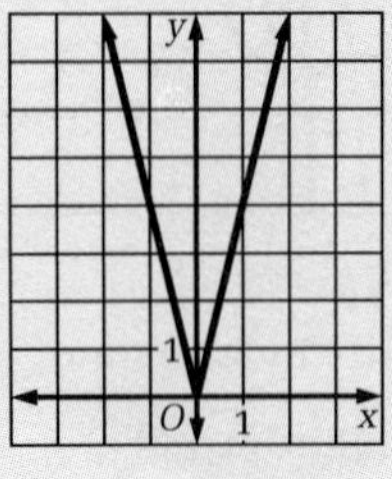

28.
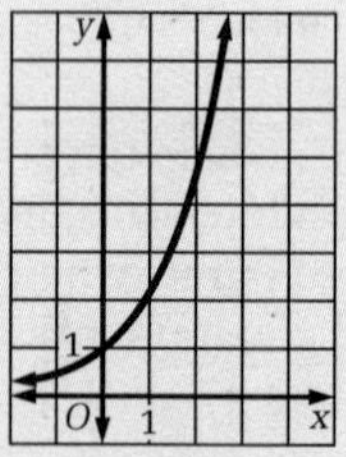

29.
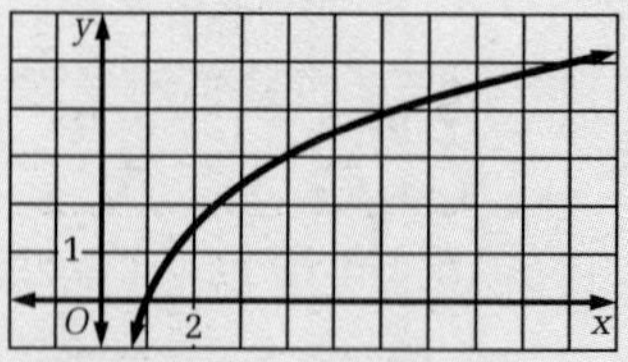

30. a, b.
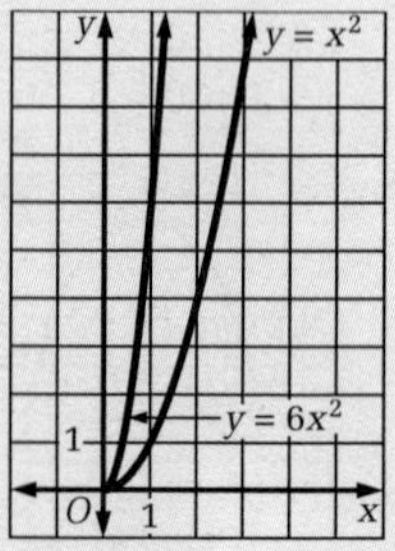

Drama A community theater company is presenting a musical. The manager wants to know how the ticket price will affect the profit. Here is the general form of the profit equation.

$$\text{profit} = \left(\text{number of shows} \times \text{average size of audience} \times \text{ticket price}\right) - \text{costs}$$

MUSICAL BUDGET	
ITEM	COST IN DOLLARS
orchestra	800
sound system	400
publicity	300
director	500
lighting/set	600
costumes	200
rental of rehearsal materials	630
royalties (5 shows at $300 each)	1500
miscellaneous	70
TOTAL	5000

The maximum seating capacity of the theater is 100 people. The theater company plans to do five shows.

31. In the worst case, suppose the average audience size is 50 people. Let x = the ticket price. Let y = the profit.

a. Use the information above and in the table to write a profit equation for this situation.

b. Graph your equation.

c. What ticket price must the company charge in order to break even?

32. Repeat Exercise 31 using an average audience size of 100 people, which is the best case.

33. Suppose the theater company wants to reserve 5% of the price of each ticket for a scholarship fund. How will this affect the equations, the graphs, and the break-even points for the best and worst case situations?

Career Note

There are many jobs and careers associated with the theater. In addition to the performers on the stage, the musical budget for Exs. 31–33 shows that many others are also involved in a production. There are the musicians in the orchestra, the director, and the various people that do the behind-the-scenes work, such as handling the sound, lighting, and costumes. Other individuals create the advertising and publicity, sell tickets, write and publish the program for the play, and rent rehearsal material. A business manager creates the budget, pays bills, and handles other financial matters. All of these jobs require special skills and offer rewarding careers in the theater.

Answers to Exercises and Problems

c. Let the graph of area of a square as a function of length of a side ($y = x^2$) be the parent graph. The graph of surface area of a cube as a function of length of a side ($y = 6x^2$) is the parent graph stretched vertically by a factor of 6.

31. a. $y = 250x - 5000$

b.

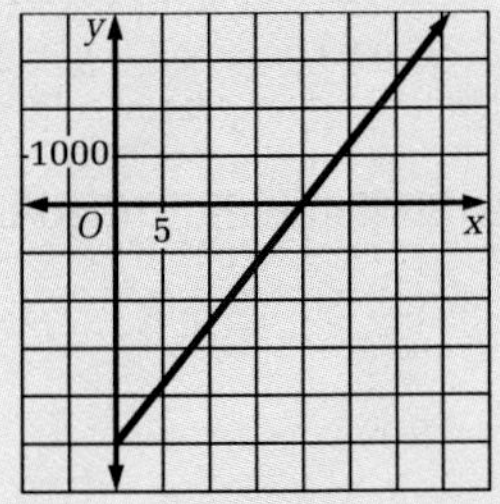

c. $20

32. a. $y = 500x - 5000$

b.

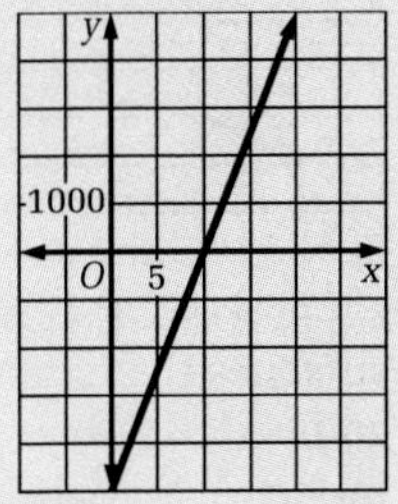

c. $10

33. In both equations, $0.95x$ will be substituted for x. Both graphs will be stretched horizontally by a factor of $\frac{1}{0.95} \approx 1.05$. Both break-even points will be increased by a factor of about 1.05.

Communication: Drawing

Exs. 34–37 provide an opportunity for students to work with transformations of four common parent functions. Students should be able to recognize the correct parent function by looking at the general shape of the transformed graph.

Assessment: Portfolio

Students should examine the parent functions in their portfolios and investigate how the stretching factor affects their graphs. For example, for $f(x) = x^2$, investigate $\frac{f(x)}{2} = x^2$ and $f(x) = \left(\frac{x}{2}\right)^2$. Students should keep the transformed equations along with their graphs in their portfolios for future reference.

Write an equation for each graph. Assume that each graph is a transformation of the graph of one of these parent functions: $y = x^2$, $y = x^3$, $y = |x|$, $y = \sqrt{x}$, $y = \log x$, $y = 10^x$.

34.

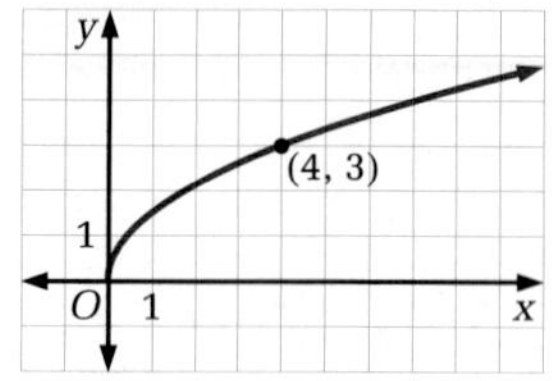

35.

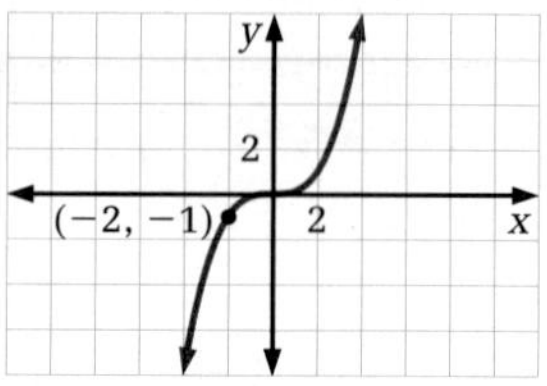

36.

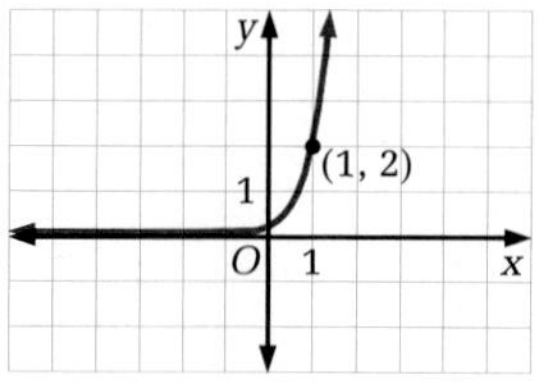

37.

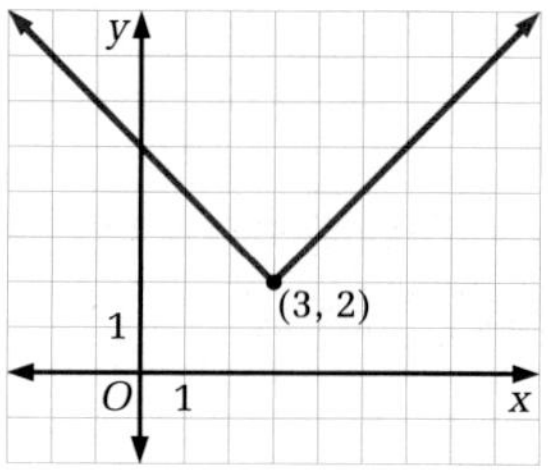

Ongoing **ASSESSMENT**

38. **Group Activity** Work with another student.
   a. Copy and complete the table.
   b. Use the values in the table to sketch the graphs of $y = \log x$ and $y = \log_2 x$ on the same set of axes.
   c. Describe the transformation that transforms the graph of $y = \log x$ into the graph of $y = \log_2 x$.
   d. Use the change-of-base property of logarithms to rewrite $y = \log_2 x$ in terms of $\log x$. Find the stretch factor to the nearest hundredth for the transformation in part (c).

x	$\log_2 x$	$\log x$
$\frac{1}{4}$	?	?
$\frac{1}{2}$	?	?
1	?	?
2	?	?
4	?	?
8	?	?

Review **PREVIEW**

39. The mean price of a pair of shoes at one store is \$30. The standard deviation is \$10. Suppose prices are reduced by 20%. How does this affect the mean and the standard deviation? *(Section 9-3)*

For each given angle, find another angle between 0° and 360° with the same cosine. *(Section 8-2)*

40. 50°   41. 135°   42. 240°   43. −25°

44. Write an equation for the graph shown. Assume that the shape matches the shape of the graph of $y = |x|$. *(Section 9-2)*

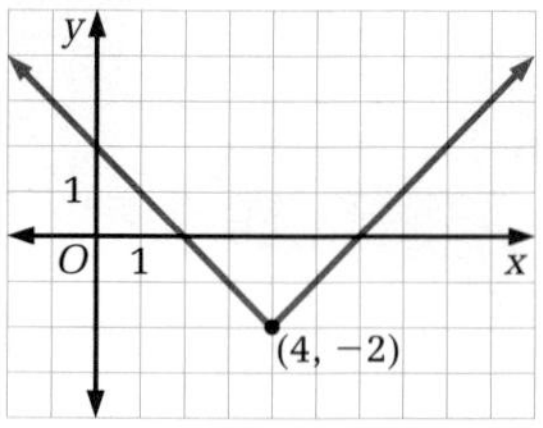

Answers to Exercises and Problems

34. $y = \frac{3}{2}\sqrt{x}$

35. $y = \left(\frac{x}{2}\right)^3$

36. $y = \frac{1}{5}10^x$

37. $y - 2 = |x - 3|$

38. a. Values of log x are given to four decimal places.

x	$\log_2 x$	$\log x$
$\frac{1}{4}$	−2	−0.6021
$\frac{1}{2}$	−1	−0.3010
1	0	0
2	1	0.3010
4	2	0.6021
8	3	0.9031

b.

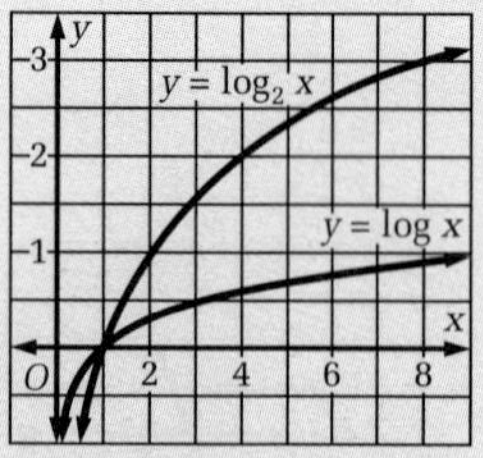

c. a transformation that stretches the parent graph vertically by a factor of $\frac{1}{\log 2}$, or about $3\frac{1}{3}$

d. $y = \log_2 x = \frac{\log x}{\log 2}$ or $\frac{y}{\frac{1}{\log 2}} = \log x$; The stretch factor is $\frac{1}{\log 2} \approx 3.32$.

39. Both are multiplied by 0.8.

40. 310°

41. 225°

42. 120°

43. 25° or 335°

44. $y + 2 = |x - 4|$

45. Stretch the graph horizontally by a factor of 34 and vertically by a factor of 1,444,000.

46. Stretch the graph horizontally by a factor of 6.6 and vertically by a factor of 280,000.

Working on the Unit Project

45. The number of farm operators and managers in the United States has been declining in recent years. The function $y = 1{,}444{,}000\left(\frac{1}{2}\right)^{x/34}$ approximates the number of jobs in this field, where x is the number of years since 1983. Describe how you can transform the graph of $y = \left(\frac{1}{2}\right)^x$ into the graph of this function.

46. The number of computer analysts/scientists in the United States has been growing in recent years. The function $y = 280{,}000 \cdot 2^{x/6.6}$ approximates the number of jobs in this field, where x is the number of years since 1983. Describe how you can transform the graph of $y = 2^x$ into the graph of this function.

Unit 9 CHECKPOINT

1. **Writing** Describe how adding a constant to each value in a data set affects the mean, the median, the range, and the standard deviation of the data set. Then describe how multiplying each data value by a positive constant affects these four statistics.

2. The mean age of workers at a company is 34. The standard deviation of the ages is 10. Suppose the same people worked for the company 5 years ago. What were the mean and the standard deviation of their ages then? 9-1

Without using graphing technology, sketch the graph of each equation. 9-2

3. $y - 2 = |x + 4|$ 4. $y + 1 = (x - 2)^2$ 5. $y = 10^{x-3}$

6. At an appliance store, the median price of small microwave ovens is $147.50. The range of the prices is $90. When you include the sales tax, the actual price of each oven is 5% higher. Find the median and the range of the actual prices with the sales tax. 9-3

Business Expenses Claudia Pyle travels for her job. Her gasoline mileage is about 24 mi/gal and she usually pays $1.20 per gallon. This is an average of $.05 per mile. 9-4

Let x = the number of miles she travels in a week. Then the cost function $y = 0.05x$ represents the amount she spends on gasoline in a week.

7. Suppose a tax increases the cost of gasoline by 5%. Find the new cost function.

8. Suppose Claudia buys a more fuel-efficient car that reduces her gasoline costs by 25%. Find the new cost function.

Working on the Unit Project

Some students may be curious about when the number of farm operators and managers will be equal to the number of computer analysts/scientists in the United States. Have them graph both functions on a graphics calculator and look for the intersection.

Quick Quiz (9-1 through 9-4)

See page 537.

See also Test 36 in the *Assessment Book.*

Practice 70 For use with Section 9-4

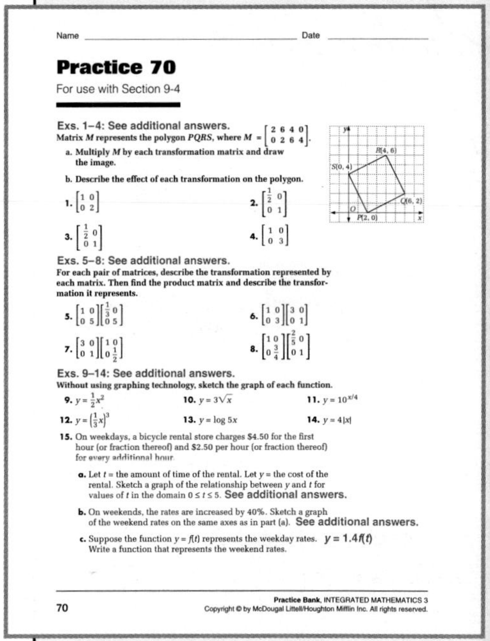

Name ______ Date ______

Practice 70

For use with Section 9-4

Exs. 1–4: See additional answers.

Matrix M represents the polygon $PQRS$, where $M = \begin{bmatrix} 2 & 6 & 4 & 0 \\ 0 & 2 & 6 & 4 \end{bmatrix}$.

a. Multiply M by each transformation matrix and draw the image.

b. Describe the effect of each transformation on the polygon.

1. $\begin{bmatrix} 1 & 0 \\ 0 & 2 \end{bmatrix}$ 2. $\begin{bmatrix} \frac{1}{2} & 0 \\ 0 & 1 \end{bmatrix}$

3. $\begin{bmatrix} \frac{1}{2} & 0 \\ 0 & 1 \end{bmatrix}$ 4. $\begin{bmatrix} 1 & 0 \\ 0 & 3 \end{bmatrix}$

Exs. 5–8: See additional answers.

For each pair of matrices, describe the transformation represented by each matrix. Then find the product matrix and describe the transformation it represents.

5. $\begin{bmatrix} 1 & 0 \\ 0 & 5 \end{bmatrix}\begin{bmatrix} \frac{1}{3} & 0 \\ 0 & 5 \end{bmatrix}$ 6. $\begin{bmatrix} 1 & 0 \\ 0 & 3 \end{bmatrix}\begin{bmatrix} 3 & 0 \\ 0 & 1 \end{bmatrix}$

7. $\begin{bmatrix} 3 & 0 \\ 0 & 1 \end{bmatrix}\begin{bmatrix} 1 & 0 \\ 0 & \frac{1}{2} \end{bmatrix}$ 8. $\begin{bmatrix} 1 & 0 \\ 0 & \frac{3}{4} \end{bmatrix}\begin{bmatrix} \frac{2}{5} & 0 \\ 0 & 1 \end{bmatrix}$

Exs. 9–14: See additional answers.

Without using graphing technology, sketch the graph of each function.

9. $y = \frac{1}{2}x^2$ 10. $y = 3\sqrt{x}$ 11. $y = 10^{x/4}$

12. $y = \left(\frac{1}{3}x\right)^3$ 13. $y = \log 5x$ 14. $y = 4|x|$

15. On weekdays, a bicycle rental store charges $4.50 for the first hour (or fraction thereof) and $2.50 per hour (or fraction thereof) for every additional hour.

a. Let t = the amount of time of the rental. Let y = the cost of the rental. Sketch a graph of the relationship between y and t for values of t in the domain $0 \le t \le 5$. See additional answers.

b. On weekends, the rates are increased by 40%. Sketch a graph of the weekend rates on the same axes as in part (a). See additional answers.

c. Suppose the function $y = f(t)$ represents the weekday rates. Write a function that represents the weekend rates. $y = 1.4f(t)$

70 **Practice Bank**, INTEGRATED MATHEMATICS 3

Answers to Checkpoint

1. When the same constant value is added to each value in a data set, the same value is added to the mean and the median. The range and standard deviation are unchanged. When each value in a data set is multiplied by the same positive constant, the mean, the median, the range, and the standard deviation are all multiplied by the same constant.

2. 29; 10

3.

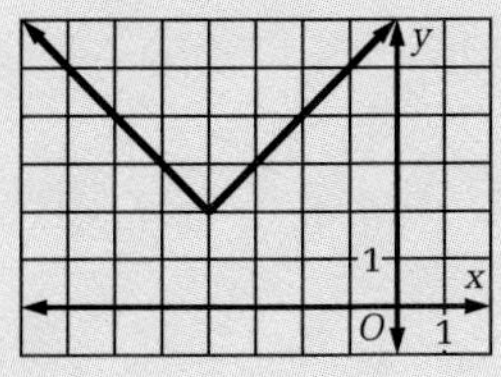

4.

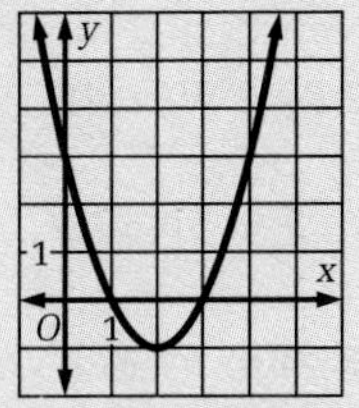

5.

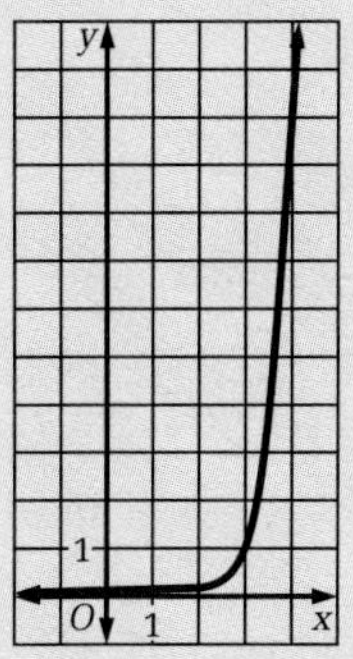

6. $154.88; $94.50

7. $\frac{y}{1.05} = 0.05x$ or $y = 0.0525x$

8. $\frac{y}{0.75} = 0.05x$ or $y = 0.0375x$

PLANNING

Objectives and Strands
See pages 480A and 480B.

Spiral Learning
See page 480B.

Materials List
- Graph paper

Recommended Pacing
Section 9-5 is a two-day lesson.

Day 1
Pages 510–511: Opening paragraph through Sample 1, *Exercises 1–9*

Day 2
Pages 512–514: Top of page 512 through Look Back, *Exercises 10–28*

Extra Practice
See pages 622–624.

Warm-Up Exercises
Warm-Up Transparency 9-5

Support Materials
- Practice Bank: Practice 71
- Activity Bank: Enrichment 63
- Study Guide: Section 9-5
- Problem Bank: Problem Set 20
- Explorations Lab Manual: Diagram Master 2
- Assessment Book: Quiz 9-5

Section 9-5

Multiple Transformations

Sliding Scale

Focus
Recognize the effects of multiple transformations on sets of data and graphs of functions.

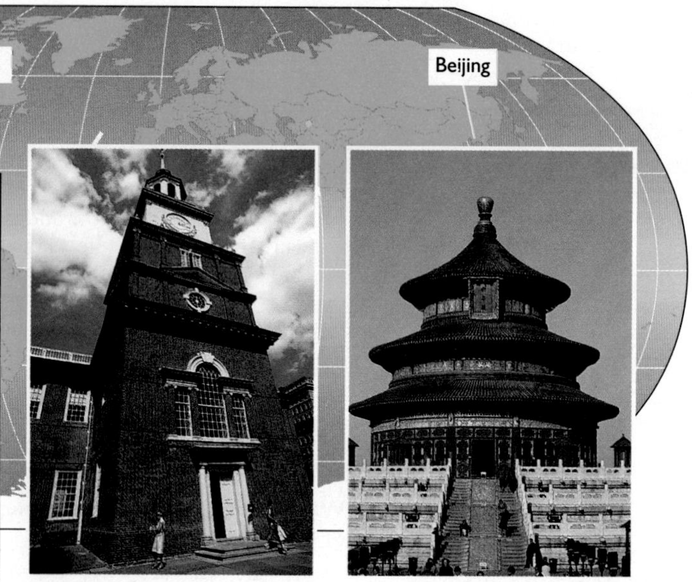

Average April Temperatures		
Year	Beijing (°C)	Phila. (°F)
1961	15.0	49.8
1962	13.3	52.0
1963	12.8	52.5
1964	11.8	50.7
1965	10.9	48.9
1966	12.6	47.8
1967	12.8	51.6
1968	14.2	54.7
1969	12.9	55.2
1970	13.6	51.4
1971	13.0	51.6
1972	13.8	49.6
1973	14.6	53.4
1974	13.6	55.8
1975	14.6	48.7
1976	11.4	56.7
1977	14.3	57.2
1978	14.0	50.5
1979	10.5	52.3
1980	10.5	54.7
Mean	13.0	52.3
Standard deviation	1.3	2.7

Philadelphia, Pennsylvania, and Beijing, China, are both located at a latitude of 40°N.

The table gives temperature data for the two cities over a period of 20 years.

Talk it Over

1. What do you need to know in order to compare the temperatures for the two cities?
2. Use the formula $C = \frac{5}{9}(F - 32)$. Find the mean April temperature in degrees Celsius for Philadelphia for the 20-year period. How does it compare to Beijing's mean April temperature for the period?
3. What factors besides latitude may account for variations in temperatures between different locations?

Answers to Talk it Over

1. how Celsius and Fahrenheit temperatures are related
2. about 11.3°C; slightly lower
3. Answers may vary. Examples are given. elevation, nearness to the ocean, prevailing wind patterns

The temperature conversion formula, $C = \frac{5}{9}(F - 32)$, is a combination of transformations.

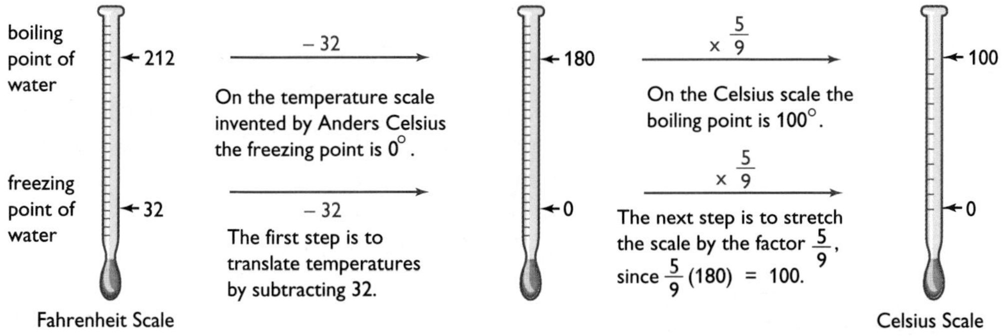

boiling point of water
212
freezing point of water
32
Fahrenheit Scale
– 32
On the temperature scale invented by Anders Celsius the freezing point is 0°.
– 32
The first step is to translate temperatures by subtracting 32.
180
0
× 5/9
On the Celsius scale the boiling point is 100°.
× 5/9
The next step is to stretch the scale by the factor 5/9, since 5/9 (180) = 100.
100
0
Celsius Scale

Sample 1

Writing Describe the effect of the transformations from Fahrenheit to Celsius on the mean and the standard deviation of the Philadelphia temperatures.

Sample Response

Michael Von Dohlen

From the table on page 510, the mean of the Philadelphia temperatures is 52.3°F and the standard deviation is 2.7°F.

① Since translating and stretching both affect the mean, you can substitute into the formula to find the mean temperature in degrees Celsius.

$C = \frac{5}{9}(52.3 - 32)$

$\approx 11.3°C$

② Since only stretching affects the standard deviation, the standard deviation in degrees Celsius is $\frac{5}{9}$ of the standard deviation in degrees Fahrenheit.

$\frac{5}{9}(2.7) = 1.5°C$

These histograms show the effect of the transformation $C = \frac{5}{9}(F - 32)$ on the distribution of April temperatures for Philadelphia.

Average Temperature for April, 1961–1980
Philadelphia

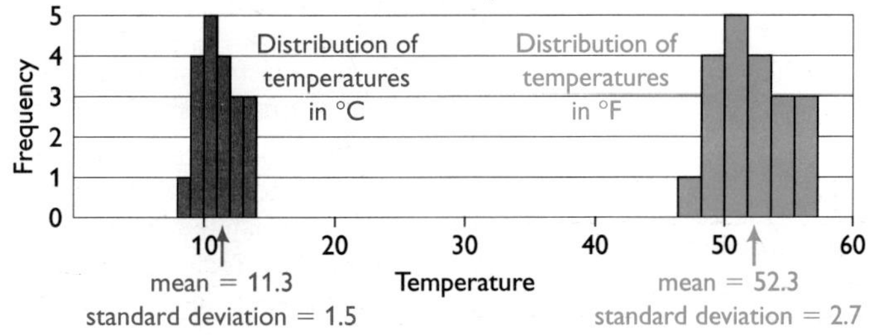

Now you are ready for: Exs. 1–9 on pp. 514–516

TEACHING

Talk it Over

For question 2, most students will probably convert each temperature to Celsius and then calculate the mean. Some students, however, may realize that they can convert the mean itself. This method leads into the discussion of Sample 1.

Additional Sample

S1 Karla Nashat has lunch at the same restaurant every Friday. She leaves a 15% tip for the waiter or waitress and a \$1.00 tip for the person who checks coats. She bases her 15% tip on the total bill for her meal (tax included). Suppose the mean of her lunch bills for one month before the tip is \$9.50, and suppose the standard deviation is \$.75. How does her method of tipping affect the mean and the standard deviation of her lunch bills for one month?

Let x be the bill for a meal. Then the equation $y = 1.15x + 1$ gives Karla's total expense for the meal including the 15% server tip and \$1.00 coat check tip. This equation is a combination of transformations: a stretch by a factor of 1.15 followed by a vertical translation of 1. Since translating and stretching both affect the mean, you can substitute the mean amount of the bill before a tip to find the mean amount of the total expenses.

$y = 1.15(9.50) + 1$
$= 11.925$

Her mean expense for one lunch is about \$11.93.

The standard deviation is affected only by the stretching that results from the 15% tip. So, the standard deviation for the expense totals is 1.15(.75), or about \$.86.

Additional Sample

S2 Transform the function $f(x) = x^3$ with a translation of 3 units to the left followed by a vertical stretch by a factor of $\frac{1}{4}$.

a. Write an equation for the transformed function.

The horizontal translation gives $y = (x + 3)^3$. The vertical stretch then gives $y = \frac{1}{4}(x + 3)^3$. The resulting function is $y = \frac{(x + 3)^3}{4}$.

b. Graph the transformed function.

Graph the parent function $y = x^3$. Then translate the graph to the left 3 units.

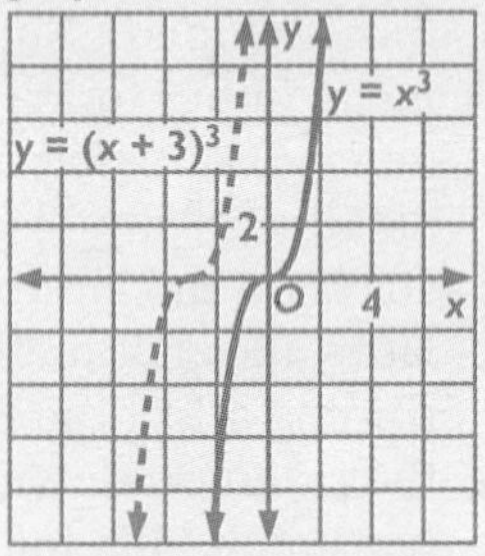

Next, stretch the translated graph by a factor of $\frac{1}{4}$.

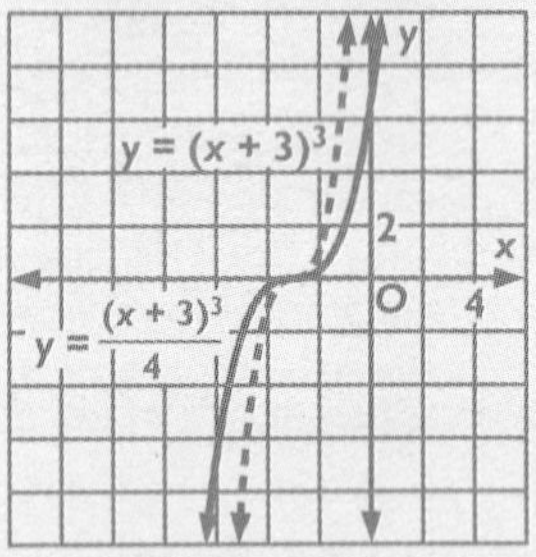

Multiple transformations can be performed on functions as well as on data sets.

Sample 2

Transform the function $f(x) = 2^x$ with a vertical stretch by a factor of 3 followed by a translation 4 units to the right.

a. Write an equation for the transformed function.

b. Graph the transformed function.

Sample Response

a. The vertical stretch gives $\frac{y}{3} = f(x)$ or $y = 3f(x)$.
The horizontal translation then gives $y = 3f(x - 4)$.
The resulting function is $y = 3(2^{x-4})$.

b. First graph the parent function $y = 2^x$.

Stretch the graph of $y = 2^x$ vertically by a factor of 3. This produces the graph of $y = 3(2^x)$.

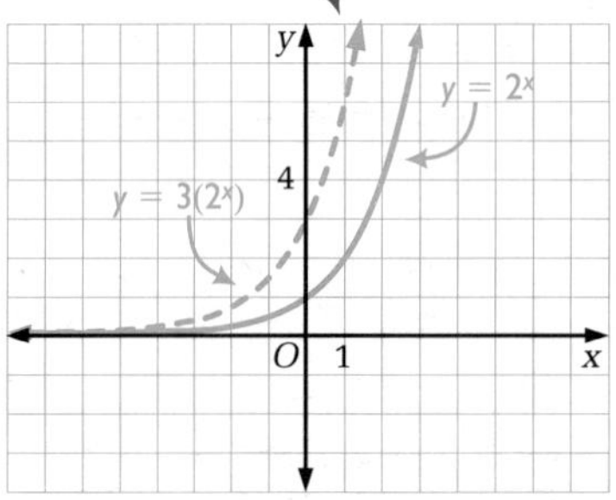

Translate the graph of $y = 3(2^x)$ 4 units to the right. This produces the graph of $y = 3(2^{x-4})$.

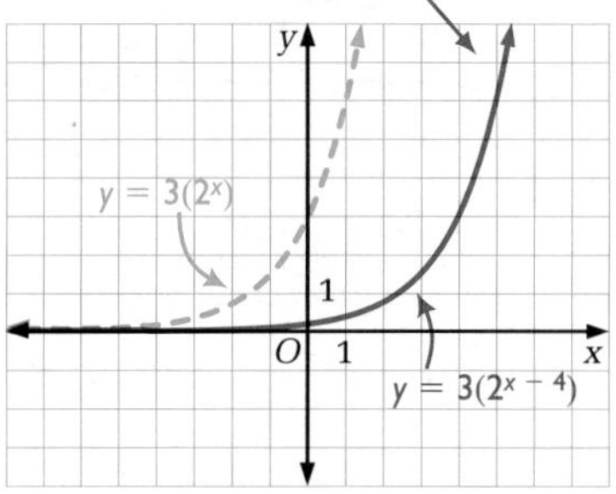

Talk it Over

4. Describe the effects of these transformations on the function $y = 2^x$:
   a vertical stretch by a factor of 2 followed by a translation 1 unit to the right
5. Find the function and sketch the graph when $y = x^3$ is first stretched by a factor of 1.5 in the horizontal direction and then translated 2 units up.

Answers to Talk it Over

4. The resulting function is $y = 2(2^{x-1}) = 2^x$. The graph is unchanged.

5. $y = \left(\frac{2}{3}x\right)^3 + 2$

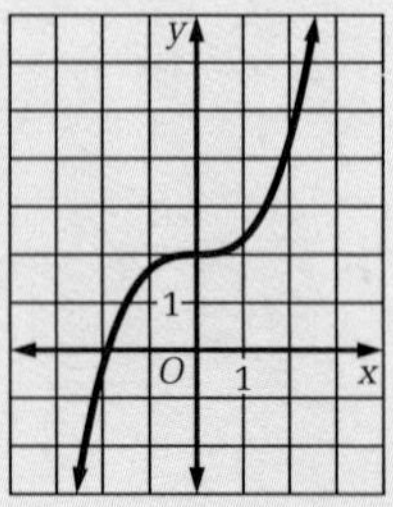

Sample 3

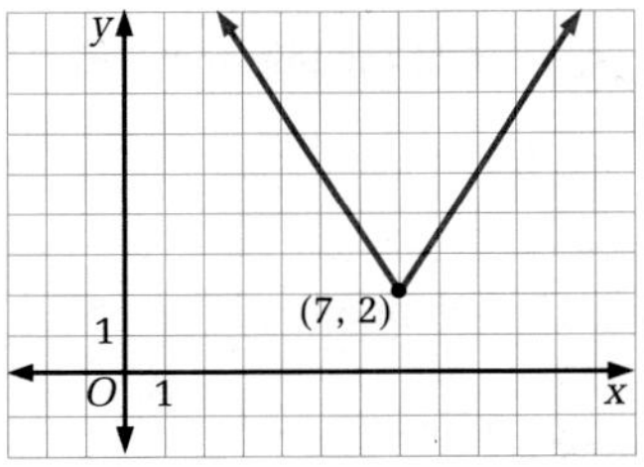

Write an equation of the graph. Assume that the parent function is one of the following:
$y = x^2$, $y = x^3$, $y = |x|$, $y = \sqrt{x}$, $y = \log x$, or $y = 10^x$

Sample Response

Step 1 Identify the parent graph.

The given graph looks like a transformation of the graph of the absolute value function, $y = |x|$.

Since the vertex of the parent graph is located at the origin, the vertex is affected only by translations.

Step 2 Find the stretch factor.

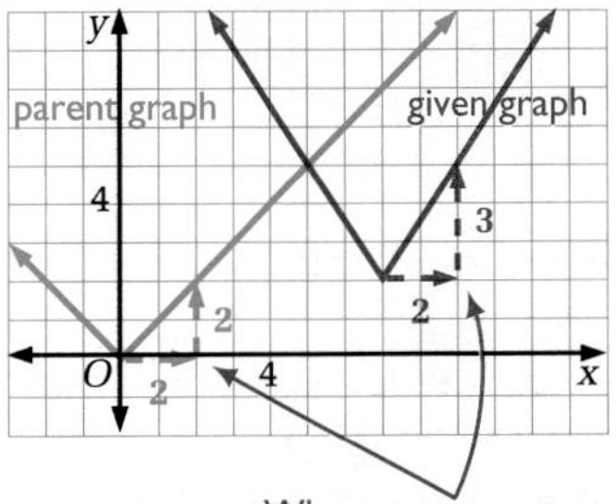

When you move two units to the right of the vertex, the given graph rises 3 units, while the parent graph rises 2 units. So the vertical stretch factor is $\frac{3}{2}$.

Step 3 Find the translation vector.

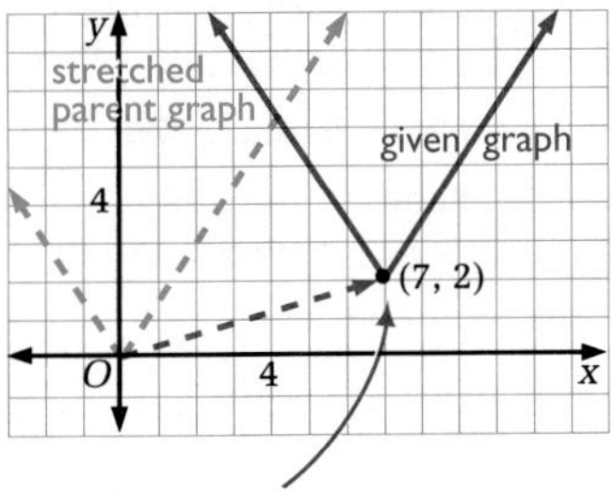

The vertex is translated by the vector (7, 2).

Step 4 Write an equation.

$$\frac{y-k}{b} = \left|\frac{x-h}{a}\right|$$

$$\frac{y-2}{\frac{3}{2}} = |x-7|$$

There is no horizontal stretch, so a is 1.
Substitute **2** for k, **7** for h, and $\frac{3}{2}$ for b.

$$y - 2 = \frac{3}{2}|x-7|$$

An equation of the graph is $y - 2 = \frac{3}{2}|x-7|$.

Additional Sample

S3 Write an equation of the graph below. Assume that the parent function is one of the following: $y = x^2$, $y = x^3$, $y = |x|$, $y = \sqrt{x}$, $y = \log x$, or $y = 10^x$.

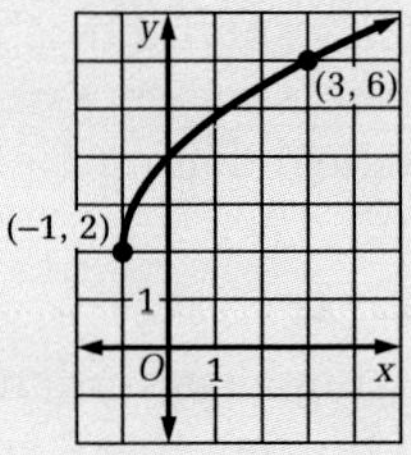

Step 1. **Identify the parent graph. The given graph looks like a transformation of the square root function, $y = \sqrt{x}$.**
Step 2. **Find the stretch factor.**

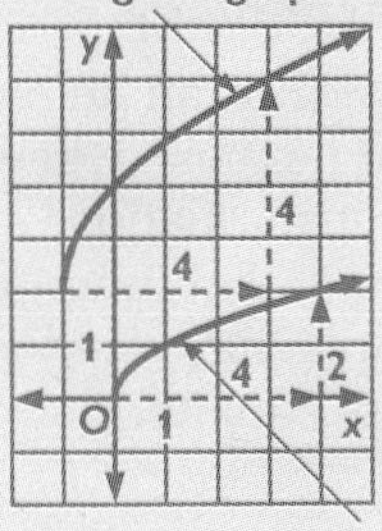

The leftmost point on the parent graph is (0, 0). If you move 4 units right on the x-axis and up 2 units, you are at (4, 2) on the parent graph. The leftmost point on the given graph is (–1, 2). If you move 4 units right and up to the graph, you rise 4 units and come to (3, 6). So, the vertical stretch factor is 2.
Step 3. **Find the translation vector.**

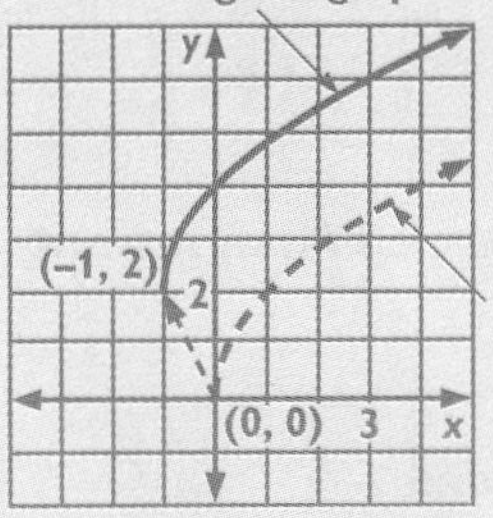

To move the leftmost point of the stretched parent graph to the leftmost point of the given graph, you use the translation vector (–1, 2).
Step 4. **Write an equation.**

$$\frac{y-k}{b} = \sqrt{\frac{x-h}{a}}$$

Since $b = 2$, $a = 1$, $h = -1$, and $k = 2$, an equation of the graph is $y - 2 = 2\sqrt{x+1}$.

Reasoning

After students complete Sample 3 on page 513, ask them if they think finding the translation vector before finding the stretch factor would affect the equation of the graph. They should be able to support their answers with a mathematical demonstration.

Talk it Over

Question 6 makes students aware of the fact that a vertical stretch factor can sometimes be replaced by a horizontal stretch factor to produce the same result.

Using Technology

Using Y-VARS can help display what happens when certain transformations are applied to parent functions. For example, have students enter these equations on the Y= list:

$Y_1=X^2$

$Y_2=0.5Y_1(X-3)$

For the second equation, the symbol Y_1 is accessed by pressing [2nd] [Y-VARS], selecting item 1 (Function) and then item 1 (Y_1) from the FUNCTION menu. After both equations have been entered, students can view their graphs and discuss how they are related.

APPLYING

Suggested Assignment

Day 1

Standard 1–4, 7–9

Extended 1–9

Day 2

Standard 10–16, 18–23, 25–28

Extended 10–23, 25–28

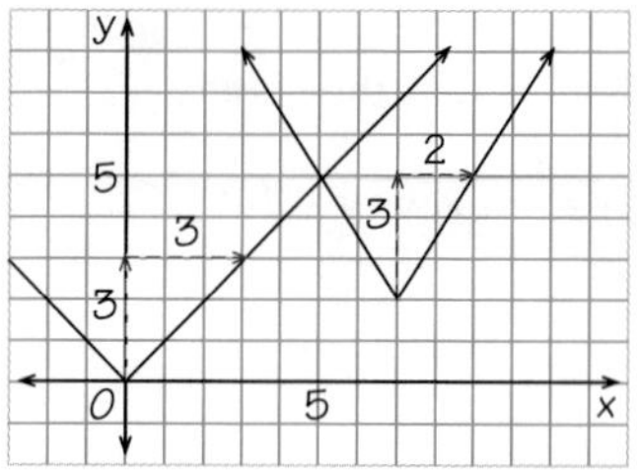

Now you are ready for: Exs. 10–28 on pp. 516–518

Talk it Over

6. a. In Sample 3, Céline assumed the stretch was horizontal and not vertical. Using her diagram at the left, find the horizontal stretch factor.
   b. What is an equation of the graph using the horizontal stretch factor?
   c. Is this equation equivalent to the one in Sample 3?

Look Back

Describe the effect of combining stretching with translating on the graphs of data sets and functions.

9-5 Exercises and Problems

1. **Reading** Which types of transformations affect the mean of a set of data? Which affect the standard deviation?
2. The school nurse takes the temperatures of a class of twelve first-grade students. The temperatures (in °F) are:

   98.6, 98.4, 98.0, 99.9, 98.9, 98.6, 98.3, 97.9, 100.2, 98.7, 99.0, 98.8

   Find the mean and the standard deviation of the data. What are the mean and the standard deviation in degrees Celsius?
3. Use the Philadelphia temperature data on page 510.
   a. In what years was the average April temperature more than one standard deviation above the mean? In what years was it more than one standard deviation below the mean?
   b. According to the 68-95-99.7 rule, about 68% of the data of a normal distribution lie within one standard deviation of the mean. Does this rule appear to hold for the Philadelphia temperature data?
   c. Repeat parts (a) and (b) for the Beijing temperature data.
4. a. Use the formula

   $$F = \frac{9}{5}C + 32$$

   to convert the mean April temperature for Beijing to degrees Fahrenheit.
   b. Describe this formula in terms of transformations.

BY THE WAY...

In Canada, the following rhyme is used to help people remember Celsius temperatures:

Thirty is hot, twenty is pleasing,
Ten is not, and zero is freezing.

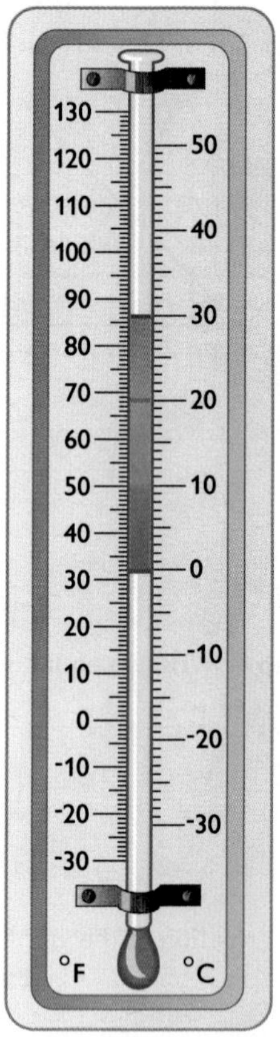

Answers to Talk it Over

6. a. $\frac{2}{3}$
   b. $y-2=\left|\frac{x-7}{\frac{2}{3}}\right|$ or $y-2=\left|\frac{3(x-7)}{2}\right|$
   c. Yes.

Answers to Look Back

Answers may vary. An example is given. Combining stretching with translating causes the graph of a function to shrink or stretch vertically or horizontally (or both) and to be translated along the *x*-axis or *y*-axis (or both). Translating a data set affects the mean, the median, and the graph, while stretching affects the mean, the median, the range, the standard deviation, and the graph.

connection to LITERATURE

In Ann C. Crispin's novel *Starbridge*, the main character, Mahree, is a 16-year-old space colonist. She and her human companion, Rob, arrive on an alien planet. Their host, Shirazz, offers to provide them with food and to take care of other needs.

> "Could we have a *bath*?" Mahree blurted... "A real *bath*?"
>
> Shirazz was unperturbed. "Of course. Sand or mud? Or if with other cleansing medium, please provide its chemical composition."
>
> Mahree stared at Rob, dismayed, but he gave her a reassuring grin, and put in, smoothly, "Liquid H_2O, Esteemed One, in quantity deep enough for immersion."...
>
> "What temperature?"
>
> Rob and Mahree stared at each other, blankly. Precise Simiu temperature measurements were still foreign to them. "This air," Rob said, finally, "is approximately ten of our temperature units less than our body temperatures, which you no doubt determined when you scanned us. We would like our liquid H_2O eight of those units *higher* than our body temperatures, if that is possible."
>
> "Entirely possible," the CLS Liaison said. "It shall be as you have requested."

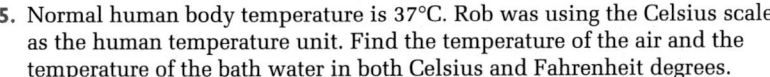

5. Normal human body temperature is 37°C. Rob was using the Celsius scale as the human temperature unit. Find the temperature of the air and the temperature of the bath water in both Celsius and Fahrenheit degrees.
6. Suppose Shirazz measures Rob's and Mahree's body temperatures at 1135.5 Simiu temperature units and the air at 1098.9 Simiu temperature units.
   a. Find the temperature of the bath water in Simiu temperature units.
   b. Find the freezing point of H_2O to the nearest whole Simiu temperature unit.

For Exercises 7 and 8, use this list of annual salaries for the 10 employees of a small business.

A. \$17,500	**B.** \$23,000	**C.** \$28,000	**D.** \$35,000	**E.** \$46,700
F. \$18,300	**G.** \$24,500	**H.** \$31,500	**I.** \$40,400	**J.** \$53,000

7. Find the mean and the standard deviation of this set of salaries.
8. Suppose each employee must pay state income tax at a rate of 4.8% on all income above \$6000. The formula $T = 0.048(S - 6000)$ gives the tax on a salary of S dollars.
   a. Describe the formula using the language of transformations.
   b. Find the amount of tax each employee pays.
   c. Find the mean and the standard deviation of the taxes.
   d. **Writing** Explain how the mean and the standard deviation of the taxes are related to the mean and the standard deviation of the salaries.

Integrating the Strands

Algebra Exs. 2–6, 8, 9, 17
Functions Exs. 10–25
Geometry Ex. 9
Statistics and Probability Exs. 1–4, 7, 8, 26–28
Discrete Mathematics Exs. 1–4, 7–9, 28
Logic and Language Exs. 1, 8, 24

Communication: Discussion

By sharing their results for Exs. 1–4, students will be able to solidify their understanding of the concepts of mean, standard deviation, normal distribution, and the effects of transformations on the mean and standard deviation of a set of data.

Students Acquiring English

Certain terms from the *Starbridge* passage may pose difficulties for students acquiring English (*alien planet, cleansing medium, Esteemed One, scanned*) and others are not typical English terms (*Simiu, CLS Liaison*). You may wish to pair students with English-proficient partners for the exercises.

Answers to Exercises and Problems

1. stretching and translating; stretching
2. about 98.78°F and about 0.66°F; about 37.1°C and about 0.37°C
3. a. 1969, 1974, 1976, and 1977; 1965, 1966 and 1975
   b. No.
   c. part (a) above: 1961, 1973, and 1975; below: 1965, 1976, 1979, and 1980; part (b) No.
4. 55.4°F; a vertical stretch by a factor of $\frac{9}{5}$ followed by a translation 32 units up
5. air: 27°C or 80.6°F; bath water: 45°C or 113°F
6. a. 1164.7 Simiu temperature units
   b. 1000 Simiu temperature units
7. \$31,790; about \$11,347
8. a. vertical stretch by a factor of 0.048 and translation 6000 units right
   b. \$552; \$816; \$1056; \$1392; \$1953.60; \$590.40; \$888; \$1224; \$1651.20; \$2256
   c. \$1237.92; \$544.66
   d. The mean is affected by both stretching and translation, so the mean for the taxes is 0.048(mean salary – 6000); the standard deviation is affected only by the stretching, so the standard deviation for the taxes is 0.048 times the standard deviation for salaries.

Integrating the Strands

Ex. 9 integrates the strands of algebra, geometry, and discrete mathematics by using a matrix equation to describe a transformation of a quadrilateral.

Research

Some students may wish to research past postal rates. They can then make a graph of the past rates and write an equation for the graph using $y = c(x)$ as the parent function. These graphs and functions can then be compared to those from Exs. 15 and 16.

9. This matrix equation shows how the polygon $ABCD$ with vertices $A(0, 0)$, $B(1, 1)$, $C(1, 5)$, and $D(0, 5)$ is transformed.

$$\begin{bmatrix} 3 & 0 \\ 0 & 1 \end{bmatrix}\begin{bmatrix} x \\ y \end{bmatrix} + \begin{bmatrix} -10 \\ 2 \end{bmatrix} = \begin{bmatrix} x' \\ y' \end{bmatrix}$$

a. Sketch the original polygon $ABCD$ and its image $A'B'C'D'$.

b. Describe the transformations used to produce the image.

For Exercises 10 and 11, apply the given transformations to the parent function. Then sketch the graphs of the parent function and the transformed function on the same set of axes.

10. $y = \sqrt{x}$; a vertical stretch by a factor of 3 followed by a translation 4 units to the right

11. $y = |x|$; a horizontal stretch by a factor of 3 followed by a translation 8 units down

12. On page 493, you saw the function $y - 0.30 = 0.03|x - 177|$, which approximates the toll, y, for a car as a function of x, the mile marker where the car exits the Kansas turnpike. Describe the multiple transformations used to produce this function from the parent function $y = |x|$.

13. A stone is dropped from a bridge above a deep, narrow canyon. The height of the bridge is 160 ft. The height in feet of the stone as it falls is given by $y = 160 - 16x^2$ where x is the elapsed time in seconds. Consider this function to be the result of multiple transformations on the parent function $y = -x^2$. Sketch the graphs of both functions and describe the transformations used to produce the height function.

For Exercises 14–16, use this postage information and the graph.

The cost of first class postage in the United States depends on the weight of the letter. In 1995 the cost of postage was \$.32 for the first ounce and \$.23 for each additional ounce or fraction of an ounce.

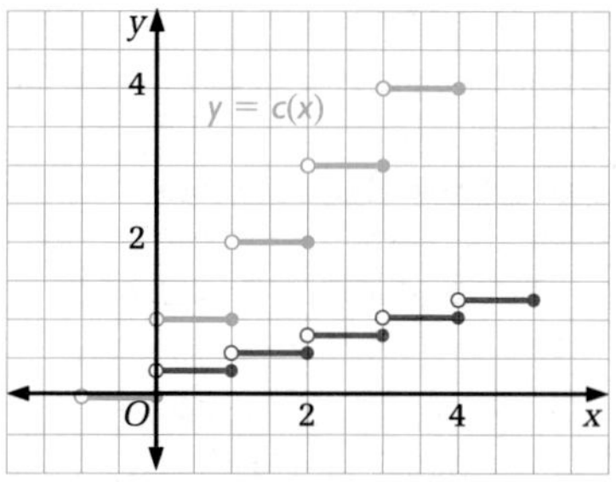

14. The blue graph is the ceiling function, $y = c(x)$, where y is the least integer greater than or equal to x. Find $c(2.7)$, $c(2.1)$, $c(-0.5)$, $c(1)$, and $c(7.4)$.

15. The red graph shows the cost of postage in dollars, y, for the weight of a letter in ounces, x, in 1995.

a. Why is there an open circle at (0, \$.32)?

b. Use the postage formula to find the vertical distance between the steps of the red graph.

c. What is the vertical distance between the steps of the graph of the ceiling function?

16. Write an equation for the red graph using $y = c(x)$ as the parent function.

Answers to Exercises and Problems

9. a.

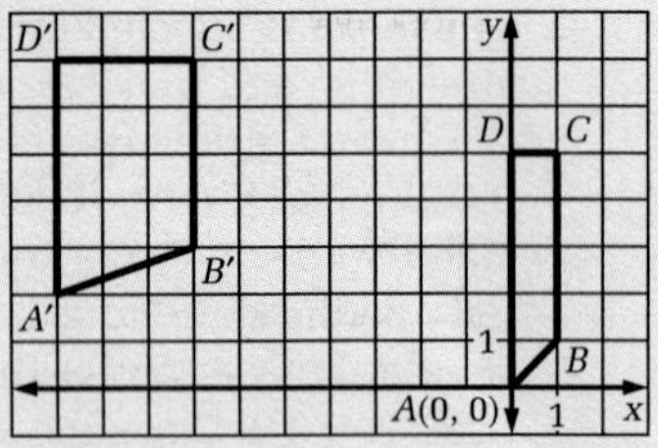

b. a horizontal stretch by a factor of 3 and translation 10 units left and 2 units up

10. $y = 3\sqrt{x - 4}$

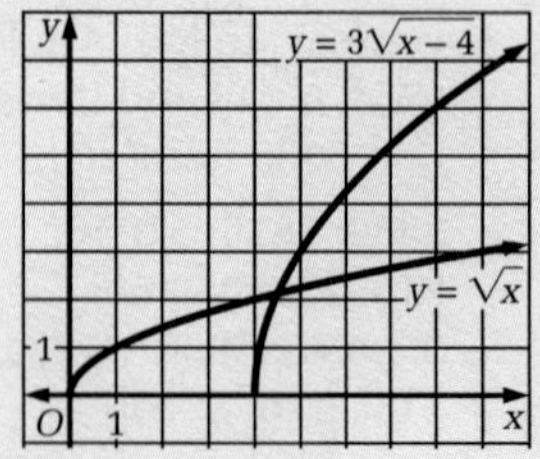

11. $y + 8 = \left|\frac{x}{3}\right|$

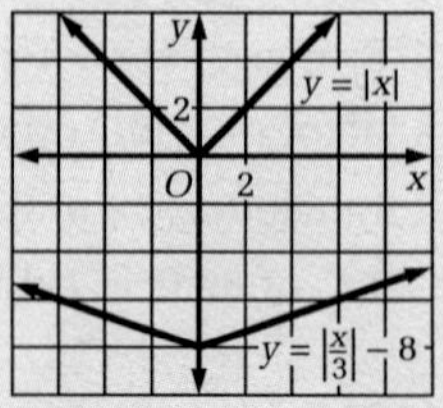

12. a translation 177 units to the right and 0.30 units up and a vertical stretch by a factor of 0.03

13. a vertical stretch by a factor of 16 and a translation 160 units up

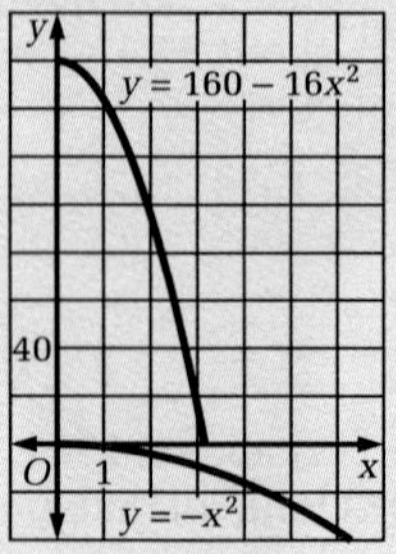

17. Think about the relationship between Fahrenheit and Celsius temperatures. Graph $y = x$ and $y = \frac{5}{9}(x - 32)$ on the same set of axes.

a. Describe how to transform the first graph to get the second.

b. Interpret the point where the second graph crosses the y-axis in terms of Fahrenheit and Celsius temperatures.

c. Find the temperature at which the Fahrenheit and Celsius readings are the same.

Write an equation for each graph. Assume that the parent graph is one of the six listed in Sample 3 on page 513.

18.

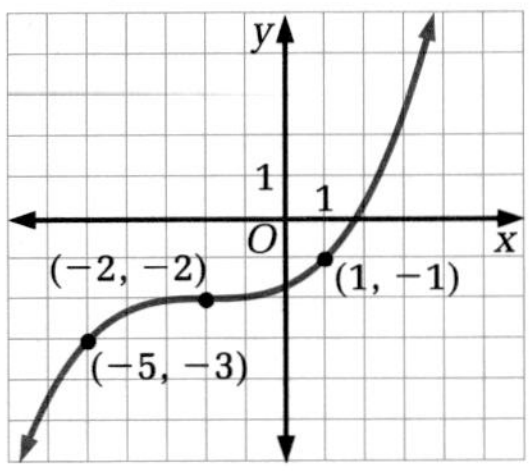

19.

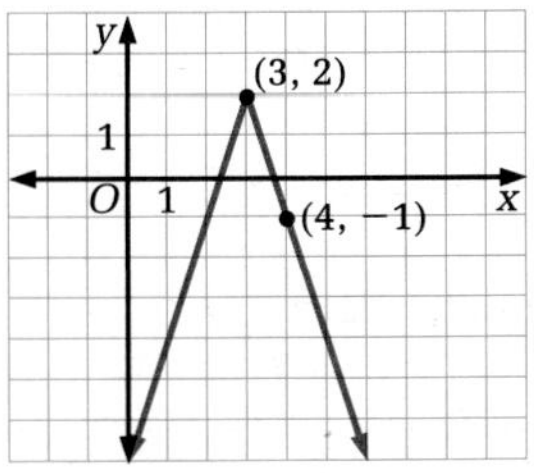

20.

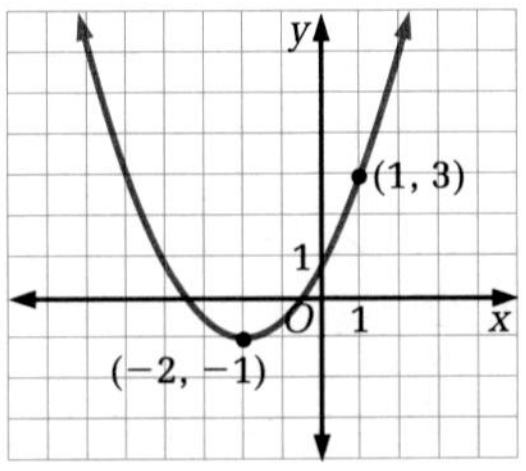

21.

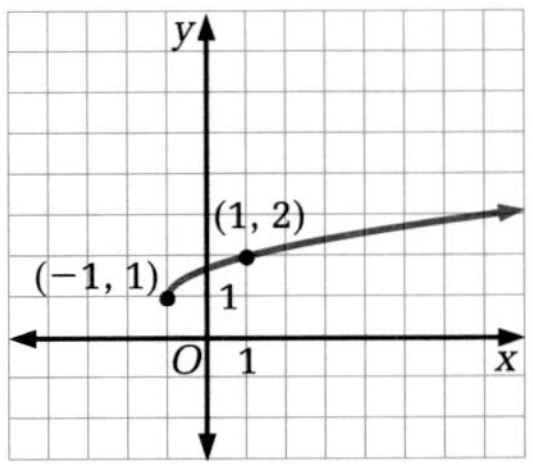

22.

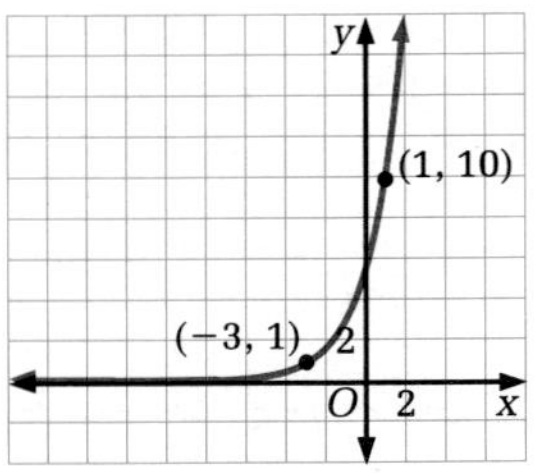

23.

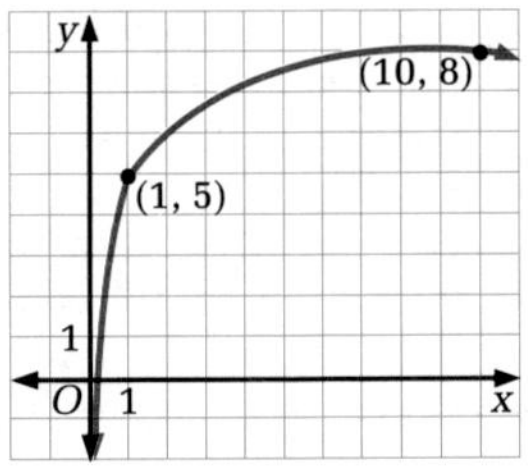

Ongoing **ASSESSMENT**

24. **Group Activity** Work in a group of four students.

a. Each of you should choose a different pair of transformations, one from column A and one from column B.

Column A	Column B
• vertical stretch by a factor of 3	• translation 5 units down
• horizontal stretch by a factor of $\frac{1}{2}$	• translation 6 units to the right

b. Each of you should choose a nonlinear parent function.

c. Find the result of performing transformation A followed by transformation B, using the parent function from part (b).

d. Then find the result of performing transformation B followed by transformation A, using the parent function from part (b).

e. Share results with your group.

f. Discuss whether composition of transformations is commutative. That is, do you always get the same results in parts (c) and (d)?

Using Technology

You can present exercises like Exs. 18–23 by entering a transformed function as the equation for Y10 on the Y= list of a TI-82 calculator. Display the graph and ask students to see if they can construct an equation for it. Students can graph their equations on their own calculators to see if they coincide with the one that you have placed on the screen.

Cooperative Learning

Allow extra time in class for students to work on Ex. 24. Each student in a group should pick a different nonlinear function for part (b). Once all groups have finished their work, they can share their results. Specific functions that each student used can be listed on the board by one member from each group.

Answers to Exercises and Problems

14. 3, 3, 0, 1, 8

15. a. 0 is not in the domain of the function.

b. 0.23

c. 1

16. $y = 0.23(c(x - 1)) + 0.32$

17.

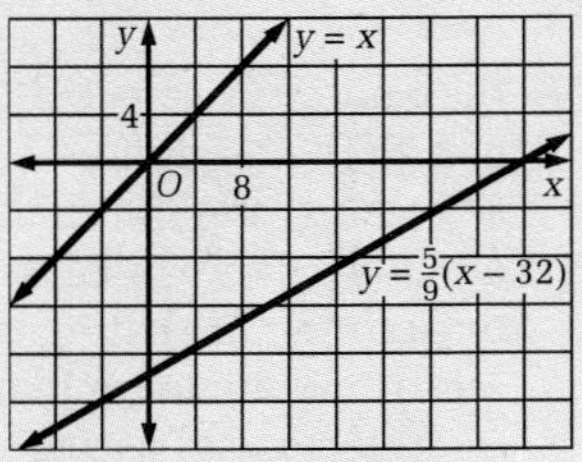

a. vertical stretch by a factor of $\frac{5}{9}$ and translation 32 units to the right

b. The y-coordinate of that point is the Celsius equivalent of 0° Fahrenheit.

c. −40°

18. $y = \left(\frac{x + 2}{3}\right)^3 - 2$

19. $y = -3|x - 3| + 2$

20. $y = \frac{4}{9}(x + 2)^2 - 1$

21. $y = \sqrt{\frac{x + 1}{2}} + 1$

22. $y = 10^{(x + 3)/4}$

23. $y = 3 \log x + 5$

24. a–e. Answers may vary. Check students' work.

f. Composition of transformations is not commutative.

Working on the Unit Project

Ex. 28 introduces students to the concept of percentiles. It may be necessary to discuss this concept in class to ensure that all students understand it.

Practice 71 For use with Section 9-5

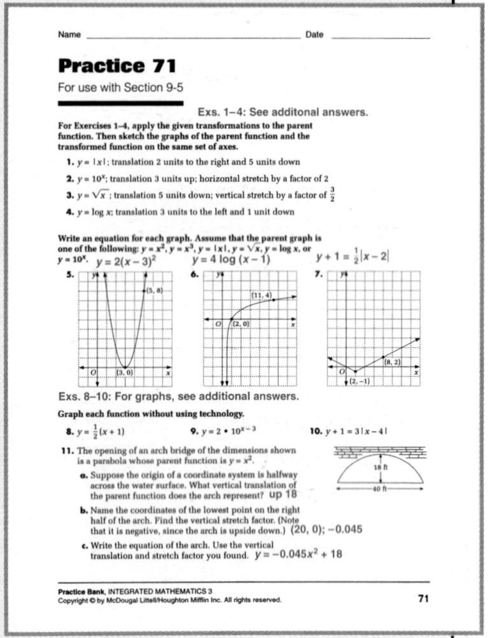
Name ______________________ Date ____________

Practice 71

For use with Section 9-5

Exs. 1–4: See additonal answers.

For Exercises 1–4, apply the given transformations to the parent function. Then sketch the graphs of the parent function and the transformed function on the same set of axes.

1. $y = |x|$; translation 2 units to the right and 5 units down
2. $y = 10^x$; translation 3 units up; horizontal stretch by a factor of 2
3. $y = \sqrt{x}$; translation 5 units down; vertical stretch by a factor of $\frac{3}{2}$
4. $y = \log x$; translation 3 units to the left and 1 unit down

Write an equation for each graph. Assume that the parent graph is one of the following: $y = x^2$, $y = x^3$, $y = |x|$, $y = \sqrt{x}$, $y = \log x$, or $y = 10^x$.

5. $y = 2(x - 3)^2$
6. $y = 4 \log (x - 1)$
7. $y + 1 = \frac{1}{2}|x - 2|$

Exs. 8–10: For graphs, see additional answers.

Graph each function without using technology.

8. $y = \frac{1}{2}(x + 1)$
9. $y = 2 \cdot 10^{x-3}$
10. $y + 1 = 3|x - 4|$

11. The opening of an arch bridge of the dimensions shown is a parabola whose parent function is $y = x^2$.
    a. Suppose the origin of a coordinate system is halfway across the water surface. What vertical translation of the parent function does the arch represent? up 18
    b. Name the coordinates of the lowest point on the right half of the arch. Find the vertical stretch factor. (Note that it is negative, since the arch is upside down.) (20, 0); −0.045
    c. Write the equation of the arch. Use the vertical translation and stretch factor you found. $y = -0.045x^2 + 18$

71

Review PREVIEW

25. Without using graphing technology, sketch the graph of the function $y = 2x(2x + 3)(2x + 3)$. Use $y = x(x + 3)(x + 3)$ as the parent function. Identify the stretch factor and the direction of the stretch. *(Section 9-4)*

26. A basketball player has a 60% free-throw average. What is the probability that the player makes at least three baskets in the next five free-throw attempts? *(Section 7-6)*

27. Each person in a group of 20 journalists and a group of 20 randomly selected people is asked to write an essay. The number of grammatical errors in each essay for each group is counted. Which group do you think has the higher mean number of errors? the higher standard deviation? Explain. *(Section 6-2)*

Working on the Unit Project

A certain career interest survey gives results in terms of *percentiles*. For example, a score in the 65th percentile for a particular category on this survey means that 65% of the general population are less interested in that category than you are.

28. The table shows Will's scores on the interest survey described above.

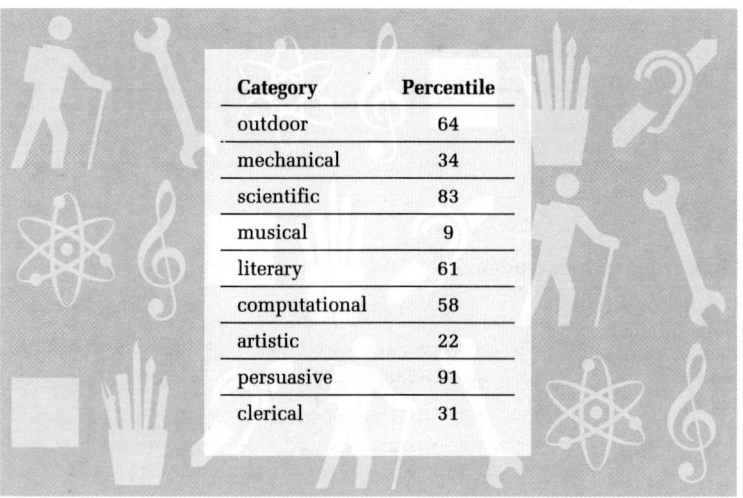

Category	Percentile
outdoor	64
mechanical	34
scientific	83
musical	9
literary	61
computational	58
artistic	22
persuasive	91
clerical	31

a. For which categories is Will's interest level above the median for the general population?

b. For which categories is Will's interest level above the upper quartile?

c. For which category is Will's interest level below that of 90% of the general population?

Answers to Exercises and Problems

25. The graph is the graph of $y = x(x + 3)(x + 3)$ with a horizontal stretch by a factor of $\frac{1}{2}$.

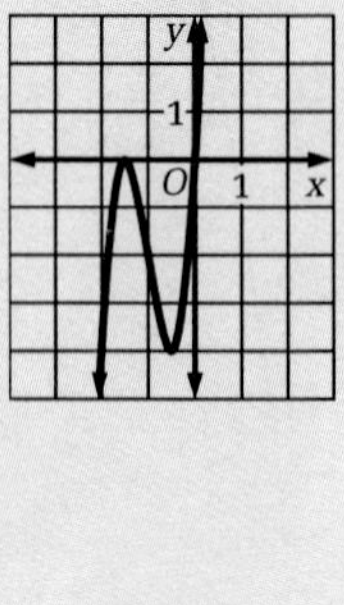

26. 0.682256

27. the randomly selected group; the randomly selected group; Journalists are more likely to have better training in writing correctly and are more likely to perform well as a group. The randomly selected people are more likely to include both people who are poor writers and people who are good writers.

28. a. outdoor, scientific, literary, computational, persuasive
    b. scientific, persuasive
    c. musical

Section

9-6 Standardizing Data

Focus

Use z-scores and the standard normal distribution to compare data.

Every April, runners from all over the world compete in the Boston Marathon. Thousands of runners gain a sense of accomplishment just from completing the race.

In 1994, 8181 runners finished the Boston Marathon. They competed in eight divisions determined by age and gender. Statistics about the women's divisions are shown. The data form a normal distribution.

Women's divisions (age in years)	Number of runners finishing	Mean time $\bar{x}$ (minutes)	Standard deviation σ (minutes)
18–39	1195	225.31	26.64
40–49	447	231.91	26.24
50–59	94	242.58	21.78
60 and over	5	249.90	27.23

Talk it Over

A 57-year-old woman from Georgia finished the race in 4:09:08 (hours : minutes : seconds). A 34-year-old woman from Texas finished the race in 4:00:15.

1. Do you think it is fair to compare the two women's times directly and conclude that the woman from Texas is a better runner? Explain.
2. a. Find the difference between each woman's time and the mean time in her division. (*Hint:* First convert the women's times to minutes.)
   b. Is each woman's time *above* or *below* the mean in her division? In a race like this, is it better to be *above* or *below* the mean? Why?
   c. Divide each difference from part (a) by the standard deviation for the division. This tells you the number of standard deviations from the mean for each time. How many standard deviations above or below the mean was each woman's time?
   d. Based on the calculations in part (c), which woman performed better relative to her division? Explain.

BY THE WAY...

In the 1994 Boston Marathon, 13 men and 2 women competed in the visually impaired category, and 71 men and 9 women competed in the wheelchair race.

PLANNING

Objectives and Strands

See pages 480A and 480B.

Spiral Learning

See page 480B.

Materials List

- Graph paper

Recommended Pacing

Section 9-6 is a two-day lesson.

Day 1

Pages 519–520: Opening paragraph through Talk it Over 4, *Exercises 1–4*

Day 2

Pages 521–523: The Standard Normal Distribution through Look Back, *Exercises 5–21*

Extra Practice

See pages 622–624.

Warm-Up Exercises

- Warm-Up Transparency 9-6

Support Materials

- Practice Bank: Practice 72
- Activity Bank: Enrichment 64
- Study Guide: Section 9-6
- Problem Bank: Problem Set 20
- Explorations Lab Manual: Diagram Master 2
- Assessment Book: Quiz 9-6, Alternative Assessment 2

Answers to Talk it Over

1. No; it is unreasonable to compare the two because of the age difference.
2. a. Georgia: 6.55 min; Texas: 14.94 min
   b. Both are above; below. A lower time indicates a faster run.
   c. Georgia: 0.30 above; Texas: 0.56 above
   d. The Georgia woman did better. She was only 0.30 standard deviation above the mean for her division, while the Texas woman was 0.56 standard deviation above the mean for her division.

TEACHING

Multicultural Note

Many nations have produced outstanding marathoners. Although he began running competitively at the late age of twenty-four, Abebe Bikila of Ethiopia was the first man to win the Olympic gold medal for the marathon twice—first in 1960 and then again in 1964. He ran the 1964 race in 2:12:11.2, at that time the fastest completion of a marathon in recorded history. Carlos Lopes, a Portuguese runner, outran Japan's Toshishiko Seko for the 1984 men's Olympic gold medal. Norwegian Grete Waitz, the woman whose running convinced Olympic Game organizers to create a woman's marathon event for the first time in 1984, is also an important figure in the history of marathons. Another great marathoner is Hwang Young-Cho from South Korea, gold-medalist in the 1992 Olympic games.

To compare the runners in different divisions, you did some calculations in *Talk it Over* question 2 on page 519 to adjust or *standardize* the times. These standardized values are called **z-scores.**

CALCULATING z-SCORES

A *z*-score for a data value is the number of standard deviations it is above or below the mean of the data set. To find a *z*-score:

- subtract the mean to find the deviation from the mean,
- then divide by the standard deviation.

$$z = \frac{x - \bar{x}}{\sigma}$$

Sample 1

A 40-year-old Louisiana woman's time was 3:50:28. Compare her time to the times of the women from Texas and Georgia. Use the statistics from page 519.

Sample Response

Step 1 Convert the woman's time to minutes.

$$x = 3{:}50{:}28 = 3(60) + 50 + \frac{28}{60} \approx 230.47$$

Step 2 Calculate the *z*-score.

The woman is in the 40–49 division. The mean time in this division is 231.91 min and the standard deviation is 26.24 min.

$$z = \frac{x - \bar{x}}{\sigma}$$

$$\approx \frac{230.47 - 231.91}{26.24}$$ ← Substitute **230.47** for *x*, **231.91** for $\bar{x}$, and **26.24** for σ.

$$\approx -0.055$$

The *z*-score for the Louisiana woman is about −0.06. Her time is about 0.06 standard deviations below the mean in her division.

The *z*-scores for the women from Georgia and Texas are positive. The *z*-score for the woman from Louisiana is negative, which represents a better-than-average performance.

The woman from Louisiana did better relative to her division than the other two women did relative to theirs.

Talk it Over

3. Does a positive *z*-score indicate a value *below* or *above* the mean? Does a negative *z*-score indicate a value *below* or *above* the mean? What does a *z*-score of 0 indicate?
4. Explain how *z*-scores allow you to compare the performance of women marathoners in different age divisions.

Now you are ready for: Exs. 1–4 on pp. 523–524

Answers to Talk it Over

3. above; below; The score and the mean are the same.
4. A lower *z*-score means a better performance relative to a particular age group, so *z*-scores for different age groups can be compared.

The Standard Normal Distribution

z-scores are often used with data that is normally distributed. The parent graph of any normal distribution is called the **standard normal curve.**

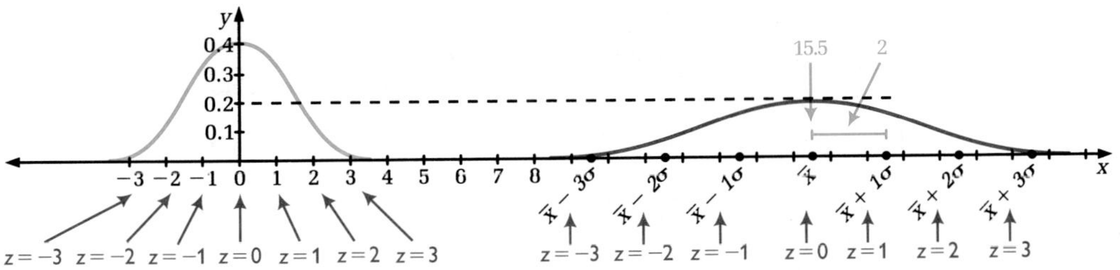

In Unit 6 you saw that in any normal distribution, the data are arranged around the mean in a predictable way. The percents of data in six regions of the distribution are shown below. These percents are estimates based on the standard normal curve.

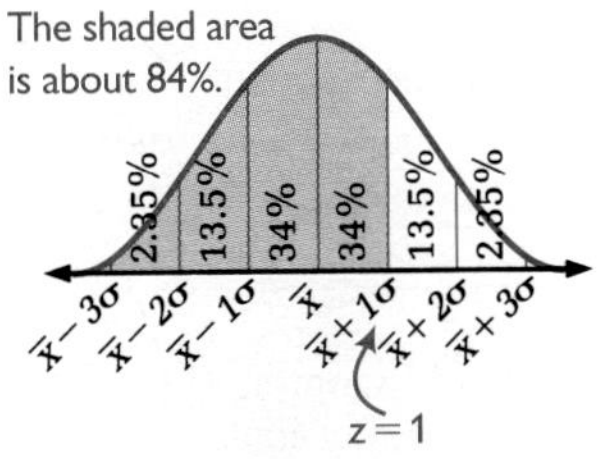

The *standard normal table,* shown below, gives these percents more accurately than the estimates you used in Unit 6. Also, you can use it with z-scores that are not integers.

If you know the z-score for a data value, you can use the table to find the percent of data that is less than the given value.

Standard Normal Table

z	.0	.1	.2	.3	.4	.5	.6	.7	.8	.9
−3	.0013	.0010	.0007	.0005	.0003	.0002	.0002	.0001	.0001	.0000+
−2	.0228	.0179	.0139	.0107	.0082	.0062	.0047	.0035	.0026	.0019
−1	.1587	.1357	.1151	.0968	.0808	.0668	.0548	.0446	.0359	.0287
−0	.5000	.4602	.4207	.3821	.3446	.3085	.2743	.2420	.2119	.1841
0	.5000	.5398	.5793	.6179	.6554	.6915	.7257	.7580	.7881	.8159
1	.8413	.8643	.8849	.9032	.9192	.9332	.9452	.9554	.9641	.9713
2	.9772	.9821	.9861	.9893	.9918	.9938	.9953	.9965	.9974	.9981
3	.9987	.9990	.9993	.9995	.9997	.9998	.9998	.9999	.9999	1.0000−

(.0000+ ← slightly more than 0; 1.0000− ← slightly less than 1)

Additional Sample

S1 A school district gave a basic arithmetic skills test to all students. Data from the test are shown in the table.

Ages	Mean number of correct answers	Standard Deviation
10–12	102.4	16.3
13–15	128.7	20.0
16–18	146.0	18.1

Gina is an 11-year-old who had 113 correct answers. Michelle is an 18-year-old who had 157 correct answers. Compare Gina's score to Michelle's.

Calculate the z-score for each student.

Gina: $z = \frac{x - \bar{x}}{\sigma} = \frac{113 - 102.4}{16.3} \approx 0.650$

Michelle: $z = \frac{x - \bar{x}}{\sigma} = \frac{157 - 146}{18.1} \approx 0.608$

Both students have scores above the means for their age groups. Since 0.650 > 0.608, Gina did better relative to her group than Michelle did to hers.

Talk it Over

Question 3 and 4 have students analyze and explain z-scores, and thus help solidify their understanding of this concept.

Reasoning

Ask students to examine the top row of entries and bottom row of entries for the standard normal table. If you add corresponding entries in these rows, the sum in each case is 1. Ask students to interpret the significance of this pattern. For which other rows of the table does a similar pattern hold? What does the answer mean in terms of a normally distributed set of data?

Additional Sample

S2 A teacher told her class that the mean score on the first test she gave was 65 with a standard deviation of 18. She said that the scores were normally distributed.

a. Miguel had a score of 88 on the test. What percent of the class had scores lower than Miguel? scores higher than Miguel?

Find Miguel's z-score.

$$z = \frac{x - \bar{x}}{\sigma}$$
$$= \frac{88 - 65}{18}$$
$$\approx 1.28$$

Next, use the table on page 521. In the table, the z-score closest to 1.28 is 1.3. The value that corresponds to 1.3 is 0.9032, or about 90%. Therefore, about 90% of the students had scores lower than Miguel's. About 10% of the students had scores higher than Miguel's.

b. Helen is a student in the class, and she had a higher score than 86% of the students. What was Helen's score?

The table entry closest to 86% is 0.8643. This corresponds to a z-score of 1.1. Substitute in the formula and solve for x.

$$z = \frac{x - \bar{x}}{\sigma}$$
$$1.1 \approx \frac{x - 65}{18}$$
$$x \approx (1.1)(18) + 65$$
$$\approx 84.8$$

Helen's score was about 85.

Sample 2

The heights of women 18 years old in the United States are normally distributed with a mean of 64 in. and a standard deviation of 2.2 in.

a. Sara is 18 years old and 60.5 in. tall. What percent of 18-year-old women in the United States are shorter than she is? taller than she is?

b. Margaret is 18 years old. 76% of 18-year-old women in the United States are shorter than Margaret. How tall is she?

Sample Response

a. **1** Find Sara's z-score.

$$z = \frac{x - \bar{x}}{\sigma}$$
$$= \frac{60.5 - 64}{2.2} \quad \leftarrow \text{Substitute } \mathbf{60.5} \text{ for } x\text{, } 64 \text{ for } \bar{x}\text{, and } 2.2 \text{ for } \sigma.$$
$$\approx -1.59$$

2 Use the table.

For $z \approx -1.6$, the value from the table is 0.0548, or about 5.5%.

z	.0	.1	.2	.3	.4	.5	.6	.7	.8	.9
−3	.0013	.0010	.0007	.0005	.0003	.0002	.0002	.0001	.0001	.0000+
−2	.0228	.0179	.0139	.0107	.0082	.0062	.0047	.0035	.0026	.0019
−1	.1587	.1357	.1151	.0968	.0808	.0668	.0548	.0446	.0359	.0287
−0	.5000	.4602	.4207	.3821	.3446	.3085	.2743	.2420	.2119	.1841

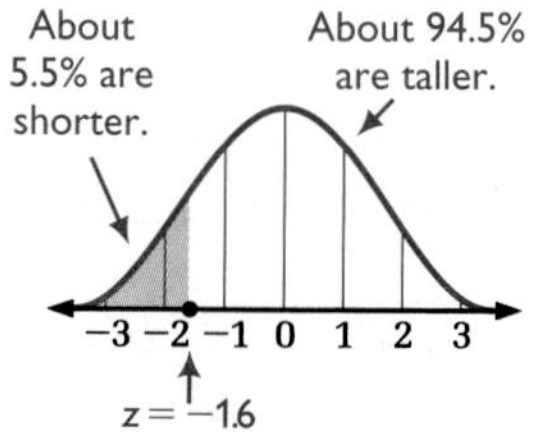

About 5.5% of the 18-year-old women in the United States are shorter than Sara. This means that about 100% − 5.5% = 94.5% are taller than Sara.

b. In the table, find the value that is closest to 0.76.

z	.0	.1	.2	.3	.4	.5	.6	.7	.8	.9
−0	.5000	.4602	.4207	.3821	.3446	.3085	.2743	.2420	.2119	.1841
0	.5000	.5398	.5793	.6179	.6554	.6915	.7257	.7580	.7881	.8159
1	.8413	.8643	.8849	.9032	.9192	.9332	.9452	.9554	.9641	.9713

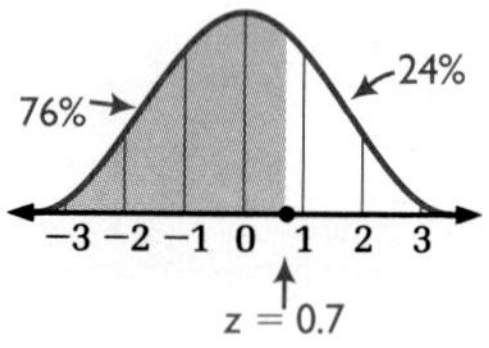

The closest value is 0.7580, which corresponds to a z-score of 0.7.

$$z = \frac{x - \bar{x}}{\sigma} \quad \leftarrow \text{Use this equation to find } x\text{, Margaret's height.}$$
$$0.7 \approx \frac{x - 64}{2.2} \quad \leftarrow \text{Substitute } \mathbf{0.7} \text{ for } z\text{, } 64 \text{ for } \bar{x}\text{, and } 2.2 \text{ for } \sigma.$$
$$x \approx (0.7)(2.2) + 64 \quad \leftarrow \text{Solve for } x.$$
$$\approx 65.54$$

Margaret is about 65.5 in. tall.

In Sample 2, Margaret's height of about 65.5 in. is said to be at the 76th *percentile* for adult female heights. A **percentile** is a whole number between 0 and 100 that tells what percent of the data in a set are less than or equal to a given value.

Talk it Over

5. Explain why Sara's height in Sample 2 is at the 5th percentile.
6. At what percentile is the median of a distribution?

Look Back

How are z-scores useful in comparing data from different distributions?

Now you are ready for: Exs. 5–21 on pp. 524–525

9-6 Exercises and Problems

1. **Reading** In what kinds of situations are z-scores useful? Why?

2. **Swimming** At a Central Masters Swimming Association meet in 1991, swimmers competed in categories by age, gender, stroke, and length of course.

   One 28-year-old man entered the competition for the 25–29 age category in both 100-yard freestyle and 100-yard backstroke.

   He swam the 100-yard freestyle in 53.76 s. The mean for this race was 56.23 s, and the standard deviation was 6.28 s. He swam the 100-yard backstroke in 62.74 s. The mean for this race was 65.04 s, and the standard deviation was 5.94 s.

   a. Find his z-score for each race.

   b. For which stroke did this swimmer do relatively better?

3. **Swimming** At the same swimming meet, 10 swimmers competed in the 50-yard butterfly for women in the 25–29 age category. Here are their times in seconds:

27.68	28.26	28.62	29.35	30.66
33.03	34.01	34.42	35.15	35.39

   a. Find the mean and the standard deviation of these times.

   b. Find the z-score for each swimmer.

   c. Find the mean and the standard deviation of the z-scores. What do you notice?

BY THE WAY...

United States Masters Swimming (USMS) events are open to male and female swimmers over the age of 19. The USMS holds competitions throughout the United States.

Talk it Over

When students answer question 6, you may wish to have them provide an explanation for the percentile they chose.

APPLYING

Suggested Assignment

Day 1
Standard 1–4
Extended 1–4

Day 2
Standard 5–13, 17–21
Extended 5–21

Integrating the Strands

Algebra Ex. 20
Functions Exs. 18–20
Statistics and Probability Exs. 1–17, 21
Discrete Mathematics Exs. 1–17
Logic and Language Exs. 1, 17

Application

The use of z-scores to compare two items of data from different distributions has applications to many areas in the real world. Exs. 2, 3, and 14 demonstrate how they can be used in sports.

Answers to Talk it Over

5. 5% of the heights are less than Sara's height.
6. 50th percentile

Answers to Look Back

Answers may vary. An example is given. Using z-scores allow you to standardize numbers from different distributions by comparing them to the means and standard deviations of their own distributions.

Answers to Exercises and Problems

1. Answers may vary. An example is given; in situations where you want to compare data from different categories; z-scores standardize results so that results from different categories can be compared.
2. a. 100-yard freestyle: −0.393; 100-yard backstroke: −0.387

   b. the 100-yard freestyle
3. Answers are given to the nearest thousandth.

   a. 31.657; 2.900

   b. −1.371; −1.171, −1.047; −0.796; −0.344; 0.473; 0.811; 0.953; 1.204; 1.287

   c. 0; 1; These are the mean and standard deviation of the normal distribution.

Assessment: Standard

For Ex. 4, ask students what Sandy would have to score to be two standard deviations above the mean. According to the chart of page 522, what percent of the class scored higher than she did? Lower than she did?

4. Sandy took two tests in her physics class.

On the first test her score was 73. The class average was 65.1, and the standard deviation was 6.8.

On the second test her score was 88. The class average was 79.3, and the standard deviation was 7.5.

a. Find her z-score for each test.

b. On which test did she do relatively better?

For each z-score:

a. Sketch the standard normal curve and shade the region under the curve for values of z less than the given score.

b. Use the table on page 521 to find the percent of data less than the given score.

5. $z = 0.7$ **6.** $z = -1.3$ **7.** $z = -0.4$ **8.** $z = 2.1$

For each percent, use the table on page 521 to find the corresponding z-score for the standard normal distribution.

9. 25% **10.** 85% **11.** 60% **12.** 0.1%

13. Two normal curves are shown.
For distribution 1, $\overline{x} = 0$ and $\sigma = 1$.
For distribution 2, $\overline{x} = 0$ and $\sigma = 2$.
What transformation transforms curve 1 into curve 2?

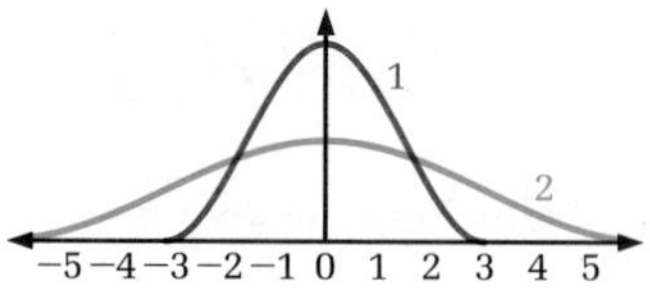

14. Running The winning times for the women's divisions in the 1994 Boston Marathon are shown.

Women's divisions (age in years)	Winning time (minutes)
18–39	141.75
40–49	153.60
50–59	200.47
60 and over	219.38

a. Find the z-score for each of the four winners based on her age group. Use the means and the standard deviations given on page 519.

b. Suppose the winner in the 40–49 division had run in the 18–39 division. Find her z-score for that division. What percent of the runners in the 18–39 division would she have beaten?

c. Suppose the winner in the 50–59 division had run in the 40–49 division. Find her z-score for that division. What percent of the runners in the 40–49 division would she have beaten?

15. Career Pediatricians monitor the height and weight of children as they grow.

Suppose the weights of one-year-old boys in the United States have a normal distribution with a mean of 11.8 kg and a standard deviation of 1.9 kg.

a. What percent of one-year-old boys weigh less than 13.1 kg?

b. What percent of one-year-old boys weigh more than 13.1 kg?

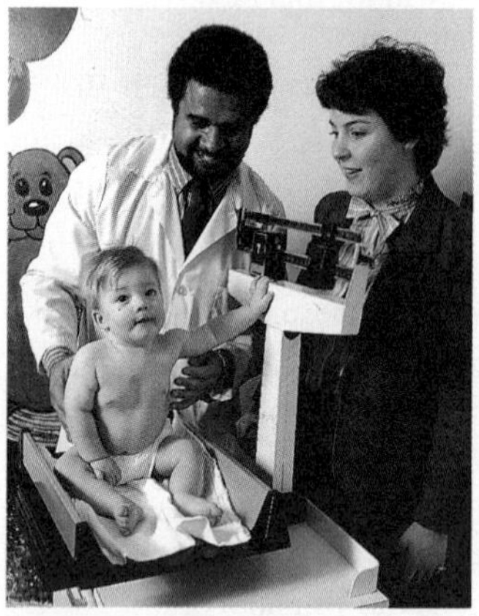

Answers to Exercises and Problems

4. a. 1.162; 1.160
   b. She did slightly better on the first test.
5. a.

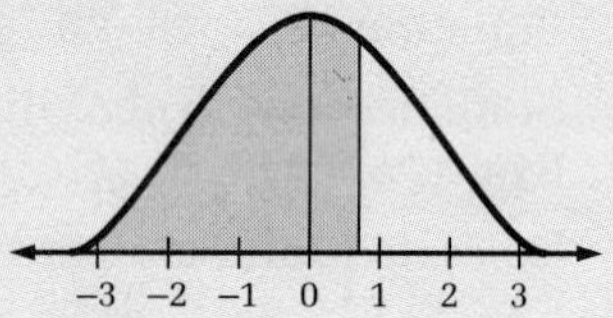

   b. about 76%
6. a.

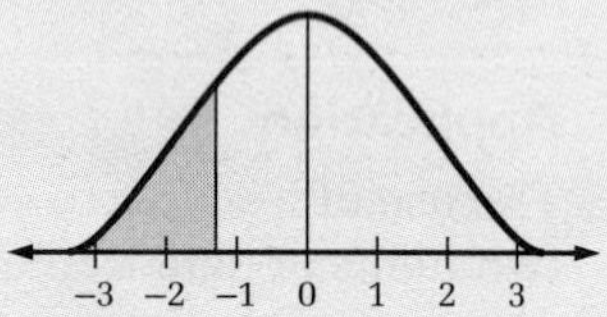

   b. about 10%
7. a.

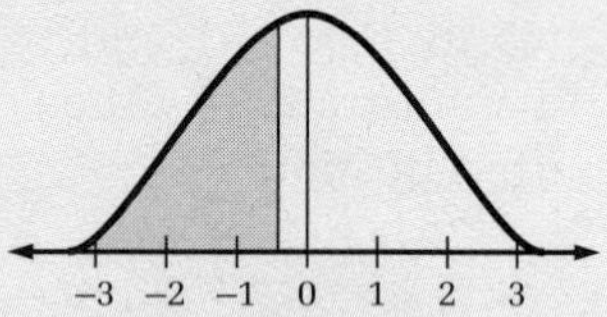

   b. about 34%
8. a.

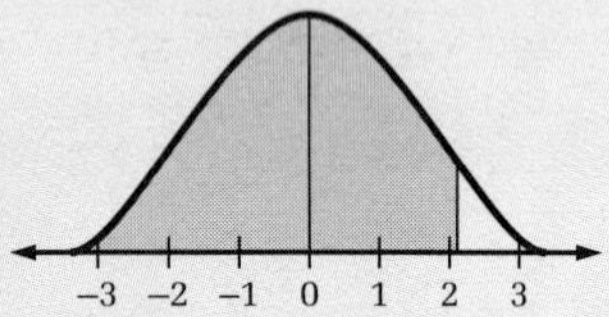

   b. about 98%
9. about −0.7
10. about 1.0
11. about 0.3
12. about −3.1
13. Stretch curve 1 horizontally by a factor of 2 and vertically by a factor of $\frac{1}{2}$.
14. Answers are given to the nearest hundredth.
   a. 18–39: −3.14; 40–49: −2.98; 50–59: −1.93; 60 and over: −1.12
   b. −2.69; about 99.65%
   c. −1.20; about 88%
15. a. about 75%
   b. about 25%
16. a. 1.65
   b. about 65 in.
17. a. 84.13%; 15.87%; 68.26%
   b. ±2 standard deviations: 97.72%; 2.28%; 95.44%; ±3 standard deviations: 99.87%; 0.13%; 99.74%
   c. Yes; the answers show that about 68% of the data is within 1 standard deviation of the mean, about 95% of the data is within 2 standard deviations of the mean, and about 99.7% of the data is within 3 standard deviations of the mean.
18.

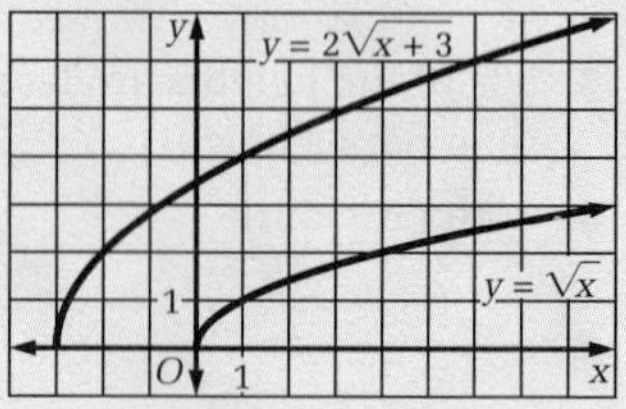

16. When he was 12 years old and in the seventh grade, Corey had a height advantage playing basketball in his school league, since he was in the 95th percentile for 12-year-old boys.

a. Find the z-score for Corey's height.

b. Suppose the mean height for 12-year-old boys is 60 in. and the standard deviation is 3 in. Find Corey's height, to the nearest inch, when he was 12 years old.

Ongoing ASSESSMENT

17. Use the standard normal table on page 521.

a. Find the percent of data less than 1 standard deviation from the mean ($z = 1$). Find the percent less than -1 standard deviation from the mean. Subtract the values. What percent is *within* 1 standard deviation of the mean?

b. Repeat part (a) for ± 2 standard deviations and for ± 3 standard deviations.

c. **Writing** Do your answers to parts (a) and (b) support the 68-95-99.7 rule? Explain.

Review PREVIEW

18. Sketch both the parent and transformed graphs when the graph of $y = \sqrt{x}$ is stretched vertically by a factor of 2 and then translated 3 units to the left. *(Section 9-5)*

19. Sketch the graphs of $y = 2^x$ and $y = \left(\frac{1}{2}\right)^x$ on the same set of axes. Explain why these functions are not inverses of each other. Describe the relationship between the two graphs. *(Section 5-5)*

20. Graph the equations $y = e^{2x}$ and $y = e^{-2x}$. Find the point of intersection of the graphs. *(Section 5-4)*

9 Working on the Unit Project

Many careers require a college degree and many colleges require applicants to submit standardized test scores. Two tests that are commonly required are the SAT and the ACT.

21. Use the data in the table.

a. Find Joan's z-score for each of the tests.

b. Find an approximate percentile for her score on each of the tests.

Test	Joan's scores	Mean score	Standard deviation
SAT verbal	490	423	113
SAT math	510	479	124
ACT composite	23	20.8	4.6

Reasoning

Ex. 17 allows students to make a connection between the percent of data within 1, 2, and 3 standard deviations of the mean and the 68-95-99.7 rule.

Working on the Unit Project

Students can answer Ex. 21 individually and then compare answers in their project groups.

Practice 72 For use with Section 9-6

Name ______ Date ______

Practice 72

For use with Section 9-6

For each z-score, use a table to find the percent of data less than the given score.

1. -1.6 ≈5.5% **2.** 1.8 ≈96.4% **3.** 2.7 ≈99.7% **4.** -0.4 ≈34.5%

For each percent, use a table to find the corresponding z-score for the standard normal distribution.

5. 69% 0.5 **6.** 46% -0.1 **7.** 8% -1.4 **8.** 97% 1.9

For Exercises 9–14, complete each row of the following table.

	Data value	Mean	Std. dev.	z-score	Percentile
9.	25	21	4	1	84th
10.	63.5	72	6.8	-1.25	11th
11.	40	44	2.5	-1.6	5th
12.	97.5	87	4.2	2.5	99th
13.	64.7	63	8.5	0.2	58th
14.	15.9	18	1.5	-1.4	8th

15. The following are scores of a group of amateur golfers on the same golf course: 82, 75, 87, 90, 92, 97, 80, 78, 86, 85.

a. Find the mean and the standard deviation of the golf scores. 85.2; ≈6.4

b. Find the z-score, to the nearest tenth, for each golf score. -0.5, -1.6, 0.3, 0.7, 1.1, 1.8, -0.8, -1.1, 0.1, 0.0

c. Find the percentile of each golf score. See bottom of page.

16. On a standardized test, Carmen scored in the 92nd percentile.

a. What was her z-score on the test? 1.4

b. Suppose the mean score on the test was 400, and Carmen's score was 720. What was the standard deviation of the test scores? about 228.6

c. Suppose Carmen's score had been 740. What percentile would she have been in? 93rd

17. ***Writing*** Suppose you are given two values in a set of normally distributed data, and suppose you know the mean and the standard deviation of the data. Describe a method of finding what percent of the data falls between the two given values. Check students' work.

15. c. 31st, 5th, 62nd, 79th, 86th, 96th, 21st, 14th, 54th, 50th

72

Answers to Exercises and Problems

19.

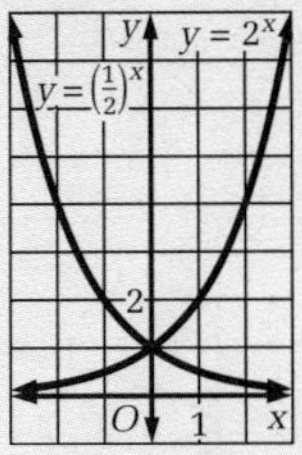

Answers may vary. An example is given. Consider the point (1, 2), which is on the graph of $y = 2^x$. If the functions were inverses, (2, 1) would be on the graph of $y = \left(\frac{1}{2}\right)^x$, which it is not, since $\left(\frac{1}{2}\right)^2 = \frac{1}{4}$, not 1. The graphs are reflections of one another over the y-axis.

20. (0, 1)

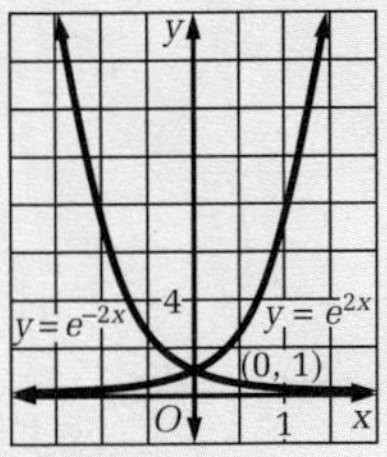

21. a. SAT verbal: 0.59; SAT math: 0.25; ACT composite: 0.48

b. SAT verbal: about the 72nd percentile; SAT math: about the 60th percentile; ACT composite: about the 69th percentile

PLANNING

Objectives and Strands

See pages 480A and 480B.

Spiral Learning

See page 480B.

Materials List

- Graphics calculator or graphing software
- Graph paper
- MIRA® transparent mirror

Recommended Pacing

Section 9-7 is a two-day lesson.

Day 1

Pages 526–527: Exploration through Sample 1, *Exercises 1–7*

Day 2

Pages 528–530: Sample 2 through Look Back, *Exercises 8–22*

Extra Practice

See pages 622–624.

Warm-Up Exercises

Warm-Up Transparency 9-7

Support Materials

- Practice Bank: Practice 73
- Activity Bank: Enrichment 65
- Study Guide: Section 9-7
- Problem Bank: Problem Set 20
- Explorations Lab Manual: Additional Exploration 11 Diagram Master 2
- Using TI-81 and TI-82 Calculators: Transforming Lines
- Using Plotter Plus: Transforming Lines
- Assessment Book: Quiz 9-7, Test 37, Alternative Assessment 3

Section **9-7**

Reflecting Graphs

On The Flip Side

Focus

Recognize the effect of a reflection on a graph and on the graph's equation.

EXPLORATION

How are the graphs of $y = -f(x)$ and $y = f(-x)$ related to the graph of $y = f(x)$?

- **Materials: graphics calculators or graphing software, graph paper**
- **Work with another student.**

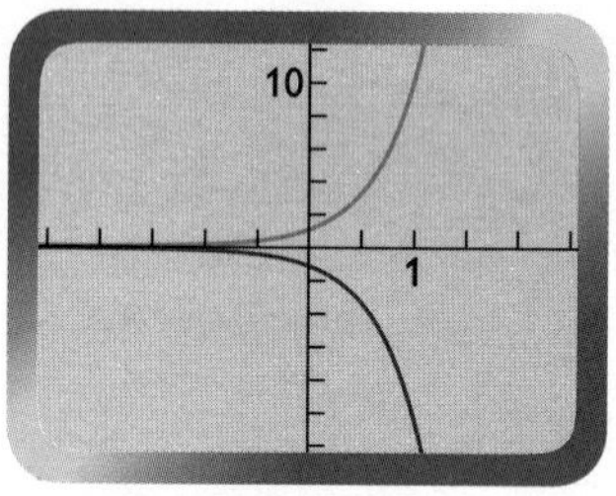

① Let $f(x) = 10^x$.

a. Write a rule for $-f(x)$.

b. Use a graphics calculator to graph $y = f(x)$ and $y = -f(x)$ on the same screen. What is the relationship between the two graphs?

② **a.** Write a rule for $f(-x)$ where f is the function in step 1.

b. Use a graphics calculator to graph $y = f(x)$ and $y = f(-x)$ on the same screen. What is the relationship between the two graphs?

③ Repeat steps 1 and 2 using the function $f(x) = \sqrt{x}$. Are your observations about the graphs of $y = f(x)$, $y = -f(x)$, and $y = f(-x)$ the same?

④ Use the function $f(x) = 4|x - 5|$.

a. Sketch the graph of $y = f(x)$.

b. Based on your results from steps 1–3, sketch the graphs of $y = -f(x)$ and $y = f(-x)$. Use a graphics calculator to check your graphs.

Answers to Exploration

1. a. $-f(x) = -10^x$

b.

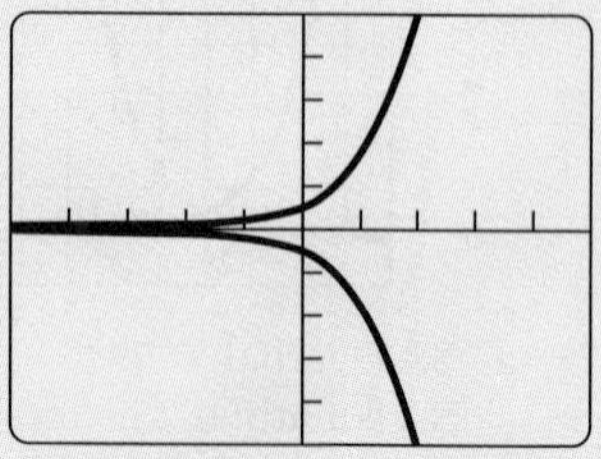

Each graph is the reflection of the other over the x-axis.

2. a. $f(-x) = 10^{-x}$

b.

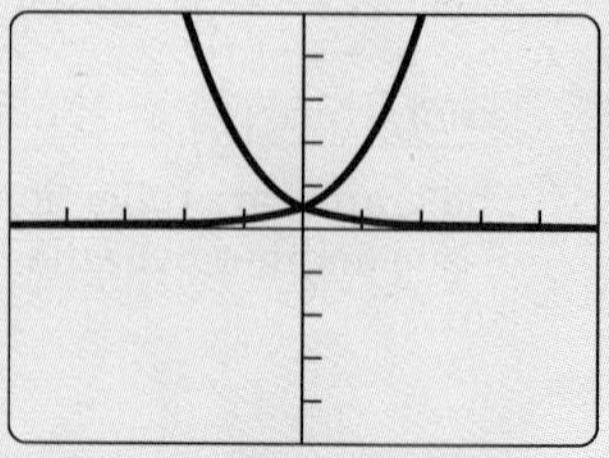

Each graph is the reflection of the other over the y-axis.

3. a. $-f(x) = -\sqrt{x}$ and $f(-x) = \sqrt{-x}$

b.

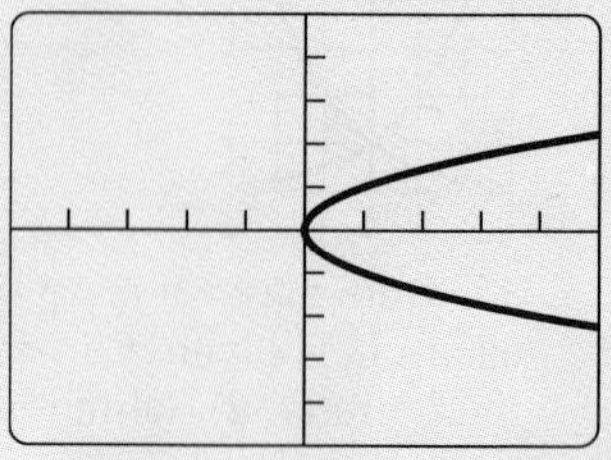

REFLECTIONS OF GRAPHS

For any function f, the graph of $y = -f(x)$ is the graph of $y = f(x)$ reflected over the x-axis.

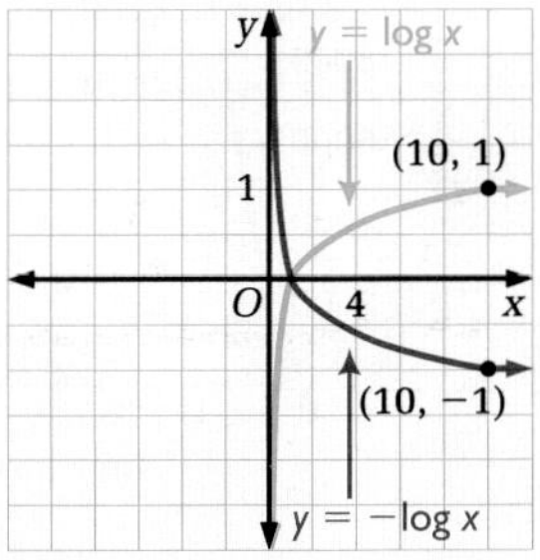

For any function f, the graph of $y = f(-x)$ is the graph of $y = f(x)$ reflected over the y-axis.

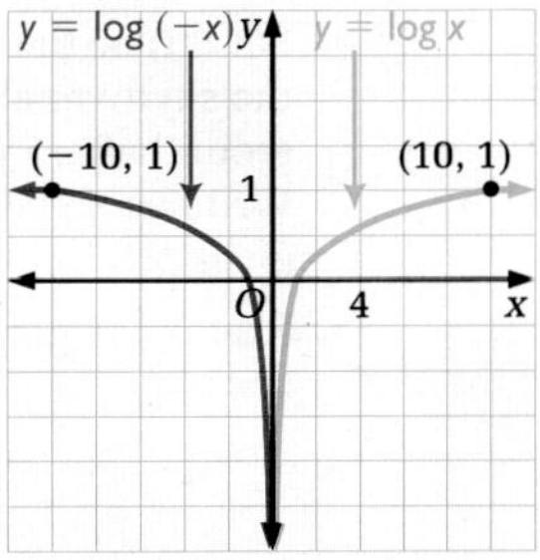

Sample 1

Sketch the graph of $y = 2\sqrt{-x+4} + 1$.

Sample Response

The parent function is $y = \sqrt{x}$. To identify the transformations, rewrite the given equation as $\frac{y-1}{2} = \sqrt{-(x-4)}$. Graph this equation using these steps:

1 Reflect the graph of $y = \sqrt{x}$ over the y-axis.

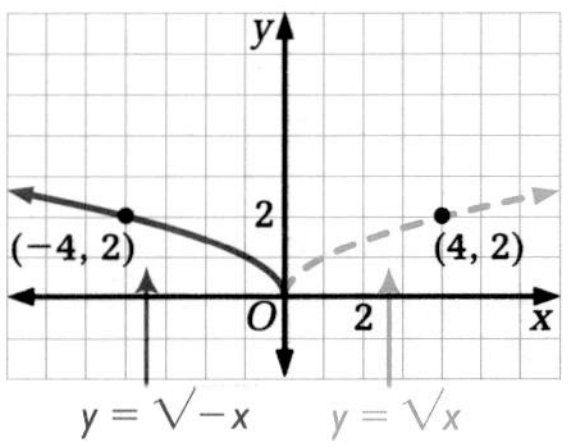

2 Stretch the graph of $y = \sqrt{-x}$ vertically by a factor of 2.

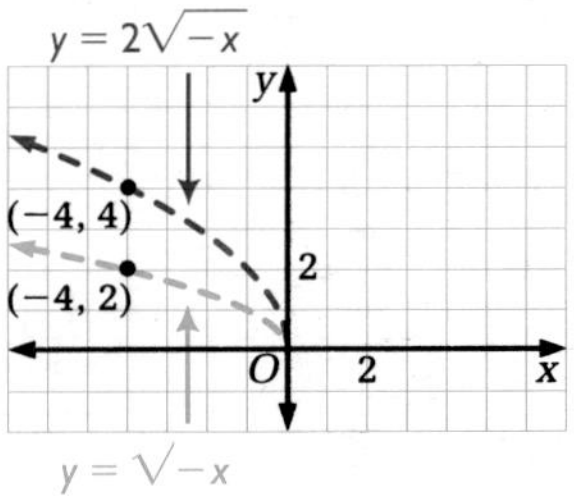

3 Translate the graph of $y = 2\sqrt{-x}$ by the vector (4, 1).

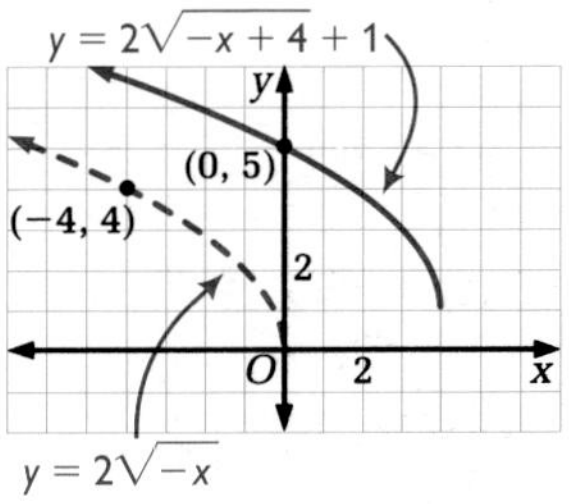

Now you are ready for: Exs. 1–7 on p. 530

TEACHING

Exploration

The goal of the Exploration is to lead students to the conclusions that $y = -f(x)$ is the reflection of $y = f(x)$ over the x-axis and that $y = f(-x)$ is the reflection of $y = f(x)$ over the y-axis.

Additional Sample

S1 Sketch the graph of $y = -\frac{(x-5)^3}{4} - 2$.

The parent function is $y = x^3$. Identify the transformation by rewriting the given equation as follows:

$\frac{y+2}{\left(\frac{1}{4}\right)} = -(x-5)^3$.

First, reflect the graph of $y = x^3$ over the x-axis.

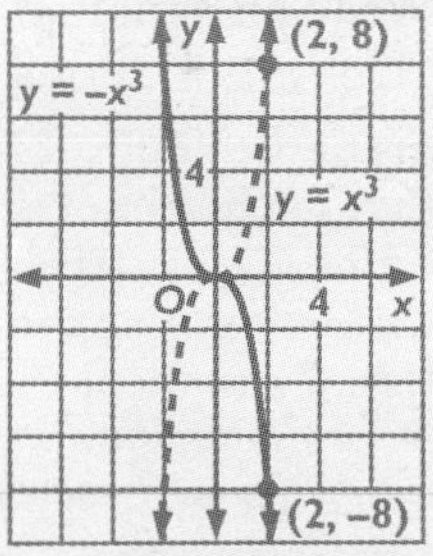

Stretch the graph of $y = -x^3$ vertically by a factor of $\frac{1}{4}$.

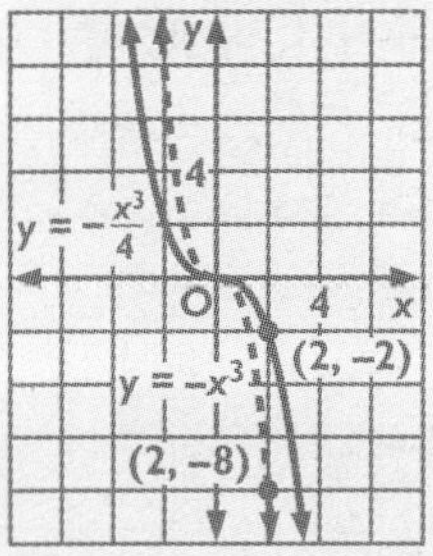

Finally, translate the graph of $y = -\frac{x^3}{4}$ by the vector (5, −2).

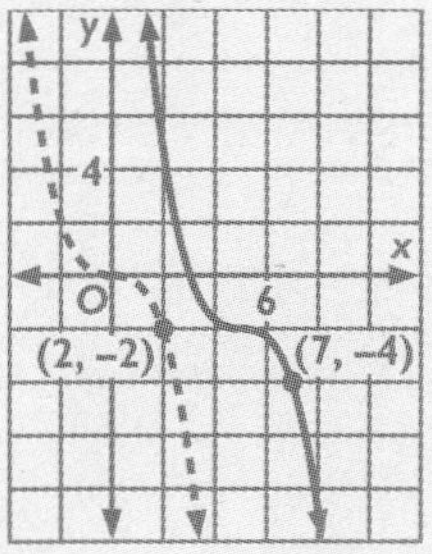

Answers to Exploration

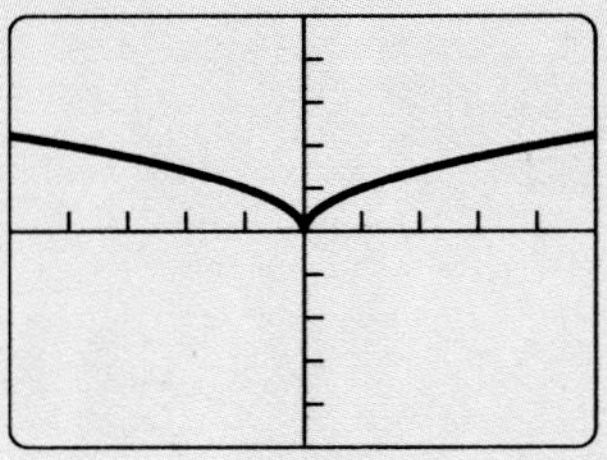

Yes.

4. a, b.

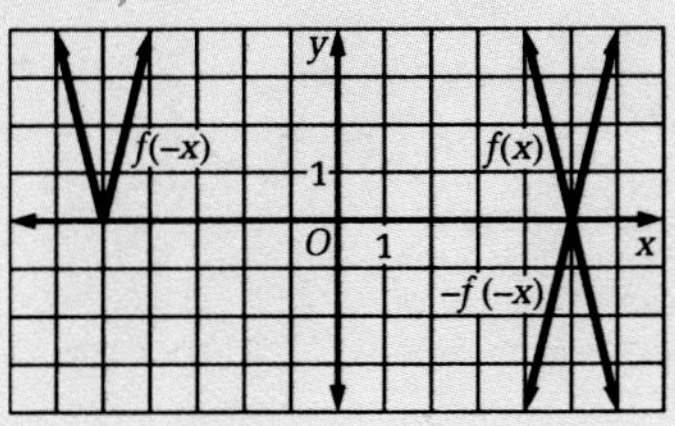

Additional Sample

S2 The graph shows the distance $f(t)$ above the floor of an insect that crawls up one side of a door frame, across the top, and down the other side. The entire trip takes 30 seconds, and the insect moves at an approximately constant rate of speed. Times are in seconds.

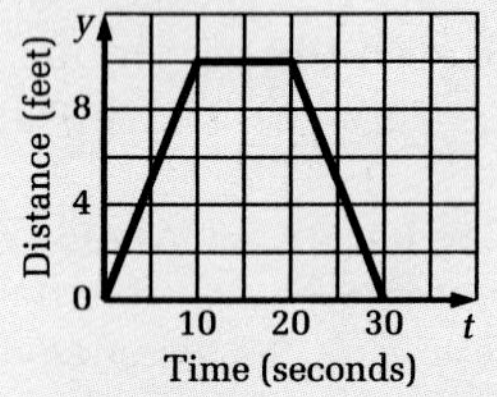

a. Let $c(t)$ be the distance of the insect from the top of the door frame t seconds from the start. Write an equation that gives $c(t)$ in terms of $f(t)$.

From the graph, it is clear that the height of the door frame is 10 ft. At any moment during the trip, the sum of the distance from the floor and the distance from the ceiling is 10 ft. Therefore, $f(t) + c(t) = 10$ and, hence, $c(t) = 10 - f(t)$.

b. Sketch the graph of $y = c(t)$.

First reflect the graph of $y = f(t)$ over the x-axis.

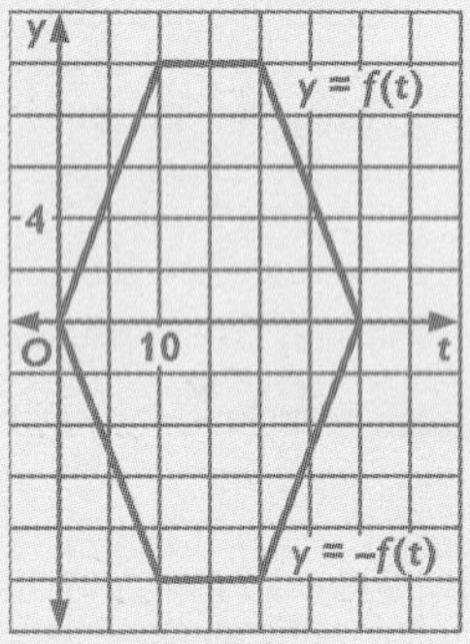

Then translate the graph of $y = -f(t)$ up 10 units.

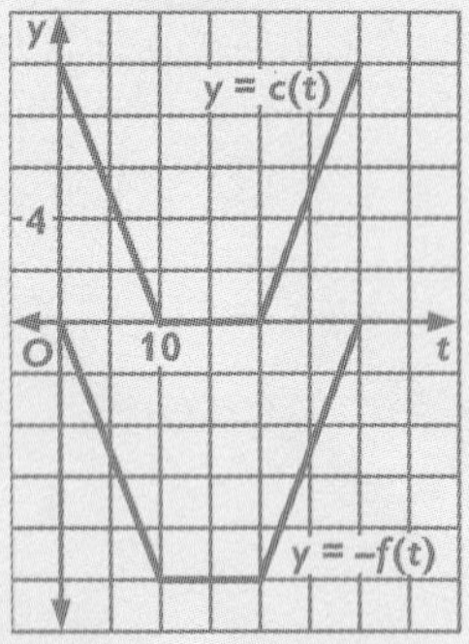

Sample 2

A skydiver jumps from an airplane at an altitude of 13,000 ft. The graph shows the approximate distance $d(t)$ that the skydiver has fallen after t minutes.

a. Let $h(t)$ be the height of the skydiver above the ground after t minutes. Write an equation that gives $h(t)$ in terms of $d(t)$.

b. Sketch the graph of $y = h(t)$.

Distance Fallen by a Skydiver

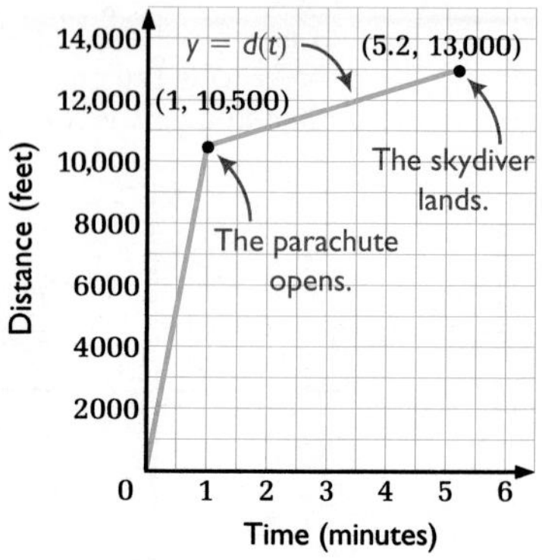

Sample Response

a. **Problem Solving Strategy:** Use a diagram.

Notice from the diagram that:

$$d(t) + h(t) = 13{,}000$$

$$h(t) = -d(t) + 13{,}000$$

An equation is $h(t) = -d(t) + 13{,}000$.

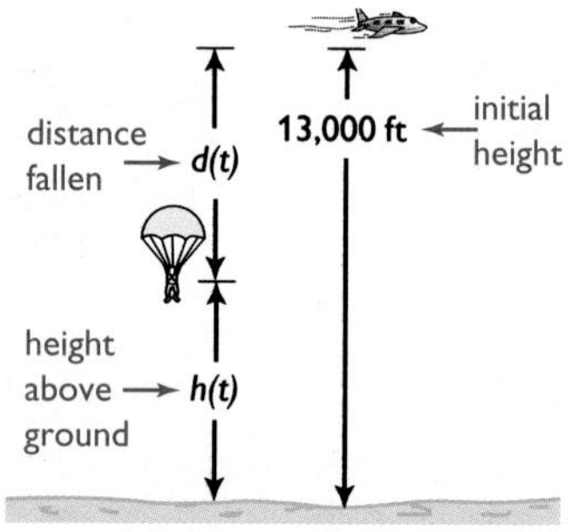

b. The equation in part (a) shows that the graph of $y = h(t)$ is a result of transforming the graph of $y = d(t)$. Follow these steps:

1 Reflect the graph of $y = d(t)$ over the t-axis.

2 Translate the graph of $y = -d(t)$ by the vector $(0, 13{,}000)$.

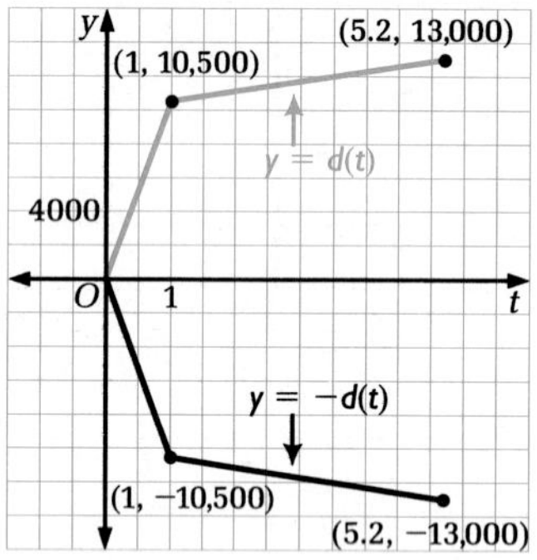

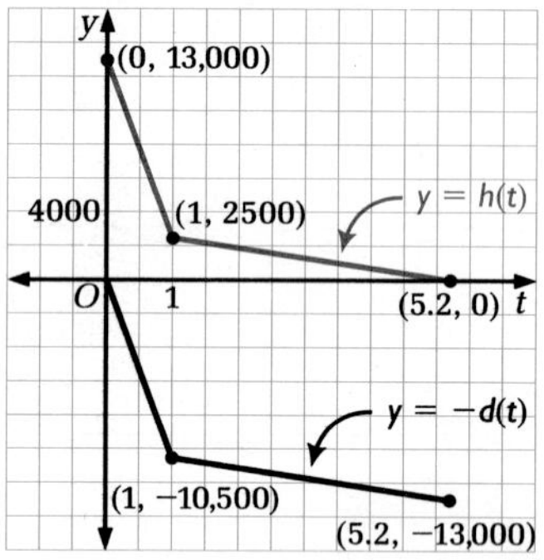

Talk it Over

1. List the steps you would use to sketch the graph of the function $y = -\frac{1}{2}|x - 6| - 3$.
2. Use the graphs of $y = d(t)$ and $y = h(t)$ in Sample 2.
   a. How far above the ground does the skydiver open the parachute?
   b. How long does it take for the skydiver to reach the ground?

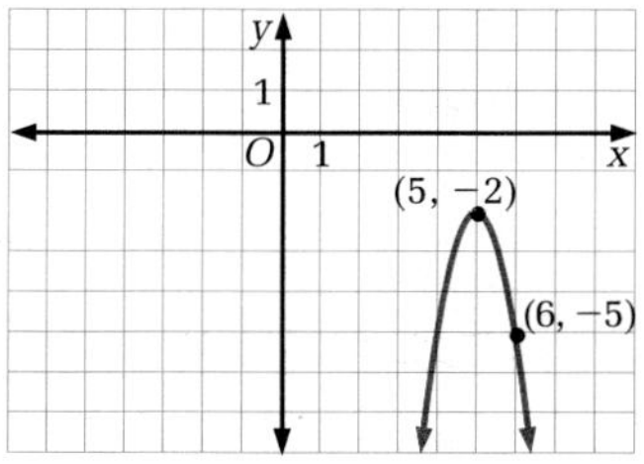

Sample 3

Find an equation for the graph shown. Assume the parent function is one of the following: $y = x^2$, $y = x^3$, $y = \sqrt{x}$, $y = |x|$, $y = 10^x$, or $y = \log x$.

Sample Response

Use the shape of the graph to identify the parent function as $y = x^2$. Apply transformations to the graph of $y = x^2$ until you produce the given graph.

1 Reflect the graph of $y = x^2$ over the x-axis.

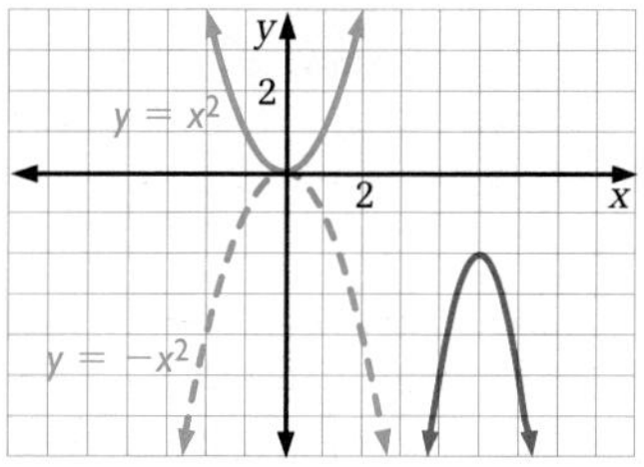

2 Stretch the graph of $y = -x^2$ vertically by a factor of $\frac{-3}{-1} = 3$.

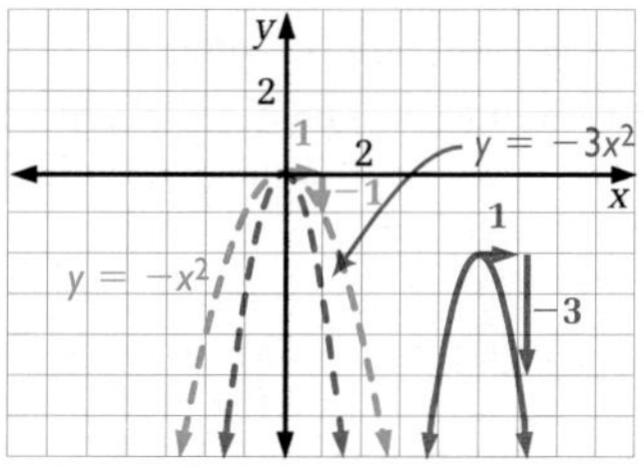

3 Translate the graph of $y = -3x^2$ by the vector $(5, -2)$. The translated graph is the given graph.

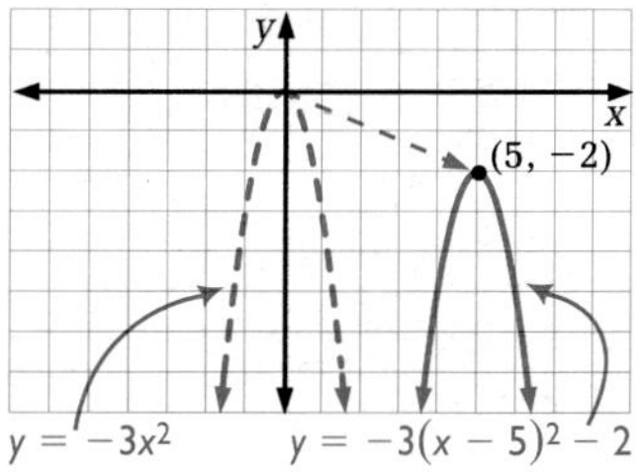

An equation of the given graph is $y = -3(x - 5)^2 - 2$.

Additional Sample

S3 Find an equation for the graph shown. Assume the parent function is one of the following: $y = x^2$, $y = x^3$, $y = \sqrt{x}$, $y = |x|$, $y = 10^x$, or $y = \log x$.

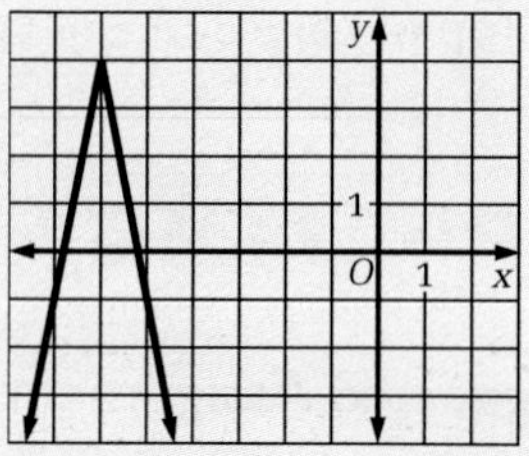

The shape of the graph indicates that the parent function is $y = |x|$. Apply transformations to the graph of $y = |x|$ until you produce the given graph.

First, reflect the graph of $y = |x|$ over the x-axis.

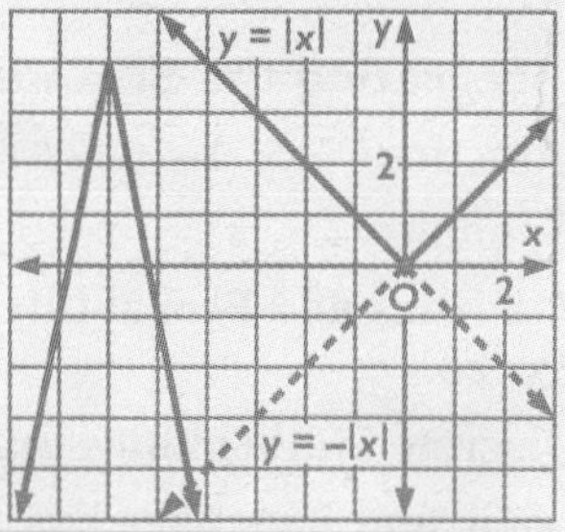

Next, stretch the graph of $y = -|x|$ vertically by a factor of 5.

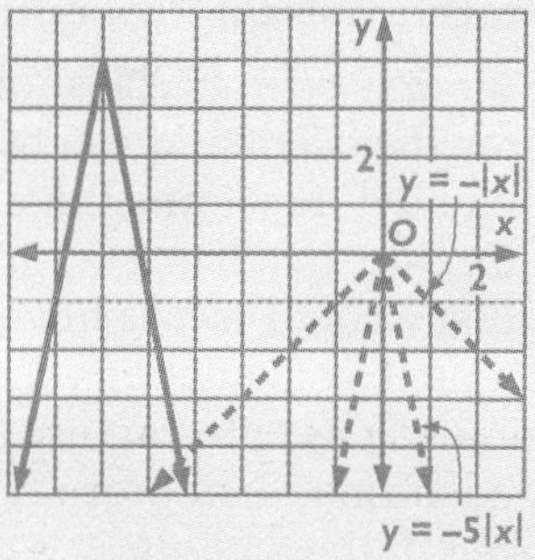

Finally, translate the graph of $y = -5|x|$ by the vector $(-6, 4)$.

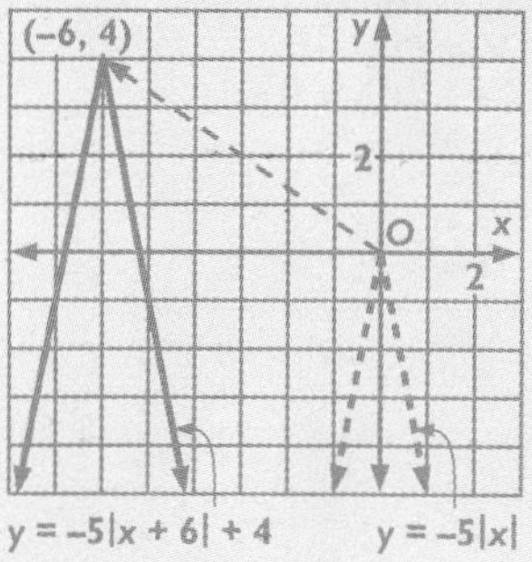

An equation of the given graph is $y = -5|x + 6| + 4$.

Answers to Talk it Over

1. Write the equation as $\frac{y + 3}{\frac{1}{2}} = -|x - 6|$. Reflect the graph of the parent function $y = |x|$ over the x-axis to get the graph of $y = -|x|$. Stretch the graph of $y = -|x|$ vertically by a factor of $\frac{1}{2}$. Finally, translate the graph by the vector $(6, -3)$.

2. a. 2500 ft
   b. 5.2 min

Look Back

You may wish to ask students to illustrate their answers to this question with a specific example of a function. Any students having trouble answering this question should refer to the top of page 527.

APPLYING

Suggested Assignment

Day 1

Standard 1–7

Extended 1–7

Day 2

Standard 12–17, 19–22

Extended 8–17, 19–22

Integrating the Strands

Functions Exs. 1–19, 22

Geometry Ex. 21

Statistics and Probability Ex. 20

Discrete Mathematics Ex. 20

Logic and Language Exs. 1, 8, 10, 22

Mathematical Procedures

You may wish to ask students how the calculator can be used to model transformational thinking. Use Sample 1 as an illustration. Discuss how the sequence of equations shown below is used to capture the transformational procedure used in the Sample.

```
Y1 = √X
Y2 = Y1(–X)
Y3 = 2Y2
Y4 = Y3(X – 4) + 1
Y5 =
Y6 =
Y7 =
Y8 =
```

Now you are ready for: Exs. 8–22 on pp. 530–532

Look Back

For any function f, how are the graphs of $y = -f(x)$ and $y = f(-x)$ related to the graph of $y = f(x)$?

9-7 Exercises and Problems

1. **Reading** What are the domain and the range of the function in Sample 1?

Sketch the graph of each function.

2. $y = -x^3$
3. $y = 10^{-x-3}$
4. $y = -3(x + 6)^2 + 4$
5. $y = -\frac{2}{3}|x - 2| - 5$
6. $y = \log(-2x - 9)$
7. $y = -4\sqrt{-x + 2} + 7$

BY THE WAY...

Chinese acrobats may have used parachute-like devices as early as 1306. The first known parachute jump was made by the French physicist Sebastian Lenormand in 1783.

8. **Skydiving** The speed of a skydiver t seconds after opening a parachute can be modeled by the function

$$s(t) = 10 + 166e^{-3.2t}$$

where $s(t)$ is measured in feet per second.

a. **Writing** What transformations can you apply to the graph of $y = e^t$ to obtain the graph of $y = s(t)$?

b. Sketch the graph of $y = s(t)$. Based on your graph, what speed does a skydiver approach as the time t increases?

c. **Research** Look up "parachute" in an encyclopedia. Write a short report on the history of parachuting.

Career A financial consultant helps people decide how to invest their money.

The graph shows what one consultant thinks is the percent of money $m(x)$ that a person x years old should invest in stocks.

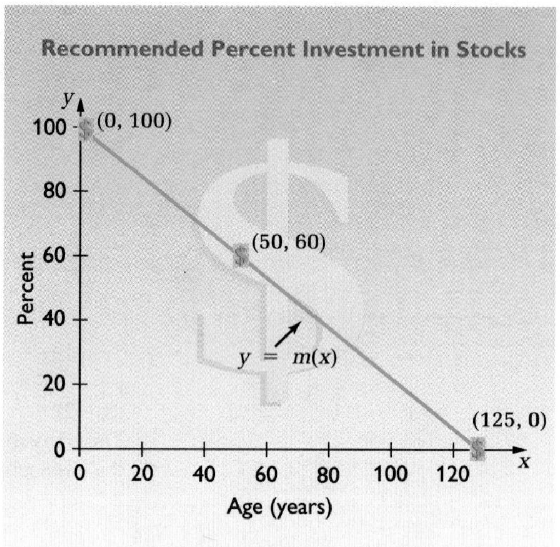

9. a. Let $b(x)$ be the percent of money that a person x years old should invest in bonds.

Write an equation that gives $b(x)$ in terms of $m(x)$. (Assume any money not invested in stocks is invested in bonds.)

b. Sketch the graph of $y = b(x)$.

10. **Writing** In general, stocks are more profitable than bonds over long periods of time. However, stocks are usually riskier investments over short periods of time.

Use these facts to explain why the graphs of $y = m(x)$ and $y = b(x)$ make sense.

Answers to Look Back

The graph of $y = -f(x)$ is the graph of $y = f(x)$ reflected over the x-axis. The graph of $y = f(-x)$ is the graph of $y = f(x)$ reflected over the y-axis.

Answers to Exercises and Problems

1. domain: all real numbers less than or equal to 4; range: all real numbers greater than or equal to 1

2.

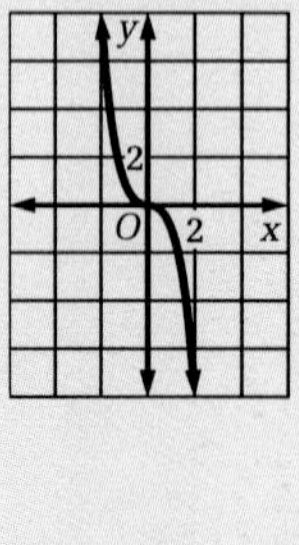

3.

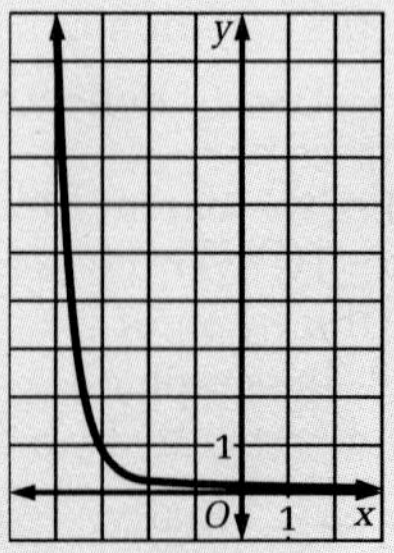

11. Scuba Diving An ocean diver is working in water 200 ft deep. The graph shows the total atmospheric and water pressure $p(d_s)$ on the diver as a function of the diver's distance d_s from the surface.

Total Atmospheric and Water Pressure on a Diver

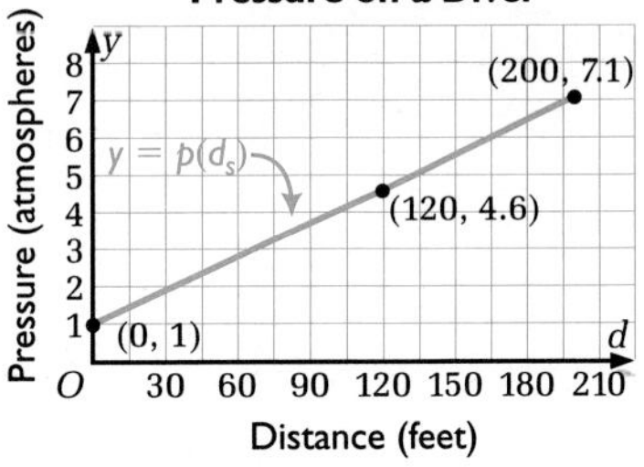

a. Let d_b be the distance of the diver from the ocean bottom. Complete this equation with an expression involving d_s:

$$p(d_b) = p(\underline{?})$$

b. Use the equation in part (a) to graph $y = p(d_b)$.

Find an equation for each graph. Assume the parent function of each graph is one of the following: $y = x^2$, $y = x^3$, $y = \sqrt{x}$, $y = |x|$, $y = 10^x$, or $y = \log x$.

12.

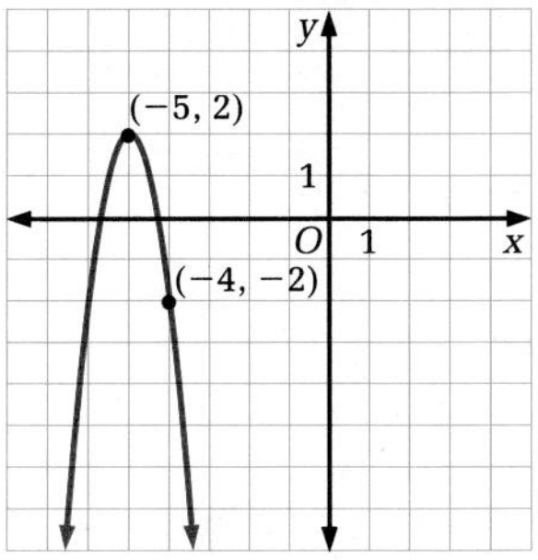

13.

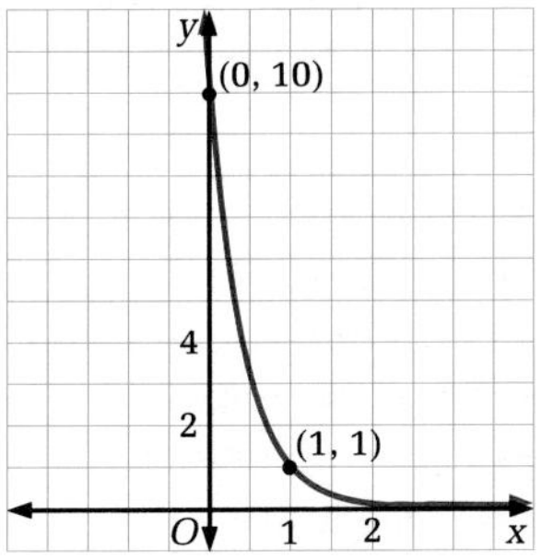

14.

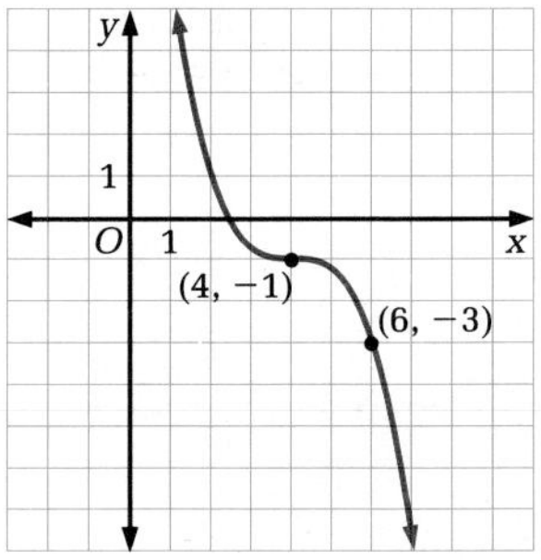

15.

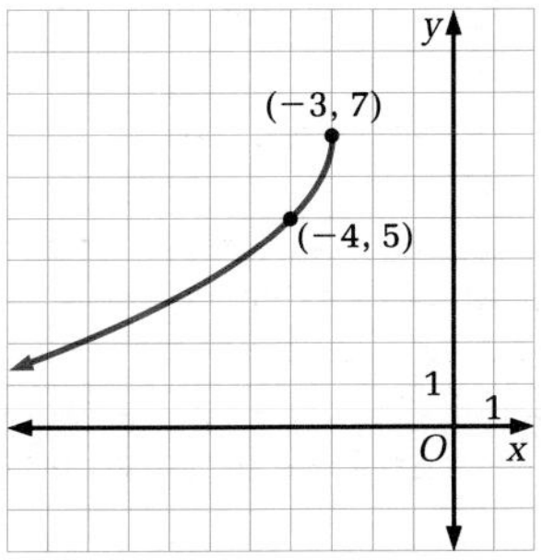

16.

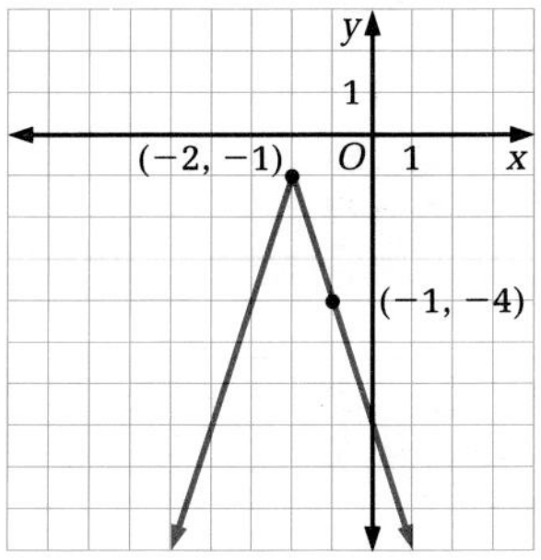

17.

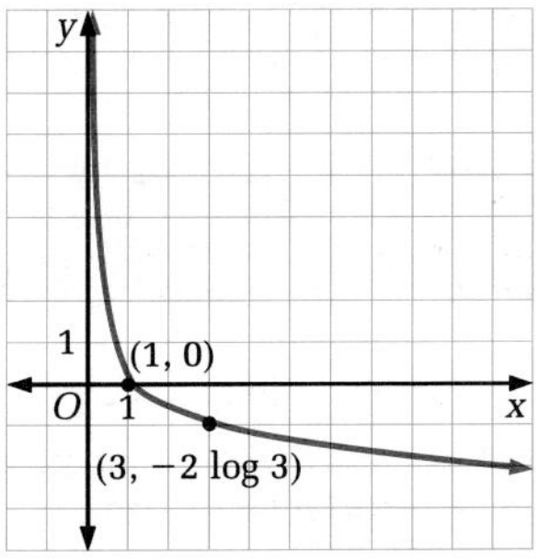

Application

Exs. 8–11 apply transformations to some real-world situations involving skydiving, financial planning, and scuba diving.

b.

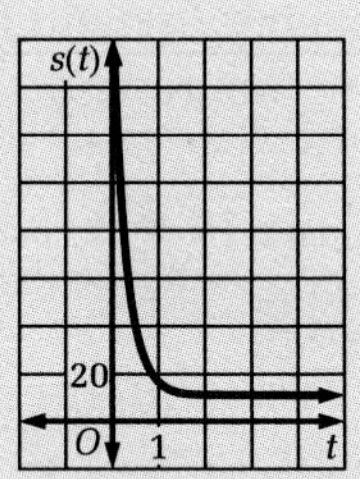

10 ft/s

c. Answers may vary. An example is given. Leonardo DaVinci sketched a parachute in 1495; the first recorded jumps were made in the late eighteenth century. A basic parachute consists of triangle-shaped panels, but many newer parachutes are rectangular. Until about 1950, parachutes were used solely to slow falling bodies but are now used to slow airplane landings, the final descent of space capsules, and race cars. Parachuting has also become a popular sport.

9. See answers in back of book.

10. Since stocks are more profitable over long periods of time, it makes sense that the percent of funds invested in stocks should decline as the investor ages. Since bonds are less risky in the short run, it makes sense that the percent of funds invested in bonds should increase as the investor ages.

11. a. $200 - d_s$

b. **Total Amtospheric and Water Pressure on a Diver**

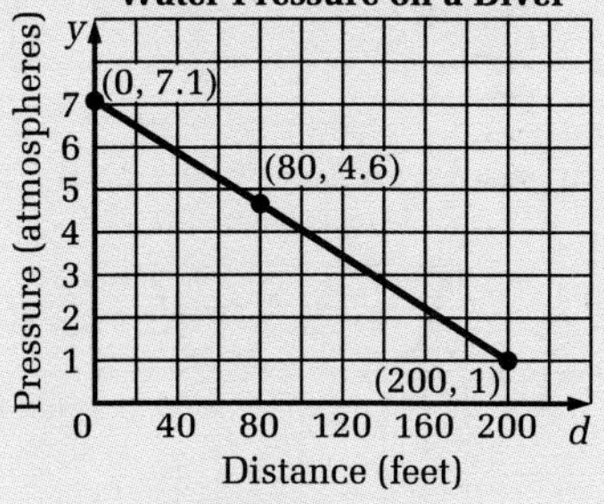

12. $y - 2 = -4(x + 5)^2$

13. $y = 10^{-(x-1)}$

14. $y = -\frac{1}{4}(x - 4)^3 - 1$

15. $y - 7 = -2\sqrt{-x - 3}$

16. $y + 1 = -3|x + 2|$

17. $y = -2 \log x$

Answers to Exercises and Problems

4.

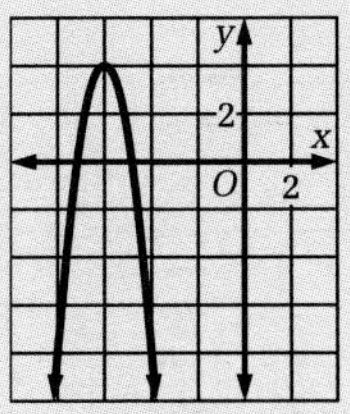

5.

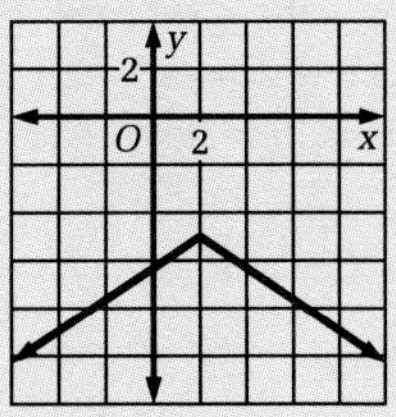

6.

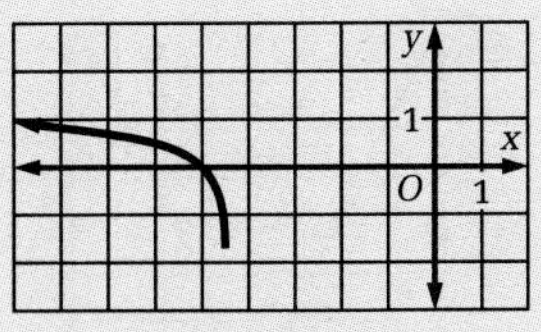

7.

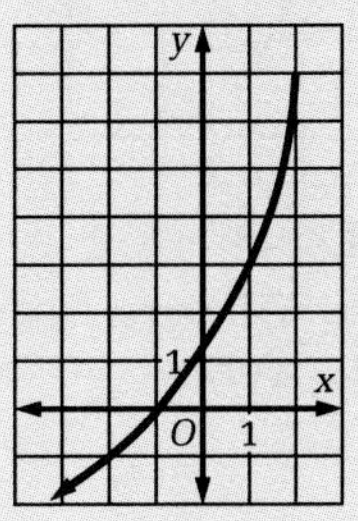

8. a. Stretch the graph of $y = e^x$ horizontally by a factor of $\frac{1}{3.2} = 0.3125$ to obtain the graph of $y = e^{3.2x}$. Reflect the graph of $y = e^{3.2x}$ over the y-axis. Stretch that graph vertically by a factor of 166, then translate the resulting graph by the vector (0, 10).

Using Manipulatives

MIRA® transparent mirrors can be used for viewing many types of transformations. You may wish to extend Ex. 18 by giving students other functions having different parent functions and asking them to perform parts (a), (b), and (c) on these new functions.

Quick Quiz (9-5 through 9-7)

See page 534.

See also Test 37 in the *Assessment Book.*

Practice 73 For use with Section 9-7

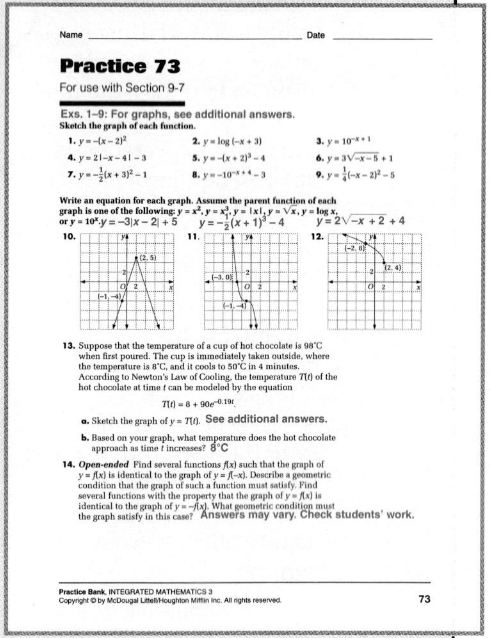
Name ____________ Date ________

Practice 73

For use with Section 9-7

Exs. 1–9: For graphs, see additional answers.

Sketch the graph of each function.

1. $y = -(x-2)^2$
2. $y = \log(-x+3)$
3. $y = 10^{-x+1}$
4. $y = 2|-x-4| - 3$
5. $y = -(x+2)^3 - 4$
6. $y = 3\sqrt{-x-5} + 1$
7. $y = -\frac{1}{2}(x+3)^2 - 1$
8. $y = -10^{-x+4} - 3$
9. $y = \frac{1}{4}(-x-2)^2 - 5$

Write an equation for each graph. Assume the parent function of each graph is one of the following: $y = x^2$, $y = x^3$, $y = |x|$, $y = \sqrt{x}$, $y = \log x$, or $y = 10^x$. $y = -3|x-2| + 5$ $y = -\frac{1}{2}(x+1)^3 - 4$ $y = 2\sqrt{-x+2} + 4$

10. (2, 5) (–1, –4)
11. (–3, 0) (–1, –4)
12. (–2, 8) (2, 4)

13. Suppose that the temperature of a cup of hot chocolate is 98°C when first poured. The cup is immediately taken outside, where the temperature is 8°C, and it cools to 50°C in 4 minutes. According to Newton's Law of Cooling, the temperature $T(t)$ of the hot chocolate at time t can be modeled by the equation

$$T(t) = 8 + 90e^{-0.19t}.$$

a. Sketch the graph of $y = T(t)$. See additional answers.

b. Based on your graph, what temperature does the hot chocolate approach as time t increases? 8°C

14. *Open-ended* Find several functions $f(x)$ such that the graph of $y = f(x)$ is identical to the graph of $y = f(-x)$. Describe a geometric condition that the graph of such a function must satisfy. Find several functions with the property that the graph of $y = f(x)$ is identical to the graph of $y = -f(x)$. What geometric condition must the graph satisfy in this case? Answers may vary. Check students' work.

Practice Bank, INTEGRATED MATHEMATICS 3

73

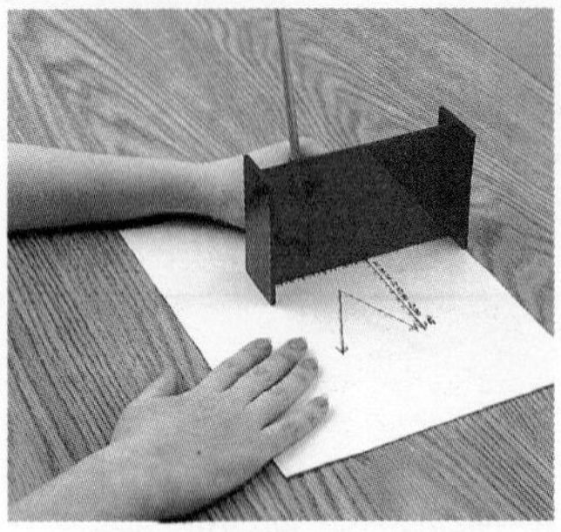

18. **Using Manipulatives** Work with another student. Use a MIRA® transparent mirror and graph paper.
    a. Graph the function $f(x) = 2|x-4| + 3$.
    b. Use the MIRA® to sketch the reflection of the graph of f over the x-axis. What is an equation of the reflected graph?
    c. Use the MIRA® to sketch the reflection of the graph of f over the y-axis. What is an equation of the reflected graph?

Ongoing ASSESSMENT

19. **Open-ended** This design was made by graphing $y = f(x)$, $y = -f(x)$, $y = f(-x)$, and $y = -f(-x)$ where $f(x) = |x-5|$. Use a different function to make your own design.

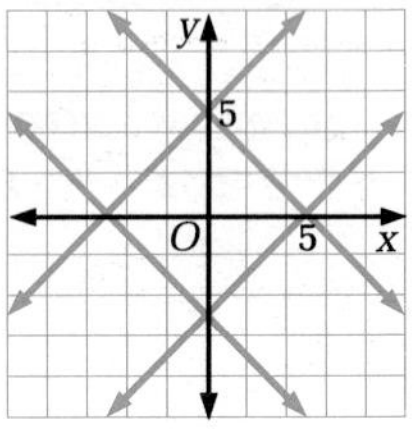

Review PREVIEW

20. Elaine's score on a chemistry test is 87. The mean of the scores in Elaine's class is 75, and the standard deviation is 16. Find Elaine's z-score. *(Section 9-6)*

21. Use the diagram. Show that $\angle PQR$ is a right angle. (*Hint:* Use the converse of the Pythagorean theorem: If $(PQ)^2 + (QR)^2 = (PR)^2$, then $\triangle PQR$ is a right triangle.) *(Toolbox Skill 31)*

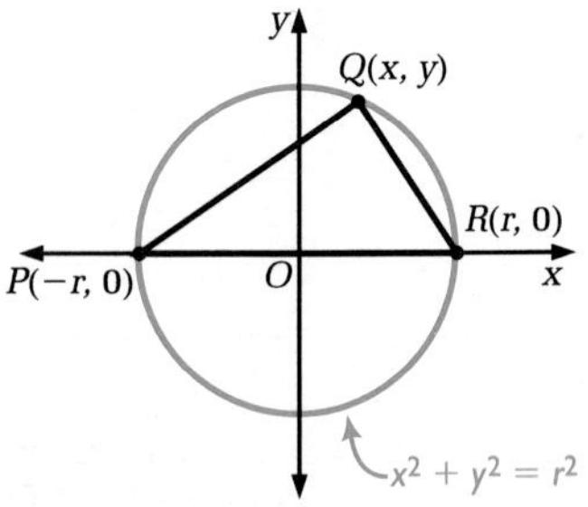

9 Working on the Unit Project

22. Claudia estimates that she will owe about \$20,000 in student loans when she finishes college. She plans to make gradually increasing monthly payments after she graduates to pay off her loans. The function

$$A(n) = (-15.3)(1.007)^n + 35.3$$

gives the amount of money (in thousands of dollars) that Claudia will still owe after making n monthly payments.

a. **Writing** What transformations can you apply to the graph of $y = (1.007)^n$ to obtain the graph of $y = A(n)$?

b. Sketch the graph of $y = A(n)$.

c. TECHNOLOGY Graph $y = A(n)$ on a graphics calculator. Use the graph to find how long it will take Claudia to pay off her student loans.

Answers to Exercises and Problems

18. a.

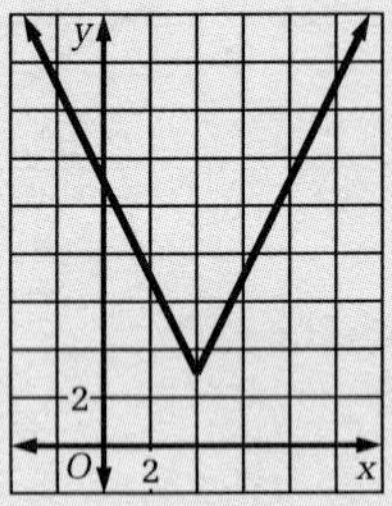

b.

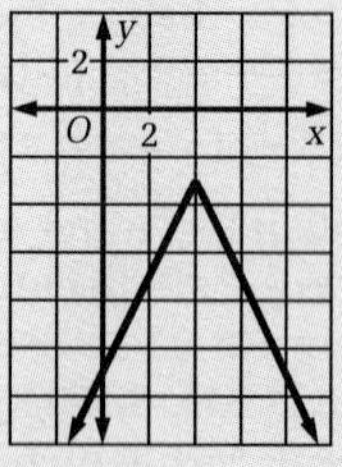

$y = -2|x-4| - 3$

c.

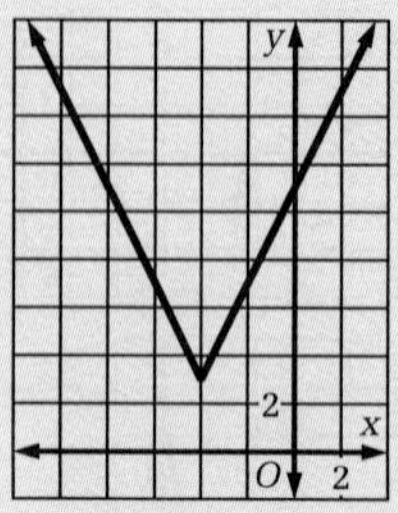

$y = 2|x+4| + 3$

19. Answers will vary. Check students' work.

20. 0.75

21. $PQ = \sqrt{x-(-r)^2 + (y-0)^2} = \sqrt{(x+r)^2 + y^2}$; $QR = \sqrt{(x-r)^2 + (y-0)^2}$; $\overline{PR}$ is a diameter, so $PR = 2r$. Then $(PQ)^2 + (QR)^2 = (x+r)^2 + y^2 + (x-r)^2 + y^2 = x^2 + 2xr + r^2 + y^2 + x^2 - 2xr + r^2 + y^2 = 2(x^2 + y^2 + r^2) = 2(2r^2) = 4r^2 = (PR)^2$. By the converse of the

Unit Project 9

Completing the Unit Project

Now you are ready to complete your list of questions and interview a career representative.

Your questions should include these topics:

- **college or other training needed for entry and advancement**
- **the role of mathematics on the job**
- **required tests and their scoring**
- **potential for job openings and job security**
- **working conditions, hours, and benefits**

The completed project should include a well-organized set of notes for reviewing and analyzing what you learned in the interview.

Look Back

Compare your interview questions with those of other groups. What additional questions would you ask if you had a second job interview?

Alternative Projects

Project 1: Re-scaling the SAT Test

In 1994 the College Board announced that beginning in 1995 scores on the Scholastic Aptitude Test (SAT) would be re-scaled. In the years leading up to the change, the mean SAT verbal score was 424 and the mean SAT math score was 478. Under the new system, the mean for both sections of the test is 500.

This decision was based on the concept of transforming a data set. Use articles from newspapers and magazines and literature from the College Board to research this question. Write a paper explaining why the College Board decided to "re-center" the scores and how students and parents should interpret the new scores.

Project 2: Transformations of Functions

You have studied linear, quadratic, polynomial, rational, exponential, logarithmic, and absolute value functions. In applications, the functions that model a real-life situation are often transformations of these parent functions. Find at least six examples of transformed functions that model real-life applications. Make a poster that presents these applications along with graphs of the transformed functions.

Assessment

A scoring rubric for the Unit Project can be found on pages 480 and 481 of this Teacher's Edition and also in the ***Project Book.***

Answers to Exercises and Problems

Pythagorean theorem, $\triangle PQR$ is a right triangle and $\angle PQR$ is a right angle.

22\. a. Stretch the graph vertically by a factor of 15.3 and reflect the resulting graph over the x-axis. Finally, translate the reflected graph by the vector (0, 35.3).

b.

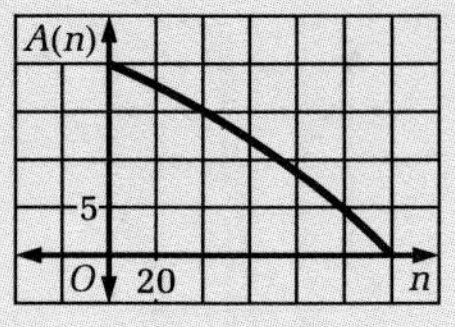

c.

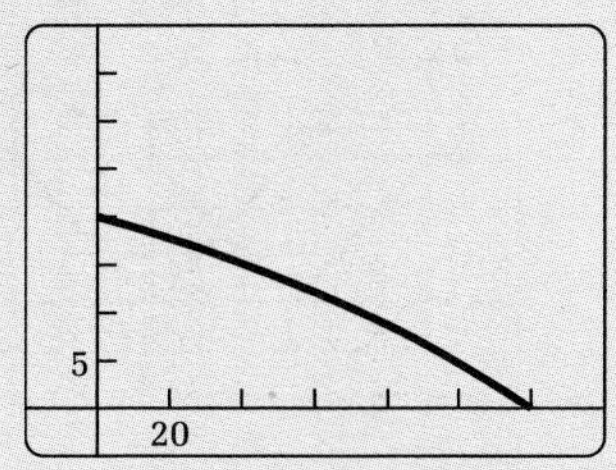

It takes about 120 months for Claudia to pay off her loans.

Unit Support Materials

- *Practice Bank:* Cumulative Practice 74
- *Study Guide:* Unit 9 Review
- Problem Bank: Unifying Problem 9
- *Assessment Book:* Unit Tests 38 and 39 Spanish Unit Tests Alternative Assessment
- *Test Generator* software with Test Bank
- *Teacher's Resources for Transfer Students*

Quick Quiz (9-5 through 9-7)

1. Juan records the temperature outside his house at noon for one week. The temperatures (in °F) are: 68.5, 72.3, 59.9, 65.7, 67.6, 70.3, 62.4. Find the mean and standard deviation of the data. What are the mean and standard deviation in degrees Celsius? [9-5] **about 66.7°F; about 4°F; about 19.3°C; about 2.2°C**
2. Find the function and sketch the graph when $y = \sqrt{x}$ is stretched by a factor of 0.5 in the vertical direction and then translated right 2 units. [9-5]
$y = 0.5\sqrt{x - 2}$

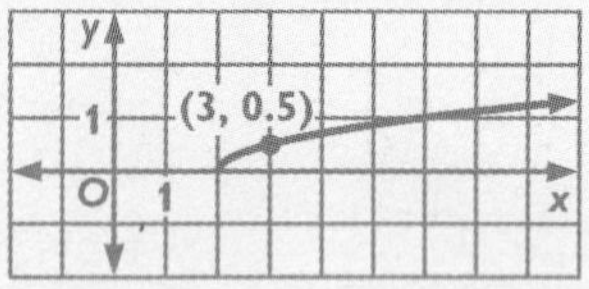

The grades on a science test in Sato's class were normally distributed with a mean of 72.5 and a standard deviation of 6.8.

3. If Sato got a 77 on the test, what percentage of the class scored better than her? [9-6] **about 24%**
4. The teacher decided she would not scale the test if 85% of the students scored better than 65. Will she need to scale the test? [9-6] **No.**

1. The table shows the heights of the ten tallest buildings in Denver, Colorado. **9-1**

Name of building	Height (ft)
Amoco Building	448
Anaconda Tower	507
Arco Tower	527
First Interstate Tower North	434
Mountain Bell Center	709
1999 Broadway	544
Republic Plaza	714
17th Street Plaza	438
Stellar Plaza	437
United Bank of Denver	698

a. Find the mean, the median, the range, and the standard deviation for the heights of the buildings.

b. Denver is about 5280 ft above sea level. Find the height of each building above sea level.

c. Use what you know about adding a constant to each value in a data set to find the mean, the median, the range, and the standard deviation for the heights of the buildings above sea level.

Sketch a parent graph and translate it to obtain a graph of each equation. **9-2**

2. $y = |x + 6|$
3. $y - 8 = 10^x$
4. $y + 1 = \sqrt{x - 3}$
5. **Writing** Danielle built a large scale model of the ten tallest buildings in Denver. The scale is 1 ft = 100 ft. Describe two ways you could use the scale to find the mean and the standard deviation of the heights for the ten buildings in Danielle's model. **9-3**

The matrix $M = \begin{bmatrix} 0 & 5 & 6 & 1 \\ 0 & 0 & 3 & 3 \end{bmatrix}$ represents the vertices of a trapezoid. **9-4**

Multiply M by each transformation matrix and then describe the effect of each transformation.

6. $\begin{bmatrix} 4 & 0 \\ 0 & 1 \end{bmatrix}$
7. $\begin{bmatrix} 1 & 0 \\ 0 & \frac{1}{3} \end{bmatrix}$
8. $\begin{bmatrix} 2 & 0 \\ 0 & 5 \end{bmatrix}$

Without using graphing technology, sketch the graph of each function.

9. $y = 3\sqrt{x}$
10. $y = \left|\frac{x}{5}\right|$
11. $y = (x - 2)^3$

Answers to Unit 9 Review and Assessment

1. See answers in back of book.
2. 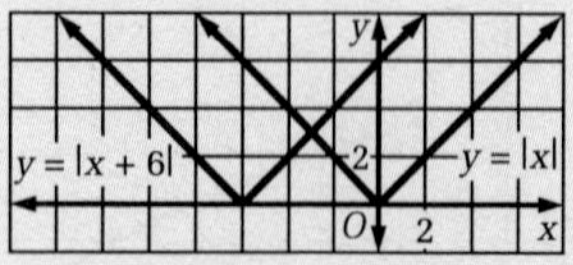

3. 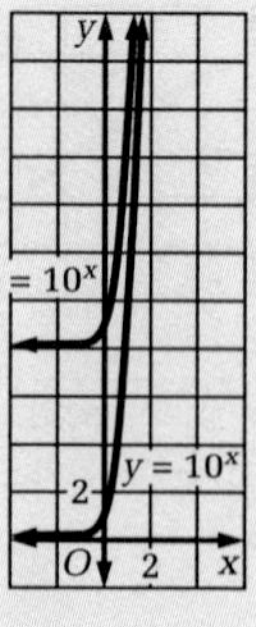

4. 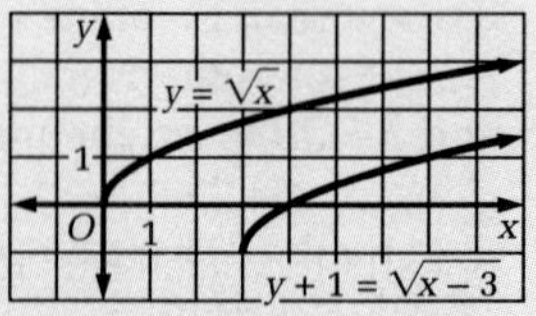

5. Answers may vary. An example is given. (1) Determine the mean and standard deviation of the model heights and multiply both by 100 to deter-

12. For 1961–1980, the mean of the average September temperatures in Beijing, China, was 66.7°F, and the standard deviation was 1.2°F. What are the mean and the standard deviation in degrees Celsius? **9-5**

13. **Open-ended** The ten students in an advanced class received the following scores on a 100-point test:

47 52 39 44 51 63 50 48 45 56

The teacher has decided to rescale the scores using a linear function $f(x) = mx + b$ where x is a student's actual score and $f(x)$ is the rescaled score. Find a function that brings all rescaled scores within the range 60–100. Explain your choice of m and b.

Find an equation for each graph. The parent function is given.

14. parent function: $y = |x|$

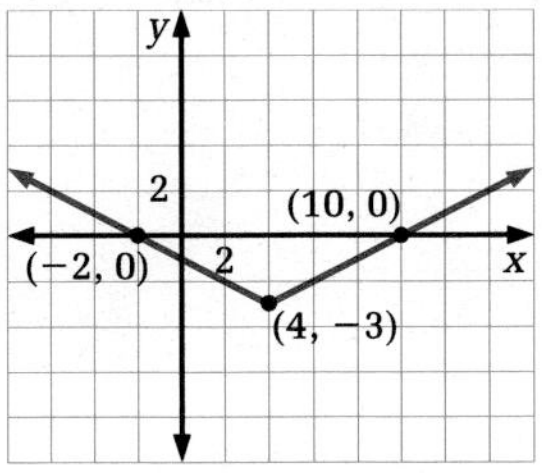

15. parent function: $y = x^3$

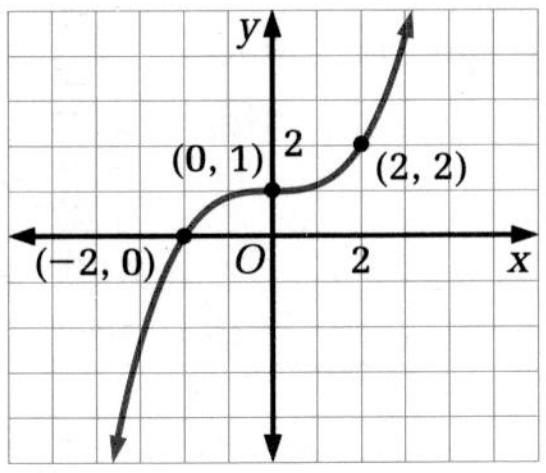

16. Use the standard normal table on page 521 to find the percent of data less than the given z-score. **9-6**

a. $z = 0.6$
b. $z = -1.3$

17. The heights of men 18 years old or older in the United States are normally distributed with a mean of 69.1 in. and a standard deviation of 2.8 in. What percent of men in the United States are taller than 72 in.?

18. Sketch the graph of $y = -3|x + 4| + 2$. **9-7**

19. Jerry drops a stone into an empty well that is 80 ft deep. The graph shows the stone's distance $f(t)$ from the top of the well after t seconds.

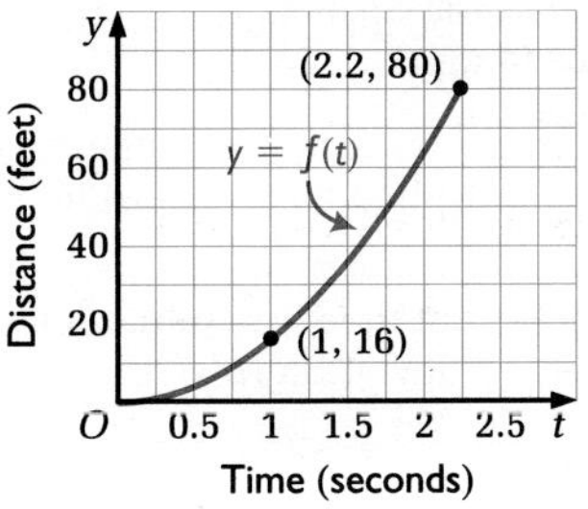

a. Let $g(t)$ be the stone's distance from the bottom of the well after t seconds. Write an equation that gives $g(t)$ in terms of $f(t)$.

b. Sketch the graph of $g(t)$.

5. Write an equation for the graph below if the parent function is $y = \log x$. [9-7]
$y = \log(-x + 1)$

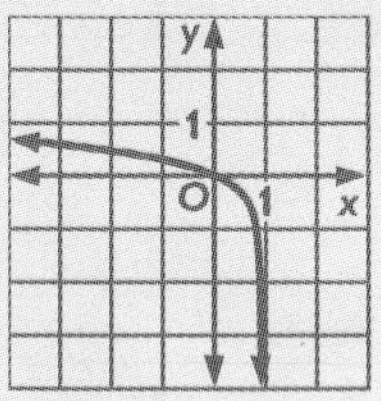

Answers to Unit 9 Review and Assessment

mine the actual mean and standard deviation.
(2) Multiply each model height by 100 and then calculate the mean and standard deviation of the actual heights.

6. $\begin{bmatrix} 0 & 20 & 24 & 4 \\ 0 & 0 & 3 & 3 \end{bmatrix}$; horizontal stretch by a factor of 4

7. $\begin{bmatrix} 0 & 5 & 6 & 1 \\ 0 & 0 & 1 & 1 \end{bmatrix}$; a vertical stretch by a factor of $\frac{1}{3}$

8. $\begin{bmatrix} 0 & 10 & 12 & 2 \\ 0 & 0 & 15 & 15 \end{bmatrix}$; a horizontal stretch by a factor of 2 and a vertical stretch by a factor of 5

9.

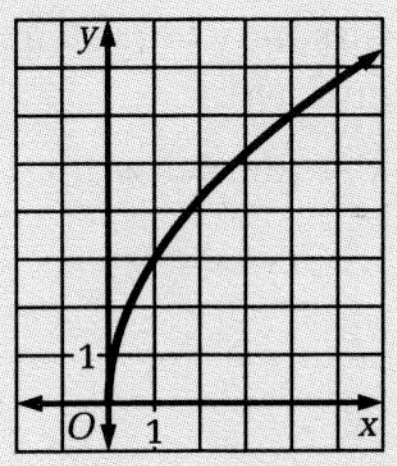

10.

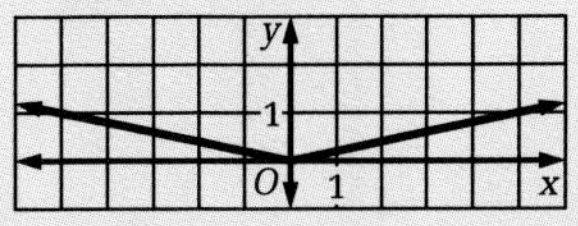

11.

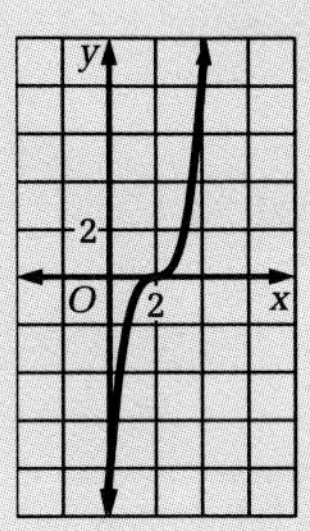

12. about 19.3°C; about 0.7°C

13. Answers may vary. An example is given. $f(x) = \frac{5}{3}x - 5$. The range of the actual scores is 24 and we want it to be $100 - 60 = 40$, so $m = \frac{40}{24} = \frac{5}{3}$. Since $\frac{5}{3} \cdot 39 = 65$ and we want a 39 to be scaled to a 60, $b = -5$.

14. $y + 3 = \frac{1}{2}|x - 4|$

15. $y - 1 = \left(\frac{x}{2}\right)^3$

16. a. about 73%
b. about 10%

17. about 16%

18.

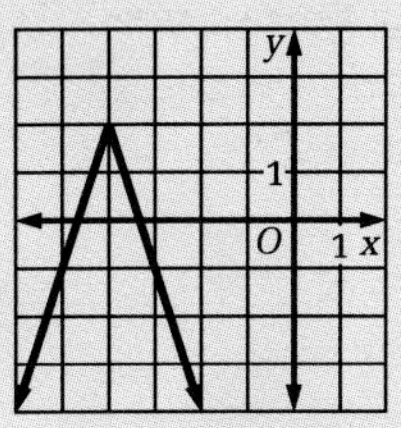

19. a. $g(t) = 80 - f(t)$

b.

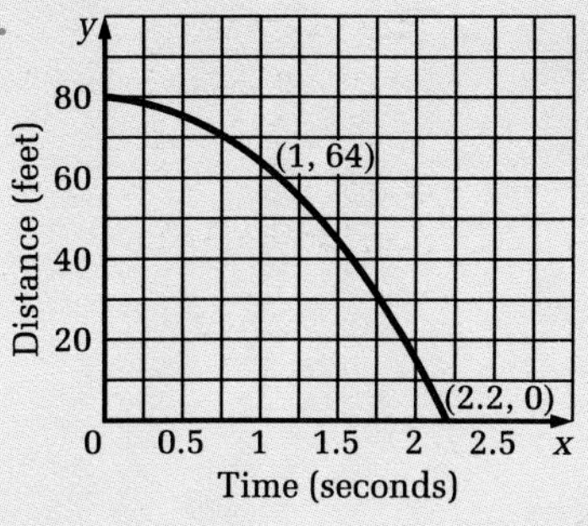

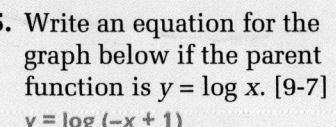

20. **Self-evaluation** Make a concept map to show your understanding of the transformations of data and functions discussed in this unit.

21. **Group Activity** Work in a group of three students. The group should sketch graphs on the same set of axes.
    a. Student 1 should choose a parent function.
    b. Student 2 should sketch the graph of the parent function from part (a). Then student 2 should decide whether student 3 should reflect the graph over the x-axis or over the y-axis.
    c. Student 3 should sketch the reflected graph. Then student 3 should choose a horizontal stretch or a vertical stretch for student 1 to graph.
    d. Student 1 should sketch the stretched graph. Then student 1 should choose a horizontal translation or a vertical translation for student 2 to graph.
    e. Student 2 should sketch the translated graph.
    f. Student 3 should write an equation of the graph sketched in part (e).

IDEAS AND (FORMULAS) = X^2 $_5P_5$

STATISTICS & PROBABILITY

- When the same constant value is added to each value in a data set:
    - the mean, the median, and the graph of the distribution are translated by the constant value. *(p. 485)*
    - the range and the standard deviation remain the same. *(p. 485)*
- When each value in a data set is multiplied by the same positive constant:
    - the graph of the distribution is stretched by the constant value. *(p. 496)*
    - the mean, the median, the range, and the standard deviation are multiplied by the constant value. *(p. 498)*
- To find the z-score for a data value, subtract the mean to find the deviation from the mean, and then divide the difference by the standard deviation *(p. 520)*:

$$z = \frac{x - \bar{x}}{\sigma}$$

- The parent curve of any normal distribution is the standard normal curve. *(p. 521)*
- The standard normal table gives the percent of data less than any given z-score. *(p. 521)*

Answers to Unit 9 Review and Assessment

20. Answers may vary. Check students' work.
21. Check students' work.

ALGEBRA

- When the graph of $y = f(x)$ is translated horizontally by the constant h, an equation of the new graph is $y = f(x - h)$. *(p. 490)*
- When the graph of $y = f(x)$ is translated vertically by the constant k, an equation of the new graph is $y - k = f(x)$. *(p. 490)*
- When the graph of $y = f(x)$ is stretched horizontally by the positive factor a:
  - the original graph crossing the y-axis remains fixed. *(p. 502)*
  - an equation of the new graph is $y = f\left(\frac{x}{a}\right)$. *(p. 502)*
- When the graph of $y = f(x)$ is stretched vertically by the positive factor b:
  - all points where the original graph crosses the x-axis remain fixed. *(p. 502)*
  - an equation of the new graph is $\frac{y}{b} = f(x)$. *(p. 502)*
- The temperature conversion formula $C = \frac{5}{9}(F - 32)$ is a combination of the transformations of translating and stretching. *(p. 511)*
- When the graph of $y = f(x)$ is reflected over the x-axis, an equation of the new graph is $y = -f(x)$. *(p. 527)*
- When the graph of $y = f(x)$ is reflected over the y-axis, an equation of the new graph is $y = f(-x)$. *(p. 527)*
- Combinations of translation, stretching, and reflection may be used to produce a new function from a given parent function. *(p. 527)*

GEOMETRY

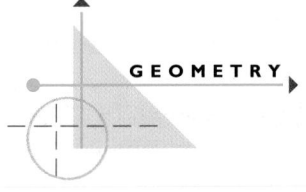

- A figure in the coordinate plane can be stretched in the horizontal or vertical direction by multiplying the coordinates of its vertices by a transformation matrix. *(pp. 501, 502)*

DISCRETE MATH

- For positive numbers a and b, the transformation matrix $\begin{bmatrix} a & 0 \\ 0 & b \end{bmatrix}$ represents stretching by a factor of a in the horizontal direction and by a factor of b in the vertical direction. *(p. 502)*

Key Terms

- **transformation matrix** (p. 501)
- **standard normal curve** (p. 521)
- **z-score** (p. 520)
- **percentile** (p. 523)

Quick Quiz (9–1 through 9–4)

The salaries at a company have a mean of $37,000, a median of $31,000, a range of $100,000, and a standard deviation of $11,500.

1. Find the mean, median, range, and standard deviation of the salaries including a $750 holiday bonus which is given to each employee. [9-1]
   $37,750; $31,750; $100,000; $11,500
2. Find the mean, median, range and standard deviation of the salaries after an across-the-board raise of 3.5%. [9-3]
   $38,295; $32,085; $103,500; $11,902.50
3. Suppose that the function that represents the original salaries is $y = f(x)$, where x is the salary step, a number assigned each employee based on their position, education, and experience. Write a function that represents the salaries after the 3.5% raise. [9-4]
   $y = 1.035\, f(x)$
4. Write an equation for the graph below. [9-2]

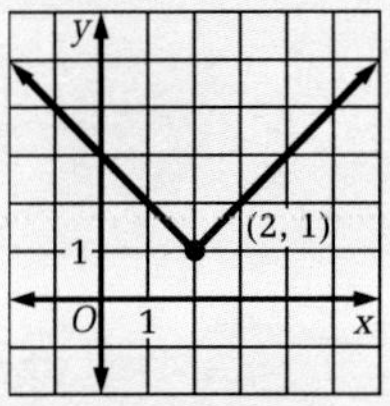

   $y = |x - 2| + 1$
5. Describe the transformation represented by $\begin{bmatrix} 2 & 0 \\ 0 & \frac{1}{3} \end{bmatrix}$. [9-4]
   a horizontal stretch of 2 and a vertical stretch of $\frac{1}{3}$

Planning the Unit

10 Periodic Models

OVERVIEW

- **Unit 10** explores periodic situations found in the real world. The features of these situations are used in examining the sine and cosine functions and their graphs. Transformations of these functions are examined for the related changes in the graphs. Sine and cosine functions are then used to model many real-life situations.
- Transformation skills are used to describe and create border patterns found in decorations and art. The periodicity of these patterns is explored. Tessellations of figures are also explored for periodicity. Students identify and create regular and semiregular tessellations.
- The **Unit Project** revolves around making a functional scrapbook. Students use the scrapbook to describe specific situations that can be modeled by graphs. Half of the siutations should be periodic. The situations chosen can be from students' lives, as well as local, national, and international topics.
- **Connections** to state lottery revenues, trampoline jumping, water power, a Ferris wheel, and the art of M. C. Escher are some of the topics contained in the teaching material and the exercises.
- **Graphics calculators** are used in Section 10-1 to compare rates of growth in linear, quadratic, and exponential functions, in Section 10-3 to graph the sine and cosine functions, and in Section 10-4 to graph transformations of sine and cosine functions.
- **Problem-solving strategies** used in Unit 10 include making a table and drawing a graph.

Unit Objectives

Section	Objectives	NCTM Standards
10-1	• Describe graphs and examine rates of change.	1, 2, 3, 4, 5, 6
10-2	• Recognize periodic functions and features of their graphs.	1, 2, 3, 4, 5, 6
10-3	• Recognize and understand characteristics of the graphs of the sine function and the cosine function.	1, 2, 3, 4, 5, 6, 9
10-4	• Recognize and understand characteristics of the graphs of sine functions and cosine functions that are transformed.	1, 2, 3, 4, 5, 6, 9
10-5	• Use sine and cosine functions to model periodic situations.	1, 2, 3, 4, 5, 6, 9, 10, 12
10-6	• Recognize periodic border patterns. • Use transformations to classify and create border patterns.	1, 2, 3, 4, 7, 8
10-7	• Recognize, classify, and create tessellations.	1, 2, 3, 4, 5, 7, 8

Topic Spiraling

Section	Connections to Prior and Future Concepts
10-1	**Section 10-1** describes graphs as constant, increasing, decreasing, periodic, saturating, and damping. Students explore rates of change of functions. Periodic functions are important in Sections 10-2 through 10-5 of Book 3.
10-2	**Section 10-2** explores periodic functions and the features of their graphs. The cycle, period, frequency, and amplitude are covered. Knowledge of these topics is necessary in graphing the sine and cosine functions in the next section.
10-3	**Section 10-3** covers graphs of the sine and cosine functions. The graphs are examined for similarities and differences. Comparisons include the topics introduced in the previous sections. Sine and cosine were first introduced in Section 6-7 of Book 1.
10-4	**Section 10-4** explores transformations on the sine and cosine graphs. Students recognize and understand characteristics of the graphs of sine and cosine functions that are transformed. Transformations were first covered in Sections 4-3, 6-6, 10-1, and 10-2 of Book 1, and were reviewed and expanded upon in Sections 3-6 and 4-2 of Book 2.
10-5	**Section 10-5** covers modeling periodic situations using the sine and cosine functions. Equations are written for these models, based on change in period, amplitude, and so on. Mathematical modeling of real-world situations was introduced in Section 5-1 of Book 1 and is used extensively throughout Books 1, 2, and 3. The use of mathematical models to solve problems is a fundamental procedure used throughout more advanced courses in mathematics, science, and many other disciplines.
10-6	**Section 10-6** explores border patterns used in decorations. Patterns are examined for translations, glide reflections, vertical reflections, and so on, applying transformation skills learned in Sections 4-3, 4-4, and 10-1 of Book 1, Sections 4-2 and 5-4 of Book 2, and Unit 9 of Book 3.
10-7	**Section 10-7** explores tessellations—regular, semiregular, and nonregular. The transformation skills referred to in Section 10-6 are applied to tiling problems.

Integrating the Strands

Strands	Sections
Algebra	10-1, 10-2, 10-3, 10-4, 10-5, 10-6, 10-7
Functions	10-1, 10-2, 10-3, 10-4, 10-5, 10-6, 10-7
Geometry	10-2, 10-3, 10-4, 10-5, 10-6, 10-7
Trigonometry	10-2, 10-3, 10-4, 10-5, 10-6
Statistics and Probability	10-5, 10-6
Discrete Mathematics	10-4, 10-5, 10-6
Logic and Language	10-1, 10-2, 10-3, 10-4, 10-5, 10-6, 10-7

Planning the Unit 10 Periodic Models

Section Planning Guide

- Essential exercises and problems are indicated in boldface.
- Ongoing work on the Unit Project is indicated in color.
- Exercises and problems that require student research, group work, manipulatives, or graphing technology are indicated in the column headed "Other."

Section	Materials	Pacing	Standard Assignment	Extended Assignment	Other
10-1	graphics calculator	Day 1	**1–3, 7–20,** 25–34, 35–40	**1–3,** 4–6, **7–20,** 21–34, 35–40	
10-2		Day 1	**1–22,** 23–28, 31–36, 37	**1–22,** 23–29, 31–36, 37	30
10-3	graphing technology	Day 1	**1–14,** 20–27, 28	**1–14,** 20–27, 28	15–19
10-4	graphing technology	Day 1 Day 2	**1, 6–17** **22–33,** 44–47, 48	**1, 6–17,** 18–21 **22–33,** 34–39, 42, 44–47, 48	2–5 40, 41, 43
10-5		Day 1 Day 2	**2–8** **9–11,** 12–20, 21, 22	1, **2–8** **9–11,** 12–20, 21, 22	8d 22
10-6		Day 1 Day 2	**1–4** **5–11, 14–18,** 20–23, 24, 25	**1–4** **5–11,** 12, 13, **14–18,** 20–23, 24, 25	 19
10-7	paper cutouts of regular polygons, geometric drawing software, scissors	Day 1 Day 2	**1–12** **14–16,** 17–23, 25, 31–34, 35	**1–12,** 13 **14–16,** 17–25, 27–34, 35	 26
Review Test		**Day 1** **Day 2**	**Unit Review** **Unit Test**	**Unit Review** **Unit Test**	

Yearly Pacing	Unit 10 Total	Units 1–10 Total	Remaining	Total
	15 days (2 for Unit Project)	154 days	0 days	154 days

Support Materials

- See **Project Book** for notes on Unit 10 Project: Make a Functional Scrapbook.
- UPP and disk refer to **Using Plotter Plus** booklet and **Plotter Plus** disk.
- TI-81/82 refers to **Using TI-81 and TI-82 Calculators** booklet.
- Warm-up exercises for each section are available on **Warm-Up Transparencies.**

Section	Study Guide	Practice Bank	Problem Bank	Activity Bank	Explorations Lab Manual	Assessment Book	Visuals	Technology
10-1	10-1	Practice 75	Set 21	Enrich 66	Master 2	Quiz 10-1		
10-2	10-2	Practice 76	Set 21	Enrich 67	Masters 1, 2	Quiz 10-2		
10-3	10-3	Practice 77	Set 21	Enrich 68	Master 2	Quiz 10-3	Folder 10	
10-4	10-4	Practice 78	Set 21	Enrich 69	Add. Expl. 12 Master 2	Quiz 10-4		TI-81/82, page 71 UPP, page 61
10-5	10-5	Practice 79	Set 21	Enrich 70	Masters 1, 2	Quiz 10-5 Test 40		
10-6	10-6	Practice 80	Set 22	Enrich 71		Quiz 10-6		
10-7	10-7	Practice 81	Set 22	Enrich 72	Masters 2, 21	Quiz 10-7 Test 41		
Unit 10	Unit Review	Practice 82	Unifying Problem 10	Family Involve 10		Tests 42–45		

Form A

Spanish versions of these tests are on pages 151–154 of the **Assessment Book.**

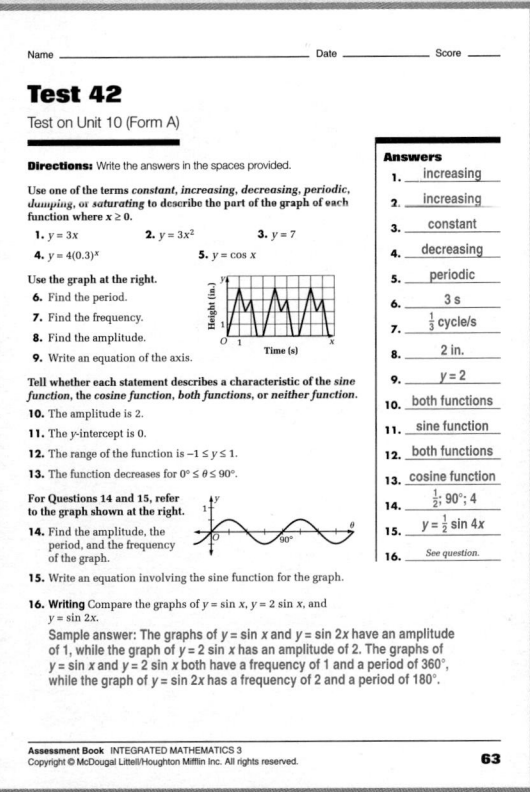

Name ________ Date ________ Score ________

Test 42

Test on Unit 10 (Form A)

Directions: Write the answers in the spaces provided.

Use one of the terms *constant, increasing, decreasing, periodic, damping,* or *saturating* to describe the part of the graph of each function where $x \geq 0$.

1. $y = 3x$ **2.** $y = 3x^2$ **3.** $y = 7$

4. $y = 4(0.3)^x$ **5.** $y = \cos x$

Use the graph at the right.

6. Find the period.

7. Find the frequency.

8. Find the amplitude.

9. Write an equation of the axis.

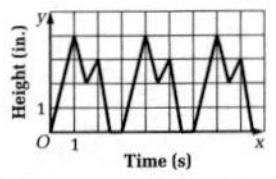

Tell whether each statement describes a characteristic of the *sine function*, the *cosine function*, *both functions*, or *neither function*.

10. The amplitude is 2.

11. The *y*-intercept is 0.

12. The range of the function is $-1 \leq y \leq 1$.

13. The function decreases for $0° \leq \theta \leq 90°$.

For Questions 14 and 15, refer to the graph shown at the right.

14. Find the amplitude, the period, and the frequency of the graph.

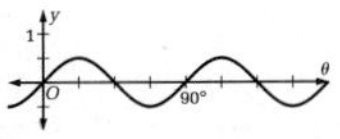

15. Write an equation involving the sine function for the graph.

16. Writing Compare the graphs of $y = \sin x$, $y = 2 \sin x$, and $y = \sin 2x$.

Sample answer: The graphs of $y = \sin x$ and $y = \sin 2x$ have an amplitude of 1, while the graph of $y = 2 \sin x$ has an amplitude of 2. The graphs of $y = \sin x$ and $y = 2 \sin x$ both have a frequency of 1 and a period of 360°, while the graph of $y = \sin 2x$ has a frequency of 2 and a period of 180°.

Answers

1. increasing
2. increasing
3. constant
4. decreasing
5. periodic
6. 3 s
7. $\frac{1}{3}$ cycle/s
8. 2 in.
9. $y = 2$
10. both functions
11. sine function
12. both functions
13. cosine function
14. $\frac{1}{2}$; 90°; 4
15. $y = \frac{1}{2} \sin 4x$
16. *See question.*

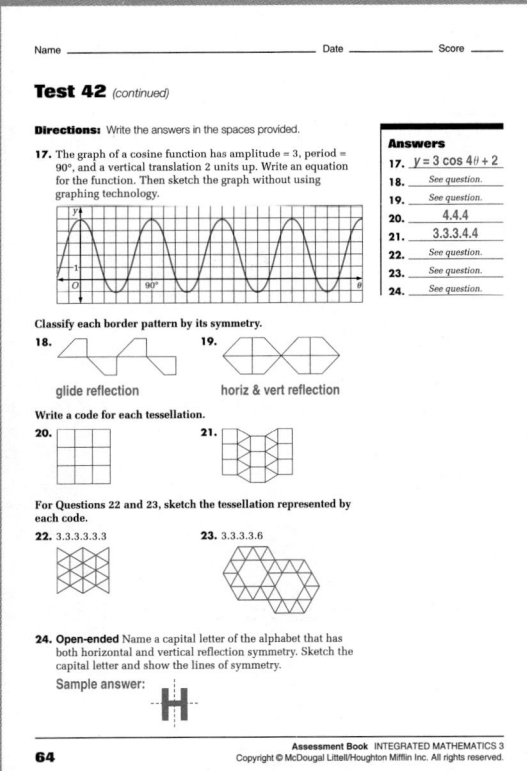

Assessment Book INTEGRATED MATHEMATICS 3

63

Name ________ Date ________ Score ________

Test 42 *(continued)*

Directions: Write the answers in the spaces provided.

17. The graph of a cosine function has amplitude = 3, period = 90°, and a vertical translation 2 units up. Write an equation for the function. Then sketch the graph without using graphing technology.

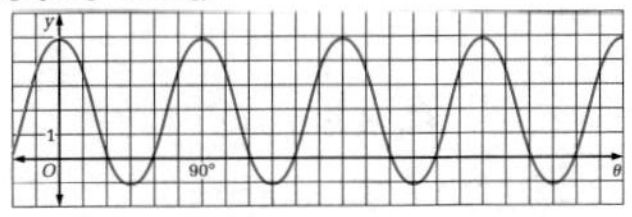

Classify each border pattern by its symmetry.

18.

glide reflection

19.

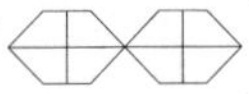

horiz & vert reflection

Write a code for each tessellation.

20.

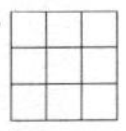

21.

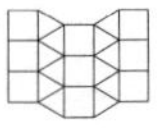

For Questions 22 and 23, sketch the tessellation represented by each code.

22. 3.3.3.3.3.3

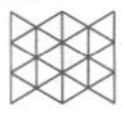

23. 3.3.3.3.6

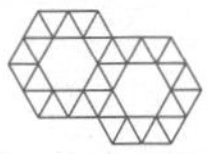

24. Open-ended Name a capital letter of the alphabet that has both horizontal and vertical reflection symmetry. Sketch the capital letter and show the lines of symmetry.

Sample answer:

Answers

17. $y = 3 \cos 4\theta + 2$
18. *See question.*
19. *See question.*
20. 4.4.4
21. 3.3.3.4.4
22. *See question.*
23. *See question.*
24. *See question.*

64

Assessment Book INTEGRATED MATHEMATICS 3

Form B

Name ________ Date ________ Score ________

Test 43

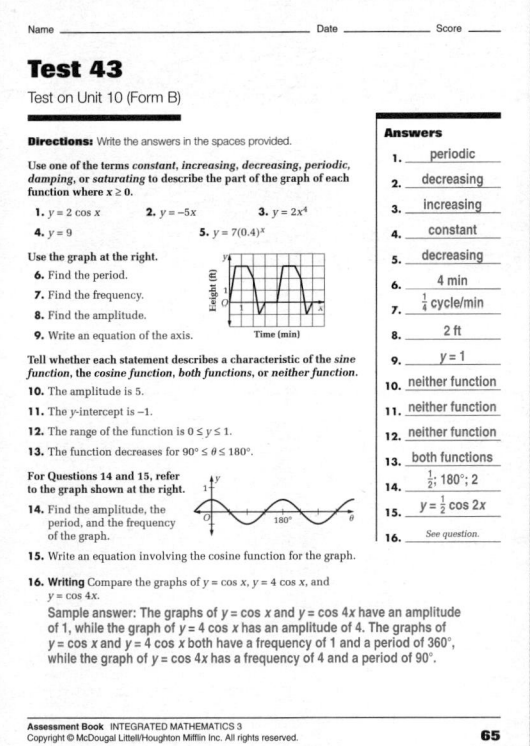

Test on Unit 10 (Form B)

Directions: Write the answers in the spaces provided.

Use one of the terms *constant, increasing, decreasing, periodic, damping,* or *saturating* to describe the part of the graph of each function where $x \geq 0$.

1. $y = 2 \cos x$ **2.** $y = -5x$ **3.** $y = 2x^4$

4. $y = 9$ **5.** $y = 7(0.4)^x$

Use the graph at the right.

6. Find the period.

7. Find the frequency.

8. Find the amplitude.

9. Write an equation of the axis.

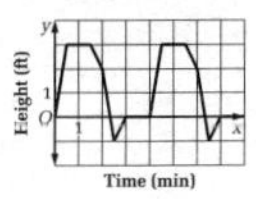

Tell whether each statement describes a characteristic of the *sine function*, the *cosine function*, *both functions*, or *neither function*.

10. The amplitude is 5.

11. The *y*-intercept is -1.

12. The range of the function is $0 \leq y \leq 1$.

13. The function decreases for $90° \leq \theta \leq 180°$.

For Questions 14 and 15, refer to the graph shown at the right.

14. Find the amplitude, the period, and the frequency of the graph.

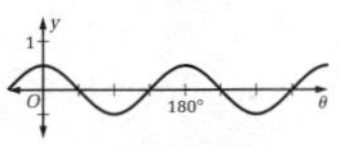

15. Write an equation involving the cosine function for the graph.

16. Writing Compare the graphs of $y = \cos x$, $y = 4 \cos x$, and $y = \cos 4x$.

Sample answer: The graphs of $y = \cos x$ and $y = \cos 4x$ have an amplitude of 1, while the graph of $y = 4 \cos x$ has an amplitude of 4. The graphs of $y = \cos x$ and $y = 4 \cos x$ both have a frequency of 1 and a period of 360°, while the graph of $y = \cos 4x$ has a frequency of 4 and a period of 90°.

Answers

1. periodic
2. decreasing
3. increasing
4. constant
5. decreasing
6. 4 min
7. $\frac{1}{4}$ cycle/min
8. 2 ft
9. $y = 1$
10. neither function
11. neither function
12. neither function
13. both functions
14. $\frac{1}{2}$; 180°; 2
15. $y = \frac{1}{2} \cos 2x$
16. *See question.*

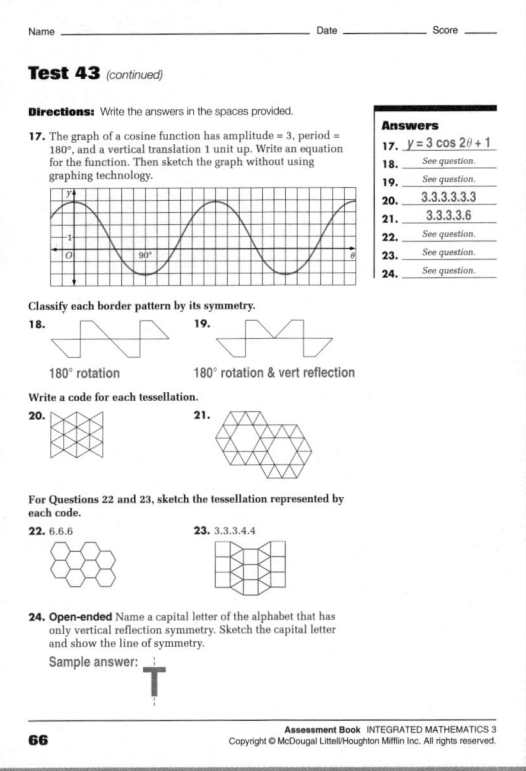

Assessment Book INTEGRATED MATHEMATICS 3

65

Name ________ Date ________ Score ________

Test 43 *(continued)*

Directions: Write the answers in the spaces provided.

17. The graph of a cosine function has amplitude = 3, period = 180°, and a vertical translation 1 unit up. Write an equation for the function. Then sketch the graph without using graphing technology.

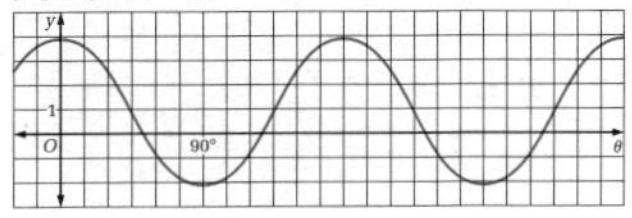

Classify each border pattern by its symmetry.

18.

180° rotation

19.

180° rotation & vert reflection

Write a code for each tessellation.

20.

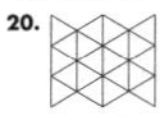

21.

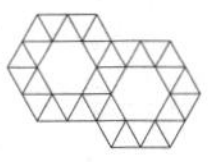

For Questions 22 and 23, sketch the tessellation represented by each code.

22. 6.6.6

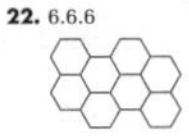

23. 3.3.3.4.4

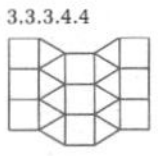

24. Open-ended Name a capital letter of the alphabet that has only vertical reflection symmetry. Sketch the capital letter and show the line of symmetry.

Sample answer:

Answers

17. $y = 3 \cos 2\theta + 1$
18. *See question.*
19. *See question.*
20. 3.3.3.3.3.3
21. 3.3.3.3.6
22. *See question.*
23. *See question.*
24. *See question.*

66

Assessment Book INTEGRATED MATHEMATICS 3

OUTSIDE RESOURCES

Books/Periodicals

Teles, Elizabeth J. "Understanding Arcsin [sin(*x*)] and Arccos [cos(*x*)]." *Mathematics Teacher* (March 1992): pp. 198–200.

Levine, Bernard S. "The Taming of the Ambiguous Case." *Mathematics Teacher* (March 1992): p. 200.

"Trigonometry and Logarithms." *Sourcebook of Applications of School Mathematics.* Prepared by Joint Committee of the MAA and NCTM: pp. 207–247. NCTM.

Activities/Manipulatives

HiMAP Module 4, *Symmetry, Rigid Motions and Patterns.* Section 5: "Two-Dimensional Patterns": pp. 22–30. COMAP, Inc.

Moody, Mally. "Trigonometry Drills." *Mathematics Teacher* (March 1992): pp. 201–203.

Activity Bank: Volume 1, Chapter 4: "Trigonometry Walk": pp. 219–224. Assembled by the Mathematics Curriculum and Teaching Program (MCTP). Curriculum Corporation.

Edwards, Thomas. "Building Mathematical Models of Simple Harmonic and Damped Motion." *Mathematics Teacher* (January 1995): pp. 18–22.

Software

TesselMania! For Macintosh LC or later, color display recommended. Key Curriculum Press.

Videos

Three-Dimensional Symmetry. Applications of symmetry in a 17-minute video. Key Curriculum Press.

PROJECT GOALS

- Students make a scrapbook describing specific situations that can be modeled using a graph.
- Students' scrapbooks include situations from their lives, as well as local, national, and international topics.
- Students describe each situation, include a sketch of a graph that models it, and explain why the situation can be modeled by the graph selected.

PROJECT PLANNING

Project Teams

Have students work on the project in groups of four. Groups can meet to discuss which functions to include in their scrapbooks and which situations they would like to model. Each group should work through all of the "Working on the Unit Project" exercises and save any graphs made. They will need them when completing the project. All members of a group should participate in suggesting functions, graphs, situations, descriptions, and in actually making the scrapbook. Members can take turns doing the writing required to collect information and complete the project.

Support Materials

The ***Project Book*** contains information about the following topics for use with this Unit Project.

- Project Description
- Teaching Commentary
- Working on the Unit Project Exercises
- Completing the Unit Project
- Assessing the Unit Project
- Alternative Projects
- Outside Resources

unit 10 Periodic Models

When a total eclipse of the sun occurred during a battle fought in ancient Mesopotamia, the armies were so frightened that they dropped all their weapons and made peace. Because eclipses occur in predictable cycles, modern astronomers have determined that the battle ended on May 28, 585 B.C. — making this historical event the earliest for which a specific date is known.

CYCLES

Every autumn millions of monarch butterflies **migrate** from the eastern United States to a small region of central Mexico — a journey of 2000 mi. In January they fly back north for the summer. The next autumn their offspring repeat the cycle.

monarchs

538

Suggested Rubric for Unit Project

4 Students make a functional scrapbook that satisfies all of the criteria listed on page 589 under Completing the Unit Project. The scrapbook is well organized and the situations chosen are interesting. Each graph is mathematically accurate and the explanation of why the graph can be used to model the situation is correct. The descriptions are clear and well written. The scrapbook is attractive and the border design is symmetrical.

3 Students make a functional scrapbook that meets almost all of the criteria for the project. Students list the functions they know how to graph, but a few of the graphs drawn are slightly inaccurate. The written descriptions of situations and the explanations of reasons for the graphs can be improved somewhat. The scrapbook itself is well made and the design is symmetrical but not altogether visually appealing.

2 Students make a functional scrapbook, but it falls short of the goals of the project. Not all of the things listed on page 589 are included. Mathematical

Unit Project 10

Make a Functional Scrapbook

Your project is to make a scrapbook describing specific situations that can be modeled using a graph. Your scrapbook should include situations from your life, as well as local, national, and international topics. Use at least one page for each situation.

Describe each situation and include a sketch of a graph that models it. Explain why the situation can be modeled by the graph you selected. Include pictures or sketches to help describe each situation or each model.

Include at least ten different types of graphs in your scrapbook. About half of the graphs you include should be from previous units and half should be from Unit 10. Describe the features of each graph. Explain what these features tell about each situation.

Between June and September, the winds blow across the Indian Ocean from the southwest instead of the northeast, causing large downpours throughout Asia. Because the rain is essential to the agriculture and economy in the area, the people of India celebrate the return of the **monsoon season** with the festival of *Teej*.

Rice fields in Goa are dry and brown in May.

The same fields turn bright green by August due to the monsoon.

The agricultural cycles of planting, growing, and harvesting once determined when school would start and end. Now **school vacations** influence economic cycles. Each year many industries focus their products or services on students — stores stock back-to-school supplies in September, movie companies release major films during school vacations, and travel agents and airlines plan trips during spring break.

Students Acquiring English

Making a functional scrapbook is an excellent activity for students acquiring English in that it emphasizes the mathematical concepts of the unit without requiring extensive written responses.

ADDITIONAL BACKGROUND

Multicultural Note

In 4000 B.C., people living in the region that is now Iraq created a calendar based on the phases of the moon, one of the first known calendars. Sumerians invented a solar calendar soon after. Neither calendar was totally accurate: the lunar Iraqi calendar did not coincide with the solar year and the Sumerian one, based on twelve months of thirty days each, required a thirteenth month every six years. Egyptians refined that calendar by adding five extra days to the twelve months, but were still not able to account for one-fourth of a day per year. About 360 B.C., the Mayans independently devised the most complex and accurate calendar of the ancient world, calculating a year to be 365.2422 days and reconciling solar and lunar years. Today calendars throughout the world have different starting dates and period divisions: for example, the year Americans and Europeans called A.D. 1900 corresponded to the Chinese year 4597 and the Muslim year 1318.

Suggested Rubric for Unit Project

errors in functions and graphs are common. Descriptions and explanations are not entirely clear and the writing itself needs substantial revision. The actual scrapbook is poorly made and the border design is not symmetrical or is missing completely.

1 Students have difficulty getting their project organized. Parts of the project are attempted, but no final scrapbook is completed. The work that is done is very poor and contains many errors. Students should be encouraged to speak with the teacher as soon as possible to review their work and to make a new start on the project.

Patterns and Symmetry

Patterns and symmetry abound in the natural world and in the cultural world created by people. Scientists observe the periodic motion of the planets in our solar system. Mathematicians study number patterns and symmetrical figures. Artists and designers use patterns and symmetry to create interesting and beautiful designs. At a very deep level, research scientists have used patterns and symmetry to make great discoveries about the physical world. This is because they believe that many of the most fundamental aspects of nature exhibit a basic simplicity that manifests itself in a periodic behavior or a symmetrical form. Physicists, for example, have used concepts of patterns and symmetry to build mathematical models that explain the most fundamental components of matter and the four basic forces that exist throughout the universe.

ALTERNATIVE PROJECTS

Project 1, page 589

Predict Future Trends

Think of a situation in nature that can be modeled by a sine wave. Gather data to develop an accurate model, graph the data, and write an equation to model the data. Use the model to make predictions about the situation.

Project 2, page 589

Penrose Tiles

Research the mathematician Roger Penrose and the tiles named for him. Describe how the dimensions of the tiles relate to the golden ratio and the kinds of nonperiodic tessellations that can be made with them.

Unit Project 10

Getting Started

For this project you should work in a group of four students. Here are some ideas to help you get started.

☞ Make a list of all the functions you can graph. Decide which of these functions to include in your scrapbook.

☞ Think about situations that you would like to model. Decide which type of graph best models these situations.

☞ Decide what materials you will use to make your scrapbook. How will you keep your scrapbook together?

Working on the Unit Project

Your work in Unit 10 will help you make your scrapbook.

Related Exercises:

Section 10-1, Exercises 35–40
Section 10-2, Exercise 37
Section 10-3, Exercise 28
Section 10-4, Exercise 48
Section 10-5, Exercises 21, 22
Section 10-6, Exercises 24, 25
Section 10-7, Exercise 35

Alternative Projects p. 589

Can We Talk CYCLES

➤ What types of things happen every day? every week? every month? every year?

➤ Halley's Comet returns about every 76 years. What events happen in regular intervals longer than one year?

➤ Meteorologists study patterns in the weather. What are some patterns that you notice in the weather?

➤ A popular saying states that "history repeats itself." Do you think history repeats itself? Why or why not? Give examples of events from history to support your position.

➤ Do you know of any annual events or holidays that are celebrated in other parts of the world? If so, when do they occur?

Answers to Can We Talk?

➤ Answers may vary. Examples are given. Every day, the sun rises and sets; every week, most people work 5 days and are off on the weekends; every month, the moon completes a complete cycle; every year contains the four seasons.

➤ Answers may vary. Examples are given. A census is taken every 10 years in the United States; presidential elections happen every 4 years.

➤ Answers may vary, depending upon which area of the country you live in.

➤ Answers may vary. An example is given. Yes, because the reasons for problems that have existed in the past also can exist in the present. For example, the American colonists who were upset because they did not have any decision-making powers about taxes imposed by the British king can be related to the college students who were upset because they did not have any decision-making powers about being drafted into the army to fight in Viet Nam.

➤ Answers may vary. An example is given. In Mexico, people celebrate "Cinco de Mayo," or "Fifth of May," as their day of independence.

Section **10-1**

Describing the Behavior of Functions

Focus
Describe graphs and examine rates of change.

From The Looks Of It

The Indianapolis 500 automobile race, first run in 1911, attracts about 300,000 spectators each year. The winner is the first driver to complete 200 laps around the track, a distance of 500 mi.

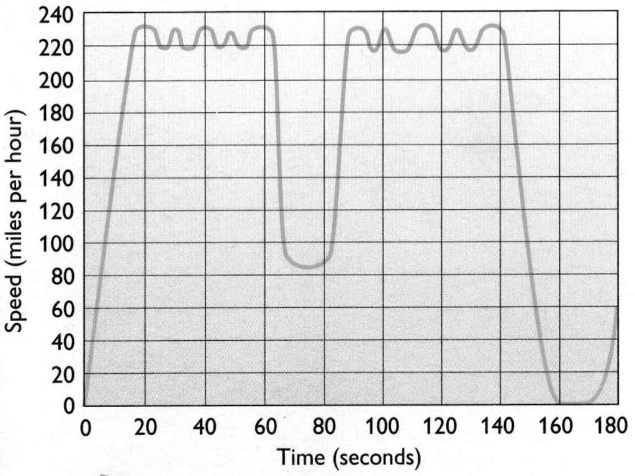

Throughout the race, drivers change speed as they enter or leave pit stops, straightaways, and curves. When a yellow caution flag is dropped, drivers must slow down to 80 mi/h.

Talk it Over

1. In the graph shown, identify the variables and any relationships they have.
2. Tell what is happening in the race as you follow the graph from left to right. Describe any increases or decreases and any patterns or breaks in patterns and what they may mean.

PLANNING

Objectives and Strands
See pages 538A and 538B.

Spiral Learning
See page 538B.

Materials List
- Graphics calculator
- Graph paper

Recommended Pacing
Section 10-1 is a one-day lesson.

Extra Practice
See pages 624–625.

Warm-Up Exercises
Warm-Up Transparency 10-1

Support Materials
- Practice Bank: Practice 75
- Activity Bank: Enrichment 66
- Study Guide: Section 10-1
- Problem Bank: Problem Set 21
- Explorations Lab Manual: Diagram Master 2
- Assessment Book: Quiz 10-1, Alternative Assessment 1

Answers to Talk it Over

1. time and speed; There appears to be little relationship between the two.
2. The car accelerates to 230 mi/h, races in the 220–230 mi/h range, slows to 80 mi/h for 30 s, accelerates to resume normal racing speed, slows and stops for 10 s, and finally accelerates back into the race. The two similar-looking sections appear to be the normal racing section of the graph, with speeds varying slightly due to turns. The slowdown from 60 s to 90 s could be due to a caution condition on the track, perhaps because of an accident. The stop from 155 s to 165 s could be a pit stop.

TEACHING

Talk it Over

Questions 1–8 start students thinking about the behavior of various types of graphs by relating the graphs and their behavior to a number of real-world situations. Question 9 introduces students to the names for various types of graphs.

Mathematical Procedures

A fundamental procedure used by applied mathematicians and other scientists who attempt to understand physical processes or other events in the real world is to create a mathematical model that describes what is happening. The questions, exercises, and problems in this section provide many excellent examples of this.

Talk it Over

Choose the letter of the graph on the next page that best matches each situation shown below. Tell what variable goes on each axis.

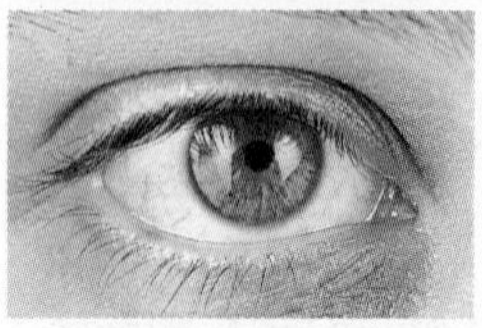

3. The iris of your eye widens as you enter a dark room.

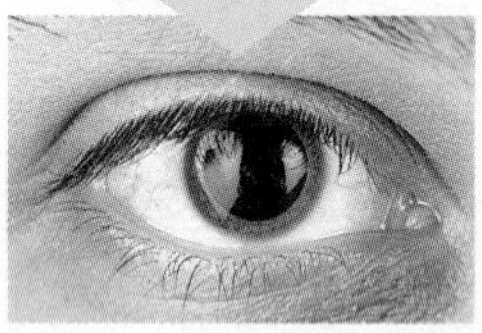

4. As a pot of water boils, the temperature of the water is recorded.

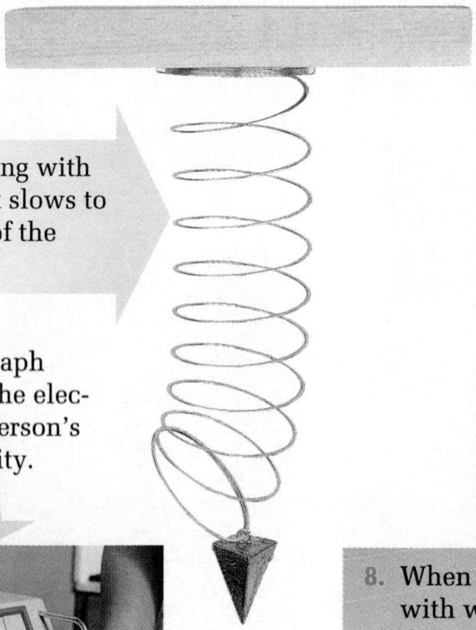

5. As a bouncing spring with an attached weight slows to a stop, the height of the weight changes.

6. A hot drink cools after it is poured into a cup.

7. An electrocardiograph traces a record of the electric currents of a person's heart muscle activity.

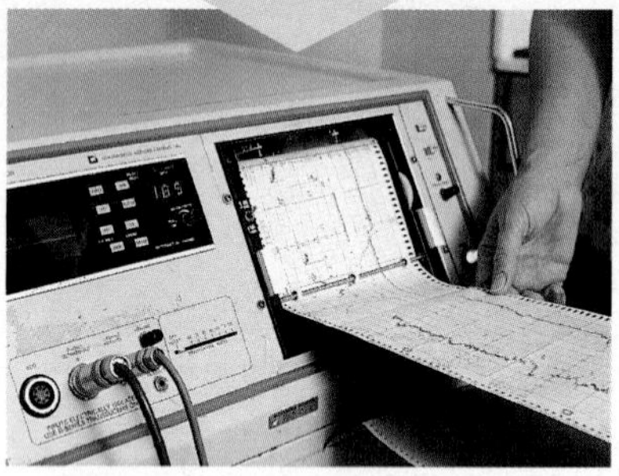

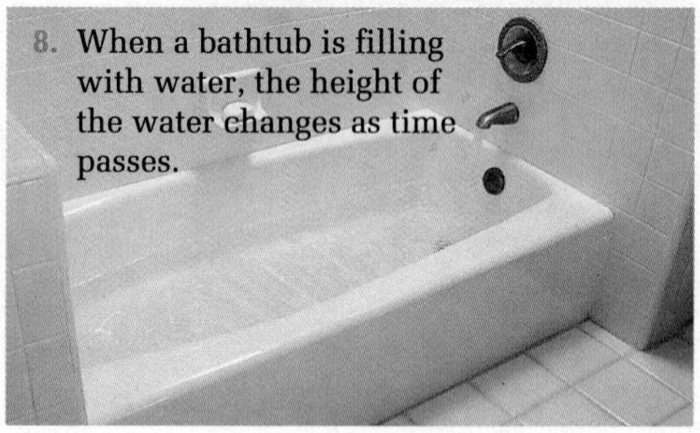

8. When a bathtub is filling with water, the height of the water changes as time passes.

Answers to Talk it Over

3. B
4. E
5. D
6. A
7. C
8. F

A.

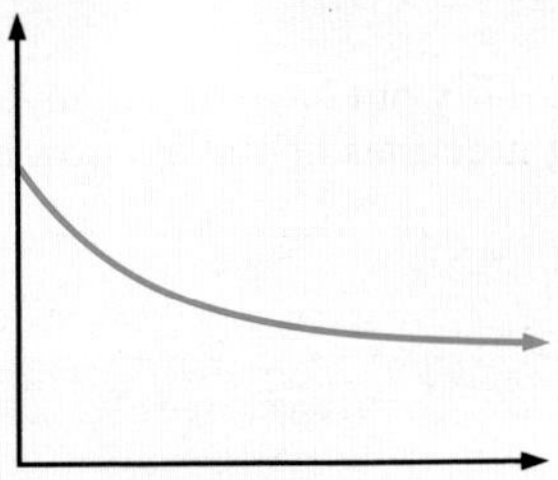

B.

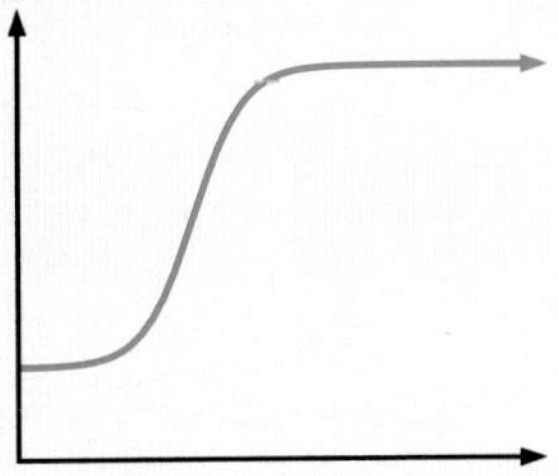

C.

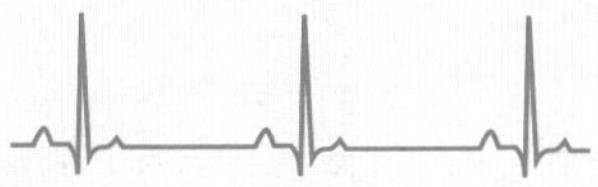

D.

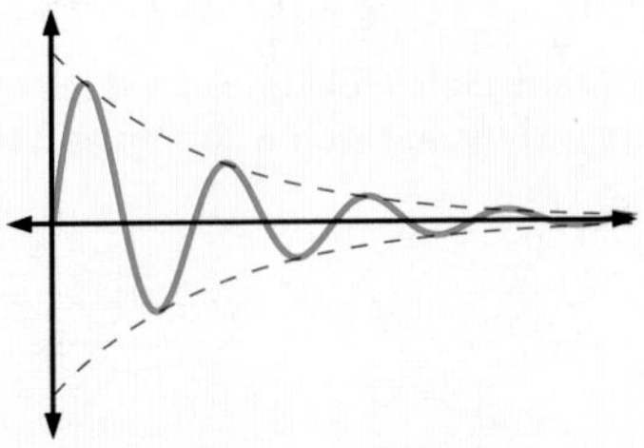

E.

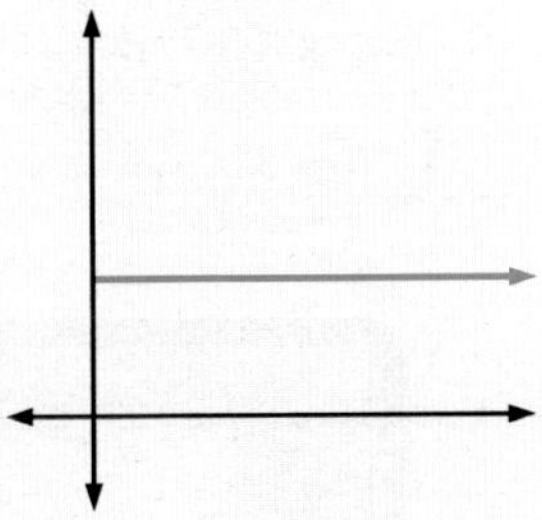

F.

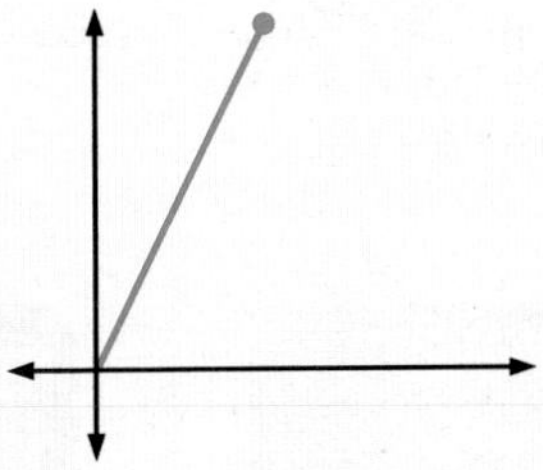

9. Which of graphs A–F can be described by each term?
   a. constant
   b. increasing
   c. decreasing
   d. **periodic**, or repeating the same pattern
   e. **saturating**, or continually increasing or decreasing toward a constant value
   f. **damping**, or moving toward a constant value by alternately increasing and decreasing around that value

Communication: Reading

Reading and interpreting a graph that represents real-life data is an important skill. For additional practice with this skill, students can work together in pairs. Each student can write down real-life situations that can be modeled by the graphs on this page. The other student can then identify which graph describes each situation.

Teaching Tip

The terms periodic, saturating, and damping discussed in Talk it Over question 9, all have meanings that can be intuitively understood by students. Ask students to suggest some situations from their own experiences in which they can use these terms in a sentence.

Answers to Talk it Over

9. Answers may vary. An example is given.
   a. E
   b. B, F
   c. A
   d. C
   e. A, B
   f. D

Additional Sample

Compare the rates of decay in the fourth quadrant of these functions: $y = -4^x$, $y = -2x^2$, $y = -5x$.

Method 1. Make tables for integer values of x. Look at the differences in the y-values to reach a conclusion about the rates of decay.

$y = -4^x$

x	y	differences
0	−1	
1	−4	−3
2	−16	−12
3	−64	−48
4	−256	−192
5	−1024	−768

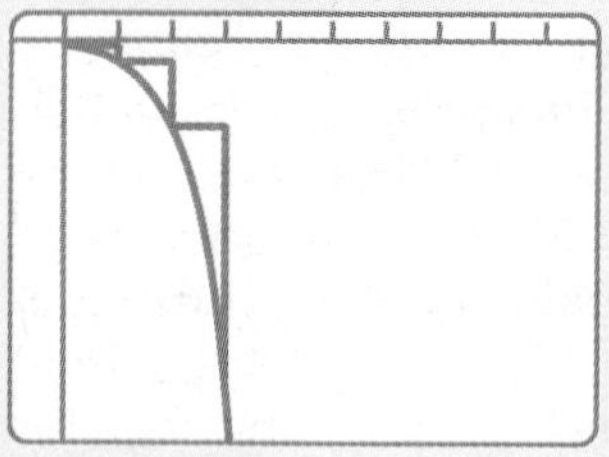

The differences are decreasing in a geometric sequence.

$y = -2x^2$

x	y	differences
0	0	
1	−2	−2
2	−8	−6
3	−18	−10
4	−32	−14
5	−50	−18

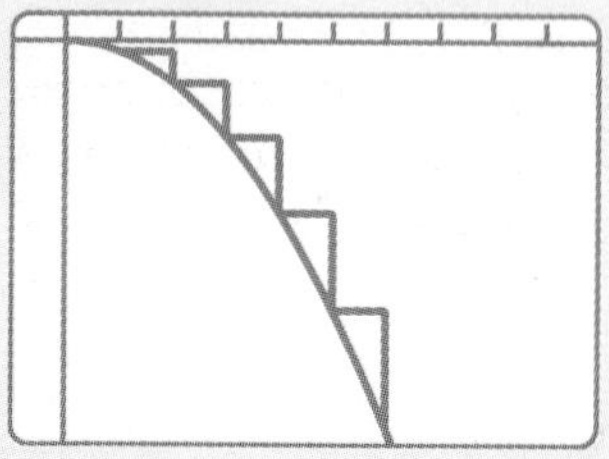

The differences are decreasing by 4 in an arithmetic sequence.

$y = -5x$.

x	y	differences
0	0	
1	−5	−5
2	−10	−5
3	−15	−5
4	−20	−5
5	−25	−5

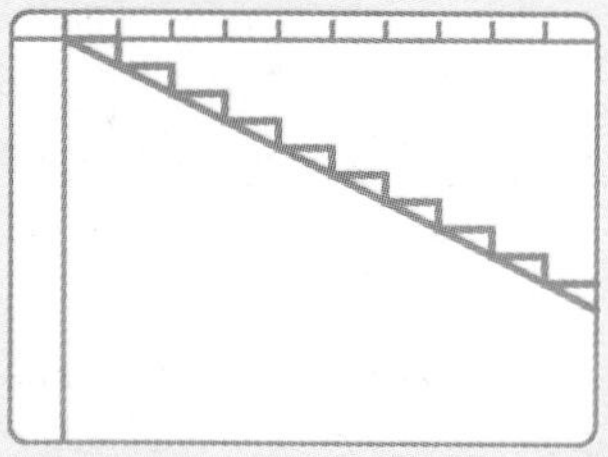

The differences are constant.

Rates of Change

Two functions may behave similarly but have important differences. For example, increasing functions do not all grow at the same rate.

Sample

Compare the rates of growth in the first quadrant of these linear, quadratic, and exponential functions:

$y = 2x \quad y = x^2 \quad y = 2^x$

Sample Response

Method 1 Make tables for integer values of x. Look at the differences in the y-values to reach a conclusion about the growth rates.

$y = 2x$

x	y	differences
0	0	
1	2	2
2	4	2
3	6	2
4	8	2
5	10	2

The differences are constant.

$y = x^2$

x	y	differences
0	0	
1	1	1
2	4	3
3	9	5
4	16	7
5	25	9

The differences are increasing by 2, in an arithmetic sequence.

$y = 2^x$

x	y	differences
0	1	
1	2	1
2	4	2
3	8	4
4	16	8
5	32	16

The differences are increasing in a geometric sequence.

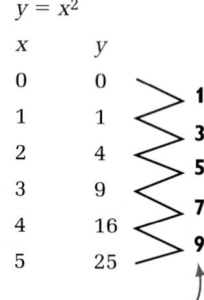

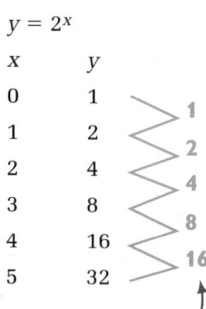

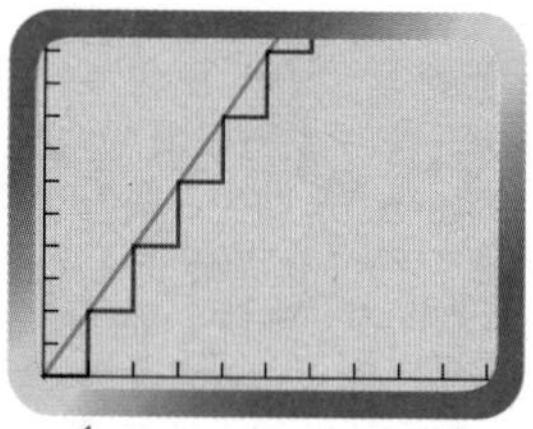

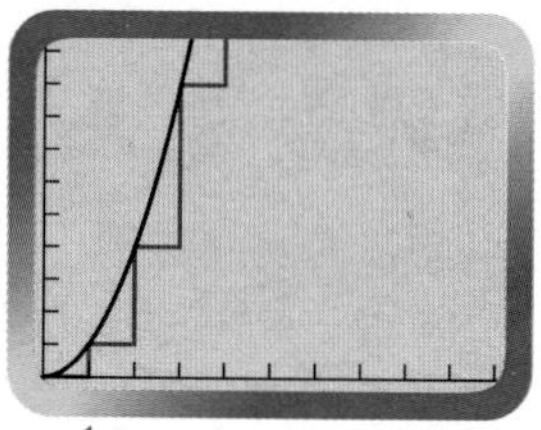

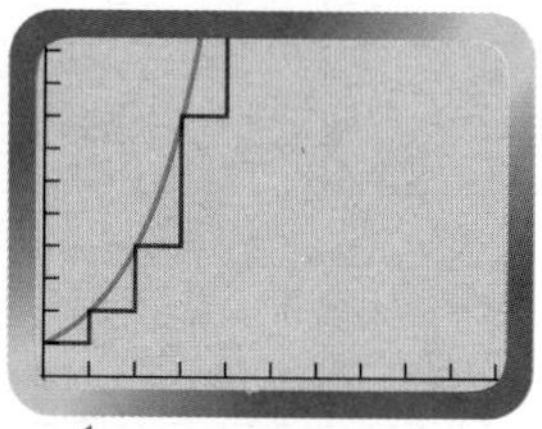

Method 2 Graph the three functions on the same set of axes.

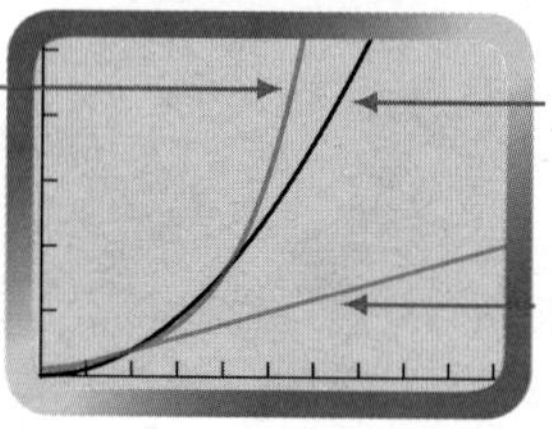

The tables and the graphs show that the exponential function $y = 2^x$ grows the fastest, followed by the quadratic function $y = x^2$, and then the linear function $y = 2x$.

Answers to Talk it Over

10. The results are not exactly the same but are similar. 4^x grows fastest, followed by x^4, then $4x$. The differences in y-values for $y = 4x$ are constant, the differences for $y = x^4$ are increasing, and the differences for $y = 4^x$ are increasing in a geometric sequence.

11. $y = \frac{1}{2}x$ increases fastest, followed by $y = \sqrt{x}$, and then $y = \log_2 x$.

12. a. slowly at the start and quickly at the end of this interval
    b. slowly over the whole interval

Answers to Look Back

Answers may vary. Examples are given.

a. heart muscle activity
b. the height of a spring that has been stretched and released
c. the temperature of a hot drink cooling in a cup
d. the number of crayons in successive boxes coming off a manufacturing assembly line

Talk it Over

10. Repeat the Sample for the functions $y = 4x$, $y = x^4$, and $y = 4^x$. Are the results the same? Explain.

11. Compare the rates of growth in the first quadrant for these linear, square root, and logarithmic functions:

$$y = \frac{1}{2}x \qquad y = \sqrt{x} \qquad y = \log_2 x$$

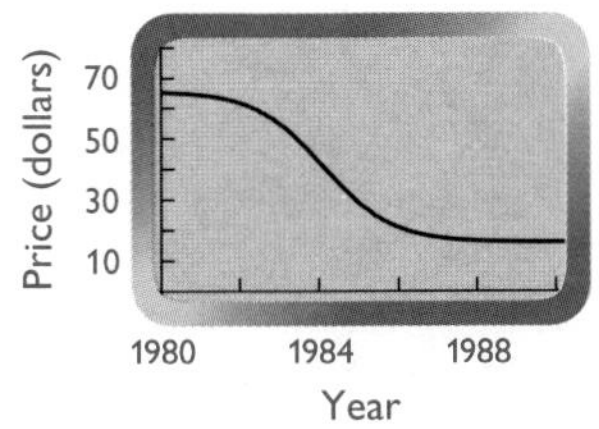

12. The graph shows the price of prerecorded videocassettes, in dollars, between 1980 and 1990. Tell whether the prices decreased quickly or decreased slowly for each interval.
   a. 1982–1986
   b. 1986–1990

Look Back

Describe a situation that can be modeled by each type of graph.

a. periodic b. damping c. saturating d. constant

10-1 Exercises and Problems

1. a. Interpret this graph of the height of an elevator in an office building.
   b. What part of the graph is periodic? What do you think this represents?
   c. When do most workers arrive at work?

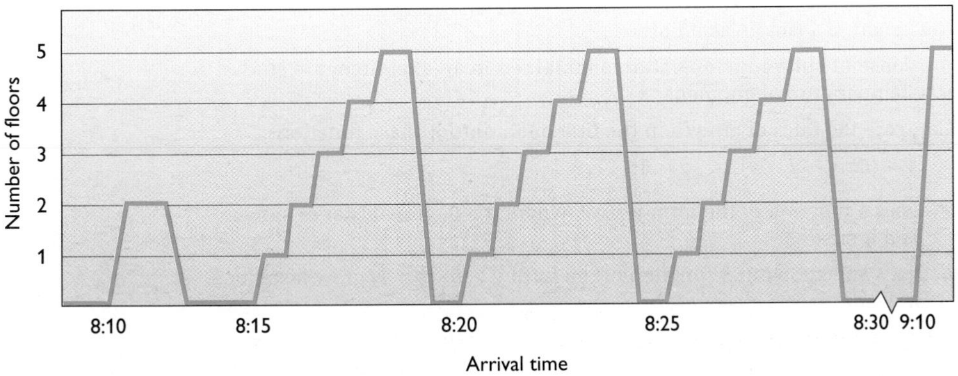

2. **Reading** How are saturating functions and damping functions alike? How are they different?

3. **Writing** Refer to the graphs in *Talk it Over* questions 5 and 7 on page 542. How are the graphs alike? How are they different?

Open-ended Describe a situation that can be modeled by each type of graph.

4. increasing 5. decreasing 6. constant

Method 2. **Graph the three functions on the same set of axes.**

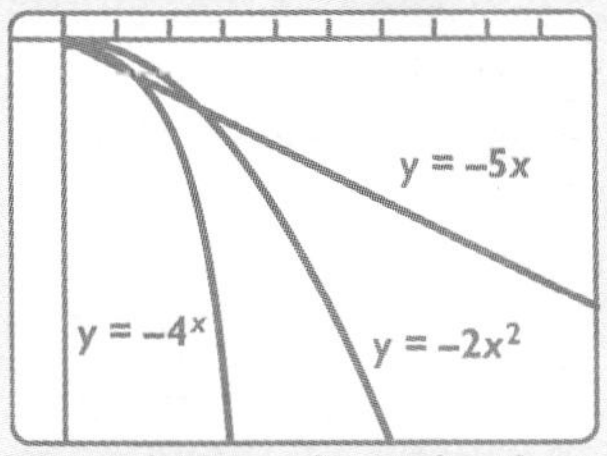

The tables and graphs show that the function $y = -4^x$ decays the fastest, followed by $y = -2x^2$, and then by $y = -5x$.

Talk it Over

Some students may have difficulty with question 11. Encourage them to graph the functions. They should observe that although the square-root function and the logarithmic function grow quickly initially, their rates of growth slow.

Teaching Tip

In discussing the Sample and Talk it Over questions 10 and 11, you may wish to ask for ways in which the functions differ, other than in their rates of change. Students should see, for example, that $y = 2x$ and $y = 2^x$ are everywhere increasing, but for negative values of x, the function decreases as x increases.

APPLYING

Suggested Assignment

Standard 1–3, 7–20, 25–40

Extended 1–40

Integrating the Strands

Algebra Exs. 17–19, 26–34

Functions Exs. 1–26, 30–40

Logic and Language Exs. 1–4, 20

Interdisciplinary Problems

Graphs are used to represent real-life situations in every discipline. For Exs. 4–6, ask students to describe situations from fields of study they especially enjoy.

Answers to Exercises and Problems

1. a. The elevator goes to the second floor and returns to the ground floor. It goes to the fifth floor, stopping at every floor, and returns to the ground floor. It does this 3 times. Then it stops at the ground floor for 40 minutes and goes to the fifth floor.
   b. 8:15 to 8:30; In each period, the elevator takes the same time to fill, go up while stopping at each floor, then return to the ground floor and await the next trip.
   c. between 8:15 and 8:25

2. They both approach a constant value. Saturating functions do it either by continuously increasing or decreasing. Damping functions do it by alternately increasing and decreasing.

3. The graphs are alike in that they alternate between high values and low values. They are different in that the graph in question 5 does not repeat values, while the graph in question 7 does repeat values.

4–6. Answers may vary. Check students' work.

Error Analysis

In Ex. 17, and in other similar exercises, students may assume that when the values of one function are greater than those of another, the rate of growth of the first function is greater than that of the second. Whether this is true or not depends on the functions and the x-values used to calculate the rates of growth. This error can be corrected by having students calculate growth rates over different-sized intervals and for different ranges of x-values. For instance, in Ex. 17, students can consider intervals of length 0.01 from $x = 0.1$ to $x = 0.2$. Then consider intervals of length 0.01 from $x = 4$ to $x = 4.1$.

Mathematical Procedures

When students examine growth rates of functions, it is helpful to graph the functions on a graphics calculator. Encourage students to experiment with different graphing windows to get a more complete picture of how the growth rates compare for different intervals of x-values.

Assessment: Standard

Ex. 21 is an excellent example for students to define rate of growth over intervals. You may wish to add $y = x^3$, $x \geq 0$, for additional comparison.

For Exercises 7–12, which of the terms *constant, increasing, decreasing, periodic, damping,* or *saturating* best describes each graph?

7.

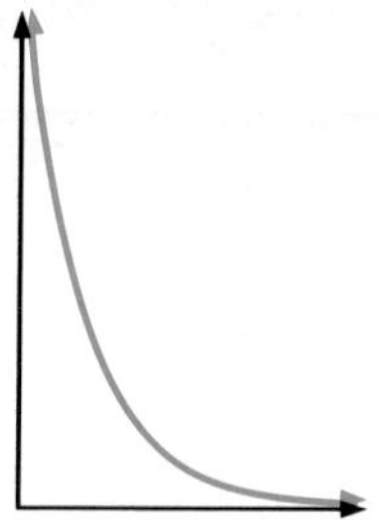

8.

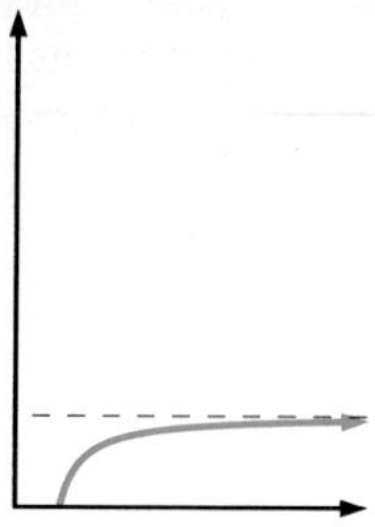

9.

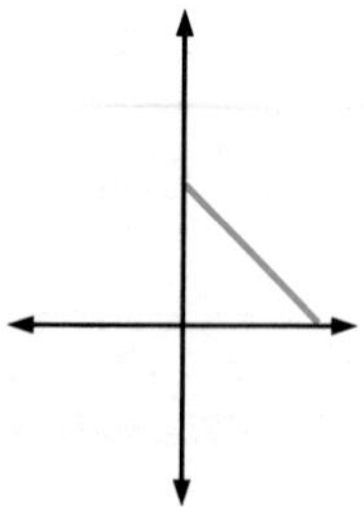

10.

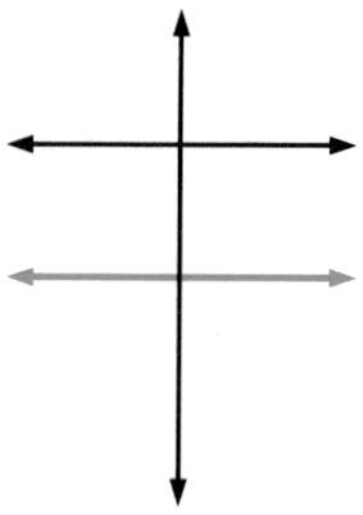

11.

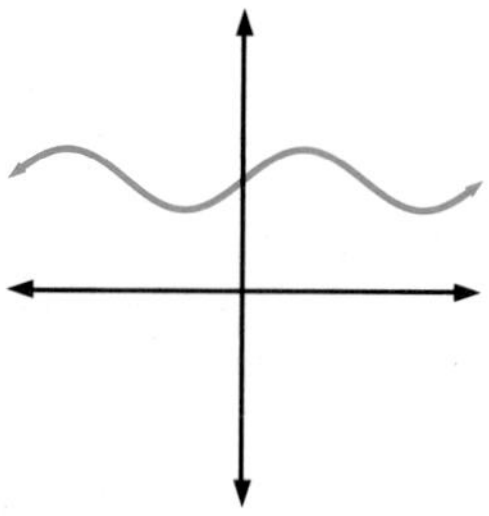

12.

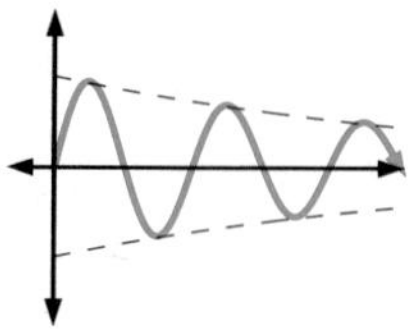

For Exercises 13–16, choose the number of the graph above that best models each situation.

13. the volume of air in your lungs as you breathe normally

14. the amount of radioactivity in a rock as years pass

15. the height of a candle as it burns

16. the constant subzero temperature maintained in an experiment to study low-temperature phenomena

17. Compare the rates of growth in the first quadrant for these functions:

$y = 3x \qquad y = x^3 \qquad y = 3^x$

18. a. Does a function of the form $y = x^n$, where $n > 0$, grow faster or slower as n increases?

b. Does an exponential function of the form $y = b^x$ $(b > 1)$ grow faster or slower as b gets close to 1?

19. Compare the rates of decay of these reciprocal, exponential decay, and linear decay functions when $x \geq 2$:

$y = \frac{2}{x} \qquad y = 2^{-x} \qquad y = 2 - x$

20. Writing Can a periodic function be constantly increasing? Explain.

21. a. In the Sample on page 544, in which interval(s) is $y = 2x$ greater than the other two functions?

b. In which interval(s) is $y = x^2$ greater than the other two functions?

Answers to Exercises and Problems

7–12. Answers may vary. Examples are given.

7. decreasing or saturating

8. increasing or saturating

9. decreasing **10.** constant

11. periodic **12.** damping

13. 11 **14.** 7

15. 9 **16.** 10

17. $y = 3^x$ grows fastest, followed by $y = x^3$, and then $y = 3x$.

18. a. faster **b.** slower

19. $y = 2 - x$ decays fastest, followed by $y = 2^{-x}$, and then $y = \frac{2}{x}$.

20. No; if it is constantly increasing, the function values cannot be repeating.

21. a. $1 < x < 2$ **b.** $2 < x < 4$

22–24. Answers may vary. An example is given.

22. from 1983 to 1985

23. from 1977 to 1979

24. The graph is increasing over its entire domain.

25. The periodic function increases and decreases in a repeating cycle. The constant function has no growth rate, and the linear function grows at a constant rate.

26. a.

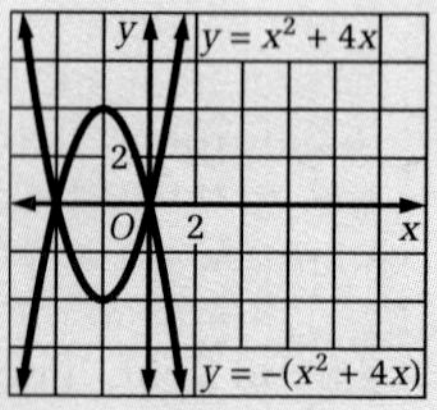

b. Reflect it over the x-axis.

27. 16

28. $5\sqrt[6]{5^5}$

29. $\frac{27}{125}$

For Exercises 22–24, use the graph of yearly revenues of a state lottery.

22. Identify an interval of the graph where the revenues increased sharply.

23. Identify an interval of the graph where the revenues increased slowly.

24. Describe the graph. Use as many of the terms *constant, increasing, decreasing, periodic, damping,* and *saturating* as you can.

Lottery Revenues

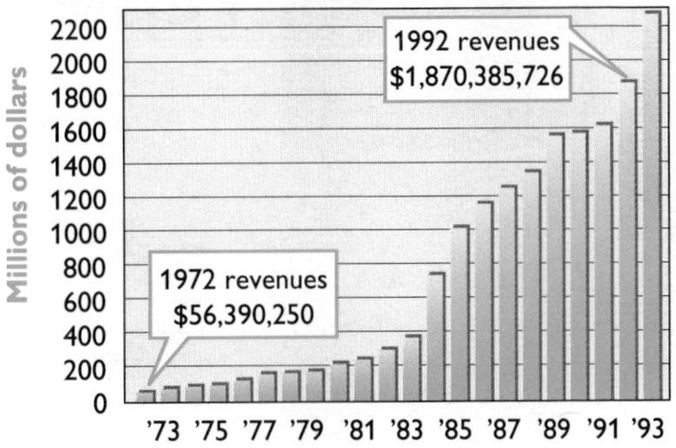

Source: Massachusetts State Lottery

Ongoing **ASSESSMENT**

25. **Writing** Compare the rates of growth of the periodic, constant, and linear functions shown.

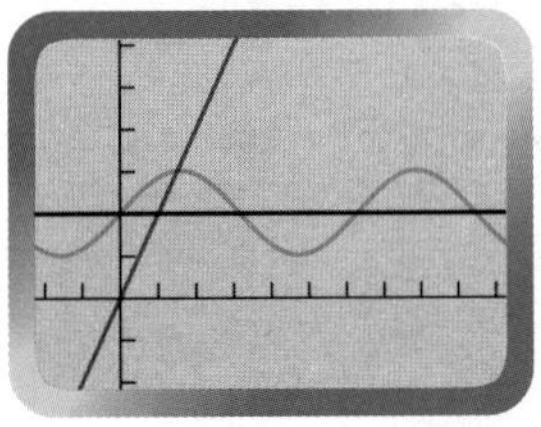

Review **PREVIEW**

26. a. Sketch the graphs of $y = x^2 + 4x$ and $y = -(x^2 + 4x)$ on the same set of axes.

    b. What transformations can you apply to the graph of $y = x^2 + 4x$ to obtain the graph of $y = -(x^2 + 4x)$? *(Section 9-7)*

Simplify. *(Section 5-3)*

27. $64^{2/3}$

28. $5^{1/2} \cdot 5^{4/3}$

29. $\left(\frac{36}{100}\right)^{3/2}$

Sketch each graph. *(Toolbox Skill 28)*

30. $y = x^2$

31. $y = (x - 2)^2$

32. $y - 2 = x^2$

33. $y = -x^2$

34. Tell what kinds of transformations are used to obtain the graphs in Exercises 31 – 33 from the original graph in Exercise 30.

10 Working on the Unit Project

35. Bob made this graph to describe the sound produced by a cymbal. What type of graph did Bob make? Do you think his graph accurately models the situation? Why or why not?

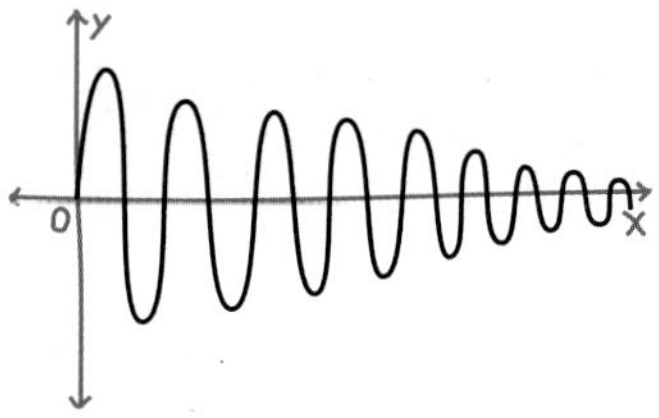

Describe a situation that can be modeled by a graph having each given characteristic. Make a graph to model the situation. Explain why each situation can be modeled by each type of graph. Include each graph on a page of your scrapbook.

36. constant

37. increasing

38. decreasing

39. saturating

40. damping

Communication: Writing

Students may want to discuss Ex. 25 with a partner before writing their individual responses. Making tables for integer values of x and looking at the differences in the y-values may help students reach a conclusion about the growth rates.

Working on the Unit Project

Students can work in their project groups to complete Exs. 35–40. All results should be recorded in a project journal for later use in completing the Unit Project.

Practice 75 For use with Section 10-1

Name ______________________ Date ____________

Practice 75
For use with Section 10-1

12. $y = \frac{1}{x}$ is greatest from 0 to 0.38; $y = 3 - x$ is greatest from 0.38 to 1.99; $y = 10^{-x} + 1$ is greatest from 1.99 to 4.

For Exercises 1–6, use one of the terms *constant, increasing, decreasing, periodic, damping,* or *saturating* to describe each graph.

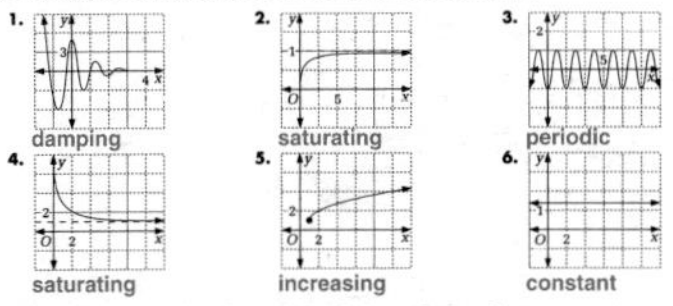

1. damping
2. saturating
3. periodic
4. saturating
5. increasing
6. constant

For Exercises 7–10, match each situation with the graph above that best models it.

7. the vertical position of the outer end of the minute hand of a clock, over time 3

8. the gravitational force on a chair, over time 6

9. the temperature of a cold drink left in a warm room 2

10. the displacement of a guitar string from its rest position after you play a note 1

11. Compare the rates of growth, for $x \geq 1$, of these functions: $y = \sqrt{x}$, $y = x$, $y = \log x$. fastest to slowest: $y = x$, $y = \sqrt{x}$, $y = \log x$

12. Use the three functions: $y = 10^{-x} + 1$, $y = \frac{1}{x}$, and $y = 3 - x$.

    Graph the functions using technology and tell on which intervals between $x = 0$ and $x = 4$ each function is greater than the other two. See top of page.

13. ***Writing*** One way to test the suspension of a car is to push down on a fender and watch the height of the car after you release the pressure. One of the graphs at the top of the page will tell you that the car's suspension needs work; another will tell you it is in good shape. Which graphs do you think these are? Explain your choice. Answers may vary. Check students' work.

 75

Answers to Exercises and Problems

30.

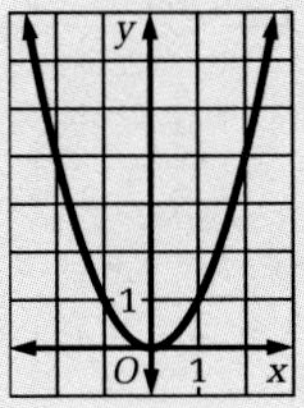

31.

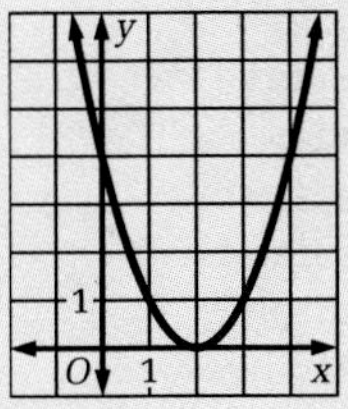

32.

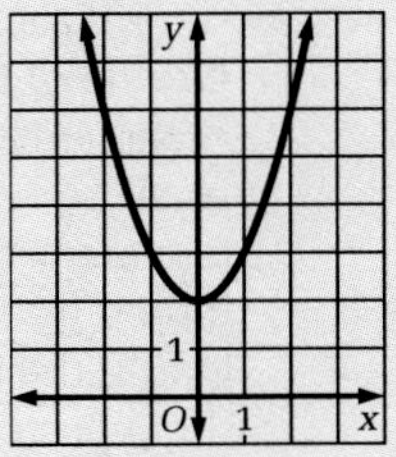

33.

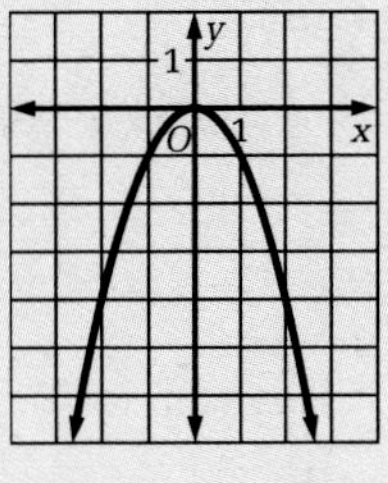

34. $y = (x - 2)^2$: transformation by (2, 0) or shift 2 units to the right; $y - 2 = x^2$: transformation by (0, 2) or shift 2 units up; $y = -x^2$: reflection over the x-axis

35. damping; Answers will vary. The actual sound one hears would fade away following a saturation curve, but the vibration of the cymbal could have this graph.

36–40. Answers may vary. Check students' work.

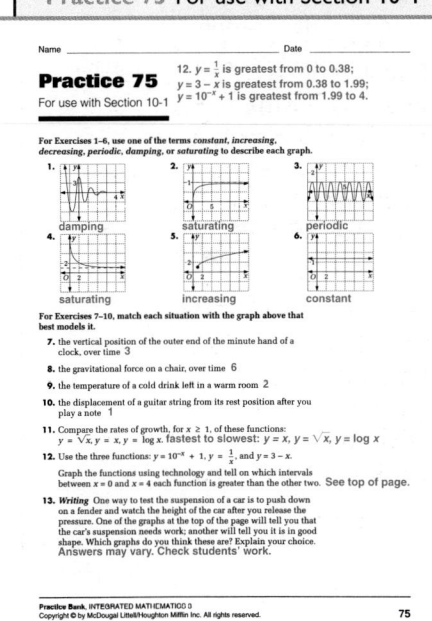

PLANNING

Objectives and Strands
See pages 538A and 538B.

Spiral Learning
See page 538B.

Materials List
- Graph paper

Recommended Pacing
Section 10-2 is a one-day lesson.

Extra Practice
See pages 624–625.

Warm-Up Exercises
- Warm-Up Transparency 10-2

Support Materials
- Practice Bank: Practice 76
- Activity Bank: Enrichment 67
- Study Guide: Section 10-2
- Problem Bank: Problem Set 21
- Explorations Lab Manual: Diagram Masters 1, 2
- Assessment Book: Quiz 10-2, Alternative Assessments 2 and 3

Section 10-2 Periodic Functions

Focus
Recognize periodic functions and features of their graphs.

Cycle Recycle

The automatic dishwasher in a busy restaurant is in constant use. The graph shows the volume of water in the dishwasher as a function of time.

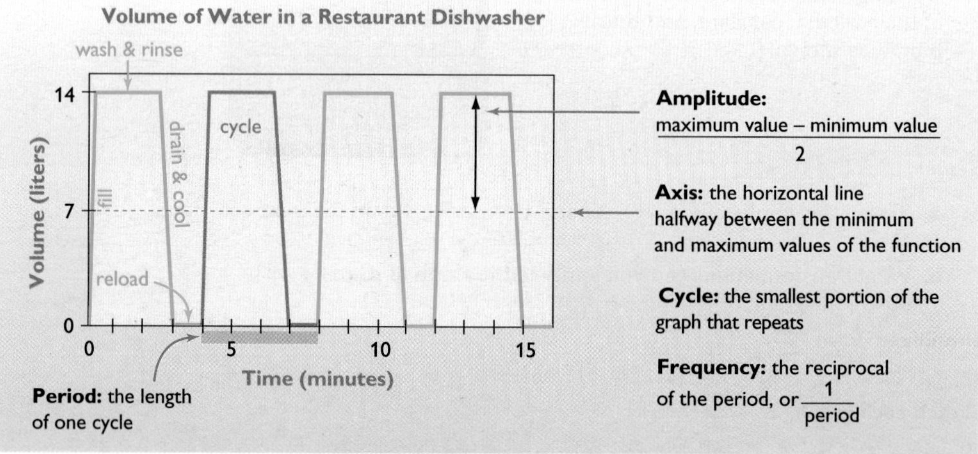

BY THE WAY...

Amplitude and *ample* are from the same Latin word meaning "large, wide, spacious, or roomy."

Talk it Over

1. How long is the period of the dishwasher graph?
2. How many washing cycles are completed in 1 h? What is this in cycles per minute? Which of the key terms does your answer describe?
3. What is the amplitude of the dishwasher graph?
4. Predict what will be happening in the dishwashing process 4 min, 5.5 min, and 18 min from the start time. Explain how you decided.

Many things, such as the dishwasher, have patterns that repeat. A repeating pattern can be represented by a *periodic graph*, which has a piece that is repeatedly translated horizontally. For a periodic function, the amplitude is the distance of the graph from a high point (or low point) to its axis.

Answers to Talk it Over

1. 4 min
2. 15; $\frac{1}{4}$; the frequency
3. 7
4. start of the second cycle; middle of the wash and rinse part of the cycle; wash and rinse; The first two answers can be determined by looking at the graph, the third by using the repeating nature of the graph to make a prediction.

Sample 1

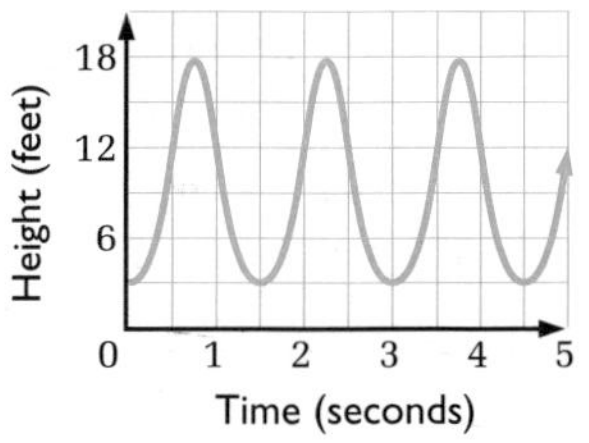

Sam is jumping on a trampoline in a national competition. The graph shows how high his feet are off the ground.

a. Find the cycle.

b. Find the period.

c. Find the frequency.

d. Find the amplitude.

e. Write an equation of the axis.

Sample Response

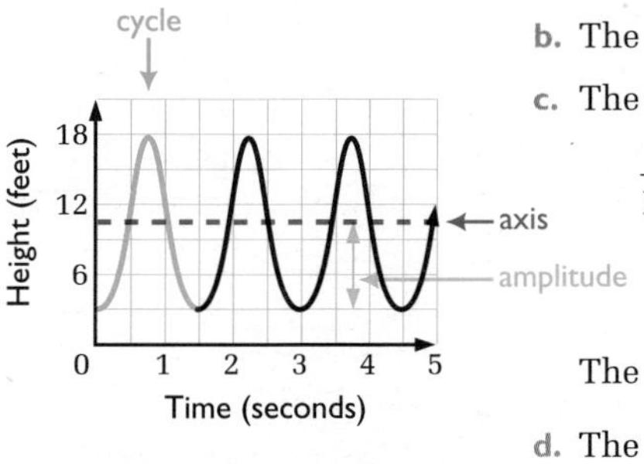

a. The cycle is the part of the graph shown in green.

b. The period is the length of one cycle, or 1.5 s.

c. The frequency is the reciprocal of the period.

$$\frac{1}{\text{period}} = \frac{1}{1.5} \quad \longleftarrow \text{Substitute 1.5 for the period.}$$

$$= \frac{2}{3}$$

The frequency is $\frac{2}{3}$ cycle per second.

d. The amplitude is half the difference between the maximum and minimum values of the function.

$$\frac{\text{maximum value} - \text{minimum value}}{2} = \frac{18 - 3}{2}$$

$$= 7.5$$

The amplitude is 7.5 ft.

e. The axis is the horizontal line halfway between the minimum and maximum values 3 and 18.

$$y = \frac{\text{maximum value} + \text{minimum value}}{2}$$

$$= 10.5$$

An equation of the axis is $y = 10.5$.

Jennifer Sans was the United States Women's National Champion Trampolinist in 1992, 1993, and 1994.

Talk it Over

5. When Jennifer jumps on the trampoline with the same frequency as Sam, she can jump 2 ft higher than Sam. How is a graph of Jennifer's jumping different from a graph of Sam's?

6. When Jennifer jumps rope, she jumps faster and not as high as when she jumps on a trampoline. How is a graph of Jennifer's jumping rope different from a graph of her jumping on a trampoline?

Answers to Talk it Over

5. The amplitude for Jennifer's graph is 8.5 and the axis is $y = 11.5$.

6. The amplitude and the period would decrease, the axis would be lower, and the frequency would increase.

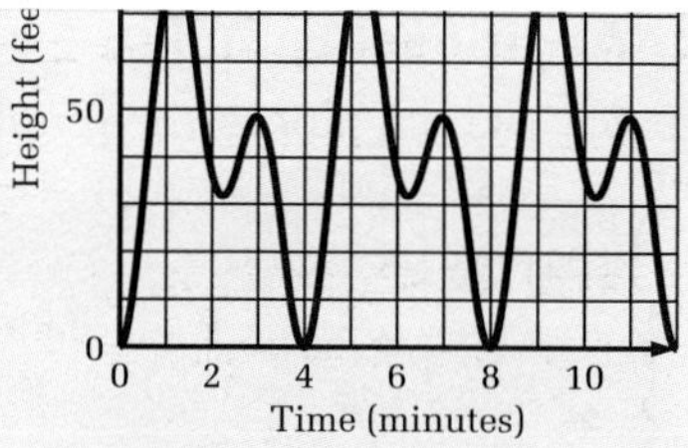

a. Find the cycle.
The cycle is the part of the graph from *A*, through *B*, to *C*.

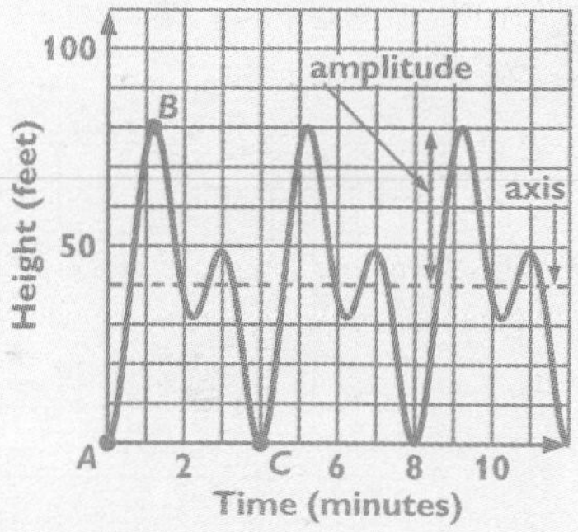

b. Find the period.
The period is the length of one cycle, or 4 min.

c. Find the frequency.
The frequency is the reciprocal of the period, or $\frac{1}{4}$.

d. Find the amplitude.
The amplitude is $\frac{80 - 0}{2}$, or 40 ft.

e. Write an equation of the axis.
The horizontal line halfway between the maximum and minimum values is the line $y = 40$.

Visual Thinking

You may wish to ask students to sketch graphs of recurring activities in their own lives. Examples might include hours of homework or sleep each day of the week, daily food intake, athletic events, music practice, and so on. Ask them to explain their sketches to the class, pointing out the cycle, the period, the frequency, and the amplitude. This activity involves the visual skills of *interpretation* and *communication.*

Additional Sample

S2 Sketch a periodic graph with the axis line $y = -1$ and frequency $\frac{1}{7}$.

Many graphs are possible. Since the frequency is the reciprocal of the period, the period must be 7. The maximum value must be as far above the line $y = -1$ as the minimum value is below that line. Two examples are shown.

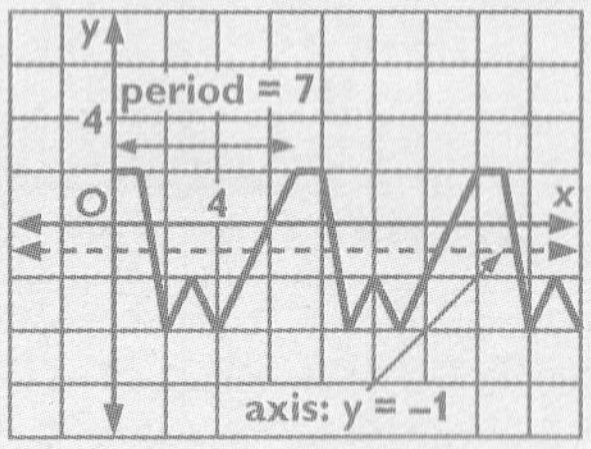

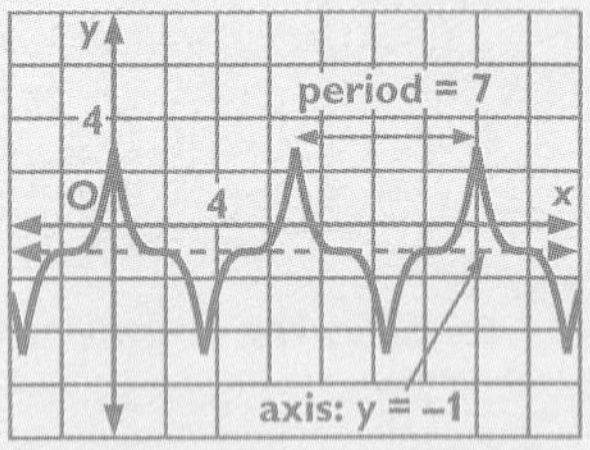

Mathematical Procedures

In applications, it is sometimes difficult to know whether a function is periodic or not. This is especially true if the function has a very large period. For this reason, a good understanding of the behavior of the variables and an equation that models the situation are often crucial in deciding whether periodic behavior is involved. Numerous examples in astronomy bear out the importance of collecting accurate data over extended periods of time.

Sample 2

Sketch a periodic graph with period 3 and amplitude 5.

Sample Response

Many graphs are possible. An amplitude of 5 means $\frac{\text{maximum} - \text{minimum}}{2} = 5$, so maximum − minimum = 10.

Here are two responses.

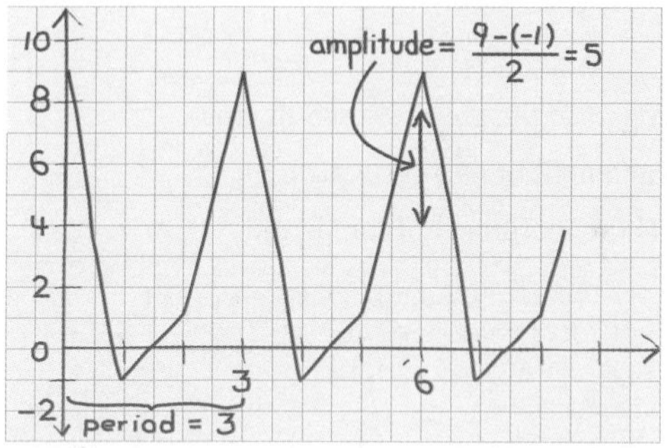

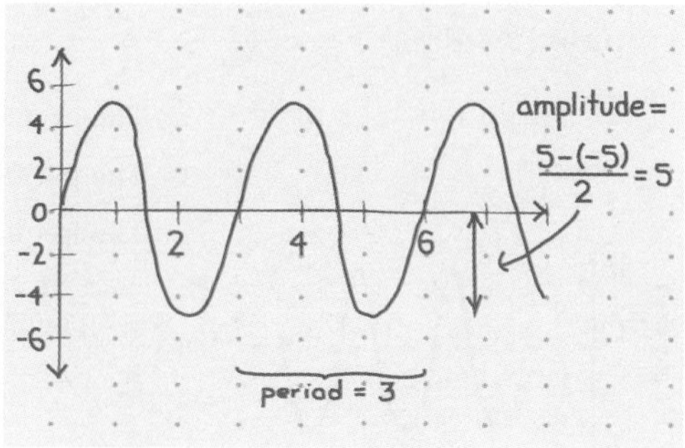

Look Back

On a periodic graph, how do you find the amplitude? the period? the frequency? an equation of the axis?

10-2 Exercises and Problems

1. **Reading** How is the period of a graph related to its frequency?

For Exercises 2–5, use the graph. It shows the way one person's body temperature changes over time.

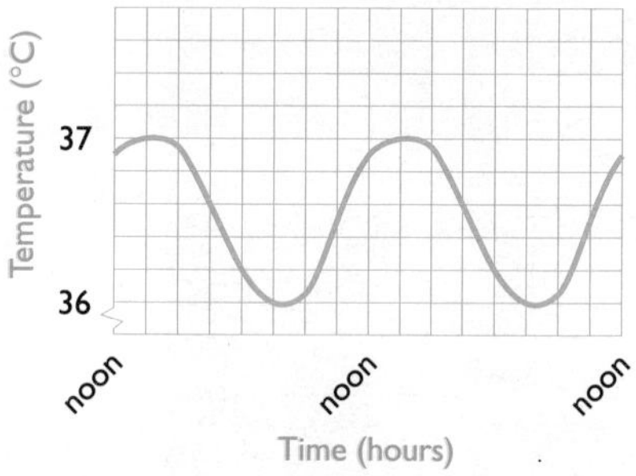

2. What is the amplitude of the function?
3. What is the period of the function?
4. Write an equation of the axis of the graph.
5. At what times of day is the temperature highest? lowest?

For Exercises 6–8, sketch a periodic graph with each set of characteristics.

6. amplitude: 4
   period: 3
   equation of axis: $y = 4$
7. amplitude: 10
   period: 5
   equation of axis: $y = -1$
8. amplitude: 32
   frequency: $\frac{1}{4}$
   equation of axis: $y = 0$

Answers to Look Back

The amplitude is $\frac{\text{maximum} - \text{minimum}}{2}$. The period is the length of one cycle. The frequency is the reciprocal of the period. The equation of the axis is $y = \frac{\text{maximum} + \text{minimum}}{2}$.

Answers to Exercises and Problems

1. The period and the frequency are reciprocals.
2. $\frac{1}{2}$
3. 24 h
4. $y = 36.5$
5. 3 P.M.; 4:30 A.M.

6–8. Answers may vary. Examples are given.

6.

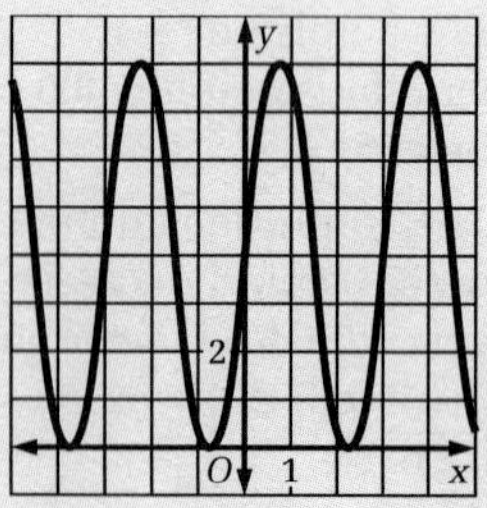

For Exercise 9–12, use this graph that shows the color of a traffic signal light over time.

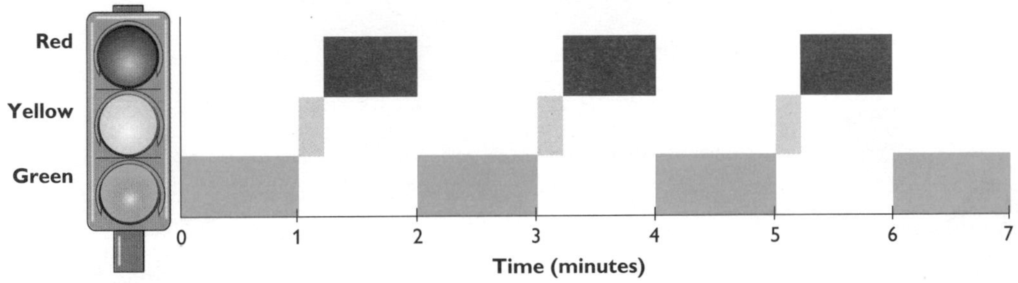

9. Describe one cycle of the traffic signal light.

10. How many times per hour does the traffic signal light repeat its pattern?

11. At rush hour, the pattern is changed so that the light stays green for 110 s. The length of the red and yellow lights are unchanged. How many times per hour does the traffic signal light repeat this pattern?

12. Explain how changing the length of the pattern changes the frequency with which it occurs.

For each graph in Exercises 13–16:

a. Find the period. **b. Find the frequency.**

c. Write an equation of the axis. **d. Find the amplitude.**

13.

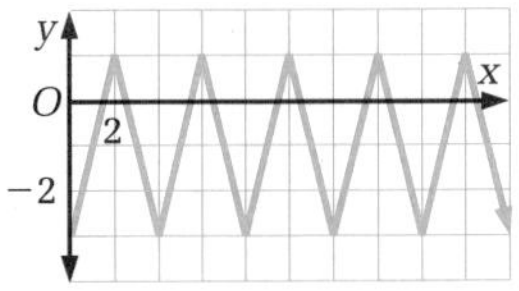

14.

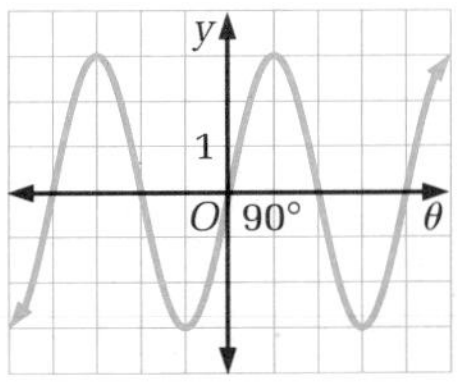

15.

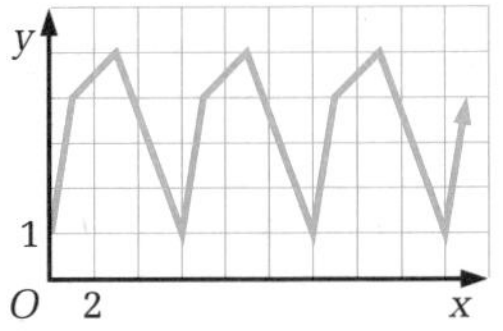

16.

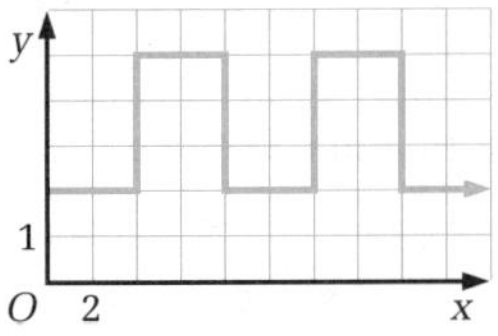

Redraw the graph in Exercise 13 using each given transformation.

17. a horizontal translation of 3

18. a vertical translation of 6

19. a period that is doubled

20. a vertical stretch by a factor of $\frac{1}{2}$

21. a horizontal stretch by a factor of 2

22. a frequency that is doubled

APPLYING

Suggested Assignment

Standard 1–28, 31–37

Extended 1–29, 31–37

Integrating the Strands

Algebra Exs. 32–36

Functions Exs. 1–31

Geometry Exs. 32–36

Trigonometry Exs. 33–36

Logic and Language Exs. 1, 28, 29

Teaching Tip

When discussing the exercises for this section, urge students to propose additional situations that involve periodic functions. Ask students if they can estimate periods and amplitudes for these situations.

Communication: Discussion

After students complete Exs. 17–22, it would be worthwhile to discuss how the different transformations discussed in Unit 9 affect the characteristics of a periodic function. For example, in Ex. 21 the horizontal stretch doubles the period of the graph. Remind students of the function notation for transformations. If the function in Ex. 13 is $y = f(x)$, then the function in Ex. 21 is $y = f\left(\frac{x}{2}\right)$.

Answers to Exercises and Problems

7.

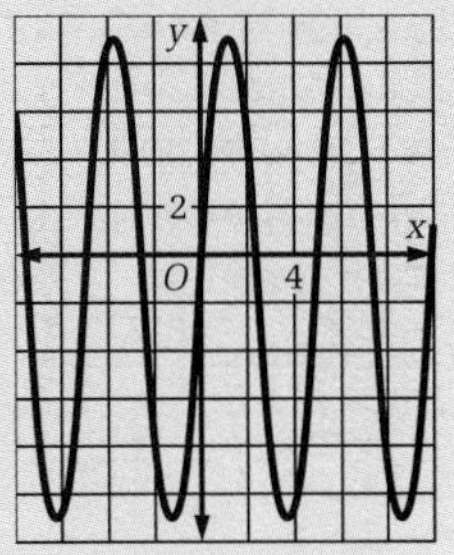

8.

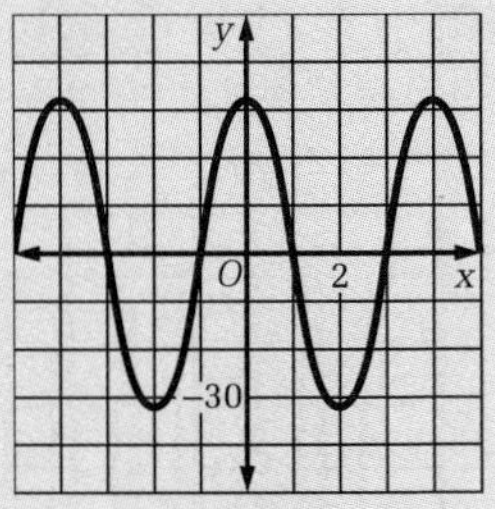

9. The light is green for 1 min, yellow for 15 s, and red for 45 s.

10. 30 times

11. about 21 times

12. Since the frequency and the period (the length of the pattern) are reciprocals, if the period increases (decreases), the frequency decreases (increases).

13. a. 4 b. $\frac{1}{4}$ c. $y = -1$ d. 2

14. a. 360° b. $\frac{1}{360°}$ c. $y = 0$ d. 3

15. a. 6 b. $\frac{1}{6}$ c. $y = 2.5$ d. 1.5

16. a. 8 b. $\frac{1}{8}$ c. $y = \frac{7}{2}$ d. $\frac{3}{2}$

17–22. See answers in back of book.

Application

Periodic functions can be used to model any cyclical phenomenon. Exs. 23–29 use periodic functions to study an ecosystem.

Assessment: Open-ended

For Exs. 23–28, students can find the peak in the hare population and examine what is happening in the willow mass at that time in terms of rate of growth. Ask students to repeat this exercise for the lowest point in the hare population. Is there a "lag" in the reaction? Why do they think this is true?

Using Technology

Students who have TI-81 or TI-82 calculators may find it interesting to examine the graph of $y = f\text{Part } x$. The function is found on the MATH NUM menu. Using a graphing window with $-1.5 \le y \le 1.5$, vary the settings for x, in the order listed below.

(1) $0 \le x < 1$

(2) $0 \le x < 6$

(3) $-6 \le x < 6$

Students will see that the graph for (3) exhibits periodic behavior for negative values of x and then changes its periodic behavior for positive values of x.

Students Acquiring English

Many of the exercises in this section are particularly suitable for students acquiring English in that they promote the understanding of mathematical concepts through drawings and explanations. You may want to pair students with English-proficient partners for Exs. 28 and 29, since both of these call for lengthy written responses.

connection to **BIOLOGY**

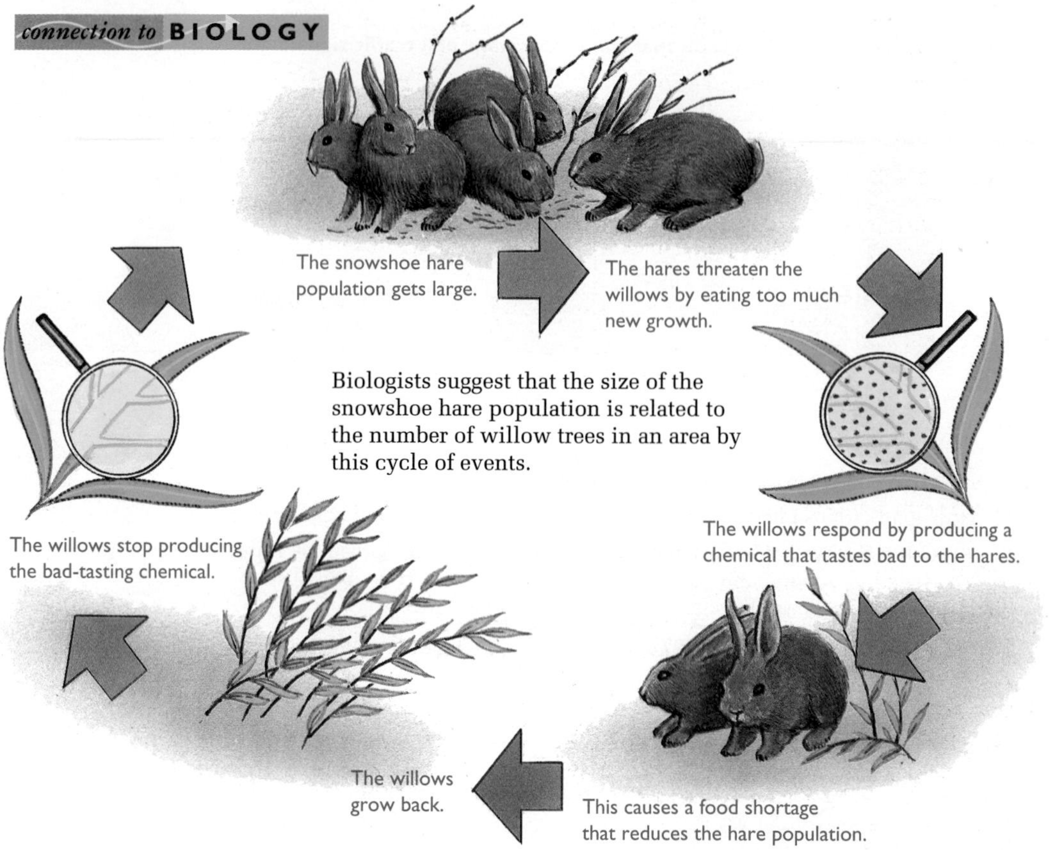

The snowshoe hare population gets large.
The hares threaten the willows by eating too much new growth.
Biologists suggest that the size of the snowshoe hare population is related to the number of willow trees in an area by this cycle of events.
The willows respond by producing a chemical that tastes bad to the hares.
This causes a food shortage that reduces the hare population.
The willows grow back.
The willows stop producing the bad-tasting chemical.

The graph shows how the numbers of hares and willows can change through several cycles.

Biomass is a measure of the weight of an organism per unit area.

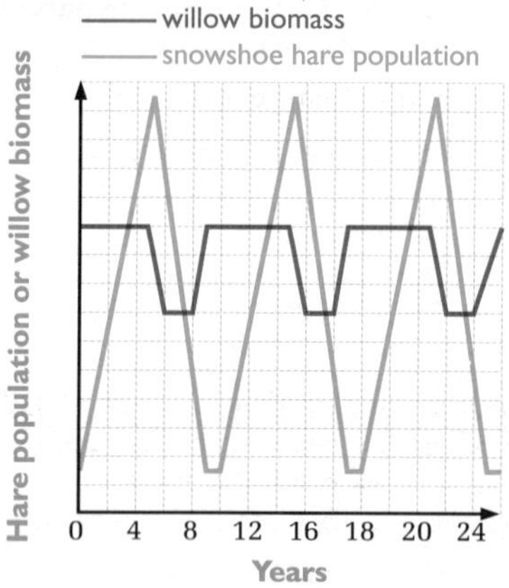

23. What is the period of the hare's population cycle?

24. What is the period of the willow's population cycle?

25. Identify the intervals when the willows produce the chemical that makes them taste bad.

26. When will the next peak in the hare population occur?

27. When will the next drop in the willow population occur?

28. **Writing** What do you think this ecosystem will be like 37 years after the "0" year on the graph?

Answers to Exercises and Problems

23. 10 years
24. 10 years
25. when the hare population is declining sharply, for example, between about 5 years and 9 years
26. at 35 years
27. at 36 years
28. The hares will be in the middle of a decline in population, while the willows will be at a low point in the size of the biomass.
29. a. 1974; 2008

b.

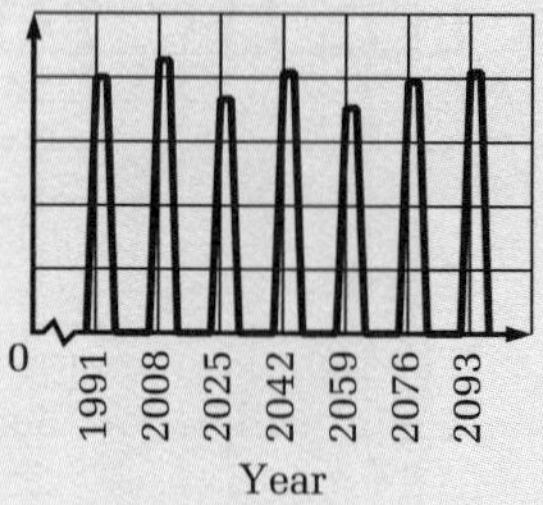

The actual number of cicadas at each appearance is not known.

c. The spikes would be closer together.

30. Check students' work.
31. $y = 5^x$ grows fastest, followed by $5x^2$, then $5x$.

29. Biology The cicada, or *Magicicada septendecim,* is an insect that develops underground for 17 years. In the spring of the year when the cicadas reach adulthood, they all leave the ground at the same time, and 20,000–40,000 may appear under one tree. Adults live for 30–40 days.

a. Indiana, Ohio, Pennsylvania, Kentucky, Tennessee, and North Carolina had adult cicada populations in 1991. What was the year of the previous adult cicada population? What will be the year of the next one?

b. Sketch a graph of the population of adult 17-year cicadas for one location for the 100 years after 1991. Explain why you cannot label the *y*-axis accurately.

c. How would the graph for 13-year cicadas be different from the graph for 17-year cicadas?

BY THE WAY...

Cicadas are sometimes called harvest flies or locusts, but they are not related to either flies or locusts. In the southeastern United States, another species of cicada has a 13-year cycle.

Ongoing **ASSESSMENT**

30. **Group Activity** Work with another student.

a. Draw a periodic graph.

b. Trade graphs with your partner. Identify the cycle, the period, the frequency, and the amplitude, and write an equation of the axis.

Review **PREVIEW**

31. Compare the growth rates in the first quadrant for these functions:

$y = 5x \quad y = 5x^2 \quad y = 5^x$ *(Section 10-1)*

32. Use right triangle *ABC.* *(Toolbox Skills 30 and 36)*

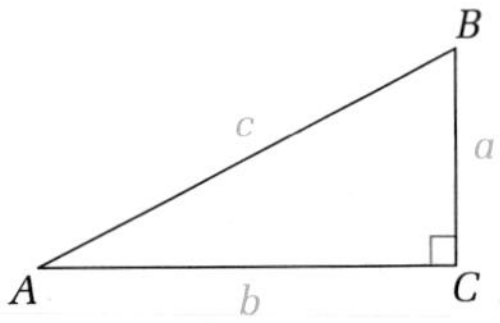

a. $m\angle A + m\angle B = \underline{?}$.

b. Find an expression for sin *A* in terms of *a* and *c*.

c. Find an expression for cos *B* in terms of *a* and *c*.

For Exercises 33–36, find the missing parts of each △*ABC* using the law of cosines or the law of sines. *(Sections 8-6, 8-7)*

33. $c = 10$ cm, $a = 5$ cm, $m\angle B = 30°$

34. $b = 15.4$ cm, $c = 2$ cm, $m\angle A = 110°$

35. $m\angle A = 30°$, $m\angle C = 50°$, $b = 45$ cm

36. $m\angle B = 22°$, $m\angle C = 95°$, $c = 17$ cm

Working on the Unit Project

37. Describe a situation that can be modeled by a periodic function. Make a graph to model the situation. Find or identify the cycle, the period, the frequency, the amplitude, and an equation of the axis of the graph. Include this graph on a page of your scrapbook.

Reasoning

Ask students to conjecture whether the sum of two periodic functions will or will not be periodic. Discuss how to sketch the graph of $f(x) + g(x)$ if the graphs of $f(x)$ and $g(x)$ are given. Invite students to test their conjectures by sketching two graphs that are periodic and using these graphs to sketch the related sum.

Working on the Unit Project

Students can work on Ex. 37 in their project groups. They should first agree on a situation that can be modeled by a periodic function and then complete the other activities.

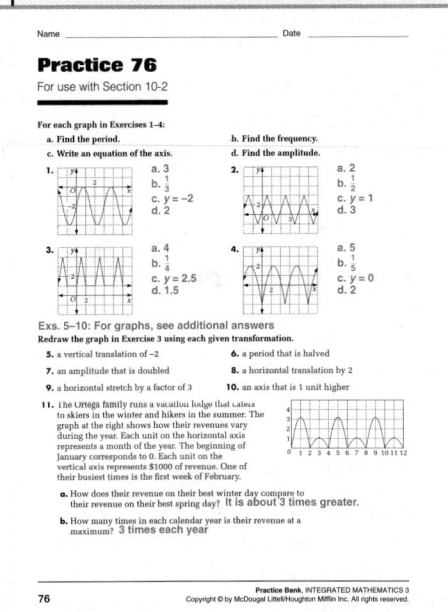

Practice 76 For use with Section 10-2

Name ______________________ Date ____________

Practice 76

For use with Section 10-2

For each graph in Exercises 1–4:

a. Find the period. **b. Find the frequency.**

c. Write an equation of the axis. **d. Find the amplitude.**

1. 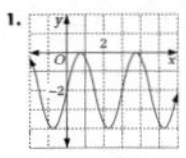 a. 3 b. $\frac{1}{3}$ c. $y = -2$ d. 2

2. 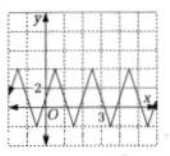 a. 2 b. $\frac{1}{2}$ c. $y = 1$ d. 3

3. 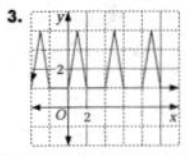 a. 4 b. $\frac{1}{4}$ c. $y = 2.5$ d. 1.5

4. 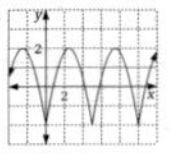 a. 5 b. $\frac{1}{5}$ c. $y = 0$ d. 2

Exs. 5–10: For graphs, see additional answers

Redraw the graph in Exercise 3 using each given transformation.

5. a vertical translation of –2

6. a period that is halved

7. an amplitude that is doubled

8. a horizontal translation by 2

9. a horizontal stretch by a factor of 3

10. an axis that is 1 unit higher

11. The Ortega family runs a vacation lodge that caters to skiers in the winter and hikers in the summer. The graph at the right shows how their revenues vary during the year. Each unit on the horizontal axis represents a month of the year. The beginning of January corresponds to 0. Each unit on the vertical axis represents $1000 of revenue. One of their busiest times is the first week of February.

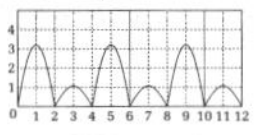

a. How does their revenue on their best winter day compare to their revenue on their best spring day? It is about 3 times greater.

b. How many times in each calendar year is their revenue at a maximum? 3 times each year

76 **Practice Bank**, INTEGRATED MATHEMATICS 3

Answers to Exercises and Problems

32. a. 90°

b. $\sin A = \frac{a}{c}$

c. $\cos B = \frac{a}{c}$

33. $b \approx 6.2$ cm, $m\angle A \approx 23.8°$, and $m\angle C \approx 126.2°$

34. $a \approx 16.2$ cm; $m\angle C \approx 6.7°$, and $m\angle B \approx 63.3°$

35. $m\angle B = 100°$, $a \approx 22.8$ cm, and $c \approx 35$ cm

36. $m\angle A = 63°$, $a \approx 15.2$ cm, and $b \approx 6.4$ cm

37. Answers may vary.

PLANNING

Objectives and Strands
See pages 538A and 538B.

Spiral Learning
See page 538B.

Materials List
- Graphics calculator or graphing software
- Graph paper

Recommended Pacing
Section 10-3 is a one-day lesson.

Extra Practice
See pages 624–625.

Warm-Up Exercises
- Warm-Up Transparency 10-3

Support Materials
- Practice Bank: Practice 77
- Activity Bank: Enrichment 68
- Study Guide: Section 10-3
- Problem Bank: Problem Set 21
- Explorations Lab Manual: Diagram Master 2
- Overhead Visuals: Folder 10
- Assessment Book: Quiz 10-3

Section 10-3

Graphs of Sine and Cosine

Focus
Recognize and understand characteristics of the graphs of the sine function and the cosine function.

As the Wheel Turns

The Ferris wheel built by George Ferris in Chicago for the World's Columbian Exposition in 1893 was the largest Ferris wheel ever built. The wheel was 250 ft in diameter and each cab could hold 60 people.

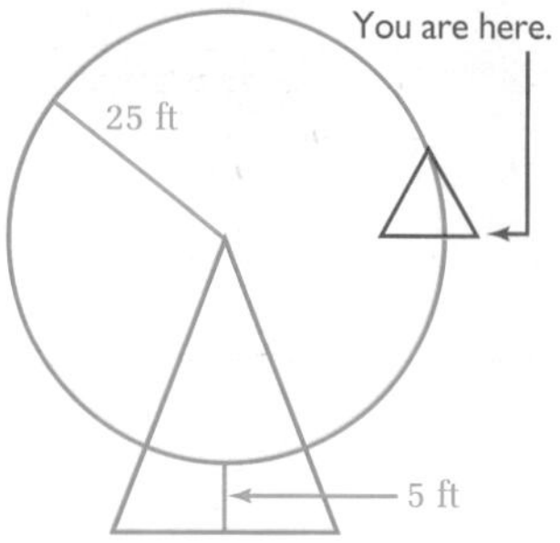

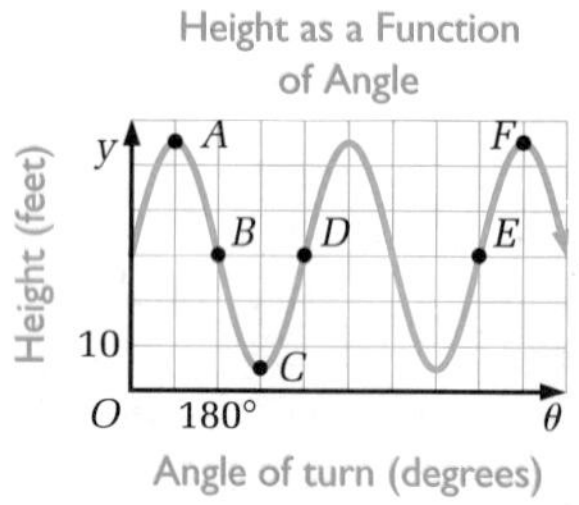

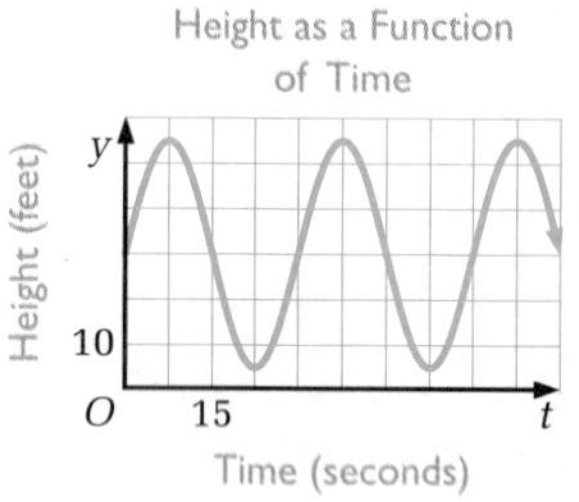

Talk it Over

1. Suppose that you are on a Ferris wheel in the seat shown. For each angle (measured counterclockwise) through which the Ferris wheel turns, make a sketch to show your position.
   a. 90° b. 180° c. 270°
   d. 360° e. 400° f. 800°
2. How does your height above the ground change as the wheel turns?
3. The graph shows your height above the ground as the Ferris wheel turns. Answer parts (a)–(c) for points A–F.
   a. What is your position on the Ferris wheel?
   b. How many degrees did you turn from the starting point?
   c. How many *revolutions*, or complete turns, did you make?
4. What is the period of the graph from question 3? What does this represent in terms of the Ferris wheel ride?
5. Answer question 4 for the amplitude of the graph.
6. The graph shows your height after t seconds.
   a. How long does it take to make one complete revolution on the Ferris wheel?
   b. How many revolutions do you make in 1 min?
   c. Suppose the Ferris wheel completes one revolution in 20 s. How would the graph change?

Answers to Talk it Over

1. a. b. c. d. e. f.

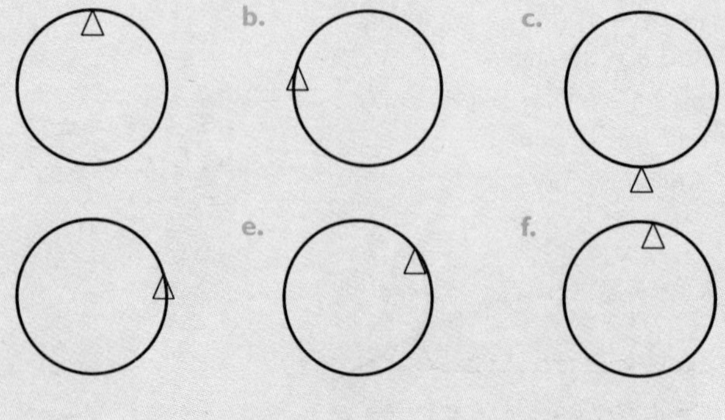

2. It increases and decreases in cycles, going from a low of 5 ft to a high of 55 ft.
3. A: a. at the top
   b. 90°
   c. $\frac{1}{4}$
   B: a. on the left side, 30 ft up
   b. 180°

The Sine Function

Many situations, like the height of a seat on a Ferris wheel, can be modeled by a periodic function whose control variable is angle measure or time. Often the graphs of these functions are transformations of the graph of the *sine function.*

A circle with a radius of 1 and its center at the origin is called a *unit circle.* Suppose you move an object around the unit circle in a counterclockwise direction starting at (1, 0). The **sine function**, $y = \sin\theta$, measures the vertical position of the object.

One revolution around the unit circle creates one cycle of the sine function.

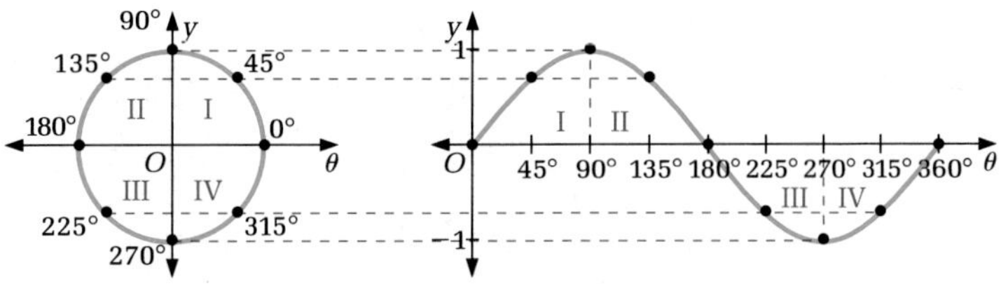

The domain of the sine function is any angle θ, positive or negative. The range of $y = \sin\theta$ is $-1 \le y \le 1$.

TECHNOLOGY NOTE

Set your calculator in degree mode. When you enter the sine function on your graphics calculator, the screen may show sin *x*, not sin *θ*.

Sample 1

a. Graph the function $y = \sin\theta$.

b. Describe the graph of $y = \sin\theta$.

Sample Response

a.

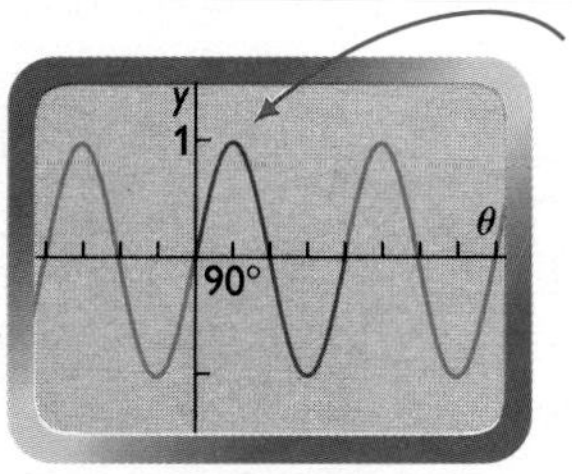

The cycle repeats every 360°.

b. The graph is periodic. The period is 360°.

The amplitude is 1.

The graph crosses the θ-axis when θ is a multiple of 180°. The θ-intercepts are . . . , −360°, −180°, 0°, 180°, 360°, ...

The y-intercept is 0.

TEACHING

Talk it Over

In questions 1–5, a Ferris wheel is used as a real-world model to help students understand the relationship between a point rotating on a circle and a sine function. Students examine how the coordinates, period, and amplitude of the graph are related to the movement of a point around the wheel.

Integrating the Strands

The introduction of the sine function integrates the strands of functions, algebra, geometry, and trigonometry.

Visual Thinking

Ask students to draw their own unit circles and make sketches of the sine function. Have them trace the circle with one finger and the sine curve with another, and have them explain the position of their fingers at various points. This activity involves the visual skills of *correlation* and *communication.*

Additional Sample

S1 a. Graph the function $y = |\sin\theta|$.

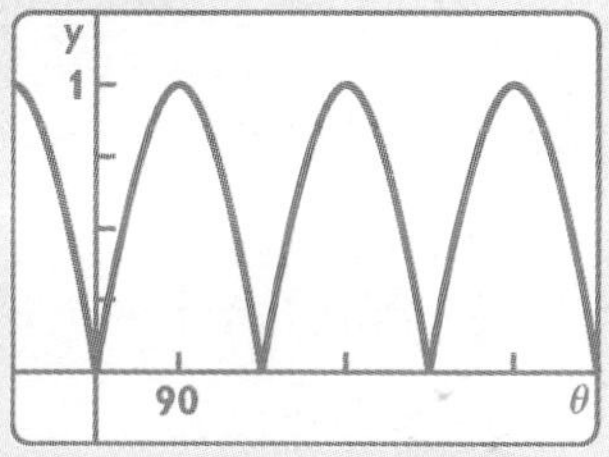

The cycle is the portion of the graph for which $0 \le \theta \le 180$.

b. Describe the graph of $y = |\sin\theta|$.

The graph is periodic. The period is 180°. The amplitude is $\frac{1}{2}$. The graph touches the θ-axis when θ is a multiple of 180°. The θ-intercepts are ..., −180°, 0°, 180°, 360°, ... The y-intercept is 0.

Answers to Talk it Over

c. $\frac{1}{2}$

C: a. at the bottom

b. 270°

c. $\frac{3}{4}$

D: a. on the right side, 30 ft up

b. 360°

c. 1

E: a. on the right side, 30 ft up

b. 720°

c. 2

F: a. at the top

b. 810°

c. $2\frac{1}{4}$

4. 360°; one turn of the wheel

5. 25 ft; the radius of the Ferris wheel

6. a. 30 s

b. 2

c. It would be stretched horizontally by a factor of $\frac{2}{3}$.

Talk it Over

Questions 7 and 8 introduce the cosine function, $y = \cos \theta$, and have students draw its graph for $0° \le \theta \le 360°$.

Communication: Discussion

Many students probably have not seen a graph generated from another graph. To prepare them for this idea, a class discussion of the sine and cosine graphs may be helpful. When drawing the cosine graph, in particular, students may be confused about the variables. The control variable on the cosine graph is the angle of the rotated object around the circle. The dependent variable is the horizontal position of the object; thus, the x-coordinate of the circle graph becomes the y-coordinate of the cosine graph. Exs. 18 and 19 on page 559 give students some experience with these ideas.

Visual Thinking

Ask students to create a sketch of the graph of the sine function of a circle. In another color, ask them to create a sketch of the cosine function of the circle on the same graph. Ask them to explain their diagram, describing both horizontal and vertical positions. This activity involves the visual skills of *exploration* and *communication.*

Reasoning

Students can explore the characteristics of sine and cosine graphs using a Venn diagram containing two overlapping circles. In the overlapping part, they can write how the graphs are alike. The distinctive characteristics are written in the nonoverlapping parts of the circles.

The Cosine Function

The **cosine function**, $y = \cos \theta$, measures the horizontal position of an object moving around a unit circle.

Talk it Over

7. Suppose an object is moving around a unit circle in a counterclockwise direction as shown. Explain why each statement describes the object's horizontal position as it moves through each of the four quadrants.

   I. decreasing from 1 to 0 **II.** decreasing from 0 to -1

   III. increasing from -1 to 0 **IV.** increasing from 0 to 1

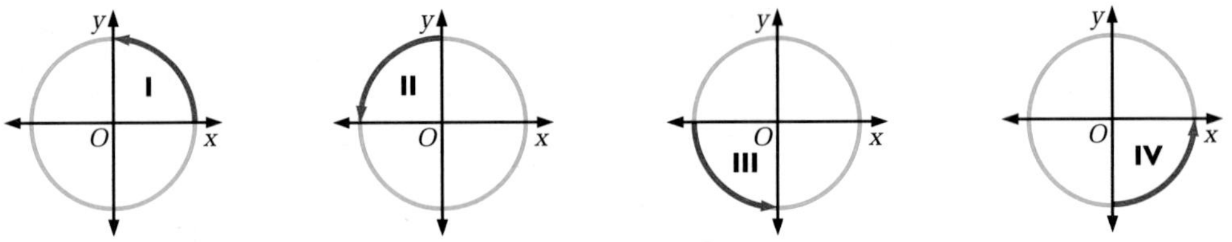

8. Use your answers to question 7 to sketch a graph of the cosine function for $0° \le \theta \le 360°$.

GRAPHS OF SINE AND COSINE

For an object moving counterclockwise (forward) or clockwise (backward) around a unit circle, the sine function gives the **vertical** position of the object.

For an object moving counterclockwise (forward) or clockwise (backward) around a unit circle, the cosine function gives the **horizontal** position of the object.

$y = \sin \theta$ **$y = \cos \theta$**

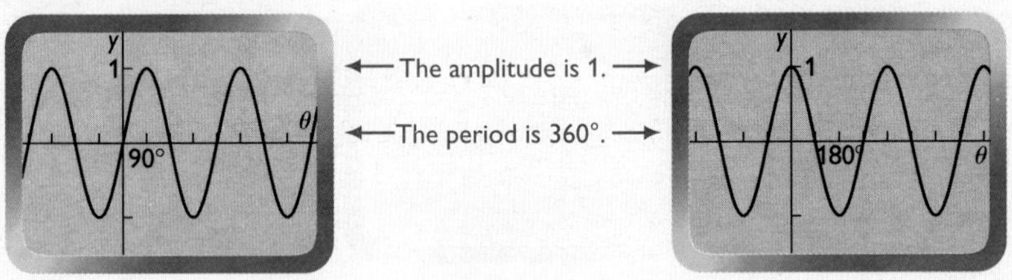

Because the cosine graph can be obtained from the sine graph through a horizontal translation, both graphs and all similarly-shaped graphs produced by transforming the sine or cosine graph can be called *sine waves.*

Answers to Talk it Over

7. I: the x-coordinate starts at 1 and ends up at 0.
   II: the x-coordinate starts at 0 and ends up at –1.
   III: the x-coordinate starts at –1 and ends up at 0.
   IV: the x-coordinate starts at 0 and ends up at 1.

8. 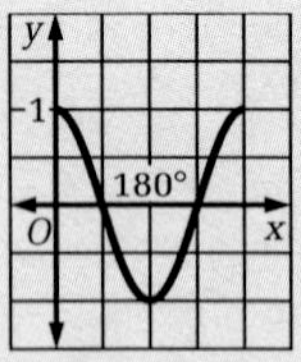

Sample 2

Writing **How is the graph of a cosine function like the graph of a sine function? How are they different?**

Sample Response

The graph of the cosine function is a 90° horizontal translation of the graph of the sine function. The two functions have the same period, range, and amplitude, but the intercepts of the two graphs are different. The y-intercept of the sine function is 0, but the y-intercept of the cosine function is 1.

Both graphs are periodic functions that have a period of 360° and an amplitude of 1. The domain of each function is all real angles. The range of each function is $-1 \le y \le 1$.

The graph of the sine function crosses the θ-axis when θ is a multiple of 180°. The graph of the cosine function crosses the θ-axis when θ is an odd multiple of 90°.

Look Back

Which function, *sine* or *cosine*, gives the horizontal position of an object moving around a unit circle? Which function gives the vertical position?

10-3 Exercises and Problems

1. **Reading** What is a unit circle? How is a unit circle related to the sine and cosine functions?

Tell whether each statement describes a characteristic of the *sine function*, the *cosine function*, *both functions*, or *neither function*.

2. The function has a period of 360°.
3. The function has an amplitude of 2.
4. The *y*-intercept is 1.
5. The *y*-intercept is 0.
6. The range of the function is $-1 \le y \le 1$.
7. The horizontal intercepts occur only at multiples of 90°.
8. The function decreases in the interval $0° \le \theta \le 90°$.
9. The function increases in the interval $0° \le \theta \le 90°$.

Additional Sample

S2 How is the graph of $y = |\cos \theta|$ like the graph of $y = |\sin \theta|$? How is it different?

Both graphs show periodic functions that have a period of 180° and an amplitude of $\frac{1}{2}$. The domain of each function is all real angles. The range of each function is $0 \le y \le 1$. The graph of $y = |\cos \theta|$ is a 90° horizontal translation of the graph of $y = |\sin \theta|$. The two functions have the same period, range, and amplitude. The θ-intercepts of $y = |\sin \theta|$ are the multiples of 180°. The θ-intercepts of $y = |\cos \theta|$ are the odd-numbered multiples of 90°. The y-intercept of $y = |\sin \theta|$ is 0, and the y-intercept of $y = |\cos \theta|$ is 1.

APPLYING

Suggested Assignment

Standard 1–14, 20–28

Extended 1–14, 20–28

Integrating the Strands

Algebra Exs. 21–28

Functions Exs. 1–28

Geometry Exs. 1, 18, 19

Trigonometry Exs. 1–19

Logic and Language Exs. 1, 18, 19

Answers to Look Back

cosine; sine

Answers to Exercises and Problems

1. The unit circle is a circle with radius 1 centered at the origin. If an object moves counterclockwise around the unit circle, the sine function indicates the vertical position and the cosine function indicates the horizontal position of the object.
2. both
3. neither
4. cosine
5. sine
6. both
7. both (The horizontal intercepts of the sine function occur at multiples of 180°; the horizontal intercepts of the cosine function occur at odd multiples of 90°.)
8. cosine
9. sine

Multicultural Note

The water wheel was probably first developed about 100 B.C. in Greece. It quickly spread as a source of power throughout the Mediterranean world, where the Greeks traded extensively. As trade routes expanded, knowledge of water wheels spread to the Middle East, northern Europe, and many other parts of the globe. Water wheels were often connected to grinding stones, and the power was used to mill grain. During the Industrial Revolution, water wheels were often used to run machinery in factories. This source of power, however, was not entirely dependable: floods would make the wheels turn too fast, and droughts sometimes left the factories without power. The more reliable steam engine, invented in the 1600s, largely replaced the water wheel in factories by the end of the nineteenth century. Water wheels are still an important source of power in some areas. The Knight Foundry in Sutter Creek, California is the last remaining foundry in the United States powered by a water wheel.

Error Analysis

In Ex. 15, students will get incorrect graphs if their calculators have been set to radian mode. This error can be corrected by having students check that the mode setting is for degrees.

Students may try to enter Y1 = sin θ on the Y= list by pressing [SIN] [ALPHA] [θ]. This will not result in a correct graph. You can correct this error by pointing out that θ is used in calculator equations mainly when using a *polar* coordinate system. All of the graphs in this unit require a *rectangular* coordinate system. In the written equations, θ is used as a reminder that it is usually an angle measure expressed in degrees. When entering equations for graphing on the calculator, change θ to x.

Water power During the Industrial Revolution, water wheels provided power to run factories. Suppose a water wheel is rotating at the rate of 3 revolutions per minute (rev/min). A bucket's height above the water level is a function of the amount of time the wheel has turned. Its graph is a sine wave.

Use the information above for Exercises 10–14.

10. Choose the most reasonable graph for the height of bucket P at t seconds. Explain.

A.
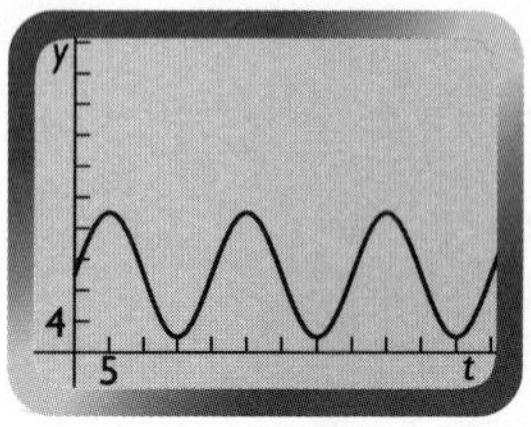

B.
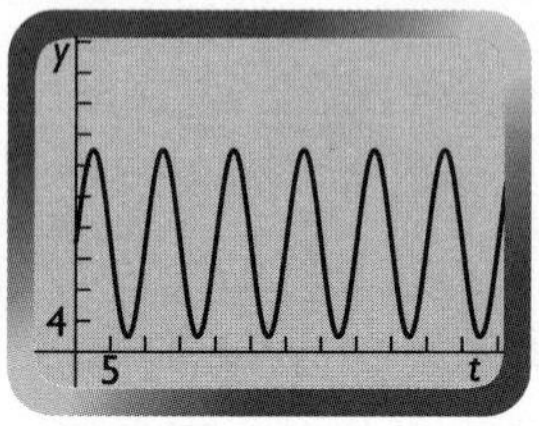

C.
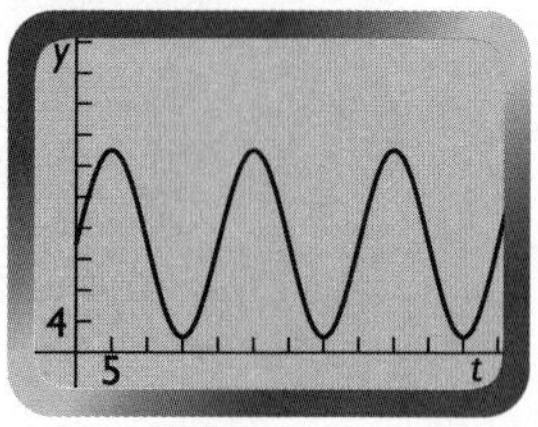

11. What is the amplitude of the function?
12. Describe an interval in which the height of the bucket changes quickly.
13. The height of the bucket at 23 s is the same as its height at 3 s. Why? When is the next time the height will be the same as the height at 3 s?
14. Suppose the speed of the water flowing in the river caused the wheel to turn at 4 rev/min. How would the graph change?
15. a. TECHNOLOGY Use a graphics calculator or graphing software to graph the function $y = \sin \theta$ for $-360° \le \theta \le 360°$. How many cycles does the graph show?
    b. For what values of θ is $\sin \theta = 0.5$?
    c. What is the maximum value of the function? What is the minimum value of the function? What values of θ give the maximum value of the function? the minimum value of the function?
    d. In what intervals are the values of the function positive? negative?
    e. When are the values of the function increasing? When are they decreasing?
16. Repeat Exercise 15 for the function $y = \cos \theta$.
17. a. What is the relationship between sin 45° and sin (−45°)? sin 210° and sin (−210°)? sin 270° and sin (−270°)?
    b. Make a conjecture based on any pattern you see in your answers to part (a).

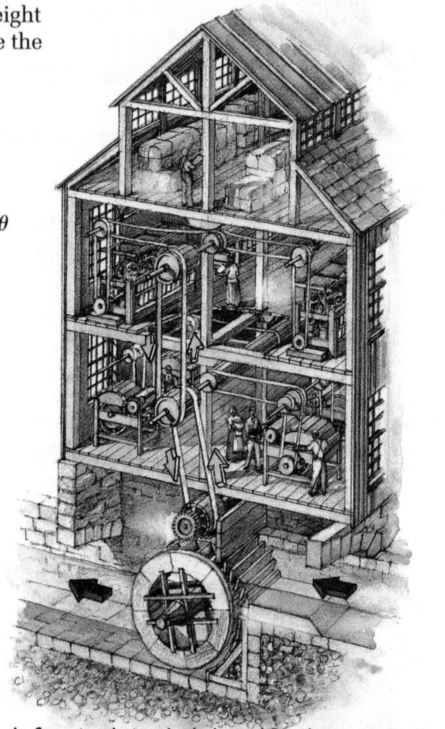

In factories during the Industrial Revolution, a spinning water wheel caused gears to turn. The gears operated pulleys and belts throughout the mill, which gave power to the machines.

Answers to Exercises and Problems

10. C; Explanations may vary. An example is given. The bucket starts at a height of 14 ft, reaches a height of 26 ft, and completes 1 revolution in 20 s. Graph C models this situation.
11. 12
12. The height of the bucket is changing most quickly every 10 s from the start.
13. The period of the function is 20 s; 27 s.
14. The period would be 15 s instead of 20 s. The graph would be stretched horizontally by a factor of $\frac{3}{4}$.
15. a.

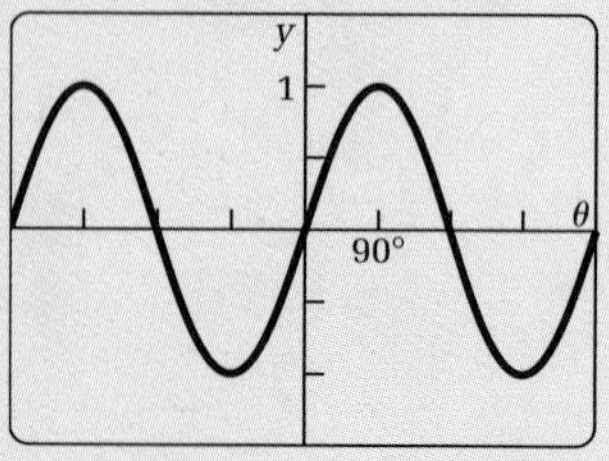

2 cycles

b. −330°, −210°, 30°, 150°

Ongoing ASSESSMENT

18. **Group Activity** Work with another student.

TECHNOLOGY You will each need a graphics calculator or graphing software. Each of you should set your calculator to degree mode and parametric mode and use the following window:

$0 \le t \le 360$, t-step $= 10$ $\quad -1.25 \le y \le 1.25$, y-scale $= 0.5$

a. One student should graph the unit circle and the other should graph the sine function. Adjust your viewing window to fit the x-values appropriate for your graph.

unit circle	sine function
$x_1 = \cos t$	$x_1 = t$
$y_1 = \sin t$	$y_1 = \sin t$
$-1.88 \le x \le 1.88$, x-scale $= 0.5$	$0 \le x \le 360$, x-scale $= 30$

b. Trace along each graph for various values of t. Select a common t-value and compare where the cursor is on each graph. What does the y-value represent on each graph? What does the x-value represent on each graph?

c. **Writing** Compare where the cursor is on each graph for other values of t. How is the position of the cursor on the graph of the unit circle related to the position of the cursor on the sine function?

19. Repeat Exercise 18 using the cosine function instead of the sine function. The person who graphed the unit circle in Exercise 18 should graph the cosine function.

Review PREVIEW

20. Sketch a graph of a periodic graph that has an amplitude of 2 and a period of 4. *(Section 10-2)*

For Exercises 21–24, use the functions $f(x) = x^2$, $g(x) = \frac{x}{2}$, and $h(x) = \sqrt{x} - 1$. Find each function rule or value. *(Section 2-8)*

21. $(f \circ g)(x)$ 22. $(h \circ f)(x)$ 23. $(f \circ g)(-8)$ 24. $(g \circ h)(25)$

Without using graphing technology, sketch the graph of each function. *(Section 9-2)*

25. $y + 2 = (x - 4)^2$ 26. $y = \sqrt{x + 3}$ 27. $y - 1 = x^3$

10 Working on the Unit Project

28. Make a list of all the types of functions you know how to graph. Include the list and examples of each type of graph in your scrapbook.

Problem Solving

Portions of certain graphs may look like portions of sine graphs without actually being so. Ask students how they can tell a graph is a sine graph without having the equation available. A TI-82 graphics calculator can be used to illustrate this situation. You can enter these equations at the bottom of the Y= list so they are not visible:

$Y_9 = \sin X$

$Y_{10} = (-0.0012344)X^2 + (0.02221987)X$

Set the graphing window for $0 \le x \le 180$ and $-1.5 \le y \le 1.5$ and show the graphs. Ask students whether they see two graphs or one. Is either a portion of a sine curve? How can you be sure?

Practice 77 For use with Section 10-3

Name ______________________ Date ____________

Practice 77
For use with Section 10-3

1. a. $90° \le \theta \le 180°$ b. $0° \le \theta \le 90°$
2. a. $270° \le \theta \le 360°$ b. $180° \le \theta \le 270°$
3. a. $0° \le \theta \le 90°$ b. $270° \le \theta \le 360°$
4. a. $180° \le \theta \le 270°$ b. $90° \le \theta \le 180°$

For Exercises 1–4, name the portion of the interval $0° \le \theta \le 360°$ in which the statement describes the graph of (a) $y = \sin \theta$, and (b) $y = \cos \theta$. Exs. 1–4: See top of page.

1. decreasing from 1 to 0
2. increasing from –1 to 0
3. increasing from 0 to 1
4. decreasing from 0 to –1

Tell whether each statement describes a characteristic of the *sine function*, the *cosine function*, *both functions*, or *neither function*.

5. The graph models the horizontal position of an object moving around a unit circle. cosine
6. The function increases throughout the interval $180° \le \theta \le 360°$. cosine
7. The domain of the function is all real angles. both
8. The graph of the function crosses the θ-axis at integer multiples of 180°. sine
9. The amplitude of the function is 1. both
10. The function has period 180°. neither

Use the graph of the sine function to give the approximate vertical position of an object at the given angle on a unit circle. Give answers to the nearest tenth.

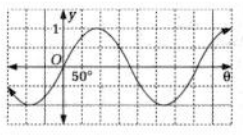

11. 37° 0.6
12. 115° 0.9
13. 235° –0.8
14. 335° –0.4

15. A vertical gear of an antique clock makes one counterclockwise revolution every 60 seconds. Suppose there is a catch on the front side of the gear that is at its rightmost position at time $t = 0$, and suppose the vertical position of the catch at this time is called 0.
    a. Suppose the vertical position of the catch after 5 seconds is h. After how many more seconds will it first return to vertical position h? 20 s
    b. Name two times when its vertical position will be $-h$. Answers may vary. A sample is given. 35 s, 55 s

16. ***Open-ended*** Describe a situation in which the rotation of a wheel or disc can be modeled by a sine function. State what the y-values represent in terms of the physical situation. Answers may vary. Check students' work.

Practice Bank, INTEGRATED MATHEMATICS 3 77

Answers to Exercises and Problems

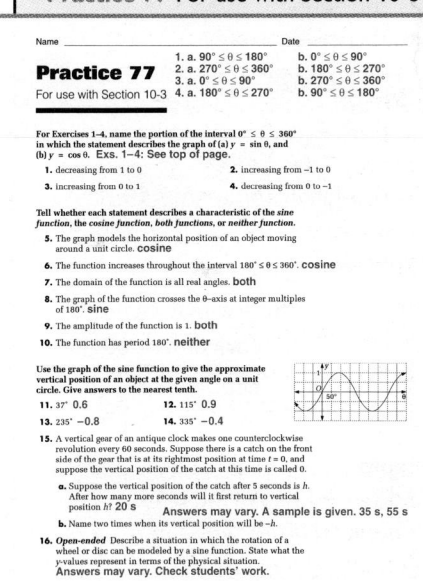

c. 1; –1; –270° and 90°; –90° and 270°

d. –360° to –180°, and 0° to 180°; –180° to 0° and 180° to 360°

e. –360° to –270°, –90° to 90°, and 270° to 360°; –270° to –90° and 90° to 270°

16. a.

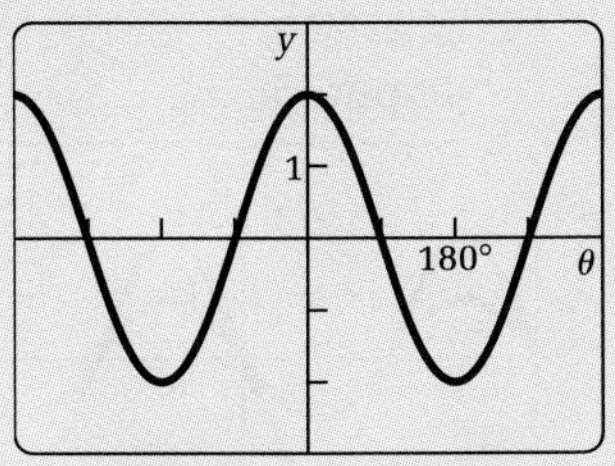

2 cycles

b. –300°, –60°, 60°, 300°

c. 1; –1; –360°, 0°, 360°; –180°, 180°

d. –360° to –270°, –90° to 90°, and 270° to 360°; –270° to –90° and 90° to 270°

e. –180° to 0° and 180° to 360°; –360° to –180° and 0° to 180°

17. a. sin (–45°) = – sin 45°; sin (–210°) = –sin 210°; sin (–270°) = –sin 270°

b. sin $(-\theta) = -\sin \theta$

18. a. Check students' graphs.

b. On both graphs, the y-value is the sine of the angle. The x-value on the unit circle is the cosine of the angle and the x-value on the sine graph is the measure of the angle.

c. The y-values are the same.

19–28. See answers in back of book.

PLANNING

Objectives and Strands
See pages 538A and 538B.

Spiral Learning
See page 538B.

Materials List
- Graphics calculator or graphing software
- Graph paper

Recommended Pacing
Section 10-4 is a two-day lesson.

Day 1

Pages 560–561: Exploration through Transformations of the Sine Graph, *Exercises 1–21*

Day 2

Pages 561–563: Sample 1 through Look Back, *Exercises 22–48*

Extra Practice
See pages 624–625.

Warm-Up Exercises
- Warm-Up Transparency 10-4

Support Materials
- Practice Bank: Practice 78
- Activity Bank: Enrichment 69
- Study Guide: Section 10-4
- Problem Bank: Problem Set 21
- Explorations Lab Manual: Additional Exploration 12 Diagram Master 2
- Using TI-81 and TI-82 Calculators: Adding Sine Graphs
- Using Plotter Plus: Adding Sine Graphs
- Assessment Book: Quiz 10-4, Alternative Assessment 4

Section **10-4**

Transforming Sine and Cosine Graphs

WAVES IN MOTION

Focus

Recognize and understand characteristics of the graphs of sine functions and cosine functions that are transformed.

EXPLORATION

What happens when sine and cosine graphs are transformed?

- **Materials: graphics calculators or graphing software**
- **Work with another student.**

Set your calculator in degree mode. Use this viewing window:
$-360 \le x \le 360$, x-scale $= 90$
$-3.6 \le y \le 3.6$, y-scale $= 0.5$

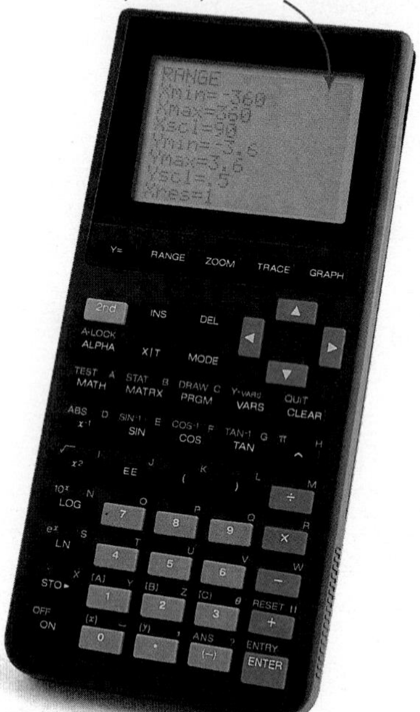

1. Graph $y = \sin\theta$. What is the amplitude? the period? the frequency? What is the y-intercept of the graph?

2. Graph each equation on the same screen as $y = \sin\theta$. (Have only two graphs on screen.) How is each graph like the graph of $y = \sin\theta$? How is each graph different?
   - **a.** $y = 2\sin\theta$
   - **b.** $y = \sin 2\theta$
   - **c.** $y = \sin(\theta - 45°)$
   - **d.** $y = \sin\theta + 1$

3. Use your answers from step 2 to sketch each graph in parts (a)–(h). To check your prediction, graph each function. How is each graph like the graph of $y = \sin\theta$? How is each graph different?
   - **a.** $y = 0.5\sin\theta$
   - **b.** $y = -\sin\theta$
   - **c.** $y = \sin 3\theta$
   - **d.** $y = \sin(-\theta)$
   - **e.** $y = \sin(\theta - 60°)$
   - **f.** $y = \sin(\theta + 90°)$
   - **g.** $y = \sin\theta + 2$
   - **h.** $y = \sin\theta - 1.5$

4. For each equation, write a general rule to tell the effect of A, B, C, or D on the graph of $y = \sin\theta$.
   - **a.** $y = A\sin\theta$
   - **b.** $y = \sin B\theta$
   - **c.** $y = \sin(\theta - C)$
   - **d.** $y = \sin\theta + D$

5. Use your rules from step 4 to tell what transformation(s) of the graph of $y = \sin\theta$ will give you the graph of each equation in parts (a)–(d). Then sketch each graph. Graph each equation using technology to check your sketch.
   - **a.** $y = 4\sin 2\theta$
   - **b.** $y = 0.5\sin(\theta - 30°)$
   - **c.** $y = \sin 0.75(\theta + 45°)$
   - **d.** $y = \sin(\theta + 180°) - 1.5$

6. For each graph sketched in step 5, find the amplitude, the frequency, and the period.

Answers to Exploration

1.

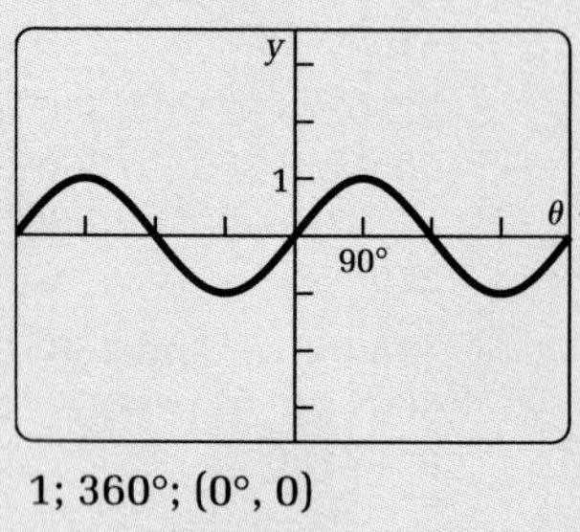

1; 360°; (0°, 0)

2. a. same period and intercepts as the parent graph; stretched vertically by a factor of 2; amplitude = 2

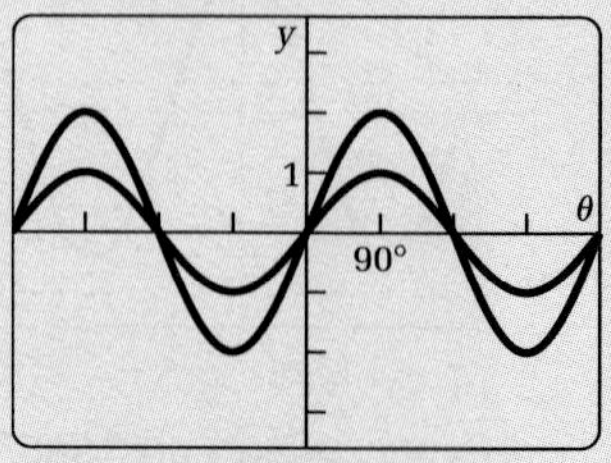

b. same amplitude and y-intercept as the parent graph; stretched horizontally by a factor of $\frac{1}{2}$; period is half that of the parent graph

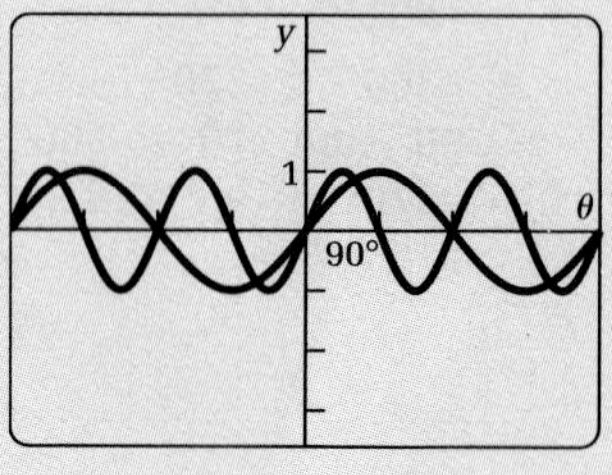

TRANSFORMATIONS OF THE SINE GRAPH

The graph of a function of the form $y = A \sin B(\theta - C) + D$ differs from the graph of $y = \sin \theta$ in the following ways.

➤ The graph is vertically stretched or shrunk by a factor of $|A|$. If $A < 0$, there is a reflection over the θ-axis.

➤ The graph is horizontally stretched or shrunk by a factor of $\frac{1}{|B|}$. The period is $\frac{360°}{|B|}$. If $B < 0$, there is a reflection over the y-axis.

➤ The graph is translated horizontally C units.

➤ The graph is translated vertically D units.

Example

$y = 3 \sin 2(\theta - 45°) + 8$

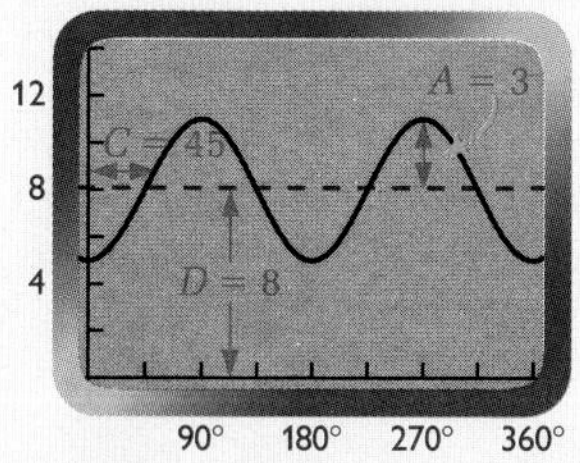

➤ Now you are ready for: Exs. 1–21 on pp. 563–564

In Exercises 2–5, you will verify that the same properties apply to the graph of the cosine function.

Sample 1

Without using graphing technology, sketch the graph of $y = -0.25 \cos 2(\theta - 90°) - 1$.

Sample Response

Sketch the graph of the parent function, $y = \cos \theta$.

The equation has the form $y = A \cos B(\theta - C) + D$.

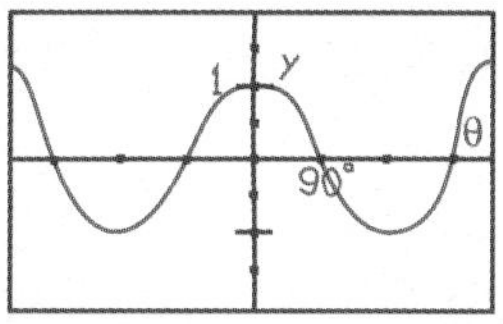

The value of A is -0.25. The amplitude is 0.25. $A < 0$, so the graph is reflected over the θ-axis.

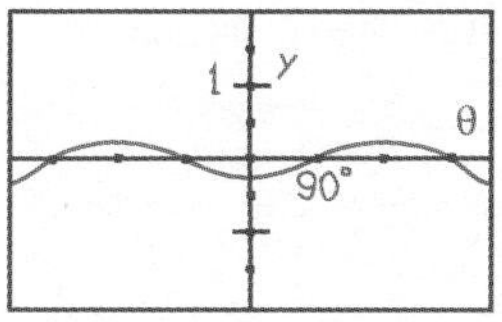

The value of B is 2. The period of the graph is $\frac{1}{2} \cdot 360° = 180°$.

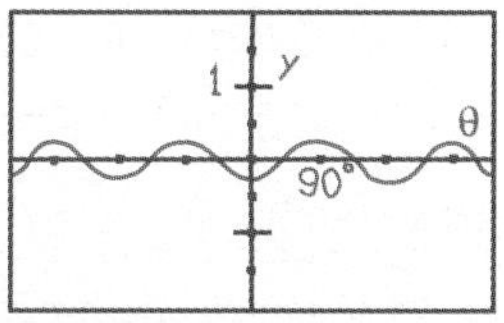

The value of C is 90°. The graph is translated 90° to the right.

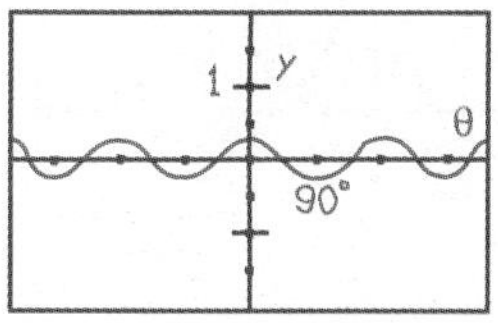

The value of D is -1. The graph is translated 1 unit down.

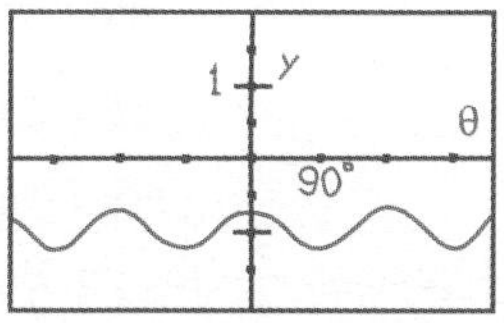

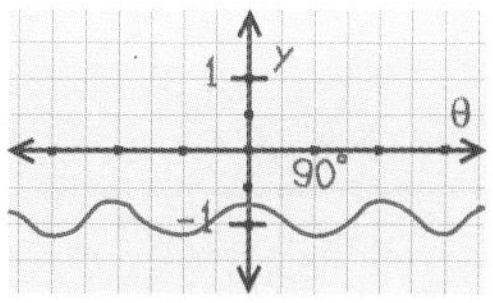

TEACHING

Exploration

The goal of the Exploration is to have students recognize the effects of transformations on the sine and cosine graphs.

Additional Sample

S1 Without using graphing technology, sketch the graph of $y = 3 \sin (-4(\theta + 90)) + 2$.

Sketch the graph of the parent function $y = \sin \theta$.

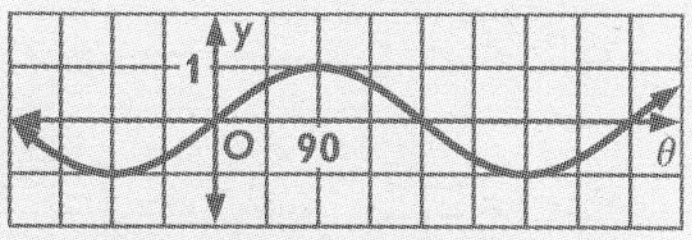

The equation has the form $y = A \sin B(\theta - C) + D$. The value of A is 3. The amplitude is 3.

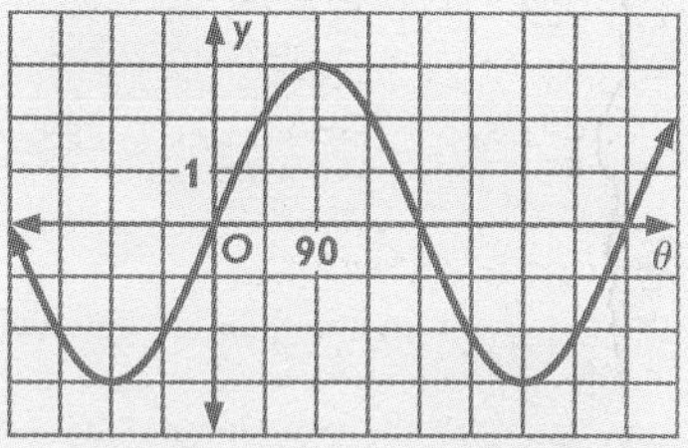

The value of B is –4. The period of the graph is $\frac{1}{4} \cdot 360° = 90°$. Since B is negative, the graph is reflected over the y-axis.

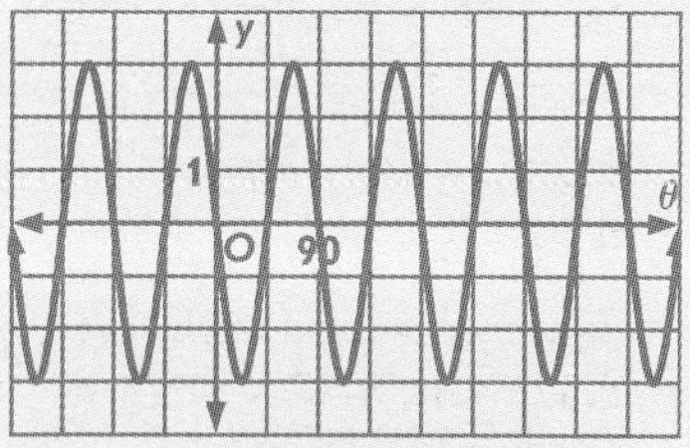

The value of C is –90°. The graph is translated 90° to the left. Since the period of the function is 90°, the appearance of the graph is not altered at this stage.

The value of D is 2. The graph is translated up 2 units.

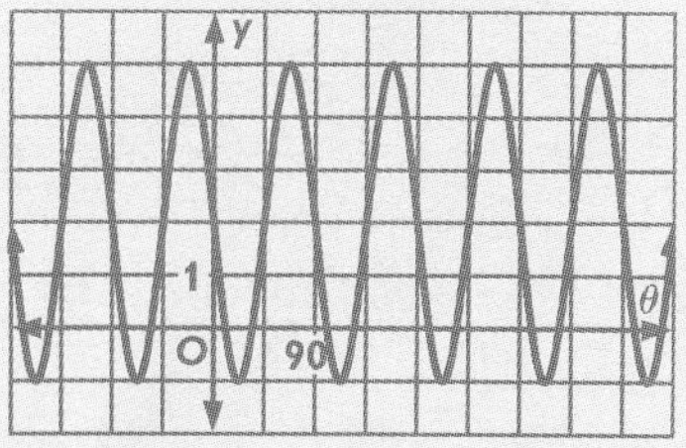

Answers to Exploration

c. same period and amplitude as the parent graph; translated 45° to the right

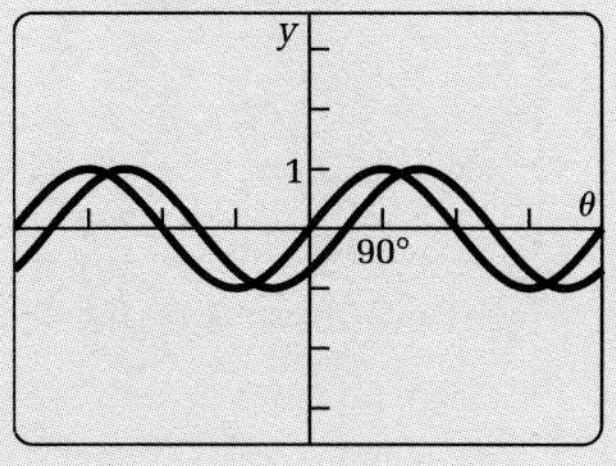

d. same amplitude and period as the parent graph; translated up 1 unit

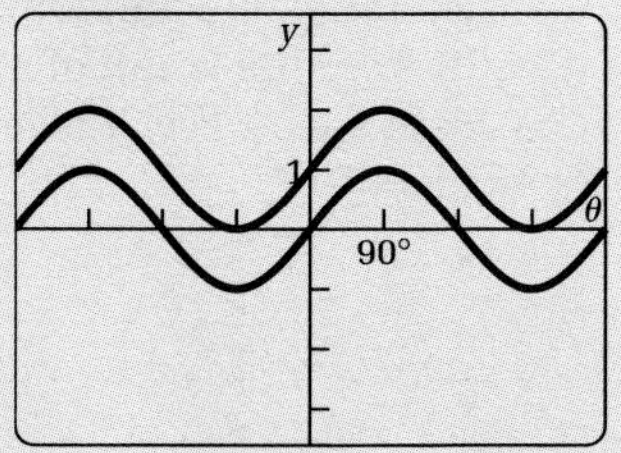

3–6. See answers in back of book.

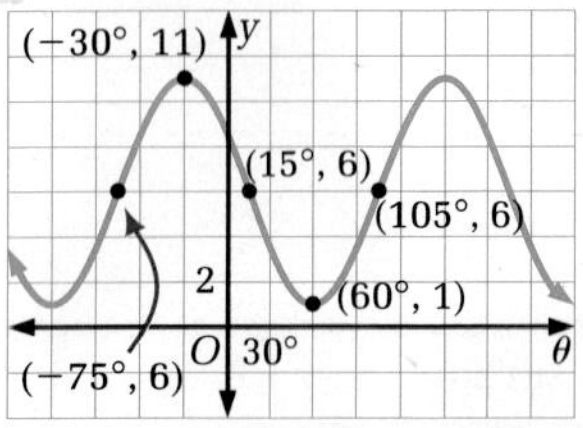

Sample 2

Write an equation for the graph shown.

Sample Response

The values of A, B, and D are the same for a sine function or a cosine function. Find these values first.

Step 1 Find the amplitude, A.

$$A = \frac{\text{maximum value} - \text{minimum value}}{2}$$
$$= \frac{11 - 1}{2}$$
$$= 5$$

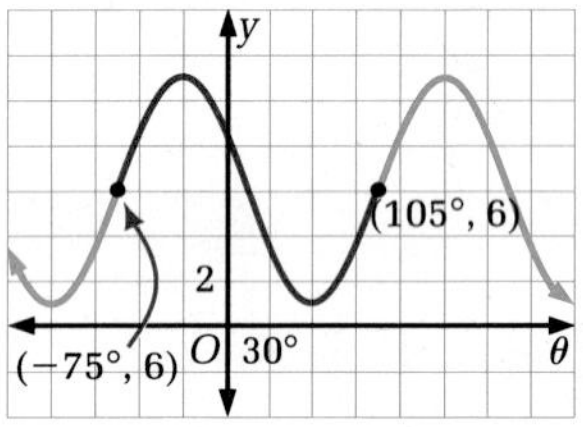

Step 2 Find the period, $\frac{360°}{B}$, and solve for B.

$$\text{period} = \frac{360°}{B}$$ ← The period is $105° - (-75°) = 180°$.

$$180° = \frac{360°}{B}$$ ← Solve for B.

$$B = \frac{360°}{180°}$$
$$= 2$$

Step 3 Find the axis, or the vertical translation, D.

$$D = \frac{\text{maximum value} + \text{minimum value}}{2}$$
$$= \frac{11 + 1}{2}$$
$$= 6$$

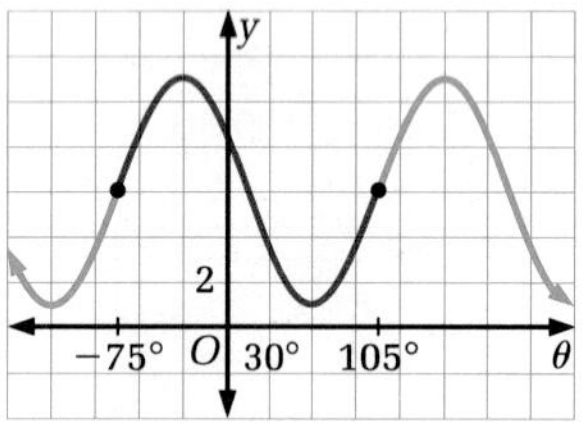

Step 4 Write an equation.

Method 1 Use a sine function.

$$y = A \sin B(\theta - C) + D$$
$$= 5 \sin 2(\theta - C) + 6$$ ← Substitute **5** for **A**, **2** for **B**, and **6** for **D**.
$$= 5 \sin 2(\theta + 75°) + 6$$ ← The sine graph is shifted left 75°, so $C = -75°$.

An equation of the graph is $y = 5 \sin 2(\theta + 75°) + 6$.

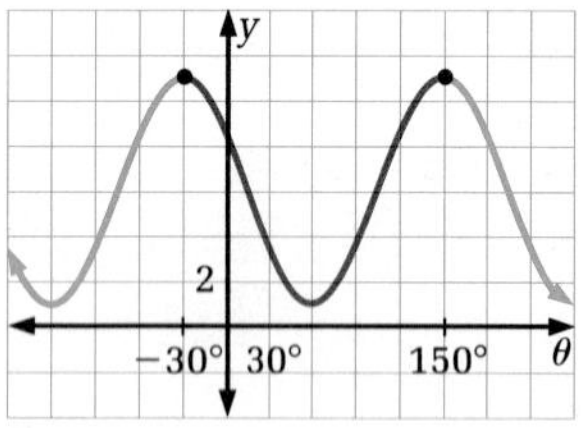

Method 2 Use a cosine function.

$$y = A \cos B(\theta - C) + D$$
$$= 5 \cos 2(\theta - C) + 6$$ ← Substitute **5** for **A**, **2** for **B**, and **6** for **D**.
$$= 5 \cos 2(\theta + 30°) + 6$$ ← The cosine graph is shifted left 30°, $C = -30°$.

An equation of the graph is $y = 5 \cos 2(\theta + 30°) + 6$.

Additional Sample

S2 Write an equation for the graph shown.

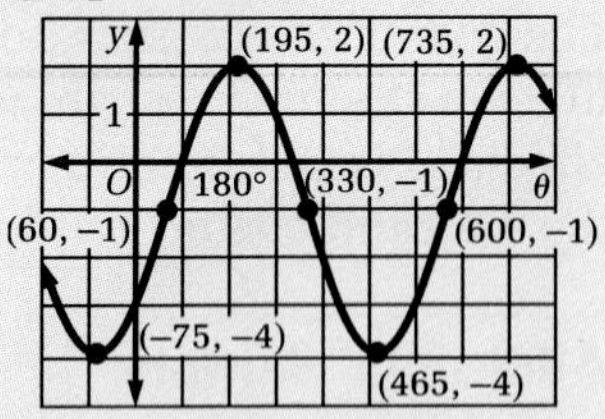

Step 1. Find the amplitude, A.

$A = \frac{\text{max. value} - \text{min. value}}{2}$

$= \frac{2 - (-4)}{2}$

$= 3$

Step 2. Find the period, $\frac{360°}{B}$, and solve for B.

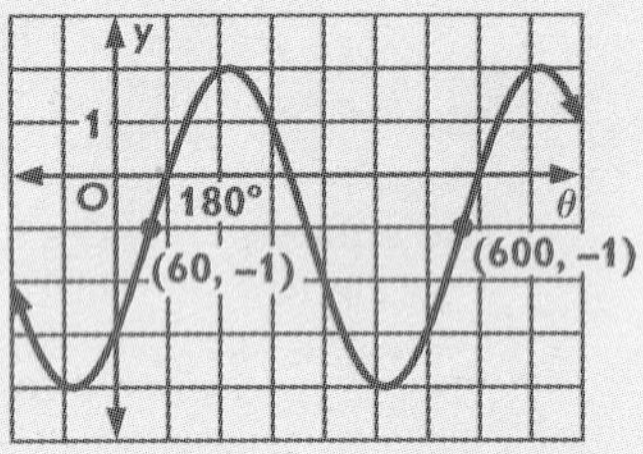

$\text{period} = \frac{360°}{B}$

$600° - 60° = 540° = \frac{360°}{B}$

$B = \frac{360°}{540°}$

$= \frac{2}{3}$

Step 3. Find the vertical translation, D.

$D = \frac{\text{max. value} + \text{min. value}}{2}$

$= \frac{2 + (-4)}{2}$

$= -1$

Step 4. Write an equation.

Method 1. Use a sine function.

$y = A \sin B(\theta - C) + D$

Substitute 3 for A, $\frac{2}{3}$ for B, and −1 for D.

$y = 3 \sin \frac{2}{3}(\theta - C) - 1$

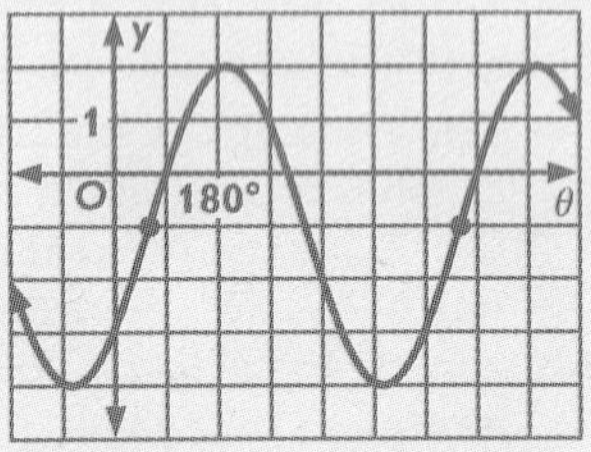

The sine graph is shifted 60° right, so $C = 60$. An equation of the graph is

$y = 3 \sin \frac{2}{3}(\theta - 60°) - 1$.

Method 2. Use a cosine function. $y = A \cos B(\theta - C) + D$

Substitute 3 for A, $\frac{2}{3}$ for B, and −1 for D.

$y = 3 \cos \frac{2}{3}(\theta - C) - 1$

Answers to Look Back

The graph of $y = A \cos B(\theta - C) + D$ is the graph of $y = \cos \theta$ stretched vertically by a factor of $|A|$, horizontally by a factor of $\frac{1}{|B|}$, translated horizontally C units and vertically D units. If $A < 0$, the graph is reflected over the θ-axis. If $B < 0$, the graph is reflected over the y-axis. The amplitude is $|A|$ times that of the parent graph, while the period is that of the parent graph divided by $|B|$.

Answers to Exercises and Problems

1. Answers may vary. An example is given. $|A|$ is the amplitude, $\frac{360°}{|B|}$ is the period, C is the horizontal shift from the graph of $y = \sin \theta$, and D is the vertical shift.

2. 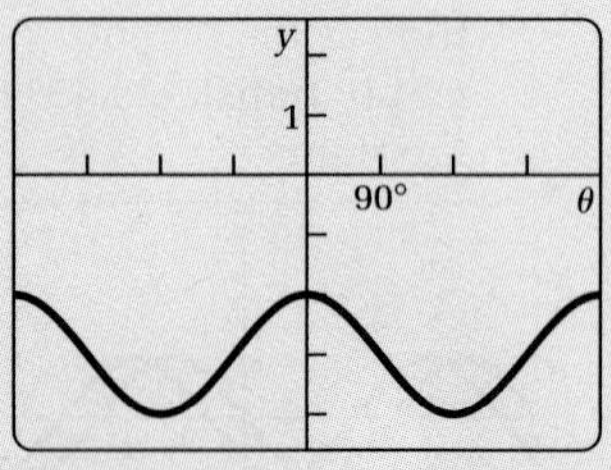

shifted down 3 units

Now you are ready for:
Exs. 22–48 on pp. 564–567

Look Back

Consider the function $y = A \cos B(\theta - C) + D$. Tell how its graph is different from the graph of $y = \cos \theta$.

10-4 Exercises and Problems

1. **Reading** On the graph of $y = A \sin B(\theta - C) + D$, what do the values of A, B, C, and D represent?

TECHNOLOGY **Use a graphics calculator or graphing software. Graph each equation. Then tell how each graph is different from the graph of $y = \cos \theta$.**

2. $y = \cos \theta - 3$
3. $y = \cos 2\theta$
4. $y = -2 \cos \theta$
5. $y = \cos (\theta + 30°)$

Choose the letter of the graph that best matches each equation.

6. $y = 4 \cos \theta$
7. $y = 2 \sin \theta$
8. $y = 2 \sin 2\theta$
9. $y = 0.25 \cos \theta - 1$
10. $y = 2 \cos (\theta - 45°)$
11. $y = \sin 2\theta + 1$

A.

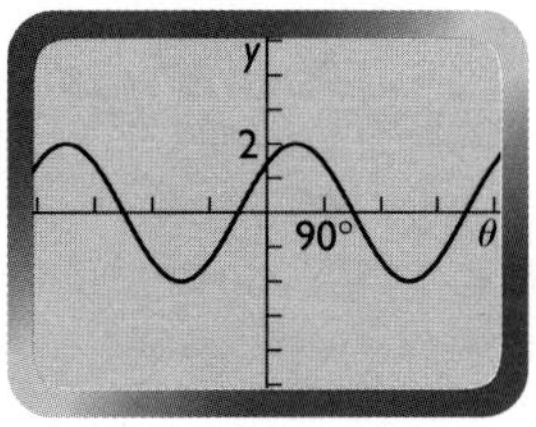

B.

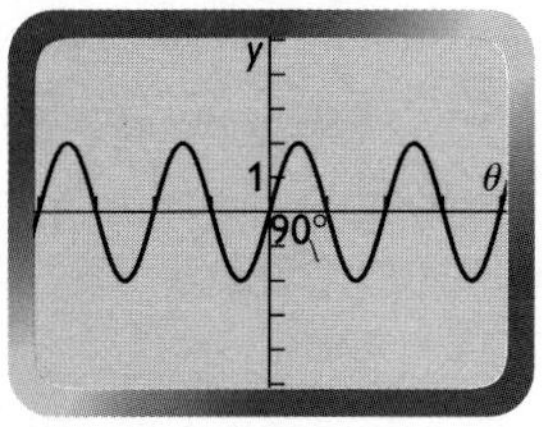

C.

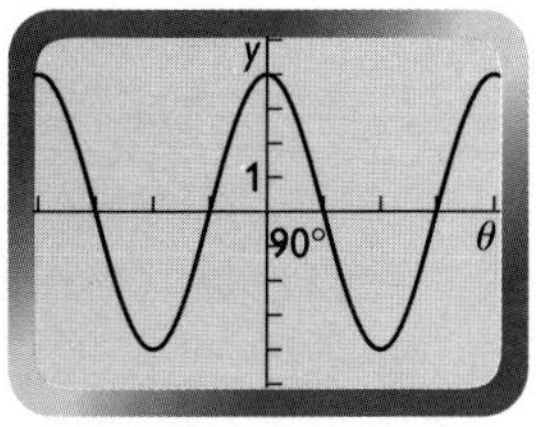

D.

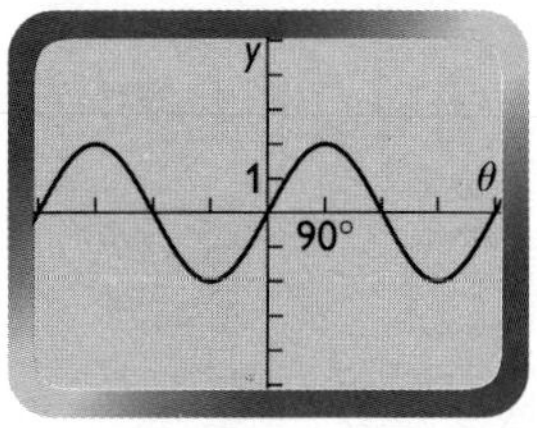

E.

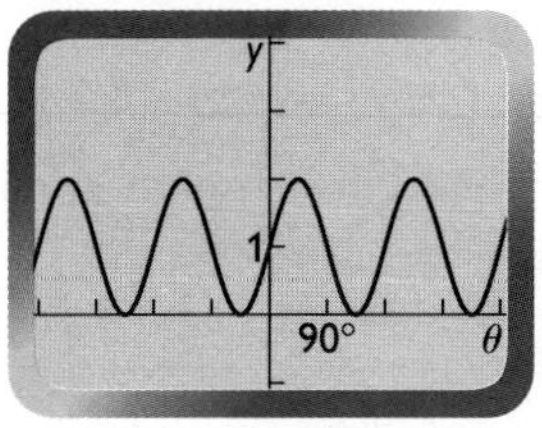

F.

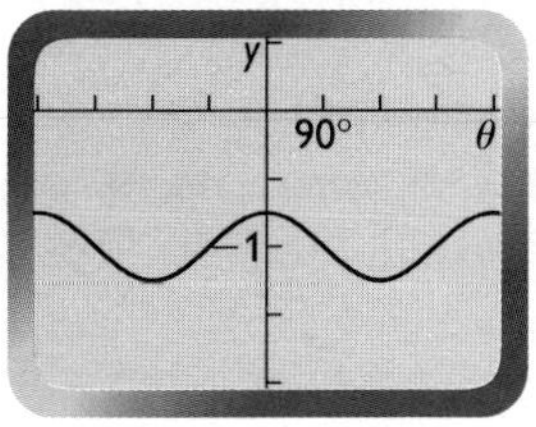

For each equation:

a. Find the amplitude, the period, and the frequency of its graph.

b. Sketch the graph without using graphing technology.

12. $y = \cos 2\theta$
13. $y = 3 \sin \theta$
14. $y = -\cos \theta + 3$
15. $y = \sin (\theta + 60°)$
16. $y = 0.5 \cos \theta - 2$
17. $y = \sin 2(\theta - 45°)$

18. **Writing** How is the graph of $y = \sin (\theta - 90°)$ like the graph of $y = \sin (\theta + 90°)$? How is it different?

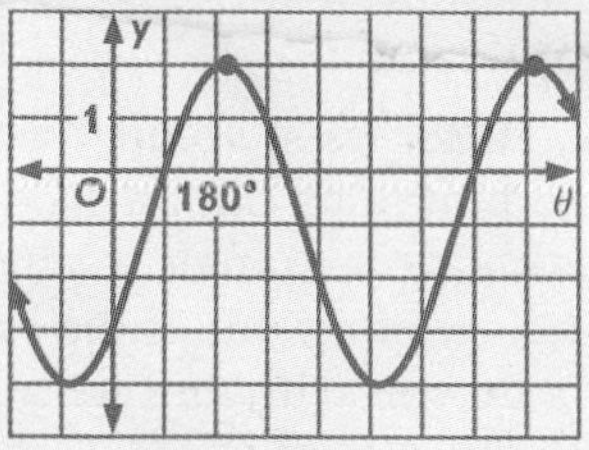

The cosine graph is shifted 195° right, so $C = -195$. An equation of the graph is $y = 3 \cos \frac{2}{3}(\theta - 195°) - 1$.

Reasoning

Ask students why every function that can be described by a function of the form $y = A \sin B(\theta - C) + D$ can also be described by a function of the form $y = A \cos B(\theta - C) + D$.

Look Back

You may wish to have individual students first describe how introducing each value A, B, C, or D affects the graph of $y = \cos \theta$. Then they can consider the function $y = A \cos B(\theta - C) + D$ and tell how its graph is different from the graph of $y = \cos \theta$.

APPLYING

Suggested Assignment

Day 1

Standard 1, 6–17

Extended 1, 6–21

Day 2

Standard 22–33, 44–48

Extended 22–39, 42, 44–48

Integrating the Strands

Algebra Exs. 1, 20, 21, 31–33, 35, 36, 42, 45, 46

Functions Exs. 1–44, 48

Geometry Ex. 47

Trigonometry Exs. 1–44, 48

Discrete Mathematics Exs. 45, 46

Logic and Language Exs. 1, 18, 21

Answers to Exercises and Problems

3.

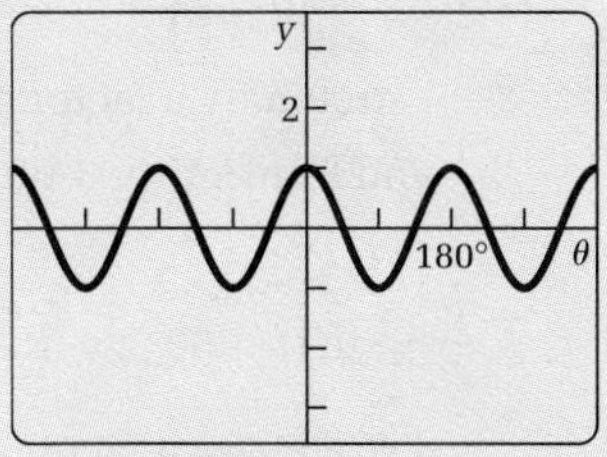

horizontal stretch by a factor of $\frac{1}{2}$; period is 180°

4.

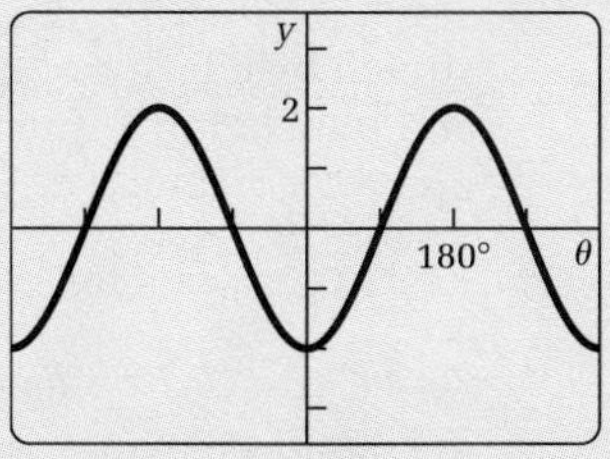

vertical stretch by a factor of 2; reflected over the θ-axis

5.

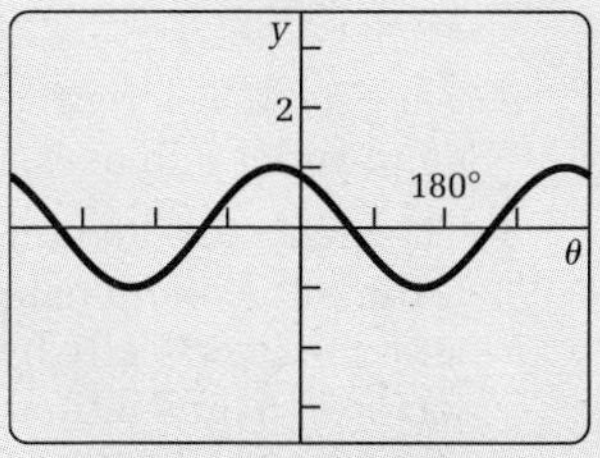

shift of 30° to the left

6–18. See answers in back of book.

Application

Ex. 19 shows students how sine functions can be used by health professionals to analyze the flow of air out of the lungs of a person with and without asthma.

Communication: Writing

Ex. 21 helps students to relate the transformed sine function to the transformations studied in Unit 9.

Problem Solving

As an open-ended activity, suggest that students use their graphics calculators to investigate sums, products, and composites of periodic functions. Ask them if the resulting functions are periodic. If they are, what can be said about their periods and amplitudes?

Using Technology

A graphics calculator can be used to obtain some surprising graphs. Students can experiment with made-up functions and keep sketches of some of their more interesting results. Two possibilities that use sine and cosine functions in combination with functions from the MATH NUM menu of the TI-82 calculator are given below. The graphs should be studied with various graphing windows.

$y = \max(\sin X, \cos X)$

$y = \max(\sin X, f\text{Part } X)$

Another equation that has an interesting graph is

$y = \sin(3X) + \cos(5X)$.

19. **Health** The graph shows the maximum flow of air out of the lungs of a person with asthma and a person without asthma.
    a. In what ways are the graphs similar?
    b. In what ways are the graphs different?
    c. How is the graph of the maximum air flow for the asthmatic patient a transformation of the graph for the non-asthmatic patient?

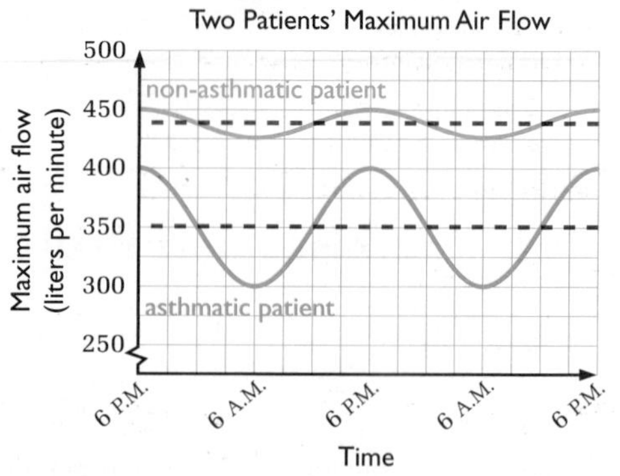

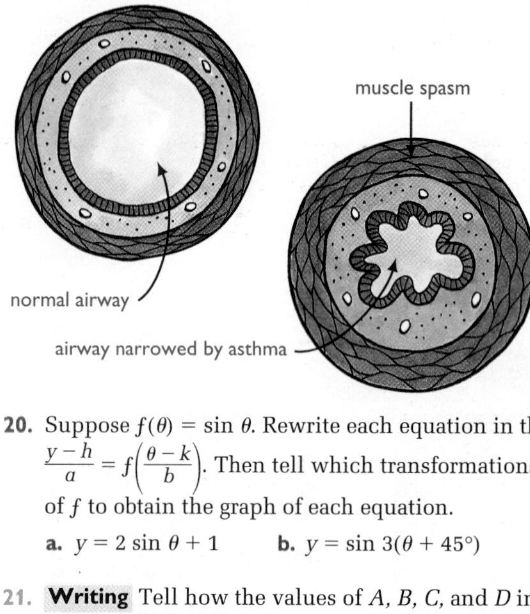

20. Suppose $f(\theta) = \sin\theta$. Rewrite each equation in the general form $\frac{y-h}{a} = f\left(\frac{\theta-k}{b}\right)$. Then tell which transformations are applied to the graph of f to obtain the graph of each equation.
    a. $y = 2\sin\theta + 1$
    b. $y = \sin 3(\theta + 45°)$
    c. $y = \sin 4\theta - 3$
    d. $y = 0.5\sin(\theta - 60°)$

21. **Writing** Tell how the values of A, B, C, and D in the equation $y = A\sin B(\theta - C) + D$ compare to the values of a, b, h, and k in the equation $\frac{y-h}{a} = f\left(\frac{x-k}{b}\right)$.

For each equation:

a. Find the amplitude, the period, and the frequency of its graph.

b. Sketch the graph without using graphing technology.

22. $y = 3\sin 4\theta - 2$
23. $y = \cos 2(\theta - 75°) + 1$
24. $y = 0.3\cos 0.5(\theta - 45°)$
25. $y = -2\sin 0.25(\theta + 30°) + 3$

Choose the letter of the graph that best matches each equation.

26. $y = 0.5\cos(\theta - 90°) - 2$
27. $y = 2\sin 0.5(\theta - 60°) - 2$
28. $y = -\cos 2(\theta + 45°) - 1$

A.

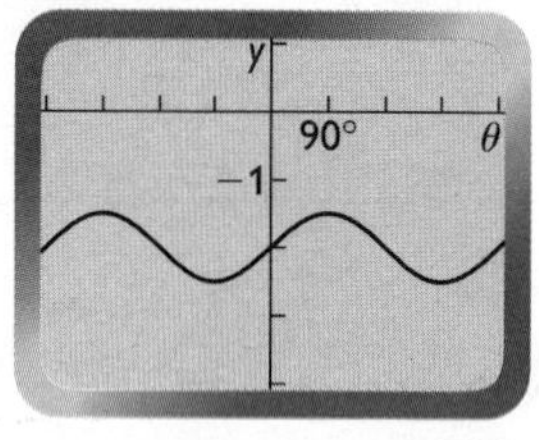

B.

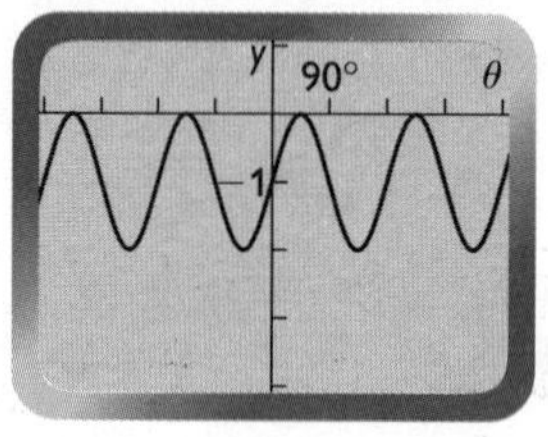

C.

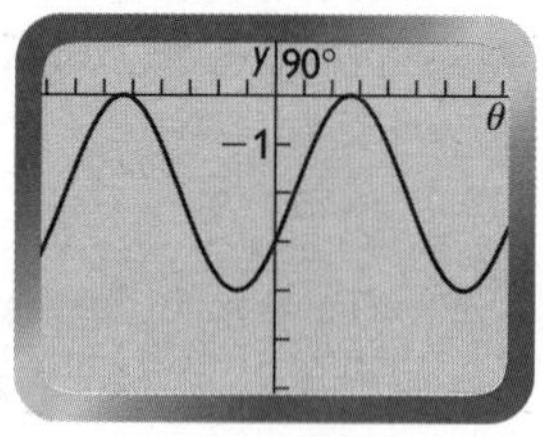

Answers to Exercises and Problems

19. a. The graphs have the same period.
    b. They have different amplitudes and axes.
    c. The graph for the non-asthmatic patient can be stretched vertically by a factor of 5 and shifted down 100 units to produce the graph for the asthmatic patient.

20. a. $\frac{y-1}{2} = \sin\theta$; vertical stretch by a factor of 2 and translation 1 unit up
    b. $y = \sin\left(\frac{\theta + 45°}{\frac{1}{3}}\right)$; horizontal stretch by a factor of $\frac{1}{3}$ and translation 45° left
    c. $y + 3 = \sin\left(\frac{\theta}{\frac{1}{4}}\right)$; horizontal stretch by a factor of $\frac{1}{4}$ and translation 3 units down
    d. $\frac{y}{\frac{1}{2}} = \sin(\theta - 60°)$; vertical stretch by a factor of $\frac{1}{2}$ and translation 60° right

For Exercises 29 and 30, the graph of a function is described. Write an equation for each function. Then sketch the graph.

29. The graph is a cosine function with amplitude = 3, period = 45°, and a horizontal translation 30° to the left.

30. The graph is a sine function with amplitude = 1, period = 60°, and a vertical translation 1.5 units up.

Write an equation for each graph.

31.

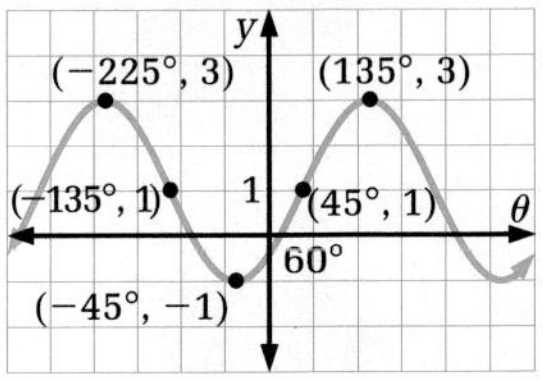

32.

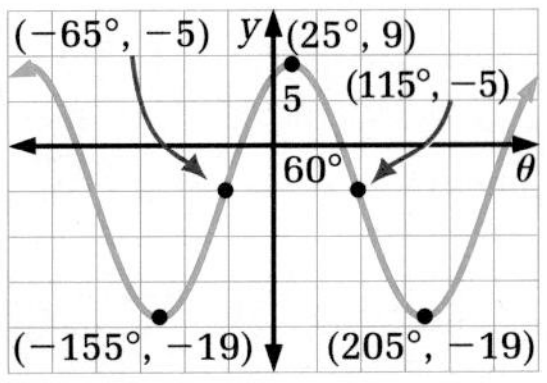

33.

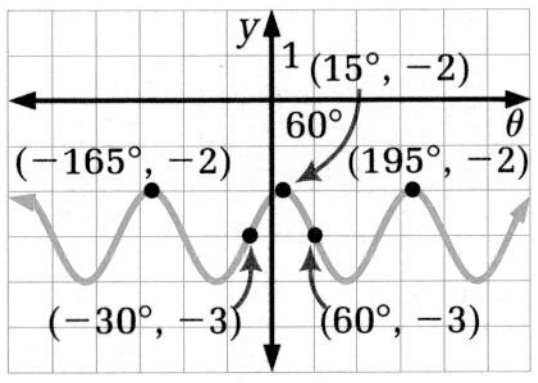

connection to EARTH SCIENCE

For a particular location on Earth, the number of hours of daylight in a 24-hour period depends on the time of year. There are always 12 hours of daylight at the equator.

The graph below shows the number of hours of daylight for two different locations north of the equator. Locations south of the equator experience the same pattern six months later than locations in the Northern Hemisphere.

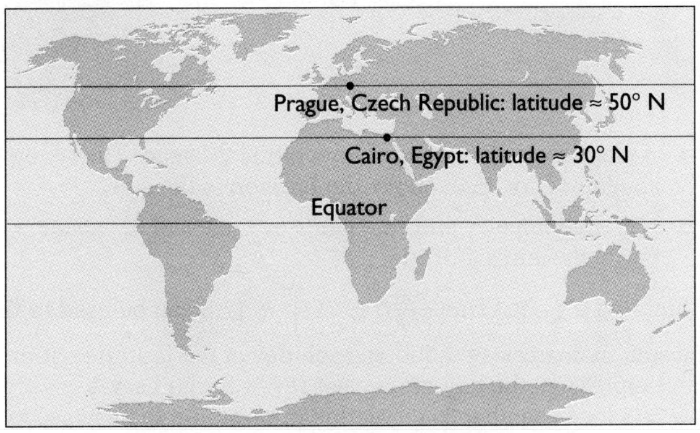

34. a. What axis do the two graphs have in common? What do you think this axis represents in this situation? Why?

b. What do the points represent where the two graphs cross their common axis?

35. The function $y = 4\sin\left(\frac{360}{365}(t-80)\right)^{\circ} + 12$ describes the number of hours of daylight as a function of t, the day of the year, for Prague. Use this function to write a function for the number of hours of daylight for Cairo.

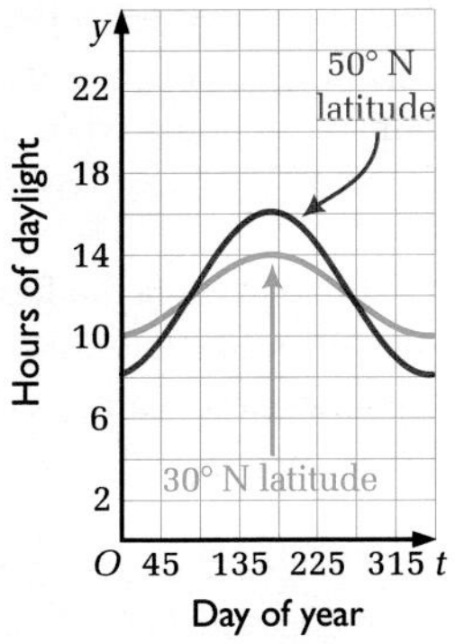

36. a. Describe two ways that you can transform the graphs to show the number of hours of daylight for regions in the Southern Hemisphere.

b. Write a function for the number of hours of daylight for Pôrto Alegre, Brazil, which has a latitude of about 30°S.

Assessment: Group Activity

For Exs. 31–33, students can work in groups of four. Two students can write an equation for each graph in terms of the sine, and the other two can write equations for the graphs in terms of the cosine. Sine pairs would compare their results, while cosine pairs compare theirs. As an entire class, students could verify that either the cosine or sine function produces equivalent results.

Answers to Exercises and Problems

21. $a = A$; $h = D$; $b = \frac{1}{B}$; $k = C$

22–25. See answers in back of book.

26. A

27. C

28. B

29, 30. See answers in back of book.

31–33. Answers may vary. Examples are given.

31. $y = 2\sin(\theta - 45°) + 1$

32. $y = 14\sin(\theta + 65°) - 5$

33. $y = \sin 2(\theta + 30°) - 3$

34. a. $y = 12$; the equinox, that is, the day when the number of hours of daylight equals the number of hours of darkness

b. the vernal and autumnal equinoxes

35. $y = 2\sin\left(\frac{360}{365}(t-80)\right)^{\circ} + 12$

36. a. Shift each graph 182.5 units to the right or reflect each graph over its axis.

b. $y = -2\sin\left(\frac{360}{365}(t-80)\right)^{\circ} + 12$

Interdisciplinary Problems

Exs. 37–40 provide students with an opportunity to make some astronomical observations based on a passage from Chaucer's *Canterbury Tales.* To do this, students use both trigonometry and a sine function.

Students Acquiring English

Understanding the passage by Chaucer may be difficult for students acquiring English. You might want to go through this passage line by line, asking students to restate the meaning of each line.

connection to LITERATURE

The Canterbury Tales were written by Geoffrey Chaucer in the late fourteenth century. The following passage is from the prologue to *The Parson's Tale.*

When the Manciple's tale was wholly ended,
From the south line the broad sun had decended
So low that it appeared now, to my sight,
Not more than twenty-nine degrees in height.
And it was four o-clock, as I should guess,
For eleven feet, or somewhat more or less,
My shadow at that time appeared to be,
Of such feet as, proportioned equally,
Would fix my height at six, by calculation.
And the moon also, in his exaltation,
Libra I mean, continued to ascend,
As we were entering a hamlet's end.

37. a. According to the passage, how tall is the man? How long is his shadow? How high above the horizon is the sun?
    b. Use trigonometry and the man's estimates of his shadow and height to verify the sun's altitude.

The function $y = 19.3 \sin\left(\frac{360}{365}(t - 71)\right)^\circ + 17.6$ can be used to find the height of the sun, in degrees, at 4:00 P.M. each day at the latitude of Canterbury, where t represents the day of the year ($t = 1$ for January 1, $t = 2$ for January 2, . . . $t = 365$ for December 31). Use this information to answer Exercises 38–40.

38. Why is the period of the function 365?
39. An *equinox* occurs when the number of hours of daylight equals the number of hours without daylight. In 1387 when scholars think Chaucer began writing *The Canterbury Tales*, the spring equinox occurred on March 12th.
    a. If January 1 is the first day of the year, what day of the year is March 12th?
    b. How is this value used in the equation? Why?
40. a. TECHNOLOGY Graph the function $y = 19.3 \sin\left(\frac{360}{365}(t - 71)\right)^\circ + 17.6$. On what days of the year is the height of the sun about 29° at 4:00 P.M.?
    b. **Research** Read the prologue to *The Canterbury Tales.* Which of your answers from part (a) represents the date referred to in the prologue to *The Parson's Tale*? Why?

Answers to Exercises and Problems

37. a. 6 ft; 11 ft; 29°
    b. $\tan\theta = \frac{6}{11}$; $\theta = \tan^{-1}\left(\frac{6}{11}\right) = 28.61°$
38. There are 365 days in a year. period $= \frac{360}{|B|} = \frac{360}{\frac{360}{365}} = 365$
39. a. day 71
    b. It is the horizontal shift. The equinox represents the axis of the graph.
40. a.

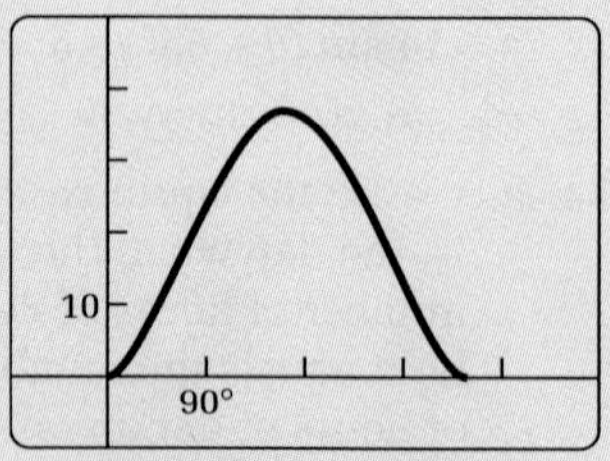

the 108th day (April 18th) or the 217th day (August 5th); The given dates assume Chaucer was not writing in a leap year, which 1387 was not.
    b. the 108th day; In the prologue, Chaucer describes how people long to go on pilgrimages in springtime, and tells how the incidents he describes "happened in that season."

41 TECHNOLOGY Graph each of the equations found in Sample 2 on page 562 to verify that they represent the same graph.

42. Suppose you are on a Ferris wheel that completes 2.5 revolutions per minute. The function $h = 25 \sin \theta + 5$ can be used to model your height, in feet, above the ground as a function of the angle the wheel has turned.

a. How many degrees are in 1 revolution?

b. Convert 2.5 revolutions per minute to degrees per second.

c. Write an equation for the angle, θ, the wheel has turned as a function of t, the amount of time the wheel has turned.

d. In part (c), you found an equation for θ as a function of t. The equation $h = 25 \sin \theta + 5$ gives the height as a function of θ. Use these equations and composition of functions to write an equation for the height as a function of time.

Ongoing ASSESSMENT

43. **Group Activity** Work with a partner. You will each need a graphics calculator or graphing technology.

a. Enter a function of the form $y = A \sin B(x - C) + D$ or the form $y = A \cos B(x - C) + D$ on your calculator. Graph the function in an appropriate window. Trade calculators with your partner.

b. Trace along the graph you receive to determine its properties. Write an equation for the graph.

c. Does your equation match the one your partner entered? Graph your equation on the same screen as the original to see if the graphs match.

d. Repeat parts (a)–(c) using other graphs.

Review PREVIEW

44. Graph the function $y = \sin \theta$ for $-360° \le \theta \le 360°$. For what values of θ is $\sin \theta \approx 0.86$? *(Section 10-3)*

Tell whether each series appears to have a sum. If it does, give its value. *(Section 4-7)*

45. $6 + 3.5 + 1 + (-1.5) + \cdots$

46. $\sum_{n=1}^{\infty} (-0.99)^{n-1}$

47. A wheel is turning at 5 revolutions per minute. How many degrees will a point on the wheel rotate in 1 min? in 1 s? *(Section 10-3)*

10 Working on the Unit Project

48. Write a function for the number of hours of daylight at a location on this map. Graph your function. Put it in your scrapbook. Include pictures and describe the location.

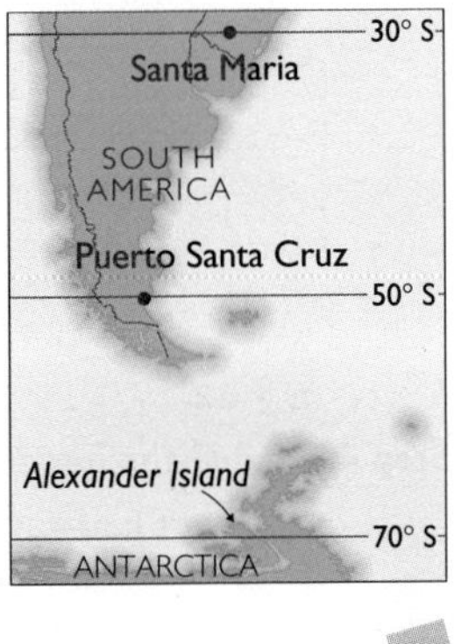

Cooperative Learning

Ex. 43 provides an excellent opportunity for students to assess their understanding of the concepts discussed in this unit.

Practice 78 For use with Section 10-4

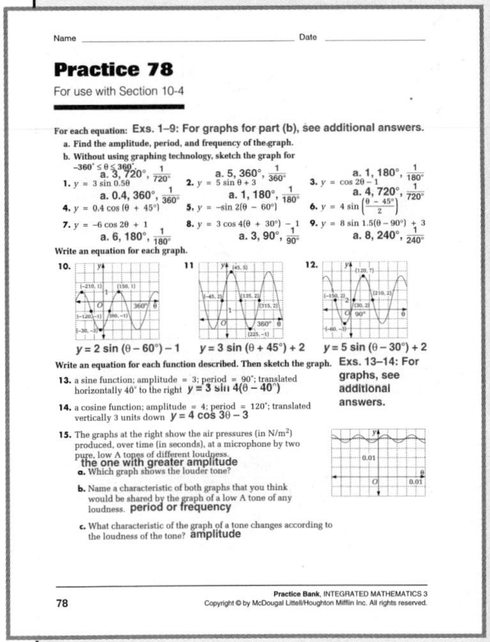

Name ______________ Date ______________

Practice 78

For use with Section 10-4

For each equation: Exs. 1–9: For graphs for part (b), see additional answers.

a. Find the amplitude, period, and frequency of the graph.

b. Without using graphing technology, sketch the graph for $-360° \le \theta \le 360°$.

1. $y = 3 \sin 0.5\theta$ a. 3, 720°, $\frac{1}{720°}$
2. $y = 5 \sin \theta + 3$ a. 5, 360°, $\frac{1}{360°}$
3. $y = \cos 2\theta - 1$ a. 1, 180°, $\frac{1}{180°}$
4. $y = 0.4 \cos(\theta + 45°)$ a. 0.4, 360°, $\frac{1}{360°}$
5. $y = -\sin 2(\theta - 60°)$ a. 1, 180°, $\frac{1}{180°}$
6. $y = 4 \sin\left(\frac{\theta - 45°}{2}\right)$ a. 4, 720°, $\frac{1}{720°}$
7. $y = -6 \cos 2\theta + 1$ a. 6, 180°, $\frac{1}{180°}$
8. $y = 3 \cos 4(\theta + 30°) - 1$ a. 3, 90°, $\frac{1}{90°}$
9. $y = 8 \sin 1.5(\theta - 90°) + 3$ a. 8, 240°, $\frac{1}{240°}$

Write an equation for each graph.

10. $y = 2 \sin(\theta - 60°) - 1$
11. $y = 3 \sin(\theta + 45°) + 2$
12. $y = 5 \sin(\theta - 30°) + 2$

Write an equation for each function described. Then sketch the graph. Exs. 13–14: For graphs, see additional answers.

13. a sine function; amplitude = 3; period = 90°; translated horizontally 40° to the right $y = 3 \sin 4(\theta - 40°)$

14. a cosine function; amplitude = 4; period = 120°; translated vertically 3 units down $y = 4 \cos 3\theta - 3$

15. The graphs at the right show the air pressures (in N/m^2) produced, over time (in seconds), at a microphone by two pure, low A tones of different loudness.

a. Which graph shows the louder tone? the one with greater amplitude

b. Name a characteristic of both graphs that you think would be shared by the graph of a low A tone of any loudness. period or frequency

c. What characteristic of the graph of a tone changes according to the loudness of the tone? amplitude

78 Practice Bank, INTEGRATED MATHEMATICS 3

Answers to Exercises and Problems

41.

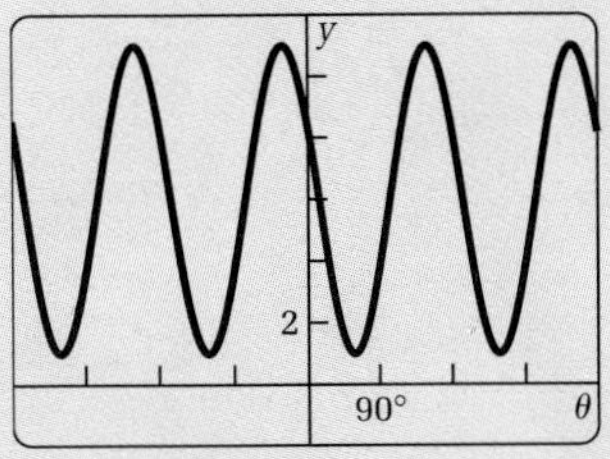

The graph shown represents both functions.

42. a. 360° b. 15°/s

c. $\theta(t) = 15t$

d. $h \circ \theta(t) = 25 \sin 15t + 5$

43. Check students' work.

44.

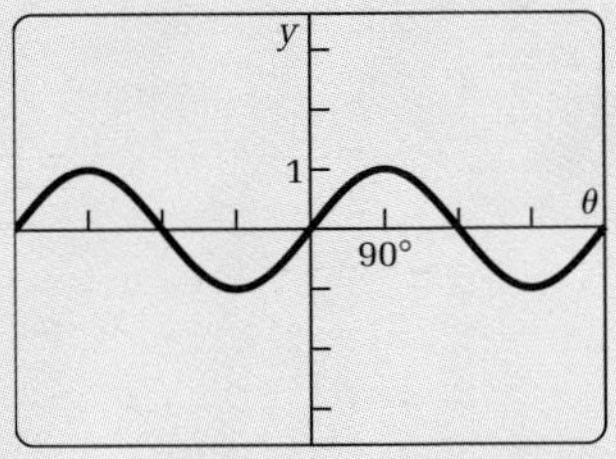

–300.7°; –239.3°; 59.3°; 120.7° (in general, $\theta = 59.3° + n \cdot 360°$ or $\theta = 120.7° + n \cdot 360°$ for any integer n)

45. no sum

46. $\frac{100}{199} \approx 0.5$

47. 1800°; 30°

48. Answers may vary.

PLANNING

Objectives and Strands

See pages 538A and 538B.

Spiral Learning

See page 538B.

Materials List

- Graph paper

Recommended Pacing

Section 10-5 is a two-day lesson.

Day 1

Pages 568–569: Opening paragraph through Talk it Over 2, *Exercises 1–8*

Day 2

Pages 569–571: Sample 2 through Look Back, *Exercises 9–22*

Extra Practice

See pages 624–625.

Warm-Up Exercises

Warm-Up Transparency 10-5

Support Materials

- Practice Bank: Practice 79
- Activity Bank: Enrichment 70
- Study Guide: Section 10-5
- Problem Bank: Problem Set 21
- Explorations Lab Manual: Diagram Masters 1, 2
- Assessment Book: Quiz 10-5, Test 40, Alternative Assessment 5

Section 10-5 Modeling with Sine and Cosine

Reinventing The Wheel

Focus

Use sine and cosine functions to model periodic situations.

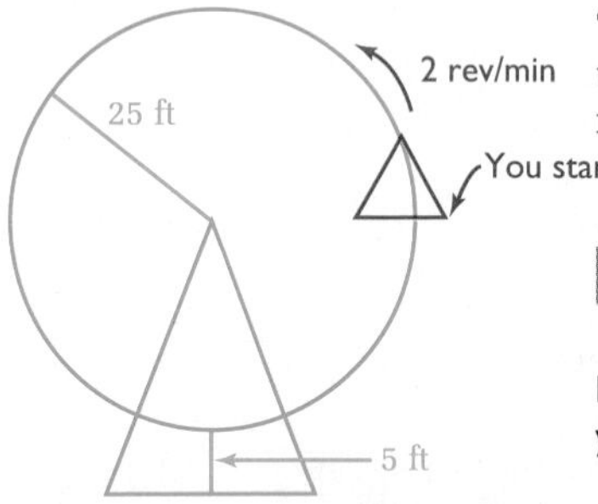

Transformations of the sine graph and the cosine graph can be used to model many periodic situations, such as the Ferris wheel ride discussed in Section 10-3.

Sample 1

Find an equation that models your height above the ground as you ride on a Ferris wheel.

Sample Response

Step 1 Choose the sine function or the cosine function.

The Sine Graph

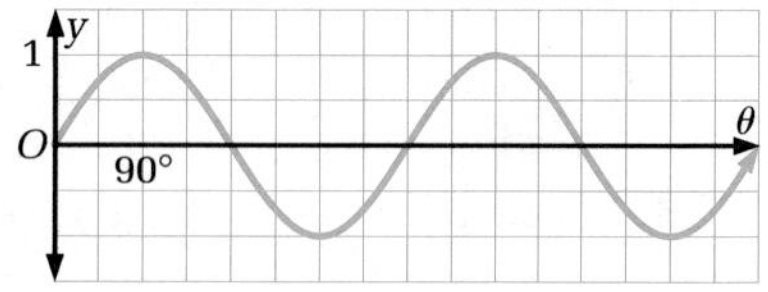

Since you know that you are halfway between the minimum height and the maximum height at time $t = 0$, choose the sine function. The general model is $h(t) = A \sin B(t - C) + D$ where h is the height, in feet, and t is the time, in seconds, since starting.

Step 2 Find A, the amplitude.

$$A = \frac{\text{maximum value} - \text{minimum value}}{2}$$

$$= \frac{55 - 5}{2}$$

← maximum value = the diameter of the wheel + the distance from the bottom of the wheel to the ground = 55

$$= 25$$

Step 3 Find B, the number of degrees the wheel turns each second.

Convert revolutions per minute to degrees per second.

$$2 \text{ rev/min} = \frac{2 \text{ revolutions}}{1 \text{ min}} \cdot \frac{360°}{1 \text{ revolution}} \cdot \frac{1 \text{ min}}{60 \text{ s}}$$

$$= \frac{720°}{60 \text{ s}}$$

$$= 12°/\text{s}$$

$B = 12$

Step 4 Find C, the horizontal translation.

You start halfway between the maximum and minimum heights, so there is no horizontal translation. $C = 0$

> **BY THE WAY...**
>
> A Ferris wheel operator places riders around the wheel so that the weight is distributed symmetrically. The order of loading may vary, depending on the operator's estimation of the size of the load.

Step 5 Find D, the vertical translation.

$$D = \frac{\text{maximum value} + \text{minimum value}}{2}$$

$$= \frac{55 + 5}{2}$$

$$= 30$$

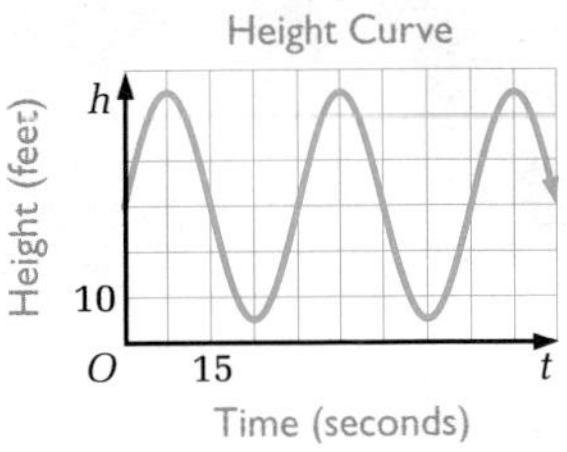

Step 6 Substitute the values of A, B, C, and D into the general model.

An equation that models your height, in feet, above the ground after t seconds is $h(t) = 25 \sin (12t)° + 30$.

Talk it Over

1. What is your height above the ground after riding the Ferris wheel for each given time?
   a. 90 s   b. 2.5 min   c. 200 s
2. Suppose you chose the cosine function. How does this change the equation that models your height?

Now you are ready for: Exs. 1–8 on p. 572

Time	Altitude
midnight	10.23°
1:00 A.M.	8.02°
2:00 A.M.	6.75°
3:00 A.M.	6.53°
4:00 A.M.	7.35°
5:00 A.M.	9.17°
6:00 A.M.	11.88°
7:00 A.M.	15.32°
8:00 A.M.	19.30°
9:00 A.M.	23.58°
10:00 A.M.	27.93°
11:00 A.M.	32.07°
noon	35.65°
1:00 P.M.	38.40°
2:00 P.M.	40.02°
3:00 P.M.	40.32°
4:00 P.M.	39.25°
5:00 P.M.	36.97°
6:00 P.M.	33.70°
7:00 P.M.	29.77°
8:00 P.M.	25.48°
9:00 P.M.	21.13°
10:00 P.M.	16.98°
11:00 P.M.	13.28°

Sample 2

In northern latitudes around the beginning of summer (June 21), there are days when the sun never sets. A region where this occurs is called "the land of the midnight sun."

a. Graph the data provided for the altitude (angle above the horizon) of the sun in Arctic Bay, Northwest Territories, Canada, for June 23, 1994. What parent function does the graph suggest?

b. Find an equation that models the altitude of the sun on this day.

c. Graph the equation and compare it to your graph for part (a).

Continued on next page.

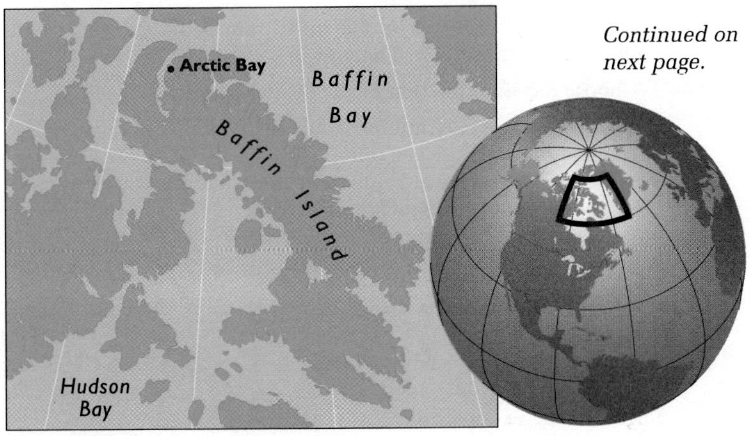

Answers to Talk it Over

1. a. 30 ft
   b. 30 ft
   c. about 8.35 ft
2. $h(t) =$
   $25 \cos (12(t - 7.5))° + 30$

TEACHING

Additional Sample

S1 A sheet of paper is wrapped around a cylinder and taped in place. Along the seam of the paper, a point P is marked 8 cm from the bottom.

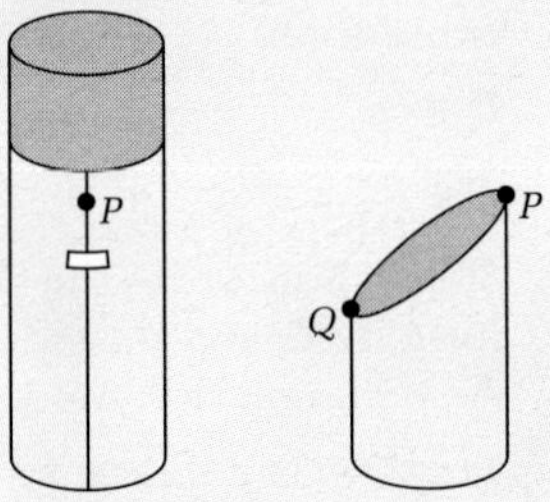

A point Q is marked 180° from P and 5 cm from the bottom. The cylinder and paper are then cut, as shown in the diagram at the right above. When the paper is removed and unwound, the edge where the cut was made is one cycle of a periodic curve.

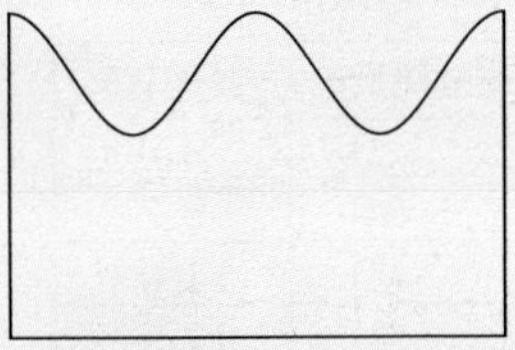

Find an equation that models the curve.

Use the left edge and bottom edge as the vertical and horizontal axes.

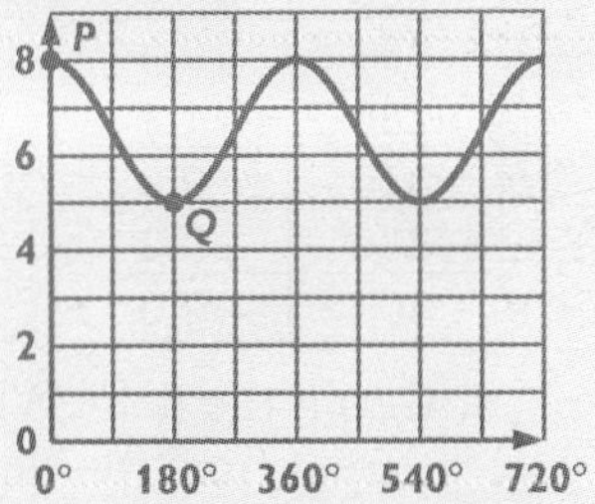

Step 1. **Choose the cosine function, since the high point is on the vertical edge at the left. The general model is $y = A \cos B(x - C) + D$, where x is the number of degrees from P where each point was on the cylinder before the paper was unwrapped.**

Step 2. **Find A, the amplitude. Since P was at a height of 8 cm and Q at 5 cm,**

$$A = \frac{\text{max. value} - \text{min. value}}{2}$$

$$= \frac{8 - 5}{2} = \frac{3}{2}$$

Sample continued on next page.

Additional Sample (continued)

Step 3. Find B, the number of revolutions around the cylinder between successive high points: $B = 1$.

Step 4. Find C, the horizontal translation. Since the curve is at a high point where $x = 0$, the value of C is 0.

Step 5. Find D, the vertical translation.

$$D = \frac{\text{max. value} + \text{min. value}}{2} = \frac{8+5}{2} = \frac{13}{2}$$

Step 6. Substitute the values for A, B, C, and D.

An equation that models the situation is $y = \frac{3}{2}\cos x + \frac{13}{2}$.

Additional Sample

S2 The table gives the approximate number of hours of daylight (hours between sunrise and sunset) for a city at a latitude of about 40° N. The numbers in the Days column are days after December 21.

Days	Hours
0	9.0
30	9.7
60	11.0
90	12.5
120	13.9
150	14.8
172	15.0
190	14.8
220	14.0
250	12.6
280	11.1
310	9.8
330	9.2
365	9.0

a. Graph the data. What parent function does the graph suggest?

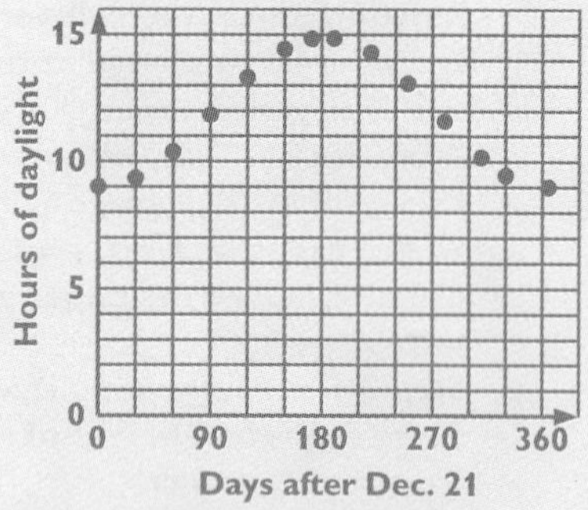

The graph looks like an inverted cosine wave.

Time-lapse photo of the midnight sun in Norway

Sample Response

a.

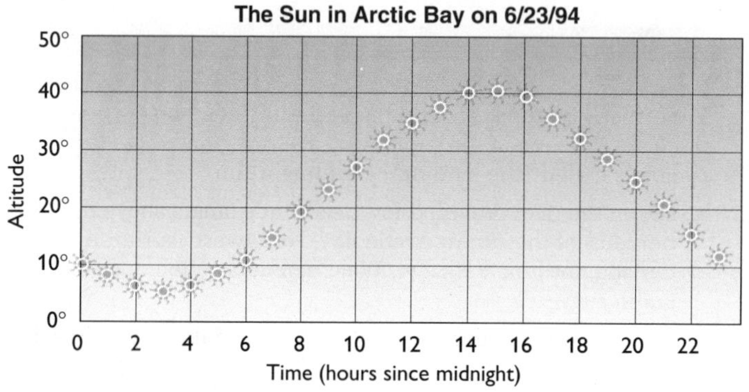

The graph looks like a sine wave. The maximum altitude is 40.32°, and the minimum altitude is 6.53°. The period is 24 h.

b. Step 1 Choose the sine function or the cosine function.

The time of the maximum altitude is known, so choose the cosine function. The general model is $h(t) = A\cos B(t - C) + D$, where h is the sun's altitude in degrees, and t is the time, in hours, since midnight.

Step 2 Find A, the amplitude.

$$A = \frac{\text{maximum value} - \text{minimum value}}{2} = \frac{40.32 - 6.53}{2} \approx 16.90$$

Step 3 Find B, the factor that relates the period of the cosine function to the period of the altitude function.

$$1 \text{ cycle in 24 hours} = \frac{360°}{24 \text{ h}}$$

$$= 15°/\text{h} = 15t° \text{ in } t \text{ hours}$$

$B = 15$

Step 4 Find C, the horizontal translation.

The maximum occurs at 3 P.M., or 15 h after midnight. $C = 15$

Step 5 Find D, the vertical translation.

$$D = \frac{\text{maximum value} + \text{minimum value}}{2}$$

$$= \frac{40.32 + 6.53}{2}$$

$$\approx 23.43$$

Step 6 Substitute the values of A, B, C, and D into the general model.

An equation that models the altitude, in degrees, of the sun in Arctic Bay on June 23, 1994, as a function of the number of hours after midnight is $h(t) = 16.90 \cos 15(t - 15)° + 23.43$.

c. The model fits the data well.

The Sun in Arctic Bay on 6/23/94

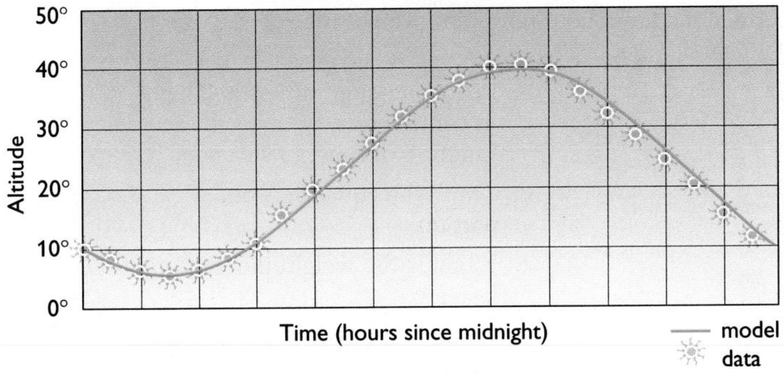

Talk it Over

3. Suppose you use the sine function as the parent graph for this model. What is the value of C?
4. Sketch the graph of the altitude of the sun when sunrise is at 6:00 A.M., the sun's altitude reaches a maximum of 60° at noon, and there are 12 hours of daylight. What happens to the graph when the sun sets?

Now you are ready for:
Exs. 9–22 on pp. 573–575

Look Back

Describe the process for fitting a sine wave to data.

b. Find an equation that models the number of hours of daylight for days throughout the year.

Step 1. There is a minimum point on the y-axis, so choose the cosine function. The general model is $y = A \cos B(x - C) + D$, where x is the number of days after December 21.

Step 2. Find A, the amplitude.

$$A = \frac{\text{max. value} - \text{min. value}}{2}$$

$$= \frac{15 - 9}{2} = 3$$

Since the graph intersects the vertical axes at a minimum rather than a maximum point, $A = -3$.

Step 3. Find B.

1 cycle in 365 days = $\frac{360°}{365} = \left(\frac{72}{73}\right)°$ per day = $\left(\frac{72}{73}\right)x°$ in x days.

So, $B = \frac{72}{73}$.

Step 4. Find C, the horizontal translation. Having a minimum point on the vertical axis indicates no horizontal translation. So, $C = 0$.

Step 5. Find D, the vertical translation.

$$D = \frac{\text{max. value} + \text{min. value}}{2}$$

$$= \frac{15 + 9}{2} = 12$$

Step 6. Substitute the values for A, B, C, and D into the general model.

$$y = -3 \cos \frac{72}{73}x + 12.$$

c. Graph the equation and compare it to part (a).

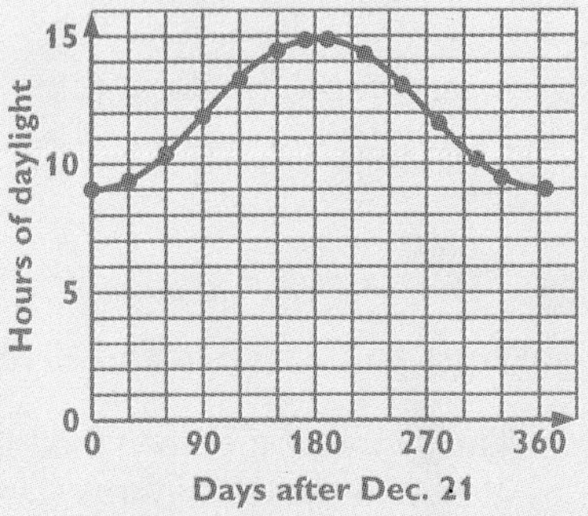

The model appears to fit the data well.

Answers to Talk it Over

3. 9
4.

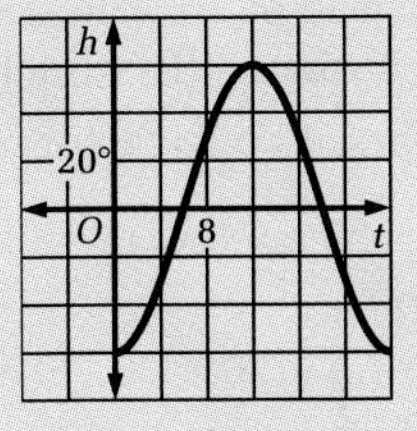

The graph crosses the horizontal axis.

Answers to Look Back

Plot the points to determine whether the graph looks like a sine wave. If so, determine the amplitude A using the formula $\frac{\text{maximum value} - \text{minimum value}}{2}$. Decide whether to use a sine or a cosine function. Find the period of the graph and use it to determine the value of B. Find the horizontal translation C. Find the vertical translation D using the formula $\frac{\text{maximum value} + \text{minimum value}}{2}$. Substitute the values of A, B, C, and D in the general equation.

APPLYING

Suggested Assignment

Day 1

Standard 2–8

Extended 1–8

Day 2

Standard 9–22

Extended 9–22

Integrating the Strands

Algebra Exs. 2–8, 10, 11, 13, 16, 21, 22

Functions Exs. 2–15, 21, 22

Geometry Exs. 16–20

Trigonometry Exs. 2–8, 10, 11, 13, 14, 21, 22

Statistics and Probability Exs. 15, 16

Discrete Mathematics Ex. 15

Logic and Language Exs. 1, 13, 14, 22

Communication: Reading

Ex. 1 directs students' attention to the procedure used to find the constant B in a sine or cosine model. This is often the most difficult step for students.

Teaching Tip

In connection with Exs. 8–12, you may wish to suggest that students consult books on physics, Earth science, meteorology, and astronomy for other situations that involve sine and cosine functions.

10-5 Exercises and Problems

1. **Reading** Explain step 3 of Sample 1.

For Exercises 2 and 3, use $h(t) = 25 \sin (12t) + 30$.

2. Find your height at each given time after the Ferris wheel starts.
   a. 10 s b. 25 s c. 100 s
3. a. Graph the equation.
   b. At what times will you be back at your starting position?
   c. Which points on the graph do these times correspond to?

For Exercises 4–6, change the dimensions of the Ferris wheel in Sample 1 as indicated.

a. How does the graph change?

b. What is the new equation?

4. The radius of the Ferris wheel is 17.5 ft.
5. The bottom of the Ferris wheel is 4 ft off the ground.
6. The Ferris wheel makes 5 revolutions in 2 min.
7. **Physics** A weight attached to a spring is pulled down so that it is 10 cm from the floor and is released so that it bounces up and down. When the effects of friction and gravity are ignored, its height can be modeled by a sine function of the time since it started bouncing. The weight reaches its first maximum height of 50 cm at 1.5 s.

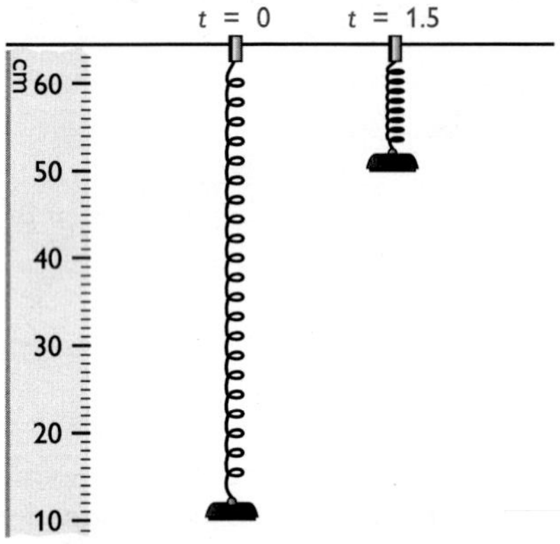

   a. Write an equation for the height, in centimeters, of the weight as a function of time, in seconds.
   b. Graph the equation from part (a).
   c. When is the weight moving up fastest?
   d. When is it moving down fastest?
   e. At what times is the weight changing direction?
8. High and low tides for Buenos Aires, Argentina, on February 17, 1993, are shown:
   a. Sketch a graph of the function. Assume that the water height can be modeled by a sine function.
   b. Find an equation for the height of the water as a function of the number of hours after midnight.
   c. Predict the time and height of the next high tide.
   d. **Research** Why do tides behave like sine waves? Why are they not exactly sine waves?

Tide	Time	Height
High	4:09 A.M.	118 cm
Low	10:52 A.M.	57 cm

Answers to Exercises and Problems

1. The period of the sine function is 360°, and the wheel turns $2 \cdot 360° = 720°$ in one minute. Since t is in seconds, the factor that relates the period of the sine function to the period of the Ferris wheel is $\frac{720°}{60 \text{ s}} = 12°/\text{s}$.

2. a. about 51.7 ft
   b. about 8.35 ft
   c. about 51.7 ft

3. a.

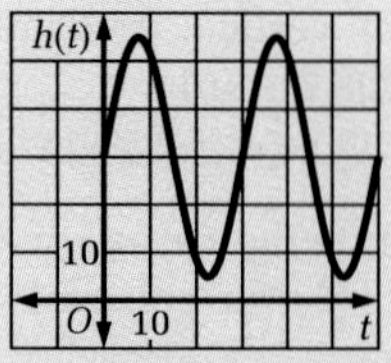

   b. every 30 s after the start
   c. $(30n, 30)$, where n is a positive integer

4. a. The amplitude is now 17.5. The graph is stretched (shrunk) vertically by a factor of $\frac{7}{10}$.
   b. $h = 17.5 \sin (12t)° + 22.5$

5. a. The axis is now $h = 29$; the graph is shifted down 1 unit.
   b. $h = 25 \sin (12t)° + 29$

6. a. The period is now 24 instead of 30. The graph is horizontally stretched (shrunk) by a factor of $\frac{4}{5}$.
   b. $h = 25 \sin (15t)° + 30$

7. a. $h(t) = 20 \sin (120(t - 0.75))° + 30$

connection to HISTORY

Thomas Jefferson recorded the temperatures at Monticello, his home near Charlottesville, Virginia, from 1810 to 1816. During this time Jefferson was experimenting with new farming techniques and new crops. (*Note:* Jefferson mistakenly switched the "max." and "min." listings.)

Jefferson began designing Monticello in 1768 at the age of 24.

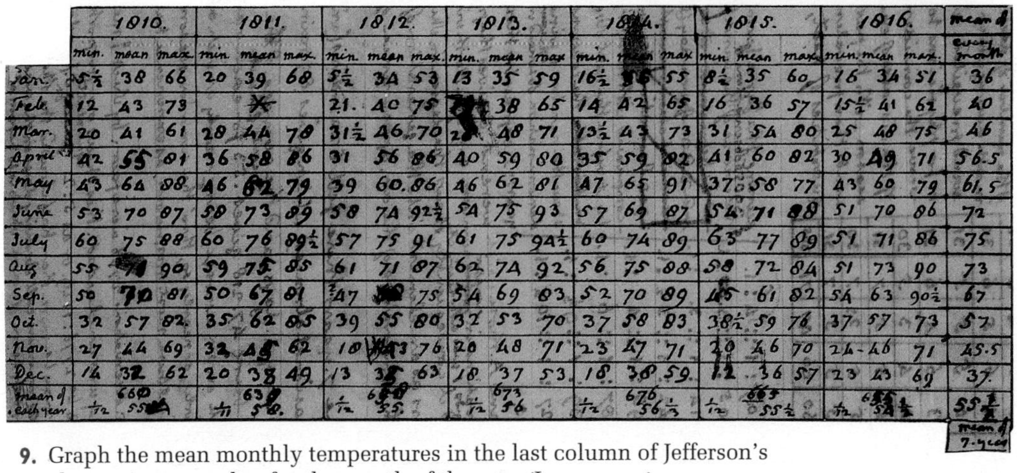

	1810.			1811.			1812.			1813.			1814.			1815.			1816.			mean of
	min.	mean	max.	min	mean	max.	min.	mean	max.	min.	mean	max	min.	mean	max	min.	mean	max.	min.	mean	max.	every month
Jan.	5½	38	66	20	39	68	5½	34	53	13	35	59	16½	[illegible]	55	8½	35	60	16	34	51	36
Feb.	12	43	73		*		21.	40	75	[illegible]	38	65	14	42	65	16	36	57	15½	41	62	40
Mar.	20	41	61	28	44	78	31½	46	70	[illegible]	48	71	13½	43	73	31	54	80	25	48	75	46
April	42	55	81	36	58	86	31	56	86	40	59	80	35	59	82	41	60	82	30	49	71	56.5
May	43	64	88	46	62	79	39	60.	86	46	62	81	47	65	91	37	58	77	43	60	79	61.5
June	53	70	87	58	73	89	58	74	92½	54	75	93	57	69	87	54	71	88	51	70	86	72
July	60	75	88	60	76	89½	57	75	91	61	75	94½	60	74	89	63	77	89	51	71	86	75
Aug.	55	[illegible]	90	59	75	85	61	71	87	62	74	92	56	75	88	58	72	84	51	73	90	73
Sep.	50	70	81	50	67	81	47	[illegible]	75	54	69	83	52	70	89	45	61	82	54	63	90½	67
Oct.	32	57	82.	35	62	85	39	55	80	32	53	70	37	58	83	38½	59	76	37	57	73	57
Nov.	27	44	69	32	45	62	10	[illegible]	76	20	48	71	23	47	71	20	46	70	24	46	71	45.5
Dec.	14	32	62	20	38	49	13	35	63	18	37	53	18	38	59	12	36	57	23	43	69	37.
mean of each year	1/12	660 55		1/11	638 58.		1/12	660 55.		1/12	673 56		1/12	676 56⅓		1/12	665 55½		1/12	[illegible]		55½

mean of 7 years

9. Graph the mean monthly temperatures in the last column of Jefferson's chart using a number for the month of the year (January = 1, February = 2, and so on).

10. a. Find an equation that models the temperature at Monticello as a function of time.

b. Graph your equation from part (a).

c. According to your equation, what was the mean temperature in May? Does this agree with the recorded temperatures?

11. Jefferson comments that many people believed "that the degrees of both cold & heat are moderated" since the country was settled.

> It is a common opinion that the climates of the several states of our union have undergone a sensible change since the dates of their first settlements; that the degrees both of cold & heat are moderated.

a. How would a moderation in temperature affect the graph?

b. Graph these monthly mean temperatures in Charlottesville, Virginia, for 1931 to 1960.

Jan.	Feb.	Mar.	Apr.	May	June	July	Aug.	Sept.	Oct.	Nov.	Dec.
37.2	38.4	45.3	56.6	66.3	73.7	77.3	76.0	69.9	59.5	48.4	38.7

c. What kind of change appears to have occurred in the mean temperatures for this area of Virginia since Jefferson's time?

Assessment: Open-ended

For Exs. 9 and 10, students should find an equation that models the monthly mean temperatures in Charlottesville for 1810 to 1816. In Ex. 11, students graph the monthly mean temperatures for 1931 to 1960. In answering Ex. 11(c), they should discuss these questions: Has the original graph of temperatures in Jefferson's time been translated, vertically stretched or horizontally stretched, and by how much? Is this what you would expect? Why?

Research

After completing Ex. 11(c), some students may think of the theory of global warming and wish to research it. After completing their research, ask them to prepare a brief verbal report to the class.

Answers to Exercises and Problems

b.

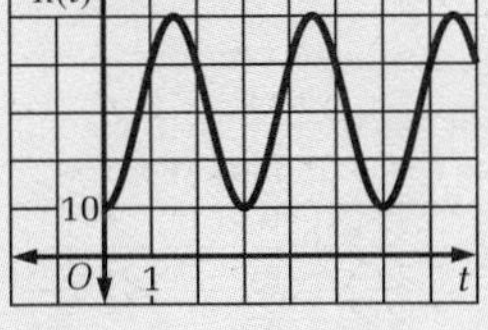

c. near 0.75 s and every 3 s after that

d. near 2.25 s and every 3 s after that

e. every 1.5 s

8. a.

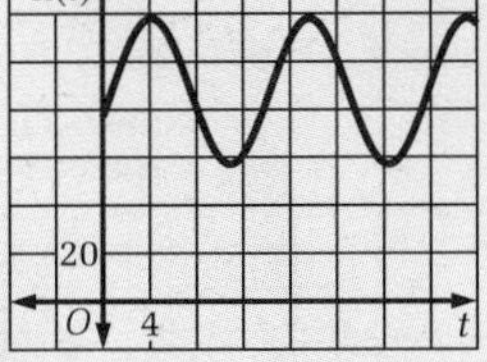

b. $h(t) = 30.5 \cos\left(\frac{360}{13.43}(t - 4.15)\right)^\circ + 87.5$

c. 118 cm at 5:35 P.M.

d. Summaries may vary. Tides behave like sine waves in that they are periodic rises and falls of the surface of the ocean. Tides are produced by gravitational force between Earth and the moon and between Earth and the sun and by rotation of Earth. The forces and their combined effect vary, so the tides are not exactly sine waves.

9.

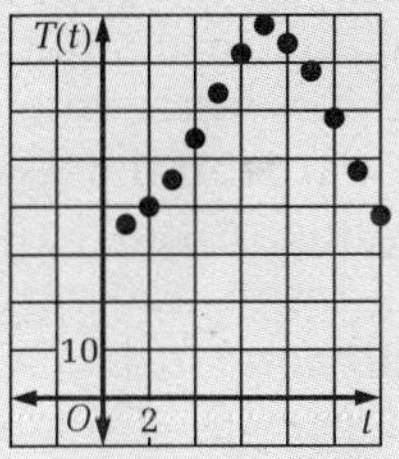

10. a. $T(t) = 19.5 \cos(30(t - 7))^\circ + 55.5$

b.

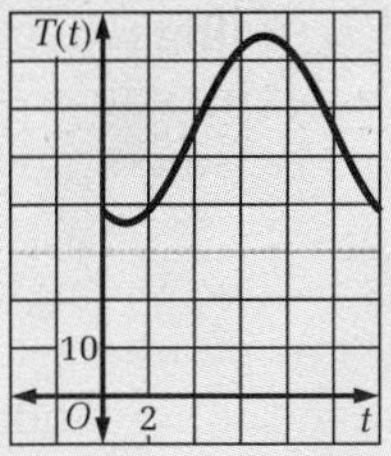

c. 65.25; It is higher than the recorded temperature.

11. a. The amplitude would be smaller.

b.

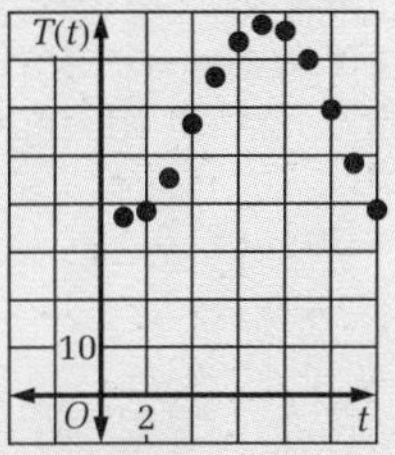

c. The temperatures are slightly higher than in Jefferson's time.

Multicultural Note

The 1883 eruption of Krakatoa is one of the most famous volcanic events in history. Krakotoa was an island near Java and Sumatra in Indonesia. Split into three smaller islands by the force of the eruption, Krakatoa no longer exists as it was in 1882. Krakatoa's eruption was so violent that the booming sounds of the explosions were heard in Rodriguez, an island 3000 miles away. Officials in Timor, 1350 miles away, thought the booms from the eruption were a ship's distress signal and sent out rescue teams. The explosion also created a tsunami that decimated villages on nearby islands, leaving thousands dead. Scientists have speculated that ash from such a volcanic eruption can block radiant energy from the sun and cause a drop in annual global temperature; and the year following the eruption of Krakatoa was particularly cold. The same phenomenon occurred in 1982 when El Chinchón erupted in Mexico. The huge ash cloud generated by El Chinchón may have played a role in the unusual weather of 1982–1983, in which California had very heavy rainfall and Australia experienced a severe drought.

Reasoning

Without discussing radian measure in detail, students can experiment with a graphics calculator to discover how radian measure is related to degree measure. They should set their calculators to radian mode and view the graph of $y = \sin x$ on a graphing window for which $0 \le x \le 10$ and $-1.2 \le y \le 1.2$. Focus on the x-intercepts. What are they and how are they related to the x-intercepts that were observed in degree mode?

12. **Earth Science** In 1815, the Indonesian volcano Tambora erupted. The volcanic ash thrown into the atmosphere from this eruption may have caused unusually cold weather in many places in 1816.

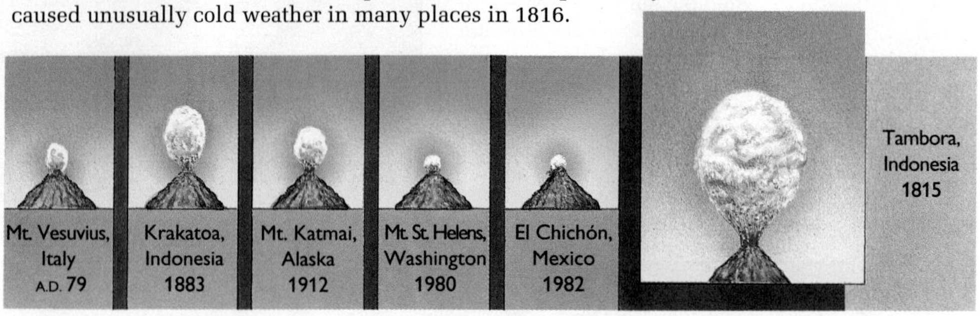

A pictorial comparison of the amounts of ash from six major volcanic eruptions

Graph Monticello's mean temperatures in 1816. Compare this graph to your graph of mean temperatures in Exercise 9 or 10. Does the eruption seem to have affected the observed temperatures?

13. a. **Open-ended** Find an equation that models the data in the graph. Let t be the number of months since January 1992.

Residential and Commercial Electricity Use in the United States from January 1992 to June 1994

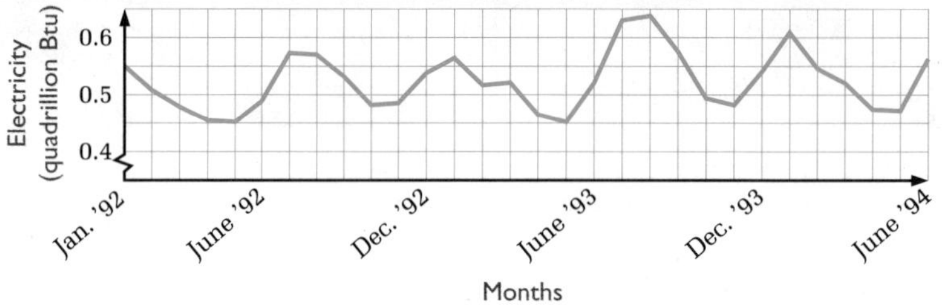

b. Graph your equation. Is it a good model?

c. **Writing** How is electricity use related to temperature? Explain why a mathematical model cannot predict the exact amount of electricity used.

Ongoing **ASSESSMENT**

14. Radio stations broadcast in one of two ways: AM, which means amplitude modulation, or FM, which means frequency modulation.

a. Which of these graphs represents AM? Which represents FM?

i.

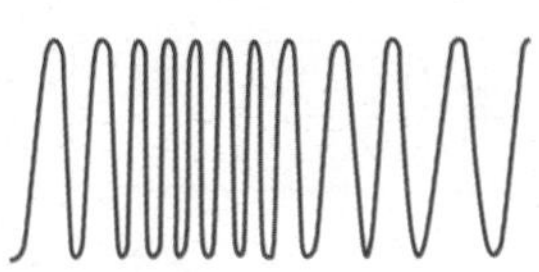

ii.

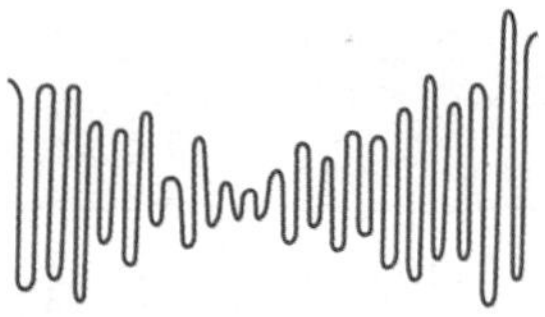

b. **Writing** Can either of these waves be modeled with a sine wave? Explain.

Answers to Exercises and Problems

12.

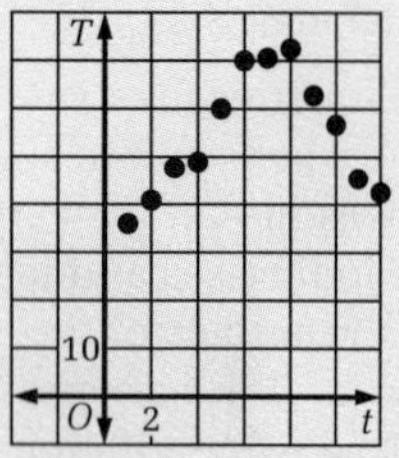

The summertime temperatures are somewhat lower than in the graphs in Exs. 9 and 10. Yes.

13. a, b. Answers may vary. An example is given.

a. $y = 0.075 \sin (72(t - 4))^\circ + 0.5$

b. **Residential and Commercial Electricity Use in the United States from January 1992 to June 1994**

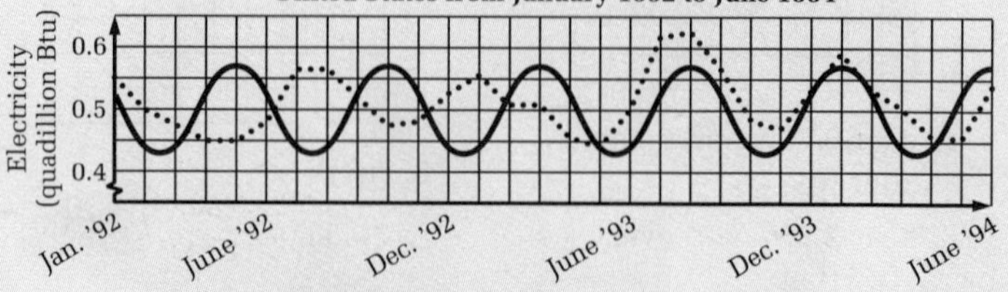

The graph is a fairly good model.

Review **PREVIEW**

15. Fit a linear model to this data. *(Section 6-4)*

Number of Women Employed in the United States Who Are at Least 20 Years Old

Year	1989	1990	1991	1992
Number Employed	49,745,000	50,455,000	50,535,000	51,180,000

16. You and a friend agree to meet for lunch between 12:10 P.M. and 12:20 P.M. You decide to wait 5 min for each other before sitting with someone else. What is the probability that you will meet? *(Section 7-3)*

For Exercises 17–20, tell whether each shape has *vertical reflection symmetry, horizontal reflection symmetry,* or *180° rotational symmetry.*
(Toolbox Skill 39)

17.

18.

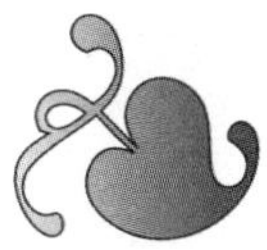

19.

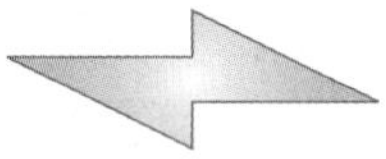

20.

10 Working on the Unit Project

21. Katherine's group gathered data on the number of cars passing through an intersection at different times during the day.

a. What function do you think best models the data? Why?

b. Write an equation to model the data. Graph your equation and the data on the same set of axes. How accurate is your model?

Daily Traffic through an Intersection

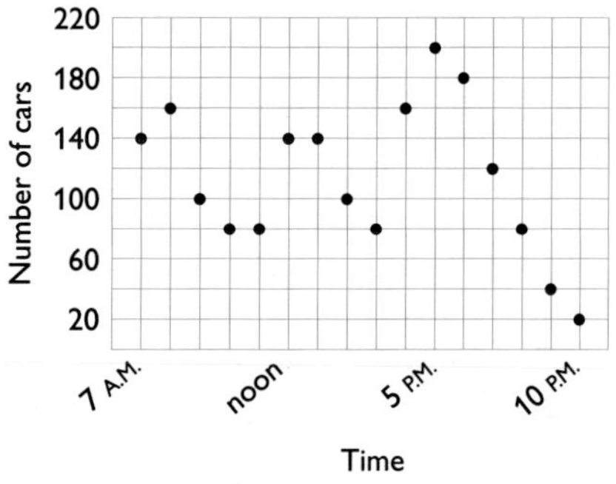

22. Include your answers to parts (a)–(e) in your scrapbook.

a. **Research** Think of a situation that can be modeled by a sine or cosine function. Gather any data you need to develop a model. Remember to include the source of your data or to describe the methods you used to gather data.

b. Make a graph of your data. Label each axis.

c. Write an equation to model your data.

d. Graph the equation you wrote in part (c) on the same set of axes as the data. How well does your model fit the data? What are some reasons why the data may not fit the model for all points?

e. **Writing** Do you think your model can be used to make accurate predictions about the future of the situation? Why or why not?

Working on the Unit Project

For Ex. 22(a), some students may have difficulty thinking of a situation to model. Remind them that sine and cosine waves model situations that are cyclical. They can start by thinking of phenomena that go through cycles every hour, day, or year.

Answers to Exericses and Problems

c. Temperature extremes may result in increased use of heating or air-conditioning appliances. Since temperature fluctuations cannot be accurately predicted, the amount of electricity used cannot be predicted exactly.

14. a. ii; i

b. No; a sine wave has a constant period and constant amplitude.

15. Answers may vary. An example is given. $N = 438.5x + 49821$, where N is the number (in thousands) of women employed and x is the number of years since 1989.

16. 0.75

17. vertical reflection symmetry

18. none of these

19. 180° rotational symmetry

20. all three

21. Answers may vary. Check students' work.

22. Answers may vary.

Quick Quiz (10-1 through 10-5)

See page 593.

See also Test 40 in the *Assessment Book.*

Practice 79 For use with Section 10-5

Name ______________________ Date __________

Practice 79

For use with Section 10-5

4. a. The graph is translated up 1 unit. b. $y = 20 \sin 9t + 26$ 5. a. The period is reduced. b. $y = 20 \sin 10t + 25$ 6. a. The amplitude is increased. b. $y = 25 \sin 9t + 30$

Your height above the ground (in ft) on a Ferris wheel of radius 20 ft that is 5 ft off the ground is modeled by the equation

$$h(t) = 20 \sin 9t + 25,$$

where t is in seconds. Use this equation for Exercises 1–3.

1. How long does it take the wheel to make one complete revolution? How many revolutions does the Ferris wheel make in 1 minute? 40 s; 1.5 revolutions
2. Find your height at each given time after the Ferris wheel starts.
   a. 15 s about 39.1 ft b. 40 s 25 ft c. 60 s 25 ft
3. a. Graph the equation. For graph, see additional answers.
   b. At what times will you be at the point closest to the ground? 30 s, 70 s, 110 s, and so on
   c. What are the corresponding points on the graph for these times? (30, 5), (70, 5), (110, 5), and so on

For Exercises 4–6, change the dimensions of the Ferris wheel described for Exercises 1–3 as indicated. Exs. 4–6: See top of page.

a. **Describe the change in the graph.** b. **Write a new equation.**

4. The Ferris wheel is 6 ft off the ground.
5. The Ferris wheel takes 36 s to make one revolution.
6. The radius of the Ferris wheel is 25 ft.

The table below gives the monthly mean temperatures in the Dallas-Ft. Worth area. Use this table for Exercises 7 and 8.

Jan.	Feb.	Mar.	Apr.	May	June	July	Aug.	Sept.	Oct.	Nov.	Dec.
43	48	57	66	73	81	85	85	77	67	56	47

7. Draw a scatter plot of the data, using 1 for January, 2 for February, 3 for March, and so on. For scatter plot, see additional answers.
8. Assume the following facts: (1) The lowest temperature occurs in January and the highest temperature occurs in July; (2) the graph is periodic with period 12; and (3) the graph can be modeled by the sine function. Using these facts, find an equation of the function that best fits the data. Then graph the equation on the same axes as the scatter plot you drew in Exercise 7.

   $y = 21 \sin 30(x - 4) + 64$; For graph, see additional answers.

Practice Bank, INTEGRATED MATHEMATICS 3

79

Unit 10 CHECKPOINT

1. **Writing** When $0° \le \theta \le 360°$, for which values of θ is $\sin \theta > \cos \theta$? Explain your reasoning.
2. Compare the rates of growth in the first quadrant of $y = 5x$, $y = x^5$, and $y = 5^x$. 10-1
3. Draw a periodic graph with period 2, amplitude 3, and axis at $y = 2$. 10-2

Tell whether each statement is true for the *sine function*, the *cosine function*, *both functions*, or *neither function*. 10-3

4. The amplitude is 1.
5. The y-intercept is 1.
6. The domain of the function is all angles.
7. The function increases in the interval $0° \le \theta \le 90°$.

For each graph, find the period, the amplitude, and the frequency. Then write an equation for each graph. 10-4

8.

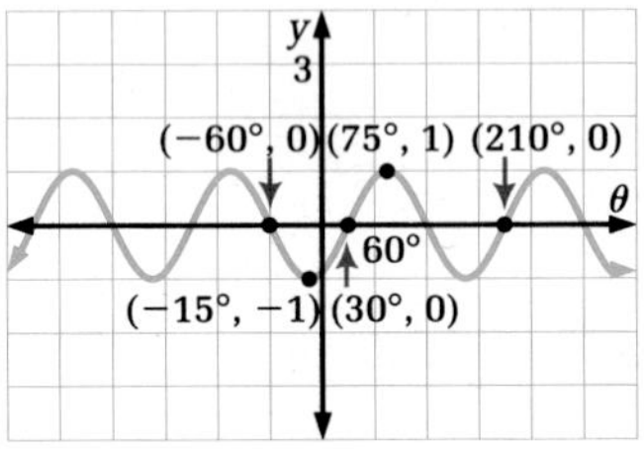

9.

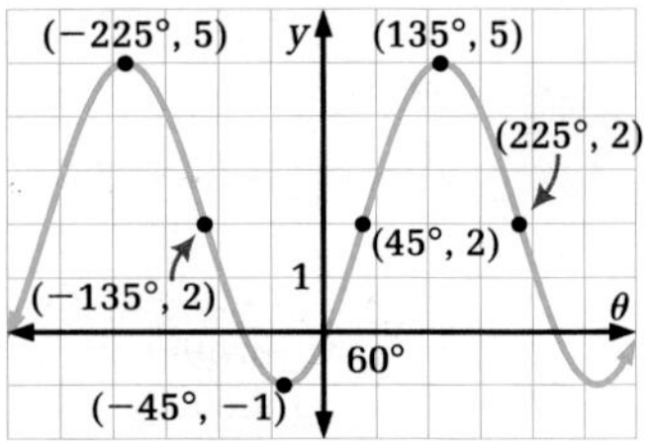

10. The graph of a transformed cosine function has period 180°, amplitude 2, and a vertical translation 3 units down. Without using graphing technology, make a sketch of the graph.
11. Use the table to answer parts (a)–(c). 10-5

Ice Cream Production in the United States (millions of gallons)

	Jan.	Feb.	Mar.	Apr.	May	June	July	Aug.	Sept.	Oct.	Nov.	Dec.
1991	56.4	60.8	72.8	73.7	81.8	91.5	88.1	83.7	72.9	68.0	57.2	58.0
1992	58.7	62.3	77.1	78.9	79.7	89.7	87.4	79.2	73.8	64.6	58.3	58.3
1993	53.7	64.3	76.7	75.9	78.0	90.3	89.7	84.4	74.2	62.8	57.9	58.3

a. Graph the mean monthly ice cream production in the United States for 1991–1993. What type of function do you think best models the data? Why?

b. Write an equation to model the data.

c. Graph your equation and the data on the same set of axes. How accurate is your model? What are some reasons why some of the data points do not fit the model?

Answers to Checkpoint

1. $45° < \theta < 225°$; In this interval, the graph of $y = \sin \theta$ is above the graph of $y = \cos \theta$.
2. $y = 5^x$ grows the fastest, followed by $y = x^5$, and then by $y = 5x$.
3. Answers may vary. An example is given.

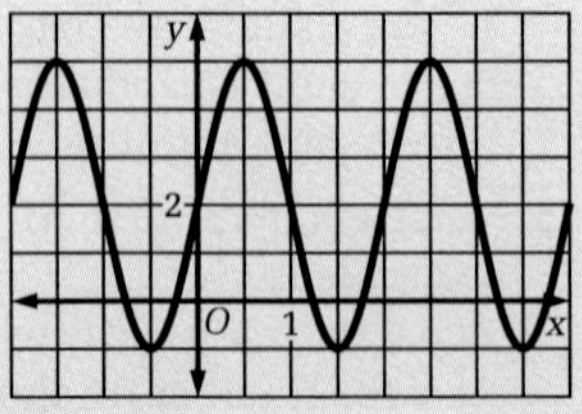

4. both
5. cosine
6. both
7. sine
8. 180°; 1; $\frac{1}{180°}$; $y = \sin 2(\theta - 30°)$
9. 360°; 1; $\frac{1}{180°}$; $y = 3 \sin (\theta - 45°) + 2$

10, 11. See answers in back of book.

Section **10-6**

Patterns and Transformations

Focus
Recognize periodic border patterns. Use transformations to classify and create border patterns.

FRIEZE FRAME

In many cultures people use border patterns to decorate edges of pottery, curtains, belts, and ribbons. Mathematicians have found that all border patterns fall into one of seven types. Examples of each type from various cultures are shown below.

A.

Native American

B.

Incan

C.

Zairean

D.

Chinese

E.

Egyptian

F.

Islamic

G.

Greek

Mosaic floor in Delos, Greece

Talk it Over

1. Explain why these patterns are periodic.
2. List all the border patterns that have each type of symmetry.
   a. vertical reflection symmetry
   b. horizontal reflection symmetry
   c. 180° rotational symmetry

Student Resources Toolbox
p. 663 *Transformations*

PLANNING

Objectives and Strands
See pages 538A and 538B.

Spiral Learning
See page 538B.

Recommended Pacing
Section 10-6 is a two-day lesson.

Day 1
Pages 577–579: Opening paragraph through top of page 579, *Exercises 1–4*

Day 2
Page 579: Sample 1 through Look Back, *Exercises 5–25*

Toolbox References
- **Toolbox Skill 39:** Recognizing Symmetry

Extra Practice
See pages 624–625.

Warm-Up Exercises
Warm-Up Transparency 10-6

Support Materials
- Practice Bank: Practice 80
- Activity Bank: Enrichment 71
- Study Guide: Section 10-6
- Problem Bank: Problem Set 22
- Assessment Book: Quiz 10-6

Answers to Talk it Over

1. Each pattern repeats the same figure over and over.
2. a. A, C, and D
   b. B and D
   c. C, D, E, and F

TEACHING

Talk it Over

Questions 1 and 2 help students make a connection between the border patterns shown on page 577 and the ideas of periodicity and symmetry.

Multicultural Note

Archaeologists and anthropologists have gained clues about the trading patterns of ancient Mexico through the discovery of pottery with distinctive regional border patterns in unexpected places. Researchers have uncovered pottery with border patterns typical of the great city of Teotihuacán as far away as present-day Guatemala. Teotihuacán, an important civic and religious center in central Mexico from about A.D. 200 to A.D. 700, was known to be influential throughout central Mexico, but finding incidences of contact between that city and the great city of Tikal in Guatemala may indicate that Teotihuacán's influence spread farther than previously thought.

Teaching Tip

When looking at a border pattern, such as the ones of the first column of the chart, some students may have trouble seeing the repeating pattern. Remind students they should check any pattern they are having trouble with against the examples in the third column of the chart.

Cooperative Learning

You may wish to have groups of students prepare a set of transparencies that could be used to illustrate the ideas about the seven types of border patterns shown on this page.

This table shows the seven types of border patterns and how each is constructed. To classify a border pattern by type, you have to identify all the types of symmetry it has.

The Seven Types of Border Patterns

Symmetry Type	How to Construct	Example
Translation 	Translate an asymmetric figure repeatedly.	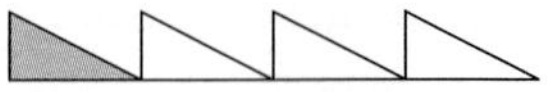
Vertical reflection 	Reflect an asymmetric figure over a vertical line. Translate the pair repeatedly.	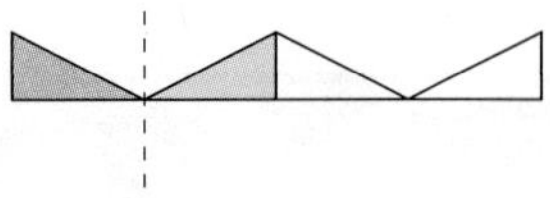
Horizontal reflection 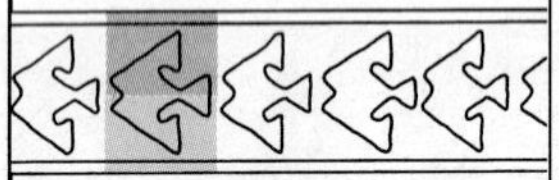	Reflect an asymmetric figure over a horizontal line. Translate the pair repeatedly.	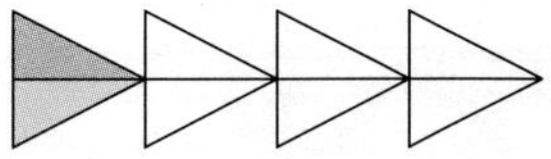
Glide reflection 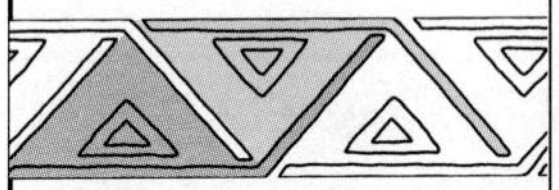	Reflect an asymmetric figure over a horizontal line, then translate the image to the next space. Translate the pair repeatedly.	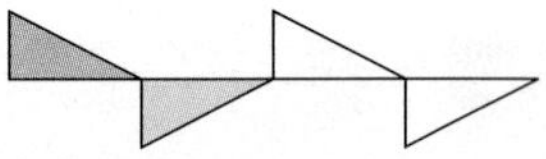
Horizontal and vertical reflection 	Reflect an asymmetric figure over a horizontal line, then reflect the pair over a vertical line. Translate all four repeatedly.	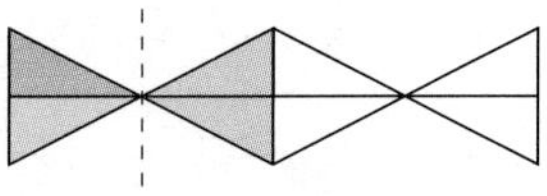
180° rotation 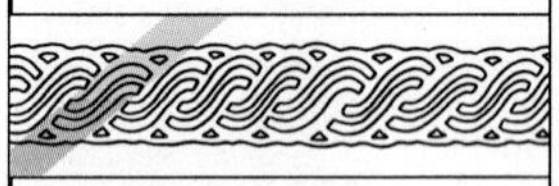	Rotate an asymmetric figure 180° into the next space. Translate the pair repeatedly.	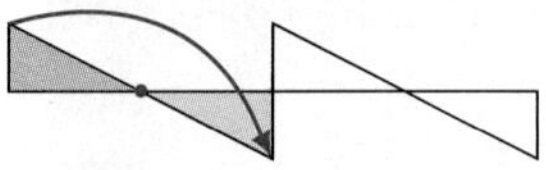
180° rotation and vertical reflection 	Rotate an asymmetric figure 180°, then reflect the pair over a vertical line. Translate all four repeatedly.	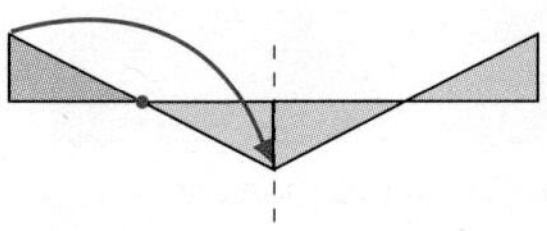

BY THE WAY...

Some cultures use only certain subsets of the seven border patterns. Archaeologists use the presence of unexpected border patterns as evidence of trading between cultures.

You can use a tree diagram to classify border patterns by the type(s) of symmetry they have. Every border pattern has translational symmetry, so begin by looking for a vertical reflection.

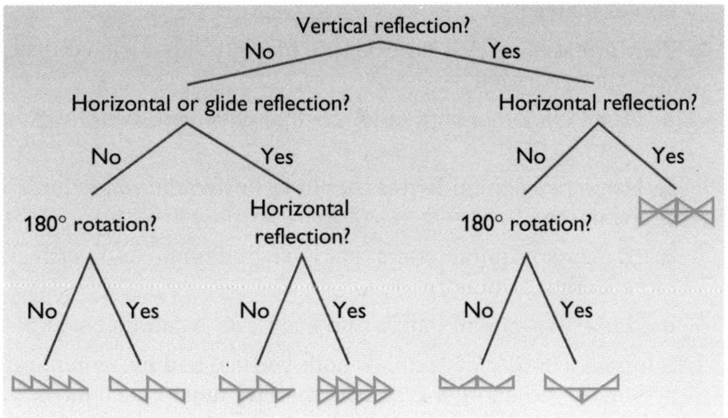

Now you are ready for: Exs. 1–4 on p. 580

Sample 1

Classify this border pattern by its symmetry.

Sample Response

Is there vertical reflection?  Yes.

Is there horizontal reflection?  No.

Is there 180° rotation?  Yes.

This border pattern has 180° rotation and vertical reflection symmetry.

Sample 2

Use this asymmetric figure. Create a border pattern that has vertical reflection symmetry.

Sample Response

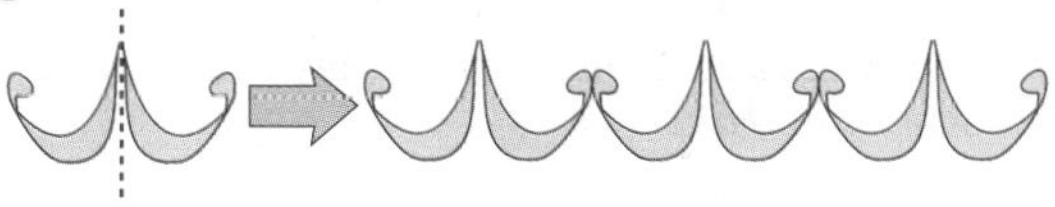

Now you are ready for: Exs. 5–25 on pp. 580–581

Look Back

What transformations are used in creating border patterns?

Integrating the Strands

A tree diagram, which is frequently used with probability and statistics problems, is applied here in a geometric situation to classify border patterns. Students should notice how a decision needs to be made at each step; thus, this diagram is essentially an algorithm for classifying a border design. Algorithms are part of the strand of discrete mathematics.

Additional Samples

S1 Classify this border pattern by its symmetry.

The pattern does not have vertical symmetry or 180° rotation symmetry. It does have translation symmetry and horizontal symmetry.

S2 Use this figure to create a border pattern that has glide reflection symmetry.

Reflect the figure over a horizontal line and translate it to the right. Then translate the pair repeatedly.

Answers to Look Back

translations, reflections, and rotations

APPLYING

Suggested Assignment

Day 1

Standard 1–4

Extended 1–4

Day 2

Standard 5–11, 14–18, 20–25

Extended 5–18, 20–25

Integrating the Strands

Algebra Ex. 20

Functions Ex. 18

Geometry Exs. 1–19, 21–25

Trigonometry Exs. 18, 21, 22

Statistics and Probability Ex. 20

Discrete Mathematics Exs. 20, 21

Logic and Language Exs. 1, 4

Career Note

There are many job opportunities in business for people who can create interesting and appealing graphic designs. All types of products sold to consumers require a design, including this textbook. The world of advertising creates messages both verbally and through graphics. Many designers, such as those who create patterns for linens, clothing, stationary, packaging and wallpaper, often use transformations to create their designs. Many designers have a college degree with a major in art or design.

Assessment: Open-ended

For Ex. 2, parts (c) and (d), students name two capital letters that have horizontal and vertical reflection symmetry and find that they also have 180° rotational symmetry. Have students determine if this is a general rule. Try $f(x) = \sin x$, $0 \le x \le 180°$.

10-6 Exercises and Problems

1. **Reading** What one transformation is used in all types of border patterns?
2. The capital letter A is an example of a letter that has vertical reflection symmetry.
   - a. Name two other capital letters that have vertical reflection symmetry.
   - b. Name two capital letters that have horizontal reflection symmetry.
   - c. Name two capital letters that have horizontal and vertical reflection symmetry.
   - d. Name two capital letters that have 180° rotational symmetry.

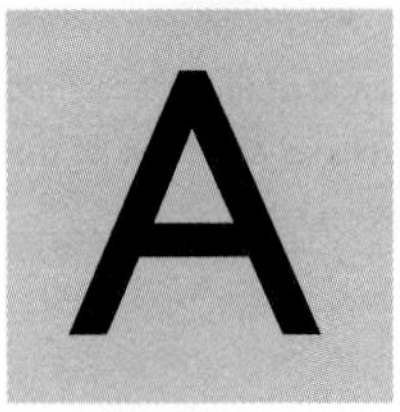

3. Suppose a border pattern has both vertical and horizontal reflection symmetry. What other type of symmetry must it also have?
4. a. How are glide reflection symmetry and horizontal reflection symmetry alike? How are they different?
   - b. **Writing** Explain why horizontal reflection is a special case of glide reflection.

Classify each border pattern by its symmetry.

5. 

Native American

6. 

Native American

7. 

Native American

8. 

Native American

9. 

Native American

10. 

Islamic

11. a. Why can a border pattern *not* have rotational symmetries other than 180°, such as 60° rotational symmetry?
    - b. Why can a border pattern *not* have other lines of symmetry besides those that are parallel or perpendicular to the edges?
12. **Open-ended** Create your own asymmetric basic element. Use it to generate two different border patterns. Tell the type of symmetry that each border pattern has.

Answers to Exercises and Problems

1. A translation is used in each pattern.
2. a. H, I, M, O, T, U, V, W, X, Y
   - b. B, C, D, E, H, I, O, X
   - c. H, I, O, X
   - d. H, I, N, O, S, X, Z
3. 180° rotational
4. a. They both involve reflection over a horizontal line. In the case of a glide reflection, the resulting figure does not have horizontal reflection symmetry.
   - b. A horizontal reflection is a glide reflection where the translation vector is (0, 0).
5. translation
6. 180° rotation
7. vertical
8. 180° rotation, horizontal and vertical
9. horizontal
10. vertical and 180° rotation
11. a. If a border pattern has a rotational symmetry of θ, the image of each edge under a rotation of θ must be the other edge. Since the edges are parallel, θ must be 180°.

13. Find all the types of border patterns in the table on page 578 that have both 180° rotational symmetry and horizontal reflection symmetry.

Suppose the asymmetric figure used to construct a border pattern contains the letter "b." In which types of border patterns constructed with this figure will each of these letters appear?

14. d

15. p

16. q

17. Find all the types of border patterns in the table on page 578 that contain glide reflection symmetry.

18. Classify this sine wave by its symmetry.

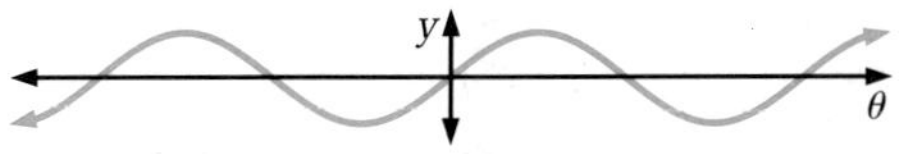

Ongoing ASSESSMENT

19. **Group Activity** Work with another student. Without showing your partner your work, choose one type of symmetry and create a border pattern. Trade papers and classify your partner's border pattern by its symmetry.

Review PREVIEW

20. Find an equation to model the mean monthly temperatures shown in the table. *(Section 10-5)*

Mean Monthly Temperatures at Yellowstone Park, WY

Jan.	Feb.	Mar.	Apr.	May	June	July	Aug.	Sept.	Oct.	Nov.	Dec.
18.3°	21.6°	26.9°	38.1°	46.9°	53.8°	62.9°	61.1°	52.4°	42.6°	29.1°	22.2°

21. Convert the vector $(-7, 4)$ to polar form. *(Section 8-4)*

22. Convert the vector $(5, -25°)$ to rectangular form. *(Section 8-4)*

23. Find the measures of each interior and each exterior angle of a regular hexagon. *(Section 3-4)*

10 Working on the Unit Project

24. Create a border design to include on the cover of your scrapbook. On the inside front cover of your scrapbook, describe the symmetry used to make the border design.

25. Make another border design and describe its symmetry. Use this design to highlight items in your scrapbook.

Error Analysis

When answering Exs. 5–10, students may omit some possibilities or suggest incorrect possibilities. Students can find and correct their errors by copying patterns or parts of patterns on tracing paper. They can then use the tracing of the patterns to perform the transformations that might have been used in each situation.

Problem Solving

Ask students to list the most basic transformations used in generating border patterns. Then challenge them to explain why combinations of these transformations result in just seven types of border patterns.

Practice 80 For use with Section 10-6

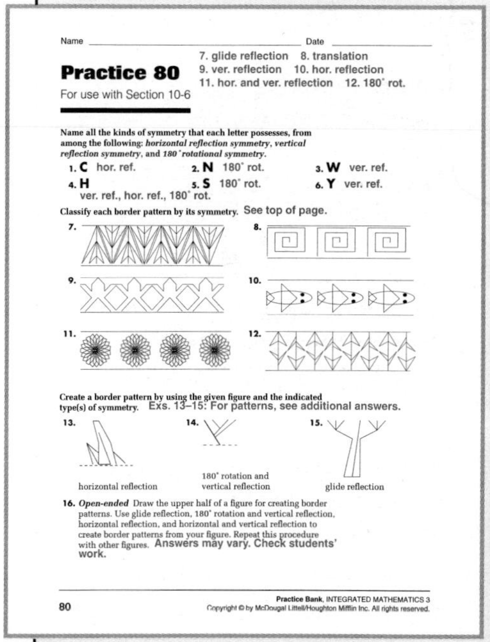
Name ____________ Date ________

Practice 80
For use with Section 10-6

7. glide reflection 8. translation
9. ver. reflection 10. hor. reflection
11. hor. and ver. reflection 12. 180° rot.

Name all the kinds of symmetry that each letter possesses, from among the following: *horizontal reflection symmetry*, *vertical reflection symmetry*, and *180° rotational symmetry*.

1. **C** hor. ref. 2. **N** 180° rot. 3. **W** ver. ref.
4. **H** ver. ref., hor. ref., 180° rot. 5. **S** 180° rot. 6. **Y** ver. ref.

Classify each border pattern by its symmetry. See top of page.

7. 8. 9. 10. 11. 12.

Create a border pattern by using the given figure and the indicated type(s) of symmetry. Exs. 13–15: For patterns, see additional answers.

13. horizontal reflection
14. 180° rotation and vertical reflection
15. glide reflection

16. ***Open-ended*** Draw the upper half of a figure for creating border patterns. Use glide reflection, 180° rotation and vertical reflection, horizontal reflection, and horizontal and vertical reflection to create border patterns from your figure. Repeat this procedure with other figures. Answers may vary. Check students' work.

80 **Practice Bank**, INTEGRATED MATHEMATICS 3

Answers to Exercises and Problems

b. If a line is a line of symmetry of a border pattern, then the image of each edge under reflection over the line must be the same edge or the other edge. Then the line of symmetry must be parallel to or perpendicular to the edges.

12. Answers may vary. Check students' work.

13. horizontal and vertical reflection

14. vertical reflection; horizontal and vertical reflection

15. glide reflection; 180° rotation and vertical reflection; horizontal and vertical reflection

16. 180° rotation; 180° rotation and vertical reflection; horizontal and vertical reflection

17. glide reflection; horizontal reflection; horizontal and vertical reflection

18. 180° rotation symmetry and horizontal and vertical symmetry

19. Answers may vary. Check students' work.

20. Answers may vary. An example is given. $y = 22.3 \cos (30(x - 7))° + 40.6$

21. $(8.06, 150.3°)$

22. $(4.53, -2.11)$

23. 120°; 60°

24, 25. Check students' work.

PLANNING

Objectives and Strands
See pages 538A and 538B.

Spiral Learning
See page 538B.

Materials List
- Paper cutouts of each type of regular polygon
- Geometric drawing software
- Scissors
- Graph paper

Recommended Pacing
Section 10-7 is a two-day lesson.

Day 1
Pages 582–583: Exploration through Sample 1, *Exercises 1–13*

Day 2
Pages 584–585: Top of page 584 through Look Back, *Exercises 14–35*

Extra Practice
See pages 624–625.

Warm-Up Exercises
Warm-Up Transparency 10-7

Support Materials
- Practice Bank: Practice 81
- Activity Bank: Enrichment 72
- Study Guide: Section 10-7
- Problem Bank: Problem Set 22
- Explorations Lab Manual: Diagram Masters 2, 21
- Assessment Book: Quiz 10-7, Test 41, Alternative Assessment 6

Section 10-7 Periodic Tessellations

Focus
Recognize, classify, and create tessellations.

It All Fits

EXPLORATION

What do regular and semiregular tessellations look like?

- **Materials: paper cutouts of 8 of each type of regular polygon — 3-, 4-, 5-, 6-, 8-, 10-, and 12-gon — with sides that are the same length, or geometric drawing software**
- **Work in a group of four students.**

1. By placing regular polygons so their vertices touch, make a tiling — a design that covers the plane without gaps or overlaps. Your design is called a **tessellation.**
2. A **regular tessellation** is a tessellation made from only one kind of regular polygon. Which regular polygons can you use to create regular tessellations?
3. A **semiregular tessellation** is a tessellation made from two or more kinds of regular polygons arranged the same way at every vertex.

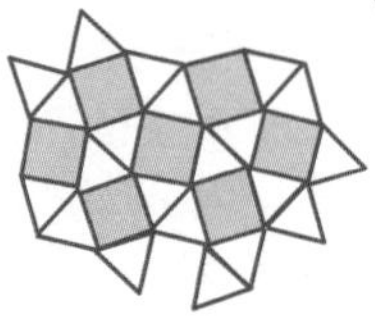

3.3.4.3.4

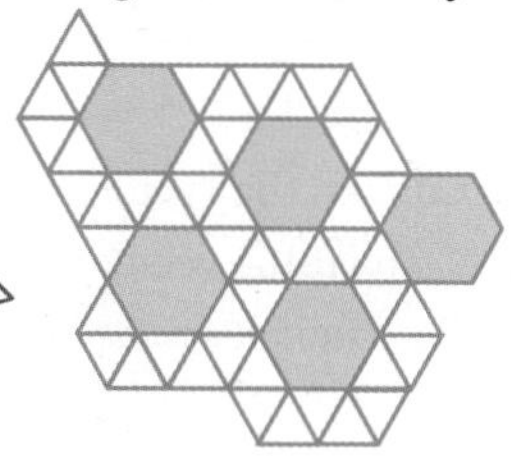

3.3.3.3.6

Explain how each number code describes the semiregular tessellation above it.
4. Make as many semiregular tessellations as you can that use *two* kinds of regular polygons. Draw a sketch and write a code for each tessellation.
5. Make a semiregular tessellation that uses *three* kinds of regular polygons.
6. Which polygons did you use in your semiregular tessellations? Which polygons did you *not* use?

BY THE WAY...
The word *tessellation* comes from the Latin word "tessera," a small square piece of stone used in mosaics.

Answers to Exploration

1. Answers may vary.
2. equilateral triangles, squares, and regular hexagons
3. The number code identifies the polygons that meet at a vertex. The code 3.3.4.3.4 indicates that at each vertex of the tessellation on the left, there are, in order, two triangles, then a square, then another triangle, then a square. The code 3.3.3.3.6 indicates that at each vertex of the tessellation on the right, there are, in order, four triangles and one hexagon.
4. Answers may vary. Examples are given.

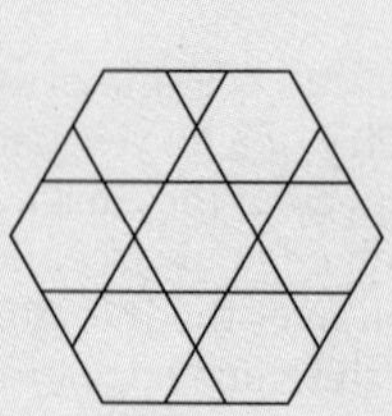

3.6.3.6

Tilings have been used by many cultures to cover floors, walls, ceilings, and columns.

Detail of Moorish-style wall tile in Córdoba, Spain

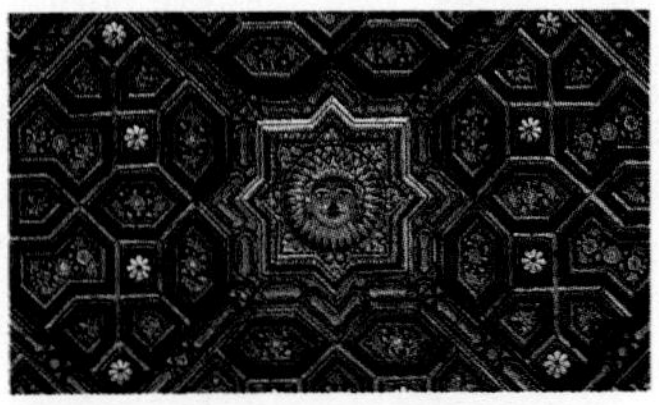

Ceiling in Junagarh Fort, Bikaner Rajasthan, India

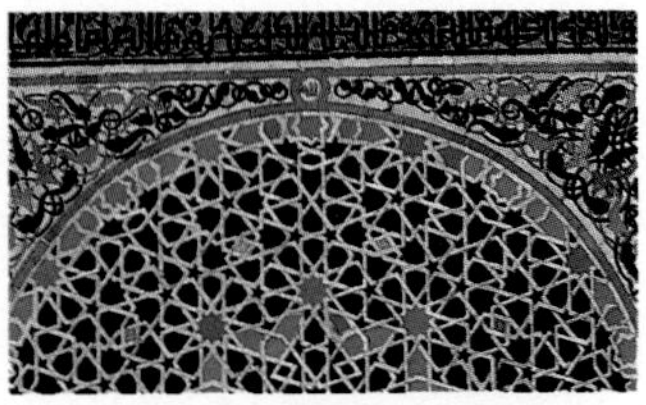

Tile dating from 1325, in Fez, Morocco

Sample 1

Use an indirect proof to show that each group of polygons cannot tessellate a plane.

a. regular pentagons

b. regular hexagons and squares

Sample Response

a. Suppose regular pentagons tessellate a plane. Let p = the number of regular pentagons that surround a vertex.

Each interior angle of a regular pentagon has a measure of 108°, so

$$108p = 360$$
$$p = 3\frac{1}{3}$$

Since p does not have an integer value, the assumption that regular pentagons tessellate a plane is false.

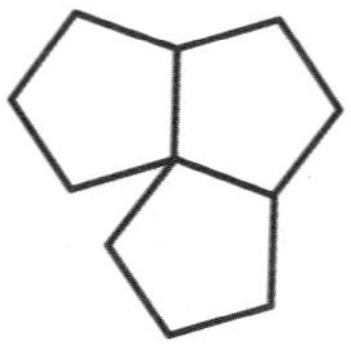

b. Suppose regular hexagons and squares tessellate a plane. Let h = the number of regular hexagons at each vertex and let s = the number of squares at each vertex.

The measures of the interior angles of regular hexagons and squares are 120° and 90°, so

$$120h + 90s = 360$$

Since the number of regular hexagons at each vertex, h, can be 0, 1, 2, or 3, the solutions to this equation include (0, 4), $\left(1, 2\frac{2}{3}\right)$, $\left(2, 1\frac{1}{3}\right)$, and (3, 0).

The only solutions with integer values are the first and last, which involve only squares or only regular hexagons, but not both shapes.

Since there is no combination of regular hexagons and squares that can surround a vertex, the assumption that regular hexagons and squares tessellate a plane is false.

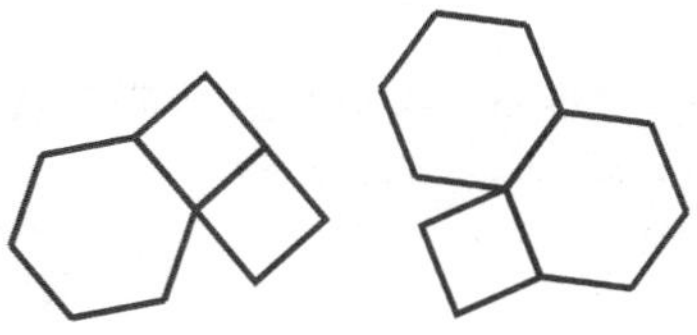

Now you are ready for: Exs. 1–13 on pp. 585–586

TEACHING

Exploration

The goal of this Exploration is to introduce students to tessellations through hands-on experience. Students work with regular and semiregular tessellations and become familiar with the number code for semiregular tessellations.

Using Technology

A computer software program called *TesselMania* does an excellent job of explaining and creating tessellations. It is available from MECC and is referenced in the Outside Resources on page 538D.

Additional Sample

S1 Use an indirect proof to show that each group of polygons cannot tessellate a plane.

a. regular nonagons

Suppose regular nonagons tessellate the plane. Let n = the number of regular nonagons that surround a vertex. Each interior angle of a nonagon has a measure of 140°. So,

$$140n = 360$$
$$n = 2\frac{4}{7}.$$

Since $2\frac{4}{7}$ is not an integer, the assumption that regular nonagons tessellate a plane is false.

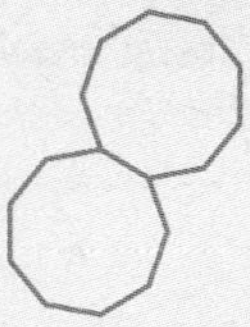

b. regular nonagons and equilateral triangles

Suppose regular nonagons and equilateral triangles tessellate the plane. Let n = the number of regular nonagons at each vertex and t = the number of equilateral triangles at each vertex. Then the equation $140n + 60t = 360$ would have to have positive whole number solutions for which $n < 3$ and

Sample continued on next page.

Answers to Exploration

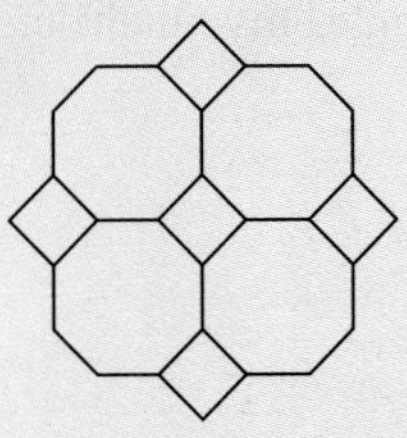

4.8.8

The figure in Ex. 3 on page 585 is a semiregular tessellation with code 3.12.12. The figure in Ex. 4 on page 585 is a semiregular tessellation with code 3.3.3.4.4.

5. Answers may vary. An example is given.

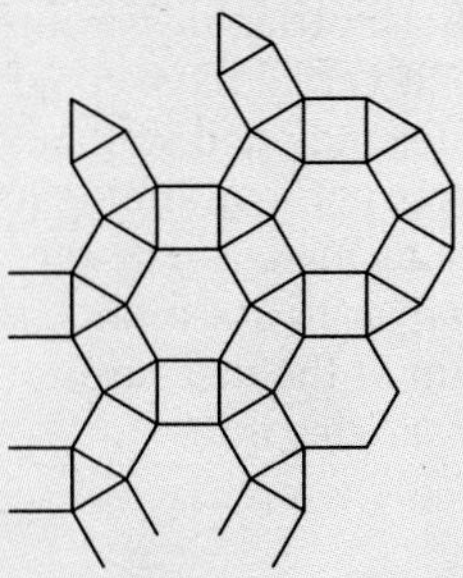

6. Answers may vary. In the sketches shown, there are triangles, squares, hexagons, and octagons. There are no pentagons, heptagons, or nonagons. The figure in Ex. 3 on page 585 includes dodecagons as well.

Additional Answer (continued)

$t \le 6$. There is no such pair (n, t) that is a solution of the equation. Therefore, regular nonagons and equilateral triangles cannot tessellate the plane.

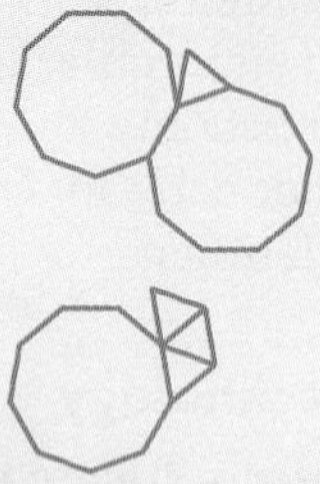

Talk it Over

To help answer questions 3 and 4, students should study the diagrams immediately above the questions. For question 4, suggest that students try to make a connection between the two diagrams.

Additional Sample

S2 Create a new tessellation by modifying a tessellation of squares.

Step 1. Start with four squares that meet at a vertex.

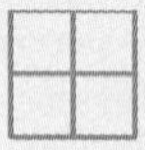

Step 2. Draw a semicircle that cuts across the two squares on the bottom and has the segment made of their top sides as its diameter.

Step 3. Using the exterior corners of the two squares on top as centers, draw quarter circles using a radius equal to the length of a side of one square.

Step 4. Translate the figure with curved sides horizontally and diagonally.

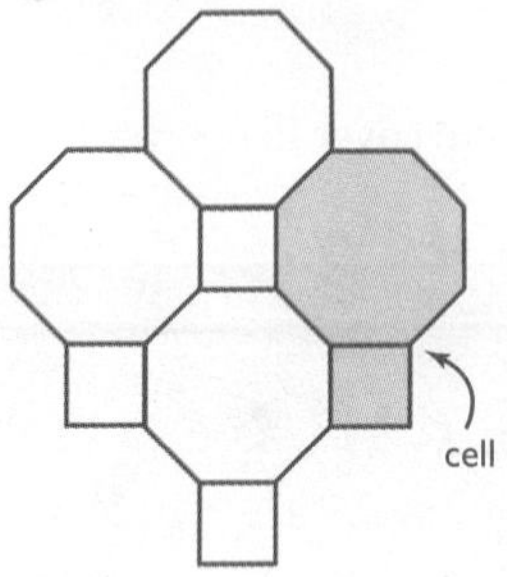

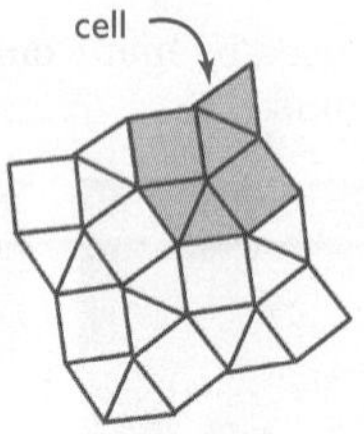

A **periodic tessellation** is a tessellation that matches itself exactly when it is translated.

Every periodic tessellation has a fundamental unit, called a **cell,** that can be copied and translated to generate the pattern.

Talk it Over

1. Choose one of the regular tessellations you made in the Exploration. Is it periodic? If so, shade a cell.
2. Choose one of the semiregular tessellations you made in the Exploration. Is it periodic? If so, shade a cell.

Tessellations with Nonregular Polygons

Regular and semiregular tessellations are made from regular polygons. You can also make tessellations with nonregular polygons.

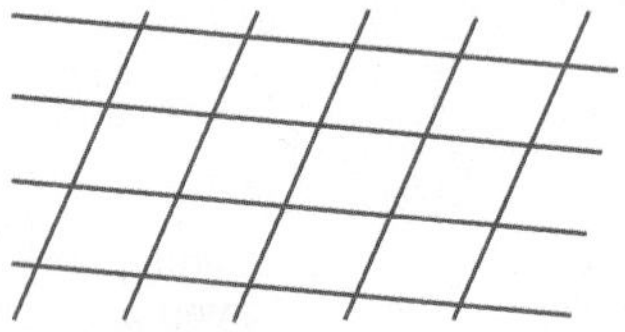

All parallelograms tessellate.

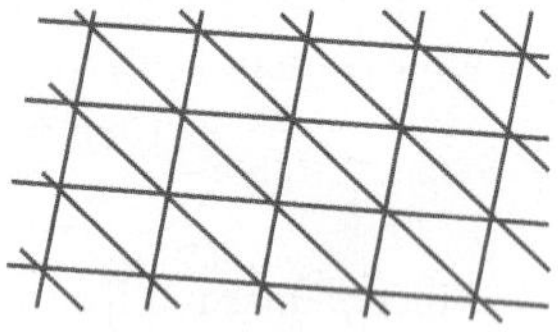

All triangles tessellate.

BY THE WAY...

The Dutch artist Maurits Cornelius Escher (1898–1972) spent much of his career creating tessellations inspired by those he admired in the Alhambra, a 14th century Moorish palace in Granada, Spain.

Talk it Over

3. Explain why the sum of the measures of the angles around each vertex is 360° in a tessellation of congruent parallelograms.
4. Explain why the sum of the measures of the angles around each vertex is 360° in a tessellation of congruent triangles.

Sample 2

Create a new tessellation by modifying a tessellation of parallelograms.

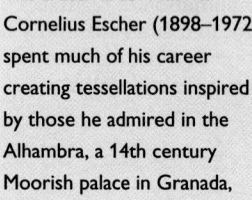

Answers to Talk it Over

1. Any regular tessellation is periodic. For a tessellation consisting of squares or hexagons, the basic figure is a cell. For a tessellation consisting of triangles, a hexagon-shaped cluster of triangles is a cell.
2. Answers may vary.
3. The four angles at each vertex are congruent to the four angles of each parallelogram. The sum of the measures of the interior angles of a parallelogram is 360°.
4. The six triangles around each vertex have congruent sides matched so that four congruent parallelograms are formed. So, by the answer to question 3, the sum of the measures of the six angles is 360°.

Answers to Look Back

any triangle, any quadrilateral

Sample Response

1 Start with a parallelogram.

2 Make a change on one side and translate the change to the opposite side.

3 Make a different change to the other pair of opposite sides.

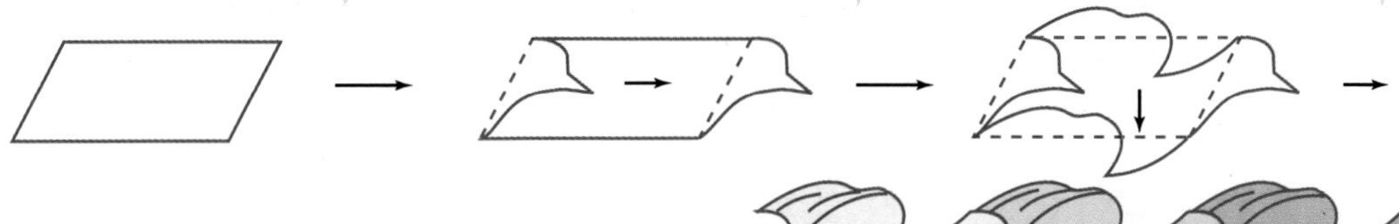

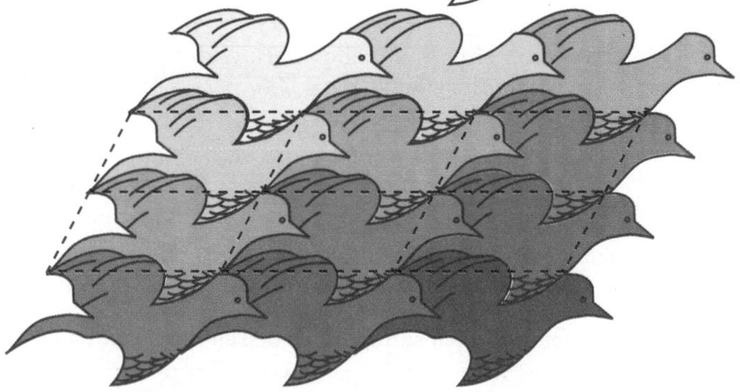

4 Draw details inside the cell if you wish.

Now you are ready for: Exs. 14–35 on pp. 586–588

Look Back

Which of these polygons will tessellate periodically — *any triangle, any quadrilateral, any pentagon*?

10-7 Exercises and Problems

1. **Reading** How does a regular tessellation differ from a semiregular one?

Write a code for each tessellation.

2. 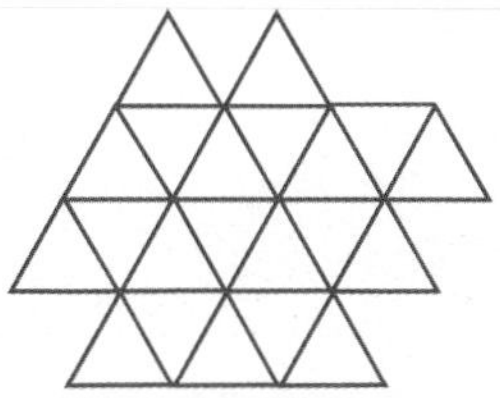

3. 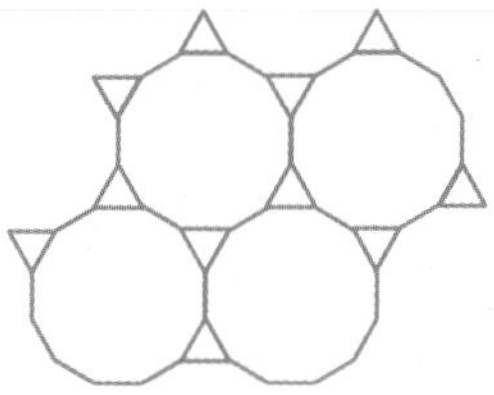

4. 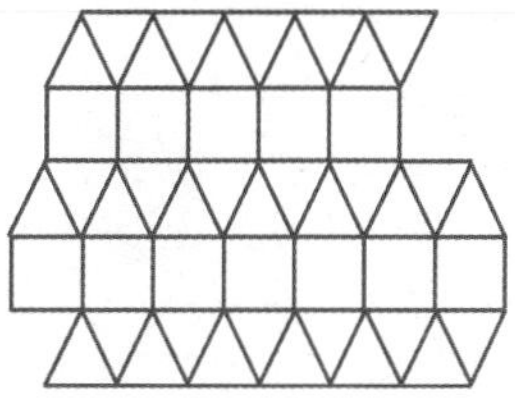

Sketch the tessellation represented by each code.

5. 3.3.3.3.3.3
6. 8.8.4
7. 6.3.6.3
8. 6.4.3.4

Use an indirect proof to show that each group of regular polygons cannot tessellate a plane.

9. regular pentagons and equilateral triangles
10. regular heptagons
11. regular decagons
12. regular octagons and equilateral triangles

Problem Solving

Ask students if they think the basic shape used in the following pattern can be used to tessellate the plane.

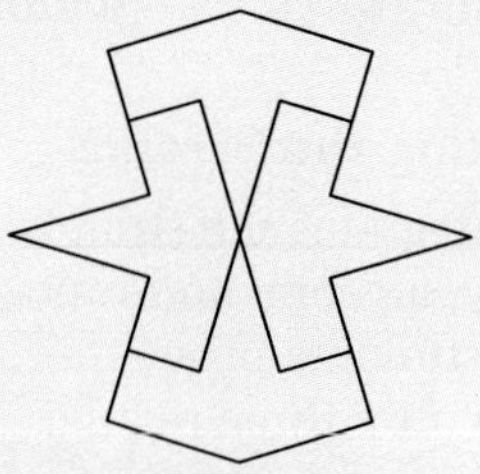

Challenge students to give a precise mathematical description of the basic shape.

Look Back

You may wish to have students discuss this Look Back in small groups. Group members can make sketches to help them decide which polygons will tessellate periodically.

APPLYING

Suggested Assignment

Day 1

Standard 1–12

Extended 1–13

Day 2

Standard 14–23, 25, 31–35

Extended 14–25, 27–35

Integrating the Strands

Algebra Exs. 13, 33, 34

Functions Exs. 13, 31, 33

Geometry Exs. 1–33, 35

Logic and Language Exs. 1, 13

Answers to Exercises and Problems

1. A regular tessellation uses only one kind of regular polygon, while a semiregular tessellation uses two or more regular polygons.
2. 3.3.3.3.3.3
3. 3.12.12
4. 3.3.3.4.4
5. 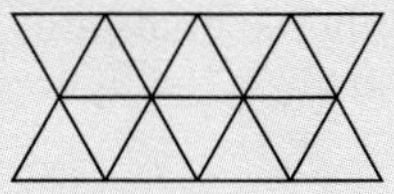
6. 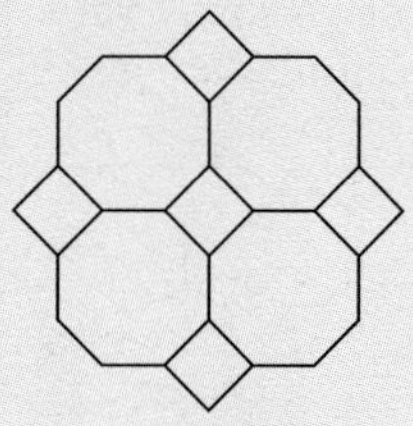
7. 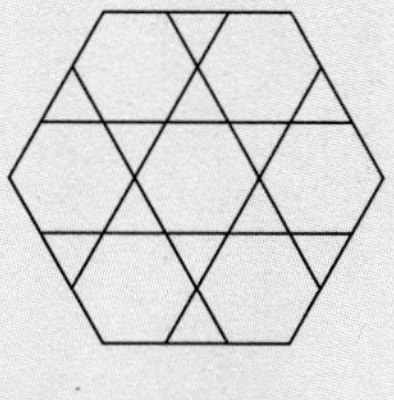
8. 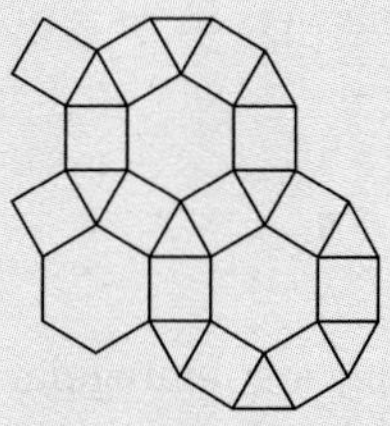
9. Suppose regular pentagons and equilateral triangles tessellate a plane. Let p = the number of regular pentagons at each vertex and let t = the number of equilateral triangles at each vertex. The measures of interior angles of regular pentagons and equilateral triangles are 108° and 60°, so $108p + 60t = 360$. The only solution with nonnegative integer values is (0, 6). Since there is no combination of regular pentagons and triangles that can surround a vertex, the assumption that regular pentagons and triangles tessellate a plane is false.

10–12. See answers in back of book.

Communication: Reading

Ex. 1 provides an opportunity to check whether students understand the difference between a regular tessellation and a semiregular tessellation.

Integrating the Strands

Ex. 13 integrates the strands of geometry, algebra, and functions as students prove that there are only three regular tessellations.

Communication: Drawing

Exs. 17–23 introduce the concept of a dual. To answer Exs. 18–23, students may wish to trace the diagrams on page 587 and work directly on the tracings.

13. You can prove that there are only three regular tessellations by considering the way regular polygons fit around a vertex. The measure of each angle of a regular n-gon is $\frac{(n-2)180°}{n}$.

a. Explain why the equation $\frac{(n-2)180q}{n} = 360$ is true for all regular tessellations in which q regular n-gons surround each vertex.

b. Use algebra to show that the equation in part (a) is equivalent to the equation $(n-2)(q-2) = 4$. Explain each step.

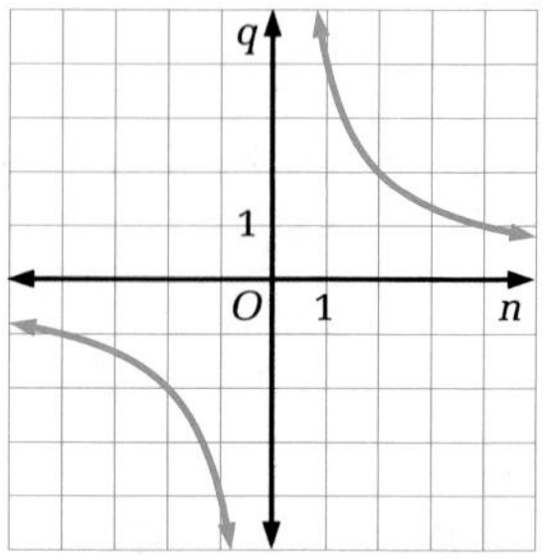

c. A graph of $nq = 4$ is shown. How should you translate it to obtain the graph of $(n-2)(q-2) = 4$? Make a careful drawing of the translated graph.

d. Name the points on the graph of $(n-2)(q-2) = 4$ whose coordinates are positive integers. How are these points related to regular tessellations?

e. Explain how the graph of $(n-2)(q-2) = 4$ shows that there are no more than three regular tessellations.

Which region in each semiregular tessellation is *not* a cell?

14.

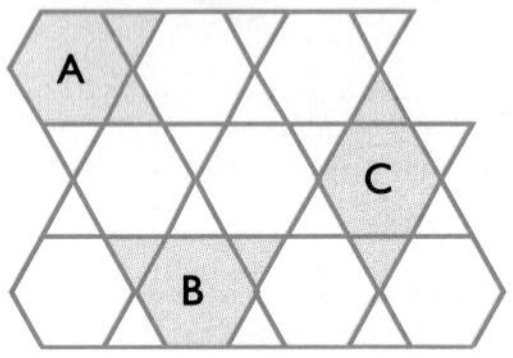

15.

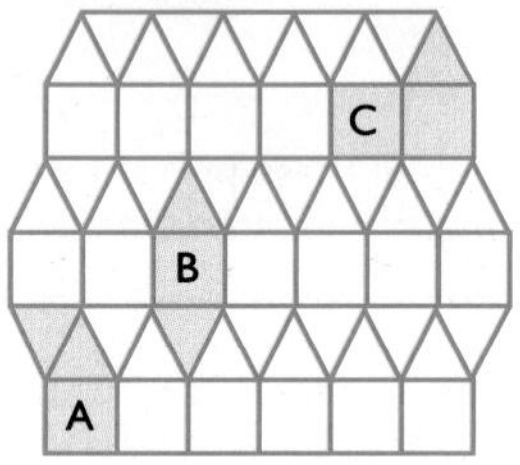

16.

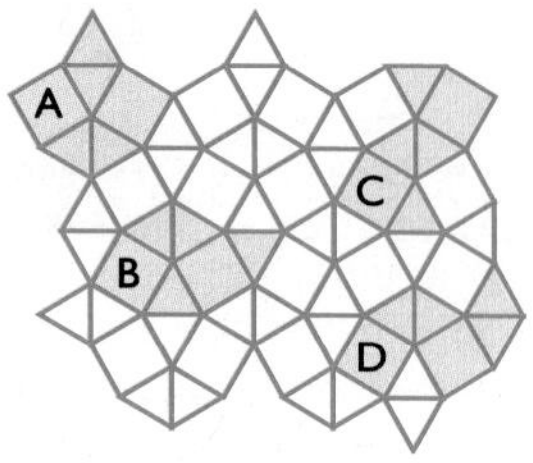

For Exercises 17–23, use this information. A *dual* is made by drawing segments that connect the centers of adjacent polygons in a regular or semiregular tessellation.

The *center of a regular polygon* is the point where the perpendicular bisectors of the sides meet.

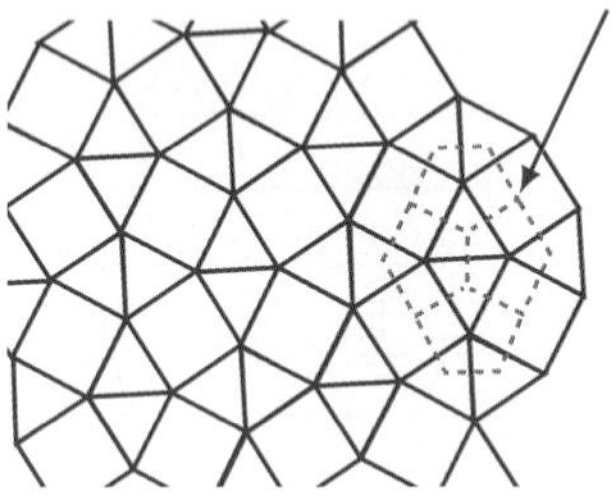

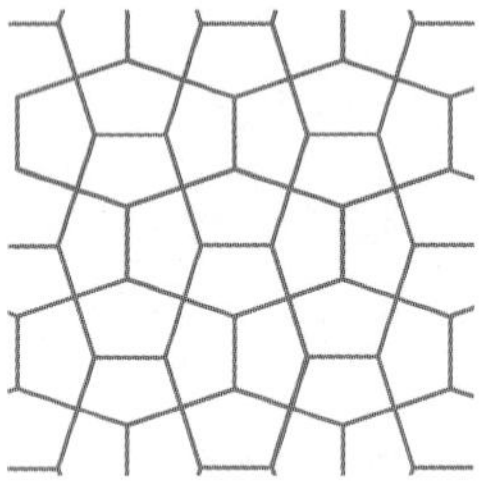

17. The 3.3.4.3.4 semiregular tessellation and its dual are shown.

a. Describe the polygons in the dual.

b. Regular pentagons cannot tessellate a plane. Explain why the polygons in the dual are able to tessellate a plane.

c. The dual is periodic. What is a cell for the dual?

BY THE WAY...

In 1975 Marjorie Rice read that mathematicians had found all possible tessellations made from nonregular pentagons. Over the next two years, without any special training in mathematics, she discovered four more nonregular pentagons that tessellate.

Answers to Exercises and Problems

13. See answers in back of book.

14. region B

15. region A

16. region A

17. a. They are nonregular pentagons.

b. Since the pentagons are not regular, it is possible for the sum of 3 angles to be 360°.

c. Answers may vary. Examples are given.

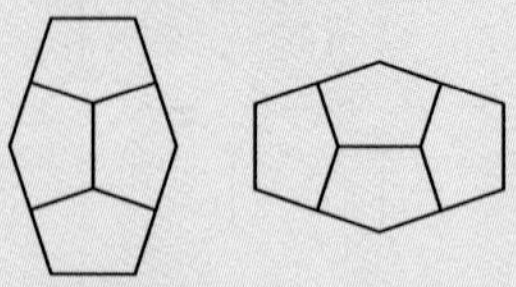

18.

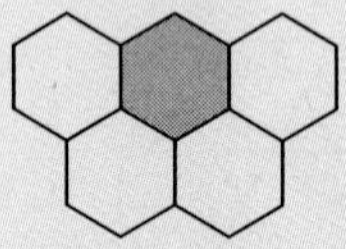

19.

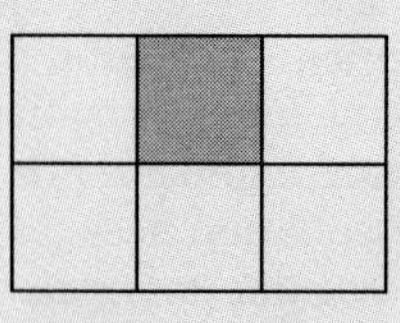

20.

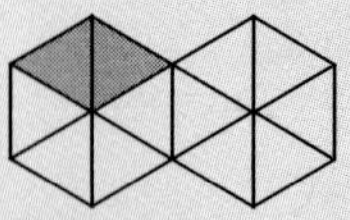

21–23. Only the duals are shown in the figures.

Draw the dual of each regular tessellation. If the dual is periodic, shade a cell.

18.

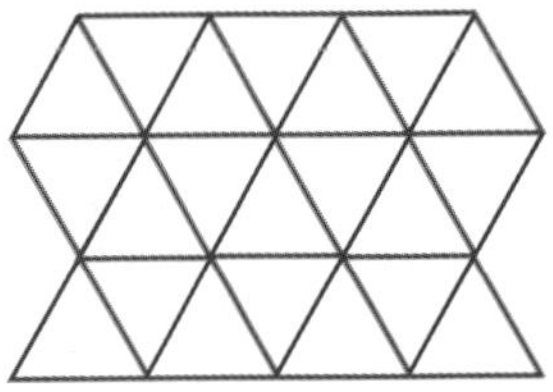

19.

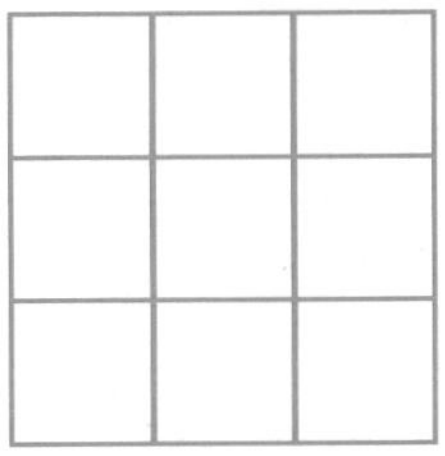

20.

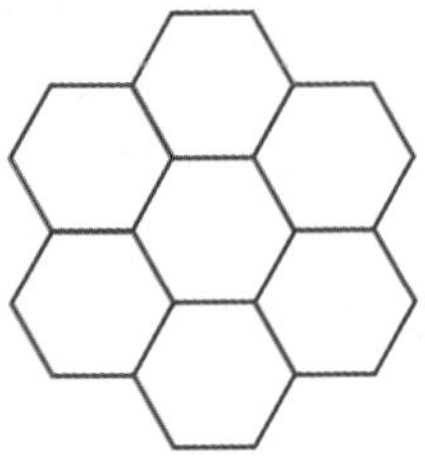

Draw each semiregular tessellation and its dual. If the dual is periodic, shade a cell.

21

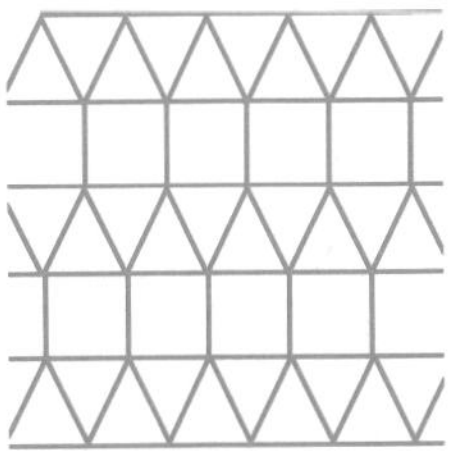

22.

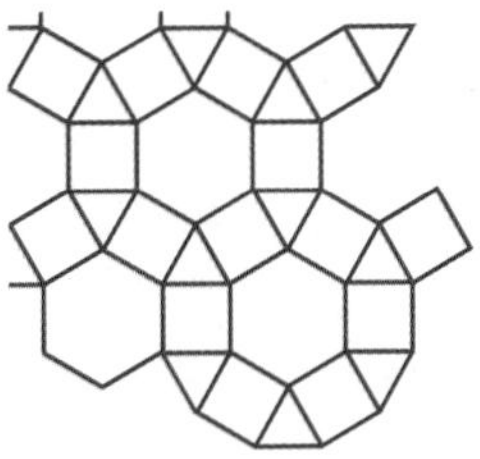

23.

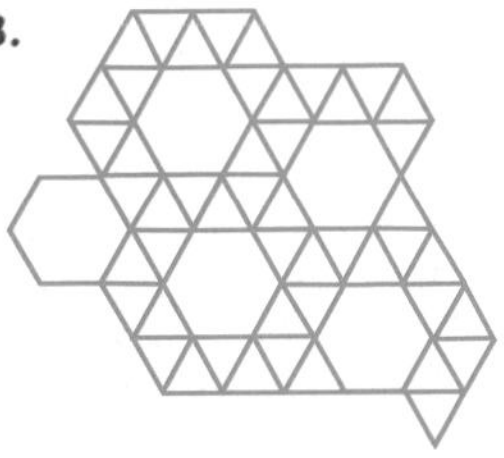

24. **Open-ended** Draw a nonperiodic tessellation using regular polygons.

25. Any quadrilateral will tessellate a plane.

a. Draw any quadrilateral and show how congruent copies can be placed around it to tessellate a plane.

b. Draw a concave quadrilateral like the one shown and show how congruent copies can be placed around it to tessellate a plane.

26. **Using Manipulatives** The three types of hexagons shown tessellate a plane.

a. Choose one of the three types of hexagons. Create an example using side lengths and angle measures that follow the rules shown for the type of hexagon you choose.

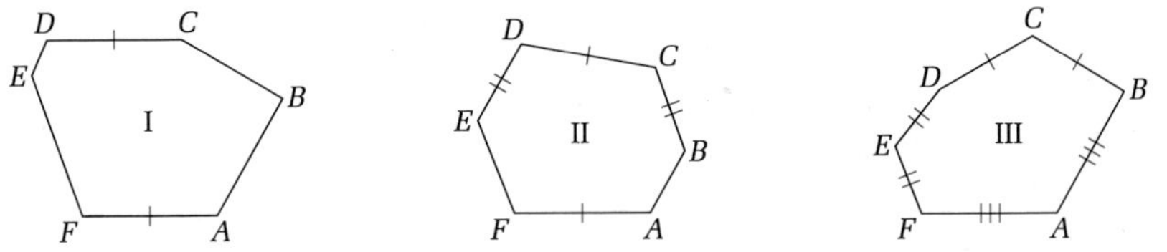

$m\angle A + m\angle B + m\angle C = 360°$ $m\angle A + m\angle B + m\angle D = 360°$ $m\angle A = m\angle C = m\angle E = 120°$

b. Cut out at least 12 copies of your hexagon from a piece of paper. Use your copies to make a periodic tessellation. Draw a sketch of it.

c. How many hexagons form a cell for your periodic tessellation?

27. **Open-ended** Use the method of Sample 2 to make a creative tessellation by modifying each pair of opposite sides of a parallelogram.

Using Manipulatives

For Ex. 26 (b), you may wish to have students cut out more than 12 copies of their hexagons. Students can use these copies to make a periodic tessellation on poster board, which can be presented to the class or turned in as an assignment.

Reasoning

Exs. 27 and 28 provide not only an opportunity for students to display their creativity, but also an opportunity to check whether they completely understand the algorithm presented in the Sample Response to Sample 2.

Answers to Exercises and Problems

21.

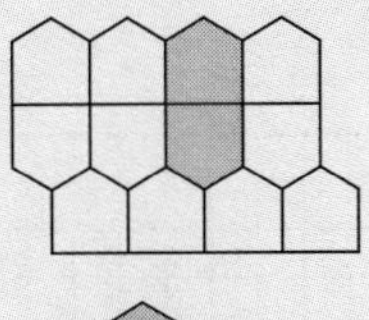

22.

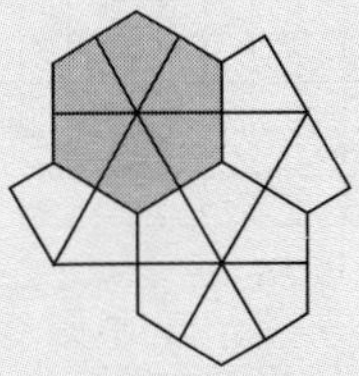

23.

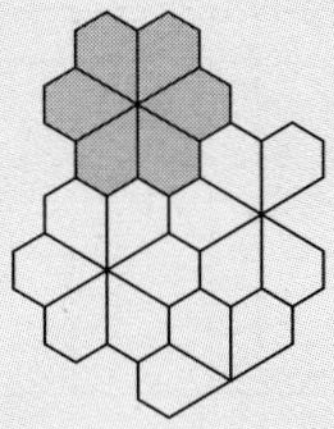

24. Answers may vary. One example is the 3.3.4.3.4 semiregular tessellation shown in Ex. 17.

25. a.

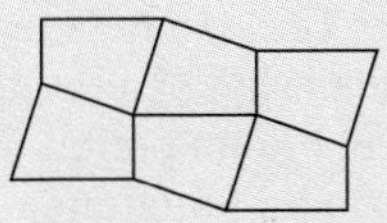

b. Drawings may vary. An example is given.

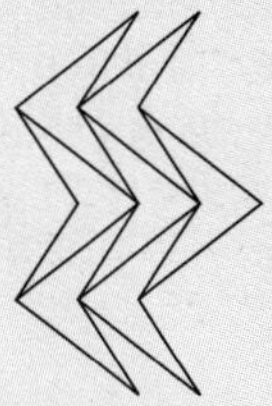

26. Answers may vary. Check students' work.

27. Answers may vary. An example is given.

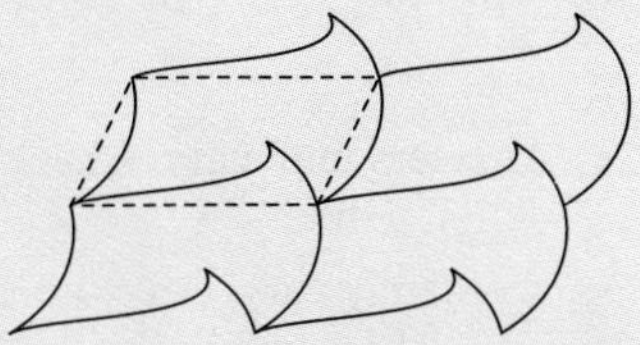

Communication: Writing

Ex. 31 gives students an opportunity to compare and contrast periodic tessellations, periodic functions, and border patterns and, thus, to tie together three of the major concepts of this unit.

Working on the Unit Project

For Ex. 35, students may want to use color to enhance their tessellations. The same tessellation can often be colored in various ways, which can make it appear quite different.

Quick Quiz (10-6 through 10-7)

See page 591.

See also Test 41 in the *Assessment Book.*

Practice 81 For use with Section 10-7

Name ______________ Date ______________

Practice 81
For use with Section 10-7

8. The sum of the angle measures at one vertex would have to be a multiple of 135°, but 360° is not evenly divisible by 135°. 9. There would have to be integer solutions of $108x + 90y = 360$, where $0 < x < 4$ and $0 < y < 4$. There are no such solutions. 10. There would have to be integer solutions of $144x + 90y = 360$, where $0 < x < 3$ and $0 < y < 4$. There are no such solutions. 11. There would have to be integer solutions of $120x + 108y = 360$, where $0 < x < 3$ and $0 < y < 4$. There are no such solutions.

For Exercises 1–3:

a. Write the code for each tessellation.

b. Tell whether each tessellation is *regular, semiregular,* or *nonregular.*

1. 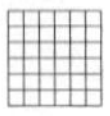 a. 4.4.4.4 b. regular

2.  a. 4.8.8 b. semiregular

3. 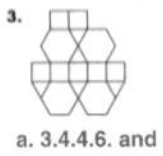 a. 3.4.4.6. and 3.6.3.6 b. nonregular

Sketch the tessellation represented by each code.

4. 3.3.3.3.3.3 **5.** 3.3.3.4.4 **6.** 3.12.12 **7.** 3.3.4.3.4

Exs. 4–7: For tessellations, see additional answers.

Use an indirect proof to show that each group of regular polygons cannot tessellate a plane. Exs. 8–11: See top of page.

8. octagons **9.** pentagons and squares

10. decagons and squares **11.** hexagons and pentagons

12. Show that the plane cannot be tessellated with triangles and n-sided polygons, where n is an integer greater than 12.

12. If $n > 12$, then meas. of one int. angle > 150°. Two such polygons could not share vertex with a triangle. One such polygon would need 3 triangles at the same vertex, but 4 triangles would not fit.

Tell whether each region in each semiregular tessellation is a cell.

13. 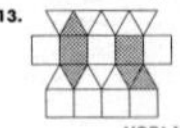 yes; yes

14.  yes; no

15. 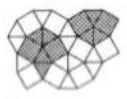 yes; no

16. ***Writing*** Suppose you are given a tessellation of the plane. Describe a procedure you might use for finding a cell. Also, describe a procedure for checking that the cluster of polygons you found is actually a cell.
Answers may vary. Check students' work.

Practice Bank, INTEGRATED MATHEMATICS 3

81

28. **Open-ended** Use the method of Sample 2 to make a creative tessellation by modifying each pair of opposite sides of a hexagon whose opposite sides are parallel and congruent.

29. Modify each side of a triangle so that the modified side has 180° rotational symmetry about the midpoint of the segment. Make a creative tessellation based on this method.

30. Modify each side of a quadrilateral so that the modified side has 180° rotational symmetry about the midpoint of the segment. Make a creative tessellation based on this method.

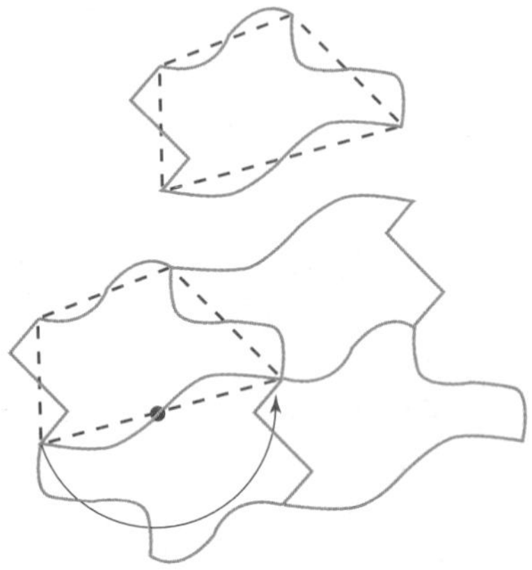

Ongoing ASSESSMENT

31. **Writing** Describe how periodic tessellations are like periodic functions and border patterns. How are they different?

Review PREVIEW

32. Use this shape. Create a border pattern having horizontal reflection symmetry. *(Section 10-6)*

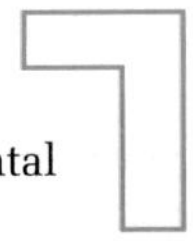

33. Transform the function $f(x) = x^2$ with a horizontal stretch by a factor of 4 followed by a translation of 5 units down. Write an equation for the transformed function and graph it. *(Section 9-5)*

34. Evaluate each logarithm. Round decimal answers to the nearest hundredth. *(Section 5-6)*

a. log 1000 **b.** log 1 **c.** ln 5

10 Working on the Unit Project

35. **a.** Make a polygonal tessellation to include in your scrapbook. Tell whether the tessellation is *regular* or *semiregular.*

b. Write a number code to describe your tessellation. Explain how the code describes the tessellation.

Answers to Exercises and Problems

28. Answers may vary. An example is given.

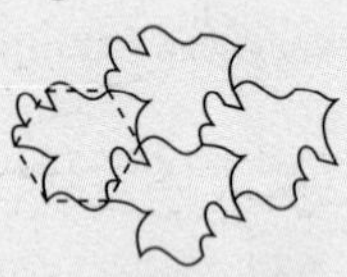

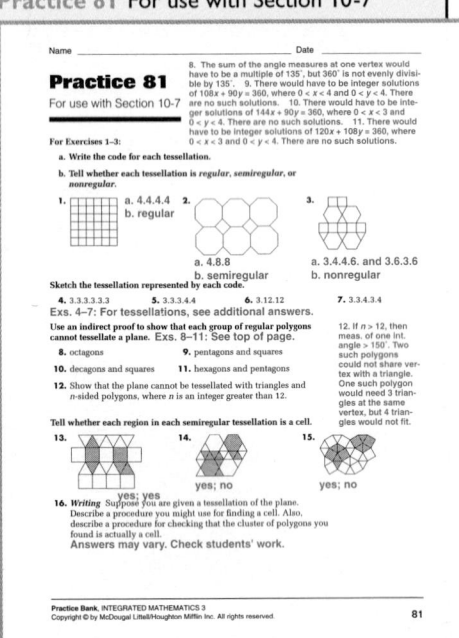

29, 30. Answers may vary. Check students' work.

31. Answers may vary. An example is given. They are similar in that they involve a repeating pattern. One cell of a tessellation, one design of a border pattern, or one period of a periodic function is repeated again and again. For border patterns and periodic functions, the patterns are translated horizontally. For tessellations, the translations are both horizontal and vertical.

32. 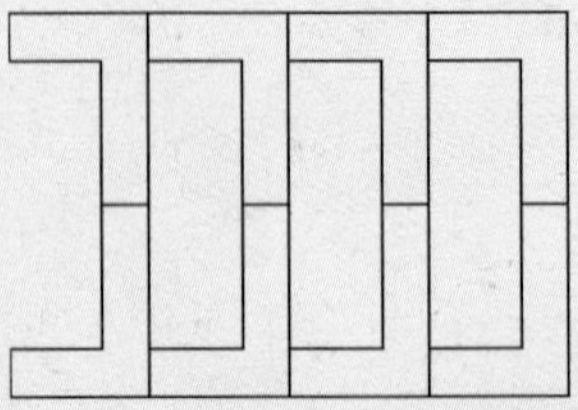

33. $y = \frac{x^2}{16} - 5$

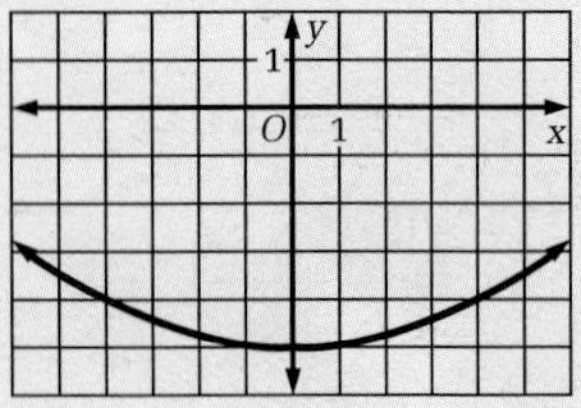

34. **a.** 3 **b.** 0 **c.** 1.61

35. Answers may vary.

Unit Project 10

Completing the Unit Project

Now you are ready to complete your scrapbook. Your project should include these things:

- **at least ten different types of graphs modeling specific situations, including the graphs you made in the "Working on the Unit Project" exercises**
- **a description and a picture of situations from your life, as well as local, national, and international topics**
- **a list of the functions you know how to graph**
- **an explanation of each graph in your scrapbook, including why the graph can be used to model a specific situation**
- **a description of how you gathered any data you needed to model a situation with a graph**
- **a description of the symmetry used to make the border design for the scrapbook cover**

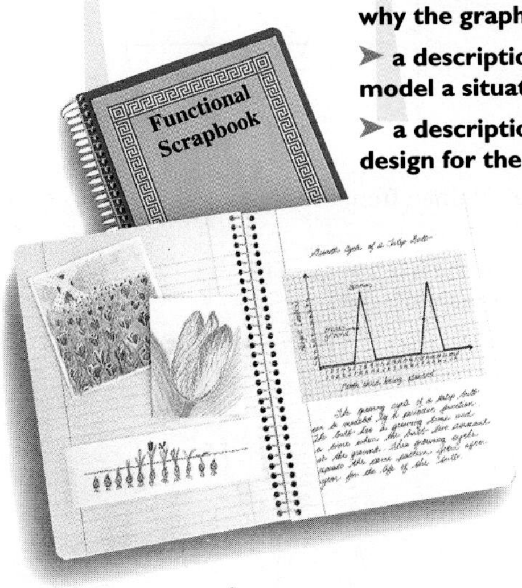

Look Back

Look at the scrapbooks made by other groups. Which graphs did they include? What situations did they describe? How would you change your scrapbook to make it better?

Alternative Projects

Project 1: **Predict Future Trends**

Think of a situation in nature that can be modeled by a sine wave. Gather data to develop an accurate model and graph your data.

Write an equation to model your data. Graph your equation and the data on the same set of axes Use your model to make predictions about the situation. Do you think your predictions are accurate? Explain.

Project 2: **Penrose Tiles**

Roger Penrose is a mathematician who worked with two kite-shaped tiles that always create nonperiodic tessellations when certain rules are followed. Research Penrose and the Penrose tiles that are named for him. How are the dimensions of the tiles related to the golden ratio? What kinds of nonperiodic tessellations can you make with Penrose tiles? Can you make periodic tessellations with the tiles if you ignore the rules Penrose used?

Assessment

A scoring rubric for the Unit Project can be found on pages 538 and 539 of this Teacher's Edition and also in the ***Project Book***.

Unit Support Materials

- *Practice Bank:* Cumulative Practice 82
- *Study Guide:* Unit 10 Review
- *Problem Bank:* Unifying Problem 10
- *Assessment Book:* Unit Tests 42 and 43; Cumulative Tests 44 and 45; Spanish Unit Tests; Alternative Assessment
- *Test Generator* software with Test Bank
- *Teacher's Resources for Transfer Students*

Quick Quiz (10-6 through 10-7)

1. Describe how to construct a border pattern with vertical reflection symmetry. [10-6] Reflect an asymmetric figure over a vertical line. Translate the pair repeatedly.
2. Use this asymmetric figure.

Create a border pattern using glide reflection. [10-6]

The common floor tile pattern shown below is made with squares and irregular octagons.

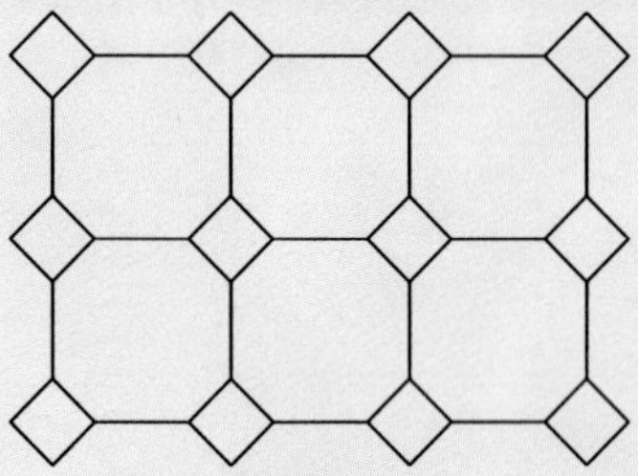

3. Write the code for this tessellation. [10-7] 8.4.8
4. Draw a cell for this tessellation. [10-7]

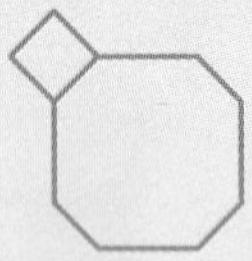

5. What are the interior angle measures of the octagons? [10-7] 135°

Unit 10 REVIEW AND ASSESSMENT

Which of the terms *constant, increasing, decreasing, periodic, damping,* or *saturating* best describes the graph of each function? 10-1

1. $y = \sin\theta$ 2. $y = 5x$ 3. $y = \frac{2}{3}$ 4. $y = (2.1)(0.6)^x$
5. Compare the rates of growth in the first quadrant of $y = 4x$, $y = x^4$, and $y = 4^x$.
6. Sketch a periodic graph with amplitude 4, period 3, and an axis with the equation $y = 4$. 10-2
7. For the graph shown, identify the following:
   a. a cycle
   b. the period
   c. the frequency
   d. the amplitude
   e. an equation of the axis

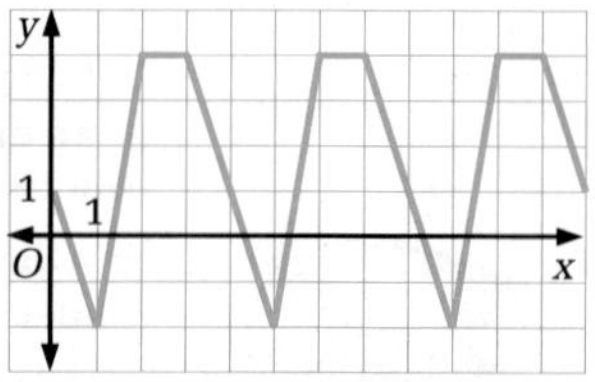

8. **Writing** What is a sine wave and how can it be obtained from the graph of the sine or the cosine function? 10-3

For each equation, find the amplitude, the period, and the frequency of its graph. Then sketch the graph without using graphing technology. 10-4

9. $y = 3\sin(\theta - 45°) + 1$ 10. $y = \cos 2\theta - 4$ 11. $y = 0.5\sin 3\theta + 2$

Write an equation for each graph.

12.

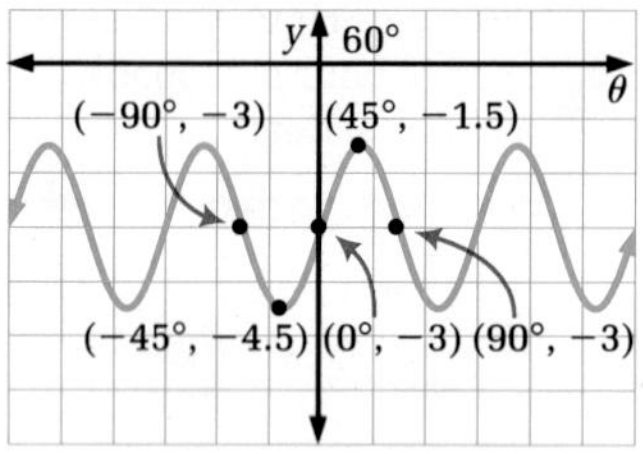

13.

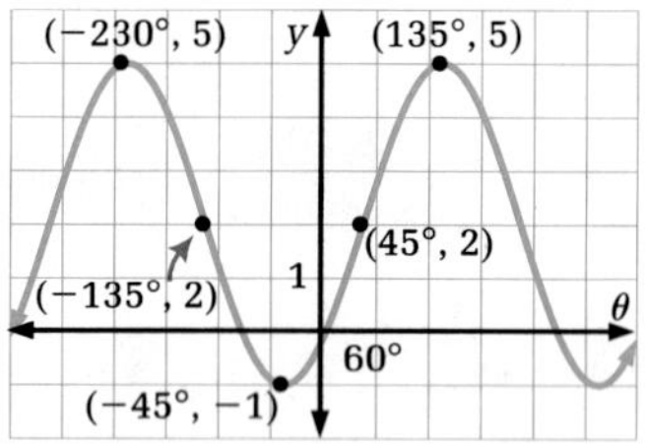

Use the table for Exercise 14. 10-5

Natural Gas Use by United States Electricity Utilities in 1992

Month	Jan.	Feb.	Mar.	Apr.	May	June	July	Aug.	Sept.	Oct.	Nov.	Dec.
Natural gas use (billion cubic ft)	169	170	208	229	236	266	334	303	274	213	189	176

14. a. Graph the data in the table.
    b. Write an equation to model the data. Let x = the month expressed as a number. (Jan. = 1, Feb. = 2, . . .)
    c. What type of function did you write an equation for in part (b)? Why did you choose this type?

Answers to Unit 10 Review and Assessment

1. periodic
2. increasing
3. constant
4. decreasing
5. $y = 4^x$ grows fastest, followed by $y = x^4$, and then by $y = 4x$.
6. Answers may vary. An example is given.

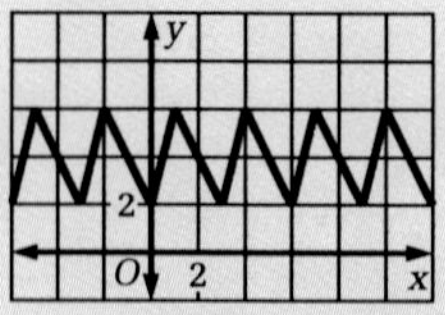

7. a. Answers may vary. An example is given. the part of the graph from 0 to 4 on the horizontal axis
   b. 4
   c. $\frac{1}{4}$
   d. 3
   e. $y = 1$
8. A sine wave is any graph that has an equation of the form $y = A\sin B(\theta - C) + D$ or $y = A\cos B(\theta - C) + D$. It can be obtained by transforming the graph of $y = \sin\theta$ or $y = \cos\theta$.

Classify each border pattern by its symmetry. 10-6

15.

16.

17.

18. **Open-ended** Draw an asymmetric figure. Use it to create three different border patterns that each have a different type of symmetry. Describe the symmetry of each pattern.

For Exercises 19 and 20, tell whether each group of polygons can tessellate a plane. 10-7

19. squares and equilateral triangles

20. regular decagons

21. Use an indirect proof to show that regular 9-gons and squares cannot tessellate a plane.

22. **Self-evaluation** Describe the process of modeling data with a sine wave. How have you helped yourself to be more successful with these steps?

23. **Group Activity** Work in a group of three students.

a. Write the names of the seven types of border patterns on separate slips of paper. Fold each slip of paper in half. Each member of the group should pick two slips of paper at random. Set aside the seventh slip.

b. Each of you should draw an asymmetric figure on a piece of paper. Pass your drawing to the person on your right.

c. Choose one of the border patterns you selected in part (a). Draw the chosen border pattern using the figure you receive. Pass your border design and the original figure to the person on your right.

d. Repeat part (c) using the other border pattern you selected.

e. Identify the types of border patterns made from your asymmetric figure. Check your answers with the members of your group.

Answers to Unit 10 Review and Assessment

9–14. See answers in back of book.

15. vertical

16. vertical and 180° rotation

17. horizontal, vertical, and 180° rotation

18. Answers may vary. Check students' work.

19. Yes.

20. No.

21. Suppose regular 9-gons and squares tessellate a plane. Let n = the number of regular 9-gons at each vertex and let s = the number of squares at each vertex. The measures of interior angles of regular 9-gons and squares are 140° and 90°, so $140n + 90s = 360$. There is no solution with nonnegative integer values. Since there is no number of regular 9-gons and squares that surround a vertex, the assumption that 9-gons and squares tessellate a plane is false.

22. Plot the points to determine whether the graph looks like a sine wave. If so, determine the amplitude A using the formula $A = \frac{\text{maximum value} - \text{minimum value}}{2}$. Decide whether to use a sine or a cosine function. Find the period of the graph and use it to determine the value of B. Find the horizontal translation C. Find the vertical translation D using the formula $D = \frac{\text{maximum value} + \text{minimum value}}{2}$. Substitute the values of A, B, C, and D in the general equation. Answers may vary.

23. Answers may vary.

IDEAS AND (FORMULAS)

ALGEBRA

- The terms *constant*, *increasing*, *decreasing*, *periodic*, *saturating*, and *damping* can be used to describe the graph of a function. *(p. 543)*
- You can use tables or graphs to compare the rates of growth of functions. *(p. 544)*
- You can find the period, the frequency, the amplitude, and the equation of the axis of a periodic graph by looking at the graph. *(p. 548)*

period = the length of one cycle

frequency = the reciprocal of the period

$$\text{amplitude} = \frac{\text{maximum value} - \text{minimum value}}{2}$$

$$\text{equation of the axis: } y = \frac{\text{maximum value} + \text{minimum value}}{2}$$

- You can make predictions about the future behavior of periodic events by analyzing a part of the graph that includes at least one complete cycle. *(p. 568)*

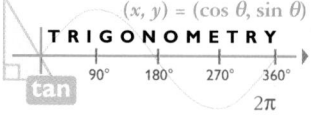

TRIGONOMETRY

- For an object moving counterclockwise (forward) or clockwise (backward) around a unit circle, the sine function gives the vertical position of the object. *(p. 555)*
- The graph of the sine function has an amplitude of 1 and a period of 360°. *(p. 556)*
- For an object moving counterclockwise (forward) or clockwise (backward) around a unit circle, the cosine function gives the horizontal position of the object. *(p. 556)*
- The graph of the cosine function has an amplitude of 1 and a period of 360°. *(p. 556)*
- The graph of the cosine function is a 90° horizontal translation of the graph of the sine function. *(p. 557)*
- The general form for the equation of a sine wave is $y = A \sin B(\theta - C) + D$ or $y = A \cos B(\theta - C) + D$. The graph of each of these functions has the following properties:
  - $|A|$ is the amplitude.
  - $\frac{360°}{|B|}$ is the period.
  - The graph of the parent function is translated horizontally C units.
  - The equation of the axis is $y = D$. *(p. 561)*

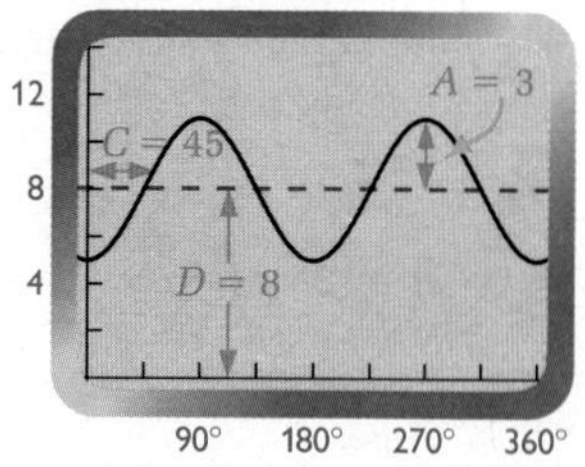

592

- The graph of a function of the form $y = A \sin B(\theta - C) + D$ can be obtained from the graph of $y = \sin \theta$ by the following transformations.
  - The graph is vertically stretched or shrunk by a factor of $|A|$. If $A < 0$, then there is also a reflection over the θ-axis.
  - The graph is horizontally stretched or shrunk by a factor of $\frac{1}{|B|}$. If $B < 0$, then there is also a reflection over the y-axis.
  - The graph is translated horizontally C units.
  - The graph is translated vertically D units.

  The same transformations apply to the graphs of $y = A \cos B(\theta - C) + D$ and $y = \cos \theta$. *(p. 561)*
- You can determine the period, the amplitude, the horizontal translation, and the vertical translation of a sine wave from its graph. *(p. 562)*
- You can write an equation for the graph of a sine wave by using the general model $y = A \sin B(\theta - C) + D$ and finding the values for A, B, C, and D. *(p. 562)*
- The general model for a sine wave can be used to write an equation that models a set of periodic data. *(p. 568)*

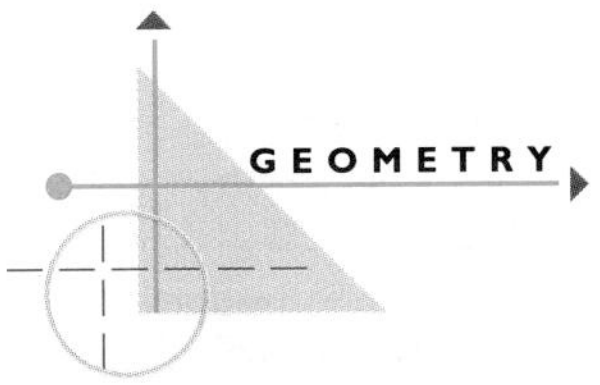

- There are seven types of border patterns. *(p. 577)*
- Border patterns are periodic. *(p. 577)*
- All border patterns are created using some combination of vertical reflection, horizontal reflection, and 180° rotation with repeated translations. *(p. 578)*
- You can use indirect proof to show that a group of polygons cannot tessellate a plane. *(p. 583)*
- Some nonregular polygons tessellate a plane, for example, all triangles and all parallelograms. *(p. 584)*
- You can make a creative tessellation by modifying the sides of a polygon that tessellates a plane. *(pp. 584–585)*

Key Terms

- **periodic** (p. 543)
- **cycle** (p. 548)
- **amplitude** (p. 548)
- **saturating** (p. 543)
- **period** (p. 548)
- **axis of a periodic function** (p. 548)
- **damping** (p. 543)
- **frequency** (p. 548)
- **sine function** (p. 555)
- **cosine function** (p. 556)
- **semiregular tessellation** (p. 582)
- **tessellation** (p. 582)
- **periodic tessellation** (p. 584)
- **regular tessellation** (p. 582)
- **cell** (p. 584)

Quick Quiz (10-1 through 10-5)

1. Compare the rates of growth in the first quadrant of the functions $y = 2$, $y = 2x$, and $y = 2 \sin x$. [10-1]
$y = 2x$ grows the fastest, $y = 2$ has a zero rate of growth, and the rate of growth of $y = 2 \sin x$ is constantly changing.

2. Find the period, frequency, amplitude, and axis of the graph below. [10-2]

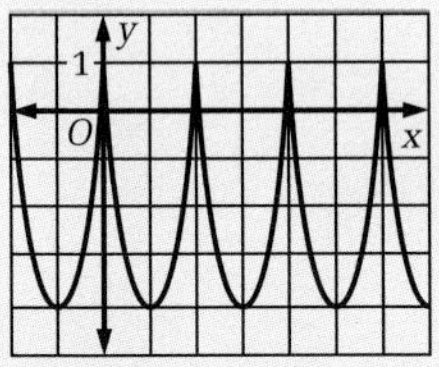

2; $\frac{1}{2}$; 2.5; $y = -1.5$

3. Why is the graph of the cosine function positive from 0° to 90° and from 270° to 360°. [10-3]
The cosine function gives the horizontal position of an object moving around the unit circle. From 0° to 90° and from 270° to 360°, the object is in Quadrants I and IV, respectively. This is where the horizontal position is positive.

4. Write an equation for the graph below using a sine function. [10-4]

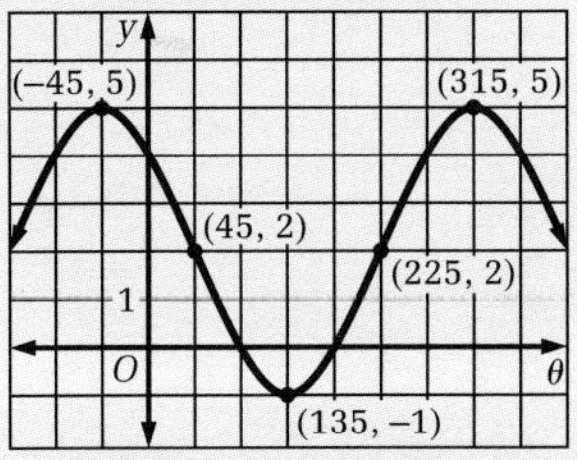

$y = -3\sin(\theta - 45) + 2$

5. A Ferris wheel ride rotates at a rate of 1.5 rev/min. When creating a model for this ride, what is the value of B in the formula $y = A \sin B(x - C) + D$? [10-5] **9**

Contents of Student Resources

Technology Handbook

This handbook introduces you to the basic features of most graphics calculators. Check your calculator's instruction manual for specific keystrokes and any details not provided here.

Performing Calculations

➤ The Keyboard

Look closely at your calculator's keyboard. Notice that most keys serve more than one purpose. Each key is labeled with its primary purpose, and labels for any secondary purposes appear somewhere near the key. You may need to press **2nd**, **SHIFT**, or **ALPHA** to use a key for a secondary purpose.

Examples of using the **X^2** key:

Press **X^2** to square a number.

Press **2nd** and then **X^2** to take a square root.

Press **ALPHA** and then **X^2** to get the letter I.

➤ The Home Screen

Your calculator has a "home screen" where you can do calculations. You can usually enter a calculation on a graphics calculator just as you would write it on a piece of paper.

Shown below are other things to remember as you enter calculations on your graphics calculator.

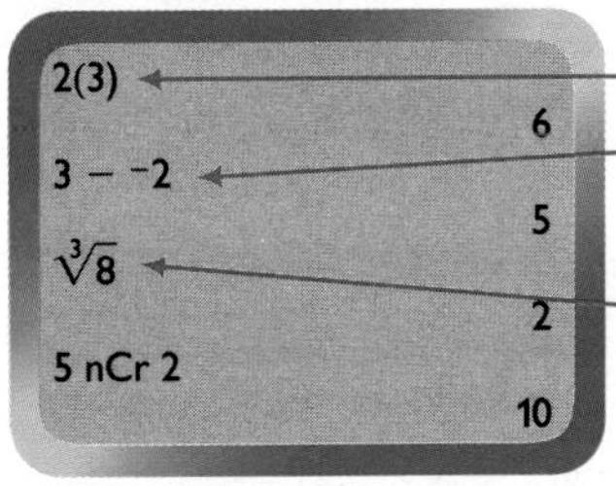

The calculator may recognize implied multiplication.

The calculator has a subtraction key, **−**, and a negation key, **(−)**. If you use these incorrectly, you will get an error message.

To perform an operation that is not often used, such as to calculate a cube root or $_nC_r$, you may need to press the **MATH** key to display a menu of operations.

Try This

1. Use your calculator to find the value of each expression.

   a. $\sqrt{12.25}$ b. 2^5 c. $\sin 75°$ d. $\cos^{-1} 0.5$

   e. $\sqrt[3]{64}$ f. 5! g. ${}_5P_2$ h. ${}_{12}C_6$

Displaying Graphs

➤ The Viewing Window

Think of the calculator screen as a "viewing window" that lets you look at a portion of the coordinate plane.

On many calculators, the standard viewing window uses values from –10 to 10 on both the *x*- and *y*-axes. You can adjust the viewing window by pressing the RANGE or WINDOW key and entering new values for the window variables.

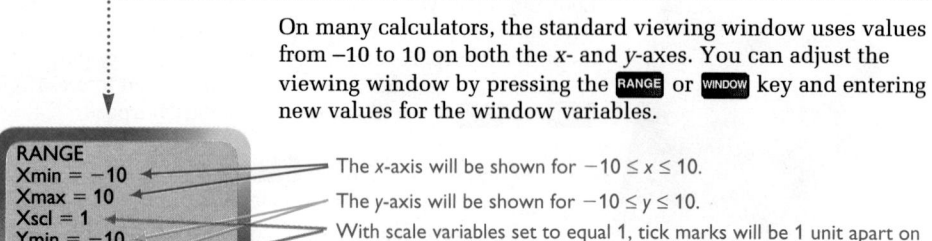

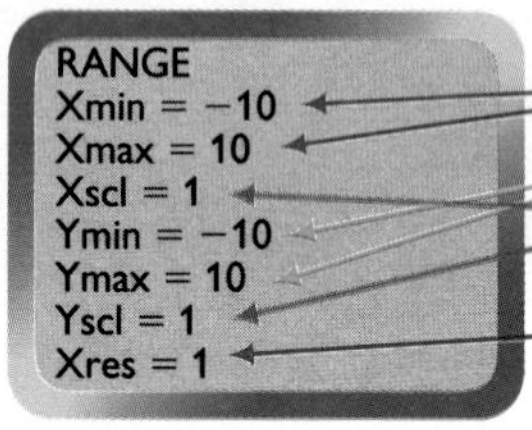

➤ Entering and Graphing a Function

To graph a function, enter its equation in the form $y = \ldots$. If the equation does not use *x* and *y*, rewrite the equation. Let x = the control variable and y = the dependent variable. Set the variables for an appropriate viewing window. The graph of $y = \frac{1}{2}x + 3$ is shown using the standard viewing window.

Use parentheses. If you enter $y = 1/2x + 3$ instead, the calculator may interpret the equation as $y = \frac{1}{2x} + 3$.

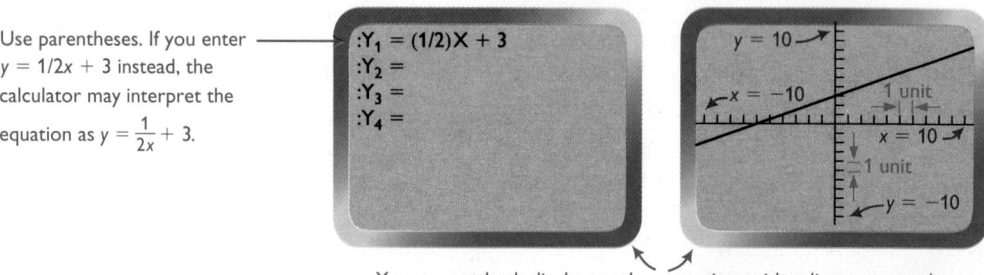

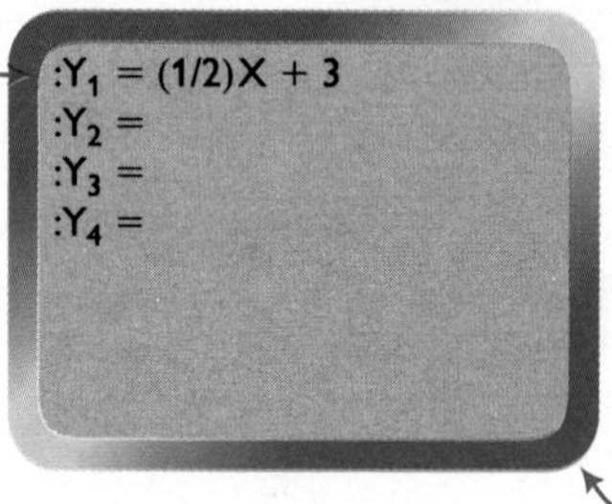

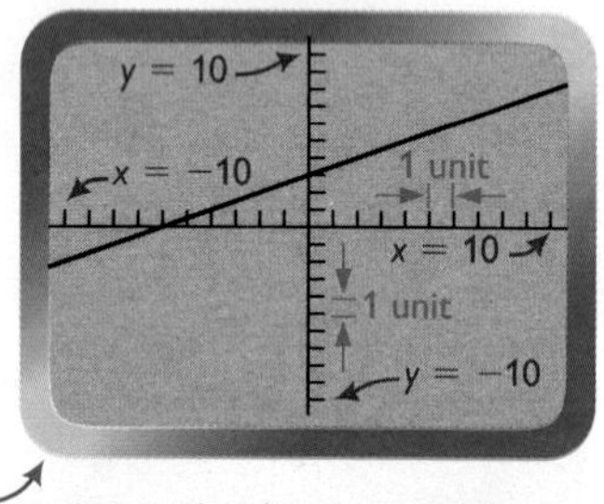

You can see both displays at the same time with split-screen mode.

To enter some radical functions, you may need to press Y= and then MATH to find and enter the menu items $\sqrt[3]{\ }$ or $\sqrt[x]{\ }$. Remember to enter $y = \sqrt[3]{x-1}$ as $y = \sqrt[3]{(x-1)}$, so the calculator does not interpret the equation as $y = \sqrt[3]{x} - 1$.

Answers to Try This

1. a. 3.5
   b. 32
   c. about 0.966
   d. 60°
   e. 4
   f. 120
   g. 20
   h. 924

➤ Squaring the Screen

A "square screen" is a viewing window with equal unit spacing on the two axes. For example, the graph of $y = x$ is shown for two different windows.

Standard Viewing Window

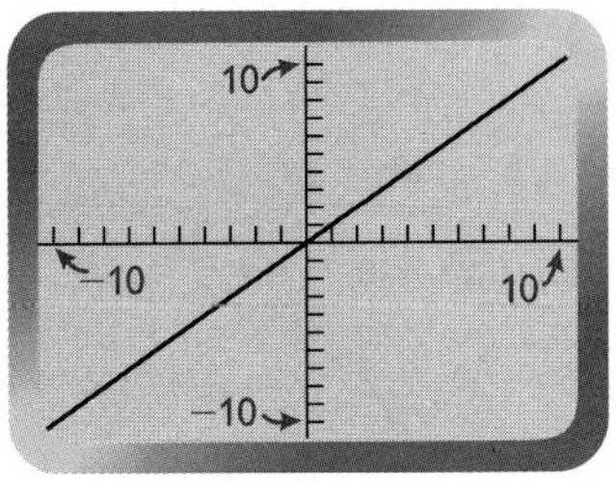

Square Screen Window

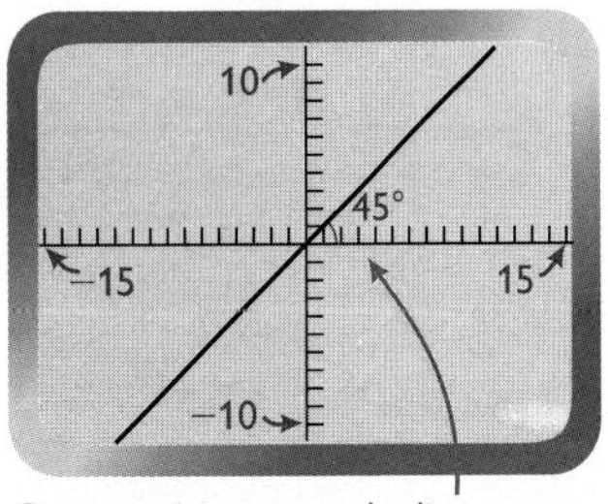

On a square screen, the line $y = x$ makes a 45° angle with the x-axis.

On many graphics calculators, the ratio of the screen's height to its width is about 2 to 3. Your calculator may have a feature that gives you a square screen. If not, choose values for the window variables that make the "length" of the y-axis about two-thirds the "length" of the x-axis:

$$(\text{Ymax} - \text{Ymin}) \approx \frac{2}{3}(\text{Xmax} - \text{Xmin})$$

Try This

2. Enter and graph each equation separately. Use the standard viewing window.

a. $y = 3x + 1$ **b.** $y = 2^x$ **c.** $y = \left|\frac{1}{x}\right|$ **d.** $y = \sqrt[3]{2x + 2}$

3. Find a good viewing window for the graph of $y = 65 - 3x$. Be sure your window shows where the graph crosses both axes.

4. Find a viewing window that will allow you to graph these two lines so that they appear to be perpendicular: $y = 2x + 1$ and $y = -0.5x - 2$.

Reading a Graph

➤ The TRACE Feature

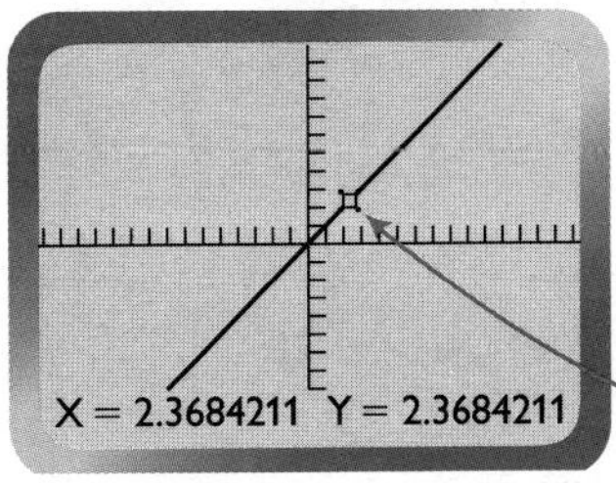

After a graph is displayed, you can use the calculator's TRACE feature. When you press TRACE, a flashing cursor appears on the graph. The x- and y-coordinates of the cursor's location are shown at the bottom of the screen. Press the left- and right-arrow keys to move the TRACE cursor along the graph.

The TRACE cursor is at the point (2.3684211, 2.3684211) on the graph of $y = x$.

Answers to Try This

2. a.

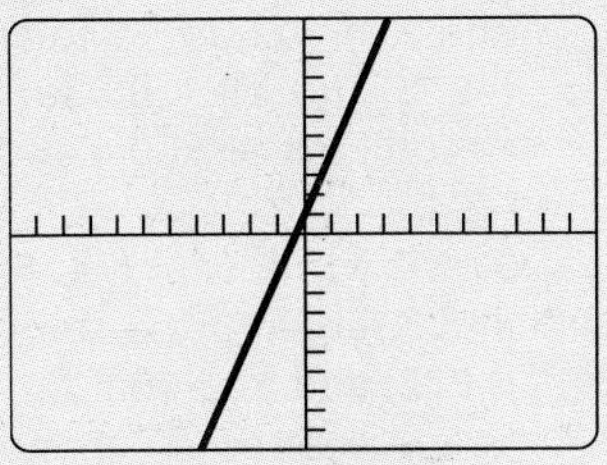

b.

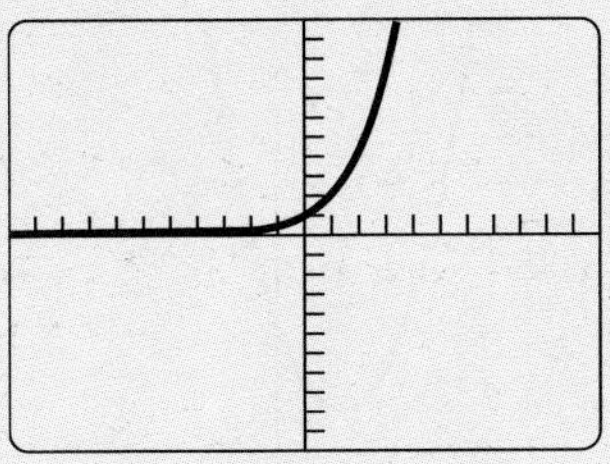

c.

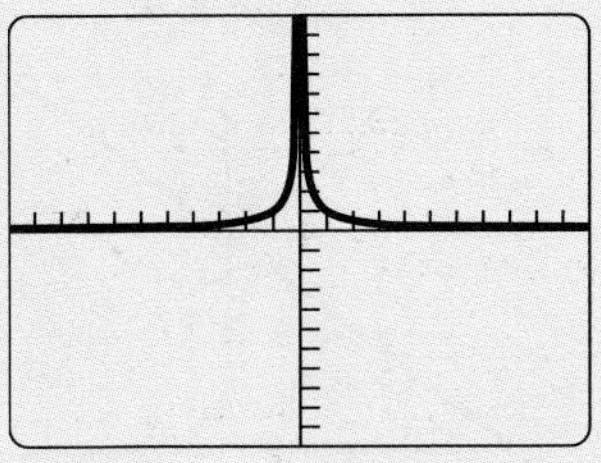

d.

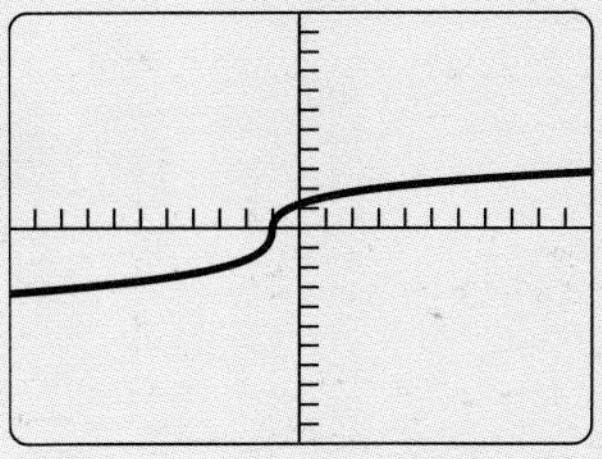

3. Answers may vary. An example is given. [–5, 30] and Xscl=5 for x, [–10, 75] and YSCL=5 for y.

4. Answers may vary. An example is given. [–7.5, 7.5] and Xscl=1 for x, [–5, 5] and YSCL=1 for y.

Suppose you want to find the radius of a circular hole with an area of 20 in². The formula for the area of a circle is $A = \pi r^2$. Rewrite the equation using x and y. Graph the equation $y = \pi x^2$.

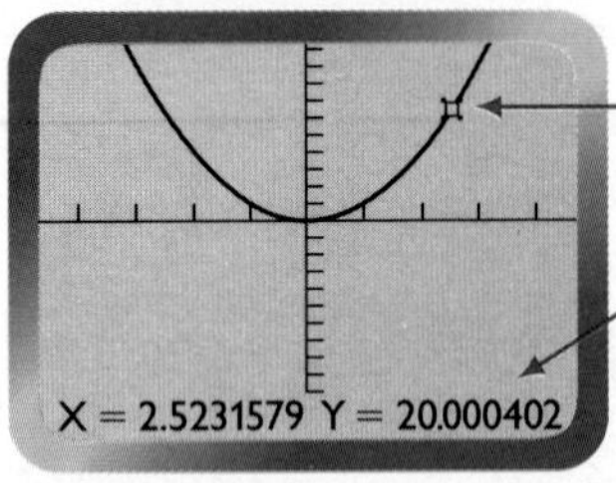

Using the TRACE key, move the cursor along the graph until the y-value is approximately equal to 20. This will give you the value of x that is the radius of the circular hole.

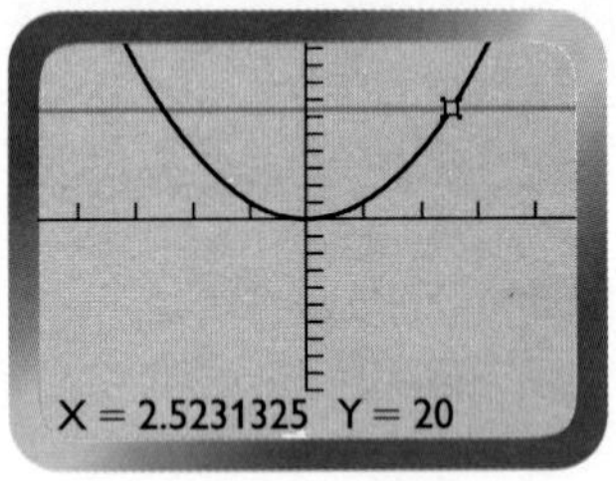

Another way to solve the problem is to graph two equations on the same screen: $y = \pi x^2$ and $y = 20$. Use TRACE to move the cursor to the intersection of the two lines to find your answer.

Some calculators have an intersection feature as a menu item that will calculate and display the coordinates of the point of intersection of two graphs.

➤ Friendly Windows

As you press the right-arrow key while tracing a graph, the x-coordinate may increase by "unfriendly" increments.

Your calculator may allow you to control the x-increment, ΔX, directly. If not, you can control it indirectly by choosing an appropriate Xmax for a given Xmin. For example, on a TI-82 graphics calculator, choose Xmax so that

$$\text{Xmax} = \text{Xmin} + 94\Delta X.$$

This number depends upon the calculator you are using.

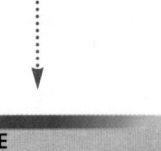

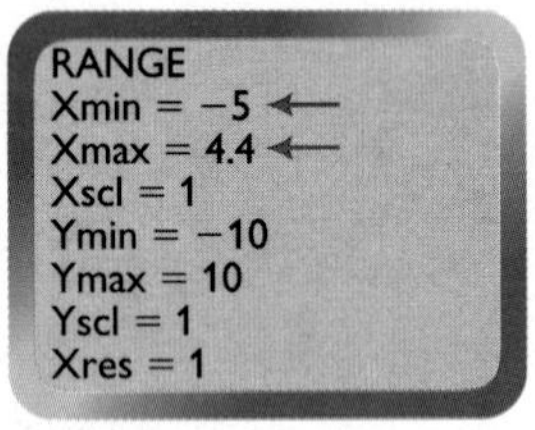

Suppose you want ΔX to equal 0.1. If Xmin $= -5$, then set Xmax equal to $-5 + 94(0.1)$, or 4.4. This gives a "friendly window" where the TRACE cursor's x-coordinate will increase by 0.1 each time you press the right-arrow key.

Try This

5. Graph $y = 2\pi x$. Find the value of x when $y = 10$.

6. Graph the equation $y = x^2 + 2x - 1$. Choose a friendly window where $\Delta X = 0.1$. Use the TRACE feature to determine, to the nearest tenth, the x-coordinate of each of the two points where the graph crosses the x-axis.

Answers to Try This

5. about 1.59

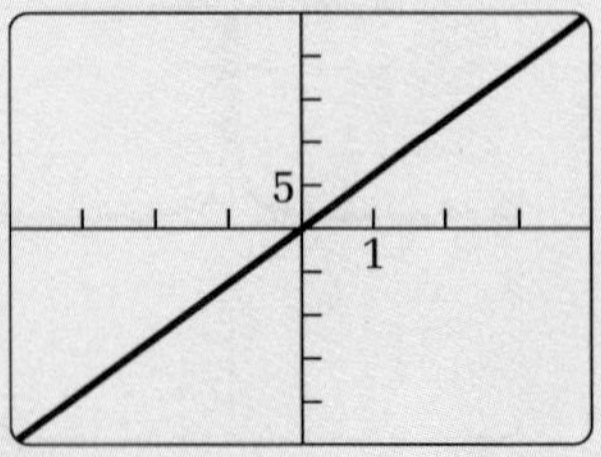

6.

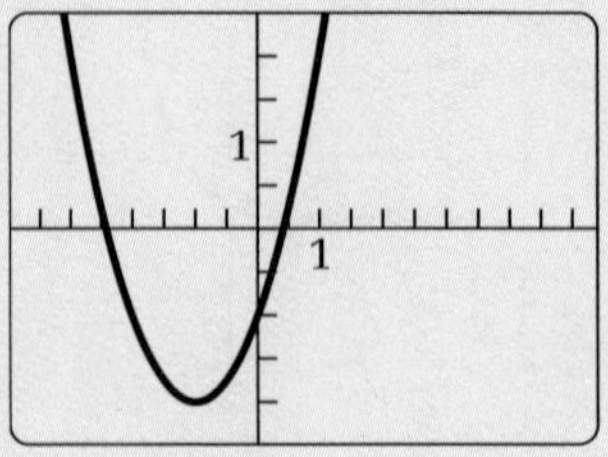

TI-82: window of [–4, 4.4] and Xscl=0.5 for x, and [–2, 2] and Yscl=0.5 for y. Graph crosses at 0.4 and –2.4.
TI-81: window of [–4, 5.5] and Xscl=0.5 for x, and [–2, 2] and Yscl=0.5 for y. Graph crosses at 0.4 and –2.4.

➤ The TABLE Feature

Instead of tracing the graph of an equation, you may wish to examine a table of values. Not all calculators have a TABLE feature. Check to see if yours does.

The screen shows a table of values for $y = x^2 + 2x - 1$. Here the value of x increases from a minimum of 0 in steps of 0.1. Some calculators have a table set-up feature that allows you to set the table minimum and the change in the control variable.

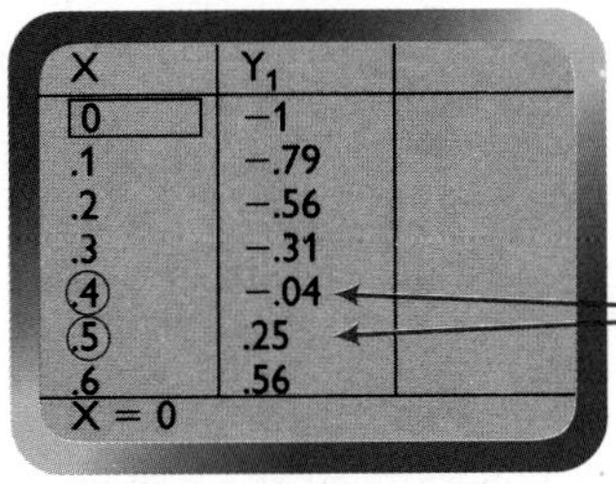

Notice that the y-values change sign between $x = 0.4$ and $x = 0.5$.

Taking a Closer Look at a Graph

➤ The ZOOM Feature

Suppose you are interested in the point where the graph of the equation $y = x^2 + 2x - 1$ crosses the positive x-axis.

To get a closer look at the point of interest, you can use the ZOOM feature. Your calculator may have more than one way to zoom. A common way is to put a "zoom box" around the point. The calculator will then draw what's inside the box at full-screen size.

Define a "zoom box" . . .

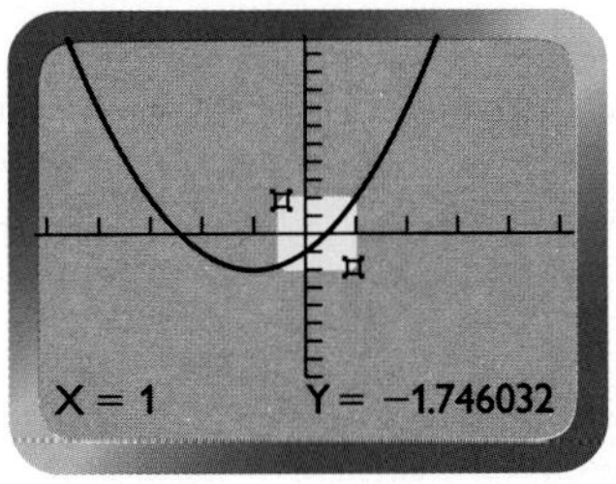

You create a "zoom box" by fixing first one corner and then the opposite corner of the box. (See your calculator's manual.)

and then zoom.

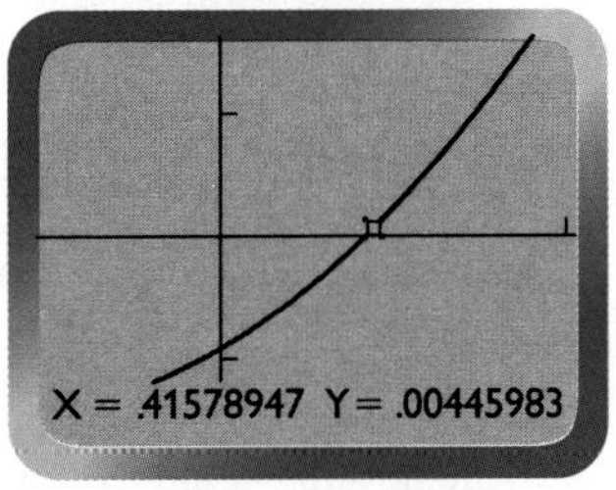

Tracing reveals that the graph crosses the x-axis between $x = 0.41$ and $x = 0.42$. Repeat zooming and tracing.

When you graph a rational equation using the standard viewing window, the display may be unclear near asymptotes of the graph.

The graph is unclear.

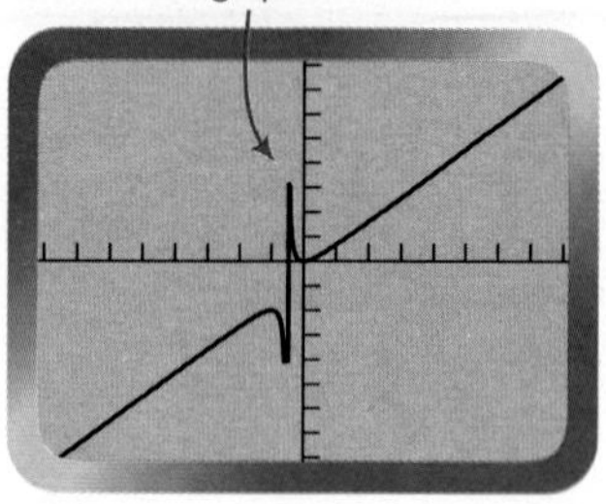

Zoom in to see the graph of $y = \frac{2x^2}{2x + 1}$ better near the origin.

$x = -\frac{1}{2}$ is an asymptote.

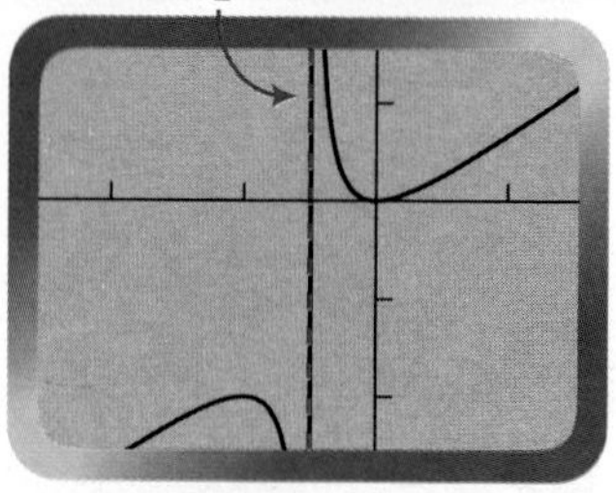

On many graphics calculators you can use ▲ and ▼ to move between graphs. This process helps you solve the system:

$$x + y = 3$$
$$x = \frac{2}{3}y - 5$$

Rewrite the equations. Enter $Y_1 = -X + 3$ and $Y_2 = \left(\frac{3}{2}\right)X + \left(\frac{15}{2}\right)$.

1 TRACE along the first line (Y_1) until you are near the intersection.

2 Press ▲

3 Now you are on the other line.

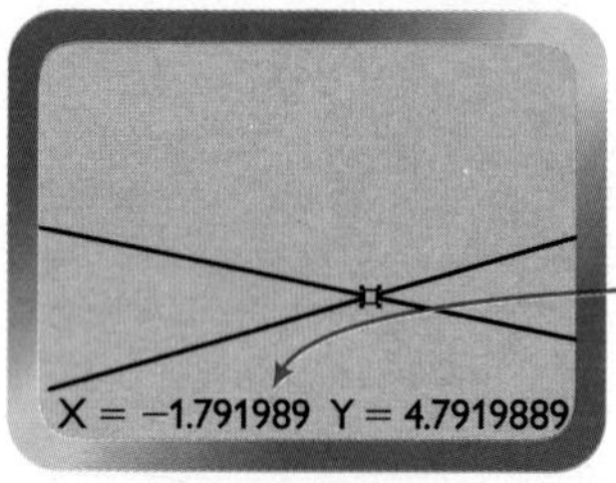

The *x*-coordinates are the same. The *y*-coordinates are close.

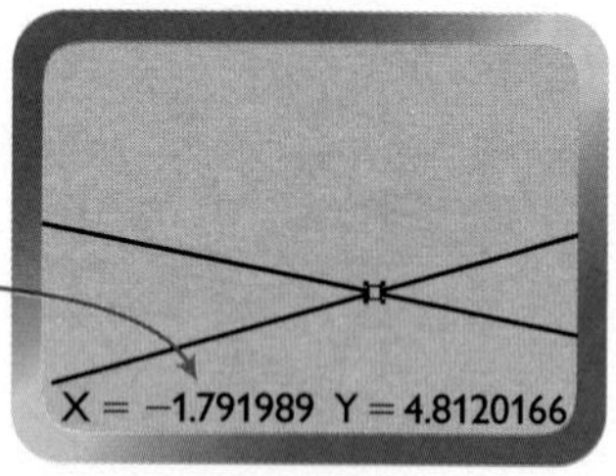

You can ZOOM in and repeat the process described above until the *y*-coordinates are the same to the nearest tenth, hundredth, or any other decimal place.

Try This

7. Try zooming in on the point where the graph of $y = x^2 + 2x - 1$ crosses the negative *x*-axis. Between what two values, to the nearest hundredth, does the *x*-coordinate of the point lie?

8. Sketch a clear graph of $y = \frac{x^2 - 2}{3x - 3}$.

9. Solve this system by graphing: $9x + 3y = 14$
$-3x + 2y = 8$

Answers to Try This

7. between −2.41 and −2.42

8.

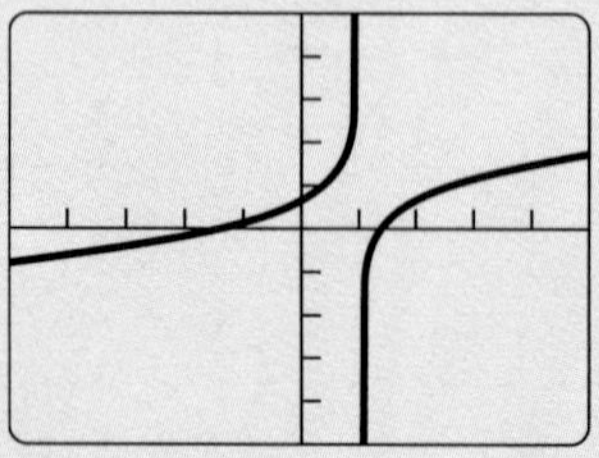

9.

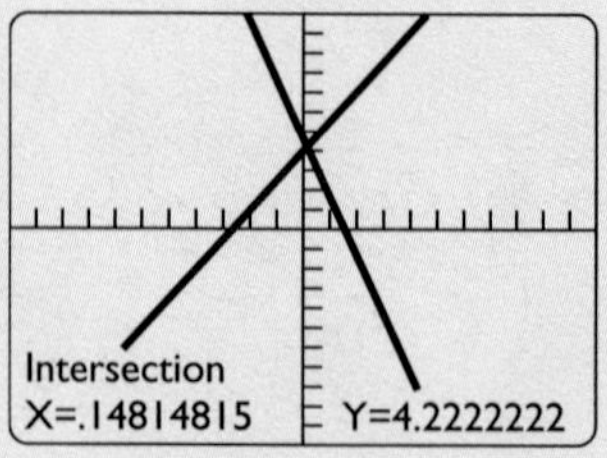

Comparing Graphs

Using a List to Graph a Family of Curves

Some calculators allow you to enter a list as an element in an expression. The calculator can then plot a function for each value in the list and graph a family of curves.

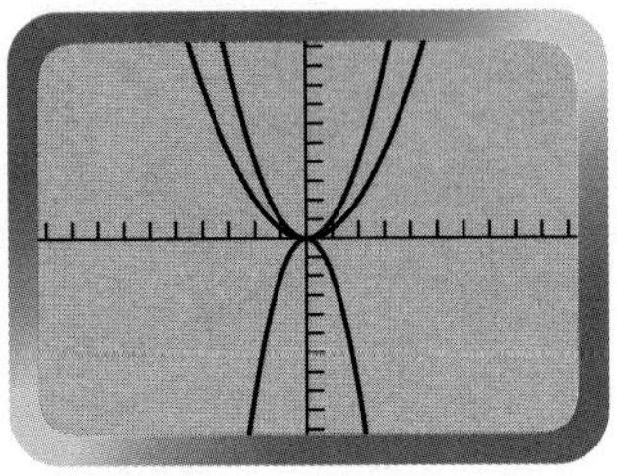

$Y_1 = \{1, -2, 0.5\}X^2$ plots the family of functions $y = x^2$, $y = -2x^2$, $y = 0.5x^2$.

$Y_2 = X^2 + \{-4, 5, 2\}$ plots the family of functions $y = x^2 - 4$, $y = x^2 + 5$, $y = x^2 + 2$.

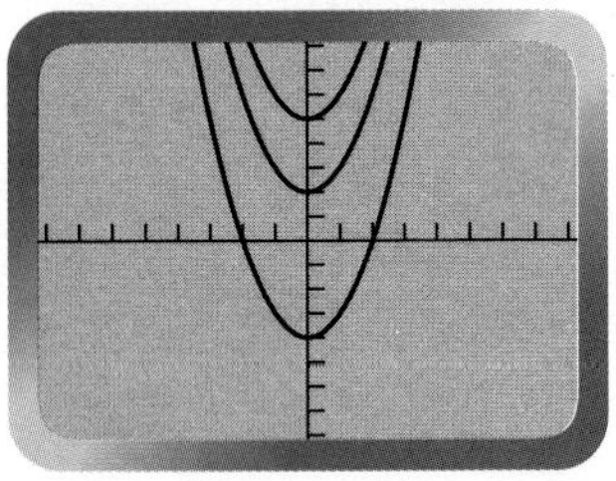

A Function and Its Inverse

When two functions are inverses of each other, their graphs are reflections of each other over the line $y = x$.

Suppose you want to graph $y = 2x + 4$ and its inverse.

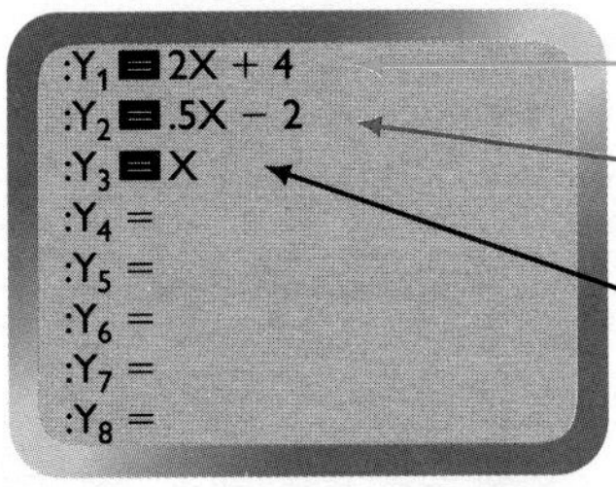

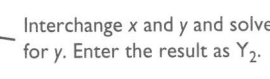

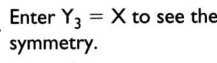

Enter the equation as Y_1.

Interchange x and y and solve for y. Enter the result as Y_2.

Enter $Y_3 = X$ to see the symmetry.

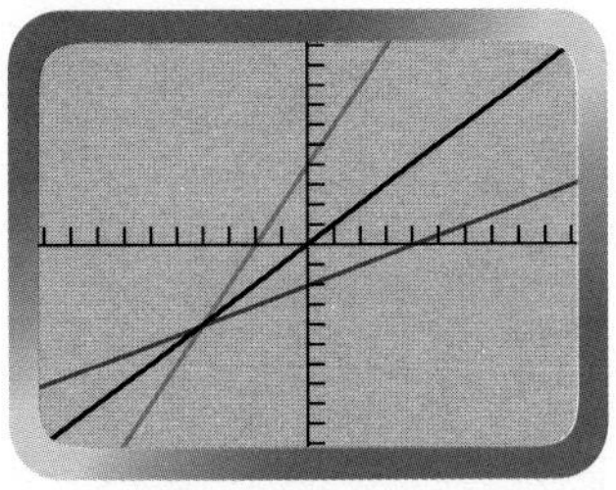

Some graphics calculators have a DRAW INV feature as a menu item. You can enter $y = 2x + 4$ as Y_1 and then enter DRAW INV Y_1 to graph Y_1 and its inverse. (See page 300, Exercise 27, for a method that uses parametric equations.)

Try This

10. Describe how to use a list to plot the family of functions $y = x$, $y = -x$, $y = \frac{1}{2}x$, $y = 2x$.

11. Graph each family of functions. (Set your calculator in degree mode. Use $-360 \le x \le 360$, x-scale $= 90$; $-4 \le y \le 4$, y-scale $= 1$.)

a. $y = \{1, -1, 0.5\} \cos x$ **b.** $y = \sin x + \{-1, 0, 1\}$

12. Graph each function and its inverse.

a. $y = x + 3$ **b.** $y = -3x - 1$ **c.** $y = 2^x$

Answers to Try This

10. Enter Y1={1,–1,0.5,2}X.

11. a.

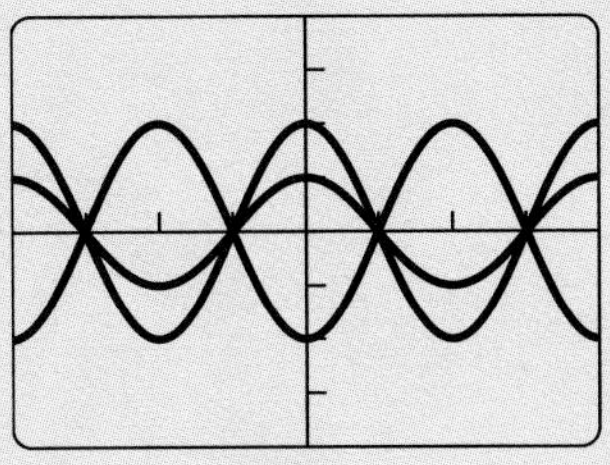

b.

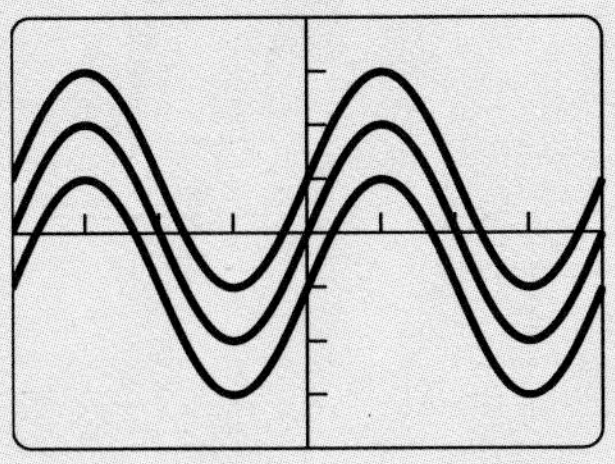

12. a.

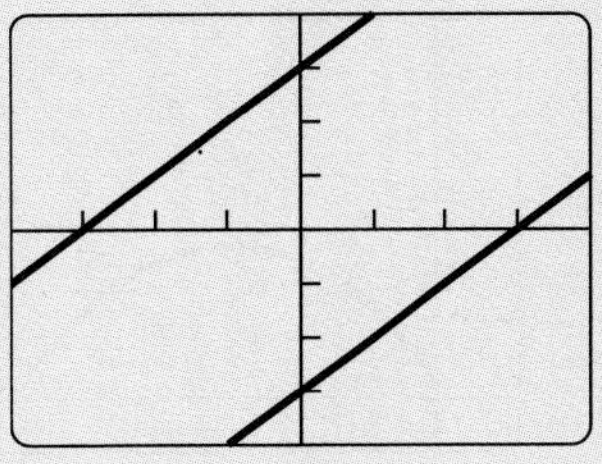

b.

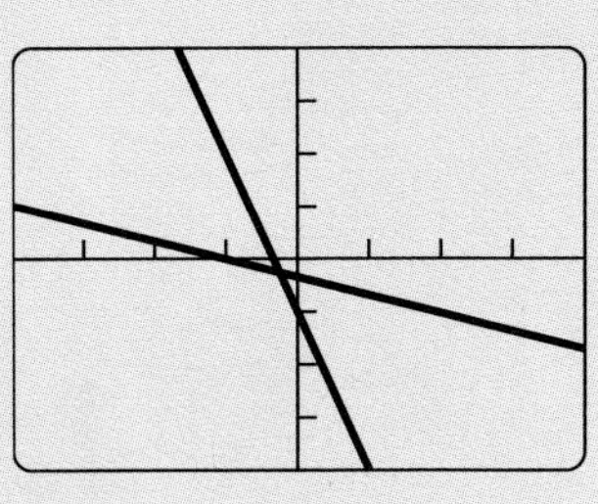

c.

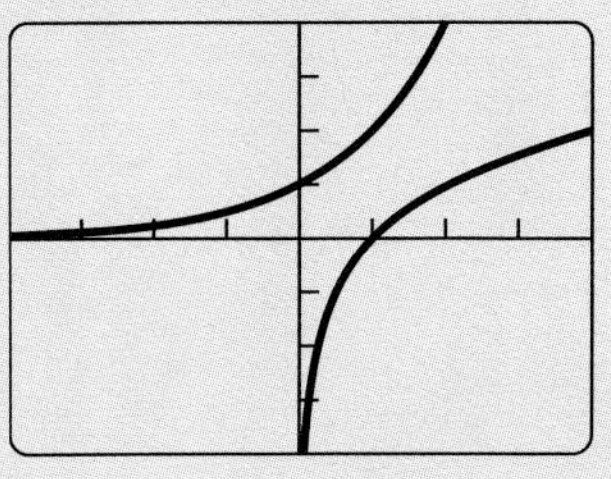

Graphing Parametric Equations

➤ Parametric Mode

When a ball is tossed up at an angle, it travels both vertically and horizontally. Suppose its path can be described by these two parametric equations:

$x = 0.3 + 6t$ ← x is the horizontal distance traveled in t seconds.

$y = 6t - 4.9t^2$ ← y is the vertical distance traveled in t seconds.

The variables x and y are both expressed in terms of a third variable, the parameter t. A graphics calculator in parametric mode will substitute a t-value in both equations and plot (x, y).

Choose parametric mode (par). In this mode, you set a range for t as well as for x and y.

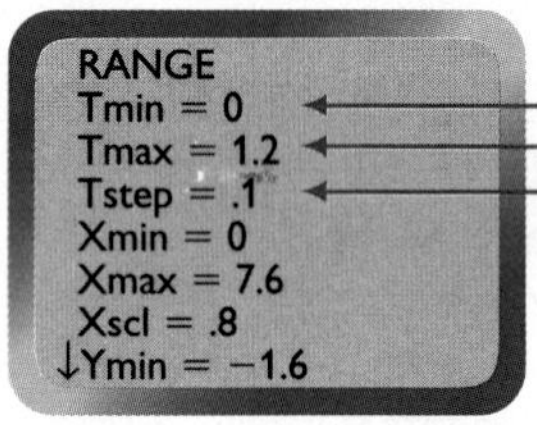

the minimum value of t in seconds
the maximum value of t in seconds
increments of time in each step from $t = 0$ to $t = 1.2$

➤ Exploring a Parametric Graph

Use these settings for the variables:

$0 \le t \le 1.2$ with increments of 0.1 for t-values,

$0 \le x \le 7.6$ with increments of 0.8 for x-values, and

$-1.6 \le y \le 3.44$ with increments of 0.8 for y-values.

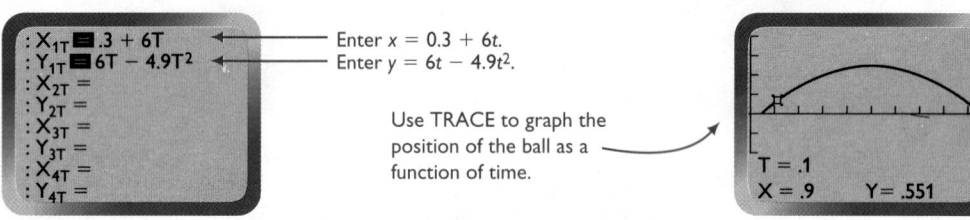

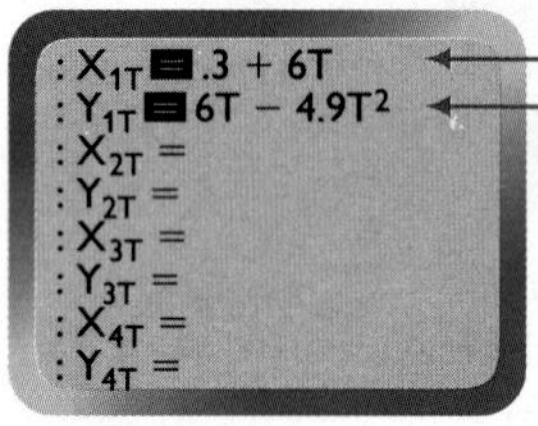

Enter $x = 0.3 + 6t$.
Enter $y = 6t - 4.9t^2$.

Use TRACE to graph the position of the ball as a function of time.

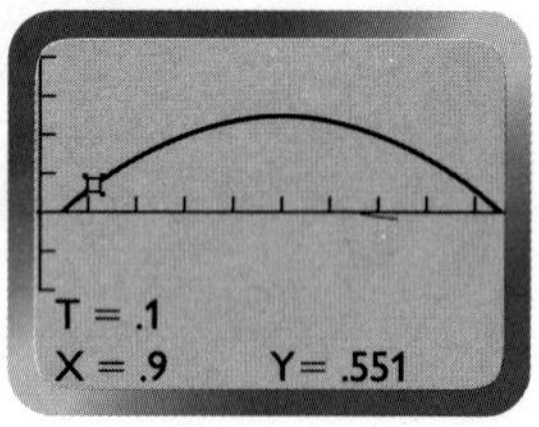

Try This

13. Suppose the path of a baseball t seconds after it is hit is described by the parametric equations $x = 100t$ and $y = -16t^2 + 40t + 3$.

a. Graph the parametric equations using these settings:

$0 \le t \le 3$ with increments of 0.05 for t-values,

$0 \le x \le 300$ with increments of 50 for x-values, and

$0 \le y \le 70$ with increments of 10 for y-values.

b. Use the TRACE feature. In how many seconds will the ball reach its maximum height?

Answers to Try This

13. a.

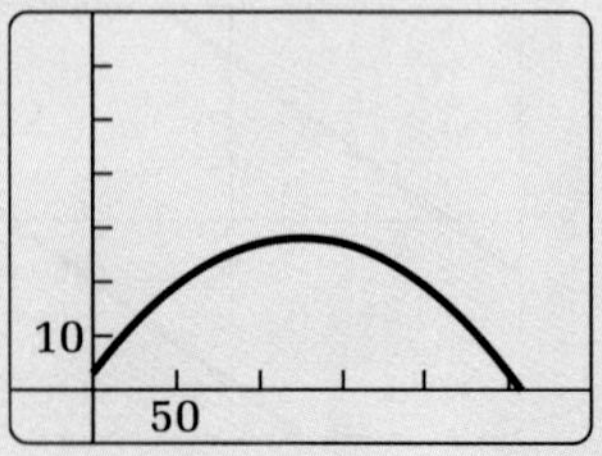

b. 1.25 seconds

Using Matrices

➤ Entering and Multiplying a Matrix by a Number

Suppose you want to enter this table of sales data as a matrix. Press the matrix function key. Select EDIT and choose matrix [A].

This Month's Sales by Age Group				
	VCRs	CD players	Faxes	Telephone answerers
Under 25	73	211	24	106
25–49	132	188	67	142
Over 49	89	55	46	98

Enter the elements. →

$$\begin{bmatrix} 73 & 211 & 24 & 106 \\ 132 & 188 & 67 & 142 \\ 89 & 55 & 46 & 98 \end{bmatrix}$$

Set the dimensions: 3 rows, 4 columns.

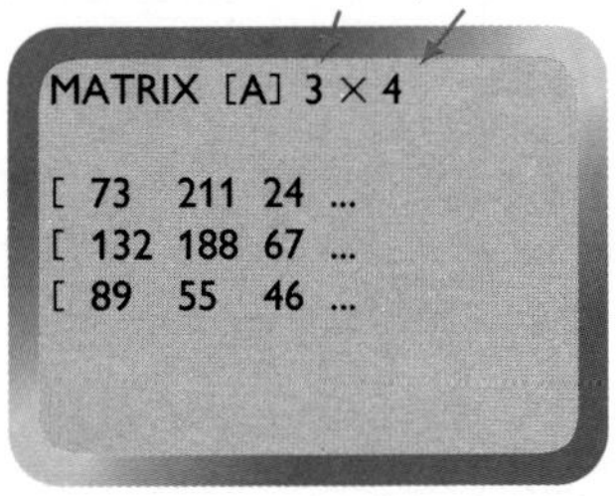

Suppose the data are this month's sales. You want to see what a projected increase of 14% in sales would be.

You have entered the data as matrix [A]. To increase each element by 14%, you can multiply [A] by 1.14.

First return to the home screen.

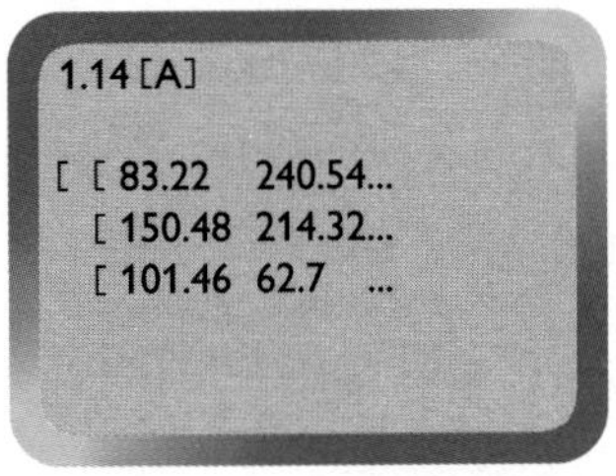

← Press 1.14 MATRX ENTER ENTER on a TI-82 graphics calculator.

➤ Adding and Subtracting Matrices

You can add or subtract two matrices that have the same dimensions. Suppose you enter two months' sales data as [A] and [B].

To compare the two data sets, you can subtract the matrices. The answer is a matrix with the same dimensions as the original matrices. Its elements are the sums or differences of elements in the same positions in [A] and in [B].

this month's sales

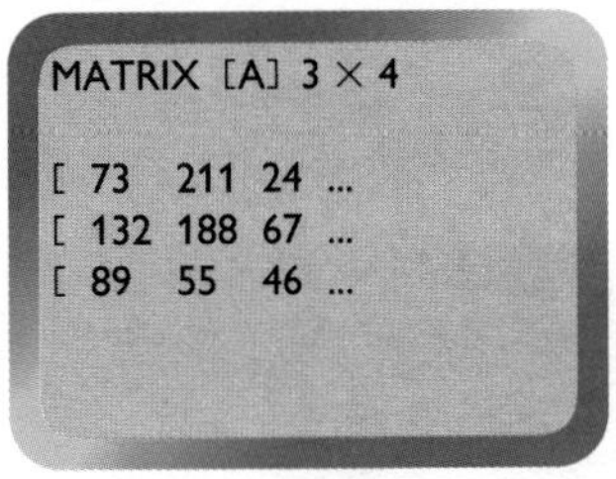

last month's sales

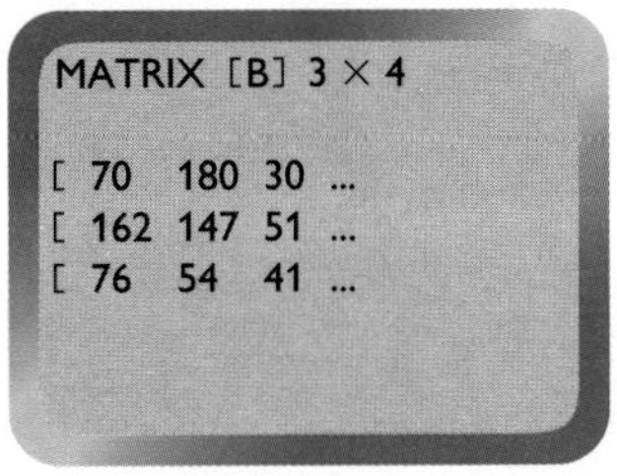

Press MATRX ENTER − MATRX 2 ENTER.

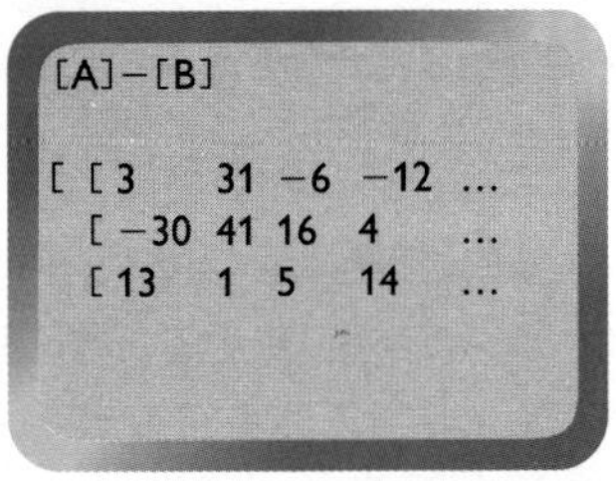

➤ Finding the Product of Two Matrices

You can multiply two matrices when the number of columns of the first matrix is the same as the number of rows of the second matrix.

Suppose you want to find this product: $\begin{bmatrix} 9 & 4 \\ 3 & 1 \\ 2 & 8 \\ 1 & 5 \end{bmatrix} \begin{bmatrix} 4 & 2 & 0 \\ 3 & 0 & 2 \end{bmatrix}$

Enter the two matrices as [A] and [B]. Their dimensions are 4×2 and 2×3. Then find the product of the matrices.

```
[A][B]

[[ 48 18 8  ]
 [ 15 6  2  ]
 [ 32 4  16 ]
 [ 19 2  10 ]]
```

Press MATRX ENTER MATRX 2 ENTER on a TI-82 graphics calculator.

➤ Finding the Inverse of a Matrix

You can use a graphics calculator to find the inverse of a matrix when the number of rows is the same as the number of columns.

Suppose you want to find the inverse of this matrix: $\begin{bmatrix} 3 & -1 \\ 4 & 0 \end{bmatrix}$

Enter the elements of the matrix as [A].

```
MATRIX [A] 2 × 2

[ 3 -1 ]
[ 4 0  ]
```

To find $[A]^{-1}$ press MATRX ENTER X^{-1} ENTER.

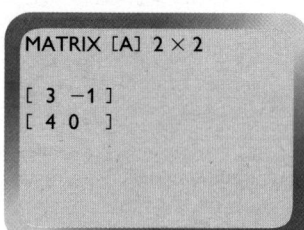

```
[A]^-1

[[ 0   .25 ]
 [ -1 .75 ]]
```

Try This

Use matrices *A*, *B*, *C*, or *D* to find each answer. If a matrix does not exist, write *not defined*.

$A = \begin{bmatrix} 3 & 4 \\ 6 & 7 \end{bmatrix}$ $B = \begin{bmatrix} 5 & 11 \\ -3 & -7 \end{bmatrix}$ $C = \begin{bmatrix} 9 & 3 \\ 12 & 4 \end{bmatrix}$ $D = \begin{bmatrix} 2 & 0 & -8 & 1 \\ 5 & 7 & 3 & 9 \end{bmatrix}$

14. $A + B$ **15.** $B - C$ **16.** $2A$ **17.** AB

18. $C + D$ **19.** CD **20.** A^{-1} **21.** D^{-1}

Answers to Try This

14.
```
[A]+[B]

[[8 15]
 [3 0 ]]
```

15.
```
[B]-[C]

[[-4   8  ]
 [-15 -11]]
```

16.
```
2[A]

[[6  8 ]
 [12 14]]
```

17.
```
[A][B]

[[3 5 ]
 [9 17]]
```

18. not defined

19.
```
[C][D]

[[33 21 -63 36]
 [44 28 -84 48]]
```

20.
```
[A]^-1

[[-2.33 1.33 ]
 [2     -1   ]]
```

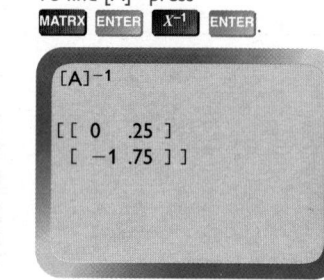

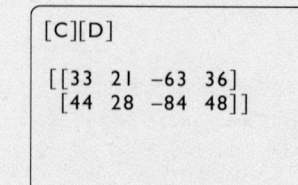

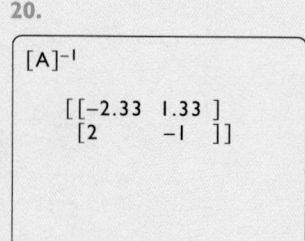

21. not defined

Working with Statistics

Histograms, Line Graphs, and Box-and-Whisker Plots

Many graphics calculators can display histograms, line graphs, and even box-and-whisker plots of data that you enter.

Readers of **galaxy** Magazine

Age group	Frequency
10–14	1110
15–19	3398
20–24	4344
25–29	3215
30–34	332
35–39	112

← The data about readers of *Galaxy* magazine are displayed in a histogram. →

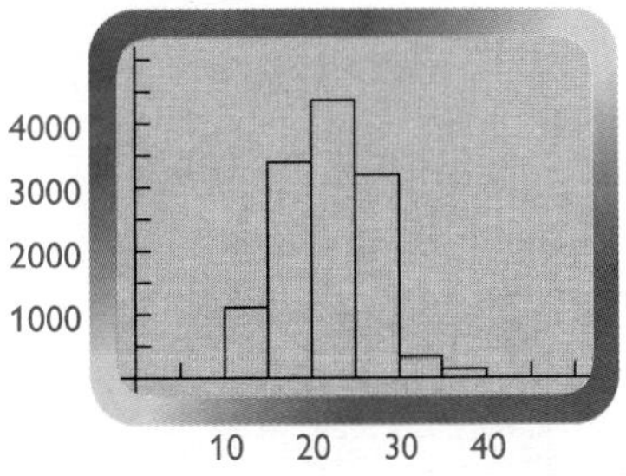

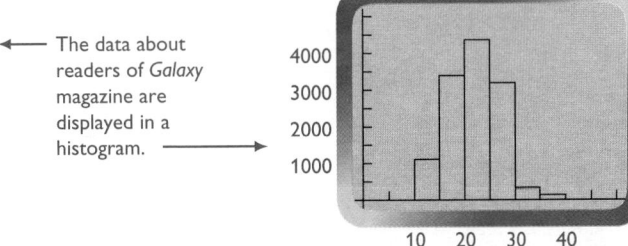

Scatter Plots and Curve Fitting

Curve fitting is the process of finding an equation that describes a set of ordered pairs. Often, the first step is to enter the two lists of data and then graph the paired data in a scatter plot.

Men's Winning Times in Olympic 400 m Freestyle Swimming		
Year	Years after 1960	Time (seconds)
1960	0	258.3
1964	4	252.2
1968	8	249.0
1972	12	240.27
1976	16	231.93
1980	20	231.31
1984	24	231.23
1988	28	226.95
1992	32	225.00

On a TI-82 graphics calculator, press the STAT key to find a menu item for editing, or entering, a list of data.

Enter the years after 1960 as List 1.
Enter the time in seconds as List 2.

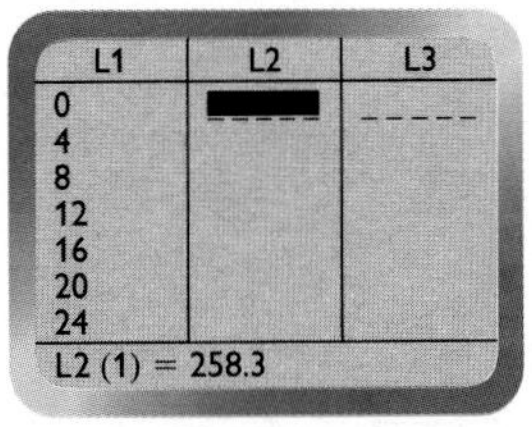

Other statistics menu items will fit lines and curves to your data plot, such as the median-median line, the least-squares line (linear regression), and the least-squares parabola (quadratic regression). Below, a least-squares line is fit to a plot of the Olympic data.

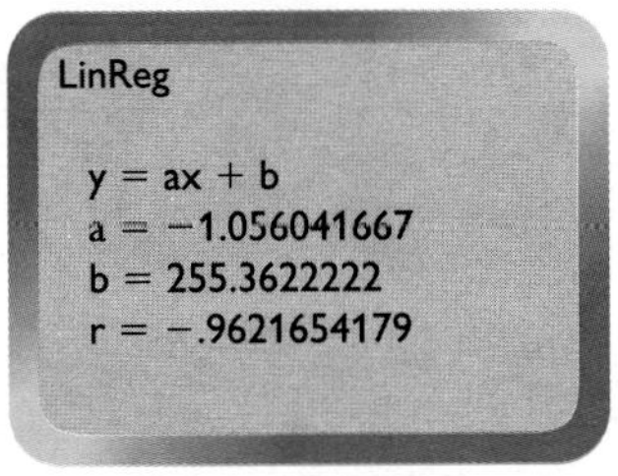

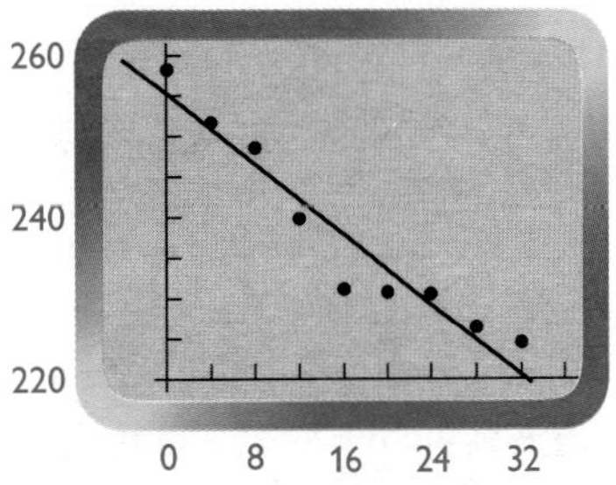

➤ Mean and Standard Deviation

Most scientific and graphics calculators will calculate and display the mean and the standard deviation of a set of data using formulas that are already programmed into the calculator.

Your calculator may have two formulas for standard deviation. This book (see page 339) uses the standard deviation σ, which uses all n data items. (The other formula uses a sample of data items.)

Suppose you have entered the table of Olympic swimming data into your graphics calculator as on page 605. You are interested in the mean and the standard deviation of the winning times.

L1	L2	L3
0	258.3	-----
4	252.2	
8	249.0	
12	240.27	
16	231.93	
20	231.31	
24	231.23	

L2 (7) = 231.23

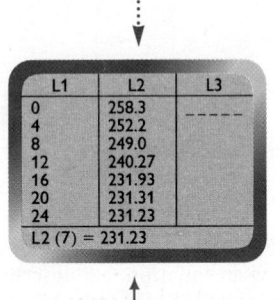

You have set up List 2 as the y-variable.

On a TI-82 graphics calculator, press the STAT key to access the menu of statistical calculations and choose the 2-variable statistics.

You will first see calculations for the x-variable, including $\bar{x}$ and σx. Press the down-arrow key until you find the statistical calculations for the y-variable.

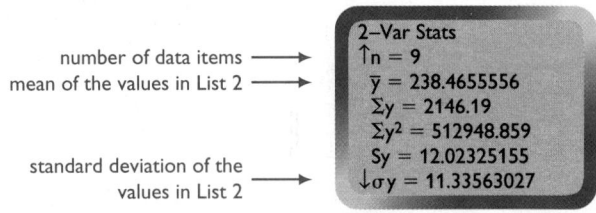

2–Var Stats
↑n = 9 ← number of data items
$\bar{y}$ = 238.4655556 ← mean of the values in List 2
Σy = 2146.19
Σy^2 = 512948.859
Sy = 12.02325155
↓σy = 11.33563027 ← standard deviation of the values in List 2

Try This

22. The table provides some data values for the eruptions of Old Faithful, a geyser in Yellowstone Park.

Length of eruption (l)	3	4.5	4	2.5	3.5	5	2
Time (in minutes) until next eruption (t)	68	89	83	62	75	92	57

a. Enter the two lists of data on a graphics calculator.

b. Graph the paired data in a scatter plot.

c. Fit a least-squares line to the data (linear regression).

d. Fit a median-median line to the data.

e. Find the mean and the standard deviation of the data in the second row of the table.

Answers to Try This

22. a. Check students' work.

b.

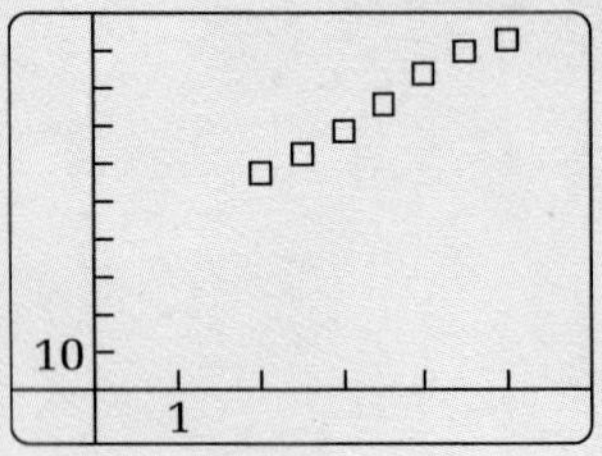

c. $y = 12.43x + 31.64$

d. $y = 12.4x + 31.6$

e. 75.14; 12.48

➤ Using a Spreadsheet

In addition to using a graphics calculator, you may want to use a computer with a spreadsheet program. A spreadsheet can help you solve a problem like this one: Suppose you want to buy a CD player that costs $195, including tax. You already have $37 and can save $9 per week. After how many weeks can you buy the player?

A spreadsheet is made up of cells named by a column letter and a row number, like A3 or B4. You can enter a label, a number, or a formula into a cell.

row numbers

column letters

CD Savings

	A	B
1	Week number	Total saved
2	0	37
3	= + A2 + 1	= + B2 + 9
4	= + A3 + 1	= + B3 + 9
5	= + A4 + 1	= + B4 + 9
6	= + A5 + 1	= + B5 + 9
7	= + A6 + 1	= + B6 + 9
8	= + A7 + 1	= + B7 + 9
9	= + A8 + 1	= + B8 + 9
10	= + A9 + 1	= + B9 + 9
11	= + A10 + 1	= + B10 + 9
12	= + A11 + 1	= + B11 + 9
13	= + A12 + 1	= + B12 + 9
14	= + A13 + 1	= + B13 + 9
15	= + A14 + 1	= + B14 + 9
16	= + A15 + 1	= + B15 + 9
17	= + A16 + 1	= + B16 + 9
18	= + A17 + 1	= + B17 + 9
19	= + A18 + 1	= + B18 + 9
20	= + A19 + 1	= + B19 + 9

Cell B1 contains the label "Total saved."

Cell B2 contains the number 37.

Cell B3 contains the formula "=+B2+9." This formula tells the computer to take the number in cell B2, add 9 to it, and put the result in cell B3. (Likewise, the formula in cell A3 tells the computer to take the number in cell A2, add 1 to it, and put the result in cell A3.)

Instead of typing a formula into each cell individually, you can use the spreadsheet's copy and fill commands.

CD Savings

	A	B
1	Week number	Total saved
2	0	37
3	1	46
4	2	55
5	3	64
6	4	73
7	5	82
8	6	91
9	7	100
10	8	109
11	9	118
12	10	127
13	11	136
14	12	145
15	13	154
16	14	163
17	15	172
18	16	181
19	17	190
20	18	199

In this spreadsheet, the computer has replaced all the formulas with calculated values. You can have the computer draw a scatter plot with a line connecting the plotted points. As you can see, you will have enough money to buy the CD player after 18 weeks.

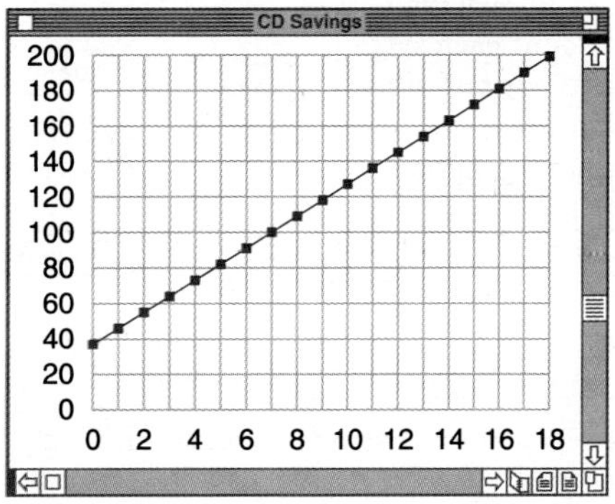

Extra Practice

Unit 1

1. Mu Lan had two positive integers, A and B. She divided B by A and found that B was not evenly divisible by A. Then she divided $2B$ by A to see if $2B$ was evenly divisible by A. She continued this process using multiples of B until she found a multiple that was evenly divisible by A. 1-1

a. What procedure did Mu Lan's algorithm perform?

b. Did Mu Lan's algorithm contain a loop? If it did, describe the loop.

Solve each equation or inequality in two ways. 1-1

2. $2x + 3 = 15$ **3.** $-4 + 7x = 31$ **4.** $12 = \frac{1}{2}x + 5$

5. $-x + 9 \geq 14$ **6.** $\frac{1}{3}x + 5 > 2$ **7.** $30 - 4x \leq 6$

Tell whether each set of measurements can be the lengths of the sides of a triangle. 1-2

8. 3 in., 5 in., 7 in. **9.** 4 ft, 15 ft, 18 ft **10.** 2.5 m, 5.5 m, 2.5 m

11. 14 cm, 6 cm, 8 cm **12.** 20 yd, 20 yd, 10 yd **13.** 25 ft, 18 ft, 6 ft

14. a. Find all the different rectangles that have sides with whole-number lengths and a perimeter of 32.

b. Which of the rectangles you found in part (a) has the largest area? 1-2

The table shows the number of books in some large city library systems in 1992 and the amount each system spent that year to buy new library materials. 1-3

15. Make a scatter plot of the relationship between number of books and amount spent.

16. Describe any trend(s) suggested by the scatter plot.

City	Books (millions)	Amount spent (million $)
Anchorage, AK	0.4	0.7
Boston, MA	6.0	2.0
Chicago, IL	1.7	3.6
Denver, CO	2.2	1.9
Houston, TX	4.0	4.0
Kansas City, MO	1.8	1.6
Los Angeles, CA	5.7	4.5
Milwaukee, WI	2.3	1.8
Philadelphia, PA	3.5	6.2
Toledo, OH	1.8	3.0

Answers to Extra Practice Unit 1

1. a. finding the least common multiple of A and B
   b. Yes; testing multiples of B until one is found that is evenly divisible by A.

2–7. Solution methods may vary.

2. $x = 6$ 3. $x = 5$

4. $x = 14$ 5. $x \leq -5$

6. $x > -9$ 7. $x \geq 6$

8. Yes. 9. Yes.

10. No.

11. No.

12. Yes.

13. No.

14. a. 1, 1, 15, 15; 2, 2, 14, 14; 3, 3, 13, 13; 4, 4, 12, 12; 5, 5, 11, 11; 6, 6, 10, 10; 7, 7, 9, 9; 8, 8, 8, 8
    b. 8, 8, 8, 8

15.

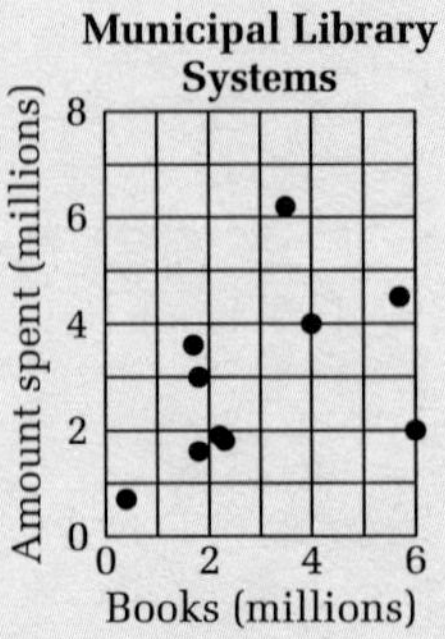

16. The more books in a library system, the greater the expenditure for new books.

17. Weefixit: $y = 16x + 25$; Appliance Rehab: $y = 19x + 18$

18. Appliance Rehab; Weefixit

19. 3

20. 90

21. 8.0

Weefixit Appliance Service charges $25 for a house call, plus $16 for each hour spent fixing an appliance. Appliance Rehab charges $18 for a house call, plus $19 per hour. 1-4

17. Write an equation that models each company's repair charges.

18. Which company charges less if a repair takes 1.5 h? 4 h?

Use algebra to find each break-even point. Round answers to the nearest tenth. 1-4

19. $E = 12 + 3.5x$
$I = 7.5x$

20. $E = 144 + 2.7x$
$I = 4.3x$

21. $E = 55 + 1.8x$
$I = 8.7x$

Solve. When necessary, round decimal answers to the nearest tenth. 1-5

22. $x + y = 8$
$3x + 4y = 22$

23. $x - y = 5$
$7x + 3y = 7$

24. $-x + 2y = 11$
$5x + y = 22$

25. $x + y = 1$
$y + z = 3$
$x + z = 6$

26. $x + y - z = 9$
$x - y + z = -3$
$x + 2y + 4z = 3$

27. $x + 2y = 6$
$3x - y + z = 9$
$x - y + 4z = 5$

Use a matrix to represent the connections among the vertices of each diagram. 1-6

28.

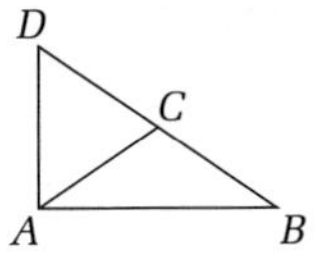

29.

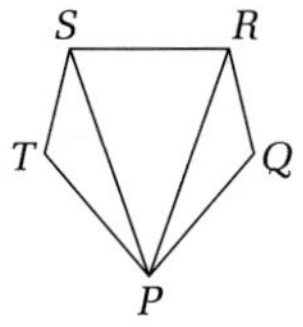

30.

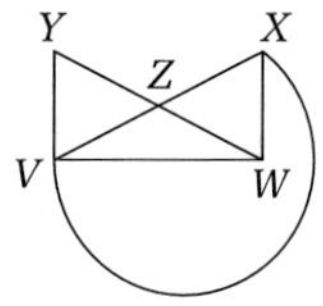

For Exercises 31 and 32, use the table of the routes of a public bus service. 1-6

Buses leaving from	Provide service to . . .		
Central Square	Veterans' Plaza	Zoo	Downtown
Downtown	Veterans' Plaza	Northside Mall	
Northside Mall	Veterans' Plaza	Downtown	
Veterans' Plaza	Central Square		
Zoo	Northside Mall		

31. Make a diagram to represent the bus routes.

32. Make a matrix to represent the bus routes.

Find the value of *x* that maximizes or minimizes each function. Find the maximum or minimum value. Tell whether it is a *maximum* or a *minimum*. 1-7

33. $y = -x^2 + 6x + 2$

34. $y = 3x^2 - 24x - 5$

35. $y = -2x^2 - 18x + 1$

Answers to Extra Practice Unit 1

22. (10, –2)
23. (2.2, –2.8)
24. (3, 7)
25. (2, –1, 4)
26. (3, 4, –2)
27. (3.2, 1.4, 0.8)
28.
$$\begin{array}{c} \\ A \\ B \\ C \\ D \end{array}\begin{array}{c} \begin{array}{cccc} A & B & C & D \end{array} \\ \begin{bmatrix} 0 & 1 & 1 & 1 \\ 1 & 0 & 1 & 0 \\ 1 & 1 & 0 & 1 \\ 1 & 0 & 1 & 0 \end{bmatrix} \end{array}$$

29.
$$\begin{array}{c} \\ P \\ Q \\ R \\ S \\ T \end{array}\begin{array}{c} \begin{array}{ccccc} P & Q & R & S & T \end{array} \\ \begin{bmatrix} 0 & 1 & 1 & 1 & 1 \\ 1 & 0 & 1 & 0 & 0 \\ 1 & 1 & 0 & 1 & 0 \\ 1 & 0 & 1 & 0 & 1 \\ 1 & 0 & 0 & 1 & 0 \end{bmatrix} \end{array}$$

30. Answers may vary. An example is given.
$$\begin{array}{c} \\ V \\ W \\ X \\ Y \\ Z \end{array}\begin{array}{c} \begin{array}{ccccc} V & W & X & Y & Z \end{array} \\ \begin{bmatrix} 0 & 1 & 1 & 1 & 1 \\ 1 & 0 & 1 & 0 & 1 \\ 1 & 1 & 0 & 0 & 1 \\ 1 & 0 & 0 & 0 & 1 \\ 1 & 1 & 1 & 1 & 0 \end{bmatrix} \end{array}$$

31.

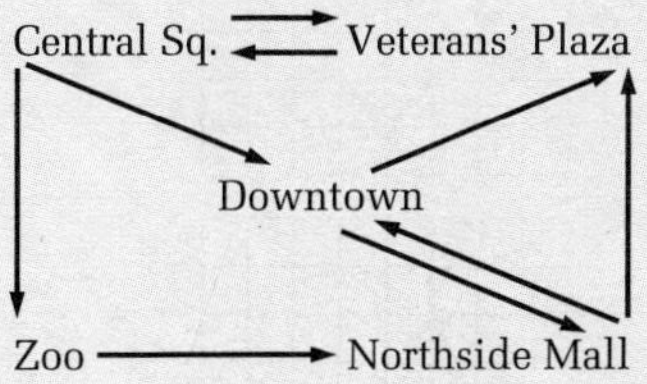

32.
$$\begin{array}{c} \\ CS \\ D \\ NM \\ VP \\ Z \end{array}\begin{array}{c} \begin{array}{ccccc} CS & D & NM & VP & Z \end{array} \\ \begin{bmatrix} 0 & 1 & 0 & 1 & 1 \\ 0 & 0 & 1 & 1 & 0 \\ 0 & 1 & 0 & 1 & 0 \\ 1 & 0 & 0 & 0 & 0 \\ 0 & 0 & 1 & 0 & 0 \end{bmatrix} \end{array}$$

33. 3; 11; maximum
34. 4; –53; minimum
35. –4.5; 41.5; maximum

At Central High's Community Day, some students will go on a walkathon and others will wash cars. At most, 15 students will take part. Each walker will spend about 4 h, and each washer will spend about 6 h. Organizers have agreed to limit the total student time to 72 h. **1-8**

36. Let x represent the number of walkers and let y represent the number of washers. Write a system of inequalities that represents all the constraints on x and y. Graph the feasible region and label each vertex with its coordinates.

37. Suppose each walker raises \$7 and each washer raises \$8. How many walkers and how many washers should there be in order to maximize the Community Day income?

Unit 2

For each function, find $f(3)$, $f(-2)$, and $f(a+b)$. **2-1**

1. $f(x) = (x+2)^2$ **2.** $f(x) = \dfrac{x+1}{x}$ **3.** $f(x) = 7 - \frac{1}{2}x$

4. $f(x) = 2x^2 - 5x + 1$ **5.** $f(x) = |x - 5|$ **6.** $f(x) = \dfrac{x}{x^2+1}$

Tell whether each graph represents a *function*. If it does, tell whether the function is *one-to-one* or *many-to-one*. **2-1**

7.

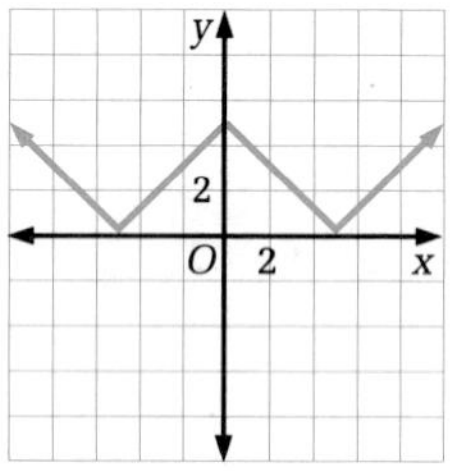

8.

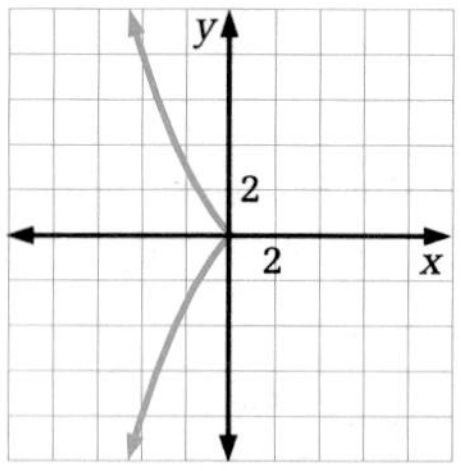

9.

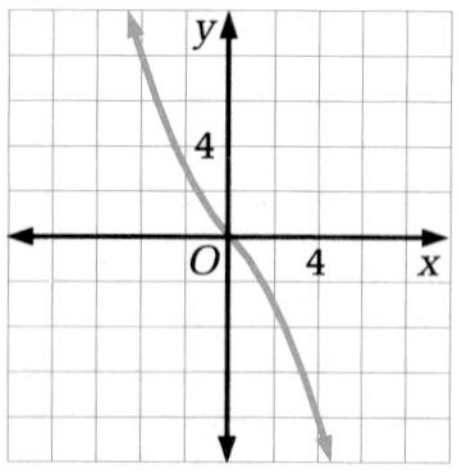

Tell whether each function is *linear*. If it is, find $f(2)$, $f(-3)$, and $f(a-b)$. **2-2**

10. $f(x) = \dfrac{1}{2x+7}$ **11.** $f(x) = -3x + 2$ **12.** $f(x) = \dfrac{x}{6} - 5$

13. $f(x) = x^2 - 2x + 1$ **14.** $f(x) = 8 - 3x$ **15.** $f(x) = 5 + 4(2 - x)$

Write an equation for each function in the form $f(x) = mx + b$. **2-2**

16. $f(0) = -6$; $m = 3$ **17.** $f(-1) = 4$; $f(5) = 7$ **18.** $m = -2$; $f(3) = 8$

19. $f(2) = 11$; $f(-3) = 1$ **20.** $f(-3) = 2$; $f(1) = -5$ **21.** $f\left(\frac{1}{2}\right) = 3$; $f\left(\frac{5}{2}\right) = 6$

Answers to Extra Practice Unit 1

36. $x \ge 0$; $y \ge 0$; $x + y \le 15$; $4x + 6y \le 72$

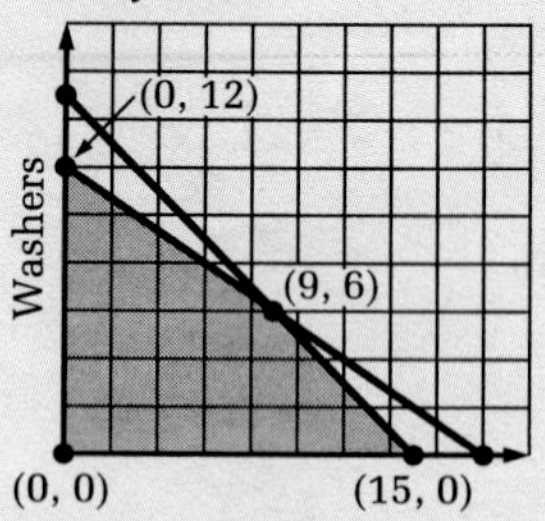

37. 9 walkers and 6 washers (\$111 raised)

Answers to Extra Practice Unit 2

1. 25; 0; $a^2 + 2ab + b^2 + 4a + 4b + 4$
2. $\frac{4}{3}$; $\frac{1}{2}$; $\dfrac{a+b+1}{a+b}$
3. $\frac{11}{2}$; 8; $7 - \dfrac{a+b}{2}$
4. 4; 19; $2a^2 + 4ab + 2b^2 - 5a - 5b + 1$
5. 2; 7; $|a + b - 5|$
6. $\frac{3}{10}$; $-\frac{2}{5}$; $\dfrac{a+b}{a^2+2ab+b^2+1}$
7. Yes; many-to-one.
8. No.
9. Yes; one-to-one.
10. No.
11. Yes; -4; 11; $-3a + 3b + 2$.
12. Yes; $-\frac{14}{3}$; $-\frac{11}{2}$; $\frac{a-b}{6} - 5$.
13. No.
14. Yes; 2; 17; $8 - 3a + 3b$.
15. Yes; 5; 25; $13 - 4a + 4b$.
16. $f(x) = 3x - 6$
17. $f(x) = \frac{1}{2}x + \frac{9}{2}$
18. $f(x) = -2x + 14$
19. $f(x) = 2x + 7$
20. $f(x) = -\frac{7}{4}x - \frac{13}{4}$
21. $f(x) = \frac{3}{2}x + \frac{9}{4}$
22.

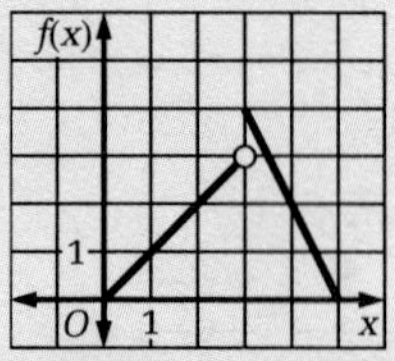

23.

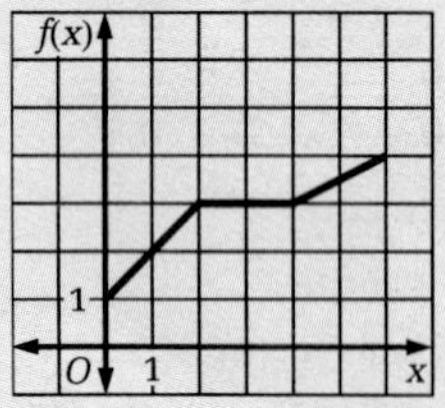

24.

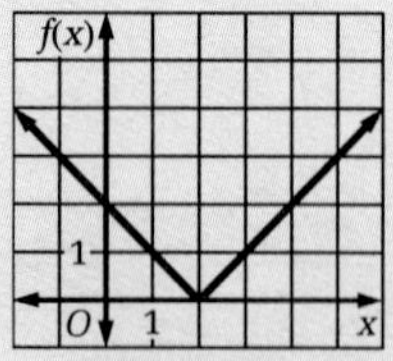

25.

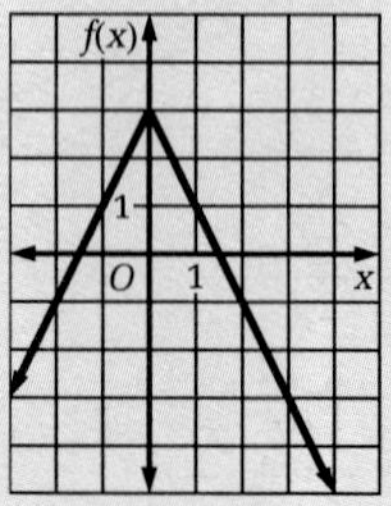

26.

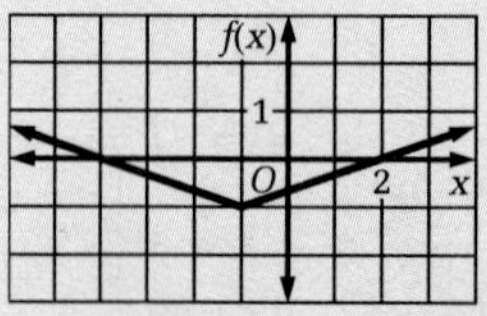

27. a. quadratic; -2; $-\frac{9}{2}$, $-\dfrac{a^2 - 2ab + b^2}{2}$

b.

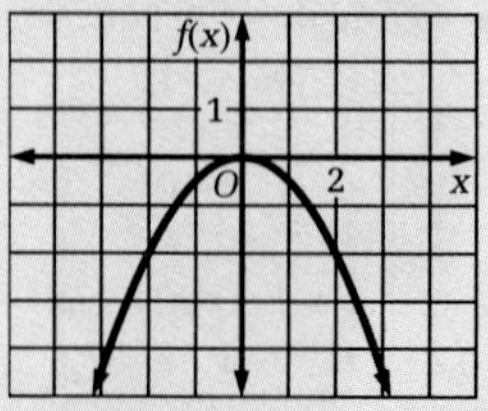

28. a. quadratic; 24; -1; $(a - b + 3)^2 - 1$

b.

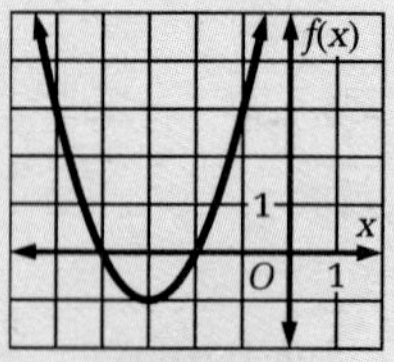

29. not quadratic
30. a. quadratic; -9; 41; $2a^2 - 4ab + 2b^2 - 8a + 8b - 1$

Graph each piecewise function or absolute value function. 2-3

22. $f(x) = \begin{cases} x & \text{if } 0 \le x < 3 \\ 10 - 2x & \text{if } 3 \le x \le 5 \end{cases}$

23. $f(x) = \begin{cases} x + 1 & \text{if } 0 \le x < 2 \\ 3 & \text{if } 2 \le x \le 4 \\ 1 + 0.5x & \text{if } 4 < x \le 6 \end{cases}$

24. $f(x) = |x - 2|$ **25.** $f(x) = -2|x| + 3$ **26.** $f(x) = \frac{|x + 1|}{3} - 1$

Tell whether each function is a quadratic function. For each quadratic function: 2-4

a. Find $f(2)$, $f(-3)$, and $f(a - b)$.

b. Graph f.

27. $f(x) = -\frac{1}{2}x^2$ **28.** $f(x) = (x + 3)^2 - 1$ **29.** $f(x) = \frac{1}{x^2 + 4}$

30. $f(x) = 2x^2 - 8x - 1$ **31.** $f(x) = \frac{(x - 1)^2}{4}$ **32.** $f(x) = x^3 + x^2 + 5$

Write an equation in the form $f(x) = kx^n$ for each direct variation function. 2-5

33. $f(1) = 3$; $n = 5$ **34.** $k = 4$; $f(3) = 36$ **35.** $f\left(\frac{1}{2}\right) = 3$; $n = 4$

36. $f(2) = 56$; $f(1) = 7$ **37.** $f(0.2) = 9$; $n = 3$ **38.** $f(10) = 500$; $f\left(\frac{1}{10}\right) = \frac{5}{100}$

Graph each polynomial function. 2-5

39. $f(x) = x(x - 1)(x + 3)$ **40.** $f(x) = x^2(x - 3)$

41. $f(x) = (x + 1)(x - 2)(x - 4)$ **42.** $f(x) = -x(x + 2)^2$

Solve. 2-6

43. $\sqrt{x + 6} = x$ **44.** $\sqrt{x} = x - 12$ **45.** $\sqrt{2x + 1} = x - 7$

46. $\sqrt{3x + 1} + 1 = x$ **47.** $\sqrt{7x - 3} + 3 = 2x$ **48.** $\sqrt[3]{4x} = x$

Find the domain and range of each function. 2-6

49. $f(x) = \sqrt[3]{x - 5}$ **50.** $f(x) = \sqrt{x + 2}$ **51.** $f(x) = -\sqrt{x - 3}$

52. $f(x) = \sqrt{5 - 2x}$ **53.** $f(x) = -\sqrt[3]{x + 10}$ **54.** $f(x) = 2\sqrt{x + 1} - 4$

For Exercises 55–60: 2-7

a. Graph the function.

b. Write the equations of all asymptotes of the graph.

c. Find the domain and range of the function.

55. $f(x) = \frac{4}{x - 2}$ **56.** $f(x) = -\frac{3}{x}$ **57.** $f(x) = \frac{x}{x + 3}$

58. $f(x) = \frac{x - 5}{x}$ **59.** $f(x) = \frac{x}{2x - 8}$ **60.** $f(x) = -\frac{x - 1}{x + 2}$

Answers to Extra Practice Unit 2

b.

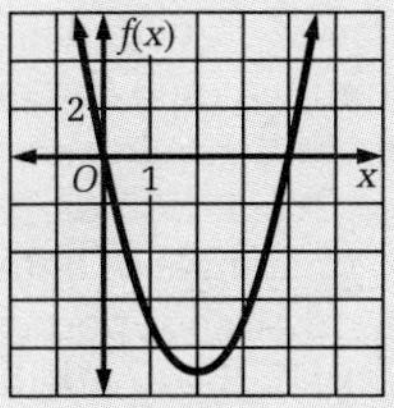

b.

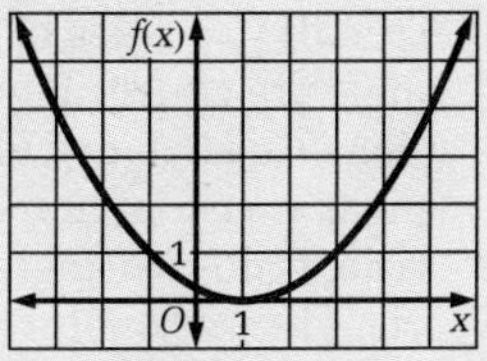

31. a. quadratic; $\frac{1}{4}$; 4; $\frac{(a - b - 1)^2}{4}$

32. not quadratic

33. $f(x) = 3x^5$

34. $f(x) = 4x^2$

35. $f(x) = 48x^4$

36. $f(x) = 7x^3$

37. $f(x) = 1125x^3$

38. $f(x) = 5x^2$

39.

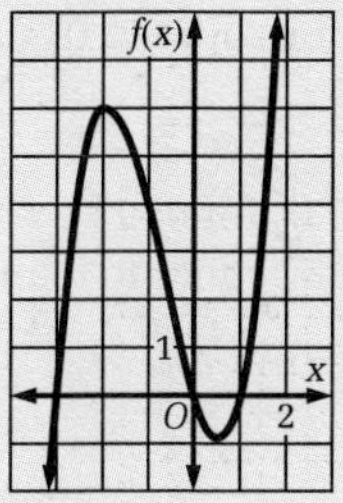

40.

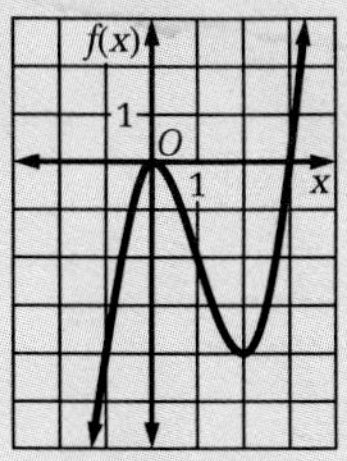

41.

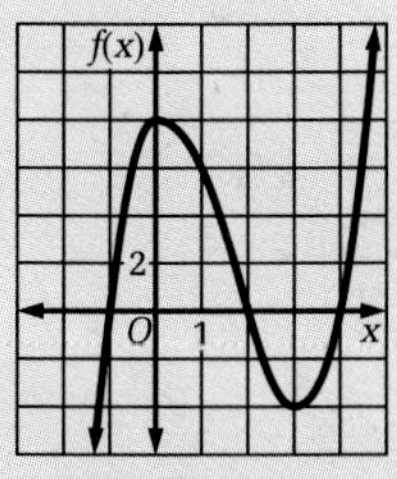

42.

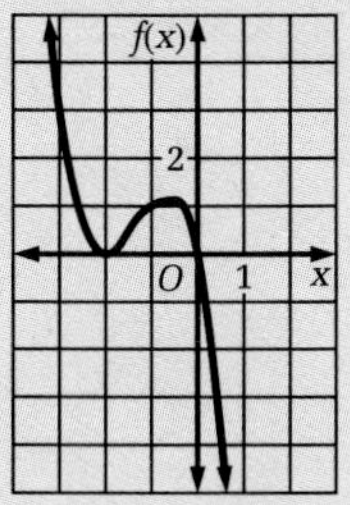

43. 3 44. 16 45. 12

46. 5 47. 4 48. −2, 0, 2

49. all real numbers; all real numbers

50. all real numbers greater than or equal to −2; all real numbers greater than or equal to 0

51. all real numbers greater than or equal to 3; all real numbers less than or equal to 0

52. all real numbers less than or equal to 2.5; all real numbers greater than or equal to 0

53. all real numbers; all real numbers

54. all real numbers greater than or equal to −1; all real numbers greater than or equal to −4

55. a.

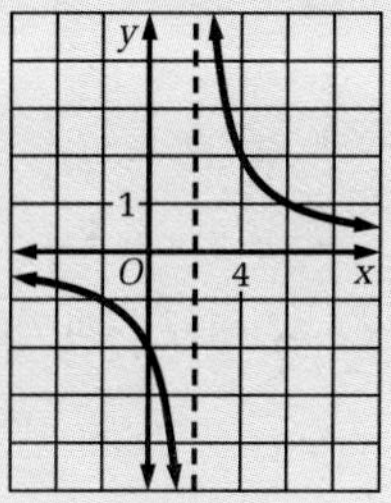

b. $x = 2$; $y = 0$

c. all real numbers except 2; all real numbers except 0

56. a.

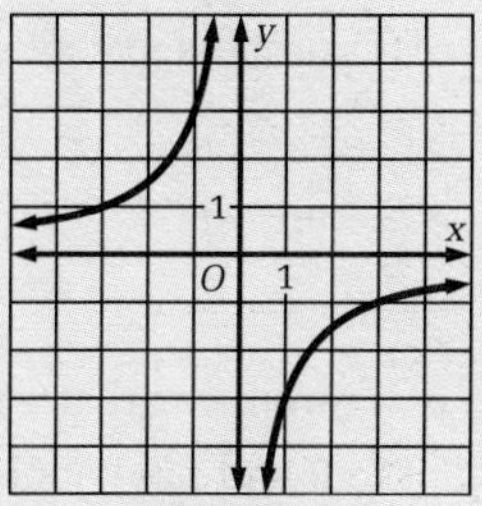

b. $x = 0$; $y = 0$

c. all real numbers except 0; all real numbers except 0

57–60. See answers in back of book.

For Exercises 61–68, use the functions $f(x) = 7x^2 + 1$, $g(x) = \sqrt{x}$, and $h(x) = \frac{x}{2x - 1}$. For each function, (a) find a rule and (b) find the domain of the function. 2-8

61. $f \cdot g$ **62.** $f + h$ **63.** $g \circ f$ **64.** $h \circ g$

65. $\frac{f}{h}$ **66.** $f - g$ **67.** $f \circ g$ **68.** $g \cdot h$

Unit 3

Determine whether each implication is *True* or *False*. Then state the converse, inverse, and contrapositive of each implication. Determine whether each is *True* or *False*. 3-1

1. If two angles of a triangle are equal in measure, then two sides of the triangle are equal in measure.

2. If the diagonals of a quadrilateral are perpendicular, then the quadrilateral is a rhombus.

3. If n is a prime number greater than 2, then n is a positive odd integer.

4. If at least one of two numbers, a and b, is positive, then their product, ab, is positive.

5. Use the diagram at the right. Write a coordinate proof of the statement: If the midpoints of the sides of a quadrilateral are joined in order, then the resulting figure is a parallelogram. 3-2

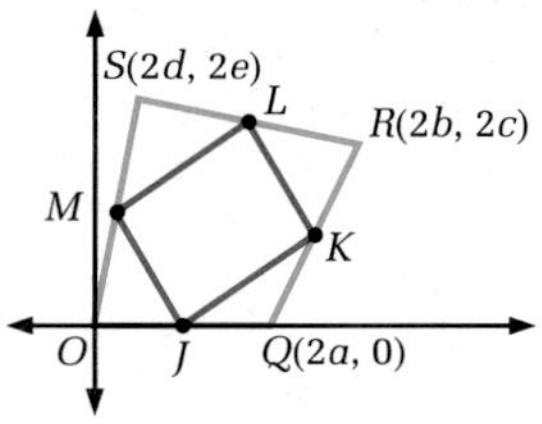

Suppose a kite is defined as a quadrilateral with at least one line of symmetry that passes through two opposite vertices. 3-3

6. Is this definition *inclusive* or *exclusive?*

7. Using this definition, is it true that if a quadrilateral is a square, then it is a kite?

8. Using this definition, is it true that if a quadrilateral is an isosceles trapezoid, then it is a kite?

9. Suppose the words *at least one* in the definition were changed to read *exactly one.* Would it then be true that if a quadrilateral is a rhombus, then it is a kite?

Find the measure of each interior angle and each exterior angle of a regular polygon with the given number of sides. 3-4

10. 5 **11.** 7 **12.** 18 **13.** 16

Answers to Extra Practice Unit 2

61. a. $(7x^2 + 1)\sqrt{x}$
 b. all real numbers greater than or equal to 0

62. a. $7x^2 + 1 + \frac{x}{2x - 1}$
 b. all real numbers except $\frac{1}{2}$

63. a. $\sqrt{7x^2 + 1}$
 b. all real numbers

64. a. $\frac{\sqrt{x}}{2\sqrt{x} - 1}$
 b. all real numbers greater than or equal to 0 and not equal to $\frac{1}{4}$

65. a. $\frac{(7x^2 + 1)(2x + 1)}{x}$
 b. all real numbers except 0 or $\frac{1}{2}$

66. a. $7x^2 + 1 - \sqrt{x}$
 b. all real numbers greater than or equal to 0

67. a. $7x + 1$
 b. all real numbers greater than or equal to 0

68. a. $\frac{x\sqrt{x}}{2x - 1}$
 b. all real numbers greater than or equal to 0 and not equal to $\frac{1}{2}$

Answers to Extra Practice Unit 3

1. True. If two sides of a triangle are equal in measure, then two angles of the triangle are equal in measure; True. If no two angles of a triangle are equal in measure, then no two sides of the triangle are equal in measure; True. If no two sides of a triangle are equal in measure, then no two angles of the triangle are equal in measure; True.

2. False. If a quadrilateral is a rhombus, then the diagonals of the quadrilateral are perpendicular; True. If the diagonals of a quadrilateral are not perpendicular, then the quadrilateral is not a rhombus; True. If a quadrilateral is not a rhombus, then the diagonals of the quadrilateral are not perpendicular; False.

ABCDEFGHIJ is a regular decagon. Find the measure of each arc or angle. 3-5

14. $\angle AOB$ **15.** $\overset{\frown}{AC}$

16. $\overset{\frown}{JFC}$ **17.** $\angle OAB$

18. $\angle G$ **19.** $\overset{\frown}{CF}$

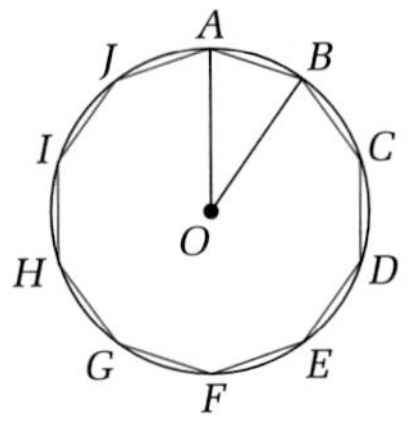

In the diagram at the right, $m\angle FCE = 55°$, $m(\overset{\frown}{FC}) = 130°$, and $\overset{\frown}{FCD}$ is a semicircle. Find the measure of each arc or angle. 3-6

20. $\angle CDF$ **21.** $\overset{\frown}{EF}$

22. $\overset{\frown}{DE}$ **23.** $\angle ECD$

24. $\overset{\frown}{CD}$ **25.** $\angle CDE$

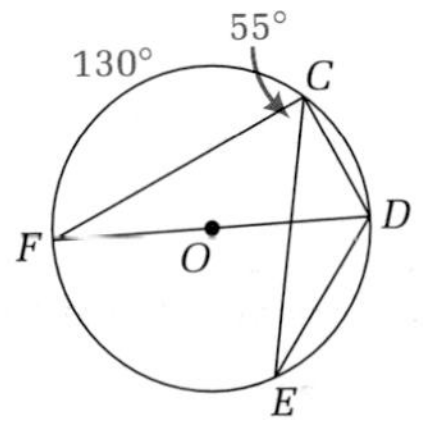

The diagram at the right shows two circles, O and P, placed so that the line *m* is tangent to *both* circles at point Q. 3-7

26. How do you know that $\overline{OQ}$ and $\overline{PQ}$ lie in the same line?

27. Assume temporarily that the circles have another point *R* in common. Draw $\overline{OQ}$, $\overline{PQ}$, $\overline{OR}$, and $\overline{PR}$. Compare the length *OP* to the sum *OR* + *PR*. Which is larger, or are they the same?

28. The triangle inequality states that the sum of the lengths of any two sides of a triangle is greater than the length of the third side. Use the triangle inequality in an indirect proof that two circles such as *O* and *P* have no other point in common other than *Q*.

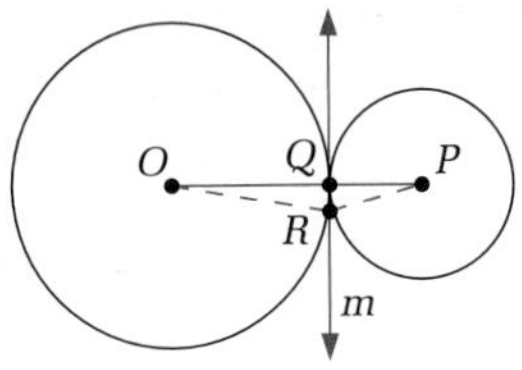

Use the diagram at the right, in which $\triangle BCD$ is an equilateral triangle circumscribed about a circle of radius 2. 3-8

29. Find $m\angle APB$.

30. Use trigonometry to find *DB*.

31. Find the area of $\triangle BCD$.

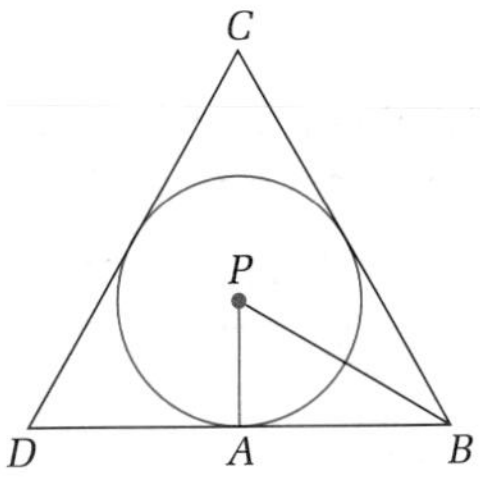

Find the area of each triangle with the given sides AB, BC, and AC, and with inscribed circle of radius *r*. 3-8

32. $AB = 17$; $BC = 25$; $AC = 26$; $r = 6$

33. $AB = 50$; $BC = 56$; $AC = 34$; $r = 12$

Write the reasoning you would use in an indirect proof of each fact. 3-9

34. A regular polyhedron cannot have 6 edges meeting at each vertex.

35. A regular polyhedron cannot have octagonal faces.

Answers to Extra Practice Unit 3

3. True. If *n* is a positive odd integer, then *n* is a prime number greater than 2; False. If *n* is not a prime number greater than 2, then *n* is not a positive odd integer; False. If *n* is not a positive odd integer, then *n* is not a prime number greater than 2; True.

4. False. If the product of two numbers *ab* is positive, then at least one of the two numbers is positive; False. If neither of two numbers *a* and *b* is positive, then their product *ab* is not positive; False. If the product of two numbers *ab* is not positive, then neither of the numbers *a* and *b* is positive; False.

5. The midpoints are $J(a, 0)$, $K(a + b, c)$, $L(b + d, c + e)$, $M(d, e)$. The slope of both $\overline{JK}$ and $\overline{ML}$ is $\frac{c}{b}$; the slope of both $\overline{KL}$ and $\overline{MJ}$ is $\frac{e}{d - a}$. Thus, by definition, *JKLM* is a parallelogram.

6. inclusive

7. Yes. 8. No.

9. No. 10. 108°; 72°

11. about 128.6°; about 51.4°

12. 160°; 20°

13. 157.5°; 22.5°

14. 36° 15. 72°

16. 252° 17. 72°

18. 144° 19. 108°

20. 65° 21. 110°

22. 70° 23. 35°

24. 50° 25. 120°

26. They are both ⊥ to line *m* at point *Q*.

27. They are the same, since each is equal to the sum of the radii of the two circles.

28. Assume there is another point, *R*, which is on both circles. Then $OQ = OR$ and $PQ = PR$. $PO = PQ + OQ$ (from Ex. 26). $PO = PR + OR$ (from Ex. 27). But this means that in $\triangle POR$, the sum of the lengths of two sides ($\overline{PR}$ and $\overline{OR}$) is equal to the length of the third side ($\overline{PO}$). This contradicts the triangle inequality. So there cannot be a point other than *Q* in common with both circles.

29. 60°

30. $4\sqrt{3} \approx 6.93$

31. $12\sqrt{3} \approx 20.8$ units^2

32. 204 units^2

33. 840 units^2

34. Assume temporarily that there is such a regular polyhedron. Then 6 faces, all congruent regular polygons, would meet at each vertex. The smallest angle a regular polygon can have is 60°. Thus, the defect at each vertex would be 0. This is impossible.

35. Assume temporarily that a regular polyhedron with octagonal faces exists. Then at each vertex at least three regular octagons meet. This is impossible, since each has interior angles that measure 135°, and $3 \cdot 135 = 405° > 360°$.

Answers to
Extra Practice Unit 4

1. a. 6, −7, 8
   b. −11
2. a. 200, 2000, 20,000
   b. 2,000,000
3. a. 8, 5, 2
   b. −7
4. a. $\frac{1}{162}, \frac{1}{486}, \frac{1}{1458}$
   b. $\frac{1}{39{,}366}$
5. a. 4, 2, 1
   b. 4
6. a. −25, 36, −49
   b. 100
7. Yes; 1.
8. No.
9. Yes; $\frac{2}{3}$.
10. −7, −4, −1, 2; 50
11. $\frac{1}{2}, \frac{4}{3}, \frac{9}{4}, \frac{16}{5}; \frac{400}{21}$
12. $\frac{1}{4}, \frac{1}{8}, \frac{1}{16}, \frac{1}{32}; \frac{1}{2{,}097{,}152}$
13. $\frac{1}{3}, \frac{3}{5}, \frac{5}{7}, \frac{7}{9}; \frac{39}{41}$
14. $a_n = \frac{1}{n^2}$
15. $a_n = 3 + 2n$
16. $a_n = 1 + \frac{1}{2}n$
17. $a_n = 1500\left(\frac{1}{10^n}\right)$
18. $a_n = \frac{n+1}{n}$
19. $a_n = 625\left(\frac{1}{5}\right)^n$
20. 79, 75, 71, 67, 63, 59
21. 80, 40, 20, 10, 5, 2.5
22. 24, 12, 4, 1, $\frac{1}{5}, \frac{1}{30}$
23. 4, $\frac{1}{2}$, 4, $\frac{1}{2}$, 4, $\frac{1}{2}$
24. 256, 16, 4, 2, $\sqrt{2}$, $\sqrt{\sqrt{2}}$
25. 10, 2, 6, $\frac{8}{3}, \frac{19}{4}, \frac{59}{19}$
26. a. arithmetic
    b. $a_1 = 12$; $a_n = a_{n-1} - 8$
27. a. geometric
    b. $a_1 = 16$; $a_n = \frac{5}{4}a_{n-1}$
28. a. geometric
    b. $a_1 = 7$; $a_n = -a_{n-1}$
29. a. geometric
    b. $a_1 = 8$; $a_n = \frac{1}{10}a_{n-1}$
30. a. neither
    b. $a_1 = 1$; $a_n = a_{n-1} + \frac{1}{10^{n-1}}$
31. a. arithmetic
    b. $a_1 = \frac{1}{2}$; $a_n = a_{n-1} - 2$

Unit 4

For each sequence in Exercises 1–6: 4-1

a. Find the next three terms.

b. Find the 10th term.

1. 2, −3, 4, −5, . . .
2. 0.02, 0.2, 2, 20, . . .
3. 20, 17, 14, 11, . . .
4. $\frac{1}{2}, \frac{1}{6}, \frac{1}{18}, \frac{1}{54}, \ldots$
5. 4, 2, 1, 4, 2, 1, . . .
6. −1, 4, −9, 16, . . .

Tell whether each infinite sequence appears to have a limit. If so, give its value. 4-1

7. 0.9, 1.1, 0.99, 1.01, . . .
8. $\frac{3}{2}, \frac{9}{4}, \frac{27}{8}, \frac{81}{16}, \ldots$
9. $\frac{3}{5}, \frac{33}{50}, \frac{333}{500}, \frac{3333}{5000}, \ldots$

Write the first four terms and the 20th term of each sequence. 4-2

10. $a_n = 3n - 10$
11. $b_n = \frac{n^2}{n+1}$
12. $a_n = \frac{1}{2^{n+1}}$
13. $t_n = \frac{2n-1}{2n+1}$

Write an explicit formula for each sequence. 4-2

14. $1, \frac{1}{4}, \frac{1}{9}, \frac{1}{16}, \ldots$
15. 5, 7, 9, 11, . . .
16. $\frac{3}{2}, 2, \frac{5}{2}, 3, \ldots$
17. 150, 15, 1.5, 0.15, . . .
18. $\frac{2}{1}, \frac{3}{2}, \frac{4}{3}, \frac{5}{4}, \ldots$
19. 125, 25, 5, 1, . . .

Write the first six terms of each sequence. 4-3

20. $a_1 = 79$
$a_n = a_{n-1} - 4$
21. $a_1 = 80$
$a_n = \frac{1}{2}a_{n-1}$
22. $a_1 = 24$
$a_n = \frac{a_{n-1}}{n}$
23. $a_1 = 4$
$a_n = \frac{2}{a_{n-1}}$
24. $a_1 = 256$
$a_n = \sqrt{a_{n-1}}$
25. $a_1 = 10$
$a_n = \frac{a_{n-1} + 10}{a_{n-1}}$

For each sequence in Exercises 26–31: 4-4

a. Decide whether the sequence is *arithmetic, geometric,* or *neither.*

b. Write a recursive formula.

26. 12, 4, −4, −12
27. 16, 20, 25, 31.25, . . .
28. 7, −7, 7, −7, . . .
29. $8, \frac{4}{5}, \frac{2}{25}, \frac{1}{125}, \ldots$
30. 1, 1.1, 1.11, 1.111, . . .
31. $\frac{1}{2}, -\frac{3}{2}, -\frac{7}{2}, -\frac{11}{2}, \ldots$

For each sequence in Exercises 32–37: 4-4

a. Find *x* when the sequence is arithmetic.

b. Find *x* when the sequence is geometric.

32. 3, *x*, 12
33. −4, *x*, −25
34. 5, *x*, 45
35. 2, *x*, 100
36. 0.5, *x*, 8
37. −3, *x*, −300

32. a. 7.5 b. ±6
33. a. −14.5 b. ±10
34. a. 25 b. ±15
35. a. 51 b. $\pm 10\sqrt{2}$
36. a. 4.25 b. ±2
37. a. −151.5 b. ±30
38. 207
39. 105
40. 27.5
41. 364
42. $\sum_{n=1}^{9} 4^n$
43. $\sum_{n=1}^{12} (6 + 10n)$
44. $\sum_{n=1}^{12} \frac{n-1}{3^n}$
45. $\sum_{n=1}^{9} (9n - 21)$
46. 461.16
47. 513
48. $\frac{665}{486}$
49. 55.55555
50. a. neither b. 63
51. a. geometric
    b. $\frac{-189}{64} = -2.953125$
52. a. arithmetic
    b. 588
53. a.

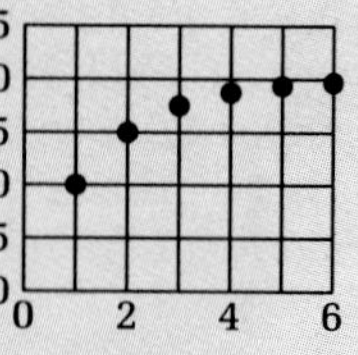

    b. Yes; 20.

Find the sum of each finite arithmetic series. 4-5

38. $(-5) + 2 + 9 + 16 + \cdots + 51$

39. $2 + 2.3 + 2.6 + 2.9 + \cdots + 8$

40. $15 + 12.5 + 10 + 7.5 + \cdots + (-10)$

41. $4 + \frac{14}{3} + \frac{16}{3} + 6 + \cdots + 22$

Write each series in sigma notation. 4-5

42. $4^2 + 4^3 + 4^4 + \cdots + 4^9$

43. $16 + 26 + 36 + 46 + \cdots + 126$

44. $\frac{4}{3^5} + \frac{5}{3^6} + \frac{6}{3^7} + \frac{7}{3^8} + \cdots + \frac{11}{3^{12}}$

45. $(-12) + (-3) + 6 + 15 + \cdots + 60$

Find the sum of each series. 4-6

46. $125 + 100 + 80 + 64 + \cdots + \frac{1024}{25}$

47. $3 + (-6) + 12 + (-24) + \cdots + 768$

48. $\frac{1}{2} + \frac{1}{3} + \frac{2}{9} + \frac{4}{27} + \cdots + \frac{16}{243}$

49. $50 + 5 + 0.5 + 0.05 + \cdots + 0.00005$

For each series in Exercises 50–52: 4-6

a. Decide whether the series is *arithmetic, geometric,* or *neither.*

b. Find the sum.

50. $1 + 2 + 4 + 7 + \cdots + 22$

51. $\sum_{n=1}^{6} -3(0.5)^n$

52. $\sum_{k=0}^{11} 8(k+1) - 3$

For each series in Exercises 53–58: 4-7

a. Graph the first six partial sums.

b. Decide whether the series has a sum. If it does, give its value.

53. $10 + 5 + 2.5 + 1.25 + \cdots$

54. $0.64 + 0.8 + 1 + 1.25 + \cdots$

55. $17 + 11 + 5 + (-1) + \cdots$

56. $\frac{5}{3} + \frac{5}{9} + \frac{5}{27} + \frac{5}{81} + \cdots$

57. $\sum_{n=1}^{\infty} 30\left(\frac{4}{5}\right)^n$

58. $\sum_{n=1}^{\infty} \left(\frac{1}{10^n} + 3\right)$

Unit 5

After reaching a peak, the concentration of medication in a person's blood stream decreases exponentially. Suppose the medication loses 18% of its concentration every hour. Suppose that the concentration y of the medication (in μg/mL) at time t (in hours) is modeled by the equation $y = (1.4)(0.82)^t$. 5-1

1. Graph the equation.
2. Find the peak concentration.
3. Find the time when half of the peak concentration of the medication remains.
4. Find the time when the concentration is less than 0.4 μg/mL.

Answers to Extra Practice Unit 5

1.
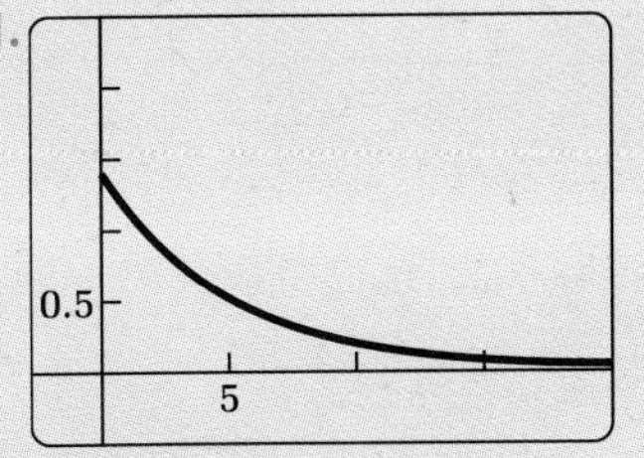

2. 1.4 μg/mL
3. 3.49 h
4. 6.32 h

Answers to Extra Practice Unit 4

54. a.
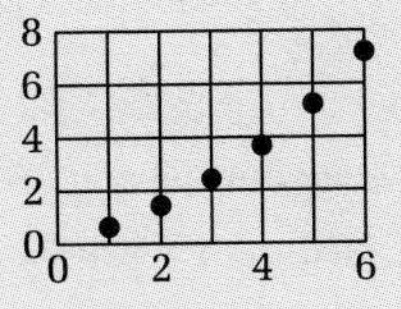

b. No.

55. a.
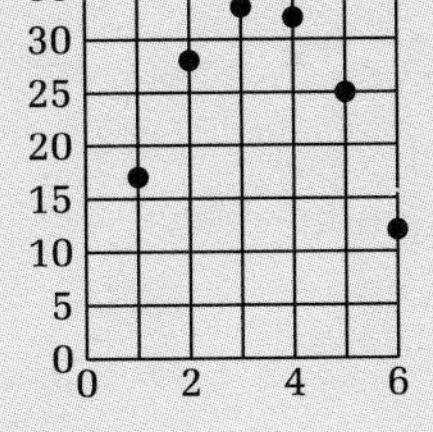

b. No.

56. a.
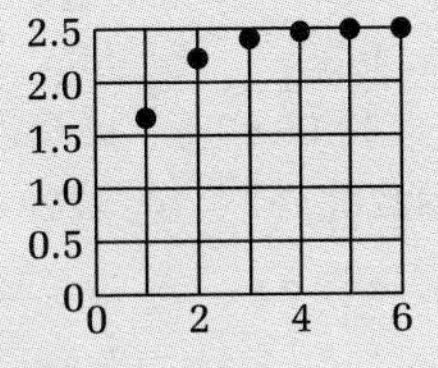

b. Yes; 2.5.

57. a.
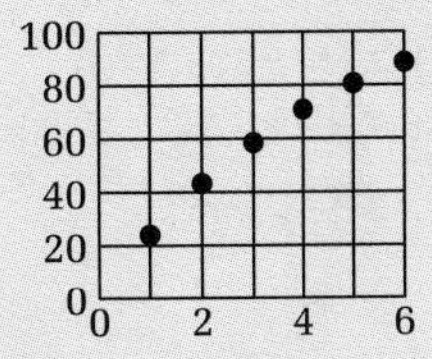

b. Yes; 120.

58. a.
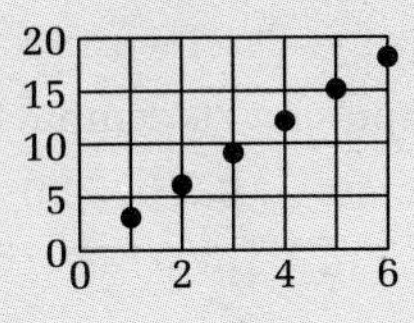

b. No.

Since 1980, when the Chavez Park public school system installed its first computers, the number of computers in the school system has been growing exponentially. There were 320 computers in the system in 1990 and 368 computers in the system in 1991. 5-2

5. Write an exponential function that models the number of computers in the Chavez Park school system.

6. How many computers were there in the school system in 1985? In 1980?

7. How many computers did the school system have in 1970?

Simplify. 5-3

8. $100^{5/2}$ **9.** $9^{1/3} \cdot 9^{1/6}$ **10.** $40^{2/3}$ **11.** $\left(\frac{25}{16}\right)^{3/4}$

Simplify. 5-3

12. $\sqrt{x^7y^8}$ **13.** $\sqrt[3]{8a^4b^6}$ **14.** $\sqrt{25p^{-8}}$ **15.** $\sqrt[3]{w^2} \cdot \sqrt[6]{w^{-2}}$

Find the value of each function, to the nearest hundredth, for (a) $x = 1.5$ and (b) $x = -0.8$. 5-4

16. $f(x) = 480e^{-2.6x}$ **17.** $f(x) = \frac{1}{1 + 14e^{-1.2x}}$ **18.** $f(x) = 22(e^{x/25} + e^{-x/25})$

Find the value of an investment of \$1500 for five years at an annual rate of 8%, if interest is compounded as specified. 5-4

19. annually **20.** monthly **21.** daily

Write an equation for the inverse of each function. 5-5

22. $y = 5 - 4x$ **23.** $y = \frac{3}{7x}$ **24.** $y = \frac{1}{8}x^3$ **25.** $y = \sqrt{x} + 2, x \geq 0$

26. $y = (x + 3)^{5/3}$ **27.** $y = |2x|, x \geq 0$ **28.** $y = (x - 3)^2, x \leq 3$ **29.** $y = x^3 + 1$

Evaluate each logarithm. Round decimal answers to the nearest hundredth. 5-6

30. $\log \frac{1}{\sqrt{10}}$ **31.** $\log_3 81$ **32.** $\log 40$ **33.** $\log_5 \frac{1}{125}$

34. $\ln 20$ **35.** $\log_6 1$ **36.** $\ln e^3$ **37.** $\log_2 \sqrt{32}$

Solve. Round decimal answers to the nearest hundredth. 5-6

38. $\log_x 1{,}000{,}000 = 3$ **39.** $\log_x 25 = -2$ **40.** $e^x = \sqrt[3]{e}$

41. $\ln x = 0$ **42.** $e^x = 17$ **43.** $\log x = -0.25$

Write each expression as a single logarithm. 5-7

44. $\log 48 - 3 \log 2$ **45.** $\log_3 14 + \frac{1}{2} \log_3 16$ **46.** $\frac{1}{3} \log 64 - 2 \log 4 + \log 2$

Answers to Extra Practice Unit 5

5. $y = 320(1.15)^x$, where x is the number of years since 1990
6. 159 computers; 79 computers
7. 0 computers
8. 100,000
9. 3
10. $4\sqrt[3]{25}$
11. $\frac{5\sqrt{5}}{8}$
12. $x^3y^4\sqrt{x}$
13. $2ab^2\sqrt[3]{a}$
14. $5p^{-4}$
15. $\sqrt[3]{w}$
16. a. 9.72 b. 3842.15
17. a. 0.30 b. 0.03
18. a. 44.08 b. 44.02
19. \$2203.99
20. \$2234.77
21. \$2237.74
22. $y = \frac{5 - x}{4}$
23. $y = \frac{3}{7x}$
24. $y = 2\sqrt[3]{x}$
25. $y = (x - 2)^2, x \geq 2$
26. $y = x^{3/5} - 3$
27. $y = \frac{x}{2}, x \geq 0$
28. $y = -\sqrt{x} + 3, x \geq 0$
29. $y = \sqrt[3]{x - 1}$
30. −0.5
31. 4
32. 1.60
33. −3
34. 3.00
35. 0
36. 3
37. 2.5
38. 100
39. 0.20
40. 0.33
41. 1
42. 2.83
43. 0.56
44. log 6
45. $\log_3 56$
46. log 0.5
47. $\log_7 A + 0.5 \log_7 B$
48. $0.5 \log_7 A - 3 \log_7 B$
49. $0.25(\log_7 A + 5 \log_7 B)$
50. 3
51. 1
52. 0
53. 1.95
54. 0.39
55. 26.40
56. 20
57. $\frac{7}{17}$
58. 6
59. 4, 20

Write each expression in terms of $\log_7 A$ and $\log_7 B$. 5-7

47. $\log_7 A\sqrt{B}$

48. $\log_7 \frac{\sqrt{A}}{B^3}$

49. $\log_7 \sqrt[4]{AB^5}$

Evaluate. 5-7

50. $\log 2000 - \log 2$

51. $\frac{1}{2}(\log_6 4 + \log_6 9)$

52. $\log_5 25 - 4 \log_5 \sqrt{5}$

Solve. 5-8

53. $4^x = 15$

54. $12^{2x} = 7$

55. $8(1.05)^x = 29$

Solve. 5-8

56. $\log x + \log (x - 15) = 2$

57. $\log_6 3x + \log_6 (x + 7) = -1$

58. $2 \log_4 x - \log_4 (x + 3) = 1$

59. $2 \log_2 x - \log_2 (3x - 10) = 3$

Unit 6

The ages, in months, of members of a mathematics class are given in the table at the right. 6-1

1. Draw a histogram for the age data.

2. Identify the type of distribution.

3. Find the mean and the median ages of the members of Jim Abbott's mathematics class.

Ages of Students in Jim Abbott's Mathematics Class (in months)

180	192	165	178	182
185	184	182	194	198
172	176	178	184	186
192	186	188	185	182
176	196	188	176	168
174	172	170	183	191

The table at the right shows intervals for the land areas of the 50 states and the number of states that fall into each interval. 6-1

4. Complete the table.

5. Draw a relative frequency histogram for the data.

6. Identify the type of distribution for the histogram.

7. What percent of the states have land areas less than 60,000 mi^2?

Land Area (1000 mi^2)	No. of states	Percent of total
0–19	9	?
20–39	6	?
40–59	15	?
60–79	7	?
80–99	5	?
100–119	3	?
120–139	1	?
140–159	2	?
160 and over	2	?

Answers to Extra Practice Unit 6

1. Histograms may vary. An example is given.

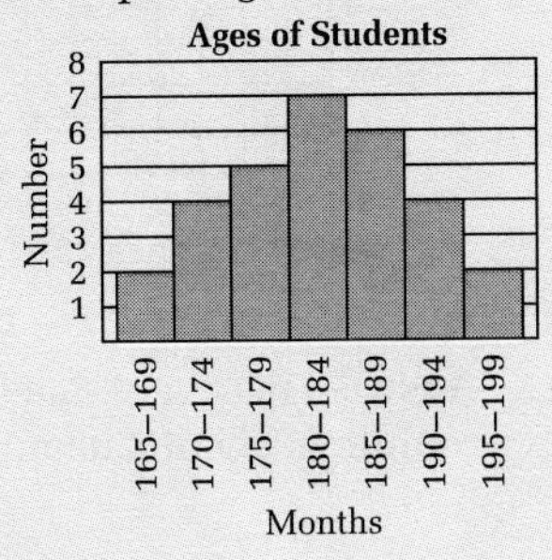

2. mound-shaped

3. 182.1; 182.5

4.

Land Area (1000 mi^2)	No. of states	Percent of total
0–19	9	18
20–39	6	12
40–59	15	30
60–79	7	14
80–99	5	10
100–119	3	6
120–139	1	2
140–159	2	4
160 and over	2	4

5.

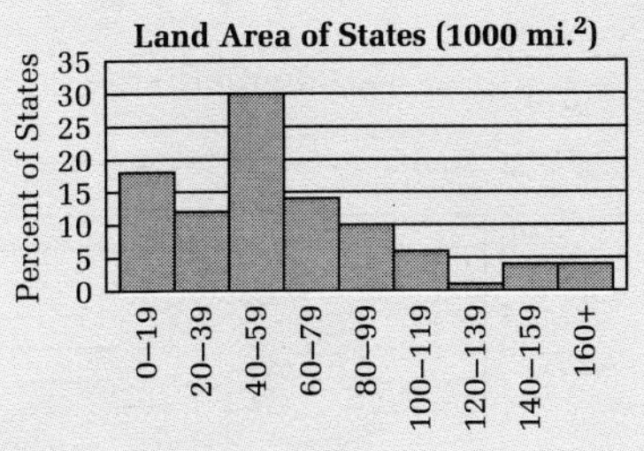

6. skewed left

7. 60%

For Exercises 8–11, the table shows a sampling of prices of regular unleaded gasoline in two communities on the same day. 6-2

	Prices of one gallon of unleaded regular gasoline ($)
Midway	1.38 1.25 1.23 1.31 1.32 1.36 1.21 1.34 1.32 1.28 1.39 1.24 1.25
Fairview	1.32 1.31 1.28 1.36 1.35 1.32 1.33 1.37 1.33 1.31 1.32 1.30 1.34

8. Find the mean, the range, and the standard deviation of the data for Midway.

9. Find the mean, the range, and the standard deviation for Fairview.

10. For each of the two data sets, find the percent of the data within one standard deviation of the mean.

11. Suppose histograms based on the two data sets were drawn on the same axes using two different shadings for the bars. Without graphing, predict how the two histograms would differ.

For Exercises 12–14: 6-3

a. Draw a histogram for each data set.

b. Find the mean and the median of each data set.

c. Decide whether the data appears to show a normal distribution.

12. times an airport shuttle bus took to reach the downtown stop (min)

23, 15, 18, 26, 17, 20, 21, 17, 14, 16, 22, 24, 18, 13, 21, 24, 22, 19, 22, 23, 24, 15, 20, 21, 19, 20, 22, 23, 19, 16

13. numbers of raisins in a sample of home-baked oatmeal cookies

8, 7, 8, 4, 9, 6, 5, 5, 6, 7, 4, 10, 8, 11, 9, 12, 10, 6, 7, 7, 9, 8, 10, 9, 9, 8, 11, 8, 6, 7

14. duration of music tracks on albums by the group Live Bait (min)

3.2, 3.3, 5.6, 2.4, 4.2, 2.4, 3.0, 2.8, 4.6, 1.8, 3.2, 4.6, 2.4, 2.0, 1.4, 4.2, 5.0, 2.0, 3.4, 3.6, 5.2, 4.2, 1.6, 3.8, 4.0

For Exercises 15–18, the table shows laboratory trials of an insect repellent. For each trial, various amounts of repellent were applied to a test patch. The numbers of insects landing on the patch during a fixed period of time were recorded. 6-4

Amount of repellent used (oz)	0.5	1.0	1.5	2.0	2.5	3.0	3.5	4.0	4.5
No. of insects landing	38	34	33	30	27	25	20	16	14

15. Make a scatter plot of the data.

16. Find an equation of the least-squares line.

17. Find an equation of the median-median line.

18. How do these two lines compare in this situation?

Answers to Extra Practice Unit 6

8. mean ≈ 1.30; range = 0.18; $\sigma \approx 0.057$

9. mean ≈ 1.33; range = 0.09; $\sigma \approx 0.024$

10. Midway: about 69%; Fairview = about 69%

11. Bars of the the second would be shifted to the right and more closely bunched around the mean.

12–14. Histograms may vary depending on intervals chosen. Examples are given.

12. a.

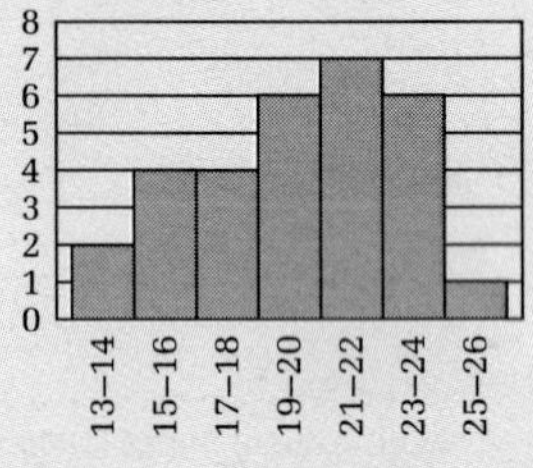

b. mean ≈ 19.8; median = 20

c. No.

13. a.

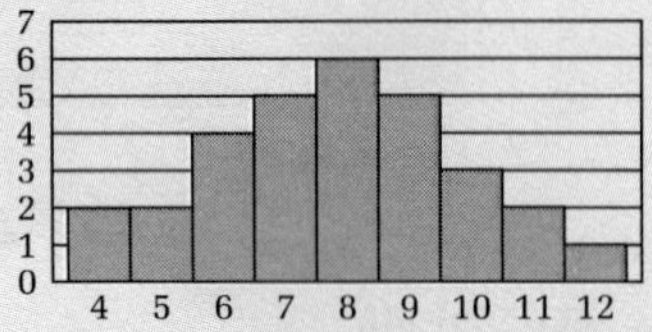

b. mean = 7.8; median = 8

c. Yes.

14. a.

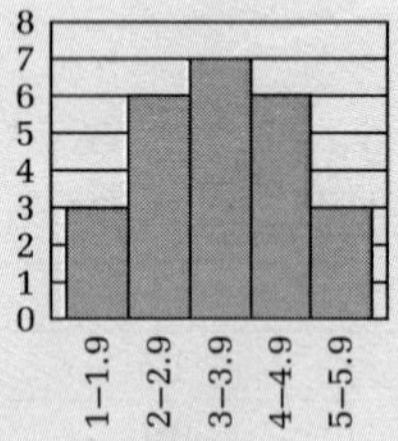

b. mean ≈ 3.4; median = 3.3

c. Yes.

15.

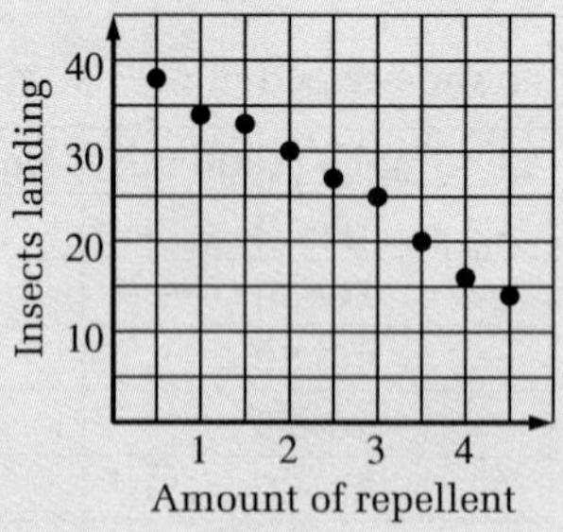

16. $y = -6.0x + 41.4$

17. $y = -6x + 40.7$

18. They are nearly identical.

19, 21.

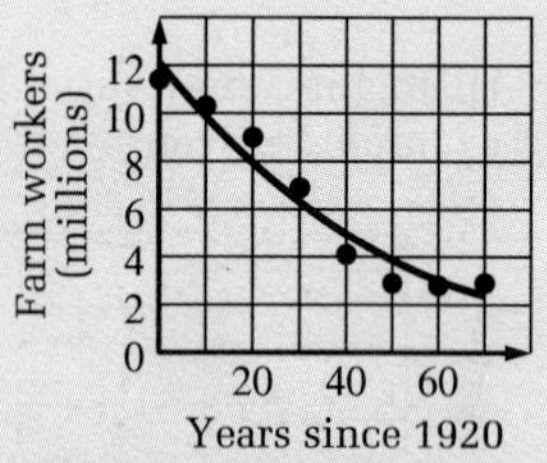

20. $y = 0.0013x^2 - 0.23x + 12.1$

22. Answers may vary due to rounding. about 2.02 million

For Exercises 19–22, the table below shows the numbers of Americans employed on farms in census years since 1920. 6-5

Year	1920	1930	1940	1950	1960	1970	1980	1990
No. of farm workers (millions)	11.4	10.3	9.0	6.9	4.1	2.9	2.8	2.9

19. Make a scatter plot of the data. (Use "Years since 1920" for the x-values.)

20. Find a quadratic model for the data.

21. Graph the quadratic model on the same set of axes as the scatter plot.

22. Use your model to predict the number of farm workers in the year 2000.

Unit 7

Design a formula for a calculator to produce each set of random numbers. 7-1

1. integers from 1 to 6

2. integers from 10 to 20

For Exercises 3 and 4, design a simulation to investigate each situation. 7-1

3. Suppose 5 cards are drawn from a standard deck of cards and each card is replaced before the next card is drawn. What is the probability that cards drawn include at least two cards of the same denomination (that is, two threes, two kings, etc.)?

4. In a certain state, lottery ticket numbers are four-digit numbers. If any three digits match digits of the winning number, the ticket holder wins a prize. Matching digits do not have to be in the same order. Suppose the winning number consists of four different digits. What is the probability of buying a winning ticket?

A card is drawn from a standard deck. Find the probability of each event. 7-2

5. It is a king or a diamond.

6. It is a face card or a red card.

You look at the second hand of a clock at two random times during the day and record the position as a real number between 0 and 60. Find the probability of each event. 7-2

7. The sum of the two readings will be greater than 80?

8. Both of the readings will be less than 20?

You roll a die twice. Find the probability of each event. 7-2

9. The sum of the numbers showing is at least 9.

10. The first number is odd or the second number is 6.

Answers to Extra Practice Unit 7

1. int(6 * rand) + 1
2. int(11 * rand) + 10
3. On a calculator, choose groups of 5 random integers from 1 to 13. Count the number of duplicate integers in each group. Perform a large number of trials. The experimental probability is $\frac{\text{number of times at least two integers match}}{\text{number of trials}}$.
4. Pick a 4-digit number to be the sample winning number. All four digits must be different. Generate 4 random number from 0 to 9. A success is defined as "3 or 4 of the random numbers match different numbers of the sample winning number." Repeat this trial many times. The experimental probability of having a winning ticket is $\frac{\text{number of successes}}{\text{number of trials}}$.
5. $\frac{4}{13}$
6. $\frac{8}{13}$
7. $\frac{2}{9}$
8. $\frac{1}{9}$
9. $\frac{5}{18}$
10. $\frac{7}{12}$

Dao Nguyen has $500 in his bank account. He wants to buy a CD player, which will cost between $100 and $300, and a pair of speakers in the same price range. 7-3

11. What is the probability that he will have enough in his bank account to cover the cost of those items?

12. What is the probability that the speakers will cost at least $50 more than the CD player?

The table shows the amount of time a sample of students spent listening to music. Use the table in Exercises 13 and 14. 7-4

Hours per week	Freshmen	Sophomores	Juniors	Seniors	Totals
0–10	8	10	20	24	62
10–20	18	14	16	12	60
20–30	22	18	10	8	58
Totals	48	42	46	44	180

13. Find $P(\text{10–20 hours per week} \mid \text{sophomore})$, $P(\text{sophomore})$, and $P(\text{10–20 hours per week and sophomore})$.

14. Find $P(\text{10–20 hours per week})$. Are the events "$X$ is a sophomore" and "X listens to music 10–20 hours per week" independent?

Box *A* contains 10 white and 10 black marbles. Box *B* contains 10 white and 20 black marbles. One of the two boxes is chosen at random. Then a marble is drawn at random from that box. Find each probability. 7-4

15. $P(\text{from Box } A \text{ and black})$ **16.** $P(\text{black})$ **17.** $P(\text{from Box } A \mid \text{black})$

In a board game, you roll a die to determine your moves. A roll of 5 or 6 advances your token 4 squares, a roll of 2, 3, or 4 advances your token 3 squares, and a roll of 1 puts your token back 1 square. 7-5

18. How many squares do you expect to advance each turn?

19. In certain situations during the game, you may pick a card, but you may choose not to. There is an equal likelihood that the card will say "Advance 6 spaces" or "Go back 1 space." Should you pick the card?

At a certain intersection, the light for northbound traffic switches between red for 15 seconds and green for 35 seconds. Find the probability of each event. 7-6

20. Out of the next 6 northbound cars, exactly 4 will hit the light red.

21. Out of the next 6 northbound cars, 4 or more will hit the light red.

22. Consider the event: "Out of the next 12 northbound cars, exactly 8 will hit the light red." Would you expect the probability of this event to be smaller or larger than the answer to Exercise 20? Calculate this probability.

Answers to Extra Practice Unit 7

11. $\frac{7}{8}$

12. $\frac{9}{32}$

13. $\frac{1}{3}; \frac{7}{30}; \frac{7}{90}$

14. $\frac{1}{3}$; Yes.

15. $\frac{1}{4}$

16. $\frac{7}{12}$

17. $\frac{3}{7}$

18. $\frac{8}{3}$

19. The average expected advance from picking a card is 2.5 spaces. You should pick the card unless rolling the die is an option. Rolling the die has a greater expected advance.

20. 0.116

21. 0.146

22. smaller; 0.030

Unit 8

Graph each point on polar graph paper. 8-1

1. (4, 150°) 2. (2, −70°) 3. (3, 230°)
4. (1.5, −110°) 5. (3.5, 110°) 6. (4, 315°)

Convert the polar coordinates to rectangular coordinates. Give answers to the nearest hundredth. 8-2

7. (7, 270°) 8. (3, 120°) 9. (4, 225°) 10. $(5\sqrt{2}, 45°)$
11. (2, −155°) 12. (6, 32°) 13. (5, 142°) 14. (8, −40°)

Convert each set of rectangular coordinates to polar coordinates. 8-2

15. $(-2, 2\sqrt{3})$ 16. $(-6\sqrt{2}, -6\sqrt{2})$ 17. (4, −3)
18. (12, 5) 19. (−4.2, 3.4) 20. (−6.5, −1)

Use the diagram at the right for Exercises 21–29. Write each vector in rectangular form. 8-3

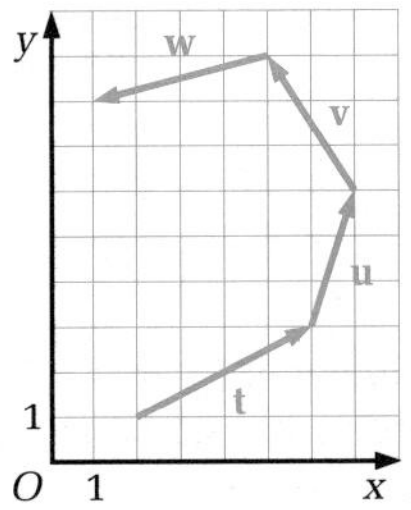

21. **t** 22. **u** 23. **v**
24. **w** 25. 2**u** 26. −**v**
27. **t** + **u** 28. **u** + **v** 29. **v** + **w**

Let t = (5, 240°) and u = (3, 90°). Use a protractor and ruler to draw each sum. Write the resultant vector in polar and rectangular form. 8-3

30. −**t** 31. 3**u** 32. **t** + **u**
33. **u** − **t** 34. **t** − 2**u** 35. 2**t** + **u**

Find each vector sum in rectangular form. 8-4

36. (2, −7) + (5, 4) 37. (−5, −8) + (3, 8) 38. (4, −2) + (11, −9)
39. (3.2, 0) + (−4.6, −1.5) 40. $(3\sqrt{2}, 4) + (-\sqrt{2}, -3)$ 41. (0.2, −1.6) + (−1.9, 0.3)

Find each vector sum in polar form. 8-4

42. (3, 40°) + (2, −80°) 43. (1, 130°) + (4, 270°) 44. (2, 15°) + (3, 148°)
45. (4, −160°) + (3, 20°) 46. (5, 112°) + (4, 310°) 47. (6, 200°) + (7, 75°)

Write parametric equations for the motion of an object thrown with a velocity vector v (with magnitude in m/s). 8-5

48. **v** = (45, 30°) 49. **v** = (63, 28°)

Answers to Extra Practice Unit 8

Answers may vary due to rounding.

1–6.

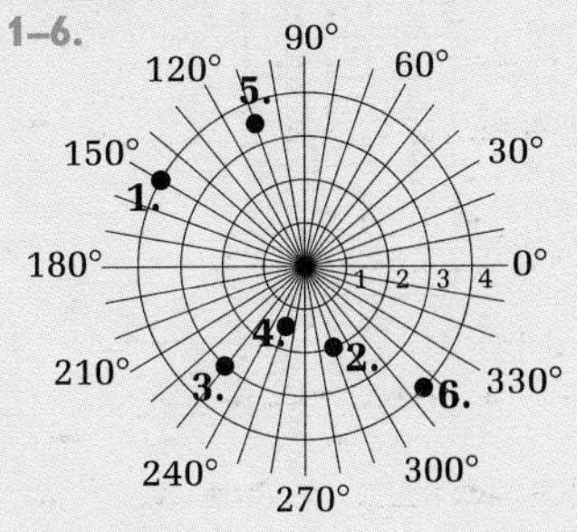

7. (0, –7)
8. (–1.5, 2.60)
9. (–2.83, –2.83)
10. (5, 5)
11. (–1.81, –0.85)
12. (5.09, 3.18)
13. (–3.94, 3.08)
14. (6.13, –5.14)
15. (4, 120°)
16. (12, 225°)
17. (5, –36.9°)
18. (13, 22.6°)
19. (5.4, 141°)
20. (6.58, –171.3°)
21. (4, 2)
22. (1, 3)
23. (–2, 3)
24. (–4, –1)
25. (2, 6)
26. (2, –3)
27. (5, 5)
28. (–1, 6)
29. (–6, 2)
30–35. Check students' diagrams.
30. (5, 120°); (–2.5, 4.33)
31. (9, 90°); (0, 9)
32. (2.83, –152°); (–2.5, –1.33)
33. (7.74, 71.2°); (2.5, 7.33)
34. (10.63, –103.6°); (–2.5, –10.33)
35. (7.55, –131.5°); (–5, –5.66)
36. (7, –3)
37. (–2, 0)
38. (15, –11)
39. (–1.4, –1.5)
40. $(2\sqrt{2}, 1)$
41. (–1.7, –1.3)
42. (2.66, –0.89°)
43. (3.30, –101.2°)
44. (2.19, 106.2°)
45. (1, –20°)
46. (1.72, 66.1°)
47. (6.07, 129.1°)
48. $x = 39.0t$; $y = 22.5t - 4.9t^2$
49. $x = 55.6t$; $y = 29.6t - 4.9t^2$

Answers to Extra Practice Unit 8

50. $y = 0.69x - 0.38x^2$
51. $y = 0.37x - 0.065x^2$
52. $(x - 4)^2 + (y + 7)^2 = 9$
53. $(x - 1)^2 + (y - 2)^2 = \frac{1}{4}$
54. $\frac{x^2}{16} + \frac{y^2}{81} = 1$
55. $\frac{(x - 3)^2}{36} + \frac{(y + 1)^2}{25} = 1$
56. 5.6
57. 19.8
58. 5.4
59. 108.6°
60. 35.5°
61. $a = 6.72$, $m\angle C = 86°$, $b = 9.69$
62. $m\angle B = 40.3°$, $m\angle C = 16.7°$, $c = 5.39$
63. $m\angle C = 43.9°$, $m\angle A = 64.1°$, $a = 8.0$
64. $b = 5.0$, $m\angle A = 99°$, $c = 8.6$
65. $m\angle A = 51.5°$, $m\angle B = 14.5°$, $b = 6.9$
66. $a = 3.8$, $m\angle B = 43°$, $c = 11.8$
67. 0
68. 1
69. 2
70. 1

Answers to Extra Practice Unit 9

1. 66.7; 62.5; 34.9
2. 76.7; 72.5, 34.9
3. $y = 10^x$; (1, –2)
4. $y = \sqrt{x}$; (–4, 6)
5. $y = \log x$; (9, 0)
6. $y = x^3$; (11, 5)
7. $y = |x|$; (0, –8)
8. $y = x^2$; (–7, –4)
9. $y - 4 = (x - 3)^3$

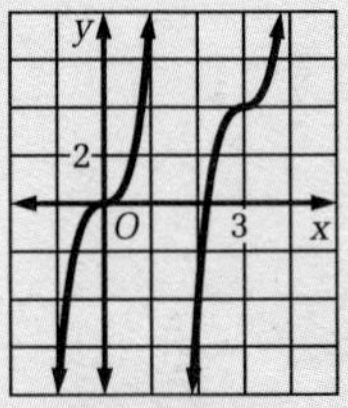

10. $y - 2 = \log (x + 4)$

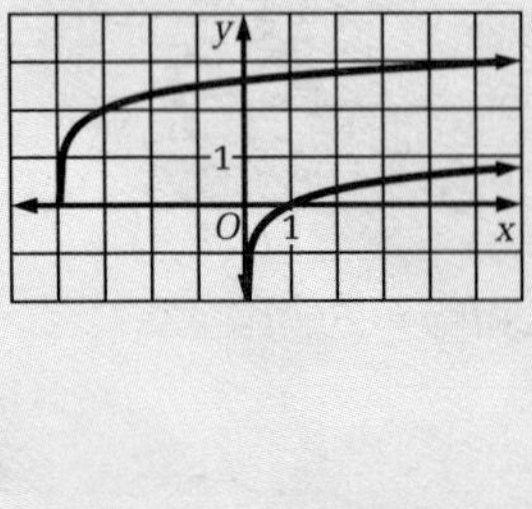

11. $y + 5 = |x - 3|$

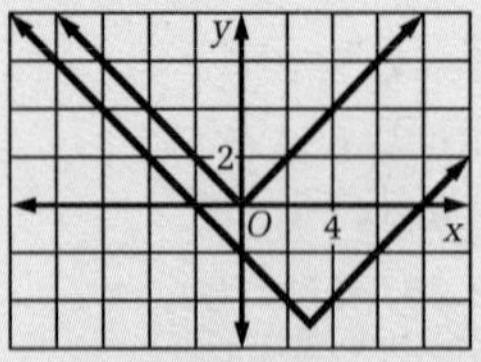

12. $y + 3 = \sqrt{x + 2}$

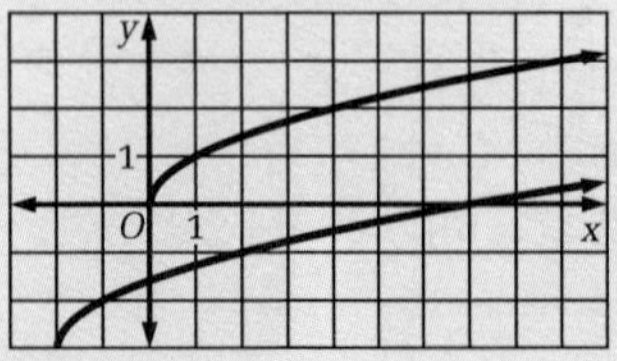

13. $y + 5 = (x - 1)^2$

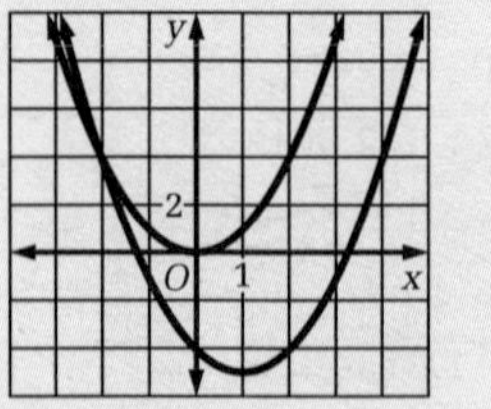

14. $y + 4 = 10^{x - 3}$

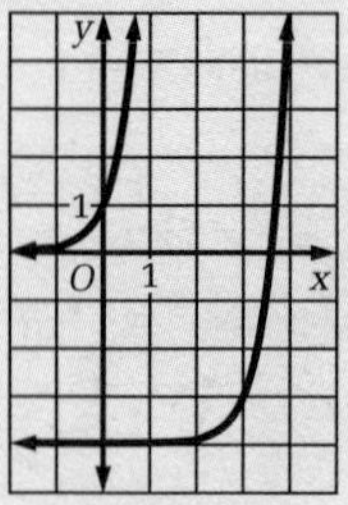

Combine each pair of parametric equations into a single equation relating y to x. 8-5

50. $x = 3.6t$, $y = 2.5t - 4.9t^2$
51. $x = 15.7t$, $y = 5.8t - 16.1t^2$
52. $x = 4 + 3 \cos t$, $y = -7 + 3 \sin t$
53. $x = 1 + \frac{1}{2} \cos t$, $y = 2 + \frac{1}{2} \sin t$
54. $x = 4 \cos t$, $y = 9 \sin t$
55. $x = 3 + 6 \cos t$, $y = -1 + 5 \sin t$

Use the law of cosines to find the length of the missing side or the measure of the missing angle of △ABC. 8-6

56. $a = 6.7$ cm, $b = 4.3$ cm, $m\angle C = 56°$, $c = ?$
57. $b = 12.8$ cm, $c = 10.6$ cm, $m\angle A = 115°$, $a = ?$
58. $a = 9.4$ cm, $c = 8.1$ cm, $m\angle B = 35°$, $b = ?$
59. $a = 12.5$ cm, $b = 5.9$ cm, $c = 9.3$ cm, $m\angle A = ?$
60. $a = 4.7$ cm, $b = 3.2$ cm, $c = 2.8$ cm, $m\angle C = ?$

Use the law of sines to solve for the missing parts of each △ABC. 8-7

61. $A = 36°$, $B = 58°$, $c = 11.4$
62. $A = 123°$, $a = 15.7$, $b = 12.1$
63. $B = 72°$, $c = 6.2$, $b = 8.5$
64. $B = 28°$, $C = 53°$, $a = 10.6$
65. $C = 114°$, $a = 21.6$, $c = 25.2$
66. $A = 16°$, $C = 121°$, $b = 9.4$

For each angle measure and side lengths, tell how many triangles can be formed. 8-7

67. $A = 49°$, $b = 12$, $a = 8$
68. $A = 37°$, $b = 7$, $a = 9$
69. $A = 43°$, $b = 6$, $a = 5$
70. $A = 112°$, $b = 10$, $a = 13$

Unit 9

Carmelita works for the fund-raising office of a hospital. Last year she received the following contributions from a group of small donors:

\$45, \$25, \$30, \$65, \$100, \$150, \$35, \$40, \$60, \$75, \$100, \$75 9-1

1. Find the mean, the median, and the standard deviation of the contributions.
2. This year Carmelita is asking each donor in the group to increase his or her contribution by \$10. Suppose every donor agrees to this plan. Find the mean, the median, and the standard deviation of the new contributions.

Each equation is in the form $y - k = f(x - h)$. For each equation, find the parent function and the translation vector. 9-2

3. $y + 2 = 10^{x-1}$
4. $y - 6 = \sqrt{x + 4}$
5. $y = \log (x - 9)$
6. $y - 5 = (x - 11)^3$
7. $y + 8 = |x|$
8. $y + 4 = (x + 7)^2$

A parent function and a translation vector are given. Write the equation of the function after translation, and graph the original function and the translated function on the same set of axes. 9-2

9. $y = x^3$; (3, 4) **10.** $y = \log x$; (−4, 2) **11.** $y = |x|$; (3, −5)

12. $y = \sqrt{x}$; (−2, −3) **13.** $y = x^2$; (1, −5) **14.** $y = 10^x$; (3, −4)

For Exercises 15 and 16: 9-3

a. Find the mean, the median, the range, and the standard deviation of the given data.

b. Repeat part (a) for the data obtained by multiplying each value by *a*.

c. Make histograms of the original and transformed data. Draw your histograms on the same axes.

15. data: 12, 14, 14, 18, 20, 22, 24, 34, 36, 38, 42, 42, 42, 44, 46
$a = 1.5$

16. data: 54, 57, 65, 66, 66, 68, 68, 71, 74, 77, 79, 83, 85, 88, 92
$a = 2$

Matrix *M* represents the triangle *ABC*. Find each product matrix *TM*. Then describe the effect on the original triangle. 9-4

17. $T = \begin{bmatrix} 4 & 0 \\ 0 & 1 \end{bmatrix}$ **18.** $T = \begin{bmatrix} 1 & 0 \\ 0 & \frac{3}{2} \end{bmatrix}$ $M = \begin{bmatrix} 4 & 6 & 2 \\ 0 & 2 & 6 \end{bmatrix}$

19. $T = \begin{bmatrix} 1 & 0 \\ 0 & \frac{1}{4} \end{bmatrix}$ **20.** $T = \begin{bmatrix} 5 & 0 \\ 0 & 1 \end{bmatrix}$

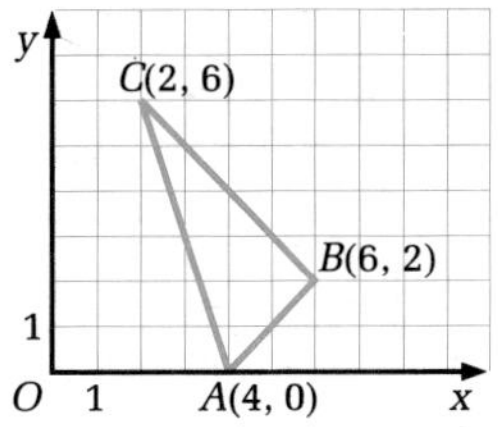

A parent function and two transformations are given. Write the equation of the transformed function and draw its graph. 9-5

21. $f(x) = \log x$; a translation of 3 units to the left followed by a vertical stretch by a factor of 2

22. $f(x) = x^2$; a translation of 2 units to the left followed by a translation of 1 unit down

23. $f(x) = |x|$; a vertical stretch by a factor of $\frac{1}{2}$ followed by a translation of 4 units to the right

24. $f(x) = 10^x$; a translation of 5 units to the left followed by a translation of 2 units down

For each z-score: 9-6

a. Sketch the standard normal curve and shade the region under the curve for values of z less than the given score.

b. Use the table on page 521 to find the percent of data less than the given score.

25. $z = 0.2$ **26.** $z = -1.0$ **27.** $z = -2.1$ **28.** $z = 1.9$

Answers to Extra Practice Unit 9

15. a. 29.9, 34, 34, 12.1
b. 44.8, 51, 51, 18.1
c.

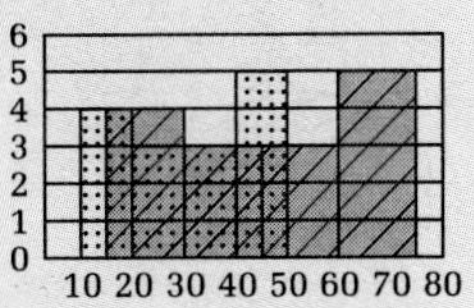

16. a. 72.9, 71, 38, 10.7 b. 145.7, 142, 76, 21.4
c.

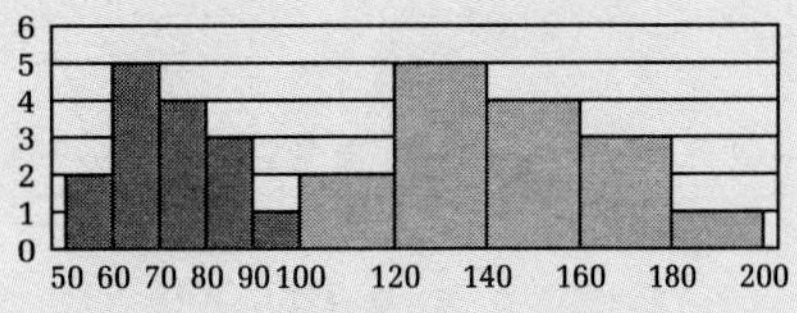

17. $\begin{bmatrix} 16 & 24 & 8 \\ 0 & 2 & 6 \end{bmatrix}$; horizontal stretch by a factor of 4

18. $\begin{bmatrix} 4 & 6 & 2 \\ 0 & 3 & 9 \end{bmatrix}$; vertical stretch by a factor of $\frac{3}{2}$

19. $\begin{bmatrix} 4 & 6 & 2 \\ 0 & 0.5 & 1.5 \end{bmatrix}$; vertical stretch by a factor of $\frac{1}{4}$

20. $\begin{bmatrix} 20 & 30 & 10 \\ 0 & 2 & 6 \end{bmatrix}$; horizontal stretch by a factor of 5

21. $y = 2 \log (x + 3)$

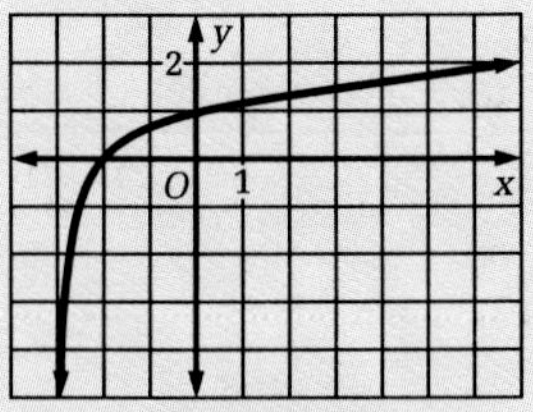

22. $y + 1 = (x + 2)^2$

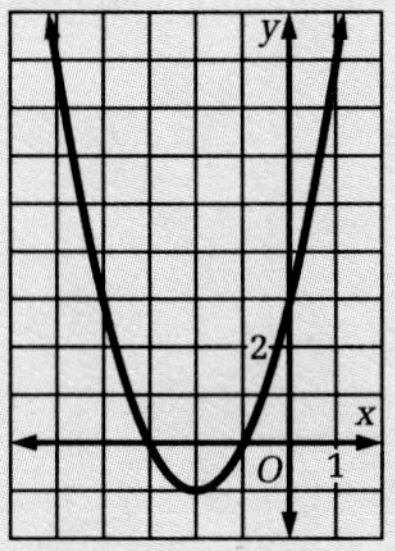

23. $y = \frac{1}{2}|x - 4|$

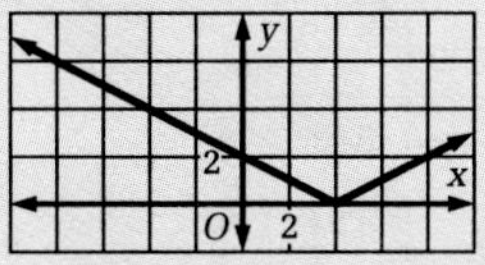

24. $y + 2 = 10^{x+5}$

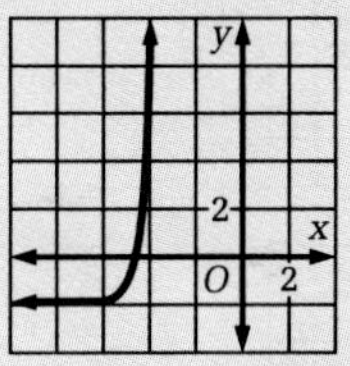

25. a.

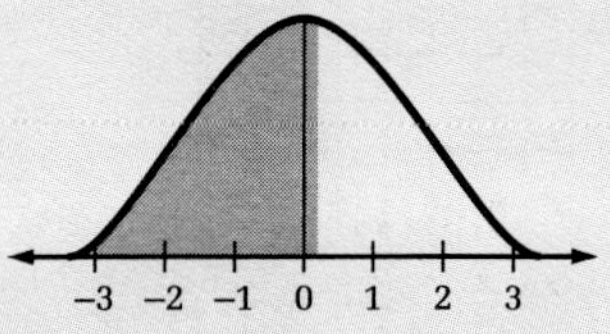

b. 57.93%

26. a.

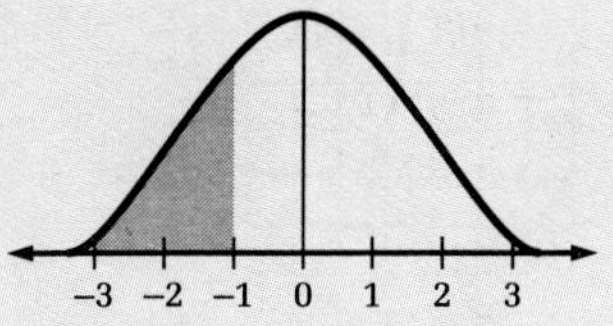

b. 15.87%

27. a.

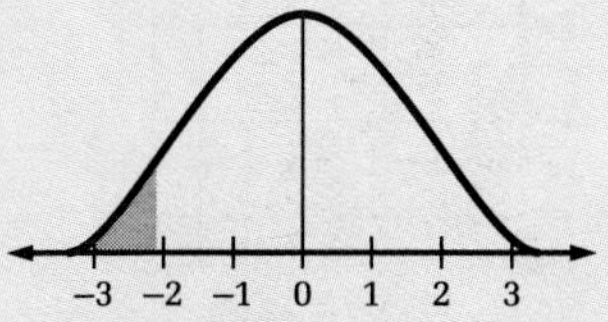

b. 1.79%

28. a.

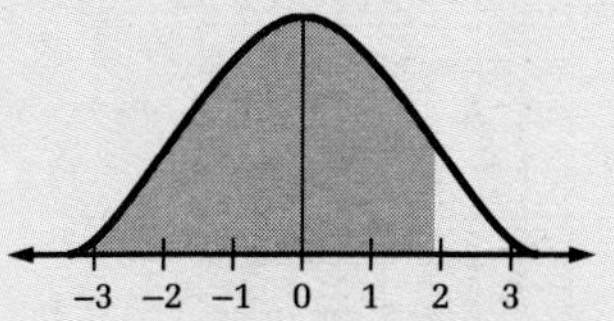

b. 97.13%

For each percent, use the table on page 521 to find the corresponding z-score for the standard normal distribution. 9-6

29. 54% **30.** 10% **31.** 1% **32.** 79%

Use the table on page 521, to find the missing information about a normal distribution from the given information. 9-6

33. $\overline{x} = 34.5$; $\sigma = 2.6$; data value = 35.2; z-score = ?; percentile = ?

34. $\overline{x} = 86.3$; $\sigma = 4.8$; percentile = 90; z-score = ?; data value = ?

35. $\sigma = 2.9$; percentile = 24; data value = 55.2; z-score = ?; $\overline{x}$ = ?

Sketch the graph of each function. 9-7

36. $y - 4 = -2|x + 3|$ **37.** $y = 3 \log(-x - 2)$ **38.** $y + 6 = -\frac{1}{5}x^3$

39. $y - 1 = 10^{-x+2}$ **40.** $y = -4\sqrt{x + 5}$ **41.** $y - 2 = -\frac{3}{4}(x - 3)^2$

Unit 10

Use one of the terms *constant, increasing, decreasing, periodic, damping,* or *saturating* to describe each graph. 10-1

1.

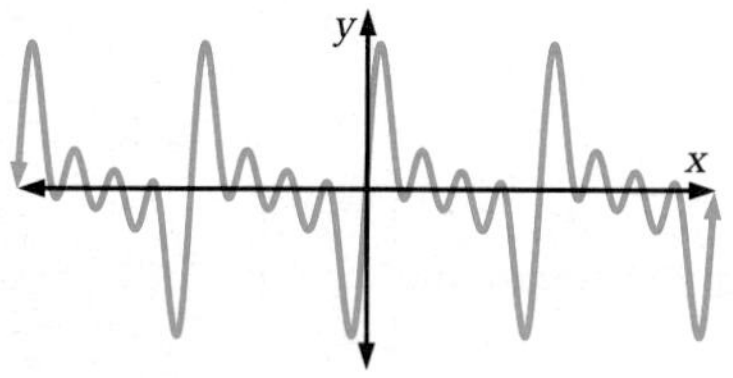

2.

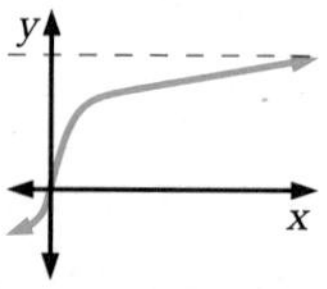

3.

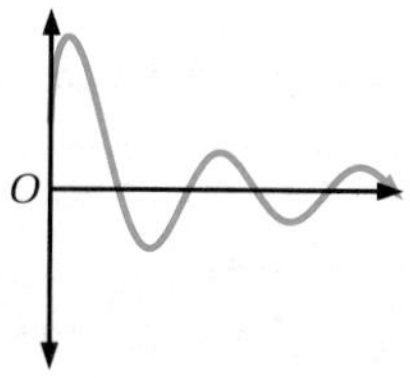

For each graph in Exercises 4 and 5: 10-2

a. Find the period. **b. Find the frequency.**

c. Write an equation of the axis. **d. Find the amplitude.**

4.

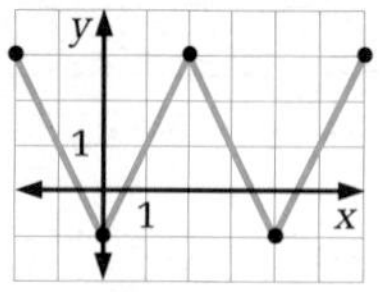

5.

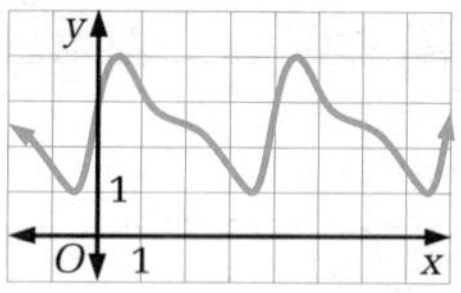

For Exercises 6–9, give the value of each characteristic of each function for $0° \le x \le 360°$. 10-3

a. sine function **b. cosine function**

6. The x-intercepts **7.** The y-intercept

8. The intervals where the function increases **9.** The range of the function

Answers to Extra Practice Unit 9

29. 0.1
30. −1.3
31. −2.3
32. 0.8
33. 0.3; 38
34. 1.3; 92.54
35. −0.7; 57.2
36.

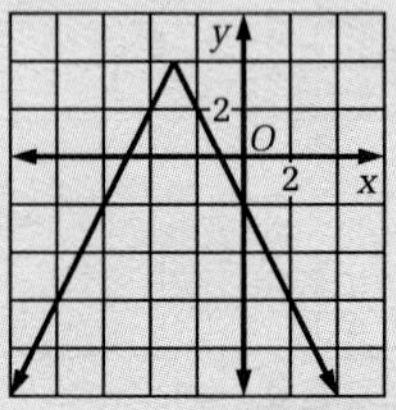

37.

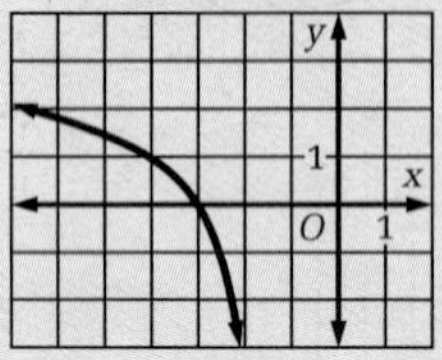

38.

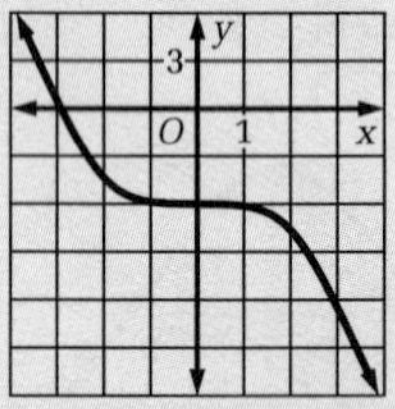

39.

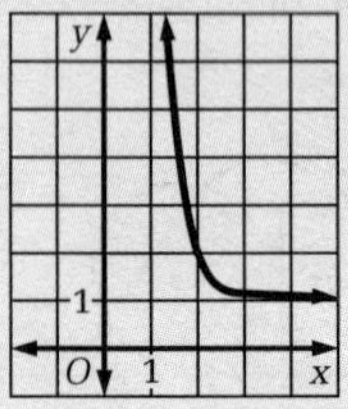

40.

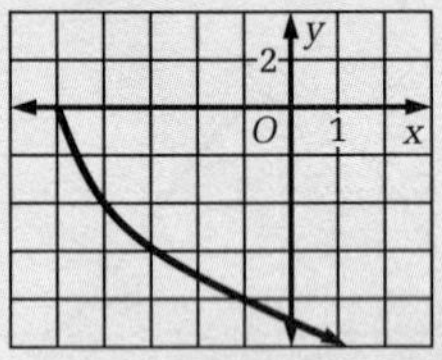

41.

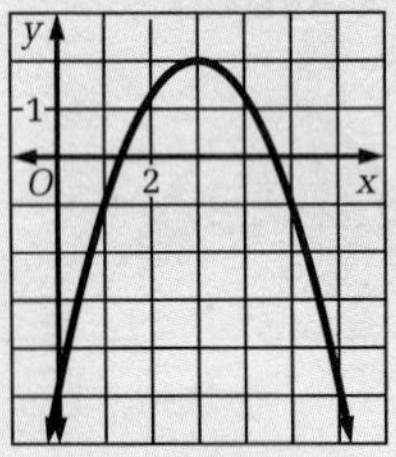

Answers to Extra Practice Unit 10

1. periodic
2. saturating
3. damping
4. a. 4
   b. $\frac{1}{4}$
   c. $y = 1$
   d. 2
5. a. 4
   b. $\frac{1}{4}$
   c. $y = 2.5$
   d. 1.5
6. a. 0, 180, 360
   b. 90, 270
7. a. 0
   b. 1
8. a. 0 to 90, 270 to 360
   b. 180 to 360
9. a. $-1 \le y \le 1$
   b. $-1 \le y \le 1$
10. 145°
11. 220°
12. 345°
13. 295°

Give another angle between 0° and 360° where the given function has the same value. 10-3

10. sin 35° **11.** cos 140° **12.** sin 195° **13.** cos 65°

For each equation: 10-4

a. Find the amplitude, period, and frequency of the graph.

b. Sketch the graph without using graphing technology, for $-360° \le \theta \le 360°$.

14. $y = 3 \sin (\theta - 60°)$ **15.** $y = 5 \sin (2\theta) + 1$ **16.** $y = -\cos (\theta + 30°) - 2$

17. $y = \cos 2(\theta + 90°) - 1$ **18.** $y = -4 \sin 0.5(\theta - 60°)$ **19.** $y = 2.5 \cos (3\theta) + 3$

Each arm of an amusement park ride called the Terrible Twister has a radius of 15 ft and each arm makes one vertical counterclockwise revolution every 20 seconds. Suppose your seat starts at the 3 o'clock position of the ride, and at its starting position your seat is 20 ft off the ground. 10-5

20. Find your height at each time.

a. 5 seconds after the ride starts **b.** 15 seconds after the ride starts

21. Write an equation of your vertical position (in feet) as a function of time (in seconds).

22. Rewrite the equation you wrote for Exercise 21 to suit each condition.

a. Each arm takes 24 s to make one complete revolution

b. You start at the highest position of the ride.

Classify each border pattern by its symmetry. 10-6

23.

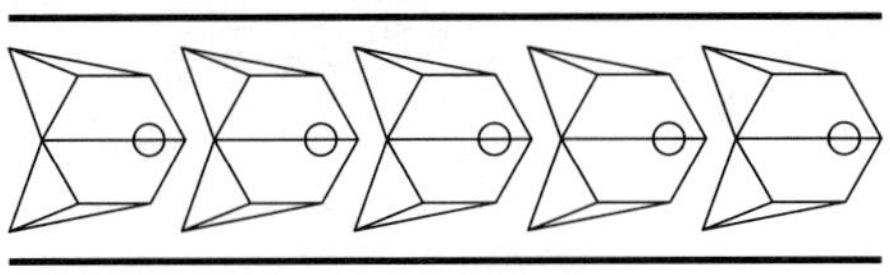

24.

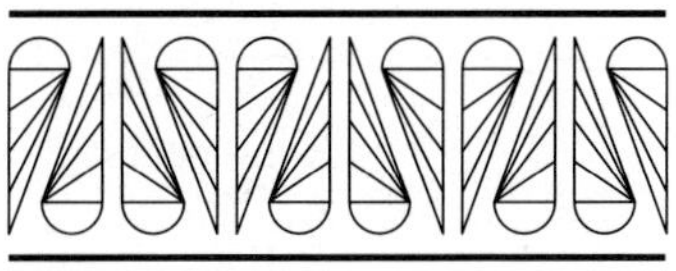

25.

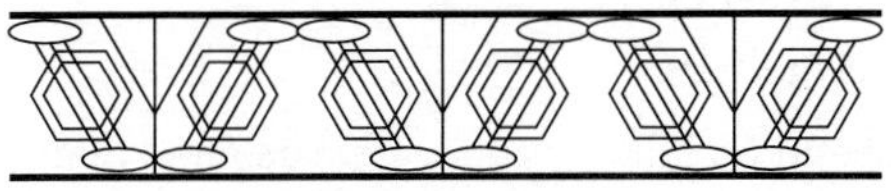

26.

Use the tessellation shown at the right. 10-7

27. Shade a cell of the tessellation.

28. Tell whether the tessellation is *regular*, *semiregular*, or *nonregular*.

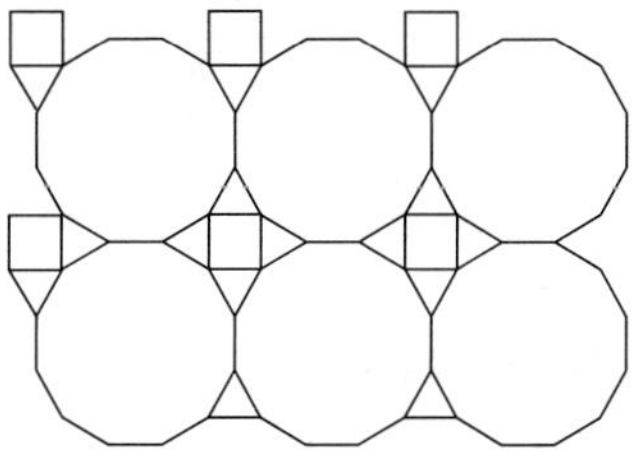

Sketch a tessellation represented by each code. 10-7

29. 4.4.4.4 **30.** 3.3.3.4.4

Answers to Extra Practice Unit 10

14. a. 3, 360, $\frac{1}{360}$

b.

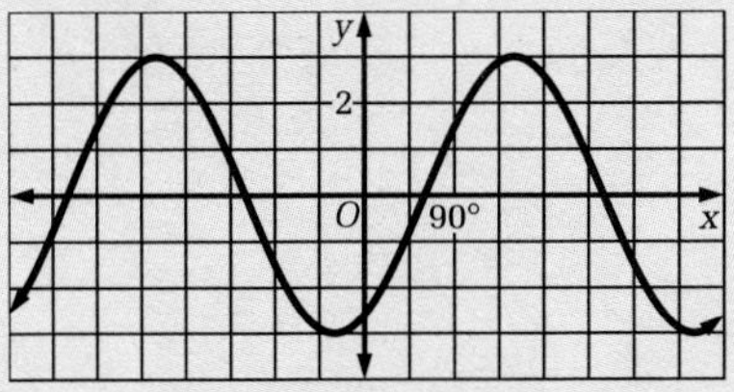

15. a. 5, 180, $\frac{1}{180}$

b.

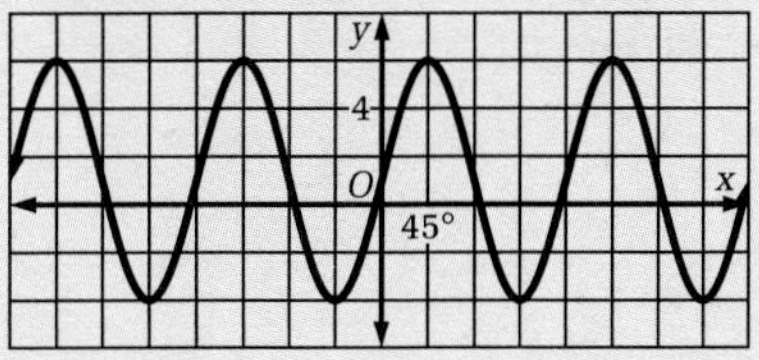

16. a. 1, 360, $\frac{1}{360}$

b.

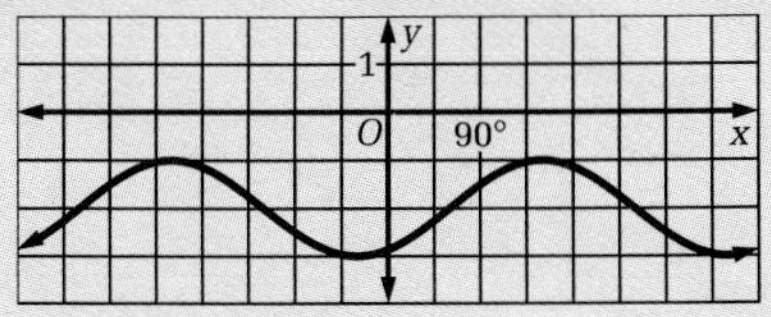

17. a. 1, 180, $\frac{1}{180}$

b.

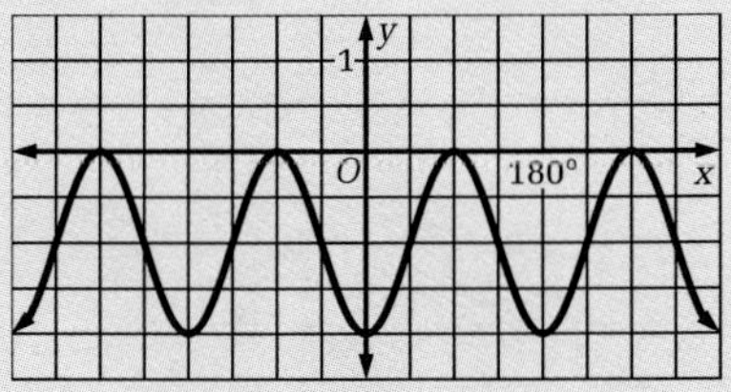

18. a. 4, 720, $\frac{1}{720}$

b.

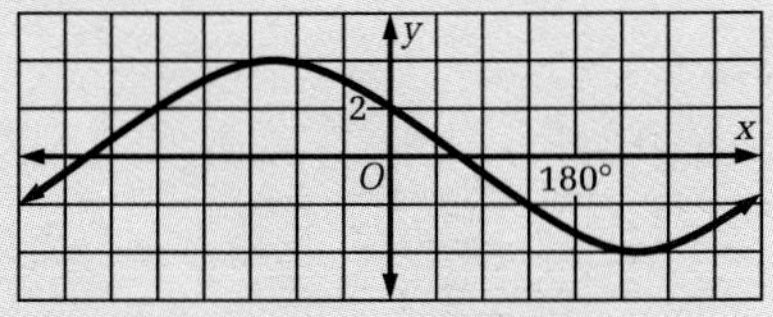

19. a. 2.5, 120, $\frac{1}{120}$

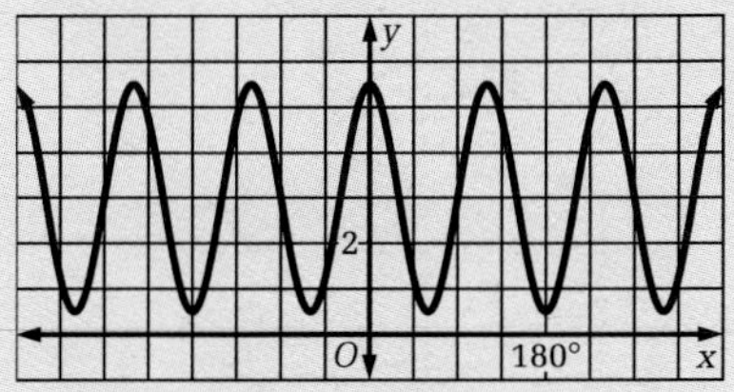

20. a. 35 ft

b. 5 ft

21. $y = 15 \sin (18t) + 20$

22. a. $y = 15 \sin (15t) + 20$

b. $y = 15 \cos (18t) + 20$

23. horizontal reflection

24. 180° rotation and vertical reflection

25. vertical reflection

26. 180° rotation

27.

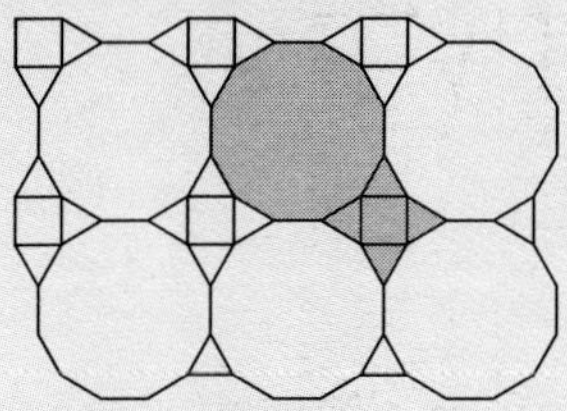

28. nonregular

29.

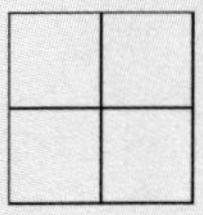

30.

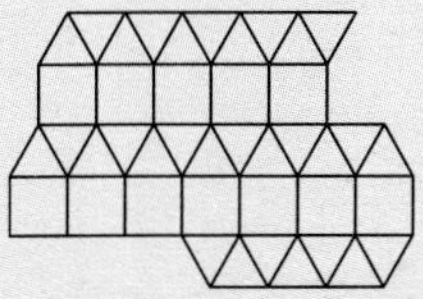

Toolbox

Discrete Mathematics

Data Measures and Displays

Skill 1 Finding Mean, Median, Mode, and Range

mean: The sum of the data in a data set divided by the number of items.	**median:** The middle number or the mean of the two middle numbers in a data set arranged in order.
mode: The data item, or items, appearing most often. There may be more than one mode or no mode.	**range:** The difference between the smallest and largest data items.

Examples The following high temperatures were recorded on days in March:

42, 38, 56, 56, 60, 40, 65, 45, 48, 42, 56.

Find the mean, the median, the mode, and the range of the data.

Mean: There are 11 temperatures. Add the numbers. Divide the sum by 11.

$$\frac{42 + 38 + 56 + 56 + 60 + 40 + 65 + 45 + 48 + 42 + 56}{11} = 49.8$$

To the nearest tenth, the mean is 49.8.

Median: List the numbers in order. Find the middle number.

38 40 42 42 45 48 56 56 56 60 65

The median is 48.

Mode: Since 56 appears three times, and no other number appears more than twice, 56 is the mode.

Range: The smallest number is 38. The largest is 65. Subtract to find the range: $65 - 38 = 27$. The range is 27.

Use the table of data, which shows protein per serving of some foods.

1. Find the mean.
2. Find the median.
3. Find the mode.
4. Find the range.

Protein per serving (g)					
24	36	42	10	32	24
10	18	16	18	12	30

Answers to Skill 1

1. about 22.7
2. 21
3. 10, 18, and 24
4. 32

Skill 2 Making a Histogram

A **histogram** displays grouped data. Each group is defined by an interval on the horizontal axis. The situation and the spread of the data determine how many intervals will be used. Five to 15 intervals are common. The height of the bar for an interval indicates the number of data items that fall in the interval. This number is called a **frequency.**

Example The frequency table and the histogram show the number of supporters for candidate Kaoru Hirakawa in different age groups in her city.

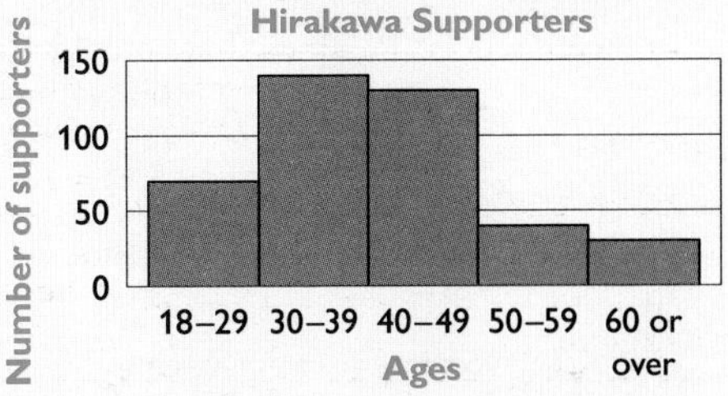

Age Group	Frequency
18–29	70
30–39	140
40–49	130
50–59	40
60 or over	30

Use the table or the histogram to answer each question.

a. How many Hirakawa supporters are between the ages of 50 and 59?

Histogram: The bar for the interval 50–59 has height 40.
Frequency table: The frequency for the age group 50–59 is 40.

There are 40 supporters between the ages of 50 and 59.

b. In which age groups does she have the greatest support and the least support?

Using the histogram, you see that the tallest bar is for the interval 30–39. She has the greatest support in this age group. The shortest bar is for the group "60 or over." She has the least support in this age group.

Use the histogram or the frequency table to answer each question.

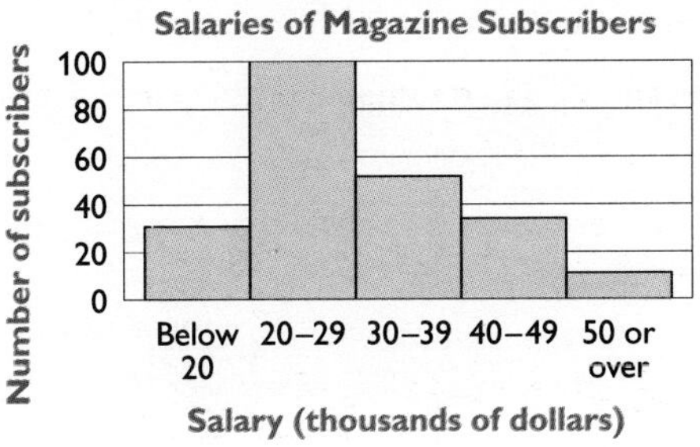

Salary Range (thousand $)	Number of subscribers
Below 20	30
20–29	100
30–39	52
40–49	34
50 or over	11

How many subscribers are in each salary range?

1. Below $20,000
2. Between $20,000 and $29,000
3. $50,000 or over
4. Which salary range has the most subscribers? the least?

Answers to Skill 2

1. 30
2. 100
3. 10 or 11
4. $20,000–29,000; $50,000 or over

Using Counting Techniques

Skill 3 Counting Permutations

A **permutation** is an arrangement of a certain number of items in a definite order. The symbol for the number of different permutations of r items chosen all possible ways from among n items is ${}_nP_r$.

Example How many different four-letter arrangements can be made from the letters of the word GRATES? Assume that a letter cannot be used more than once.

The number of four-letter arrangements is ${}_6P_4$, since there are 6 letters to choose from.

6 choices for the first letter; 4 choices for the third letter

$$6 \cdot 5 \cdot 4 \cdot 3 = 360$$ ← number of 4-letter arrangements

5 choices for the second letter; 3 choices for the fourth letter

The symbol $n!$ (read "n factorial") means the product of the positive integers through n.

$$n! = n(n-1)(n-2)\cdots 3 \cdot 2 \cdot 1.$$

The example above suggests a formula for ${}_nP_r$.

Number of permutations of *r* items selected from *n* items:

$${}_nP_r = \underbrace{n(n-1)(n-2)\cdots}_{r \text{ factors}} = \frac{n!}{(n-r)!}$$

n items; arranged r at a time

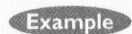

Example To solve the example above using the formula, you write:

$${}_6P_4 = \frac{6!}{(6-4)!} = \frac{6!}{2!} = \frac{6 \cdot 5 \cdot 4 \cdot 3 \cdot 2 \cdot 1}{2 \cdot 1} = 6 \cdot 5 \cdot 4 \cdot 3 = 360$$

Evaluate each expression.

1. $5!$ **2.** $7!$ **3.** $\frac{6!}{3!}$ **4.** $\frac{10!}{2!}$

5. ${}_7P_3$ **6.** ${}_8P_5$ **7.** ${}_9P_4$ **8.** ${}_6P_6$

Find the number of four-letter arrangements that can be made from the letters of each word.

9. COMPUTER **10.** HARVEST **11.** NUMERATOR

Answers to Skill 3

1. 120
2. 5040
3. 120
4. 1,814,400
5. 210
6. 6720
7. 3024
8. 720
9. 1680
10. 840
11. 3024

Skill 4 Counting Combinations

A **combination** is a selection of a definite number of items from a group of items, without regard to order. The symbol for the number of different combinations of r items selected all possible ways from a group of n items is ${}_nC_r$.

As you saw in Skill 3, each selection of r items can be arranged to form ${}_rP_r$ (or $r!$) different *permutations.* Therefore, ${}_nP_r = {}_rP_r \cdot {}_nC_r$. When you solve this equation for ${}_nC_r$, you get the following formula.

Number of combinations of *r* items selected from *n* items:

$$ {}_nC_r = \frac{{}_nP_r}{{}_rP_r} = \frac{n!}{(n-r)!\,r!} $$

Example How many different groups of 3 actresses can be chosen to play the witches in *Macbeth* from these 6 actresses who audition: Val, Carla, Susan, Pat, Mimi, and Rose?

Order does not matter in this situation. You have to find the number of combinations of 6 items selected 3 at a time.

Method 1

Problem Solving Strategy: Make a systematic list.

Represent each actress with one letter.

VCS	VSP	VPM	VMR	CSP	CPM	CMR	SPM	SMR	PMR
VCP	VSM	VPR		CSM	CPR		SPR		
VCM	VSR			CSR					
VCR									

20 different groups of 3 actresses can be chosen from 6 actresses.

Method 2

Problem Solving Strategy: Use a formula.

$$ {}_6C_3 = \frac{6!}{(6-3)!3!} = \frac{6!}{3!3!} = \frac{6 \cdot 5 \cdot 4 \cdot 3 \cdot 2 \cdot 1}{3 \cdot 2 \cdot 1 \cdot 3 \cdot 2 \cdot 1} = 20 $$

20 different groups of 3 actresses can be chosen from 6 actresses.

Evaluate each expression.

1. ${}_6C_4$
2. ${}_8C_3$
3. ${}_7C_5$
4. ${}_9C_4$
5. How many different subcommittees of 5 can be selected from a committee of 11?
6. How many different ways can you select 7 CDs out of 12 to bring to a party?

Answers to Skill 4

1. 15
2. 56
3. 21
4. 126
5. 462
6. 792

Skill 5 Using Pascal's Triangle

Pascal's triangle shows that there is a simple pattern to the numbers ${}_nC_r$.

Example Use Pascal's triangle to find ${}_4C_2$.

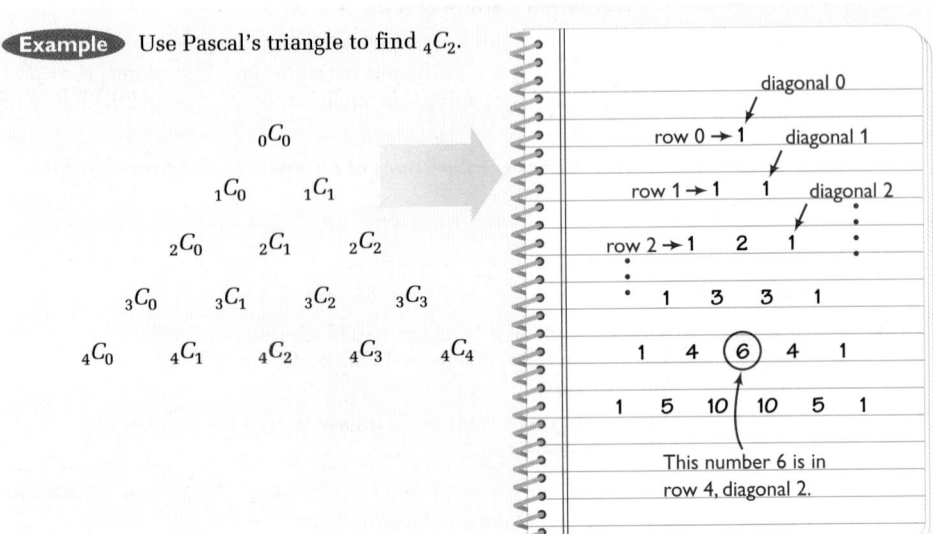

$${}_0C_0$$
$${}_1C_0 \quad {}_1C_1$$
$${}_2C_0 \quad {}_2C_1 \quad {}_2C_2$$
$${}_3C_0 \quad {}_3C_1 \quad {}_3C_2 \quad {}_3C_3$$
$${}_4C_0 \quad {}_4C_1 \quad {}_4C_2 \quad {}_4C_3 \quad {}_4C_4$$

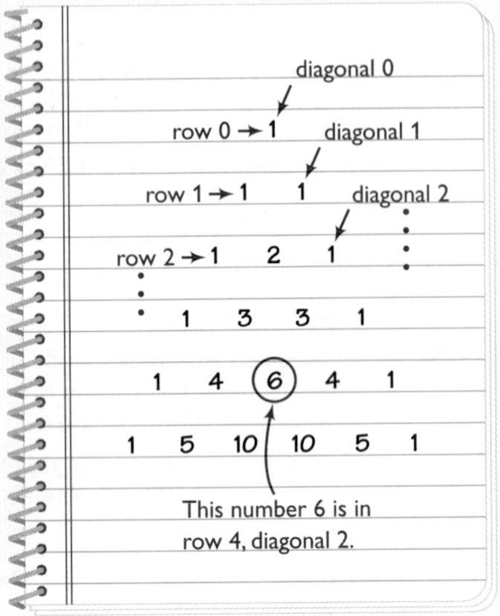

Some Properties of Pascal's Triangle

Each number (except the 1's) is the sum of the two numbers just above it.

The number in row n, diagonal r, is ${}_nC_r$.

The sum of the numbers in row n is 2^n.

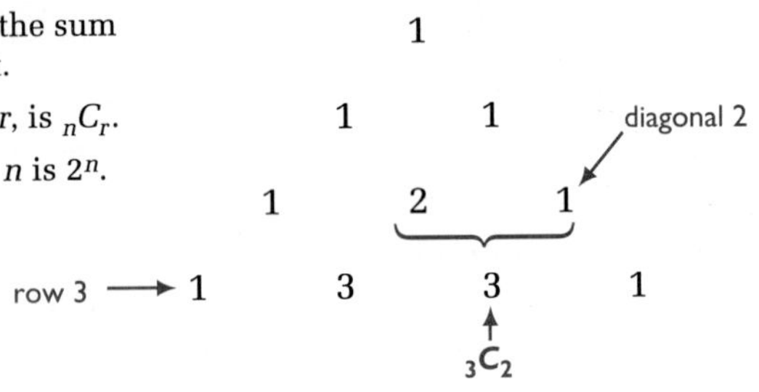

Find each value using Pascal's triangle.

1. ${}_4C_2$ **2.** ${}_5C_3$ **3.** ${}_6C_3$ **4.** ${}_6C_4$

5. How could you find ${}_7C_4$ from your answers to Questions 3 and 4 above, using Pascal's triangle?

6. The numbers in diagonal 2 of Pascal's triangle are called *triangular numbers*. Why do you think this name was chosen? (*Hint*: Draw an equilateral triangle made up of dots arranged in a grid.)

7. Make a conjecture about the sums of pairs of consecutive numbers in diagonal 2 of Pascal's triangle.

Answers to Skill 5

1. 6
2. 10
3. 20
4. 15
5. Add the two numbers together.
6. The numbers in diagonal 2 of Pascal's triangle are called triangular numbers because they can be represented by dots arranged to form equilateral triangles.
7. The sum of any two consecutive numbers in diagonal 2 of Pascal's triangle is a perfect square. These numbers are called square numbers because they can be represented by dots arranged in a square.

Probability

Skill 6 Finding the Probability of an Event

The **probability** of an event is a number between 0 and 1 that tells how likely the event is to happen. The closer the probability is to 1, the more likely the event. If the same experiment is performed over and over, you can find the **experimental probability** $P(E)$ of an event E as follows:

$$P(E) = \frac{\textbf{number of times the event } E \textbf{ happened}}{\textbf{number of times the experiment was performed}}$$

Example Out of a crate of 400 light bulbs, 8 were found to be defective. What is the experimental probability of finding a defective bulb?

$$P(\text{defective bulb}) = \frac{\text{number of defective bulbs found}}{\text{total number of bulbs checked}} = \frac{8}{400} = 0.02\text{, or } 2\%$$

When all the outcomes of an experiment are equally likely, you can find the **theoretical probability** of an event by calculation. You can think of an event as the group of outcomes that are considered favorable. Theoretical probability is the ratio:

$$P(E) = \frac{\textbf{number of favorable outcomes}}{\textbf{number of possible outcomes}}$$

Example Suppose you pick a card at random from the 13 hearts of a standard deck. What is the probability that it is a face card?

There are 13 possible outcomes:

There are 3 favorable outcomes:

$$P(\text{face card}) = \frac{3}{13} \approx 0.23\text{, or about } 23\%$$

The table shows how many of each different size of a particular style of sweatshirt were sold at one store last month. Find the probability that a sweatshirt sold will be each size.

Size	Number Sold
Small	15
Medium	32
Large	44

1. Small **2.** Medium **3.** Large

A card is chosen at random from the 13 hearts of a standard deck. Find each probability.

4. P(ace of hearts) **5.** P(not a face card) **6.** P(a diamond) **7.** P(not a diamond)

Answers to Skill 6

1. about 16%
2. about 35%
3. about 48%
4. about 8%
5. about 77%
6. 0
7. 100%

Skill 7 Finding the Probability of Mutually Exclusive Events

Two events that cannot happen at the same time are said to be **mutually exclusive.** When two events are mutually exclusive, you can find the probability that either one or the other will occur by adding the probabilities of the two events.

For mutually exclusive events A and B: $\mathbf{P(A \text{ or } B) = P(A) + P(B)}$

Example A spinner numbered 1 through 12 is spun. What is the probability that the spinner lands on 12 or an odd number?

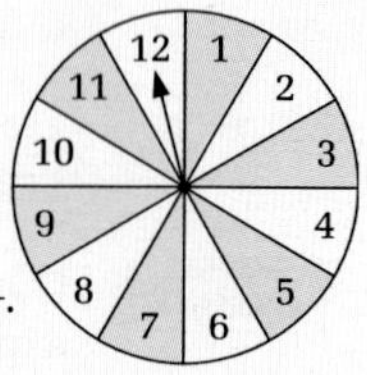

Since 12 is not an odd number, these two events are mutually exclusive. Therefore,

$P(\text{odd number or } 12) = P(\text{odd number}) + P(12) = \frac{6}{12} + \frac{1}{12} = \frac{7}{12}.$

Use the spinner from the example above. Find the probability of each event.

1. 3 or 7
2. 8 or a multiple of 3
3. number less than 4 or greater than 8
4. a multiple of 4 or a prime number

Operations with Matrices

Skill 8 Representing a Data Set with a Matrix

A **matrix** (plural: *matrices*) is an arrangement of numbers, called **elements,** in rows and columns. The number of rows followed by the number of columns are the **dimensions** of the matrix. The price matrix below has dimensions 3 × 2, read "three by two."

Example What is the daily rental price for a compact car on a weekend?

Daily Car Rental Prices

	Weekday	Weekend
Subcompact	32.50	44.00
Compact	38.00	56.00
Full-size	52.50	68.00

three rows; two columns

The daily rental for a compact car on a weekend is $56.00.

Use the matrix at the right.

1. Give the dimensions of the matrix.
2. How much is a $\frac{1}{4}$ lb burger special? How much is a dinner with a plain burger?

Burger Prices

	$\frac{1}{4}$ lb	$\frac{1}{2}$ lb	dinner
Plain	3.50	4.00	5.50
Special	4.50	5.00	7.00

Answers to Skill 7

1. $\frac{1}{6}$
2. $\frac{5}{12}$
3. $\frac{7}{12}$
4. $\frac{2}{3}$

Answers to Skill 8

1. 2 × 3
2. $4.50; $5.50

Skill 9 Multiplying a Matrix by a Number

To multiply a matrix P by a number, you multiply each entry of P by the number. Multiplying a matrix by a number is called **scalar multiplication.**

Example The matrix $P = \begin{bmatrix} 3.50 & 4.00 & 5.50 \\ 4.50 & 5.00 & 7.00 \end{bmatrix}$ represents burger prices, as in the exercises on page 632. Suppose burger prices increase by 8%. Write the matrix for the new pricing scheme.

Each new price will be 100% + 8% of the current price.
The new matrix will be:

$$(1.08)P = \begin{bmatrix} (1.08)\,3.50 & (1.08)\,4.00 & (1.08)\,5.50 \\ (1.08)\,4.50 & (1.08)\,5.00 & (1.08)\,7.00 \end{bmatrix}$$

$$= \begin{bmatrix} 3.78 & 4.32 & 5.94 \\ 4.86 & 5.40 & 7.56 \end{bmatrix}$$

Multiply each matrix by the given number.

1. $\begin{bmatrix} 1 & 4 \\ 7 & 0 \\ 6 & 2 \end{bmatrix}$; 5

2. $\begin{bmatrix} 1 & 1 & 0 & 1 \\ 3 & 2 & 4 & 6 \end{bmatrix}$; 0.5

3. $\begin{bmatrix} 24 & 36 & 8 \\ 12 & 48 & 4 \end{bmatrix}$; 1.25

Skill 10 Adding and Subtracting Matrices

When two matrices have the same dimensions, you can add or subtract the matrices by adding or subtracting the elements in corresponding positions.

Example

$$\begin{bmatrix} 25 & 46 & 53 \\ 0 & 38 & 22 \end{bmatrix} + \begin{bmatrix} 15 & 16 & -5 \\ 25 & 32 & -12 \end{bmatrix} = \begin{bmatrix} 25 + 15 & 46 + 16 & 53 + (-5) \\ 0 + 25 & 38 + 32 & 22 + (-12) \end{bmatrix}$$

$$= \begin{bmatrix} 40 & 62 & 48 \\ 25 & 70 & 10 \end{bmatrix}$$

Example

$$\begin{bmatrix} 5 & 6 \\ -4 & 9 \\ 2 & 3 \end{bmatrix} - \begin{bmatrix} 3 & 6 \\ 5 & 2 \\ -1 & 8 \end{bmatrix} = \begin{bmatrix} 5 - 3 & 6 - 6 \\ -4 - 5 & 9 - 2 \\ 2 - (-1) & 3 - 8 \end{bmatrix} = \begin{bmatrix} 2 & 0 \\ -9 & 7 \\ 3 & -5 \end{bmatrix}$$

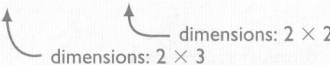

Example You cannot add $\begin{bmatrix} 3 & 2 & 4 \\ 6 & 5 & 1 \end{bmatrix}$ and $\begin{bmatrix} 8 & 3 \\ 2 & 5 \end{bmatrix}$, because they have different dimensions.

dimensions: 2×2

dimensions: 2×3

Answers to Skill 9

1. $\begin{bmatrix} 5 & 20 \\ 35 & 0 \\ 30 & 10 \end{bmatrix}$

2. $\begin{bmatrix} 0.5 & 0.5 & 0 & 0.5 \\ 1.5 & 1 & 2 & 3 \end{bmatrix}$

3. $\begin{bmatrix} 30 & 45 & 10 \\ 15 & 60 & 5 \end{bmatrix}$

Add or subtract as indicated.

1. $\begin{bmatrix} 4 & -1 & 6 \\ 9 & 3 & -2 \end{bmatrix} + \begin{bmatrix} 5 & 3 & -7 \\ 6 & -2 & 8 \end{bmatrix}$

2. $\begin{bmatrix} 12 & 15 \\ 22 & 0 \\ 17 & 25 \\ 8 & 16 \end{bmatrix} - \begin{bmatrix} -5 & 6 \\ 15 & 9 \\ 12 & -4 \\ 8 & 20 \end{bmatrix}$

3. $\begin{bmatrix} 1.5 & 3.8 \\ 6.2 & 4.6 \end{bmatrix} - \begin{bmatrix} 2.5 & 1.4 \\ 4.3 & 5.9 \end{bmatrix}$

4. $[62 \;\; 112 \;\; 58] + [48 \;\; 68 \;\; 12]$

Skill 11 Multiplying Matrices

The matrix product $L \cdot R$ is defined only when the number of columns of matrix L equals the number of rows of matrix R. If the dimensions of L are $m \times n$ and those of R are $n \times p$, the product will have dimensions $m \times p$.

To multiply a row of L by a column of R, you multiply each element of the row of L by the corresponding element of the column of R and then add.

Example

Multiply the first elements.

Multiply the second elements.

$$[3 \;\; -1]\begin{bmatrix} 4 \\ 5 \end{bmatrix} = [3 \cdot 4 + (-1)5] = [7]$$

Add.

To multiply a matrix L by a matrix R, you multiply each *row* of L by each *column* of R, and enter the result in the position whose coordinates are (*row, column*).

Example When matrix A has dimensions 2×4 and matrix B has dimensions 4×1, the dimensions of the product matrix are 2×1.

Multiply each row of A by the one column of B.

$$\overset{A}{\begin{bmatrix} 1 & 0 & 5 & 7 \\ 2 & 12 & -2 & 10 \end{bmatrix}} \overset{B}{\begin{bmatrix} 8 \\ 12 \\ -3 \\ 4 \end{bmatrix}} = \begin{bmatrix} 1(8) + 0(12) + 5(-3) + 7(4) \\ 2(8) + 12(12) + (-2)(-3) + 10(4) \end{bmatrix}$$

$$= \begin{bmatrix} 21 \\ 206 \end{bmatrix} \begin{matrix} \leftarrow \text{(row 1 of A)} \times \text{(column 1 of B)} \\ \leftarrow \text{(row 2 of A)} \times \text{(column 1 of B)} \end{matrix}$$

Multiply.

1. $[5 \;\; -2]\begin{bmatrix} 3 \\ 6 \end{bmatrix}$

2. $\begin{bmatrix} 4 & -1 \\ 3 & 5 \\ 2 & -2 \end{bmatrix}\begin{bmatrix} 4 & 7 \\ 0 & 1 \end{bmatrix}$

3. $\begin{bmatrix} 6 & 1 \\ 0 & 4 \\ 3 & 2 \end{bmatrix}\begin{bmatrix} 1 & -1 & 2 & 0 \\ 3 & 4 & -5 & -2 \end{bmatrix}$

Answers to Skill 10

1. $\begin{bmatrix} 9 & 2 & -1 \\ 15 & 1 & 6 \end{bmatrix}$

2. $\begin{bmatrix} 17 & 9 \\ 7 & -9 \\ 5 & 29 \\ 0 & -4 \end{bmatrix}$

3. $\begin{bmatrix} -1 & 2.4 \\ 1.9 & -1.3 \end{bmatrix}$

4. $[110 \;\; 180 \;\; 70]$

Answers to Skill 11

1. $[3]$

2. $\begin{bmatrix} 16 & 27 \\ 12 & 26 \\ 8 & 12 \end{bmatrix}$

3. $\begin{bmatrix} 9 & -2 & 7 & -2 \\ 12 & 16 & -20 & -8 \\ 9 & 5 & -4 & -4 \end{bmatrix}$

➤Algebra and Graphing

Algebraic Expressions

Skill 12 Evaluating Expressions

You evaluate an algebraic expression, such as $7x^2 - 3x + 5$, by substituting a given value for the variable.

Example Evaluate $7x^2 - 3x + 5$ when $x = -2$.

$7x^2 - 3x + 5 = 7(-2)^2 - 3(-2) + 5$ ← Substitute −2 for *x*.

$= 7 \cdot 4 - 3(-2) + 5$ ← $(-2)^2$ means $(-2)(-2)$.

$= 28 - (-6) + 5$ ← Do multiplication next.

$= 28 + 6 + 5$ ← To subtract −6, add its opposite.

$= 39$ ← Do addition last.

Evaluate each expression when $x = 3$.

1. $x^2 - 7x + 12$ **2.** $4x^3 - 5x + 1$ **3.** $15 - 4x^2$

4. $-2x^2 + 6x - 4$ **5.** $2x^5 - 4(x - 1)$ **6.** $(x - 5)^3 - 4(x + 1)^2$

Evaluate each expression when $a = -10$ and $b = 2$.

7. $3(a + b)$ **8.** $-a - b^2$ **9.** $-a^2b$

10. $-4a^2 + b^2$ **11.** $-4(a^2 + b^2)$ **12.** $-4(a + b)^2$

Skill 13 Adding and Subtracting Polynomials

A **monomial** is a number, a variable, or the product of a number and one or more variables. A **polynomial** is a monomial or a sum of monomials.

Terms having the same variables raised to the same powers are called **like terms.** To add two polynomials, **combine like terms.**

Example

$(2x^2 + 4xy) + (-5x^2 + 7xy) = 2x^2 + 4xy - 5x^2 + 7xy$ ← Rewrite without parentheses.

$= (2x^2 - 5x^2) + (4xy + 7xy)$ ← Group like terms.

$= (2 - 5)x^2 + (4 + 7)xy$ ← Use the distributive property.

$= -3x^2 + 11xy$

Answers to Skill 12

1. 0
2. 94
3. −21
4. −4
5. 478
6. −72
7. −24
8. 6
9. −200
10. −396
11. −416
12. −256

To subtract one polynomial from another, you can remove parentheses and rewrite the difference as the sum of the first polynomial and -1 times the second. Then use the distributive property.

Example

$$\begin{aligned}(3x^2 - 2xy + 8y^2) - (7x^2 - 5xy + 2y^2) &= (3x^2 - 2xy + 8y^2) + (-1)(7x^2 - 5xy + 2y^2)\\ &= 3x^2 - 2xy + 8y^2 - 7x^2 + 5xy - 2y^2\\ &= (3 - 7)x^2 + (-2 + 5)xy + (8 - 2)y^2\\ &= -4x^2 + 3xy + 6y^2\end{aligned}$$

Add or subtract, as indicated.

1. $(4a^2 - 3a + 2) + (-5a^2 + 2a)$

2. $(x^2 - xy + y^2) + (2x^2 + xy + y^2)$

3. $(7x^3 - 4x + 9) - (5x^3 + 3x^2 - 4x + 10)$

4. $(3c^2 + cd - 4d^2) - (5c^2 + 7cd + d^2 + 1)$

Skill 14 Using Rules of Exponents

For all real numbers x and all positive integers n,

$$x^n \text{ means } \underbrace{x \cdot x \cdot x \cdot \cdots \cdot x}_{n \text{ factors}}.$$

Product of Powers Rule: $a^m \cdot a^n = a^{m+n}$, for all real numbers a.

Quotient of Powers Rule: $\frac{a^m}{a^n} = a^{m-n}$, for all real numbers $a \neq 0$.

Example $3^5 \cdot 3^4 = 3^{5+4}$ ← Use the product of powers rule.

$= 3^9$

Example $\frac{2^7}{2^4} = 2^{7-4}$ ← Use the quotient of powers rule.

$= 2^3$

Example $\frac{6.7 \times 10^9}{3.2 \times 10^4} = \frac{6.7}{3.2} \times \frac{10^9}{10^4}$

$= \frac{6.7}{3.2} \times 10^{9-4}$ ← Use the quotient of powers rule.

$\approx 2.1 \times 10^5$

Answers to Skill 13

1. $-a^2 - a + 2$
2. $3x^2 + 2y^2$
3. $2x^3 - 3x^2 - 1$
4. $-2c^2 - 6cd - 5d^2 - 1$

To extend the domain of exponents to include 0 and the negative integers in a way that is consistent with the rules of exponents on page 636, you make the following definitions:

Zero Exponent Rule: For any real number $a \neq 0$, $a^0 = 1$.

Negative Exponent Rule: For any real number $a \neq 0$, $a^{-n} = \frac{1}{a^n}$.

Example Simplify $6a^0b^{-4}$. Write the answer without negative exponents.

$$6a^0b^{-4} = 6 \cdot 1 \cdot b^{-4}$$ ← Use the zero exponent rule.

$$= 6\left(\frac{1}{b^4}\right)$$ ← Use the negative exponent rule.

$$= \frac{6}{b^4}$$

If a, b, c, and d are real numbers with $b \neq 0$ and $d \neq 0$, then

$$\frac{ac}{bd} = \frac{a}{b} \cdot \frac{c}{d}.$$

Example Simplify $\frac{2x^{12}y^{-2}}{6x^3y^4}$. Write your answer without negative exponents.

$$\frac{2x^{12}y^{-2}}{6x^3y^4} = \frac{2}{6} \cdot \frac{x^{12}}{x^3} \cdot \frac{y^{-2}}{y^4}$$

$$= \frac{2}{6} \cdot x^{12-3} \cdot y^{-2-4}$$ ← Use the quotient of powers rule.

$$= \frac{1}{3} \cdot x^9 \cdot y^{-2}$$ ← Simplify.

$$= \frac{x^9}{3y^2}$$ ← Use the negative exponent rule.

Simplify. Write your answer without negative exponents.

1. $(3t^4)(9t^9)$

2. $\frac{6a^{10}}{8a^2}$

3. $(5r^2)(-4r^{-3})$

4. $5a^{-2}b^0$

5. $\frac{45u^{-5}}{9}$

6. $(4m^{\ 2})(3n^{\ 4})$

7. $(-2x^2y^5)(3x^{-2}y^{-7})$

8. $\frac{4v^0w^3}{6v^{-2}w^4}$

9. $\frac{14x^{-5}y^{-7}}{7xy^2}$

Answers to Skill 14

1. $27t^{13}$
2. $\frac{3}{4}a^8$
3. $\frac{-20}{r}$
4. $\frac{5}{a^2}$
5. $\frac{5}{u^5}$
6. $\frac{12}{m^2n^4}$
7. $\frac{-6}{y^2}$
8. $\frac{2v^2}{3w}$
9. $\frac{2}{x^6y^9}$

Skill 15 Multiplying Polynomials

To multiply two polynomials, you use the distributive property as often as needed.

Example $(2a + 3)(5a + 7) = (2a + 3) \cdot 5a + (2a + 3) \cdot 7$ ← Use the distributive property.

$= 2a \cdot 5a + 3 \cdot 5a + 2a \cdot 7 + 3 \cdot 7$ ← Use the distributive property again.

$= 10a^2 + 15a + 14a + 21$

$= 10a^2 + 29a + 21$ ← Combine like terms.

Example Expand $(2p + 3)(4p^2 - 5p + 8)$.

$(2p + 3)(4p^2 - 5p + 8) = (2p + 3) \cdot 4p^2 + (2p + 3) \cdot (-5p) + (2p + 3) \cdot 8$

$= 8p^3 + 12p^2 - 10p^2 - 15p + 16p + 24$

$= 8p^3 + 2p^2 + p + 24$ ← Combine like terms.

There are patterns that help you expand certain special products.

Special Products

$(a + b)^2 = a^2 + 2ab + b^2$
$(\text{first} + \text{second})^2 = (\text{first})^2 + 2(\text{first})(\text{second}) + (\text{second})^2$

$(a - b)^2 = a^2 - 2ab + b^2$
$(\text{first} - \text{second})^2 = (\text{first})^2 - 2(\text{first})(\text{second}) + (\text{second})^2$

$(a + b)(a - b) = a^2 - b^2$
$(\text{first} + \text{second})(\text{first} - \text{second}) = (\text{first})^2 - (\text{second})^2$

Example Expand $(3x + 5y)^2$.

You can use the distributive property, as in the Examples above, or you can use the first pattern above, with a replaced by $3x$ and b replaced by $5y$:

$(3x + 5y)^2 = (3x)^2 + 2 \cdot 3x \cdot 5y + (5y)^2$

$= 9x^2 + 30xy + 25y^2$

Expand.

1. $(a + 2)(a - 2)$
2. $(a + 2)^2$
3. $(a - 2)^2$
4. $(3k + 4)(7k - 2)$
5. $(5x + 2y)^2$
6. $(7c + 6d)(7c - 6d)$
7. $(9w - 4)^2$
8. $(10x - 3y)(2x - 5y)$
9. $(a + 2)(a^2 - 3a + 1)$
10. $(3x + 2y)^2$
11. $(x + 3)(x^2 - 2x + 4)$
12. $(2n - 5)(2n^2 - 3n - 2)$

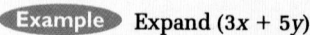

Answers to Skill 15

1. $a^2 - 4$
2. $a^2 + 4a + 4$
3. $a^2 - 4a + 4$
4. $21k^2 + 22k - 8$
5. $25x^2 + 20xy + 4y^2$
6. $49c^2 - 36d^2$
7. $81w^2 - 72w + 16$
8. $20x^3 - 56xy + 15y^2$
9. $a^3 - a^2 - 5a + 2$
10. $9x^2 + 12xy + 4y^2$
11. $x^3 + x^2 - 2x + 12$
12. $4n^3 - 16n^2 + 11n + 10$

Skill 16 Using the Binomial Theorem

You can use Pascal's triangle (see Skills 4 and 5) to help you expand a power of a binomial. Notice the pattern of the coefficients in the expansions below.

$$(a+b)^0 = 1$$
$$(a+b)^1 = \mathbf{1}a + \mathbf{1}b$$
$$(a+b)^2 = \mathbf{1}a^2 + \mathbf{2}ab + \mathbf{1}b^2$$
$$(a+b)^3 = \mathbf{1}a^3 + \mathbf{3}a^2b + \mathbf{3}ab^2 + \mathbf{1}b^3$$

This pattern suggests that if n is a positive integer, the coefficients in the expansion of $(a+b)^n$ can be found in row n of Pascal's triangle.

The Binomial Theorem

If n is a positive integer, then this is the expanded form of $(a+b)^n$:

$$({}_nC_0)a^nb^0 + ({}_nC_1)a^{n-1}b^1 + ({}_nC_2)a^{n-2}b^2 + \cdots$$
$$\cdots + ({}_nC_{n-2})a^2b^{n-2} + ({}_nC_{n-1})a^1b^{n-1} + ({}_nC_n)a^0b^n$$

The coefficients of the form $({}_nC_r)$ are numbers in the nth row of Pascal's triangle.

Example Write $(x+3)^4$ in expanded form.

Use the binomial theorem with $n = 4$.

For coefficients, use the fourth row of Pascal's triangle.

1
1 1
1 2 1
1 3 3 1
1 4 6 4 1 → $(a+b)^4 = 1a^4b^0 + 4a^3b^1 + 6a^2b^2 + 4a^1b^3 + 1a^0b^4$

Substitute x for a and 3 for b. Simplify.

$$\begin{aligned}(x+3)^4 &= 1x^4(3)^0 + 4x^3(3)^1 + 6x^2(3)^2 + 4x^1(3)^3 + 1x^0(3)^4\\ &= 1x^4(1) + 4x^3(3) + 6x^2(9) + 4x^1(27) + 1x^0(81)\\ &= x^4 + 12x^3 + 54x^2 + 108x + 81\end{aligned}$$

Expand.

1. $(x+y)^5$
2. $(2r+3)^4$
3. $(p^2+q)^6$
4. $(2n-m)^3$
5. $(c+10d)^5$
6. $(v^2+2w^3)^4$

Answers to Skill 16

1. $x^5 + 5x^4y^1 + 10x^3y^2 + 10x^2y^3 + 5xy^4 + y^5$
2. $16r^4 + 96r^3 + 216r^2 + 216r + 81$
3. $p^{12} + 6p^{10}q + 15p^8q^2 + 20p^6q^3 + 15p^4q^4 + 6p^2q^5 + q^6$
4. $8n^3 - 12n^2m + 6nm^2 - m^3$
5. $c^5 + 50c^4d + 1000c^3d^2 + 10{,}000c^2d^3 + 50{,}000cd^4 + 100{,}000d^5$
6. $v^8 + 8v^6w^2 + 24v^4w^6 + 32v^2w^9 + 16w^{12}$

Skill 17 Factoring Polynomials

One way to factor a trinomial is to try different combinations of binomial factors.

Example Factor $2x^2 + 5x + 3$ by trial and error.

You want an answer in this form: $(\underline{?}\ x + \underline{?})(\underline{?}\ x + \underline{?})$

These terms will be factors of $2x^2$. For example: $2x$, x

These terms will be factors of 3. For example: 1, 3; -1, -3

Try different combinations. Focus on getting the correct middle term.

$(2x + 1)(x + 3)$	$(2x + 3)(x + 1)$	$(2x - 1)(x - 3)$	$(2x - 3)(x - 1)$
x	$3x$	$-x$	$-3x$
$+\ 6x$	$+\ 2x$	$+\ -6x$	$+\ -2x$
$7x$	$5x$	$-7x$	$-5x$

The second pair of factors gives the correct middle term.

$2x^2 + 5x + 3 = (2x + 3)(x + 1)$

To factor certain trinomials, use the patterns from Skill 15.

Perfect Square Trinomials: $a^2 + 2ab + b^2 = (a + b)^2$
$a^2 - 2ab + b^2 = (a - b)^2$

Difference of Two Squares: $a^2 - b^2 = (a + b)(a - b)$

Example $49x^2 - 4y^2 = (7x)^2 - (2y)^2$
$= (7x + 2y)(7x - 2y)$ ← Use the pattern for $a^2 - b^2$.

You can factor some polynomials by finding the greatest common (monomial) factor, or **GCF,** of the terms.

Example $6x^5 - 8x^7 + 10x^2 = 6(x^2 \cdot x^3) - 8(x^2 \cdot x^5) + 10x^2$ ← Find the GCF of the variable parts.
$= 2x^2(3x^3) - 2x^2(4x^5) + 2x^2(5)$ ← The GCF of the coefficients is 2. The GCF of the variable parts is x^2.
$= 2x^2(3x^3 - 4x^5 + 5)$

Factor.

1. $x^2 - 20x + 100$
2. $25y^2 - 81z^2$
3. $a^2 - 5a - 14$
4. $2c^2 + 9c - 5$
5. $3x^2 - 7xy + 2y^2$
6. $49m^2 + 70mn + 25n^2$
7. $4x^5 - 2x^4 + 16x^2$
8. $9p^3q + 15p^2$
9. $45x^3y - 18x^2y^5$

Answers to Skill 17

1. $(x - 10)(x - 10)$
2. $(5y + 9z)(5y - 9z)$
3. $(a - 7)(a + 2)$
4. $(c + 5)(2c - 1)$
5. $(3x - y)(x - 2y)$
6. $(7m + 5n)(7m + 5n)$
7. $2x^2(2x^3 - x^2 + 8)$
8. $3p^2(3pq + 5)$
9. $9x^2y(5x - 2y^4)$

Skill 18 Simplifying Rational Expressions

A **rational expression** is a quotient of two polynomials. You can often simplify rational expressions by factoring both the numerator and the denominator of the expression.

Example Simplify $\frac{x^2 + 5x + 6}{x + 2}$.

$\frac{x^2 + 5x + 6}{x + 2} = \frac{(x + 3)\cancel{(x + 2)}}{\cancel{x + 2}}$ ← Factor both the numerator and the denominator.

$= x + 3$ ← Look for common factors in numerator and denominator to simplify.

Example Simplify $\frac{x^2 - 6x + 9}{x^2 - 2x - 3}$.

$\frac{x^2 - 6x + 9}{x^2 - 2x - 3} = \frac{(x - 3)^2}{(x - 3)(x + 1)}$ ← Factor both the numerator and the denominator.

$= \frac{x - 3}{x + 1}$ ← Look for common factors in numerator and denominator to simplify.

Sometimes an algebraic expression includes several rational expressions as terms. You can simplify such an expression by finding the least common multiple (LCM) of the denominators, just as you would when adding fractions.

Example Simplify $\frac{3}{4x^2} - \frac{1}{8x}$.

$\frac{3}{4x^2} - \frac{1}{8x} = \frac{2 \cdot 3}{2 \cdot 4x^2} - \frac{x \cdot 1}{x \cdot 8x}$ ← The LCM is $8x^2$.

$= \frac{6}{8x^2} - \frac{x}{8x^2}$ ← Simplify.

$= \frac{6 - x}{8x^2}$ ← Subtract the numerators and write the result over the common denominator.

Simplify.

1. $\frac{y + 5}{y^2 + 10y + 25}$
2. $\frac{x^2 - x - 2}{x^2 - 4}$
3. $\frac{a^2 - 7a + 12}{a^2 - a - 12}$
4. $\frac{2}{3x} + \frac{4}{x}$
5. $\frac{2}{x - 5} - \frac{3}{2x - 10}$
6. $\frac{2}{x^2 + x} - \frac{1}{x + 1}$

Answers to Skill 18

1. $\frac{1}{y + 5}$
2. $\frac{x + 1}{x + 2}$
3. $\frac{a - 3}{a + 3}$
4. $\frac{14}{3x}$
5. $\frac{1}{2x - 10}$
6. $\frac{2 - x}{x^2 + x}$

Skill 19 Simplifying Radical Expressions

When a is a nonnegative number, $\sqrt{a}$ is a real number. When a is negative, $\sqrt{a}$ is not a real number.

You may be able to simplify $\sqrt{a}$ by finding the largest perfect-square factor of a, and then using the fact that for all nonnegative numbers a and b,

$$\sqrt{ab} = \sqrt{a} \cdot \sqrt{b}.$$

Example $\sqrt{18} = \sqrt{9 \cdot 2}$ ← 9 is the largest perfect-square factor of 18. Write 18 as $9 \cdot 2$.
$= \sqrt{9} \cdot \sqrt{2}$ ← Use the fact that $\sqrt{ab} = \sqrt{a} \cdot \sqrt{b}$.
$= 3\sqrt{2}$ ← Write $\sqrt{9}$ as 3.

Note: It would *not* have been helpful to write $\sqrt{18}$ as $\sqrt{6 \cdot 3}$, because neither 6 nor 3 is a perfect square.

To simplify a product of radical factors, you use the fact that for all nonnegative numbers a and b,

$$\sqrt{a} \cdot \sqrt{b} = \sqrt{ab}.$$

Example $5\sqrt{3} \cdot \sqrt{2} = 5\sqrt{3 \cdot 2}$ ← Use the fact that $\sqrt{a} \cdot \sqrt{b} = \sqrt{ab}$.
$= 5\sqrt{6}$

Example $4\sqrt{5} \cdot 2\sqrt{15} = 4 \cdot 2 \cdot \sqrt{5} \cdot \sqrt{15}$ ← Group radical terms.
$= 8\sqrt{5 \cdot 15}$ ← Use the fact that $\sqrt{a} \cdot \sqrt{b} = \sqrt{ab}$.
$= 8\sqrt{75}$ ← Simplify.
$= 8\sqrt{25 \cdot 3}$ ← 25 is the largest perfect-square factor of 75.
$= 8\sqrt{25} \cdot \sqrt{3}$ ← Use the fact that $\sqrt{ab} = \sqrt{a} \cdot \sqrt{b}$.
$= 8 \cdot 5\sqrt{3}$ ← Write $\sqrt{25}$ as 5.
$= 40\sqrt{3}$

Simplify.

1. $\sqrt{45}$ **2.** $\sqrt{300}$ **3.** $\sqrt{76}$ **4.** $5\sqrt{12}$ **5.** $12\sqrt{125}$

6. $\sqrt{6} \cdot \sqrt{7}$ **7.** $3\sqrt{5} \cdot \sqrt{2}$ **8.** $4.2\sqrt{11} \cdot \sqrt{3}$ **9.** $8\sqrt{2} \cdot 3\sqrt{3}$

10. $\sqrt{6} \cdot \sqrt{6}$ **11.** $(3\sqrt{3})^2$ **12.** $2\sqrt{6} \cdot 3\sqrt{2}$ **13.** $4\sqrt{5} \cdot 3\sqrt{10}$

Answers to Skill 19

1. $3\sqrt{5}$
2. $10\sqrt{3}$
3. $2\sqrt{19}$
4. $10\sqrt{3}$
5. $60\sqrt{5}$
6. $\sqrt{42}$
7. $3\sqrt{10}$
8. $4.2\sqrt{33}$
9. $24\sqrt{6}$
10. 6
11. 27
12. $12\sqrt{3}$
13. $60\sqrt{2}$

Skill 20 Operating with Complex Numbers

The equation $x^2 = -1$ has no real-number solution. The **imaginary unit** $\boldsymbol{i}$, which is not a real number, is defined to be a solution of this equation:

$$i = \sqrt{-1} \quad \text{and} \quad i^2 = -1$$

The square root of a negative number is defined as follows for all real numbers $a > 0$:

$$\sqrt{-a} = \sqrt{-1} \cdot \sqrt{a} = i\sqrt{a}$$

Example **a.** $\sqrt{-81} = i\sqrt{81} = 9i$ **b.** $\sqrt{-33} = i\sqrt{33} \approx 5.7i$

In order to create a system that includes both the real numbers and the **pure imaginary numbers,** such as i, $2i$, and $i\sqrt{3}$, the **complex numbers** are defined. These are numbers of the form $\boldsymbol{a + bi}$, where a and b are real numbers.

You can use what you know about adding like terms and multiplying binomials to do operations with complex numbers.

Example Simplify.

a. $(9i)(3i) = 27i^2$

$= 27(-1)$ ← Substitute -1 for i^2.

$= -27$

b. $(2 + 6i) - (1 - 3i) = 2 + 6i - 1 + 3i$ ← Rewrite without parentheses.

$= (2 - 1) + (6 + 3)i$ ← Group real parts and group imaginary parts.

$= 1 + 9i$

c. $(5 + 7i)(4 + 8i) = (5 + 7i)4 + (5 + 7i)8i$ ← Use the distributive property.

$= 20 + 28i + 40i + 56i^2$ ← Use the distributive property again.

$= 20 + 68i + 56(-1)$ ← Combine like terms. Substitute -1 for i^2.

$= 20 + 68i - 56$ ← Simplify.

$= -36 + 68i$ ← Combine like terms.

Simplify.

1. $\sqrt{-100}$
2. $\sqrt{-0.25}$
3. $\sqrt{-17}$
4. $\sqrt{-83}$
5. $(5i)(7i)$
6. $(4i)(-3i)(0.5i)$
7. $9i(1 - 8i)$
8. $-7i(12 + 3i)$
9. $(8 - 14i) + (10 - 5i)$
10. $(-4 + 2i) - (9 - 9i)$
11. $(3 + 4i) - (-7 + 11.5i)$
12. $(5 + 4i)(6 - 12i)$
13. $(-2 + 8i)(1 + 0.2i)$
14. $(4 - 10i)(4 + 10i)$

Answers to Skill 20

1. $10i$
2. $0.5i$
3. $i\sqrt{17} \approx 4.12i$
4. $i\sqrt{83} \approx 9.11i$
5. -35
6. $6i$
7. $72 + 9i$
8. $21 - 84i$
9. $18 - 19i$
10. $-13 + 11i$
11. $10 - 7.5i$
12. $78 - 36i$
13. $-3.6 + 7.6i$
14. 116

Solving Equations

Skill 21 Solving a Linear System

Solving a linear system of equations means finding one pair of numbers (x, y) that satisfies both equations. One method of solving a linear system is **substitution.**

Example Solve the system: $x - 2y = -11$
$x + y = 4$

$x + y = 4$ →	$y = 4 - x$	← Solve the second equation for y.
$x - 2y = -11$ →	$x - 2(4 - x) = -11$	← Substitute $4 - x$ for y in the other equation.
	$x - 8 + 2x = -11$	← Remove parentheses.
	$3x = -3$	← Combine like terms.
	$x = -1$	← Solve for x.
	$-1 + y = 4$	← Substitute the value of x back into either equation.
	$y = 5$	

The solution (x, y) is $(-1, 5)$.

You can also use **addition-or-subtraction** to solve a system.

Example Solve by the addition-or-subtraction method.

a. $3y = 2x + 2$
$3y = x - 2$

b. $5a + 2b = 23$
$7a - 2b = 13$

a. The coefficients of y are *the same.* *Subtract* to eliminate the y-terms.

$$\begin{array}{rl} & 3y = 2x + 2 \\ - & 3y = x - 2 \\ \hline & 0 = x + 4 \end{array}$$

$0 = x + 4$

$-4 = x$

$3y = -4 - 2$

$3y = -6$

$y = -2$

The solution (x, y) is $(-4, -2)$.

Solve the new equation.

Substitute in one of the original equations to find the other variable.

b. The coefficients of b are *opposites.* *Add* to eliminate the b-terms.

$$\begin{array}{rl} & 5a + 2b = 23 \\ + & 7a - 2b = 13 \\ \hline & 12a + 0 = 36 \end{array}$$

$12a = 36$

$a = 3$

$5(3) + 2b = 23$

$15 + 2b = 23$

$2b = 8$

$b = 4$

The solution (a, b) is $(3, 4)$.

Some linear systems contain no like terms with the same or opposite coefficients. You can change the coefficients by multiplying both sides of either equation by a constant.

Example Solve the system: $4x - 3y = 11$
$-5x + 2y = -12$

$5(4x - 3y) = 5(11)$ ← Multiply both sides of the first equation by **5**.

$4(-5x + 2y) = 4(-12)$ ← Multiply both sides of the second equation by 4.

$20x - 15y = 55$
$-20x + 8y = -48$ ← The new system has *x*-terms that are opposites.

Now you can solve the system as in the preceding example.

Solve each system.

1. $y = 4x$
$y = -x - 15$

2. $b = a + 5$
$b = 3a - 1$

3. $x - y = 1$
$x + y = -1$

4. $3a + 5b = 17$
$3a + 8b = 5$

5. $p - 6q = 14$
$2p + 6q = 1$

6. $-2c + 9d = 35$
$6c + 7d = 65$

Skill 22 Solving a Quadratic Equation

A **quadratic** equation is one that contains an x^2-term. Some quadratic equations can be solved by factoring and using the *zero-product property.*

Example Solve $4x^2 - x = 5$.

$4x^2 - x = 5$

$4x^2 - x - 5 = 0$ ← Rewrite the equation in standard form.

$(4x - 5)(x + 1) = 0$ ← Factor the trinomial by trial and error.

$4x - 5 = 0$ *or* $x + 1 = 0$ ← Use the zero-product property. When a product of factors is equal to 0, one or more of the factors equals 0.

$4x = 5$ $x = -1$

$x = 1.25$

The solutions are 1.25 and −1.

Answers to Skill 21

1. $x = -3, y = -12$
2. $a = 3, b = 8$
3. $x = 0, y = -1$
4. $a = \frac{37}{3}, b = -4$
5. $p = 5, q = -\frac{3}{2}$
6. $c = 5, d = 5$

Quadratic Formula: If $a \neq 0$, the solutions of the equation $ax^2 + bx + c = 0$ are

$$\mathbf{x = -\frac{b}{2a} + \frac{\sqrt{b^2 - 4ac}}{2a}} \text{ and } \mathbf{x = -\frac{b}{2a} - \frac{\sqrt{b^2 - 4ac}}{2a}}.$$

← The discriminant, $b^2 - 4ac$.

Two different real roots when $b^2 - 4ac > 0$.
One real root when $b^2 - 4ac = 0$.
No real roots when $b^2 - 4ac < 0$.

Example Solve $2x^2 + 3x = 4$.

$2x^2 + 3x - 4 = 0$ ← Rewrite the equation in standard form: $ax^2 + bx + c = 0$.

$x = -\frac{b}{2a} \pm \frac{\sqrt{b^2 - 4ac}}{2a}$ ← Use the formula.

$= -\frac{3}{2(2)} \pm \frac{\sqrt{3^2 - 4(2)(-4)}}{2(2)}$ ← Substitute **2** for ***a***, 3 for *b*, −4 for *c*.

$= -\frac{3}{4} \pm \frac{\sqrt{41}}{4} \approx -0.75 \pm 6.40$

$x \approx 5.65$ or $x \approx 7.15$

Solve.

1. $x^2 - 4x - 5 = 0$
2. $2x^2 - 3x = 5$
3. $0 = 9x^2 + 6x - 3$
4. $x^2 - 4x + 1 = 0$
5. $2x^2 + 3x = 7$
6. $5x^2 - 10x = -2$

Skill 23 Solving Rational Equations

A **rational equation** is one that involves rational expressions.

Example Solve $\frac{1}{x-4} = \frac{2}{x^2 - 16}$.

$x^2 - 16 = 2(x - 4)$ ← Use cross products.

$x^2 - 16 = 2x - 8$ ← Use the distributive property.

$x^2 - 2x - 8 = 0$ ← Write the equation in standard form.

$(x + 2)(x - 4) = 0$ ← Factor.

$x + 2 = 0$ *or* $x - 4 = 0$ ← Use the zero-product property.

$x = -2$ $x = 4$

Check Substitute the results in the original equation.

$\frac{1}{-2-4} \stackrel{?}{=} \frac{2}{(-2)^2 - 16}$ ← Substitute −2 for *x*.

$\frac{1}{-6} = \frac{1}{-6}$ ✓ ← This is a valid statement. −2 is a solution.

$\frac{1}{4-4} \stackrel{?}{=} \frac{2}{4^2 - 16}$ ← Substitute 4 for *x*.

$\frac{1}{0} = \frac{2}{0}$ ← Division by zero is undefined. 4 is **not** a solution.

The solution is −2. (4 is an **extraneous solution.**)

Answers to Skill 22

1. 5, −1
2. $\frac{5}{2}$, −1
3. $\frac{1}{3}$, −1
4. about 3.7, about 0.27
5. about −2.8, about 1.25
6. about 1.77, about 0.23

To solve rational equations, you can eliminate denominators by multiplying both sides by the LCM of all the denominators.

Example Solve $\frac{3}{5} + \frac{4}{x-3} = 7$.

First look for a common multiple of the denominators. A common multiple is an expression that has both denominators as factors.

$$\frac{3}{5} + \frac{4}{x-3} = 7 \quad \longleftarrow \text{ 5(x − 3) is a common multiple.}$$

> **Watch Out!**
> A common multiple of $\frac{3}{5}$ and $\frac{4}{x-3}$ is $5(x-3)$, not $5x-3$.

Then multiply both sides of the equation by this common multiple.

$$5(x-3)\left(\frac{3}{5} + \frac{4}{x-3}\right) = 5(x-3)(7)$$

$$5(x-3)\left(\frac{3}{5}\right) + 5(x-3)\left(\frac{4}{x-3}\right) = 35x - 105 \quad \longleftarrow \text{ Use the distributive property.}$$

$$(x-3)3 + 5(4) = 35x - 105 \quad \longleftarrow \text{ Simplify.}$$

$$3x + 11 = 35x - 105$$

$$116 = 32x$$

$$3.625 = x$$

Solve.

1. $\frac{1}{x} = \frac{4}{x-2}$
2. $\frac{3}{y+1} = \frac{y}{2}$
3. $\frac{2a+5}{a-1} = \frac{3a}{a-1}$
4. $\frac{6}{t} - \frac{2}{t-1} = 1$
5. $\frac{x-1}{x} + \frac{9}{4x} = 6$
6. $\frac{x}{x-2} + \frac{30}{x+2} = 9$

Skill 24 Solving Polynomial Equations

Some polynomial equations can be solved using the zero-product property (see Skill 22), even if they are not quadratic.

Example Solve $x^3 + 3x^2 + 2x = 0$

$x(x^2 + 3x + 2) = 0 \quad \longleftarrow$ Factor out the monomial factor x.

$x(x + 2)(x + 1) = 0 \quad \longleftarrow$ Factor the trinomial $x^2 + 3x + 2$ by trial and error.

$x = 0$ *or* $x + 2 = 0$ *or* $x + 1 = 0 \quad \longleftarrow$ Use the zero-product property.

$x = -2 \qquad x = -1$

The solutions are 0, −2, and −1.

Solve.

1. $x^3 - 5x^2 + 6x = 0$
2. $x^3 + 4x^2 + 3x = 0$
3. $(x-1)(x+3)(3x-1) = 0$
4. $x^3 + 3x^2 - 4x = 0$
5. $x^3 - 4x = 0$
6. $(x+6)(x-2)^2 = 0$

Answers to Skill 23

1. $-\frac{2}{3}$
2. 2, −3
3. 5
4. 3, 2
5. $\frac{1}{4}$
6. 3

Answers to Skill 24

1. 0, 2, 3
2. 0, −1, −3
3. 1, −3, $\frac{1}{3}$
4. 0, −4, 1
5. 0, 2, −2
6. 2, −6

Answers to Skill 25

1. $y = 4x + 7$
2. $y = -0.5x - 2$
3. $y = -3x + 11$
4. $y = -\frac{2}{3}x - \frac{1}{3}$
5. $y = -\frac{1}{2}x + \frac{3}{2}$

Answers to Skill 26

1. Tables may vary. Examples are given.

x	y	(x, y)
−2	9	(−2, 9)
−1	7	(−1, 7)
0	5	(0, 5)
1	3	(1, 3)
2	1	(2, 1)

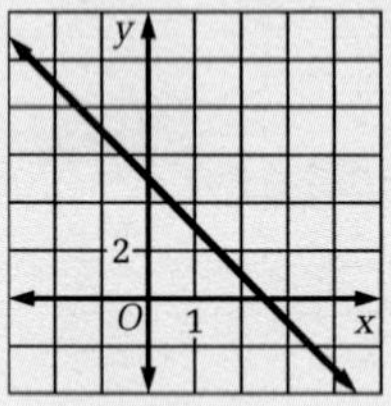

2.

x	y	(x, y)
−2	−2	(−2, −2)
−1	$-\frac{3}{2}$	$\left(-1, -\frac{3}{2}\right)$
0	−1	(0, −1)
1	$-\frac{1}{2}$	$\left(-1, -\frac{1}{2}\right)$
2	0	(2, 0)

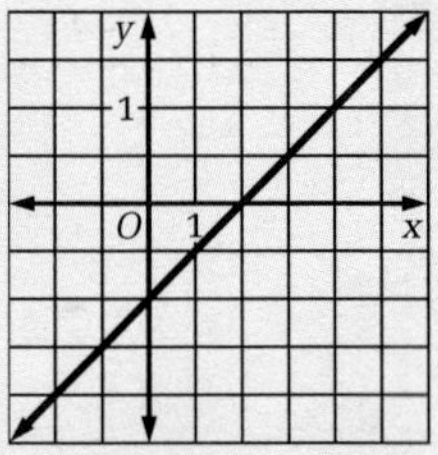

3.

x	y	(x, y)
−2	−1	(−2, −1)
−1	−4	(−1, −4)
0	−5	(0, −5)
1	−4	(1, −4)
2	−1	(2, −1)

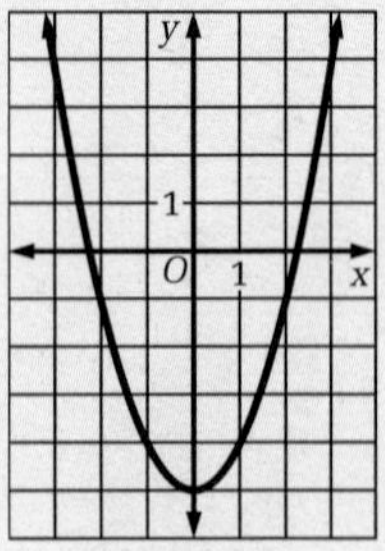

4.

x	y	(x, y)
−2	0	(−2, 0)
−1	−2	(−1, −2)
0	−2	(0, −2)
1	0	(1, 0)
2	4	(2, 4)

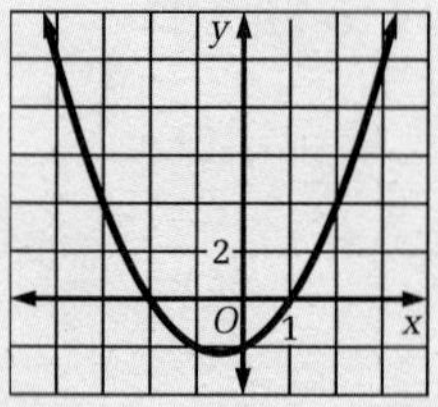

Graphing

Skill 25 Writing an Equation of a Line

The **slope** of a nonvertical line in a coordinate plane is the ratio

$$m = \frac{\text{change in } y}{\text{change in } x}, \text{ or } \frac{\text{rise}}{\text{run}}.$$

The ***y*-intercept** of such a line is the *y*-coordinate of the point where the line crosses the *y*-axis.

A nonvertical line in a coordinate plane always has an equation that can be expressed in **slope-intercept form,** where m is the slope of the line and b is the *y*-intercept:

$$\boldsymbol{y = mx + b}$$

Example Write an equation of the line with slope 3 and *y*-intercept −5.

Since $m = 3$ and $b = -5$, an equation is $y = 3x + (-5)$, or $y = 3x - 5$.

Example Write an equation of the line through the points (−3, 9) and (2, −1).

Step 1 Find the slope of the line: $m = \frac{\text{change in } y}{\text{change in } x} = \frac{-1 - 9}{2 - (-3)} = \frac{-10}{5} = -2.$

Step 2 Since $m = -2$, the equation $y = mx + b$ is $y = -2x + b$. Substitute the coordinates of either of the two given points into this equation to find the value of b:

$9 = -2(-3) + b$ ⟵ Substitute the coordinates (−3, 9) into $y = -2x + b$.

$9 = 6 + b$ ⟵ Simplify.

$3 = b$ ⟵ Solve for b.

An equation is $y = -2x + 3$.

You can find the equation of a graphed line by choosing two points on the line and using them to find the slope. If the *y*-intercept can easily be read off the graph, there is no need to calculate the value of b.

Example Write an equation of the line shown in the graph.

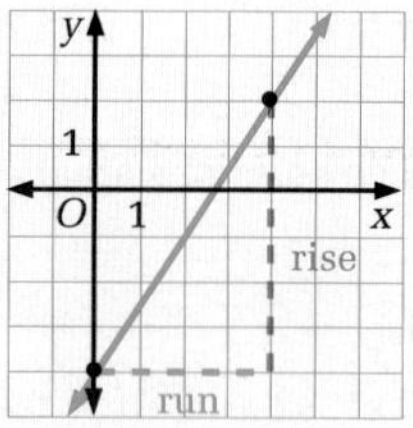

Choose two points on the graph to find m and b: (0, −4) and (4, 2).

Using these two points, you find:

$m = \frac{\text{change in } y}{\text{change in } x} = \frac{2 - (-4)}{4 - 0} = \frac{6}{4} = 1.5$

Since the graph crosses the *y*-axis at (0, −4), you know that $b = -4$.

Therefore, an equation is $y = 1.5x - 4$.

Find an equation of each line.

1. with slope 4, through (0, 7)
2. with slope -0.5, through $(2, -3)$
3. through $(6, -7)$ and (2, 5)
4. through $(-2, 1)$ and $(7, -5)$
5. the line shown at the right

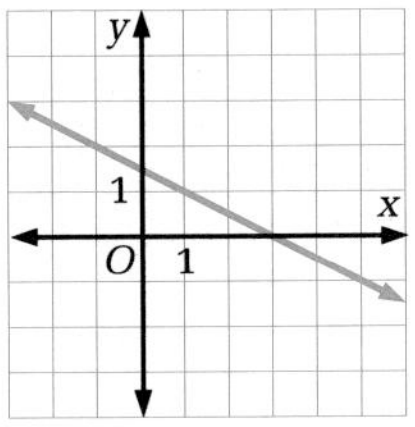

Skill 26 Making a Table of Values to Graph

To graph a function without using graphing technology, you need to make a table of values.

Example Make a table and graph the function $y = x^2 - 4x - 3$.

Choose any convenient x-values. The values 0, 1, and -1 often make calculation of y easy.

List x-values, y-values, and ordered pairs in three columns.

$y = x^2 - 4x - 3$		
x	y	(x, y)
-1	2	$(-1, 2)$
0	-3	$(0, -3)$
1	-6	$(1, -6)$
2	-7	$(2, -7)$
3	-6	$(3, -6)$
4	-3	$(4, -3)$

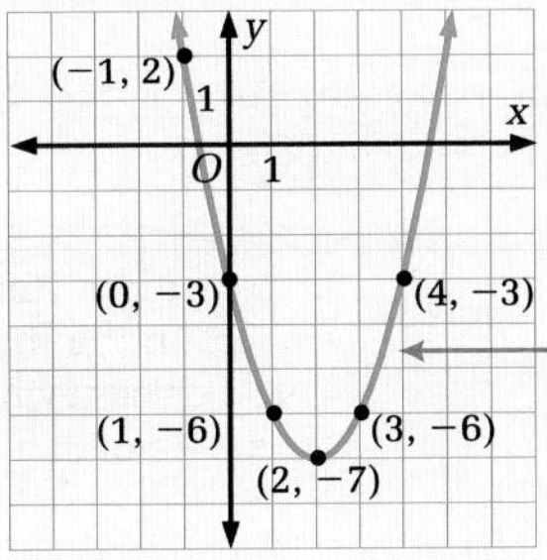

Make a table and graph each function.

1. $y = -2x + 5$
2. $y = \frac{x}{2} - 1$
3. $y = x^2 - 5$
4. $y = x^2 + x - 2$
5. $y = |x - 2|$
6. $y = 3 - \frac{|x|}{2}$
7. $y = 2|x + 2| - 3$
8. $y = \frac{x^3}{4}$
9. $y = x^3 - x$

Answers to Skill 26

5.

x	y	(x, y)
−2	4	(−2, 4)
−1	3	(−1, 3)
0	2	(0, 2)
1	1	(1, 1)
2	0	(2, 0)
3	1	(3, 1)
4	2	(4, 2)

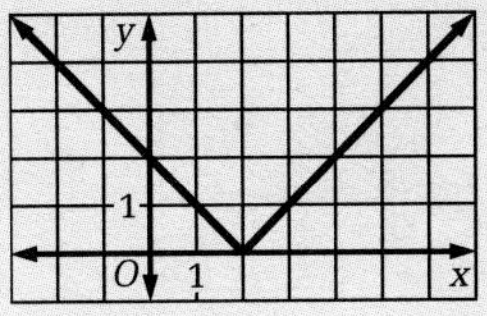

6.

x	y	(x, y)
−2	2	(−2, 2)
−1	2.5	(−1, 2.5)
0	3	(0, 3)
1	2.5	(1, 2.5)
2	2	(2, 2)

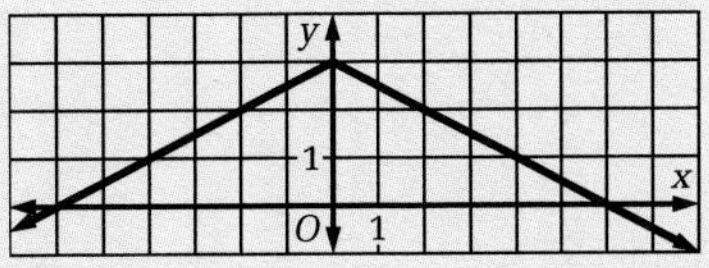

7.

x	y	(x, y)
−4	1	(−4, 1)
−3	−1	(−3, −1)
−2	−3	(−2, −3)
−1	−1	(−1, −1)
0	1	(0, 1)
1	3	(1, 3)
2	5	(2, 5)

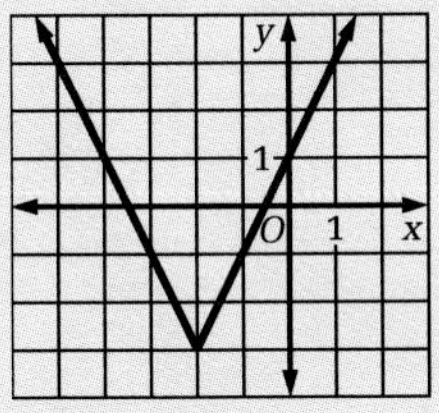

8.

x	y	(x, y)
−3	−6.75	(−3, −6.75)
−2	−2	(−2, −2)
−1	−0.25	(−1, −0.25)
0	0	(0, 0)
1	0.25	(1, 0.25)
2	2	(2, 2)
3	6.75	(3, 6.75)

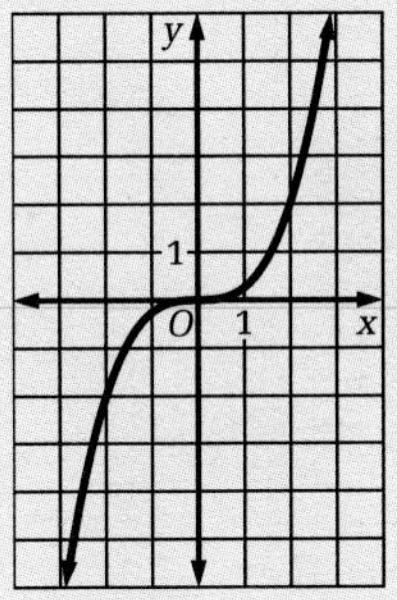

9.

x	y	(x, y)
−3	−24	(−3, −24)
−2	−6	(−2, −6)
−1	0	(−1, 0)
−0.5	0.325	(−0.5, 0.325)
0	0	(0, 0)
0.5	−0.325	(0.5, −0.325)
1	0	(1, 0)
2	6	(2, 6)
3	24	(3, 24)

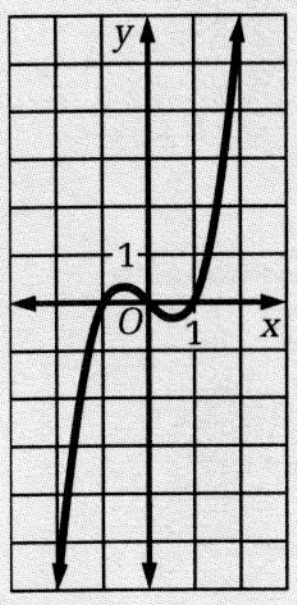

Skill 27 Graphing Inequalities

To graph an inequality in one variable on a number line, you shade the part of the number line that contains those *x*-values that make the inequality true.

Example **a.** Graph $x < 4$.

Draw a ray to the left to include all points less than 4.

Use an open circle to show that 4 is not included.

b. Graph $x \geq 5$.

Use a closed circle to show that −5 is included.

Draw a ray to the right to include all points greater than −5.

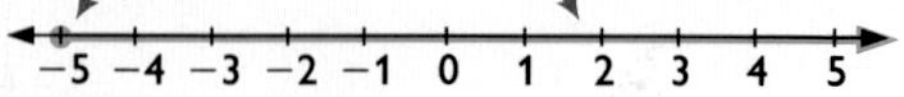

To graph a system of inequalities in two variables, you shade the region whose points make *both* inequalities true.

Example Graph this system of inequalities: $y \geq -x + 4$
$2x - y > -2$

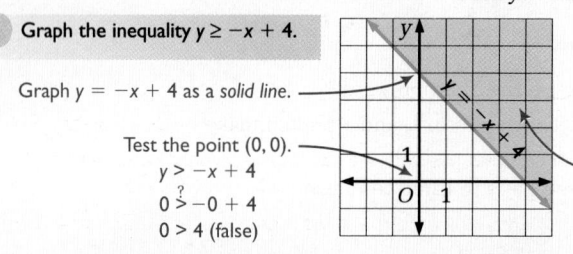

1 Graph the inequality $y \geq -x + 4$.

Graph $y = -x + 4$ as a *solid line.*

Test the point (0, 0).
$y > -x + 4$
$0 \overset{?}{>} -0 + 4$
$0 > 4$ (false)

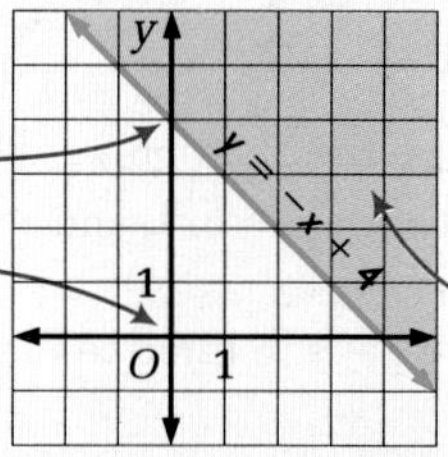

(0, 0) is not a solution. Shade the region on the *opposite side* of the line from (0, 0).

2 Graph the inequality $2x - y > -2$ on the same set of axes.

Graph $2x - y = -2$ as a *dashed line.*

Test the point (0, 0).
$2x - y > -2$
$2(0) - 0 \overset{?}{>} -2$
$0 > -2$ (true)

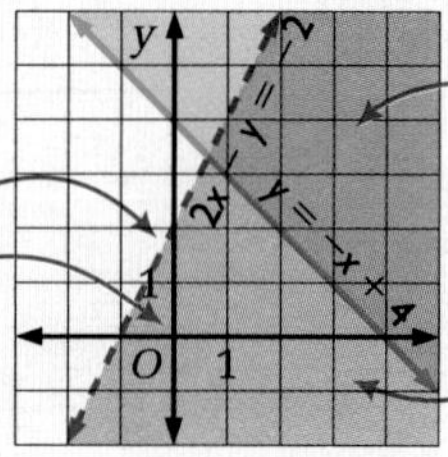

You may want to darken the region of overlap.

(0, 0) is a solution. Shade the region on the *same side* of the line as (0, 0).

3 Find the solution region for the system. The points inside the region of overlap make both inequalities true.

Graph each inequality.

1. $x > -3$ **2.** $x \leq 4$ **3.** $x < 1$ **4.** $x \geq -2$

Graph each system of inequalities in a coordinate plane.

5. $y \leq 3x - 3$
$y > -\frac{1}{2}x + 2$

6. $y < 3x$
$3y - 2x \geq 9$

7. $x < 2$
$x + 2y > -2$

8. $y - 3x \geq 5$
$2y + 5x < 7$

Answers to Skill 27

1. 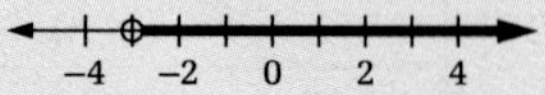

2. 

3. 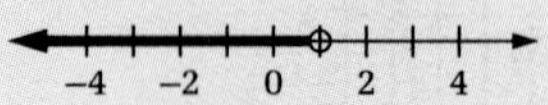

4. 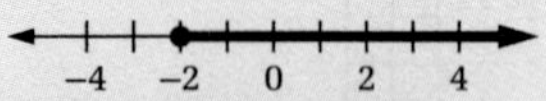

5. 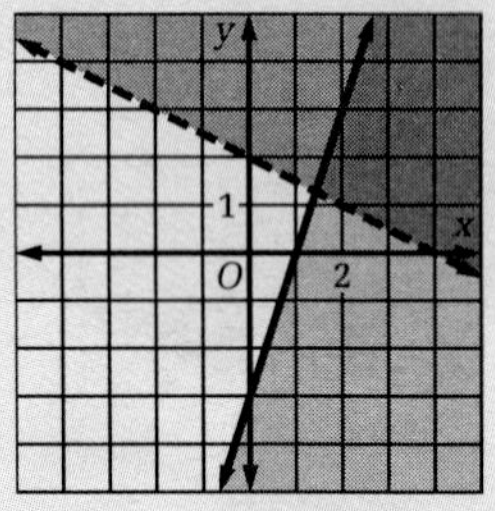

6. 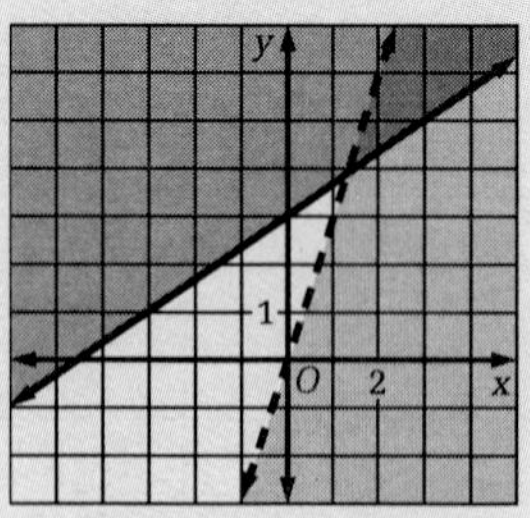

7. 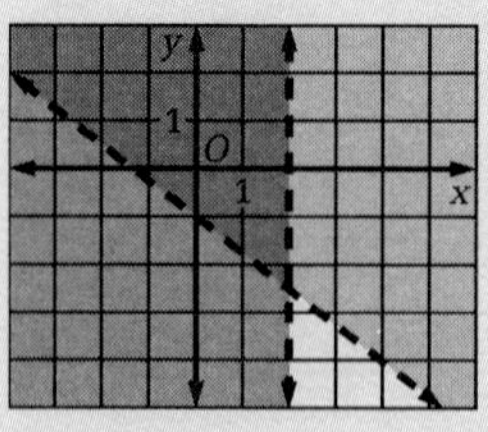

8. 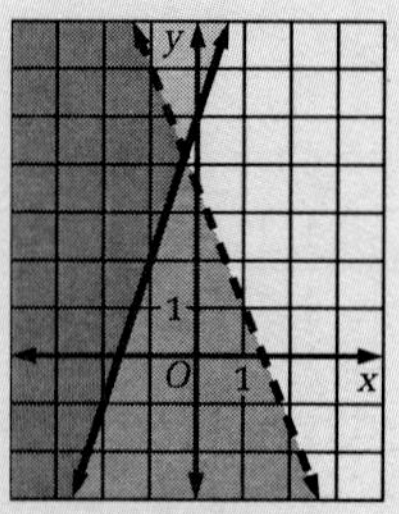

Skill 28 Graphing Quadratic Functions

A **quadratic function** is a function of the form $f(x) = ax^2 + bx + c$, where $a \neq 0$.

The graph of a quadratic function is always a U-shaped curve called a **parabola.** The coefficients a, b, and c help you draw the parabola in a coordinate system.

y-intercept of the parabola: the number c

line of symmetry of the parabola: the vertical line whose equation is $x = -\frac{b}{2a}$

Example Find the y-intercept and the equation of the line of symmetry of the graph of $f(x) = 3x^2 - 5x + 2$.

In the given equation, $c = 2$, so the y-intercept is 2.

In the given equation, $a = 3$ and $b = -5$, so the equation of the line of symmetry is

$$x = -\frac{b}{2a} = -\frac{-5}{2(3)}, \text{ or } x = \frac{5}{6}.$$

The parabola determined by a quadratic function $f(x) = ax^2 + bx + c$ can open upward or downward.

If the number a is positive, the parabola opens upward, and the function has a minimum value.

If a is negative, the parabola opens downward, and the function has a maximum value.

Example Graph the function $y = -1.5x^2 + 6x + 3$.

In this function, $a = -1.5$, $b = 6$, and $c = 3$.

The y-intercept is 3.

The equation of the line of symmetry is

$$x = -\frac{6}{2(-1.5)}, \text{ or } x = 2.$$

Since $a < 0$, the graph opens downward and y has a maximum value.

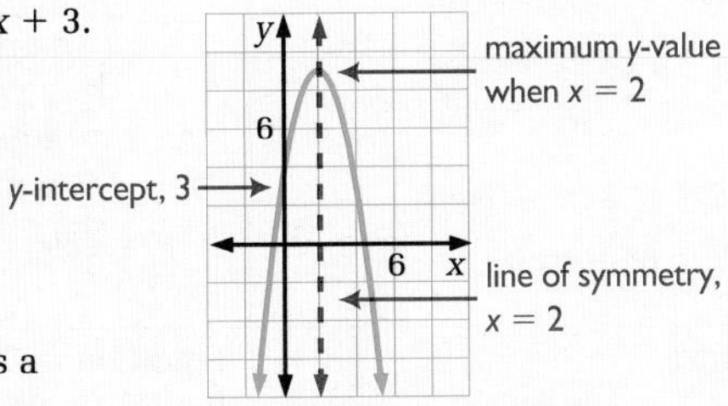

Find the y-intercept and the equation of the line of symmetry for each quadratic function. Then sketch the graph.

1. $y = x^2 - 4$

2. $y = -2x^2 + 7$

3. $y = -0.5x^2 + 3x$

4. $y = 0.25x^2 - 2x - 1$

5. $y = 3x^2 - 12x - 10$

6. $y = -0.75x^2 + 4.5x + 6$

Answers to Skill 28

1. y-intercept: -4; $x = 0$

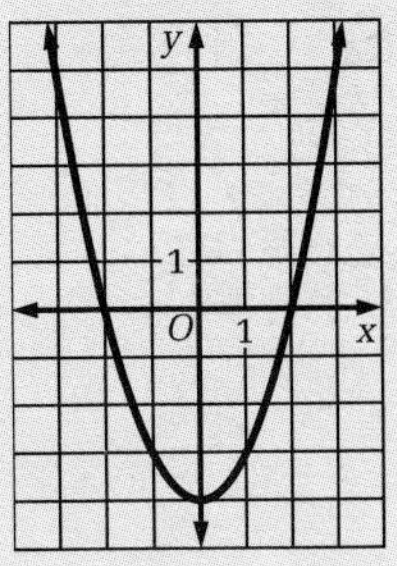

2. y-intercept: 7; $x = 0$

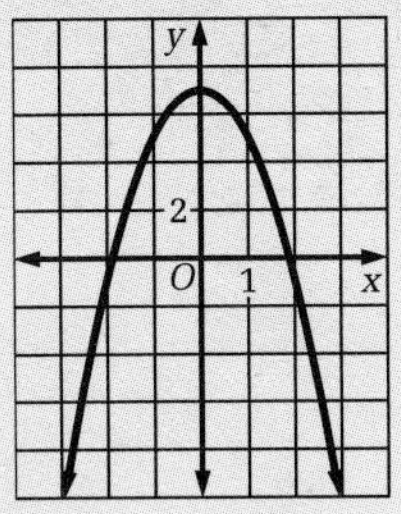

3. y-intercept: 0; $x = 3$

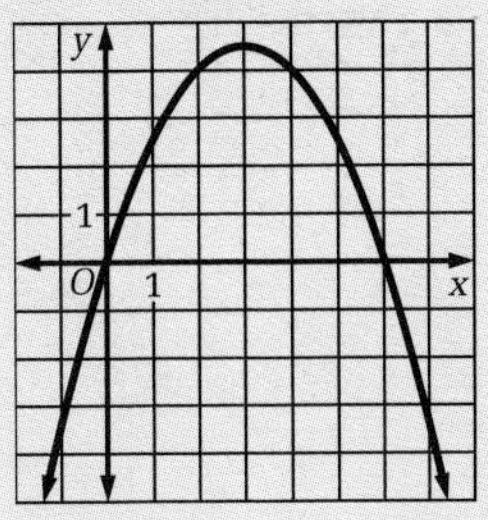

4. y-intercept: -1; $x = 4$

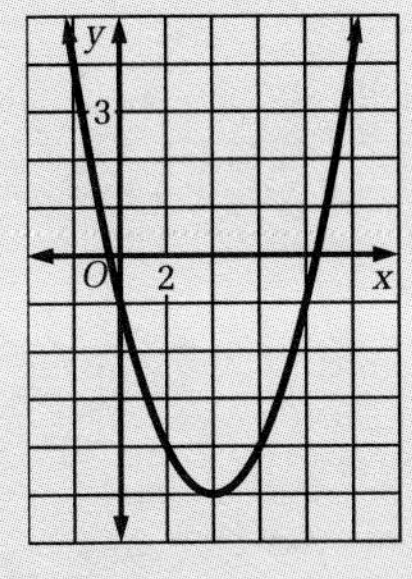

5. y-intercept: -10; $x = 2$

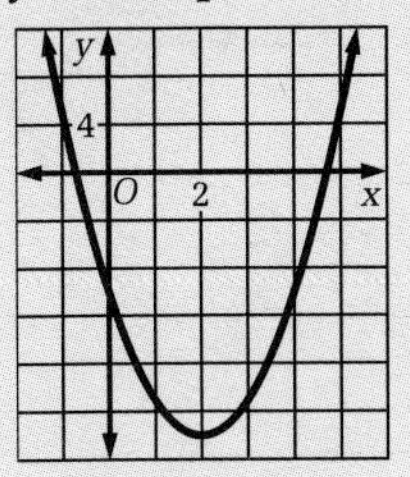

6. y-intercept: 6; $x = 3$

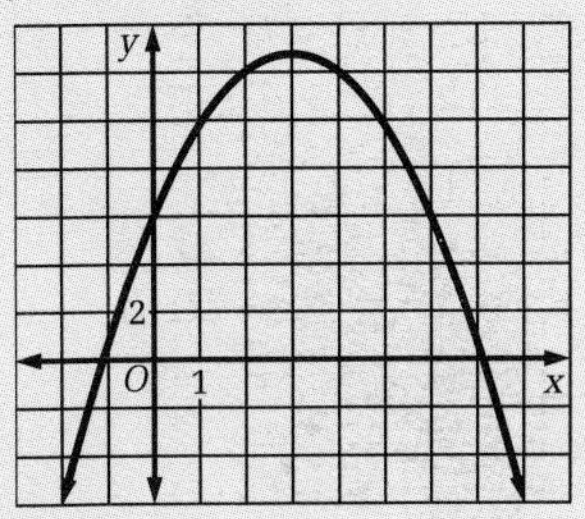

Geometry and Measurement

Formulas and Relationships

Skill 29 Using Formulas from Geometry

A = area B = area of base S.A. = surface area V = volume

Parallelogram

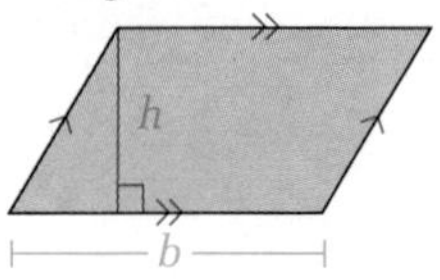

$A = bh$

Triangle

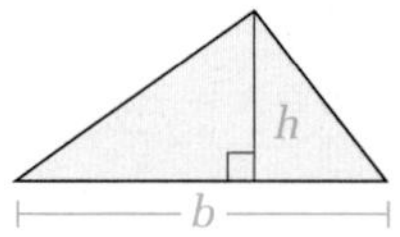

$A = \frac{1}{2}bh$

Trapezoid

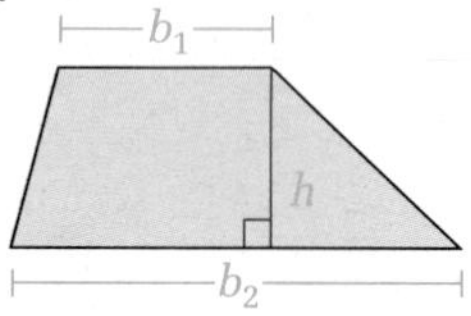

$A = \frac{1}{2}(b_1 + b_2)h$

Circle

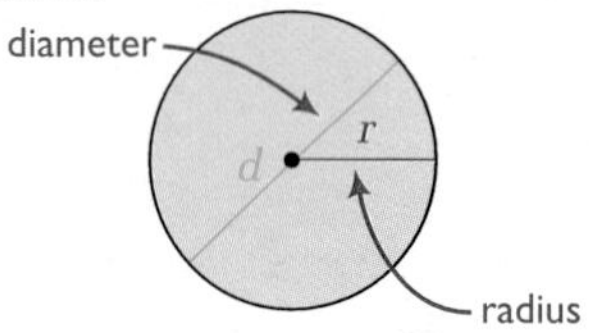

Circumference $= 2\pi r = \pi d$
$A = \pi r^2$

Sector

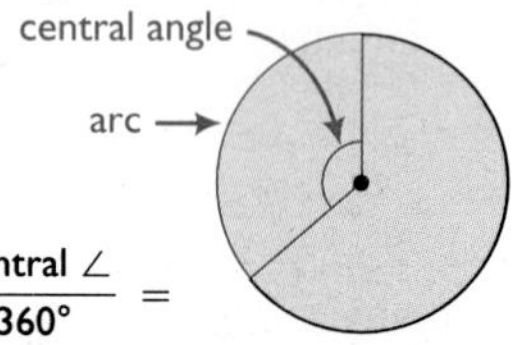

$\frac{\text{central } \angle}{360°} = \frac{\text{arc length}}{\text{circumference}} = \frac{\text{area of sector}}{\text{area of circle}}$

Prism

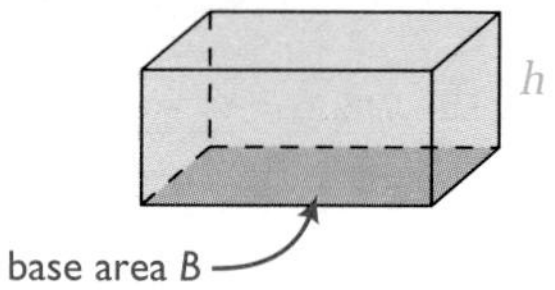

S.A. = areas of two bases + areas of faces
$V = Bh$

Cylinder

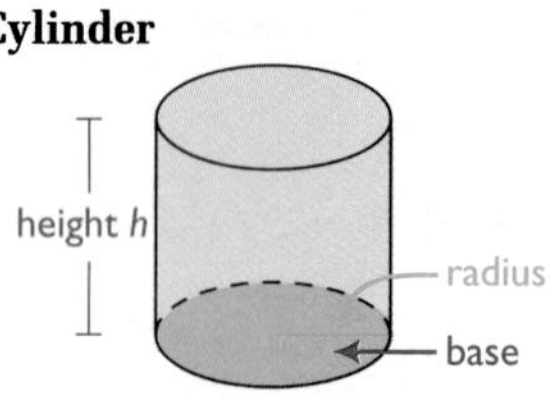

S.A. $= 2\pi r^2 + 2\pi rh$
$V = Bh = \pi r^2 h$

Pyramid

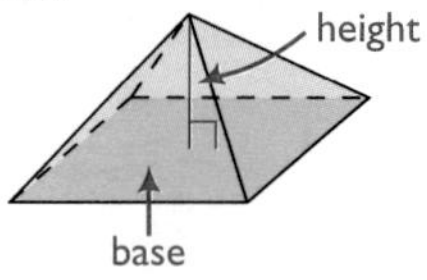

S.A. = area of base + areas of faces
$V = \frac{1}{3}Bh$

Cone

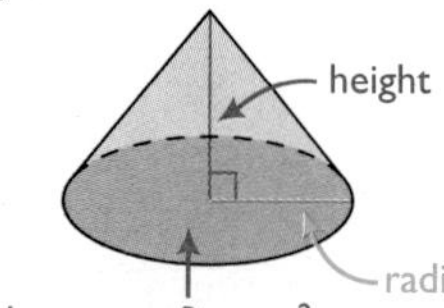

base area $B = \pi r^2$

$V = \frac{1}{3}Bh = \frac{1}{3}\pi r^2 h$

Sphere

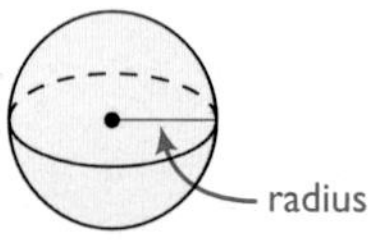

S.A. $= 4\pi r^2$ $V = \frac{4}{3}\pi r^3$

Equation of Circle with Center (0, 0)

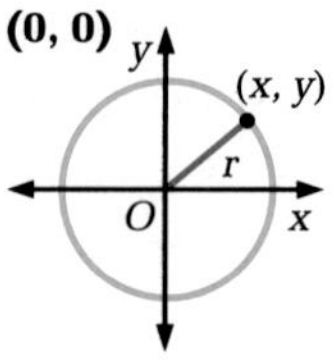

$x^2 + y^2 = r^2$

Equation of Sphere with Center (0, 0, 0)

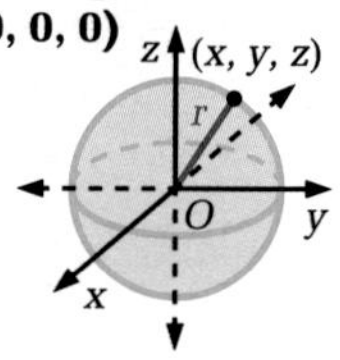

$x^2 + y^2 + z^2 = r^2$

Example Find the area of the trapezoid shown.

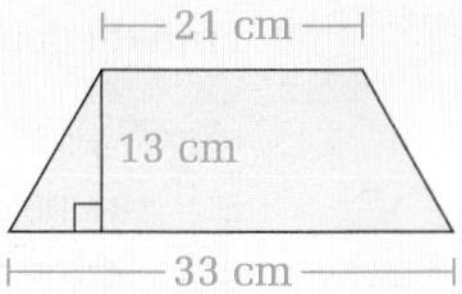

$A = \frac{1}{2}(b_1 + b_2)h$ ← Write the formula for the area of a trapezoid.

$A = \frac{1}{2}(21 + 33)13$ ← Substitute **21** for b_1, **33** for b_2, and **13** for h.

$= \frac{1}{2}(54)13 = 351$

The area is 351 cm^2. ← $cm \times cm = cm^2$

Find the area of each figure.

1. a parallelogram of height 10 ft with a base of length 16 ft
2. a trapezoid of height 3 in. with bases of length 5 in. and 8 in.
3. a sector with central angle 40° and radius 9 in.

Find the surface area of each space figure.

4. a rectangular prism with dimensions 18 in., 12 in., and 4.5 in.
5. a cylinder of height 20 cm and with a base of radius 5 cm

Find the volume of each space figure.

6. a triangular prism whose height is 8 cm and whose base area is 25 cm^2
7. a cylinder of height 8.5 m and with a base of radius 2 m
8. a pyramid of height 12 in., whose base is a square with sides of length 10 in.

Skill 30 Using Special Triangle Relationships

Triangle Sum Theorem

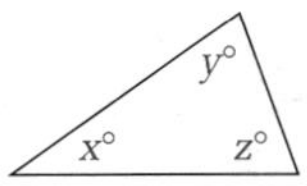

$x + y + z = 180$

Pythagorean Theorem

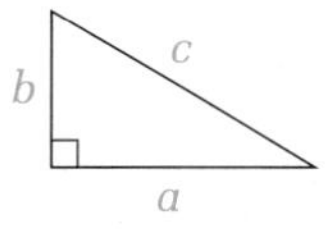

$a^2 + b^2 = c^2$

45-45-90 Right Triangle

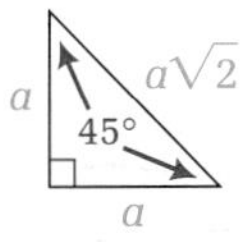

30-60-90 Right Triangle

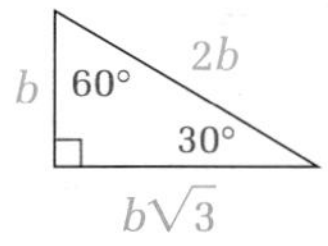

Answers to Skill 29

1. 160 ft^2
2. 19.5 $in.^2$
3. about 28.27 $in.^2$
4. 702 $in.^2$
5. about 785.4 cm^2
6. 200 cm^3
7. about 106.8 m^3
8. 400 $in.^3$

Example Find the length of the hypotenuse in the right triangle.

$c^2 = 6.4^2 + 12^2$ ← Use the Pythagorean theorem: $c^2 = a^2 + b^2$.

$c^2 = 184.96$

$c = \sqrt{184.96}$ ← Find the positive square root.

$c = 13.6$

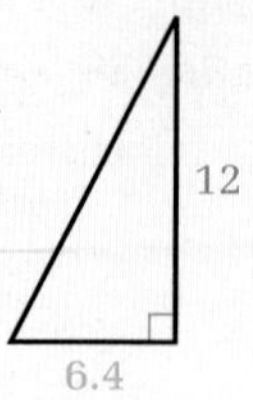

Example Find the missing sides of the right triangle.

$QR = a = 10$ ← Use the pattern for 30-60-90 triangles.

$PR = a\sqrt{3} = 10\sqrt{3}$

$PQ = 2a = 2 \cdot 10 = 20$

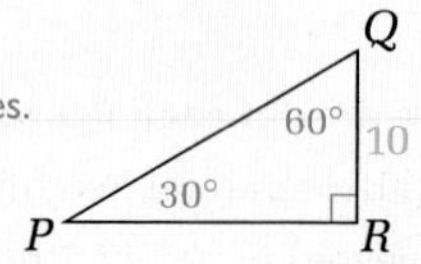

Find the length of the missing side of a right triangle with legs of length *a* and *b* and hypotenuse of length *c*.

1. $a = 3, b = 4, c = \underline{?}$ **2.** $a = 7, c = 25, b = \underline{?}$ **3.** $b = 8, c = 11, a = \underline{?}$

4. Find the lengths of the missing sides of a 30-60-90 triangle when the side opposite the 60° angle is 12 cm long.

Skill 31 Using the Distance and Midpoint Formulas

When you know the coordinates of two points (x_1, y_1) and (x_2, y_2), you can find the distance between the points by using the Pythagorean theorem or the **distance formula:**

$$d = \sqrt{(\text{change in } x)^2 + (\text{change in } y)^2} = \sqrt{(x_2 - x_1)^2 + (y_2 - y_1)^2}$$

Example Find the distance between the points $(-4, 6)$ and $(0, 3)$.

$\text{distance} = \sqrt{(x_2 - x_1)^2 + (y_2 - y_1)^2}$ ← Use the distance formula.

$= \sqrt{(0 - (-4))^2 + (3 - 6)^2}$ ← Substitute $(-4, 6)$ for (x_1, y_1) and $(0, 3)$ for (x_2, y_2).

$= \sqrt{4^2 + (-3)^2}$

$= \sqrt{25}$

$= 5$

The distance between the two points is 5 units.

Answers to Skill 30

1. 5
2. 24
3. about 7.55
4. $b \approx 6.9$ cm, $c \approx 13.9$ cm

The Midpoint Formula

The midpoint of a segment with endpoints (x_1, y_1) and (x_2, y_2) has coordinates

$$\left(\frac{x_1 + x_2}{2}, \frac{y_1 + y_2}{2}\right).$$

the mean of the x's — the mean of the y's

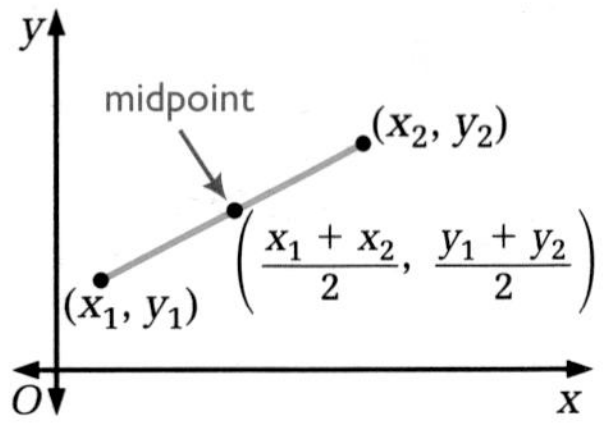

Example Find the coordinates of the midpoint of the segment whose endpoints are (9, −14) and (3, 4).

Let (x_1, y_1) be (9, −14) and let (x_2, y_2) be (3, 4).

The x-coordinate of the midpoint is $\frac{x_1 + x_2}{2} = \frac{9 + 3}{2} = 6$.

The y-coordinate of the midpoint is $\frac{y_1 + y_2}{2} = \frac{-14 + 4}{2} = -5$.

The coordinates of the midpoint are (6, −5).

Find the distance between each pair of points.

1. (2, 7) and (5, 3)
2. (4, −1) and (−8, 4)
3. (−2, 7) and (3, 9)

Find the coordinates of the midpoint of a segment with the given endpoints.

4. (3, 8) and (7, 4)
5. (−2, 1) and (6, −9)
6. (3, −4) and (8, −5)

Skill 32 Recognizing Similar Figures

Two polygons are **similar** if and only if their vertices can be matched up so that (1) corresponding angles are equal in measure and (2) corresponding sides are in proportion.

For *triangles*, condition (1) alone is enough to guarantee similarity. In fact, all you need are *two* pairs of angles that are equal in measure.

Example $m\angle C = m\angle F$ and $m\angle B = m\angle E$, so the triangles are similar.

Corresponding sides: AB and DE
BC and EF
CA and FD

Equal ratios: $\frac{AB}{DE} = \frac{BC}{EF} = \frac{CA}{FD}$.

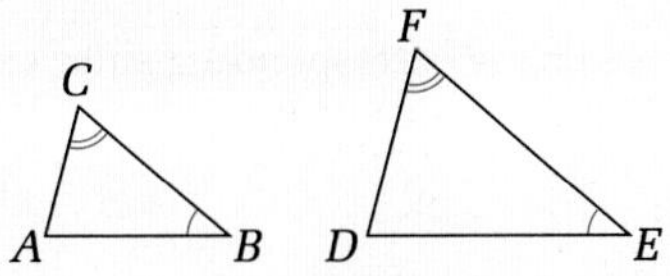

$\triangle ABC \sim \triangle DEF$

Answers to Skill 31

1. 5
2. 13
3. $\sqrt{29} \approx 5.4$
4. (5, 6)
5. (2, −4)
6. (5.5, −4.5)

Explain why the triangles in each pair are similar. Name all pairs of corresponding sides and angles.

1. 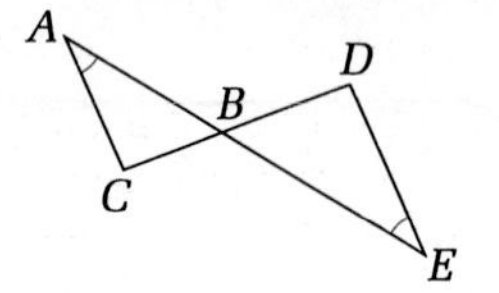

2. 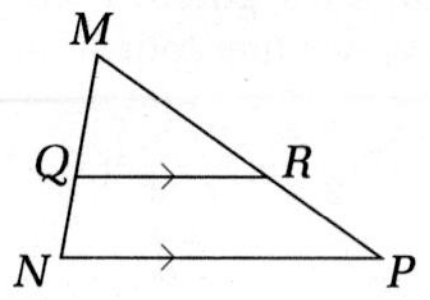

3. $\triangle MQR$ is similar to $\triangle XYZ$. $QR = 9$, $RM = 8$, $MQ = 12$, and $ZX = 16$. Find YZ.

Skill 33 Recognizing Congruent Figures

Two polygons are **congruent** if and only if their vertices can be matched up so that the *corresponding parts* (angles and sides) are equal in measure.

For *triangles,* you can prove congruence when only three pairs of corresponding parts are known to be equal in measure. See the list of postulates and theorems about triangles on page 670.

Example Name the congruent triangles. Explain why they are congruent.

$\triangle ABC \cong \triangle DEF$ by ASA.

Two pairs of angles and the included sides are equal in measure:

$m\angle B = m\angle E$, $m\angle A = m\angle D$, $AB = DE$.

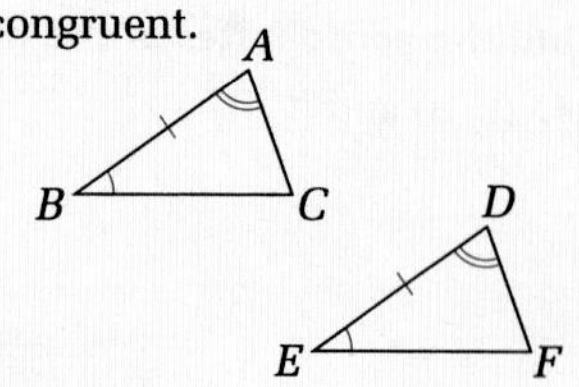

For each pair of triangles, tell whether there is enough information to prove that the triangles are congruent. Write *Yes* or *No*. Explain.

1. 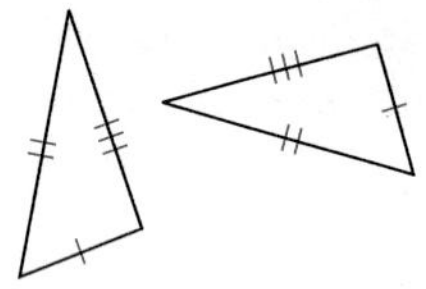

2. 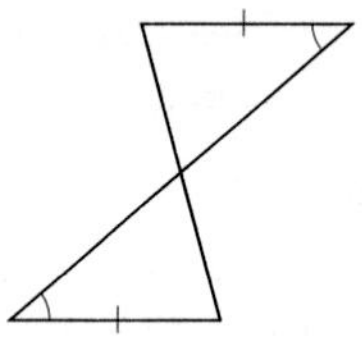

3. 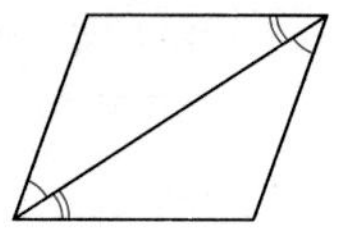

4. 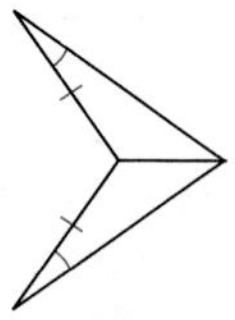

5. Name the corresponding sides and angles. Find the value of x.

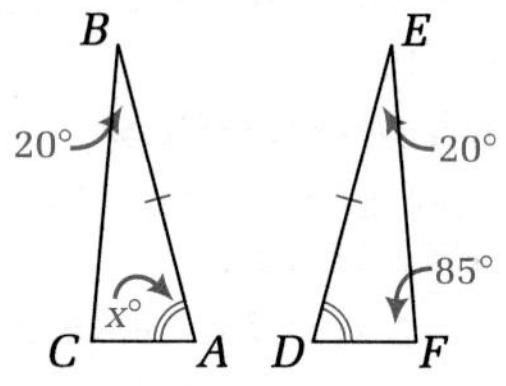

Answers to Skill 32

1. $m\angle CAB = m\angle DEB$ and $m\angle CBA = m\angle DBE$, so the triangles are similar. Corresponding angles are $\angle CAB$ and $\angle DEB$, $\angle CBA$ and $\angle DBE$, and $\angle BCA$ and $\angle BDE$; corresponding sides are $\overline{AC}$ and $\overline{ED}$, $\overline{BC}$ and $\overline{BD}$, and $\overline{AB}$ and $\overline{EB}$.
2. $m\angle PNM = m\angle RQM$ and $m\angle NMP = m\angle QMR$, so the triangles are similar. Corresponding angles are $\angle PNM$ and $\angle RQM$, $\angle NMP$ and $\angle QMR$, and $\angle M$ and $\angle M$; corresponding sides are $\overline{NP}$ and $\overline{QR}$, $\overline{MN}$ and $\overline{MQ}$, and $\overline{MP}$ and $\overline{MR}$.
3. 18

Answers to Skill 33

1. Yes; by SSS.
2. Yes; by AAS.
3. Yes; by ASA.
4. No.
5. $CB = FE$, $AC = DF$, $AB = DE$, $m\angle A = m\angle D$, $m\angle C = m\angle F$, $m\angle B = m\angle E$; 75

Logical Reasoning

Skill 34 Understanding Implications

A statement of the form $p \to q$ (read "p implies q" or "if p, then q") is called an **implication** or **conditional.** The statement represented by p is the **hypothesis** of the implication, and the statement represented by q is the **conclusion.**

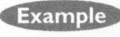

Example The statement "All prairie dogs are rodents" can be rewritten in the following equivalent ways:

- *Every* prairie dog is a rodent.
- The fact that an animal is a prairie dog *implies* that it is a rodent.
- *If* an animal is a prairie dog, *then* it is a rodent.

(hypothesis: "an animal is a prairie dog"; conclusion: "it is a rodent")

The **converse** of an implication is formed by reversing the hypothesis and the conclusion.

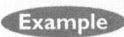

Example The converse of the statement "If an animal is a prairie dog, then it is a rodent" is the statement: "If an animal is a rodent, then it is a prairie dog."

Since mice are also rodents, the converse is *false*. ← The converse of a true inplication may be false.

The conjunction of a statement and its converse is called a **biconditional.** Biconditionals are often stated using "if and only if."

Example The biconditional "Today is Tuesday if and only if yesterday was Monday" can be rewritten as the conjunction: "If today is Tuesday, then yesterday was Monday, and if yesterday was Monday, then today is Tuesday."

Rewrite each implication in an equivalent form using the key word(s) given.

1. If a figure is a rhombus, then its diagonals are perpendicular. (*all*)
2. A figure is a square only if it is a rhombus. (*implies*)
3. Whenever it is Saturday, it is a day of the weekend. (*every*)
4. If a figure is a cylinder, then the formula for its volume is $V = \pi r^2 h$. (*only if*)

Write each conjunction as a biconditional using "if and only if."

5. A polygon is a quadrilateral if it has four sides, and a polygon has four sides if it is a quadrilateral.
6. If $5x = 20$, then $x = 4$, and if $x = 4$, then $5x = 20$.

Answers to Skill 34

1. All rhombuses have perpendicular diagonals.
2. The fact that a figure is a square implies that it is a rhombus.
3. Every Saturday is a day of the weekend.
4. A figure is a cylinder only if the formula for its volume is $V = Bh$.
5. A polygon is a quadrilateral if and only if it has four sides.
6. $5x = 20$ if and only if $x = 4$.

Skill 35 Using Valid Arguments

A **rule of logic** consists of two given statements, called **premises,** and a statement, called a **conclusion,** that follows from them.

RULES OF LOGIC

Direct Argument

If p is true, then q is true.	$p \rightarrow q$
p is true.	p
Therefore, q is true.	$\therefore q$ ← Read "Therefore q."

Indirect Argument

If p is true, then q is true.	$p \rightarrow q$
q is not true.	not q
Therefore, p is not true.	$\therefore$ not p

Chain Rule

If p is true, then q is true.	$p \rightarrow q$
If q is true, then r is true.	$q \rightarrow r$
Therefore, if p is true, then r is true.	$\therefore p \rightarrow r$

***Or* Rule**

p is true or q is true.	p or q
p is not true.	not p
Therefore, q is true.	$\therefore q$

Example What conclusion can you reach when both premises are true?

a. Premise 1: If an animal is an insect, then it has exactly six legs.
Premise 2: A spider does not have exactly six legs.

b. Premise 1: If a polygon is a square, then it is a rectangle.
Premise 2: If a polygon is a rectangle, then it is a parallelogram.

Translate each premise into its symbolic form.

a. *insect* → *six legs*
not *six legs*

These are the premises of an indirect argument.

$p \rightarrow q$
not q
$\therefore$ not p

Conclusion: A spider is not an insect.

b. *square* → *rectangle*
rectangle → ***parallelogram***

These are the premises of a chain rule argument.

$p \rightarrow q$
$q \rightarrow r$
$\therefore p \rightarrow r$

Conclusion: If a polygon is a square, it is a parallelogram.

658

Arguments that do not use rules of logic are considered errors, or **invalid arguments.**

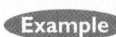

Example Explain why each of the following arguments is invalid.

a. *Argument:* All multiples of 20 are multiples of 5. Sam's locker number is a multiple of 5. Therefore, Sam's locker number is a multiple of 20.

b. *Argument:* Every member of YHC likes to travel. Indira is not a member of YHC. Therefore, Indira does not like to travel.

Write the premises as if-then statements. Use symbols to examine the argument:

a. If a number is a multiple of 20, then it is a multiple of 5. $p \to q$

Sam's locker number is a multiple of 5. $\underline{q}$

Therefore, Sam's locker number is a multiple of 20. $\therefore p$

This argument does not use a rule of logic. If the first premise were the converse, "$q \to p$," instead of "$p \to q$," this argument would be valid. This is called a *converse error.*

b. If a person is a member of YHC, then the person likes to travel. $p \to q$

Indira is not a member of YHC. $\underline{\text{not } p}$

Therefore, Indira does not like to travel. $\therefore$ not q

This argument would be valid if the first premise were the inverse, "not $p \to$ not q," instead of "$p \to q$." This is an example of *inverse error.*

Tell which rule of logic is used in these valid arguments.

1. John is in his room or John is in the kitchen. John is not in the kitchen. Therefore, John is in his room.
2. If I take the Number 10 bus, then I can get to the mall. If I can get to the mall, then I can go to the movies. Therefore, if I take the Number 10 bus, then I can go to the movies.
3. If Manuela has guacamole, then she has tasted avocado. Manuela has eaten guacamole. Therefore, she has tasted avocado.

What conclusion can you reach when both premises are true?

4. If an animal has a backbone, it is a vertebrate. A horse has a backbone.
5. If a triangle is isosceles, then it has at least two angles equal in measure. $\triangle ABC$ does not have at least two angles equal in measure.
6. If $2x + 3 = 21$, then $2x = 18$. If $2x = 18$, then $x = 9$.
7. If $P(A) = 0$, then A is an impossible event. A is not an impossible event.

Answers to Skill 35

1. *or* rule
2. chain rule
3. direct argument
4. A horse is a vertebrate.
5. $\triangle ABC$ is not isosceles.
6. If $2x + 3 = 21$, then $x = 9$.
7. $P(A) \neq 0$.

Trigonometry

Skill 36 Finding Sine, Cosine, and Tangent Ratios

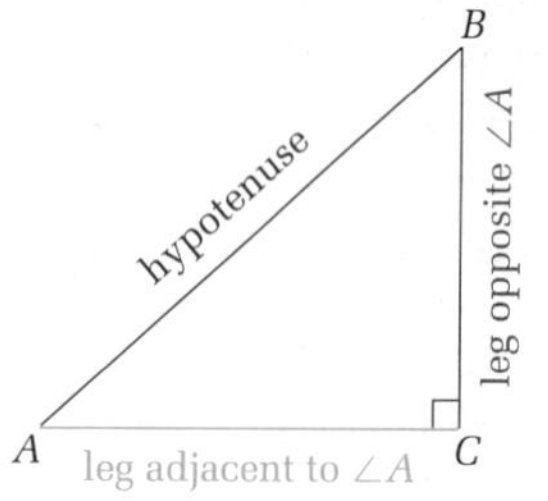

Suppose $\angle A$ is one of the acute angles of a right triangle. The **trigonometric ratios** for $\angle A$ are ratios of the measures of the sides of the triangle.

Sine of $\angle A$ $\quad \sin A = \dfrac{\text{leg opposite } \angle A}{\text{hypotenuse}} = \dfrac{BC}{AB}$

Cosine of $\angle A$ $\quad \cos A = \dfrac{\text{leg adjacent to } \angle A}{\text{hypotenuse}} = \dfrac{AC}{AB}$

Tangent of $\angle A$ $\quad \tan A = \dfrac{\text{leg opposite } \angle A}{\text{leg adjacent to } \angle A} = \dfrac{BC}{AC}$

Example Write each trigonometric ratio as a fraction and as a decimal rounded to the nearest hundredth.

a. $\sin P$ **b.** $\cos P$ **c.** $\tan P$

d. $\sin Q$ **e.** $\cos Q$ **f.** $\tan Q$

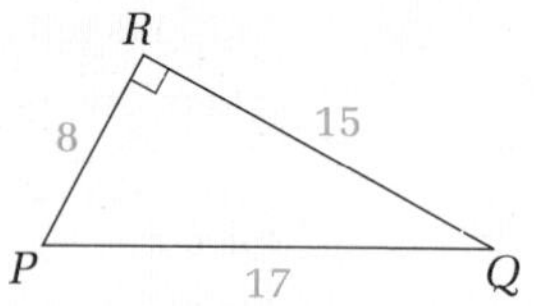

a. $\sin P = \dfrac{\text{leg opposite } \angle P}{\text{hypotenuse}} = \dfrac{15}{17} \approx 0.88$

b. $\cos P = \dfrac{\text{leg adjacent to } \angle P}{\text{hypotenuse}} = \dfrac{8}{17} \approx 0.47$

c. $\tan P = \dfrac{\text{leg opposite } \angle P}{\text{leg adjacent to } \angle P} = \dfrac{15}{8} \approx 1.88$

d. $\sin Q = \dfrac{\text{leg opposite } \angle Q}{\text{hypotenuse}} = \dfrac{8}{17} \approx 0.47$

e. $\cos Q = \dfrac{\text{leg adjacent to } \angle Q}{\text{hypotenuse}} = \dfrac{15}{17} \approx 0.88$

f. $\tan Q = \dfrac{\text{leg opposite } \angle Q}{\text{leg adjacent to } \angle Q} = \dfrac{8}{15} \approx 0.53$

Use the triangle at the right. Find the value of each trigonometric ratio.

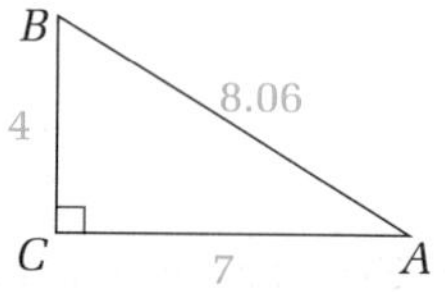

1. $\sin A$ **2.** $\cos A$ **3.** $\tan A$

4. $\sin B$ **5.** $\cos B$ **6.** $\tan B$

Draw and label a right triangle that fits each description.

7. $\triangle ABC$ with $\sin A = \dfrac{\text{leg opposite } \angle A}{\text{hypotenuse}} = \dfrac{3}{5}$

8. $\triangle XYZ$ with $\cos X = \dfrac{\text{leg adjacent to } \angle X}{\text{hypotenuse}} = \dfrac{5}{13}$

Answers to Skill 36

1. $\dfrac{4}{8.06} \approx 0.50$
2. $\dfrac{7}{8.06} \approx 0.87$
3. $\dfrac{4}{7} \approx 0.57$
4. $\dfrac{7}{8.06} \approx 0.87$
5. $\dfrac{4}{8.06} \approx 0.50$
6. $\dfrac{7}{4} \approx 1.75$
7. 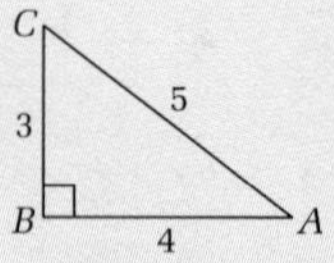

8. 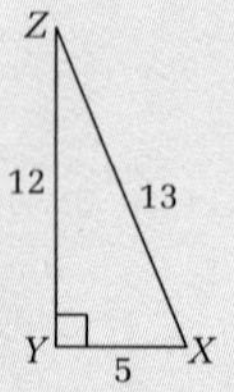

Skill 37 Finding Missing Lengths in Right Triangles

You can use the trigonometric ratios to find missing lengths in right triangles. Scientific calculators are programmed to find these ratios. A table of trigonometric ratios is on page 667.

Example Find the missing length in the triangle at the right.
Use the tangent ratio.

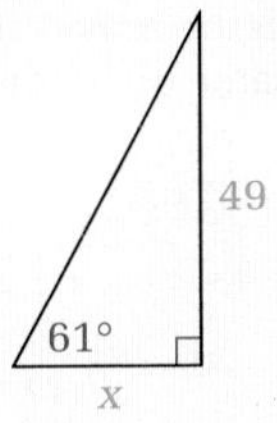

$$\tan 61° = \frac{49}{x}$$

$$1.804 \approx \frac{49}{x}$$ ← Use the **tan** key on your calculator.

$$1.804 \cdot x \approx \frac{49}{x} \cdot x$$ ← Multiply both sides by *x*.

$$1.804x \approx 49$$

$$\frac{1.804x}{1.804} \approx \frac{49}{1.804}$$ ← Divide both sides by 1.804.

$$x \approx \frac{49}{1.804} \approx 27.2$$

Example Find the altitude to side $\overline{AB}$ of $\triangle ABC$.
Let h = the altitude to $\overline{AB}$.

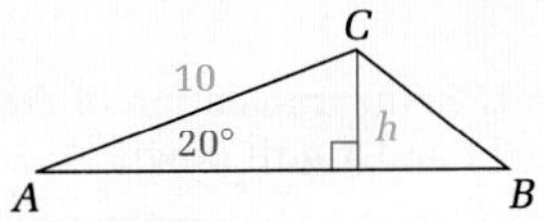

$$\sin 20° = \frac{h}{10}$$

$$h = 10 \sin 20°$$ ← Multiply both sides by 10.

$$\approx 10(0.3420)$$ ← Use the **sin** key on your calculator.

$$\approx 3.42$$

Find each missing length.

1.

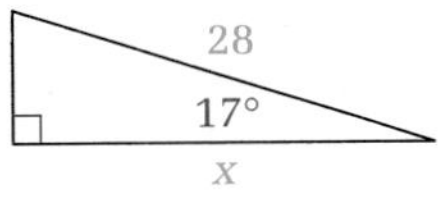

2.

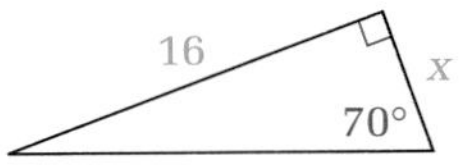

3.

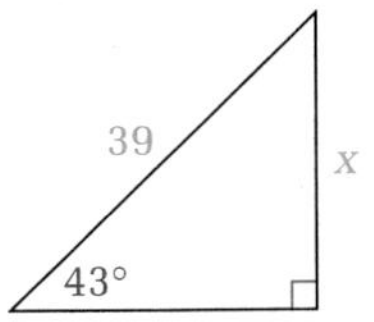

4. Find the length of the altitude to side $\overline{PQ}$ in $\triangle PQR$ at the right.

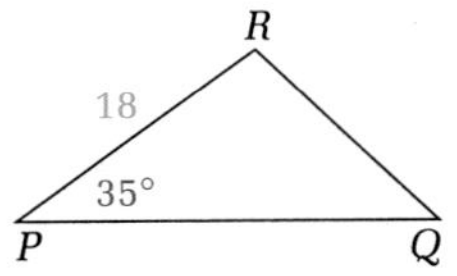

Answers to Skill 37

1. $x \approx 26.8$
2. $x \approx 5.8$
3. $x \approx 26.6$
4. $h \approx 10.3$

Skill 38 Transforming a Figure

A **transformation** is a procedure by which a new geometric figure is produced from a given figure according to a definite rule.

Translation

A **translation** slides a figure vertically and/or horizontally without changing its size or shape and without turning it. Each point of the figure moves the same distance and in the same direction.

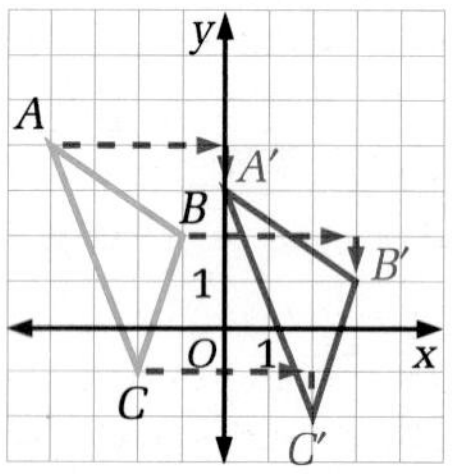

$\triangle A'B'C'$ is the translation of $\triangle ABC$ 4 units to the right and 1 unit down.

Rotation

A **rotation** moves each point of a figure through a circular arc about a common point, called the **center of rotation.** Each point of the figure is rotated by the same number of degrees.

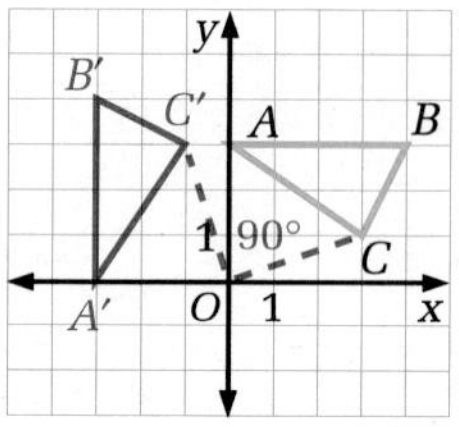

$\triangle A'B'C'$ is the 90° counterclockwise rotation of $\triangle ABC$ about the origin.

Dilation

A **dilation** transforms a figure into a similar figure. Each point of the figure is transformed along a line connecting it to the **center of dilation.** The ratio of any length in the transformed figure to the corresponding length in the original figure is the **scale factor.**

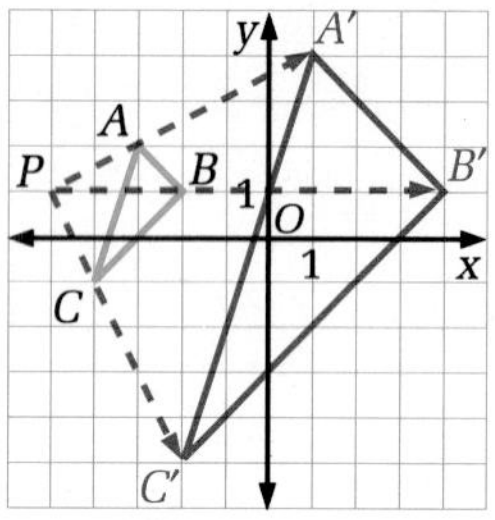

$\triangle A'B'C'$ is the dilation of $\triangle ABC$ with center P and scale factor 3.

Reflection

A **reflection** is a flip over a line. The line is called the **line of reflection.** The reflected figure is congruent to the original figure, but its orientation is reversed.

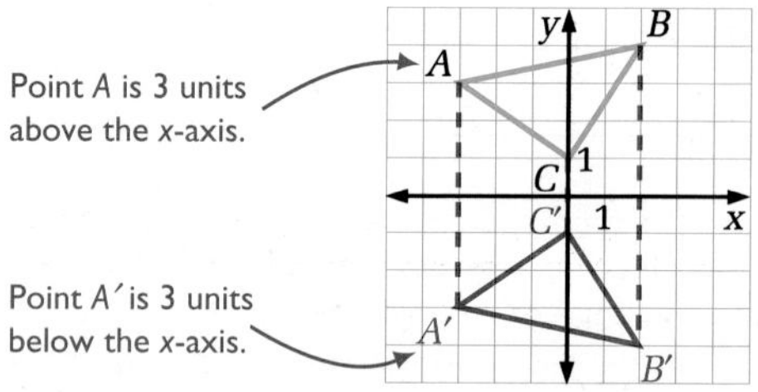

$\triangle A'B'C'$ is the reflection of $\triangle ABC$ over the x-axis.

Answers to Skill 38

1. 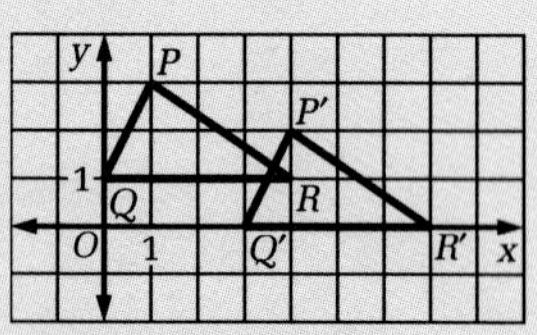

2. 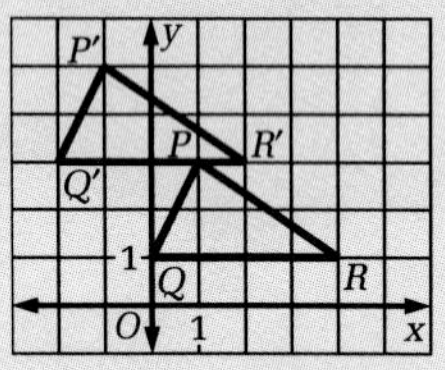

3. 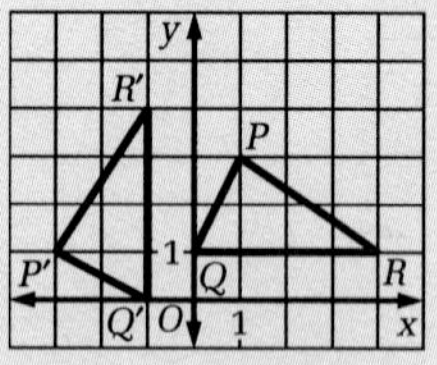

4. 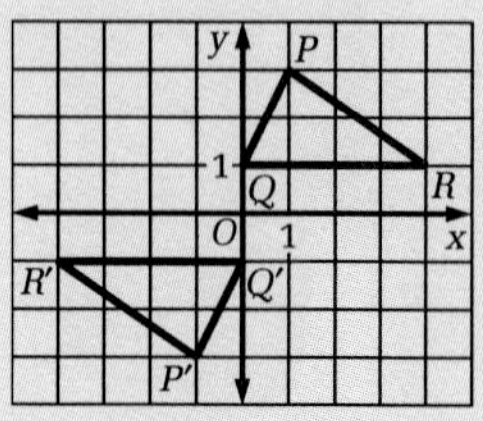

5. 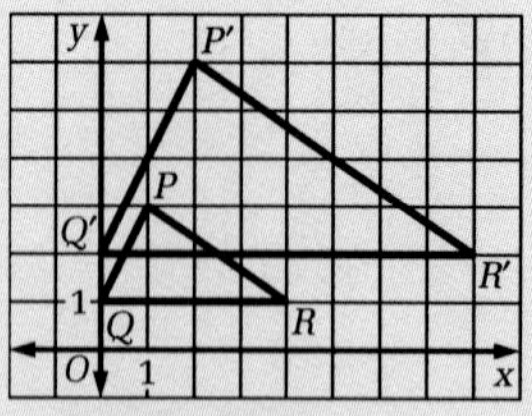

6. 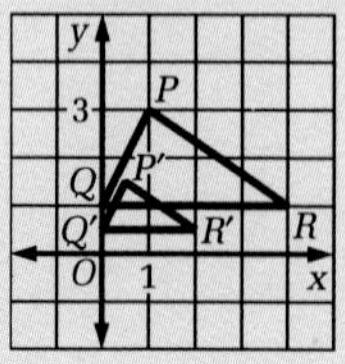

7. 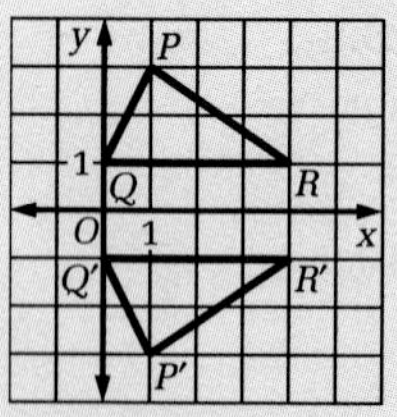

8. 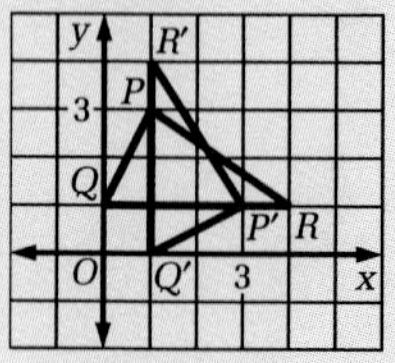

Draw the figure produced by each transformation of the figure at the right.

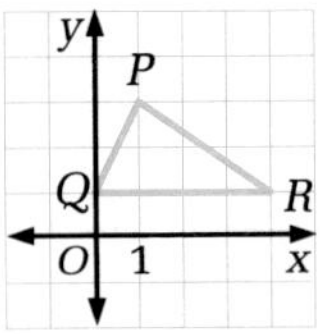

1. translation by 3 units to the right and 1 unit down
2. translation by 2 units to the left and 2 units up
3. a 90° counterclockwise rotation about the origin
4. a 180° rotation about the origin
5. a dilation about the origin with a scale factor 2
6. a dilation about the origin with a scale factor 0.5
7. a reflection over the x-axis
8. a reflection over the line $y = x$

Skill 39 Recognizing Symmetry

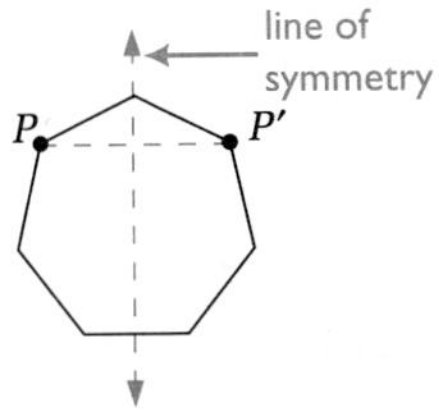

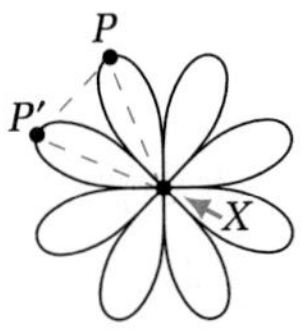

$m\angle PXP' = 45°$

A figure that looks the same after reflection over a line has **reflection symmetry.** The line over which the figure is reflected is called the **line of symmetry.**

Each point P corresponds to a point P' in such a way that the line of symmetry is the perpendicular bisector of the segment $\overline{PP'}$.

A figure that looks the same after a rotation through an angle, clockwise or counterclockwise, has **rotational symmetry.** The center of rotation is called the **point of symmetry.**

Each point P corresponds to a point P' in such a way that all the triangles PXP' are similar isosceles triangles, when X is the point of symmetry.

Example The regular hexagon at the right has three lines of symmetry: l, m, and n, as well as 60° rotational symmetry about point X.

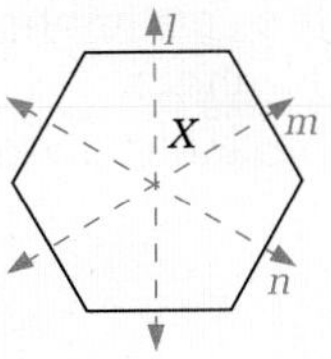

Name the type(s) of symmetry each figure has. Copy the figure and indicate lines and points of symmetry. For rotational symmetry, name the angle of the symmetry.

1. 

2. 

3. 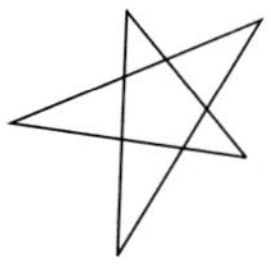

Answers to Skill 39

1. 72° rotational symmetry about point X

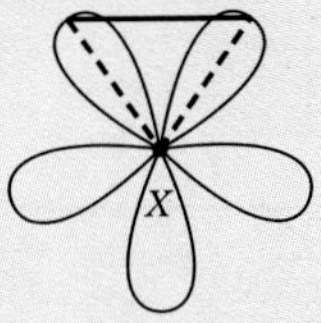

2. reflection symmetry

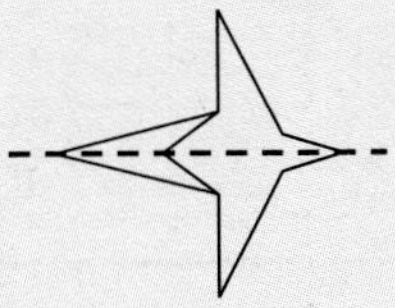

3. reflection symmetry

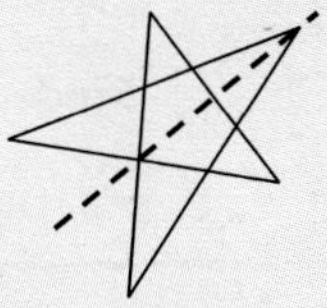

Table of Measures

Time

60 seconds (s)	=	1 minute (min)
60 minutes	=	1 hour (h)
24 hours	=	1 day
7 days	=	1 week
4 weeks (approx.)	=	1 month
365 days; 52 weeks (approx.); 12 months	=	1 year
10 years	=	1 decade
100 years	=	1 century

Metric

Length

10 millimeters (mm)	=	1 centimeter (cm)
100 cm; 1000 mm	=	1 meter (m)
1000 m	=	1 kilometer (km)

Area

100 square millimeters (mm^2)	=	1 square centimeter (cm^2)
10,000 cm^2	=	1 square meter (m^2)
10,000 m^2	=	1 hectare (ha)

Volume

1000 cubic millimeters (mm^3)	=	1 cubic centimeter (cm^3)
1,000,000 cm^3	=	1 cubic meter (m^3)

Liquid Capacity

1000 milliliters (mL)	=	1 liter (L)
1000 L	=	1 kiloliter (kL)

Mass

1000 milligrams (mg)	=	1 gram (g)
1000 g	=	1 kilogram (kg)
1000 kg	=	1 metric ton (t)

Temperature — Degrees Celsius (°C)

0°C	=	freezing point of water
37°C	=	normal body temperature
100°C	=	boiling point of water

United States Customary

Length

12 inches (in.)	=	1 foot (ft)
36 in.; 3 ft	=	1 yard (yd)
5280 ft; 1760 yd	=	1 mile (mi)

Area

144 square inches ($in.^2$)	=	1 square foot (ft^2)
9 ft^2	=	1 square yard (yd^2)
43,560 ft^2; 4840 yd^2	=	1 acre (A)

Volume

1728 cubic inches ($in.^3$)	=	1 cubic foot (ft^3)
27 ft^3	=	1 cubic yard (yd^3)

Liquid Capacity

8 fluid ounces (fl oz)	=	1 cup (c)
2 c	=	1 pint (pt)
2 pt	=	1 quart (qt)
4 qt	=	1 gallon (gal)

Weight

16 ounces (oz)	=	1 pound (lb)
2000 lb	=	1 ton (t)

Temperature — Degrees Fahrenheit (°F)

32°F	=	freezing point of water
98.6°F	=	normal body temperature
212°F	=	boiling point of water

Table of Symbols

Symbol		Page
$-a$	opposite of a	5
$\geq$	is greater than or equal to	7
$\leq$	is less than or equal to	7
$\approx$	is approximately equal to	7
a^n	nth power of a	8
$\circ$	degree(s)	15
(x, y)	ordered pair	18
$\cdot$	$\times$ (times)	25
$\sqrt{a}$	nonnegative square root of a	28
A^{-1}	inverse of matrix A	31
(x, y, z)	ordered triple	32
a^{-n}	$\frac{1}{a^n}$, $a \neq 0$	36
$f(x)$	f of x, or the value of f at x	49
π	pi, about 3.14	50
$P(E)$	probability of event E	60
$\frac{1}{a}$	reciprocal of a, $a \neq 0$	69
$\lvert a \rvert$	absolute value of a	74
m	slope	78
tan	tangent	91
$\neq$	is not equal to	95
sin	sine	96
$\pm$	plus-or-minus sign	97
$\sqrt[n]{a}$	nth root of a	113
$f \circ g$	$f(g(x))$, or f of g of x	127
AB	the length of $\overline{AB}$	140
$\triangle ABC$	triangle ABC	141
$m\angle 1$	measure of angle 1	141
$\parallel$	is parallel to	141
$\overline{AB}$	segment AB	141
$\angle$s	angles	141
$\triangle$s	triangles	141
$\cong$	congruent, is congruent to	141
$p \rightarrow q$	*p implies q*, or *if p, then q*	142
$m\,\widehat{AB}$	measure of arc AB	167
$\sim$	is similar to	177
$\perp$	is perpendicular to	200
i	$\sqrt{-1}$	221
Σ	summation sign	238
∞	infinity	255
$a^{1/n}$	nth root of a, $n \neq 0$	282
e	irrational number about 2.718	288
$f^{-1}(x)$	inverse function of $f(x)$	298
$\log_b x$	base-b logarithm of x	302
$\log_{10} x$	base-10 logarithm of x	302
$\ln x$	base-e logarithm of x	302
$\overline{x}$	mean of data set $x_1, x_2, \cdots, x_n$	339
σ	standard deviation	339
$P(B \mid A)$	probability of B given A	397
!	factorial	411
${}_nC_r$	combination	411
(r, θ)	polar coordinates	425
cos	cosine	431
$\cos^{-1}$	inverse cosine	433
$\sin^{-1}$	inverse sine	433
v	vector	441
$A \rightarrow A'$	point A is transformed to point A'	490
${}_nP_r$	permutation	628

665

Table of Squares and Square Roots

No.	Square	Sq. Root	No.	Square	Sq. Root	No.	Square	Sq. Root
1	1	1.000	**51**	2,601	7.141	**101**	10,201	10.050
2	4	1.414	**52**	2,704	7.211	**102**	10,404	10.100
3	9	1.732	**53**	2,809	7.280	**103**	10,609	10.149
4	16	2.000	**54**	2,916	7.348	**104**	10,816	10.198
5	25	2.236	**55**	3,025	7.416	**105**	11,025	10.247
6	36	2.449	**56**	3,136	7.483	**106**	11,236	10.296
7	49	2.646	**57**	3,249	7.550	**107**	11,449	10.344
8	64	2.828	**58**	3,364	7.616	**108**	11,664	10.392
9	81	3.000	**59**	3,481	7.681	**109**	11,881	10.440
10	100	3.162	**60**	3,600	7.746	**110**	12,100	10.488
11	121	3.317	**61**	3,721	7.810	**111**	12,321	10.536
12	144	3.464	**62**	3,844	7.874	**112**	12,544	10.583
13	169	3.606	**63**	3,969	7.937	**113**	12,769	10.630
14	196	3.742	**64**	4,096	8.000	**114**	12,996	10.677
15	225	3.873	**65**	4,225	8.062	**115**	13,225	10.724
16	256	4.000	**66**	4,356	8.124	**116**	13,456	10.770
17	289	4.123	**67**	4,489	8.185	**117**	13,689	10.817
18	324	4.243	**68**	4,624	8.246	**118**	13,924	10.863
19	361	4.359	**69**	4,761	8.307	**119**	14,161	10.909
20	400	4.472	**70**	4,900	8.367	**120**	14,400	10.954
21	441	4.583	**71**	5,041	8.426	**121**	14,641	11.000
22	484	4.690	**72**	5,184	8.485	**122**	14,884	11.045
23	529	4.796	**73**	5,329	8.544	**123**	15,129	11.091
24	576	4.899	**74**	5,476	8.602	**124**	15,376	11.136
25	625	5.000	**75**	5,625	8.660	**125**	15,625	11.180
26	676	5.099	**76**	5,776	8.718	**126**	15,876	11.225
27	729	5.196	**77**	5,929	8.775	**127**	16,129	11.269
28	784	5.292	**78**	6,084	8.832	**128**	16,384	11.314
29	841	5.385	**79**	6,241	8.888	**129**	16,641	11.358
30	900	5.477	**80**	6,400	8.944	**130**	16,900	11.402
31	961	5.568	**81**	6,561	9.000	**131**	17,161	11.446
32	1,024	5.657	**82**	6,724	9.055	**132**	17,424	11.489
33	1,089	5.745	**83**	6,889	9.110	**133**	17,689	11.533
34	1,156	5.831	**84**	7,056	9.165	**134**	17,956	11.576
35	1,225	5.916	**85**	7,225	9.220	**135**	18,225	11.619
36	1,296	6.000	**86**	7,396	9.274	**136**	18,496	11.662
37	1,369	6.083	**87**	7,569	9.327	**137**	18,769	11.705
38	1,444	6.164	**88**	7,744	9.381	**138**	19,044	11.747
39	1,521	6.245	**89**	7,921	9.434	**139**	19,321	11.790
40	1,600	6.325	**90**	8,100	9.487	**140**	19,600	11.832
41	1,681	6.403	**91**	8,281	9.539	**141**	19,881	11.874
42	1,764	6.481	**92**	8,464	9.592	**142**	20,164	11.916
43	1,849	6.557	**93**	8,649	9.644	**143**	20,449	11.958
44	1,936	6.633	**94**	8,836	9.695	**144**	20,736	12.000
45	2,025	6.708	**95**	9,025	9.747	**145**	21,025	12.042
46	2,116	6.782	**96**	9,216	9.798	**146**	21,316	12.083
47	2,209	6.856	**97**	9,409	9.849	**147**	21,609	12.124
48	2,304	6.928	**98**	9,604	9.899	**148**	21,904	12.166
49	2,401	7.000	**99**	9,801	9.950	**149**	22,201	12.207
50	2,500	7.071	**100**	10,000	10.000	**150**	22,500	12.247

666

Table of Trigonometric Ratios

Angle	Sine	Cosine	Tangent
1°	.0175	.9998	.0175
2°	.0349	.9994	.0349
3°	.0523	.9986	.0524
4°	.0698	.9976	.0699
5°	.0872	.9962	.0875
6°	.1045	.9945	.1051
7°	.1219	.9925	.1228
8°	.1392	.9903	.1405
9°	.1564	.9877	.1584
10°	.1736	.9848	.1763
11°	.1908	.9816	.1944
12°	.2079	.9781	.2126
13°	.2250	.9744	.2309
14°	.2419	.9703	.2493
15°	.2588	.9659	.2679
16°	.2756	.9613	.2867
17°	.2924	.9563	.3057
18°	.3090	.9511	.3249
19°	.3256	.9455	.3443
20°	.3420	.9397	.3640
21°	.3584	.9336	.3839
22°	.3746	.9272	.4040
23°	.3907	.9205	.4245
24°	.4067	.9135	.4452
25°	.4226	.9063	.4663
26°	.4384	.8988	.4877
27°	.4540	.8910	.5095
28°	.4695	.8829	.5317
29°	.4848	.8746	.5543
30°	.5000	.8660	.5774
31°	.5150	.8572	.6009
32°	.5299	.8480	.6249
33°	.5446	.8387	.6494
34°	.5592	.8290	.6745
35°	.5736	.8192	.7002
36°	.5878	.8090	.7265
37°	.6018	.7986	.7536
38°	.6157	.7880	.7813
39°	.6293	.7771	.8098
40°	.6428	.7660	.8391
41°	.6561	.7547	.8693
42°	.6691	.7431	.9004
43°	.6820	.7314	.9325
44°	.6947	.7193	.9657
45°	.7071	.7071	1.0000

Angle	Sine	Cosine	Tangent
46°	.7193	.6947	1.0355
47°	.7314	.6820	1.0724
48°	.7431	.6691	1.1106
49°	.7547	.6561	1.1504
50°	.7660	.6428	1.1918
51°	.7771	.6293	1.2349
52°	.7880	.6157	1.2799
53°	.7986	.6018	1.3270
54°	.8090	.5878	1.3764
55°	.8192	.5736	1.4281
56°	.8290	.5592	1.4826
57°	.8387	.5446	1.5399
58°	.8480	.5299	1.6003
59°	.8572	.5150	1.6643
60°	.8660	.5000	1.7321
61°	.8746	.4848	1.8040
62°	.8829	.4695	1.8807
63°	.8910	.4540	1.9626
64°	.8988	.4384	2.0503
65°	.9063	.4226	2.1445
66°	.9135	.4067	2.2460
67°	.9205	.3907	2.3559
68°	.9272	.3746	2.4751
69°	.9336	.3584	2.6051
70°	.9397	.3420	2.7475
71°	.9455	.3256	2.9042
72°	.9511	.3090	3.0777
73°	.9563	.2924	3.2709
74°	.9613	.2756	3.4874
75°	.9659	.2588	3.7321
76°	.9703	.2419	4.0108
77°	.9744	.2250	4.3315
78°	.9781	.2079	4.7046
79°	.9816	.1908	5.1446
80°	.9848	.1736	5.6713
81°	.9877	.1564	6.3138
82°	.9903	.1392	7.1154
83°	.9925	.1219	8.1443
84°	.9945	.1045	9.5144
85°	.9962	.0872	11.4301
86°	.9976	.0698	14.3007
87°	.9986	.0523	19.0811
88°	.9994	.0349	28.6363
89°	.9998	.0175	57.2900

Postulates and Theorems

Postulates of Algebra

Postulate 1 **Addition Property of Equality** If the same number is added to equal numbers, then the sums are equal. *(Book 2)*

Postulate 2 **Subtraction Property of Equality** If the same number is subtracted from equal numbers, then the differences are equal. *(Book 2)*

Postulate 3 **Multiplication Property of Equality** If equal numbers are multiplied by the same number, then the products are equal. *(Book 2)*

Postulate 4 **Division Property of Equality** If equal numbers are divided by the same nonzero number, then the quotients are equal. *(Book 2)*

Postulate 5 **Reflexive Property of Equality** A number is equal to itself. *(Book 2)*

Postulate 6 **Substitution Property** If values are equal, then one value may be substituted for the other. *(Book 2)*

Postulate 7 **Distributive Property** An expression of the form $a(b + c)$ is equivalent to $ab + ac$. *(Book 2)*

Postulates and Theorems of Geometry

Angles

Theorem If two angles are supplements of the same angle, then they are equal in measure. *(Book 2)*

Theorem If two angles are complements of the same angle, then they are equal in measure. *(Book 2)*

Postulate If the sides of an angle form a straight line, then the angle is a straight angle with measure 180°. *(Book 2)*

Postulate For any segment or angle, the measure of the whole is equal to the sum of the measures of its non-overlapping parts. *(Book 2)*

Theorem Vertical angles are equal in measure. *(Book 2)*

Theorem The sum of the measures of the angles of a triangle is 180°. *(Book 2)*

Theorem An exterior angle of a triangle is equal in measure to the sum of the measures of its two remote interior angles. *(Book 2)*

Theorem If two sides of a triangle are equal in measure, then the angles opposite those sides are equal in measure. *(Book 2)*

Theorem If two angles of a triangle are equal in measure, then the sides opposite those angles are equal in measure. *(Book 2)*

Theorem If a triangle is equilateral, then it is also equiangular, with three 60° angles. *(Book 2)*

Theorem If a triangle is equiangular, then it is also equilateral. *(Book 2)*

Theorem 3.4 The sum of the angle measures of an n-gon is given by the formula $S(n) = (n - 2)180°$. *(p. 161)*

Theorem 3.5 The sum of the exterior angle measures of an n-gon, one angle at each vertex, is 360°. *(p. 161)*

Lines

Postulate If two parallel lines are intersected by a transversal, then corresponding angles are equal in measure. *(Book 2)*

Theorem If two parallel lines are intersected by a transversal, then alternate interior angles are equal in measure. *(Book 2)*

Theorem If two parallel lines are intersected by a transversal, then co-interior angles are supplementary. *(Book 2)*

Postulate If two lines are intersected by a transversal and corresponding angles are equal in measure, then the lines are parallel. *(Book 2)*

Theorem If two lines are intersected by a transversal and alternate interior angles are equal in measure, then the lines are parallel. *(Book 2)*

Theorem If two lines are intersected by a transversal and co-interior angles are supplementary, then the lines are parallel. *(Book 2)*

Theorem If two lines are perpendicular to the same transversal, then they are parallel. *(Book 2)*

Theorem If a transversal is perpendicular to one of two parallel lines, then it is perpendicular to the other one also. *(Book 2)*

Postulate Through a point not on a given line, there is one and only one line parallel to the given line. *(Book 2)*

Theorem If a point is the same distance from both endpoints of a segment, then it lies on the perpendicular bisector of the segment. *(Book 2)*

Postulate A segment can be drawn perpendicular to a given line from a point not on the line. The length of this segment is the shortest distance from the point to the line. *(p. 182)*

Postulates and Theorems of Geometry (cont.)

Triangles

Postulate If two angles of one triangle are equal in measure to two angles of another triangle, then the two triangles are similar. (AA Similarity) *(Book 2)*

Theorem If a line is drawn from a point on one side of a triangle parallel to another side, then it forms a triangle similar to the original triangle. *(Book 2)*

In a triangle, a segment that connects the midpoints of two sides is parallel to the third side and half as long. *(Book 2)*

Theorem If two angles and the included side of one triangle are equal in measure to the corresponding angles and side of another triangle, then the triangles are congruent. (ASA) *(Book 2)*

Theorem If two angles and a non-included side of one triangle are equal in measure to the corresponding angles and side of another triangle, then the triangles are congruent. (AAS) *(Book 2)*

Postulate If two sides and the included angle of one triangle are equal in measure to the corresponding sides and angle of another triangle, then the triangles are congruent. (SAS) *(Book 2)*

Postulate If three sides of one triangle are equal in measure to the corresponding sides of another triangle, then the triangles are congruent. (SSS) *(Book 2)*

Theorem If the altitude is drawn to the hypotenuse of a right triangle, then the two triangles formed are similar to the original triangle and to each other. *(Book 2)*

Theorem In any right triangle, the square of the length of the hypotenuse is equal to the sum of the squares of the lengths of the legs. *(Book 2)*

Theorem If the altitude is drawn to the hypotenuse of a right triangle, then the measure of the altitude is the geometric mean between the measures of the parts of the hypotenuse. *(Book 2)*

The sum of the lengths of any two sides of a triangle is greater than the length of the third side. *(p. 9)*

Theorem 3.2 In an isosceles triangle, the medians drawn to the legs are equal in measure. *(p. 150)*

Quadrilaterals

In a parallelogram, the diagonals have the same midpoint. *(Book 2)*

In a kite, the diagonals are perpendicular to each other. *(Book 2)*

In a rectangle, the diagonals are equal in measure. *(Book 2)*

In a parallelogram, opposite sides are equal in measure. *(Book 2)*

670

Theorem If a quadrilateral is a parallelogram, then consecutive angles are supplementary. *(Book 2)*

Theorem If a quadrilateral is a parallelogram, then opposite angles are equal in measure. *(Book 2)*

Theorem The sum of the measures of the angles of a quadrilateral is 360°. *(Book 2)*

Theorem If both pairs of opposite angles of a quadrilateral are equal in measure, then the quadrilateral is a parallelogram. *(Book 2)*

Theorem 3.1 If the two diagonals of a quadrilateral bisect each other, then the quadrilateral is a parallelogram. *(p. 141)*

Theorem 3.3 In an isosceles trapezoid, (1) the legs are equal in measure, (2) the diagonals are equal in measure, and (3) the two angles at each base are equal in measure. *(p. 150)*

Circles

Theorem 3.6 The perpendicular bisector of a chord of a circle passes through the center of the circle. *(p. 166)*

Theorem 3.7 The measure of an inscribed angle of a circle is equal to half the measure of its intercepted arc. *(p. 174)*

Theorem 3.8 An inscribed angle whose intercepted arc is a semicircle is a right angle. *(p. 174)*

Theorem 3.9 If two inscribed angles in the same circle intercept the same arc, then they are equal in measure. *(p. 174)*

Theorem 3.10 If a line is tangent to a circle, then the line is perpendicular to the radius drawn from the center to the point of tangency. *(p. 183)*

Theorem 3.11 If a line in the plane of a circle is perpendicular to a radius at its outer endpoint, then the line is tangent to the circle. *(p. 183)*

Theorem 3.12 If two tangent segments are drawn from the same point to the same circle, then they are equal in measure. *(p. 183)*

Polyhedra

Postulate In any convex polyhedron, the sum of the measures of the angles at each vertex is less than 360°. *(p. 194)*

Theorem 3.13 There are exactly five regular polyhedra: tetrahedron, octahedron, icosahedron, hexahedron, and dodecahedron. *(p. 195)*

Visual Glossary of Functions

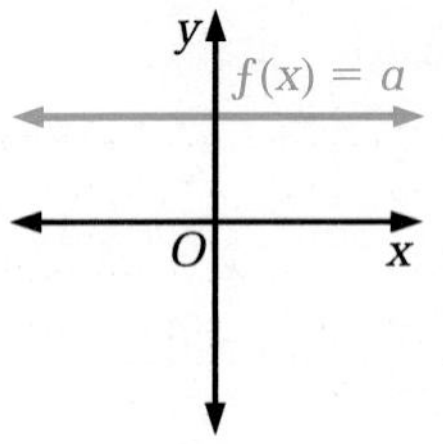

Constant

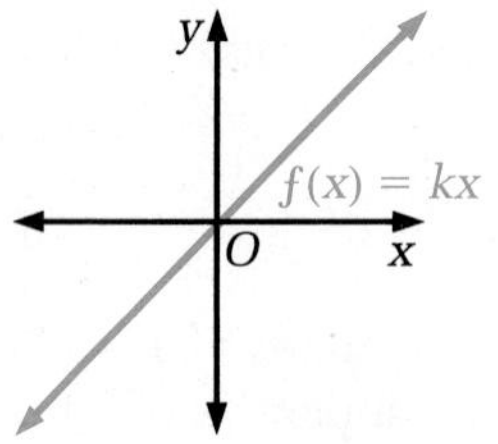

Direct Variation

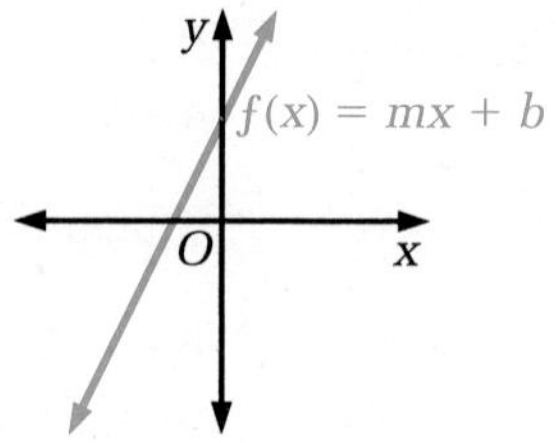

Linear

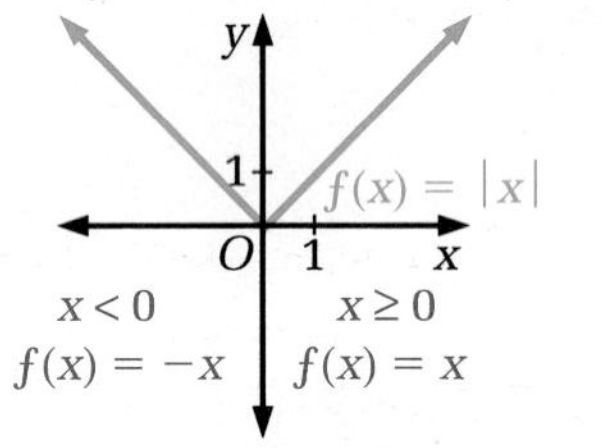

Absolute Value (Piecewise)

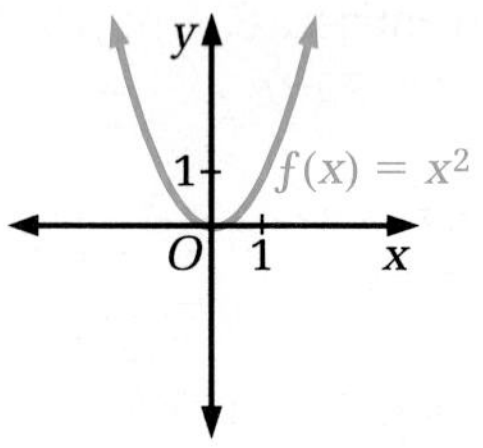

Quadratic

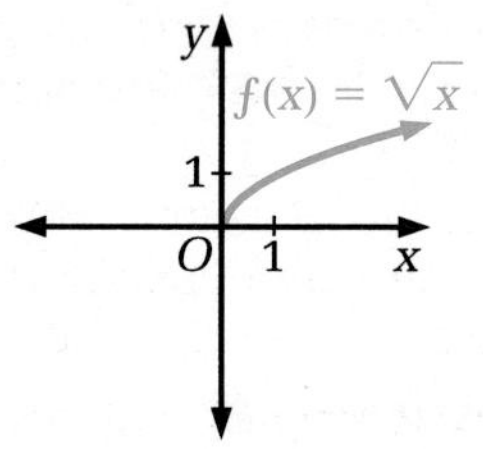

Square Root (Radical)

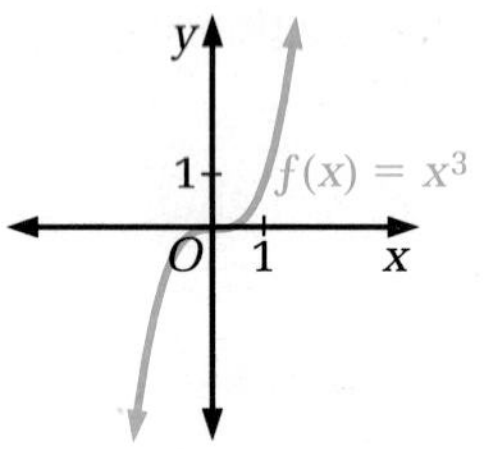

Cubic

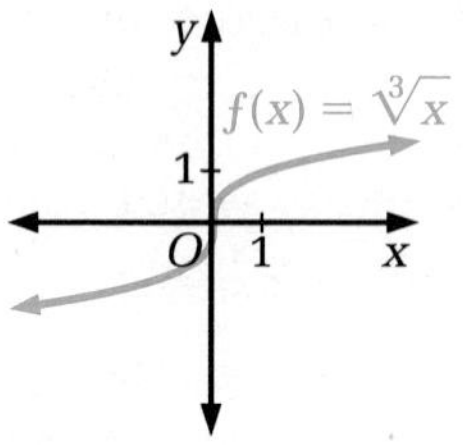

Cube Root (Radical)

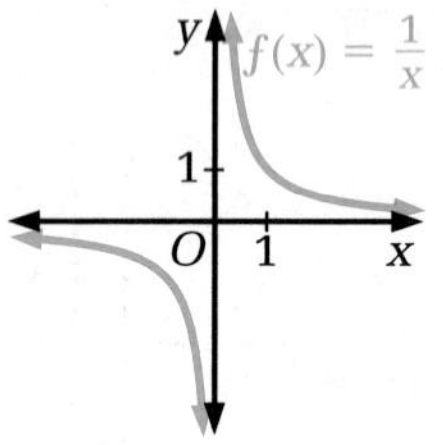

Reciprocal (Rational)

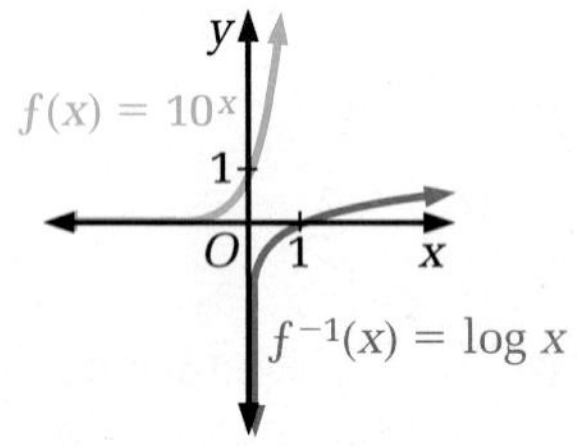

Exponential Logarithmic
(Inverse Functions)

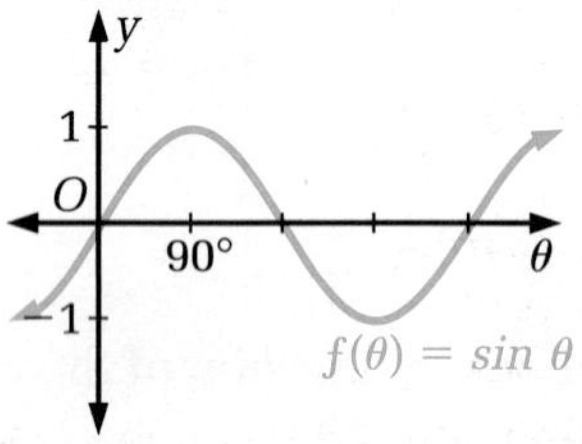

Sine (Periodic)

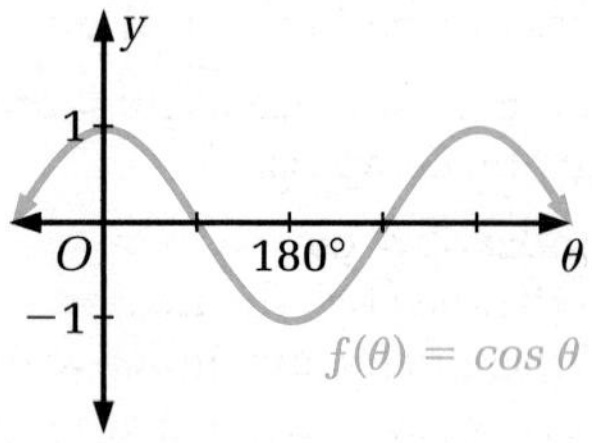

Cosine (Periodic)

Glossary

absolute value (p. 87) The distance that a number is from zero on a number line.

$$|x| = \begin{cases} x & \text{if } x \geq 0 \\ -x & \text{if } x < 0 \end{cases}$$

absolute value function (p. 88) A function that has a variable within the absolute value symbol.

algorithm (p. 4) A set of step-by-step directions for a process.

ambiguous case (p. 470) A situation where two triangles can be formed from the same side-side-angle information.

amplitude (p. 548) One half the difference between the maximum and minimum values of a periodic function.

arithmetic sequence (p. 229) A sequence in which the difference between any term and the term before it is a constant.

arithmetic series (p. 238) A series whose terms form an arithmetic sequence.

asymptote (p. 123) A line that a graph approaches more and more closely.

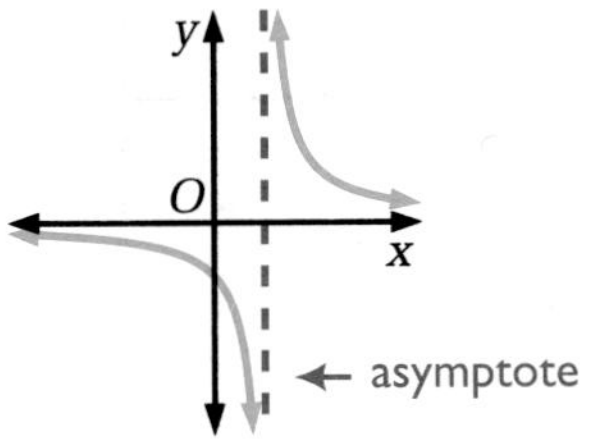

axis of a periodic function (p. 548) The horizontal line halfway between the minimum and maximum values of a periodic function.

bearing (p. 425) An angular direction given by a magnetic compass.

bimodal distribution (p. 332) A frequency distribution that has two distinct intervals or groups of intervals that contain the greatest frequencies.

binomial experiment (p. 411) An experiment with a fixed number of independent trials. For each trial there are two mutually exclusive outcomes, success and failure. Each trial has the same *P*(success), and *P*(success) + *P*(failure) = 1.

box-and-whisker plot (p. 17) A method for displaying the median, quartiles, and extremes of a data set.

cell of a periodic tessellation (p. 584) A fundamental unit of a periodic tessellation that can be copied and translated to generate the pattern.

central angle (p. 167) An angle with its vertex at the center of a circle.

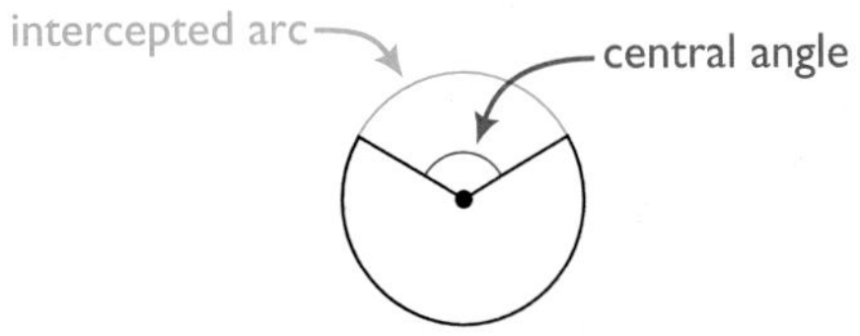

chord (p. 166) A segment joining two points on a circle.

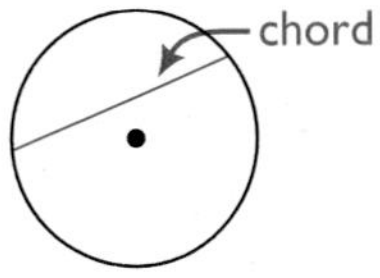

circumscribed circle (p. 166) A circle is *circumscribed* about a polygon when each vertex of the polygon lies on the circle. The polygon is *inscribed* in the circle.

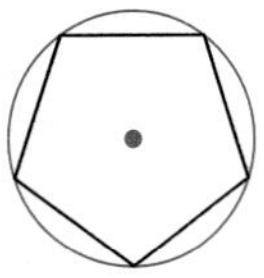

circumscribed polygon (p. 187) A polygon is *circumscribed* about a circle when each side of the polygon is tangent to the circle. The circle is *inscribed* in the polygon.

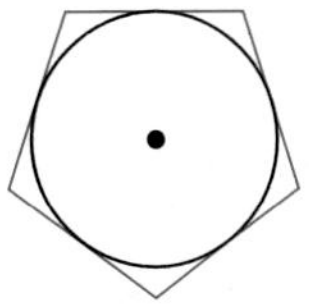

common difference (p. 229) The value $a_n - a_{n-1}$ in an arithmetic sequence.

common logarithm (p. 302) A logarithm with base 10. It is usually written log *x*.

common ratio (p. 229) The value $\frac{a_n}{a_{n-1}}$ in a geometric sequence.

complementary angles (p. 668) Two angles whose measures have the sum 90°.

complementary events (p. 386) Two mutually exclusive events that account for all possibilities.

composite of two functions (p. 126) The function whose value at x is $f(g(x))$ is the composite of the functions f and g. Also written as $(f \circ g)(x)$.

conclusion of an implication (p. 140) The *then* part of an *if-then* statement.

conclusion of a logical argument (p. 658) A statement resulting from the premises of a logical argument.

conditional probability (p. 396) The probability that an event will occur given that another event has occurred.

congruent (p. 656) Having the same size and shape.

conjecture (p. 139) A statement, opinion, or conclusion based on observation.

constraint (p. 54) Any condition that must be met by a variable or by a linear combination of variables.

contrapositive (p. 142) A statement obtained by interchanging and adding "not" to the *if* and *then* parts of an *if-then* statement.

converse (p. 142) A statement obtained by interchanging the *if* and *then* parts of an *if-then* statement.

coordinate proof (p. 147) A method of proof based on a coordinate system in which all points are represented by ordered pairs of numbers.

correlation coefficient (p. 352) A number between 1 and −1, inclusive, that describes how well a set of data points can be modeled by a line.

cosine function (p. 556) A function that measures the horizontal position of an object moving around a unit circle.

cycle (p. 548) The smallest portion of the graph that repeats.

damping function (p. 543) A function whose graph moves towards a constant value by alternately increasing and decreasing around that value.

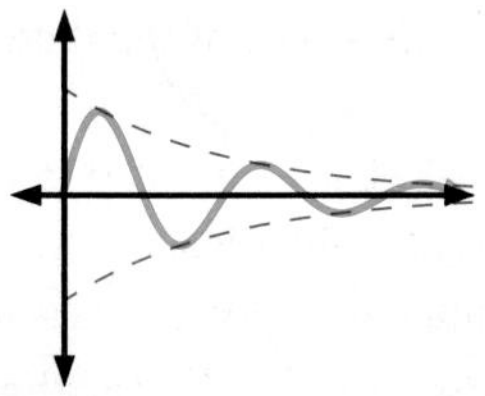

defect (p. 194) The angle measure of the gap in a net for a polyhedron.

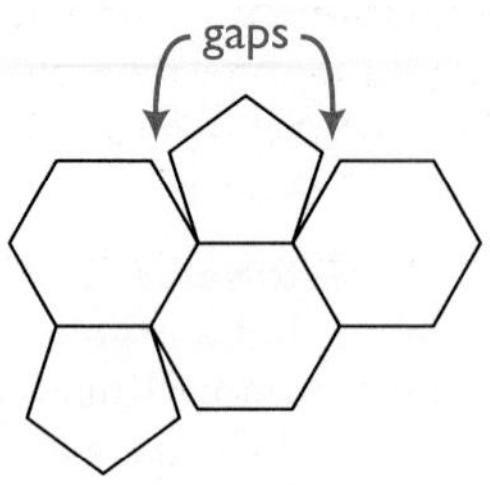

degree of a polynomial function (p. 103) The largest exponent, n, of a polynomial function.

direct variation function (p. 102) A function of the form $f(x) = kx^n$ where n is a positive integer and $k \neq 0$.

domain of a function (p. 70) All values of the control variable of a function.

edge of a network (p. 37) A line connecting two vertices of a network.

exclusive definition (p. 154) A definition that excludes some possibilities.

expected value (p. 404) If a situation has a gain or loss x_1 with probability $P(x_1)$, a gain or loss x_2 with probability $P(x_2)$, and so on, then the expected value is

$$x_1 \cdot P(x_1) + x_2 \cdot P(x_2) + \cdots + x_n \cdot P(x_n).$$

explicit formula (p. 217) A formula for a sequence that expresses the value of any term a_n using n, the position number of the term in the sequence $a_1, a_2, a_3, \ldots$.

exponential decay function (p. 270) An exponential function $y = ab^x$, where $0 < b < 1$.

exponential function (p. 267) A function of the form $y = ab^x$.

exponential growth function (p. 270) An exponential function $y = ab^x$, where $b > 1$.

extraneous solution (p. 112) A solution of a simplified equation that is not a solution of the original equation.

fair (p. 405) A game or situation is considered fair if it has an expected value of 0.

feasible region (p. 54) The graph of the solution of a system of inequalities that meets all given constraints.

flow proof (p. 144) A proof written as a diagram using arrows to show the connections between statements. Numbers written over the arrows refer to a numbered list of the justifications for the statements.

frequency distribution (p. 332) A display of data that shows the number of occurrences of each data item.

frequency of a periodic function (p. 548) The reciprocal of the period.

function (p. 70) A relationship where there is only one value of the dependent variable for each value of the control variable. All the values of the control variable are the *domain.* All the values of the dependent variable over the domain are the *range.*

function notation (p. 71) The notation $y = f(x)$, which means that y represents the value of the function f for a value of x.

geometric mean (p. 232) The *geometric mean* of any two positive numbers a and b is $\sqrt{ab}$, since the sequence $a, \sqrt{ab}, b$ is geometric.

geometric sequence (p. 229) A sequence in which the ratio of any term to the term before it is a constant.

geometric series (p. 245) A series whose terms form a geometric sequence.

histogram (p. 331) A type of bar graph that displays frequencies.

hyperbola (p. 123) The graph of a rational function where the degree of the polynomial in the numerator is zero or 1 and the degree of the polynomial in the denominator is 1.

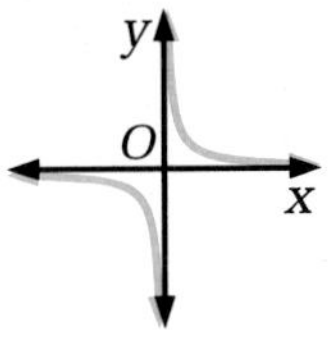

hypothesis (p. 140) The *if* part of an *if-then* statement.

implication (p. 142) A statement with an *if* part and a *then* part. The *if* part is the *hypothesis* and the *then* part is the *conclusion.* Also called a *conditional.*

inclusive definition (p. 154) A definition that includes all possibilities.

independent events (p. 398) Two events are independent if and only if the occurrence of one does not affect the probability of the occurrence of the other.

indirect proof (p. 180) A method of proof in which you temporarily assume the opposite of what you want to prove and then show that this leads to an impossible situation or a contradiction of a known fact.

inscribed angle (p. 172) An angle formed by two chords intersecting at a point on a circle.

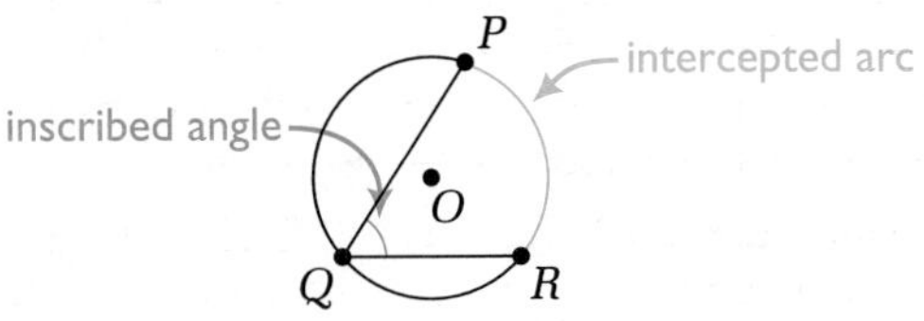

inscribed circle (p. 187) *See* circumscribed polygon.

inscribed polygon (p. 166) *See* circumscribed circle.

intercepted arc (pp. 167, 172) The arc that lies within an inscribed angle or central angle. *See also* inscribed angle and central angle.

inverse (p. 142) A statement obtained by adding "not" to both the *if* and *then* parts of an *if-then* statement.

inverse functions (p. 297) Two functions f and g are inverse functions if $g(b) = a$ whenever $f(a) = b$. The inverse of a function is written as $f^{-1}(x)$ or f^{-1}.

inverse matrices (p. 31) Two 2×2 matrices whose product is the matrix $\begin{bmatrix} 1 & 0 \\ 0 & 1 \end{bmatrix}$. The symbol A^{-1} is used to represent the inverse of matrix A.

isosceles trapezoid (p. 150) A trapezoid with a line of symmetry that passes through the midpoints of the bases.

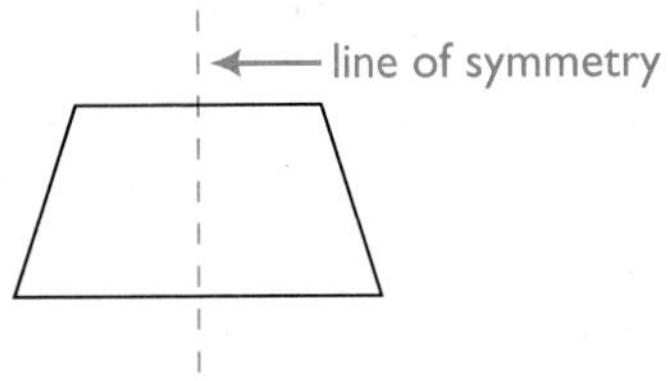

kite (p. 155) A quadrilateral with two pairs of consecutive sides equal in measure. These pairs do not have a side in common.

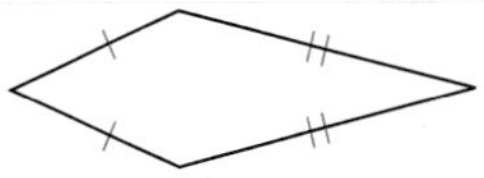

law of cosines (p. 461) In any $\triangle ABC$:

$$a^2 = b^2 + c^2 - 2bc \cos A$$

$$b^2 = a^2 + c^2 - 2ac \cos B$$

$$c^2 = a^2 + b^2 - 2ab \cos C$$

law of sines (p. 467) In any $\triangle ABC$:

$$\frac{a}{\sin A} = \frac{b}{\sin B} = \frac{c}{\sin C}$$

least-squares line (p. 351) A linear model used to fit a line to a data set. The model minimizes the sum of the squares of the distances of the data points from the line.

least-squares parabola (p. 360) A quadratic model used to fit a curve to a data set.

limit of a sequence (p. 210) When the terms of a sequence get closer and closer to a single fixed number L, then L is called the *limit* of the sequence.

675

linear combination (p. 53) An expression in the form $ax + by$, where a and b are constants.

linear equation with three variables (p. 32) A linear equation of the form $ax + by + cz = d$.

linear function (p. 78) A function that has an equation of the form $f(x) = mx + b$, where m is the slope of its linear graph and b is the vertical intercept.

linear programming (p. 56) A process that uses a system of linear inequalities called *constraints* to solve certain decision-making problems where a linear combination is to be maximized or minimized.

logarithm (p. 302) The exponent a when a positive number x is written as a power of a positive base b, $(b \neq 1)$: $\log_b x = a$ when $x = b^a$.

logarithmic function (p. 302) The inverse of the exponential function $f(x) = 10^x$ and written as $f^{-1}(x) = \log_{10}x$.

logistic growth function (p. 290) A function whose rate of growth slows down after increasing or decreasing exponentially.

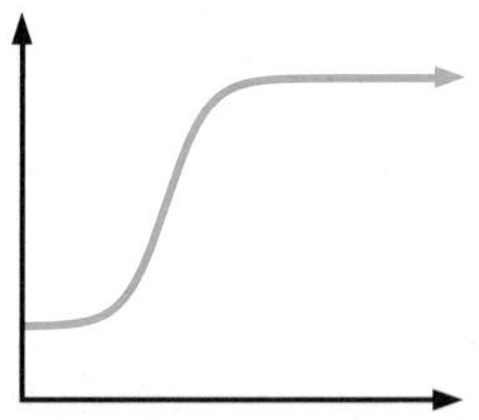

loop (p. 4) A group of steps that are repeated for a certain number of times or until some condition is met.

lower extreme (p. 17) In a data set, the smallest data item.

lower quartile (p. 17) In a data set, the median of the lower half of the data set.

magnitude (p. 441) The length of a vector.

major arc (p. 167) An arc of a circle with a measure greater than 180° and less than 360°.

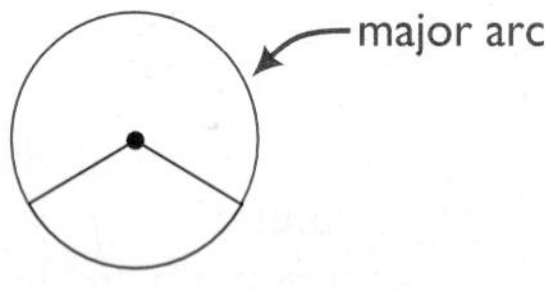

many-to-one function (p. 72) A function in which a member of the range may be paired with more than one member of the domain.

matrix (p. 632) An arrangement of numbers, called *elements*, in rows and columns. The number of rows and columns are the *dimensions* of the matrix.

matrix equation (p. 31) For a system of linear equations in two variables, a matrix equation has the form $A\begin{bmatrix} x \\ y \end{bmatrix} = B$.

maximum (p. 45) The greatest value.

mean (p. 16) The sum of the data in a data set divided by the number of items.

median (p. 17) In a data set, the middle number or the average of the two middle numbers when the data are arranged in numerical order.

median-median line (p. 352) A linear model used to fit a line to a data set. The line is fit only to summary points, key points calculated using medians.

median of a triangle (p. 147) A segment joining a vertex of a triangle to the midpoint of the opposite side.

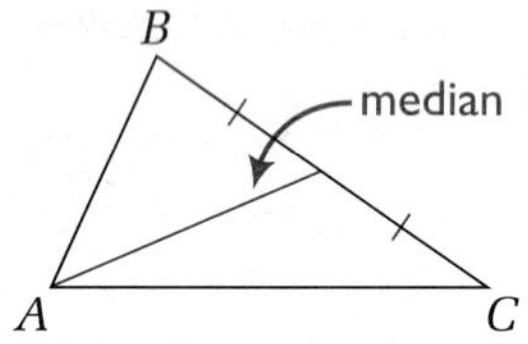

minimum (p. 45) The least value.

minor arc (p. 167) An arc of a circle with a measure less than 180°.

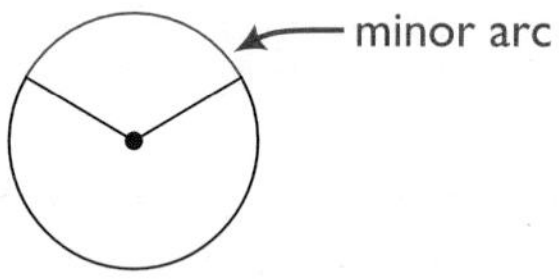

mound-shaped distribution (p. 332) A frequency distribution approximately symmetrical about a line passing through the interval or between the intervals with the greatest frequencies.

mutually exclusive events (p. 386) Events that cannot occur at the same time.

natural logarithm (p. 302) A logarithm with base e. It is usually written $\ln x$.

net (p. 193) A set of polygons that can be used to make a polyhedron. *See also* defect.

normal distribution (p. 345) A distribution of data with a histogram that follows a bell-shaped curve, or *normal curve*, that reaches its maximum height at the mean.

one-to-one function (p. 72) A function in which each member of the range is paired with exactly one member of the domain.

opposite vectors (p. 443) Two vectors that have the same magnitude but opposite directions.

outlier (p. 332) In a data set, a value whose distance from the center of the data is much greater than the distances for other data values.

parabola (p. 95) The graph of $y = ax^2 + bx + c$, $a \neq 0$. The point where the curve turns is either the highest point or the lowest point and is called the *vertex.*

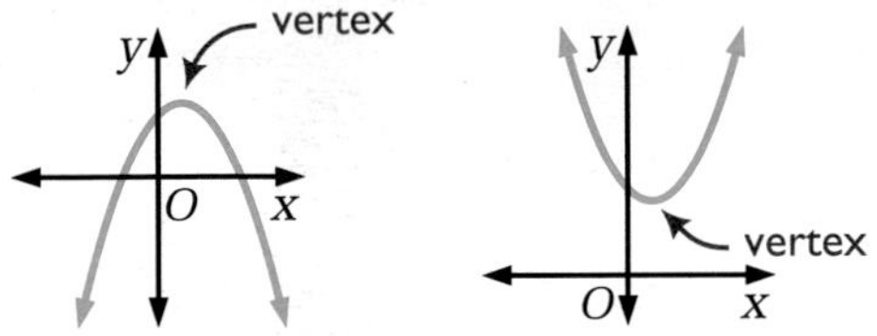

paragraph proof (p. 166) A proof whose statements and justifications are written in paragraph form.

parallelogram (p. 155) A quadrilateral with two pairs of parallel sides.

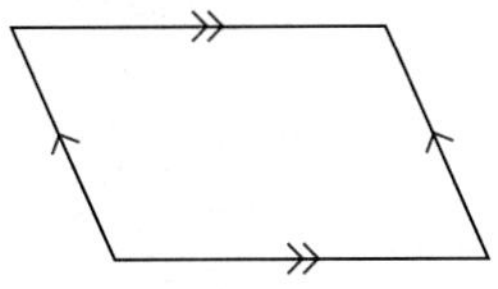

parametric equations (p. 453) Equations that express two variables in terms of a third variable, called the *parameter.*

partial sum (p. 252) The sum of the first n terms of an infinite series.

percentile (p. 523) A whole number between 1 and 100 that tells what percent of the data in a set are less than a given value.

period (p. 548) The length of one cycle.

periodic function (p. 543) A function whose graph repeats a pattern.

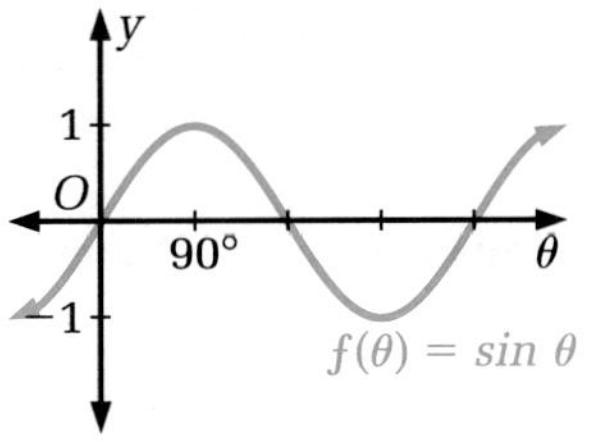

periodic tessellation (p. 584) A tessellation that matches itself exactly if it is translated.

perpendicular bisector (p. 165) A line, ray, or segment that bisects a segment and forms right angles with it.

piecewise function (p. 86) A function defined by two or more equations. Each equation applies to a different part of the function's domain.

polar coordinates (p. 425) A set of coordinates that use a distance from a reference point and an angle to describe a location.

polyhedron (p. 193) A space figure whose faces are polygons.

polynomial function (p. 103) A function of the form $P(x) = a_nx^n + a_{n-1}x^{n-1} + a_{n-2}x^{n-2} + \cdots + a_1x + a_0$ where the coefficients are real numbers and n is a nonnegative integer.

premise (p. 658) A given statement in an argument.

probability distribution (p. 376) A histogram in which the relative frequencies on the vertical axis represent probabilities.

Pythagorean identity (p. 435) The relationship $\cos^2 \theta + \sin^2 \theta = 1$.

quadratic function (p. 95) A function of the form $f(x) = ax^2 + bx + c$ where a, b, and c are constants and $a \neq 0$.

quadrilateral (p. 155) A polygon with four sides. *See also* kite, parallelogram, rectangle, rhombus, square, and trapezoid.

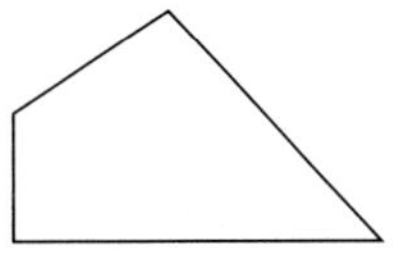

radical function (p. 113) A function that has a variable under a radical symbol.

random numbers (p. 374) A sequence of numbers that has no pattern, and each number has an equally likely chance of appearing in the sequence.

range of a data set (p. 338) The difference between the greatest and least data values.

range of a function (p. 70) All values of the dependent variable.

rational expression (p. 120) The quotient of two polynomials.

rational function (p. 120) A function that is defined by a rational expression.

rectangle (p. 155) A quadrilateral with four right angles.

recursive formula (p. 223) A formula for a sequence that expresses the value of the nth term of a sequence, a_n, using the term before it, a_{n-1}.

regression (p. 351) The process of fitting a curve to a set of data.

regular polygon (p. 161) A polygon with all sides equal in length and all angles equal in measure.

677

regular polyhedron (p. 193) A space figure whose faces are all regular polygons of one type and which has the same number of faces at each vertex.

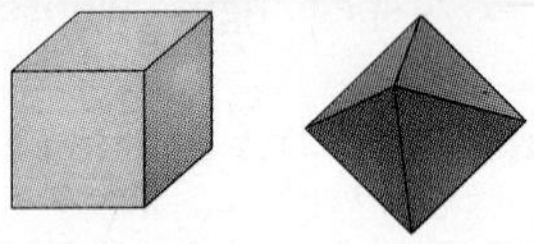

relative frequency histogram (p. 334) A histogram whose vertical scale shows the fraction or percent of the data values in each interval.

resultant vector (p. 442) The sum of two vectors. A vector from the tail of the first vector to the head of the second vector.

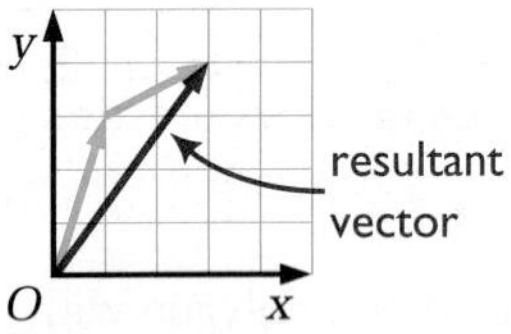

rhombus (p. 155) A quadrilateral with four sides of equal measure.

saturating function (p. 543) A function whose graph continually increases or decreases toward a constant value.

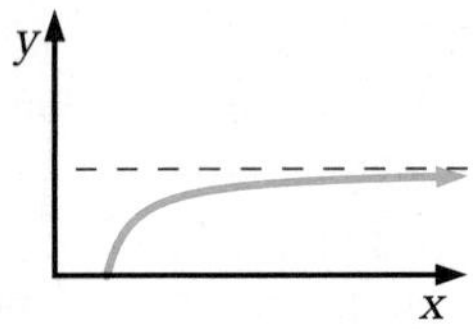

scalar multiple of a vector (p. 443) A vector $\mathbf{v} = k\mathbf{u}$ is called a *scalar multiple* of $\mathbf{u}$.

scatter plot (p. 18) A graph that shows the relationship between two data sets.

secant (p. 181) A line intersecting a circle in two points.

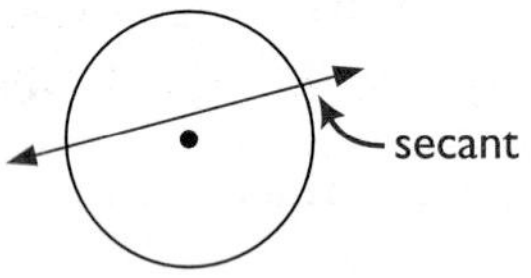

semicircle (p. 167) An arc of a circle with a measure of 180°.

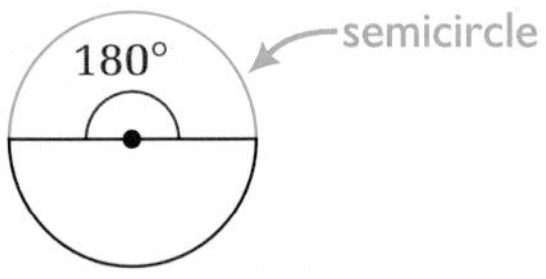

semiperimeter of a polygon (p. 188) Half the perimeter.

semiregular polyhedron (p. 193) A space figure whose faces are all regular polygons and which has the same number of faces of each type at each vertex.

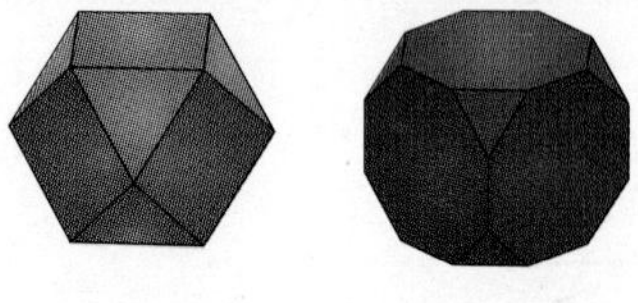

sequence (p. 208) An ordered list of numbers called the *terms* of the sequence. A *finite sequence* has a last term. An *infinite sequence* has no last term.

series (p. 238) The indicated sum of the terms of a sequence. A *finite series* has a last term. An *infinite series* has no last term.

sigma notation (p. 238) A form of notation using the summation symbol Σ to express a series.

significance level (p. 413) In a probability model, a probability level that would be considered unusual if the results were due to chance.

similar triangles (p. 655) Two triangles whose vertices can be matched up so that corresponding angles are equal in measure and corresponding sides are in proportion.

simulation (p. 374) An experiment that models an actual event.

sine function (p. 555) A function that measures the vertical position of an object moving around a unit circle. (*See also* periodic function.)

skewed distribution (p. 332) A frequency distribution that has an interval or group of intervals that contains the greatest frequencies near one end.

slope (p. 78) The measure of the steepness of a line given by the ratio of rise to run for any two points on the line.

square (p. 155) A quadrilateral with four right angles and four sides of equal measure.

standard deviation (p. 339) A measure of the variability of a data set. The formula for the standard deviation σ (read *sigma*), where $\bar{x}$ is the mean and n is the number of data values, is

$$\sigma = \sqrt{\frac{\sum_{i=1}^{n} (x_i - \bar{x})^2}{n}}.$$

standard normal curve (p. 521) The parent graph of any normal distribution.

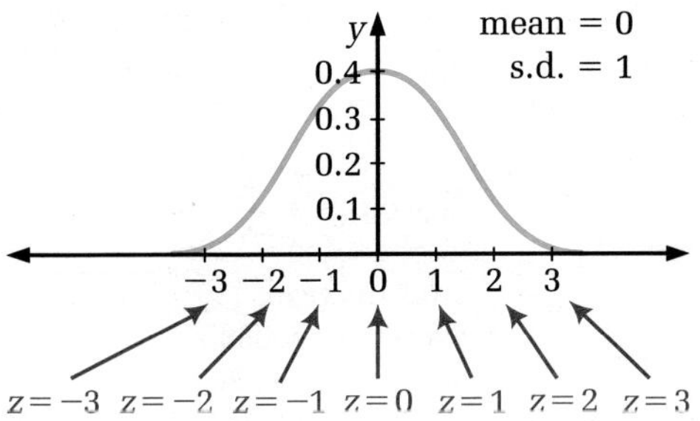

supplementary angles (p. 161) Two angles whose measures have the sum 180°.

synthetic proof (p. 140) A method of proof based on a system of postulates and theorems in which the properties of figures are studied, but not their actual measurements.

system of equations (p. 30) Two or more equations involving the same variables.

tangent (p. 181) A line in the plane of the circle that intersects the circle in exactly one point.

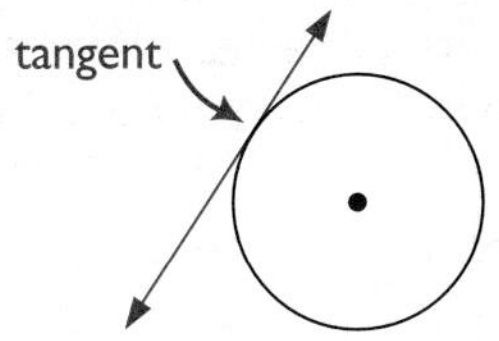

tessellation (p. 582) A design made up of regular polygons placed so their vertices touch without gaps or overlaps to cover a plane. A *regular* tessellation has only one kind of polygon. A *semiregular* tessellation has two or more kinds of polygons, and the arrangement at each vertex is the same.

transformation (p. 662) A change in the size or the position of a figure.

transformation matrix (p. 501) A matrix used to represent a transformation.

translation (p. 662) A transformation that moves each point of a figure the same distance in the same direction.

trapezoid (p. 154) A quadrilateral with at least one pair of parallel sides.

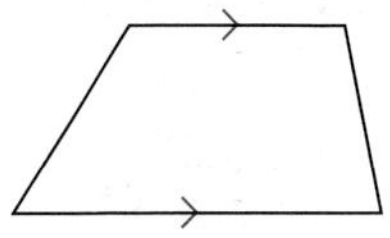

two-column proof (p. 141) A proof written in two columns. Statements are listed in one column and justifications in the other.

uniform distribution (p. 332) A frequency distribution where all the intervals have approximately equal frequencies.

upper extreme (p. 17) In a data set, the largest data item.

upper quartile (p. 17) In a data set, the median of the upper half of the data.

valid argument (p. 658) An argument that uses rules of logic.

vector (p. 441) A quantity that has both a length, or magnitude, and a direction.

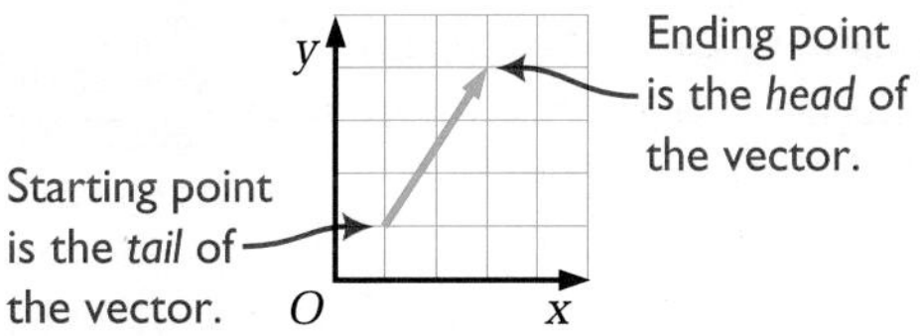

vertex of a network (p. 37) In a network, the point where two edges meet.

x-intercept (p. 648) The x-coordinate of a point where a graph intersects the x-axis (where $y = 0$). Also called a *horizontal intercept*.

y-intercept (p. 648) The y-coordinate of a point where a graph intersects the y-axis (where $x = 0$). Also called a *vertical intercept*.

z-score (p. 520) The distance of a data value from the mean of a data set, measured in standard deviations.

zero of a function (p. 104) A value of x that makes $f(x)$ equal to zero.

Index

689

Credits

DESIGN

Book Design: Two Twelve Associates, Inc.
Cover Design: Two Twelve Associates, Inc. Photographic collage by Susan Wides; student photography by Ken Karp.
Electronic Technical Art: American Composition & Graphics, Inc.

ACKNOWLEDGMENTS

20 Data from "The Best and Worst According to Nielsen," *TV Guide*, June 11, 1994. **248** From *The Mythology of All Races*, Volume 1: Greek and Roman, by William Sherwood Fox, A.M., Ph.D. Originally published in 1916 by Marshall Jones Company, copyright renewed 1944 by Macmillan Publishing Company. **249** From *The Oxford Dictionary of Nursery Rhymes*, edited by Iona and Peter Opie, Oxford University Press, 1983. **321** From *The Andromeda Strain*, by Michael Crichton. Copyright ©1969 by Centesis Corporation. Reprinted by permission of Alfred A. Knopf, Inc. **348** From *Jurassic Park* by Michael Crichton. Copyright ©1990 by Michael Crichton. Reprinted by permission of Alfred A. Knopf, Inc. **408** From *The Promise*, by Chaim Potok. Copyright ©1969 by Chaim Potok. Reprinted by permission of Alfred A. Knopf, Inc. **466** Cipher from *The Gold-Bug*, by Edgar Allen Poe. **515** From *Starbridge*, by A.C. Crispin. Copyright ©1989 by A.C. Crispin. Reprinted by permission of The Putnam Berkley Group. **566** From "The Parson's Tale" from *The Canterbury Tales*, by Geoffrey Chaucer, translated by Nevill Coghill, published by Penguin Books, 1977.

STOCK PHOTOGRAPHY

viii © Richard Megna/Fundamental Photographs (tl); **viii** Superstock (tr); **vii** The *Polyhedron Earth* found on page vii was designed and invented by R. Buckminster Fuller. © 1938, 1967, 1982 & 1992 Buckminster Fuller Institute, Santa Barbara, CA. All Rights Reserved.; **ix** Courtesy of the Playing Card Museum, The United States Playing Card Company, Cincinnati, Ohio; **xi** Murray Alcosser/The Image Bank; **xiii** Richard E. Smalley (b); **xiii** Lester Lefkowitz/Tony Stone Images/Chicago Inc. (t); **xxvi** Shinichi Kanno/FPG International Corp. (t); **xxvi** Superstock (bl); **xxvi** The Bettmann Archive (br); **xxvi** Portrait by William A. Smith, courtesy of Silvio Bedini and Landmark Enterprises (br); **1** Terry Qing/FPG International Corp. (bl); **1** Fort Worth Star-Telegram/Steve Gariepy (br); **2** © Bob Daemmrich (tl); **3** Dallas & John Heaton/Westlight; **5** NASA (l); **7** From *The Living Legend of Maria Martinez*, by Susan Peterson, Kodansha International © 1977, © 1989 (tl, tc, tr, bl, bc, br); **7** © Jerry Jacka. All Rights Reserved (br); **14** © 1994 Reneé Stockdale (tr); **14** Blair Seitz/Photo Researchers, Inc. (tl); **16** © The Stock Market/Otto Rogge (r); **16** Superstock (l); **16** © The Stock Market/Tom Van Sant 1993 (c); **19** Superstock; **20** NASA; **21** Bob Daemmerich/Stock Boston; **23** Tony Stone Images/Chicago Inc. (b); **23** The Bettmann Archive (t); **28** Copyright © David Smart/Stock South/Atlanta; **29** Henley & Savage/Uniphoto; **34** Ken Ross/FPG International Corp. (inset); **34** Copyright Tom Van Sant/Geosphere Project, Santa Monica/Science Photo Library/Photo Researchers, Inc. (b); **34** NASA (b); **36** Alex MacLean/Landslides (br); **36** © Kindra Clineff (bl); **37** FPG International Corp.; **37** Reproduced by permission of the Massachusetts Bay Transportation Authority, John J. Haley, Jr., General Manager (r); **39** Terry Donnelly/Tony Stone Images/Chicago Inc. (c); **43** Jerry Austin/The Picture Cube; **48** Martin Rogers/Uniphoto (bl); **49** Vance Tucker (r); **50** Per Breiehagen/Black Star (r); **52** David J. Sams/Tony Stone Images/Chicago Inc. (b); **53** Alan Becker/The Image Bank (r); **53** Mike Mazzaschi/Stock Boston (bl); **53** Don Smetzer/Tony Stone Images/Chicago Inc. (tr); **56** AT&T Archives (t); **57** Stamp designs © 1995 United States Postal Service. All Rights Reserved.; **59** Andy Sacks/Tony Stone Images/Chicago Inc. (r); **59** David R. Frazier/Tony Stone Images/Chicago Inc. (l); **66** Superstock (t); **66** The Bettmann Archive (bfr); **66** FPG International Corp. (bcr); **66** Ackerman/Mary Evans Picture Library (bfl); **66** Mary Evans Picture Library (bcl); **66** © Rob Crandall (m); **67** Eric Roth/The Picture Cube (bfl); **67** Courtesy Major Taylor Velodrome (m); **67** Indiana State Museum (t); **67** Mark Richards/PhotoEdit (bfr); **67** Duomo (bcr); **67** Lori Adamski Peek/Tony Stone Images/Chicago Inc. (bcl); **69** Superstock; **73** Jane Reed; **74** J. Goell/The Picture Cube (l); **74** Peter Pearson/Tony Stone Images/Chicago Inc. (r); **77** F. St. Clair-Renard/Photo Researchers; **83** Robert Caputo/Stock Boston (br); **83** From *The Albert N'yanza, Great Basin of the Nile, and exploration of the Nile Sources,* vol.1, by Samuel White Baker, © 1866, London: Macmillan and Co. (tr, mr); **85** Bill Bachman/Stock Boston; **91** Cathlyn Melloan/Tony Stone Images/Chicago Inc. (tr); **92** Chris Noble/Tony Stone Images/Chicago Inc.; **94** © Arthur & Elaine Morris/BIRDS AS ART (t); **94** Carr Clifton/Tony Stone Images/Chicago Inc. (b); **96** Superstock; **97** Reproduced by permission of the Lacrosse Foundation and Hall of Fame Museum, Balimore, MD and Time-Life Books Inc.; From *American Indians: Realm of the Iroquois.* Photograph by Larry Sherer © 1993 Time-Life Books Inc. (l); **99** John Running/Stock Boston; **102** © Animals Animals/Zig Leszczynski; **106** ABB Traction, Inc.; **107** Professors P.M. Motta & S. Correr/Science Photo Library/Photo Researchers; **108** Elena Rooraid/PhotoEdit; **109** Tim Davis/Tony Stone Images/Chicago Inc.; **110** Art Resource, NY (m); **110-111** © The Stock Market/Darrell Jones 1987; **116** © The Stock Market/Joe Towers 1991 (tl); **117** © Allsport USA/Ken Levine, 1992; **118** Gerard Vandystadt/Photo Researchers, Inc.; **125** NASA; **126** Solar Design Associates, Harvard MA; **130** © Animals Animals/Arthur Gloor (tl); **130** © Animals Animals/Joe McDonald (ml); **130** © The Stock Market/Michele Burgess 1991 (mr); **130** Stan Osolinski/Tony Stone Images/Chicago Inc. (br); **130** Chris Harvey/Tony Stone

Images/Chicago Inc.; **142** Superstock; **145** Ivan Massar/Black Star; **146** Peter Menzel/Stock Boston (br); **146** Peter Menzel/Stock Boston (bl); **147** Courtesy of Tai Soo Kim Partners (l); **147** Tai Soo Kim, FAIA (tr); **150** Superstock; **152** © Sally Humphrey (tr); **155** Courtesy The Institute of Architectural Design, Shenzhen University (tl); **161** Photo by George Gould, courtesy of John Hanou and Purdue Research Foundation; **162** Arthur Friedberg/The Coin & Currency Institute, Clifton NJ (tl, tr, bl, br); **170** Cherokee Historical Association, Council House replica at Oconaluftee Indian Village, Cherokee, NC (l); **170** Reproduced by permission of Cherokee Publications, Cherokee, NC (r); **171** David Young Wolff/Tony Stone Images/Chicago Inc.; **172** Superstock; **177** Paul Chesley/Tony Stone Images/Chicago Inc.; **178** Dennis MacDonald/PhotoEdit; **185** Foto Marburg/Art Resource; **186** Courtesy the Aerospace Corporation (l); **187** Hughes Leroux and Jules Gardner, *Acrobats and Mountebanks.* London: Chapman & Hall, 1890. (l); **191** Drawing by Shanahan; © 1993, The New Yorker Magazine, Inc.; **192** © Richard Pasley/Stock Boston (l); **192** Courtesy Richard E. Smalley (c); **194** Sylvia Wiegand; **196** Courtesy Richard E. Smalley; **197** The *Polyhedron Earth* was designed and invented by R. Buckminster Fuller. © 1938, 1967, 1982, & 1992 Buckminster Fuller Institute, Santa Barbara, CA. All Rights Reserved. (br); **204** Superstock (t); **204** Gary Buss/FPG International Corp. (l); **204** Superstock (r); **205** The Bettmann Archive (tr); **212** © The Stock Market/Tom Sanders 1990; **216** Cosmo Condina/Tony Stone Images/Chicago Inc.; **222** Rod Planck/Tony Stone Images/Chicago Inc.; **222** Michael F. Barnsley, *Fractals Everywhere*, Academic Press (1989, 1992) (inset); **226** © Animals Animals/Oxford Scientific Films; **234** Grant Heilman Photography, Inc./Grant Heilman (l); **234** © Animals Animals/Oxford Scientific Films/G.I. Bernard (inset); **236** Shelburne Museum, Shelburne, Vermont, Photograph by Ken Burris; **242** Grant Heilman Photography, Inc./Grant Heilman; **244** Steven E. Sutton/Duomo; **256** © 1994 M.C. Escher/Cordon Art—Baarn—Holland. All Rights Reserved.; **264** Alain Evrard/Gamma-Liaison (tl); **264** Tony Stone Images/Chicago Inc. (t); **264** Edith Munroe/Catgut Acoustical Society, Inc. (bl); **264** Michael Neveux/Westlight (br); **264** Superstock (tr); **265** Mangbetu Peoples, Zaire. Hampton University Museum, Hampton, Virginia (t); **265** Mark Harmel/FPG International Corp. (bl); **265** Marc Romanelli/The Image Bank (br); **267** Dick Luria/FPG International Corp.; **268** Eric Sander/Gamma-Liaison; **271** Walker Collection of A'Leilia Bundles (tr); **271** Photograph by Madame C.J. Walker, Collection of Indiana Historical Society Library (bl); **272** "The March to Valley Forge," by William B.T. Trego, 1883, courtesy of the Valley Forge Historical Society; **273** Courtesy of Arbor Drugs, Inc.; **274** Superstock (t); **274** © Richard Megna/Fundamental Photographs (b); **275** National Baseball Library, Cooperstown, NY; **285** University of Massachusetts Photo Services; **289** The Bettmann Archive (b); **292** Frans Halsmuseum, Haarlem, The Netherlands (t); **292** AP/Wide World Photos Inc. (b); **293** © Dallas & John Heaton/Westlight (t); **293** Superstock (b); **294** Steve Dunwell/The Image Bank (r); **294** © The Stock Market/Gabe Palmer (l); **296** © Duomo/Al Tielemans 1994; **299** "Girl amid autumn leaves from Lucky Days (decorative calendar, 1844) by Kunisada Utagawa (1786-1864)"/ Bridgeman Art Library; **301** © 1989 Scott Robinson/Stock South; **304** © Jerry Jacka. All Rights Reserved.; **304** From *Eyewitness Books: Volcanoes and Earthquakes* by Susanna van Rose, Copyright © 1992 by Dorling Kindersley Ltd. Reprinted by permission of Alfred A. Knopf, Inc.; **306** Photograph by Ansel Adams, Copyright © 1995 by the Trustees of the Ansel Adams Publishing Rights Trust. All rights reserved. (b); **306** Barbara Alper/Stock Boston (t); **307** © Tom McHugh, 1978 Steinhart Aquarium/Photo Researchers; **311** © Leonard Lessin/Peter Arnold, Inc. (t); **311** © Leonard Lessin/Peter Arnold, Inc. (b); **314** James Simon/The Picture Cube (t); **314** Uniphoto (b); **314** Chris Noble/Tony Stone Images/Chicago Inc. (b inset); **316** Superstock; **317** James King-Holmes/Science Photo Library/Photo Researchers (inset); **317** © 1995 Sarah Leen/Matrix International, Inc.; **318** Sygma (r); **318** Kenneth Garrett © National Geographic Society (l); **320** © Charles O'Rear/Westlight; **321** Dr. Tony Brain/Science Photo Library/Photo Researchers; **322** Superstock; **328-329** Artwork provided by American Honda Finance Corporation; **329** Ted Spiegel/Black Star (br); **333** The Bettmann Archive; **335** © Animals Animals/William D. Griffin; **338** Tony Freeman/Photo Edit (t); **338** Henley & Savage/Uniphoto (b); **342** Superstock (b); **342** Official U.S. Mint Photograph (t); **348** Steve Kirk/Marshall Editions; **350** Nigel Press/US Department of the Interior/Tony Stone Images, Chicago Inc.; **351** John Tlumacki/The Boston Globe Photo; **355** Courtesy Terry Pimsleur, president International Pumpkin Association; **356** Courtesy West Virginia Department of Education; **359** David Lissy/FPG International Corp.; **362** Superstock; **367** Lonny Kalfus/Tony Stone Images/Chicago Inc.; **370** Chris Cole/Tony Stone Images/Chicago Inc. (m); **370** Victor De Palma/Black Star (tr); **370** Archive Photos (tl); **370** © Johnny Stockshooter, International Stock Photo (b); **371** Benn Mitchell/The Image Bank (tl); **371** © The Stock Market/Paul Steel 1991 (br); **371** Tony Stone Images/Chicago Inc. (m); **371** Alan Bergman/FPG International Corp. (bl); **371** Will & Deni McIntyre/Tony Stone Images/Chicago Inc. (tr); **373** © The Stock Market/Richard Steedman (fr); **373** Hilarie Kavanagh/Tony Stone Images/Chicago Inc. (cr); **373** Michael MacIntyre/The Hutchinson Library–London (cl); **373** David Young-Wolff/PhotoEdit (bl); **381** Canadian Museum of Civilization, Negative S93-9725 (l); **381** Cranbrook Institute of Science (r); **387** Courtesy of the Playing Card Museum, The United States Playing Card Company, Cincinnati, Ohio; **390** Spencer Grant/The Picture Cube; **399** David Ryan/Uniphoto; **400** © The Stock Market/Peter A. Simon 1987; **401** © Heather Angel; **402** Stock Montage; **403** Jim Pickerell/Tony Stone Images/Chicago Inc. (tl); **403** Superstock (tr); **403** Photo: Wan-go Weng, Collection of Wan-go H.C. Weng (bl); **408** Marvin E. Newman/The Image Bank; **409** © Bob Daemmrich; **410** © The Stock Market/Masahiro Sano 1991; **410-413** © The Stock Market/Masahiro Sano 1991 (c); **415** © Animals Animals/Richard Kolar; **416** Lester Lefkowitz/Tony Stone Images/Chicago Inc.; **422** Stephen Dalton/Photo Researchers, Inc. (bl); **422** British Museum (br); **423** Greenland National Museum and Archives (br); **423** National Maritime Museum Picture Library (tl);

423 The Bridgeman Art Library (bl); **425** Alcosser Murray/The Image Bank (br); **425** Courtesy Colorado Historical Society F13,238 (tl); **428** Map adapted from THE FELLOWSHIP OF THE RING by J.R.R. Tolkien. Copyright © 1954, 1965 by J.R.R. Tolkien. Copyright © renewed 1982 by Christopher R. Tolkien, Michael H.R. Tolkien, John F.R. Tolkien and Priscilla M.A.R. Tolkien. Reprinted by permission of Houghton Mifflin Co. All rights reserved. (b); **436** © The Stock Market/H.P.Merten 1991; **437** Archive Photos; **438** Alex MacLean/The Picture Cube; **442** Alex MacLean/Landslides; **446** Copyright © Jack Gardner/Stock South/Atlanta; **451** Jerome Shaw/FPG International Corp.; **451** Phoebe Hearst Museum of Anthropology, University of California at Berkeley (tr); **452** John Terence Turner/FPG International Corp.; **453** © Richard Megna, Fundamental Photographs, NY; **456** NASA (lc); **457** © Kirk Schlea/Sports Photo Masters, Inc.; **459** Superstock; **464** Eureka! ® Tent; **465** Roddy Mims/Uniphoto (tl); **465** Everett C. Johnson/Folio, Inc. (tr); **467** David Ball/The Picture Cube (tl); **476** © Animals Animals/Oxford Scientific Films/Stephen Dalton; **477** Alfred Gescheidt/The Image Bank (tr); **480** David Joel/Tony Stone Images/Chicago Inc. (t); **480** Philip & Karen Smith/Tony Stone Images/Chicago Inc. (l); **481** Eddie Stangler/The Picture Cube (r); **481** Robert Frerck/Odyssey Productions/Chicago (br, bl, m); **487** Courtesy NFL Properties, Inc. (b); **494** NASA; **497** J.M. Charles/Black Star/Rapho (l); **497** Jeff Dunn/The Picture Cube (r); **507** Bob Daemmrich/Stock Boston (l); **507** Donald Dietz/Stock Boston (r); **510** Steve Weinrebe/The Picture Cube (l); **510** Jean-Marc Truchet/Tony Stone Images/Chicago Inc. (r); **519** Steve Dunwell/The Image Bank; **523** Superstock; **524** Lee Balterman/The Picture Cube; **534** James Balog/Tony Stone Images/Chicago Inc.; **538** E. Carlin/The Picture Cube (t); **538** © Animals Animals/Patti Murray (bl); **538** © Animals Animals/Patti Murray (br); **539** Steve McCurry/Magnum Photos (tl); **539** Steve McCurry/Magnum Photos (tr); **539** David Young-Wolff/PhotoEdit (b); **540** © Bob Daemmrich; **541** Harlen Hunter/Indiana Motor Speedway (tl); **541** Ron McQueeny/Indiana Motor Speedway (bl); **541** Frank Cezus/FPG International Corp. (br); **542** Ann Purcell/Uniphoto (mr); **542** David Young-Wolff/PhotoEdit (br); **542** David Young-Wolff/PhotoEdit (tr); **542** Diane Schiumo/Fundamental Photo (tl); **542** Stuart Krasner/Uniphoto (bl); **549** Courtesy Vicki Sans; **553** Superstock; **554** The Bettmann Archive; **566** Reproduced by permission of The Huntington Library, San Marino, California; **570** Arnulf Husmo/Tony Stone Images/Chicago Inc.; **573** Richard Nowitz/FPG International Corp.; **573** Courtesy of the Massachusetts Historical Society; **575** © Bob Daemmrich; **577** Eleni Mylonas/Viesti Associates, Inc. (b); **583** Robert Frerck/Odyssey; **584** © 1995 M.C. Escher/Cordon Art—Baarn—Holland. All Rights Reserved.;

ASSIGNMENT PHOTOGRAPHERS

Ken Karp **vi**
Richard Haynes **v, vi** (b), **x, xiv, xv, 9, 10, 11, 12, 14** (l), **27, 32, 35, 51, 57, 58, 61, 64, 68, 69** (b), **70, 100, 101, 115, 132, 139, 153, 154, 164, 165, 175, 176, 190, 198, 199, 206, 207, 230, 241, 251, 259, 266, 269, 282, 289** (t), **295, 323, 330, 331, 364, 365, 372, 382, 393, 395, 418, 424, 432, 439, 440, 450, 451** (bl), **462, 469, 475, 480** (r), **482, 483, 485, 504, 526, 532, 533, 542** (m), **560, 582, 589, 591**
Banta Digital Group **162** (tc, bc), **179, 192-193**

ILLUSTRATIONS

Jana Brenning **388, 422**
Chris Costello **4, 17, 15, 91** (m), **95, 98, 126, 138** (b), **151, 152** (l), **182, 220, 237, 248, 264-266, 274, 293, 379, 427** (b), **450, 466** (mr), **476, 493, 564, 577, 578, 579** (m), **580, 591**
Steve Cowden **xii, 6, 76, 84, 116, 454**
Christine Czernota **161**
Dartmouth Publishing, Inc. **16, 18, 20, 22, 23, 29, 30, 33, 35** (r), **37, 38,** (t, c, b), **40, 42** (t, b), **44, 45, 52, 53, 58, 62, 137** (t), **138** (t), **157, 159, 163, 187** (r), **196** (m), **281, 287, 288, 295, 331, 333, 340, 342, 344, 346, 348, 350** (bl, tr, br), **352, 373, 374, 376** (tl), **387** (t), **408** (tl), **425** (m), **426, 427** (tr, tl), **429, 439, 449, 451, 456, 457, 459** (l), **460, 464, 465** (ml, mr), **466** (tr), **471, 472, 473** (m), **474, 477, 511, 514, 528, 530, 541, 543** (lm), **545, 547, 548, 550, 551, 554, 558** (t), **561, 568, 570, 571, 572, 574** (b), **575, 579** (t, b) **582, 583, 584, 585, 586, 587, 588**
DLF Group **3, 46, 80, 111** (t), **119** (b), **142, 144, 152** (r), **184** (b), **278-279, 286, 320, 430, 437, 565, 567, 569**
Bruce Holloway **291, 315, 322, 552, 574** (t)
Piotr Kaczmareck **169, 186** (r), **201, 213, 428** (t), **473** (tr, br), **499**
Aggie Kenny **136-137**
Joe Klim **91** (b), **308, 408** (m)
Ellen Kuzdro **119** (m), **325, 440, 444, 445, 467, 468**
Andrew Myer **180, 277** (r), **480**
Steve Patricia **558** (b)
Neil Pinchin **267, 277** (l), **296, 301, 321**
Photo manipulations by Lisa Rahon **296, 317**
Glenn Reid **51, 82, 350** (tl), **354, 359, 362, 357**
Michael Smith **1, 2, 61**

TYPOGRAPHIC TITLES

Chris Costello **136**
Frank Loose Design **0, 3, 9, 16, 23, 29, 37, 45, 53, 69, 77, 85, 94, 101, 110, 119, 126, 139, 147, 153, 159, 165, 172, 179, 187, 192, 207, 216, 222, 228, 237, 244, 251, 267, 275, 281, 288, 296, 301, 309, 317, 331, 338, 344, 350, 359, 373, 382, 390, 395, 403, 410, 425, 430, 440, 446, 453, 459, 467, 483, 489, 496, 501, 510, 519, 526, 541, 548, 560, 568, 577, 582**
Piotr Kaczmareck **422, 424**

Selected Answers

Student Resources

Unit 1

Pages 3–6 Talk it Over

1. Answers may vary. An example is given. Walk south on Horikawa-dori Ave. to Shichijo-dori Ave. Turn left at Shichijo-dori Ave. and walk east. **2.** Answers may vary. Examples are given. **a.** Michiko's **b.** Michiko's **c.** Jiro's **d.** (1) Jiro's; Michiko's directions give no idea of how long to walk along Horikawa-dori Ave. and I'd think I missed the temple. (2) Michiko's; I wouldn't have to make decisions at as many intersections as with Jiro's directions. **3.** Steps 1, 6, and 7 **4.** End the first loop when a space is found that the car will fit in and in which it may be legally parked. End the second loop when there is enough space for the car to be backed in parallel to the curb. End the third loop when the car is parallel to the curb and less than 1 foot from it. **5.** In Step 4, turn the front wheels all the way to the left. In Step 6, turn the steering wheel all the way to the right. **6.** Answers may vary. An example is given for solving an inequality of the form $ax + b < c$. **Step 1:** If $b = 0$, go to Step 3. **Step 2:** Add the opposite of b to both sides of the inequality. **Step 3:** If $a = 1$, go to Step 6. **Step 4:** Multiply both sides of the inequality by the reciprocal of a. **Step 5:** If $a < 0$, reverse the inequality symbol. **Step 6:** Simplify if necessary. **7.** Yes; **Step 1:** Graph $y = 4$. **Step 2:** Graph $y = -3x + 9$. **Step 3:** Estimate the x-coordinate of the intersection point.

Pages 6–8 Exercises and Problems

5. E, F, C, A, B, B **7, 9.** Solution methods may vary. Solutions are given. **7.** $x > 1274$ **9.** $x \le -4\frac{2}{5}$

11. a. Answers may vary. An example is given. I prefer Ryan's algorithm because it requires fewer steps and calculations. If the server didn't add the sales tax to the check, then I'd use Susan's algorithm. **b.** Divide the bill, with tax, by 7. Divide the bill, without tax, by 6. **c.** Answers may vary. An example is given. I prefer Ryan's algorithm because multiplying is easier than dividing. My algorithm is less complicated than Susan's. **17.** $(3x + 7)(x - 1)$ **18.** $3(x + 3)(x + 2)$ **19.** $(3x - 8)(3x + 8)$ **20.** $(4x + 7)(4x + 7)$ **21.** 24 **22.** 360 **23.** 15,120 **24.** 21 **25.** 792 **26.** 5

Pages 10–12 Talk it Over

1. If one side is 1 unit long, the other two sides have lengths x and $11 - x$ for some integer x. Because of the triangle inequality, all of the following must be true: $x + 1 > 11 - x$, $x + (11 - x) > 1$, and $1 + (11 - x) > x$. For all three to be true, x must be greater than 5 and less than 6. There is no such integer x. **2.** The algorithm begins with the smallest possible length for one side and considers all possible lengths of each of the other two sides in order (from least to greatest for the second side and greatest to least for the third). **3.** No; first, there is no relationship among letters similar to inequality with numbers. If you were to use "≤" to mean "is the same or comes before in alphabetical order" the algorithm would still not be successful. In the given algorithm, order does not matter. It would in producing different ordering of letters. Also, in Sample 1, side lengths can repeat in a single triangle; in determining possible orders of letters, each letter would be used only once in each grouping. **4.** Answers may vary. An example is given. Beginning with 10 for the length of side 1, list all possible lengths for sides 2 and 3 that add up to 12 and also satisfy this inequality: length of side 1 ≥ length of side 2 ≥ length of side 3. Cross out all the combinations that do not satisfy the triangle inequality. **5.** Assign the first task to a mechanic. The second task would take that mechanic past 8 h, so assign it to a new mechanic. Similarly, assign the third task to a third mechanic. For each succeeding task, determine if any of the first three mechanics (in order) can complete the task. If not, assign another mechanic. **6.** Even if Paula Hernandez does not need to use the least number of mechanics, the second algorithm may be more reasonable, because each mechanic's time is used more efficiently. Three of the four will be working all day. With the first algorithm, four are working 6 h or less and two are working less than 5 h. **7.** Answers may vary. An example is given. Look for the most time-consuming job and assign it to mechanic 1. Then find a smaller job to fill his or her 8 hours. Then take the next most time-consuming job and assign it to mechanic 2. Then find smaller jobs to fill his or her 8 hours. Continue this until all the jobs are assigned. This method should result in the least number of mechanics. **8.** If some customers need to pick up their cars by noon, Paula would have to assign those tasks so that they could be completed in time, then schedule the others. If two or more mechanics are needed for some tasks, they would have to be assigned first. Then the remaining tasks would have to be assigned to individual mechanics as they become available.

Pages 13–15 **Exercises and Problems**
1. The sum of the lengths of any two sides must be greater than the length of the third side. This relationship is called the triangle inequality. **3.** Yes. **5.** Yes. **7.** No.

13. a.

Side 1	Side 2	Side 3
8	8	14
9	9	12
10	10	10
11	11	8
12	12	6
13	13	4
14	14	2

b. the triangle with sides lengths of 10, 10, and 10
15. Answers may vary. An example is given. (1) 57 lb; (2) 43 lb 9 oz, 26 lb 7 oz; (3) 38 lb 12 oz, 30 lb 7 oz; (4) 34 lb 8 oz, 30 lb 6 oz; (5) 25 lb 5 oz, 22 lb 8 oz, 18 lb 14 oz **17. a.** Six breaks are needed. Plans may vary. An example is given. Run two pairs of 60 s commercials, and two pairs of one 90 s and one 30 s commercials. Run the remaining three 30 s and the two 15 s commercials together. Run the 40 s, 20 s, and 45 s commercials together. **b.** Answers may vary. An example is given. I grouped numbers with a sum of 120. There were five such groups. I grouped the remaining commercials together, since their running time was less than 2 min. **19.** She needs four boards. Plans may vary. An example is given. She could cut two 12 ft boards into three pieces each, one $6\frac{1}{2}$ ft long and the other two $2\frac{1}{2}$ ft long. The third board could be cut into four pieces, two 4 ft long and two 2 ft long. She would need a $6\frac{1}{2}$ ft piece from the fourth board. **21.** $x < 4\frac{1}{4}$ **22.** $x > -48$ **23.** $x \leq -2$ **24.** $x \geq 2\frac{1}{2}$ **25.** 45 **26.** 110 **27.** 7 **28.** 12 **29.** If there is heavy traffic in the city, Corey will have to wait until the first act ends to be seated.

Pages 16–18 **Talk it Over**
1. Caracas **2.** Answers may vary. Examples are given. Caracas in both January and July; the temperature is at or near 70°F during both months in Caracas, but not in Alice Springs. **3. a.** No; it means that, on average, Caracas is rainier than Alice Springs.
b. Answers may vary. An example is given. Yes; if it rains a lot, then my activities could be limited.
4. No; the graph would still rise to the right, indicating the same relationship between increasing employment percentages and increasing tuition.
5. Estimates may vary. about 90%

Pages 19–21 **Exercises and Problems**
3. Mexico City: about 16.3°; Shanghai: about 16.6°
5. a, b. Answers may vary. An example is given.
a. Of the 13 points on the scatter plot, 9 appear to lie near the line $y = 80$, which indicates that the annual cost of operating a top-freezer refrigerator is fairly constant, about \$80, no matter what the cost of the refrigerator. **b.** No; four of the data points lie fairly far from the line, indicating that four of the refrigerators cost somewhat more or less than \$80 per year to operate. **7.** Fox **9.** ABC; ABC has the highest upper extreme as well as the highest median.

16.

Side 1	Side 2	Side 3
1	6	6
2	5	6
3	4	6
3	5	5
4	4	5

17.

Side 1	Side 2	Side 3
2	6	6
3	5	6
4	4	6
4	5	5

18.

Side 1	Side 2	Side 3
2	8	8
3	7	8
4	6	8
4	7	7
5	5	8
5	6	7
6	6	6

19.

Side 1	Side 2	Side 3
2	12	12
3	11	12
4	10	12
4	11	11
5	9	12
5	10	11
6	8	12
6	9	11
6	10	10
7	7	12
7	8	11
7	9	10
8	8	10
8	9	9

20. a. $x = -\frac{3}{8}$ **b.** 8 **21. a.** $x = 1\frac{5}{6}$ **b.** 0 **22. a.** $x = \frac{5}{14}$ **b.** 1 **23.** $\left(-5\frac{1}{5}, -13\frac{4}{5}\right)$ **24.** $\left(-2\frac{4}{7}, -7\frac{5}{7}\right)$ **25.** $\left(-\frac{10}{23}, 1\frac{6}{23}\right)$ **26.** $\left(14\frac{5}{7}, 2\frac{1}{7}\right)$

Page 22 Checkpoint

1. Answers may vary. An example is given. Box-and-whisker plots let you compare the extremes and the quartiles. For example, you can determine which of the data sets contains the largest or smallest number. You can also compare which has half of its data grouped more closely about the median, or which is spread out more. You can also tell how the upper and lower quartiles of the data compare. **2.** Answers may vary. An example is given. **Step 1:** Decide which movies are showing at the theater you plan to go to. **Step 2:** Decide which movie you would all like to see. If you do not agree on a movie, discuss your preferences, toss a coin , or compromise until a movie is selected. **Step 3:** Decide what times the movie is showing and whether any one time is suitable for all of you. If not, go back to the previous step to choose another movie. **3.** garden salad, small cole slaw, small herbal tea **4.** about 75% both years
5. Answers may vary. An example is given. Generally, usage is up slightly. Every specific point on the graph is higher in 1990 than it was in 1980.

Pages 23–26 Talk it Over

1. There were more workers employed in agriculture than in any of the other three sectors. **2. a.** They intersect. **b.** about 1905 **3. a.** the Industrial Age **b.** about 1954 **4.** the Information Age; More workers are employed in Information than in any of the other three sectors. **5.** The x-coordinates represent the number of checks Camille will write. The y-coordinates indicate the total monthly charge for the number of checks indicated by the x-coordinate. **6.** At that point, both the graph of $y = 3 + 0.75x$ and the graph of $y = 10$ are above the graph of $y = 6 + 0.35x$. The monthly charges at Main St. Bank and Co-op Bank would be higher for that number of checks than for the same number of checks at Central Bank.
7. Both graphs involve the same control variable and independent variable. **8.** Find the solutions of the three pairs of equations $y = 6 + 0.35x$ and $y = 3 + 0.75x$, $y = 6 + 0.35x$ and $y = 10$, and $y = 3 + 0.75x$ and $y = 10$. Check which bank is least expensive on each of the four intervals determined by the x-values of the three solutions. Answers may vary. Examples are given. The graphing solution to Sample 1 is easier to interpret. The algebra solution to Sample 2 is quicker.

Pages 26–28 Exercises and Problems

1. You can determine the intervals where the values of one function are less than, greater than, or equal to the values of another function.
3. B **5.** \$25,000 per year **7.** Physicians' Services
9, 11. Estimates may vary.
9. about 333.3

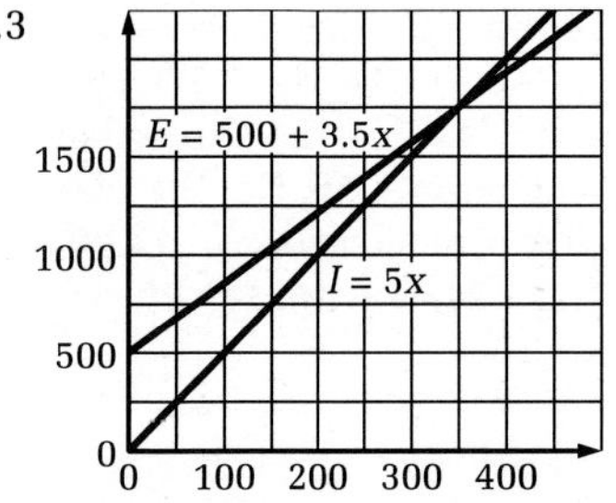

11. about 26

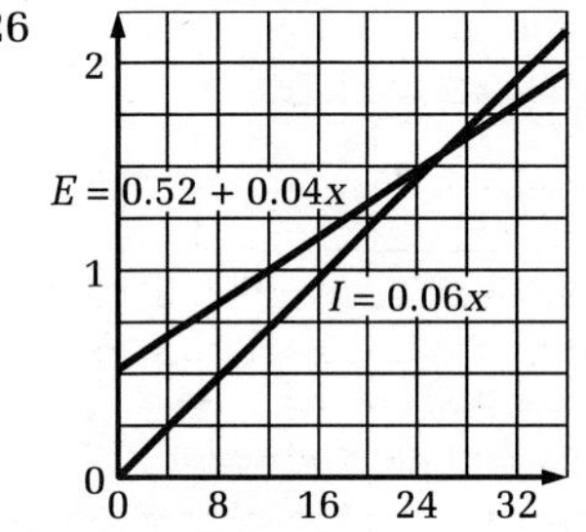

13. 22.7 **15.** 131.8 **24.** lower extreme: 25; lower quartile: 38; median: 48; upper quartile: 63.5; upper extreme: 83 **25.** lower extreme: 80; lower quartile: 90; median: 98.5; upper quartile: 115; upper extreme: 140 **26.** $TR = 4$; $RS = 4\sqrt{3}$ **27.** $DE = 11$; $DF = 11\sqrt{2}$ **28.** $KL = 10$; $JL = 5$

29. $XY = ZY = 11\sqrt{2}$ **30.** $\begin{bmatrix} 12.6 & 78 & 28 \\ 10 & 70 & 9 \end{bmatrix}$

Pages 29–33 Talk it Over

1. the seating capacity of the auditorium **2.** premier; These seats are closer to the stage. **3.** Answers may vary. Examples: average ticket prices in the area; local sales of the band's music; how much of a profit the promoter can reasonably expect on each ticket **4. a.** Yes; when the given values of x and y are substituted in the equation, the result is a true statement. $3(-1) + 4(2) = -3 + 8 = 5$✓ **b.** Yes; when the given values of x, y, and z are substituted in the equation, the result is a true statement. $3(-1) + 4(2) + 2(3) = -3 + 8 + 6 = 11$✓ **c.** No; when the given values of x, y, and z are substituted in the equation, the result is not a true statement. $3(-1) + 4(2) + 2(-3) = -3 + 8 + (-6) = -1$; $-1 \neq 11$ **5. a.** Yes; (–1, 2) is a solution of both equations. **b.** No; (–1, 2, 3) is not a solution of the second equation or the third. **6. a.** Yes; methods may vary. An example is given. Substitute $y + z$ for x in the first equation to get $2y + 2z = 10{,}000$. Next, substitute $y + z$ for x in the second equation to get $16y + 20z = 88{,}000$. Then solve the system of equations in y and z to get $z = 2000$. Use this value to get $y = 3000$ in one of the two equations. Finally, use $z = 2000$ and $y = 3000$ in $x = y + z$ to get $x = 5000$. **b.** No; it is only possible to graph equations in two variables on a graphics calculator.

Pages 34–36 Exercises and Problems

3. (–1, 2) **5.** (–0.5, 1.5) **7.** (5000, 2000) **9.** (4.7, 1.7) **11.** (112, –170) **13.** No; the planes have no common point of intersection. **15.** No; the planes have no common point of intersection. **17.** (0, 7, 1) **19.** (–2, 4, 1.5) **21.** (–3, 10, 1) **29.** 311.7 **30.** 59.9 **31.** 56.1 **32.** 27

33.
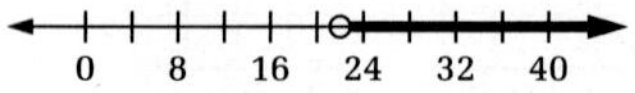

34.
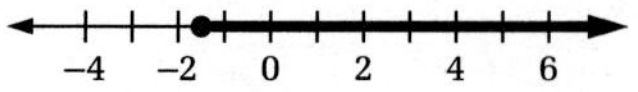

35.
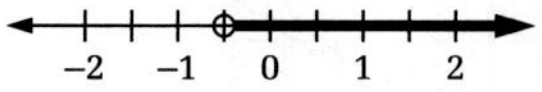

36. $\frac{2}{c^{1/2}}$ **37.** 0 **38.** $\frac{6}{n^5}$ **39.** $-3r^4$

Pages 37–41 Talk it Over

1. The transportation map is easier to read. Summaries may vary. The labels are clearly written in horizontal lines. The labels on the street map are written within boundary lines in varying directions. **2.** stations; There is a transportation line connecting the two stations. There is no transportation line directly connecting the two stations. **3.** Answers may vary. An example is given. I would take the Green Line from the Museum station to Government Center, then take the Blue Line to the Airport. The Airport would be the twelfth stop. **4.** Kenmore, Hynes, Copley, Arlington, Boylston, Park St., Government Center, Haymarket, North Station, Science Park; 9 edges **5.** No. **6.** Summaries may vary. Both maps indicate locations in the city of Boston. The street map shows the actual layout of the streets, with distances to scale. The transportation map shows transportation routes around the city. Distances are not to scale and streets are not shown. The street map would be helpful if you were walking or driving around the city. The transportation map would be of little help when walking and driving, other than indicating relative positions of some locations. **7.** 2 nonstop flights; 4 nonstop flights; Count the arrows coming out of the vertex for the indicated city; add the 1's in the row for the indicated city. **8.** Answers may vary. An example is given. It is very easy to read the number of nonstop departing flights from either the diagram or the matrix. It is easier to read the number of roundtrip nonstop flights from the diagram because you only have to look for arrows going both ways between two vertices.

9. Answers may vary. An example is given.

Meeting time A	Meeting time B	Meeting time C
band	student council	science club
honor society	language club	school newspaper
drama club		

10. Answers may vary. An example is given. Use each vertex as a starting point and find all possible colorings for the vertex. Record all resulting schedules and eliminate any duplicates.

Pages 41–44 Exercises and Problems

1. The vertices labeled "Cárdenas" and "volleyball" each have four edges connected to them. The number of edges connected to a story indicates the number of people who can cover it. The number of edges connected to a name indicate the number of stories the reporter can cover.

3.

	A	B	C	D	E	F	G
A	0	1	0	0	0	0	1
B	1	0	1	0	0	1	1
C	0	1	0	1	1	1	0
D	0	0	1	0	1	0	0
E	0	0	1	1	0	1	1
F	0	1	1	0	1	0	1
G	1	1	0	0	1	1	0

5.
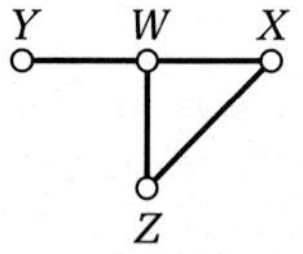

7.
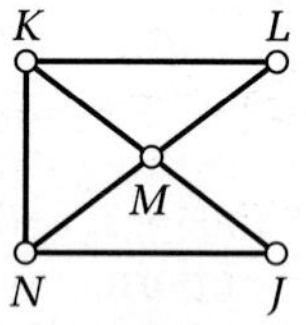

9. Answers may vary. Examples are given. children's hospital: Joyce, Becka; food pantry: Miguel, Hannah; homeless shelter: Desmond, Marissa; community center: Kirby, John **19.** Schedules may vary. An example is given. scheduling sequence: math; chemistry and history; economics and American literature; French and Spanish **22.** (–2, 1, –1) **23.** (–1, –3, 5) **24.** (–1, 0, –4) **25.** $2\sqrt{5} \approx 4.5$ **26.** $2\sqrt{5} \approx 4.5$ **27.** $5\sqrt{2} \approx 7.1$ **28.** $\sqrt{73} \approx 8.5$ **29.** $2\sqrt{26} \approx 10.2$ **30.** $\sqrt{5} \approx 2.2$

Pages 45–48 Talk it Over

1. It increases initially, reaches a maximum at about 35 mi/h, and then begins to decrease. **2.** Yes; there must be a maximum. For example, there could be no speed at which a car achieves 200 MPG. The maximum appears to occur at about 35 mi/h. **3.** Yes; clearly the MPG cannot be less than 0. **4. a.** St. Louis to Effingham to Champaign to Danville; 201 mi **b.** Answers may vary. An example is given. I used trial and error. **5.** 201 mi; They are equal. **6.** St. Louis is the starting point. There is no edge leading into this vertex. **7.** 7 times **8.** Yes. Yes. In both cases, you are able to do so, but it is unnecessarily complicated, since the shortest routes are fairly apparent.

Pages 49–52 Exercises and Problems

3. a. minimum **b.** 2 **c.** −11 **5. a.** maximum **b.** −3 **c.** 25.5 **7. a.** maximum **b.** $\frac{1}{8}$ **c.** $\frac{49}{96}$
13. $A - D - E - Z$; 24

16.

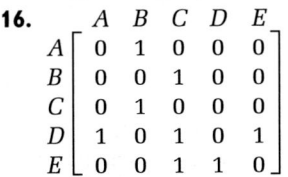

$$\begin{array}{c} \\ A \\ B \\ C \\ D \\ E \end{array}\begin{array}{c} \begin{array}{ccccc} A & B & C & D & E \end{array} \\ \begin{bmatrix} 0 & 1 & 0 & 0 & 0 \\ 0 & 0 & 1 & 0 & 0 \\ 0 & 1 & 0 & 0 & 0 \\ 1 & 0 & 1 & 0 & 1 \\ 0 & 0 & 1 & 1 & 0 \end{bmatrix} \end{array}$$

17.
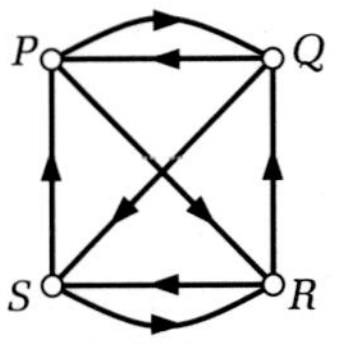

18.
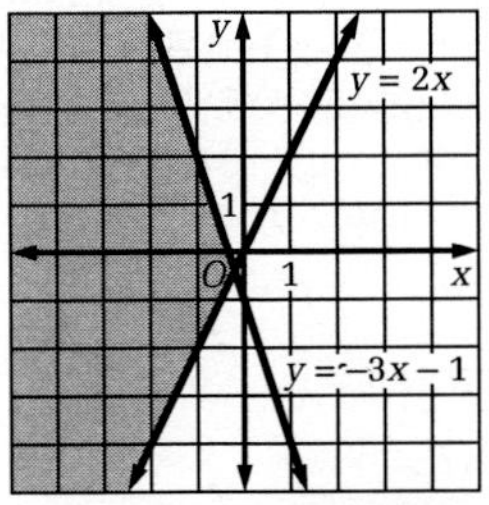

19.
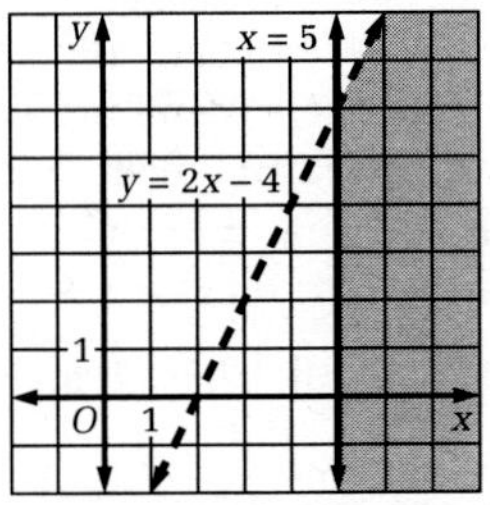

20.
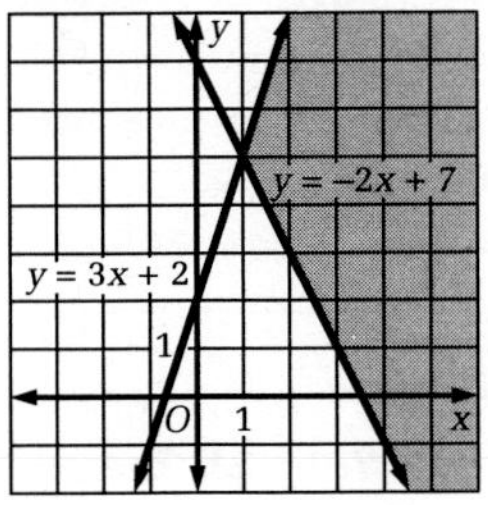

21.
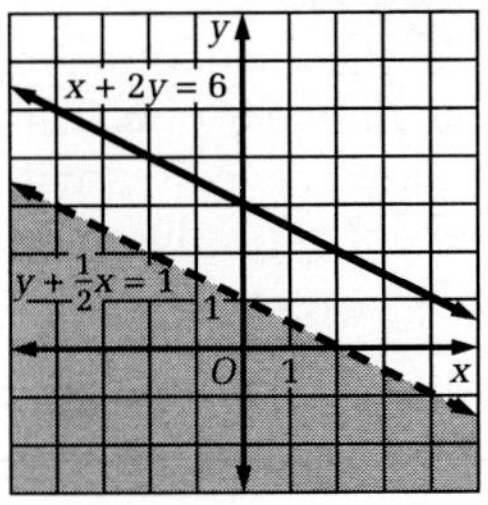

Pages 53–55 Talk it Over

1. Answers may vary. An example is given. I think more people listen to the radio in the morning than in the afternoon or evening, because people listen while getting ready for and going to work or school. **2.** Answers may vary. **3. a.** 22 ads; \$2600 **b.** 16 ads; \$1700 **c.** 18 ads; \$2400 **d.** 22 ads; \$1400 **4. a.** the total number of ads **b.** the total cost of x A.M. ads and y P.M. ads **5.** $200x + 50y$; $x + y$ **6.** b, c, and d; b **7.** No; x and y represent numbers of ads, which are always nonnegative. **8.** O: 0 ads, \$0; A: 20 ads, \$1000; B: 20 ads, \$2200; C: 11 ads; \$2200 **9.** No; only points with both coordinates whole numbers. **10. a.** 300,000 people **b.** 300,000 people **c.** 300,000 people **11. a.** the total number of listeners **b.** on the line labeled $T = 300{,}000$ in the figure (the line with equation $90{,}000x + 30{,}000y = 300{,}000$) **c.** (3, 1) **12. a–c.** Answers may vary. Examples are given. **a.** 4 A.M. ads, 8 P.M. ads **b.** 8 A.M. ads, 9 P.M. ads **c.** 8 A.M. ads, 12 P.M. ads **d.** The line $T = 1{,}260{,}000$ does not intersect the feasible region. None of the combinations will reach that many listeners. **13. a.** It moves to the right and both intercepts increase. **b.** All other points in the feasible region fall below the graph of $90{,}000x + 30{,}000y = 1{,}080{,}000$. That is, for all other points in the feasible region, $T < 1{,}080{,}000$. **c.** 8 A.M. ads and 12 P.M. ads; 1,080,000 listeners

Pages 57–60 Exercises and Problems

1. a. $0.45a + 0.30p$ **b.** $0.45a + 0.30p \leq 8.10$ **c.** $a + p \geq 8$ **d.** $a \geq 0$; $p \geq 0$

e.
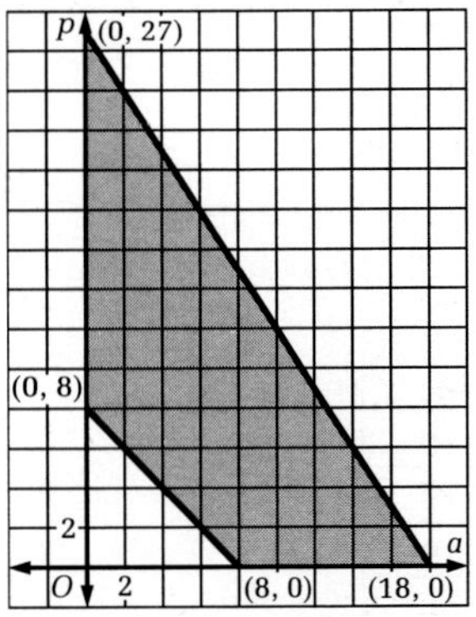

f. Answers may vary. An example is given. For (10, 6), the total cost is \$6.30.

3.
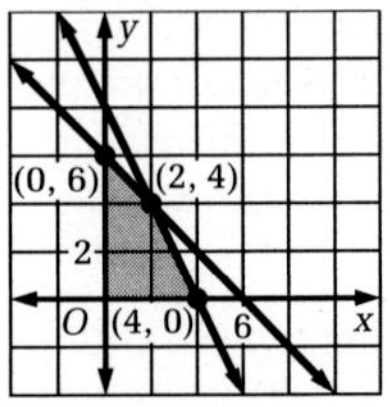

5.
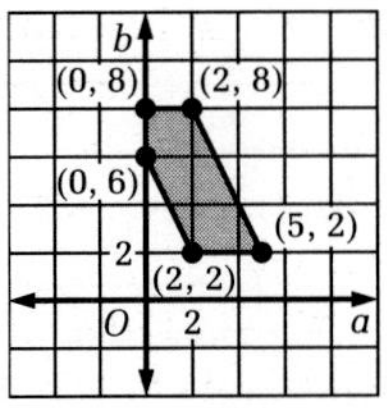

11. 35 kg of silicon carbide and 150 kg of silicon tetrachloride; \$48,700 **15.** maximum height: about 13 ft; distance traveled: about 14 ft **16.** $\frac{1}{36} \approx 0.03$ **17.** $\frac{1}{12} \approx 0.08$ **18.** $\frac{1}{4} \approx 0.25$ **19.** S.A. ≈ 254.5 cm^2; $V \approx 381.7$ cm^3 **20.** S.A. $\approx 16{,}286$ in.^{2}; $V \approx 195{,}432$ in.^{3} **21.** S.A. ≈ 1017.9 m^2; $V \approx 3053.6$ m^3 **22.** S.A. ≈ 2123.7 ft^2; $V \approx 9202.8$ ft^3

Pages 62–63 Unit 1 Review and Assessment

1. Answers may vary. Examples are given for two equations in x and y. **a.** *Method 1:* **Step 1:** Solve for y in the first equation. **Step 2:** Substitute the resulting value for y in the second equation. **Step 3:** Solve the resulting equation for x. **Step 4:** Substitute the value of x in the first equation. **Step 5:** Solve the resulting equation for y. *Method 2:* **Step 1:** Graph both equations on a graphics calculator. **Step 2:** Use TRACE to determine the point of intersection and find x and y. **b.** The results will be the same (although the first may result in a fractional answer, while the second gives a decimal approximation). Neither algorithm is significantly easier to use.

2. *Method 1:*

$3x > 7 + 10x$

$-7x > 7$ ← Subtract $10x$ from both sides.

$x < -1$ ← Divide both sides by -7, and reverse the inequality sign.

Method 2: Graph $y = 3x$ and $y = 7 + 10x$ on the same set of axes.

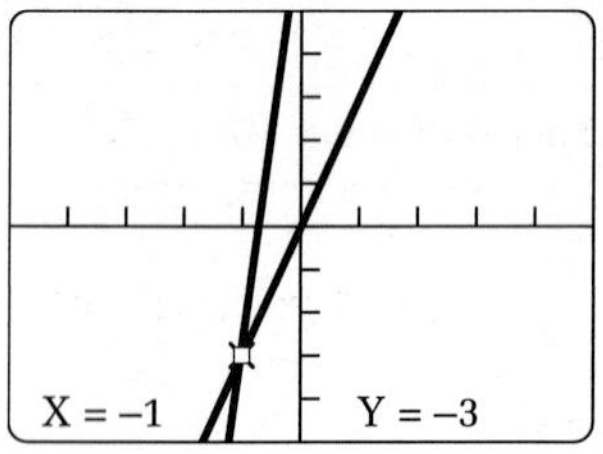

The solution of the inequality is all values of x for which the graph of $y = 7 + 10x$ is below the graph $y = 3x$, that is $x < -1$.

3.

Side 1	Side 2	Side 3
2	10	10
3	9	10
4	8	10
4	9	9
5	7	10
5	8	9
6	6	10
6	7	9
6	8	8
7	7	8

4. Answers may vary. Examples are given. **a.** In general, as the weight of an SLR Camera zoom lens increases, the price increases. **b.** No; although the points lie close to the line $y = 3x$, they do not all lie on the line. So, for example, there is a 17 oz lens that costs less than a 16 oz lens. **5. a.** less than about 110 min or between about 128 min and about 238 min **b.** less than about 178 min **c.** If Robert expects to spend less than about 110 min or between about 128 min and about 238 min on local calls, he should get Measured Service. If he expects to spend between about 110 min and about 128 min on local calls, he should get Measured Circle Service. If he plans to spend more than 238 min on local calls, then he should get Unlimited Service. **6.** (2, 1, –2) **7.** (–2, –4, 5) **8.** (–3, 0, 2) **9. a.** Answers may vary. Example: The access graph might represent five towns and their connections by direct train service.

b.

$$\begin{array}{c} \\ A \\ B \\ C \\ D \\ E \end{array}\begin{array}{c} \begin{array}{ccccc} A & B & C & D & E \end{array} \\ \begin{bmatrix} 0 & 1 & 1 & 0 & 0 \\ 1 & 0 & 1 & 1 & 0 \\ 1 & 1 & 0 & 0 & 0 \\ 0 & 1 & 0 & 0 & 1 \\ 0 & 0 & 0 & 1 & 0 \end{bmatrix} \end{array}$$

10. three colors

11. a. minimum **b.** $1\frac{1}{4}$ **c.** $7\frac{7}{8}$ **12. a.** maximum **b.** $1\frac{1}{2}$ **c.** $\frac{1}{4}$ **13. a.** maximum **b.** about –2 **c.** about 6.9 **14.** She should spend 50 days in Arizona and 120 days in New Mexico for a sales volume of $486,000.

Unit 2

Pages 70–72 Talk it Over

1. walking speed **2.** all real numbers except 0 **3.** all 366 days of the year **4.** $C(r) = 2\pi r$ **5.** Answers may vary. An example is given. B(George Washington) = February 22 **6. a.** Answers may vary. An example is given. Let m be the minimum stopping distance function; $m(x) = x + \frac{x^2}{20}$. **b.** 0; The point (0, 0) is on the graph. That is, for $x = 0$, $m(x) = 0$.

Pages 73–76 Exercises and Problems

1. a. the relation that pairs words and meanings in the dictionary **b.** The word *friendly* on page 69, for example, is paired with several meanings. **3.** No; the domain values Giles and Kalisha are both paired with two range values. **9. a.** $\frac{1}{3}$ **b.** $-\frac{1}{3}$ **c.** $\frac{1}{a + b + 1}$ **11. a.** 2 **b.** 4 **c.** $|a + b|$ **13. a.** 11 **b.** 77 **c.** $4a^2 + 8ab + 4b^2 - 3a - 3b + 1$ **17.** 1 **19.** 1 **21.** Yes. **23.** Yes. **25.** B and C are functions, since no member of either domain is paired with more than one member of the range. B is many-to-one, since two members of the domain are paired with the same member of the range. C is one-to-one, since every member of the range is paired with exactly one member of the domain. A is not a function, since 0 is paired with three different members of the range. **27.** one-to-one; No horizontal line intersects the graph in more than one point. **29.** one-to-one; Each value y of the range is paired with exactly one x-value. **31.** all real numbers; all real numbers greater than or equal to 1

34.

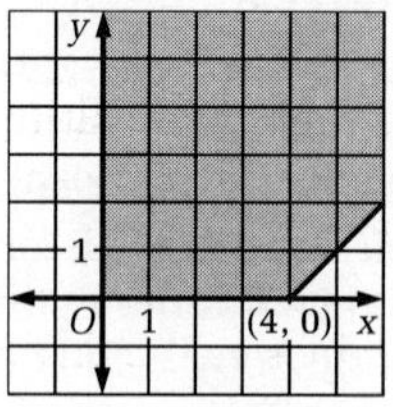

35.

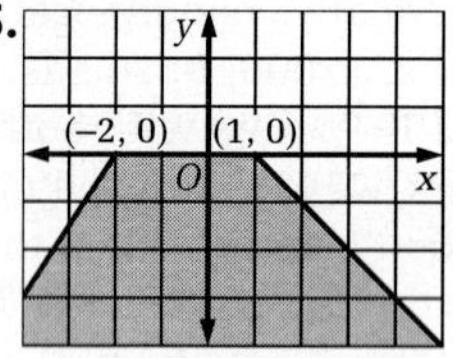

36.

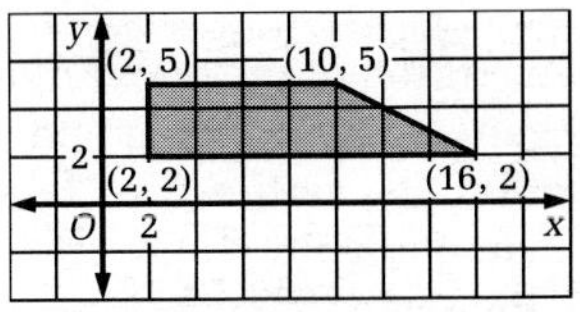

37. Answers may vary. An example is given. **Step 1:** I make sure I have my homework assignment. **Step 2:** I read over my notes if necessary. **Step 3:** I do my math homework. If I can't complete it, I reread my notes and my math text and work through sample problems. **Step 4:** I complete my homework. If there are problems I still can't do, I ask for help at home, or make a note to ask my teacher for help. **Step 5:** I put my homework, my notebook and textbook, a pencil and, if necessary, a calculator together in one place so I will be ready for the next math class.
38. $x > 6$ **39.** -3.75

Pages 77–79 Talk it Over

1. The points appear to lie in a line. As altitude increases, the boiling temperature of water decreases.

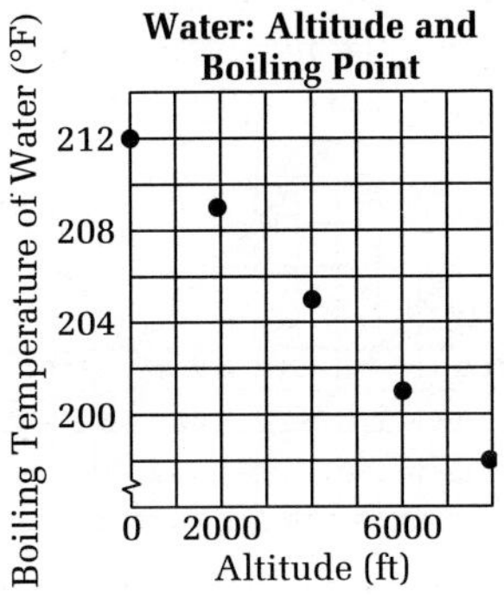

2. The graph appears to represent a function. It appears that no two points on the graph lie on the same vertical line. **3. a.** Estimates may vary. Example: about 200°F **b.** Estimates may vary. Example: about 2800 ft

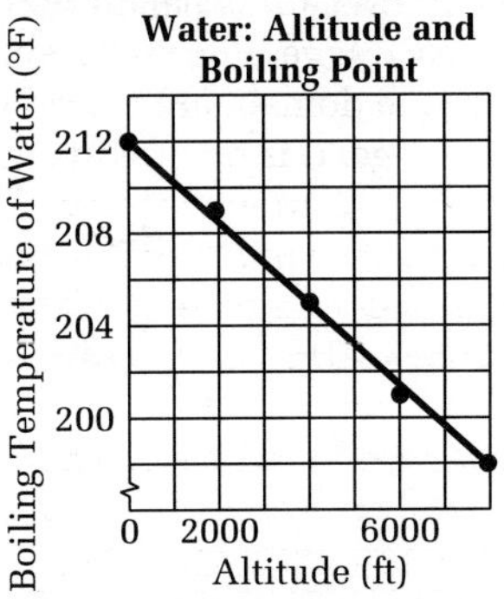

4. 212; the boiling temperature of water at an altitude of 0 ft, or sea level **5.** The coefficient of a is negative. The slope of the graph is negative, so as a increases, $T(a)$ decreases. As the values in the second column (altitudes) increase, the values in the third column (temperatures) decrease. **6.** Cathie used the form of a linear function with the calculated value of m and substituted the a and $T(a)$ values from one of the pairs in the table. She solved the resulting equation for b. **7.** Answers may vary. An example is given. For Kimberley and Colorado Springs, the value of m is $\frac{201-205}{6012-4013} = -\frac{4}{1999} \approx -0.002$ and the equation is $T(a) = -0.002a + 212$. The equation will vary only slightly, since for any two points in the table, the value of m rounded to three decimal places is -0.002. **8.** Yes; the equation is written in the form $f(x) = mx + b$ with $m = \frac{1}{2}$ and $b = -13$. **9.** Yes; the equation is written in the form $f(x) = mx + b$ with $m = 4$ and $b = 0$. **10.** No; the equation cannot be written in the form $f(x) = mx + b$. **11.** Substitute 10,000 for a; $T(10{,}000) = 212 + (-0.0018)(10{,}000) = 194$; 194°. **12.** Substitute 185 for $T(a)$; $185 = 212 + (-0.0018)a$; $a = 15{,}000$; 15,000 ft.

Pages 81–84 Exercises and Problems

1. Its graph is a line. Its equation can be written in the form $y = mx + b$. **3.** Yes. **5.** Yes. **7.** No. **9.** Yes; 24, -16, $4 + 10a + 10b$ **11.** Yes; 7, 1, $\frac{3}{2}a + \frac{3}{2}b + 4$ **17.** $f(x) = -5x - 6$ **19.** $f(x) = \frac{1}{2}x - 1$ **21. a.** $C(T) = 4T - 156$ **b.** 204; the number of times per minute a snowy tree cricket chirps at 90°F **c.** $39 \le T \le 105$; $0 \le C(T) \le 264$ **27. a.** $g(y) + g(z) = 3y + 2 + 3z + 2 = 3y + 3z + 4$; $g(y + z) = 3(y + z) + 2 = 3y + 3z + 2$ **b.** There are values of y and z for which $g(y) \cdot g(z) = g(y \cdot z)$. However, the statement is not true in general. Consider $y = 1$ and $z = 2$: $g(1) \cdot g(2) = (3(1) + 2)(3(2) + 2) = 40$; $g(1 \cdot 2) = g(2) = 3(2) + 2 = 8$ **30.** 120; 0; $8a^3 - 2a$ **31.** When a linear programming problem is represented by a system of linear inequalities in x and y, any maximum or minimum value of a linear combination $ax + by$ will occur at one of the vertices of the feasible region. **32.** $-10 \le x < 5$ **33.** $x \ge 0$ **34.** $x < 2$

Pages 85–88 Talk it Over

1. A person's heart rate decreases steadily over the first 20 years of life, then remains constant after the age of 20. **2.** Yes; no two points have the same first coordinate and different second coordinates. **3.** No; the first part of the graph clearly has an equation of the form $f(x) = mx + b$, where m is negative and b is positive. The second part of the graph clearly has an equation of the form $g(x) = c$, for a constant c.
4. 104 beats/min; 82 beats/min; 72 beats/min
5. a. $90 = -2a + 112$ **b.** 11; Etenia is 11 years old.
6. a.

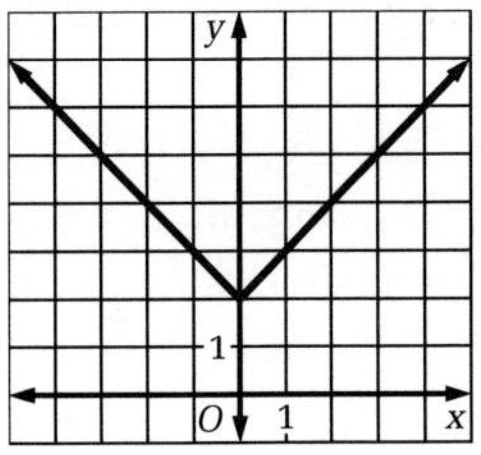

The graphs have the same shape; the graph of $y = |x| + 2$ is the graph of $y = |x + 2|$ shifted 2 units to the right and up 2 units. **b.** all x greater than or equal to 0

7. 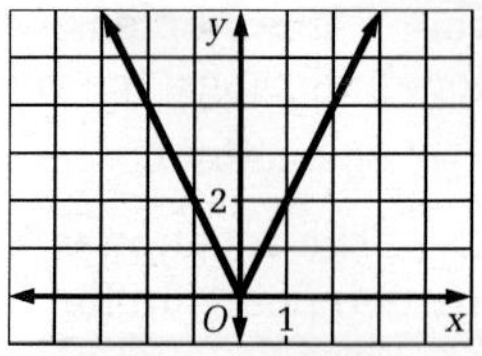

The graphs are identical. Each of the two sections of the graph of $y = |2x|$ and $y = 2|x|$ has slope twice that of the corresponding section of the graph of $y = |x|$.

8. 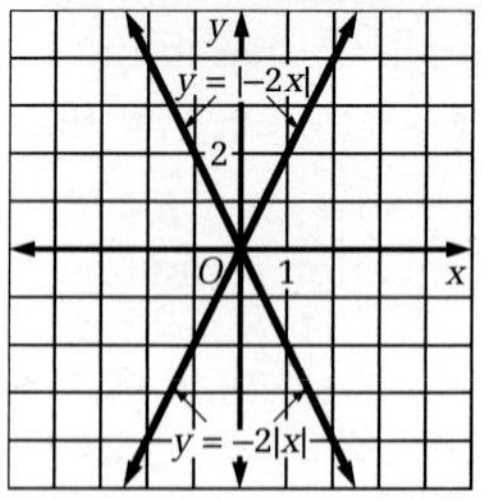

Each graph is the image of the other reflected across the x-axis. Each of the two sections of the graph of $y = |-2x|$ has slope twice that of the corresponding section of the graph of $y = |x|$. Each of the two sections of the graph of $y = -2|x|$ has slope twice that of the corresponding section of the graph of $y = -|x|$, which is the graph of $y = |x|$ reflected across the x-axis.

Pages 89–92 Exercises and Problems

1. Heart rate is a piecewise function of age. If a person's age is 20 or less, the heart rate is given by $h(a) = -2a + 112$. If his or her age is greater than 20, the heart rate is given by $h(a) = 72$.

3. $y = \begin{cases} -\frac{3}{2}x + 3 & \text{if } 0 \le x \le 2 \\ x - 4 & \text{if } x > 2 \end{cases}$

5. 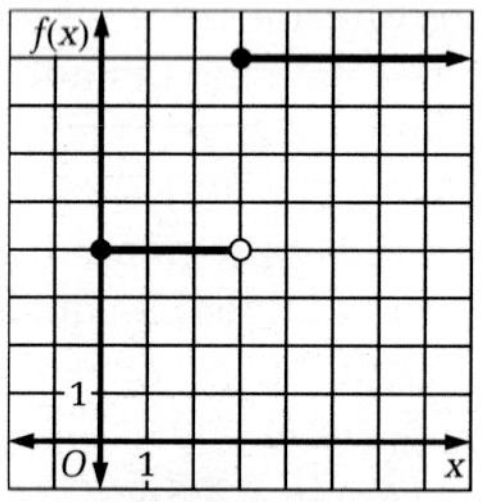

7. 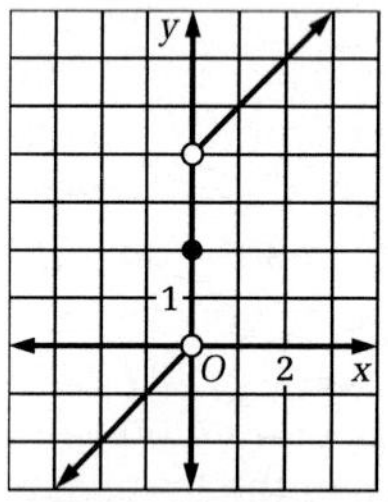

13. 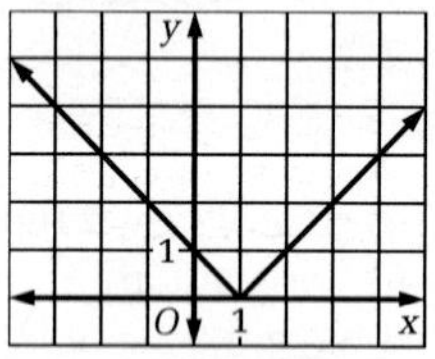

15. 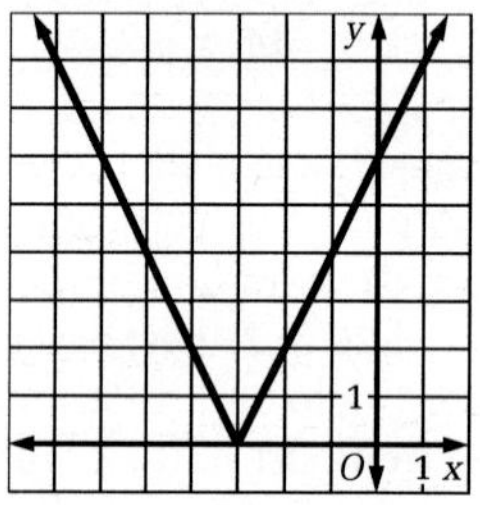

17. 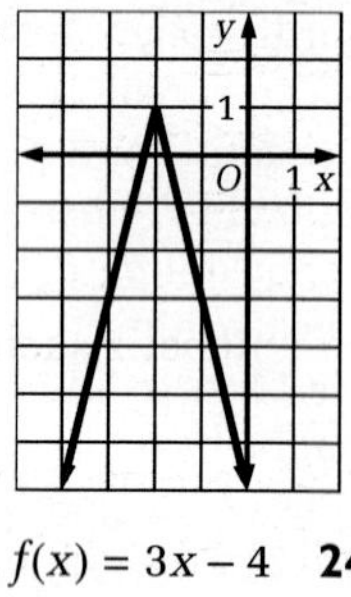

23. $f(x) = 3x - 4$ **24.** $f(x) = -\frac{1}{2}x + 5$ **25.** kite **26.** rectangle **27.** parallelogram **28.** about −0.38; about −2.6 **29.** about 1.6; about −1.1

Page 93 Checkpoint

1. If any vertical line drawn through the graph intersects the graph in more than one point, the graph does not represent a function. If a vertical line intersects the graph in more than one point, there are two points on the graph with the same first coordinate and different second coordinates. Then the graph does not represent a function. **2.** Yes. The domain consists of the numbers 12, 20, and 32. The range consists of the numbers 0.75, 1, and 1.5. **3. a.** $V(r) = 3\pi r^2$ **b.** about 59 units^3; about $462x^2$ units^3 **4.** many-to-one; For every real number x except 0, $x \neq -x$ but $f(x) = f(-x)$. **5. a.** Let n be the number of movies ordered and $C(n)$ the cost in dollars; $C(n) = 4.99n + 24.28$. **b.** \$49.23 **c.** 4 movies **d.** He will be able to order 3 movies per month. (If he carries over the \$2.75 he has left to the next month, he can order 4 movies every other month.) **6.** $f(x) = -\frac{1}{3}x + 5$ **7.** Let w be the weight of the chicken in pounds and $c(w)$ be the cost in dollars;

$$c(w) = \begin{cases} 2.5w & \text{if } 0 < w \le 5 \\ 2w & \text{if } w > 5 \end{cases}$$

8. 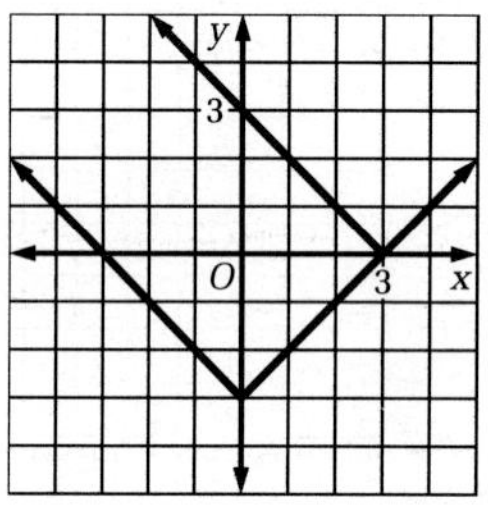

The graphs are translations of the graph of $y = |x|$. The graph of $y = |x - 3|$ is a translation 3 units to the right; the graph of $y = |x| - 3$ is a translation 3 units down.

Pages 94–97 Talk it Over

1. a. 4 ft; 12 ft; 20 ft **b.** No. If d were a linear function of t, the clam would have fallen the same distance in every 0.5-second period. **c.** Yes; the graph is a curve, not a line.

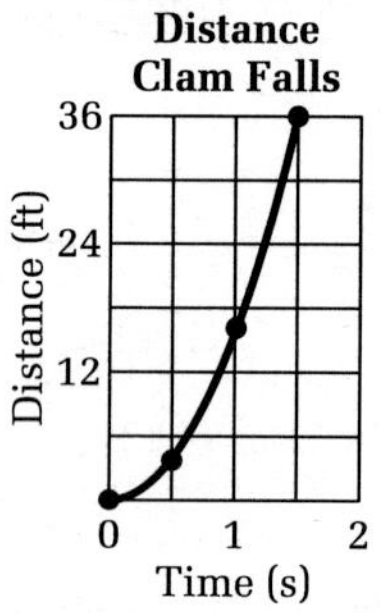

2. a. $d(t) = kt^2$ **b.** Using the point (1, 16) gives $d(1) = 16$; $k \cdot 1^2 = 16$; $k = 16$. **3.** Answers may vary. An example is given. I like using a graphics calculator because it is simpler than using the quadratic formula. **4.** after about 1.4 s; about 38.6 ft **5.** No; it shows the height of the ball as a function of time and does not give any information about the angle of the ball with respect to the horizontal. **6.** No; every height but the maximum is paired with two times.

Pages 97–100 Exercises and Problems

1. 36 ft

3. a. quadratic; 25; 7; $2a^2 + 4ab + 2b^2 + 3a + 3b - 2$

b.

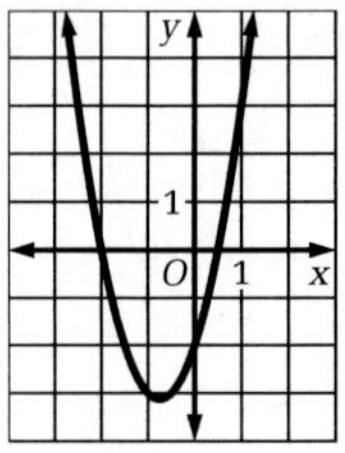

5. a. quadratic; −58; −58; $5 - 7a^2 - 14ab - 7b^2$

b.

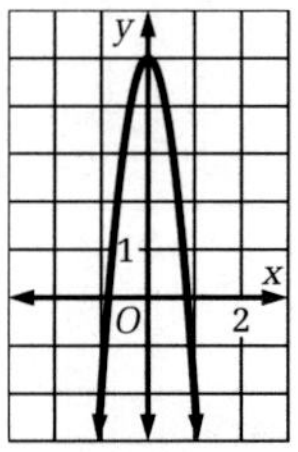

7. a. quadratic; 1; 49; $a^2 + 2ab - 8a + b^2 - 8b + 16$

b.

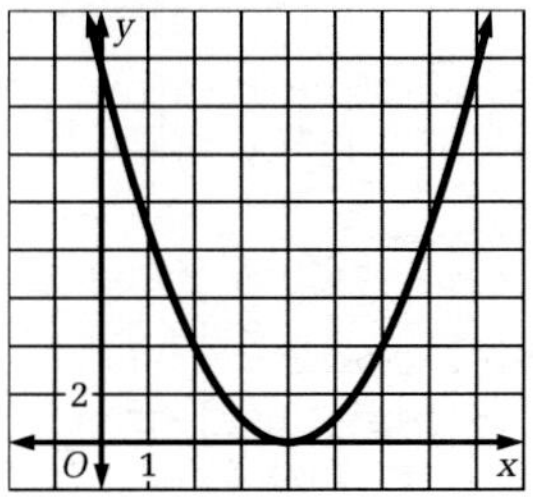

18. 80.3; 82.5; none

19.

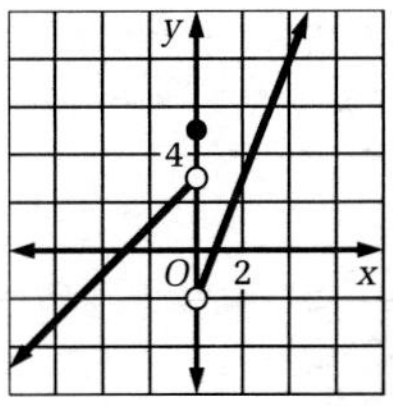

20.

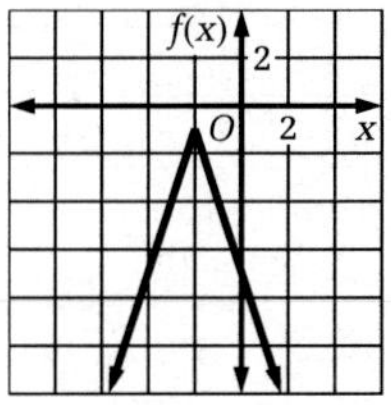

21. $(x + 6)(x - 4)$ **22.** $(3a - 5)(2a + 3)$
23. $25x(x + 2)(x - 2)$

Pages 102–105 Talk it Over

1. Each function has the form $f(x) = kx^n$; the degrees of the equations (from left to right) are 1, 2, 3, 4, and 5. **2.** $y = x$; $y = x^3$; $y = x^5$ **3.** $y = x$, $y = x^3$, and $y = x^5$ have rotational symmetry of 180° about the origin; $y = x^2$ and $y = x^4$ have symmetry about the y-axis. **4.** They are one-to-one and have 180° rotational symmetry about the origin. Their domain and range are all real numbers. **5.** They are many-to-one and are symmetric about the y-axis. Their domain is all real numbers, and their range is all nonnegative real numbers. **6. a.** Yes; 1; linear. **b.** 6 **7. a.** Yes; 2; quadratic. **b.** 0, 3 **8.** Answers may vary. An example is given. You could graph the function and see that there is no value for which $P(x) = -1$. **9.** the points in the first quadrant **10.** Draw the graph of $y = 40$ on the same axes and determine the x-coordinate of the point where the graphs intersect. ($x = 2$)

Pages 105–109 Exercises and Problems

3. No. **5.** $f(x) = 0.8x^3$ **7.** $f(x) = \frac{1}{2}x^4$ **9.** $w(d) = 0.41d^3$ **13.** Yes. **15.** No; in a polynomial function, the exponents are positive whole numbers. **17. a.** 2; polynomial function of degree 2 and quadratic function **b.** 10; $1\frac{3}{4}$ **c.** 0, −3 **19.** 4; 1 **21.** 4; 2

23. a. 1, 3

b.

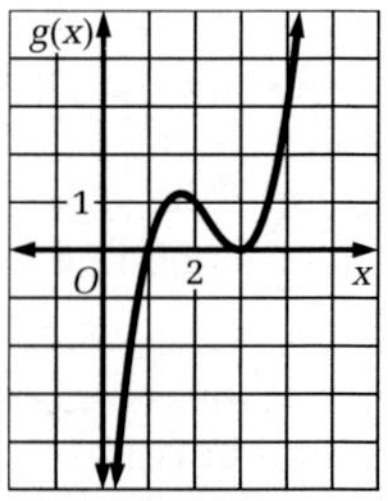

25. a. 0, 2

b.

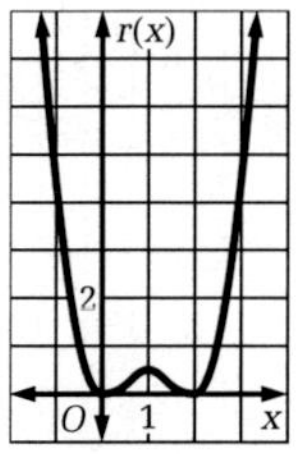

30. Let t be the time in seconds after the fall and $h(t)$ the height in feet of the pebble; $h(t) = -16t^2 + 75$.
31. No. **32.** Yes. **33.** Yes. **34.** True.

Pages 110–114 Talk it Over

1. a.

Depth d (ft)	Speed $s(d)$ (mi/h)
100	38.6
200	54.6
346	71.8
5000	272.9
10,000	386
20,000	545.9
34,600	718

b. The speed at a depth of 10,000 ft is 10 times that at 100 ft. The speed at a depth of 34,600 ft is 10 times that at 346 ft. **2. a.** $s(100d) = 3.86\sqrt{100d} = 38.6\sqrt{d}$ **b.** The speed is 10 times greater at a depth 100 times as deep. **c.** Yes. **3.** The graphs do not intersect the x-axis at $x = -3$. **4.** $\sqrt{x + 7}$ is not a real number if $x < -7$. **5.** The domain of the function is all possible values of the control variable. Jane knows that $x + 1$ must be nonnegative for $\sqrt{x + 1}$ to be a real number. **6.** The first function's domain is limited by the fact that for $x < -1$, $\sqrt{x + 1}$ is not a real number. The range is limited by the fact that $\sqrt{x}$ is nonnegative for every real number x. For the second function, $\sqrt[3]{x + 1}$ is a real number for every real number x, and for every real number y, there is a real number z such that $y^3 = z$.

Pages 115–117 Exercises and Problems

1. 6.3 **3.** 5 **5.** 1.2 **7.** Table values may vary. The functions are the same. $\sqrt{25x} = \sqrt{25} \cdot \sqrt{x} = 5\sqrt{x}$
9. False. The statement is only true if $a = 0$ or $b = 0$.

11. 0, $\frac{1}{4}$ **13.** 4, 3 **15.** −2, −1 **19.** $8|x|$ **21.** $7p$
23. 2 **25.** $5\sqrt{3}|x|$

27. 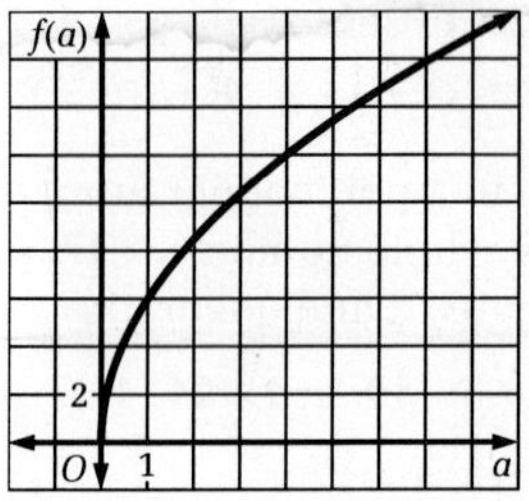

domain: all real numbers greater than or equal to 0; range: all real numbers greater than or equal to 0

29. 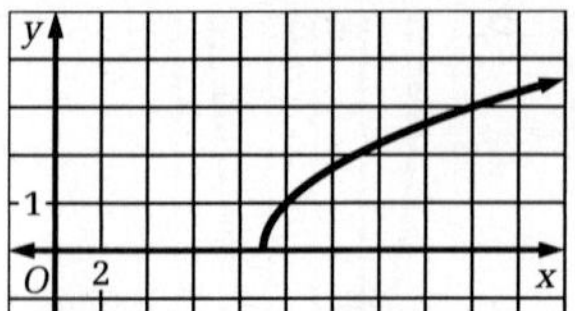

domain: all real numbers greater than or equal to 9; range: all real numbers greater than or equal to 0

31. 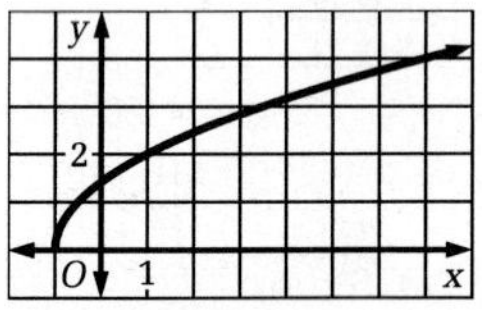

domain: all real numbers greater than or equal to −1; range: all real numbers greater than or equal to 0

33. all real numbers less than or equal to 0 **35.** all real numbers; all real numbers **37.** all real numbers greater than or equal to 0; all real numbers less than or equal to 1 **39.** all real numbers; all real numbers

45. a. 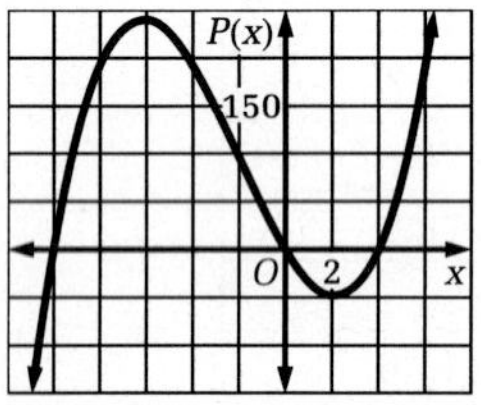

b. −10, 0, 4

46. lower extreme: 0.171; lower quartile: 0.190; median: 0.224; upper quartile: 0.2735; upper extreme: 0.325 **47.** $x = 1$; If $x = 1$, $1 - x = 0$ and $\frac{1}{1-x}$ is not defined.

Page 118 Checkpoint

1. Summaries may vary. An example is given. The graphs of quadratic functions and square root functions are both curves. The graph of a quadratic function is a parabola. The graph of a square root function is a curve that resembles half of a parabola. **2. a.** Let t be the time in seconds and $h(t)$ the height in feet. $h(t) = -16t^2 + 60$

b. 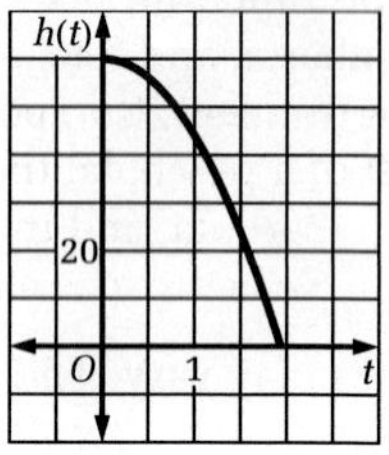

3. a. $h(t) \approx -16t^2 + 3.14t + 4$ **b.** about 0.5 s

4. a. −162; $64b^3 - 144b^2 + 72b$

b. 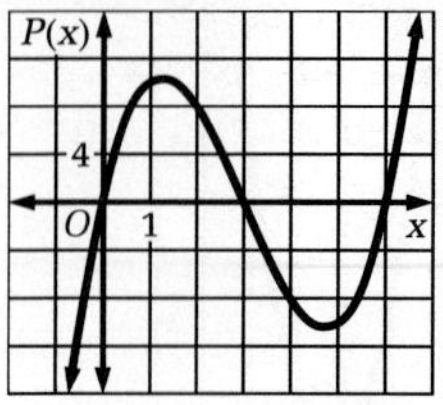

5. about 2267.57 **6.** 1 **7.** 3

8. 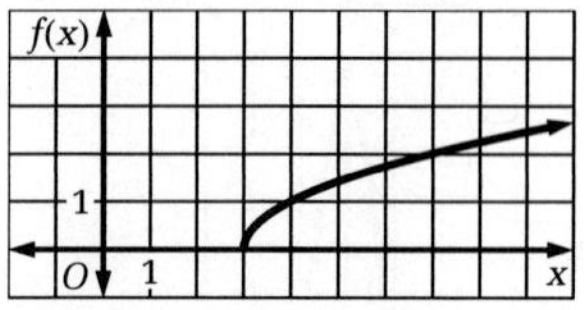

domain: all real numbers greater than or equal to 3; range: all real numbers greater than or equal to 0

9. all real numbers; all real numbers

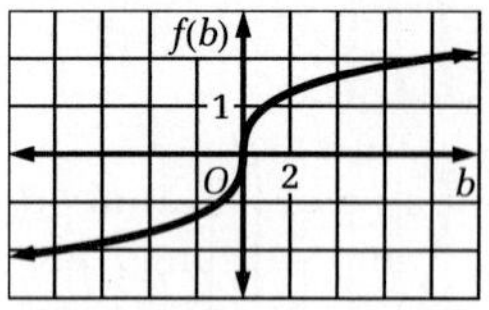

Pages 119–123 Talk it Over

1.

Speed s (miles per hour)	Time t (hours)
25	12
40	7.5
55	5.45
65	4.62

2. a. distance = speed × time or $d = st$ **b.** $t = \frac{d}{s}$; the rate of speed **c.** In a polynomial function, exponents of the control variable are whole numbers in $t = ds^{-1}$, the exponent of s is a negative number. **3.** to produce a quadratic equation in standard form **4.** You get a fraction with denominator 0. **5.** The average is the sum divided by 2, so the sum is $2 \cdot 63$. **6.** The function becomes $a(t) = \frac{74 + 100t}{t + 1}$. **7.** No fraction with numerator 1 is equal to 0. **8.** Note that $\frac{1}{1-x}$ is not defined when $1 - x = 0$, that is, when $x = 1$. **9.** $y = 0$ **10.** A vertical asymptote $x = k$ indicates that the domain of the function is all real numbers except k. A horizontal asymptote $y = n$ indicates that the range is all real numbers except n.

Pages 124–125 Exercises and Problems

1. to show that the graph may not cross those lines, they are not part of the graph **3.** $\frac{11}{3}$ **5.** 4 **7.** 9, −1 **9.** 3 tests

11. a.

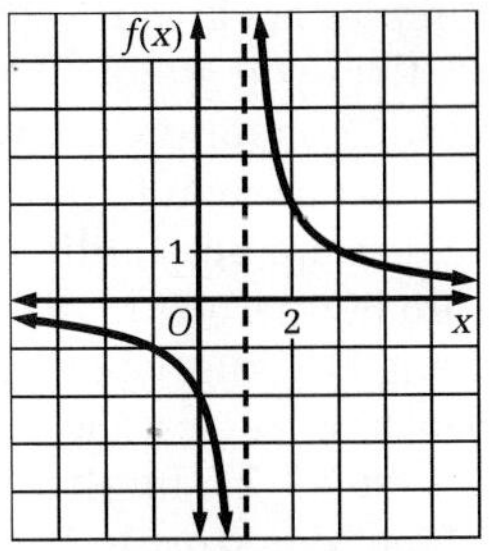

b. $x = 1$; $y = 0$
c. domain: all real numbers except 1; range: all real numbers except 0

13. a.

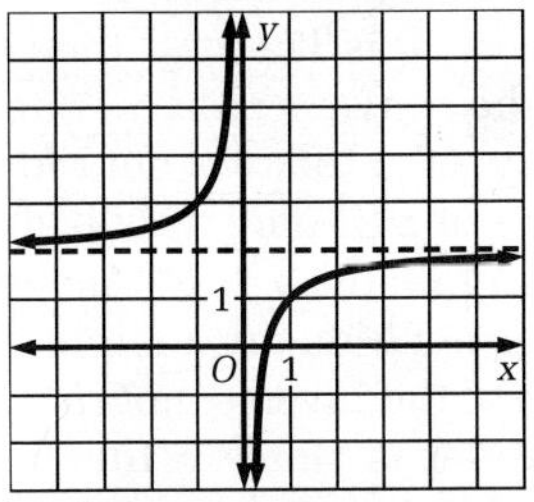

b. $x = 0$; $y = 2$
c. domain: all real numbers except 0; range: all real numbers except 2

15. a.

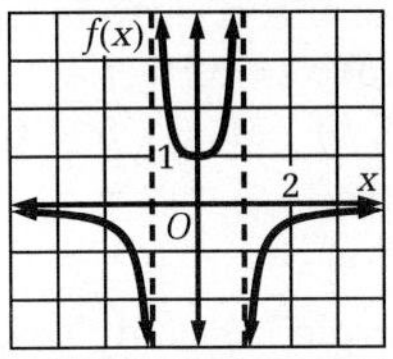

b. $x = 1$, $x = -1$, $y = 0$
c. domain: all real numbers except 1 and –1; range: all real numbers greater than or equal to 1 or less than 0

17. a.

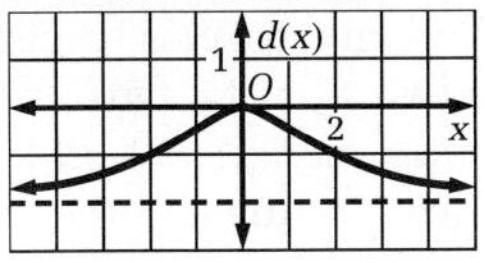

b. $y = -2$ **c.** domain: all real numbers; range: all real numbers greater than –2 and less than or equal to 0

23. 5, 4 **24.** 1 **25.** $f(x) = \begin{cases} 3x - 12 & \text{if } x \geq 4 \\ -3x + 12 & \text{if } x < 4 \end{cases}$

26. $a + 1$

Pages 126–128 Talk it Over

1. $s(A) = 1000A$ **2.** $p(s) = 0.12s$ **3.** $p(A) = 120A$ **4.** 240 W **5.** The domain of $(h \circ k)(x)$ is those numbers in the domain of $h(x)$ that are also in the range of $k(x)$. **6.** Sandra found the value of $k\left(\frac{1}{3}\right)$ and substituted that value in the equation for h. She found $k\left(\frac{1}{3}\right) = 2$, then $h(2) = 6$. **7. a.** $(f \circ g)(x) = 4x^2 - 4x$ **b.** $(g \circ f)(x) = 3 - 2x^2$ **c.** No; the given functions $f(x)$ and $g(x)$ provide a counterexample, since $4x^2 - 4x \neq 3 - 2x^2$ for all x. **8.** the number of bottles and the cost; the number of people and the number of bottles; the number of people and the cost; $b(p)$ and $(C \circ b)(p)$ **9.** Answers may vary. An example is given. I disagree. The range of the number of bottles function is even numbers greater than 10. The domain of the cost function is positive whole numbers. However, every number in the range of the number of bottle functions is in the domain of the cost function.

Pages 128–131 Exercises and Problems

1. $\sqrt{2x}$ **3.** $12x^2 - 2$ **5.** 20 **7.** f of g of x; the composite of f and g **15.** $x^2 - 3x + 2$ **17.** $\frac{x-4}{2}$, $x \neq 1$ **20.** domain: all real numbers except 2; range: all real numbers except 0 **21.** 2; $4a^2 - 6a + 2$ **22.** SSS, ASA, AAS, SAS

Pages 133–134 Unit 2 Review and Assessment

1. Yes; for every number in the domain, there is only one value in the range. **2. a.** $C(g) = 1.17g$ **b.** $0 \leq g \leq 12$; $0 \leq C(g) \leq 14.04$ **3.** 1; $\frac{2}{a - b + 4}$ **4.** $f(x) = \frac{1}{9}x + 4$ **5.** $f(x) = -\frac{3}{2}x + 6$

6. $f(x) = \begin{cases} -3 & \text{if } -4 < x \leq -1 \\ 2 & \text{if } -1 < x < 2 \\ x + 1 & \text{if } x \geq 2 \end{cases}$

7. a. $h(t) = -16t^2 + 4$ **b.** about 3 ft **c.** after about 0.35 s **8.** 0; –20 **9.** $V = \frac{\pi}{6}d^3$ **10. a.** 4 **b.** –9 **11.** $2\sqrt{3}|a|$ **12.** 1 **13. a.** $-\frac{10}{19}$ **b.** about –1.93; about 52.93 **14.** No; she would need to score 46 points. It is unlikely she could do so, with an average of 22 points per game. **15.** $(f \circ g)(x) = \frac{3}{x}$

16. a.

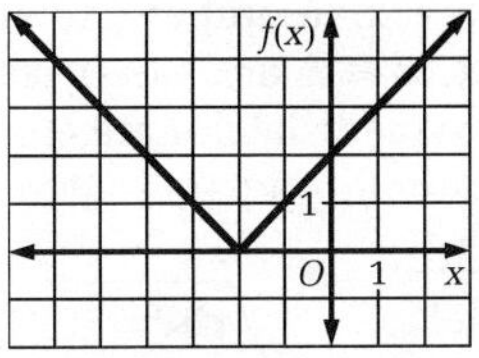

domain: all real numbers; range: all real numbers greater than or equal to 0

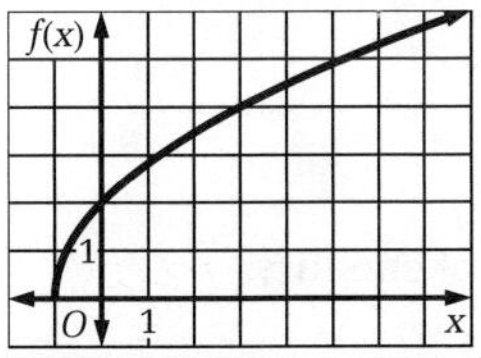

domain: all real numbers greater than or equal to –1; range: all real numbers greater than or equal to 0

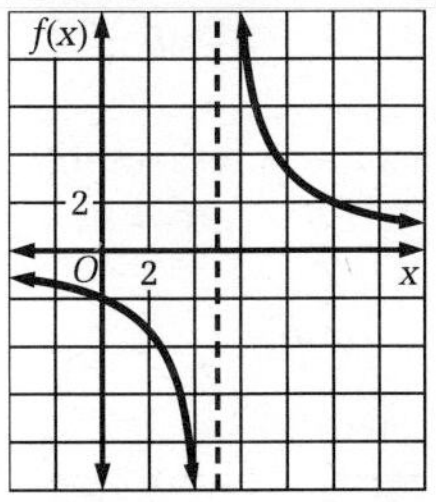

domain: all real numbers except 5; range: all real numbers except 0

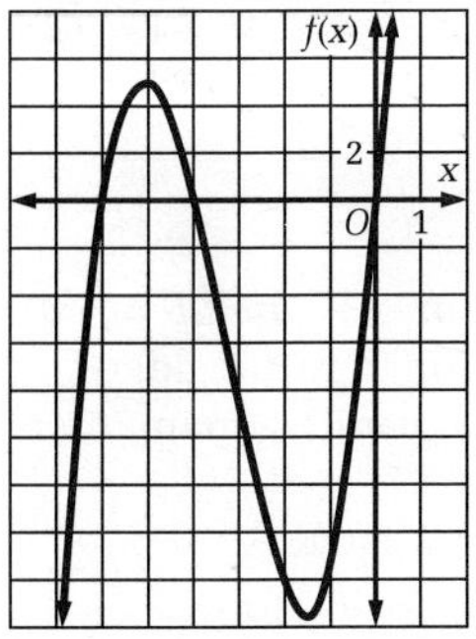

domain: all real numbers; range: all real numbers

b. Answers may vary. Examples are given. The first consists of two rays, the second is half of a parabola, the third is a hyperbola, and the fourth is an S-shaped curve.

17. Answers may vary. Examples are given. Every direct variation function is a polynomial function. No absolute value function is a direct variation function. Some rational functions are polynomial functions.

Pages 140–141 Talk it Over

1. If the diagonals of a quadrilateral bisect each other, then the quadrilateral is a parallelogram; True. Since vertical angles are equal in measure, two pairs of congruent triangles are formed (SAS). If two lines are intersected by a transversal and alternate interior angles are equal in measure, the lines are parallel. This fact can be used with the pairs of congruent triangles to show that both pairs of opposite sides of the quadrilateral are parallel. **2.** If the diagonals of a quadrilateral are equal in length, then the quadrilateral is a rectangle; False. The diagonals must also bisect each other. **3.** Steps 1–5 **4.** Steps 1 and 6–9

Pages 143–146 Exercises and Problems

5. a. Yes. **b.** SSS **7. a.** No. **9. a.** If two parallel lines are intersected by a transversal, then alternate interior angles are equal in measure. **b.** True.
11. ❶ definition of parallelogram; ❷ If two ∥ lines are intersected by a transversal, then alternate interior ∠s are equal in measure; ❸ Reflexive property of equality; ❹ ASA; ❺ Corres. parts of ≅ △s are = in measure.
13. a. If opposite sides of a quadrilateral are equal in measure, then the quadrilateral is a parallelogram.
b. *Given:* Quadrilateral *ABCD* with $AB = CD$ and $BC = AD$ *Prove: ABCD* is a parallelogram.

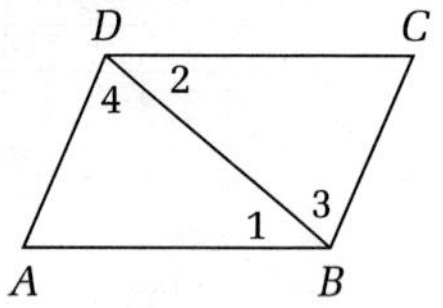

Statements (Justifications)
1. $AB = CD$ and $BC = AD$ (Given)
2. $DB = DB$ (Reflexive property of equality)
3. $\triangle ABD \cong \triangle CDB$ (SSS)
4. $m\angle 1 = m\angle 2$; $m\angle 3 = m\angle 4$ (Corres. parts of ≅ △s are = in measure.
5. $\overline{AB} \parallel \overline{CD}$ and $\overline{BC} \parallel \overline{AD}$ (If two lines are intersected by a transversal and alternate interior ∠s are = in measure, then the lines are ∥.)
6. *ABCD* is a parallelogram. (Definition of parallelogram)

15. *Given:* Parallelogram *ABCD* is a rectangle. *Prove:* $AC = BD$

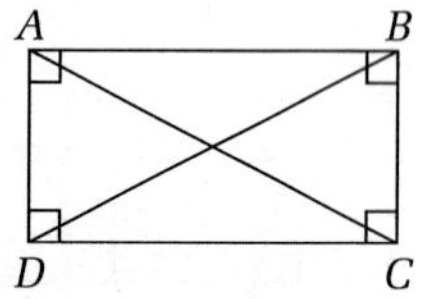

Statements (Justifications)
1. *ABCD* is a rectangle. (Given)
2. *ABCD* is a parallelogram and $m\angle A = m\angle B = m\angle C = m\angle D = 90°$. (Definition of rectangle)
3. $AD = BC$ (If a quadrilateral is a parallelogram, then opposite sides are = in measure.)
4. $AB = AB$ (Reflexive property of equality)
5. $\triangle DAB \cong \triangle CBA$ (SAS [Steps 2, 3, and 4])
6. $AC = BD$ (Corres. parts of ≅ △s are = in measure.)

17. Answers may vary. Examples are given. **a.** If you are in Queensland, then you are in Australia. **b.** If you are not in Australia, then you are not in Queensland. **19.** Converse: If two lines form right angles, then the lines are perpendicular; True. Inverse: If two lines are not perpendicular, then the lines do not form right angles; True. Contrapositive: If two lines do not form right angles, then the lines are not perpendicular; True. **27.** Implication: $p \to q$ (If today is Monday, then tomorrow is Tuesday.) Converse: $q \to p$ (If tomorrow is Tuesday, then today is Monday.) Inverse of the converse: not $q \to$ not p (If tomorrow is not Tuesday, then today is not Monday.) Contrapositive: not $q \to$ not p (If tomorrow is not Tuesday, then today is not Monday.) **29.** Inverse: not $p \to$ not q (If today is not Monday, then tomorrow is not Tuesday.) Inverse of the inverse: not (not p) $\to$ not (not q) or $p \to q$ (If today is Monday, then tomorrow is Tuesday.) The inverse of the inverse is the original implication. **31.** $(f \circ g)(x) = x^2 + 1$
32. $(-4, 6)$ **33.** $(3, 0)$ **34.** $(-a, b)$ **35.** $3\sqrt{5} \approx 6.7$

Page 147 Talk it Over

1. Answers may vary. An example is given. Yes. I think *ADEC* should be called an isosceles trapezoid because it appears to have two sides that are equal in measure, just like an isosceles triangle. **2.** $E(2, 3.5)$; $A(-5, 0)$ **3.** Use the distance formula and the coordinates of *A*, *C*, *D*, and *E*. **4.** Use the midpoint formula and the coordinates of *A* and *B* to find the coordinates of the midpoint of $\overline{AB}$ to determine if they are the same as the coordinates of *D* (they are not). Repeat for $\overline{CB}$ and *E*. $\overline{DC}$ and $\overline{AE}$ are not medians.

Pages 150–152 Exercises and Problems

3. $\overline{DC}$ has endpoints $(-2, 3.5)$ and $(5, 0)$ and length $\sqrt{(5-(2))^2 + (0-3.5)^2} = \sqrt{61.25}$. *AE* has endpoints $(2, 3.5)$ and $(-5, 0)$ and length $\sqrt{(2-(-5))^2 + (3.5-0)^2} = \sqrt{61.25}$. The supports are the same length. **5.** ❶ Division property of equality; ❷ Definition of median; ❸ Substitution property; ❹ Base ∠s of an isosceles triangle are = in measure; ❺ Reflexive property of equality; ❻ SAS; ❼ Corres. parts of ≅ △s are = in measure. **7.** One side has endpoints $(-9, 56)$ and $(-34, 0)$ and length $\sqrt{(-9-(-34))^2 + (56-0)^2} = \sqrt{3761}$. The other side has endpoints $(9, 56)$ and $(34, 0)$ and length $\sqrt{(9-34)^2 + (56-0)^2} = \sqrt{3761}$. The sides are the same length. **9.** It would make it difficult, if not impossible, to get into the swing.
11. a. $PS = \sqrt{(-b-(-a))^2 + (c-0)^2} = \sqrt{(-b+a)^2 + c^2}$ $= \sqrt{(a-b)^2 + c^2}$; $QR = \sqrt{(b-a)^2 + (c-0)^2} = \sqrt{(b-a)^2 + c^2}$; since $(a-b)^2 = (b-a)^2$, $PS = QR$.

b. $PR = \sqrt{(b-(-a))^2 + (c-0)^2} = \sqrt{(b+a)^2 + c^2}$;
$QS = \sqrt{(-b-a)^2 + (c-0)^2} = \sqrt{(-b-a)^2 + c^2}$; since $(b+a)^2 = (-b-a)^2$, $PR = QS$.
c. Form of proof may vary. An example is given.

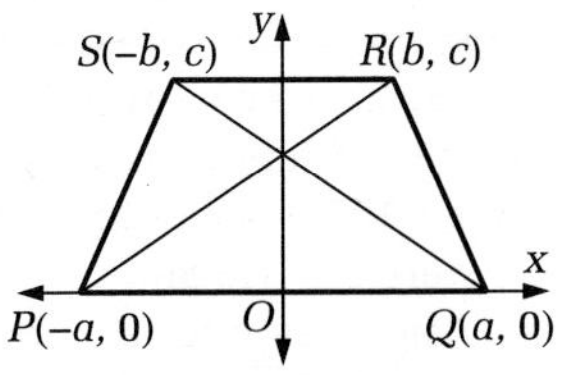

Given: $PQRS$ is an isosceles trapezoid. The y-axis is a line of symmetry of $PQRS$, passing through the midpoints of its bases.
Prove: $m\angle SPQ = m\angle RQP$ and $m\angle PSR = m\angle QRS$

Statements (Justifications)

1. $PQRS$ is an isosceles trapezoid. (Given)
2. $PS = QR$ (Definition of isosceles trapezoid)
3. $PR = QS$ (In an isosceles trapezoid, the diagonals are = in measure.)
4. $SR = SR$ and $PQ = PQ$ (Reflexive property of equality)
5. $\triangle SPQ = \triangle RQP$ and $\triangle QRS = \triangle PSR$ (SSS)
6. $m\angle SPQ = m\angle PQR$ and $m\angle QRS = m\angle RSP$ (Corres. parts of $\cong$ ⧌ are = in measure.)

13.

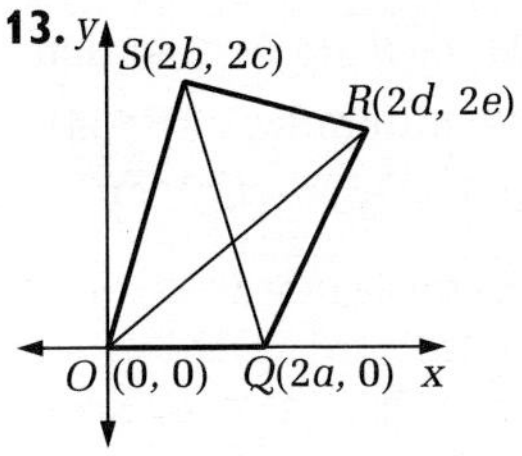

a. $\left(\frac{2d+0}{2}, \frac{2e+0}{2}\right) = (d, e)$
b. $\left(\frac{2a+2b}{2}, \frac{2c}{2}\right) = (a+b, c)$
c. $a + b$; c

d.

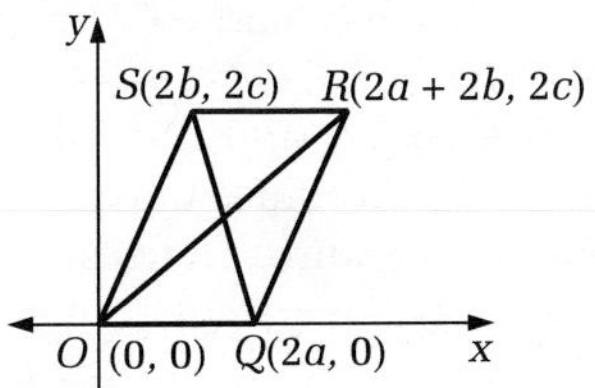

The coordinates of R are $(2a + 2b, 2c)$. $\overline{OQ}$ is on the x-axis, so the slope of $\overline{OQ}$ is 0. The slope of $\overline{RS}$ is $\frac{2c-2c}{2a+2b-2b} = \frac{0}{2a} = 0$, so $\overline{OQ} \parallel \overline{RS}$. The slope of $\overline{SO}$ is $\frac{2c-0}{2b-0} = \frac{c}{b}$. The slope of $\overline{RQ}$ is $\frac{2c-0}{2a+2b-2b} = \frac{c}{b}$. Then $\overline{SO} \parallel \overline{RQ}$ and $OQRS$ is a parallelogram. **17.** If a quadrilateral is not a rectangle, then it is not a square. **18. a.** All squares are rhombuses. **b.** All parallelograms are quadrilaterals. **c.** All rectangles are parallelograms. **19.** Answers may vary. An example is given. A quadrilateral is a parallelogram if and only if both pairs of opposite sides are parallel.

Pages 153–154 Talk it Over

1–3. Answers may vary. Examples are given.
1. Among the terms that may need to be defined in order to interpret the city ordinance are: *commercial establishment, meals,* and *premises.* **2.** A commercial establishment could be defined as any place of business that offers goods or services for sale. The bookstore is a commercial establishment. A meal could be defined as any food consumed in one sitting or as a substantial serving of food as opposed to a snack or light refreshment. Premises could be defined as land owned by a business and the buildings on the land, or as the building occupied by a business. **3.** If a meal is defined as any food consumed in one sitting and premises are defined as land owned by a business and the buildings on the land, then the bookstore manager must apply for a license. If a meal is defined as a substantial serving of food as opposed to a snack or light refreshment or if premises are defined as only the building occupied by a business, then the owner does not have to apply for a license.
4. Answers may vary. **5–7.** Answers may vary. Examples are given. **5.** Yes; $MNRS$ fits the given definition of trapezoid. **6.** Yes; the definition does not exclude the possibility of two pairs of parallel sides. **7.** disagree; I don't think a quadrilateral can be both a parallelogram and a trapezoid.
8. True. **9.** True. **10.** True. **11.** False.

Pages 156–158 Exercises and Problems

3. exactly one **5.** at least one **7. a.** I, II, III, V **b.** I, III **11.** True. **13.** False. **15.** In a rectangle, either pair of opposite sides may be considered the bases. A rectangle is a parallelogram, so each pair of opposite sides is equal in measure. The second property was proved in Ex. 15 on page 144. (If a parallelogram is a rectangle, then its diagonals are equal in measure.) All four angles of a rectangle are equal in measure, so the two angles at each base are equal in measure. **17.** Answers may vary. An example is given. A quadrilateral is an isosceles trapezoid if and only if it has at least one pair of parallel sides and a line of symmetry that passes through the midpoints of the bases. **21.** $PQ = \sqrt{(-1-0)^2 + (-2-0)^2} = \sqrt{5}$;
$QR = \sqrt{(2-0)^2 + (1-0)^2} = \sqrt{5}$;
$RS = \sqrt{(1-2)^2 + (-1-1)^2} = \sqrt{5}$;
$PS = \sqrt{(1-(-1))^2 + (-1-(-2))^2} = \sqrt{5}$;
Since $PQ = QR = RS = PS$, $PQRS$ is a rhombus.
22. Answers may vary. An example is given.
Step 1: Determine the author's last name. **Step 2:** Look in the card catalog under the author's last name and find the reference number. **Step 3:** If you know the general location of the book, go to Step 5.
Step 4: Find the general location of the book. If necessary, ask the librarian. **Step 5:** Use the reference number to locate the book.

23.

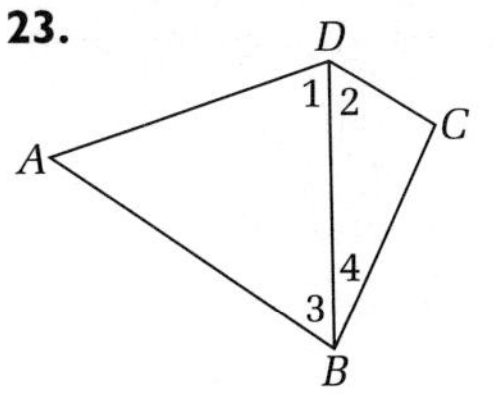

a. 180°; 180°; $m\angle 2$; $m\angle ABC$

b. $m\angle A + m\angle ABC + m\angle C + m\angle CDA = m\angle A + (m\angle 3 + m\angle 4) + m\angle C + (m\angle 1 + m\angle 2)$ (Substitution property) $= (m\angle A + m\angle 1 + m\angle 3) + (m\angle C + m\angle 2 + m\angle 4)$ (Associative and commutative properties of addition) $= 180° + 180°$ (Substitution property) $= 360°$ **c.** The sum of the measures of the angles of a quadrilateral is 360°.

Page 158 Checkpoint

1. In the inverse of an implication, the "if" and the "then" are in the same order as in the original implication. In the contrapositive, they are in reverse order. In both cases, the "if" and the "then" are negated. **2.** If a quadrilateral is not a rectangle, then its diagonals are not equal in measure; False.

3. $FD = \sqrt{(0 - (-6))^2 + (3 - 0)^2} = 3\sqrt{5}$;
$FE = \sqrt{(0 - 5)^2 + (3 - 0)^2} = \sqrt{34}$;
$ED = \sqrt{(5 - (-6))^2 + (0 - 0)^2} = 11$. $\triangle DEF$ is not isosceles, since no two sides are equal in measure. $\triangle DEF$ has no line of symmetry, so $\triangle DEF$ is not isosceles.

4. $JK = \sqrt{(-4 - 0)^2 + (5 - (-2))^2} = \sqrt{65}$;
$LK = \sqrt{(4 - 0)^2 + (5 - (-2))^2} = \sqrt{65}$; Since two sides of $\triangle JKL$ are equal in measure, $\triangle JKL$ is isosceles. $\triangle JKL$ is an isosceles triangle because it has a line of symmetry, the y-axis. **5.** True. **6.** False.

Pages 159–161 Talk it Over

1. 360°; 540°; 720°; 900°

2.

Name of polygon	Number of sides, n	Sum of angle measures, $S(n)$
Triangle	3	180° = (180°)
Quadrilateral	4	360° = 2(180°)
Pentagon	5	540° = 3(180°)
Hexagon	6	720° = 4(180°)
Heptagon	7	900° = 5(180°)
Octagon	8	1080° = 6(180°)

3. $S(n) = (n - 2)(180°)$ Answers may vary. An example is given. Every time another side is added, another triangle can be formed, so 180° is added to the sum of the angle measures. For example, the formula is true for octagons, $S(8) = (8 - 2)(180°) = 6(180°) = 1080°$. When a ninth side is added, 180° is added and the sum of the measures of the angles is $1260° = 7(180°) = (9 - 2)(180°)$. **4.** $(k - 2)180° + 180° = (k - 2)180° + (1)180° = (k - 2 + 1)180°$ **5.** Answers may vary. An example is given. I know the formula is true for $n = 3$. Since I can show that whenever it is true for n, it is true for $n + 1$, I am convinced that it is true for all values of n greater than 3.

Pages 162–164 Exercises and Problems

1. Yes. All the sides of a square are equal in measure, and all four angles are right angles and thus equal in measure. So a square is a regular polygon. **3. a.** Yes; a square is a rectangle that is a regular polygon. **b.** Yes; a square is a rhombus that is a regular polygon. **c.** Yes; a square is a trapezoid that is a regular polygon. **5.** about 128.6°; about 51.4° **7.** 140°; 40° **9.** about 147.3°; about 32.7° **11. a.** At each vertex, the interior angle and one exterior angle form a linear pair. Angles that form a linear pair are supplementary. **b.** $n(180°)$ **c.** $(n - 2)180° + E(n) = n(180°)$; $E(n) = n(180°) - (n - 2)180° = 2(180°) = 360°$

13. a. $X(n) = \frac{360°}{n}$; $I(n) = \frac{(n-2)180°}{n}$ **b.** rational

c.

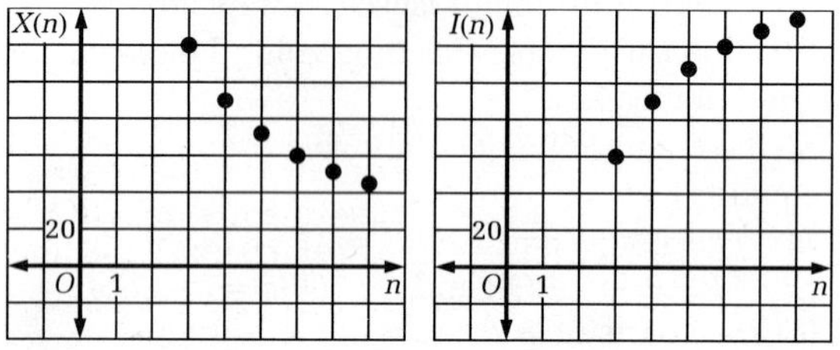

18. True. **19.** True. **20.** 4 **21.** b ($\overleftrightarrow{RS}$ is perpendicular to $\overline{PQ}$ and passes through its midpoint); c ($\overleftrightarrow{RS}$ is perpendicular to $\overline{PQ}$ and its intersection with $\overline{PQ}$ forms two congruent triangles (HL), so $\overleftrightarrow{RS}$ bisects $\overline{PQ}$.)

Page 167 Talk it Over

1. $\triangle ODE$, $\triangle OEF$, $\triangle OFG$, and $\triangle OGH$, and $\triangle OHD$ ($\triangle OED$, $\triangle OFE$, $\triangle OGF$, $\triangle OHG$, and $\triangle ODH$ are all congruent to the five given triangles as well.) All ten triangles are congruent by SSS. The sides that are radii of the circle are equal in measure. The bases that are sides of the regular polygon are equal in measure as well. **2.** 72°

Pages 168–171 Exercises and Problems

1.

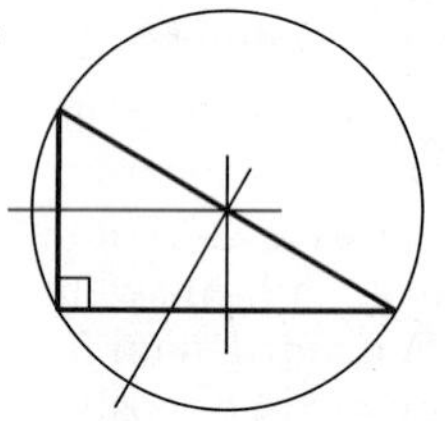

3. agree; The statement is true because every triangle has a circumscribed circle, the center of which is equidistant from all three vertices.

5. a. (0, 4); (3, 1) **b.** $y = 4$ **c.** $y = -3x + 10$ **d.** (2, 4) **e.** $2\sqrt{5} \approx 4.5$; $(x - 2)^2 + (y - 4)^2 = 20$ **f.** $y = x + 2$; $4 = 2 + 2$, so the perpendicular bisector of $\overline{ST}$ passes through H. **7.** 45° **9. a.** 21,600 nautical miles **b.** about 24,840 statute miles; about 7907 statute miles **13. a.** 120° **b.** 120° **c.** 30° **d.** 120° **19.** about 152.3° **20.** Corresponding angles are equal in measure and corresponding sides are in proportion. **21.** 43°; 86°; 51°

Pages 172–173 Talk it Over

1. A; C **2.** Yes; let X and Y be the endpoints of the stage. Consider a circle of which $\overline{XY}$ is a chord and B a point on the circle. The points with the same viewing angle are on $\overset{\frown}{XBY}$. **3.** $\overset{\frown}{AB}$ **4.** a semicircle **5.** They are inscribed angles that intercept the same arc, so each has measure $\frac{1}{2}\overset{\frown}{KL}$.

Pages 175–178 Exercises and Problems

1. The vertex of an inscribed angle of a circle is on the circle, so the sides of an inscribed angle are chords of the circle. The vertex of a central angle is at the center of the circle, so the sides of a central angle are radii of the circle. **3.** $m\,\overset{\frown}{KF} = 180°$; $m\,\overset{\frown}{MF} = 130°$; $m\,\angle K = 65°$; $m\,\angle F = 25°$

7.

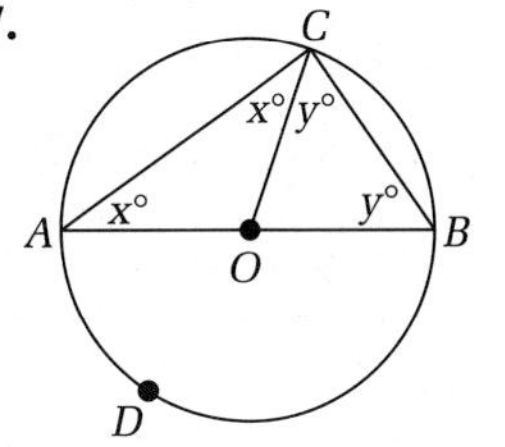

In $\triangle ABC$, $2x + 2y = 180°$. In $\triangle BOC$, $2y + m\,\angle BOC = 180°$. So $m\,\angle BOC = 2x$. Consequently, $m\,\angle BOC = 2m\,\angle BAC$. Since the measure of an arc is equal to the measure of its central angle, $m\,\overset{\frown}{BC} = m\,\angle BOC = 2m\,\angle BAC$. Therefore, $m\,\angle BAC = \frac{1}{2}m\,\overset{\frown}{BC}$.

11. The slope of $\overline{PR}$ is $\frac{y-0}{x-(-r)} = \frac{y}{x+r}$. The slope of $\overline{QR}$ is $\frac{y-0}{x-r} = \frac{y}{x-r}$. Since $x^2 - r^2 = -y^2$, the product of the slopes is $\left(\frac{y}{x+r}\right)\left(\frac{y}{x-r}\right) = \frac{y^2}{x^2-r^2} = \frac{y^2}{-y^2} = -1$. Then $\overline{PR} \perp \overline{QR}$ and $\angle PRQ$ is a right angle. **17.** 45° **18.** Tanya eats taco salad for lunch. **19.** Akio used a black pen or a blue pen to fill out the application. **20.** The odd integers can be written $2k + 1$ and $2n + 1$ for integers k and n. Their sum is $2k + 1 + 2n + 1 = 2k + 2n + 2 = 2(k + n + 1)$, which is even.

Page 178 Checkpoint

1. a regular 19-gon; a regular 18-gon; The measure of an interior angle of a regular n-gon is $\frac{(n-2)180°}{n} = 180° - \frac{360°}{n}$, which increases as n increases. The measure of an exterior angle of a regular n-gon is $\frac{360°}{n}$, which decreases as n increases.
2. Construct the perpendicular bisectors of two of the sides. The point where the bisectors intersect is the center of the circumscribed circle. The radius is the distance from the center to any vertex of the triangle.
3. $m\,\overset{\frown}{AC} = 150°$; $m\,\angle A = 45°$; $m\,\angle B = 75°$; $m\,\angle C = 60°$
4. $m\,\overset{\frown}{DGE} = 180°$; $m\,\overset{\frown}{DF} = 90°$; $m\,\angle D = 45°$; $m\,\angle E = 45°$
5. $m\,\angle I = 70°$; $m\,\overset{\frown}{JI} = 150°$; $m\,\overset{\frown}{HJ} = 140°$; $m\,\overset{\frown}{HI} = 70°$

Pages 179–182 Talk it Over

1. 62 squares; 32 black squares, 30 red squares
2. 31 dominoes **3.** red **4.** 31 black; 31 red **5.** No; there are only 30 red squares to be covered. There are 32 black squares. **6.** the point of tangency **7.** C

Pages 183–186 Exercises and Problems

1. Sample 1: Line j is not parallel to line k; Sample 2: All three numbers in a Pythagorean triple are odd; Sample 3: $\overline{OP}$ is perpendicular to line t. **3.** Suppose quadrilateral $ABCD$ has four acute angles. Then $m\,\angle A < 90°$, $m\,\angle B < 90°$, $m\,\angle C < 90°$, and $m\,\angle D < 90°$. Then $m\,\angle A + m\,\angle B + m\,\angle C + m\,\angle D < 360°$. This contradicts the fact that the sum of the measures of the interior angles of a quadrilateral is 360°. So the assumption that a quadrilateral can have four acute angles must be false. Therefore, no quadrilateral can have four acute angles. **7.** Suppose there is a Pythagorean triple with two even numbers and one odd number. There are two possibilities. One, either a or b is odd, and two, c is odd. Suppose a is odd and b and c are even. Then a^2 is odd and b^2 is even, so $a^2 + b^2 = c^2$ is odd, since it is the sum of an even number and an odd number. This contradicts the assumption that c is even. Suppose a and b are even and c is odd. Then a^2 and b^2 are even, so $a^2 + b^2 = c^2$ is even, since it is the sum of two even numbers. This contradicts the assumption that c is odd. The assumption that there is a Pythagorean triple with two even numbers and one odd number must be false. Therefore, there is no Pythagorean triple with two even numbers and one odd number. **11.** Since $\overline{OA} \perp m$, $\angle A$ is a right angle. Since $\triangle OAB$ is isosceles, its base angles are equal in measure and $m\,\angle B = 90°$. Then since $m\,\angle AOB > 0$, $m\,\angle AOB + m\,\angle A + m\,\angle B > 180°$. This contradicts the triangle sum theorem. The assumption that m is not tangent to the circle must be false. Therefore, m is tangent to circle O.
19. Figures may vary. An example is given.

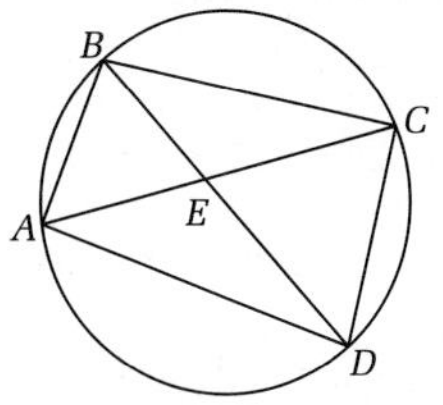

$\triangle ABE \sim \triangle DCE$ and $\triangle BEC \sim \triangle AED$. Both triangles can be proved to be similar using Theorem 3.9 (If two inscribed angles in the same circle intercept the same arc, then they are equal in measure) and AA. $\angle BAC$ and $\angle CDB$ both intercept $\overset{\frown}{BC}$, so they are equal in measure and $\angle ABD$ and $\angle DCA$ both intercept $\overset{\frown}{AD}$, so they are equal in measure. Then $\triangle ABE \sim \triangle DCE$ by AA. $\angle CBD$ and $\angle DAC$ both intercept $\overset{\frown}{CD}$, so they are equal in measure and $\angle BCA$ and $\angle ADB$ both intercept $\overset{\frown}{AB}$, so they are equal in measure. Then $\triangle BEC \sim \triangle AED$ by AA. **20. a.** $-\frac{1}{3}$ **b.** 1, −2
21. circumference: 9π m, or about 28.3 m; area: 20.25π m^2, or about 63.6 m^2

Pages 187–189 Talk it Over
1. 45°; all arcs whose endpoints are consecutive points on the circle O **2.** Answers may vary. An example is given. $OR = OR$ (The measure of a segment is equal to itself.) $OF = OE$ (All radii of a circle are equal in measure.) $FR = FE$ (If two tangent segments are drawn from the same point to the same circle, then they are equal in measure.) Then $\triangle OER \cong \triangle OFR$ by SSS. **3.** $22\frac{1}{2}°$; If $\overline{OG}, \overline{OH}, \overline{OI}, \overline{OJ}, \overline{OK}$, and $\overline{OD}$ were drawn, the sixteen central angles produced would all have measure $22\frac{1}{2}°$. **4.** $\triangle EOR \cong \triangle FOR$, so $FR = ER$ and $m\angle ORF = m\angle ORE$. (Corresponding parts of congruent triangles are equal in measure.) $OR = OR$ (Reflexive property of equality) Then $\triangle SOR \cong \triangle ROQ$ by ASA. **5.** Using the same methods as in questions 1–4, you can prove that $\triangle SOR \cong \triangle ROQ \cong \triangle QOP \cong \triangle POW \cong \triangle WOV \cong \triangle VOU \cong \triangle UOT \cong \triangle TOS$. Then the sides of $PQRSTUVW$ are corresponding parts of congruent triangles and are equal in measure. Using the same property along with addition of angle measures, you can show that all the angles of $PQRSTUVW$ are equal in measure. **6.** about 3.312 m **7.** The octagon consists of eight congruent triangles. Let p be the perimeter of the octagon. Each triangle has height h (which is also the radius of the circle) and base $\frac{p}{8}$. The area of one triangle is $\frac{1}{2}h\left(\frac{p}{8}\right) = h \cdot \frac{p}{16}$. The area of the octagon is $8 \cdot h\frac{p}{16} = h \cdot \frac{p}{2}$. So the area of the octagon is equal to the radius times one half of the perimeter, which is the semiperimeter. **8.** The area of the circle is about 3.14 m^2; the area of the circle is slightly less than that of the octagon. **9. a.** $s = \pi r$ **b.** $A = sr$ or $\frac{s^2}{\pi}$

Pages 189–191 Exercises and Problems
1. A semiperimeter of a figure is half the perimeter. **3.** 36°; 54° **5.** Method 1: $A = rs = 9\left(\frac{25 + 63 + 52}{2}\right) = 630$ ft^2; Method 2: Draw $\overline{OA}, \overline{OB}$, and $\overline{OC}$; A = area of $\triangle AOB$ + area of $\triangle AOC$ + area of $\triangle BOC = \frac{1}{2} \cdot 9 \cdot 25 + \frac{1}{2} \cdot 9 \cdot 63 + \frac{1}{2} \cdot 9 \cdot 52 = 630$ ft^2 **13.** Suppose that $m\angle MOA = 25°$. Then the figure is an n-gon where $\frac{180}{n} = 25$ and $n = 7.2$. This contradicts the fact that n must be a whole number. The assumption that $m\angle MOA = 25°$ must be false. Therefore, $m\angle MOA$ can never be 25°. **14.** cylinder; cone; pentagonal prism **15.** Yes; –6; –2; 14

Pages 192–195 Talk it Over
1. two hexagons, one pentagon **2.** two 120° angles, one 108° angle **3.** 348° **4.** 12° **5.** The tessellation of hexagons has no defects; the sum of the measures of the angles at each vertex is 360°. **6.** Yes; Yes; No; No. The sum of the measures of the angles at each vertex must be less than 360°, so there can be no more than five equilateral triangles at each vertex. **7.** No; No. The sum of the measures of the angles at each vertex must be less than 360°, so there can be no more than three squares or three pentagons at each vertex. **8.** No. The measure of each angle of a regular hexagon is 120°. The sum of the measures of three such angles is 360°.

Pages 195–198 Exercises and Problems
1. a space figure with faces that are all regular polygons of at least two types, with the same number of faces of each type at each vertex

5.

	Number of Faces (F)	Number of Vertices (V)	Number of Edges (E)
tetrahedron	4	4	6
hexahedron	6	8	12
octahedron	8	6	12
dodecahedron	12	20	30
icosahedron	20	12	30

tetrahedron: $4 + 4 = 6 + 2$; hexahedron: $6 + 8 = 12 + 2$; octahedron: $8 + 6 = 12 + 2$; dodecahedron: $12 + 20 = 30 + 2$; icosahedron: $20 + 12 = 30 + 2$ **9.** 720° **11.** 720° **13.** Suppose that there is a regular polyhedron with faces that have more than five sides. That is, the faces are regular n-gons with $n \geq 6$. The measure of each angle of the n-gon is $\frac{(n-2)180°}{n} = 180° - \frac{360°}{n}$. Since $n \geq 6$, the measure of each angle is greater than or equal to 120°. There are at least three n-gons at each vertex of the polyhedron, so the sum of the measures of the angles at each vertex is greater than or equal to 360°. This contradicts the convex polyhedron postulate. The assumption that there is a regular polyhedron with faces that have more than five sides must be false. Therefore, there is no regular polyhedron with faces that have more than five sides. **15.** at least **17.** at least **19. a.** $r = 1$ m; $R = \sqrt{3}$ m $\approx$ 1.7 m **b.** The inner sphere has surface area 4π m^2 or about 12.6 m^2. The outer sphere has surface area 12π m^2 or about 37.7 m^2. The cube has surface area 24 m^2. **c.** The inner sphere has volume $\frac{4}{3}\pi$ m^3 or about 4.2 m^3. The outer sphere has volume $4\sqrt{3}\pi$ m^3 or about 21.8 m^3. The cube has volume 8 m^3. **23. a.** prism: $500\sqrt{3}$ cm$^3 \approx 866$ cm^3; cylinder: 250π cm$^3 \approx 785.4$ cm^3 **b.** Its volume is greater than that of the cylinder but less than that of the hexagonal prism. **24.** 13 **25.** $x < -4$ **26.** 2; 1.5

27. a. 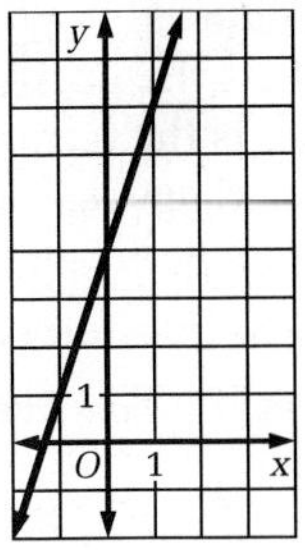

b. 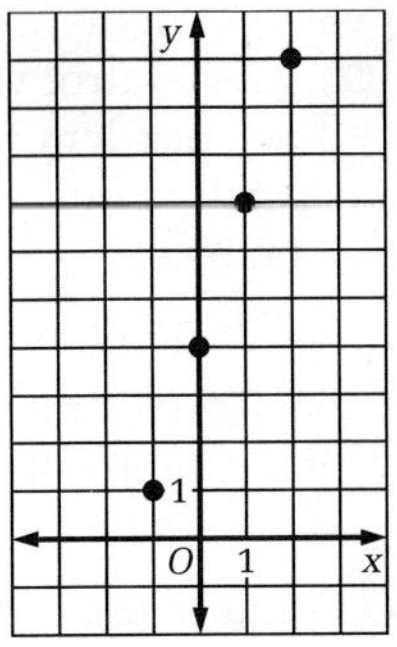

c. Yes; answers may vary. An example is given. Neither graph includes two points with the same first coordinate and different second coordinates.

Pages 200–201 Unit 3 Review and Assessment

1. a. *ABCD* is a parallelogram and $\overline{AC} \perp \overline{BD}$ (Given); $\angle DEC$ and $\angle BEC$ are right angles (Def. of ⊥ lines); $\angle DEC$ and $\angle BEC$ are equal in measure (Def. of right angles); $\overline{AC}$ and $\overline{BD}$ bisect each other (The diagonals of a parallelogram bisect each other.); $DE = BE$ (Def. of segment bisector); $CE = CE$ (Reflexive property of equality); $\triangle DEC \cong \triangle BEC$ (SAS) **b.** Answers may vary. An example is given. I would use the proof to show that $\overline{DC}$ and $\overline{BC}$ are corresponding parts of congruent triangles and are equal in measure. Since *ABCD* is a parallelogram, we already know that $\overline{DA}$ and $\overline{CB}$ are equal in measure and $\overline{AB}$ and $\overline{DC}$ are equal in measure. It would then follow that all four sides of *ABCD* are equal in measure and *ABCD* is a rhombus. **2. a.** Converse: If the diagonals of a quadrilateral are perpendicular, then the quadrilateral is a kite. Inverse: If a quadrilateral is not a kite, then its diagonals are not perpendicular. Contrapositive: If the diagonals of a quadrilateral are not perpendicular, then the quadrilateral is not a kite. **b.** The converse and the inverse are false. Counterexample: Consider a quadrilateral whose diagonals are perpendicular, but whose intersection is not the midpoint of either diagonal. The contrapositive is true.
3. a. $Q(6, -2)$; $S(4, 6)$; $U(-7, 2)$ **b.** *QRSTUV* can be formed by joining *QVUR* and *UTSR*, which are isosceles trapezoids, since both have one pair of parallel sides ($\overline{TS}$, $\overline{UR}$, and $\overline{VQ}$ are all parallel to the *x*-axis) and both have a line of symmetry (the *y*-axis).
4. There are no longer segments joining parallelograms to trapezoids or rectangles to trapezoids and isosceles trapezoids. **5.** Answers may vary. Among the terms or phrases which may need to be defined in order to interpret the by-laws are *member, in good standing, appropriate account,* and *reliable financial institution.* Sample definitions are given. A member in good standing is one who has paid dues for the last membership year and has been absent from fewer than half of all meetings (unless those absences were due to illness or for other causes found reasonable by the executive board). A reliable financial institution is one which is licensed by the state to do business and is insured by the Federal Deposit Insurance Corporation. **6.** 8 sides **7.** 12 sides **8.** 8 cm; 8 cm; 17 cm **9.** $m\,\overset{\frown}{GF} = 142°$; $m\,\overset{\frown}{EF} = 54°$; $m\angle G = 27°$; $m\angle F = 82°$ **10.** 22.5° (for example, $\angle BAC$); 45° ($\angle BAD$); 67.5° ($\angle BAE$); 90° ($\angle BAF$); 112.5° ($\angle BAG$); 135° ($\angle BAH$) **11.** Suppose that *P* is a regular *n*-gon with an exterior angle measuring 150°. *P* has *n* such exterior angles and their sum is 360°. (The sum of the exterior angles of an *n*-gon is 360°.) Then $150n = 360$ or $n = \frac{360}{150} = 2.4$. This contradicts the fact that *n* must be a whole number. So the assumption that a regular polygon can have an exterior angle measuring 150° must be false. Therefore, no regular polygon has an exterior angle measuring 150°. **12. a.** about 900 mi **b.** If a line is tangent to a circle, then it is perpendicular to the radius from the center to the point of tangency. In any right triangle, the square of the length of the hypotenuse is equal to the sum of the squares of the lengths of the legs. **c.** Let *h* be the distance from *R* to *H*; $h = \sqrt{(4000 + d)^2 - (4000)^2} = \sqrt{8000d + d^2}$. **13.** 50 in.^{2}; 100 in.^{2}; the circumscribed square **14. a.** In a regular polyhedron, only one type of regular polygon is used for the faces. **b.** 720° (There are 4 vertices with defects of 150° and one vertex with a defect of 120°.

Unit 4

Pages 209–211 Talk it Over

1. a. Each term is the cube of the position number. **b.** 1000 **c.** Method 1 **2.** Yes. Answers may vary. An example is given. The graphs pass the vertical-line test. **3.** Yes; the positive integers; ranges may vary. **4.** The graph of a sequence is discrete because the domain is a set of distinct points. **5. a.** (a) The sequence does not get close to any number. (b) 0 **b.** (a) The terms will increase steadily as you go farther out in the sequence. The points lie on the line with equation $y = 3x - 1$. (b) The graph will get closer and closer to, but never intersect, the *x*-axis as you go farther out in the sequence. The points lie on the curve with equation $y = \frac{1}{x}$.

6. 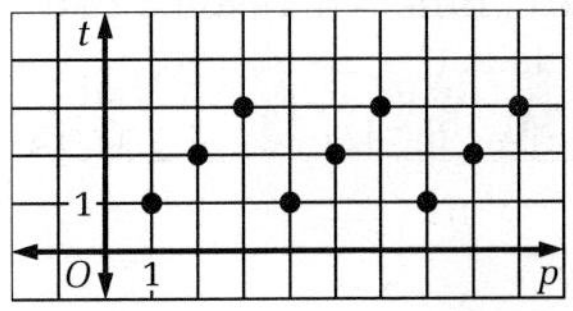

There is no single number to which the terms get close.

7. Answers may vary. An example is given. The limit of a sequence is a number that the sequence approaches but never reaches. The speed limit is a speed that the car approaches, reaches, and exceeds.

Pages 212–215 Exercises and Problems
3. a. 48, 96, 192 **b.** 6144 **c.** Answers may vary. An example is given. I noticed that each term is twice the term before it. I continued the pattern to find the 12th term. **5. a.** 360, 2160, 15,120 **b.** 1,437,004,800 **c.** Answers may vary. An example is given. I noticed that each term is the product of the term before it and the position number. I continued the pattern to find the 12th term. **7. a.** heptagon, octagon, nonagon **b.** 14-gon **c.** I noticed each term is the name of a figure with $n + 2$ sides, where n is the position number.
13. a.

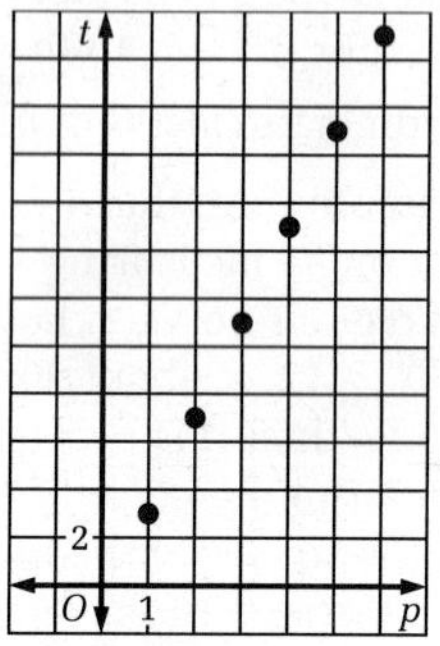

b. No; each term is 4 more than the term before it. The terms increase without limit.

15. a.

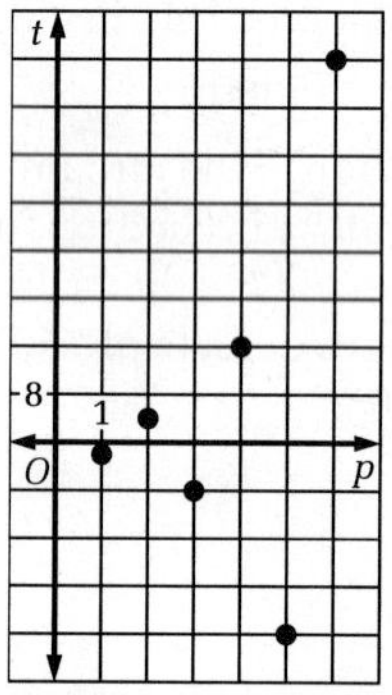

b. No; each term is –2 times the term before it. Odd-numbered terms decrease without limit, while even-numbered terms increase without limit.

17. a.

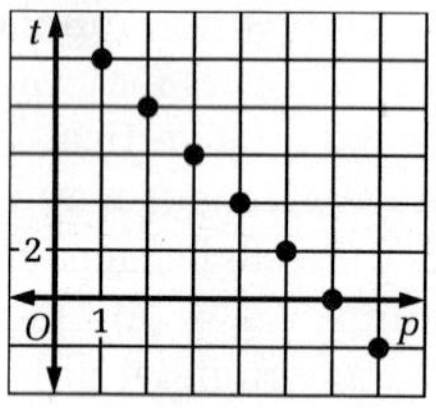

b. No; each term is 2 less than the term before it. The terms decrease without limit.

19. Yes; the terms decrease toward 0. **21.** Yes; odd-numbered terms increase toward 0, while even-numbered terms decrease toward 0. **27.** Each face of an octahedron is an equilateral triangle. Four faces meet at each vertex. Therefore, the sum of the measures of the angles at each vertex is 240°. The defect at each vertex is 360° – 240° = 120°. Since there are 6 vertices, the sum of the defects is 6 • 120° = 720°.
28. $\frac{1}{32}$ **29.** 1 **30.** $\frac{1}{343}$ **31.** –5; –17; 43

Pages 216–219 Talk it Over
1. $16\frac{2}{3}$ ft; 25 ft **2.** 48 ft; the width of the first terrace is 90 ft; that is, $a_1 = 90$. Each succeeding terrace is 7 ft narrower, so the width of the seventh terrace is 42 ft less, that is, $a_7 = 90 - 42 = 48$. **3.** eight times **4.** For each value of n, there is a single value of $90 - (n - 1)(7)$. The control variable is the position number and the dependent variable is the term. The domain is the positive integers. The range is the set of numbers given by the sequence 90, 83, 76, 69, ... **5.** $b_n = \frac{25n}{3}$ **6.** The domain of the explicit formula is the nonnegative integers rather than the positive integers. **7.** about \$1161.62

Pages 219–221 Exercises and Problems
1. when the initial value of a sequence is the value before any changes take place **3.** $0, \frac{1}{2}, 1, \frac{3}{2}; \frac{23}{2}$
5. $3, \frac{1}{2}, \frac{1}{9}, \frac{1}{36}; \frac{1}{3^{22} \cdot 24} \approx 0.000000000001328$
7. $a_n = 3^n$ **9.** $d_n = n - 3$, where n is the number of vertices of the polygon
16. a.

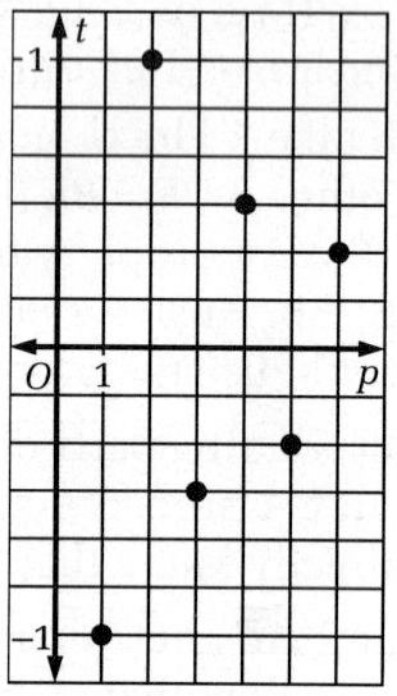

b. Yes; odd-numbered terms increase toward 0, while even-numbered terms decrease toward 0.
17. 15 ways **18.** $11 + 3i$ **19.** –6 **20.** $24 + 10i$ **21.** $4i + 8$

Pages 222–225 Talk it Over
1. Summaries may vary. A fractal tree begins with a single branch; a tree has a trunk, which can be thought of as the first single branch. For both, new branches grow from existing branches. When you magnify any part of the fractal tree, what you see looks like the whole tree. This is not true of a real tree. **2.** 1, 3, 7, ..., 63 **3.** Each term is a power of 2 added to the previous term. The power is the position number of the previous term. **4. a.** $a_n = 1 \cdot 2 \cdot 3 \cdot \ldots \cdot n = n!$ **b.** Answers may vary. An example is given. I think the recursive formula was easier to find in this case. I don't think this will always be the case. In general, I think no one method will always be easier. **5.** the sixth dose; 30 h **6.** The first term would be 1000 rather than 650. The recursion equation would be the same.

Pages 225–227 Exercises and Problems
1. the value of the first term and a recursion equation, a formula showing how to find each term from the term(s) before it. **3.** $a_1 = -3$; $a_n = -3a_{n-1}$ **5.** $a_1 = 8.32$; $a_n = a_{n-1} + 0.12$ **7.** $a_1 = \frac{1}{2}$; $a_n = (2a_{n-1})\left(\frac{n}{n+1}\right)$
9. 0.5, 0.1, 0.02, 0.004, 0.0008, 0.00016

11. $5, \frac{5}{2}, \frac{5}{6}, \frac{5}{24}, \frac{1}{24}, \frac{1}{144}$ **13.** 1, 2, 2, 4, 8, 32 **19.** $\frac{3}{6}, \frac{9}{7}, \frac{27}{8}, \frac{81}{9}; \frac{19{,}683}{14}$ **20.** $\frac{1}{2}$ **21. a.** 80, 160, 320 **b.** 2560
c. Answers may vary. An example is given. I used an explicit formula, $a_n = 5 \cdot 2^{n-1}$; $a_{10} = 5 \cdot 2^9 = 2560$.
22. a. 25, 30, 35 **b.** 50 **c.** Answers may vary. An example is given. I used an explicit formula, $a_n = 5 \cdot n$; $a_{10} = 5 \cdot 10 = 50$.

Pages 228–233 Talk it Over

1. a. $a_1 = 1$; $a_n = a_{n-1} + 7$ **b.** $a_1 = 1$; $a_n = a_{n-1} - \frac{1}{4}$ **c.** $a_1 = 2$; $a_n = a_{n-1} + 2$ **2. a.** $a_1 = 32$; $a_n = \frac{3}{4}a_{n-1}$ **b.** $a_1 = 1$; $a_n = \frac{1}{2}a_{n-1}$ **c.** $a_1 = 2$; $a_n = 2a_{n-1}$
3. Answers may vary. An example is given. The recursion equations for all the sequences in Group 1 involve adding the same number to the previous term. The recursion equations for all the sequences in Group 2 involve multiplying the previous term by the same number. The graphs of Group 1 functions are linear; the graphs of Group 2 functions are curves.
4. a. $a_n = 12 - 2.5(n - 1)$ or $a_n = 14.5 - 2.5n$ **b.** $(n - 1)$ times **5. a.** $a_n = 27\left(\frac{1}{3}\right)^{n-1}$ or $a_n = \frac{81}{3^n}$ **b.** $(n - 1)$ times
6. a.

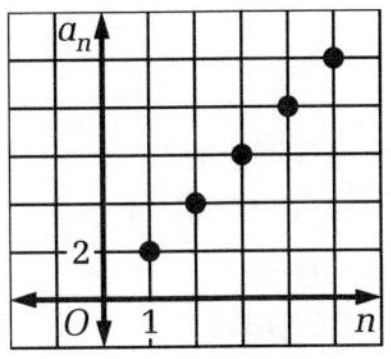
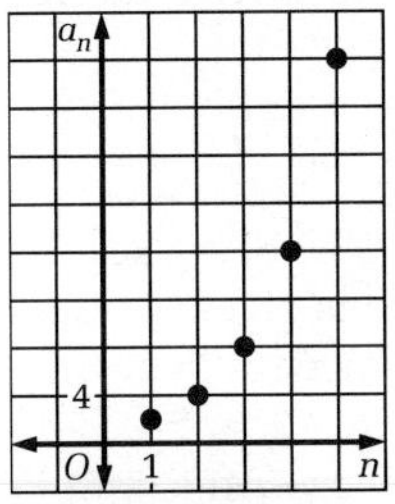

The graph of the first sequence is linear; the graph of the second is a curve.

b.

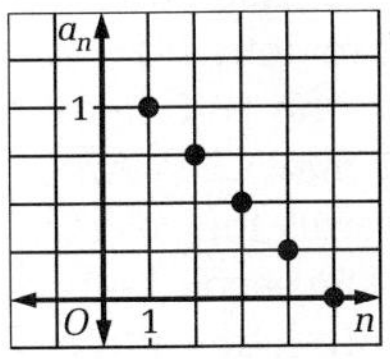
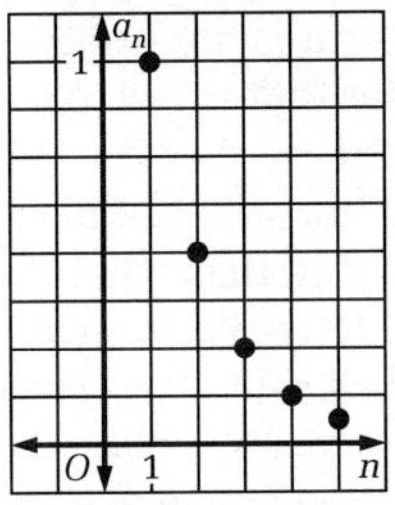

The graph of the first sequence is linear; the graph of the second is a curve.

c. The graph of an arithmetic sequence is part of a line. The graph of a geometric sequence is part of a curve. **7. a.** 2 sequences; 3, $3\sqrt{6}$, 18; 3, $-3\sqrt{6}$, 18 **b.** $\sqrt{6}$; $-\sqrt{6}$ **c.** $18\sqrt{6}$; $-18\sqrt{6}$ **8. a.** 10.5 **b.** 1 sequence; 7.5 **c.** 10.5 **d.** The arithmetic mean and the mean are the same.

Pages 233–235 Exercises and Problems

1. arithmetic **3.** neither **5.** neither **7.** arithmetic **9.** neither **11.** arithmetic **13. a.** geometric **b.** $a_1 = 64$; $a_n = \frac{1}{4}a_{n-1}$ **15. a.** arithmetic **b.** $a_1 = 5$; $a_n = a_{n-1} - 2$ **17. a.** arithmetic **b.** $a_1 = 550$; $a_n = a_{n-1} + 550$ **19.** 18 terms **21.** 33 terms

27. a. $\sqrt{35}$ or $-\sqrt{35}$ **b.** $\frac{\sqrt{35}}{5}$ or $-\frac{\sqrt{35}}{5}$ **c.** $\frac{7\sqrt{35}}{5}, \frac{49}{5}, \frac{49\sqrt{35}}{25}$ or $-\frac{7\sqrt{35}}{5}, \frac{49}{5}, -\frac{49\sqrt{35}}{25}$ **29. a.** -12 **b.** $4\sqrt{5}$ or $-4\sqrt{5}$ **31. a.** 10 **b.** 8 or -8 **33.** $8\sqrt{3} \approx 13.9$ **36.** 15, 18, 23, 30, 39, 50 **37.** $\frac{68}{7} \approx 9.7$ **38.** 42

Page 236 Checkpoint

1. a. If the inside ring is considered the first ring, the number of parallelograms in each ring is 8 times the ring number. The number of parallelograms in the rings forms the sequence 8, 16, 24, 32, ... **b.** 40 parallelograms for the fifth ring and 48 parallelograms for the sixth ring, for a total of 88 parallelograms; If the color begins to repeat, Kiyoski will need 40 golden and 48 violet parallelograms.

2. a.

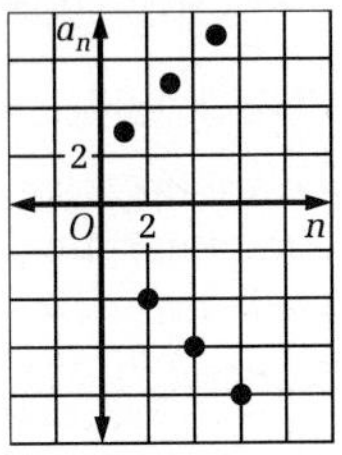

b. No.

3. a.

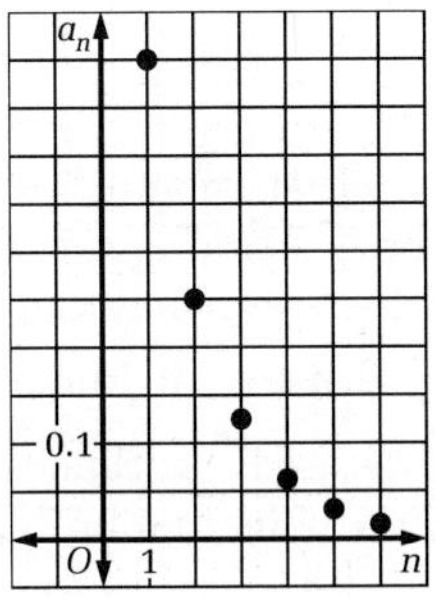

b. Yes; 0.

4. $a_n = \frac{3n-2}{4n-3}$ **5.** $a_n = \frac{3n-5}{n}$ **6.** $a_1 = 6$; $a_n = a_{n-1} + 7$ **7.** $a_1 = 1$; $a_n = a_{n-1} + 2n - 1$ **8.** geometric **9.** neither **10.** arithmetic **11.** neither

Pages 237–240 Talk it Over

1. 1, 2, 3, 4, ..., n **2.** 21 reeds **3.** It is the sum of the first six terms of the sequence. **4.** $1 + 2 + 3 + 4 + \ldots + n$ **5. a.** 101; 101; 101; 101 **b.** 50 sums; There are 100 numbers altogether and they are grouped in pairs. **c.** $\frac{100}{2} \cdot 101 = 5050$ **6. a.** Yes; Yes; Yes; Yes, except that the middle term will not be paired with another term. **7.** Answers may vary. An example is given. I think it was easier to use the formula. I think deciding which method is easier depends on several things, including the number of terms in the series and the types of numbers. For example, if the terms involved fractions or both positive and negative numbers, I think it would be easier to use the formula. **8.** The series is not arithmetic.

Pages 241–243 Exercises and Problems

1. the first term, the last term, and the number of terms **3.** 1248 **5.** 119 **13.** 144 **15.** -21
17. $\sum_{n=1}^{8}[28 + (n-1)(-5)] = \sum_{n=1}^{8}(33 - 5n)$ **19.** $\sum_{n=3}^{10}\frac{1}{2^n}$

24. $2\sqrt{3}$ or $-2\sqrt{3}$ **25.** $AB = \begin{bmatrix} 14 & 2 \\ 7.5 & 1 \end{bmatrix}$;

$BA = \begin{bmatrix} -5 & -8.5 & -6 \\ 2 & 16 & -6 \\ 2 & 1 & 4 \end{bmatrix}$ **26. a.** geometric **b.** 189

Pages 244–246 Talk it Over

1. a. 32, 16, 8, 4, 2, 1; geometric **b.** 32 + 16 + 8 + 4 + 2 + 1 **c.** 63 games **2. a.** She multiplied each term of the series by the common ratio, 2. The new series has the same terms as the original series except that it does not include the first term of the original and has a new last term, 192. When she subtracts, all but the first term of the original series and the last term of the new series cancel out, leaving $S - rS = a_1 - ra_n$. She then divides both sides by $1 - r$. Her answer is correct. **b.** $S = 32 + 16 + 8 + 4 + 2 + 1$; $\frac{1}{2}S = 16 + 8 + 4 + 2 + 1 + \frac{1}{2}$; $S - \frac{1}{2}S = 32 - \frac{1}{2}$; $S = 64 - 1 = 63$; Yes.
3. Alternating terms of the sequence are negative.
4. the first and last terms and the common ratio
5. a. Answers may vary. An example is given. 3 + 3 + 3 + 3 + 3 + 3 **b.** Since $r = 1$, $1 - r = 0$ and $\frac{a_1 - a_n r}{1 - r}$ is not defined. **c.** Since all the terms are the same, multiply the first term by the number of terms. $3 \cdot 6 = 18$

Pages 247–250 Exercises and Problems

1. the number of terms and an explicit formula for the related geometric sequence, for which you need to know the first term and the common ratio **3.** 684
5. about 28.24 **7.** 102 **9.** about 5.99
11. a. arithmetic **b.** 105 **13. a.** geometric
b. 0.134375 **15. a.** geometric **b.** 182 **22.** 935
23. $AB = BC = 5$ **24.** $EF = 3$; $DF = 3\sqrt{3}$ **25.** No; there is no single number to which the terms of the sequence get close. **26.** Yes; the odd-numbered terms decrease toward 0 and the even-numbered terms increase toward 0.

Pages 253–254 Talk it Over

1. Yes; each partial sum S_n is defined in terms of S_{n-1}.
2. a. Answers may vary. An example is given. For the sequence 1, 0, –1, –2, –3, ..., the common difference is –1 and the sequence of partial sums is 1, 1, 0, –2, –5, ... The series has no sum. (This will be true for any such example.) **b.** No. No arithmetic sequence with $d \neq 0$ will have a sum. The sequence of partial sums will increase (if $d > 0$) or decrease (if $d < 0$) without limit. **3.** $\frac{3}{1 - \left(-\frac{1}{2}\right)} = 2$; Yes. **4. a.** Answers may vary. An example is given. For the sequence 2, 4, 8, 16, ..., the common ratio is 2 and the sequence of partial sums is 2, 6, 14, 30, ... The series has no sum. (This will be true for any such example.) **b.** No. No geometric series with $|r| > 1$ will have a sum. The sequence of partial sums behaves as did the one in Sample 1(b) if $r < -1$ and increases without limit if $r > 1$.

Pages 255–258 Exercises and Problems

1. a.

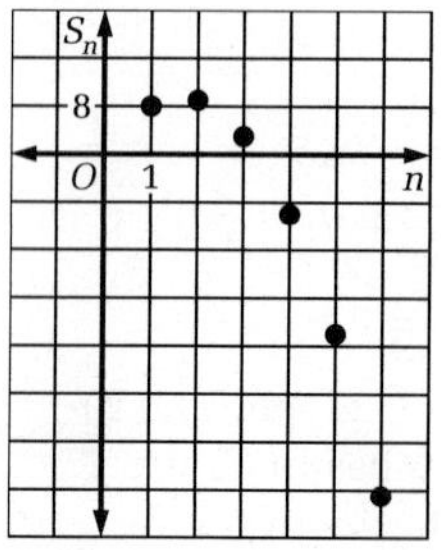

b. No.

3. a.

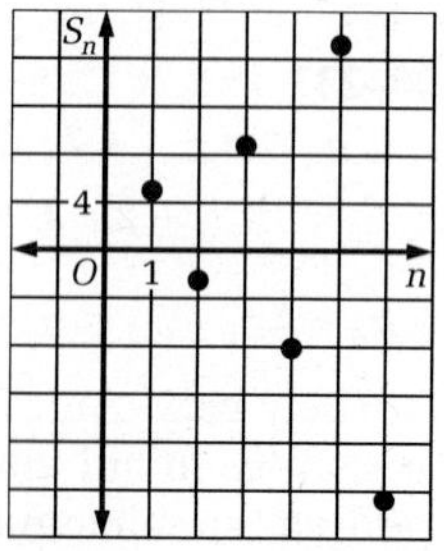

b. No.

5. Yes; 8. **7.** No. **11.** those for which $|r| < 1$
13. Yes; 40.5. **15.** Yes; –2. **17.** No. **19.** 66 **21.** no sum **23.** 12.5 **29.** 555.55555 **30.** 51 **31.** 8 cm; The longest chord of a circle is a diameter. The length of a diameter is twice the length of a radius.
32. a. No vertical line intersects the graph in more than one point. **b.** all real numbers; all positive real numbers

Pages 260–261 Unit 4 Review and Assessment

1. a. 121, 364, 1093 **b.** 29,524 **c.** Answers may vary. An example is given. I noticed that the terms increase by powers of 3. I continued the pattern to find the 10th term. **2. a.** 78, 158, 318 **b.** 2558
c. Answers may vary. An example is given. I noticed that the second term is 5 more than the first, and each increase after that is twice the previous increase. I continued the pattern to find the 10th term.
3. a. AE, AF, AG **b.** AJ **c.** Answers may vary. An example is given. I noticed that the terms follow a pattern. The first letter is always A. The second letters are in alphabetical order. The tenth letter of the alphabet is J.

4. a.

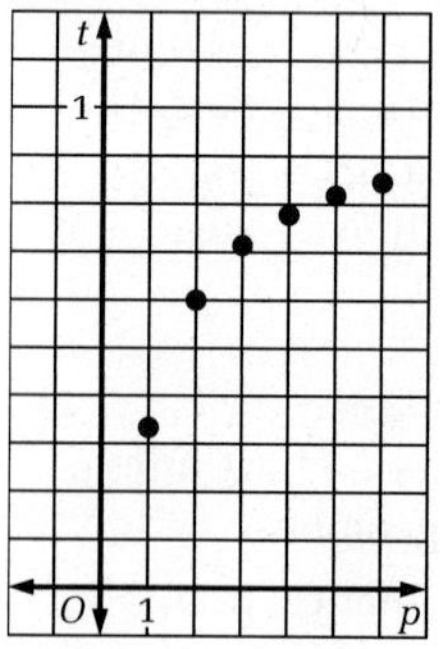

b. Yes; the terms appear to approach 1.

5. a.

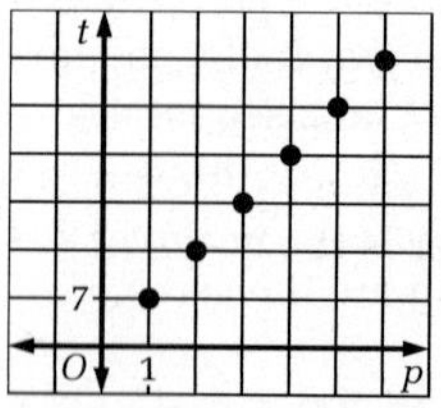

b. No. The terms appear to increase without limit.

6. a.

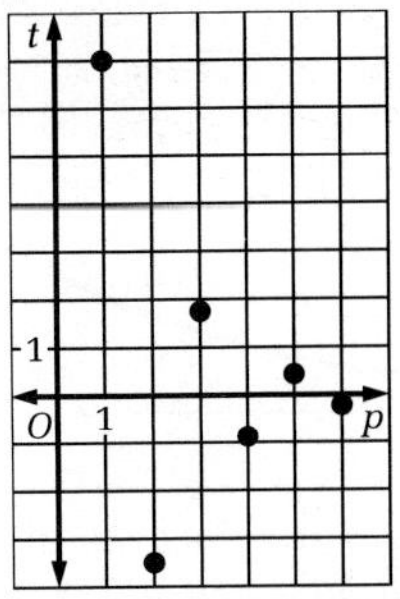

b. Yes. the terms appear to approach 0.

7. $a_n = 3^{3-n}$ **8.** $a_n = \frac{1}{n+2}$

9. 2, 1, 0, 1; 14

10. 17, 8, 5, $\frac{7}{2}$; $\frac{1}{17}$

11. a. $a_n = 200 + 30n$
b. \$60

12. 3, 4, 3, 4, 3 **13.** 6, –3.6, 2.16, –1.296, 0.7776 **14.** $a_1 = 11$; $a_n = a_{n-1} + 13$ **15. a.** 48π, 80π, 112π, 144π, 176π, 208π, 240π, 272π, 304π **b.** recursive: $a_1 = 48\pi$, $a_n = a_{n-1} + 32\pi$; explicit: $a_n - 48\pi + 32\pi(n-1)$ **c.** 304π cm$^2 \approx 955$ cm^2 **16. a.** neither **b.** $a_1 = 10$; $a_n = 2a_{n-1} + 1$ **17. a.** neither **b.** $a_1 = \frac{2}{3}$; $a_n = (a_{n-1})^2$ **18. a.** arithmetic **b.** $a_1 = \frac{1}{8}$; $a_n = a_{n-1} + \frac{3}{8}$ **19.** 1643 **20.** 84 **21.** $\sum_{n=1}^{20}(100 - 5n)$

22. Maps will vary. An example is given.

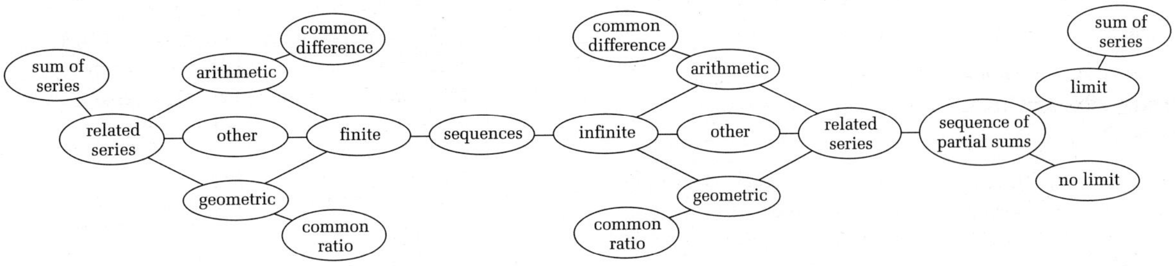

23. 333.33333 **24.** about 5.33

25. a.

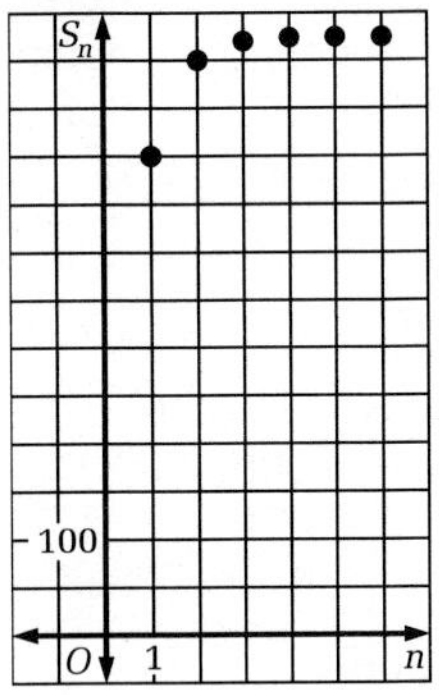

b. Yes; 625.

26. a.

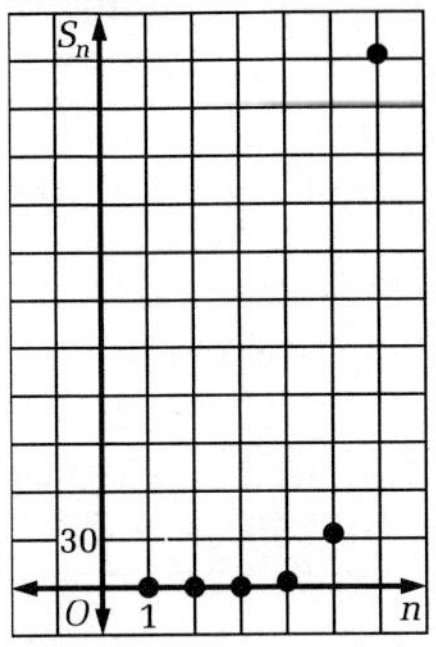

b. No.

27. no sum **28.** –8 **29.** Answers may vary. Examples are given. $\sum_{n=1}^{\infty}\left(\frac{1}{2}\right)^n$ has a sum. (The sum is 1.)

$\sum_{n=1}^{\infty} 2^n$ has no sum.

Unit 5

Pages 267–269 Talk it Over

1. a. 32 ancestors **b.** the positive integers; all numbers 2^n for n an integer and $n \geq 1$ **c.** The number of generations and the number of ancestors must both be positive whole numbers. **2.** Estimates may vary. An example is given; about 4 billion people. **3.** The model assumes that the growth rate will remain unchanged during the time period in question. The growth rate may not stay the same. For example, it might increase (for example, in response to an economic boom in the state) or decrease (for example, in response to a decline in available natural resources). **4.** For the same value of a, the greater the value of b, the steeper is the graph of $y = ab^x$.

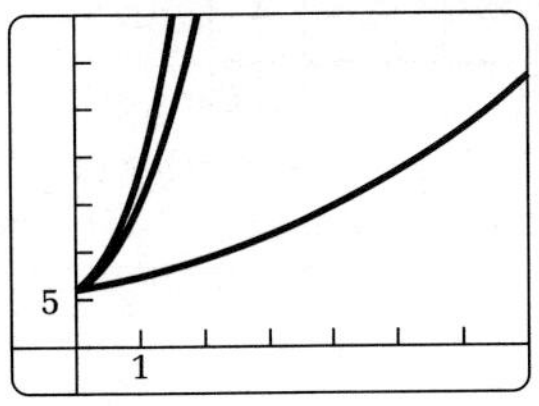

Pages 271–274 Exercises and Problems

1. a represents the initial amount, b represents the rate of change, x represents the time interval, and y represents the amount after x time intervals. **3.** A **11. a.** discrete **b.** $y = 48{,}000(1.05)^x$ **c.** in 2027 **15.** about \$104,245,000,000 **19. a.** Both functions are exponential functions, that is, both have the form $y = ab^x$. For the first equation, $b > 1$, so the first function models exponential growth. For the second equation, $0 < b < 1$, so the second function models exponential decay.

b.

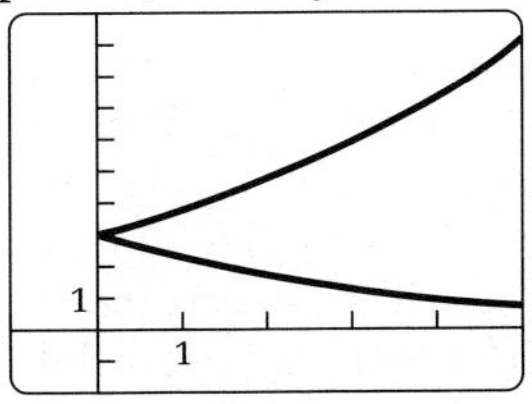

Both curves are exponential. The first models exponential growth; the second models exponential decay. The graphs have the point (0, 3) in common.

21. a.

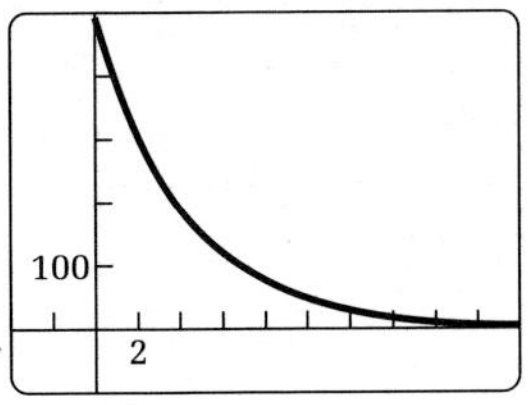

b. 492.5 ng/mL
c. about 2.9 h
d. about 14.5 h

25. The sum does not exist. **26.** 1 **27.** $-\frac{5}{0.997} \approx -5.015$ **28.** No; if such a polygon existed, the number, n, of sides of the polygon would satisfy the equation $(n-2)180° = 655$. The solution to this equation is not a whole number. **29.** $\frac{15}{x}$ **30.** $5s^2t^2$ **31.** $\frac{p^2}{q^2}$

Page 275 Talk it Over

1. 1984; 1987; 1992 **2.** 7 **3.** about \$1,014,000 **4.** -1 **5.** about \$114,000 **6.** about \$38,000 **7.** about \$12,750

Pages 277–280 Exercises and Problems

3. $y = (0.5)(0.125)^x$ **5. a.** 1985; 1980; 2000 **b.** The graph declines and moves toward the x-axis; the graph rises and moves away from the x-axis. **c.** y-values increase; Yes; a percent can be any real number. **7. a.** about 162,132 **b.** 0 **c.** For years represented by negative values of x, there was no cellular phone system, and so no subscribers.
16. $y = (132{,}000{,}000)(1.013)^x$, where x is the number of years since 1940 **17.** -2 **18.** 1 **19.** $3d^3$ **20.** $2|x|$ **21. a.** ± 123 **b.** 3 **c.** ± 1107, 3321, ± 9963

Pages 282–283 Talk it Over

1. $f(x) = (40)\left(\frac{1}{2}\right)^x$, where x is the number of 2-hour periods since the beverage was consumed **2.** It takes 5780 years for half of a given amount of carbon 14 to decay. **3.** $6^{2/5}$ **4.** $18^{3/2} = (2 \cdot 3^2)^{3/2} = 2^{3/2} \cdot 3^{2 \cdot 3/2} = 2^{1+1/2} \cdot 3^3 = 2 \cdot 2^{1/2} \cdot 3^3 = 54\sqrt{2}$

Pages 285–287 Exercises and Problems

1. For integers m and n with $n > 0$ and any positive real number a, $a^{1/n} = \sqrt[n]{a}$, and $a^{m/n} = (\sqrt[n]{a})^m = \sqrt[n]{a^m}$. Examples may vary. **3.** $\sqrt[5]{13}$ **5.** $\sqrt[3]{15^5}$ or $\sqrt[3]{759{,}375}$ or $15\sqrt[3]{225}$ **7.** $12^{1/4}$ **9.** $21^{4/3}$ **11.** $4\sqrt[4]{2}$ **13.** $\sqrt[6]{243}$ **15.** $\frac{512}{125} = 4\frac{12}{125}$ **17.** $800\sqrt{5}$ **19. a.** $f(t) = 0.1(1.35)^t$ **b.** about 0.11 g; about 0.15 g **c.** nonnegative real numbers; Yes. **d.** after about 2.3 h **23.** $|x|\sqrt{x}$ **25.** $7v^2\sqrt{v}$ **27.** $6z^{-2}$ **29.** 1 **37.** 1,000,000,000 **39.** $\frac{16}{25} = 0.64$ **42.** $\left(\frac{1}{b}\right)^x = b^{-x}$; $\left(\frac{1}{b}\right)^x = \frac{1^x}{b^x} = \frac{1}{b^x}$
43. Methods may vary. An example is given.
Step 1: Find the product of the two first terms.
Step 2: Find the product of the two outside terms.
Step 3: Find the product of the two inside terms.
Step 4: Find the product of the two last terms.
Step 5: Add your results from Steps 1 through 4.
Step 6: Simplify if necessary.
44. exponential decay **45.** exponential growth
46. exponential decay **47.** exponential growth

Page 289 Talk it Over

1. about \$35.63
2. Answers may vary. An example is given. Since the rate of change is greater than 1, the function models exponential growth. The graph lies completely above the x-axis, approaching the x-axis as x decreases and increasing exponentially as x increases. It has a y-intercept of 2.62.

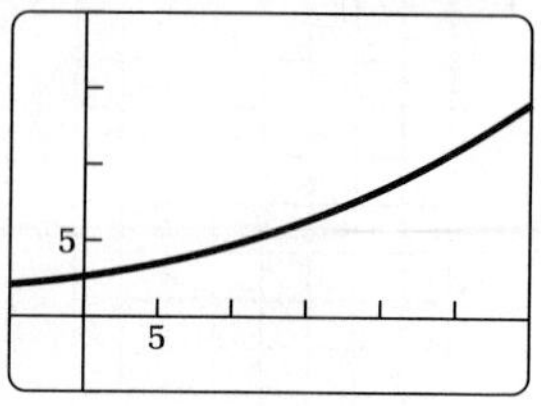

Pages 291–294 Exercises and Problems

1. irrational; about 2.718281828 **3.** C **5.** A **7.** about $2.95 \cdot 10^{-6}$ **13.** \$106.17 **15. a.** \$7430.90 **b.** \$7404.56 **c.** He will earn more in the account in which interest is compounded continuously; the added \$10 in the other account does not make up for the interest lost by compounding quarterly rather than continuously. **26.** 5 **27.** 10,000 **28.** $7|x|\sqrt{2x}$ **29.** $4|b^3|c^2\sqrt[4]{c^2}$ or $4|b^3|c^2\sqrt{|c|}$

30.

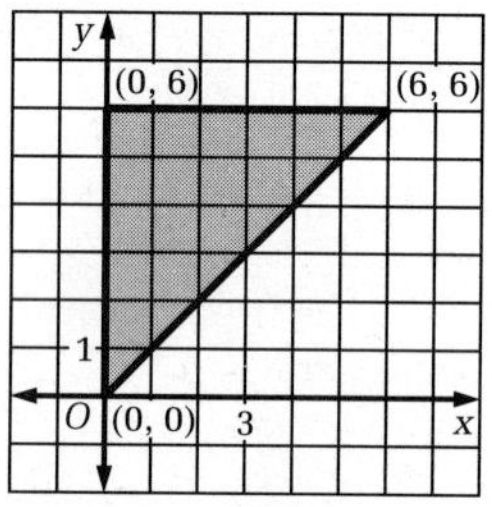

31.

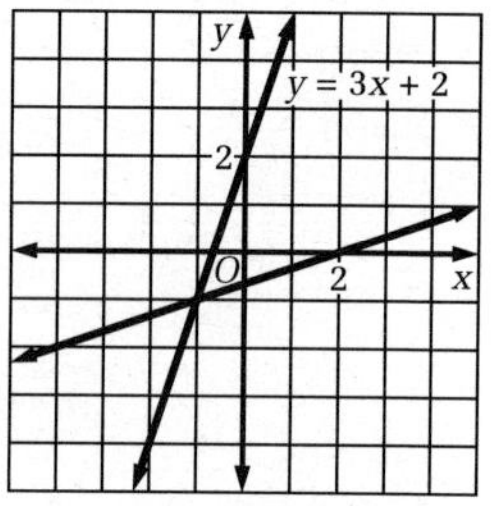

32.

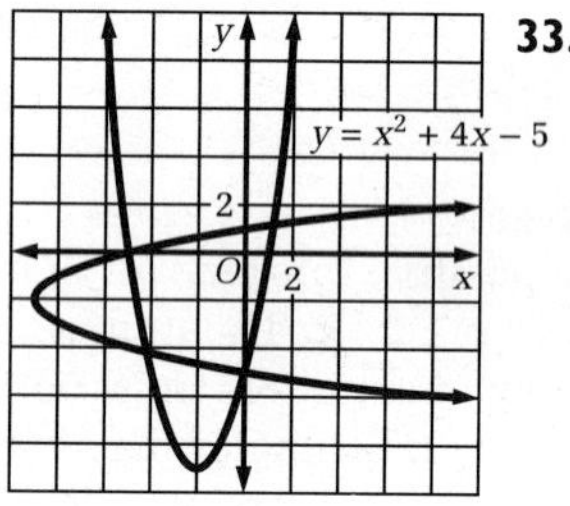

33.

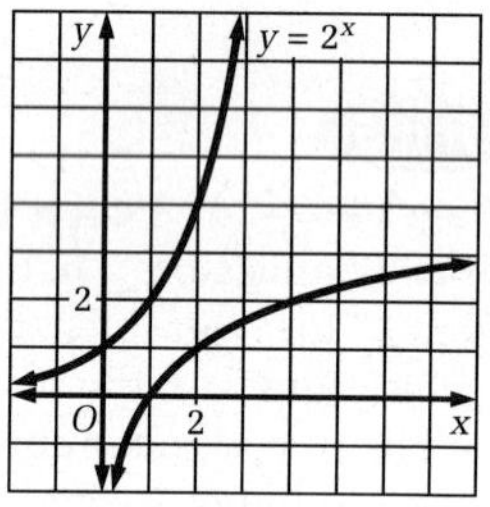

Page 295 Checkpoint

1. Exponential growth functions and exponential decay functions both have the form $y = ab^x$. For an exponential growth function, $b > 1$. For an exponential decay function, $0 < b < 1$.
2. a. $y = 15{,}112{,}000(1.16)^x$, where x is the number of 5-year periods after 1980 **b.** about 27,360,000 **c.** about 11,231,000 **3.** 9 **4.** 243 **5.** 8 **6.** \$286.23 **7.** \$287.45 **8.** \$287.57

Pages 296–298 Talk it Over

1.

Distance (km)	Number of laps
10	25
5	12.5
3	7.5
1.5	3.75
1	2.5
0.5	1.25

2. $L(D) = \frac{D}{0.4} = 2.5D$

3. $D(L) = 0.4L$

4. The domain of L consists of the numbers 0.5, 1, 1.5, 3, 5, and 10, while the range consists of the numbers 1.25, 2.5, 3.75, 7.5, 12.5, and 25. The domain of D consists of the numbers 1.25, 2.5, 3.75, 7.5, 12.5, and 25, while the range consists of the numbers 0.5, 1, 1.5, 3, 5, and 10. The domain of L is the range of D and the range of L is the domain of D.

5.

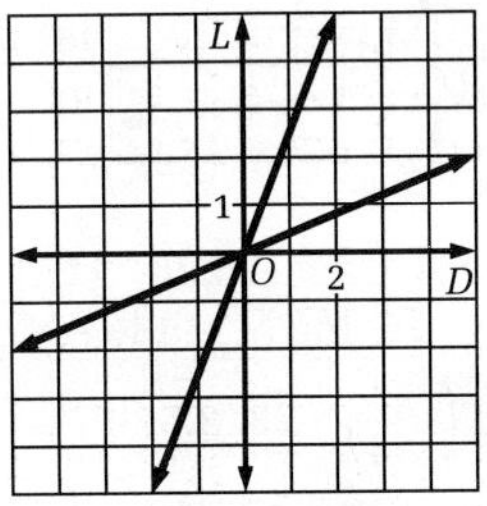

Each graph is the image of the other after reflection over the line $L = D$. **6.** Yes; No; the inverse does not pass the horizontal line test. **7.** Yes; Yes; the inverse passes the horizontal line test.

8. Yes; No; the inverse does not pass the horizontal line test. **9. a.** nonnegative real numbers; nonnegative real numbers **b.** the portion of the graph in the first quadrant **c.** Yes; the function $y = x^2$ with the domain restricted to $x \geq 0$ has an inverse. Its graph passes the horizontal line test. The graph of the inverse is the first-quadrant portion of the inverse function shown in Sample 1(b).

Pages 299–300 Exercises and Problems

1. The domain of $f(x)$ is the range of $f^{-1}(x)$ and the range of $f(x)$ is the domain of $f^{-1}(x)$.

3. Yes.

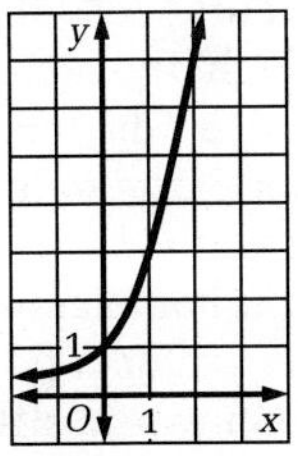

5. No. Answers may vary. An example is given.

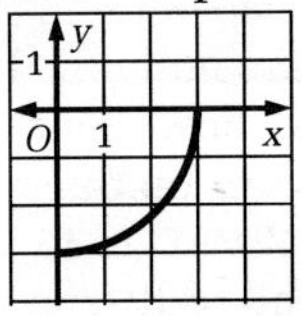

7. Yes.

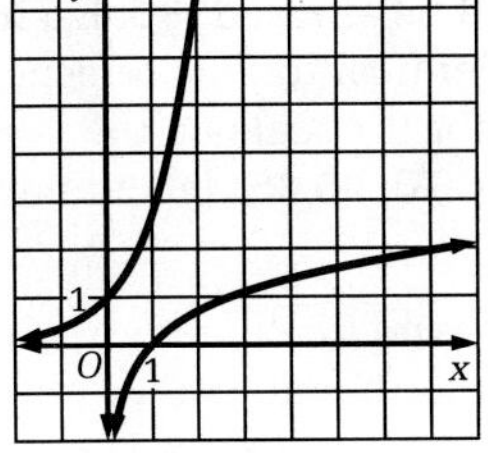

11. $y = -x + 6$ **13.** $y = x$ **15.** $y = \sqrt[5]{\frac{x+3}{2}}$ **17.** $y = -\frac{x}{2}, x \geq 2$ **19.** $y = x - b$ **21.** $y = x^{b/a}$ **23.** No. **25.** No.

29. about 6.2 billion transactions **30.** Answers may vary. An example is given. Use the distance formula to find OR, OP, RQ, and QP.
$OR = \sqrt{(a-0)^2 + (b-0)^2} = \sqrt{a^2 + b^2}$,
$OP = \sqrt{(a-0)^2 + (b-0)^2} = \sqrt{a^2 + b^2}$,
$RQ = \sqrt{(c-a)^2 + (0-b)^2} = \sqrt{(c-a)^2 + b^2}$, and
$QP = \sqrt{(c-a)^2 + (0+b)^2} = \sqrt{(c-a)^2 + b^2}$. Therefore, $OR = OP$ and $RQ = QP$. By the definition of a kite, $OPQR$ is a kite. **31.** $2.91 \cdot 10^{-6}$ **32.** $4.7 \cdot 10^7$ **33.** $4.829 \cdot 10^5$

Pages 301–304 Talk it Over

1. The scale number is the same as the number of zeroes in the relative size. **2.** about 125,893 **3.** No; the relative size of an earthquake that measures 8 on the Richter scale is 100,000,000. The relative size of an earthquake that measures 2 on the Richter scale is 100. The size of the larger earthquake is a million times the size of the smaller quake. **4.** arithmetic; geometric **5.** Yes; the graph passes the horizontal line test. **6.** because 1 raised to any power is 1

7. a. $5^2 = 25$ **b.** $4^1 = 4$ **c.** $10^{-2} = 0.01$

8. a. $\log (100{,}000) = 5$ **b.** $\log_4 8 = \frac{3}{2}$ **c.** $\log_7 1 = 0$

9. more acidic **10.** less acidic

Pages 305–307 Exercises and Problems

1. For any positive number b, $f(x) = \log_b x$ and $g(x) = b^x$ are inverse functions. **3.** $\log_7 y = x$ **5.** $\ln e = 1$ **7.** $\log_2 \left(\frac{1}{8}\right) = -3$ **9.** $\log_8 1 = 0$ **11.** $6^0 = 1$

13. $10^4 = 10{,}000$ **15.** $e^1 \approx 2.718$ **17.** 2 **19.** 1 **21.** 2.30 **23.** 0 **25. a.** Estimates may vary; about 3.25. **b.** Answers may vary. An example is given. 10 is about one-quarter of the way between 8 and 16. **27.** $\frac{1}{2}$ **29.** 1.32 **31.** 2.30 **40.** $y = \frac{1}{4}(x - 1)$ **41.** $\frac{16x^4}{y}$ **42.** $5t$ **43.** $\frac{9}{4a^6b^{12}}$ **44.** $\frac{1}{m^3n^6}$ **45.** Answers may vary. An example is given. The graph of $y = -16x^2 + 20$ is a parabola that opens down. Its vertex is at (0, 20).

Pages 311–313 Talk it Over

1. No; as the sample shows, doubling the intensity of a sound increases the loudness by about 3 decibels. The loudness would double only if the original loudness were 3 decibels. **2.** increase in loudness =
$L_2 - L_1 = 10 \log \frac{2I}{I_0} - 10 \log \left(\frac{I}{I_0}\right) = 10\left(\log \frac{2I}{I_0} - \log \frac{I}{I_0}\right) =$
$10\left(\log\left(\frac{2I}{I_0} \div \frac{I}{I_0}\right)\right) = 10 \log \left(\frac{2I}{I_0} \cdot \frac{I}{I_0}\right) = 10 \log 2 \approx 3$

3. Find $3^{1.893}$. **4.** $\log_3 8 = \frac{\ln 8}{\ln 3} \approx \frac{2.0794}{1.0986} \approx 1.893$

5. Graph $y = \frac{\ln x}{\ln 6}$. This graph is identical to the graph of $\frac{\log x}{\log 6}$ from Sample 5.

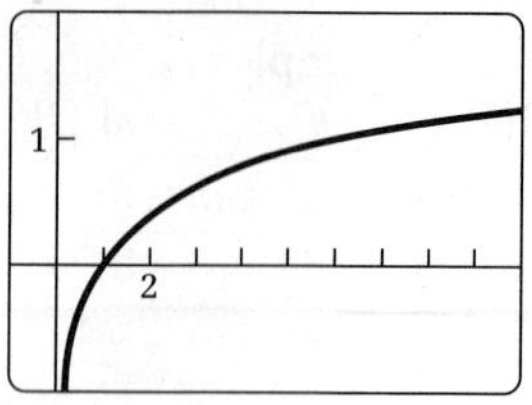

Pages 313–316 Exercises and Problems

3. $3 \log M$ **5.** $7 \log M - 5 \log N$ **7.** $\frac{4}{5} \log M$ **9.** $\ln \frac{2}{343}$ **11.** $\log_4 35$ **13.** $\ln 4$ **15.** 1 **17.** −2 **19.** $-3x$ **23.** 3.459 **25.** 0.921 **27.** −0.102 **33.** Let $\log_b M = x$. Then $M = b^x$, so that $\log_b M^k = \log_b (b^x)^k = \log_b b^{kx} = kx = k \log_b M$. **35.** Let $\log_b M = x$. Then $M = b^x$, so that $\log_c M = \log_c b^x = x \log_c b = \log_b M \cdot \log_c b$; so $\log_b M = \frac{\log_c M}{\log_c b}$. **39.** $3^4 = 81$ **40.** $e^{1.61} \approx 5$ **41.** $2^6 = 64$ **42.** 145 **43.** 14 **44.** $7\frac{3}{4}$ **45.** $0; \frac{1}{9}$ **46.** 1 **47.** 3

Pages 318–319 Talk it Over

1. Take the common logarithms of both sides of the equation to express the equation in logarithmic form. Use the power property of logarithms to express $\log (0.883)^t$ as $t \log 0.883$. Finally, divide both sides of the equation by $\log 0.883$. **2.** Yes. Using logarithms with any base will result in the same answer. **3.** Take the common logarithms of both sides of the equation. Use the power property of logarithms to express $\log 5^x$ as $x \log 5$. Divide both sides of the equation by $\log 5$; $x \approx 2.05$. **4.** Graph the equation $y = 0.883^t$ and use TRACE to find the t-value when $y = 0.37$. **5.** Graph $y = \frac{\log x + \log (x - 2)}{\log 2}$ and $y = 3$ on the same screen and use TRACE to find the x-coordinate of the intersection point. **6.** Graph both equations on the same screen and use TRACE to find the x-coordinate of the intersection point; $x \approx 1.9$, $y \approx 5.2$.

Pages 320–322 Exercises and Problems

1. $\log_2 x + \log_2 (x - 2)$ is not defined for $x = -2$. **3, 5, 7.** Answers are given to the nearest hundredth. **3.** 0.56 **5.** 1.82 **7.** −0.90 **11.** 8 **13.** 2 **15.** no solution **25.** $\log_5 9$ **26.** $\log_b \frac{p^{17}}{q}$ **27.** $w^3 - w$ **28.** $4a^2 + 2a - 23$ **29.** $x^2 - x$ **30.** mean: 76; median: 81 **31.** mean: 3.05; median: 2.85

Pages 324–325 Unit 5 Review and Assessment

1. The growth factor would be 1.05, so the equation would be $y = a(1.05)^x$. The decay factor would be 0.95, so the equation would be $y = a(0.95)^x$. **2.** about 22 billion dollars **3.** Answers may vary. Examples are given. value of an investment at a fixed rate of interest over a period of time; amount of carbon 14 left in organic material over time **4.** A **5.** B **6.** C **7. a.** $y = 10586000(1.485)^x$, where x is the number of decades since 1950 **b.** about 1,466,000 **8.** $y = (20)(5^x)$ **9.** 125 **10.** $10\sqrt[3]{2}$ **11.** $2x^4\sqrt[4]{4}$ **12.** 27 **13. a.** 0.94 L/min **b.** about 18 mi/h **14.** $y = 0$; $y = 5$

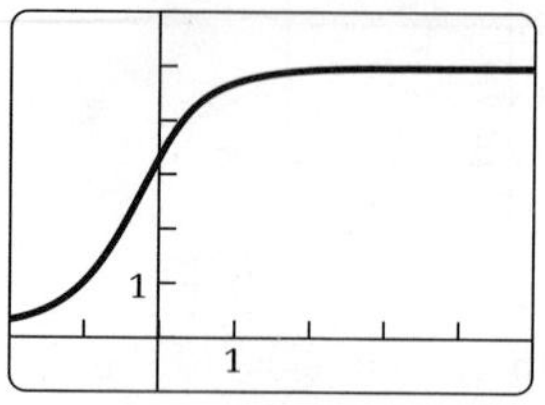

15. Yes.

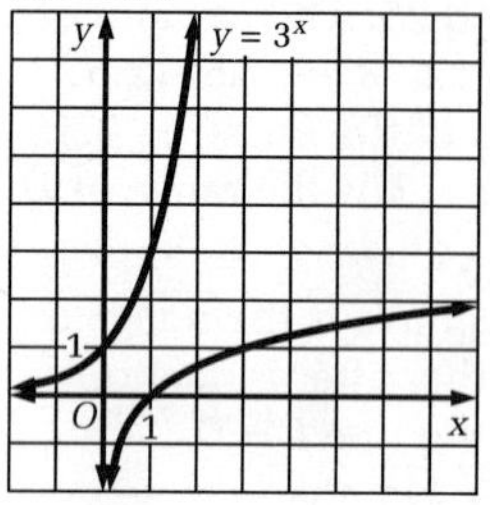

16. No.

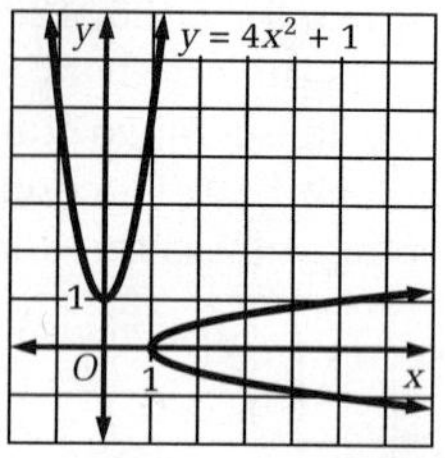

17. Yes; $y = 4x - 8$. **18.** −5 **19.** 1.79 **20.** 3 **21.** 11 **22.** −4 **23.** 1.54 **24.** $7 \log N$ **25.** $\log M + 5 \log N$ **26.** $\frac{1}{2} \log M - 6 \log N$ **27.** $\log \left(13\frac{1}{3}\right)$ **28.** 2.579 **29.** 8 **30.** about 2.65 billion dollars **31.** about 3.7%

Unit 6

Pages 332–334 Talk it Over

1. Answers may vary. An example is given. The telephone-digit data set shows a uniform distribution. **2.** Estimate the total heights of all the bars and divide by 2. The median will be in the bar that contains the item of data that corresponds to this quotient. **3.** Answers may vary. An example is given. I think measures of central tendency describe how numbers relate to a central, or average, value. **4. a.** An outlier might show as a bar separated from the rest of the bars. If the intervals are large enough, the histogram might not show the outliers at all. **b.** The median is determined by the numerical order of the numbers in a data list, not by how large or small they are, so an outlier has no effect on the median. For example, for the data 3, 5, 6, 7, 7, 7, 8, 12, if the last data point were 1750 instead of 12, the median would still be 7. Since the outlier may significantly change the sum of the data, the mean would be affected. **5.** Each histogram in Sample 2 is identical to the corresponding one in Sample 1 except for the scale on the vertical axis. **6.** 38.7% (about 40%); 90.3% (about 90%)

Pages 335–337 Exercises and Problems

1. In a skewed distribution, the interval or intervals that contain the greatest frequencies are at one end. In a mound-shaped distribution, the interval with the greatest frequency is in the middle of the histogram. **3.** Answers may vary. As the widths of the intervals change, the shape of the histogram will change. **5. a.** Estimates may vary. Monday, Wednesday, and Thursday: about 48; Tuesday: 50; Friday: about 45; Saturday: about 42; Sunday: 40 **b.** Yes; it seems reasonable that there is no particular day of the week on which more births occur. Therefore, it is reasonable to expect a uniform distribution.

11. a.

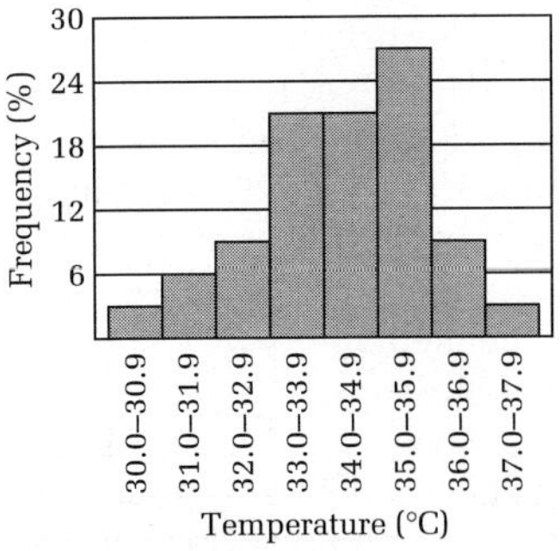

b. about 39%

13. a.

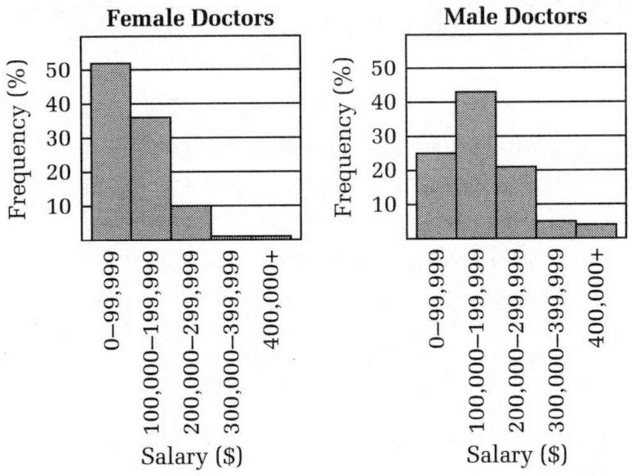

b. Both distributions are basically skewed. **c.** female doctors: 46%; male doctors: 64% **d.** Answers may vary. Examples are given. It appears that the salaries of male doctors are significantly higher than those of female doctors. Over half of the female doctors earn less than $100,000 per year, while 75% of all male doctors earn more than $100,000 per year. Also, while only 2% of all female doctors earn more than $300,000 per year, 11% of male doctors earn that much. Both the mean and median salaries are higher for male doctors. **15.** about 2.16 **16.** No. **17.** –63

Pages 338–340 Talk it Over

1. The San Francisco distribution is fairly uniform, while the Kansas City distribution is basically mound-shaped. **2.** about 56.8° for both cities; No. Although the histograms show that the monthly mean temperatures for the two cities are very different, the mean annual temperature is the same for both cities. **3.** San Francisco: 11.3°; Kansas City: 49.8° **4.** Answers may vary. An example is given. I think the mean and range together are better than the mean alone. Given both, I can tell that there is a greater variation in the temperature in Kansas City. **5.** San Francisco: –6.1, –3.8, –2.1, –1.1, 0.6, 2.3, 2, 2.6, 5.2, 4.6, 0.6, –4.3; Kansas City: –25.1, –21, –13.5, –1.1, 8.8, 19.1, 24.7, 23, 14.5, 3.4, –12.2, –21; San Francisco average: about 0.04; Kansas City average: about –0.03 **6.** Kansas City; The standard deviation for Kansas City, about 17.4, is greater than that for San Francisco, about 3.4. Also, the range for Kansas City, 49.8, is greater than the range for San Francisco, 11.3. **7.** San Francisco: 60.2, 53.4; 58.3% of the temperatures are within one standard deviation of the mean; Kansas City: 74.2, 39.4; 50% of the temperatures are within one standard deviation of the mean.

Pages 341–343 Exercises and Problems

1. 2, the least and the greatest; all the data items **3. a.** 282.4 **b.** 13 **c.** 3.9 **d.** 60% **5. a.** 8.2 **b.** 7 **c.** 1.7 **d.** 84.6% **11.** Answers may vary depending on intervals chosen. In the example shown, the data are not grouped into intervals.

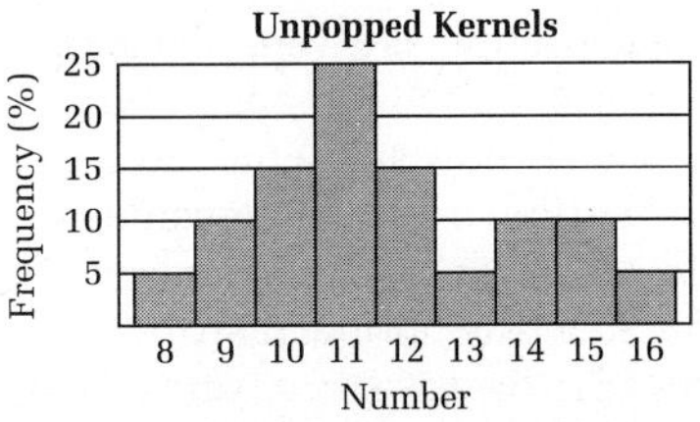

12. –3, –5 **13. a.** mound-shaped **b.** Estimates may vary; about 42%.

Pages 345–346 Talk it Over

1. about 68.5 in. **2.** about 71% **3.** normal; Most heights and weights approximate normal distributions. **4.** Both distributions have the same mean, but the distribution described by the shorter, wider bell-shaped curve has a greater standard deviation than the distribution described by the narrower curve. **5.** The 68-95-99.7 rule applies only to normal distributions. **6.** 5%

Pages 347–349 Exercises and Problems

1, 3. Answers may vary depending on interval size chosen. Examples are given.

1. a, b.

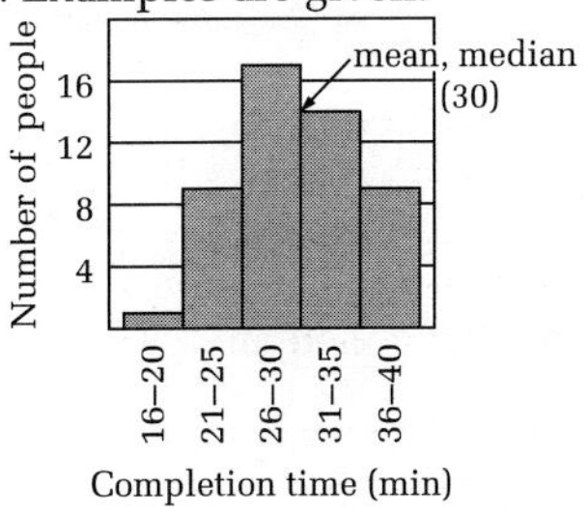

c. No.

3. a, b.

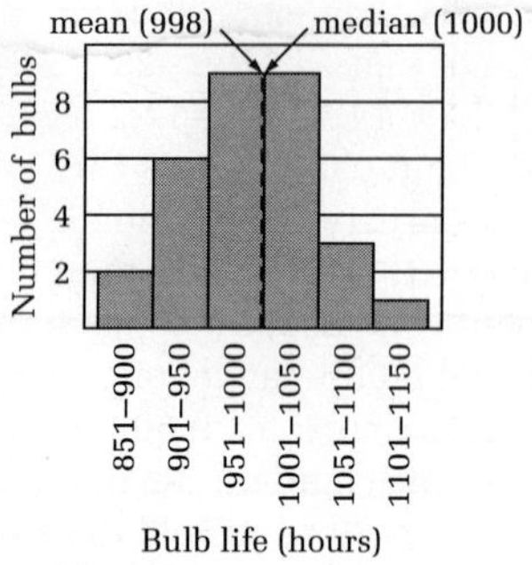

c. Yes.

7. a. 144 and 200 **b.** about 34% **c.** about 2.5% **9.** about 2.5% **13.** 13.4°; 10.5° **14.** $y = 0.5x - 1$ **15.** $y = 4x + 3$ **16.** $\begin{bmatrix} 2\frac{2}{3} \\ -5\frac{2}{3} \end{bmatrix}$

Page 349 Checkpoint

1. Answers may vary. An example is given. The graph of a normal distribution has the same shape as the data set gets larger, no matter how small the interval size. For a normal distribution, a known percentage of the data falls within one, two, or three standard deviations of the mean. In a true normal distribution, the mean and median of the data are equal and fall at the line of symmetry of the distribution's bell-shaped curve. **2. a.** Answers may vary depending on the intervals chosen. Examples are given. See histograms below. **b.** financial: basically bimodal; communications: basically mound-shaped

c. financial: 50%; about 42%; communications: about 44.5%; about 55.5%

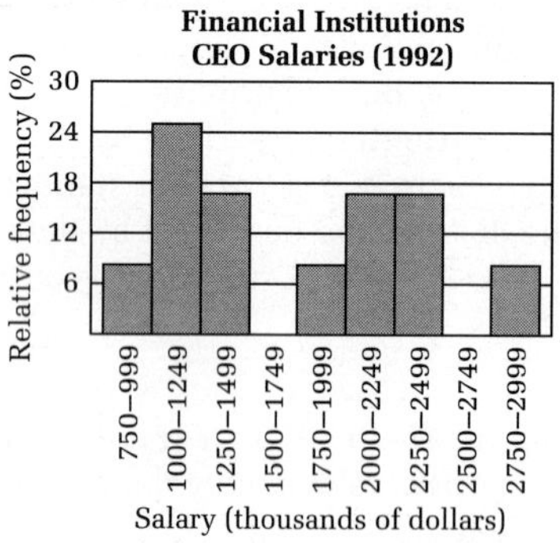

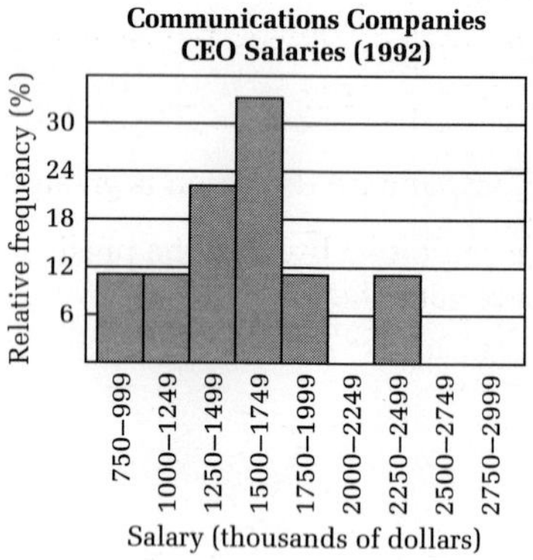

d. The range of salaries for financial institutions is greater than that for communications companies, so there is more variation for financial institutions. The two groups appear to have similar median salaries, but it appears that, overall, salaries are higher in financial institutions. **3.** about 3601; 3031; about 775 **4.** between 67 and 83

Pages 350–353 Talk it Over

1. distance from NYC vs. airfare; Jan. low temp. vs. population **2.** latitude vs. Jan. low temp.; distance from NYC vs. airfare **3.** The point is not near either the least-squares line or the median-median line. **4.** the least-squares line; The slope of the least-squares line is determined using actual data points, while the slope of the median-median line is determined using points whose coordinates are means of the coordinates of summary points. **5.** Answers may vary. An example is given. I think the point will be closer to the median-median line. That line is less affected by the point (11, 914) than is the least-squares line. **6.** the median-median line

Pages 354–358 Exercises and Problems

1. No; if the correlation coefficient of a data set is close to 0, the distribution of the data set is nearly random. A line would not model the data well.

3. a. $\frac{1}{2} < r \leq 1$ **b.** Yes; the data show a general upward trend, but the data points are somewhat scattered.

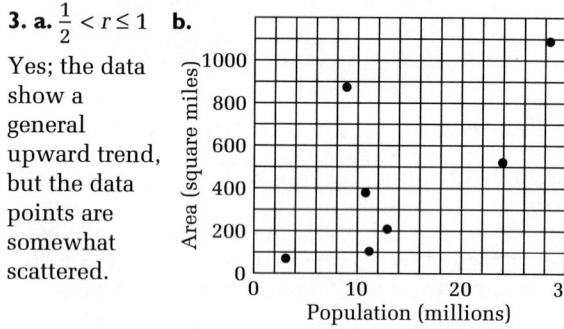

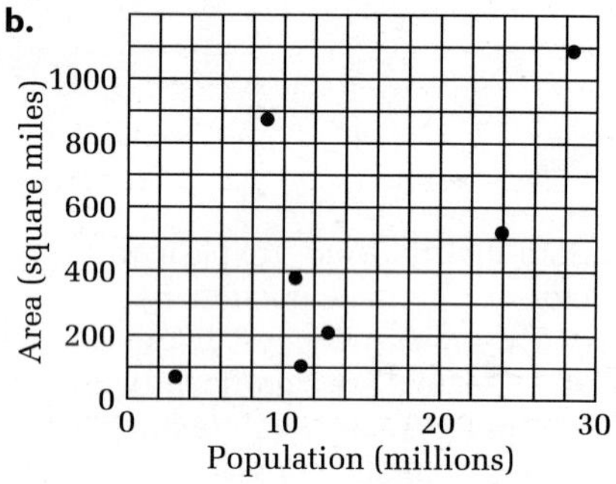

11. a–c.

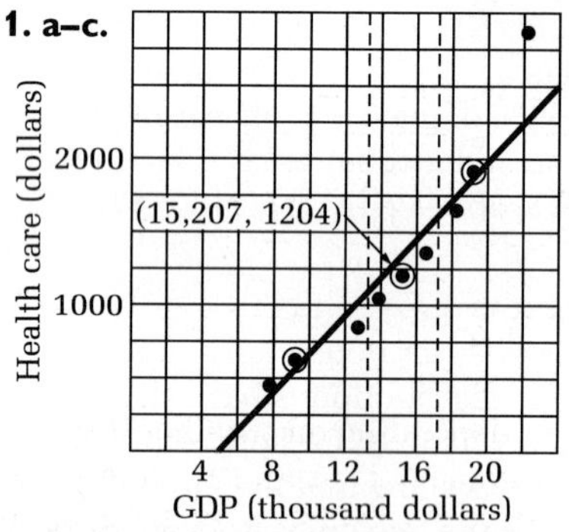

d. Answers may vary due to rounding; $y = 0.13x - 639.54$. **e.** Answers may vary. An example is given. I think the median-median line is a better fit than the least-squares line. The data point for the United States appears to be an outlier.

13. a–c.

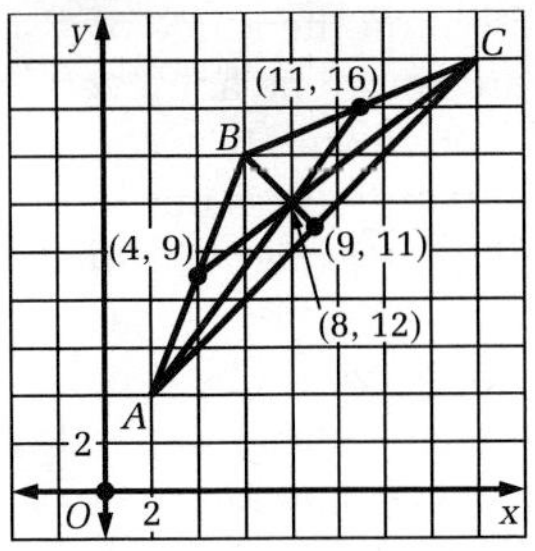

d. $\left(\frac{4 + 11 + 9}{3}, \frac{9 + 16 + 11}{3}\right) =$ (8, 12)

15. about 81.5%

16. −3, −1, 3, 11, 27

17. $\begin{bmatrix} 2 & 3 & -1 \\ -1 & 5 & 8 \\ 5 & -1 & 0 \end{bmatrix} \begin{bmatrix} x \\ y \\ z \end{bmatrix} = \begin{bmatrix} 16 \\ 16 \\ -7 \end{bmatrix}$

Pages 359–361 Talk it Over

1. a. Answers may vary due to rounding; $y = -1.16x + 121.6$.

b–d.

x	y-values Actual	Model	Deviation d	d^2
20	108	98.4	9.6	92.16
25	91	92.6	−1.6	2.56
30	81	86.8	−5.8	33.64
35	74	81	−7	49
40	71	75.2	−4.2	17.64
45	70	69.4	0.6	0.36
50	71	63.6	7.4	54.76

e. 250.12 **2.** Yes; a parabola. **3.** There are three unknowns: a, b, and c. **4.** In Sample 1, all the points were used to find the least-squares parabola. In Sample 2, a curve that passes through 3 of the points was found. **5.** No; only if all the points actually are on the same curve as the points in Sample 2. **6.** Choose the quadratic equation for which the sum of the squares of the distances from the curve is the least.

Pages 362–364 Exercises and Problems

1. Graph points to see if they appear to lie on a line or on a curve.

3. a.

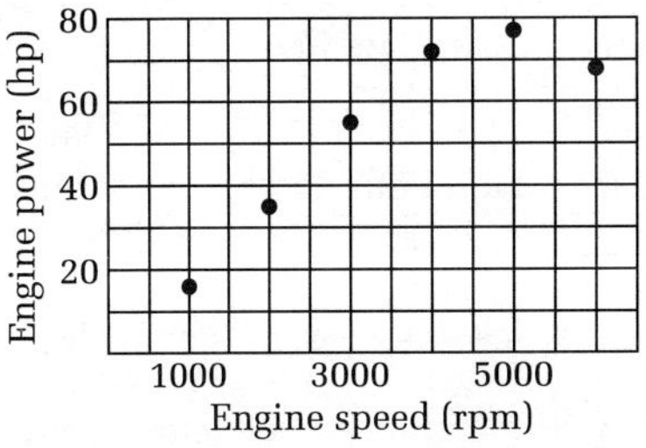

b. Answers may vary depending on whether the matrix method or technology is used. An example is given; $y = (-4.13 \times 10^{-6})x^2 + 0.039x - 19.2$. **c.** about 4718 rpm **5. a.** $4a + 2b + c = 20$; $25a + 5b + c = 56$; $9a - 3b + c = 0$ **b.** $\begin{bmatrix} 4 & 2 & 1 \\ 25 & 5 & 1 \\ 9 & -3 & 1 \end{bmatrix} \cdot \begin{bmatrix} a \\ b \\ c \end{bmatrix} = \begin{bmatrix} 20 \\ 56 \\ 0 \end{bmatrix}$

c. $y = x^2 + 5x + 6$

7. a. $a + b + c = 16$; $a - b + c = 16$; $4a - 2b + c = -32$

b. $\begin{bmatrix} 1 & 1 & 1 \\ 1 & -1 & 1 \\ 4 & -2 & 1 \end{bmatrix} \cdot \begin{bmatrix} a \\ b \\ c \end{bmatrix} = \begin{bmatrix} 16 \\ 16 \\ -32 \end{bmatrix}$ **c.** $y = -16x^2 + 32$

14. a.

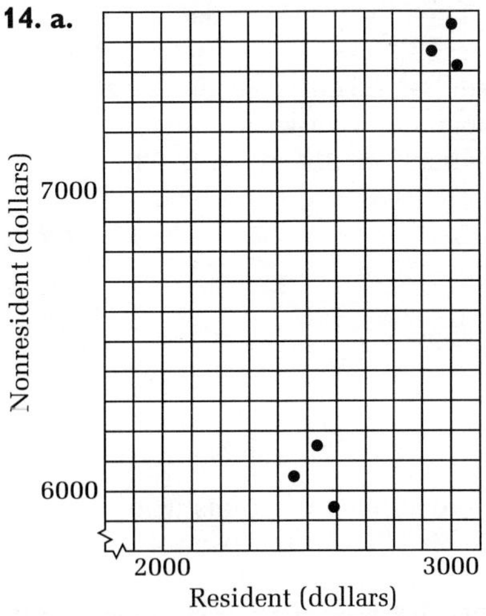

b. linear; The points appear to lie near a line.

c. $y = 2.6814x - 608.36$

15. Suppose that $\angle B$ is a right angle. Then, since $m\angle C > 0$, $m\angle A + m\angle B + m\angle C = 100° + 90° + m\angle C > 190°$. This is not possible, since the sum of the measures of the angles of a triangle is 180°. So the assumption that $\angle B$ is a right angle is false. Therefore, $\angle B$ is not a right angle. **16.** 0.5

Pages 366–368 Unit 6 Review and Assessment

1. a. A: mound-shaped; B: skewed; C: uniform

b. A: 7.55, 8; B: 8.85, 9; C: 7.77, 8

2. a. Intervals may vary; bimodal.

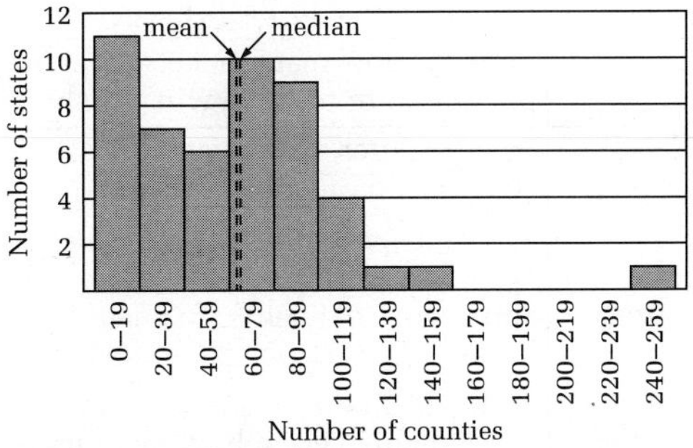

b. 254 **c.** mean: 61.64; median: 62.5; See graph.

d. 18% **3. a.** sons: 2.15, 8; daughters: 1.46, 6

b. 2.05; 1.47 **c.** The mean is the sum of the means; $\frac{\text{number of children}}{\text{number of presidents}} = \frac{\text{number of daughters}}{\text{number of presidents}} + \frac{\text{number of sons}}{\text{number of presidents}}$. The standard deviation is greater when the numbers are combined because the presidents had so many more sons than daughters.

4. a. 43 yr to 67 yr **b.** 47.5%; 34%

5. a–c. Sketches may vary.

a, b.

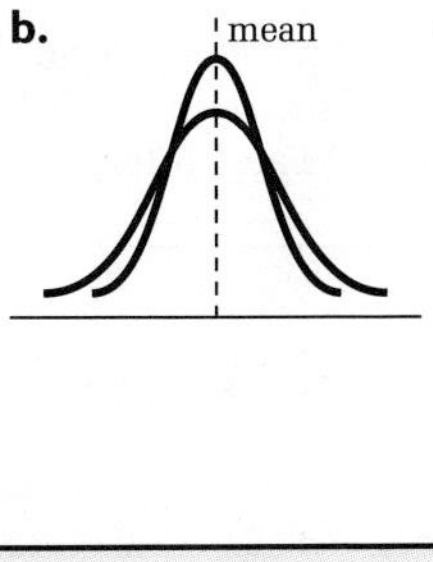

c.

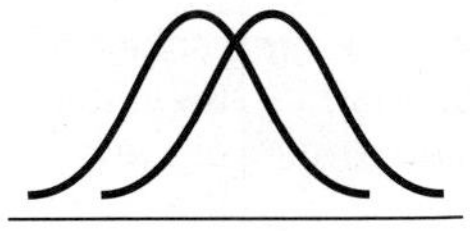

d. If the standard deviation is unchanged and the mean is increased (or decreased), the curve retains the same shape but shifts to the right (or left). If the mean is unchanged and the standard deviation increases (or decreases), the line of symmetry of the curve is unchanged, but the curve becomes flatter and wider (or steeper and narrower).
6. a. $P = 0.0723A + 0.23$ **b.** 314 in.^{2}; \$22.93
7. Answers may vary. An example is given.
a. $y = -186.14x^2 + 2201.7x + 23{,}895$, where x is the number of years since 1986 **b.** \$23,895

Unit 7

Pages 373–378 Talk it Over

1. 20%, or $\frac{1}{5}$ **2.** 5; If an infinite number of tickets were printed, then no number of tickets would guarantee winning a T-shirt. But in this example, only a finite number of tickets will be printed. Since it is equally likely that each letter will appear, there must be an equal number of tickets, n, with each letter. Since there are 5 different letters, buying $4n + 1$ tickets guarantees getting one of each letter, thus winning a T-shirt. **3.** Answers may vary. **4.** The random sequence produced will consist of integers between 0 and 9. **5.** The random sequence produced will consist of integers between 3 and 11. **6.** int (20*rand) + 10 **7.** int (2*rand) + 1 **8.** Yes; 15 of the random numbers represent passengers who do not take the flight, and 85 represent passengers who do.

9.

General model	Model applied to T-shirt problem
State the problem clearly.	Find out how many tickets must be purchased in order to win a T-shirt.
Define the key question.	Select a letter. Does that letter show up on one of your tickets?
State the underlying assumptions.	The probability that any given letter appears on a ticket is 20%. The appearance of a letter on one ticket has no effect on any other ticket.
Choose a model to the answer the key question.	Generate random numbers from 1 to 5 inclusive. Each of the numbers represents one of the 5 letters.
Define and conduct a trial.	Generate a sequence of random numbers until each of the digits 1–5 has appeared once.
Record the observation of interest.	How many numbers had to be generated before each of the digits 1–5 appeared once?
Repeat the last two steps several times.	Generate a sequence as described above several times.
Summarize the information and draw conclusions.	Make a table or graph; determine the measures of central tendency of the data. Analyze and interpret the results.

Pages 378–381 Exercises and Problems

3. No. Since there is no restriction on the range of numbers, a random distribution cannot be achieved. Responses would tend to be positive one- or two-digit numbers. Also, by saying the numbers out loud, some students' responses might be affected by others' responses. **5.** Yes; a sequence of the letter "A" and random numbers between 2 and 10 would be produced (or 1–10 if aces are counted as ones). Each of the numbers occurs with the same frequency in the deck. **7.** int(100*rand) + 1 **9.** int(19*rand) + 11
11. a. It will increase the number of tickets needed. **b.** Answers may vary. An example is given. Use the function int(25*rand) + 1 to generate a sequence of random numbers from 1 to 25. Assign, say, the numbers 1–6 to the letter U, 7–12 to the letter S, and so on, and the number 25 to the letter M. For each trial, generate random numbers until each letter is represented once. **13.** Estimates may vary; about 45%.
15. Methods of simulation and results may vary. Use the function int(12*rand) + 1. Let the numbers 1–12 represent the months in order. (1 = January, and so on) For each trial, generate random numbers until the number for your birth month appears. Analyze the data. **22.** $y = x^2 + 2x - 8$ **23.** 160°; 20° **24.** $\frac{1}{8}$, or 12.5%

Pages 382–386 Talk it Over

1. a. die: 1, 2, 3, 4, 5, 6; dial: all real numbers greater than 0 and less than 6 **b.** die: discrete (exactly six possible outcomes); dial: continuous (the number of outcomes cannot be counted) **2.** $\frac{5}{6} \approx 0.83$; $\frac{\text{number of favorable outcomes}}{\text{number of possible outcomes}} = \frac{5}{6}$ **3.** Since the outcomes are continuous, measurements can be used to compare the ratio of favorable outcomes to possible outcomes. **4. a.** $\frac{1}{2} = 0.5$ **b.** $\frac{1}{3} \approx 0.33$ **c.** $\frac{1}{6} \approx 0.17$
5. To find $P(A \text{ or } B)$ for the given situation, you can add $P(A)$ and $P(B)$ and subtract $P(A \text{ and } B)$ so no out comes are counted twice. $P(A \text{ or } B) = \frac{1}{2} + \frac{1}{3} - \frac{1}{6} = \frac{2}{3} \approx 0.67$ **6.** the square region bounded by the x- and y-axes and the lines $x = 5$ and $y = 5$ **7. a.** $\frac{1}{6} \approx 0.17$ **b.** $\frac{3}{6} = \frac{1}{2} = 0.5$ **c.** $\frac{4}{6} = \frac{2}{3} \approx 0.67$ **d.** 0 **8.** Part (b) becomes $\frac{5}{6}$, or about 0.83; part (c) becomes 1; part (d) remains 0. **9.** If A and B are mutually exclusive events, then $P(A \text{ and } B) = 0$, so $P(A \text{ or } B) = P(A) + P(B) - P(A \text{ and } B) = P(A) + P(B) - 0 = P(A) + P(B)$.
10. A and not A are complementary events, so A and not A are mutually exclusive, which means that $P(A \text{ and not } A) = 0$. Also, A and not A account for all possibilities, so $P(A \text{ or not } A) = 1$. Then $P(A \text{ or not } A) = P(A) + P(\text{not } A) - P(A \text{ and not } A)$. Then $1 = P(A) + P(\text{not } A) - 0$ and $P(\text{not } A) = 1 - P(A)$.

Pages 387–389 Exercises and Problems

When necessary, probabilities are rounded to the nearest hundredth. **1.** Use measurements such as length or area to determine the sizes of favorable outcomes and possible outcomes. **3. a.** discrete **b.** $\frac{1}{12} \approx 0.083$ **5. a.** continuous **b.** $\frac{1}{12} \approx 0.083$ **7. a.** continuous **b.** $\frac{8}{48} = \frac{1}{6} \approx 0.17$ **9.** $\frac{67}{72} \approx 0.93$ **11.** $\frac{32}{52} = \frac{8}{13} \approx 0.62$ **20.** int(13*rand) **21.** int(35*rand) + 1 **22.** int(16*rand) + 47

23.

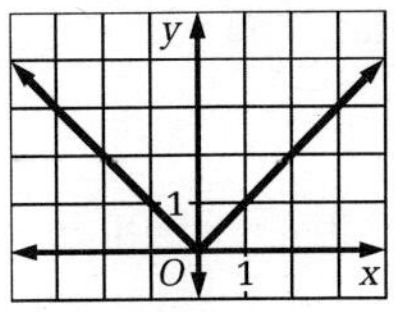

24.

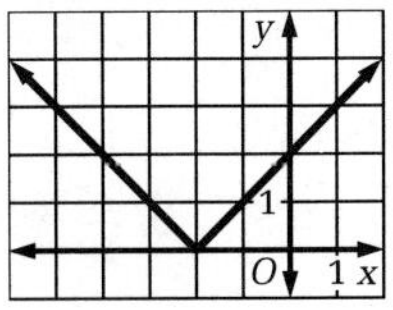

25.

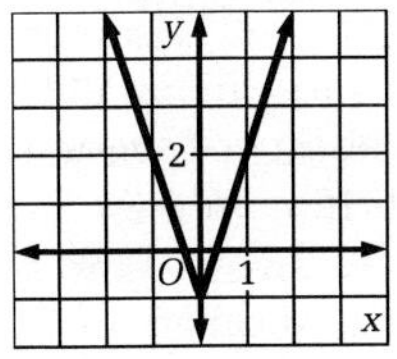

26. a. between 326 and 442 **b.** about 34%

Page 392 Talk it Over

1. Answers may vary. **2.** May is right. She describes $y = x$ and the two boundary lines. The region *does* represent the friends arriving at the same time ($y = x$) and all times *within* 10 minutes of that event, for the region inside the box. **3.** Answers may vary. An example is given. Divide the shaded region into a rectangle and two congruent triangles. Then the area of the shaded region is $10\sqrt{2}(20\sqrt{2}) + 2\left(\frac{1}{2}\right)(10)(10)$ or 500 square units. **4.** The region of the graph would be bounded by the inequalities $y < x + 20$ and $y > x - 20$. The probability would be about 0.89.

Pages 392–394 Exercises and Problems

3. $\frac{275}{900} \approx 0.31$

5.

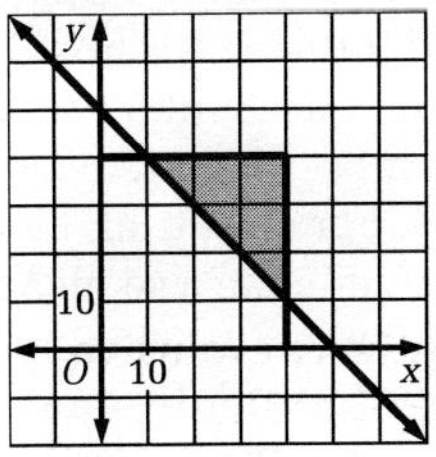

x is the amount of one sister's check; y is the amount of the other sister's check; the square represents all possible sums of the two checks written; the account is overdrawn if $x + y > 50$, that is, if the sum of the checks is more than was deposited.

$P(\text{overdrawn check}) = \frac{450}{1600} = \frac{9}{32} \approx 0.28$

7.

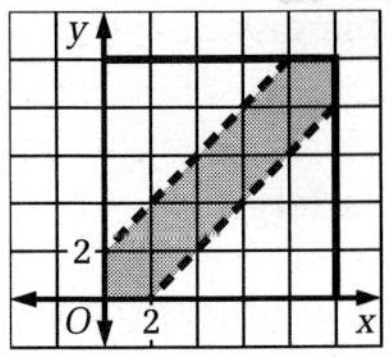

area of square = 100
area of shaded region = 36

Let x be time in minutes after the beginning of the ten-minute period when Janelle starts ringing up the sale and y the time in minutes when Abby starts ringing up the sale. The sales overlap if $x - 2 < y < x + 2$. The rectangle represents all possible times within the 10-minute interval when one of the clerks starts using the register for the indicated sale. The shaded area represents the possible times when the clerks would be using the register at the same time, that is, the time when one clerk would have to wait. $P(\text{one of them will have to wait}) = \frac{36}{100} = 0.36$.

10. $\frac{10}{36} = \frac{5}{18} \approx 0.28$ **11.** $\frac{13}{36} \approx 0.36$ **12.** $\frac{8}{36} = \frac{2}{9} \approx 0.22$ **13.** $\frac{10}{36} = \frac{5}{18} \approx 0.28$ **14.** $f(x) = -2x + 11$ **15.** $f(x) = 0$ **16. a.** $\frac{4}{12} = \frac{1}{3} \approx 0.33$ **b.** $\frac{9}{12} = \frac{3}{4} = 0.75$

Page 394 Checkpoint

1. Answers may vary. Examples are given. I used a formula in Exercise 10 on page 387 to find the probability that one of two events that are not mutually exclusive will occur. The formula $P(A \text{ or } B) = P(A) + P(B) - P(A \text{ and } B)$ is used by finding the individual probabilities and subtracting the probability that both events occur. I also used the formula for the area of a triangle in Exercise 6 on page 393. Areas are used when finding probability of an event with continuous outcomes. **2.** int(20*rand) **3.** int(7*rand) + 6 **4.** For each trial, use the function int(20*rand) + 1 to generate a sequence of 60 random numbers between 1 and 20. Let the number 1 represent a cancelled order and the numbers 2–20 represent orders that are not cancelled. Analyze the number of sequences for which the number 1 appears three or more times. **5.** $\frac{1}{2} = 0.5$ **6.** $\frac{16}{32} = \frac{4}{13} \approx 0.31$ **7.** $P(\text{one person will have to wait}) = \frac{1.75}{4} = 0.4375$

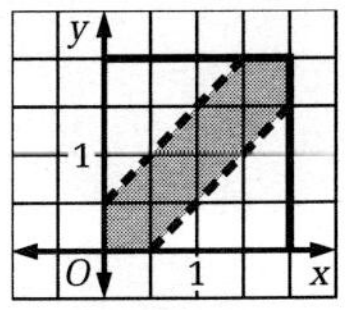

area of square = 4
area of shaded region = 1.75

Pages 396–398 Talk it Over

1. It was given that the student selected is male. There are 19 male students. **2.** $\frac{9}{17} \approx 0.53$ **3.** Tree 1 determines probabilities such as $P(\text{allergic} \mid \text{male})$, while Tree 2 determines probabilities such as $P(\text{male} \mid \text{allergic})$. **4. a.** $P(\text{male and allergic}) = P(\text{male}) \cdot P(\text{allergic} \mid \text{male})$; $P(\text{male and allergic}) = P(\text{allergic}) \cdot P(\text{male} \mid \text{allergic})$

b. Answers may vary. An example is given. $P(\text{female}) = \frac{17}{36}$; $P(\text{not allergic} \mid \text{female}) = \frac{8}{17}$; $P(\text{not allergic}) = \frac{15}{36}$; $P(\text{female} \mid \text{not allergic}) = \frac{8}{15}$; $P(\text{female and not allergic}) = \frac{8}{36}$. The relationship holds. **5.** with replacement: $\frac{4}{25} = 0.16$; without replacement: 0.1 **6.** situation (b), in which the first ball was not replaced

Pages 399–402 Exercises and Problems
3. $\frac{160}{400} = \frac{2}{5} = 0.4$ **5.** $\frac{25}{110} = \frac{5}{22} \approx 0.23$ **7.** $\frac{175}{400} = \frac{7}{16} = 0.4375$ **11.** $\frac{1}{3}$ **17.** $\frac{30}{50} = \frac{3}{5} = 0.6$ **19.** $\frac{12}{20} = \frac{3}{5} = 0.6$ **21.** $\frac{21}{50} = 0.42$ **30.** $\frac{2475}{8100} \approx 0.31$ **31.** $2 \log M + 3 \log N$ **32.** 1540 groups

Page 404 Talk it Over
1. a. Answers may vary. An example is given. In both cases, each average claim is multiplied by the number of such claims and divided by 10,000 and the results are added. The order of the operations is different for the two methods. **b.** Answers may vary. **c.** $\left(\frac{1}{10{,}000}\right)[\$200{,}000(1) + (\$20{,}000)(10) + (\$2000)(200) + (\$0)(9789)] = \frac{\$200{,}000(1)}{10{,}000} + \frac{(\$20{,}000)(10)}{10{,}000} + \frac{(\$2000)(200)}{10{,}000} + \frac{(\$0)(9789)}{10{,}000}$
2. Answers may vary. An example is given. I think it makes sense to share the cost evenly unless a policyholder has repeated claims. Then I think that policyholder should pay more, being more of a risk.

Pages 406–409 Exercises and Problems
5. \$3.50 **7.** No; the person running the game. **15.** \$3.30 **17.** No; if the game were fair, the expected value would be 0. **19.** $\frac{2}{3}$ **20.** 90° **21. a.** $(a + b)^3 = 1a^3 + 3a^2b + 3ab^2 + 1b^3$; $(a + b)^4 = 1a^4 + 4a^3b + 6a^2b^2 + 4ab^3 + 1b^4$ **b.** Answers may vary. Examples are given. The coefficients of the expanded form of $(a + b)^n$ are the numbers in row n of Pascal's triangle. The powers of a in the expanded form of $(a + b)^n$ begin at n and decrease to 0, while the powers of b begin at 0 and increase to n. The sum of the powers of a and b in each term equals n.

Pages 412–413 Talk it Over
1. Answers may vary. Responses are given based on the sample results from step 2 of the Exploration. The results predict a 4% probability, while theoretically the probability is 7.6%. **2.** Answers may vary. Examples are given. Use a systematic list, or use Pascal's triangle. The number in row 10 and diagonal 8 of Pascal's triangle is ${}_{10}C_8$. **3.** 0.55^{10}, or about 0.25% **4. a.** step 2 and step 3 since $P(\text{man}) = 0.3$ and $P(\text{woman}) = 0.7$ **b.** about 0.14%; This probability is about $\frac{1}{50}$ of the probability found in Sample 1. **5.** Six men are hired. **6.** mound-shaped **7. a.** about 0.17 **b.** about 0.265 **c.** The events are mutually exclusive. **8.** No. **9. a.** Yes. **b.** Yes. **c.** Yes. **d.** No.

Pages 415–417 Exercises and Problems
3. about 0.00000525 **5.** about 0.00042 **7.** There are exactly two outcomes for each trial. The trials are independent. Each trial has the same probability of success. The experiment has a fixed number of trials. **9. a.** $\frac{11}{20} = 0.55$ **b.** $\frac{10}{19} \approx 0.53$ if the first person hired is male, and $\frac{11}{19} \approx 0.58$ if the first person hired is female **c.** The trials are not independent, so the trials do not all have the same probability of success. **11. a.** 0.3125 **b.** 0.15625 **c.** 0.03125 **d.** 0.50 **15, 17, 19.** Estimates may vary. **15.** about 0.29; about 0.23; The values are different because $P(\text{senior has a job}) > P(\text{junior has a job})$. **17.** about 0.33 **19.** about 0.96 **25.** –25¢ **26.** 170.8; 37.7 **27.** 10 **28.** 13 **29.** $5\sqrt{2} \approx 7.07$

Pages 419–420 Unit 7 Review and Assessment
1. If the numbers given to the students were originally assigned in a random manner (for example, by putting all the names in a hat and assigning 1 to the first name, and so on), then the procedure described would give a random sampling. If the numbers were not originally assigned at random (for example, if the numbers were given out in sequence by someone who simply visited all the areas of the school), then the sequence produced would not be random, since it is likely that friends would have numbers assigned at the same time and would arrive together, as would groups of athletes, band members, and so on. **2.** int(100*rand) **3.** int(10*rand) + 1 **4. a.** Add the frequencies; divide each frequency by the sum. **b.** Answers may vary. Examples are given. If a call is selected at random, the probability that it was received between 4 P.M. and midnight is about 55%. If a call is selected at random, it is most likely to be in the 8 P.M.–midnight category. **5.** For each trial, use the function int(100*rand) + 1 to generate a sequence of 10 random numbers. Let the numbers 1–60 represent people with previous experience and the numbers 61–100 represent people with no previous experience. Count the number of sequences generated and the number in which 7 numbers greater than 60 appear. Analyze the data. **6.** $\frac{5}{18} \approx 0.28$ **7.** $\frac{1}{6} \approx 0.17$ **8.** $\frac{5}{12} \approx 0.42$ **9.** Answers may vary. Examples are given. Two dice are rolled. Their sum is less than 5 or

greater than 7. Both events cannot occur at the same time, so they are mutually exclusive. However, they account for only 21 of the 36 possible outcomes, so the events are not complementary. **10. a.** True. **b.** False. **11.** $\frac{7}{16} = 0.4375$ **12. a.** Two events are independent if the occurrence of one of the events does not affect the probability of the occurrence of the other. **b.** Events A and B are independent if the probability that B occurs given that A has occurred is exactly the same as the probability of B alone.
13. $\frac{7}{12} \approx 0.58$ **14.** $\frac{7}{15} \approx 0.47$ **15.** No. $P(\text{female}) = \frac{20}{32} = \frac{5}{8}$; $P(\text{female} \mid \text{in the band}) = \frac{7}{15}$; $\frac{5}{8} \neq \frac{7}{15}$

16. a.

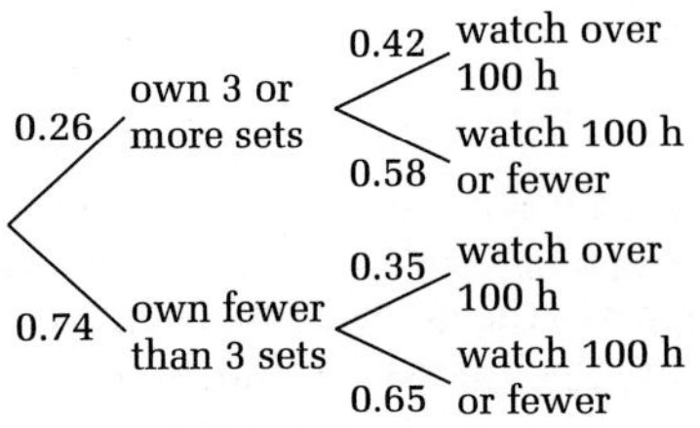

	Watch over 100 h	Watch 100 h or less	Total
Own 3 or more sets	0.109	0.151	0.26
Own fewer than 3 sets	0.259	0.481	0.74
Total	0.368	0.632	1

b. No; $P(\text{over 100 hours per week} \mid \text{three or more televisions}) \approx 0.11$; $P(\text{over 100 hours per week}) \approx 0.37$
17. –40¢ **18. a.** 278 **b.** 278 Calories **19.** ≈ 0.16
20. ≈ 0.00000018

Unit 8

Pages 426–427 Talk it Over

1. For each pair, $r = 0.4$ and $\theta = 330° + n(360°)$ for some integer n. For –30°, $n = -1$; for 690°, $n = 1$.
2. Answers may vary. Examples are given. (0.8, –185°); (0.8, 535°) **3.** Answers may vary. An example is given. Use the same r-coordinate and find $\theta + n(360°)$ for integer values of n. **4.** Answers may vary. An example is given. I think it is easier to use negative angles when the negative angles are small.
5. You can measure distances using a ruler or compass and measure angles using a protractor. You must first designate a reference point.

Pages 427–429 Exercises and Problems

3. 180°; 270° **5.** 45°; 45° **7.** 225°; 225°

17, 19, 21.

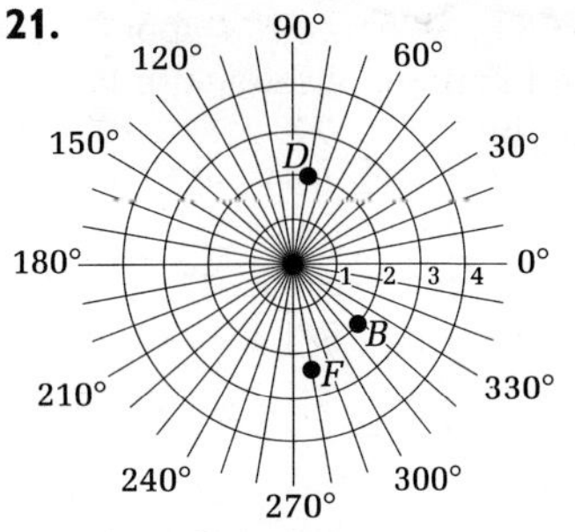

23. Mt. Erebor **25, 27.** Estimates may vary.
25. (about 260, about 92°)
27. (about 560, about –30°) **36.** about 0.38
37. $\frac{1}{2} = 0.5$ **38.** Measure the length of the hypotenuse. To find the cosine of an acute angle, measure the length of the leg adjacent to the angle and divide by the length of the hypotenuse. To find the sine of the acute angle, measure the length of the leg opposite the angle and divide by the length of the hypotenuse.

Pages 431–435 Talk it Over

1. $r = 2.5$ and $\theta = 53.1°$; $x = r\cos\theta \approx 2.5(0.6) = 1.5$ and $y = r\sin\theta \approx 2.5(0.8) = 2$ **2. a.** The three pairs of polar coordinates describe the same point.
b. (5.3, 37.6); (5.3, 37.6) **3.** 90°; 270° **4.** 0°; 180°; 360° **5.** 116.6°; 296.6° **6.** $\sin^{-1}\frac{y}{r} = \sin^{-1}\frac{-7}{7.616} = \sin^{-1}(-0.9191) = -66.8°$; The polar angle in Quadrant III with this sine value is 246.8°.
7. $\tan^{-1}\frac{y}{x} = \tan^{-1}\frac{-7}{-3} = \tan^{-1}(2.333) = 66.8°$; The polar angle in Quadrant III with this tangent value is 246.8°. **8.** Method 1: Use the Pythagorean theorem to find r and use the inverse trigonometric feature of a calculator to find θ. $r^2 = \sqrt{5^2 + (-8)^2}$; $r \approx 9.434$; $\cos^{-1}\frac{x}{r} = \cos^{-1}\frac{5}{9.434} \approx 58°$. The polar angle 57.99° has the correct cosine value but is in the first quadrant. The polar angle in Quadrant IV with this cosine is 302.01°. The polar coordinates of (5, –8) are about (9.43, 302.01). Method 2: Use a graphics calculator.

Pages 436–439 Exercises and Problems

1. If a perpendicular is constructed from the point with polar coordinates (r, θ) to the x-axis and a line is drawn from the origin to the point, a right triangle is formed with acute angle θ and hypotenuse r. The horizontal leg has length x and the vertical leg has length y, so $\cos\theta = \frac{x}{r}$ and $\sin\theta = \frac{y}{r}$. Then $x = r\cos\theta$ and $y = r\sin\theta$. **3.** (1, 2); Drive 1.5 mi east on Five Mile Rd., 2 mi north on Farmington, and then 0.5 mi west on Seven Mile Rd. **5.** (1, 0) **7.** (0.71, 0.71)
9. (16.2, 22.5°) and (16.2, 67.5°)
13. $\tan\theta = \frac{y}{x} = \frac{y/r}{x/r} = \frac{\sin\theta}{\cos\theta}$

15. about (–482, 688); Belgrano II is about 482 mi "west" and 688 mi "north" of the South Pole.
23. opposite **25.** the same **27.** 215° **29.** 10°, 350° **31.** 195°, 345 **33.** 155° **35.** 350° **37.** 100°, 280° **39.** 152.87°, 207.13° **41.** 311.41°, 228.59° **43.** (5.7, 45°) **45.** (28.0, 173.84°) **47.** (4.3, 180°) **49.** (2, 120°) **51.** $C(1.4, 0°)$, $B(3.4, 49.8°)$, $A(0, 90°)$, $E(3.4, 130.2°)$, $D(1.4, 180°)$ **53.** $-\frac{4}{5}$ **55.** –0.9772 **57.** (2, 0°) or (2, 120°) **58.** 2000, 4140, 6429.8, 8879.886, 11,501.47802, 14,306.58148 **59.** 5

Pages 441–442 Talk it Over

1. The (–4, –3) on the left of the diagram indicates the rectangular coordinates of the point in the plane that is 4 units to the left of the *y*-axis and 3 units below the *x*-axis. The (–4, –3) on the right indicates the vector with horizontal shift –4 and vertical shift –3.
2. a. (2, 3) **b.** (5, 126.9°) **3. a. u** and **w** **b. t** and **v**
4. a. No. **b.** Answers may vary. An example is given. The figure shows the vectors added in both orders, beginning at the same point. The resulting figure is a parallelogram whose diagonal is the resultant vector from both additions, **a** + **b** and **b** + **a**. **5. a.** The magnitude of **u** is twice that of **v**; the direction is the same. **b. w** has magnitude three times that of **v** and the same direction as **v**; **t** has magnitude half that of **v** and the same direction as **v**. **c.** (–1)**v** has the same magnitude as **v**. Its direction is opposite that of **v**.

Pages 444–445 Exercises and Problems

5, 7, 9.

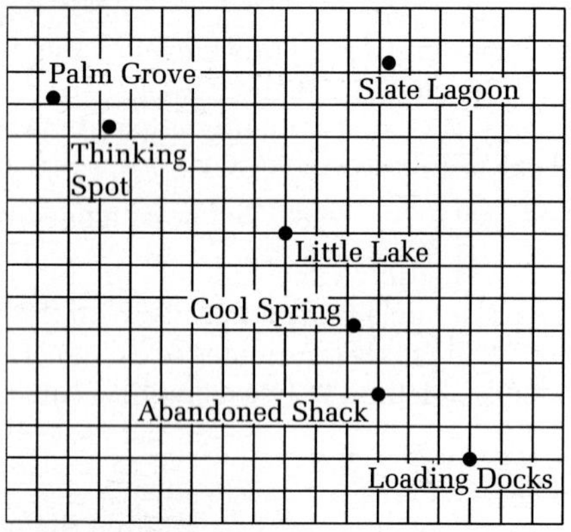

11.

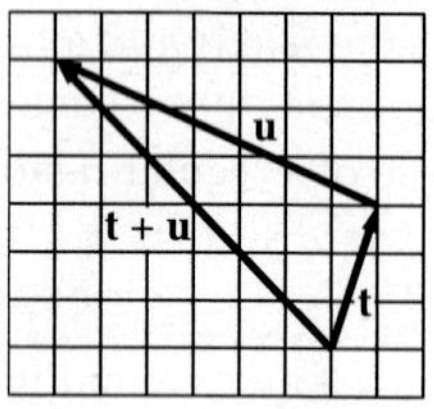

(8.5, 135°); (–6, 6)

13.

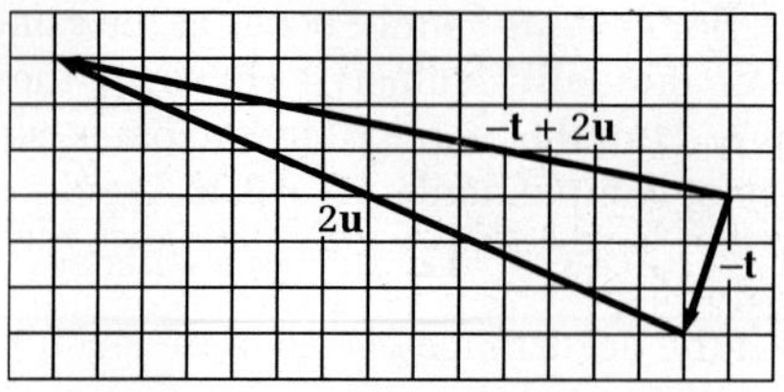

(15.3, 168.7°); (–15, 3)

15.

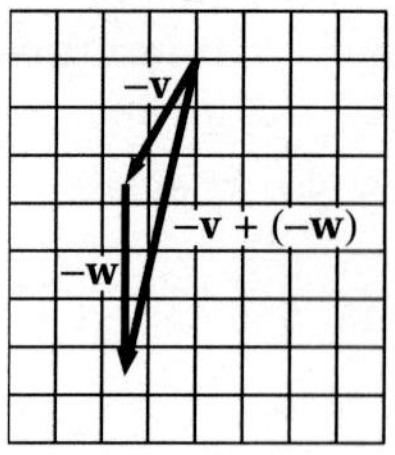

(6.8, –102.8°); (–1.5, –6.6)

21. (8.7, –33.69°) **22.** 0.588 **23.** (1.56, –12.70)

Pages 447–449 Talk it Over

1. The horizontal component of the airplane's speed is its speed relative to the ground, or *ground speed.* The vertical component of the airplane's speed is the rate at which it is climbing, or *rate of climb.*
2. a. (200, 15°); its actual speed through the air and its angle of ascent **b.** (193.19, 51.8); the horizontal and vertical distances of the jet from its takeoff point after 1 h **3. a.** $\sqrt{34}$; 60; $\sqrt{34}$; 60 **b.** 5; 3; $\frac{3}{5}$ **4.** The methods are the same. **5.** so the vectors could be added by adding coordinates **6.** to find the bearing of the airplane **7. a.** decreases **b.** Yes; with a bearing of 30°, the wind would increase the airplane's speed. **c.** 40°; 220°

Pages 449–452 Exercises and Problems

3. (6.32, 161.57°) **5.** (3.5, 6.06) **7.** (2.05, –1.43) **9.** (4, 3) **11.** $(2\sqrt{2}, 1)$ **13.** (85.10, –93.96°) **15.** about 246.2 mi/h; about 43.41 mi/h **22.** (–6, 15) **23.** about 11.9 **24.** 11 **25. a.** $h(t) = -16t^2 + 34t + 6$ **b.** after about 2.3 s

Page 452 Checkpoint

1. No; the magnitude of **u** + **v** will be $r_1 + r_2$ only if **u** and **v** have the same direction. **2.** *C* **3.** *F* **4.** *G* **5.** (6.4, 128.66°) **6.** (4.24, –45°) **7.** (4, 0°) **8.** (–2.74, 7.52) **9.** (5.83, –59.04°) **10.** 51.37°; 308.63° **11.** 0.8722 **12.**

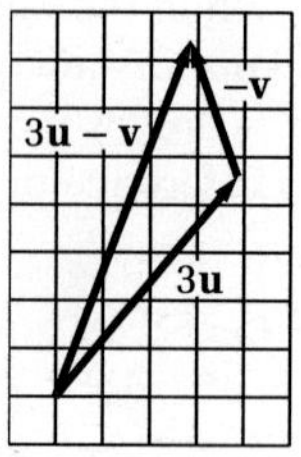

13. (3, 14) **14.** (6.0, –68.4°)

Pages 453–455 Talk it Over
1. a. The golf ball on the left has no horizontal motion. The golf ball on the right moves horizontally at a steady rate as it falls. **b.** The golf balls move vertically at the same rate. **c.** No; the vertical motion of the two golf balls is the same. **2. a.** $a \approx -4.9$ **b.** $x = 2t$ **3. a.** -3 **b.** after about 0.71 s
4. Yes.

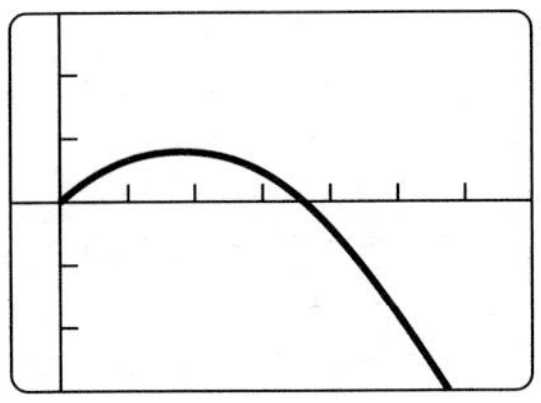

5. Answers may vary. An example is given. The skateboard will not achieve the same height and will hit the ground slightly sooner.

Pages 456–458 Exercises and Problems
5. $y = 1.36x$ **7.** $y = 0.35x - 0.06x^2$ **9.** $x^2 + y^2 = 12.96$ or $y = \pm\sqrt{12.96 - x^2}$ **11. a.** $x = 33.88t$; $y = 26.47t - 4.9t^2 + 1$

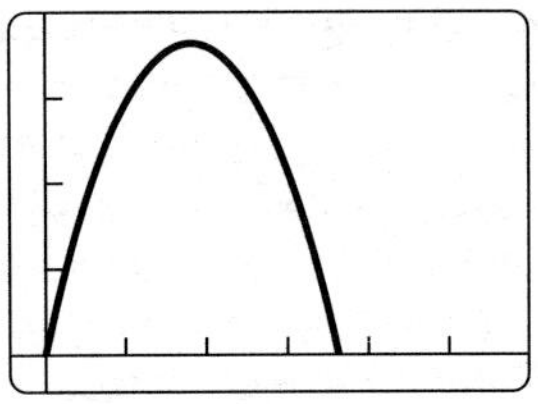

b. $y = 0.78x - 0.0043x^2 + 1$
c.

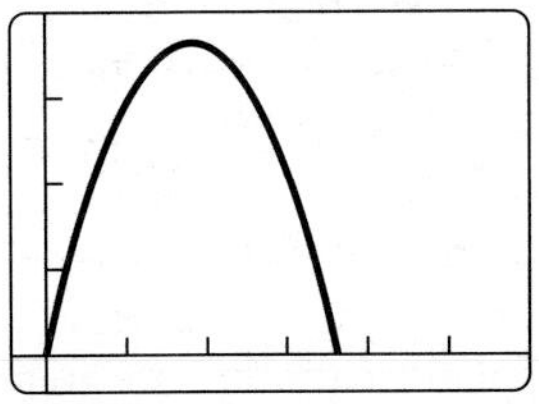

d. The graphs are essentially the same. They may not be identical due to rounding.

e. Answers may vary. An example is given. The parametric equations allow you to determine the horizontal and vertical positions of the ball at any given time t. The single equation allows you to determine the height of the ball at any given horizontal distance x from the point where the ball was hit. **29.** If $a \neq c$, the graph is an ellipse centered at (b, d) with one axis $2a$ units high on the line $x = b$, and one axis $2c$ units wide on the line $y = d$. If $a = c$, the graph is a circle with radius a and center (b, d). **31.** $x = 6 \cos t + 3$; $y = 6 \sin t - 2$ **35. a.** $(3, -2)$ **b.** $(1, -12)$
36. a. $(3, -3)$ **b.** $(7, -9)$ **37.** $5\sqrt{5} \approx 11.18$ cm
38. about 7.28 units

Page 461 Talk it Over
1. because of the Pythagorean identity, which states that for every angle θ, $\cos^2\theta + \sin^2\theta = 1$ **2.** Answers may vary due to measurement or rounding; 1: 1230 ft; 2: 1140 ft; 3: 940 ft; They are the same. **3.** Yes; since the term $2ab \cos C = 0$, the formula simplifies to $a^2 + b^2 = c^2$. The Pythagorean theorem is a special case of the law of cosines for $\theta = 90°$.

Pages 463–466 Exercises and Problems
1. the lengths of two sides and the measure of the included angle **3.** Measurements may vary slightly. Calculated value is given to the nearest tenth. 9.9 cm **11.** Answer is given to nearest tenth. 3.5 in. **17.** The angle at the vertex marked by the Lincoln Memorial is 44.4°; the third angle is 10.4°. **23.** Answer is given to the nearest tenth. 109.5° **29.** $x = 5 \cos \theta$; $y = 5 \sin \theta$ **30. a.** True. **b.** True. **31.** 14

Page 469 Talk it Over
1. from Aisyālukharka: about 29,030 ft; from Mayām: about 28,720 ft **2.** SAS

Pages 471–474 Exercises and Problems
3, 5, 7, 9. Measurements are given to the nearest tenth. **3.** $m\angle Z = 57°$; $x = 16.7$ m; $y = 21.3$ m
5. $m\angle A = 104°$; $a = 40.3$ m; $b = 21.4$ m

	$m\angle A$	$m\angle B$	$m\angle C$	a	b	c
7.	109°	46°	25°	58	44.1	25.9
9.	90°	38°	52°	30.9	19	24.3

11. a. 39.249° **b.** 7.359 km; 8.124 km **c.** 2.518 km **d.** 2.520 km **e.** 2.519 km **f.** (a) 40° (b) 7.21 km; 8.000 km (c) 2.443 km (d) 2.483 km (e) 2.463 km; The estimate is decreased about 2.2%.
17.

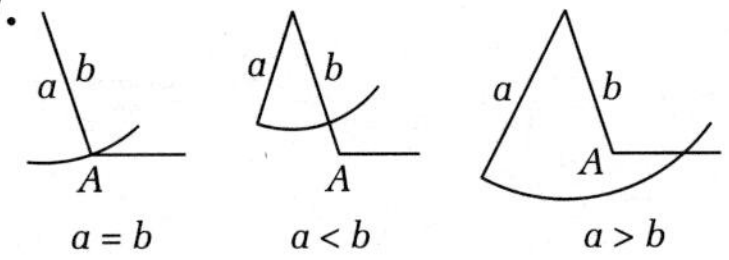

If $a \leq b$, no triangle is formed. If $a > b$, one triangle is formed.

19. 1 triangle **21.** no triangle **23.** no triangle
25, 27. Answers are given to the nearest tenth. Answers may vary slightly due to rounding or to solution method. **25.** $m\angle B = 24.6°$; $m\angle C = 44.4°$; $b = 5.4$ **27.** not possible **29, 31, 33.** Solution methods may vary; examples are given. Answers are given to the nearest tenth. Answers may vary slightly due to rounding or to solution method. **29.** $m\angle S = 69°$; $r = 12.3$ km; $s = 10.5$ km; law of sines **31.** $m\angle R = 46°$; $r = 11.1$ in.; $s = 13.5$ in.; law of sines
33. $m\angle S = 72°$; $s = 10.4$ km; $t = 10.9$ km; law of sines
35. $x^2 = y^2 + z^2 - 2yz \cos X$; $y^2 = x^2 + z^2 - 2xz \cos Y$; $z^2 = x^2 + y^2 - 2xy \cos Z$ **36.** Methods may vary. An example is given. Suppose a regular n-sided polygon has an interior angle whose measure is 80°. Then the measure of an exterior angle is $180° - 80° = 100°$. The measure of each exterior angle of a regular n-sided polygon is $\frac{360°}{n}$, so $100° \cdot n = 360°$ and $n = 3.6$. This contradicts the fact that n must be an integer. The assumption that there is a regular polygon with an interior angle whose measure is 80° is false. Therefore, no regular polygon has an interior angle whose measure is 80°. **37.** Answers may vary. An example is given. data: 75, 66, 84, 82, 88, 71, 72, 87; standard deviation: 7.67

Pages 476–477 Unit 8 Review and Assessment
1. (180, 180°) **2.** (180, 202.5°) **3.** (180, 225°)
4. (−180, 0) **5.** about (−166.3, −68.9)
6. about (−127.3, −127.3)
7.
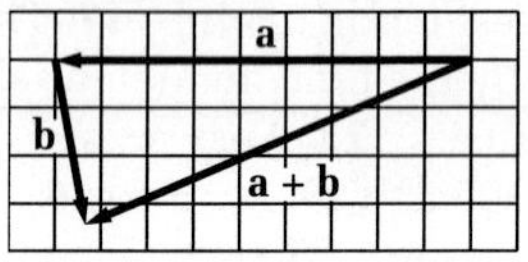

8.
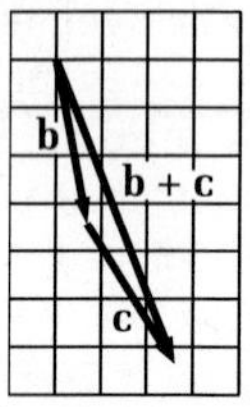

9.
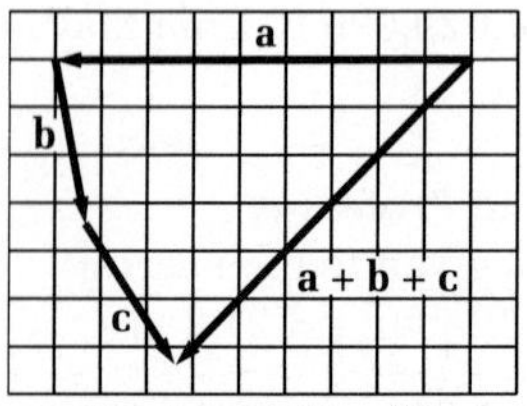

10.
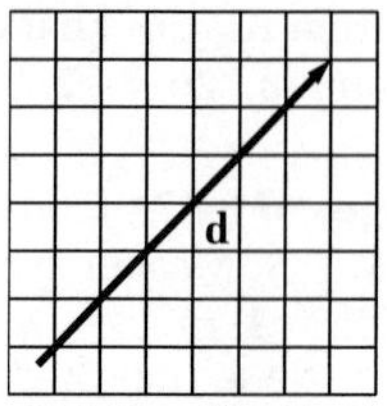

11. (−166.3, −68.9) **12.** (−127.3, −127.3) **13.** (0, 0)
14. a circle with radius 1 centered at the origin
15. a circle with radius 1 centered at (4, −5)
16. an ellipse with its center at the origin, 8 units wide, 10 units high **17.** about 78.6 in.
18.
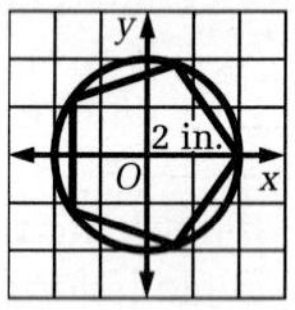

2.35 in.

19.
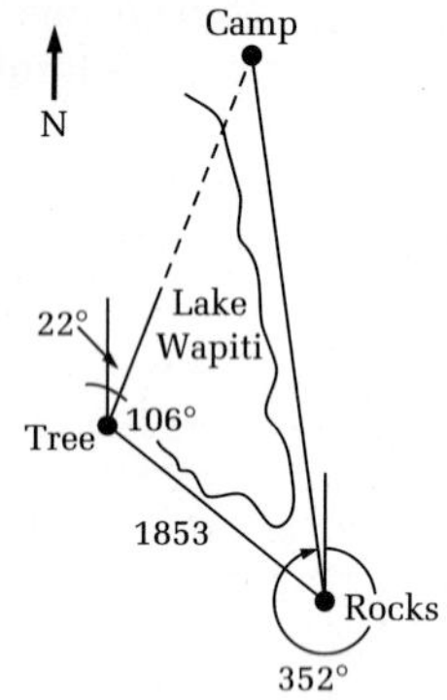

20. about 3562 paces
21. about 2574 paces
22. about 2:28

Unit 9

Pages 483–485 Talk it Over
1. The number in each row of the third column is 14 more than the number in the corresponding row of the second column. **2.** about 66.7; about 80.7; The mean of the heights with stilts is 14 more than the mean of the heights without stilts. **3.** 66; 80; The median of the heights with stilts is 14 more than the median of the heights without stilts. **4.** 12; 12; The ranges are the same. **5.** The histograms have exactly the same shape. The second is the first shifted to the right. **6.** Answers may vary. An example is given. The standard deviation is a measure of spread. The data in both cases are spread out in exactly the same manner. Let x_i represent the first data points and X_i represent the second. For every i, $X_i = x_i + 14$, so $X_i - \overline{X} = (x_i + 14) - (\overline{x} + 14) = x_i - \overline{x}$). Since the sum of the deviations is the same, the variance and standard deviation are the same. **7. a.** decrease; When the same constant value is added to each value in a data set, the mean is translated by the constant value, in this case, a negative number. **b.** decrease; When the same constant value is added to each value in a data set, the median is translated by the constant value, in this case, a negative number. **c.** stay the same; When the same constant value is added to each value in a data set, the range remains the same. **d.** stay the same; When the same constant value is added to each value in a data set, the standard deviation remains the same. **8.** left; When the same constant value is added to each value in a data set, the graph of the distribution is translated by the constant value, in this case, a negative number.

Pages 486–488 Exercises and Problems
3. The median, lower extremes, and upper extremes for the heights with 18-in. stilts are each 18 more than the corresponding values for the heights without stilts. The ranges are the same. **5.** 160; about 40.2
7. 156.5; 40.2 **20.** about 20.78 cm **21.** (7, −1)
22. (15, −4) **23.** (17, −8)
24.
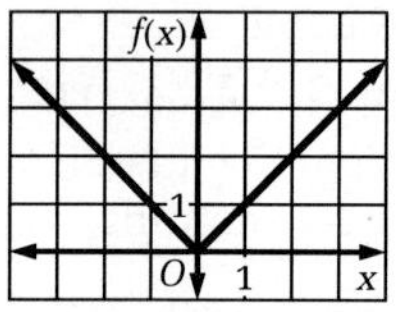

domain: all real numbers; range: all real numbers greater than or equal to 0
25.
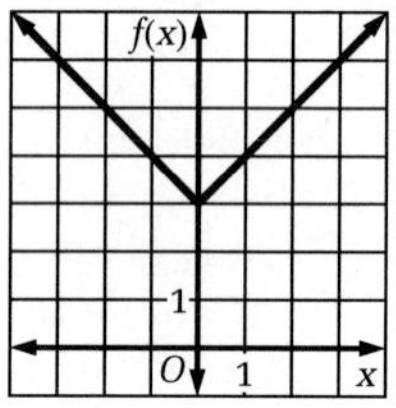

domain: all real numbers; range: all real numbers greater than or equal to 3
26.
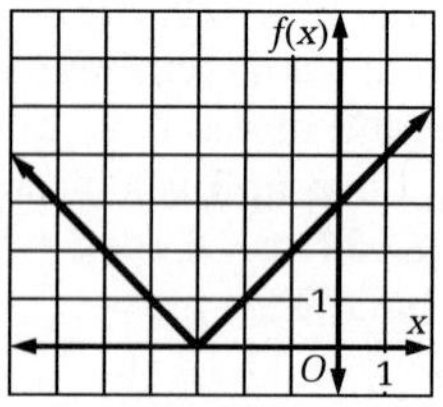

domain: all real numbers; range: all real numbers greater than or equal to 0

Pages 490–492 Talk it Over
1. (0, 0); (2, 1); (2, 1) **2.** b and c; a and c; The graph of $y - k = f(x - h)$ is the graph of $y = f(x)$ translated k units up if $k > 0$, k units down if $k < 0$, h units to the right if $h > 0$, and h units to the left if $h < 0$.
3. a. $x + h$; $y + k$ **b.** $\begin{bmatrix} h \\ k \end{bmatrix}$ **4.** $9 + 1 = 10^{4-3}$, $10 = 10$✓
5. The origin is not part of the graph of $y = 10^x$.

6. The domain of the parent function is all real numbers greater than or equal to 0. The domain of the function in part (a) is all real numbers greater than or equal to −4. If you think of the domain of the parent function as an interval of the x-axis, the domain of the second function is the interval with the endpoint shifted 4 units to the left. **7.** The asymptote of the graph in part (b) is the asymptote of the parent graph translated 1 unit down.

10.

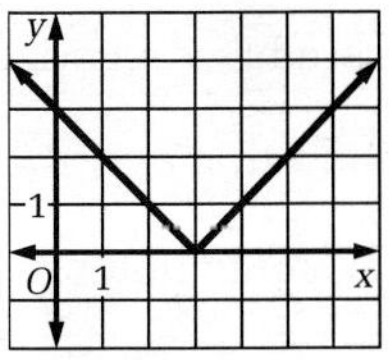

11.

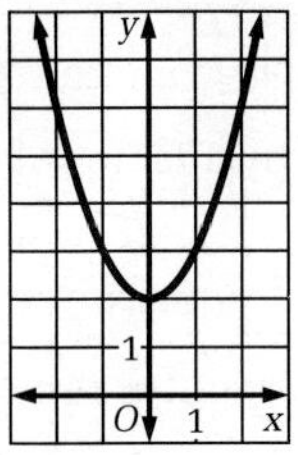

12.

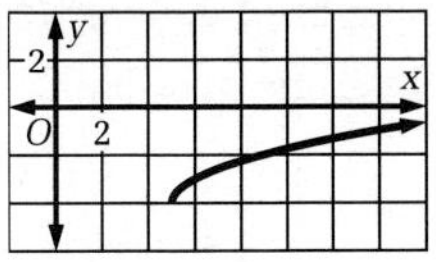

13. 2.322 **14.** 0.565 **15.** −2.066 **16.** −0.120

17. $y \approx 2.8x^2 - 13.5x + 18.7$ or $y = \frac{17}{6}x^2 - \frac{27}{2}x + \frac{56}{3}$

18. $y \approx -0.08x^2 + 0.2x + 1.1$ or $y = \frac{-7}{88}x^2 + \frac{19}{88}x + \frac{47}{44}$

Pages 492–495 Exercises and Problems

1. $y = x^2$; $(-2, 7)$ **3.** $y = 10^x$; $(-9, 0)$

5.

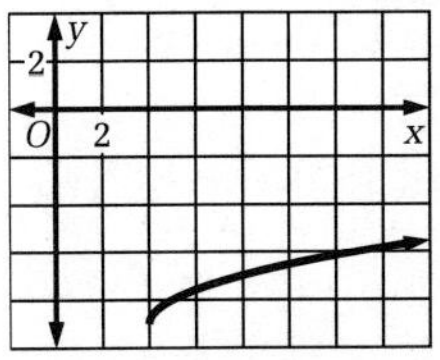

7.

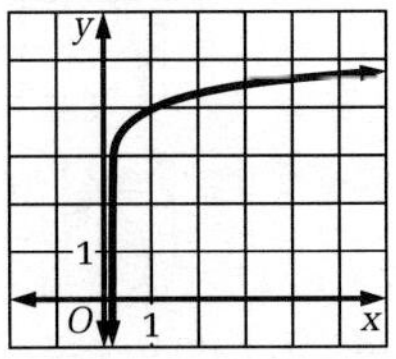

15. $y + 3 = x^3$ **17.** $y - 2 = \sqrt{x - 1}$ **19.** $y - 2 = 10^x$
21. a. $h = 3$; $k = 1$ **b.** $x = 3$; $y = 1$ **26.** 82; 9

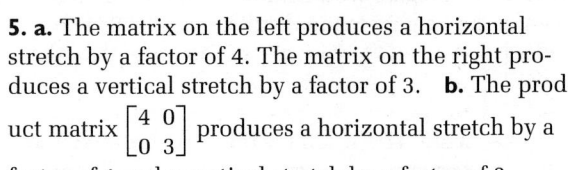

27. about 1.6 **28.** $\begin{bmatrix} 5 & 0 \\ 0 & 8 \end{bmatrix}$ **29.** $\begin{bmatrix} 1\frac{3}{5} & 0 \\ 0 & \frac{1}{2} \end{bmatrix}$

30. $\begin{bmatrix} 6 & 9 & 6 & 12 \\ 2 & 0 & 4 & 2 \end{bmatrix}$

Pages 496–497 Talk it Over

1. 0.2; 0.2 dollars to 1 franc **2. a.** 1500 **b.** 300 **c.** 0.2 **3. a.** 1550 **b.** 310 **c.** 0.2 **4. a.** 750 **b.** 150 **c.** 0.2 **5.** The columns are the same height. The columns in the second histogram are five times as wide as the columns in the first. The endpoints of each column in the second histogram are 5 times the endpoints of the corresponding column in the first. **6.** 0.2 **7.** Predictions will vary. Each will be multiplied by 4. For the original data, the mean is 8, the median is 7, the range is 15, and the standard deviation is about 4.94. For the transformed data, the mean is 32, the median is 28, the range is 60, and the standard deviation is about 19.76.

Pages 498–500 Exercises and Problems

3. a. 7.3; 6.7; 4.5; about 1.8 **b.** 21.9; 20.1; 13.5; about 5.4 **c.** Data intervals may vary.

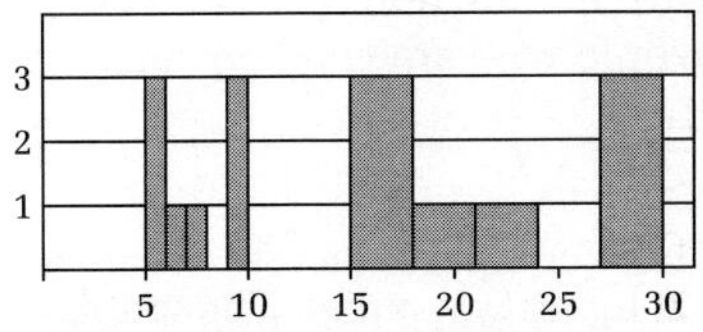

5. a. $d > 1$ **b.** $d < 1$ **7. a.** about 0.000126
b. Answers are rounded to the nearest tenth. 3.5; 1.0; 11.0; 3.9 **c.** Aaron used the fact that the value of d for Earth is the median; Yolanda used the fact that the value of d for Earth is significantly less than the mean.

Page 503 Talk it Over

1. a. The parent graph is stretched horizontally by a factor of 2. **b.**

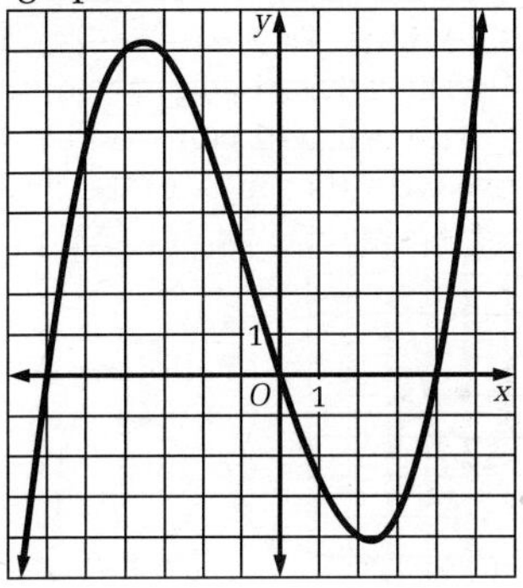

2. a. $\frac{y}{3} = 2x(2x - 2)(2x + 3)$ **b.** the origin

Pages 504–508 Exercises and Problems

1. a. $\begin{bmatrix} 0 & 8 & 8 & 4 & 0 \\ 0 & 0 & 3 & 5 & 3 \end{bmatrix}$

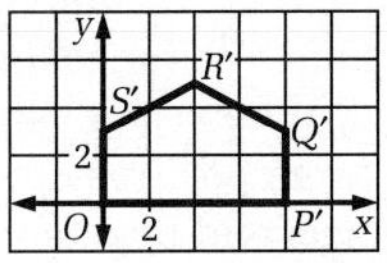

b. The polygon is stretched vertically by a factor of $\frac{1}{2}$.

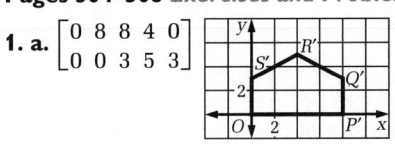

3. a. $\begin{bmatrix} 0 & 8 & 8 & 4 & 0 \\ 0 & 0 & 12 & 20 & 12 \end{bmatrix}$

b. The polygon is stretched vertically by a factor of 2.

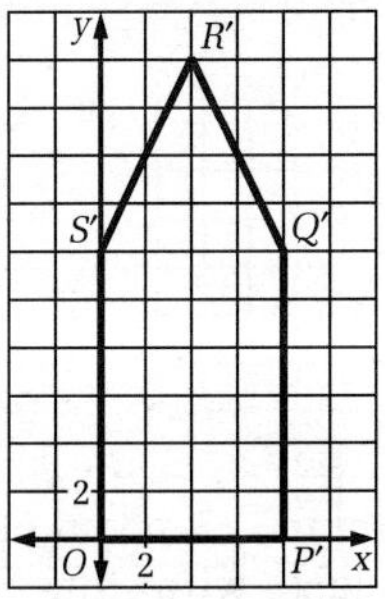

5. a. The matrix on the left produces a horizontal stretch by a factor of 4. The matrix on the right produces a vertical stretch by a factor of 3. **b.** The product matrix $\begin{bmatrix} 4 & 0 \\ 0 & 3 \end{bmatrix}$ produces a horizontal stretch by a factor of 4 and a vertical stretch by a factor of 3.

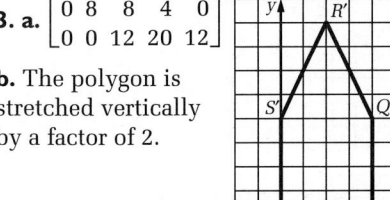

9. $\begin{bmatrix} 9 & 0 & -9 \\ 0 & 9 & 0 \end{bmatrix}$; The transformation produces a horizontal stretch by a factor of 3 and a reflection over the y-axis. **11.** $\begin{bmatrix} 9 & 0 & -9 \\ 0 & -4\frac{1}{2} & 0 \end{bmatrix}$; The transformation produces a horizontal stretch by a factor of 3, a vertical stretch by a factor of $\frac{1}{2}$, and reflections over both axes.

17. a.

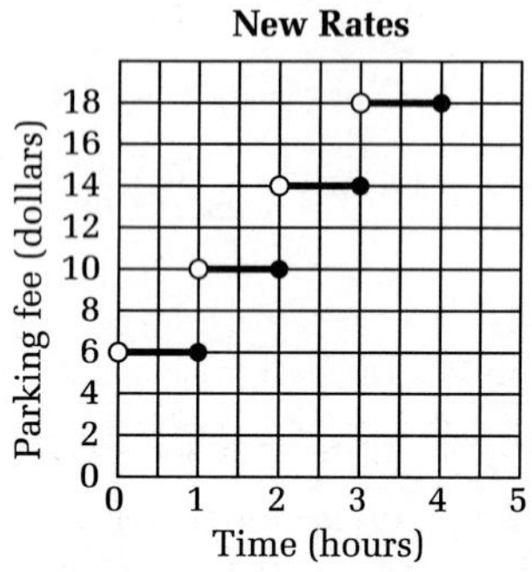

b. $\frac{y}{2} = f(x)$ or $y = 2(f(x))$ **23.** A **25.** D

27.

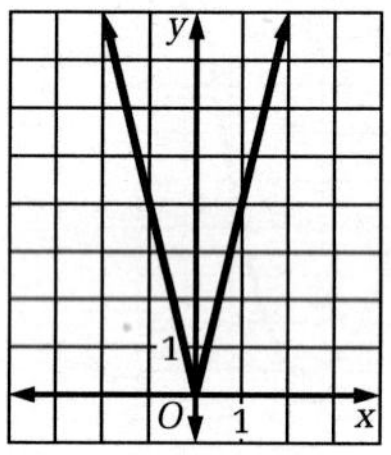

35. $y = \left(\frac{x}{2}\right)^3$ **37.** $y - 2 = |x - 3|$ **39.** Both are multiplied by 0.8. **40.** 310° **41.** 225° **42.** 120° **43.** 25° or 335° **44.** $y + 2 = |x - 4|$

Page 509 Checkpoint

1. When the same constant value is added to each value in a data set, the same value is added to the mean and the median. The range and standard deviation are unchanged. When each value in a data set is multiplied by the same positive constant, the mean, the median, the range, and the standard deviation are all multiplied by the same constant. **2.** 29; 10

3.

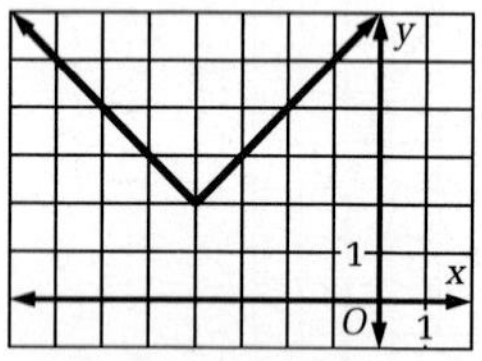

4.

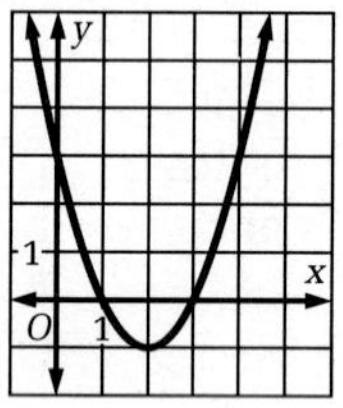

5.

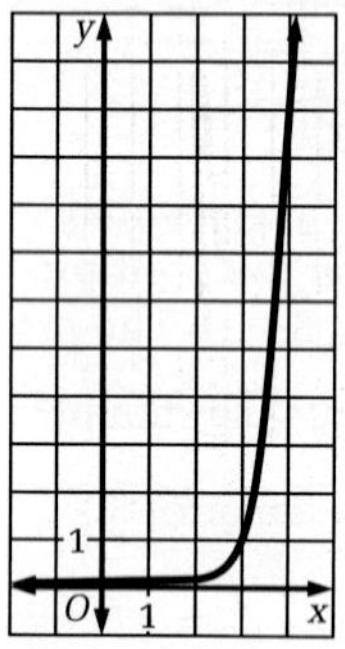

6. \$154.88; \$94.50

7. $\frac{y}{1.05} = 0.05x$ or $y = 0.0525x$

8. $\frac{y}{0.75} = 0.05x$ or $y = 0.0375x$

Pages 510–514 Talk it Over

1. how Celsius and Fahrenheit temperatures are related **2.** about 11.3°C; slightly lower **3.** Answers may vary. Examples are given. elevation, nearness to the ocean, prevailing wind patterns **4.** The resulting function is $y = 2(2^{x-1}) = 2^x$. The graph is unchanged.

5. $y = \left(\frac{2}{3}x\right)^3 + 2$

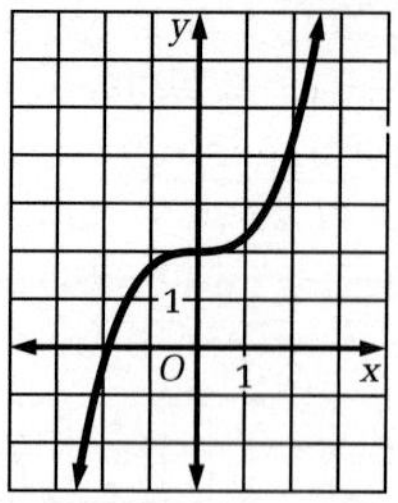

6. a. $\frac{2}{3}$ **b.** $y - 2 = \left|\frac{x - 7}{\frac{2}{3}}\right|$ or $y - 2 = \left|\frac{3(x - 7)}{2}\right|$ **c.** Yes.

Pages 514–518 Exercises and Problems

1. stretching and translating; stretching **3. a.** 1969, 1974, 1976, and 1977; 1965, 1966 and 1975 **b.** No. **c.** part (a) above: 1961, 1973, and 1975; below: 1965, 1976, 1979, and 1980; part (b) No. **7.** \$31,790; about \$11,347 **13.** a vertical stretch by a factor of 16 and a translation 160 units up

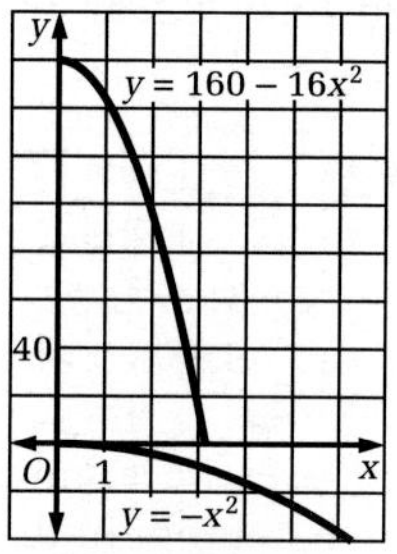

19. $y = -3|x - 3| + 2$

21. $y = \sqrt{\frac{x + 1}{2}} + 1$

23. $y = 3 \log x + 5$

25. The graph is the graph of $y = x(x + 3)(x + 3)$ with a horizontal stretch by a factor of $\frac{1}{2}$.

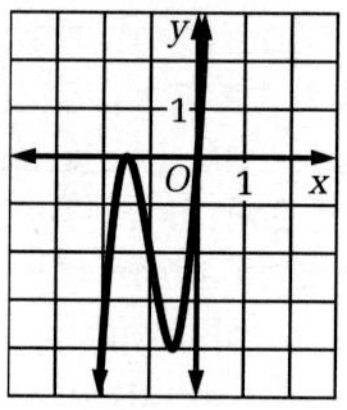

26. 0.682256

27. the randomly selected group; the randomly selected group; Journalists are more likely to have better training in writing correctly and are more likely to perform well as a group. The randomly selected people are more likely to include both people who are poor writers and people who are good writers.

Page 519–523 Talk it Over

1. No; it is unreasonable to compare the two because of the age difference. **2. a.** Georgia: 6.55 min; Texas: 14.94 min **b.** Both are above; below. A lower time indicates a faster run. **c.** Georgia: 0.30 above; Texas: 0.56 above **d.** The Georgia woman did better. She was only 0.30 standard deviation above the mean for her division, while the Texas woman was 0.56 standard deviation above the mean for her division.

3. above; below; The score and the mean are the same. **4.** A lower z-score means a better performance relative to a particular age group, so z-scores for different age groups can be compared. **5.** 5% of the heights are less than Sara's height. **6.** 50th percentile

Pages 523–525 Exercises and Problems

1. Answers may vary. An example is given; in situations where you want to compare data from different categories; z-scores standardize results so that results from different categories can be compared.
3. Answers are given to the nearest thousandth.
a. 31.657; 2.900 **b.** –1.371; –1.171, –1.047; –0.796; –0.344; 0.473; 0.811; 0.953; 1.204; 1.287 **c.** 0; 1; These are the mean and standard deviation of the normal distribution.

5. a.

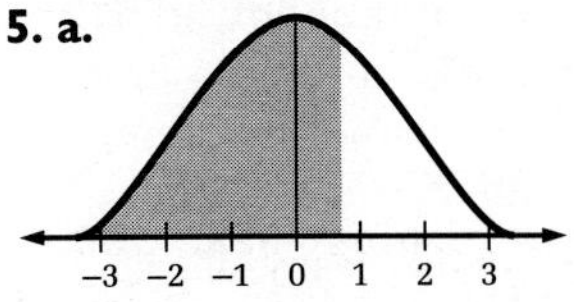

b. about 76%

7. a.

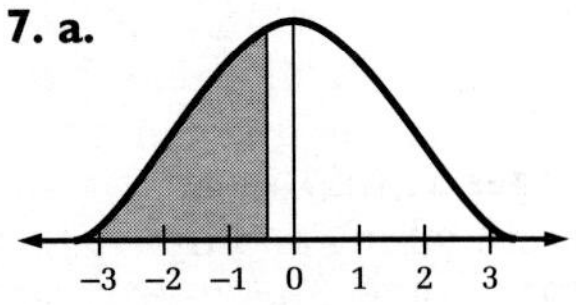

b. about 34%

9. about –0.7 **11.** about 0.3 **13.** Stretch curve 1 horizontally by a factor of 2 and vertically by a factor of $\frac{1}{2}$.

18.

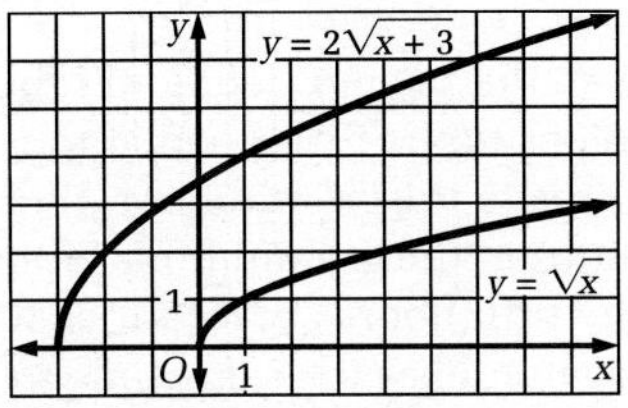

19.

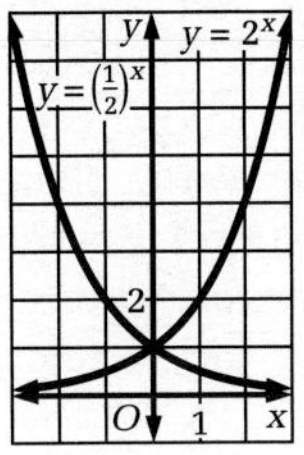

Answers may vary. An example is given. Consider the point (1, 2), which is on the graph of $y = 2^x$. If the functions were inverses, (2, 1) would be on the graph of $y = \left(\frac{1}{2}\right)^x$, which it is not, since $\left(\frac{1}{2}\right)^2 = \frac{1}{4}$, not 1. The graphs are reflections of one another over the y-axis **20.** (0, 1)

Page 529 Talk it Over

1. Write the equation as $\frac{y+3}{\frac{1}{2}} = -|x - 6|$. Reflect the graph of the parent function $y = |x|$ over the x-axis to get the graph of $y = -|x|$. Stretch the graph of $y = -|x|$ vertically by a factor of $\frac{1}{2}$. Finally, translate the graph by the vector (6, –3). **2. a.** 2500 ft **b.** 5.2 min

Pages 530–532 Exercises and Problems

1. domain: all real numbers less than or equal to 4; range: all real numbers greater than or equal to 1

3.

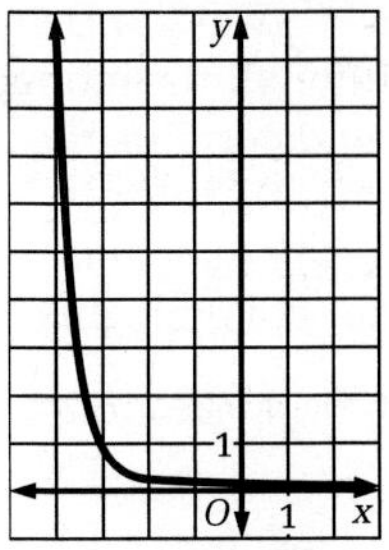

5.

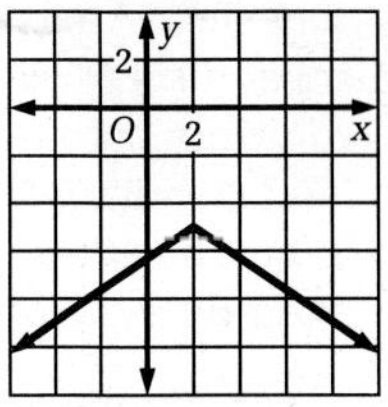

7.

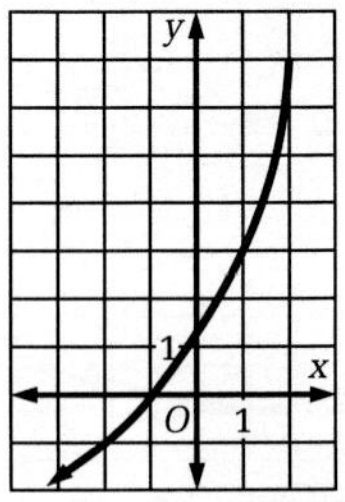

13. $y = 10^{-(x-1)}$
15. $y - 7 = -2\sqrt{-x-3}$
17. $y = -2 \log x$ **20.** 0.75

21. $PQ = \sqrt{x-(-r)^2 + (y-0)^2} = \sqrt{(x+r)^2 + y^2}$; $QR = \sqrt{(x-r)^2 + (y-0)^2}$; $\overline{PR}$ is a diameter, so $PR = 2r$. Then $(PQ)^2 + (QR)^2 = (x+r)^2 + y^2 + (x-r)^2 + y^2 = x^2 + 2xr + r^2 + y^2 + x^2 - 2xr + r^2 + y^2 = 2(x^2 + y^2 + r^2) = 2(2r^2) = 4r^2 = (PR)^2$. By the converse of the Pythagorean theorem, $\triangle PQR$ is a right triangle and $\angle PQR$ is a right angle.

Pages 534–535 Unit 9 Review and Assessment

1. a. 545.6; 517; 280; about 112

b.

Name of building	Height (ft)
Amoco Building	5728
Anaconda Tower	5787
Arco Tower	5807
First Interstate Tower North	5714
Mountain Bell Center	5989
1999 Broadway	5824
Republic Plaza	5994
17th Street Plaza	5718
Stellar Plaza	5717
United Bank of Denver	5978

c. 5825.6; 5797; 280; about 112

2.

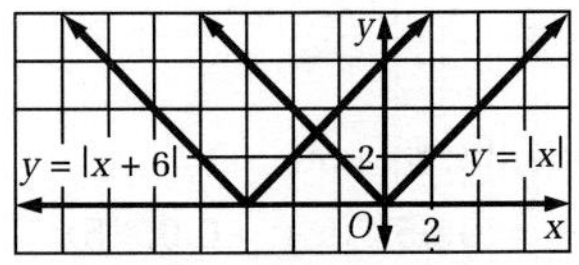

3.

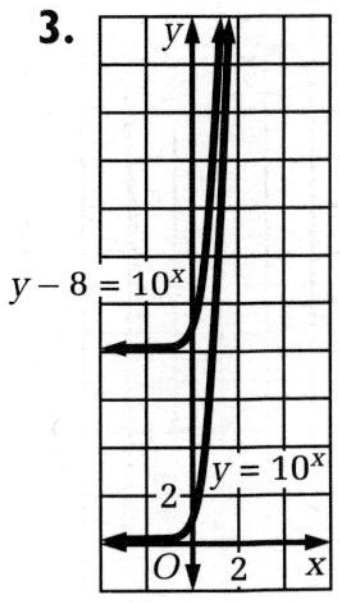

4.

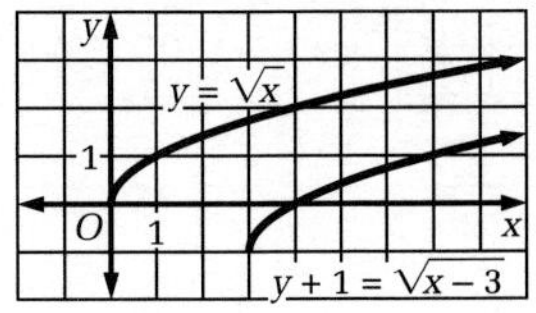

5. Answers may vary. An example is given. (1) Determine the mean and standard deviation of the model heights and multiply both by 100 to determine the actual mean and standard deviation. (2) Multiply each model height by 100 and then calculate the mean and standard deviation of the actual heights.

6. $\begin{bmatrix} 0 & 20 & 24 & 4 \\ 0 & 0 & 3 & 3 \end{bmatrix}$; horizontal stretch by a factor of 4

7. $\begin{bmatrix} 0 & 5 & 6 & 1 \\ 0 & 0 & 1 & 1 \end{bmatrix}$; a vertical stretch by a factor of $\frac{1}{3}$

8. $\begin{bmatrix} 0 & 10 & 12 & 2 \\ 0 & 0 & 15 & 15 \end{bmatrix}$; a horizontal stretch by a factor of 2 and a vertical stretch by a factor of 5

9.

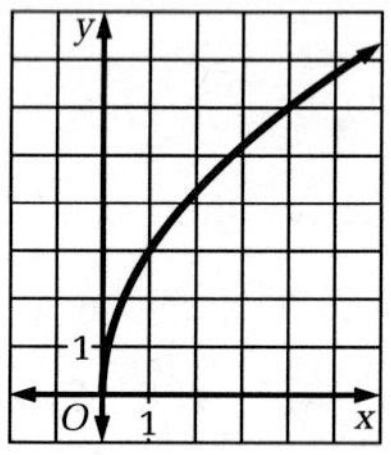

10.

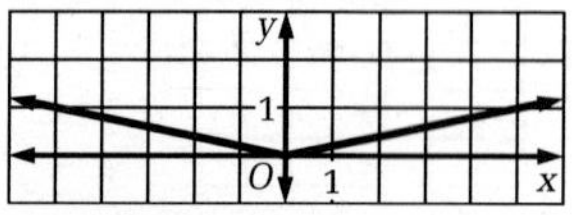

11.

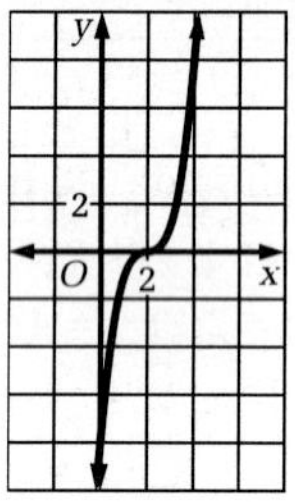

12. about 19.3°C; about 0.7°C **13.** Answers may vary. An example is given. $f(x) = \frac{5}{3}x - 5$. The range of the actual scores is 24 and we want it to be 100 – 60 = 40, so $m = \frac{40}{24} = \frac{5}{3}$. Since $\frac{5}{3} \cdot 39 = 65$ and we want a 39 to be scaled to a 60, $b = -5$. **14.** $y + 3 = \frac{1}{2}|x - 4|$

15. $y - 1 = \left(\frac{x}{2}\right)^3$ **16. a.** about 73% **b.** about 10%

17. about 16% **18.**

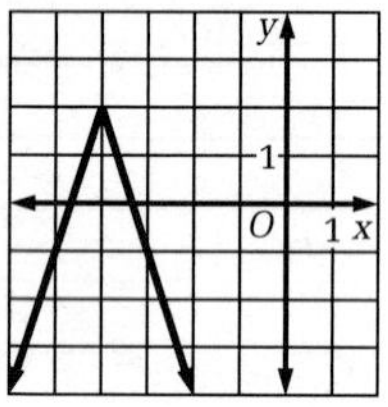

19. a. $g(t) = 80 - f(t)$

b.

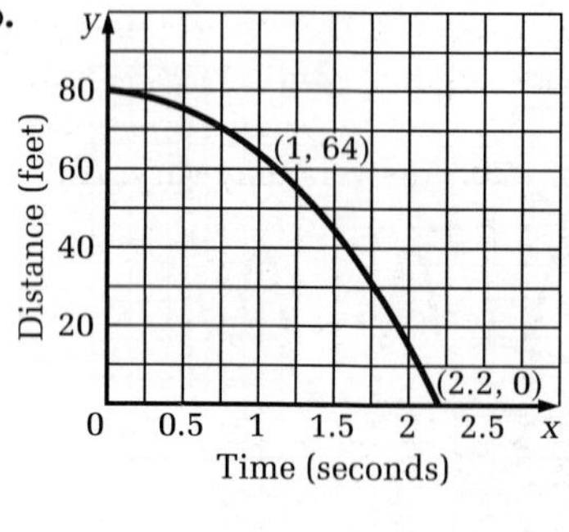

Unit 10

Pages 541–545 **Talk it Over**

1. time and speed; There appears to be little relationship between the two. **2.** The car accelerates to 230 mi/h, races in the 220–230 mi/h range, slows to 80 mi/h for 30 s, accelerates to resume normal racing speed, slows and stops for 10 s, and finally accelerates back into the race. The two similar-looking sections appear to be the normal racing section of the graph, with speeds varying slightly due to turns. The slowdown from 60 s to 90 s could be due to a caution condition on the track, perhaps because of an accident. The stop from 155 s to 165 s could be a pit stop. **3.** B **4.** E **5.** D **6.** A **7.** C **8.** F **9.** Answers may vary. An example is given. **a.** E **b.** B, F **c.** A **d.** C **e.** A, B **f.** D **10.** The results are not exactly the same but are similar. 4^x grows fastest, followed by x^4, then $4x$. The differences in y-values for $y = 4x$ are constant, the differences for $y = x^4$ are increasing, and the differences for $y = 4^x$ are increasing in a geometric sequence. **11.** $y = \frac{1}{2}x$ increases fastest, followed by $y = \sqrt{x}$, and then $y = \log_2 x$.
12. a. slowly at the start and quickly at the end of this interval **b.** slowly over the whole interval

Pages 545–547 **Exercises and Problems**

1. a. The elevator goes to the second floor and returns to the ground floor. It goes to the fifth floor, stopping at every floor, and returns to the ground floor. It does this 3 times. Then it stops at the ground floor for 40 minutes and goes to the fifth floor. **b.** 8:15 to 8:30; In each period, the elevator takes the same time to fill, go up while stopping at each floor, then return to the ground floor and await the next trip. **c.** between 8:15 and 8:25 **3.** The graphs are alike in that they alternate between high values and low values. They are different in that the graph in question 5 does not repeat values, while the graph in question 7 does repeat values. **7, 9, 11.** Answers may vary. Examples are given. **7.** decreasing or saturating **9.** decreasing **11.** periodic **13.** 11 **15.** 9 **17.** $y = 3^x$ grows fastest, followed by $y = x^3$, and then $y = 3x$.

19. $y = 2 - x$ decays fastest, followed by $y = 2^{-x}$, and then $y = \frac{2}{x}$. **26. a.**

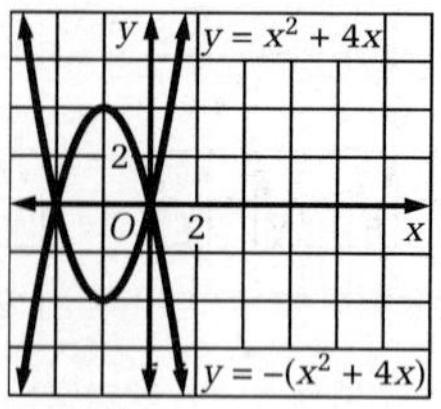

b. Reflect it over the x-axis.

27. 16 **28.** $5\sqrt[6]{5^5}$ **29.** $\frac{27}{125}$

30. 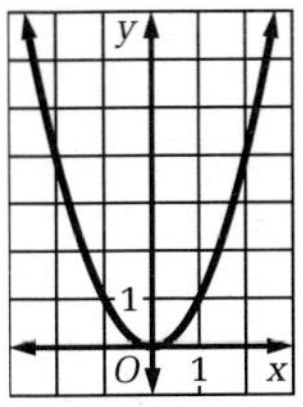
31. 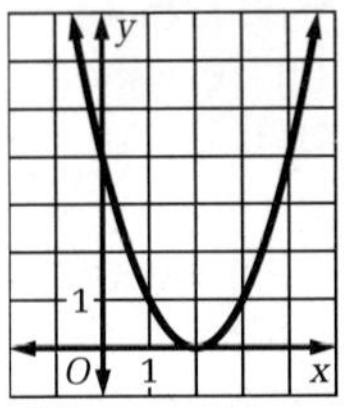
32. 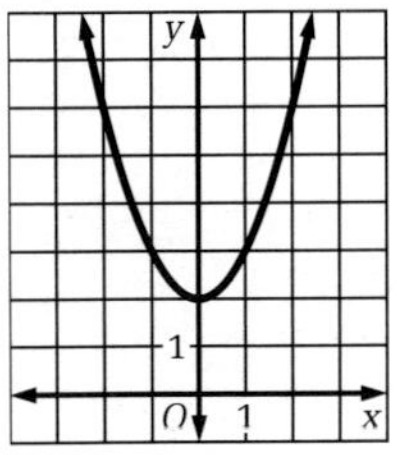
33. 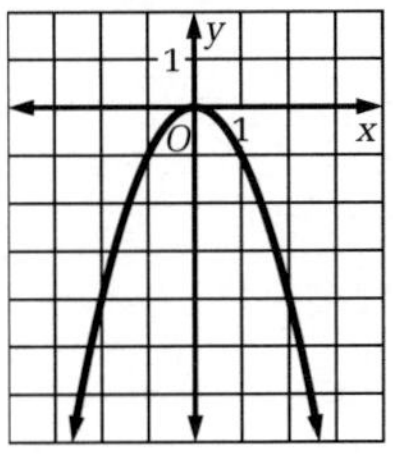

34. $y = (x - 2)^2$: transformation by (2, 0) or shift 2 units to the right; $y - 2 = x^2$: transformation by (0, 2) or shift 2 units up; $y = -x^2$: reflection over the x-axis

Pages 548–549 Talk it Over

1. 4 min **2.** 15; $\frac{1}{4}$; the frequency **3.** 7 **4.** start of the second cycle; middle of the wash and rinse part of the cycle; wash and rinse **5.** The amplitude for Jennifer's graph is 8.5 and the axis is $y = 11.5$. **6.** The amplitude and the period would decrease, the axis would be lower, and the frequency would increase.

Pages 550–553 Exercises and Problems

1. The period and the frequency are reciprocals. **3.** 24 h **5.** 3 P.M.; 4:30 A.M. **7.** Answers may vary. An example is given.

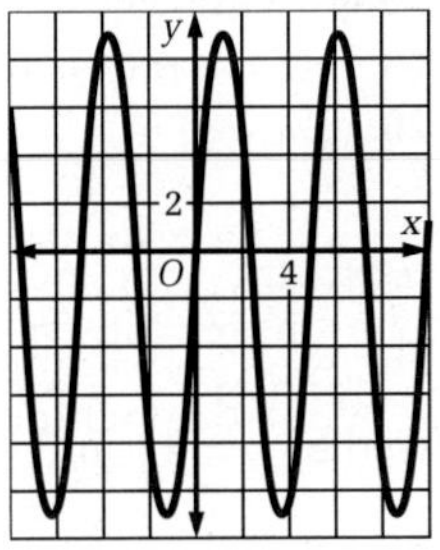

11. about 21 times **13. a.** 4 **b.** $\frac{1}{4}$ **c.** $y = -1$ **d.** 2 **15. a.** 6 **b.** $\frac{1}{6}$ **c.** $y = 2.5$ **d.** 1.5

17. 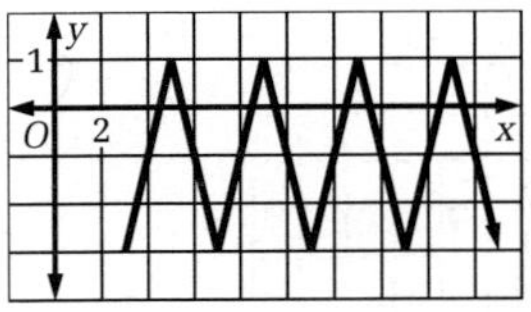

19. 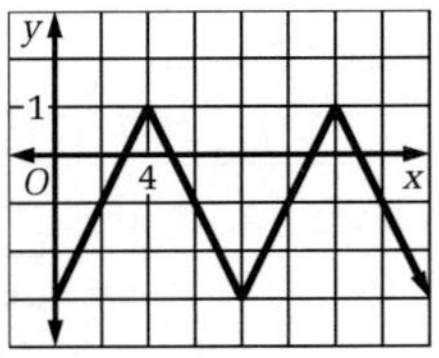

31. $y = 5^x$ grows fastest, followed by $5x^2$, then $5x$.

32. a. 90° **b.** $\sin A = \frac{a}{c}$ **c.** $\cos B = \frac{a}{c}$

33. $b \approx 6.2$ cm, $m\angle A \approx 23.8°$, and $m\angle C \approx 126.2°$

34. $a \approx 16.2$ cm; $m\angle C \approx 6.7°$, and $m\angle B \approx 63.3°$

35. $m\angle B = 100°$, $a \approx 22.8$ cm, and $c \approx 35$ cm

36. $m\angle A = 63°$, $a \approx 15.2$ cm, and $b \approx 6.4$ cm

Pages 554–556 Talk it Over

1. a. 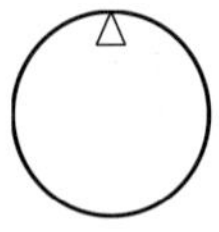 **b.** 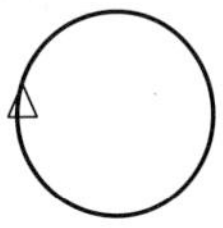 **c.** 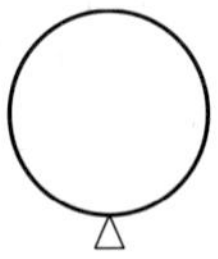

d. 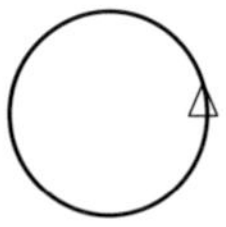 **e.** 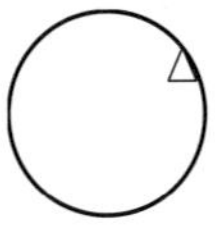 **f.** 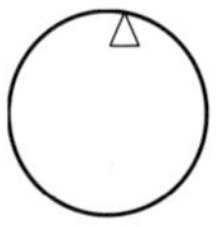

2. It increases and decreases in cycles, going from a low of 5 ft to a high of 55 ft.
3. A: a. at the top **b.** 90° **c.** $\frac{1}{4}$ **B: a.** on the left side, 30 ft up **b.** 180° **c.** $\frac{1}{2}$ **C: a.** at the bottom **b.** 270° **c.** $\frac{3}{4}$ **D: a.** on the right side, 30 ft up **b.** 360° **c.** 1 **E: a.** on the right side, 30 ft up **b.** 720° **c.** 2 **F: a.** at the top **b.** 810° **c.** $2\frac{1}{4}$ **4.** 360°; one turn of the wheel **5.** 25 ft; the radius of the Ferris wheel
6. a. 30 s **b.** 2 **c.** It would be stretched horizontally by a factor of $\frac{2}{3}$. **7.** I: the x-coordinate starts at 1 and ends up at 0. II: the x-coordinate starts at 0 and ends up at –1. III: the x-coordinate starts at –1 and ends up at 0. IV: the x-coordinate starts at 0 and ends up at 1.
8. 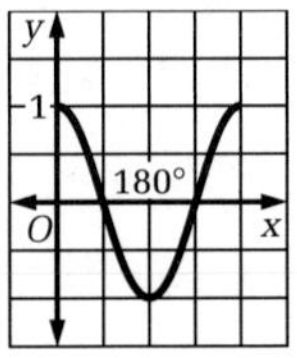

Pages 557–559 Exercises and Problems

1. The unit circle is a circle with radius 1 centered at the origin. If an object moves counterclockwise around the unit circle, the sine function indicates the vertical position and the cosine function indicates the horizontal position of the object. **3.** neither **5.** sine **7.** both (The horizontal intercepts of the sine function occur at multiples of 180°; the horizontal intercepts of the cosine function occur at odd multiples of 90°.) **9.** sine **11.** 12 **13.** The period of the function is 20 s; 27 s. **20.** Answers may vary. An example is given.

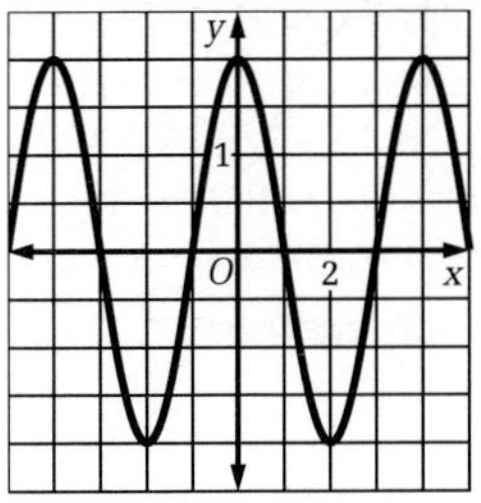

21. $f(g(x)) = \left(\frac{x}{2}\right)^2 = \frac{x^2}{4}$ **22.** $h(f(x)) = \sqrt{x^2} - 1$ or $|x| - 1$

23. 16 **24.** 2

25.

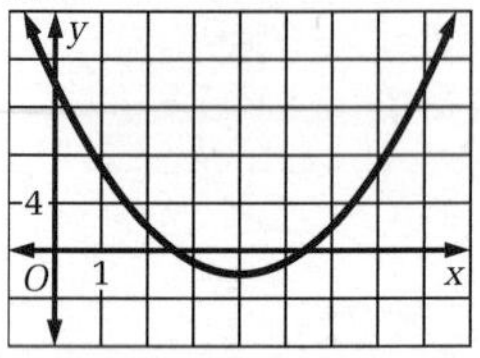

27.

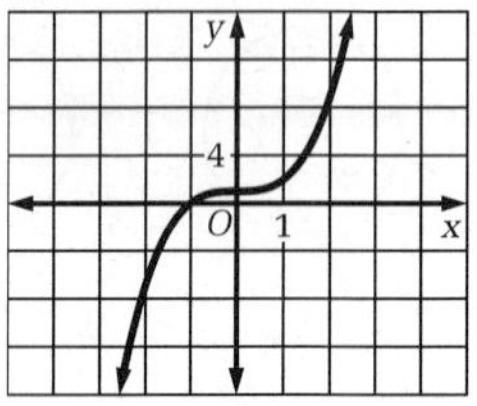

26.

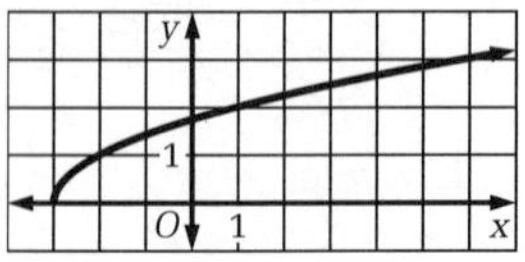

Pages 563–567 Exercises and Problems

1. Answers may vary. An example is given. $|A|$ is the amplitude, $\frac{360°}{|B|}$ is the period, C is the horizontal shift from the graph of $y = \sin\theta$, and D is the vertical shift.

7. D **9.** F **11.** E **13. a.** 3; 360°; $\frac{1}{360°}$

b.

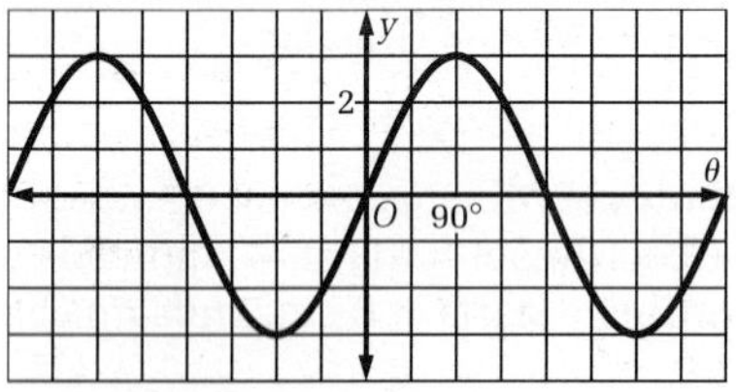

15. a. 1; 360°; $\frac{1}{360°}$

b.

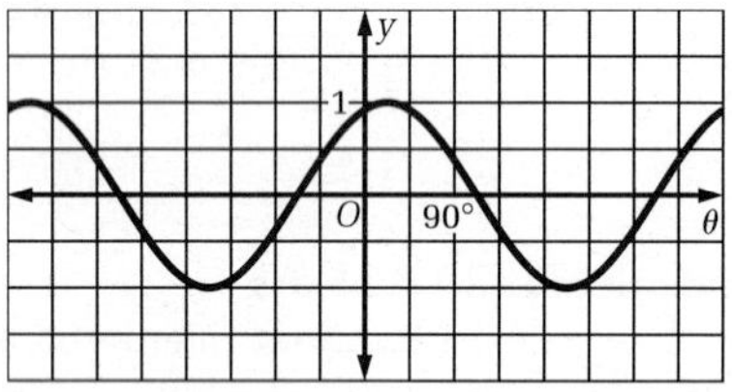

23. a. 1; 180°; $\frac{1}{180°}$

b.

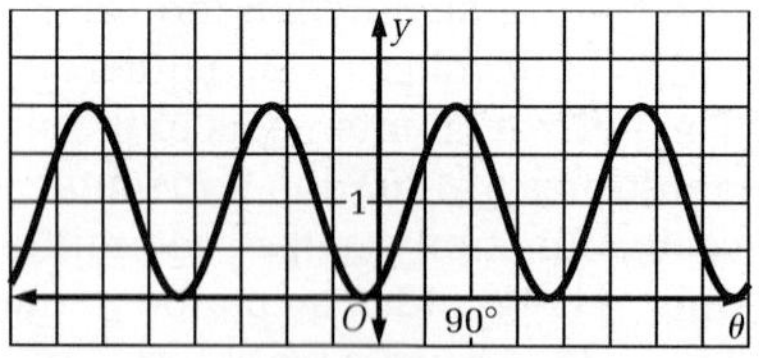

27. C **31, 33.** Answers may vary. Examples are given. **31.** $y = 2\sin(\theta - 45°) + 1$
33. $y = \sin 2(\theta + 30°) - 3$ **44.** –300.7°; –239.3°; 59.3°; 120.7° (in general, $\theta = 59.3° + n \cdot 360°$ or $\theta = 120.7° + n \cdot 360°$ for any integer n) **45.** no sum
46. $\frac{100}{199} \approx 0.5$ **47.** 1800°; 30°

Pages 569–571 Talk it Over

1. a. 30 ft **b.** 30 ft **c.** about 8.35 ft
2. $h(t) = 25\cos(12(t - 7.5))° + 30$ **3.** 9

4.

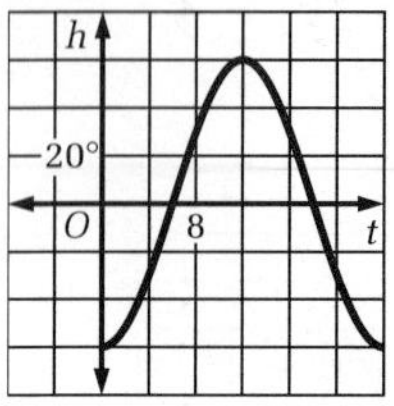

The graph crosses the horizontal axis.

Pages 572–575 Exercises and Problems

3. a.

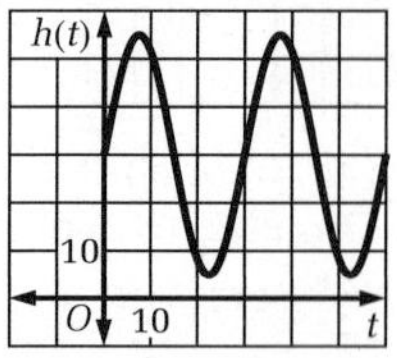

b. every 30 s after the start **c.** $(30n, 30)$, where n is a positive integer

5. a. The axis is now $h = 29$; the graph is shifted down 1 unit. **b.** $h = 25\sin(12t)° + 29$
7. a. $h(t) = 20\sin(120(t - 0.75))° + 30$

b.

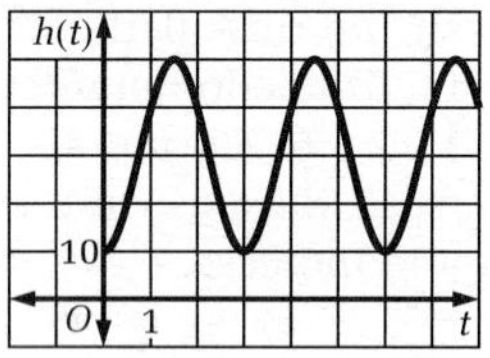

c. near 0.75 s and every 3 s after that **d.** near 2.25 s and every 3 s after that **e.** every 1.5 s

15. Answers may vary. An example is given. $N = 438.5x + 49821$, where N is the number (in thousands) of women employed and x is the number of years since 1989. **16.** 0.75 **17.** vertical reflection symmetry **18.** none of these **19.** 180° rotational symmetry **20.** all three

Page 576 Checkpoint

1. $45° < \theta < 225°$; In this interval, the graph of $y = \sin\theta$ is above the graph of $y = \cos\theta$. **2.** $y = 5^x$ grows the fastest, followed by $y = x^5$, and then by $y = 5x$.
3. Answers may vary. An example is given.

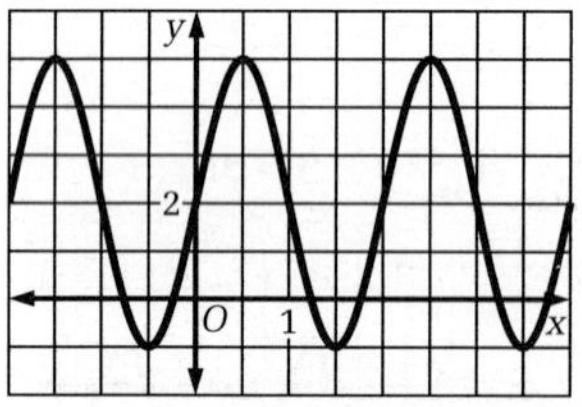

4. both **5.** cosine **6.** both **7.** sine
8. 180°; 1; $\frac{1}{180°}$; $y = \sin 2(\theta - 30°)$
9. 360°; 1; $\frac{1}{180°}$; $y = 3\sin(\theta - 45°) + 2$

10.

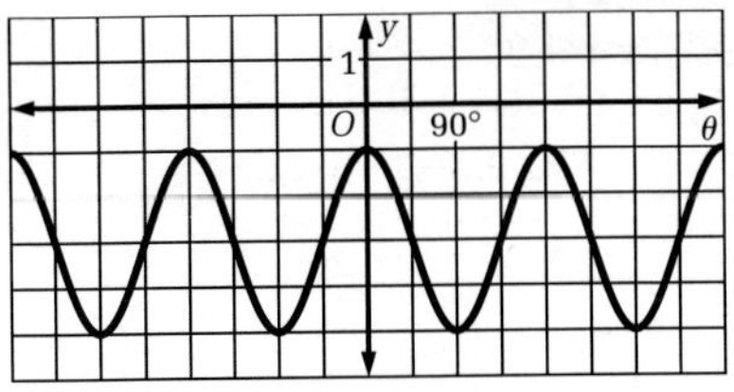

11. a.

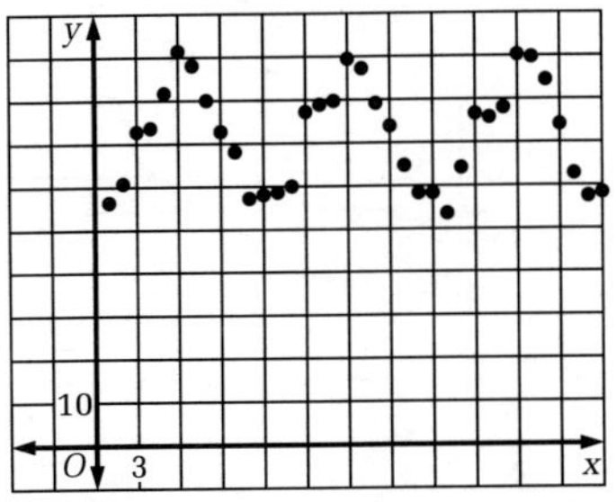

a sine curve; The points appear to lie on a sine curve.

b, c. Answers may vary. An example is given.
b. $y = 17.2 \cos (30(x - 6))° + 73.3$
c.

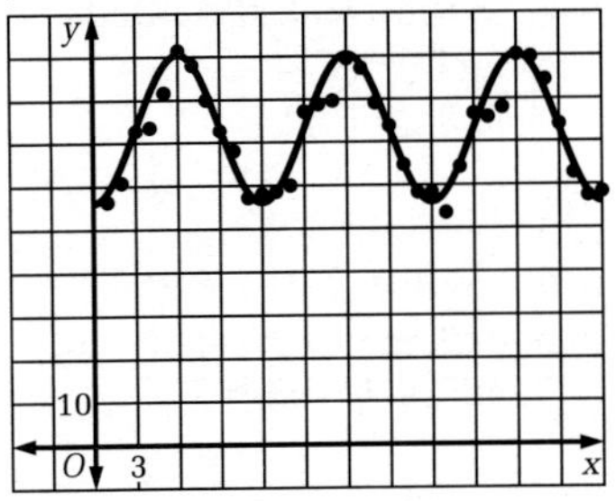

My graph fits the data fairly well. I think no graph will fit all the data points exactly because ice cream production reflects sales and sales are dependent on human behavior, which is not entirely predictable.

Page 577 Talk it Over

1. Each pattern repeats the same figure over and over.
2. a. A, C, and D **b.** B and D **c.** C, D, E, and F

Pages 580–581 Exercises and Problems

1. A translation is used in each pattern. **3.** 180° rotational **5.** translation **7.** vertical **9.** horizontal
11. a. If a border pattern has a rotational symmetry of θ, the image of each edge under a rotation of θ must be the other edge. Since the edges are parallel, θ must be 180°. **b.** If a line is a line of symmetry of a border pattern, then the image of each edge under reflection over the line must be the same edge or the other edge. Then the line of symmetry must be parallel to or perpendicular to the edges. **15.** glide reflection; 180° rotation and vertical reflection; horizontal and vertical reflection **17.** glide reflection; horizontal reflection; horizontal and vertical reflection
20. Answers may vary. An example is given.
$y = 22.3 \cos (30(x - 7))° + 40.6$ **21.** (8.06, 150.3°)
22. (4.53, –2.11) **23.** 120°; 60°

Page 584 Talk it Over

1. Any regular tessellation is periodic. For a tessellation consisting of squares or hexagons, the basic figure is a cell. For a tessellation consisting of triangles, a hexagon-shaped cluster of triangles is a cell.
2. Answers may vary. **3.** The four angles at each vertex are congruent to the four angles of each parallelogram. The sum of the measures of the interior angles of a parallelogram is 360°. **4.** The six triangles around each vertex have congruent sides matched so that four congruent parallelograms are formed. So, by the answer to question 3, the sum of the measures of the six angles is 360°.

Pages 585–588 Exercises and Problems

1. A regular tessellation uses only one kind of regular polygon, while a semiregular tessellation uses two or more regular polygons. **3.** 3.12.12

5.

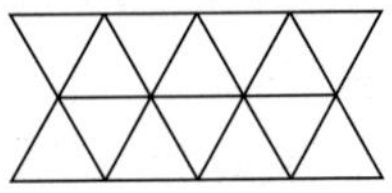

7.

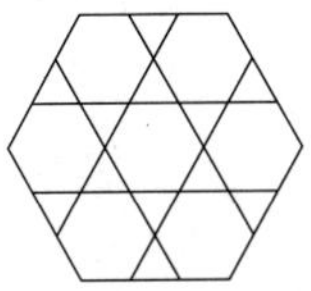

9. Suppose regular pentagons and equilateral triangles tessellate a plane. Let p = the number of regular pentagons at each vertex and let t = the number of equilateral triangles at each vertex. The measures of interior angles of regular pentagons and equilateral triangles are 108° and 60°, so $108p + 60t = 360$. The only solution with nonnegative integer values is (0, 6). Since there is no combination of regular pentagons and triangles that can surround a vertex, the assumption that regular pentagons and triangles tessellate a plane is false. **15.** region A **32.**

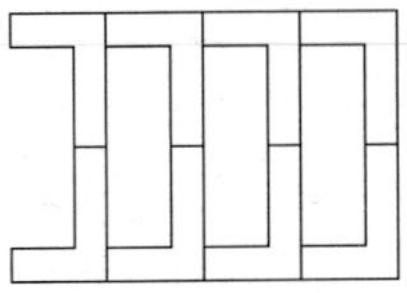

33. $y = \frac{x^2}{16} - 5$

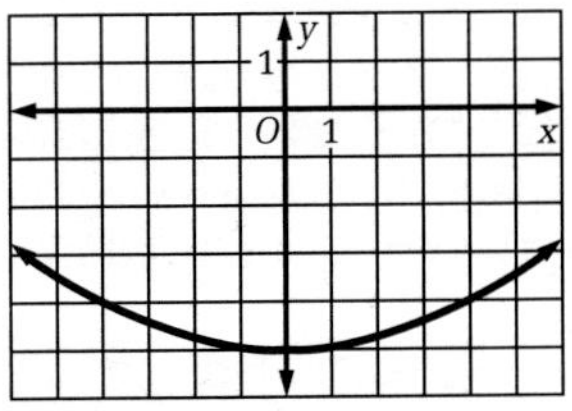

34. a. 3 **b.** 0 **c.** 1.61

Pages 590–591 Unit 10 Review and Assessment

1. periodic **2.** increasing **3.** constant
4. decreasing **5.** $y = 4^x$ grows fastest, followed by $y = x^4$, and then by $y = 4x$.
6. Answers may vary. An example is given.

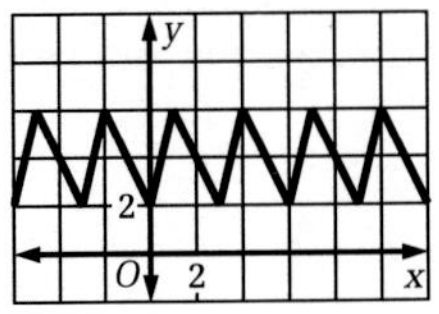

7. a. Answers may vary. An example is given. the part of the graph from 0 to 4 on the horizontal axis **b.** 4 **c.** $\frac{1}{4}$ **d.** 3 **e.** $y = 1$ **8.** A sine wave is any graph that has an equation of the form $y = A \sin B(\theta - C) + D$ or $y = A \cos B(\theta - C) + D$. It can be obtained by transforming the graph of $y = \sin \theta$ or $y = \cos \theta$.

9. 3; 360°; $\frac{1}{360°}$

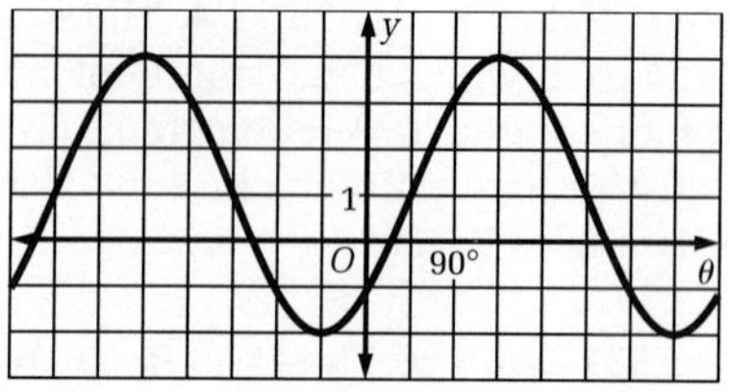

10. 1; 180°; $\frac{1}{180°}$

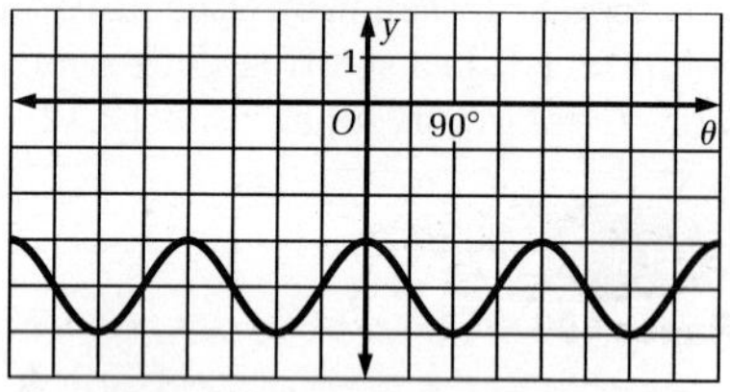

11. $\frac{1}{2}$; 120°; $\frac{1}{120°}$

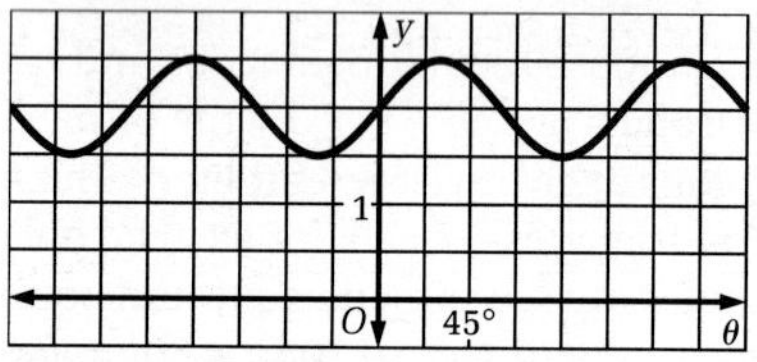

12, 13. Answers may vary. Examples are given. **12.** $y = 1.5 \cos 2\theta - 3$ **13.** $y = 3 \sin(\theta - 45°) + 2$

14. a.

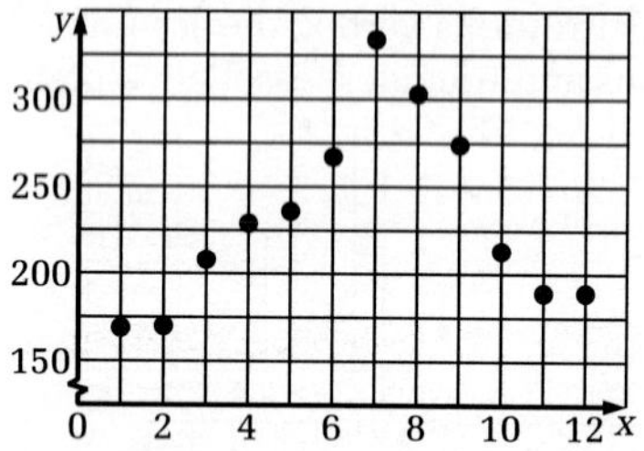

b. Answers may vary. An example is given. $y = 75 \cos(30(x - 7))° + 235$ **c.** A periodic function that can be modeled by a sine wave. The given data are periodic. **15.** vertical **16.** vertical and 180° rotation **17.** horizontal, vertical, and 180° rotation **19.** Yes. **20.** No. **21.** Suppose regular 9-gons and squares tessellate a plane. Let n = the number of regular 9-gons at each vertex and let s = the number of squares at each vertex. The measures of interior angles of regular 9-gons and squares are 140° and 90°, so $140n + 90s = 360$. There is no solution with non-negative integer values. Since there is no number of regular 9-gons and squares that surround a vertex, the assumption that 9-gons and squares tessellate a plane is false.

Technology Handbook

Page 596 Try This

1. a. 3.5 **b.** 32 **c.** about 0.966 **d.** 60° **e.** 4 **f.** 120 **g.** 20 **h.** 924

Page 597 Try This

2. a.

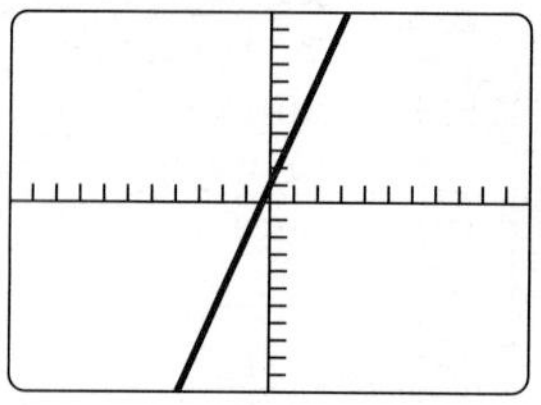

b.

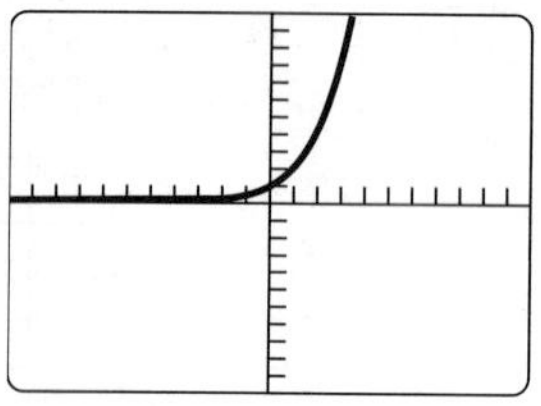

c.

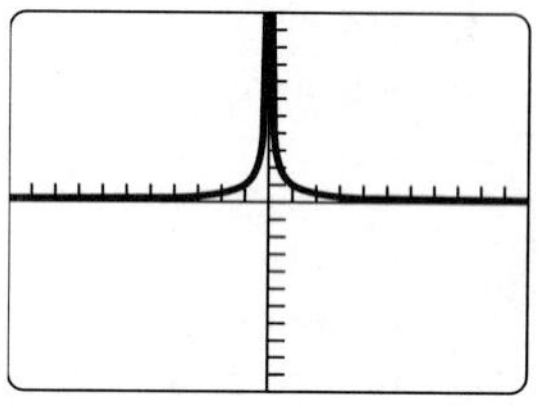

d.

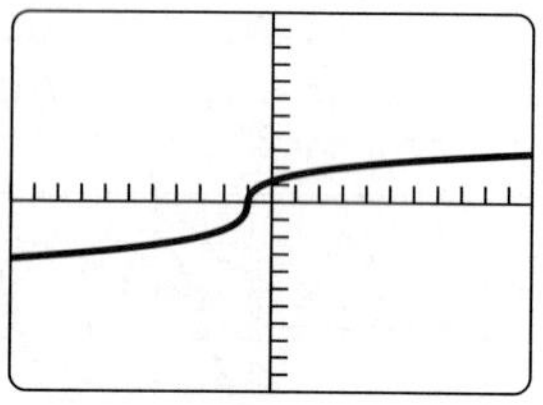

3. Answers may vary. An example is given. [–5, 30] and Xscl=5 for x, [–10, 75] and Yscl=5 for y.
4. Answers may vary. An example is given. [–7.5, 7.5] and Xscl=1 for x, [–5, 5] and Yscl=1 for y.

Page 598 Try This

5. about 1.59

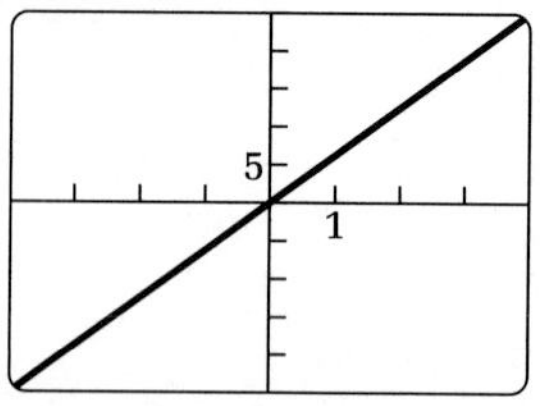

6.

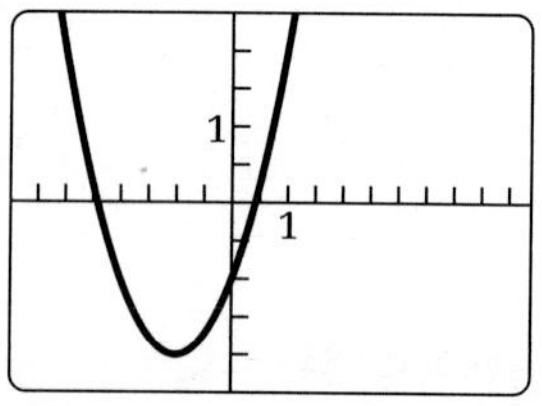

TI-82: window of [–4, 4.4] and Xscl=0.5 for x, and [–2, 2] and Yscl=0.5 for y. Graph crosses at 0.4 and –2.4. TI-81: window of [–4, 5.5] and Xscl=0.5 for x, and [–2, 2] and Yscl=0.5 for y. Graph crosses at 0.4 and –2.4.

Page 600 Try This

7. between –2.41 and –2.42

8.

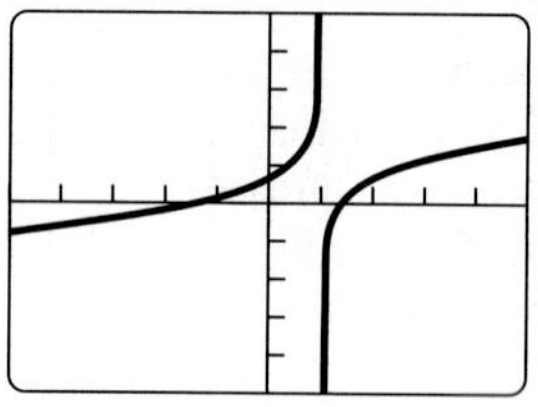

9.

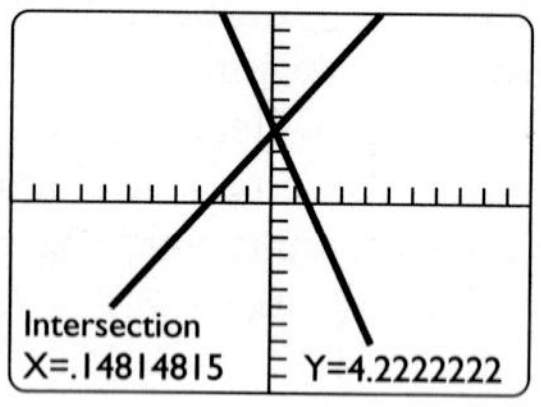

Page 601 Try This

10. Enter Y1={1,–1,0.5,2}X.

11. a.

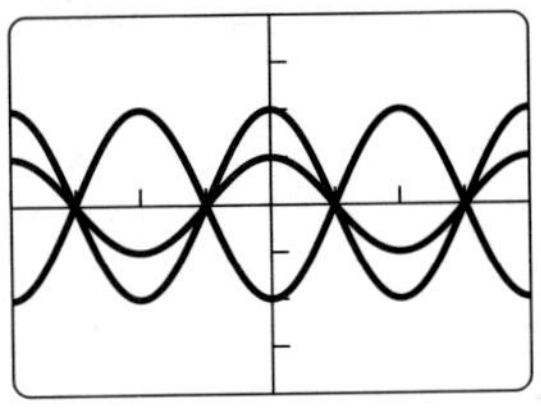

b.

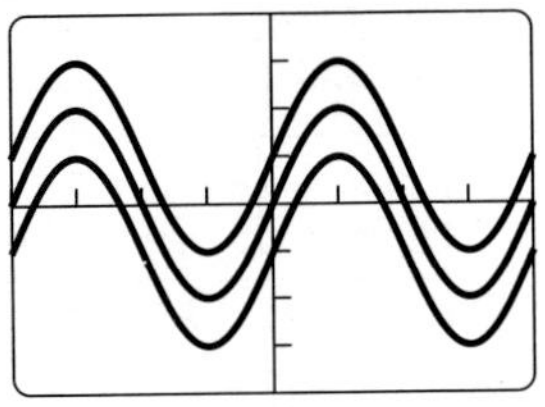

12. a.

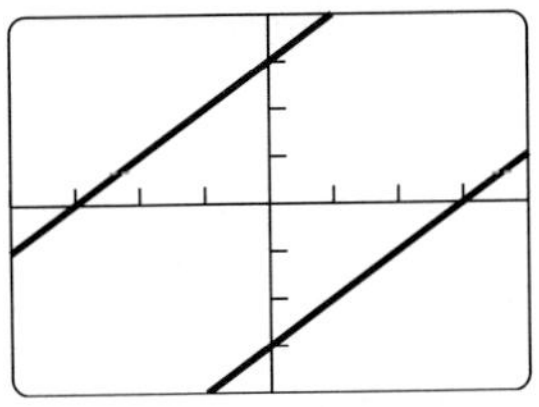

b.

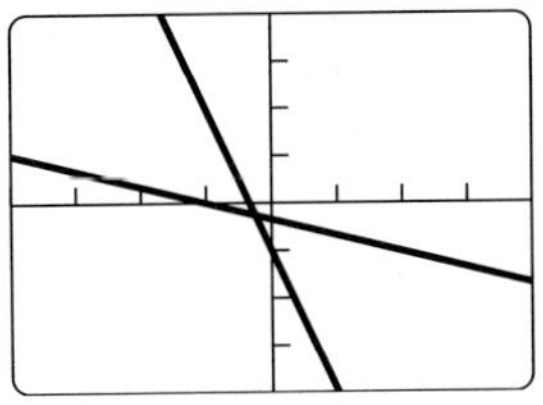

c.

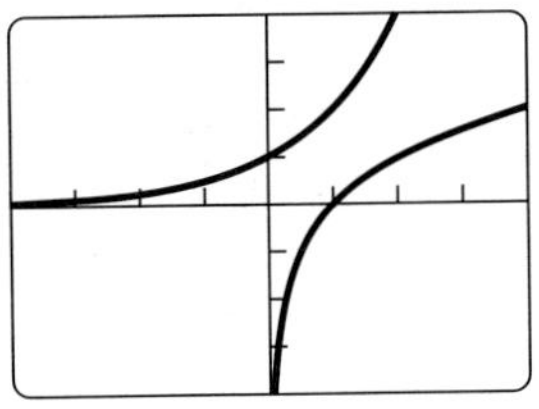

Page 602 Try This

13. a.

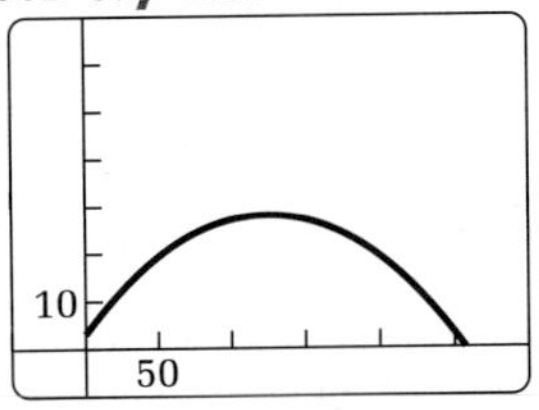

b. 1.25 seconds

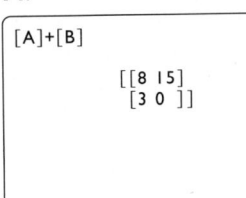

Page 604 Try This

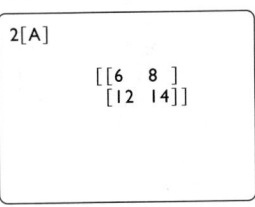

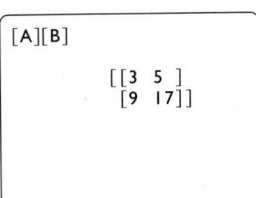

14.

```
[A]+[B]
        [[8 15]
         [3 0 ]]
```

15.

```
[B]-[C]
        [[-4   8  ]
         [-15  -11]]
```

16.

```
2[A]
        [[6   8 ]
         [12  14]]
```

17.

```
[A][B]
        [[3  5 ]
         [9  17]]
```

18. not defined

19.

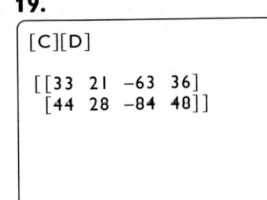

```
[C][D]
 [[33  21  -63  36]
  [44  28  -84  48]]
```

20.

```
[A]-1
 [[-2.33  1.33 ]
  [2      1    ]]
```

21. not defined

Page 606 Try This

22. b.

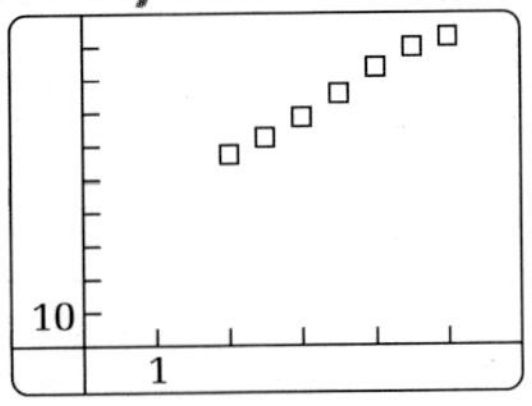

c. $y = 12.43x + 31.64$
d. $y = 12.4x + 31.6$
e. 75.14; 12.48

Extra Practice

Pages 608–610 Extra Practice Unit 1

1. a. finding the least common multiple of *A* and *B* **b.** Yes; testing multiples of *B* until one is found that is evenly divisible by *A*. **3, 5, 7.** Solution methods may vary. **3.** $x = 5$ **5.** $x \le -5$ **7.** $x \ge 6$ **9.** Yes. **11.** No. **13.** No. **15.**

Municipal Library Systems

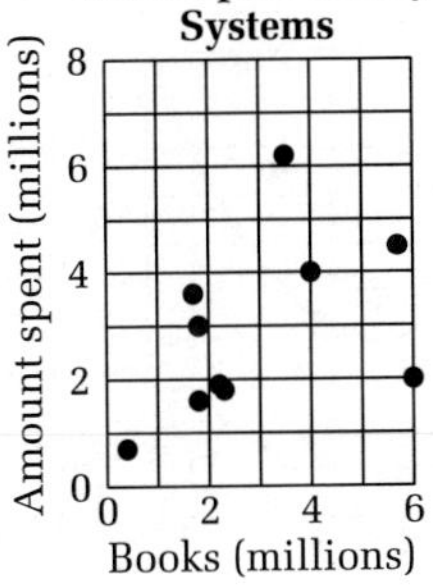

17. Weefixit: $y = 16x + 25$; Appliance Rehab: $y = 19x + 18$ **19.** 3 **21.** 8.0 **23.** (2.2, –2.8) **25.** (2, –1, 4) **27.** (3.2, 1.4, 0.8)

29.

	P	*Q*	*R*	*S*	*T*
P	0	1	1	1	1
Q	1	0	1	0	0
R	1	1	0	1	0
S	1	0	1	0	1
T	1	0	0	1	0

31.

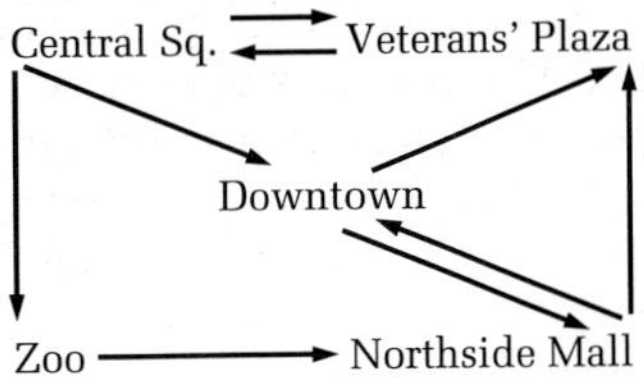

33. 3; 11; maximum **35.** –4.5; 41.5; maximum **37.** 9 walkers and 6 washers ($111 raised)

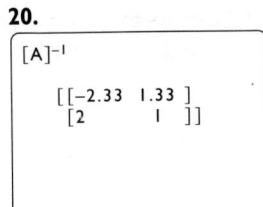

Pages 610–612 Extra Practice Unit 2

1. 25; 0; $a^2 + 2ab + b^2 + 4a + 4b + 4$ **3.** $\frac{11}{2}$; 8; $7 - \frac{a+b}{2}$ **5.** 2; 7; $|a + b - 5|$ **7.** Yes; many-to-one. **9.** Yes; one-to-one. **11.** Yes; –4; 11; $-3a + 3b + 2$. **13.** No. **15.** Yes; 5; 25; $13 - 4a + 4b$. **17.** $f(x) = \frac{1}{2}x + \frac{9}{2}$ **19.** $f(x) = 2x + 7$ **21.** $f(x) = \frac{3}{2}x + \frac{9}{4}$

23.

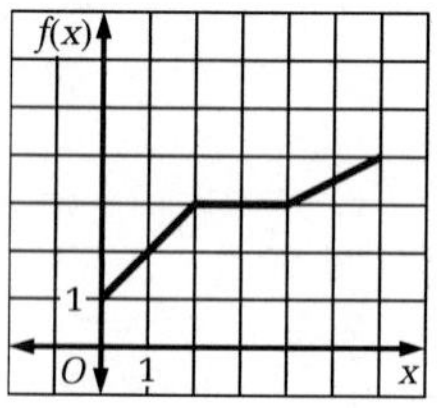

25.

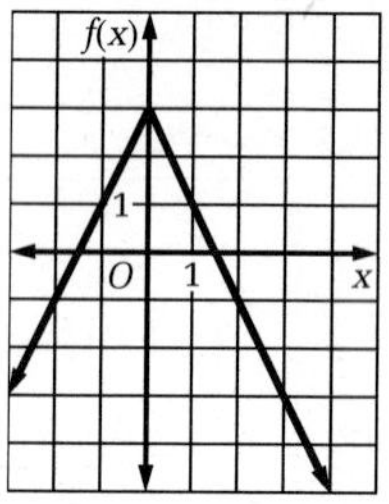

27. a. quadratic; –2; $-\frac{9}{2}$, $-\frac{a^2 - 2ab + b^2}{2}$

b.

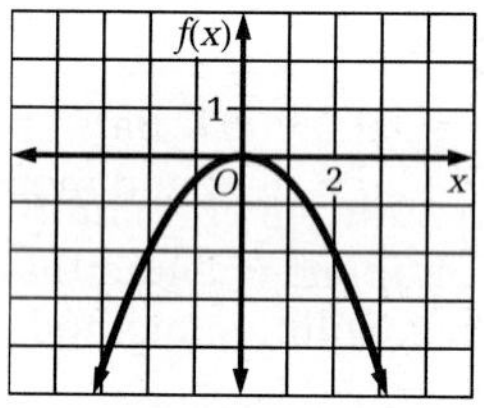

29. not quadratic **31. a.** quadratic; $\frac{1}{4}$; 4; $\frac{(a-b-1)^2}{4}$

b.

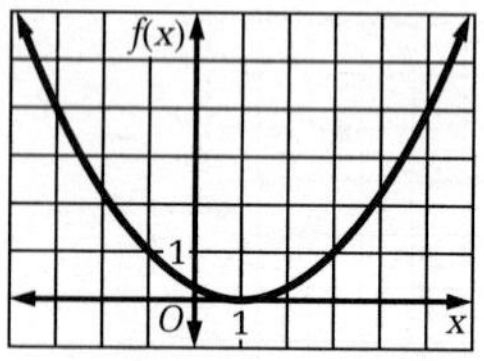

33. $f(x) = 3x^5$ **35.** $f(x) = 48x^4$ **37.** $f(x) = 1125x^3$

39.

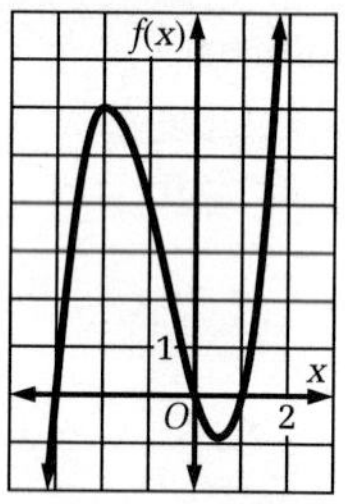

41.

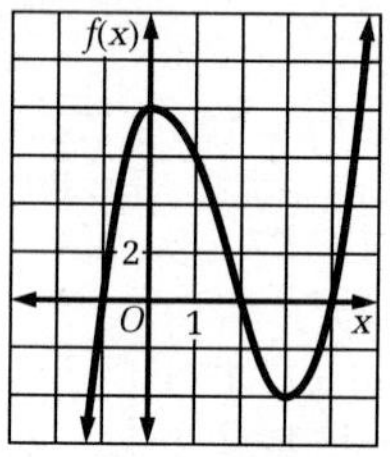

43. 3 **45.** 12 **47.** 4 **49.** all real numbers; all real numbers **51.** all real numbers greater than or equal to 3; all real numbers less than or equal to 0 **53.** all real numbers; all real numbers

55. a.

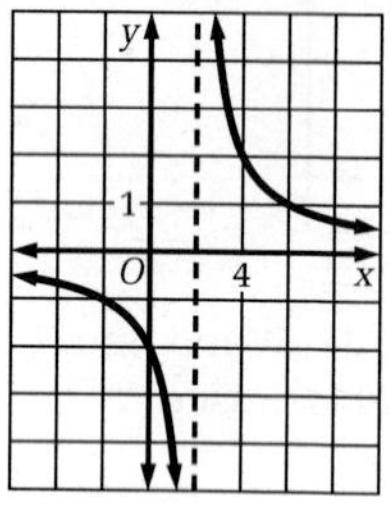

b. $x = 2$; $y = 0$
c. all real numbers except 2; all real numbers except 0

57. a.

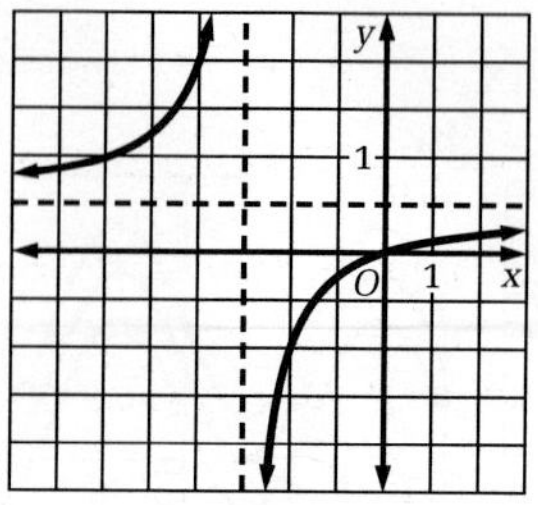

b. $x = -3$; $y = 1$
c. all real numbers except –3; all real numbers except 1

59. a.

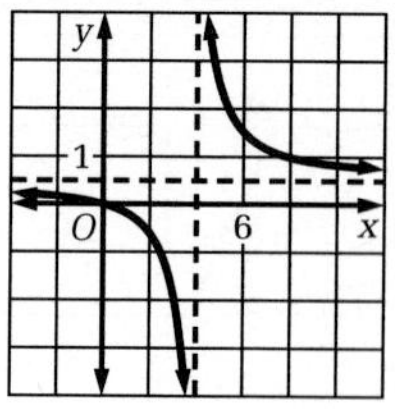

b. $x = 4$; $y = 0.5$
c. all real numbers except 4; all real numbers except 0.5

61. a. $(7x^2 + 1)\sqrt{x}$ **b.** all real numbers greater than or equal to 0 **63. a.** $\sqrt{7x^2 + 1}$ **b.** all real numbers

65. a. $\frac{(7x^2 + 1)(2x + 1)}{x}$ **b.** all real numbers except 0 or $\frac{1}{2}$

67. a. $7x + 1$ **b.** all real numbers greater than or equal to 0

Pages 612–613 Extra Practice Unit 3

1. True. If two sides of a triangle are equal in measure, then two angles of the triangle are equal in measure; True. If no two angles of a triangle are equal in measure, then no two sides of the triangle are equal in measure; True. If no two sides of a triangle are equal in measure, then no two angles of the triangle are equal in measure; True. **3.** True. If n is a positive odd integer, then n is a prime number greater than 2; False. If n is not a prime number greater than 2, then n is not a positive odd integer; False. If n is not a positive odd integer, then n is not a prime number greater than 2; True. **5.** The midpoints are $J(a, 0)$, $K(a + b, c)$, $L(b + d, c + e)$, $M(d, e)$. The slope of both $\overline{JK}$ and $\overline{ML}$ is $\frac{c}{b}$; the slope of both $\overline{KL}$ and $\overline{MJ}$ is $\frac{e}{d-a}$. Thus, by definition, $JKLM$ is a parallelogram. **7.** Yes. **9.** No. **11.** about 128.6°; about 51.4° **13.** 157.5°; 22.5° **15.** 72° **17.** 72° **19.** 108° **21.** 110° **23.** 35° **25.** 120° **27.** They are the same, since each is equal to the sum of the radii of the two circles. **29.** 60° **31.** $12\sqrt{3} \approx 20.8$ units^2 **33.** 840 units^2 **35.** Assume temporarily that a regular polyhedron with octagonal faces exists. Then at each vertex at least three regular octagons meet. This is impossible, since each has interior angles that measure 135°, and $3 \cdot 135 = 405° > 360°$.

Pages 614–615 Extra Practice Unit 4

1. a. 6, –7, 8 **b.** –11 **3. a.** 8, 5, 2 **b.** –7 **5. a.** 4, 2, 1 **b.** 4 **7.** Yes; 1. **9.** Yes; $\frac{2}{3}$. **11.** $\frac{1}{2}, \frac{4}{3}, \frac{9}{4}, \frac{16}{5}; \frac{400}{21}$ **13.** $\frac{1}{3}, \frac{3}{5}, \frac{5}{7}, \frac{7}{9}; \frac{39}{41}$ **15.** $a_n = 3 + 2n$ **17.** $a_n = 1500\left(\frac{1}{10^n}\right)$

19. $a_n = 625\left(\frac{1}{5}\right)^n$ **21.** 80, 40, 20, 10, 5, 2.5 **23.** 4, $\frac{1}{2}$, 4, $\frac{1}{2}$, 4, $\frac{1}{2}$ **25.** 10, 2, 6, $\frac{8}{3}$, $\frac{19}{4}$, $\frac{59}{19}$ **27. a.** geometric **b.** $a_1 = 16$; $a_n = \frac{5}{4}a_{n-1}$ **29. a.** geometric **b.** $a_1 = 8$; $a_n = \frac{1}{10}a_{n-1}$ **31. a.** arithmetic **b.** $a_1 = \frac{1}{2}$; $a_n = a_{n-1} - 2$

33. a. −14.5 **b.** ±10 **35. a.** 51 **b.** $\pm 10\sqrt{2}$
37. a. −151.5 **b.** ±30 **39.** 105 **41.** 364

43. $\sum_{n=1}^{12}(6 + 10n)$ **45.** $\sum_{n=1}^{9}(9n - 21)$ **47.** 513

49. 55.55555 **51. a.** geometric **b.** $\frac{-189}{64} = -2.953125$

53. a.

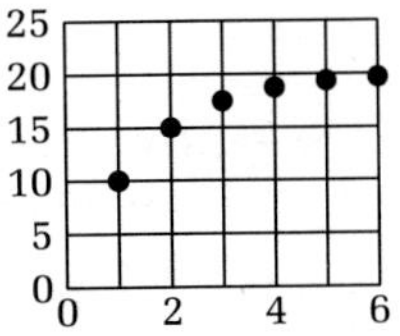

b. Yes; 20.

55. a.

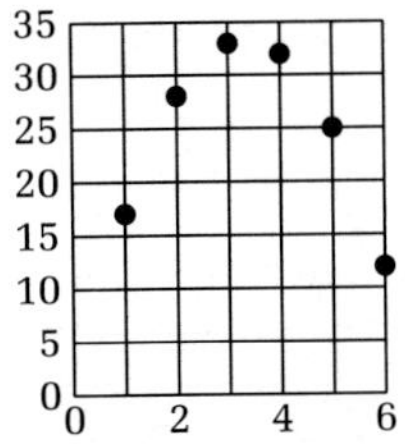

b. No.

57. a.

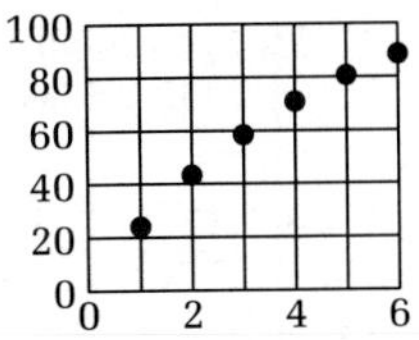

b. Yes; 120.

Pages 615–617 Extra Practice Unit 5

1.

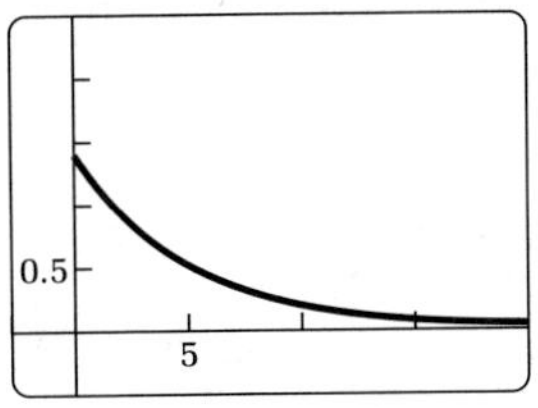

3. 3.49 h **5.** $y = 320(1.15)^x$, where x is the number of years since 1990 **7.** 0 computers **9.** 3 **11.** $\frac{5\sqrt{5}}{8}$
13. $2ab^2\sqrt[3]{a}$ **15.** $\sqrt[3]{w}$ **17. a.** 0.30 **b.** 0.03
19. \$2203.99 **21.** \$2237.74 **23.** $y = \frac{3}{7x}$

25. $y = (x - 2)^2$, $x \geq 2$ **27.** $y = \frac{x}{2}$, $x \geq 0$ **29.** $y = \sqrt[3]{x - 1}$

31. 4 **33.** −3 **35.** 0 **37.** 2.5 **39.** 0.20 **41.** 1

43. 0.56 **45.** $\log_3 56$ **47.** $\log_7 A + 0.5 \log_7 B$
49. $0.25(\log_7 A + 5 \log_7 B)$ **51.** 1 **53.** 1.95

55. 26.40 **57.** $\frac{7}{17}$ **59.** 4, 20

Pages 617–619 Extra Practice Unit 6

1. Histograms may vary. An example is given.

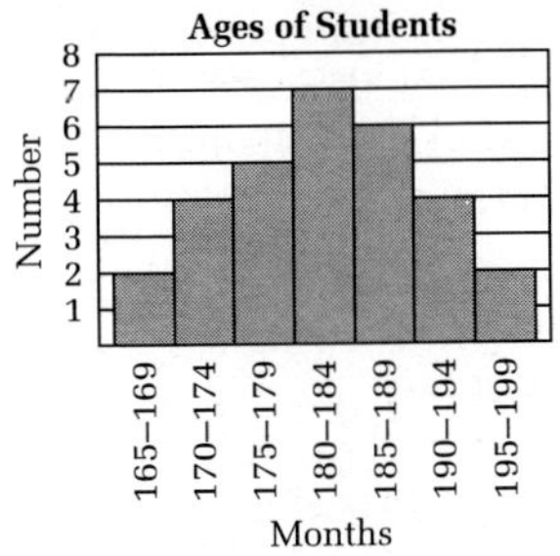

3. 182.1; 182.5

5.

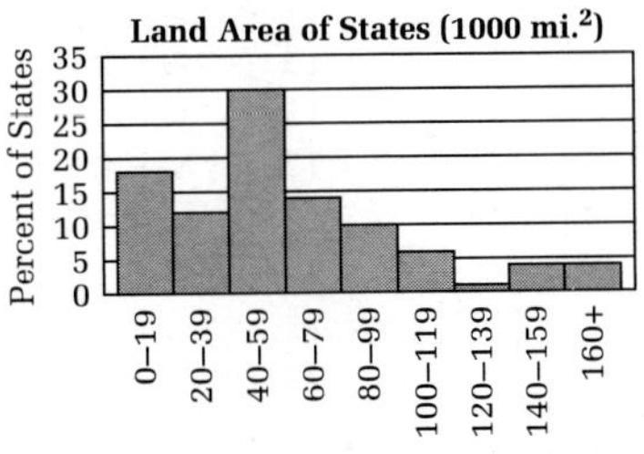

7. 60%

9. mean ≈ 1.33; range = 0.09; $\sigma \approx 0.024$ **11.** Bars of the the second would be shifted to the right and more closely bunched around the mean. **13. a.** Histogram may vary depending on intervals chosen. An example is given.

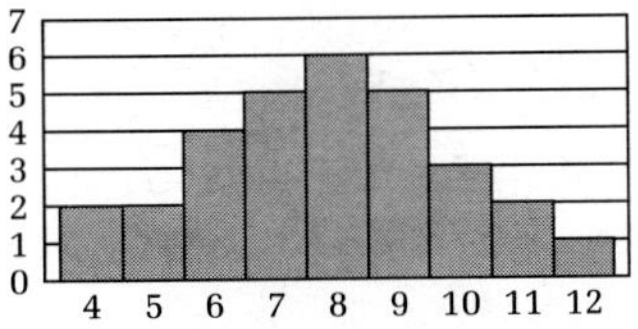

b. mean = 7.8; median = 8
c. Yes.

15.

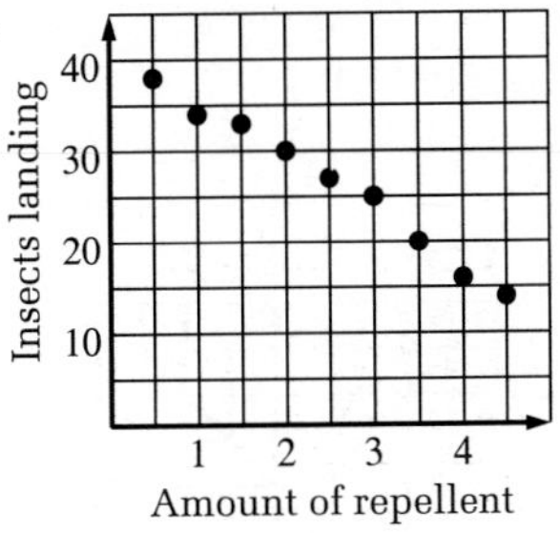

17. $y = -6x + 40.7$

19, 21.

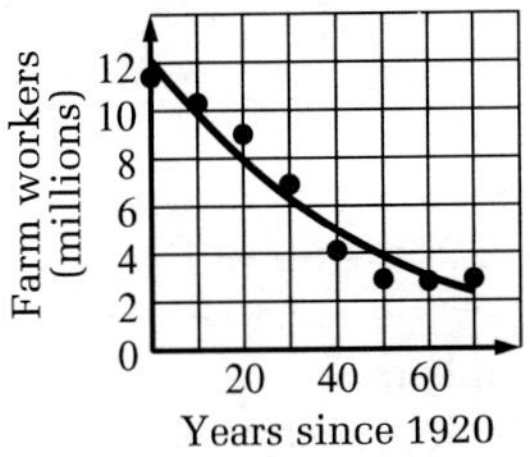

Pages 619–620 Extra Practice Unit 7

1. int(6 * rand) + 1 **3.** On a calculator, choose groups of 5 random integers from 1 to 13. Count the number of duplicate integers in each group. Perform a large number of trials. The experimental probability is $\frac{\text{number of times at least two integers match}}{\text{number of trials}}$.

5. $\frac{4}{13}$ **7.** $\frac{2}{9}$ **9.** $\frac{5}{18}$ **11.** $\frac{7}{8}$ **13.** $\frac{1}{3}; \frac{7}{30}; \frac{7}{90}$ **15.** $\frac{1}{4}$ **17.** $\frac{3}{7}$

19. The average expected advance from picking a card is 2.5 spaces. You should pick the card unless rolling the die is an option. Rolling the die has a greater expected advance. **21.** 0.146

Pages 621–622 Extra Practice Unit 8
Answers may vary due to rounding.

1, 3, 5.

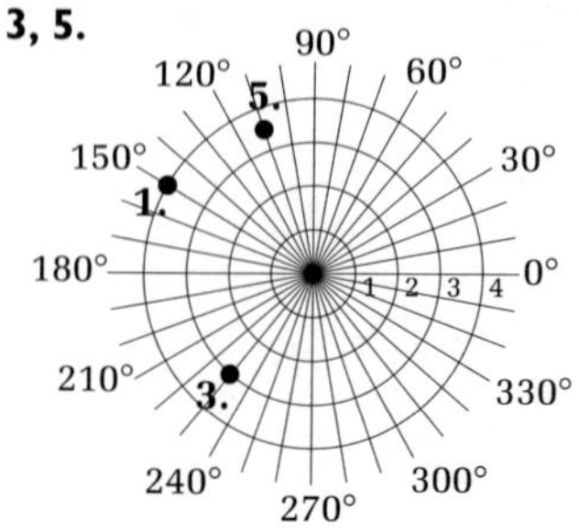

7. (0, −7)
9. (−2.83, −2.83)
11. (−1.81, −0.85)
13. (−3.94, 3.08)
15. (4, 120°)
17. (5, −36.9°)
19. (5.4, 141°)
21. (4, 2)
23. (−2, 3)
25. (2, 6)

27. (5, 5) **29.** (−6, 2) **31.** (9, 90°); (0, 9) **33.** (7.74, 71.2°); (2.5, 7.33) **35.** (7.55, −131.5°); (−5, −5.66) **37.** (−2, 0) **39.** (−1.4, −1.5) **41.** (−1.7, −1.3) **43.** (3.30, −101.2°) **45.** (1, −20°) **47.** (6.07, 129.1°) **49.** $x = 55.6t$; $y = 29.6t - 4.9t^2$

51. $y = 0.37x - 0.065x^2$ **53.** $(x-1)^2 + (y-2)^2 = \frac{1}{4}$

55. $\frac{(x-3)^2}{36} + \frac{(y+1)^2}{25} = 1$ **57.** 19.8 **59.** 108.6°

61. $a = 6.72$, $m\angle C = 86°$, $b = 9.69$ **63.** $m\angle C = 43.9°$, $m\angle A = 64.1°$, $a = 8.0$ **65.** $m\angle A = 51.5°$, $m\angle B = 14.5°$, $b = 6.9$ **67.** 0 **69.** 2

Pages 622–624 Extra Practice Unit 9
1. 66.7; 62.5; 34.9 **3.** $y = 10^x$; (1, −2) **5.** $y = \log x$; (9, 0) **7.** $y = |x|$; (0, −8)

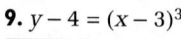

9. $y - 4 = (x-3)^3$

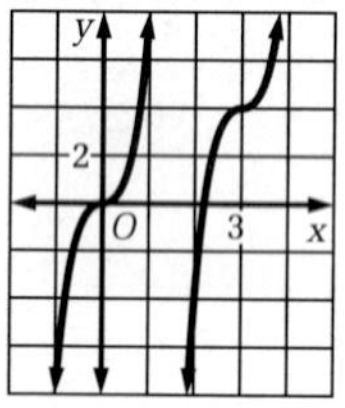

11. $y + 5 = |x - 3|$

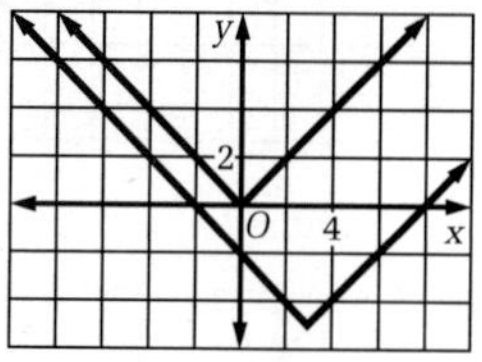

13. $y + 5 = (x-1)^2$

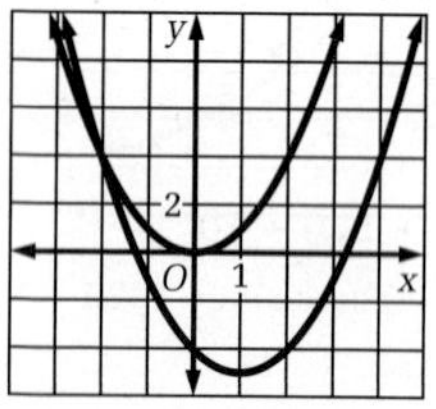

15. a. 29.9, 34, 34, 12.1 **b.** 44.8, 51, 51, 18.1
c.

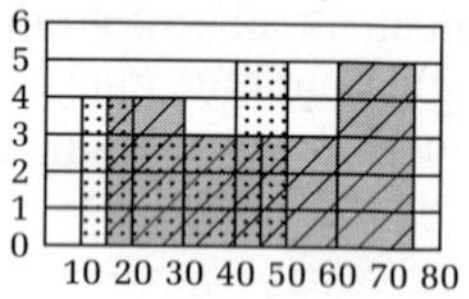

17. $\begin{bmatrix} 16 & 24 & 8 \\ 0 & 2 & 6 \end{bmatrix}$; horizontal stretch by a factor of 4

19. $\begin{bmatrix} 4 & 6 & 2 \\ 0 & 0.5 & 1.5 \end{bmatrix}$; vertical stretch by a factor of $\frac{1}{4}$

21. $y = 2\log(x+3)$

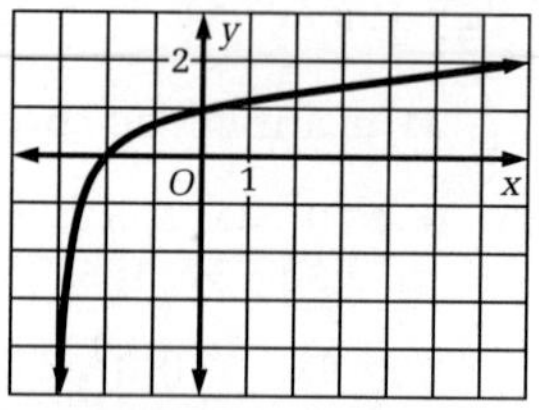

23. $y = \frac{1}{2}|x - 4|$

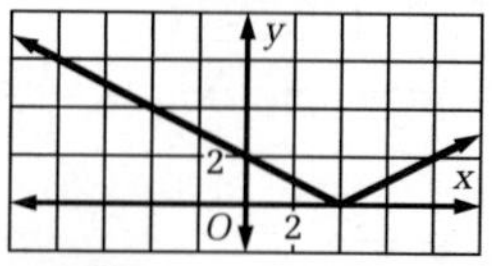

25. a.

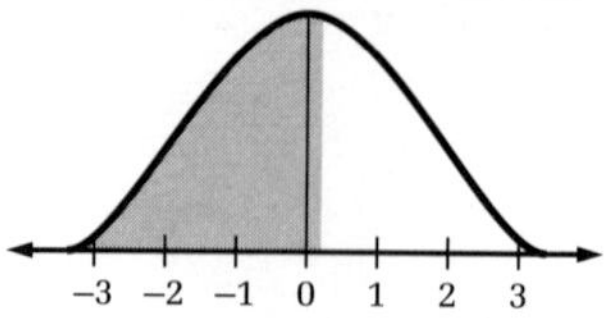

b. 57.93%

27. a.

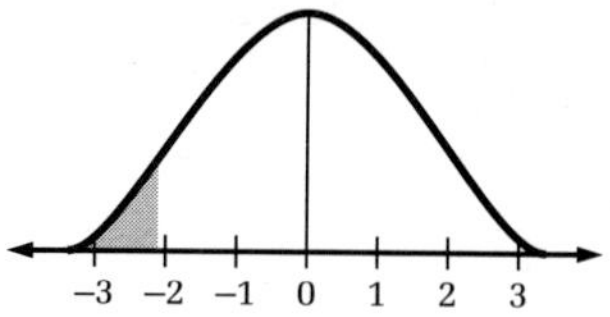

b. 1.79%

29. 0.1 **31.** −2.3 **33.** 0.3; 38 **35.** −0.7; 57.2

37.

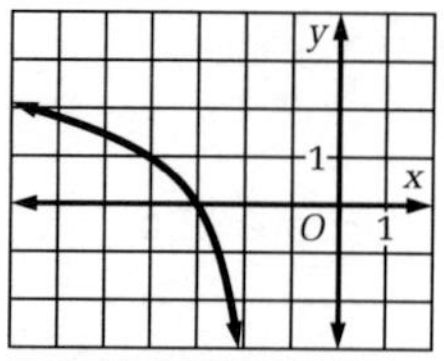

39.

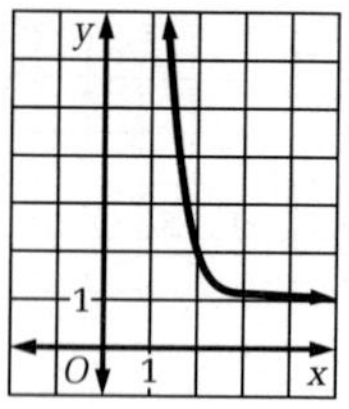

41.

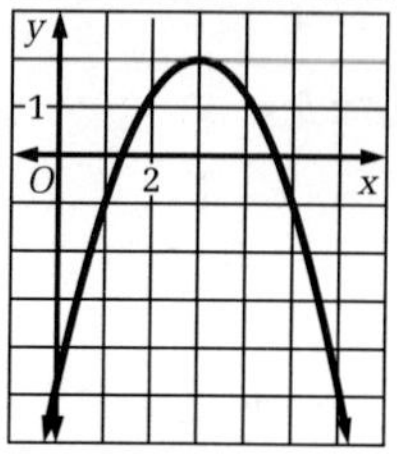

Pages 624–625 Extra Practice Unit 10
1. periodic **3.** damping **5. a.** 4 **b.** $\frac{1}{4}$ **c.** $y = 2.5$ **d.** 1.5 **7. a.** 0 **b.** 1 **9. a.** $-1 \le y \le 1$ **b.** $-1 \le y \le 1$ **11.** 220° **13.** 295° **15. a.** 5, 180, $\frac{1}{180}$

b.

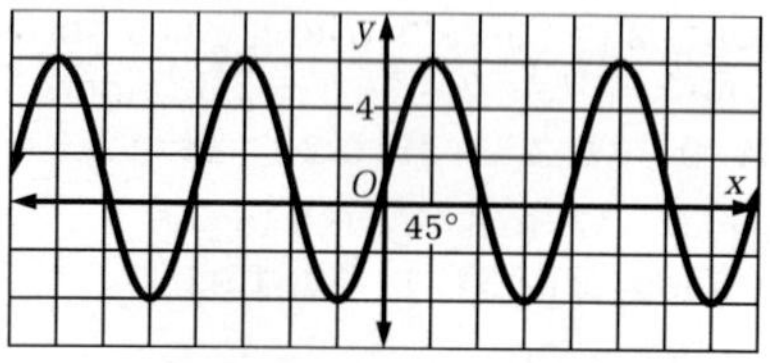

17. a. 1, 180, $\frac{1}{180}$

b.

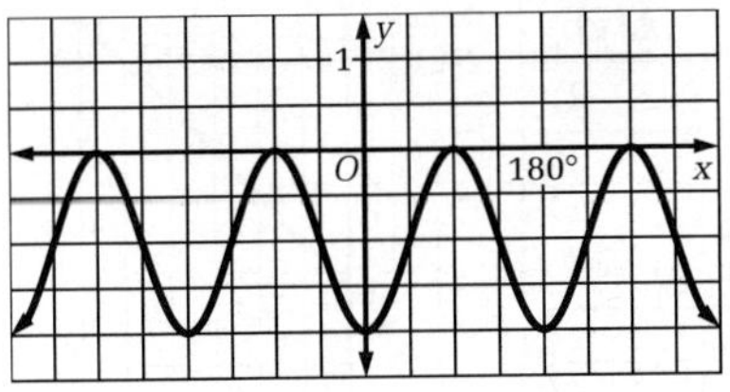

19. a. 2.5, 120, $\frac{1}{120}$

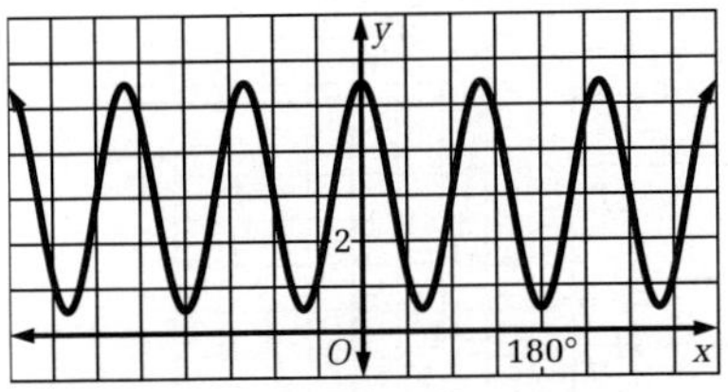

21. $y = 15 \sin (18t) + 20$ **23.** horizontal reflection

25. vertical reflection

27.

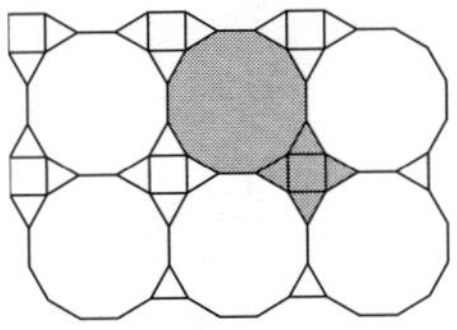

29.

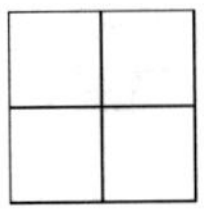

Toolbox Skills

Page 626 Skill 1

1. about 22.7 **2.** 21 **3.** 10, 18, and 24 **4.** 32

Page 627 Skill 2

1. 30 **2.** 100 **3.** 10 or 11 **4.** \$20,000–29,000; \$50,000 or over

Page 628 Skill 3

1. 120 **2.** 5040 **3.** 120 **4.** 1,814,400 **5.** 210 **6.** 6720 **7.** 3024 **8.** 720 **9.** 1680 **10.** 840 **11.** 3024

Page 629 Skill 4

1. 15 **2.** 56 **3.** 21 **4.** 126 **5.** 462 **6.** 792

Page 630 Skill 5

1. 6 **2.** 10 **3.** 20 **4.** 15 **5.** Add the two numbers together. **6.** The numbers in diagonal 2 of Pascal's triangle are called triangular numbers because they can be represented by dots arranged to form equilateral triangles. **7.** The sum of any two consecutive numbers in diagonal 2 of Pascal's triangle is a perfect square. These numbers are called square numbers because they can be represented by dots arranged in a square.

Page 631 Skill 6

1. about 16% **2.** about 35% **3.** about 48% **4.** about 8% **5.** about 77% **6.** 0 **7.** 100%

Page 632 Skill 7

1. $\frac{1}{6}$ **2.** $\frac{5}{12}$ **3.** $\frac{7}{12}$ **4.** $\frac{2}{3}$

Page 632 Skill 8

1. 2×3 **2.** \$4.50; \$5.50

Page 633 Skill 9

1. $\begin{bmatrix} 5 & 20 \\ 35 & 0 \\ 30 & 10 \end{bmatrix}$ **2.** $\begin{bmatrix} 0.5 & 0.5 & 0 & 0.5 \\ 1.5 & 1 & 2 & 3 \end{bmatrix}$ **3.** $\begin{bmatrix} 30 & 45 & 10 \\ 15 & 60 & 5 \end{bmatrix}$

Pages 633–634 Skill 10

1. $\begin{bmatrix} 9 & 2 & -1 \\ 15 & 1 & 6 \end{bmatrix}$ **2.** $\begin{bmatrix} 17 & 9 \\ 7 & -9 \\ 5 & 29 \\ 0 & -4 \end{bmatrix}$ **3.** $\begin{bmatrix} -1 & 2.4 \\ 1.9 & -1.3 \end{bmatrix}$

4. [110 180 70]

Page 634 Skill 11

1. [3] **2.** $\begin{bmatrix} 16 & 27 \\ 12 & 26 \\ 8 & 12 \end{bmatrix}$ **3.** $\begin{bmatrix} 9 & -2 & 7 & -2 \\ 12 & 16 & -20 & -8 \\ 9 & 5 & -4 & -4 \end{bmatrix}$

Page 635 Skill 12

1. 0 **2.** 94 **3.** –21 **4.** –4 **5.** 478 **6.** –72 **7.** –24 **8.** 6 **9.** –200 **10.** –396 **11.** –416 **12.** –256

Pages 635–636 Skill 13

1. $-a^2 - a + 2$ **2.** $3x^2 + 2y^2$ **3.** $2x^3 - 3x^2 - 1$

4. $-2c^2 - 6cd - 5d^2 - 1$

Pages 636–637 Skill 14

1. $27t^{13}$ **2.** $\frac{3}{4}a^8$ **3.** $\frac{-20}{r}$ **4.** $\frac{5}{a^2}$ **5.** $\frac{5}{u^5}$ **6.** $\frac{12}{m^2n^4}$ **7.** $\frac{-6}{y^2}$

8. $\frac{2v^2}{3w}$ **9.** $\frac{2}{x^6y^9}$

Page 638 Skill 15

1. $a^2 - 4$ **2.** $a^2 + 4a + 4$ **3.** $a^2 - 4a + 4$

4. $21k^2 + 22k - 8$ **5.** $25x^2 + 20xy + 4y^2$

6. $49c^2 - 36d^2$ **7.** $81w^2 - 72w + 16$

8. $20x^3 - 56xy + 15y^2$ **9.** $a^3 - a^2 - 5a + 2$

10. $9x^2 + 12xy + 4y^2$ **11.** $x^3 + x^2 - 2x + 12$

12. $4n^3 - 16n^2 + 11n + 10$

Page 639 Skill 16

1. $x^5 + 5x^4y^1 + 10x^3y^2 + 10x^2y^3 + 5xy^4 + y^5$

2. $16r^4 + 96r^3 + 216r^2 + 216r + 81$ **3.** $p^{12} + 6p^{10}q + 15p^8q^2 + 20p^6q^3 + 15p^4q^4 + 6p^2q^5 + q^6$

4. $8n^3 - 12n^2m + 6nm^2 - m^3$ **5.** $c^5 + 50c^4d + 1000c^3d^2 + 10{,}000c^2d^3 + 50{,}000cd^4 + 100{,}000d^5$

6. $v^8 + 8v^6w^2 + 24v^4w^6 + 32v^2w^9 + 16w^{12}$

Page 640 Skill 17

1. $(x-10)(x-10)$ **2.** $(5y+9z)(5y-9z)$
3. $(a-7)(a+2)$ **4.** $(c+5)(2c-1)$ **5.** $(3x-y)(x-2y)$
6. $(7m+5n)(7m+5n)$ **7.** $2x^2(2x^3-x^2+8)$
8. $3p^2(3pq+5)$ **9.** $9x^2y(5x-2y^4)$

Page 641 Skill 18

1. $\frac{1}{y+5}$ **2.** $\frac{x+1}{x+2}$ **3.** $\frac{a-3}{a+3}$ **4.** $\frac{14}{3x}$ **5.** $\frac{1}{2x-10}$
6. $\frac{2-x}{x^2+x}$

Page 642 Skill 19

1. $3\sqrt{5}$ **2.** $10\sqrt{3}$ **3.** $2\sqrt{19}$ **4.** $10\sqrt{3}$ **5.** $60\sqrt{5}$
6. $\sqrt{42}$ **7.** $3\sqrt{10}$ **8.** $4.2\sqrt{33}$ **9.** $24\sqrt{6}$ **10.** 6
11. 27 **12.** $12\sqrt{3}$ **13.** $60\sqrt{2}$

Page 643 Skill 20

1. $10i$ **2.** $0.5i$ **3.** $i\sqrt{17} \approx 4.12i$ **4.** $i\sqrt{83} \approx 9.11i$
5. -35 **6.** $6i$ **7.** $72+9i$ **8.** $21-84i$ **9.** $18-19i$
10. $-13+11i$ **11.** $10-7.5i$ **12.** $78-36i$
13. $-3.6+7.6i$ **14.** 116

Pages 644–645 Skill 21

1. $x=-3,\ y=-12$ **2.** $a=3,\ b=8$ **3.** $x=0,\ y=-1$
4. $a=\frac{37}{3},\ b=-4$ **5.** $p=5,\ q=-\frac{3}{2}$ **6.** $c=5,\ d=5$

Pages 645–646 Skill 22

1. 5, −1 **2.** $\frac{5}{2}$, −1 **3.** $\frac{1}{3}$, −1 **4.** about 3.7, about 0.27
5. about −2.8, about 1.25 **6.** about 1.77, about 0.23

Pages 646–647 Skill 23

1. $-\frac{2}{3}$ **2.** 2, −3 **3.** 5 **4.** 3, 2 **5.** $\frac{1}{4}$ **6.** 3

Page 647 Skill 24

1. 0, 2, 3 **2.** 0, −1, −3 **3.** 1, −3, $\frac{1}{3}$ **4.** 0, −4, 1
5. 0, 2, −2 **6.** 2, −6

Pages 648–649 Skill 25

1. $y=4x+7$ **2.** $y=-0.5x-2$ **3.** $y=-3x+11$
4. $y=-\frac{2}{3}x-\frac{1}{3}$ **5.** $y=-\frac{1}{2}x+\frac{3}{2}$

Page 649 Skill 26

1. Tables may vary. Examples are given.

x	y	(x, y)
−2	9	(−2, 9)
−1	7	(−1, 7)
0	5	(0, 5)
1	3	(1, 3)
2	1	(2, 1)

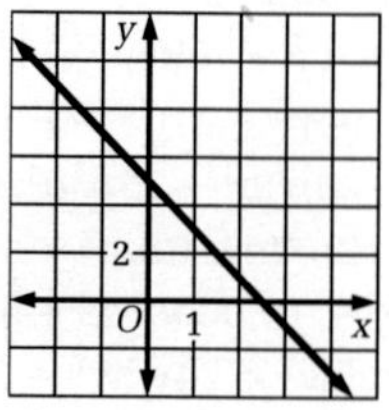

2.

x	y	(x, y)
−2	−2	(−2, −2)
−1	$-\frac{3}{2}$	$\left(-1, -\frac{3}{2}\right)$
0	−1	(0, −1)
1	$-\frac{1}{2}$	$\left(-1, -\frac{1}{2}\right)$
2	0	(2, 0)

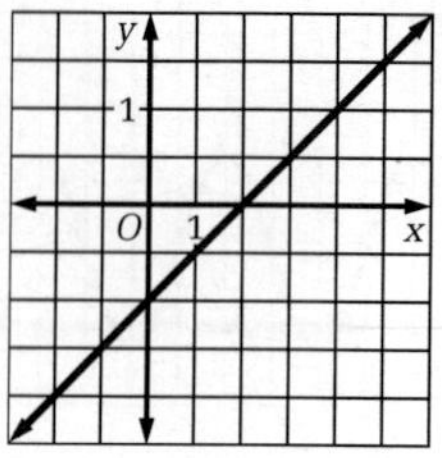

3.

x	y	(x, y)
−2	−1	(−2, −1)
−1	−4	(−1, −4)
0	−5	(0, −5)
1	−4	(1, −4)
2	−1	(2, −1)

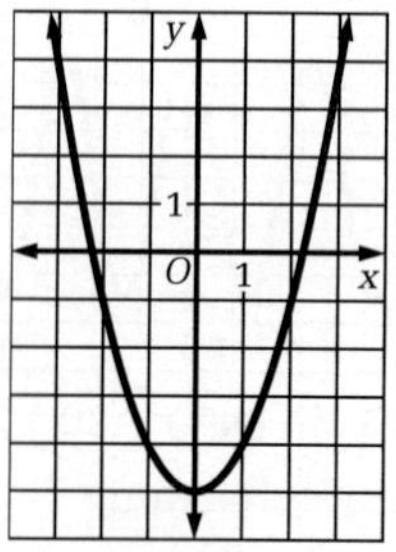

4.

x	y	(x, y)
−2	0	(−2, 0)
−1	−2	(−1, −2)
0	−2	(0, −2)
1	0	(1, 0)
2	4	(2, 4)

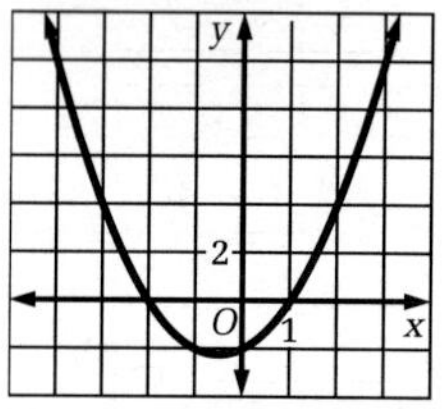

5.

x	y	(x, y)
−2	4	(−2, 4)
−1	3	(−1, 3)
0	2	(0, 2)
1	1	(1, 1)
2	0	(2, 0)
3	1	(3, 1)
4	2	(4, 2)

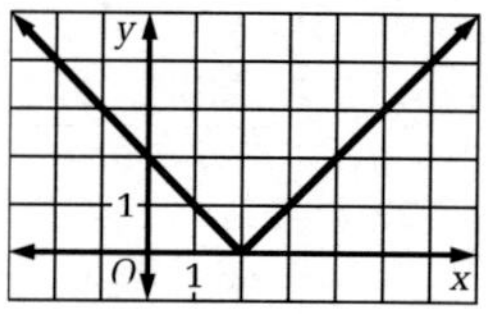

6.

x	y	(x, y)
−2	2	(−2, 2)
−1	2.5	(−1, 2.5)
0	3	(0, 3)
1	2.5	(1, 2.5)
2	2	(2, 2)

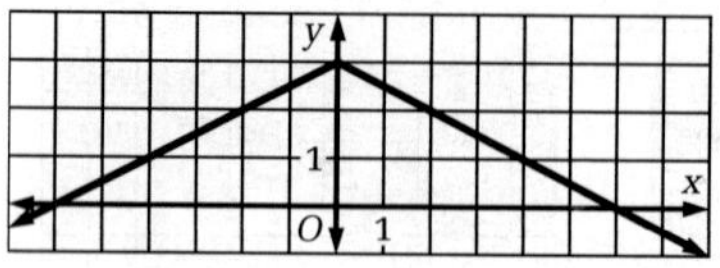

7.

x	y	(x, y)
–4	1	(–4, 1)
–3	–1	(–3, –1)
–2	–3	(–2, –3)
–1	–1	(–1, –1)
0	1	(0, 1)
1	3	(1, 3)
2	5	(2, 5)

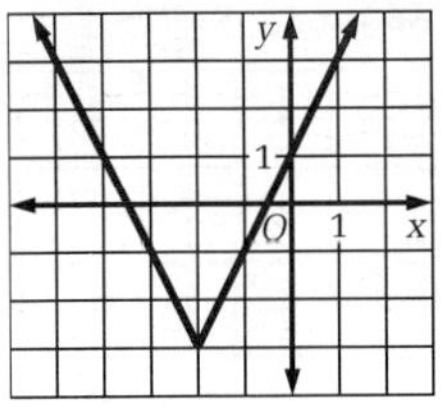

8.

x	y	(x, y)
–3	–6.75	(–3, –6.75)
–2	–2	(–2, –2)
–1	–0.25	(–1, –0.25)
0	0	(0, 0)
1	0.25	(1, 0.25)
2	2	(2, 2)
3	6.75	(3, 6.75)

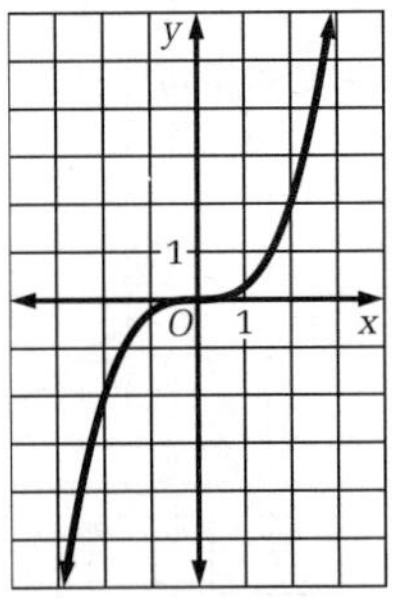

9.

x	y	(x, y)
–3	–24	(–3, –24)
–2	–6	(–2, –6)
–1	0	(–1, 0)
–0.5	0.325	(–0.5, 0.325)
0	0	(0, 0)
0.5	–0.325	(0.5, –0.325)
1	0	(1, 0)
2	6	(2, 6)
3	24	(3, 24)

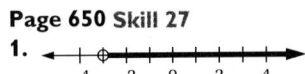

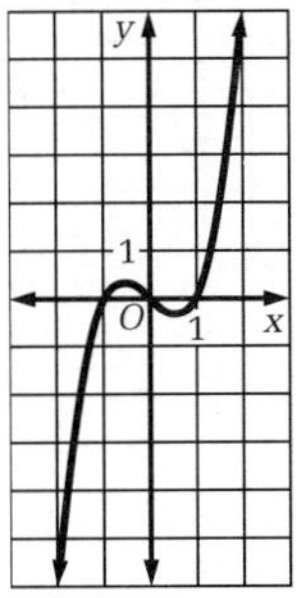

Page 650 Skill 27

1. 

2. 

3. 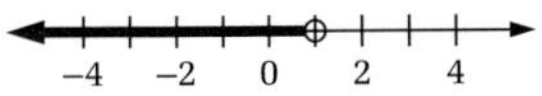

4. 

5. 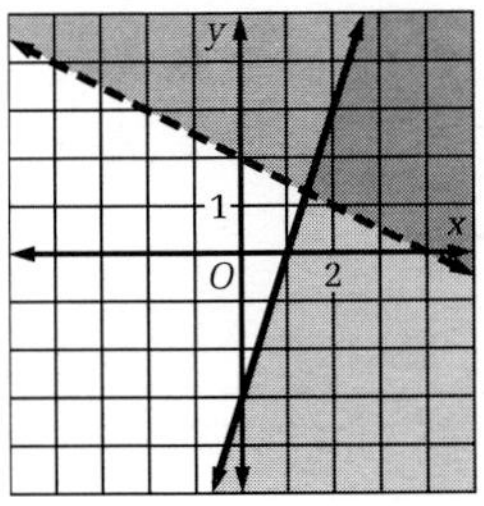

6. 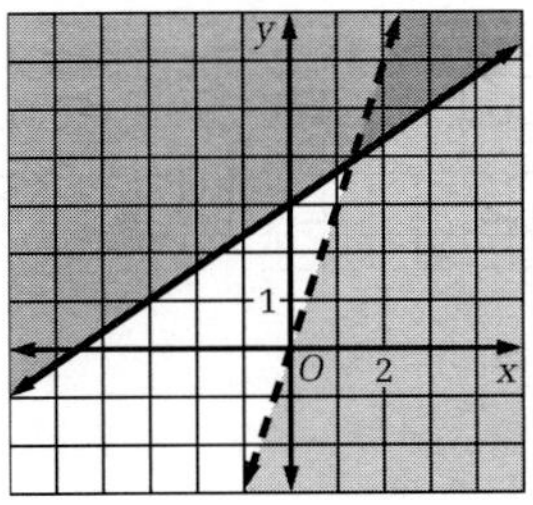

7. 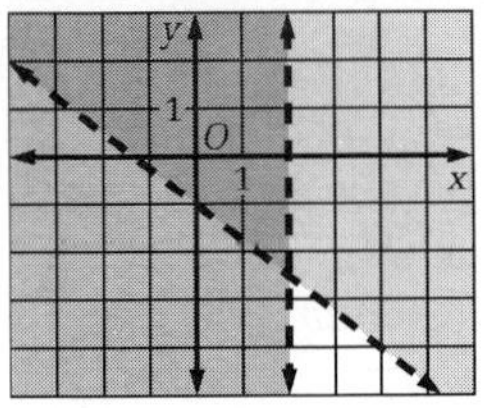

8. 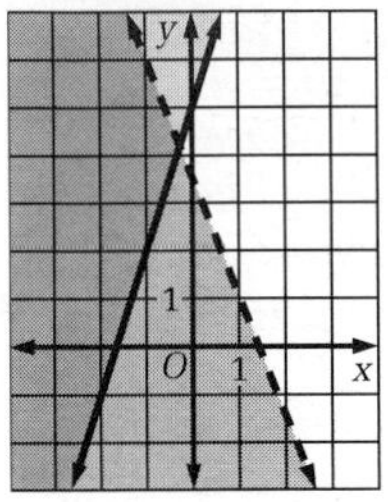

Page 651 Skill 28

1. *y*-intercept: –4; *x* = 0

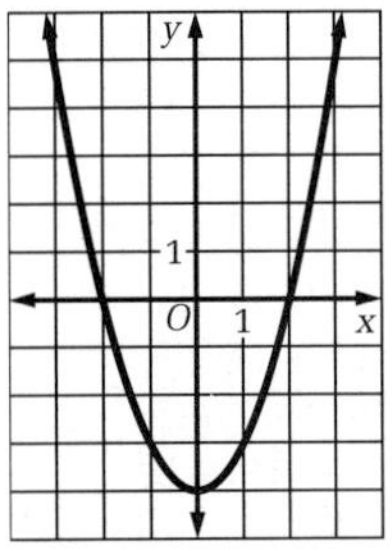

2. *y*-intercept: 7; *x* = 0

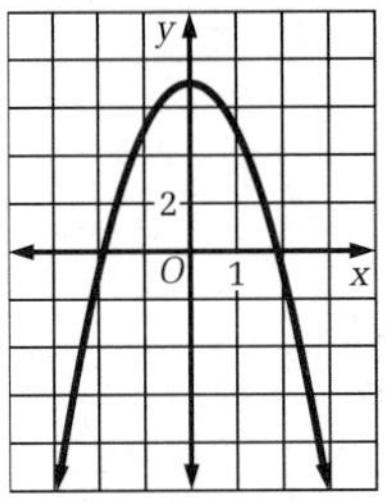

3. *y*-intercept: 0; *x* = 3

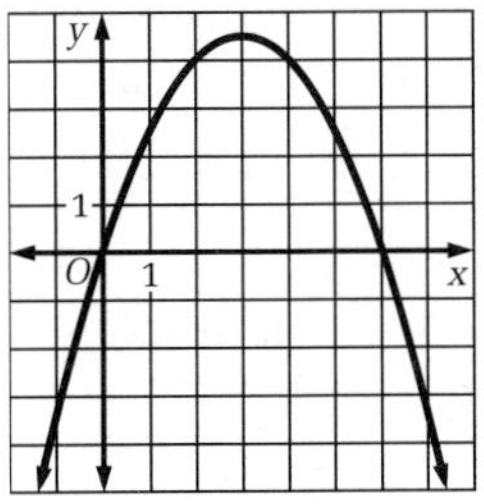

4. *y*-intercept: –1; *x* = 4

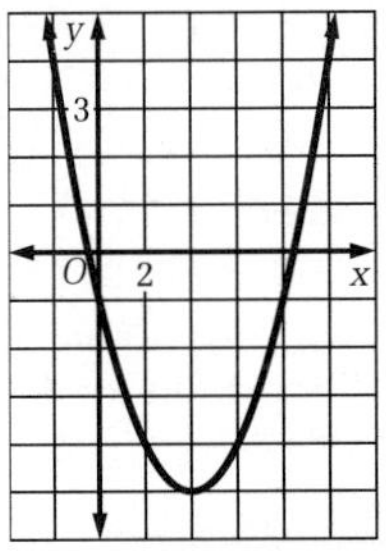

5. *y*-intercept: –10; *x* = 2

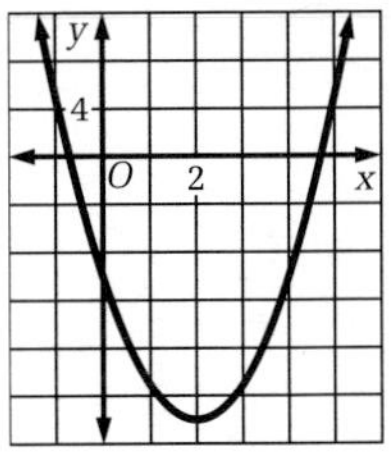

6. *y*-intercept: 6; *x* = 3

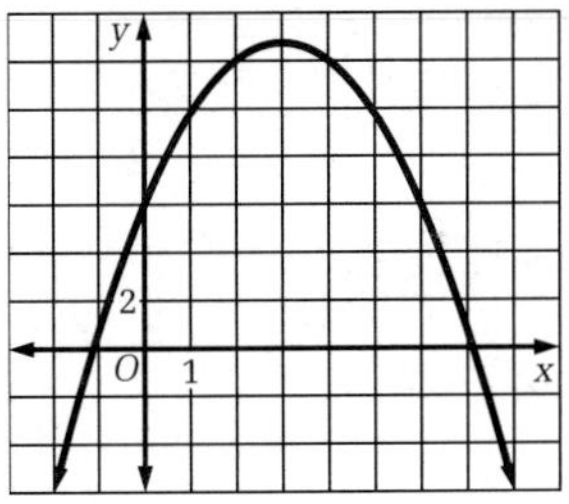

Pages 652–653 Skill 29

1. 160 ft^2 **2.** 19.5 $in.^2$ **3.** about 28.27 $in.^2$ **4.** 702 $in.^2$ **5.** about 785.4 cm^2 **6.** 200 cm^3 **7.** about 106.8 m^3 **8.** 400 $in.^3$

Pages 653–654 Skill 30

1. 5 **2.** 24 **3.** about 7.55 **4.** $b \approx 6.9$ cm, $c \approx 13.9$ cm

Pages 654–655 Skill 31

1. 5 **2.** 13 **3.** $\sqrt{29} \approx 5.4$ **4.** (5, 6) **5.** (2, –4) **6.** (5.5, –4.5)

Pages 655–656 Skill 32

1. $m\angle CAB = m\angle DEB$ and $m\angle CBA = m\angle DBE$, so the triangles are similar. Corresponding angles are $\angle CAB$ and $\angle DEB$, $\angle CBA$ and $\angle DBE$, and $\angle BCA$ and $\angle BDE$; corresponding sides are $\overline{AC}$ and $\overline{ED}$, $\overline{BC}$ and $\overline{BD}$, and $\overline{AB}$ and $\overline{EB}$. **2.** $m\angle PNM = m\angle RQM$ and $m\angle NMP = m\angle QMR$, so the triangles are similar. Corresponding angles are $\angle PNM$ and $\angle RQM$, $\angle NMP$ and $\angle QMR$, and $\angle M$ and $\angle M$; corresponding sides are $\overline{NP}$ and $\overline{QR}$, $\overline{MN}$ and $\overline{MQ}$, and $\overline{MP}$ and $\overline{MR}$. **3.** 18

Page 656 Skill 33

1. Yes; by SSS. **2.** Yes; by AAS. **3.** Yes; by ASA. **4.** No. **5.** $CB = FE$, $AC = DF$, $AB = DE$, $m\angle A = m\angle D$, $m\angle C = m\angle F$, $m\angle B = m\angle E$; 75

Page 657 Skill 34

1. All rhombuses have perpendicular diagonals.
2. The fact that a figure is a square implies that it is a rhombus. **3.** Every Saturday is a day of the weekend. **4.** A figure is a cylinder only if the formula for its volume is $V = Bh$. **5.** A polygon is a quadrilateral if and only if it has four sides. **6.** $5x = 20$ if and only if $x = 4$.

Pages 658–659 Skill 35

1. *or* rule **2.** chain rule **3.** direct argument
4. A horse is a vertebrate. **5.** $\triangle ABC$ is not isosceles.
6. If $2x + 3 = 21$, then $x = 9$. **7.** $P(A) \neq 0$.

Page 660 Skill 36

1. $\frac{4}{8.06} \approx 0.50$ **2.** $\frac{7}{8.06} \approx 0.87$ **3.** $\frac{4}{7} \approx 0.57$

4. $\frac{7}{8.06} \approx 0.87$ **5.** $\frac{4}{8.06} \approx 0.50$ **6.** $\frac{7}{4} \approx 1.75$

7.

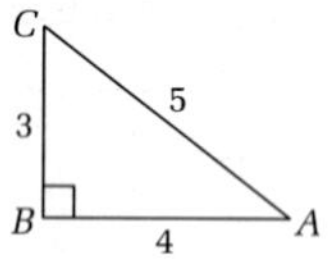

8.

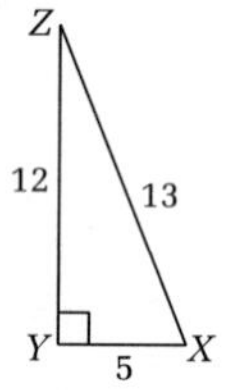

Page 661 Skill 37

1. $x \approx 26.8$ **2.** $x \approx 5.8$ **3.** $x \approx 26.6$ **4.** $h \approx 10.3$

Pages 662–663 Skill 38

1.

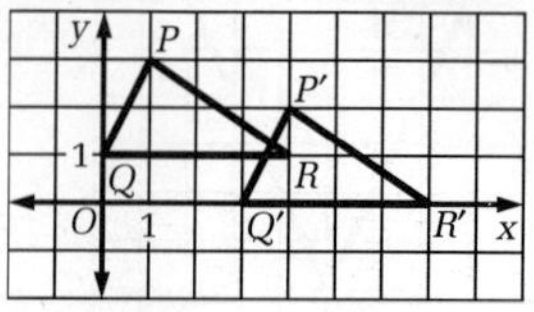

2.

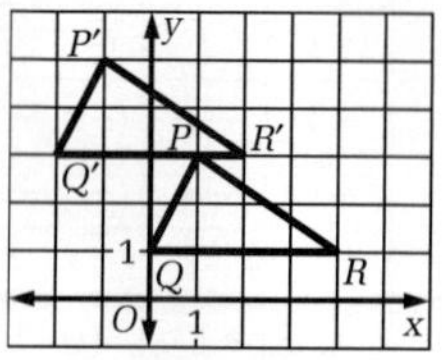

3.

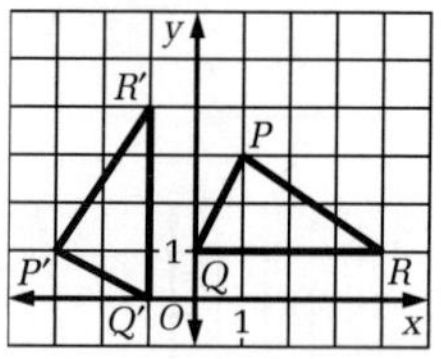

4.

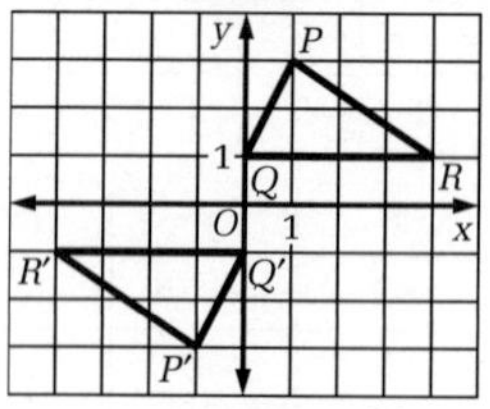

5.

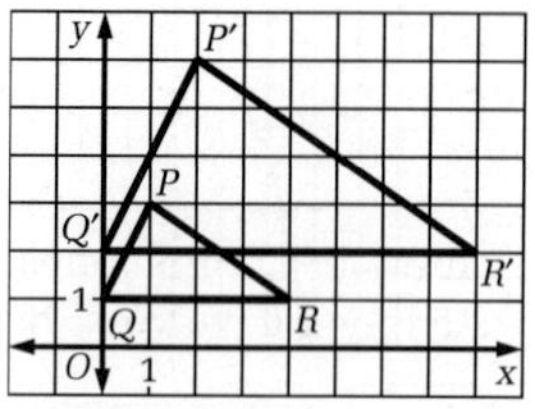

6.

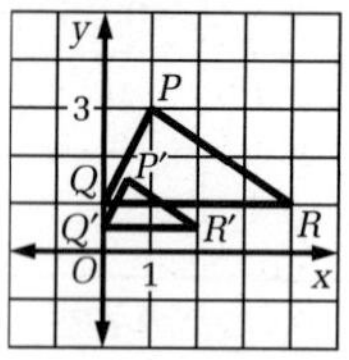

7.

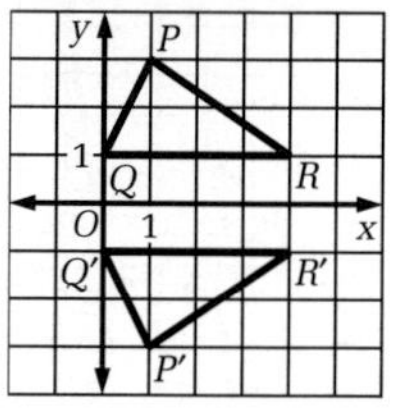

8.

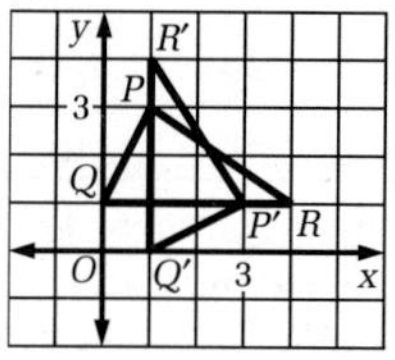

Page 663 Skill 39

1. 72° rotational symmetry about point X

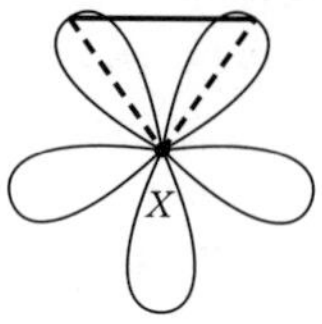

2. reflection symmetry

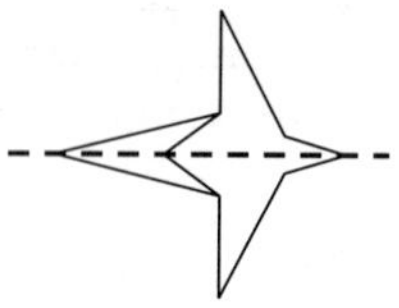

3. reflection symmetry

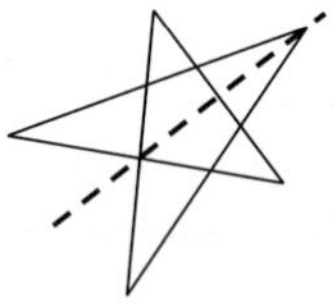

Additional Answers

Unit 1

Page 8 Exercises and Problems

15. a. It computes the mean of the numbers in the list. **b.** At Step 6, it either repeats Steps 3–5 or, if all the numbers in the list have been read, it moves on to Step 7. **16. a.** Answers may vary. An example is given. **Step 1:** Start with the amount of the sale. **Step 2:** Count the number of coins (pennies, nickels, dimes, quarters) needed to add up to the next dollar. **Step 3:** Count the number of bills (ones, fives, and so on) needed to equal the amount the buyer gave you. **b.** Answers will vary.

Page 12 Talk it Over

7. Answers may vary. An example is given. Look for the most time-consuming job and assign it to mechanic 1. Then find a smaller job to fill his or her 8 hours. Then take the next most time-consuming job and assign it to mechanic 2. Then find smaller jobs to fill his or her 8 hours. Continue this until all the jobs are assigned. This method should result in the least number of mechanics. **8.** If some customers need to pick up their cars by noon, Paula would have to assign those tasks so that they could be completed in time, then schedule the others. If two or more mechanics are needed for some tasks, they would have to be assigned first. Then the remaining tasks would have to be assigned to individual mechanics as they become available.

Page 13 Exercises and Problems

8.

0	0	0	16
0	0	1	15
0	0	2	14
0	0	3	13
0	0	4	12
0	0	5	11
0	0	6	10
0	0	7	9
0	0	8	8
0	1	1	14
0	1	2	13
0	1	3	12
0	1	4	11
0	1	5	10
0	1	6	9
0	1	7	8
0	2	2	12
0	2	3	11
0	2	4	10
0	2	5	9
0	2	6	8
0	2	7	7

0	3	3	10
0	3	4	9
0	3	5	8
0	3	6	7
0	4	4	8
0	4	5	7
0	4	6	6
0	5	5	6
1	1	1	13
1	1	2	12
1	1	3	11
1	1	4	10
1	1	5	9
1	1	6	8
1	1	7	7
1	2	2	11
1	2	3	10
1	2	4	9
1	2	5	8
1	2	6	7
1	3	3	9

1	3	4	8
1	3	5	7
1	3	6	6
1	4	4	7
1	4	5	6
1	5	5	5
2	2	2	10
2	2	3	9
2	2	4	8
2	2	5	7
2	2	6	6
2	3	3	8
2	3	4	7
2	3	5	6
2	4	4	6
2	4	5	5
3	3	3	7
3	3	4	6
3	3	5	5
3	4	4	5
4	4	4	4

9.

Side 1	Side 2	Side 3
1	8	8
2	7	8
3	6	8
3	7	7
4	5	8
4	6	7
5	5	7
5	6	6

10.

Side 1	Side 2	Side 3	Side 1	Side 2	Side 3
2	11	11	6	7	11
3	10	11	6	8	10
4	9	11	6	9	9
4	10	10	7	7	10
5	8	11	7	8	9
5	9	10	8	8	8

11.

Side 1	Side 2	Side 3	Side 1	Side 2	Side 3
2	15	15	7	12	13
3	14	15	8	9	15
4	13	15	8	10	14
4	14	14	8	11	13
5	12	15	8	12	12
5	13	14	9	9	14
6	11	15	9	10	13
6	12	14	9	11	12
6	13	13	10	10	12
7	10	15	10	11	11
7	11	14			

12. a.

quarters	dimes	nickels	quarters	dimes	nickels
4	0	0	1	2	11
3	2	1	1	1	13
3	1	3	1	0	15
3	0	5	0	10	0
2	5	0	0	9	2
2	4	2	0	8	4
2	3	4	0	7	6
2	2	6	0	6	8
2	1	8	0	5	10
2	0	10	0	4	12
1	7	1	0	3	14
1	6	3	0	2	16
1	5	5	0	1	18
1	4	7	0	0	20
1	3	9			

b. all combinations but two: 4 quarters, 10 dimes **c.** Answers may vary. Examples are given. (a) First I listed the number of quarters, beginning with the greatest possible number. Then I listed possible numbers of dimes for that number of quarters, beginning with the greatest possible number. Then I determined the correct number of nickels to total $1. (b) Using my list from part (a), I determined whether each entry gave the correct change.

13. a.

Side 1	Side 2	Side 3
8	8	14
9	9	12
10	10	10
11	11	8
12	12	6
13	13	4
14	14	2

b. the triangle with sides lengths of 10, 10, and 10

14.

fixed income fund	mutual fund	company stock
0	0	3%
0	0	4%
0	0	5%
1%	0	3%
0	1%	3%
1%	1%	3%
1%	0	4%
0	1%	4%
2%	0	3%
0	2%	3%
0	2%	2%
1%	1%	2%
2%	0	2%
0	3%	2%
3%	0	2%
2%	1%	2%
1%	2%	2%
4%	0	1%
0	4%	1%
2%	2%	1%
3%	1%	1%
1%	3%	1%

Page 15 Exercises and Problems

30. a.

Swings	Slides	Whirligigs	Sandboxes
3	4	4	4
4	3	4	4
4	4	3	4
4	4	4	3
5	4	3	3
5	3	4	3
5	3	3	4
4	5	3	3
3	5	4	3
3	5	3	4
3	3	5	4
3	4	5	3
4	3	5	3
3	3	4	5
3	4	3	5
4	3	3	5
6	3	3	3
3	6	3	3
3	3	6	3
3	3	3	6

b. The designer should include six swings, three slides, three whirligigs, and three sandboxes.

Page 19 Exercises and Problems

2. lower extreme: 66; lower quartile: 93; median: 121; upper quartile: 189; upper extreme: 254 **3.** Mexico City: about 16.3°; Shanghai: about 16.6°

4. a. **Mean Monthly Temperature (°C)**

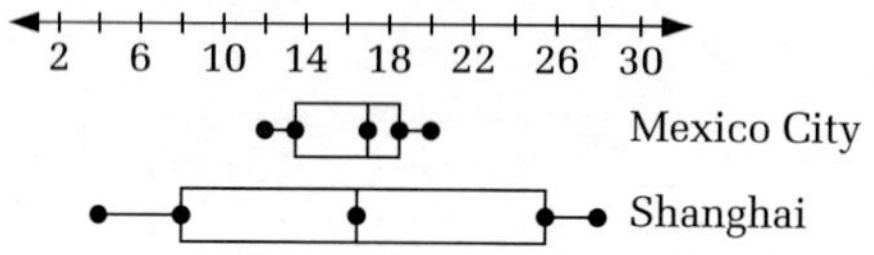

b. Answers may vary. Examples are given. Since the plot for Shanghai is much wider than that for Mexico City, the temperature varies much more in Shanghai than in Mexico City. Since the boxes have nearly the same midpoint, the median temperatures of the two cities are close. The lower extreme for Mexico City is higher than the first quartile for Shanghai, so more than 25% of Shanghai's mean monthly temperatures are below Mexico City's lowest mean monthly temperature. **c.** Answers may vary. An example is given. Which graph you use would depend on what type of information you want to convey. For example, if you are interested in the kind of information given in part (b) above, a box-and-whisker plot is helpful. If you want to compare the mean temperatures for given months or compare trends over given periods, a line graph is more helpful.

Page 21 Exercises and Problems

18.

Side 1	Side 2	Side 3
2	8	8
3	7	8
4	6	8
4	7	7
5	5	8
5	6	7
6	6	6

19.

Side 1	Side 2	Side 3
2	12	12
3	11	12
4	10	12
4	11	11
5	9	12
5	10	11
6	8	12
6	9	11
6	10	10
7	7	12
7	8	11
7	9	10
8	8	10
8	9	9

Page 22 Exercises and Problems

27. Answers may vary. Examples are given. The examples are based on the assumption that the population of the area where the park will be located is similar to the population on which the chart is based.

People Who Play Selected Recreational Sports in the United States (1991)

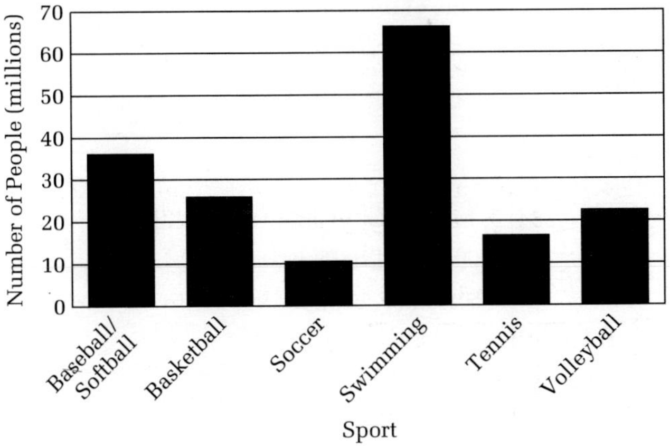

People Who Play Selected Recreational Sports in the United States (1991)

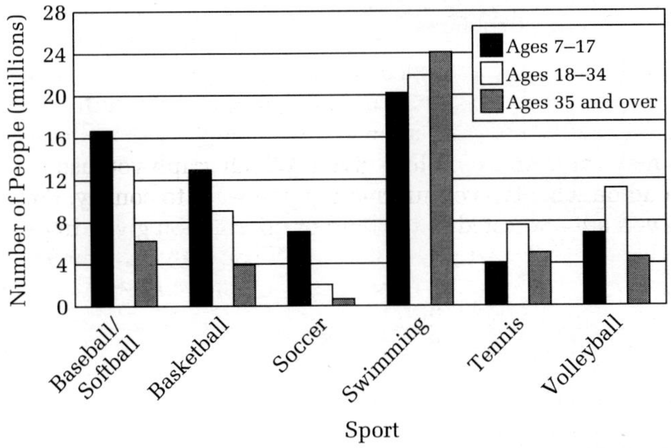

I would choose swimming, baseball/softball, and basketball. Swimming and baseball/softball are first and second on the overall chart, and first and second with each of the age groups. To choose the third facility, I considered how the remaining four sports ranked with each age group. Weighting each according to how it ranked with the age groups produced a tie between basketball and volleyball. Since the numbers are greater for basketball, I chose that.

Page 27 Exercises and Problems

18. a. Let n be the number of miles and t the total cost in dollars. U-Rent: $t = 19.95 + 0.59n$; MacRents: $t = 44.95 + 0.39n$; Wiley's: $t = 79.95 + 0.22n$; Zeke's: $t = 150$ **b.** If she will drive less than about 128 mi, she should choose U-Rent. If she will drive more than about 128 mi but less than about 205 mi, she should choose MacRents. If she will drive more than about 205 mi but less than about 318 mi, she should choose Wiley's. If she will drive more than about 318 mi, she should choose Zeke's. All of these determinations are based only on total cost. There may be other factors involved, however, such as convenience and availability of equipment. **19. a.** $F = 2C + 30$

b.

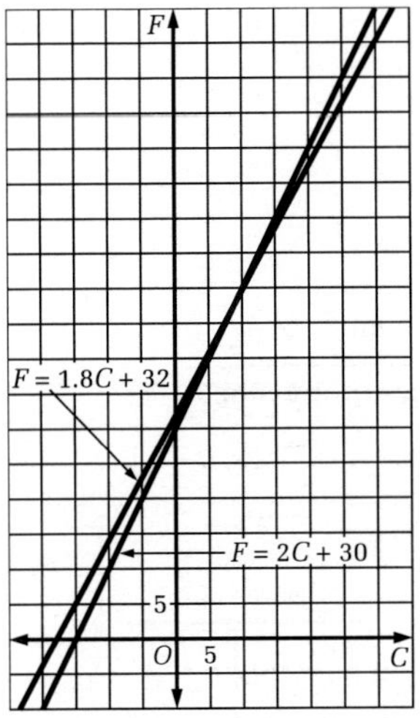

c. 10°C **d.** $-15° \le C \le 35°$ **e.** Answers may vary. An example is given. For an estimation error of 1°, $5° \le C \le 15°$.

20. Let x be the number of hours Alejandro works. **a.** $E = 559 + 1.25x$ **b.** $I = 12x$ **c.** about 11 weeks (52 h) **21.** Answers may vary. An example is given. A Booster Club pays $52.50 for button-making equipment. The material for each "We're #1!" button costs $1.50. If each button sells for $3.25, how many buttons must the club sell to break even?

Page 35 Exercises and Problems

16. Answers may vary. An example is given. Substitute $-x - y + 2$ for z in the second and third equations. The result will be a system of two equations in two variables. The second equation becomes $x - y + (-x - y + 2) = 6$. When you simplify, you have $-2y + 2 = 6$, so $y = -2$. The third equation becomes $x - y - (-x - y + 2) = 8$. When you simplify, you have $2x - 2 = 8$, so $x = 5$. Now substitute 5 for x and -2 for y in one of the original equations to find z. $x - y - z = 8$; $5 - (-2) - z = 8$; $z = -1$. The solution is $(5, -2, -1)$.

Page 41 Exercises and Problems

2.

	A	B	C	D	E
A	0	2	0	1	1
B	2	0	1	0	1
C	0	1	0	1	1
D	1	0	1	0	1
E	1	1	1	1	0

3.

	A	B	C	D	E	F	G
A	0	1	0	0	0	0	1
B	1	0	1	0	0	1	1
C	0	1	0	1	1	1	0
D	0	0	1	0	1	0	0
E	0	0	1	1	0	1	1
F	0	1	1	0	1	0	1
G	1	1	0	0	1	1	0

4.

	A	B	C	D	E	F	G	H	I
A	0	1	0	0	0	0	0	0	1
B	1	0	1	0	1	0	0	1	0
C	0	1	0	1	0	0	0	0	0
D	0	0	1	0	1	0	0	0	0
E	0	1	0	1	0	1	0	1	0
F	0	0	0	0	1	0	1	0	0
G	0	0	0	0	0	1	0	1	0
H	0	1	0	0	1	0	1	0	1
I	1	0	0	0	0	0	0	1	0

Page 42 Exercises and Problems

11. a.

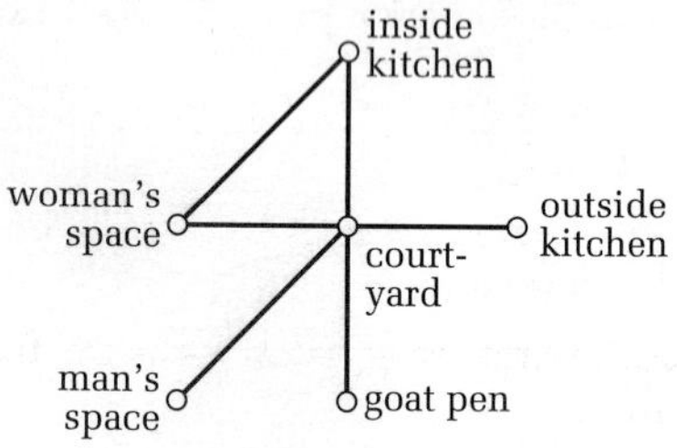

12.

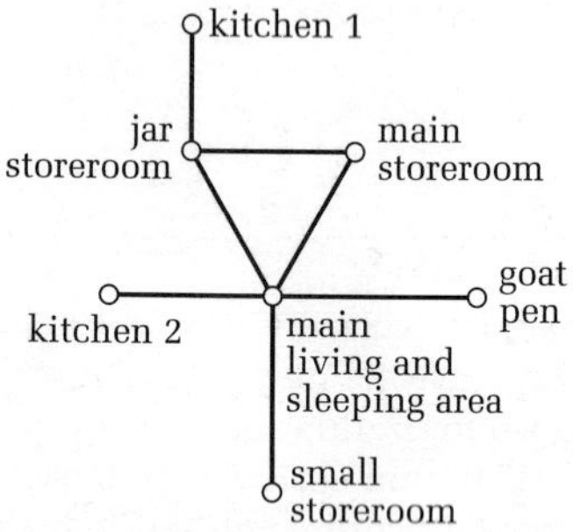

13.

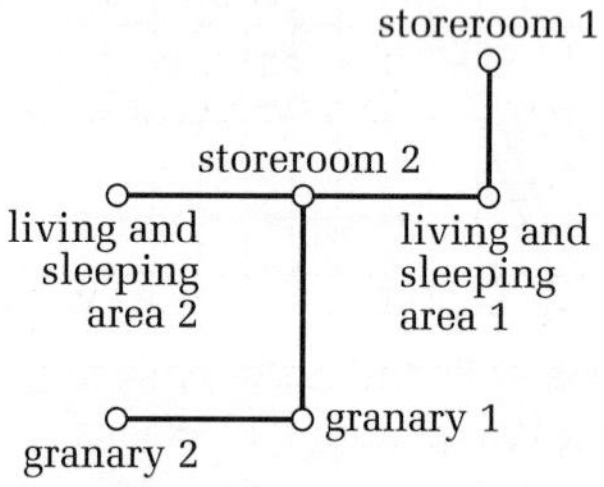

Page 43 Exercises and Problems

17. a.

	Alb.	Asp.	Chey.	Den.	SLC
Alb.	0	0	0	1	1
Asp.	0	0	0	1	0
Chey.	0	1	0	1	0
Den.	0	1	1	0	1
SLC	1	0	1	1	0

b.

	Alb.	Asp.	Chey.	Den.	SLC
Alb.	1	1	2	1	1
Asp.	0	1	1	0	1
Chey.	0	1	1	1	1
Den.	1	1	1	3	0
SLC	0	2	1	2	2

c. Albuquerque–Denver–Aspen, Albuquerque–Denver–Cheyenne, Albuquerque–Salt Lake City–Cheyenne, Albuquerque–Salt Lake City–Denver, Albuquerque–Denver–Salt Lake City, Aspen–Denver–Cheyenne, Aspen–Denver–Salt Lake City, Cheyenne–Aspen–Denver, Cheyenne–Denver–Aspen, Cheyenne–Denver–Salt Lake City, Denver–Cheyenne–Aspen, Denver–Salt Lake City–Albuquerque, Denver–Salt Lake City–Cheyenne, Salt Lake City–Albuquerque–Denver, Salt Lake City–Cheyenne–Aspen, Salt Lake City–Cheyenne–Denver, Salt Lake City–Denver–Aspen, Salt Lake City–Denver–Cheyenne

Unit 2

Page 76 Exercises and Problems

37. Answers may vary. An example is given. **Step 1:** I make sure I have my homework assignment. **Step 2:** I read over my notes if necessary. **Step 3:** I do my math homework. If I can't complete it, I reread my notes and my math text and work through sample problems. **Step 4:** I complete my homework. If there are problems I still can't do, I ask for help at home, or make a note to ask my teacher for help. **Step 5:** I put my homework, my notebook and textbook, a pencil and, if necessary, a calculator together in one place so I will be ready for the next math class.

Page 77 Talk it Over

1. The points appear to lie in a line. As altitude increases, the boiling temperature of water decreases.

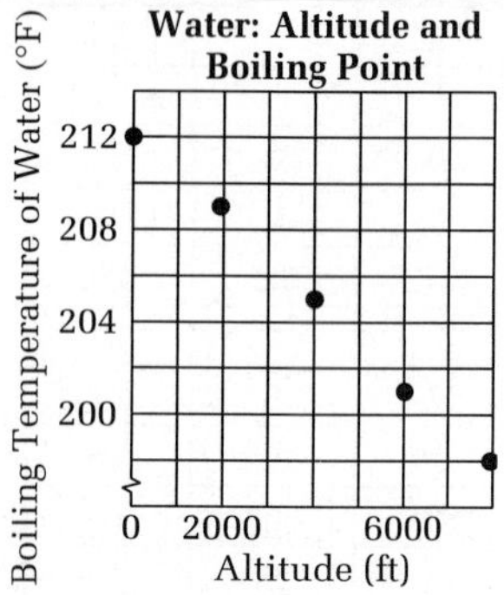

Page 88 Talk it Over

8.

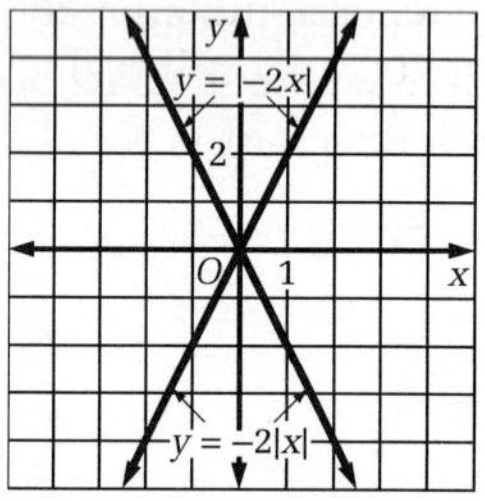

Each graph is the image of the other reflected across the x-axis. Each of the two sections of the graph of $y = |-2x|$ has slope twice that of the corresponding section of the graph of $y = |x|$. Each of the two sections of the graph of $y = -2|x|$ has slope twice that of the corresponding section of the graph of $y = -|x|$, which is the graph of $y = |x|$ reflected across the x-axis.

Page 89 Exercises and Problems

8.

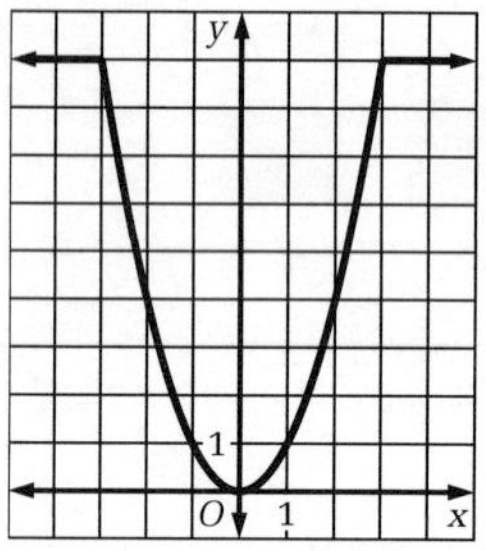

9. a. Let t represent time in hours and $C(t)$ cost in dollars. $C(t) = \begin{cases} 6 & \text{if } 0 \le t \le 1 \\ 6.6t - 0.6 & \text{if } t > 1 \end{cases}$

b. \$9.30 **c.** about 3 h 7 min

Page 91 Exercises and Problems

12.

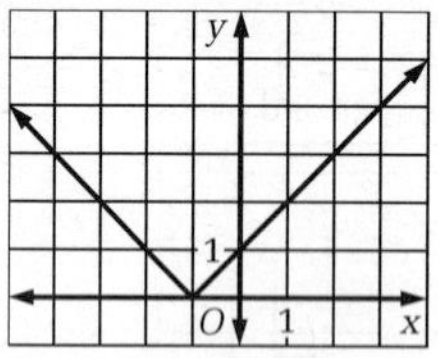

Page 117 Exercises and Problems

45. a.

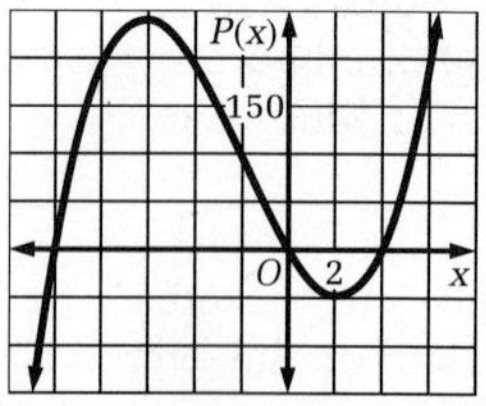

b. –10, 0, 4

46. lower extreme: 0.171; lower quartile: 0.190; median: 0.224; upper quartile: 0.2735; upper extreme: 0.325 **47.** $x = 1$; If $x = 1$, $1 - x = 0$ and $\frac{1}{1 - x}$ is not defined.

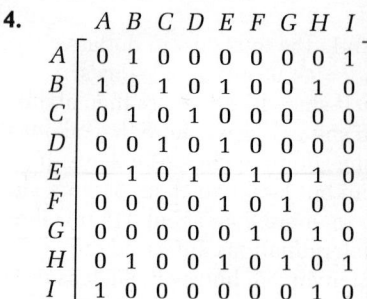

Page 124 Exercises and Problems

11. a.

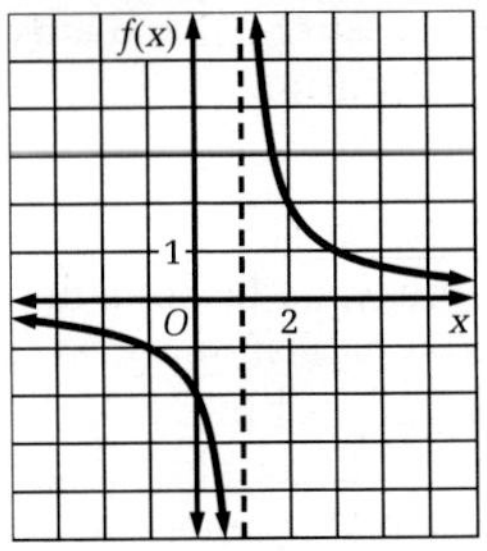

b. $x = 1$; $y = 0$ **c.** domain: all real numbers except 1; range: all real numbers except 0

12. a.

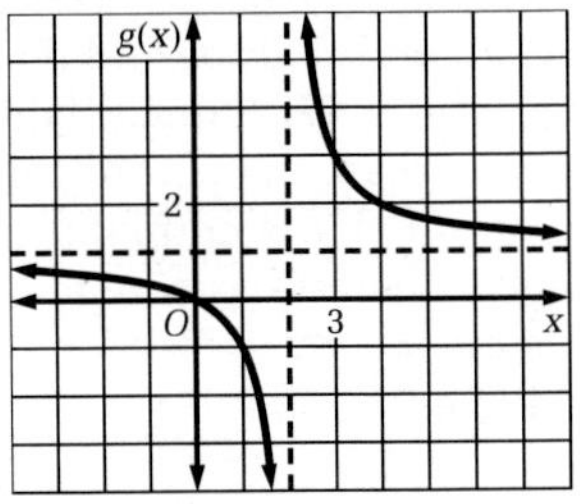

b. $x = 2$; $y = 1$ **c.** domain: all real numbers except 2; range: all real numbers except 1

13. a.

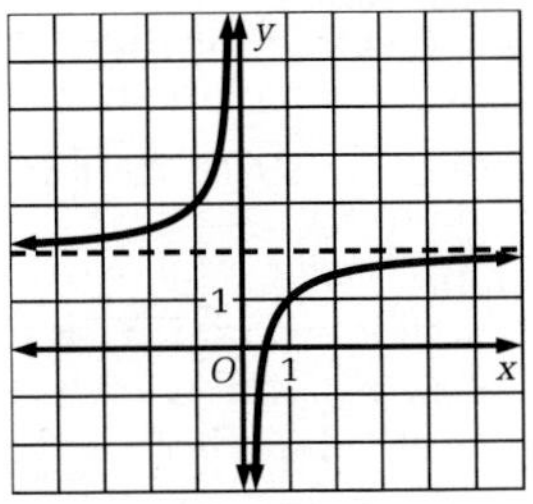

b. $x = 0$; $y = 2$ **c.** domain: all real numbers except 0; range: all real numbers except 2

14. a.

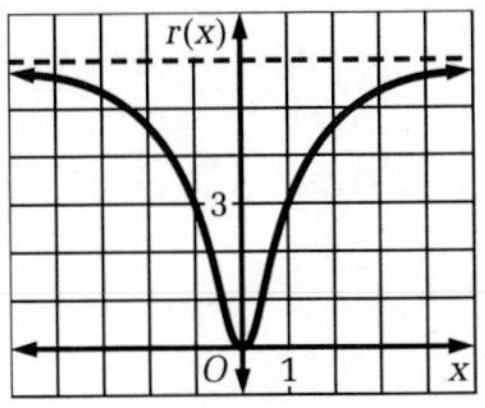

b. $y = 6$ **c.** domain: all real numbers; range: all real numbers greater than or equal to 0 or less than 6

15. a.

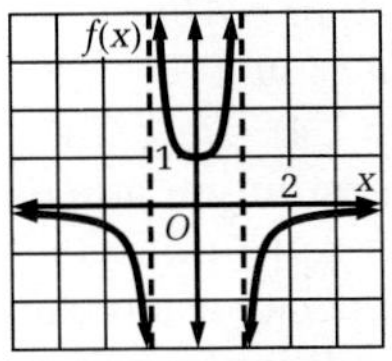

b. $x = 1$, $x = -1$, $y = 0$ **c.** domain: all real numbers except 1 and –1; range: all real numbers greater than or equal to 1 or less than 0

16. a.

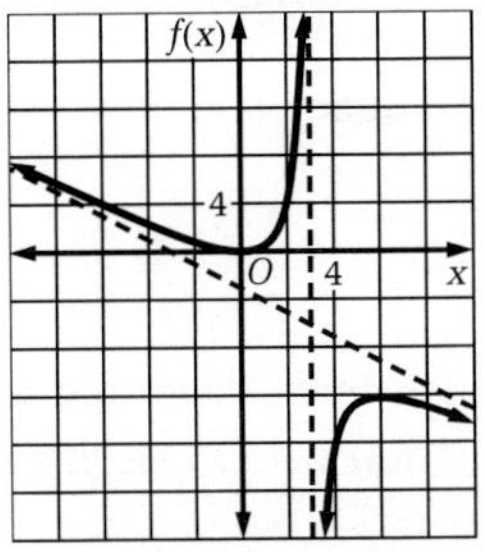

b. $x = 3$, $y = -x - 3$ **c.** domain: all real numbers except 3; range: all real numbers greater than or equal to 0 or less than or equal to –12

17. a.

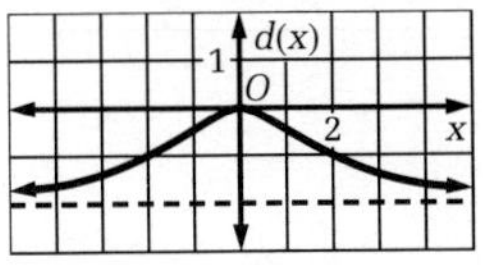

b. $y = -2$ **c.** domain: all real numbers; range: all real numbers greater than –2 and less than or equal to 0

Page 129 Exercises and Problems

11. a. the cost after the \$10 discount from the coupon; the cost with the regularly advertised discount **b.** Let f be the price Lily pays. $f(x) = 0.95x - 10$ **c.** Let g be the price Lily pays. $g(x) = 0.95(x - 10)$ **d.** $(c \circ d)(x) = 0.95x - 10$; $(d \circ c)(x) = 0.95x - 9.50$; $(c \circ d)(x)$ is the function for part (b) because this function represents taking the discount before using the coupon; $(d \circ c)(x)$ is the function for part (c) because this function represents using the coupon before taking the discount.

Unit 3

Page 143 Exercises and Problems

12. *Given:* $ABCD$ is a parallelogram.
Prove: $AB = CD$ and $BC = AD$

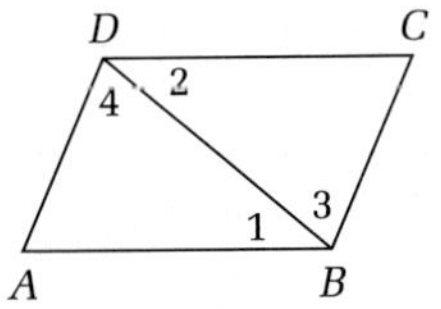

Statements	Justifications
1. $ABCD$ is a parallelogram.	1. Given
2. $\overline{AB} \parallel \overline{CD}$ and $\overline{BC} \parallel \overline{AD}$	2. Definition of parallelogram
3. $m\angle 1 = m\angle 2$ $m\angle 3 = m\angle 4$	3. If two ∥ lines are intersected by a transversal, then alternate interior ∠s are = in measure.
4. $DB = DB$	4. Reflexive property of equality
5. $\triangle ABD \cong \triangle CDB$	5. ASA
6. $AB = CD$ and $BC = AD$.	6. Corres. parts of ≅ △s are = in measure.

13. a. If opposite sides of a quadrilateral are equal in measure, then the quadrilateral is a parallelogram.

b. *Given:* Quadrilateral $ABCD$ with $AB = CD$ and $BC = AD$
Prove: $ABCD$ is a parallelogram.

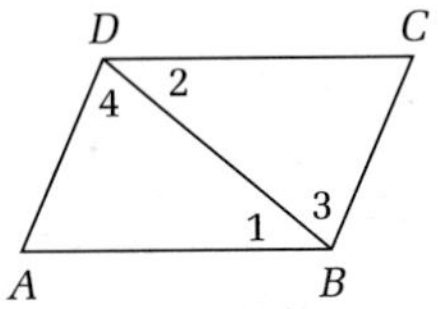

Statements	Justifications
1. $AB = CD$ and $BC = AD$	1. Given
2. $DB = DB$	2. Reflexive property of equality
3. $\triangle ABD \cong \triangle CDB$	3. SSS
4. $m\angle 1 = m\angle 2$ $m\angle 3 = m\angle 4$	4. Corres. parts of ≅ △s are = in measure.
5. $\overline{AB} \parallel \overline{CD}$ and $\overline{BC} \parallel \overline{AD}$	5. If two lines are intersected by a transversal and alternate interior ∠s are = in measure, then the lines are ∥.
6. $ABCD$ is a parallelogram.	6. Definition of parallelogram

14. *Given:* Parallelogram $ABCD$ with diagonals intersecting at E
Prove: $AE = EC$ and $BE = ED$

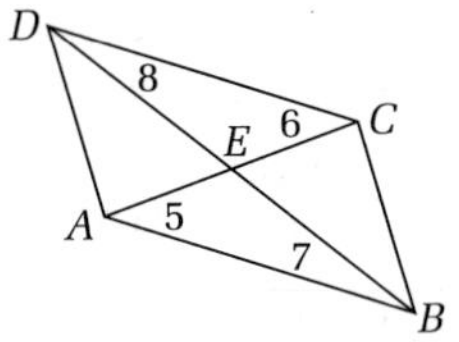

Statements	Justifications
1. $ABCD$ is a parallelogram.	1. Given
2. $\overline{AB} \parallel \overline{CD}$	2. Definition of parallelogram
3. $m\angle 5 = m\angle 6$ and $m\angle 7 = m\angle 8$	3. If two ∥ lines are intersected by a transversal, then alternate interior ∠s are = in measure.
4. $AB = CD$	4. If a quadrilateral is a parallelogram, then opposite sides are = in measure.
5. $\triangle AEB \cong \triangle CED$	5. ASA
6. $AE = EC$ and $BE = ED$	6. Corres. parts of ≅ △s are = in measure.

15. *Given:* Parallelogram *ABCD* is a rectangle.
Prove: $AC = BD$

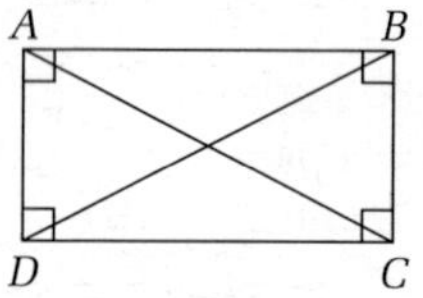

Statements	Justifications
1. *ABCD* is a rectangle.	1. Given
2. *ABCD* is a parallelogram and $m\angle A = m\angle B = m\angle C = m\angle D = 90°$.	2. Definition of rectangle
3. $AD = BC$	3. If a quadrilateral is a parallelogram, then opposite sides are = in measure.
4. $AB = AB$	4. Reflexive property of equality
5. $\triangle DAB \cong \triangle CBA$	5. SAS [Steps 2, 3, and 4]
6. $AC = BD$	6. Corres. parts of $\cong$ ⚠ are = in measure.

Page 151 Exercises and Problems

11. a. $PS = \sqrt{(-b-(-a))^2 + (c-0)^2} = \sqrt{(-b+a)^2 + c^2} = \sqrt{(a-b)^2 + c^2}$; $QR = \sqrt{(b-a)^2 + (c-0)^2} = \sqrt{(b-a)^2 + c^2}$; since $(a-b)^2 = (b-a)^2$, $PS = QR$. **b.** $PR = \sqrt{(b-(-a))^2 + (c-0)^2} = \sqrt{(b+a)^2 + c^2}$; $QS = \sqrt{(-b-a)^2 + (c-0)^2} = \sqrt{(-b-a)^2 + c^2}$; since $(b+a)^2 = (-b-a)^2$, $PR = QS$.

c. Form of proof may vary. An example is given.

Given: *PQRS* is an isosceles trapezoid. The *y*-axis is a line of symmetry of *PQRS*, passing through the midpoints of its bases.

Prove: $m\angle SPQ = m\angle RQP$ and $m\angle PSR = m\angle QRS$

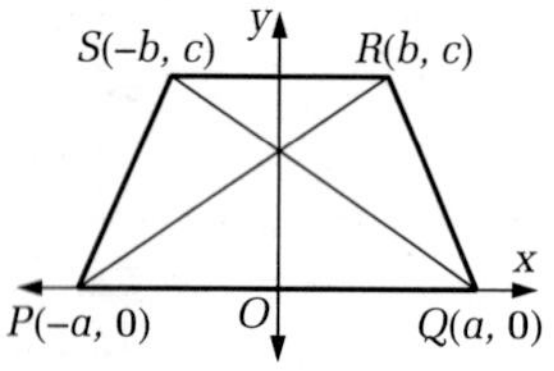

Statements	Justifications
1. *PQRS* is an isosceles trapezoid.	1. Given
2. $PS = QR$	2. Definition of isosceles trapezoid
3. $PR = QS$	3. In an isosceles trapezoid, the diagonals are = in measure.
4. $SR = SR$ and $PQ = PQ$	4. Reflexive property of equality
5. $\triangle SPQ = \triangle RQP$ and $\triangle QRS = \triangle PSR$	5. SSS
6. $m\angle SPQ = m\angle PQR$ and $m\angle QRS = m\angle RSP$	6. Corres. parts of $\cong$ ⚠ are = in measure.

12. Answers may vary. An example is given. There is no simple method for expressing the measure of an angle in terms of coordinates. You could use trigonometric ratios if you could express the coordinates of the intersection of the diagonals in terms of *a*, *b*, and *c*. Even then, a proof would be unwieldy.

Page 152 Exercises and Problems

15. a. If the diagonals of a quadrilateral are equal in measure and bisect each other, then the quadrilateral is a rectangle.
b. Form of proof may vary. An example is given.
Given: $AC = DB$ and the diagonals of *ABCD* bisect each other.
Prove: *ABCD* is a rectangle.

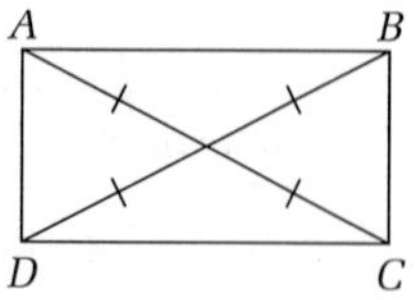

Since the diagonals of *ABCD* bisect each other, *ABCD* is a parallelogram (bisecting diagonals theorem). Since opposite sides of a parallelogram are equal in measure and the measure of a segment is equal to itself, $\triangle BAD \cong \triangle ABC \cong \triangle ADC \cong \triangle BCD$ by SSS. Then $m\angle BAD = m\angle ABC = m\angle ADC = m\angle BCD$ (Corres. parts of $\cong$ ⚠ are = in measure). Since all four angles of the quadrilateral are equal in measure and the sum of their measures is 360°, the measure of each angle is 90° and the quadrilateral is a rectangle by definition.

Page 157 Exercises and Problems

15. In a rectangle, either pair of opposite sides may be considered the bases. A rectangle is a parallelogram, so each pair of opposite sides is equal in measure. The second property was proved in Ex. 15 on page 144. (If a parallelogram is a rectangle, then its diagonals are equal in measure.) All four angles of a rectangle are equal in measure, so the two angles at each base are equal in measure. **16–18.** Answers may vary. Examples are given. **16.** A quadrilateral is a trapezoid if and only if it has at least one pair of parallel sides. **17.** A quadrilateral is an isosceles trapezoid if and only if it has at least one pair of parallel sides and a line of symmetry that passes through the midpoints of the bases. **18.** A quadrilateral is a kite if and only if it has two pairs of consecutive sides that are equal in measure.

19. *Given:* Quadrilateral *ABCD* is a rectangle.
Prove: *ABCD* is a parallelogram.

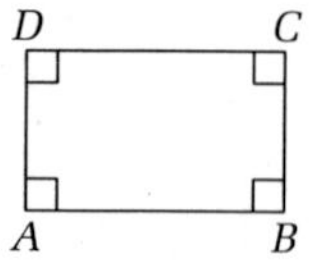

Since *ABCD* is a rectangle, we know by definition that $\angle A$, $\angle B$, $\angle C$, and $\angle D$ are all right angles. So $\overline{AD} \perp \overline{CD}$ and $\overline{AD} \perp \overline{AB}$. Since lines $\perp$ to the same line are $\parallel$, $\overline{CD} \parallel \overline{AB}$. Similarly, $\overline{AD} \perp \overline{CD}$ and $\overline{BC} \perp \overline{CD}$, so $\overline{AD} \parallel \overline{BC}$. So *ABCD* is a parallelogram.

20. *Given:* Quadrilateral *ABCD* is a rectangle.
Prove: *ABCD* has four right angles.

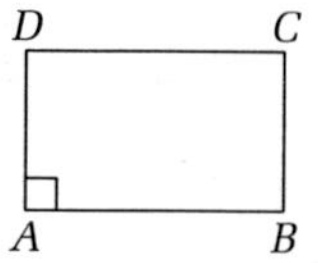

Since *ABCD* is a rectangle, it has a right angle, say, $\angle A$. Also, *ABCD* is a parallelogram, so $\overline{AB} \parallel \overline{CD}$ and $\overline{AD} \parallel \overline{BC}$. Since $\angle A$ is a right $\angle$, $\overline{AD} \perp \overline{AB}$. Then $\overline{AD} \perp \overline{CD}$ also, since a line $\perp$ to one of two $\parallel$ lines is $\perp$ to the other as well. So $\angle D$ is a right $\angle$. Then $\overline{CD} \perp \overline{BC}$ (since $\overline{AD} \parallel \overline{BC}$ and $\overline{AD} \perp \overline{CD}$), so $\angle C$ is a right $\angle$. Finally, since $\overline{AD} \parallel \overline{BC}$ and $\overline{AB} \perp \overline{AD}$, $\overline{AB} \perp \overline{BC}$ and so $\angle B$ is a right $\angle$.

Page 158 Exercises and Problems

23.

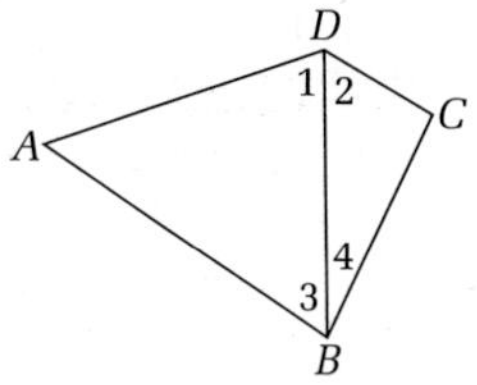

a. 180°; 180°; $m\angle 2$; $m\angle ABC$
b. $m\angle A + m\angle ABC + m\angle C + m\angle CDA = m\angle A + (m\angle 3 + m\angle 4) + m\angle C + (m\angle 1 + m\angle 2)$ (Substitution property) $= (m\angle A + m\angle 1 + m\angle 3) + (m\angle C + m\angle 2 + m\angle 4)$ (Associative and commutative properties of addition) $= 180° + 180°$ (Substitution property) $= 360°$
c. The sum of the measures of the angles of a quadrilateral is 360°.
24. Answers may vary.

Page 164 Exercises and Problems

21. b ($\overleftrightarrow{RS}$ is perpendicular to $\overline{PQ}$ and passes through its midpoint); c ($\overleftrightarrow{RS}$ is perpendicular to $\overline{PQ}$ and its intersection with $\overline{PQ}$ forms two congruent triangles (HL), so $\overleftrightarrow{RS}$ bisects $\overline{PQ}$).

Page 186 Exercises and Problems

19. Figures may vary. An example is given.

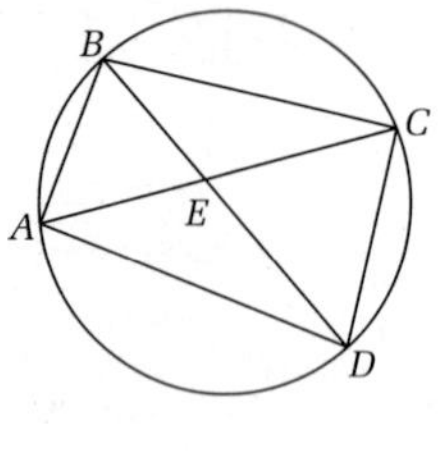

$\triangle ABE \sim \triangle DCE$ and $\triangle BEC \sim \triangle AED$. Both triangles can be proved to be similar using Theorem 3.9 (If two inscribed angles in the same circle intercept the same arc, then they are equal in measure) and AA. $\angle BAC$ and $\angle CDB$ both intercept $\overparen{BC}$, so they are equal in measure and $\angle ABD$ and $\angle DCA$ both intercept $\overparen{AD}$, so they are equal in measure. Then $\triangle ABE \sim \triangle DCE$ by AA. $\angle CBD$ and $\angle DAC$ both intercept $\overparen{CD}$, so they are equal in measure and $\angle BCA$ and $\angle ADB$ both intercept $\overparen{AB}$, so they are equal in measure. Then $\triangle BEC \sim \triangle AED$ by AA.

Page 191 Exercises and Problems

11. a. Where necessary, angle measures are rounded to the nearest tenth. Other measures are given to 4 decimal places.

n	$m\angle MOA$	$\tan\angle MOA$	Length of side	Semiperimeter	Area
6	30°	0.5774	1.1547	3.4641	3.4641
7	25.7°	0.4816	0.9631	3.3710	3.3710
8	22.5°	0.4142	0.8284	3.3137	3.3137
9	20°	0.3640	0.7279	3.2757	3.2757
10	18°	0.3250	0.6498	3.2492	3.2492
11	16.4°	0.2936	0.5873	3.2299	3.2299
12	15°	0.2679	0.5359	3.2154	3.2154
13	13.8°	0.2465	0.4930	3.2042	3.2042
14	12.9°	0.2282	0.4565	3.1954	3.1954
15	12°	0.2126	0.4251	3.1883	3.1883
16	11.3°	0.1989	0.3978	3.1826	3.1826
17	10.6°	0.1869	0.3739	3.1779	3.1779
18	10°	0.1763	0.3527	3.1739	3.1739
19	9.5°	0.1669	0.3337	3.1705	3.1705
20	9°	0.1584	0.3168	3.1677	3.1677
21	8.6°	0.1507	0.3015	3.1652	3.1652
22	8.2°	0.1438	0.2876	3.1631	3.1631
23	7.8°	0.1374	0.2749	3.1613	3.1613
24	7.5°	0.1317	0.2633	3.1597	3.1597
25	7.2°	0.1263	0.2527	3.1582	3.1582

Answers may vary. An example is given. As n increases, the area of the n-gon approaches π.

b. Semiperimeters and areas are given to 6 decimal places.

n	$m\angle MOA$	$\tan\angle MOA$	Length of side	Semiperimeter	Area
6	30°	0.5774	1.1547	3.464102	3.464102
12	15°	0.2679	0.5359	3.215390	3.215390
24	7.5°	0.1317	0.2633	3.159660	3.159660
48	3.75°	0.0655	0.1311	3.146086	3.146086
96	1.875°	0.0327	0.0655	3.142715	3.142715
192	0.9375°	0.0164	0.0327	3.141873	3.141873
384	0.4688°	0.0082	0.0164	3.141663	3.141663
768	0.2344°	0.0041	0.0082	3.141610	3.141610
1536	0.1172°	0.0020	0.0041	3.141597	3.141597
3072	0.0586°	0.0010	0.0020	3.141593	3.141593

$n \approx 3072$ **c.** Both ratios decrease and approach 1 as n increases.

Page 196 Exercises and Problems

5.

	Number of Faces (F)	Number of Vertices (V)	Number of Edges (E)
tetrahedron	4	4	6
hexahedron	6	8	12
octahedron	8	6	12
dodecahedron	12	20	30
icosahedron	20	12	30

tetrahedron: 4 + 4 = 6 + 2; hexahedron: 6 + 8 = 12 + 2; octahedron: 8 + 6 = 12 + 2; dodecahedron: 12 + 20 = 30 + 2; icosahedron: 20 + 12 = 30 + 2

Page 198 Exercises and Problems

28. Summaries may vary. An example is given. A legal brief is like a mathematical proof in that both use a series of statements accepted to be or previously shown to be true to demonstrate the truth of a statement. In a legal brief, however, there may be no absolute measure of whether a statement is true. Also, any reasonable person with sufficient background should be able to determine if a mathematical proof is successful; that is, the theorem is either proved or not. However, two sufficiently knowledgeable people responding to a legal brief may disagree on whether the issue has been proven or not. Answers may vary.

Unit 4

Page 207 Exploration

1. a. Check students' work. **b.** Answers may vary. An example is given. I added a square at each end of the existing rows of squares and added a single square at the top and bottom.

c.

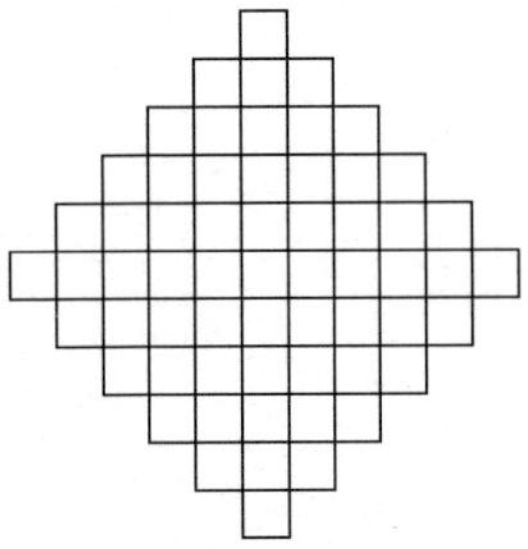

d.

Stage number	1	2	3	4	5
Number of squares	5	13	25	41	61

e. 221 squares; Answers may vary. An example is given. I noticed that for the nth stage, the number of squares is $4n$ plus the number of squares at the previous stage.

Page 214 Exercises and Problems

25. a. $\frac{1}{9}; \frac{1}{16}; \frac{1}{25}; \frac{1}{36}$ **b.** 0; The differences between the terms and 1 appear to get closer and closer to 0, which means that 1 is the limit of the sequence. **c.** Answers may vary. An example is given. In Exercise 24, the limit was determined by considering what the terms would look like as the position number increased. In part (a), the limit was determined by considering how the terms were related to a particular number. I prefer the method in Exercise 24, because to use the method in part (a), you have to have an idea of what the limit is in the first place. **26.** Answers may vary. Several examples are given. 2, 4, 6, 8, ... (Each increase is 2.); 2, 4, 8, 16, ... (Each term is twice the term before it.); 2, 4, 7, 11, ... (Each increase is 1 more than the one before it.); 2, 4, 8, 14, ... (Each increase is 2 more than the one before it.)
27. Each face of an octahedron is an equilateral triangle. Four faces meet at each vertex. Therefore, the sum of the measures of the angles at each vertex is 240°. The defect at each vertex is 360° – 240° = 120°. Since there are 6 vertices, the sum of the defects is 6 • 120° = 720°.

Page 236 Checkpoint

2. a.

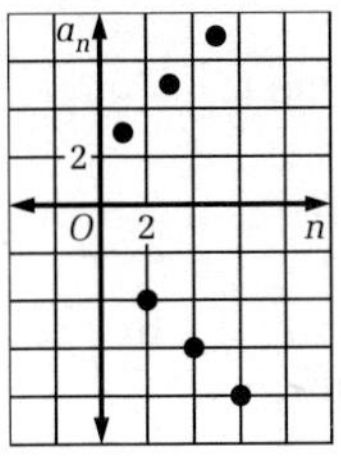

b. No. **3. a.**

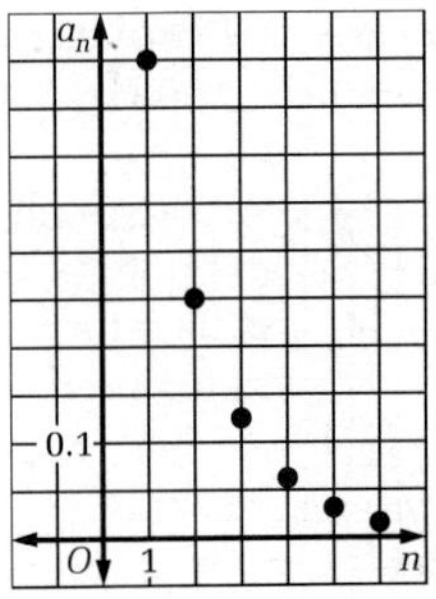

b. Yes; 0.

4. $a_n = \frac{3n-2}{4n-3}$ **5.** $a_n = \frac{3n-5}{n}$ **6.** $a_1 = 6$; $a_n = a_{n-1} + 7$

7. $a_1 = 1$; $a_n = a_{n-1} + 2n - 1$ **8.** geometric

Page 255 Exercises and Problems

2. a. 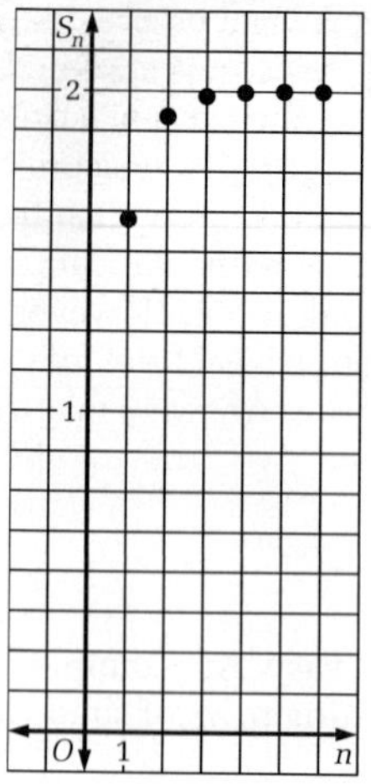

b. Yes; 2

Unit 5

Page 273 Exercises and Problems

20. a. The y-intercept values represent the peak concentration of medication.
b. weakest strength: about 12 h; medium strength: about 16 h; greatest strength: about 20 h

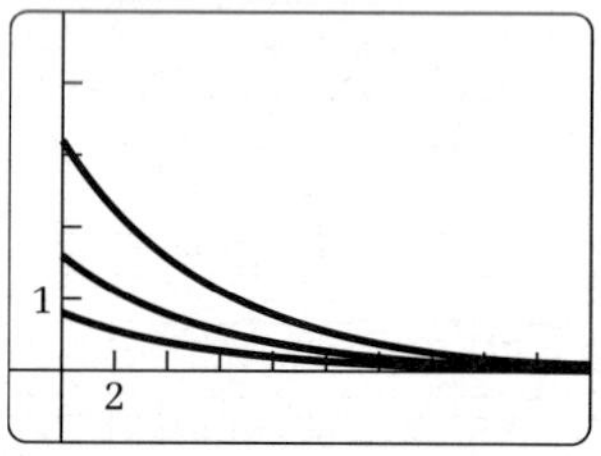

21. a. 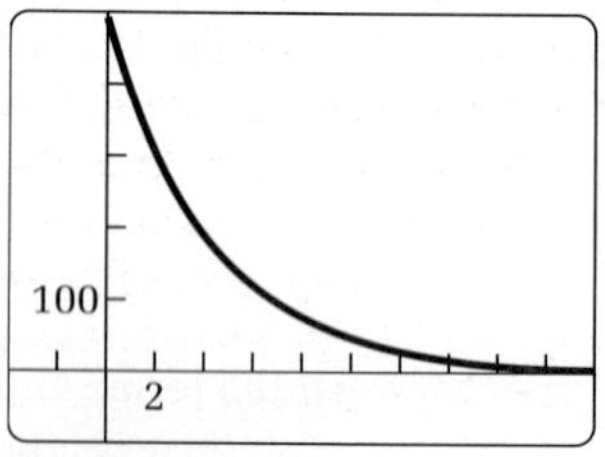

b. 492.5 ng/mL
c. about 2.9 h
d. about 14.5 h

22. a. 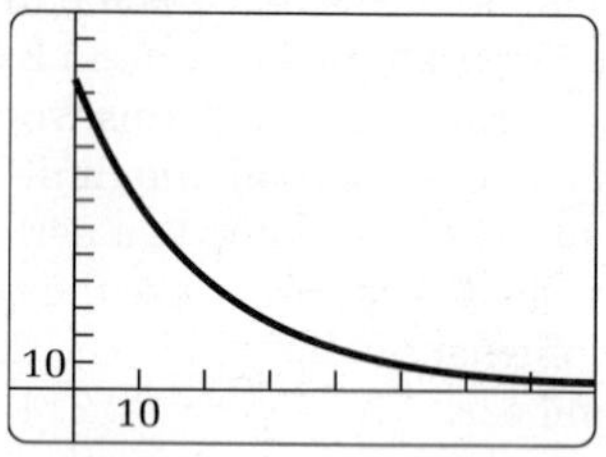

b. 115 ng/mL
c. about 13.5 h
d. about 67.5 h

23. a. 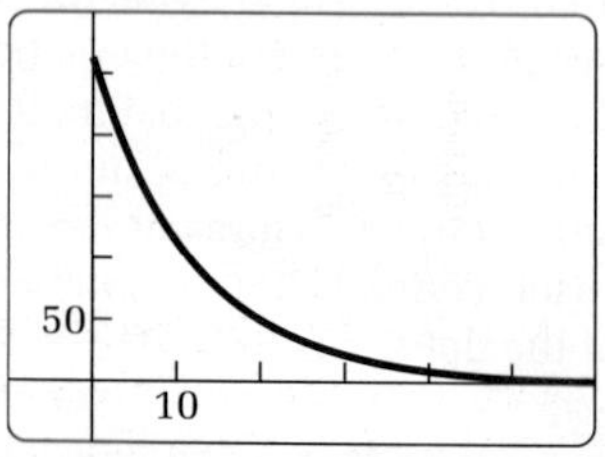

b. 263 ng/mL
c. about 8.3 h
d. about 41.5 h

Page 274 Exercises and Problems

32. a. 40 Hz **b.** 80 Hz; 160 Hz; 320 Hz; 640 Hz; 1280 Hz; 2560 Hz; 5120 Hz

c. 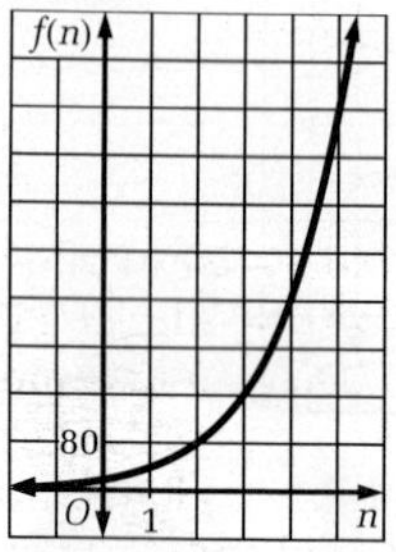

As the pitch increases, the frequency grows exponentially.
d. The human ear can hear only the first nine notes produced by the instrument.

33. 261.6 Hz

Page 276 Exploration

2. $f(x) = 3 \cdot 2^{-x}$

3. 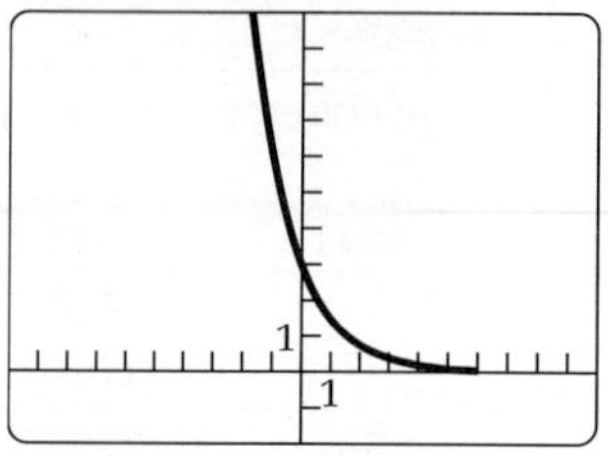

$g(x) = h(x)$ for every value of x.

4. $(2)\left(\frac{1}{4}\right)^x = (2)(4^{-1})^x = (2)(4^{-1 \cdot x}) = (2)(4)^{-x}$

Page 280 Exercises and Problems

15. The graphs of the functions are images of each other after reflection over the y-axis. Let (p, q) be any point on the graph of $y = ab^x$; $ab^p = q$, so $b^p = \frac{q}{a}$. Then $ab^{-p} = \frac{a}{b^p} = \frac{a}{\frac{q}{a}} = q$, so $(-p, q)$ is on the graph of $y = ab^{-x}$.

Page 288 Exploration

2.

Frequency of compounding	Number of times interest is calculated in a year (n)	$\left(1 + \frac{1}{n}\right)^n$	Dollar value A
annual	$n = 1$	$\left(1 + \frac{1}{1}\right)^1$	2.00
semiannual	$n = 2$	$\left(1 + \frac{1}{2}\right)^2$	2.25
quarterly	$n = 4$	$\left(1 + \frac{1}{4}\right)^4$	2.44141
monthly	$n = 12$	$\left(1 + \frac{1}{12}\right)^{12}$	2.61304
weekly	$n = 52$	$\left(1 + \frac{1}{52}\right)^{52}$	2.69260
daily	$n = 365$	$\left(1 + \frac{1}{365}\right)^{365}$	2.71457
hourly	$n = 8760$	$\left(1 + \frac{1}{8760}\right)^{8760}$	2.71813
every minute	$n = 525,600$	$\left(1 + \frac{1}{525,600}\right)^{525,600}$	2.71828
every second	$n = 31,536,000$	$\left(1 + \frac{1}{31,536,000}\right)^{31,536,000}$	2.71828

Page 292 Exercises and Problems

23. 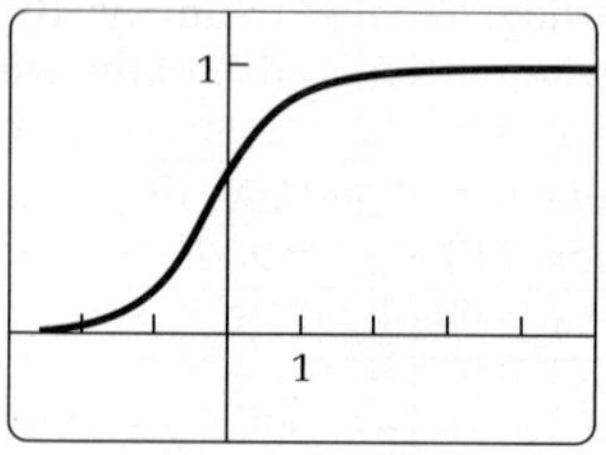

$y = 0$; $y = 1$

24. a. 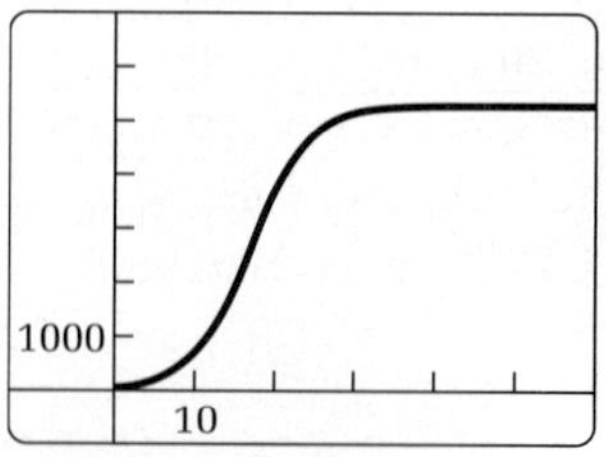

about 53; about 2260 (Estimates may vary.)

b. between about 1960 and 1985; Answers may vary. An example is given. In the early 1960s, there was a rise in the feminist movement helped in part by civil rights legislation. **c.** about 1988 (Estimates may vary.)

d. (a) 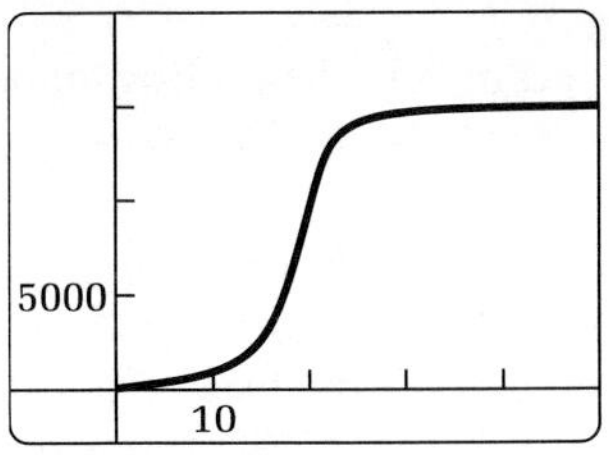

about 37; about 4900 (Estimates may vary.) (b) between about 1960 and 1985; see above. (c) about 1990

e. 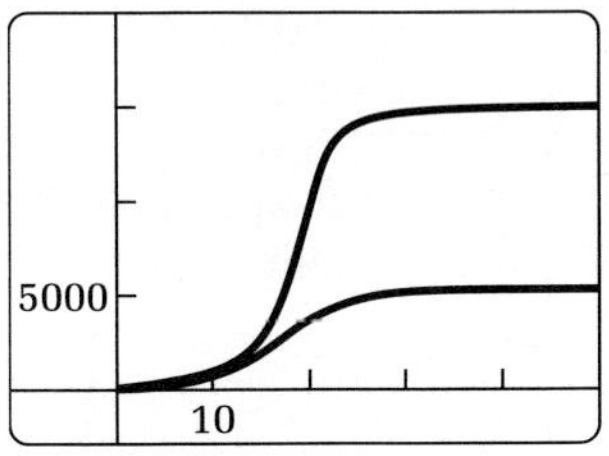

from about 1960 to 1965; about 1965

f. about 9560

Page 299 Exercises and Problems

7. Yes. 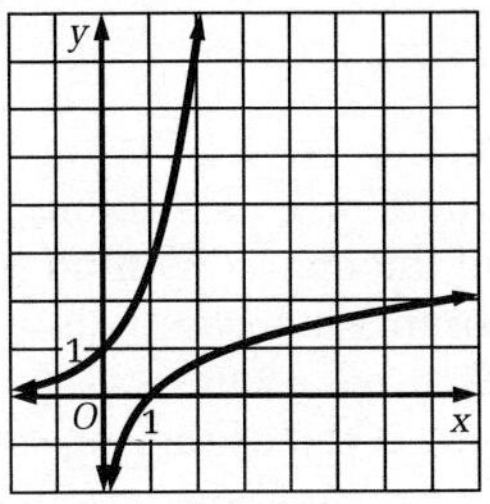

8. No. 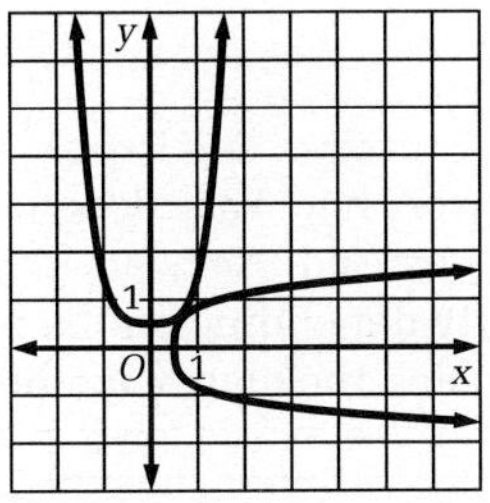

9. a. $j(e) = e + 660$ **b.** $e(j) = j - 660$ **c.** 1 **d.** No; the slopes are equal so the graphs are parallel. The Japanese calendar is always 660 years ahead, so there can be no point at which the Japanese year and the European year are the same. **10.** $y = 3x + 15$ **11.** $y = -x + 6$
12. $y = x + 3\frac{1}{2}$ **13.** $y = x$ **14.** $y = \sqrt{-\frac{x}{2}}, x \le 0$ **15.** $y = \sqrt[5]{\frac{x+3}{2}}$
16. $y = x^{5/2}$ **17.** $y = -\frac{x}{2}, x \ge 2$ **18.** $y = \frac{x-b}{m}, m \ne 0$ **19.** $y = x - b$
20. $y = \sqrt[b]{\frac{x}{a}}$ **21.** $y = x^{b/a}$ **22. a.** $r = \frac{d}{31.15}$; No; the rule for exchanging rupees for dollars is $d = \frac{r}{31.15}$. **b.** domain and range of function: nonnegative real numbers; domain and range of inverse: all real numbers

Unit 6

Page 335 Exercises and Problems

3. Answers may vary. As the widths of the intervals change, the shape of the histogram will change. **4.** Estimates may vary; about 325.
5. a. Estimates may vary. Monday, Wednesday, and Thursday: about 48; Tuesday: 50; Friday: about 45; Saturday: about 42; Sunday: 40
b. Yes; it seems reasonable that there is no particular day of the week on which more births occur. Therefore, it is reasonable to expect a uniform distribution. **6.** No. The graph shows that most students have 1 or 2 brothers and sisters. **7.** Answers may vary. Examples are given. I think the mean and the median of the number of minutes it takes to get to school occur in the interval from 15–19, so I think the average is 15–19 minutes. Similarly, I think the average number of brothers and sisters is about 2.

8. a. 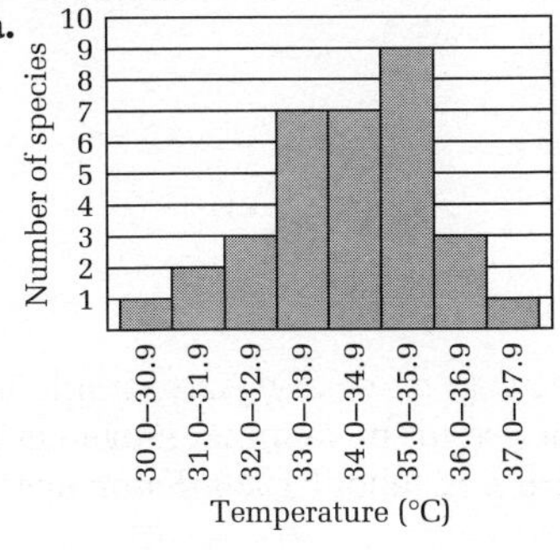

b. Since the interval that contains the greatest frequency is close to one end, the distribution is skewed.

Page 336 Exercises and Problems

10. The scale on a histogram tells you how many data values are in each interval. The scale on a relative frequency histogram tells you what fraction or what percent of the data values are in each interval.

11. a. 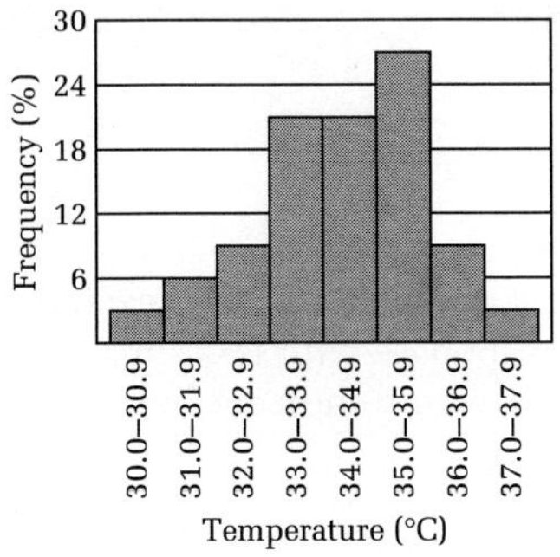

b. about 39%

12. Intervals may vary. Answers are based on the histograms shown in the figures.

a. 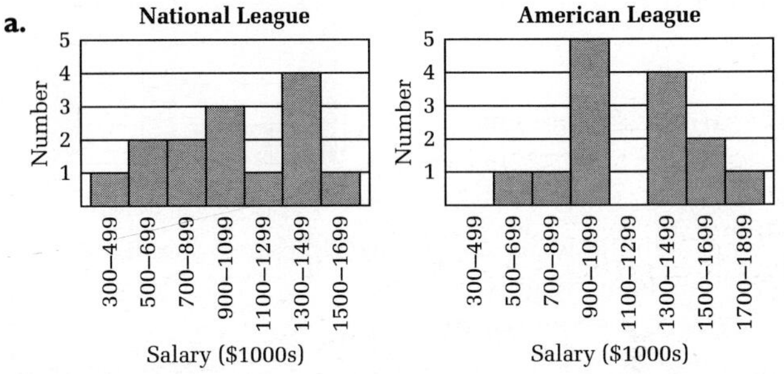

b. National League: basically uniform; American League: basically bimodal **c.** National League: about 1,061,000; about 1,056,000; American League: about 1,188,000; about 1,192,000; For the National League, the median is to the left of the mean in the bar with frequency three; for the American League, the median is to the right of the mean, and both fall in an interval with zero frequency. **d.** Summaries may vary. American League salaries are on average slightly higher than National League salaries. Only 2 American League teams have average salaries less than $900,000 per year, while 5 National League teams have salaries in this category.

13. a. 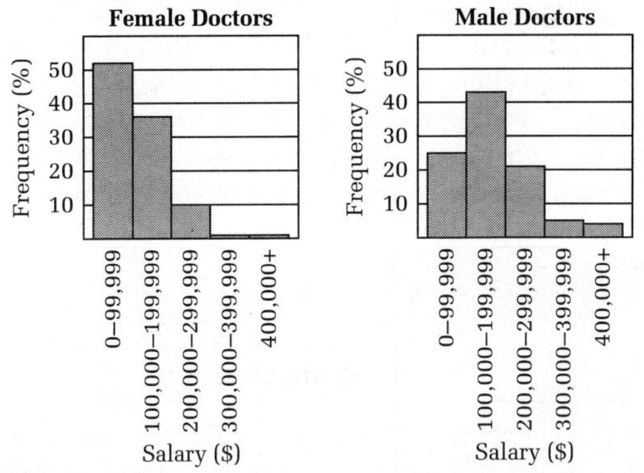

b. Both distributions are basically skewed. **c.** female doctors: 46%; male doctors: 64% **d.** Answers may vary. Examples are given. It appears that the salaries of male doctors are significantly higher than those of female doctors. Over half of the female doctors earn less than $100,000 per year, while 75% of all male doctors earn more than $100,000 per year. Also, while only 2% of all female doctors earn more than $300,000 per year, 11% of male doctors earn that much. Both the mean and median salaries are higher for male doctors.

Page 341 Exercises and Problems

7. a. South Atlantic: 26.59, 14.18; New England: 26.82, 4.90; While the means are the same, the standard deviation for the South Atlantic region is greater. The number of physicians per 10,000 people is more uniform in the New England region than in the South Atlantic region. **b.** 21.85, 4.90; The mean is lower for the South Atlantic region, while the standard deviations are identical. The number of physicians per 10,000 people is, on average, lower for the South Atlantic region than for New England, but the distribution of the data is similar for both groups. The conclusions are not the same as in part (a). **c.** Answers may vary. An example is given. An outlier seems to have a greater effect on the standard deviation than on the mean. In the example, omitting the outlier resulted in a decrease of 17.8% of the mean, while the standard deviation decreased by 65.4%.

Page 343 Exercises and Problems

13. a. mound-shaped **b.** Estimates may vary; about 42%.
14. a, b. Answers may vary. **c.** Answers may vary. An example is given. I think the mean of a set of responses with less spread would make the list of features because this would indicate a more uniformly positive response to such a feature.

Page 347 Exercises and Problems

4. a. 8, 8, 8, 0, 5, 5, 5, 5, 5, 0, 2, 3, 0, 5, 5, 5, 0, 0, 0, 0, 5, 0, 0, 2, 0

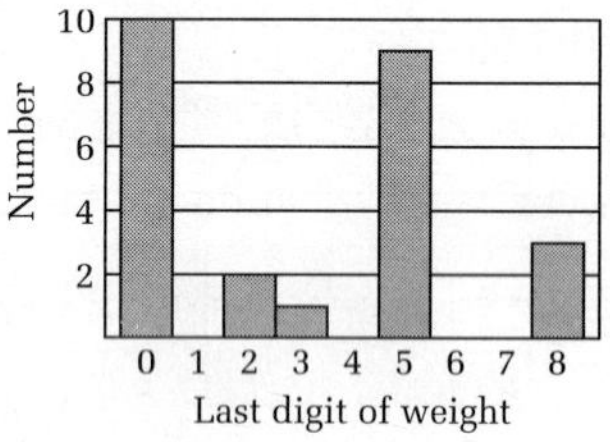

b. If the men were actually weighed, you would expect the distribution of the last digits to be roughly uniform.
5. Drawings may vary. Examples are given.

a.

b.

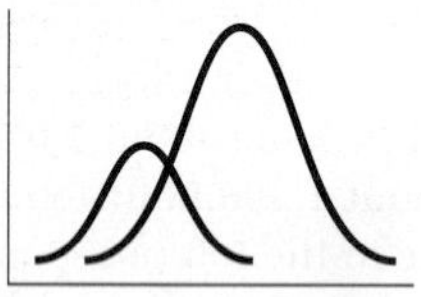

6. In any normal distribution, about 68% of the data is within one standard deviation of the mean, about 95% of the data is within two standard deviations of the mean, and about 99.7% of the data is within three standard deviations of the mean.

Page 349 Checkpoint

1. Answers may vary. An example is given. The graph of a normal distribution has the same shape as the data set gets larger, no matter how small the interval size. For a normal distribution, a known percentage of the data falls within one, two, or three standard deviations of the mean. In a true normal distribution, the mean and median of the data are equal and fall at the line of symmetry of the distribution's bell-shaped curve.

Page 355 Exercises and Problems

5. a, b.

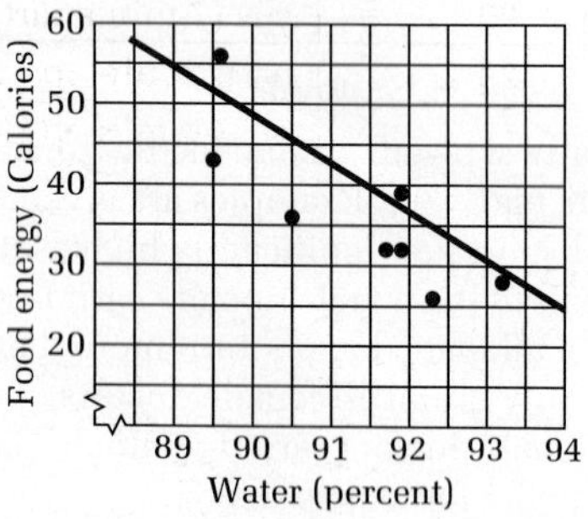

b. $y = -6x + 588.7$
c. As the percent of water increases, the number of Calories decreases.
d. about 76 Calories

Page 362 Exercises and Problems

4. Answers may vary as a result of rounding and depending on methods used to determine models. **a.** $y = 0.222865x - 359.04$
b. $y = 0.00002487x^2 - 0.02865x + 193.3$

c.

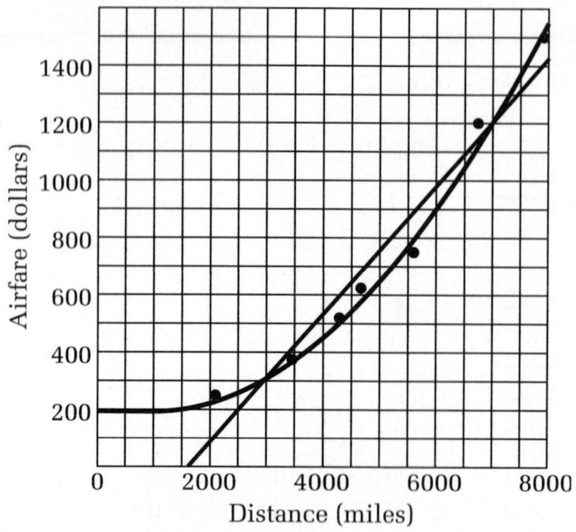

d. Although the correlation coefficient for the least-squares line is 0.98, the parabola appears to fit the data better. For the line, Σd^2 is about 60,134. For the parabola, Σd^2 is about 10,461.

Page 363 Exercises and Problems

11. a. Answers may vary depending on whether the matrix method or technology is used. Answers derived from the matrix method will vary slightly depending on the points chosen. An example is given based on using the matrix method: $y = -35.57x^2 + 2365.4x + 33{,}078$.

b.

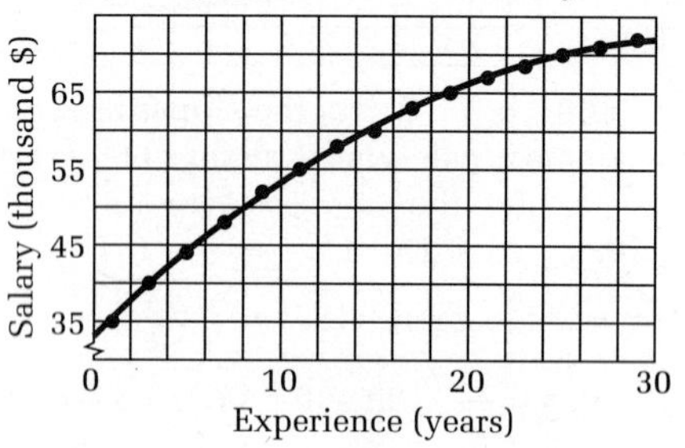

c. the beginning salary of an engineer with no previous experience

Page 364 Exercises and Problems

14. a.

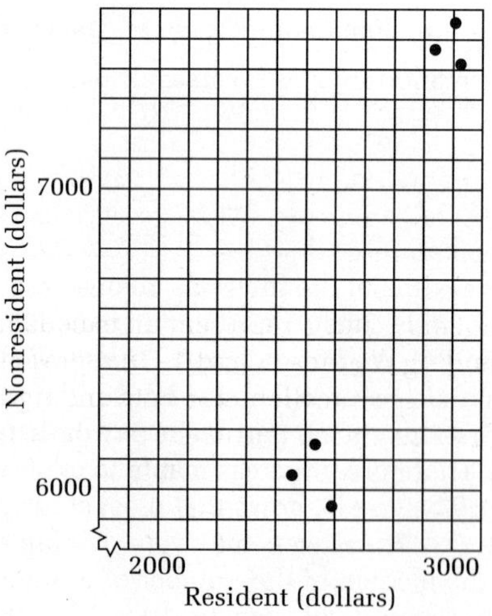

b. linear; The points appear to lie near a line.
c. $y = 2.6814x - 608.36$

Page 378 Talk it Over

9.

General model	Model applied to T-shirt problem
State the problem clearly.	Find out how many tickets must be purchased in order to win a T-shirt.
Define the key question.	Select a letter. Does that letter show up on one of your tickets?
State the underlying assumptions.	The probability that any given letter appears on a ticket is 20%. The appearance of a letter on one ticket has no effect on any other ticket.
Choose a model to answer the key question.	Generate random numbers from 1 to 5 inclusive. Each of the numbers represents one of the 5 letters.
Define and conduct a trial.	Generate a sequence of random numbers until each of the digits 1–5 has appeared once.
Record the observ-ation of interest.	How many numbers had to be generated before each of the digits 1–5 appeared once?
Repeat the last two steps several times.	Generate a sequence as described above several times.
Summarize the information and draw conclusions.	Make a table or graph; determine the measures of central tendency of the data. Analyze and interpret the results.

Pages 378 and 379 Exercises and Problems

5. Yes; a sequence of the letter "A" and random numbers between 2 and 10 would be produced (or 1–10 if aces are counted as ones). Each of the numbers occurs with the same frequency in the deck. **6.** int(10*rand) **7.** int(100*rand) + 1 **8.** int(4*rand) + 5 **9.** int(19*rand) + 11 **10.** Answers may vary. Examples are given. You do not need to find a sales source and either travel there or make phone orders, so you will save both time and money. You will not end up with tickets you do not need. **11. a.** It will increase the number of tickets needed. **b.** Answers may vary. An example is given. Use the function int(25*rand) + 1 to generate a sequence of random numbers from 1 to 25. Assign, say, the numbers 1–6 to the letter U, 7–12 to the letter S, and so on, and the number 25 to the letter M. For each trial, generate random numbers until each letter is represented once. **12.** 30–39 **13.** Estimates may vary; about 45%. **14.** The bars would all be the same height. **15–18.** Methods of simulation and results may vary. **15.** Use the function int(12*rand) + 1. Let the numbers 1–12 represent the months in order. (1 = January, and so on) For each trial, generate random numbers until the number for your birth month appears. Analyze the data. **16.** Use the function int(10*rand) + 1. Let the numbers 1 and 2 represent an excellent candidate, 3 represent an outstanding candidate, and 4–10 represent candidates that are qualified but neither excellent nor outstanding. For each trial, generate random numbers until one of the numbers 1–3 appears. Record the number of terms in each sequence. Analyze the data. **17.** Use the function int(50*rand) + 1. Let each of the numbers 1–30 represent a man and each of the numbers 31–50 represent a woman. For each trial, generate a sequence of five random numbers. Record the number of sequences generated and the number that include either four or five numbers greater than 30. Analyze the data. **18.** You need to find out how often 11, 12, 13, or 14 people show up for the plane. Use the function int(100*rand) + 1. Let each of the numbers 1–25 represent a person who does not show up, and the remaining numbers represent people who do show up. For each trial, generate a sequence of 14 random numbers. Record the number of sequences generated and the number that include 3 or fewer numbers which are less than 26. Analyze the data.

Page 388 Exercises and Problems

16. a. Answers may vary. **b.** In each case, let the outcome be represented by an ordered sequence of numbers, with the position number representing the student and the term representing the jacket selected. That is, (2, 1) indicates that there are two students and Student 1 selects Student 2's jacket and vice versa. If there are two students, either they both pick up the right jacket, or neither does. The two possible outcomes (1, 1), which is the same as (2, 2), and (1, 2), which is the same as (2, 1), are equally likely. Then *P*(at least one student picks up the right jacket) = 0.5. Suppose there are three students. The possible outcomes are (1, 2, 3), (1, 3, 2), (3, 2, 1), (2, 1, 3), (2, 3, 1), and (3, 1, 2). *P*(at least one student picks up the right jacket) = $\frac{2}{3} \approx 0.67$

c. Answers may vary. **d.** For any number n of students, *P*(at least one student picks up the right jacket) = 1 – *P*(all students pick up the wrong jacket). *P*(all four students pick up the wrong jacket) is the ratio of the number of ways the first 4 whole numbers can be arranged in a sequence with no digit in the right numerical position to the number of ways the first 4 whole numbers can be arranged in a sequence. For 4 students, that ratio is $\frac{9}{24} = \frac{3}{8}$, so *P*(at least one student among 4 picks up the right jacket) = $1 - \frac{3}{8} = \frac{5}{8} \approx 0.63$.

Page 390 Exploration

2, 3. Answers may vary. Example results are given.

Trial	One friend's arrival time	The other friend's arrival time	Did they meet?
1	24	16	yes (8 min apart)
2	14	22	yes (8 min apart)
3	21	7	no (14 min apart)
4	5	28	no (23 min apart)
5	12	25	no (13 min apart)
6	5	28	no (23 min apart)
7	2	10	yes (8 min apart)
8	28	7	no (21 min apart)
9	9	17	yes (8 min apart)
10	7	18	no (11 min apart)
11	26	10	no (16 min apart)
12	20	7	no (13 min apart)
13	9	13	yes (4 min apart)
14	28	0	no (28 min apart)
15	7	14	yes (7 min apart)
16	12	2	yes (10 min apart)
17	7	4	yes (3 min apart)
18	13	20	yes (7 min apart)
19	0	23	no (23 min apart)
20	2	28	no (26 min apart)

Page 393 Exercises and Problems

6.

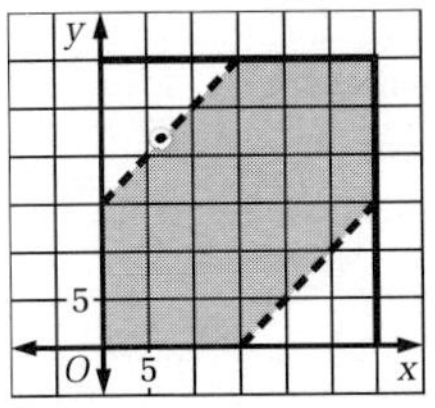

area of square = 900
area of shaded region = 675

Let x be the time in minutes after 7:00 P.M. that one friend calls and y the time in minutes after 7:00 P.M. that the other friend calls. The calls overlap if $x - 15 < y < x + 15$. The rectangle represents all possible times for at least one caller to be on the phone. The shaded area represents the possible times when the calls overlap.

P(calls will overlap) = $\frac{675}{900} \approx 0.75$

7. 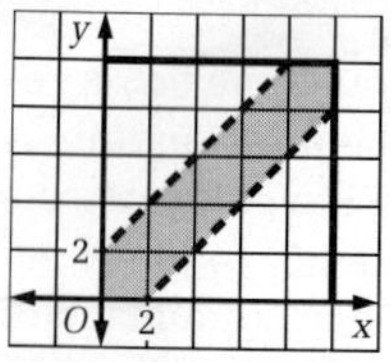

area of square = 100
area of shaded region = 36

Let x be time in minutes after the beginning of the ten-minute period when Janelle starts ringing up the sale and y the time in minutes when Abby starts ringing up the sale. The sales overlap if $x - 2 < y < x + 2$. The rectangle represents all possible times within the 10-minute interval when one of the clerks starts using the register for the indicated sale. The shaded area represents the possible times when the clerks would be using the register at the same time, that is, the time when one clerk would have to wait.

$P(\text{one of them will have to wait}) = \frac{36}{100} = 0.36$.

Page 394 Exercises and Problems

17. Answers may vary. An excample is given. Many versions of the game referred to by Europeans as "Dish" exist as part of the cultural traditions of various Native American groups. The various versions are similar to da-un-dah-qua as described in Section 7-1 in that two players shake two-sided disks in a dish to generate random outcomes. The number and size of the disks used, the material employed for the actual disk, and the scoring system sometimes differ. According to custom, the game was played as part of a festival or some other community gathering or ceremony. Large crowds of spectators would add to the enthusiasm and excitement of the event by cheering on the participants.

Page 394 Checkpoint

5. $\frac{1}{2} = 0.5$ **6.** $\frac{16}{32} = \frac{4}{13} \approx 0.31$

7. $P(\text{one person will have to wait}) = \frac{1.75}{4} = 0.4375$

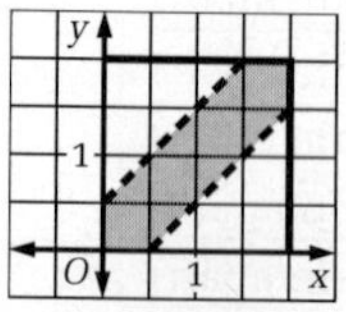

area of square = 4
area of shaded region = 1.75

Page 399 Exercises and Problems

10. The second formula is derived by dividing both sides of the first formula by $P(A)$. (This assumes that $P(A) \neq 0$.) **11.** $\frac{1}{3}$ **12.** Check students' work. **13.** 0.4

Page 400 Exercises and Problems

16. a.

	Likes Film	Dislikes Film	Total
Females	21	9	30
Males	12	8	20
Total	33	17	50

b.

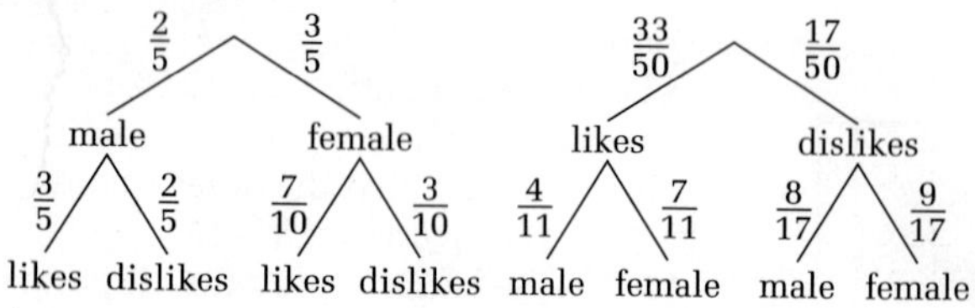

Unit 8

Page 429 Exercises and Problems

34. a.

Chess piece	Standard chess board	Circular chess board
king	can move one square in any direction including sideways, forward, backward, and diagonally as long as the move does not place it under attack	can move one rank, one file, or one space diagonally in any direction
queen	can move in a straight line in any direction as long as its path is not blocked	can move any number of ranks, any number of files, or along a spiral path
rook	can move any distance, forward, backward, or sideways, provided the line of motion is not blocked by any other piece	can move any number of ranks or any number of files
bishop	moves diagonally (forward or backward) for any unblocked distance	can move along a spiral
knight	makes an L-shaped move, two squares forward, backward, or sideways, then one square at a right angle to that move, or one square forward, backward, or sideways, then two squares at a right angle to that move. Pieces in its way do not block its path.	can move one rank and two files or two ranks and one file
pawn	can move forward one or two squares on the opening move, one square on subsequent moves; move diagonally forward one square to capture an opposing piece on that square	can move forward one or two ranks on the first move, one on subsequent moves, spirally forward one space to capture an opposing piece

b. Answers may vary. Examples are given. Move the pawn at (2, 157.5°) to (3, 157.5°); move the queen to (2, 112.5°); move the bishop at (1, 157.5°) to (2, 112.5°). Of these moves, the best is the third. If Black's bishop takes White's bishop, White's queen may then take Black's bishop. **c.** Check students' work.

Page 439 Exercises and Problems

56. a.

$m\angle\theta$	$\cos\theta = x$	$\sin\theta = y$	(x, y)
0°	1	0	(1, 0)
10°	0.9848	0.1736	(0.9848, 0.1736)
20°	0.9397	0.3420	(0.9397, 0.3420)
30°	0.8660	0.5	(0.8660, 0.5)
40°	0.7660	0.6428	(0.7660, 0.6428)
50°	0.6428	0.7660	(0.6428, 0.7660)
60°	0.5	0.8660	(0.5, 0.8660)
70°	0.3420	0.9397	(0.3420, 0.9397)
80°	0.1736	0.9848	(0.1736, 0.9848)
90°	0	1	(0, 1)
100°	–0.1736	0.9848	(–0.1736, 0.9848)
110°	–0.3420	0.9397	(–0.3420, 0.9397)
120°	–0.5	0.8660	(–0.5, 0.8660)
130°	–0.6428	0.7660	(–0.6428, 0.7660)
140°	–0.7660	0.6428	(–0.7660, 0.6428)
150°	–0.8660	0.5	(–0.8660, 0.5)
160°	–0.9397	0.3420	(–0.9397, 0.3420)
170°	–0.9848	0.1736	(–0.9848, 0.1736)
180°	–1	0	(–1, 0)
190°	–0.9848	–0.1736	(–0.9848, –0.1736)
200°	–0.9397	–0.3420	(–0.9397, –0.3420)
210°	–0.8660	–0.5	(–0.8660, –0.5)
220°	–0.7660	–0.6428	(–0.7660, –0.6428)
230°	–0.6428	–0.7660	(–0.6428, –0.7660)
240°	–0.5	–0.8660	(–0.5, –0.8660)
250°	–0.3420	–0.9397	(–0.3420, –0.9397)
260°	–0.1736	–0.9848	(–0.1736, –0.9848)
270°	0	–1	(0, –1)
280°	0.1736	–0.9848	(0.1736, –0.9848)
290°	0.3420	–0.9397	(0.3420, –0.9397)
300°	0.5	–0.8660	(0.5, –0.8660)
310°	0.6428	–0.7660	(0.6428, –0.7660)
320°	0.7660	–0.6428	(0.7660, –0.6428)
330°	0.8660	–0.5	(0.8660, –0.5)
340°	0.9397	–0.3420	(0.9397, –0.3420)
350°	0.9848	–0.1736	(0.9848, –0.1736)
360°	1	0	(1, 0)

b, c.

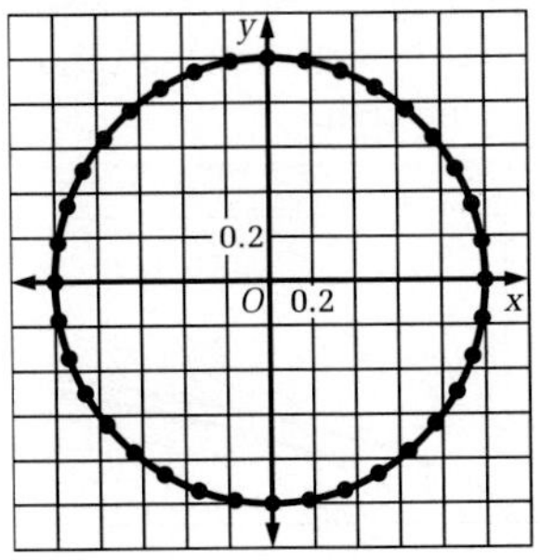

a circle centered at the origin with radius 1 **d.** No. **e.** Answers may vary. An example is given. This activity extends the work done in the Exploration, showing that $(x, y) = (\cos\theta, \sin\theta)$, and visually demonstrating that $\cos^2\theta + \sin^2\theta = 1$, for all θ.

Page 444 Exercises and Problems

5–10.

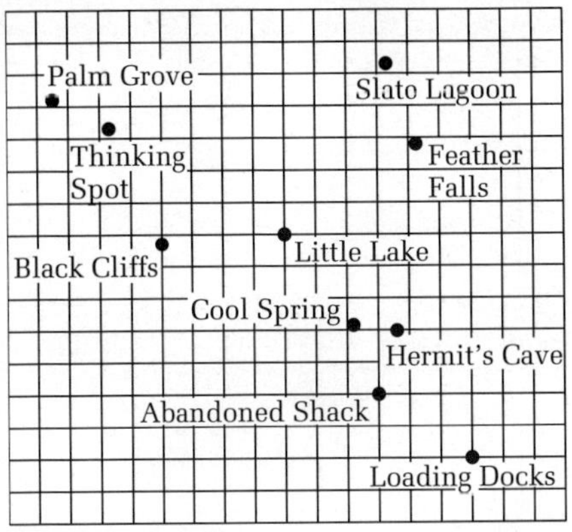

11.

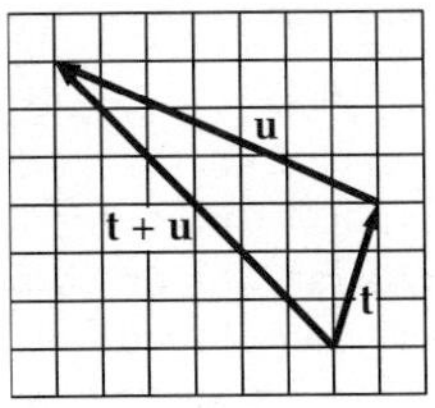

(8.5, 135°); (–6, 6)

12.

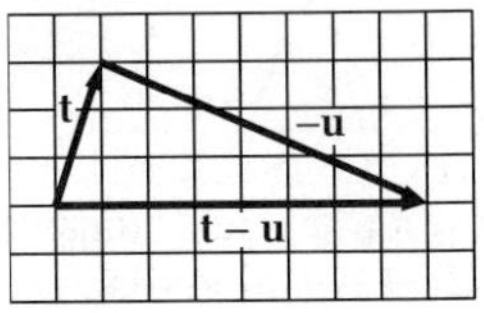

(8, 0°); (8, 0)

13.

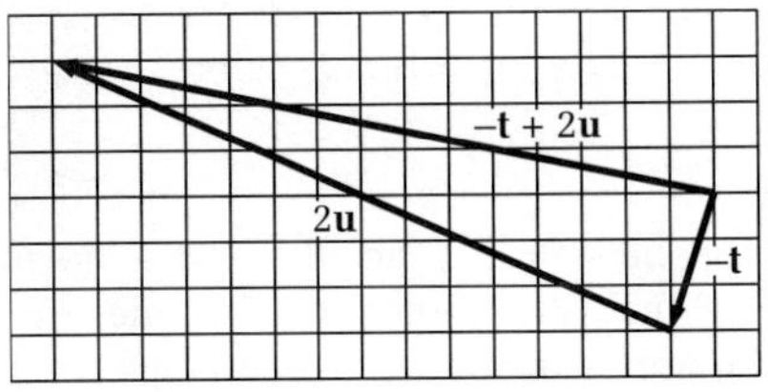

(15.3, 168.7°); (–15, 3)

14.

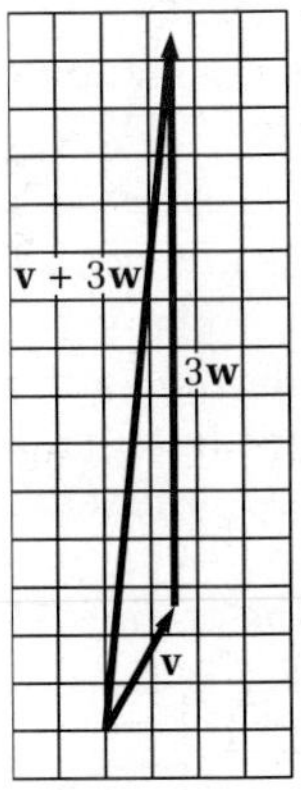

(14.7, 84.1°); (1.5, 14.6)

Page 456 Exercises and Problems

3. a. $x = 3t$; $y = -4.9t^2$

b.

Time (seconds)	Horizontal position x	Vertical position y	Ordered pair
0.0	0	0	(0, 0)
0.1	0.3	–0.049	(0.1, –0.049)
0.2	0.6	–0.196	(0.2, –0.196)
0.3	0.9	–0.441	(0.3, –0.441)
0.4	1.2	–0.784	(1.2, –0.784)
0.5	1.5	–1.225	(0.5, –1.225)

c. about 0.45 s; The ball hits the floor when $y = -1$, which occurs sometime between $t = 0.4$ and $t = 0.5$.

Page 457 Exercises and Problems

12. a, b. 30°: $x = 6.75t$; $y = 3.90t - 4.9t^2 + 1.5$

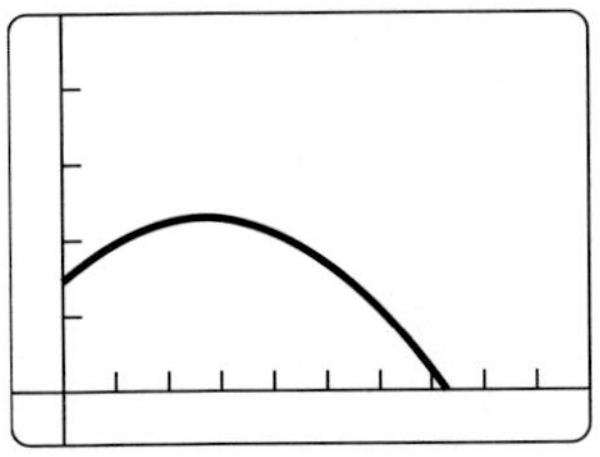

$40°$: $x = 5.98t$; $y = 5.01t - 4.9t^2 + 1.5$

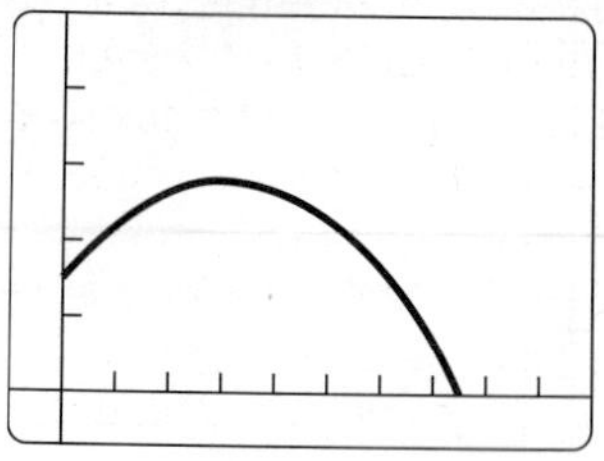

$50°$: $x = 5.01t$; $y = 5.98t - 4.9t^2 + 1.5$

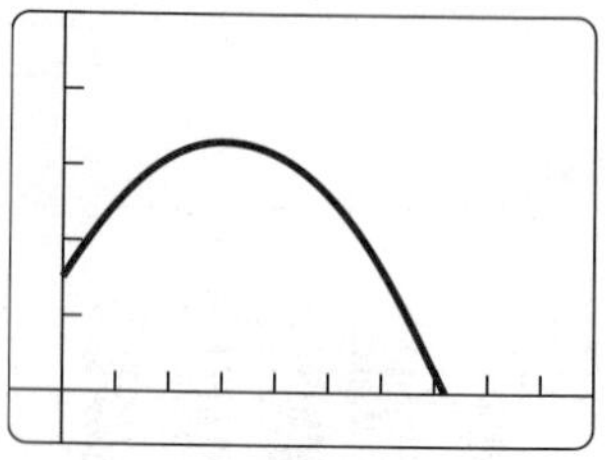

$60°$: $x = 3.90t$; $y = 6.75t - 4.9t^2 + 1.5$

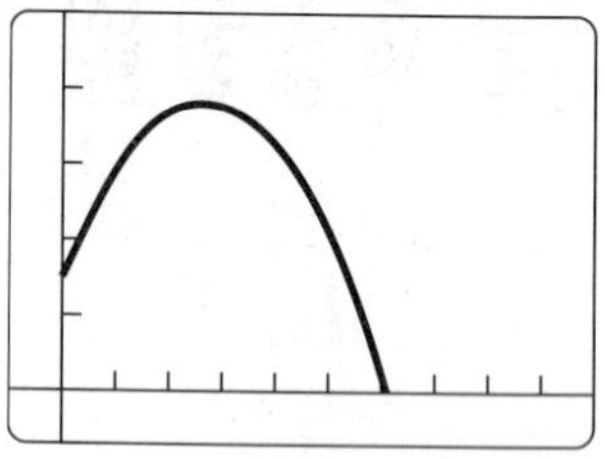

c. Yes; Yes, it does matter where she stands. Since the best range for watering her plants is less than 8 feet, she cannot stand at one corner of her garden and still reach the opposite corner, which is 9.4 ft away. If she stands in the middle of one long side, she is about 6.4 ft from the farthest corner and can water the whole garden. **13.** a circle with radius 8 and its center at the origin **14.** a circle with radius 5 and its center at the origin **15.** a circle with radius 7.5 and its center at the origin **16.** $x = 3 \cos t$; $y = 3 \sin t$

17.

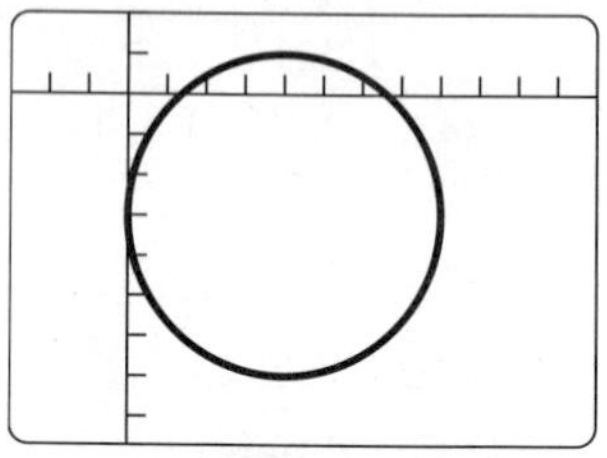

18. $x = 7 \cos t - 3$; $y = 7 \sin t - 2$ **19.** The graph is a circle with radius a and its center at (b, c).

20–22.

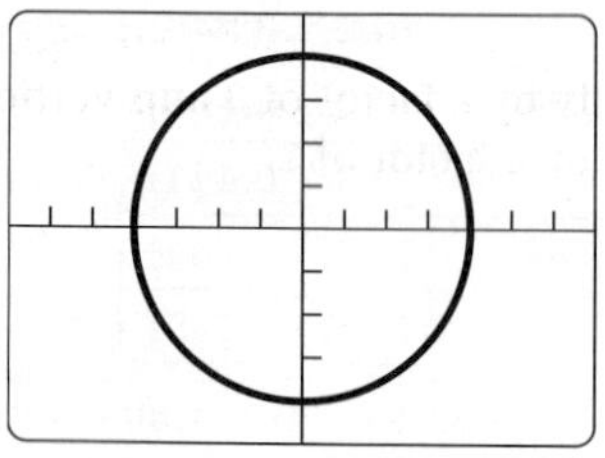

20. a circle with radius 8 and its center at the origin **21.** a 270° arc of the circle with radius 8 and center (0, 0) in quadrants II, III, and IV **22.** a 195° arc of the circle with radius 8 and center (0, 0) in quadrants II, III, and IV **23.** Answers may vary. An example is given. Graph $x = 6 \cos t$, $y = 6 \sin t$ with the following settings: $180° \le t \le 360°$ and t-step = 1. **24.** Answers may vary. An example is given. Graph $x = 7 \cos t$, $y = 7 \sin t$ with the following settings: $115° \le t \le 162°$ and t-step = 1. **25.** Answers may vary. Check students' work.

Page 473 Exercises and Problems

15. a. $m\angle V_1 = 29.329°$; $m\angle ESV_1 = 129.524°$; $EV_1 = 1.548$ a.u.; $m\angle V_2 = 150.671°$; $m\angle ESV_2 = 8.182°$; $EV_2 = 0.286$ a.u.
b. $m\angle M_1 = 25.746°$; $m\angle ESM_1 = 146.048°$; $EM_1 = 1.264$ a.u.; $m\angle M_2 = 154.254°$; $m\angle ESM_2 = 17.54°$; $EM_2 = 0.682$ a.u. **c.** V_1; M_2
d.

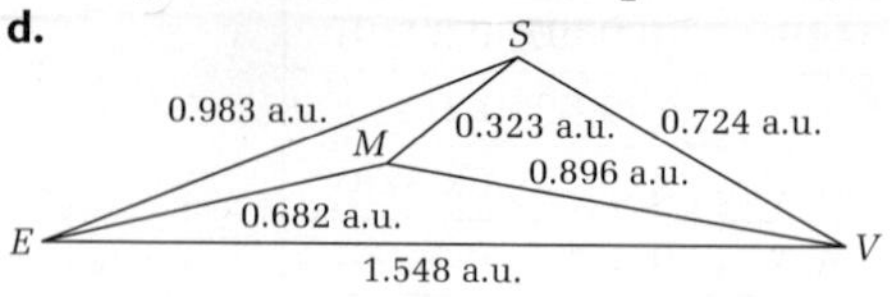

Page 474 Exercises and Problems

34. Answers may vary. Examples are given.

Given	To find sides, use ...	To find angles, use ...	Easier to find—first
SAS	law of cosines	law of sines or law of cosines	sides
SSS	all sides known	law of cosines or law of sines	—
ASA	law of sines	triangle sum theorem	angles
AAS	law of sines	triangle sum theorem	angles
SSA	law of sines	law of cosines or law of sines	sides

Unit 9

Page 489 Exploration

3. step 1:

a.

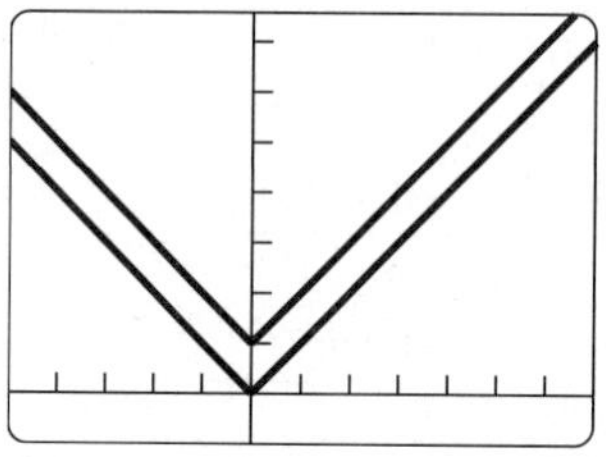

b.

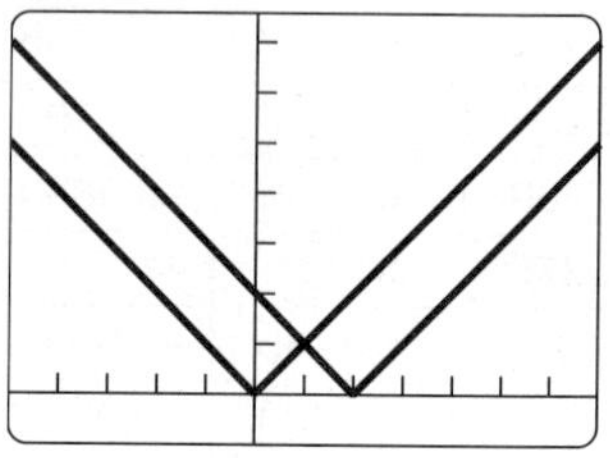

c.

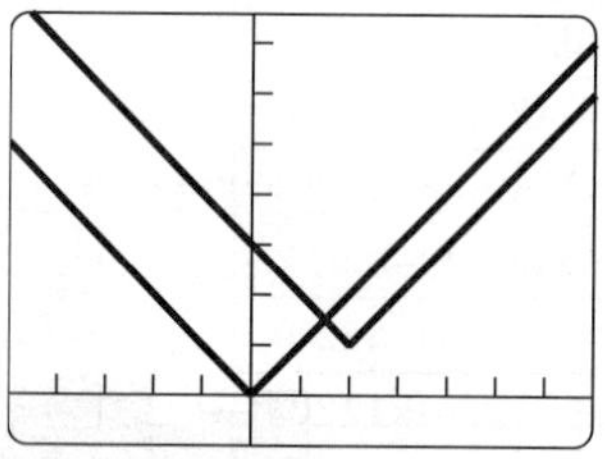

step 2: **a.** The graph of $y - 1 = |x|$ is the graph of $y = |x|$ translated 1 unit up. **b.** The graph of $y = |x - 2|$ is the graph of $y = |x|$ translated 2 units to the right. **c.** The graph of $y - 1 = |x - 2|$ is the graph of $y = |x|$ translated 1 unit up and 2 units to the right.

4. step 1:

a.

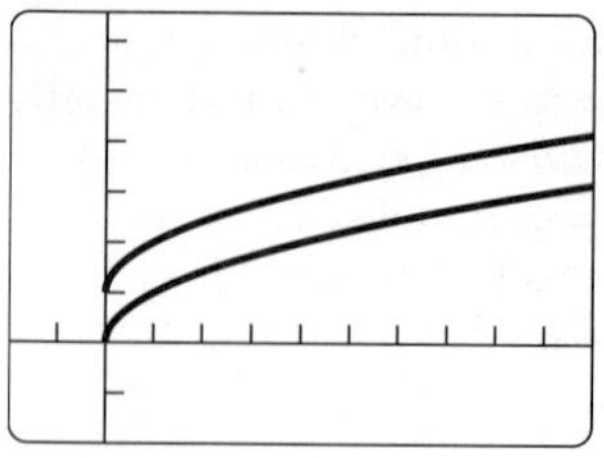

b.

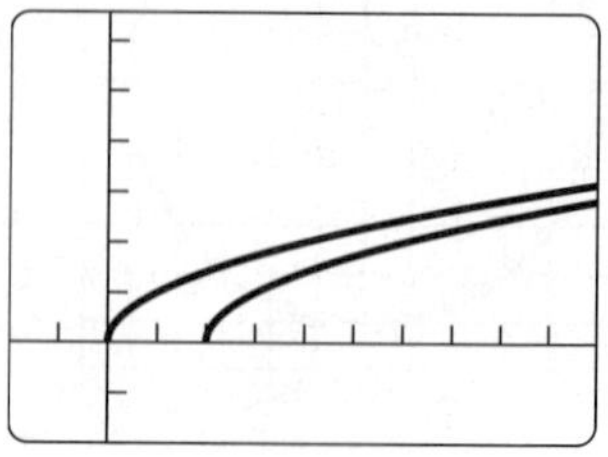

c.

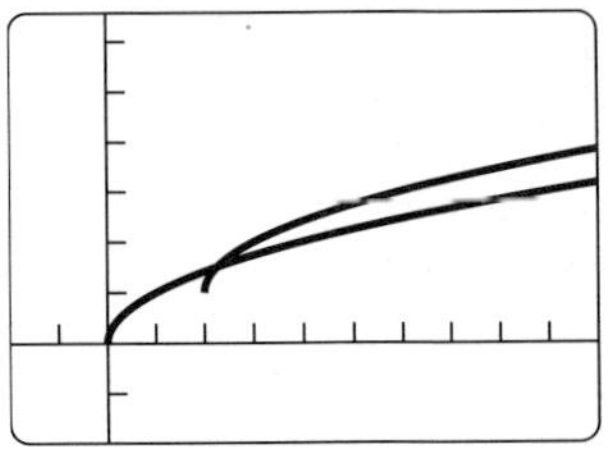

step 2: **a.** The graph of $y - 1 = \sqrt{x}$ is the graph of $y = \sqrt{x}$ translated 1 unit up. **b.** The graph of $y = \sqrt{x - 2}$ is the graph of $y = \sqrt{x}$ translated 2 units to the right. **c.** The graph of $y - 1 = \sqrt{x - 2}$ is the graph of $y = \sqrt{x}$ translated 1 unit up and 2 units to the right.

5. step 1:

a.

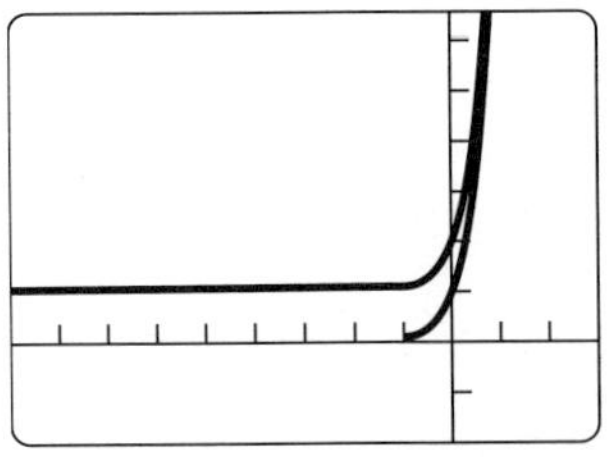

b.

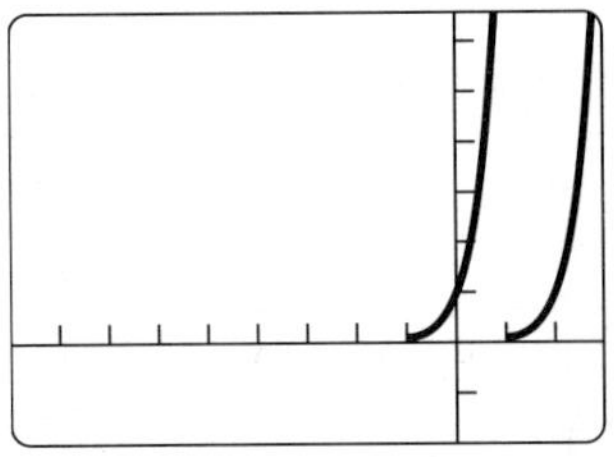

c.

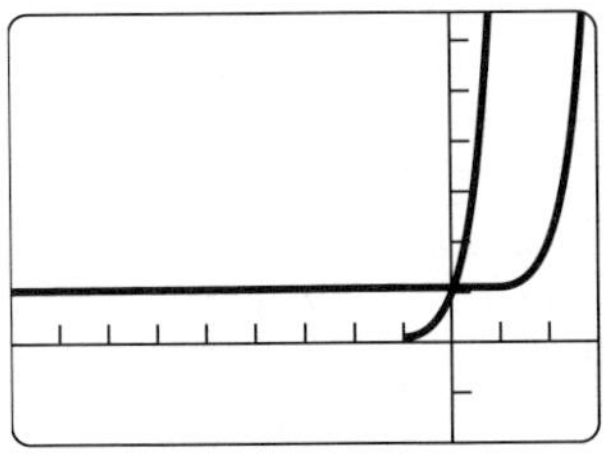

step 2: **a.** The graph of $y - 1 = 10^x$ is the graph of $y = 10^x$ translated 1 unit up. **b.** The graph of $y = 10^{x-2}$ is the graph of $y = 10^x$ translated 2 units to the right. **c.** The graph of $y - 1 = 10^{x-2}$ is the graph of $y = 10^x$ translated 1 unit up and 2 units to the right. **6.** Answers may vary. An example is given. The graph of $y - k = f(x - h)$ is the graph of $y = f(x)$ translated k units up if $k > 0$, k units down if $k < 0$, h units to the right if $h > 0$, and h units to the left if $h < 0$.

Page 488 Exercises and Problems

27. The mean of the n data values with k added is

$$\frac{(x_1 + k) + (x_2 + k) + \dots + (x_n + k)}{n} = \frac{(x_1 + x_2 + \dots x_n) + nk}{n} = \frac{x_1 + x_2 + \dots x_n}{n} + k.$$

This is $\bar{x} + k$.

28.

a.

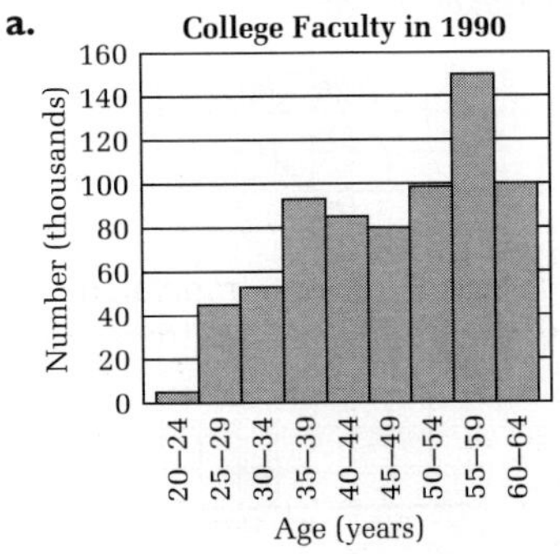

b.

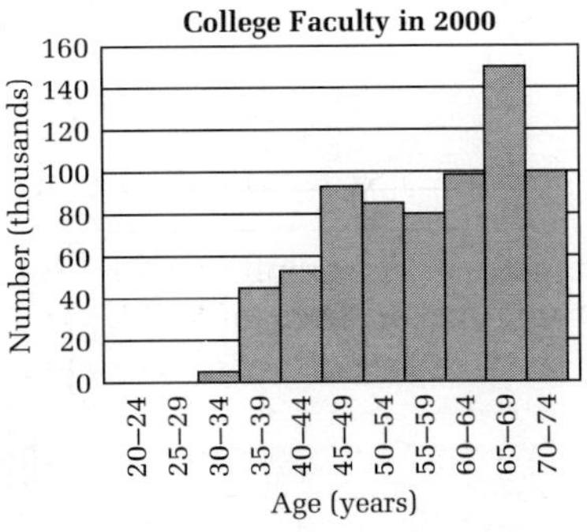

c. The histogram in part (b) is the histogram from part (a) shifted 10 units to the right. **d.** 250,000 faculty members **e.** Answers may vary. Examples are given: student enrollment. changes in educational trends, changes in the economy leading teachers into industry or vice versa, new developments producing new fields of study or growth or decline in existing fields

Page 493 Exercises and Problems

14. a, b.

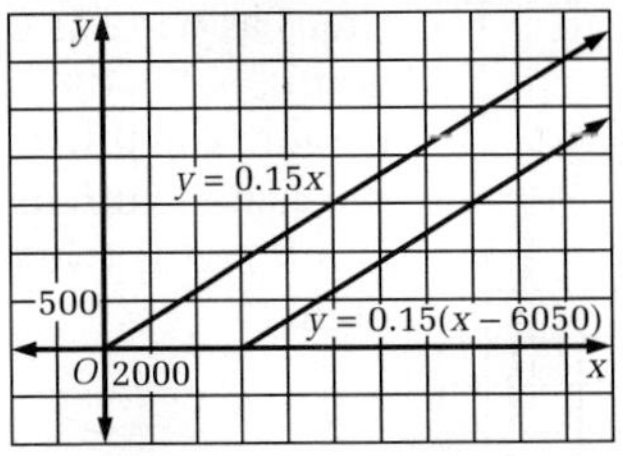

c. The parent graph is translated 6050 units to the right. **d.** One graph gives the tax as a function of taxable income, the other as a function of earned income. **e.** The graph of $y = 0.15x$ applies; for earned income of \$3500, the taxable income is 0 and the tax is $0.15(0) = 0$.

Page 500 Exercises and Problems

19. a.

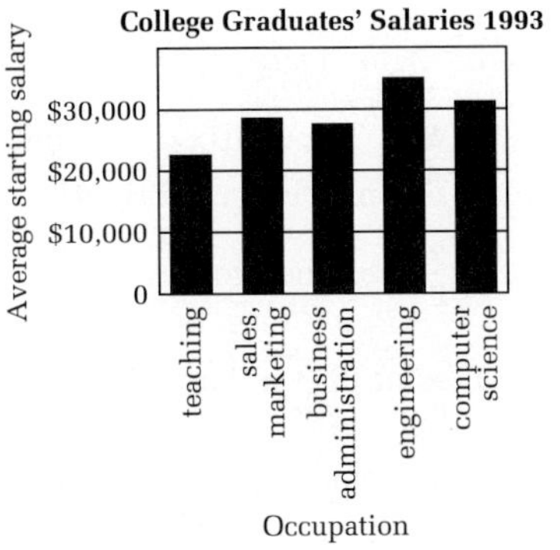

b.

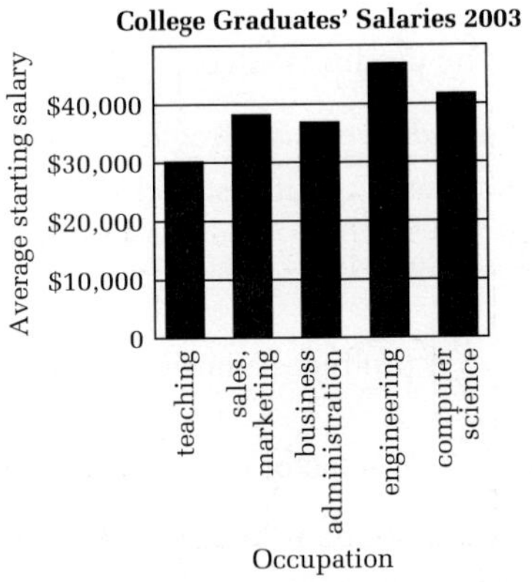

c. Each data item in the second graph is obtained by multiplying the corresponding data item in the first graph by 1.34.

Page 501 Exploration

3. a. step 1: $\begin{bmatrix} -3 & 0 & 3 \\ 0 & 18 & 0 \end{bmatrix}$

step 2:

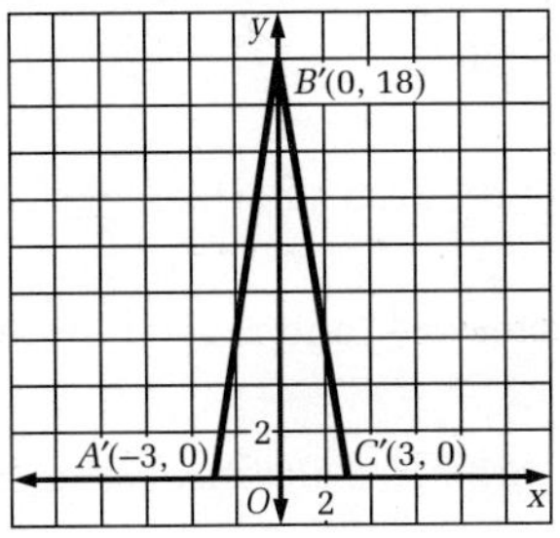

The graph is stretched vertically by a factor of 2.

b. step 1: $\begin{bmatrix} -9 & 0 & 9 \\ 0 & 18 & 0 \end{bmatrix}$

step 2:

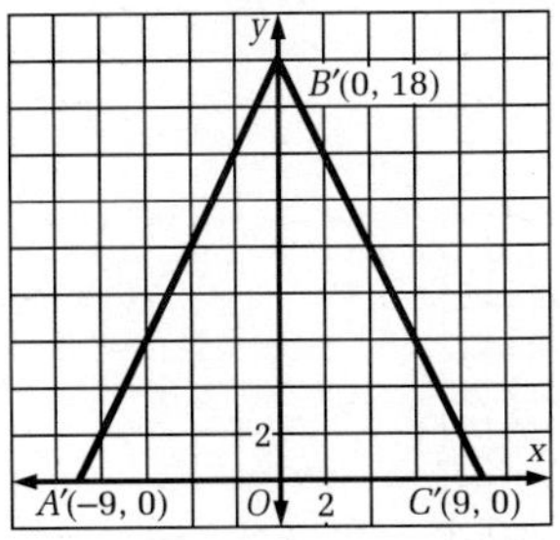

The graph is stretched horizontally by a factor of 3 and vertically by a factor of 2.

c. step 1: $\begin{bmatrix} -1 & 0 & 1 \\ 0 & 4\frac{1}{2} & 0 \end{bmatrix}$

step 2:

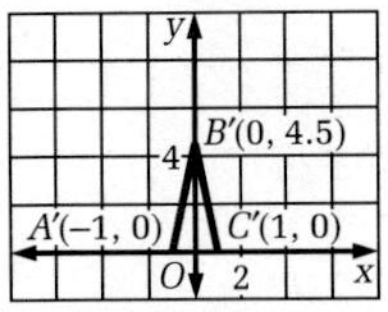

The graph is stretched horizontally by a factor of $\frac{1}{3}$ and vertically by a factor of $\frac{1}{2}$.

6. a. $\frac{y}{1} = -\left(\frac{x}{3}\right)^2 + 9$; $a = 3$; $b = 1$ **b.** $\frac{y}{2} = -x^2 + 9$; $a = 1$; $b = 2$

c. 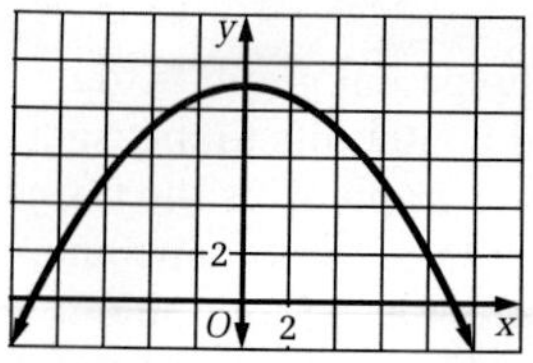

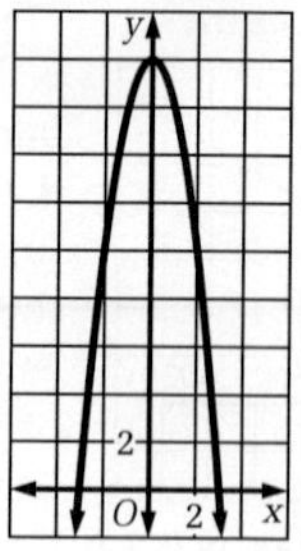

The second graph is three times as wide as the first graph, but is the same height. The third graph is the same width as the first graph, but is twice as high.

Page 504 Exercises and Problems

5. a. The matrix on the left produces a horizontal stretch by a factor of 4. The matrix on the right produces a vertical stretch by a factor of 3.
b. The product matrix $\begin{bmatrix} 4 & 0 \\ 0 & 3 \end{bmatrix}$ produces a horizontal stretch by a factor of 4 and a vertical stretch by a factor of 3. **6. a.** The matrix on the left produces a vertical stretch by a factor of $\frac{1}{3}$. The matrix on the right produces a horizontal stretch by a factor of $\frac{3}{2}$. **b.** The product matrix $\begin{bmatrix} \frac{3}{2} & 0 \\ 0 & \frac{1}{3} \end{bmatrix}$ produces a horizontal stretch by a factor of $\frac{3}{2}$ and a vertical stretch by a factor of $\frac{1}{3}$. **7.** Answers may vary. Check students' work.

8. a. all points where the original graph intersects the *y*-axis
b. all points where the original graph intersects the *x*-axis

9. $\begin{bmatrix} 9 & 0 & -9 \\ 0 & 9 & 0 \end{bmatrix}$; The transformation produces a horizontal stretch by a factor of 3 and a reflection over the *y*-axis.

10. $\begin{bmatrix} -3 & 0 & 3 \\ 0 & -4\frac{1}{2} & 0 \end{bmatrix}$; The transformation produces a vertical stretch by a factor of $\frac{1}{2}$ and a reflection over the *x*-axis.

11. $\begin{bmatrix} 9 & 0 & -9 \\ 0 & -4\frac{1}{2} & 0 \end{bmatrix}$; The transformation produces a horizontal stretch by a factor of 3, a vertical stretch by a factor of $\frac{1}{2}$, and reflections over both axes.

12. $\begin{bmatrix} -3 & 0 & 3 \\ 0 & -27 & 0 \end{bmatrix}$; The transformation produces a vertical stretch by a factor of 3 and a reflection over the *x*-axis.

Page 505 Exercises and Problems

14. a. $\begin{bmatrix} 15 & 12 & 0 & -12 & -15 & 0 \\ 0 & 3 & 5 & 3 & 0 & -5 \end{bmatrix}$

b. 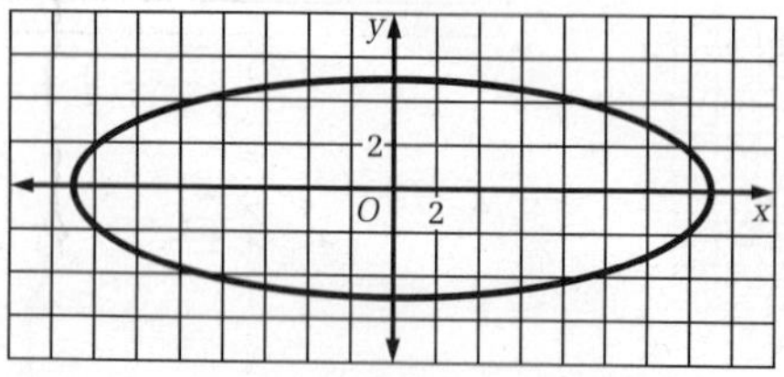

c. $(15, 0)$: $\left(\frac{15}{3}\right)^2 + 0^2 = 25$; $(12, 3)$: $\left(\frac{12}{3}\right)^2 + 3^2 = 16 + 9 = 25$; $(0, 5)$: $\left(\frac{0}{3}\right)^2 + 5^2 = 25$; $(-12, 3)$: $\left(\frac{-12}{3}\right)^2 + 3^2 = 16 + 9 = 25$; $(-15, 0)$: $-\left(\frac{-15}{3}\right)^2 + 0^2 = 25$; $(0, -5)$: $\left(\frac{0}{3}\right)^2 + (-5)^2 = 25$

15. 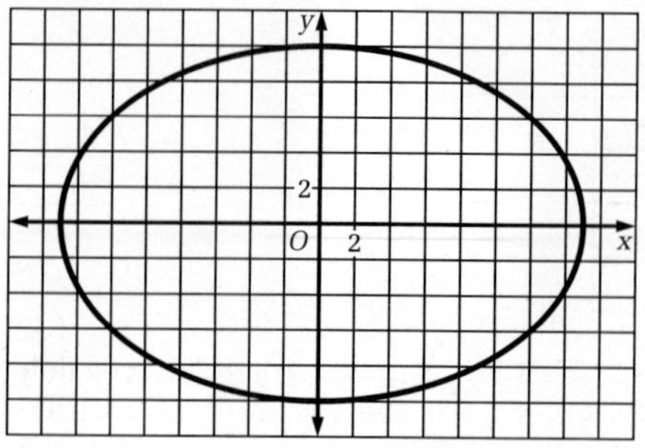

21. a. 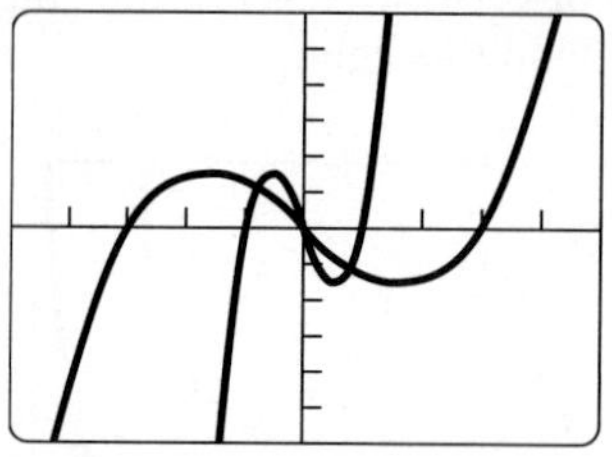

b. The graph is stretched horizontally by a factor of 3.

Page 506 Exercises and Problems

22. a, b.

Telephone Rates

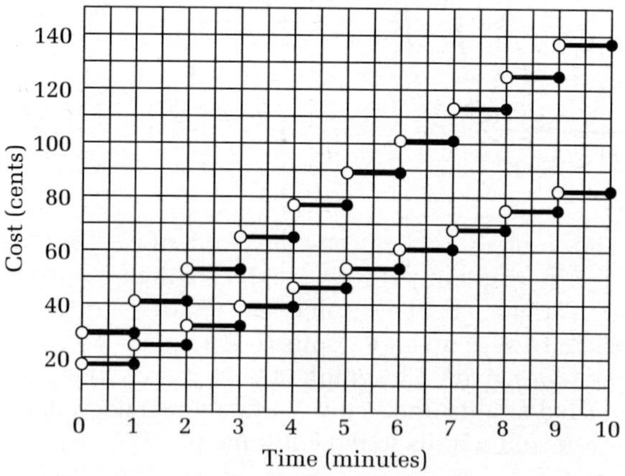

c. The parent graph is stretched vertically by a factor of 0.6.
d. $\frac{y}{0.6} = f(x)$ or $y = 0.6(f(x))$

Page 530 Exercises and Problems

9. a. $b(x) = 100 - s(x)$ **b.**

Recommended Percent of Money in Bonds

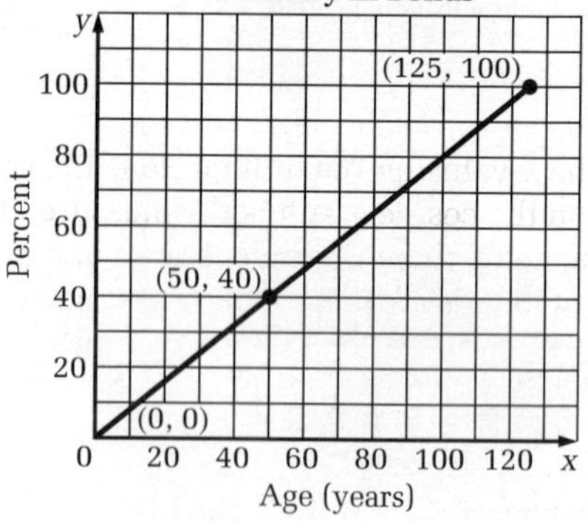

Page 534 Unit 9 Review and Assessment

1. a. 545.6; 517; 280; about 112

b.

Name of building	Height (ft)
Amoco Building	5728
Anaconda Tower	5787
Arco Tower	5807
First Interstate Tower North	5714
Mountain Bell Center	5989
1999 Broadway	5824
Republic Plaza	5994
17th Street Plaza	5718
Stellar Plaza	5717
United Bank of Denver	5978

c. 5825.6; 5797; 280; about 112

Unit 10

Page 551 Exercises and Problems

17.

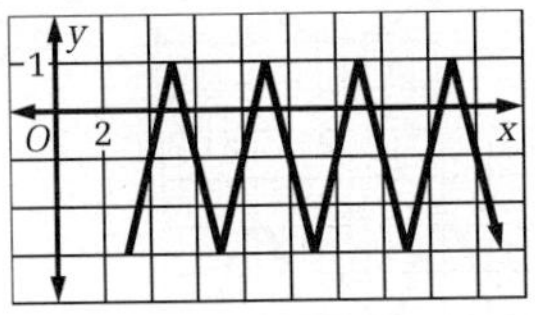

18.

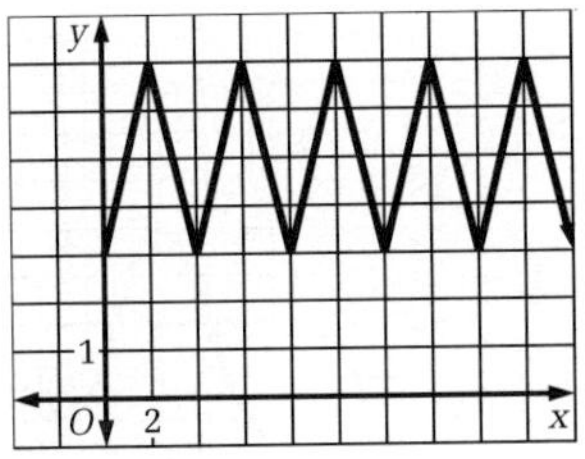

19.

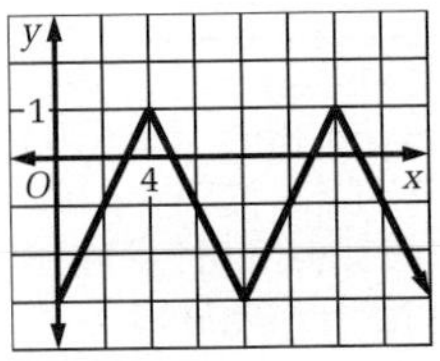

20.

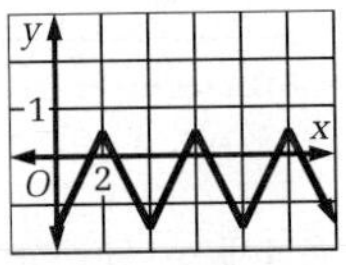

21.

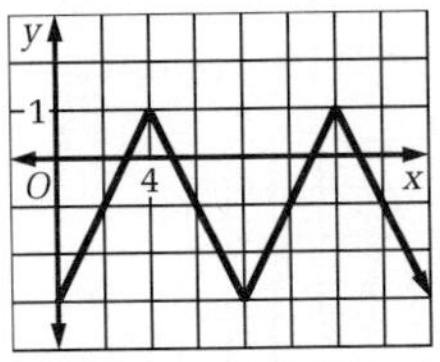

22.

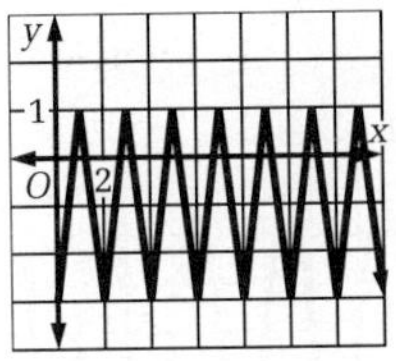

Page 559 Exercises and Problems

19. a. Check students' graphs. **b.** The *y*-value on the unit circle is the sine of the angle. The *y*-value on the cosine graph is the cosine of the angle. The *x*-value on the unit circle is the cosine of the angle and the *x*-value on the cosine graph is the measure of the angle. **c.** The *x*-coordinate on the unit circle is the same as the *y*-coordinate on the cosine graph. **20.** Answers may vary. An example is given.

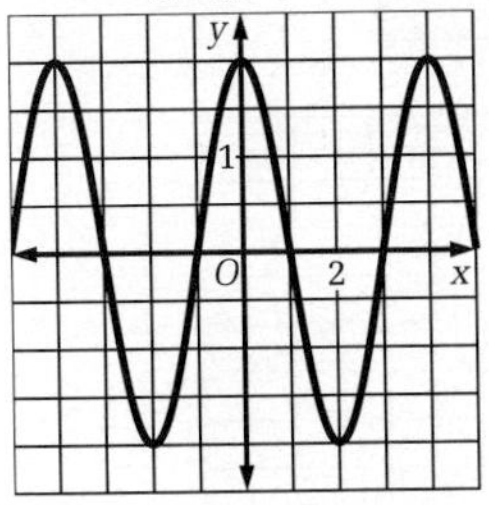

21. $f(g(x)) = \left(\frac{x}{2}\right)^2 = \frac{x^2}{4}$ **22.** $h(f(x)) = \sqrt{x^2} - 1$ or $|x| - 1$ **23.** 16 **24.** 2

25.

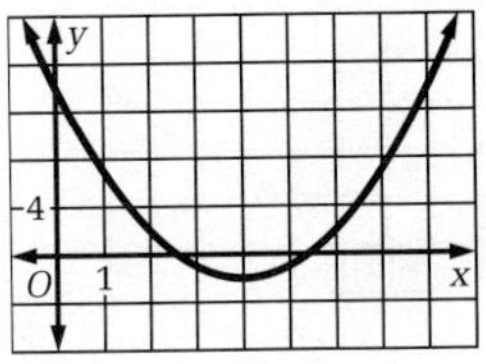

26.

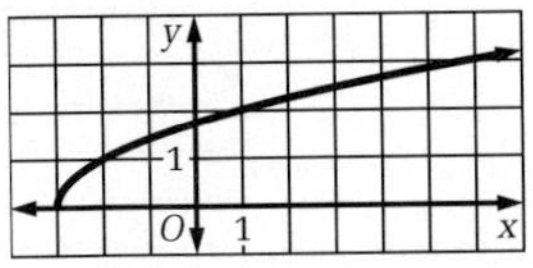

27.

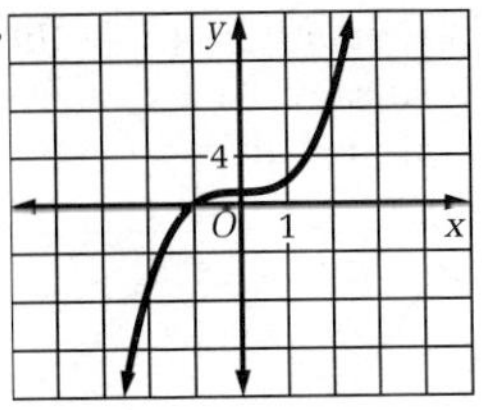

28. Answers will vary. Check students' work.

Page 560 Exploration

3. a. same period and intercepts as the parent graph; amplitude is half that of the parent graph.

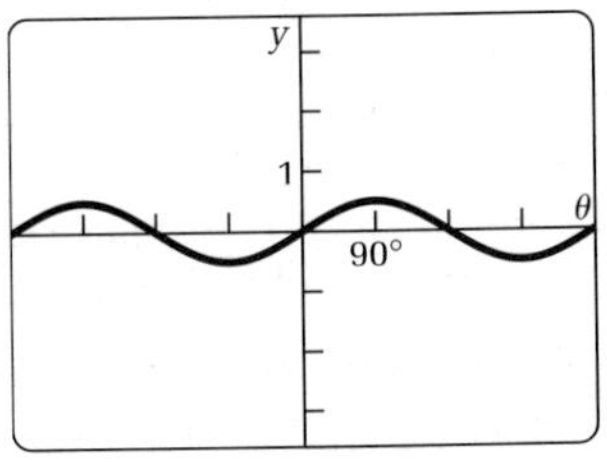

b. same period, amplitude, and intercepts as the parent graph; reflected over the θ-axis

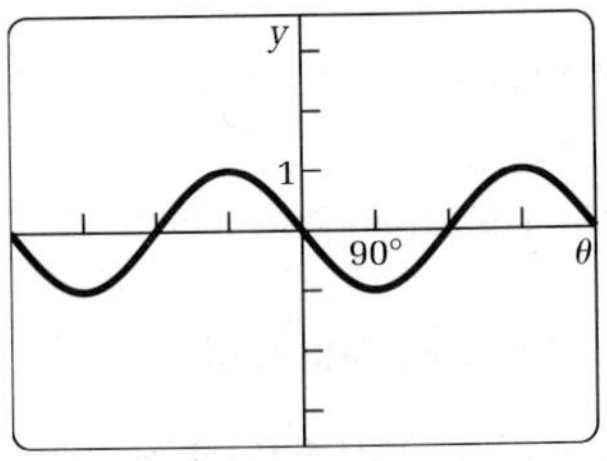

c. same amplitude and *y*-intercept as the parent graph; period is one-third that of the parent graph

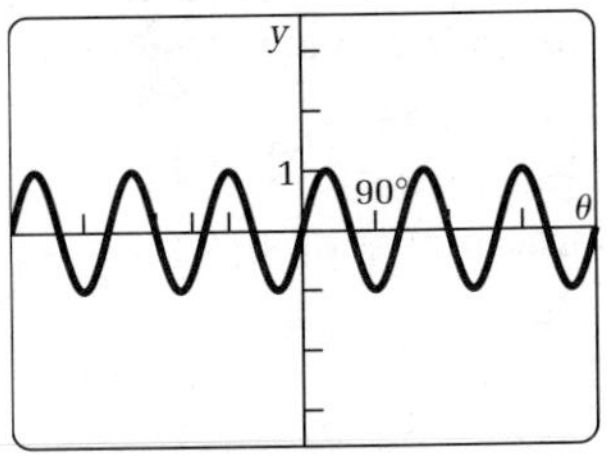

d. same period, amplitude, and intercepts as the parent graph; reflected over the *y*-axis

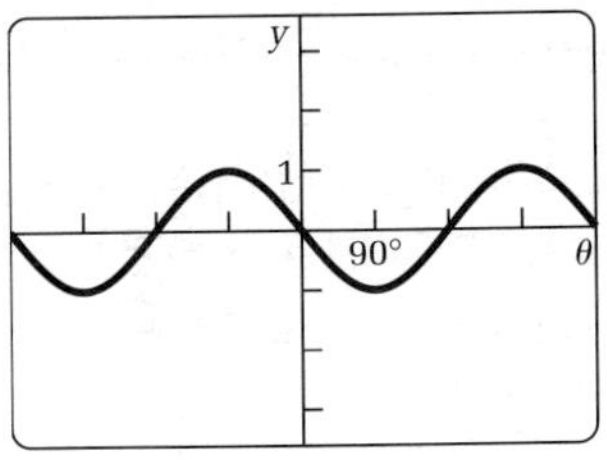

e. same period and amplitude as the parent graph; translated 60° to the right

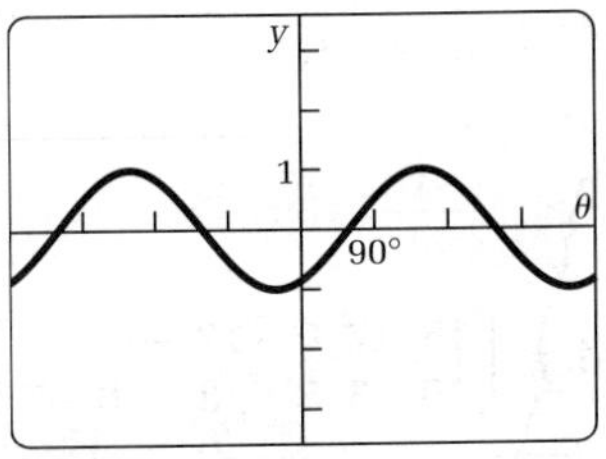

f. same period and amplitude as the parent graph; translated 90° to the left

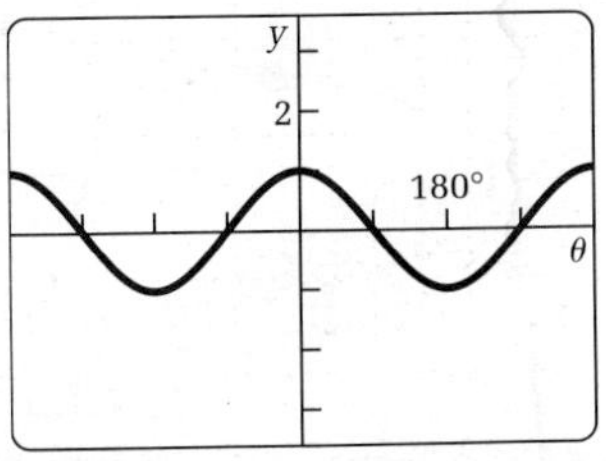

g. same period and amplitude as the parent graph; translated up 2 units

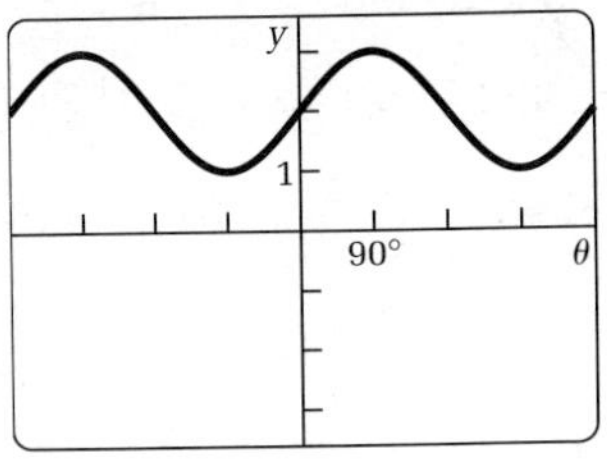

h. same period and amplitude as the parent graph; translated down 1.5 units

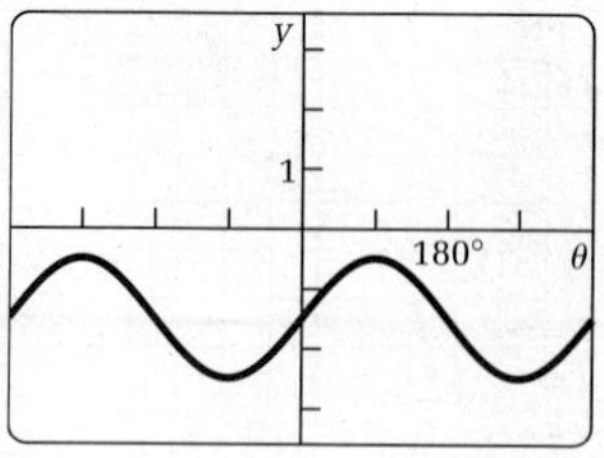

4. a. vertical stretch by a factor of $|A|$; If $A < 0$, the graph is reflected over the θ-axis. **b.** horizontal stretch by a factor of $\frac{1}{|B|}$; If $B < 0$, the graph is reflected over the y-axis.
c. horizontal translation of C units **d.** vertical translation of D units

5. a vertical stretch by a factor of 4, horizontal stretch by a factor of $\frac{1}{2}$

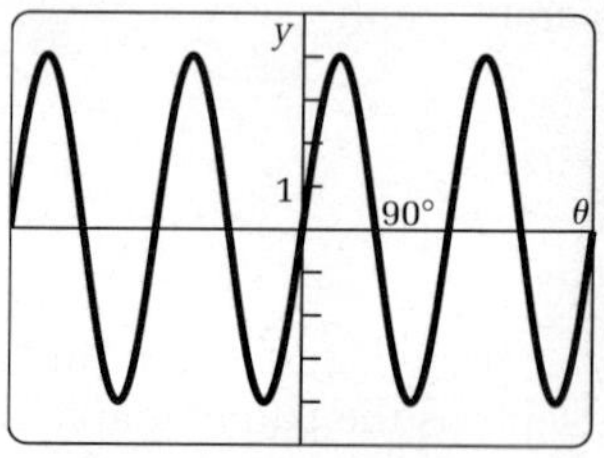

b. vertical stretch by a factor of 0.5; horizontal shift 30° to the right

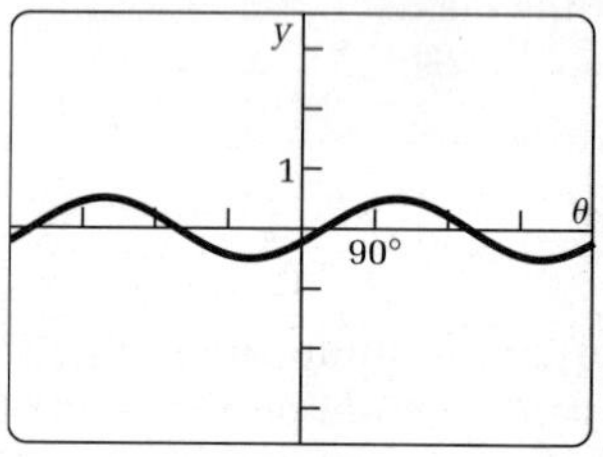

c. horizontal stretch by a factor of $\frac{4}{3}$; horizontal shift 45° to the left

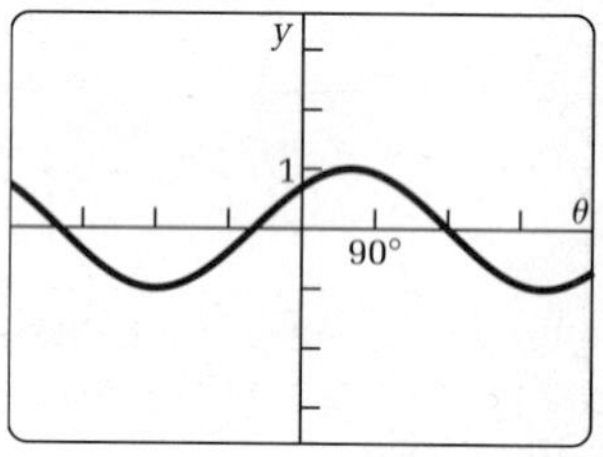

d. horizontal shift 180° to the left; vertical shift 1.5 units down

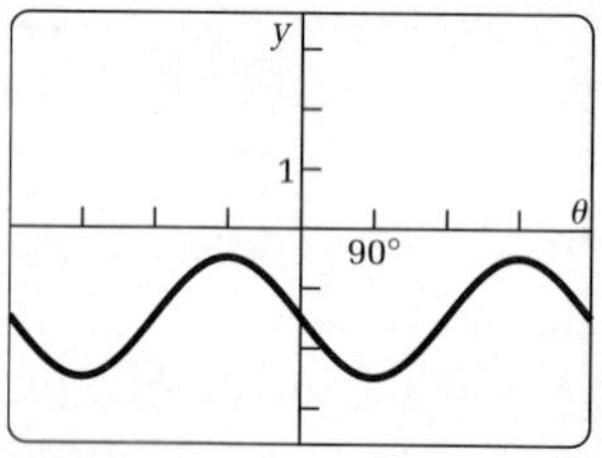

6. a. 4; $\frac{1}{180°}$; 180° **b.** 0.5; $\frac{1}{360°}$; 360° **c.** 1; $\frac{1}{480°}$; 480° **d.** 1; $\frac{1}{360°}$; 360°

Page 563 Exercises and Problems

6. C **7.** D **8.** B **9.** F **10.** A **11.** E

12. a. 1; 180°; $\frac{1}{180°}$ **b.**

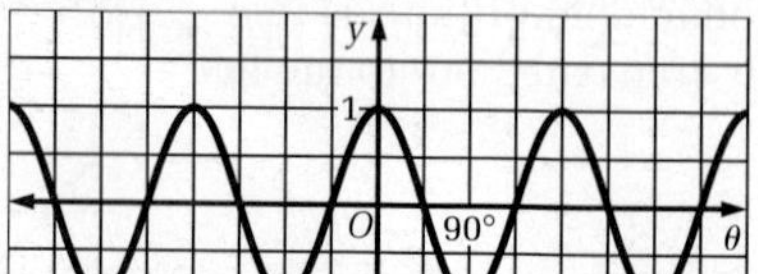

13. a. 3; 360°; $\frac{1}{360°}$ **b.**

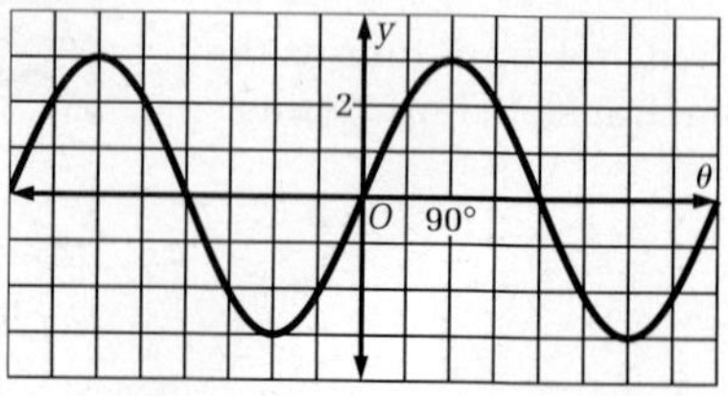

14. a. 1; 360°; $\frac{1}{360°}$ **b.**

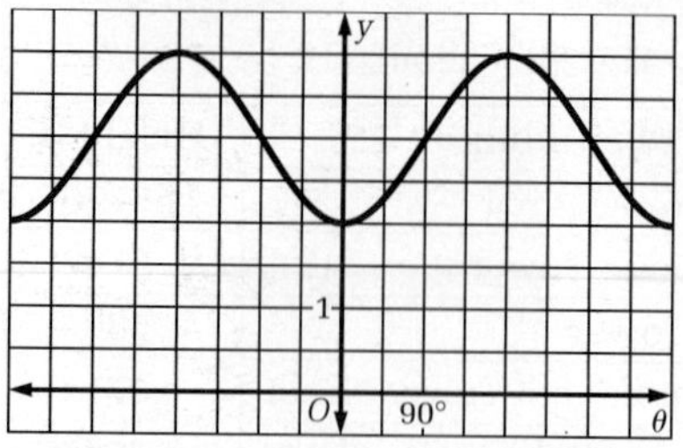

15. a. 1; 360°; $\frac{1}{360°}$ **b.**

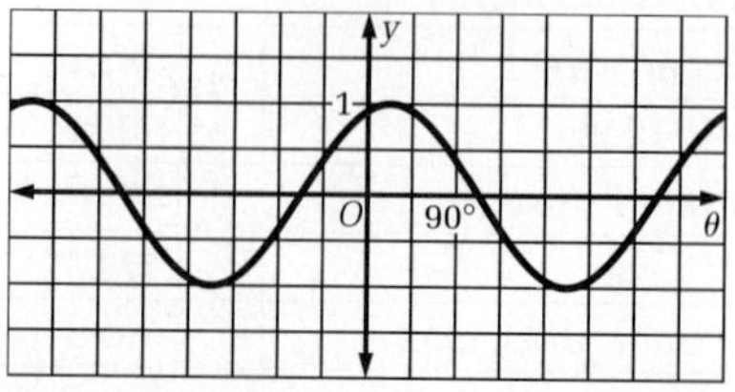

16. a. $\frac{1}{2}$; 360°; $\frac{1}{360°}$ **b.**

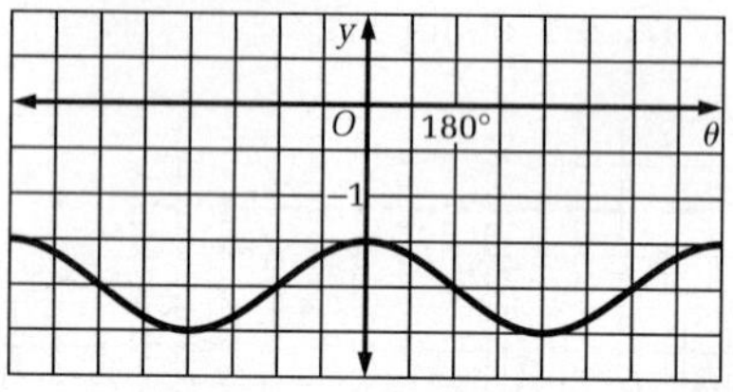

17. a. 1; 180°; $\frac{1}{180°}$ **b.**

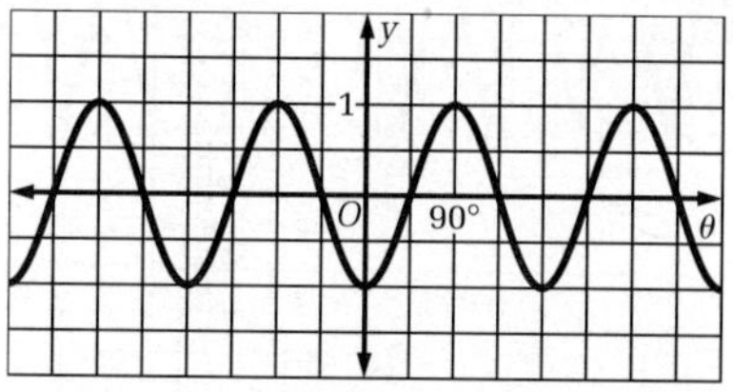

18. The graphs have the same shape, the same θ-intercepts, the same amplitude, and the same period. The graphs have different y-intercepts. The graph of $y = \sin(\theta - 90°)$ is the graph of $y = \sin(\theta + 90°)$ shifted 180° to the right. (Each graph can also be obtained by shifting the other 180° to the left or reflecting the other in the θ-axis.)

Page 564 Exercises and Problems

22. a. 3; 90°; $\frac{1}{90°}$ **b.**

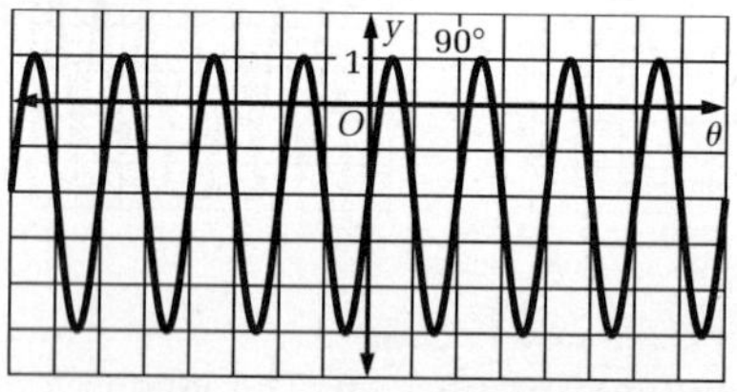

23. a. 1; 180°; $\frac{1}{180°}$ **b.**

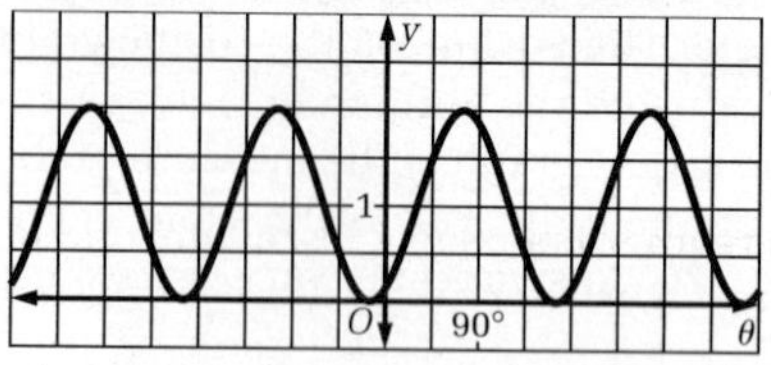

24. a. 0.3; 720°; $\frac{1}{720°}$ **b.**

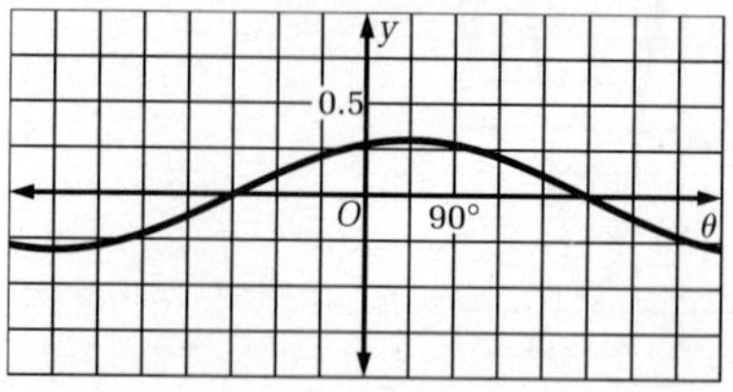

25. a. 2; 1440°; $\frac{1}{1440°}$ **b.**

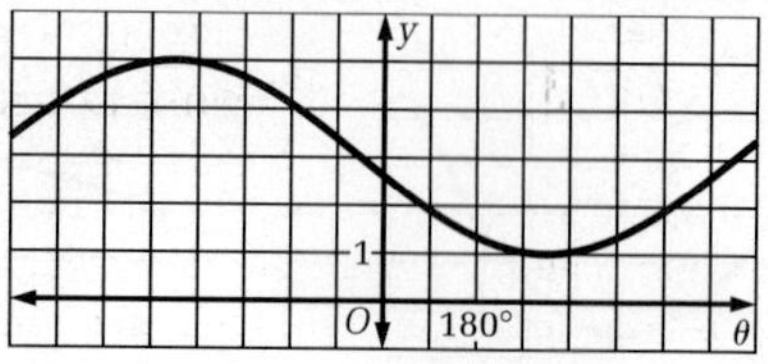

Page 565 Exercises and Problems

29, 30. Equations may vary. Examples are given.

29. $y = 3 \cos 8(\theta + 30°)$ **b.**

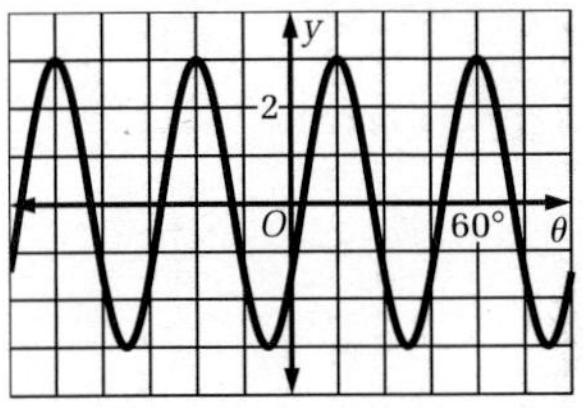

30. $y = \sin 6\theta + 1.5$ **b.**

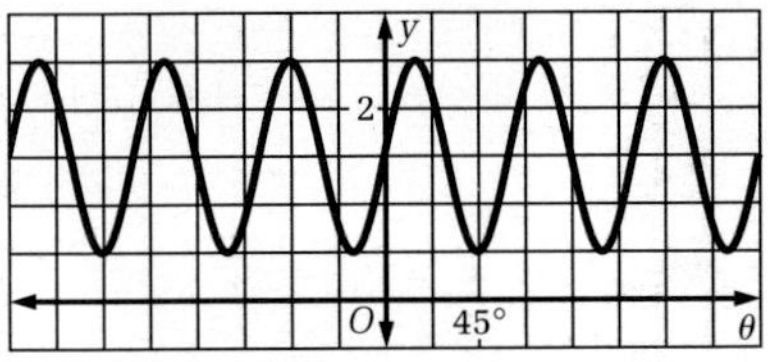

Page 576 Checkpoint

10.

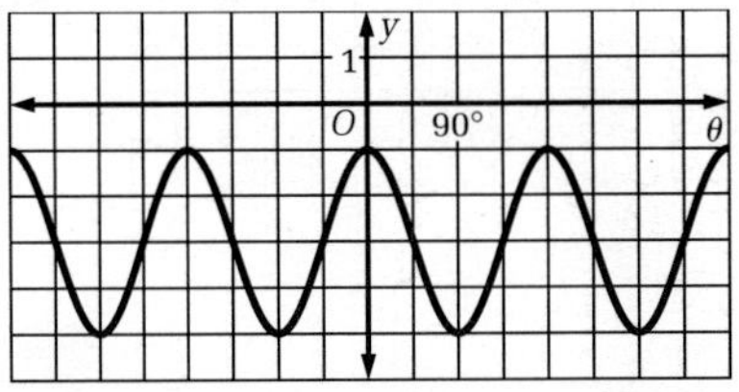

11. a.

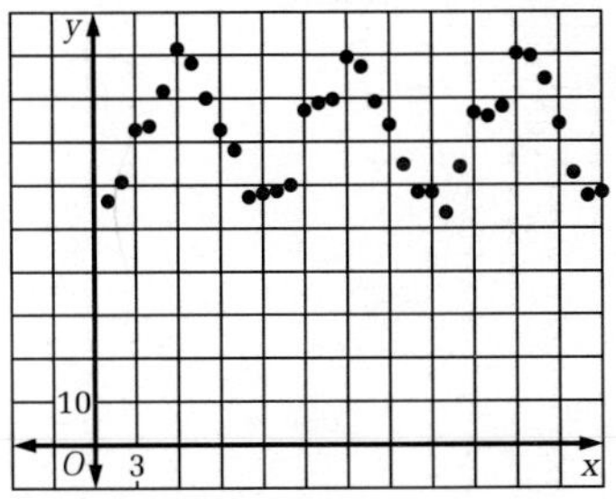

a sine curve; The points appear to lie on a sine curve.

b, c. Answers may vary. An example is given.

b. $y = 17.2 \cos (30(x - 6))° + 73.3$

c.

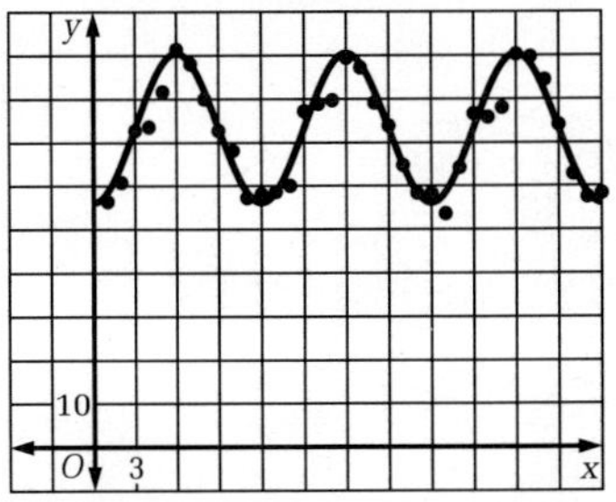

My graph fits the data fairly well. I think no graph will fit all the data points exactly because ice cream production reflects sales and sales are dependent on human behavior, which is not entirely predictable.

Page 585 Exercises and Problems

10. Suppose regular heptagons tessellate a plane. Let h = the number of regular heptagons at each vertex. The measures of interior angles of regular heptagons are $128\frac{4}{7}°$, so $128\frac{4}{7}h = 360$. There is no solution with integer values. Since there is no number of regular heptagons that surround a vertex, the assumption that heptagons tessellate a plane is false. **11.** Suppose regular decagons tessellate a plane. Let d = the number of regular decagons at each vertex. The measures of each interior angle of a regular decagon is 144°, so $144d = 360$. There is no solution with integer values. Since there is no number of regular decagons that surround a vertex, the assumption that decagons tessellate a plane is false. **12.** Suppose regular octagons and equilateral triangles tessellate a plane. Let x = the number of regular octagons at each vertex and let t = the number of equilateral triangles at each vertex. The measures of the interior angles of regular octagons and equilateral triangles are 135° and 60°, so $135x + 60t = 360$. The only solution with nonnegative integer values is (0, 6). Since there is no combination of regular octagons and triangles that surround a vertex, the assumption that octagons and triangles tessellate a plane is false.

Page 586 Exercises and Problems

13. a. The measure of each interior angle of the regular n-gon is $\frac{(n-2)180}{n}$. There are q at each vertex of the tessellation, so the sum is $\frac{(n-2)180q}{n}$, which must equal 360°. **b.** Methods may vary. An example is given. Divide both sides of the equation by 180 to produce the equivalent equation $\frac{(n-2)q}{n} = 2$. Multiply both sides of the new equation by n. Simplify the resulting equation and subtract 2 from both sides to produce the equivalent equation $nq - 2q - 2n = 0$. Add 4 to both sides. Use the distributive property to rewrite the resulting equation: $q(n-2) - 2(n-2) = 4$; $(n-2)(q-2) = 4$.
c. Translate it by the vector (2, 2).

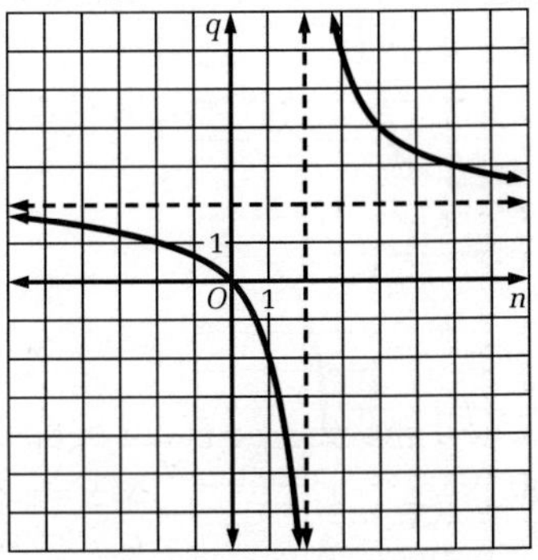

d. (3, 6), (4, 4), and (6, 3); The first coordinate is the number of sides of the n-gon in a regular tessellation and the second coordinate is the number of n-gons at each vertex. The only regular tessellations of a plane consist of regular 3-gons (equilateral triangles), regular 4-gons (squares), and regular 6-gons (regular hexagons). **e.** There are no other points on the graph with both coordinates positive integers.

Page 590 Review and Assessment

9. 3; 360°; $\frac{1}{360°}$

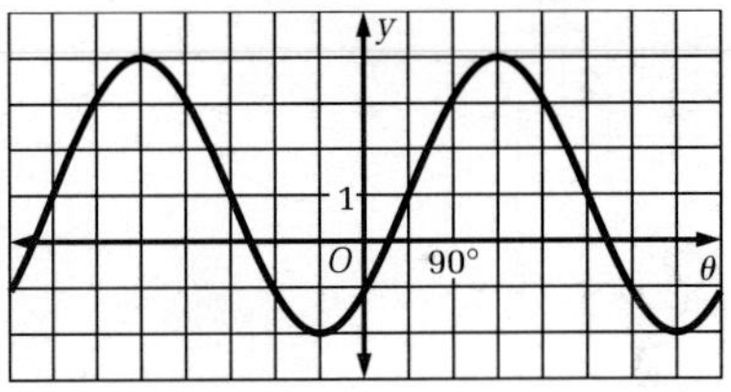

10. 1; 180°; $\frac{1}{180°}$

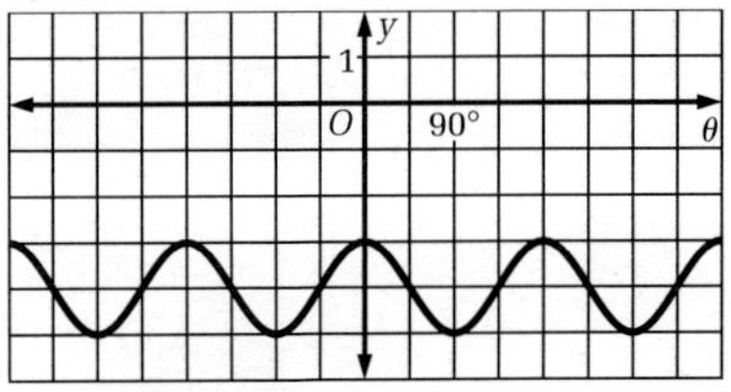

11. $\frac{1}{2}$; 120°; $\frac{1}{120°}$

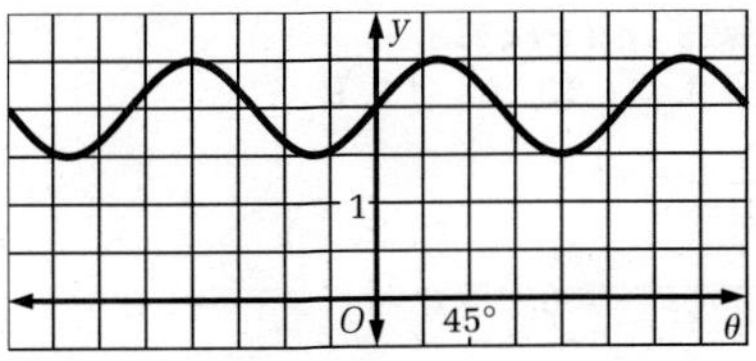

12, 13. Answers may vary. Examples are given. **12.** $y = 1.5 \cos 2\theta - 3$
13. $y = 3 \sin (\theta - 45°) + 2$

14. a.

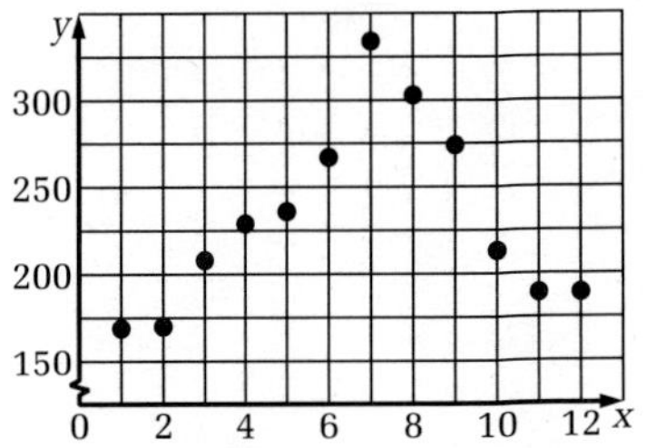

b. Answers may vary. An example is given. $y = 75 \cos (30(x - 7))° + 235$
c. A periodic function that can be modeled by a sine wave. The given data are periodic.

Extra Practice

Page 611 Extra Practice Unit 2

57. a.

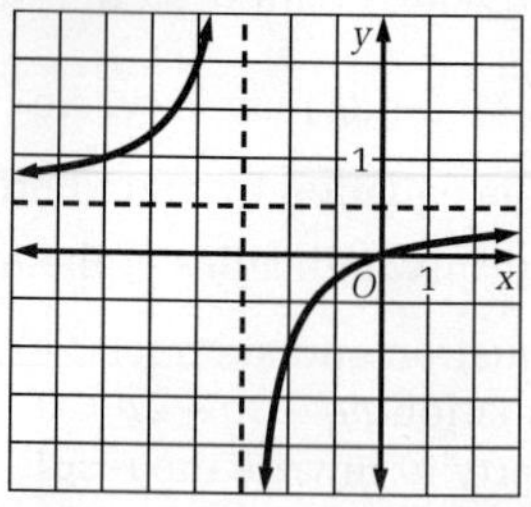

b. $x = -3$; $y = 1$
c. all real numbers except −3; all real numbers except 1

58. a.

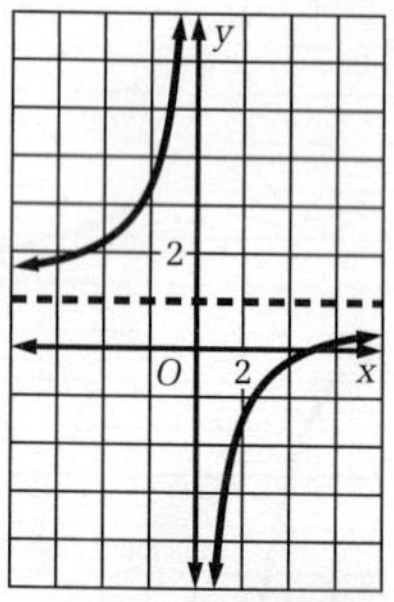

b. $x = 0$; $y = 1$
c. all real numbers except 0; all real numbers except 1

59. a.

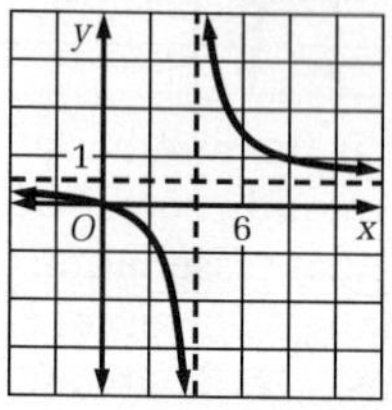

b. $x = 4$; $y = 0.5$
c. all real numbers except 4; all real numbers except 0.5

60. a.

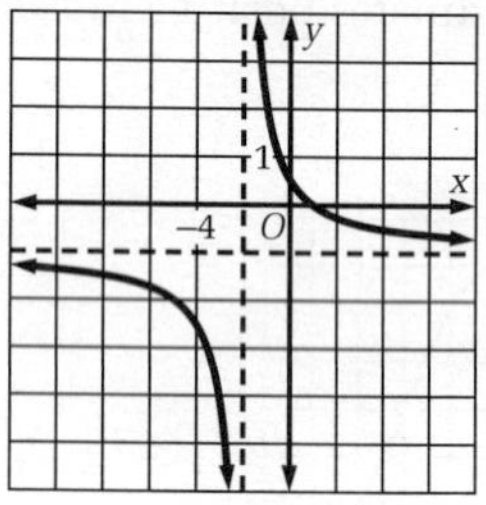

b. $x = -2$; $y = -1$
c. all real numbers except −2; all real numbers except −1